Biogeochemist

Citations
Please use the following example for citations:
Nisbet E.G. and Fowler C.M.R. (2003) The early history of life, pp. 1–39. In *Biogeochemistry* (ed. W.H. Schlesinger) Vol. 8 *Treatise on Geochemistry* (eds. H.D. Holland and K.K. Turekian), Elsevier–Pergamon, Oxford.

Cover photo: A high-elevation lake in the Sierra Nevada of California, showing the juxtaposition of air, land and wetland ecosystems that provide an arena for biogeochemistry. (Photograph provided by William H. Schlesinger)

Biogeochemistry

Edited by

W. H. Schlesinger
Duke University, Durham, NC, USA

TREATISE ON GEOCHEMISTRY
Volume 8

Executive Editors

H. D. Holland
Harvard University, Cambridge, MA, USA

and

K. K. Turekian
Yale University, New Haven, CT, USA

ELSEVIER

2005

AMSTERDAM – BOSTON – HEIDELBERG – LONDON – NEW YORK – OXFORD
PARIS – SAN DIEGO – SAN FRANCISCO – SINGAPORE – SYDNEY – TOKYO

Elsevier
The Boulevard, Langford Lane, Kidlington, Oxford OX5 1GB, UK
Radarweg 29, PO Box 211, 1000 AE Amsterdam, The Netherlands

First edition 2005
Reprinted 2006

British Library Cataloguing in Publication Data
A catalogue record for this book is available from the British Library

Library of Congress Cataloging-in-Publication Data
A catalog record for this book is available from the Library of Congress

ISBN–13: 978-0-08-044642-4

For information on all Elsevier publications
visit our website at books.elsevier.com

Transferred to digital print 2008
Printed and bound by CPI Antony Rowe, Eastbourne

DEDICATED
TO

G. EVELYN HUTCHINSON
(1903–1991)

Photograph provided by Images of Yale Individuals, circa 1750–2001 (RU684). Manuscripts and Archives, Yale University.

Contents

Executive Editors' Foreword

H. D. Holland

Harvard University, Cambridge, MA, USA

and

K. K. Turekian

Yale University, New Haven, CT, USA

Geochemistry has deep roots. Its beginnings can be traced back to antiquity, but many of the discoveries that are basic to the science were made between 1800 and 1910. The periodic table of elements was assembled, radioactivity was discovered, and the thermodynamics of heterogeneous systems was developed. The solar spectrum was used to determine the composition of the Sun. This information, together with chemical analyses of meteorites, provided an entry to a larger view of the universe.

During the first half of the twentieth century, a large number of scientists used a variety of methods to determine the major-element composition of the Earth's crust, and the geochemistries of many of the minor elements were defined by V. M. Goldschmidt and his associates using the then new technique of emission spectrography. V. I. Vernadsky founded biogeochemistry. The crystal structures of most minerals were determined by X-ray diffraction techniques. Isotope geochemistry was born, and age determinations based on radiometric techniques began to define the absolute geologic timescale. The intense scientific efforts during World War II yielded new analytical tools and a group of people who trained a new generation of geochemists at a number of universities. But the field grew slowly. In the 1950s, a few journals were able to report all of the important developments in trace-element geochemistry, isotopic geochronometry, the exploration of paleoclimatology and biogeochemistry with light stable isotopes, and studies of phase equilibria. At the meetings of the American Geophysical Union, geochemical sessions were few, none were concurrent, and they all ranged across the entire field.

Since then the developments in instrumentation and the increases in computing power have been spectacular. The education of geochemists has been broadened beyond the old, rather narrowly defined areas. Atmospheric and marine geochemistry have become integrated into solid Earth geochemistry; cosmochemistry and biogeochemistry have contributed greatly to our understanding of the history of our planet. The study of Earth has evolved into "Earth System Science," whose progress since the 1940s has been truly dramatic.

Major ocean expeditions have shown how and how fast the oceans mix; they have demonstrated the connections between the biologic pump, marine biology, physical oceanography, and marine sedimentation. The discovery of hydrothermal vents has shown how oceanography is related to economic geology. It has revealed formerly unknown oceanic biotas, and has clarified the factors that today control, and in the past have controlled the composition of seawater.

Seafloor spreading, continental drift and plate tectonics have permeated geochemistry. We finally understand the fate of sediments and oceanic crust in subduction zones, their burial and their exhumation. New experimental techniques at temperatures and pressures of the deep Earth interior have clarified the three-dimensional structure of the mantle and the generation of magmas.

Moon rocks, the treasure trove of photographs of the planets and their moons, and the successful search for planets in other solar systems have all revolutionized our understanding of Earth and the universe in which we are embedded.

Geochemistry has also been propelled into the arena of local, regional, and global anthropogenic problems. The discovery of the ozone hole came as a great, unpleasant surprise, an object lesson for optimists and a source of major new insights into the photochemistry and dynamics of the atmosphere. The rise of the CO_2 content of the atmosphere due to the burning of fossil fuels and deforestation has been and will continue to be at the center of the global change controversy, and will yield new insights into the coupling of atmospheric chemistry to the biosphere, the crust, and the oceans.

The rush of scientific progress in geochemistry since World War II has been matched by organizational innovations. The first issue of *Geochimica et Cosmochimica Acta* appeared in June 1950. The Geochemical Society was founded in 1955 and adopted *Geochimica et Cosmochimica Acta* as its official publication in 1957. The International Association of Geochemistry and Cosmochemistry was founded in 1966, and its journal, *Applied Geochemistry*, began publication in 1986. *Chemical Geology* became the journal of the European Association for Geochemistry.

The Goldschmidt Conferences were inaugurated in 1991 and have become large international meetings. Geochemistry has become a major force in the Geological Society of America and in the American Geophysical Union. Needless to say, medals and other awards now recognize outstanding achievements in geochemistry in a number of scientific societies.

During the phenomenal growth of the science since the end of World War II an admirable number of books on various aspects of geochemistry were published. Of these only three attempted to cover the whole field. The excellent *Geochemistry* by K. Rankama and Th.G. Sahama was published in 1950. V. M. Goldschmidt's book with the same title was started by the author in the 1940s. Sadly, his health suffered during the German occupation of his native Norway, and he died in England before the book was completed. Alex Muir and several of Goldschmidt's friends wrote the missing chapters of this classic volume, which was finally published in 1954.

Between 1969 and 1978 K. H. Wedepohl together with a board of editors (C. W. Correns, D. M. Shaw, K. K. Turekian and J. Zeman) and a large number of individual authors assembled the *Handbook of Geochemistry*. This and the other two major works on geochemistry begin with integrating chapters followed by chapters devoted to the geochemistry of one or a small group of elements. All three are now out of date, because major innovations in instrumentation and the expansion of the number of practitioners in the field have produced valuable sets of high-quality data, which have led to many new insights into fundamental geochemical problems.

At the Goldschmidt Conference at Harvard in 1999, Elsevier proposed to the Executive Editors that it was time to prepare a new, reasonably comprehensive, integrated summary of geochemistry. We decided to approach our task somewhat differently from our predecessors. We divided geochemistry into nine parts. As shown below, each part was assigned a volume, and a distinguished editor was chosen for each volume. A tenth volume was reserved for a comprehensive index:

(i) *Meteorites, Comets, and Planets*: Andrew M. Davis

(ii) *Geochemistry of the Mantle and Core*: Richard Carlson

(iii) *The Earth's Crust*: Roberta L. Rudnick

(iv) *Atmospheric Geochemistry*: Ralph F. Keeling

(v) *Freshwater Geochemistry, Weathering, and Soils*: James I. Drever

(vi) *The Oceans and Marine Geochemistry*: Harry Elderfield

(vii) *Sediments, Diagenesis, and Sedimentary Rocks*: Fred T. Mackenzie

(viii) *Biogeochemistry*: William H. Schlesinger

(ix) *Environmental Geochemistry*: Barbara Sherwood Lollar

(x) *Indexes*

The editor of each volume was asked to assemble a group of authors to write a series of chapters that together summarize the part of the field covered by the volume. The volume editors and chapter authors joined the team enthusiastically. Altogether there are 155 chapters and 9 introductory essays in the Treatise. Naming the work proved to be somewhat problematic. It is clearly not meant to be an encyclopedia. The titles *Comprehensive Geochemistry* and *Handbook of Geochemistry* were finally abandoned in favor of *Treatise on Geochemistry*.

The major features of the Treatise were shaped at a meeting in Edinburgh during a conference on Earth System Processes sponsored by the Geological Society of America and the Geological Society of London in June 2001. The fact that the Treatise is being published in 2003 is due to a great deal of hard work on the part of the editors, the authors, Mabel Peterson (the Managing Editor), Angela Greenwell (the former Head of Major Reference Works), Diana Calvert (Developmental Editor, Major Reference Works),

Bob Donaldson (Developmental Manager), Jerome Michalczyk and Rob Webb (Production Editors), and Friso Veenstra (Senior Publishing Editor). We extend our warm thanks to all of them. May their efforts be rewarded by a distinguished journey for the Treatise.

Finally, we would like to express our thanks to J. Laurence Kulp, our advisor as graduate students at Columbia University. He introduced us to the excitement of doing science and convinced us that all of the sciences are really subdivisions of geochemistry.

Contributors to Volume 8

R. Amundson
University of California, Berkeley, CA, USA

P. Brimblecombe
University of East Anglia, Norwich, UK

J. J. Brocks
Harvard University, Cambridge, MA, USA

F. S. Chapin III
University of Alaska, Fairbanks, AK, USA

V. T. Eviner
Institute of Ecosystem Studies, Millbrook, NY, USA

P. G. Falkowski
Rutgers University, New Brunswick, NJ, USA

C. M. R. Fowler
Royal Holloway, University of London, Egham, UK

J. N. Galloway
University of Virginia, Charlottesville, VA, USA

M. E. Hines
University of Massachusetts Lowell, MA, USA

R. A. Houghton
Woods Hole Research Center, MA, USA

A. H. Jahren
Johns Hopkins University, Baltimore, MD, USA

J. P. Megonigal
Smithsonian Environmental Research Center, Edgewater, MD, USA

K. H. Nealson
University of Southern California, Los Angeles, CA, USA

E. G. Nisbet
Royal Holloway, University of London, Egham, UK

S. T. Petsch
University of Massachusetts, Amherst, MA, USA

K. C. Ruttenberg
University of Hawaii at Manoa, Honolulu, HI, USA

R. Rye
University of Southern California, Los Angeles, CA, USA

J. Sanderman
University of California, Berkeley, CA, USA

H. C. W. Skinner
Yale University, New Haven, CT, USA

R. E. Summons
Massachusetts Institute of Technology, Cambridge, MA, USA

E. T. Sundquist
US Geological Survey, Woods Hole, MA, USA

P. T. Visscher
University of Connecticut/Avery Point, Groton, CT, USA

K. Visser
US Geological Survey, Woods Hole, MA, USA

Volume Editor's Introduction

W. H. Schlesinger
Duke University, Durham, NC, USA

I am uncertain of what milepost should mark the origin of biogeochemistry as a scientific discipline. I am tempted to pick Priestley's (1772) discovery that plants produce oxygen as the first recognition that life affects the chemistry of our environment. Nearly a century later, Darwin (1881) wrote an entire book on the role of worms in the decomposition of plant material, realizing that plant growth flourished only by their good deeds. The first use of the term "biogeochemistry" is often credited to Vernadsky (1926), who recognized that at the surface of the Earth there is no geochemistry that is not affected by the presence of life on Earth (Gorham, 1991). This volume of *The Treatise on Geochemistry* is dedicated to G. Evelyn Hutchinson, whose 1950 paper on the biogeochemistry of vertebrate excretion provides a masterful description of the role of birds in recycling nitrogen and phosphorus between land and sea.

A space traveler arriving at Earth would find that its chemical composition is most unusual—low concentrations of reduced gases such as methane (CH_4) persist in an atmosphere with unusually high concentrations of O_2 (Sagan *et al.*, 1993). The ocean shows a remarkably constant ratio between dissolved nitrogen and phosphorus—about 16—despite processes that add and remove nitrogen and phosphorus from seawater much more rapidly than it mixes each year (Redfield *et al.*, 1963). And the land surface stores remarkably little organic carbon despite a large annual production of it by land plants (Schlesinger, 1990). All these characteristics reflect the presence of life on Earth, which has determined the chemical composition of the environment of life today.

An examination of the sedimentary record shows fluctuations in the activity of the biosphere, which we will define here as the total mass of living organisms on Earth. The chemical composition of the atmosphere and oceans has also changed through geologic time, strongly affected by life, which appeared ca. 3.8 billion years ago (Schidlowski, 2001). Some characteristics of the Earth's surface chemistry seem to show cyclic behavior, while others, such as the rise of oxygen in the atmosphere, seem to have progressed from low initial values to the higher, but relatively stable, levels of today (Chapter 8.11). As we examine the composition of molecules, including genetic materials left in the sedimentary record, we can read the history of life on Earth—its diversity, its metabolic sophistication, and its imprint on Earth's surface chemistry (Benner *et al.*, 2002, Newman and Banfield, 2002; Chapter 8.03).

In models of Earth's biogeochemistry, showing the movement of materials between the oceans, atmosphere and land surface, we see that today a single species, *Homo sapiens*, dominates the flow of many chemical elements on our planet. Humans use roughly half of the available freshwater on Earth (Postel *et al.*, 1996), and future shortages of good-quality freshwater will be critical to much of our population globally (Vorosmarty *et al.*, 2000). Humans mobilize nitrogen and phosphorus into their global biogeochemical cycles at a rate that roughly doubles the natural availability of these elements to the biosphere each year (Schlesinger, 1997). The resulting impacts are well known, ranging from the eutrophication of lakes with excessive phosphorus loading to the enrichment of entire watersheds and their coastal waters by excessive loads of nitrogen fertilizer (Chapter 8.12). Humans have nearly doubled the annual mobilization of sulfur in its global cycle, and our effect is manifest in the acid deposition that scrubs SO_2 from the atmosphere in regions downwind of industrial activity (Chapter 8.14).

Much biogeochemical research has focused on the carbon cycle, in which, by the extraction of carbon-based fossil fuels from the Earth's crust, humans add >22 Gt of carbon dioxide (more than 6×10^{15} g C) to the atmosphere each year (Chapter 8.10). This global flux of carbon from the Earth's crust is equivalent to 5% of the total annual exchange of CO_2 to and from the atmospheric reservoir by photosynthesis. Looking broadly at the periodic table, we see that humans have affected the global cycle of nearly every chemical element with economic value, dominating the mobilization of metallic elements from the Earth's crust (Bertine and Goldberg, 1971), their movement through the atmosphere (Lantzy and MacKenzie, 1979), and their content in river flow (Martin and Meybeck, 1979).

This volume of the *Treatise on Geochemistry* assesses our knowledge of Earth-system function at the beginning of the new millennium. We look at the Earth's past, in its sedimentary records, for an index of how life has affected Earth's chemistry through its history (Chapter 8.01). We look at the present to assess the main processes maintaining the relatively stable conditions of Earth's chemistry during the Holocene and the extent of current human impact. We look at the future, using our best ability to model how the Earth system works and to make predictions of where we may be headed as we enter a new era, the *anthropocene*...

Biogeochemistry is an interdisciplinary science. It has its roots in biochemistry—a science now informed by genomics—to understand the metabolic reactions of organisms, which obtain raw materials from the environment and cast their chemical wastes into nature. Cycles of reduction and oxidation—i.e., the flow of electrons in metabolism—control Earth's biogeochemistry (Chapter 8.02). Recognizing that past life has affected the chemical characteristics and mineral deposits of sedimentary rocks, the science of geomicrobiology grew as an early subdiscipline of biogeochemistry (Chapter 8.04). We now know that life can also affect the composition of rocks deep in the Earth—long regarded as the academic province only of "hard-rock" geologists (Fisk *et al.*, 1998). Conversely, geobotanical prospecting at the Earth's surface uses the distribution and chemistry of today's plants to indicate mineral deposits in the underlying crust (Brooks, 1972). To the extent that human activities control the chemistry on Earth's surface, biogeochemistry is now informed by studies of economics, politics, and human nature.

Dating back to the early studies by Redfield *et al.* (1963), biogeochemists have often focused on the linkage between chemical elements—both in biochemistry and in their global biogeochemical cycles (Reiners, 1986, Sterner and Elser, 2002). Signal transduction mediated by phosphorus seems to determine the activation of N-fixation in bacteria (Stock *et al.*, 1990). Available iron determines the rate of pyritization in marine sediments, hence the O_2 content in Earth's atmosphere (Boudreau and Westrich, 1984). And the rate of net primary production by plants is closely regulated by the availability of nitrogen and phosphorus that are used to build plant tissues (Chapters 8.05 and 8.06). Expected ratios between elements determine the basic stoichiometry for the biosphere, allowing us to predict its response to alterations of nutrient availability, such as changes in N-fixation in lakes (Howarth *et al.*, 1988). The production of trace gases (e.g., CH_4) in anaerobic sediments is determined by the relative amount of other chemical species (e.g., SO_4^{2-}) that may be present (Chapter 8.08). Thus, biogeochemistry is inherently interdisciplinary—one cannot study biogeochemistry one element at a time.

Biogeochemistry has matured as a scientific discipline as we have come to recognize that the current human impact on our planet may disrupt the stable chemistry of our evolutionary environment, which is at least partially determined by the diversity of species that occupy this planet with us. The level of CO_2 in the atmosphere, the amount of precipitation that falls on land, the content of nitrogen and phosphorus in rivers, and the silicon that is deposited in ocean sediments are all determined by biota—ranging from bacteria to higher plants. Rare as well as ubiquitous species are involved; for example, sea turtles, which transport nutrients from marine to terrestrial ecosystems, provide important inputs to coastal maritime ecosystems (Bouchard and Bjorndal, 2000). Life's diversity—the *bio* in biogeochemistry—performs a great service to us, which we must understand better if we are to preserve its function.

Among the tasks facing today's biogeochemists is to understand fully the global biogeochemical cycles of water and the various chemical elements and the human impacts on each of them. A study of political history shows compelling motivation for doing so. When we were able to articulate clear and complete budgets for the natural and perturbed global cycles of lead (Pb; see Chapter 9.03), the need to remove lead from gasoline appeared starkly before policy makers who heard of the effects of excessive lead on human health. We need to develop the same level of understanding of the global carbon cycle, and the nagging persistence of the "missing sink" in atmospheric CO_2 budgets weakens our case for immediate policy actions to curb CO_2 emissions and stem global warming. Policy makers will also delay action on global problems concerning nitrogen, phosphorus, and sulfur, until their global cycles are better known. Biogeochemistry must emerge as the critical scientific discipline that informs planetary stewardship through the rest of this century. This volume shows where we are and the challenges that lie before us.

REFERENCES

Benner S. A., Caraco M. D., Thomson J. M., and Gaucher E. A. (2002) Planetary biology—paleontological, geological, and molecular histories of life. *Science* **296**, 864–868.

Bertine K. K. and Goldberg E. D. (1971) Fossil fuel combustion and the major sedimentary cycle. *Science* **173**, 233–235.

Bouchard S. S. and Bjorndal K. A. (2000) Sea turtles as biological transporters of nutrients and energy from marine to terrestrial ecosystems. *Ecology* **81**, 2305–2313.

Boudreau B. P. and Westrich J. T. (1984) The dependence of bacterial sulfate reduction on sulfate concentration in marine sediments. *Geochim. Cosmochim. Acta* **48**, 2503–2516.

Brooks R. R. (1972) *Geobotany and Biogeochemistry in Mineral Exploration.* Harper and Row, New York.

Darwin C. (1881) *The Formation of Vegetable Mould, through the Action of Worms.* John Murray, London.

Fisk M. R., Giovannoni S. J., and Thorseth I. H. (1998) Alteration of oceanic volcanic glass: textural evidence for microbial activity. *Science* **281**, 978–980.

Gorham E. (1991) Biogeochemistry: its origins and development. *Biogeochemistry* **13**, 199–239.

Howarth R. W., Marino R., and Cole J. J. (1988) Nitrogen fixation in freshwater, estuarine, and marine ecosystems: 2. Biogeochemical controls. *Limnol. Oceanogr.* **33**, 688–701.

Hutchinson G. E. (1950) The biogeochemistry of vertebrate excretion. *Bulletin of the American Museum of Natural History*, Number 96, 554pp.

Lantzy R. J. and MacKenzie F. T. (1979) Atmospheric trace metals: global cycles and assessment of man's impact. *Geochim. Cosmochim. Acta* **43**, 511–525.

Martin J.-M. and Meybeck M. (1979) Elemental mass-balance of material carried in major world rivers. *Mar. Chem.* **7**, 173–206.

Newman D. K. and Banfield J. F. (2002) Geomicrobiology: how molecular-scale interactions underpin biogeochemical systems. *Science* **296**, 1071–1077.

Postel S. L., Daily G. C., and Ehrlich P. R. (1996) Human appropriation of renewable fresh water. *Science* **271**, 785–788.

Priestley J. (1772) Observations son different kinds of air. *Phil. Trans. Roy. Soc. London* **62**, 147–152.

Redfield A. C., Ketchum B. H., and Richards F. A. (1963) The influence of organisms on the composition of seawater. In *The Sea* (ed. M. N. Hill). Wiley, New York, vol. 2, pp. 26–77.

Reiners W. A. (1986) Complementary models for ecosystems. *Am. Natural.* **127**, 59–73.

Sagan C., Thompson W. R., Carlson R., Gurnett D., and Hord C. (1993) Search for life on Earth from the Galileo spacecraft. *Nature* **365**, 715–721.

Schidlowski M. (2001) Carbon isotopes as biogeochemical recorders of life over 3.8 Ga of Earth's history: evolution of a concept. *Precamb. Res.* **106**, 117–134.

Schlesinger W. H. (1990) Evidence from chronosequence studies for a low carbon-storage potential of soils. *Nature* **348**, 232–234.

Schlesinger W. H. (1997) *Biogeochemistry: An Analysis of Global Change.* Academic Press, San Diego.

Sterner R. W. and Elser J. J. (2002) *Ecological Stoichiometry: The Biology of Elements from Molecules to the Biosphere.* Princeton University Press, Princeton.

Stock J. B., Stock A. M., and Mottonen J. M. (1990) Signal transduction in bacteria. *Nature* **344**, 395–400.

Vernadsky V. (1926) *The Biosphere* (English translation). Copernicus, Springer, New York.

Vorosmarty C. J., Green P., Salisbury J., and Lammers R. B. (2000) Global water resources: vulnerability from climate change and population growth. *Science* **289**, 284–288.

8.01
The Early History of Life

E. G. Nisbet and C. M. R. Fowler
Royal Holloway, University of London, Egham, UK

8.01.1 INTRODUCTION

8.01.1.1 Strangeness and Familiarity—The Youth of the Earth

The youth of the Earth is strange to us. Many of the most fundamental constraints on life may have been different, especially the oxidation state of the surface. Should we suddenly land on its Hadean or early Archean surface by some sci-fi accident, we would not recognize our home. Above, the sky may have been green or some other unworldly color, and above that the weak young Sun might have been unrecognizable to someone trying to identify it from its spectrum. Below, seismology would show a hot, comparatively low-viscosity interior, possibly with a magma ocean in the deeper part of the upper mantle (Drake and Righter, 2002; Nisbet and Walker, 1982), and a core that, though present, was perhaps rather smaller than today. The continents may have been small islands in an icy sea, mostly frozen with some leads of open water, (Sleep *et al.*, 2001). Into these icy oceans, huge protruding Hawaii-like volcanoes would have poured out vast far-spreading floods of komatiite lavas in immense eruptions that may have created sudden local hypercane storms to disrupt the nearby icebergs. And meteorites would rain down.

Or perhaps it was not so strange, nor so violent. The child is father to the man; young Earth was mother to Old Earth. Earth had hydrogen, silicate rock below and on the surface abundant carbon, which her ancient self retains today. Moreover, Earth was oxygen-rich, as today. Today, a tiny part of the oxygen is free, as air; then the oxygen would have been in the mantle while the surface oxygen was used to handcuff the hydrogen as dihydrogen monoxide. Oxygen dihydride is dense, unlikely to fly off to space, and at the poles, rock-forming. Of all the geochemical features that make Earth unique, the initial degassing (Genesis 2:b) and then the sustained presence of liquid water is the defining oddity of this planet. Early Earth probably also kept much of its carbon, nitrogen, and sulfur as oxide or hydride. And, after the most cataclysmic events had passed, ~4.5 Ga ago, for the most part the planet was peaceful. Even the most active volcanoes are mostly quiet; meteorites large enough to extinguish all dinosaurs may have hit as often as every few thousand years, but this is not enough to be a nuisance to a bacterium (except when the impact boiled the ocean); while to the photosynthesizer long-term shifts in the solar spectrum may be less of a problem than cloudy hazy days. Though, admittedly, green is junk light to biology, the excretion from the photosynthetic antennae, nevertheless even a green sky would have had other wavelengths also in its spectrum.

Most important of all, like all good houses, this planet had location: Earth was just in the right spot. Not too far from the faint young Sun (Sagan and Chyba, 1997), it was also far enough away still to be in the comfort zone (Kasting *et al.*, 1993) when the mature Sun brightened. As many have pointed out, when Goldilocks arrived, she found everything just right. But what is less obvious is that as she grew and changed, and the room changed too, she commenced to rearrange the furniture to make it ever righter for her. Thus far, the bears have not arrived, though they may have reclaimed Mars from Goldilocks's sister see (Figure 1).

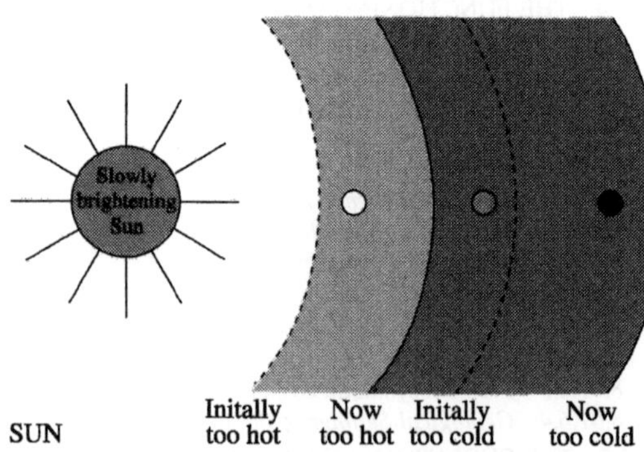

Figure 1 The habitable zone (Kasting *et al.*, 1993). Too close to the Sun, a planet's surface is too hot to be habitable; too far, it is too cold. Early in the history of the solar system, the Sun was faint and the habitable zone was relatively close; 4.5 Ga later, with a brighter Sun, planets formerly habitable are now too hot, and the habitable zone has shifted out. Note that boundaries can shift. By changing its albedo and by altering the greenhouse gas content of the air, the planet can significantly widen the bounds of the habitable zone (Lovelock, 1979, 1988).

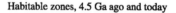

8.01.1.2 Evidence in Rocks, Moon, Planets, and Meteorites—The Sources of Information

The information about the early history of life comes from several sources: ancient relics, modern descendants, and models. The ancient material is in the rocks, in meteorites, in what we can learn from other planets, and in solar system and stellar science. The Lucretian view of a planet, ramparts crumbling with age, may apply to Mars, but as Hutton realized, virtually all of the surface of the Earth is renewed every few hundred million years, and if it were not so, life would die from lack of resources. But in the tiny fragment that is not renewed, relics of early life remain. Some of these relics are direct—specks of carbon, or structures of biogenic origin. Other relics are indirect: changes in the isotopic ratio of inorganic material or oxidation states of material that is of inorganic origin. Yet other information is simply scene-setting: evidence, for example, that water was present, or that volcanism was active.

Extraterrestrial sources of evidence are also important. From Venus there is evidence that a planet can have water oceans and then lose all its hydrogen (Donahue *et al.*, 1982; Watson *et al.*, 1984). From Mars there is evidence that planets can die geologically, and become unable to renew their surface by tectonics and volcanism. Perhaps they can also die biologically. From moons of outer planets comes evidence that a wide variety of early conditions was possible. Meteorites (Ahrens, 1990; Taylor, 2001) provide clear signs that in the early part of the history of the solar system there could have been significant exchange of surface material between the inner planets. Study of the Sun and of sun-like stars demonstrates that even stable stars do change, and over the past 4.6 billion years the Sun has significantly increased in power (Sagan and Chyba, 1997), and altered in spectrum.

8.01.1.3 Reading the Palimpsests—Using Evidence from the Modern Earth and Biology to Reconstruct the Ancestors and their Home

"Ontogeny," the old saying went, "recapitulates phylogeny." We each start as a couple of lengths of DNA, one loose with a few attendants, the other comfortable in a pleasant container full of goodies, itself held in a warm and safe maternal universe. The DNA-, the RNA-based processing, and information-transfer systems, and the protein machinery of the cell all carry historical information. Every human cell lives in its own seawater, the blood—we came from a warm kindly ocean. Every oxygen-handling blood cell carries iron: we learned this trick somewhere that our ancestors could acquire iron, surely without the sophisticated metal-gathering equipment that is provided by modern biochemistry. At the very heart of the information transfer in the cell is the ribosome: a massive (compared to other enzymes) RNA-based super-enzyme that in a strange way is both chicken and egg and, though much modified by evolution, is surely of the very greatest antiquity.

Modern life comes in very many forms: animals, plants, and single-celled eukaryotes in the *Eucarya* domain; prokaryotes in two great domains, *Archea* and *Bacteria* (Woese, 1987; Woese *et al.*, 1990), and *not-life* viruses. Some not-life is even anthropogenic: the wild-type polio virus that used to be found in water bodies is now replaced in the pools and rivers of America and Europe by the altered vaccine-type virus. From all this information, deductions can be made. Clearly, multicelled life came from single-celled life; less obviously but most probably each of our cells carries mitochondria that are descended from symbiotic purple bacteria. Plants, in addition, carry chloroplasts that are descended from partner cyanobacteria.

There is an enormous wealth of this type of information that is only just beginning to be deciphered. Indeed, deciphering the molecular record (Zuckerkandl and Pauling, 1965) may be the best route to understanding of Archean palaeontology. Geological study interacts with this, both by calibrating the timing of the evolutionary steps (e.g., by dating the arrival of the multicelled organisms), and secondly by identifying the impact of each step (e.g., the onset of oxygenic photosynthesis). Both molecular and rock-based studies are needed: without the rock information, molecular evidence can lead to (and has done so) very erroneous deduction; equally, the geological evidence cannot of itself give much detail about major steps. But there is also a danger of circularity of reasoning: just because something looks plausible biochemically, it is possible to reinterpret the geological evidence to fit, but wrongly; conversely, weakly supported geological models can on occasion unduly sway interpretation of complex and nonunique molecular evidence.

8.01.1.4 Modeling—The Problem of Taking Fragments of Evidence and Rebuilding the Childhood of the Planet

Model building is part of all science: lovely falsifiable hypotheses are built and then broken on the cold facts. Certain key components are common to all models of life's origins—water (though not necessarily in an ocean: aerosols are possible hosts of proto-life); inorganic supplies of thermodynamic drive (i.e., interface settings where two or more different conditions are accessible); ambient temperatures in the 0–110 °C range.

Models of the early history of life come in two broad categories—models of the origin and first development of life itself and models of the environmental settings of that life. The geologist can contribute much more to the second class of model than to the first: from the geological evidence it is possible to make reasonable models of the early planets, including their surface condition and the supply of chemicals and nutrients from the interior and from space.

All information is fragmental, and the further back in time, the less the information. But enough is left that reasonable guesses may be made about the surface conditions of the four inner planets and the Moon as they evolved in the first two billion years of the solar system's history. These models set the scene for the biochemists: without them, the biochemical deductions are unconstrained and can be wrong (e.g., the primaeval "soup" is geologically unlikely). Thus, the debate over the various models of origin is avoided here: for that, seek out google.com. The focus instead is on what geologists and geochemists can usefully contribute.

8.01.1.5 What Does a Planet Need to be Habitable?

Venus may have been in the right place once, and is the right size, but in the long term it was too close to the Sun. Mars may be in a tolerable place, and on occasion with liquid water (Baker, 2001; Carr, 1996), but has a small heart, almost dead of cold. Only Earth had long-term location and is large enough to keep an active interior.

What are the requirements? First, liquid water. It is difficult to imagine biology that does not include water. It may be possible (indeed, someday computers may produce some sort of water-free inorganic sentience that achieves genetic take-over from organic life) but not in Nature as we know it. Externally, the planet needs to be far enough from the Sun not to overheat, close enough not to freeze entirely. The allowable bounds of the habitable zone (Kasting *et al.*, 1993) are wide at any one moment, given the range of temperature control provided by changes in atmospheric greenhouse heat trapping, but these bounds progressively shift outwards as the Sun evolves and brightens. Thus, while Venus was probably within the bounds of the habitability zone early in the history of the solar system with a faint young Sun, and covered by oceans, it may today be too close to the bright old Sun to allow life. Even if water were added it would be difficult to sustain liquid water on the planet with any plausible life-supporting atmosphere. However, that is not to say it is uninhabitable: some day humanity may well add water supplies from an outer solar system source and hang aluminum foil mirrors around the planet to reduce sunlight input. Mars, alternatively, is too cold to sustain liquid water, but could in future be warmed by chlorofluorocarbon and methane greenhouse gases, such that it sustained puddle oceans. These thought-experiments with Venus and Mars demand teleological action, not possible in Darwinian evolution, but there are persuasive arguments that feedbacks from non-teleological life have carried out very similar processes on Earth over 4 Ga (Lovelock and Margulis, 1974; Lovelock, 1979, 1988).

Planets also need to be geologically active to sustain life over long periods. Nature needs to renew her face continuously, or the chemical and thermodynamic resources behind life, especially early life, are rapidly exhausted. For example, DNA-based life is built on phosphates. If the available surface phosphorus supply is exhausted, and not renewed continually by volcanism and tectonics, then life must become hungry for phosphorus and eventually die out. Life depends on a small number of essential house-keeping proteins, and many of these proteins use metals: if the geological metal supply ended, the proteins would not be formed and life would be unsustainable. This need for "supply" places constraints on the physical evolution of a planet, if it is to be capable of sustaining life over many aeons. Planets vary (Taylor, 2001). The Moon is too small. It was once active, but now has died. Mars is just about dead. Mercury may once have been larger but now seems to be a barren metal-rich relict of the innards of a planet. Jupiter and the outer gas giants are too large. Some of the tidally heated great moons (some with radii comparable to Mars) of the gas giants do remain very active geologically and offer possible homes. But of the internally warm bodies that have a Sun-warm outer surface, only Earth and Venus are just the right size.

8.01.1.6 The Power of Biology: The Infinite Improbability Drive

What is just right in one moment becomes wrong in the next. The porridge that Goldilocks tasted would have been perfect in the first mouthful, but a little later, especially if she ate slowly as her mother would have taught her, it would have cooled.

Biology has the power to sustain, to draw out, its environmental conditions (Lovelock, 1979, 1988), and indeed to remake them in an improbable path. Swiss travelers do not descend peaks by jumping over the cliffs. Instead they use cable cars, and as they descend they help others to ascend: only a small input of energy is needed to overcome frictional losses. Indeed, consider a hypothetical cable car that had an attached snowtank, filled from a snowfield at the top. At the bottom the snow would be dropped off, so that the rising car was always somewhat lighter than the descending car. This system could work without extra input perpetually carrying tourists up to the peak and down again, as the potential energy transfer would make up for the frictional losses. Indirectly, this is solar power: the Sun lifts the water, evaporating it from the bottom and replacing it back on the top as snow.

Most microbial processes are like that—they move enormous numbers of traveling chemical species on cogways up and down the thermodynamic peaks and valleys with only small extra inputs of externally sourced energy. Moreover, at the intermediate stations part-way up (or down) the peaks, the microbial processes link with innumerable smaller cable car systems that scatter metabolic tourists around the ecological mountain sides in a complex web of ascents, descents, and lateral movements. Thus, biology creates local order, primarily by using the high quality of sun-given energy, to exploit and create redox contrast between the surface of the Earth and its interior.

8.01.2 THE HADEAN (~4.56–4.0 Ga AGO)

8.01.2.1 Definition of Hadean

The Hadean is the first of the four aeons of Earth history (Nisbet, 1991). Aeons are the largest divisions of geological time: Hadean, Archean, Proterozoic, Phanerozoic. The first and last aeons are "short" (relatively, if 560 Ma can be called short); the middle two are billions of years long. The Hadean was the period of the formation of the Earth, from the first accretion of planetesimals at the start of the Hadean, to the end of the aeon, when the Earth was an ordered, settled planet, with a cool surface under oceans and atmosphere, and with a hot active interior mantle and core.

The bounds of the Hadean have never been properly defined. The birth of the Earth is the start of the Hadean, but is this the moment of the beginning of the solar system, or the moment when the first significant planetesimals collected to begin the accretion? Fortunately, this is not much more than an academic discussion—the time of formation of the oldest material in meteorites is usually taken to be representative of the start of accretion. Very roughly, 4.56 Ga is taken as the start, where 1 Ga is one thousand million (10^9) years.

The end of the Hadean is more difficult to define, though the definition is more useful to the geologist, as this is within the terrestrial geological record. For the interim, as a rough guide, 4 Ga is used as a working definition of the Hadean/Archean boundary. But this is unsatisfactory—a random number. The choice of the start of life as the defining moment for the boundary between the Hadean and Archean has great appeal and should surely be preferred. Fortunately, at present the best guess for the origin of life is also "somewhen around 4 Ga ago."

8.01.2.2 Building a Habitable Planet

The solar system accreted from a dust cloud, formed after a supernova explosion. From this primitive solar nebula condensed the Sun and the planets. Some of the oldest objects in the solar system yet found are Ca–Al-rich inclusions in meteorites, ~4.566 Ga old (Allegre *et al.*, 1995). It is possible that these grains predate the solar nebula and may have been formed in the expanding envelope of the supernova explosion (Cameron, 2002).

The formation of planetesimals may have been very rapid after the initial formation of the solar nebula. Objects as large as Mars would have grown within 10^5 yr (Weatherill, 1990). The core of the asteroid Vesta may have formed within only 3–4 Myr, and lavas flows on its surface may have occurred at this time also (Yin *et al.*, 2002). Bodies like Vesta would have collided rapidly, aggregating their cores to form larger planetoids and then planets. The date of core formation in the Earth remains controversial but may have been as little as 30 Myr or less after the birth of the solar system (Kleine *et al.*, 2002). Yin *et al.* (2002) suggest that the aggregation of the Earth's core took place within 29 Myr. The core of Mars may have formed as early as within 13 Myr.

The special events of this planet's accretion (Newsom and Jones, 1990; Weatherill, 1990; Ahrens, 1990; Taylor, 2001) were crucial in making Earth habitable over billions of years. Segregation of the core physically separated reduction power in the iron-rich center of the Earth, from a more oxidized mantle. Simultaneously, the early events controlling the surface environment made possible the development of a habitable ocean/atmosphere system.

The most important single physical event took place roughly 4.5 Ga ago, 25–30 Myr after the birth of the solar system. At this stage, Earth was probably a substantial fraction of its present mass, with a segregated core. Sunwards of Earth, Venus and Mercury had formed; outwards, were Mars-like planets. Then, the Earth suffered its largest collision: a defining moment in habitability.

One model is that a planet at least double the mass of Mars hit a half-formed Earth in a double collision (Cameron, 2002); an alternative model is that a Mars-sized body hit the 90%-formed Earth (Canup and Asphaug, 2001). When the impact took place, the Earth was transformed. Internally, it would have been melted, even if primordial radiogenic and infall heat had not already melted it. The already-formed iron core of the impactor would have crashed to the center to join the core of the Earth. This large core, with its solid center and molten outer region, gave Earth its distinctive magnetic field, and life-protecting van Allen belts. Arguably, planets without a strong protective field (e.g., Venus), are initially uninhabitable as the surface environment may be too severe for unstable early genomes.

The surface of Earth was completely changed. Any deep primordial atmosphere/ocean, possibly rich in noble gases, would have been removed by the impact. Presumably, the event was followed by further cometary infall and further degassing from the interior to produce our present thin water-dominated inventory of volatiles. By this stage, the inner solar system was probably swept clear of volatiles and was a relatively gas-poor environment. Volatile influx would have come perhaps in larger planetesimals infalling from the outer solar system.

The mechanical effect of the impact was that the Earth was tilted, creating winter and summer. This is very important in distributing heat evenly across the surface, as the intensity of solar radiation falling on any particular place varies in the annual cycle. Even more important for the habitability of the Earth is the spin: much faster immediately after the impact but now slowed by aeons of tidal friction to give the 24 h day. Thus there is no hot day-side and cold night-side, but an even illumnation. Moreover, the night-day cycle allows a variety of photosynthetic/respiratory cycles in cells, and contributes greatly to the diversity of biota.

The Moon was created from the mantle-derived ejecta. Physically, over the aeons this may have played a useful sheltering role in protecting the planet from some meteorite impacts. Arguably more important, the presence of the Moon leads to the tides. These create the intertidal and near-subtidal habitat with rapidly varying geochemical settings, from wet submarine to dry subaerial, in which sediment is repeatedly flushed with fluid. Such cyclically varying habitats may have been vital in the early evolution of microbial biofilms and eventually microbial mats.

Other planets had varying histories (Taylor, 2001). Mercury also had a major collision, possibly being hit by an object ~0.2 Mercury masses, removing much of its silicate mantle and leaving a planet of high intrinsic density, with a major core and a thin rocky mantle, an uninhabitable planet. Mars and Venus had kinder gentler histories. On neither did a great impact eject splat; neither planet gained a significant Moon. Though subject to geological or impact catastrophe, both planets evolved sustainable systems within the constraints of kinetics and the thermodynamics of equilibrium; only Earth produced an intrinsically unsustainable disequilibrium system.

On Mars (a tenth of Earth mass and 38% of its radius), the present water inventory is much less, enough to cover the planet to a few tens of meters: puddle oceans (Carr, 1996). On Venus, which must have been very nearly Earth's twin prior to the giant impact on Earth (0.815 modern Earth mass, 95% of its radius), the atmosphere evolved to its present runaway CO_2 greenhouse. There has been much speculation about early Venusian oceans, perhaps some kilometers deep, but possibly only a few meters if Venus formed too close to the Sun to inherit a large water inventory (see Taylor, 2001 for a brief summary of this dispute).

The main part of the accretion of the Earth can be considered complete by ~4.45 Ga. By this stage most of what now makes up the Earth was in place. The then much nearer Moon orbited close by. The Earth would have been molten except for a thin rocky outer carapace, possibly of broadly basaltic composition (komatiitic basalt; or even komatiite?). A large magma ocean may have persisted in the mantle for some time in the Hadean or even longer. Within the Earth, ongoing late precipitation of the core may have continued, with reaction between water in the mantle and infalling iron, adding oxygen to the iron, and giving a mantle source of hydrogen that may have made its way eventually to the surface via mantle plumes and thence volcanoes.

The composition of the Earth is unique, subtly different from the other rocky planets, and this suggests that different parts of the material of the inner solar system went to make each planet (Drake and Righter, 2002). The origin of Earth's water is particularly interesting (Yung *et al.*, 1989). A significant fraction of Earth's early hydrogen endowment may have been lost to space in short-lived steam greenhouse events. Seawater has a D/H ratio of 150×10^{-6} in contrast to Mars water which has D/H of 300×10^{-6}. Perhaps Mars lost more hydrogen to space, enriching D, but it also may be possible that the Martian interior has water of very different D/H, since cooler Mars has outgassed less than Earth. One possibility is that temperatures were high in the inner part of the accretion disk: thus the Earth may have accreted as a dry planet, with water and carbon compounds delivered after the main accretion by comets and meteorites. Alternately, Earth did indeed accrete with a significant water content, and some geochemical evidence suggests the early magma ocean was hydrous (Drake and Righter, 2002).

Late Hadean Earth (say 4.2 Ga to 4 Ga ago) was thus very unusual among the inner rocky planets, in its Moon, spin, tilt, likely magnetic field, and especially in its inventory of water and its location in the "habitable band." Such a planet is not improbable, given the allowable common accidents of accreting planets by collision, but perhaps may be found to be rare as knowledge of distant extra-solar planetary systems increases.

8.01.2.3 The Hadean Record

Jack Hills and Mt. Narryer, Western Australia. Some of the oldest material found on Earth consists of a few crystals of detrital zircon that are now preserved in quartzites in the Mt. Narryer and nearby Jack Hills area, Western Australia (Compston and Pidgeon, 1986, Wilde *et al.*, 2001, Halliday, 2001). The host sediment is ~3.3–3.5 Ga old, but the some of the zircons themselves are up to 4.2–4.4 Ga old (Figure 2). There are several implications of the discovery. First (also shown in many other successions) that by 3.3 Ga ago, in the mid-Archean, there was already old continental crust being eroded and redeposited by water. Second, abundant zircons are typical of rocks broadly characteristic of continental crust. This line of reasoning thus suggests granitoid rocks and continental crust were present in the Hadean. Intuitive reasoning would suggest komatiitic and basaltic rocks would be expected to be typical of 4.4 Ga ago crust, rather than granitoids, but the existence of Hadean zircon implies otherwise, at least locally in what is now Western Australia. Moreover, to form granitoid nowadays, subduction of old hydrated oceanic plate is needed: water is needed to make granites, and subduction is needed to supply the water (Campbell and Taylor, 1983). Did subduction occur as early as 4.4 Ga ago, and did oceans of water exist to hydrate the crust? Oxygen isotope evidence (Wilde *et al.*, 2001; Mojzsis *et al.*, 2001) supports the deduction that oceans of liquid water were indeed present. The zircons contain isotopically heavy oxygen: suggesting derivation from liquid surface water. This is speculation, and just as one swallow does not make a summer, one zircon does not make either a continent or an ocean of water (Moorbath, 1983). Yet the question remains open: did Hadean continents exist, and oceans, and were hydrothermal systems present on continental land surfaces around andesitic volcanoes, fed by water-mediated subduction?

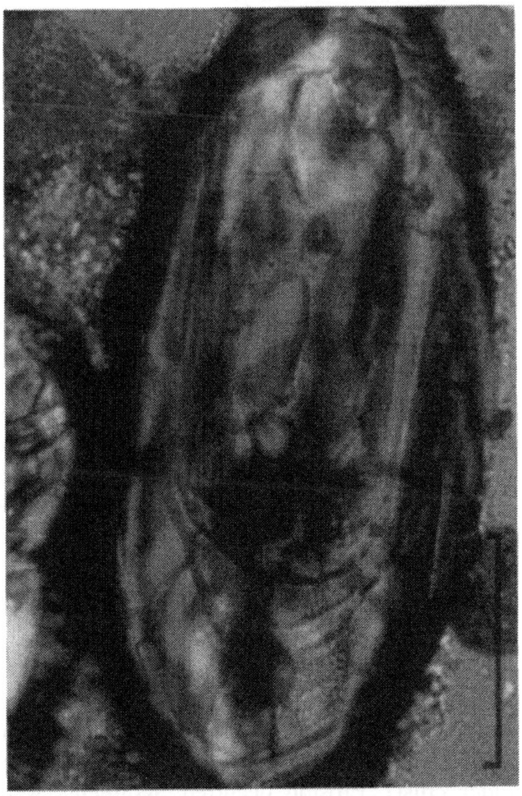

Figure 2 Zircon grain, in part ~4.3 Ga old (Compston and Pidgeon, 1986). Jack Hills, Western Australia. Scale bar is 100 μm long.

Acasta Gneiss, Canada. Next oldest is the Acasta Gneiss, close to 4 Ga old (Bowring *et al.*, 1989). This is a rock, of sorts, though highly deformed and metamorphically recrystallized. The oldest rocks form a small part of a 20 km^2 terrain of old rocks. There are various such terrains worldwide: examples include the Nain province in Labrador (~3.9 Ga); the Napier complex in Antarctica (up to ~3.7 Ga); and the Narryer complex, Australia (host rock up to ~3.7 Ga, hosting the older zircons). Some of these terrains are up to several thousand square kilometers, though the datable older rocks may only be a small proportion of the whole. The implication is that massifs of continental crust at least up to the size of, say, Luxembourg or Rhode Island existed in the latest Hadean and earliest Archean.

There is no evidence for the existence of life before 4 Ga ago. Even if a living organism had appeared, life would probably have been obliterated within a few million years, killed in the intense Hadean bombardment. This was a time when from time to time (say every few million to tens of millions of years, large meteorite impact events would have occurred that so heated the oceans and the atmosphere as to make the Earth briefly uninhabitable, sterilized at several hundred °C (Sleep *et al.*, 2001).

8.01.2.4 When and Where Did Life Start?

Enough has been said of the origin of life to show that the problem is as far from solution as it was in Charles Darwin's time. The debate continues (Line, 2002). The geologist can make little contribution to this debate, except to point out possible habitats where the first life could have been born.

There are many possibilities: in the air, in the sea, on the shallow seafloor, on the deep seafloor, near on-land hydrothermal systems around andesite volcanoes (variable, intermediate to low pH), near on-land hydrothermal systems around komatiite volcanoes and hot ultramafic rocks (alkaline), near deep-water hydrothermal systems (acid), near carbonatite-driven hydrothermal systems (which could be phosphorus-rich), in hydrothermal systems under ice caps, in shallow-water tidal muds, anywhere else that is fancied.

There are also five planets on which life could have begun (Nisbet and Sleep, 2001, 2003). Earth is the most likely, as it is the only place where Cartesian logic suggests life exists today. Next most likely on the list is Mars, which could at one stage have had an early wet environment under a strong greenhouse. Mars would have been hit by many impacts capable of ejecting relatively unshocked rocks that could have carried a living cell to Earth, surviving the transit frozen in space. There would have been a numerically vast early flux of such rocks in the Hadean, and it is thus very reasonable to infer that *if* life had begun on Mars, it would have been transferred to Earth. The logic that applies to Mars also applies to Venus, except that it is a very much deeper gravity well, and thus the outward flux of ejecta would have been much less, and those ejecta would be more shocked. The Moon is a possible though unlikely candidate, early on. Finally, a candidate is the impactor planet that hit the Earth. This Mars-sized object could have hosted life. On the great impact, ejected cells could have gone into space, either seeding Mars or much later falling back to Earth or new Moon. The most likely first homes are Earth or Mars; the other candidates are varying shades of improbable, only entertained because life is itself so improbable.

8.01.3 THE ARCHEAN (~4–2.5 Ga AGO)

8.01.3.1 Definition of Archean

The inter-aeon boundary between the Hadean and Archean is presently not defined (Nisbet, 1991). There are various options: (i) the date of the first life on Earth; (ii) the date of the last common ancestor; (iii) a "round" number, such as exactly 4 Ga—4,000,000,000 years ago; (iv) the oldest record of a terrestrial rock (~4 Ga ago); (v) the oldest record of a terrestrial mineral crystal (~4.3–4.4 Ga ago).

Each option has attractions and problems. The choice of a "round number" goes sharply against long-held stratigraphic logic, which firmly maintains any definition should be "in the rock." Dating calibrations shift when decay constants are remeasured and can be made more precise: such changes would reclassify material across the boundary. But a definition rooted in rock does not shift. The choice of a particular "oldest" rock or "oldest" mineral has more logic, but inevitably the candidate would be supplanted as a new "oldest" is discovered.

Life-based definitions are more satisfying. After all, the word "Archean" comes from the Greek for beginning: St. John's gospel starts with the words "In the Archae…." One option is the start of life: it is not clear when this was, yet, but given life's impact on carbon isotopes, it is perhaps not over-optimistic to hope that the geological record may eventually provide some insight into when life began. A second option—perhaps better—is suggested by phylogenetic studies that infer a *last common ancestor* of life—the cell or group of cells from which all modern cells are descended (Woese, 1987, 1999). Any such successful cell would spread rapidly across the globe to inhabit all accessible habitats within a geological moment—and thus there is a hope that a global signature of its metabolism could be found. Moreover, there are clocks in the genetic divergence, and the rRNA record has already been used for this. The clocks may not be very accurate at present, but there is the hope that they can be calibrated better. The date of the last common ancestor is thus perhaps the most attractive candidate for the definition of the Hadean/Archean boundary.

Once life had begun, the early Archean bombardment during later phases of accretion would have imposed a major constraint on its survival (Sleep *et al.*, 2001; Gogarten-Boekels *et al.*, 1995).

8.01.3.2 The Archean Record

8.01.3.2.1 Greenland

The most informative old sequence is the Itsaq gneiss complex of southern West Greenland (Nutman *et al.*, 1996). This complex includes a wide variety of rocks older than 3.6 Ga and ranging up to 3.9 Ga (early Archean): components are the Isua Belt, the Amitsoq gneisses, and the Akilia association. The Isua belt is especially interesting because it is supracrustal: it was laid down on the surface of the planet. The rocks include mafic pillow lavas, felsic volcanics, and volcaniclastic rocks, some of which were deposited from turbidity currents. The ensemble is reminiscent of material deposited today in volcanic island arcs, for example, in the western Pacific volcanic island chains. The implications are profound. There was clearly an ocean present, and land masses (at least volcanoes, possibly other older crust). Erosion occurred, sediments were deposited; volcanic eruptions must have been normal features of the geological setting. Moreover, this was a time early enough that the Earth was still under heavy bombardment by meteorites (the face of the Moon, like a ravaged battlefield, dates from this time). There is good evidence from Isua of a meteoritic component in sediment (Schoenberg *et al.*, 2002).

With volcanoes come hydrothermal systems, and there is good evidence for these in Isua. Localized low-strain zones in ~3.75 Ga rocks show many primary features (Appel *et al.*, 2001), including mafic lavas with fine-grained cooling rims, and in pillow breccias, quartz globules occur. These globules are interpreted as former gas vesicles, infilled with quartz from hydrothermal veins that formed during and immediately after volcanism. These quartz infills contain rare fluid inclusions. Appel *et al.* (2001) describe inclusions containing remnants of two independent fluid/mineral systems, comprising pure methane and highly saline (25% NaCl) aqueous fluids, and co-precipitating calcite. These fluids strongly resemble modern sea-floor hydrothermal fluids. The conclusion reached by Appel *et al.* (2001) is thus that methane-brine hydrothermal systems operated 3.75 Ga ago, in the early Archean. If correct, the implications are twofold: that, as common sense already tells us, hydrothermal systems existed, and that they emitted methane, useful for metabolism.

There have been various claims of evidence for life in the rocks of west Greenland. These have been reviewed by Myers and Crowley (2000), and also studied by van Zuilen *et al.* (2002) and Fedo and Whitehouse (2002). Significantly, they contest claims (Mojzsis *et al.*, 1996) for evidence of very early life at Akilia island. Fedo and Whitehouse (2002) showed that the rock studied by Mojzsis *et al.* was not sedimentary but an ultramafic igneous rock. They further considered that the isotopic ratios of the carbon particles at Akilia recorded high temperature metamorphic processes, not life, and yielding abiotic hydrocarbons. Thus the Akilia rock, though interesting, is not a guide to early life.

Rosing (1999) reported carbon microparticles from >3,700 Ma rocks in Isua that are strongly depleted in ^{13}C relative to bulk Earth. $\delta^{13}C$ in these particles is in the range of −10‰ to −20‰, strongly indicative of organic fractionation though inorganic processes can also fractionate carbon isotopes (Pavlov *et al.*, 2001). This work is not contested by Fedo and Whitehouse (2002). The carbon is present as 2–5 μm graphite globules, that appear to be biogenic detritus. They are hosted in turbiditic sediments and in pelagic muds. The simplest interpretation is that these carbon particles were originally (before deformation and metamorphism) organic remains, and represent the bodies of settled planktonic organisms. The implication is that plankton, and hence mesothermophilic organisms, were present globally before 3.7 Ga ago. Currently, this is the oldest claimed evidence for life on Earth that has as yet withstood critical scepticism.

8.01.3.2.2 Barberton

Evidence for early life comes from the Barberton Mountain land of South Africa (Byerly *et al.*, 1986), in material from the 3.3 Ga to 3.5 Ga Swaziland Supergroup.

Byerly *et al.* (1986) described probable stromatolites in the Fig Tree Group, preserved in grey-black finely laminated chert. The structures are made primarily of microcrystalline chert, forming low-relief laterally linked domes and in places pseudo-columnar structures. Byerly *et al.* did not find evidence of microfossils but inferred an organic origin from the morphology of the structures. However, Lowe (1994) disputed this evidence and concluded that the structures were not demonstrably of biotic origin.

Elsewhere in the Barberton Mountain Land is a wide array of mid-Archean volcanic and sedimentary rocks, ranging up to >3.5 Ga old. Some material is clearly biogenic (Westall *et al.*, 2001), with highly fractionated carbon isotopes ($\delta^{13}C$ −27‰), but may be of non-Archean age. Thus the case for mid-Archean biotic material in Barberton remains open.

8.01.3.2.3 Western Australia

Rocks of similar age to Barberton occur in the 3.4–3.5 Ga Warrawoona Group, Pilbara, Western Australia. A wide range of rock types is present, both lavas and sediments. There is strong controversy as to whether or not microfossils are present in the Apex cherts of the Warrawoona Group (Buick *et al.*, 1981): this controversy is summarized by the debate between Schopf *et al.* (2002) and Brasier *et al.* (2002) (see also Gee (2002) and Kerr (2002) for excellent reporting on the debate, and Buick, 1990). Lowe (1994) also dismisses claims that structures described as stromatolites in the Warrawoona Group are actually of organic origin.

Schopf *et al.* (2002) and earlier work cited therein, found evidence for microbial fossils in Pilbara and Barberton material. The laser-Raman imagery reported by Schopf *et al.* (2002) demonstrated that the material was made of kerogen and they interpreted this as evidence for remains microbial life. Brasier *et al.* disputed the earlier work by Schopf and Packer (1987) and Schopf (1993) on Warrawoona material, constructing a detailed case in which they reinterpreted the supposed microfossils of the earlier study as secondary artifacts of graphite in hydrothermal veins. However, Brasier *et al.* (2002) did report C isotopic results that are most easily (though not conclusively) interpreted as microbial. Thus although the "microfossils" earlier reported by Schopf may not be organic, there is isotopic evidence suggesting biological activity, though of uncertain age (possibly later than the host country rock).

Several notable pieces of evidence for early life come from Western Australia. Shen *et al.* (2001) found isotopic evidence for microbial sulfate reduction in 3.47 Ga barites from North Pole in the Pilbara. Intuitively, sulfate reduction would be expected to be very old: this confirmatory evidence is strong. Also notable is the discovery by Rasmussen (2000) of filamentous microfossils in a 3.235 Ga old volcanogenic massive sulfide deposit, a type of deposit that only forms under deep water. The implication is that hyperthermophile microbial life was certainly present on Earth by this date, and in deep water. One diversion is of interest here. The abundant microbial life around mid-ocean ridge vents would have meant that considerable amounts of reduced carbon were preserved under the lava flows. This would have affected the net balance of the atmosphere, leaving an excess of oxygen. It would also have introduced reduced carbon down subduction zones. Interestingly, some diamonds have light carbon isotopes that may have "organic" ratios prior to metamorphism, and also contain "ophiolite like" inclusions, palimpsests of a mid-ocean ridge origin. Just possibly, some diamonds may be carbon from ancient microbial colonies (Nisbet *et al.*, 1994).

There is also evidence for the presence of methanotrophs in the ~2.8 Ga old Mount Roe palaeosol. This contains highly fractionated organic carbon, probably recording the activity of methanotrophs living near ephemeral ponds: this implies that significant biological methane sources existed in the late Archean (Rye and Holland, 2000). Oil is also present in some Archean sandstone (Dutkeiwicz *et al.*, 1998, Rasmussen and Buick, 2000).

In the late Archean of Western Australia, there is much evidence of life, both macroscopic and microscopic. Of particular interest are stromatolites from the Tumbiana Formation, in the 2.7 Ga Fortescue Group (Buick, 1992). These have diverse morphology and occur in lacustrine sediments. Texturally, they closely resemble younger microbialites, and they are most probably the product of phototrophic microbial life, living by oxygenic photosynthesis in shallow water with negligible sulfate concentrations. Slightly younger, the 2.5 Ga Mt. McRae shale yields bitumens that contain biomolecules characteristic of cyanobacteria (Summons *et al.*, 1999). This evidence strongly supports the notion that cyanobacterial oxygenic photosynthesis was fully established.

The late Archean of Australia contains many carbonate rocks with $\delta^{13}C \sim 0‰$. This is strong circumstantial evidence for global oxygenic photosynthesis. The logic depends on the strong fractionation imposed by rubisco as it selects carbon from the ocean/atmosphere system to incorporate it into living organisms (Schidlowski and Aharon, 1992; Schidlowski, 2002). Though some rubisco-using cells are not photosynthetic, most are, and the energy that allows rubisco to incorporate carbon into life is photosynthesis. Carbon emitted from the mantle has $\delta^{13}C \sim -5‰$ to $-7‰$. This is emitted into the air and ocean mainly as carbon dioxide. From this mantle-derived carbon, carbon is acquired into organic matter by rubisco, using the harvest of thermodynamic reduction power from the apparatus of oxygenic photosynthesis in the presence of abundant ambient atmospheric CO_2. This carbon chosen by life is strongly selected for ^{12}C and thus has $\delta^{13}C \sim -28‰$ to $-30‰$. Thus, the residue left in the air/sea system is enriched in ^{13}C. In modern-day carbonates, $\delta^{13}C \sim 0‰$, implying by balance ($-7‰$ source, partitioning into $-28‰$ organic life and $0‰$ inorganic sinks) that about a quarter to a fifth of primitive carbon is captured by organic matter, and three-quarters to four-fifths is left as carbonate with $\delta^{13}C \sim 0‰$. Because carbon dioxide is globally mixed, the presence of carbonate with $\delta^{13}C \sim 0‰$ implies a global fractionation of carbon by oxygenic photosynthesis. This indeed is what is recorded in the late Archean.

8.01.3.2.4 Steep Rock, Ontario, and Pongola, South Africa

The evidence from the 3.0 Ga sequence at Steep Rock, Northwest Ontario, Canada, is very different (Wilks and Nisbet, 1985; Nisbet, 1987) (Figure 3). Here is a large limestone reef, some kilometers long, displaying a wide variety of structures interpreted as formed by life, and also with a range of isotopic evidence that is not greatly dissimilar to modern sequences. The structures vary from large stromatolites (several meters long) to smaller (1–20 cm) stromatolitic structures (sadly some of the loveliest of these have been fractured recently by unknown collectors), deposited close to a major unconformity. These are among the oldest unchallenged examples of stromatolites: claims of older examples have been strongly criticized (Lowe, 1994). Isotopic evidence from Steep Rouch (Abell, Grassineau, and Nisbet, unpublished) indicates that rubisco-mediated carbon capture (i.e., oxygenic photosynthesis) controlled the global carbon partitioning between carbon dioxide and carbonate: this is some of the oldest evidence for global oxygenic photosynthesis.

The Pongola sequence in South Africa (Matthews and Scharrer, 1968; von Brunn and Hobday, 1976) also includes stromatolites above a major unconformity, and is uncannily like Steep

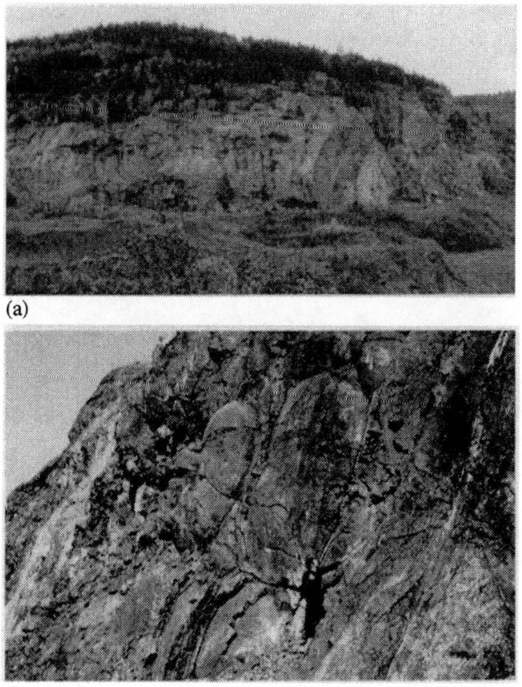

(a)

(b)

Figure 3 (a) The surface of the 3 Ga Earth, Steep Rock, NW Ontario, Canada. The hill-face is very close to a 3 Ga unconformity surface, and the rocks (granitoids and mafic dikes) exposed on the hill-face are immediately below the unconformity. Above them are assorted sediments, including thick stromatolitic limestones. (b) Stromatolitic limestone, Steep Rock, Ontario, Canada (ca. 3 Ga old). The palaeohorizontal surface dips ~70°. Stromatolitic domes are up to 4–5 m long and 2 m high.

Rock both in age and sequence: it is tantalizing to wonder if they were once contiguous before the vagaries of continental breakup and re-assembly.

8.01.3.2.5 Belingwe

The evidence for life in the sediments of the Belingwe belt, Zimbabwe, has been described by Martin *et al.* (1980), Nisbet (1987), Grassineau *et al.* (2001, 2002), and Nisbet (2002). The Belingwe Greenstone Belt has a wide and diverse array of evidence for late Archean life. In this it is not unique—many Australian and South African sequences also have abundant evidence of life. What makes the Belingwe belt fascinating is the range of features outcropping in a small area, coupled with some extremely well-preserved igneous rocks (Bickle *et al.*, 1975; Nisbet, 1987).

The rocks of the Belingwe belt span a range of ages, but the sequence that carries the most detailed evidence for life (Figures 4 and 5), the Ngesi Group, is 2.7 Ga old. The base of the Group includes shallow-water sediment locally rich in carbon and sulfur that is highly fractionated isotopically, suggesting the original presence of methanogens, as well as the operation of complex sulfur fractionating processes (Grassineau *et al.*, 2001, 2002). Oil is present in some rocks (Grassineau and Nisbet, own observations). Locally associated with this stratigraphic unit are stromatolites made of calcite with $\delta^{13}C \sim 0‰$, with kerogen that contains carbon which is strongly fractionated isotopically, implying the selection of carbon by rubisco (see Section 8.01.7.2). Immediately above the basal sediments are komatiite pillow lavas and flows (Figure 5). Close to the contact with the lavas, in the uppermost sediments, are sediments that in places are very rich in kerogen and sulfides, with highly variable fractionated carbon and sulfur isotopes, different in very small physical distances. The simplest interpretation of this (Grassineau *et al.*, 2002; Nisbet, 2002) is that the complex isotopic fractionation is a record of consortia of prokaryotes, some reducing sulfate and some perhaps oxidizing sulfur, others generating methane, some photosynthesizing and capturing carbon by using rubisco, and perhaps carrying out other microbial biochemistry (using metal enzymes). Both in shallow-water photosettings and in deeper water below the photic zone, microbial mats may have cycled sulfur in sulfureta, as in modern parallels (Fenchel and Bernard, 1995).

Above the komatiites are thick basalt pillows and flows. At the top of the sequence is a further sequence of shallow-water sediments, including limestones that locally have extensive and very well preserved stromatolites (Figure 4). These too have evidence for rubisco fractionation (see Section 8.01.7.2), both in kerogen carbon, and in carbonate with $\delta^{13}C \sim 0‰$.

8.01.4 THE FUNCTIONING OF THE EARTH SYSTEM IN THE ARCHEAN

8.01.4.1 The Physical State of the Archean Planet

The map of the surface of the Archean planet remains largely blank, populated by imagined beasts and perhaps some features seen dimly but truly (Macgregor, 1949). The main input from the mantle to the surface is via volcanism. Late Hadean and early Archean volcanism would have provided thermodynamic contrast, placing material that had equilibrated with the mantle in contact with the ocean-atmosphere system that was open at the top to space and light. In the latest Hadean and earliest Archean this contrast would have been most likely thermodynamic basis of life.

Early Archean volcanism was probably largely basaltic or komatiitic, but perhaps with some

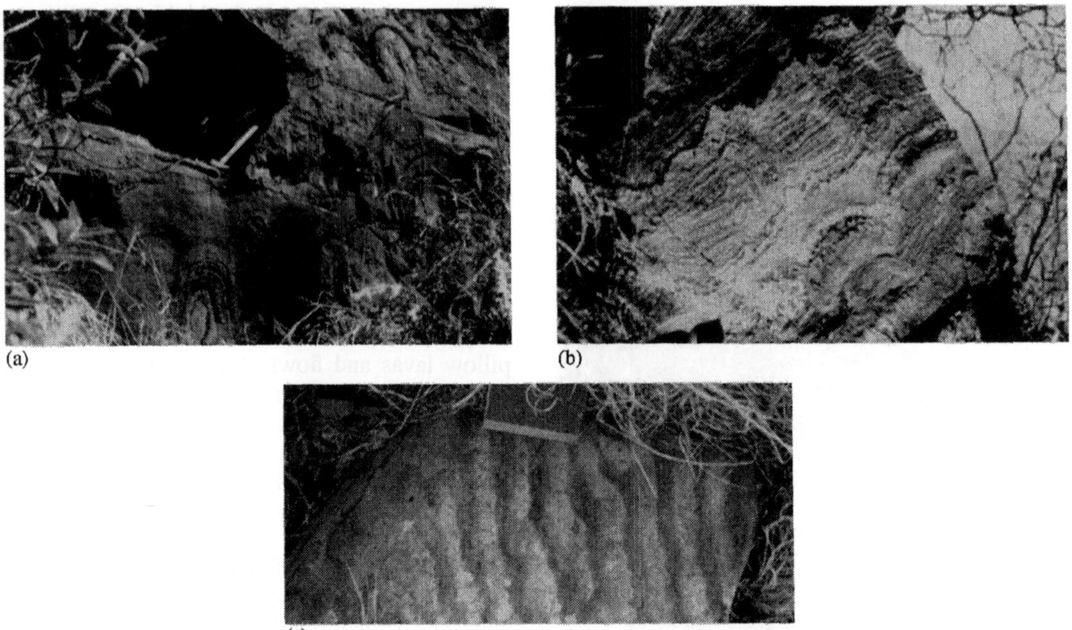

(a) (b) (c)

Figure 4 Stromatolitic limestone, Cheshire Formation, Belingwe belt, Zimbabwe (2.6–2.7 Ga old): (a) outcrop surface—structures occur on a variety of scales, from microscopic to metre relief; (b) detail of one outcrop (from Nisbet, 1987); and (c) shallow-water shale associated with Cheshire stromatolites.

andesitic and alkaline centers. The mantle may have been somewhat hotter than today (Nisbet *et al.*, 1993), and thus the primary melt at mid-ocean ridges would likely have been more magnesian than today. Moreover, a hotter mantle would likely have sourced more plume volcanoes than today. These volcanoes would have been comparable to modern Hawaii but may have ranged up to much larger sizes. The plumes would have emitted komatiite lava flows. These are less viscous than basalt, and would have flowed long distances on relatively flat surfaces, creating huge flat shields, perhaps as large wide islands emerging as the upper fraction of enormous volcanic platforms resting on oceanic plate.

Komatiite lava flows are very rich in MgO. They contain significant iron oxide, and are typically associated with nickel sulfides and chromite. Hydrothermal systems in highly magnesian rocks can be very alkaline, with very high pH. Thus, it would be expected that rain falling onto komatiite flows, or flows into shallow (low-pressure) seawater, would generate very alkaline outflows of hot or warm water.

The zircon evidence and the existence of 4 Ga gneiss provide evidence for the existence of continental crust, but this may have been of limited areal extent. Significantly, by the early Archean there had probably been inadequate time for deep continental lithosphere to develop, yet by 2.7 Ga, late Archean diamonds are known in the Witwatersrand record (Nisbet, 1987). Diamonds imply lithosphere at least 150 km or so thick, and suggest kimberlite and probably a spectrum of alkali volcanism on land. Alkaline volcanism is indeed known to have occurred, a source of high pH and perhaps phosphatic environments. There is a small but significant record of Archean alkali volcanics (Nisbet, 1987), for example, in the Timiskaming Group in Northern Ontario (Cooke and Moorhouse, 1969), which includes leucitic flows and pyroclasts. Just possibly phosphatic volcanics did occur—arguably the most likely setting for constructing sugar–phosphate chains in an inorganic process.

Early Archean continents were subject to erosion. Rocks from Isua include sediments, implying the action of rain and the existence of subaerial exposure, as well as the presence of wide oceans capable of evaporating the rainwater. The nature of the sediment was different from today, however. Nowadays, most surface rock is actually recycled previous sediment, and aluminous clay-containing muds (mature sediments) are common. Most of what little there is of the early Archean sedimentary record is not mature: primary volcanic terrains were being eroded. Clays would have been widely present, but were probably mainly magnesium-rich clays derived from weathering of volcanic rock, not aluminum-rich material. This scarcity of mud may be important in considering likely biological host environments.

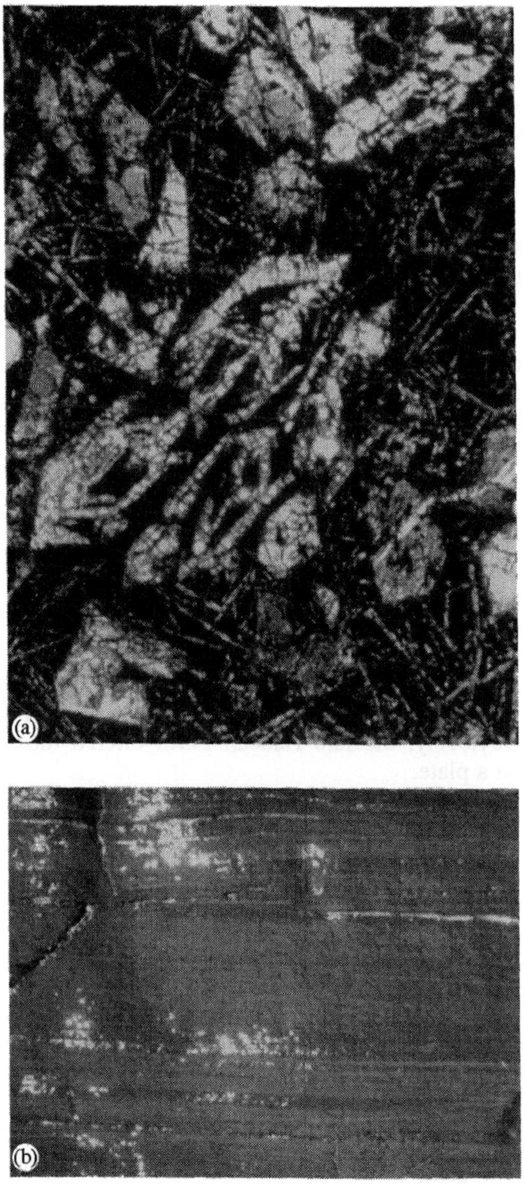

Figure 5 (a) Thin section of 2.7 Ga komatiite lava, Reliance Fm, Belingwe belt, Zimbabwe. About 6.5 mm across photo. Olivine crystals set in fine grained to once-glassy groundmass. For details see Nisbet *et al.* (1987). Photo W. E. Cameron. (b) Alternating iron-rich and carbon-rich shales. White bands are chert: this lithology is transitional to banded ironstone. Approximately, 20 cm across picture. Belingwe belt, Zimbabwe.

8.01.4.2 The Surface Environment

The sedimentary evidence implies the existence of oceans. Although the initial deep volatile inventory of the planet would have been removed by the late great impact that formed the Moon, much of the water presently in Earth's oceans would have degassed from the hot mantle or infallen as comets soon after that great impact, and the ongoing volcanism would have added more. However, at ridges water is rapidly returned to ocean crust by serpentinization and metamorphic hydration of basalt. As soon as old oceanic plate developed, cold plate and hence crustal water would begin to fall back in to the interior down subduction zones, returning more water than mid-ocean ridge volcanism emitted. Given the high mantle temperature, subduction zone volcanism probably rapidly restored the subducted water to the surface. Nisbet and Sleep (2003) suggest that in effect the Earth's mantle is self-fluxing. The net annual contribution of primary new water to from the deep interior the surface (ocean) would thus be set by the inputs of volcanism at ridges and plumes, plus infall of cometary material, minus net loss back into the interior from the small net amount of water that was carried down into the deep interior, and net loss by loss of hydrogen to space.

The controls on carbon dioxide would have been somewhat different. Today, carbon dioxide is stored in carbonate minerals in the ocean floor and on the continental shelf. Subduction, followed by volcanism, cycles the carbon dioxide to the mantle and then restores the CO_2 to the air. Metamorphic decarbonation of the lower crust also returns carbon dioxide. The carbon dioxide is then cycled back to the water, some via rain, some dissolved via wave bubbles. Erosion provides calcium and magnesium, eventually to precipitate the carbonate. In the earliest Archean, parts of this cycle may have been inefficient. The continental supply of calcium may have been limited; however, subseafloor hydrothermal systems would have been vigorous and abundant, exchanging sodium for calcium in spilitization reactions, and hence providing calcium for *in situ* precipitation in oceanic crust.

Before significant thicknesses of lithosphere had cooled over large areas, the subduction may have been limited, and hence the return of carbonate-held carbon dioxide to air, via subduction volcanoes, would have been hindered in the earliest Hadean: by late Hadean subduction should have become the general fate of old oceanic plate. Cooling of plate depends on having a cool surface. The temperature of the late Hadean Earth's surface is unknown, but Sleep and Zahnle (2001) and Sleep *et al.* (2001) have made an excellent circumstantial case that the ambient surface environment was glacial, ice over cold ocean. The crustal Urey cycle buffers carbon dioxide in the air. In the Urey cycle, if global warming occurs, silicate weathering is speeded up, more calcium, magnesium, strontium cations are released and hence carbonate is formed: thus the carbon dioxide greenhouse is reduced, ending the warming. Carbon dioxide is also cycled via the mantle: outgassing at the mid-ocean ridges adds carbon dioxide to the air, while alteration of ocean floor basalt precipitates carbonate, and the subsequent subduction of carbonated oceanic crust returns carbon dioxide to the Earth's interior.

To return now to the carbon dioxide question, early in Earth's history, degassing would have been vigorous but so would have been the return of carbon to the interior, and it is likely that the mantle cycle would have dominated (Sleep and Zahnle, 2001). Moreover, frequent meteorite impacts would have created vast quantities of basalt ejecta that would also have reacted with carbon dioxide to precipitate carbonate. Sleep and Zahnle (2001) concluded that so much carbon dioxide would have been held in the mantle that the greenhouse warming would have been small: the Earth was probably heavily glaciated—the Hadean was probably a Norse ice-hell. Possibly early Hadean Earth risked loss of atmospheric carbon dioxide to the interior more than dehydration by hydrogen loss to space, though this would depend on how much methane was in the air.

If so, the likely ambient conditions (Sleep *et al.*, 2001) would have included a dry troposphere with little water vapor (and hence little OH) in the low temperatures, and wide ice cover only locally broken by water leads on the sea surface (Figure 9(a)). The air would have had very high dust content from the volcanic eruptions and meteorite impacts, possibly being so dusty as to inhibit rainfall (especially given the dryness). Continents would have been covered by dirty ice or perhaps dry permafrost (given the very low humidity). If conditions were cold enough and CO_2 concentration high enough, possibly carbon dioxide was present in polar ice. From time to time, perhaps millions of years apart, massive meteorite impacts would have ejected huge amounts of water and dust, melting ice and changing albedo (Sleep *et al.*, 2001). Brief warm episodes would result, with water-aided greenhouse conditions, and then slowly the ice cover would return. In this oscillating climate there would be many local oases of warmth around volcanic hydrothermal systems. Some of these would operate under ice cover (as in Iceland today) offering an interesting and very diverse variety of chemical settings very closely juxtaposed in space and perhaps repeatedly replacing each other in time as the hydrothermal systems fluctuated. These settings would include all possible phases: warm rock surfaces in warm or hot water/brine; fumaroles with various vapor phases; and locations in ice; and in warm water (Nisbet and Sleep, 2001a,b).

In addition to plate boundary and plume-related hydrothermal systems, the chemistry of the prebiotic world would have had strong redox contrasts in the restricted areas that had tidal coasts, and perhaps within the oceans where differing water masses interacted, or under ice. These redox contrasts were ultimately driven by photolysis in the atmosphere/ocean (presumably made of water and carbon gases) and escape from the top of the system, and also by magmatic interaction at the bottom (around komatiitic vents), where reduced species such as H_2 would have been generated. Likely terrestrial sources of redox contrast included: hydrogen emitted from serpentinization reactions when water reached hot ultramafic rock; sulfates in air and water versus sulfides in hydrothermal deposits; carbon dioxide in air versus methane or CO in hydrothermal systems; nitrogen oxides in air and water versus ammonia in hydrothermal emissions; as well as the contrast between water and magmatic hydrogen. Meteorites would have provided reduced iron and carbon particles. Hot iron, falling in to water, could generate hydrogen.

A world is only interesting to biology if it offers a way of making a living. The first life must have been unskilled, not equipped to search out the necessities of life: Thus it must have existed where a strong redox contrast was accessible, either spatially (over a few microns) or temporally (in a fluctuating setting, where regular variation took place between one redox regime and another, within hours or even minutes (e.g., as in a geyser). Obviously, the late Hadean Earth offered these on a plate.

8.01.5 LIFE: EARLY SETTING AND IMPACT ON THE ENVIRONMENT

8.01.5.1 Origin of Life

Over the origin of life, Nature has chosen to draw a veil. A basic criterion in science is that the result should be reproducible, falsifiable. Not one of the notions of the origin of life has led to reproduction, yet not one can be falsified. No doubt success will soon come in the effort to understand the detailed step-by-step molecular controls of reproduction. There are many notions about the origin of life. Where there is little fact, imagination is allowable and profitable, but where there is no fact, then even imagination is best left unimagined here. Similarly, the question "what is life?" is perhaps best left to the consideration of Hades by trouser-role opera singers, of uncertain reproductive ability, seeking Eurydice. Life is more than reproduction, which clay minerals also achieve. Defining the boundary between life and nonlife is, to quote N. H. Sleep, like searching for the world's smallest giant.

Nevertheless, despite these warnings, the questions are of supreme interest. Given that life bends the rules, a slight digression is warranted. A definition of life is perhaps best approached via thermodynamics (Nisbet, 1987). Life is growth—it is always in disequilibrium with its surroundings, and its actions are such as to increase that disequilibrium. Sustainable, equilibrium molecules are dead molecules. In practice the boundary is set between the cell and the virus: the cell can in principle reproduce and thus increase the scale of the disequilibrium, while in contrast the virus can crystallize and thus set itself in a fixed point on the entropy scale.

There are several favorite notions of the site of the origin of life (Nisbet, 1987). The best known is the Marxist hypothesis of the "primaeval soup"—that the early ocean was a soup of organic molecules that had fallen in from meteorites (which frequently contain complex carbon-chain compounds: organic chemicals, but made by prebiotic inorganic processes). In this soup, lipid blobs somehow evolved into living cells. The discovery of hydrothermal systems led to the realization that early oceans would have pervasively reacted with basalt, both in hydrothermal systems and also with basalt ejecta after impacts. Thus, the late Hadean ocean was most unlikely to be a festering broth, but more likely a cool clean ocean not greatly dissimilar to the modern ocean: exit the primaeval soup.

Other hypotheses note the properties of minerals, especially clay minerals (Bernal, 1951, 1967), iron oxides and zeolites. Hooker, in a letter to Darwin that provoked the "warm little pond" hypothesis (Darwin, 1959), noted the characteristics of modern hydrothermal systems: abiotic formation of hydrocarbons may occur today in mid-ocean ridge systems (Holm and Charlou, 2001). An interesting variant is the idea of "genetic take-over" (Cairns-Smith, 1982). This is based on the notion that some minerals are not greatly different from viruses—as Schrodinger (1944) pointed out, life is based on molecules that can be crystallized as aperiodic crystals. Mineral crystals reproduce, in a sense, when they grow—each crystal surface seeds new copies of itself. In one version of the genetic takeover hypothesis, the earliest replicating structures were simply minerals, that replicated just as clays minerals grow. These structures bound proteins, which helped in the reproduction. Then nucleic acid took over the role of the mineral template, and occupied the central direction of the reproducing body (Figure 6).

The "panspermia" hypothesis is simple (Crick, 1981)—Earth was seeded by little green men from outer space, who spread life cells by sending rockets throughout the galaxy. This hypothesis has the attraction of avoiding the impossible task of elucidating how life began on Earth by transferring the problem to another planet far away and long ago; it also achieves a happy congruence with Star Trek's DNA-based universe. However, it is not discussed why the men were green, or why they were men: pan-oo would be perhaps more likely than pan-sperm.

8.01.5.2 RNA World

Of the many origin of life ideas, the "RNA-first" idea (Gilbert, 1986) is worth noting in more detail: the idea that prior to DNA, the genetic code was held in RNA. This does not necessarily mean that life began as RNA (a takeover is possible), but at some stage it seems likely that life was RNA-based. All cells today use ribosomes—a giant RNA enzyme—to read the DNA tape, and RNA retains the key role of carrying messages in the cell. It may be that at one stage life was a few self-replicating RNA molecules.

If so, how did these RNA molecules exist? Possibly they were sophisticated enough already to have outer bags and thus containers for the protein they made. But it is also possible to imagine an early RNA world (Gilbert, 1986; Nisbet, 1986) in vesicles in a rock, where the container was provided either by the rock itself, or by minerals with large tubular shapes, such as faujasitic zeolites or some of the iron oxide minerals. Chemicals and redox drive would be provided by fluids flushing through the setting. Any RNA molecules that accidentally managed to self-replicate would be protected and would propagate; one might next accidentally develop the ability to synthesize proteins that could be assembled to act as enzymes aiding replication, increasing the population. Volcanic accident could spread the molecules from the first container into other parts of the system. Finally, any molecule that accidentally acquired the ability to enclose itself with a lipid bag would be pre-adapted to life in the open environment, away from the rock vesicle. But this is a notion—many other notions have equal or greater validity.

Geologically, some inferences can be made. The setting of the first life to use nucleic acids would presumably have had abundant local phosphate sources and accessible phosphorus, as well as sugars and nitrogen bases. Here the evidence of the existence of komatiite plumes and the antiquity of continents is just possibly relevant. Alkali volcanism is a feature of plume volcanoes (e.g., Mauna Kea in Hawaii). Carbonatite volcanism and associated very unusual rocks (such as phosphatites) occur today mainly on ancient continental crust. Whether phosphate-rich volcanism could have been possible as early as the Hadean is a moot point. Then the lithosphere may have been thin and limited to a segregated cooled–melt earliest crust, plus giant plume volcanic centers, fractionated in their upper stages. Assuming phosphate-rich igneous rocks did exist, then phosphorus-rich hydrothermal systems may have occurred.

More generally, alkaline hydrothermal systems would have occurred around the widespread

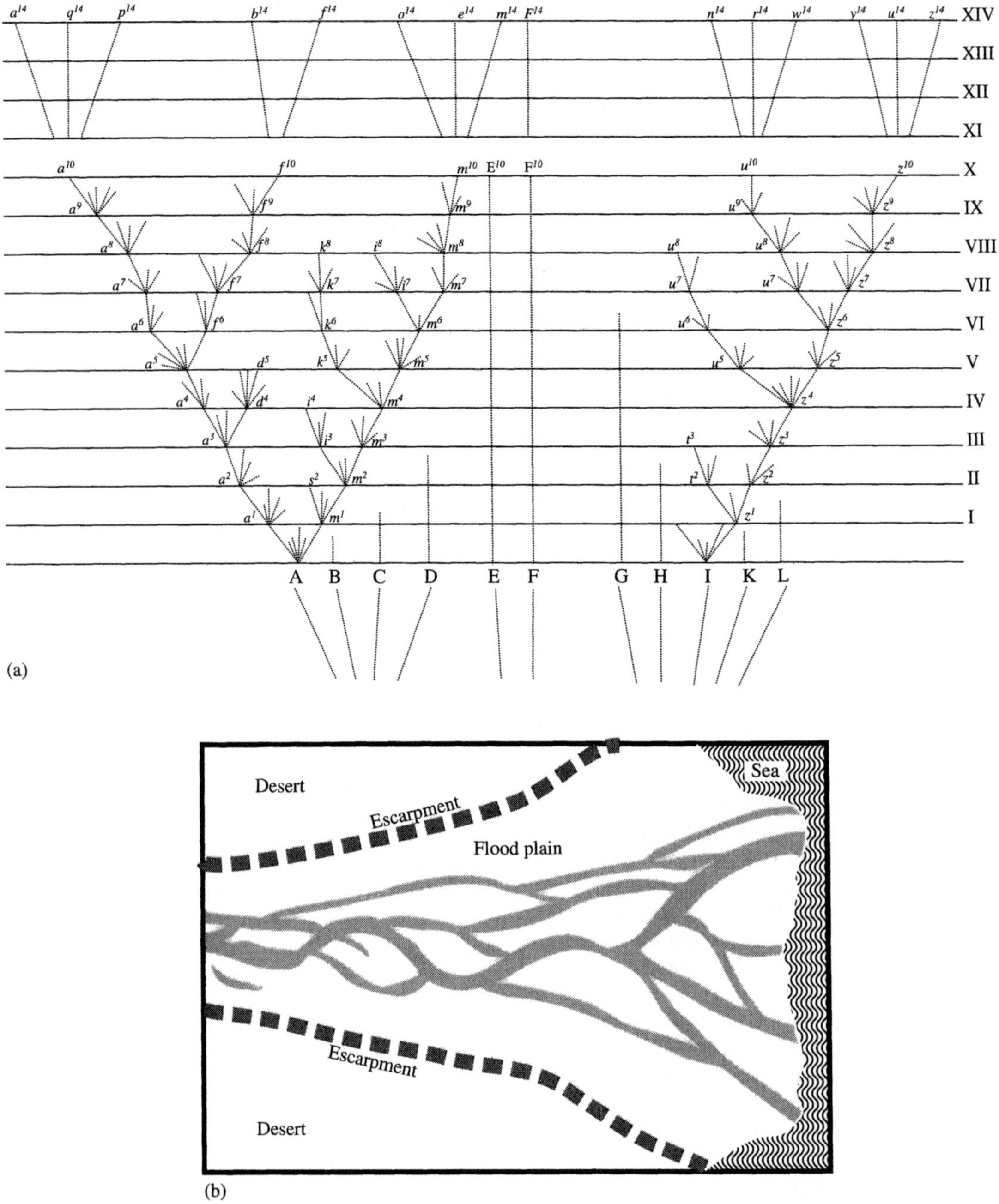

Figure 6 Models of the descent of life: (a) after Darwin's single illustration in Origin of Species (chapter IV) (Darwin, 1859, 1872) and (b) braided delta model, assuming large-scale lateral gene transfers and boundaries of nonviability.

cooling ultramafic rocks, such as the enormous komatiite flows that would have issued from komatiitic plume volcanoes, and also at distal sites near early mid-ocean ridges (themselves possibly fed by komatiitic basalt liquid). These hydrothermal systems would emit high-pH hot fluids. Here ammoniacal hydrothermal systems (Hall, 1989, Hall and Alderton, 1994) would probably have occurred. Under such high-pH conditions metal atoms (e.g., iron, copper) can form compounds within cages of four nitrogen atoms. Possibly the cytochrome family of proteins, which is clearly very ancient, may have had its origins in such a setting. These proteins have at their heart a metal surrounded by four nitrogen atoms: haem with iron and four nitrogens; chlorophyll with magnesium surrounded by four nitrogens.

8.01.5.3 The Last Common Ancestor

The *last* common ancestor is more accessible to geology and molecular biology than the first ancestor. Though not less controversial than the first ancestor, it is at least the subject of testable hypotheses.

The *last* common ancestor is the notional cell, or population of cells, from which all modern living cells are descended (Woese, 1999). One definition of the Hadean/Archean boundary is the date of the last common ancestor. This last ancestor would have been a DNA-based organism, already complex, with many of the so-called housekeeping proteins that are broadly common to nearly all modern types of cell. Note however, that viruses, especially RNA viruses, may (or may not) be separately descended from an earlier ancestor.

There is much debate about the habitat—and hence metabolic processes—of the last common ancestor. The majority view is that the root was a prokaryote, more bacterial than anything else, from which diverged the sister domains of Archea and Eucarya (Woese, 1987) (Figure 7). In this view, complex eukaryotes evolved from simple prokaryotes. This interpretation also leads to the inference that the last common ancestor was a hyperthermophile, living in hot conditions (>85 °C) probably in close proximity to a hydrothermal system (Stetter, 1996; Nisbet and Fowler, 1996a,b; Miyazaki *et al.*, 2001). In standard microbial phylogenies (e.g., Woese, 1987; Barnes *et al.*, 1996; Pace, 1997), the most deeply rooted organisms all appear to live in high-temperature settings. This view makes abundant geological sense, as the diversity and fluctuation of chemical settings in hydrothermal systems offers readily accessible thermodynamic drive for prephotosynthetic life; while the deep involvement of metals in the ubiquitous (and thus presumably ancient) enzymes responsible for the housekeeping biochemistry of cells strongly suggests hydrothermal supply. Moreover, heat shock proteins are integral to protein shaping, suggesting the speculation that heat shock was a general problem for early life (Figure 8).

However, the argument in favor of a mesophile last common ancestor is equally strong (Forterre, 1995; Forterre and Philippe, 1999; Galtier *et al.*, 1999). Heat is a threat to life: it cooks it, and cells have heat-shock proteins to restore them if slightly cooked. It seems counterintuitive to imagine that life started in a place so risky, before it could evolve protective mechanisms. Forterre suggested that life began in milder mesophile settings, with an initially poorly organized and complex structure. Then, when bacteria and archaea spread to the more dangerous but thermodynamically advantageous hyperthermophile settings, those that prospered were cells descended from lines that had evolved more efficient, streamlined genomes ("thermoreduction"). Forterre (1995) considered the RNA-world idea incompatible with the notion (e.g., Stetter, 1996; Nisbet and Fowler, 1996a) that early life was hyperthermophile. RNA is unstable at very high temperatures. Moreover, modern hyperthermophiles have very sophisticated mechanisms to sustain them in hot environments: unlikely in very primitive cells. Forterre (1995) suggested that early cells were complex mesophiles, and those that strayed into hotter settings slowly adapted to the conditions by selection for reduced and streamlined genotypes, to produce the hyperthermophiles.

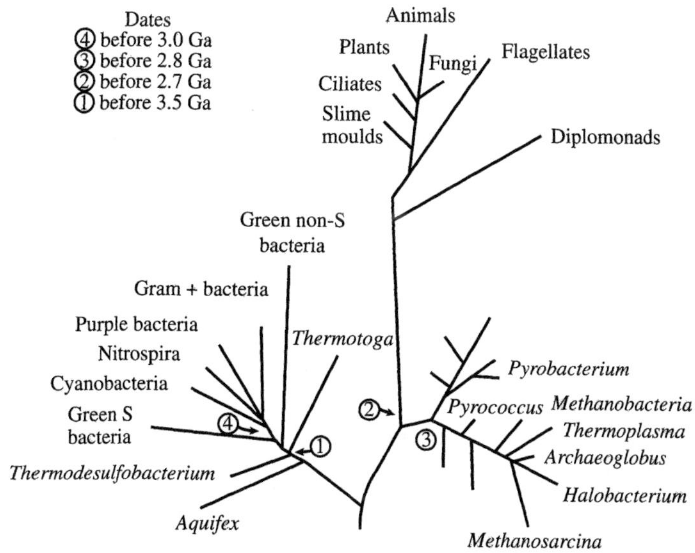

Figure 7 Model of the descent of life following the "standard" model of Woese (1987), as calibrated by the geological evidence (source Shen *et al.*, 2001, and other evidence). See Figure 11 for alternative model.

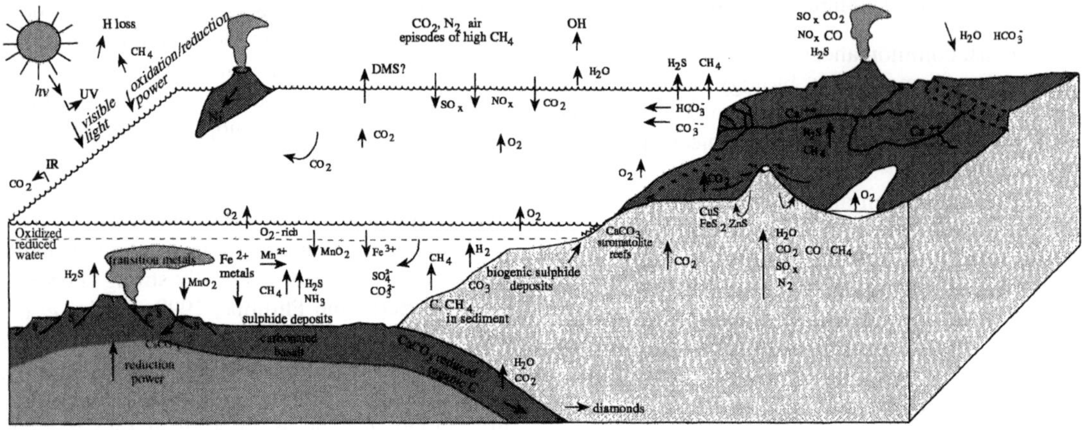

Figure 8 Sources of disequilibrium—possible geochemical (redox) resources for life in the early to mid-Archean.

An analogy would be the comparison between geometrically complex early multiple-winged aircraft such as Sopwith biplanes of the 1914–1918 era, the streamlined Hawker Hurricane monoplanes of the Battle of Britain with protruding propellers and tail wheel, and the Hawker Harrier still today in service: all built under the direction of Thomas Sopwith. These airplanes simplified in shape as they became more powerful and internally complex. Yet they are all part of a single line. Or, continuing the analogy, the officer commanding those Battle of Britain fighters, Marshal of the RAF Sholto Douglas, was brother-in-law of author J. D. Salinger: close relations, utterly different careers.

Derivation of molecular phylogeny from rRNA suffers from various mathematical pitfalls, especially the difficulty of dealing with branches of the tree of life that evolve especially rapidly. Any model that assumes uniform rates of evolution will make these branches appear inaccurately ancient. Moreover, there is massive evidence for multiple gene transfer between distinct lines within domains, and across domain boundaries. For example, up to 18% of *E. coli's* genes may be relatively recent foreign acquisitions (Martin, 1999). This complicates interpretation enormously (Doolittle, 1999, 2000), and leads to models not so much of "trees" of descent but of "mangrove roots" (Martin, 1999), or analogies with braided deltas (Nisbet and Sleep, 2001; Figure 6). Woese (1999) concluded that the communal ancestor was not so much a single discrete organism but a diverse community of cells that evolved together as a biological unit. In this view, the universal phylogenetic tree is not an organismal tree at its base, but becomes one as the peripheral branchings emerge.

The choice between explanations suggesting (i) shared ancestry between the lines, rather than (ii) lateral transfer of information between contemporary but unrelated lines, is not easy. Thus the molecular record is very "noisy" and the interpretation of descent is ambiguous. It is very difficult to be sure of the limited number of positions in an amino acid or nucleotide sequence that actually record true antiquity.

Initially, it was thought that derivation of phylogeny from molecular information is intrinsically superior to phenotypic information. Forterre and Philippe (1999), Penny and Poole (1999), and Glansdorff (2000) point out that this is not necessarily true. The microsporidae, for example, were originally misclassified as very ancient. More recently, these have been shown to be closely related to fungi, a much younger line (Hirt *et al.*, 1999). The discovery of the error in placing the microsporidae increases awareness that massive lateral gene transfer has occurred between the three domains of life. Each domain is distinct and monophyletic, but members of each domain have obtained genetic information from other domains.

In this view, the eukaryotes may well preserve some very primitive characteristics that are not seen in prokaryotes. Glansdorff (2000) reappraised claims for lateral gene transfer and concluded that the extent of transfer was overemphasized; moreover, Glansdorff inferred that the last common ancestor was probably nonthermophilic and perhaps a protoeukaryote, from which the thermophilic archaea may have been the first divergent branch.

Conceivably, if the last common ancestor were mesophile, the majority of bacteria (except perhaps planctomycetes (Brochier and Philippe, 2002)) may descend from an early mesophile prokaryote, perhaps via a genetically streamlined descendant that occupied a hyperthermophile setting. Archea too may descend from the last common ancester via a streamlined cell that had evolved to inhabit hyperthermophile settings. In contrast, the Eucarya may be directly descended from a mesophile, as may the planctomycetes.

A possible geological scenario for this process may be that the last common ancestor lived on the warm (~40 °C) but not hot periphery of a hydrothermal system, on a glaciated planet. Descendants of the last common ancestor may have evolved to occupy hyperthermophile habitats, with sophisticated biochemical processes to ensure their survival. Other descendants may have spread to occupy planktonic mesophile habitats. Large meteorite impacts, capable of heating the oceans to near 100 °C, would have occurred occasionally prior to 3.8 Ga ago. Such impacts would have destroyed all life except two types of organism: those forms capable of living in high-temperature conditions, and perhaps also those organisms that had been accidentally preserved in especially thick ice caps. Modern organisms can survive up to half a million years or more in ice (Reeve, 2002), and there are cells preserved in ice that has crystallized from Lake Vostok, the great ancient lake under the Antarctic ice cap. Just possibly, early relatives of the planctomycetes, a bacterial branch which may be of the greatest antiquity, may have been distributed in the glacial oceans, and would have been subject to freezing in thick ice cap, and thus preferentially likely to survive a global heating event after a meteorite impact.

8.01.5.4 A Hyperthermophile Heritage?

Whatever the setting of the last common ancestor, there are many aspects of modern cells that have a possible or likely hyperthermophile origin. To possess such a heritage, it is not necessary that a cell's primary ancestral line once occupied a hyperthermophile habitat. There has been much genetic exchange between organisms both within lines and even massively between domains (Figure 6(b)).

Candidates for biochemical processes or molecules with hyperthermophile origins include the heat shock proteins, and the metal enzymes (Nisbet and Fowler, 1996b). Heat-shock proteins are ubiquitous in all domains of life. They help repair damage after heat shock, but more generally they help to shape new protein molecules so they can carry out their proper functions. The heat-shock proteins are clearly of the greatest antiquity, given their involvement in very basic housekeeping processes. Their role as heat-shock repairers may of course simply be a relatively late adaptation to life in hot settings. Alternately, however, heat-shock proteins may indeed descend from an original function evolved to enable life to enter hyperthermophile settings around hot-water vents.

Like the heat-shock proteins, the metal enzymes are central to many very basic cell functions. The Metal-4N and Ni proteins have already been mentioned. Many other metal proteins involve metals such as iron, copper, or zinc, often associated with four sulfur atoms. Such metals are characteristic of hydrothermal systems hosted by basaltic and andesitic volcanism. More generally, easily available metals in hydrothermal systems play a key role in many vital housekeeping proteins, often but not always associated with four sulfurs. Examples include zinc in carbonic anhydrase, alcohol dehydrogenase, and RNA and DNA polymerases; copper in proteins used in respiration, such as cytochrome *c* oxidase; cobalt in transcarboxylase; Mo in many enzymes participating in the nitrogen cycle, in sulfite oxidase, in some dehydrogenases, and in Dimethylsulfoxide-trimethylamine oxide reductase (which may have had an important role in early methane-linked atmospheric chemistry); selenium in hydrogenases; and iron in a wide range of catalases, peroxidases, ferredoxins, oxidases, and all nitrogenases.

Nickel, in particular, is interesting to the geologist. For example, carbon monoxide dehydrogenase, which is at the center of the acetyl-coA pathway of reducing carbon dioxide, characteristically contains nickel, zinc, iron, and molybdenum. Both coenzyme F_{430} of methanogens and hydrogenase contain nickel. Consequently, nickel is essential to methanogens. Moreover, urease, a key part of the nitrogen cycle, converting urea to carbon dioxide and ammonia, is based on nickel. The most obvious supply of nickel in nature is komatiite: highly magnesian high-temperature lavas that would have been widespread in the late Hadean and early Archean. Around komatiites nickel sulfide would have been freely available. It could be that it was in this setting that nickel metal proteins evolved: perhaps it was around komatiite flows that hydrogenases, carbon monoxide dehydrogenase, and urease began. It may be that it was in such settings that methanogens first appeared, exploiting the hydrogen made from serpentinization reactions (see Section 8.01.6.5) (Figure 9). It is interesting to wonder if the cytochromes, methanogens, and the nitrogen cycle all first evolved on the flanks of komatiite volcanoes.

Today, metals are scavenged from water by extremely sophisticated biochemical processes (Morel and Price, 2003). Thus, seawater can have very low ambient levels of metal ions. Early Archean seawater would likely have been much richer in trace metals. But given that early organisms presumably had very unsophisticated processes for capturing metals, even in seawater rich in metal it would have been difficult to access the metal. Perhaps the earliest distribution of organisms was very restricted, with few cells living away from locations such as volcanoes that had readily accessible metals. Only the evolution of effective metal-gaining siderophores would have allowed the spread of life. There is thus reason to believe that, even if the last common ancestor was not hyperthermophile but lived

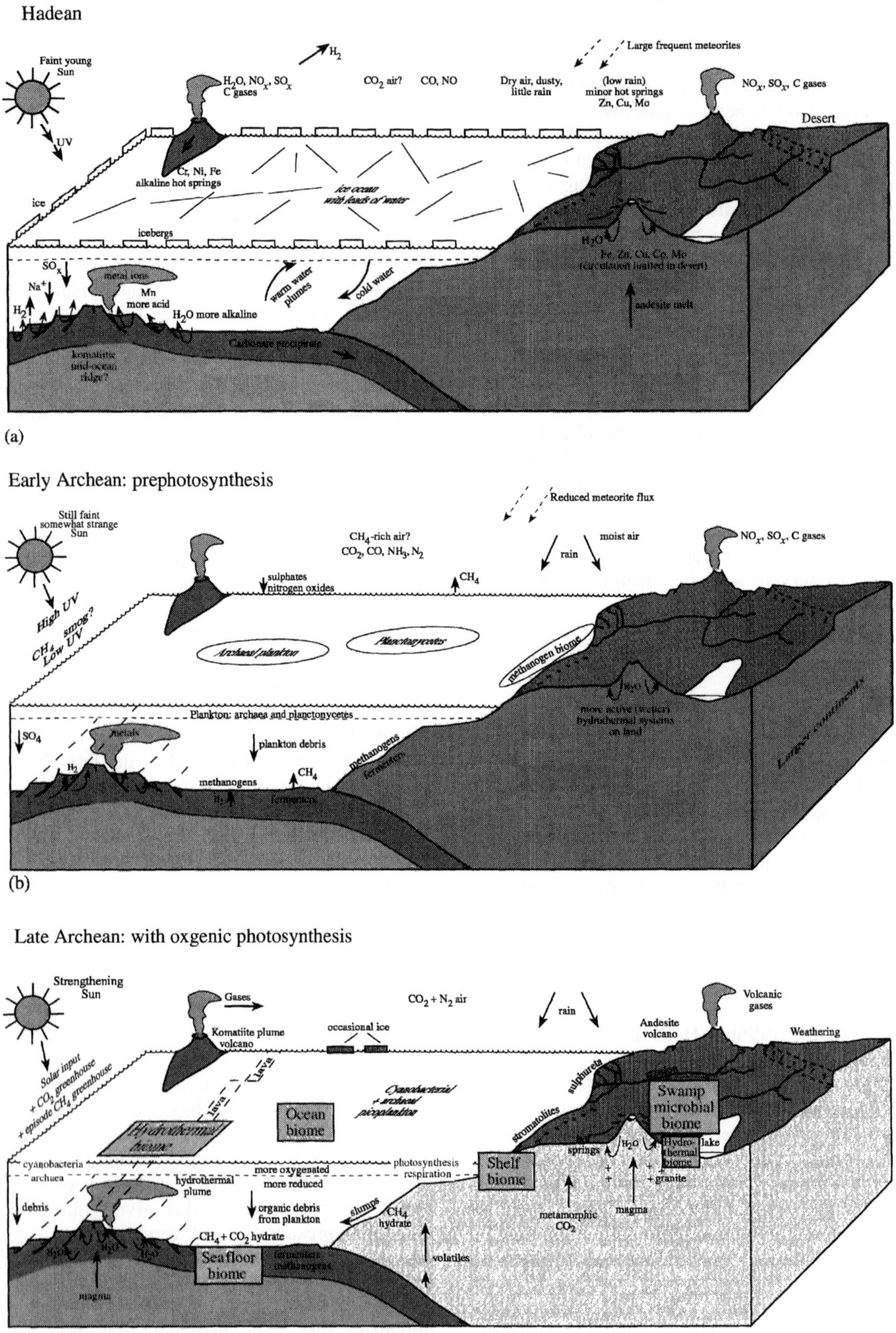

Figure 9 Model of the evolution of the planetary surface: (a) Hadean surface, possibly glacial (apart from rare very hot events after major meteorite impacts); (b) early Archean surface, before the onset of photosynthetic processing of the air; and (c) late Archean surface, assuming that the major biochemical pathways had evolved, and that the main groups of prokaryotes had evolved.

in somewhat cooler conditions, from it came volcano-hosted hyperthermophile ancestral lines, living in and around hydrothermal systems, that led to the Archeal domain and perhaps also to most bacteria (excluding perhaps the planctomycetes). There has been much gene exchange since then, and consequently enzymes of hyperthermophilic origin are ubiquitous in the housekeeping chemistry of all cell lines. The volcanic signature is written deeply into all life.

8.01.5.5 Metabolic Strategies

It is likely that the oldest organisms were not photosynthetic (see discussion in Nisbet *et al.*, 1995; Nisbet and Sleep, 2001). Prephotosynthetic organisms would have depended on natural redox contrasts, and would thus have lived in habitats where such contrasts were accessible, either spatially or temporally (in fluctuating conditions). Air is never in chemical equilibrium, and always contains both reduced and oxidized species. At the top of the atmosphere and in the higher levels there would have been radiation-induced sources of oxidation power: the oxygen left after loss of hydrogen knocked out by UV, cosmic rays or solar wind; and also OH formed from water vapor in the lower air. The flux of UV in particular has major biological impact (Cockell, 2000). In addition, sulfate and nitrate from volcanic eruptions would have been present (Kasting *et al.*, 1989; see also Alt and Shanks, 1998). Such transient species would have contributed vital oxidation power to the oceans; simultaneously reduced species such as CO, H_2, and perhaps NH_3 would have been present also. The chief source of reduction power would be hydrothermal exchange with magma, providing reduced sulfur species, H_2, methane, CO, and ammonia.

Major reactions supporting prephotosynthetic life (Reysenbach and Shock, 2002) may have included a series of processes that depended on molecular hydrogen that was formed inorganically. Water/rock reaction at high temperature (Stevens and McKinley, 1995; Kaiser, 1995) produces molecular hydrogen when circulating groundwater reacts with ferromagnesian minerals (FeO silicate), producing iron oxide (e.g., Fe_2O_3) and quartz. Deeper in the earliest Earth hydrogen may also have been formed by water reaction with iron, as the iron precipitated to the core, producing an oxide component to the core and releasing molecular hydrogen to the mantle.

Such inorganically released hydrogen would have been available to be exploited by microbial life. Some archaea and bacteria use the "knallgas" reaction:

$$\tfrac{1}{2}O_2(\text{aqueous}) + H_2(\text{aqueous}) \rightarrow H_2O$$

Others reduce sulfur:

$$S + H_2(\text{aqueous}) \rightarrow H_2S(\text{aqueous})$$

Methanogenesis (Thauer, 1998) is another process that involves H_2; in this,

$$CO_2(\text{aqueous}) + 4H_2(\text{aqueous}) \rightarrow CH_4(\text{aqueous}) + 2H_2O$$

These processes then allow sulfate chemistry to give many microbial possibilities: for example, an extreme option is

$$CH_4 + SO_4^{-} \rightarrow HCO_3^{-} + HS^{-} + H_2O$$

Similarly, some planctomycetes can exploit ammonium and nitrogen oxides, likely to be found around volcanoes, to make dinitrogen:

$$NO_2^{-} + NH_4^{+} \rightarrow N_2 + 2H_2O$$

Nature evolves by processing waste dumps. It is possible to imagine, for example, an early community of cells, as a single biofilm located on the site of a redox contrast, making its living from one of the hydrogen-using reactions. This would produce a waste of dead cells—reduced carbon—and also the by-products of metabolism such as sulfur or bicarbonate. Various specialist cells would evolve to tap into the new opportunities afforded either by oxidizing the dead carbon, or using the by-product. These new cells would form a substrate, thickening the biofilm. Then, in turn, the waste of the new cells would be utilized, until the whole network resembled a complex clock with innumerable wheels cycling and recycling the thermodynamic possibilities provided by the basic metabolic redox-driven winding of the spring.

8.01.6 THE EARLY BIOMES

8.01.6.1 Location of Early Biomes

Replication of the last common ancestor would lead to mutation: in turn, mutation would create accidental pre-adaptation to life in diverse new habitats. Whatever the habitat of the last common ancestor, the spread of life across the more accessible other locations on Earth was probably rapid, when compared to a geological timescale.

The habitats available were disparate. Examples include: hydrothermal high-temperature (>85 °C) settings; moderate thermophile settings (~40 °C) on the fringes of hot areas, or in cooler, probably more alkaline springs; very cool (~0 °C) water distal to hydrothermal vents but in the flux of metals and geochemical contrasts from hydrothermal plumes; in tidal waters where currents create a flux of nutrient; around terrestrial volcanoes; under ice; or even in air in dust clouds in frequent eruptions and meteorite impacts.

The first organisms to replicate in each habitat would immediately create new habitat by their very existence. Nature spreads on its own ordure. Dead cells would provide reduced organic matter that could be exploited by reoxidation by other cells, and specialist cells would rapidly evolve. Within a number of generations, mutation would lead to a diversified biofilm, relatively reduced at its base and relatively oxidized above (assuming that the redox gradient is between more reduced rock substrate and more oxidized air/ocean system).

This diversification would lead to distinct types of biofilms in specific habitats—the earliest biomes. They may have been only a few microns thick, but these would have been the first complex communities: the ancestors of interdependent ecologies.

8.01.6.2 Methanogenesis: Impact on the Environment

Life operates on a global scale. On a geological timescale, once the first cell had replicated, all habitats on the planet would immediately be filled. This would rapidly have consequences for the atmosphere. In particular, methanogens are likely very ancient, and may long predate methanotrophic bacteria. Methanogens most probably predate photosynthesizers if the evolutionary lengths in the standard models of molecular palaeontology (Woese, 1987; Barnes *et al.*, 1996; Pace, 1997) have value. Possibly they also predate the methane oxidizing archaea. These operate by anaerobic oxidation of methane against sulfate, to produce bicarbonate, HS^- and water: their impact would have been limited by the supply of sulfate oxidant. Once methanogens had evolved, they would have occupied proximal and distal hydrothermal habitats, and then perhaps wider habitats such as open ocean (Sansone *et al.*, 2001) and tidal habitats. Possibly methanotrophs evolved quickly following the arrival of methanogens, to exploit the new opportunity: but, in the likely absence of abundant free molecular oxygen, they would have been severely limited by the supply of oxidant.

Methanogenesis on a scale large enough to affect the atmosphere would have been possible if the hydrogen supply from inorganic and organic sources (and hence methanogenesis) had been adequate: given the likely abundance of ultramafic rock near the surface, interacting with hydrothermal water, it is not unreasonable to suppose a major flux of inorganic hydrogen. If so, and there was a surplus of methane, then much of the methane formed by the first methanogens would have been emitted directly to atmosphere. In the dry air on a cold glacial planet, this methane might rapidly overwhelm the OH. Over a few tens of millennia, the atmospheric methane burden would build up and have a major greenhouse impact (see Pavlov *et al.*, 2000), until enough ice melted to permit OH in air and thus control the methane.

Methane may have played a crucial role in allowing the early Earth to be habitable (Pavlov *et al.*, 2000, 2001). Methane emitted by organisms would have had a substantial greenhouse effect, and if the methane/carbon dioxide ratio in the air were high, methane could have fostered an organic smog that protected shallow-level life against ultraviolet radiation in sunlight (Lovelock, 1988). Thus, there is a possible progression here, from the first methanogens, few in total number and confined to the immediate vicinity of hydrothermal systems on a very cold planet, then a warming trend, then development of planktonic life and much more widely spread methanogens, increasing the warming.

Catling *et al.* (2001) pointed out that in the early Archean, biogenic methane may have saved the Earth from permanent glaciation. On the modern Earth, on a 20 yr timescale, emission of methane has an incremental greenhouse impact nearly 60 times, weight-for-weight, or 21 times, molecule-for-molecule, that of carbon dioxide. On the Archean planet, this ratio would have been very different, and the difference is nonlinear with burden. But whatever the greenhouse impact was, methane is a very powerful greenhouse gas. Indeed, unless abundant methane existed in the air it is difficult to imagine how intense global glaciation was avoided. Thus geologically likely models of the early Archean atmosphere, that are consistent with the Isua evidence for water-eroded and water-transported sediment, would be expected to invoke high methane concentrations (10^2–10^3 ppmv—compared to modern air with less than 2 ppmv CH_4 and ~375 ppmv of CO_2). Such high levels of methane would lead to hydrogen escape by photolysis and loss from the top of the atmosphere, and hence irreversible net oxidation of the planetary surface environment (Catling *et al.*, 2001), though not necessarily to significant ambient O_2 at any particular time.

Methanogenesis may have had the interesting consequence of triggering the evolution of nitrogen fixation (Navarro-Gonzalez *et al.*, 2001; Kasting and Siefert, 2001). On an early planet with CO_2 present in the air, nitrogen fixation would have occurred in lightning strikes, which would have used oxygen atoms from the carbon dioxide (or from water) to form NO. However, if CO_2 levels declined and CH_4 rose, the oxygen supply would be reduced, limiting the synthesis of NO. This would have created a crisis for the biosphere as usable nitrogen is essential. Out of this crisis, Navarro-Gonzalez *et al.* suggested, may have come what now appears to be the essentially "altruistic" process of nitrogen fixation, which is very expensive in energy.

Another, not necessarily incompatible hypothesis is that nitrogenase first evolved as a manager of excess ammonia in the lower, anaerobic part of microbial mats, where hydrogen is present. The product, dinitrogen, could be safely bubbled away. Had a crisis occurred, in which there was a shortage of fixed nitrogen, any cell or consortium of cells able to reverse the process would have been advantaged. It is perhaps notable that in nitrogenase the N_2 is bound to a cluster of Mo–3Fe–3S. Molybdenum, iron and sulfur are likely to be abundant together at hydrothermal systems, especially around andesite volcanoes, and this may be a protein with a hydrothermal heritage. Falkowski (1997) points out that the requirement for iron, and the need for anoxia, would have put severe limits on nitrogen fixation, such that fixed nitrogen supply (and hence the availability of iron), not phosphorus, may be the chief limitation on the productivity of the biosphere. Indeed, the vast scale of human fixation of nitrogen, and perhaps the pH change of the ocean, may some day be seen as the greatest peril of global climate change: not the greenhouse.

8.01.6.3 Prephotosynthetic Ecology

Early life most likely depended on exploiting the transient redox contrasts available from two sources: within the inorganic geological system—especially at hydrothermal vents (Reysenbach and Shock, 2002); and secondly from inorganic light-driven reactions, such as the formation of transient oxidizing and reducing species in the atmosphere by incident radiation.

These sources of redox contrast would have been limited. The hydrothermal contrasts depend on local thermally driven juxtaposition (e.g., in vent fluids) of chemical species from differing environments. From the vents would come H_2, H_2S, and probably CH_4. The size and activity of the hydrothermal biosphere, and hence its impact, would have been considerable, as early Archean volcanism was probably much more common than today, with a higher heat flow out of the Earth. Nevertheless, the total potential productivity of an early hydrothermal biosphere would have been small on a global scale compared to the modern photosynthetically driven biosphere. Moreover, modern biota at hydrothermal vents depend on the supply of sulfate, oxidized in the photosynthetic biosphere: before photosynthesis, the sulfate supply may have been limited. Thus, as a first guess, with a planetary heat flow higher than today but not massively so, and with a limited supply of oxidation power, it is unlikely that the early Archean chemolithotrophic biosphere would have been vastly greater than the sum of today's hydrothermal communities.

In addition, there would have been redox input from transient chemical species formed in the air. The solar radiation, acting on an atmosphere containing water vapor, would likely have produced OH, and probably some O_2. Volcanic gases, taking part in atmospheric chemistry, would produce a small but important supply of sulfur oxides—and hence sulfate and sulfide in the sea, as well as nitrates and nitrites. Moreover, H_2 and CO would have been present. Together, the inorganic sources of redox contrast probably would have been capable of sustaining a small global biological community.

Life must be continuous—it must always have habitat. Volcanoes, however, become extinct. Thus, life must either have been able to live in the open ocean or must have hopped from dying volcano to new volcano. Volcanic vents were probably abundant enough, close enough and accessible enough (especially to cells capable of floating in cool water, or blowing in wind) that they could host gypsy-like cells that were perpetually seeking a new home as the old one was exhausted.

Nonphotosynthetic plankton are abundant today (Karl, 2002). Many of these are eukaryote zooplankton, but there is also a massive population of planktonic archaea, that live near the base of the photic zone. Indeed, in the Pacific, the archaea dominate the deeper waters below ~1,000 m depth, where pelagic crenarchaeota are abundant (Karner *et al.*, 2001). In the early Archean, there may have been a significant boundary between deeper, more reduced water, and shallower water in sunlight. This boundary, as it shifted diurnally, would provide a fluctuating redox contrast for organisms that could exploit it. For example, the planctomycetes, form macroscopic aggregates (>0.5 mm) of detritus, in which they create tiny microaerobic or microanaerobic habitats in otherwise aerobic environments (Fuerst, 1995). They can thus exploit local redox contrast. Among the diverse and interesting properties of the planctomycetes is their ability to react nitrate with ammonia, evolving dinitrogen (the anammox process: Jetten *et al.*, 2001; Fuerst, 1995). This too may be of the greatest antiquity.

Most intriguing of all, they are bacteria that have babies.

8.01.6.4 Geological Settings of the Early Biomes

Geological evidence for the early distribution of life is fragmentary. In the early Archean of the Isua belt, Rosing (1999) reported isotopic and textural evidence of planktonic life, presumably occupying mesophile or cool, even near-freezing habitats, from prior to 3.7 Ga. A possible (though not robust) inference is that from the last common ancestor, fairly early in Earth's history, came the occupation of a diversity of habitats. If Rosing's evidence is correctly interpreted, by ~3.7 Ga, mesophile plankton existed. On the modern Earth, archaeal plankton are abundant in the deeper parts of the upper ocean, in the deep photic zone and below. Though ill-studied, the planctomycetes have marine examples. Thus, a marine biome, occupied by free-living cells, was probably well established and diversified by the mid-Archean.

The geological evidence for the presence of sulfur-processing microbial life and for methanogens goes back at least as far as the late Archean, and probably earlier. Rocks containing highly fractionated sulfur isotopes, closely spatially associated with highly fractionated carbon, are known from many localities (Goodwin *et al.*, 1976). For example, in the late Archean 2.7 Ga sediments of the Belingwe belt, Grassineau *et al.* (2001, 2002) describe what is interpreted as evidence for a complex biological sulfur cycle. Fractionated pyrite, implying sulfur-processing bacteria, is also known from 3.4 Ga Barberton rocks in South Africa (Ohmoto *et al.*, 1993).

Strong evidence for Archean methanogens comes from highly fractionated carbon isotopes. As mentioned above, these have been found in 2.7 Ga material from Belingwe (Grassineau *et al.*, in press), and also from similarly aged rocks in Australia (Rye and Holland, 2000).

Standard rRNA molecular phylogeny (Woese, 1987; Barnes *et al.*, 1996; Stetter, 1996; Pace, 1997) implies the antiquity of hyperthermophile organisms. Though there has been much dispute about the rRNA interpretation, there is some consensus that, whether or not it is the very most ancient, life around hot-water vents is certainly of great antiquity. The implication is that by mid-Archean, hyperthermophile habitats around hot vents were populated by microbial mats, and the waters around hot vents were occupied by free-swimming cells. Mesophile prephotosynthetic plankton probably existed in the open seas, and, distal to the thermophile life in the surroundings of vents, the mesophile habitats further from the hot springs were also occupied.

The reactions that involve sulfur oxidation states leave isotopically fractionated sulfur and hence sulfide, a target for investigation by the geologist. Though there is controversy about sulfur isotope fractionation (Farquhar *et al.*, 2000), the strong fractionation of $\delta^{34}S$ seen in the best-preserved Archean organo-sedimentary rocks can only be biological. Sulphate reducers are probably very old, present 3.5 Ga ago in the early Archean (Shen *et al.*, 2001), and may have provided sulfur deposits, which in turn supported an increase in the supply of HS and H_2S at the bottom of the biofilm: the biofilm would have thickened, diversified, and turned to a microbial mat, created by structured consortia of prokaryotes (Fenchel and Bernard, 1995; Nisbet and Fowler, 1999). Such mats could have had a large impact on the production of reduced gases added to the air (Hoehler *et al.*, 2001), and could have had a global significance in keeping the planet warm (Kasting and Siefert, 2002). Methane generated at the bottom may have been recycled nearer the top of the mat, in processes such as those described in the modern ocean by Boetius *et al.* (2000) in which archaea and sulfate-reducing bacteria consort.

The evolution of photosynthetic oxidation of sulfur compounds permitted the development of the full microbial sulfur cycle in sulfureta. In this cycle, some bacteria and archaea reduce oxidized sulfur compounds, pumping them downward in the microbial mat, while other bacteria reoxidize them photosynthetically. The development of this cycle, coupled with the use of stored sulfur as a redox bank balance that could be exploited either way the redox budget swung during tidal and diurnal cycles, would have greatly expanded the thermodynamic power of the biosphere.

The thermodynamic drive for this life would have come from various sources. In hot-spring settings, reduced species such as CH_4, H_2S, and H_2 would have emanated from inorganic reactions around hot magma. These could have provided the basis of methanogenic life; quickly the supply of dead biomass would provide opportunity for other organisms to generate H_2 organically, thus multiplying the opportunities of the methanogens. At the top of the biofilms, sulfate was probably available in water. In the open seas, prephotosynthetic archaeal and planctomycete planktonic life probably spread ubiquitously even before the advent of photosynthesis—it is a small evolutionary hop from a cell loosely bound to a microbial biofilm and a cell that lives in the sea, floating up and down between redox setting. Possible sources of life support, though limited in total flux, would have been widespread. They would have come from volcanic sources, especially in plumes of hot water, creating the contrast between, above, SO_x and NO_y chemical species dissolved in seawater from the atmosphere, and below, reduced chemical species emanating from hydrothermal vents on the seafloor. Structured consortia of archaea and sulfate-reducing bacteria (Boetius *et al.*, 2000) may have had global distribution.

8.01.7 THE EVOLUTION OF PHOTOSYNTHESIS

8.01.7.1 The Chain of Photosynthesis

Photosynthesis is the source of the redox power that allowed life to escape from the very restricted early settings where inorganic redox contrast existed, and occupy the planet. Without access to light energy, life would have been permanently restricted to a few narrow settings, probably as thin biofilms, and as plankton near upwellings.

Photosynthesis involves a complex chain of events, each of which must have its roots in the remote Archean (Blankenship, 2001). The chain is of great interest, as each unit presents a separate puzzle in explaining its evolutionary history. Light is captured by pigments, such as chlorophylls (in oxygenic photosynthesis by eukaryotes and cyanobacteria) or bacteriochlorophylls (in other bacteria), as well as accessory pigments such as phycobiliproteins. The light is harvested by an array of chlorophyll molecules (say 300) that form an antenna, around a light-harvesting complex. This array passes the energy of the absorbed photon from molecule to molecule until it reaches a photosynthetic reaction center. In purple bacteria, the photosynthetic reaction center consists of special bacteriochlorophyll molecules, linked to other molecules and a central Fe(II) atom. In the overall process in purple bacteria, the net result of two photons hitting the reaction center is the transfer of four H^+ from the interior cytoplasm to the external medium.

In oxygenic photosynthesis, in cyanobacteria and chloroplasts in plants, there are two linked reaction centers. One (photosystem II; PSII) is similar to that in purple bacteria. At PSII, an oxygen evolving complex based on manganese oxide splits two water molecules into $4H^+$ and dioxygen, O_2, which is evolved as waste. The other, PSI, is electrically connected to the PSII production of H^+, and, with two further electrons, generates NADPH; in addition, ADP synthesis occurs on the membrane, driven by proton flow turning the ADP synthase motor. Thus, the products of light capture are NADPH and ATP.

Then in the biosynthesis reactions, the NADPH and ATP are used to capture carbon from the environment, for use in biology. Three ATP and two NADPH, with two H^+ combine with a water and a CO_2 molecule to form carbohydrate. In sum, a dozen quanta of light energy are needed to incorporate one molecule of CO_2. This process is accomplished by the enzyme ribulose-1,5-bisphosphate carboxylase oxygenase, or rubisco, which can in effect work both ways, either capturing carbon dioxide from the air, or oppositely to return it, depending on the $O_2 : CO_2$ ratio it is exposed to (Lorimer and Andrews, 1973, Lorimer, 1981). On rubisco hangs the balance of the atmosphere (Tolbert, 1994).

8.01.7.2 The Rubisco Fingerprint

Geologically, photosynthesis presents several quarries to be hunted down in the geological record. The distinctive isotopic fingerprint of rubisco, which presumably must predate oxygenic photosynthesis, is the most obvious target—it is very selective in the carbon atoms it accepts and hence the organic molecules it creates are highly depleted in ^{13}C. There are two main types of rubisco. Rubisco I is used in oxygenic photosynthesis, it operates in aerobic or microaerobic conditions, not anaerobic settings. Rubisco II is characteristic of organisms that fix CO_2 anaerobically. It may be more ancient, and is today found in deep-sea rent organisms (Elsaied and Nagunama, 2001). The oxygen-evolving complex is also a target for the geologist, as it is based on manganese oxide, as are the transition metal isotopes that are likely to have been fractionated by capture in key enzymes.

More subtly, the isotopic signatures of photosynthesis in inorganic sediment are also valuable. Rubisco depletes the environment of ^{12}C. Hence, inorganic carbonate is enriched in ^{13}C if rubisco operates on a planetary scale. Carbon dioxide emitted from the mantle is about $\delta^{13}C - 5‰$ to $-7‰$, on the arbitrary PDB scale. About a quarter to a fifth of carbon in the environment is captured by rubisco to make organic matter: kerogen (rubisco-fractionated organic matter) has about $\delta^{13}C \sim -28‰$ to $-30‰$ when fractionated by rubisco I, but around $-11‰$ when fractionated by rubisco II (e.g., Guy *et al.*, 1993; Robinson *et al.*, 1998). Three-quarters to four-fifths is residue precipitated as carbonate at $\delta^{13}C \sim 0‰$. The presence of $\delta^{13}C \sim 0‰$ carbonate is thus testimony that rubisco I was capturing carbon on a global scale, in aerobic conditions: this is known as the rubisco fingerprint.

8.01.7.3 The Evolutionary Chain

Respiration most probably evolved before photosynthesis (Xiong and Bauer, 2002). Each step in this chain must have a long and complex evolutionary history (Pierson, 1994): the puzzle is similar to Darwin's puzzle—what use is half of an eye? And half of photosynthesis? The debate is vigorous and is addressed by Blankenship (2001), and references cited therein. How did the full chain evolve, given that half a chain is useless? The challenge to the geologist is to identify the small steps of pre-adaptive advantage on which evolutionary change worked, to date those steps, and to explain the way the individual links in the chain were incorporated.

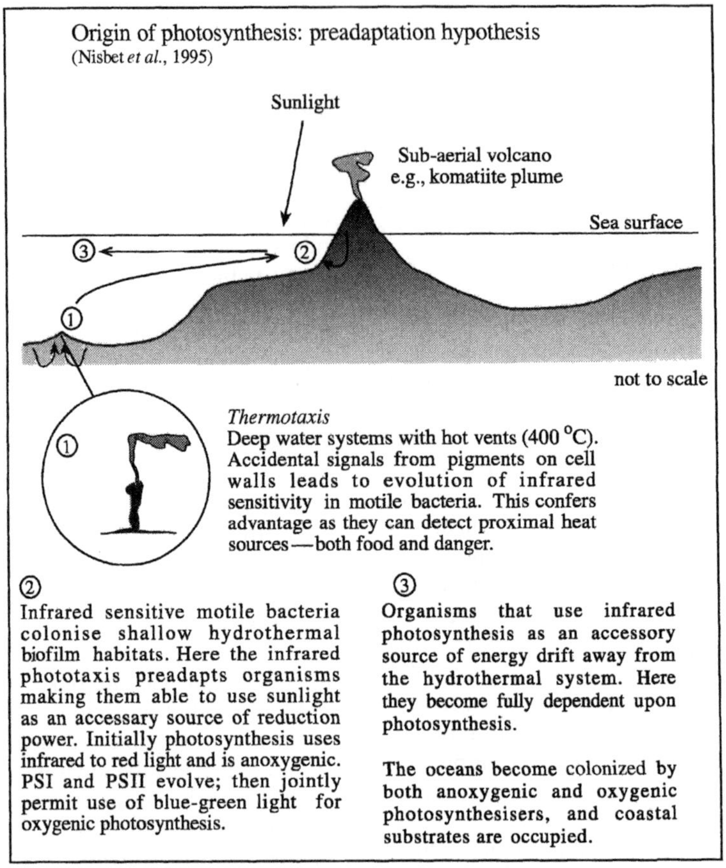

Origin of photosynthesis: preadaptation hypothesis
(Nisbet *et al.*, 1995)
Sunlight
Sub-aerial volcano
e.g., komatiite plume
Sea surface
③
②
①
not to scale
①
Thermotaxis
Deep water systems with hot vents (400 °C). Accidental signals from pigments on cell walls leads to evolution of infrared sensitivity in motile bacteria. This confers advantage as they can detect proximal heat sources—both food and danger.
②
Infrared sensitive motile bacteria colonise shallow hydrothermal biofilm habitats. Here the infrared phototaxis preadapts organisms making them able to use sunlight as an accessary source of reduction power. Initially photosynthesis uses infrared to red light and is anoxygenic. PSI and PSII evolve; then jointly permit use of blue-green light for oxygenic photosynthesis.
③
Organisms that use infrared photosynthesis as an accessory source of energy drift away from the hydrothermal system. Here they become fully dependent upon photosynthesis.
The oceans become colonized by both anoxygenic and oxygenic photosynthesisers, and coastal substrates are occupied.

Figure 10 Possible evolutionary chain leading to photosynthesis: hypothesis of pre-adaptation for infrared thermotaxis (Nisbet *et al.*, 1995).

There is much debate about the origin of photosynthesis, and little agreement. Among the many hypotheses, Nisbet *et al.* (1995) suggested that photosynthesis began in organisms that were pre-adapted by their ability to use IR thermotaxis to detect hot sources (Figure 10). This hypothesis offers a set of small incremental steps, each immediately advantageous, each depending on accidental pre-adaptation that led to the very sophisticated electron management that occurs in photosynthesis. The steps begin with accidental IR sensitivity in cells that had pigments in their outer surfaces. Deep-water hot vents emit IR radiation at around 350–400 °C, slightly below the temperature of a hot plate on a kitchen cooker before it becomes a visible cherry-red. Detection of this radiation would have been very advantageous to motile organisms, and such organisms that possessed IR detection and the ability to move towards the source (or away if it became too powerful) would have gained survival advantage. Evolutionary survival would then have favored those cells that were increasingly finer-tuned to the IR. Then, in the next step, organisms that had spread to a shallow-water vent would be pre-adapted to use solar IR as a supplementary energy source; finally, full dependence on abundant and energetic visible light energy would follow. The hypothesis suggested by Nisbet *et al.* (1995) invokes IR phototaxis: the cells that use IR depend on bacteriochlorophyll and are anoxygenic and usually act in anaerobic settings.

However, other hypotheses suggest that oxygenic photosynthesis came first, depending on chlorophyll, and that from chlorophyll evolved bacteriochlorophyll. Bacteriochlorophyll absorbs into the IR. Chlorophyll *a* is green in color as it absorbs red and blue light and reflects green: thus green, though much loved, is waste light: the biosphere's chief excretion. If chlorophyll came first, the hypothesis of "evolution via IR thermotaxis" would be invalid.

Which came first—bacteriochlorophyll or chlorophyll? Much of the debate centers on the long-held Granick hypothesis (Granick, 1965). The steps to the synthesis of chlorophyll being simpler, it would intuitively be expected to have come first. Recent work supports the notion that bacteriochlorophyll predates chlorophyll (Xiong *et al.*, 2000), refuting the Granick hypothesis, so the "IR thermotaxis" hypothesis remains tenable.

Chlorophyll and bacteriochlorophyll are closely related, and both center around a porphyrin ring that contains an magnesium atom surrounded by four nitrogen atoms (see Section 8.01.5.2 for the argument that these originated in alkaline fluids from hydrothermal systems in ultramafic lavas such as komatiites). Similar porphyrin rings lie at the heart of haem (where the central metal is iron) and the enzyme catalase, that helps split hydrogen peroxide to water and dioxygen (thereby allowing the excretion of the poison, either to the external environment or to attack neighboring cells), as well as in the cytochromes. Many of these must be of the very greatest antiquity and probably predate the last common ancestor. They were clearly exploited in the ancestry of photosynthesis, which may have been via evolutionary tweaking of respiratory processes. Xiong and Bauer (2002) concluded that cytochrome *b* may have been the ancestor of type II photosynthetic reaction centers. Inorganically, linking metal with nitrogen occurs at very high pH.

8.01.7.4 Anoxygenic Photosynthesis

Anoxygenic photosynthesis is carried out by a wide range of bacteria. The chief groups are green sulfur bacteria, such as *Chlorobium* (which do not use rubisco), green nonsulfur bacteria, such as *Chloroflexus*, purple sulfur bacteria, such as *Thiospirillum*, and purple nonsulfur bacteria (e.g., *Rhodobacter*). The purple bacteria (proteobacteria) are classified by 16S rRNA study into several major evolutionary groups (Woese, 1987). Green sulfur bacteria are strict anaerobes and obligate phototrophs, using hydrogen sulfide, hydrogen, or elemental sulfur, and are unable to respire in the dark. Some have gas vesicles that allow them to float up and down in lakes, adjusting their level with the movement of the redox boundary. Green nonsulfur bacteria are thermophiles, and *Chloroflexus* is typically found as gliding bacteria in mats in hot springs. Purple sulfur bacteria are strict anaerobes, oxidizing hydrogen sulfide to sulfur and often eventually to sulfate. They typically inhabit deeper, anaerobic, parts of the photic layer of lakes, where IR light penetrates. Purple nonsulfur bacteria (many of which are nonphototrophic) are very flexible in life. Normally anaerobic photosynthesizers that use organic molecules as electron acceptors and carbon sources, some species can also oxidize low (nontoxic) levels of sulfide to sulfate. In dark, most purple nonsulfur bacteria can grow in aerobic or microaerobic conditions.

The linking characteristic between these groups is the use of various types of bacteriochlorophyll in a single stage process, involving either photosynthetic reaction center II (e.g., purple bacteria) or photosynthetic reaction center I (e.g., green sulfur bacteria). This photosynthetic process uses electron donors such as H_2, H_2S, S, or organic matter, and does not, as a consequence, evolve waste oxygen. Many green and purple bacteria can grow phototrophically using H_2 as the sole electron donor and CO_2 as the carbon source, using hydrogenase (a nickel enzyme) for CO_2 reduction.

The two photosystems are structurally related, and Xiong *et al.* (2000) concluded from a study of sequence information in photosynthesis genes that green sulfur and green nonsulfur bacteria are each other's most closely related groups. Phototrophic purple bacteria use the Calvin cycle, and utilize rubisco, with its characteristic (and geologically identifiable) strong fractionation of carbon. Green bacteria, however, do it differently and do not produce the same isotopic signature: *Chlorobium* uses the reverse citric acid cycle, and *Chloroflexus* the hydroxyproprionate pathway. Geologically, these should be distinguishable in the kerogen record from rubisco-captured carbon.

When the first photosynthetic sulfur-compound oxidizers first appeared, the development of full sulfureta would have been possible. Sulfate reducers would take sulfate from the external environment and eventually produce H_2S. Then the photosynthetic oxidizers would reverse the steps, e.g.,

$$6CO_2 + 12H_2S \rightarrow C_6H_{12}O_6 + 6H_2O + 12S^0$$

depositing the sulfur either outside the cells (as in phototrophic green bacteria) or inside them (in most of the purple bacteria). This is a trail the geologist can hunt.

The isotopic evidence of Shen *et al.* (2001) is not inconsistent with a full sulfur cycle but does not prove it. The wider fractionation observed in 2.7 Ga material by Grassineau *et al.* (2002) (which also includes carbon isotope support) is very suggestive of a full S cycle. Grassineau *et al.* (2002) reported abundant evidence for highly fractionated carbon in kerogen, the signature of rubisco. Such evidence suggests either the presence of cyanobacteria, or anoxygenic photosynthesizers: by extrapolation, this implies the presence of bacteria capable of oxidizing sulfur compounds.

Once anoxygenic photosynthesis had evolved, the planet would have become widely habitable and the biosphere much more productive. Tidal and shallow-water environments around the globe would have been immediately occupied by life. Sulfureta would have cycled sulfur, derived from oceanic volcanogenic sulfate, between the upper more oxidized layers of mats and the lower H_2S-rich layers, creating a complex microbial mat habitat.

Plankton today are very diverse, including archaea, bacteria, and eukaryotes (Beja *et al.*, 2002). It is very likely that anoxygenic photosynthesizers would rapidly have spread as plankton, limited only by the availability of reducing chemicals. It is possible to imagine a microbial biosphere dependent on anoxygenic photosynthesis, with widespread abundance of oxidized sulfur and nitrogen chemical species in the uppermost few tens of meters of the sea, above an anoxic deeper mass of the seas and oceans. At this stage, the oceanic biomes may have been stratified with anoxygenic purple and green photosynthesizers, as well as planctomycetes (Fuerst, 1995), which may be a very ancient prephotosynthetic branch of the bacteria (Brochier and Philippe, 2002), as discussed earlier.

Complex global-scale nitrogen cycles would also have become possible at this stage. The planctomycetes, if presumed ancient, are capable of emitting N_2 by reacting NH_3 with NO_2. In the reverse direction, supplying nitrogen from the air, the inorganic sources are mainly volcanoes and lightning. But this source could have been restricted (e.g., see Navarro-Gonzalez *et al.*, 2001). The modern nitrogen cycle is dominated by bacteria. Some bacteria such as *Pseudomonas* release N_2. Many purple and green bacteria can fix nitrogen, using the Fe–Mo enzyme nitrogenase. Anoxic nitrification can occur, coupled with manganese reduction (Hulth *et al.*, 1999). A nitrogen cycle may have become possible very early on, some species emitting gaseous nitrogen, others capturing nitrogen from air/ocean. Nitrogen fixation may be closely connected with hydrogen emission: in reducing N_2 to NH_3, eight electrons are consumed, six for producing $2NH_3$ and two to make H_2: hydrogen production and nitrogen fixation appear closely linked. Possibly nitrogenase originally evolved to manage ammonia in close association with methanogenesis using H_2.

With the evolution of the anoxygenic bacteria, the global-scale biosphere would have been greatly enriched. It would have been capable of cycling sulfur, carbon, and nitrogen on a global, scale, and presumably with fluxes that were on a much greater scale than the inorganic volcanogenic fluxes—over a geologically brief time, bacterial emissions would thus have used photosynthetic energy to reconstruct the atmosphere. From this date also N_2 has been a biological product in the main, produced and consumed by organisms.

8.01.7.5 Oxygenic Photosynthesis

The development of oxygenic photosynthesis created the modern biosphere. The use of ubiquitous ingredients, water, carbon dioxide, and light, to capture carbon into life, was the final metabolic step that made the entire planet habitable by life. The waste product was simply dumped—indeed, it may originally have been a deliberate toxic by-product in toxin warfare between cyanobacteria and their neighbors. Cyanobacteria achieve this by a multicomponent system. Most likely (though not if the Granick hypothesis is correct), oxygenic photosynthesis came *after* the development of photosynthetic reaction system II in purple bacteria and reaction system I in green sulfur bacteria (Nisbet and Fowler, 1999).

There are many notions about how oxygenic photosynthesis evolved (e.g., Blankenship, 2001; Nisbet and Fowler, 1999). All photosystems are basically alike and must have had a common origin (Jordan *et al.*, 2001; Kuhlbrandt, 2001). Heliobacteria, which are anoxygenic phototrophs living in tropical soils, utilize a modified form, bacteriochlorophyll *g*, that is related to cyanobacterial chlorophyll *a*. They may be the microbial branch with the photosynthetic genes, which are most closely related to the ancestral cyanobacteria (Xiong *et al.*, 2000). Perhaps the photosynthetic reaction system in green gliding bacteria, such as *Chloroflexus*, is ancestral to both. What of the host cell, apart from the photosynthetic process? One possibility is that the cyanobacterial cells themselves are chimaera, created by genetic transfer (perhaps lunch) between close-living or symbiotic purple and heliobacteria. This would imply that purple bacteria evolved before cyanobacteria. Perhaps a primitive reaction system evolved first, in the mutation that produced the common ancestor of the purple bacteria, then a further mutation led to the ancestor of the green sulfur bacteria and of the heliobacteria. It is possible that the first O_2-evolving photoreaction center originated in green nonsulfur bacteria, and that this was later incorporated into cyanobacteria (Dismukes *et al.*, 2001). Then, to speculate further, possibly the two lines formed a symbiotic partnership across a redox boundary and eventually became so close that the genes for PSI and PSII were incorporated into the cell. Another possibility is that, following the development of photosynthesis in the purple bacteria, transfer of Mg-tetrapyrrole genes occurred to the line leading to the cyanobacteria occurred, plus gene duplication, to produce the cyanobacterial reaction center II in the ancestral cyanobacterium (Xiong and Bauer, 2002). The puzzle remains open.

The evolution of the cyanobacteria massively changed the ability of the biosphere by harvesting sunlight, and using it to sequester reduced chemical species from the waste oxidation power dumped into the air. These cells would be able, in a single cell, to photosynthesize with the most available of ingredients, water, light and air, and to fix nitrogen (e.g., Zehr *et al.*, 2001), and even to grow anaerobically if need be.

Nitrogen supply is a key limitation on productivity (Falkowski, 1997). Cyanobacteria fix nitrogen. This is a process that needs low oxygen tension in heterocysts, yet the cyanobacteria can also use the oxygen-evolving complex to excrete waste oxygen. The formation of nitrate from ammonium needs molecular oxygen (Falkowski, 1997): it is reasonable to suppose that this could not have evolved until oxygenic photosynthesis appeared. But, conversely, productive oxygenic photosynthesis could not have become global unless there was a good supply of biologically accessible nitrogen. Cyanobacterial plankton still today occupy the tropical oceans in vast numbers (Capone *et al.*, 1997), and the chloroplast in a modern plant is in effect a cyanobacterium in a space suit. Given that respiration today is still carried out by mitochondria, which are in effect proteobacteria also in space suits, the modern cycle of life had begun.

When did this occur? The key signature (see Section 8.01.7.2) is in the $\delta^{13}C \sim 0‰$ isotopic signature of rubisco I in carbonate rocks (Schidlowski and Aharon, 1992; Schidlowski, 1988, 2002; Nisbert, 2002). This is the modern fingerprint imposed by the chloroplast, still a member of the cyanobacterial line. Carbon dioxide in the atmosphere and ocean is well mixed. For the $\delta^{13}C \sim 0‰$ fingerprint to occur, carbon dioxide must have been managed by rubisco I on a global scale. The only process that could perform this is photosynthesis: Although purple bacteria use rubisco, arguably only oxygenic photosythesis can drive the Calvin cycle to capture carbon dioxide on a scale large enough to create the isotopic signature.

The evidence for the $\delta^{13}C \sim 0‰$ signature is strong around 2.7 Ga (e.g., Grassineau *et al.*, 2002). Buick (1992) in the 2.7 Ga Tumbiana formation in Western Australia presents strong textural evidence for oxygenic photosynthesis in stromatolites growing in shallow lakes. However, older evidence for oxygenic photosynthesis is problematic. The ca. 3 Ga Steep Rock carbonates have $\delta^{13}C$ not far from 0‰ (Abell, Grassineau and Nisbet, unpublished), but in older material there is strong controversy (e.g., Brasier *et al.*, 2002; see also Schopf *et al.*, 2002).

8.01.7.6 Archean Oxygen

By 2.7 Ga ago, the modern carbon cycle was in operation: the oxygen production must have been considerable. Did it build up in the air? For contrasting views on this vexed problem, see Holland (1999) and Ohmoto (1997). Catling *et al.* (2001) argue persuasively for a high-methane atmosphere, or Earth would have frozen over. Towe (2002), commenting on Catling *et al.* (2001) presented strong arguments that it would be very difficult for the Earth system to scavenge back the free dioxygen released by the cyanobacteria, and argued equally persuasively for a low-O_2 but oxic atmosphere in the late Archean. Catling *et al.* in response (see Towe, 2002), with somewhat different assumptions, defended the methane-rich model of the air, though agreeing that local high-O_2 "oases" (presumably water masses rich in dissolved oxygen) and high-O_2 events could occur just as today methane accumulates in swamps despite the O_2-rich air. Phillips *et al.* (2001), in a careful review of the actual rock evidence, based on much field knowledge, consider that some of the mineralogical and field evidence can be interpreted as supporting an oxidized Archean atmosphere but conclude that the geological evidence for a reducing atmosphere remains ambiguous. In particular, postdepositional processes may need far more examination. Similar conclusions can be drawn from the rocks of Steep Rock and Belingwe.

Kasting (2001) argues in support of the view of Farquhar *et al.* (2000) (but see also Ohmoto *et al.*, 2001) that sulfur isotope fractionation changed around 2.3 Ga. This opinion is based on the claim, from comparison of sulfur isotopes, that so-called "mass independent" fractionation occurred as a result of gas-phase photochemical reactions, particularly photolysis of SO_2. Such fractionation would be much more likely to occur in a low-O_2 atmosphere in which sulfur was present in a variety of oxidation states. Thus, the claim that fractionation changed around 2.3 Ga ago can be seen as supporting the notion that there was a substantial rise in O_2 around this time. This, however, raises the question: if cyanobacterial oxygen production had been sufficient to create the rubisco fingerprint in carbonates as early as 2.7–3.0 Ga ago, why did the rise of free O_2 only occur 400–700 Myr later?

The implications of the Catling *et al.* (2001) suggestion that the air had high methane concentrations (>0.1%) in the late Archean are worth further thought. If so, then consequently, as methane mixed into the stratosphere and upwards through the mesosphere, the Earth would have lost much hydrogen through the thermosphere at the top of the atmosphere. Loss of hydrogen from biologically produced methane equates to surplus of oxygen. This would have produced a substantial net accumulation of oxygen, consumed by oxidation of crust and perhaps by the creation of an upside-down biosphere (Walker, 1987), in which the sediment was more oxidized than the water or air above. The debate continues.

8.01.8 MUD-STIRRERS: ORIGIN AND IMPACT OF THE EUCARYA

8.01.8.1 The Ancestry of the Eucarya

The origin of the *Eucarya* remains deep mystery. Some (e.g., Forterre, 1995, 1996) would place it very early indeed; yet it has also been ascribed to a time as recently as 850 Ma ago, in the later Proterozoic (Cavalier- Smith, 2002).

The geological evidence for Archean eukaryotes can be dealt with swiftly. Brocks *et al.* (1999) found organic molecules (sterols) in Archean sediment, that they ascribed to the presence of eukaryotes. This is permissive but not necessarily persuasive evidence, as some bacteria (e.g., methanotrophs, planctomycetes) may leave similar molecular records; thus the interpretation by Brocks *et al.* (1999) is contested (Cavalier-Smith, 2002). Nevertheless, the simplest interpretation is that this is a just-plausible record of Archean eukaryotes.

The molecular evidence for the descent of the eukaryotes (Hartman and Fedorov, 2002) is deeply controversial (see Section 8.01.6.3). Standard models (Woese, 1987) suggest an ancestral line among the Archea, with massive transfers and symbioses from the bacteria. The standard model (e.g., see summaries in Pace (1997); Nisbet and Fowler, 1996a,b) is that early archaea and bacteria diverged from a hyperthermophile last common ancestor. Then, a sequence of symbiotic events took place between a stem–cell line, among the archaea, that developed partnerships with symbiotic purple and cyanobacteria, either in separate events, or in a single moment of fusion. This produced the eukaryote cell, with the mitochondria derived (Bui *et al.*, 1996 from within the α-proteobacteria such as *Rickettsia*. The other great acquisition of the eukaryotes, the chloroplast, is clearly related to the cyanobacteria (Figure 11).

Much discussion followed on the timing of the event or events, especially as some eukaryotes lack mitochondria. Could they be more primitive? At first it was thought so, but recently it has become clear that even eukaryotes, such as the microsporidae, that are no longer capable of aerobic respiration still have relict mitochondrial proteins (Williams *et al.*, 2002; Roger and Silberman, 2002). The ancestral eukaryote did probably possess mitochondria, and the amitochondrial eukaryotes lost them. Even simple eukaryotes that today do not have mitochondria (e.g., some parasites) appear to have once had them and then lost them; moreover, mitochondria and hydrogenosomes (distinctive hydrogen-producing organelles in some amitochondrial organisms) appear to have had a common ancestor (Bui *et al.*, 1996). Thus it appears that the ancestors of all modern eukaryotes diverged after the mitochondrion symbiosis. Likewise, animals may have descended from an ancestral photosynthesizer by loss of the chloroplast.

Whatever the explanation of the stem eukaryote, the eukaryote organelles, both mitochondria and chloroplasts, are best explained as symbiont bacteria. Explanations of the mitochondrial symbiosis mostly invoke an early Archean stem that incorporated a bacterial symbiont. One explanation of the mitochondrion is that the origin of the mitochondrion was simultaneous with the origin of the eukaryote nucleus (Grey *et al.*, 1999). In the "hydrogen hypothesis" (Martin and Muller, 1998), the symbiosis is seen as the end product of a tight physical association between anaerobic

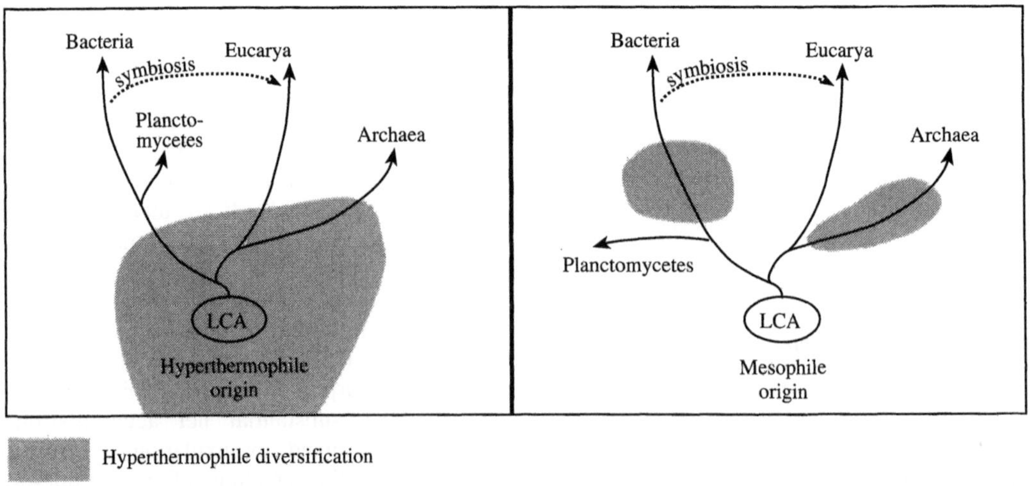

Figure 11 Standard and alternative models of eukaryote evolution (see Figure 7) (with thanks to J. Fuerst). Alternative model assumes a last common ancestor that was mesophile, and that the divergence of the planctomycetes was very ancient (see Brochier and Philippe, 2002).

archaea and heterotrophic proteobacteria capable of producing molecular H_2 through anaerobic fermentation. In another version of the close-association idea, anaerobic archaea may have evolved the ability to survive in oxidizing settings by incorporating respiring proteobacteria (Vellai and Vida, 1999).

In contrast, Penny and Poole (1999) suggested that the last universal common ancestor may have been a mesophile with many features of the eukaryote genome, and the first distinct eukaryote may thus also have been mesophile—a distal inhabitant of a hydrothermal system, or a planktonic form.

However, if the "eukaryote-like" view of the last common ancestor (Forterre and Philippe, 1999; Glansdorff, 2000) is correct, then the sequence of events may have been greatly different. In this view, an ancient common ancestor may have been a fairly complex organism, living in mesophile conditions, possibly some hundreds of meters distal to a shallow hydrothermal system, in water between, say, 35–45 °C (blood-temperature, optimal for DNA-based life), perhaps in water with pH around 7 or more alkaline (to account for the cytochromes).

From this, in Forterre's thermoreduction hypothesis (1995), came several lines of descendants. Some became colonists of much hotter environments. Of these lines, some necessarily streamlined both their genomes and their physiology, in order to survive, while others, also with reduced physiology, developed heat shock proteins to correct damage. This led to the distinct domains, the *Bacteria* and *Archea*. In contrast, other descendants retained the complex more primitive physiology—biplanes or triplanes, as opposed to prokaryote monoplanes. This third line became the *Eucarya*. This third line may share some characteristics with the planctomycetes (Fuerst, 1995; Lindsay *et al.*, 2001).

At some stage in this hypothesis came the key acquisition by the ancestor of the modern *Eucarya* of the mitochondrion. Possibly: (i) this was very early—perhaps not so much an acquisition as a primitive characteristic; alternately (ii) it may have been a later product of a symbiosis between a mesophile eukaryote stem-organism directly descended from a mesophile eukaryote-like last common ancestor and a proteobacterium that had evolved from a line that had passed through a hyperthermophile bottleneck; or, (iii) it could have been a later product of a symbiosis between two organisms that had both been through a hyperthermophile stage, an achaea-like host and a proteobacterial symbiont.

Chloroplasts may have been acquired at the same time as the mitochondrion, or much later. Some lines of dinoflagellates appear to have had multiple gains and losses of plastids (Saldarriga *et al.*, 2001), although perhaps from a single ancient endosymbiotic origin (Fast *et al.*, 2001). The same hypotheses apply. It is not clear whether all modern eukaryotes are descended from organisms that possessed chloroplasts, but the hypothesis is attractive. Some have multiply acquired and lost chloroplasts.

At many stages in their evolution each of the three domains gave and received genetic material with the other two lines, so that major innovations were acquired by sharing between all three domains.

As for the antiquity of the *Eucarya* there is no consensus. Those who support a eukaryote-like last common ancestor, of course, propose that the eukaryotes date to the very start of the Archean and end of the Hadean. It is not improbable to those who consider that the *Eucarya* were the last domain to appear, that Eukaryotes first evolved in the Archean aeon. There is, however, little support in the rock record for the hypothesis of a "very late" origin of both the archaea and eukaryotes, proposed by Cavalier-Smith (2002), especially as the evidence for early methanogens is strong (Grassineau *et al.*, 2002; Rye and Holland, 2000). However, a proterozoic origin of the eukaryotes is not yet excluded, as the sterols found by Brocks *et al.* (1994) could be of prokaryote origin.

8.01.8.2 Possible Settings for the Eukaryote Endosymbiotic Event

What was the purpose of symbiosis? And were the organelles incorporated simultaneously or sequentially? The answers are not known, but an argument can be made that simultaneous acquisition of both organelles took place, as they in effect balance.

The most likely setting of symbiosis is a microbial mat community, in which a complex community of cells is clustered across a redox boundary, cycling and recycling redox power (Nisbet and Fowler, 1999; Nisbet, 2002). The aerobic top of the mat would include photosynthetic cyanobacteria, above photosynthesizing purple bacteria. There would be a very sharply focused redox boundary. Below would be the green photosynthetic bacteria, and at the base the methanogens and the hydrogen producers.

In the Archean, such prokaryotic mats would be limited to some extent by diffusion gradients, in the absence of multicelled organisms like worms capable of physical movement of fluid on a large scale. However, microbes are motile and, moreover, they can move fluid, so the thickness of the mat would be substantial compared to the dimensions of a single cell, despite lack of physical power.

In such a setting there is great benefit from being very close to the redox boundary between aerobic and anaerobic conditions, where the greatest thermodynamic power is to be had. Any cluster of cells that straddled this boundary, or incorporated it within itself, would possess great advantage. To some extent, some cyanobacteria already do so within their cell, as they include heterocysts, which protect nitrogenase, which enzyme needs to function within the cell in anaerobic conditions despite the emission of molecular oxygen from the cell during photosynthesis. Any cluster of cells that carried out oxygenic photosynthesis and yet managed to control redox levels by respiration also, would be greatly advantaged. Oxygen is dangerous, and mitochondria may have evolved to manage it (Abele, 2002). There would be much advantage to a symbiotic association, located just above the redox boundary, of a host cell that linked cyanobacteria and purple bacteria, alternately providing useful redox waste to each.

Photosynthesis by a symbiont cyanobacterium would produce reactive oxygen species in the fluid. The buildup of oxidation power in the near environment would be a nuisance to the cyanobacterium as its rubisco would begin to work in reverse. Thus, it would have to wait until the oxygen diffused away before continuing photosynthesis. Moreover, the oxygen would be damaging to the nearby symbiont cell. However, if nearby, on the redox boundary, there were an α-proteobacterium, this would mop up the excess oxidation power immediately, allowing the cyanobacterium to keep photosynthesizing. Thus, the host cell would be protected, the rubisco in the cyanobacterium would operate, and the respiring α-proteobacterium would flourish. Such an arrangement is beneficial to all: thus it is tempting to imagine that both incorporations took place by single lucky improbable accident. Arguably, however, it is more likely that initially only one partner was incorporated, then the second.

8.01.8.3 Water and Mud Stirring—Consequences

Multi-celled eukaryotes have some unique advantages and some disadvantages. The disadvantage is that they evolve by Darwinian evolution. Genetic change can only occur when parent organisms have a number of different offspring, some of which are better suited than others to the environment in which they find themselves: these are more likely to survive and in turn have offspring, so natural selection chooses the genes most suited. Only females reproduce, so the ability to "bloom" is slightly restricted by one generation time. For single-celled eukaryotes, rapidly passing through the generations, the evolutionary process of adapting to a changed environment (e.g., the arrival of a virus) can be quick and population recovery fast, but for an elephant that lives for decades, or a tree that lasts for centuries, adaptation can be slow and the population can be brought dangerously near extinction before it can respond to the new challenge.

Human cultures evolve as bacteria do, by swapping genetic information among living individuals. This is a rapid and highly advantageous method of adaptation. Most readers of this (except Scots) are likely to wear some variety of undergarment, but the habit is only a few generations old: prior to that it was thought unclean. The change was non-Darwinian. Those human families (probably most) who did not wear such attire a few lifetimes ago did not become extinct—they mysteriously acquired the habit from contemporaries by a hidden process of cultural infection. Eukaryotes do, to some extent, adopt such quasi-Lamarkian evolution, in the immune system. This has an extraordinarily bacterial-like ability to learn in life—perhaps it alone has ensured the domain's survival in the face of viral challenge and microbial attack.

The advantages of the eukaryotes are that they can mix genomes over long distances (males wander), and also create striking cellular architecture, and thereby link together colonies of cells so that they form a single unit with distributed tasks. This may have major consequences in the Proterozoic when multicelled eukaryotes became capable of moving water and stirring mud. Bacteria do this to some extent, but only slightly. By stirring mud and water, the eukaryotes expand the range of the biosphere. The bacterial biosphere is at most only a few millimeters thick—the growing biofilm in a microbial mat. The physical structure of a prokaryote mat may include a debris layer a meter or more thick, but most of the action is close to the redox boundary. In contrast, eukaryotes can move redox power up and down, and widen the environment—they become capable of more reducing power, more photosynthesis, limited only by nutrients such as iron. Eventually, they even become able to send roots down into the soil and rock to extract the nutrient, or, in humans, to dig for potash to put on fields, or to fix nitrogen directly.

The expansion of the productivity of the biosphere by the eukaryotes must have begun slowly, but it probably started in late Archean. Around 2,300 Ma, much evidence suggests (but does not conclusively prove: see Section 8.01.8.6) that oxygen levels rose sharply. Possibly the eukaryotes were beginning to muscle the world.

8.01.9 THE BREATH OF LIFE: THE IMPACT OF LIFE ON THE OCEAN/ATMOSPHERE SYSTEM

8.01.9.1 The Breath of Life

The modern atmosphere is the breath of life—a biological construction. Excepting the argon, the balance of the exosphere is entirely fluxed by life: the gases are emitted and taken up by living organisms. This does not necessarily mean that life manages the air, as life may simply be fast-tracking inorganic processes that would happen anyway, but there is also the possibility that life is maintaining thermodynamic disequilibrium (e.g., the sharp contrast between reducing sediment and oxidizing air; or on an even finer scale, the kinetically improbable presence of ammonia and methane in oxygen-rich lower tropospheric air).

The controls on the atmosphere's operation (Walker, 1977) are complex and poorly understood, yet have been robust enough to keep the planet habitable over 4 Ga. The greenhouse effect adds ~33 °C to the temperature (Lewis and Prinn, 1984). Without the atmosphere, the temperature of the planet would be ~−18 °C; with the atmosphere it is a pleasant +15 °C. But is the control pure inorganic chance, or is it somehow implemented because the Earth, uniquely, is inhabited? And when did the control begin?

8.01.9.2 Oxygen and Carbon Dioxide

Since the Archean the oxygen that has been emitted into the atmospheric reserve by the oxygen-evolving complex as the waste product of oxygenic photosynthesis, has been taken up again by respiration. The carbon is the other side of this coin: it is stored for the most part in the biosphere and crustal reserves, forming a well of reduction power that matches the surficial oxidation power, with just enough carbon dioxide is sustained in the air to allow rubisco to operate in balance (Tolbert, 1994). There are complex inorganic controls and buffers in the carbon dioxide content of the air and the partitioning of carbon between air/water and surface/crust (Walker, 1994), but the extent to which biological processes exert the fundamental control remains very controversial. One argument that may be made that the most basic control may lie within the cell itself, in within-cell controls (Joshi and Tabita, 1996). The control in photosynthetic cells would lie in the balance between chloroplasts and mitochondria in eukaryote cells. The debate remains open.

8.01.9.3 Nitrogen and Fixed Nitrogen

Dinitrogen, like dioxygen and carbon, is almost entirely a biological product: there is roughly 3.8×10^{21} g presently in air, and each year $\sim 3 \times 10^{14}$ g are added and subtracted from this reservoir by denitrifying and nitrifying bacteria (lifetime in air ~10–100 Ma). Nitrogen is emitted to the air by guilds of denitrifying bacteria (e.g., *Pseudomonas, Bacillus*), which reduce nitrate to N_2 as an alternative respiration process in anaerobic settings when oxygen is absent. This raises the inference that this process may be very old, and could have evolved before oxygenic photosynthesis, predating the time of abundant oxygen supply. As argued above (Section 8.01.6.3), the anammox process may also be of great antiquity, perhaps long predating oxygenic photosynthesis.

The nitrogen cycle has a controlling role on the carbon dioxide/oxygen cycle (Falkowski, 1997). Were all fixed nitrogen to be evolved as dinitrogen, the biosphere would rapidly be reduced to the nitrogen oxides and ammonia emitted by volcanoes and hydrothermal systems. Nitrogen fixation reverses this, and is carried out by cyanobactria, by free-living bacteria such as *Azotobacter* and *Clostridium*, by archaea such as *Methanococcus*, and (perhaps since late Palaeozoic) by plant symbionts. Fixation is very expensive energetically, requiring a large ATP price. At least six electrons and 12 ATP are required to fix one dinitrogen molecule. The use of nitrogen fixation by cyanobacteria is especially interesting geologically, as cyanobacterial picoplankton are very important today and presumably were from the geological moment the cyanobacteria evolved. Given the clear evidence for cyanobacteria (albeit bound in mats) in the Archean (Summons *et al.*, 1999), it is reasonable to assume that cyanobacterial plankton were presumably ubiquitous in the Archean oceanic photic zone. Thus, there is a reasonable presumption that in the Archean the dinitrogen atmospheric burden was organically fluxed.

8.01.9.4 Methane

Of the lesser gases, methane is the most interesting. Today, the natural methane sources are primarily archaea, but operating in eukaryote hosts (e.g., archaea symbiotically cooperating with plants in wetland, termite stomachs, cows, and sheep). In the Archean, methane was probably managed by complex microbial communities, comparable to those in modern oceans (Boetius *et al.*, 2000). Walker's (1987) surprising hypothesis that the Archean biosphere at times may have been inverted, with a relatively reducing atmosphere and a relatively oxidizing sediment is not as absurd as it seems. Today the oxygen-rich air is sustained by photosynthesis. Prior to oxygenic photosynthesis, the air would have contained relatively oxidized species (carbon dioxide and water, as well as dinitrogen), but also substantial methane and probably ammonia emissions occurred, that would have had multiyear atmospheric lifetimes. In the continental slope sediments, huge methane hydrate reserves would have built up, as they do today (Kvenvolden, 1988), In these circumstances, episodes of major atmospheric methane burden could occur, as perhaps happened at the Archean–Proterozoic transition (Hayes, 1994). For example, this could occur after massive release of geological methane stores (e.g., see Kvenvolden, 1988; Harvey and Huang, 1995).

Today a large part of the biosphere is reducing—much of the soft sediment. It is possible, especially if methanotrophy were absent or ineffective in the absence of abundant oxidant, that Archean "Walkerworld" events may have occurred, when the biosphere was inverted: relatively reducing air and oxidizing sediment. Once such an event was established, it might be stable for long periods, until reversed by the combined impact of volcanic degassing of carbon dioxide and nitrogen oxides, and of methanotrophy. With methane, other reduced gases such as ammonia would build up in the air, reversing the nitrogen cycle also.

8.01.9.5 Sulfur

The oxidation states of sulfur may have been the core tool by which life bootstrapped its way to a global biosphere (Kasting *et al.*, 1989; Kasting, 1993, 2001). Sulfur offers a wide range, from H_2S through HS and sulfur to the oxidized species up to SO_3 (H_2SO_4). Moreover, dimethyl-sulfide is two methyls linked with one sulfur.

For bacteria living close to a redox boundary, sulfur is a marvellous reservoir. Should conditions become reducing, they can tap it and make H_2S. Conversely, if conditions become strongly oxidizing, they can make SO_x species. Thus, the bacteria can sequester sulfur rather as in a piggy bank, saved for a needful day: it becomes a redox currency. Even better, sulfur-bearing chemical species are common components of hydrothermal fluids—readily available!

In the inorganic world, sulfur would have been available in a variety of oxidation states. Even in a reduced atmosphere, transient SO_x would have been present from volcanic sources, supplemented by interaction between sulfur-bearing aerosols and oxidants produced by photolytic chemistry in the early UV flux, or from escape of hydrogen to space. Reduced sulfur species would have been widely available in lavas and volcanic vents. Thus, for the early organisms, shuffling sulfur between various oxidation states would have been the best way of exploiting redox ratchets. When anoxygenic photosynthesis started, a full sulfuretum cycle would have been possible in sediments, fluxing sulfur endlessly up and down to capture a living from oxidizing decaying organic matter, or reducing available oxidant.

Once abundant oxygenic photosynthesis began, the sulfureta would have become much more productive, and sulfur would have become the chief currency of redox transactions on the bottom. Finally, as in most piggy banks, the contents are lost, buried in the mass of reduced sediment as pyrite or sulfide mineral, or even as sulfur, eventually to return via the plate cycle to the volcanoes or groundwater as oxidized sulfur.

8.01.10 FEEDBACK FROM THE BIOSPHERE TO THE PHYSICAL STATE OF THE PLANET

The planet shapes life, but life also shapes the planet (Nisbet, 2002). The maintenance of surface temperature is managed by the air: hence, as life controls the composition of the air and the atmospheric greenhouse, then life sets the surface temperature (Lovelock and Margulis, 1974; Lovelock, 1979, 1988).

What would be the nature of the air if life did not exist? If for the past 4 Ga, life had not captured carbon and sequestered it, and cycled nitrogen back from soluble compounds, returning it as atmospheric dinitrogen, and evolved oxygen and hence permitted ozone to form in the stratosphere, what would the atmosphere be? It is almost impossible to say. Reasonable guesses include a nitrogen–carbon dioxide atmosphere; or perhaps a nitrogen atmosphere over ice, with the carbon dioxide removed as carbonate after volcanic paroxysm. One possibility is that over time the air would have evolved as Venus's air may have evolved, first as a steam greenhouse, then after hydrogen loss, to a dry hot carbon dioxide greenhouse over a dehydrated planet. Alternately, the surface could have become very cold and icy. This would have interesting consequences, as it may have changed the operation of the erosional cycle and the plate system, perhaps leading to periods of long quiescence, followed by volcanic resurfacing. The persistence of oceans of liquid water is closely interwoven with the long-term history of the continents and oceans (Hess, 1962), and the controls on water depth may be linked to the physical properties of water (Kasting and Holm, 1992) as well as to the nature of the atmosphere and its greenhouse impact. In an inanimate planet, whether or not liquid water would have persisted for as long as 4 Ga is a moot point. If it had disappeared, would Earth have had plate tectonics? Perhaps not: perhaps the life has shaped the face of the Earth.

ACKNOWLEDGMENTS

Thanks to the community of Archean field geologists in Zimbabwe and Canada for many years of discussions, especially Mike Bickle and Tony Martin, Jim Wilson, John Orpen, and Tom Blenkinsop; to Freeman Dyson, Jim Lovelock, Norm Sleep, Crispin Tickell, and Kevin Zahnle for much thought; and to the late Preston Cloud, Teddy Bullard, Harold Jeffreys, and Drum Matthews for tuition and insight. Finally, thanks to Mabel of Yale for e-mail encouragement during the task of writing this, other matters being more pressing.

REFERENCES

Abele D. (2002) The radical life-giver. *Nature* **420**, 27.

Ahrens T. J. (1990) Earth accretion. In *Origin of the Earth* (eds. H. E. Newsom and J. H. Jones). Oxford University Press, Oxford, pp. 211–227.

Allegre C., Manhes G., and Gopel C. (1995) The age of the Earth. *Geochim. Cosmochim. Acta* **59**, 1445–1456.

Alt J. C. and Shanks W. C. (1998) Sulfur in serpentinized oceanic peridotites: serpentinization processes and microbial sulfate reduction. *J. Geophys. Res.* **103**, 9917–9929.

Appel P. W. U., Rollinson H. R., and Touret J. L. R. (2001) Remnants of an early Archean (>3.75 Ga) sea-floor hydrothermal system in the Isua greenstone belt. *Precamb. Res.* **112**, 27–49.

Baker V. R. (2001) Water and the Martian landscape. *Nature* **412**, 228–236.

Barnes S. M., Delwiche C. F., Palmer J. D., and Pace N. R. (1996) Perspectives on archaeal diversity, thermophyly and monophyly from environmental rRNA sequences. *Proc. Natl. Acad. Sci. USA* **93**, 9188–9193.

Beja O., Suzuki M. T., Heidelberg J. F., Nelson W. C., Preston C. M., Hamada T., Eisen J. A., Fraser C. M., and deLong E. F. (2002) Unsuspected diversity among marine aerobic anoxygenic phototrophs. *Nature* **415**, 630–632.

Bernal J. D. (1951) *The Physical Basis of Life*. Routledge and Kegan Paul, London.

Bernal J. D. (1967) *The Origin of Life*. Weidenfeld and Nicholson, London.

Bickle M. J., Martin A., and Nisbet E. G. (1975) Basaltic and peridotitic komatiites, stromatolites, and a basal unconformity in the Belingwe Greenstone belt, Rhodesia. *Earth Planet. Sci. Lett.* **27**, 155–162.

Blankenship R. E. (2001) Molecular evidence for the evolution of photosynthesis. *Trends Plant Sci.* **6**, 4–6.

Boetius A., Ravenschlag K., Schubert C., Rickert D., Widdel F., Gieseke A., Amann R., Jorgensen B. B., Witte U., and Pfannkuche O. (2000) A marine microbial consortium apparently mediating anaerobic oxidation of methane. *Nature* **407**, 623–626.

Bowring S. A., Williams I. S., and Compston W. (1989) 3.96 Ga gneisses from the Slave province, Canada. *Geology* **17**, 760–764.

Brasier M. D., Green O. R., Jephcoat A. P., Kleppe A., van Kranedonk M. J., Lindsay J. F., Steele A., and Grassineau N. V. (2002) Questioning the evidence for Earth's oldest fossils. *Nature* **416**, 76–81.

Brochier C. and Philippe H. (2002) A non-hyperthermophilic ancestor for Bacteria. *Nature* **417**, 244.

Brocks J. J., Logan G. A., Buick R., and Summons R. E. (1999) Archean molecular fossils and the early rise of eukaryotes. *Science* **285**, 1033–1036.

Bui E. T. N., Bradley P. J., and Johnson P. J. (1996) A common evolutionary origin for mitochondria and hydrogenosomes. *Proc. Natl. Acad. Sci. USA* **93**, 9651–9656.

Buick R. (1990) Microfossil recognition in Archean rocks: an appraisal of spheroids and filaments form a 3,500 MY old chert-barite unit at North Pole, Western Australia. *Palaios* **5**, 441–459.

Buick R. (1992) The antiquity of oxygenic photosynthesis: evidence from stromatolites in sulphate-deficient Archean lakes. *Science* **255**, 74–77.

Buick R., Dunlop J. S. R., and Groves D. I. (1981) Stromatolite recognition in ancient rocks: an appraisal of irregularly laminated structures in an Early Archean chert-barite unit from North Pole, Western Australia. *Alcheringa* **5**, 161–181.

Byerly G. R., Lowe D. R., and Walsh M. M. (1986) Stromatolites from the 3,300–3,500 Myr Swaziland supergroup, Barberton Mountain Land, South Africa. *Nature* **319**, 489–491.

Cairns-Smith A. G. (1982) *Genetic Take-over and the Mineral Origins of Life*. Cambridge University Press, Cambridge.

Cameron A. G. W. (2002) Birth of a solar system. *Nature* **418**, 924–925.

Campbell I. H. and Taylor S. R. (1983) No water, no granites, no oceans, no continents. *Geophys. Res. Lett.* **10**, 1061–1064.

Canup R. M. and Asphaug E. (2001) The lunar-forming giant impact. *Nature* **412**, 708–712.

Capone D. G., Zehr J. P., Paerl H. W., Bergman B., and Carpenter E. J. (1997) *Trichodesmium*, a globally significant marine cyanobacterium. *Science* **276**, 1221–1229.

Carr M. (1996) *Water on Mars*. Cambridge University Press, Cambridge.

Catling D. C., Zahnle K. J., and McKay C. P. (2001) Biogenic methane, hydrogen escape, and the irreversible oxidation of the Earth. *Science* **293**, 839–843. (see also Towe below).

Cavalier-Smith T. (2002) The phagotrophic origin of eukaryotes and phylogenetic classification of Protozoa. *Int. J. Sys. Evol. Microbiol.* **52**, 297–354.

Cockell C. S. (2000) The ultraviolet history of the terrestrial planets—implications for biological evolution. *Planet. Space Sci.* **48**, 203–214.

Compston W. and Pidgeon R. T. (1986) Jack Hills, evidence of more very old detrital zircons in Western Australia. *Nature* **321**, 766–769.

Cooke D. L. and Moorhouse W. W. (1969) Timiskaming volcanism in the Kirkland Lake area, Ontario, Canada. *Can. J. Earth Sci.* **6**, 117–132.

Crick F. H. C. (1981) *Life Itself: Its Origin and Nature*. Simon and Schuster, Touchstone, New York.

Darwin C. (1859, 1872) *On the Origin of Species by Means of Natural Selection or the Preservation of Favoured Races in the Struggle for Life,* diagram from 1872 5th edn. D. Appleton and Co., New York.

Darwin C. (1959). *Some Unpublished Letters* (1871) (ed. Sir Gavin de Beer). Notes and Records of the Royal Society, London, **14**, pp. 1.

Dismukes G. C., Klimov V. V., Baranov S., Kozlov Yu. N., DasGupta J., and Tryshkin A. (2001) The origin of atmospheric oxygen on Earth: the innovation of oxygenic photosynthesis. *Proc. Natl. Acad. Sci. USA* **98**, 2170–2175. (www.pnas.org/cgi/doi/10.1073/pnas.061514798).

Donahue T. M., Hoffman J. H., Hodges R. R., Jr., and Watson A. J. (1982) Venus was wet: a measurement of the ratio of deuterium to hydrogen. *Science* **216**, 630–633.

Doolittle W. F. (1999) Phylogenetic classification and the universal tree. *Science* **284**, 2124–2128.

Doolittle W. F. (2000) Uprooting the tree of life. *Sci. Am.* (February), 72–77.

Drake M. J. and Righter K. (2002) Determining the composition of the Earth. *Nature* **416**, 39–44.

Dutkeiwicz A., Rasmussen B., and Buick R. (1998) Oil preserved in fluid inclusions in Archean sandstone. *Nature* **395**, 885–888.

Elsaied H. and Nayunama T. (2001) Phylogenetic diversity of Ribulose-1,5-Birphosphate Carboxylase/Oxygenase large-subunit genes from deep sea microorganisms. *Appl. Environ. Microbiol.* **67**, 1751–1765.

Falkowski P. G. (1997) Evolution of the nitrogen cycle and its influence on the biological sequestration of CO_2 in the ocean. *Nature* **387**, 272–275.

Farquhar J., Bao H., and Thiemans M. (2000) Atmospheric influence of Earth's earliest sulfur cycle. *Science* **289**, 756–758, see also comment by Ohmoto *et al.* (2001), and response by Farquhar *et al.*(1959a), *Science*, **292**.

Fast N. M., Kissinger J. C., Roos D. S., and Keeling P. J. (2001) Nuclear-encoded, plastid targeted genes suggest a single common origin for apicomplexan and dinoflagellate plastids. *Mol. Biol. Evol.* **18**, 418–426.

Fedo C. M. and Whitehouse M. J. (2002) Metasomatic origin of quartz-pyroxene rock, Akilia, Greenland, and implications for Earth's earliest life. *Science* **296**, 1448–1452.

Fenchel T. and Bernard C. (1995) Mats of colourless sulphur bacteria: I. Major microbial processes. *Mar. Ecol. Prog. Ser.* **128**, 161–170.

Forterre P. (1995) Thermoreduction, a hypothesis for the origin of prokaryotes. *C. R. Acad. Sci. Paris: Sci. de la vie* **318**, 415–422.

Forterre P. (1996) A hot topic: the origin of hyperthermophiles. *Cell* **85**, 789–792.

Forterre P. and Philippe H. (1999) Where is the root of the universal tree of life? *BioEssays* **21**, 871–879.

Fuerst J. A. (1995) The planctomycetes: emerging models for microbial ecology, evolution, and cell biology. *Microbiology* **141**, 1493–1506.

Galtier N., Tourasse N., and Gouy M. (1999) A non-hyperthermophile common ancestor to extant life forms. *Science* **283**, 220–221.

Gee H. (2002) That's life? *Nature* **416**, 28.

Gilbert W. (1986) The RNA world. *Nature* **319**, 618.

Glansdorff N. (2000) About the last common ancestor, the universal life-tree and lateral gene-transfer: a reappraisal. *Mol. Microbiol.* **38**, 177–185.

Goodwin A. M., Monster J., and Thode H. G. (1976) Carbon and sulphur isotope abundances in iron-formations and early Precambrian life. *Econ. Geol.* **71**, 870–891.

Gogarten-Boekels M., Hilario E., and Gogarten J. P. (1995) The effects of heavy meteorite bombardment on the early evolution the emergence of the three domains of Life. *Origins Life Evol. Biosphere* **25**, 251–264.

Granick S. (1965) *Evolving Genes and Proteins* (eds. V. Bryson and H. J. Vogel). Academic Press, New York.

Grassineau N. V., Nisbet E. G., Bickle M. J., Fowler C. M. R., Lowry D., Mattey D. P., Abell P., and Martin A. (2001) Antiquity of the biological sulphur cycle: evidence from sulphur and carbon isotopes in 2700 million year old rocks of the Belingwe belt, Zimbabwe. *Proc. Roy. Soc. London* **B268**, 113–119.

Grassineau N. V., Nisbet E. G., Fowler C. M. R., Bickle M. J., Lowry D., Chapman H. J., Mattey D. P., Abell P., Yong J., and Martin A. (2002) Stable isotopes in the Archean Belingwe belt, Zimbabwe: evidence for a diverse prokaryotic mate ecology. In *The Early Earth: Physical, Chemical and Biological Development*, Geological Society London, Special Publication 199, (eds. C. M. R. Fowler, *et al.*), pp. 309–328.

Grey M. W., Burger G., and Lang B. F. (1999) Mitochondrial evolution. *Science* **283**, 1476–1481.

Guy R. D., Fogel M., and Berry J. A. (1993) Photosynthetic fractionation of the stable isotopes of oxygen and carbon. *Plant Physiol.* **101**, 37–47.

Hall A. (1989) Ammonium in spilitized basalts of southwest England and its implications for the recycling of nitrogen. *Geochem. J.* **23**, 19–23.

Hall A. and Alderton D. H. M. (1994) Ammonium enrichment associated with hydrothermal activity in the granites of south-west England. *Proc. Ussher Soc.* **8**, 242–247.

Halliday A. N. (2001) In the beginning. *Nature* **409**, 144–145.

Hartman H. and Fedorov A. (2002) The origin of the eukaryotic cell: a genomic investigation. *Proc. Natl. Acad. Sci.* **99**, 1420–1425. www.pnas.org/cgi/doi/10.1073/pnas.032658599.

Hayes J. M. (1994) Global methanotrophy at the Archean-Proterozoic transition. In *Early Life on Earth* (ed. S. Bengtson). Columbia University Press, New York, pp. 220–236.

Hirt R. P., Logsdon J. M., Healy B., Dorey M. W., Doolittle W. F., and Embley T. M. (1999) Microsporidae are related to fungi: evidence from the largest subunit of RNA polymerase II and other proteins. *Proc. Natl. Acad. Sci. USA* **96**, 580–585.

Harvey L. D. D. and Huang Z. (1995) Evaluation of the potential impact of methane clathrate destabilisation on future global warming. *J. Geophys. Res.* **100**, 2905–2926.

Hess H. H. (1962) History of ocean basins. In *Petrological Studies: A Volume in Honour of A. F. Buddington* (eds. A. E. J. Engel, H. L. James, and B. F. Leonard). Geological Society of America, pp. 599–620.

Hoehler T. M., Bebout B. M., and Des Marais D. J. (2001) The role of microbial mats in the production of reduced gases on the early earth. *Nature* **412**, 324–327.

Holland H. D. (1999) When did the Earth's atmosphere become oxic? A reply. *Geochem. News* **100**, 20–22.

Holm N. G. and Charlou J. L. (2001) Initial indications of abiotic formation of hydrocarbons in the Rainbow ultramafic hydrothermal system, mid-Atlantic ridge. *Earth Planet. Sci. Lett.* **191**, 1–8.

Hulth S., Aller R. C., and Gilbert F. (1999) Coupled anoxic nitrification/manganese reduction in marine sediments. *Geochim. Cosmochim. Acta* **63**, 49–66.

Jetten M. S. M., Wagner M., Fuerst J., van Loosdrecht M., Kuenen G., and Strous M. (2001) Microbiology and application of the anaerobic ammonium oxidation (anammox) process. *Curr. Opinion Biotechnol.* **12**, 283–288.

Jordan P., Fromme P., Tobias Witt H., Klukas O., Saenger W., and Krauss N. (2001) Three dimensional structure of cyanobacterial photosystem: I at 2.5. A resolution. *Nature* **411**, 909–917.

Joshi H. M. and Tabita F. R. (1996) A global two-way component signal transduction system that integrates the control of photosynthesis, carbon dioxide assimilation and nitrogen fixation. *Proc. Natl. Acad. Sci. USA* **93**, 14515–14520.

Kaiser J. (1995) Can deep bacteria live on nothing but rocks and water? *Science* **270**, 377.

Karl D. M. (2002) Hidden in a sea of microbes. *Nature* **415**, 590–591.

Karner M. B., DeLong E. F., and Karl D. M. (2001) Archeal dominance in the mesopleagic zone of the Pacific Ocean. *Nature* **409**, 507–510.

Kasting (1993) Earth's early atmosphere. *Science* **259**, 920–926.

Kasting J. F. (2001) The rise of atmospheric oxygen. *Science* **293**, 819–820.

Kasting J. F. and Holm N. G. (1992) What determines the volume of the oceans? *Earth Planet. Sci. Lett.* **109**, 507–515.

Kasting J. F. and Siefert J. L. (2001) The nitrogen fix. *Nature* **412**, 26–27.

Kasting J. F. and Siefert J. L. (2002) Life and the evolution of the Earth's atmosphere. *Science* **296**, 1066–1067.

Kasting J. F., Zahnle K. J., Pinto J. P., and Young A. T. (1989) Sulfur, ultraviolet radiation, and the early evolution of life. *Origins Life Evol. Biosphere* **19**, 95–108.

Kasting J. F., Whitmire D. P., and Reynolds R. T. (1993) Habitable zones around main sequences stars. *Icarus* **101**, 108–128.

Kerr R. A. (2002) Reversals reveal pitfalls in spotting ancient and E. T. life. *Science* **296**, 1384–1385.

Kleine T., Munker C., Mezger K., and Palme H. (2002) Rapid accretion and early core formation on asteroids and the terrestrial planets from Hf–W chronometry. *Nature* **418**, 952–955.

Kuhlbrandt W. (2001) Chlorophylls galore. *Nature* **411**, 896–898.

Kvenvolden K. (1988) Methane hydrate—a major reservoir of carbon in the shallow geosphere? *Chem. Geol.* **29**, 159–162.

Lewis J. S. and Prinn R. G. (1984) *Planets and their Atmospheres*. Academic Press, Orlando.

Lindsay M. R., Webb R. I., Strous M., Jetten M. S. M., Butler M. K., Forde R. J., and Fuerst J. A. (2001) Cell compartmentalisation in plactomycetes: novel types of structural organization for the bacterial cell. *Arch. Microbiol.* **175**, 413–429.

Line M. A. (2002) The engima of the origin of life and its timing. *Microbiology* **148**, 21–27.

Lorimer G. H. (1981) The carboxylation and oxygenation of ribulose 1,5-Bisphosphate: the primary events in photosynthesis and photorespiration. *Ann. Rev. Plant Physiol.* **32**, 349–383.

Lorimer G. H. and Andrews T. J. (1973) Plant photorespiration-an inevitable consequence of the existence of atmospheric oxygen. *Nature* **243**, 359.

Lovelock J. E. (1979) *Gaia: A New Look at Life on Earth.* Oxford University Press, Oxford.

Lovelock J. E. (1988) *Ages of Gaia.* Norton, London.

Lovelock J. E. and Margulis L. (1974) Homeostatic tendencies of the Earth's atmosphere. *Origins Life* **5**, 93–103.

Lowe D. R. (1994) Abiological origin of described stromatolites older than 3.2 Ga. *Geology* **22**, 387–390.

Macgregor A. M. (1949) The influence of life on the face of the Earth. In *Presidential Address. Rhodesia Scientific Association.* Proceedings and Transactions, XLII, 5–11 (now Zimbabwe Scientific Association).

Martin A., Nisbet E. G., and Bickle M. J. (1980) Archean stromatolites of the Belingwe greenstone belt, Zimbabwe (Rhodesia). *Precamb. Res.* **13**, 337–362.

Martin W. (1999) Mosaic bacterial chromosomes: a challenge en route to a tree of genomes. *BioEssays* **21**, 99–104.

Martin W. and Muller M. (1998) The hydrogen hypothesis for the first eukaryote. *Nature* **392**, 37–41.

Matthews P. E. and Scharrer R. H. (1968) A graded unconformity at the base of the early Precambrian Pongola system. *Trans. Geol. Soc. S. Afr.* **71**, 257–272.

Miyazaki J., Nakaya S., Suzuki T., Tamakoshi M., Oshima T., and Yamagishi A. (2001) Ancestral residues stabilising 3-isopropylmalate dehydrogenase of an extreme thermophile; experimental evidence supporting the thermophilic common ancestor hypothesis. *J. Biochem.* **129**, 777–782.

Mojzsis S. J., Arrhenius G., McKeegan K. D., Harrison T. M., Nutman A. P., and Friend C. R. L. (1996) Evidence for life on Earth before 3,800 million years ago. *Nature* **384**, 55–59.

Mojzsis S. J., Harrison T. M., and Pidgeon R. T. (2001) Oxygen-isotope evidence from ancient zircons for liquid water at the Earth's surface 4,300 Myr ago. *Nature* **409**, 178–181.

Morel F. M. M. and Price N. M. (2003) The biogeochemical cycles of trace metals in the oceans. *Science* **300**, 944–947.

Moorbath S. (1983) The most ancient rocks. *Nature* **304**, 585–586.

Myers J. S. and Crowley J. L. (2000) Vestiges of life in the oldest Greenland rocks? a review of early Archean geology in the Godthabsfjord region, and reappraisal of field evidence for the >3,850 Ma life on Akilia. *Precamb. Res.* **103**, 101–124.

Navarro-Gonzalez R., McKay C. P., and Mvondo D. N. (2001) A possible nitrogen crisis for Archean life due to reduced nitrogen fixation by lightning. *Nature* **412**, 61–64.

Newsom H. E. and Jones J. H. (1990) *Origin of the Earth.* Oxford University Press, New York.

Nisbet E. G. (1986) RNA and hydrothermal systems. *Nature* **322**, 206.

Nisbet E. G. (1987) *The Young Earth.* George Allen and Unwin, London, 402pp.

Nisbet E. G. (1991) Of clocks and rocks the four aeons of Earth. *Episodes* **14**, 327–331.

Nisbet E. G. (1995) Archean ecology: a review of evidence for the early development of bacterial biomes, and speculations on the development of a global scale biosphere. In *Early Precambrian Processes*, Geological Society of London Spec. Publ. 95 (eds. M. P. Coward and A. C. Ries). Geological Society of London, London, pp. 27–51.

Nisbet E. G. (2002) The influence of life on the face of the Earth: garnets and moving continents, Geol. Soc. London Fermor lecture. In *The Early Earth: Physical, Chemical and Biological Development*, Geological Society of London Spec. Publ. 199 (ed. C. M. R. Fowler, *et al.*). Geological Society of London, London, pp. 275–307.

Nisbet E. G. and Fowler C. M. R. (1996a) Some liked it hot. *Nature* **382**, 404–405.

Nisbet E. G. and Fowler C. M. R. (1996b) The hydrothermal imprint on life: did heat shock proteins, metalloproteins and photosynthesis begin around hydrothermal vents? In *Tectonic, Magmatic, Hydrothermal, and Biological Segmentation of Mid-Ocean Ridges,* Geological Society of London Spec. Publ. 118 (eds. C. J. MacLeod, P. A. Tyler, and C. L. Walker). Geological Society of London, London, pp. 239–251.

Nisbet E. G. and Fowler C. M. R. (1999) Archean metabolic evolution of microbial mats. *Proc. Roy. Soc. London* **B266**, 2375–2382.

Nisbet E. G. and Sleep N. H. (2001) The habitat and nature of early life. *Nature* **409**, 1083–1091.

Nisbet E. G. and Sleep N. H. (2003) The early earth. In *The Physical Setting for Early Life* (eds. A. Lister and L. Rothschild). Academic Press, San Diego, pp. 3–24.

Nisbet E. G. and Walker D. (1982) Komatiites and the structure of the Archean mantle. *Earth Planet. Sci. Lett.* **60**, 105–113.

Nisbet E. G., Arndt N. T., Bickle M. J., Cameron W. E., Chauvel C., Cheadle M., Hegner E., Kyser T. K., Martin A., Renner R., and Roedder E. (1987) Uniquely fresh 2.7 Ga komatiites from the Belingwe greenstone belt, Zimbabwe. *Geology* **15**, 1147–1150.

Nisbet E. G., Cheadle M. J., Arndt N. T., and Bickle M. J. (1993) Constraining the potential temperature of the Archean mantle: a review of the evidence from komatiites. *Lithos* **30**, 291–307.

Nisbet E. G., Mattey D. P., and Lowry D. (1994) Can diamonds be dead bacteria? *Nature* **367**, 694.

Nisbet E. G., Cann J. R., and van Dover C. L. (1995) Origins of photosynthesis. *Nature* **373**, 479–480.

Nutman A. P., Macgregor V. R., Friend C. R. L., Bennett V. C., and Kinny P. D. (1996) The Itsaq geniss complex of southern West Greenland: the world's most extensive record of early crustal evolution (3,900–3,600 Ma). *Precamb. Res.* **78**, 1–39.

Ohmoto H. (1997) When did the Earth's atmosphere become oxic? *Geochem. News* **93**, 12–13.

Ohmoto H., Kakagawa T., and Lowe D. R. (1993) 3.4 billion year old pyrites from Barberton, South Africa: sulfur isotope evidence. *Science* **262**, 555–557.

Ohmoto H., Yamaguchi K. E., and Ono S. (2001) Questions regarding Precambrian sulfur fractionation, and Response by Farquhar *et al. Science* **292**, 1959a.

Pace N. R. (1997) A molecular view of biodiversity and the biosphere. *Science* **276**, 734–740.

Pavlov A. A., Kasting J. F., Brown L. L., Rages K. A., and Freedman R. (2000) Greenhouse warming by CH_4 in the atmosphere of early Earth. *J. Geophys. Res.* **105**, 11981–11990.

Pavlov A. A., Kasting J. F., Eigenbrode J. L., and Freeman K. H. (2001) Organic haze in Earth's early atmosphere: source of low ^{13}C kerogens? *Geology* **29**, 1003–1006.

Phillips G. N., Law J. D. M., and Myers R. E. (2001) Is the redox state of the Archean atmosphere constrained? *SEG Newslett.: Soc. Econ. Geol.*, Oct. 2001, No. 47, 1–19.

Pierson B. K. (1994) The emergence, diversification, and role of photosynthetic eubacteria. In *Early Life on Earth*, Nobel Symposium 84 (ed. S. Bengtson). Columbia University Press, New York, pp. 161–180.

Rasmussen R. (2000) Filamentous microfossils in a 3, 235 million year old volcanogenic massive sulphide deposit. *Nature* **405**, 676–679.

Rasmussen R. and Buick R. (2000) Oily old ores, evidence for hydrothermal petroleum generation in an Archean volcanogenic massive sulpide deposit. *Geology* **27**, 115–118.

Reeve J., Christner B. C., Kvitko B. H., Mosley-Thompson E., and Thompson L. G. (2002) Life in glacial ice. In *Extremophiles 2002: 4th International Congress on Extremophiles, Naples*, L5, 27.

Reysenbach A.-L. and Shock E. (2002) Merging genomes with geochemistry in hydrothermal ecosystems. *Science* **296**, 1077–1082.

Robinson J. J., Stein J. L., and Cavanaugh C. M. (1998) Cloning and sequencing of a Form II Ribulose-1,5-Bisphosphate Carboxylase/Oxygenase from the bacterial symbiont of the hydrothermal vent tubeworm *Riftia pachyptila*. *J. Bacteriol.* **180**, 1596–1599.

Roger A. J. and Silberman J. D. (2002) Mitochondria in hiding. *Nature* **418**, 827–829.

Rosing M. T. (1999) ^{13}C-depleted carbon in >3700 Ma seafloor sedimentary rocks from West Greenland. *Science* **283**, 674–676.

Rye R. and Holland H. D. (2000) Life associated with a 2.76 Ga ephemeral pond? *Geology* **28**, 483–486.

Sagan C. and Chyba C. (1997) The early Sun paradox: organic shielding of ultraviolet-labile greenhouse gases. *Science* **276**, 1217–1221.

Saldarriaya J. F., Taylor F. J. R., Keeling P. J., and Cavalier-Smith T. (2001) Dinoflagellate nuclear SSU rRNA phylogeny suggests multiple plastid losses and replacements. *J. Mol. Evol.* **53**, 204–213.

Sansone F. J., Popp B. N., Gasc A., Graham A. W., and Rust T. M. (2001) Highly elevated methane in the eastern tropical North Pacific and associated isotopically enriched fluxes to the atmosphere. *Geophys. Res. Lett.* **28**, 4567–4570.

Schidlowski M. (1988) A 3,800 million year record of life from carbon in sedimentary rocks. *Nature* **333**, 313–318.

Schidlowski M. (2002) Sedimentary carbon isotope archives as recorders of early life: implications for extraterrestrial scenarios, chap. 11. In *Fundamentals of Life*, Editions Scientifiques et Medicales (eds. G. Palyi, C. Zucchi, and L. Caglioti). Elsevier, Amsterdam, pp. 307–329.

Schidlowski M. and Aharon P. (1992) Carbon cycle and carbon isotopic record: geochemical impact of life over 3.8 Ga of earth history. In *Early Organic Evolution: Implications for Mineral and Energy Resources* (ed. M. Schidlowski, *et al.*). Springer, Berlin, pp. 147–175.

Schoenberg R., Kambeer B. S., Collerson K. D., and Moorbath S. (2002) Tungsten isotope evidence from ~3.8-Gyr metamorphosed sediments for early meteorite bombardment of the Earth. *Nature* **418**, 403–405.

Schopf J. W. (1993) Microfossils of the early Archean Apex chert: new evidence of the antiquity of life. *Science* **260**, 640–646.

Schopf J. W. and Packer B. M. (1987) Early Archean (3.3 billion to 3.5 billion-year-old) microfossils from Warrawoona group, Australia. *Science* **237**, 70–73.

Schopf J. W., Kudryavtsev A. B., Agresti D., Wdowiak T., and Czaja A. D. (2002) Laser-Raman imagery of Earth's earliest fossils. *Nature* **416**, 73–76.

Schrodinger E. (1944) *What is Life?* Cambridge University Press, Cambridge.

Shen Y., Buick R., and Canfield D. E. (2001) Isotopic evidence for microbial sulphate reduction in the early Archean era. *Nature* **410**, 77–81.

Sleep N. H. and Zahnle K. (2001) Carbon dioxide cycling and implications for climate on ancient Earth. *J. Geophys. Res.* **106**, 1373–1399.

Sleep N. H., Zahnle K., and Neuhoff P. S. (2001) Initiation of clement surface conditions on the earliest Earth. *Proc. Natl. Acad. Sci. USA* **98**, 3666–3672.

Stetter K. O. (1996) Hyperthermophiles in the history of life. In *Evolution of Hydrothermal Systems on Earth and Mars?* Ciba Foundation Symposium 202 (eds. G. R. Bock and J. A. Goode). Wiley, Chichester, pp. 1–10.

Stevens T. O. and McKinley J. P. (1995) Lithoautotrophic microbial ecosystems in deep basalt aquifers. *Science* **270**, 450–454.

Summons R. E., Jahnke L. L., Hope J. M., and Logan G. A. (1999) 2-Methylhopanoids as biomarkers for cyanobacterial oxygenic photosynthesis. *Nature* **400**, 554–557.

Taylor S. R. (2001) Solar System Evolution. In *A New Perspective*, 2nd edn. Cambridge University Press, Cambridge.

Thauer R. K. (1998) Biochemistry of methanogenesis: a tribute to Marjory Stephenson. *Microbiology* **144**, 2377–2406.

Tolbert N. E. (1994) Role of photosynthesis and photorespiration in regulating atmospheric CO_2 and O_2. In *Regulation of Atmospheric CO_2 and O_2 by Photosynthetic Carbon Metabolism* (eds. N. E. Tolbert and J. Preiss). Oxford University Press, Oxford, pp. 8–33.

Towe K. M. (2002) The problematic rise of Archean oxygen. *Science* **295**, 1419a (see also reply by Catling *et al.*, p1419b).

van Zuilen A., Lepland A., and Arrhenius G. (2002) Reassessing the evidence for the earliest traces of life. *Nature* **418**, 627–630.

Vellai T. and Vida G. (1999) The origin of eukaryotes: the difference between eukaryotic and prokaryotic cells. *Proc. Roy. Soc. London* **B266**, 1571–1577.

von Brunn V. and Hobday D. K. (1976) Early Precambrian tidal sedimentation in the pongola supergroup of South Africa. *J. Sedim. Petrogr.* **46**, 670–679.

Walker J. C. G. (1977) *Evolution of the Atmosphere*. Macmillan, New York.

Walker J. C. G. (1987) Was the Archean biosphere upside down? *Nature* **329**, 710–712.

Walker J. C. G. (1994) Global geochemical cycles of carbon. In *Regulation of Atmospheric CO_2 and O_2 by Photosynthetic Carbon Metabolism* (eds. N. E. Tolbert and J. Preiss). Oxford University Press, Oxford, pp. 75–89.

Watson A. J., Donahue T. M., and Kuhn W. R. (1984) Temperatures in a runaway greenhouse on the evolving Venus. *Earth Planet. Sci. Lett.* **68**, 1–6.

Weatherill G. W. (1990) Formation of the Earth. *Ann. Rev. Earth Planet. Sci.* **18**, 205–256.

Westall F., de Wit M. J., Dann J., van der Gaast S., de Ronde C. E. J., and Gerneke D. (2001) Early Archean fossil bacteria and biofilms in hydrothermally-influenced sediments from the Barberton greenstone belt, South Africa. *Precamb. Res.* **106**, 93–116.

Wilde S. A., Valley J. W., Peck W. H., and Graham C. M. (2001) Evidence from detrital zircons for the existence of continental crust and oceans on the Earth 4.4 Gyr ago. *Nature* **409**, 175–178.

Wilks M. E. and Nisbet E. G. (1985) Archean stromatolites from the Steep Rock group NW Ontario. *Can. J. Earth Sci.* **22**, 792–799.

Williams B. A. P., Hirt R. P., Lucocq J. M., and Embley T. M. (2002) A mitochondrial remnant in the microsporidian *Trachipleistophora hominis*. *Nature* **418**, 865–869.

Woese C. R. (1987) Bacterial evolution. *Microbiol. Rev.* **51**, 221–271.

Woese C. R. (1999) The universal ancestor. *Proc. Natl. Acad. Sci.* **95**, 6854–6859.

Woese C. R., Kandler O., and Wheelis M. I. (1990) Towards a natural system of organisms: proposals for the domains Archea, Bacteria, and Eucarya. *Proc. Natl. Acad. Sci. USA* **87**, 4576–4579.

Xiong J. and Bauer C. E. (2002) A cytochrome *b* origin of photosynthetic reaction centers: an evolutionary link between respiration and photosynthesis. *J. Molecul. Biol.* **322**, 1025–1037.

Xiong J., Fischer W. M., Inoue K., Nakahara M., and Bauer C. E. (2000) Molecular evidence for the early evolution of photosynthesis. *Science* **289**, 1724–1730.

Yin Q., Jacobsen S. B., Yamashita K., Blichert-Toft J., Telouk P., and Albarede F. (2002) A short timescale for terrestrial planet formation from Hf–W chronometry of meteorites. *Nature* **418**, 949–952.

Yung Y., Wen J.-S., Moses J. I., Landry B. M., and Allen M. (1989) Hydrogen and deuterium loss from the terrestrial atmosphere: a quantitative assessment of non-thermal escape fluxes. *J. Geophys. Res.* **94**, 14971–14989.

Zehr J. P., Waterbury J. B., Turner P. J., Montoya J. P., Omoregie E., Steward G. F., Hansen A., and Karl D. M. (2001) Unicellular cyanobacteria fix N_2 in the subtropical North Pacific Ocean. *Nature* **412**, 635–638.

Zuckerkandl E. and Pauling L. (1965) Molecules as documents of evolutionary history. *J. Theor. Biol.* **8**, 357–366.

8.02 Evolution of Metabolism

K. H. Nealson and R. Rye

University of Southern California, Los Angeles, CA, USA

8.02.1 INTRODUCTION

This chapter is devoted to the discussion of the evolution of metabolism, with a particular focus towards redox metabolism and the utilization of redox energy by life. We will deal with various aspects of metabolism that involve direct interaction with, and the extraction of energy from, the environment (catabolic metabolism) and will talk briefly of the reactions that affect mineral formation and dissolution. However, we will de-emphasize the aspects related to the formation of complex molecules and organisms. To some, it will be refreshingly brief; to others, somewhat superficial. This is unavoidable, as our knowledge of the details of the evolution of metabolism is at best slim. However, by piecing together aspects of the properties and history of the Earth and coupling these with what we know of today's metabolism, it is possible to at least frame several different hypotheses that, with time, should be possible to test and modify so that the next writing of this chapter might contain some intellectual entrees and not just the appetizers. Any discussion of metabolic evolution must occur in concert with a consideration of the Earth—the understanding of the forces that drove the co-evolution of life and Earth can be achieved only by considering them together. This theme will pervade this chapter, and any real understanding of the evolution of metabolism must be inexorably coupled to, and consistent with, the geological record of the Earth.

The first aspect of evolution concerns the metabolic participants as we know them now (i.e., a definition of metabolic diversity), and the second concerns the sequence of events that have led to this remarkable metabolic diversity. The first part is fairly straightforward: a discussion of the domains of life, and the metabolic achievements that are expressed in the various domains, and relating metabolism to biogeochemical processes whenever possible. The second part is much more problematic. While it is possible to make up nearly any story regarding the evolution of metabolism (and nearly all have been attempted!), the starting point of life is not known (great debates still rage as to the nature and origin of the first living systems), and it is not a trivial matter to specify the sequence and timing of metabolic innovations. As will be discussed below, genetic and genomic data have revealed that genetic exchange between organisms has been so pervasive that it has essentially uncoupled the evolution of taxonomic groups from the evolution of metabolic processes, thus, obscuring the evolutionary trail with blurred signals. Given these challenges, it may be prudent at this time to admit what we do not know, and lay out the challenges for the coming years.

8.02.2 THE DOMAINS OF LIFE

Prior to the early 1980s, life was considered to reside in five kingdoms, four of which were eukaryotic, and the Monera, containing the very small, anucleate cells, called prokaryotes (Figure 1(a)). However, with the advent of molecular taxonomy (sequence comparisons of the genes coding for 16S ribosomal RNA), our view of life on Earth was radically changed. The kingdom Eukarya was established to accommodate what had previously been four kingdoms (animals, plants, fungi, and protists), while two kingdoms (Archea and Bacteria) were created to account for major differences between RNA sequences from organisms that had previously been grouped into the Monera (Figure 1(b); Woese, 1987). As will be seen, this division has some interesting implications with regard to metabolism, and especially with regard to metabolism important in biogeochemical cycles. The living world can be divided into two metabolic groups, with the prokaryotes possessing rather remarkable metabolic plasticity, and the eukaryotes being much more monolithic in their approach to metabolism (i.e., metabolizing only a few different carbon compounds for energy, and respiring only oxygen). For the most part, this division also holds with regard to mineral metabolism, separating the prokaryotes, which harvest energy from minerals (and inadvertently dissolve or precipitate them) from the eukaryotes, which invest energy in biomineral synthesis.

8.02.3 LIFE AND ROCKS

The interactions between life and minerals are many (Banfield and Nealson, 1997; Nealson and

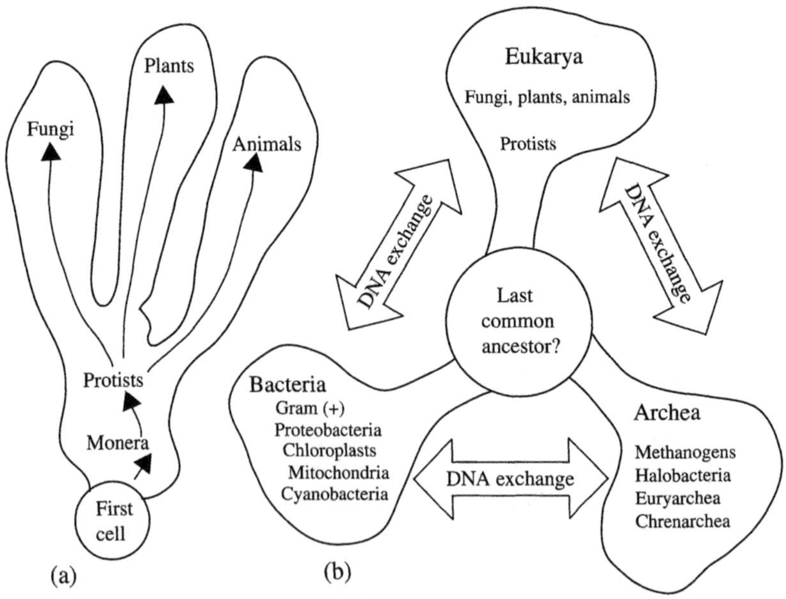

Figure 1 The domains of life. A diagrammatic representation of life as it can be viewed by classical (left) and molecular (right) taxonomists and phylogeneticists. Prior to the advent of molecular (16S rRNA sequence comparison) methods, four eukaryotic kingdoms and one prokaryotic kingdom, referred to as the Monera, were postulated. Molecular methods (Pace, 1996; Woese, 1987) have given rise to a dramatic revision of this view to one in which the four eukaryotic kingdoms are combined in the Eukarya, and the prokaryotes are expanded into two kingdoms, or domains, called the Bacteria and the Archea.

Ghiorse, 2001), including both energy-yielding and energy-consuming processes (Figure 2). With regard to the evolution of metabolism, life can be divided by consideration of the "use" of the mineral metabolism by the organism (Figure 2). On the one hand, the prokaryotes have developed a wide array of abilities to utilize geochemical compounds either as energy sources (electron donors), or oxidants (electron acceptors) for respiration, many of which are mineral-forming and/or geochemically important elements. As we like to say, they "eat and breathe" anything, including the rocks on this planet, perhaps an expression of the importance of the geological energy sources in their past evolution. While it is very difficult to ascribe to any given prokaryote a time of appearance, and especially hard to prove that when it appeared it possessed a given metabolic ability, it seems clear that the prokaryotes (Archea and Bacteria) have enjoyed a long presence on Earth, and that the development of much of their metabolism preceded the formation of the multicellular eukaryotic mineral-forming organisms.

On the other hand, the Eukarya do not indulge in the redox metabolism of minerals; their interactions involve structure or behavior—synthesizing minerals and "rocks" for structures involved primarily with predation or the escape from it (bones, teeth, shells, frustules, etc.). In general, these cases do not involve a change of redox state, and do require the input of energy for the synthesis. As far as is known, none of "rock syntheses" can be observed in the fossil record prior to ~800 Ma (million years ago) so, while we have these processes to thank for our fossil record, our thanks must be limited to the relatively recent past. Probably the earliest examples are the biomineralization of silica found in ancient fossils of testate amoebae (Porter and Knoll, 2000) and primitive sponge-like cells (Li *et al.*, 1998), with the invention of both apatite- and carbonate-based structures following soon after (see Lipps, 1993).

While this review is focused on the energy-related metabolism of the prokaryotes, we note that the more recently evolved eukaryotic abilities have also effected major changes on planetary chemistry through the precipitation of massive carbonates. Many of these abilities produce "hard parts" providing us with an eloquent fossil record in contrast with the far more cryptic record in older sediments.

8.02.4 MECHANISMS FOR ENERGY CONSERVATION

The mechanism (or some slight variations of it) whereby life garners useful energy from the environment appears to be nearly universal, and involves using redox energy to pump ions (usually protons) across an impermeable membrane, thus establishing a charge separation that can be used to drive the synthesis of adenosine triphosphate (ATP), the energy "currency" of living

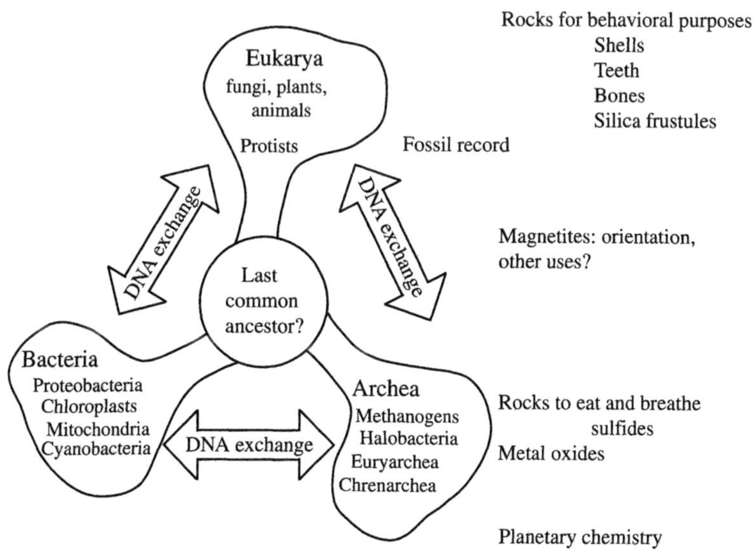

Figure 2 Biogeochemistry and the domains of life. This figure demonstrates that the relationships between biogeochemical reactions and life are divided rather nicely between those that are involved with energy (redox chemistry), which are done primarily by prokaryotes, and those that are involved with behavior (protective structures, etc.), which are done primarily by eukaryotes. The prokaryotic reactions involve the use of the geological world as either electron donors (energy sources) or electron acceptors (oxidants), while the eukaryotic reactions do not in general involve redox chemistry, and almost always require energy. Magnetite, which is a structural (perhaps behaviorally used) mineral, is unique, being synthesized and used by all groups.

systems. The wider the redox separation, the more ions can be "pumped," and ultimately, the more ATP can be made (Figure 3; see Nealson, 1997; Nealson and Stahl, 1997).

The fact that virtually all life utilizes a similar strategy for energy production makes it relatively easy to understand the situation mechanistically, and strongly suggests that it was invented very early—as Nisbet and Sleep (2001) say, the rest is "tinkering with the available equipment, adapting existing organs to new purposes." To this end, the basic metabolism of life, which is, for the most part, redox chemistry and enzymology, can be viewed as the ancient invention, and the tinkering and adaptations of the unknown challenges. In order to accomplish energy conversion, cellular life utilizes only a few components: electron carriers and hydrogen carriers of different potentials, enzymes to catalyze oxidation and reduction, membranes to separate charges, and enzymes (ATPases) for conversion of the membrane potential to useful chemical energy (i.e., ATP).

For the sake of discussion, we have divided the elements of metabolism into three parts, labeled (a)–(c) in Figure 3. Part (a) includes functions devoted to harvesting environmental energy, be it sunlight, inorganics, or organics, and transferring this energy to biological (electron or hydrogen) carriers. The electron carriers or "energy movers" used by living cells are, for the most part, the same molecules. Thus, nicotine adenine dinucleotide (NAD), flavins, and quinones are found throughout life's domains, and are either synthesized (by bacteria, fungi, and plants) or obtained by feeding on other organisms. The universality of these compounds either reflects their very early invention, or an overwhelming advantage that allowed them to supplant other molecules after later interdomain transfer. Thus, with the exception of some electron carriers involved with one-carbon metabolism (Schäfer *et al.*, 1999), a small group of compounds has been adopted by virtually all life, ranging from aerobic animals, to strictly anaerobic Bacteria and Archea. The universality of these molecules and the genes and proteins associated with them is being strongly reinforced as genomic data are compiled, showing that the pathways used to synthesize these compounds and catalyze their reactions are reflected in genes that can be easily recognized across domain boundaries.

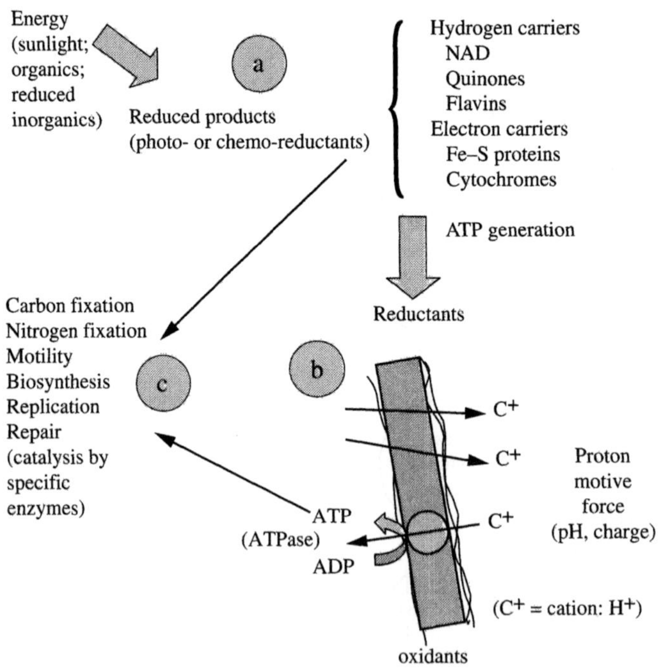

Figure 3 Fundamental metabolic processes of life. This diagram illustrates the features of the mode of energy utilization that is nearly universally used by life on Earth. (a) Formation of reduced products: whether the energy source is sunlight, reduced inorganics, or reduced carbon in the form of organics, life interacts to produce reduced hydrogen or electron carriers that are then used by the cell. These reduced "carriers" can then be used either directly for anabolic reactions (biosynthesis), for reduction of carbon or nitrogen (C or N fixation), or for energy generation. (b) Energy generation: energy in the form of ATP is the usual product of metabolism, and it is obtained via pumping of cations (usually protons) across a semipermeable membrane to establish a gradient. A specific enzyme complex (ATPase) then takes advantage of the flow of ions back across the membrane, and couples this flow to ATP synthesis. (c) Functions: many functions of the cell require reducing equivalents and/or ATP, as listed here.

The universality of the mechanism and of some of the electron carriers, however, does not mean that there are no major differences between metabolic groups. Clearly, one of the major evolutionary diversions with regard to niche occupation must have been the development of the cellular equipment to take advantage of various electron donors and acceptors. Thus, while the overall mechanism for energy harvest is similar, the individual enzymes (reductases, iron–sulfur proteins, etc.) are highly specialized for their individual metabolic functions. Furthermore, eukaryotic cells have similar mechanisms for energy harvesting, but in fact these similarities exist because of the symbiotic acquisition of prokaryotic cells. This is the ultimate in lateral gene transfer: the acquisition of an entire genome. Given that virtually all eukaryotic cells possess mitochondria (Margulis, 1981), or evidence of relict mitochondria (Roger and Silberman, 2002), it can be assumed that the acquisition of mitochondria was a very early event, and that surely the invention of aerobic respiration occurred prior to the symbiotic event.

The second part of metabolism (b) represents perhaps one of the most important and fundamental inventions of life—the use of redox energy to "pump" ions across semipermeable biological membranes to establish an ionic charge gradient (because most prokaryotes pump protons, it is often referred to as the PMF, or proton motive force, consisting of both a pH and a charge component), and then to utilize this charged membrane to produce chemical energy. It should be noted here that other ions can be used for the purpose of establishing gradients, notably Na^+ in marine and alkaliphilic microorganisms.

Once an electron is extracted from an energetic substrate, it flows "downhill" energetically until the circuit is closed by the reduction of the electron acceptor. Through these "electron transport" reactions, the energy is obtained for the pumping of the ions across the membrane. One of the fundamental inventions of life is expressed in this part of metabolism, namely, the use of certain classes of molecules as electron or hydrogen carriers. These include the iron-containing cytochromes and iron–sulfur proteins as the major components. Also critical are cyclic double-bonded carbon-containing quinones.

The charged membrane (effectively, a biological capacitor) is used for a variety of purposes (transport, motility, etc.), but is best known for its ability to synthesize ATP (adenosine triphosphate), the universal high-energy compound of life. The ATPase class of enzymes (and genes that code for them) can be identified in abundance in all domains of life, and have adapted to accomplish a variety of functions (usually having nothing to do with ATP), such as transport (Saier, 2000). With the exception of a few fermentative bacteria, such as the *Lactobacilli*, the chemistry and enzymology of the reactions in this category are universal. One imagines they were invented and selected, found to be advantageous very early, and then distributed, both horizontally and vertically through all the domains of life.

The energy transducing reactions of category (b) are found as integral parts of both Archea and Bacteria, and are found in the Eukarya as a result of symbiosis. Once again, with regard to evolution, the fundamental inventions common to all life were ancient—virtually all forms of life contain these traits. By the time the three domains of life had emerged, this ability was established as a metabolic paradigm. However, as opposed to the mitochondria, which use only oxygen, the prokaryotic world has developed the accessory materials to utilize many other electron acceptors (and donors).

Finally, part (c) represents that part of metabolism in which energetic molecules (ATP and reductants) are used to synthesize new cells, or repair existing cells. As with the other categories, the enzymes and genes involved with these processes are easily recognizable across domain boundaries. In this category lie the anabolic reactions that lead to complex cells, multicellular organisms, tissues, and "behavioral" biominerals, all characteristic of the Eukarya.

Looking at metabolism makes it easier to understand the unifying features of energy transduction, but presents certain difficulties with regard to unraveling the evolutionary details. All three domains of life utilize the same approaches to gain energy, and the same or very similar molecules to accomplish this end. As far as can be ascertained, the last common ancestor of the three domains had achieved similar approaches. Another option is that upon lateral transfer of these abilities, they were so superior to any other that they simply outcompeted them. Thus, the evolution of redox chemistry and diversity probably occurred prior to the deposition of easily discernible macrofossils.

8.02.5 EXTANT PATTERNS OF METABOLISM

Figure 4 shows that at the surface of the Earth there are only two major energy sources, light from the sun, and geothermal energy from the interior: all other sources are quantitatively trivial in comparison (although man has learned how to harvest most of them). Light is used directly by the phototrophs, while hydrothermal energy is utilized mainly via the heat-catalyzed

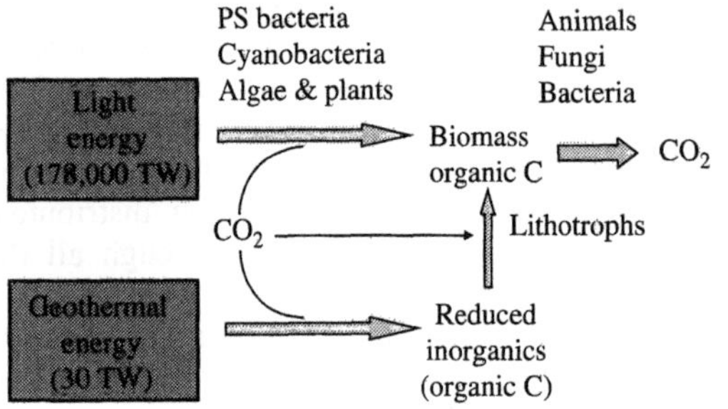

Figure 4 Energy flow on Earth. This diagram illustrates the general flow of energy on Earth, including the relative levels of energy for the only two significant sources: solar (light) energy and geothermal energy. The energy flux into the biosphere via photosynthetic harvest is ~100 TW of energy, or ~1 part in 2,000 of the incoming solar flux. For reference, we note that humans use renewable energy sources, most of which are related in some way to solar energy input, at a rate of ~1 TW. As shown, primary producers harvest the physical energy of light, converting it into organic carbon in the form of biomass. The rest of the biota then recycles this biomass. A poorly quantified but smaller fraction of the geothermal energy is converted into reduced inorganics (and perhaps some directly to organic carbon by abiotic carbon fixation). The inorganics (H_2S, H_2, Fe(II), CO, CH_4) in this small pool are then used by lithotrophic bacteria to fix carbon and contribute to the organic food chain.

Table 1 Metabolic divisions of life.

Metabolic types	*Energy source*	*Organisms*
Phototrophs	Sunlight	Bacteria[a]
		Archea[b]
		Eukarya
Lithotrophs	Inorganics	Bacteria
		Archea
Heterotrophs	Organic carbon	Bacteria
		Archea
		Eukarya[c]

[a] Includes anoxygenic photosynthetic bacteria and cyanobacteria.
[b] Only the halobacteria, and for energy only, not carbon fixation.
[c] Includes animals, fungi, protists, and plants in the dark.

production of reduced inorganics (H_2, CH_4, CO, H_2S), and perhaps the direct fixation of carbon via thermally catalyzed reactions (Cody *et al.*, 2000; Shock *et al.*, 1995). On the early Earth, geothermal energy was thought to be significantly higher, while the sun was perhaps 25% less intense than it is today, but still by far the dominant energy source. As shown in Figure 4, both sunlight and reduced inorganics are used to produce organic products: organic carbon thus becomes both the energy source to drive the abundant and diverse life via food chains, and the mechanism for storage of energy for later use. We thus divide metabolic life into groups based on energy usage: phototrophs, lithotrophs, and organotrophs (Table 1).

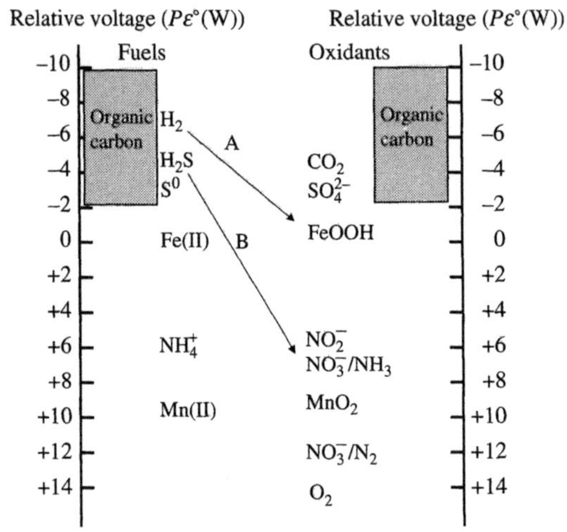

Figure 5 Energetics of biologically useful energy sources. We have ranked, on the left side of the diagram, the available fuels for life from most energetic (top) to the least energetic (bottom), using a scale of $P\varepsilon^{0}$(W) (the log of the electron potential adjusted to pH 7 and water). On the right side, using the same scale, we have ranked the oxidants that life uses to oxidize these fuels. If an arrow is drawn from a fuel to an oxidant, and the slope is negative, then energy is available in this reaction (i.e., arrow A illustrates the amount of energy available when hydrogen is oxidized by goethite; arrow B, when sulfide is oxidized by nitrate).

Figure 5 is an energy diagram that includes the major energy sources and oxidants on Earth: an attempt to put this information into an energetic and biogeochemical context. As will be discussed, life takes advantage of virtually all the existing voltage separations between the various energy sources (left side of diagram), and the oxidants (right side of the diagram). If one connects an energy source on the left, with an oxidant on the right, and the connecting arrow has a downward slope, then energy is available. For almost every combination that yields a downward arrow, an organism is known that lives by that metabolism. The strongest reductants are generated by photoreduction—reductants that can be used in the cell to generate chemical reductants for carbon fixation or biosynthesis. Hydrogen, a common by-product of many geochemical activities, is one of the most common metabolic fuels.

8.02.5.1 Kinds of Phototrophs

In Table 2, the phototrophs have been grouped with regard to which electron donors (reductants) are used during photosynthesis. When a reductant is utilized, it leaves behind an oxidized product that can have substantial biogeochemical impact. Thus, as shown in Table 2, for the anaerobic (anoxygenic) photosynthetic bacteria, a wide range of electron donors are used, each of which

Table 2 Phototrophs grouped according to electron donors.

Types of phototrophs	*Specific groups*	*Electron donors*	*Oxidized products*
Oxygenic	Cyanobacteria	H_2O	O_2
	Algae		
	Green plants		
Anoxygenic	Cyanobacteria[a]	H_2S	SO_4^{2-}
	Bacteria[b,c]	H_2	H_2O
		H_2S	S^0, SO_4^{2-}
		S^0	SO_4^{2-}
		Fe(II)	Fe(III)
		Organic carbon	Oxidized organics

[a] Some cyanobacteria can grow via anoxygenic photosynthesis (Cohen *et al.*, 1975; Padan, 1979). [b] There are four groups of anoxygenic phototrophs, all in the domain Bacteria: (i) filamentous green bacteria, (ii) purple bacteria, (iii) green sulfur bacteria, and (iv) Gram positive heliobacillus (Blankenship, 2002; Xiong *et al.*, 2000). [c] Another group of prokaryotic phototrophs has been known for many years, but not yet well characterized. These are two marine and six freshwater genera, all members of the alpha proteobacteria (see Yurkov and Beatty, 1998).

is oxidized during the process, leading to the fixation of reduced carbon on the one hand, and the oxidation of an electron donor, on the other. If organic carbon is ultimately buried, then these oxidizing equivalents will contribute to the oxidation of Earth's surface including the oceans and atmosphere. Of particular note are those bacteria that use hydrogen sulfide, sulfur, or ferrous iron as electron donors. As shown, these bacteria, under strict anaerobic conditions, photosynthetically generate copious quantities of sulfur, sulfate, or iron. The latter group, the anaerobic iron oxidizers, were discovered only a few years ago (Widdel *et al.*, 1993), and have been suggested as potential players in the formation of banded iron formations (BIFs) (Ehrenreich and Widdel, 1994; Konhauser *et al.*, 2002).

The oxygenic photosynthetic organisms include cyanobacteria, eukaryotic algae, and green plants (Table 2). These organisms utilize water as the hydrogen donor, generating oxygen as the product. The ability of eukaryotic algae and plants to photosynthesize is due to an early symbiotic event in which a cyanobacterial symbiont combined with some other kind of cell to create a cell fusion capable of oxygenic photosynthesis (Margulis, 1981). Molecular genetic data clearly show that these symbionts have come from the domain Bacteria, via organisms closely related to extant cyanobacteria (Woese, 1987).

As mentioned above, because of their ability to produce oxidized products under strictly anaerobic conditions, both types of photosynthetic processes are of potential importance in the chemical evolution of Earth, and understanding the timing of their appearance and metabolic success on Earth represent major challenges for understanding the evolution of metabolism. The timing of the appearance, expansion, and distribution of such abilities over time is of great interest with regard to the stepwise oxidation of the Earth, as proposed by DesMarais (1994) and DesMarais *et al.* (1992) (see also Cloud (1968) for an earlier perspective on this issue). Clearly, the energy from the Sun, in addition to supplying photoreductants for the fixation of carbon, drives the production of oxidants under anoxic conditions, activities that have had far-reaching effects on the evolution of both the geosphere and the biosphere.

Molecular analyses of the five groups of prokaryotic phototrophs indicate that all of these photosynthetic systems are genetically and functionally related. The invention of photosynthesis as a process has not resulted in a wide variety of types, but a group of closely related systems that have enjoyed extensive gene exchange (Blankenship, 2002; Blankenship and Hartman, 1998; Raymond *et al.*, 2002; Xiong *et al.*, 2000). Because the anoxygenic photosynthetic systems are structurally and chemically simpler, it is assumed that they evolved first, and that the complex oxygen-emitting systems followed: molecular genetic data support such a view. To this end, some cyanobacteria can exist both anaerobically (using sulfide as an electron donor for photosynthesis), and aerobically, using water as an electron donor and emitting oxygen (Cohen *et al.*, 1975; Padan, 1979). This co-occurrence of anoxygenic and oxygenic photosynthesis lifestyles in the cyanobacteria, while it is consistent with the current ideas of the evolution of oxygenic photosynthesis, creates some potential problems in the interpretation of the ancient-rock record (see below).

8.02.5.2 Lithotrophic Energy Sources

A wide range of prokaryotic genera, distributed among the Bacteria and Archea, are capable of lithotrophic growth, using inorganic molecules as their sole or major energy source. Many different lithotrophic energy sources are available in the environment, virtually all of which can be utilized by prokaryotes (Table 3 and Figure 4). As we view prokaryotic diversity today, we see that nearly any energy source that is available and abundant is being utilized, either aerobically, or with other electron acceptors.

As will be discussed below, respiration requires electron acceptors, and many of the lithotrophic modes of metabolism require oxygen, making many believe that they were rather late metabolic "inventions," occurring only after the world was oxic, or at least after oxygen respiration was invented. This is not necessarily the case, for a variety of reasons: first, the "invention" of oxygenic photosynthesis, and the resulting

Table 3 Lithotropic metabolism and organisms.

Metabolic type	*Substrate utilized*	*Product(s) produced*	*Distribution of abilities*[e]
Hydrogenotrophs	H_2	H_2O	B, A
Methanogens[a]	H_2	CH_4	A
Carboxydotrophs	CO	CO_2, H_2	B
Sulfur oxidizers	S^0	SO_4^{2-}	B, A
Sulfide oxidizers	H_2S	S^0, SO_4^{2-}	B, A
Iron oxidizers[b]	Fe(II)	Fe(III)	B
Ammonia oxidizers[c]	NH_4^+	NO_2^-, N_2	B
Nitrite oxidizers	NO_2^-	NO_3^-	B
Phosphite oxidizers[d]	PO_3^{2-}	PO_4^{2-}	B
Arsenite oxidizers	AsO_3^{3-}	AsO_4^{3-}	B

[a] While methanogens are not usually thought of as lithotrophs, many of them utilize H_2 as their sole energy source, using CO_2 as the oxidant, producing CH_4. [b] The iron oxidizers fall into four groups: acidiphilic iron oxidizers, using oxygen or nitrate as electron acceptors; neutrophilic iron oxidizers, using oxygen or nitrate as electron acceptors. [c] The ammonia oxidizers include both the classical aerobic nitrifying bacteria that produce nitrite from ammonia, and a newly discovered group of organisms that accomplish ammonia oxidation anaerobically, a process is called ANAMOX (Strous *et al.*, 1999). [d] Phosphite oxidation at the expense of sulfate reduction was recently described (Schink and Friedrich, 2000; Schink *et al.*, 2002). [e] A = Archea; B = Bacteria.

Table 4 Metabolic divisions based on carbon source utilization.

Metabolic type	*Carbon source*	*Representative groups*[a]	*Challenges*[b]
Heterotrophs	Organic carbon	E, B, A	Break down polymers to monomers; reassemble
One-carbon	Formate	B, A	Synthesize all C—C bonds: make monomers; assemble polymers
	Methanol[c]	B, A	
	Methane[d]	B, A	
	CO	B, A	
	CN^-	A	
Autotrophs	CO_2[e]	E, B, A	Reduce CO_2 to CH_2O; synthesize all C—C bonds; make monomers; assemble polymers

[a] E = Eukarya; B = Bacteria; A = Archea. [b] The metabolic requirements and challenges faced by the various groups. [c] One group of eukaryotes, the methanol-utilizing yeasts are known (Higgins *et al.*, 1981). [d] The archeal methane utilizers are methanogens apparently running "backwards," which live in syntrophy with sulfate-reducing bacteria, which remove hydrogen. [e] The Eukarya include green plants and algae; the Bacteria include cyanobacteria, lithotrophic bacteria, and photosynthetic bacteria; the Archea include lithotrophs.

appearance of oxygen, perhaps confined to microenvironments, may have occurred long before it was abundant (i.e., detectable) in the geological record; second, many lithotrophs utilize other electron acceptors (Table 3), including methanogens, which utilize hydrogen as the energy source, and CO_2 as the oxidant; finally, when the symbiotic event took place to form the first respiratory eukaryotic mitochondrial organelle, the antecedent was almost certainly an oxygen-respiring organism, as virtually all known eukaryotic cells today have active mitochondria or remnants of former mitochondria that can be detected by genetic and genomic analyses (Roger and Silberman, 2002). Thus, even oxygen-utilizing lithotrophic metabolism cannot be constrained to be "recent" process.

8.02.5.3 Carbon Sources for Life

Life can also be categorized according to the carbon sources utilized. Organisms deriving their carbon from pre-made organic carbon are called heterotrophs, while those that utilize inorganic carbon (CO_2 or bicarbonate) are referred to as autotrophs. As shown in Table 4, there is a third group we consider here, namely the one-carbon organisms, which share many of the same metabolic challenges faced by the autotrophs. Heterotrophs utilize energetically rich organic carbon, breaking down the polymers to monomers, and resynthesizing polymers, using the energy obtained from metabolism. One-carbon organisms begin with reduced carbon, but still have the formidable challenge of assembling an entire cell from a starting point containing no C–C bonds, building the entire complexity of the cell from the bottom up. Autotrophs face the same challenge, but with the added difficulty of the need to convert inorganic carbon in the form of CO_2 to a reduced form before beginning the synthesis of the new cell. We have added the one-carbon organisms to this list because, along with the autotrophs, they tend to be very important in terms of isotopic signatures. In general, any organism that uses a one-carbon compound for a carbon source has the likelihood of leaving an easily discernible (strongly fractionated) carbon isotopic signature.

The most abundant autotrophs on Earth are, not surprisingly, oxygenic photoautotrophs, as there is almost an unlimited amount of energy available to support the carbon fixation and biosynthesis, and a limitless supply of the electron donor, water. In contrast, lithotrophs are limited both by the fact that the free energy available from the energy-yielding reactions is small, and the availability of the reduced species is often limited and confined to anaerobic (subsurface) environments. Thus, both lithotrophs and anoxygenic phototrophs are often confined to areas isolated from the oxic atmosphere, where their electron donors are stable and available.

8.02.5.4 Fermentative and Respiratory Metabolism

Virtually all energy-yielding reactions used by life are the result of oxidation–reduction (redox) reactions, with a reductant (electron donor) as fuel, being respired via an oxidant (electron acceptor). The one seeming exception is the process of fermentation, wherein microbes (prokaryotes and some yeasts) utilize organic carbon as both oxidant and reductant. Often a single carbon source is enzymatically split into two entities, one more oxidized than the other, so that one becomes the oxidant and the other the reductant. Since both parts of this reaction are derived from reduced organic carbon, the differences in redox state are small, and the reactions are, by necessity, very inefficient, yielding minimal biomass and maximal waste products. Entire industries are built around these inefficient processes and the resulting abilities of fermentative organisms to produce large amounts of by-products such as lactic acid (yogurt, cheese, etc.), alcohol, and solvents such as butyrate.

In marked contrast, respiration involves the utilization of an electron acceptor (usually, but not always inorganic) that is utilized in a dissimilatory fashion (i.e., its reduction is coupled to respiration and growth, but it is neither taken up nor incorporated into cellular material). Respiration can be aerobic, in which oxygen is the electron acceptor and maximal energy is yielded, or it can be any of a variety of other compounds ranging from nitrate or Mn(IV), which have redox potentials near that of oxygen, to CO_2, with a potential so low that very little energy can be obtained. Some of these, along with the common names for such metabolic types, are shown in Table 5. These are not all of the known dissimilatory groups, but serve to demonstrate the metabolic diversity: diversity that does not extend to the eukaryotes, which, almost without exception, are limited to oxygen as an electron acceptor (Table 5).

The full range of prokaryotic respiratory versatility is not yet fully known, as demonstrated by the fact that just a few years ago it was established that dissimilatory metal-ion reducing bacteria could grow at the expense of iron or manganese oxide reduction (see Nealson and Saffarini, 1994). It is probably safe to say that any electron acceptor that is in abundance, and that has a redox potential electronegative enough to allow an organism to use it for an energy-yielding reaction, has been (or will be) discovered by the microbial world. A good example of this is that of arsenate (As^{5+}), which in Mono Lake, CA is sufficiently abundant as to have enriched for arsenic-respiring microbes that account for ~10% of the recycling of organic carbon in the lake (Stolz and Oremland, 1999; R. Oremland, personal communication). In addition, new types of organisms, such as those reported by Coates *et al.* (1999) that live via the respiration of perchlorate, are being frequently described.

Table 5 Types of respiration.

Type of respirer	*Electron acceptor*	*Product*	*Distribution of ability*[a]
Aerobes	O_2	H_2O	E, B, A
Nitrate reducers	NO_3^-	N_2, NH_3	B, A
Nitrite reducers	NO_2^-	N_2, NH_3	B, A
Mn reducers[b]	Mn(III), Mn(IV)	Mn(II)	B, A
Iron reducers[b]	Fe(III)	Fe(II)	B, A
Fumarate reducers	Fumarate	Succinate	B, A
Sulfur reducers	S^0	H_2S	B, A
Sulfate reducers	SO_4^{2-}	H_2S	B, A
Arsenate reducers[c]	AsO_4^{3-}	AsO_3^{3-}	B
Selenate reducers[c]	SeO_4^{3-}	SeO_3^{3-}	B
Perchlorate reducers[d]	ClO_4^-	ClO_3^-	B
Methanogens	CO_2	CH_4	A

[a] E = Eukarya; B = Bacteria; A = Archea. [b] Nealson and Saffarini (1994). [c] Oremland *et al.* (2002) and Stolz and Oremland (1999). [d] Coates *et al.* (1999).

8.02.6 RECONSTRUCTING THE EVOLUTION OF METABOLISM

8.02.6.1 Approaches Employing Genomics and Molecular Genetics

With the advent of gene sequencing it became clear that the relationships of proteins could be easily studied either by the comparison of the DNA sequences of genes coding for them, or by conversion of the DNA sequence into a protein sequence and studying the similarities of protein sequences directly. As the techniques improved, it became apparent that genes catalyzing the same reaction were similar across all the domains of life. Thus, the genes that were functionally similar were often structurally related, suggesting that life clings to a good idea once it has been "invented." Thus, genes were categorized as orthologs—those that share descent from a common ancestor and presumably have the same function, or paralogs—those that arose by gene duplication (and thus share sequence similarity) and have diverged to many different functions. However, as is now well known, there is no sure way to tell from sequence data alone which is which. The usual practice is that highly homologous genes are labeled as orthologs and ascribed a function on the basis of that assignment.

Now, with more than 100 microbial genomes sequenced or nearly completed, we are presented with an immense amount of information concerning single genes, multiple gene groups, and entire genomes. With the advent of such genomic analysis, there was great enthusiasm that it might be possible to tease from comparative genomics some of the keys to the evolution of life, especially some of the metabolic details. However, to do this, several things need to be true. First, the rate of evolution needs to be sufficiently slow so that the information is not lost in geological time. Any mutation will eventually act to "hide" the information about the past, and if the rates of mutation are such that the information is randomized over geological timescales of hundreds of millions of years, then looking back for billions of years becomes a hopeless endeavor. In a nice discussion of this, Benner *et al.* (2002) conclude that, when the rate of substitution for eukaryotic genes is taken into account, the maximum time that might be hoped for will be ~600 Myr. Thus, such approaches may well work for processes occurring over the last 500 Myr. This is particularly encouraging, as over this same time period, fossilization has occurred and been preserved, so that outside markers can be used to calibrate the molecular clock.

Second, the rate of evolution needs to be constant. This is a much harder issue to deal with, especially in prokaryotes, where growth rates are highly variable, symbiotic relationships that alter growth needs are common, and microbial communities are notorious for slow rates of growth and metabolism.

Finally, horizontal (lateral) gene transfer (HGT or LGT) needs either to represent a minor component with regard to genomic information, or to be easily documented (i.e., to tell the time and extent of LGT). For example, while exchange among the domains of life may have been "easy" when all were unicells, heritable exchange is much more difficult now, given the need to transfer genetic equivalents into the germ cells of specialized eukarya.

When genes, such as the 16S ribosomal RNA, are used, the results can be encouraging—the rates of mutation and evolution of these genes are slow, and potentially useful for those who would study evolution. To this end, Doolittle and colleagues have been able to use very slowly evolving proteins to estimate some of the major events in the history of life, such as the time of separation of major eukaryotic biological group, and even estimated the time of the last shared ancestor as being ~2 Ga. This is an interesting extrapolation, as it implies that oxygen respiration had evolved prior to this time, an assumption needed to account for the universal presence of mitochondria in eukaryotic cells. However, when more metabolically relevant genes are examined, the situation gets less encouraging. As stated by DeLong (1998), "Despite utility for identification, molecular sequence comparisons do not provide much detail on phenotype"—to paraphrase this, one might say, "between rapid and unpredictable rates of mutational substitution on one hand, and horizontal gene transfer on the other, trying to use molecular data to specify ancient metabolic evolution is at best an effort fraught with many intellectual traps, and at worst, impossible." This is a very important perspective, as it has become almost commonplace for statements to be made relating the presence of phenotypes in modern bacteria to conclusions regarding their antiquity.

Such statements are intellectual landmines that serve little purpose other than to mislead the uninformed. An example of this situation is that of iron reduction by bacteria, now supposed to be an "ancient" property based on "microbiological evidence" (Vargas *et al.*, 1998). While on the surface this seems to be a reasonable assumption, there is danger in proclaiming it to be so, based on genomic data. All organisms studied today are far removed in time from any ancient phenotypes, and assumptions that define any given function as ancient should be made with great caution. Vargas *et al.* state on the basis of the measurement of iron reduction by hyperthermophiles in the laboratory, that iron reduction was the first type of respiration; in fact, there are many reasons to question this conclusion. One of their central assumptions is that hyperthermophiles are ancient, a contentious point of view (Brochier and Philippe, 2002; Galtier *et al.*, 1999). Perhaps more relevant, there is virtually no biochemistry or physiology to back up the assumption. Further work on the iron reductase from *Archeoglobus fulgidus* (Vadas *et al.*, 1999) has revealed that it functions only with chelated iron, similar to many other assimilatory iron reductases, and may well have been acquired via LGT. In aqueous environments, chelation of iron is presumably necessary only under oxidizing conditions, conditions that most likely did not exist on Earth until after oxygenic photosynthesis had been developed. As always, time and quality data will resolve this issue, but it is but one of many examples that serve to demonstrate the potential hazards of trying to infer the evolution of metabolism from inadequate data.

The bottom line, with regard to genomics and genetics, is that while there may be great promise in such an approach, available data suggest that LGT is much more of a major process than was imagined a few years ago. Doolittle (2002) has characterized this as "molecular Lego," and noted that, "no one had anticipated that prokaryotic genomes would be so extensively mosaic, and so obviously pieced together by assembly of genes from so many antecedents." While this is fine for genomicists, who want to understand how prokaryotes have gone about their evolution, it muddies the concept of prokaryotic species, and greatly obscures the relationship between taxonomy, phylogeny, and function.

With regard to the discussion here, it is very important to grasp the importance of LGT, and discuss the evolution of metabolism apart from the evolution of bacterial species, and perhaps apart from bacterial systematics altogether. Eventually, these two discussions may reunite, but for now the discussion of metabolic evolution is probably better done independently.

The above discussion suggests that, at least for the present, molecular genomic methods have sufficient limitations to negate their value for specifying the evolution of ancient metabolism. This statement stands in apparent defiance to the work of Benner *et al.* (2002), who have used the molecular record to establish major events when gene duplication occurred. Using these approaches, the authors have tracked the development of genes for ethanol synthesis in yeast, showing that their appearance correlates with the emergence of flowering plants, and fructose synthesis. Usually, such work is difficult to correlate with timed geological or paleontological events, but if it can be done, it allows for the possibility of relating the evolution of one pathway to the exploitation of resources evolving elsewhere. Such approaches represent the hope and promise of genomics and proteomics. However, as the authors point out, there are substantial limitations to the approach. Given the substitution rates for eukaryotic genes, the authors predict a maximum extent for these methods to be ~600 Ma. Thus, while it is easy to agree with the statement, "correlation of events in the molecular, paleontological, and geological records, and the molecular dissection of historical events occurring in the past, offers a paradigm to dissect the function of the planetary proteome," it is also very important to recognize the limitations, as Benner and colleagues have done.

For the moment, taking the conservative approach, it is probably fair to say first, that it is difficult, if not impossible, to tell from current molecular systematics data, using the 16S rRNA methods, the actual time of divergence of almost any group of prokaryotes, and second, that it is probably equally challenging to assign a definite physiology to any group of ancient organisms based on this approach. While this may seem overly pessimistic, it is prudent to err on the side of caution, and to anticipate that as genomic methods mature, the situation may improve.

8.02.6.2 Approaches Employing Geochemical and Geophysical Methods

The above discussion strongly suggests that genomic and molecular genetic approaches have little to tell us of the precise timing and/or exact sequence of events that resulted in the metabolic diversity life enjoys today. However, there are a number of other methods available, all of which involve the analysis of the rock record for: (i) physical fossils of past life; (ii) signature molecules (usually organic) derived from past life; (iii) geochemical (inorganic) signatures of past chemical evolution; and (iv) isotopic fractionations indicative of life. All of these techniques involve examination of the rock record, usually layered sediments, the ages of which must be known, and which, to a best approximation, must be understood in terms of their physical and chemical history. These requirements make the task very difficult. As samples age, they are profoundly affected by biological and chemical processes, by burial and resulting heating, and by tectonic processes leading to mixing, melting, uplifting, etc. All this conspires to obscure the rock record. Samples are physically and chemically altered, and sometimes displaced in the geological record. Finding ancient samples that have not been profoundly altered is a great challenge: well-preserved ancient material is rare and highly prized by the paleontological community (Bottjer *et al.*, 2002; Schopf and Klein, 1992).

8.02.6.2.1 Physical fossils

Physical fossils are the standard tools of paleobiology. For the larger eukaryotic organisms, both plants and animals, the recognizable fossil remains provide markers through which we can calibrate past evolutionary events: the emergence of carbonate, silicate, and phosphate "hard parts," and the emergence of major groups of invertebrates, vertebrates, grasses, plants, etc. For example, the "Cambrian explosion" of eukaryotic species is obvious in the rock record, and can be documented in samples from around the world. The record from shortly before this explosion to the present allows one to calibrate species appearances with a geological timescale, and to reconstruct a picture of organismal evolution over relatively recent (800 Myr) timescales. With regard to the evolution of metabolism, however, this record does little for us. By the time larger eukaryotes emerged, the atmosphere was highly oxygenated (Figure 6). Basic metabolic evolution was essentially complete. All of these complex eukaryotes are metabolically similar, utilizing mitochondria for "standard" aerobic heterotrophic metabolism, and/or chloroplasts for oxygenic photosynthesis. However, as one ventures further back in time, the fossil record becomes more cryptic in every sense. First, the absence of hard parts results in less obvious and more ambiguous preserved samples. Second, the organisms are smaller, usually single-celled, and generally lack distinctive markers that allow them to be identified. Third, even if it can be established that prokaryotic fossils of a particular morphology are present, given the above discussions of genetic exchange, it is nearly impossible to say with certainty that a given type of metabolism was operating.

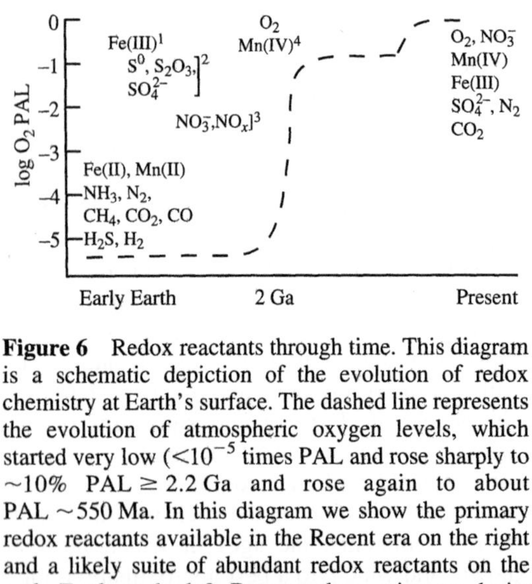

Figure 6 Redox reactants through time. This diagram is a schematic depiction of the evolution of redox chemistry at Earth's surface. The dashed line represents the evolution of atmospheric oxygen levels, which started very low ($<10^{-5}$ times PAL and rose sharply to $\sim$10% PAL $\geq$ 2.2 Ga and rose again to about PAL $\sim$550 Ma. In this diagram we show the primary redox reactants available in the Recent era on the right and a likely suite of abundant redox reactants on the early Earth on the left. Between these suites we depict the rough order and timing of the appearance of those reactants that are abundant today but were probably largely absent on the very early Earth. [1]Given the widespread appearance of the banded iron formations in the Archean, Fe(III) may have been an abundant oxidant present from these early times. [2]Isotopic evidence suggests that some oxidized sulfur compounds were present from 3.5 Ga, though S^0 may have been the primary species undergoing reduction at that time. [3]Isotopic evidence suggests that oxidized forms of nitrogen became abundant enough to leave a signal between 2.7 Ga and 2.0 Ga, though the timing is difficult to pin down, [4]while Mn oxides appear only later (2.3 Ga), nearly coincident with the rise of atmospheric oxygen.

Nevertheless, microfossils are one of the major tools of the trade, and visible microfossils, when correlated with other kinds of data, provide convincing evidence for past communities, and some hints as to their activities (Barghoorn and Tyler, 1965; Rasmussen, 2000; Schopf, 1999; Walsh and Lowe, 1985). Perhaps the most compelling fossil evidence from the ancient rock record is that of the so-called stromatolites: layered structures composed of carbonates that are commonly ascribed to the cyclic activities of photosynthetic microbes, namely, cyanobacteria (Bertrand-Sarfati and Walter, 1981; Schopf and Klein, 1992; Semikhatov *et al.*, 1979). These structures have been consistently used as "the marker" for ancient life, and the time of origin for the advent of photosynthesis. Modern analogues for stromatolite formation are documented (Laval *et al.*, 2000; Reid *et al.*, 2000). However, it has been pointed out that such structures can be formed abiotically (Grotzinger and Knoll, 1999; Grotzinger and Rothman, 1996; Pope *et al.*, 2000): as with almost all biomarkers, there is some ambiguity. Furthermore, recent studies of both deep-sea carbonate deposits from serpentinization emanations (Kelley *et al.*, 2001), or due to pH changes in methane oxidation environments (Michaelis *et al.*, 2002), may provide other options for carbonate formation. The latter example is particularly interesting, as it involves symbiosis between anaerobic methane-oxidizing archeal cells and sulfate-reducing bacteria. The sulfate reducers rapidly consume the hydrogen produced by the methane oxidizers, promoting further methane oxidation and generating alkalinity. In concert, these prokaryotes oxidize methane, precipitate carbonate into large reef-like structures, and leave a decided isotopic signature behind that we will discuss below.

8.02.6.2.2 Signature molecules

Signature molecules are commonly used to infer the presence of certain types of organisms. These approaches are the basis of community taxonomic analyses commonly done in modern sediments (White, 1993). These molecules are excellent for laboratory studies, and valuable for field analyses of extant biota. However, most biological molecules serve as excellent food sources for other organisms, and thermal and abiological chemical alterations take their toll during diagenesis and metamorphism, often leaving little to be analyzed in old samples.

Despite these obstacles, many signature molecules have been developed and utilized for the study of ancient samples (see Schopf and Klein, 1992). Of these molecules, perhaps the most interesting group recently utilized has been the methylhopanoids, breakdown products of 2-methyl-hopanepolyols, which are produced by cyanobacteria, and are common in microbial mats. The discovery of methylhopanoids in organic-rich sediments from 2.7 Ga (Brocks *et al.*, 1999; Summons *et al.*, 1999) provided the single best piece of evidence for the presence of cyanobacteria at this time.

Given that all contemporary cyanobacteria are capable of oxygenic photosynthesis, Summons *et al.* (1999) conclude that these observations also provide the strongest evidence for the evolution of oxygenic photosynthesis by 2.7 Ga. However, Cohen *et al.* (1975) and many others (see Padan, 1979, for review) have reported that many extant cyanobacteria grow well under strictly anoxic conditions with sulfide as the electron donor, utilizing only photosystem I, and releasing no oxygen. If these anaerobically grown cyanobacteria also produce hopanepolyols (they are reported to do so under aerobic conditions), then the compelling evidence for cyanobacteria cannot be in itself compelling evidence for oxygen generation.

In general, this exciting approach and the results it has yielded are remarkable. The survival of indicator organics for gigayears is encouraging. Certainly, these results will motivate the search for more well-preserved samples with the hope of seeing how far back in time various signature molecules can be found.

8.02.6.2.3 *Inorganic chemistry of sediments*

The study of the inorganic chemistry of sediments has provided a broad-brush picture of the evolution of the Earth. Riverine sediments from the deepest past contain minerals (e.g., uraninite, pyrite, and siderite) that decompose rapidly under oxic conditions whereas such minerals are absent from more recent sediments (Kirkham and Roscoe, 1993; Ramdohr, 1958; Rasmussen and Buick, 1999; Robb *et al.*, 1992). Other sedimentary features associated with highly oxic conditions (e.g. red beds) are absent in the earliest sediments and present in more recent ones (Rye and Holland, 1998). The sedimentary record could reflect localized geochemical environments or regional conditions rather than the global average, depending on many different things. Thus, one must integrate all of the data and try to understand the evolution of the planet in this context. For such integration, adding the study of paleosol chemistry to the mix has proven to be quite valuable in unraveling the evolution of the Earth and, more importantly for our purposes, the corresponding metabolic processes catalyzed by living systems. Paleosols are ancient soils that, when retrieved in a relatively unaltered state, can be used to infer the conditions of the environment in which they formed. Investigation of the mineralogy and redox chemistry (usually focusing on redox sensitive components like iron) of accurately dated paleosols has led to the general picture shown in Figure 6 (modeled after Rye and Holland, 1998), in which a large increase in atmospheric oxygen is inferred at ~2.2 Ga, presumably due to the global success of oxygenic photosynthetic metabolism, and the titration of reducing equivalents. Using these approaches, it has been possible to estimate the atmospheric levels of oxygen (Holland, 1984; Holland and Rye, 1997; Rye and Holland, 1998; Yang *et al.*, 2002) and of CO_2 (Rye and Holland, 2000a; Rye *et al.*, 1995). Of course, relating such data to the origin and evolution of metabolism is difficult. With regard to understanding the evolution of metabolism, paleosol chemical data supply the definitive times at which a given metabolism left a measurable trace—almost certainly temporally uncoupled from the origin and early evolution of this process. However, as will be discussed below, while we now have some confidence with regard to ancient and modern oxygen levels, neither the rapidity of the rise once the necessary metabolic innovations were first made, nor the corresponding appearance of other oxidants is easy to specify with confidence.

8.02.6.2.4 *Stable isotope fractionation*

The fractionation of stable isotopes often occurs as a result of interaction with biological systems. Enzymatically catalyzed reactions tend to favor the chemical transformations of the lighter isotopes. Consequently, such reactions generally produce biological material that is isotopically lighter than their source material. Carbon and sulfur isotopes (and, to a much lesser extent, nitrogen isotopes) have been extensively utilized to look for traces of metabolic activity (see Schopf and Klein, 1992) in diverse settings ranging from modern sediments to some of the oldest known rock formations. Because it is a function of atomic mass ratios, fractionation is easy to detect between ^{12}C and ^{13}C. Usually, this is described in terms of parts per thousand fractionation (‰) or $\delta^{13}C$ relative to a standard (all values in this chapter are given relative to the Pee Dee Belemnite of PDB standard). Negative numbers connote lighter carbon, and are taken to indicate the possibility of biological activity. When dealing with one-carbon compounds, the fractionation is particularly large, so that the major signals that are seen in organic carbon in the fossil record are those due to carbon fixation by photosynthesis (usually to −13‰ to −25‰), and those due to the production and/or oxidation of methane (ranging from −45‰ to −90‰ or more).

In a similar way, mass-dependent sulfur isotopic fractionation has been used for tracking sulfur metabolism through time, with light sulfur, ^{32}S, being favored over the heavier, ^{33}S, ^{34}S, and ^{36}S. As with carbon isotopes, differences in sulfur isotope mass ratio ($^{33}S/^{32}S$, $^{34}S/^{32}S$, or $^{36}S/^{32}S$) are described in terms of parts per thousand fractionation, $\delta^{33}S$, $\delta^{34}S$, or $\delta^{36}S$ relative to a standard. In all known aqueous, solid-state and UV-shielded atmospheric processes $\delta^{33}S = 0.515\delta^{34}S$ and $\delta^{36}S = 1.90\delta^{34}S$; the fractionation is thus said to be mass dependent. Recently (Farquhar *et al.*, 2000a, 2002; Pavlov and Kasting, 2002), mass-independent fractionation of sulfur, an atmospheric process that occurs only in the presence of a high UV-flux, has been used as a tool to help resolve the debate concerning the rise of stratospheric ozone and atmospheric oxygen.

The stable isotopes of iron have been recognized as a potential tracer for iron metabolism. Iron isotopes, like those of carbon, nitrogen, and sulfur, are fractionated by a variety of biological processes (Table 6). The isotopic signatures left by such processes may be more robust than carbon, nitrogen, and sulfur isotopic signatures, inasmuch as iron-bearing compounds are typically less volatile than those bearing carbon, nitrogen, and sulfur. However, our understanding of the biological and abiological processes that fractionate iron is still immature and it is unclear whether evidence in the rock record for any given fractionation of iron isotopes will emerge as compelling evidence for biological processes.

8.02.6.2.5 Carbon isotopes

Carbon isotopes have been extensively used as evidence for the appearance of life on Earth (see Schopf and Klein, 1992), and simultaneously for the "invention" of carbon fixation. Mojzsis *et al.* (1996) used these approaches to push the earliest evidence for life back to 3.8 Ga. However, as often occurs with such ancient samples, the age of the samples has been debated (Kamber and Moorbath, 1998; Mojzsis and Harrison, 2002), the sedimentary origin of at least some of the samples is in doubt (Fedo and Whitehouse, 2002), and it now seems possible that the graphite that was the source of carbon for the measurements could have been there via a geological secondary injection (van Zuilen *et al.*, 2002). Irrespective of the resolution of this claim, it is widely accepted that there is abundant examples of carbon isotopic data supporting the presence of life back to ~3.5 Ga (Figure 7; see Schopf and Klein, 1992) though the origin and biogenicity of some of the oldest samples is still being debated (Brasier *et al.*, 2002; Schopf *et al.*, 2002). Though life almost certainly was present over 3 Ga and probably by 3.5 Ga, we are certain of neither the mechanism of carbon fixation (including the source of oxidized carbon), nor the organisms responsible. Carbon monoxide is an appropriate candidate for reduction to organic carbon, and may well have been abundant on the ancient Earth as a result of the photolysis of carbon dioxide. Many organisms are known today that can utilize CO as their carbon source for autotrophic growth (Maness and Weaver, 2001; Svetlitchnyi *et al.*, 2001).

When carbon isotopic values fall below −40 per mil, the involvement of methane producing and consuming microbes is suspected. In modern environments the isotopic composition

Table 6 Possible processes for biological iron isotope fractionations.

Process	*Description*	*Location*	*Conditions*
Assimilation	Uptake of Fe(II)	Cell membrane	Anoxic
	Uptake of Fe(III)	Cell membrane	Oxic
	Reduction of Fe(III)-chelates	Cytoplasm	Oxic
	Biosynthesis of organo-iron	Cytoplasm	Anoxic and oxic
Scavenging	Binding by siderophores	Extracellular	Oxic
Storage	Ferritins	Intracellular	Anoxic and oxic
Transfer	Transferrins	Intracellular	Oxic
Oxidation	Lithotrophic iron oxidation	Cell membrane	Anoxic or oxic
	Photosynthetic iron oxidation	Cell membrane	Anoxic[a,c]
Dissimilation	Enzymatic iron reduction	Cell membrane	Anoxic[b,c]
	Reduction via electron shuttles	Extracellular	Anoxic

[a] Isotopic fractionation between ferric and ferrous species over a wide range of oxidation rates is +1.3 per mil to +1.5 per mil (Croal *et al.*, 2003). [b] Isotopic fractionation between ferric and ferrous species over wide range of reduction rates is +1.3 per mil (Beard *et al.*, 1999, 2003). [c] Biological fractionation nearly half that in inorganic ferric–ferrous system (Johnson *et al.*, 2002).

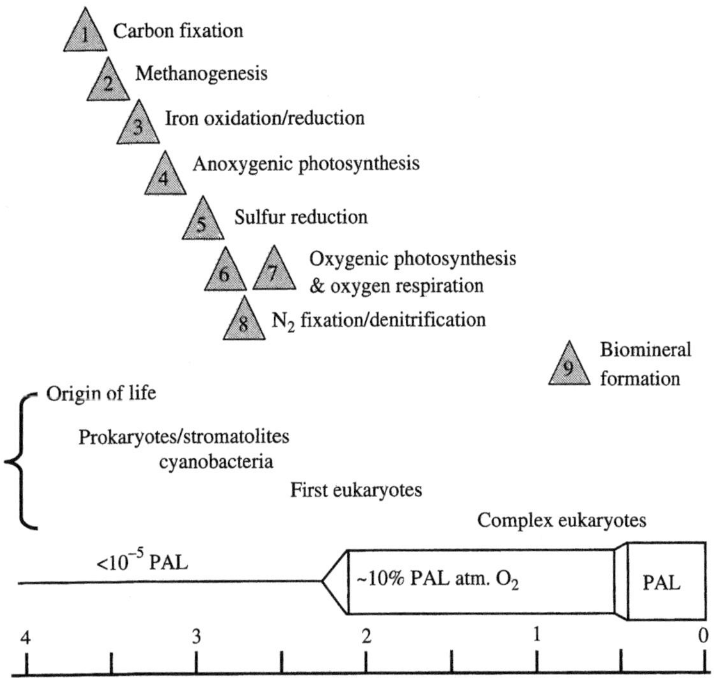

Figure 7 Time of appearance of major metabolic abilities. We list here some of the major metabolic events, and the times at which we first have evidence for their appearance. As noted in the text, this is a minimum time for their "invention," as it is not known how long a given process was present before it left a recognizable imprint in the rock record. Carbon fixation, methanogenesis, and methane oxidation can be estimated through carbon isotope signatures (Buick, 1992; DesMarais, 1994; DesMarais *et al.*, 1992; Mojzsis *et al.*, 1996), while iron oxidation is inferred from the presence of BIFs. Iron reduction is inferred from the presence of abundant iron, but to date no direct evidence to support iron reduction as an ancient process is available. Sulfur reduction is inferred from sulfur isotopic data (Canfield *et al.*, 2000; Shen *et al.*, 2001). Anoxygenic photosynthesis in put here because it is regarded as an antecedent to oxygenic photosynthesis. Oxygen respiration is placed after the invention of oxygenic photosynthesis. Of interest is the late invention of biominerals, which occurred only a few tens or hundreds of millions of years before the Cambrian explosion and the rise to the PAL of oxygen.

of carbon in methane gas can be used to assess biological production, with values reaching below −100 per mil (Schoell, 1980). In ancient environments, methane is not preserved; so ancient signals are usually not so negative. Thus, when values below −40 per mil are observed, they are usually attributed to the action of methanotrophic organisms (Schoell and Wellmer, 1981). Such light values have been observed as far back as 2.75 Ga (Buick, 1992; Hayes, 1994; Rye and Holland, 2000b), suggesting that methanotrophy is at least that old. Traditionally, evidence for ancient methanotrophy has also been taken as evidence for ancient oxygenic photosynthesis (Buick, 1992) as all cultured methanotrophs require molecular oxygen (Higgins *et al.*, 1981). However, recently new discoveries in the Black Sea have opened up potentially new interpretations of carbon isotopes (Michaelis *et al.*, 2002). Massive microbial reefs with carbonate depletion signatures of −25 per mil to −32 per mil are seen. It is proposed by the authors that the reefs are formed via a microbial community including methanogenic bacteria (that catalyze the anaerobic oxidation of methane) working symbiotically with sulfate-reducing bacteria whose function is the removal of hydrogen. Keeping the hydrogen concentration very low makes the anaerobic oxidation of methane energetically favorable. Such symbioses, in which product removal by one organism allows another partner to accomplish an otherwise unfavorable process are well known in the microbial world (see Jackson and McInerney, 2002). The symbiosis accounting for anaerobic methane oxidation has only recently been discovered and documented (Boetius *et al.*, 2000; Orphan *et al.*, 2001). These results show that ancient isotopic evidence of methanotrophy in marine systems does not necessarily show that either oxygenic photosynthesis or classical chemolithotrophy were functioning then. Alternatively, some samples of organic carbon from 2.75 Ga that formed in lacustrine sediments (Buick, 1992) and in sediments associated with soil environments (Rye and Holland, 2000b) have very negative $\delta^{13}C$ values (<−40 per mil). Neither of these environments appears to have contained significant amounts of sulfate. It is tempting then to conclude that the light carbon in these nonmarine samples still demonstrate that oxygenic photosynthesis was extant 2.75 Ga but it seems likely that other syntrophic pathways are possible as well.

8.02.6.2.6 Sulfur isotopes

Isotopes of sulfur are likewise extremely valuable for tracing metabolism, and have been extensively used for the study of ancient sediments. The fractionation most commonly measured is between ^{32}S and the heavier ^{34}S. The sulfate reducers have long been considered to be the organisms responsible for the major signals seen in the rock record, a logical assumption considering their ubiquity in anaerobic environments today. In particular, Canfield and his colleagues have done extensive work in unraveling the details of sulfate-reducing bacteria, and their relationship to isotope fractionation. The elegant studies of this group have led to several major conclusions, which include the following:

(i) Low sulfate (<200 μM) in the early Archean ocean—inferred from fractionations of less than 10 per mil between seawater sulfate and sedimentary sulfide (Canfield, 1998; Habicht *et al.*, 2002).

(ii) Metabolic sulfate reduction in evaporitic environments at 3.47 Ga—indicated by fractionations of ~20 per mil between microscopic sulfide and coexisting sulfate (Shen *et al.*, 2001).

(iii) Marine sulfate levels begin to rise (>200 μM) at 2.4 Ga, but remain low (<1 mM)—indicated by low sulfur fractionation in sediments deposited in nonevaporitic settings (Habicht *et al.*, 2002).

(iv) The deep ocean was finally oxidized after 1.05 Ga—indicated by large fractionations between sulfate and coexisting sulfide (Canfield and Teske, 1996).

The same group points out that during this time major events, such as the formation of BIFs, were occurring, and could have secondary effects on phosphate concentration, and thus net productivity (Bjerrum and Canfield, 2002). The above conclusions result from elegant arguments, are very convincing, and show the importance of correlating isotopic data with other major geological events, and linking both of these to known properties of metabolic groups. For example, the argument that sulfate levels have risen, and are correlated with levels of isotope fractionation observed is supported by physiological studies of sulfate reducers in which chemostat studies with various sulfate levels show that low sulfate levels are consistent with low fractionation of sulfur (Habicht *et al.*, 2002).

This being said, Canfield and his colleagues recognize (and comment upon) the fact that even if these events account for the sequence of events observed in the rock record, there is almost certainly an uncoupling in time between this record, and the "invention" of the processes. As they say, "... the geochemical expression of both oxygenic photosynthesis and sulfate reduction are separated in time from their biological origin" (Canfield *et al.*, 2000). Thus, as is so often the case we seek to reconstruct the evolution of metabolism but are offered evidence with regard to the evolution of biogeochemical cycles.

Despite our enthusiasm for this model (Canfield *et al.*, 2000), some notes of caution should be added. First, Detmers *et al.* (2001) noted that when 32 different sulfate-reducing bacteria were examined under optimal conditions for sulfate reduction, the observed fractionation was in the range 2–42 ppm. There was no apparent correlation with any variables tested, and the conclusion was that previous models correlating sulfate reduction with the rate of the process were apparently too simple. These results demand explanation, and create problems with interpreting the rock record. We need to know why some bacteria strongly fractionate the isotopes of sulfur, whereas others do not.

In addition, the report by Smock *et al.* (1998) that when thiosulfate is used as the oxidant (instead of sulfate), the fractionation of sulfur is much lower (on the order of 15 per mil), reminds us of two critical items. First, there are many organisms now in culture that contribute to the sulfur cycle, and many of them (and their various chemistries) are not well characterized with regard to sulfur fractionation. Perhaps it would be wise to revisit the fractionation that occurs with the oxidation of various sulfur species, both photosynthetic and nonphotosynthetic. Second, we are reminded that many of the organisms that contribute to the environment today have not been grown in culture.

Mass-independent sulfur fractionation, which occurs exclusively in the presence of high UV-fluxes (Farquhar *et al.*, 2000a,b, 2002; Pavlov and Kasting, 2002), has recently been reported in samples from the rock record, and its abrupt disappearance at between ~2.45 Ga and 2.32 Ga has been used by these authors to infer the presence of molecular oxygen in the atmosphere from this time until the present. Pavlov and Kasting (2002) have modeled this process, and concluded that the oxygen levels prior to 2.2 Ga must have been 10^{-5} PAL (present atmospheric level) or less. While the debate surrounding the history of oxygen in the atmosphere has raged for many years (Beukes *et al.*, 2002; Dimroth and Kimberley, 1976; Lasaga and Ohmoto, 2002; Ohmoto, 1997), it is hard to escape the significance of these findings with regard to the early history of Earth's atmosphere.

8.02.6.2.7 *Nitrogen isotopes*

Isotopes of nitrogen, like those of other biologically important elements, have potential as recorders of metabolic history. Few minerals retain enough nitrogen to allow for such a record to be preserved, so that kerogens of biological origin may be disproportionately represented among ancient samples that contain preserved nitrogen. Beaumont and Robert (1998, 1999) have found evidence of a shift in the isotopic composition of nitrogen in kerogens from ~3.4 Ga to 2.1 Ga. Average values shifted from about 0 per mil to +6 per mil. This shift is consistent with a transition from a system in which marine nitrogen was predominantly ammonium ion to one in which nitrate was the main form. If their interpretation is correct, this suggests that fixing nitrogen as nitrate is a capability life had developed by 2.7 Ga. Furthermore, the spread in values seen in samples from about 2.7 Ga and certainly in those from 2.0 Ga is consistent with progressive denitrification of a pool of nitrate. It should be noted though that while no other simple explanation for the shift in values seems likely, this conclusion is based on a very limited data set.

8.02.6.2.8 *Iron isotopes*

Isotopes of iron are tools of potential importance for the understanding of the evolution of metabolism—when organisms interact with iron it is usually with the result of altering the redox state of the iron—it is the nearly universal electron carrier of life, being a component of cytochromes and iron–sulfur proteins. The time at which the ability to oxidize and reduce iron evolved is difficult to tell, but nearly every organism that has been investigated seems to contain an iron reductase of some kind. Given the almost universal importance of iron in metabolic systems, it is a reasonable assumption that it was one of the first, if not *the* first, metal to be "adopted" by life for electron transfer. However, upon the invention of oxygenic photosynthesis, life apparently played a cruel trick on itself, as the solubility of iron is quite low in the presence of oxygen at pH values of 5 or higher. Thus, present-day organisms have invented a wide array of oxygen binding compounds to scavenge (siderophores; Butler, 1998; Neilands, 1981), store (ferritins; Ford *et al.*, 1984), and transfer (transferrins; Weinberg, 1984) iron. Thus, there are many opportunities for iron isotope fractionation, as shown in Table 6. Of the processes shown in Table 6, only a few have been investigated with regard to iron fractionation, and clearly, before this approach is valuable for understanding the evolution of metabolism, a great deal of work will need to be done. Nonbiological processes must also be further studied as some are known to fractionate iron isotopes (Anbar *et al.*, 2000). That being said, as pointed out recently by Anbar (2001), there is great potential here, not only for iron, but for other metals as well.

One of the potential advantages that iron isotopes offer over those of carbon and sulfur is their resistance to volatilization during heating, thus raising expectations that they might be stable to diagenetic activities. To this end, recent reports by Beard and colleagues have shown measurable isotopic anomalies in the iron in BIFs (Beard *et al.*, 2003). While there are presently no definitive interpretations of these isotopic anomalies, the fact that other iron-containing rocks show no such variations suggests that biological factors may be involved in BIF genesis, and that further understanding of the processes in Table 6 may lead to insights into this important and ancient geological process.

8.02.7 OVERVIEW

It may seem from the above discussion that we have very little hope of ever understanding the details of the evolution of metabolism, and to some extent we are. Two major negative forces appear to be at work. First, it appears that the origin and evolution of metabolic innovation are temporally uncoupled from the appearance of the activities in the rock record. Thus, while we may well be able to identify certain types of biogeochemical cycles resulting from successful metabolic innovation, this gives one only a very rough sketch of metabolic evolution. Second, LGT, along with unknown conditions on early Earth, work together to add great uncertainty to the origin of metabolic abilities. It is nevertheless instructive to assemble what is known, or at least what is considered likely.

Figure 6 presents the generalized plot (after Rye and Holland, 1998) of atmospheric oxygen during the Earth's last 4 Gyr. Shown on the left are some of the major components of the atmosphere at the onset of life, and on the right are the major redox components seen presently. Using the results of the mass-independent sulfur fractionation, it is possible to specify with some confidence that between 2.45 Ga and 2.32 Ga, oxygen moved from very low to a level of greater than 10^{-5} PAL. Furthermore, paleosol data strongly indicate that oxygen was at least 10% of PAL by ~2.0 Ga (Holland and Beukes, 1990; Rye and Holland, 1998; Yang *et al.*, 2002). However, it is not so easy to specify the other components of today's biogeochemical cycles. Iron oxidation almost certainly occurred early on, as the massive BIFs of the Early Archean indicate, but the mechanism underlying this oxidation is difficult to specify. Ancient oxygen-emitting cyanobacteria are tempting speculative culprits. However, it could well be that the iron was oxidized anaerobically via either photosynthetic iron oxidation (Ehrenreich and Widdel, 1994; Hartman, 1984; Widdel *et al.*, 1993), or lithotrophic iron oxidization (Ehrenreich and Widdel, 1994). Finally, iron oxidation may have proceeded entirely abiotically, relying neither on biogenic oxygen nor on iron metabolism. Irrespective of the mechanism(s), it is clear from the rock record that iron oxidation and reduction were important processes very early in Earth history, and that massive accumulations of oxidized iron in the form of BIFs were the result. As many have previously noted (Cloud, 1973, 1988; Nealson and Myers, 1990; Vargas *et al.*, 1998; Walker, 1984, 1987) the presence of abundant oxidized iron in a solid form could offer an energetic opportunity for respiratory bacteria in the form of some of the first stable and abundant oxidants present on early Earth. As mentioned above, preliminary work by Beard *et al.* (2003) has revealed iron isotopic fractionation between various layers in the BIF deposits, leading to speculation that one or more active biological processes (oxidation, reduction, or both) are involved in the formation and alteration of these deposits.

Of note in Figure 6 is the fact that, while some confidence can be given that a rise in oxygen (to a level 10% or greater of PAL) occurred around 2 Ga, with the exception of iron (in the BIF deposits), the same confidence cannot be achieved for the other electron acceptors. Canfield has speculated that sulfate reduction occurred on ancient Earth, and became a major process at ~2.5 Ga. However, given the uncertainties with other forms of sulfur reduction (Smock *et al.*, 1998), and the wide range of fractionation seen in modern sulfate reducers (Detmers *et al.*, 2001), this is hard to say with certainty. Manganese oxides become abundant for the first time around 2 Ga, suggesting that their accumulation may be biologically mediated, and/or that molecular oxygen may have been required, while nitrogen oxides leave little in the way of a mineral record (almost no stable minerals contain nitrogen as a major component) though there are tantalizing suggestions of nitrate formation and denitrification between 2.7 Ga and 2.0 Ga.

In Figure 7 we put together our best estimates of the times of detection of various metabolic abilities. We stress here that this has no necessary connection with the time of origin of any given process, but rather is the time that it is first seen in the rock record. However, this serves as the time at which the process has certainly been present. The message that emerges from this presentation is that the evolution of respiration may have been rather mature, perhaps "complete" prior to the rise in oxygen at 2.2 Ga, with sulfur and iron reduction probably in place, and carbon reduction via either CO_2 or CO almost certainly accomplished. If the estimates of the last common ancestor are correct (Doolittle *et al.*, 1996), then oxygen respiration was also in place at this time.

We end with some cautious notes of optimism amid our frustrating attempt to pin down the origins of earthly metabolism. First, while it seems unlikely that the molecular clocks of any of the metabolic genes will be sufficiently slow to reveal the times of their appearance, or even divergence from other pathways, it may well be that as genomic analyses become more mature, there will be better ways to subtract the effects of lateral gene transfer, clock speed variations, and other things that obscure the record. As this happens, the time of divergence of major metabolic groups may well be possible to discern. Second, the fossil record will almost certainly yield new and better samples for analysis of major functions, using both organic geochemistry and stable isotope fractionation. Finally, with some luck, the isotopes of iron, as well as of other metals, will prove to be a valuable diagnostic tool for the dissection of past metabolic activities.

REFERENCES

Anbar A. D. (2001) Iron isotope biosignatures: promise and progress. *EOS, Trans., AGU* **82**, 173–179.

Anbar A. D., Roe J. E., Barling J., and Nealson K. H. (2000) Nonbiological fractionation of iron isotopes. *Science* **288**(5463), 126–128.

Banfield J. F. and Nealson K. H. (1997) Geomicrobiology: interactions between microbes and minerals. In *Reviews in Mineralogy*. Mineralogical Society of America, vol. 35, 448pp.

Barghoorn E. S. and Tyler S. A. (1965) Microorganisms from the Gunflint Chert. *Science* **147**(3658), 563–577.

Beard B. L., Johnson C. M., Cox L., Sun H., Nealson K. H., and Aguilar C. (1999) Iron isotope biosignatures. *Science* **285**(5435), 1889–1892.

Beard B. L., Johnson C. M., Skulan J. L., Nealson K. H., Cox L., and Sun H. (2003) Application of Fe isotopes to tracing the geochemical and biological cycling of Fe. *Chem. Geol.* (published online 1/8/03).

Beaumont V. and Robert F. (1998) Nitrogen isotopic composition of organic matter from Precambrian cherts: new keys for nitrogen cycle evolution? *Bull. Soc. Geol. Fr.* **169**(2), 211–220.

Beaumont V. and Robert F. (1999) Nitrogen isotope ratios of kerogens in Precambrian cherts: a record of the evolution of atmosphere chemistry? *Precamb. Res.* **96**(1–2), 63–82.

Benner S. A., Caraco M. D., Thomson J. M., and Gaucher E. A. (2002) Evolution—Planetary biology—Paleontological, geological, and molecular histories of life. *Science* **296**(5569), 864–868.

Bertrand-Sarfati J. and Walter M. R. (1981) Stromatolite biostratigraphy. *Precamb. Res.* **15**(3–4), 353–371.

Beukes N. J., Doland H., Gutzmer J., Nedachi M., and Ohmoto H. (2002) Tropical laterites, life on land, and the history of atmospheric oxygen in the Paleoproterozoic. *Geology* **30**(6), 491–494.

Bjerrum C. J. and Canfield D. E. (2002) Ocean productivity before about 1.9 Gyr ago limited by phosphorus adsorption onto iron oxides. *Nature* **417**(6885), 159–162.

Blankenship R. E. (2002) *Molecular Mechanisms of Photosynthesis*. Blackwell, London.

Blankenship R. E. and Hartman H. (1998) The origin and evolution of oxygenic photosynthesis. *Trends Biochem. Sci.* **23**(3), 94–97.

Boetius A., Ravenschlag K., Schubert C. J., Rickert D., Widdel F., Gieseke A., Amann R., Jorgensen B. B., Witte U., and Pfannkuche O. (2000) A marine microbial consortium apparently mediating anaerobic oxidation of methane. *Nature* **407**(6804), 623–626.

Bottjer D. J., Etter W., Hagadorn J. W., and Tang C. M. (2002) *Exceptional Fossil Preservation: A Unique View on the Evolution of Marine Life*. Columbia University Press, 403pp.

Brasier M. D., Green O. R., Jephcoat A. P., Kleppe A. K., Van Kranendonk M. J., Lindsay J. F., Steele A., and Grassineau N. V. (2002) Questioning the evidence for Earth's oldest fossils. *Nature* **416**(6876), 76–81.

Brochier C. and Philippe H. (2002) Phylogeny—a non-hyperthermophilic ancestor for bacteria. *Nature* **417**(6886), 244–244.

Brocks J. J., Logan G. A., Buick R., and Summons R. E. (1999) Archean molecular fossils and the early rise of eukaryotes. *Science* **285**(5430), 1033–1036.

Buick R. (1992) The antiquity of oxygenic photosynthesis: evidence from stromatolites in sulphate-deficient Archaean lakes. *Science* **255**, 74–77.

Butler A. (1998) Acquisition and utilization of transition metal ions by marine organisms. *Science* **281**(5374), 207–210.

Canfield D. E. (1998) A new model for Proterozoic ocean chemistry. *Nature* **396**(6710), 450–453.

Canfield D. E. and Teske A. (1996) Late Proterozoic rise in atmospheric oxygen concentration inferred from phylogenetic and sulphur-isotope studies. *Nature* **382**(6587), 127–132.

Canfield D. E., Habicht K. S., and Thamdrup B. (2000) The Archean sulfur cycle and the early history of atmospheric oxygen. *Science* **288**(5466), 658–661.

Cloud P. (1968) Atmospheric and hydrospheric evolution on the primitive Earth. *Science* **160**, 729–736.

Cloud P. (1973) Paleoecological significance of the banded iron-formation. *Econ. Geol.* **68**(7), 1135–1143.

Cloud P. (1988) *Oasis in Space: Earth History from the Beginning*. W. W. Norton and Company.

Coates J. D., Michaelidou U., Bruce R. A., O'Connor S. M., Crespi J. N., and Achenbach L. A. (1999) Ubiquity and diversity of dissimilatory (per)chlorate-reducing bacteria. *Appl. Environ. Microbiol.* **65**(12), 5234–5241.

Cody G. D., Boctor N. Z., Filley T. R., Hazen R. M., Scott J. H., Sharma A., and Yoder H. S. (2000) Primordial carbonylated iron–sulfur compounds and the synthesis of pyruvate. *Science* **289**(5483), 1337–1340.

Cohen Y., Jorgensen B. B., Padan E., and Shilo M. (1975) Sulfide-dependent anoxygenic photosynthesis in cyanobacterium oscillatoria-limnetica. *Nature* **257**(5526), 489–492.

Croal L. R., Johnson C. M., Beard B. L., and Newman D. K. (2003) Iron isotope fractionation by anoxygenic Fe(II)-phototrophic bacteria. *Geochim. Cosmochim. Acta* (in press).

DeLong E. (1998) Microbiology—Archaeal means and extremes. *Science* **280**(5363), 542–543.

DesMarais D. J. (1994) Tectonic control of the crustal organic-carbon reservoir during the Precambrian. *Chem. Geol.* **114**(3–4), 303–314.

DesMarais D. J., Strauss H., Summons R. E., and Hayes J. M. (1992) Carbon isotope evidence for the stepwise oxidation of the Proterozoic environment. *Nature* **359**(6396), 605–609.

Detmers J., Bruchert V., Habicht K. S., and Kuever J (2001) Diversity of sulfur isotope fractionations by sulfate-reducing prokaryotes. *Appl. Environ. Microbiol.* **67**(2), 888–894.

Dimroth E. and Kimberley M. M. (1976) Precambrian atmospheric oxygen—evidence in sedimentary distributions of carbon, sulfur, uranium, and iron. *Can. J. Earth Sci.* **13**(9), 1161–1185.

Doolittle R. F., Feng D. F., Tsang S., Cho G., and Little E. (1996) Determining divergence times of the major kingdoms of living organisms with a protein clock. *Science* **271**(5248), 470–477.

Doolittle W. F. (2002) Lateral DNA transfer: mechanisms and consequences. *Nature* **418**(6898), 589–590.

Ehrenreich A. and Widdel F. (1994) Anaerobic oxidation of ferrous iron by purple bacteria, a new-type of phototrophic metabolism. *Appl. Environ. Microbiol.* **60**(12), 4517–4526.

Farquhar J., Bao H. M., and Thiemens M. (2000a) Atmospheric influence of Earth's earliest sulfur cycle. *Science* **289**(5480), 756–758.

Farquhar J., Savarino J., Jackson T. L., and Thiemens M. H. (2000b) Evidence of atmospheric sulphur in the martian regolith from sulphur isotopes in meteorites. *Nature* **404**(6773), 50–52.

Farquhar J., Wing B. A., McKeegan K. D., Harris J. W., Cartigny P., and Thiemens M. H. (2002) Mass-independent sulfur of inclusions in diamond and sulfur recycling on early earth. *Science* **298**(5602), 2369–2372.

Fedo C. M. and Whitehouse M. J. (2002) Metasomatic origin of quartz-pyroxene rock, Akilia, Greenland, and implications for Earth's earliest life. *Science* **296**(5572), 1448–1452.

Ford G. C., Harrison P. M., Rice D. W., Smith J. M. A., Treffry A., White J. L., and Yariv J. (1984) Ferritin—design and formation of an iron-storage molecule. *Phil. Trans. Roy. Soc. London Ser. B-Biol. Sci.* **304**(1121), 551–565.

Galtier N., Tourasse N., and Gouy M. (1999) A nonhyperthermophilic common ancestor to extant life forms. *Science* **283**(5399), 220–221.

Grotzinger J. P. and Knoll A. H. (1999) Stromatolites in Precambrian carbonates: evolutionary mileposts or environmental dipsticks? *Ann. Rev. Earth Planet. Sci.* **27**, 313–358.

Grotzinger J. P. and Rothman D. H. (1996) An abiotic model for stromatolite morphogenesis. *Nature* **383**(6599), 423–425.

Habicht K. S., Gade M., Thamdrup B., Berg P., and Canfield D. E. (2002) Calibration of sulfate levels in the Archean Ocean. *Science* **298**(5602), 2372–2374.

Hartman H. (1984) The evolution of photosynthesis and microbial mats: a speculation on the banded iron formations. In *Microbial Mats: Stromatolites* (eds. Y. Cohen, R. W. Castenholz, and H. O. Halvorson). Alan R. Liss, pp. 449–453.

Hayes J. M. (1994) Global methanotrophy at the Archean–Proterozoic transition. In *Early Life on Earth* (ed. S. Bengtson). Columbia University Press, vol. 84, pp. 220–236.

Higgins I. J., Best D. J., Hammond R. C., and Scott D. (1981) Methane-oxidizing microorganisms. *Microbiol. Rev.* **45**(4), 556–590.

Holland H. D. (1984) *The Chemical Evolution of the Atmosphere and Oceans*. Princeton University Press.

Holland H. D. and Beukes N. J. (1990) A paleoweathering profile from Griqualand West, South-Africa—evidence for a dramatic rise in atmospheric oxygen between 2.2 and 1.9 By BP. *Am. J. Sci.* **290A**, 1–34.

Holland H. D. and Rye R. (1997) Evidence in pre-2.2 Ga paleosols for the early evolution of atmospheric oxygen and terrestrial biota: Comment. *Geology* **25**(9), 857–858.

Jackson B. E. and McInerney M. J. (2002) Anaerobic microbial metabolism can proceed close to thermodynamic limits. *Nature* **415**(6870), 454–456.

Johnson C. M., Skulan J. L., Beard B. L., Sun H., Nealson K. H., and Braterman P. S. (2002) Isotopic fractionation between Fe(III) and Fe(II) in aqueous solutions. *Earth Planet. Sci. Lett.* **195**(1–2), 141–153.

Kamber B. S. and Moorbath S. (1998) Initial Pb of the Amitsoq gneiss revisited: implication for the timing of early Archaean crustal evolution in West Greenland. *Chem. Geol.* **150**(1–2), 19–41.

Kelley D. S., Karson J. A., Blackman D. K., Fruh-Green G. L., Butterfield D. A., Lilley M. D., Olson E. J., Schrenk M. O., Roe K. K., Lebon G. T., and Rivizzigno P. (2001) An off-axis hydrothermal vent field near the mid-Atlantic ridge at 30 degrees N. *Nature* **412**(6843), 145–149.

Kirkham R. V. and Roscoe S. M. (1993) Atmospheric evolution and ore deposit formation. *Resour. Geol. Spec. Issue* **15**, 1–17.

Konhauser K. O., Hamade T., Raiswell R., Morris R. C., Ferris F. G., Southam G., and Canfield D. E. (2002) Could bacteria have formed the Precambrian banded iron formations? *Geology* **30**(12), 1079–1082.

Lasaga A. C. and Ohmoto H. (2002) The oxygen geochemical cycle: dynamics and stability. *Geochim. Cosmochim. Acta* **66**(3), 361–381.

Laval B., Cady S. L., Pollack J. C., McKay C. P., Bird J. S., Grotzinger J. P., Ford D. C., and Bohm H. R. (2000) Modern freshwater microbialite analogues for ancient dendritic reef structures. *Nature* **407**(6804), 626–629.

Li C. W., Chen J. Y., and Hua T. E. (1998) Precambrian sponges with cellular structures. *Science* **279**(5352), 879–882.

Lipps J. (1993) *Fossil Prokaryotes and Protists*. Blackwell, London, 342pp.

Maness P. C. and Weaver P. F. (2001) Evidence for three distinct hydrogenase activities in *Rhodospirillum rubrum*. *Appl. Microbiol. Biotechnol.* **57**(5–6), 751–756.

Margulis L. (1981) *Symbiosis in Cell Evolution*. W. H. Freeman.

Michaelis W., Seifert R., Nauhaus K., Treude T., Thiel V., Blumenberg M., Knittel K., Gieseke A., Peterknecht K., Pape T., Boetius A., Amann R., Jorgensen B. B., Widdel F., Peckmann J. R., Pimenov N. V., and Gulin M. B. (2002) Microbial reefs in the Black Sea fueled by anaerobic oxidation of methane. *Science* **297**(5583), 1013–1015.

Mojzsis S. J. and Harrison T. M. (2002) Establishment of a 3.83-Ga magmatic age for the Akilia tonalite (southern West Greenland). *Earth Planet. Sci. Lett.* **202**(3–4), 563–576.

Mojzsis S. J., Arrhenius G., McKeegan K. D., Harrison T. M., Nutman A. P., and Friend C. R. L. (1996) Evidence for life on Earth before 3,800 million years ago. *Nature* **384**(6604), 55–59.

Nealson K. H. (1997) Sediment bacteria: Who's there, what are they doing, and what's new? *Ann. Rev. Earth Planet. Sci.* **25**, 403–434.

Nealson K. H. and Ghiorse W. C. (2001) *Geobiology: Exploring the Interface between the Biosphere and the Geosphere*. American Academy of Microbiology.

Nealson K. H. and Myers C. R. (1990) Iron reduction by bacteria—a potential role in the genesis of banded iron formations. *Am. J. Sci.* **290A**, 35–45.

Nealson K. H. and Saffarini D. (1994) Iron and manganese in anaerobic respiration—environmental significance, physiology, and regulation. *Ann. Rev. Microbiol.* **48**, 311–343.

Nealson K. H. and Stahl D. A. (1997) Microorganisms and biogeochemical cycles: What can we learn from layered microbial communities?. In *Geomicrobiology: Interactions between Microbes and Minerals*, vol. 35, pp. 5–34.

Neilands J. B. (1981) Microbial iron compounds. *Ann. Rev. Biochem.* **50**, 715–731.

Nisbet E. G. and Sleep N. H. (2001) The habitat and nature of early life. *Nature* **409**(6823), 1083–1091.

Ohmoto H. (1997) Evidence in pre-2.2 Ga paleosols for the early evolution of atmospheric oxygen and terrestrial biota: Reply. *Geology* **25**(9), 858–859.

Oremland R. S., Hoeft S. E., Santini J. A., Bano N., Hollibaugh R. A., and Hollibaugh J. T. (2002) Anaerobic oxidation of arsenite in Mono Lake water and by facultative, arsenite-oxidizing chemoautotroph, strain MLHE-1. *Appl. Environ. Microbiol.* **68**(10), 4795–4802.

Orphan V. J., House C. H., Hinrichs K. U., McKeegan K. D., and DeLong E. F. (2001) Methane-consuming archaea revealed by directly coupled isotopic and phylogenetic analysis. *Science* **293**(5529), 484–487.

Pace N. R. (1996) New perspective on the natural microbial world: molecular microbial ecology. *ASM News* **62**(9), 463–470.

Padan E. (1979) Facultative anoxygenic photosynthesis in cyanobacteria. *Ann. Rev. Plant Physiol. Plant Molecul. Biol.* **30**, 27–40.

Pavlov A. A. and Kasting J. F. (2002) Mass-independent fractionation of sulfur isotopes in Archean sediments: strong evidence for an anoxic Archean atmosphere. *Astrobiology* **2**, 27–41.

Pope M. C., Grotzinger J. P., and Schreiber B. C. (2000) Evaporitic subtidal stromatolites produced by *in situ* precipitation: textures, facies associations, and temporal significance. *J. Sedim. Res.* **70**(5), 1139–1151.

Porter S. M. and Knoll A. H. (2000) Testate amoebae in the Neoproterozoic Era: evidence from vase-shaped microfossils in the Chuar Group, Grand Canyon. *Paleobiology* **26**(3), 360–385.

Ramdohr P. (1958) New observations on the ores of the Witwatersrand in South Africa and their genetic significance. *Trans. Geol. Soc. S. Afr.* **61**, 1–50.

Rasmussen B. (2000) Filamentous microfossils in a 3,235-million-year-old volcanogenic massive sulphide deposit. *Nature* **405**(6787), 676–679.

Rasmussen B. and Buick R. (1999) Redox state of the Archean atmosphere: evidence from detrital heavy minerals in ca. 3, 250–2,750 Ma sandstones from the Pilbara Craton, Australia. *Geology* **27**(2), 115–118.

Raymond J., Zhaxybayeva O., Gogarten J. P., Gerdes S. Y., and Blankenship R. E. (2002) Whole-genome analysis of photosynthetic prokaryotes. *Science* **298**(5598), 1616–1620.

Reid R. P., Visscher P. T., Decho A. W., Stolz J. F., Bebout B. M., Dupraz C., Macintyre L. G., Paerl H. W., Pinckney J. L., Prufert-Bebout L., Steppe T. F., and DesMarais D. J. (2000) The role of microbes in accretion, lamination and early lithification of modern marine stromatolites. *Nature* **406**(6799), 989–992.

Robb L. J., Davis D. W., Kamo S. L., and Meyer F. M. (1992) Ages of altered granites adjoining the Witwatersrand Basin with implications for the origin of gold and uranium. *Nature* **357**, 677–680.

Roger A. J. and Silberman J. D. (2002) Cell evolution: mitochondria in hiding. *Nature* **418**(6900), 827–829.

Rye R. and Holland H. D. (1998) Paleosols and the evolution of atmospheric oxygen: a critical review. *Am. J. Sci.* **298**(8), 621–672.

Rye R. and Holland H. D. (2000a) Geology and geochemistry of paleosols developed on the Hekpoort basalt, Pretoria Group, South Africa (vol. 300, pp. 85, 2000). *Am. J. Sci.* **300**(4), 344–345.

Rye R. and Holland H. D. (2000b) Life associated with the 2.76 billion year old Mt. Roe #2 paleosol, Western Australia. *Geology* **28**, 483–486.

Rye R., Kuo P. H., and Holland H. D. (1995) Atmospheric carbon-dioxide concentrations before 2.2-billion years ago. *Nature* **378**(6557), 603–605.

Saier M. H. (2000) A functional-phylogenetic classification system for transmembrane solute transporters. *Microbiol. Molecul. Biol. Rev.* **64**(2), 354–411.

Schäfer G., Engelhard M., and Muller V. (1999) Bioenergetics of the archaea. *Microbiol. Molecul. Biol. Rev.* **63**(3), 570–620.

Schink B. and Friedrich M. (2000) Bacterial metabolism—phosphite oxidation by sulphate reduction. *Nature* **406**(6791), 37–37.

Schink B., Thiemann V., Laue H., and Friedrich M. W. (2002) Desulfotignum phosphitoxidans sp. nov., a new marine sulfate reducer that oxidizes phosphite to phosphate. *Arch. Microbiol.* **177**(5), 381–391.

Schoell M. (1980) The hydrogen and carbon isotopic composition of methane from natural gases of various origins. *Geochim. Cosmochim. Acta* **44**, 649–661.

Schoell M. and Wellmer F.-W. (1981) Anomalous ^{13}C depletion in early Precambrian graphites from Superior Province, Canada. *Nature* **290**, 696–699.

Schopf J. W. (1999) *Cradle of Life: The Discovery of Earth's Earliest Fossils*. Princeton University Press, Princeton.

Schopf J. W. and Klein C. (1992) *The Proterozoic Biosphere: A Multidisciplinary Study*. Cambridge University Press, Cambridge, 1348pp.

Schopf J. W., Kudryavtsev A. B., Agresti D. G., Wdowiak T. J., and Czaja A. D. (2002) Laser-Raman imagery of Earth's earliest fossils. *Nature* **416**(6876), 73–76.

Semikhatov M. A., Gebelein C. D., Cloud P., Awramik S. M., and Benmore W. C. (1979) Stromatolite morphogenesis—progress and problems. *Can. J. Earth Sci.* **16**(5), 992–1015.

Shen Y. A., Buick R., and Canfield D. E. (2001) Isotopic evidence for microbial sulphate reduction in the early Archaean era. *Nature* **410**(6824), 77–81.

Shock E. L., McCollom T., and Schulte M. D. (1995) Geochemical constraints on chemolithoautotrophic reactions in hydrothermal systems. *Origins Life Evol. Biosphere* **25**(1–3), 141–159.

Smock A. M., Bottcher M. E., and Cypionka H. (1998) Fractionation of sulfur isotopes during thiosulfate reduction by Desulfovibrio desulfuricans. *Arch. Microbiol.* **169**(5), 460–463.

Stolz J. F. and Oremland R. S. (1999) Bacterial respiration of arsenic and selenium. *Fems Microbiol. Rev.* **23**(5), 615–627.

Strous M., Fuerst J. A., Kramer E. H. M., Logemann S., Muyzer G., van de Pas-Schoonen K. T., Webb R., Kuenen J. G., and Jetten M. S. M. (1999) Missing lithotroph identified as new planctomycete. *Nature* **400**(6743), 446–449.

Summons R. E., Jahnke L. L., Hope J. M., and Logan G. A. (1999) 2-Methylhopanoids as biomarkers for cyanobacterial oxygenic photosynthesis. *Nature* **400**(6744), 554–557.

Svetlitchnyi V., Peschel C., Acker G., and Meyer O. (2001) Two membrane-associated NiFeS-carbon monoxide dehydrogenases from the anaerobic carbon-monoxide- utilizing eubacterium Carboxydothermus hydrogenoformans. *J. Bacteriol.* **183**(17), 5134–5144.

Vadas A., Monbouquette H. G., Johnson E., and Schroder I. (1999) Identification and characterization of a novel ferric reductase from the hyperthermophilic Archaeon Archaeoglobus fulgidus. *J. Biol. Chem.* **274**(51), 36715–36721.

van Zuilen M. A., Lepland A., and Arrhenius G. (2002) Reassessing the evidence for the earliest traces of life. *Nature* **418**(6898), 627–630.

Vargas M., Kashefi K., Blunt-Harris E. L., and Lovley D. R. (1998) Microbiological evidence for Fe(III) reduction on early Earth. *Nature* **395**(6697), 65–67.

Walker J. C. G. (1984) Suboxic diagenesis in banded iron formations. *Nature* **309**(5966), 340–342.

Walker J. C. G. (1987) Was the Archean biosphere upside down. *Nature* **329**(6141), 710–712.

Walsh M. M. and Lowe D. R. (1985) Filamentous microfossils from the 3,500-Myr-old Onverwacht Group, Barberton Mountain Land, South-Africa. *Nature* **314**(6011), 530–532.

Weinberg E. D. (1984) Iron withholding—a defense against infection and neoplasia. *Physiol. Rev.* **64**(1), 65–102.

White D. C. (1993) *In-situ* measurement of microbial biomass, community structure and nutritional-status. *Phil. Trans. Roy. Soc. London Ser. A: Math. Phys. Eng. Sci.* **344**(1670), 59–67.

Widdel F., Schnell S., Heising S., Ehrenreich A., Assmus B., and Schink B. (1993) Ferrous iron oxidation by anoxygenic phototrophic bacteria. *Nature* **362**(6423), 834–836.

Woese C. R. (1987) Bacterial evolution. *Microbiol. Rev.* **51**(2), 221–271.

Xiong J., Fischer W. M., Inoue K., Nakahara M., and Bauer C. E. (2000) Molecular evidence for the early evolution of photosynthesis. *Science* **289**(5485), 1724–1730.

Yang W. B., Holland H. D., and Rye R. (2002) Evidence for low or no oxygen in the late Archean atmosphere from the ~2.76 Ga Mt. Roe #2 paleosol, Western Australia: Part 3. *Geochim. Cosmochim. Acta* **66**(21), 3707–3718.

Yurkov V. V. and Beatty J. T. (1998) Aerobic anoxygenic phototrophic bacteria. *Microbiol. Molecul. Biol. Rev.* **62(3)**, 695–724.

8.03
Sedimentary Hydrocarbons, Biomarkers for Early Life

J. J. Brocks
Harvard University, Cambridge, MA, USA

and

R. E. Summons
Massachusetts Institute of Technology, Cambridge, MA, USA

8.03.1 INTRODUCTION

Molecular biological markers, or biomarkers, are natural products that can be assigned to a particular biosynthetic origin. For environmental and geological studies, the most useful molecular biomarkers are organic compounds with high taxonomic specificity and potential for preservation. In other words, the most effective biomarkers have a limited number of well-defined sources; they are recalcitrant against geochemical changes and easily analyzable in environmental samples. Accordingly, biomarkers can be proxies in modern environments as well as chemical fossils that afford a geological record of an organism's activities. One of the first significant outcomes of biomarker research was Treibs' (1936) recognition of unquestionable biological signatures in sedimentary organic matter. Subsequent research (Eglinton and Calvin, 1967; Eglinton *et al.*, 1964) pursued the concept that biomarkers can provide information about the nature of early life in the absence of recognizable fossils and that petroleum is composed of biological remains (Whitehead, 1973). As of early 2000s, thirty years of accumulated facts about sedimentary organic matter are clearly commensurate with the aforesaid and falsify the hypotheses (e.g., Gold, 2001) about the primordial origins of petroleum and natural gas.

Largely as a result of efforts to understand the detail of the transformation of biogenic organic matter into petroleum (Hunt, 1996; Tissot and Welte, 1984) and individual chemical fossils, geochemists began to appreciate the value of biomarkers as tools for environmental research (e.g., Brassell *et al.*, 1986) and their potential for elucidating biogeochemical processes (e.g., Hinrichs *et al.*, 1999; Kuypers *et al.*, 2003). The structural and isotopic information in biomarkers allows them to be distinguished from abiogenic organic compounds that are widely distributed throughout the cosmos (e.g., Cronin and Chang, 1993; Engel and Macko, 1997). Consequently, biomarkers will be an important tool in the search for extraterrestrial life. A thorough review of recent biomarker research is not possible within the limitations of this chapter. Instead, this chapter introduces some of the general principles, provides examples of their use for discerning the identities and physiologies of microbes in contemporary environments and summarizes biomarker research aimed at elucidating aspects of biological and environmental evolution in the Precambrian.

8.03.2 BIOMARKERS AS MOLECULAR FOSSILS

Molecular fossils that are stable under geological conditions mostly originate from biological lipids. These biomarkers encode information about ancient biodiversity, trophic associations, and environmental conditions. They are recorders of element cycling, sediment and water chemistry, redox conditions, and temperature histories. Most importantly, however, hydrocarbon biomarkers are stable for billions of years if they are enclosed in intact sedimentary rocks that have only suffered a mild thermal history. Therefore, biomarkers offer a powerful means to study life and its interaction with the environment as recorded in rocks of Precambrian age. In sedimentary environments, and under appropriate diagenetic conditions, functionalized biolipids are reduced to hydrocarbon skeletons (e.g., (**11**) to (**10**), or (**31**) to (**32**)). During this process, much of the biological information is retained and it is thereby possible to assign specific hydrocarbon skeletons to specific taxa (Figure 1) wherever their biosynthetic pathways are exclusive. For example, pentacyclic terpanes of the C_{31} to C_{35} extended hopane series (**55**) are diagnostic biomarkers for the domain Bacteria. The biological precursors of extended hopanes (**55**), the bacteriohopanepolyols (**56**), probably have the physiological function of membrane rigidifiers, a role in Eukarya fulfilled by sterols. Important hydrocarbon fossils of eukaryotic sterols such as cholesterol (**65**) are steranes (e.g., (**66**)) and aromatic steroids (e.g., (**68**)). Although some Bacteria are capable of synthesizing a limited variety of sterols, including lanosterol and 4-methylsterols, the wide structural range of

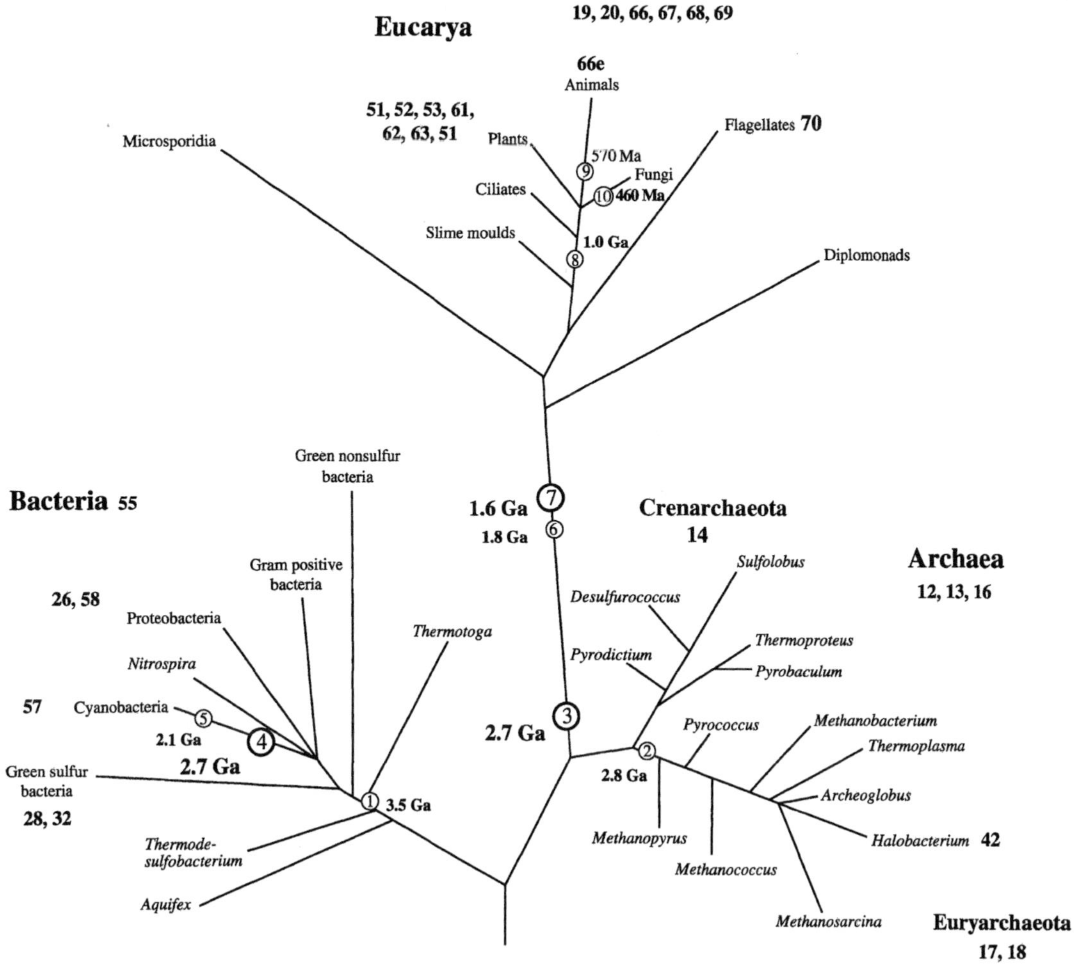

Figure 1 The Universal Phylogenetic Tree annotated with structure numbers of selected diagnostic biomarkers for some taxonomic groups. Ages refer to minimum ages of selected branches based on biomarkers (large fonts) and inorganic geochemical and paleontological data. ① Sulfur-isotopic evidence for mesothermophilic sulfate-reducing bacteria from North Pole, Pilbara Craton, Western Australia (Shen *et al.*, 2001). ② Circumstantial evidence for the activity of methanogens from a global carbon-isotopic excursion of kerogen to very light values between ~2.8 Ga and ~2.5 Ga (Hayes, 1983; Hayes, 1994). ③ Oldest probably syngenetic sterane biomarkers diagnostic of Eukarya from the ~2.7 Ga Fortescue Group, Hamersley Basin, Western Australia (Brocks *et al.*, 1999). ④ 2α-methylhopanes with an extended side-chain diagnostic for oxygenic cyanobacteria from the ~2.7 Ga Fortescue Group, Hamersley Basin, Western Australia (Brocks *et al.*, 2003b). ⑤ Oldest fossils with diagnostic cyanobacterial morphology from the 2.15 Ga Belcher Supergroup, Canada (Hofmann, 1976). ⑥ Oldest known probably eukaryotic fossils from the ~1.8 Ga to 1.9 Ga Chuanlinggou Formation, China (Hofmann and Chen, 1981). ⑦ Oldest known occurrence of certainly syngenetic eukaryotic biomarkers from the 1.64 Ga Barney Creek Formation, McArthur Basin, Northern Territory (Summons *et al.*, 1988b). ⑧ Oldest known eukaryotic fossils that confidently belong to an extant phylum (Rhodophyta) from the 1.26 Ga to 0.95 Ga Hunting Formation, Somerset Island, Canada (Butterfield, 2001; Butterfield *et al.*, 1990). ⑨ Phosphatized embryonic metazoans from the 555–590 Ma Doushantuo Formation, southern China (Xiao *et al.*, 1998). ⑩ Oldest fossils with diagnostic fungal morphology from the Ordovician 460 Ma Guttenberg Formation, Wisconsin (Redecker *et al.*, 2000). Branch lengths and branching order are based on SSU rRNA from Shen *et al.* (2001) and Canfield and Raiswell (1999).

fossil steranes typically found in oils and bitumens is diagnostic for organisms of the domain Eukarya. Similarly, a range of structurally distinctive acyclic and cyclic isoprenoids found in sedimentary rocks can be assigned exclusively to the domain Archaea (Figure 1). Their precursor lipids are hydrocarbon chains bound to glycerol through ether linkages with varied chain lengths, branching patterns and modes of cyclization (e.g., (**12**)–(**14**)). Other biomarkers are evidently diagnostic for taxonomic groups below domain level. These include extended 2α-methylhopanes (**57**) for cyanobacteria, 24-*n*-propylcholestanes (**66d**) for pelagophyte algae, 24-isopropylcholestane (**66e**) for certain sponges, and a large number of very distinctive polycyclic compounds characteristic of various plant taxa (e.g., (**53**) and (**61**)–(**63**)). Botryococcanes (e.g., (**20**)) are hydrocarbon fossils that appear to be diagnostic for a single taxon, the alga *Botryococcus braunii*. The study of biomarkers in sedimentary rocks thus allows the existence of a taxonomic group to be established for a given geological period. This capability is especially useful prior to the Cambrian where diagnostic body fossils are mostly absent and the affinities of microfossil are less certain.

8.03.2.1 The Fate of Dead Biomass: Diagenesis, Catagenesis, and Metagenesis

Organic matter from defunct organisms is almost quantitatively remineralized back to carbon dioxide in aquatic environments. However, a small fraction of total biomass, on an average less than 0.1% (Holser *et al.*, 1988), escapes remineralization, and eventually accumulates in sediments. As compounds with rapid biological turnover rates—including carbohydrates, proteins, and nucleic acids—are most prone to recycling, more resistant molecules such as lipids and recalcitrant structural biopolymers become concentrated (Tegelaar *et al.*, 1989). During transport through the water column, and subsequently in the unconsolidated sediment, this organic matter is further altered by a variety of chemical and biological processes commonly referred to as diagenesis (e.g., Hedges and Keil, 1995; Hedges *et al.*, 1997; Rullkötter, 1999). During diagenesis a large fraction of the lipid and other low-molecular weight components react via condensation and sulfur vulcanization reactions (e.g., Sandison *et al.*, 2002) and combine with degradation-resistant macromolecules to form kerogen (e.g., de Leeuw and Largeau, 1993; Derenne *et al.*, 1991). Formally, kerogen is defined as the fraction of large chemical aggregates in sedimentary organic matter that is insoluble in solvents. In contrast, the fraction of organic matter that can be extracted from sediments with organic solvents such as dichloromethane and methanol, is defined as bitumen (pyrobitumen and radiobitumen are residues of migrated petroleum that was cross linked and immobilized by heat and radioactivity, respectively). Bitumen in fresh sediments is predominantly composed of functionalized lipids. During diagenesis, these lipids undergo oxidation, reduction, sulfurization, desulfurization, and rearrangement reactions, generating an array of partly or entirely defunctionalized breakdown products that can have different stereo- and structural isomers. Analysis of these alteration products often yields valuable information about prevailing chemical conditions in the sediment during and after deposition because the extent and relative speed of diagenetic reactions is dependent on environmental conditions such as redox state, pH, and availability of catalytic sites on mineral surfaces. Where reducing conditions prevail in the sediment, biolipids eventually lose all functional groups but remain identifiable as geologically stable hydrocarbon skeletons.

Diagenetic reactions in the presence of reduced sulfur species have a profound effect on the sedimentary fate of lipids and other biological debris (Sinninghe Damsté and de Leeuw, 1990) and the preservation of diagnostic carbon skeletons (e.g., Adam *et al.*, 1993; Kenig *et al.*, 1995a; Kohnen *et al.*, 1992, 1993, 1991a,b; Schaeffer *et al.*, 1995; Wakeham *et al.*, 1995) in complex, sulfur-rich macromolecules. The sequestration and subsequent release of these skeletons upon burial provides one of the most important mechanisms for preserving the structural integrity of organism-specific biomarkers.

With increasing burial over millions of years, geothermal heat will initiate catagenesis, the thermal degradation of kerogen and bitumen. Kerogen is cracked into smaller fragments, releasing increasing volumes of bitumen that might eventually be expelled from its source rock as crude oil. Weaker chemical bonds, such as S—S and S—C, are cleaved at relatively low temperatures with the result that sulfur-rich kerogens might commence oil generation at lower temperatures (e.g., Koopmans *et al.*, 1997; Lewan, 1985). Hydrocarbon chains attached to kerogen via stronger C—O and C—C bonds are sequentially released at higher temperatures. Also, with increasing heat flux, biomarkers and other components in the bitumen undergo thermal rearrangement and cracking reactions. By measuring the relative abundances of these thermal products, it is possible to assess the maturity of an oil or bitumen (Section 8.03.3). With continuing burial, and at temperatures and pressures that initiate low-grade metamorphism of the host rock, most or all of residual bitumen is expelled or cracked to gas and the kerogen becomes progressively depleted in hydrogen to form a partly crystalline, highly aromatic carbon phase (metagenesis). The exact temperature and time constraints of metagenesis and the preservation of hydrocarbon biomarkers are much debated (e.g., Mango, 1991; Price, 1997) and are briefly reviewed later in Section 8.03.3.2.

8.03.2.2 Compound-specific Stable Isotopes

The carbon-isotopic content of organic matter carries information about the immediate environment of an organism, its primary carbon assimilation pathways and subsequent processing of its metabolic products in the environment. While isotopic measurements of bulk organic materials (e.g., biomass, kerogen, bitumen, petroleum) allow some correlations between precursor and product, measurements at the molecular or intramolecular level reveal stunningly detailed information about the biosynthetic pathways and organismic sources of individual carbon skeletons. The feasibility of developing a tool such as gas chromatography–isotope ratio mass spectrometry for routine natural-abundance isotopic measurements of individual organic compounds was first demonstrated by Matthews and Hayes (1978). Subsequent improvements in sensitivity, precision, calibration, analysis software, and general ease of use then enabled a wide range of biogeochemical applications for compound-specific carbon isotope analysis to be explored and exploited (e.g., Freeman *et al.*, 1990; Hayes *et al.*, 1990; Jahnke *et al.*, 1999). Hydrogen, nitrogen, and oxygen isotopes are also amenable to analysis, and multiple isotope ratios for the same compound offers a precise way to determine its provenance (e.g., Engel and Macko, 1997; Hinrichs *et al.*, 2001). Compound-specific carbon-isotopic patterns also reveal much about fractionation during carbon assimilation and biosynthesis (e.g., Grice *et al.*, 1998a; Hayes, 1993, 2001; Jahnke *et al.*, 1999; Kohnen *et al.*, 1992; Rieley *et al.*, 1993; Schouten *et al.*, 1998a; Summons *et al.*, 1994a; van der Meer *et al.*, 2001). Measurements of the individual radiocarbon ages of different organic compounds add a further dimension to studies of recent sediments (e.g., Eglinton *et al.*, 1997). It is now clear that compound-specific isotope analysis has completely revolutionized biomarker research.

8.03.3 THERMAL STABILITY AND MATURITY OF BIOMARKERS

8.03.3.1 Biomarkers as Maturity Indicators

One of the most widely used applications for biomarkers is for the measurement of thermal maturity of organic matter to estimate the petroleum-generation potential and temperature history of sedimentary basins (e.g., Mackenzie, 1984; Radke *et al.*, 1997). A large number of biomarker parameters that are sensitive to different stages of maturity have been developed and are reviewed in Peters and Moldowan (1993). Two examples are described further below. For the interpretation of maturity parameters in the literature, it is important to note that the thermal evolution of biomarkers, and organic matter in general, might be widely different in rocks of different lithological compositions and from different basins and formations. Clay minerals, for example, provide catalytic sites for degradation and isomerization reactions and strongly influence the type and extent of isomer conversion (e.g., Moldowan *et al.*, 1991a). Moreover, the range of biological inputs, presence of organic sulfur compounds and a host of other factors might cause disparate maturity values from the outset. Therefore, considerable caution has to be applied when comparing maturity parameters across disparate sample sets. Similar caution is necessary for the interpretation of conventional organic geochemical nomenclature for hydrocarbon thermal maturity (Figure 2). The generation and maturation process of petroleum can follow

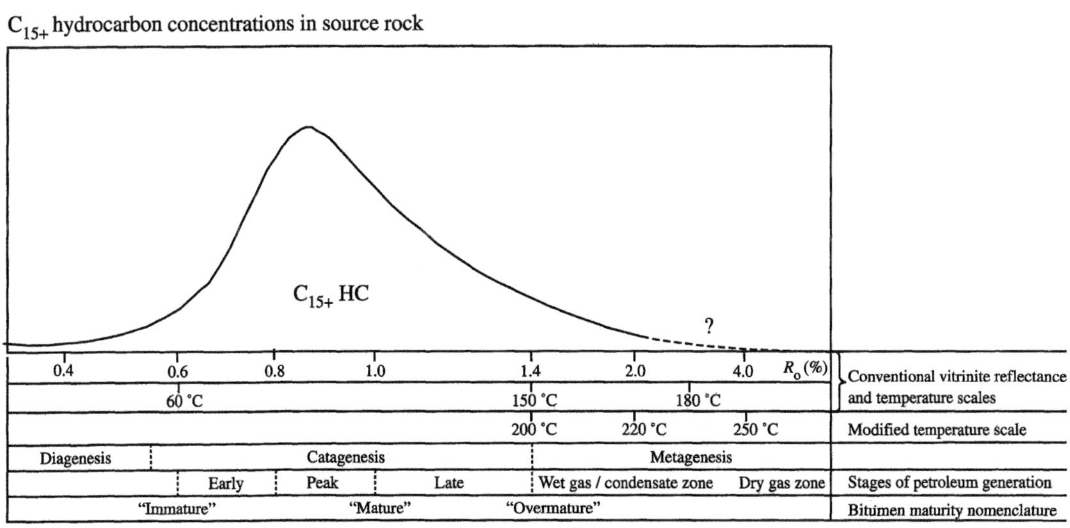

Figure 2 Terminology for bitumen maturity commonly used in the literature (e.g., Peters and Moldowan, 1993). The "modified temperature scale" pertains to hydrocarbon preservation under ideal conditions and was derived from data in Table 1.

markedly different pathways between different samples and between different components of the same sample (e.g., Radke *et al.*, 1997). Thus, the description of the preservation state of petroleum and bitumen using such terminology as "peak oil generation" or "overmature" is quite vague and clearly qualitative. It does not necessarily correlate with kerogen maturity data (e.g., vitrinite reflectance) or absolute temperatures unless calibrated for each sample set. Figure 2 should therefore only be used as a visual indicator of the relationships between bitumen descriptions expressed in words, temperatures, and vitrinite reflectance data.

As an example of a typical biomarker maturity parameter, the ratio of 20S/(20S + 20R) isomers in a sterane measures the relative abundance of the S and R configurations at C-20 of sterane hydrocarbons with 5α, 14α, 17α (H) stereochemistry (Figure 3; for sterane nomenclature see Section 8.03.5.11). In living organisms, sterols exclusively possess the 20R configuration, but during diagenesis and catagenesis steranes are gradually transformed to a mixture of 20R and 20S isomers. The thermal equilibrium value of ~0.55 for the 20S/(20S + 20R) ratio is apparently reached close to the peak of oil generation (Peters and Moldowan, 1993). Ratios based on triaromatic steroids (TA) (**68a**) and (**68b**) are an example of parameters sensitive at higher thermal maturities, i.e., the late stage of petroleum generation (Riolo *et al.*, 1985). Triaromatic steroids (**68**) form apparently predominantly by aromatization of monoaromatic steroid precursors (Mackenzie *et al.*, 1981). Thermal cleavage of the side chain of intact C_{26} to C_{28}-TAs (**68b**) (TA-II) leads to the generation of degradation products with 20 to 21 carbon atoms (TA-I (**68a**)). Consequently, in the transition from immature through mature to overmature petroleum, the ratio TA-I/(TA-I + TA-II) increases from <5% to close to 100% (Figure 4).

Hydrocarbons with typical "overmature" compositions and isomer distributions are characteristically found in rocks with deep-burial history, e.g., in Archean sequences (Brocks *et al.*, 2003a). Adamantanes and diamantanes, for example, are diagnostic classes of "diamondoid" hydrocarbons that persist and become concentrated at extreme levels of thermal maturity (Chen *et al.*, 1996;

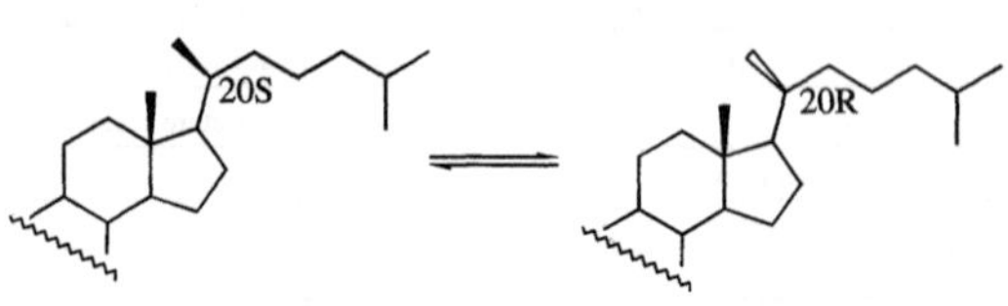

Figure 3 Equilibration between the biological 20R epimer and the geological 20S epimer of cholestane **66a**.

Dahl *et al.*, 2002, 1999). In contrast to burial metamorphism, hydrocarbons may also be generated over short timescales at very high temperatures such as those prevailing at recent hydrothermal vents (e.g., Simoneit and Fetzer, 1996; Simoneit *et al.*, 1992), in shales proximal to centers of hydrothermal ore formation (e.g., Brocks *et al.*, 2003d; Chen *et al.*, 2003; Gize, 1999; Landais and Gize, 1997; Püttmann *et al.*, 1988) or near-volcanic intrusions (e.g., Farrimond *et al.*, 1999; George, 1992). Bitumens that form in these extreme environments also have very distinctive hydrocarbon distribution patterns.

8.03.3.2 The Survival of Biomarkers with Increasing Temperature and Time

The number of sedimentary units that contain indigenous bitumen drastically decreases with increasing age of the rock. However, time alone is not the driver of degradation of organic molecules. For example, amino acids and other highly sensitive compounds may survive in carbonaceous meteorites for many billion years (Engel and Macko, 1997), and hydrocarbon biomarkers have endured in sedimentary rocks with little alteration for as long as 1.7 Gyr (Jackson *et al.*, 1986). Instead, the main factor driving molecular degradation, next to oxidation and erosion of the host rock, is thermal cleavage of covalent bonds during catagenesis and metamorphism. All known sedimentary successions older than ~1.7 Ga have suffered burial metamorphism to at least prehnite-pumpellyite facies at temperatures between 175 °C and 280 °C. Unfortunately, it is not entirely clear whether these temperatures, if experienced over geological periods of time, are consistent with preservation of biomarker hydrocarbons (Price, 1997). However, it is possible to obtain reliable minimum estimates of molecular preservation by observing biomarkers in deep-subsurface petroleum reservoirs (Table 1). Some of these reservoirs produce gas condensate and oil at present-day temperatures up to 200 °C, but still contain intact C_{15+} biomarkers (Brigaud, 1998; Knott, 1999; McNeil and BeMent, 1996; Pepper and Dodd, 1995). So far, the highest reliable temperature was observed in the rapidly subsiding Los Angeles Basin that contains moderately mature kerogens and biomarkers at 223 °C (Price, 2000; Price *et al.*, 1999). Earlier reports of relatively immature bitumens allegedly preserved at even higher temperatures (226–296 °C) lack credible information about syngeneity and require reconsideration (Price, 1982, 1983, 1993, 1997; Price *et al.*, 1981). These observations are consistent with kinetic models of petroleum degradation that predict

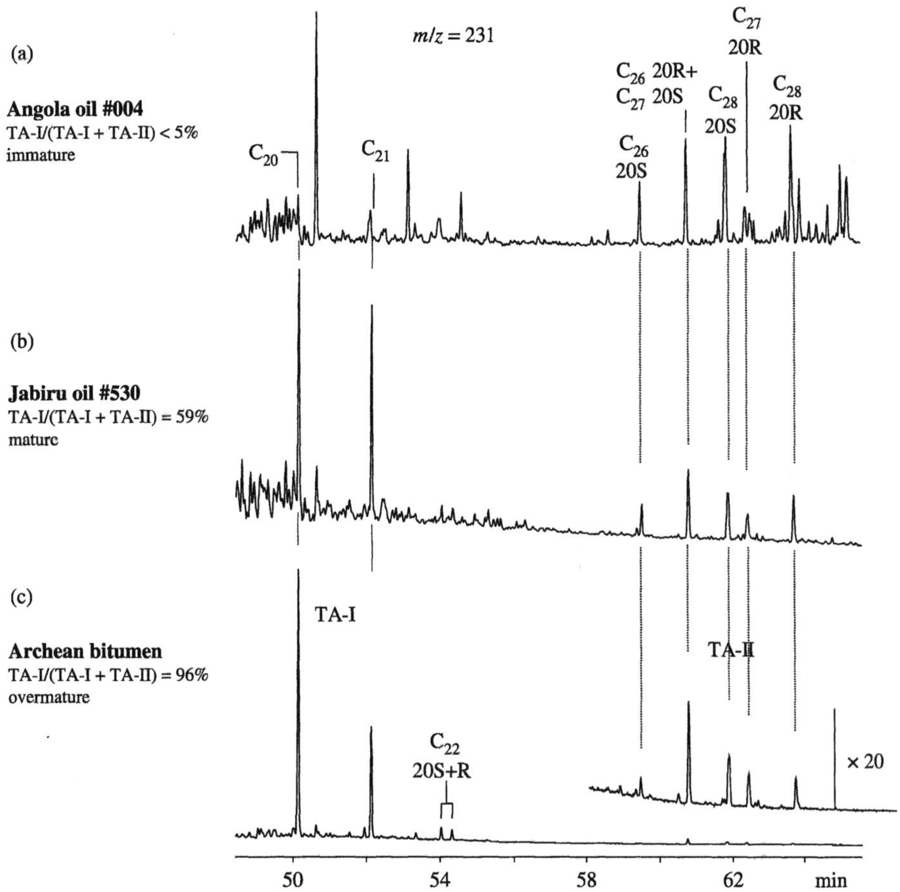

Figure 4 Distribution of triaromatic steroids (**68**) in GC-MS $m/z = 231$ selected ion chromatograms in (a) a Phanerozoic oil of low thermal maturity, (b) a mature Phanerozoic oil, and (c) an overmature bitumen from the late Archaean Fortescue Group in Western Australia. The inset in (c) is a 20× magnification of the elution range of C_{26} to C_{28} triaromatic steroids (**68b**) (Brocks *et al.*, 2003a,b) (reproduced by permission of Elsevier from *Geochim. Cosmochim. Acta* **2003**, in press).

persistence of aliphatic hydrocarbons over geological periods of time at 250 °C (Burnham *et al.*, 1997; Dominé *et al.*, 2002; Pepper and Dodd, 1995). Therefore, the existence of residual biomarkers in the lowest grade of metasedimentary rocks and, therefore, in units older than 1.7 Ga is at least theoretically possible.

The preservation of commercial quantities of oil at reservoir temperatures of 200 °C and the existence of moderately mature bitumen in source rocks at temperatures of 200–220 °C, though evidently real, are certainly not normal. Such excellent thermal preservation therefore requires exceptional conditions. One favorable condition is rapid heating caused by fast-basin subsidence and/or high geothermal gradients increasing maximum temperatures for hydrocarbon preservation by up to 30 °C (Waples, 2000). A second favorable condition is increased fluid or gas pressure that has the potential to retard petroleum generation, biomarker maturation and hydrocarbon thermal degradation to a significant degree (Fang *et al.*, 1995; Lewan, 1997; Price and Wenger, 1992). The absence of catalysts that promote the degradation of hydrocarbons is a third factor influencing petroleum preservation at high temperatures. Some organometallic complexes and active mineral surfaces have the capacity to induce hydrocarbon cracking at considerably reduced temperatures (Mango, 1990; Mango and Elrod, 1999; Mango and Hightower, 1997; Mango *et al.*, 1994). Mango (1987) has even argued that purely thermally induced, uncatalyzed cracking of higher hydrocarbons to gas is generally an insignificant process. Furthermore, hydrocarbon degradation is strongly retarded and shifted to higher temperatures in the absence of sulfur and organic sulfur compounds that are known to initiate radical chain reactions (Lewan, 1998). Some or all of these exceptional conditions might have prevailed in some Paleoproterozoic and Archean

Table 1 Deep subsurface petroleum reservoirs and bituminous source rocks existing at extreme present-day temperatures.

Location	*Preservation*	*Temp.*[a] (°C)	*References*
Word North Field, South Texas; 4.1 km depth	Gas condensate; ~3 wt.% oil, 97% gas; ~30 Ma at peak temperature; C_{35}-biomarkers	175	McNeil and BeMent (1996)
Elgin Field, North Sea; 5.25 km depth	Oil reservoir; ~60 wt.% oil, 40% gas; 1,100 bar reservoir pressure	185	Knott (1999)
Central Graben, North Sea	Oil reservoir; ~70 wt.% oil, 30% gas; steranes and hopanes; no evidence for in-reservoir cracking	195	Pepper and Dodd (1995)
Franklin Field, North Sea; 5.3 km depth	Oil reservoir; ~45 wt.% oil, 55% gas; 1,090 bar reservoir pressure	196	Knott (1999)
Well C403, California; 3.2 km depth	Source rock; 1,800–3,100 ppm[b] C_{15+} bitumen; TOC ≈ 2–5%; HI ≈ 390–560[c]; H/C = 1.2; T_{max} ≈ 440 °C[d]	198	Price (2000) and Price *et al.* (1999)
Sweethome Field, South Texas; 4.1 km depth	Gas condensate; ~20 wt.% oil, 80% gas; ~30 Ma at peak temperature; C_{35}-biomarkers	200	McNeil and BeMent (1996)
Franklin Field, North Sea	Condensate containing 9% C_{15+} and 32% C_6–C_{15} hydrocarbons	203	Brigaud (1998) and Dominé *et al.* (1998)
Well Makó-2, Hungary; Miocene; 4.8 km depth	Oil reservoir; the deeper lying source rocks are apparently at 210–215 °C	208	Sajgó (2000)
Well KCL-A 72-4, California; 6.4 km depth	Source rock; 550–2,100 ppm[b] C_{15+} bitumen; HI ≈ 35 to 96[c]; T_{max} ≈ 440 °C[d]	214	Price (2000) and Price *et al.* (1999)
Central Graben, North Sea	Source rocks in petroleum generation stage	220	Pepper and Dodd (1995)
Well Apex-1, California; 6.3 km depth	Source rock; 1,400 ppm[b] C_{15+} bitumen; HI ≈ 40 to 80[c]; R_o = 1.5%[e]; T_{max} ≈ 450 °C[d]	223	Price (2000) and Price *et al.* (1999)

[a] Present-day reservoir or source rock temperature. [b] 1 ppm = 1 μg g^{-1} of rock. [c] Hydrogen index HI = mg hydrocarbons produced per gram of TOC by ROCK EVAL pyrolysis of pre-extracted rock powder. [d] ROCK EVAL parameter. [e] Vitrinite reflectance.

sedimentary basins and might be used as a guide to find preserved biomarkers in low-grade metasedimentary rocks.

8.03.4 EXPERIMENTAL APPROACHES TO BIOMARKER AND KEROGEN ANALYSIS

Analysis of organic matter in sediments generally begins with determination of total organic carbon content (TOC) and an approximate evaluation of thermal maturity and organic matter type using a screening tool such as Rock-Eval pyrolysis (e.g., Espitalié *et al.*, 1977; Peters and Moldowan, 1993). This may be accompanied by organic petrographic analysis, palynology, determination of elemental carbon, hydrogen, oxygen, nitrogen, and sulfur contents and bulk isotope analyses. Bitumen is isolated from the sediment by extraction with solvents such as dichloromethane and methanol and further separated into components of different molecular sizes and polarities by liquid chromatography. Saturated hydrocarbons, aromatic hydrocarbons and a polar fraction with organic oxygen, nitrogen, and sulfur compounds are readily separated from macromolecular material of the asphaltene fraction in this way. The insoluble organic component, kerogen, is then obtained from the rock residue after demineralization with hydrochloric and hydrofluoric acids.

Gas chromatography (GC) and combined gas chromatography-mass spectrometry (GC-MS) are the primary instrumental means for identifying and quantifying biomarkers in the saturated and aromatic hydrocarbon fractions of bitumen. Mass spectrometers operated in full scan mode provide detailed information of fragmentation pathways and the identity of compounds. When operated to detect selected fragment ions (selected ion monitoring or SIM) there is considerable enhancement of signal to noise and more sensitive detection of trace components. Instruments with tandem mass analyzers (GC-MS-MS) allow compounds to be detected through their most-specific fragmentation reactions (multiple reaction monitoring (MRM)) with further enhancements in detection, provided one knows the structure of the compound being sought and some details of its mass spectrum (e.g., Philp and Oung, 1992; Summons, 1987). Improvements in chromatographic resolution (e.g., multidimensional GC; Reddy *et al.*, 2002) and the high mass-spectrometer scan rates of time-of-flight (TOF) mass analyzers are creating the means to identify more of the thousands of components that occur in fossil-hydrocarbon mixtures. Liquid chromatography and mass spectrometry is opening new windows on the structures and compositions of intact polar lipid mixtures from cultured organisms and environmental samples (e.g., Hopmans *et al.*, 2000; Rutters *et al.*, 2001; Talbot *et al.*, 2001). This new direction has important consequences for improved knowledge of biomarker sources and the identification of high molecular weight compounds that are inaccessible by GC-MS.

Obtaining information about the overall structures and biomarker contents of kerogen and other kinds of macromolecular organic matter is complex and best accomplished with a combination of controlled chemical and pyrolytic degradation techniques and solid-state spectroscopic methods such as Nuclear Magnetic Resonance (NMR) (e.g., Cody *et al.*, 2002; Wilson, 1987; Wilson *et al.*, 1994) and Fourier Transform Infrared Spectroscopy (FT-IR) (e.g., Ganz and Kalkreuth, 1987; Marshall *et al.*, 2001; Solomon and Carangelo, 1987). Successful approaches that target the structures of component biomolecules have been based on analytical pyrolysis techniques (Larter and Horsfield, 1993, and references therein), catalytic hydropyrolysis (Love *et al.*, 1995) and various types of chemical degradation (e.g., Kohnen *et al.*, 1991b).

8.03.5 DISCUSSION OF BIOMARKERS BY HYDROCARBON CLASS

8.03.5.1 Advantages and Limitations of the Biomarker Approach

In contrast to provenance and authenticity issues faced by paleontologists studying Proterozoic and Archean rocks (e.g., Brasier *et al.*, 2002; Schopf *et al.*, 2002) it is generally straightforward, by virtue of their chemical structures, to recognize when complex hydrocarbon molecules are genuine biogenic remains. However, a more detailed interpretation of these molecular structures is complicated by three factors. The first is the fragmentary knowledge of biomarker distributions across the wide range of extant organisms, second is the presence of compounds that are obviously biogenic but which have no known precursor organism, and third are the uncertainties associated with the extrapolation of the biomarker relationships of extant organisms back in time over hundreds of millions to billions of years. The biological interpretation of molecular fossils is almost exclusively based on the distribution of biolipids in living organisms. However, the full repertoire of lipid biosynthetic capabilities is only known for a small fraction of microorganisms that have been cultured (Volkman *et al.*, 1993). Therefore, it is probable that some biomarkers have a broader taxonomic distribution and less diagnostic value than is currently accepted. Moreover, pathways for the biosynthesis of particular lipids might have evolved independently in different lineages or could have been acquired by horizontal gene transfer between lineages. Lastly, lipids believed to be diagnostic for specific taxonomic groups might also be representative of unrelated extinct clades.

Some tests are available to verify the accuracy of biomarker assignments. In many cases, unusual biomarkers are associated with organisms known for a particular physiology or biochemical capacity. Examples of this would be an uncommon carbon fixation pathway (e.g., van der Meer *et al.*, 2001), the consumption of methane or capacity to survive anoxia or hypersalinity. These physiologies may be associated with specific isotopic fractionations or with specific geological settings. In such cases, isotopic ratios are particularly valuable in determining whether or not to assign a specific source organism, biogeochemical process or environmental niche to a particular compound (e.g., Jahnke *et al.*, 1999; Summons and Powell, 1986).

8.03.5.2 *n*-Alkanes, Algaenans, and other Polymethylenic Biopolymers

n-Alkanes, such as hexadecane (**1**),

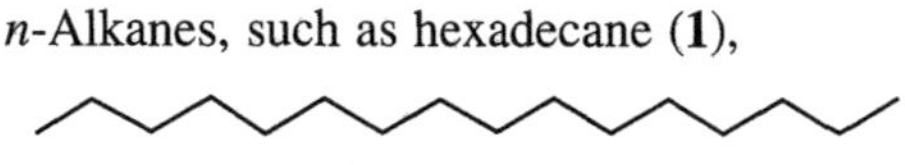

(**1**) Hexadecane (n-C_{16})

are the most abundant hydrocarbons in all nonbiodegraded oils and mature bitumens. Their potential biological precursors can be found in virtually all extant organisms. The bulk of *n*-alkanes in most Phanerozoic bitumens is derived from membrane components such as phospholipids and sphingolipids produced by bacteria and algae, polymethylenic biopolymers such as algaenans biosynthesized by microalgae (Tegelaar *et al.*, 1989), and waxes introduced by vascular plant debris (Hedberg, 1968). However, despite the ubiquity of straight-chain lipids in the biosphere, some *n*-alkane profiles can be environmentally and taxonomically diagnostic (Table 2), especially if combined with micropaleontological and carbon-isotopic analysis (Hoffmann *et al.*, 1987; Rieley *et al.*, 1991). For example, elevated concentrations of *n*-alkanes with odd carbon numbers between n-C_{15} and n-C_{19} in Ordovician rocks point to the presence of the marine cyanobacterium or alga *Gloeocapsomorpha prisca* (Fowler, 1992; Hoffmann *et al.*, 1987; and references therein). Long chain *n*-alkanes with more than ~27 carbon atoms and a predominance of odd-over-even carbon numbers (OEP) are frequently derived from plant waxes indicating a post-Silurian age and organic matter input from terrestrial sources (Hedberg, 1968; Tissot and Welte, 1984).

Also abundant in most crude oils, coals, and bitumens are *n*-alkanes with more than 40 and up to 110 carbon atoms (del Rio and Philp, 1999; Hsieh *et al.*, 2000; Killops *et al.*, 2000; Mueller and Philp, 1998). In coals and oils from dominantly terrigenous sources, high-molecular weight alkanes are likely to be the diagenetic alteration products of cuticular waxes and plant-derived aliphatic macromolecules such as cutan (McKinney *et al.*, 1996; Nip *et al.*, 1986a,b) and suberan (Tegelaar *et al.*, 1995). However, in sedimentary rocks with little terrigenous organic matter input, algaenans are probably the most important sources for high-molecular weight aliphatic hydrocarbons.

Algaenans are insoluble, nonhydrolyzable, and highly aliphatic macromolecules that serve as a structural component in the cell wall of several marine (Derenne *et al.*, 1992; Gelin *et al.*, 1996) and freshwater (Blokker *et al.*, 1998) green algae (chlorophytes), and marine eustigmatophytes (Gelin *et al.*, 1996, 1999). Chemists are still uncertain about the precise structures of algaenans. Elucidation with various chemical and thermal degradation techniques suggests that the biopolymers comprise mainly linear, long-chain aliphatic building blocks, derived from even-carbon-numbered C_{30} to C_{34} mono- and di-unsaturated ω-hydroxy fatty acids (Blokker *et al.*, 1998), and C_{28} to C_{36} diols and alkenols (Gelin *et al.*, 1997) that are apparently intermolecularly cross-linked by mid-chain ether bridges (Blokker *et al.*, 2000, 1998; Gelin *et al.*, 1997). In contrast, Allard *et al.* (2002), studying the same organisms, mainly observed ester linked, extremely long-chain, linear alcohols and acids with more than 36 and up to 80 carbon atoms, and an absence of ether cross-linking. Best known is the structure of algaenan biosynthesized by the chlorophyte *Botryococcus braunii*. Its algaenan is not based on ester-linked monomers but has a polyacetal structure constructed from linear di-unsaturated C_{32} α,ω-dialdehydes via an aldolization-dehydration mechanism (Bertheas *et al.*, 1999; Gelin *et al.*, 1994; Metzger *et al.*, 1993).

Algaenan likely plays an important role in the marine carbon cycle (Derenne and Largeau, 2001; Volkman *et al.*, 1998). Although the full extent to which algaenans are present in photosynthetic microorganisms of marine and lacustrine ecosystems is not known, their resistance against chemical and biological degradation leads to selective preservation during diagenesis (Tegelaar *et al.*, 1989). This recalcitrance also ensures that algaenan is one form of organic matter which may be quantitatively exported from the surface ocean and eventually into sediments where further selectivity in its preservation leads to accumulation in kerogen. Degradation-resistant qualities probably make algaenan and related materials one of the major sedimentary sinks for organic carbon (Derenne and Largeau, 2001; Gelin *et al.*, 1996). With burial of the host rock, and upon cracking of kerogen, algaenans become an important source of crude oil hydrocarbons and are, therefore, relevant to understanding petroleum occurrence (Tegelaar *et al.*, 1989). It is probably not just coincidence that the world's oldest known commercial deposits of petroleum in Oman and Siberia are from rocks of Late Neoproterozoic to Early Cambrian age (Grantham *et al.*, 1988), which correspond in age to the rising prominence of marine planktonic algae as suggested by a major diversification in acritarchs (Knoll, 1992; Mendelson, 1993; Zang and Walter, 1989). These old oils are geochemically distinctive with high abundances of long-chain methyl-alkanes with chain length and branching patterns consistent with thermal cracking of a specific type of aliphatic biopolymer such as algaenan (Fowler and Douglas, 1987; Höld *et al.*, 1999; Klomp, 1986). An exceptionally high content of 24-ethylcholestanes in the same oils suggests an overwhelming input of green algal (Chlorophycean) biomass to the source kerogen. Resolution of the exact structures of various algaenans might yield a new biomarker tool that could help to recognize major algal groups that contributed to organic matter in sedimentary rocks (Blokker *et al.*, 2000). Different algal groups biosynthesize algaenans with different monomeric units and various modes of linking,

Table 2 Aliphatic and monocyclic saturated hydrocarbons in the molecular fossil record and their paleobiological interpretation.

Biomarkers	*Biological and environmental interpretation*	*References*
n-Alkanes		
Outstanding concentrations of n-C_{15}, n-C_{17}, and n-C_{19} in early Paleozoic rocks	*Gloeocapsomorpha prisca,* marine phytoplankton of uncertain affinity, probably an alga; identified in Cambrian–Devonian sediments but most prominent in Ordovician. Estonian kukersite is a typical source.	Blokker *et al.* (2001) and Fowler (1992)
$>n$-C_{27} with OEP[a]	Waxes derived from higher plants; terrestrial input; post-Silurian age.	Hedberg (1968) and Tissot and Welte (1984)
$>n$-C_{40}	Predominantly degradation products of aliphatic macromolecules such as algaenan (marine, lacustrine), cutan and suberan (terrestrial, plant derived).	Allard *et al.* (2002) and Killops *et al.* (2000)
Branched alkanes and acyclic isoprenoids		
Monomethylalkanes and dimethylalkanes (MMA and DMA)	Cyanobacteria both cultured and in mat communities from hypersaline and hydrothermal environments.	For example, Dembitsky *et al.* (2001), Kenig *et al.* (1995b), Köster *et al.* (1999), and Shiea *et al.* (1990)
5,5-diethylalkanes with OEP[a] (wrongly reported as 3,7- or 3,ω7-dimethylalkanes)	These structures widely and incorrectly assigned. Chemical synthesis of a 5,5-diethylalkane indicates this is a major series. Often occurs with other alkanes with quaternary carbon centers (BAQC's). Source organisms not known but commonly found in association with benthic microbial mats.	Arouri *et al.* (2000a,b), Kenig *et al.* (2002), Logan *et al.* (1999), Logan *et al.* (2001) and Simons *et al.* (2002)
Pristane (**9**) (Pr) and phytane (**10**) (Ph)	From chlorophylls of cyanobacteria, algae and plants; bacteriochlorophylls *a* and *b* of phototrophic bacteria; tocopherols; Ph: archaeal membrane lipids.	Peters and Moldowan (1993)
Regular acyclic isoprenoids (**6**) i_{21} to i_{30}	probable source is halophilic Archaea; abundant in evaporitic environments.	Grice *et al.* (1998b)

(continued)

Table 2 (continued).

Biomarkers	*Biological and environmental interpretation*	*References*
Squalane (**15**) (tail–tail C_{30} acyclic isoprenoid)	All organisms produce some squalene; most sedimentary squalane probably from Archaea.	Grice *et al.* (1998b)
Crocetane (**17**)	Archaea (anaerobic methane oxidizers); associated with sub-sea gas, gas hydrate, and mud volcanoes.	Bian *et al.* (2001) and Thiel *et al.* (1999)
PMI (**18**) (2,6,10,15, 19-pentamethylicosane)	Methanogenic and methanotrophic archaea.	Elvert *et al.* (1999), Schouten *et al.* (1997), and Thiel *et al.* (1999)
TMI (2,6,15,19-tetramethylicosane)	Only reported from a mid-Cretaceous oceanic anoxic event; nonhyperthermophilic marine Crenarchaeota?	Kuypers *et al.* (2001)
C_{20}, C_{25}, C_{30} and C_{35} highly branched isoprenoids (**19**)	Unsaturated and polyunsaturated isoprenoid hydrocarbons are prominent biochemicals in some diatom taxa such as *Rhizosolenia*, *Haslea*, *Pleurosigma*, and *Navicula*.	Sinninghe Damsté *et al.* (1999a), Volkman *et al.* (1994), Belt *et al.* (2000), and Rowland *et al.* (2001)
Botryococcenes and botryococcanes (**20**), cyclobotryococcenes, polymethylsqualenes	The unsaturated, sometimes cyclic, biogenic hydrocarbons and their saturated fossil counterparts are diagnostic markers of the chlorophyte *B. braunii* and their preferred habitat of fresh to brackish water.	Huang *et al.* (1988), Metzger and Largeau (1999), and Summons *et al.* (2002)
Monocyclic saturated hydrocarbons		
C_{42}–C_{46} cyclopentylalkanes (**3**) with OEP	Oils from marine environments; unknown biological source.	Carlson *et al.* (1993), and Hsieh and Philp (2001)
C_{42}–C_{46} cyclopentylalkanes (**3**) with no distinct carbon preference	Oils from freshwater lacustrine settings; unknown biological source.	
C_{42}–C_{46} cyclopentylalkanes (**3**) with strong EOP[b]	Oils from saline lacustrine settings; unknown biological source.	
Cyclohexylalkanes (**4**) without carbon number preference	Formed during pyrolysis of biopolymers with long aliphatic carbon chains suggesting an origin from acyclic polymethylenic precursors.	For example, Gelin *et al.* (1994)
Macrocyclic alkanes C_{15}–C_{34}	Bitumens extracted from torbanites containing remains of *B. braunii;* fresh to brackish water.	Audino *et al.* (2002)

[a] Odd-over-even carbon number predominance. [b] Even-over-odd carbon number predominance.

and the resulting structures are resistant and often preserved in sedimentary rocks with minor alterations. Thus, analysis of the structures of sedimentary algaenans and comparison with their counterparts in extant organisms might eventually enable organic matter from different algae to be distinguished, possibly down to the family level (Blokker *et al.*, 2000). Prior to the Neoproterozoic, and the major algal diversifications, aliphatic algaenans probably played a relatively minor role in organic matter accumulation and oil generation (Brocks *et al.*, 2003c).

8.03.5.3 Methyl and Ethyl Alkanes

Acyclic alkanes with one or more sites of branching are notably abundant components of Archean, Proterozoic, and Early Paleozoic bitumens with most reported occurrences being low-molecular weight (C_{14}–C_{19}) monomethylalkanes (e.g., Hoering, 1976, 1981; Summons and Walter, 1990). Microbial mat communities, particularly those where cyanobacteria are the predominant organism, are well known for having high abundances and distinctive patterns of short-chain (C_{15}–C_{20}) methyl alkanes and are considered to be one of the major sources since these same hydrocarbons have been identified in cyanobacterial cultures (Dembitsky *et al.*, 2001; Köster *et al.*, 1999), as well as modern and ancient sediments with actual or remnant cyanobacterial mat assemblages (Kenig *et al.*, 1995b; Robinson and Eglinton, 1990; Shiea *et al.*, 1990, 1991; Summons and Walter, 1990). Hydrolysis and decarboxylation of branched fatty acyl bacterial lipids is another possible origin for C_{15}–C_{20} methyl alkanes.

As mentioned above (Section 8.03.5.2), a striking feature of some Neoproterozoic to Early Cambrian oils from Oman and Siberia are C_{20+} methylalkanes with the locus of branching located towards the centers of the chains and a marked reduction on their abundances above C_{24} (Höld *et al.*, 1999). These compounds were attributed to C_{28+} precursor lipids with alkyl substituents at C-12 or C-13.

Other Proterozoic sediments contain abundant pseudohomologous series of odd carbon-numbered C_{19}–C_{33} branched alkanes that were originally and mistakenly assigned as 3,7-dimethylalkanes on the basis of similar GC and MS data to published literature (Logan *et al.*, 1999, 2001; Mycke *et al.*, 1988). These hydrocarbons have been reported as major components of 1,640 Ma microbial mat sediments in the Barney Creek Formation, Australia (Logan *et al.*, 2001) and the Tanana Formation and correlatives of the Centralian Superbasin, Australia (Arouri *et al.*, 2000a; Arouri *et al.*, 2000b; Logan *et al.*, 1999), while a separate series of branched alkanes, consisting of predominantly even carbon numbers ranging from C_{22} to C_{36} were also found in Barney Creek sediments in association with assemblages of large filamentous microfossils (Logan *et al.*, 2001). Some uncertainties and errors concerning the exact structures of these odd- or even-carbon numbered series of branched alkanes have recently been resolved. Kenig *et al.* (2001) identified a series of even-numbered carbon monoethylalkanes in a Mesozoic black shale by comparisons with reference mass spectra of ethylalkanes earlier identified by Wharton *et al.* (1997). Chemical synthesis of a member of another series, namely 5,5-diethylalkanes (**2**)

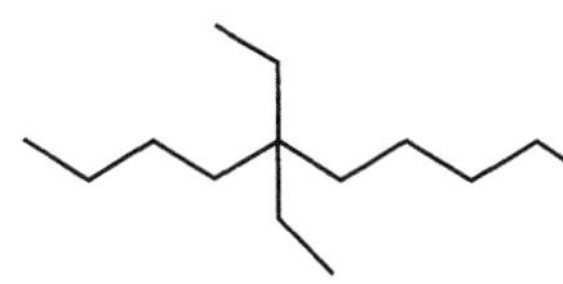

(**2**) 5,5-Diethylalkanes

(Kenig *et al.*, 2002), has led to an appreciation of their widespread occurrence. Mass-spectral and gas chromatographic analysis of related series suggests that branched alkanes with quaternary carbon centers (BAQCs) may be ubiquitous and, although the full range of structures and their biological sources are not established, they appear to be especially abundant in ancient sediments and associated with microbial mats (Kenig *et al.*, 2002; Logan *et al.*, 2001; Simons *et al.*, 2002).

Audino *et al.* (2001) have reported a unique distribution of branched alkanes ranging from C_{23} to C_{31+} in the extractable organic matter and kerogen of several Permian torbanites. Every series begins with the 2-methylalkane. Each member of a particular homologous series has a common alkyl group and each series differs from the next by two carbon atoms. These components were assigned either to an origin from the A-race of *Botryococcus braunii* based on structural similarities to the botryals biosynthesized by these organisms or by subsequent heterotrophic organisms reworking the *Botryococcus braunii* biomass.

8.03.5.4 Alkyl Cyclohexanes and Cyclopentanes

Although specific biological sources for alkylcyclopentanes (**3**) are unknown, the distribution

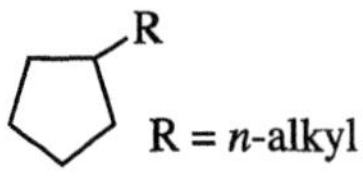

(3) Alkylcyclopentanes

of high-molecular weight homologs in the C_{41} to C_{46} range may be a useful tool to obtain information about the depositional environment (Carlson *et al.*, 1993; Hsieh and Philp, 2001) (Table 2). A predominance of odd-over-even (OEP) carbon numbers in the above range seems to indicate petroleum from marine sources, while petroleum hydrocarbons with no distinct carbon number preference or a low even-over-odd (EOP) predominance might have a freshwater origin. A strong EOP of C_{41} to C_{46} alkylcyclopentanes may be a useful indicator for oils sourced from saline lake sediments. However, the statistical basis for the above interpretations is still limited and requires a study of a larger set of oils and bitumens from different depositional environments.

n-Alkylcyclohexanes (**4**),

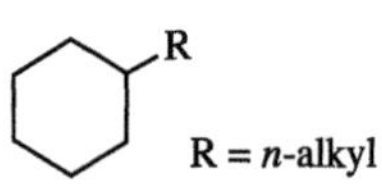

(4) Alkylcyclohexanes

methyl-*n*-alkylcyclohexanes and related compounds such as alkyl phenols have long been recognized as important components of sedimentary hydrocarbon assemblages. Potential precursors are cyclohexyl fatty acids that are known from some thermophilic and nonthermophilic bacteria (e.g., De Rosa *et al.*, 1971; Suzuki *et al.*, 1981). However, the limited carbon number distributions of these biological lipids compared to the long chain lengths of the cyclohexanes in sediments suggest there are less exotic sources. A wide variety of alkylcyclohexanes has been reported in pyrolysis products of fatty acids, aliphatic polyaldehydes, and algaenans (e.g., Fowler *et al.*, 1986; Gelin *et al.*, 1994; Rubinstein and Strausz, 1979). Moreover, a homologous series of *n*-alkylcyclohexanes was identified in pyrolysis products of microbial mats (Kenig, 2000) suggesting that they can arise from chemical or thermal alteration of acyclic precursors.

8.03.5.5 Isoprenoids

Hydrocarbons formally constructed from repeating C_5 isoprene (**5**) units, are ubiquitous in ancient sediments and petroleum. The most common and abundant of these are the C_{19} and C_{20} regularly branched (head-to-tail linking of isoprene units (**6**)) compounds pristane (**9**) and phytane (**10**) which are widely viewed as transformation products of phytol (**11**), the esterifying alcohol of cyanobacterial and green-plant chlorophylls (e.g., Chlorophyll *a* (**43**)) (Didyk *et al.*, 1978). Tocopherols are additional plant and phytoplanktonic sources of pristane (Goosens *et al.*, 1984). Archaeol (**12**) (diphytanylglycerol)

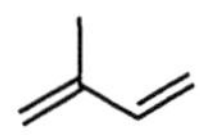

(5) Isoprene

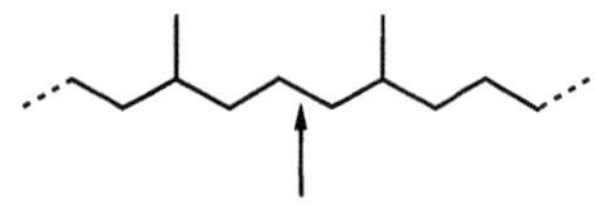

(6) Head-to-tail

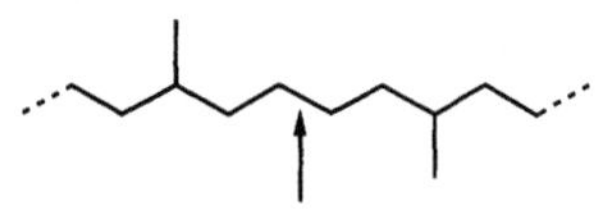

(7) Tail-to-tail

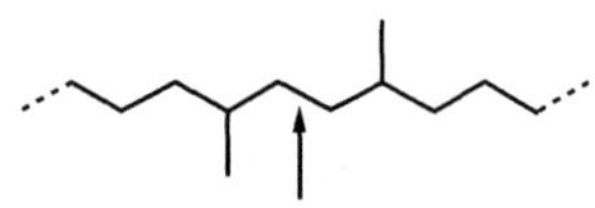

(8) Head-to-head

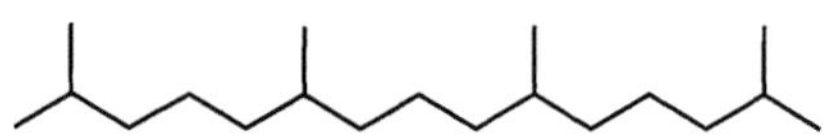

(9) Pristane

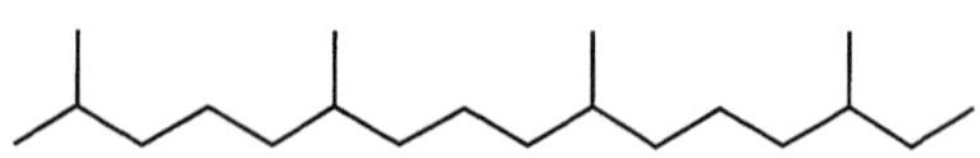

(10) Phytane

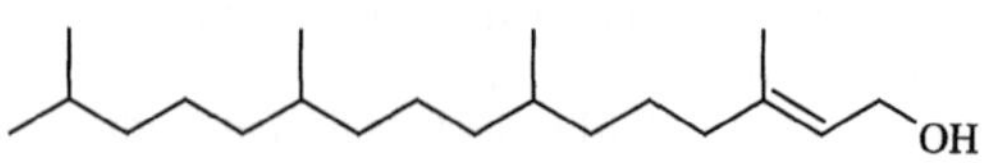

(11) Phytol

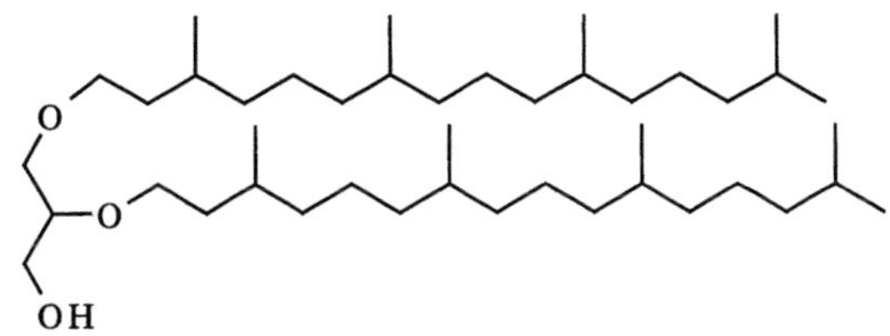

(12) Archaeol

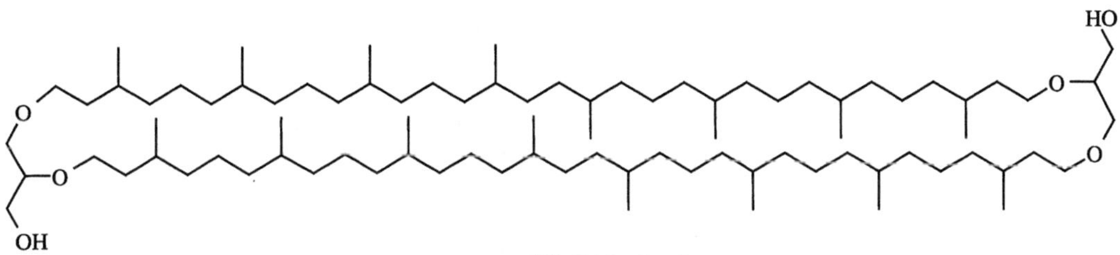

(13) Caldarchaeol

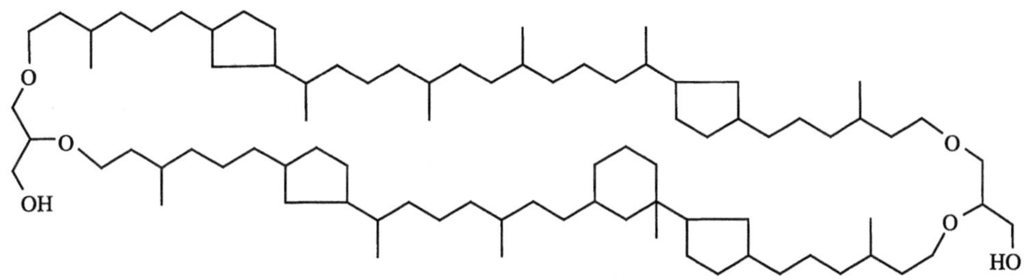

(14) A polycyclic caldarcheol charactresitic of nonthermophilic crenarcheota

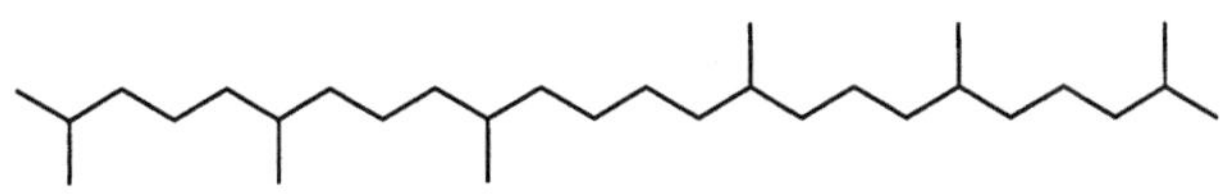

(15) Squalane

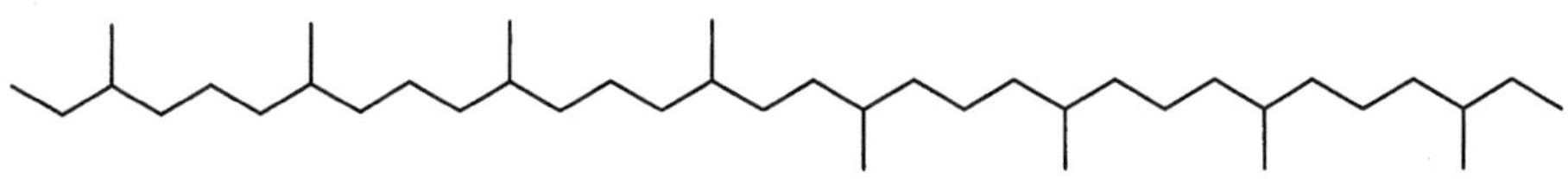

(16) Biphytane

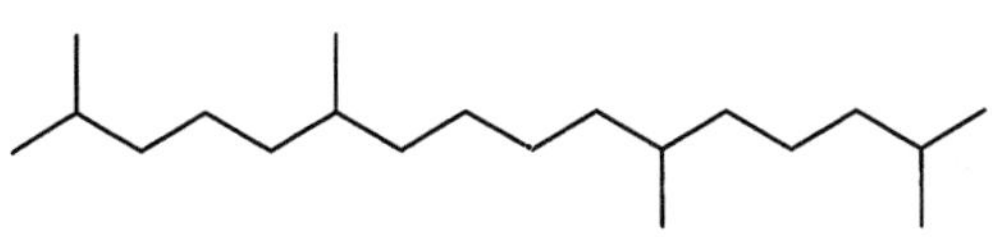

(17) Crocetane

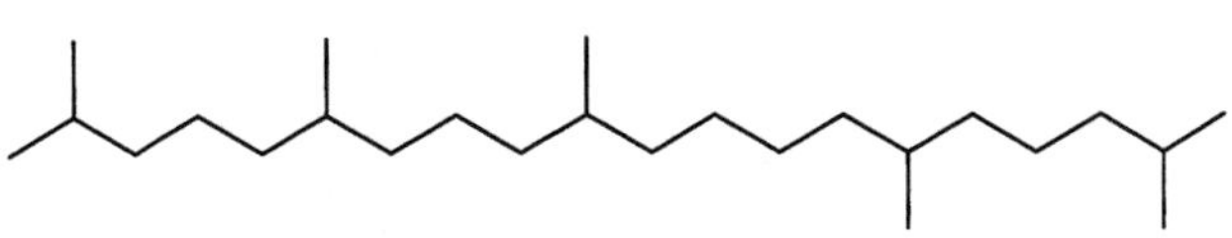

(18) 2,6,10,15,19-Pentamethylicosane (PMI)

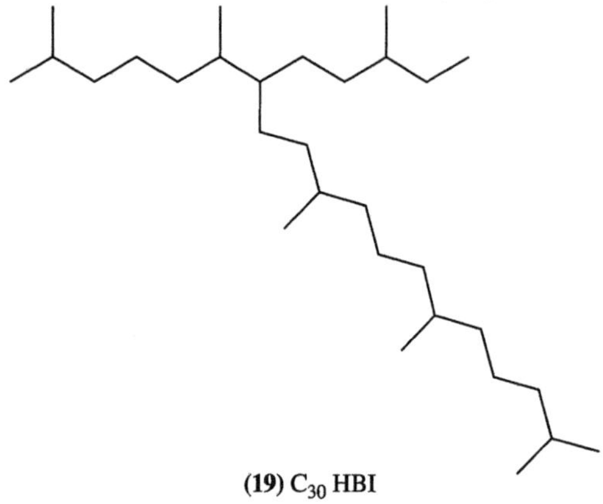

(19) C_{30} HBI

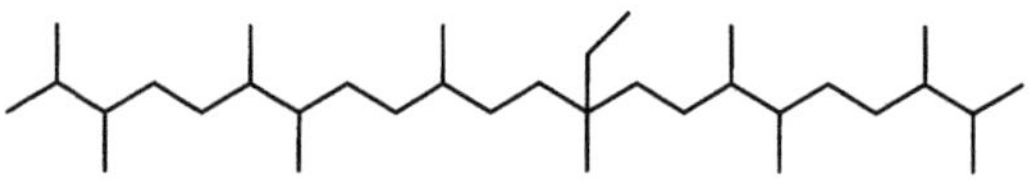

(20) Botryoccocane

is the most commonly reported core lipid in Archaea, occurring in both major kingdoms, the Euryarchaeota and Crenarchaeota (Kates, 1993; Koga *et al.*, 1993) and is also an important source of sedimentary phytane (**10**), especially in samples from extreme environments. Many oils and bitumens, however, also contain varying abundances of C_{21+} regularly branched acyclic isoprenoids (**6**) which must have originated from $>C_{20}$ precursors. Archaea are presumed to be the major source of compounds of this type. Although Albaiges (1980) has reported extended regular isoprenoids with chains as long as C_{45} in oils, there is limited knowledge of their occurrence in cultured organisms. Langworthy *et al.* (1982) have cited the presence of regular isoprenoid chains as long as C_{30} in the neutral lipid fractions of thermoacidophiles but this does not explain the wider range of carbon numbers in fossil assemblages. The polar ether lipids of extreme halophiles have often been reported to contain the C_{20}–C_{25} and C_{25}–C_{25} diether analogues of archaeol (**12**) and are, therefore, a logical source of the C_{25} and lower regular acyclic isoprenoid hydrocarbons that are invariably prominent in bitumens and oils from saline lakes (e.g., McKirdy *et al.*, 1982). Other C_{21+} regular isoprenoids in extant organisms might have remained unnoticed as the majority of lipid profiling studies to date have focused on compounds that are able to be made volatile for analysis by GC-MS. Many of the recently identified archaeal lipids have irregular C_{40} isoprenoid chains and it is quite probable that other, presently unknown, high-molecular weight polar lipid precursors exist but have escaped detection through conventional analytical windows.

Irregularly branched isoprenoids are also prominent sedimentary hydrocarbons. Squalane (**15**), comprising two tail–tail (**7**) linked C_{15} isoprenoid moieties, is a very common component of bitumens and oils and, although its logical precursor squalene occurs in most organisms, Archaea are likely to be the predominant sources. Squalane and a variety of unsaturated derivatives are present in the neutral lipid fractions of many Archaea and their abundances are highest in environmental samples with overall elevated acyclic isoprenoid content such as those from saline lakes (McKirdy *et al.*, 1986; ten Haven *et al.*, 1988). The tail-to-tail (**7**) linked C_{40} isoprenoid lycopane (**22**) is often detected in lacustrine and marine sediments (e.g., Freeman *et al.*, 1990, 1994; Wakeham *et al.*, 1993) and in particulate organic matter from anoxic water columns (e.g., Wakeham *et al.*, 1993). Feasible precursors include carotenoids of the lycopene (**21**) family that occur, for example, in anoxygenic phototrophic bacteria (Section 8.03.6.1.4), or the lycopadiene-like precursors produced by algae such as *Botryococcus braunii* (Derenne *et al.*, 1990; Wakeham *et al.*, 1993).

A major source of biphytane (**16**), the C_{40} isoprenoid with head-to-head (**8**) branching, is caldarchaeol (**13**) (dibiphytanyl—diglycerol–tetraether) which is a prominent core lipid in methanogens (Koga *et al.*, 1993) and members of the kingdom Crenarchaeota (e.g., Kates, 1993). Crenarchaeotes and some methanogens are known to produce polar lipids with variants of the caldarchaeol core with cyclopentane, and, occasionally, cyclohexane rings (e.g., (**14**)). These complex lipids have been discovered in abundance in filtrates from open ocean waters attesting to the probability that these Archaea are an important component of ocean plankton (DeLong *et al.*, 1998; Sinninghe Damsté *et al.*, 2002a). Biphytane (**16**) has long been recognized as a prominent sedimentary hydrocarbon (Moldowan and Seifert, 1979) and can have both crenarchaeote and euryarchaeote origins.

Lower molecular weight irregularly branched isoprenoids are also sometimes prominent in sediments and oils. The irregular tail-to-tail (**7**) linked C_{20} isoprenoid hydrocarbon 2,6,11,15-tetramethylhexadecane (**17**) (crocetane) and its C_{25} counterpart 2,6,10,15,19-pentamethylicosane (**18**) (often referred to as PMI, or in older literature PME) are considered diagnostic markers for

Archaea that are central to the methane cycle. These compounds have been detected in various cultured organisms, microbial communities and sediments comprising methanogenic (e.g., Brassell *et al.*, 1981; Koga *et al.*, 1993; Risatti *et al.*, 1984; Schouten *et al.*, 2001a,b,c, 1997) and methanotrophic Archaea (e.g., Bian *et al.*, 2001; Elvert *et al.*, 1999; Hinrichs *et al.*, 2000; Pancost *et al.*, 2000; Thiel *et al.*, 1999). Crocetane has also recently been reported in crude oils (Barber *et al.*, 2001; Barber *et al.*, 2002; Greenwood and Summons, 2003). PMI appears to be confined to Mesozoic and younger rocks, whereas crocetane probably has a much longer geological record since it has been detected in Triassic, Devonian, and Proterozoic rocks (Greenwood and Summons, 2003).

There are further distinctive classes of isoprenoids which are thought to have quite restricted biological origins. Compounds referred to as "highly branched isoprenoids" or HBIs with C_{20}, C_{25}, and C_{30} (e.g., (**19**)) members are biosynthesized by some diatoms (Volkman *et al.*, 1994) and are therefore considered very specific biomarkers for these organisms (Allard *et al.*, 2001; Belt *et al.*, 2000; Robson and Rowland, 1986; Rowland *et al.*, 2001; Sinninghe Damsté *et al.*, 1999a,b; Summons *et al.*, 1993; Volkman *et al.*, 1994). The biomarker botryococcane (**20**) and related compounds are derived from botryococcenes, C_{30}–C_{37} polymethylated, and polyunsaturated derivatives of an irregularly constructed isomer of squalene, and are only known to be biosynthesized by the green alga *Botryococcus braunii* (Metzger and Largeau, 1999). Certain strains biosynthesize cyclobotryococcenes (e.g., Metzger *et al.*, 1985) and polymethylsqualenes which occur as their saturated hydrocarbon analogs in ancient sediments and oils (Summons *et al.*, 2002).

8.03.5.6 Carotenoids

Carotenoids are usually yellow to red colored lipids formally derived from the C_{40} isoprenoid lycopene (**21**) carbon skeleton by varied hydrogenation, dehydrogenation, cyclization and oxidation reactions. In excess of 600 different carotenoid structures have been identified (Britton, 1995). They are biosynthesized *de novo* by all photosynthetic bacteria, eukaryotes and halophilic archaea, but also occur in a large variety of nonphotosynthetic organisms. Vertebrates and invertebrates have to incorporate carotenoids through their diet, but often have the capacity to generate structurally modified products from ingested precursors (Liaaen-Jensen, 1979). Carotenoids function most commonly as accessory pigments in phototrophs, as pigments for photoprotection, as photoreceptors for phototropism and phototaxis, and as pigments for the coloration of plants and animals (Liaaen-Jensen, 1979). Several hundred natural carotenoids have been described that are distinguished by different cyclic and linear end-groups, and a large variety of functionalities in various positions such as keto, aldehyde, ester, hydroxy, methoxy, and glycoside groups. Many of the functionalized carotenoids extracted from living organisms and recent sediments have been used to obtain information about biological origins, evolution, and ecology (e.g., Britton *et al.*, 1995; Frank *et al.*, 1999; Liaaen-Jensen, 1979; Watts *et al.*, 1977; Xiong *et al.*, 2000). However, the large variety of biological carotenoids, such as (**21**), (**23**), and (**25**) is based on a limited number of different carbon skeletons. Thus, most carotenoids, lose their diagnostic value during diagenesis by reduction of all functional groups and generation of much less specific-fossil hydrocarbons such as lycopane (**22**) and β-carotane (**24**).

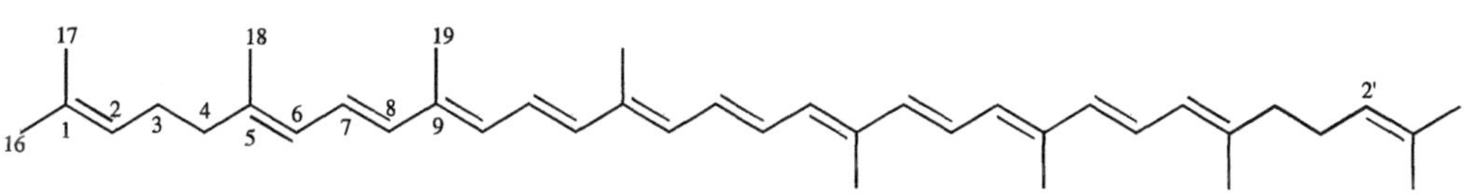

(**21**) Lycopene

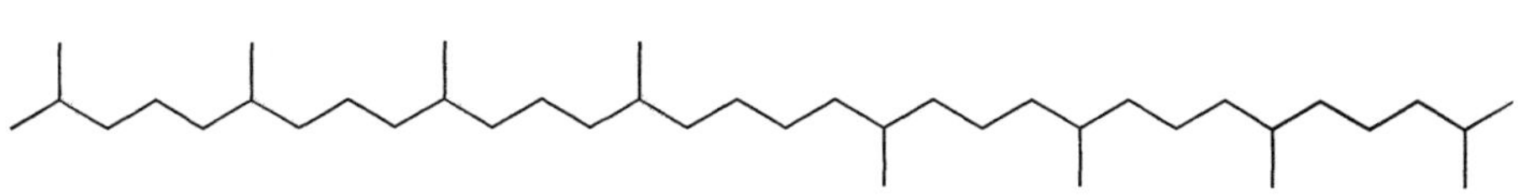

(**22**) Lycopane

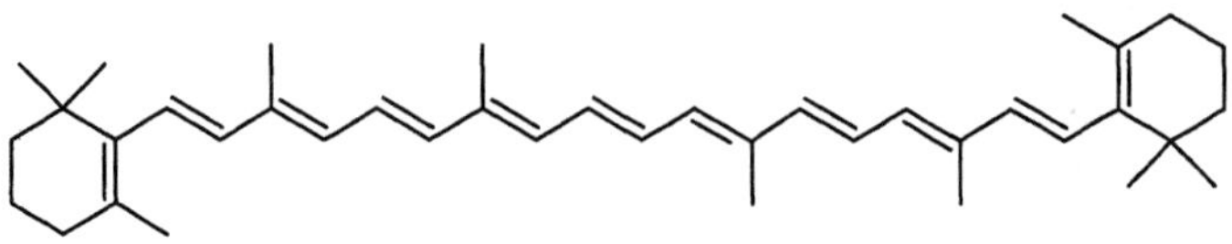

(23) β-Carotene

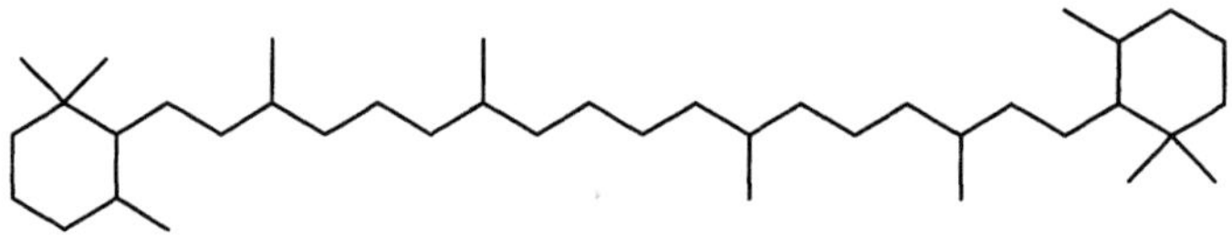

(24) β-Carotane

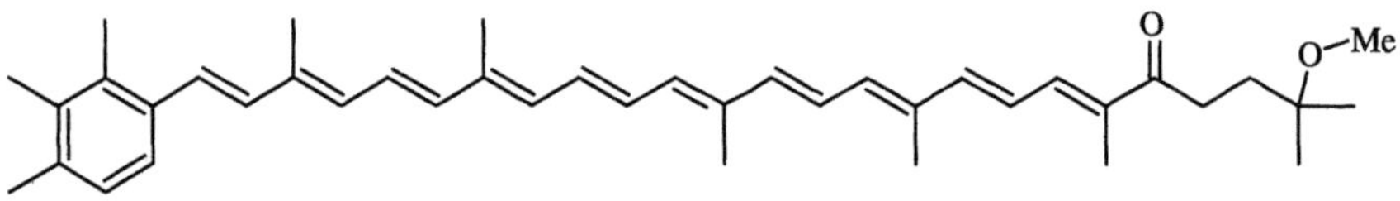

(25) Okenone

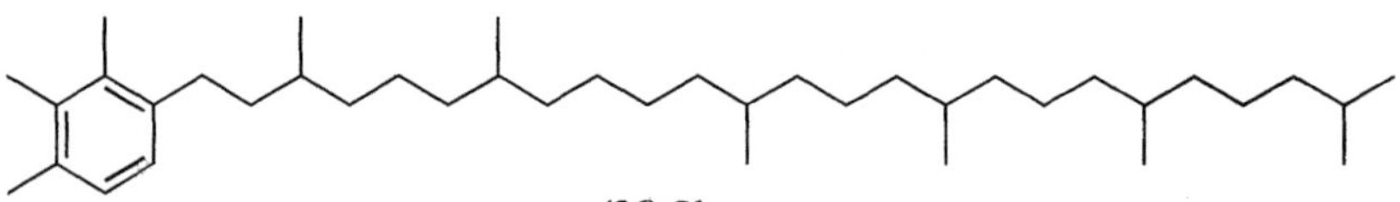

(26) Okenane

However, some carotenoids retain a taxonomically diagnostic structure during diagenesis and belong to the most important biomarkers for paleoenvironmental reconstructions as discussed below.

8.03.5.6.1 Aromatic carotenoids and arylisoprenoids

The only significant biological source for aromatic carotenoids in aquatic sedimentary environments are phototrophic green (Chlorobiaceae) and purple (Chromatiaceae) sulfur bacteria (Table 3). The growth of most phototrophic sulfur bacteria requires the presence of light and reduced sulfur species in the absence of oxygen. Thus, aromatic carotenoids are often applied as biomarkers for photic-zone euxinia (Koopmans *et al.*, 1996a; Requejo *et al.*, 1992; Summons and Powell, 1986). Okenone (**25**), the potential precursor of yet undiscovered okenane (**26**), is exclusively known from planktonic species of Chromatiaceae, while chlorobactane (**28**), the fossil equivalent of chlorobactene (**27**) and hydroxychlorobactene, is a biomarker for planktonic as well as benthic mat-forming green pigmented species of Chlorobiaceae. Brown pigmented species of Chlorobiaceae, in contrast, predominantly contain the carotenoids isorenieratene (**31**) and β-isorenieratene (**29**), the precursors for sedimentary isorenieratane (**32**) and β-isorenieratane (**30**) (Liaaen-Jensen, 1965). As carbon assimilation in Chlorobiaceae follows the reductive or reversed tricarboxylic acid cycle (TCA), their biomass is often distinguished by a strong carbon-isotopic enrichment in ^{13}C by more than ~10‰ relative to that of oxygenic phototrophs (e.g., Kohnen *et al.*, 1992). The distinctive carbon-isotopic composition of Chlorobiaceae, and the ecology of phototrophic sulfur bacteria are further discussed in Section 8.03.6.1.4.

A second source for aromatic carotenoids are some genera of actinomycetes, such as *Mycobacterium* and *Streptomyces* (Krügel *et al.*, 1999). However, the contribution of carotenoids from these organisms to organic matter in aquatic sediments is probably insignificant. A larger variety of aromatic carotenoids also occurs in selected species of marine sponges (Liaaen-Jensen *et al.*, 1982) suggesting the presence of bacterial symbionts. These carotenoids include isorenieratene (**31**) and β-isorenieratene (**29**) also found in Chlorobiaceae, but in addition two aromatic structures

Table 3 Aromatic carotenoids and maleimides as indicators for photic zone euxinia.

Geological carotenoid	*Possible biological precursors*	*Biological sources*	*References*
Okenane (**26**)	Okenone (**25**)	Chromatiaceae	Schaeffer *et al.* (1997)[a]
Chlorobactane (**28**)	Chlorobactene (**27**); hydroxychlorobactene	Green pigmented Chlorobiaceae	Grice *et al.* (1998c)
Isorenieratane[b] (**32**)	Isorenieratene (**31**)	Brown pigmented Chlorobiaceae	Bosch *et al.* (1998), Grice *et al.* (1996b), Hartgers *et al.* (1993), Koopmans *et al.* (1996a), Pancost *et al.* (1998), Putschew *et al.* (1998), Simons and Kenig (2001), and Sinninghe Damsté *et al.* (2001)
β-isorenieratane[c] (**30**)	β-isorenieratene (**29**); β-carotene[d] (**23**)	Brown pigmented Chlorobiaceae	Grice *et al.* (1998c)
Renieratane (**34**)	Renieratene (**33**)	Sponges or sponge symbionts? phototrophic sulfur bacteria?	Hartgers *et al.* (1993), and Schaeflé *et al.* (1977)
Renierapurpurane[e] (**36**)	Renierapurpurin (**35**)	Sponges or sponge symbionts? phototrophic sulfur bacteria?	Schaeflé *et al.* (1977)
Palaerenieratane[f] (**37**)	Unknown	Chlorobiaceae?	Hartgers *et al.* (1993), Koopmans *et al.* (1996a), and Requejo *et al.* (1992)
2,3,6-TMAs[g] (**38**)	Chlorobactene (**27**); hydroxychlorobactene; isorenieratene (**31**); β-isorenieratene (**29**); β-carotene (**23**)[d] and similar structures	Mostly Chlorobiaceae[c]	Hartgers *et al.* (1993), Requejo *et al.* (1992), Summons and Powell (1986), Summons and Powell (1987, 1992)
2,3,4-TMAs[g] (**39**)	Okenone (**25**); renieratene (**33**) renierapurpurin (**35**)	Chromatiaceae; Chlorobiaceae?	Summons and Powell (1987)
3,4,5-TMAs[g] (**40**)	Precursor of palaerenieratane (**37**)	Chlorobiaceae?	Hartgers *et al.* (1993), Requejo *et al.* (1992), Summons and Powell (1987)
Me *i*-Bu maleimide (**49d**)	BChl *c*, *d*, and *e* (**46**)–(**48**)	Chlorobiaceae, Chloroflexaceae	Grice *et al.* (1996a, 1997), and Pancost *et al.* (2002)

[a] Report of okenane as a hydrogenation product of okenone after H_2/PtO_2 treatment of a polar fraction extracted from a Recent lake sediment. Okenane is unknown as a hydrocarbon biomarker from sedimentary rocks. [b] Also including a large variety of di- to pentacyclic early diagenetic cyclization and rearrangement products of isorenieratene (Grice *et al.*, 1996b; Koopmans *et al.*, 1996a). [c] Only diagnostic for Chlorobiaceae if the carbon isotopic composition of individual arylisoprenoids shows an enrichment in ^{13}C diagnostic of the reversed tricarboxylic acid cycle. [d] According to Koopmans *et al.* (1996b), β-carotene can undergo aromatization during diagenesis to β-isorenieratane and further degrade to 2,3,6-TMAs. [e] New trivial name suggested here (= perhydrorenierapurpurin). [f] New trivial name suggested here. [g] TMA = trimethylarylisoprenoids.

unknown from other organisms: renieratene (**33**) and renierapurpurin (**35**). However, sponges are not capable of *de novo* carotenoid biosynthesis. Therefore, these biomarkers are either generated by the sponge by modification of dietary carotenoids, or they are derived from sponge symbionts, possibly phototrophic sulfur bacteria (Liaaen-Jensen *et al.*, 1982). Although phototrophic sulfur bacteria have not yet been reported as symbionts in sponges, determination of the carbon-isotopic composition of the aromatic carotenoids could confirm their presence. The diagenetic products renieratane (**34**) and reniera-purpurane (**36**) (= perhydrorenierapurpurin), are rare in the geological record. In the Upper Devonian Duvernay Formation of the Western Canada Basin, renieratane (**34**) occurs together with isorenieratane (**32**) (Hartgers *et al.*, 1993). In these particular samples isorenieratane was enriched in ^{13}C by up to 15‰ relative to aliphatic hydrocarbons and, therefore, clearly derived from Chlorobiaceae. Although the carbon-isotopic composition of renieratane (**34**) was not reported, its co-occurrence with isorenieratane (**32**) in the Duvernay Formation indirectly suggests that it is also product of phototrophic sulfur bacteria (Hartgers *et al.*, 1993). The Duvernay Formation also contains a third diaromatic carotenoid (**37**), here named palaerenieratane, unknown from extant organisms (Hartgers *et al.*, 1993; Requejo *et al.*, 1992). Palaerenieratane (**37**)

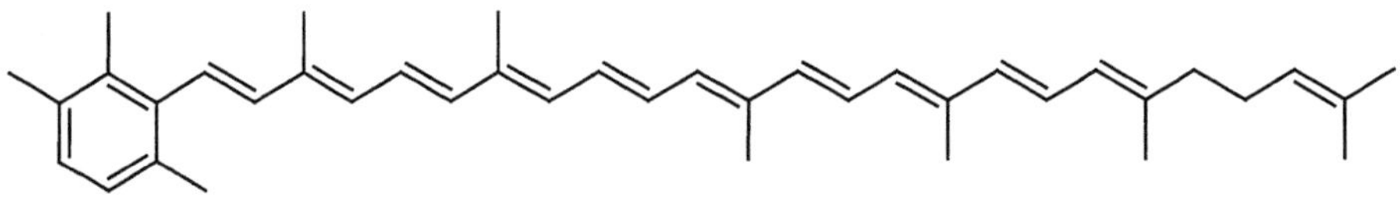

(27) Chlorobactene

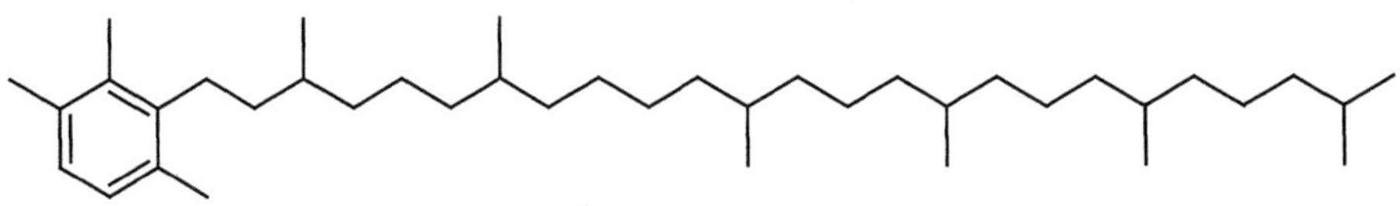

(28) Chlorobactane

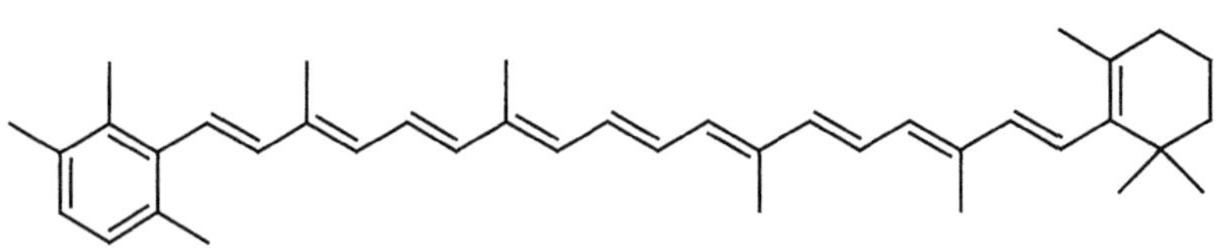

(29) *β*-Isorenieratene

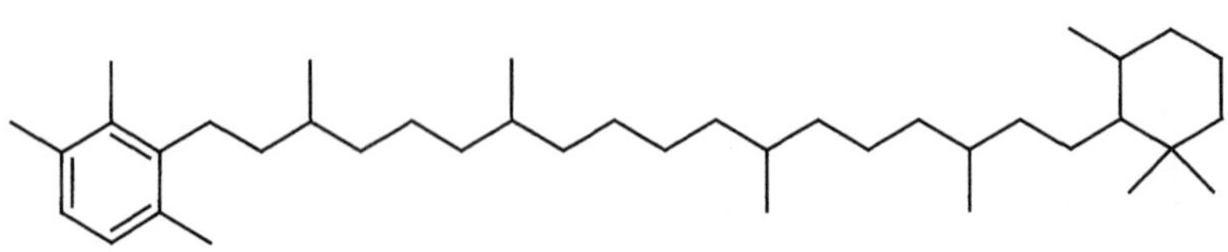

(30) *β*-Isorenieratane

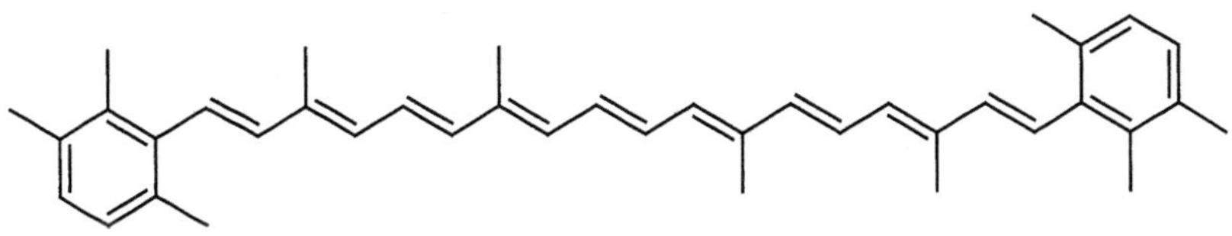

(31) Isorenieratene

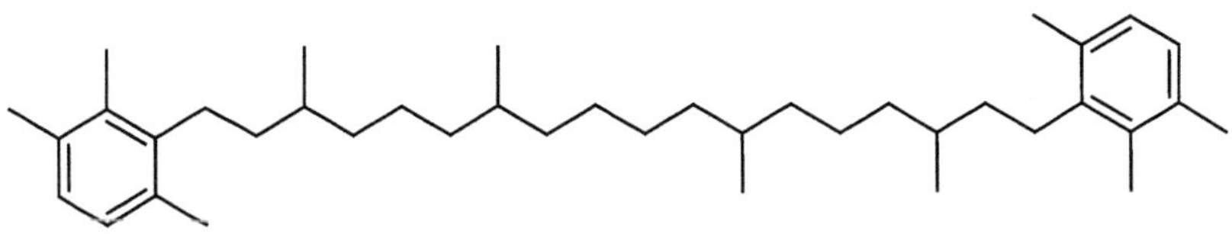

(32) Isorenieratane

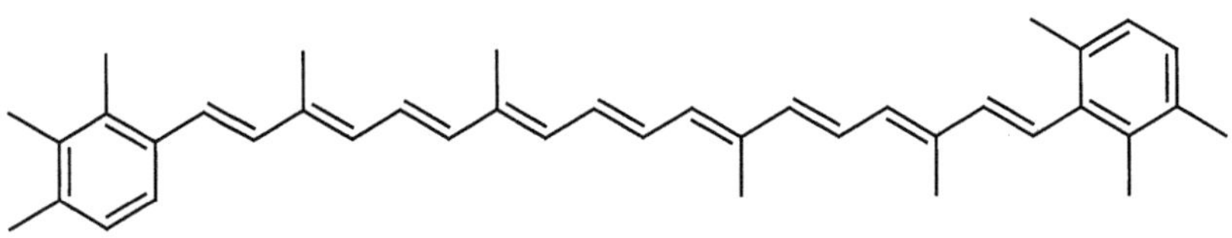

(33) Renieratene

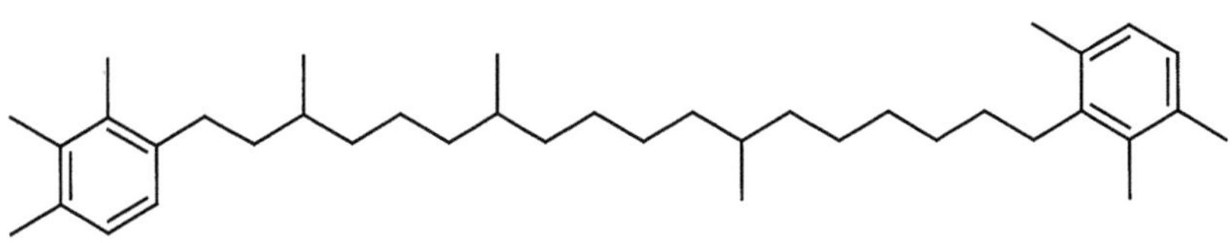

(34) Renieratane

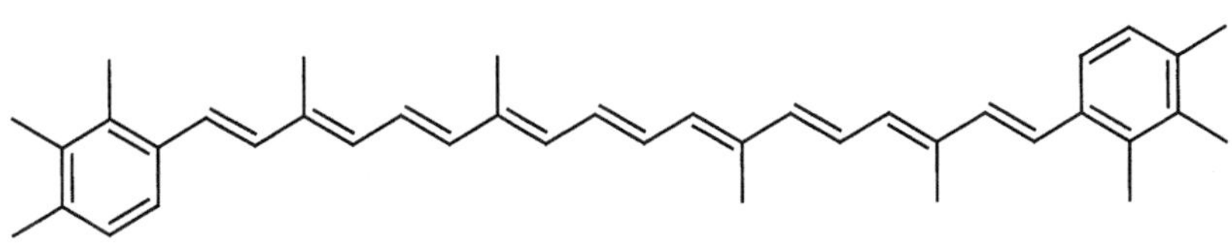

(35) Renierapurpurin

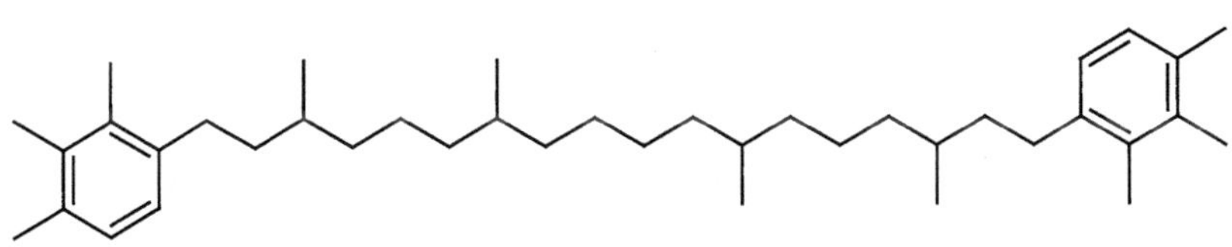

(36) Renierapurpurane

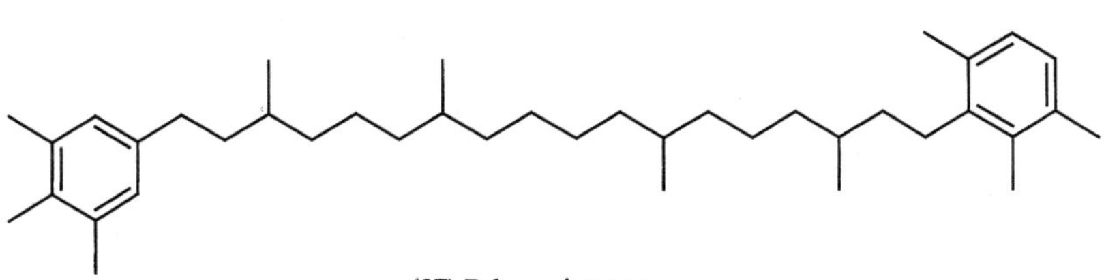

(37) Palaerenieratane

is formally derived from renieratane (**34**) by dislocation of one methyl group at one of the terminal aromatic rings. In the Duvernay samples, palaerenieratane (**37**) is strongly enriched in ^{13}C just as isorenieratane (**32**). As the enrichment is diagnostic for carbon assimilation via the reversed TCA cycle (Section 8.03.6.1.4), palaerenieratane (**37**) is also almost certainly derived from either an extinct species, or an as yet undetected species, of extant Chlorobiaceae (Hartgers *et al.*, 1993).

Arylisoprenoids with the 2,3,6- (**38**), 2,3,4- (**39**), and 3,4,5-trimethyl (**40**)

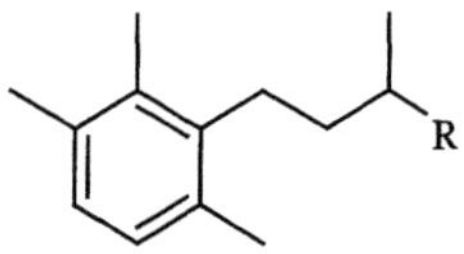

(**38**) 2,3,6-Trimethylarylisoprenoids

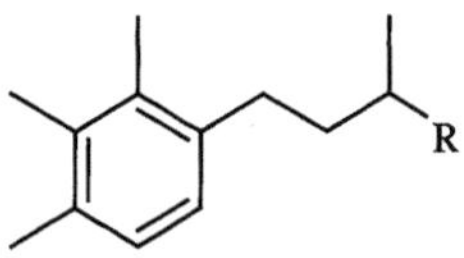

(**39**) 2,3,4-Trimethylarylisoprenoids

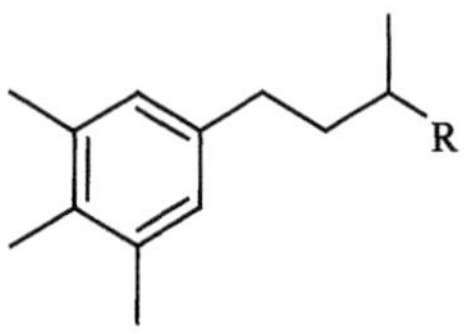

(**40**) 3,4,5-Trimethylarylisoprenoids

substitution patterns are diagenetic and catagenetic cracking products of the above C_{40} aromatic carotenoids (Hartgers *et al.*, 1993; Summons and Powell, 1986, 1987). Therefore, arylisoprenoids are also applied as biomarkers for phototrophic sulfur bacteria. However, 2,3,6-trimethyl aromatic arylisoprenoids (**38**) are also purported to form by diagenetic aromatization and rearrangement of cyclic, nonaromatic carotenoids (Koopmans *et al.*, 1996b). (**38**) are therefore more clearly biomarkers for Chlorobiaceae if they also show the carbon-isotopic enrichment typical for the reductive TCA cycle. Aromatic carotenoids might also form a large variety of other rearrangement, cyclization, and degradation products that have diagnostic value for phototrophic sulfur bacteria (Grice *et al.*, 1996b; Koopmans *et al.*, 1996a; Sinninghe Damsté *et al.*, 2001).

8.03.5.6.2 Bacterioruberin

Pigments of the bacterioruberin group (**41**) are an example for uncommon C_{50} carotenoids. The unique carbon skeleton (**42**) is biosynthesized by addition of two C_5 isoprenoid units to the 2 and 2′ positions of the C_{40} carotenoid lycopene (**21**) (Kushwaha and Kates, 1979; Kushwaha *et al.*, 1976). Bacterioruberin is a ubiquitous and abundant, red-orange pigment in moderately (Rønnekleiv and Liaaen-Jensen, 1995) to extremely halophilic archaea (Halobacteria) (Liaaen-Jensen, 1979) (Section 8.03.6.2.3). Located in the membrane of Halobacteria, it plays a role in the photoprotection system (Cockell and Knowland, 1999), but might also be important for the adaptation of membrane fluidity to changing osmotic conditions (D'Souza *et al.*, 1997). Carotenoid pigments with the bacterioruberin skeleton have also been detected in several species of the class Actinobacteria. These include the plant pathogen *Curtobacterium flaccumfaciens* (Häberli *et al.*, 2000), the psychrotrophic Micrococcaceae *Micrococcus roseus* (Strand *et al.*, 1997) and *Arthrobacter agilis* found in Antarctic soil and ice (Fong *et al.*, 2001), and the highly radioresistant *Rubrobacter radiotolerans* (Saito *et al.*, 1994). In psychrotrophic species, C_{50} carotenoids play an adaptive role in membrane stabilization at low temperature (Fong *et al.*, 2001).

The fossil equivalent of bacterioruberin (**41**), perhydro bacterioruberin (**42**), has yet to be discovered in geological samples. However, the abundance and ubiquity of bacterioruberin in Halobacteria, and the occurrence of dense blooms in salt lakes and pools of evaporating seawater, makes (**42**) a potential, highly diagnostic biomarker for Halobacteria and moderate-to-extreme hypersaline conditions. It is worth bearing in mind that some high-molecular weight biomarkers may have escaped detection because they are difficult to analyze by conventional GC-MS methods.

8.03.5.7 Chlorophylls and Maleimides

The major chlorophyll (Chl) found in all oxygenic photosynthetic organisms, i.e., prochlorophytes, cyanobacteria, and photosynthetic Eukarya, is Chl *a* (**43**). However, the partially defunctionalized diagenetic products of Chl *a* can usually not be distinguished from products of bacteriochlorophyll (BChl) *a* (**44**) and *b* (**45**), which are mostly derived from anoxygenic phototrophic purple sulfur bacteria. However, BChl *c*, *d*, and *e* (**46**)–(**48**) are highly specific. BChl *c* (**46**) and *d* (**47**) are restricted to green

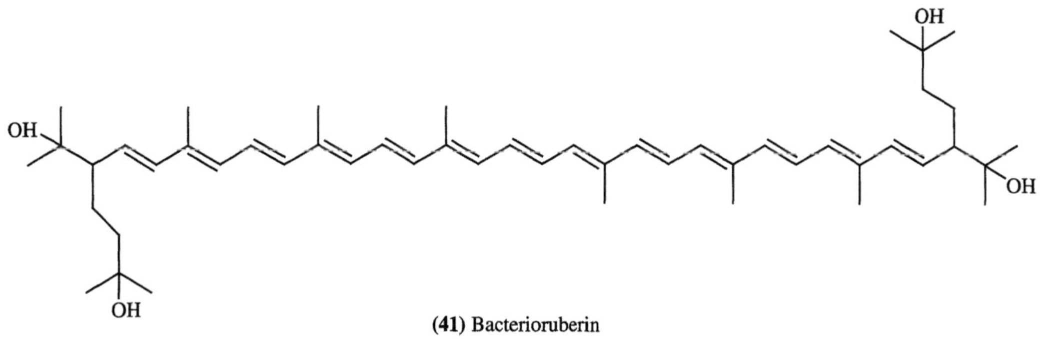

(41) Bacterioruberin

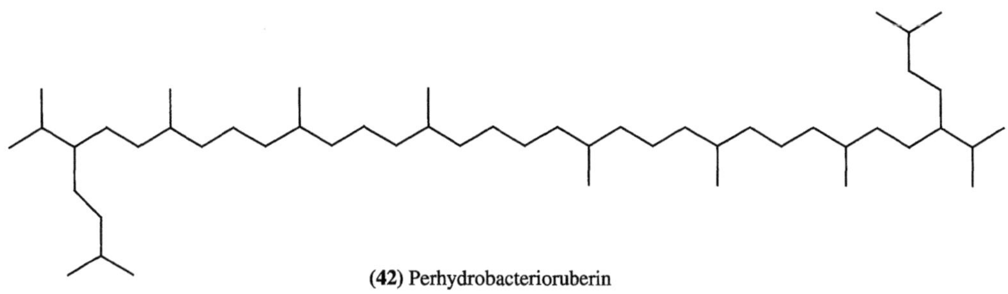

(42) Perhydrobacterioruberin

filamentous bacteria (Chloroflexaceae) and green sulfur bacteria (Chlorobiaceae), while BChl *e* (**48**) is found as a major component only in brown pigmented strains of the Chlorobiaceae.

An elegant methodology to study the input of BChl *c*, *d*, or *e* into ancient sediments was developed by Grice *et al.* (1996a). The tetrapyrrole structure of Chl and BChl is only rarely preserved in thermally mature sedimentary rocks. Grice *et al.* (1996a) observed that Chl and BChl might undergo oxidative degradation to maleimides (**49**) (1*H*-pyrrole-2,5-diones), possibly induced by enzymatic activity or light. As the major distinguishing structural characteristics of BChl *c*, *d*, and *e* (**46**)–(**48**), in comparison to (**43**)–(**45**), are additional carbon atoms in positions C-8, C-12, and C-20; their oxidative degradation will generate a distinctive suite of maleimides. While the major products of Chl *a* (**43**) degradation are indistinct 3,4-dimethyl (**49(a)**) and 3-ethyl-4-methylmaleimide (**49(b)**), the oxidation of BChl *c*, *d*, and *e* ((**46**)–(**48**)) of Chlorobiaceae additionally generates the 3-isobutyl-4-methylmaleimide (**49(d)**). The diagnostic value of 3-isobutyl-4-methylmaleimide (**49(d)**) in the Permian Kupferschiefer was confirmed by determination of the carbon-isotopic composition of individual maleimides. 3-methyl-4-propyl- (**49(c)**) and 3-isobutyl-4-methylmaleimide (**49(d)**) were enriched in ^{13}C by 10–11‰ relative to 3-ethyl-4-methylmaleimide (**49(b)**) (Grice *et al.*, 1996b). This isotopic enrichment is typical for the reductive TCA cycle, the pathway of CO_2 fixation followed by Chlorobiaceae (Section 8.03.6.1.4).

8.03.5.8 Sesquiterpanes (C_{15}) and Diterpanes (C_{20})

Bicyclic terpanes are common in oils and bitumens and can have separate origins in bacteria and plants. Compounds of the drimane (**50**) series, which are ubiquitous and occur in rocks of all ages, are thought to be degradation products of bacteriohopanoids (Alexander *et al.*, 1983). Diterpanes with a far more restricted distribution appear to be derived from vascular plant precursors such as abietic acid. Prominent sedimentary hydrocarbons include beyerane, kaurane, phyllocladane (**51**), and isopimarane. These compounds and structurally related aromatic hydrocarbons regularly co-occur with resins and other remains of conifers and are therefore considered biomarkers for vascular plants and, more specifically, for gymnosperms (e.g., Noble *et al.*, 1985). Compound-specific isotopic data support the gymnosperm-diterpane relationships (e.g., Murray *et al.*, 1998).

Another important class of diagnostic plant terpenoids is the cadinane group derived from cadinene-based polymers of resinous tropical

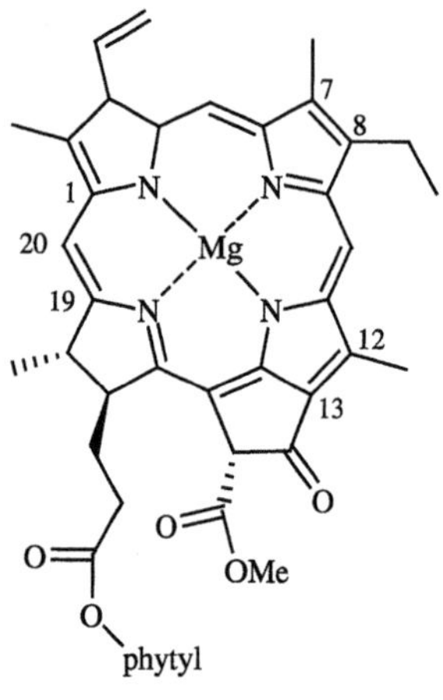

(**43**) Chl *a*

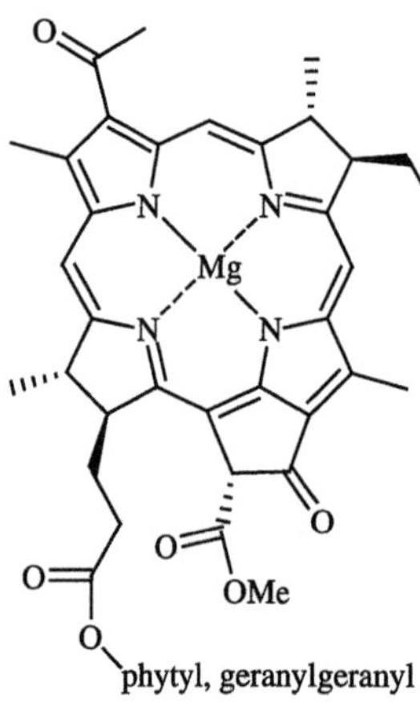

(**44**) BChl *a*

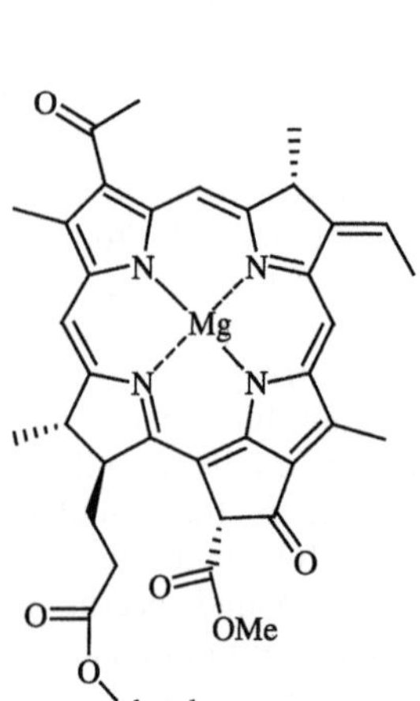

(**45**) BChl *b*

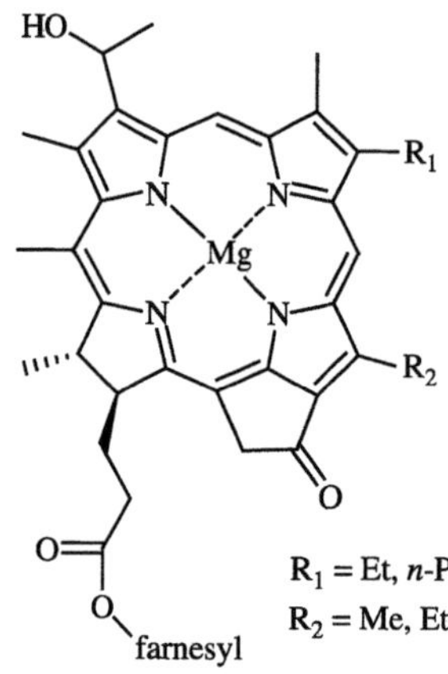

R_1 = Et, *n*-Pr, isobutyl
R_2 = Me, Et

(**46**) BChl *c*

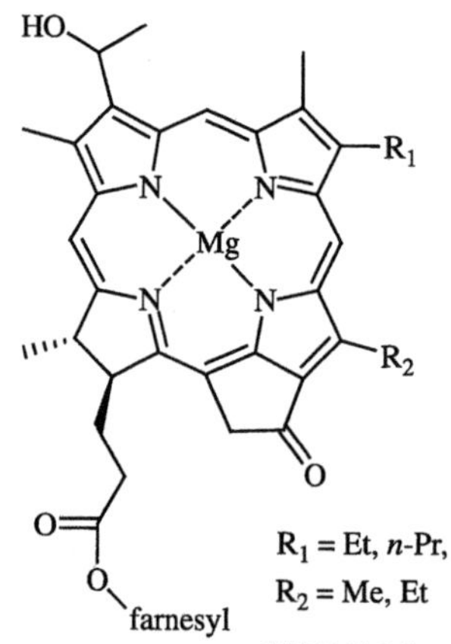

R_1 = Et, *n*-Pr, isobutyl, neopentyl
R_2 = Me, Et

(**47**) BChl *d*

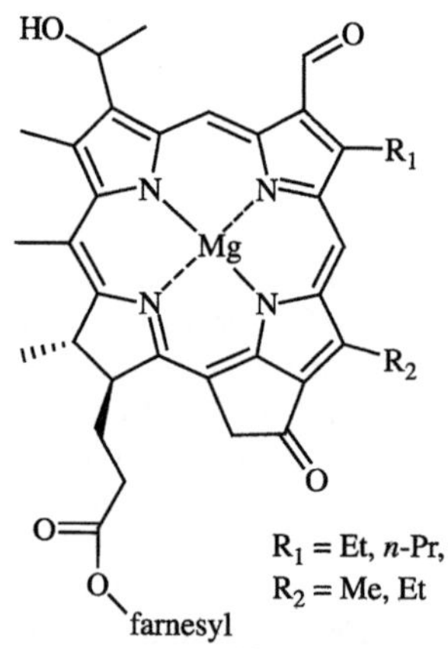

R_1 = Et, *n*-Pr, isobutyl
R_2 = Me, Et

(**48**) BChl *e*

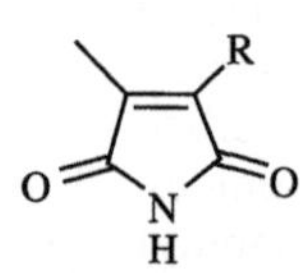

(**49**) Maleimides
(**a**) R = Me; (**b**) R = Et; (**c**) R = *n*-Pr; (**d**) R = isobutyl

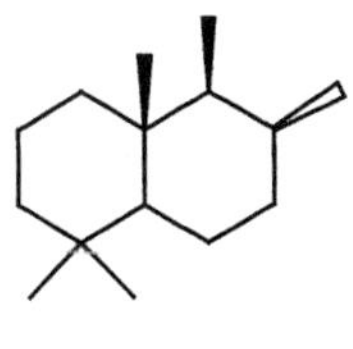

(50) Drimane

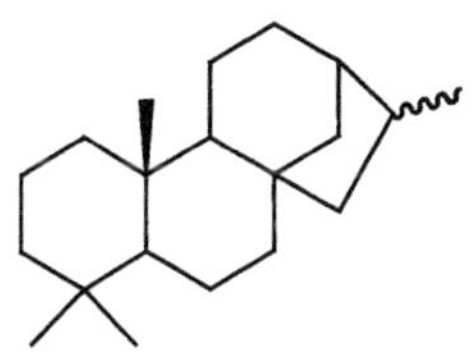

(51) Phyllocladane

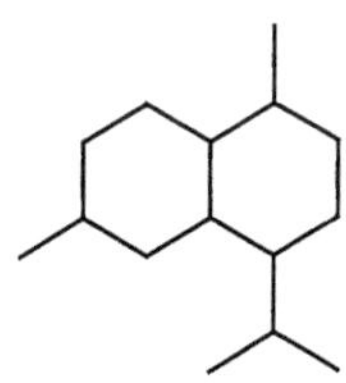

(52) Cadinane

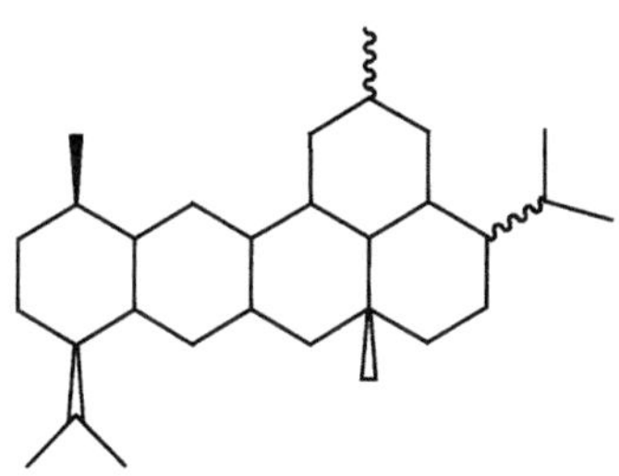

(53) Bicadinane

angiosperms. Cadinane (**52**), isomeric bicadinanes (**53**), and tricadinanes are generated by thermal alteration of polycadinene resins and are putative biomarkers for the Dipterocarpaceae (van Aarssen *et al.*, 1990; Sorsowidjojo *et al.*, 1996).

8.03.5.9 Cheilanthanes and other Tricyclic Polyprenoids

The most common compounds in this class are cheilanthanes (**54**) (= 13-methyl, 14-alkylpodocarpanes), tricyclic terpanes that extend from C_{19}

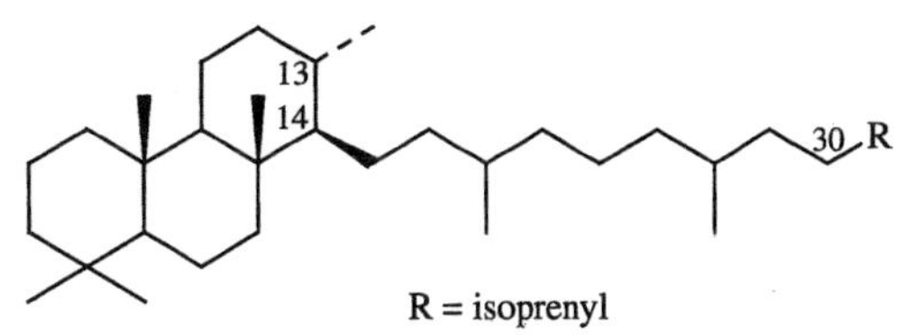

(54) Cheilanthane

to at least C_{45} (Moldowan and Seifert, 1983). These compounds are, on theoretical grounds, derived by cyclization of regular polyprenol precursors (Aquino Neto *et al.*, 1983). The only known natural products that have the cheilanthane skeleton are very unlikely precursors for the ubiquitous hydrocarbon counterparts found in bitumens and petroleum: for example cytotoxins with exotic structures found in sponges (e.g., Gomez Paloma *et al.*, 1997; Manes *et al.*, 1988) and nudibranchs (Miyamoto *et al.*, 1992), and cheilanthatriol, extracted from the fern *Cheilanthes*, the organism that gave the compound class its name (Khan *et al.*, 1971). Cheilanthanes are "orphan biomarkers" because their actual source remains unknown. While Bacteria have been hypothesized as their precursors, they have been found to occur abundantly in association with tasmanite algae, and are cogenerated with related monoaromatic to triaromatic tricyclic hydrocarbons during pyrolysis of *Tasmanites* kerogen (e.g., Aquino Neto *et al.*, 1992; Greenwood *et al.*, 2000; Revill *et al.*, 1994). Therefore, cheilanthanes could be derived from an unusual algal biopolymer. This could explain why feasible precursors have not been found in the extractable lipids of extant organisms using conventional techniques. Tricyclic terpanes occur widely in the geological record but are most abundant in mature shales and their derived oils. More highly cyclized structures based on the same regular polyprenol carbon chain also occur in sediments and have been positively identified by comparison with standards produced by chemical synthesis (Grosjean *et al.*, 2001).

8.03.5.10 Hopanoids and other Pentacyclic Triterpanes

It is often said that hopanoids are "the most abundant natural products on Earth" (Ourisson and Albrecht, 1992) and a major body of work exists on their distributions in sediments, in prokaryotes and in plants. Most commonly, hopanoids are found in select groups of Bacteria, all of which are aerobic (Farrimond *et al.*, 1998; Rohmer *et al.*, 1984). In fact, hopanoids were recognized as chemical fossils (e.g., pentakishomohopane (**55**)) well before their bacterial origins were established. Hopanoids are ubiquitous components in sedimentary organic matter and petroleum of all geological eras. The functional forms of hopanoids in bacteria are the amphiphilic bacteriohopanepolyols (**56**)

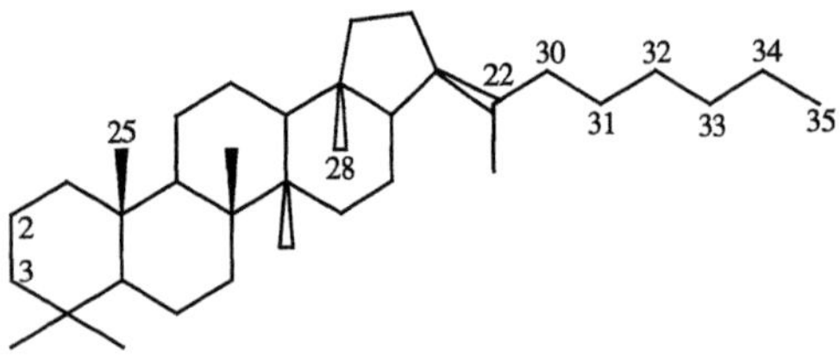

(**55**) Pentakishomohopane

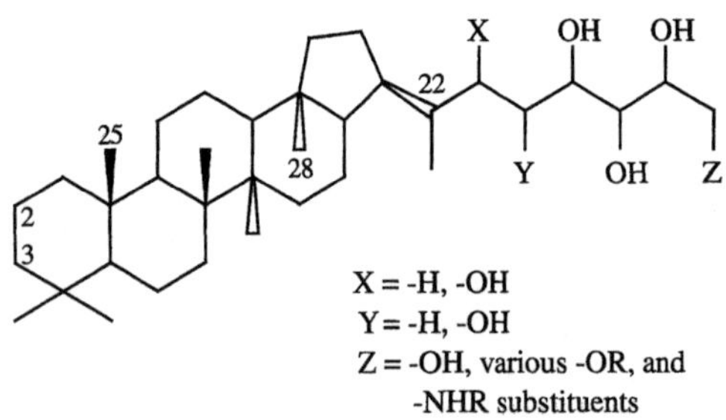

(**56**) Bacteriohopanepolyols (BHP)

(BHP) where a five carbon sugar-derived moiety is C-bound to the C_{30} pentacyclic hopane skeleton. This C_5 unit may have additional sugar, amino acid, or other polar groups attached. It is hypothesized that BHP are the bacterial surrogates of sterols which perform a role as membrane modifiers in eukaryotic cells (Ourisson and Albrecht, 1992; Ourisson *et al.*, 1987).

Although they are known to be synthesized by a wide variety of cultured aerobic bacteria there does not appear to be any obligate requirement for oxygen in their biosynthesis. The biosynthesis and cyclization of squalene to a pentacyclic triterpenoid with a hopane skeleton does not seem to require oxygen and, therefore, hopanoid synthesis might also be possible in anaerobes. For instance, analysis of microbial mats at methane seeps under anoxic Black Sea water revealed the presence of ^{13}C-depleted ($\delta^{13}C = -78‰$) hopanoids with an unusual stereochemistry. This isotopic depletion indicates *in situ* production and, therefore, suggests that anaerobes are responsible (Thiel *et al.*, 2003).

Besides the apparent paradox of finding BHP only in cultured aerobic bacteria, specific precursor-to-hopane product relationships are very poorly constrained. The major problem in elucidating their sources lies in the huge variety of potential contributing organisms, the low number of these that have been cultured for screening and the relatively low number of individual compounds that have been so far identified in both cultures and in environmental samples. For the vast majority of natural situations, the hopanoid content of a particular sediment- or water-column sample cannot be reliably attributed to any specific source without additional information. Such information might include the amounts of a specific hopanoid known to be contained in different bacteria versus their quantitative importance in a particular setting. Or, it might be the presence of characteristic chemical attributes or stable carbon isotopic compositions as in the case of (**57**) and (**58**) (Table 4).

The presence of alkyl substituents on the hopanoid skeleton, for example, A-ring methyl groups, appears to be limited to specific physiological types. For example, methanotrophic bacteria and acetic acid bacteria biosynthesize a range of 3β-hopanoids (Summons and Jahnke, 1992; Zundel and Rohmer, 1985a,b,c). The corresponding 3β-methylhopane hydrocarbons (**58**) could be derived from either group of bacteria but a profound ^{13}C depletion that has been observed in several of their sedimentary occurrences points to methanotrophic sources being more important (e.g., Burhan *et al.*, 2002; Collister *et al.*, 1992). 2β-Methylhopanoids are produced by many cyanobacteria and have few other demonstrated sources (Bisseret *et al.*, 1985) and, accordingly, it is hypothesized that the corresponding sedimentary 2α-methylhopane (**57**) hydrocarbons are biomarkers for cyanobacteria (Summons *et al.*, 1999).

It also appears that further clues about hopanoid origins can be drawn from the polar side-chains which carry different numbers and types of substituents. This, in turn, affects their subsequent diagenesis and the types of hopane hydrocarbon, ketone, and other products that are recorded in sediments. In addition to the diagnostic 3β-methyl substituents, hexafunctionalized side-chains are prevalent in Type 1 methanotrophic bacteria (Neunlist and Rohmer, 1985; Zundel and Rohmer, 1985a). The hydroxy substituent at C-31 of these compounds appears to assist oxidative loss of this carbon or the one at C-30, resulting in a predominance of C_{30} hopane and 30-norhopane products where methanotrophs are prevalent or even dominant (e.g., Burhan *et al.*, 2002; Rohmer *et al.*, 1992). Anomalous ^{13}C-depletion of these hopanoids often observed in sediments and oils is quite consistent with this interpretation (e.g., Summons *et al.*, 2002).

28,30-Dinorhopane (a.k.a. 28,30-bisnorhopane) and 25,28,30-trinorhopane (**59**) are often very prominent hydrocarbons in sediments from euxinic environments and their derived oils

Table 4 Common steranes, hopanes and other polycyclic terpanes in the molecular fossil record and their paleobiological interpretation.

Biomarkers	*Biological and environmental interpretation*	*References*
Hopanoids		
C_{30}-hopanes	Diverse bacterial lineages; few eukaryotic species (e.g., some cryptogams, ferns, mosses, lichens, filamentous fungi, protists).	Rohmer *et al.* (1984)
Extended C_{31} to C_{35} hopanes (a.k.a. homohopanes) e.g., (**55**)	Diagnostic for Bacteria; biosynthesis appears to be restricted to lineages that are not strictly anaerobic (with a possible exception (Thiel *et al.*, 2003)).	Ourisson and Albrecht (1992), Rohmer *et al.* (1984)
Extended C_{32} to C_{36} 2α-methylhopanes (**57**)	Diagnostic for cyanobacteria and prochlorophytes.	Bisseret *et al.* (1985), Summons *et al.* (1999)
Extended C_{32} to C_{36} 3β-methylhopanes (**58**)	Diagnostic for some microaerophilic proteobacteria (certain methylotrophs, methanotrophs, acetic acid bacteria).	Zundel and Rohmer (1985a,b), (1985c), Summons and Jahnke (1992)
28,30-Dinorhopane; 25,28,30-trinorhopane TNH (**59**)	Often prominent in sediments from euxinic environments.	Grantham *et al.* (1980), Peters and Moldowan (1993)
Steranes and steroids		
24-Norcholestane (C_{26})	Possible diatom origin; high concentrations relative to 27-norcholestane indicate Cretaceous or younger crude oil.	Holba *et al.* (1998a,b)
Cholestane (**66a**)	In aquatic sources probably almost exclusively derived from diverse eukaryotes; in organic matter from terrestrial sources (e.g., paleosols) input from soil bacteria of the order Myxococcales conceivable.	Volkman (2003), Bode *et al.* (2003), Kohl *et al.* (1983)
Ergostane (**66b**), stigmastane (**66c**)	Exclusively eukaryotic; but usually no distinct sources discernible.	Volkman (2003)
24-*n*-propylcholestane (**66d**)	Pelagophyte algae; a biomarker for marine conditions with few exceptions.	Moldowan *et al.* (1990)
24-Isopropylcholestane (**66e**)	Sponges and possibly the sponge-related stromatoporids.	McCaffrey *et al.* (1994b)
2- and 3-Alkylsteranes	Ubiquitous in bitumens of all ages; possibly heterotrophic alteration products of sedimentary steroids.	Summons and Capon (1991)
4-Methylcholestane (**69a**); 4,4-dimethylcholestane	Diverse eukaryotic sources; high concentrations likely indicate a dinoflagellate origin.	Volkman (2003)
	If strongly depleted in ^{13}C indicative for methylotrophic bacteria (Methylococcaceae).	Summons *et al.* (1994a)
4-Methylergostane (**69b**); 4-methylstigmastane (**69c**)	Diverse eukaryotic sources; high concentrations likely indicate a dinoflagellate origin.	Volkman (2003)
Dinosterane (**70**)	In the Mesozoic and Cenozoic specific for dinoflagellates (with possible minor diatom contribution); in Paleozoic and Neoproterozoic samples probably derived from protodinoflagellates.	Moldowan and Talyzina (1998), Robinson *et al.* (1984), Volkman *et al.* (1993)

(Grantham *et al.*, 1980; Schoell *et al.*, 1992). The association between these compounds and evidence of sulfidic water columns is very strong. Furthermore, it is not easy to rationalize how these compounds could arise by diagenesis of known BHP precursors. Thus, their occurrence is an indicator that some, presently unidentified, bacteria specific to these environments are the ultimate source and we predict they will be found in due course.

The pentacyclic triterpenoid gammacerane (**60**) occurs in trace amounts in almost all bitumens and oils, but is often abundant in sediments that were deposited under a stratified water column, a condition often observed in lacustrine and hypersaline settings (Sinninghe Damsté *et al.*, 1995). The most likely diagenetic precursor of gammacerane (**60**) is tetrahymanol (ten Haven *et al.*, 1989). Tetrahymanol has multiple sources. It has been isolated from a fern (Zander *et al.*, 1969), a fungus (Kemp *et al.*, 1984) and the ubiquitous phototrophic purple nonsulfur bacterium *Rhodopseudomonas palustris* (Kleemann *et al.*, 1990). However, the most likely source for abundant tetrahymanol in sediments is bacterivorous ciliates (e.g., Harvey and McManus, 1991). Predatory ciliates can thrive under oxic and anoxic conditions and are known to graze microorganisms across the oxic–anoxic interface, where they might feed on phototrophic sulfur bacteria or methanotrophs. In some cases it was possible to reconstruct this diverse bacterial diet by measuring the carbon-isotopic composition of ciliate lipids. Gammacerane enriched in ^{13}C relative to most other lipids in the Miocene Gessos-solfifera Formation suggests that ciliates were partially feeding on green sulfur bacteria (Sinninghe Damsté *et al.*, 1995); and tetrahymanol, strongly depleted in ^{13}C, extracted from cold-seep sediments from Kazan mud volcano in the eastern Mediterranean Sea indicates that ciliates were probably grazing on methane metabolizing prokaryotes (Werne *et al.*, 2002).

Plant-derived triterpenoids are overtly abundant in sediments from the late Mesozoic onwards. These include oleanane (**61**), lupane (**62**), and taraxastane (**63**). There is a clear relationship between these compounds and triterpenoid precursors such as β-amyrin (**64**)

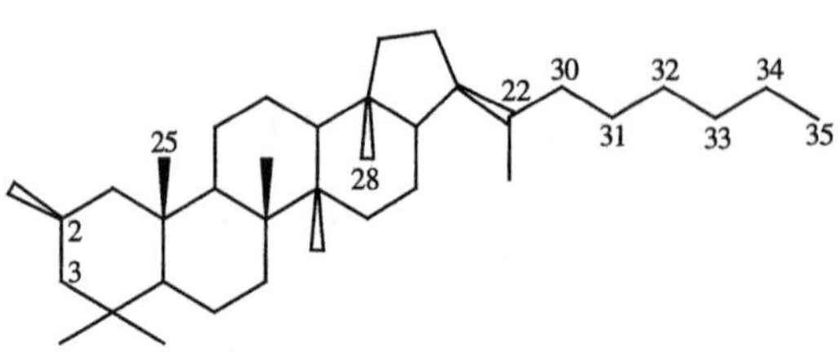

(**57**) 2α-Methylpentakishomohopane

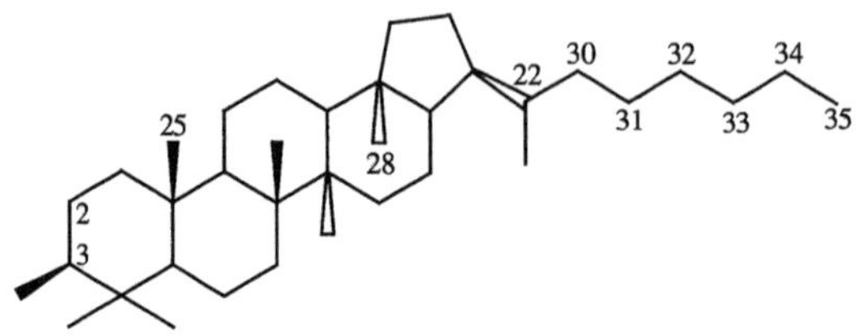

(**58**) 3β-Methylpentakishomohopane

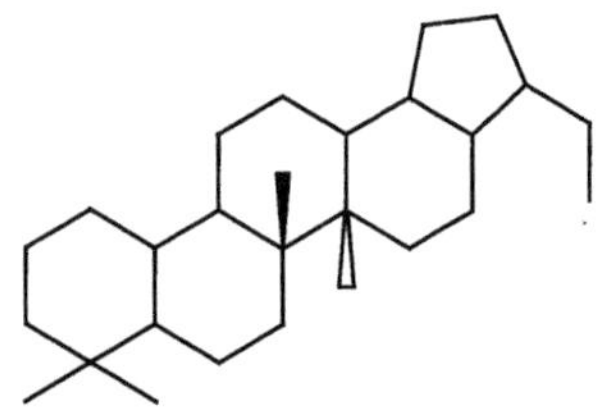

(**59**) 25,28,30-Trinorhopane (TNH)

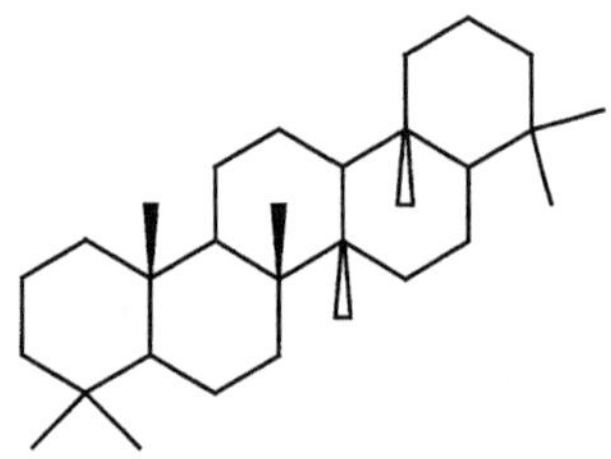

(**60**) Gammacerane

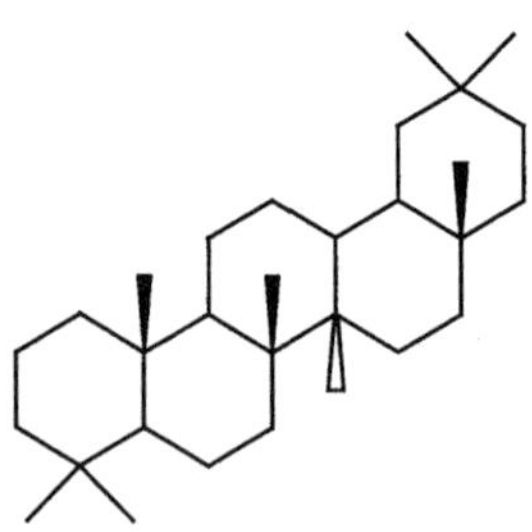

(**61**) Oleanane

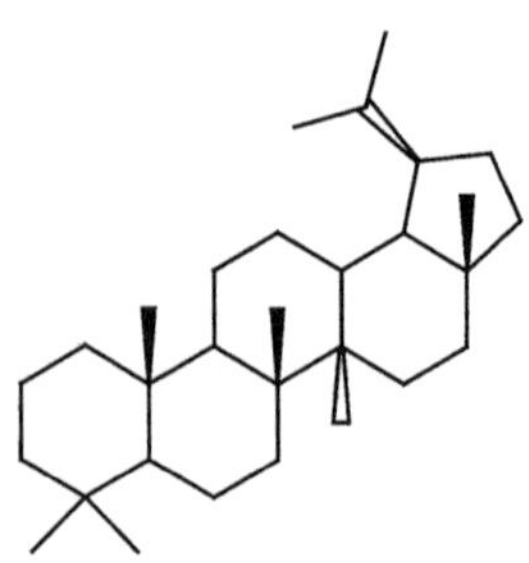

(**62**) Lupane

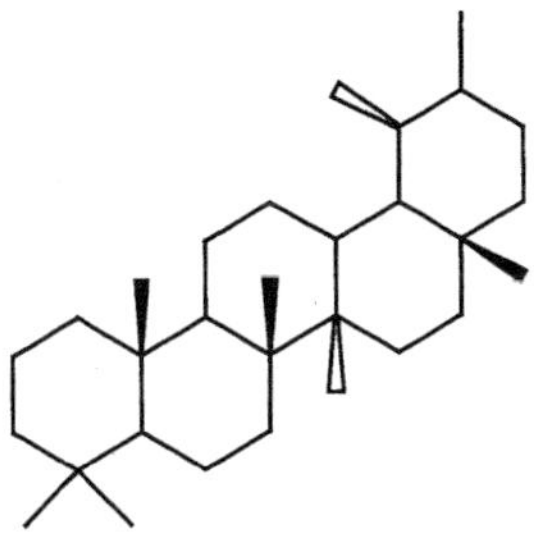

(63) Taraxastane

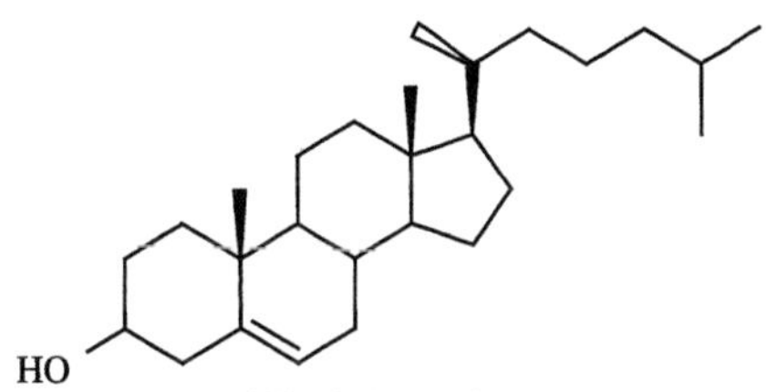

(65) Cholesterol

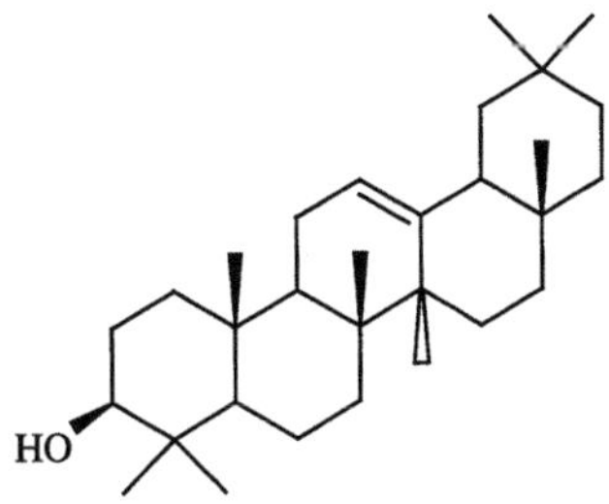

(64) β-Amyrin

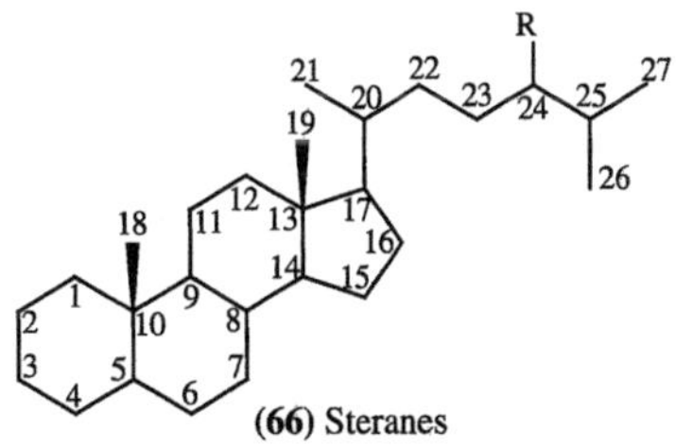

(66) Steranes

(a) R = H (cholestane); (b) R = Me (ergostane);
(c) R = Et (stigmastane); (d) R = *n*-Pr (24-*n*-proplycholestane);
(e) R = i-Pr (24-isopropylcholestane).

in angiosperms and, consequently, they are considered excellent biomarkers. The appearance of oleanane (**61**) and its increasing abundance in the Cenozoic is clearly related to the radiation of flowering plants (Moldowan *et al.*, 1994), although there are also numerous diagenetic controls on its occurrence and preservation (e.g., Murray *et al.*, 1997; ten Haven *et al.*, 1992). In fact, the sedimentary distributions of vascular plant triterpanes reflects not only the existence of precursor triterpenoids but the outcome of numerous, kinetically controlled diagenetic reactions. The ultimate preservation of just a few of the most thermodynamically stable isomers probably masks much more diverse contributions from the original biological precursors (Rullkötter *et al.*, 1994; ten Haven *et al.*, 1992).

8.03.5.11 Steroid Hydrocarbons

Sterols, such as cholesterol (**65**), are essential lipids in all eukaryotic organisms (Table 4). They are often quantitatively important components in membranes where they control membrane permeability and rigidity. Recent sediments contain an extensive variety of different functionalized sterols characterized by the position and number of double bonds, hydroxy groups, alkyl and various other substituents. Many are widespread among eukaryotes, but a considerable number is diagnostic for certain taxonomic groups (e.g., Volkman, 2003). Although double bonds and heteroatomic groups are commonly lost during diagenesis, it is still possible in mature sedimentary rocks to distinguish fossil steroids with different alkyl substituents (e.g., (**66a**)–(**66e**)). Furthermore, it has been established that zooplankton feeding on phytoplankton do not alter the stable carbon-isotopic compositions of lipids such as the sterols (Grice *et al.*, 1998b), therefore stable carbon-isotopic composition of steranes in sedimentary material is assumed to be unaltered. There is also evidence that phytoplankton grazing by zooplankton has only a minor impact on the composition of sterols present in fecal pellets (e.g., Méjanelle *et al.*, 2003). Steranes (**66**), diasteranes (**67**) and aromatic steroids (e.g., (**68**)) with 26 to 30 carbon atoms are abundant in most oils and bitumens from the Cenozoic to the Paleoproterozoic (Summons and Walter, 1990) and possibly the Archaean (Section 8.03.9.2).

During diagenesis and catagenesis the biological stereospecificity of sterols, particularly at C-5, C-14, C-17, and C-20, is usually lost (see structure (**66**) for numbering) and a diverse range of isomers is generated. The nomenclature for the structural and stereoisomers used in the literature, and also in this review, requires a short explanation. The term $\alpha\beta\beta$ sterane (sometimes just $\alpha\beta$) is commonly used as short-hand to denote steranes with the 5α(H), 14β(H), 17β(H) configuration, while $\alpha\alpha\alpha$ sterane refers to those with 5α(H), 14α(H), 17α(H) stereochemistry. The notation 14α(H) indicates that the hydrogen is located below the plane of the paper whereas in 14β(H) it is above the plane. In steranes, if no other carbon number is cited, S and R always refer to the stereochemistry at C-20. The prefix "nor", as for example, in 27-norcholestane, indicates that the molecule is formally derived from the parent structure by loss of the indicated carbon atom, i.e., in the above example, C-27 is removed from cholestane (**66a**). The term "desmethylsteranes" is sometimes used to distinguish steranes that do not possess an additional alkyl group at ring A, i.e., at carbon atoms C-1 to C-4. Diasteranes refers to hydrocarbons with the distinctive structure (**67**). Diasteranes have no direct biological precursors (Ourisson, 1994) and form by diagenetic rearrangement of sterols or sterenes (Sieskind *et al.*, 1979). The rearrangement is probably catalyzed by clay minerals, regularly leading to elevated concentrations of diasteranes in petroleum derived from clay-rich source rocks (van Kaam-Peters *et al.*, 1998) (Section 8.03.7.4). Finally, monoaromatic and triaromatic steroids (**68**) form either by diagenetic alteration of unsaturated and polyunsaturated steroids or by dehydrogenation of steranes during catagenesis (de Leeuw and Baas, 1986; Moldowan and Fago, 1986).

Desmethylsteranes with 26–30 carbon atoms have a large number of different sources. C_{26} steranes are ubiquitous in sedimentary rocks, although usually in relatively low concentrations. Moldowan *et al.* (1991b) have identified three series of C_{26} steranes. The 21- and 27-norcholestanes have apparently no direct biological precursors and are probably degradation products of steroids with higher carbon numbers. In contrast, the third series, 24-norcholestanes, probably has a direct biological source as corresponding sterols are commonly found in recent marine sediments. Circumstantial evidence points to diatoms, or at least to organisms or diagenetic processes associated with diatom blooms (Holba *et al.*, 1998a). A marine algal origin for these sterols is corroborated by the compound-specific radiocarbon ages of C_{26}–C_{29} sterols in shallow marine sediments (Pearson *et al.*, 2000, 2001). The abundance of 24-norcholestanes relative to 27-norcholestanes in crude oils increases considerably from the Jurassic to the Cretaceous and again in the Tertiary, a distribution that appears to coincide with diatom radiation and deposition of major diatomaceous sediments (Holba *et al.*, 1998b). Therefore, the abundances of 24-norcholestanes relative to the more common 27-nor isomers is considered to be an age-diagnostic marker for post-Jurassic oils and bitumens.

Desmethylsteranes with 27–29 carbon atoms are the most abundant steranes and occur in virtually all bitumens and oils that are not overmature. Biological precursors of cholestane (**66a**) (C_{27}) are common in animals and red algae (Rhodophyceae), while precursors of ergostane (C_{28}) (**66b**) are frequently found in yeast and fungi, diatoms (Bacillariophyceae), and several other classes of microalgae (Volkman, 2003). Sterols with the stigmastane skeleton (C_{29}) (**66c**) typically occur in higher plants (Volkman, 1986), but are also the major sterols in many microalgae, such as several freshwater eustigmatophytes and chrysophytes, and green algae of the class Chlorophyceae. Unfortunately, the C_{27} to C_{29} desmethylsteranes are not characteristic for any specific taxon, because the precursors are widely distributed in the domain Eukarya. Even related species within the same class may contain major sterols with different carbon numbers or even mixtures of all three carbon skeletons (Volkman, 1986, 2003, 1980).

Highly specific, on the other hand, are the C_{30} desmethylsteranes (**66d**) and (**66e**). 24-*n*-propylcholestane (**66d**) is even regarded as one of the most specific indicators for marine conditions (Moldowan *et al.*, 1985). Its potential biological precursors have only been detected in five marine algae of the class Pelagophyceae. These include the "brown tide" algae *Aureoumbra* (Giner and Li, 2000; Giner *et al.*, 2001) and *Aureococcus* (Giner and Boyer, 1998) and three species of the order Sarcinochrysidales (Moldowan *et al.*, 1990; Raederstorff and Rohmer, 1984) (Sarcinochrysidales were previously grouped with the class Chrysophyceae but were reclassified into the new class Pelagophyceae (Saunders *et al.*, 1997)). Sterols with the 24-isopropylcholestane skeleton (**66e**) are only abundant in extant demosponges. Therefore, 24-isopropylcholestane (**66e**) in sedimentary rocks is generally attributed to the contribution of sponges (McCaffrey *et al.*, 1994b). The ratio 24-isopropylcholestane/24-*n*-propylcholestane is high in the terminal Proterozoic to Ordovician but low in all following periods. This distribution might reflect the radiation of early sponges or sponge-related organisms that were the dominant reef builders during this time (McCaffrey *et al.*, 1994b).

Steranes ((**66a**)–(**66c**)) with alkyl substituents at C-2 or C-3 are ubiquitous in oils and bitumens of all ages (Summons and Capon, 1988). 2- and 3-methylsteranes are usually most abundant, but many oils also contain alkyl substituents at C-3 with up to seven and possibly more than ten carbon atoms (Dahl *et al.*, 1995). Biological steroids with an alkyl substituent in 2- or 3- position have not been observed in extant organisms, and a direct biological source seems unlikely. Instead, 2- and 3-alkylsteranes probably form by addition of a substituent to diagenetically-formed Δ^2-sterenes, possibly mediated by heterotrophic organisms (Summons and Capon, 1991). It is possible that pentose and hexose sugars are important reactants in this process, as substituents with five and six carbon atoms are particularly abundant in some samples (Dahl *et al.*, 1992; Schouten *et al.*, 1998b). Moreover, by desulfurization of the polar fraction of oils it was possible to show that the diagenetic precursors of these alkyl substituents originally carried multiple functionalities, consistent with a sugar origin (Dahl *et al.*, 1992). The analysis of the sedimentary processes that lead to the formation of steroids functionalized at C-2 or C-3 could, in principle, lead to the discovery of heterotrophic organisms that mediate the reaction. In this case, 2-and 3-alkylsteranes might eventually gain biomarker status.

A third series of alkylsteranes (**69**) common in bitumens and oils carries a methyl group at C-4. Sterols with the corresponding carbon structure are ubiquitous in eukaryotic organisms because 4-methylsterols and 4,4-dimethylsterols (e.g., lanosterol and cycloartenol) are intermediates in the biosynthesis of all other sterols (Volkman, 2003). However, the usually low concentration of these reaction intermediates suggests that their contribution to sedimentary organic matter is not significant (Volkman *et al.*, 1990). The most important source for sedimentary 4-methylsteranes (**69**) appear to be dinoflagellates. Dinoflagellates contain relatively high concentrations of sterols with the 4-methyl cholestane (**69a**), 4-methylergostane (**69b**) and 4-methylstigmastane (**69c**) skeletons (e.g., Piretti *et al.*, 1997; Robinson *et al.*, 1984; Volkman *et al.*, 1999). Although dinoflagellates are probably the only significant origin of 4-methylsteranes (**69**) in the majority of sedimentary rocks, multiple other potential sources are known (Volkman, 2003). Sterols with the 4-methylergostane (**69b**) and 4-methylstigmastane (**69c**) skeletons have been isolated from a slime mold (Nes *et al.*, 1990) and further potential precursors for 4-methylstigmastanes (**69c**) occur in the Pavlovales order of haptophyte algae (Volkman *et al.*, 1990). Methylotrophic bacteria of the family Methylococcaceae biosynthesize sterols with the (**69a**) structure (Bird *et al.*, 1971; Schouten *et al.*, 2000), and potential (**69**)-precursors were also detected in red algae (Beastall *et al.*, 1974), higher plants (Menounos *et al.*, 1986; Yano *et al.*, 1992) and fungi (Méjanelle *et al.*, 2000). Therefore, low relative concentrations of regular 4-methylsteranes (**69**) are not specific for any particular taxon, but high concentrations likely indicate biomarker contribution from dinoflagellates.

A distinct group of 4-methylsteranes, the dinosteranes (4α,23,24-trimethylcholestanes (**70**)), possess a unique side-chain alkylation pattern with an additional methyl group at C-23. Dinosteranes are regarded as very sound biomarkers for dinoflagellates (Robinson *et al.*, 1984; Summons *et al.*, 1987). Their biological source, dinosterol and related compounds, (Robinson *et al.*, 1984) are the most abundant sterols in the majority of dinoflagellates species (Volkman, 2003, and references therein). The only other organism that is known to contain sterols with the dinosterane skeleton is a single diatom species (Nichols *et al.*, 1990; Volkman *et al.*, 1993). Although dinosteranes (**70**) and triaromatic dinosteroids

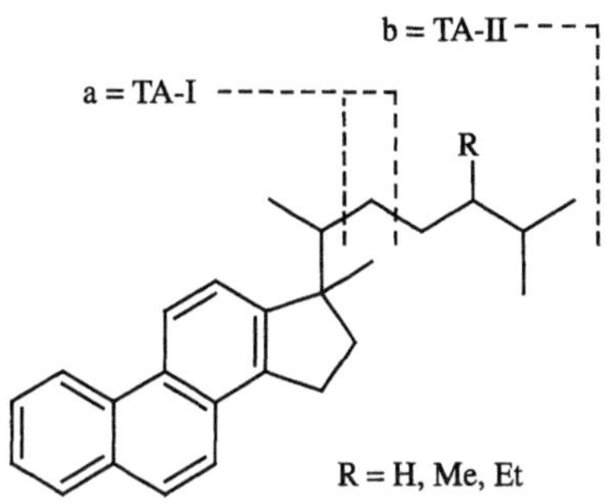

(**68**) (a) C_{20}–C_{21} Triaromatic steroids (TA-I)
(b) C_{26}–C_{28} Triaromatic steroids (TA-II)

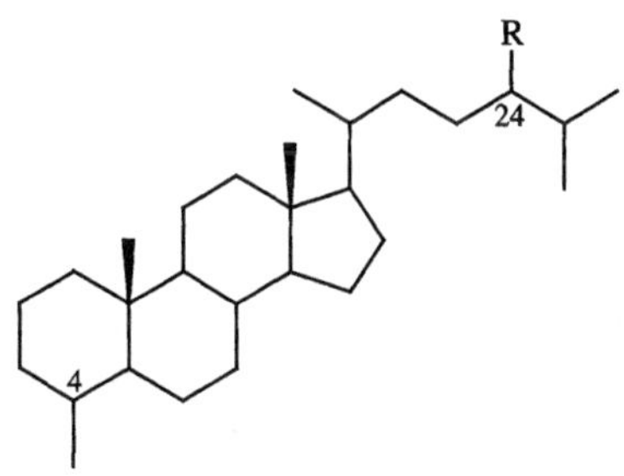

(**69**) 4-Methylsteranes

(**a**) R = H (4-methylcholestane);
(**b**) R = Me (4-methylergostane);
(**c**) R = Et (4-methylstigmastane)

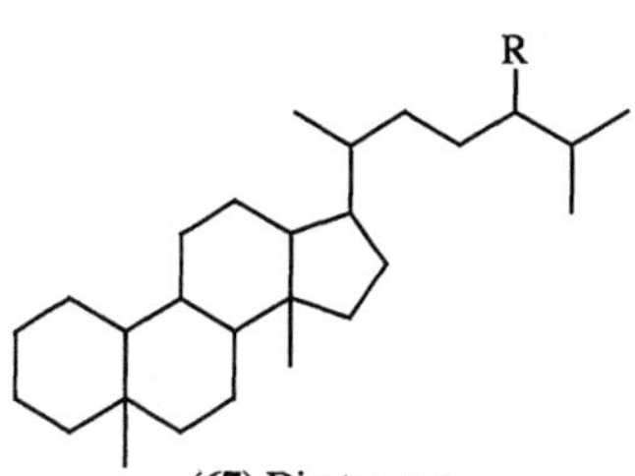

(**67**) Diasteranes

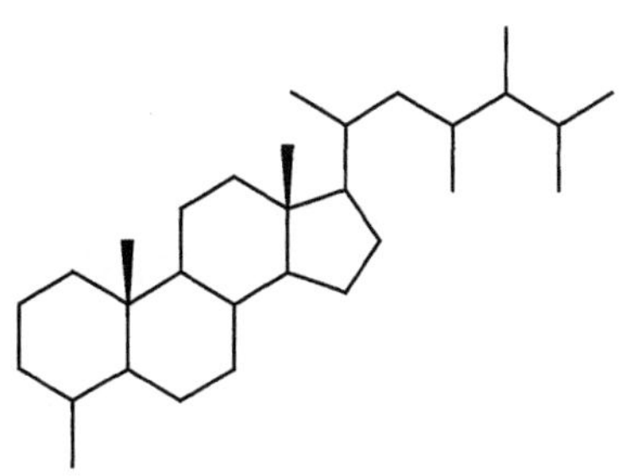

(**70**) Dinosterane

occur in Precambrian and Paleozoic oils and bitumens, they are rare. However, they become abundant in the Mesozoic to Cenozoic, possibly reflecting the first appearance of recognizable dinoflagellates in the fossil record (Moldowan *et al.*, 1996; Moldowan and Talyzina, 1998; Summons *et al.*, 1992, 1987). There is some good evidence that dinosteroids in many Paleozoic and Precambrian rocks might be the product of ancestral dinoflagellates (protodinoflagellates) (Moldowan *et al.*, 1996; Talyzina *et al.*, 2000).

The oldest known, probably syngenetic sterane biomarkers were detected in 2.7 Ga old rocks from the Hamersley Basin in Western Australia (Brocks *et al.*, 2003b; Brocks *et al.*, 1999) (Section 8.03.9.2). These ancient biomarkers include the full range of isomers of C_{26} to C_{30} desmethyl steranes, diasteranes, ring-A methylated steranes, and aromatic steroids also found in other rocks of Precambrian age. It was argued that the wide structural range of steranes in the Archean rocks, their relative abundance like those of younger bitumens, and in particular the presence of steranes methylated at C-24 ((**66b**)–(**66d**)) is convincing evidence for the existence of ancestral eukaryotes 2.7 Ga. This conclusion was challenged by Cavalier-Smith (2002) who rejected steranes as biomarkers for Eukarya pointing to the discovery of sterols in several lineages of Bacteria. However, the presence of sterols in bacteria was critically reviewed by Volkman (2003) and Brocks *et al.* (2003b). The only known bacteria with the unequivocal capacity for *de novo* sterol biosynthesis appear to be the methylotrophic bacteria *Methylococcus* and *Methylosphaera* (both Methylococcaceae) (Bird *et al.*, 1971; Schouten *et al.*, 2000), and several species of soil bacteria of the order Myxococcales, for example, *Nannocystis exedens* (Bode *et al.*, 2003; Kohl *et al.*, 1983). However, the Methylococcaceae synthesize exclusively 4-methyl and 4,4-dimethyl sterols with an uncommon unsaturation pattern and do not have the capacity to alkylate the sterol side chain at C-24. Some Myxococcales appear to generate C_{27}-cholesteroids, but they also do not have the biosynthetic capacity to alkylate the side-chain. Sterols have also been detected repeatedly in cyanobacterial cultures, although only in trace amounts. It appears now likely that these low quantities were introduced by eukaryotic culture contamination, probably fungi (Summons *et al.*, 2001). Moreover, the complete DNA sequence data of several cyanobacterial lineages are available now, and they do not indicate that cyanobacteria possess the genes required for full sterol biosynthesis (Volkman, 2003). The same criticism (Volkman, 2003) applies to sterols allegedly biosynthesized by mycobacteria.

In conclusion, sterol biosynthesis in Bacteria is probably limited to a small number of taxa that either have an incomplete sterol biosynthetic pathway or lack the capacity to alkylate the side chain. Therefore, steranes ((**66b**)–(**66e**) and (**69b**) and (**69c**)) in bitumens and oils can be reliably attributed to the activity of eukaryotic organisms (contra Cavalier-Smith, 2002). Additionally, some steranes with a diagnostic alkylation pattern ((**66d**)–(**66e**), and (**70**)) have taxonomic value below domain level.

8.03.6 RECONSTRUCTION OF ANCIENT BIOSPHERES: BIOMARKERS FOR THE THREE DOMAINS OF LIFE

8.03.6.1 Bacteria

8.03.6.1.1 Hopanoids as biomarkers for bacteria

The first extensive survey of biohopanoids in bacteria (Rohmer *et al.*, 1984) indicated that biosynthesis of this important class of biomarkers was the province of aerobic bacteria. Subsequent research, using new screening approaches such as liquid chromatography-mass spectrometry (LC-MS) to measure intact polar lipid structures, has verified their widespread occurrence in cultured aerobic bacteria and environmental samples (e.g., Farrimond *et al.*, 1998; Talbot *et al.*, 2001). It also showed that the isoprenoid building block of the hopanoid skeleton was produced by a biosynthetic pathway, the methylerythritol phosphate "MEP" pathway, new to science (Rohmer *et al.*, 1993) as well as other distinctive biochemistries (Rohmer, 1993). These discoveries illustrate the extent to which specific aspects of lipid biosynthetic pathways can also function as a biomarker. Since it would be an impossible task to measure the hopanoid contents of all bacteria growing under the full diversity of natural situations, we have to look to other methods to extend our knowledge of these biomarkers. One of the most promising approaches flows from studies of the DNA that codes for enzymes of key biosynthetic pathways and making use of the genomes of cultured organisms and sequences of DNA cloned from natural environmental samples. Although triterpenoids with a hopane skeleton occur in some plants, hopanoids with an extended side chain (i.e., C_{35} bacteriohopanes (**55**)) have only ever been found in the Bacteria (Section 8.03.5.10).

8.03.6.1.2 Cyanobacteria

Many, but not all, cyanobacteria biosynthesize bacteriohopanepolyol (**56**) (BHP) (e.g., Rohmer *et al.*, 1984). As discussed above (Section 8.03.5.10), cyanobacterial hopanoids have a number of distinctive attributes such as specific polar side-chain groups and sometimes an additional methyl substituent at position 2 of the hopane skeleton (Bisseret *et al.*, 1985) that makes them readily distinguished from other hopanoids (e.g., Summons and Jahnke, 1990). It is these 2-methylhopanoids (**57**) that are recognized in ancient sediments and oils as being largely of cyanobacterial origin (Summons *et al.*, 1999). A survey of cultured cyanobacteria indicates that biosynthesis of both BHP and 2-Me-BHP is widely and evenly distributed through cyanobacterial phylogeny (L. Jahnke, personal communication). However, compared to their freshwater counterparts, cyanobacteria from saline and hypersaline environments are poorly studied in this respect and this is an obvious target for further research.

Monomethyl and dimethylalkanes in the range $C_{16}-C_{20}$ are prominent in many cultured cyanobacteria as well as most cyanobacterial mat communities that have been studied (Section 8.03.5.3). No specific physiological role has been assigned to these hydrocarbons. Because they have probably multiple origins in ancient sediments and petroleum, these monomethyl and dimethylalkanes alone probably have limited chemotaxonomic specificity. However, they may be very useful in multivariate approaches for linking isotopic and molecular-structure data for a less ambiguous identification of sedimentary cyanobacterial lipids.

8.03.6.1.3 Methanotrophs, methylotrophs, and acetic acid bacteria

These are further groups of aerobic bacteria that produce distinctive hopanoids in abundance. In this case the distinctive features are an additional methyl group at C-3 of the hopane skeleton or the degree of functionality of the polar side chain (Farrimond *et al.*, 2000; Zundel and Rohmer, 1985b) and these hopanoids are also easily distinguished from other series on the basis of their GC-MS or LC-MS behavior (e.g., Summons and Jahnke, 1992; Talbot *et al.*, 2001). In the case of hopanoids from methanotrophic bacteria, an additional signature for their physiology can be a depletion in ^{13}C content compared to co-occurring compounds (Jahnke *et al.*, 1999; Summons *et al.*, 1994a). This isotopic characteristic is preserved along with the diagnostic carbon skeleton in sedimentary hydrocarbons (**58**) from communities supported by methane oxidation (e.g., Burhan *et al.*, 2002). Some methylotrophic bacteria are also unusual in having the capacity to simultaneously biosynthesize hopanoids along with 4-methyl and 4,4-dimethylsterols (Section 8.03.5.11) with both groups of compounds recording comparable isotopic depletion (Ourisson *et al.*, 1987; Summons and Capon, 1988).

8.03.6.1.4 Phototrophic sulfur bacteria

Anoxygenic phototrophic bacteria are a taxonomically very heterogeneous group. Based on phenotypic criteria they are divided into heliobacteria, purple nonsulfur bacteria, green filamentous bacteria (Chloroflexaceae), green sulfur bacteria (Chlorobiaceae), and purple sulfur bacteria (Chromatiaceae and Ectothiorhodospiraceae; Imhoff, 1995). Among these groups, diagnostic and geologically stable hydrocarbon biomarkers are known for the Chromatiaceae and Chlorobiaceae (Sections 8.03.5.6.1 and 8.03.5.7). As purple and green sulfur bacteria are highly specialized organisms, these biomarkers provide important paleoenvironmental tools. To form blooms, purple and green sulfur bacteria require reduced sulfur species and light. They represent the only known indicators for euxinic conditions in the photic zone of ancient lacustrine and marine environments.

The Chlorobiaceae form a monophyletic group, separated from other phototrophs (Figure 1). Brown pigmented strains contain bacteriochlorophyll *e* (**48**) (BChl *e*) and the major specific carotenoids isorenieratene (**31**) and β-isorenieratene (**29**). Green strains obtain their distinctive color from BChl *c* (**46**) or *d* (**47**) (Section 8.03.5.7) and the diagnostic carotenoids chlorobactene (**27**) and hydroxychlorobactene (Imhoff, 1995) (Section 8.03.5.6.1). Chlorobiaceae are strictly anaerobic, obligate phototrophs that utilize only photosystem I (PS I). In contrast to cyanobacteria that have the capacity to oxidize water, green sulfur bacteria require sulfide or other reduced sulfur species as the electron donor. CO_2 is the sole carbon source and is assimilated via the reductive or reversed TCA cycle. This mode of carbon fixation gives biomarkers of Chlorobiaceae a diagnostic isotopic fingerprint. Isorenieratane (**32**), chlorobactane (**28**), and other biomarkers are often enriched in ^{13}C by ~10‰ relative to organic matter from co-occurring oxygenic phototrophs.

According to 16S rRNA (ribosomal RNA) analyses, purple sulfur bacteria form a well separated group in the γ-subgroup of Proteobacteria. Several genera of the family Chromatiaceae contain the taxonomically diagnostic monoaromatic carotenoid okenone (**25**). Although okenone has been extracted from recent sediments (Schaeffer *et al.*, 1997), the equivalent fossil hydrocarbon okenane (**26**) has surprisingly not been reported from sedimentary rocks. However, if okenane is discovered it should be possible to establish its specific biological origin by measuring the carbon-isotopic composition. CO_2 fixation in Chromatiaceae, as in cyanobacteria and algae, follows the Calvin-Benson cycle. However, the CO_2 utilized by the Chromatiaceae originates in the anoxic zone of the water column and usually carries the distinct carbon isotopic depletion of remineralized organic matter. Accordingly, the biomass of Chromatiaceae should be depleted in ^{13}C relative to organic matter derived from oxygenic phototrophs. Moreover, biomarkers of Chlorobiaceae, using the same source of ^{13}C depleted CO_2 but following the reversed TCA cycle, should be strongly enriched in ^{13}C relative to biomarkers of Chromatiaceae. Accordingly, it has been observed that okenone from recent sediments is depleted in ^{13}C by ~20‰ relative to isorenieratene derived from Chlorobiaceae (Schaeffer *et al.*, 1997). A similar depletion is predicted for okenane (**26**) relative to isorenieratane (**32**) extracted from sedimentary rocks.

8.03.6.2 Archaea

The Archaea is often considered Life's "extremist" domain because of their overwhelming presence in volcanic vent systems, strongly acidic and alkaline springs, evaporitic settings, and in deep-subsurface sediments (e.g., Rothschild and Mancinelli, 2001). However, recent research is showing that archaeans are also quite abundant in the picoplankton of the open ocean (e.g., DeLong, 1992; DeLong *et al.*, 1998). The two broad metabolic themes of archaea, both of which rely on molecular hydrogen as an energy source, make them an important driving force in biogeochemical cycles. In the Euryarchaeota, CO_2 is the predominant electron acceptor and methane the predominant product. In the Crenarchaeota, there is a strong bias toward the oxidation of molecular hydrogen using sulfur compounds as electron acceptors. Given the importance of these processes in biogeochemical cycling it is not surprising, therefore, that biomarkers from archaea are widely and abundantly present in environmental samples (e.g., Sinninghe Damsté *et al.*, 2002a) as well as bitumen and petroleum (e.g., Moldowan and Seifert, 1979).

8.03.6.2.1 Methanogens

Methanogen lipids have been intensively studied and characterized due to their structures being one of the most remarkable features that distinguish the Archaea from all other organisms (Woese *et al.*, 1990). The polar lipids of methanogens comprise both di- and tetra-ethers of glycerol and isoprenoid alcohols with most compounds being based on the core lipids archaeol (**12**) or caldarchaeol (**13**). Minor core lipids are *sn*-2- and *sn*-3-hydroxyarchaeol and macrocyclic archaeol (Koga *et al.*, 1993). As discussed earlier (Section 8.03.5.5), nonpolar lipids are also distinctive with many methanogens having high contents of hydrocarbons including the characteristic irregularly branched compound PMI (**18**) and structurally related analogs (e.g., Risatti *et al.*, 1984; Schouten *et al.*, 1997; Tornabene *et al.*, 1979).

8.03.6.2.2 Biomarkers and ecology at marine methane seeps

The advent of compound-specific isotope analysis (CSIA) has forever altered the way geochemists approach the analysis of sediments. One example of this is the capacity to screen environmental samples for the carbon-isotopic signatures of a process as opposed to the traditional mode of analyzing for the diagnostic molecular structures. CSIA of lipids from near-surface sediments of the Kattegat Strait draining the Baltic Sea showed extreme ^{13}C-depletion of a chromatographic peak, normally attributed to phytane, within a zone corresponding to high rates of sulfate reduction and concomitant methane oxidation. Closer inspection revealed the localized occurrence of the hydrocarbon crocetane (**17**), hitherto rarely reported isomer of phytane (Bian *et al.*, 2001) and pointed to a relationship between organisms biosynthesizing crocetane and the oxidation of methane with sulfate as the terminal electron acceptor. This had been hypothesized on the basis of other geochemical indicators (Hoehler *et al.*, 1994). The introduction of gene surveys of small subunit ribosomal RNA (16S rRNA) in concert with data from lipid biomarkers and their individual isotopic compositions further demonstrated that the anaerobic oxidation of methane (AOM) was conducted by archaea in close association with sulfate-reducing bacteria (Hinrichs *et al.*, 1999) and that the process was characterized by distinctive assemblages of lipids such as *sn*-2 hydroxyarchaeol, crocetane (**17**), PMI (**18**) from the Archaea and nonisoprenoid branched fatty acids and ether lipids from the sulfate-reducing partners (Elvert *et al.*, 1999; Hinrichs *et al.*, 1999, 2000; Pancost *et al.*, 2000; Thiel *et al.*, 1999). Recent lipid work suggests thermophilic archaea can mediate anaerobic oxidation of methane in environments with steep geothermal gradients (Schouten *et al.*, 2003).

The above studies established a precedent for the combined use of microbiological, genomic, and isotopic methods to study important biogeochemical processes. Furthering the revolution of culture-independent methods for studying these processes on natural samples has been the visualization of the active microbes through fluorescence *in situ* hybridization (Boetius *et al.*, 2000), the determination of sites and patterns of isotopic fractionation by active microbial consortia by ion-microprobe analysis and, overall, the direct linking of microorganisms from nature with biogeochemical cycles (Orphan *et al.*, 2001). Methane venting is now recognized as widespread and sometimes spectacular process at the sediment–water interface (e.g., Michaelis *et al.*, 2002) with likely implications for rapid climate change (Hinrichs *et al.*, 2003) and with directly associated biomarkers that facilitate studies of its occurrence in the geological past (e.g., Greenwood and Summons, 2003; Thiel *et al.*, 1999).

8.03.6.2.3 Halobacteria

Halophiles are chemo-organotrophic Euryarchaeota that are often the predominant organisms in salt lakes, pools of evaporating seawater, solar salterns and other hypersaline environments with salt concentrations as high as halite saturation (e.g., Oren, 2002). Their lipids closely resemble those of methanogens with the principal difference being the occasional substitution of regular C_{25} isoprenoid chains in place of C_{20} in the archaeol lipid cores. A unique lipid abundant in Halobacteria is the C_{50} carotenoid bacterioruberin (**41**). It is likely that under diagenetic conditions bacterioruberin is reduced to perhydrobacterioruberin (**42**), a biomarker potentially diagnostic for Halobacteria but that has, so far, eluded discovery in sedimentary rocks (Section 8.03.5.6.2).

8.03.6.2.4 Marine Crenarchaeota

An amazing example of a massive occurrence of archaea was reported by Kuypers *et al.* (2001). They discovered that black shales from the Mid-Cretaceous Oceanic Anoxic Event OAE1b contained an unusual assemblage of cyclic and acyclic isoprenoids including a lipid (**14**) diagnostic for nonthermophilic Crenarchaeota. Kuypers *et al.* (2001) calculated that up to 80% of the sedimentary organic matter deposited during this event was derived from nonthermophilic Crenarchaeota. The archaeal biomarkers were enriched in ^{13}C by more than 10‰ relative to algal lipids. This isotopically heavy biomass is not only responsible for the positive carbon-isotopic excursion of organic matter during OEB1b, but also suggests that the marine Crenarchaeota did not live heterotrophically but followed a chemoautotrophic metabolism (e.g., Hoefs *et al.*, 1997; Kuypers *et al.*, 2001; Pearson *et al.*, 2001; Sinninghe Damsté *et al.*, 2002a). Further work on elucidating the precise structures of lipids from Crenarchaeota is underway (e.g., Sinninghe Damsté *et al.*, 2002b) and will likely result in profound new insights into the paleobiology and biogeochemistry of this group of Archaea.

8.03.6.3 Eukarya

Eukaryotes are an ancient clade. The oldest acritarchs that are clearly eukaryotic come from shales of the 1.49–1.43 Ga Roper Group, McArthur Basin, Australia (Javaux *et al.*, 2001) and the oldest body fossils believed to have eukaryotic affinity have been found in rocks ~1.8–1.9 Gyr old (Figure 5) (Hofmann and Chen, 1981; Zhang, 1986; Han and Runnegar, 1992; see Schneider *et al.*, 2002 for an up-to-date age of the Negaunee Iron-Formation). Biomarker evidence for eukaryotes comes from steranes with diagnostic alkylation patterns in the side chain ((**66b**)–(**66d**)) extracted from ~1.64 Gyr old rocks of the Barney Creek Formation (Summons *et al.*, 1988b) and possibly the ~2.7 Ga Fortescue Group (Brocks *et al.*, 1999) (Section 8.03.9.3), both in Australia (Figure 5).

Extant eukaryotes contain thousands of natural products that are only found in members of their domain. Although much information in these molecules is lost in the diagenetic transition to hydrocarbon fossils, many retain a structure based on a specific carbon skeleton (Figure 1). Diagnostic hydrocarbon skeletons include ergostane (**66b**) and stigmastane (**66c**) for Eukarya as a whole, 24-*n*-propylcholestane (**66d**) for pelagophyte algae, dinosteranes (**70**) for dinoflagellates and, possibly, a few diatoms (Section 8.03.5.11), and botryococcane (**20**) for the chlorophyte *Botryococcus* (Section 8.03.5.5).

The vast majority of biomarkers that can be traced to distinct branches in the eukaryotic tree belong to higher plants, for example, oleanane (**61**) (e.g., Moldowan *et al.*, 1994; Murray *et al.*, 1997), taraxastane (**63**) (e.g., Perkins *et al.*, 1995), and bicadinane (**53**) (e.g., Cox *et al.*, 1986; van Aarssen *et al.*, 1992). Bicyclic and tricyclic diterpenoid compounds such as abietic acid are major components of conifer resins (e.g., Simoneit, 1977). These are the proposed biological precursors of sedimentary diterpane biomarkers retene, simonellite, phyllocladane (**51**), kaurane, bayerane, and many others (e.g., Alexander *et al.*, 1988, 1992, 1987; Simoneit, 1977; Noble *et al.*, 1985, 1986; Otto and Simoneit, 2001, 2002). For example, retene has been detected in high relative abundance in Tertiary carbonaceous shales and has been attributed to Podocarpaceae and Araucariaceae conifer resins (Villar *et al.*, 1988). The biomarker cadalene occurs widely in recent and ancient sediments (e.g., Noble *et al.*, 1991; Wang and Simoneit, 1990). Cadinenes and cadinols in plants, bryophytes, fungi, and extant and fossils plant resins (e.g., Grantham and Douglas, 1980; van Aarssen *et al.*, 1990) are the proposed precursors for cadalene (Simoneit *et al.*, 1986). Surprisingly, there appears to be only one

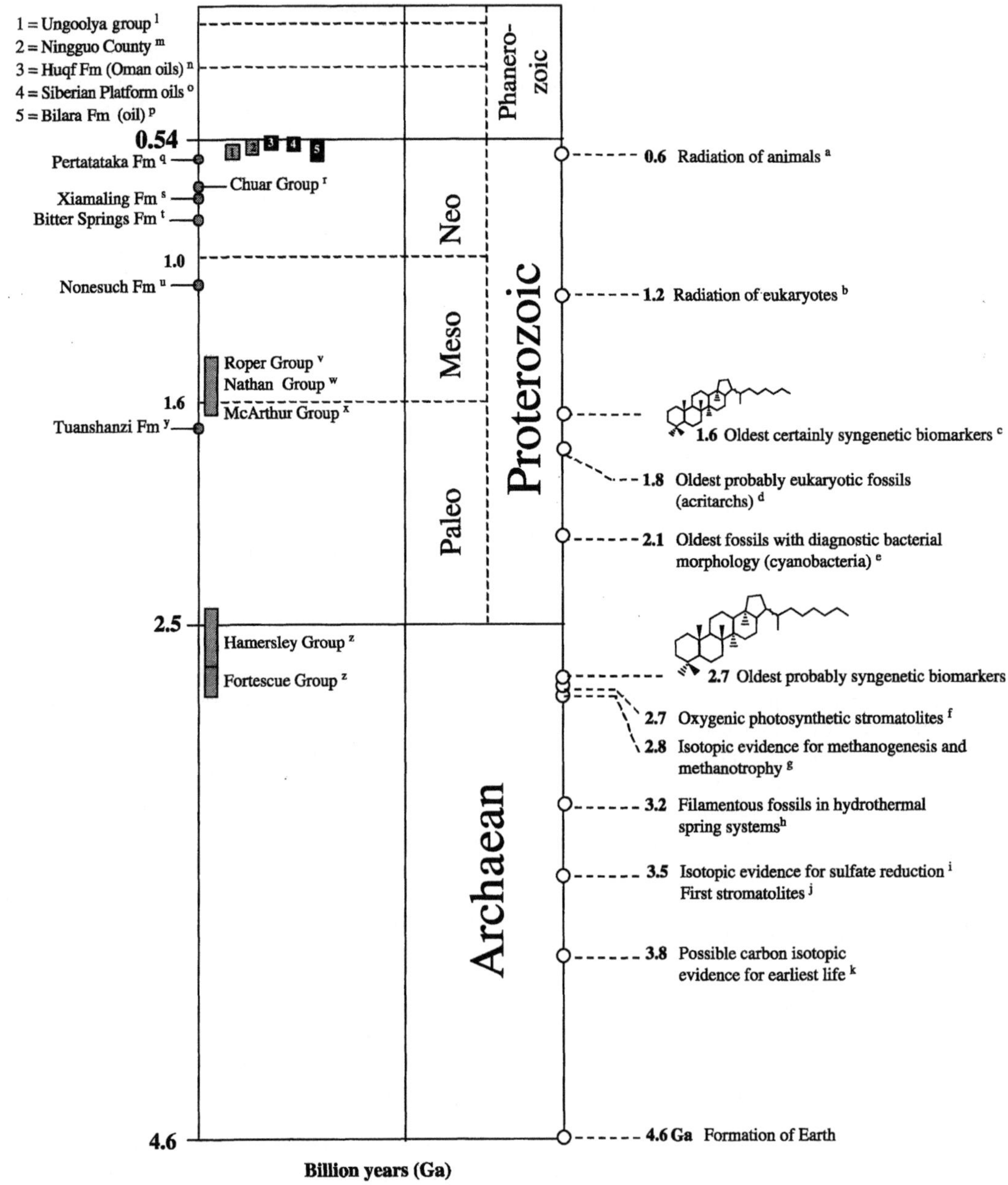

Figure 5 Geological timescale with important biological events, and observations of well-preserved Precambrian biomarkers (gray) and crude oil (black). (a) Xiao *et al.* (1998); (b) Knoll (1992); (c) Jackson *et al.* (1986); (d) Hofmann and Chen (1981); (e) Hofmann (1976); (f) Buick (1992); (g) Hayes (1983); (h) Rasmussen (2000); (i) Shen *et al.* (2001); (j) Buick *et al.* (1981), Walter *et al.* (1980); (k) Rosing (1999); (l) Arouri *et al.* (2000a,b); Logan *et al.* (1997)); (m) Jiang *et al.* (1995); (n) e.g., Grantham (1986); Klomp (1986); McCaffrey *et al.* (1994b); Summons *et al.* (1999); (o) e.g., Fowler and Douglas (1987); McCaffrey *et al.* (1994b); Summons *et al.* (1988b); (p) Peters *et al.* (1995); (q) Logan *et al.* (1999, 1997); Summons and Powell (1991); (r) Summons *et al.* (1988a, 1999); Höld *et al.* (1997); (s) Wang (1991); Wang and Simoneit (1995); (t) Logan *et al.* (1997); McCaffrey *et al.* (1994b); Summons *et al.* (1988a); Summons *et al.* (1999); Summons and Powell (1991); (u) Ho *et al.* (1990); Pratt *et al.* (1991); (v) Brocks *et al.* (2003c); Crick *et al.* (1988); George and Ahmed (2003); George and Jardine (1994); Summons *et al.* (1999, 1988b, 1994b); Taylor *et al.* (1994); (w) Summons *et al.* (1999); (x) Crick *et al.* (1988); Greenwood and Summons (2003); Logan *et al.* (2001); McCaffrey *et al.* (1994b); Summons *et al.* (1999, 1988b), Jackson *et al.* (1986); (y) Peng *et al.* (1998); (z) Brocks *et al.* (2003a,b,c,d, 1999); Arouri *et al.* (2000a,b).

hydrocarbon fossil in oils and bitumens that is clearly derived from an animal: 24-isopropylcholestane diagnostic for sponges (Section 8.03.5.11), although cholestane is likely to have significant contributions from the cholesterol of animals.

8.03.7 BIOMARKERS AS ENVIRONMENTAL INDICATORS

Organic matter can provide important clues for paleoenvironmental assessments (Table 5) (de Leeuw *et al.*, 1995). Because some biomarkers point to specific taxa, they can also act as indicators of specific habitats. Paleoenvironmental conditions that are often readily inferred from the presence and distribution patterns of biomarkers are marine (e.g., (**66d**)), terrestrial (e.g., (**61**)), and deltaic environments where plant and algal hydrocarbons are mixed or show stratigraphy-related fluctuations in abundance.

8.03.7.1 Marine versus Lacustrine Conditions

As discussed above (Section 8.03.5.11), 24-*n*-propylcholestane (**66d**) is considered an unambiguous indicator of marine depositional environments. Additionally, marine conditions can often be inferred from high abundances and the compositions of organo-sulfur compounds as the prevalence of sulfide in euxinic marine environments strongly affects the diagenetic pathways and preservation of many classes of lipids (e.g., Kohnen *et al.*, 1992, 1993, 1991a; Schouten *et al.*, 2001a; Wakeham *et al.*, 1995). Organosulfur compounds are usually less abundant in sediments that were deposited in freshwater. However, freshwater environments are often indicated by the presence of biomarkers of typical freshwater organisms such as *Botryococcus braunii*. Lacustrine conditions are often indicated by preponderances of algal steroids (e.g., Chen and Summons, 2001), biomarkers for aerobic methanotrophs (Collister *et al.*, 1992) and, very often, by the presence of certain C_{30} tetracyclic polyprenoid hydrocarbons (Holba *et al.*, 2003). The Cenozoic lacustrine basins of China provide numerous examples of biomarker patterns that are characteristic of nonmarine (freshwater and saline) depositional systems (e.g., Chen and Summons, 2001; Chen *et al.*, 1989; Li *et al.*, 2003; Philp *et al.*, 1992; Ping'an *et al.*, 1992).

Table 5 Biomarkers as environmental indicators.

Depositional environment	*Typical biomarker patterns*	*Reference example*
Marine	24-*n*-Propylcholestane (**66d**)	Moldowan *et al.* (1985)
Lacustrine	Botryococcane (**20**) and other biomarkers of *Botryococcus* (fresh to brackish water).	Metzger and Largeau (1999)
	Elevated concentrations of C_{30} tetracyclic polyprenoids (fresh to brackish water).	Holba *et al.* (2003)
Hypersaline	C_{21} to C_{25} regular isoprenoids **6** enriched in ^{13}C relative to biomarkers of phytoplanktonic origin.	Grice *et al.* (1998b)
	High gammacerane[a] (**60**).	Sinninghe Damsté *et al.* (1995)
Terrestrial organic matter input	Diverse biomarkers of higher plants	Section 8.03.6.3
Strongly anoxic conditions (water column anoxia?)	28,30-Dinorhopane; 25,28,30-trisnorhopane (**59**).	Peters and Moldowan (1993)
	Gammacerane[a] (**60**)	Sinninghe Damsté *et al.* (1995)
Photic zone euxinia[b]	Isorenieratane (**32**); 2,3,6- (**38**) and 2,3,4- (**39**) trimethylarylisoprenoids; chlorobactane (**28**); Me *i*-Bu maleimide (**49d**)	Grice *et al.* (1996a), Hartgers *et al.* (1993), Koopmans *et al.* (1996a), Summons and Powell (1987)
Carbonates and evaporites	Low diasterane (**67**)/sterane (**66**) ratios[a]	van Kaam-Peters *et al.* (1998)
	High 2α-methylhopane (**57**) concentrations[a]	Summons *et al.* (1999)
	High 30-norhopanes[a]	Subroto *et al.* (1991)

[a] Typical for, but not necessarily restricted to, this depositional environment. [b] This might include environments with an anoxic and sulfidic water column that persists into the photic zone, or microbial mats in very shallow water settings that become anoxic within millimeters below the sediment water interface.

8.03.7.2 Hypersaline Conditions

Halophiles are found in all three domains of life with a wide diversity of metabolisms such as aerobic heterotrophy and fermentation, sulfate reduction, denitrification, methanogenesis, and anoxygenic and oxygenic phototrophy. Many sediments deposited under hypersaline conditions contain abundant biomarkers probably derived from archaeal Halobacteria (Section 8.03.6.2.3). For instance, Miocene/Pliocene halite deposits from the Dead Sea Basin in Israel contain pristane, phytane and C_{21} to C_{25} regular isoprenoids as the dominant lipids of the hydrocarbon fraction (Grice *et al.*, 1998b). These isoprenoids are enriched in ^{13}C by up to 7‰ relative to biomarkers of presumed phytoplanktonic origin, consistent with Halobacteria as the dominant source. Hypersaline lakes and ponds often develop anoxic conditions if saline deep water is covered with water of lower density. Sedimentary rocks that were deposited under these conditions often contain high relative concentrations of gammacerane (**60**), a biomarker generally associated with water column stratification (Section 8.03.5.10) (Sinninghe Damsté *et al.*, 1995). However, as water column stratification occurs under other conditions as well, gammacerane is also often abundant in freshwater sediments (e.g., Grice *et al.*, 1998c).

8.03.7.3 Anoxic and Euxinic Conditions

Biomarker analysis is one of the best paleoenvironmental tools to identify anoxic and euxinic conditions in the water column. Biomarkers of phototrophic sulfur bacteria such as isorenieratane (**32**) (Section 8.03.5.6.1) and 3-isobutyl-4-methylmaleimide (**49d**) (Section 8.03.5.7) unambiguously indicate euxinic conditions within the photic zone of the water–sediment system. Other biomarkers that are often associated with sediments deposited beneath anoxic waters are 28,30-dinorhopane, 25,28,30-trisnorhopane (**59**) and gammacerane (**60**) (Section 8.03.5.10).

Nitrogen cycling in anoxic water columns and sediments is another prominent biogeochemical process that can have direct molecular and isotopic indicators. Sinninghe Damsté *et al.* (2002c) discovered that organisms (Planctomycetales) capable of oxidizing ammonia with nitrate (anammox) biosynthesize unprecedented glycerol ester and ether lipids with hydrocarbon chains comprising concatenated cyclobutane rings, or ladderanes (e.g., (**71**)).

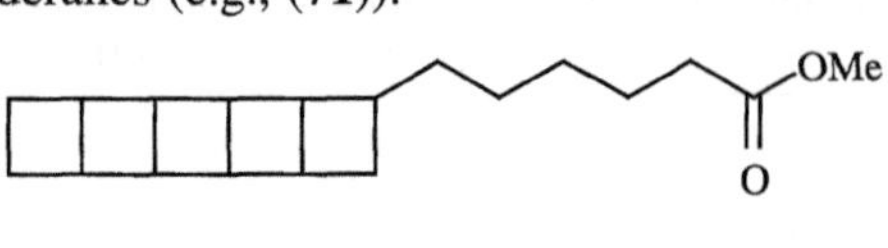

(71) A [5]-ladderane

A combination of water column nutrient profiles, fluorescently labeled RNA probes, the vertical distribution of specific "ladderane" membrane lipids, and experiments with ^{15}N labeled ammonium and nitrate demonstrate that anammox organisms are presently active in anaerobic oxidation of ammonia below the oxic zone of the Black Sea (Kuypers *et al.*, 2003). These observations suggest that the anammox pathway of nitrogen cycling may be widespread in suboxic environments in the modern marine realm. 16S rRNA sequences indicate that the Planctomycetales are a distinct and ancient lineage within the bacterial domain (Brochier and Philippe, 2002). Further, the anammox reaction may have been even more significant in times past when ocean-water columns were largely anaerobic (Anbar and Knoll, 2002; Canfield, 1998). If ladderane-like lipids evolve diagenetically to recognizable chemical fossils, we expect this would be prominently recorded in Proterozoic sediments.

8.03.7.4 Carbonates versus Clay-rich Sediments

Acid-catalyzed rearrangement reactions are promoted during diagenesis of organic matter adsorbed to clay particles (e.g., Rubinstein *et al.*, 1975). Accordingly, rearranged steranes (diasteranes) are relatively more abundant in clastic sediments than in carbonates (van Kaam-Peters *et al.*, 1998). Hopanoids appear to be similarly affected so that diahopanes and neohopanes are relatively more prominent in bitumens and oils derived from shales as opposed to carbonates (Peters and Moldowan, 1993). However, increasing thermal maturity is also a key factor in the conversion of biomarkers to their rearranged forms.

Carbonate-dominated sediments tend to be deposited in low-latitude environments and, therefore, biomarkers for organisms that preferentially colonize warm waters tend to be important signatures in these sediments. Cyanobacterial 2α-methylhopanes (**57**) (Summons *et al.*, 1999) and 30-norhopanes (Subroto *et al.*, 1991) are generally elevated in bitumens from carbonates and marls.

8.03.7.5 Paleotemperature and Paleolatitude Biomarkers

Paleotemperature might be reconstructed in ancient sediments through biomarker signals if cold-adapted and warm-adapted organisms produced distinctive lipids. The best-known example of such a signal is the long-chain ketones produced by haptophytes that carry patterns of unsaturation determined by sea-surface temperature (Brassell *et al.*, 1986).

Water temperature is one factor that can influence the concentration of dissolved CO_2 and, thereby, the isotopic fractionation encoded during photosynthetic carbon assimilation. This has been suggested as a means through which paleolatitude could be reconstructed from the carbon-isotopic composition of petroleum hydrocarbons sourced from rocks laid down during time intervals when significant pole to equator temperature gradients prevailed (Andrusevich *et al.*, 2001).

8.03.8 AGE DIAGNOSTIC BIOMARKERS

It is now well known that the hydrocarbon composition of petroleum has evolved over geological time reflecting a corresponding evolution in sedimentary organic matter and, hence, biology. Unusual oils with high abundances of branched alkanes appeared to be exclusively associated with "Infracambrian" source rocks of Siberia and Oman (e.g., Fowler and Douglas, 1987; Grantham *et al.*, 1988) while oils with strong odd carbon number predominances at n-C_{15}, n-C_{17}, and n-C_{19}, and extremely low abundances of acyclic isoprenoids are often found in Ordovician strata (e.g., Hoffmann *et al.*, 1987; Reed *et al.*, 1986). These are features of bulk hydrocarbon composition that denote an overwhelming input from a single organism such as *Gloeocapsomorpha prisca,* in the case of Ordovician oils and kukersite oil shales (Section 8.03.5.2). Further major changes in petroleum composition accompanied vascular plant radiations in the Late Paleozoic and again during the Cenozoic, with organic matter contributions from leaf waxes, resins, and other terpene-based biopolymers. The isotopic composition of marine organic carbon has changed over geological time (Hayes *et al.*, 1999) and there is a concomitant secular variation in the isotopic compositions of petroleum from marine source rocks (Andrusevich *et al.*, 1998).

More subtle changes occur in the distribution of hydrocarbons that reflect the radiation of specific taxa and their distinctive biochemicals. Algal steroids show particularly strong age-related trends and can be used in a forensic sense to constrain the age of organic sedimentary matter, including petroleum. Prime examples are the heightened occurrences of 24-isopropylcholestanes (**66e**) in Proterozoic and Early Paleozoic sediments (McCaffrey *et al.*, 1994a), dinosteroids (e.g., (**70**)) in the Mesozoic and Cenozoic (e.g., Moldowan and Talyzina, 1998; Summons *et al.*, 1992) and 24-norcholestanes (Holba *et al.*, 1998a) in the Cenozoic. Triterpenoids from angiosperms are another class of compounds that show very strong age-related patterns of occurrence (e.g., Moldowan *et al.*, 1994).

8.03.9 BIOMARKERS IN PRECAMBRIAN ROCKS

8.03.9.1 Biomarkers in the Proterozoic (0.54–2.5 Ga)

There are numerous sedimentary sequences in the Proterozoic that contain abundant and well-preserved organic matter. Characterization of this organic matter, and especially the establishment of its age, has provided a major challenge for geochemists, and much of this progress has been reviewed by Hayes *et al.* (1983), Summons and Walter (1990), and Brocks *et al.* (2003a). Organic matter in the form of distinctive, morphologically diverse, organic-walled microfossils abounds in otherwise organic-lean shales and carbonates (e.g., Butterfield *et al.*, 1988) and, as with other paleoflora, carry information about biota and environments. Rocks with the high contents of organic matter tend to have amorphous kerogen which is difficult to study by optical methods but may be amenable to pyrolysis and chemical degradation studies for paleoenvironmental and paleobiological reconstruction (e.g., Arouri *et al.*, 2000a, 1999). For this, identification and selection of sediments with very mild thermal histories and freedom from the damaging effects of ionizing radiation (e.g., Dahl *et al.*, 1988) is essential for making accurate assessments of these issues.

Studies of bitumens have been far more extensive than studies of kerogen due to the relative ease with which extractable and volatile hydrocarbons can be analyzed (e.g., Summons and Walter, 1990). The carbon skeletons found prominently include algal steroids, bacterial hopanoids, and archaeal polyisoprenoids.

Distinctive biomarker distribution patterns are common in the Neoproterozoic. Prime examples include organic-rich shales and marls within the Chuar Group, Grand Canyon, USA, (Summons *et al.*, 1988a), Rodda Beds, Bitter Springs and Pertatataka Formations of Central Australia (Hayes *et al.*, 1992), and the Terminal Proterozoic of Oman and the Siberian Platform (Figure 5) (e.g., Fowler and Douglas, 1987; Grantham, 1986; Klomp, 1986; Summons and Powell, 1992). It is in these sediments that one finds unprecedented predominances of a single-sterane homolog, either C_{27} (cholestanes (**66a**) or C_{29} (stigmastanes (**66c**), signals that might be related to the radiation and massive occurrence of specific algal clades. Similarly, 24-isopropylcholestane (**66e**) shows a unique predominance in the Neoproterozoic to Ordovician, and this is hypothesized to be a consequence of the radiation of sponges and their archaeocyathid or stromatoporid relatives (McCaffrey *et al.*, 1994b).

Heightened relative abundances of monomethyl, dimethyl, and other branched acyclic alkanes is another distinctive feature of Proterozoic bitumens (e.g., Höld *et al.*, 1999; Klomp, 1986; Logan *et al.*, 1999). This is most often seen in clastic lithologies and only rarely in carbonates. Some patterns of highly branched alkanes appear to be specific to benthic microbial mats and, on the basis of carbon and sulfur isotope anomalies, are hypothesized to be associated with sulfide-oxidizing microbial communities (Kenig *et al.*, 2002; Logan *et al.*, 1999).

Evidence from the age-distribution of mineral deposits, sedimentary patterns of redox-sensitive trace metals combined with sulfur- and carbon-isotope systematics point to a profound evolution of the ocean redox structure during the Proterozoic eon (e.g., Anbar and Knoll, 2002; Canfield, 1998; Des Marais *et al.*, 1992). In particular, it is hypothesized that the oceans were sulfide-rich and sulfate-poor after the cessation of deposition of banded iron formations (BIF) in the Paleoproterozoic roughly 1.8 Ga and prior to the existence of ventilated oceans, possibly as early as the end of the Mesoproterozoic (1.0 Ga) (Canfield, 1998; Shen *et al.*, 2003) or at the end of the Neoproterozoic (Logan *et al.*, 1995). Analyses of organic matter provide some evidence for unusual diagenetic pathways and support the hypothesis that the biogeochemical carbon cycle in the Proterozoic was fundamentally different from that of the Phanerozoic. Studies of kerogens indicate that Proterozoic sedimentary organic matter, despite having high elemental hydrogen to carbon ratios, tends to be unusually aromatic in nature and yields relatively low amounts of aliphatic hydrocarbons during burial maturation (Summons *et al.*, 1994b) and also during catalytic hydropyrolysis (Brocks *et al.*, 2003c). Carbon-isotopic compositions of kerogens and co-occurring individual hydrocarbons in sediments throughout the Proterozoic show a different order to those observed in the Cambrian (Logan *et al.*, 1997). This was hypothesized to be a hallmark of a major re-organization of the biogeochemical carbon and sulfur cycles at the Proterozoic-Phanerozoic transition (Logan *et al.*, 1995). Rothman *et al.* (2003) analyzed fluctuations in the isotopic records of sedimentary organic and inorganic carbon through the Neoproterozoic and found evidence for non-steady-state behavior of the carbon cycle at this time. Thus, there are numerous clues pointing to an evolution in carbon cycle and in the type of organic matter that was being buried. While the actual compounds that are found in Proterozoic sediments tend to be the same ones that are encountered in younger rocks, their relative abundances, distribution patterns, and isotopic characteristics can be quite different. It is in this regard that studies of kerogen composition, biomarkers and compound-specific isotope data may prove to be most useful for evaluating environmental and ecological evolution during the Proterozoic and especially across the Proterozoic–Phanerozoic transition.

8.03.9.2 Biomarkers Extracted from Archean Rocks (>2.5 Ga)

An example that illustrates the difficulties that might be associated with establishing the age of solvent extractable organic matter are biomarkers detected in 2.7–2.5 Ga rocks from the Hamersley Basin, Western Australia (Brocks *et al.*, 1999). The host rocks from the Hamersley and Fortescue Group, although exceptionally well preserved by Archean standards, have suffered low-grade metamorphism at temperatures between 175 °C and 300 °C (Brocks *et al.*, 2003a). Yet, solvent extraction of kerogen-rich shales unexpectedly yielded 1 ppm–1,000 ppm *n*-alkanes, methylalkanes, acyclic isoprenoids, adamantanes, tri- to penta-cyclic terpanes, steranes, and polyaromatic hydrocarbons (PAH). One sample, a black shale from a hydrothermally altered iron mine in the Hamersley Group, exclusively contained adamantanes, parent PAH and minor concentrations of methylated PAH, patterns indicating extremely high thermal maturity and possibly hydrothermal alteration. Moreover, PAH with the same over-mature pattern were also released by pyrolytic degradation of isolated kerogens from other iron deposits in the Hamersley Basin (Brocks *et al.*, 2003c). The unusual composition, extreme thermal maturity and covalent bonding to kerogen rank these adamantanes and PAH as the by far oldest known "certainly syngenetic" bitumens in terrestrial rocks.

However, most samples from the Hamersley Basin additionally contain aliphatic hydrocarbons and polycyclic biomarkers in mixture with the certainly syngenetic adamantanes and PAH. The origin and age of these thermally less-stable components is less well constrained. Arguments against their syngeneity are a pronounced carbon-isotopic difference between bitumen and kerogen (Brocks *et al.*, 2003a), the absence of saturated hydrocarbons in kerogen pyrolysates (Brocks *et al.*, 2003c) and, most significantly, a strong inhomogeneous distribution of bitumen in individual drill core samples that is potentially consistent with surficial staining and migration of hydrocarbons into the rock (Brocks *et al.*, 2003a). However, the samples come from eight independent drill cores, drilled by several different companies, stored several hundred kilometers apart, collected by different workers over several years and analyzed in two laboratories with consistent results. All samples contain bitumen with a typical earlier Precambrian composition: absence of plant biomarkers, predominance of C_{27}-steranes, high C_{31}-2α-methylhopane indices

(8–20%), and phytane isotopically depleted relative to $n\text{-}C_{18}$. Moreover, the thermal maturity of the biomarkers is within the wet-gas zone of petroleum generation, younger petroleum source rocks are absent within the basin and were never deposited over the top, and the shales were collected from diamond drill core over an area of several hundred kilometers (Brocks *et al.*, 2003a). Therefore, the biomarkers are characterized "probably syngenetic." A less ambiguous classification might become available when fresh material is collected in the Hamersley Basin under controlled conditions as part of the *Deep Time Drilling Project* (Dalton, 2001).

However, if the biomarkers are in fact syngenetic, then they provide new insights into Archean biodiversity and ecology (Brocks *et al.*, 2003b; Brocks *et al.*, 1999). The presence of hopanes confirms the antiquity of the domain bacteria, and biomarkers of the 3β-methylhopane series suggest that microaerophilic Proteobacteria, probably methanotrophs, were active in Late Archean marine environments. High relative abundances of C_{30} to C_{36} 2α-methylhopanes indicate that cyanobacteria were important primary producers in the Late Archean. Therefore, oxygenic photosynthesis probably evolved before 2.7 Ga. High relative concentrations of cyanobacterial biomarkers were also detected in thin layers of Late Archean shales interbedded with oxide-facies banded iron formations (BIF) suggesting that, although some Archean BIF might have been formed by anoxygenic phototrophic bacteria or nonbiological photochemical processes, those in the Hamersley Group formed as a direct consequence of biogenic oxygen production. As chlorophyll biosynthesis in cyanobacteria probably succeeded the evolution of bacteriochlorophylls in anoxygenic phototrophic bacteria (Xiong *et al.*, 2000), the 2α-methylhopanes also give indirect evidence that all lineages of anoxygenic phototrophs—heliobacteria, purple bacteria, green sulfur bacteria, and green nonsulfur bacteria—evolved before 2.7 Ga (Des Marais, 2000). Steranes, including 4-methylsteranes (**69**), desmethylsteranes alkylated at C-24 ((**66b**)–(**66d**)), and aromatic steroids (**68**), occur in relative abundances similar to those from other Precambrian sources, providing evidence that ancestral eukaryotes existed ~900 Ma before the earliest microfossil evidence indicates that the lineage arose (Hofmann and Chen, 1981; Zhang, 1986). Sterol biosynthesis in extant eukaryotes requires dissolved molecular oxygen in concentrations equivalent to ~1% of the present atmospheric level (Jahnke and Klein, 1979; Jahnke and Klein, 1983). Therefore, it is likely that oxygen concentrations in Archean surface waters were high enough to support aerobic respiration to some extent.

8.03.10 OUTLOOK

In recent years, the discovery of new biomarkers and their sources has been greatly aided by the combination of molecular and compound-specific isotopic analysis methods (e.g., Hinrichs *et al.*, 2000), the advent of genomic tools to screen natural samples for the identities of dominant taxa (e.g., Boetius *et al.*, 2000; Hinrichs *et al.*, 1999) and the advent of culture-independent methods for studying important biogeochemical processes (e.g., Orphan *et al.*, 2001). The extensive screening of cultured extant organisms in the past has shown very general connections between taxa and their diagnostic markers. While this will continue, access to important organisms that are difficult or impossible to grow in the laboratory can be accomplished by studies of their genomes. Moreover, these genomes also encode an evolutionary history so that it may eventually be possible to reconstruct genetic information (paleogenomics) about extinct ancestors and their biochemical capacities (Benner, 2001). Accurate timing of evolutionary events can only be accomplished by studies of the rock record, and exploration of the fossil biomarkers will continue to be an important activity in the search for life's early history on Earth.

ACKNOWLEDGMENTS

The manuscript was greatly improved by helpful comments on various sections of this chapter by Alison Cohen, Jennifer Eigenbrode, David Fike, Kliti Grice, John Hayes, Fabien Kenig, Ann Pearson, and John Volkman. We thank Ann Pearson for interesting discussions about the occurrence of sterols in bacteria. Roger Summons is supported by grants from the NASA Exobiology program and the NASA Astrobiology Institute and Jochen Brocks gratefully acknowledges the Harvard Society of Fellows for providing support during the preparation of this work.

REFERENCES

Adam P., Schmid J. C., Mycke B., Strazielle C., Connan J., Huc A., Riva A., and Albrecht P. (1993) Structural investigation of non-polar sulfur cross-linked macromolecules in petroleum. *Geochim. Cosmochim. Acta.* **57**, 3395–3419.

Albaiges J. (1980) Identification and geochemical significance of long chain acyclic isoprenoids in crude oils. In *Advances in Organic Geochemistry 1979* (eds. A. G. Douglas and J. R. Maxwell). Pergamon, Oxford, pp. 19–28.

Alexander R., Kagi R. I., and Noble R. (1983) Identification of the bicyclic sesquiterpenes drimane and eudesmane in petroleum. *Chem. Commun.*, 226–228.

Alexander R., Noble R., and Kagi R. I. (1987) Fossil resin biomarkers and their application in oil to source rock correlation, Gippsland Basin, Australia. *APEA J.* **27**, 63–71.

Alexander R., Larcher A. V., Kagi R. I., and Price P. L. (1988) The use of plant-derived biomarkers for correlation of oils with source rocks in the Cooper/Eromanga basin system, Australia. *APEA J.* **28**, 310–324.

Alexander R., Larcher A. V., Kagi R. I., and Price P. L. (1992) An oil-source correlation study using age-specific plant-derived biomarkers. In *Biological Markers in Sediments and Petroleum* (eds. J. M. Moldowan, P. Albrecht, and R. P. Philp). Prentice Hall, Engelwood Cliff, NJ, pp. 210–221.

Allard B., Rager M.-N., and Templier J. (2002) Occurrence of high molecular weight lipids (C_{80+}) in the trilaminar outer cell walls of some freshwater microalgae: a reappraisal of algaenan structure. *Org. Geochem.* **33**, 789–801.

Allard W. G., Belt S. T., Massé G., Naumann R., Robert J.-M., and Rowland S. J. (2001) Tetra-unsaturated sesterterpenoids (Haslenes) from *Haslea ostrearia* and related species. *Phytochemistry* **56**, 795–800.

Anbar A. D. and Knoll A. H. (2002) Proterozoic ocean chemistry and evolution: a bioinorganic bridge? *Science* **297**, 1137–1142.

Andrusevich V. E., Engel M. H., Zumberge J. E., and Brothers L. A. (1998) Secular, episodic changes in stable isotopic composition of crude oils. *Chem. Geol.* **152**, 59–72.

Andrusevich V. E., Engel M. H., and Zumberge J. E. (2001) Applications of paleogeographic reconstructions for evaluating secular, isotopic trends exhibited by crude oils. *Abstr. 20th Int. Meet. Org. Geochem.* Nancy, France, 243–244.

Aquino Neto F. R., Trendel J. M., Restle A., Connan J., and Albrecht P. A. (1983) Occurrence and formation of tricyclic and tetracyclic terpanes in sediments and petroleums. In *Advances in Organic Geochemistry 1981* (ed. M. Bjorøy). Wiley, New York, pp. 659–667.

Aquino Neto F. R., Trigüis J., Azevedo D. A., Rodrigues R., and Simoneit B. R. T. (1992) Organic geochemistry of geographically unrelated Tasmanites. *Org. Geochem.* **18**, 791–803.

Arouri K., Greenwood P. F., and Walter M. R. (1999) A possible chlorophycean affinity of some Neoproterozoic acritarchs. *Org. Geochem.* **30**, 1323–1337.

Arouri K., Conaghan P. J., Walter M. R., Bischoff G. C. O., and Grey K. (2000a) Reconnaissance sedimentology and hydrocarbon biomarkers of Ediacaran microbial mats and acritarchs, lower Ungoolya Group, Officer Basin. *Precamb. Res.* **100**, 235–280.

Arouri K., Greenwood P. F., and Walter M. R. (2000b) Biological affinities of Neoproterozoic acritarchs from Australia: microscopic and chemical characterisation. *Org. Geochem.* **31**, 75–89.

Audino M., Grice K., Alexander R., Kagi R. I., and Boreham C. J. (2001) Unusual distribution of monomethylalkanes in *Botryococcus braunii*-rich samples: origin and significance. *Geochim. Cosmochim. Acta* **65**, 1995–2006.

Audino M., Grice K., Alexander R., and Kagi R. I. (2002) Macrocyclic alkenes in crude oils from the algaenan of *Botryococcus braunii. Org. Geochem.* **33**, 979–984.

Barber C. J., Grice K., Bastow T. P., Alexander R., and Kagi R. I. (2001) The identification of crocetane in Australian crude oils. *Org. Geochem.* **32**, 943–947.

Barber C. J., Grice K., Bastow T. P., Alexander R., and Kagi R. I. (2002) Corrigendum to: the identification of crocetane in Australian crude oils. *Org. Geochem.* **33**, 89.

Beastall G. H., Tyndall A. M., Rees H. H., and Goodwin T. W. (1974) Sterols of the *Porphyridium* series. 4α-Methyl-5α-cholesta-8, 22-dien-3β-ol and 4, 24-dimethyl-5α-cholesta-8, 22-dien-3β-ol: two novel sterols from *Porphyridium cruentum. Euro. J. Biochem.* **41**, 301–309.

Belt S. T., Allard W. G., Massé G., Robert J.-M., and Rowland S. J. (2000) Highly branched isoprenoids (HBIs): identification of the most common and abundant sedimentary isomers. *Geochim. Cosmochim. Acta* **64**, 3839–3851.

Benner S. A. (2001) Natural progression. *Nature* **409**, 459.

Bertheas O., Metzger P., and Largeau C. (1999) A high molecular weight complex lipid, aliphatic polyaldehyde tetraterpenediol polyacetal from *Botryococcus braunii* (L race). *Phytochemistry* **50**, 85–96.

Bian L., Hinrichs K.-U., Xie T., Brassell S. C., Iversen N., Fossing H., Jørgensen B. B., Sylva S. P., and Hayes J. M. (2001) Algal and archaeal polyisoprenoids in a recent marine sediment: molecular isotopic evidence for anaerobic oxidation of methane. *Geochem. Geophys. Geosys.* **2**, 2000GC000112.

Bird C. W., Lynch J. M., Pirt F. J., Reid W. W., Brooks C. J. W., and Middleditch B. S. (1971) Steroids and squalene in *Methylococcus capsulatus* grown on methane. *Nature* **230**, 473–474.

Bisseret P., Zundel M., and Rohmer M. (1985) 2β-Methylhopanoids from *Methylobacterium organophilum* and *Nostoc muscorum*, a new series of prokaryotic triterpenoids. *Euro. J. Biochem.* **150**, 29–34.

Blokker P., Schouten S., van den Ende H., de Leeuw J. W., Hatcher P. G., and Sinninghe Damsté J. S. (1998) Chemical structure of algaenans from the fresh water algae *Tetraedron minimum, Scenedesmus communis* and *Pediastrum boryanum. Org. Geochem.* **29**, 1453–1468.

Blokker P., Schouten S., de Leeuw J. W., Sinninghe Damsté J. S., and van den Ende H. (2000) A comparative study of fossil and extant algaenans using ruthenium tetroxide degradation. *Geochim. Cosmochim. Acta* **64**, 2055–2065.

Blokker P., van Bergen P. F., Pancost R. D., Collinson M. E., Sinninghe Damsté J. S., and de Leeuw J. W. (2001) The chemical structure of *Gloeocapsamorpha prisca* microfossils: implication for their origin. *Geochim. Cosmochim. Acta* **65**, 885–900.

Bode H. B., Zeggel B., Silakowski B., Wenzel S. C., Hans R., and Müller R. (2003) Steroid biosynthesis in prokaryotes: identification of myxobacterial steroids and cloning of the first bacterial 2,3(S)-oxidosqualene cyclase from the myxobacterium *Stigmatella aurantiaca. Mol. Microbiol.* **47**, 471–481.

Boetius A., Ravenschlag K., Schubert C. J., Rickert D., Widdel F., Gieseke A., Amann R., Jørgensen B. B., Witte U., and Pfannkuche O. (2000) A marine microbial consortium apparently mediating anaerobic oxidation of methane. *Nature* **407**, 623–626.

Bosch H.-J., Sinninghe Damsté J. S., and de Leeuw J. W. (1998) Molecular palaeontology of eastern Mediterranean sapropels: evidence for photic zone euxinia. *Proc. ODP.: Sci. Res.* **160**, 285–295.

Brasier M. D., Green O. R., Jephcoat A. P., Kleppe A. K., van Kranendonk M. J., Lindsay J. F., Steele A., and Grassineau N. V. (2002) Questioning the evidence for the Earth's oldest fossils. *Nature* **416**, 76–81.

Brassell S. C., Wardroper A. M. K., Thomson I. D., Maxwell J. R., and Eglinton G. (1981) Specific acyclic isoprenoids as biological markers of methanogenic bacteria in marine sediments. *Nature* **290**, 693–696.

Brassell S. C., Eglinton G., Marlowe I. T., Pflaufmann U., and Sarnthein M. (1986) A new tool for climatic assessment. *Nature* **320**, 129–133.

Brigaud F. (1998) *HP–HT Petroleum System Prediction from Basin to Prospect Scale.* Final report, project OG/211/94FR/UK. Commission of the European Community, Directorate General for Energy, Brussel.

Britton G. (1995) History: 175 years of carotenoid chemistry. In *Carotenoids* (eds. G. Britton, H. Pfander, and S. Liaaen-Jensen). Birkhauser, vol. 1a, pp. 13–26.

Britton G., Liaaen-Jensen S., and Pfander H. (1995) Carotenoids today and challenges for the future. In *Carotenoids: Isolation and Analysis*, Birkhäuser, vol. 1a, pp. 13–26.

Brochier C. and Philippe H. (2002) A non-thermophilic ancestor for Bacteria. *Nature* **417**, 244.

Brocks J. J., Logan G. A., Buick R., and Summons R. E. (1999) Archean molecular fossils and the early rise of eukaryotes. *Science* **285**, 1033–1036.

Brocks J. J., Buick R., Logan G. A., and Summons R. E. (2003a) Composition and syngeneity of molecular fossils from the 2.78–2.45 billion year old Mount Bruce Supergroup, Pilbara Craton, Western Australia. *Geochim. Cosmochim. Acta* (in press).

Brocks J. J., Buick R., Summons R. E., and Logan G. A. (2003b) A reconstruction of Archean biological diversity based on molecular fossils from the 2.78–2.45 billion year old Mount Bruce Supergroup, Hamersley Basin, Western Australia. *Geochim. Cosmochim. Acta* (in press).

Brocks J. J., Love G. D., Snape C. E., Logan G. A., Summons R. E., and Buick R. (2003c) Release of bound aromatic hydrocarbons from late Archean and Mesoproterozoic kerogens via hydropyrolysis. *Geochim. Cosmochim. Acta* **67**(8), 1521–1530.

Brocks J. J., Summons R. E., Buick R., and Logan G. A. (2003d) Origin and significance of aromatic hydrocarbons in giant iron ore deposits of the late Archean Hamersley Basin in Western Australia. *Org. Geochem.* **34**, 1161–1175.

Buick R. (1992) The antiquity of oxygenic photosynthesis: evidence from stromatolites in sulphate-deficient Archaean lakes. *Science* **255**, 74–77.

Buick R., Dunlop J. S. R., and Groves D. I. (1981) Stromatolite recognition in ancient rocks: an appraisal of irregularly laminated structures in an early Archean chert-barite unit from North Pole, Western Australia. *Alcheringa* **6**, 161–181.

Burhan R. Y. P., Trendel J. M., Adam P., Wehrung P., Albrecht P., and Nissenbaum A. (2002) Fossil bacterial ecosystem at methane seeps: origin of organic matter from Be'eri sulfur deposit, Israel. *Geochim. Cosmochim. Acta* **66**, 4085–4101.

Burnham A. K., Gregg H. R., Ward R. L., Knauss K. G., Copenhaver S. A., Reynolds J. G., and Sanborn R. (1997) Decomposition kinetics and mechansism of *n*-hexadecane-1, 2-$^{13}C_2$ and dodec-1-ene-1, 2-$^{13}C_2$ doped in petroleum and *n*-hexadecane. *Geochim. Cosmochim. Acta* **61**, 3725–3737.

Butterfield N. J. (2001) Paleobiology of the late Mesoproterozoic (ca. 1,200 Ma) hunting formation, Somerset Island, arctic Canada. *Precamb. Res.* **111**, 235–256.

Butterfield N. J., Knoll A. H., and Swett K. (1988) Exceptional preservation of fossils in an Upper Proterozoic shale. *Nature* **334**, 424–427.

Butterfield N. J., Knoll A. H., and Swett K. (1990) A bangiophyte red alga from the Proterozoic of arctic Canada. *Science* **250**, 104–107.

Canfield D. E. (1998) A new model for Proterozoic ocean chemistry. *Nature* **396**, 450–453.

Canfield D. E. and Raiswell R. (1999) The evolution of the sulfur cycle. *Am. J. Sci.* **299**, 697–723.

Carlson R. M. K., Teerman S. C., Moldowan J. M., Jacobson S. R., Chan E. I., Dorrough K. S., Seetoo W. C., and Mertani B. (1993) High temperature gas chromatography of high-wax oils. In *Indonesian Petroleum Association, 22nd Annual Convention Proceedings, Jakarta, Indonesia*, 483–507.

Cavalier-Smith T. (2002) The neomuran origin of archaebacteria, the negibacterial root of the universal tree and bacterial megaclassification. *Int. J. Syst. Evol. Microbiol.* **52**, 7–76.

Chen J. and Summons R. E. (2001) Complex patterns of steroidal biomarkers in Tertiary lacustrine sediments of the Biyang Basin, China. *Org. Geochem.* **32**, 115–126.

Chen J., Fu J., Sheng G., Liu D., and Zhang J. (1996) Diamondoid hydrocarbon ratios: novel maturity indices for highly mature crude oils. *Org. Geochem.* **25**, 179–190.

Chen J., Walter M. R., Logan G. A., Hinman M. C., and Summons R. E. (2003) The Paleoproterozoic McArthur river (HYC) Pb/Zn/Ag deposit of northern Australia: organic geochemistry and ore genesis. *Earth. Planet. Sci. Lett.* **210**, 467–479.

Chen J. H., Philp R. P., Fu F. M., and Sheng G. Y. (1989) The occurrence and identification of C_{30}–C_{32} lanostanes: a novel series of tetracyclic triterpenoid hydrocarbons. *Geochim. Cosmochim. Acta* **53**, 2775–2779.

Cockell C. S. and Knowland J. (1999) Ultraviolet radiation screening compounds. *Biol. Rev.* **74**, 311–345.

Cody G., Alexander C. M. O., and Tera F. (2002) Solid-state (^1H and ^{13}C) nuclear magnetic resonance spectroscopy of insoluble organic residue in the Murchison meteorite: a self-consistent quantitative analysis. *Geochim. Cosmochim. Acta* **66**, 1851–1865.

Collister J. W., Summons R. E., Lichtfouse E., and Hayes J. M. (1992) An isotopic biogeochemical study of the Green River oil shale. *Org. Geochem.* **19**, 265–276.

Cox H. C., de Leeuw J. W., Schenk P. A., van Konigsveld H., Jansen J. C., van de Graaf B., van Geerstein V. J., Kanters J. A., Kruk C., and Jans A. W. H. (1986) Bicadinane, a C_{30} pentacyclic isoprenoid hydrocarbon found in crude oil. *Nature* **319**, 316–318.

Crick I. H., Boreham C. J., Cook A. C., and Powell T. G. (1988) Petroleum geology and geochemistry of middle Proterozoic McArthur Basin, Northern Australia: II. Assessment of source rock potential. *AAPG Bull.* **72**, 1495–1514.

Cronin J. R. and Chang S. (1993) Organic matter in meteorites: molecular and isotopic analyses of the Murchison meteorite. In *The Chemistry of Life's Origin* (eds. J. M. Greenberg, C. X. Mendoza-Gómez, and V. Pirronello). Kluwer Academic, pp. 209–258.

D'Souza S. E., Altekar W., and D'Souza S. F. (1997) Adaptive response of *Haloferax mediterranei* to low concentrations of NaCl (<20%) in the growth medium. *Arch. Microbiol.* **168**, 68–71.

Dahl J. E. P., Hallberg R., and Kaplan I. R. (1988) Effects of irradiation from uranium decay on extractable organic matter in the Alum Shales of Sweden. *Org. Geochem.* **12**, 559–571.

Dahl J. E. P., Moldowan J. M., McCaffrey A. M., and Lipton P. (1992) A new class of natural products revealed by 3β-alkyl steranes in petroleum. *Nature* **355**, 154–157.

Dahl J. E. P., Moldowan J. M., Summons R. E., McCaffrey A. M., Lipton P., Watt D. S., and Hope J. M. (1995) Extended 3-alkyl steranes and 3-alkyl triaromatic steroids in crude oils and rock extracts. *Geochim. Cosmochim. Acta* **59**, 3717–3729.

Dahl J. E. P., Moldowan J. M., Peters K. E., Claypool G. E., Rooney M. A., Michael G. E., Mello M. R., and Kohnen M. L. (1999) Diamondoid hydrocarbons as indicators of natural oil cracking. *Nature* **399**, 54–57.

Dahl J. E. P., Liu S. G., and Carlson R. M. K. (2002) Isolation and structure of higher diamondoids, nanometer-sized diamond molecules. *Science* **299**, 96–99.

Dalton R. (2001) Cores set to unearth whole picture of evolution. *Nature* **414**, 476.

de Leeuw J. W. and Baas M. (1986) Early-stage diagenesis of steroids. In *Biological Markers in the Sedimentary Record* (ed. R. B. Johns). Elsevier, Amsterdam, vol. 24, pp. 101–123.

de Leeuw J. W. and Largeau C. (1993) A review of macromolecular organic compounds that comprise living organisms and their role in kerogen, coal, and petroleum formation. In *Organic Geochemistry* (eds. M. H. Engel and S. A. Macko). Plenum, New York and London, pp. 23–72.

de Leeuw J. W., Frewin N. L., Van Bergen P. F., Sinninghe Damsté J. S., and Collinson M. E. (1995) Organic carbon as a palaeoenvironmental indicator in the marine realm. In *Geol. Soc. Spec. Publ.* (eds. D. W. J. Bosence and P. A. Allison). Geological Society of London, vol. 83, pp. 43–71.

De Rosa M., Gambacorta A., Minale L., and Bu'Lock J. D. (1971) Cyclohexane fatty acids from a thermophilic bacterium. *Chem. Commun.* 1334.

del Rio J. C. and Philp R. P. (1999) Field ionization mass spectrometric study of high molecular weight hydrocarbons in a crude oil and a solid bitumen. *Org. Geochem.* **30**, 279–286.

DeLong E. F. (1992) Archaea in coastal marine environments. *Proc. Natl. Acad. Sci. USA* **89**, 5685–5689.

DeLong E. F., King L. L., Massana R., Cittone H., Murray A., Schleper C., and Wakeham S. G. (1998) Dibiphytanyl ether lipids in nonthermophilic Crenarchaeotes. *Appl. Environ. Microbiol.* **64**, 1133–1138.

Dembitsky V. M., Dor I., Shkrob I., and Aki M. (2001) Branched alkanes and other apolar compounds produced by the cyanobacterium *Microcoleus vaginatus* from the Negev Desert. *Russian J. Bioorg. Chem.* **27**, 110–119.

Derenne S. and Largeau C. (2001) A review of some important families of refractory macromolecules: composition, origin and fate in soils and sediments. *Soil Sci.* **166**, 833–847.

Derenne S., Largeau C., Casadevall E., and Sellier N. (1990) Direct relationship between the resistant biopolymer and the tetraterpenic hydrocarbon in the lycopadiene race of *Botryococcus braunii. Phytochemistry* **28**, 2187–2192.

Derenne S., Largeau C., Casadevall E., Berkaloff C., and Rousseau B. (1991) Chemical evidence of kerogen formation in source rocks and oil shales via selective preservation of thin resistant outer walls of microalgae: origin of ultralaminae. *Geochim. Cosmochim. Acta* **55**, 1041–1050.

Derenne S., Le Berre F., Largeau C., Hatcher P. G., Connan J., and Raynaud J.-F. (1992) Formation of ultralaminae in marine kerogens via selective preservation of thin resistant outer walls of microalgae. *Org. Geochem.* **19**, 345–350.

Des Marais D. J. (2000) When did photosynthesis emerge on Earth? *Science* **289**, 1703–1705.

Des Marais D. J., Strauss H., Summons R. E., and Hayes J. M. (1992) Carbon isotope evidence for the stepwise oxidation of the Proterozoic environment. *Nature* **359**, 605–609.

Didyk B. M., Simoneit B. R. T., Brassell S. C., and Eglinton G. (1978) Organic geochemical indicators of palaeoenvironmental conditions of sedimentation. *Nature* **272**, 216–222.

Dominé F., Dessort D., and Brévart O. (1998) Towards a new method of geochemical kinitic modelling: implications for the stability of crude oils. *Org. Geochem.* **28**(9–10), 597–612.

Dominé F., Bounaceur R., Scacchi G., Marquaire P.-M., Dessort D., Pradier B., and Brevart O. (2002) Up to what temperature is petroleum stable? New insights from a 5200 free radical reaction model. *Org. Geochem.* **33**, 1487–1499.

Eglinton G. and Calvin M. (1967) Chemical fossils. *Sci. Am.* **261**, 32–43.

Eglinton G., Scott P. M., Belsky T., Burlingame A. L., and Calvin M. (1964) Hydrocarbons of a biological origin from a one-billion-year-old sediment. *Science* **145**, 263–264.

Eglinton T. I., Benitez-Nelson B. C., Pearson A., McNichol A. P., Bauer J. E., and Druffel E. R. M. (1997) Variability in radiocarbon ages of individual organic compounds from marine sediments. *Science* **277**, 796–799.

Elvert M., Suess E., and Whiticar M. J. (1999) Anaerobic methane oxidation associated with marine gas hydrates: superlight C-isotopes from saturated and unsaturated C_{20} and C_{25} irregular isoprenoids. *Naturwissenschaften* **31**, 1175–1187.

Engel M. and Macko S. (1997) Isotopic evidence for extraterrestrial non-racemic amino acids in the Murchison meteorite. *Nature* **389**, 265–268.

Espitalié J., Laporte J. L., Madec M., Marquis F., Leplat P., Paulet J., and Boutefeu A. (1977) Méthode rapide de caractérisation des roches mères, de leur potentiel pétrolier et de leur degre d'évolution. *Revue de l'Institut Français du Pétrole* **32**, 23–42.

Fang H., Yongchuan S., Sitian L., and Qiming Z. (1995) Overpressure retardation of organic-matter maturation and petroleum generation: a case study from the Yinggehai and Qiongdongnan Basins, South China Sea. *AAPG Bull.* **79**, 551–562.

Farrimond P., Fox P. A., Innes H. E., Miskin I. P., and Head I. M. (1998) Bacterial sources of hopanoids in recent sediments: improving our understanding of ancient hopane biomarkers. *Ancient Biomol.* **2**, 147–166.

Farrimond P., Bevan J. C., and Bishop A. N. (1999) Tricyclic terpane maturity parameters: response to heating by an igneous intrusion. *Org. Geochem.* **30**, 1011–1019.

Farrimond P., Head I. M., and Innes H. E. (2000) Environmental influence on the biohopanoid composition of recent sediments. *Geochim. Cosmochim. Acta* **64**, 2985–2992.

Fong N. J. C., Burgess M. L., Barrow K. D., and Glenn D. R. (2001) Carotenoid accumulation in the psychrotrophic bacterium *Arthrobacter agilis* in response to thermal and salt stress. *Appl. Microbiol. Biotechnol.* **56**, 750–756.

Fowler M. G. (1992) The influence of *Gloeocapsomorpha prisca* on the organic geochemistry of oils and organic-rich rocks of Late Ordovician age from Canada. In *Early Organic Evolution: Implications for Mineral and Energy Resources* (eds. M. Schidlowski, S. Golubic, M. M. Kimberly, and P. A. Trudinger). Springer, Berlin, pp. 336–356.

Fowler M. G. and Douglas A. G. (1987) Saturated hydrocarbon biomarkers in oils of late Precambrian age from eastern Siberia. *Org. Geochem.* **11**, 201–213.

Fowler M. G., Abolins P., and Douglas A. G. (1986) Monocyclic alkanes in Ordovician organic matter. *Org. Geochem.* **10**, 815–823.

Frank H. A., Young A. J., Britton G., and Cogdell R. J. (1999) The photochemistry of carotenoids. In *Advances in Photosynthesis.* Kluwer Academic, vol. 8, 420pp.

Freeman K. H., Hayes J. M., Trendel J.-M., and Albrecht P. (1990) Evidence from carbon isotope measurements for diverse origins of sedimentary hydrocarbons. *Nature* **343**, 254–256.

Freeman K. H., Wakeham S. G., and Hayes J. M. (1994) Predictive isotopic biogeochemistry: hydrocarbons from anoxic marine basins. *Org. Geochem.* **21**, 629–644.

Ganz H. and Kalkreuth W. (1987) Application of infrared spectroscopy to the classification of kerogen-types and the evaluation of source rock and oil shale potentials. *Fuel* **66**, 708–711.

Gelin F., De Leeuw J. W., Sinninghe Damsté J. S., Derenne S., and Metzger P. (1994) The similarity of chemical structures of soluble aliphatic polyaldehyde and insoluble algaenan in the green microalga *Botryococcus braunii* race A as revealed by analytical pyrolysis. *Org. Geochem.* **21**, 423–435.

Gelin F., Boogers I., Noordeloos A. A. M., Sinninghe Damsté J. S., de Leeuw J. W., and Hatcher P. G. (1996) Novel, resistant microalgal polyethers: an important sink of organic carbon in the marine environment? *Geochim. Cosmochim. Acta* **60**, 1275–1280.

Gelin F., Boogers I., Noordeloos A. A. M., Sinninghe Damsté J. S., Riegman R., and de Leeuw J. W. (1997) Resistant biomacromolecules in marine microalgae of the classes eustigmatophyceae and chlorophyceae: geochemical implications. *Org. Geochem.* **26**, 659–675.

Gelin F., Volkman J. K., Largeau C., Derenne S., Sinninghe Damsté J. S., and de Leeuw J. W. (1999) Distribution of aliphatic, nonhydrolyzable biopolymers in marine microalgae. *Org. Geochem.* **30**, 147–159.

George S. C. (1992) Effect of igneous intrusion on the organic geochemistry of a siltstone and an oil shale horizon in the midland valley of Scotland. *Org. Geochem.* **18**, 705–723.

George S. C. and Ahmed M. (2003) Use of aromatic compound distributions to evaluate organic maturity of the proterozoic middle velkerri formation, McArthur Basin, Australia. In *The Sedimentary Basins of Western Australia 3* (eds. M. Keep and S. Moss). PESA, Perth, pp. 253–270.

George S. C. and Jardine D. R. (1994) Ketones in a Proterozoic dolerite Sill. *Org. Geochem.* **21**(8/9), 829–839.

Giner J.-L. and Boyer G. L. (1998) Sterols of the brown tide alga *Aureococcus anophagefferens. Phytochemistry* **48**, 475–477.

Giner J.-L. and Li X. (2000) Stereospecific synthesis of 24-propylcholesterol isolated from the Texas brown tide. *Tetrahedron* **56**, 9575–9580.

Giner J.-L., Li X., and Boyer G. L. (2001) Sterol composition of *Aureoumbra lagunensis*, the Texas brown tide alga. *Phytochemistry* **57**, 787–789.

Gize A. P. (1999) Organic alteration in hydrothermal sulfide ore deposits. *Econ. Geol.* **94**, 967–980.

Gold T. (2001) *The Deep Hot Biosphere: The Myth of Fossil Fuels*. Freeman Dyson.

Gomez Paloma L., Randazzo A., Minale L., Debitus C., and Roussakis C. (1997) New cytotoxic sesterterpenes from the New Caledonian marine sponge *Petrosaspongia nigra* (Bergquist). *Tetrahedron* **53**, 10451–10458.

Goosens H., de Leeuw J. W., Schenck P. A., and Brassell S. C. (1984) Tocopherols as likely precursors of pristane in ancient sediments and crude oils. *Nature* **312**, 440–442.

Grantham P. J. (1986) The occurrence of unusual C_{27} and C_{29} sterane predominances in two types of Oman crude oil. *Org. Geochem.* **9**, 1–10.

Grantham P. J. and Douglas A. G. (1980) The nature and origin of sesquiterpenoids in some tertiary fossil resins. *Geochim. Cosmochim. Acta* **44**, 1801–1810.

Grantham P. J., Posthuma J., and DeGroot K. (1980) Variation and significance of the C_{27} and C_{28} triterpane content of a North Sea core and various North Sea crude oils. In *Advances in Organic Geochemistry 1979* (eds. A. G. Douglas and J. R. Maxwell). Pergamon, Oxford, pp. 29–38.

Grantham P. J., Lijmbach G. W. M., Postuma J., Hughes-Clark M. W., and Willink R. J. (1988) Origin of crude oils in Oman. *J. Petrol. Geol.* **11**, 61–80.

Greenwood P. F. and Summons R. E. (2003) GC–MS detection and significance of crocetane and pentamethylicosane in sediments and crude oils. *Org. Geochem.* **34**, 1211–1222.

Greenwood P. F., Arouri K. R., and George S. C. (2000) Tricyclic terpenoid composition of *Tasmanites* kerogen as determined by pyrolysis GC–MS. *Geochim. Cosmochim. Acta* **64**, 1249–1263.

Grice K., Gibbison R., Atkinson J. E., Schwark L., Eckardt C. B., and Maxwell J. R. (1996a) Maleimides (1*H*-pyrrole-2, 5-diones) as molecular indicators of anoxygenic photosynthesis in ancient water columns. *Geochim. Cosmochim. Acta* **60**, 3913–3924.

Grice K., Schaeffer P., Schwark L., and Maxwell J. R. (1996b) Molecular indicators of palaeoenvironmental conditions in an immature Permian shale (Kuperschiefer, Lower Rhine Basin, north-west Germany) from free and S-bound lipids. *Org. Geochem.* **25**, 131–147.

Grice K., Schaeffer P., Schwark L., and Maxwell J. R. (1997) Changes in palaeoenvironmental conditions during deposition of the Permian Kupferschiefer (Lower Rhine Basin, northwest Germany) inferred from molecular and isotopic compositions of biomarker components. *Org. Geochem.* **26**, 677–690.

Grice K., Klein Breteler W. C. M., Schouten S., Grossi V., de Leeuw J. W., and Sinninghe Damsté J. S. (1998a) The effects of zooplankton herbivory on biomarker proxy records. *Paleoceanography* **13**, 686–693.

Grice K., Schouten S., Nissenbaum A., Charrach J., and Sinninghe Damsté J. S. (1998b) Isotopically heavy carbon in the C_{21} to C_{25} regular isoprenoids in halite-rich deposits from the Sdom Formation, Dead Sea Basin, Israel. *Org. Geochem.* **28**, 349–359.

Grice K., Schouten S., Peters K. E., and Sinninghe Damsté J. S. (1998c) Molecular isotopic characterisation of hydrocarbon biomarkers in Palaeocene-eocene evaporitic, lacustrine source rocks from the Jianghan Basin, China. *Org. Geochem.* **29**, 1745–1764.

Grosjean E., Poinsot J., Charrié-Duhaut A., Tabuteau S., Adam P., Trendel J., Schaeffer P., Connan J., Dessort D., and Albrecht P. (2001) Synthesis and NMR characterization of novel highly cyclised polyprenoid hydrocarbons from sediments. *J. Chem. Soc. Perkin Trans.* 1, 711–719.

Häberli A., Bircher C., and Pfander H. (2000) Isolation of a new carotenoid and two new carotenoid glycosides from *Curtobacterium flaccumfaciens pvar poinsettiae*. *Helvetica Chimica Acta* **83**, 328–335.

Han T.-M. and Runnegar B. (1992) Megascopic eukaryotic algae from the 2.1-billion-year-old Negaunee Iron-formation, Michigan. *Science* **257**, 232–235.

Hartgers W. A., Sinninghe Damsté J. S., Requejo A. G., Allan J., Hayes J. M., Ling Y., Tiang-Min X., Primack J., and de Leeuw J. W. (1993) A molecular and carbon isotopic study towards the origin and diagenetic fate of diaromatic carotenoids. *Org. Geochem.* **22**, 703–725.

Harvey H. R. and McManus G. B. (1991) Marine ciliates as a widespread source of tetrahymanol and hopan-3β-ol in sediments. *Geochim. Cosmochim. Acta* **55**, 3387–3390.

Hayes J. M. (1983) Geochemical evidence bearing on the origin of aerobiosis, a speculative hypothesis. In *Earth's Earliest Biosphere, its Origin and Evolution* (ed. J. W. Schopf). Princeton University Press, Princeton, pp. 291–301.

Hayes J. M. (1993) Factors controlling ^{13}C contents of sedimentary organic compounds: principles and evidence. *Mar. Geol.* **113**, 111–125.

Hayes J. M. (1994) Global methanotrophy at the Archean-proterozoic transition. In *Early life on Earth* (ed. S. Bengtson). Columbia University Press, New York, vol. 84, pp. 220–236.

Hayes J. M. (2001) Fractionation of the isotopes of carbon and hydrogen in biosynthetic processes. In *Stable Isotope Geochemistry*, Reviews in Mineralogy and Geochemistry. Mineralogical Society of America, Washington, DC, vol. 43, pp. 225–278.

Hayes J. M., Kaplan I. R., and Wedeking K. W. (1983) Precambrian organic geochemistry, preservation of the record. In *Earth's Earliest Biosphere, its Origin and Evolution* (ed. J. W. Schopf). Princeton University Press, Princeton, pp. 93–134.

Hayes J. M., Freeman K. H., Popp B. N., and Hoham C. H. (1990) Compound-specific isotope analysis: a novel tool for reconstruction of ancient biogeochemical processes. In *Advances in Organic Geochemistry 1989* (eds. B. Durand and F. Behar). Pergamon, Oxford, pp. 1115–1128.

Hayes J. M., Summons R. E., Strauss H., Des Marais D. J., and Lambert I. B. (1992) Proterozoic biogeochemistry. In *The Proterozoic Biosphere: A Multidisciplinary Study* (eds. J. W. Schopf and C. Klein). Cambridge University Press, Cambridge, pp. 81–133.

Hayes J. M., Strauss H., and Kaufman A. J. (1999) The abundance of ^{13}C in marine organic matter and isotopic fractionation in the global biogeochemical cycle of carbon during the past 800 Ma. *Chem. Geol.* **16**, 103–125.

Hedberg H. D. (1968) Significance of high-wax oils with respect to genesis of petroleum. *Am. Assoc. Petrol. Geol. Bull.* **52**, 736–750.

Hedges J. I. and Keil R. G. (1995) Sedimentary organic matter preservation: an assessment and speculative synthesis. *Mar. Chem.* **49**, 81–115.

Hedges J. I., Keil R. G., and Benner R. (1997) What happens to terrestrial organic matter in the ocean? *Org. Geochem.* **27**, 195–212.

Hinrichs K.-U., Hayes J. M., Sylva S. P., Brewer P. G., and DeLong E. F. (1999) Methane-consuming archaebacteria in marine sediments. *Nature* **398**, 802–805.

Hinrichs K.-U., Summons R. E., Orphan V., Sylva S. P., and Hayes J. M. (2000) Molecular and isotopic analysis of anaerobic methane-oxidising communities in marine sediments. *Org. Geochem.* **31**, 1685–1701.

Hinrichs K.-U., Eglinton G., Engel M. H., and Summons R. E. (2001) Exploiting the multivariate isotopic nature of organic compounds. *Geochem. Geophys. Geosys.* **1**, Paper number 2001GC000142.

Hinrichs K. U., Hmelo L. R., and Sylva S. P. (2003) Molecular fossil record of elevated methane levels in late pleistocene coastal waters. *Science* **299**, 1214–1217.

Ho E. S., Meyers P. A., and Mauk J. L. (1990) Organic geochemical study of mineralization in the Keweenawan Nonesuch formation at White Pine, Michigan. *Org. Geochem.* **16**, 229–234.

Hoefs M. J. L., Schouten S., King L., Wakeham S. G., de Leeuw J. W., and Sinninghe Damsté J. S. (1997) Ether lipids of planktonic archaea in the marine water column. *Appl. Environ. Microbiol.* **63**, 3090–3095.

Hoehler T. M., Alperin M. J., Albert D. B., and Martens C. S. (1994) Field and laboratory studies of methane oxidation in an anoxic marine sediment: evidence for a methanogen-sulfate reducer consortium. *Global Biogeochem. Cycles* **8**, 451–463.

Hoering T. C. (1976) Molecular fossils from the Precambrian Nonesuch Shale. *Carnegie Inst. Wash. Yearb.* **75**, 806–813.

Hoering T. C. (1981) Monomethyl acyclic hydrocarbons in petroleum and rock extracts. *Carnegie Inst. Wash. Yearb.* **80**, 389–393.

Hoffmann C. F., Foster C. B., Powell T. G., and Summons R. E. (1987) Hydrocarbon biomarkers from ordovician sediments and the fossil alga *Gloeocapsomorpha prisca* Zalessky 1917. *Geochim. Cosmochim. Acta* **51**, 2681–2697.

Hofmann H. J. (1976) Precambrian microflora, Belcher Islands, Canada: significance and systematics. *J. Paleontol.* **50**, 1040–1073.

Hofmann H. J. and Chen J. (1981) Carbonaceous megafossils from the Precambrian (1,800 Ma) near Jixian, northern China. *Can. J. Earth Sci.* **18**, 443–447.

Holba A. G., Dzou L. I. P., Masterson W. D., Hughes W. B., Huizinga B. J., Singletary M. S., Moldowan J. M., Mello M. R., and Tegelaar E. (1998a) Application of 24-norcholestanes for constraining source age of petroleum. *Org. Geochem.* **29**, 1269–1283.

Holba A. G., Tegelaar E. W., Huizinga B. J., Moldowan J. M., Singletary M. S., McCaffrey M. A., and Dzou L. I. P. (1998b) 24-norcholestanes as age-sensitive molecular fossils. *Geology* **26**, 783–786.

Holba A. G., Dzou L. I. P., Wood G. D., Ellis L., Adam P., Schaeffer P., Albrecht P., Green T., and Hughes W. B. (2003) Application of tetracyclic polyprenoids as indicators of input from fresh-brackish water environments. *Org. Geochem.* **34**, 441–469.

Höld I.M., Brussee N.J., Schouten S., and Sinninghe Damsté J. S. (1997) Occurrence of bound monoterpenoids in Palaeozoic and Proterozoic marine kerogens. *Abstr. 18th Int Meet. Org. Geochem.* Maastricht, 493–494.

Höld I. M., Schouten S., Jellema J., and Sinninghe Damsté J. S. (1999) Origin of free and bound mid-chain methyl alkanes in oil, bitumens and kerogens of the marine, Infracambrian Huqf Formation (Oman). *Org. Geochem.* **30**, 1411–1428.

Holser W. T., Schidlowski M., Mackenzie F. T., and Maynard J. B. (1988) Biogeochemical cycles of carbon and sulfur. In *Chemical Cycles in the Evolution of the Earth* (eds. C. B. Gregor, R. M. Garrels, F. T. Mackenzie, and J. B. Maynard). Wiley, New York, pp. 105–173.

Hopmans E. C., Schouten S., Pancost R. D., van der Meer M. J. T., and Sinninghe Damsté J. S. (2000) Analysis of intact tetraether lipids in archaeal cell material and sediments using high performance liquid chromatography/atmospheric pressure ionization mass spectrometry. *Rapid Commun. Mass. Spectrom.* **14**, 585–589.

Hsieh M. and Philp R. P. (2001) Ubiquitous occurrence of high molecular weight hydrocarbons in crude oils. *Org. Geochem.* **32**, 955–966.

Hsieh M., Philp R. P., and del Rio J. C. (2000) Characterization of high molecular weight biomarkers in crude oils. *Org. Geochem.* **31**, 1581–1588.

Huang Z., Poulter C. D., Wolf F. R., Somers T. C., and White J. D. (1988) Braunicene, a novel cyclic C_{32} isoprenoid from *Botryococcus braunii*. *J. Am. Chem. Soc.* **110**, 3959–3964.

Hunt J. M. (1996) *Petroleum Geochemistry and Geology*. Freeman, New York.

Imhoff J. F. (1995) Taxonomy and physiology of phototrophic purple bacteria and green sulfur bacteria. In *Anoxygenic Photosynthetic Bacteria* (eds. R. E. Blankenship, M. T. Madigan, and C. E. Bauer). Kluwer Academic, Dordrecht, pp. 1–15.

Jackson M. J., Powell T. G., Summons R. E., and Sweet I. P. (1986) Hydrocarbon shows and petroleum source rocks in sediments as old as 1.7 billion years. *Nature* **322**, 727–729.

Jahnke L. L. and Klein H. P. (1979) Oxygen as a factor in eukaryote evolution: some effects of low levels on *Saccharomyces cerevisiae*. *Origins Life Evol. Biosphere* **9**, 329–334.

Jahnke L. L. and Klein H. P. (1983) Oxygen requirements for formation and activity of the squalene epoxidase in *Saccharomyces cerevisiae*. *J. Bacteriol.* **155**, 488–492.

Jahnke L. L., Summons R. E., Hope J. M., and Des Marais D. J. (1999) Carbon isotopic fractionation in lipids from methanotrophic bacteria: II. The effects of physiology and environmental parameters on the biosynthesis and isotopic signatures of biomarkers. *Geochim. Cosmochim. Acta* **63**, 79–93.

Javaux E., Knoll A. H., and Walter M. R. (2001) Morphological and ecological complexity in early eukaryotic ecosystems. *Nature* **412**, 66–69.

Jiang N., Tong Z., Ren D., Song F., Yang D., Zhu C., and Yijun G. (1995) The discovery of retene in Precambrian and lower Paleozoic marine formations. *Chin. J. Geochem.* **14**, 41–51.

Kates M. (1993) Membrane lipids of Archaea. In *The Biochemistry of Archaea (Archaebacteria)* (eds. M. Kates, D. J. Kushner, and A. T. Matheson). Elsevier, Amsterdam, pp. 261–295.

Kemp P., Lander D. J., and Orpin C. G. (1984) The lipids of the rumen fungus *Piromonas communis*. *J. Gen. Microbiol.* **130**, 27–37.

Kenig F. (2000) C_{16}–C_{29} homologous series of monomethylalkanes in the pyrolysis products of a Holocene microbial mat. *Org. Geochem.* **31**, 237–241.

Kenig F., Sinninghe Damsté J. S., Frewin N. L., Hayes J. M., and de Leeuw J. W. (1995a) Molecular indicators for palaeoenvironmental change in a Messinian evaporitic sequence (Vena del Gesso, Italy): II. High-resolution variations in abundances and ^{13}C contents of free and sulphur-bound carbon skeletons in a single marl bed. *Org. Geochem.* **23**, 485–526.

Kenig F., Sinninghe Damsté J. S., Kock-van Dalen A. C., Rijpstra W. I. C., Huc A. Y., and de Leeuw J. W. (1995b) Occurrence and origin of mono-, di-, and trimethylalkanes in modern and Holocene cyanobacterial mats from Abu Dhabi, United Arab Emirates. *Geochim. Cosmochim. Acta* **59**, 2999–3015.

Kenig F., Simons D.-J. H., and Anderson K. B. (2001) Distribution and origin of ethyl-branched alkanes in a Cenomanian transgressive shale of the western interior seaway. *Org. Geochem.* **32**, 949–954.

Kenig F., Simons D.-J. H., Critch D., Cowen J. P., Ventura G. T., Brown T. C., and Rehbein T. (2002) Alkanes with a quaternary carbon centre: a 2,200 Myr record of sulfide oxidizing bacteria. *Geochim. Cosmochim. Acta* **66**, A393 (abstr.).

Khan H., Zaman A., Chetty G. L., Gupta A. S., and Dev S. (1971) Cheilanthatriol a new fundamental type in sesterterpenes. *Tetrahedron Lett.* **12**, 4443–4446.

Killops S. D., Carlson R. M. K., and Peters K. E. (2000) High-temperature GC evidence for the early formation of C_{40+} *n*-alkanes in coals. *Org. Geochem.* **31**, 589–597.

Kleemann G., Poralla K., Englert G., Kjøsen H., Liaaen-Jensen S., Neunlist S., and Rohmer M. (1990) Tetrahymanol from the phototrophic bacterium *Rhodopseudomonas palustris*: first report of a gammacerane triterpene from a prokaryote. *J. Gen. Microbiol.* **136**, 2551–2553.

Klomp U. C. (1986) The chemical structure of a pronounced series of iso-alkanes in South Oman crudes. *Org. Geochem.* **10**, 807–814.

Knoll A. H. (1992) The early evolution of eukaryotes: a geological perspective. *Science* **256**, 622–627.

Knott D. (1999) Elf UK expands HP–HT expertise with Elgin-Franklin development. *Oil Gas J.* **June 21**, 18–22.

Koga Y., Nishihara M., Morii H., and Akagawa-Matsushita M. (1993) Ether polar lipids of methanogenic bacteria: structures, comparative aspects, and biosynthesis. *Microbiol. Rev.* **57**, 164–182.

Kohl W., Gloe A., and Reichenbach H. (1983) Steroids from the myxobacterium *Nannocystis exedens*. *J. Gen. Microbiol.* **129**, 1629–1635.

Kohnen M. E. L., Sinninghe Damsté J. S., and de Leeuw J. W. (1991a) Biases from natural sulphurization in palaeoenvironmental reconstruction based on hydrocarbon biomarker distributions. *Nature* **349**, 775–778.

Kohnen M. E. L., Sinninghe Damsté J. S., Kock-van Dalen A. C., and de Leeuw J. W. (1991b) Di- or polysulfide-bound biomarkers in sulfur-rich geomacromolecules as revealed by selective chemolysis. *Geochim. Cosmochim. Acta* **55**, 1375–1394.

Kohnen M. E. L., Schouten S., Sinninghe Damsté J. S., de Leeuw J. W., Merrit D. A., and Hayes J. M. (1992) Recognition of paleobiochemicals by a combined molecular sulphur and isotope geochemical approach. *Science* **256**, 358–362.

Kohnen M. E. L., Sinninghe Damsté J. S., Baas M., Kock-van Dalen A. C., and de Leeuw J. W. (1993) Sulphur-bound steroid and phytane carbon skeletons in geomacromolecules: implications for the mechanism of incorporation of sulphur into organic matter. *Geochim. Cosmochim. Acta* **57**, 2515–2528.

Koopmans M. P., Köster J., van Kaam-Peters H. M. E., Kenig F., Schouten S., Hartgers W. A., de Leeuw J. W., and Sinninghe Damsté J. S. (1996a) Diagenetic and catagenetic products of isorenieratene: molecular indicators for photic zone anoxia. *Geochim. Cosmochim. Acta* **60**, 4467–4496.

Koopmans M. P., Schouten S., Kohnen M. E. L., and Sinninghe Damsté J. S. (1996b) Restricted utility of aryl isoprenoids for photic zone anoxia. *Geochim. Cosmochim. Acta* **60**, 4873–4876.

Koopmans M. P., de Leeuw J. W., Lewan M. D., and Sinninghe Damsté J. S. (1997) Impact of dia- and catagenesis on sulphur and oxygen sequestration of biomarkers as revealed by artificial maturation of an immature sedimentary rock. *Org. Geochem.* **25**, 391–426.

Köster J., Volkman J. K., Rullkötter J., Scholz-Böttcher B. M., Rethmeier J., and Fischer U. (1999) Mono-, di- and trimethyl-branched alkanes in cultures of the filamentous cyanobacterium *Calothrix scopulorum*. *Org. Geochem.* **30**, 1367–1379.

Krügel H., Krubasik P., Weber K., Saluz H. P., and Sandmann G. (1999) Functional analysis of genes from *Streptomyces griseus* involved in the synthesis of isorenieratene, a carotenoid with aromatic end groups, revealed a novel type of carotenoid desaturase. *Biochim. Biophys. Acta* **1439**, 57–64.

Kushwaha S. C. and Kates M. (1979) Studies of the biosynthesis of C_{50} carotenoids in *Halobacterium cutirubrum*. *Can. J. Microbiol.* **25**, 1292–1297.

Kushwaha S. C., Kates M., and Porter J. W. (1976) Enzymatic synthesis of C_{40} carotenes by cell-free preparation from *Halobacterium cutirubrum*. *Can. J. Biochem.* **54**, 816–823.

Kuypers M. M. M., Blokker P., Erbacher J., Kinkel H., Pancost R. D., Schouten S., and Sinninghe Damsté J. S. (2001) Massive expansion of marine Archaea during a Mid-Cretaceous Oceanic anoxic event. *Science* **293**, 92–94.

Kuypers M. M. M., Sliekers O. A., Lavik G., Schmid M., Jørgensen B. B., Kuenen J. G., Sinninghe Damsté J. S., Strous M., and Jetten M. S. M. (2003) Anaerobic ammonium oxidation by anammox bacteria in the Black Sea. *Nature* **422**, 608–611.

Landais P. and Gize A. P. (1997) Organic matter in hydrothermal ore deposits. In *Geochemistry of Hydrothermal Ore Deposits* (ed. H. L. Barnes). Wiley, Chichester, pp. 613–655.

Langworthy T. A., Tornabene T. G., and Holzer G. (1982) Lipids of Archaebacteria. *Zbl. Bakt. Hyg. I. Abt. Orig.* **C3**, 228–244.

Larter S. R. and Horsfield B. (1993) Determination of structural components of kerogens by the use of analytical pyrolysis methods. In *Organic Geochemistry* (eds. M. H. Engel and S. A. Macko). Plenum, New York and London, pp. 271–288.

Lewan M. D. (1985) Evaluation of petroleum generation by hydrous pyrolysis experimentation. *Phil. Trans. Roy. Soc. London Ser. A* **315**, 123–134.

Lewan M. D. (1997) Experiments on the role of water in petroleum formation. *Geochim. Cosmochim. Acta* **61**, 3691–3723.

Lewan M. D. (1998) Sulphur-radical control on petroleum formation rates. *Nature* **391**, 164–166.

Li S., Pang X., Li M., and Jin Z. (2003) Geochemistry of petroleum systems in the Niuzhuang south slope of Bohai Bay Basin: Part 1. Source rock characterization. *Org. Geochem.* **34**, 389–412.

Liaaen-Jensen S. (1965) Bacterial carotenoids XVIII: Aryl-carotenes from *Phaeobium*. *Acta Chem. Scand.* **19**, 1025–1030.

Liaaen-Jensen S. (1979) Marine carotenoids. In *Marine Natural Products, Chemical and Biological Perspectives* (ed. P. Scheuer). Academic Press, vol. 2, pp. 1–73.

Liaaen-Jensen S., Renstrøm B., Ramdahl T., Hallenstvet M., and Bergquist P. (1982) Carotenoids of marine sponges. *Biochem. Sys. Ecol.* **10**, 167–174.

Logan G. A., Hayes J. M., Hieshima G. B., and Summons R. E. (1995) Terminal proterozoic reorganisation of biogeochemical cycles. *Nature* **376**, 53–56.

Logan G. A., Summons R. E., and Hayes J. M. (1997) An isotopic biogeochemical study of Neoproterozoic and early Cambrian sediments from the Centralian Superbasin, Australia. *Geochim. Cosmochim. Acta* **61**, 5391–5409.

Logan G. A., Calver C. R., Gorjan P., Summons R. E., Hayes J. M., and Walter M. R. (1999) Terminal proterozoic mid-shelf benthic microbial mats in the Centralian Superbasin and their environmental significance. *Geochim. Cosmochim. Acta* **63**, 1345–1358.

Logan G. A., Hinman M. C., Walter M. R., and Summons R. E. (2001) Biogeochemistry of the 1640 Ma McArthur river (HYC) lead-zinc ore and host sediments, northern territory, Australia. *Geochim. Cosmochim. Acta* **65**, 2317–2336.

Love G. D., Snape C. E., Carr A. D., and Houghton R. C. (1995) Release of covalently-bound alkane biomarkers in high yields from kerogen via catalytic hydropyrolysis. *Org. Geochem.* **23**, 981–986.

Mackenzie A. S. (1984) Application of biological markers in petroleum geochemistry. In *Advances in Petroleum Geochemistry* (eds. J. Brooks and D. H. Welte). Academic Press, vol. 1, pp. 115–214.

Mackenzie A. S., Hoffmann C. F., and Maxwell J. R. (1981) Molecular parameters of maturation in the Toarcian shales, Paris Basin, France: III. Changes in aromatic steroid hydrocarbons. *Geochim. Cosmochim. Acta* **45**, 1345–1355.

Manes L. V., Crews P., Kernan M. R., Faulkner D. J., Fronczek F. R., and Gandour R. D. (1988) Chemistry and revised structure of suvanine. *J. Org. Chem.* **53**, 570–575.

Mango F. D. (1987) An invariance in the isoheptanes of petroleum. *Science* **237**, 514–517.

Mango F. D. (1990) The origin of light hydrocarbons in petroleum: a kinetic test of the steady-state catalytic hypothesis. *Geochim. Cosmochim. Acta* **54**, 1315–1323.

Mango F. D. (1991) The stability of hydrocarbons under the time-temperature condition of petroleum genesis. *Nature* **352**, 146–148.

Mango F. D. and Elrod L. W. (1999) The carbon isotopic composition of catalytic gas: a comparative analysis with natural gas. *Geochim. Cosmochim. Acta* **63**(7/8), 1097–1106.

Mango F. D. and Hightower J. (1997) The catalytic decomposition of petroleum into natural gas. *Geochim. Cosmochim. Acta* **61**, 5347–5350.

Mango F. D., Hightower J. W., and James A. T. (1994) Role of transition-metal catalysis in the formation of natural gas. *Nature* **268**, 536–538.

Marshall C. P., Wilson M. A., Hartung-Kagi B., and Hart G. (2001) Potential of emission Fourier transform infrared spectroscopy for *in situ* evaluation of kerogen in source rocks during pyrolysis. *Chem. Geol.* **175**, 623–633.

Matthews D. E. and Hayes J. M. (1978) Isotope-ratio-monitoring gas chromatography mass spectrometry. *Anal. Chem.* **50**, 1465–1473.

McCaffrey A. M., Dahl J. E. P., Sundararaman P., Moldowan J. M., and Schoell M. (1994a) Source rock quality determination from oil biomarkers: II. A case study using tertiary-reservoired Beauford Sea oils. *AAPG Bull.* **78**, 1527–1540.

McCaffrey M. A., Moldowan J. M., Lipton P. A., Summons R. E., Peters K. E., Jeganathan A., and Watt D. S. (1994b) Paleoenvironmental implications of novel C_{30} steranes in Precambrian to Cenozoic age petroleum and bitumen. *Geochim. Cosmochim. Acta* **58**, 529–532.

McKinney D. E., Bortiatynski J. M., Carson D. M., Clifford D. J., de Leeuw J. W., and Hatcher P. G. (1996) Tetramethylammonium hydroxide (TMAH) thermochemolysis of the aliphatic biopolymer cutan: insights into the chemical structure. *Org. Geochem.* **24**, 641–650.

McKirdy D. M., Aldridge A. K., and Ypma P. J. M. (1982) A geochemical comparison of some crude oils from pre-Ordovician carbonate rocks. In *Advances in Organic Geochemistry 1981* (ed. M. Bjorøy *et al.*). Wiley, pp. 99–107.

McKirdy D. M., Cox R. E., Volkman J. K., and Howell V. J. (1986) Botryococcane in a new class of Australian non-marine crude oils. *Nature* **320**, 57–59.

McNeil R. I. and BeMent W. O. (1996) Thermal stability of hydrocarbons: laboratory criteria and field examples. *Energy Fuels* **10**, 60–67.

Méjanelle L., Lòpez J. F., Gunde-Cimerman N., and Grimalt J. O. (2000) Sterols of melanized fungi from hypersaline environments. *Org. Geochem.* **31**, 1031–1040.

Méjanelle L., Sanchez-Gargallo A., Bentaleb I., and Grimalt J. O. (2003) Long chain *n*-alkyl diols, hydroxy ketones and sterols in a marine eustigmatophyte, *Nannochloropsis gaditana*, and in *Brachionus plicatilis* feeding on the algae. *Org. Geochem.* **34**, 527–538.

Mendelson C. V. (1993) Acritarchs and prasinophytes. In *Fossil Prokaryotes and Protists* (ed. J. H. Lipps). Blackwell, pp. 77–104.

Menounos P., Staphylakis K., and Gegiou D. (1986) The sterols of *Nigella sativa* seed oil. *Phytochemistry* **25**, 761–763.

Metzger P. and Largeau C. (1999) Chemicals of *Botryococcus braunii*. In *Chemicals from Microalgae* (ed. Z. Cohen). Taylor and Francis, pp. 205–260.

Metzger P., Casadevall E., Pouet M. J., and Pouet Y. (1985) Structures of some botryococcenes: branched hydrocarbons from the B-race of the green alga *Botryococcus braunii*. *Phytochemistry* **24**, 2995–3002.

Metzger P., Pouet Y., Bischoff R., and Casadevall E. (1993) An aliphatic polyaldehyde from *Botryococcus braunii* (A race). *Phytochemistry* **32**, 875–883.

Michaelis W., Seifert R., Nauhaus K., Treude T., Thiel V., Blumenberg M., Knittel K., Gieseke A., Peterknecht K., Pape T., Boetius A., Amann R., Jørgensen B. B., Widdel F., Peckmann J., Pimenov N. V., and Gulin M. B. (2002) Microbial reefs in the Black Sea fueled by anaerobic oxidation of methane. *Science* **297**, 1013–1015.

Miyamoto T., Sakamoto K., Amano H., Higuchi R., Komori T., and Sasaki T. (1992) Three new cytotoxic sesterterpenoids, inorolide A, B, and C from the nudibranch *Chromodoris inomata*. *Tetrahedron Lett.* **33**, 5811–5814.

Moldowan J. M. and Fago F. J. (1986) Structure and significance of a novel rearranged monoaromatic steroid hydrocarbon in petroleum. *Geochim. Cosmochim. Acta* **50**, 343–351.

Moldowan J. M. and Seifert W. K. (1979) Head-to-head linked isoprenoid hydrocarbons in petroleum. *Science* **204**, 169–171.

Moldowan J. M. and Seifert W. K. (1983) Identification of an extended series of tricyclic terpanes in petroleum. *Geochim. Cosmochim. Acta* **47**, 1531–1534.

Moldowan J. M., Seifert W. K., and Gallegos E. J. (1985) Relationship between petroleum composition and depositional environment of petroleum source rocks. *AAPG Bull.* **69**, 1255–1268.

Moldowan J. M., Fago F. J., Lee C. Y., Jacobson S. R., Watt D. S., Slougui N.-E., Jeganathan A., and Young D. C. (1990) Sedimentary 24-*n*-propylcholestanes, molecular fossils diagnostic of marine algae. *Science* **247**, 309–312.

Moldowan J. M., Fago F. J., Carlson R. M. K., Young D. C., Van Duyne G., Clardy J., Schoell M., Pillinger C. T., and Watt D. S. (1991a) Rearranged hopanes in sediments and petroleum. *Geochim. Cosmochim. Acta* **55**, 3333–3353.

Moldowan J. M., Lee C. Y., Watt D. S., Jeganathan A., Slougui N. E., and Gallegos E. J. (1991b) Analysis and occurrence of C_{26}-steranes in petroleum and source rocks. *Geochim. Cosmochim. Acta* **55**, 1065–1081.

Moldowan J. M., Dahl J. E. P., Huizinga B. J., Fago F. J., Hickey L. J., Peakman T. M., and Taylor D. W. (1994) The molecular fossil record of oleanane and its relation to angiosperms. *Science* **265**, 768–771.

Moldowan J. M., Dahl J. E. P., Jacobson S. R., Huizinga B. J., Fago F. J., Shetty R., Watt D. S., and Peters K. E. (1996) Chemostratigraphic reconstruction of biofacies: molecular evidence linking cyst-forming dinoflagellates with Pre-Triassic ancestors. *Geology* **24**, 159–162.

Moldowan J. M. and Talyzina N. M. (1998) Biogeochemical evidence for dinoflagellate ancestors in the Early Cambrian. *Science* **281**, 1168–1170.

Mueller E. and Philp R. P. (1998) Extraction of high molecular weight hydrocarbons from source rocks: an example from the Green River Formation, Uinta Basin, Utah. *Org. Geochem.* **28**, 625–632.

Murray A. P., Sosrowidjojo I. B., Alexander R., Kagi R. I., Norgate C. M., and Summons R. E. (1997) Oleananes in oils and sediments: evidence of marine influence during early diagenesis? *Geochim. Cosmochim. Acta* **61**, 1261–1276.

Murray A. P., Edwards D., Hope J. M., Boreham C. J., Booth W. E., Alexander R. A., and Summons R. E. (1998) Carbon isotope biogeochemistry of plant resins and derived hydrocarbons. *Org. Geochem.* **29**, 1199–1214.

Mycke B., Michaelis W., and Degens E. T. (1988) Biomarkers in sedimentary sulfides of Precambrian age. *Org. Geochem.* **13**, 619–625.

Nes W. D., Norton R. A., Crumley F. G., Madigan S. J., and Katz E. R. (1990) Sterol phylogenesis and algal evolution. *Proc. Natl. Acad. Sci.* **87**, 7565–7569.

Neunlist S. and Rohmer M. (1985) Novel hopanoids from the methylotrophic bacteria *Methylococcus capsulatus* and *Methylomonas methanica*. *Biochem. J.* **231**, 635–639.

Nichols P. D., Volkman J. K., Palmisano A. C., Smith G. A., and C. W. D. (1988) Occurrence of an isoprenoid C_{25} di-unsaturated alkene and high neutral lipid content in Antarctic sea-ice diatom communities. *J. Phycol.* **24**, 90–96.

Nichols P. D., Palmisano A. C., Rayner M. S., Smith G. A., and White D. C. (1990) Occurrence of novel C_{30} sterols in Antarctic sea-ice diatom communities during a spring bloom. *Org. Geochem.* **15**, 503–508.

Nip M., Tegelaar E. W., Brinkhuis H., de Leeuw J. W., Schenck P. A., and Holloway P. J. (1986a) Analysis of modern and fossil plant cuticles by Curie-point Py–GC and Curiepoint Py–GC–MS: recognition of a new highly aliphatic and resistant biopolymer. *Org. Geochem.* **10**, 769–778.

Nip M., Tegelaar E. W., de Leeuw J. W., Schenck P. A., and Holloway P. J. (1986b) A new non-saponifable highly aliphatic and resistant biopolymer in plant cuticles: evidence from pyrolysis and ^{13}C NMR analysis of present day and fossil plants. *Naturwissenschaften* **73**, 579–585.

Noble R. A., Alexander R., Kagi R. I., and Knox J. (1985) Tetracyclic diterpenoid hydrocarbons in some Australian coals, sediments and crude oils. *Geochim. Cosmochim. Acta* **49**, 2141–2147.

Noble R. A., Alexander R., and Kagi R. I. (1986) Identification of some diterpenoid hydrocarbons in petroleum. *Org. Geochem.* **10**, 825–829.

Noble R. A., Wu C. H., and Atkinson C. D. (1991) Petroleum generation and migration from Talang Akar coals and shales offshore NW Java, Indonesia. *Org. Geochem.* **17**, 363–374.

Oren A. (2002) Molecular ecology of extremely halophilic Archaea and Bacteria. *FEMS Microbiol. Ecol.* **39**, 1–7.

Orphan V. J., House C. H., Hinrichs K. U., McKeegan K. D., and DeLong E. F. (2001) Methane-consuming archaea revealed by directly coupled isotopic and phylogenetic analysis. *Science* **293**, 484–487.

Otto A. and Simoneit B. R. T. (2001) Chemosystematics and diagenesis of terpenoids in fossil conifer species and sediment from the Eocene Zeitz formation, Saxony, Germany. *Geochim. Cosmochim. Acta* **65**, 3505–3527.

Otto A. and Simoneit B. R. T. (2002) Biomarkers of Holocene buried conifer logs from Bella Coola and north Vancouver, British Columbia, Canada. *Org. Geochem.* **33**, 1241–1251.

Ourisson G. (1994) Biomarkers in the Proterozoic record. In *Early Life on Earth, Nobel Symposium* (ed. S. Bengtson). Columbia University Press, New York, vol. 84, pp. 259–269.

Ourisson G. and Albrecht P. (1992) Hopanoids: 1. Geohopanoids: the most abundant natural products on Earth? *Acc. Chem. Res.* **25**, 398–402.

Ourisson G., Rohmer M., and Poralla K. (1987) Prokaryotic hopanoids and other polyterpenoid sterol surrogates. *Ann. Rev. Microbiol.* **41**, 301–333.

Pancost R. D., Freeman K. H., Patzkowsky E., Wavrek D. A., and Collister J. W. (1998) Molecular indicators of redox and marine photoautotroph composition in the late Middle Ordovician of Iowa USA. *Org. Geochem.* **29**, 1649–1662.

Pancost R. D., Sinninghe Damsté J. S., De Lint S., van der Maarel M. J., Gottschal J. C., and the MEDINAUT Shipboard Scientific Party (2000) Biomarker evidence for widespread anaerobic methane oxidation in Mediterranean sediments by a consortium of methanogenic archaea and bacteria. *Appl. Environ. Microbiol.* **67**, 1126–1132.

Pancost R. D., Crawford N., and Maxwell J. R. (2002) Molecular evidence for basin-scale photic zone euxinia in the Permian Zechstein sea. *Chem. Geol.* **188**, 217–227.

Pearson A., Eglinton T. I., and McNichol A. P. (2000) An organic tracer for surface ocean radiocarbon. *Paleoceanography* **15**, 541–550.

Pearson A., McNichol A. P., Benitez-Nelson B. C., Hayes J. M., and Eglinton T. I. (2001) Origins of lipid biomarkers in Santa Monica Basin surface sediment: a case study using compound-specific ^{14}C analysis. *Geochim. Cosmochim. Acta* **65**, 3123–3137.

Peng P., Sheng G., Fu J., and Yan Y. (1998) Biological markers in 1.7 billion year old rock from the Tuanshanzi formation, Jixian strata section, North China. *Org. Geochem.* **29**, 1321–1329.

Pepper A. S. and Dodd T. A. (1995) Simple kinetic models of petroleum formation: Part II. oil-gas cracking. *Mar. Petrol. Geol.* **12**, 321–340.

Perkins G. M., Bull I. D., Ten Haven H. L., Rullkötter J., Smith Z. E. F., and Peakman T. M. (1995) First positive identification of triterpanes of the taraxastane family in petroleums and oil shales: 19α(H)-taraxastane and 24-nor-19α(H)-taraxastane. Evidence for a previously unrecognised diagenetic alteration pathway of lup-20(29)-ene derivatives. In *17th International Meeting on Organic Geochemistry. Organic Geochemistry: Developments and Applications to Energy, Climate, Environments and Human History*, 247–279.

Peters K. E. and Moldowan J. M. (1993) *The Biomarker Guide*. Prentice Hall, Engelwood Cliffs, NJ.

Peters K. E., Clark M. E., Das Gupta U., McCaffrey A. M., and Lee C. Y. (1995) Recognition of an infracambrian source rock based on biomarkers in the Baghewala-1 oil, India. *AAPG Bull.* **79**, 1481–1494.

Philp R. P. and Oung J.-N. (1992) Biomarker distributions in crude oils as determined by tandem mass spectrometry. In *Biomarkers in Sediments and Petroleum* (eds. J. M. Moldowan, P. Albrecht, and R. P. Philp). Prentice Hall, Engelwood Cliffs, NJ, pp. 106–123.

Philp R. P., Chen J. H., Fu F. M., and Sheng G. Y. (1992) A geochemical investigation of crude oils and source rocks from Biyang Basin, China. *Org. Geochem.* **18**, 933–945.

Ping'an P., Eglinton G., Jiamo F., Guoying S., and Jiayou X. (1992) Biological markers in Chinese ancient sediments: 2. Alkanes, cycloalkanes, and geoporphyrins in paleoenvironmental assessment. *Energy Fuels* **6**, 225–235.

Piretti M., Pagliuca G., Boni L., Pistocchi R., Diamante M., and Gazzotti T. (1997) Investigation of 4-methyl sterols from cultured dinoflagellate algal strains. *J. Phycol.* **33**, 61–67.

Pratt L. M., Summons R. E., and Hieshima G. B. (1991) Sterane and triterpane biomarkers in the Precambrian Nonesuch Formation, North American Midcontinent Rift. *Geochim. Cosmochim. Acta* **55**, 911–916.

Price L. C. (1982) Organic geochemistry of core samples from an ultra-deep hot well (300°C, 7 km). *Chem. Geol.* **37**, 215–228.

Price L. C. (1983) Geologic time as a parameter in organic metamorphism and vitrinite reflectance as an absolute paleogeothermometer. *J. Petrol. Geol.* **6**, 5–38.

Price L. C. (1993) Thermal stability of hydrocarbons in nature: limits, evidence, characteristics, and possible controls. *Geochim. Cosmochim. Acta* **57**, 3261–3280.

Price L. C. (1997) Minimum thermal stability levels and controlling parameters of methane, as determined by C_{15+} hydrocarbon thermal stabilities. In *Geologic Controls of Deep Natural Gas Resources in the United States* (eds. T. S. Dyman, D. D. Rice, and P. A. Westcott). USGS, vol. 2146-K, pp. 139–176.

Price L. C. (2000) Organic metamorphism in the California petroleum basins: Chapter B. Insights from extractable bitumen and saturated hydrocarbons. *US Geol. Surv. Bull.* **B2174-B**, 33pp.

Price L. C. and Wenger L. M. (1992) The influence of pressure on petroleum generation and migration as suggested by aqueous pyrolysis. *Org. Geochem.* **19**, 141–159.

Price L. C., Clayton J., and Rumen L. L. (1981) Organic geochemistry of the 9.6 km Bertha No. 1, well, Oklahoma. *Org. Geochem.* **3**, 59–77.

Price L. C., Pawlewicz M. J., and Daws T. (1999) Organic metamorphism in the California petroleum basins: Chapter A. Rock Eval and vitrinite reflectance. *US Geol. Surv. Bull.* **B 2174-A**, 34pp.

Putschew A., Schaeffer P., Schaeffer-Reiss C., and Maxwell J. R. (1998) Carbon isotope characteristic of the diaromatic carotenoid, isorenieratene (intact and sulfide bound) and a novel isomer in sediments. *Org. Geochem.* **28**, 1849–1856.

Püttmann W., Hagemann H. W., Merz C., and Speczik S. (1988) Influence of organic material on mineralization processes in the Permian Kupferschiefer Formation, Poland. *Org. Geochem.* **13**, 357–363.

Radke M., Horsfield B., Littke R., and Rullkötter J. (1997) Maturation and petroleum generation. In *Petroleum and Basin Evolution* (eds. D. H. Welte, B. Horsfield, and D. R. Baker). Springer, Berlin, pp. 169–229.

Raederstorff D. and Rohmer M. (1984) Sterols of the unicellular algae *Nematochrysopsis roscoffensis* and *Chrysotila lamellosa*: isolation of (24E)-24-*n*-propylidenecholesterol and 24-*n*-propylcholesterol. *Phytochemistry* **23**, 2835–2838.

Rasmussen B. (2000) Filamentous microfossils in a 3,235-million-year-old volcanogenic massive sulphide deposit. *Nature* **405**, 676–679.

Reddy C. M., Eglinton T. I., Hounshell A., White H. K., Xu L., Gaines R. B., and Frysinger G. S. (2002) The West Falmouth oil spill: the persistence of petroleum hydrocarbons in marsh sediments. *Environ. Sci. Technol.* **36**, 4754–4760.

Redecker D., Kodner R., and Graham L. E. (2000) Glomalean fungi from the Ordovician. *Science* **289**, 1920–1921.

Reed J. H., Illich H. A., and Horsfield B. (1986) Biochemical evolutionary significance of Ordovician oils and their sources. *Org. Geochem.* **10**, 347–358.

Requejo A. G., Creaney S., Allan J., Gray N. R., and Cole K. S. (1992) Aryl isoprenoids and diaromatic carotenoids in Paleozoic source rocks and oils from the western Canada and Williston Basins. *Org. Geochem.* **19**, 245–264.

Revill A. T., Volkman J. K., O'Leary T., Summons R. E., Boreham C. J., Banks M. R., and Denwer K. (1994) Depositional setting, hydrocarbon biomarkers and thermal maturity of tasmanite oil shales from Tasmania, Australia. *Geochim. Cosmochim. Acta* **58**, 3803–3822.

Rieley G., Collier R. J., Jones D. M., Eglinton G., Eakin P. A., and Fallick A. E. (1991) Sources of sedimentary lipids inferred from carbon isotopic analysis of individual compounds. *Nature* **352**, 425–427.

Rieley G., Collister J. W., Stern B., and Eglinton G. (1993) Gas chromatography/isotope ratio mass spectrometry of leaf wax *n*-alkanes from plants of differing carbon dioxide metabolisms. *Rapid Commun. Mass. Spectrom.* **7**, 488–491.

Riolo J., Hussler G., Albrecht P., and Connan J. (1985) Distribution of aromatic steroids in geological samples: their evaluation as geochemical parameters. *Org. Geochem.* **10**, 981–990.

Risatti J. B., Rowland S. J., Yon D. A., and Maxwell J. R. (1984) Stereochemical studies of acyclic isoprenoids: XII. Lipids of methanogenic bacteria and possible contributions to sediments. *Org. Geochem.* **6**, 93–104.

Robinson N. and Eglinton G. (1990) Lipid chemistry of Icelandic hot spring microbial mats. *Org. Geochem.* **15**, 291–298.

Robinson N., Eglinton G., and Brassell S. C. (1984) Dinoflagellate origin for sedimentary 4α-methylsteroids and 5α(H)-stanols. *Nature* **308**, 439–442.

Robson J. N. and Rowland S. J. (1986) Identification of novel widely distributed sedimentary acyclic sesterpenoids. *Nature* **324**, 561–563.

Rohmer M. (1993) The biosynthesis of triterpenoids of the hopane series in the Eubacteria: mine of new enzyme reactions. *Pure Appl. Chem.* **65**, 1293–1298.

Rohmer M., Bouvier-Navé P., and Ourisson G. (1984) Distribution of hopanoid triterpenes in prokaryotes. *J. Gen. Microbiol.* **130**, 1137–1150.

Rohmer M., Bisseret P., and Neunlist S. (1992) The hopanoids, prokaryotic triterpenoids and precursors of ubiquitous molecular fossils. In *Biological Markers in Sediments and Petroleum* (eds. J. M. Moldowan, P. Albrecht, and R. P. Philp). Prentice Hall, Englewood Cliffs, NJ, pp. 1–17.

Rohmer M., Knani M., Simonin P., Sutter B., and Sahm H. (1993) Isoprenoid biosynthesis in bacteria: a novel pathway for the early steps leading to isopentenyl diphosphate. *Biochem. J.* **295**, 517–524.

Rønnekleiv M. and Liaaen-Jensen S. (1995) Bacterial carotenoids 53, C_{50}–carotenoids 23, Carotenoids of *Haloferax volcanii* versus other halophilic bacteria. *Biochem. Sys. Ecol.* **23**, 627–634.

Rosing M. T. (1999) ^{13}C-depleted carbon microparticles in >3700-Ma sea-floor sedimentary rocks from West Greenland. *Science* **283**, 674–676.

Rothman D. H., Hayes J. M., and Summons R. E. (2003) Dynamics of the Neoproterozoic carbon cycle. *PNAS* **100**, 8124–8129.

Rothschild L. and Mancinelli R. L. (2001) Life in extreme environments. *Nature* **409**, 1092–1101.

Rowland S. J., Allard W. G., Belt S. T., Massé G., Robert J.-M., Blackburn S. I., Frampton D., Revill A. T., and Volkman J. K. (2001) Factors influencing the distributions of polyunsaturated terpenoids in the diatom, *Rhizosolenia setigera*. *Phytochemistry* **58**, 717–728.

Rubinstein I. and Strausz O. P. (1979) Geochemistry of the thiourea adduct fraction from an Alberta petroleum. *Geochim. Cosmochim. Acta* **43**, 1387–1392.

Rubinstein I., Sieskind O., and Albrecht P. (1975) Rearranged steranes in a shale: occurrence and simulated formation. *J. Chem. Soc. Perkin Trans.* 1, 711–719.

Rullkötter J. (1999) Organic matter: the driving force for early diagenesis. In *Marine Geochemistry* (eds. H. D. Schulz and M. Zabel). Springer, Berlin, pp. 129–172.

Rullkötter J., Peakman T. M., and ten Haven H. L. (1994) Early diagenesis of terrigenous terpenoids and its implications for petroleum geochemistry. *Org. Geochem.* **21**, 215–233.

Rutters H., Sass H., Cypionka H., and Rullkötter J. (2001) Monoalkylether phospholipids in the sulfate-reducing bacteria *Desulfosarcina variabilis* and *Desulforhabdus amnigenus*. *Arch. Microbiol.* **176**, 435–442.

Saito T., Terato H., and Yamamoto O. (1994) Pigments of *Rubrobacter radiotolerans*. *Arch. Microbiol.* **162**, 414–421.

Sajgó C. (2000) Assessment of generation temperatures of crude oils. *Org. Geochem.* **31**, 1301–1323.

Sandison C. M., Alexander R., Kagi R. I., and Boreham C. J. (2002) Sulfurisation of lipids in a marine-influenced lignite. *Org. Geochem.* **33**, 1053–1077.

Saunders G. W., Potter D., and Andersen R. A. (1997) Phylogenetic affinities of the Sarcinochrysidales and Chrysomeridales (Heterokonta) based on analysis of molecular and combined data. *J. Phycol.* **33**, 310–318.

Schaeffer P., Reiss C., and Albrecht P. (1995) Geochemical study of macromolecular organic matter from sulfur-rich sediments of evaporitic origin (Messinian of Sicily) by chemical degradations. *Org. Geochem.* **23**, 567–581.

Schaeffer P., Adam P., Wehrung P., and Albrecht P. (1997) Novel aromatic carotenoid derivatives from sulfur photosynthetic bacteria in sediments. *Tetrahedron Lett.* **38**, 8413–8416.

Schaeflé J., Ludwig B., Albrecht P., and Ourisson G. (1977) Hydrocarbures aromatiques d'origine géologique: II. Nouveaux carotenoïdes aromatiques fossiles. *Tetrahedron Lett.* **18**, 3673–3676.

Schneider D. A., Bickford M. E., Cannon W. F., Schulz K. J., and Hamilton M. A. (2002) Age of volcanic rocks and syndepositional iron formations, Marquette range supergroup: implications for the tectonic setting of Paleoproterozoic iron formations of the Lake Superior region. *Can. J. Earth Sci.* **39**, 999–1012.

Schoell M., McCaffrey A. M., Fago F. J., and Moldowan J. M. (1992) Carbon isotopic compositions of 28,30-bisnorhopanes and other biological markers in a Monterey crude oil. *Geochim. Cosmochim. Acta* **56**, 1391–1399.

Schopf J. W., Kudryavtsev A. B., Agresti D. G., Wdowlak T. J., and Czaja A. D. (2002) Laser-Raman imagery of Earth's earliest fossils. *Nature* **416**, 73–76.

Schouten S., van der Maarel M. J., Huber R., and Sinninghe Damsté J. S. (1997) 2,6,10,15,19-pentamethylicosenes in *Methanolobus bombayensis*, a marine methanogenic archaeon, and in *Methanosarcina mazei*. *Org. Geochem.* **26**, 409–414.

Schouten S., Klein Breteler W. C. M., Blokker P., Schogt N., Rijpstra W. I. C., Grice K., Baas M., and Sinninghe Damsté J. S. (1998a) Biosynthetic effects on the stable carbon isotopic composition of algal lipids: implications for deciphering the carbon isotopic biomarker record. *Geochim. Cosmochim. Acta* **62**, 1397–1406.

Schouten S., Sephton S., Baas M., and Sinninghe Damsté J. S. (1998b) Steroid carbon skeletons with unusually branched C-3 alkyl side chains in sulphur-rich sediments. *Geochim. Cosmochim. Acta* **62**, 1127–1132.

Schouten S., Bowman J. P., Rijpstra W. I. C., and Sinninghe Damsté J. S. (2000) Sterols in a psychrophilic methanotroph, *Methylosphaera hansonii*. *FEMS Microbiol. Lett.* **186**, 193–195.

Schouten S., De Loureiro M. R. B., Sinninghe Damsté J. S., and de Leeuw J. W. (2001a) Molecular biogeochemistry of Monterey sediments, Naples Beach, California: I. Distributions of hydrocarbons and organic sulfur compounds. In *The Monterey Formation: from Rocks to Molecules* (eds. C. M. Isaacs and J. Rullkötter). Columbia University Press, New York, pp. 150–174.

Schouten S., Rijpstra W. I. C., Kok M., Hopmans E. C., Summons R. E., Volkman J. K., and Sinninghe Damsté J. S. (2001b) Molecular organic tracers of biogeochemical processes in a saline meromictic lake (Ace Lake). *Geochim. Cosmochim. Acta* **65**, 1629–1640.

Schouten S., Schoell M., Sinninghe Damsté J. S., Summons R. E., and de Leeuw J. W. (2001c) Molecular biogeochemistry of Monterey sediments, Naples Beach, California: II. Stable carbon isotopic compositions of free and sulphur-bound carbon skeletons. In *The Monterey Formation: From Rocks to Molecules* (eds. C. M. Isaacs and J. Rullkötter). Columbia University Press, New York, pp. 175–188.

Schouten S., Wakeham S. G., Hopmans E. C., and Sinninghe Damsté J. S. (2003) Biogeochemical evidence that thermophilic archaea mediate the anaerobic oxidation of methane. *Appl. Environ. Microbiol.* **69**, 1680–1686.

Shen Y., Buick R., and Canfield D. E. (2001) Isotopic evidence for microbial sulphate reduction in the early Archaean era. *Nature* **410**, 77–81.

Shen Y., Knoll A. H., and Walter M. R. (2003) Evidence for low sulfate and anoxia in a mid-proterozoic marine basin. *Nature* **423**, 632–635.

Shiea J., Brassell S. C., and Ward D. M. (1990) Mid-chain branched mono- and dimethyl alkanes in hot spring cyanobacterial mats: a direct biogenic source for branched alkanes in ancient sediments? *Org. Geochem.* **15**, 223-231.

Shiea J., Brassell S. C., and Ward D. M. (1991) Comparative analysis of extractable lipids in hot spring microbial mats and their component photosynthetic bacteria. *Org. Geochem.* **17**, 309–319.

Sieskind O., Joly G., and Albrecht P. (1979) Simulation of the geochemical transformation of sterols: superacid effect of clay minerals. *Geochim. Cosmochim. Acta* **43**, 1675–1680.

Simoneit B. R. T. (1977) Diterpenoid compounds and other lipids in deep-sea sediments and their geochemical significance. *Geochim. Cosmochim. Acta* **41**, 463–476.

Simoneit B. R. T. and Fetzer J. C. (1996) High molecular weight polycyclic aromatic hydrocarbons in hydrothermal petroleums from the Gulf of California and Northeast Pacific Ocean. *Org. Geochem.* **24**, 1065–1077.

Simoneit B. R. T., Grimalt J. O., Hwang T. G., Cox R. E., Hatcher P. G., and Nissenbaum A. (1986) Cyclic terpenoids of contemporary resinous plant detritus and of fossil woods, ambers and coals. *Org. Geochem.* **10**, 877–889.

Simoneit B. R. T., Kawka O. E., and Wang G.-M. (1992) Biomarker maturation in contemporary hydrothermal systems, alteration of immature organic matter in zero geological time. In *Biomarkers in Sediments and Petroleum* (eds. J. M. Moldowan, P. Albrecht, and R. P. Philp). Prentice Hall, Englewood Cliffs, NJ, pp. 124–141.

Simons D.-J. H. and Kenig F. (2001) Molecular fossil constraints on the water column structure of the Cenomanian-turonian western interior seaway, USA. *Palaeogeogr. Palaeoclimatol. Palaeoecol.* **169**, 129–152.

Simons D.-J. H., Kenig F., Critch D., and Schröder-Adams C. J. (2002) Significance of novel branched alkanes with quaternary carbon centers in black shales. *Geochim. Cosmochim. Acta* **66**, A718 (abstr).

Sinninghe Damsté J. S. and de Leeuw J. W. (1990) Analysis, structure and geochemical significance of organically-bound sulphur in the geosphere: state of the art and future research. *Org. Geochem.* **16**, 1077–1101.

Sinninghe Damsté J. S., Kenig F., Koopmans M. P., Köster J., Schouten S., Hayes J. M., and De Leeuw J. W. (1995) Evidence for gammacerane as an indicator of water column stratification. *Geochim. Cosmochim. Acta* **59**, 1895–1900.

Sinninghe Damsté J. S., Rijpstra W. I. C., Schouten S., Peletier H., van der Maarel M. J., and Gieskes W. W. C. (1999a) A C_{25} highly branched isoprenoid alkene and C_{25} and C_{27} *n*-polyenes in the marine diatom *Rhizosolenia setigera*. *Org. Geochem.* **30**, 95–100.

Sinninghe Damsté J. S., Schouten S., Rijpstra W. I. C., Hopmans E. C., Peletier H., Gieskes W. W. C., and Geenevasen J. A. J. (1999b) Structural identification of the C_{25} highly branched isoprenoid pentaene in the marine diatom *Rhizosolenia setigera*. *Org. Geochem.* **30**, 1581–1583.

Sinninghe Damsté J. S., Schouten S., and van Duin A. C. T. (2001) Isorenieratene derivatives in sediments: possible controls on their distribution. *Geochim. Cosmochim. Acta* **65**, 1557–1571.

Sinninghe Damsté J. S., Rijpstra W. I. C., Hopmans E. C., Prahl F. G., Wakeham S. G., and Schouten S. (2002a) Distribution of membrane lipids of planktonic Crenarchaeota in the Arabian Sea. *Appl. Environ. Microbiol.* **68**, 2997–3002.

Sinninghe Damsté J. S, Schouten S., Hopmans E. C., van Duin A. C. T., and Geenevasen J. A. J. (2002b) Crenarchaeol: the characteristic core glycerol dibiphytanyl glycerol tetraether membrane lipid of cosmopolitan pelagic crenarchaeota. *J. Lipid Res.* **43**, 1641–1651.

Sinninghe Damsté J. S., Strous M., Rijpstra W. I. C., Hopmans E. C., Geenevasen J. A. J., van Duin A. C. T., van Niftrik L. A., and Jetten M. S. M. (2002c) Linearly concatenated cyclobutane lipids form a dense bacterial membrane. *Nature* **419**, 708–712.

Solomon P. R. and Carangelo R. M. (1987) FT-IR analysis of coal. *Fuel* **67**, 949–959.

Sosrowidjojo I. B., Murray A. P., Alexander R., Kagi R. I., and Summons R. E. (1996) Bicadinanes and related compounds as maturity indicators for oils and sediments. *Org. Geochem.* **24**, 43–55.

Strand A., Shivajib S., and Liaaen-Jensen S. (1997) Bacterial carotenoids 55: C_{50}–carotenoids 25: revised structures of carotenoids associated with membranes in psychrotrophic *Micrococcus roseus*. *Biochem. Sys. Ecol.* **25**, 547–552.

Subroto E. A., Alexander R., and Kagi R. I. (1991) 30-Norhopanes: their occurrence in sediments and crude oils. *Chem. Geol.* **93**, 179–192.

Summons R. E. (1987) Branched alkanes from ancient and modern sediments: isomer discrimination by GC/MS with multiple reaction monitoring. *Org. Geochem.* **11**, 281–289.

Summons R. E. and Capon R. J. (1988) Fossil steranes with unprecedented methylation in ring-A. *Geochim. Cosmochim. Acta* **52**, 2733–2736.

Summons R. E. and Capon R. J. (1991) Identification and significance of 3-ethyl steranes in sediments and petroleum. *Geochim. Cosmochim. Acta* **55**, 2391–2395.

Summons R. E. and Jahnke L. L. (1990) Identification of the methylhopanes in sediments and petroleum. *Geochim. Cosmochim. Acta* **54**, 247–251.

Summons R. E. and Jahnke L. L. (1992) Hopenes and hopanes methylated in ring-A: correlation of the hopanoids from extant methylotrophic bacteria with their fossil analogues. In *Biological Markers in Sediments and Petroleum* (eds. J. M. Moldowan, P. Albrecht, and R. P. Philp). Prentice Hall, Englewood Cliffs, NJ, pp. 182–200.

Summons R. E. and Powell T. G. (1986) Chlorobiaceae in Paleozoic seas revealed by biological markers, isotopes and geology. *Nature* **319**(6056), 763–765.

Summons R. E. and Powell T. G. (1987) Identification of aryl isoprenoids in source rocks and crude oils: biological markers for the green sulphur bacteria. *Geochim. Cosmochim. Acta* **51**, 557–566.

Summons R. E. and Powell T. G. (1991) Petroleum source rocks of the Amadeus Basin. In *The Amadeus Basin Central Australia,* BMR Bulletin (eds. R. J. Korsch and

J. M. Kennard). Bureau of Mineral Resources, Canberra, vol. 236, pp. 511–524.

Summons R. E. and Powell T. G. (1992) Hydrocarbon composition of the late Proterozoic oils of the Siberian platform: implications for the depositional environment of source rocks. In *Early Organic Evolution: Implications for Mineral and Energy Resources* (eds. M. Schidlowski, S. Golunic, M. M. Kimberley, D. M. McKirdy, and P. A. Trudinger). Springer, Berlin, pp. 296–307.

Summons R. E. and Walter M. R. (1990) Molecular fossils and microfossils of prokaryotes and protists from Proterozoic sediments. *Am. J. Sci.* **290-A**, 212–244.

Summons R. E., Volkman J. K., and Boreham C. J. (1987) Dinosterane and other steroidal hydrocarbons of dinoflagellate origin in sediments and petroleum. *Geochim. Cosmochim. Acta* **51**, 3075–3082.

Summons R. E., Brassell S. C., Eglinton G., Evans E., Horodyski R. J., Robinson N., and Ward D. M. (1988a) Distinctive hydrocarbon biomarkers from fossiliferous sediments of the late Proterozoic Walcott Member, Chuar Group, Grand Canyon, Arizona. *Geochim. Cosmochim. Acta* **52**, 2625–2637.

Summons R. E., Powell T. G., and Boreham C. J. (1988b) Petroleum geology and geochemistry of the Middle Proterozoic McArthur Basin, northern Australia: III. Composition of extractable hydrocarbons. *Geochim. Cosmochim. Acta* **52**, 1747–1763.

Summons R. E., Thomas J., Maxwell J. R., and Boreham C. J. (1992) Secular and environmental constraints on the occurrence of dinosterane in sediments. *Geochim. Cosmochim. Acta* **56**, 2437–2444.

Summons R. E., Barrow R. A., Capon R. J., Hope J. M., and Stranger C. (1993) The structure of a new C_{25} isoprenoid alkane biomarker from diatomaceous microbial communities. *Aust. J. Chem.* **46**, 907–915.

Summons R. E., Jahnke L. L., and Roksandic Z. (1994a) Carbon isotope fractionation in lipids from methanothrophic bacteria: relevance for interpretation of the geological record of biomarkers. *Geochim. Cosmochim. Acta* **58**, 2853–2863.

Summons R. E., Taylor D., and Boreham C. (1994b) Geochemical tools for evaluating petroleum generation in Middle Proterozoic sediments of the McArthur Basin, Northern Territory, Australia. *APEA J.* **34**, 692–706.

Summons R. E., Jahnke L. L., Hope J. M., and Logan G. A. (1999) 2-methylhopanoids as biomarkers for cyanobacterial oxygenic photosynthesis. *Nature* **400**, 554–557.

Summons R. E., Jahnke L. L., Cullings K. W., and Logan G. A. (2001) Cyanobacterial biomarkers: triterpenoids plus steroids? *EOS, Trans., AGU: Fall Meet. Suppl.* **82**, Abstract B22D-0184.

Summons R. E., Metzger P., Largeau C., Murray A. P., and Hope J. M. (2002) Polymethylsqualanes from *Botryococcus braunii* in lacustrine sediments and oils. *Org. Geochem.* **33**, 99–109.

Suzuki K.-I., Saito K., Kawaguchi A., Okuda S., and Komagata K. (1981) Occurrence of ω-cyclohexyl fatty acids in *Curtobacterium pusillum* strains. *J. Gen. Appl. Microbiol.* **27**, 261–266.

Talbot H. M., Watson D. F., Murrell J. C., Carter J. F., and Farrimond P. (2001) Analysis of intact bacteriohopanepolyols from methanotrophic bacteria by reversed phase high performance liquid chromatography-atmospheric pressure chemical ionisation-mass spectrometry. *J. Chromatogr.* **A 921**, 175–185.

Talyzina N. M., Moldowan J. M., Johannisson A., and Fago F. J. (2000) Affinities of early Cambrian acritarchs studied by using microscopy, fluorescence flow cytometry and biomarkers. *Rev. Palaeobot. Palynol.* **108**(1–2), 37–53.

Taylor D., Kontorovich A. E., Larichev A. I., and Glikson M. (1994) Petroleum source rocks in the Roper Group of the McArthur Basin: source characterisation and maturity determinations using physical and chemical methods. *APEA J.* **34**, 279–296.

Tegelaar E. W., de Leeuw J. W., Derenne S., and Largeau C. (1989) A reappraisal of kerogen formation. *Geochim. Cosmochim. Acta* **53**, 3103–3106.

Tegelaar E. W., Hollman G., van der Vegt P., de Leeuw J. W., and Holloway P. J. (1995) Chemical characterization of the periderm tissue of some angiosperm species: recognition of an insoluble, non-hydrolyzable, aliphatic biomacromolecule (Suberan). *Org. Geochem.* **23**, 239–251.

ten Haven H. L., de Leeuw J. W., Sinninghe Damsté J. S., Schenck P. A, Palmer S. E., and Zumberge J. E. (1988) Application of biological markers in the recognition of palaeo-hypersaline environments. In *Lacustrine Petroleum Source Rocks* (eds. K. Kelts, A. Fleet, and M. Talbot). Blackwell, vol. 40, pp. 123–130.

ten Haven H. L., Rohmer M., Rullkötter J., and Bisseret P. (1989) Tetrahymanol, the most likely precursor of gammacerane, occurs ubiquitously in marine sediments. *Geochim. Cosmochim. Acta* **53**, 3073–3079.

ten Haven H. L., Peakman T. M., and Rullkötter J. (1992) Early diagenetic transformation of higher-plant triterpenoids in deep-sea sediments of Baffin Bay. *Org. Geochem.* **56**, 2001–2024.

Thiel V., Peckmann J., Seifert R., Wehrung P., Reitner J., and Michaelis W. (1999) Highly isotopically depleted isoprenoids: molecular markers for ancient methane venting. *Geochim. Cosmochim. Acta* **63**, 3959–3966.

Thiel V., Blumenberg M., Pape T., Seifert R., and Michaelis W. (2003) Unexpected occurrence of hopanoids at gas seeps in the Black Sea. *Org. Geochem.* **34**, 81–87.

Tissot B. P. and Welte D. H. (1984) *Petroleum Formation and Occurrence*. Springer, Berlin.

Tornabene T. G., Langworthy T. A., Holzer G., and Oro J. (1979) Squalenes, phytanes and other isoprenoids as major neutral lipids of methanogenic and thermoacidophilic archaebacteria. *J. Mol. Evol.* **13**, 73–83.

Treibs A. (1936) Chlorophyll and hemin derivatives in organic mineral substances. *Angew. Chem.* **49**, 682–686.

van Aarssen B. G. K., Cox H. C., Hoogendoorn P., and de Leeuw J. W. (1990) A cadinene biopolymer in fossil and extant dammar resins as a source for cadinanes and bicadinanes in crude oils from South East Asia. *Geochim. Cosmochim. Acta* **54**, 3021–3031.

van Aarssen B. G. K., Quanxing Z., and de Leeuw J. W. (1992) An unusual distribution of bicadinanes, tricadinanes and oligocadinanes in sediments from the Yacheng gasfield, China. *Org. Geochem.* **18**, 805–812.

van der Meer M. T. J., Schouten S., van Dongen B. E., Rijpstra W. I. C., Fuchs G., Sinninghe Damsté J. S., de Leeuw J. W., and Ward D. M. (2001) Biosynthetic controls on the ^{13}C contents of organic components in the photoautotrophic bacterium *Chloroflexus aurantiacus*. *J. Biol. Chem.* **276**, 10971–10976.

van Kaam-Peters H. M. E., Köster J., van der Gaast S. J., Dekker M., de Leeuw J. W., and Sinninghe Damsté J. S. (1998) The effect of clay minerals on diasterane/sterane ratios. *Geochim. Cosmochim. Acta* **62**, 2923–2929.

Villar H. J., Püttmann W., and Wolf M. (1988) Organic geochemistry and petrography of Tertiary coals and carbonaceous shales from Argentina. *Org. Geochem.* **13**, 1011–1021.

Volkman J. K. (1986) A review of sterol markers for marine and terrigenous organic matter. *Org. Geochem.* **9**, 83–99.

Volkman J. K. (2003) Sterols in microorganisms. *Appl. Microbiol. Biotechnol.* **60**, 496–506.

Volkman J. K., Eglinton G., and Corner E. D. S. (1980) Sterols and fatty acids of the marine diatom *Biddulphia sinensis*. *Phytochemistry* **19**, 1809–1813.

Volkman J. K., Kearney P., and Jeffrey S. W. (1990) A new source of 4-methyl sterols and 5α(H)-stanols in sediments: prymnesiophyte microalgae of the genus *Pavlova*. *Org. Geochem.* **15**, 489–497.

Volkman J. K., Barrett S. M., Dunstan G. A., and Jeffrey S. W. (1993) Geochemical significance of the occurrence of dinosterol and other 4-methyl sterols in a marine diatom. *Org. Geochem.* **20**, 7–15.

Volkman J. K., Barrett S. M., and Dunstan G. A. (1994) C_{25} and C_{30} highly branched isoprenoid alkanes in laboratory cultures of two marine diatoms. *Org. Geochem.* **21**, 407–413.

Volkman J. K., Barrett S. M., Blackburn S. I., Mansour M. P., Sikes E. L., and Gelin F. (1998) Microalgal biomarkers: a review of recent research developments. *Org. Geochem.* **29**, 1163–1179.

Volkman J. K., Rijpstra W. I. C., de Leeuw J. W., Mansour M. P., Kackson A. E., and Blackburn S. I. (1999) Sterols of four dinoflagellates from the genus *Prorocentrum*. *Phytochemistry* **42**, 659–668.

Wakeham S. G., Freeman K. H., Pease T., and Hayes J. M. (1993) A photoautotrophic source for lycopane in marine water columns. *Geochim. Cosmochim. Acta.* **57**, 159–165.

Wakeham S. G., Sinninghe Damsté J. S., Kohnen M. E. L., and de Leeuw J. W. (1995) Organic sulfur compounds formed during early diagenesis in Black Sea sediments. *Geochim. Cosmochim. Acta* **59**, 521–533.

Walter M. R., Buick R., and Dunlop J. S. R. (1980) Stromatolites 3,400–3,500 Myr old from the North Pole area, Western Australia. *Nature* **284**, 443–445.

Wang T.-G. (1991) A novel tricyclic terpane biomarker series in the upper Proterozoic bituminous sandstone, eastern Yanshan Region. *Sci. China Ser. B* **34**, 479–489.

Wang T.-G. and Simoneit B. R. T. (1990) Organic geochemistry and coal petrology of Tertiary brown coal in the Zhoujing mine, Baise Basin, South China: 2. Biomarker assemblage and significance. *Fuel* **69**, 12–20.

Wang T.-G. and Simoneit B. R. T. (1995) Tricyclic terpanes in Precambrian bituminous sandstone from the eastern Yanshan region, North China. *Chem. Geol.* **120**, 155–170.

Waples D. W. (2000) The kinetics of in-reservoir oil destruction and gas formation: constraints from experimental and empirical data, and from thermodynamics. *Org. Geochem.* **31**, 553–575.

Warton B., Alexander R., and Kagi R. I. (1997) Identification of some single branched alkanes in crude oils. *Org. Geochem.* **27**, 465–476.

Watts C. D., Maxwell J. R., and Kjøsen H. (1977) The potential of carotenoids as environmental indicators. *Adv. Org. Geochem.: Proc. 7th meet Org. Geochem.* 391–413.

Werne J. P., Baas M., and Sinninghe Damsté J. S. (2002) Molecular isotopic tracing of carbon flow and trophic relationships in a methane-supported benthic microbial community. *Limnol. Oceanogr.* **47**, 1694–1701.

Whitehead E. V. (1973) Molecular evidence for the biogenesis of petroleum and natural gas. In *Proceedings of Symposium on Hydrogeochemistry and Biogeochemistry* (ed. E. Ingerson). Clarke Co., vol. 2, pp. 158–211.

Wilson M. A. (1987) *NMR Techniques and Applications in Geochemistry and Soil Chemistry*. Pergamon.

Wilson M. A., La Fargue E., and Gizachew D. (1994) Solid state ^{13}C NMR for characterising source rocks. *APEA J.* **34**, 210–215.

Woese C. R., Kandler O., and Wheelis M. L. (1990) Towards a natural system of organisms: proposal for the domains Archaea, Bacteria, and Eucarya. *Proc. Natl. Acad. Sci. USA* **87**, 4576–4579.

Xiao S., Zhang Y., and Knoll A. H. (1998) Three-dimensional preservation of algae and animal embryos in a Neoproterozoic phosphorite. *Nature* **391**, 553–558.

Xiong J., Fischer W. M., Inoue K., Nakahara M., and Bauer C. E. (2000) Molecular evidence for the early evolution of photosynthesis. *Science* **289**, 1724–1730.

Yano K., Akihisa T., Tamura T., and Matsumoto T. (1992) Four 4α-methylsterols and triterpene alcohols from *Neolitsea aciculata*. *Phytochemistry* **31**, 2093–2098.

Zander J. M., Caspi E., Pandey G. N., and Mitra C. R. (1969) The presence of tetrahymanol in *Oleandra wallichii*. *Phytochemistry* **8**, 2265–2267.

Zang W. L. and Walter M. R. (1989) Latest Proterozoic plankton from the Amadeus Basin in central Australia. *Nature* **337**, 642–645.

Zhang Z. (1986) Clastic facies microfossils from the Chuanlinggou formation (1800 Ma) near Jixian, North China. *J. Micropalaeontol.* **5**, 9–16.

Zundel M. and Rohmer M. (1985a) Hopanoids of the methylotrophic bacteria *Methylococcus capsulatus* and *Methylomonas sp.* as possible precursors for the C_{29} and C_{30} hopanoid chemical fossils. *FEMS Microbiol. Lett.* **28**, 61–64.

Zundel M. and Rohmer M. (1985b) Prokaryotic triterpenoids: 1. 3-methylhopanoids from *Acetobacter sp.* and *Methylococcus capsulatus*. *Euro. J. Biochem.* **150**, 23–27.

Zundel M. and Rohmer M. (1985c) Prokaryotic triterpenoids: 3. The biosynthesis of 2β-methylhopanoids and 3β-methylhopanoids of *Methylobacterium organophilum* and *Acetobacter pasteurianus* spp. *pasteurianus*. *Euro. J. Biochem.* **150**, 35–39.

8.04
Biomineralization

H. C. W. Skinner
Yale University, New Haven, CT, USA
and
A. H. Jahren
Johns Hopkins University, Baltimore, MD, USA

8.04.1 INTRODUCTION

Biomineralization is the process by which living forms influence the precipitation of mineral materials. The process creates heterogeneous accumulations, composites composed of biologic (or organic) and inorganic compounds, with inhomogeneous distributions that reflect the environment in which they form. The products are, however, disequilibrium assemblages, created and maintained during life by dynamic metabolism and which, on death, may retain some of the original characteristics. The living forms discussed in this chapter produce carbonate, phosphate, oxalate, silica, iron, or sulfur-containing minerals illustrating the remarkable range of biomineralization chemistries and mechanisms. Biomineralization, in the broadest use of the term, has played a role in Earth cycles since water appeared on the surface. The general perception that the onset of biomineralization coincides with the appearance of fossils that left "hard parts" amenable to analysis is marked geologically as the dawn of the Cambrian. At least for some of us, the origins of life, and possibly biomineralization, go back 3.8 Gyr. The startling chemical and biological range encompassed by the term biomineralization implies that life forms have adapted and altered geoenvironments from the beginning, and will continue to do so, an essential ingredient for establishing and maintaining Earth's environment in the future as in the past (Figure 1).

8.04.1.1 Outline of the Chapter

It is only with the advent of more sensitive, higher-resolution techniques that we can identify the exact mineral components, and appreciate the precision control exercised by the life forms on their mineralized structures, and the potential that their formation and evolution are responses to the environment and climatic, or any other pervasive geochemical changes. We first present the crystal chemistry, or mineralogy, of some of the common minerals that are encountered in biomineralization (Section 8.04.2) and then provide examples from the range of life forms. We start with the Archea and Bacteria although we are just beginning to investigate their mineralization processes. However, these primitive forms were probably the first to generate the mechanisms that led to the accumulation of elements, the precursor to the formation of biominerals. The details of the mechanisms and the ranges of possible creatures in these classes remain to be fully defined and understood. However, we consider them an essential base to

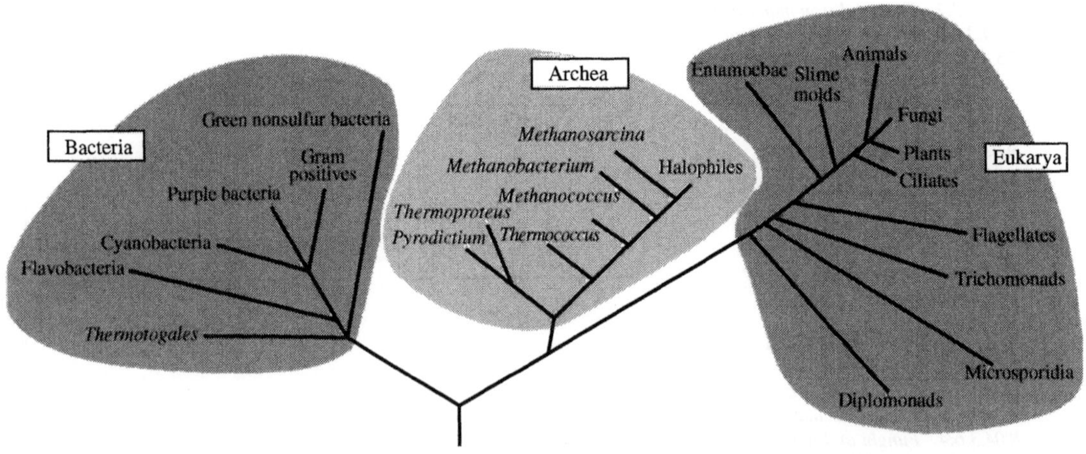

Figure 1 The "tree of life" (source Raven *et al.*, 1999, figure 13.8, p. 270).

any biomineralization discussions. Further, and perhaps most intriguing, many other biomineralizing forms incorporate them as symbionts. Following these opening sections, we move onto $CaCO_3$ deposition associated with cyanophytes (photosynthetic cyanobacteria) that marks the end of the Precambrian period, ~0.6 Gyr ago (Riding, 1982). At the start of the Cambrian period (~570 Myr ago), calcareous skeletons appear; rapidly and dramatically (within 40–50 Myr) some form of biomineralized structures appear in all existing phyla. Since that time only corals, some algae, and the vertebrates have developed new skeletons in the marine habitat (Simkiss and Wilbur, 1989); thus, the obvious developments of biomineralization are concentrated into ~1% of Earth's history, along with all other major diversifications of life. We present information, and include references to works we think will be most helpful to geochemists, on selected invertebrate and vertebrate skeletons many of which have a vast literature available. We select a few structures, such as teeth (in chitons and humans), because they are examples of different mineralizing systems whose tissue textures, and mechanisms, are unique, and because they may be, or have been, important to geochemical studies. In looking to future opportunities we include brief introductions to otoliths and antlers, because they offer novel sampling sites to test geochemical variations in the present environment.

The importance of the survival of land-based communities on plants suggested that our purview must include plant biomineralization, the materials, mechanisms, and strategies. For our summary we ask "why biomineralize?" and offer a few suggestions based predominantly on plant researches. The reasons for biomineralization we have outlined are appropriate to other mineral-producing life forms, and have often been discussed. However, in the process of reviewing the evolutionary development, some novel approaches, if not answers, to this basic question are preferred.

8.04.1.2 Definitions and General Background on Biomineralization

We classify biomineralization in our examples either as extracellular or as intracellular (Pentecost, 1985a) and include the specific cell types if known. We follow the standard definitions of Borowitzka (1982) that extracellular biomineralization involves inorganic, often crystalline, materials forming on the outer wall of the cell, within the cell wall, or in the immediate surrounding tissue areas, and is the usual type of biomineralization. Intracellular biomineralization is mineral formation within the cell such as the calcite formation for the group of algae the coccolithophoridae (Section 8.04.3.4.3). Another illustration of the diversity of intracellular biomineralization is found in the freshwater green alga *Spirogyra hatillensis* T. that contains calcium oxalate inclusions. The inclusions are not associated with the central vacuole, but instead are in cytoplasmic strands (Pueschel, 2001).

With the advent of eukaryotes, subdivisions within the cell, or compartments, were created. Within these subcellular compartments, or specialized anatomical sites, mineralization may be facilitated, with the result that biomineralization became more extensive and diverse. By creating lipid membranes, the eukaryotes could selectively "pump" ions and bioaccumulate them in a small volume. In plants, this subcellular compartment is usually a vacuole (Matile, 1978), whose membrane may serve both as a pre-existing surface for nucleation and as the ultimate determinant of mineral shape, as the crystal(s) grow to fill the vacuole (Simkiss and Wilbur, 1989). Most ion pumps translocate ions against electrochemical gradients (Carapoli and Scarpa, 1982) accomplished either by attaching the ion to a carrier molecule that is moving with an electrochemical gradient, or by directly using ATP as an energy source for the translocation. Often described for Ca^{2+} transport, similar transport mechanisms have been suggested to supply the anions that control the onset of mineral deposition in cells (Simkiss and Wilbur, 1989). Like cations, anions are involved in a wide variety of cell activities. Primary among them is the ability of anionic complexes, e.g., carbonate and phosphate, to act as inorganic pH buffers. In support of the anion-supply hypothesis, it has long been recognized that there is a positive relationship between photosynthetic rate (acquisition of CO_2) and rate of calcite biomineralization in algae. Specific studies have shown that when *Corallina officianalis* algae achieve a certain level of photosynthesis, the relationship between calcite biomineralization rate and CO_2 acquisition is roughly linear (Pentecost, 1978).

An organic matrix or pre-existing nucleation surface is usually considered to be the determining feature in many systems, especially the higher biomineralizing systems, such as the vertebrates (Section 8.04.3.7). Organic matrices within biomineralizing plant vacuoles (Webb *et al.*, 1995) provide the sites where "seed" cations bind as loose chelates (Tyler and Simkiss, 1958), and can be alternately soluble and insoluble (Wheeler *et al.*, 1981), or a combination of the two (Degens, 1976). Several nucleation centers may be present within a matrix, and each may grow independently, and perhaps produce similar crystallographically oriented biominerals (Simkiss and Wilbur, 1989). Although animal cell membranes are becoming well known, a limited number of studies have investigated the protein and lipid compositions of membranes within plant vacuoles. The crystal-associated organics in most biomineralizing systems contain a complex assortment of polypeptides (Miller, 1984; Webb *et al.*, 1995) and fatty acids (Jahren *et al.*, 1998) that could aid nucleation. In the invertebrates and plants the vacuole effectively separates the crystal from the sap, allowing for chemical control over the environment of crystallization (Webb, 1999). As growing crystals fill the vacuole, the matrix is compressed between neighboring crystallites giving rise to the mosaic or "honeycomb" structure of many biominerals.

Within multicellular organisms, biomineralization is determined by specific tissues and cells, which we describe for the calcium phosphate biomineralization in human teeth (Section 8.04.3.7.4). Within *Thalassia testudinum* (Turtlegrass) silica exists only as isolated deposits within leaves (Brack-Hanes and Greco, 1988), while within *Rheum officinale* M. (Chinese rhubarb), calcium oxalate is confined to the roots (Duke, 1992). Some biomineralized structures, such as the silica "hairs" found on *Phalaris canariensis* (Annual canarygrass) (Perry *et al.*, 1984), are associated exclusively with the defense structures of the plant. Similarly, some plants only biomineralize around reproductive tissues, such as the silica-aragonite endocarp surrounding the endosperm within the *Celtis* spp. (Hackberry) fruit (Cowan *et al.*, 1997). It has been shown that in sugar cane, different cell types accumulate silicon at very different rates (Sakai and Sanford, 1984).

The favored hypothesis regarding why cells within the same organism biomineralize with different minerals, e.g., calcite and aragonite in bivalves, is that there are different organic moieties which lead to changed nucleation opportunities. For example, permeability of a vacuole membrane in plants may lead to different concentrations of the ions and elements within, or to the inclusion of inhibitors that prevent nucleation of a specific mineral forms (Goodwin and Mercer, 1982).

The biomineralization mechanisms are far from fully investigated and understood for most species, but come close for the coccolithophorids. What is certainly true is that there is not a single or grand scheme or mechanism that is established throughout living forms. What can be taken from this brief overview is that biomineralization entails cooperative efforts involving cells, producing and controlling organic and inorganic molecules that combine, in structurally distinct ways, the ultimate and unique expression of a species. Biomineralization which added inorganic to the usual organic moieties necessary for life increased the possibility for survival in new environments and extended the range of potential ecological niches available on Earth.

8.04.2 BIOMINERALS

8.04.2.1 Calcium Carbonates

8.04.2.1.1 Calcite

The most widely known biomineral is calcite, which forms the "hard part" of many common invertebrates such as corals and mollusks. The anhydrous calcium carbonate $CaCO_3$ is one of the eight mineral end-members with identical crystal structure but different divalent cations, that form the calcite group (Gaines *et al.*, 1997, p. 427). These minerals, many almost as common as calcite, all crystallize in the hexagonal/rhombohedral space group, R3c. Figure 2(A), a projection of the three-dimensional crystal structure down the unique "*c*" axis shows the array of the planar and trigonal carbonate ions (CO_3^{2-}). The trigonal species (CO_3^{2-}) is composed of a carbon atom at the center and coplanar with three equidistant oxygen ions. Carbonate ions form layers along the *c*-axis and alternate with the cations in the calcite sequence along the *c*-axis $-Ca-CO_3-Ca-CO_3-$ and every other layer of CO_3 ions points in the opposite direction. Calcium is in octahedral, or sixfold, coordination with six oxygen ions from six different carbonate groups, i.e., the divalent calcium ion has hexagonal distribution. However, the superposition of the alternating directions of the trigonal ions in the third dimension is a distinguishing feature of the structure, and influences the morphology of the crystalline form. Rhombs are the dominant morphology expressed during inorganic crystallization of calcite, and the shape often encountered when the crystals are fractured, or cleaved (Figure 2(B)).

Perfect, clear calcite crystals that showed either hexagonal, or rhombohedral, characteristics fascinated seventeenth-century scientists, and were instrumental in advancing physics and materials science. Cleavage fragments were used in the discovery of double refraction, and the polarization of light in the mid-seventeenth to early nineteenth centuries. The X-ray diffraction studies by Bragg in the twentieth century confirmed the relationship between the optical character and the morphology. Calcite shows amazingly high birefringence, a property directly related to the anisotropy of the structure. The ability to generate plane polarized light from the insertion of an oriented calcite crystal in the beam led to the polarizing microscope, a tool still providing the most useful technique for an initial discrimination of minerals and other crystalline materials, and a quick way to identify calcite. The properties of anisotropy in crystals and the use of polarizing microscopy for

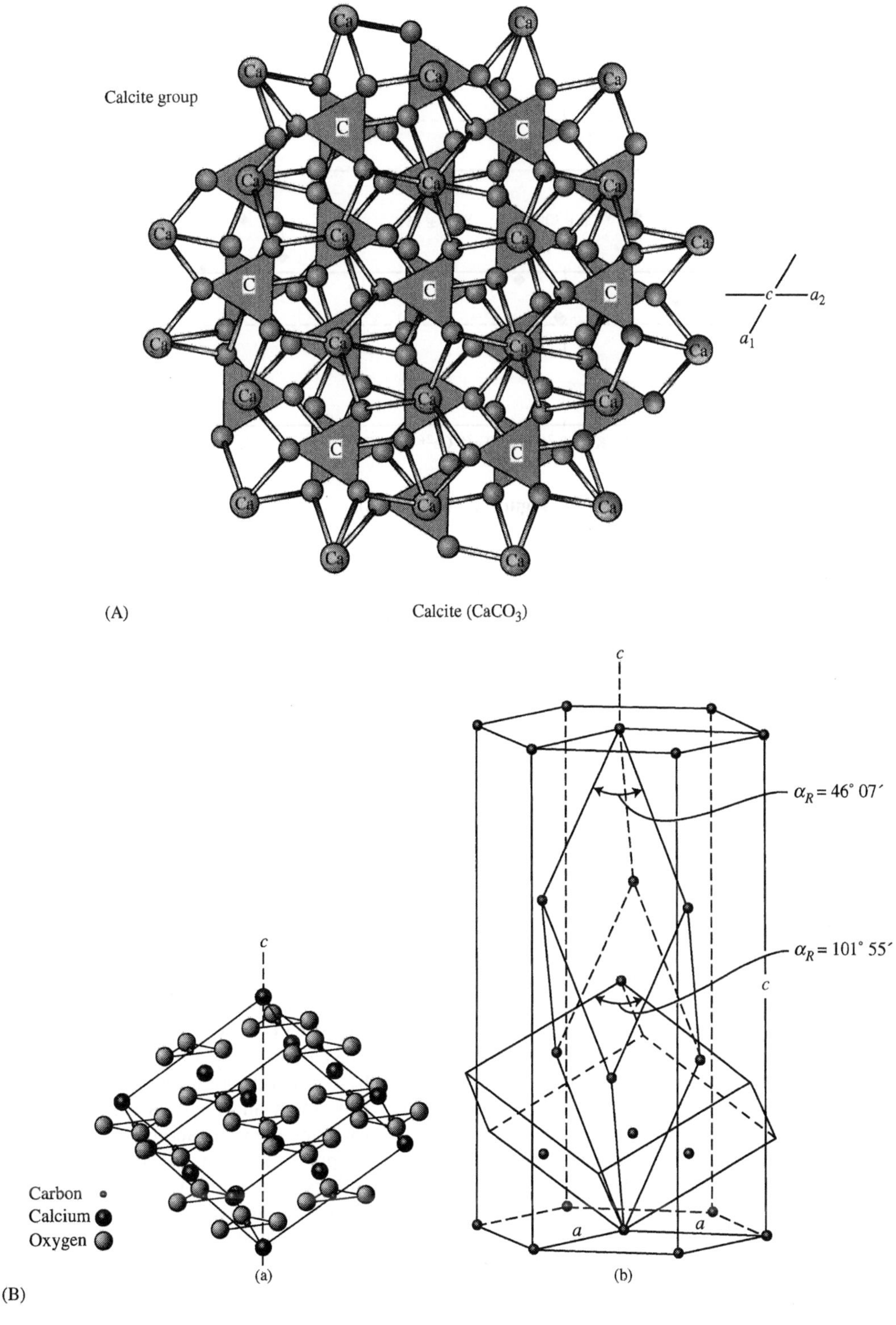

Figure 2 (A) Calcite. The crystal structure projected down the unique *c*-axis showing the hexagonal disposition of Ca and CO_3 ions (Gaines *et al.*, 1997, p. 427). (B) Morphological relationships: (a) the arrangement of Ca and CO_3 groups relative to the calcite cleavage rhomb; (b) the true rhombohedral unit cell (steep rhomb) to the cleavage rhomb and the hexagonal cell where $c/a = 3.42$ (Gaines *et al.*, 1997, p. 429). (C) Plot of the possible "best fit," 15%, and next best fit, 30%, of cations that could take the place of Ca in sixfold coordination in the calcite structure (courtesy of Stefan Nicolescu, Department of Geology and Geophysics, Yale University).

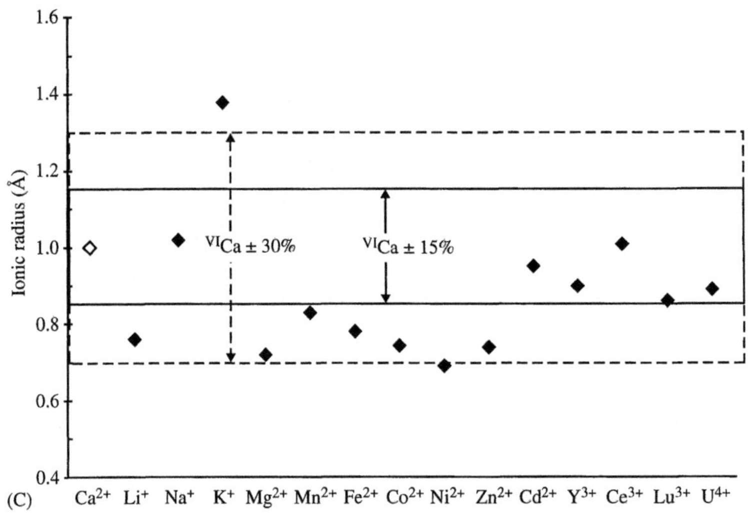

Figure 2 (continued).

complete identification of any crystalline solid is reviewed in Klein and Hurlbut (2000) and in most optical mineralogy texts.

The isostructural calcite group consists of several minerals with the following elements in place of calcium: magnesium, iron, manganese, cobalt, zinc, cadmium, and nickel. All these mineral carbonates have different names and are considered end-members in the series with an ideal composition expressed, e.g., as $MgCO_3$, for the mineral magnesite. These are the elements most likely to be incorporated in calcite whether formed strictly inorganically or precipitated when the elements are bioaccumulated. It should be pointed out that there is virtually no substitution for the carbonate group in calcite group structures (Reeder, 1983).

Figure 2(C) is the result of a calculation that illustrates which cations might fit into sixfold coordination position in the calcite group structures. It is an interesting insight as both light and heavy rare earth elements are possible substitutes for calcium in the calcium carbonate structure, i.e., they plot within ±15% of the calcium ionic size. However, some of the end-members incorporate elements into this crystal structure and are outside this deviation but within ±30%, an expression of the potential physical expansion for this layered crystal structure. These are ionic charge differences important in whether a stable crystalline structure can be produced. Trace amounts of all these ions can be incorporated in calcite and may dictate the morphology of the crystallites. Therefore, the presence and amount of any ions in the environment in which carbonate crystallization occurs may possibly be recorded. However, in spite of the predominance of sodium and potassium in the solutions where biomineralization takes place, neither element is found in calcite to any extent. These single charged species more readily associate with Cl^-, and together with their size (larger than calcium ionic size), mean that they are not usually accommodated in the rhombohedral calcite structure. The formation of any biomineral indicates the composition of the ions available in the surrounding media and habitat with the specific mineral species and form dictated by an input of energy, to overcome the nucleation barrier. The continued availability of ions is essential for growth of the mineral species consistent with the biological needs.

Calcite can easily be identified without a microscope by its ready dissolution in 1 N HCl, often fizzing or forming bubbles indicating the liberation of CO_2. The other isostructural members of the calcite group are less soluble, but since all of these minerals are usually well crystallized, X-ray diffraction and analysis can identify them as a group member. In addition, the diffraction maxima positions can be used to estimate the quantities of different elements incorporated. The diffraction analysis relies on a physical shift in the crystal structure parameters based on the size of the cation, and is not a chemical analysis.

8.04.2.1.2 Aragonite

An equally important carbonate mineral is aragonite, the common polymorph of calcite. Polymorphism of minerals implies the same chemical composition but distinct crystal structure. The aragonite structure also has alternate layers of carbonate groups and cations with the triangular CO_3 pointing in opposite directions along the *c*-axis, but the cations are in ninefold coordination catering to larger size elements such as strontium, barium, and lead. Aragonite crystallizes in the orthorhombic space group Pmcn and is one of the members of an isostructural group that includes these other cations (Gaines *et al.*, 1997, pp. 440–448). Figure 3(a) projected down the crystallographic *c*-axis allows comparison with the calcite structure and the different polygonal arrangement of calcium. The carbon of the carbonate groups in aragonite is very slightly out of the plane (0.026 Å), somewhat displaced along the *b*-axis (0.20 Å), and the plane defined by the three oxygens is tilted ~2.5°. In addition, the cations are puckered, alternating ~0.05 Å above and below the cation plane. Six bonds are made between the calcium with two oxygens in each of three carbonate groups and three shorter bonds with the corner oxygens of three other carbonate groups. The pseudohexagonal array has stacking order along "*c*" of $A(CO_3)_1$–$B(CO_3)_2$–$A(CO_3)_1$–$B(CO_3)_2$, which results in orthorhombic symmetry. Crystals have a prismatic habit, elongated and flattened along the *c*-axis appearing acicular. Most aragonite that appears to be in the form of single crystals is usually twinned, a mosaic composed of discrete differently oriented

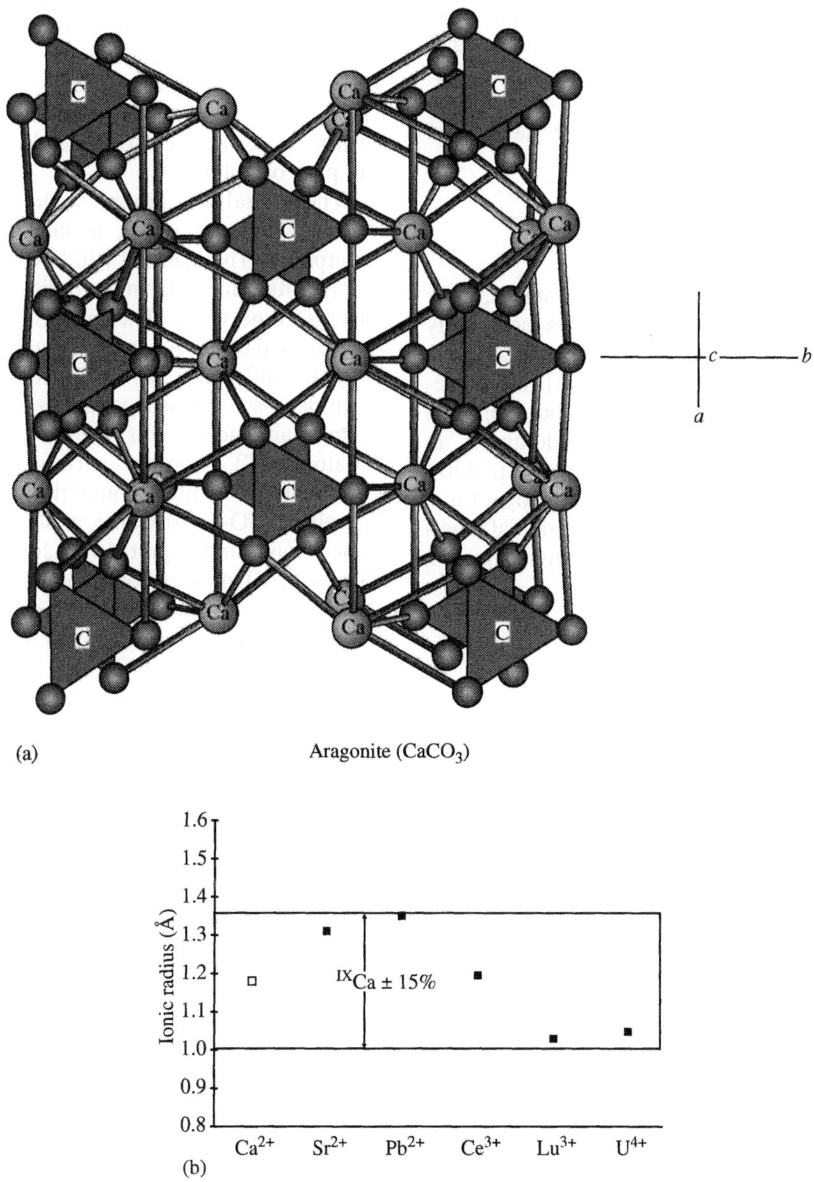

Figure 3 (a) Aragonite. The crystal structure projected down the *c*-axis (Gaines *et al.*, 1997, p. 441). (b) Plot of the "best fit," 15%, of cations that could take the place of Ca in eightfold coordination in the aragonite structure (courtesy of Stefan Nicolescu, Department of Geology and Geophysics, Yale University).

portions of the same mineral (with identical crystallographic parameters) and related by strict geometric laws. These polysynthetically twinned aragonites produce lamellae oriented parallel to the *c*-axis that can be detected as fine striations in large crystals.

Some substitution of strontium (up to 14 mol.%), of lead (2 mol.% reported) but no barium has been reported in aragonite, although investigations at elevated temperatures and pressures show almost complete miscibility of these elements in the structure (Gaines *et al.*, 1997, p. 442), and $SrCO_3$ (strontionite), $BaCO_3$ (witherite), and $PbCO_3$ (cerussite) are common minerals. A calculated plot (Figure 3(b)) for cations in ninefold coordination shows that this coordination theoretically allows trivalent rare earth elements and quadravalent U^{4+}, and many other elements to be substituents in the structure. Ytterbium, europium, samarium, and radium carbonates with aragonite structure have been synthesized (Spear, 1983).

8.04.2.1.3 Vaterite

A third polymorph of $CaCO_3$, vaterite, may form as an inorganic precipitate, but is most likely to be encountered as the mineral forming in the nacre of mollusks (Section 8.04.3.4.6). The mineral crystallizes in the hexagonal/rhombohedral space group $P6_3/mmc$, and the calcium is in sixfold coordination (Gaines *et al.*, 1997, p. 440). It is unstable and often reverts to calcite over time. Elevated temperatures or water high in NaCl will accelerate the transition; in dry environments, vaterite will slowly revert to calcite.

8.04.2.2 Silica

Anhydrous SiO_2, silica, is one of the most common rock-forming minerals, quartz (Frondel, 1960). Large (inches to feet), transparent, and clear, hexagonal crystals (rock crystal) are found in some locations, but the mineral also occurs in a variety of crystalline forms, and may be clear to milky, transparent or opaque, and from white to black, with all shades of colors in between (Rossman, 1994). Quartz is the dominant mineral phase in some rocks, as it is in sands, and in soils, and many of the colored varieties have been used as decorations, and in jewelry, for as long as humans have existed. The earliest tools were made from flint and chert, microcrystalline varieties of quartz, that had all the qualities desired: a hardness of 7 and sharp edges on fracture. Mineralogists have studied all these aspects, and have assiduously determined variations in crystal structure brought about under different temperature and pressure conditions (Graetsch, 1994; Gaines *et al.*, 1997, pp.1568–1586). However, it is the low-temperature hydrated variety of silica, opal ($SiO_2 \cdot nH_2O$) which is a biomineral.

8.04.2.2.1 Opal

Opal is composed of differing amounts and arrangements of structural units of amorphous SiO_2, water, and the crystalline polymorphs of quartz, crystobalite, and tridymite (Gaines *et al.*, 1997, pp. 1587–1592). A disordered solid, the mineral may display colors due to patchy areas of short-term crystalline order (Figures 4(A) and (B)). Close-packed silica spheres, typically 0.25 μm in diameter, about half the wavelength of visible light, randomly associate producing vacancies and stacking faults containing variable amounts of water in the aggregated solid (Levin and Ott, 1933). Jewelry made from opaline material is cut *en cabachon* to maximize the play of colors.

When totally random, without regularized crystal structural order, a substance is called amorphous. The material will have no discernable X-ray diffraction pattern, an indication of the lack of crystalline order. The biological material has very limited order. A variety known as opal-A with only one diffraction band is identified in samples from deep-sea deposits which are the remains of diatoms and radiolarians (there are also freshwater "diatomites"). Another variety is opal-CT, a discontinuous combination of the crystalline polymorphs of SiO_2, cristobalite, and/ or tridymite, and water (Jones and Segnit, 1971; Jones *et al.*, 1966; Graetsch, 1994). Crystobalite and tridymite have multiple modifications of their crystal structures known as high- and low-temperature forms (see Gaines *et al.*, 1997, pp. 1568–1586) for more details). Many names have been applied to the different varieties of "amorphous hydrated silica" as the solid responds to the source of the silica and the environment, and takes up locally dictated morphologies. Some silica deposits will change over time with compaction or diagenesis. For example, diatomaceous earth is soft and fine grained with individual diatoms visible with some magnification, while "tripolite" is a variety of diatomaceous earth showing little evidence of diatom remains.

Opalline materials may not show any texture at the level of a polarizing microscope but may appear platy, or fibrous, at the higher resolution of transmission or scanning electron microscopy (De Jong *et al.*, 1987). Using magic angle spinning of ^{29}Si, a nuclear magnetic resonance technique, opal-CT was shown to be effectively amorphous compared to the anhydrous silica species, crystobalite and trydimite. However, opal-CT is birefringent with an index of refraction that varies

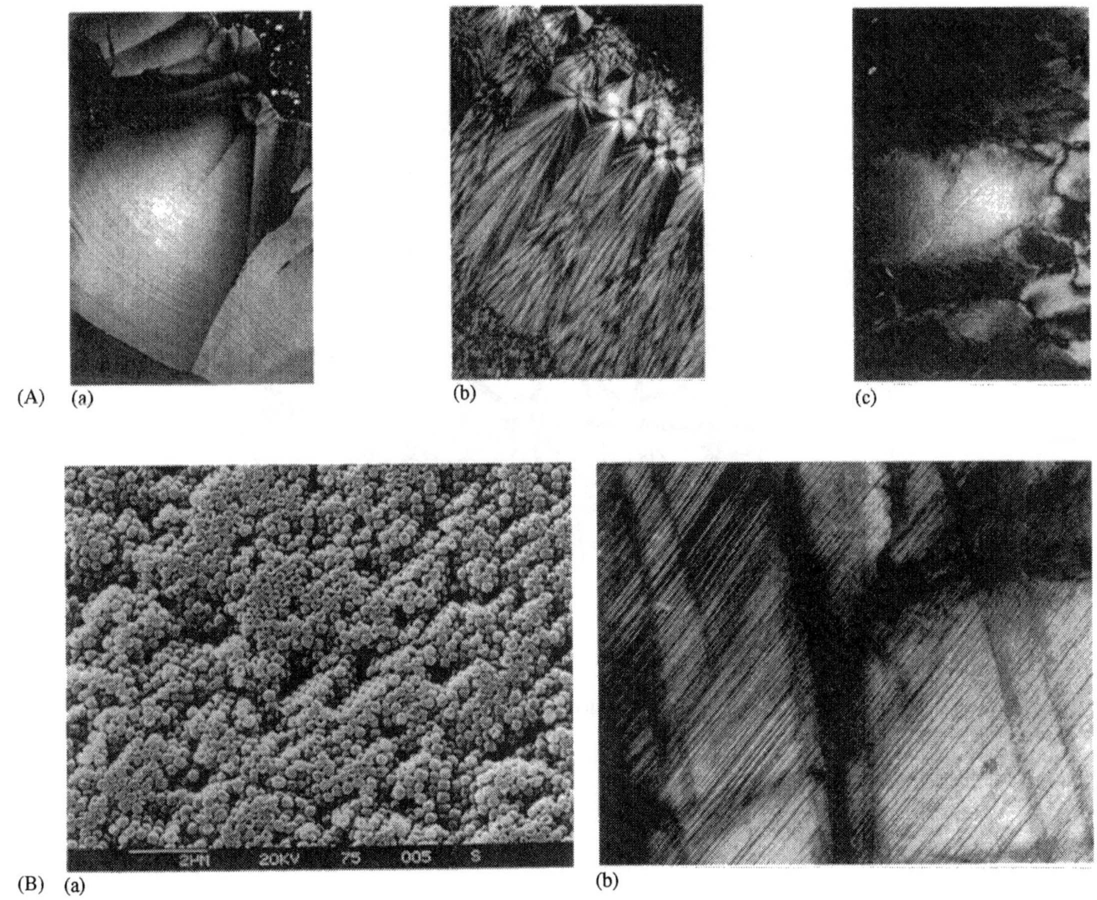

Figure 4 (A) Microradiographs using crossed polars of three forms of opal: (a) opal-C, (b) opal-CT, (c) opal-AN. The long edge of the micrographs is 1.11 mm. (B) Precious opal: (a) scanning electron micrograph and (b) petrographic thin section micrograph, crossed polars. The long edge of the micrograph is 1.11 mm.

from $N = 1.43$ to 1.46. The X-ray diffraction pattern, which shows broadened maxima due to very fine grain size as well as packing variations, will reflect the amount of SiO_2 polymorphs relative to water content. Opal-CT may also be formed in volcanic rocks.

A combination of several different opaline materials may be found juxtaposed. For example, in fossilized wood both tridymite and crystobalite may be determined in a sample while adjacent portions of the sample may be composed of opal-A. The silicified wood may be transparent or translucent, clear and colorless, or white, yellow, red, brown, and black, indicating inclusions of other, usually iron-containing, complexes during precipitation of the colloid or gel. The faithfully preserved structures of fossilized wood suggest that the replacement phenomena are molecule-by-molecule processes that take place under low temperatures and pressures, and require concomitant removal of nonsiliceous compounds but do not disrupt the cellularity of the woody tissues. Alternatively, primary biomineralization by living forms of hydrated silica can produce amazingly delicate structures. The morphology and mechanisms in diatoms, radiolarians, and the phytoliths that occur in grasses and plants are described below.

8.04.2.3 Bioapatite

Apatite is the name for a group of minerals with the general formula $A_5(TO_4)_3Z$ (Gaines *et al.*, 1997, pp. 854–868) in which A is usually calcium, but may be strontium, lead, or barium; T is usually P, but As and V forms crystallize with the same hexagonal or pseudohexagonal crystal structural symmetry, and Z may be OH or one of the halogens. Fluorapatite, $Ca_5(PO_4)_3F$ (Figure 5(a)), is the most common form occurring in trace quantities in igneous, metamorphic, and sedimentary rocks. Often selected for analyses because of the wide range of other elements, specifically the rare earths, uranium, and thorium, that may be incorporated (Skinner *et al.*, 2003).

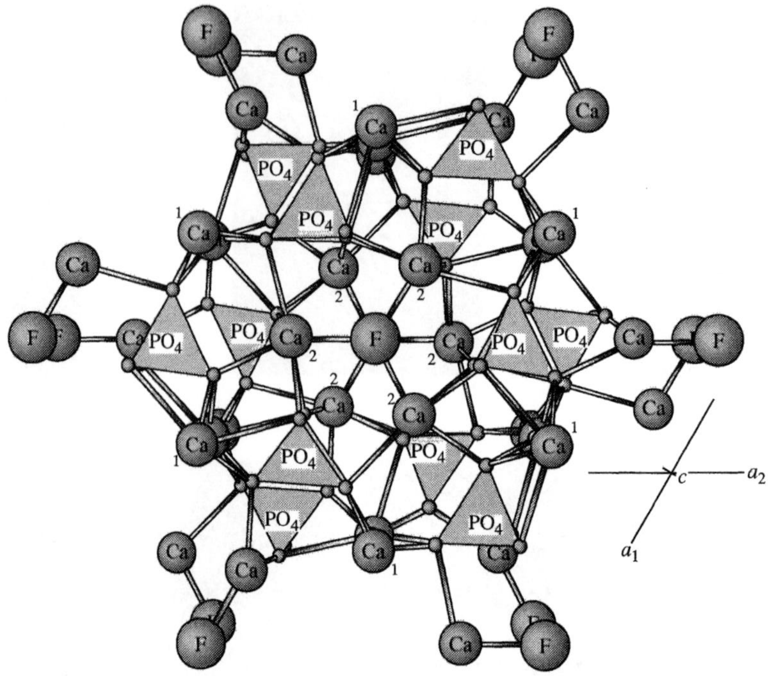

(a) Fluorapatite ($Ca_5(PO_4)_3F$)

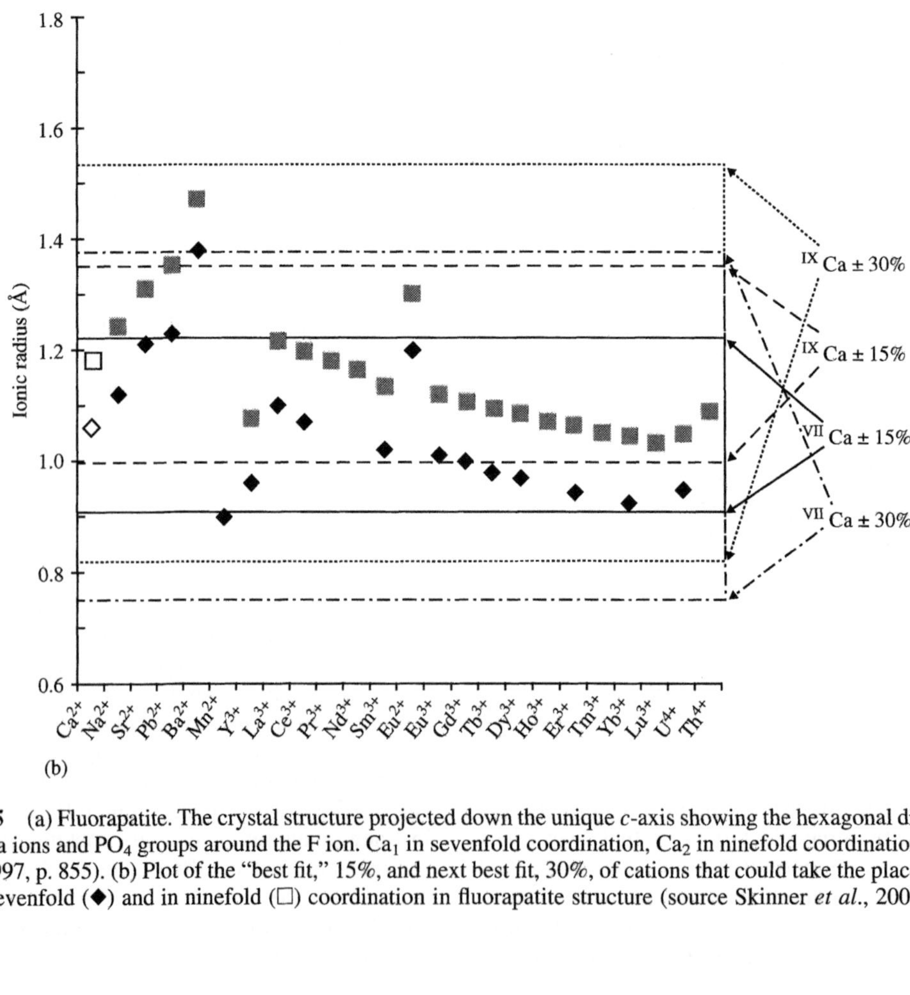

(b)

Figure 5 (a) Fluorapatite. The crystal structure projected down the unique *c*-axis showing the hexagonal disposition of the Ca ions and PO_4 groups around the F ion. Ca_1 in sevenfold coordination, Ca_2 in ninefold coordination (Gaines *et al.*, 1997, p. 855). (b) Plot of the "best fit," 15%, and next best fit, 30%, of cations that could take the place of Ca in sevenfold (◆) and in ninefold (□) coordination in fluorapatite structure (source Skinner *et al.*, 2003).

It is easily identified by its optical or X-ray diffraction characteristics as it is always well crystallized. The apatite formed at low temperatures, however, is not well crystallized. For example, the phosphate species in the Phosphoria Formation of western US, though a calcium apatite containing a wide range of other elements, is very fine grained and associated with many other fine-grained or poorly crystalline materials, i.e., clays (Gulbrandson, 1966). It is this low-temperature, poorly crystalline form of calcium apatite that is found in biomineralized tissues. The bioapatite, predominantly a hydroxylapatite, has a formula that can be written as $Ca_5(PO_4,CO_3)_3$ (OH, F, Cl, CO_3).

In this formula CO_3 is shown as a substituent in the PO_4 structural site and in the OH or halogen site. Biological apatites may contain up to 6 wt.% CO_2. For a particular bioapatite sample, it is difficult to determine the actual quantity of CO_2 or how it is incorporated. Bioapatites are so fine grained that spectroscopic and wide and small angle X-ray techniques have been applied in efforts to interpret the possible crystallographic location (Elliott, 2002). In addition, the mineral is usually intimately mixed with bioorganic molecules, and extraction is required before analysis (Skinner *et al.*, 1972). The exact location of carbonate in bioapatites remains a question. If carbonate substitutes for even a small portion of the phosphate in the structure, there must be some lattice disruption as the planar CO_3 ion has a double negative charge compared to the tetrahedral phosphate group that has a triple negative charge. Further, it is the phosphate framework and the associated calcium ions that are responsible for the characteristic hexagonal structure and morphology of the crystals. The spatial and charge differences may give rise to disruptions and vacancies in the structure if CO_3 is incorporated. However, the usual suggestion is that the carbonate is balanced by concomitant substitution of other elements for calcium. Calcium occurs in two different lattice sites: Ca_1 is in sevenfold coordination with the oxygens on the phosphate groups, while Ca_2 is in ninefold coordination with phosphate oxygens and the OH or F. Figure 5(b) compares the possible cationic substitutions for calcium. It shows that virtually all cations whether single, double, or triply charged are of appropriate size to fit into the metal ion sites associated with the PO_4 oxygens. Triply charged species could compensate for carbonate inclusion maintaining local charge balance in the structure. If the carbonate inserts in the OH or halogen site in the structure, similar disruptions, vacancies, or substitutions would be necessary to maintain a balanced apatite lattice.

What is clear is that bioapatites are poorly crystalline, and carbonate may be a major factor affecting crystallinity. Whether the carbonate ion is actually within the lattice in small amounts interfering with crystal structure formation, or adsorbed on the very large surface area of the tiny crystallites, it will affect the stability of the phase and the crystallite size. The crystallites in bone are less than 1,000 Å along the *c*-axis or the long dimension of apatite, and only a few unit cells (<50 Å) in the *a*-dimensions (Elliott, 2002). These very small sizes are typical of biological apatites. Such size and morphology may be an advantage when the mineral is deposited intra-cellularly, or inter-biomolecularly, as it is in bone (Section 8.05.3.7.2). Since mineral resorption is an essential part of normal tissue metabolism maintaining the stability of bone, this mineral/chemical system for biomineralization has been well chosen (Skinner, 2002).

Although essential and predominant in bones and teeth, bioapatites have been found in all living forms including bacteria, invertebrates, and plants (e.g., Macintyre *et al.*, 2000). Bioapatite is also the mineral that occurs in some aberrant or pathological deposits (Skinner, 2000).

8.04.2.4 Iron Oxides and Hydroxides

The most visible iron-containing minerals are the ubiquitous subaerial rusts on rocks. They are ferric oxide or oxyhydroxide flocculants of exceedingly fine grain size that change over time. They may be considered abiotic, but more likely their localization and formation are expressions of the presence of microorganisms whose metabolism depends on iron (Section 8.04.3.1). All organisms from microbes to humans require iron with one exception: fermenting lactic streptoccoci (Ehrlich, 2002, p. 347). Iron is one of the elements utilized for the transfer of electrons in oxidation and reduction reactions. Special proteins, the cytochromes, which have iron-bearing heme groups, or the ferrodoxins, which are Fe–S organic complexes, are prominent cell constituents in aerobic or anaerobic microorganisms and are detected in creatures and plants on up the phylogenetic tree. We will not address the potentially biologically mediated rust mineral formation; instead, we focus on the mineral magnetite, because its formation and properties play important roles in several life forms.

8.04.2.4.1 Magnetite

Magnetite, ideal formula Fe_3O_4, is an intriguing mineral, and not only because it is magnetic. The formula could be written $Fe^{2+}Fe_2^{3+}O_4$, which more precisely designates one of its peculiarities. It contains both ferrous (Fe^{2+}) and ferric (Fe^{3+}) ions implying synthesis, growth, and stability within an environment where oxidized and reduced states of iron are present and maintained. Magnetite is a member of the spinel mineral group that crystallizes in the isometric space group Fd3m. In the basic unit (unit cell) that completely describes the chemistry and crystal structure, there are 32 oxygen ions and 24 cations with eight of the cations in tetrahedral (fourfold or A site) coordination and the remaining 16 in octahedral (sixfold or B site) coordination with the oxygen ions. A view of the structure (Figure 6) oriented with the vertical [111] crystallographic direction shows layers of oxygen atoms alternating with layers of cations in this projection of the three-dimensional array. Layers of octahedrally coordinated iron alternate with layers in which the cations are distributed among two possible metal sites, A and B positions in the ratio A/B = 2/1. There are two types of spinel structures, normal and inverse, depending on the trivalent and divalent metal ion distribution. The *normal* structure would have eight ferrous (Fe^{2+}) ions in the A site and 16 ferric (Fe^{3+}) ions in the B site. *Inverse* would have eight ferric (Fe^{3+}) ions in the A site and for the remaining 16 cations, one-half or eight Fe^{2+} and eight Fe^{3+} are in the B site. Magnetite has the inverse spinel form with predominance of Fe^{3+}. In natural materials there may be slight excess or a deficiency of total cations relative to oxygen (Flint, 1984). Another mineral species, maghemite, also magnetic, formula γ-$Fe_{2.67}O_4$, indicates lower amounts of iron relative to oxygen, and is also known to be biologically formed (Gaines *et al.*, 1997, p. 229)

The predominance of ferric cations in magnetite is mandatory, but magnesium, manganese, zinc, nickel, copper, and germanium may substitute for some of the iron within the spinel structure, occurring in solid solution in a magnetite, or maghemite. When these elements are dominant cations, they are the end-members in the spinel group of minerals with separate names (Gaines *et al.*, 1997, pp. 293–294). As might be suspected, these elements in addition to iron that are likely to be detected when magnetites are chemically analyzed.

Magnetite is widely distributed usually with octahedral, or spherical, morphology, in igneous and metamorphic rocks. Magnetite is also found in limestones, in fumerolic deposits, and in living forms. For the latter the mineral plays some very important role, which we highlight in Sections 8.04.3.3.2 and 8.04.4.6.3. Black magnetite, with a specific gravity greater than 5, is easily detected and separated from sediments and tissues by its magnetic properties and the identification can be confirmed by X-ray diffraction.

Studies using high-resolution transmission electron microscopy coupled with Mossbauer spectroscopy on experimental systems (Tamaura *et al.*, 1981, 1983) and on flocculants, the precipitates associated with bacterial activity (Schwertmann and Fitzpatrick, 1992), often detect the presence of other iron minerals such as the more hydrated iron

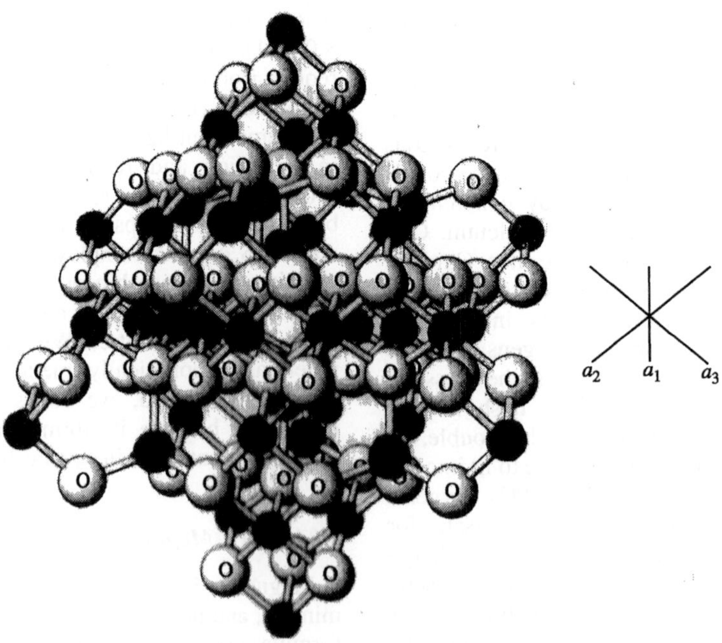

Figure 6 Magnetite. The crystal structure of magnetite, a member of the spinel group of minerals. The [111] crystallographic direction is vertical to show the horizontal appearance of cubic-close-packed oxygen (O) atoms. Fe^{2+} and Fe^{3+} are in tetrahedral and in octahedral interstices (both dark colored) (after Gaines *et al.*, 1997, p. 293).

oxide phases—ferrihydrite, $Fe_5O_8H{\cdot}4H_2O$, lepidocrocite, γ-FeO(OH), and goethite, α-FeO(OH), as well as hematite, Fe_2O_3. A calculated Eh–pH diagram indicates that there is a much larger field for biologically mediated magnetite formation from the metastable ferrihydrite phase compared with the inorganic formation of common goethite (Skinner and Fitzpatrick, 1992). The relationships between oxidized iron mineral phases and iron sulfides (e.g., pyrite, FeS_2), in natural environments and in mine waste sites, so visibly obvious in the environment, have sparked interest in the roles of microbes, and their biomineralization utilizing iron and the deposition of many different iron minerals (Rabenhorst and James, 1992; Bigham *et al.*, 1992).

8.04.3 EXAMPLES OF BIOMINERALIZATION

8.04.3.1 Introduction

Archea and Bacteria, two of the three domains that make up the Tree of Life as presently understood (see Figure 1), are grouped together as prokaryotes. The classes under the Archea and Bacteria are relatively simple. They are unicellular organisms that are only beginning to be explored. Their genetic information is not packaged within a membrane, and therefore the cells do not have a true nucleus. Archea and Bacteria lack mitochondria and chloroplasts, organelles that are prominent in the third domain, the Eukarya, and assist in respiration or photosynthesis. These functions are carried out in, or on, the membranes of the cell walls, or in some prokaryotes with the aid of symbionts. Archea are distinguished from Bacteria by the composition and structure of the cell wall, and by the ribosomes and the enzymes produced (Brock and Madigan, 1988; Ehrlich, 2002).

Within the Archea are methanogens and extremophiles (hypersaline and/or thermotolerant species), whereas Bacteria include species that can metabolize hydrogen in aerobic and anaerobic environments, can nitrify or denitrify, reduce sulfate or oxidize sulfur, or utilize iron and manganese in either their oxidized or reduced forms to gain energy. The number and diversity of bacterial species is enormous and still growing. A relatively new arena, characterizing the range of Archea, has captured scientific interest (Woese *et al.*, 1990). In future, the largest collection of living forms with the greatest diversity may well be Archea.

Both Archea and Bacteria include heterotrophs, i.e., creatures that can derive energy and carbon needed for the construction of their essential bioorganic molecules through the assimilation and oxidation of organic compounds. We know of aerobic and anaerobic species and facultative organisms that use oxygen as a terminal electron acceptor when it is available, and when not, they may use reducible inorganic (e.g., nitrate or ferric ions) or organic compounds in the metabolic activities. There are fermentors, species that enzymatically control the breakdown of energy-rich organic compounds, such as carbohydrates, to carbon dioxide and water, in the process of securing other nutrients, and chemolithotrophs, microorganisms that derive energy from oxygenation of inorganic compounds. These latter assimilate carbon from CO_2, HCO_3^-, or CO_3^{2-} without the benefit of light, and therefore survive in deep-sea hydrothermal vent sites. The discovery of this vast range of "bugs" has caused us to rethink some very basic questions on alternate sources of energy for creating and maintaining living creatures and the origin of life.

These "simple" forms accumulate, actively or passively, elements from their environment to construct metabolically essential components: nucleic acids, proteins, lipids, including a range of cations. Such anabolic activities require catabolic machinery and the ability to adapt to the surroundings. In the process of producing, maintaining, and reproducing, they maximize the selection and transport of inorganic and organic complexes, and are some of the most important geological agents and catalysts (Ehrlich, 2002, pp.119–151; Krumbein, 1986).

While they are about sustaining themselves, they change their environment. Some accumulations can rightfully be labeled biomineralization in that these creatures are involved with the precipitation of a variety of minerals. The mineralization may take place adjacent to the metabolizing form, associated with, but appearing nonessential to, their life cycle. Since eukaryotes incorporate a wide range of these living forms as symbionts to facilitate the same functions, a significant portion of future research in biomineralization will probably be devoted to investigating their range of activities, determining the details of the processes involved, and integrating that information into geoenvironments and monitoring ecological changes. They are some of the most important, if incompletely charted, agents in the geochemistry of the Earth and we would be remiss if we did not include these most primitive organisms in this chapter on biomineralization.

8.04.3.2 Sulfur Biomineralization

There are two large groups of prokaryotic organisms: those that oxidize and those that reduce sulfur compounds, sulfur being an element with one of the widest possible ranges of oxidation states ($-$II to $+$VI). There are chemolithotrophic bacteria that are able to disproportionate sulfur all the way from H_2S to SO_4^{2-} but only in the presence of sulfide scavengers such as FeOOH, $FeCO_3$, or MnO_2 (Janssen *et al.*, 1996), but most organisms are either sulfate reducers or sulfur and hydrogen sulfide oxidizers. There are aerobic, anaerobic, and facultative forms that perform these reactions, from the Bacteria and Archea kingdoms that occur in marine or freshwaters, on land, in soils, and in sediments (Barns and Nierzwicki-Bauer, 1997), and perhaps most importantly, the reducers and oxidizers are often juxtaposed, occur together and act in consort. Morphologically these creatures can be cocci, rodlike, or filamentous, and many of the bacteria show gram-negative cell types.

8.04.3.2.1 Sulfur oxidizers

Strictly anaerobic sulfur oxidizers are found in the photosynthetic purple (*Chromatiascea*) and green bacteria (*Chjlorobiaceae*) and in the cyanobacteria (Ehrlich, 2002, table 18.3, p. 557); there are also chemosynthetic autotrophs that use hydrogen as an energy source. Some species operate under extreme environments (at high salinities, at high or low pH), or are thermophylic (up to 110 °C); many identified from hot springs or ore deposits. The oxidizing activity is coupled to the reduction of available CO_2, with the carbon fixed into bioorganic molecules. In addition, these anaerobes may use nitrate as the terminal electron acceptor reducing it, via nitrite, to NO, to N_2O to N_2, meaning that these microbes can exist in virtually all surficial environments. Thiobaccilli, the family Beggiaatoaceae, and in the Archea, *Sulfolobus* and *Acidianus*, are among the most widely studied (Ehrlich, 2002, table 18.2, p. 554).

Natural bacterial oxidation processes are exploited industrially. The beneficiation of metal sulfide mineral ores and waste materials via bioleaching has been used to recover very small quantities of gold intimately sequestered within pyrite, FeS_2. Acidic pH and a consortium of bacteria including thiobacillii and *Metallogenium* are utilized. A lixiviate, ferric iron (Fe^{3+}), which is consumed in the reaction, has been shown to be essential (Mustin *et al.*, 1992). For specifics on the many sulfur-oxidizing microorganisms used on copper, iron, manganese, and mercury, mineral sulfides, see Ehrlich (2002, pp. 642–657).

The thiobacilli are autotrophs that produce sulfate (or H_2SO_4) directly from the oxidation of H_2S, the gas that may be produced by other bacteria or by volcanic emanations. Other bacteria, and including some Archea, may accumulate elemental S^0 when H_2S is in short supply, and some sulfur oxidizers can start with S^0 rather than H_2S, and produce SO_4 irrespective of the oxygen tension (London and Rittenberg, 1967), whereas species operating in partially reduced environment only oxidize to S^0. When the sources of sulfur are unlimited, most photoautotrophic species oxidize to S^0, which may accumulate intracellularly (purple sulfur bacteria) or extracellularly (green sulfur bacteria). The elemental sulfur accumulated by one species, *Chlorobium*, is readily available to the cell that produced it but not to other individuals in the population nor other species. Scanning electron microscope studies on this bacteria showed that the sulfur was concentrated in hollow proteinaceous protruding tubes on the cell surface, an example of the construction of an organic edifice to contain inorganic, by which we mean "mineral," deposits! (Douglas and Douglas, 2000). Polysulfides ($S_{n-1}SH^-$), but not free sulfur, have also been detected, and are probably an intermediate, in the course of sulfide oxidation to SO_3 and onto SO_4 in some species (Aminuddin and Nicholas, 1974).

8.04.3.2.2 Sulfate reducers

Sulfate reducers, originally identified as the bacterial species—*desulfovibrio, desulfomaculum,* and *desulfomonas*—have restricted nutrition unable to degrade organic materials below acetate. Recent investigations have vastly expanded the number of species that anaerobically use aliphatic, aromatic, and heterocyclic organic molecules, and some use H_2 as an energy source to reduce sulfate.

Archea sulfate reducers can also use simple organic molecules (glucose) as well as more complex substrates, adding to the previously identified methanogenic activity possible in these cells. Some varieties were isolated from the hot (to 110 °C) vent areas in the Pacific (Jannasch and Mottl, 1985; Jorgansen *et al.*, 1992), although low-temperature (10 °C) species occur in sediments (Ghiorse and Balkwell, 1983). For more details on the metabolism and range of all these microbes, see Ehrlich (2002, pp. 549–620) and Banfield and Nealson (1997).

8.04.3.2.3 Formation of elemental sulfur

There is widespread deposition of elemental sulfur from H_2S or SO_4 sources by bacteria. Native sulfur in the cyrenaician lakes of Libya, North Africa, is part of a cycle in which sulfate-reducing bacteria utilize the SO_4 available in the waters, and reduce it to H_2S, and associated photosynthetic bacteria oxidize the H_2S to S^0 (Butlin and Postgate, 1952). Sulfur-containing nodules covered with crystalline gypsum, $CaSO_4{\cdot}2H_2O$, are found in the shallow edge waters of Lake Eyre, South Australia (Ivanov, 1968 pp. 146–150). These nodules and the waters contain active sulfate-reducing bacteria and thiobacilli (Bass-Becking and Kaplan, 1956), and it is assumed that the sulfurogenesis is aided, if not actively the product of bacterial activity. Fumerolic hot springs with H_2S are also localities where oxidization to S^0, or in some cases to H_2SO_4, is bacterially mediated. The result is not only sulfur production but adjacent rock and any other materials, dissolve promptly in such acidic environments (Ehrlich, 1996).

8.04.3.2.4 Sulfate biomineralization

There are a few species that utilize sulfate minerals to form their hard parts. Spangenberg (1976) describes the epidermal intracellular (vesicle) formation of varying size statoliths of gypsum, $CaSO_4{\cdot}2H_2O$, in the jellyfish *Aurelia arita*. This citation is a modern follow-up to the identification by Fischer in 1884. Statoliths of barite, $BaSO_4$, in *Charandother protoctista* (Schroter *et al.*, 1975) are recorded by Lowenstam and Weiner (1989, table 4.1, p. 52, table 5.1, p. 76), who suggest that the mineral enabled the animal to respond to gravity by aiding location determination, which makes sense as the specific gravity values for these two sulfate minerals are: 2.3 for gypsum, and 4.5 for barite.

Acantharia. The acantharia use $SrSO_4$, the mineral celestite, or celestine (Butschli, 1906), to biomineralize a remarkably regular geometric test consisting of 20 spines. These creatures were described as early as 1858 by Muller (commented on in Thompson (1942) and illustrated in Schewiakoff (1926). The spines are radially distributed: 10 pairs of spines, each one a single crystal with the elongation of the spine parallel to the mineral crystallographic axes. The spines may grow to 1.5 mm long with a diameter of 38 μm (Schreiber *et al.*, 1959). The ranges of geometric patterns in the hard tissue of acantharia were presented by Popofsky (1940) and Reshertnjak (1981) with the most recent compilation listing over 150 species (Bernstein *et al.*, 1999).

Acantharians are exclusively marine, planktonic, single-celled protozoa, 0.05–5 mm in diameter, and show a wide range and often multiple types of symbionts (Michaels, 1988, 1991). Common in tropical and subtropical waters worldwide, they do occur in lower numbers in temperate and polar seas (Caron and Swanberg, 1990). Originally detected by dragging nets of fine mesh in shallow (200 m depth) open ocean, the early reports on distribution gave conflicting results: patchy low distribution, less than 1% of the total mineralized biomass, but up to 70% in the Ligurian Sea, and occasionally exceeding the concentrations of radiolarians, foraminifera, and diatoms (Bottazzi *et al.*, 1971). The different assessments were at least partially the result of dissolution of the acantharia skeleton that requires special preservation techniques (Michaels, 1988). $SrSO_4$ is highly undersaturated in the oceans (Whitfield and Watson, 1983), and early estimations were affected by the collection system since a particular catch might be analyzed only after month-long storage (Bottazzi *et al.*, 1971). Further, as Odum (1951) commented, strontium, as the major constituent of acantharia, is also found in trace amounts in the carbonate minerals of other life forms. The growth of acantharia is a prime example of biologic accumulation against a gradient, since the level of strontium is 8–10 mg (8,000 μg) per liter in seawater. It is now known that barium, and a variety of other elements (calcium, manganese, nickel, zinc, lead, arsenic, and bromine), can be detected in acantharia (Brass, 1980). The average Ba/Sr molar ratio is 3×10^{-3} (Brass and Turekian, 1974), but the value varies for different samples: those from the Pacific were 2% lower than those collected from the Atlantic Ocean. It has been suggested that deceased acantharia undergoing dissolution at depth would release more barium than strontium to the waters, and these creatures may play a larger role in barium than in strontium cycling.

It should be noted that celestine has been found in hydrothermal deposits and in evaporates (Skinner, 1963), although these occurrences may not necessarily be biologically mediated. The mineral also has a rather high specific gravity, 3.98.

8.04.3.3 Iron Biomineralization

The essential element iron is not only utilized by all living systems but bacteria materially assist the nucleation of minerals that contain ferric and ferrous forms. After a brief consideration of the roles of iron in bacterial systems, we discuss the magnetotactics, the bacteria that biomineralize with euhedral nanosized particles of magnetite, Fe_3O_4, and greigite, Fe_3S_4.

8.04.3.3.1 Roles of iron in Archea and Bacteria

Fe^{2+} will auto-oxidize to Fe^{3+} in air above pH 5. However, at lower pH, where Fe^{2+} is the dominant form, many Bacteria and Archea species hasten the reaction with the energy gained in the transformation, Fe^{2+} to Fe^{3+}, to fix carbon. Under anaerobic conditions there would be no oxidation without bacteria in light or in the dark. Some anaerobes can use nitrate as the electron acceptor coupled with Fe^{2+} oxidation (Benz *et al.*, 1998). These prokaryotes can facilitate mineral deposition, and are considered by some researchers as prime candidates to explain the widespread banded iron formations found in the Precambrian (Widdel *et al.*, 1993).

Fe^{3+}, common in surficial oxic environments, forms hydroxides, complexes formerly called "limonite" (a collection of rust-like precipitates, chemically designated as $Fe(OH)_3$). The aggregate has now been shown to contain several oxyhydroxide species, FeOOH, and the oxides, hematite, Fe_2O_3, and magnetite, Fe_3O_4, and some new minerals have been identified (Schwertmann and Fitzpatrick, 1992, tables 1–3, pp. 10–11).

The ferric forms are stable in neutral or slightly alkaline solutions, soluble in acids, but can only be reduced within a reasonable time frame to Fe^{2+} via bacterial action. Therefore, in oxic environments iron is usually present in Fe^{3+}-containing minerals, i.e., hematite, and unavailable to living forms, thus presenting a limit to growth. To provide for such situations many microbes produce chelators known as siderophores with very high binding constants for Fe^{3+} (10^{22}–10^{50}) relative to those for Fe^{2+} (10^8) (Reid *et al.*, 1993). Note that plants have similar needs and utilize similar products and mechanisms. Fungi and microorganisms located on the roots of plants are symbionts that aid in the selection of essential elements from the environment and make them available to the plant. Cells convert the transported ferric compounds into ferrous ions and into heme as well as nonheme complexes. Many microorganisms are involved with iron reduction and oxidation (Lovley, 1987) but only a few biomineralize with magnetite, a mixture of the two oxidation states, in spite of the widespread availability of iron.

8.04.3.3.2 Bacterial iron mineral formation

Bacteria are major mediators in the deposition of many nonmagnetic oxyhydroxide iron minerals such as ferrihydrite and goethite, but some form the iron oxide magnetite, Fe_3O_4 (see Figure 6), or the magnetic iron sulfides, greigite, Fe_3S_4, and pyrrhotite, Fe_7S_8. The location of these minerals may be intracellular (biologically controlled), or extracellular (biologically mediated) (Mann *et al.*, 1992) (Figure 7). The best-studied extracellular producers are *Geobacter metallireducens* and *Shewanella putrefaciens.* Both are rod shaped, gram negative, freshwater chemoheterotrophic forms, meaning that they utilize organic carbon compounds from a range of sources (Lovley, 1987). A halotolerant iron-reducing bacterium has been shown to produce nonstoichiometric magnetitic particles with the mineral composition between magnetite and maghemite (Hanzlik *et al.*, 1996).

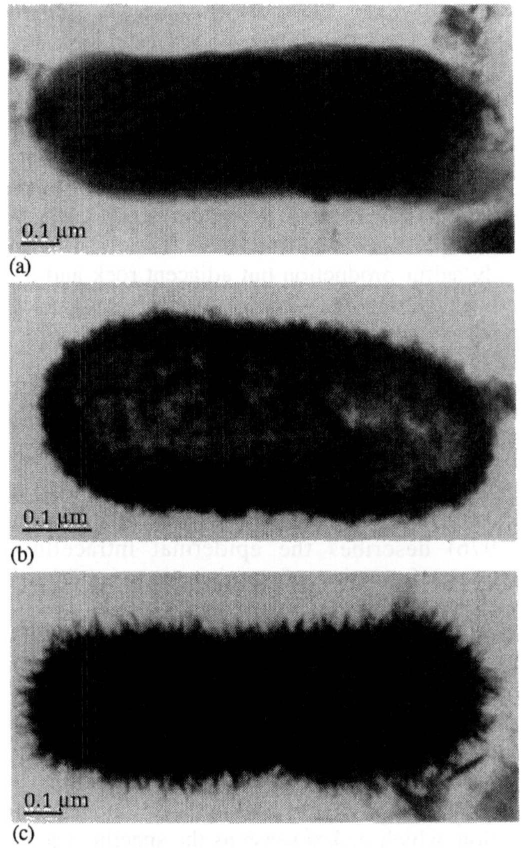

Figure 7 Progressive stages of microbially mediated iron oxide mineralization of a rod-shaped bacteria. Scale bars = 0.1 μm: (a) original bacteria, (b) early deposition of amorphous hydrated iron oxide (ferrihydrite), and (c) intense deposits of acicular magnetite (source Mann *et al.*, 1992, photo 2, p. 118).

Some anaerobic sulfur-reducing bacteria can also use iron as the sole electron acceptor. Since all anaerobic sulfur-reducing bacteria respire H_2S, this product together with a source of iron from an iron-rich environment could aid in the extracellular nucleation and production of greigite, pyrrhotite, or nonmagnetic sulfides, such as the two polymorphs of FeS_2, pyrite, and marcasite. Intracellular magnetite production studies detected temperature-dependent fractionation of oxygen in the formation for Fe_3O_4 and water consistent with that observed with extracellular magnetite produced by thermophyllic Fe^{3+}-reducing bacteria (Mandernach *et al.*, 1999). The minerals formed will depend on the pH, Eh, temperature, and the source and availability of iron, sulfur, and oxygen (Freke and Tate, 1961; Rickard, 1969).

8.04.3.3.3 Magnetotactic bacteria

The processes that create the magnetic particles within bacteria, the magnetotactics, are

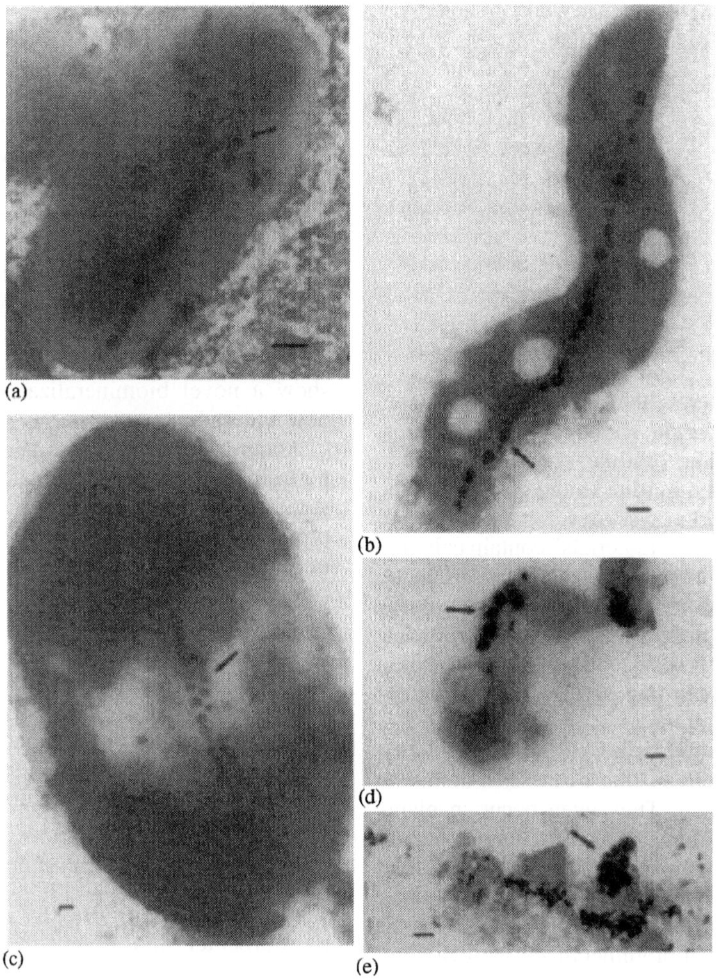

Figure 8 TEM of the magnetotactic bacteria of the Pettaquamescatt estuary: (a) coccoid from 3.6 m, (b) vibrio from 7.2 m, (c) large rods with double row of magnetite crystallites, (d) moribund bacteria on surface of the sediment, (e) particles from 1 m depth. Scale bars = 100 nm (source Stolz, 1992, p. 138).

biologically controlled (R. P. Blakemore and N. A. Blakemore, 1990; Bazylinski, 1996). The particles ranging from 30 nm to 120 nm in diameter are well crystallized, and each is a single crystal with distinctive crystal morphology, and occasionally several crystallites appear within one bacterium and usually become aligned (Mann, 1985). Although all magnetotactics described so far are gram-negative motile bacteria, it is possible that a magnetotactic Archea may exist (Bazylinski and Moskowitz, 1997). Found in highest numbers where an oxic environment changes to anoxic in fresh, brackish, or ocean waters, the magnetotactics show species specificity for geographic as well as depth locations.

The crystallites grow within an organic membrane bound vesicle, or sac, the magnetosome. First a precursor iron oxide forms that matures into a single magnetic domain crystal of a magnetic iron mineral (Balkwill *et al.*, 1980; Mann, 1985). The morphology of the individual crystallites is maintained within a species (Figure 8). Cubes, parallelepipeds, tooth, and arrow-shaped crystallites, some with truncated hexagonal or cubo-octahedral faces as well as the crystallographic growth patterns of the grains have been investigated using high-resolution transmission and scanning electron microscopy. The crystals are usually magnetite or greigite, but there is one magnetotactic species that has both minerals and another that contains nonmagnetic pyrite (Mann *et al.*, 1990). The size and linear arrangement of the crystallites into a chain of magnetic particles, or two chains in some cases, orient so as to enhance the dipole moment of the bacterium. The enhancement probably overcomes any thermal or other forces that might interfere and prevent detection of the Earth's geomagnetic field. Bacteria orient themselves toward the sediments, rather than the oxidized surface waters, with the direction of orientation distinct for northern and southern hemispheres. Alignment is not so simple as this might suggest. Magnetotactics swim up and down the geomagnetic field using their flagella to find optimum oxygen concentration locations, and probably move when the oxygen concentration changes. The biological control of internal mineral formation and the choice of the mineral species allow the bacteria to use both magnetotaxis and aerotaxis to maximize their habitat (Frankel *et al.*, 1997).

8.04.3.4 Carbonate Biomineralization

Of all the minerals that have been associated with biomineralization, carbonates are the most obvious. From the coralline materials found in atolls to the shells of mollusks and gastropods, the average person knows these creatures, and possibly that the "hard materials" contain calcium, if not that the mineral is calcium carbonate, $CaCO_3$. It is the carbonate minerals that provide the distinctive shapes allowing immediate recognition, and the precise compositions are often the means by which one designates a particular invertebrate species.

Lowenstam and Weiner (1989, pp. 8–11, table 2.1) clearly show that carbonates dominate in biomineralization. They even occur in plants and fungi. The volume percent of limestone and marbles is well documented from the Precambrian to the present. Whether these rocks are inorganic precipitates or festooned with fossils, many of the living creatures had biomineralized with calcium carbonate, is usually clear. Indeed, some strata are composed entirely of calcium carbonate shells. We present examples of carbonate mineral deposition in cyanobacteria, corals, coccoliths, foraminifera, mollusks, echinoids, and the arthropods.

Although many calcium carbonate deposits are well known geologically, and the intimate association of bioorganic molecules with the mineral materials fully appreciated for geochemical investigations on stable isotopes, identification of the precise mineral species has often been ignored. Bulk inorganic chemical analysis in early investigations documented the presence of calcium and the term "calcification" was, and is, widely used. This designation does provide important chemical information on the cation but not on the anion. There are many calcium-containing biomineralized tissues that are not carbonates, e.g., calcium sulfates, and oxalates, and a chemical analysis cannot distinguish between the several possible calcium carbonate mineral species, calcite, aragonite, and vaterite (see Section 8.04.2.1.3). Further, because the mineral crystallites in biomineralized systems are usually tiny, and optical microscopy was the only technique generally available until the twentieth century, accurate identification was often impossible. Biominerals, difficult to assess, were labeled "amorphous." It was not until the 1920s that they were adequately evaluated with X-ray diffraction analyses, and the mineral species were accurately determined. The more sensitive spectrographic methods (IR, Raman, ICP-MS) allow for more specific investigations. The stretching and bending modes of the carbonate anions can be evaluated by identifying the species of calcium carbonate mineral from its structural character, and the polymorph determined. The latter may show a novel biomineralization, and perhaps a new invertebrate species.

Another important point about calcium carbonate biomineralization is that more than one of the three polymorphs may be present in close proximity within an individual sample and each may be a different calcium carbonate polymorph (mollusks provide such an example), or the mineral may change over time. In many species the larval stage is aragonitic, while the adult biomineral is calcite. The different minerals in biomineralized tissues play specific roles in the proper functioning of the organism.

The uptake and incorporation of other elements, magnesium or strontium, in the calcite have been investigated and shown the bioavailability of these elements. In modern ocean waters, the content and range of magnesium in biotic and abiotic deposited minerals were virtually identical in studies by Carpenter and Lohmann (1992). However, there was a more rapid strontium uptake in the calcium-rich biomineral. The composition of the calcium carbonate mineral deposited is a function of kinetics and related to the metabolism of the organism.

8.04.3.4.1 Cyanobacteria

Cyanobacteria, one of the earliest creatures to biomineralize, are also one of the most pervasive and abundant life forms on the planet. Although they usually occur in the upper oceans where they form mats, and hundreds of species are well known for the deposition of calcium carbonate minerals, they are also found in lakes, in the oxygenated layer in the upper horizons of soils, and associated with fungi and other creatures that bore into limestones or other rock masses (Riding and Voronova, 1982; Ehrlich, 2002). Together with calcifying algae and a number of other mineralizing taxa, cyanobacteria markedly expand at the end of the Precambrian (Lowenstam and Weiner, 1989, p. 233).

The calcifying varieties of cyanobacteria do so extracellularly in and on a mucilaginous and fibrous polysaccharide sheath (Figure 9(A)).

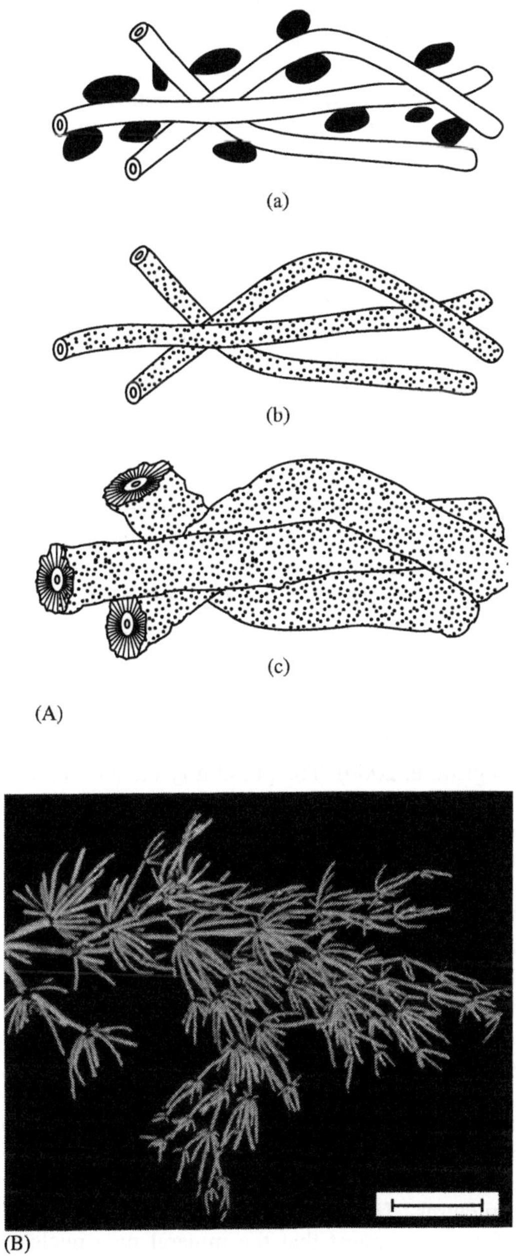

Figure 9 (A) Sketch of the three stages of $CaCO_3$ deposition associated with filamentous bacteria: (a) trapping sediment, (b) calcification of the sheath, and (c) encrustation of the sheath (source Pentecost and Riding, 1986, figure 51, p. 75). (B) Articulated coralline algae. Scale bar 1 cm (source Simkiss and Wilbur, 1989, figure 7.5A, p. 94).

The nucleation of a calcitic or aragonitic form is species specific, and they grow where conditions are also favorable for abiotic carbonate deposition. The filamentous cyanobacterial mats may passively accumulate and bind local sedimentary particles that become entombed through cementation with additional calcium carbonate. Distinguishing between calcium carbonate precipitates adjacent to actively metabolizing cyanobacteria versus cell directed mineralization, i.e., via bioorganic molecular templates, is often moot. It is probable that both processes take place intermittently if not sequentially with shifts in temperature and composition, alkalinity or salinity, of the surrounding fluid media, especially when the waters are effectively saturated with respect to carbonate and calcium ions. Where the waters are in rapid motion, as in springs and waterfalls, $CaCO_3$ deposition may be aided by evaporation while bioprecipitation of carbonates is also taking place (Golubic, 1973, 1983).

Tuffaceous carbonate deposition ascribed to actions of cyanobacteria is known from the arctic to the tropics, and even in caves where light is restricted (Pentecost, 1978, 1985b). The mineral in freshwater mats is usually of lower magnesian content than that in the oceans possibly responding to species specificity but certainly to a difference in the availability of magnesium (Folk, 1974). Studies of the isotopic variability of carbon and oxygen in coralline algae samples indicate that some species are at equilibrium with the oceanic carbonate while others are not (Lee and Carpenter, 2001). The same may well be true of cyanobacteria. The contributions toward fractionation of stable element isotopes by any biomineralizing form during sequestration of carbonate minerals are not obvious nor straightforward.

Oriented calcite crystals of different morphologies enhanced through epitaxy (growth of the mineral continues in crystallographic register on a preformed inorganic template), as well as completely unstructured and disordered mineralized mats have been described. The mineral expression suggests a range from total to no biological direction or control (Pentecost and Riding, 1986). Secondary, and perhaps even tertiary, precipitation of calcium carbonate and/or silicate from the activities of accompanying biomineralizing creatures would lead to the varied microfabrics that have been described for ancient sediments. The textures are undoubtedly related to multiple stages of deposition, even if predominantly due to cyanobacteria (Wray, 1977). Modern investigations of bacterial biomineralization can determine more precisely the species of cyanobacteria by employing the latest advances in genetic techniques and matching methods (Dale, 1998).

The appearance of marine calcified cyanobacteria marking the base of the Cambrian suggests a major change on the Earth, either in the environmental availability of appropriate elements, or a shift in biological capacities that enhanced mineral precipitation. Mineralized reef deposits fortuitously provide a permanent geological record of the existence and activity of cyanobacteria. The reduction of extensive carbonate deposition post-Cretaceous and the variations recorded in today's marine environments have made cyanobacteria one of the foci for marine geochemical analysis. Wary researchers anticipate diagenetic changes in ancient carbonate mineral deposits when stromatolites were abundant (Walter, 1976; Riding, 1977; Golubic, 1983), but similar care should be exercised in modern reef investigations. Whenever and wherever living creatures expire, their mineralized materials become exposed to a different environment. Addition or subtraction, or modification, of both organic and inorganic constituents, should be anticipated and the evidence sought for in the samples. Using high-resolution techniques (TEM, SEM), the possible morphological or chemical effects attributable to local conditions and diagenesis should be investigated before geochemical analysis to assure accurate and meaningful assays.

Because of the widespread geographic locations and geologic persistence of cyanobacteria, volumes of work have been published on this group of organisms. The reader may find the following references of some advantage in searching for particulars on species and environments of mineral deposition: Brock and Madigan (1988); Carr and Whitten (1982); Rai (1997); and Sebald (1993).

8.04.3.4.2 Cnidaria (coelenterates)

The group Cnidaria includes jellyfish, anemones, and hydras, mobile invertebrates, most of which sparingly mineralize at only a few anatomical sites with specialized functions. For example, the gravity detectors, or statoliths, in the hydrozoa mineralize with magnesian–calcium phosphates or calcium sulfate. However, another group of organisms in this phylum form the reefs, fringing many islands and continents today as they have in the past. These obvious, large geostructures are the mineralized product of sessile, or fixed, cnidarians—the corals. The corals create massive carbonate deposits in the relatively shallow waters of the circum-equatorial regions where salinities are at least 27 ppt and temperatures remain at or above 15 °C year round. Some thousands of miles in extent, the coral reefs are a unique ecological niche that contains many other biomineralizing invertebrates such as sponges, mollusks, and gastropods.

Corals. Among the most important members of the class Anthozoa are the scleratinian corals each of which secrete their own calcified exoskeleton. Most are colonial but some deeper-water (and lower-temperature) species are isolated individuals. Their skeletons mineralize with calcite or aragonite, and sometimes both calcium carbonate polymorphs occur in different anatomic sites in the same animal (Lowenstam, 1964). One member of this class, Octocoralia, produces individual mineralized spicules, or spicular aggregates whose distinctive morphologies are the basis for separating species (see below), while another produces massive or branching but coherent structures (Figure 9(B)). The massive reef building forms contain symbionts within their endodermal tissues. Known as zooanthellae, these unicellular dinoflagellate algae are essential to the metabolism of the coral, including mineralization of their skeletons. Trench and Blank (1987) identified *Symbiodinium*, the gymnodinoid dinoflagellate, and showed that this intimate, indeed essential, relationship was pervasive throughout all reef corals and other benthic marine cnidarians. Using modern genetic techniques on extracted ribosomal DNA, the diversity of symbionts is being confirmed (Gast and Caron, 1996). In sea anemones, the concentrations of different symbionts show a dependence on habitat, illustrating how they could be used to signal specific benthic conditions (Huss *et al.*, 1994; Secord and Augustine, 2000). The possibility for interference with the viability of coral reefs by global climate change is a relatively new arena of investigation (e.g., Lindquist, 2002).

The functional parts in the body of a reef coral have a protective surface of ectoderm which is external but adjacent to the specialized cell layer that gives rise to the mineralized structures (Johnson, 1980). Once the coralline larvae (free-swimming stage) settle onto a surface, a base plate of bioorganic molecules is formed (Wainwright, 1963) and calcite deposits as spherulite interspersed with small rodlike granules (Vandermeulen and Watabe, 1973). The granules, all about the same size, do not coalesce to form larger structures. The two morphological aggregations of calcite suggest that the mineral may nucleate and be retained in spaces defined by an organic membrane. In the second stage of larval mineralization, large aragonite crystals similar to those seen in the adult skeleton are deposited on top of the base plate.

The mineralized portions of an adult scleratinian coral are composed of bunches of bladed aragonite crystallites elongated along their *c*-axes forming spherulites. The spherulites show a dense center, presumably the site of initial mineral nucleation, and on electron microscopic examination the centers were shown to be calcite for the species *Mussa angulosa* (Constantz and Mielke, 1988). The aragonite crystals have distinct morphologies used to define distinct genera. It appears that the mineral deposition is under local control, possibly the result of chemically specific bioorganic molecules produced by the cell that later disappear. Lipids, phospholipids, and proteins have been detected on analysis of the coral body (~0.02 wt.% of each of these compounds relative to the skeletal weight), and may assist with mineralization. The proteins are characterized by high concentrations (~50 mol.%) of acidic aminoacids, especially aspartic acid, glutamic acid, and γ-carboxyglutamic acid, the latter biomolecular species that is also detected in vertebrate tissues (Hamilton *et al.*, 1982).

Other taxa that biomineralize. Spicules formed from magnesian calcitic (calcite that incorporates some magnesium into the structure, Mg-calcite) are found in the Octocoralia and Gorgoniacea. They may occur in the axis, or in the cortex, of the skeletons of these colonial corals as well as in the tentacles with morphologies distinct from the anatomical sites. The orientation of the Mg-calcite crystallites can differ within a particular spicular aggregate or the mineral may form as a single crystal, similar to the calcite mineral deposited in echinoderms. There are examples of functional adaptations of spicular biomineralization. The sea whip (*Leptogorgia virgulata*) has a plant-like appearance and a skeleton composed of the protein collagen reinforced by Mg-calcite spicules. The sea fan (*Melithacea ochracea*) has anatomizing branches of jointed segments that are dense fused aggregates of calcitic spicules alternating with protein segments rich in collagen. It would appear that the design of the mineralized skeleton enables them to withstand the forces of tidal currents with flexible as well as mineralized portions, securing their survival. In the Gorgoniidae family of the Gorgoniacea, calcium phosphate has been identified in the axial spines and holdfasts (Macintyre *et al.*, 2000) in addition to the Mg-calcite spicules throughout the coral. In *Pocillipora damicornis*, the skeletal organic constituent is chitin, a polysaccharide (Wainwright, 1963).

The mineralization mechanisms displayed by Cnidaria are extraordinarily varied and, although a relatively primitive invertebrate group, they present highly specialized biological systems, including the incorporation of symbionts that show genetic distinctions related to the ecologic niche occupied by the coral.

8.04.3.4.3 Coccoliths

The most abundant $CaCO_3$ depositing organisms are algae, members of the phylum Haptophyta, which produce a flagellum used during their motile stage, and surface scales that cover their cells. Some species mineralize the outer portion of the scale, especially the coccolithophorids. They account for most of the oceanic carbonate sedimentations and chalk deposits worldwide (Lowenstam and Weiner, 1989, p. 68). The discoid fenestrated double-rimmed scales, less than 10 μm in diameter, of these unicellular photosynthetic planktonic algae are mineralized with calcite. *Emiliania huxleyi* is probably the best-known species. The calcitic mineral deposits on an organic template intracellularly. Similar to inorganic calcite the $[10\bar{1}4]$ faces of the usual calcite cleavage rhomb may be displayed in some coccoliths, while other species do not show any calcite crystal forms, and a few species mineralize with aragonite or silica. This range of minerals with different compositions, and different habits expressed by the minerals, is used to delineate species and suggests that the biomineralization process expresses different degrees of intracellular as well as extracellular control (Green, 1986).

Emiliana huxleyi: intracellular calcification. Using *E. huxleyi* as the example, we outline the sequence of intracellular, intravesicle calcification. The first stage is the formation of a flattened vesicle adjacent to the Golgi apparatus in the cell. The vesicle accumulates polysaccharides and proteins (De Vrind-de Jong *et al.*, 1986) forming a base plate, the site of mineral nucleation and deposition that may commence before the scale is extruded. The remarkable mineralized scale structures formed by *E. huxleyi* are composed of a fitted collection of upwards of 20 calcitic segments with anvil-like shapes that are smooth surfaced and behave as a single calcite crystal (Wilbur and Watabe, 1963) (Figures 10(A)–(C)).

To further emphasize the exquisite specificity of mineral deposition, each initial 40 nm rhomb of calcite formed in the plate is oriented crystallographically with the *c*-axis parallel to the base of the segment and aligned radially. The *a*-axes of the calcite are aligned vertically to the plane of the coccolith plate (Watabe, 1967; Mann and Sparks, 1988). Protrusions develop as the mineral is deposited and the calcite segments thicken and fuse. The extracellular coccoliths festooning the surface of the cell display anvil forms (Figure 10(B)) that precess in a clockwise direction with the calcite {104} face always to the right-hand side; this means that there is chirality expressed in the construction. The anvil-shaped units overlap in the completely formed ring (see developmental scheme in Figures 10(A) and (B) and micrograph of the amazing final product; Figure 10(C)).

In other coccoliths species the intracellular vesicles may be mineralized and completed after the scale has been extruded onto the cell surface but still covered by an organic membrane. In the genus *Chyrsotila*, nonmotile cells produce a mucilaginous sheath in which spherulitic aragonite crystallites form, which eventually encapsulate the cell (Green and Course, 1983). There are

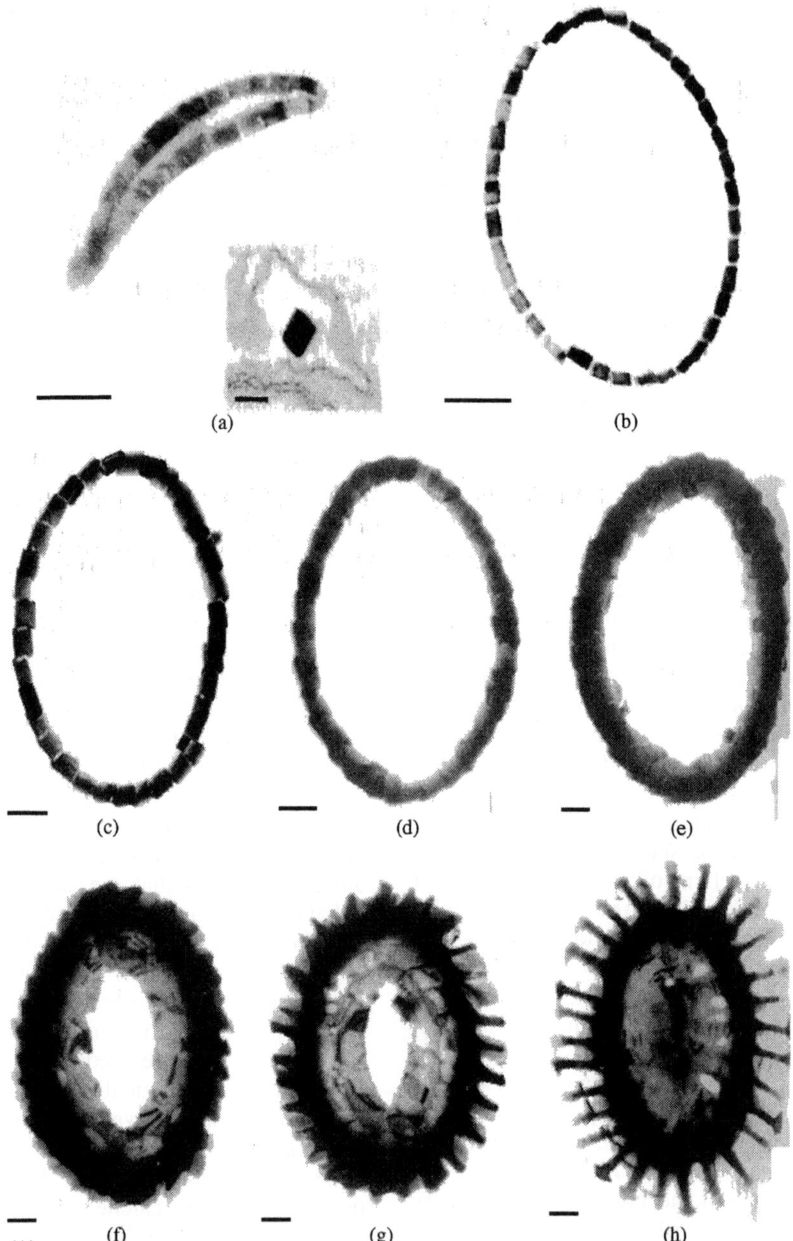

Figure 10 (A) Stages in the formation of *E. huxleyi*: (a) precoccolith ring, (b) mineral growth stages, (c)–(e) continued growth and then overlap of individual segments, (f)–(h) development of extensions from the initial ring. Scale bars for (a)–(d), (g), (h) = 0.2 μm, inset (a) = 0.5 μm. Scale bars for (e) and (f) = 0.1 μm (source De Vrind-de Jong and De Vrind, 1997, figure 17, p. 286). (B) Sketch of a cross-section through the *E. huxleyi* coccolith cell. The dark colored mineralized sections are produced internally and then moved externally (source Simkiss and Wilbur, 1989, figure 6.3, p. 68). (C) An SEM of a complete adult coccolith showing the juxtaposition of the coccolith plates into a sphere (source Mann, 2001, figure 8.9, p. 147).

species that form silica instead of calcitic cysts (Green *et al.*, 1982). Delicate structures with deliberate morphology and crystallographic orientations displayed by the mineral and a precision growth sequence are also characteristic of echinoderm exoskeletons.

The complicated intracellular process means that the *E. Huxleyi* cell must accumulate Ca, CO_2, or HCO_3^-, for the inorganic phase, as well as the carbon and nitrogen to create the organic molecules by a regularized, programmed cycle. The coccolithophorids are remarkable dynamos that illustrate the complexity of biological mineralization we only partially understand (see Mann (2001) for more details on calcification mechanisms).

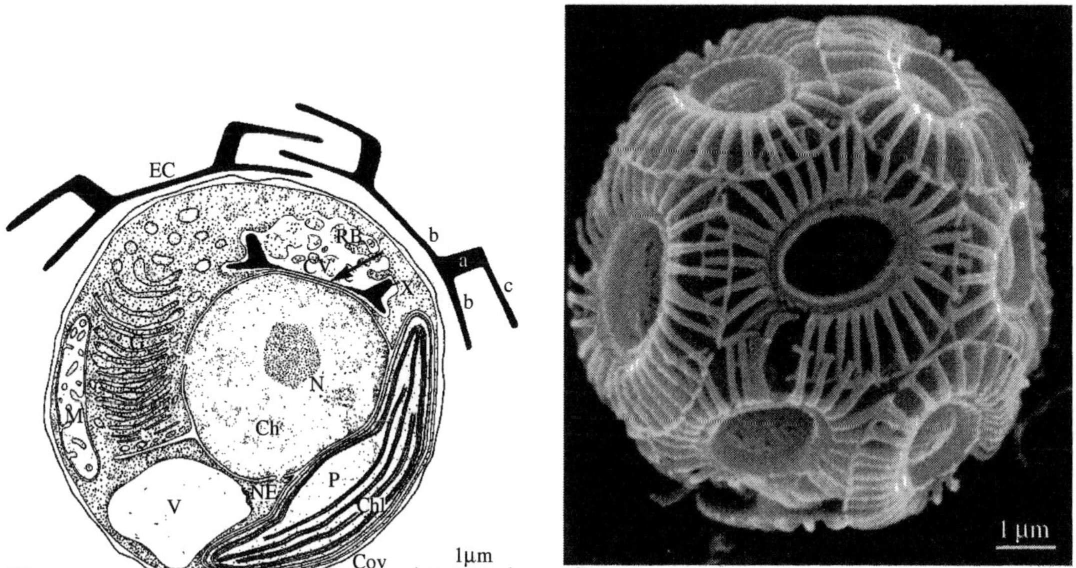

(B) (C)

Figure 10 (continued).

8.04.3.4.4 Foraminifera

Planktonic species that occur with the coccolithophorids and nektonic mollusks, foraminifera, are another biogenic carbonate-forming species in open oceans. A significant constituent of the biomaterials that formed the rocks typical of Early Mesozoic to the present, and because of their preservation and morphologic diversity, foraminifera have played a prominent role in biostratigraphy, providing both age and environmental information. The *Catalogue of Foraminifera* (Ellis and Messina, 1940) (now upwards of 70 volumes) describes over 3.5×10^4 species with over 4,000 living today, most of which are benthic, not planktonic. Divisions within the foraminifera are based on the materials used to construct their shells.

One group, the Allogrominia, has membranous shells composed of organic macromolecules and inorganic material picked up from their environment and attached on the exterior surface. Some of these species are so selective that they choose a particular mineral. Another group actually cements whatever it can find with acid mucopolysaccharide and proteins, using its pseudopodia to strategically place the accumulations.

Some foraminifera form calcitic, others aragonitic mineralized tests. The ancient, now extinct, Fusulinia constructed its shells out of microcrystalline calcite with unique morphologies (Green *et al.*, 1980). The Miliolina form magnesian-calcite crystals, 1–2 μm long and 0.1 μm wide within vesicles, which are then haphazardly arranged on their exterior surface, while another species makes a shell from opal (Resig *et al.*, 1980). Most of the Rotaliina group of foraminifera mineralizes with magnesian calcite. The magnesium content may vary but the *c*-axes of the crystals are oriented perpendicular to the inside surface of the shell wall.

Constructed in a modular manner, each chamber of a foraminifera is preformed with an organic layer that subsequently mineralizes. In some cases a differentiating protoplasmic bulge produces two layers and calcite crystals nucleate on both surfaces of the membrane, first in patches that laterally fuse within a few hours and after tens of hours the final dimension of the chamber is achieved. Each new chamber is of larger dimensions and the chamber walls are a composite of organic and inorganic compounds. The growth of the mineral crystallites may be epitaxial (continuation of crystal growth in preordained crystallographic directions) once the signals for nucleation were inculcated through proteinaceous macromolecules generated by the cell (Addadi *et al.*, 1987). One other insight into the range of biomineralization in this phyla comes from the test formed by Spirillinacea which, in polarized light or with X-ray diffraction, appears to be as a single magnesian-calcite crystal (Towe *et al.*, 1977). Other foraminifera taxa utilize aragonite to mineralize their structures.

Of the many different minerals and mechanisms of mineralization expressed in the foraminifera, we must add the fact that they contain photosynthesizing symbionts, dinoflagellates, and chlorophytes, whether they are carnivorous or herbivorous forms. Further, during growth some mineral crystals show dissolution, suggesting that individual crystallites do not grow in isolated compartments, and there might be some "diagenesis" taking place during their lifetimes. Any resorption or secondary deposition will depend on the bioavailability of specific elements, and the relationships with the symbionts. The organic constituents, although of exceedingly small amount (0.2 wt.%), in the foraminifera are predominantly acidic macromolecules and resemble those found in many mineralized tissues (King and Hare, 1972). They may function at different biochemical levels as: (i) nucleators, attracting cations such as Ca^{2+}, or (ii) merely directing, the organization of mineral crystallites into an appropriate texture. By studying the mineral and patterns of growth relative to the organic frameworks, the distinctions between taxa and species can be delineated (Towe and Cifelli, 1967).

8.04.3.4.5 Echinoids

There are five major subdivisions of the echinodermata with distinct anatomical expressions, but characteristically all show a fivefold symmetry. In each subdivision there are forms that biomineralize their tissues; some forms have rigid tests while others have articulated structures. The mineral, magnesian-calcite, appears as granules, spicules, or spines with the $MgCO_3$ content ranging from 5% to 15% (Chave, 1952, 1954; Raup, 1966). The calcified skeletons are usually smooth externally but spongy and riddled with cavities which may become secondarily filled with mineral and therefore appear solid. There is no obvious internal organic sheath around the individual crystallites, so the mineral behaves as a single crystal (Towe, 1967) even with its high magnesia content (Schroeder *et al.*, 1969).

Studies of the mineralization of the larvae of sea urchins (they can be conveniently grown in synchronous culture) have made it possible to utilize molecular biologic techniques to determine gene expression, protein synthesis, and macromolecular organization as well as the sequences of mineral deposition (Benson *et al.*, 1987). These are some of the earliest forms to be studied and illustrate the connections between genetics and biomineralization. Stem cells responsible for spicule formation can be isolated and produce normal spicules *in vitro* (Kitajima and Okazaki, 1980); since spines regenerate the secondary mineralization process can also be studied (Ebert, 1967).

Spicule formation in the larval stage of the sea urchin commences when mesenchyme cells migrate into special locations where they fuse forming a syncytium that produces a membrane-bound vacuole. The spicule morphology is directed by the size and orientation of the vacuole, which aggregate and connect through stalks (Beverlander and Nakahara, 1960). Several sites of mineral deposition start but resorb and ultimately only one granule remains per vacuole. As it enlarges, a tri-radiate form is produced that reflects the crystallography of the magnesian-calcite: the long axis of the spicule is aligned with the *c*-axis of the mineral. Cross-sections through the spicule show concentric layers of organic glycoproteins in concentrations only 0.1% by weight of the spicule that effectively compartmentalize the mineral deposition (Benson *et al.*, 1983).

The adult mineralized spicules use the larval spicules as a nucleus extending the arms into branches that join into fenestrated plates. At the macrostructural level the spine plate segments show infilling of secondary calcite (Figure 11). The entire skeleton is covered by epithelia and any space within the mineralized portions contains a variety of cells, such as phagocytes and sclerocytes, which are capable of resorbing and reprecipitating mineralized tissue when the skeleton is broken (Loven, 1982).

The crystals forming the plates are aligned with their *c*-axes either perpendicular or parallel to the plane of the plate depending on the species and appear to be single crystals (Raup, 1959). As well, spines behave as single crystals with *c*-axes aligned along the length (Raup, 1966) and, if broken, regenerate to their former length by elongating, connecting, and fusing to restore the characteristic structural pattern (Davies *et al.*, 1972).

The regeneration of biomineralized skeletal tissues by echinoids using magnesian-calcite is reminiscent of that observed for the precipitation of calcium phosphate in regenerating fractured vertebrate bones (Section 8.05.3.7.2). These organisms and their cell systems are completely

Figure 11 An SEM image of the cross-section through the spine of a sea urchin showing secondary mineral infilling. Diameter of spine is ~1 mm (source Lowenstam and Weiner, 1989, figure 8.2.A, p. 125).

different from vertebrates, but they act in similar fashion and the processes have the same end: a mineralized form that continues to function.

8.04.3.4.6 Mollusks

The biomineralized shells of mollusks, especially the common bivalves, oysters, clams, and mussels, have been well studied possibly because they are easily found in shallow marine waters, where they become a food source for humans as well as animals. One of the largest groups of invertebrates originated in the earliest Cambrian and had a rapid expansion in the Ordivician (Stanley, 1973, 1976). Of the five out of seven taxa in the phylum that mineralize 17 different mineral species have been identified from a range of anatomic sites (Lowenstam and Weiner, 1989, table 6.1, pp. 90–93). Mostly they mineralize with calcite or aragonite, but the third polymorph of $CaCO_3$, vaterite, is found in the freshwater snail egg capsules of gastropods, which also precipitate goethite and opal in their radula. Further, weddellite, $CaC_2O_4{\cdot}2H_2O$, fluorite, CaF_2, and amorphous calcium phosphate are found in the gizzard plates of some species of gastropods. Only a general overview of the aragonite and calcite shell mineralization in bivalves will be presented, along with the interesting anatomical biomineralized structure—the formation of magnetite teeth in chitons, another member of the mollusk family.

Bivalves. The physical expressions of the mineralized portions of modern and fossil bivalves are similar, and faithfully record incremental growth stages. Mollusks, therefore, have been used to study not only the history of the life form, but also the effects of environmental influences, including climatic changes on their growth and development. Any exoskeletal alteration in general shape or size, and interruptions in mineralization due to variations in salinity, temperature, or tidal cycles, can be detected, and correlated with some external event (Seed, 1980). Whether burrowing in soft sediment or attached to a hard surface by the byssus, the macro- and micromorphology and chemical constituents of bivalve shells from far-flung reaches of the oceans have been sampled. The early workers were cognizant of the adaptation strategies of living forms to the environment (Clarke and Wheeler, 1922; Vinogradov, 1953). For example, in the 1920s it was known that the mineralogy of the shells was one of the ways they responded to the physical environment: forms that grew in the colder waters, or at higher salinity, deposited calcite rather than aragonite.

Was it possible that the physical milieu dictated the genetic determinants, and benefited the choice of mineral phase, or was the geographic expansion to different environments the instigator in shifting the mineral precipitated an example of convergent adaptation? Today we know that the mineral phase dictates the uptake of other elements: magnesium is a prominent element in solid solution in calcite, whereas strontium is more readily incorporated in aragonite. However, we note that all mineralogical variations in bivalves are associated with differences in shell architecture.

Bivalve shell architecture. The simple double-capped shell so typical of these species with macroscopically visible incremental growth lines hides a complicated mineralization system (Seed, 1980). There are different patterns of ultrastructure, distinctive to species, and to freshwater forms (Saluddin and Petit, 1983). Whatever the mineralization patterns, they arise through the actions of an organic structure produced by epithelial cells, the mantle. The mantle in the bivalves is composed of a gel-like organic molecular complex with three folds and two grooves. The groove closest to the shell surface is the site where specialized macromolecules, mostly proteins, are secreted, polymerize and form the periostracum (the outer shell covering that contains in addition to water insoluble proteins and chitin) (Waite, 1983). The periostracum varies in thickness, ultrastructure, color, and texture between species; the inner portions are sites of mineralization, i.e., calcitic or aragonitic deposition takes place interior to an organic layer that protects the animal from the external world. The shell is thus a layered structure, with interior cells the site of mineralization, and the calcium and carbonate for the mineral is transported either through or between the cells that produce the periostracum (Figure 12(A)). The cells actively involved in mineralization are elongated with obvious and numerous organelles: mitochondria, golgi, and endoplasmic reticula (Wilbur and Jodrey, 1952; Crenshaw, 1980). The shell thickens vertically and enlarges (horizontally) as the animal grows. Growth hormones and calcium transport proteins have been identified as important in the formation of mollusk shells, whether the process takes place on land or in water (Doterton and Doderer, 1981).

A cross-section through the growing shell edge shows several prismatic layers of $CaCO_3$ separated by organic layers. When the two polymorphs of $CaCO_3$ are present in the same shell, they are separated by ultrathin organic sheets with different compositions; in addition, each mineralite may be enveloped by biomolecules with different compositions (Watabe, 1965). The myostracum, site of muscle attachments in bivalves, is always mineralized with aragonite, adding to the complexities of accurate determination of the mineral in different

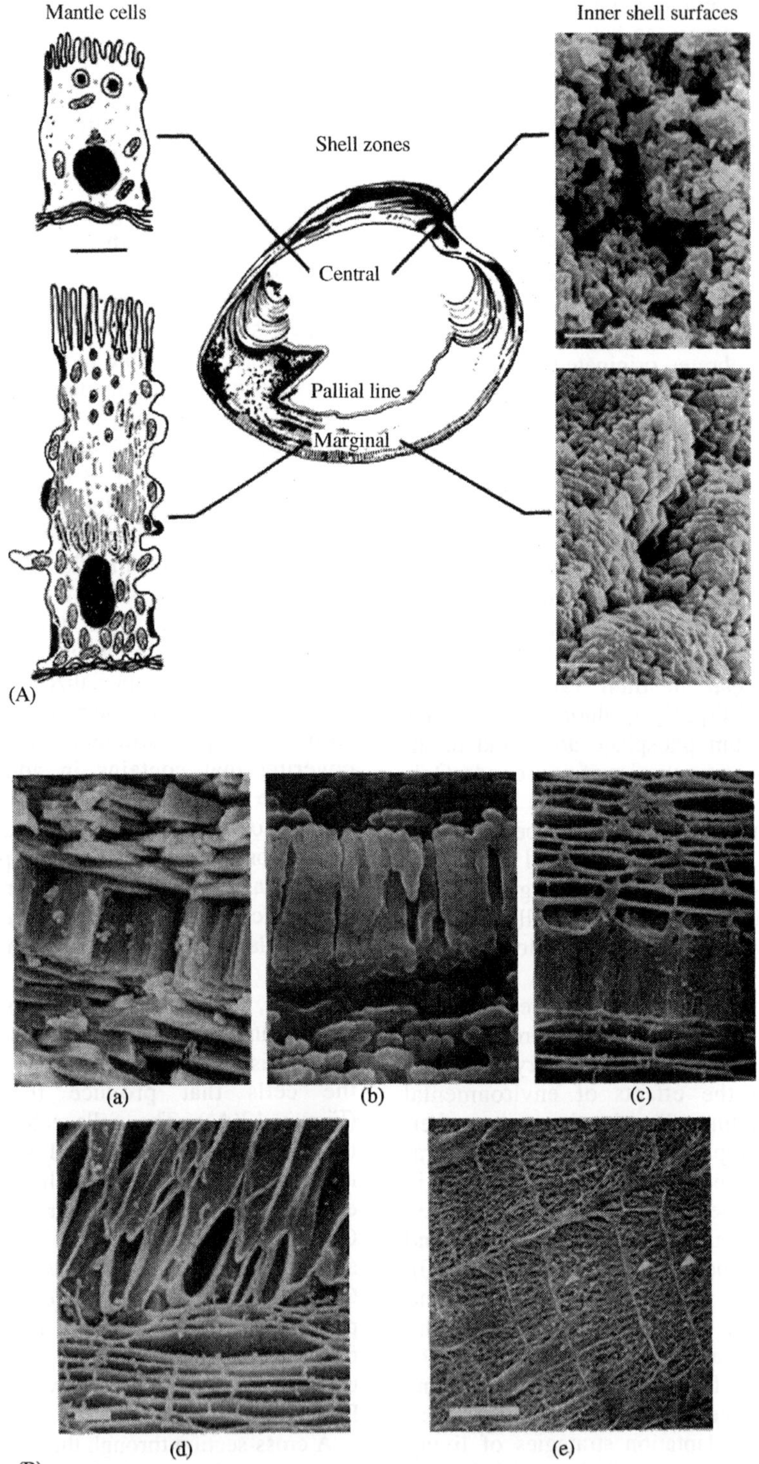

Figure 12 (A) Zones, cells, and appearance of the mineralized portions of the shell of *Mercinaria mercinaria*. Scale bar with sketches of cells 3 μm. SEM images after removal of organic with sodium hypochlorite solution, upper = 0.5 μm, lower = 0.1 μm (source Crenshaw, 1980, figure 2, p. 118). (B) SEM images of the various layers in bivalves, *Mytilus californianus* air dried and fractured sections: (a) pallial myostratum; (b) polished and etched section in which the organic matrix exposed by etching has collapsed over the dried surface; (c) polished and etched section obtained by critical point drying showing organic matrix in high relief above etched surface; (d) transition from prismatic (top) to nacreous layer (scale bar (a)–(d) = 1 μm); and (e) prismatic layer of *Mercinaria mercinaria* showing prominent growth lines (arrows) formed by concentrations of organic matter (scale bar = 10 μm) (source Clark, 1980, plate 1, p. 608).

portions of the shell as it is retained during the shell enlargement.

Over 50 types of shell ultrastructures have been catalogued with different size layers, patterns, and orientations of calcite or aragonite in the layers depicted (Carter, 1980; Carter and Clark, 1985; Boggild, 1930) Figure 12(B). The biomineralization processes appear to be directly responsive to an anatomical site. Such specificity suggests sophisticated genetic control not only on the production of the bioorganic molecules but also for the mineral phase, its site, and orientation. For example, tablet crystals of aragonite in nacre in one bivalve species may have uniform height in all the layers but show variable or different heights and arrangements in another species. The length, height, and shape of the aragonite tablets may vary in a third species from one layer to another layer. The implications of such refinements have been used to gain some general understanding of the mechanisms of biomineralization in invertebrates (Weiner, 1986).

Chitons. Another class of mollusk is the chitons. The shell is not solid as in the bivalves but composed of a series of eight overlapping plates mineralized by aragonite. The plates arise from a mantle-like mass that also produces the girdle and aragonitic spicules. The spicules initially form within vesicles of the epithelial cell system and protrude beyond these enclosures as they enlarge. The mineralized chitons have an additional layer between the periostracum and the inner shell layer, the tegument, in which spherulites and other organs, e.g., light sensing (Haas, 1976), are aligned parallel to the plate face. For details of the anatomy of an adult mineralized shell, see Beedham and Truemann (1967).

The chitons have another and quite different anatomical mineralized site, teeth; these teeth contain magnetite. Since these creatures inhabit the intertidal and subtidal zones on tropical limestone coasts, or other rocky areas, they utilize the teeth for obtaining food, the algae that cling to the rock surfaces. There are other organisms that grow magnetite-containing teeth (Lowenstam and Kirschvink, 1985), and in order to provide a unique insight into a novel biomineralizing system, we describe the chiton teeth that illustrate specialized invertebrate structures.

Teeth of chitons. Chiton teeth are arranged in transverse rows consisting of a central tooth and eight pairs of flanking teeth along the radula, a ribbon-like structure in the mouth portion of the chiton (Figure 13). One end of the radula moves providing an area of 8–10 rows of teeth for scraping. As teeth abrade they are discarded, the rate dependent on species, and within a couple of days new magnetite teeth move into place having been produced at the same rate at the other end of the radula (Nesson, 1968; Kirschvink and Lowenstam, 1979).

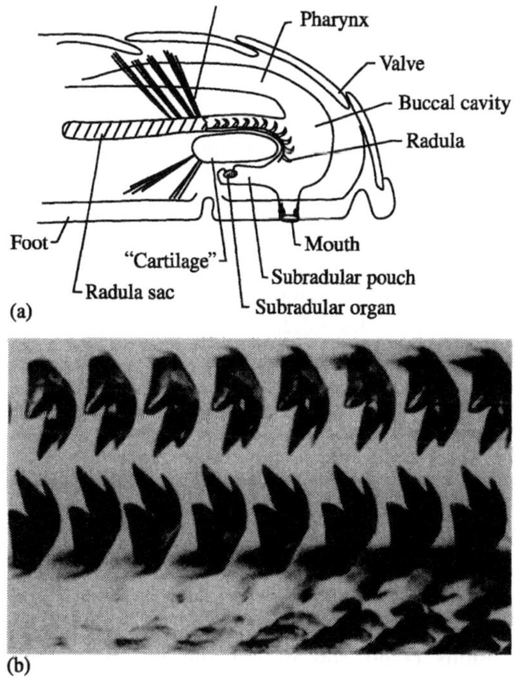

Figure 13 Chiton teeth. A sketch of the radular organ in the chiton mouth and a micrograph showing the shape and rows of magnetite teeth. Each tooth is less than 20 μm in length (source Nesson and Lowenstam, 1985, figure 1, p. 336).

Each tooth with its mineralized cap has a flexible stalk through which it is attached to the radula. Studies in *Chionida* showed that the original organ was defined, and the mineralization sequence orchestrated, by epithelial cells. The cells first create the housing of a mature tooth, then the framework of α-chitin, organic molecules on which iron hydroxides precipitate. The source of iron is ferritin, an iron–protein complex found in the blood. In which the protein encloses the iron oxyhydroxide ferrihydrite ($5Fe_2O_3{\cdot}9H_2O$). Ferritin is brought to the tooth location and stored until iron is extracted and transported in a reduced, soluble form to the mineralizing sites (Nesson, 1968). The first mineral to be precipitated in the tooth is ferrihydrite (Towe and Lowenstam, 1967), a poorly crystalline phase with light brown color, that is a precursor to the nucleation and growth of magnetite. A thin layer of another iron hydroxide phase, lepidocrocite (γ-FeOOH), forms on the inner surface of the enlarging magnetite layer. Note that magnetite occurs on the rasping and pointed edge, as well as on the concave portion of the tooth surface, while the central core and most of the remainder of the tooth is filled with dahlite, carbonate hydroxylapatite or francolite, the fluorine-containing carbonate apatite. A similar mineral sequence occurs in another chiton, *Cryptochiton*, except that amorphous hydrous ferric phosphate with trace amounts of amorphous silica form the tooth interior. The selection of magnetite, and/or apatites to produce a strong structure is appropriate for the rasping function of this specialized organ compared to the soft calcium carbonate mineralized tests.

For additional information on magnetite, and other iron mineral species see Section 8.04.2.4, and for iron biomineralization and mechanisms see Section 8.04.3.3.2.

8.04.3.4.7 Arthropods

The arthropod domain includes insects, the largest group of living creatures but a group that has not so far been known to be obviously biomineralized. It is certainly true that specialized anatomical sites within this group of organisms contain mineral materials and future investigations could add to our present list. Mobile arthropods may be found anywhere, on land, in freshwaters and ocean waters, and have need of bodily protection, as they are the prey of other invertebrates and vertebrates. Some arthropods, e.g., crayfish or lobsters, like mollusks, and fish, have become important constituents of the human diet. The life cycle and nutritional requirements of these creatures have become well documented so that they can be economically "farmed" (Nowak, 1972; Marx, 1986).

The Crustacea show a wide range of biominerals in specialized organs within their segmented bodies (Lowenstam and Weiner, table 7.1, pp. 112–113). The cuticle, or carapace, the external covering that is shed, or molted, as the creatures age and grow in size, may be mineralized with calcite, aragonite, or calcium phosphate mineral materials. These creatures also produce gastroliths, a much more heavily mineralized organ that functions as an internal storage site for the calcium required in the seasonal regeneration of the carapace. Usually composed of amorphous calcium carbonate (calcite in crabs and lobsters), in insects other calcium minerals, such as calcium oxalates and calcium citrate (in addition to the carbonate or phosphate species), are also found. Copepods show some opal in their mandibles, presumably an advantage for crushing and grinding prey, and several crustacea, including the trilobites, perhaps had calcite lenses. Bees are assumed to harbor magnetite for navigation (Gould *et al.*, 1978).

Crustacea exoskeleton, the carapace. The carapace of Crustacea strengthens their exoskeleton through mineral deposition (calcium carbonate or calcium phosphate) in, or on, a fibrous organic template. The first stage is the secretion of protein fibrils, 2.5–5 nm in width, that aggregate in a special manner with chitin, an aliphatic polysaccharide, β(1-4)-N-acetyl-D-glucosamine, which closely resembles cellulose, and is also found in fungi. β-Chitin was recognized as early as 1811 in mushrooms and in insects, and by mid-century the name "Chitine" was applied to arthropod cuticle materials. A relatively recent book summarizes the chemistry and molecular characteristics of chitin including industrial and medical applications (Muzzarelli, 1977), while the molecular biology of chitin has since been investigated by Blackwell and Weih (1980) and Atkins (1985). Chitin, the organic constituent in nacre biomineralization sites in mollusks, is illustrated by Mann (2001, p. 104). In arthropods, the fibrous protein polysaccharide mixture forms layers which stack with the fiber direction 180° out of phase between adjacent lamellar arrays forming a very tough, amazingly resilient "skin".

The total amount of mineral within the matrix layers varies with species (Greenaway, 1985) and from place to place within the cuticle of an individual (Huner *et al.*, 1978). Prenant (1927) measured the ratio of calcium phosphate to amorphous or crystalline carbonate in the tissues, and suggested that phosphate may act as an inhibitor to the formation of a crystalline carbonate phase. Hegdahl *et al.* (1977) showed that the cuticle of the crab *Cancer pagurus* contained three distinct layers, each with different ultrastructural organizations and percentages of calcite. It would appear that the mineral selectively associates, and may be nucleated and localized by the specific proteins or the chitinous polysaccharides that characterize each of the species (Roer and Dilliman, 1985; Degens *et al.*, 1967).

Gastroliths. The initiation of the crustacean molt cycle takes place below the cuticle (Greenaway, 1985). However, before shedding a new organic molecular composite is formed and some or all of the mineral in the old carapace is resorbed. The calcium released during the molt is retained in a separate storage site and structure. A mineralized button, known as a gastrolith, is created in the cardiac stomach of crabs and lobsters from where the element will be mobilized to mineralize the next and larger shell. Since the amount and availability of calcium depends on habitat, terrestrial and freshwater crustaceans maximize calcium recycling. These forms, when they consume their cuticle, must have at their disposal some "hardening" materials, at least for their mouth parts, as they lose every means of protection, including their ability to feed, when the cuticle disappears. Gastroliths in freshwater crayfish are composed of spherical aggregates of amorphous $CaCO_3$. This ultrafhin, poorly crystalline, solid form probably proves advantageous as a more crystalline aggregate would not be as easily dissolved with rapid mobilization of the calcium (Travis, 1976).

Other members of the crustacea sequester calcium as calcium phosphate in a gland, the hepatopancreas, where needle-shaped rods up to 300 Å or concentric granules to 4 μm diameter have been detected (Chen *et al.*, 1974; Becker *et al.*, 1974). In one species the granules have a high concentration of lead, and it has been suggested that these gastroliths may also play a role as a detoxification site (Hopkin and Nott, 1979). Calcium pyrophosphate (amorphous) was identified in the hepatopancreas of gastropods (Howard *et al.*, 1981) and yet another biomineralization form and system, extracellular phosphatic concretions, is found in the crustacean *Orchestia cavimana*. The calcium phosphate form may be a problem. Not only is this mineral less soluble but its localization requires that this crustacean create a customized extracellular transport system to move the calcium to the newly forming epithelial cuticle; however, the phosphate is a more durable (hard) mineral than carbonate. Perhaps this species is not a biological oddity, but a trial run for biomineralization in bone where mineral dissolution and reprecipitation is essential to the proper functions of vertebrates.

Calcium is not usually stored in the blood in these creatures, although a freshwater land crab *Holthusiana transversa* shows hemolymph cloudy with microspheres of calcium carbonate after shedding, presumably awaiting new skeletal deposition sites. Specialized storage mechanisms are not unique to crustaceans. A calcium phosphate compound is found in an insect, the larval form of *Musca autumnalis*. Mineral containing granules accumulate in tubules, and resorb, coincidental with the pupation of a fully developed larva (Grondowitz and Broce, 1983; Grondowitz *et al.*, 1987).

The arthropods present us with an intriguing range of biomineralization mechanisms that suits their peculiar bodily needs and functions. Biomineralization activities change on demand during their life cycle and clearly have been modulated to maximize a particular form and function. Sequestering and recycling of calcium for "hardening" their carapaces, an essential part of their defense mechanism.

8.04.3.5 Silica Biomineralization

Silica biomineralization is the incorporation of the extremely fine-grained, and virtually amorphous varieties of opal, a hydrated form of the common, but highly insoluble species that has the formula $SiO_2{\cdot}nH_2O$ (see Section 8.05.2.2.1). Opal is related to the very common SiO_2 mineral species, quartz. Oceans are at present undersaturated with respect to opal (Broecker, 1971) possibly because of the biological formation of animals with silicified skeletons such as the diatoms. These delicate structured creatures, which proliferate in the upper photic zone, dissolve at depth. Therefore, only robust siliceous skeletons such as sponge spicules are retained in sediments that accumulate in deep waters, although some diatoms survive on the continental shelf under zones with high productivity. The initial deposition of the amorphous hydrated silica, opal, converts first to opal-CT and eventually to crystalline quartz (Kastner, 1981).

The oldest sponge spicule accumulation has been found at the Precambrian–Cambrian boundary (Sdzuy, 1969) (Figure 14). The beginning of extensive siliceous deposits of radiolarian skeletons marks the mid-Ordivician. Cherts derived from shallow-water deposits of radiolarian skeletons, and sponge spicules are found throughout the Paleozoic (Lowenstam and Weiner, 1989, p. 245) and suggest that seawater may have been silica saturated at this time. Diatoms become quantitatively important constituents of the marine planktonic community in mid-Cretaceous along with the first appearance of silicoflagellates ~100 Myr ago, while dinoflagellates evolved in the Paleocene just after the massive Cretaceous extinction. The huge explosion of diatoms extracting silica from ocean waters, especially in the Cenozoic, has been cited as the reason for the decrease in the level of siliceous mineralization in radiolarians and siliceous sponges (Harper and Knoll, 1975), but reduction of silica biomineralization has not been detected in the silicoflagellates, and some other families (Tappan, 1980). The expansion of these geologically modern biomineralizers may affect the global ocean silica saturation levels, which may account for a decrease in: (i) the accumulation of siliceous marine deposits, (ii) the number of species of radiolarians, and (iii) the concentration of mineral per individual radiolarian skeleton (Moore, 1969).

Freshwater diatom accumulations, diatomites, are part of the radiation of this family of siliceous mineralizing species in the Cenozoic (Round, 1981). Along with members of the plant kingdom, such as the angiosperms with silica in their seeds, and the grasses (see Section 8.05.3.6), silica biomineralization expanded to the land more or less at the same time as diatoms proliferated in the oceans.

8.04.3.5.1 Radiolarians

The extensive sampling during the Deep Sea Drilling Project (1969–1979) has expanded

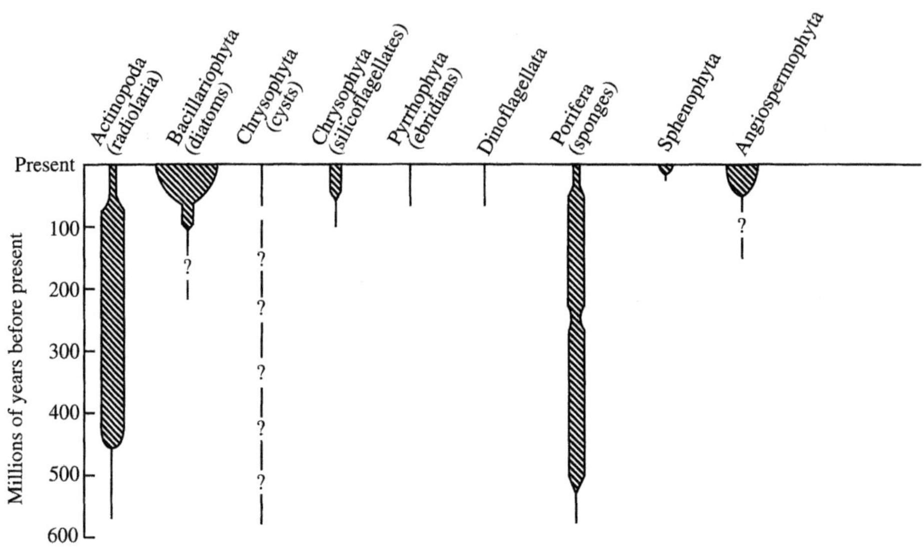

Figure 14 The stratigraphic ranges of invertebrates with silica mineralized hard parts (source Lowenstam and Weiner, 1989, figure 12.4, p. 245).

the number of varieties of the beautiful and delicate structures formed by the radiolarians. Originally depicted by Haeckel (1899–1904b) from the Challenger expedition, the range of structures has delighted lay and scientific observers ever since (Figure 15(a)). Similar to other invertebrates radiolarian mineralized remains, especially the intricate structural detail has been, and is, used to distinguish between families and species. Planktonic photoautotrophs, most radiolarian species, live in the top few meters in normal composition seawater. Their well-preserved skeletons have been part of cherts since mid-Ordivician and throughout the Paleozoic. These deposits are the sites not only of the silica skeletons but they were also often cemented by silica, suggesting that the oceans may have been saturated with silicon during that period (Lowenstam, 1972; Broecker, 1974; Whitfield and Watson, 1983).

The frothy appearance of a live radiolarian (Anderson, 1986) (Figure 15(b)) belies the mineralizing material, opal. Produced by the uptake of monomeric silicic acid, followed by dehydration and polymerization, the bioprocessing that creates these creatures was difficult to study because of the differential hardness of the silica mineral and the organic components. However, employing $Ge(OH)_4$, as a substitute for $Si(OH)_4$, has now elucidated the rates of uptake of silicon as well as the mechanisms of biomineralization. With its longer-lived isotope ^{68}Ge (half-life 282 d) relative to ^{31}Si (156 min), the transport, sequence, skeletal production, and insight into other biologically controlled processes have been investigated (Azam and Volcani, 1981). An outline of these mechanisms is presented under diatoms (Section 8.04.3.5.2); here we merely reiterate that growth and proliferation of all the organisms that mineralize with silica have customized systems that present some interesting chemical conundrums.

The modern radiolarian skeletal frameworks appear less robust. Typically their tests show dissolution and dissociation. This is diagenesis taking place within a living form, and though often encountered it may not be recorded; it is difficult to document that which is missing. However, examination with scanning or high-resolution microscopy of the sample, and attention to the individual test morphology, or alteration, should be a regular part of sample preparation before any geochemical analysis.

Through examination of the recently obtained Cenozoic deposits, the record of radiolarians is sufficiently complete to provide us with a remarkable picture of morphologic diversity. The amount of information now available affords insight into the evolution as well as extinction of a few taxa (Figure 15(c)) and the ability to study their distinctive morphologic variance, growth, and development, as well as biomineralization processes, all useful to paleontologists, biostratigraphers and to those interested in biomineralization.

8.04.3.5.2 Diatoms

Of all the biomineralized species that are used as geobiological indicators, the best studied are the diatoms. As the base of the food chain for many eukaryotic organisms, the group also presents opportunities to geochemists investigating environmental issues. Diatoms are abundant today in hypersaline marine to freshwater

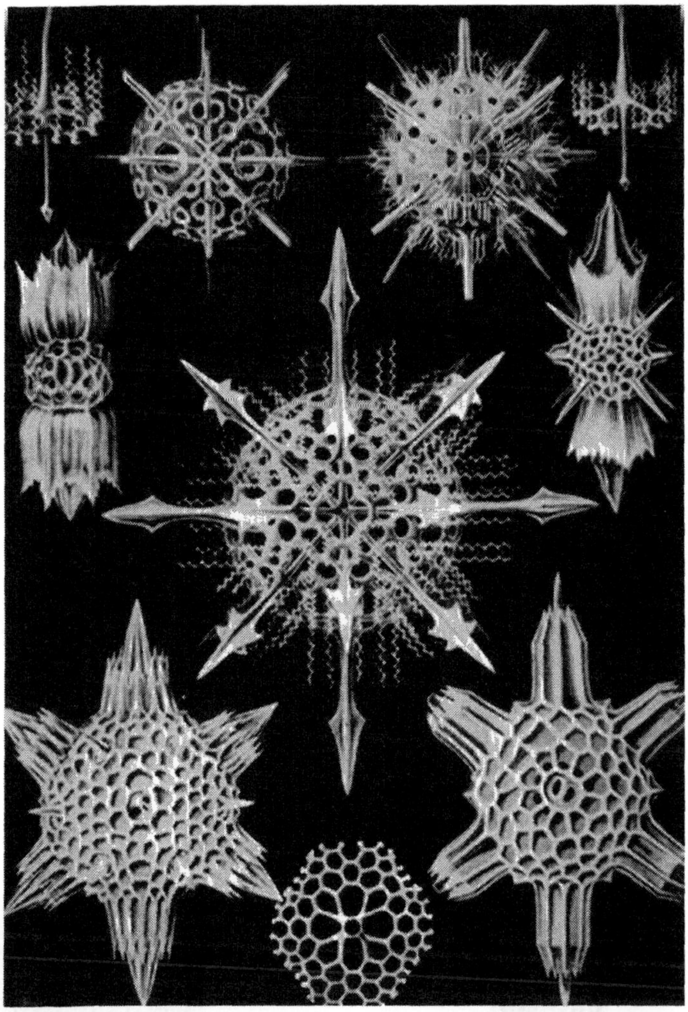

(a)

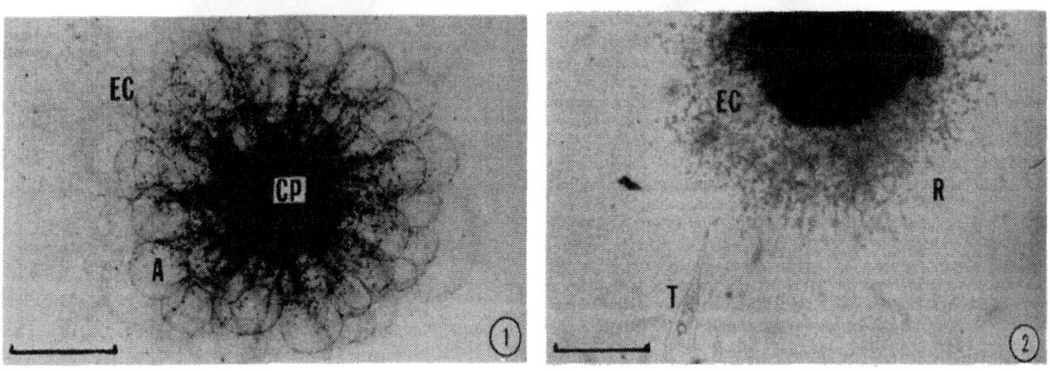

(b)

Figure 15 (a) Radiolarian morphologies as depicted by Haeckel (1899–1904b). (b) Light micrographs of fresh radiolarians showing frothy appearance: (1) *Thalassicolla nucleata* A = cytoplasmic bubbles which are penetrated by algal symbionts and enclose captured algal prey, EC = extracellular capsular cytoplasm, and CP = central capsular region; (2) a member of the Astrospaeorida family of radiolarians showing the spherical organization of the cytoplasm (scale bar = 0.5 mm) (source Anderson, 1981, figure 13.1, p. 348). (c) Evolution of radiolarian forms over time: *Lithocycla* (1–4) and *Didymocryrtis* (5–11) (source Reidel and Sanfilippo, 1981, figure 12.6, p. 332).

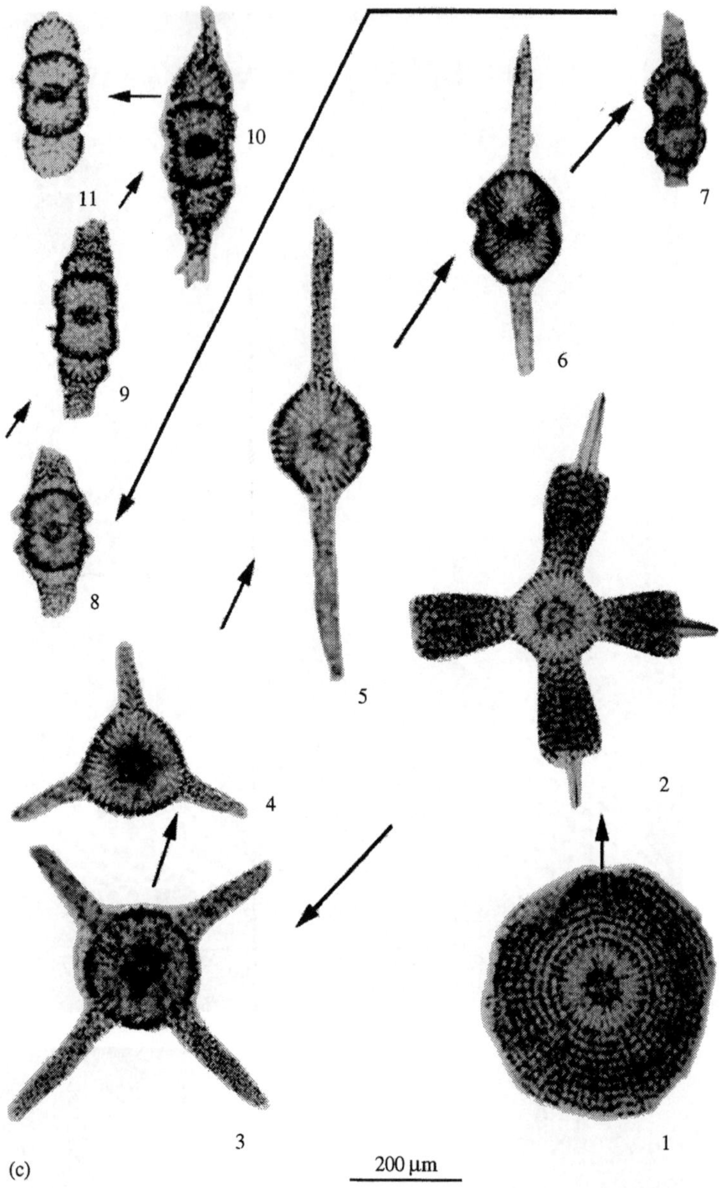

Figure 15 (continued).

environments. Phototrophic diatoms mineralize with opal similar to radiolarians. Accumulations of diatom-rich sediments may be a source of silica-rich solutions which, through diagenesis and metamorphism, formed siliceous veins, layers, or pods in sediments, a process analogous to the accumulation of plant materials that give rise to oil and gas deposits.

The diatom's delicate structures, similar to radiolarians, were originally presented by Haeckel (1899–1904a) (Figure 16(a)) and over 10^4 species, modern and fossil, have been described (Locker, 1974). The variety of the delicate mineralized structures not only identify species but have also been used to determine the ecology of their depositional environment, freshwater or salt water (Crawford and Schmid, 1986).

The mineralized structures of tiny silica diatoms (<50 μm in diameter) appear smooth even at the highest magnifications reflecting the amorphous state of the mineral. Diatom tests are multipartite, a combination of specialized forms described by Li and Volcani (1984, p. 519) as "box-and-cover-like structures" with the parts known as valves and girdles (Figure 16(b)). The valves are often elaborately patterned with ribs, processes (tubes), and pores, while the girdles form bands that may vary in number. There are two major groups of diatoms based on the pattern of the perforations on the valves: pinnate diatoms

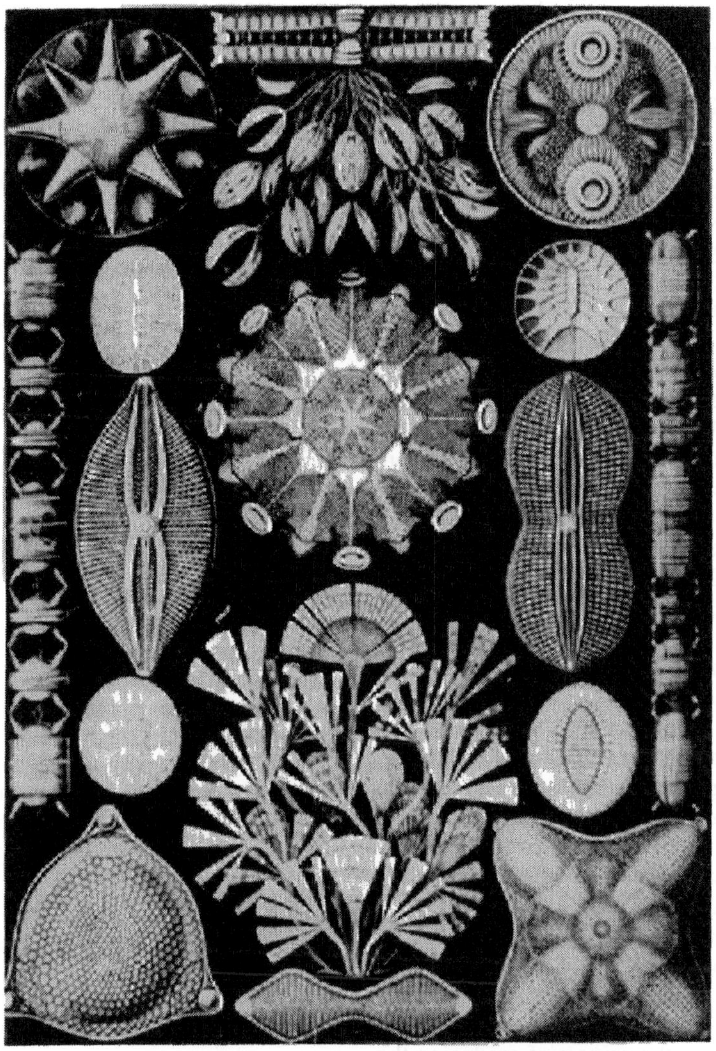

(a)

(b)

Figure 16 (a) Diatom morphologies as depicted by Haeckel (1899–1904a). (b) Whole cell of the centric diatom *Stephanodiscus* showing the mineralized "box and cover" structure with a protruding ring of spines, 1,000× (source Round, 1981, figure 5.1, p. 100).

that show bilateral symmetry, and centric diatoms with tri- or circular form. Silicification of the valves varies with species from almost none, i.e., the structure is purely organic, to heavily silicified, while the girdle portion characteristically contains mineral.

There is a complicated morphogenesis of these siliceous structures. The initial uptake of silicate, probably as silicic acid, moves into the cell and becomes associated with the endoplasmic reticulum and most prominently with a specialized membrane-bound vesicle known as the silicalemma. This vesicle occurs in all organisms that form siliceous structures (Sullivan and Volcani, 1981; Round, 1981). Mann (2001) constructed a simplified schematic for the diatom *Conscinodiscus wailesii* to illustrate the interplay of specialized and sequential cell activities: elongated vesicles are secreted but stay attached to the cell wall. Close-packed and distributed with regular polygonal symmetry, silica deposition takes place not within but between the vesicles on the vesicle membranes. The initially isolated silica deposits fuse into a silica web with pores representing the opening, or internal void of the vesicles. The three-dimensional silica scaffold is directed through precise placement of the vesicles until the full diameter of the diatom is a series of concentric fenestrated shells with specific nanoscale geometric patterns (Mann, 2001, pp. 134–135).

Diatoms not only have a siliceous mineralized skeleton but also require silicon for synthesizing their DNA, and transcriptional proteins, making their entire metabolism dependent on silicon. It would appear that in the process of producing the silica mineralized structures, sufficient silicon becomes sequestered for all requirements. The original silica pool created within the organism (up to 500 mM silicon has been measured in the solution within cells, considerably above the 3.5 mM L^{-1} in inorganic solutions which autopolymerize) is distributed as granules along the preformed vesicle membrane. The membranes, composed of polycarbonates and acid polyprotein molecules, play several roles: they act as the templates aiding the placement of appropriate amounts of silica in precise arrays and thereby prevent premature silica precipitation (Lowenstam and Weiner, 1989). The biomineralization process in diatoms is an exercise in control on the uptake, localization, and deposition of opaline silica. Studies comparing the water chemistry surrounding diatoms show that biological activity keeps the mineralizing system out of equilibrium with the surrounding environment. Freshwater and oceanic diatoms may also incorporate locally available aluminum and iron as they produce their delicate, distinctive tests.

8.04.3.5.3 Sponges

Found in all aquatic environments these filter-feeding organisms may occur on soft or on hard substrates at a variety of depths. Some members are entirely organic, others form mineralized skeletons that are calcareous, found in association with other $CaCO_3$ biomineralizers, i.e., corals; and there are siliceous forms, the demosponges (Finks, 1970). The mineral portions form as spicules, which are distinctive to a species, and some sponges produce secondary mineral deposits, which leads to dense and rigid structures, generally typical of deeper water forms. Modern sponge lineages, mostly soft sponges, derive from Early Carboniferous as few of the Paleozoic families survived the Frasne/Famenene (Devonian) extinction (Reitner and Keupp, 1991, p. v). Before sponges with rigid calcareous skeletons were rediscovered in the 1960s (Hartman and Goreau, 1970), there had been differences of opinion on an appropriate inclusive classification to cover the entire range of such a diverse group of organisms (see discussions by van Soest (1991); see also Reitner and Keupp (1991) and Hartman (1981)).

The oldest heteractinid calcareous genus *Eiffelia* was described from the well-preserved spicules in the Burgess shale of British Columbia (Cambrian age) by Walcott (1920) (Figure 17(a)). These skeletons were composed of three different sizes of hexaform spicules, regularly and spherically distributed but other tri-stellate, Y-shaped, tetra- and octaform spicule species were described. By the Pennsylvanian marked knobby calcareous overgrowths that occasionally became fused and obscured, the initial calcareous spicules in a grossly tubular form with a variety of substructures were described (Rigby and Webby, 1988) (Figure 17(b)).

The carbonate spicules, initiated in the outer or dermal part of the organism, become interior in later forms of these organisms. The process of secondary mineralization leads to a diversity of macrostructures, many centimeters in height (Rigby, 1991). Localities in the Silurian of North America and the Devonian of Australia show an explosion of such structures, but the morphological expressions are different suggesting ecological specificity. Similar structures are seen in modern localities (Rigby and Webby, 1988).

In siliceous sponges the spicule-forming site is known as the desma. An organic biomolecule defined structure located in the ectosomal regions near the surface of the growing sponge; it had been defined in 1888 (Sollas, 1888). Successive layers of silica (opal) are deposited as rods that become decorated with symmetric or asymmetric distributed sidepieces. These spicules may fuse joining several mineralized desmas to produce

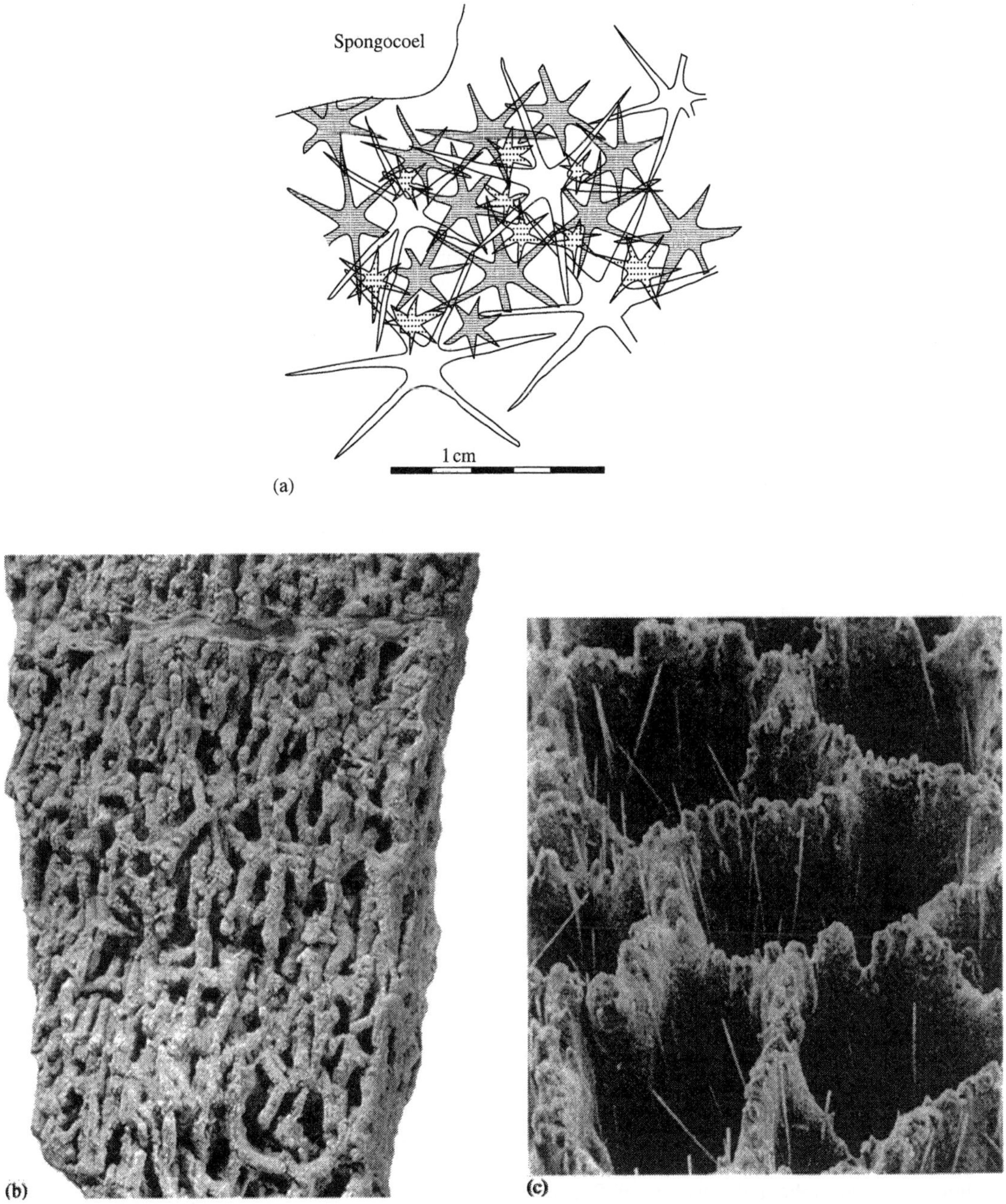

Figure 17 (a) Drawing of the skeleton of a sponge, *Eiffelia globosa* from the Middle Cambrian Burgess shale of Canada illustrated in Walcott (1920). Three ranks of spicules are shown: the first order is unpatterned, the second is loosely stippled, while the third is dark stippled (source Rigby, 1991, figure 2, p. 84). (b) Photomicrographs of *Wewokella solida* Girty 1911 from the Pennsylvanian Deese formation, Oklahoma, 4× (source Rigby, 1991, figure 7B, p. 88). (c) SEM of a cleaned surface of a sclerosponge skeleton showing silicious needles embedded in, and protruding from, the aragonitic skeleton, 100× (source Hartman, 1981, figure 16–28, p. 488).

a complex rigid skeletal mass (Hartman and Goreau, 1970; Hartman, 1981). A similar process occurs in carbonate-mineralizing sponges. In the sclerosponges, a modern group, there is a species that has very thin silica needles in a massive calcareous (aragonitic) structure (Hartman, 1981) (Figure 17(c)).

Studies of sponge lineages show evolutionary patterns with convergent adaptation and mimicry across groups in both the calcareous and siliceous varieties. The amazing reappearance of the same basic structural elements, after a lengthy hiatus, even by the geologic timescale, and of diverse spicule morphology that enlarges and may fuse, producing rigid forms, suggests that sponges, like many other invertebrate mineralized species, adapt to exploit ecologic niches today as they did in remote times.

8.04.3.6 Plant Biomineralization

8.04.3.6.1 Introduction

Biomineralization, or the precipitation of mineral as a result of the metabolic functioning of a living organism, is a process present in all five kingdoms of life, and is widely practiced by plants (Figure 18). Both unicellular and multicellular plants biomineralize producing structures known as "phytoliths," utilizing a range of chemical compounds with various mineralogies. The morphology of these structures varies as well: some plant biomineral structures are intricate and ornate, while others are simply massive. Investigations of the mineral fractions that sequester in plants are sparse, and to date they are mostly related to anthropological interests for identifying specific agricultural varieties.

8.04.3.6.2 Plant biominerals

Plant-tissue biominerals are overwhelmingly of three chemical types: calcium oxalate, silica, and calcium carbonate. Of these types, calcium oxalate is thought to be the most prevalent and widespread (Webb, 1999). Opaline silica (Geis, 1978) and calcium oxalate (Finley, 1999) have been reported within plant tissues from hundreds of families in the plant kingdom, while calcium carbonate is widely described in many plant species. In addition to these compounds, which form common minerals, several oxides have been reported in isolated plant species, although the number of plant species is small, i.e., <4% of total mass (Duke, 1992). They include magnesium oxide throughout the tissues of *Stellaria media* (chickweed), calcium oxide in the leaf of *Corylus avellana* (hazel), iron oxide in the seeds of *Cannabis sativa* (marijuana), and aluminum oxide as well as manganese oxide in the seeds of *Gossypium* spp. (cotton) (Duke, 1992). A few additional compounds have been noted, but <0.5% of total mass: potassium nitrate is found in the leaves of *Piper betel* (betel pepper); calcium sulfate is present in the roots of *Echinacea* spp. (coneflower); and potassium chloride can be found in the calyx of *Physalis peruviana* (ground cherry) (Duke, 1992). In addition, some metal–organic associations have been noted. Phytoferritin, a central iron-containing core surrounded by a polypeptide shell, has been found in the plastids of some higher plants and in the seeds of legumes (Webb and Walker, 1987). Aluminum and silicon were found to be co-deposited in the needles of *Pinus strobus L.* (Eastern white pine) growing on acid soils in Muskoka, Canada, with aluminum accumulation increasing toward the needle tip (Hodson and Sangster, 2002), the areas of heaviest silica biomineralization (Figure 19).

Calcium oxalates in plants. Calcium oxalate sequestration in plants is found in two chemical states related to different levels of hydration, and the stable formation of two distinct minerals: whewellite, the monohydrated calcium oxalate, $CaC_2O_4{\cdot}H_2O$, and weddellite, calcium oxalate dihydrate, $CaC_2O_4{\cdot}2H_2O$ (Gaines *et al.*, 1997, p.1011). These minerals were found as well-formed isolated crystals in *Opuntia microdasys* (Pricklypear) and *Chamaecereus silvestrii* (Unnamed cactus), respectively (Monje and Baran, 1996, 1997) and both minerals are also commonly found in human urinary calculi (Gaines *et al.*, 1997, p. 1011). Calcium carbonate occasionally occurs as isolated deposits, such as within the seeds of *Magnifera indica* (Mango) (Duke, 1992), and both calcite and aragonite have been known as co-precipitates with silica in plants. For example, Cystoliths in the Acanthacea plant *Beloperone californica* contain both calcium carbonate and silica (Hiltz and Pobeguin, 1949), and the prickly emergences of *Urtica dioica* (Stinging nettle) are silica structures upon which calcite has been bioprecipitated (Thurston, 1974). Similarly, the endocarp of the *Celtis occidentalis* (Hackberry) fruit is composed of a reticulate opaline network, within which aragonite has been precipitated (Jahren, 1996) (Figures 20(A) and (B)), and developmental studies showed that vilification is complete by the time calcification begins (Jahren *et al.*, 1998), indicating that the plant exerts strategic control over the timing and the composition of biomineral precipitation (Table 1).

Silica in plants. Mineral precipitation is a fundamental cellular activity for many photosynthesizers and the vast majority of the dry mass of a biomineralizing organism is in the mineral, which is dense, having a higher specific gravity than the organic constituents, or the fluids that make up the great volume of the tissues. Diatoms are a striking example of this, with concentrations of 95% (dry weight) silica in their walls (Kaufman *et al.* (1981); see Section 8.04.2.2); another extreme example is *Bambusa* spp. (Bamboo) (Figures 21(A) and (B)), which may consist of 70% silica (dry weight) (Jones *et al.*, 1966). The dry mass of rice husks at harvest contains up to 20% silica (Garrity *et al.*, 1984), and the amount of calcium oxalate is in excess of 10% (dry weight) within the bark of *Quillaja sapona* M. (Soapbark) (Duke, 1992).

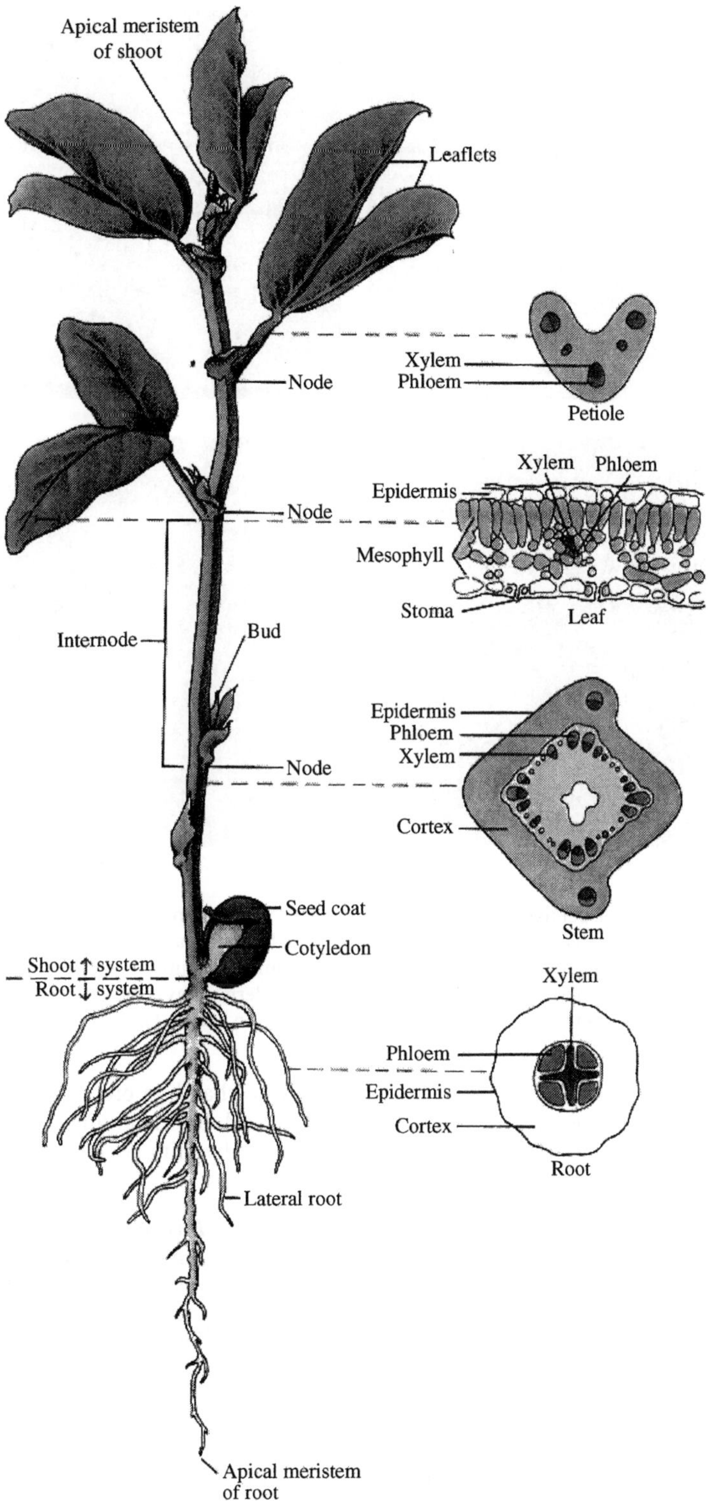

Figure 18 Diagram of a young broad bean illustrating the anatomy (source Raven *et al.*, 1999, p. 8).

The amount of mineral offers structural advantages. Among the *Bambusa* whose hollow stems are reinforced by silica are the tallest members of the Poaceae. Commonly viewed as grasses, some bamboo achieve tree-like stature, while the vast majority of grasses are less heavily silicified, and commonly are less than 1 m in height. Throughout much of Asia, heavily silicified bamboo is strong enough to use as a building material, possibly a long-standing practice. One interesting illustration

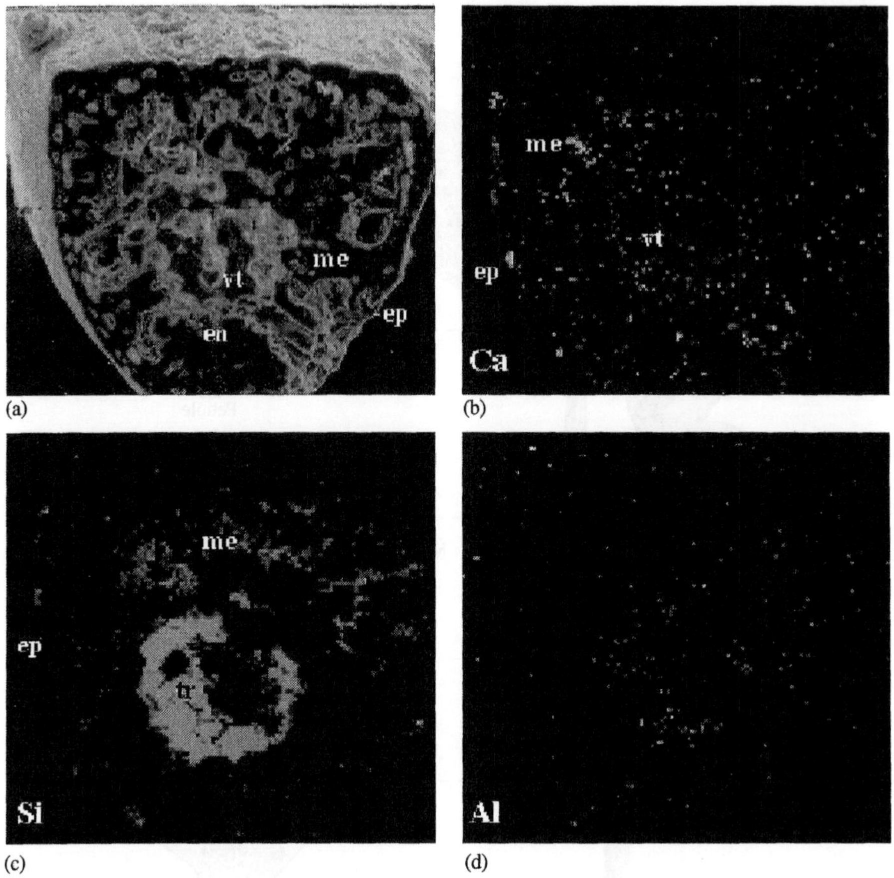

Figure 19 SEM and energy dispersive analysis micrographs (175×) showing mineral distribution, Ca, Si, and Al, localization in a frozen, planar transverse section 1 mm behind the tip in a second year needle of *Pinus strobus* (Eastern white pine): (a) secondary electron image, (b) calcium distribution, (c) Si distribution, and (d) Al distribution. Abbreviations: endodermis (en), epidermis (ep), hypodermis (hy), mesophyll (me), transfusion tissue (tr), vascular tissue (vt), and xylem wall (xw) (courtesy of M. J. Hodson and A. G. Sangster, unpublished collection).

of this concerns the "Movius Line," or the imaginary geographical line drawn for the Lower Paleolithic which divides Africa, the Near East and Western Europe with their developing hand axe technologies from Eastern Europe and Southeastern Asia, where "chopping-tool" industries and use of casually retouched flint and chert flakes were dominant. Several suggestions have been offered for this discontinuity. Eastern regions were subject to chronological, geographical, and other barriers, and there were different requirements between the separated peoples. One provocative explanation includes the favored use of bamboo as a raw material (Schick and Toth, 1994). Bamboo was, and is, plentiful in these regions and its high silica content gives crude tools fashioned from splitting bamboo glass-like sharpness. Under this hypothesis, stone tools take on a secondary role, their main purpose being the initial processing of bamboo. Because the biomineralized silica bundles found in bamboo are recognizable by their morphology and elemental composition, it may be possible to place the early phases of bamboo processing in the archaeological record by examining crude stone tools for bamboo phytolith residue (Jahren *et al.*, 1997).

8.04.3.6.3 *Phytoliths: indicators of the environment and paleoenvironment*

The eventual decay of organic tissues releases plant biominerals into the soil, where they can be recognized as evidence of plants by paleobotanists, paleontologists, archaeologists, and other environmental historians (Pearsall and Piperno, 1993; Piperno, 1988). Extensive genetic influence is exerted over all aspects of plant morphology, rendering different species of plants distinct in their shape, size, and collection of accruements. As an extension of this, most plant phytoliths are distinctive of plant family, and many can be used to determine plant genus or even species.

Phytoliths can be grouped broadly by their morphology (Figures 22(A) and (B)) with great accuracy: monocotyledon (herbaceous annual

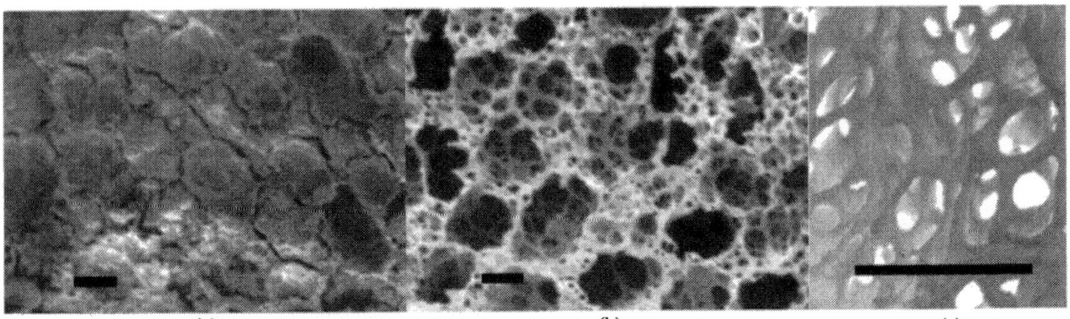

(A) (a) (b) (c)

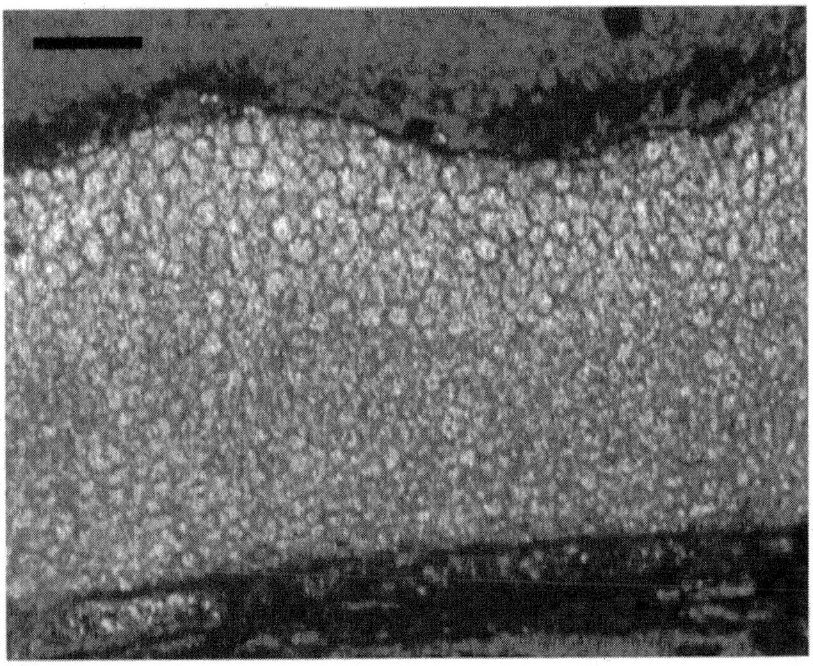

(B)

Figure 20 (A) SEM images of *Celtis Occidentalis* (Hackberry) endocarp. Aragonite deposited in a: (a) honeycomb pattern, (b) the inner silica scaffolding, and (c) organic matter occluded within the silica (scale bar = 10 μm) (source A. H. Jahren, unpublished collection). (B) Petrographic thin section of *Celtis Occidentalis* (Hackberry) endocarp, note honeycomb pattern (scale bar = 100 μm). Photograph taken with crossed polars and quartz plate inserted (source A. H. Jahren, unpublished collection).

plants) are recognizable and distinct from dicotyledon (angiosperm tree) phytoliths. Furthermore, both groups are distinct from gymnosperm phytoliths, which include conifer trees. Recognition of changes in the relative abundance of each group in archaeological and geological records has illuminated continental-scale changeovers from grassland to forestland through geological time.

The diversity and abundance of phytolith assemblages can be used to determine the composition of paleoplant communities. A study across the Great Plains of the US showed that modern phytolith assemblages displayed a geographic pattern consistent with modern grassland composition (Fredlund and Tieszen, 1997a). Because the species composition of grasslands is observed to change systematically with environmental parameters such as temperature, the study also suggested that phytolith assemblages from prehistoric grasslands could be used to reconstruct paleoclimate conditions.

For example, phytoliths from the Black Hills, South Dakota (complemented with stable isotope data) show that regional vegetation changed from C3 grasslands to mixed C3/C4 grasslands between 11 kya and 9 kya, and that the presence of C4 vegetation was stable into the Late Holocene (Fredlund and Tieszen, 1997b). Use of phytolith assemblages reveals that grassland and steppe ecosystems were widespread in Washington State

Table 1 Biomineralization in plants.

Mineral	*General distribution*	*Specific occurrences*[a]
Calcium oxalate ($CaC_2O_4 \cdot nH_2O$)	215 Families, including both gymnosperms and angiosperms (Mcnair, 1932)	*Chamaecereus silvestrii* (Unnamed cactus) (Monje and Baran, 1996) *Opuntia microdasys* (Pricklypear) (Monje and Baran, 1997) *Phaseolus vulgaris* (Common bean) (Zindler *et al.*, 2001) *Quillaja saponaria* M. (Soapbark) (Duke, 1992) *Sida rhombifolia* (Teaweed) (Molano, 2001) *Vitis* spp. (Grape) (Webb *et al.*, 1995)
Silica ($SiO_2 \cdot nH_2O$)	>100 Families of tropical plants, including both gymnosperms, angiosperms and pteridophytes (Piperno, 1991)	*Arundo donax* (Giant reed) (Mulholland, 1990) *Bambusa* spp. (Bamboo) (Jones *et al.*, 1966) *Bouteloua curtipendula* and *Panicum virgatum* (Sideoats and Switchgrass) (Ball and Brotherson, 1992) *Phalaris canariensis* (Annual canarygrass) (Perry *et al.*, 1984) *Thalassia testudinum* (Turtlegrass) (Brack-Hanes and Greco, 1988) *Zea* spp. (Corn) (Doebley and Iltis, 1980)

[a] Not a complete list.

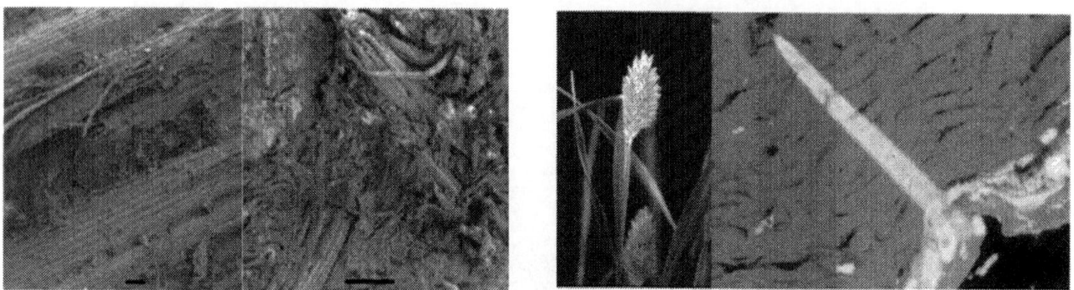

(A) (a) (b) (B) (a) (b)

Figure 21 (A) SEM images of silica phytoliths from *Bambusa indigena* (Bamboo) (a), and of the surface of a chert tool that had been used as a wedge during the chopping of Bamboo stems (b). Note the silica phytoliths fused to the tool edge (scale bar = 30 μm) (source A. H. Jahren, unpublished collection). (B) Inflorescence of the grass *Phalaris canariensis* (Canarygrass) (a), and a siliceous hair from the inflorescence of this grass species penetrating the skin of a mouse (b). The hair is stained with fluorescein and is 50 μm long with a midpoint diameter of 3 μm; there is a fracture near the tip of the hair; the tissue is stained in acid fuschin (courtesy of M. J. Hodson, R. J. Smith, A. Van Blaaderen, T. Crafton and C. H. O'Neill, unpublished collection).

during the Pleistocene, compared with the forested ecosystems of the region today (Blinnikov *et al.*, 2002).

Phytolith assemblages show a clear turnover from savanna to forest ecosystems in south central Brazil ~4 kya (Alexandre *et al.*, 1999) and indicate that the closed forest in the Amazon Basin has existed since 4.6 kya, but prior to this time the forest composition and species abundance fluctuated widely (Piperno and Becker, 1996), leading researchers to hypothesize that climate change stabilized the vegetation during the middle Holocene.

Occasionally, phytolith analyses suggest a reinterpretation of the evolutionary effects of vegetation change. The changes in tooth morphology of herbivores during the Miocene are often attributed to expanding grasslands and the development of phytolith-resistant dentition. However, phytolith assemblages from northwestern Nebraska showed that open C3 grasslands dominated the landscape between 25 Ma and 17 Ma, millions of years before the observed changes in tooth morphology of regional herbivores (Stromberg, 2002).

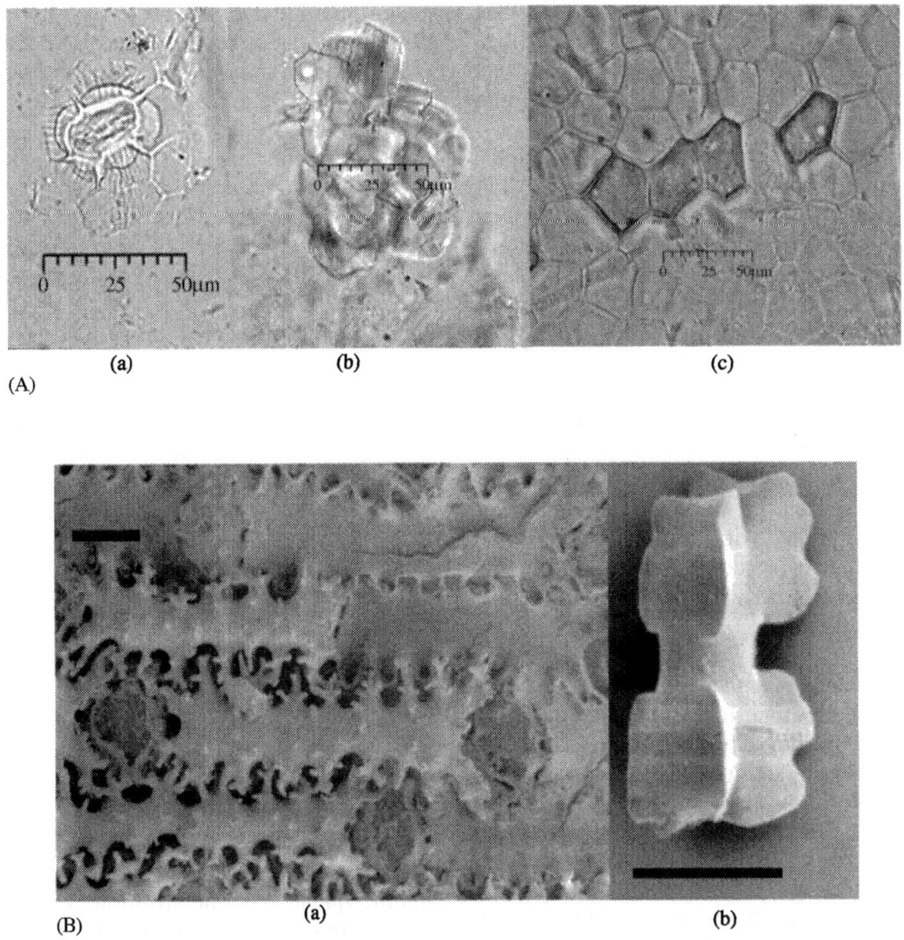

Figure 22 (A) Light microscope images of plant phytoliths (scale bar = 50 μm). Silicified stomate from *Cordia lutea* (Flor de overo) (a), irregularly sized blocky crystals of calcium carbonate from *Brownea grandiceps* (Rose of Venezuela) (b), and silicified epidermal cells (dark cells near center) from *Licania longistyla* (Licania) (c) (courtesy of D. M. Pearsall, unpublished collection). (B) SEM images of silica phytoliths from common crop plants. *In situ* dendriforms from *Triticum monococcum* (Wheat) inflorescence (a), and a bilobate cross-body phytolith from *Zea mays* (Corn) (b) (scale bar = 10 μm long) (courtesy of T. B. Ball, unpublished collection).

Variability in critical environmental parameters has been shown to affect the total height, biomass, and leaf area (among other aspects) of a growing plant. However, extreme variations in soil texture, light levels, and water availability did not alter the morphometry of phytoliths in two grass species: *Bouteloua curtipendula* (Sideoats grama) and *Panicum virgatum* (Switch grass) (Ball and Brotherson, 1992). Further, crop plant phytoliths (Figure 22(B)) can be differentiated from wild plant phytoliths by recognition of genus- and species-specific morphologies (Cummings, 1992). One example is the suite of recognizable traits in *Zea Mays* (Corn) phytoliths compared to those of wild grasses (Piperno, 1984). The morphology and taxonomy of New World domesticated plants has been extensively documented (Piperno, 1985) and used to explore the history and significance of various crop species to the peoples of this region (Piperno, 1991). Phytolith material extracted from residues on prehistoric pottery, and from human dental calculus, presented an opportunity to evaluate cooking practices and nourishment levels, allowing archaeologists to characterize and understand past civilizations in terms of agrarian lifestyles (Pearsall and Trimble, 1984). Much of what is known about the importance of *Zea Mays* (corn) in the New World has been interpreted from thousands of years of phytolith-bearing sites (Bush *et al.*, 1989). The mixed record of phytoliths from both wild and cultivated plants was used to interpret environmental change as the result of agricultural activities. On another continent Kealhofer and Penny (1998) found that the Late Holocene transition to rice agriculture in Thailand was accompanied by a reduction in dry-land forest and the subsequent establishment of secondary-growth forests. Plant biominerals are incorporated into the geologic record when plant communities are subjected to natural fires, or by slash-and-burn agriculture practices.

Some authors have warned against overinterpretation of the phytolith record: Carter (2002) compared phytolith assemblages with pollen extracted from sediment cores and showed that trends in vegetation through time were not only disparate, but completely opposite. Additional contextual information led Carter (2002) to conclude that the phytoliths reflected a very local vegetation signal, whereas widely distributed pollen was more representative of a larger geographical signal. Also, phytoliths, like most minerals, are subject to diagenesis. This may be a particular problem in highly weathered and actively weathering soils. Boyd *et al.* (1998) have developed a specialized methodology for preparing and interpreting phytolith assemblages from wet tropical sites.

Radiocarbon dating using plant biominerals. Carbon associated with plant biominerals has been explored as a substrate for radiocarbon dating in Quaternary phytolith samples. The ^{14}C content of calcium carbonate biomineralized in Hackberry endocarps over the past 120 yr parallels the observed ^{14}C variations of atmospheric CO_2 during that time span, indicating that these common phytoliths faithfully record the year in which they formed (Wang *et al.*, 1997). When compared to other ^{14}C dating substrates from Quaternary archaeological and geological sites, Hackberry endocarp carbonate yielded ages that compared favorably with those obtained by more established means (Jahren *et al.*, 2001; Wang *et al.*, 1997). Additional studies have focused upon carbon in the organic matrix occluded within the plant biomineral as a potential substrate for radiocarbon dating. Pigment from rock art near Catamarca, Argentina has been sampled and analyzed for its mineral content. These analyses revealed that plant material was used as a pigment binder, resulting in calcium oxalate and calcium carbonate from local cacti incorporated into the paint. By extracting calcium oxalate phytoliths from the pigments and using occluded organic carbon for ^{14}C analyses, Hedges *et al.* (1998) were able to hypothesize a Quaternary age for the rock paintings.

Stable isotope investigations on plant biominerals. In contrast, only a limited number of studies have investigated the stable isotope composition of plant biominerals as a source of paleoenvironmental information. Silica extracted from the Eocene paleosols of Axel Heiberg Island in arctic Canada were found to have notably high stable oxygen isotope value, leading to the suggestion that the mineral represented an accumulation of plant phytoliths, which presumably formed from isotopically enriched leaf water (Kodama *et al.*, 1992). Jahren *et al.* (2001) found that the oxygen stable-isotope composition of aragonite in Hackberry endocarps from 101 North American sites was well correlated with the oxygen isotope composition of environmental water at the sites where the trees grew. They used this correlation to interpret the oxygen isotope composition of fossilized endocarps from Pintwater Cave, Nevada as resulting from summer influx of isotopically heavy monsoon precipitation during the Early Holocene. Organic matter occluded with silica phytoliths has also been used as a substrate for carbon isotope analysis. Stable carbon isotope composition of phytolith-occluded organics was found to be in agreement with values from paleosol organic matter, and isotopic trends through time reflected expanded C4 plant communities and regionally warmer conditions in the Great Plains during the middle Holocene, relative to the present (Kelly *et al.*, 1998).

8.04.3.6.4 Funghi and lichen biomineralization

Plant biomineralization influences the fungal communities that coexist with plants, and often confers advantages to the decomposers. There are several reports of fungal biomineralization that mimic plant biomineralization, occurring only when fungi are grown directly on plants. In compost fungi, much of the hyphae actually consist of calcium oxalate crystals that originate within cell walls (Arnott and Webb, 1983). Calcium oxalate crystals encrust the hyphae of the fungi, *Cyathus striatus* and *Cyathus olla*, common decomposers of plant debris (Tewari *et al.*, 1997). Calcium oxalate crystals have been found on the aerial hyphae of the fungus *Agaricus bisporus* when it was grown on natural plant substrates (Whitney and Arnott, 1987). Lichens (a symbiotic association between a fungus and an alga) may have the ability to biomineralize as well: within lichen communities, the lichens with the greatest capacity to weather volcanogenic sediment showed weddellite (dihydrate calcium oxalate) and calcite, as well as many bacteria (Sancho and Rodriguez, 1990; Easton, 1997).

8.04.3.7 Vertebrate Biomineralization

8.04.3.7.1 Introduction

In all vertebrates, including humans, the calcium phosphate mineralized endoskeleton is not only the basis of species identification, but plays two distinct major roles in this phylum. The skeleton is the scaffold with sites for muscle attachment permitting vertebrates their peculiar stance and mobility, and bones act as storehouses for the elements required for all facets of metabolism (Skinner, 2000). In forensics, or in paleontological studies, distinctive features of each bone or tooth tells its own story of growth, development, aging, trauma, and diagenesis.

There are 32 teeth and over 200 bones in an adult human, each a separate organ with its own essential role. These organs are composed of mineralized tissues, composites of bioorganic, and mineral that record the chemistry of what has been ingested, i.e., the C4 or C3 characteristics of the diet, as well as the overabundance of certain, perhaps hazardous cations, e.g., copper (Pyatt and Grattan, 2001), or shortfall of essential elements, i.e., calcium, that may lead to rickets, usually expressed in the abnormal appearance and function of the long bones.

There is at present worldwide concern with personal and public health. Coupling genetic and nutrition information it is possible to discriminate the predisposition, or susceptibility, of some disorders of the vertebrate hard tissues. For example, using animal models the aberrant production of tooth enamel, amelogenesis imperfecta (Dong *et al.*, 2000), is now traceable, and osteoporosis (Avioli, 2000), which may have previously been attributed solely to nutrition or inactivity, is being re-examined in light of molecular genetic predisposition. Mindful that genetic abnormalities are not confined to current populations of humans or animals, a major caveat in evaluating samples for geochemical investigation is to discriminate between "normal" and diseased or "pathologic" tissues. The distribution and composition of the familiar, and common, calcium bioapatite mineral in teeth and bones, relatively easily obtained, is usually extracted from the organic constituents before being examined. Teeth, because of their distinctive forms, functions, and high degree of mineralization, are an important source of samples, while bones present some drawbacks. The background on human mineralized tissues, briefly outlined below, should allow us to make more informed choices for geochemical investigations.

8.04.3.7.2 Bones and bone tissues

The formation of a mature skeleton composed of a series of individual bone organs of various shapes ultimately depends on cellular control and the deposition of mineral within an organic matrix, and the creation of specialized tissues. The mineral, a calcium phosphate, closely approximates hydroxylapatite, ideal formula $Ca_5(PO_4)_3(OH)$, described in Section 8.04.2.3.

The mineral in skeletal tissues (see Section 8.04.2.3) is found as part of several different textures, e.g., woven, lamellar, haversian, trabecular and cortical bone that can be discriminated optically, or histologically, and at higher resolutions using TEM or SEM (Figure 23). The most dense, most highly mineralized tissues are found in the external portions of bones, and especially the shaft of the long bones in tetrapods, whether warm or cold blooded. The textural term for these tissues is cortex or cortical bone (Figures 24(a) and (b)). Interior to the cortex is an area of less dense tissue which becomes more porous grading into the marrow cavity where only mineralized spicules, or trabeculae, are located. This mineralized tissue is known as trabecular bone. The marrow, largely fat, is where the blood tissues are produced.

The fact that bones have hollow centers is an expression of the clever mechanical design of these organs. Bones and bone tissues fulfill their structural function while protecting the marrow where essential cellular material is generated (Wainwright *et al.*, 1976). Sufficient bone stiffness is achieved by increasing the amount of calcium phosphate mineral relative to the organic components in the tissue. Each bone is an independent structure in which the amount and distribution of mineralized tissue suits its size and functional needs while minimizing the organ weight. Programmed construction of each mineralized organ conserves mass and provides a mechanical advantage, the strength required, for example, to prevent buckling in the long bones of land-based vertebrates (Albright, 1987). As animals increase in size, the skeleton becomes a greater percentage of body weight. The skeleton of a mouse (20 g) is 5% of its body weight, that of a 70 kg man is 14% of its body weight, while a 1,000 kg elephant has a skeleton 27% of its body weight. The largest mammals that exist today are aquatic, and dwell in buoyant environments. One of the largest captured, a blue whale, weighed 203 t, with a skeleton that was 15% of its body weight (Kayser and Heusner, 1964).

Bones are successful living structures, because they are composed of tissues, biomolecules, and mineral, maintained by a highly integrated system containing several different cell types with specialized functions.

Biomineralization of bone: the bone morphogenic unit. Bone tissues have an auto-inductive cell cycle that keeps these essential organs viable throughout the life of the animal. Tissues are resorbed and regenerated within restricted sites in the cortex or trabeculae, known as bone morphogenic units (BMUs), a few micrometers in diameter (Urist and Strates, 1971). Three specialized cells are part of the unit; they interact responding to constantly changing conditions so that small portions of the tissue resorb and re-form without jeopardizing the function of the organ.

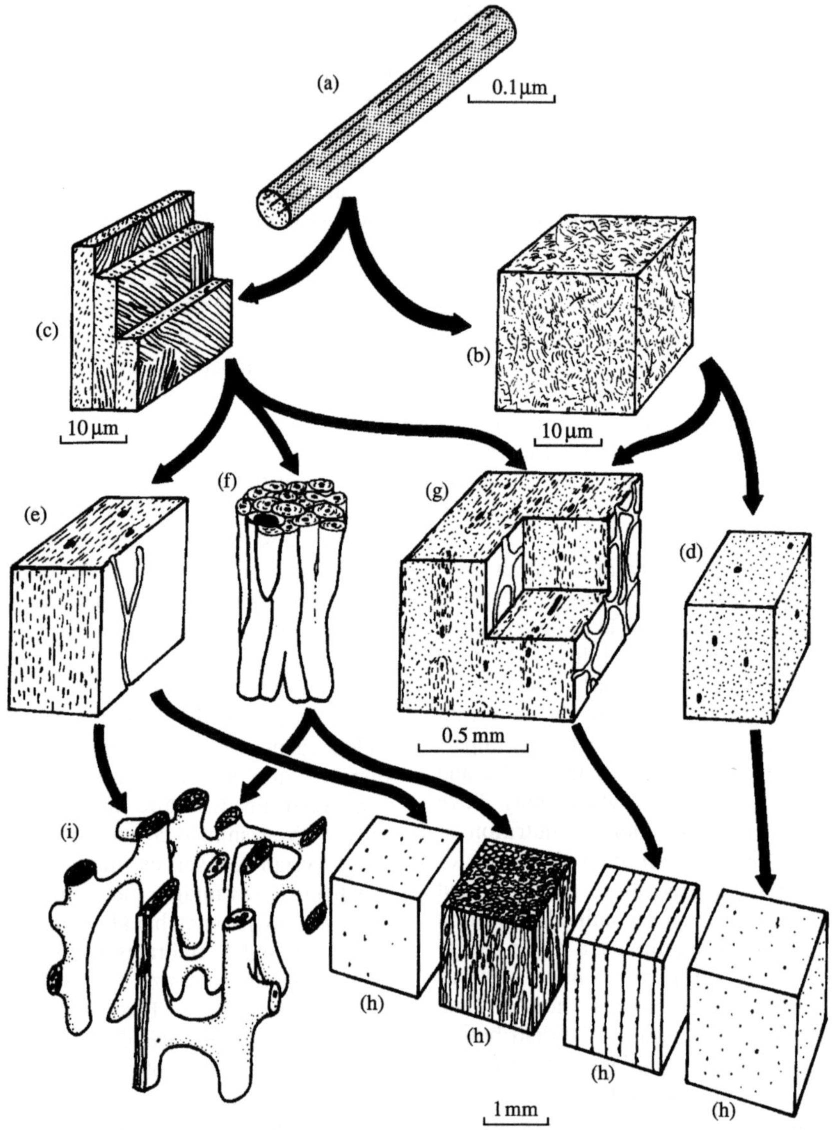

Figure 23 Mammalian bone at different levels of resolution: (a) Collagen fibril with associated mineral. (b) Woven bone (random collagen distribution). (c) Lamellar bone showing separate lamellae with collagen organized in domains with preferred orientation alternating in adjacent lamellae. (d) Woven bone with blood channels shown as dark spots, woven bone stippled. (e) Primary lamellar bone orientation indicated by dashes. (f) Haversian bone, a collection of haversian systems are shown as a longitudinal structure. Each system has concentric lamellae around a central blood channel. Darkened area represents an empty (eroded) portion of the section which will be reconstituted with new bone. (g) Alternation of woven and lamellar bone. (h) Various orientations of heavily mineralized (cortical, or compact) bone. (i) Trabecular, or cancellous, bone (Wainwright *et al.*, 1976) (reproduced by permission of Hodder Arnold from *Mechanical Design in Organisms*, **1976**).

The cells are connected via the cardiovascular system that permeates each organ and transfers nutrients and wastes as required.

A BMU is generated, and a cycle commences, when circulating hematopoetic stem cells from the marrow are recruited, fuse, and differentiate into multinucleate osteoclasts (Figure 25(a)). Enhanced by circulating hormones, osteoclasts resorb a portion of mineralized tissue a few micrometers in size, creating lacunae, or depressions, which subsequently fill in with organic matrix produced by cells called osteoblasts (Figure 25(b)). These cells originate from stromal mesenchymal cells and migrate to the sites vacated by the relocation or death of the osteoclasts. Mineralization of the refilled area takes time depending on the animal species and, in some disease states, may take months (Mundy, 1999). There is another, and perhaps the most important, cell in the bone morphometric

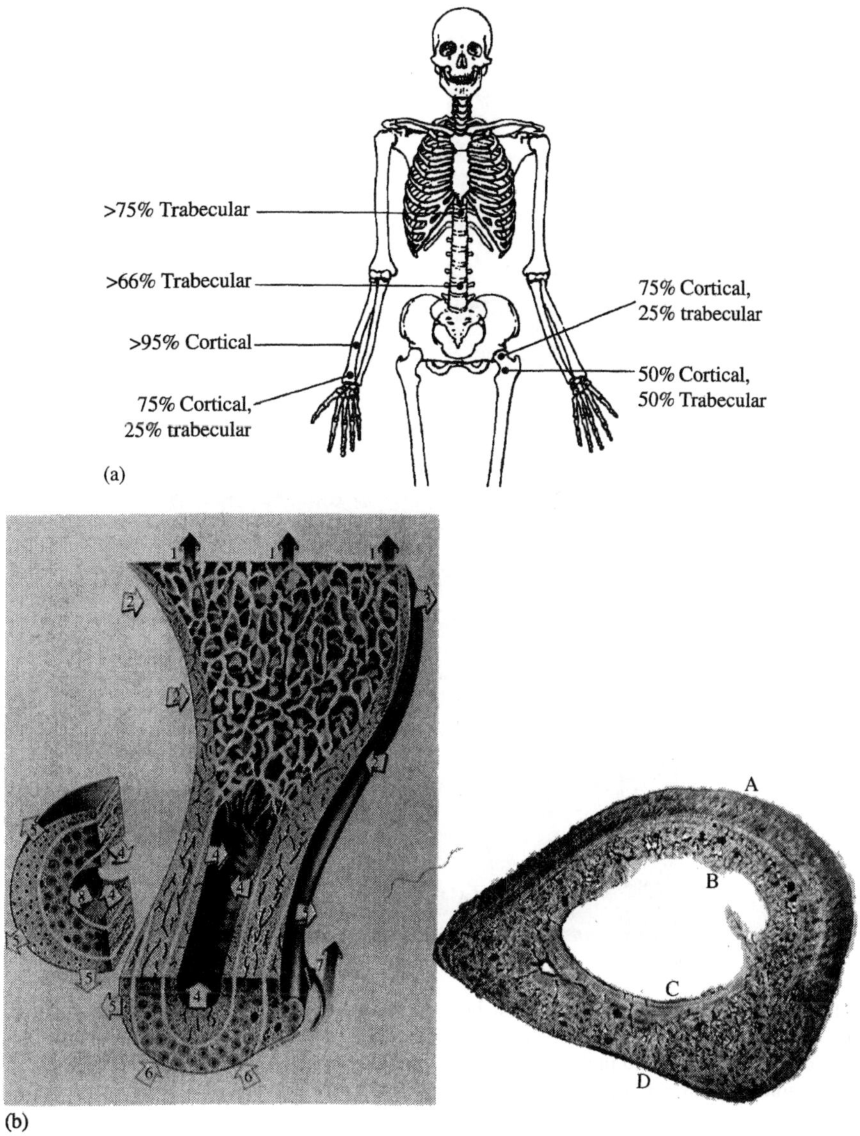

Figure 24 (a) Diagram of the human skeleton showing the percentages of cortical and trabecular bone in different bone organs (source Mundy, 1999, figure 2, p. 31). (b) Composite diagram through a long bone illustrating the tissue distribution and the changes that will take place as a bone grows and remodels. The longitudinal cross-section shows diagramatically areas of cortical and trabecular bone with directions of growth and/or resorbtion (numbers with arrows show the directions). A cross-section of an actual long bone is on the right-hand side of the diagram. It presents the appearances one observes of bone tissue changes: A = at the periosteum where there is active bone deposition, B = in the marrow cavity showing the response to bone resorption, C = at the opposite side of the marrow cavity showing lamellar bone deposition, and D = external planing of a portion of the bone in response to organ shape changes (source Enlow, 1963, figure 53, p. 111).

unit. Osteocytes (Figure 25(c)) are osteoblasts that become embedded in mineralized matrix. They remain viable, connected through pseudopoda and small blood vessels to each other, and the cardiovascular system to obtain nutrients, circulating small molecules, including hormones, and output any wastes. The detection of external signals, and the production of specific chemical species in the life span, days to weeks, of the osteocyte, the ultimate gateway for maintaining living and metabolizing bone tissue, is critical to the processes involved in bone biomineralization (Skinner, 1987).

A similar cycle acts when bone organs restructure during growth and development from the fetus to the adult except the tissue resorption and deposition are displaced to accommodate revisions in the shape of the organ (Enlow, 1963; Figure 24(b)). To achieve the adult size of the long bones, a separate mineralizing system, known as endochondral ossification, is generated. The system has specialized cells and organic

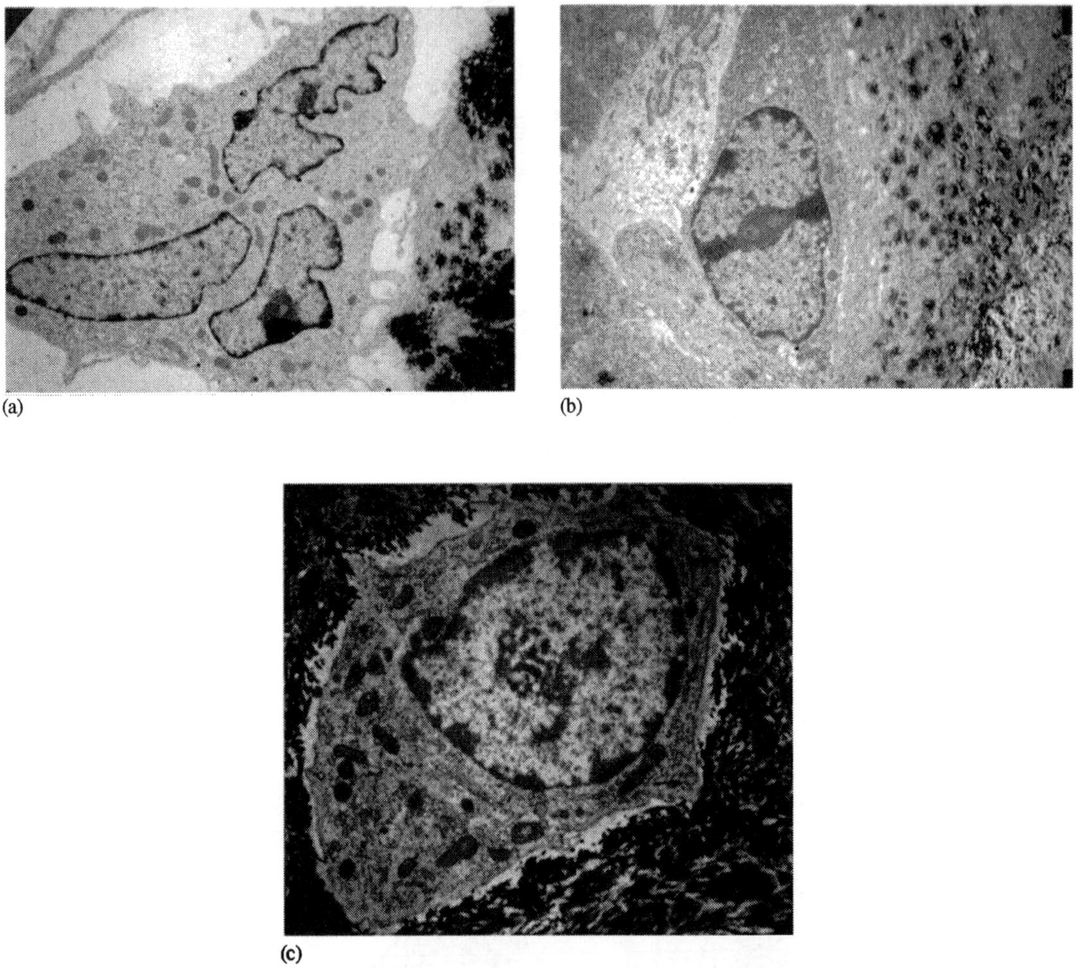

Figure 25 TEM microradiographs of individual bone cells: (a) osteoclast, a multinucleate cell; (b) osteoblast, mononucleate with surrounding matrix only partially mineralized (dark spots in fibrous matrix); (c) osteocyte, mononucleate, completely embedded in darkly colored mineralized tissue. Scale: the nuclei of these cells are ~5 μm in diameter (courtesy of Lynn Ann Neff, Department of Orthopaedics, Yale University).

components, and a growth plate with distinctive morphology to accommodate the extension and calcium phosphate deposition necessary to achieve the adult size of the organ (see Section 8.04.3.7.3).

When mineralized tissues are dissolved, both the organic and inorganic fractions become available for redeposition or redistribution. Some of the breakdown products (elements) will immediately reincorporate into new bone tissue, but others may move elsewhere to other body systems and cycles. Tissue reworking, the bone remodeling required during growth of the organ or the consistent tissue turnover throughout life via the BMU, provides many different elements stored in the apatitic mineral to the general circulation. The crystal structure of apatite allows the incorporation of a wide variety of cations and ions beyond the essential calcium and phosphate needed for apatitic bone and tooth mineral formation. Indeed, most of the elements in the periodic table can be found in solid solution within the precipitated bioapatite or adsorbed on the exceptionally large surface area that characterizes bone crystallites (Table 2; Section 8.04.2.3). The chemical composition of bone mineral can be used to evaluate some of the element concentrations in the environment in which it formed (Leeman, 1970; Larsen *et al.*, 1992; Leethorp *et al.*, 1994; Skinner *et al.*, 2003).

Bone cells and products: collagen. With the advent of high-resolution transmission and scanning electron microscopy, the intimate relationships between the cells (Figures 25(a)–(c)) that form (osteoblasts), maintain (osteocytes), and remodel (osteoclasts) mineralized bone tissues have been depicted and their products and reactions studied. The first extracellular products of the osteoblast are bioorganic molecules dominated by the fibrous asymmetric protein

Table 2 Major element chemistry of geoapatite and bioapatite.

	1	*2*	*3*	*4*	*5*
SiO_2	0.55	0.34	11.9	bd/br	bd/br
Al_2O_3	0.04	0.07	1.7	0.02–0.14	0.00–1.63
Fe_2O_3	0.14	0.06	1.1	(bd, incl.)	(bd, incl.)
FeO	0.24	0.00	bd/br	0.22–1.90	0.00–0.01
MnO	0.08	0.01	bd/br	bd/br	bd/br
MgO	bd/br	0.01	0.03	0.12–1.07	0.00–0.39
CaO	53.91	54.02	44.0	49.80–56.70	60–62.1
SrO	bd/br	0.07	Bd/br	2.20–2.50	0.00–0.17
Na_2O	0.10	0.23	0.06	0.86–7.56	0.00–0.46
K_2O	0.01	0.01	0.05	0.06–1.04	0.00–0.01
REE_2O_3	1.69	1.43	3.75	0.00–0.23	0.00–0.37
P_2O_5	41.13	40.78	30.5	29.20–33.30	36.3–37.6
SO_3	bd/br	0.37	1.8	bd/br	bd/br
CO_2	bd/br	0.05	2.2	5.84–5.98	bd/br
F	2.93	3.53	3.1	bd/br	bd/br
Cl	0.71	0.41	0.04	bd/br	bd/br
Net	101.66	101.40	103.23	NA	NA
-O = F,Cl	1.23	1.58		NA	NA
Total	100.43	99.82		NA	NA

Source: Skinner *et al.* (2003). 1-Granitoid rock, Cougar Canyon, Elko County, Nevada (Chang *et al.*, 1996, table 3), 2-Fe-apatite ore, Cerro de Mercado, Durango, Mexico (Chang *et al.*, 1996, table 4), 3-Phosphorite average, Phosphoria Fm., CO (Gulbrandson, 1966), 4-"Typical" ranges for bone bioapatite composition (Skinner *et al.*, 1972 and Rink and Schwarcz, 1995), 5-"Typical" ranges for enamel bioapatite composition (Cavalho *et al.*, 2001 and Brown *et al.*, 2002). "bd/br" = below detection limit or below rounding (two significant figures). "bd,incl") = Fe_2O_3 may not be reported separated, included within FeO value.

known as type I collagen (Miller, 1984). One of 11 different composition fibrous proteins in the collagen family (Kuhn, 1987), the molecular biology of type I collagen has been thoroughly investigated (Skinner (1987, pp. 207–208) presents a sketch of these details). It is composed one-third of the amino acid glycine (GLY), significant amounts of the imino acids proline and hydroxyproline, with many molecular repeats of GLY–X–Y where X and Y represent other amino acids on the initial strand. Some globular end regions assist, as does the subsequent hydroxylation of proline and lysine, in the cross-linking of individual protein chains into a triple stranded helical array of uniform size, roughly 1.5 μm wide × 280 μm long, known as a fibril (Miller, 1984). Note that the collagen fibril has a morphology reminiscent of a DNA molecule but is composed of amino acids not nucleic acids, is triple rather than double stranded, and contains no phosphate groups. The triple stranded collagen fibrils are extruded and aggregate extracellularly with a very special 1/4 stagger into fibers. The stagger leads to "holes," which may become sites for the deposition of apatitic mineral. Fibrillar collagen can be detected by tissue staining as the fibril aggregation appears to maximize the association of regions along the chains of positive and negative amino acids (Kuhn, 1987). Figure 26 illustrates, at the ultramicroscopic level, the typical banded pattern of osmium-stained collagen-containing tissue samples dotted with fine mineral deposits. The figure contains an electron diffraction pattern that confirms the presence and identity of the bone mineral as apatitic (Skinner, 1968). The grain size of the mineral deposits range from less than 300 Å in young animals up to 20 nm in maximum dimension for mature cortical bone (Elliott, 2002). The mineral permeates throughout, within as well as on, the collagen and protein–polysaccharide matrix produced by the osteoblasts. So thoroughly and intimate is the association that even after extraction of greater than 95% of the organic fraction, the gross morphology of the bone and texture are retained (Skinner *et al.*, 1972).

Matrix mediated biomineralization and other possibilities. The collagen family of proteins is the dominant constituent of most connective tissues in the human, other vertebrates (Mayne and Burgess, 1987), and some invertebrates (Bairati and Garrone, 1985). Collagens other than the types I and II (cartilage), but always with high glycine content, are typically found in articular cartilage, and the many other nonmineralized connective tissues throughout the vertebrate body. Most of the other collagen species are smaller fibrous molecules that form networks with glycoproteins. Highly resistant to proteolysis, and only slowly degraded after death even under nonbiological conditions, several techniques have been developed to evaluate, and to date, the organic and

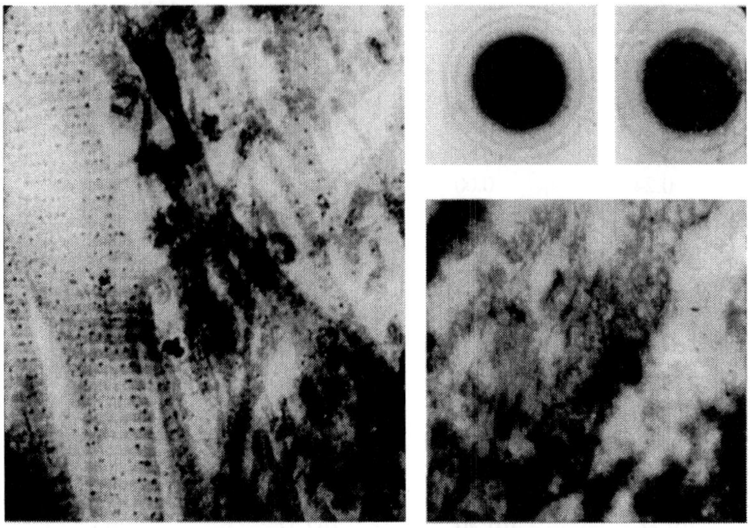

Figure 26 Electron micrograph of partially mineralized collagen. Magnification is $\sim 6 \times 10^4$. An electron diffraction pattern is shown in the inset that identifies the mineral as apatite (after Glimcher, 1959; reproduced in Glimcher, 1984, figure 13, plate 2).

mineral materials in fossils (Hare and Abelson, 1965; Kohn and Cerling, 2002; Trueman and Tuross, 2002).

Biomolecules, such as collagen, with spatial as well as charged surficial sites could overcome the nucleation barrier (heterogeneous nucleation) required for mineral deposition, whether the mineral is hydroxylapatite or one of the many polymorphs of calcium carbonate. The strategy, matrix-mediated biomineralization, has been espoused for a long time, although the specificity for consistent and regularized mineral deposition in bone tissues remains elusive (Mann, 2001). A hypothesis that offered such an advantageous scenario relied on the bone cells to control the chemistry of the biomolecular species: the amino acids serine and threonine in collagen, especially if located within the hole regions could became phosphorylated. As phosphate monoesters, they could attract calcium ions providing the bridge between the organic and inorganic (Glimcher, 1959, 1960, 1984). Another proposal for inducing extracellular biomineralization, the type found in bone tissues, invoked small noncollagenous protein molecules produced and secreted by the bone cells. The roles of exogenous molecules to stimulate the resorption, formation, and mineralization of bone tissues are beyond the scope of this review, but reports on their potential are part of the search for treatment of osteoporosis (Nordin, 1971; Wasnich, 1999). Using cell culture techniques and animal model systems, the effects of osteonectin, osteocalcin, calcitonin, and many growth factors, for example bone morphogenic protein #7, or osteogenic protein #1 (OP-1), on the production and regeneration of bone, especially the healing of fractures, are under investigation (Deftos *et al.*, 1999; Friedlaender *et al.*, 1999). It should be pointed out that many small molecules originally described from bone studies have been detected in other tissues and organs where they may play roles in the general metabolism of the body (Hollinger, 1997). Another suggestion of how calcium phosphate mineral is provided to growing bone tissues was by Anderson (1973). He suggested that apatite could become localized within lipid bound sacs or vesicles. Vesicles with lipid membranes will self-assemble in aqueous environments, and are not confined to the eukaryotes. They are prominent at sites of intracellular nucleation of iron oxides and sulfides in magnetotactic bacteria (Section 8.04.3.3.2; Mann, 2001, pp. 70–71). The lipid bilayer of cells, and of vesicles, often incorporates proteins that enhance the transfer of elements across these membranes, i.e., calcium, with the possibility that apatite crystallites could be prefabricated within vesicles. If assembled in the extracellular environment, the mineral could be released at the mineralizing tissue site, and deposit on the collagen. Mitochondria are also regarded as possible sources of mineral for biomineralization (Lehninger, 1983). The possibilities that any or all of these mechanisms act in the mineralization of bone tissues are still niether fully evaluated nor understood (Mann, 2001, pp. 143–144).

In addition to the molecular localization of apatite in vertebrate tissues, we hasten to point out that the quest to understand and control, or at least modulate, the mechanisms of bone biomineralization with its many facets and levels of mineral deposition continue. The intricate processes that lead to calcium phosphate deposition and specialized tissue formation have been refined over most biogeologic time not only since the first mammals (Halstead, 1969; Tuross and Fischer, 1989). Enlow, one of the early researchers into the histological characteristics of bone tissues, found counterparts of the human textures in the long extinct dinosaurs (Enlow and Brown, 1956–1958) (Figures 27(a) and (b)).

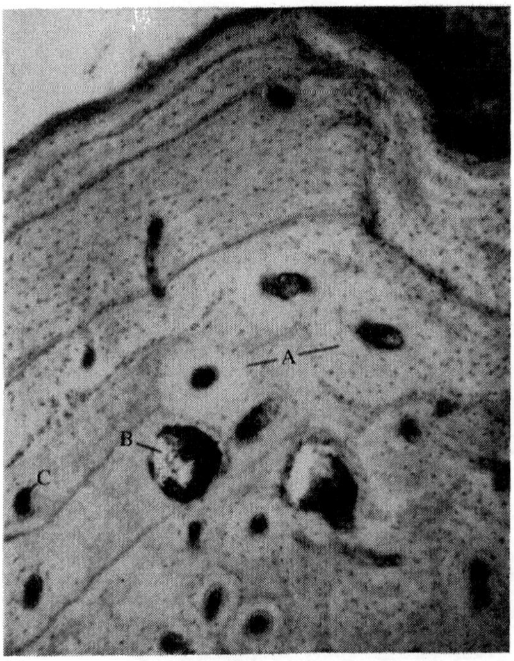

(a)

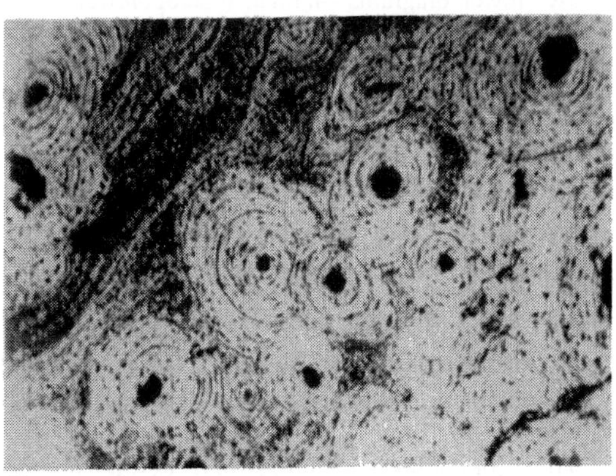

(b)

Figure 27 (a) Thin section observed with optical microscopy of lamellar bone from a femur of a Rhesus monkey through an area of muscle attachment showing resorption spaces (B), and haversian systems (A) and vascular canals (C), 150× (source Enlow, 1963, figure 32, p. 68). (b) Thin section of a dinosaur (Ceretops) bone showing lamellar and haversian bone similar to the tissue characteristics seen in (a) (source Enlow and Brown, 1956, p. 410).

Analyses of bone. The regeneration of mineralized tissues within all bone organs continues throughout the life of the organism. Therefore, though bones are durable and appear unchanged, the tissues are dynamic during the lifetime of the individual, and the integration of several textures during sampling may contain locally distinct, as well as with variable composition, organic and inorganic materials. A particular BMU will reflect the elements and molecules circulating during the time spanning the deposition phase. By employing markers specific for mineral, such as tetracycline, or fluorine, the sites, and the rates, of mineral deposition have been investigated in a variety of animals, and in human disease (Skinner and Nalbandian, 1975; Figure 28).

Because the external appearance of a bone gives no indication of this continual, internal reworking sampling may not be straightforward. Cortical bone, the more heavily mineralized and homogeneous tissue with slower turnover (species related), is often chosen, but any sample will contain several bone morphogenic units with heterogeneous organic and mineral deposition expressing responses to short-term, i.e., dietary, changes. Analyses of the mineral using multiple samples even from one individual may, therefore, show considerable variability. The results are an aggregate, a summary of the formation, resorption, and redeposition from several bone morphometric units. With such dynamic activity effectively hidden within bone tissue samples, it is not surprising that bone is considered less suitable than enamel, one of the tissues in teeth (see Section 8.04.3.7.5), for evaluating the elemental composition of bioapatites. If the object of the research effort is to establish the range of ingestion and uptake of certain natural, or anthropogenically provided, elements or compounds in a population, bone from known age individuals could indicate bioavailability, and some pollution or contamination (Kohn and Cerling, 2002; Trueman and Tuross, 2002). In the past skeletal investigations focused on the physical expression of bones comparing "normal" bones with those that might have experienced trauma, metabolic disturbances, or disease showing obvious differences (Ortner, 1992). With the advent of the more sensitive analytical techniques, such as scanning electron microcopy with energy dispersive analysis, there are many more investigations of the chemical variations of bone tissues (Skinner *et al.*, 2003; Ceruti *et al.*, 2003).

8.04.3.7.3 Cartilage

Cartilage is a distinctive skeletal tissue that also mineralizes with calcium phosphate.There are multiple locations in the vertebrate skeleton

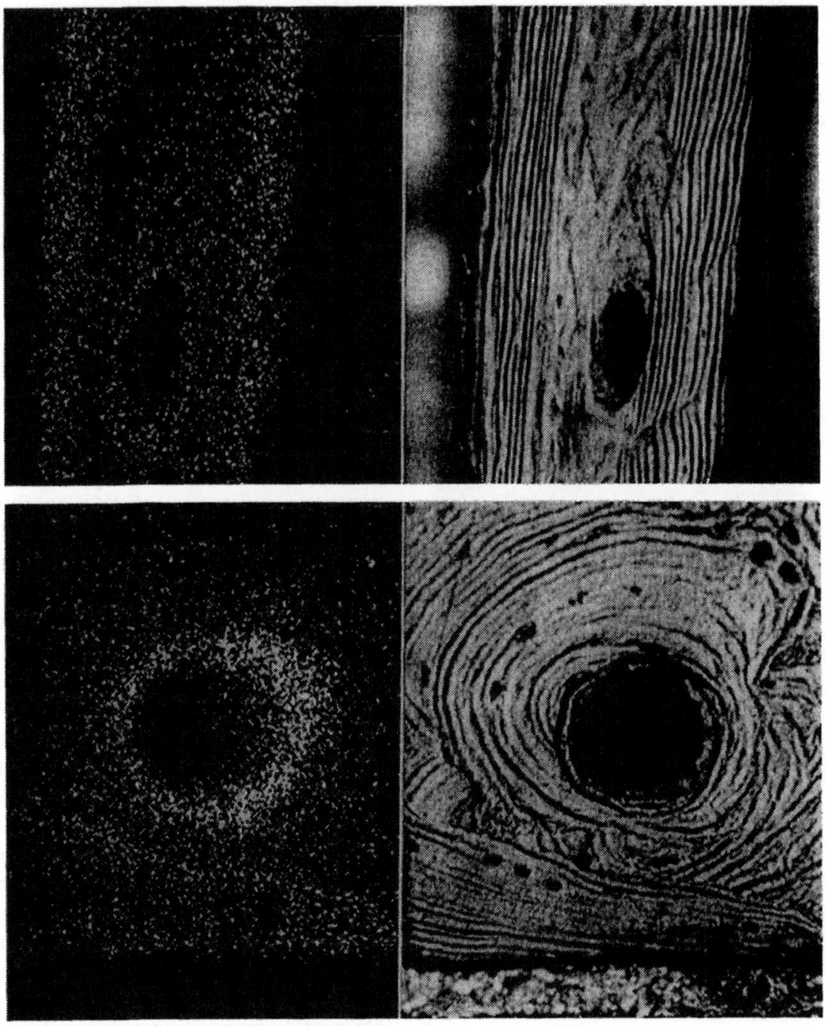

Figure 28 Fluorine uptake in bones. Upper diagrams—mouse; lower diagrams—human (osteoporotic) bones. White dots in left two plates indicate the location of F using electron microprobe emission analysis selecting the Kα wavelength of F. Companion plates on the right are photomicrographs of the same area using general illumination, and show the mineralized tissue structures, lamellar bone mineral deposition in the mouse, and mineral deposition at the center of a haversian system in the human coinciding with the concentration of white dots indicating F in the associated diagram (source Vischer, 1970, figures 1 and 2, p. 28).

that are predominantly cartilage but only some sites biomineralize, e.g., *in utero*, where an aggregation of cells, chondrocytes, proliferate and create a preformed model of a bone. The cartilage will eventually be replaced as part of the normal gestation and maturity sequence of the organ (Ogden and Grogan, 1987; Ogden *et al.*, 1987). This precursor bone formation system, known as endochondral ossification, is typical of vertebrate long bones where elongation of the organ takes place at the growth plate and the cells as they are produced assemble in columns (Figure 29). The chondroblasts produce a water-rich (up to 80%) gel-like aggregate (anlage) of two organic molecular species: proteoglycans, or protein–polysaccharides, and type II collagen, which becomes mineralized with apatite (Lowenstam and Weiner, 1989, pp. 167–175). Sharks and some fish keep such cartilaginous "hard" tissues to maturity (McLean and Urist, 1968; Moss and Moss-Salentijn, 1983), whereas in humans this "cancellous bone" is replaced by "membranous bone."

The other sites where cartilage occurs are not meant to mineralize. Articular cartilage is the shiny slippery textured material found at the ends of many bones, a tissue that facilitates the motion between two bones at the joints. The deposition of mineral in articular cartilage and joints is pathological, and briefly discussed in Skinner (2000). Cartilagenous tissues that occur in the ear, epiglottis, and intervertebral disks are normally not mineralized, and the organic components at these sites are distinct chemically and

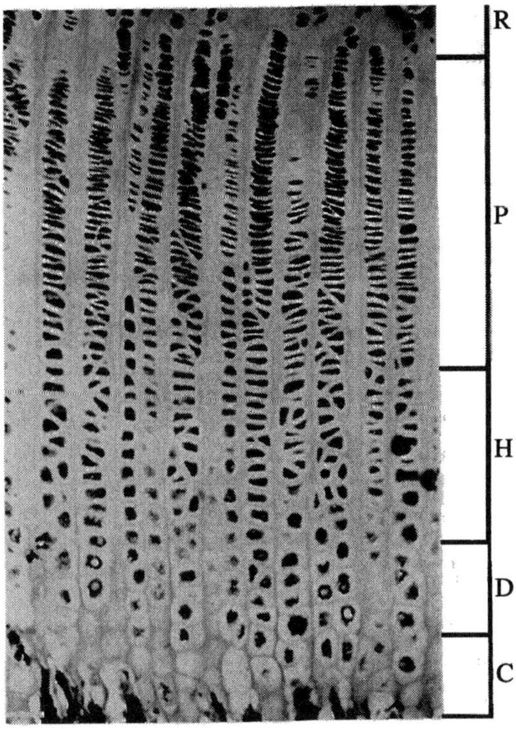

Figure 29 Optical micrograph of epiphyseal cartilage in the femur (large leg bone) of a rabbit showing columns of chondrocytes in the growth plate. The growth plate is divided into zones (C–R) that reflect changes in the chondrocytes during the elongation and mineralization of the endochondral bone (Source Lowenstam and Weiner, 1989, figure 9.12, p.172).

histologically from the mineralized cartilage sites. Miller (1985) compares the collagen compositions and the associated proteoglycan molecules that have molecular weights upwards of 200 kDa. These molecules have a core of hyaluronic acid (a polymer of glucuronic acid and *N*-acetylglucosamine), ~1,500 nm in length. This aliphatic protein base has upwards of 100 protein-linked monomers with noncovalently linked "side chains." Each of the monomers, formed intracellularly by chondroblasts, also has a protein backbone ~300 nm long with tens of negatively charged glycosaminoglycan chains covalently attached through serine and threonine residues. The assembly of such very large molecules, some of which contain sulfated species, e.g., chondroitin sulfate, results in extracellular, negatively charged species which, like bone, provides sites where mineralization commences and continues.

Biomineralization of cartilage. Prior to mineralization, mitochondria in the chondrocytes, cells equivalent to osteocytes, buried in the protein-glycosamino-gel, load up with calcium and phosphorus. Using SEM/EDAX analyses and microdissection, a timed efflux of calcium and phosphorus from mitochondria was shown to be coordinated with mineral deposition (Shapiro and Boyde, 1984). The mineralized tissue produced contained platy hydroxylapatite crystallites similar in size and composition to that in bone, but not uniformly associated with the cartilage collagen (type II). The extracellular cartilaginous matrix with its highly charged anionic polysaccharide chains may attract additional calcium and aid mineral nucleation (Hunter, 1987). However, the matrix also contains abundant lipid bound vesicles (Bonucci, 1967). At the earliest stages of cartilage matrix formation, the vesicles are without mineral but over time accumulate apatite crystallites, which coalesce, and the vesicles disappear leaving a haphazardly mineralized tissue (Ali *et al.*, 1970, 1977). There are slight differences in cartilage mineralization from that of bone: the mineral is not uniquely associated with the type II collagen, and the large amount of proteoglycans provides multiple opportunities for nucleation in the extracellular environment. The vesicles, acidic proteoglycans, and the procollagen type II molecules appear before mineralization and each could be, or could become, dominant as the mechanism of biomineralization, or they could behave cooperatively (Lowenstam and Weiner, 1989, p. 175). Vesicles have been identified as transport packets of mineral during biomineralization in invertebrates (Addadi *et al.*, 1987). In cartilage, acidic phospholipids on the vesicle membrane may act as a site of calcium accumulation (Wuthier, 1984), or nucleate apatite crystallites (Vogel and Boyan-Salyers, 1976). Vesicles could transport calcium and/or mineral to the mineralization front. Studies of biomineralization mechanisms in model animal systems use isotopically labeled elements and molecules, such as ^{45}Ca, ^{31}P, or the sulfur, in the sulfated organic complexes to elucidate the process.

8.04.3.7.4 Antlers

There is one other "bone" which could be useful to geochemists interested in assessing environmental exposures. Antlers, but not horns, are shed annually. Both these organs have a bony core but horn is covered with dead keratinous tissues while antlers are covered initially by velvet, an epidermal tissue with separate blood and neural supply. Once the velvet has been rubbed off, the bony superstructure that has formed and mineralized very rapidly provides the animal with a remarkable headdress. After the breeding season osteoclasts resorb at the base where the antler is attached to the frontal bone of the skull and antlers may be relatively easily obtained for analyses. Male deer, reindeer, and caribou regrow their membranous bone excrescences each year with an increase in the complexity and size of these unique structures

(Halstead, 1974, pp. 98–100). Frank (2003) investigated an odd disease of moose in southern Sweden showing that a molybdenum deficiency in their forage leads to diabetes in these animals.

8.04.3.7.5 Teeth

Introduction. The evolution of dentitions as independent mineralized organs shows that from the earliest vertebrates to modern mammals, there has been a diversity of biomineralized structures and processes (Halstead, 1974). The following discussion focuses on human dentition as an example of the many teeth and oral cavity arrangements found in warm or cold-blooded vertebrates.

The adult human normally has 32 teeth (Figure 30(a)) with distinctive shapes related to function, a system that parallels the morphological expression of bones in the skeleton. Each tooth should be considered an independent organ. The significance of the macrostructural variety and the biomechanical contributions, especially tooth wear, has become important to those interested in the evolution of vertebrates, particularly the hominids. The relationships of dentition to the skull bones, the mandible and the maxilla, and the size and shape reflect the habitat, and diet, of the species. Oral anatomy, tooth development, mineralization, and demineralization (caries) remain active research areas by a host of dental professionals, as well as biologists, anthropologists, and geologists. In fact, studies of calcification, elucidation of the composition and structure of bioapatites, and biomolecular species, especially collagen, are and have been supported in the US by the National Institute of Dental Health and Craniofacial Anomalies. Obvious malformations such as cleft palate provided some of the first information on genetic abnormalities and hazardous drug ingestion (Goose and Appleton, 1982). Early studies into tooth biomineralization came from Europe (Schmidt, 1921) and markedly expanded in the US, around fluoridation of domestic waters to minimize the formation of cavities (caries), especially in young children (Vischer, 1970). This review of the biomineralization of teeth affords insight into different sites of calcium phosphate deposition and some novel geochemical applications.

Tooth biology and mineralogy. There are three mineralized tissues in a tooth each with distinct cell systems and different histological expressions. Table 3 compares enamel, dentine, and bone, the relative amount of calcium phosphate to organics and the composition range of the mineral material. Most analyses of the mineral are on samples that have been chemically extracted, or ashed, to remove organic components and water (Skinner *et al.*, 1972, 2003) and reported on the dry-fat-free basis. Many papers report only calcium and phosphorus, calculate the Ca/P ratio, and compare the results with the ideal for hydroxylapatite (2.15 wt.%, 1.67 mol.%). Some analyses include magnesium (always less than 2.5 wt.%), sodium (usually less than 1 wt.%), and chlorine, always less than 0.1 wt.%. Occasionally CO_2 is reported (between 3 wt.% and 6 wt.%) with other elements mentioned when related to specific research activities. Comprehensive elemental analyses that add to 100% are not usual (Skinner *et al.*, 1972).

The tooth structure is created when ectodermal cells, ameloblasts, situated in apposition to the mesodermal dentinoblasts secrete organic matrices, and direct calcium phosphate deposition. The third tissue, cementum, covers the dentine below the gum line in the tooth root (Figure 30(b)). Teeth are anchored into the jaw bones via fibers that start in the cementum (below the enamel cap) and end in the bone; known as periodontal ligaments, these essential supporting guy wires must be discarded when deciduous teeth are replaced by the permanent dentition. Each tooth in its socket represents a multiple mineralization site, forming and forcing resorption of adjacent bone (remodeling) over time. In humans "baby teeth" are evulsed after resorption of the root portion as a new larger tooth is created with new enamel, dentine, and cementum, and erupts into the oral cavity through a jawbone enlarged to the adult size.

Mature enamel is a cell-free tissue while cementum though mineralized is avascular, similar to some bone tissues, and may have developed to serve the local needs, not a primitive, but a derived and modulated tissue system (Poole, 1967). The bulk of the mineralized tooth is dentine, also a vital tissue that contains passageways known as tubules that are intermineralized with apatite. The tubules reach from the pulp cavity with its supply of nerves and blood vessels that enter at the base of the root(s) and extend to the enamel. Dentine has the possibility to repair itself.

The incremental growth of a human tooth is demonstrated in Figure 30(c). The bright nested lines are generated from sequential doses of tetracycline. An antibiotic, tetracycline, interacts with the mineral, recording the sites and rates of deposition at the mineralizing front during tooth gestation of this molar. Tetracycline incorporation can be detected because of its characteristic fluorescence in UV (Skinner and Nalbandian, 1975). Mineralization starts at the dentine–enamel junction, fills in toward the central pulp canal and tapers towards the root(s), while the enamel grows in the opposite direction forming a cap over the entire upper surface. The central canal reduces in size as the organ completes its maturation and erupts into the oral cavity.

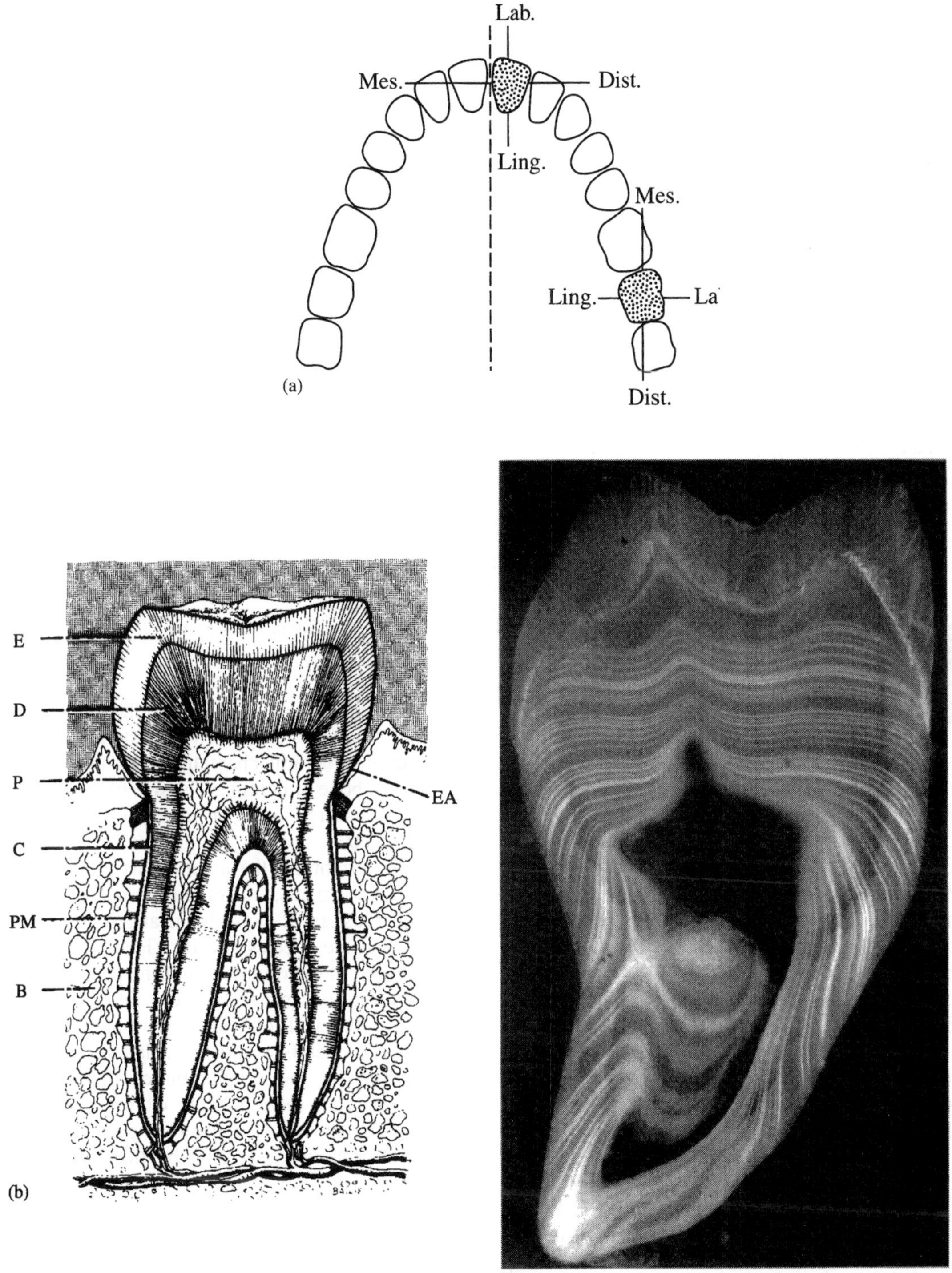

Figure 30 (a) Schematic of the maxillar (upper jaw) dentition of the adult human (source Peyer, 1968, figure 10, p. 16). (b) Diagram through a molar: E = enamel, D = dentin, P = pulp chamber, C = cementum, PM = pericementum muscle attachments, B = bone, and EA = enamel cap finishes with the cementum the mineralized tissue below the gumline (source Goose and Appleton, 1982, figure 8.1, p. 126). (c) Cross-section of molar, comparable to Figure 30(b), showing tetracycline uptake as a series of nested bright lines indicating where and when the antibiotic was ingested, separated by darker areas when no antibiotic was circulating in the system, and therefore was not deposited with the mineralized tissue of the growing tooth (Skinner and Nalbandian, 1975, fig. 4A, p. 386).

Table 3 Bulk composition, density and crystallite size of mineral in bone, dentine, enamel, and the composition of the extracted bioapatites.

	Bone[a]		*Dentine*[a]		*Enamel*[a]	
	wt.%	vol.%	wt.%	vol.%	wt.%	vol.%
Inorganic	70	49	70	50	96	90
Water	6	13	10	20	3	8
Organic	24	38	20	30	1	2
Density (avg) (g cm^{-3})	2.35		2.51		2.92	
Crystal size (maximum)[b] (angstroms, Å)						
Length	300 Å	1,600 Å				
Width	60 Å	410 Å				
Hydroxylapatite composition (wt.% on a dry, fat-free basis)[c]						
Ash	57.1		70.0		95.7	
Ca	22.5		25.9		35.9	
P	10.3		12.6		17.0	
Ca/P	2.18		2.06		2.11	
Mg	0.26		0.62		0.42	
Na	0.52		0.25		0.55	
K	0.089		0.09		0.17	
CO_2	3.5		3.19		2.35	
Cl	0.11		0.0		0.27	
F	0.054		0.02		0.01	

[a] Driessens and Verbeeck (1990: table 8.2, p. 107; table 9.4, p. 165; table 10.5, p. 183). [b] Elliott (2002, table 4, p. 441).
[c] Zipkin (1970, table 17, p. 72).

Once erupted, the cells that produced enamel disappear; hence, there is no possibility of inducing new enamel formation, should cavities develop. Enamel is the only tissue observed in the mouth. Enamel is a hard, composite, tissue, and the most highly mineralized tissue in humans, consisting of over 95% mineral. There is virtually no organic component in enamel (Table 3), although it is essential to creating the patterns distinctive to this mineralized tissue. The ameloblasts secret small (2.5×10^4 MW) hydrophobic proteins known as amelogenins (Fincham and Simmer, 1997) which have high concentrations of proline, glutamine, leucine, and histadine, but no regular repeating [Gly–X–Y] aminoacid sequence typical of the collagen fiber series. There is a self-assembly of 20 nm nanospheres of ~100 amino acid (AA) residues, that appear to assist in the orientation of the crystals parallel to the *c*-axis or long axis of the apatite. Enamelins, acidic glycoproteins (β-pleated sheath structure; Deutsch *et al.* (1991)), enclose each growing crystallite binding to specific crystallite surfaces.

Enamel contains well-oriented apatite crystallites, of sizes up to 1 nm in maximum dimension (Elliott, 2002). Ultra-high-resolution photos of thin sections of developing enamel show the unmistakable hexagonal outlines of apatite (Figure 31(A)). Powder X-ray diffraction analysis which can be used to accurately identify the crystal structural characteristics of any solid shows that the material is clearly apatite and that there is preferred orientation of the crystallites. These analyses reinforce the fact that the mineral of enamel mineral is much better crystallized (more regular in structural character) than the mineral portion of bone or dentine. The size of the crystals can be measured using a variety of techniques (Elliott, 2002).

In all vertebrates the ameloblasts create packets, called prisms, of organic materials with aligned clusters of crystals, arranged in arcuate arrays that are species specific (Halstead, 1974). Differences in the orientation of the prisms and in the degree of mineralization determine the textures observed optically, or with TEM, in enamel (G. Gustafson and A. G. Gustafson, 1967). Depending on the orientation of the mineral-containing thin section being examined, optical microscopy reveals (i) narrow stripes (called the Lines of Retzius) generated by the relationship between adjacent prisms during successive periods of growth in human enamel, or (ii) Hunter–Schrager bands that extend from the dentine–enamel junction ~2/3 of the way toward the surface of the enamel and result from the changing directions of the prisms. Disruptions or variations in amounts of mineral within or in adjacent prisms can also be observed (G. Gustafson and A. G. Gustafson, 1967).

Dentine is less mineralized than enamel with a lower Ca/P, and shows variable composition individual to individual (Rowles, 1967), similar to the other mineralized tissues. Some of the chemical variations reported on teeth may be the result of the pretreatment, the level of analytical

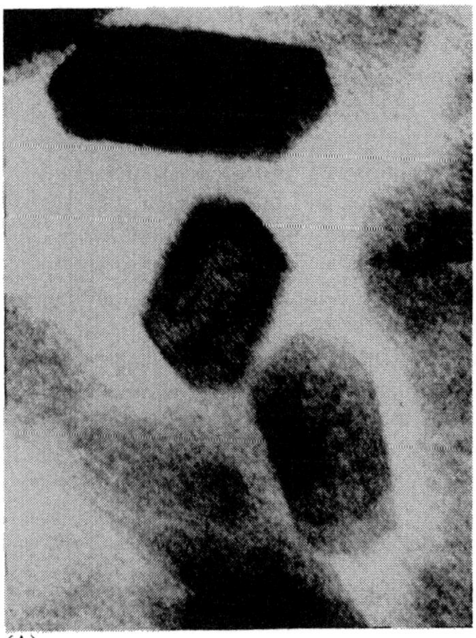

(A)

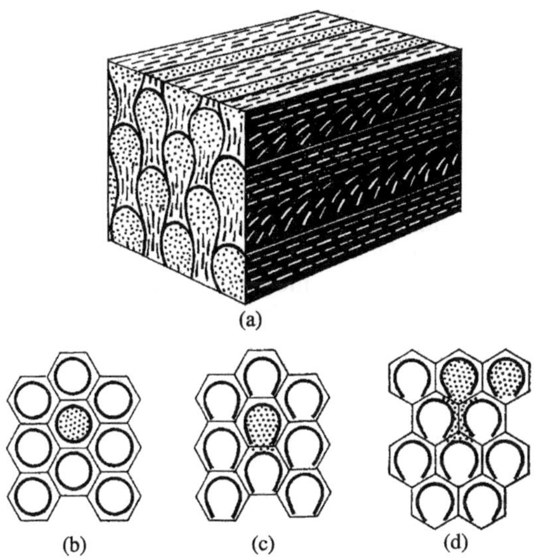

(B)

Figure 31 (A) Ultra-high-resolution (1,109,000×) TEM photo of apatite crystallites in enamel (early developing rat enamel) (source Helmcke, 1968, figure 8, p. 143). (B) Diagram showing the distribution of enamel crystallites within prisms and prism arrangements formed by ameloblasts in enamel of vertebrates illustrating the variety of arrays in the tissue depending on the angle of light microscopic observation. Note: different vertebrate species have distinct enamel patterns depending on the sites where crystallites are deposited and the number of ameloblasts producing the prisms. The textures observed not only vary with the orientation of the thin section examined but are distinctive for the ameloblasts contribution to crystallite orientation. For example in "d" the deposition of crystallites in a prism is contributed by four ameloblasts (source Halstead, 1974, figure 12.3, p. 89).

sensitivity, as well as relate to the particular tooth, or portion of the tooth, examined. For example, ingestion of fluoridated water and the age of the individual or of the sample may increase the mineral content postextraction due to dehydration. Using transmission microradiography increased mineralization around dentine tubules, and the formation of secondary dentine after tooth trauma, are detectable. These expressions of mineralization post the original organ mineralization may be large and could contribute to analytical variations. The apatitic mineral in dentine, and cementum, is similar in size and composition range and variations depending on the stage and diet during gestation, plus whatever uptake (usually minor) after maturity of the organs. Fluorine content in human dentine, for example, may increase over time with constant ingestion of fluoridated water reaching a maximum at ~55 yr of age. Enamel, alternatively, is completely formed, in a short time period for an individual tooth, with little, if any, change postcompletion and emplacement of the whole tooth into the oral cavity, including uptake of elements (except as might be briefly adsorbed when flushed by saliva). The post-mortem uptake (passive) of fluorine in both teeth and bone from groundwater is well known. It is an expression of the stability of fluorapatite over hydroxylapatite in Earth environments (see fluorine uptake in bone, Figure 28). Besides, the fluorine content of fossil samples helps determine whether there has been obvious diagenesis. During formation of bioapatites in humans, however, the mineral is a hydroxylapatite with trace quantities of fluorine even in areas where high fluorine waters are ingested.

Over geological time, vertebrates have generated a variety of mineralized structures. Their multiple forms of dental organs and tissues, and replacement modes reflect genetic diversity, probably modulated through adaptations or conversions relative to the habitat and diet of the species. Some of the tissues which can be examined today in lampreys, bony fish, and amphibians are histologically similar to tissues known as aspidin, and enameloid that may have functioned as armor in the earliest Devonian vertebrates (Orvig, 1967; Halstead, 1974). The morphological expressions of these tissues distinctive to vertebrates may be unique, but the mineral matter is invariably the calcium phosphate, apatite.

8.04.3.7.6 Otoliths

There are other mineralized structures that have been used for dating and investigations of the ecology of vertebrates—the otoconia or statoliths—that occur in the ears of mammals, birds, and fish. The predominant mineral in these structures is $CaCO_3$ in the form of aragonite, but as shown in Figure 32(a) there are some families which possess calcium phosphate otoliths, and many otoliths are a combination of two mineral species.

Otoliths are tiny, often barely visible, mineralized structures that are unattached and do not form part of the skeleton *per se*. They are highly mineralized single, or multiple and fused crystals, of one of the three polymorphs of $CaCO_3$, aragonite, calcite, or vaterite. The crystallites (Figure 32(b)) grow in the canals or labyrinths of the ear ducts and may have several functions: sound detection, as sensors of gravity, or as part of the system used to determine orientation equilibrium of the animal body, making it essential that they be mineralized structures. Otoliths show distinctive morphology that is used to make taxonomic decisions (Nolf, 1985).

Large otolith accumulations in sedimentary horizons led Cuvier in 1836 to suggest that the taxonomy of fish species, e.g., teleosts, could be used to determine the ecological parameters for the marine environment. The otoliths became markers for deciphering ocean sediments comparable to the use of mammal teeth in terrestrial environments. In the ensuing years otoliths or "buttons" from the Jurassic to the present in localities as diverse as the Viennese Basin, Sumatra, Nigeria, New Zealand, the Mediterranean, and England were investigated. Past studies and applications are summarized by Nolf (1985). Nolf, along with Milton and Cheney (1998), have all been concerned with diagenetic alteration of the specimens. The chemical analyses of recent fish otoliths from marine and freshwater sites for trace elements that include both stable and radioactive species provide an independent source of data on migration, maturation, as well as being indicators of pollution (Adami *et al.*, 2001; Gillanders *et al.*, 2001; Hanson and Zdanowicz, 1999; Spencer *et al.*, 2000; Volk *et al.*, 2000). Analyses for specific elements, e.g., lead and strontium, can go beyond bulk determinations as otoliths may record a time series. Their growth rings (Figure 32(b)) are similar to tree rings documenting seasonal changes. Laser-ablation inductively coupled plasma mass spectroscopy (ICPMS) analyses on cross-sections through the collection of rings in the larger, up to centimeter-sized, oval to circular otoliths typical of some species, is a most effective analytic technique.

The deposition of mineral matter in fish otoliths may take place in three paired canals designated as saccular, utricular, and lagenar, for the actinopterygians by Norman and Greenwood (1975), or saggita, lapillus, and asteriscus for other species (Nolf, 1985, p. 3). Although there are multiple crystallites, the otoliths may be no more than a 0.01 mm in diameter depending on the species, age, and size of the fish. However, since there will be two of each sort for intact fish, sufficient sample of these biogenic carbonates may not be a problem (Arslan and Paulson, 2002). The rings, a light colored and usually thicker deposit, consisting predominantly of inorganic materials formed during summer and fall alternate with dark, and thin, winter rings that may contain up to 10% organic materials. The organic portion in both light and dark areas is up to 85% otolin, a collagen-like protein.

Interpretations of the otolith data to document climate change or pollution have been careful to consider the bioavailability of specific elements, variations in habitat, or ingestion idiosyncrasies of the specific fish species (Campana, 1999).

8.04.4 SUMMARY: WHY BIOMINERALIZE?

It appears that the capacity for biomineralization has been an evolutionarily widespread and an enduring trait, because it conferred strong and obvious selective advantage to organisms that possessed it. Biomineralization, in this analysis, is a fundamental life process by which animals and plants gain structure and mass, virtually without tissue maintenance cost, drawing on the chemistry of the environment to find strategies for maintenance and defense. Mineralization increases the opportunities for organisms: they can become mobile, producing endoskeletons on which to attach appropriate muscles (vertebrates), and therefore move out of hostile environments, or, if sessile, adapt by strengthening their internal skeletons (gorgonid corals), or produce external protection such as shells (mollusks). Biomineralization also contributes to the classic adaptation to land environments by plants.

Beneficial attributes conferred on biological forms by biomineralization and biominerals are actually twofold: (i) physical, or macrocontributions, the production of skeletal structures that provide integrity and specificity on the organism, and (ii) chemical, or micro- or sub-micro-contributions. In the latter, biomineralization provides a personal storage system from which ions are mobilized to other portions of the living creature if they are necessary: the nutrients that assist or regulate growth or, at the other extreme, ions may be permanently sequestered to avoid toxicity, if they are hazardous. Availability of stored nutrient ions is crucial to the buffering of body fluids and to skeletal and tissue repair activities central to the evolutionary survival and competitive status of an organism. The generation and systemic circulation of hormones or other special molecules could facilitate these activities and probably represent evolutionary advances or adaptations.

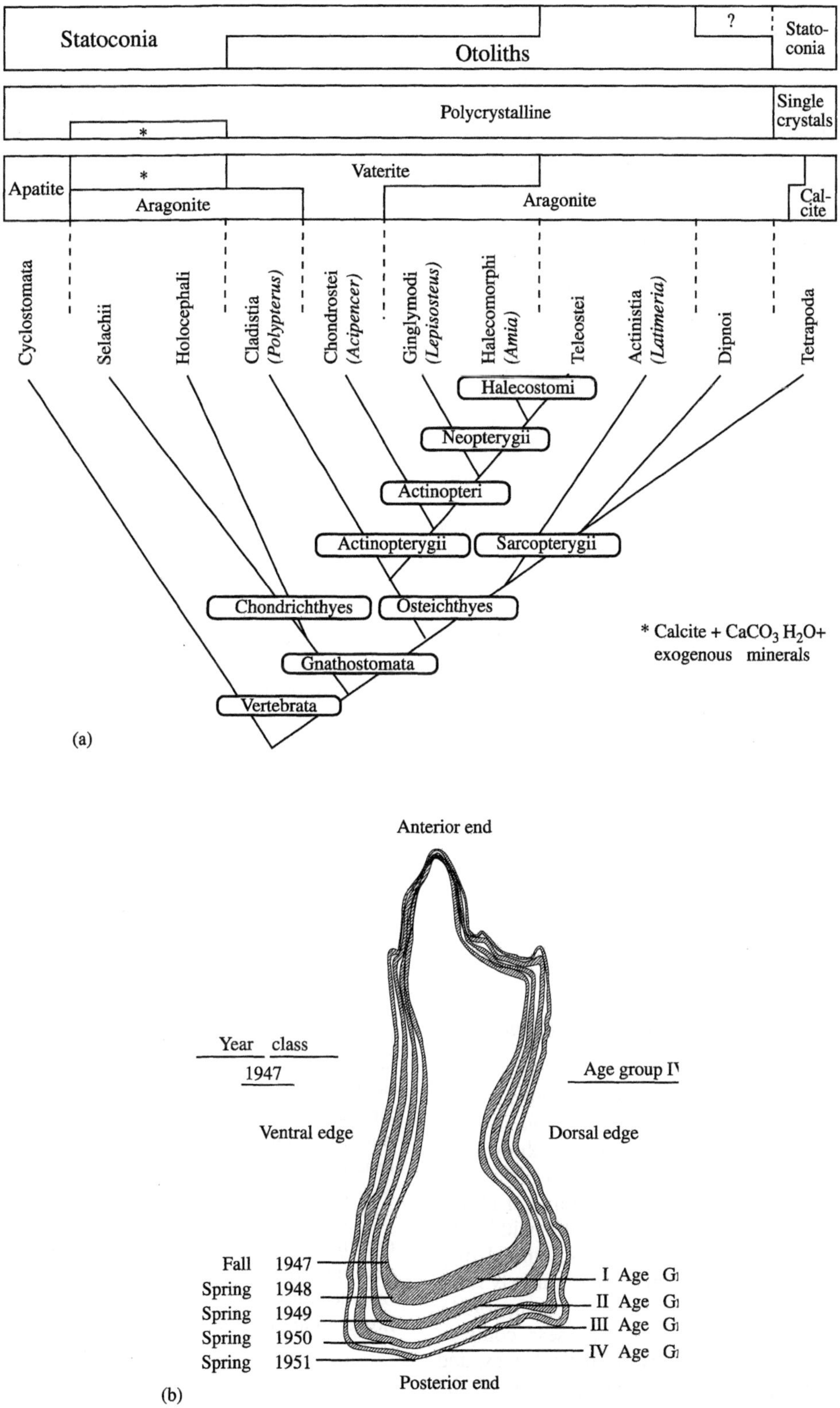

Figure 32 (a) Mineral materials found in the statoconia and otoliths of vertebrates (source Nolf, 1985, figure 1, p. 3). (b) An otolith showing incremental growth stages (source Nolf, 1985, figure 5, p. 5).

8.04.4.1 Physical or Macrobiomineralization Contributions

As animals and plants came to colonize land surfaces, gravity (cf. buoyancy) became a primary force; organisms required a structure, or framework, in order to stand up under the earthward pull of gravity. Minerals became the ideal choice. Incorporating them into the biological structures provided the stiffening agents. Throughout evolution the design and engineering of the composite (inorganic and organic), tissues became integrated into macroscopic, but relatively low-density structures. Produced internally or externally by the animals or plants, biomineralization did not reduce, but rather aided flexibility, if not mobility, for some forms. Once mobile, for example, animals could migrate to more benign habitats, seek different food sources, and avoid becoming prey. Sparsely mineralized carapaces of arthropods protect soft tissues but allow for mobility through an articulated exoskeleton. These creatures can live under water, in the air, as well as on land; mollusks burrow into mud flats to depths commensurate with the mineralization level of their shells. Mineralized internal structures characteristic of coccoliths and diatoms establish their domains, while silica reinforcement of plant stem tissues gives them structural integrity that could never be achieved using soft organic tissues alone. To remain rigid is critical for plants and it provides competitive advantages. Structural support during growth periods increases the plant's ability to compete for light, and provides resilience against heavy rain, strong winds, or trampling by animals.

8.04.4.1.1 Defense and protection through biomineralization

Many defense structures and mechanisms exist in plants, living forms that must survive in fixed locations. Sharp hairs or spikes known as "trichomes" are extensively biomineralized. The stiffness and sharpness of the biomineral allows the structure to effectively stick, lodge within and irritate an animal, even though the animal may possess several times the mass and infinitely more mobility than the plant (Hodson *et al.*, 1994). Stinging nettles (*Urtica* spp.) possess complex trichomes on the surfaces of their leaves and stems. The proximal silicious portion of the trichome attaches to a distal calcified portion with a terminal bulb. Upon contact with an herbivore, the bulb is dislodged leaving a beveled silicious hollow that functions like a hypodermic needle. This sharp needle, filled with various irritant chemicals including histamine, acetylcholine, and 5-hydroxytryptamine, may be released into the animal's skin sometimes causing extreme irritation or allergic reaction (Kulze and Greaves, 1988). Silicious spines or sharp needle-like crystals could damage soft mouthparts with instantaneous effect, preventing herbivores from specializing in, and thus obliterating, the biomineralized plant species and communities.

Many plant fruits contain a hard covering for their inner endosperm contents, providing structural protection for the reproductive tissues that will grow into a new plant, while allowing dispersers to enjoy enticing fruit. The hard covering is usually hard because of the incorporation of biomineral: peach-pits contain aragonite, and walnut shells are heavily silicified. Some biominerals may be designed to protect the endosperm as it travels through the acidic gastro-intestinal tract of a dispersing herbivore. Hackberry carbonate endocarps dissolve in low pH solutions, leaving a reticulate silica network, porous enough to let the endosperm germinate after it passes through the herbivore (Cowan *et al.*, 1997). Immature stamens of *Philodendron megalophyllum* (Philodendron) are reinforced with calcium oxalate (Barabe and Lacroix, 2001), which may serve to keep these delicate structures intact until pollen is ready for dispersal. In developing fruits, biomineralization shows specific timing (Jahren *et al.*, 1998), with the bulk of the mass added at the same time that the endosperm forms. After the mineralized endocarp is essentially complete, the edible fleshy portion of the fruit begins to develop. Studies of dual-mineral plant phytoliths have shown that vilification is complete by the time calcification begins (Jahren *et al.*, 1998; Thurston, 1974).

Horses who graze sandy-quartz soils suffer from silica enteroliths, either as a result of phytolith intake, or direct soil intake. Geis (1978) studied three species of opal-bearing Gramineae and found that the amount of opal deposited to the soil annually by root systems and aboveground parts of plants was approximately equal in magnitude, indicating that biomineralization in grasses is not specific to leaves, but results in considerable contribution of phytolith material directly to the rhizosphere. The evolution of horse dental morphology through geologic time is thought to be partially controlled by the impact of phytolith-bearing grasses on horse enamel (MacFadden *et al.*, 1999).

8.04.4.2 Chemical or Microbiomineralization Contributions

Biominerals because of their crystal chemistry naturally incorporate certain elements via solid solution in the crystal lattice, and their habits may present morphological (size or twinning) advantages for the adsorption of specific elements, or molecular species. The result is that the mineral may act as a storehouse for useful or hazardous ions.

Silica may deter animals from eating or contacting (and possibly crushing or damaging) plant tissues, but silica biomineralization may help to alleviate the toxicity of aluminum in plants growing on contaminated soils (Hodson and Sangster, 2000). Plants relieve potential Ca^{2+} toxicity via biomineralization of calcium oxalate (Webb, 1999), but calcium oxalates are also well known for preventing herbivory through chemical restraints.

Calcium oxalate results from the reaction of oxalic acid and calcium ions (McNair, 1932), and oxalic acid itself is an effective deterrent to herbivores in its own right (Arnott and Webb., 1983). At least four different biochemical pathways have been shown to result in the synthesis of oxalate in plants (Raven *et al.*, 1982). Utilizing radiocarbon-labeled ascorbic acid administered to roots of *Yucca torreyi* L. (Torrey's yucca); the incorporation into vacuole crystal bundles demonstrated that ascorbate is an important precursor to oxalate in this plant (Horner *et al.*, 2000).

The number of calcium oxalate crystals was found to be highest in young leaves and lowest in mature leaves of five tropical plants, indicating that plant organisms use biominerals to preferentially defend their most vulnerable tissues (Finley, 1999). Plants are also able to respond to ongoing herbivory by increasing the amount of biomineralized calcium oxalate crystals: leaves from seedlings of *Sida rhombifolia* (Teaweed) subjected to herbivory had a greater crystal density than those grown under protection (Molano, 2001). However, many herbivores have developed strategies to deal with plant biominerals. Microscopic spherulites of calcium carbonate found in the dung of herbivores may represent transformed plant-derived calcium oxalate (Canti, 1997), suggesting that the original biomineralized material can be transformed and excreted.

To discuss the range of possibilities of "why biomineralize?" we utilized plants, but there are parallels when assessing the possible reasons for biomineralization in other life forms, some of which have been mentioned or alluded to in the body of those sections. Our purpose has been to accentuate the cross-overs between the Earth and the biological sources for nutrients (and hazards) for all forms of life. We underscored the case of biomineralization in plants, because we and all land-based and most marine life must feed upon them. There are, of course, many agricultural examples which have not been included herein. The fact is that biomineralized tissues do reflect the environment and have become one of the important areas for future geochemical research. Each biological form mentioned in this chapter has mineral components, storage sites of elements, and molecules that deserve further investigation.

Biomineralization strategies and the biominerals are, for many good reasons, important to all who inhabit the Earth. The crossovers between the biomineralizers, the chemistry and the morphology of the minerals, the tissue textures, the cells, and the mechanisms that characterize some life forms offer intriguing insights to the physical–chemical laws of nature.

ACKNOWLEDGMENTS

Catherine Skinner wishes to acknowledge the sustaining contributions and assistance of Mabel Peterson, Yale Department of Geology and Geophysics, at many levels in the preparation of this manuscript, and to thank the Department of Geological and Environmental Sciences, School of Earth Sciences, Stanford University for the opportunity to collate and complete the chapter as the Allan Cox Visiting Professor, January–March 2003.

REFERENCES

Adami G., *et al.* (2001) Metal contents in tench otoliths: relationships to the aquatic environment. *Ann. Chim.* **91**, 401–408.

Addadi L., *et al.* (1987) A chemical model for the cooperation of sulfates and carboxylates in calcite crystal nucleation: relevance to biomineralization. *Proc. Natl. Acad. Sci. USA* **84**, 2732–2736.

Albright J. A. (1987) Bone: physical properties. In *The Scientific Basis of Orthopaedics*, 2nd edn. (eds. J. A. Albright and R. A. Brand). Appleton and Lange, Norwalk, CT, pp. 213–240.

Albright J. A. and Skinner H. C. W. (1987) Bone: structural organization and remodeling dynamics. In *The Scientific Basis of Orthopaedics,* 2nd edn. (eds. J. A. Albright, and R. A. Brand). Appleton and Lange, Norwalk, CT, pp. 161–198.

Alexandre A., *et al.* (1999) Late Holocene phytolith and carbon-isotope record from a latosol at Salitre, south-central Brazil. *Quat. Res.* **51**(2), 187–194.

Ali S. Y., Sajdera S. W., and Anderson H. C. (1970) Isolation and characterization of calcifying matrix vesicles from epiphseal cartilage. *Proc. Natl. Acad. Sci. USA* **67**, 1513–1520.

Ali S. Y., Wisby A., Evans L., and Craig-Gray J. (1977) The sequence of calcium and phosphorus accumulation by matrix vesicles. *Calcified Tiss. Res.* **225**, 490–493.

Aminuddin M. and Nicholas D. J. D. (1974) Electron transfer during sulfide to sulfite oxidation in *Thiobacillis denitrificans. J. Gen. Microbiol.* **82**, 115–123.

Anderson H. (1973) Calcium-accumulating vesicles in the intercellular matrix of bone. In *Hard Tissue Growth, Repair, and Remineralization*, CIBA Fondation Symposium (eds. K. Elliott and D. W. Fitzsimons). Elsevier, Amsterdam, pp. 213–246.

Anderson O. R. (1981) Radiolarian fine structure and silica deposition. In *Silicon and Silicious Structures in Biological Systems* (eds. T. L. Simpson and B. E. Volcani). Springer, New York, pp. 347–380.

Anderson O. R. (1986) Silicification in radiolaria—deposition and ontogenetic origins of form. In *Biomineralization in Lower Plants and Animals* (eds. B. S. C. Leadbeater and R. Riding). Clarendon Press, Oxford, UK, pp. 375–392.

Arnott H. J. and Webb M. A. (1983) The structure and formation of calcium oxalate crystal deposits on the hyphae of a wood rot fungus. *Scan. Electr. Micros.* **3**, 1747–1750.

Arslan Z. and Paulson A. J. (2002) Analysis of biogenic carbonates by inductively coupled plasma-mass spectroscopy (ICP-MS). Flow injection on-line solid-phase preconcentration for trace element determination in fish otoliths. *Anal. Bioanal. Chem.* **372**, 776–785.

Atkins E. (1985) Conformations in polysaccharides and complex carbohydrates. Proc. Int. Symp. Biomol. Struct. Interact. Suppl. *J. Biosci.* **8**, 375–387.

Avioli L. V. (2000) *The Osteoporotic Syndrome: Detection, Prevention and Treatment,* 4th edn. Academic Press, New York.

Azam F. and Volcani B. E. (1981) Germanium-silicon interaction in biological systems. In *Silicon and Siliceous Structures in Biological Systems* (eds. T. L. Simpson and B. E. Volcani). Springer, New York, pp. 43–67.

Bairati A. and Garrone R. (eds.) (1985) *Biology of Invertebrate and Lower Vertebrate Collagens.* NATO Advanced Research Workshop on the Biology of Invertebrate and Vertebrate Collagens (1984). Plenum, New York.

Balkwill D. L., Maratea D., and Blakemore R. P. (1980) Ultrastructure of a magnetotactic bacterium. *J. Bacteriol.* **141**, 1399–1408.

Ball T. B. and Brotherson J. D. (1992) The effect of varying environmental conditions on phytolith morphometries in two species of grass (*Bouteloua curtipendula* and *Panicum virgatum*). *Scan. Micros.* **6**(4), 1163–1181.

Banfield J. F. and Nealson K. H. (eds.) (1997) *Geomicrobiology: Interactions between Microbes and Minerals*, Rev. Mineral. 35. Mineralogical Society of America, Washington, DC.

Barabe D. and Lacroix C. (2001) Aspects of floral development in *Philodendron Grandifolium* and *Philodendron megalophyllum* (Araceae). *Int. J. Plant Sci.* **162**(1), 47–57.

Barns S. W. and Nierzwicki-Bauer S. (1997) Microbial diversity in modern subsurface, ocean, surface environments. In *Geomicrobiology: Interactions between Microbes and Minerals*, Rev. Mineral. 35 (eds. J. F. Banfield and K. H. Nealson). Mineralogical Society of America, Washington, DC, pp. 35–80.

Bass-Becking L. G. M. and Kaplan I. R. (1956) The microbiological origin of the sulfur nodules of Lake Eyre. *Trans. Roy. Soc. Austral.* **79**, 52–65.

Bazylinski D. A. (1996) Controlled biomineralization of magnetic minerals by magnetotactic bacteria. *Chem. Geol.* **132**, 191–198.

Bazylinski D. A. and Moskowitz B. M. (1997) Microbial biomineralization of magnetic iron minerals: microbiology, magnetism, and environmental significance. In *Geomicrobiology: Interactions between Microbes and Minerals*, Rev. Mineral. 35 (eds. J. F. Banfiled and K. H. Nealson). Mineralogical Society of America, Washington, DC, pp. 181–223.

Becker G. L., Chen C. H., Greenwalt J. W., and Lehninger A. L. (1974) Calcium phosphate granules in the hepatopandreas of the Blue Crab, *Callinectes sapidus*. *J. Cell. Biol.* **61**, 310–326.

Beedham G. E. and Truemann E. R. (1967) The relationship of the mantle and shell of the Polyplacophora in comparison with that of other mollusks. *J. Zool. London* **151**, 215–231.

Benson S. C., *et al.* (1983) Morphology of the organic matrix of the spicule of the seas urchin larvae. *Exp. Cell Res.* **148**, 249–253.

Benson S. C., *et al.* (1987) Lineage specific gene encoding a major matrix protein of the sea urchin emrbyo spicule: 1. Authentification of its development expression. *Dev. Biol.* **120**, 499–506.

Benz M., Brune A., and Schink B. (1998) Anaerobic and aerobic oxidation of ferrous iron at neutral pH by chemoheterotrophic nitrate-reducing bacteria. *Arch. Microbiol.* **169**, 159–165.

Bernstein R., Kling S. A., and Boltovskoy D. (1999) Acantharia. In *South Atlantic Zooplankton*, Volume 1 (ed. D. Boltovskoy). Backhuys Publishers, Leiden, The Netherlands, pp. 75–147.

Beverlander G. and Nakahara H. (1960) Development of the skeleton of the sand dollar. In *Calcification in Biological Systems* (ed. R. F. Soggnaes). Am. Assoc. Adv. Science, Washington DC, pp. 41–56.

Bigham J. M., Schwertmann U., and Carlson L. (1992) Mineralogy of precipitate formed by the biogeochemical oxidation of Fe(II) in mine drainage. In *Biomineralization Processes of Iron and Manganese, Modern and Ancient Environments*, suppl. 21 (eds. H. C. W. Skinner and R. W. Fitzpatrick). Catena-Verlag, Cremlingen-Destedt, Germany, pp. 219–232.

Blackwell J. and Weih M. A. (1980) Structure of chitin-protein complexes: ovipositor of the ichneumon fly, *Megarhyssa. J. Mol. Biol.* **137**, 49–60.

Blakemore R. P. and Blakemore N. A. (1990) Magnetotactic magnetogens. In *Iron Biominerals* (eds. R. B. Frankel and R. P. Blakemore). Plenum, New York, pp. 51–67.

Blinnikov M., Busacca A., and Whitlock C. (2002) Reconstruction of the Late Pleistocene grassland of the Columbia Basin, Washington, USA, based on phytolith records in loess. *Palaeogeogr. Palaeoclimatol. Palaeoecol.* **177**(1–2), 77–101.

Boggild O. B. (1930) The shell structure of the mollusks. K. Dan. *Vidensk. Selsk. Skr. Naturvidensk. Math Afd.* **9**, 233–326.

Bonucci E. (1967) Fine structure of early cartilage calcification. *J. Ultrastruct. Res.* **20**, 33–50.

Borowitzka M. A. (1982) Morphological and cytological aspects of algal calcification. *Int. Rev. Cytol.* **74**, 127–162.

Bottazzi E. M., Schreiber B., and Bowen V. Y. (1971) Acantharia in the Atlantic Ocean, their abundance and preservation. *Limnol. Oceanogr.* **16**, 677–684.

Boyd W. E., Lentfer C. J., and Torrence R. (1998) Phytolith analysis for a wet tropics environment: methodological issues and implications for the archaeology of Garua Island, West New Britain, Papua New Guinea. *Palynology* **22**, 213–228.

Brack-Hanes S. D. and Greco A. M. (1988) Biomineralization in *Thalassia Testudimun liliopsida hydrocharitaceae* and an Eocene seagrass. *Trans. Am. Microsc. Soc.* **107**(3), 286–292.

Brass G. W. (1980) Trace elements in acantharia skeletons. *Limnol. Oceanogr.* **25**, 146–149.

Brass G. W. and Turekian K. K. (1974) Strontium distribution in GEOSECS oceanic profiles. *Earth Planet. Sci. Lett.* **23**, 141–148.

Brock T. D. and Madigan M. T. (1988) *Biology of Microorganisms*, 2nd edn. Prentice-Hall, Engelwood Cliffs, NJ.

Broecker W. (1971) A kinetic model for the chemical composition of sea water. *Quat. Res.* **1**, 188–207.

Broecker W. (1974) *Chemical Oceanography*. Harcourt Brace Jovanovich, New York.

Brown C. J., *et al.* (2002) A sampling and analytical methodology for dental trace element analysis. *Analyst* **127**, 319–323.

Bush M. B., Piperno D. R., and Colinvaux P. A. (1989) A 6,000 year history of an Amazonian maize cultivation. *Nature* **340**(6231), 303–305.

Butlin K. R. and Postgate J. R. (1952) The microbial formation of sulfur in the Cyrenaikan Lakes. In *Biology of Deserts* (ed. J. Cloudsley-Thompson). Institute of Biology, London, pp. 112–122.

Butschli O. (1906) Ueber die chemische Natur der skeletsubstanz der Acantharier. *Zool. Anz.* **30**, 784.

Campana S. E. (1999) Chemistry and composition of fish otoliths: pathways, mechanisms and applications. *Mar. Ecol. Prog. Ser.* **188**, 263–297.

Canti M. G. (1997) An investigation of microscopic calcareous spherulites from herbivore dungs. *J. Archaeol. Sci.* **24**(3), 219–231.

Carapoli E. and Scarpa A. (1982) Transport ATPases. *Ann. NY Acad. Sci.* **402**, 1–604.

Caron D. A. and Swanberg N. R. (1990) The ecology of planktonic sarcodines. In *Aquatic Science*. CRC Press, Boca Raton, vol. 3, pp. 147–180.

Carpenter S. J. and Lohmann K. C. (1992) Sr/Mg ratios of modern marine calcite: empirical indicators of ocean chemistry and precipitation rate. *Geochim. Cosmochim. Acta* **56**, 1837–1849.

Carr N. G. and Whitten B. A. (eds.) (1982) *The Biology of Cyanobacteria*. Blackwell, Oxford.

Carter J. A. (2002) Phytolith analysis and paleoenvironmental reconstruction from Lake Poukawa core, Hawkes Bay, New Zealand. *Global Planet. Change* **33**(3–4), 257–267.

Carter J. G. (1980) Guide to bivalvue shell microstructures. In *Skeletal Growth in Aquatic Organisms* (eds. D. C. Rhoads and R. A. Lutz). Plenum, New York, pp. 243–627.

Carter J. G. and Clark G. R. (1985) Classification and phylogenetic significance of molluscan shell structure. In *Mollusks, Notes for a Short Course* (ed. T. W. Broadhead). University of Tennessee, Dept. of Geol. Sci. Studies in Geology, Knoxville, TN, pp. 50–71.

Cavalho M. L., *et al.* (2001) Human teeth elemental profiles measured by synchrotron x-ray fluorescence: dietary habits and environmental influence. *X-Ray Spectrometry* **30**, 190–193.

Cerling T. E., Wang Y., and Quade J. (1993) Expansion of C4 ecosystems as an indicator of global ecological change in the late Miocene. *Nature* **361**, 344–345.

Ceruti P. O., Fey M., and Pooley J. (2003) Soil nutrient deficiencies in an area of endemic osteoarthritis (Mseleni joint disease) and dwarfism in Maputoland, South Africa. In *Geology and Health: Closing the Gap* (eds. H. C. W. Skinner and A. R. Berger). Oxford University Press, New York, pp. 151–154.

Chang L. L. Y., *et al.* (1996) *Rock Forming Minerals. Non Silicates: Carbonates, Phosphates, Halides*. Longmans, UK, vol. 5B.

Chave K. E. (1952) A solid solution between calcite and dolomite. *J. Geol.* **60**, 190–192.

Chave K. E. (1954) Aspects of the biogeochemistry of magnesium: 1. *J. Geol.* **62**, 266–283.

Chen C. H., Greenaway J. W., and Lehninger A. L. (1974) Biochemical and ultrastructural aspects of Ca^{2+} by mitochondria and the hepatopancreas of the blue crab, *Callinectes sapidus*. *J. Cell Biol.* **61**, 301–305.

Clark G. A., II (1980) Techniques for observing the original matrix of mollusc shells. In *Skeletal Growth of Aquatic Organisms* (eds. D. C. Rhoads and R. A. Lutz). Plenum, New York, pp. 607–612.

Clarke F. W. and Wheeler W. C. (1922) The inorganic constituents of marine invertebrates. *US Geol. Surv. Prof. Pap.* **124** (Washington, DC).

Clementz M. T. and Koch P. L. (2001) Differentiating aquatic mammal habitat and foraging ecology with stable isotopes in tooth enamel. *Oecologia* **129**(3), 461–472.

Cocker K. M., Evans D. E., and Hodson M. J. (1998) The amelioration of aluminium toxicity by silicon in higher plants: solution chemistry or an in planta mechanism? *Physiol. Planta.* **104**, 608–614.

Constantz B. R. and Mielke A. (1988) Calcite centers of calcification in *Mussa angulosa* (scleractinia). In *Origin, Evolution and Modern Aspects of Biomineralization in Plants and Animals* (ed. R. E. Crick). Plenum, New York, pp. 201–208.

Cowan M. R., *et al.* (1997) Growth and biomineralization of *Celtis occidentalis* (Ulmaceae) pericarps. *Am. Midland Natural.* **137**, 266–273.

Crawford R. M. and Schmid A. M. (1986) Ultrastructure of silica deposition in diatoms. In *Biomineralization in Lower Plants and Animals* (eds. B. S. C. Leadbeater and R. Riding). Clarendon Press, Oxford, UK, pp. 291–314.

Crenshaw M. A. (1980) Mechanisms of shell formation and dissolution. In *Skeletal Growth of Aquatic Organisms* (eds. D. C. Rhoads and R. A. Lutz). Plenum, New York, pp. 115–132.

Cummings L. S. (1992) Illustrated phytoliths from assorted food plants. In *Phytolith Systematics*, Emerging Issues (eds. G. J. Rapp and S. C. Mulholland). Plenum, pp. 175–192.

Dale J. (1998) *Molecular Genetics of Bacteria*, 3rd edn. Wiley, Chichester, UK.

Davies T., Crenshaw A., and Healthfield M. (1972) The effect of temperature on the chemistry and structure of echinoid spine regeneration. *J. Paleontol.* **46**, 833–874.

Deftos L. J., Roos B. A., and Oates E. L. (1999) Calcitonin. In *Primer on the Metabolic Bone Diseases and Disorders of Mineral Metabolism*, 4th edn. (ed. M. J. Favus). Lippincott Williams and Wilkins, Philadelphia, pp. 99–104.

Degens E. T. (1976) Molecular mechanisms of calcium phosphate and silica deposition in the living cell. *Top. Curr. Chem.* **64**, 1–112.

Degens E. T., Carey F. G., and Spencer D. W. (1967) Amino-acids and amino-sugars in calcified tissues of portunid crabs. *Nature (London)* **216**, 601–603.

Degens E. T., Kasmierczak J., and Ittekkot U. (1985) Cellular responses to $Ca2^{+}$ stress and its geological implications. *Palaeontologica* **30**, 115–135.

De Jong E. W., *et al.* (1987) X-ray diffraction and ^{29}Si magic angle spinning NMR of opals: incoherent long- and short-range order in opal CT. *Am. Min.* **72**, 1195–1203.

Dettman D. L., Reische A. K., and Lohmann K. C. (1999) Controls on the stable isotope composition of seasonal growth bands in aragonitic fresh-water bivalves (unionidae). *Geochim. Cosmochim. Acta* **63**(7–8), 1049–1057.

Deutsch D., *et al.* (1991) Sequencing bonvine enamelin (tuftelin): a novel acidic enamel protein. *J. Bio. Chem.* **266**, 16021–16028.

De Vrind-de Jong E. W, *et al.* (1986) Calcification in the coccolithophorids *Emiliania huxleyi* and *Pleurochrysis caterae*: II. Biochemical aspects. In *Biomineralization in Lower Plants and Animals* (eds. B. S. C. Leadbeater and R. Riding). Clarendon Press, Oxford, pp. 205–217.

De Vrind-de Jong E. W. and De Vrind J. P. M. (1997) Algal deposition of carbonates and silicates. In *Geomicrobiology: Interactions between Microbes and Minerals*, Rev. Mineral. 35 (eds. J. F. Banfield and K. H. Nealson). Mineralogical Society of America, Washington, DC, pp. 267–307.

Doebley J. F. and Iltis H. H. (1980) Taxonomy of Zea (Gramineae): I. A subspecific classification with key to taxa. *Am. J. Bot.* **67**, 982–993.

Dong J., Gu T. T., Simmons D., and MacDougall M. (2000) Enamelin maps to human chromosome 4q21 within the autosomal dominant amelogenesis imperfecta locus. *Eur. J. Oral. Sci.* **108**, 353–358.

Doterton A. A. and Doderer A. (1981) A hormone dependent calcium-binding protein in the mantle edge of the freshwater snail *Lymnae stagnalis*. *Calc. Tiss. Int.* **33**, 505–508.

Douglas and Douglas (2000) Environmental scanning electron microscopy studies of colloidal sulfur deposition in a natural microbial community from a cold sulfide spring near Ancaster, Ontario, Canada. *Geomicrobiol. J.* **17**, 275–289.

Driessens F. C. M. and Verbeek R. M. H. (1990) *Biominerals*. CRC Press, Boca Raton, FL.

Duke J. A. (1992) *Handbook of Phytochemical Constituents of GRAS Herbs and other Economic Plants*. CRC Press, Boca Raton.

Easton R. M. (1997) Lichen-rock-mineral interactions:an overview. In *Biological-mineralogical Interactions*, Short Course, 25. Canadian Mineralogical Association, Ottawa.

Ebert T. A. (1967) Growth and repair of spines in the sea urchin *Strongylcentrotus purpuratus* (Stimpson). *Biol. Bull.* **133**, 141–147.

Ehrlich H. L. (1996) How microbes influence mineral growth and dissolution. *Chem. Geol.* **132**, 5–9.

Ehrlich H. L. (2002) *Geomicrobiology*, 4th edn. Dekker, New York.

Elliott J. (2002) Calcium phosphate biominerals. In *Phosphates: Geochemical, Geobiological and Materials Importance*, Reviews in Mineralogy and Geochemistry (eds. M. J. Kohn, J. Hughes, and J. Rakovan). Mineralogical Society of America, Washington, DC, vol. 48, pp. 427–454.

Ellis B. F. and Messina A. (1940) *The Catalogue of Foraminifera.* American Museum of Natural History, New York.

Enlow D. H. (1963) *Principles of Bone Remodeling.* Charles C. Thomas, Springfield, Ill.

Enlow D. H. and Brown S. O. (1956–1958) A comparative histological study of fossil and recent bone tissues: Part 1. *Texas J. Sci.* **VII**(4), 405–430, (1956); Part II. *Texas J. Sci.* **IX**(2), 186–214, (1957); Part III *Texas J. Sci.* **X**(2), 187–230, (1958).

Fincham A. G. and Simmer J. F. (1997) Amelogenin proteins of developing dental enamel. In *Dental Enamel.* CIBA Symposium #205. Wiley, Chichester, pp. 118–134.

Finks R. M. (1970) The evolution and ecologic history of sponges during Palaeozoic times. In *The Biology of the Portifera* (ed. W. G. Fry). Symp. Zool. Soc. London Academic Press, London, vol. 25.

Finley D. S. (1999) Patterns of calcium oxalate crystals in young tropical leaves: a possible role as an anti-herbivory defense. *Rev. Biol. Trop.* **47**(1–2), 27–31.

Flint M. E. (1984) The structure of magnetite: two annealed natural magnetites $Fe_{3.005}O_4$ and $Fe_{2.96}Mg_{0.04}O_4$. *Acta Cryst.* **C40**, 1491–1493.

Folk R. I. (1974) The natural history of crystalline calcium carbonate: effect of magnesium content and salinity. *J. Sediment. Petrol.* **44**, 40–53.

Frank A. (2003) Molybdenosis in Swedish moose. In *Geology and Health, Closing the Gap* (eds. H. C. W. Skinner and A. R. Berger). Oxford University Press, New York, pp. 73–76.

Frankel R. B., *et al.* (1997) Magneto-aerotaxis in marine coccoid bacteria. *Biophys. J.* **73**, 994–1000.

Fredlund G. G. and Tieszen L. L. (1997a) Calibrating grass phytolith assemblages in climatic terms: application to late Pleistocene assemblages from Kansas and Nebraska. *Palaeogeogr. Palaeoclimatol. Palaeoecol.* **136**(1–4), 199–211.

Fredlund G. G. and Tieszen L. L. (1997b) Phytolith and carbon isotope evidence for Late Quaternary vegetation and climate change in the southern Black Hills, South Dakota. *Quat. Res.* **47**(2), 206–217.

Freke A. M. and Tate D. (1961) The formation of magnetic iron sulfide by bacterial reduction of iron solutions. *J. Biochem. Micicrobiol. Technol. Eng.* **3**, 29–39.

Fricke H. C. and Rogers R. R. (2000) Multiple taxon-multiple locality approach to providing oxygen isotope evidence for warm-blooded theropod dinosaurs. *Geology* **28**(9), 799–802.

Fricke H. C., Clyde W. C., and O'Neil J. R. (1998) Intra-tooth variations in delta O-18 (PO4) of mammalian tooth enamel as a record of seasonal variations in continental climate variables. *Geochim. Cosmochim. Acta* **62**(11), 1839–1850.

Friedlaender G. E., *et al.* (1999) Osteogenic protein-1 (bone morphogenic protein-7) in the treatment of tibial nonunions. *J. Bone Joint Surg.* **83-A**, S1-151–S1-158.

Frondel C. (1960). *The System of Mneralogy of James Dwight Dana.* Wiley, New York, vol. 3.

Gaines R. V., Skinner H. C. W., Foord E. E., Mason B., and Rosensweig A. (1997) *Dana's New Mineralogy.* Wiley, New York.

Garrity D. P., Vidal E. T., and O'Toole J. C. (1984) Genotypic variation in the thickness of silica deposition on flowering rice spikelets. *Ann. Bot.* **54**, 413–421.

Gast R. J. and Caron D. A. (1996) Molecular phylogeny of symbiotic dinoflagellates from plankton formainifera and radiolaria. *Mol. Biol. Evol.* **13**(9), 1192–1197.

Geis J. W. (1978) Biogenic opal in three species of Gramineae. *Ann. Bot.* **42**, 1119–1129.

Ghiorse W. C. and Balkwell D. L. (1983) Enumeration and morphological characterization of bacteria indigenous to subsurface environments. *Dev. Ind. Microbiol.* **24**, 213–224.

Ghiorse W. C. and Wilson J. T. (1988) Microbial ecology of the terrestrial subsurface. *Adv. Appl. Microbiol.* **33**, 107–171.

Gillanders C., *et al.* (2001) Trace elements in otoliths of the two banded bream from a coastal reion in the south-west Mediterranean: are there differences among locations? *J. Fish Biol.* **59**, 350–363.

Glimcher M. J. (1959) Molecular biology of mineralized tissues with particular reference to bone. *Rev. Mod. Phys.* **31**, 359–393.

Glimcher M. J. (1960) Specificity of the molecular structure of organic matricies in mineralization. In *Calcification in Biological Systems* (ed. R. F. Sognaes). AAAS, Washington, DC, pp. 421–487.

Glimcher M. J. (1984) Recent studies of the mineral phase in bone and its possible linkage to the organic matrix by protein-bound phosphate bonds. *Proc. Trans. Roy. Soc. London B.* **304**, 479–508.

Golubic S. (1973) The relationship between blue-green algae and carbonate deposits. In *The Biology of Blue Green Algae* (eds. N. G. Carr and B. A. Whitton). Blackwell, London, pp. 434–472.

Golubic S. (1983) Stromatolites, fossil and recent, a case history. In *Biomineralization and Biological Metal Accumulation* (eds. P. Westbroek and E. W. de Jong). D. Reidel, Dordrecht, Germany, pp. 313–326.

Goodwin P. W. and Mercer E. T. (1982) *Introduction to Plant Biochemistry.* Pergamon.

Goose D. H. and Appleton J. (1982) *Human Dentofacial Growth.* Pergamon, Oxford, UK.

Gould J. L., Kirschvink J. L., and Defeffeyes K. S. (1978) Bees have magnetic remanence. *Science* **201**, 1026–1028.

Graetsch H. (1994) Microcrystalline silica minerals. In *Silica Physical Behavior, Geochemistry and Materials Applications*, Rev. Mineral. 29 (eds. P. J. Heaney, C. R. Prewitt, and G. V. Gibbs). Mineralogical Society of America, Washington, DC, pp. 209–232.

Green H. W., Lipps J. H., and Showers W. J. (1980) Test structure of fusulinid Foraminifera. *Nature (London)* **283**, 853–855.

Green J. C. (1986) Biomineralization in the algal class Pymnesiophyceae. In *Biomineralization in Lower Plants and Animals* (eds. B. S. C. Leadbeater and R. Riding). Clarendon Press, Oxford, pp. 173–188.

Green J. C. and Course P. A. (1983) Extracellular calcification in *Chryosotila lamellose* (Prymnesiophyceae). *Br. Phycol. J.* **18**, 367–382.

Green J. C., Hibbard D. J., and Pienaar R. N. (1982) The taxonomy of Pyrmnesium (Pyrmesiophycea) including a description of a new cosmopolitan species *P. patellifera* sp. and further observations on *P. parvum* M. Carter. *Br. Phycol. J.* **17**, 363–382.

Greenaway P. (1985) Calcium balance and moulting in the crustacea. *Biol. Rev.* **50**, 425–454.

Grondowitz M. J. and Broce A. B. (1983) Calcium storage in face fly (Dipter: Muscidae) larvae for puparium formation. *Ann. Entomol. Sci. Am.* **76**, 418–424.

Grondowitz M. J., Broce A. B., and Kramer K. J. (1987) Morphology and biochemical composition of mineralized granules form the Malphigian tubules of *Musca autumnalis* De Geer larvae (*Diptera muscidae*). *Insect Biochem.* **17**, 335–345.

Grossman E. L. (1987) Stable isotopes in modern benthic foraminifera—a study of vital effect. *J. Foraminiferal Res.* **17**(1), 48–61.

Gulbranson R. A. (1966) Chemical composition of phosphorite in the Phosphoria Formation. *Geochim. Cosmochim. Acta* **30**, 769–778.

Gustafson G. and Gustafson A. G. (1967) Microanatomy and histochemistry of enamel. In *Structural and Chemical Organization of Teeth*, Volume II (ed. A. E. W. Miles). Academic Press, New York, pp. 76–134.

Haas W. (1976) Observations on the shell and mantle of the Placophora. In *The Mechanisms of Mineralization in the Invertebrates and Plants* (eds. N. Watabe and K. M. Wilbur). University of S. Carolina Press, Columbia, SC, pp. 389–402.

Haeckel E. (1899–1904a) Diatoms From *Kunstformen der Natur* originally published by Haeckel and reproduced in *Art forms in nature* (1974), Dover publications, New York, and in *Silicon and Siliceous Structures in Biological Systems* (eds. T. L. Simpson and B. E. Volcani 1978). Figs. 1–2, 5p.

Haeckel E. (1899–1904b) Radiolarians. From *Kunstformen der Natur* based on Report on radiolaria collected by H. M. S. Challenger during the years 1873–1876 by (eds. C. W. Thompson and J. Murray), vol. 18, pp. 1–1760, HMSO, London. Plate reproduced from *Art Forms in Nature* (1974) Dover Publications, New York and in *Silicon and Siliceous Structures in Biological Systems* (eds. T. L. Simpson and B. E. Volcani 1978).

Halstead L. H. (1969) *The Pattern of Vertebrate Evolution*. Oliver and Boyd, London.

Halstead L. H. (1974) *Vertebrate Hard Tissues*. Wykeham Publications, Science Series, London.

Hamilton S. E., *et al.* (1982) γ-carboxyglutamic acid in invertebrates: its identification in hermatypic corals. *Biochem. Biophys. Res. Commun.* **108**, 610–613.

Hanson P. J. and Zdanowicz V. S. (1999) Elemental composition of otoliths from Atlantic croaker along an estuarine pollution gradient. *J. Fish Biol.* **54**, 656–668.

Hanzlik M. M., *et al.* (1996) Electron microscopy and ^{57}Fe Mossbauer spectra of 10 nm particles, intermediate in composition between Fe_3O_4 and X Fe_2O_3 produced by bacteria. *Geophys. Res. Lett.* **23**, 479–482.

Hare P. E. and Abelson P. H. (1965) Amino acid composition of some calcified proteins. *Geophys. Lab. Ann. Report 1964–1965*. Carnegie Institute of Washington, pp. 223–235.

Harper H. E. and Knoll A. H. (1975) Silica, diatoms and Cenazoic radiolarian evolution. *Geology* **3**, 175–177.

Hartman W. D. (1981) Form and distribution of silica in sponges. In *Silicon and Siliceous Structures in Biological Systems* (eds. T. L. Simpson and B. E. Volcani). Springer, New York, pp. 294–453.

Hartman W. D. and Goreau T. F. (1970) Jamaican coralline sponges: their morphology, ecology and fossil relatives. In *The Biology of the Porifera*. Symp. 25 (ed. W. G. Fry). Zool. Soc. London, Academic Press, London, pp. 205–243.

Hedges R. E. M., *et al.* (1998) Methodological issues in the ^{14}C dating of rock paintings. *Radiocarbon* **40**(1), 35–44.

Hegdahl T. L., Gustavsen F., and Silness J. (1977) The structure and mineralization of the carapace of the crab (*Cancer pagurus L.*): 2. The exocuticle. *Zool. Scripta* **6**, 215–220.

Helmcke J.-G. (1968) Ultrastructure of enamel. In *Structural and Chemical Organization of Teeth* (ed. A. E. W. Miles). Academic Press, New York, vol. II, pp. 135–164.

Hiltz P. and Pobeguin T. (1949) On the constitution of cystoliths of Ficus elastica C. R. *Hebd. Seances. Acad. Sci.* **228**, 1049–1051.

Hodson M. J. and Sangster A. G. (1999) Aluminum/silicon interactions in conifers. *J. Inorg. Biochem.* **76**, 89–98.

Hodson M. J. and Sangster A. G. (2000) Aluminum localization in conifers growing on highly acidic soils in Ontario, Canada. In *International Symposium on Impact of Potential Tolerance of Plants on the Increased Productivity Under Aluminum Stress*. Research Institute for Bioresources, Okayama University, pp. 103–106.

Hodson M. J. and Sangster A. G. (2002) X-ray microanalytical studies of mineral localization in the needles of white pine (*Pinus strobus* L.). *Ann. Bot.* **89**, 367–374.

Hodson M. J., Smith R. J., Van Blaaderen A., Crafton T., and O'Neill C. H. (1994) Detecting plant silica fibres in animal tissue by confocal fluorescence microscopy. *Ann. Occupat. Hyg.* **38**, 149–160.

Hollinger J. O. (1997) What's new in bone biology? *J. Histotechnol.* **20**, 235–240.

Hopkin S. P. and Nott J. A. (1979) Some observations on concentrically structured intracellular granules in the hepatopancreas of the shore crab *Carinus maenas* (L.). *J. Mar. Biol. Assoc. UK* **59**, 867–877.

Horner H. T., Kausch A. P., and Wagner B. L. (2000) Ascorbic acid: a precursor of oxalate in crystal idioblasts of Yucca torreyi in liquid root culture. *Int. J. Plant Sci.* **161**(6), 861–868.

Howard B., Mitchell P. C. H., Ritchie A., Simkiss K., and Taylor M. (1981) The composition of intracellular granules from the metal-accumulating cells of the common garden snail (*Helix aspersa*). *Biochem. J.* **194**, 507–511.

Huner J. V., Kowalczuk J. G., and Avault J. W. (1978) Postmolt calcification in subadult red swamp crayfish, *Procambarus clarkiii* (Girard) (Decapoda, Cambridae). *Crustaceana* **34**, 275–280.

Hunter G. K. (1987) An ion-exchage mechanism of cartilage calcification. *Conn. Tissue Res.* **16**, 111–120.

Huss V., Holweg C., Seidel B., Reich V. R., and Kessler E. (1994) There is an ecological basis for host symbiont specificity in Chlorella Hydra symbiosis. *Endocytobiosis Cell Res.* **10**, 35–46.

Ivanov M. V. (1968) Microbial processes in the formation of sulfur deposits. Israel Program for Scientific Translations, Washington DC, US Dept. of Agricultrue and Natl. Sci. Foundation.

Jahren A. H. (1996) The Stable Isotope Composition of the Hackberry (*Celtis*) and its use as a Paleoclimate Indicator. PhD, University of California at Berkeley.

Jahren A. H., *et al.* (1997) Distinguishing the nature of tool residue: chemical and morphological analysis of bamboo and bone tool residues. *J. Archaeol. Sci.* **24**(3), 245–250.

Jahren A. H., Gabel M. L., and Amundson R. (1998) Biomineralization in seeds: developmental trends in isotopic signatures of hackberry. *Palaeogeogr. Palaeoclimatol. Palaeoecol.* **138**, 259–269.

Jahren A. H., *et al.* (2001) Paleoclimatic reconstruction using the correlation in $\delta^{18}O$ of hackberry carbonate and environmental water, North America. *Quat. Res.* **56**(2), 252–263.

Jannasch H. W. and Mottl M. J. (1985) Geomicrobiology of the deep-sea hydrothermal vents. *Science* **229**, 717–725.

Janssen P. H., *et al.* (1996) Disproportionation of inorganic sulfur compounds by the sulfate reducing bacterium *Desulfocapsa thiozymogenes*, gen. nov., spec. nov. *Arch. Microbiol.* **166**, 184–192.

Johnson I. S. (1980) The ultrastructure of skelatogenesis in hermatypic corals. *Int. Rev. Cytol.* **67**, 171–214.

Jones J. B. and Segnit E. R. (1971) Nature of opal: I. Nomenclature and constituent phases. *J. Geol. Soc. Austral.* **18**, 57–68.

Jones L. H. P., Milne A. A., and Sanders J. V. (1966) Tabashir: an opal of plant origin. *Science* **151**, 464–466.

Jorgansen B. B., Isaksen M. F., and Jannasch H. W. (1992) Bacterial sulfate reduction above 100°C. in deep sea hydrothermal cent sediments. *Science* **258**, 1756–1757.

Kastner M. (1981) Authigenic silicates in deep sea sediments: formation and diagenesis. In *The sea* (ed. C. Emiliani). Wiley, New York, pp. 915–980.

Kaufman P. B., *et al.* (1981) Silica shoots of higher plants. In *Silicon and Siliceous Structures in Biological Systems* (eds. T. L. Simpson and B. E. Volcani). Springer, New York, pp. 409–449.

Kayser C. H. and Heusner A. (1964) Etude comparative du metabolisme energetique dans la serie animale. *J. Physiol. Paris* **56**, 489.

Kealhofer L. and Penny D. (1998) A combined pollen and phytolith record for fourteen thousand years of vegetation change in northeastern Thailand. *Rev. Palaeobot. Palynol.* **103**(1–2), 83–93.

Kelly E. F., *et al.* (1998) Stable isotope composition of soil organic matter and phytoliths as paleoenvironmental indicators. *Geoderma* **82**(1–3), 59–81.

King K. and Hare P. E. (1972) Amino acid composition of the test as a taxonomic character for living and fossil plankton foraminifera. *Micropaleont.* **18**, 285–293.

Kirschvink J. L. and Lowenstam H. A. (1979) Mineralization and magnetism of chiton teeth: paleomagnetic, sedimentologic and biologic implications of organic magnetite. *Earth Planet. Sci. Lett.* **44**, 193–204.

Kitajima T. and Okazaki K. (1980) Spicule formation *in vitro* by the descendents of precocious micromere formed at the 8-cell stage of sea urchin embryo. *Dev. Growth Differ.* **22**, 266–279.

Klein C. and Hurlbut C. (2000) *Manual of Mineralogy*. Wiley, New York.

Koch P. L., Fogel M. F., and Tuross N. (1994) Tracing the diets of fossil animals using stable isotopes. In *Stable Isotopes in Ecology and Environmental Science* (eds. K. Lajtha and R. H. Michener). Blackwell, Oxford, pp. 63–92.

Kodama H., *et al.* (1992) Platy quartz phytoliths found in the fossil forest deposits, Axel Heiberg Island, Northwest Territories, Canada. *Zeitschrift für Pflanzenareale* **155**, 401–406.

Kohn M. J. and Cerling T. E. (2002) Stable isotope compositions of biological apatite. In *Phosphates: Geochemical, Geobiological, and Materials Importance*, Rev. Mineral. Geochem. 48 (eds. M. J. Kohn, J. Rakovan, and J. M. Hughes). Mineralogical Society of America, Washington, DC, pp. 455–488.

Krumbein W. E. (1986) Biotransfer of minerals by microbes and microbial mats. In *Biomineralization in Lower Plants and Animals* (eds. B. S. C. Leadbeater and R. Riding), Clarendon Press, Oxford, UK, pp. 55–72.

Kuhn K. (1987) The classical collagens. In *Structure and Function of Collagen Types* (eds. R. Mayne and R. E. Burgess). Academic Press, New York, pp. 1–42.

Kulze A. and Greaves M. (1988) Contact urticaria caused by stinging nettles. *Br. J. Dermatol.* **119**, 169–270.

Larsen C. S., *et al.* (1992) Carbon and nitrogen stable isotopic signatures of human dietary change in the Georgia Bight. *Am. J. Phys. Anthropol.* **89**(2), 197–214.

Lee D. and Carpenter S. J. (2001) Isotopic disequilibrium in marine calcareous algae. *Chem. Geol.* **172**, 307–329.

Leeman W. (1970) Fluorosis in cattle. In *Fluoride in Medicine* (ed. T. L. Vischer). Hans Huber Publishers, Bern, Switzerland, pp. 130–135.

Leethorp J. A., vanderMerwee N. J., and Brain C. K. (1994) Diet of australopithecus-robustus at Swartkrans from stable carbon isotopic analysis. *J. Human Evol.* **27**(4), 361–372.

Lehninger A. L. (1983) The possible role of mitochondria and phosphocitrate in biological calcification. In *Biomineralization and Biological Metal Accumulation* (eds. P. Westbroek and E. W. de Jong). D. Reidel, Dordrecht, Holland, pp. 107–121.

Levin I. and Ott E. (1933) X-ray study of opals, silica glass and silica gel. *Z. Kristallogr.* **85**, 305–318.

Li C. W. and Volcani B. E. (1984) Aspects of silicification in wall morphogenesis of diatoms. *Phil. Trans. Roy. Soc. London B* **304**, 510–528.

Lindquist N. (2002) Chemical defence of early life stages of benthic marine invertebrates. *J. Chem. Ecol.* **28**, 1987–2000.

Locker S. (1974) Revision der silicoflagellaten sus der Mikrogeologischen Sammlung von C. G. Ehrenberg. *Ecologae Geol. Helv.* **67**, 631–646.

London J. and Rittenberg S. C. (1967) Path of sulfur in sulfide and thiosulfate oxidation by thiobacillus. *Arch. Microbiol.* **59**, 218–225.

Loven S. (1982) Echinologica. *Svenska Vetensk-Acad. Handl.* **18**, 1–73.

Lovley D. R. (1987) Organic matter mineralization with the reduction of ferric iron: a review. *Geomicrobiology* **5**, 375–399.

Lowenstam H. (1964) Coexisting calcites and aragonites from skeletal carbonates of marine organisms and their strontium and magnesium contents. In *Recent Researches in the Fields of Hydrosphere, Atmosphere, and Nuclear Geochemistry* (eds. Y. Miyake and T. Koyama). Maruzen, Tokyo, pp. 373–404.

Lowenstam H. A. (1972) Biogeochemistry of hard tissues: their depth and possible pressure relations. In *Barobiology and the Experimental Biology of the Deep Sea* (ed. R. W. Brauer). University of N. Carolina, Chapel Hill, NC, pp. 19–32.

Lowenstam H. A. and Kirschvink J. L. (1985) Iron biomineralization: a geobioloical perspective. In *Magnetite Biomineralization and Magnetoreception in Organisms, A New Biomagnetism* (eds. J. L. Kirschvink, D. S. Jones, and B. J. MacFadden). Plenum, New York, pp. 3–15.

Lowenstam H. L. and Weiner S. (1989) *On Biomineralization*. Oxford University Press, New York.

Lutz R. A. and Rhoads D. C. (1980) Growth patterns within molluscan shells. In *Skeletal Growth in Aquatic Organisms* (eds. D. C. Rhoads and R. A. Lutz). Plenum, New York, pp. 203–254.

MacFadden B. J. (2000) Cenozoic mammalian herbivores from the Americas: reconstructing ancient diets and terrestrial communities. *Ann. Rev. Ecol. Syst.* **31**, 33–59.

MacFadden B. J., Solounias N., and Cerling T. E. (1999) Ancient diets, ecology, and extinction of 5-million-year-old horses from Florida. *Science* **283**, 824–827.

Macintyre I. G., *et al.* (2000) Possible vestige of early phosphatic biomineralization in gorgonian octocorals (Coelenterata). *Geology* **28**, 453–458.

Mandernach, *et al.* (1999) Oxygen isotope studies of magnetite produced by magnetotactic bacteria. *Science* **285**, 1892–1896.

Mann H. *et al.* (1992) Microbial accumulation of ion and manganese in different aquatic environments, an elecgtron optical study. In *Biomineralization of Iron and Manganese, Modern and Ancient Environments* (eds. H. C. W. Skinner and R. W. Fitzpatrick). Catena-Verlag, Cremlingen-Destadt, Germany, suppl. 21, pp. 115–132.

Mann S. (1985) Structure, morphology, and crystal growth of bacterial magnetite. In *Magnetite Biomineralization and Magnetoreception in Organisms* (eds. J. L. Kirschvink, D. S. Jones, and B. J. MacFadden). Plenum, New York, pp. 311–332.

Mann S. (2001) *Biomineralization: Principles and Concepts in Bioinorganic Materials Chemistry*. Oxford University Press, Oxford, UK.

Mann S. and Sparks N. H. C. (1988) Single crystalline nature of coccolith elements of the marine alga *Emiliania huxlei* as determined by electron diffraction and high resolution transmission electron microscopy. *Proc. Roy. Soc. London B* **234**, 441–453.

Mann S., *et al.* (1990) Biomineralization of ferrimagneic griegite and iron pyrite (FeS_2) in a magnetotactic bacterium. *Nature* **343**, 258–260.

Marx J. M. (1986) Species profiles: life histories and environmental requirements of coastal fishes and invertebrates (S. Florida). Fish and Wildlife Serv., US Interior Dept. Vicksburg, Ms., Coastal Ecology Gr., Waterways Exp. Stn., US Army Corps of Engineers, Washington, DC.

Matile P. (1978) Biochemistry and function of vacuoles. *Ann. Rev. Plant Physiol. Plant Mol. Biol.* **29**, 193–213.

Mayne R. and Burgess R. E. (1987) *Structure and Function of Collagen Types*. Academic Press, New York.

McLean F. C. and Urist M. R. (1968) *Bone: Fundamentals of the Physiology of Skeletal Tissue*. University of Chicago Press, Chicago.

McNair J. B. (1932) The interrelation between substances in plants: essential oils and resins, cyanogen and oxalate. *Am. J. Bot.* **19**, 255–272.

Michaels A. F. (1988) Vertical distribution and abundance of Acantharia and their symbionts. *Marine Biol.* **97**, 559–569.

Michaels A. F. (1991) Acantharian abundance and symbiont productivity at the VERTEX seasonal station. *J. Planton Res.* **13**, 399–418.

Miller E. J. (1984) Collagen: the organic matrix of bone. *Phil. Trans. Roy. Soc. London B* **304**, 455–477.

Miller E. J. (1985) Recent information on the chemistry of collagens. In *The Chemistry and Biology of Mineralized Tissues* (ed. W. T. Butler). EBSCO Media, Birmingham, Alabama, pp. 80–93.

Milton D. A. and Cheney S. R. (1998) The effect of otolith storage methods on the concentration of elements detected by laser-ablation ICPMS. *J. Fish Biol.* **53**, 785–794.

Molano F. B. (2001) Herbivory and calcium concentrations affect calcium oxalate crystal formation in leaves of *Sida* (Malvaceae). *Ann. Bot. (London)* **88**(3), 387–391.

Monje P. V. and Baran E. J. (1996) On the formation of weddellite in Chamaecereus silvestrii, a Cactaceae species from Northern Argentina. *Zeitschrift fuer Naturforshung (Section C): J. Biosci.* **51**(5–6), 426–428.

Monje P. V. and Baran E. J. (1997) On the formation of whewellite in the cactaceae species Opuntia microdasys. *Zeitschrift fuer Naturforshung (Section C): J. Biosci.* **52**(3–4), 267–269.

Moore T. C. (1969) Radiolaria: change in skeletal weight and resistance to solution. *Geol. Spec. Am. Bull.* **80**, 2103–2108.

Moss M. L. and Moss-Salentijn L. (1983) Verterate cartilages. In *Cartilage, Structure, Function and Biochemistry* (ed. B. K. Hall). Academic Press, New York, pp. 1–30.

Mulholland S. C. (1990) *Arundo Donax* phytolith assemblages. *The Phytolitharien* **6**, 3–9.

Mundy G. R. (1999) Bone remodeling. In *Primer on the Metabolic Diseases and Disorders of Mineral Metabolism*, 4th edn. (ed. M. J. Favus). Lippincott Williams and Wilkins, Philadelphia, pp. 30–38.

Mustin C., *et al.* (1992) Corrosion and electrochemical oxidation of a pyrite by *Thiobacillus ferrooxidans*. *Appl. Environ. Microbiol.* **58**, 1175–1182.

Muzzarelli R. A. (1977) *Chitin*. Pergamon, New York.

Nesson M. H. (1968) Studies on radula tooth mineralization in the polyplacophora. PhD Thesis, California Institute of Technology, Pasadena, CA.

Nesson M. H. and Lowenstam H. A. (1985) Biomineralization processes of the radula teeth of chitons. In *Magnetite Biomineralization and Magnetoreception in Organisms* (eds. J. L. Kirschvink, D. S. Jones, and MacFadden). Plenum, New York, pp. 333–363.

Nolf D. (1985) *Otolith Piscium*. Handbook of Paleoichthyology (ed. H. P. Schultze). Gustave Fischer Verlag, Stuttgart, Germany, vol. 10.

Nordin B. E. (1971) Clinical significance and pathogenosis of osteoporosis. *Br. Med J.* **1**, 571–576.

Norman J. R and Greenwood P. H. (1975) *A History of Fishes*. Ernst Benn, London, UK.

Nowak W. S. W. (1972) *The Lobster (Homaridae) and the Lobster Fisheries: An Interdisciplinary Bibliography*. Memorial University of Newfoundland, St. John's, Newfoundland, Canada.

Odum H. T. (1951) Notes on the strontium content of seawater, celestite radiolaria and strontianite snail shells. *Science* **114**, 211–213.

Ogden J. A. and Grogan D. P. (1987) Prenatal development and growth of the musculoskeletal system. In *The Scientific Basis of Orthopaedics*, 2nd edn. (eds. J. A. Albright and R. A. Brand). Appleton and Lange, Norwalk, CT, pp. 47–90.

Ogden J. A., Grogan D. P., and Light T. A. (1987) Postnatal development and growth of the musculoskeletal system. In *The Scientific Basis of Orthopaedics*, 2nd edn. (eds. J. A. Albright and R. A. Brand). Appleton and Lange, Norwalk, CT, pp. 91–160.

Ortner D. (1992) Skeletal paleopathology. In *Disease and Demography in the Americas* (eds. J. W Verano and D. H. Uberlaker). Smithsonian Institution Press, Washington, DC, pp. 5–13.

Orvig T. (1967) Phylogeny of tooth tissues: evolution of some calcified tissues in early vertebrates. In *Structural and Chemical Organization of Teeth* (ed. A. E. W. Miles). Academic Press, New York, vol. I, pp. 25–110.

Pearsall D. M. and Piperno D. R. (1993) Current research in phytolith analysis: applications in archaeology and paleoecology. In *MASCA Research Papers in Science and Archaeology*. The University Museum of Archaeology and Anthropology, University of Pennsylvania, Philadelphia, vol. 10, 212pp.

Pearsall D. M. and Trimble M. K. (1984) Identifying past agricultural activity through soil phytolith analysis: a case study from the Hawaiian Islands. *J. Archaeol. Sci.* **11**(2), 119–133.

Pentecost A. (1978) Blue green algae and freshwater carbonate deposits. *Proc. Roy. Soc. London B* **200**, 43–61.

Pentecost A. (1985a) Association of cyanobacteria and tufa deposits:identity, enumeration and nature of the sheath material revealed by histochemistry. *Geomicrobiol. J.* **4**, 285–298.

Pentecost A. (1985b) Photosynthetic plants as intermediary agents between environmental HCO_3^- and carbonate deposition. In *Inorganic Carbon Uptake by Aquatic Photosynthetic Organisms* (eds. W. J. Lucas and J. A. Berry). American Society of Plant Physiology, pp. 459–480.

Pentecost A. and Riding R. (1986) Calcification in cyanobacteria. In *Biomineralization in Lower Plants and Animals* (eds. B. S. C. Leadbeater and R. Riding). Clarendon Press, Oxford UK, pp. 73–90.

Perry C. C., *et al.* (1984) *Proc. Roy. Soc. London B, Biol. Sci.* **222**, 438–445.

Peyer B. (1968) *Comparative Odontology* (translated by Rainer Zangerl). University of Chicago Press, Chicago.

Piperno D. R. (1984) A comparison and differentiation of phytoliths from maize and wild grasses: use of morphological criteria. *Am. Antiq.* **49**(2), 361–383.

Piperno D. R. (1985) Phytolith analysis and tropical paleoecology: production and taxonomic significance of siliceous forms in new world plant domesticates and wild species. *Rev. Palaeobot. Palynol.* **45**, 185–228.

Piperno D. R. (1988) *Phytolith Analysis: An Archaeological and Geological Perspective*. Academic Press, New York.

Piperno D. R. (1991) The status of phytolith analysis in the American tropics. *J. World Prehist.* **5**(2), 155–189.

Piperno D. R. and Becker P. (1996) Vegetational history of a site in the central Amazon Basin derived from phytolith and charcoal records from natural soils. *Quat. Res.* **45**(2), 202–209.

Poole D. F. G. (1967) Phylogeny of tooth tissues: enameloid and enamel in recent vertebrates, with a note on the history of cementum. In *Structural and Chemical Organization of Teeth* (ed. A. E. W. Miles). Academic Press, New York, pp. 111–150.

Popofsky A. W. K. (1940) *Die Acantharia der Plankton-expedition. Ergebnisse der Plankton-expedition der Humboldt-stiftung*. Lipsius and Tischer, Kiel, Germany.

Prenant M. (1927) Les formes mineralogiques du calcaire chez les etres vivants, et le probleme de leur derterminisme. *Bio. Rev.* **2**, 365–393.

Pueschel C. M. (2001) Calcium oxalate crystals in the green alga *Spirogyra hatillensis* (Zygnematales, Chlorophyta). *Int. J. Plant Sci.* **162**(6), 1337–1345.

Pyatt F. B. and Grattan J. P. (2001) Consequences of ancient mining activities on the health of ancient and modern human populations. *J. Publ. Health Med.* **23**, 235–236.

Rabenhorst M. C. and James B. R. (1992) Iron sulfidization in tidal marsh soils. In *Biomineralization Processes of Iron and Manganese Modern and Ancient Environments* (eds. H. C. W. Skinner and R. W. Fitzpatrick). Catena-Verlag, Cremlingen-Dested, Germany, suppl. 21, pp. 218–263.

Rai A. K. (1997) *Cyanobacterial Nitrogen Metabolism and Environmental Biotechnology*. Springer, New York.

Raup D. M. (1959) Crystallography of echinoid calcite. *J. Geol.* **67**, 661–674.

Raup D. M. (1966) The endoskeleton. In *Physiology of Echinodermata* (ed. R. A. Boolootian). Interscience, New York, pp. 379–395.

Raven J. A., Griffiths H., Glidewell S. M., and Preston T. (1982) The mechanism of oxalate biosynthesis in higher plants: investigating with the stable isotopes ^{18}O and ^{13}C. *Proc. Roy. Soc. London B, Biol. Sci.* **216**, 87–101.

Raven J. A., Evert R. F., and Eichorn S. E. (1999) *Biology of the Plants*, 6th edn. Worth Pub. W. H. Freeman, New York.

Reeder R. R. (1983) Crystal chemistry of the rhobohedral carbonates. In *Carbonates: Mineralogy and Chemistry*, Rev. Mineral. (ed. R. R. Reeder). Mineralogical Society of America, Washington, DC, vol. 11, pp. 1–48.

Reid R. T., Live D. T., Faulkner D. J., and Butler A. (1993) A siderophore from a marine bacterium with an exceptional ferric iron affinity constant. *Nature (London)* **366**, 455–458.

Reidel W. R. and Sanfilippo A. (1981) Evolution and diversity of form in radiolaria. In *Silicon and Siliceous Structures in Biological Systems* (eds. T. L. Simpson and B. E. Volcani). Springer, New York, pp. 323–346.

Reitner J. and Keupp H. (1991) (eds.) *Fossil and Recent Sponges*. Springer, Berlin.

Reshertnjak V. V. (1981) *Acantharia Fauna SSSR* 123-1-210. Akad Nauk. SSSR Zool. Inst, Nauka, Leningrad.

Resig J. M., Lowenstam H. A., Echols R. J., and Weiner S. (1980) An extant opaline foraminifera: test ultrastructure, mineralogy and taxonomy. *Cushman Foundation Spec. Publ.* **19**, 205–214.

Rhoads D. C. and Lutz R. A. (1980) *Skeletal Growth of Aquatic Organisms*. Plenum, New York.

Rickard D. T. (1969) The microbiological formation of iron sulfides. *Stockholm Contrib. Geol.* **20**, 50–66.

Riding R. (1977) Skeletal stromatolites. In *Fossil Alga Recent Results and Developments* (ed. E. Flugel). Springer, Berlin, pp. 57–60.

Riding R. (1982) Cyanophyte calcification and changes in ocean chemistry. *Nature* **299**, 814–815.

Riding R. and Voronova L. (1982) Recent freshwater oscillatorian analogue of the Lower Paleozoic calcareous alga *Angulocellularia. Lethaia* **15**, 105–114.

Rigby J. K. (1991) Evolution of Paleozoic heterocalcareous sponges and demosponges-patterns and records. In *Fossil and Recent Sponges* (eds. J. Reitner and H. Keupp). Springer, Berlin, pp. 83–101.

Rigby J. K. and Webby R. D. (1988) Late Ordivician sponges from the malonfullit formation of central New South Wales, Australia. *Paleontogr. Am.* **56**, 1–47.

Rink W. J. and Schwarcz H. P. (1995) Tests for diagenesis in tooth enamel-ESR dating signals and carbonate contents. *J. Archalaeol. Sci.* **22**, 251–255.

Roer R. and Dilliman R. (1985) The structure and calcification of the crustacean cuticle. *Am. Zool.* **24**, 893–909.

Rossman G. R. (1994) Colored varieties of the Silica mineral. In *Silica: Physical Behavior, Geochemistry and Materials Applications*, Rev. Mineral. (eds. P. J. Heaney, C. T. Prewitt, and G. V. Gibbs). Mineralogical Society of America, Washington, DC, vol. 29, pp. 433–468.

Round F. E. (1981) Morphology and phyletic relationships of the silicified algae and the archetypical diatom—monophyly or polyphyly. In *Silicon and Siliceous Structures in Biological Systems* (eds. T. L. Simpson and B. E. Vocani). Springer, New York, pp. 97–128.

Rowles S. L. (1967) Chemistry of the mineral phase of dentine. In *Structural and Chemical Organization of Teeth* (ed. A. E. W. Miles). Academic Press, New York, vol. II, pp. 201–246.

Sakai W. S. and Sanford W. G. (1984) A developmental study of silicification in the abaxial epidermal cells of sugarcane leaf blades using scanning electron microscopy and energy dispersive X-ray analysis. *Am. J. Bot.* **71**, 1315–1322.

Saluddin A. S. M. and Petit H. P. (1983) The mode of formation and structure of the periostracum. In *The Mollusca* (ed. K. M. Wilbur). Plenum, New York, vol. 4, pp. 199–234.

Sancho L. G. and Rodriguez P. C. (1990) The weathering action of saxicolous lichens in maritime antarctica. *Polar Biol.* **11**(1), 33–40.

Schewiakoff W. (1926) Die Acantharia des Golfs von Nepal. *Fauna e Flora del Golfo di Napoli* **37**, 1–755.

Schick K. D. and Toth N. (1994) *Making Silent Stones Speak*. Simon and Schuster, New York.

Schmidt W. J. (1921) Uber den kristallofraphischen charackter der prismen in den muschelschalen. *Z. Allegegem Physiol.* **19**, 191.

Schreiber B., Cavalca L., and Bottazzi E. M. (1959) Ecologia degli Acantari e circulazione dello Sr nel mare. *Boll. Zool.* **24**, 213.

Schroeder J. H., Dwornik E. J., and Papike J. J. (1969) Primary protodolomite in echinoid skeletons. *Bull. Geol. Soc. Am.* **80**, 1613–1616.

Schroter K., Lachli A., and Sievers A. (1975) Mikroanalyztische identifikation von bariumsulfat-kristallen in den statolithen der rhizoide von *Chara fragilis* Desv. *Planta* **122**, 213–225.

Schwertmann U. and Fitzpatrick R. W. (1992) Iron minerals in surface environments. In *Biomineralization, Processes of Iron and Manganese Modern and Ancient Environments* (eds. H. C. W. Skinner and R. W. Fitzpatrick). Catena Verlag, Cremingen-Destedt, Germany, suppl. 21, pp. 7–30.

Sdzuy K. (1969) Unter-und mittelkambrische Porifera (Chancelloriida und Hexactinellida). *Pallaeontol Z.* **43**, 115–147.

Sebald M. (1993) *Genetics and Molecular Biology of Anaerobic Bacteria*. Springer, New York.

Secord D. and Augustine L. (2000) Biogeography and microhabitat variation in temperate algal-invertebrate symbioses: zooxanthellae and zoochlorellae in two Pacific intertidal sea anemones, Anthopleura elegantissima and A. xanthogrammica. *Invertebr. Biol.* **119**, 139–146.

Seed R. (1980) Shell growth and form in the Bivalvia. In *Skeletal Growth of Aquatic Organisms* (eds. D. C. Rhoads and R. A. Lutz). Plenum, New York, pp. 23–68.

Shapiro I. M. and Boyde A. (1984) Microdissection-elemental analysis of the mineralizing growth cartilage of the normal and rachitic chick. *Metabolic Bone Dis.* **5**, 317–326.

Simkiss K. and Wilbur K. M. (1989) *Biomineralization*. Academic Press, New York.

Simpson T. L. and Volcani B. E. (1981) Introduction. In *Silicon and Siliceous Structures in Biological Systems* (eds. T. L. Simpson and B. E. Volcani). Springer, New York, pp. 3–11.

Skinner H. C. W. (1963) Precipitation of calcium dolomite and magnesian calcites in the Southeast of South Australia. *Am. J. Sci.* **261**, 449–472.

Skinner H. C. W. (1968) X-ray diffraction analysis to monitor composition fluctuations within the mineral group Apatite. *Appl. Spectrosc.* **22**, 412–414.

Skinner H. C. W. (1987) Bone: mineralization. In *The Scientific Basis of Orthopaedics*, 2nd edn. (eds. J. A. Albright and R. A. Brand). Appleton and Lange, Norwalk, CT, pp. 199–212.

Skinner H. C. W. (2000) Minerals and human health. In *Environmental Mineralogy*. European Mineralogical Union Notes in Mineralogy (eds. D. Vaughan and R. Wogelius). Eotvos University Press, Budapest, vol. 2, pp. 383–412.

Skinner H. C. W. (2002) In praise of phosphates or why vertebrates chose apatite to mineralize their skeletal tissues. In *Frontiers in Geochemistry: Organic, Solution and Ore*

Deposit Chemistry (ed. W. G. Ernst). Geological Society of America, Boulder, CO, vol. 2, pp. 41–49.

Skinner H. C. W. and Fitzpatrick R. W. (1992) Iron and manganese biomineralization. In *Biomineralization Processes of Ion and Manganese, Modern and Ancient Environments* (eds. H. C. W. Skinner and R. W. Fitzpatrick). Catena-Verlag, Cremlingen-Destedt, Germany, suppl. 21 pp. 1–6.

Skinner H. C. W. and Nalbandian J. (1975) Tetracycline and mineralized tissues: review and perspectives. *Yale J. Biol. Med.* **48**, 377–397.

Skinner H. C. W., Kempner E., and Pak C. Y. C. (1972) Preparation of the mineral phase of bone using ethylene diamine extaction. *Calc. Tiss. Res.* **10**, 257–268.

Skinner H. C. W., Nicolescu S., and Raub T. D. (2003) A tale of two apatites. Studia Ambientum Universitatis Babes Bolyai, Rumania (in press).

Sollas W. J. (1888) Report on the Tetractinellida. Collected by H. M. S. Challenger during the yeatrs 1873–1876. *Rep. Sci Res. HMS Challenger (Zool.)* **25**, 1–458.

Spangenberg D. B. (1976) Intracellular statolith synthesis *Auralia Aurita*. In *Mechanisms of Biominerization in the Invertebrates and Plants* (eds. N. J. Watabe and K. M. Wilbur). University of S. Carolina Press, Columbia, SC, pp. 231–418.

Spear J. A. (1983) Crystal chemistry and phase relations pf orthorhombic carbonates. In *Carbonates: Mineralogy, and Chemistry*. Rev. Mineral. (ed. R. R. Reeder). Mineralogical Society of America, Washington, DC, vol. 11, pp. 145–225.

Spencer K., *et al.* (2000) Stable lead isotope ratios from distinct anthropogenic sources in fish otoliths: a potential nursery ground stock marker. *Comp. Biochem. Physiol.: Part A. Molec. Integr. Physiol.* **127**, 273–284.

Stanley S. M. (1973) An ecological theory for the sudden origin of multicellular life in the late Precambrian. *Proc. Natl. Acad. Sci. USA* **70**, 1486–1489.

Stanley S. M. (1976) Fossil data and the Precambrian–Cambrian evolutionary transition. *Am. J. Sci.* **276**, 56–76.

Stolz J. F. (1992) Magnetotactic bacteria: biomineralization, ecology, sediment magnetism, environmental indicator. In *Biomineralization: Processes of Iron and Manganese-Modern and Ancient Environments* (eds. H. C. W. Skinner and R. W. Fitzpatrick). CATENA, Catena-Verag, Cremlingen-Destedt, Germany, suppl. 21, pp. 133–146.

Stromberg C. A. E. (2002) The origin and spread of grass-dominated ecosystems in the late Tertiary of North America; preliminary results concerning the evolution of hypsodonty. *Palaeogeogr. Palaeoclimatol. Palaeoecol.* **177**(1–2), suppl. 21, 59–75.

Sullivan C. W. and Volcani B. E. (1981) Silicon in the cellular metabolism of diatoms. In *Silicon and Siliceous Structures in Biological Systems* (eds. T. L. Simpson and B. E. Volcani). Springer, New York, pp. 15–42.

Tamaura Y., Buduan P. V., and Katsura T. (1981) Studies on the oxidation of iron(II) ion during formation of Fe_3O_4 and a-FeOOH by oxidation of $Fe(OH)_2$suspensions. *J. Chem Soc. Dalton Trans.*, 1807–1811.

Tamaura T., Ito K., and Katsura T. (1983) Transformation of g-FeOOH to Fe_3O_4 by adsorbtion of iron(II) ion on g-FeOOH. *J. Chem. Soc. Dalton Trans.*, 189–194.

Tappan H. (1980) *The Paleobiology of Plant Protests*. W. H. Freeman, San Francisco.

Tewari J. P., Shinners T. C., and Briggs K. G. (1997) Production of calcium oxalate crystals by two species of Cyathus in culture and infested plant debris. *Zeitschrift fuer Naturforshung (Section C): J. Biosci.* **52**(7–8), 421–425.

Thompson D. W. (1942) *On Growth and Form*. Cambridge University Press, Cambridge, UK.

Thurston E. L. (1974) Morphology, fine structure and ontogeny of the stinging emergence of *Urtica dioica*. *Am. J. Bot.* **61**, 809–817.

Towe K. M. (1967) Echinoderm calcite: single crystal or polycrystalline aggregate. *Science* **157**, 1047–1050.

Towe K. M. and Cifelli R. (1967) Wall ultrastructure in the calcareous foraminifera: crystallographic aspects and a model for calcification. *J. Paleontol.* **41**, 742–762.

Towe K. M. and Lowenstam H. A. (1967) Ultrastructure and development of iton mineralization in the radular teeth of *Crytochiton stelleri* (Mollusca). *J. Ultrastruc. Res.* **17**, 1–13.

Towe K. M., Berthold W. U., and Appleman D. E. (1977) The crystallography of *Patellina corrugata* Williamson: 'a' axis preferred orientation. *J. Foraminiferal Res.* **7**, 58–61.

Travis D. F. (1976) Structural features of mineralizarion from tissue to macromolecular levels of organization in the decapod Crustacaea. *Ann. NY Acad. Sci.* **109**, 177–245.

Trench R. K. and Blank R. J. (1987) *Symbiodiunium microadriatacum* Freudenthal *S. Goreauii* Sp. Nov. *S. kawagutti* Sp. Nov. and *S. pilosum* Sp. Nov., Gymnodinioid dinoflagellate symbionts of marine invertebrates. *J. Phycol.* **23**, 469–481.

Trueman M. and Tuross N. (2002) Trace elements in recent and fossil bone In *Phosphates: Geochemical, Geobiological and Material Importance*. Rev. Mineral. Geochem. (eds. M. J. Kohn, J. Rakovan, and J. M. Hughes). Mineralogical Society of America, Washington, DC, vol. 48, pp. 489–522.

Tuross N. and Fisher L. W. (1989) The proteins in the shell of *Lingula*. In *Origin, Evolution and Modern Aspects of Biomineralization in Plants and Animals* (ed. R. Crick). Plenum, New York, pp. 325–328.

Tyler C. and Simkiss K. (1958) Reactions between eggshell organic matrix and metallic cations. *Q. J. Microsc. Sci.* **99**, 5–13.

Urist M. R. and Strates B. S. (1971) Bone mophogenic protein. *J. Dent. Res.* **50**(6), 1392–1406.

Van Soest R. W. M. (1991) Demosponge higher taxa classification re-examined. In *Fossil and Recent Sponges* (eds. J. Reitner and H. Keupp). Springer, Berlin, pp. 54–71.

Vandermeulen J. H. and Watabe N. (1973) Studies on reef corals: I. Skeleton formation by newly settled plaula larva of *Poccillopora damicornis*. *Mar. Biol.* **23**, 47–57.

Vaughn D. J. (1984) The influence of bone cells on the formation, and mineralization, of bone matrix. *Phil. Trans. Roy. Soc. London B* **304**, 453–454.

Vinogradov A. P. (1953) *The Elementary Chemical Composition of Marine Organisms*. Memoir Sears Found. For Marine Res.: II. Yale university, New Haven, CT.

Vischer T. L. (1970) *Fluoride in Medicine*. Hans Huber, Bern, Switzerland.

Vogel J. J. and Boyan-Salyers B. D. (1976) Acidic lipids associated with the local mechanism of calcification. *Clinic. Orthop. Relat. Res.* **118**, 230–241.

Volk *et al.* (2000) Otolith chemistry reflects migratory characteristics of Pacific salmonids: using otolith core chemistry to distinguish maternal associations with sea and freshwaters. *Fish. Res.* **46**, 251–266.

Wainwright S. A. (1963) Skeletal organization in the coral, *Pocillopora damicornis*. *Quart. J. Microsc. Sci.* **104**, 160–183.

Wainwright S. A., Biggs W. D., Currey J. D., and Gosline J. M. (1976) *Mechanical Design in Organisms*. Arnold, New York.

Waite J. H. (1983) Quinone–tanned scleroproteins. In *The Mollusca* (ed. K. M. Wilbur). Academic Press, New York, pp. 467–504.

Walcott C. D. (1920) Middle Cambrian Spongiae, Cambrian geology and paleontology. *Smithson. Misc. Collect.* **67**(6), 261–364.

Walter M. (1976) Introduction. In *Stromatolites*. Developments in Sedimentology (ed. M. Walter). Elsevier, Amsterdam, vol. 20, pp. 1–3.

Wang Y., Jahren A. H., and Amundson R. (1997) Potential for ^{14}C dating of biogenic carbonate in hackberry (*Celtis*) endocarps. *Quat. Res.* **47**, 337–343.

Wasnich R. D. (1999) Epidemiology of osteoporosis. In *Primer on the Metabolic Bone Diseases and Disorders of Mineral Metabolism*, 4th edn. (ed. M. J. Favus). Lippincott, Williams and Wilkins, Philadelphia, pp. 257–259.

Watabe N. (1965) Studies on shell formation: XI. Crystal-matrix relationships in the inner layers of mollusks. *J. Ultrastuct. Res.* **12**, 351–360.

Watabe N. (1967) Crystallographic analysis of the coccoliths of *Coccolithus huxleii. Calc. Tiss. Res.* **1**, 114–121.

Webb J. and Walker C. D. (1987). In *Thalassemia: Pathophysiology and Management* (eds. S. Fucharoen, P. T. Rowley, and N. W. Paul). Alan R. Liss, vol. B, pp. 147–154.

Webb M. A. (1999) Cell-mediated crystallization of calcium oxalate in plants. *The Plant Cell* **11**, 751–761.

Webb M. A., *et al.* (1995) The intravacuolar organic matrix associated with calcium oxalate crystals in leaves of *Vitis. Plant J.* **7**(4), 633–648.

Weiner S. (1986) Organization of extracellularly mineralized tissues: a comparative study of biological crystal growth. *CRC Crit. Rev. Biochem.* **20**, 365–408.

Wheeler A. P., George J. W., and Evans C. R. (1981) Control of calcium carbonate nucleation and crystal growth by soluble matrix of oyster shell. *Science* **212**, 1397–1398.

Whitfield M. and Watson A. J. (1983) The influence of biomineralization on the composition of seawater. In *Biomineralization and Biological Metal Accumulation* (eds. P. Westbroek and E. W. deJong). D. Reidel, Dordrecht, Germany, pp. 57–72.

Whitney K. D. and Arnott H. J. (1987) Calcium oxalate crystal morphology and development in *Agaricus bisporus. Mycologia* **79**(2), 180–187.

Widdel F., *et al.* (1993) Ferrous iron oxidation in anoxygenic photoptrophic bacteria. *Nature (London)* **362**, 834–836.

Wilbur K. M. and Jodrey L. H. (1952) Studies on shell formation: I. Measurement of the rate of shell formation using ^{45}Ca. *Biol. Bull.* **103**, 269–276.

Wilbur K. M. and Watabe N. (1963) Experimental studies on calcification in mollusks and the alga *Coccolithus huxleii. Ann. NY Acad. Sci.* **109**, 82–112.

Woese C. R., Kandler O., and Wheelis M. L. (1990) Toward a natural selection of organisms: proposal for the domains Archea, Bacteria, and Eucarya. *Proc. Natl. Acad. Sci. USA* **87**, 4576–4579.

Wray J. L. (1977) *Calcareous Algae*. Elsevier, Amsterdam.

Wuthier R. E. (1984) Calcification of Vertebrate Hard Tissues. In *Calcium and its Role in Biology* (ed. H. Sigel). Dekker, New York, pp. 411–472.

Zindler F. E., Hoenow R., and Hesse A. (2001) Calcium and oxalate content of the leaves of *Phaseolus vulgaris* at different calcium supply in relation to calcium oxalate crystal formation. *J. Plant Physiol.* **158**(2), 139–144.

Zipkin I. (1970) The inorganic composition of bones and teeth. In *Biological Calcification: Cellular and Molecular Aspects* (ed. H. Schraer) Appleton-Century-Crofts, New York, pp. 69–104.

8.05
Biogeochemistry of Primary Production in the Sea

P. G. Falkowski

Rutgers University, New Brunswick, NJ, USA

As the present condition of nations is the result of many antecedent changes, some extremely remote and others recent, some gradual, others sudden and violent, so the state of the natural world is the result of a long succession of events, and if we would enlarge our experience of the present economy of nature, we must investigate the effects of her operations in former epochs.

Charles Lyell,
Principles of Geology, 1830

8.05.1 INTRODUCTION

Earth is the only planet in our solar system that contains vast amounts of liquid water on its surface and high concentrations of free molecular oxygen in its atmosphere. These two features are not coincidental. All of the original oxygen on Earth arose from the photobiologically catalyzed splitting of water by unicellular photosynthetic organisms that have inhabited the oceans for at least 3 Gyr. Over that period, these organisms have used the hydrogen atoms from water and other substrates to form organic matter from CO_2 and its hydrated equivalents. This process, the *de novo* formation of organic matter from inorganic carbon, or primary production, is the basis for all life on Earth. In this chapter, we examine the evolution and biogeochemical consequences of primary production in the sea and its relationship to other biogeochemical cycles on Earth.

8.05.1.1 The Two Carbon Cycles

There are two major carbon cycles on Earth. The two cycles operate in parallel. One cycle is slow and abiotic. Its effects are observed on multimillion-year timescales and are dictated by tectonics and weathering (Berner, 1990). In this cycle, CO_2 is released from the mantle to the atmosphere and oceans via vulcanism and seafloor spreading, and removed from the atmosphere and ocean primarily by reaction with silicates to form carbonates in the latter reservoir. Most of the carbonates are subsequently subducted into the mantle, where they are heated, and their carbon is released as CO_2 to the atmosphere and ocean, to carry out the cycle again. The chemistry of this cycle is dependent on acid–base reactions, and would operate whether or not there was life on the planet (Kasting *et al.*, 1988). This slow carbon cycle is a critical determinate of the concentration of CO_2 in Earth's atmosphere and oceans on timescales of tens and hundreds of millions of years (Kasting, 1993).

The second carbon cycle is dependent on the biologically catalyzed reduction of inorganic carbon to form organic matter, the overwhelming majority of which is oxidized back to inorganic carbon by respiratory metabolism (Schlesinger, 1997). This cycle, which is observable on timescales of days to millenia, is driven by reduction—oxidation (redox) reactions that evolved over ~2 Gyr, first in microbes, and subsequently in multicellular organisms (Falkowski *et al.*, 1998). A very small fraction of the reduced carbon escapes respiration and becomes incorporated into the lithosphere. In so doing, some of the organic matter is transferred to the slow carbon cycle. In this chapter, we will focus primarily on this fast, biologically mediated carbon cycle in the sea, and the supporting biogeochemical processes and feedbacks.

8.05.1.2 A Primer on Redox Chemistry

The biologically mediated redox reactions cycle carbon through three mobile pools: the atmosphere, the ocean, and the biosphere. Of these, the ocean is by far the largest (Table 1); however, more than 98% of this carbon is found in its oxidized state as CO_2 and its hydrated equivalents, HCO_3^- and CO_3^{2-}. To form organic molecules, the inorganic carbon must be chemically reduced, a process that requires the addition of hydrogen atoms (not just protons, but protons plus electrons) to the carbon atoms. Broadly speaking, these biologically catalyzed reduction reactions are carried out by two groups of organisms, chemoautotrophs and photoautotrophs, which are collectively called primary

Table 1 Carbon pools in the major reservoirs on Earth.

Pools	*Quantity* ($\times 10^{15}$ g)
Atmosphere	720
Oceans	38,400
Total inorganic	37,400
Surface layer	670
Deep layer	36,730
Dissolved organic	600
Lithosphere	
Sedimentary carbonates	>60,000,000
Kerogens	15,000,000
Terrestrial biosphere (total)	2,000
Living biomass	600–1,000
Dead biomass	1,200
Aquatic biosphere	1–2
Fossil fuels	4,130
Coal	3,510
Oil	230
Gas	140
Other (peat)	250

producers. The organic carbon they synthesize fuels the growth and respiratory demands of the primary producers themselves and all remaining organisms in the ecosystem.

All redox reactions are coupled sequences. Reduction is accomplished by the addition of an electron or hydrogen atom to an atom or molecule. In the process of donating an electron to an acceptor, the donor molecule is oxidized. Hence, redox reactions require pairs of substrates, and can be described by a pair of partial reactions, or half-cells:

$$A_{ox} + n(e^-) \leftrightarrow A_{re} \quad (1a)$$

$$B_{red} - n(e^-) \leftrightarrow B_{ox} \quad (1b)$$

The tendency for a molecule to accept or release an electron is therefore "relative" to some other molecule being capable of conversely releasing or binding an electron. Chemists scale this tendency, called the redox potential, E, relative to the reaction

$$H_2 \leftrightarrow 2H^+ + 2e^- \quad (2)$$

which is arbitrarily assigned an E of 0 at pH 0, and is designated E_0. Biologists define the redox potential at pH 7, 298 K (i.e., room temperature) and 1 atm pressure (=101.3 kPa). When so defined, the redox potential is denoted by the symbols E_0' or sometimes E_{m_7}. The E_0' for a standard hydrogen electrode is -420 mV.

8.05.2 CHEMOAUTOTROPHY

Organisms capable of reducing sufficient inorganic carbon to grow and reproduce in the dark without an external organic carbon source are called chemoautotrophs (literally, "chemical self-feeders"). Genetic analyses suggest that chemoautotrophy evolved very early in Earth's history, and is carried out exclusively by prokaryotic organisms in both the Archea and Bacteria superkingdoms (Figure 1).

Early in Earth's history, the biological reduction of inorganic carbon may have been directly coupled to the oxidation of H_2. At present, however, free H_2 is scarce on the planet's surface. Rather, most of the hydrogen on the surface of Earth is combined with other atoms, such as sulfur or oxygen. Activation energy is required to break these bonds in order to extract the hydrogen. One source of energy is chemical bond energy itself. For example, the ventilation of reduced mantle gases along tectonic plate subduction zones on the seafloor provides hydrogen in the form of H_2S. Several types of microbes can couple the oxidation of H_2S to the reduction of inorganic carbon, thereby forming organic matter in the absence of light.

Ultimately all chemoautotrophs depend on a nonequilibrium redox gradient, without which there is no thermodynamic driver for carbon fixation. For example, the reaction involving the oxidation of H_2S by microbes in deep-sea vents described above is ultimately coupled to oxygen

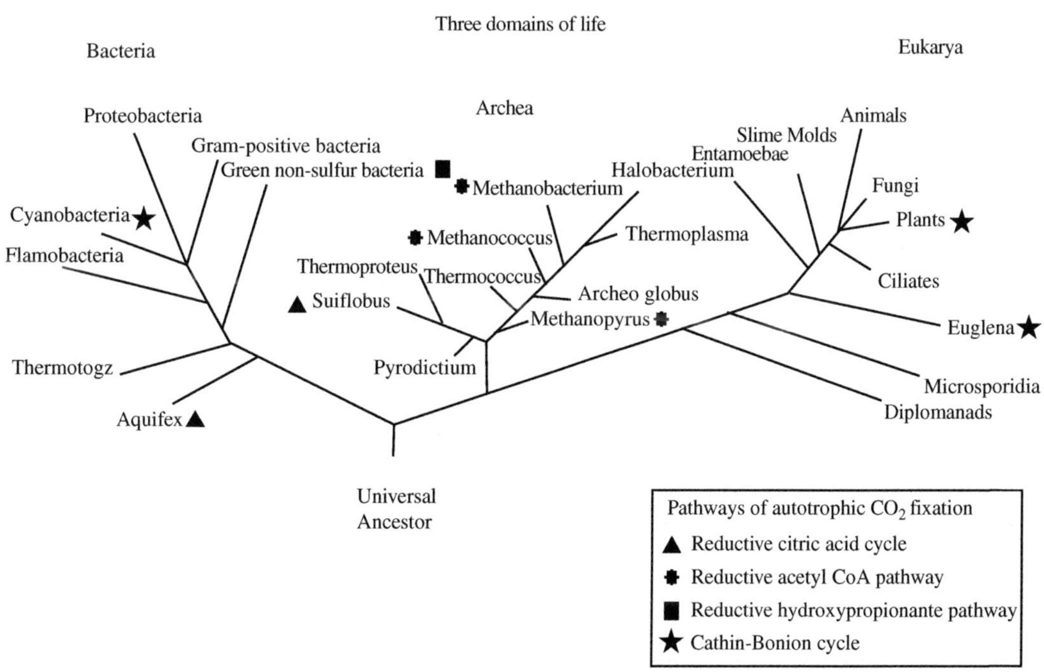

Figure 1 The distribution of autotrophic metabolic pathways among taxa within the three major domains of life (as inferred from ^{16}S ribosomal RNA sequences (Pace, 1997)). Specific metabolic pathways are indicated.

in the ocean interior. Hence, this reaction is dependent on the chemical redox gradient between the ventilating mantle plume and the ocean interior that thermodynamically favors oxidation of the plume gases. Maintaining such a gradient requires a supply of energy, either externally, from radiation (solar or otherwise), or internally, via planetary heat and tectonics, or both.

The overall contribution of chemoautotrophy in the contemporary ocean to the formation of organic matter is relatively small, accounting for <1% of the total annual primary production in the sea. However, this process is critical in coupling reduction of carbon to the oxidation of low-energy substrates, and is essential for completion of several biogeochemical cycles.

8.05.3 PHOTOAUTOTROPHY

The oxidation state of the ocean interior is a consequence of a second energy source: light, which drives photosynthesis. Photosynthesis is a redox reaction of the general form:

$$2H_2A + CO_2 + \text{light} \rightarrow (CH_2O) + H_2O + 2A \qquad (3)$$

where A is an atom, e.g., S. In this formulation, light is specified as a substrate, and a fraction of the light energy is stored as chemical bond energy in the organic matter. Organisms capable of reducing inorganic carbon to organic matter by using light energy to derive the source of reductant or energy are called photoautotrophs. Analyses of genes and metabolic sequences strongly suggest that the machinery for capturing and utilizing light as a source of energy to extract reductants was built on the foundation of chemoautotrophic carbon fixation; i.e., the predecessors of photoautotrophs were chemoautotrophs. The evolution of a photosynthetic process in a chemoautotroph forces consideration of both the selective forces responsible (why) and the mechanism of evolution (how).

8.05.3.1 Selective Forces in the Evolution of Photoautotrophy

Reductants for chemoautotrophs are generally deep in the Earth's crust. Vent fluids are produced in magma chambers connected to the Athenosphere. As such, the supply of vent fluids is virtually unlimited. While the chemical disequilibria between vent fluids and bulk seawater provides a sufficient thermodynamic gradient to continuously support chemoautotrophic metabolism in the contemporary ocean, in the early Earth the oceans would not have had a sufficiently large thermodynamic energy potential to support a pandemic outbreak of chemoautotrophy. Moreover, magma chambers, vulcanism and vent fluid fluxes are tied to tectonic subduction and spreading regions, which are transient features of Earth's crust and hence only temporary habitats for chemoautotrophs. In the Archean and early Proterozoic oceans, the chemoautotrophs would have already been dispersed throughout the oceans by physical mixing and helping to colonize new vent regions. This same dispersion process would have also helped ancestral chemoautotrophs exploit solar energy near the ocean surface.

Although the processes that selected the photosynthetic reactions as the major energy transduction pathway remain obscure, central hypotheses have emerged based on our understanding of the evolution of Earth's carbon cycle, the evolution of photosynthesis, biophysics, and molecular phylogeny. Photoautotrophs are found in all three major superkingdoms (Figure 1); however, there are very few known Archea capable of this form of metabolism. Efficient photosynthesis requires harvesting solar radiation, and hence the evolution of a light harvesting system. While some Archea and Bacteria use the pigment-protein rhodopsin, by far, the most efficient and ubiquitous light harvesting systems are based on chlorins. The metabolic pathway for the synthesis of porphyrins and chlorins is one of the oldest in biological evolution, and is found in all chemoautotrophs (Xiong *et al.*, 2000). Mulkidjanian and Junge (1997) proposed that the chlorin-based photosynthetic energy conversion apparatus originally arose from the need to prevent UV radiation from damaging essential macromolecules such as nucleic acids and proteins. The UV excitation energy could be transferred from the aromatic amino acid residues in the macromolecule to a blue absorption band of membrane-bound chlorins to produce a second excited state which subsequently decays to the lower-energy excited singlet. This energy dissipation pathway can be harnessed to metabolism if the photochemically produced, charge-separated, primary products are prevented from undergoing a back-reaction, but rather form a biochemically stable intermediate reductant. This metabolic strategy was selected for the photosynthetic reduction of CO_2 to carbohydrates, using reductants such as S^{2-} or Fe^{2+}, which have redox potentials that are too positive to reduce CO_2 directly.

The synthesis of reduced (i.e., organic) carbon and the oxidized form of the electron donor permits a photoautotroph to use "respiratory" metabolism, but operate them in reverse. However, not all of the reduced carbon and oxidants remain accessible to the photoautotrophs. In the oceans, cells tend to sink, carrying with them organic carbon. The oxidation of Fe^{2+} forms insoluble Fe^{3+} salts that precipitate. The sedimentation and subsequent burial of organic carbon and Fe^{3+}

removes these components from the water column. Without replenishment, the essential reductants for anoxygenic photosynthesis would eventually become depleted in the surface waters. Thus, the necessity to regenerate reductants potentially prevented anoxygenic photoautotrophs from providing the major source of fixed carbon on Earth for eternity. Major net accumulation of reduced organic carbon in Proterozoic sediments implies local depletion of reductants such as S^{2-} and Fe^{2+} from the euphotic zone of the ocean. These limitations almost certainly provided the evolutionary selection pressure for an alternative electron donor.

8.05.3.2 Selective Pressure in the Evolution of Oxygenic Photosynthesis

H_2O is a potentially useful biological reductant with a vast supply on Earth relative to any redox-active solute dissolved in it. Liquid water contains ~100 kmol of H atoms per m^3, and, given $>10^{18}$ m^3 of water in the hydrosphere and cryosphere, $>10^{20}$ kmol of reductant are potentially accessible. Use of H_2O as a reductant for CO_2, however, requires a larger energy input than does the use of Fe^{2+} or S^{2-}. Indeed, to split water by light energy requires 0.82 eV at pH 7 and 298 K. Utilizing light at such high energy levels required the evolution of a new photosynthetic pigment, chlorophyll *a*, which has a red (lowest singlet) absorption band that is 200–300 nm blue shifted relative to bacteriochlorophylls. Moreover, stabilization of the primary electron acceptor to prevent a back-reaction necessitates thermodynamic inefficiency that ultimately requires two light-driven reactions operating in series. This sequential action of two photochemical reactions is unique to oxygenic photoautotrophs and presumably involved horizontal gene transfer through one or more symbiotic events (Blankenship, 1992).

In all oxygenic photoautotrophs, Equation (3) can be modified to:

$$2H_2O + CO_2 + \text{light} \xrightarrow{\text{Chl } a} (CH_2O) + H_2O + O_2 \quad (4)$$

where Chl *a* is the pigment chlorophyll *a* exclusively utilized in the reaction. Equation (4) implies that somehow chlorophyll *a* catalyzes a reaction or a series of reactions whereby light energy is used to oxidize water:

$$2H_2O + \text{light} \xrightarrow{\text{Chl } a} 4H^+ + 4e + O_2 \quad (5)$$

yielding gaseous, molecular oxygen. Hidden within Equation (5) are complex suites of biological innovations that have heretofore not been successfully mimicked *in vitro* by humans. At the core of the water splitting complex is a quartet of manganese atoms, that sequentially extract electrons, one at a time, from $2H_2O$ molecules, releasing gaseous O_2 to the environment, and storing the reductants on biochemical intermediates.

The photochemically produced reductants generated by the reactions schematically outlined in Equation (5) are subsequently used in the fixation (fixation is an archaic term meaning to make non-volatile, as in the chemical conversion of a gas to a solid phase) of CO_2 by a suite of enzymes that can operate *in vitro* in darkness and, hence, the ensemble of these reactions are called the dark reactions. At pH 7 and 25 °C, the formation of glucose from CO_2 requires an investment of 915 cal mol^{-1}. If water is the source of reductant, the overall efficiency for photosynthetic reduction of CO_2 to glucose is ~30%; i.e., 30% of the absorbed solar radiation is stored in the chemical bonds of glucose molecules.

8.05.4 PRIMARY PRODUCTIVITY BY PHOTOAUTOTROPHS

When we subtract the costs of all other metabolic processes by the chemoautotrophs and photoautotrophs, the organic carbon that remains is available for the growth and metabolic costs of heterotrophs. This remaining carbon is called *net* primary production (NPP) (Lindeman, 1942). From biogeochemical and ecological perspectives, NPP provides an upper bound for all other metabolic demands in an ecosystem. If NPP is greater than all respiratory consumption of the ecosystem, the ecosystem is said to be net autotrophic. Conversely, if NPP is less than all respiratory consumption, the system must either import organic matter from outside its bounds, or it will slowly run down—it is net heterotrophic.

It should be noted that NPP and photosynthesis are not synonymous. On a planetary scale, the former includes chemoautrophy, the latter does not. Moreover, photosynthesis *per se* does not include the integrated respiratory term for the photoautotrophs themselves (Williams, 1993). In reality, that term is extremely difficult to measure directly, hence NPP is generally approximated from measurements of photosynthetic rates integrated over some appropriate length of time (a day, month, season, or a year) and respiratory costs are either assumed or neglected.

8.05.4.1 What are Photoautotrophs?

In the oceans, oxygenic photoautotrophs are a taxonomically diverse group of mostly single-celled, photosynthetic organisms that drift with currents. In the contemporary ocean, these organisms, called phytoplankton (derived from Greek, meaning to wander), are comprised of $\sim 2 \times 10^4$ species distributed among at least eight taxonomic divisions or phyla (Table 2). By comparison, higher plants are comprised of $>2.5 \times 10^5$ species, almost all of which are contained within one class in one division. Thus, unlike terrestrial plants, phytoplankton are represented by relatively few species but they are

Table 2 The taxonomic classification and species abundances of oxygenic photosynthetic organisms in aquatic and terrestrial ecosystems. Note that terrestrial ecosystems are dominated by relatively few taxa that are species rich, while aquatic ecosystems contain many taxa but are relatively species poor.

Taxonomic group	*Known species*	*Marine*	*Freshwater*
Empire: Bacteria (=Prokaryota)			
Kingdom: Eubacteria			
Subdivision: Cyanobacteria (*sensu strictu*) (=Cyanophytes, blue-green algae)	1,500	150	1,350
Subdivision: Chloroxybacteria (=Prochlorophyceae)	3	2	1
Empire: Eukaryota			
Kingdom: Protozoa			
Division: Euglenophyta	1,050	30	1,020
Class: Euglenophyceae			
Division: Dinophyta (Dinoflagellates)			
Class: Dinophyceae	2,000	1,800	200
Kingdom: Plantae			
Subkingdom: Biliphyta			
Division: Glaucocystophyta			
Class: Glaucocystophyceae	13		
Division: Rhodophyta			
Class: Rhodophyceae	6,000	5,880	120
Subkingdom: Viridiplantae			
Division: Chlorophyta			
Class: Chlorophyta	2,500	100	2,400
Prasinophyceae	120	100	20
Ulvophyceae	1,100	1,000	100
Charophyceae	12,500	100	12,400
Division: Bryophyta (mosses, liverworts)	22,000		1,000
Division: Lycopsida	1,228		70
Division: Filicopsida (ferns)	8,400		94
Division: Magnoliophyta (flowering plants)	(240,000)		
Subdivision: Monocotyledoneae	52,000	55	455
Subdivision: Dicotyledoneae	188,000		391
Kingdom: Chromista			
Subkingdom: Chlorechnia			
Division: Chlorarachniophyta			
Class: Chlorarachniophyceae	3–4	3–4	0
Subkingdom: Euchromista			
Division: Crytophyta			
Class: Crytophyceae	200	100	100
Division: Haptophyta			
Class: Prymensiophyceae	500	100	400
Division: Heterokonta			
Class: Bacillariophyceae (diatoms)	10,000	5,000	5,000
Chrysophyceae	1,000	800	200
Eustigmatophyceae	12	6	6
Fucophyceae (brown algae)	1,500	1,497	3
Raphidophyceae	27	10	17
Synurophyceae	250		250
Tribophyceae (Xanthophyceae)	600	50	500
Kingdom: Fungi			
Division: Ascomycontina (lichens)	13,000	15	20

Source: Falkowski (1997).

phylogenetically diverse. This deep taxonomic diversity is reflected in their evolutionary history and ecological function (Falkowski, 1997).

Within this diverse group of organisms, three basic evolutionary lineages are discernable (Delwiche, 2000). The first contains all prokaryotic oxygenic phytoplankton, which belong to one class of bacteria, namely, the cyanobacteria. Cyanobacteria are the only known oxygenic photoautotrophs that existed prior to ~2.5 Gyr BP (Ga) (Lipps, 1993; Summons *et al.*, 1999). These prokaryotes numerically dominate the photoautotrophic community in contemporary marine ecosystems and their continued success bespeaks an extraordinary adaptive capacity. At any moment in time, there are $\sim 10^{24}$ cyanobacterial cells in the contemporary oceans. To put that into perspective, the number of cyanobacterial cells in the oceans is two orders of magnitude more than all the stars in the sky.

The evolutionary history of cyanobacteria is obscure. The first microfossils assigned to this group were identified in cherts from 3.1 Ga by Schopf (Schopf, 1993). Macroscopic stromabolites, which are generally of biological (oxygenic photoautotrophic) origin, are first found in strata a few hundred million years younger. However, much of the fossil evidence provided by Schopf (e.g., Schopf, 1993) has been questioned (Brasier *et al.*, 2002), and many researchers believe in a later origin. The origin of this group is critical to establishing when net O_2 production (and hence, an oxidized atmosphere) first occurred on the planet. Although photodissociation of H_2O vapor could have provided a source of atmospheric O_2 in the Archean, the UV absorption cross-section of O_2 constrains the reaction, and theoretical calculations supported by geochemical evidence suggest that prior to ca. 2.4 Ga atmospheric O_2 was less than 10^{-5} of the present level (Holland and Rye, 1998; Pavlov and Kasting, 2002). There was a lag between the first occurrence of oxygenic photosynthesis and a global buildup of O_2 possibly due to the presence of alternative electron acceptors, especially Fe^{2+} and S^{2-} in the ocean. Indeed, the dating of oxidation of Earth's oceans and atmosphere is, in large measure, based on analysis of the chemical precipitation of oxidized iron in sedimentary rocks (the "Great Rust Event (Holland and Rye, 1998) and the mass-independent (Farquhar *et al.*, 2002) and mass-dependent (Habicht *et al.*, 2002) fractionation of sulfur isotopes. The ensemble of these analyses indicate that atmospheric oxygen rose sharply, from virtually insignificant levels, to between 1% and 10% of the present atmospheric concentration over a 100 Myr period beginning ca. 2.4 Ga. Thus, there may be as much as a 1 Gyr or as little as a 100 Myr gap between the origin of the first oxygenic photoautotrophs and oxygenation of Earth's oceans and atmosphere.

All other oxygen-producing organisms in the ocean are eukaryotic, i.e., they contain internal organelles, including a nucleus, one or more chloroplasts, one or more mitochondria, and, in some cases, a membrane-bound storage compartment, the vacuole. Within the eukaryotes, we can distinguish two major groups, both of which appear to have descended from a common ancestor thought to be the endosymbiotic appropriation of a cyanobacterium into a heterotrophic host cell (Delwiche, 2000). The appropriated cyanobacterium became a chloroplast.

8.05.4.1.1 *The red and green lineages*

In one group of eukaryotes, chlorophyll *b* was synthesized as a secondary pigment; this group forms the "green lineage," from which all higher plants have descended. The green lineage played a major role in oceanic food webs and the carbon cycle from ca. 1.6 Ga until the end-Permian extinction, ~250 Ma (Lipps, 1993). Since that time however, a second group of eukaryotes has risen to ecological prominence in the oceans; that group is commonly called the "red lineage" (Figure 2). The red lineage is comprised of several major phytoplankton divisions and classes, of which the diatoms, dinoflagellates, haptophytes (including the coccolithophorids), and the chrysophytes are the most important. All of these groups are comparatively modern organisms; indeed, the rise of dinoflagellates and coccolithophorids approximately parallels the rise of dinosaurs on land, while the rise of diatoms approximately parallels the rise of mammals in the Cenozoic. The burial and subsequent diagenesis of organic carbon produced primarily by members of the red lineage in shallow seas in the Mesozoic era provide the carbon source for many of the petroleum reservoirs that have been exploited for the past century by humans.

8.05.4.2 Estimating Chlorophyll Biomass

As implied in Equation (4), given an abundance of the two physical substrates, CO_2 and H_2O, primary production is, to first order, dependent on the concentration of the catalyst Chl *a* and light. The distribution of Chl *a* in the upper ocean can be discerned from satellite images of ocean color. The physical basis of the measurement is straightforward; making the measurements is technically challenging. Imagine two small parcels of water that are adjacent to each other. As photons from the sun enter the water column, they are either absorbed or scattered. Water itself absorbs red wavelengths of light, at shorter wavelengths of the visible spectrum, light is not

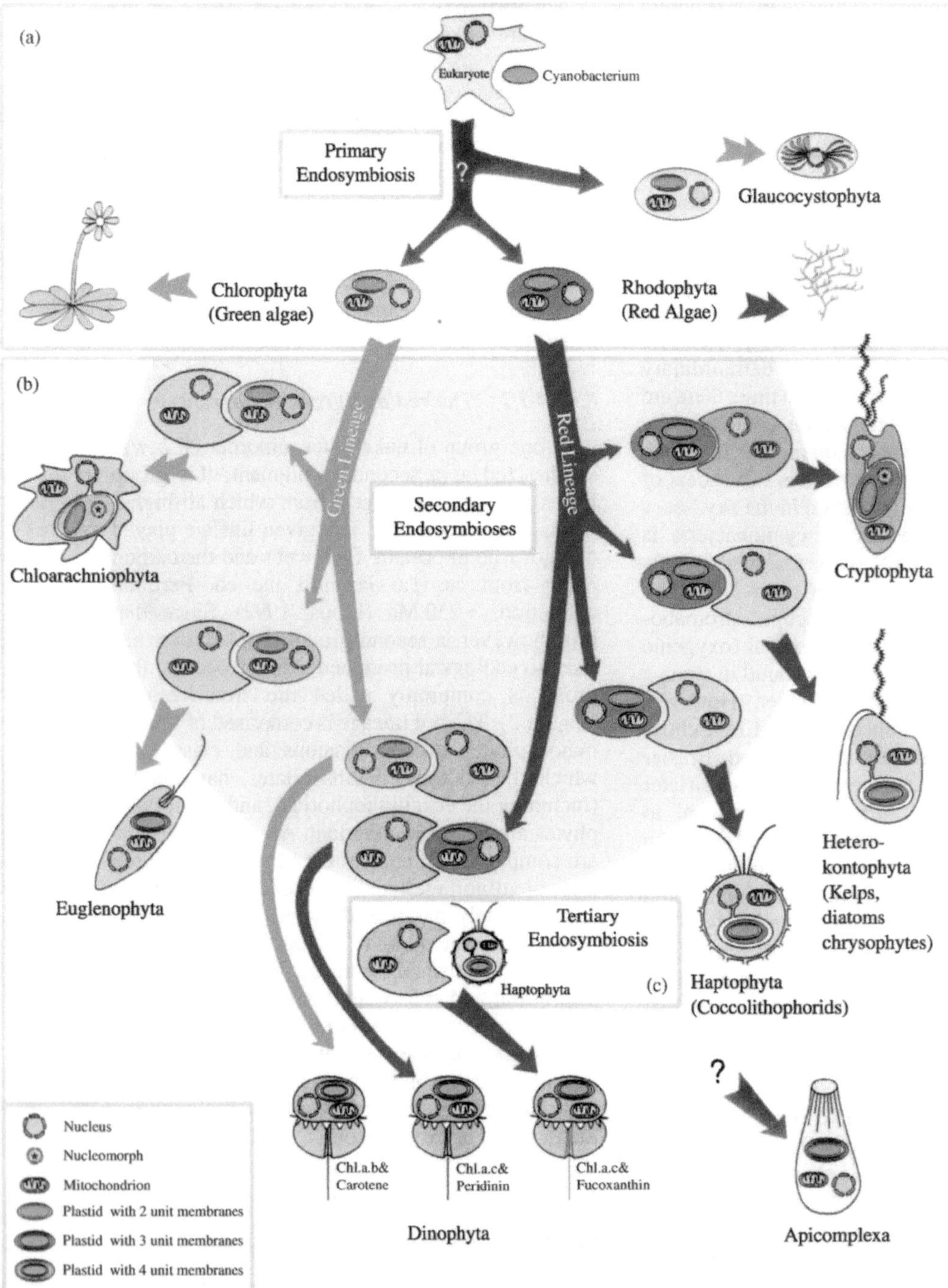

Figure 2 The basic pathway leading the evolution of eukaryotic algae. The primary symbiosis of a cyanobacterium with a apoplastidic host gave rise to both chlorophyte algae and red algae. The chlorophyte line, through secondary symbioses, gave rise to the "green" line of algae, one division of which was the predecessor of all higher plants. Secondary symbioses in the red line with various host cells gave rise to all the chromophytes, including diatoms, cryptophytes, and haptophytes (after Delwiche, 2000).

as efficiently absorbed. However, because water molecules can randomly move from one adjacent parcel to another, there are continuous minor changes in density and hence in the refractive index of the water parcels. These minor changes in refractive index lead to incoherence in the downwelling light stream. The incoherence, in turn, increases the probability of photon scattering (a process called "fluctuation density scattering"), such that light in the shorter wavelengths is more likely to be scattered back to space (Einstein, 1910; Morel, 1974). If the ocean contained sterile, pure seawater, an observer looking at the surface from space would see the oceans as blue. However, chlorophyll *a* has a prominent absorption band in the blue portion of the spectrum. Hence, in the presence of chlorophyll some of the downwelling and upwelling photons from the sun are absorbed by the phytoplankton themselves and the ocean becomes optically darker. As chlorophyll concentrations increase even further, blue wavelengths are largely eliminated from the outbound reflectance, and the ocean appears optically dark green (Morel, 1988).

8.05.4.2.1 Satellite based algorithms for ocean color retrievals

Empirically, satellite sensors that measure ocean color utilize a number of wavelengths. In addition to the blue and green region, red and far-red spectra are determined to derive corrections for scattering and absorption of the outbound or reflected radiation from the ocean by the atmosphere. In fact, only a very small fraction (~5%) of the light leaving the ocean is observed by a satellite; the vast majority of the photons are scattered or absorbed in the atmosphere. However, based on the ratio of blue-green light that is reflected from the ocean, estimates of photosynthetic pigments are derived. It should be pointed out that the blue-absorbing region of the spectrum is highly congested; it is virtually impossible to derive the fraction of absorption due solely to chlorophyll *a* as opposed to other photosynthetic pigments that absorb blue light. The estimation of chlorophyll *a* is based on empirical regression of the concentration of the pigment to the total blue-absorbing pigments (Gordon and Morel, 1983). Water-leaving radiances (L_W) at specific wavelengths are corrected for atmospheric scattering and absorption, and the concentration of chlorophyll is calculated from the ratios of blue and green light reflected from the water body. The calibration of the sensors is empirical and specifically derived for individual satellites. Examples of such algorithms for five satellites are given in Table 3.

One limitation of satellite images of ocean chlorophyll is that they do not provide information about the vertical distribution of phytoplankton. The water-leaving radiances visible to an observer outside of the ocean are confined to the upper 20% of the euphotic zone (which is empirically defined as the depth to which 1% of the solar radiation penetrates). In the open ocean there is almost always a subsurface chlorophyll maximum that is not visible to satellite ocean color sensors. A number of numerical models have been developed to estimate the vertical distribution of chlorophyll based on satellite color data (Berthon and Morel, 1992; Platt, 1986). The models rely on statistical parametrizations and require numerous *in situ* observations to obtain typical profiles for a given area of the world ocean (Morel and Andre, 1991; Platt and Sathyendranath, 1988). In addition, large quantities of phytoplankton associated with the bottom of ice flows in both the Arctic and Antarctic are not visible to satellite sensors but do contribute significantly to the primary production in the polar seas (Smith and Nelson, 1990). Despite these deficiencies, the satellite data allow high-resolution, synoptic observations of the temporal and spatial changes in phytoplankton chlorophyll in relation to the physical circulation of the atmosphere and ocean on a global scale.

The global distribution of phytoplankton chlorophyll in the upper ocean for winter and summer, derived from a compilation of satellite images, is shown in Figure 3. To a first order, the images reveal how the horizontal and temporal distribution of phytoplankton is related to the physical circulation of the oceans, especially the major features of the basin-scale gyres. For example, throughout most of the central ocean basins, between 30° N and 30° S, phytoplankton biomass is extremely low, averaging 0.1–0.2 mg chlorophyll *a* m^{-3} at the sea surface. In these regions the vertical flux of nutrients is generally extremely low, limited by eddy diffusion through the thermocline. Most of the chlorophyll biomass is associated with the thermocline. Because there is no seasonal convective overturn in this latitude band, there is no seasonal variation in

Table 3 Algorithms used to calculate chlorophyll *a* (*C*) from remote sensing reflectance. *R* is determined as the maximum of the values shown. Sensor algorithms are for Sea-viewing Wide Field of view Sensor (SeaWiFS), Ocean Color and Temperature Scanner (OCTS), Moderate Resolution Imaging Spectroradiometer (MODIS), Coastal Zone Color Scanner (CZCS), and Medium Resolution Imaging Spectrometer (MERIS).

Sensor	*Equation*	*R*
SeaWiFS/OC2	$C=10.0^{(0.341-3.001R+2.811R^2-2.041R^3)}-0.04$	490/555
OCTS/OC4O	$C=10.0^{(0..405-2.900R+1.690R^2-0.530R^3-1.144R^4)}$	443>490>520/565
MODIS/OC3M	$C=10.0^{(0.2830-2.753R+1.457R^2-0.659R^3-1.403R^4)}$	443>490/550
CZCS/OC3C	$C=10.0^{(0.362-4.066R+5.125R^2-2.645R^3-0.597R^4)}$	443>520/550
MERIS/OC4E	$C=10.0^{(0.368-2.814R+1.456R^2+0.768R^3-1.292R^4)}$	443>490>510/560
SeaWiFS/OC4v4	$C=10.0^{(0.366-3.067R+1.930R^2+0.649R^3-1.532R^4)}$	443>490>510/555
SeaWiFS/OC2v4	$C=10.0^{(0.319-2.336R+0.879R^2-0.135R^3)}-0.071$	490/555

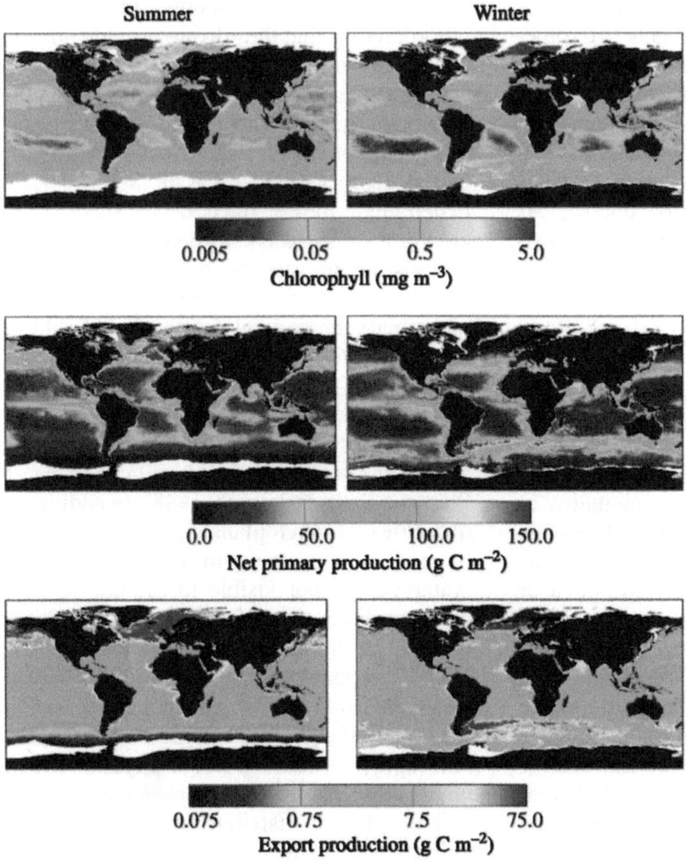

Figure 3 Composite global images for winter and summer of upper-ocean chlorophyll concentrations (top panels) derived from satellite-based observations of ocean color, net primary production (middle panels) calculated based on the algorithms of Behrenfeld and Falkowski (1997a), and export production (lower panels) calculated from the model of Laws *et al.* (2000).

phytoplankton chlorophyll. The chlorophyll concentrations are slightly increased at the equator in the Pacific and Atlantic Oceans, and south of the equator in the Indian Ocean. In the equatorial regions the thermocline shoals laterally as a result of long-range wind stress at the surface (Pickard and Emery, 1990). The wind effectively piles up water along its fetch, thereby inclining the upper mixed layer. This results in increased nutrient fluxes, shallower mixed layers, and higher chlorophyll concentrations on the eastern end of the equatorial band, and decreased nutrient fluxes, deeper mixed layers, and lower chlorophyll concentrations on the western end. This effect is most pronounced in the Pacific. The displacement of the band south of the equator in the Indian Ocean is primarily a consequence of basin scale topography.

At latitudes above ~30°, a seasonal cycle in chlorophyll can occur (Figure 3). In the northern hemisphere, areas of high chlorophyll are found in the open ocean of the North Atlantic in the spring and summer. The southern extent and intensity of the North Atlantic phytoplankton bloom are not found in the North Pacific. The North Atlantic bloom is associated with deep vertical convective mixing, which allows resupply of nutrients to the upper mixed layer of the ocean. This phenomenon does not occur in the Pacific due to a stronger vertical density gradient in that basin (driven by the hydrological cycle). The North Atlantic bloom leads to a flux of organic matter into the ocean interior that is observed even at the seafloor.

In the southern hemisphere, phytoplankton chlorophyll is generally lower at latitudes symmetrical with the northern hemisphere in the corresponding austral seasons. For example, in the austral summer (January–March), phytoplankton chlorophyll is slightly lower between 30° S and the Antarctic ice sheets than in the northern hemisphere in July to September (Yoder *et al.*, 1993).

8.05.4.3 Estimating Net Primary Production

8.05.4.3.1 Global models of net primary production for the ocean

Using satellite data to estimate upper-ocean chlorophyll concentrations, satellite-based observations of incident solar radiation, atlases of seasonally averaged sea-surface temperature, and models that incorporate a temperature response function for photosynthesis, it is possible to estimate global net photosynthesis in the world oceans (Antoine and Morel, 1996; Behrenfeld and Falkowski, 1997a; Longhurst *et al.*, 1995). Although estimates vary between models, based on how the parameters are derived, for illustrative purposes we use a model based on empirical parametrization of the daily integrated photosynthesis profiles as a function of depth. The physical depth at which 1% of irradiance incident on the sea surface remains is called the euphotic zone. This depth can be calculated from surface chlorophyll concentrations, and defines the base of the water column at which net photosynthesis can be supported. Given such information, net primary production can be calculated following the general equation:

$$PP_{eu} = C_{sat} \cdot Z_{eu} \cdot P^{b}_{opt} \cdot DL \cdot F \qquad (6)$$

where PP_{eu} is daily net primary production integrated over the euphotic zone, C_{sat} is the satellite-based (upper water column; i.e., derived from Table 3) chlorophyll concentration, P^{b}_{opt} is the maximum daily photosynthetic rate within the water column, Z_{eu} is the depth of the euphotic zone, DL is the photoperiod (Behrenfeld and Falkowski, 1997a), and F is a function describing the shape of the photosynthesis depth profile. This general model can be both expanded (differentiated) and collapsed (integrated) with respect to time and irradiance; however, the global results are fundamentally similar (Behrenfeld and Falkowski, 1997). The models predict that NPP in the world oceans amounts to 40–50 Pg per annum (Figure 3 and Table 4).

In contrast to terrestrial ecosystems, the fundamental limitation of primary production in the ocean is not irradiance *per se*, but temperature and the concentration of chlorophyll in the upper ocean. The latter is a negative feedback; i.e., the more the chlorophyll in the water column, the shallower is the euphotic zone. Hence, to double NPP requires nearly a fivefold increase in chlorophyll concentration.

8.05.4.4 Quantum Efficiency of NPP

The photosynthetically available radiation (400–700 nm) for the world oceans is 4.5×10^{18} mol of photons per annum, which is $\simeq 9.8 \times 10^{20}$ kJ yr^{-1}. The average energy stored by photosynthetic organisms amounts to ~39 kJ per gram of carbon fixed (Platt and Irwin, 1973). Given an annual net production of 40 Pg C for phytoplankton, and an estimated production of 4 Pg yr^{-1} by benthic photoautotrophs, the photosynthetically stored radiation is equal to $\sim 1.7 \times 10^{18}$ kJ yr^{-1}. The fraction of photosynthetically available solar energy conserved by photosynthetic reactions in

Table 4 Annual and seasonal net primary production (NPP) of the major units of the biosphere.

	Ocean NPP		*Land NPP*
Seasonal			
April–June	10.9		15.7
July–September	13.0		18.0
October–December	12.3		11.5
January–March	11.3		11.2
Biogeographic			
Oligotrophic	11.0	Tropical rainforests	17.8
Mesotrophic	27.4	Broadleaf deciduous forests	1.5
Eutrophic	9.1	Broadleaf and needleleaf forests	3.1
Macrophytes	1.0	Needleleaf evergreen forests	3.1
		Needleleaf deciduous forest	1.4
		Savannas	16.8
		Perennial grasslands	2.4
		Broadleaf shrubs with bare soil	1.0
		Tundra	0.8
		Desert	0.5
		Cultivation	8.0
Total	48.5		56.4

Source: Field *et al.* (1998). After Field *et al.* (1998). All values in GtC. Ocean color data are averages from 1978 to 1983. The land vegetation index is from 1982 to 1990. Ocean NPP estimates are binned into three biogeographic categories on the basis of annual average C_{sat} for each satellite pixel, such that oligotrophic = $C_{sat} < 0.1$ mg m^{-3}, mesotrophic = $0.1 < C_{sat} < 1$ mg m^{-3}, and eutrophic = $C_{sat} > 1$ mg m^{-3} (Antoine *et al.*, 1996). This estimate includes a 1 GtC contribution from macroalgae (Smith, 1981). Differences in ocean NPP estimates between Behrenfeld and Falkowski (1997) and those in the global annual NPP for the biosphere and this table result from: (i) addition of Arctic and Antarctic monthly ice masks; (ii) correction of a rounding error in previous calculations of pixel area; and (iii) changes in the designation of the seasons to correspond with Falkowski *et al.* (1998).

the world oceans amounts to 1.7×10^{18} kJ/ $9.8 \times 10^{20} =$ 0.0017 or 0.17%. Thus, on average, in the oceans, 0.0007 mol C is fixed per mole of incident photons; this is equivalent to an effective quantum requirement of 1,400 quanta per CO_2 fixed. This value is less than 1% of the theoretical maximum quantum efficiency of photosynthesis; the relatively small realized efficiency is due to the fact that photons incident on the ocean surface have a small probability of being absorbed by phytoplankton before they are either absorbed by water or other molecules (e.g., organic matter), or are scattered back to space.

8.05.4.4.1 Comparing efficiencies for oceanic and terrestrial primary production

The average chlorophyll concentration of the world ocean is 0.24 mg m^{-3} and the average euphotic zone depth is 54 m; thus the average integrated chlorophyll concentration is ~13 mg m^{-2}. Carbon to chlorophyll ratios of phytoplankton typically range between 40:1 and 100:1 by weight (Banse, 1977). Given the total area of the ocean of 3.1×10^8 km^2, the total carbon biomass in phytoplankton is 0.25–0.65 Pg. If NPP is ~40 Pg per annum, and assuming the ocean is in steady state (a condition we will discuss in more detail), the living phytoplankton biomass turns over 60–150 times per year, which is equivalent to a turnover time of 2–6 d. In contrast, terrestrial plant biomass amounts to ~600–1,000 Pg C, most of which is in the form of wood (Woodwell *et al.*, 1978). Estimates of terrestrial plant NPP are in the range of 50–65 Pg C per annum, which gives an average turnover time of ~12–20 yr (Field *et al.*, 1998). Thus, the flux of carbon through aquatic photosynthetic organisms is about 1,000-fold faster than terrestrial ecosystems, while the storage of carbon in the latter is about 1,000-fold higher than the former. Moreover, the total photon flux to terrestrial environments amounts to ~2×10^{18} mol yr^{-1}, which gives an effective quantum yield of ~0.002. In other words, on average one CO_2 molecule is fixed for every 500 incident photons. The results of these calculations suggest that terrestrial vegetation is approximately three times more efficient in utilizing *incident* solar radiation to fix carbon than are aquatic photoautotrophs. This situation arises primarily from the relative paucity of aquatic photoautotrophs in the ocean and the fact that they must compete with the media (water) for light.

This comparison points out a fundamental difference between the two ecosystems in the context of the global carbon cycle. On timescales of decades to centuries, carbon fixed in terrestrial ecosystems can be temporarily stored in organic matter (e.g., forests), whereas most of the carbon fixed by marine phytoplankton is rapidly consumed by grazers or sinks and is transferred from the surface ocean to the ocean interior. Upon entering the ocean interior, virtually all of the organic matter is oxidized by heterotrophic microbes, and in the process is converted back to inorganic carbon. Elucidating how this transfer occurs, what controls it, how much carbon is transferred via this mechanism, on what timescales, and whether the process is in steady state was a major focus for research in the latter portion of the twentieth century.

8.05.5 EXPORT, NEW AND "TRUE NEW" PRODUCTION

We can imagine that NPP produced by photoautotrophs in the upper, sunlit regions of the ocean (the euphotic zone) is consumed in the same general region by heterotrophs. In such a case, the basic reaction given by Equation (4) is simply balanced in the reverse direction due to respiration by heterotrophic organisms, and no organic matter leaves the ecosystem. This very simple "balanced state" model, also referred to as the microbial loop (e.g., Azam, 1998), accounts for the fate of most of the organic matter in the oceans (and on timescales of decades, terrestrial ecosystems as well). In marine ecology, this process is sometimes called "regenerated production"; i.e., organic matter produced by photoautotrophs is locally regenerated to inorganic nutrients (CO_2, NH_4^+, PO_3^{2-}) by heterotrophic respiration. It should be noted here that with the passage of organic matter from one level of a marine food chain to the next (e.g., from primary producer to heterotrophic consumer), a metabolic "tax" must be paid in the form of respiration, such that the net metabolic potential of the heterotrophic biomass is always less than that of the primary producers. This does not mean, *a priori*, that photoautotrophic biomass is always greater, as heterotrophs may grow slowly and accumulate biomass; however, as heterotrophs grow faster, their respiratory rates must invariably increase. The rate of production (i.e., the energy flux) of heterotrophic biomass is always constrained by NPP.

Let us imagine a second scenario. Some fraction of the primary producers and/or heterotrophs sink below a key physical gradient, such as a thermocline, and for whatever reason cannot ascend back into the euphotic zone. If the water column is very deep, sinking organic matter will most likely be consumed by heterotrophic microbes in the ocean interior. The flux of organic carbon from the euphotic zone is often called *export production*, a term coined by Wolfgang Berger. Export production is an important conduit for the exchange of carbon between the upper ocean and the ocean interior (Berger *et al.*, 1987).

This conduit depletes the upper ocean of inorganic carbon and other nutrients essential for photosynthesis and the biosynthesis of organic matter. In the central ocean basins, export production is a relatively small fraction of total primary production, amounting to 5–10% of the total carbon fixed per annum (Dugdale and Wilkerson, 1992). At high latitudes and in nutrient-rich areas, however, diatoms and other large, heavy cells can form massive blooms and sink rapidly. In such regions, export production can account for 50% of the total carbon fixation (Bienfang, 1992; Campbell and Aarup, 1992; Sancetta *et al.*, 1991; Walsh, 1983). The subsequent oxidation and remineralization of the exported production enriches the ocean interior with inorganic carbon by ~200 μM in excess of that which would be supported solely by air–sea exchange (Figure 4 and Table 5). This enrichment is called the *biological pump* (Broecker *et al.*, 1980; Sarmiento and Bender, 1994; Volk and Hoffert, 1985). The biological pump is crucial to maintaining the steady-state levels of atmospheric CO_2 (Sarmiento *et al.*, 1992; Siegenthaler and Sarmiento, 1993).

8.05.5.1 Steady-state versus Transient State

The concepts of new, regenerated, and export production are central to understanding many aspects of the role of aquatic photosynthetic organisms in biogeochemical cycles in the oceans. In steady state, the globally averaged fluxes of new nutrients must match the loss of the nutrients contained in organic material. If this were not so, there would be a continuous depletion of nutrients in the euphotic zone and photoautotrophic biomass and primary production would slowly decline (Eppley, 1992). Thus, in the steady state, the sinking fluxes of organic nitrogen and the production of N_2 by denitrifying bacteria must equal the sum of the upward fluxes of inorganic nitrogen, nitrogen fixation, and the atmospheric

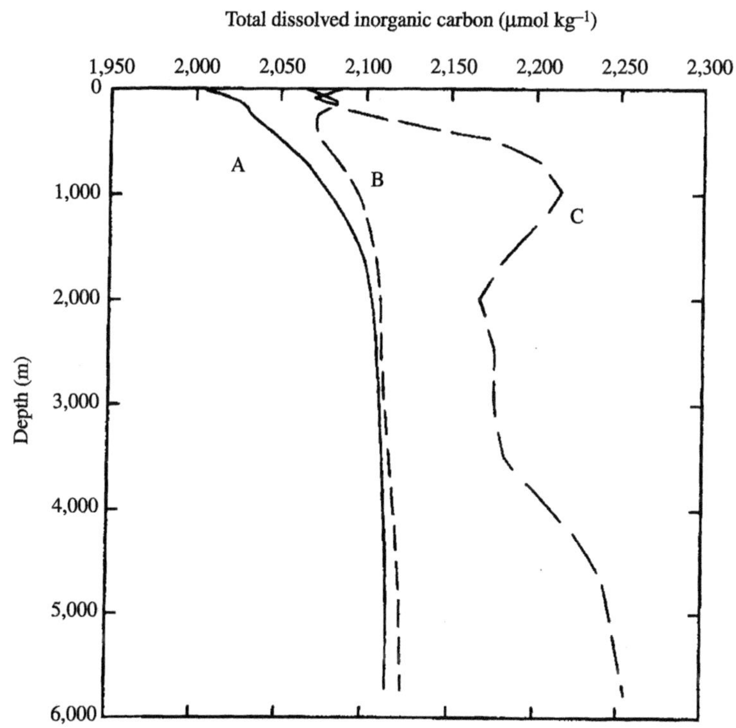

Figure 4 Vertical profiles of total dissolved inorganic carbon (TIC) in the ocean. Curve A corresponds to a theoretical profile that would have been obtained prior to the Industrial Revolution with an atmospheric CO_2 concentration of 280 $\mu mol\ mol^{-1}$. The curve is derived from the solubility coefficients for CO_2 in seawater, using a typical thermal and salinity profile from the central Pacific Ocean, and assumes that when surface water cools and sinks to become deep water it has equilibrated with atmospheric CO_2. Curve B corresponds to the same calculated solubility profile of TIC, but in the year 1995, with an atmospheric CO_2 concentration of 360 $\mu mol\ mol^{-1}$. The difference between these two curves is the integrated oceanic uptake of CO_2 from anthropogenic emissions since the beginning of the Industrial Revolution, with the assumption that biological processes have been in steady state (and hence have not materially affected the net influx of CO_2). Curve C is a representative profile of measured TIC from the central Pacific Ocean. The difference between curve C and B is the contribution of biological processes to the uptake of CO_2 in the steady state (i.e. the contribution of the "biological pump" to the TIC pool.) (courtesy of Doug Wallace and the World Ocean Circulation Experiment).

Table 5 Export production and ef ratios calculated from the model of Laws *et al.* (2000).

	Export (GtC y^{-1})	*ef*
Ocean basin		
Pacific	4.3	0.19
Atlantic	4.3	0.25
Indian	1.5	0.15
Antarctic	0.62	0.28
Arctic	0.15	0.56
Mediterranean	0.19	0.24
Global	11.1	0.21
Total production		
Oligotrophic (chl $a < 0.1$ mg m^{-3})	1.04	0.15
Mesotrophic ($0.1 \leq$ chl $a < 1.0$ mg m^{-3})	6.5	0.18
Eutrophic (chl $a \geq 1.0$ mg m^{-3})	3.6	0.36
Ocean depth		
0 − 100 m	2.2	0.31
100 m−1 km	1.4	0.33
>1 km	7.4	0.18

deposition of fixed nitrogen in the form of aerosols (the latter is produced largely as a consequence of air pollution and, to a lesser extent, from lightning).

8.05.6 NUTRIENT FLUXES

Primary producers are not simply sacks of organic carbon. They are composed of six major elements, namely, hydrogen, carbon, oxygen, nitrogen, phosphorus, and sulfur, and at least 54 other trace elements and metals (Schlesinger, 1997). In steady state, the export flux of organic matter to the ocean interior must be coupled to the upward flux of several of these essential nutrients. The fluxes of nutrients are related to the elemental stoichiometry of the organic matter that sinks into the ocean interior. This relationship, first pointed out by Alfred Redfield in 1934, was based on the chemistry of four of the major elements in the ocean, namely, carbon, nitrogen, phosphorus, and oxygen.

8.05.6.1 The Redfield Ratio

In the ocean interior, the ratio of fixed inorganic nitrogen (in the form of NO_{3-}) to PO_4 in the dissolved phase is remarkably close to the ratios of the two elements in living plankton. Hence, it seemed reasonable to assume that the ratio of the two elements in the dissolved phase was the result of the sinking and subsequent remineralization (i.e., oxidation) of the elements in organic matter produced in the open ocean. Further, as carbon and nitrogen in living organisms are largely found in chemically reduced forms, while remineralized forms are virtually all oxidized, the remineralization of organic matter was coupled to the depletion of oxygen. The relationship could be expressed stoichiometrically as

$$[106(CH_2O)16NH_3 1PO_4^{2-}] + 138O_2 \rightarrow 106CO_2 + 122H_2O + 16NO_3^- + PO_4^{2-} + 16H^+] \quad (7)$$

Hidden within this balanced chemical formulation are biochemical redox reactions, which are contained within specific groups of organisms. (In the oxidation of organic matter, there is some ambiguity about the stoichiometry of O_2/P. Assuming that the mean oxidation level of organic carbon is that of carbohydrate (as is the case in Equation (9)), then the oxidation of that carbon is equimolar with O. Alternatively, some organic matter may be more or less reduced than carbohydrate, and therefore require more or less O for oxidation. Note also that the oxidation of NH_3 to NO_3^- requires four atoms of O, and leads to the formation of one H_2O and one H^+.) When the reactions primarily occur at depth, Equation (7) is driven to the right, while when the reactions primarily occur in the euphotic zone, they are driven to the left. Note that in addition to reducing CO_2 to organic matter, formation of organic matter by photoautotrophs requires reduction of nitrate to the equivalent of ammonia. These two forms of nitrogen are critically important in helping to quantify "new" and export production.

8.05.7 NITRIFICATION

NO_3^- is produced via oxidation of NH_3 by a specific group of eubacteria, the nitrifiers, that are oblibate aerobes found primarily in the water column. The oxidation of NH_3 is coupled to the reduction of inorganic carbon to organic matter; hence nitrification is an example of a chemoautotrophic process that couples the aerobic nitrogen cycle to the carbon cycle. However, because the thermodynamic gradient is very small, the efficiency of carbon fixation by nitrifying bacteria is low and does not provide an ecologically significant source of organic matter in the oceans. In the contemporary ocean, global CO_2 fixation by marine nitrifying bacteria only amounts to ~0.2 Pg C per annum, or ~0.5% of marine photoautotrophic carbon fixation.

There are two major sources of nutrients in the euphotic zone. One is the local regeneration of simple forms of combined elements (e.g., NH_4^+, HPO_4^{2-}, SO_4^{2-}) resulting from the metabolic activity of metazoan and microbial degradation. The second is the influx of distantly produced, "new" nutrients, imported from the deep ocean, the atmosphere (i.e., nitrogen fixation, atmospheric pollution), or terrestrial runoff from streams, rivers, and estuaries (Dugdale and Goering, 1967). In the open ocean, these two sources can be usefully related to the form of inorganic nitrogen assimilated by phytoplankton. Because biological nitrogen fixation is relatively low in the ocean (see below) and nitrification in the upper mixed layer is sluggish relative to the assimilation of nitrogen by photoautotrophs, nitrogen supplied from local regeneration is assimilated before it has a chance to become oxidized. Hence, regenerated nitrogen is primarily in the form of ammonium or urea. In contrast, the fixed inorganic nitrogen in the deep ocean has sufficient time (hundreds of years) to become oxidized, and hence the major source of new nitrogen is in the form of nitrate. Using $^{15}NH_4$ and $^{15}NO_3$ as tracers, it is possible to estimate the fraction of new nitrogen that fuels phytoplankton production (Dugdale and Wilkerson, 1992). This approach provides an estimate of both the upward flux of nitrate required to sustain the $^{15}NO_3^-$ supported production, as well as the downward flux of organic carbon, which is required to maintain a steady-state balance (Dugdale and Wilkerson, 1992; Eppley and Peterson, 1979).

8.05.7.1 Carbon Burial

On geological timescales, there is one important fate for NPP, namely, burial in the sediments. By far the largest reservoir of organic matter on Earth is locked up in rocks (Table 1). Virtually all of this organic carbon is the result of the burial of exported marine organic matter in coastal sediments over literally billions of years of Earth's history. On geological timescales, the burial of marine NPP effectively removes carbon from biological cycles, and places most (not all) of that carbon into the slow carbon cycle. A small fraction of the organic matter escapes tectonic processing via the Wilson cycle and is permanently buried, mostly in continental rocks. The burial of organic carbon is inferred not from direct measurement, but rather from indirect means. One of the most common proxies used to derive burial on geological timescales is based on isotopic fractionation of carbonates. The rationale for this analysis is that the primary enzyme responsible for inorganic carbon fixation is ribulose 1,5-bisphosphate carboxylase/oxygenase (RuBisCO), which catalyzes the reaction between ribulose 1,5-bisphosphate and CO_2 (not HCO_3^-), to form two molecules of 3-phosphoglycerate. The enzyme strongly discriminates against ^{13}C, such that the resulting isotopic fractionation amounts to ~27‰ relative to the source carbon isotopic value. The extent of the actual fractionation is somewhat variable and is a function of carbon availability and of the transport processes for inorganic carbon into the cells, as well as the specific carboxylation pathway (Laws *et al.*, 1997). However, regardless of the quantitative aspects, the net effect of carbon fixation is an enrichment of the inorganic carbon pool in ^{13}C, while the organic carbon produced is enriched in ^{12}C.

8.05.7.2 Carbon Isotope Fractionation in Organic Matter and Carbonates

The isotopic fractionation in carbonates mirrors the relative amount of organic carbon buried. It is generally assumed that the source carbon, from vulcanism (the so-called "mantle" carbon) has an isotopic value of approximately −5‰. As mass balance must constrain the isotopic signatures of carbonate carbon and organic carbon with the mantle carbon, then

$$f_{org} = \frac{\delta_w - \delta_{carb}}{\Delta_B} \qquad (8)$$

where f_{org} is the fraction of organic carbon buried, δ_w is the average isotopic content of the carbon weathered, δ_{carb} is the isotopic signature of the carbonate carbon, and Δ_B the isotopic difference between organic carbon and carbonate carbon deposited in the ocean. Equation (8) is a steady-state model that presumes the source of carbon from the mantel is constant over geological time. This basic model is the basis of nearly all estimates of organic carbon burial rates (Berner *et al.*, 1983; Kump and Arthur, 1999).

Carbonate isotopic analyses reveal positive excursions (i.e., implying organic carbon burial) in the Proterozioc, and more modest excursions throughout the Phanaerozic (Figure 5). Burial of organic carbon on geological timescales implies that export production must deviate from the steady state on ecological timescales. Such a deviation requires changing one or more of (i) ocean nutrient inventories, (ii) the utilization of unused nutrients in enriched areas, (iii) the average elemental composition of the organic material, or (iv) the "rain" ratios of particulate organic carbon to particulate inorganic carbon to the seafloor.

8.05.7.3 Balance between Net Primary Production and Losses

In the ecological theater of aquatic ecosystems, the observed photoautotrophic biomass at any moment in time represents a balance between the rate of growth and the rate of removal of that trophic level. The burial of organic carbon in

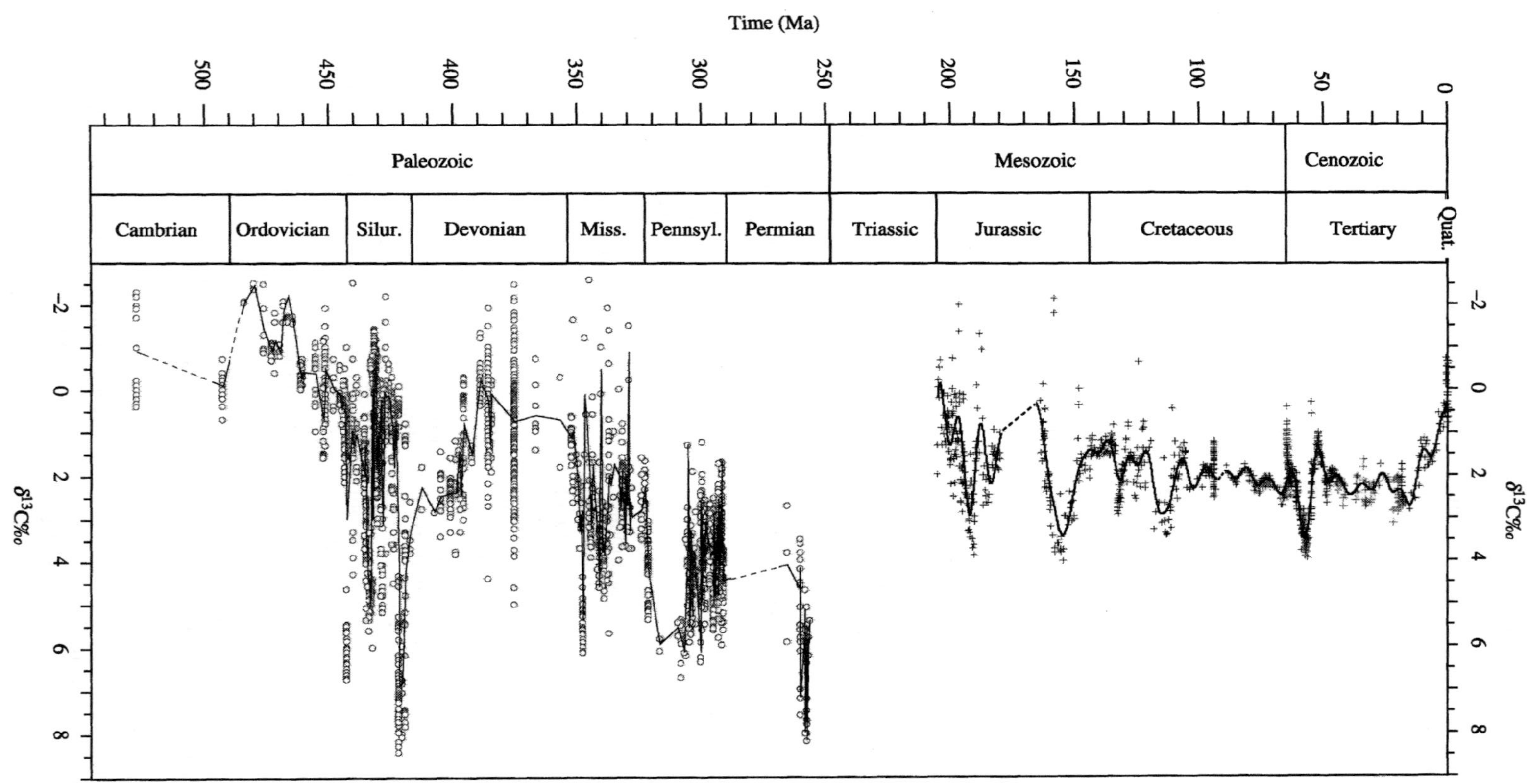

Figure 5 Phanerozoic $\delta^{13}C_{carb}$ record. The Jurassic through the Cenozoic record was generated from bulk sediment carbonates primarily from open ocean Atlantic Deep Sea Drilling Project boreholes; Lower Jurassic samples were used from the Mochras Borehole (Wales) (Katz *et al.*, in review). Dashed intervals indicate data gaps. Singular spectrum analysis was used to generate the Mesozoic–Cenozoic curve. The Paleozoic record was generated from brachiopods (Veizer *et al.*, 1999; note that the timescale has been adjusted from the original reference by interpolating between period boundaries). Data were averaged for each time slice to obtain the Paleozoic curve. The timescales of Berggren *et al.* (1995; Cenozoic), Gradstein *et al.* (1995; Mesozoic), and GSA (Paleozoic) were used.

the lithosphere requires that the ecological balance between NPP and respiration diverge; i.e., the global ocean must be net autotropic. For simplicity, we can express the time-dependent change in photoautrophic biomass by a linear differential equation:

$$dP/dt = [P](\mu - m) \quad (9)$$

where $[P]$ is photoautrophic biomass (e.g., organic carbon), μ is the specific growth rate (units of 1/time), and m is the specific mortality rate (units of 1/time). In this equation, we have lumped all mortality terms, such as grazing and sinking together into one term, although each of these loss processes can be given explicitly (Banse, 1994). Two things should be noted regarding Equation (9). First, μ and m are independent variables; i.e., changes in P can be independently ascribed to one or the other process. Second, by definition, a steady state exists when dP/dt is zero.

8.05.7.4 Carbon Burial in the Contemporary Ocean

The burial of organic carbon in the modern oceans is primarily confined to a few regions where the supply of sediments from terrestrial sources is extremely high. Such regions include the Amazon outfall and Indonesian mud belts. In contrast, the oxidation of organic matter in the interior of the contemporary ocean is extremely efficient; virtually no carbon is buried in the deep sea. Similarly, on most continental margins, organic carbon that reaches the sediments is consumed by microbes within the sediments, such that very little is actually buried in the contemporary ocean (Aller, 1998). The solution to Equation (9) must be close to zero; and consequently, in the absence of human activities, the oxygen content of Earth's atmosphere is very close to steady state and has been zso for tens, if not hundreds, of millions of years.

8.05.7.5 Carbon Burial in the Precambrian Ocean

In contrast, carbon burial in the Proterozoic ocean must have occurred as oxygen increased in the atmosphere; i.e., the global solution to Equation (9) must have been >0. Photoautotrophic biomass could have increased until some element became limiting. Thus, the original feedback between the production of photoautotrophic biomass in the oceans and the atmospheric content of oxygen was determined by an element that limited the crop size of the photoautotrophs in the Archean or Proterozoic ocean. What was that element, and why did it become limiting?

8.05.8 LIMITING MACRONUTRIENTS

A general feature of aquatic environments is that because the oxidation of organic nutrients to their inorganic forms occurs below the euphotic zone where the competing processes of assimilation of nutrients by photoautotrophs do not occur, the pools of inorganic nutrients are much higher at depth. As the only natural source of photosynthetically active radiation is the sun, the gradients of light and nutrients are from opposite directions. Thermal or salinity differences in the surface layers produce vertical gradients in density that effectively retard the vertical fluxes of soluble nutrients from depth. Thus, in the surface layers of a stratified water column, nutrients become depleted as the photoautotrophs consume them at rates exceeding their rate of vertical supply. Indeed, throughout most of the world oceans, the concentrations of dissolved inorganic nutrients, especially fixed inorganic nitrogen and phosphate, are exceedingly low, often only a few nM. One or the other of these nutrients can limit primary production. However, the concept of limitation requires some discussion.

8.05.8.1 The Two Concepts of Limitation

The original notion of limitation in ecology was related to the *yield* of a crop. A limiting factor was the substrate least available relative to the requirement for synthesis of the crop (Liebig, 1840). This concept formed a strong underpinning of agricultural chemistry and was used to design the elemental composition of fertilizers for commercial crops. This concept subsequently was embraced by ecologists and geochemists as a general "law" (Odum, 1971).

Nutrients can also limit the *rate* of growth of photoautotrophs (Blackman, 1905; Dugdale, 1967). Recall that if organisms are in balanced growth, the rate of uptake of an inorganic nutrient relative to the cellular concentration of the nutrient defines the growth rate (Herbert *et al.*, 1956). The uptake of inorganic nutrients is a hyperbolic function of the nutrient concentration and can be conveniently described by a hyperbolic expression of the general form

$$V = (V_{max}(U, \ldots))/(K_s + (U, \ldots)) \quad (10)$$

where V is the instantaneous rate of nutrient uptake, V_{max} is the maximum uptake rate, $(U, \ldots)$ represent the substrate concentration of nutrient U, etc., and K_s is the concentration supporting half the maximum rate of uptake (Dugdale, 1967; Monod, 1942). There can be considerable variation between species with regard to K_s and V_{max} values and these variations are potential sources of competitive selection (Eppley *et al.*, 1969; Tilman, 1982).

It should be noted that Liebig's notion of limitation was not related *a priori* to the intrinsic rate of photosynthesis or growth. For example, photosynthetic rates can be (and often are) limited by light or temperature. The two concepts of limitation (yield and rate) are often not understood correctly: the former is more relevant to biogeochemical cycles, the latter is more critical to selection of species in ecosystems.

8.05.9 THE EVOLUTION OF THE NITROGEN CYCLE

Globally, nitrogen and phosphorus are the two elements that immediately limit, in a Liebig sense, the biologically mediated carbon assimilation in the oceans by photoautotrophs. It is frequently argued that since N_2 is abundant in both the ocean and the atmosphere, and, in principle, can be biologically reduced to the equivalent of NH_3 by N_2-fixing cyanobacteria, nitrogen cannot be limiting on geological timescales (Barber, 1992; Broecker *et al.*, 1980; Redfield, 1958). Therefore, phosphorus, which is supplied to the ocean by the weathering of continental rocks, must ultimately limit biological productivity. The underlying assumptions of these tenets should, however, be considered within the context of the evolution of biogeochemical cycles.

By far, the major source of fixed inorganic nitrogen for the oceans is via biological nitrogen fixation. Although in the Archean atmosphere, electrical discharge or bolide impacts may have promoted NO formation from the reaction between N_2 and CO_2, the yield for these reactions is extremely low. Moreover, atmospheric NH_3 would have photodissociated from UV radiation (Kasting, 1990), while N_2 would have been stable (Kasting, 1990; Warneck, 1988). Biological N_2 fixation is a strictly anaerobic process (Postgate, 1971), and the sequence of the genes encoding the catalytic subunits for nitrogenase is highly conserved in cyanobacteria and other eubacteria, strongly suggesting a common ancestral origin (Zehr *et al.*, 1995). The antiquity and homology of nitrogen fixation capacity also imply that fixed inorganic nitrogen was scarce prior to the evolution of diazotrophic organisms; i.e., there was strong evolutionary selection for nitrogen fixation in the Archean or early Proterozoic periods. In the contemporary ocean, N_2 is still catalyzed solely by prokaryotes, primarily cyanobacteria (Capone and Carpenter, 1982).

While apatite and other calcium-based and substituted solid phases of phosphate minerals precipitated in the primary formation of crustal sediments, secondary reactions of phosphate with aluminum and transition metals such as iron are mediated at either low salinity, low pH, or high oxidation states of the cations (Stumm and Morgan, 1981). Although these reactions would reduce the overall soluble phosphate concentration, the initial condition of the Archean ocean probably had a fixed low N:P ratio in the dissolved inorganic phase. As N_2 fixation proceeded, that ratio would have increased with a buildup of ammonium in the ocean interior. The accumulation of fixed nitrogen in the oceans would continue until the N:P ratio of the inorganic elements reached equilibrium with the N:P ratio of the sedimenting particulate organic matter (POM). Presumably, the latter ratio would approximate that of extant, nitrogen-fixing marine cyanobacteria, which is ~16:1 by atoms (Copin-Montegut and Copin-Montegut, 1983; Redfield, 1958; Quigg *et al.*, 2003) or greater (Letelier *et al.*, 1996) and would ultimately be constrained by the availability of phosphate (Falkowski, 1997; Tyrell, 1999).

The formation of nitrate from ammonium by nitrifying bacteria requires molecular oxygen; hence, nitrification must have evolved following the formation of free molecular oxygen in the oceans by oxygenic photoautotrophs. Therefore, from a geological perspective, the conversion of ammonium to nitrate probably proceeded rapidly and provided a substrate, NO_3^-, that eventually could serve both as a source of nitrogen for photoautotrophs and as an electron acceptor for a diverse group of heterotrophic, anaerobic bacteria, the denitrifiers.

In the sequence of the three major biological processes that constitute the nitrogen cycle, denitrification must have been the last to emerge. This process, which permits the reduction of NO_3^- to (ultimately) N_2, occurs in the modern ocean in three major regions, namely, continental margin sediments, areas of restricted circulation such as fjords, and oxygen minima zones of perennially stratified seas (Christensen *et al.*, 1987; Codispoti and Christensen, 1985; Devol, 1991; Nixon *et al.*, 1996). In all cases, the process requires hypoxic or anoxic environments and is sustained by high sinking fluxes of organic matter. Denitrification appears to have evolved independently several times; the organisms and enzymes responsible for the pathway are highly diverse from a phylogenetic and evolutionary standpoint.

With the emergence of denitrification, the ratio of fixed inorganic nitrogen to dissolved inorganic phosphate in the ocean interior could only be depleted in nitrogen relative to the sinking flux of the two elements in POM. Indeed, in all of the major basins in the contemporary ocean, the N:P ratio of the dissolved inorganic nutrients in the ocean interior is conservatively estimated at 14.7

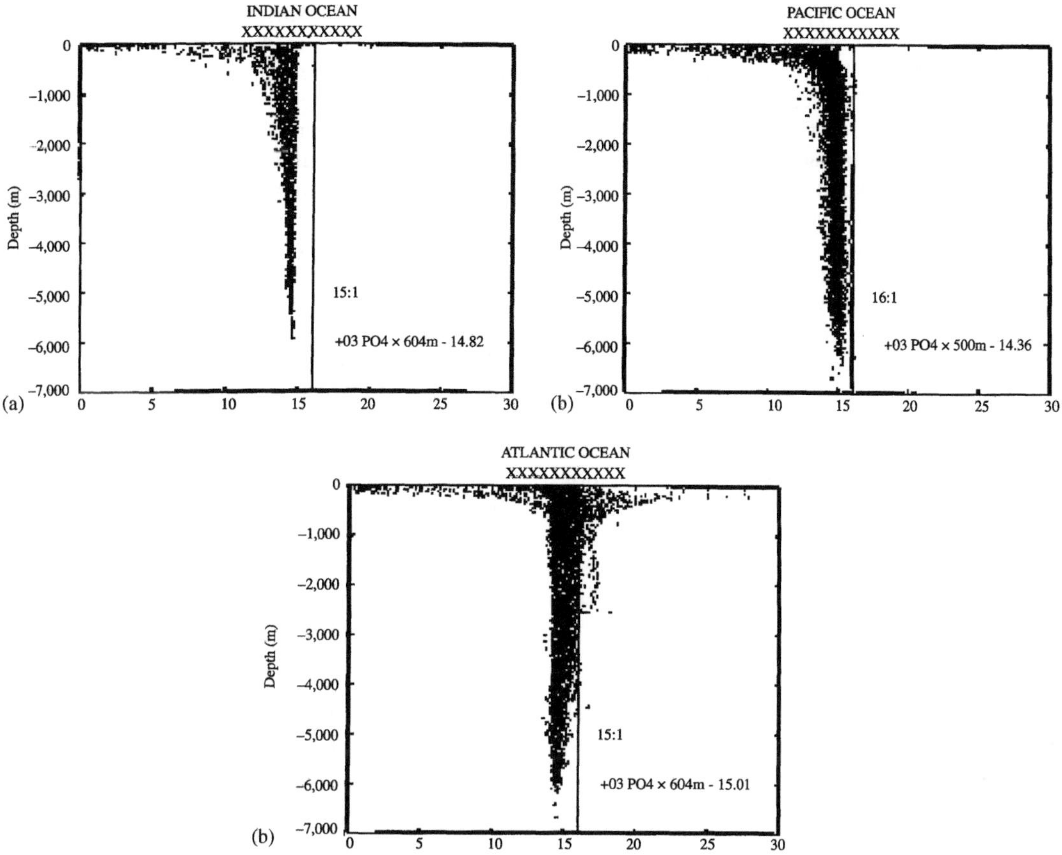

Figure 6 (a)–(c) Vertical profiles of NO_3^-/HPO_4^{3-} ratios in each of the three major ocean basins. The data were taken from the GEOSECS database. In all three basins, the N:P ratio converges on an average value that is significantly lower than the 16:1 ratio predicted by Redfield. The deficit in nitrogen relative to phosphorus is presumed to be a result of denitrification. Note that in the upper 500 m of the water column, NO_3^-/HPO_4^{3-} ratios generally decline except in a portion of the Atlantic that corresponds to the eastern Mediterranean.

by atoms (Fanning, 1992) or less (Anderson and Sarmiento, 1994) (Figure 6).

There are three major conclusions that may be drawn from the foregoing discussion:

(i) Because the ratio of the sinking flux of particulate organic nitrogen and particulate phosphorus exceeds the N:P ratio of the dissolved pool of inorganic nutrients in the ocean interior, the average upward flux of inorganic nutrients must be slightly enriched in phosphorus relative to nitrogen as well as to the elemental requirements of the photoautotrophs (Gruber and Sarmiento, 1997; Redfield, 1958). Hence, although there are some exceptions (Kromer, 1995; Wu *et al.*, 2000), dissolved, inorganic fixed nitrogen generally limits primary production throughout most of the world's oceans (Barber, 1992; Falkowski *et al.*, 1998).

(ii) The N:P ratio of the dissolved pool of inorganic nutrients in the ocean interior was established by biological processes, not vice versa (Redfield *et al.*, 1963, 1934). The elemental composition of marine photoautotrophs has been conserved since the evolution of the eukaryotic phytoplankton (Lipps, 1993). The Redfield N:P ratio of 16:1 for POM (Codispoti, 1995; Copin-Montegut and Copin-Montegut, 1983; McElroy, 1983; Redfield *et al.*, 1963, 1958) is an upper bound, which is not observed for the two elements in the dissolved inorganic phase in the ocean interior. The deficit in dissolved inorganic fixed nitrogen relative to soluble phosphate in the ocean represents a slight imbalance between nitrogen fixation and denitrification on timescales of $\sim 10^3 - 10^4$ yr (Codispoti, 1995).

(iii) If dissolved inorganic nitrogen rather than phosphate limits productivity in the oceans, then it follows that the ratio of nitrogen fixation/denitrification plays a critical role in determining the net biologically mediated exchange of CO_2 between the atmosphere and ocean (Codispoti, 1995).

8.05.10 FUNCTIONAL GROUPS

As we have implied throughout the foregoing discussion, the biologically mediated fluxes of elements between the upper ocean and the ocean interior are critically dependent upon key groups of organisms. Fluxes between the atmosphere and ocean, as well as between the ocean and the lithosphere, are mediated by organisms that catalyze phase state transitions from either gas to solute/solid or from solute to solid/gas phases. For example, autotrophic carbon fixation converts gaseous CO_2 to a wide variety of organic carbon molecules, virtually all of which are solid or dissolved solids at physiological temperatures. Respiration accomplishes the reverse. Nitrogen fixation converts gaseous N_2 to ammonium and thence to organic molecules, while denitrification accomplishes the reverse. Calcification converts dissolved inorganic carbon and calcium to solid phase calcite and aragonite, whereas silicification converts soluble silicic acid to solid hydrated amorphous opal. Each of these biologically catalyzed processes is dependent upon specific metabolic sequences (i.e., gene families encoding a suite of enzymes) that evolved over hundreds of millions of years of Earth's history, and have, over corresponding periods, led to the massive accumulation of oxygen in the atmosphere, and opal, carbonates, and organic matter in the lithosphere. Presumably, because of parallel evolution as well as lateral gene transfer, these metabolic sequences have frequently co-evolved in several groups of organisms that, more often than not, are not closely related from a phylogenetic standpoint (Falkowski, 1997). Based on their biogeochemical metabolism, these homologous sets of organisms are called functional groups or biogeochemical guilds; i.e., organisms that are related through common biogeochemical processes rather than a common evolutionary ancestor affiliation.

8.05.10.1 Siliceous Organisms

In the contemporary ocean, the export of particulate organic carbon from the euphotic zone is highly correlated with the flux of particulate silicate. Most of the silicate flux is a consequence of precipitation of dissolved orthosilicic acid by diatoms to form amorphous opal that makes up the cell walls of these organisms. These hard-shelled cell walls presumably help the organisms avoid predation, or if ingested, increase the likelihood of intact gut passage through some metazoans (Smetacek, 1999). In precipitating silicate, diatoms simultaneously fix carbon. Upon depleting the euphotic zone of nutrients, the organisms frequently sink *en masse*, and while some are grazed en route, many sink as intact cells. Ultimately, either fate leads to the gravitationally driven export flux of particulate organic carbon into the ocean interior.

Silica is supplied to the oceans from the weathering of continental rocks. Because of precipitation by silicious organisms, however, the ocean is relatively depleted in dissolved silica. Although diatom frustules (their silicified cell walls) tend to dissolve and are relatively poorly preserved in marine sediments, enough silica is buried to keep the seawater undersaturated, throughout the ocean. As the residence time of silica in the oceans is $\sim 10^4$ yr (i.e., about an order of magnitude longer than the mean deep-water circulation), one can get an appreciation for the silicate demands and regeneration rates by following the concentration gradients of dissolved silica along isopycnals. While these demands are generally attributed to diatoms, radiolarians (a group of nonphotosynthetic, heterotrophic protists with silicious tests that are totally unrelated to diatoms) are not uncommon, and radiolarian shells are abundant in the sediments of Southern Ocean. Silica is also precipitated by various sponges and other protists. As a functional group, the silicate precipitators are identified by their geochemical signatures in the sediments and in the silica chemistry of the oceans.

8.05.10.2 Calcium Carbonate Precipitation

Like silica precipitation, calcium carbonate is not confined to a specific phylogenetically distinct group of organisms, but evolved (apparently independently) several times in marine organisms. Carbonate sediments blanket much of the Atlantic basin, and are formed from the shells of both coccolithophorids and foraminifera (Milliman, 1993). (In the Pacific, the carbon compensation depth is generally higher than the bottom, and hence, in that basin carbonates tend to dissolve rather than become buried.) As the crystal structure of the carbonates in both groups is calcite (as opposed to the more diagenetically susceptible aragonite), the preservation of these minerals and their co-precipitating trace elements provides an invaluable record of ocean history. Although on geological timescales huge amounts of carbon are removed from the atmosphere and ocean and stored in the lithosphere as carbonates, on ecological timescales, carbonate formation leads to the formation of CO_2. This reaction can be summarized by the following:

$$2HCO_3^- + Ca^{2+} \rightarrow CaCO_3 + CO_2 + H_2O \qquad (11)$$

Unlike silicate precipitation, calcium carbonate precipitation leads to strong optical signatures that can be detected both *in situ* and remotely (Balch *et al.*, 1991; Holligan and Balch, 1991). The basic principle of detection is the large, broadband (i.e., "white") scattering cross-sections of calcite. The high scattering cross-sections are detected by satellites observing the upper ocean as relatively highly reflective properties (i.e., a "bright" ocean).

Using this detection scheme, one can reconstruct global maps of planktonic calcium carbonate precipitating organisms in the upper ocean. *In situ* analysis can be accompanied by optical rotation properties (polarization) to discriminate calcite from other scattering particles. *In situ* profiles of calcite can be used to construct the vertical distribution of calcium carbonate-precipitating planktonic organisms that would otherwise not be detected by satellite remote sensing because they are too deep in the water column.

Over geological time, the relative abundances of key functional groups change. For example, relative coccolithophorid abundances generally increased through the Mesozoic, and underwent a culling at the Cretaceous Tertiary (K/T) boundary, followed by a general waning throughout the Cenozoic. The changes in the coccolithophorid abundances appear to follow eustatic sea-level variations, suggesting that transgressions lead to higher calcium carbonate deposition. In contrast, diatom sedimentation increases with regressions and, since the K/T impact, diatoms have generally replaced coccolithophorids as ecologically important eukaryotic phytoplankton. On much finer timescales during the Pleistocene, it would appear that interglacial periods favor coccolithophorid abundance, while glacial periods favor diatoms. The factors that lead to glacial–interglacial variations between these two functional groups are relevant to elucidating their distributions in the contemporary ecological setting of the ocean (Tozzi, 2001).

8.05.10.3 Vacuoles

In addition to a silicic acid requirement, diatoms, in contrast to dinoflagellates and coccolithophores, have evolved a nutrient storage vacuole (Raven, 1987). The vacuole, which occupies ~35% of the volume of the cell, can retain high concentrations of nitrate and phosphate. Importantly, ammonium cannot be (or is not) stored in a vacuole. The vacuole allows diatoms to access and hoard pulses of inorganic nutrients, thereby depriving potentially competing groups of these essential resources. Consequently, diatoms thrive best under eutrophic conditions and in turbulent regions where nutrients are supplied with high pulse frequencies.

The competition between diatoms and coccolithophorids can be easily modeled by a resource acquisition model based on nutrient uptake (Equation (9)). In such a model, diatoms dominate under highly turbulent conditions, when their nutrient storage capacity is maximally advantageous, while coccolithophorids dominate under relatively quiescent conditions (Tozzi *et al.*, 2003).

The geological record during the Pleistocene reveals a periodicity of opal/calcite deposition corresponding to glacial/interglacial periods. Such alterations in mineral deposition are probably related to upper ocean turbulence; i.e., the sedimentary record is a "fax" machine of mixing (Falkowski, 2002). Glacial periods appear to be characterized by higher wind speeds and a stronger thermal contrast between the equator and the poles. These two factors would, in accordance with the simple nutrient uptake model, favor diatoms over coccolithophores. During interglacials, more intense ocean stratification, weaker winds, and a smaller thermal contrast between the equator and the poles would tend to reduce upper-ocean mixing and favor coccolithophores (Iglesias-Rodriguez *et al.*, 2002). While other factors such as silica availability undoubtedly also influenced the relative success of diatoms and coccolithophores on these timescales, we suggest that the climatically forced cycle, played out on timescales of 40 kyr and 100 kyr (over the past 1.9 Myr), can be understood as a long-term competition that never reaches an exclusion equilibrium condition (Falkowski *et al.*, 1998).

Can the turbulence argument be extended to even longer timescales to account for the switch in the dominance from coccolithophorids to diatoms in the Cenozoic? The fossil record of diatoms in the Mesozoic is obscured by problems of preservation; however several species are preserved in the late Jurassic (Harwood and Nikolaev, 1995), suggesting that the origins were in the early Jurassic or perhaps as early as the Triassic. It is clear, however, that this group did not contribute nearly as much to export production during Mesozoic times. We suggest that the ongoing successional displacement of coccolithophores by diatoms in the Cenozoic is, to first order, driven by tectonics (i.e., the Wilson cycle). The Mesozoic period was relatively warm and was characterized by a two-cell Hadley circulation, with obliquity greater than 37°, resulting in a thoroughly mixed atmosphere with nearly uniform temperatures over the surface of the Earth. The atmospheric meridional heat transport decreased the latitudinal thermogradients; global winds and ocean circulation were both sluggish (Huber *et al.*, 1995). This relatively quiescent period of Earth's history was ideal for coccolithophorids. Following the K/T impact, and more critically, the onset of polar ice caps about 32 Ma, the Hadley circulation changed dramatically. Presently, there are six Hadley cells, and the atmosphere has become drier. The net result is more intense thermohaline circulation, greater wind mixing and decreased stability (Barron *et al.*, 1995; Chandler *et al.*, 1992). Associated with this decreased stability is the rise of the diatoms.

Over the past 50 Myr, both carbon and oxygen isotopic records in fossil foraminifera suggest that there has been a long-term depletion of CO_2 in the ocean–atmosphere system and a decrease in temperature in the ocean interior. The result has been increased stratification of the ocean, which has, in turn, led to an increased importance for wind-driven upwelling and mesoscale eddy turbulence in providing nutrients to the euphotic zone. The ecological dominance of diatoms under sporadic mixing conditions suggests that their long-term success in the Cenozoic reflects an increase in event-scale turbulent energy dissipation in the upper ocean. But, was the Wilson cycle the only driver?

Although weathering of siliceous minerals by CO_2 (the so-called "Urey reactions"; but see (Berner and Maasch, 1996)) contributed to the long-term flux of silica to the oceans (Berner, 1990), and potentially fostered the radiation of diatoms in the Cenozoic, by itself, orogeny cannot explain the relatively sharp increase in diatoms at the Eocene/Oligocene boundary. Indeed, the seawater strontium isotope record does not correspond with these radiations in diatoms (Raymo and Ruddiman, 1992). We must look for other contributing processes.

Shortly after, or perhaps coincident with Paleocene thermal maximum (55 Ma), was a rise in true grasses (Retallack, 2001). This group, which rapidly radiated in the Eocene, rose to prominence in Oligocene, a period coincident with a global climatic drying. During this period, however, there was a rapid co-evolution of grazing ungulates that displaced browsers (Janis and Damuth, 1990). Grasses contain up to 10% dry weight of silica, which forms micromineral deposits in the cell walls; phytoliths (Conley, 2002). Indeed, the selection of hypsodont (high crown) dentition in ungulates from the brachydont (leaf eating) early-appearing browsing mammals, coincides with the widespread distribution of phytoliths and grit in grassland forage. It is tempting to suggest that the rise of grazing ungulates, which spurred the radiation of grasses, was, in effect, a biologically catalyzed silicate weathering process. The deep-root structure of Eocene grasses certainly facilitated silicate mobilization into rivers and groundwaters (Conley, 2002). Additionally, upon their annual death and decay, the phytoliths of many temperate grasses are potentially transported to the oceans via wind.

The feedback between the co-evolution of mammals and grasses and the supply of silicates to the ocean potentially explains the rapid radiation of diatoms, and their continued dominance in the Cenozoic. There is another potential feedback at play, however, which "locked in" the diatom preeminence. It is likely that the increase in diatom dominance, and the associated increase in the efficiency of carbon burial, played a key role in decreasing atmospheric CO_2 over the past 32 Myr. That biological selection may influence climate is clearly controversial; however, the trends in succession between taxa on timescales of tens of millions of years, and cycles in dominance on shorter geological timescales beg for explanation.

8.05.11 HIGH-NUTRIENT, LOW-CHLOROPHYLL REGIONS—IRON LIMITATION

On ecological timescales, the biologically mediated net exchange of CO_2 between the ocean and atmosphere is limited by nutrient supply and the efficiency of nutrient utilization in the euphotic zone. There are three major areas of the world ocean where inorganic nitrogen and phosphate are in excess throughout the year, yet the mixed-layer depth appears to be shallower than the critical depth; these are the eastern equatorial Pacific, the subarctic Pacific, and Southern (i.e., Antarctic) Oceans. In the subarctic North Pacific, it has been suggested that there is a tight coupling between phytoplankton production and consumption by zooplankton (Miller *et al.*, 1991). This grazer-limited hypothesis has been used to explain why the phytoplankton in the North Pacific do not form massive blooms in the spring and summer like their counterparts in the North Atlantic (Banse, 1992). In the mid-1980s, however, it became increasingly clear that the concentration of trace metals, especially iron, was extremely low in all three of these regions (Martin, 1991). Indeed, in the eastern equatorial Pacific, for example, the concentration of soluble iron in the euphotic zone is only 100–200 pM. Although iron is the most abundant transition metal in the Earth's crust, in its most commonly occurring form, Fe^{3+}, it is virtually insoluble in seawater. The major source of iron to the euphotic zone is Aeolian dust, originating from continental deserts. In the three major areas of the world oceans with high inorganic nitrogen in the surface waters and low chlorophyll concentrations, the flux of Aeolian iron is extremely low (Duce and Tindale, 1991). In experiments in which iron was artificially added on a relatively large scale to the waters in the equatorial Pacific, Southern Ocean, and subarctic Pacific, there were rapid and dramatic increases in photosynthetic energy conversion efficiency and phytoplankton chlorophyll (Abraham *et al.*, 2000; Behrenfeld *et al.*, 1996; Kolber *et al.*, 1994; Tsuda *et al.*, 2003). Beyond doubt, NPP and export production in all three regions are limited by the availability of a single micronutrient—iron.

8.05.12 GLACIAL–INTERGLACIAL CHANGES IN THE BIOLOGICAL CO_2 PUMP

In the modern (i.e., interglacial) ocean, two major factors affect iron fluxes. First, changes in land-use patterns and climate over the past several thousand years have had, and continue to have, marked effects on the areal distribution and extent of deserts. At the height of the Roman Empire some 2,000 years ago, vast areas of North Africa were forested, whereas today these same areas are desert. These changes were climatologically induced. Similarly, the Gobi Desert in North Central Asia has increased markedly in modern times. The flux of Aeolian iron from the Sahara Desert fuels photosynthesis for most of the North Atlantic Ocean; that from the Gobi is deposited over much of the North Pacific (Duce and Tindale, 1991). The primary source of iron for the Southern Ocean is Australia, but the prevailing wind vectors constrain the delivery of the terrestrial dust to the Indian Ocean. Consequently, the Southern Ocean is iron limited in the modern epoch (Martin, 1990).

The second factor in this climatological feedback is that the major wind vectors are driven by atmosphere–ocean heat gradients. Changes in thermal gradients between the equator and poles lead to changes in wind speed and direction. Wind vectors prior to glaciations appear to have supported high fluxes of iron to the Southern Ocean, thereby presumably stimulating phytoplankton production, the export of carbon to depth, and the drawdown of atmospheric CO_2 appears to have accompanied glaciations in the recent geological past (Berger, 1988).

8.05.13 IRON STIMULATION OF NUTRIENT UTILIZATION

The enhancement of *net* export production (i.e., "true" new production) requires the addition of a limiting nutrient to the ocean, an increase in the efficiency of utilization of preformed nutrients in the upper ocean, and/or a change in the elemental stoichiometry of primary producers (Falkowski *et al.*, 1998; Sarmiento and Bender, 1994). Indeed, an analysis of ice cores from Antarctica, reconstruction of Aeolian iron depositions and concurrent atmospheric CO_2 concentrations over the past 4.2×10^5 yr (spanning four glacial–interglacial cycles) suggests that, when iron fluxes were high, CO_2 levels were low and vice versa (Martin, 1990; Petit *et al.*, 1999). Variations in iron fluxes were presumably a consequence of the areal extent of terrestrial deserts and wind vectors. It is hypothesized that increased fluxes of iron to the high-nutrient, low chlorophyll Southern Ocean stimulated phytoplankton photosynthesis and led to a drawdown of atmospheric CO_2. Model calculations suggest that the magnitude of this drawdown could have been cumulatively significant, and accounted for the observed variations in atmospheric CO_2 recorded in gases trapped in the ice cores. However, the sedimentary records reveal large glacial fluxes of organic carbon in low- and mid-latitude regions; areas that are presumably nutrient impoverished. Was there another factor besides iron addition to high-nutrient, low-chlorophyll regions that contributed to a net export of carbon during glacial times?

Given that N:P ratios in the ocean interior are lower than for the sinking flux, an increase in the net delivery of fixed inorganic nitrogen to the ocean would also potentially contribute to a net drawdown of CO_2. The Antarctic ice core records suggest that atmospheric CO_2 declined from $\sim$290 μmol mol^{-1} to $\sim$190 μmol mol^{-1} over a period of $\sim 8 \times 10^4$ yr between the interglacial and glacial maxima (Sigman and Boyle, 2000). Assuming a C:N ratio of about 6.5 by atoms for the synthesis of new organic matter in the euphotic zone, a simple equilibrium, three-box model calculation suggests that 600 Pg of inorganic carbon should have been fixed by marine photoautotrophs to account for the change in atmospheric CO_2. This amount of carbon is approximately threefold greater than that released to the atmosphere from the cumulative combustion of fossil fuels since the beginning of the Industrial Revolution. The calculated change in atmospheric CO_2 would have required an addition of $\sim$1.5 Tg fixed nitrogen per annum; this is $\sim$2% of the global mean value in the contemporary ocean.

8.05.14 LINKING IRON TO N_2 FIXATION

Iron also appears to exert a strong constraint on N_2 fixation in the modern ocean. Nitrogen-fixing cyanobacteria, like most cyanobacteria, require relatively high concentrations of iron (Berman-Frank *et al.*, 2001). The high-iron requirements come about because these organisms generally have high requirements for this element in their photosynthetic apparatus (Fujita *et al.*, 1990), as well as for nitrogen fixation and electron carriers that are critical for providing the reductants for CO_2 and N_2 fixation *in vivo*. Increased Aeolian flux of iron to the oceans during glacial periods may have therefore not only stimulated the utilization of nutrients in high-nutrient, low-chlorophyll regions, but also stimulated nitrogen fixation by cyanobacteria, and hence indirectly provided a significant source of new nitrogen. Both effects would have led to increased photosynthetic carbon fixation, and a net drawdown of atmospheric CO_2 (Falkowski, 1997; Falkowski *et al.*, 1998).

8.05.15 OTHER TRACE-ELEMENT CONTROLS ON NPP

Is iron the only limiting trace element in the sea? Prior to the evolution of oxygenic photosynthesis, the oceans contained high concentrations (~1 mM) of dissolved iron in the form of Fe^{2+} and manganese (>1 mM) in the form of Mn^{2+}, but essentially no copper or molybdenum; these and other elements would have been precipitated as sulfides. Thus, both iron and manganese were readily available to the early photoautotrophs. The availability of these two elements permitted the evolution of the photosynthetic apparatus and the oxygen-evolving system that ultimately became the genetic template for all oxygenic photoautotrophs (Blankenship, 1992). Indeed, the availability of these transition metals, which is largely determined by the oxidation state of the environment, appears to account for their use in photosynthetic reactions. However, over geological time, oxygenic photoautotrophs themselves altered the redox state of the ocean, and hence the availability of the elements in the soluble phase (Anbar and Knoll, 2002).

As photosynthetic oxygen evolution proceeded in the Proterozoic oceans, singlet oxygen ($^1O_2$), peroxide (H_2O_2), superoxide anion radicals ($O_2^{\cdot-}$), and hydroxide radicals ($^{\cdot}OH$) were all formed as by-products (Kasting, 1990; Kasting *et al.*, 1988). These oxygen derivatives can oxidize proteins and photosynthetic pigments as well as cause damage to reaction centers (Asada, 1994). A range of molecules evolved to scavenge or quench the potentially harmful oxygen by-products; these include superoxide dismutase (which converts $O_2^{\cdot-}$ to O_2 and H_2O_2), peroxidase (which reduces H_2O_2 to H_2O by oxidizing an organic cosubstrate for the enzyme), and catalase (which converts $2H_2O_2$ to H_2O and O_2). The oldest superoxide dismutases contained iron and/or manganese, while the peroxidases and catalases contained iron (Asada *et al.*, 1980). These transition metals facilitate the electron transfer reactions that are at the core of the respective enzyme activity, and their incorporation into the proteins undoubtedly occurred because the metals were readily available (Williams and Frausto da Silva, 1996). As O_2 production proceeded, the oxidation of Fe^{2+}, Mn^{2+}, and S^{2-} eventually led to the virtual depletion of these forms of the elements in the euphotic zone of the oceans. The depletion of these elements had profound consequences on the subsequent evolution of life. In the first instance, a number of enzymes were selected that incorporated alternative transition metals that were available in the oxidized ocean. For example, a superoxide dismutase evolved in the green algae, and hence higher plants (and many non-photosynthetic eukaryotes) that utilized copper and zinc (Falkowski, 1997). Similar metal substitutions occurred in the photosynthetic apparatus and in mitochondrial electron transport chains.

The overall consequence of the co-evolution of oxygenic photosynthesis and the redox state of the ocean is a relatively well-defined trace-element composition of the bulk phytoplankton. Analogous to Redfield's relationship between the macronutrients, trace-element analyses of phytoplankton reveals a relation for trace elements normalized to cell phosphorus of $(C_{125}N_{16}P_1S_{1.3}K_{1.7}Mg_{0.56}Ca_{0.4})_{1000}Sr_{4.4}Fe_{7.5}Mn_{3.8}Zn_{0.80}Cu_{0.38}Co_{0.19}Cd_{0.21}Mo_{0.03}$. The relative composition of transition trace elements in phytoplankton is reflected in that of black shales (Figure 7). Thus, while there is a strong correlation between N:P ratios in phytoplankton and seawater (Anderson and Sarmiento, 1994; Lenton and Watson, 2000), there is no parallel correlation for trace elements (Whitfield, 2001; Morel and Hudson, 1985) (Figure 7 inset). There are two underlying explanations for the lack of a strong relationship. First, low abundance cations that have a valance state of two or higher are often assimilated with particles, whether or not they are metabolically required. Hence, while zinc, manganese, and cobalt are depleted in surface waters and are used in metabolic cycles, mercury, lanthanum, and other rare-earth elements are also depleted and have no known biological function. The profiles of these elements are dictated, to first order, by particle fluxes and their ligands. Second, the absolute abundance of transition metals is critically dependent on the solubility of the source minerals, which is, in turn, regulated by redox chemistry. Most key metabolic pathways evolved prior to the oxidation of Earth's atmosphere and ocean, and hence many of the transition metals selected for catalyzing biological redox reactions reflect their relative abundance under anoxic or suboxic conditions. For example, although the contemporary ocean is oxidized, no known metal can substitute for manganese in the water splitting complex in the photosynthetic apparatus. Similarly, the photosynthetic machinery has maintained a strict requirement for iron for over 2.8 Ga, and all nitrogenases require at least 16 iron atoms per enzyme complex. Some key biological processes do not have the flexibility to substitute trace elements based simply on their availability (Williams and Frausto da Silva, 1996). Hence, while oxygenic photoautotrophs indirectly determine the distribution of the major trace elements in the ocean interior, the distribution of these elements in the soluble phase does not reflect with the composition of the organisms.

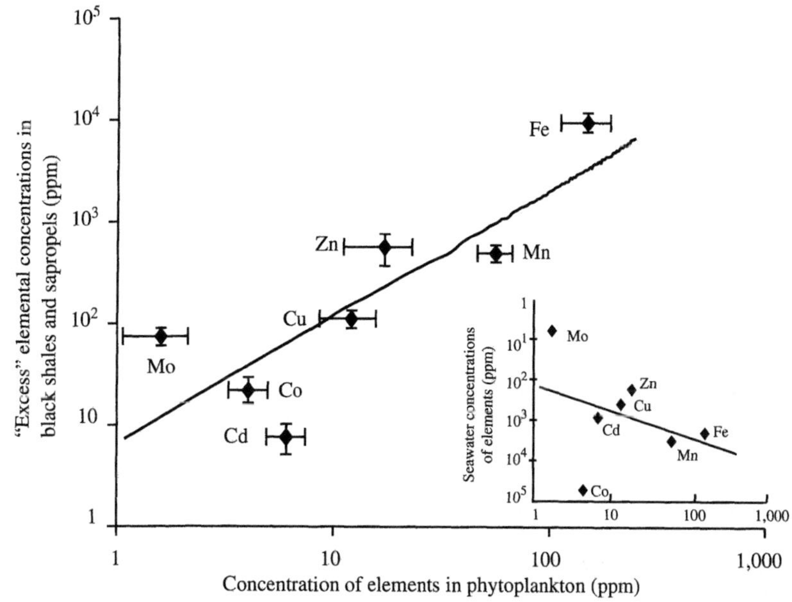

Figure 7 The excess trace element composition in black shales compared with that of calculated marine phytoplankton, and the relationship between trace elements in seawater and phytoplankton (inset) (source Quigg *et al.*, 2003).

On long timescales, seawater concentrations of trace elements reflect the balance between their sources and sinks. For trace elements, the sink term is tightly coupled with the extent of redox conditions throughout the ocean, where periods of extended reducing conditions result in greater partitioning of organic carbon into the deep ocean and sediments. This, in turn, leads to enhanced sequestering of redox-sensitive trace elements into sediments, thereby decreasing their seawater concentrations. These sedimentary rocks have a high content of marine fractions (i.e., organic matter, apatite, biogenic silica and carbonates), and so are enriched by 1–2 orders of magnitude in several trace elements: zinc, copper, nickel, molybdenum, chromium, and vanadium (Piper, 1994). The positive correlation between trace-element ratios in phytoplankton and sediments is consistent with the notion that phytoplankton have imprinted their activities on the lithosphere.

8.05.16 CONCLUDING REMARKS

The evolution of primary producers in the oceans profoundly changed the chemistry of the atmosphere, ocean, and lithosphere of Earth. The photosynthesis processes catalyzed by ensemble of these organisms not only influences the six major light elements, but directly and indirectly affect every major soluble redox-sensitive trace element and transition metal on Earth's surface. These processes continue to provide, primarily through the utilization of solar radiation, a disequilibrium in geochemical processes, such that Earth maintains an oxidized atmosphere and ocean. This disequilibrium prevents atmospheric oxygen from being depleted, maintains a lowered atmospheric CO_2 concentration, and simultaneously imprints on the ocean interior and lithosphere elemental composition that reflect those of the bulk biological material from which it is derived.

While primary producers in the ocean comprise only ~1% of Earth's biomass, their metabolic rate and biogeochemical impact rivals the much larger terrestrial ecosystem. On geological timescales, these organisms are the little engines that are essential to maintaining life as we know it on this planet.

ACKNOWLEDGMENTS

The author's research is supported by grants from the US National Science Foundation, the National Aeronautical and Space Administration, the US Department of Energy, and the US Department of Defense.

REFERENCES

Abraham E. R., Law C. S., Boyd P. W., Lavender S. J., Maldonado M. T., and Bowie A. R. (2000) Importance of stirring in the development of an iron-fertilized phytoplankton bloom. *Nature* **407**(6805), 727–730.

Aller R. C. (1998) Mobile deltaic and continental shelf muds as fluidized bed reactors. *Mar. Chem.* **61**, 143–155.

Anbar A. D. and Knoll A. H. (2002) Proterozoic ocean chemistry and the evolution: a bioinorganic bridge? *Science* **297**, 1137–1142.

Anderson L. and Sarmiento J. (1994) Redfield ratios of remineralization determined by nutrient data analysis. *Global Biogeochem. Cycles* **8**, 65–80.

Antoine D., Andre J. M., and Morel A. (1996) Oceanic primary production 2. Estimation at global-scale from satellite (coastal zone color scanner) chlorophyll. *Global Biogeochem. Cycles* **10**, 57–69.

Asada K. (1994) Mechanisms for scavenging reactive molecules generated in chloroplasts under light stress. In *Photoinhibition of Photosynthesis: from Molecular Mechanisms to the Field* (eds. N. Baker and J. Bowyer). Bios Scientific, Cambridge, pp. 129–142.

Asada K., Kanematsu S., Okada S., and Hayakawa T. (1980) Phytogenetic distribution of three types of superoxide dismutases in organisms and cell organelles. In *Chemical and Biochemical Aspects of Superoxide Dismutase* (eds. J. V. Bannister and H. A. O. Hill). Elsevier, Amsterdam, pp. 128–135.

Azam F. (1998) Micriobial control of oceanic carbon flux: the plot thickens. *Science* **280**, 694–696.

Balch W. M., Holligan P. M., Ackleson S. G., and Voss K. J. (1991) Biological and optical properties of mesoscale coccolithophore blooms in the Gulf of Maine. *Limnol. Oceanogr.* **36**(4), 629–643.

Banse K. (1977) Determining the carbon-to-chlorophyll ratio of natural phytoplankton. *Mar. Biol.* **41**, 199–212.

Banse K. (1992) Grazing, temporal changes of phytoplankton concentrations, and the microbial loop in the open sea. In *Primary Productivity and Biogeochemical Cycles in the Sea* (ed. P. G. Falkowski). Plenum, New York and London, pp. 409–440.

Banse K. (1994) Grazing and zooplankton production as key controls of phytoplankton production in the open ocean. *Oceanogr.* **7**, 13–20.

Barber R. T. (1992) Geological and climatic time scales of nutrient availability. In *Primary Productivity and Biogeochemical Cycles in the Sea* (eds. P. G. Falkowski and A. Woodhead). Plenum, New York, pp. 89–106.

Barron E. J., Fawcett P. J., Peterson W. H., Pollard D., and Thompson S. L. (1995) A simulation of midcretaceous climate. *Paleoceanography* **10**(5), 953–962.

Beaumont V. I., Jahnke L. L., and Des Marais D. J. (2000) Nitrogen isotopic fractionation in the synthesis of photosynthetic pigments in Rhodobacter capsulatus and Anabaena cylindrica. *Org. Geochem.* **31**(11), 1075–1085.

Behrenfeld M. J. and Falkowski P. G. (1997a) Photosynthetic rates derived from satellite-based chlorophyll concentration. *Limnol. Oceanogr.* **42**, 1–20.

Behrenfeld M. and Falkowski P. (1997b) A consumer's guide to phytoplankton productivity models. *Limnol. Oceanogr.* **42**, 1479–1491.

Behrenfeld M., Bale A., Kolber Z., Aiken J., and Falkowski P. (1996) Confirmation of iron limitation of phytoplankton photosynthesis in the equatorial Pacific. *Nature* **383**, 508–511.

Beja O., Spudich E. N., Spudich J. L., Leclerc M., and DeLong E. F. (2001) Proteorhodopsin phototrophy in the ocean. *Nature* **411**(6839), 786–789.

Berger A. (1988) Milankovitch theory and climate. *Rev. Geophys.* **26**, 624–657.

Berger W. H., Smetacek V. S., *et al.* (eds.) (1989) *Productivity of the Ocean: Present and Past.* Wiley, New York, 471pp.

Berggren W. A., Kent D. V., Swisher C. C., and Aubry M.-P. (1995) A revised Cenozoic geochronology and chronostratigraphy. In *Geochronology, Time Scales, and Global Stratigraphic Correlations: A Unified Temporal Framework for an Historical Geology*, Spec. Vol. Soc. Econ. Paleontol. Mineral. 54 (eds. W. A. Berggren, D. V. Kent, and J. Hardenbol). pp. 129–212.

Berman-Frank I., Cullen J. T., Shaked Y., Sherrell R. M., and Falkowski P. G. (2001) Iron availability, cellular iron quotas, and nitrogen fixation in Trichodesmium. *Limnol. Oceanogr.* **46**, 1249–1260.

Berner R. A. (1990) Atmospheric carbon dioxide levels over phaneroic time. *Science* **249**, 1382–1386.

Berner R. A. and Maasch K. (1996) Chemical weathering and controls on atmospheric O_2 and CO_2: fundamental principles were enunciated by J. J. Ebelmen in 1845. *Geochem. Geophys. Geosys.* **60**, 1633–1637.

Berner R. A., Lasaga A., and Garrels R. (1983) The carbonate-silicate geochemical cycle and its effect of atmospheric carbon dioxide over the past 100 million years. *Am. J. Sci.* **283**, 641–683.

Berthon J. F. and Morel A. (1992) Validation of a spectral light-photosynthesis model and use of the model in conjunction with remotely sensed pigment observations. *Limnol. Oceanogr.* **37**(4), 781–796.

Bienfang P. K. (1992) The role of coastal high latitude ecosystems in global export production. In *Primary Productivity and Biogeochemical cycles in the Sea* (eds. P. G. Falkowski and A. Woodhead). Plenum, New York, pp. 285–297.

Blackman F. F. (1905) Optima and limiting factors. *Ann. Bot.* **19**, 281–298.

Blankenship R. E. (1992) Origin and early evolution of photosynthesis. *Photosyn. Res.* **33**, 91–111.

Brasier M. D., Green O. R., Jephcoat A. P., Kleppe A. K., Van Kranendonk M. J., Lindsay J. F., Steele A., and Grassinau N. V. (2002) Questioning the evidence for Earth's oldest fossils. *Nature* **416**, 76–81.

Bricaud A. and Morel A. (1987) Atmospheric corrections and interpretation of marine radiances in CZCS imagery: use of a reflectance model. In *Oceanography From Space: Proceedings Of The Atp Symposium On Remote Sensing, Brest, France* vol. 7, pp. 33–50.

Broecker W. S., Peng T.-H., and Engh R. (1980) Modeling the carbon system. *Radiocarbon* **22**, 565–598.

Brumsack H.-J. (1986) The inorganic geochemistry of Cretaceous black shales (DSDP Leg 41) in comparison to modern upwelling sediments from the Gulf of California. In *North Atlantic Palaeoceanography*, Geological Society Special Publication. No. 21, (eds. C. P. Summerhayes and N. J. Shackleton), pp. 447–462.

Campbell J. W. and Aarup T. (1992) New production in the North Atlantic derived from the seasonal patterns of surface chlorophyll. *Deep-Sea Res.* **39**, 1669–1694.

Capone D. G. and Carpenter E. J. (1982) Nitrogen fixation in the marine environment. *Science* **217**(4565), 1140–1142.

Chandler M. A., Rind D., and Ruedy R. (1992) Pangean climate during the Early Jurassic: GMC simulations and the sedimentary record of paleoclimate. *Geol. Soc. Am. Bul.* **104**, 543–559.

Christensen J. P., Murray J. W., Devol A. H., and Codispoti L. A. (1987) Denitrification in continental shelf sediments has major impact on the oceanic nitrogen budget. *Global Biogeochem. Cycles* **1**, 97–116.

Codispoti L. (1995) Is the ocean losing nitrate? *Nature* **376**, 724.

Codispoti L. A. and Christensen J. P. (1985) Nitrification, denitrification and nitrous oxide cycling in the eastern tropical south Pacific Ocean. *Mar. Chem.* **16**, 277–300.

Conley D. J. (2002) Terrestrial ecosystems and the global biogeochemical cycle. *Glob. Biogeochem. Cycles* **16**, doi: 10.129/2002GB001894.

Copin-Montegut C. and Copin-Montegut G. (1983) Stoichiometry of carbon, nitrogen, and phosporus in marine particulate matter. *Deep-Sea Res.* **30**, 31–46.

Delwiche C. (2000) Tracing the thread of plastid diversity through the tapestry of life. *Am. Nat.* **154**, S164–S177.

Des Marais D. J. (2000) When did photosynthesis emerge on Earth? *Science* **289**, 1703–1705.

Devol A. H. (1991) Direct measurement of nitrogen gas fluxes from continental shelf sediments. *Nature* **349**, 319–321.

Duce R. A. and Tindale N. W. (1991) Atmospheric transport of iron and its deposition in the ocean. *Limnol. Oceanogr.* **36**, 1715–1726.

Dugdale R. C. (1967) Nutrient limitation in the sea: dynamics, identification and significance. *Limnol. Oceanogr.* **12**, 685–695.

Dugdale R. C. and Goering J. J. (1967) Uptake of new and regenerated forms of nitrogen in primary productivity. *Limnol. Oceanogr.* **12**, 196–206.

Dugdale R. and Wilkerson F. (1992) Nutrient limitation of new production in the sea. In *Primary Productivity and Biogeochemical Cycles in the Sea.* (eds. P. G. Falkowski). Plenum, New York and London, pp. 107–122.

Einstein A. (1910) Therioe der Opaleszenz von homogenen Flussigkeiten un Flussigkeitsgemischen in der Nahe des kritischen Zustandes. *Ann. Physik.* **33**, 1275.

Eppley R. W. (1992) Towards understanding the roles of phytoplankton in biogeochemical cycles: personal notes. In *Primary Productivity and Biogeochemical Cycles in the Sea* (eds. P. G. Falkowski and A. D. Woodhead). Plenum, New York and London, pp. 1–7.

Eppley R. W. and Peterson B. J. (1979) Particulate organic matter flux and planktonic new production in the deep ocean. *Nature* **282**, 677–680.

Eppley R. W., Rogers J. N., and McCarthy J. J. (1969) Half-Saturation constant for uptake of nitrate and ammonium by marine phytoplankton. *Limnol. Oceanogr.* **14**, 912–920.

Falkowski P. (1997) Evolution of the nitrogen cycle and its influence on the biological sequestration of CO_2 in the ocean. *Nature* **387**, 272–275.

Falkowski P. G. (2002) On the evolution of the carbon cycle. In *Phytoplankton Productivity: Carbon Assimilation in Marine and Freshwater Ecosystems* (eds. P. I. B. Williams, D. Thomas, and C. Renyolds). Blackwell, Oxford, pp. 318–349.

Falkowski P. G. and Raven J. A. (1997) *Aquatic Photosynthesis.* Blackwell, Oxford, 375pp.

Falkowski P., Barber R., and Smetacek V. (1998) Biogeochemical controls and feedbacks on ocean primary production. *Science* **281**, 200–206.

Fanning K. (1992) Nutrient provinces in the sea: concentration ratios, reaction rate ratios, and ideal covariation. *J. Geophys. Res.* **97C**, 5693–5712.

Farquhar J., Wing B. A., McKeegan K. D., Harris J. W., Cartigny P., and Thiemens M. H. (2002) Mass-independent sulfur of inclusions in diamond and sulfur recycling on early Earth. *Science* **298**(5602), 2369–2372.

Field C., Behrenfeld M., Randerson J., and Falkowski P. (1998) Primary production of the biosphere: integrating terrestrial and oceanic components. *Science* **281**, 237–240.

Fujita Y., Murakami A., and Ohki K. (1990) Regulation of the stoichiometry of thylakoid components in the photosynthetic system of cyanophytes: model experiments showing that control of the synthesis or supply of Ch A can change the stoichiometric relationship between the two photosystems. *Plant. Cell. Physiol.* **31**, 145–153.

Gordon H. R. and Morel A. (1983) *Remote Sensing of Ocean Color for Interpretation of Satellie Visible Imagery: A Review.* Springer, New York.

Gradstein F. M., Agterberg F. P., Ogg J. G., Hardenbol H., van Veen P., Thierry J., and Huang Z. A. (1995) A Triassic, Jurassic, and Cretaceous time scale. In *Geochronology, Time Scales, and Global Stratigraphic Correlations: A Unified Temporal Framework for an Historical Geology* (eds. W. A. Berggren, D. V. Kent, and J. Hardenbol). Spec. Vol.- Soc. Econ. Paleontol. Mineral. **54**, 95–126.

Gruber N. and Sarmiento J. (1997) Global patterns of marine nitrogen fixation and denitrification. *Global Biogeochem. Cycles* **11**, 235–266.

Habicht K. S., Gade M., Thamdrup B., Berg P., and Canfield D. E. (2002) Calibration of sulfate levels in the Archean Ocean. *Science* **298**, 2372–2374.

Harrison W. G. (1980) Nutrient regeneration and primary production in the sea. In *Primary Production in the Sea* (ed. P. G. Falkowski). Plenum, New York, pp. 433–460.

Harwood D. M. and Nikolaev V. A. (1995) Cretaccous diatoms: morphology, taxonomy, biostratigraphy. In *Siliceous Microfossils* (eds. C. D. Blome, P. M. Whalen, and R. Katherine). Paleontological Society, Number 8, pp. 81–106.

Herbert D., Elsworth R., and Telling R. C. (1956) The continuous culture of bacterial a theoretical and experimental study. *J. Gen. Microbiol.* **14**, 601–622.

Holland H. and Rye R. (1998) *Am. J. Sci.* **298**, 621–672.

Holligan P. M. and Balch W. M. (1991) From the ocean to cells: coccolithophore optics and biogeochemistry. In *Particle Analysis in Oceanography* 27 (eds. S. Demers). Springer, Berlin, pp. 301–324.

Huber B. T., Hodell D., and Hamilton C. (1995) Middle–Late Cretaceous climate of the southern high Latitudes—stable isotopic evidence for minimal equator-to-pole thermal gradients. *Geol. Soc. Am. Bull.* **107**, 1164–1191.

Iglesias-Rodriguez D. M., Brown C. W., Doney S. C., Kleypas J. A., Kolber D., Kolber Z., Hayes P. K., and Falkowski G. P. (2002) Representing key phytoplankton functional groups in ocean carbon cycle models: coccolithophorids. *Global Biogeochem. Cycles* **16** (in press).

Janis N. and Damuth J. (1990) Mammals. In *Evolutionary Trends* (ed. K. McNammara). Belknap, London, pp. 301–345.

Johnston A. M. and Raven J. A. (1987) The C{-4}-like characteristics of intertidal macroalga Ascophyllum nedosum (Fucales, Phaeophyta). *Phycologia* **26**(2), 159–166.

Kasting J. F. (1990) Bolide impacts and the oxidation state of carbon in the Earth's early atmosphere. *Origins Life Evol. Biosphere* **20**, 199–231.

Kasting J. F. (1993) Earth's early atmosphere. *Science* **259**, 920–926.

Kasting J. F., Toon O. B., and Pollack J. B. (1988) How climate evolved on the terrestrial planets. *Sci. Am.* **258**, 90–97.

Katz M. E., Wright J. D., Miller K. G., Cramer B. S., Fennel K., and Falkowski P. G. Biological overprint of the geological carbon cycle. *Nature* (in review).

Kolber Z. S., Barber R. T., Coale K. H., Fitzwater S. E., Greene R. M., Johnson K. S., Lindley S., and Falkowski P. G. (1994) Iron limitation of phytoplankton photosynthesis in the Equatorial Pacific Ocean. *Nature* **371**, 145–149.

Kromer S. (1995) Respiration during photosynthesis. *Ann. Rev. Plant Physiol. Plant Mol. Biol.* **0046**, 00045–00070.

Kump L. and Arthur M. (1999) Interpreting carbon-isotope excursions: carbonates and organic matter. *Chem. Geol.* **161**, 181–198.

Laws E. A., Popp B. N., and Bidigare R. (1997) Effect of growth rate and CO_2 concentration on carbon siotoic fractionation by the marine diatom *Phaeodactylum tricornutum. Limnol. Oceanogr.* (in press).

Lenton T. and Watson A. (2000) Redfield revisited: 2. What regulates the oxygen content of the atmosphere. *Global Biogeochem. Cycles* **14**, 249–268.

Letelier R. M., Dore J. E., Winn C. D., and Karl D. M. (1996) Seasonal and interannual variations in photosynthetic carbon assimilation at Station ALOHA. *Deep-Sea Res.* **43**, 467–490.

Liebig J. (1840) *Chemistry and its Application to Agriculture and Physiology.* Taylor and Walton, London.

Lindeman R. (1942) The trophic-dynamic aspect of ecology. *Ecology* **23**, 399–418.

Lipps J. H. (ed.) (1993) *Fossil Prokaryotes and Protists.* Blackwell, Oxford.

Martin J. H. (1990) Glacial-interglacial CO_2 change: the iron hypothesis. *Paleoceangraphy* **5**, 1–13.

Martin J. H. (1991) Iron, Liebig's Law and the greenhouse. *Oceanography* **4**, 52–55.

McElroy M. (1983) Marine biological controls on atmospheric CO_2 and climate. *Nature* **302**, 328–329.

Miller C. B., Frost B. W., Wheeler P. A., Landry M. R., Welschmeyer N., and Powell T. M. (1991) Ecological dynamics in the subarctic Pacific, a possibly iron-limited ecosystem. *Limnol. Oceanogr.* **36**, 1600–1615.

Milliman J. (1993) Production and accumulation of calcium carbonate in the ocean: budget of a nonsteady state. *Global Biogeochem. Cycles* **7**, 927–957.

Monod J. (1942) *Recheres sur la croissance des cultures bacteriennes*. Hermann & Cie, Paris.

Morel A. (1974) Optical properties of pure water and pure seawater. In *Optical Aspects of Oceanography* (eds. N. G. Jerlov and E. S. Nielsen). Academic Press, London, pp. 1–24.

Morel A. (1988) Optical modeling of the upper ocean in relation to it's biogenous matter content (case one waters). *J. Geophys. Res.* **93**, 10749–10768.

Morel F. M. M. and Hudson R. J. (1985) The geobiological cycle of trace elements in aquatic systems: Redfield revisited. In *Chemical Processes in Lakes* (ed. W. Strumm). Wiley-Interscience, pp. 251–281.

Morel A. and Andre J. M. (1991) Pigment distribution and primary production in the western Mediterranean as derived and modeled from coastal zone color scanner observations. *J. Geophys. Res. C. Oceans* **96**(C7), 12685–12698.

Mulkidjanian A. and Junge W. (1997) On the origin of photosynthesis as inferred from sequence analysis—a primordial UV-protector as common ancestor of reaction centers and antenna proteins. *Photosyn. Res.* **51**, 27–42.

Nixon S. W., Ammerman J. W., Atkinson L. P., Berounsky V. M., Billen G., Boicourt W. C., Boynton W. R., Church T. M., Ditoro D. M., Elmgren R., Garber J. H., Giblin A. E., Jahnke R. A., Owens N. J. P., Pilson M. E. Q., and Seitzinger S. P. (1996) The fate of nitrogen and phosphorus at the land-sea margin of the North Atlantic Ocean. *Biogeochemistry* **35**, 141–180.

Odum E. P. (1971) *Fundamentals of Ecology*. Philadelphia.

Pace N. R. (1997) A molecular view of microbial diversity and the biosphere. *Science* **276**(5313), 734–740.

Pavlov A. and Kasting J. (2002) Mass-independent fractionation of sulfur isotopes in Archean sediments: strong evidence for an anoxic Archean atmosphere. *Astrobiology* **2**, 27–41.

Petit J. R., Jouzel J., Raynaud D., Barkov N. I., Barnola J.-M., and Basile I. (1999) Climate and atmospheric history of the past 420,000 years from the Vostok ice core, Antarctica. *Nature* **399**, 429–436.

Pickard G. L. and Emery W. J. (1990) *Descriptive Physical Oceanography*. Pergamon, Oxford.

Piper D. Z. (1994) Seawater as the source of minor elements in black shales, phosphorites and other sedimentary rocks. *Chem. Geol.* **114**, 95–114.

Platt T. (1986) Primary production of the ocean water column as a function of surface light intensity: algorithms for remote sensing. *Deep-Sea Res.* **33**, 149–163.

Platt T. and Irwin B. (1973) Caloric content of phytoplankton. *Limnol. Oceanogr.* **18**, 306–310.

Platt T. and Sathyendranath S. (1988) Oceanic primary production: estimation by remote sensing at local and regional scales. *Science* **241**, 1613–1620.

Postgate J. R. (ed.) (1971) *The Chemistry and Biochemistry of Nitrogen Fixation*. Plenum, New York.

Quigg A., Finkel Z. V., Irwin A. J., Rosenthal Y., Ho T.-Y., Reinfelder J. R., Schofield O., More F., and Falkowski P. (2003) Plastid inheritance of elemental stoichiometry in phytoplankton and its imprint on the geological record. *Nature* **425**, 291–294.

Ramus J. (1992) Productivity of seaweeds. In *Primary Productivity and Biogeochemical Cycles in the Sea* (ed. P. G. Falkowski). Plenum, New York and London, pp. 239–255.

Raven J. A. (1987) The role of vacuoles. *New Phytol.* **106**, 357–422.

Raymo M. and Ruddiman W. (1992) Tectonic forcing of the late Cenozoic climate. *Nature* **359**, 117–122.

Redfield A. C. (1934) *On the Proportions of Organic Derivatives in Sea Water and Their Relation to the Composition of Plankton*. Liverpool, James Johnstone Memorial Volume, Liverpool Univ. Press, pp. 176–192.

Redfield A. C. (1958) The biological control of chemical factors in the environment. *Am. Sci.* **46**, 205–221.

Redfield A., Ketchum B., and Richards F. (1963) The influence of organisms on the composition of sea-water. In *The Sea* (ed. M. Hill). Interscience, New York, **2**, 26–77.

Retallack G. (2001) Cenozoic expansion of grasslands and climatic cooling. *J. Geol.* **109**, 407–426.

Sancetta C., Villareal T., and Falkowski P. G. (1991) Massive fluxes of rhizosolenoid diatoms: a common occurrence? *Limnol. Oceanogr.* **36**, 1452–1457.

Sarmiento J. L. and Bender M. (1994) Carbon biogeochemistry and climate change. *Photosyn. Res.* **39**, 209–234.

Sarmiento J. L., Orr J. C., and Siegenthaler U. (1992) A perturbation simulation of CO_2 uptake in an ocean general circulation model. *J. Geophys. Res.* **94**, 3621–3645.

Schlesinger W. H. (1997) *Biogeochemistry: An Analysis of Global Change*. Academic Press, New York.

Schopf J. (1993) Microfossils of the early Archean Apex Chert: new evidence of the antiquity of life. *Science* **260**, 640–646.

Siegenthaler U. and Sarmiento J. L. (1993) Atmospheric carbon dioxide and the ocean. *Nature* **365**, 119–125.

Sigman D. and Boyle E. (2000) Glacial/interglacial variations in atmospheric carbon dioxide. *Nature* **407**, 859–869.

Smetacek V. (1999) Diatoms and the ocean carbon cycle. *Protist* **150**, 25–32.

Smith D. F. (1981) Tracer kinetic analysis applied to problems in marine biology. In *Physiological Bases of Phytoplankton Ecology* (ed. T. Platt) Can. Bull. Fish Aquat. Sci. **170**, Ottawa, pp. 113–129.

Smith W. O., Jr. and Nelson D. M. (1990) The importance of ice-edge phytoplankton production in the Southern Ocean. *BioScience* **36**, 151–157.

Stumm W. and Morgan J. J. (1981) *Aquatic Chemistry*. Wiley, New York.

Summons R., Jahnke L., Hope J., and Logan G. (1999) 2-Methylhopanoids as biomarkers for cyanobacterial oxygenic photosynthesis. *Nature* **400**, 55–557.

Tilman D. (1982) *Resource Competition and Community Structure*. Princeton University Press, Princeton.

Tozzi S. (2001) Competition and succession of key marine phytoplankton functional groups in a variable environment. *IMCS*. New Brunswick, Rutgers University, 79.

Tozzi S., Schofield O., and Falkowski P. (2003) Turbulence as a selective agent of two phytoplankton functional groups. *Global Change Biology* (in press).

Tsuda A. and Kawaguchi S., *et al.* (2003) Microzooplankton grazing in the surface-water of the Southern Ocean during an Austral Summer. *Polar Biology*.

Tyrell T. (1999) The relative influences of nitrogen and phosphorus on oceanic primary production. *Nature* **400**, 525–531.

Veizer J., Ala D., Azmy K., Bruckschen P., Buhl D., Bruhn F., Carden G. A. F., Diener A., Ebneth S., Godderis Y., Jasper T., Korte C., Pawellek F., Podlaha O. G., and Strauss H. (1999) $^{87}Sr/^{86}Sr$, $\delta^{13}C$ and $\delta^{18}O$ evolution of Phanerozoic seawater. *Chem. Geol.* **161**, 59–88.

Volk T. and Hoffert M. I. (1985) Ocean carbon pumps: analysis of relative strengths and efficiencies in ocean-driven atmospheric CO_2 exchanges. In *The Carbon Cycle and Atmospheric CO_2: Natural Variations Archean to Present* (eds. E. T. Sunquist and W. S. Broeker). American Geophysical Union. **32**, Washington, DC, pp. 99–110.

Whitfield M. (2001) Interactions between phytoplankton and trace metals in the ocean. *Adv. Mar. Biol.* **41**, 3–128.

Walsh J. J. (1983) Death in the sea: enigmatic phytoplankton losses. *Prog. Oceanogr.* **12**, 1–86.

Warneck P. (1988) *Chemistry of the Natural Atmosphere.* Academic Press, New York.

Williams P. J. L. (1981) Incorporation of microheterotrophic processes into the classical paradigm of the planktonic food web. *Kieler Meeresforsch. Sonderh.* **5**, 1–27.

Williams P. J. L. (1993) On the definition of plankton production terms. *ICES Mar. Sci. Symp.* **197**, 9–19.

Williams R. and Frausto da Silva J. (1996) *The Natural Selection of the Chemical Elements.* Clarendon Press, Oxford.

Woodwell G. M., Whittaker R. H., Reiners W. A., Likens G. E., Delwiche C. C., and Botkin D. B. (1978) The biota and the world carbon budget. *Science* **199**, 141–146.

Wu J., Sunda W., Boyle E. A., and Karl D. M. (2000) Phosphate depletion in the western North Atlantic ocean. *Science* **289**(5480), 759–762.

Xiong J., Fischer W. M., Inoue K., Nakahara M., and Bauer C. E. (2000). Molecular evidence for the early evolution of photosynthesis. *Science* **289**(5485), 1724–1730.

Yoder J. A., McClain C. R., Feldman G. C., Esaias W. E., (1993) Annual cycles of phytoplankton chlorophyll concentrations in the global ocean: a satellite view. *Global Biogeochem. Cycles* **7**, 181–193.

Zehr J. P., Mellon M., Braun S., Litaker W., Steppe T., and Paerl H. W. (1995) Diversity of heterotrophic nitrogen-fixation genes in a marine cyanobacterial mat. *Appl. Environ. Microbiol.* **0061**, 02527–02532.

8.06
Biogeochemistry of Terrestrial Net Primary Production

F. S. Chapin, III

University of Alaska, Fairbanks, AK, USA

and

V. T. Eviner

Institute of Ecosystem Studies, Millbrook, NY, USA

8.06.1 INTRODUCTION

Net primary production (NPP) is the amount of carbon and energy that enters ecosystems. It provides the energy that drives all biotic processes, including the trophic webs that sustain animal populations and the activity of decomposer organisms that recycle the nutrients required to support primary production. NPP not only sets the baseline for the functioning of all ecosystem components but also is the best summary variable of ecosystem processes, being the result of numerous interactions among elements, organisms, and environment. This dual role makes NPP the key integrative process in ecosystems (McNaughton *et al.*, 1989) and thus a critical component in our understanding of ecosystem responses to the many changes that are occurring in the global environment. In this chapter, we explain the mechanisms that control NPP, including the environmental constraints on plant growth and the ways in which plants adjust to and alter these constraints.

8.06.2 GENERAL CONSTRAINTS ON NPP

8.06.2.1 What is NPP?

NPP is the net carbon gain by vegetation over a particular time period—typically a year. It is the balance between the carbon gained by photosynthesis and the carbon released by plant respiration. NPP includes the new biomass produced by plants, the soluble organic compounds that diffuse or are secreted by roots into the soil (root exudation), the carbon transfers to microbes that are symbiotically associated with roots (e.g., mycorrhizae and nitrogen-fixing bacteria), and the volatile emissions that are lost from leaves to the atmosphere (Clark *et al.*, 2001).

"Measured" NPP is more of an index of net primary production than a true value. Most field measurements of NPP document only the new plant biomass produced and therefore probably underestimate the true NPP by at least 30% (Table 1). There are many sources of error to this estimate. Some biomass above and below ground dies or is removed by herbivores before it can be measured, so even the new biomass measured in field studies is an underestimate of biomass production. Root exudates are rapidly taken up and respired by microbes adjacent to roots and are generally measured in field studies as a portion of root respiration (i.e., a portion of carbon lost from plants), rather than a component of carbon gain. Volatile emissions are also rarely measured, but are generally a small fraction (<5%) of NPP and thus are probably not a major source of error (Guenther *et al.*, 1995; Lerdau, 1991). For some purposes, these errors may not be too important. A frequent objective of measuring NPP, for example, is to estimate the rate of biomass accumulation. Root exudates, transfers to symbionts, losses to herbivores, and volatile emissions are lost from plants and therefore do not contribute directly to biomass accumulation. Consequently, failure to measure these components of NPP does not bias estimates of biomass accumulation rates. However, these losses of NPP from plants fuel other ecosystem processes such as nitrogen fixation, herbivory, decomposition, and nutrient turnover, so they are important components of the overall carbon dynamics of ecosystems and strongly influence the rates of and interactions among element cycles.

Some components of NPP, such as root production, are particularly difficult to measure and have sometimes been assumed to be some constant ratio (e.g., 1:1) of aboveground production (Fahey *et al.*, 1998). Fewer than 10% of the studies that report total ecosystem NPP actually measure components of belowground production (Clark *et al.*, 2001). Estimates of aboveground NPP sometimes include only large plants (e.g., trees in forests) and exclude understory shrubs or mosses, which can account for a

Table 1 Major components of NPP and typical relative magnitudes[a].

Components of NPP	*% of NPP*
New plant biomass	40–70
Leaves and reproductive parts (fine litterfall)	10–30
Apical stem growth	0–10
Secondary stem growth	0–30
New roots	30–40
Root secretions	20–40
Root exudates	10–30
Root transfers to mycorrhizae	10–30
Losses to herbivores and mortality	1–40
Volatile emissions	0–5

[a] Seldom, if ever, have all of these components been measured in a single study (Chapin *et al.*, 2002).

substantial proportion of NPP in some ecosystems. Most published summaries of NPP do not state explicitly which components of NPP have been included (or sometimes even whether the units are grams of carbon or grams of biomass). For these reasons, considerable care must be used when comparing data on NPP or biomass among studies. These limitations suggest that the large number of NPP estimates that are available globally may not be a valid indication of our understanding of the process.

8.06.2.2 The General Biochemistry of NPP

NPP is the carbon gained by photosynthesis after taking into account the respiratory costs associated with growth and maintenance. Thus, the basic recipe for NPP is simply a function of the resources required for photosynthesis (light, CO_2, nutrients, water), coupled with the environmental factors that influence the rate at which these ingredients are assembled. NPP requires the proper balance of resources and is constrained by the resource in least abundance, relative to plant demand. Increasing the availability of the most limiting resource will increase NPP up to the point that another resource becomes limiting. Because NPP ultimately depends on a *balance* of resources, one of the simplest approaches to understanding controls over NPP is a stoichiometric approach. In marine systems, it has been established that the cytoplasm of primary producers has a certain ratio of elements (the Redfield ratio) that supports optimal metabolism (Redfield, 1958). Similar ratios are observed in terrestrial vegetation, with land plants having an average C : N : S : P of 790 : 7.6 : 3.1 : 1 (Bolin *et al.*, 1983). Departures from this ratio of plant nutrients can be used as an indicator of nutrient limitation (Koerselman and Mueleman, 1996). We will discuss the specific stoichiometry of terrestrial NPP later in this chapter, but for now, we will base these discussions on the premise that plants need a balance of these photosynthetic ingredients, and NPP is limited by the resources that are in lowest supply relative to plant demand.

The relative importance of the resources and environmental conditions that limit NPP vary by scale and ecosystem. At the global scale, total NPP varies 14-fold among mature stands of the major terrestrial biomes (Table 2). This variation correlates strongly with climate. In ecosystems where moisture is favorable, NPP increases exponentially with temperature. Where temperature is favorable, NPP increases to a maximum in tropical rainforests with moderately high precipitation (2–3 m annual precipitation) and declines at extremely high precipitation, due to anaerobic conditions and/or depletion of soil minerals by rapid weathering (Schuur, in press) (Figure 1). The global pattern of NPP reflects patterns of precipitation more strongly than patterns of temperature (Foley *et al.*, 1996; Gower, 2002; Kucharik *et al.*, 2000; New *et al.*, 1999) (Figure 2) because most of the terrestrial surface receives an order of magnitude less precipitation than is optimal for NPP.

Much of the variation in NPP simply reflects the length of the growing season. NPP that is averaged over the time that plants actively produce new biomass varies only fourfold among biomes (Table 2). When NPP is normalized by both growing-season length and the quantity of leaf area available to fix carbon, there is no consistent relationship between NPP and climate (Chapin *et al.*, 2002). Biome differences in NPP per unit leaf area and time probably reflect uncertainty in

Table 2 Productivity per day and per unit leaf area.

Biome	*Total NPP* ($g\ m^{-2}\ yr^{-1}$)[a]	*Season length*[b] (days)	*Daily NPP per ground area* ($g\ m^{-2}\ d^{-1}$)	*Total LAI*[c] ($m^2\ m^{-2}$)	*Daily NPP per leaf area* ($g\ m^{-2}\ d^{-1}$)
Tropical forests	2,500	365	6.8	6.0	1.14
Temperate forests	1,550	250	6.2	6.0	1.03
Boreal forests	380	150	2.5	3.5	0.72
Mediterranean shrublands	1,000	200	5.0	2.0	2.50
Tropical savannas and grasslands	1,080	200	5.4	5.0	1.08
Temperate grasslands	750	150	5.0	3.5	1.43
Deserts	250	100	2.5	1.0	2.50
Arctic tundra	180	100	1.8	1.0	1.80
Crops	610	200	3.1	4.0	0.76
Range of values	14-fold	3.7-fold	3.8-fold	6-fold	3.3-fold

[a] NPP is expressed in units of dry mass (Saugier *et al.*, 2001). [b] Estimated. [c] Data from Gower (2002).

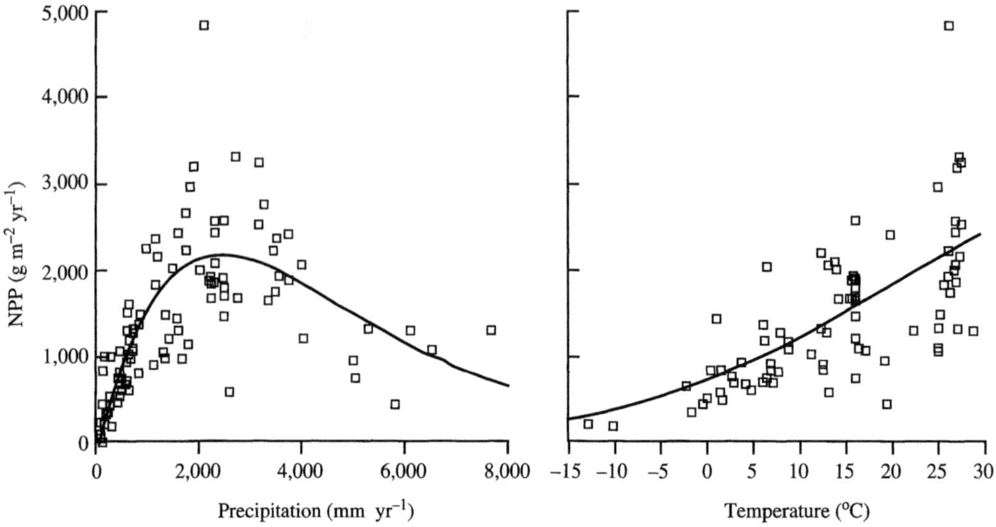

Figure 1 Correlation of NPP (in units of biomass) with temperature and precipitation (Schuur, 2003) (reproduced by permission of Springer from *Principles of Terrestrial Ecosystem Ecology*, **2002**).

the data at least as much as any underlying climatic influence. The climatic controls over NPP of mature stands can therefore be viewed as a combination of the climatic constraints on the length of growing season and the capacity of vegetation to produce and maintain leaf area. Fine root length may be just as important as leaf area in governing the productive potential of vegetation (Craine *et al.*, 2001), but fewer comparative data are available for roots.

At a global scale, water is the most limiting resource to NPP, and nutrient limitation becomes an important limiting factor at more local scales. Broad global patterns of nutrient limitation exist, with phosphorus being the most commonly limiting nutrient to NPP in wet tropical systems, and nitrogen being limiting in most temperate systems. Beyond these broad patterns, it is necessary to consider environmental conditions, resource availability, and their interactions to understand the constraints on NPP at different scales. To do this, we must first consider the overall constraints of *potential* NPP within an ecosystem (state factors), and then within these constraints, to determine the interactions that occur within ecosystems to determine the conditions that directly influence NPP (interactive controls).

Dokuchaev (1879) and Jenny (1941) proposed that five independent state factors (climate, parent material, topography, time, and potential biota) govern the properties of soils and ecosystems (Amundson and Jenny, 1997). These state factors represent the overall constraints on NPP within an ecosystem. On broad geographic scales, climate is the state factor that most strongly influences ecosystem structure and functioning and determines the global patterns of NPP (Figure 2). Within this broad climatic context, parent material influences the types of soils that develop and the availability of some nutrients, both of which explain much of the regional variation in ecosystem processes. Limestone, granite, and marine sands support radically different patterns of biogeochemistry within a climate zone. Patterns of ecosystem development over time lead to shifts in the relative availability of different nutrients, causing long-term changes in an ecosystem's potential NPP. Topography influences both microclimate and soil development at a local scale, causing additional fine-scale variation in biogeochemical processes. If NPP were determined by a fixed stoichiometry of resources in all terrestrial plants, these first four state factors would be sufficient to predict overall potential patterns of NPP. Functional types of plants differ dramatically, however, in their potentials for growth under different limiting conditions. Potential biota governs the types and diversity of organisms that actually occupy a site. The resulting species composition then determines the observed response of NPP to other state factors because plant species differ in their stoichiometry of NPP.

Figure 2 Global patterns of mean annual temperature and precipitation (New *et al.*, 1999) and of modeled NPP (Foley *et al.*, 1996; Kucharik *et al.*, 2000) (reproduced by permission of Atlas of the Biosphere http://atlas.sage.wisc.edu).

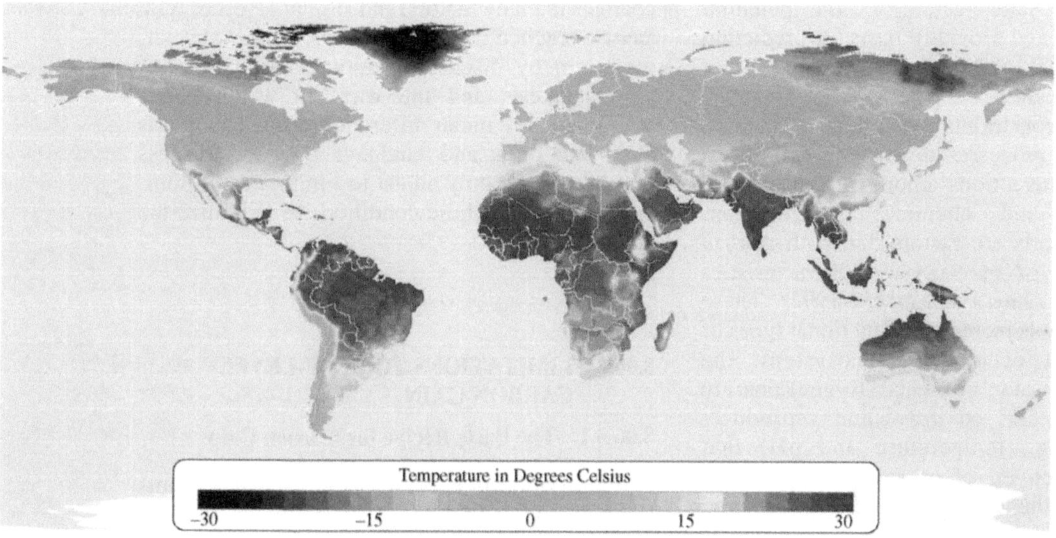
Temperature in Degrees Celsius
−30
−15
0
15
30

(a)

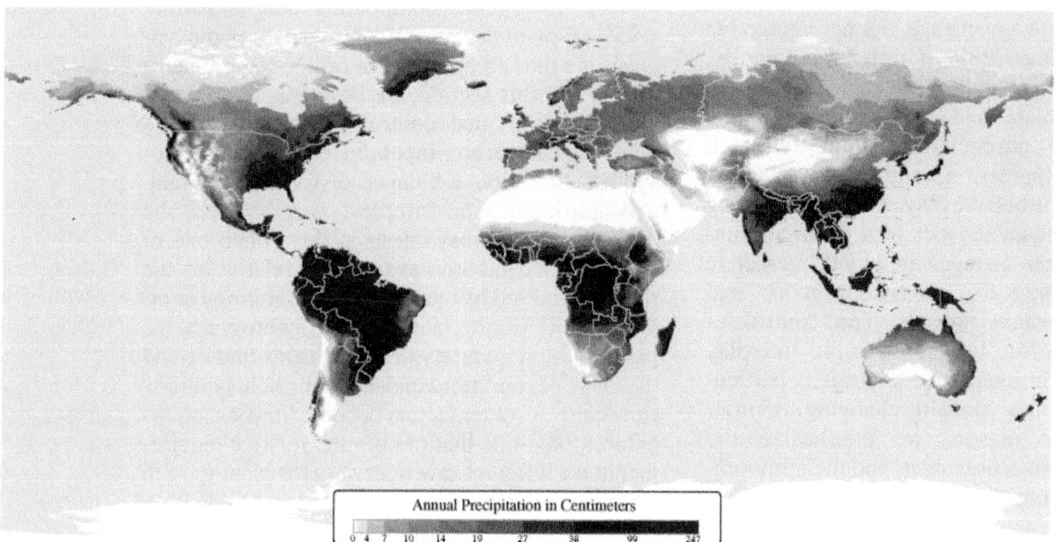
Annual Precipitation in Centimeters
0
4
7
10
14
19
27
38
99
247

(b)

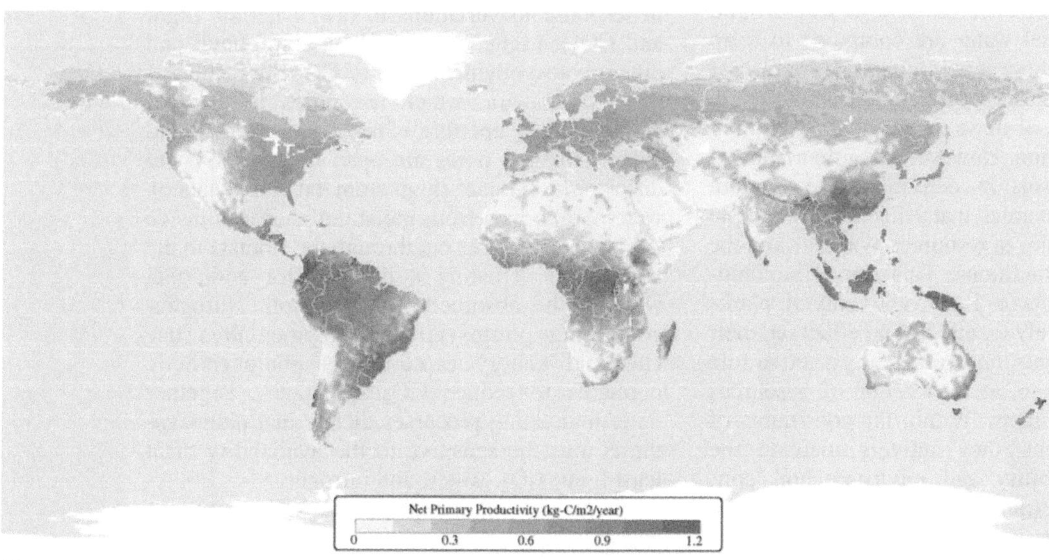
Net Primary Productivity (kg-C/m2/year)
0
0.3
0.6
0.9
1.2

(c)

Together these state factors set the potential patterns of NPP and provide a basis for predicting local and global patterns of NPP and other ecosystem processes.

Within the constraints set by state factors, biogeochemical processes are strongly influenced by a web of interactions among organisms and the physical and chemical environment. Interactive controls are factors that both *control* and *are controlled by* ecosystem characteristics (Chapin *et al.*, 1996; Field *et al.*, 1992). These interactive controls include the functional types of organisms that occupy the ecosystem; the resources (e.g., water, nutrients, oxygen) that are used by organisms to grow and reproduce; modulators (e.g., temperature and pH) that influence the activity of organisms but are not consumed by them; disturbance regime; and human activities. These interactive controls respond dynamically to any external change in state factors and to any change in other interactive controls. The composition of a plant community, for example, is influenced both by the global changes in climate and regional biota (state factors) and by nitrogen deposition, livestock density, fire suppression, and timber harvest (interactive controls). Many of the resulting changes in the characteristics of a plant community cause further changes in other interactive controls, including the ecosystem goods and services that benefit society. The control of ecosystem processes by the dynamic interplay among changes in interactive controls is particularly important in a globally changing environment. For this reason, we emphasize the *interactions* between organisms and their environment in describing the biogeochemical controls over NPP.

To revisit our general recipe for NPP, environmental factors influence the rate at which light, CO_2, nutrients and water are combined to form NPP. Any of these environmental factors or resources may constrain NPP, and it is ultimately a proper balance of these factors that is required for plant production. However, the importance of interacting controls in determining ecosystem processes demonstrates that NPP is not a simple function of the ratio of resources available and the environmental conditions. This simple stoichiometric approach would be valid only if plants responded passively to, and had no effect on, their environment. Plants, however, play an active role in their response to, and mediation of, resources and their environment. Within the constraints of their environment, they actively mediate the resource availability and environmental conditions that constrain NPP. Ultimately, biogeochemical cycling is driven by the interactions between organisms and their physical and chemical environment. NPP is therefore sensitive to changes in many factors, and similar levels of NPP can be reached in multiple ways. In this chapter, we begin by discussing resource limitations of photosynthesis and the ways in which plants maximize NPP under different limiting conditions at the leaf, plant, and stand levels. We then discuss how plants not only adjust to limiting conditions, but also modify these conditions to minimize the limitations.

8.06.3 LIMITATIONS TO LEAF-LEVEL CARBON GAIN

8.06.3.1 The Basic Recipe for Carbon Gain

Photosynthesis is the process by which plants use light energy to reduce carbon dioxide (CO_2) to sugars, which are subsequently converted to a variety of organic compounds that constitute ~95% of plant dry mass. Controls over photosynthesis are thus a key regulator of the stoichiometry of NPP. In this section, we describe the environmental factors that control photosynthesis and therefore the carbon inputs to vegetation. Photosynthesis requires a balance of CO_2, H_2O, light, and nutrients. The simplest way to describe limitation of photosynthesis is that, when one of these factors has low availability relative to the ratio of required resources, this is a limiting factor. When any single factor limits photosynthesis, plants exhibit a variety of adjustments that extend the range of conditions under which photosynthesis can occur. As other factors become limiting, plants exhibit trade-offs that modify the relative requirements for different raw materials for plant growth and therefore alter the stoichiometry of NPP.

This principal of adjustments and trade-offs is illustrated by changes in photosynthesis that occur in response to variations in raw materials (light and CO_2). Light is captured by chlorophyll and other photosynthetic pigments. CO_2 enters the leaf through stomata, which are pores in the leaf surface whose aperture is regulated by the plant. When stomatal pores are open to allow CO_2 to diffuse into the leaf (high stomatal conductance), water evaporates from moist cell surfaces inside the leaf and diffuses out through the stomata to the atmosphere, creating a demand for additional water to be absorbed from the soil. Nitrogen-containing photosynthetic enzymes then use chemical energy captured by photosynthetic pigments to reduce CO_2 to sugars. Together these interacting processes dictate that photosynthesis must be sensitive to the availability of at least light, CO_2, water, and nitrogen.

Plants are not exposed to the resources necessary for photosynthesis in optimal proportions, but under a wide variety of circumstances, plants adjust the components of photosynthesis so all these components are about equally limiting to photosynthesis (Farquhar and Sharkey, 1982). Plants make this adjustment by altering the size of stomatal openings, which alters the rate of diffusion of CO_2 and water vapor, or by changing the concentrations of light-harvesting pigments or photosynthetic enzymes, which alters the nitrogen requirement for carrying out the biochemistry of photosynthesis.

The general principle of colimitation of photosynthesis by biochemistry and diffusion provides the basis for understanding most of the adjustments by individual leaves to minimize the environmental limitations of photosynthesis.

8.06.3.2 Light Limitation

When light is the only factor limiting photosynthesis, net photosynthesis increases linearly with increasing light. The slope of this line (the quantum yield of photosynthesis) is a measure of the efficiency with which plants use absorbed light to produce sugars. The quantum yield is similar among all C_3 plants at low light in the absence of environmental stress. In other words, all C_3 plants have a relatively constant photosynthetic light-use efficiency (~6%) of converting absorbed visible light into chemical energy under low-light conditions. At high irradiance, photosynthesis becomes light saturated, i.e., it no longer responds to changes in light supply, due to the finite capacity of the light-harvesting reactions to capture light. As a result, light energy is converted less efficiently into sugars at high light. As described later, leaves at the top of a plant canopy and species that characteristically occur in high-light habitats saturate at higher light intensities than do leaves and plants characteristic of low-light environments.

In response to short-term environmental variation, individual leaves minimize light limitation by adjusting stomatal conductance and photosynthetic capacity to maximize carbon gain in different light environments (Chazdon and Field, 1987; Chazdon and Pearcy, 1991; Pearcy, 1988; Pearcy, 1990). Stomatal conductance increases in high light, when CO_2 demand is high, and decreases in low light, when photosynthetic demand for CO_2 is low. These stomatal adjustments result in a relatively constant CO_2 concentration inside the leaf, as expected from the hypothesis of colimitation of photosynthesis by biochemistry and diffusion. It allows plants to conserve water under low light and to maximize carbon uptake at high light, thus regulating the trade-off between carbon gain and water loss.

Over longer timescales (days to months) plants acclimate to variations in light availability by producing leaves with different photosynthetic properties. Sun leaves at the top of the canopy have more cell layers, are thicker, and therefore have greater photosynthetic capacity per unit leaf area than do shade leaves (Terashima and Hikosaka, 1995; Walters and Reich, 1999). The respiration rate of a tissue depends on its protein content, as described later, so the low photosynthetic capacity and protein content of shade leaves are associated with a lower respiration rate per unit area than in sun leaves. For this reason, shade leaves maintain a more positive carbon balance (photosynthesis minus respiration) under low light than do sun leaves. The changes in photosynthetic properties as a result of genetic adaptation are similar to patterns observed with acclimation. Species that are adapted to high light and are intolerant of shade typically have a higher photosynthetic capacity per unit mass or area and higher respiration rate than do shade-tolerant species, even in the shade (Walters and Reich, 1999). The net effect of acclimation or adaptation to variation in light availability is to extend the range of light availability over which vegetation maintains a relatively constant light-use efficiency, i.e., a relatively constant relationship between absorbed photosynthetically active radiation and net photosynthesis (Chapin *et al.*, 2002).

8.06.3.3 CO_2 Limitation

When CO_2 is the only factor limiting photosynthesis, net photosynthesis increases linearly with increasing CO_2 concentration, until other factors limit photosynthesis, at which point the curve saturates, much as described for the photosynthetic response to light. Most plants operate at the upper end of the linear portion of the CO_2–response curve, where CO_2 and biochemical processes are about equally limiting to photosynthesis (Farquhar and Sharkey, 1982).

The free atmosphere is so well mixed that its CO_2 concentration varies globally by only 4%. Consequently, spatial variation in CO_2 concentration does not explain much of the global variation in photosynthetic rate (Field, 1991). Nonetheless, the continued worldwide increases in atmospheric CO_2 concentration could cause a general increase in carbon gain by ecosystems. A doubling of the CO_2 concentration to which leaves are exposed, for example, leads to a 30–50% increase in photosynthetic rate over the short term (Curtis and Wang, 1998). The long-term enhancement of photosynthesis by addition of CO_2 is, however, uncertain. Herbaceous plants and deciduous trees (but not conifers) sometimes acclimate to increased CO_2 concentration by reducing photosynthetic capacity and stomatal conductance (Ellsworth, 1999; Mooney *et al.*, 1999). This reduces the nitrogen and water required to fix a given amount of carbon, as expected from the hypothesis of colimitation of photosynthesis by biochemistry and diffusion. In other cases acclimation has no effect on photosynthetic rate and stomatal conductance (Curtis and Wang, 1998). The downregulation of CO_2 uptake in response to elevated CO_2 causes photosynthesis to respond less strongly to elevated CO_2 than we might expect from a simple extrapolation of a CO_2-response curve of photosynthesis.

Over the long term, indirect effects of elevated CO_2 often have an important influence on trade-offs between CO_2 uptake and requirements for water and nitrogen. In dry environments, for example, the reduced stomatal conductance caused by elevated CO_2 leads to a decline in transpiration, which reduces evapotranspiration and increases soil moisture, which can affect nitrogen mineralization (Curtis *et al.*, 1996; Díaz *et al.*, 1993; Hungate *et al.*, 1997). Elevated CO_2 often has a greater effect on plant growth through changes in moisture and nutrient supply than through a direct stimulation of photosynthesis by elevated CO_2 (Hungate *et al.*, 1997; Owensby *et al.*, 1993). Given that the atmospheric CO_2 concentration has increased 30% (by 90 parts per million by volume; ppmv) since the beginning of the Industrial Revolution, it is important to understand and predict these indirect effects of elevated CO_2 on carbon gain by ecosystems.

8.06.3.4 Nitrogen Limitation

Photosynthetic capacity, i.e., the photosynthetic rate per unit leaf mass measured under favorable conditions of light, moisture, and temperature, increases linearly with leaf nitrogen concentration over almost the entire range of nitrogen concentrations found in natural ecosystems (Evans, 1989; Field and Mooney, 1986; Poorter, 1990; Reich *et al.*, 1999, 1992, 1997). This relationship exists because photosynthetic enzymes account for a large proportion of the nitrogen in leaves. Only at extremely high nitrogen concentrations or under conditions where other factors limit photosynthesis is there an accumulation of nitrate and other forms of nitrogen unrelated to photosynthetic capacity (Bloom *et al.*, 1985). Many ecological factors can lead to a high leaf nitrogen concentration and therefore a high photosynthetic capacity. Plants growing in high-nitrogen soils, for example, have higher tissue nitrogen concentrations and photosynthetic rates than do the same species growing on less fertile soils. This acclimation of plants to a high nitrogen supply contributes to the high photosynthetic rates in agricultural fields and other ecosystems with a rapid nitrogen turnover. Many species differ in their nitrogen concentration, even when growing in the same soils. Species adapted to productive habitats usually produce leaves that are short-lived and have high tissue nitrogen concentrations and high photosynthetic rates. Nitrogen-fixing plants also typically have high leaf nitrogen concentrations and correspondingly high photosynthetic rates. Environmental stresses that cause plants to produce leaves with a low leaf nitrogen concentration result in low photosynthetic capacity. In summary, regardless of the cause of variation in leaf nitrogen concentration, there is always a strong positive correlation between leaf nitrogen concentration and photosynthetic capacity (Field and Mooney, 1986; Reich *et al.*, 1999, 1997). Thus, as with adjustment to variation in light availability, plants adjust to variation in nitrogen supply by the same physiological mechanism within species (acclimation) as between species (adaptation), in this case by increasing the concentration of photosynthetic enzymes and pigments.

Plants with a high photosynthetic capacity have a high stomatal conductance, in the absence of environmental stress (Reich *et al.*, 1999, 1997). This enables plants with a high photosynthetic capacity to gain carbon rapidly, at the cost of high rates of water loss. Conversely, species with a low photosynthetic capacity conserve water as a result of their lower stomatal conductance. This illustrates the trade-off between water and nitrogen in response to variation in nitrogen supply. As described later, water stress induces the same trade-off. Plants acclimated and adapted to low water availability have a low stomatal conductance to conserve water and a low tissue nitrogen concentration, which reduces photosynthetic capacity. The net effect of these trade-offs is to maintain colimitation of photosynthesis by diffusive and biochemical processes.

There appears to be an unavoidable trade-off between traits that maximize photosynthetic rate and traits that maximize leaf longevity (Reich *et al.*, 1999, 1997). Many plant species that grow in low-nutrient environments produce long-lived leaves because there are insufficient nutrients to support rapid leaf turnover (Chapin, 1980). Shade-tolerant species also produce longer-lived leaves than do shade-intolerant species (Walters and Reich, 1999). Long-lived leaves typically have a low leaf nitrogen concentration and a low photosynthetic capacity; they must therefore photosynthesize for a relatively long time to break even in their lifetime carbon budget (Chabot and Hicks, 1982; Gulmon and Mooney, 1986; Reich *et al.*, 1997). To survive, long-lived leaves must have sufficient structural rigidity to withstand drought and/or winter desiccation. These structural requirements cause leaves to be dense, i.e., to have a small surface area per unit of biomass, termed specific leaf area (Chapin, 1993; Lambers and Poorter, 1992). Long-lived leaves must also be well defended against herbivores and pathogens, if they are to persist. This requires substantial allocation to lignin, tannins, and other compounds that deter herbivores, but also contribute to tissue mass and a low specific leaf area (Coley *et al.*, 1985; Gulmon and Mooney, 1986). Many woody plants in dry environments also produce long-lived leaves. For the same reasons, these leaves typically have a low specific leaf area and a low photosynthetic capacity (Reich *et al.*, 1999).

The broad relationship among species with respect to photosynthetic rate and leaf life span is similar in all biomes; a 10-fold decrease in leaf life span gives rise to about a fivefold increase in photosynthetic capacity (Reich *et al.*, 1999). Species with long-lived leaves, low photosynthetic capacity, and low stomatal conductance are common in all low-resource environments, including those that are dry, infertile, or shaded.

Plants in productive environments, in contrast, produce short-lived leaves with a high tissue nitrogen concentration and a high photosynthetic capacity; this allows a large carbon return per unit of biomass invested in leaves, if sufficient light is available. These leaves have a high specific leaf area, which maximizes the quantity of leaf area displayed and the light captured per unit of leaf mass. The resulting high rates of carbon gain support a high maximum relative growth rate in the absence of environmental stress or competition from other plants but render plants more vulnerable to environmental stresses such as drought (Schulze and Chapin, 1987). Many early successional habitats, such as recently abandoned agricultural fields or post-fire sites, have sufficient light, water, and nutrients to support high growth rates and are characterized by species with short-lived leaves, high tissue nitrogen concentration, high specific leaf area, and high photosynthetic rates. Even in late succession, environments with high water and nutrient availability are characterized by species with relatively high nitrogen concentrations and photosynthetic rates. Plants in these habitats can grow quickly to replace leaves removed by herbivores or to fill canopy gaps produced by death of branches or individuals.

The changes in tissue nitrogen, and therefore in the C : N ratio of tissues, that occur in response to variation in nitrogen supply constitute an important change in element stoichiometry. This occurs through changes in the ratio of cytoplasm to cell wall and changes in compounds such as tannins and nitrate that are stored in vacuoles. Aquatic phytoplankton have no cell walls and limited capacity for storing compounds in vacuoles and therefore exhibit a much smaller range of variation in C : N ratio than do terrestrial plants (Elser *et al.*, 2000). This variation in stoichiometry enables plants to maximize carbon gain under favorable conditions and maximize efficiency of using other resources to fix carbon, when these resources are limiting to plant growth.

In summary, plants produce leaves with a continuum of photosynthetic characteristics, ranging from short-lived thin leaves with a high nitrogen concentration and high photosynthetic rate to long-lived dense leaves with a low nitrogen concentration and low photosynthetic rate. These correlations among traits are so consistent that specific leaf area (leaf area per unit leaf mass) is often used in ecosystem comparisons as an easily measured index of photosynthetic capacity.

8.06.3.5 Water Limitation

Water limitation reduces the capacity of individual leaves to match CO_2 supply with light availability. Water stress is often associated with high light because sunny conditions correlate with low precipitation (low water supply) and with low humidity (high rate of water loss). High light also increases leaf temperature and water vapor concentration inside the leaf, leading to greater water loss by transpiration. The high-light conditions in which a plant would be expected to increase stomatal conductance to minimize CO_2 limitations to photosynthesis are therefore often the same conditions in which the resulting transpirational water loss is greatest and most detrimental to the plant. When water supply is abundant, leaves typically open their stomata in response to high light, despite the associated high rate of water loss. As leaf water stress develops, stomatal conductance declines to reduce water loss. This decline in stomatal conductance reduces photosynthetic rate and the efficiency of using light to fix carbon below levels found in unstressed plants.

Plants that are acclimated and adapted to dry conditions reduce their photosynthetic capacity and leaf nitrogen content toward a level that matches the low stomatal conductance that is necessary to conserve water in these environments (Wright *et al.*, 2001). A high photosynthetic capacity provides little benefit if the plant must maintain a low stomatal conductance to conserve water. Conversely, low nitrogen availability or other factors that constrain leaf nitrogen concentration result in leaves with low stomatal conductance. This strong correlation between photosynthetic capacity and stomatal conductance maintains the balance between photosynthetic capacity and CO_2 supply, i.e., the colimitation of photosynthesis by diffusional and biochemical processes. In addition to their low photosynthetic capacity and low stomatal conductance, plants in dry areas minimize water stress by reducing leaf area (by shedding leaves or producing fewer new leaves). Some drought-adapted plants produce leaves that minimize radiation absorption; their leaves reflect most incoming radiation or are steeply inclined toward the sun (Ehleringer and Mooney, 1978; Forseth and Ehleringer, 1983). The low leaf area, the reflective nature of leaves, and the steep angle of leaves are the main factors accounting for the low absorption of radiation and low carbon inputs in dry environments. In other words, plants adjust to dry environments primarily by altering leaf area and radiation absorption rather than by altering photosynthetic capacity per unit leaf area. By altering their coarse-scale allocation to biomass (leaves versus roots), plants maintain photosynthetic capacity and associated variation in stomatal conductance within a range in which normal physiological regulation can continue to occur.

Water-use efficiency of photosynthesis is defined as the carbon gain per unit of water lost. Water use is quite sensitive to the size of stomatal openings, because stomatal conductance has slightly different effects on the rates of CO_2 entry and water loss. Water leaving the leaf encounters two resistances to flow: the stomata and the boundary layer of still air on the leaf surface. Resistance to CO_2 diffusion from the bulk air to the site of photosynthesis includes the same stomatal and boundary-layer resistances *plus* an additional internal resistance associated with diffusion of CO_2 from the cell surface into the chloroplast and any biochemical resistances associated with carboxylation. Because of this additional resistance to CO_2 movement into the leaf, any change in stomatal conductance has a *proportionately* greater effect on water loss than on carbon gain. In addition, water diffuses more rapidly than does CO_2 because of its smaller molecular mass and because of the steeper concentration gradient that drives diffusion across the stomata. For all these reasons, as stomata close, water loss declines to a greater extent than does CO_2 absorption. The low stomatal conductance of plants in dry environments results in less photosynthesis per unit of time but greater carbon gain per unit of water loss, i.e., greater water-use efficiency. Plants in dry environments also enhance water-use efficiency by maintaining a somewhat higher photosynthetic capacity than would be expected for their stomatal conductance, thereby drawing down the internal CO_2 concentration and maximizing the diffusion gradient for CO_2 entering the leaf (Wright *et al.*, 2001).

8.06.3.6 Summary of Leaf-level Carbon Gain

The individual leaves of plants exhibit a similar response to photosynthetic limitation by any single environmental factor, whether it is CO_2, light, nitrogen, or water. Photosynthesis initially increases linearly in response to increases in the limiting factor, until some point at which other environmental factors become limiting. Because photosynthetic capacity is geared to match the typical availability of resources that the leaf experiences, there is a limit to which photosynthesis can instantaneously respond to changes in availability of a single limiting factor. Over a longer time period, plants acclimate (physiological adjustment), change their distribution (changes in community composition), or adapt (genetic adjustment). In general, both acclimation and adaptation to low availability of an environmental resource occur by the same physiological mechanism. These adjustments extend the range of environmental conditions over which carbon gain occurs in ecosystems. Many of these adjustments involve changes in photosynthetic capacity, which entail changes in C:N ratio. This variation in element stoichiometry enables plants to maximize carbon gain under favorable environmental conditions. Under unfavorable conditions the increased C:N ratio associated with reduced photosynthetic capacity maximizes the efficiency of using other resources to gain carbon, primarily by prolonging leaf longevity and by shifting allocation to production of other tissues such as wood or roots that have lower tissue nitrogen concentrations than leaves.

8.06.4 STAND-LEVEL CARBON GAIN

8.06.4.1 Scaling of Carbon Gain

Gross primary production (GPP) is the sum of the net photosynthesis by all leaves measured at the ecosystem scale. Modeling studies and field measurements suggest that most conclusions derived from leaf-level measurements of net photosynthesis also apply to GPP. In most closed-canopy ecosystems, photosynthetic capacity decreases exponentially through the canopy in parallel with the exponential decline in irradiance (Field, 1983; Hirose and Werger, 1987). This matching of photosynthetic capacity to light availability maintains the colimitation of photosynthesis by diffusion and biochemical processes in each leaf. The matching of photosynthetic capacity to light availability occurs through the preferential transfer of nitrogen to leaves at the top of the canopy, as a result of at least three processes:

(i) Sun leaves at the top of the canopy develop more cell layers than shade leaves and therefore contain more nitrogen per unit leaf area (Terashima and Hikosaka, 1995).

(ii) New leaves are produced primarily at the top of the canopy, causing nitrogen to be transported to the top of the canopy (Field, 1983; Hirose and Werger, 1987).

(iii) Leaves at the bottom of the canopy senesce when they become shaded to the point that they no longer maintain a positive carbon balance, i.e., they consume more energy in respiration than they produce in photosynthesis.

Much of the nitrogen resorbed from these senescing leaves is transported to the top of the canopy to support the production of young leaves with high photosynthetic capacity. The accumulation of nitrogen at the top of the canopy is most pronounced in dense canopies, which develop under circumstances of high water and nitrogen availability (Field, 1991). In environments in which leaf area is limited by water, nitrogen, or time since disturbance, there is less advantage to concentrating nitrogen at the top of the canopy, because light is abundant throughout the canopy. In these canopies, light availability, nitrogen concentrations, and photosynthetic rates are more uniformly distributed through the canopy.

Canopy-scale relationships between light and nitrogen appear to occur even in multispecies communities (Hirose *et al.*, 1995; Hikosaka and Hirose, 2001). In a single individual, there is an obvious selective advantage to optimizing nitrogen distribution within the canopy because this provides the greatest carbon return per unit of nitrogen invested in leaves. We know less about the factors governing carbon gain in multispecies stands. In such stands, the individuals at the top of the canopy account for most of the photosynthesis and may be able to support greater root biomass to acquire more nitrogen, compared to smaller subcanopy or understory individuals (Hikosaka and Hirose, 2001; Hirose and Werger, 1994). This specialization and competition among individuals probably contributes to the vertical scaling of nitrogen and photosynthesis that is observed in multispecies stands.

Vertical gradients in other environmental variables reinforce the maximization of carbon gain near the top of the canopy. In addition to irradiance, the canopy modifies wind speed, temperature, relative humidity, and CO_2 concentration. The most important of these effects is the exponential decrease in wind speed from the free atmosphere to the ground surface. This vertical reduction in wind speed is most pronounced in smooth canopies, characteristic of crops or grasslands, whereas rough canopies, characteristic of many forests, create more friction and turbulence that increases the vertical mixing of air within the canopy (McNaughton and Jarvis, 1991). Wind speed is important because it reduces the thickness of the boundary layer of still air around each leaf, producing steeper gradients in temperature and in concentrations of CO_2 and water vapor from the leaf surface to the atmosphere. This speeds the diffusion of CO_2 into the leaf and the loss of water from the leaf. The net effect of wind on photosynthesis is generally positive at moderate wind speeds and adequate moisture supply, enhancing photosynthesis at the top of the canopy. When low soil moisture or a long pathway for water transport from the soil to the top of the canopy reduces water supply to the uppermost leaves, as in tall forests, the uppermost leaves reduce their stomatal conductance, causing the zone of maximum photosynthesis to shift farther down in the canopy (Landsberg and Gower, 1997). Although multiple environmental gradients within the canopy have complex effects on photosynthesis, they probably enhance photosynthesis near the top of canopies in ecosystems with sufficient water and nutrients to develop dense canopies.

Canopy properties extend the range of light availability over which the light-use efficiency of the canopy remains constant. The light-response curve of canopy photosynthesis, measured in closed canopies (total leaf area index (LAI)—the leaf area per unit ground area—is larger than ~3), saturates at higher irradiance than does photosynthesis by a single leaf (Jarvis and Leverenz, 1983). The canopy increases the efficiency of converting light energy into fixed carbon for several reasons. The more vertical orientation of leaves at the top of the canopy reduces the likelihood that they become light-saturated and increases light penetration deeper into the canopy. The clumped distribution of leaves in shoots, branches, and crowns also increases light penetration into the canopy, particularly in conifer canopies in which needles are clumped around stems. This could explain why conifer forests frequently support a higher LAI than deciduous forests. The light compensation point also decreases from the top to the bottom of the canopy, so lower leaves maintain a positive carbon balance, despite the relatively low light availability. In crop canopies, where water and nutrients are highly available, the linear relationship between canopy carbon exchange and irradiance (i.e., constant light-use efficiency) extends up to irradiance typical of full sunlight. In other words, there is no evidence of light saturation, and light-use efficiency remains constant over the full range of natural light intensities (Figure 3) (Ruimy *et al.*, 1996). In most natural canopies, however, canopy photosynthesis becomes light-saturated at high irradiance.

8.06.4.2 Scaling of Controls over GPP

As described in the previous section, stand-level photosynthesis (GPP) responds to limiting

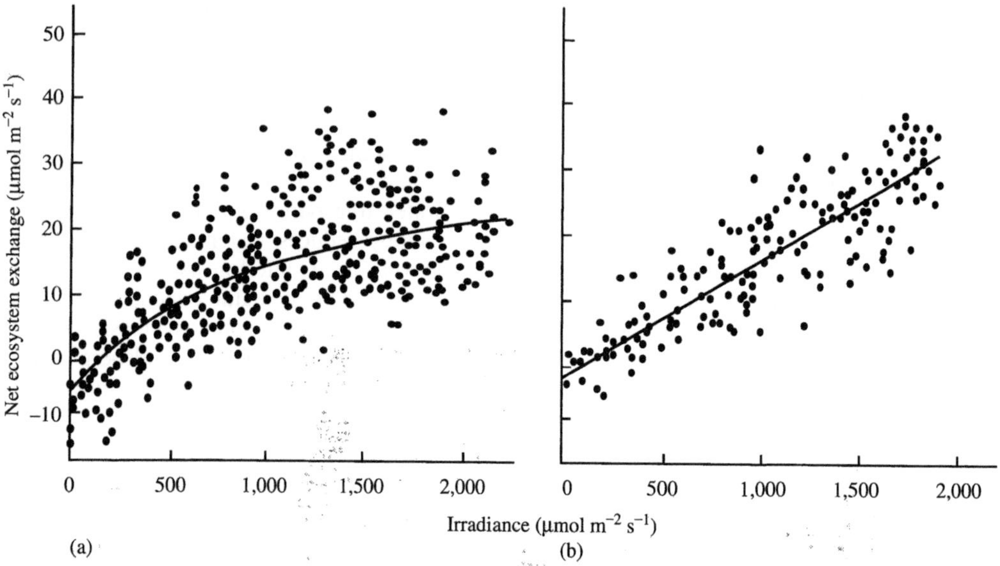

Figure 3 Effect of vegetation and irradiance on net ecosystem exchange in (a) forests and (b) crops (reproduced by permission of Academic Press from *Adv. Ecol. Res.*, **1996**, *26*, 1–68).

factors in a way that qualitatively matches the responses of individual leaves. This occurs because the leaves at the top of the canopy are exposed to the highest irradiance and have primary access to plant nitrogen. In addition, these leaves experience the highest wind speed and therefore have a thin boundary layer, so the gradients in CO_2 concentration, water vapor concentration, and temperature between the leaf and the air are similar to patterns measured on individual leaves.

The major differences between leaf-level and stand-level responses of photosynthesis to environmental constraints relate to differences in leaf area and its control. LAI is both a cause and a consequence of ecosystem differences in NPP. It is governed primarily by the availability of soil resources (water and nutrients) and by the time for recovery from past disturbances and other processes (e.g., herbivory) that remove leaves from vegetation.

Variation in soil resource supply accounts for much of the spatial variation in leaf area and GPP among ecosystem types. Analysis of satellite imagery shows that ~70% of the ice-free terrestrial surface has relatively open canopies (Graetz, 1991) (Figure 4). GPP correlates closely with leaf area below a total LAI of ~8 (projected LAI of 4) (Schulze *et al.*, 1994), suggesting that leaf area is a critical determinant of GPP on most of Earth's terrestrial surface. GPP saturates with increasing LAI in dense canopies, because the leaves in the middle and bottom of the canopy contribute relatively little to GPP. The availability of soil resources, especially water and nutrient supply, is

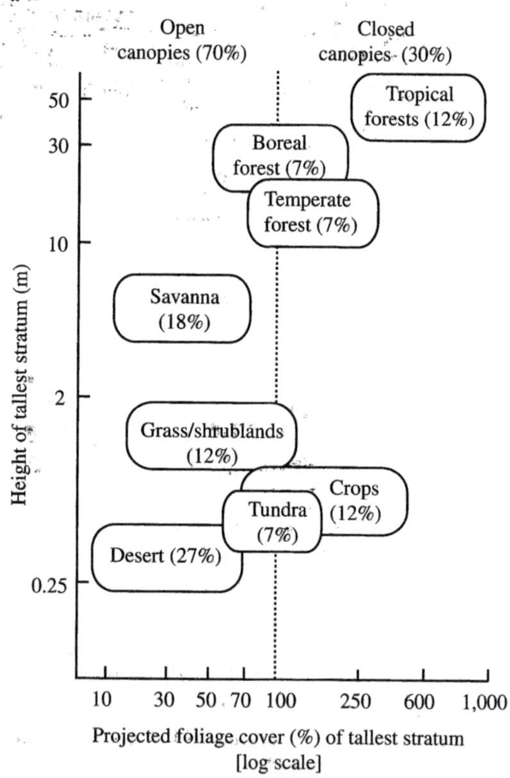

Figure 4 Projected foliage cover and canopy height of the major biomes. Typical values for that biome and the percentage of the terrestrial surface that it occupies are shown. The vertical line shows 100% canopy cover (reproduced by permission of Kluwer from *Climatic Change*, **1991**, *18*, 147–173).

a critical determinant of LAI for two reasons: (i) Plants in high-resource environments produce a large amount of leaf biomass; and (ii) leaves produced in these environments have a high SLA, i.e., a large leaf area per unit of leaf biomass. As discussed earlier, a high specific leaf area maximizes light capture and therefore carbon gain per unit of leaf biomass (Lambers and Poorter, 1992; Reich *et al.*, 1997). In low-resource environments, plants produce fewer leaves, and these leaves have a lower specific leaf area. Ecosystems in these environments have a low LAI and therefore a low GPP.

Soil resources and light extinction through the canopy determine the upper limit to the leaf area that an ecosystem can support. However, many factors regularly reduce leaf area below this potential LAI. Drought and freezing are climatic factors that cause plants to shed leaves. Other causes of leaf loss include physical disturbances (e.g., fire and wind) and biotic agents (e.g., herbivores and pathogens). After major disturbances the remaining plants may be too small, have too few meristems, or lack the productive potential to produce the leaf area that could potentially be supported by the climate and soil resources of a site. For this reason, LAI tends to increase with time after disturbance to an asymptote.

8.06.5 RESPIRATION

All controls on NPP that we have discussed so far have focused on production side, but NPP is also a function of carbon loss through respiration. The environmental controls over plant respiration are quite similar to the controls over GPP because respiration, like photosynthesis, is tightly linked to environmental factors that regulate plant activity. The mechanistic basis for this relationship can be understood by separating plant respiration into three functional components: growth respiration, maintenance respiration, and the respiratory cost of ion uptake.

The carbon expended in plant growth consists of the carbon incorporated into new tissue plus the respiration required to produce the ATPs necessary to carry out this synthesis (Penning de Vries, 1975). This carbon cost can be calculated from the chemical composition of tissues and an estimate from biochemical pathways of the carbon required to synthesize each class of chemical compound (Chapin, 1989; Merino *et al.*, 1982; Penning de Vries, 1975; Williams *et al.*, 1987). Although there is a threefold range in the carbon cost of synthesis among the major classes of chemical compounds in plants, the carbon cost per gram of tissue is surprisingly similar across species, tissue types, and ecosystems (Chapin, 1989; Poorter, 1994). All plant parts contain some expensive constituents. For example, metabolically active tissues, such as leaves, have high concentrations of proteins, tannins, and lipids (primarily lipophilic substances such as terpenes that defend protein-rich tissues from herbivores and pathogens) (Bryant and Kuropat, 1980; Coley *et al.*, 1985), whereas structural tissue is rich in lignin. Similar chemical correlations are observed within a tissue type across species or growing conditions. Leaves of rapidly growing species with high protein concentration, for example, have higher tannin and lower lignin concentrations than leaves with low protein concentrations. Consequently, most plant tissues contain some expensive constituents, although the nature of these constituents differs among plant parts and species. Given that the carbon cost of growth is nearly constant, we expect that growth respiration should be a relatively constant fraction of NPP. Gas exchange and modeling studies support this hypothesis: growth respiration is ~25% of the carbon incorporated into new tissues (Waring and Running, 1998). In summary, the rates of growth and therefore of growth respiration measured at the ecosystem scale (g C m^{-2} d^{-1}) increase when temperature and moisture favor growth, but growth respiration is always a nearly constant fraction of NPP, regardless of environmental conditions.

Ion transport across membranes may account for 25–50% of root respiration (Bloom, 1986; Lambers *et al.*, 1996, 1998). This large requirement for respiratory energy is not well quantified in field studies but may correlate with NPP because the quantity of nutrients absorbed is greatest in productive environments. Several factors cause this cost of ion uptake to differ among ecosystems. The respiratory cost of nitrogen uptake and use depends on the form of nitrogen absorbed, because nitrate must be reduced to ammonium (an exceptionally expensive process) before it can be incorporated into proteins or other organic compounds. The cost of nitrate reduction is also variable among plant species and ecosystems, depending on whether the nitrate is reduced in the leaves, where it may be supported by excess reducing power from the light reaction, or in the roots, where it depends on carbohydrates transported to roots. In general, we expect respiration associated with ion uptake to correlate with the total quantity of ions absorbed and therefore to show a positive relationship with NPP. However, there are few data available to evaluate this hypothesis.

All live cells, even those that are not actively growing, require energy to maintain ion gradients across cell membranes and to replace proteins, membranes, and other constituents. Maintenance respiration provides the ATP for these maintenance and repair functions. Laboratory experiments suggest that ~85% of maintenance respiration is associated with the turnover of proteins (~6% turnover per day), explaining why there is a strong correlation between protein concentration and whole-tissue respiration rate in nongrowing tissues (Penning de Vries, 1975; Ryan and Waring, 1992; van der Werf *et al.*, 1992). We therefore expect maintenance respiration to be greatest in ecosystems with high tissue nitrogen concentrations and/or a large plant biomass and thus to be greatest in productive ecosystems. Simulation models suggest that maintenance respiration may account for about half of total plant respiration; the other half is associated with growth and ion uptake (Lambers *et al.*, 1998). These proportions may vary with environment and plant growth rate and are difficult to estimate precisely.

Maintenance respiration depends on environment as well as tissue chemistry. It increases with temperature because proteins and membrane lipids turn over more rapidly at high temperatures. Drought also imposes short-term metabolic costs associated with synthesis of osmotically active organic solutes. These effects of environmental stress on maintenance respiration are the major factors that alter the partitioning between growth and respiration and therefore are the major sources of variability in the efficiency of converting GPP into NPP. Maintenance respiration increases during times of environmental change but, following acclimation, maintenance respiration returns to values close to those predicted from biochemical composition (Semikhatova, 2000). Over the long term, therefore, maintenance respiration may not be strongly affected by environmental stress.

Plant respiration is a relatively constant proportion of GPP, when ecosystems are compared. Although the respiration rate of any given plant increases exponentially with ambient temperature, acclimation and adaptation counterbalance this direct temperature effect on respiration. Plants from hot environments have lower respiration rates at a given temperature than do plants from cold places (Billings *et al.*, 1971; Billings and Mooney, 1968; Mooney and Billings, 1961). The net result of these counteracting temperature effects is that plants from different thermal environments have similar respiration rates, when measured at their mean habitat temperature (Semikhatova, 2000).

In summary, studies of the basic components of respiration associated with growth, ion uptake, and maintenance suggest that total plant respiration should be a relatively constant fraction of GPP. These predictions are consistent with the results of model simulations of plant carbon balance. These modeling studies indicate that total plant respiration is about half (48–60%) of GPP, when a wide range of ecosystems is compared (Landsberg and Gower, 1997; Ryan *et al.*, 1994). Variation in maintenance respiration is the most likely cause for variability in the efficiency of converting GPP into NPP. There are too few detailed studies of ecosystem carbon balance to know how variable this efficiency is among seasons, years, and ecosystems.

8.06.6 PHOTOSYNTHESIS, RESPIRATION, AND NPP: WHO IS IN CHARGE?

Knowing that NPP is the balance of carbon gained by photosynthesis and the carbon lost by respiration does not tell us which is the cause and which is the effect. Do the conditions governing photosynthesis dictate the amount of carbon that is available to support growth or do conditions influencing growth rate determine the potential for photosynthesis? On short timescales (seconds to days), environmental controls over photosynthesis (e.g., light and water availability) strongly influence photosynthetic carbon gain. Leaf carbohydrate concentrations increase during the day and decline at night, allowing plants to maintain a relatively constant supply of carbohydrates to nonphotosynthetic organs. Similarly, carbohydrate concentrations increase during periods (hours to weeks) of sunny weather and decline under cloudy conditions. Over these short timescales, the conditions affecting photosynthesis are the primary determinants of the carbohydrates available to support growth.

On weekly to annual timescales, however, plants adjust leaf area and photosynthetic capacity, so carbon gain matches the soil resources that are available to support growth. Plant carbohydrate concentrations are usually lowest when environmental conditions favor rapid growth (i.e., carbohydrates are drawn down by growth) and tend to accumulate during periods of drought or nutrient stress or when low temperature constrains NPP (Chapin, 1991b). If the products of photosynthesis directly controlled NPP, we would expect high carbohydrate concentrations to coincide with rapid growth or to show no consistent relationship with growth rate.

Results of growth experiments also indicate that growth is not simply a consequence of the controls over photosynthetic carbon gain. Plants respond to low availability of water, nutrients, or oxygen in their rooting zone by producing hormones that reduce growth rate. The decline in growth subsequently leads to a decline in photosynthesis (Chapin, 1991b; Davies and Zhang, 1991; Gollan *et al.*, 1985). The general conclusion from these experiments is that plants actively sense the resource supply in their environment and adjust their growth rate accordingly. These changes in growth rate then change the sink strength (demand) for carbohydrates and nutrients, leading to changes in photosynthesis and nutrient uptake (Chapin, 1991b; Lambers *et al.*, 1998). The resulting changes in growth and nutrition determine the LAI and photosynthetic capacity, which, as we have seen, largely account for ecosystem differences in carbon input (Gower *et al.*, 1999).

8.06.7 ALLOCATION OF NPP

In general, plants allocate production preferentially to those plant parts that are necessary to acquire the resources that most strongly limit growth. Plants allocate new biomass preferentially to roots when water or nutrients limit growth. They allocate new biomass preferentially to shoots when light is limiting (Reynolds and Thornley, 1982). Plants can increase acquisition of a resource by producing more biomass of the appropriate tissue, by increasing the activity of each unit of biomass, or by retaining the biomass for a longer time (Garnier, 1991). A plant can, for example, increase carbon gain by increasing leaf area or photosynthetic rate per unit leaf area or by retaining the leaves for a longer time before they are shed. Similarly, a plant can increase nitrogen uptake by altering root morphology or by increasing root biomass, root longevity, nitrogen uptake rate per unit root, or extent of mycorrhizal colonization. Changes in allocation and root morphology have a particularly strong impact on nutrient uptake. It is the integrated activity (mass × acquisition rate per unit mass × time) that must be balanced between shoots and roots to maximize growth and NPP (Garnier, 1991). These allocation rules are key features of all simulation models of NPP. Observations in ecosystems are generally consistent with allocation theory. Tundra, grasslands, and shrublands, for example, allocate a larger proportion of NPP below ground than do forests (Gower *et al.*, 1999; Saugier *et al.*, 2001).

The balance between NPP and biomass loss determines the annual increment in plant biomass. Plants retain only part of the biomass that they produce. Some biomass loss is physiologically regulated by the plant—e.g., the senescence of leaves and roots. Senescence occurs throughout the growing season in grasslands and during autumn or at the beginning of the dry season in many trees. Other losses occur with varying frequency and predictability and are less directly controlled by the plant, such as the losses to herbivores and pathogens, wind throw, and fire. The plant also influences these tissue loss rates through the physiological and chemical properties of the tissues it produces. Still other biomass transfers to dead organic matter result from mortality of individual plants. Given the substantial, although incomplete, physiological control over tissue loss, why do plants dispose of the biomass in which they have invested so much carbon, water, and nutrients to produce?

Tissue loss is an important mechanism by which plants balance resource requirements with resource supply from the environment. Plants depend on regular large inputs of carbon, water, and, to a lesser extent, nutrients to maintain vital processes. For example, once biomass is produced, it must have continued carbon inputs to support maintenance respiration. If the plant (or organ) cannot meet these carbon demands, the plant (or organ) dies. Similarly, if the plant cannot absorb sufficient water to replace the water that is inevitably lost during photosynthesis, it must shed transpiring organs (leaves) or die. The plant must therefore shed biomass whenever resources decline below some threshold needed for maintenance. Senescence is just as important as production in adjusting to changes in resource supply and is the only mechanism by which plants can reduce biomass when resources decline in abundance.

8.06.8 NUTRIENT USE

Given the importance of nutrients in controlling NPP, it is important to understand the relationship between nutrient supply and NPP. Plants respond to increased supply of a limiting nutrient in laboratory experiments primarily by increasing plant growth, giving a linear relationship between rate of nutrient accumulation and plant growth rate (Ingestad and Ågren, 1988). Plants also respond to increased nutrient supply in the field primarily through increased NPP, with proportionately less increase in tissue nutrient concentration. Tissue nutrient concentrations increase substantially only when other factors begin to limit plant growth. The sorting of species by habitat also contributes to the responsiveness of nutrient uptake and NPP to variations in nutrient supply observed across habitats. Species such as trees that use large quantities of nutrients dominate sites with high nutrient supply rates, whereas infertile habitats are dominated by species with lower capacities for nutrient absorption and growth. Despite these physiological and species adjustments, tissue nutrient concentrations in the field generally increase with an increase in nutrient supply.

Nutrient-use efficiency is greatest where production is nutrient-limited. Differences among plants in tissue nutrient concentration provide insight into the quantity of biomass that an ecosystem can produce per unit of nutrient. Nutrient-use efficiency is the ratio of nutrients to biomass lost in litterfall (i.e., the inverse of nutrient concentration in plant litter)

(Vitousek, 1982). This ratio is highest in unproductive sites, suggesting that plants are more efficient in producing biomass per unit of nutrient acquired and lost, when nutrients are in short supply. Several factors contribute to this pattern (Chapin, 1980). First, tissue nutrient concentration tends to decline as soil fertility declines, as described earlier. Individual plants that are nutrient-limited also produce tissues more slowly and retain these tissues for a longer period of time, resulting in an increase in average tissue age. Older tissues have low nutrient concentrations, causing a further decline in concentration (i.e., increased nutrient-use efficiency). Finally, the dominance of infertile soils by species with long-lived leaves that have low nutrient concentrations further contributes to the high nutrient-use efficiency of ecosystems on infertile soils.

There are at least two ways in which a plant might maximize biomass gained per unit of nutrient (Berendse and Aerts, 1987): through (i) a high nutrient productivity (a_n), i.e., a high instantaneous rate of carbon uptake per unit nutrient or (ii) a long residence time (t_r), i.e., the average time that the nutrient remains in the plant:

$$\mathrm{NUE} = a_n \times t_r$$

Species characteristic of infertile soils have a long residence time of nutrients but a low nutrient productivity (Chapin, 1980; Lambers and Poorter, 1992), suggesting that the high nutrient-use efficiency in unproductive sites results primarily from traits that reduce nutrient loss rather than traits promoting a high instantaneous rate of biomass gain per unit of nutrient. Shading also reduces tissue loss more strongly than it reduces the rate of carbon gain (Walters and Reich, 1999).

There is an innate physiological trade-off between nutrient residence time and nutrient productivity. This occurs because the traits that allow plants to retain nutrients reduce their capacity to grow rapidly (Chapin, 1980; Lambers and Poorter, 1992). Plants with a high nutrient productivity grow rapidly and have high photosynthetic rates, which are associated with thin leaves, a high specific leaf area, and a high tissue nitrogen concentration. Conversely, a long nutrient residence time is achieved primarily through slow rates of replacement of leaves and roots. Leaves that survive a long time have more structural cells to withstand unfavorable conditions and higher concentrations of lignin and other secondary metabolites that deter pathogens and herbivores. Together these traits result in dense leaves with low tissue nutrient concentrations and therefore low photosynthetic rates per gram of biomass. The high nutrient-use efficiency of plants on infertile soils therefore reflects their capacity to retain tissues for a long time rather than a capacity to use nutrients more effectively in photosynthesis.

The trade-off between nutrient-use efficiency and rate of resource capture explains the diversity of plant types along resource gradients. Low-resource environments are dominated by species that conserve nutrients through low rates of tissue turnover, high nutrient-use efficiency, and the physical and chemical properties necessary for tissues to persist for a long time. These stress-tolerant plants outcompete plants that are less effective at nutrient retention in environments that are dry, infertile, or shaded (Chapin, 1980; Walters and Field, 1987). A high nutrient-use efficiency and associated traits constrain the capacity of plants to capture carbon and nutrients. In high-resource environments species with high rates of resource capture, rapid growth rates, rapid tissue turnover, and consequently low nutrient-use efficiency therefore outcompete plants with high nutrient-use efficiency. In other words, neither a rapid growth rate nor a high nutrient-use efficiency is universally advantageous, because there are inherent physiological trade-offs between these traits. The relative benefit to the plant of efficiency versus rapid growth depends on environment.

8.06.9 BALANCING NUTRIENT LIMITATIONS

8.06.9.1 Nutrient Requirements

Thus far, we have focused on the mechanisms by which plants minimize the constraints on NPP by balancing the limitations of water, CO_2, light, and nutrients. Photosynthesis and productivity require a balanced proportion of these resources, and plants can adjust their physiology to maximize NPP across a range of limiting factors. Nutrients need further explanation. Unlike light, CO_2, and water, which are relatively homogenous in quality, the nutrient category includes many chemical elements, each with different functions and controls.

Because each nutrient performs a different function in plants (Table 3), the relative amount of each nutrient required and plant response to limitation by these nutrients vary. Primary macronutrients are the nutrients needed in the largest amounts and are most commonly limiting to plant growth. These include nitrogen, phosphorus and potassium. Secondary macronutrients include calcium, magnesium, and sulfur. These are also required in large quantities but are less frequently limiting to growth. Micronutrients are essential for plant growth but are only needed in small quantities. These include boron, chloride, copper, iron, manganese, molybdenum, and zinc. All macro- and micronutrients are essential for plant

Table 3 Nutrients required by plants and their major functions.

Nutrient	*Role in plants*
Macronutrients[a]	Required by all plants in large quantities
Primary	
Nitrogen (N)	Component of proteins, enzymes, phospholipids, and nucleic acids
Phosphorus (P)	Component of proteins, coenzymes, nucleic acids, oils, phospholipids, sugars, starches
	Critical in energy transfer (ATP)
Potassium (K)	Component of proteins
	Role in disease protection, photosynthesis, ion transport, osmotic regulation, enzyme catalyst
Secondary	
Calcium (Ca)	Component of cell walls
	Regulates structure and permeability of membranes, root growth
	Enzyme catalyst
Magnesium (Mg)	Component of chlorophyll
	Activates enzymes
Sulfur (S)	Component of proteins and most enzymes
	Role in enzyme activation, cold resistance
Micronutrients[b]	Required by all plants in small quantities
Boron (B)	Role in sugar translocation and carbohydrate metabolism
Chloride (Cl)	Role in photosynthetic reactions, osmotic regulation
Copper (Cu)	Component of some enzymes Role as a catalyst
Iron (Fe)	Role in chlorophyll synthesis, enzymes, oxygen transfer
Manganese (Mn)	Activates enzymes
	Role in chlorophyll formation
Molybdenum (Mo)	Role in N fixation, NO_3 enzymes, Fe adsorption, and translocation
Zinc (Zn)	Activates enzymes, regulates sugar consumption
Beneficial nutrients[c]	Required by certain plant groups, or by plants under specific environmental conditions
Aluminum (Al)	
Cobalt (Co)	
Iodine (I)	
Nickel (Ni)	
Selenium (Se)	
Silicon (Si)	
Sodium (Na)	
Vanadium (V)	

[a] Macronutrients: Primary—usually most limiting because used in largest amounts. Secondary—major nutrients but less often limiting.
[b] Micronutrients: essential for plant growth, but only needed in small quantities. [c] Beneficial nutrients—often aid plant growth, but not essential.

growth and metabolism, and other elements cannot substitute for their function. However, certain functions, such as maintenance of osmotic pressure, can be accomplished by various elements. A fourth class of mineral nutrients, "beneficial nutrients" can enhance plant growth, are required by plants under very specific conditions, or are necessary for very specific groups of plants (Marschner, 1995). For example, aluminum is required by ferns, cobalt by *Fabales* with symbionts, and sodium by *Chenopodiaceae* (Larcher, 1995).

8.06.9.2 Limitations by Different Nutrients

Although all of these mineral nutrients are necessary for plant growth, the particular nutrient that limits plant production may vary in space and time. The primary macronutrients, nitrogen, phosphorus, and potassium, are used by plants in the greatest amounts, and tend to most frequently limit plant production. Nitrogen is the most commonly limiting nutrient to plant growth in terrestrial systems, particularly in the temperate zone (Vitousek and Howarth, 1991). Phosphorus generally limits plant growth in the lowland wet tropics (Tanner *et al.*, 1998), on very old soils (Vitousek and Farrington, 1997), on some Mediterranean soils (Cowling, 1993; Specht and Rundel, 1990), and on glacial and aeolian sandy soils in European heathlands (Aerts and Heil, 1993). Sites that would naturally be nitrogen-limited can become phosphorus-limited under certain conditions. Phosphorus limitation, for example, occurs in areas with high nitrogen deposition (Aerts and Berendse, 1988; Aerts and Bobbink, 1999) and in European fens that have lost substantial phosphorus over time through long-term mowing treatments (Verhoeven and Schmitz, 1991). Vegetation composition can also influence whether a site is limited by nitrogen or phosphorus. In California grasslands, for example, grass-dominated sites are nitrogen-limited, but these same sites can be sulfur- and phosphorus-limited if legumes are present (Jones *et al.*, 1970; Jones and Martin, 1964; Jones *et al.*, 1983). The limitation of nitrogen versus phosphorus also changes over successional time, with soils being nitrogen-limited early in primary succession, then becoming phosphorus-limited with time (Chapin *et al.*, 1994; Vitousek and Farrington, 1997; Walker and Syers, 1976). Calcium, magnesium, and potassium also virtually disappear due to leaching in old soils, but are frequently not limiting to plant growth due to atmospheric inputs (Chadwick *et al.*, 1999). There are, however, instances when these nutrients do limit NPP. Potassium is taken up by plants in larger amounts than any element except for nitrogen (Marschner, 1995). It tends to be limiting in ecosystems with high precipitation and very late in soil development, particularly on sandy soils (Tisdale *et al.*, 1993), but its limitation is relatively infrequent compared to nitrogen and phosphorus. Highly weathered tropical soils with high leaching rates can also be limiting in calcium, although calcium is more frequently found in excess of plant demand (Barber, 1984; Chapin, 1991a; Marschner, 1995). Base cations such as calcium and magnesium have also been found to be limiting in areas with high cation leaching associated with high nitrogen deposition (Aber *et al.*, 1998; Driscoll *et al.*, 2001; Schulze, 1989). Limitation by other essential nutrients is rare, but does occur (e.g., manganese (Goransson, 1994), iron (Goransson, 1993), molybdenum (Tisdale *et al.*, 1993)).

Although certain ecosystems can be characterized as being limited by a particular mineral nutrient, changes in the environment, such as rain storms or pulses of litter inputs, can rapidly alter the relative abundance of nutrients, shifting limitation from one nutrient to another at different times during plant growth. Thus, plants must be flexible in taking up different nutrients.

8.06.9.3 Stoichiometry of NPP

A proper balance of nutrients is required for plant growth. In marine systems, the stoichiometry of primary production is determined by the ratio of elements in the cytoplasm (Redfield ratio) that supports optimal metabolism of phytoplankton (Redfield, 1958). The C:N:P ratio is fairly constant in marine phytoplankton, and this ratio in primary producers constrains the cycling of all elements (Elser *et al.*, 2000). The amount and proportions of nitrogen and phosphorus available determine the amount of carbon fixed by phytoplankton. Limitation by either of these elements constrains any further accumulation of carbon or other nutrients by phytoplankton. The carbon and nutrients in phytoplankton in turn determine the recycling of nutrients and the N:P in the deep sea and upwelling waters, so biotic demand for nitrogen and phosphorus closely match their availability.

In terrestrial systems, similar ratios are observed in vegetation, with a general C : N : S : P of land plants being 790 : 7.6 : 3.1 : 1 (Bolin *et al.*, 1983). Such generalizations have been used to guide fertilizer application in agricultural systems. The widespread use of "fixed formulas" of nutrients, such as Hoaglands solution, in controlled environments is an indicator of the robustness of such a stoichiometric relationship (Ingestad and Ågren, 1988). Departures from such ratios have been used as indicators of nutrient limitation in plants (Jones and Martin, 1964; Koerselman and Mueleman, 1996; Ulrich and Hills, 1973). However, ratios of nutrients in tissues are not necessarily an indicator of nutrient limitation in land plants, because uptake of nutrients in terrestrial vegetation is less constrained by nutrient balances than in marine phytoplankton (Marschner, 1995).

In order to extend the simple stoichiometric control implied by the Redfield ratio to terrestrial systems, the element that most constrains NPP must define the quantities of all elements cycled through vegetation. We have already seen, however, that the nutrient-use efficiency of plants differs among growing conditions and among species. In addition, to be truly comparable to marine systems, the input and recycling of nutrients in dead plant material must also approximately equal the nutrient ratio required for plant growth. Observed dynamics in terrestrial systems are far from this simple formula because of several mechanisms that decouple nutrient and carbon cycles in terrestrial ecosystems. Let us start with a simple contrast to marine systems. The Redfield ratio is based on an optimal cytoplasmic stoichiometry of single-celled organisms. Terrestrial plants are both multicellular and have different tissue types and compounds with dramatically different stoichiometries (Bazzaz, 1997; Lambers *et al.*, 1998). In this chapter, we have already described many situations in which plants shift their relative allocation among tissues in response to a change in environment. Allocation can also differ among species. Thus, even assuming that plants receive an ideal ratio of resources, plant species have inherently different allocation strategies and even different nutrient ratios within the same tissues, leading to substantial variation in the stoichiometry of NPP (Eviner and Chapin, in press). A more dramatic departure from the simple marine stoichiometric model occurs with the recycling of nutrients in terrestrial ecosystems. There are many reasons why the supply of resources does not equal demand, as is hypothesized in marine systems. The tight coupling of nitrogen and phosphorus cycling in marine systems does not occur in terrestrial systems, where nitrogen and phosphorus differ dramatically in the controls over mineralization and availability (McGill and Cole, 1981). In addition, litter inputs have a dramatically different stoichiometry from plant demand due to resorption of nutrients from senescing litter (Aerts and Chapin, 2000). Finally, unlike the well-mixed nutrient return through upwelling in marine systems, nutrient availability in the soil is extremely heterogeneous (Caldwell *et al.*, 1996). So unlike marine systems, terrestrial cycling involves significant storage in plants and soils and slow turnover of nutrients, so the stocks of available nutrients have little relation to the fluxes.

8.06.9.4 Uncoupling Mechanisms

NPP in terrestrial systems is not a simple function of the ratio of available nutrients because there are many ways in which carbon and different nutrients become uncoupled in terrestrial ecosystems (Eviner and Chapin, in press) (Figure 5). In the following sections, we discuss those uncoupling mechanisms that cause NPP in terrestrial ecosystems to depart from a simple stoichiometric model.

8.06.9.4.1 Litterfall and leaching inputs

During the transition from live tissue to litter, the ratios and concentrations of nutrients undergo dramatic changes due to both resorption and leaching (Marschner, 1995) (Aerts and Chapin, 2000) (Figure 5). Plants resorb approximately half of their leaf nitrogen and phosphorus during senescence, with a larger percentage of phosphorus than of nitrogen, tending to be resorbed (Aerts, 1995; Aerts and Chapin, 2000; Chapin and Kedrowski, 1983). In contrast, only ~35% of sulfur is resorbed (Quilchano *et al.*, 2002). Calcium and iron cannot be resorbed because they are immobile in the phloem of plants (Gauch, 1972). During resorption, there is a high potential for cations such as potassium, calcium, magnesium, and sodium to leach from leaves in plant-available forms. In fact, up to 80% of leaf

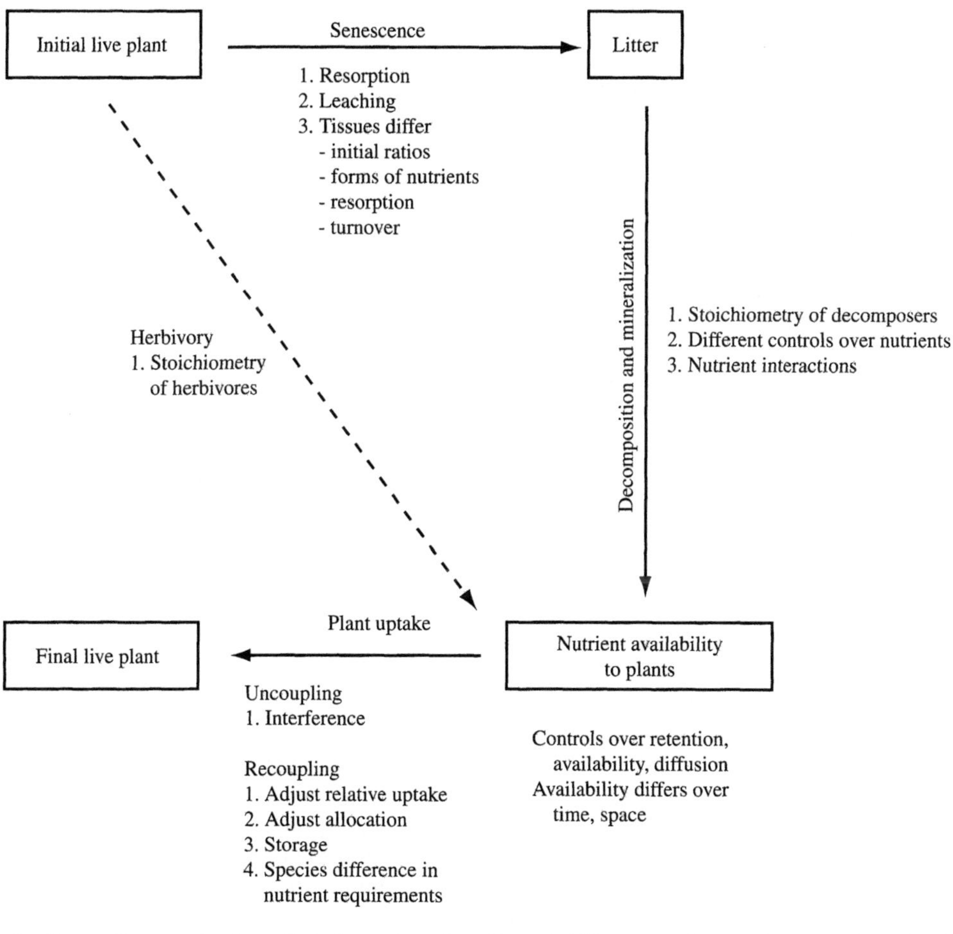

Figure 5 Uncoupling mechanisms that alter rates of cycling of different nutrients relative to the ratios initially present in live plants.

potassium, 50% of leaf calcium, but only ~15% of leaf nitrogen and phosphorus are lost through leaching (Chapin, 1991a). Thus, plant senescence results in a significant decoupling among nutrients returned to the soil in soluble and particulate forms. This causes the stoichiometry of these element inputs from the plants to soil to be extremely different than the ratio required for plant growth. This is very different from the scenario with marine phytoplankton, in which the ratios of nutrients absorbed and lost are similar to the ratios found in plankton (Elser *et al.*, 2000).

Plant species differ in the magnitude of decoupling among nutrients because of differences in allocation to, and turnover of, tissues with different element ratios (Eviner and Chapin, in press). Roots, for example, have low-nutrient-to-carbon ratios, as does wood, which also has a very high concentration of calcium. Roots and leaves, with their high enzyme concentrations, have higher N : P ratios than does wood. The types of nutrient-containing compounds also differ among tissue types (Chapin and Kedrowski, 1983) and can substantially affect recycling rates. The turnover rates of these different tissues differ due to both environmental conditions and plant species identity (Poorter and Villar, 1997). These tissues also differ in their effectiveness in resorption. Leaves resorb about half of their nitrogen and phosphorus; stems have much lower resorption (Aerts and Chapin, 2000), whereas there is no evidence for nutrient resorption from roots (Gordon and Jackson, 2000; Nambiar, 1987).

Disturbances such as hurricanes can result in large inputs of unsenesced plant tissue that contains nutrients in roughly the ratios required to produce living material. However, these inputs occur infrequently and do not govern recycling of nutrients most of the time. Herbivores also harvest plant matter before plant tissues senesce; this is often viewed as a "short circuit" in nutrient recycling. However, because the stoichiometry of herbivores differs from that of plants (Elser and Urabe, 1999), herbivores incorporate nutrients and carbon in different ratios than the plants supply and therefore excrete the nutrients in a ratio that differs from the ideal plant demand. The supply of nutrients recycled by herbivores is also spatially and temporally variable. So unlike marine systems, where phytoplankton sink to the deep ocean, and nutrients are then supplied in upwelling zones in the same nutrient ratio, the inputs of terrestrial litter have their nutrient stoichiometry decoupled from that of live leaves.

8.06.9.4.2 Nutrient mineralization

The release of nutrients from litter is further decoupled through decomposition and mineralization processes, because elements differ in controls over their cycling (McGill and Cole, 1981). From a simple stoichiometric perspective, it is instructive to first consider which organisms are doing the recycling. Soil microbes break down organic matter to meet their energy (C) and nutrient requirements for growth. Because carbon is often limiting to the microbial community and is a common currency for growth and biomass in both plants and microbes, we will express the stoichiometric relationships per unit of C. The average plant has a C:N:S:P of 1,000:9.6:3.9:1.3 (Bolin *et al.*, 1983). Assuming that roughly half of nitrogen and phosphorus (Aerts and Chapin, 2000), and 35% of sulfur (Quilchano *et al.*, 2002) is resorbed from aboveground litter, this would imply an average plant litter ratio of 1,000:4.8:2.5:0.65. Soil bacterial biomass has a ratio of 1,000:100:4.7:23.3, whereas fungi have a ratio of 1,000:62:4.3:5.3 (Bolin *et al.*, 1983). Both groups require ~40% more carbon than the stoichiometric ratios in their biomass would suggest because of the carbon expended in respiration. They also require additional nitrogen for the production of exoenzymes. If the growth efficiency of bacteria and fungi is similar (i.e., the same respiratory carbon requirement for growth), and the nitrogen requirement for exoenzyme production is similar, these stoichiometric ratios suggest that bacteria require nearly twice as much nitrogen and more than four times more phosphorus per unit of growth than do fungi, i.e., bacteria have a higher phosphorus requirement than do fungi.

Because the nutrient demands of soil microbes differ from the ratios of elements available in litter inputs, the decomposition and mineralization processes decouple the cycling of these nutrients from one another, and the nature of this decoupling depends, in part, on the identity of the decomposing organisms (Paul and Clark, 1996). In forests, for example, which are dominated by fungal activity, nitrogen tends to be immobilized by microbes, whereas phosphorus may be more readily mineralized. In bacterially dominated grasslands and agricultural systems, in contrast, there may be greater tendency to immobilize phosphorus and to mineralize or immobilize nitrogen.

The nature of chemical bonds, which bind nutrients to dead organic matter, also influences the patterns of element decoupling that occur during decomposition. Nitrogen and some of the sulfur are bonded directly to the carbon skeleton of organic matter, so nitrogen and sometimes sulfur can be mineralized to plant-available forms as "waste products" of the breakdown of organic compounds during oxidation of carbon for energy (McGill and Cole, 1981; Paul and Clark, 1996). This accounts for the strong relationship between litter C:N and rates of decomposition

(Mafongoya *et al.*, 2000; Mueller *et al.*, 1998) and net nitrogen mineralization (Maithani *et al.*, 1991; Steltzer and Bowman, 1998). Alternatively, if microbes are nitrogen-limited, decomposition may lead to immobilization of nitrogen and mineralization of sulfur. The form of inorganic nitrogen in the soil is governed by a series of redox reactions, which are influenced by soil carbon availability, oxygen, pH, and several other factors.

In contrast to nitrogen, phosphorus is mineralized from dead organic matter by extracellular phosphatases at a rate that is controlled by microbial and plant phosphorus demand, rather than by microbial demand for energy. This occurs because phosphorus is bound to organic matter through ester bonds, which can be broken without disrupting the carbon skeleton. Phosphorus tends to accumulate in microbial biomass, which accounts for 30% of organic phosphorus in the soil (versus 2% of C, 4% of N, and 3% of S) (Jonasson *et al.*, 1999; Paul and Clark, 1996). The size and turnover of this large microbial phosphorus pool is therefore the main biotic control of phosphorus availability to plants. Phosphorus availability to plants is further influenced by its chemical reactions with soil minerals, as discussed in the next section. Unlike nitrogen, phosphorus is not an energy source to microbes and is not involved in redox reactions in the soil.

The control of sulfur release is intermediate between that of nitrogen and phosphorus, because sulfur occurs in organic matter in both carbon-bonded and ester-bonded forms. The mineralization of organic sulfur is therefore responsive to microbial demands for both sulfur and energy. The ester-bonded forms are sulfur-storage compounds produced under conditions of high-sulfur availability. Under sulfur-limiting conditions, plants produce primarily carbon-bonded forms of sulfur, so its mineralization is determined mainly by the carbon demand of microbes (McGill and Cole, 1981). Because ester-bonded sulfur can be mineralized based on microbial sulfur demand, it tends to be a more important source for plant needs under high-sulfur conditions. In summary, controls over sulfur cycling are similar to those of phosphorus cycling under high-sulfur conditions and similar to those of nitrogen cycling under low-sulfur conditions.

Less work has focused on the controls of recycling of other nutrients. Decomposition dynamics are a critical determinant of calcium availability, because calcium is part of cell walls that are difficult to decompose. In contrast, potassium occurs mostly in the cell cytoplasm and is largely lost through leaching, so decomposition dynamics are less important than controls over soil availability in determining plant supply. The importance of decomposition to magnesium and manganese availability is intermediate between calcium and potassium (Chapin *et al.*, 2002).

Although similar environmental factors can limit both NPP and decomposition, these two processes are differentially affected by these constraints, so it is unlikely that the timing and amount of nutrient supply will coincide with plant demand. For example, in some ecosystems, a substantial amount of nutrient mineralization occurs underneath the snow pack and is released in spring thaw before plants actively take up nutrients (Bilbrough *et al.*, 2000; Hobbie and Chapin, 1996). Nutrients are often released from organic matter in pulses associated with the initial stages of decomposition or with wet–dry or freeze–thaw events (Haynes, 1986; Schimel and Clein, 1996; Venterink *et al.*, 2002). Timing of element release also differs among elements. Soluble elements like potassium are immediately available when they enter the soil, whereas the release of nitrogen and calcium depend on microbial demands for energy, and the release of phosphorus depends on microbial phosphorus demands and factors governing microbial turnover.

8.06.9.4.3 Nutrient availability

The ratios at which nutrients are released in their mineral form through decomposition and mineralization does not directly determine the ratio of their availability. Nutrient availability is a function of the presence of nutrients in soil solution, their diffusion rates through soil, and their chemical interactions with soil minerals. Mobile nutrients can be lost from the system through leaching, whereas nitrogen can also be lost through gaseous pathways. Less mobile nutrients, such as phosphorus, can be lost in erosion. Retention mechanisms include microbial immobilization and bonds of varying strength with soil particles and soil organic matter. These retention mechanisms can enhance nutrient availability by minimizing nutrient loss, but also decrease plant access to these nutrients.

As with mineralization dynamics, the factors governing availability of mineralized nutrients differ among nutrients. The two inorganic forms of nitrogen in soil solution behave quite differently. NH_4 diffuses slowly through the soil because its positive charge interacts with the negatively charged soil particles. NO_3 diffuses rapidly, but is also prone to leaching or gaseous loss. Organic nitrogen exhibits a variety of retention mechanisms (Neff *et al.*, 2003). Microbial immobilization of nitrogen can compete with plant nitrogen uptake but can also be important in retaining pulses of nitrogen release, particularly when the pulses do not coincide with periods of plant growth.

Phosphorus availability is determined largely by chemical interactions with soil. Complexes with other elements can remove PO_4 from soil solution. PO_4 precipitates with calcium, aluminum, iron, or manganese, forming insoluble compounds. Charged organic compounds can compete with PO_4 on the binding surface and decrease chelation with metals, increasing PO_4 availability to plants. The microbial phosphorus pool may be the main reservoir of plant-available phosphorus in the soil because it protects phosphorus from chemical reactions with soil minerals (Paul and Clark, 1996). Soil pH can greatly influence phosphorus availability, as well as the availability of manganese, copper, magnesium, and iron. Water-logged soils can limit manganese and zinc availability, and iron availability can decrease with enhanced concentrations of phosphorus, manganese, zinc, or copper (Marschner, 1995).

8.06.9.4.4 Element interactions

As discussed above, there are not only different controls on the recycling of these different nutrients, but these elements can also interact to influence one another's dynamics. Nitrogen cycling, for example, is very sensitive to availability of phosphorus and sulfur. Phosphorus and sulfur limit nitrogen fixation (Bromfield, 1975; Jones *et al.*, 1970). Phosphorus also stimulates nitrification and net nitrogen mineralization (Cole and Heil, 1981). Phosphorus availability, in turn, is often enhanced by sulfur, because sulfur can acidify rock phosphorus, and SO_4 leaching enhances leaching of cations that precipitate with phosphorus. Sulfur-releasing enzymes can be inhibited by PO_4 and stimulated or inhibited by nitrogen availability (McGill and Christie, 1983). Inorganic nitrogen additions can increase mineralization of sulfur (Ghani *et al.*, 1992). All of these element interactions modify the ratios of nitrogen, phosphorus, and sulfur availability, causing the degree of coupling of these nutrients to be sensitive to environment.

8.06.9.4.5 Plant uptake

Clearly, the ratio of nutrients available to plants does not necessarily correlate with plant needs. For example, soil solutions usually contain lower concentrations of potassium and PO_4 than plants need, and excess calcium and magnesium (reviewed in Larcher, 1995 and Marschner, 1995). This imbalance in nutrient supply can interfere with uptake of limiting nutrients. In general, uptake of cations stimulate anion uptake and vice versa. However, at low external concentrations of nutrients, such as commonly occur in ecosystems, anion and cation uptake are not necessarily coupled. The relative uptake of cations and anions also shifts with pH. In general, cation uptake decreases at low pH, when H^+ concentrations are high relative to mineral cation concentrations, although high concentrations of calcium can mitigate this effect for potassium. In contrast, low pH stimulates or has no effect on anion uptake because of low OH^- concentrations in the soil solution. At high external concentrations, there is nonspecific competition between ions of the same charge. For example potassium can inhibit calcium and magnesium uptake because they have lower transport rates through the plasma membrane. This interference in uptake of certain elements is particularly pronounced when they are supplied in ratios that are unbalanced with respect to plant demand. Plant uptake of nutrients is selective based on the physicochemical characteristics of the elements, and there can be competition for binding sites at the plasma membrane between elements with similar properties. Excess ratios of certain nutrients can inhibit the uptake of others. Ammonium, for example, decreases uptake of potassium, calcium, and magnesium; and high SO_4 decreases molybdenum uptake. High concentrations of magnesium or potassium can inhibit calcium uptake, whereas high calcium levels inhibit potassium uptake. NO_3 and chlorine can inhibit one another, and potassium and calcium can strongly inhibit magnesium uptake. In fact, high fertilization of either of these leads to magnesium deficiency in soils. There are many other negative interactions between elements during uptake. For example, boron is limited by high calcium; iron is limited by high phosphorus, copper, and manganese; and calcium requirement increases with high external concentrations of heavy metals, aluminum, NaCl and at low pH. NO_3 uptake is also inhibited by the presence of NH_4.

Relatively high concentrations of certain nutrients can also increase the uptake of other elements. For example, NH_4 and sodium enhance potassium uptake; magnesium and manganese enhance uptake of one another; calcium enhances potassium uptake; and zinc enhances uptake of both magnesium and manganese (Larcher, 1995). It is clear that the stoichiometry of elements available in soil solution can substantially decouple the stoichiometry of plant uptake from supply.

8.06.9.5 Recoupling Mechanisms

In the previous sections, we showed that many mechanisms can uncouple the stoichiometry of nutrients from their ratios in live plants, so the stoichiometry of available nutrients is very different from demand. If plant growth were dependent on the relative availability of these nutrients at any one time, NPP would be constrained by a shifting balance of nutrients. Conversely, if plants simply took up nutrients in proportion to their availability, the nutrient imbalance within the plant could interfere with its metabolic function, for example through toxicity effects (Marschner, 1995). Over time and space, plants can "recouple" nutrients in ratios needed for growth.

In general, we have seen that plants respond to nutrient limitation by increasing their root : shoot ratio, increasing their nutrient-use efficiency, and by allocating to protective compounds that increase life span. Plants also adjust their physiology to respond to limitations of specific nutrients. Just as water, CO_2, light, and nutrients need to be balanced, plants must also balance the acquisition of different nutrients to grow. Nutrient limitation is strongly determined by the balance of nutrients and cannot necessarily be predicted from the concentration of a single limiting nutrient (Koerselman and Mueleman, 1996; Larcher, 1995). Any nutrient not in balance can limit plant growth as well as plant investment in absorption of these nutrients. There are different controls over the availability of, and plant access to, these different nutrients. For example, enhancing root length is a common response to nutrient limitation, but it does not equally relieve limitation of all nutrients. NO_3 diffuses rapidly in the soil and its uptake will substantially increase with a given increase in root length. In contrast, it takes 6–10 times the root length increase to produce an equivalent increase in PO_4 or NH_4 uptake because the diffusion zones around the roots are much smaller for these nutrients (Marschner, 1995). Mass flow is usually sufficient to supply micronutrients to plants, but macronutrients require additional nutrient movement to the root by diffusion in order to attain the proper balance of these nutrients. Even among the macronutrients, up to 80% of nitrogen can be supplied to crops by mass flow, while only 5% of phosphorus is supplied this way due to lower mobility in the soil (Barber, 1984; Chapin, 1991a; Lambers *et al.*, 1998).

Balancing the supply of these multiple nutrients requires many different strategies. On an individual plant level, plants adjust their relative uptake of these different nutrients through changes in ion transporters in the roots, shifts in enzyme allocation, and by forming associations with mycorrhizal fungi. Active transport is a major mechanism by which plants absorb potentially limiting nutrients. Plants are able to greatly enhance nutrient absorption of a limiting element (Chapin, 1991a; Lee, 1982; Lee and Rudge, 1986, 1987) by increasing the transport proteins specific to that nutrient, while decreasing uptake capacity of other nutrients that do not limit growth (Chapin, 1980; Lambers *et al.*, 1998). This is particularly important for the nutrients that most frequently limit plant growth because NH_4, NO_3, potassium, and SO_4 are transported by different membrane proteins that are individually regulated (Clarkson, 1985). This preferential uptake by increasing specific carriers is seen even among the different forms of nitrogen. NH_4, NO_3 and amino acids are all absorbed by different carriers, and the relative availability of these forms of nitrogen in the soil solution influences the capacity of a plant to absorb these different nitrogen forms.

Plants can also balance their uptake of different nutrients through their production of enzymes and other compounds that help to make specific nutrients more available. Nitrate reductase is required to assimilate NO_3 into plant biomass, and its production is triggered by the presence of NO_3 in soil solution. Phosphorus limitation induces production of root phosphatase enzymes that cleave organically bound PO_4, or siderophores, which solubilize mineral phosphorus by chelating with other minerals that bind to PO_4, such as iron.

Associations with soil microbes such as mycorrhizal fungi can relieve limitation by certain nutrients. Since these fungi dramatically increase the effective surface area of absorption of nutrients, they particularly enhance uptake of nutrients that diffuse slowly in soil, so they greatly enhance uptake of PO_4, and of NH_4–N in soils with low nitrifying potential. Arbuscular mycorrhizae can help to relieve phosphorus limitation, whereas ectomycorrhizae enhance both phosphorus and nitrogen uptake. In fact the presence of arbuscular mycorrhizae can relieve phosphorus limitation, to the extent that ecosystems become nitrogen-limited (Grogan and Chapin, 2000). Associations with plant-growth-promoting rhizobacteria often stimulate growth more under low-nutrient conditions (Belimov *et al.*, 2002).

Although plants have several mechanisms by which they can balance uptake of these multiple nutrients, the supply of these nutrients is rarely in balance, and many of these nutrients are available in short pulses, or mostly at certain times of the year. Plants can balance nutrient availability over time by accumulating each nutrient at times of high availability and storing it to support growth at another time (Chapin *et al.*, 1990). In fact, in many cases, much of the nutrient uptake occurs before plant growth begins (Aerts and Chapin, 2000; Larcher, 1995). Stored nutrients can then be transported to sites of growth to achieve balanced nutrient ratios in growing tissues

(Chapin *et al.*, 1990). Nutrient storage is particularly important for nitrogen, phosphorus, potassium, sulfur, copper, and zinc, but cannot occur for calcium, which is not mobile in the phloem (Nambiar, 1987).

In summary, the many mechanisms by which plants adjust to unbalanced supplies of CO_2, H_2O, nutrients, and light enable plants to maximize NPP in situations where the ratio of supply of essential nutrients is far from balanced.

8.06.10 COMMUNITY-LEVEL ADJUSTMENTS

In the previous sections, we showed that plants can adjust on a leaf and whole-plant level to maintain NPP under limiting conditions. We have also seen that these responses can often be scaled to the stand or community level. For example, allocation of nitrogen to leaves at the top of the canopy occurs both within a plant and within a stand. Changes in community composition are another important mechanism by which vegetation can maximize NPP under limiting conditions. Plants differ in their tolerances of resource and environmental limitations, and the flexibility provided by a diversity of plant species with different traits mirrors the flexibility of traits within an individual plant in its response to limiting conditions. There are many parallels between acclimation of an individual plant and shifts in plant community composition along resource gradients.

A balance of nutrients is critical to support growth of any plant, but the specific proportions of nutrients required can differ among species. For example, species can differ dramatically in the amount of phosphorus they require (Larcher, 1995); dicots contain twice as much calcium as do monocots, and forbs contain more magnesium than do grasses (Lambers *et al.*, 1998). Due to species differences in nutrient requirements, different nutrients can simultaneously limit production, and shifts in community composition can alter the NPP attained at a given nutrient supply. For example, productivity of California grasslands can be enhanced by nitrogen additions, or alternatively, phosphorus and sulfur additions can stimulate legume growth and enhance overall ecosystem productivity beyond the stimulation by nitrogen fertilizer (Jones and Winans, 1967).

Deep-rooted species tap a larger volume of soil than do shallow-rooted species and therefore access more water and nutrients to support production. In California, the deep-rooted *Eucalyptus* trees access a deeper soil profile than do annual grasses, so the forest absorbs more water and nutrients. In dry, nutrient-limited ecosystems, this substantially enhances NPP and nutrient cycling (Robles and Chapin, 1995). Similarly, the introduction of deep-rooted phreatophytes in deserts increases the productivity in watercourses (Berry, 1970). Deep-rooted species can also tap nutrients that are available only at depth. A deep-rooted tundra sedge, for example, is the only species in arctic tussock tundra that accesses nutrients in the groundwater that flows over permafrost. By tapping nutrients at depth, the productivity of this sedge increases 10-fold in sites with abundant groundwater flow, whereas productivity of other species is unaffected by deep resources (Chapin *et al.*, 1988). In the absence of this species, NPP would be greatly reduced. Species with deep roots, and particularly with high fine root biomass in the lower soil profiles, can pump calcium up to the surface layers and enhance overall calcium availability in the system (Andersson, 1991; Dijkstra and Smits, 2002).

Phenological specialization could increase resource capture by increasing the total time available for plants to acquire resources from their environment. This is most evident when coexisting species differ in the timing of their maximal activity. In mixed grasslands, for example, C_4 species are generally active in the warmer, drier part of the growing season than are C_3 species. Consequently C_3 species account for most early-season, and C_4 species account for most late-season production. Similarly, in the Sonoran desert, there is a different suite of annuals that becomes active following winter versus summer rains, and in California grasslands a mixture of early season annuals and late season perennials enhance productivity (Eviner and Chapin, 2001). In all these cases, phenological specialization probably enhances NPP and nitrogen cycling. In mixed-cropping agricultural ecosystems, phenological specialization is more effective in enhancing production than are species differences in rooting depth (Steiner, 1982). The ecosystem consequences of phenological specialization to exploit the extremes of the growing season are less clear. Evergreen forests, for example, have a longer photosynthetic season than deciduous forests, but most carbon gain occurs in midseason in both forest types, when conditions are most favorable (Schulze *et al.*, 1977). Phenological specialization is an area where species effects on ecosystem processes could be important but these effects have been well documented primarily in agricultural ecosystems.

8.06.11 SPECIES EFFECTS ON INTERACTIVE CONTROLS

Plants do much more than simply adjust to the limitations imposed by state factors, they also actively mediate most of the resource and environmental conditions that constrain growth. Some of the most important effects of plant characteristics on NPP operate indirectly through the effects of plants on interactive controls, which are the factors that directly regulate ecosystem processes.

8.06.11.1 Vegetation Effects on Resources

Plant traits that influence the supply of limiting resources (e.g., light, water, and nutrients) have strong feedback effects on NPP. The introduction of a strong nitrogen-fixer into a community that lacks such species can substantially enhance nitrogen availability and cycling. Invasion by the exotic nitrogen-fixing tree, *Myrica faya*, in Hawaii, for example, increased nitrogen inputs, litter nitrogen concentration, and nitrogen availability (Vitousek *et al.*, 1987). A nitrogen-fixing invader is most likely to be successful in ecosystems that are nitrogen-limited, have no strong nitrogen fixers, and have adequate phosphorus, micronutrients, and light (Vitousek and Howarth, 1991).

8.06.11.1.1 Decomposition and nitrogen mineralization

Traits that govern plant growth rate and NPP also determine the microbial processing of carbon and nitrogen in soils. When plant leaves senesce, they resorb approximately half of their nitrogen and phosphorus pool and very little of the initial carbon pool, regardless of the environment in which they grow (Aerts and Chapin, 2000; Chapin and Kedrowski, 1983). The quality of leaf litter, as measured by litter C : N ratio and carbon quality, therefore correlates with corresponding parameters in live leaves. Chemical properties that promote high physiological activity and growth in plants (e.g., high tissue nitrogen concentration) and low lignin content (reflecting less sclerified leaves with a high ratio of cytoplasm to cell wall) also promote rapid decomposition (Hobbie, 1992; Melillo *et al.*, 1982). Litter from species typical of productive environments (e.g., herbs and deciduous species) typically decomposes more rapidly than those from less productive environments (e.g., evergreens) (Cornelissen, 1996; Perez-Harguindeguy *et al.*, 2000).

The quantity of litter input provides the second critical link between NPP and decomposition because NPP governs the quantity of organic matter inputs to decomposers. When biomes are compared at steady state, heterotrophic respiration (i.e., the carbon released by processing of dead plant material by decomposer organisms and animals) is approximately equal to NPP. In other words, net ecosystem production (NEP), the rate of net carbon sequestration, is approximately zero at steady state, regardless of climate or ecosystem type. This indicates that the quantity and quality of organic matter inputs to soils, as determined by plant traits, are the major determinants of decomposition, when ecosystems are compared. Environment exerts important additional controls on decomposition through effects on both NPP (quantity and quality of litter inputs) and the activity of decomposer organisms. Other factors that influence decomposition rate include pH and the composition of the microbial community. Any plant effects on these factors will also influence decomposition.

Litter properties that promote NPP and decomposition also facilitate net nitrogen mineralization. The activity of decomposer organisms, which depends strongly on the carbon quality of substrates and the nitrogen status of microbes (a function of litter nitrogen concentration) are the major effects of plant litter quality on net nitrogen mineralization (Paul and Clark, 1996). Microbes mineralize nitrogen more slowly from litter with high concentrations of lignin or other recalcitrant compounds than from litter with more labile carbon compounds. High-nitrogen litter shows greater net mineralization of nitrogen than does low-nitrogen litter because microbes are seldom nitrogen-limited below a C : N ratio of 25 : 1; the nitrogen in excess of microbial demands for growth is released into the soil, where it becomes available to plants. As with decomposition, traits governing NPP strongly influence the annual net nitrogen mineralization, because productive ecosystems produce large quantities of high-quality litter.

Species differences in litter quality magnify site differences in soil fertility. Differences among plant species in tissue quality strongly influence litter decomposition rates. Litter from low-nutrient-adapted species decomposes slowly because of the negative effects on soil microbes of low concentrations of nitrogen and phosphorus and high concentrations of lignin, tannins, waxes, and other recalcitrant or toxic compounds. This slow decomposition of litter from species characteristic of nutrient-poor sites reinforces the low nutrient availability of these sites (Hobbie, 1992; Wilson and Agnew, 1992). Species from high-resource sites, in contrast, produce rapidly decomposing litter due to its higher nitrogen and phosphorus content and fewer recalcitrant compounds, enhancing rates of nutrient turnover in nutrient-rich sites.

Species differences in labile C inputs from root exudation also influence rates of decomposition and nutrient cycling. Plant carbon inputs to the rhizosphere can increase the size and activity of microbial biomass (Newman, 1985) and have large effects on nitrogen cycling (Flanagan and Van Cleve, 1983; Schimel *et al.*, 1992). More than 70% of the total soil biomass of microbes and grazing fauna are found in the rhizosphere (Ingham *et al.*, 1985). Plant species differ in their effects on the labile carbon pool (Vinton and Burke, 1995). This is one of the key regulators of plant species effects on nitrogen cycling (Wedin and Pastor, 1993), because, beyond the initial flush of labile compounds from litter, litter is unlikely to be the major source of labile carbon. Even though labile carbon is a relatively small component of the total soil carbon pool, species effects on labile carbon are responsible for up to 10-fold differences in nitrogen cycling, with this effect disappearing relatively quickly once plants are removed from the soil (Wedin and Pastor, 1993). Labile carbon inputs provided by growing plants can also accelerate decomposition rates of both recalcitrant litter and soil organic matter (Bottner *et al.*, 1999; Mueller *et al.*, 1998; Sallih and Bottner, 1988).

8.06.11.1.2 Water dynamics

Plant species can also dramatically influence the distribution of available water through space and time. Although there are many examples indicating that the amount of water used by different plants can profoundly influence soil water availability (Gordon and Rice, 1993; Gordon *et al.*, 1989; van Vuuren *et al.*, 1992), there are also examples in which particular species can profoundly alter overall ecosystem dynamics and productivity through their unique capacity to capture water sources that are unavailable to most vegetation. Two such examples are hydraulic lift and collection of fog by vegetation. In some deep-rooted species, soil water is taken up from deep layers of soil, and then is passively released into surface soils at night, when transpiration ceases. These plants supply an appreciable amount of moisture to the surface soil that can enhance the overall productivity of the plant community (Caldwell *et al.*, 1998; Horton and Hart, 1998). Aboveground plant structure may also play a critical role in supplying water to the entire ecosystem. Species with canopies that are tall and have high surface area collect water from fog in many coastal and montane ecosystems (Weathers, 1999). This fog can dramatically enhance water availability for the species responsible for fog collection, but also for the entire ecosystem. Redwood trees in California provide 34% of the annual water input to these systems, primarily at a time of minimal precipitation. This fog water can account for up to 66% of the water use by understory plants and between 13–45% of water use by redwood trees themselves, thus dramatically enhancing the production of this water-limited system (Dawson, 1998).

8.06.11.2 Vegetation Effects on Climate

Species effects on microclimate influence ecosystem processes most strongly in extreme environments. This occurs because ecosystem processes are particularly sensitive to climate in extreme environments (Hobbie, 1995; Wilson and Agnew, 1992). Boreal mosses, for example, form thick mats that insulate the soil from warm summer air temperatures. The resulting low soil temperature retards decomposition, contributing to the slow rates of nutrient cycling that characterize these ecosystems (Van Cleve *et al.*, 1991). Some mosses such as *Sphagnum* effectively retain water, as well as insulating the soil, leading to cold anaerobic soils that reduce decomposition rate and favor peat accumulation. The sequestration of nitrogen and phosphorus in undecomposed peat reduces growth of vascular plants. The shading of soil by plants is an important factor governing soil microclimate in hot environments. Establishment of many desert cactuses, for example, occurs primarily beneath the shade of "nurse plants" (Turner *et al.*, 1966) (Nobel, 1984).

Large-scale shifts in vegetation can even influence regional or global patterns of climate. Conversion of the Amazonian rain forest to pasture, for example, is predicted to result in dramatic reductions in regional precipitation, which could be irreversible since the re-establishment of a tropical forest would be impossible under these drier conditions (Shukla *et al.*, 1990). Similarly, deforestation of the boreal forest can lead to summer cooling and prevent regrowth of trees, whereas spread of the boreal forest due to climatic warming can significantly enhance both regional and global warming (Bonan *et al.*, 1995; Bonan *et al.*, 1992).

8.06.11.3 Species Effects on Disturbance Regime

Plants that alter disturbance regimes change the balance between equilibrium and nonequilibrium processes. Following disturbance, there are substantial changes in most ecological processes, including increased opportunities for colonization by new individuals and often an imbalance between inputs to, and outputs from, ecosystems. Plants that colonize following disturbance, in turn, affect the capacity of the ecosystem to gain carbon and retain nutrients.

Most disturbances produce a pulse of nutrient availability because disturbance-induced changes in environment and litter inputs increase mineralization of dead organic matter and reduce plant biomass and nutrient uptake. Anthropogenic disturbances create a wide range of initial nutrient availabilities. Some disturbances, such as mining, can produce an initial environment that is even less favorable than most primary successional habitats for initiation of succession. Some agricultural lands are abandoned to secondary succession after erosion or (in the tropics) formation of laterite soils, reducing the nutrient-supplying power of soils. Soils from some degraded lands have concentrations of aluminum and other elements that are toxic to many plants.

When initial nutrient availability is high after disturbance, early successional species typically have high relative growth rates, supported by high rates of photosynthesis and nutrient uptake. These species reproduce at an early age and allocate a large proportion of NPP to reproduction. Their strategy is to grow quickly under conditions of high resource supply, and then disperse to new disturbed sites. These early successional species include many weeds that colonize sites disturbed by people. As succession proceeds, there is a gradual shift in dominance to species that have lower resource requirements and grow more slowly. In ecosystems with low initial availability of soil resources, succession proceeds more slowly and follows patterns similar to those in primary succession, with initial colonization by light-seeded species that colonize from outside the disturbed area.

8.06.12 SPECIES INTERACTIONS AND ECOSYSTEM PROCESSES

Plant traits that influence the interactions among species in an ecosystem are among the most profoundly important ways in which plants influence the resources and environmental conditions that control NPP. Interactions such as competition, mutualism, and predation govern the abundance of species in an ecosystem, and therefore the extent to which the traits of a species are represented in the ecosystem. These combinations of species can provide unique functions to ecosystems or increase the efficiency of resource use through niche separation that leads to complementarity of resource use. The effects of species interactions are ubiquitous, but often highly situation-specific and idiosyncratic, so it is difficult to predict *a priori* the full range of ecosystem changes caused by the introduction or loss of a species (Carpenter and Kitchell, 1993). Development of a framework for predicting the effects of species interactions is an emerging challenge that will improve our capacity to predict and mitigate the effects of global changes in species composition and biodiversity (Chapin *et al.*, 2000).

Mutualistic species interactions contribute directly to many essential ecosystem processes, such as nutrient inputs through nitrogen fixation and mycorrhizal associations that govern phosphorus and organic nitrogen uptake by plants (Read, 1991). Other mutualisms, such as pollination and seed dispersal, have indirect effects, influencing the presence or absence of species that may have strong ecosystem effects.

Plant traits that influence herbivory affect virtually all ecosystem processes. In general, plants that characterize low-fertility soils produce chemical defenses that reduce the frequency of herbivory in these habitats; these compounds also retard decomposition, nutrient cycling, and therefore subsequent NPP. In contrast, plants characteristic of high-fertility soils tend to invest preferentially in growth rather than chemical defense (Bryant *et al.*, 1983), and herbivores are an important avenue of carbon and nutrient transfer from plants to soils. In this way, herbivores magnify inherent differences in soil fertility among ecosystems (Chapin, 1993). Herbivory has a major impact on ecosystem processes for several reasons. Herbivores transfer plant tissue to soils before nutrient resorption can occur, so approximately twice as much nitrogen and phosphorus is transferred per unit of plant biomass than would occur through litterfall. Secondly, herbivores preferentially select nutrient-rich tissues, further enhancing nutrient transfer to soils. Finally, animal digestion, especially in homeotherms, uses much of the energy from ingested plant matter to support animal metabolism, resulting in the excretion of nutrients in readily available forms. In these ways, herbivory short-circuits the decomposition process and speeds rates of nutrient cycling (Kielland and Bryant, 1998).

Competitive interactions among plant species obviously influence the relative abundance of species in an ecosystem and therefore the traits that are expressed at an ecosystem scale. The importance of plant combinations is not only through competitive interactions, but also in their ability to coexist. Species with different traits can differ in their resource utilization (in space, time, or the specific form of the resource), leading to an increased use of resources, and thus enhanced NPP (Tilman *et al.*, 1996). However, this can be idiosyncratic, depending on the specific species in the combinations (Hooper, 1998).

8.06.13 SUMMARY

NPP is not a simple function of the resources available at one moment. Although plant growth depends on a balance of resources, these resources are rarely available in the ratios required for growth. Plants, as individuals and communities, can maintain production under limiting conditions. They make many adjustments to maintain the balance of limiting resources imposed by state factors through shifts in physiological traits, or by plant mixtures that enhance access to resources. These adjustments extend the range of environmental conditions over which carbon gain occurs in ecosystems. Many of these adjustments involve changes in photosynthetic capacity, which entail changes in C:N ratio. This variation in element stoichiometry enables plants to maximize carbon gain under favorable environmental conditions. Under unfavorable conditions the increased C:N ratio associated with reduced photosynthetic capacity maximizes the efficiency of using other resources to gain carbon, primarily by prolonging leaf longevity and by shifting allocation to production of other tissues such as wood or roots that have lower tissue nitrogen concentrations than leaves.

Alternatively, plants can enhance the availability of limiting factors through their effects on interactive controls. This can extend the range of habitats that provide adequate resources for plant growth. These multiple processes maximize the NPP that is possible in sites with strongly limiting conditions. There are therefore many ways to achieve a similar level of NPP within any environment.

Clearly, NPP is the product of numerous biogeochemical interactions, environmental conditions, and organisms, making NPP a key summary variable that depends on many ecosystem processes. Because NPP is the basis for sustaining all life on earth, it is critical to understand the mechanisms that determine it. This is particularly true because the biotic and environmental conditions that determine NPP are subject to dramatic changes due to human impacts on ecosystems.

Substantial decreases in NPP are occurring in many ecosystems. Forest decline and dieback are observed in many areas (Huettl, 1993), with particularly large decreases in the northeastern United States, and in 20–25% of European forests (Schulze, 1989). These declines could be due to many interacting factors, including soil acidification, sulfur and nitrogen deposition, ozone pollution, and disease, but the end result of these multiple factors is a decline in NPP. Many arid lands are experiencing desertification, or a permanent loss in productive capacity, due to climate changes and land use practices. With a doubling of atmospheric CO_2, we can expect a further 17% increase in the world area of desert (Schlesinger *et al.*, 1996). The degradation of the productive capacity of terrestrial ecosystems is also widespread through loss of topsoil. Soil erosion is one of the world's most pressing environmental problems, occurring in agricultural lands, managed forests, and natural systems (Pimentel and Kounang, 1998). Each year, six million hectares of land worldwide is lost to production through erosion or salinization (Pimentel *et al.*, 1993). This is a particularly strong trend in agricultural land. In the last forty years, almost one-third of the world's arable land has been lost to production and abandoned as farmland (Pimentel *et al.*, 1995). Eighty percent of the world's agricultural land is undergoing moderate to severe erosion due to both cropping and grazing practices (Pimentel and Kounang, 1998). This has lead to a 15–30% decrease in the productivity of rain-fed agriculture land in the last 25 years, with 8–100% decrease in production at any given site. In the mean time, deforestation is occurring to replace agricultural land at a very large scale. This land is often abandoned once it has lost its productive potential (Pimentel *et al.*, 1995). The world's productive potential is also declining due to the conversion of fertile agricultural lands to suburban and urban land uses. We know from yearly fluctuations in climate that climatic warming can have dramatic effects on NPP (Knapp and Smith, 2001). These projected shifts in climate will impact NPP, but the direction of these changes will likely vary regionally (Watson *et al.*, 1996).

NPP is the basis of life on earth, and such large changes at a global scale not only indicate the presence of significant changes in the earth's biogeochemistry, but also will likely affect many species, and ultimately, human society.

ACKNOWLEDGMENTS

We thank Bill Schlesinger, Pamela Matson, and Hal Mooney for helpful suggestions and feedback. Ideas developed in this review evolved from research funded by the Bonanza Creek Long-Term Ecological Research program (USFS grant number PNW01-JV11261952-231 and NSF grant number DEB-0080609) to FSC and by a National Science Foundation dissertation improvement grant and predoctoral fellowship, a NASA Earth System Science Fellowship, and a grant from California's Sustainable Agriculture Research and Education Program to VTE. This is a contribution to the program of the Institute of Ecosystem Studies and the institute of Arctic Biology.

REFERENCES

Aber J., McDowell W., Nadelhoffer K., Magill A., Bernstson G., Kamakea M., McNulty S., Currie W., Rustad L., and Fernandez I. (1998) Nitrogen saturation in temperate forest ecosystems. *Bioscience* **48**, 921–934.

Aerts R. (1995) Nutrient resorption from senescing leaves of perennials: are there general patterns? *J. Ecol.* **84**, 597–608.

Aerts R. and Berendse F. (1988) The effect of increased nutrient availability on vegetation dynamics in wet heathlands. *Vegetatio* **76**, 63–69.

Aerts R. and Bobbink R. (1999) The impact of atmospheric nitrogen deposition on vegetation processes in terrestrial non-forest ecosystems. In *The Impact of Nitrogen Deposition on Natural and Semi-natural Ecosystems* (ed. S. Langan). Kluwer, Dordrecht, pp. 85–122.

Aerts R. and Chapin F. S., III (2000) The mineral nutrition of wild plants revisited: a re-evaluation of processes and patterns. *Adv. Ecol. Res.* **30**, 1–67.

Aerts R. and Heil G. E. (1993) *Heathlands, Patterns and Processes in a Changing Environment.* Kluwer, Dordrecht.

Amundson R. and Jenny H. (1997) On a state factor model of ecosystems. *Bioscience* **47**, 536–543.

Andersson T. (1991) Influence of stemflow and throughfall from common oak (*Quercus robur*) on soil chemistry and vegetation patterns. *Can. J. Forest Res.* **21**, 917–924.

Barber S. A. (1984) *Soil Nutrient Bioavailability.* Wiley, New York.

Bazzaz F. (1997) Allocation and resources in plants: state of the science and critical questions. In *Plant Resource Allocation* (eds. F. Bazzaz and J. E. Grace). Academic Press, San Diego, CA, pp. 1–37.

Belimov A. A., Safronova V. I., and Mimura T. (2002) Response of spring rape (*Brassica napus* var. *oleifera* L.) to inoculation with plant growth promoting rhizobacteria containing 1-aminocyclopropane-1-carboxylate deaminase depends on nutrient status of the plant. *Can. J. Microbiol.* **48**, 189–199.

Berendse F. and Aerts R. (1987) Nitrogen-use efficiency: a biologically meaningful definition? *Funct. Ecol.* **1**, 293–296.

Berry W. L. (1970) Characteristics of salts secreted by *Tamarix aphylla. Am. J. Bot.* **57**, 1226–1230.

Bilbrough C. J., Welker J. M., and Bowman W. D. (2000) Early spring nitrogen uptake by snow-covered plants: a comparison of arctic and alpine plant function under the snowpack. *Arct. Antarct. Alp. Res.* **32**, 404–411.

Billings W. D. and Mooney H. A. (1968) The ecology of arctic and alpine plants. *Biol. Rev.* **43**, 481–529.

Billings W. D., Godfrey P. J., Chabot B. F., and Bourque D. P. (1971) Metabolic acclimation to temperature in arctic and alpine ecotypes of *Oxyria digyna. Arct. Alp. Res.* **3**, 277–289.

Bloom A. J. (1986) Plant economics. *Trends Ecol. Evol.* **1**, 98–100.

Bloom A. J., Chapin F. S., III, and Mooney H. A. (1985) Resource limitation in plants—an economic analogy. *Ann. Rev. Ecol. Syst.* **16**, 363–392.

Bolin B., Crutzen P., Vitousek P., Woodmansee R., Goldberg E., and Cook R. (1983) Interactions of biogeochemical cycles. In *The Major Biogeochemical Cycles and their Interactions* (eds. B. Bolin and R. Cook). Wiley, New York, pp. 1–39.

Bonan G. B., Pollard D., and Thompson S. L. (1992) Effects of boreal forest vegetation on global climate. *Nature* **359**, 716–718.

Bonan G. B., Chapin F. S., III, and Thompson S. L. (1995) Boreal forest and tundra ecosystems as components of the climate system. *Climat. Change* **29**, 145–167.

Bottner P., Pansu M., and Sallih Z. (1999) Modelling the effect of active roots on soil organic matter turnover. *Plant Soil* **216**, 15–25.

Bromfield A. (1975) Effect of ground rock phosphate-sulphur mixture on yield and nutrient uptake of ground nuts (*Arachis hypogaea*) in northern Nigeria. *Exp. Agri.* **11**, 265–272.

Bryant J. P. and Kuropat P. J. (1980) Selection of winter forage by subarctic browsing vertebrates: the role of plant chemistry. *Ann. Rev. Ecol. Syst.* **11**, 261–285.

Bryant J. P., Chapin F. S., III, and Klein D. R. (1983) Carbon/nutrient balance of boreal plants in relation to vertebrate herbivory. *Oikos* **40**, 357–368.

Caldwell M. M., Manwaring J. H., and Durham S. L. (1996) Species interactions at the level of fine roots in the field: influence of soil nutrient heterogeneity and plant size. *Oecologia* **106**, 440–447.

Caldwell M. M., Dawson T. E., and Richards J. (1998) Hydraulic lift: consequences of water efflux from the roots of plants. *Oecologia* **113**, 151–161.

Carpenter S. R. and Kitchell J. F. (1993) *The Trophic Cascade in Lakes.* Cambridge University Press, Cambridge, 384pp.

Chabot B. F. and Hicks D. J. (1982) The ecology of leaf life spans. *Ann. Rev. Ecol. Syst.* **13**, 229–259.

Chadwick O. A., Derry L. A., Vitousek P. M., Huebert B. J., and Hedin L. O. (1999) Changing sources of nutrients during 4 million years of soil and ecosystem development. *Nature* **397**, 491–497.

Chapin F. S., III (1980) The mineral nutrition of wild plants. *Ann. Rev. Ecol. Syst.* **11**, 233–260.

Chapin F. S., III (1989) The cost of tundra plant structures: evaluation of concepts and currencies. *Am. Nat.* **133**, 1–19.

Chapin F. S., III (1991a) Effects of multiple environmental stresses on nutrient availability and use. In *Response of Plants to Multiple Stresses* (eds. H. A. Mooney, W. E. Winner, and E. J. Pell). Academic Press, San Diego, CA, pp. 67–88.

Chapin F. S., III (1991b) Integrated responses of plants to stress. *Bioscience* **41**, 29–36.

Chapin F. S., III (1993) Functional role of growth forms in ecosystem and global processes. In *Scaling Physiological Processes: Leaf to Globe* (eds. J. R. Ehleringer and C. B. Field). Academic Press, San Diego, CA, pp. 287–312.

Chapin F. S., III and Kedrowski R. A. (1983) Seasonal changes in nitrogen and phosphorus fractions and autumn retranslocation in evergreen and deciduous taiga trees. *Ecology* **64**, 376–391.

Chapin F. S., III, Fetcher N., Kielland K., Everett K. R., and Linkins A. E. (1988) Productivity and nutrient cycling of Alaskan tundra: enhancement by flowing soil water. *Ecology* **69**, 693–702.

Chapin F. S., III, Schulze E.-D., and Mooney H. A. (1990) The ecology and economics of storage in plants. *Ann. Rev. Ecol. Syst.* **21**, 423–448.

Chapin F. S., III, Walker L. R., Fastie C. L., and Sharman L. C. (1994) Mechanisms of primary succession following deglaciation at Glacier Bay, Alaska. *Ecol. Monogr.* **64**, 149–175.

Chapin F. S., III, Torn M. S., and Tateno M. (1996) Principles of ecosystem sustainability. *Am. Nat.* **148**, 1016–1037.

Chapin F. S., III, Zaveleta E. S., Eviner V. T., Naylor R. L., Vitousek P. M., Lavorel S., Reynolds H. L., Hooper D. U., Sala O. E., Hobbie S. E., Mack M. C., and Diaz S. (2000) Consequences of changing biotic diversity. *Nature* **405**, 234–242.

Chapin F. S., III, Matson P. A., and Mooney H. A. (2002) *Principles of Terrestrial Ecosystem Ecology.* Springer, New York.

Chazdon R. L. and Field C. B. (1987) Determinants of photosynthetic capacity in six rainforest *Piper* species. *Oecologia* **73**, 222–230.

Chazdon R. L. and Pearcy R. W. (1991) The importance of sunflecks for forest understory plants. *Bioscience* **41**, 760–766.

Clark D. A., Brown S., Kicklighter D. W., Chambers J. Q., Thomlinson J. R., and Ni J. (2001) Measuring net primary production in forests: concepts and field methods. *Ecol. Appl.* **11**, 356–370.

Clarkson D. T. (1985) Factors affecting mineral nutrient acquisition by plants. *Ann. Rev. Plant Physiol.* **36**, 77–115.

Cole C. V. and Heil R. (1981) Phosphorus effects on terrestrial nitrogen cycling. In *Terrestrial Nitrogen Cycles* (eds. F. Clark and T. Rosswall). Ecological Bulletins, Stockholm.

Coley P. D., Bryant J. P., and Chapin F. S., III (1985) Resource availability and plant anti-herbivore defense. *Science* **230**, 895–899.

Cornelissen J. H. C. (1996) An experimental comparison of leaf decomposition rates in a wide range of temperate plant species and types. *J. Ecol.* **84**, 573–582.

Cowling R. (1993) *The Ecology of Fynbos, Nutrients, Fire and Diversity.* Oxford University Press, Oxford.

Craine J. M., Froehle J., Tilman D. G., Wedin D. A., and Chapin F. S., III (2001) The relationships among root and leaf traits of 76 grassland species and relative abundance along fertility and disturbance gradients. *Oikos* **93**, 274–285.

Curtis P. S. and Wang X. (1998) A meta-analysis of elevated CO_2 effects on woody plant mass, form, and physiology. *Oecologia* **113**, 299–313.

Curtis P. S., Zak D. R., Pregitzer K. S., Lussenhop J., and Teeri J. A. (1996) Linking above- and belowground responses to rising CO_2 in northern deciduous forest species. In *Carbon Dioxide and Terrestrial Ecosystems* (eds. G. W. Koch and H. A. Mooney). Academic Press, San Diego, CA, pp. 41–51.

Davies W. J. and Zhang J. (1991) Root signals and the regulation of growth and development of plants in drying soil. *Ann. Rev. Plant Physiol. Plant Mol. Biol.* **42**, 55–76.

Dawson T. E. (1998) Fog in the California redwood forest: ecosystem inputs and use by plants. *Oecologia* **117**, 476–485.

Díaz S., Grime J. P., Harris J., and McPherson E. (1993) Evidence of a feedback mechanism limiting plant response to elevated carbon dioxide. *Nature* **364**, 616–617.

Dijkstra F. A. and Smits M. M. (2002) Tree species effects on calcium cycling: the role of calcium uptake in deep soils. *Ecosystems* **5**, 385–398.

Dokuchaev V. V. (1879) Abridged historical account and critical examination of the principal soil classifications existing. *Trans. Petersburg Soc. Nat.* **1**, 64–67.

Driscoll C. T., Lawrence G. B., Bulger A. J., Butler T. J., Cronan C. S., Eagar C., Lambert K. F., Likens G. E., Stoddard J. L., and Weathers K. C. (2001) Acidic deposition in the northeastern United States: sources and inputs, ecosystem effects and management strategies. *Bioscience* **51**, 180–198.

Ehleringer J. R. and Mooney H. A. (1978) Leaf hairs: effects on physiological activity and adaptive value to a desert shrub. *Oecologia* **37**, 183–200.

Ellsworth D. S. (1999) CO_2 enrichment in a maturing pine forest: Are CO_2 exchange and water status in the canopy affected? *Plant Cell Environ.* **22**, 461–472.

Elser J. J. and Urabe J. (1999) The stoichiometry of consumer-driven nutrient recycling: theory, observations, and consequences. *Ecology* **80**, 735–751.

Elser J. J., Fagan W. F., Denno R. F., Dobberfuhl D. R., Folarin A., Huberty A., Interlandl S., Kilham S. S., McCauley E., Schulz K. L., Siemann E. H., and Sterner R. W. (2000) Nutritional constraints in terrestrial and freshwater food webs. *Nature* **408**, 578–580.

Evans J. R. (1989) Photosynthesis and nitrogen relationships in leaves of C_3 plants. *Oecologia* **78**(1), 9–19.

Eviner V. T. and Chapin F. S., III (2001) Plant species provide vital ecosystem functions for sustainable agriculture, rangeland management and restoration. *Calif. Agri.* **55**(6), 54–59.

Eviner V. T. and Chapin F. S., III Biogeochemical interactions and biodiversity. In *Element Interactions: Rapid Assessment Project of SCOPE* (eds. J. M. Melillo, C. B. Field, and M. Moldan). Island Press (in press).

Fahey T., Bledsoe C., Day R., Ruess R., and Smucker A. (1998) *Fine Root Production and Demography*. CRC Press, Boca Raton, FL.

Farquhar G. D. and Sharkey T. D. (1982) Stomatal conductance and photosynthesis. *Ann. Rev. Plant Physiol.* **33**, 317–345.

Field C. (1983) Allocating leaf nitrogen for the maximization of carbon gain: leaf age as a control on the allocation program. *Oecologia* **56**, 341–347.

Field C. and Mooney H. A. (1986) The photosynthesis-nitrogen relationship in wild plants. In *On the Economy of Plant Form and Function* (ed. T. J. Givnish). Cambridge University Press, Cambridge, pp. 25–55.

Field C., Chapin F. S., III, Matson P. A., and Mooney H. A. (1992) Responses of terrestrial ecosystems to the changing atmosphere: a resource-based approach. *Ann. Rev. Ecol. Syst.* **23**, 201–235.

Field C. B. (1991) Ecological scaling of carbon gain to stress and resource availability. In *Integrated Responses of Plants to Stress* (eds. H. A. Mooney, W. E. Winner, and E. J. Pell). Academic Press, pp. 35–65.

Flanagan P. W. and Van Cleve K. (1983) Nutrient cycling in relation to decomposition and organic matter quality in taiga ecosystems. *Can. J. Forest Res.* **13**, 795–817.

Foley J. A., Prentice I. C., Ramankutty N., Levis S., Pollard D., Sitch S., and Haxeltine A. (1996) An integrated biosphere model of land surface processes, terrestrial carbon balance, and vegetation dynamics. *Global Biogeochem. Cycles* **10**, 603–628.

Forseth I. N. and Ehleringer J. R. (1983) Ecophysiology of two solar tracking desert winter annuals: IV. Effects of leaf orientation on calculated daily carbon gain and water use efficiency. *Oecologia* **58**, 10–18.

Garnier E. (1991) Resource capture, biomass allocation and growth in herbaceous plants. *Trends Ecol. Evol.* **6**(4), 126–131.

Gauch H. G. (1972) *Inorganic Plant Nutrition*. Dowden, Hutchinson, and Ross, Stroudsburg, PA.

Ghani A., McLaren R. G., and Swift R. S. (1992) Sulfur mineralization and transformations in soils as influenced by additions of carbon, nitrogen, and sulfur. *Soil Biol. Biochem.* **24**, 331–341.

Gollan T., Turner N. C., and Schulze E. D. (1985) The responses of stomata and leaf gas exchange to vapor pressure deficits and soil water content: III. In the sclerophyllous woody species *Nerium oleander*. *Oecologia* **65**, 356–362.

Goransson A. (1993) Growth and nutrition of small *Betula-pendula* plants at different relative addition rates of iron. *Trees-Struct. Funct.* **8**, 31–38.

Goransson A. (1994) Growth and nutrition of small *Betula-pendula* plants at different relative addition rates of manganese. *Tree Physiol.* **14**, 375–388.

Gordon D. and Rice K. (1993) Competitive effects of grassland annuals on soil water and blue oak (*Quercus douglasii*) seedlings. *Ecology* **74**, 68–82.

Gordon D., Welker J. M., Menke J., and Rice K. (1989) Competition for soil water between annual plants and blue oak (*Quercus douglasii*) seedlings. *Oecologia* **79**, 533–541.

Gordon W. S. and Jackson R. B. (2000) Nutrient concentrations in fine roots. *Ecology* **81**, 275–280.

Gower S. T. (2002) Productivity of terrestrial ecosystems. In *Encyclopedia of Global Change* (eds. H. A. Mooney and J. Canadell). Blackwell, Oxford, vol. 2, pp. 516–521.

Gower S. T., Kucharik C. J., and Norman J. M. (1999) Direct and indirect estimation of leaf area index, F_{APAR}, and net primary production of terrestrial ecosystems. *Remote Sens. Environ.* **70**, 29–51.

Graetz R. D. (1991) The nature and significance of the feedback of change in terrestrial vegetation on global atmospheric and climatic change. *Clim. Change* **18**, 147–173.

Grogan P. and Chapin F. S., III (2000) Nitrogen limitation of production in a Californian annual grassland: the contribution of *Arbuscular mycorrhizae*. *Biogeochemistry* **49**, 37–51.

Guenther A., Hewitt C., Erickson D., Fall R., Geron C., Graedel T., Harley P., Klinter L., Lerdau M., McKay W., Pierce T., Scholes B., Steinbrecher R., Tallamraju R., Taylor R., and Zimmerman P. (1995) A global model of natural volatile organic compound emissions. *J. Geophys. Res.* **100D**, 8873–8892.

Gulmon S. L. and Mooney H. A. (1986) Costs of defense on plant productivity. In *On the Economy of Plant Form and Function* (ed. T. J. Givnish). Cambridge University Press, Cambridge, pp. 681–698.

Haynes R. J. (1986) The decomposition process: mineralization, immobilization, humus formation, and degradation. In *Mineral Nitrogen in the Plant-Soil System* (ed. R. J. Haynes). Academic Press, Orlando, pp. 52–126.

Hikosaka K. and Hirose T. (2001) Nitrogen uptake and use by competing individuals in a *Xanthium canadense* stand. *Oecologia* **126**, 174–181.

Hirose T. and Werger M. J. A. (1987) Maximizing daily canopy photosynthesis with respect to the leaf nitrogen allocation pattern in the canopy. *Oecologia* **72**, 520–526.

Hirose T. and Werger M. J. A. (1994) Photosynthetic capacity and nitrogen partitioning among species in the canopy of a herbaceous plant community. *Oecologia* **100**, 203–212.

Hirose T. and Werger M. J. A. (1995) Canopy structure and photon flux partitioning among species in a herbaceous plant community. *Ecology* **76**, 466–474.

Hobbie S. E. (1992) Effects of plant species on nutrient cycling. *Trends Ecol. Evol.* **7**, 336–339.

Hobbie S. E. (1995) Direct and indirect effects of plant species on biogeochemical processes in arctic ecosystems. In *Arctic and Alpine Biodiversity: Patterns, Causes and Ecosystem Consequences* (ed. F. S. Chapin, III and C. Körner). Springer, Berlin, pp. 213–224.

Hobbie S. E. and Chapin F. S., III (1996) Winter regulation of tundra litter carbon and nitrogen dynamics. *Biogeochemistry* **35**, 327–338.

Hooper D. U. (1998) The role of complementarity and competition in ecosystem responses to variation in plant diversity. *Ecology* **79**, 704–719.

Horton J. and Hart S. (1998) Hydraulic lift: a potentially important ecosystem process. *Trends Ecol. Evol.* **13**, 232–235.

Huettl R. (1993) Summary and concluding remarks. In *Forest Decline in the Atlantic and Pacific Region* (eds. R. Huettl and D. Mueller-Dumbois). Springer, New York, pp. 351–358.

Hungate B. A., Chapin F. S., III, Zhong H., Holland E. A., and Field C. B. (1997) Stimulation of grassland nitrogen cycling under carbon dioxide enrichment. *Oecologia* **109**, 149–153.

Ingestad T. and Ågren G. I. (1988) Nutrient uptake and allocation at steady-state nutrition. *Physiol. Plant.* **72**, 450–459.

Ingham R. E., Trofymow J. A., Ingham E. R., and Coleman D. C. (1985) Interactions of bacteria, fungi, and their nematode grazers: effects on nutrient cycling and plant growth. *Ecol. Monogr.* **55**, 119–140.

Jarvis P. G. and Leverenz J. W. (1983) Productivity of temperate, deciduous and evergreen forests. In *Encyclodedia of Plant Physiology*, new series (eds. O. L. Lange, P. S. Nobel, C. B. Osmond, and H. Ziegler). Springer, Berlin, vol. 12D, pp. 233–280.

Jenny H. (1941) *Factors of Soil Formation.* McGraw-Hill, New York.

Jonasson S., Michelsen A., and Schmidt I. K. (1999) Coupling of nutrient cycling and carbon dynamics in the Arctic, integration of soil microbial and plant processes. *Appl. Soil Ecol.* **11**, 135–146.

Jones M. and Martin W. (1964) Sulfate–sulfur concentration as an indicator of sulfur status in various California dryland pasture species. *Soil Sci. Soc. Am. Proc.* **28**, 539–541.

Jones M. and Winans S. (1967) Subterranean clover versus nitrogen fertilized annual grasslands: botanical composition and protein content. *J. Range Manage.* **20**, 8–12.

Jones M., Lawler P., and Ruckman J. (1970) Differences in annual clover response to phosphorus and sulfur. *Agron. J.* **62**, 439–442.

Jones M., Williams W., and Vaughn C. (1983) Soil characteristics related to production on subclover-grass range. *J. Range Manage.* **36**, 444–446.

Kielland K. and Bryant J. (1998) Moose herbivory in taiga: effects on biogeochemistry and vegetation dynamics in primary succession. *Oikos* **82**, 377–383.

Knapp A. K. and Smith M. D. (2001) Variation among biomes in temporal dynamics of aboveground primary production. *Science* **291**, 481–484.

Koerselman W. and Mueleman A. F. M. (1996) The vegetation N:P ratio: a new tool to detect the nature of nutrient limitation. *J. Appl. Ecol.* **33**, 1441–1450.

Kucharik C. J., Foley J. A., Delire C., Fisher V. A., Coe M. T., Lenters J., Young-Molling C., Ramankutty N., Norman J. M., and Gower S. T. (2000) Testing the performance of a dynamic global ecosystem model: water balance, carbon balance and vegetation structure. *Global Biogeochem. Cycles* **14**, 795–825.

Lambers H. and Poorter H. (1992) Inherent variation in growth rate between higher plants: a search for physiological causes and ecological consequences. *Adv. Ecol. Res.* **23**, 187–261.

Lambers H., Atkin O. K., and Scheurwater I. (1996) Respiratory patterns in roots in relation to their functioning. In *Plant Roots: The Hidden Half* (eds. Y. Waisel, A. Eshel, and U. Kafkaki). Dekker, New York, pp. 323–362.

Lambers H., Chapin F. S., III, and Pons T. (1998) *Plant Physiological Ecology.* Springer, Berlin.

Landsberg J. J. and Gower S. T. (1997) *Applications of Physiological Ecology to Forest Management.* Academic Press, San Diego, CA.

Larcher W. (1995) *Physiological Plant Ecology.* Springer, Berlin.

Lee R. B. (1982) Selectivity and kinetics of ion uptake by barley plant following nutrient deficiency. *Ann. Bot.* **50**, 429–449.

Lee R. B. and Rudge K. A. (1986) Effects of nitrogen deficiency on the absorption of nitrate and ammonium by barley plants. *Ann. Bot.* **57**, 471–486.

Lee R. B. and Rudge K. A. (1987) Effects of nitrogen deficiency on the absorption of nitrate and ammonium by barley plants. *Ann. Bot.* **57**, 471–486.

Lerdau M. T. (1991) Plant function and biogenic terpene emission. In *Trace Gas Emissions by Plants* (eds. T. D. Sharkey, E. A. Holland, and H. A. Mooney). Academic Press, San Diego, CA, pp. 121–134.

Mafongoya P., Barak P., and Reed J. (2000) Carbon, nitrogen, and phosphorus mineralization of tree leaves and manure. *Biol. Fertil. Soils* **30**, 298–305.

Maithani G. P., Bahuguna V. K., and Lal P. (1991) Seed germination behaviour of *Desmodium tiliaefolium* G. Don: an important shrub species of Himalayas. *Indian For.* **117**, 593–595.

Marschner H. (1995) *Mineral Nutrition in Higher Plants.* Academic Press, London.

McGill W. and Christie E. (1983) Biogeochemical aspects of nutrient cycle interactions in soils and organisms. In *The Major Biogeochemical Cycles and their Interactions* (eds. B. Bolin and R. Cook). Wiley, New York, pp. 271–301.

McGill W. and Cole C. V. (1981) Comparative aspects of cycling of organic C, N, S, and P through soil organic matter. *Geoderma* **26**, 267–286.

McNaughton K. G. and Jarvis P. G. (1991) Effects of spatial scale on stomatal control of transpiration. *Agri. Forest Meteorol.* **54**, 279–302.

McNaughton S. J., Oesterheld M., Frank D. A., and Williams K. J. (1989) Ecosystem-level patterns of primary productivity and herbivory in terrestrial habitats. *Nature* **341**, 142–144.

Melillo J. M., Aber J. D., and Muratore J. F. (1982) Nitrogen and lignin control of hardwood leaf litter decomposition dynamics. *Ecology* **63**, 621–626.

Merino J., Field C., and Mooney H. A. (1982) Construction and maintenance costs of mediterranean-climate evergreen and deciduous leaves: I. Growth and CO_2 exchange. *Oecologia* **53**, 208–213.

Mooney H. A. and Billings D. W. (1961) The physiological ecology of arctic and alpine populations of *Oxyria digyna. Ecol. Monogr.* **31**, 1–29.

Mooney H. A., Canadell J., Chapin F. S., III, Ehleringer J. R., Körner C., McMurtrie R. E., Parton W. J., Pitelka L. F., and Schulze E.-D. (1999) Ecosystem physiology responses to global change. In *The Terrestrial Biosphere and Global Change: Implications for Natural and Managed Ecosystems* (eds. B. Walker, W. Steffen, J. Canadell, and J. Ingram). Cambridge University Press, Cambridge, pp. 141–189.

Mueller T., Jensen L., Nielsen E., and Magid J. (1998) Turnover of carbon and nitrogen in a sandy loam soil following incorporation of chopped maize plants, barley straw and blue grass in the field. *Soil Biol. Biochem.* **30**, 561–571.

Nambiar E. K. S. (1987) Do nutrients retranslocate from fine roots? *Can. J. Forest Res.* **17**, 913–918.

Neff J. C., Chapin F. S., III, and Vitousek P. M. (2003) Breaks in the cycle: dissolved organic nitrogen in terrestrial ecosystems. *Front. Ecol. Environ. Sci.* **1**, 205–211.

New M. G., Hulme M., and Jones P. D. (1999) Representing 20th century space-time climate variability: I. Development of a 1961–1990 mean monthly terrestrial climatology. *J. Climate* **12**, 829–856.

Newman E. I. (1985) The rhizosphere: carbon sources and microbial populations. In *Ecological Interactions in Soil* (eds. A. H. Fitter, D. Atkinson, D. J. Read, and M. Busher). Blackwell, Oxford, pp. 107–121.

Nobel P. S. (1984) Extreme temperatures and thermal tolerances for seedlings of desert succulents. *Oecologia* **62**, 310–317.

Owensby C. E., Coyne P. I., Ham J. M., Auen L., and Knapp A. K. (1993) Biomass production in a tallgrass prairie ecosystem exposed to ambient elevated CO_2. *Ecol. Appl.* **3**(4), 644–653.

Paul E. A. and Clark F. E. (1996) *Soil Microbiology and Biochemistry*. Academic Press, San Diego, CA.

Pearcy R. W. (1988) Photosynthetic utilisation of lightflecks by understory plants. *Austral. J. Plant Physiol.* **15**, 223–238.

Pearcy R. W. (1990) Sunflecks and photosynthesis in plant canopies. *Ann. Rev. Plant Physiol.* **41**, 421–453.

Penning de Vries F. W. T. (1975) The cost of maintenance processes in plant cells. *Ann. Bot.* **39**, 77–92.

Perez-Harguindeguy N., Diaz S., Cornelissen J. H. C., Vendramini F., Cabido M., and Castellanos A. (2000) Chemistry and toughness predict leaf litter decomposition rates over a wide spectrum of functional types and taxa in central Argentina. *Plant Soil* **218**, 21–30.

Pimentel D. and Kounang N. (1998) Ecology of soil erosion in ecosystems. *Ecosystems* **1**, 416–426.

Pimentel D., Allen J., Beers A., Guinand L., Hawkins A., Linder R., McLaughlin P., Meer B., Musonda D., Perdue D., Poisson S., Salazar R., Siebert S., and Stoner K. (1993) Soil erosion and agricultural productivity. In *World Soil Erosion and Conservation* (ed. D. Pimentel). Cambridge University Press, Cambridge, pp. 277–292.

Pimentel D., Harvey C., Resosudarmo P., Sinclair K., Kurz D., McNair M., Crist S., Shpritz L., Fitton L., Saffouri R., and Blair R. (1995) Environmental and economic costs of soil erosion and conservation benefits. *Science* **267**, 1117–1123.

Poorter H. (1990) Interspecific variation in relative growth rate: on ecological causes and physiological consequences. In *Causes and Consequences of Variation in Growth Rate and Productivity in Higher Plants* (eds. H. Lambers, M. L. Cambridge, H. Konings, and T. L. Pons). SPB Academic Publishing, The Hague, pp. 45–68.

Poorter H. (1994) Construction costs and payback time of biomass: a whole-plant perspective. In *A Whole-Plant Perspective on Carbon–Nitrogen Interactions* (eds. J. Roy and E. Garnier). SPB Academic Publishing, The Hague, pp. 111–127.

Poorter H. and Villar R. (1997) Chemical composition of plants: causes and consequences of variation in allocation of C to different plant compounds. In *Resource Allocation in Plants* (eds. F. Bazzaz and J. E. Grace). Academic Press, San Diego, CA, pp. 39–72.

Quilchano C., Haneklaus S., Gallardo J. F., Schnug E., and Moreno G. (2002) Sulphur balance in a broadleaf, non-polluted, forest ecosystem (central-western Spain). *Forest Ecol. Manage.* **161**, 205–214.

Read D. J. (1991) Mycorrhizas in ecosystems. *Experientia* **47**, 376–391.

Redfield A. C. (1958) The biological control of chemical factors in the environment. *Am. Sci.* **46**, 205–221.

Reich P. B., Walters M. B., and Ellsworth D. S. (1992) Leaf life span in relation to leaf, plant and stand characteristics among diverse ecosystems. *Ecol. Monogr.*, **62**, 365–392.

Reich P. B., Walters M. B., and Ellsworth D. S. (1997) From tropics to tundra: global convergence in plant functioning. *Proc. Natl. Acad. Sci. USA.* **94**, 13730–13734.

Reich P. B., Ellsworth D. S., Walters M. B., Vose J. M., Gresham C., Volin J. C., and Bowman W. D. (1999) Generality of leaf trait relationships: a test across six biomes. *Ecology* **80**, 1955–1969.

Reynolds J. F. and Thornley J. H. M. (1982) A shoot:root partitioning model. *Ann. Bot.* **49**, 585–597.

Robles M. and Chapin F. S., III (1995) Comparison of the influence of two exotic species on ecosystem processes in the Berkeley Hills. *Madroño* **42**, 349–357.

Ruimy A., Jarvis P. G., Baldocchi D. D., and Saugier B. (1996) CO_2 fluxes over plant canopies and solar radiation: a review. *Adv. Ecol. Res.* **26**, 1–68.

Ryan M. G. and Waring R. H. (1992) Maintenance respiration and stand development in a subalpine lodgepole pine forest. *Ecology* **73**(6), 2100–2108.

Ryan M. G., Linder S., Vose J. M., and Hubbard R. M. (1994) Respiration of pine forests. *Ecol. Bull.* **43**, 50–63.

Sallih Z. and Bottner P. (1988) Effect of wheat (*Triticum aestivum*) roots on mineralization rates of soil organic matter. *Biol. Fertil. Soils* **7**, 67–70.

Saugier B., Roy J., and Mooney H. A. (2001) Estimations of global terrestrial productivity: converging toward a single number? In *Terrestrial Global Productivity* (eds. J. Roy, B. Saugier, and H. A. Mooney). Academic Press, San Diego, CA, pp. 543–557.

Schimel J., Helfer S., and Alexander I. (1992) Effects of starch additions on N turnover in Sitka spruce forest floor. *Plant Soil* **139**, 139–143.

Schimel J. P. and Clein J. S. (1996) Microbial response to freeze-thaw cycles in tundra and taiga soils. *Soil Biol. Biochem.* **28**, 1061–1066.

Schlesinger W. H., Raikes J. A., Hartley A. E., and Cross A. F. (1996) On the spatial pattern of soil nutrients in desert ecosystems. *Ecology* **77**, 364–374.

Schulze E.-D. (1989) Air pollution and forest decline in a spruce (*Picea abies*) forest. *Science* **244**, 776–783.

Schulze E.-D. and Chapin F. S., III (1987) Plant specialization to environments of different resource availability. In *Potentials and Limitations in Ecosystem Analysis* (eds. E. D. Schulze and H. Zwolfer). Springer, Berlin, pp. 120–148.

Schulze E.-D., Fuchs M., and Fuchs M. I. (1977) Spatial distribution of photosynthetic capacity and performance in a mountain spruce forest of northern Germany: III. The significance of the evergreen habit. *Oecologia* **30**, 239–248.

Schulze E.-D., Kelliher F. M., Körner C., Lloyd J., and Leuning R. (1994) Relationship among maximum stomatal conductance, ecosystem surface conductance, carbon assimilation rate, and plant nitrogen nutrition: a global ecology scaling exercise. *Ann. Rev. Ecol. Syst.* **25**, 629–660.

Schuur E. A. G. (2003) Productivity and global climate revisited: the sensitivity of tropical forest growth to precipitation. *Ecology* **84**, 1165–1170.

Semikhatova O. A. (2000) Ecological physiology of plant dark respiration: its past, present and future. *Bot. Zh.* **85**, 15–32.

Shukla J., Nobre C., and Sellers P. (1990) Amazon deforestation and climate change. *Science* **247**, 1322–1325.

Specht R. and Rundel P. (1990) Sclerophylly and foliar nutrient status of mediterranean-climate plant communities in southern Australia. *Austral. J. Bot.* **38**, 459–474.

Steiner K. (1982) *Intercropping in Tropical Smallholder Agriculture with Special Reference to West Africa.* German Agency for Technical Cooperation (GTZ), Eschborn, Germany.

Steltzer H. and Bowman W. D. (1998) Differential influence of plant species on soil nitrogen transformations in moist meadow alpine tundra. *Ecosystems* **1**, 464–474.

Tanner E. V. J., Vitonsek P. M., and Cuevas E. (1998) Experimental investigation of nutrient limitation of forest growth on wet tropical mountains. *Ecology* **79**, 10–22.

Terashima I. and Hikosaka K. (1995) Comparative ecophysiology of leaf and canopy photosynthesis. *Plant Cell Environ.* **18**, 1111–1128.

Tilman D., Wedin D., and Knops J. (1996) Productivity and sustainability influenced by biodiversity in grassland ecosystems. *Nature* **379**, 718–720.

Tisdale S., Nelson W., Beaton J., and Javlin J. (1993) *Soil Fertility and Fertilizers*. Macmillan, New York.

Turner R. M., Alcorn S. M., Olin G., and Booth J. A. (1966) The influence of shade, soil, and water on saguaro seedling establishment. *Bot. Gaz.* **127**, 95–102.

Ulrich A. and Hills F. J. (1973) Plant analysis as an aid in fertilizing sugar crops: Part I. Sugar beets. In *Soil Testing and Plant Analysis* (eds. L. M. Walsh and J. D. Beaton). Soil Science Society of America, Madison, Wisconsin, pp. 271–288.

Van Cleve K., Chapin F. S., III, Dyrness C. T., and Viereck L. A. (1991) Element cycling in taiga forest: state-factor control. *Bioscience* **41**, 78–88.

van der Werf A., van den Berg G., Ravenstein H. J. L., Lambers H., and Eising R. (1992) Protein turnover: a significant component of maintenance respiration in roots. In *Molecular, Biochemical, and Physiological Aspects of Plant Respiration* (eds. H. Lambers and L. H. W. van der Plas). SPB Academic Publishing, The Hague.

van Vuuren M. M. I., Aerts R., Berendse F., and de Visser W. (1992) Nitrogen mineralization in heathland ecosystems dominated by different plant species. *Biogeochemistry* **16**, 151–166.

Venterink H. O., Davidsson T. E., Kiehl K., and Leonardson L. (2002) Impact of drying and re-wetting on N, P, and K dynamics in a wetland soil. *Plant Soil* **243**, 119–130.

Verhoeven J. and Schmitz M. (1991) Control of plant growth by nitrogen and phosphorus in mesotrophic fens. *Biogeochemistry* **12**, 135–148.

Vinton M. A. and Burke I. C. (1995) Interactions between individual plant species and soil nutrient status in shortgrass steppe. *Ecology* **76**, 1116–1133.

Vitousek P. M. (1982) Nutrient cycling and nutrient use efficiency. *Am. Nat.* **119**, 553–572.

Vitousek P. M. and Farrington H. (1997) Nitrogen limitation and soil development: experimental test of a biogeochemical theory. *Biogeochemistry* **37**, 63–75.

Vitousek P. M. and Howarth R. W. (1991) Nitrogen limitation on land and in the sea: how can it occur? *Biogeochemistry* **13**, 87–115.

Vitousek P. M., Walker L. R., Whiteaker L. D., Mueller-Dombois D., and Matson P. A. (1987) Biological invasion by *Myrica faya* alters ecosystem development in Hawaii. *Science* **238**, 802–804.

Walker T. W. and Syers J. K. (1976) The fate of phosphorus during pedogenesis. *Geoderma* **1**, 1–19.

Walters M. B. and Field C. B. (1987) Photosynthetic light acclimation in two rainforest *Piper* species with different ecological amplitudes. *Oecologia* **72**, 449–456.

Walters M. B. and Reich P. B. (1999) Low-light carbon balance and shade tolerance in the seedlings of woody plants: do winter deciduous and broad-leaved evergreen species differ? *New Phytol.* **143**, 143–154.

Waring R. H. and Running S. W. (1998) *Forest Ecosystems: Analysis at Multiple Scales*. Academic Press, San Diego, CA.

Watson R. T., Zinowera M. C., Moss R. H., and Dokken D. J. (1996) *Climate Change 1995. Impacts, Adaptations, and Mitigation of Climate Change: Scientific-technical Analyses*. Cambridge University Press, Cambridge.

Weathers K. C. (1999) The importance of cloud and fog in the maintenance of ecosystems. *Trends Ecol. Evol.* **14**, 214–215.

Wedin D. and Pastor J. (1993) Nitrogen mineralization dynamics in grass monocultures. *Oecologia* **96**, 186–192.

Williams K., Percival F., Merino J., and Mooney H. A. (1987) Estimation of tissue construction cost from heat of combustion and organic nitrogen content. *Plant Cell Environ.* **10**, 725–734.

Wilson J. B. and Agnew D. Q. (1992) Positive-feedback switches in plant communities. *Adv. Ecol. Res.* **23**, 263–336.

Wright I. J., Reich P. B., and Westoby M. (2001) Strategy shifts in leaf physiology, structure and nutrient content between species of high- and low-rainfall and high- and low-nutrient habitats. *Funct. Ecol.* **15**, 423–434.

8.07
Biogeochemistry of Decomposition and Detrital Processing

J. Sanderman and R. Amundson
University of California, Berkeley, CA, USA

8.07.1 INTRODUCTION

Decomposition is a key ecological process that roughly balances net primary production in terrestrial ecosystems and is an essential process in resupplying nutrients to the plant community. Decomposition consists of three concurrent processes: communition or fragmentation, leaching of water-soluble compounds, and microbial catabolism. Decomposition can also be viewed as a sequential process, what Eijsackers and Zehnder (1990) compare to a Russian matriochka doll.

Soil macrofauna fragment and partially solubilize plant residues, facilitating establishment of a community of decomposer microorganisms. This decomposer community will gradually shift as the most easily degraded plant compounds are utilized and the more recalcitrant materials begin to accumulate. Given enough time and the proper environmental conditions, most naturally occurring compounds can completely be mineralized to inorganic forms. Simultaneously with mineralization, the process of humification acts to transform a fraction of the plant residues into stable soil organic matter (SOM) or humus. For reference, Schlesinger (1990) estimated that only ~0.7% of detritus eventually becomes stabilized into humus.

Decomposition plays a key role in the cycling of most plant macro- and micronutrients and in the formation of humus. Figure 1 places the roles of detrital processing and mineralization within the context of the biogeochemical cycling of essential plant nutrients. Chapin (1991) found that while the atmosphere supplied 4% and mineral weathering supplied no nitrogen and <1% of phosphorus, internal nutrient recycling is the source for >95% of all the nitrogen and phosphorus uptake by tundra species in Barrow, Alaska. In a cool temperate forest, nutrient recycling accounted for 93%, 89%, 88%, and 65% of total sources for nitrogen, phosphorus, potassium, and calcium, respectively (Chapin, 1991).

The second major ecosystem role of decomposition is in the formation and stabilization of humus. The cycling and stabilization of SOM in the litter–soil system is presented in a conceptual model in Figure 2. Parallel with litterfall and most root turnover, detrital processing is concentrated

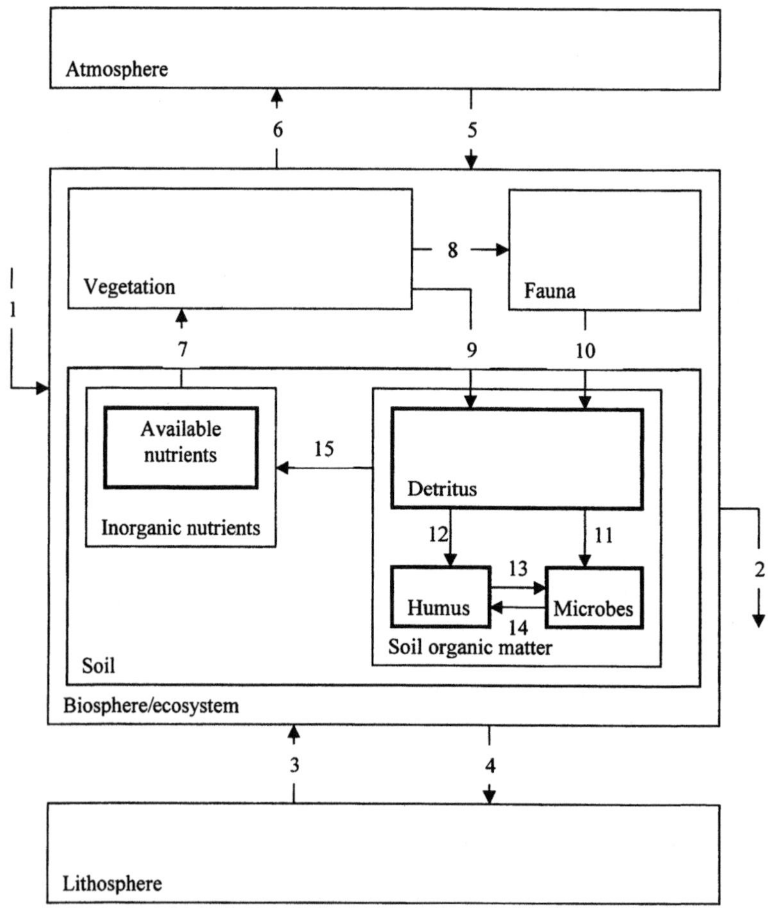

Figure 1 A decomposition-centric biogeochemical model of nutrient cycling. Although there is significant external input (1) and output (2) from neighboring ecosystems (such as erosion), weathering of primary minerals (3), loss of secondary minerals (4), atmospheric deposition and N-fixation (5) and volatilization (6), the majority of plant-available nutrients are supplied by internal recycling through decomposition. Nutrients that are taken up by plants (7) are either consumed by fauna (8) and returned to the soil through defecation and mortality (10) or returned to the soil through litterfall and mortality (9). Detritus and humus can be immobilized into microbial biomass (11 and 13). Humus is formed by the transformation and stabilization of detrital (12) and microbial (14) compounds. During these transformations, SOM is being continually mineralized by the microorganisms (15) replenishing the inorganic nutrient pool (after Swift *et al.*, 1979).

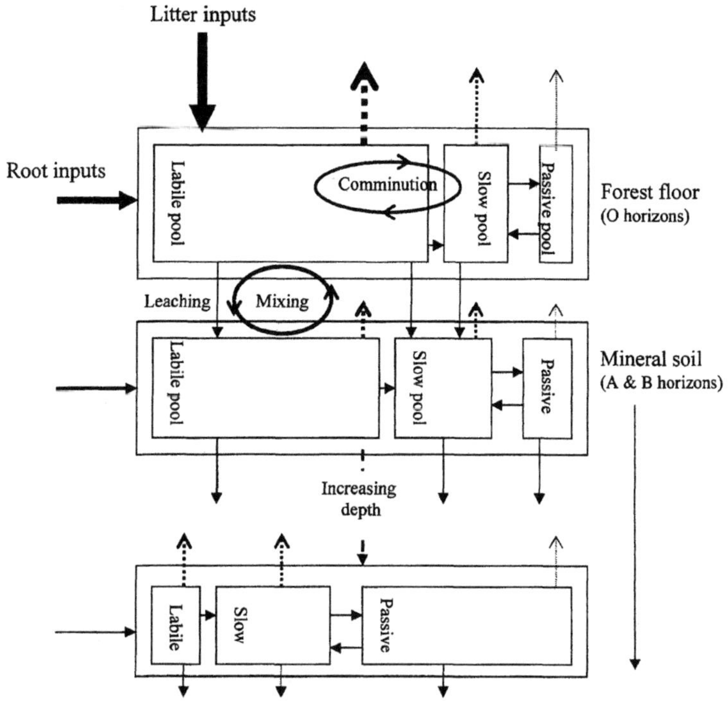

Figure 2 Conceptual model of carbon cycling in the litter–soil system. In each horizon or depth increment, SOM is represented by three pools: labile SOM, slow SOM, and passive SOM. Inputs include aboveground litterfall and belowground root turnover and exudates, which will be distributed among the pools based on the biochemical nature of the material. Outputs from each pool include mineralization to CO_2 (dashed lines), humification (labile → slow → passive), and downward transport due to leaching and physical mixing. Communition by soil fauna will accelerate the decomposition process and reveal previously inaccessible materials. Soil mixing and other disturbances can also make physically protected passive SOM available to microbial attack (passive → slow).

at or near the soil surface. As labile SOM is preferentially degraded, there is a progressive shift from labile to passive SOM with increasing depth. There are three basic mechanisms for SOM accumulation in the mineral soil: bioturbation or physical mixing of the soil by burrowing animals (e.g., earthworms, gophers, etc.), *in situ* decomposition of roots and root exudates, and the leaching of soluble organic compounds. In the absence of bioturbation, distinct litter layers often accumulate above the mineral soil. In grasslands where the majority of net primary productivity (NPP) is allocated belowground, root inputs will dominate. In sandy soils with ample rainfall, leaching may be the major process incorporating carbon into the soil.

There exists an amazing body of literature on the subject of decomposition that draws from many disciplines—including ecology, soil science, microbiology, plant physiology, biochemistry, and zoology. In this chapter, we have attempted to draw information from all of these fields to present an integrated analysis of decomposition in a biogeochemical context. We begin by reviewing the composition of detrital resources and SOM (Section 8.07.2), the organisms responsible for decomposition (Section 8.07.3), and some methods for quantifying decomposition rates (Section 8.07.4). This is followed by a discussion of the mechanisms behind decomposition (Section 8.07.5), humification (Section 8.07.6), and the controls on these processes (Section 8.07.7). We conclude the chapter with a brief discussion on how current biogeochemical models incorporate this information (Section 8.07.8).

8.07.2 COMPOSITION OF DECOMPOSER RESOURCES

A wide range of substrates are available to the decomposer organisms. Table 1 highlights the large range in biomass and production values within and across three major biomes. Not only are the absolute amounts of litter important in determining decomposition dynamics, but its spatial distribution, molecular composition and physical properties will also have a large controlling effects on both its degradation and stabilization as SOM (see Section 8.07.7). The microbial biomass is an important secondary resource and its varied composition and properties are now believed to be a significant starting material

Table 1 Range of biomass and productivity for temperate forests, temperate grasslands, and moist to wet tropical forests.

	Temperate forest	*Temperate grassland*	*Tropical forest*
Biomass (tC ha^{-1})			
Total living	13–281	2.2–27	209–473
Aboveground	5.7–220	0.5–1.8	146–431
Belowground	7.0–41	1.7–25	11–120
Litter layer	2.8–75	0.5–4.2	1.8–16.5
Soil organic matter	27–158	17–207[a]	50–599
Production (tC ha^{-1} yr^{-1})			
NPP	4.3–24	2.3–14.3	11.1–20.8
Aboveground NPP	1.6–15	0.5–5.2	9.6–18.2
Litterfall	0.4–2.7	0.3–3.1	6.4–15.3
Belowground NPP	1.9–6.4	1.5–6.4	1.5–5.5
Root turnover	0.6–6.0	0.5–2.2	1.0–3.6[b]

Sources: temperate forests—Vogt (1991); temperate grasslands—Sims and Singh (1978a,b); and tropical forests—Brown and Lugo (1982).
[a] ±1 SD of mean listed for cool temperate steppe ecosystems in Post *et al.* (1982). [b] Estimated from root turnover rates listed for tropical forests in Gill and Jackson (2000).

for SOM formation (see Section 8.07.5). In this section, we will describe the physical and molecular composition of detritus. For additional information on this subject, see reviews by De Leeuw and Largeau (1993) and Kogel-Knabner (2002).

8.07.2.1 Plant Litter

Leaves, whether broadleaf foliage or coniferous needles, are the dominant aboveground inputs to most soils. Other aboveground inputs—such as branches, bark, and fruits—account for only 20% in temperate deciduous forests (Jensen, 1984) and 20–40% in coniferous forests (Millar, 1974) of total litterfall. Herbaceous understory vegetation contributes only a small fraction to the total litterfall in most temperate forests. Whether or not a forest is managed for timber production will have a major impact on the composition of the aboveground litter. In mature natural forests, coarse woody debris may contribute 40–60% or more to the total detrital biomass (Vogt, 1991). Aboveground input in grasslands and other semi-arid ecosystems can be an order of magnitude lower than in forests (Table 1). Additionally, grazing will have a significant impact on how much of the aboveground NPP is returned to be decomposed (Sims and Singh, 1978a,b).

The timing of litterfall will also have a big impact on detrital processing and carbon dynamics in terrestrial ecosystems. Deciduous forests have a single annual pulse of fresh inputs in the fall, while evergreen coniferous and tropical forests have a somewhat continuous rate of litterfall which tends to fluctuate with water availability. Plants in grasslands and other semi-arid to arid ecosystems often turn over in response to drought stress.

Plants are predominantly composed of parenchyma and woody tissues. Parenchyma cells dominate the green tissues in leaves and are composed of a protein-rich protoplast surrounded by a cellulose wall. Woody plant cells dominate all support (sclerenchyma) and transport (xylem and phloem) structures in a plant. They are composed of several layers (middle lamella, primary wall, secondary wall, and tertiary wall) with varying proportions of cellulose, hemicellulose, and lignin (Fengel and Wegener, 1984).

In general, intracellular spaces are comprised of easily degraded energy-rich proteins and starches, while plant cell walls contain macromolecular polysaccharides (cellulose and hemicellulose), lignin, lipids, and various polyphenols with varying degrees of decomposability (Kogel-Knabner, 2002). Table 2 shows the distribution of these major organic components in several primary and secondary resources as determined by wet chemical techniques. The structure and degradation pathways of each of these compounds are discussed in Section 8.07.5.4.

8.07.2.2 Roots

Plants allocate a significant portion of their C and energy resources to root production. Vogt (1991) estimated that belowground net primary production (BNPP) accounts for 30–50% of total NPP in temperate forests, while temperate grasslands can allocate as much as 86% of NPP to belowground production (Sims and Singh, 1978b). Jackson *et al.* (1996) found that as a global average, 50% and 75% of total root biomass occurs in the top 20 cm and 40 cm of soil, respectively. Reviewing the root literature, Gill and Jackson (2000) found that annual root

Table 2 Major organic components of plant and microbial resources determined by wet chemical methods.

	Deciduous leaf Quercus	*Conifer needle* Pinus	*Grass leaf* D. flexuosa	*Grass root* Loudetia simplex	*Deciduous wood*	*Bacteria*	*Fungi*
Lipid (ether soluble)	8	24	2	11	2–6	10–35	1–42
Storage/metabolic carbohydrate (cold and hot water soluble)	22	7	13	35	1–2	5–30	8–60 (chitin)
Cell wall polysaccharide, hemicellulose (alkali soluble)	13	19	24	16	19–24	4–32	2–15
Cellulose (strong acid)	16	16	33	11	45–48	0	0
Lignin (residue)	21	23	14	34	17–26	0	0
Protein ($N \times 6.25$)	9	2	2	2		50–60	14–52
Ash (incineration)	6	2		5	0.3–1.1	5–15	5–12

Adapted from Swift *et al.* (1979) and Lavelle and Spain (2001).

turnover or mortality, defined as BNPP/maximum belowground standing crop, was 10% for entire tree root systems, 34% for total shrubland roots, 53% for grassland fine roots (<5 mm diameter), 55% for wetland fine roots (<5 mm), and 56% for forest fine roots (<5 mm). In addition, minirhizotron studies have shown the existence of a highly dynamic pool of roots (generally <1 mm) with a lifespan of days to weeks (Hendrick and Pregitzer, 1997; Tingey *et al.*, 2000). However, Gaudinski *et al.* (2001) noted the possibility of long-lived fine roots (<2 mm) based on radiocarbon measurements. Reconciling these results is an area of active research with major ramifications on terrestrial carbon cycling dynamics.

Despite its importance in detrital processing and long-term carbon storage, the biochemical composition of belowground carbon inputs has received much less attention. Most current biogeochemical models assume that the chemical composition of roots is similar to that of aboveground litter. Root diameter will have a large impact on the ratio of woody to nonwoody tissues, with the proportion of woody tissues increasing with diameter (Gill and Jackson, 2000).

Rhizodeposition represents a small but significant carbon substrate available to the decomposer organisms. Rhizodeposition, a term that describes the loss of carbon from a living root (Whipps and Lynch, 1985), includes: (1) water-soluble exudates such as sugars, amino acids, and organic acids which passively leak from roots; (2) higher-molecular-weight secretions such as complex carbohydrates and enzymes which are metabolically moved out of the root; (3) lysates from the autolysis of cells within the root; (4) mucilage coating on roots which is composed of polysaccharides and polygalacturonic acids; and (5) gases such as ethylene and CO_2 (Grayston *et al.*, 1996; Lynch and Whipps, 1990). The amount of net fixed carbon lost as nongaseous rhizodeposition has been estimated to range from 1% to 10% in most perennial plants (Grayston *et al.*, 1996), with annual plants ranging from 6% to 17% (Lynch and Whipps, 1990). The majority of this carbon is readily utilized by the highly active microbial community in the rhizosphere and only a small fraction will end up incorporated into SOM (Cheng *et al.*, 1994; Grayston, 2000).

Readily assimilable root exudates are believed to have a large impact on plant nutrient availability (Uren and Reisenauer, 1988). Exuded organic acids can increase the solubility of inorganic nutrients such as phosphorus and manganese. However, the largest effect rhizodeposition has on nutrient availability is indirectly through the stimulation of microorganisms and their subsequent acceleration of the biogeochemical cycling of most essential plant nutrients (Grayston *et al.*, 1996). Additionally, it has been hypothesized that root exudates act as a primer for the decomposition of bulk SOM through the stimulation of microorganisms in the rhizosphere (Bottner *et al.*, 1988; Sallih and Bottner, 1988).

8.07.2.3 Secondary Resources

Microbial biomass only contributes 1–5% to the total SOM pool. However, this highly dynamic protein-rich carbon pool represents a major nutrient sink and as such many organisms are capable of degrading most microbially synthesized materials. Microbial biomass is usually measured indirectly by monitoring CO_2 production (Anderson and Domsch, 1978; Jenkinson and Powlson, 1986) or levels of specific enzymes and unique biochemical compounds (Tunlid and White, 1992; West *et al.*, 1987). The chloroform ($CHCl_3$) fumigation–incubation technique, developed by Jenkinson and Powlson (1986), measures the extra CO_2 evolved (assumes that dead microorganisms are rapidly decomposed) from a soil following fumigation relative to a control soil. First proposed by Anderson and Domsch (1978), the substrate-induced respiration method measures the initial respiratory response of the microbial population following amendment with an excess carbon and energy source.

Fungal cell walls are composed primarily of the highly crystalline polysaccharides, chitin and β-glucan, and various nonhydrolyzable melanins (Peberdy, 1990). Section 8.07.5.4 describes the structure and decomposition of chitin. While chitin has been well studied, very little is known about the structure or *in situ* decomposition of melanins (Butler and Day, 1998). However, melanins are thought to be a possible humus precursor because of their similarities with humic acids (Saiz-Jimenez, 1996).

Peptidoglucan, a heteropolymer with carbohydrate and amino acid subunits, is the dominant material in bacterial cell walls (Rogers *et al.*, 1980). Bacteria and a number of algae also contain considerable amounts of insoluble, nonhydrolyzable aliphatic compounds, termed bacteran and algaenan (Kogel-Knabner, 2002). Although more research is needed, these highly recalcitrant macromolecules also have the potential to be humus precursors (Augris *et al.*, 1998).

8.07.2.4 Soil Organic Matter

SOM is composed of a continuum of organic resources from fresh plant residues to stabilized organic matter (OM) or humus (Stevenson, 1994). Although this definition of SOM includes intact plant litter, in this review we will often distinguish between decomposition of litter on the soil surface and decomposition/stabilization of OM within the mineral soil. Within the mineral soil, SOM is often divided into four categories: the "light" fraction, microbial biomass (discussed above), dissolved organic matter (DOM), and stable humic substances.

Maintaining adequate levels of humus in agriculture soils is recognized as one of the key factors in sustaining productivity (Weil, 1992). Humus imparts the typical dark color of many soils; greatly increases water retention; facilitates cementation of clay particles into aggregates; acts as a large pool of nutrients for plant growth; enhances micronutrient availability by forming stable complexes with many polyvalent cations; helps buffer the soil from pH changes; and greatly increases the cation exchange capacity of the soil (Stevenson, 1994).

Light fraction. The light fraction, so termed because it is the material that floats in liquids with densities ranging from 1.6 g cm^{-3} to 2.0 g cm^{-3}, consists of partially decomposed plant residues. In an undisturbed forest and grassland soils, the light fraction can constitute as much as 30% of the total SOM (Stevenson, 1994). However, in agriculture soils, the light fraction will typically contribute only a few percent to total SOM. Under a microscope, the majority of the light fraction can still be recognized as belonging to plant litter. As determined by ^{13}C nuclear magnetic reasonance (NMR) spectroscopy, the gross chemical composition of the light fraction has been found to be comparable to that of plant litter (Skjemstad *et al.*, 1986). The light fraction combined with the microbial biomass is what many researchers have termed the labile or active SOM fraction—the highly dynamic, rapidly cycling component of SOM responsible for supplying most of the nutrients for plant growth (Stevenson, 1994; Townsend *et al.*, 1997).

Dissolved organic matter. The water-soluble matter or DOM is a small but important component of SOM. Typical concentrations of dissolved organic carbon (DOC) in forest soil solutions are 10–50 mg L^{-1} with annual fluxes decreasing from 10–80 g C m^{-2} yr^{-1} in the surface horizons (0–20 cm) to <10 g C m^{-2} yr^{-1} in the subsurface (20–100 cm) (Neff and Asner, 2001; Qualls *et al.*, 1991). The dissolved phase of SOM is a major source of plant-available nutrients (i.e., Qualls and Haines, 1991) and it has been implicated in the alleviation of metal toxicities (i.e., Pohlman and McColl, 1988) and in movement of organic carbon into the mineral soil (i.e., Guggenberger, 1992). DOM fluxes leaving the litter layer consist mostly of simple carbohydrates and amino acids, while fluxes through the mineral horizons often contain complex biopolymers that are thought to be of microbial origin (see Section 8.07.5.3 for more details).

Humic substances. In most soils, humic substances comprise ~60–80% of SOM. Humic substances are distinguished from nonhumic substances in that they are unique to the soil or sediment environment and are composed of relatively high-molecular-weight compounds that can neither be characterized as biopolymers of microorganisms nor higher plants (Stevenson, 1994).

Researchers have devised numerous extraction and fractionation schemes to deal with the heterogeneous nature of humic substances. Traditionally, the operational definition of humic substances as used by the International Humic Substances Society (Hayes *et al.*, 1989) is based on the solubility in a series of acids and bases. In this scheme, humic substances are classified into three chemical groupings: (1) fulvic acid, soluble in both alkali and acid solutions, has the lowest molecular weight and is generally considered the most susceptible to microbial degradation; (2) humic acid, soluble in alkali but not in acid, is intermediate in molecular weight and decomposability; and (3) humin, insoluble in both alkali and acid solutions, is the most complex and recalcitrant humus fraction in this scheme (Brady and Weil, 2002). Although widely applied, this fractionation scheme has been criticized because the resulting fractions are still heterogeneous mixtures of compounds that will differ between soils and do not necessarily correlate with the biochemical processes of decomposition and humification (Kogel-Knabner, 1993; Oades, 1988).

Methods of characterization. Other physical and chemical fractionation schemes attempt to separate humus in terms of biodegradability or degree of humification. With decreasing particle size or increasing density, the recalcitrance, as defined by numerous methods, of the associated organic material often increases (Christensen, 1992). Additional chemical treatments can be applied to a sample, such as HCl or H_2SO_4 hydrolysis which removes labile carbohydrates and proteinaceous materials. Trumbore *et al.* (1996) combined density separation with hydrolysis to separate bulk SOM into three pools with increasing mean residence times based on ^{14}C analysis: a low-density ($<2.0\ g\ cm^{-3}$) fraction with rapid turnover; a high-density hydrolyzable fraction with intermediate turnover; and a high-density nonhydrolyzable fraction with very slow turnover times.

Kogel-Knabner (2002) stressed that conventional chemical degradative techniques, such as those used to produce Table 2, can only account for 50–60% of the total organic material in plant litter and SOM (Kogel *et al.*, 1988) and that these methods are often not specific for single compound classes (Preston *et al.*, 1998; Ryan *et al.*, 1990). Although these conventional chemical techniques have limitations, the majority of decomposition studies have relied on them. Recently, numerous advances have been made in the study of decomposition and humification utilizing solid-state ^{13}C NMR spectroscopy, pyrolysis in combination with gas chromatography (Py-GC) and mass spectrometry, and other advanced analytical techniques (Baldock *et al.*, 1997; Kogel-Knabner, 1997).

In NMR spectroscopy, certain nuclei (e.g., ^{1}H, ^{13}C, ^{15}N, and ^{31}P) will precess about their axis parallel with an externally applied magnetic field. When a second magnetic field is applied at a right angle, the nuclei will resonate, thereby inducing a measurable voltage change. Each nucleus is influenced by its neighbors and will only resonate in a specific magnetic field strength, so by varying the strength of the magnetic field a spectrum of electromagnetic radiation is produced. To make quantitative comparisons, results are always expressed as a "chemical shift" (δ) relative to a reference standard, usually tetramethylsilane (TMS). Solid organic samples often show broad overlapping peaks due to dipolar interactions between ^{13}C and the much more common ^{1}H nuclei and interference due to paramagnetic materials such as iron oxides in the soil. The techniques of cross-polarization and magic-angle spinning (CPMAS) have been developed to greatly reduce line broadening due to ^{1}H (Wilson *et al.*, 1981). Treatment with HF will both concentrate the organic material and remove paramagnetic materials (Schmidt *et al.*, 1997; Skjemstad *et al.*, 1994). For much more detail on theory and application of NMR spectroscopy to OM studies, see Malcolm (1989), Wilson (1989), Kogel-Knabner (1997) and Veeman (1997), and others.

Solid-state CPMAS ^{13}C NMR spectroscopy yields information on the abundance of different functional groups in a sample. In general, the four most commonly used chemical shift regions are: alkyl carbon (0–45 ppm) associated with lipids, cutin, suberin, proteins, aliphatic biopolymers, and several uncharacterized compounds; O-alkyl carbon (45–110 ppm) associated with carbohydrates and methoxyl groups; aromatic carbon (110–160 ppm) associated with lignin and tannins; and carboxyl C (160–210 ppm) associated with numerous compounds (Zech *et al.*, 1992). The ^{13}C NMR spectrum of humic acid extracted from a grassland soil along with the chemical shift regions for selected functional groups is shown in Figure 3. As this example illustrates, humic acid is a heterogeneous mixture of organic materials dominated by aromatic carbon ($\delta = 130$ ppm), followed by near equal contributions of carboxyl (178 ppm), aliphatic (25–32 ppm) and carbohydrate (75 ppm) carbon, and slightly smaller quantities of proteinaceous (58 ppm) and phenolic carbon (152 ppm).

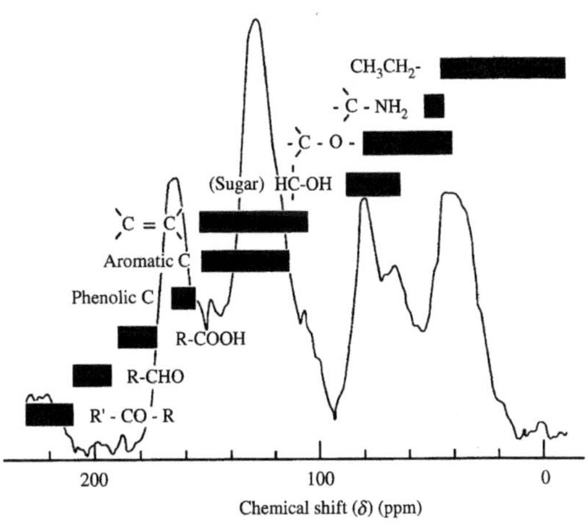

Figure 3 Solid-state CPMAS ^{13}C NMR spectrum of humic acid extracted from a grassland soil with the chemical shift ranges for select functional groups relative to TMS (after Schnitzer, 1990; Wilson, 1990).

While ^{13}C NMR studies reveal the gross chemical composition of OM, this technique cannot identify specific compounds. However, the thermal degradative technique of analytical pyrolysis, especially when coupled with mass spectrometry, can provide information on the specific compounds present in a sample. Bracewell *et al.* (1989), Saiz-Jimenez (1994), Kogel-Knabner (2000), and others have reviewed the application of pyrolysis techniques to OM studies.

8.07.2.5 Summary

The quantity and biochemical composition of aboveground litter inputs has been well characterized by wet chemical analyses such as the classical Klason lignin or van Soest analysis of proximate fractions. However, there is a dearth of data on belowground carbon inputs (i.e. roots, root exudates, and microbial biomass). Due to its importance in ecosystem functioning and in long-term carbon storage, many aspects of belowground carbon inputs are being actively researched.

Despite the fact that the classical degradative techniques have potentially serious analytical problems (Kogel *et al.*, 1988; Zech *et al.*, 1987), they are still commonly employed in decomposition studies and the majority of biogeochemical models have been parametrized using these data. Significant advances in the understanding of the macromolecular composition of detritus are being made using spectroscopic, thermolytic, and other analytical techniques (Kogel-Knabner, 2002). Where appropriate, we have attempted to integrate some recent ^{13}C NMR spectroscopy data into this review.

8.07.3 THE DECOMPOSER ORGANISMS

When examining decomposition at a large scale, we tend to obscure the fact that decomposition and detrital recycling occurs as a result of the direct and indirect actions of a myriad of soil flora and fauna. A well-developed soil may contain a thousand species of soil animals with a much greater number of microbial taxa. In fact, soil microorganisms may constitute upwards of 98% of all life (Pace, 1996), with only a fraction of a percent having been classified. Even more impressive than the shear diversity of soil dwelling organisms is their abundance. In a single gram of soil, there may be one hundred million bacterial cells supporting a population of one hundred thousand predatory protozoa (Table 3).

This taxonomic diversity emphasizes the challenges facing any analysis of the decomposer community and highlights the need to simplify the classification of the roles these organisms play in the process of decomposition (Swift *et al.*, 1979). In this section, we will first outline a functional ecology for decomposer organisms and then highlight some details of several of the major groups of soil organisms. For more detailed information on the biology and ecology of soil organisms, the reader is referred to the texts of Burges and Raw (1967), Dindal (1990), Dix and Webster (1995), and Reddy (1995b).

8.07.3.1 Functional Ecology

Depending on the interests of the particular investigator, soil organisms regardless of taxonomic identity can be classified by body size or by various physiological traits related to trophic function. In studies of decomposition, the functional division of the decomposer community

Table 3 Relative abundance and biomass of soil fauna and flora commonly found in the surface 15 cm of soil.

Organisms	*Number*		*Biomass*[b]	
	m^{-2}	g^{-1}	$kg\ ha^{-1}$	$g\ m^{-2}$
Microflora				
Bacteria	$10^{13}-10^{14}$	$10^{8}-10^{9}$	400–5,000	40–500
Actinomycetes	$10^{12}-10^{13}$	$10^{7}-10^{8}$	400–5,000	40–500
Fungi	$10^{10}-10^{11}$	$10^{5}-10^{6}$	1,000–15,000	100–1,500
Algae	$10^{9}-10^{10}$	$10^{4}-10^{5}$	10–500	1–50
Soil fauna				
Protozoa	$10^{9}-10^{10}$	$10^{4}-10^{5}$	20–200	2–20
Nematodes	$10^{6}-10^{7}$	$10-10^{2}$	10–150	1–15
Mites	$10^{3}-10^{6}$	1–10	5–150	0.5–1.5
Collembola	$10^{3}-10^{6}$	1–10	5–150	0.5–1.5
Earthworms[a]	$10-10^{3}$		100–1,500	10–150
Other fauna	$10^{2}-10^{4}$		10–100	1–10

Source: Brady and Weil (2002).
[a] A greater depth was used for earthworms. [b] Biomass values are on a liveweight basis. Dry weights are ~20–25% of these values.

based on body size has become common since Edwards and Heath (1963) introduced the mesh litter bag as a field tool for studying detrital processing (details of the litter bag technique can be found in Section 8.07.4). On the basis of what organisms can or cannot enter a specific mesh size, the appropriate dimension to consider would be body width (Figure 4) not length as researchers have traditionally done (e.g., Wallwork, 1970). Under this scheme bacteria and fungi are considered *microflora*. The protozoa, nematodes, rotifers, tardigrades, and the smallest mites (Acari), and Collembola are classified as *microfauna*. The major functional role of the microfauna is as bacterial and fungal predators. The organisms generally responsible for communition and redistribution of plant litter most often are categorized as *macrofauna*. The macrofauna include several groups which when present have inordinate impacts on the decomposition subsystem, mainly the earthworms (Oligochaeta) and the termites (Isoptera).

Alternately, the decomposer organisms can be subdivided by their respective trophic positions. The traditionally well-described food chains of herbivore–carnivore systems for higher animals often fail to capture the complexity and diversity of roles in the decomposer community. Swift *et al.* (1979) describe the food web of the decomposer community as "a more fluid, interactive structure with individual species operating on several levels which might be distinguished as trophically different." The trophic-level classification of fungus and food source devised by Lewis (1973) has been widely adopted to describe the general decomposer community (e.g., Swift *et al.*, 1979). Lewis (1973) defines three main trophic roles: *necrotrophs*, *biotrophs*, and *saprotrophs*.

The *necrotrophs* exploit other living organisms resulting in the rapid death of the food resource. They include parasitic microorganisms (individuals of which can feed on and kill both plant tissue and higher soil fauna) and the major microflora predators (i.e., the protozoa and nematodes). The *biotrophs* form a more mutualistic relationship whereby their continued existence depends on the health of the host (many root-feeders and the mycorrhizal fungi fall into this category). The majority of the decomposer organisms are classified as *saprotrophs*, organisms that utilize dead OM as their main food source. The saprotrophs can further be divided into the type of OM they

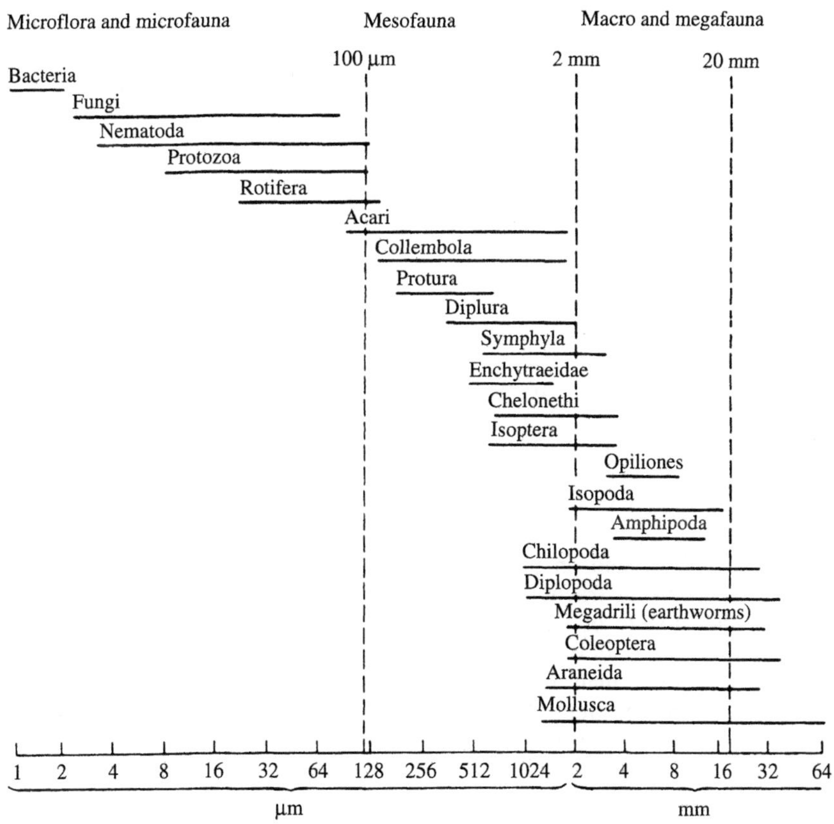

Figure 4 Size classification of organisms in decomposer food webs by body width (Swift *et al.*, 1979) (reproduced by permission of University of California Press from *Decomposition in Terrestrial Ecosystems*, **1979**).

consume, i.e., *primary* or *secondary* (Swift *et al.*, 1979). In terms of ecological strategies (Macarthur and Wilson, 1967), the necrotrophs are primarily opportunistic *r*-selected species (short lifespan, early reproduction, low biomass, and the potential to produce large numbers of offspring in a short period of time), the biotrophs are dominated by long-lived *K*-strategists (in fact, the largest and oldest known organism may be mycelial fungi), and the saprotrophs are represented by the full spectrum of ecological strategies.

Table 4 Microbial tolerance to matric-controlled (Ψ_m) water stress.

Water potential (MPa)	*Water film thickness*	*Microbial activity limited (example of genus)*
−0.03	4.0 μm	movement of protozoa, zoospores, and bacteria
−0.1	1.5 μm	
−0.5	0.5 μm	
−1.5	3.0 nm	nitrification; sulfur oxidation
−4.0	<3.0 nm	bacterial growth (*Bacillus*)
−10.0	<1.5 nm	fungal growth (*Fusarium*)
−40.0	<0.9 nm	fungal growth (*Penicillium*)

Source: Hartel (1999).

8.07.3.2 Soil Microorganisms

The diversity of microbial life is, in part, a result of the interactions between an extremely heterogeneous and complex habitat (the soil), variable nutrition sources (both in structure and availability), numerous trophic interactions (including competition and predation), and multiple methods of gene flow. For additional information on soil microorganisms the reader is referred to the texts of Paul and Clark (1996) and Sylvia *et al.* (1999).

The soil habitat. Soil is a dynamic and diverse medium—the result of a complex interplay between the local climate (especially temperature and precipitation), parent material (bedrock, alluvium, aeolian sands, etc.), living biota (especially the native vegetation), topography (slope and relief), and time (Jenny, 1941). Soils vary horizontally and continually across the landscape as a result of variations in these soil-forming factors. Vertical heterogeneity in a soil profile results from variations in chemical weathering and leaching of weathering products down a profile, asymmetric inputs of OM as litter at the surface or as root exudates at localized positions within the soil, and turbation by both biotic and abiotic factors.

A given volume of soil will contain ~50% solid particles of which less than 5% is typically defined as OM and 50% as pore space which can then be further divided into water- and air-filled pores. The ratio of water/air will depend on the water content of the soil. The mineral fraction can be defined by texture—the mixture of sand (diameters from 0.05 mm to 2.0 mm), silt (0.002 mm to 0.5 mm), and clay size (less than 0.002 mm) particles. The soil texture, especially the clay percentage due to its very high surface area, is critical in determining many properties relevant to soil microorganisms and detrital processing: (1) soil structure and the distribution of pore sizes; (2) water-holding capacity of the soil; (3) soil aeration; and (4) the ability to retain nutrient elements (cation exchange capacity).

Perhaps the single most important factor determining the distribution of microorganisms in a soil is the water potential. Nearly all microorganisms and many of the smallest soil animals rely on the thin film of water surrounding soil aggregates and in soil pores for the basic necessities of life. The soil water potential (Ψ) is a measure of the potential energy of water in the soil relative to the potential energy of pure, free water. Ψ is always negative in a soil due primarily to attractive forces of soil particles (matric potential, Ψ_m) and solute ions (osmotic potential, Ψ_π). Table 4 shows the relationship between matric potential, water film thickness, and the limitations on microbial activity. If Ψ falls too low, microorganisms must either enter a quiescent resting phase or perish from desiccation. Just as drying has dramatic effects on microbial populations, rapid rewetting can cause massive mortality and turnover (discussed in detail in Section 8.07.7).

Although representing less than 5% and often less than 1% of the soil, OM plays several critical roles in the soil system. It is the substrate for all heterotrophic microbial life. The high cation exchange capacity of OM acts as a store of plant and microbe available nutrients. OM, especially fungal muscigels and root exudates, in combination with mycelial microorganisms act to bind mineral particles into soil aggregates. The microhabitat within soil aggregates can be extremely different than that of the aggregate surface. Diffusion is limited by the increased bulk density and anaerobic interiors to aggregates are found often only a few millimeters from the aggregate surface (Tokunaga *et al.*, 2001). Anaerobic interiors offer a refuge for strict anaerobic bacteria and also play a major role in protecting and preserving OM in soils. The diversity of microhabitats created by soil aggregates is highlighted in Figure 5.

Methods of quantification. A major hindrance to the study of soil microorganisms and particularly to bacteria is the lack of appropriate methods for characterizing the community

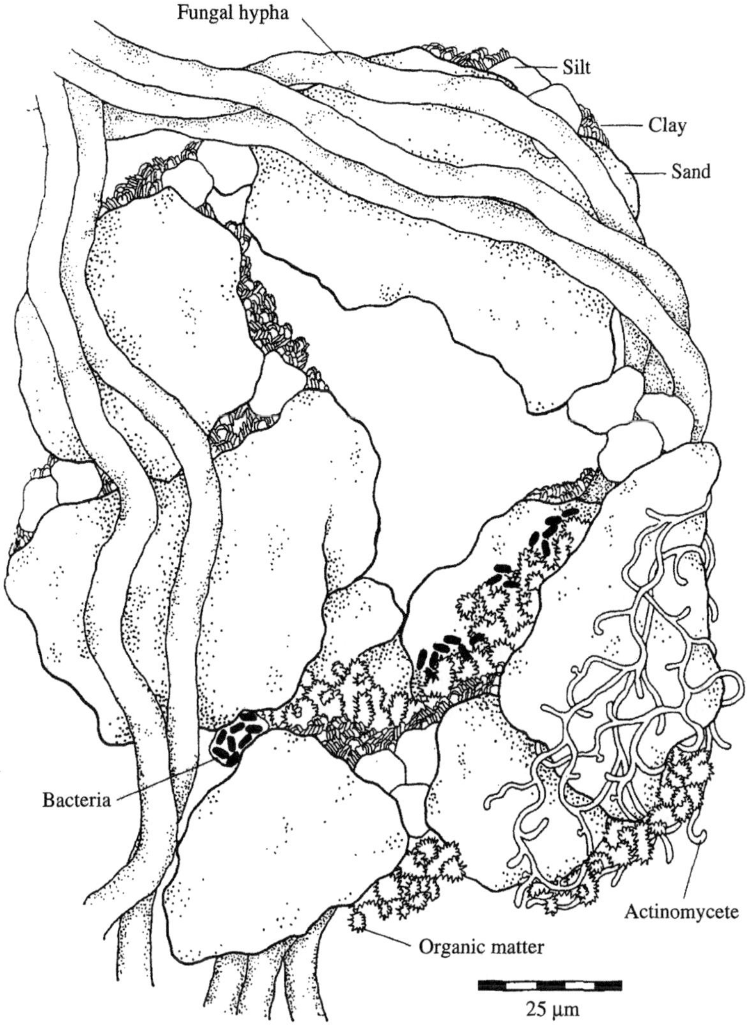

Figure 5 A typical soil aggregate. Sand, silt, and clay particles, cemented by organic matter, precipitated inorganic materials, and microorganisms, bind the soil particles together to form an aggregate. Original drawing by Kim Luoma (Fuhrmann, 1999) (reproduced by permission of Prentice Hall from *Principles and Applications of Soil Microbiology*, **1999**).

structure (Hill *et al.*, 2000). Torsvik *et al.* (1996) estimate that less than 0.1% of the microbes found in a typical agricultural soil can be cultured using current culturing techniques. A quote from Zak *et al.* (1994) expresses the difficulties inherent in studying microbial biodiversity: "we understand little about the degree to which genetic diversity is translated into taxonomic diversity, and even less about the manner in which genetic and taxonomic diversity affects functional diversity or ecosystem properties."

One culture-dependent method that is now being widely utilized in analysis of community-level physiological profiles (Garland and Mills, 1991; Zak *et al.*, 1994) is available commercially as the BIOLOG® system. This system is based on the microbial utilization (detected by the reduction of a tetrazolium dye) of a suite of 95 different carbon sources that can be categorized into several substrate guilds: carbohydrates, carboxylic acids, polymers, amines/amides, amino acids, and miscellaneous (Zak *et al.*, 1994). A major drawback of the commercially available substrates on the BIOLOG® plates is that the substrates are often not ecologically relevant and likely do not reflect the diversity of substrates found in the natural environment (Insam, 1997; Konopka *et al.*, 1998). With this and several other methodological considerations (Hill *et al.*, 2000) in mind, the BIOLOG® system has been extensively and successfully employed to characterize the functional diversity between both microbial communities and individual isolates (Garland, 1996; Lehman *et al.*, 1995; Zak *et al.*, 1994).

Culture-independent methods of soil microbial community analysis include methods based on: (1) the extraction, quantification, and identification of molecules (most commonly, phospholipids fatty acids (PLFAs) and nucleic acids) from the soil that are unique to certain microorganisms or microbial groups; or (2) fluorescence microscopic techniques where DNA segments are hybridized with labeled taxon-specific oligonucleotide probes and then scanned for cells that have incorporated the probe (Hill *et al.*, 2000). Fluorescent *in situ* hybridization (FISH) has proven to be a useful technique for direct identification and quantification of prokaryotic communities within their natural habitat (Amann *et al.*, 1995; Macnaughton *et al.*, 1996). A particularly powerful technique for assessing the decomposer community in terms of what organisms are using which substrate is to combine ^{13}C isotopic analysis with the PLFA analysis (Petersen *et al.*, 1997). These methods and applications are reviewed by Zarda *et al.* (1997), Atlas and Bartha (1998), Hill *et al.* (2000), and Ward *et al.* (1992).

Bacteria. The bacteria are single-celled prokaryotic organisms representing two of the three phylogenetic domains of life: the Bacteria and the Archea (Woese, 1987). Bacteria are a ubiquitous feature of terrestrial ecosystems—from the dry deserts of the Antarctic to the tropical rainforest soils of the Amazon. For a perspective on the diversity and abundance of prokaryotic life, see Whitman *et al.* (1998). Perhaps, most importantly, it is the variety of metabolic capabilities and the ability to rapidly reproduce and exploit a food source which make bacteria so crucial in the processes of OM decomposition, nutrient cycling, and soil formation (Sylvia *et al.*, 1999). In general, the Bacteria are mesophiles while the Archea are tolerant of harsh environments, with both utilizing an array of nutrient and energy sources (Paul and Clark, 1996). Several of the more important soil dwelling members of the Bacteria domain are the subgroups (Paul and Clark, 1996): purple bacteria (diverse group of gram negative bacteria), green sulfur bacteria (obligate anaerobic photolithotrophs), actinomycetes (slender, filamentous growth form), sporogenic bacilli (capable of forming an endospore), and the cyanobacteria (obligate phototrophs). Within the Archea, we find the extreme halophiles, methanogens, extreme thermophiles, and the thermoacidophiles (Noll, 1992).

Bacteria, as well as fungi, can only transport simple monomers and some dimers across their cell walls and as such must excrete a suite of extracellular enzymes to break down large and complex organic molecules (discussed in detail in Section 8.07.5). The production of extracellular enzymes by an individual microbe can be seen as a metabolically costly and potentially wasteful process. However, many bacteria have developed mechanisms for intercellular communication that allow homogenous populations to coordinate enzyme production and in a sense behave as a multicellular organism (Dworkin, 1998). Bacteria constantly excrete diffusible signals or "pheromones" into the environment which when at a critical concentration can induce or repress gene expression for a number of important traits (Gray, 1997; Kaiser and Losick, 1993). This density-dependent or "quorum" sensing mechanism allows for more efficient exploitation of heterogeneous carbon sources by timing the production of exoenzymes to times of high bacterial population density and substrate availability (Benedik and Strych, 1998; Chernin *et al.*, 1998). New research in the field of quorum sensing shows that plants can even excrete substances that mimic the bacterial signals (Teplitski *et al.*, 2000) which can have profound impacts on how we picture nutrient cycling.

Fungi. The fungi comprise a diverse group of multicellular eukaryotic organisms that exhibit an amazing array of morphologies, both vegetative and reproductive, and of life cycles (Carlile and Watkinson, 1994; Cooke and Rayner, 1984; Dix and Webster, 1995). On a wet biomass basis, fungi are the most abundant soil organism with biomasses ranging from 1,000 kg wet biomass ha^{-1} to 15,000 kg wet biomass ha^{-1} (see Table 3). Additionally, fungi are the dominant agents of OM decomposition in most soils (Paul and Clark, 1996). As with the bacteria, the diversity and ubiquity of fungi cannot be overemphasized. Cornejo *et al.* (1994) found over 500 morphospecies from litter of only five different tropical tree species using a single culture medium. Perhaps the most striking contrast to the bacteria is the mycelial growth form that allows a fungus to grow indefinitely towards nutrient sources and provide some protection from adverse environmental conditions (Sylvia *et al.*, 1999). Additionally, the presence of unique reproductive structures in many taxa of fungi has allowed for simple isolation and determination of basic ecological characteristics. In fact, many of these general trends in fungal decomposer community dynamics first outlined in an excellent review by Hudson (1968), still hold today.

The decomposer fungi can be classified functionally by substrate-utilization/enzyme-production and by position, both spatially and temporally, along the SOM continuum. There exists an active biotrophic community of fungi on photosynthesizing leafs, such as the common leaf-surface fungus, *Aureobasidium pullulans*, which often become the first wave of primary saprotrophs as a leaf senesces (Swift *et al.*, 1979). After contact with the soil surface, the litter-dwelling primary saprophytic fungi, which can rapidly utilize simple and readily available carbohydrates, begin to colonize the leaf. These fungi are often referred to as the "sugar fungi" and are commonly members of the class Zygomycetes. After assimilation of these readily available energy sources, the cellulose decomposers (numerous genera of the ascomycetes) and associated secondary, often soil-borne, saprophytic sugar fungi that are dependent upon the hydrolytic products of the cellulose decomposers for nutrition become dominant (Garrett, 1981). Researchers often further distinguish the primary and secondary colonizers as surface or interior colonizers. Osono and Takeda (1999) summarize the differences between these groups: "The interior colonizers grow over the mesophyll and utilize such substrates as lignocellulose or readily available hexoses and pentoses. However, surface colonizers utilize two kinds of nutrients: endogenous substrates originating from relatively recalcitrant cuticular components above the epidermal cell walls, and exogenous substrates that deposit on the surface." The final stage in this generalized succession on leaf litter is the establishment of the lignin degrading and associated fungal communities (Garrett, 1981). This generalized succession has been demonstrated on many litter types including grain sorghum litter (Beare *et al.*, 1993), black alder litter (Rosenbrock *et al.*, 1995), and pine litter (Tokumasu, 1998).

Another important class of fungi are the Basidiomycetes which contain the dominant wood-rot species responsible for the decay of coarse woody debris (Paul and Clark, 1996). The wood-rot fungi are commonly divided into the white-rot and brown-rot fungi. The brown-rot fungi lack the ability to degrade lignin thus leaving behind a brown residue, while the white rots can simultaneously degrade lignin and complex carbohydrates (cellulose and hemicellulose) (Griffin, 1994). Paul and Clark (1996) also note that many of the basidiomycetes are facultatively parasitic—attacking and killing living trees and then growing as a saprotroph on the dead wood.

8.07.3.3 Soil Fauna

Soil animals exist.
I like soil animals.
They respire too little.
Ergo, they must CONTROL something!

Faunophilic logic by O. Andrén

The term soil fauna encompasses a broad range of organisms from single-celled protozoa to burrowing earthworms to highly social insects such as the termites to mammals, e.g., pocket gophers and ground squirrels. In this review, we will limit the discussion to select groups of soil invertebrates. Although playing a small role in the direct mineralization of OM to CO_2, soil fauna greatly affect the functioning of the decomposer microorganisms both directly and indirectly as a result of their feeding habits (Seastedt, 1984). There are four basic roles that individual soil fauna play in detrital processing: (1) increasing microbial access to litter through communition and fragmentation; (2) ameliorating harsh environmental conditions that would hinder microbial processing; (3) mineralizing immobilized nutrients by grazing on microflora and fauna; and (4) stimulation and dispersal of microorganisms.

Protozoa. These single-celled eukaryotic organisms, lacking adaptations for a truly terrestrial existence, are restricted in habitat to the thin films of water in the soil matrix. Their distribution in soil is largely influenced by pore space size with flagellates and small amoebae occupying the smallest and most abundant pores. Small pores often are inaccessable to predators and stay moist due to presence of capillary water, protecting these smallest protozoa from desiccation. Larger pores, home to most amoebae, ciliates, and testecea, are subject to increased moisture variations leading to periods of low activity and encystment (Bamforth and Lousier, 1995). Typical population densities range from $10^2\ g^{-1}$ soil for desert sites to $>10^6\ g^{-1}$ soil in humid forest soils (Lousier and Bamforth, 1990).

Soil-dwelling protozoa obtain most of their nutrition through predation on bacteria. As a result of this feeding habit, protozoa play two important roles in detrital processing: (1) increased mineralization of nutrients through bacterial predation and (2) transformation of bacterial biomass into higher trophic levels (Bamforth and Lousier, 1995). As an example, Lousier and Parkinson (1984) found that the mean biomass of testacea in an aspen woodland was $72\ g\ m^{-2}$, with an annual production of $206\ g\ m^{-2}$ resulting from the consumption of $1,377\ g\ m^{-2}$ of bacteria per year. These protozoa consumed 60 times the annual standing crop of bacteria in this system.

Nematodes. The importance of nematodes in decomposition and detrital processing is exemplified by the quote from B. G. Chitwood and M. B. H. Chitwood (1974): "Soil nematodes are of chief importance in the destruction of dead plant material, but they also take part in the decomposition of animal matter. Their inter-relationships with other organisms, both dead and alive, might easily be so great that if tomorrow they all disappear, a few weeks hence the foul odour of death might pervade the whole earth as the balance of life was destroyed." Nematodes occur in nearly all soils and feed on an incredibly diverse range of food. Sharma and Sharma (1995) identify three roles in decomposition: (1) the dispersal of microorganisms by carrying them on their body surface, by ingestion and subsequent defecation of bacteria and fungal spores, and by development of microflora in their faeces; (2) mineralization of nutrients by predation on microbes; and (3) their bodies act as stores of energy and nutrients for consumption by higher trophic levels.

The nematodes can be divided into groups based on general feeding habits. Figure 6 highlights the diversity in mouth parts and digestive systems found in the soil nematodes. R. Sharma and S. B. Sharma (1995) recognized five functional groups of soil nematodes: (1) Plant feeders (order Tylenchida) are obligate parasites that feed using a stylet to pierce plant cell walls, predominately root tissue, and to suck sap. Wasilewska (1991) found that phytophagous nematodes can consume 3–20% of net root production in a drained peat soil. (2) Fungal feeders also use a stylet to pierce and withdraw hyphae contents. Many fungal feeders can also parasitize roots. (3) Bacterial feeders, such as Rhabditida, ingest material of suitable size and shape, but will only digest bacteria and some decaying OM. (4) Omnifeeders consume whole diatoms, blue-green algae, bacteria, and protozoa by sucking food particles into their oesophagus. (5) Predatory nematodes (e.g., order Mononchida) possess powerful teeth, jaws, or denticles with which they puncture, shred, or bite prey. Prey include protozoa, other nematodes, rotifers, tradigrads, and even small oligochaetes.

Nematode distribution follows the distribution of organic carbon sources in the soil. In an evergreen oak forest, fungal and bacterial feeders dominate in the rapidly decomposing surface litter with populations decreasing with increasing depth in the soil (Tanaka *et al.*, 1978). In this study, Tanaka *et al.* (1978) also found that densities of nematodes in the surface litter varied by an order of magnitude between the wet and dry seasons. Actively feeding nematodes cannot tolerate desiccation and as such will be forced to either move away from dry regions or shift into a quiescent resting state. Starvation can also induce this anhydrobiotic state (Anderson *et al.*, 1981).

Oligochaeta. This class includes the Enchytraeidae (the white worms) and the diverse group commonly referred to as the earthworms. The main differences between these two groups can be considered in terms of ecological strategies. The enchytraeids are largely *r*-selected species with a small body size, rapid development, a short life span, and high reproductive output (Dash, 1995). Alternatively, the earthworms can be generalized as *K*-selected species with a relatively large body size and long life span.

The enchytraeids through their grazing on litter and fungal populations play an important role in decomposition. Besides the indirect effects of microbial grazing listed above, the enchytraeids possess gut enzymes, including cellulase, indicating that they also play a direct role in the decomposition of plant litter (Dash *et al.*, 1981). Populations reach maximums of 2×10^4 individuals m^{-2} in tropics and 10 times that number in many temperate regions (Dash, 1995). In a typical soil, an enchytraeid biomass of $\sim 6\ g\ m^{-2}$ can consume 3–20% of the annual energy input to the forest floor based on oxygen consumption data reviewed by Dash (1990).

Charles Darwin (1881) wrote: "It may be doubted whether there are many other animals which have played so important a part in the history of the world, as have these lowly organized creatures." The importance of earthworms in

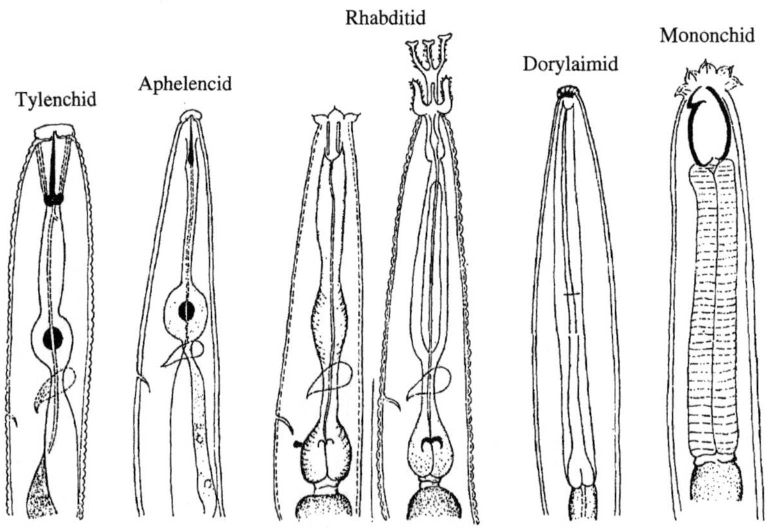

Figure 6 Diagram of mouth parts and oesophageal regions of selected soil nematodes (Sharma and Sharma, 1995) (reproduced by permission of Westview Press from *Soil Organisms and Litter Decomposition in the Tropics*, **1995**, pp. 75–88).

ecosystem functioning has indeed been recognized for centuries. Gilbert White (1789) wrote: "worms seem to eat earth, or perhaps rotten vegetables turning to earth... they perforate, loosen, and meliorate the soil, rendering it pervious to rains, the fibres of plants, and... without worms, perhaps vegetation would go on but lamely."

Edwards and Bohlen (1996) identified five major terrestrial earthworm families, with the Megoscolides (important in subtropical and tropical environments) and the Lumbricidae (temperate systems) being the most common. Earthworms occur across the globe, but rarely in extreme environments—deserts, permafrost, high mountains, and other regions lacking a well-developed soil. Most terrestrial earthworms are intolerant of salts, sensitive to very low or high pH, high temperatures, low soil moisture, and very clayey soil textures (Lee, 1985)—thus creating limits on their range. In many temperate regions earthworms dominate the soil faunal biomass (see Table 3). Fragoso and Lavelle (1992) compared soil faunal communities across a range of tropical forests and found that earthworms account for 51% of total biomass, followed by termites at 13%. A comprehensive review of earthworm biology and ecology is beyond the scope of this chapter. The interested reader is referred to the texts of Lee (1985) and Edwards and Bohlen (1996). Here we focus our review on the importance of earthworms in detrital processing.

Bouché (1977) identified three dominant ecological groups for the European lumbricid (Table 5), which have since been widely adopted to describe earthworms in all regions of the world. The epigenic worms live in the litter or upper organic horizons and feed almost exclusively on plant litter. These litter feeders are primarily agents of communition and fragmentation. The endogenic worms live in the mineral soil and feed by indiscriminately ingesting mineral soil. The anecic earthworms live in the mineral soil in permanent vertical burrows and consume partially decomposed litter after pulling the litter into their burrows. Anecic earthworms can comprise 50–75% of worm biomass in temperate zone, while often less than 10% in dry tropics (Edwards and Bohlen, 1996). Anderson and Swift (1983) state that no tropical earthworms are known to draw leaves down into the soil, but Fragoso and Lavelle (1992) showed that some earthworm communities are dominated by epigenic and anecic worms in tropical rainforests. Lavelle (1988) further divided the endogenic category into polyhumic (ingest soil with high OM content), mesohumic (feed indiscriminately on both mineral and organic particles), and oligohumic (feed on deep soil poor in OM) feeders.

During a normal wet year (1,250 mm rainfall) at Lamto on the Ivory Coast of Africa, earthworm

Table 5 General diagnostic features of the major ecological groups of European lumbricid earthworms as described by Bouché (1977).

Diagnostic feature	*Epigeic species*	*Anecic species*	*Endogeic species*
Food	Decomposing litter on the soil surface; little or no soil ingested	Decomposing litter on soil surface, some of which is pulled into burrows; some soil ingested	Mineral soil with preference for material rich in organic matter
Pigmentation	Heavy, usually both ventrally and dorsally	Medium–heavy, usually only dorsally	Unpigmented or lightly pigmented
Size of adults	Small–medium	Large	Medium
Burrows	None; some burrowing in upper few cm of soil by intermediate species	Large, permanent, vertical burrows extending into mineral soil horizon	Continuous, extensive, subhorizontal burrows, usually in the upper 10–15 cm of soil
Mobility	Rapid movement in response to disturbance	Rapid withdrawal into burrow but more sluggish than epigeics	Generally sluggish
Longevity	Relatively short lived	Relatively long lived	Intermediate
Generation time	Shorter	Longer	Shorter
Drought survival	Survives drought in the cocoon stage	Becomes quiescent during drought	Enters diapause in response to drought
Predation	Very high, particularly from birds, mammals, and predatory arthropods	High, especially when they are at the surface; somewhat protected in their burrows	Low; some predation by ground-dwelling mammals and predatory arthropods

Source: Edwards and Bohlen (1996).

communities in this savannah contain 2,000–4,100 individuals ha^{-1}, weighing 350–550 kg fresh weight, which ingest 800–1,250 Mg dry soil ha^{-1} containing 14–15 Mg OM ha^{-1} (Lavelle and Martin, 1992). At this site, Lavelle (1978) estimated that upwards of 60% of the humic pool in the upper 10 cm of the soil passes through the gut of earthworms annually (Martin, 1991). However, these high activity levels only result in 5–6% of the total soil heterotrophic respiration due to low assimilation of the worms (2–18%) (Edwards and Bohlen, 1996). Scheu (1991) found that 37% of the total carbon loss from a geophagous earthworm, *Octolasion lacteum*, was respired and 63% was excreted as high-energy, water-soluble mucous compounds. Most of the ingested soil material is excreted in discrete casts enriched in OM relative to the bulk soil. The sizes of the casts range from a few millimeters to several centimeters in length (Lee, 1985). The implications of these castings on detrital processing are discussed in detail in Section 8.07.7.

Most endogenic species form mutualistic relationships with microflora enabling digestion of low-quality substrates (Barois and Lavelle, 1986; Martin *et al.*, 1992). Gut microorganisms are generally the same species found in soils; however, many earthworms also produce cellulose and chitin degrading enzymes (Edwards and Bohlen, 1996). Kristufek *et al.* (1992) examined the bacterial counts in the fore-, mid-, and hindguts of a detritivore *Lumbricus rubellus* and a geophagous earthworm *Aporrectodea caliginosa* and found that bacterial counts increased down the digestive pathway in the detritivore while they decreased in the geophage. Kristufek *et al.* (1992) concluded that the digestive canal could be considered as a fermentor of fresh plant residues for *L. rubellus*, while *A. caliginosa* cannot support a gut community due to the low quality of SOM and that this earthworm may, in fact, utilize the microbes as nutrient sources. Casts generally have higher fungi, actinomycetes, bacteria populations, and enzyme activity than surrounding soil (Dkhas and Mishra, 1986; Tiwari and Mishra, 1993). Microbial populations show a similar trend in burrows. Bhatnagar (1975) in a French grassland found 42% of soil aerobic nitrogen fixers, 13% of anaerobic nitrogen fixers, and 16% of denitrifiers associated with burrows.

Earthworm activity can influence microbial decomposition for a number of reasons (Lavelle and Martin, 1992): (i) there is a greater nitrogen concentration in casts; (ii) gut passage can stimulate dormant microbes; and (iii) increased bacteria/fungal ratio due to increase in soluble OM from mucus secretions. Trigo and Lavelle (1993) describe this relationship between earthworms and microflora as a mutualistic digestive system—the earthworms secrete mucus for the microorganisms in return for increased decomposition and nutrient release within the gut. A similar mutualistic relationship is believed to exist between plants and rhizosphere bacteria whereby the plants supply high-energy root exudates that the bacteria can rapidly mineralize, thus resupplying nutrients to the plant (Anderson *et al.*, 1981).

Litter arthropods. In general, the dominant groups of litter arthropods are the mites (order Acari) and Collembola, with the Acarina dominating in monocotyledonous and deciduous leaf litter and the Collembola dominating in coniferous litters (Reddy, 1995a). The dominance of these two groups of mesofauna is diminished in the tropics, where the influence of several groups of macroarthropods, notably the termites, are greater (Swift *et al.*, 1979). We will discuss the termites (order Isoptera) separately due to their disproportionate impact on decomposition processes in ecosystems where they are abundant.

Population densities of most litter arthropods are often correlated with substrate availability and microclimate. Many arthropods cannot tolerate dry conditions. Reddy and Venkataiah (1989) found that Collembola and Acarina were almost completely absent from *Eucalyptus* leaf litter during the hot dry summer months in Warrangal, India. Ecologically, the litter arthropods can be divided by food preference: litter or microbial feeders. Rihani *et al.* (1995a) studied the food preferences of three oribatid mites on a decomposition gradient of beech litter from fresh litter to fungal mycelium. They found that the mite *Damaeus verticillipes* preferred fungal hyphae in the leaves (microphytophage); *Steganacarus magnus* preferred the easily digestible parenchyma tissue in the leaves (macrophytophage); and *Achipteria coleoptrata* was intermediate in food preference (panphytophage). These feeding habits are supported by the gut enzyme activities (see Section 8.07.5 for details on enzyme activities) of these three species—*S. magnus* showed only cellulase activity, while *D. verticillipes* showed chitinase and trehalase activity (Luxton, 1982).

Termites. The termites (order Isoptera), can be divided into six families (Kambhampati and Eggleton, 2000): Hodotermitidae (harvester termites); Termopsidae (rottenwood termites); Mastotermitidae (primitive Australian termites); Kalotermitidae (dampwood and drywood termites); Rhinotermitidae (subterranean termites); and Termitidae (higher termites). Approximately 85% of all known termites are considered higher termites which are differentiated phylogenetically because they harbor only bacteria in their gut instead of a mix of protozoa and bacteria (Kambhampati and Eggleton, 2000). Figure 7 highlights the dominantly tropical and subtropical distribution of termite diversity.

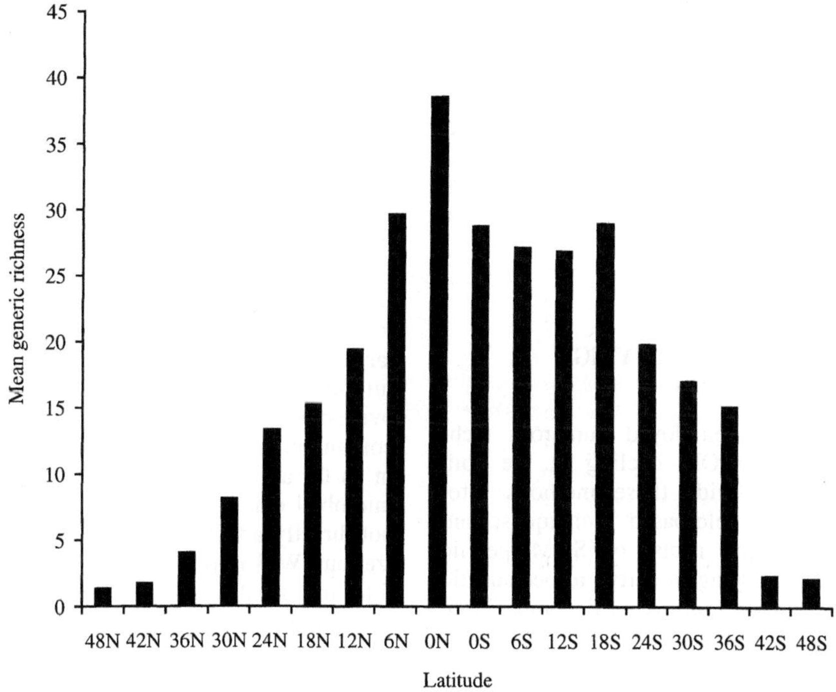

Figure 7 Latitudinal gradients of termite generic richness north and south of the equator (source Eggleton, 2000).

The Rhinotermitidae can range 1,000 km north of all other genera, because they make shelter tubes that keep them warm and moist (Eggleton, 2000). Termites are important agents in the decomposition of litter, wood, and soil humus.

The primary reason for the importance of termites in detrital processing stems from their symbiotic relationships with bacteria, fungi, and protozoa. Direct microscopic counts reveal densities of 10^9–10^{11} cells ml^{-1} of gut fluid, where often greater than 90% of the gut volume is occupied by protozoa in the lower termites (Breznak, 2000; Krasil'nikov and Satdykov, 1969). A very interesting and poorly understood observation is that these protozoa contain their own bacterial symbionts, often methanogens consuming the CO_2 and H_2 produced by the protozoa (Breznak, 2000; Lee *et al.*, 1987). These gut symbioses are discussed in length in the text of Abe *et al.* (2000). Globally, the fermenting activity of termite-gut symbionts may add 10–50 Tg CH_4 to the atmosphere annually (Houghton, 1996).

The higher termites, especially the Macrotermitinae, have evolved symbiotic relationships where they cultivate a fungus (*termitomyces* spp.) on elaborate combs within more or less developed mounds or subterranean nests. Jones (1990) described this cultivation of fungus as a "short-circuiting" of the nutrient cycling and soil food web, resulting in greatly enhanced decomposition rates. These termite-fungal systems can process an extraordinary amount of litter. For example, Buxton (1981) observed that Macrotermitinae reprocessed greater than 90% of the dry wood in a semi-arid savanna in Kenya; Collins (1981) reported that 25% of the total litter fall (leaf + wood) was consumed annually by termites in a Nigerian savanna; and Whitford *et al.* (1982) estimated that termites may be responsible for upwards of 100% of all litter consumption in many dry regions of the world. Collins (1981) concluded that mound structure determines seasonality of feeding habits, with the more complex mounds having the most stable microclimate. *Macrotermes bellicosus* builds the most complex nest in the Nigerian savanna: the individual colonies of *M. bellicosus* can cultivate a fungal comb 3 m in diameter with a mound reaching 6 m in height (Collins, 1979).

8.07.3.4 Interactions

It is often tempting to describe decomposition as a purely biochemical process where enzymes catalyze the chemical breakdown of complex organic molecules. By ignoring the soil organisms, we would not only be missing a beautifully complex ecological system but also be missing important interactions that can have a profound effect on decomposition and nutrient cycling. In terms of carbon mineralization, microorganisms are the primary agents of decomposition and are often responsible for greater than 90% of the total heterotrophic respiration (Foissner, 1987). Soil fauna, alternatively, play a minor direct role in carbon mineralization but can have major indirect impacts on decomposition rates through their actions, including—communition and fragmentation of leaf litter; mixing of litter into the mineral soil; dramatically changing soil physical properties; predation on life plant tissue and on microorganisms; stimulation and dispersal of microbial communities; and mineralization of immobilized nutrients (see references within).

8.07.4 METHODS FOR STUDYING DECOMPOSITION

Researchers have employed numerous techniques for studying OM cycling in the soil. We can broadly divide these methods into: (1) laboratory- or field-based techniques; and (2) litter- (including roots) or SOM-specific techniques. When trying to derive a mechanistic understanding of the controls on decomposition, Kirschbaum (2000) considered laboratory incubations to give a truer indication of the response to a manipulation of a single variable. By working in a highly controlled environment, laboratory studies will minimize the indirect effects of that manipulation which can often swamp the direct response. A simple example of this is that in Mediterranean environments soil moisture levels will often vary inversely with temperature (Xu and Qi, 2001). However, due to the numerous indirect controls on decomposition and SOM stabilization (Lavelle *et al.*, 1993; Sollins *et al.*, 1996), the true microbial response to a single variable measured in the laboratory may never be realized. The choice of field or laboratory methods will thus depend on the scope of the study: for microbial physiological studies, laboratory studies may be more appropriate, while field-based techniques may be more appropriate for deriving empirically based biogeochemical models of detrital processing.

8.07.4.1 Litter Techniques

In regions of the world where litter accumulates on the forest floor, annual decay rates can be estimated simply by measuring the litter fall (L) and the standing stock of recognizable litter (Olsen, 1963):

$$k = \frac{L}{X_{ss}} \qquad (1)$$

where X_{ss} is the steady-state forest floor mass. More precisely, the mass/nutrient loss of a single cohort of litter can be followed through time and k can be estimated based on the shape of the decay curve. This is the basic premise behind both litterbag and isotopic labeling techniques for studying litter decomposition both in the field and in laboratory microcosms.

Confining a known mass of litter in a mesh bag and periodically measuring the mass loss and nutrient concentrations has proven to be an easy and cost effective method in litter decay studies. In addition, confining litter in bags with varying mesh sizes is an effective method for measuring the impacts of soil fauna on detrital processing (Seastedt, 1984). Seastedt (1984) presents a nice conceptual model of faunal influences on mass loss in a litterbag (Figure 8). A drawback of using large mesh sizes is that undigested fragments of litter can be lost from the bag, thus leading to an overestimate of the decomposition rate. Although communition is a critical aspect of decomposition in as far as it exposes more surface area to the microbial community, communition *per se* does not directly affect mass loss or nutrient mineralization (Witkamp and Olson, 1963). Confining litter in a small mesh bag may also affect fungal productivity (St. John, 1980); however, decay rates obtained from litterbag studies usually agree with results from laboratory microcosms (Seastedt, 1984).

Isotope techniques for studying litter decomposition include both natural ^{13}C abundance experiments (Balesdent *et al.*, 1987) and ^{14}C and ^{15}N labeling experiments (Durall *et al.*, 1994; Harmon *et al.*, 1990b; Scheu, 1992; Van Veen *et al.*, 1985). Natural abundance experiments take advantage of the 12–15‰ difference in the $\delta^{13}C$ value of vegetation from different photosynthetic pathways. $\delta^{13}C$ (‰) is defined as

$$\delta^{13}C = \left(\frac{R_{sample}}{R_{standard}} - 1\right) \times 1{,}000 \qquad (2)$$

where R is the ratio of $^{13}C/^{12}C$ and the standard is Pee Dee Belemnite (PDB). For example,

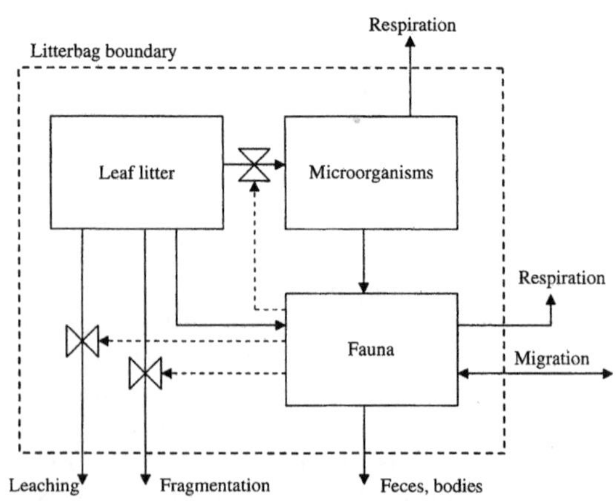

Figure 8 Conceptual model of mass loss of litter contained in litterbags. Solid arrows represent direct flows and indirect regulation by soil fauna is indicated by the dashed arrows (after Seastedt, 1984).

unconfined C4 grass litter ($\delta^{13}C \approx -13‰$) can be incubated with soil derived from a C3 forested site ($\delta^{13}C \approx -26‰$) and the CO_2 evolved can be divided into litter- and native SOM-derived CO_2 using a two-member mixing model:

$$f_{\text{litter}} = \frac{\delta_{CO_2} - \delta_{\text{soil}}}{\delta_{\text{litter}} - \delta_{\text{soil}}} \quad (3)$$

where f_{litter} is the fraction of litter-derived CO_2 and δ_{CO_2}, δ_{soil}, and δ_{litter} are the $\delta^{13}C$ of the evolved CO_2, the soil, and the litter, respectively (Balesdent *et al.*, 1990). ^{14}C-labeling experiments follow the same premise, but the ^{14}C-CO_2 evolved can be measured directly by scintillation counting (Dalias *et al.*, 2001b).

In studies of mass loss through time decomposition rate constants are not measured directly, necessitating the modeling of mass loss versus time to calculate a k value. The simplest model for this relationship is a zero-order decomposition model where mass loss progressively decreases with time. However, this model often does not capture observed trends and will yield unrealistic predictions as time gets large (Andren and Paustian, 1987). Most often, decomposition dynamics are represented as a first-order process—i.e., decomposition is proportional to the amount of material present:

$$M_t = M_0 e^{-kt} \quad (4)$$

Many authors find that a better fit is achieved with a parallel first-order model where the carbon substrate is divided into pools of similar decomposability (Dalias *et al.*, 2001a). For example, a two-pool model with a labile, l, and recalcitrant, r, component:

$$M_t = M_0 \alpha e^{-k_l t} + M_0(1 - \alpha)e^{-k_r t} \quad (5)$$

where α is the empirically derived labile carbon fraction of M. It is important to keep in mind that the α values obtained using a model such as Equation (5) to fit mass loss data do not necessarily represent any distinct pool of carbon that can be isolated in the laboratory.

Litterbags have become the primary tool for studying litter decomposition and nutrient dynamics in field studies. Although both litterbags and labeling studies are used in the laboratory, isotope labeling methods offer several advantages over litterbags. The decomposition products can be followed as they are incorporated into the mineral soil as more humified materials (Christensen and Sorenson, 1985). Additionally, researchers have labeled individual plant components, such as lignin (Scheu, 1992) and cellulose (Amato and Ladd, 1980).

8.07.4.2 SOM Techniques

Given that soils only contain a few percent OM by mass, direct measurement of SOM mass loss is much more difficult than for litter studies. The four most common techniques for measuring soil organic carbon (SOC) turnover are: (1) laboratory incubations; (2) *in situ* soil respiration measurements; (3) stable isotope measurements; and (4) radiocarbon methods. We will introduce the basic methodology and highlight some of the strengths and weaknesses of each of these four techniques. For much more detailed discussions of these methods, see the references cited within.

Laboratory incubations. Incubating sieved and homogenized soils in the laboratory under controlled moisture and temperature conditions and measuring the CO_2 evolution over a period of time is a simple and effective means of calculating the decomposition rate of SOM (Townsend *et al.*, 1997). CO_2 evolution can be measured either by trapping the CO_2 released in an airtight chamber using an alkaline absorber such as granular soda lime (Edwards, 1982) or NaOH solution (Leiros *et al.*, 1999), or by periodically measuring the CO_2 efflux rate via an infrared gas analyzer (IRGA) (Fang and Moncrieff, 2001). Decomposition rates (k) can then be calculated either by following the drop in respiration rates with time or by assuming steady-state conditions and dividing the respiration rate by the stock of carbon in the incubation (similar to Equation (1)). Theoretically, most of the initial CO_2 flux in an incubation experiment should be dominated by mineralization of the labile pool of SOM. This pool will be rapidly depleted without continued plant inputs, leaving a background mineralization rate of the much larger stabilized SOM pool (Townsend *et al.*, 1997). This conceptual model is shown in Figure 9.

Kirschbaum (2000) believes that laboratory incubations give the truest microbial response to manipulations of temperature and moisture,

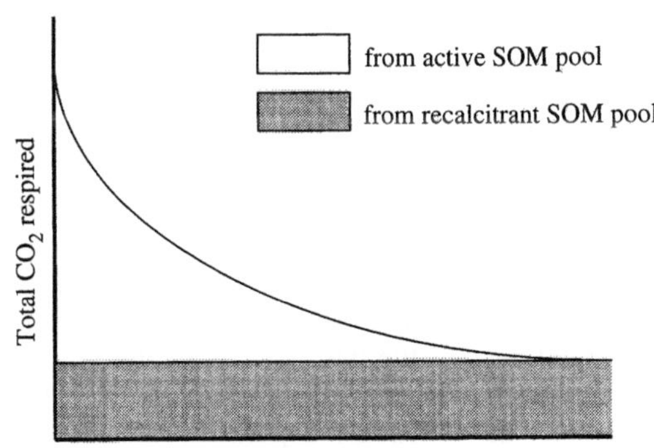

Figure 9 Theoretical contribution of active and recalcitrant SOM pools to total CO_2 respiration during a long-term laboratory incubation (after Townsend *et al.*, 1997).

because by sieving the soil roots have been removed and soil macrostructure, which can limit microbial access to carbon (see Section 8.07.7), has been destroyed. In fact, Kirschbaum (2000) found that the temperature dependence of SOM decomposition was much more sensitive for laboratory techniques than for other techniques. However, this high-temperature sensitivity may never be realized *in situ* because of the complex array of direct and indirect controls on decomposition rates.

In situ *soil respiration measurements.* Respiration can be measured directly in the field to avoid the methodological artifacts discussed in the preceding section. A number of techniques have been used to measure the CO_2 efflux rate, including the soda lime and infrared gas analysis techniques mentioned above. Each of these techniques has its own strengths and weaknesses which have been compared and reviewed by several authors (Janssens *et al.*, 2000, 2001; Pongracic *et al.*, 1997). In order to get at the SOM decomposition rate, root respiration must be separated from microbial metabolism (i.e., heterotrophic respiration). This can be done analytically by assuming that roots contribute a fixed percentage to total soil respiration (Raich and Schlesinger, 1992) or experimentally by digging a trench around a plot and removing all aboveground vegetation (Boone *et al.*, 1998; Melillo *et al.*, 1996). Once the influence of roots is removed, decomposition rates can again be calculated using the simple steady-state model (heterotrophic respiration divided by the stock of carbon in the soil).

Stable isotopes studies. Numerous researchers have taken advantage of the difference in carbon isotope fractionation during photosynthesis of plants with C3 and C4 photosynthetic pathways as a natural tracer to study soil carbon dynamics (Balesdent *et al.*, 1987). If a C3 forest is converted to a C4 pasture or C4 agricultural field, the whole soil carbon turnover time (τ) can be calculated as

$$\tau = \frac{-X}{\ln((\mathrm{C3\text{-}C_{present}})/(\mathrm{C3\text{-}C_{initial}}))} \qquad (6)$$

where X is the time since conversion, $\mathrm{C3\text{-}C_{present}}$ and $\mathrm{C3\text{-}C_{initial}}$ are quantities of C3 forested-derived carbon at the present sampling (derived from a simple two-member mixing model similar to Equation (3)) and at time of conversion (usually an adjacent forested stand is used as a surrogate) (Giardina and Ryan, 2000). This technique can be expanded beyond the bulk soil by examining the amount of C4-C that has been incorporated into different size or density separates (Christensen, 1992). This method for calculating turnover time assumes that there is no fractionation as the new carbon is incorporated into the soil and that the soils are at steady state (i.e., the soils have not accumulated or lost any carbon since conversion).

An additional consideration is that if X is large enough (perhaps only 20–50 years depending on the climate), then much, if not all, of the original C3-derived carbon will have been mineralized long before the *present* sampling date (Neill *et al.*, 1996), thus leading to an overestimation of the turnover time.

Radiocarbon approaches. Carbon-14 has proven to be a particularly powerful tool for studying the dynamics of C cycling in soils (Amundson *et al.*, 1998; Trumbore, 1993). In particular, the "bomb-spike" of ^{14}C released to the atmosphere due to nuclear weapons testing has been utilized to study nearly all aspects of the biogeochemical cycling of carbon. In order to model the turnover of SOM using ^{14}C, several assumptions must be met (Gaudinski *et al.*, 2000): (i) each SOM fraction is homogeneous with respect to decomposition; (ii) newly fixed photosynthate reflects the ratio of $^{14}C/^{12}C$ of that year's atmosphere; and (iii) ^{14}C does not fractionate during respiration. Based on these assumptions, the fraction of modern carbon in the soil (post-1950) normalized to a standard, F_s, which would be expected given a specific decay constant, k, in any given year, can be modeled as (Trumbore *et al.*, 1996)

$$F_{s(t)} = \frac{C_{s(t-1)}F_{s(t-1)}(1 - k - \lambda) + (I)F_{\mathrm{atm}(t)}}{C_{s(t-1)}} \qquad (7)$$

where C_s is the carbon stock in a given pool at year t, λ is the radioactive decay of ^{14}C, and I is the annual carbon inputs to the whole soil. Figure 10 shows the atmospheric record of ^{14}C as well as the expected values for soil carbon pools with turnover times of 5 yr and 50 yr based on Equation (7). To better constrain this model, ^{14}C measurements of archived soils can be compared to modern samples (Trumbore *et al.*, 1996).

8.07.5 DETRITAL PROCESSING

In this section, we will highlight the major biogeochemical pathways effecting detritus throughout the decomposition process. Swift *et al.* (1979) identified three distinct processes that combine to result in what is generally termed decomposition: comminution, leaching, and catabolism. The relative importance of these three processes will depend on numerous factors, including the location and type of litter, the soil faunal community, and the climate regime. These controls on detrital processes will be discussed in length in Section 8.07.7. Here we will examine the rates of decomposition, the roles of comminution, leaching, and catabolism in detrital processing, and the fate of nutrients during the decay process.

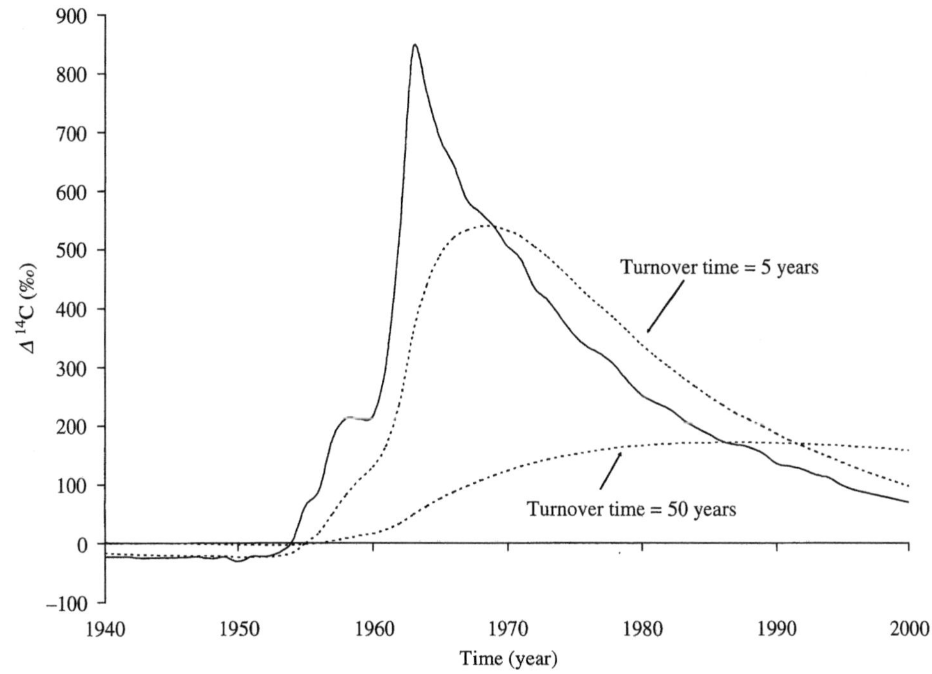

Figure 10 The atmospheric ^{14}C record in the Northern Hemisphere (solid line). Also shown are the modeled $\Delta^{14}C$ content of a homogenous, steady-state carbon pool with turnover times of 5 years and 50 years (Equation (7)) (dashed line).

8.07.5.1 Time Course of Litter Decomposition

Long-term studies of litter decomposition have been primarily conducted in temperate forested ecosystems (Aber *et al.*, 1990; Berg, 1984b; Berg and Staaf, 1980; Melillo *et al.*, 1989). In a 77-month litterbag study at the Harvard Forest, MA, Melillo and co-workers (Aber *et al.*, 1990; Melillo *et al.*, 1989) found that for two litters (red pine needles and paper birch leaves), decomposition can be described by a two-phase system: (i) a relatively constant fractional mass loss until ~20% of the original mass remains and (ii) a second phase of negligible rates of mass loss (Figure 11). Decay in phase I was best described by an exponential curve for the more palatable paper birch foliage, while the lignin-rich red pine needle was best fit by a simple linear function (see Section 8.07.7.2 for the influence of resource quality on decomposition rates). Melillo *et al.* (1989) developed a simple decomposition model where the initial litter quality (i.e., nutrient concentrations and lignin to nutrient ratios) controls the decay rate until a point where the lignocellulose index (LCI) (ratio of lignin to lignin + cellulose) reaches 0.7–0.8, at which time the factors governing the decay of lignin will control the overall decomposition rate. Berg (1984b) found that decomposition of Scots pine root litter also followed this general two-phase model.

When litter accumulates to form a well-stratified organic layer on the soil surface (mor soils), litter depth can be used as a surrogate for time of decay (Gourbière, 1982; Scheu and Parkinson, 1995). This allows for practical study of the products of late stages of decay. Gourbière (1982) was able to identify 10 years worth of decaying Spruce needles in the organic horizons of a cool temperate forest (Figure 12). An initial phase of rapid decomposition occurred in years 0–2.5. This was followed by a long phase of relative inhibition of decomposition (years 2.5–8.5). Finally, a third stage consisting of rapid mass loss occurred as a result of the establishment of a community of white-rot fungi. If edaphic conditions favor bacteria over fungi, this third phase may not be present in a particular ecosystem.

Baldock *et al.* (1992) presented a model describing decomposition based on CPMAS ^{13}C NMR spectra obtained for decreasing particle size fractions (Figure 13). Decreasing particle size fractions have commonly been employed as a surrogate for extent of degradation, with fresh plant residues being associated with the largest size fractions (>20 μm), partially degraded residues in the intermediate fractions (2–20 μm), and the most humified material associated with the finest fractions (<2 μm) (Christensen, 1992). As shown in Figure 13, the O-alkyl carbon peak at 73 ppm

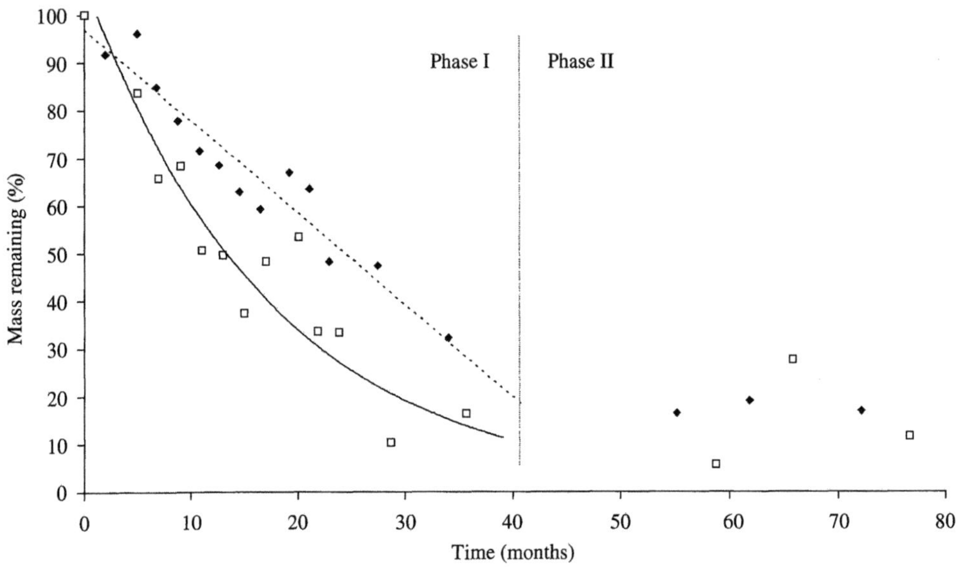

Figure 11 Percent of original mass remaining for red pine needle litter (filled diamonds) and paper birch foliage litter (open squares) versus time in months (sources Melillo *et al.*, 1989; Aber *et al.*, 1990).

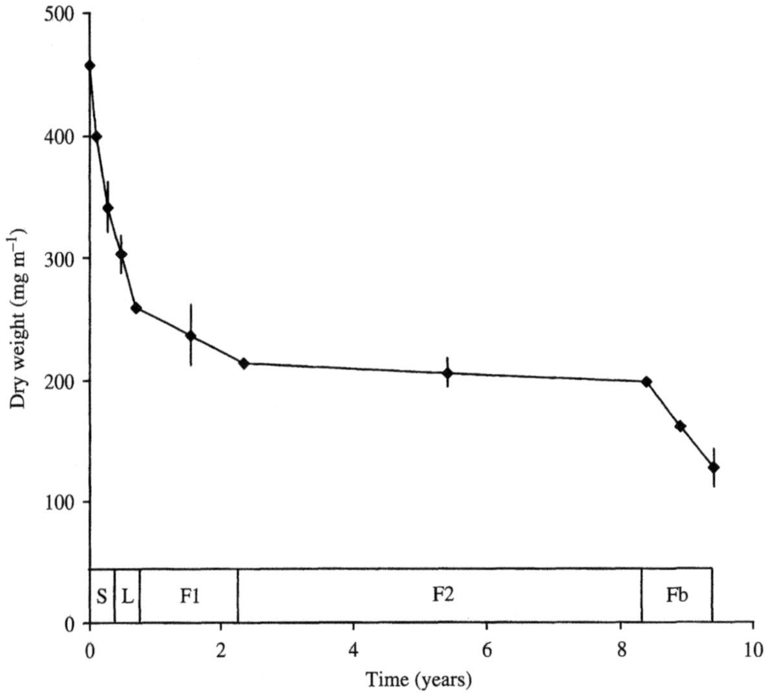

Figure 12 Decomposition dynamics of Spruce needles as indicated by increasing depth in the litter layers (S, L, F1, F2, and Fb). Fb = F layer that has been invaded by white-rot fungi (after Gourbiére, 1982).

(simple carbohydrates, cellulose, and hemicellulose) dominated the ^{13}C NMR spectra for the largest size fraction. Baldock *et al.* (1992) proposed that the relative increase in alkyl (32 ppm) and aromatic carbon (132 ppm) in the intermediate size fraction was due to the preferential degradation of carbohydrates and proteins with the selective preservation of more recalcitrant plant polymers. Microbial biomass was thought to be reason for the continued presence of O-alkyl carbon in this fraction. In the finest size fraction, aromatic carbon was lost indicating substantial lignin degradation. Baldock *et al.* (1992) proposed that the dominance of alkyl carbon in this fraction was due to the microbial synthesis of highly resistant compounds.

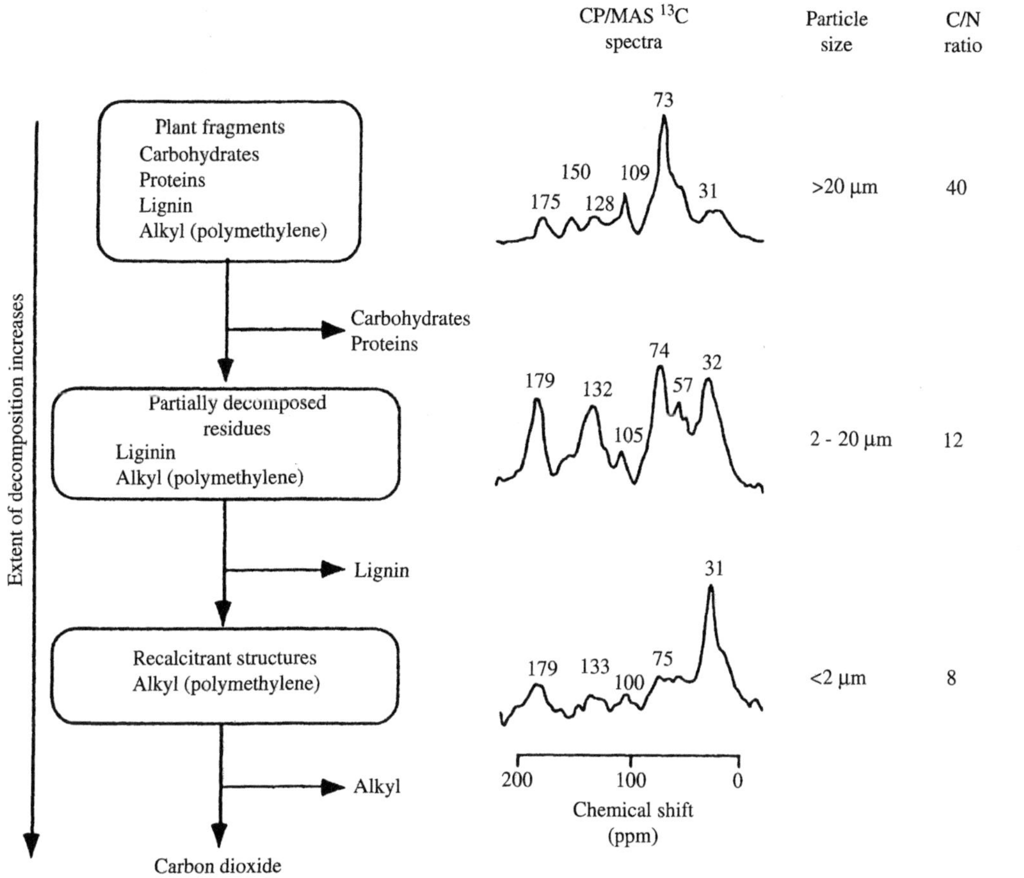

Figure 13 Decomposition model of Baldock *et al.* (1992) based on CPMAS ^{13}C NMR spectroscopy of a Mollisol (reproduced by permission of Kluwer Academic Publishers from *Biogeochemistry*, **1992**, *16*, 1–42).

8.07.5.2 Comminution

The physical fragmentation of coarse litter by the feeding habits of soil invertebrates can significantly enhance the accessibility of organic compounds to leaching (Reddy and Venkataiah, 1989) and microbial attack (Wolters, 2000). Comminution also stimulates microbes by increasing the surface area of litter available for colonization (Kheirallah, 1990). Maraun and Scheu (1995) found a large increase in the maximum initial respiratory response of the microbial community when beech litter was fragmented to $<25\ mm^2$. For a wide variety of deciduous litters, Seastedt (1984) found that microarthropods increased the average decay rate by 23% (litterbag exclusion studies lasting from 9 months to 30 months). The grinding of soil aggregates by geophagous fauna can also increase SOM accessibility by disrupting the bonds between soil particles (Lavelle and Martin, 1992; Wolters, 2000). Additional examples of soil faunal controls through comminution on decomposition can be found in Section 8.07.7.1.

8.07.5.3 Leaching

The infrequent sampling intervals in long-term litter decomposition experiments shown in Figure 11 missed a small, but significant, fraction of the mass loss by leaching which occurs in the first few days to weeks of decomposition. Leaching not only contributes to the initial phases of decomposition (Ibrahima *et al.*, 1995; Reddy and Venkataiah, 1989), but also transports organic C to underlying mineral horizons (Guggenberger and Zech, 1994). The degree to which leaching is a significant factor in the decomposition and stabilization processes depends on the local soil moisture regime, hydrologic properties, and the nature of the litter and its reaction products.

Leaching is often implicated as a reason for the faster than predicted loss based on exponential models (MacLean and Wein, 1978; Parsons *et al.*, 1990). These models otherwise yield reasonable fits with most litter decomposition data (Taylor and Parkinson, 1988b). Mass loss attributable to leaching and that due to the microbial metabolism of water-soluble compounds are often hard to separate (Andren and Paustian, 1987). Laboratory studies where fresh litter is submerged in deionized water provide data on leaching losses (Nykvist, 1959, 1961; Taylor and Parkinson, 1988b). Water absorption and leaching loss curves are shown in Figure 14 for aspen and pine litter. The striking differences in the curves for aspen and pine litter can be best explained in terms of the complexity of leaf architecture. In aspen leaves, water only has to penetrate the cuticle to access hydrophilic compounds, while in pine needles, lignified tissues protect much of the potentially soluble materials (Taylor and Parkinson, 1988b).

Because low-molecular-weight compounds such as sugars, simple carbohydrates, and ionic nutrients are preferentially leached from fresh litter (Berg and Wessen, 1984), the remaining chemical components become more recalcitrant. This will alter the decomposability of the remaining litter (Figure 15). Parsons *et al.* (1990) found that the preleached litter had lost 32% of total mass and nearly 60% of labile materials, and decomposed at a significantly reduced rate relative to the intact litter. Using a single exponential model, the underestimate in the intercept for the unleached aspen litter corresponded closely with the amount of mass loss that Parsons *et al.* (1990) attributed to leaching.

Physical fragmentation of litter also influences the leaching rates. By comparing mass and nutrient loss rates from litter in fine- and coarse-meshed litterbags that were both suspended and on the ground, Reddy and Venkataiah (1989) found that leaching losses were only significant in litter that was comminuted by soil fauna. Potassium is generally considered to be a highly mobile element that is not required by microorganisms to mineralize OM (Witkamp and Crossley, 1966). Potassium concentrations decreased nearly 50% in the first month for the litter altered

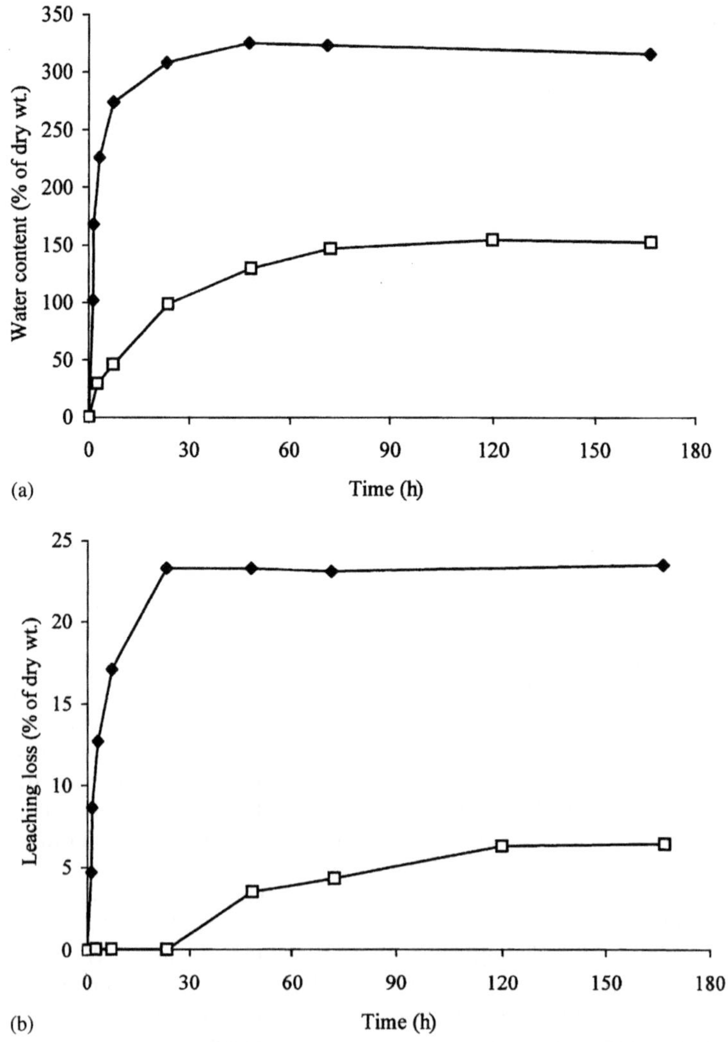

Figure 14 Rates of: (a) water absorption and (b) leaching loss from 5 g of pine needles (open squares) and aspen leaves (filled diamonds) immersed in 1.5 L of deionized water (after Taylor and Parkinson, 1988b).

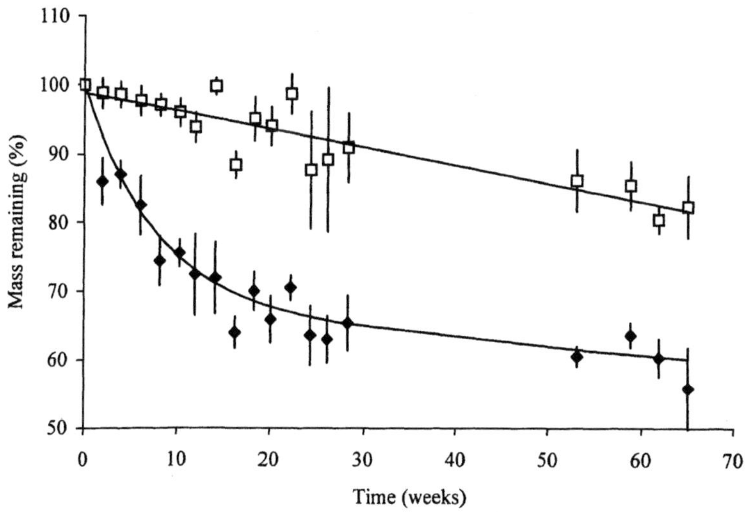

Figure 15 Mass loss (% original mass remaining) from intact (filled diamonds) and preleached (open squares) aspen leaf litter. Preleached litter had lost 31.7% of original mass before onset of incubation. A linear regression best described the mass loss of the preleached litter ($r^2 = 0.57$ $P < 0.05$), while a double exponential model (i.e., Equation (5)) best described the mass loss of the intact litter ($r^2 = 0.79$ $P < 0.01$) (after Parsons *et al.*, 1990).

by soil fauna, while litter in suspended litterbags only lost ~15% of the initial potassium concentration (Reddy and Venkataiah, 1989).

On longer timescales, leaching of DOM from the organic layer to the underlying mineral soil plays an important role in soil development and long-term carbon storage by transporting carbon deep within a profile. For three forest soils, Guggenberger (1992) estimated that between 66% and 91% of the annual carbon inputs to the mineral soil are due to DOC fluxes from the organic horizons (Guggenberger and Zech, 1994). Neff and Asner (2001) estimated that the annual carbon flux from the organic horizons in a temperate forest was 10–40 g DOC m^{-2}, which contributed 25% of the total soil profile carbon. Based on a review of DOC chemical characterization literature, Currie and Aber (1997) found that the majority of the forest floor leachate is comprised of fairly recalcitrant lignocellulose-associated materials and that only 15% could be considered labile materials. This DOC leaving the organic horizons is comprised mainly of partially degraded plant compounds and microbial biooxidation products, and can be considered important humus precursors (Guggenberger and Zech, 1994).

DOC transport through the soil and its concentration leaving a soil profile depends on abiotic sorption and desorption reactions with mineral surfaces. The tendency for organics to be strongly sorbed to soil particles through a variety of bonds can explain the order of magnitude drop in DOC fluxes in subsurface horizons (Neff and Asner, 2001; Ugolini *et al.*, 1977). For example, at the Harvard Forest, Massachusetts, Currie *et al.* (1996) found that greater than 50% of the total dissolved nitrogen (TDN) and DOC leached from the organic horizons are retained in the underlying mineral soil (Table 6). Table 6 also highlights the fact that DON is the dominant form of nitrogen leached from this system. In a mixed hardwood forest in North Carolina, Qualls *et al.* (1991) found that although throughfall consisted of 50% inorganic nitrogen, leachate from the base of the Oa horizon consisted of >90% DON. Additionally, across a range of unpolluted South American forests, Perakis and Hedin (2002) showed that nitrogen losses were dominated by DON.

8.07.5.4 Catabolism

There are very few materials, natural or man-made, that microorganisms cannot degrade (Ratledge, 1994). Upwards of 95% of total heterotrophic respiration is derived from microbial mineralization of OM (Lavelle and Spain, 2001). However, fungi and bacteria are only capable of ingesting monomeric and some dimeric compounds. These organisms digest macromolecular complexes *in situ* by secreting numerous extracellular enzymes (Griffin, 1994). Extracellular enzymes catalyze the cleavage of large molecules into monomeric subunits which can then be transported into the cell where intracellular enzymes will help complete the metabolic process. Table 7 lists the major classes of enzymes and the reactions that each catalyze. Of these enzyme classes, the hydrolases are especially important because they catalyze many of the reactions involved in the biogeochemical

Table 6 Annual solute fluxes for a red pine and mixed hardwood stand in the Harvard Forest, MA.

Flux or flux difference	*NO_3-N* ($g\ m^{-2}\ yr^{-1}$)	*NH_4-N* ($g\ m^{-2}\ yr^{-1}$)	*DON* ($g\ m^{-2}\ yr^{-1}$)	*TDN*[a] ($g\ m^{-2}\ yr^{-1}$)	*DOC* ($g\ m^{-2}\ yr^{-1}$)
Red pine stand					
Throughfall	0.696	0.223	0.348	1.27	13.9
From Oa	0.604	0.138	0.953	1.70	39.8
From subsurface	<0.001	0.010	0.536	0.549	16.7
Illuviated and retained[b]	0.604	0.128	0.417	1.15	23.1
Hardwood stand					
Throughfall	0.488	0.181	0.268	0.938	11.7
From Oa	0.199	0.104	0.611	0.915	22.5
From subsurface	0.002	0.003	0.319	0.324	12.3
Illuviated and retained[b]	0.197	0.101	0.292	0.591	10.2

Source: Currie *et al.* (1996).
[a] TDN = total dissolved nitrogen [b] Material illuviated into mineral soil and retained there, calculated as flux from Oa minus flux from subsurface.

Table 7 Major classes and subclasses of enzymes and the corresponding types of reactions catalyzed.

Class	*Representative subclasses*	*Types of reactions catalyzed by enzyme class*
Oxireductases	Dehydrogenases Oxidases Reductases Oxygenases Peroxidases Catalases	Catalyze oxidation–reduction reactions. Important in fermentation and respiration pathways.
Transferases	Aminotransferases Kinases	Catalyze the transfer of molecular constituents among molecules.
Hydrolases	Glycosidases Peptidases Phosphatases Ribonucleases	Catalyze the hydrolytic cleavage of chemical bonds.
Lyases	Decarboxylases Synthases Lyases	Catalyze the addition or removal of chemical groups such as carbon dioxide, ammonia, and water.
Isomerases	Racemases Isomerases	Catalyze inversions at asymmetric carbon atoms and the intramolecular transfer of molecular constituents.
Ligases	Synthetases Carboxylases	Catalyze the binding of two molecules with the expenditure of ATP. Important in anabolic pathways.

Source: Fuhrmann (1999).

transformations of C, N, P, and S, and are an integral component of detrital processing (Taylor *et al.*, 2002).

The major biochemical constituents of detritus (i.e., water-soluble carbohydrates, cellulose, hemicellulose, lignin, proteins, phenols, lipids, waxes, and other secondary plant compounds) differ in their ease of microbial degradation (Minderman, 1968). Microorganisms degrading fresh litter will preferentially metabolize the easily accessible, high energy-yielding compounds first. Numerous field studies have found that lignin degradation is often delayed by several years (Berg *et al.*, 1982; Gourbière, 1982; Scheu and Parkinson, 1995) until the appropriate fungal communities become established in the litter and begin producing the enzymes required to degrade these complex substrates. Figure 16 illustrates this fact: crude proteins and soluble components are the first materials to be degraded and their relative proportions decrease rapidly; polysaccharides that comprise the semicrystalline cellulose (glucans) and hemicellulose (most of the nonglucan polysaccarides) remain in their initial abundances for over one year; and the absolute amount of lignin did not change for over two years (Berg *et al.*, 1982).

Microbial metabolism can be divided into two general functions, the anabolic building of structural and functional components of the organism and the catabolic extraction of energy through the breaking of chemical bonds. Swift *et al.* (1979) defined catabolism as "the biochemical term which describes an energy-yielding enzymatic reaction, or chain of reactions, usually

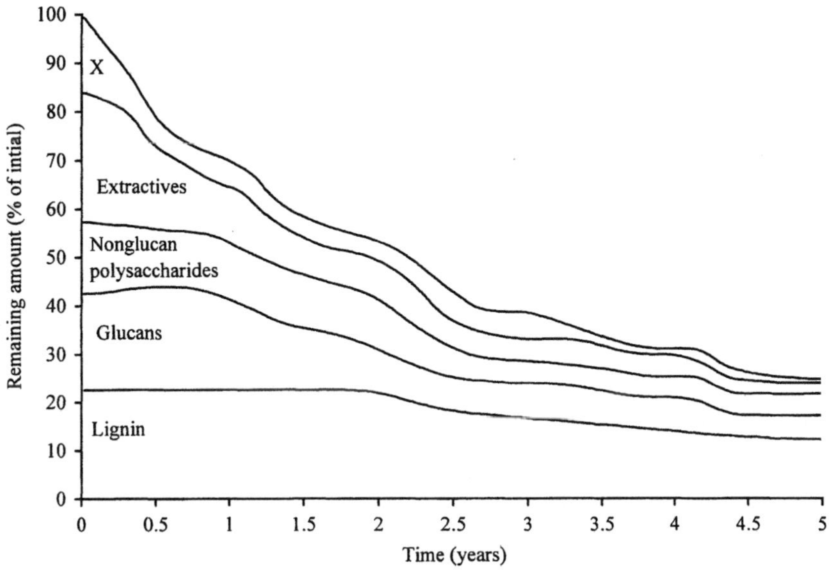

Figure 16 Changes in the composition of Scots pine needle litter during five years of decomposition. X corresponds to crude proteins, polyurinodes, ash, and unspecified products (after Berg *et al.*, 1982).

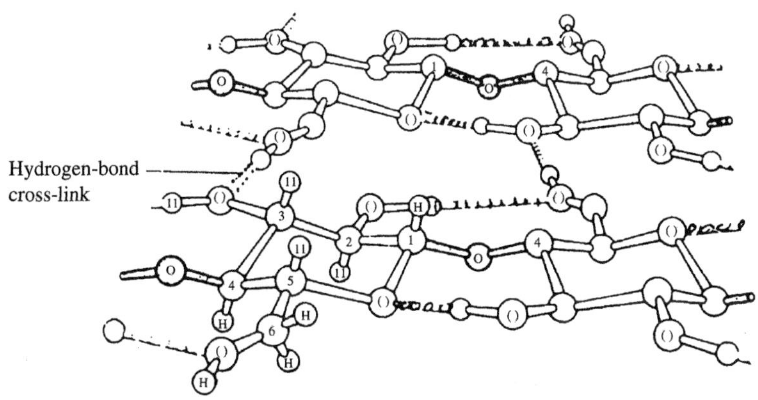

Figure 17 The structure of cellulose (Paul and Clark, 1996) (reproduced by permission of Elsevier from *Soil Microbiology and Biochemistry*, **1996**).

involving the transformation of complex organic compounds to smaller and simpler molecules." Anabolism, the synthesis of cellular materials from simpler metabolic and nutritive sources, is dependent upon catabolism for both energy in the form of ATP, NAPH, and NADPH, and the production of critical reaction intermediates (Griffin, 1994). More information on microbial metabolism and the biochemistry of microbial catabolism is available in the texts of Griffin (1994) and Ratledge (1994). In this section, we highlight some important molecular aspects of the biochemical degradation of the major plant and microbial materials.

Water-soluble organic compounds such as simple sugars, free amino acids, and organic acids are readily available for utilization by the vast majority of microorganisms. Root exudates also contribute significantly to the water-soluble pool of rapidly metabolized organic materials. Simple sugars and free amino acids can directly be transported into the cell and undergo glycolysis and then enter the appropriate metabolic pathway (Ramsh *et al.*, 1994; Wagner and Wolf, 1999). Depending on microbial species, disaccharides can be transported directly into the cell to undergo hydrolysis or the molecule can be hydrolyzed outside the cell and then transported (Griffin, 1994). For example, Sutton and Lampen (1962) found that the yeast *Saccharomyces cerevisae* converted sucrose to glucose and fructose at the cell surface then transported the monomeric compounds into the cell. In a similar manner, peptidase will catalyze the cleavage of the peptide bonds of polymeric proteins and then the individual amino acids can be transported into the cell.

The structural components of most plant cells contain most of their carbon in the form of complex carbohydrates, the most common being cellulose (Figure 17). Cellulose consists of a pseudocrystalline array of linear β-1, 4-glucan molecules, interspersed with amorphously linked regions, with hydrogen bond cross-links between chains (Paul and Clark, 1996). Individual cellulose molecules may contain upwards of 10^4 glucose units (Wagner and Wolf, 1999). Microbial cellulolysis is a complex process involving a number of extracellular enzymes that will vary with species (Wood and Garcia-Campayo, 1994). The aerobic soft-rot and white-rot fungi, and some aerobic bacteria, use a three-enzyme system consisting of an exoglucanase (normally cellobiohydrolase), endoglucanase, and β-glucosidase (Eriksson and Wood, 1985). Exoglucanase cleaves disaccharide cellobiose units from the nonreducing ends of cellulose chains and is mainly responsible for breaking interchain hydrogen bonds. Endoglucanase hydrolyzes the internal bonds of cellulose chains, while β-glucosidase hydrolyzes cellobiose to glucose (Griffin, 1994). Brown-rot fungi do not produce exoglucanases and likely use a decomposition mechanism involving H_2O_2 (Wood and Garcia-Campayo, 1994). The anaerobic bacteria use a membrane-bound multicomponent protein–enzyme complex called a cellulosome, of which the detailed biochemistry is just beginning to be worked out (Schwarz, 2001).

While the composition of cellulose is consistent between groups of plants, the composition of hemicellulose and lignin can vary dramatically. Generally found in association with cellulose in the secondary walls of plants, hemicelluloses are branched heteropolymers containing several simple sugars, including xylose, mannose, glucose, galacatose, arabanose, and glucuronic acid (Griffin, 1994; Wagner and Wolf, 1999). Hardwood hemicellulose is often comprised of a glucoxylan backbone with uronic acid branches (Figure 18), softwood hemicellulose is comprised of glucomannan backbone with uronic acid branches, and grasses (graminaceous plants) can have either of these general configurations (Dekker, 1985; Jeffries, 1994). Numerous endo- and exo-xylanases and mannanases acting on specific cross-linkages have been characterized (Jeffries, 1994). Several accessory enzymes including esterases, arabinofuranosidases, and glucuranosidases are also required for hemicellulose biodegradation (Jeffries, 1994). Despite the complex array of enzymes required to degrade hemicellulose, hemicellulose is often observed to decay at a rapid rate (e.g., Figure 16).

Lignin is synthesized by a polycondensation process involving free radicals in plant cells from three phenol propane building blocks: coumaryl alcohol, coniferyl alcohol, and sinapyl alcohol (Jeffries, 1994). Due to the random nature of its synthesis, lignin does not show a specific structure (Paul and Clark, 1996) and the relative proportions of the phenol propanoid units will vary according to plant source (Griffin, 1994). However, a generalized structure highlighting the propanoid units and the various linkages and side chains can be drawn (Figure 19). In plant cells, lignin is the impregnating material that protects the cellulose and hemicellulose matrix from enzymatic attack (Griffin, 1994). Lignin decomposition differs greatly from the biodegradation of all other plant materials for several reasons. First, only a particular group of basidiomycete fungi called the white-rot fungi have been shown to completely mineralize lignin to CO_2. Second, lignin is degraded by oxidative rather than hydrolytic attack. Third, lignin degradation requires an accompanying cometabolizable substrate to provide the carbon and energy for fungal growth (Buswell, 1991). Additionally, strict aerobic conditions and nitrogen limitations have been shown to be required for active lignin degradation (Buswell, 1991; Fog, 1988). Because of the energy costs involved in its decomposition, lignin decomposition probably only occurs because it is protecting a substantial amount of simpler carbon- and nitrogen-containing materials. Before lignin degradation can begin, the cellulose-rich secondary wall must be at least partially degraded in order for the fungi to attack the lignin. Rihani *et al.* (1995b) found that lignin loss rates in beech litter increased significantly after two weeks of incubation when ~20% of the cellulose content had disappeared.

Three families of extracellular enzymes have been implicated in lignin degradation: lignin peroxidases (ligninases), manganese-dependent peroxidases, and phenol oxidases (Griffin, 1994). The ligninase enzyme system consists of a hemeprotein (iron containing) that interacts with H_2O_2 in cyclic oxidation–reduction reactions that results in the oxidative cleavage and removal of a variety of functional groups (Buswell, 1991; Kirk and Farrell, 1987). The manganese peroxidases are distinguished from lignin peroxidases by their requirement of Mn^{2+} and by their ability to catalyze the demethoxylation of aromatic methyl esters, to act as a methyl esterase, and to oxidize several phenols (Buswell, 1991). Lignin peroxidase uses the iron in the heme-protein as an electron donor, while manganese peroxidase, which also contains the heme-protein, uses manganese as the electron donor. White-rot fungi also produce several phenol oxidases, including the copper-containing laccase (Griffin, 1994). The exact biodegradative role of laccase and other phenol oxidases has been difficult to ascertain, because polymerization reactions dominate *in vitro* experiments (Buswell, 1991). *In vivo*, quinine-oxidoreductases are thought to limit phenol polymerization reactions (Buswell, 1991; Chung *et al.*, 2000).

In soils, microbial cell walls make up a significant portion of the organic nutrient reserves.

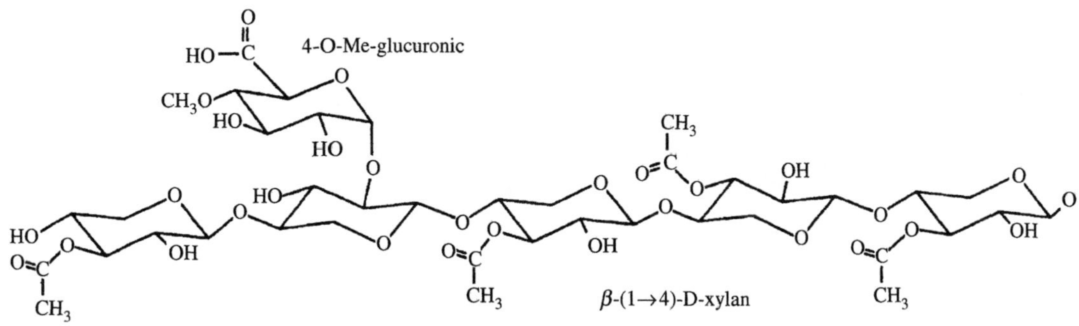

Figure 18 *O*-Acetyl-4-*O*-methyl-D-glucuronoxylan, a type of hemicellulose, structure from angiosperms (Dekker, 1985) (reproduced by permission of Academic Press from *Biosynthesis and Biodegradation of Wood Components*, **1985**).

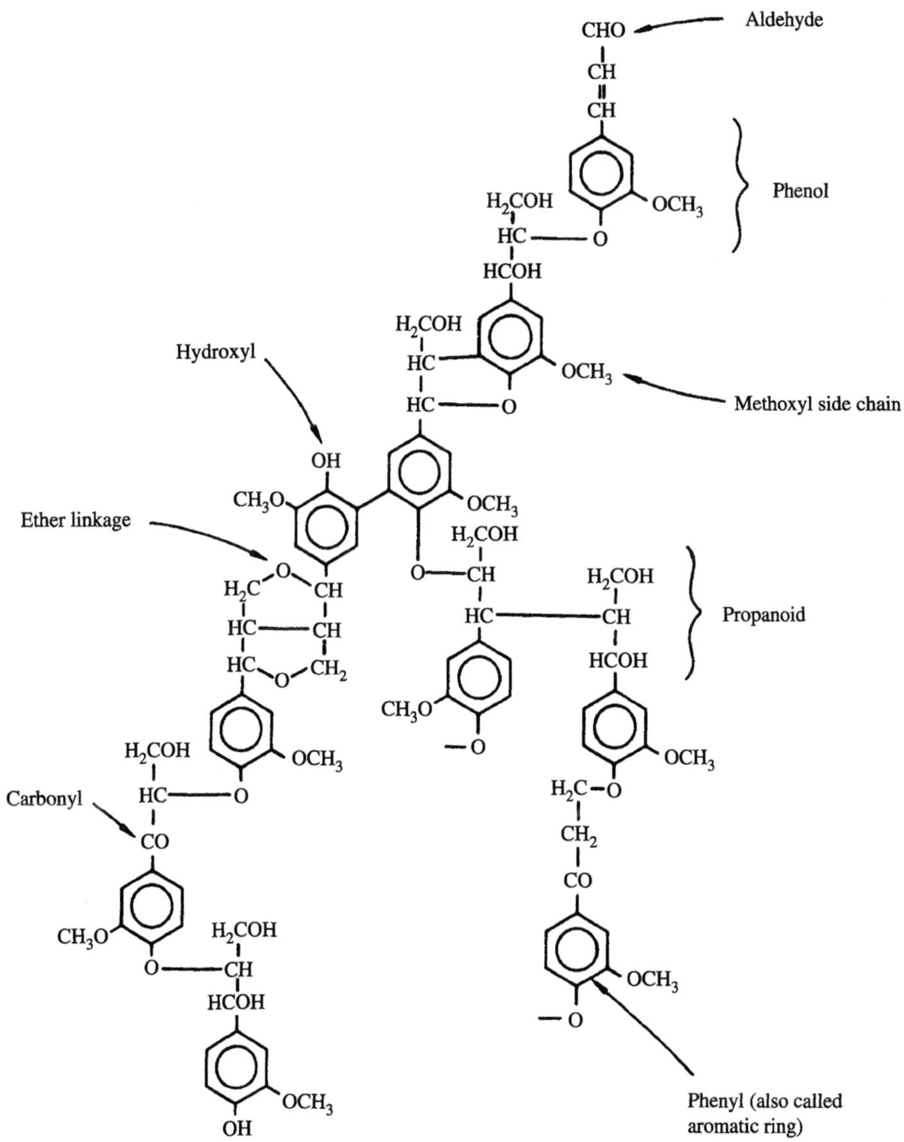

Figure 19 Generalized lignin structure, showing the common functional groups (Paul and Clark, 1996) (reproduced by permission of Elsevier from *Soil Microbiology and Biochemistry*, **1996**).

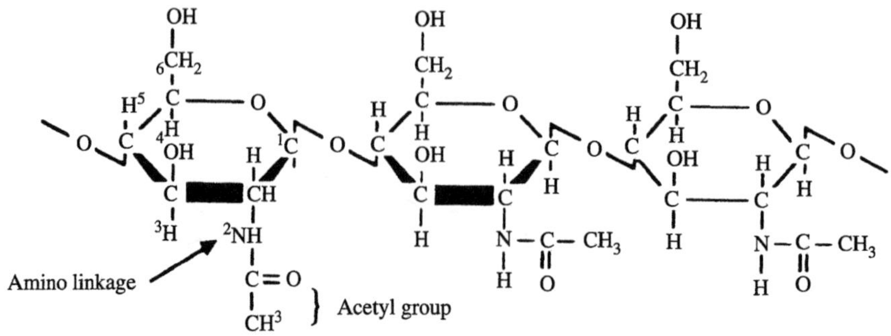

Figure 20 The structure of chitin (Paul and Clark, 1996) (reproduced by permission of Elsevier from *Soil Microbiology and Biochemistry*, **1996**).

Bacteria have a peptidogylcan cell wall composed of repeating units of *N*-acetylglucosamine and *N*-acetylmuramic acid joined by amino acids through peptide linkages (Paul and Clark, 1996). Fungal cell walls also contain significant amounts of chitin, a crystalline chain of *N*-acetylglucosamine with $\beta(1 \rightarrow 4)$ linkages (Figure 20). Many bacteria and fungi are capable of degrading microbial cell walls producing both glucanase and chitinase degrading enzyme systems. Most commonly, chitin is degraded by the hydrolysis of glycosidic bonds (Gooday, 1994): exochitinase cleaves diacetylchitobiose from the ends of chitin chains; endochitinase cleaves glycosidic linkages randomly along chitin chains; and the diacetylchitobiose will be hydrolyzed to *N*-acetylglucosimane by β-*N*-acetylglucosaminidase. An alternative pathway utilized by some microorganisms is via deacetylation of chitin to chitosan which is then hydrolyzed by chitosanase to chitobiose, which, in turn, is cleaved by glucosaminidase to yield individual glucosamine units (Gooday, 1994).

8.07.5.5 Change in Nutrient Status

The combined action of comminution, leaching, and catabolism results in patterns of mass loss such as shown in Figures 11 and 12. As the dominant element in plant tissues, carbon loss will mirror the mass loss. However, the rates at which other important nutrient elements are lost may vary significantly. Changes in nutrient status during decomposition can be generalized into three phases: initial leaching, bioaccumulation/immobilization, and final mineralization (Berg and Staaf, 1981; Cornejo *et al.*, 1994; Gosz *et al.*, 1973). Nutrients that are nonlimiting to microbial growth such as potassium, magnesium, and sodium will rapidly be leached and mineralized (Bubb *et al.*, 1998), while limiting nutrients such as nitrogen and phosphorus will follow this three-phase model (Staaf and Berg, 1982). Figure 21 highlights the differential behavior of nutrient loss for Scots pine litter decomposition in central Sweden. Gosz *et al.* (1973) hypothesized that when nutrient loss is greater than mass loss leaching must be the dominant removal pathway, and when nutrient loss is less than or equal to mass loss mineralization will dominate the eventual release of the nutrient.

Numerous researchers have reported a mobility series for major nutrients released during the decomposition of different litters (Table 8). In the chaparral of southern California, Schlesinger and Hasey (1981) found that after one year, potassium was almost completely lost from both *Salvia* and *Ceanothus* litter, phosphorus showed a 50% decline in the phosphorus-rich *Salvia* but actually bioaccumulated in the phosphorus-poor *Ceanothus* litter, magnesium and calcium were lost at a slightly greater rate than overall mass loss for both species, and nitrogen showed a net immobilization over the one-year study period. In a seasonally dry tropical forest, Cornejo *et al.* (1994) found that for several deciduous litter species potassium and phosphorus were rapidly leached, nitrogen was bioaccumulated, and magnesium and calcium concentrations increased because of abiotic exchange reactions. These authors concluded that the divalent cations, magnesium and calcium, were preferentially absorbed as the more mobile monovalent cations were leached from exchange sites on clays and OM. Calcium is also known to accumulate as calcium oxalate in fungal biomass reducing its relative mobility (Cromack *et al.*, 1975; Palm and Sanchez, 1990).

Gosz *et al.* (1973) found that carbon-to-nutrient ratios are critical in determining both the percent increase or decrease in a particular nutrient's concentration as well as the initiation of nutrient release. Figure 22 highlights the importance of the initial carbon-to-nutrient ratios in determining the final nutrient content after two years of decomposition (Rutigliano *et al.*, 1998). Researchers recognize a critical C : N ratio in forest soils

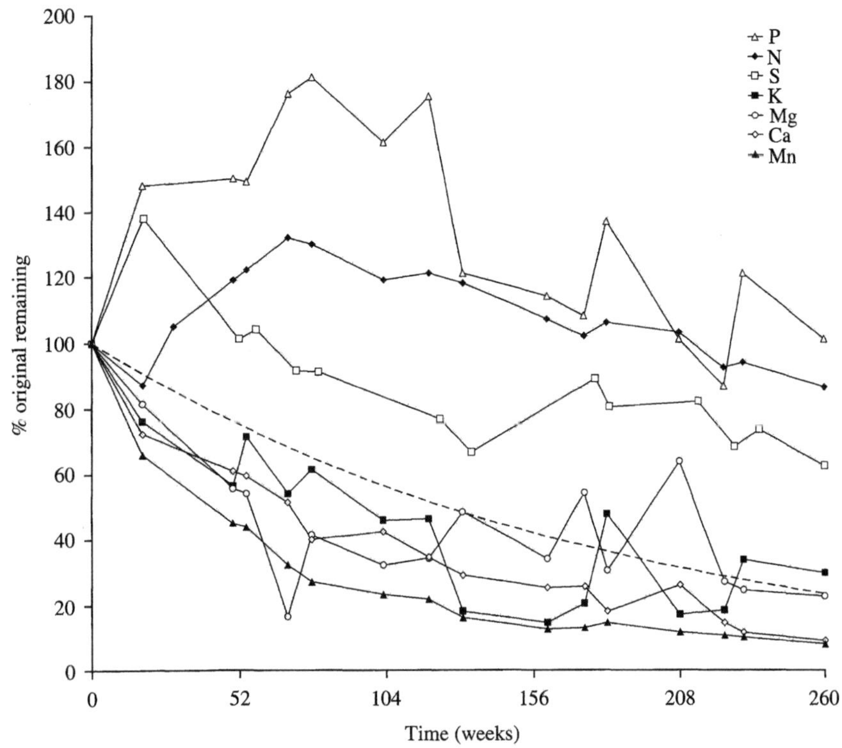

Figure 21 Changes in absolute amounts of plant nutrients for Scots pine needle litter during five years of decomposition. The dashed line shows the fitted exponential decay for mass loss of organic matter (after Staaf and Berg, 1982).

Table 8 Nutrient mobility series for decomposition of various litters.

Litter species	*Ecosystem*	*Ref.*[a]	*Mobility series*
Scots pine	Boreal forest	1	Mn > Ca > K > Mg > S > N > P
Mixed hardwood	Temperate forest	2	K > Mg > P > Ca ≈ S > C > N
Beech and fir	Temperate forest	3	K > P > Mg > Ca ≈ N
Ceanothus	Chaparral	4	K > Mg > Ca ≥ N > P
Salvia	Chaparral	4	K > P > Mg > Ca > N
Evergreen oak and stone pine	Mediterranean	5	K > P ≈ Mg ≈ C > Ca > N
Hoop pine	Subtropical forest	6	K > Na > C > Mg > P > N > Ca > Mn
Three legume species	Tropical plantation	7	K > P ≈ N ≈ Mg > C > Ca

[a] References: 1. Staaf and Berg (1982); 2. Gosz *et al.* (1973); 3. Rutigliano *et al.* (1998); 4. Schlesinger and Hasey (1981b); 5. Regina (2001); 6. Bubb *et al.* (1998); and 7. Palm and Sanchez (1990).

between 20 and 30 (Lutz and Chandler, 1946). At higher ratios nitrogen is immobilized, while at lower ratios nitrogen is released or mineralized. In most ecosystems, the preferential degradation of simple carbohydrates in the initial phases of decomposition leads to an increase in the concentration of nitrogen (Berg and Staaf, 1987; Lousier and Parkinson, 1978; Rustad, 1994; Rutigliano *et al.*, 1998). However, researchers have also found that the *absolute* quantity of nitrogen increases (Berg and Staaf, 1980; Gosz *et al.*, 1973; Lousier and Parkinson, 1978; Schlesinger, 1985). This indicates that the litter-degrading microorganisms are accessing and immobilizing an exogenous supply of nitrogen.

As decomposition proceeds, the C : N ratio declines as organic carbon is lost from the system as CO_2, and the C : N ratio of stabilized SOM approaches the C : N ratio of the microbial biomass. In a forest soil, Scheu and Parkinson (1995) found that the C : N ratio dropped from 37.5 in fresh litter to 20.7 in the organic horizons and to 9.3 in the first mineral horizon.

Results from fertilization experiments generally support the contention that nitrogen is often a limiting nutrient to microbial growth during the

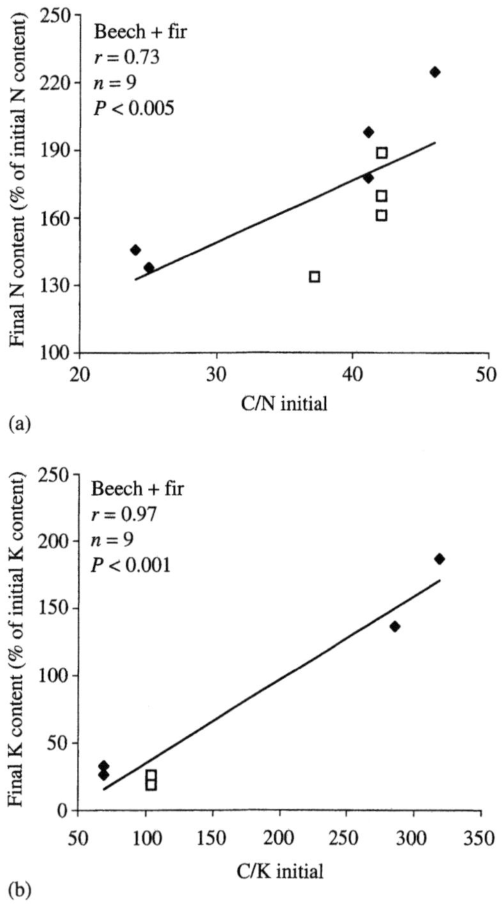

Figure 22 Relationship between: (a) final N and (b) K content (as percent of initial nutrient content) and initial C/nutrient ratio for beech (filled diamonds) and fir (open squares) litters (source Rutigliano *et al.*, 1998).

initial phase of litter decomposition (French, 1988). Additionally, the form of applied nitrogen can exert a strong influence on the microbial response (Rastin *et al.*, 1990; Scheu and Parkinson, 1995). In a temperate forest, Scheu and Parkinson (1995) found that ammonium, not nitrate, stimulated microbial respiration.

Phosphorus is often a limiting nutrient for decomposition in old highly weathered soils (Crews *et al.*, 1995; Vitousek and Sanford, 1986). In contrast to nitrogen, the dominant source of phosphorus is from the chemical weathering of parent material (Schlesinger, 1991) or from atmospheric deposition (Chadwick *et al.*, 1999). Phosphate, PO_4^-, the biologically available anionic form, will be tightly bound to sesquioxide clays in well-developed soils and allophane in young volcanic soils that have a high affinity for anionic compounds (Vitousek and Sanford, 1986). Along the Hawaiian archipelago chronosequence, Crews *et al.* (1995) found that during a 1.5-year *Metrosideros polymorpha* litter decomposition study, phosphorus was immobilized at the youngest (an Andisol) and oldest landforms (an Oxisol), but was lost at a similar rate to the overall mass loss at the nutrient-rich intermediate-aged sites (Table 9). The rapid mass and nutrient loss observed at the intermediate-aged sites was likely the combined result of the soils having more available nutrients and the leaf litter having higher nutrient contents (Crews *et al.*, 1995).

Several researchers have also found that root and leaf litter decomposition often have contrasting patterns of immobilization and release of limiting nutrients (Moretto *et al.*, 2001; Ostertag and Hobbie, 1999; Prescott *et al.*, 1993; Seastedt *et al.*, 1992). For *M. polymorpha* along the Hawaiian chronosequence, Ostertag and Hobbie (1999) found that although fine roots (<2 mm) decomposed faster than leaves possibly due to lower lignin : N and lignin : P ratios in the roots, root decomposition immobilized nitrogen regardless of site fertility, while nitrogen was released from leaf litter. For phosphorus, these authors found that only the roots at the old phosphorus-limited site showed significant phosphorus immobilization. In a semi-arid grassland in Argentina, Moretto *et al.* (2001) found that roots released nitrogen and phosphorus regardless of initial nutrient content; while leaf litter of the unpalatable grass (low nitrogen and phosphorus and high lignin content) immobilized phosphorus and the more palatable grasses released both nitrogen and phosphorus. Summarizing these studies, it would seem that differences in nutrient release patterns between leaf and root litter are primarily driven by those in initial resource quality and in the decomposition environment.

8.07.5.6 Priming Effect on Native SOM

Experimental additions of carbon- and nitrogen-labeled plant residues often show an increase in the mineralization rates of nonlabeled or native carbon and nitrogen relative to a control treatment (Bottner *et al.*, 1988; Cheng and Johnson, 1998; Fu and Cheng, 2002; Sallih and Bottner, 1988). Kuzyakov *et al.* (2000) define priming effects as "strong short-term changes in the turnover of SOM caused by comparatively moderate treatments of the soil." In considering the biogeochemistry of decomposition, the priming effects that would be of most interest involve additions of easily available organic substances (i.e., fresh litter and rhizodeposition). Positive priming effects following additions of organic material are attributed to an increase in microbial activity and an acceleration of SOM mineralization through co-metabolism (Asmar *et al.*, 1994; Breland and Hansen, 1998). Negative priming effects have also been observed following the

Table 9 Percent of initial C, N, and P remaining in *M. polymorpha* leaf litter decomposed *in situ* for 1.5 years along a chronosequence in Hawaii.

Elapsed time (yr)	*Thurston* (300 yr)			*Laupahoehoe* (20 kyr)			*Kohala* (150 kyr)			*Kokee* (4.1 Myr)		
	C	*N*	*P*	*C*	*N*	*P*	*C*	*N*	*P*	*C*	*N*	*P*
0.083	97	75	94	98	90	88	97	100	87	98		
0.25	94	81	91	87	97	87	91	96	82	96	88	78
0.50	88	90	97	62	95	96	74	83	81	92	80	90
1.00	74	89	143	29	43	55	37	64	58	85	119	146
1.50	64	81	146	18	39	36	31	69	66	80	104	194

After Crews *et al.* (1995).

addition of organic materials (Cheng, 1996; Sparling *et al.*, 1982). Kuzyakov *et al.* (2000) largely attribute the negative effects to a switch in microbial metabolism of SOM to the newly added readily available substrates.

The accumulating literature on priming effects (reviewed by Kuzyakov *et al.*, 2000) indicates that both the direction of and the mechanism driving the priming effect may depend on the study system. For example, Fu and Cheng (2002) found that in soybeans potted in soil dominated by C4 OM, native SOM decomposition increased by 70%, but when sorghum was planted in soil with C3 OM, SOM decomposition decreased by 9% relative to control soil. Priming effects can have large quantitative impacts on belowground nutrient cycling which are not included in most existing models of carbon and nitrogen dynamics (Kuzyakov *et al.*, 2000).

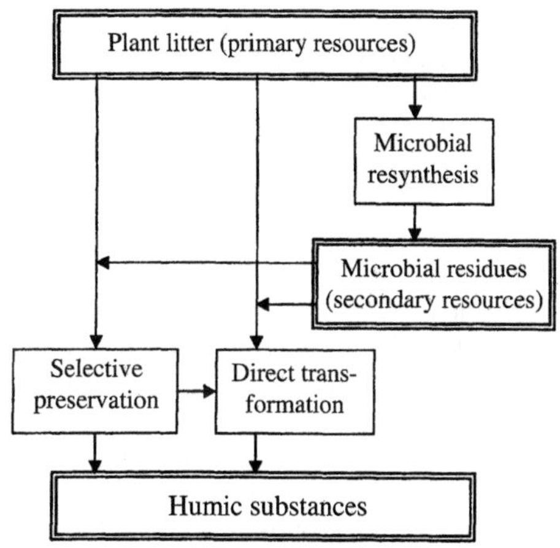

Figure 23 Different humification processes operating in the transformation of litter to humic compounds (after Kogel-Knabner, 1993).

8.07.6 HUMIFICATION

The transformation of plant detritus into stabilized humic substances is one of the most complex and least understood biogeochemical processes in the carbon cycle (Stevenson, 1994). Traditionally, decomposition and humification of plant residues was thought to be dominated by the mineralization of labile materials, while more recalcitrant aromatic compounds accumulate in the soil. The application of modern analytical techniques—including solid-state ^{13}C NMR spectroscopy, pyrolysis gas chromatography, and degradative chemical techniques—to the study of decomposition and humification has significantly altered this simple view of carbon transformation in the soil (Baldock *et al.*, 1997; Kogel-Knabner, 1997).

The formation of humus in most soils is likely the result of a combination of several processes, which Kogel-Knabner (1993) has summarized as *selective preservation* of recalcitrant plant and microbial biopolymers, *direct transformation* and *microbial resynthesis* (Figure 23). The relative importance of each of these processes will vary with resource type and soil environmental conditions. In this section, we will highlight the leading theories of humus synthesis with emphasis on some recent advances arising from the use of modern chromatographic and spectroscopic techniques. For more detailed information, the reader is referred to several excellent reviews (Baldock *et al.*, 1997; Hatcher and Spiker, 1988; Hedges, 1988; Kogel-Knabner, 1993; Kononova, 1966; Zech *et al.*, 1997).

8.07.6.1 Selective Preservation

Certain recalcitrant plant and microbial biopolymers are thought to be incorporated into stable SOM with only minor modification by microorganisms (Hatcher and Spiker, 1988; Lichtfouse, 1999; Waksman, 1932). Early investigators held the view, popularized by the Waksman (1932) lignin-protein theory, that humus was derived primarily from the incomplete utilization of lignin by microorganisms. In this theory, lignin is progressively altered by microbial attack on the exposed methyl groups and terminal side chains. Demethylation and oxidation of side chains results in products enriched in acidic functional groups (COOH and phenolic OH) which can then undergo various condensation reactions with nitrogen-containing compounds to form humic acids (Stevenson, 1994). Much of the evidence that Waksman cited in support of this theory stems from the similar biochemistry between lignin and humic acid.

Based on detailed analyses of the chemical nature of SOM, Hatcher and Spiker (1988) have extended this humification model to include other resistant biopolymers, including plant cutin and suberin, and microbial melanins and paraffinic macromolecules. During decomposition, these biopolymers are selectively preserved and modified to become part of what can be operationally defined as humin (acid and alkali insoluble component of humus) (Hatcher and Spiker, 1988; Rice, 2001). The humin becomes progressively enriched in acidic groups leading to the formation of first humic acids and then fulvic acids, which under this "degradative" scheme of SOM formation would be regarded as the most humified of humic substances (Stevenson, 1994).

Under aerobic conditions, the long-term preservation of relatively unaltered lignin seems unlikely. Several groups of fungi have been identified with the ability to completely mineralize lignin to CO_2 (Griffin, 1994). Solid-state ^{13}C NMR studies have found that the aromatic carbon (predominantly, lignin and tannins) content often decreases with depth (Baldock *et al.*, 1997) and with decreasing particle size (Baldock *et al.*, 1992). Additionally, CuO oxidation studies often find an increase in the acid/aldehyde ratio of individual phenylpropane units with depth (Kogel, 1986), indicating increased biodegradation of lignin via ring cleavage and side chain oxidation (Kogel-Knabner, 1993). These studies suggest that the selective preservation of lignin may not be as important as previously thought.

The relatively recent work combining isotopic and structural information (pyrolysis gas chromatography with mass spectrometry) has revealed several biopolymers that appear to be selectively preserved in soils (Kracht and Gleixner, 2000). Working in a *Spagnum* moss bog, Kracht and Gleixner (2000) found that the relative amount of several plant biopolymers increased with increasing depth in the bog, while the $\delta^{13}C$ of the individual components remained constant relative to fresh moss samples. These authors concluded that no change in the isotopic ratio indicated that these biopolymers were being selectively preserved rather than being synthesized by microorganisms. In a maize agricultural soil, Lichtfouse *et al.* (1998) found evidence for the selective preservation of a highly aliphatic, straight-chain biopolymer which was thought to be of microbial origin. Other researchers have found and ascribed these highly aliphatic polymers to the cutans and suberans of higher plants (Augris *et al.*, 1998; Nierop, 1998). In contrast, Poirier *et al.* (2000) found that aliphatic entities contributed only a minor amount to the refractory (nonhydrolyzable during drastic chemical treatment) organic fraction.

Charcoal or "black carbon" resulting from the incomplete combustion of plant residues during fires is another potentially important preservation pathway for SOM (Goldberg, 1985). Several researchers have shown that black carbon can account for a significant fraction of the total soil organic carbon, especially in fire-dominated landscapes (Glaser *et al.*, 1998; Golchin *et al.*, 1997; Skjemstad *et al.*, 1996). Applying ^{13}C NMR spectroscopy to UV-photo-oxidized SOM from the $<53\ \mu m$ size fraction of five different Australian soils, Knicker and Skjemstad (2000) found that charred material comprised most of the physically protected "passive" OM. Additionally, Poirier *et al.* (2000, 2002) found that the refractory organic fraction in both a temperate cultivated soil and a deep tropical savanna soil consisted of substantial amounts of black carbon. The black carbon in the tropical savanna soil was characteristic of forest vegetation which was thought to be replaced by savanna over 3,000 years BP (Poirier *et al.*, 2002).

8.07.6.2 Condensation Models

In contrast to the selective preservation theory, the condensation pathway proposes that humic substances are derived from the polymerization and condensation of low-molecular-weight molecules that are products of the partial microbial degradation of organic residues (Kogel-Knabner, 1993). Under this scheme of increasing complexation, fulvic acids would be the first humic substances synthesized, followed by humic acids and then humin (Stevenson, 1994). The two commonly accepted condensation models are the polyphenol theory and the sugar-amine or melanoidin theory.

The polyphenol theory views humus as a result of enzymatic conversion of polyphenols to quinones, which polymerize in the company or absence of amino compounds (Stevenson, 1994). Sources of polyphenol humic-precursors are thought to include the phenylpropane structural units of lignin released as a result of lignin biodegradation (Flaig *et al.*, 1975) and the synthesis products of microorganisms during xylem cellulose degradation (Kononova, 1966). Martin and Haider (1971) found that, when several different *Imperfecti* fungi were grown on cultures of plant residues, significant amounts of humic-acid-like compounds were produced. Upon structural determination, these phenol polymers were found to be composed of both biodegraded lignin subunits (i.e., syringyl and guaiacyl acids and their derivatives) and microbially derived products (i.e., flavinoids). Adding support to observations that peat derived from lignin-free mosses contained substantial amounts of humus, Martin and Haider (1969) and Martin *et al.* (1972) found that fungi grown on lignin-free cultures synthesized appreciable quantities of humic-acid-like substances.

Polyphenols derived from lignin biodegradation and by microbial synthesis can be viewed as reaction intermediates in the polyphenol theory of humus formation. The oxidation of phenols to quinones is mostly likely catalyzed by the extracellular polyphenoloxidase enzyme produced by numerous decomposer microorganisms (Kononova, 1966). There is also evidence that this reaction can occur spontaneously in an alkaline media (Stevenson, 1994). Quinones, being relatively unstable in soil, will undergo condensation and polymerization reactions with other quinones and amino compounds to form humic substances (Hedges, 1988; Kononova, 1966). In cultures of microorganisms, Flaig *et al.* (1975) found that the lignin degradation products, vanillin and vanillic acid, in the presence of various amino acids formed brown nitrogenous polymers with properties similar to natural humic acids. Additionally, Bondiett *et al.* (1972) found that similar humic-acid-like polymers could be produced from the reaction of phenol derivatives with amino sugars.

Formed by the condensation of simple carbohydrates and amino acids, melanoidins are complex brown nitrogenous macromolecules that are insoluble and resistant to chemical degradation. This reaction, commonly observed during food dehydration, was initially proposed by Maillard (1917) to be important during humus formation. The substrates for the melanoidin model, simple sugars and amino acids, are available in large quantities in plant residues; however, they are also readily metabolized by most microorganisms leading to low abundances in the mineral soil where most humus is found (Kogel-Knabner, 1993; Stevenson, 1994). Additionally, synthetically derived melanoidins have different structural characteristics from naturally occurring humic substances, as determined by solid-state ^{13}C NMR spectroscopy (Hedges, 1988; Kogel- Knabner, 1993). Despite these and other criticisms, the melanoidin theory has been supported by several researchers (Ikan *et al.*, 1986; Poirier *et al.*, 2000; Van Bergen *et al.*, 1998). In fact, Poirier *et al.* (2000) revised earlier results (Augris *et al.*, 1998) that indicated selective preservation of highly aliphatic macromolecules in the refractory humus pool to invoke the participation of melanoidins.

The field of humus research is an exciting and rapidly developing field. Newly developed methods combining structural and isotopic information with compound specific analyses hold the greatest potential in elucidating the various humification pathways. Despite these advances, there has been little conclusive evidence supporting one theory of humification over another. Humification likely involves both selective preservation and condensation reactions. The soil environment (via temperature and moisture regimes and interactions with clay minerals) will have a large impact upon the degree to which either of these processes operates.

8.07.7 CONTROL OF DECOMPOSITION AND STABILIZATION

Decomposition and stabilization of detritus as SOM are regulated by three general categories of driving variables (Swift *et al.*, 1979): the physicochemical environment, the resource quality, and the decomposer organisms. The physicochemical environment can be further divided into climatic and edaphic components. Communition and catabolism of detritus are influenced by all three variables, while leaching is primarily controlled by the climate and resource quality of the litter. Swift *et al.* (1979) considered these driving variables as three points on a triangle with each variable interacting and influencing the other.

In experiments designed to test these controls on decomposition, researchers often find that not all of these factors are equally important in all ecosystems. Heneghan *et al.* (1999) comparing a single substrate between a temperate and two tropical sites found that soil fauna are not important in controlling decomposition in the temperate forest, but soil fauna are important and could explain the differences between the two tropical sites. In a three-factorial experiment (climate, litter quality, and biota), Gonzalez and Seastedt (2001) found that climate and litter quality had dominant effects at a temperate and dry tropical site, while all three factors were important in the wet tropical site.

In a review of the decomposition literature, Lavelle *et al.* (1993) suggest a general model where decomposition is controlled by a hierarchy of factors which regulate microbial activity at decreasing spatial and temporal timescales (Figure 24). This hierarchical model best integrates seemingly disparate results—acutal evapotranspiration (AET) can explain over 90% of the variation in decomposition rates across large geographic regions (Berg *et al.*, 1993a); litter

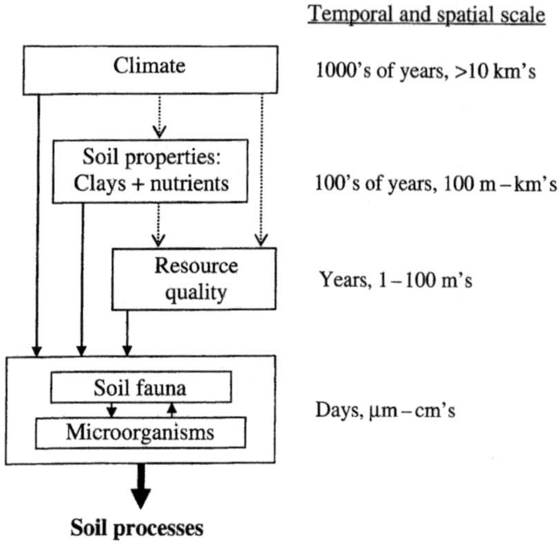

Figure 24 A hierarchical model of the factors controlling many soil processes in terrestrial ecosystems. Solid arrows represent direct regulation of biological processes and dashed arrows represent indirect controls (after Lavelle *et al.*, 1993).

quality, as measured by lignin and nitrogen content, can explain 88% of the variation in decomposition between four Rocky Mountain coniferous forests (Taylor *et al.*, 1991); and at a given site, differences in the abundance and assemblage of soil fauna can lead to threefold increases in decomposition rates (Whitford *et al.*, 1982). A corollary to this model is that when climatic and edaphic factors are ideal or at least not limiting, litter quality and mutualistic relationships between macro- and microorganisms become much more important in governing decomposition rates (Lavelle *et al.*, 1993). In this section, we will use this hierarchical model to describe the controls on decomposition beginning with the proximal controls of the decomposer organisms and ending with the more distal climatic factors.

8.07.7.1 Decomposer Organisms

Section 8.07.3 outlined the basic biology and ecology of the major decomposer organisms and Section 8.07.5 reviewed how these organisms breakdown and metabolize detritus. Here, we will highlight some of the more interesting controls that decomposer organisms have on the decomposition/stabilization process.

Microorganisms. Microorganisms are the primary agent of carbon mineralization in the detritus–soil continuum. Greater than 90% of total heterotrophic respiration is attributable to the metabolic activity of the microflora (Foissner, 1987). While acknowledging that microorganisms play a paramount role in decomposition, most biogeochemical models only implicitly include the microbial ecology (Schimel, 2001). According to Schimel (2001), two key assumptions in most biogeochemical models are: (1) microbial physiologies are global, i.e., they have an equivalent response across a range of environmental conditions and (2) microbial population sizes are never limiting and they rapidly adjust to stresses. These assumptions usually hold for processes that are aggregates of several microbial processes (i.e., soil respiration); however, as processes get more specific (i.e., nitrification or CH_4 production), these assumptions may not hold (Schimel, 2001).

Although microbial degradation is constrained by external factors (i.e., climate, edaphic conditions, etc.), the interactions between various microorganisms can have significant impacts on decomposition rates. Cox *et al.* (2001) showed the importance of individual species versus functional communities in the decomposition of pine litter; initially, a group of "sugar fungi" soon outcompeted *Marasmius androsaceus* (basidiomycetes, lignocellulose degrader), but once the most labile substrates were consumed *M. androsaceus* regained dominance. An antagonistic relationship between microorganisms was also demonstrated by Møller *et al.* (1999) where mineralization of one-year-old beech leaves was reduced by 50% when bacteria was added to a soil containing a cellulolytic fungi (*Humicola* sp.) due to carbon limitation and competition. Cox *et al.* (2001) also demonstrated that the celluolytic antagonistic fungi *Trichoderma viride* can prevent any other species from entering a volume of soil for six months.

Microorganisms also affect decomposition indirectly through the adhesive qualities of metabolic products, and the entanglement of soil particles by filamentous fungi can lead to a significant increase in soil aggregation (Caesar-TonThat and Cochran, 2000; Tisdall *et al.*, 1997). In a laboratory study on sterile sandy soil, Caesar-TonThat and Cochran (2000) demonstrated that water-stable aggregates (WSAs) increased when saprolytic fungi were present, and that carbon amendments greatly increased WSA formation because the fungus was carbon limited. Scanning electron microscopy by Caesar-TonThat and Cochran (2000) showed the extensive hyphal binding of soil particles (Figure 25). Tisdall *et al.* (1997), working with both saprophytic and mycorrhizal fungi, provided support for Miller and Jastrow's (1992) hypotheses that vesicular–arbuscular mycorrhizae (VAM) hyphae bring mineral soil and organic material together to from small microaggregates and the hyphae enmesh and bind microaggregates into larger macroaggregates (>50 μm) with the help of root exudates.

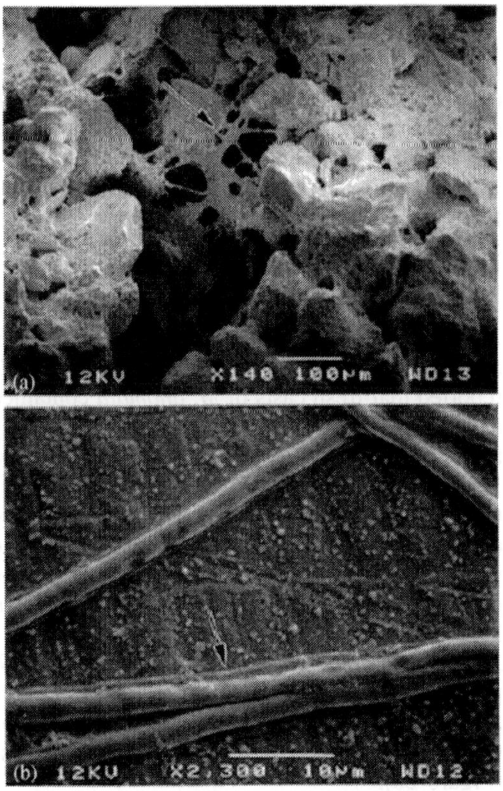

Figure 25 Low-temperature scanning electron microscopy of: (a) a portion of a soil macroaggregate with fungal material bridging soil particles (*arrow*) and (b) mucigel produced along fungal hyphae (*arrow*) (Caesar-TonThat and Cochran, 2000) (reproduced by permission of Springer Verlag from *Biol. Fert. Soils*, **2000**, *32*, 374–380).

Soil fauna. In a review of soil invertebrate control on SOM stability, Wolters (2000) noted that soil fauna affect the recalcitrance of, microbial accessibility to, and interactions with SOM. Direct effects of soil fauna include the communition, incorporation, and redistribution of organic materials (Seastedt, 1984). Soil fauna indirectly effect decomposition through numerous interactions with the microbial community (Shaw, 1992; Wolters, 2000). The mixing, aggregating, and channeling of soil material are another class of important indirect effects that can affect SOM stability and turnover. The summation of these direct and indirect effects of soil fauna can result in substantially increased decomposition rates. For a wide variety of deciduous litters, Seastedt (1984) found that microarthropods, predominately mites and collembolans, increased the average decay rate by 23% (median value of 17%) for litterbag exclusion studies lasting 9–30 months.

Anderson *et al.* (1981) showed that CO_2 production increased in the presence of nematodes because grazing increased the turnover of the bacterial population (i.e., more rapid growth). Anderson *et al.* (1981) also found that although decomposition was similar with or without nematodes after 65 days, greater nitrogen and phosphorus mineralization rates were observed with nematodes present because nematodes have low production efficiencies, ranging from 15% to 40% (Sohlenius, 1980). Net mineralization of nitrogen and phosphorus due to bacterial grazing by nematodes can also be argued based on C:N and C:P ratios—nematodes have higher carbon-to-nutrient ratios than bacteria; thus, they cannot assimilate all of the nitrogen and phosphorus in the bacteria food source (Anderson *et al.*, 1981). This observation suggests that secondary saprophages (i.e., microbial consumers) facilitate efficient nutrient cycling in soils (Bardgett *et al.*, 1999; Savin *et al.*, 2001).

A study by Rihani *et al.* (1995a) highlighted the synergistic interaction between the direct decomposition effects of faunal ingestion and the indirect effects it causes by simulating the microbial population. These authors found that the litter feeding mite *Steganacarus magnus* digested 8% of beech leaf litter alone, a white-rot fungus decomposed 24% of litter, while together the two organisms decomposed 37% of the litter. Maraun and Scheu (1996) found that fragmentation of beech leaf litter by the millipede *Glomeris marginata* resulted in initial increases in microbial biomass due to increased access to carbon, but later microbial biomass was depressed, relative to control, due to reduced carbon availabilty. *G. marginata* also significantly altered the microbial community directly through the preferential digestion of fungi, as shown by changes in ergosterol levels in the soil (Maraun and Scheu, 1996). In the same study, Maraun and Scheu (1996) also demonstrated that microbial growth in the faecal pellets of *G. marginata* is carbon limited while microbial growth in the original beech litter is nutrient limited, thereby changing the limiting factor of decomposition.

Perhaps no other soil invertebrate has such well-documented controls on decomposition and stabilization of SOM as the burrowing anecic earthworms, such as *Lumbricus terrestris*. These organisms are capable of directly incorporating surface litter into the soil, and often the absence of a surface organic horizon is due to the presence of these organisms (Edwards and Bohlen, 1996). Langmaid (1964) reported that it took only three to four years after invasion of worms to thoroughly mix a well-developed spodosol in New Brunswick, Canada. Clements *et al.* (1991) found that after earthworms had been absent from a grassland soil for 20 years, there was a significant increase in the depth of the litter layer and a large reduction in the OM content of the underlying soil. Based on radiocarbon measurements, O'Brian and Stout (1978) estimated that the introduction of earthworms led to an increase in the annual carbon flux from 300 kg ha^{-1} to 1,000 kg ha^{-1}, and the residence time of SOM decreased from 180 years to 67 years in a New Zealand pasture.

Earthworms only assimilate a low proportion of ingested OM. Lavelle and Martin (1992) reported that *Millsonia anomala* assimilates only 2–6% of the OM that passes through its gut, although the communition of litter and particulation of soil aggregates can have important secondary effects. Kanyonyo (1984), cited within Lavelle and Martin (1992), found that in the tropical anecic *Millsonia lamtoiana*, communition plus assimilation decreased the recognizable fragments of tree leaves, roots, grasses, and seeds by 83%, 71%, 41%, and 31%, respectively. Free-living microbial populations that have been ingested along with soil and litter are greatly stimulated during gut passage by the addition of water and labile intestinal mucus secretions. In this respect carbon cycling is stimulated in the drilosphere, the region of soil influenced by earthworms (i.e., burrow and casts), much like that is known to occur in the rhizosphere due to the exudation of labile root exudates that stimulate nutrient turnover (Anderson *et al.*, 1981). During gut transport, by the time material reaches the second half of the gut, most of the added mucus has been metabolized and the now active microbial community begins to degrade soil OM into assimilable compounds which are used by both the earthworms and the microorganisms (Barois and Lavelle, 1986; Edwards and Bohlen, 1996).

Microbial metabolism of soil SOM during gut transport and the excretion of urine by earthworms into their guts both result in the release of available nutrients which may be expelled in casts. The volume of casts in earthworm-dominated soils can be enormous. In a tropical savanna at Lamto, Ivory Coast, Blanchart *et al.* (1993) found that 50% of the soil volume and 65% of the soil weight consisted of high bulk density (1.97 g cm^{-3}) casts, while the remaining soil volume had a bulk density near 1.0 g cm^{-3}. In a temperate cultivated soil, *L. terrestris* casts contained 21.9 μg N-NO_3^- g^{-1} soil and 150 μg P g^{-1} soil while the surrounding bulk soil contained only 4.7 μg N-NO_3^- g^{-1} soil and 20.8 μg P g^{-1} soil (Lunt and Jacobson, 1944). In a tropical savanna, Lavelle and Martin (1992) reported that fresh casts of *M. anomala* contained an average of 26.4 μg N-NH_4^+ g^{-1} soil and only traces of NO_3^-, while the control soil contained 1.5–7.5 μg N-NH_4^+ g^{-1} soil and no NO_3^-. Generally, fresh fecal material shows elevated decomposition relative to bulk SOM due to increased accessibility of labile carbon and nutrients, and the stimulation of bacterial populations during gut passage (Wolters, 2000). In a 30-day laboratory incubation, Cortez *et al.* (1989) found that the presence of the earthworm *Nicodrilus longus* increased CO_2 production threefold over soil without earthworms (Figure 26).

As earthworm casts age, mineralization is often reduced relative to uningested soil. Martin (1991) found that carbon in the aged casts of the tropical endogenic *M. anomala* mineralized at a rate of 3% yr^{-1}, while the control soil mineralized at 11% yr^{-1}. McInerney and Bolger (2000a) showed that earthworm-cast microaggregate structure was responsible for the decreased decomposition rates in temperate oak litter casts due to fewer macropores and more micropores, which reduce

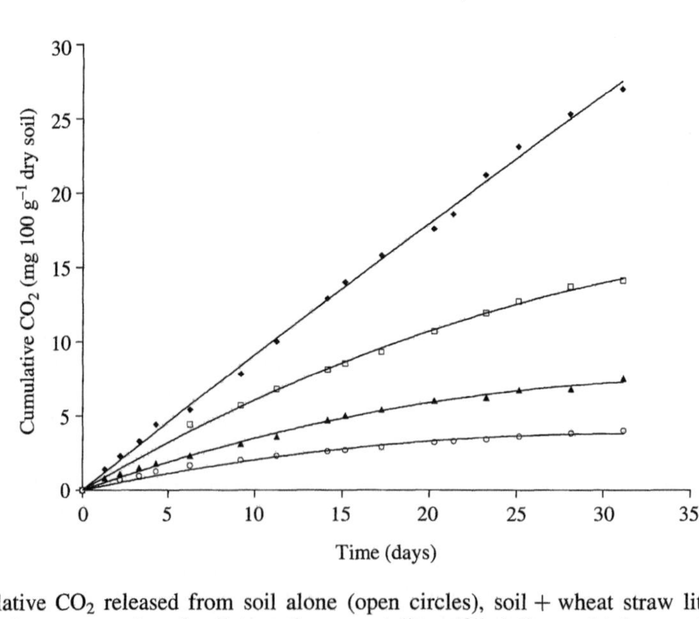

Figure 26 Cumulative CO_2 released from soil alone (open circles), soil + wheat straw litter (filled triangles), soil + earthworms (open squares), and soil + earthworms + litter (filled diamonds) (source Cortez *et al.*, 1989).

accessibility by most microorganisms. Blanchart *et al.* (1993) concluded that water re-adsorption in the hindgut pulls fine particles to the outside of the cast creating a crust (~25 μm thick) that has micropores and which prevents water movement. Shipitalo and Protz (1989) discussed the cast formation process, and suggested that aggregate stability in the casts increases via cation bridges and coordinated complexes between clay minerals and SOM.

Martin *et al.* (1992) found that the tropical geophage *M. anomala* assimilated both fresh plant material (coarse OM) and fine soil OM (resistant pool) into biomass by measuring changes in $\delta^{13}C$ of the earthworms switched from a C3 forest to a C4 grassland (or vice versa). Studies of earthworm invasions have shown that worms can metabolize recalcitrant SOM (Burtelow *et al.*, 1998; O'Brian and Stout, 1978). These studies raise the question: in earthworm-dominated systems is there really any resistant SOM? The digestion of humified OM during earthworm invasions would result in a significant release of nitrogen due to the low C/N ratios of stable SOM. Martin (1991) estimated that 12–17 kg N ha^{-1} yr^{-1} is incorporated into casts, 60% of which originated from the clay-associated OM. However, increased mineralization of the humic pool during earthworm digestion may be balanced by decreased mineralization of the remaining carbon in the resulting casts. While there is ample evidence to suggest that earthworms dramatically alter the cycling dynamics and structure of OM in soil, these competing factors make it hard to predict whether earthworms act to increase or decrease the overall storage of carbon (Blaire *et al.*, 1994).

Many of the effects on carbon decomposition and stabilization discussed above for earthworms can be generalized to many groups of soil fauna. Table 10 summarizes the mechanisms by which soil invertebrates control decomposition and stabilization of OM in soils. Wolters (2000) hypothesized that soil invertebrates externally influence SOM destabilization by microorganisms in two ways: (1) directly, by selectively grazing on

Table 10 Overview of the mechanisms by which invertebrates control the turnover of humic substances.

Process	*Level of control*[a]		*Mechanism*
Humus formation	Internal	Direct	Production of gut enzymes that favor the condensation reactions Biosynthesis of waxes, etc.
		Indirect	Production of enzymes that favor humification by gut symbionts
	External	Indirect	Facilitation of the microbial production of polymers and extracellular enzymes
			Altering distal factors affecting SOM stability by: (i) affecting the condensation of intermediates and (ii) stimulating the formation of stable organometallic complexes
Humus degradation	Internal	Direct	Disturbance of soil structure and associated disaggregation
		Indirect	Degradation of recalcitrant organic compounds aided by obligate or facultative symbionts
	External	Indirect	Increase in the competitive capacity of humus-degrading microorganisms by: (i) alterations in carbon and nutrient availability and (ii) selective grazing of fast-growing fungi and on microorganisms capable of degrading recalcitrant compounds
Other effects	Internal	Direct	Selective ingestion of less recalcitrant compounds leads to the enrichment of recalcitrant SOM
		Direct and indirect	Degradation of more labile compounds in the gut gradually increases the average recalcitrance of nonassimilated carbon
	External	Direct	Selective translocation of SOM leads to enrichment either of less or of more recalcitrant SOM
		Indirect	Rapid depletion of nutrients and labile carbon compounds shortly after defecation leads to an increase in recalcitrant components

After Wolters, (2000).

[a] Internal control refers to effects of ingestion and digestion, while external control refers to any process occurring outside of the invertebrate's body.

fast-growing fungi, thereby allowing the slower growing lignolytic fungi to gain a competitive advantage (Doube and Brown, 1998; Moody, 1993) and (2) indirectly, by altering the availability of nutrients in a way which is advantageous to the microorganisms capable of metabolizing recalcitrant compounds (Scheu, 1992). Humus formation is directly promoted through decreased access to carbon in aged casts (Shipitalo and Protz, 1989) and increased soil aggregation via the excretion of polysaccharides (Sollins *et al.*, 1996). Indirect controls on humus formation include a range of processes that favor condensation of humic products and humic precursors (Hartenstein, 1982; Wolters, 2000).

8.07.7.2 Resource Quality

The primary carbon sources, originating from plants, available to the decomposer community comprise a wide variety of tissues that differ in both physical and chemical properties (Swift *et al.*, 1979). The aboveground components (leaves, stems, and reproductive organs) and the belowground components (roots) will each show characteristic patterns of decomposition that vary between species and possibly within species in differing ecosystems. Swift *et al.* (1979) showed that decay rates, calculated as litter fall divided by standing crop, from both a temperate and a tropical forest followed the same pattern—fleshy reproductive organs decayed faster than leaves, which decayed much faster than woody parts. Differences in the apparent decay rates of these plant parts are due not only to the palatability to the decomposers, but also to the loss of water-soluble compounds (Ibrahima *et al.*, 1995; Kuiters and Sarink, 1986; Nykvist, 1961; Taylor and Parkinson, 1988b).

One of the first detailed studies to highlight the importance of the chemical nature of detritus upon decomposition was reported in a series of papers by Tenney and Waksman (1929) and Waksman and Tenney (1927, 1928). Since then, differences in decay rates between plant species have been shown to be controlled by a wide variety of chemical properties (Aerts, 1997), including the lignin concentration or lignin-to-nutrient ratios (Aerts and De Caluwe, 1997; Berg, 1984a; Meentemyer, 1978; Melillo and Aber, 1982; Tian *et al.*, 1992b; Van Vuuren *et al.*, 1993), the nitrogen concentration or the C/N ratio (Coulson and Butterfield, 1978; Perez-Harguindeguy *et al.*, 2000; Taylor *et al.*, 1989; Tian *et al.*, 1992a), phosphorus concentrations or C/P ratios (Berg *et al.*, 1987; Coulson and Butterfield, 1978; Schlesinger and Hasey, 1981a; Staaf and Berg, 1982; Vitousek *et al.*, 1994), and polyphenol content (Northup *et al.*, 1998; Yu *et al.*, 1999).

Plant physical traits that protect against biotic attack and harsh environmental conditions (Herms and Mattson, 1992) produce tissues with higher lignin, polyphenol, and wax contents and higher lignin/N and C/N ratios that decompose more slowly (Perez-Harguindeguy *et al.*, 2000). Grime *et al.* (1996) presented evidence that the least palatable litters to two herbivores also decomposed at the slowest rates. Perez-Harguindeguy *et al.* (2000) showed that litter mass loss is highly correlated ($r = -0.86$, $p = 0.014$, $n = 7$) to leaf tensile strength across a wide range of plant functional types in central Argentina (Figure 27). Although not a direct measure of the chemical nature of the litter, tensile strength is a relative measure of leaf toughness (Cornelissen *et al.*, 1999; Cornelissen and Thompson, 1997). When these same litters were moved to the temperate climate of Sheffield, UK (mean temperature of experiment = 7.7 °C versus 20 °C in Argentina), Perez-Harguindeguy *et al.* (2000) found the same trend of mass loss versus tensile strength, but that there was a 25% reduction in overall mass loss.

Table 11 summarizes the effect of initial litter quality on decay rates, *k*, for selected studies (Aber *et al.*, 1990; Fioretto *et al.*, 1998; Harmon *et al.*, 1990a; McClaugherty *et al.*, 1985; Melillo and Aber, 1982; Taylor *et al.*, 1989, 1991; White *et al.*, 1988). Included are forested sites that span a range of climates from mediterranean to temperate rainforest and that include a variety of litter sources (leaf, needle, and fine root) with greatly contrasting chemistry. For these sites, neither the nitrogen concentration nor the C/N ratio predicted *k*. However, a simple exponential model explained most of the variance between *k* and lignin content across all sites. The addition of nitrogen either in the form of the lignin/N ratio or in a multiple linear regression significantly improved the prediction for the Hubbard Brook, Kananaskis Valley, and Olympic National Park sites (Table 11). Aber *et al.* (1990) suggested that the ratio of lignin to lignin + cellulose (the lignocellulose index (LCI)) is a good predictor of *k*. For the data in Table 11, the LCI significantly increases the coefficient of determination to 0.73 and 0.94 for the Olympic National Park and Blackhawk Island sites, respectively. These results indicate that at a given site, lignin is the primary control on the decomposition rate, and that nutrient content is a secondary influence (Taylor *et al.*, 1991).

For a range of lignin concentrations wider than evaluated in Table 11; Taylor *et al.* (1991) found a rapid decrease in mass loss with increasing lignin up to a critical content of 28% lignin and a lignin-to-nitrogen ratio of 50 : 1, after which constant loss rates occurred. However, this interpretation may be confounded by the fact that all of the high lignin substrates were large woody litters

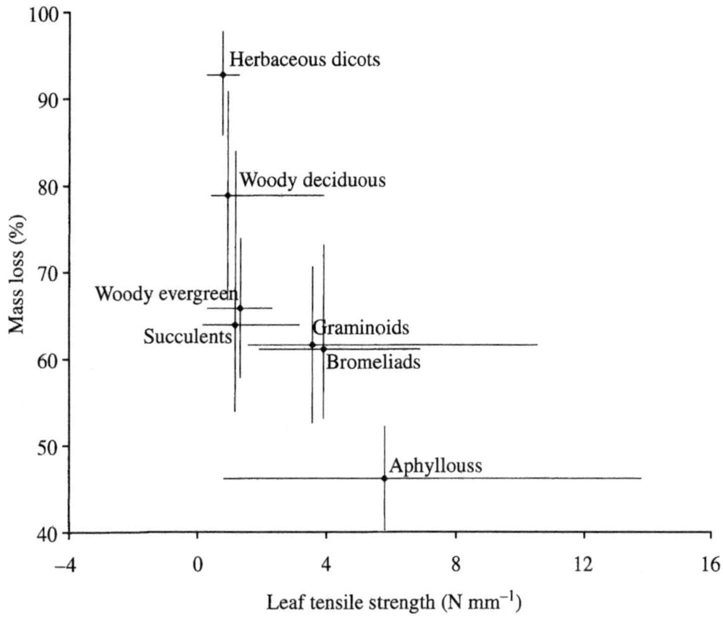

Figure 27 Relationship between leaf tensile strength and mass loss for different functional types of central Argentina. Points represent median values and error bars reprsent the quartiles 25% and 85% (source Perez-Harguindeguy *et al.*, 2000).

(branches, cones, and roots) which may exhibit more complex decay patterns due to their three-dimensional structure (Edmonds *et al.*, 1986). In the Pacific Northwest, Edmonds (1988) found no correlation between five-year mass loss and lignin content for branch, twig, and cone litter.

In a review of global patterns of root decomposition, Silver and Miya (2001) found that the substrate quality parameters, root Ca^{++} concentration and C/N ratio, could explain the greatest proportion of the variance in decay rates across a range of sites and species ($r^2 = 0.89$, $n = 17$):

$$\ln(k) = 3.79 + 0.74 \times \ln(\mathrm{Ca}) - 1.22 \times \ln(\mathrm{C:N}) \quad (8)$$

This result is in accord with Lavelle *et al.*'s (1993) hierarchical model—when climate is ameliorated as a limiting variable, resource quality becomes much more important in determining decomposition dynamics. In the soil, roots and the associated decomposer organisms are sheltered from climatic extremes relative to the surface litter system. Additionally, Silver and Miya (2001) proposed two explanations for the strong role of root calcium in predicting decay rates: (1) high levels of calcium in the root may indicate high levels of mycorrhizal fungal associations which upon root death may act as a readily available carbon substrate for heterotrophic organisms and (2) high root calcium levels may indicate nutrient-rich soil conditions that in turn would promote accelerated decomposition rates.

At the San Dimas Experimental Forest in southern California, a large lysimeter installation established in 1937 is an excellent experiment on the effects of differing monocultures of native vegetation (chamise, ceanothus, scrub oak, and Coulter pine) on soil properties and processes without the confounding influence of other factors (Colman and Hamilton, 1947). The lysimeters are large (5.3 m × 5.3 m horizontally and 2.1 m deep) earthen-walled pits that were filled with a homogenized fine sandy loam (58% sand, 31% silt, 11% clay) with only 0.2% organic carbon at time of filling (Colman and Hamilton, 1947). Previous research has addressed the influence of vegetation on soil morphological development (Graham and Wood, 1991), aggregate stability (Graham *et al.*, 1995), soil nutrient content (Ulery *et al.*, 1995), and mineralogy (Tice *et al.*, 1996). Perhaps the most striking finding is that the soil under oak has developed a dark, 7 cm thick "A" horizon overlain by a thin (6 cm) litter layer while the soil under pine has a thick "O" horizon (10 cm), minimal darkening in the A horizon, and an argillic horizon (a subsurface concentration of clay by elluvial processes) had developed in only four decades (Graham and Wood, 1991). Graham and Wood (1991) report that while worms are completely absent from the pine soil, there is substantial earthworm activity in the oak soil and that the A horizon is comprised of 95% earthworm casts.

Quideau and co-workers have compared the soil carbon cycling dynamics of the oak and pine lysimeters using physical fractionation (Quideau *et al.*, 1998), radiocarbon (Quideau *et al.*, 2001),

Table 11 Regression summary of litter chemistry controls on the decay rates of leaf and fine root (<1 mm) litter for selected sites spanning a range of climates (as indicated by MAT, MAP, AET, DEFAC, and mean *k* values for each site). Values are the coefficients of determination for a linear regression between the natural log (*k*) and the litter chemistry variable in each column. Also shown is the results of a multiple linear regression between ln (*k*) and %N and %lignin.

Site	*Ref.*[a]	*MAT* (°C)	*MAP* (mm)	*AET* (mm)	*DEFAC*[b]	*n*	*mean k*	Multiple Regression parameters					
								%N	*%lignin*	*C:N*	*LCI*	*lignin:N*	*%N, %lignin*
Kananaskis Valley, Alberta	5	2.1	660	385	0.13	10	0.14	0.09	0.81***[c]	0.03	ND[d]	0.87***	0.90*,***
Hubbard Brook, NH	4	5.0	1,300	552	0.24	6	0.25	0.00	0.55*	ND	ND	0.99***	0.99**,***
Blackhawk Island, WI	1	7.0	810	605	0.26	8	0.41	0.29	0.90***	ND	0.94***	0.06	0.92-,**
Harvard Forest, MA	3	7.0	1,120	578	0.26	4	0.55	0.26	0.96*	ND	0.65**	0.69	0.998-,*
Olympic National Park, WA	6	8.9	3,550	525	0.33	9	0.81	0.33	0.44*	ND	0.73**	0.55*	0.55-,-
Coweeta, NC	2	13.0	1,800	702	0.37	6	0.94	0.27	0.96***	0.25	0.85**	0.44	0.98-,**

[a] References: (1) McClaugherty *et al.* (1985), (2) White *et al.* (1988), (3) Aber *et al.* (1990), (4) Melillo and Aber (1982) (5) Taylor *et al.* (1989, 1991), and (6) Harmon *et al.* (1990a). [b] DEFAC = synthetic climate variable used in the Century model (see text for explanation). [c] Significance of each regression is denoted by astericks (for the multiple regression: significance of %N term, %lignin term): *$P < 0.1$, **$P < 0.01$, ***$P < 0.001$. [d] ND = not determined.

and ^{13}C NMR techniques (Quideau *et al.*, 2000). Based on radiocarbon measurements, Quideau *et al.* (2001) found that the decomposition rates were much faster under oak than under pine for both the litter layer and for nearly all SOM fractions in the A horizon (Table 12). Not only are the turnover rates faster under oak, but significantly more OM is being incorporated in the mineral soil, mainly a result of bioturbation by earthworms (Graham and Wood, 1991). Results from Quideau *et al.*'s (2000) NMR analysis suggest two separate patterns of decomposition (Figure 28). The small increase in the aromatic/O-alkyl-C ratio from the litter to floatable fraction of the A horizon under oak indicates little microbial metabolism of the labile carbon components, while substantial degradation has occurred under pine. Fragmentation and mixing of the litter into the A horizon may dominate the initial stages of decomposition under oak, followed by a more traditional pattern (Baldock *et al.*, 1992) of microbial degradation of labile (O-alkyl carbon rich), and then more recalcitrant (high in aromatic-C and alkyl-C) components. In the pine soil, without the influence of earthworms, much more microbial degradation occurs in the litter layer and transport into the mineral soil is much slower, resulting in more recalcitrant SOM as indicated by the higher alkyl/O-alkyl-C ratios in all size fractions (Figure 28). Consistent with this pattern, particle size analysis of the OM in the A horizon showed that most of the carbon was

Table 12 Carbon storage, radiocarbon content, and the estimated decomposition constant (*k*) for the O and A horizons under oak and pine vegetation.

Horizon[a]	*Depth* (cm)	*C* ($g\ m^{-2}$)	$\Delta^{14}C$ (‰)	k (yr^{-1})
Oak lysimeter				
Oi	6	493	212	0.61
A-WS	7	2,093	259	0.08
FL		963	319	0.07
SA		492	154	0.09
FS		331	274	0.08
CL		308	225	0.13
Pine lysimeter				
Oi1	4	107	216	0.51
Oi2	2	65	238	0.27
Oe	4	347	343	0.03
A-WS	1	172	90	0.04
FL		56	220	0.08
SA		49	−90	0.005
FS		31	108	0.07
CL		37	115	0.02

After Quideau *et al.* (2001).
[a] WS = whole soil (<2 mm), FL = floatables (50–2,000 μm), SA = sand + coarse and medium silt (5–2,000 μm), FS = fine silt (2–5 μm), CL = clay (<2 μm).

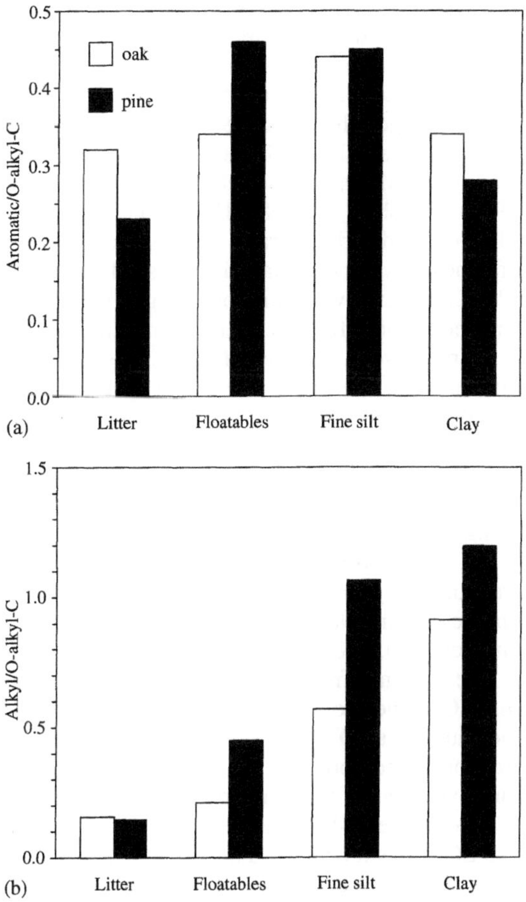

Figure 28 (a) Aromatic/O-alkyl-C and (b) alkyl/O-alkyl-C ratios in litter and particle size fractions from A horizon samples under oak and pine vegetation in the San Dimas lysimeters experiments (source Quideau *et al.*, 2000).

associated with the sand fraction under oak, while most of the carbon was associated with the silt and clay fractions under pine vegetation (Quideau *et al.*, 1998).

8.07.7.3 Soil Characteristics

In this section, we will consider the effects of soil texture and soil nutrient status on decomposition. The initial stages of leaf litter decomposition will be at least partially decoupled from control by edaphic properties of the soil environment. For example, Scott *et al.* (1996) found that while SOM decomposition varies significantly with soil texture, the CO_2 evolution from surface litter does not. However, as partially decomposed litter is incorporated into the soil both through abiotic and biotic means, the physical characteristics of the soil begin to play an important role in the overall degradation and stabilization of the organic inputs.

Christensen (2001) suggested a conceptual model for SOM dynamics organized around three levels of structural complexity: (1) primary organomineral complexes, (2) soil aggregates, and (3) the whole soil. In terms of SOM, the organo-mineral complexes associated with the sand-, silt- and clay-sized soil fractions represent the basic structural and functional units. At the scale of a soil particle, surface area and charge properties will be the dominant influences stabilizing OM from microbial degradation. Within soil aggregates, water and gas diffusion become important limiting controls on rates of carbon mineralization. At the scale of the whole soil, nutrient availability, macroporosity, the activity of soil fauna and roots, and exogenous disturbance become important influences on decomposition processes (Christensen, 2001).

Soil texture. In many biogeochemical models, soil texture, in the form of the clay + silt fraction, is suspected of being one of the key variables influencing both the decomposition rate of the active SOM pool and the efficiency of stabilizing active SOM into slow SOM (e.g., Parton *et al.*, 1987).

In a laboratory incubation of Danish soils of varying texture, Sørensen (1981) found that the proportion of labeled cellulose that decomposed after 90 days varied inversely with clay content ($r^2 = 0.99$, $P = 0.006$, $n = 4$) and that the amount of labeled carbon remaining in the soil after 1,600 days was twice as large in the high clay soil compared to the sandy soil. McInerney and Bolger (2000b) showed that CO_2 produced per gram soil carbon was 15% higher in a loam versus a clayey soil, indicating a reduction in decomposition rates. Schimel *et al.* (1985), working on a toposequence in southwestern North Dakota, also report that the amount of CO_2-C respired, normalized to carbon content, decreases with increasing clay. Giardina *et al.* (2001) found a strong, but nonsignificant ($r^2 = 0.51, P = 0.11, n = 6$) trend of decreasing CO_2 efflux with increasing clay content in lodgepole pine forest soils but not for soils under an aspen stand ($r^2 = 0.11, P = 0.46, n = 6$). When the Scott *et al.* (1996) data on carbon mineralization rates are normalized by carbon content, we find that the clay-rich soils decompose slower than the sandy soil. However, other studies have found weak (Motavalli *et al.*, 1994) or nonexistent trends (Giardina and Ryan, 2000; Thomsen *et al.*, 1999). It is not surprising that conflicting results are reported since clay content is affected by state factors (i.e., climate and parent material) which in turn influence decay rates and affect a number of soil properties (i.e., porosity, water-holding capacity, and aggregate stability). The latter can also significantly alter decay rates (Sollins *et al.*, 1996).

Analyses of SOM dynamics on particle size and density separates (Christensen, 1992) support the general concept that the clay- and silt-associated OM is a more recalcitrant pool with slower turnover times (Stevenson, 1994). Christensen (1987) found that the sand-associated OM decomposed faster than the clay fraction which, in turn, decomposed faster than the silt fraction. Overall decomposition rates were higher in the finer textured soil, because most of the OM was associated with the clay fraction which decomposed faster than silt-associated OM (Christensen, 1987). Several other particle size fraction studies support the finding that the silt-SOM is the most stable fraction (Amato and Ladd, 1980; Ladd *et al.*, 1977); however, turnover times derived from ^{14}C data usually show that the clay fraction is the most stable pool (Anderson and Paul, 1984; Christensen, 1992). During particle size fractionation, without rigorous pretreatment, silt-sized aggregates of clay particles can settle out with the silt-sized fraction, possibly reconciling these two disparate results.

Geologists are also beginning to recognize the importance of the role of clay minerals in the long-term burial and preservation of OC in shale deposits. Kennedy *et al.* (2002) showed that 85% of the variance in total OC can be explained by mineral surface area in two Cretaceous black shale deposits.

The importance of clays in stabilizing OM comes from results of labeling studies which find that the clay fraction is often most enriched, relative to the whole soil (Chichester, 1970). This enrichment is relatively more important in sandy than in clayey soils (Figure 29). However, the turnover of clay-associated SOM in sandy soils is faster (Gregorich *et al.*, 1989), suggesting that these soils are less efficient in stabilizing and storing SOM than clay-rich soils (Christensen, 1992). Cheshire and Mundie (1981) found that plant-derived carbohydrates (glucose, xylose, and arabinose) were dominant in the sand separates (20–2,000 μm), while microbial sugars (mannose, rhamnose, and fucose) were found in greatest concentration in the clay fraction. As the OM ages and becomes more humified, structural complexity and size can increase, possibly explaining the relatively greater role silt plays in holding old (native) SOM in long-term labeling studies (Christensen and Sorenson, 1985). Several studies using solid-state ^{13}C NMR spectroscopy on size separates found that the aromaticity of silt-SOM fraction is higher than the clay fraction (Catroux and Schnitzer, 1987; Oades *et al.*, 1987; Schnitzer and Schuppli, 1989).

The evidence from studies that track the change in $\delta^{13}C$ values following a vegetation shift also indicates that OM associated with the coarse fractions decomposes at the fastest rate. Twenty-three years after the conversion of a pine forest (C3) to maize (C4), Balesdent *et al.* (1987) found that SOM in silt contained the least maize-derived carbon (12%), and thus the slowest turnover, while the coarse sand had 61% maize-SOM. However, 97 years after a prairie was converted to

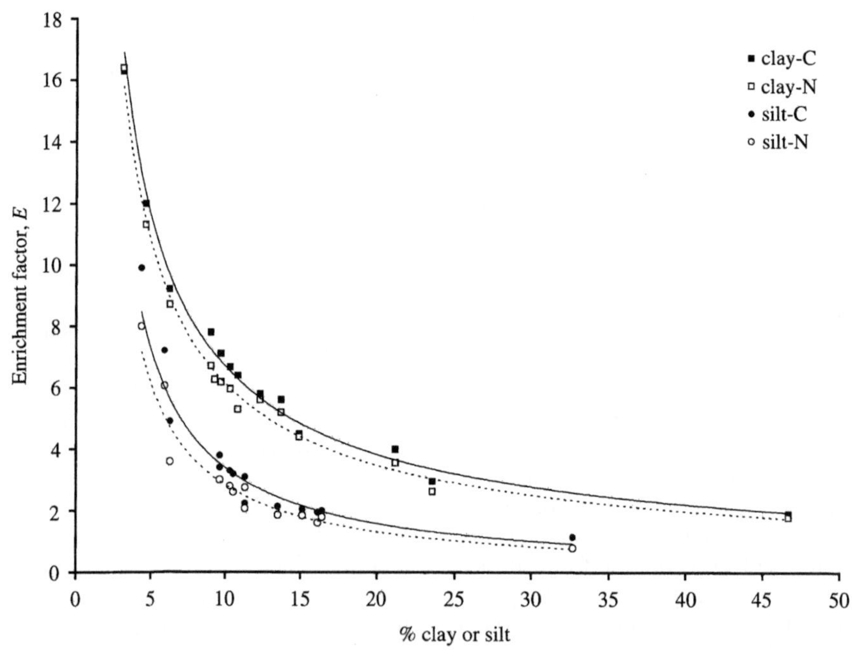

Figure 29 Relationship between fraction size and C and N enrichment factors (E = percent of C or N in fraction to percent in whole soil) for clay (<2 μm) and silt (2–20 μm) isolated from a range of Danish agriculture soils (source Christensen, 1992 and references within).

Table 13 Surface characteristics of various clay minerals.

Clay mineral	*Size* (μm)	*Surface area* ($m^2\ g^{-1}$)	*Negative charge* ($cmol_c\ kg^{-1}$)	*Positive charge* ($cmol_c\ kg^{-1}$)
Smectite	0.01–1.0	600–800	80–120	0
Vermiculite	0.1–5.0	600–800	100–180	0
Illite	0.2–2.0	70–100	10–40	0
Kaolinite	0.1–5.0	10–30	3–15	2
Gibbsite (Al oxide)	0.1–0.2	5–20	0–4	0–5
Goethite (Fe oxide)	0.01–0.2	25–95	0–4	0–5
Allophane		70–300	0–30	0–15

Sources: Brady and Weil (2002); Stevenson (1994), and others.

wheat cultivation, Balesdent *et al.* (1988) found that the fine clay-SOM (<0.2 μm) fraction had the slowest turnover and that the major losses, occurring during the first 27 years, were from the macro-OM in separates >25 μm and the fine silt-SOM (2–25 μm). For an Oxisol in the humid tropics, whose vegetation had shifted from grass savanna (C4) to dense woodland (C3), Martin *et al.* (1990) found turnover times increased from clay (70% savanna-derived C) to fine silt (56% savanna-C) to coarse silt (28% savanna-C) to the >50 μm fraction (<15% savanna-C).

Although there is ample observational evidence for soil textural effects on SOM decomposition and stabilization, mechanisms for the observations are difficult to verify. Perhaps the most straightforward hypothesis is that clays have a high surface area and possess the majority of the surface charges of the soil particles, thereby creating more opportunities to hold organic additions to the soil. Oades (1988) has suggested that SOM stabilization to clay is due to adsorption reactions (i.e., electrostatic bonding, hydrogen bonding, van der Waals forces, hydrophobic bonding, coordination, and ligand exchange (Stevenson, 1994)) of organics onto surfaces of clays and other organic complexes. It would, therefore, seem that the type of clay present, due to differences in particle size, surface area, and charge density, will have a large effect on the carbon dynamics (Table 13). Nelson *et al.* (1997) found that the ability to retain OM was less in an illite- and kaolinite-dominated soil than in a smectite-dominated soil, even though no difference in the structural stability of aggregates occurred between the two soils (Barzegar *et al.*, 1997).

Across a chronosequence of soils on the Hawaiian islands (Crews *et al.*, 1995), Torn *et al.* (1997) found that both the quantity of stored carbon and its turnover time correlated with the noncrystalline (allophane, imogolite, and ferrihydrite) mineral content of the soil (Figure 30). These amorphous minerals possess a unique geometry with a very high surface area (Table 13) which facilitates the formation of highly stable bonds with SOM (Oades, 1988).

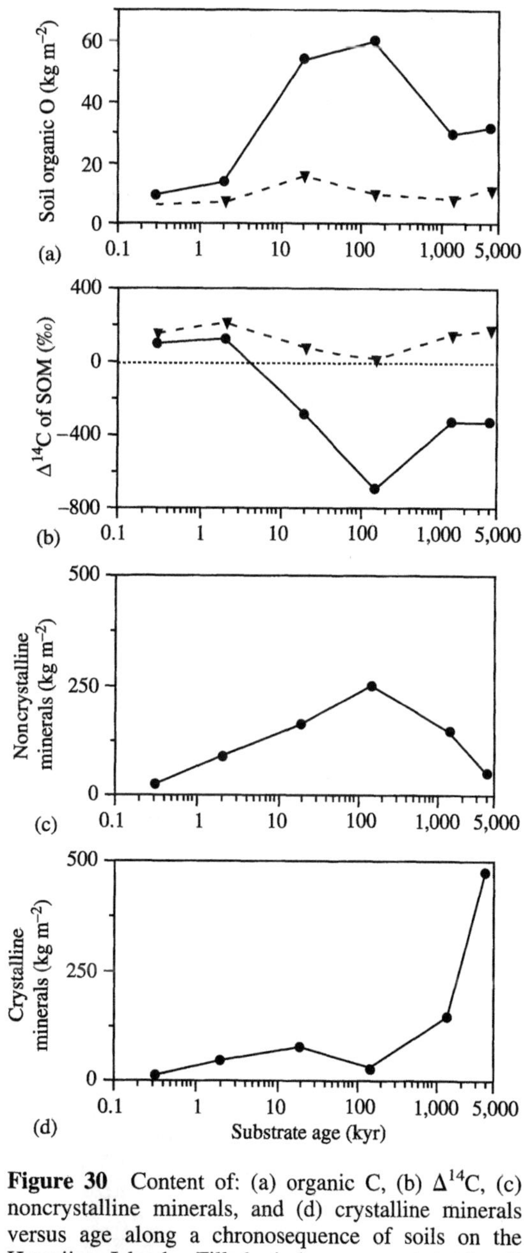

Figure 30 Content of: (a) organic C, (b) $\Delta^{14}C$, (c) noncrystalline minerals, and (d) crystalline minerals versus age along a chronosequence of soils on the Hawaiian Islands. Filled circles represent total soil profile and filled triangles represent surface (O and A) horizons only (Torn *et al.*, 1997) (reproduced by permission of Nature Publishing Group from *Nature*, **1997**, *389*, 170–173).

Textural differences may influence microbial activity and decomposition rates indirectly by modifying the physiochemical environment (Nelson *et al.*, 1997). Indirect mechanisms include entrapment of organic particles in the interiors of aggregates (Van Veen and Kuikman, 1990) and in micropores (Hassink *et al.*, 1993) where microbes are physically excluded. Van Veen *et al.* (1985) suggest that in fine textured soils, products released from dead bacterial cells are retained in the vicinity of surviving bacteria which minimizes the leaching of labile carbon. A similar mechanism has been proposed for the observed decrease in decomposition in earthworm casts (Martin, 1991) where increased organo-clay bonding (Zhang and Schrader, 1993) can reduce the amount of leachate in casts versus bulk soil (McInerney and Bolger, 2000b).

Soil texture is a major control on the distribution of pore sizes in a volume of soil, which, in turn, largely determines both the water holding capacity and the soil water potential (Ψ, MPa) for a given volumetric water content (θ, m^3 water m^{-3} soil). The water potential limits to microbial activity are listed in Table 4. Studying native SOM mineralization rates across a range of soils, Scott *et al.* (1996) found a significant interaction between soil texture and Ψ (Figure 31), which could best be explained in terms of the percentage water-filled pore space (WFPS). As clay increased from 7% to 20%, SOM mineralized per kg of soil increased linearly with increasing WFPS ($r^2 = 0.71$, $P < 0.01$). However as WFPS increases towards 100%, it might be expected that O_2 limitations would begin to decrease the activity of the microorganisms. Thomsen *et al.* (1999) incorporated the soil textural interactions with water availability by defining microbially accessible water (MAW) as the difference between volumetric water content (θ) and the volume of inaccessible water (IW). IW was defined as the water content at the permanent wilting point ($\Psi = -1.5$ MPa), which increases with increasing clay and SOM content. In a ^{14}C-labeled residue addition experiment, MAW significantly improved predictions of $^{14}CO_2$ evolution from a series of soils with increasing clay content from $r^2 = 0.55$ for θ to 0.88 for MAW (Thomsen *et al.*, 1999).

Soil structure. Mature soils often exhibit well-developed macrostructure as a result of physical binding between organo-metal–mineral complexes (Tisdall and Oades, 1982) and polysaccharides excreted by roots (Traore *et al.*, 2000) and soil organisms (Tisdall, 1994) and through the enmeshment of soil particles by fine roots and fungal hyphae (Tisdall *et al.*, 1997). Although Tisdall and Oades (1982) propose that clay particle binding to bacterial colonies is a major factor in promoting microaggregate formation, Bossuyt *et al.* (2001) found that fungi, not bacteria, were the dominant agents in macroaggregate (>2,000 μm) formation. Factors which affect fungal biomass—such as resource quality, predation, and nutrient availability—will impact the formation of macroaggregates in the soil (Bossuyt *et al.*, 2001).

The effect of cultivation on aggregate stability, and carbon cycling, is important for carbon storage, and the disruption of this structure may be the fundamental mechanism behind the rapid and large loss of soil carbon that occurs following

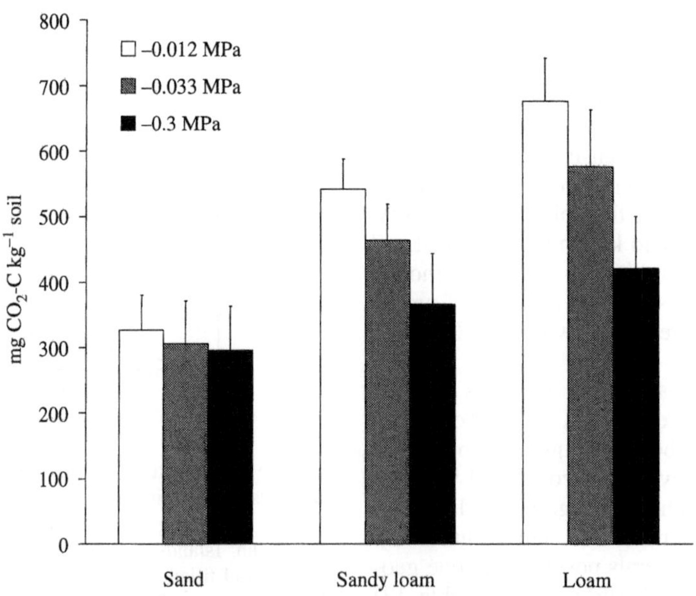

Figure 31 Interaction between soil texture and water pressure on total native soil C mineralization during a 91-day incubation. Error bars represent 1 standard error (source Scott *et al.*, 1996).

the initiation of agriculture (Paul *et al.*, 1997). Beare *et al.* (1994b) found that WSAs were larger, more abundant, and more stable in no-till agriculture versus conventional tillage, resulting in a fourfold increase in the protected nitrogen pool. An important finding of this study was that the macroaggregate-protected OM was more labile than the unprotected inter-aggregate OM (Beare *et al.*, 1994a). These authors suggested that the unprotected inter-aggregate OM has been more processed by microorganisms. Six *et al.* (2001), using multiple methods of SOM characterization, confirmed that inter-aggregate particulate OM is more decomposed than coarse intra-aggregate particulate OM.

Following the ^{13}C content after six years of continuous maize production, Puget *et al.* (2000) found that stable macroaggregates (2–3 mm diameter) contained 62% young carbon (maize derived) mostly in the form of particulate OM, whereas aggregates of medium stability contained 38% young carbon, and unstable aggregates contained only 5% young carbon. At a site that had been under maize for 23 years, the differences in maize-derived or young carbon between stability classes were less pronounced, although similar percentages of particulate OM were found relative to the six-year site, indicating that the macroaggregates are transient in nature (Puget *et al.*, 2000). These differences in aggregate structure and stability across cropping regimes result in substantially increased decomposition rates in fields under conventional tillage. For example, Balesdent *et al.* (1990) found that ^{13}C-derived turnover times decreased from 127 years for no-tillage to 68 years for superficial-tillage to 56 years for conventional-tillage after only 17 years of maize production.

Nutrient availability. Nutrient-deficient soils can limit decomposition rates directly due to nutrient limitation of the microbial community, and indirectly through the production of lower quality litter by nutrient-stressed plants. Gosz (1981) found that when climate was held constant, plants growing on nitrogen-rich sites produce leaf litter with greater nutrient concentrations that decompose more quickly than litter from nitrogen-poor sites. On a long-term age gradient of soils on the Hawaiian islands, Ostertag and Hobbie (1999) found that one-year mass loss for leaf and fine root litter is least for the old phosphorus-limited site, intermediate for the young nitrogen-limited site, and greatest at the intermediate-aged fertile site (neither nitrogen nor phosphorus limiting). This indirect effect of fertility on resource quality can often explain most of the variation in decay rates between sites of differing fertility (Hobbie, 1992; Ostertag and Hobbie, 1999; Prescott, 1995).

If nutrient availability was limiting decomposition especially of low-quality litters (high C/N and lignin/N ratios), fertilization should relieve this microbial limitation and decomposition rates will increase. Several studies report increases in k following nutrient additions to soil, indicating that decomposer organisms are nutrient limited (Ostertag and Hobbie, 1999; Prescott *et al.*, 1992). However in a review of nitrogen effects on decomposition, Fog (1988) cited more than 60 papers that reported neutral or negative effects on decomposition following nitrogen fertilization. As an explanation for this observation, microbiologists often attribute the lack of a decomposition response after fertilization to the fact that microbes are primarily limited by substrate (carbon) not by nutrient availability (McClaugherty *et al.*, 1985; Ostertag and Hobbie, 1999; Prescott *et al.*, 1992). In line with this, Fog (1988) found that positive effects of fertilization were most commonly reported for easily degradable substrates, which in effect were not substrate limited.

Several mechanisms have been proposed to explain the negative effects of nitrogen fertilization on decomposition rates (Fog, 1988): (1) the competitive balance in the microbial community is shifted to bacteria and fungi which rapidly utilize the additional nutrients at the expense of the slower growing lignolytic basidomycetes; (2) the production of ligninase is repressed in the presence of available nitrogen and recalcitrant lignocellulose accumulates preferentially; and (3) polymerization reactions with polyphenols are enhanced by the presence of amino compounds resulting in the formation of more recalcitrant compounds. These proposed mechanisms leading to decreased decomposition in the presence of nitrogen fertilizers are very similar to several theories on humification (see Section 8.07.6).

While the influence of nutrient availability on litter decomposition has been well documented, its influence on SOM turnover has received much less attention. Working along the same age gradient as Ostertag and Hobbie (1999), Torn *et al.* (in review) found that the slow SOM pool (turnover times of decades) in the O and A horizons turned over two to three slower in the young nitrogen-limited and old phosphorus-limited sites than in the non-nutrient limiting sites of intermediate age. Additionally, experimental fertilization at the young and old sites had only small and inconsistent effects on SOM turnover (Torn *et al.*, in review). These results in combination suggest that nutrient availability does not directly influence microorganisms ability to degrade SOM (consistent with the assumption that microorganisms are primarily carbon-substrate limited, not nutrient limited), but, rather, indirectly through the quality of the litter resource being incorporated into the soil (similar to the conclusions for leaf and root litter from Ostertag and Hobbie, 1999).

Soil texture also plays a major indirect role in regulating decomposition dynamics by effecting nutrient availability. The exchange capacity due to clay mineralogy will have a big impact on the nutrient cycling in a soil. Soils with a high cation exchange capacity will better retain nutrient cations (i.e., K^+, NH_4^+, Ca^{2+}, Mg^{2+}, etc.), whereas soils with appreciable anion exchange capacity will better retain nutrient anions (i.e., NO_3^-, PO_4^-, SO_4^{2-}, etc.) (Table 13).Working with a kaolonite-, Al- and Fe-oxide dominated textural gradient in Para, Brazil, Silver *et al.* (2000) found that total phosphorus increased fourfold from a sandy to clayey soil and that extractable (available) phosphorus decreased threefold. A similar pattern of phosphorus availability was observed by Tiessen *et al.* (1994).

8.07.7.4 Climate

Beginning with the work of Tenny and Waksman (1929), studies of detrital processing have often emphasized the importance of the climate of the decomposition environment. In developing their generalized conceptual decomposition model (Figure 24), Lavelle *et al.* (1993) recognized that the temperature and moisture regimes usually exert the dominant control on decay.

Meentemyer (1995) asked the question: "Is the climate of decay processes measured at weather stations?" The microclimate at the scale of an individual leaf may be nearly fully decoupled from what a nearby weather station measures. For field studies following decomposition processes at short time intervals, this decoupling may have important consequences. However, for studies comparing annual mass losses across large geographic regions, the average climate at the weather station will probably suffice. Additionally, as Lavelle *et al.* (1993) noted, climate is not an equally important constraint across all ecosystems. Whitford (1989) hypothesized that abiotic controls on decomposer food webs are least important in moist, closed-canopy forests and most important in hot deserts.

Most research has shown that temperature and moisture regimes are the relevant climatic controls on decomposition. How these "regimes" are represented as measurable variables is not universally agreed upon and will often depend on the scale of investigation. Temperature of air or soil can be represented as an average, maximum, or minimum for any given temporal scale. It can also be incorporated into time-dependent models by regressing mass loss against degree-day data (Andren and Paustian, 1987). Moisture can be represented simply by the amount of precipitation or more directly as the volumetric water content (θ) or soil water potential (Ψ). These variables can be combined with knowledge of the water-holding capacity of the soil into the AET, what many researchers (Berg *et al.*, 1993a; Meentemyer, 1978) consider to be a more robust measure of climate. In this section, we will consider how each of these groups of variables controls both litter and SOM processing across a range of scales.

Temperature. In controlled laboratory incubations of single substrates where only temperature varies and moisture is nonlimiting, investigators often find a strong positive correlation between mineralization rate and temperature (Moore, 1986; Nicolardot *et al.*, 1994; Sorensen, 1981). Similar observations have been made for fresh litter and mineral-associated OM (e.g., Kirschbaum, 1995). Swift *et al.* (1979) proposed that litters of differing quality have consistent differences in decay rates regardless of temperature or moisture regimes. However, Taylor and Parkinson (1988a) observed that pine and aspen litters have different initial (16-week) responses to temperature (Figure 32).

This differential response is generally not seen in laboratory studies of SOM mineralization (Fang and Moncrieff, 2001; Katterer *et al.*, 1998; Kirschbaum, 1995) or in an analysis of field studies conducted in nonmoisture limiting systems (Lloyd and Taylor, 1994). For example, Katterer *et al.* (1998) empirically fit two-component exponential decay models (Equation (5)) to 25 sets of incubation data and found that a single nonlinear model could explain 96% of the variance in the SOM decay rate response (r) factor to temperature (Figure 33). The r-factor is simply a scalar that adjusts all k_1 and k_2 values to a common temperature ($r = 1$ at 30 °C), i.e.,

$$k_1 = rk_1^{T=30} \qquad (9)$$

In fitting their model, Katterer *et al.* (1998) assumed that temperature affects the decay constants of the labile and recalcitrant fractions equally. There is evidence from litter decomposition studies that this assumption may not be valid. Studying the decomposition of a ^{14}C-labeled standard plant litter along an altitudinal gradient in Venezuela, Couteaux *et al.* (2002) found that the labile fraction decayed rapidly regardless of temperature, whereas $k_{recalcitrant}$ was strongly correlated with temperature. Nicolardot *et al.* (1994) also found nonproportional responses of glucose and holocellulose carbon and nitrogen mineralization rates to increasing temperature in a laboratory microcosm study.

Researchers have fit a variety of empirical models to describe the temperature dependence of soil respiration and OM decomposition. Lloyd

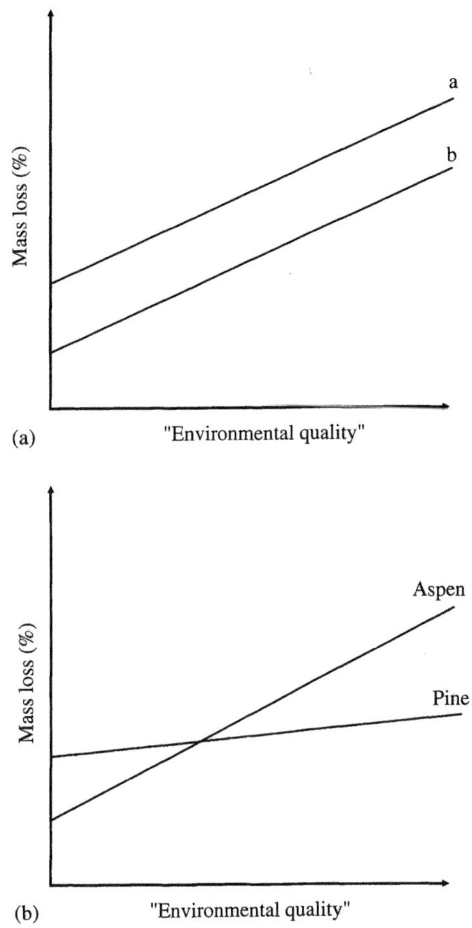

Figure 32 Graphical depiction of the response of litter decomposition rate to changes in climate and edaphic factors that favor decomposition ("environmental quality"). (a) Theoretical response that substrate "a" decomposes faster than substrate "b" under all conditions (b) The observed response of pine needles and aspen leaf litter mass loss to changing conditions. (after Taylor and Parkinson, 1988a).

and Taylor (1994) and Fang and Moncrieff (2001) have excellent discussions of many of these models. Figure 34 illustrates the relationship between the mean residence times of SOM with mean annual temperature (MAT) for a wide range of forested sites (Sanderman *et al.*, 2003). We will use this data set to discuss the fit and appropriateness of several common temperature-dependent decomposition models. Linear and exponential equations are simple empirical expressions relating turnover time with increasing temperature, although they lack any theoretical basis (Fang and Moncrieff, 2001). While some researchers have found a good linear fit for decomposition of specific substrates (Nicolardot *et al.*, 1994), a cursory examination of Figure 34 reveals its inappropriateness. A common log-transform of the data produces a more reasonable fit ($r^2 = 0.77$, $P < 0.0001$, $n = 21$). However, as Fang and Moncrieff (2001) found, an examination of the distribution of residuals reveals that turnover times are underestimated at low temperatures and overestimated at high temperatures.

The exponential equation is given as

$$\tau = \tau_0 e^{\beta(T-\bar{T})} \tag{10}$$

where β is a fitted parameter and τ_0 is the turnover time at the MAT, yields a very good fit to the data in Figure 34 ($r^2 = 0.81$, $P < 0.0001$, $n = 21$). The rate increase of any reaction to a 10 °C increase in temperature is termed the Q_{10} value (Kirschbaum, 2000). We calculated Q_{10} for these data (where $Q_{10} = e^{10\beta}$) to be 2.9 for this regression at a MAT of 281 K. Although the model produces an excellent fit with the data, fitting a constant Q_{10} is not biologically realistic (Fang and Moncrieff, 2001; Kirschbaum, 1995). In a review of laboratory incubation data, Kirschbaum (1995) found a greater temperature sensitivity of decomposition at lower temperatures. To accommodate this, the Arrhenius equation can be expressed in terms of turnover time at 10 °C (as illustrated in Lloyd and Taylor, 1994):

$$\tau = \tau_{10} \exp\left[\frac{E^*}{R}\left(\frac{1}{283.15} - \frac{1}{T}\right)\right] \tag{11}$$

where E^* is the activation energy and R is the universal gas constant (8.314 J mol^{-1} K^{-1}). Although the empirical fit of Equation (11) in Figure 34 is similar to a simple exponential model ($r^2 = 0.81$), this equation is theoretically more justifiable because it is derived from basic biochemical principles (Moore, 1986). A plot of the natural logarithm of decay rates versus the inverse of absolute temperature (the "Arrhenius graph") should reveal a linear range whose slope is the activation energy. Beyond this range, decomposition rates drop sharply due to microbial limitations in extreme cold and hot environments.

Lloyd and Taylor (1994) suggested that the assumption of a constant activation energy across a range of ecosystems, that likely have vastly different microbial communities, may not be valid. Plotting the data of both Lloyd and Taylor (1994) and Sanderman *et al.* (2003) reveals a nonlinear decrease in the ln(k) with decreasing T. Based on this observation, Lloyd and Taylor modified Equation (11) to account for the effect of a nonconstant activation energy:

$$\tau = \tau_{10} \exp E_0\left(\frac{1}{283.15 - T_0} - \frac{1}{T - \bar{T}_0}\right) \tag{12}$$

where E_0 and T_0 are fitted parameters and τ_{10} is the turnover time at 10 °C. This equation is no longer directly related to biochemical principles but

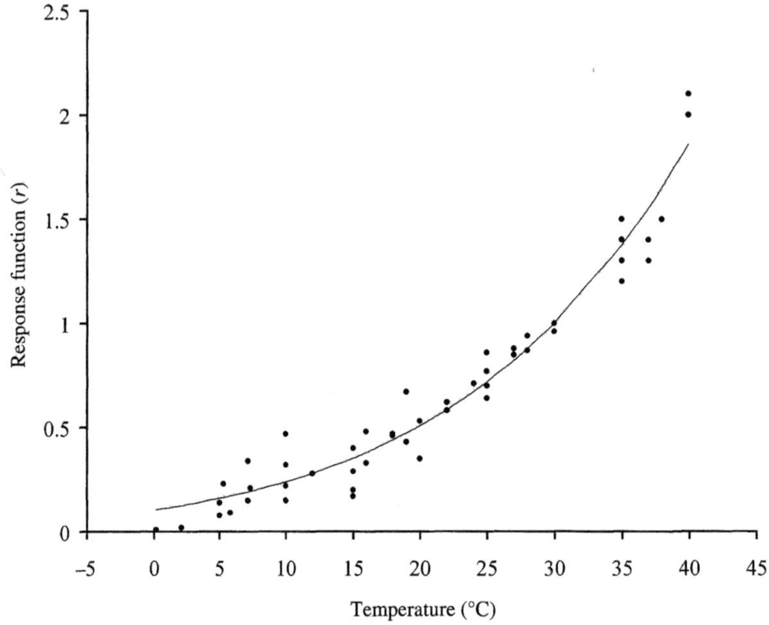

Figure 33 The temperature response function r of the first-order decomposition rate constants k_l and k_r. The r-function is a scalar that relates k at any given temperature to its maximal rate at a reference temperature ($T_{ref} = 30$ °C, see text for details). Also shown is the modeled fit of the Lloyd and Taylor equation (Equation (12), $r^2 = 0.96$). (after Katterer *et al.*, 1998).

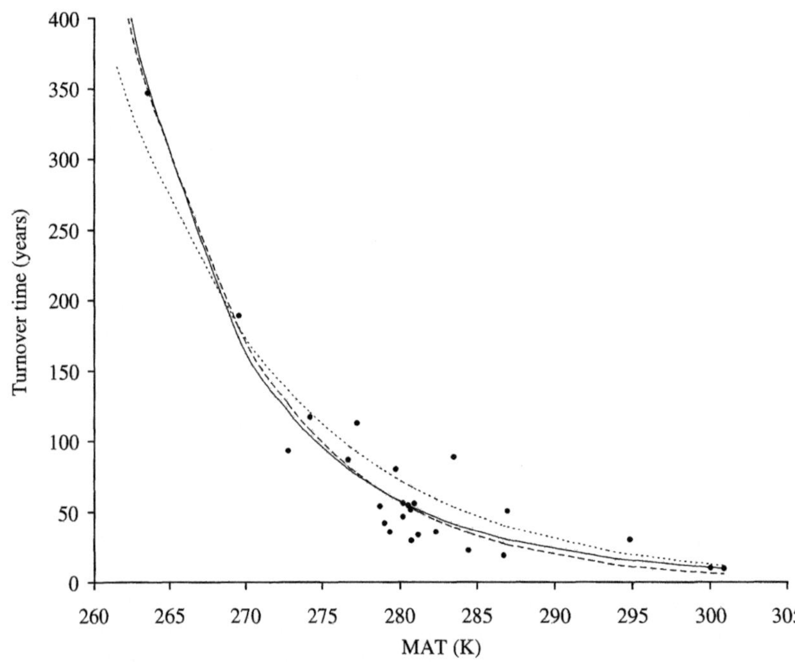

Figure 34 Steady-state SOM mean residence times derived from eddy covariance data plotted against MAT. Bold solid curve is Lloyd and Taylor fit; bold dashed curve is the Q_{10} best fit; and light dashed curve represents a linear regression wit, the common log of turnover time (source Sanderman *et al.*, 2003).

Lloyd and Taylor (1994) suggested that this "semi-empirical formulation effectively gives a decrease in activation energy with increasing temperature." Equation (12) fit to the data in Figure 34 produces a similar, albeit more gradual, temperature response than reported by Kirschbaum (1995) where the calculated Q_{10} decreases from 4 to 2 across a 36 °C temperature range.

Dalias *et al.* (2001b) incubated soils from different latitudes with a common ^{14}C-labeled substrate and found that the temperature response optima (similar to the *r*-factor of Katterer *et al.*, 1998) varied with latitudinal origin of soil and decreased with time for some soils. This finding is consistent with the notion that the native microbial community is adapted to function optimally within the normal range of temperatures they are acclimated to. Additionally, these researchers found that the proportion of litter acting as labile carbon (α in Equation (8)) decreased with increasing temperature. In a companion study, Dalias *et al.* (2001a) confirmed this temperature-dependent stabilization by moving soils, initially incubated at different temperatures until a common percent mass loss was achieved, to a common temperature. The litter conditioned at the highest temperature had the slowest mass loss rate when incubated at a common temperature, indicating a greater fraction of recalcitrant carbon. Finding similar results, Zogg *et al.* (1997) concluded that this response to temperature may be due to a shift in the microbial community away from lignin degraders which cannot compete with fast-growing bacteria at the higher temperatures. Berg *et al.* (1993b) and Johansson *et al.* (1995) also found that in long-term litter studies, a greater percentage of remaining mass was lignified at the warmer site. Dalias *et al.* (2001a) cautioned that results such as these may invalidate models that assume constant pool sizes with different temperatures. Additionally, this observed shift in the temperature optima with mass loss of litter has important implications for decay studies in general: generating thermal response functions from a single cohort or input of litter followed through time may be misleading, because there will be a progressive shift in recalcitrance and hence a non-steady-state response with temperature. In this case, steady-state methods for determining decomposition rates may be of more general use.

Most simply, steady-state annual decomposition rates can be calculated by dividing the standing stock of carbon in the forest floor or mineral soil by the annual flow in (litter production; Olsen, 1963) or out (heterotrophic respiration; Raich and Schlesinger, 1992), respectively. Although single-pool models are gross simplifications of complex systems, they offer the advantage of being inexpensive and easy ways to compare turnover times at large spatial scales. Based on estimates of the steady-state turnover of SOM using the shift in $\delta^{13}C$ following vegetation change across a range of sites, Giardina and Ryan (2000) found no relationship between mean residence time and MAT. This data set contained study sites that were forests converted to pasture, forests converted to various agricultural practices, and pastures that were re-invaded by forests. The varying disturbance histories between these sites may have masked any real trends in climate. By limiting the data set to one land-use type (i.e., forest converted to pasture), Sanderman *et al.* (2003) found that the $\delta^{13}C$-derived turnover times do indeed decrease exponentially with increasing MAT ($r^2 = 0.59$, $P = 0.001$, $n = 14$). By calculating steady-state turnover times from ^{14}C measurements of density separates of paired pre/post atomic bomb testing soils along an altitudinal transect in the Sierra Nevada, Trumbore *et al.* (1996) reported a very similar trend to that depicted in Figure 34 for the mass-weighted mean of the low-density ($<2.0\ g\ cm^{-3}$) and hydrozylable portion of the dense fraction ($>2.0\ g\ cm^{-3}$) of the surface horizons.

Moisture. Microorganisms rely on the thin film of water surrounding soil particles and in soil pores for the basic necessities of life. As water potential drops, microorganisms must either enter a quiescent resting phase or perish (Table 4). In laboratory incubations at constant temperatures of both litter (Donnelly *et al.*, 1990; Taylor and Parkinson, 1988a) and SOM (Orchard and Cook, 1983; Thomsen *et al.*, 1999), carbon mineralization rates are strongly correlated with soil moisture levels. For example, Donnelly *et al.* (1990) found that at 12 °C increasing θ from $0.2\ m^3\ m^{-3}$ to $0.6\ m^3\ m^{-3}$ increased microbial biomass fourfold and increased cellulose degradation by over 60-fold and lignin degradation by sixfold. In a 77-day incubation of a silt loam pasture soil at 25 °C, Orchard and Cook (1983) found that microbial activity measured as CO_2 evolution increased linearly with the common logarithm of water potential (Figure 35). Unfortunately, many soil moisture experiments only report volumetric water content (θ), which is hard to compare physiologically across different soils (Walse *et al.*, 1998). For any value of θ, the soil water potential (Ψ), which may be of greater physiological significance, will be higher for a fine textured soil than a coarse textured soil.

In field studies, soil moisture can affect decomposition dynamics in a number of ways. In systems where soil moisture is nonlimiting, researchers find little, if any, correlation between soil respiration and soil water potential (Davidson *et al.*, 1998). However in the same study, Davidson *et al.* (1998) found that including matrix potential with soil temperature could significantly improve soil flux predictions at sites that are affected by seasonally low Ψ (Figure 36). Xu and Qi (2001) found a similar seasonal pattern in a ponderosa pine plantation in the Mediterranean climate of northern California—soil CO_2 efflux increases sharply with increasing spring temperatures until the soil dries out a few months later and flux rates drop sharply. Bryant *et al.* (1998)

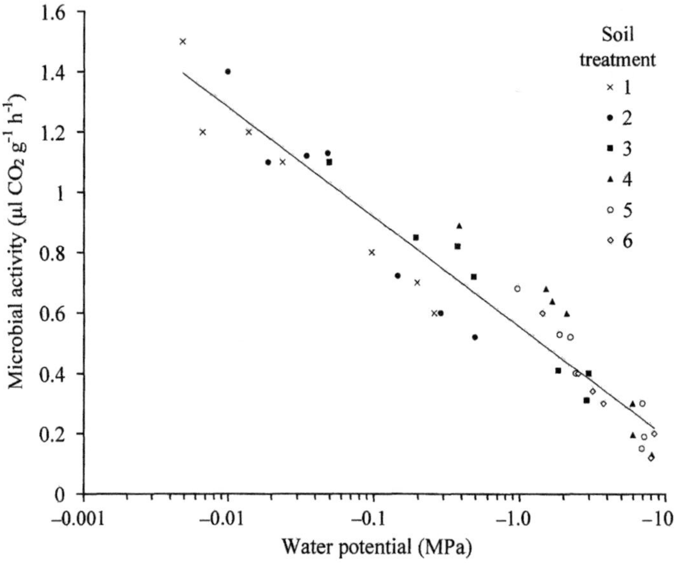

Figure 35 Relationship between water potential (MPa) and microbial respiration during a 77-day incubation of a silt loam with six different initial water potentials: −0.005 MPa, −0.01 MPa, −0.05 MPa, −0.4 MPa, −1.0 MPa, and −1.5 MPa, for treatments 1–6, respectively (source Orchid and Cook, 1983).

concluded that soil moisture was the primary control on surface litter decomposition in an alpine tundra ecosystem.

In the wet tropics, rapid forest floor (O horizon) decomposition begins with leaf fall, and nutrient releases reflect leaf fall seasonality (Anderson and Swift, 1983). In contrast, in dry tropics, the nutrient flush is delayed until the onset of the rainy season resulting in a strong, highly seasonal, pulse of nutrients to the ecosystem (Cornejo *et al.*, 1994). In Cornejo *et al.*'s study, mass loss was greater under irrigation during the dry season than in the control (no irrigation) due primarily to leaching, and secondarily to microbial enhancement. Similarly, Swift *et al.* (1981) found that during the dry season, substantial amounts of phosphorus, nitrogen, and other nutrients accumulated on the forest floor in undecomposed leaf litter, and within four weeks of the onset of rains all of the accumulated phosphorus and half of the nitrogen was released through decomposition (reported in Anderson and Swift, 1983). In cool temperate forest soils, West *et al.* (1992) found that drying produced only a small reduction in microbial biomass, but rewetting caused basal respiration and carbon mineralization to increase strongly. West *et al.* (1992) attributed this strong increase in mineralization to enhanced microbial turnover and subsequent release of immobilized nutrients. Rapid rewetting imposes osmotic stresses that many microorganisms cannot tolerate resulting in ruptured of the cell walls, and a large turnover of the microbial biomass (Kieft *et al.*, 1987). Consistent with general microbial ecology, the bacterial populations are often observed to recover faster from drying/rewetting events, than the slower growing fungal communities (Scheu and Parkinson, 1994; Van Gestel *et al.*, 1993). Studying the effects of wetting/drying cycles on earthworm casts, McInerney and Bolger (2000b) found that supersaturation of casts can swell clays enough to break the organo-clay bonds and reduce the structural stability of the cast, leading to a reduction in the protection of the SOM within.

Working along a well-controlled precipitation gradient on the island of Hawaii, Austin and Vitousek (2000) found a clear increase in *M. polymorpha* litter decomposition rates with mean annual precipitation (MAP) increasing from 500 mm to 5,500 mm. To tease apart direct versus indirect effects of increasing MAP, these authors studied the decomposition of *in situ* litter, common substrates across sites, and litter taken from each site and decomposed at the common site. The decay rates for the common substrates of varying quality all increased with MAP; as did the *in situ* litters despite the fact that palatability, as measured by mass loss at the common site, decreased with increasing MAP of the site where the litter was obtained (Austin and Vitousek, 2000). Along this gradient, climate exerts a direct control on microbial metabolism, and indirect effects on nutrient (phosphorus) availability in the soils (Austin and Vitousek, 1998), which are manifest in the decreasing litter quality from the dry to wet sites. A second important indirect effect of soil moisture that several investigators have also observed is that as soil moisture content declines the temperature sensitivity or Q_{10} value decreases in both lab-based litter incubations

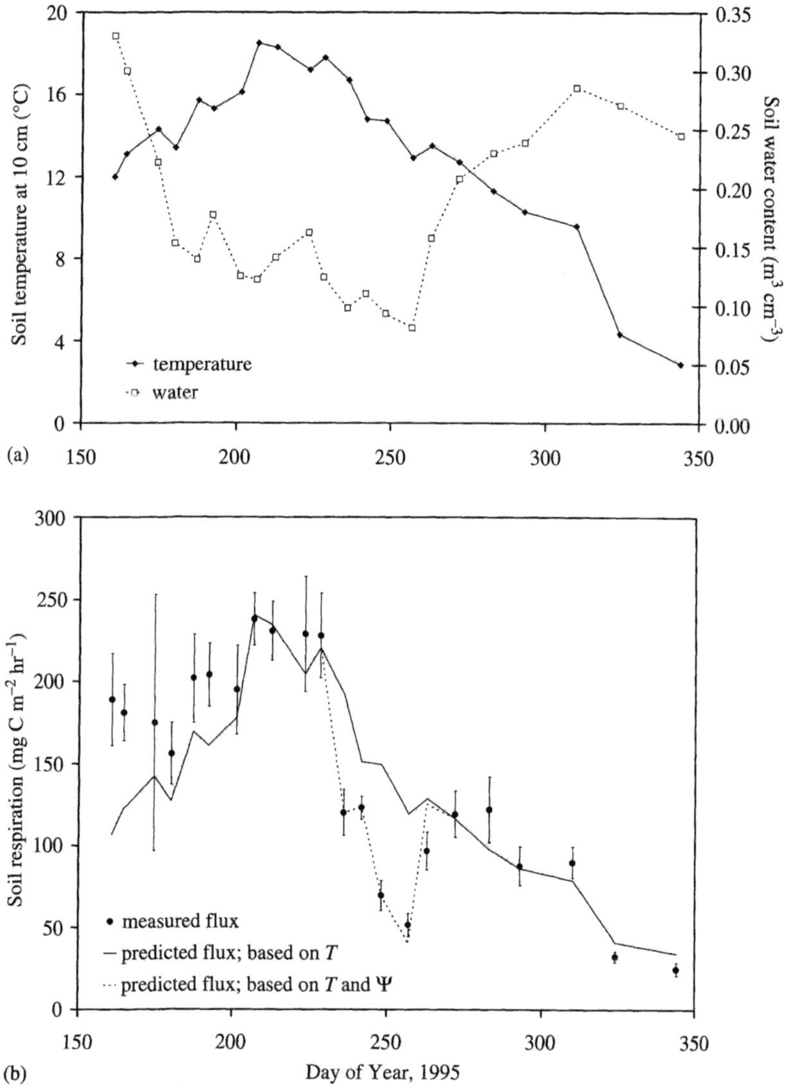

Figure 36 Seasonal variation in mean soil temperature and soil water content for well-drained soils at the Harvard Forest, MA. Also shown is the measured flux (error bars = 1 standard error) and predicted flux based on temperature alone (solid line) and based on temperature and soil water potential (dashed line) (after Davidson *et al.*, 1998).

(Moore, 1986) and field-based soil respiration studies (Xu and Qi, 2001).

AET. Combining temperature and moisture into a multivariate model is often complicated by the fact that these two factors co-vary in many ecosystems. Many researchers believe that AET best combines these interactions between temperature and precipitation along with soil physical parameters in predicting decomposition dynamics (Berg *et al.*, 1993a; Gonzalez and Seastedt, 2001; Meentemyer, 1977). Studying the first-year mass loss of a common pine needle substrate, Meentemyer and Berg (1986) found that 80% ($P < 0.0001$, $n = 14$) of the variance between pine forests along a latitudinal transect in Sweden could be explained by AET. Berg *et al.* (1993a) expanded this study to include pine forests in the eastern United States and across Europe and again found a strong linear trend between first-year mass loss and AET (adjusted $R^2 = 0.5$, $P < 0.0001$, $n = 39$). However, Whitford *et al.* (1981) suggested that this simple AET model is not appropriate in very low-AET (i.e., desert) ecosystems and in very disturbed systems (i.e., clear-cut forests) because of biotic adaptations and marked differences in microclimate, respectively.

In cross-site studies designed to examine climatic controls on litter decomposition, a common practice is to use a "standardized litter" from one forest and place it in all of the treatment sites (Berg *et al.*, 1993a; Gholz *et al.*, 2000; Gonzalez and Seastedt, 2001). This experimental design raises the question: Is there a "home field advantage" (Gholz *et al.*, 2000) for native litters?

Based on a comparison of a short-grass prairie, a mountain meadow, and a lodgepole pine ecosystem, Hunt *et al.* (1988) suggested that the decomposer community in a particular ecosystem may be adapted to the native litter of that ecosystem. Berg *et al.* (1993b) attempted to limit this treatment effect by restricting their study to a variety of coniferous forests. In a comparison of long-term (five-year) litter dynamics across 28 long-term ecological research (LTER) sites, Gholz *et al.* (2000) found that when k was normalized by AET, leaf litter from the tropical hardwood *Drypetes glauca* decomposed nearly 3 times faster in broadleaf forests than in coniferous forests ($k = 1.37$ and 0.43, respectively) and needle litter from *Pinus resinous* decomposed 30% faster in coniferous forests than in the broadleaf forests.

With this qualification in mind, Gholz *et al.* (2000) reported that the synthetic climate variable DEFAC, a composite of temperature and moisture influences on decay rates (see Section 8.07.8), developed for predicting the climate influence on litter, and SOM decomposition in the Century model (Parton *et al.*, 1993) is a better predictor of litter decay rates than AET across the LTER sites (Figure 37). Gholz *et al.* (2000) concluded that DEFAC is applicable over a wider range of ecosystems because it "places a primary emphasis on temperature over a relatively broad range of moisture availability, and still maintains some decomposition at very low precipitation." For the studies listed in Table 11, DEFAC (Pearson's $r = 0.94$, $P < 0.01$, $n = 6$) does a better job in predicting the mean k across a range of litter types than AET (Pearson's $r = 0.70$, $P > 0.2$, $n = 6$).

The studies presented in this section indicate that not only does climate play a large direct role in litter decomposition and SOM turnover, but the indirect effects of climate on the structure of the microbial community (Zogg *et al.*, 1997), litter chemistry (Alvarez and Lavado, 1998), soil texture (Alvarez and Lavado, 1998), and nutrient status (Vitousek *et al.*, 1994) combine to make climate the overriding influence on detrital processing when comparing across large spatial and temporal scales. This is especially evident in studies that calculate turnover times from an aggregate of substrates with varying quality (e.g., CO_2 efflux studies and isotopic studies on bulk samples).

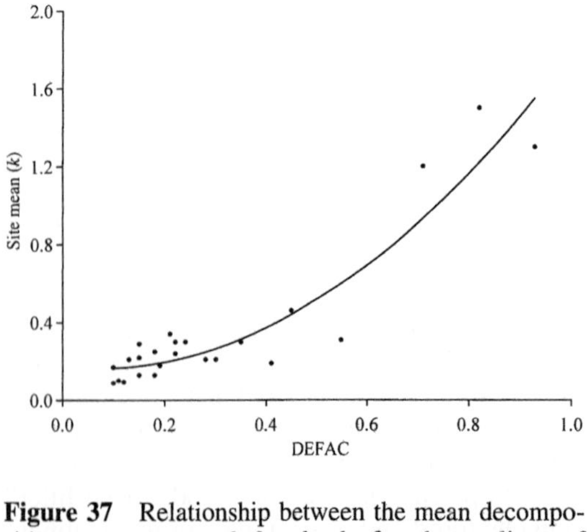

Figure 37 Relationship between the mean decomposition rate constant k for the leaf and root litter of *D. glauca* and *Pinus elliotii*, and DEFAC, a synthetic climate variable (see text for explanation), across a range of N. American sites. Solid line is a quadratic fit with the data ($r^2 = 0.88$, $n = 28$) (source Gholz *et al.*, 2000).

8.07.7.5 Multiple Constraints

In developing a model that incorporates both AET and litter quality (i.e., lignin concentration), Meentemyer (1978) found that mass loss is more sensitive to changes in AET when the litter is of higher quality. However, in nature litter quality will often co-vary with climate, making it harder to tease apart the relative strength of climate versus resource quality controls on decomposition. In a review of 44 litter decomposition data sets, Aerts (1997) was able to separate the relative contributions of direct and indirect controls of climate and litter quality (lignin/N) on decomposition rates using the multiple regression technique of path analysis (Figure 38). In this "triangular relationship," AET is significantly correlated with both k and the lignin/N ratio which, in turn, is also significantly correlated with k. More importantly, ~20% of the relationship between k and AET can be explained by the indirect effects of the lignin/N ratio (Aerts, 1997).

The lignin concentration increase rate (LCIR) as defined by Berg *et al.* (1993b) is the slope of a plot of lignin concentration versus mass loss. The LCIR for Scots pine litter is highest under favorable climate conditions (and/or soil nutrient status) which can actually lead in the long term

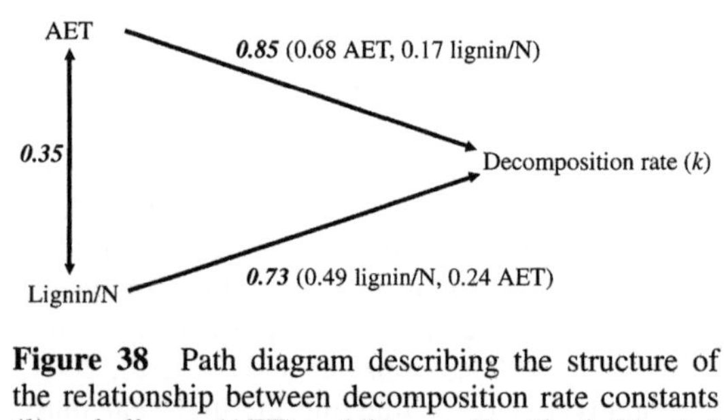

Figure 38 Path diagram describing the structure of the relationship between decomposition rate constants (k) and climate (AET) and litter quality (lignin/N ratio) for first-year decomposition from 44 locations. Numbers in bold type are the Pearson correlation coefficients among variables and the numbers in parentheses partition the coefficients into direct and indirect effects of the predictor variables (AET and lignin/N) (source Aerts, 1997).

to the formation of more recalcitrant SOM. In colder/dryer conditions, LCIR is lower because mass loss of labile components is slower leading to an accumulation of easily decomposable SOM. This line of reasoning can help explain why SOM at colder sites is relatively more sensitive to changes in temperature. These results also support the model of Berg and Staaf (1980) that in early stages of litter decay, climate, or nutrient status controls mass loss of labile components; then in later stages, lignin loss rate is rate limiting factor.

The models of Meentemyer (1978), Parton *et al.* (1987), and Aerts (1997) consider climate and litter quality as the dominant controls on decomposition rates. Of these studies, only Aerts (1997) has included tropical sites in developing his regressions. For the tropical sites only, no significant regressions could be developed with AET and the lignin/N ratio could only explain a little more than 50% of the variance in k. Lavelle *et al.* (1993) and others have suggested that in the humid tropics where macroinvertebrate populations are large, biological regulation of decomposition becomes much more important. In Section 8.07.7.1, we have reviewed numerous studies where soil fauna play an important, if not dominant, role in detrital processing. Gonzalez and Seastedt (2001) designed a three-factorial (climate, litter quality, and macrofauna) experiment to test the hypothesis that the soil biota become significant factors in controlling litter decay rates. These authors found that climate and litter quality were significant at all three sites (temperate, dry tropical, and wet tropical), while the faunal exclusion only produced significant effects in the wet tropical site.

Detrital processing can be thought of as a continuum from fresh litter to stabilized SOM (Agren and Bosatta, 2002). At different stages in this continuum, the relative importance of each of these environmental and biological factors that have been identified as controlling decomposition dynamics will likely vary. The initial stages of mass loss are characteristically most affected by climate, resource quality, and, when abundant, soil macrofauna. The physical soil environment also needs to be considered as an important control on the turnover of more humified SOM in the mineral horizons. It is also evident from this literature review that observed correlations between decay rates and "decomposition" factors are often attributable to both the direct effects of that factor on microbial metabolism and to the indirect interactions with other factors.

8.07.8 MODELING APPROACHES

Modeling has been critical to the development and advancement of biogeochemistry. Models are tools for both synthesizing and extrapolating to broader spatial and temporal scales, and for testing and refining hypotheses. Decomposition and detrital processing in terrestrial ecosystems has been represented in a number of biogeochemical-process-based simulation models, including Century (Parton *et al.*, 1987, 1993), Introductory Carbon Balance Model (ICBM) (Katterer and Andren, 2001), G'Day (Comins and McMurtrie, 1993), Linkages (Pastor and Post, 1986), and Terrestrial Ecosystem Model (TEM) (McGuire *et al.*, 1992). The use of first-order kinetics to describe decomposition rates and division of SOM into a number of pools based on decomposability are traits common to all of these models with roots dating back to the original work of Tenney and Waksman (1929). These simulation models differ from the empirical models presented in the previous sections in that they can be potentially parametrized for application across a range of ecosystems.

Complex bottom-up modeling (e.g., food webs) of decomposition and detrital processing is often impossibly difficult at the ecosystem level. To quote Andrén *et al.* (1999): "How do we construct an ecosystem out of a square meter harboring 50,000 microorganism species, 50 mite species, 10 enchytraied species, 1000 insect species, 100 plant species, etc.? And the adjacent square meter, where careful sampling revealed a slightly, but statistically significant, different species composition, is that another ecosystem?" To deal with this amazing genetic diversity, most ecosystem models use top–down or process-based approaches where the organismal physiology and their population size are only implicitly included (see Schimel (2001) for an excellent discussion on this topic). When applying a process-based model to environmental change, Andrén *et al.* (1999) suggested five crucial questions that should be addressed to determine if implicitly including the biota is sufficient: (1) Are there any keystone species? (2) Are there species-poor functional groups? (3) Are there functional groups with low dispersal abilities? (4) Are there important "narrow-physiology" microbes? and (5) Are there significant interactions that may vanish or appear?

The underlying assumption in implicitly including decomposer organisms, and microorganisms in particular, is that microbes are primarily carbon limited and the microbial community can rapidly expand and adapt to handle even extreme environmental changes (Finlay *et al.*, 1997; Schimel, 2001). As shown in numerous examples throughout this chapter, these assumptions seem to hold in most situations. However, as Schimel (2001) highlights, there are numerous situations where these assumptions may not hold and an explicit characterization of the microbial biomass and physiology would result in a more accurate model.

The basic representation of decomposition using first-order kinetics has already been discussed. Here we will highlight how the widely used Century model represents decomposition, SOM cycling, and the controls on these processes (Parton *et al.*, 1987, 1993). We choose this model due to its intensive development, and its wide use and availability to the scientific community. For a detailed description, instructions, and to download the version 5 of Century (latest as of early 2003) see www.nrel.colostate.edu/projects/century5/. A box model representation of the SOM dynamics in Century version 5 is shown in Figure 39. Litter, both aboveground and belowground, is divided into structural and metabolic carbon based on its lignin/N ratio. Surface metabolic carbon is incorporated into the microbial carbon pool, whereas the belowground metabolic carbon pool flows into the active carbon (~0.5 yr turnover time). However, as McGill (1996) noted, the microbial carbon pool is essentially indistinguishable from the active carbon pool. Most structural and active carbon is incorporated into the slow carbon pool (turnover of 20–40 yr), which can further be transformed into passive carbon (200–1500 yr). Both slow and passive carbon can be transformed back onto active carbon. During each microbially mediated transformation, a percentage of the carbon is mineralized to CO_2. For each carbon pool, there is an analogous nitrogen, phosphorus, and sulfur pool, which is linked directly to the carbon flows by the carbon-to-nutrient ratios.

Decomposition of each pool is calculated as

$$\frac{dC_i}{dt} = K_i L_C A T_m C_i \tag{13}$$

where C_i = carbon in the *i*th pool, K_i = maximum decomposition rate for the *i*th pool, L_C is the impact of lignin on structural decomposition, A is the combined effect of temperature and moisture on decomposition, and T_m is the effect of soil texture (defined as the silt + clay content of the soil) (Parton *et al.*, 1994). All of the effects on K_i are scaled from 0 to 1, so that if any of the variables are suboptimal, the maximum decomposition rate will not be realized. As shown in Figure 39, most transformations are controlled by only a subset of these scalars. The shapes of the

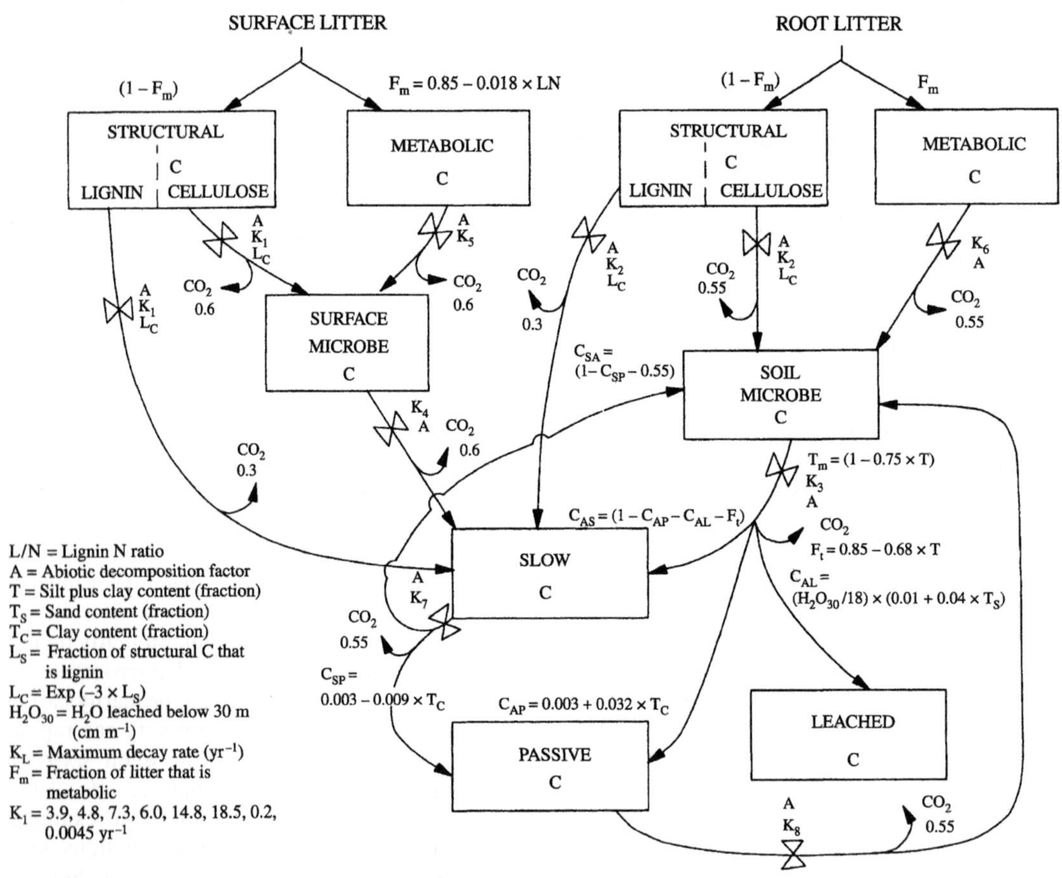

Figure 39 Diagram of carbon pools and flows in the Century version 5 agroecosystem model (Parton *et al.*, 1993) (reproduced by permission of the American Geophysical Union from *Global Biogeochem. Cycles*, **1993**, *7*, 785–809). See text for discussion.

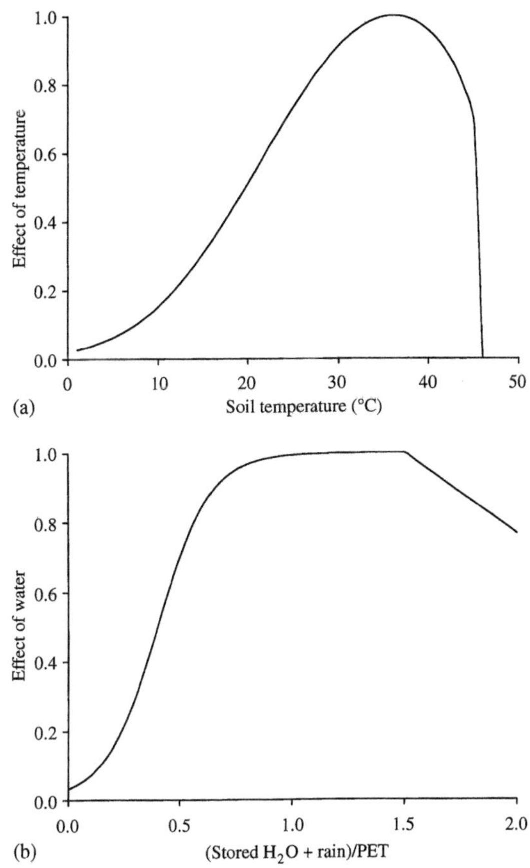

Figure 40 The effect of: (a) soil temperature and (b) soil water on decomposition in the Century model.

soil temperature and moisture effect curves are shown in Figure 40.

The general structure of most terrestrial biogeochemistry models is similar to what we have described for the Century model; however, these models do vary significantly in how they treat the controls on decomposition (Burke *et al.*, 2002). For example, TEM uses an exponentially increasing temperature scalar ($Q_{10} = 2.0$), while Century uses a generalized Poisson function that drops off rapidly above 40 °C (shown in Figure 40(a)). Similar differences between models are found when comparing the moisture, textural, and litter quality controls. These differences become strikingly apparent when several models are used to simulate ecosystem dynamics in highly contrasting ecosystems (Moorhead *et al.*, 1999).

Part of these discrepancies can be attributed to the fact that many of these models are based on a limited number of empirical studies. In Century, the temperature response function that is *the same for all pools* of carbon was derived from the Sørensen (1981) 90-day laboratory incubation study on the decomposition of labeled cellulose at three different temperatures (10 °C, 20 °C, and 30 °C) (Parton *et al.*, 1987). It is unlikely that microbial physiology developed to digest cellulose will respond to temperature the same way as physiologies designed to digest lignin and other recalcitrant polymers. In fact, Taylor and Parkinson (1988a) found that pine and aspen litters had different initial responses to changing temperature, with the more palatable aspen leaves being more sensitive to changes in temperature (Figure 32).

In Century, as well as other models, soil texture has a significant influence on the turnover rate of the active carbon pool (i.e., $T_m = 1 - 0.75T$ where T is the silt + clay content of the soil) and the rate of stabilization into the slow and passive carbon pools. This formulation of T_m is based on data from the Sørensen (1981) study in Danish agricultural soils and validated with a series of semi-arid grassland soils (Gregorich *et al.*, 1991; Schimel *et al.*, 1985; Van Veen *et al.*, 1985). The applicability of this relationship to soils with a different mineralogy, especially variably charged clays found in many tropical regions, is questionable (e.g., Giardina and Ryan, 2000, 2001). Additionally, the disruption of soil aggregates in the original laboratory experiments (Sørensen, 1981) can result in significantly different carbon dynamics than in an intact soil (see Section 8.07.7.3).

Despite these and other potential limitations, Century and other biogeochemistry models are able to reasonably simulate the carbon and nitrogen dynamics in many ecosystems (see Parton *et al.*, 1994 for example). The success of these models in simulating current conditions, given proper parametrization, gives scientists and policy makers greater confidence in believing model predictions under climate change and differing management regimes. However, caution should be used in interpreting these predictions given the uncertainties in the data used for model development, such as those described above. Greater attention should be given to designing experiments that test these models' assumptions regarding the controls on decomposition (Burke *et al.*, 2002) and on the underlying microbiology (Schimel, 2001).

8.07.9 CONCLUSIONS

Decomposition is essentially a microbial process whose rate is regulated by a suite of physical and biological controls. At the microscopic level, many of the biochemical details of microbial degradation of specific carbon and nitrogen compounds have been well documented. However, beyond this scale, most of our understanding of decomposition becomes empirical in nature. As we have highlighted in this review, there exists a wealth of data on the decomposition of plant litter while root and soil OM have received significantly less attention. The composition and activity of soil fauna, initial resource quality, soil properties such as texture and nutrient availability, and climatic factors such as temperature and moisture must all be considered. These factors can directly affect a microorganism's ability to degrade a resource or indirectly affect biodegradation by altering one or more of the above factors in a way that affects microbial activity. Most biogeochemical models incorporate only a subset of these controls and rarely incorporate the numerous indirect effects. As carbon cycling and sequestration becomes more and more of a sociopolitical issue, the demand for more precise biogeochemical models that incorporate all of these direct and indirect effects will likely increase.

REFERENCES

Abe T., Bignell D. E., and Higashi M. (2000) *Termites: Evolution, Sociality, Symbioses, Ecology*. Kluwer Academic.

Aber J. D., Melillo J. M., and McClaugherty C. A. (1990) Predicting long-term patterns of mass loss, nitrogen dynamics, and soil organic matter formation from initial fine litter chemistry in temperate forest ecosystems. *Can. J. Botany* **68**(10), 2201–2208.

Aerts R. (1997) Climate, leaf litter chemistry and leaf litter decomposition in terrestrial ecosystems: a triangular relationship. *Oikos* **79**(3), 439–449.

Aerts R. and De Caluwe H. (1997) Nutritional and plant-mediated controls on leaf litter decomposition of Carex species. *Ecology (Washington, DC)* **78**(1), 244–260.

Agren G. I. and Bosatta E. (2002) Reconciling differences in predictions of temperature response of soil organic matter. *Soil Biol. Biochem.* **34**(1), 129–132.

Alvarez R. and Lavado R. S. (1998) Climate, organic matter and clay content relationships in the Pampa and Chaco soils, Argentina. *Geoderma* **83**(1–2), 127–141.

Amann R. I., Ludwig W., and Schleifer K.-H. (1995) Phylogenetic identification and *in situ* detection of individual microbial cells without cultivation. *Microbiol. Rev.* **59**(1), 143–169.

Amato M. and Ladd J. N. (1980) Studies of nitrogen immobilization and mineralization in calcareous soils: V. Formation and distribution of isotope-labelled biomass during decomposition of 14C- and 15N-labelled plant material. *Soil Biol. Biochem.* **12**, 405–411.

Amundson R., Stern L., Baisden T., and Wang Y. (1998) The isotopic composition of soil and soil-respired CO_2. *Geoderma* **82**(1–3), 83–114.

Anderson D. W. and Paul E. A. (1984) Organo-mineral complexes and their study by radiocarbon dating. *Soil Sci. Soc. Am. J.* **48**, 298–301.

Anderson J. M. and Swift M. J. (1983) Decomposition in tropical forests. In *Tropical Rain Forest: Ecology and Management* (eds. S. L. Sutton, T. C. Whitmore, and A. C. Chadwick). Blackwell, pp. 210–287.

Anderson J. P. E. and Domsch K. H. (1978) A physiological method for the quantitative measurement of microbial biomass in soils. *Soil Biol. Biochem.* **10**, 215–221.

Anderson R. V., Coleman D. C., Cole C. V., and Elliott E. T. (1981) Effect of nematodes *Acrobeloides* sp. and *Mesodiplogaster lheritieri* on substrate utilization and nitrogen and phosphorus mineralization in soil. *Ecology* **62**, 549–555.

Andrén O., Brussaard L., and Clarholm M. (1999) Soil organism influence on ecosystem-level processes bypassing the ecological hierarchy? *Appl. Soil Ecol.* **11**(2–3), 177–188.

Andren O. and Paustian K. (1987) Barley straw decomposition in the field A comparison of models. *Ecology (Tempe)* **68**(5), 1190–1200.

Asmar F., Eiland F., and Nielsen N. E. (1994) Effect of extracellular-enzyme activities on solubilization rate of soil organic nitrogen. *Biol. Fertility Soils* **17**(1), 32–38.

Atlas R. M. and Bartha R. (1998) *Microbial Ecology: Fundamentals and Applications*. Benjamin/Cummings.

Augris N., Balesdent J., Mariotti A., and Derenne S. (1998) Structure and origin of insoluble and non-hydrolysable, aliphatic organic matter in a forest soil. *Org. Geochem.* **28**, 119–124.

Austin A. T. and Vitousek P. M. (1998) Nutrient dynamics on a precipitation gradient in Hawai'i. *Oecologia (Berlin)* **113**(4), 519–529.

Austin A. T. and Vitousek P. M. (2000) Precipitation, decomposition and litter decomposability of Metrosideros polymorpha in native forests on Hawai'i. *J. Ecol.* **88**(1), 129–138.

Baldock J. A., Oades J. M., Waters A. G., Peng X., Vassallo A. M., and Wilson M. A. (1992) Aspects of the chemical structure of soil organic materials as revealed by solid-state carbon-13 NMR spectroscopy. *Biogeochemistry (Dordrecht)* **16**(1), 1–42.

Baldock J. A., Oades J. M., Nelson P. N., Skene T. M., Golchin A., and Clarke P. (1997) Assessing the extent of decomposition of natural organic materials using solid-state 13C NMR spectroscopy. *Austral. J. Soil Res.* **35**(5), 1061–1083.

Balesdent J., Mariotti A., and Guillet B. (1987) Natural carbon-13 abundance as a tracer for studies of soil organic matter dynamics. *Soil Biol. Biochem.* **19**(1), 25–30.

Balesdent J., Wagner G. H., and Mariotti A. (1988) Soil organic matter turnover in long-term field experiments as revealed by carbon-13 natural abundance. *Soil Sci. Soc. Am. J.* **52**(1), 118–124.

Balesdent J., Mariotti A., and Boisogntier D. (1990) Effect of tillage on soil organic carbon mineralization estimated from carbon-13 abundance in maize fields. *J. Soil Sci.* **41**(4), 587–596.

Bamforth S. S. and Lousier J. D. (1995) Protozoa in tropical litter decomposition. In *Soil Organisms and Litter Decomposition in the Tropics* (ed. M. V. Reddy). Westview Press, pp. 59–74.

Bardgett R. D., Cook R., Yeates G. W., and Denton C. S. (1999) The influence of nematodes on below-ground processes in grassland ecosystems. *Plant Soil* **212**(1), 23–33.

Barois I. and Lavelle P. (1986) Changes in respiration rate and some physicochemical properties of a tropical soil during transit through Pontoscolex corethrurus (Glossoscolecidae, Oligochaeta). *Soil Biol. Biochem.* **18**(5), 539–542.

Barzegar A. R., Nelson P. N., Oades J. M., and Rengasamy P. (1997) Organic matter, sodicity, and clay type: influence on soil aggregation. *Soil Sci. Soc. Am. J.* **61**(4), 1131–1137.

Beare M. H., Pohlad B. R., Wright D. H., and Coleman D. C. (1993) Residue placement and fungicide effects on fungal communities in conventional and no-tillage soils. *Soil Sci. Soc. Am. J.* **57**(2), 392–399.

Beare M. H., Cabrera M. L., Hendrix P. F., and Coleman D. C. (1994a) Aggregate-protected and unprotected organic matter pools in conventional- and no-tillage soils. *Soil Sci. Soc. Am. J.* **58**(3), 787–795.

Beare M. H., Hendrix P. F., and Coleman D. C. (1994b) Water-stable aggregates and organic matter fractions in conventional- and no-tillage soils. *Soil Sci. Soc. Am. J.* **58**(3), 777–786.

Benedik M. J. and Strych U. (1998) Serratia marcescens and its extracellular nuclease. *FEMS Microbiol. Lett.* **165**(1), 1–13.

Berg B. (1984a) Decomposition of moss litter in a mature Scots pine forest. *Pedobiologia* **26**(5), 301–308.

Berg B. (1984b) Decomposition of root litter and some factors regulating the process—long-term root litter decomposition in a Scots pine forest. *Soil Biol. Biochem.* **16**(6), 609–617.

Berg B. and Staaf H. (1980) Decomposition rate and chemical changes of Scots pine needle litter: II. Influence of chemical composition. *Ecol. Bull. (Stockholm)* **32**, 363–372.

Berg B. and Staaf H. (1981) Leaching, accumulation and release of nitrogen in decomposing litter. In *Terrestrial Nitrogen Cycles* (eds. F. E. Clark and T. Rosswall). Ecological Bulletins, vol. 33, pp. 163–178.

Berg B. and Staaf H. (1987) Release of nutrients from decomposing white birch leaves and Scots pine needle litter. *Pedobiologia* **30**(1), 55–63.

Berg B. and Wessen B. (1984) Changes in organic-chemical components and ingrowth of fungal mycelium in decomposing birch leaf litter as compared to pine needles. *Pedobiologia* **26**, 285–298.

Berg B., Hannus K., Popoff T., and Theander O. (1982) Changes in organic chemical components of needle litter during decomposition: long-term decomposition in a Scots pine forest: I. *Can. J. Botany* **60**, 1310–1319.

Berg B., Muller M., and Wessen B. (1987) Decomposition of red clover (*Trifolium pratense*) roots. *Soil Biol. Biochem.* **19**, 589–594.

Berg B., Berg M. P., Bottner P., Box E., Breymeyer A., Calvo De Anta R., Couteaux M., Escudero A., Gallardo A., and Kratz W. (1993a) Litter mass loss rates in pine forests of Europe and Eastern United States: some relationship with climate and litter quality. *Biogeochemistry (Dordrecht)* **20**(3), 127–159.

Berg B., McClaugherty C., and Johansson M.-B. (1993b) Litter mass-loss rates in late stages of decomposition at some climatically and nutritionally different pine sites: long-term decomposition in a Scots pine forest: VIII. *Can. J. Botany* **71**(5), 680–692.

Bhatnagar T. (1975) Lombriciens et humification: un aspect noveau de l'incorporation microbienne d'azote induite par les vers de terre. In *Biodegradation et Humification* (eds. G. Kilbertus, O. Reisinger, A. Mourney, and J. A. Cancela Da Fonseca). Pierron, pp. 169–182.

Blaire J. M., Parmelee R. W., and Lavelle P. (1994) Influences of earthworms on biogeochemistry. In *Earthworm Ecology and Biogeography in North America* (ed. P. F. Hendrix). Lewis Publishers, pp. 127–158.

Blanchart E., Bruand A., and Lavelle P. (1993) The physical structure of casts of Millsonia anomala (Oligochaeta: Megascolecidae) in shrub savanna soils (Cote d'Ivoire). *Geoderma* **56**(1–4), 119–132.

Bondiett E., Martin J. P., and Haider K. (1972) Stabilization of amino sugar units in humic-type polymers. *Soil Sci. Soc. Am. Proc.* **36**(4), 597–602.

Boone R. D., Nadelhoffer K. J., Canary J. D., and Kaye J. P. (1998) Roots exert a strong influence on the temperature sensitivity of soil respiration. *Nature (London)* **396**(6711), 570–572.

Bossuyt H., Denef K., Six J., Frey S. D., Merckx R., and Paustian K. (2001) Influence of microbial populations and residue quality on aggregate stability. *Appl. Soil Ecol.* **16**(3), 195–208.

Bottner P., Sallih Z., and Billes G. (1988) Root activity and carbon metabolism in soils. *Biol. Fertility Soils* **7**(1), 71–78.

Bouché M. B. (1977) Strategies lombriciennes. *Ecol. Bull. (Stockholm)* **25**, 122–132.

Bracewell J. M., Haider K., Larter S. R., and Schulten H. R. (1989) Thermal degradation relevant to structural studies of humic substances. In *Humic Substances II: In Search of Structure* (eds. M. H. B. Hayes, P. MacCarthy, R. L. Malcolm, and M. J. Swift). Wiley, pp. 181–222.

Brady N. C. and Weil R. R. (2002) *The nature and property of soils*. Prentice Hall.

Breland T. A. and Hansen S. (1998) Comparison of the difference method and 15N technique for studying the fate of nitrogen from plant residues in soil. *Biol. Fertility Soils* **26**(3), 164–168.

Breznak J. A. (2000) Ecology of prokaryotic microbes in the guts of wood- and litter-feeding termites. In *Termites: Evolution, Sociality, Symbioses, Ecology* (eds. T. Abe, D. E. Bignell, and M. Higashi). Kluwer Academic, pp. 209–231.

Brown S. and Lugo A. E. (1982) The storage and production of organic matter in tropical forests and their role in the global carbon cycle. *Biotropica* **14**, 161–187.

Bryant D. M., Holland E. A., Seastedt T. R., and Walker M. D. (1998) Analysis of litter decomposition in an alpine tundra. *Can. J. Botany: Rev. Can. Botanique* **76**(7), 1295–1304.

Bubb K. A., Xu Z. H., Simpson J. A., and Saffigna P. G. (1998) Some nutrient dynamics associated with litterfall and litter decomposition in hoop pine plantations of southeast Queensland. Australia. *Forest Ecol. Manage.* **110**(1–3), 343–352.

Burges A. and Raw F. (1967) *Soil Biology*. Academic Press.

Burke I. C., Kaye J. P., Bird S. P., Hall S. A., McCulley R. L., and Sommerville G. L. (2002) In *Evaluating and Testing Models of Terrestrial Biogeochemistry: the Role of Temperature in Controlling Decomposition* (eds. C. Canham, J. Cole, and W. Lauenroth) (in press).

Burtelow A. E., Bohlen P. J., and Groffman P. M. (1998) Influence of exotic earthworm invasion on soil organic matter, microbial biomass and denitrification potential in forest soils of the northeastern United States. *Appl. Soil Ecol.* **9**, 197–202.

Buswell J. A. (1991) Fungal degradation of lignin. In *Handbook of Applied Mycology* (eds. D. K. Arora, B. Rai, K. G. Mukerji, and G. R. Knudsen). Dekker, vol. 1, pp. 425–480.

Butler M. J. and Day A. W. (1998) Fungal melanins: a review. *Can. J. Microbiol.* **44**(12), 1115–1136.

Buxton R. D. (1981) Termites and turn-over of dead wood in an arid tropical environment. *Oecologia* **51**, 379–384.

Caesar-TonThat T. C. and Cochran V. L. (2000) Soil aggregate stabilization by a saprophytic lignin-decomposing basidiomycete fungus: I. Microbiological aspects. *Biol. Fertility Soils* **32**(5), 374–380.

Carlile M. J. and Watkinson S. C. (1994) *The Fungi*. Academic Press.

Catroux G. and Schnitzer M. (1987) Chemical spectroscopic and biological characteristics of the organic matter in particle size fractions separated from an Aquoll. *Soil Sci. Soc. Am. J.* **51**(5), 1200–1207.

Chadwick O. A., Derry L. A., Vitousek P. M., Huebert B. J., and Hedin L. O. (1999) Changing sources of nutrients during four million years of ecosystem development. *Nature (London)* **397**(6719), 491–497.

Chapin F. S., III (1991) *Effects of Multiple Environmental Stresses on Nutrient Availability and Use*. Academic Press.

Cheng W. (1996) Measurement of rhizosphere respiration and organic matter decomposition using natural 13C. *Plant Soil* **183**(2), 263–268.

Cheng W. and Johnson D. W. (1998) Elevated CO_2, rhizosphere processes, and soil organic matter decomposition. *Plant Soil* **202**(2), 167–174.

Cheng W. X., Coleman D. C., Carroll C. R., and Hoffman C. A. (1994) Investigating short-term carbon flows in the rhizospheres of different plant species, using isotopic trapping. *Agron. J.* **86**(5), 782–788.

Chernin L. S., Winson M. K., Thompson J. M., Haran S., Bycroft B. W., Chet I., Williams P., and Stewart G. S. A. B. (1998) Chitinolytic activity in *Chromobacterium violaceum*: substrate analysis and regulation by quorum sensing. *J. Bacteriol.* **180**(17), 4435–4441.

Cheshire M. V. and Mundie C. M. (1981) The distribution of labelled sugars in soil particle size fractions as a means of distinguishing plant and microbial carbohydrate residue. *J. Soil Sci.* **32**, 605–618.

Chichester F. W. (1970) Tranformations of fertilizer nitrogen in soil: II. Total and N15-labelled nitrogen of soil organo-mineral sedimentation fractions. *Plant Soil* **33**, 437–456.

Chitwood B. G. and Chitwood M. B. H. (1974) *Introduction to Nematology*. University Park Press.

Christensen B. T. (1987) Decomposability of organic matter in particle size fractions from field soils with straw incorporation. *Soil Biol. Biochem.* **19**(4), 429–436.

Christensen B. T. (1992) Physical fractionation of soil and organic matter in primary particle size and density separates. *Adv. Soil Sci.* **20**, 1–90.

Christensen B. T. (2001) Physical fractionation of soil and structural and functional complexity in organic matter turnover. *Euro. J. Soil Sci.* **52**(3), 345–353.

Christensen B. T. and Sørensen L. H. (1985) The distribution of native and labelled carbon between soil particle size fractions isolated from long-term incubation experiments. *J. Soil Sci.* **36**, 219–229.

Chung N. H., Lee I. S., Song H. S., and Bang W. G. (2000) Mechanisms used by white-rot fungus to degrade lignin and toxic chemicals. *J. Microbiol. Biotechnol.* **10**(6), 737–752.

Clements R. O., Murray P. J., and Sturdy R. G. (1991) The impact of 20 years' absence of earthworms and three levels of nitrogen fertilizer on a grassland soil environment. *Agri. Ecosys. Environ.* **36**, 75–86.

Collins N. M. (1979) The nests of *Macrotermes bellicosus* (Smeathman) from Mokwa, Nigeria. *Insectes Soc.* **26**, 240–246.

Collins N. M. (1981) The role of termites in the decomposition of wood and leaf litter in the southern guinea savanna of Nigeria. *Oecologia* **51**, 389–399.

Colman E. A. and Hamilton E. L. (1947) The San Dimas lysimeters. Research Note 47, US Forest Service, Forest Range Experiment Station..

Comins H. N. and McMurtrie R. E. (1993) Long-term response of nutrient-limited forests to CO_2 enrichment—equilibrium behavior of plant-soil models. *Ecol. Appl.* **3**(4), 666–681.

Cooke R. C. and Rayner A. D. M. (1984) *Ecology of Saprotrophic Fungi.* Longman.

Cornejo F. H., Varela A., and Wright S. J. (1994) Tropical forest litter decomposition under seasonal drought: nutrient release, fungi and bacteria. *Oikos* **70**(2), 183–190.

Cornelissen J. H. C. and Thompson K. (1997) Functional leaf attributes predict litter decomposition rate in herbaceous plants. *New Phytol.* **135**(1), 109–114.

Cornelissen J. H. C., Perez-Harguindeguy N., Diaz S., Grime J. P., Marzano B., Cabido M., Vendramini F., and Cerabolini B. (1999) Leaf structure and defence control litter decomposition rate across species and life forms in regional floras on two continents. *New Phytol.* **143**(1), 191–200.

Cortez J., Hameed R., and Bouche M. B. (1989) Carbon and nitrogen transfer in soil with or without earthworms fed with carbon-14 and nitrogen-15 labelled wheat straw. *Soil Biol. Biochem.* **21**(4), 491–498.

Coulson J. C. and Butterfield J. (1978) Investigation of biotic factors determining rates of plant decomposition on blanket bog. *J. Ecol.* **66**(2), 631–650.

Couteaux M. M., Sarmiento L., Bottner P., Acevedo D., and Thiery J. M. (2002) Decomposition of standard plant material along an altitudinal transect (65–3,968 m) in the tropical Andes. *Soil Biol. Biochem.* **34**(1), 69–78.

Cox P., Wilkinson S. P., and Anderson J. M. (2001) Effects of fungal inocula on the decomposition of lignin and structural polysaccharides in Pinus sylvestris litter. *Biol. Fertility Soils* **33**(3), 246–251.

Crews T. E., Kitayama K., Fownes J. H., Riley R. H., Herbert D. A., Mueller-Dombois D., and Vitrousek P. M. (1995) Changes in soil phosphorus fractions and ecosystem dynamics across a long chronosequence in Hawaii. *Ecology (Washington, DC)* **76**(5), 1407–1424.

Cromack K., Jr, Todd R. L., and Monk C. D. (1975) Patterns of basidiomycete nutrient accumulation in conifer and deciduous forest litter. *Soil Biol. Biochem.* **7**, 265–268.

Currie W. S. and Aber J. D. (1997) Modeling leaching as a decomposition process in humid montane forests. *Ecology (Washington, DC)* **78**(6), 1844–1860.

Currie W. S., Aber J. D., McDowell W. H., Boone R. D., and Magill A. H. (1996) Vertical transport of dissolved organic C and N under long-term N amendments in pine and hardwood forests. *Biogeochemistry (Dordrecht)* **35**(3), 471–505.

Dalias P., Anderson J. M., Bottner P., and Couteaux M. M. (2001a) Long-term effects of temperature on carbon mineralisation processes. *Soil Biol. Biochem.* **33**(7–8), 1049–1057.

Dalias P., Anderson J. M., Bottner P., and Couteaux M. M. (2001b) Temperature responses of carbon mineralization in conifer forest soils from different regional climates incubated under standard laboratory conditions. *Global Change Biol.* **7**(2), 181–192.

Darwin C. (1881) *The Formation of Vegetable Mould, through the Action of Worms: with Observations on their Habits.* J. Murray.

Dash M. C. (1990) Oligochaeta: Enchytraeidae. In *Soil Biology Guide* (ed. D. L. Dindal). Wiley, pp. 311–340.

Dash M. C. (1995) The Enchytraeidae. In *Soil Organisms and Litter Decomposition in the Tropics* (ed. M. V. Reddy). Westview Press, pp. 89–102.

Dash M. C., Nanda B., and Mishra P. C. (1981) Digestive enzymes in three species of tropical Enchytraeidae (Oligochaeta). *Oikos* **36**, 316–318.

Davidson E. A., Belk E., and Boone R. D. (1998) Soil water content and temperature as independent or confounded factors controlling soil respiration in a temperate mixed hardwood forest. *Global Change Biol.* **4**(2), 217–227.

Dekker R. F. H. (1985) Biodegradation of the hemicelluloses. In *Biosynthesis and Biodegradation of Wood Components* (ed. T. Higuchi). Academic Press.

De Leeuw J. W. and Largeau C. (1993) A review of macromolecular organic compounds that comprise living organisms and their role in kerogen, coal, and petroleum formation. In *Organic Geochemistry* (eds. M. H. Engel and S. A. Macko). Plenum, pp. 23–72.

Dindal D. L. (1990) *Soil Biology Guide.* Wiley.

Dix N. J. and Webster J. (1995) *Fungal Ecology.* Chapman and Hall.

Dkhas M. C. and Mishra R. R. (1986) Microflora in earthworm casts. *J. Soil Biol. Ecol.* **6**, 24–31.

Donnelly P. K., Entry J. A., Crawford D. L., and Cromack K., Jr. (1990) Cellulose and lignin degradation in forest soils: response to moisture, temperature, and acidity. *Microbial Ecol.* **20**(3), 289–296.

Doube B. M. and Brown G. G. (1998) Life in a complex community: functional interactions between earthworms, organic matter, microorganisms, and plants. In *Earthworm Ecology* (ed. C. A. Edwards). Lucie Press, Boca Raton, FL, pp. 179–211.

Durall D. M., Todd A. W., and Trappe J. M. (1994) Decomposition of 14C-labelled substrates by ectomycorrhizal fungi in association with Douglas fir. *New Phytol.* **127**(4), 725–729.

Dworkin M. (1998) Multiculturalism versus the single microbe. In *Bacteria as Multicellular Organisms* (eds. J. A. Shapiro and M. Dworkin). Oxford University Press, pp. 3–13.

Edmonds R. L. (1988) Decomposition rates and nutrient dynamics in small-diameter woody litter in four forest ecosystems in Washington, USA. *Can. J. Forest Res.* **17**(6), 499–509.

Edmonds R. L., Vogt D. J., Sandberg D. H., and Driver C. H. (1986) Decomposition of Douglas-fir [Pseudotsuga menziesii] and red alder [Alnus rubra] wood in clear-cuttings. *Can. J. Forest Res.* **16**(4), 822–831.

Edwards C. A. and Bohlen P. J. (1996) *Biology and Ecology of Earthworms.* Chapman and Hall.

Edwards C. S. and Heath G. (1963) The role of soil animals in breakdown of leaf material. In *Soil Organisms* (eds. J. Doeksen and J. van der Drift). North Holland, pp. 76–84.

Edwards N. T. (1982) The use of soda-lime for measuring respiration rates in terrestrial systems. *Pedobiologia* **23**, 321–330.

Eggleton P. (2000) Global patterns of termite diversity. In *Termites: Evolution, Sociality, Symbioses, Ecology* (eds. T. Abe, D. E. Bignell, and M. Higashi). Kluwer Academic, pp. 25–51.

Eijsackers H. and Zehnder A. J. B. (1990) Litter decomposition—a Russian Matriochka doll. *Biogeochemistry* **11**(3), 153–174.

Eriksson K. E. and Wood T. M. (1985) Biodegradation of cellulose. In *Biosynthesis and Biodegradation of Wood Components* (ed. T. Higuchi). Academic Press, pp. 469–504.

Fang C. and Moncrieff J. B. (2001) The dependence of soil CO_2 efflux on temperature. *Soil Biol. Biochem.* **33**(2), 155–165.

Fengel D. and Wegener G. (1984) *Wood: Chemistry, Ultrastructure, Reactions*. de Gruyter.

Finlay B. J., Maberly S. C., and Cooper J. I. (1997) Microbial diversity and ecosystem function. *Oikos* **80**(2), 209–213.

Fioretto A., Musacchio A., Andolfi G., and De Santo V. (1998) Decomposition dynamics of litters of various pine species in a Corsican pine forest. *Soil Biol. Biochem.* **30**(6), 721–727.

Flaig W., Beutelspacher H., and Rietz E. (1975) Chemical composition and physical properties of humic substances. In *Soil Components: Vol. I. Organic Components* (ed. J. E. Gieseking). Springer, pp. 1–211.

Fog K. (1988) The effect of added nitrogen on the rate of decomposition of organic matter. *Biol. Rev. Cambridge Phil. Soc.* **63**(3), 433–462.

Foissner W. (1987) Soil protozoa fundamental problems ecological significance adaptations in ciliates and Testaceans Bioindicators and Guide to the Literature. *Prog. Protist.* **2**, 69–212.

Fragoso C. and Lavelle P. (1992) Earthworm communities of tropical rain forests. *Soil Biol. Biochem.* **24**(12), 1397–1408.

French D. D. (1988) Some effects of changing soil chemistry on decomposition of plant litters and cellulose on a Scottish [UK] moor. *Oecologia (Berlin)* **75**(4), 608–618.

Fu S. and Cheng W. (2002) Rhizosphere priming effects on the decomposition of soil organic matter in C_4 and C_3 grassland soils. *Plant Soil* **238**, 289–294.

Fuhrmann J. J. (1999) Microbial metabolism. In *Principles and Applications of Soil Microbiology* (eds. D. M. Sylvia, J. J. Fuhrmann, P. G. Hartel, and D. A. Zuberer). Prentice Hall, pp. 189–217.

Garland J. L. (1996) Patterns of potential C source utilization by rhizosphere communities. *Soil Biol. Biochem.* **28**(2), 223–230.

Garland J. L. and Mills A. L. (1991) Classification and characterization of heterotrophic microbial communities on the basis of patterns of community-level sole-carbon-source utilization. *Appl. Environ. Microbiol.* **57**(8), 2351–2359.

Garrett S. D. (1981) *Soil Fungi and Soil Fertility: An Introduction to Soil Mycology*. Pergamon.

Gaudinski J. B., Trumbore S. E., Davidson E. A., and Zheng S. (2000) Soil carbon cycling in a temperate forest: radiocarbon-based estimates of residence times, sequestration rates and partitioning of fluxes. *Biogeochemistry (Dordrecht)* **51**(1), 33–69.

Gaudinski J. B., Trumbore S. E., Davidson E. A., Cook A. C., Markewitz D., and Richter D. D. (2001) The age of fine-root carbon in three forests of the eastern United States measured by radiocarbon. *Oecologia (Berlin)* **129**(3), 420–429.

Gholz H. L., Wedin D. A., Smitherman S. M., Harmon M. E., and Parton W. J. (2000) Long-term dynamics of pine and hardwood litter in contrasting environments: toward a global model of decomposition. *Global Change Biol.* **6**(7), 751–765.

Giardina C. P. and Ryan M. G. (2000) Evidence that decomposition rates of organic carbon in mineral soil do not vary with temperature. *Nature (London)* **404**(6780), 858–861.

Giardina C. P., Ryan M. G., Hubbard R. M., and Binkley D. (2001) Tree species and soil textural controls on carbon and nitrogen mineralization rates. *Soil Sci. Soc. Am. J.* **65**(4), 1272–1279.

Gill R. A. and Jackson R. B. (2000) Global patterns of root turnover for terrestrial ecosystems. *New Phytol.* **147**(1), 13–31.

Glaser B., Haumaier L., Guggenberger G., and Zech W. (1998) Black carbon in soils: the use of benzenecarboxylic acids as specific markers. *Org. Geochem.* **29**(4), 811–819.

Golchin A., Clarke P., Baldock J. A., Higashi T., Skjemstad J. O., and Oades J. M. (1997) The effects of vegetation and burning on the chemical composition of soil organic matter in a volcanic ash soil as shown by C-13 NMR spectroscopy: 1. Whole soil and humic acid fraction. *Geoderma* **76**(3–4), 155–174.

Goldberg E. D. (1985) *Black Carbon in the Environment*. Wiley.

Gonzalez G. and Seastedt T. R. (2001) Soil fauna and plant litter decomposition in tropical and subalpine forests. *Ecology* **82**(4), 955–964.

Gooday G. W. (1994) Physiology of microbial degradation of chitin and chitosan. In *Biochemistry of Microbial Degradation* (ed. C. Ratledge). Kluwer Academic, pp. 279–312.

Gosz J. R. (1981) Nitrogen cycling in coniferous ecosystems. In *Terrestrial Nitrogen Cycles* (eds. F. E. Clark and T. Rosswall). Ecological Bulletin, pp. 405–426.

Gosz J. R., Likens G. E., and Bormann F. H. (1973) Nutrient release from decomposing lead and branch litter in the Hubbard Brook Forest, New Hamsphire. *Ecol. Monogr.* **43**, 173–191.

Gourbière F. (1982) White rot of Abies-alba mill litter: 1. Modification of litter during the activity of basidiomycetes of the genus collybia. *Revue D Ecologie Et De Biologie Du Sol* **19**(2), 163–175.

Graham R. C. and Wood H. B. (1991) Morphologic development and clay redistribution in lysimeter soils under chaparral and pine. *Soil Sci. Soc. Am. J.* **55**(6), 1638–1646.

Graham R. C., Ervin J. O., and Wood H. B. (1995) Aggregate stability under oak and pine after four decades of soil development. *Soil Sci. Soc. Am. J.* **59**(6), 1740–1744.

Gray K. M. (1997) Intercellular communication and group behavior in bacteria. *Trends Microbiol.* **5**(5), 184–188.

Grayston S. J. (2000) Rhizodeposition and its impact on microbial community structure and function in trees. *Phyton (Horn)* **40**(4), 27–36.

Grayston S. J., Vaughan D., and Jones D. (1996) Rhizosphere carbon flow in trees, in comparison with annual plants: the importance of root exudation and its impact on microbial activity and nutrient availability. *Appl. Soil Ecol.* **5**(1), 29–56.

Gregorich E. G., Kachanoski R. G., and Voroney R. P. (1989) Carbon mineralization in soil size fractions after various amounts of aggregate disruption. *J. Soil Sci.* **40**(3), 649–660.

Gregorich E. G., Voroney R. P., and Kachanoski R. G. (1991) Turnover of carbon through the microbial biomass in soils with different textures. *Soil Biol. Biochem.* **23**(8), 799–805.

Griffin D. H. (1994) *Fungal Physiology*. Wiley-Liss.

Grime J. P., Cornelissen J. H. C., Thompson K., and Hodgson J. G. (1996) Evidence of a causal connection between anti-herbivore defence and the decomposition rate of leaves. *Oikos* **77**(3), 489–494.

Guggenberger G. (1992) Eigenschaften und Dynamik geloster organischer Substanzen (DOM) auf unterschiedlich immissionsgeschadigten Fictenstandorten. *Bayreauther Bodenkundl. Ber.* **26**, 1–164.

Guggenberger G. and Zech W. (1994) Dissolved organic carbon in forest floor leachates: simple degradation products or humic substances? *Sci. Total Environ.* **152**(1), 37–47.

Harmon M. E., Baker G. A., Spycher G., and Greene S. E. (1990a) Leaf litter decomposition in the Picea, Tsuga forests of Olympic National Park, Washington, USA. *Forest Ecol. Manage.* **31**(1–2), 55–66.

Harmon M. E., Ferrell W. K., and Franklin J. F. (1990b) Effects on carbon storage of conversion of old-growth forests to young forests. *Science (Washington, DC)* **247**(4943), 699–702.

Hartel P. G. (1999) The soil habitat. In *Principles and Applications of Soil Microbiology* (eds. D. M. Sylvia, J. J. Fuhrmann, P. G. Hartel, and D. A. Zuberer). Prentice Hall, pp. 21–43.

Hartenstein R. (1982) Soil macroinvertebrates, aldehyde oxidase, catalase, cellulase and peroxidase. *Soil Biol. Biochem.* **14**, 387–391.

Hassink J., Bouwman L. A., Zwart K. B., and Brussaard L. (1993) Relationships between habitable pore space, soil biota and mineralization rates in grassland soils. *Soil Biol. Biochem.* **25**(1), 47–55.

Hatcher P. G. and Spiker E. C. (1988) Selective degradation of plant biomolceules. In *Humic Substances and their Role in the Environment* (eds. F. H. Frimmel and R. F. Christman). Wiley, pp. 59–74.

Hayes M. H. B., MacCarthy P., Malcolm R. L., and Swift R. S. (1989) The search for structure: setting the scene. In *Humic Substances II: In Search if Structure* (eds. M. H. B. Hayes, P. MacCarthy, R. L. Malcolm, and R. S. Swift). Wiley, vol. 2, pp. 3–31.

Hedges J. I. (1988) Polymerization of humic substances in natural environments. In *Humic Substances and their Role in the Environment* (eds. F. H. Frimmel and R. F. Christman). Wiley, pp. 45–58.

Hendrick R. L. and Pregitzer K. S. (1997) The relationship between fine root demography and the soil environment in northern hardwood forests. *Ecoscience* **4**(1), 99–105.

Heneghan L., Coleman D. C., Zou X., Crossley D. A., and Haines B. L. (1999) Soil microarthropod contributions to decomposition dynamics: tropical-temperate comparisons of a single substrate. *Ecology* **80**(6), 1873–1882.

Herms D. A. and Mattson W. J. (1992) The dilemma of plants to grow or defend. *Quart. Rev. Biol.* **67**(3), 283–335.

Hill G. T., Mitkowski N. A., Aldrich-Wolfe L., Emele L. R., Jurkonie D. D., Ficke A., Maldonado-Ramirez S., Lynch S. T., and Nelson E. B. (2000) Methods for assessing the composition and diversity of soil microbial communities. *Appl. Soil Ecol.* **15**(1), 25–36.

Hobbie S. E. (1992) Effects of plant species on nutrient cycling. *Trends Ecol. Evol.* **7**(10), 336–339.

Houghton J. T. (1996) *Climate Change 1995: The Science of Climate Change*. Cambridge University Press.

Hudson H. J. (1968) The ecology of fungi on plant remains above the soil. *New Phytol.* **67**, 837–874.

Hunt H. W., Ingham E. R., Coleman D. C., Elliott E. T., and Reid C. P. P. (1988) Nitrogen limitation of production and decomposition in prairie, mountain meadow, and pine forest. *Ecology (Tempe)* **69**(4), 1009–1016.

Ibrahima A., Joffre R., and Gillon D. (1995) Changes in litter during the initial leaching phase—an experiment on the leaf-litter of Mediterranean species. *Soil Biol. Biochem.* **27**(7), 931–939.

Ikan P., Rubinsztain Y., Ioselis P., Aizenshtat Z., Pugmire R., Anderson L. L., and Ishiwatari R. (1986) Carbon-13 cross polarized magic-angle spinning nuclear magnetic resonance of melanoidins. *Org. Geochem.* **9**(3), 199–212.

Insam H. (1997) A new set of substrates proposed for community characterization in environmental samples. In *Microbial Communities: Functional versus Structural Approaches* (eds. H. Insam and A. Rangger). Springer, pp. 259–260.

Jackson R. B., Canadell J., Ehleringer J. R., Mooney H. A., Sala O. E., and Schulze E. D. (1996) A global analysis of root distributions for terrestrial biomes. *Oecologia (Berlin)* **108**(3), 389–411.

Janssens I. A., Kowalski A. S., Longdoz B., and Ceulemans R. (2000) Assessing forest soil CO_2 efflux: an *in situ* comparison of four techniques. *Tree Physiol.* **20**(1), 23–32.

Janssens I. A., Kowalski A. S., and Ceulemans R. (2001) Forest floor CO_2 fluxes estimated by eddy covariance and chamber-based model. *Agri. Forest Meteorol.* **106**(1), 61–69.

Jeffries T. W. (1994) Biodegradation of lignin and hemicelluloses. In *Biochemistry of Microbial Degradation* (ed. C. Ratledge). Kluwer Academic, pp. 233–277.

Jenkinson D. S. and Powlson D. S. (1986) The effects of biocidal treatments on metabolism in soil: V. A method for measuring soil biomass. *Soil Biol. Biochem.* **8**, 209–213.

Jenny H. (1941) *Factors of Soil Formation: A System of Quantitative Pedology*. McGraw-Hill.

Jensen V. (1984) Decomposition of angiosperm tree leaf litter. In *Biology of Plant Litter Decomposition* (eds. C. H. Dickinson and G. J. F. Pugh). Academic Press, vol. 1, pp. 69–104.

Johansson M.-B., Berg B., and Meentemyer V. (1995) Litter mass-loss rates in late stages of decomposition in a climatic transect of pine forests: long-term decomposition in a Scots pine forest: IX. *Can. J. Botany* **73**(10), 1509–1521.

Jones J. A. (1990) Termites, soil fertility and carbon cycling in dry tropical Africa: a hypothesis. *J. Tropical Ecol.* **6**(3), 291–305.

Kaiser D. and Losick R. (1993) How and why bacteria talk to each other. *Cell* **73**(5), 873–885.

Kambhampati S. and Eggleton P. (2000) Taxonomy and phylogeny of termites. In *Termites: Evolution, Sociality, Symbioses, Ecology* (eds. T. Abe, D. E. Bignell, and M. Higashi). Kluwer Academic, pp. 1–23.

Kanyonyo K. K. (1984) Ecologie alimentaire du ver de terre anecique africain *Millisonia lamtoiana* (Acanthodrilidae, Oligochetes) dans la savane de Lamto (Cote d'Ivoire). Doctoral, Paris VI University.

Katterer T. and Andren O. (2001) The ICBM family of analytically solved models of soil carbon, nitrogen, and microbial biomass dynamics: descriptions and application examples. *Ecol. Model.* **136**(2–3), 191–207.

Katterer T., Reichstein M., Andren O., and Lomander A. (1998) Temperature dependence of organic matter decomposition: a critical review using literature data analyzed with different models. *Biol. Fertility Soils* **27**(3), 258–262.

Kennedy M. J., Pevear D. R., and Hill R. J. (2002) Mineral surface control of organic carbon in black shale. *Science* **295**, 657–660.

Kheirallah A. M. (1990) Fragmentation of leaf litter by a natural population of the millipede Julus scandinavius (Latzel 1884). *Biol. Fertility Soils* **10**(3), 202–206.

Kieft T. L., Soroker E., and Firestone M. K. (1987) Microbial biomass response to a rapid increase in water potential when dry soil is wetted. *Soil Biol. Biochem.* **19**(2), 119–126.

Kirk T. K. and Farrell R. L. (1987) Enzymatic combustion—the microbial-degradation of lignin. *Annu. Rev. Microbiol.* **41**, 465–505.

Kirschbaum M. U. F. (1995) The temperature-dependence of soil organic-matter decomposition, and the effect of global warming on soil organic-C storage. *Soil Biol. Biochem.* **27**(6), 753–760.

Kirschbaum M. U. F. (2000) Will changes in soil organic carbon act as a positive or negative feedback on global warming? *Biogeochemistry* **48**(1), 21–51.

Knicker H. and Skjemstad J. O. (2000) Nature of organic carbon and nitrogen in physically protected organic matter of some Australian soils as revealed by solid-state C-13 and N-15 NMR spectroscopy. *Austral. J. Soil Res.* **38**(1), 113–127.

Kogel I. (1986) Estimation and decomposition pattern of the lignin component in forest soils. *Soil Biol. Biochem.* **18**, 589–594.

Kogel I., Hempfling R., Zech W., Hatcher P. G., and Schulten H. R. (1988) Chemical composition of the organic matter in forest soils: 1. Forest litter. *Soil Sci.* **146**(2), 124–136.

Kogel-Knabner I. (1993) *Biodegradation and Humification Processes in Forest Soils*. Dekker.

Kogel-Knabner I. (1997) 13C and 15N NMR spectroscopy as a tool in soil organic matter studies. *Geoderma* **80**(3–4), 243–270.

Kogel-Knabner I. (2000) Analytical approaches for characterizing soil organic matter. *Org. Geochem.* **31**(7–8), 609–625.

Kogel-Knabner I. (2002) The macromolecular organic composition of plant and microbial residues as inputs to soil organic matter. *Soil Biol. Biochem.* **34**, 139–162.

Kononova M. M. (1966) *Soil Organic Matter: Its Nature, Its Role in Soil Formation and in Soil Fertility*. Pergamon.

Konopka A., Oliver L., and Turco R. F. (1998) The use of carbon substrate utilization patterns in environmental and ecological microbiology. *Microbial Ecol.* **35**(2), 103–115.

Kracht O. and Gleixner G. (2000) Isotope analysis of pyrolysis products from Sphagnum peat and dissolved organic matter from bog water. *Org. Geochem.* **31**(7–8), 645–654.

Krasil'nikov N. A. and Satdykov S. I. (1969) Estimation of the total bacteria in the intestines of termites. *Mikrobiologiya* **38**, 346–350.

Kristufek V., Ravasz K., and Pizl V. (1992) Changes in densities of bacteria and microfungi during Gut transit in Lumbricus-Rubellus and Aporrectodea-Caliginosa Oligochaeta Lumbricidae. *Soil Biol. Biochem.* **24**(12), 1499–1500.

Kuiters A. T. and Sarink H. M. (1986) Leaching of phenolic-compounds from leaf and needle litter of several deciduous and coniferous trees. *Soil Biol. Biochem.* **18**(5), 475–480.

Kuzyakov Y., Friedel J. K., and Stahr K. (2000) Review of mechanisms and quantification of priming effects. *Soil Biol. Biochem.* **32**(11–12), 1485–1498.

Ladd J. N., Parsons J. W., and Amato M. (1977) Studies of nitrogen immobilization and mineralization in calcareous soils: II. Mineralization of immobilized nitrogen from soil fractions of different particle size and density. *Soil Biol. Biochem.* **9**, 319–325.

Langmaid K. K. (1964) Some effects of earthworm invasion in virgin podsols. *Can. J. Soil Sci.* **44**, 34–37.

Lavelle P. (1978) Les vers de terre de la savane de Lamanto (Cote d'Ivoire): peuplements, populations et fonctions dans l'ecosysteme. PhD Dissertation, University of Paris.

Lavelle P. (1988) Earthworm activities and the soil system. *Biol. Fertility Soils* **6**(3), 237–251.

Lavelle P. and Martin A. (1992) Small-scale and large-scale effects of endogeic earthworms on soil organic matter dynamics in soils of the humid tropics. *Soil Biol. Biochem.* **24**(12), 1491–1498.

Lavelle P. and Spain A. (2001) *Soil Ecology*. Kluwer Academic.

Lavelle P., Blanchart E., Martin A., Martin S., Spain A., Toutain F., Barois I., and Schaefer R. (1993) A hierarchical model for decomposition in Terrestrial ecosystems—application to soils of the humid tropics. *Biotropica* **25**(2), 130–150.

Lee K. E. (1985) *Earthworms: Their Ecology and Relationships with Soils and Land Use*. Academic Press.

Lee M. J., Schreurs P. J., Messer A. C., and Zinder S. H. (1987) Association of methanogenic bacteria with flagellated protozoa from a termite hindgut. *Current Microbiol.* **15**(6), 337–342.

Lehman R. M., Colwell F. S., Ringelberg D. B., and White D. C. (1995) Combined microbial community-level analyses for quality assurance of terrestrial subsurface cores. *J. Microbiol. Meth.* **22**(3), 263–281.

Leiros M. C., Trasar-Cepeda C., Seoane S., and Gil-Sotres F. (1999) Dependence of mineralization of soil organic matter on temperature and moisture. *Soil Biol. Biochem.* **31**(3), 327–335.

Lewis D. H. (1973) Concepts in fungal nutrition and the origin of biotrophy. *Biol. Rev.* **48**, 261–278.

Lichtfouse E. (1999) A novel model of humin. *Analusis* **27**(5), 385–386.

Lichtfouse E., Chenu C., Baudin F., Leblond C., DaSilva M., Behar F., Derenne S., Largeau C., Wehrung P., and Albrecht P. (1998) A novel pathway of soil organic matter formation by selective preservation of resistant straight-chain biopolymers: chemical and isotope evidence. *Org. Geochem.* **28**(6), 411–415.

Lloyd J. and Taylor J. A. (1994) On the temperature dependence of soil respiration. *Funct. Ecol.* **8**(3), 315–323.

Lousier J. D. and Bamforth S. S. (1990) Soil protozoa. In *Soil Biology Guide* (ed. D. L. Dindal). Wiley, pp. 97–136.

Lousier J. D. and Parkinson D. (1978) Chemical element dynamics in decomposing leaf litter. *Can. J. Botany* **56**, 2795–2812.

Lousier J. D. and Parkinson D. (1984) Annual population dynamics and production ecology of testacea (Protozoa, Rhizopoda) in an aspen woodland soil. *Soil Biol. Biochem.* **16**, 103–114.

Lunt H. A. and Jacobson G. M. (1944) The chemical composition of earthworm casts. *Soil Sci.* **58**, 367–374.

Lutz H. J. and Chandler R. F. J. (1946) *Forest Soils*. Wiley.

Luxton M. (1982) Studies on the oribatid mites of a Danish beech wood soil. *Pedobiologia* **12**, 434–463.

Lynch J. M. and Whipps J. M. (1990) Substrate Flow in the Rhizosphere. *Plant Soil* **129**(1), 1–10.

Macarthur R. H. and Wilson E. D. (1967) *The Theory of Island Biogeography*. Princeton University Press.

MacLean D. A. and Wein R. W. (1978) Weight loss and nutrient changes in decomposing litter and forest floor material in New Brunswick forest stands. *Can. J. Botany* **56**, 2730–2749.

Macnaughton S. J., Booth T., Embley T. M., and O'Donnell A. G. (1996) Physical stabilization and confocal microscopy of bacteria on roots using 16S rRNA targeted, fluorescent-labeled oligonucleotide probes. *J. Microbiol. Meth.* **26**(3), 279–285.

Maillard L. C. (1917) Identité des matières humiques de synthèse avec les materières humiques naturelles. *Annales de Chimie (Paris)* **7**, 113–152.

Malcolm R. L. (1989) Application of solid-state 13C NMR spectroscopy to geochemical studies of humic substances. In *Humic Substances II: In Search of Structure* (eds. M. H. B. Hayes, P. MacCarthy, R. L. Malcolm, and M. J. Swift). Wiley, pp. 339–372.

Maraun M. and Scheu S. (1995) Influence of beech litter fragmentation and glucose concentration on the microbial biomass in three different litter layers of a beechwood. *Biol. Fertility Soils* **19**(2–3), 155–158.

Maraun M. and Scheu S. (1996) Changes in microbial biomass, respiration and nutrient status of beech (Fagus sylvatica) leaf litter processed by millipedes (Glomeris marginata). *Oecologia (Berlin)* **107**(1), 131–140.

Martin A. (1991) Short-term and long-term effects of the endogenic earthworm Millsonia anomala (Omodeo) (Megascolecidae, Oligochaeta) of tropical savannas on soil organic matter. *Biol. Fertility Soils* **11**(3), 234–238.

Martin J. P. and Haider K. (1969) Phenolic polymers of Stachybotrys Atra Strachybotrys Chartarum and Epicoccum Nigrum in relation to humic acid formation. *Soil Sci.* **107**(4), 260–270.

Martin J. P. and Haider K. (1971) Microbial activity in relation to soil humus formation. *Soil Sci.* **111**(1), 54–59.

Martin A., Mariotti A., Balesdent J., Lavelle P., and Vuattoux R. (1990) Estimate of organic matter turnover rate in a savanna soil by carbon-13 natural abundance measurements. *Soil Biol. Biochem.* **22**(4), 517–524.

Martin A., Mariotti A., Balesdent J., and Lavelle P. (1992) Soil organic matter assimilation by a geophagous tropical earthworm based on carbon-13 measurements. *Ecology (Tempe)* **73**(1), 118–128.

Martin J. P., Wolf D., and Haider K. (1972) Synthesis of phenols and phenolic polymers by Hendersonula Toruloidea in relation to humic acid formation. *Soil Sci. Soc. Am. Proc.* **36**(2), 311–315.

McClaugherty C. A., Pastor J., Aber J. D., and Melillo J. M. (1985) Forest litter decomposition in relation to soil nitrogen dynamics and litter quality. *Ecology (Tempe)* **66**(1), 266–275.

McGill W. B. (1996) Review and classification of ten soil organic matter (SOM) models. In *Evaluation of Soil Organic Matter Models* (eds. D. S. Powlson, P. Smith, and J. U. Smith). Springer, pp. 111–132.

McGuire A. D., Melillo J. M., Joyce L. A., Kicklighter D. W., Grace A. L., Moore B., III, and Vorosmarty C. J. (1992) Interactions between carbon and nitrogen dynamics in estimating net primary productivity for potential vegetation in North America. *Global Biogeochem. Cycles* **6**(2), 101–124.

McInerney M. and Bolger T. (2000a) Decomposition of Quercus petraea litter: influence of burial, comminution and earthworms. *Soil Biol. Biochem.* **32**(14), 1989–2000.

McInerney M. and Bolger T. (2000b) Temperature, wetting cycles and soil texture effects on carbon and nitrogen dynamics in stabilized earthworm casts. *Soil Biol. Biochem.* **32**(3), 335–349.

Meentemyer V. (1977) Macroclimate regulation of decomposition rates of organic matter in terrestrial ecosystems. In *Environmental Chemistry and Cycling Processes,* US DOE Symposium Series CONF-760429 (eds. D. C. Adriano and I. L. Brisbin) US DOE, pp. 779–789.

Meentemyer V. (1978) Macroclimate and lignin control of litter decomposition rates. *Ecology* **59**(3), 465–472.

Meentemyer V. (1995) Meteorologic control of litter decomposition with an emphasis on tropical environments. In *Soil Organisms and Litter Decomposition in the Tropics* (ed. M. V. Reddy). Westview Press, pp. 153–182.

Meentemyer V. and Berg B. (1986) Regional variation in rate of mass loss of Pinus sylvestris needle litter in Swedish pine forests as influenced by climate and litter quality. *Scand. J. Forest Res.* **1**(2), 167–180.

Melillo J. M. and Aber J. D. (1982) Nitrogen and lignin control of hardwood leaf litter decomposition dynamics. *Ecology* **63**(3), 621–626.

Melillo J. M., Aber J. D., Linkins A. E., Ricca A., Fry B., and Nadelhoffer K. J. (1989) Carbon and nitrogen dynamics along the decay continuum plant litter to soil organic matter. *Plant Soil* **115**(2), 189–198.

Melillo J. M., Newkirk K. M., Catricala C. E., Steudler P. A., Aber J. B., Nadelhoffer K. J., and Boone R. D. (1996) The soil warming experiment at Harvard forest 1991–1996. *Bull. Ecol. Soc. Am.* **77**(3 suppl. Part 2), 300.

Millar C. S. (1974) Decomposition of coniferous leaf litter. In *Biology of Plant Leaf Litter Decomposition* (eds. C. H. Dickinson and G. J. F. Pugh). Academic Press, vol. 1, pp. 105–128.

Miller R. M. and Jastrow J. D. (1992) The role of mycorrhizal fungi in soil conservation. In *Mycorrhizae in Sustainable Agriculture* (eds. C. J. Bethlenfalvay and R. G. Linderman). Soil Science Society of America, pp. 29–44.

Minderman G. (1968) Addition, decomposition and accumulation of organic matter in forests. *J. Ecol.* **56**, 355–362.

Møller J., Miller M., and Kjoller A. (1999) Fungal-bacterial interaction on beech leaves: influence on decomposition and dissolved organic carbon quality. *Soil Biol. Biochem.* **31**(3), 367–374.

Moody S. A. (1993) Selective consumption of decomposing wheat straw by earthworms. *Soil Biol. Biochem.* **27**, 1209–1213.

Moore A. M. (1986) Temperature and moisture dependence of decomposition rates of hardwood and coniferous leaf litter. *Soil Biol. Biochem.* **18**(4), 427–435.

Moorhead D. L., Currie W. S., Rastetter E. B., Parton W. J., and Harmon M. E. (1999) Climate and litter quality controls on decomposition: an analysis of modeling approaches. *Global Biogeochem. Cycles* **13**(2), 575–589.

Moretto A. S., Distel R. A., and Didone N. G. (2001) Decomposition and nutrient dynamic of leaf litter and roots from palatable and unpalatable grasses in a semi-arid grassland. *Appl. Soil Ecol.* **18**(1), 31–37.

Motavalli P. P., Palm C. A., Parton W. J., Elliott E. T., and Frey S. D. (1994) Comparison of laboratory and modeling simulation methods for estimating soil carbon pools in tropical forest soils. *Soil Biol. Biochem.* **26**(8), 934–944.

Neff J. C. and Asner G. P. (2001) Dissolved organic carbon in terrestrial ecosystems: synthesis and a model. *Ecosystems* **4**(1), 29–48.

Neill C., Fry B., Melillo J. M., Steudler P. A., Moraes J. F. L., and Cerri C. C. (1996) Forest- and pasture-derived carbon contributions to carbon stocks and microbial respiration of tropical pasture soils. *Oecologia (Berlin)* **107**(1), 113–119.

Nelson P. N., Barzegar A. R., and Oades J. M. (1997) Sodicity and clay type: influence on decomposition of added organic matter. *Soil Sci. Soc. Am. J.* **61**(4), 1052–1057.

Nicolardot B., Fauvet G., and Cheneby D. (1994) Carbon and nitrogen cycling through soil microbial biomass at various temperatures. *Soil Biol. Biochem.* **26**(2), 253–261.

Nierop K. G. J. (1998) Origin of aliphatic compounds in a forest soil. *Org. Geochem.* **29**, 1000–1019.

Noll K. M. (1992) Archaebacteria Archaea. In *Encyclopedia of Microbiology* (ed. J. Lederberg). Academic Press, vol. 1, pp. 149–160.

Northup R. R., Dahlgren R. A., and McColl J. G. (1998) Polyphenols as regulators of plant-litter-soil interactions in northern California's pygmy forest: a positive feedback? *Biogeochemistry* **42**(1–2), 189–220.

Nykvist N. (1959) Leaching and decomposition of litter: I. Experiments on leaf litter of *Fraxinus Excelsior. Oikos* **10**, 190–211.

Nykvist N. (1961) Leaching and decomposition of litter: III. Experiments on leaf litter of *Betula Verrucosa. Oikos* **12**(2), 249–263.

Oades J. M. (1988) The retention of organic matter in soils. *Biogeochemistry (Dordrecht)* **5**(1), 35–70.

Oades J. M., Vassallo A. M., Waters A. G., and Wilson M. A. (1987) Characterization of organic matter in particle size and density fractions from a red-brown earth by solid-state carbon-13 NMR. *Austral. J. Soil Res.* **25**(1), 71–82.

O'Brian B. J. and Stout J. D. (1978) Movement and turnover of soil organic matter as indicated by carbon isotope measurements. *Soil Biol. Biochem.* **10**, 309–317.

Olsen J. S. (1963) Energy storage and the balance of producers and decomposers in ecological systems. *Ecology* **44**(2), 322–331.

Orchard V. A. and Cook F. (1983) Relationship between soil respiration and soil moisture. *Soil Biol. Biochem.* **15**, 447–453.

Osono T. and Takeda H. (1999) Decomposing ability of interior and surface fungal colonizers of beech leaves with reference to lignin decomposition. *Euro. J. Soil Biol.* **35**(2), 51–56.

Ostertag R. and Hobbie S. E. (1999) Early stages of root and leaf decomposition in Hawaiian forests: effects of nutrient availability. *Oecologia (Berlin)* **121**(4), 564–573.

Pace N. R. (1996) New perspective on the natural microbial world: molecular microbial ecology. *Am. Soc. Microbiol. News* **62**, 463–470.

Palm C. A. and Sanchez P. A. (1990) Decomposition and nutrient release patterns of the leaves of 3 tropical legumes. *Biotropica* **22**(4), 330–338.

Parsons W. F. J., Taylor B. R., and Parkinson D. (1990) Decomposition of aspen (Populus tremuloides) leaf litter modified by leaching. *Can. J. Forest Res.* **20**(7), 943–951.

Parton W. J., Schimel D. S., Cole C. V., and Ojima D. S. (1987) Analysis of factors controlling soil organic matter levels in great plains grasslands USA. *Soil Sci. Soc. Am. J.* **51**(5), 1173–1179.

Parton W. J., Scurlock J. M. O., Ojima D. S., Gilmanov T. G., Scholes R. J., Schimel D. S., Kirchner T., Menaut J. C., Seastedt T., Moya E. G., Kamnalrut A., and Kinyamario J. I. (1993) Observations and modeling of biomass and soil organic matter dynamics for the grassland biome worldwide. *Global Biogeochem. Cycles* **7**(4), 785–809.

Parton W. J., Ojima D. S., Cole C. V., and Schimel D. S. (1994) A general model for soil organic matter dynamics:

sensitivity to litter chemistry, texture and management. In *Quantitative Modeling of Soil Forming Processes*, SSSA Special Publication (eds. R. B. Bryant, and R. W. Arnold). SSSA, vol. 39, pp. 147–167.

Pastor J. and Post W. M. (1986) Influence of climate, soil moisture, and succession of forest carbon and nitrogen cycles. *Biogeochemistry* **2**, 3–27.

Paul E. A. and Clark F. E. (1996) *Soil Microbiology and Biochemistry*. Academic Press.

Paul E. A., Paustian K., Elliott E. A., and Cole C. V. (1997) *Soil Organic Matter in Temperate Agroecosystems: Long-term Experiments in North America*. CRC Press.

Peberdy J. F. (1990) Fungal cell walls—a review. In *Biochemistry of Cell Walls and Membranes in Fungi* (eds. P. J. Kuhn, A. P. J. Trinci, M. J. Jung, M. W. Goosey, and L. G. Copping). Springer, pp. 5–30.

Perakis S. S. and Hedin L. O. (2002) Nitrogen loss from unpolluted South American forests mainly via dissolved organic compounds. *Nature (London)* **415**(6870), 416–419.

Perez-Harguindeguy N., Diaz S., Cornelissen J. H. C., Vendramini F., Cabido M., and Castellanos A. (2000) Chemistry and toughness predict leaf litter decomposition rates over a wide spectrum of functional types and taxa in central Argentina. *Plant Soil* **218**(1–2), 21–30.

Petersen S. O., Debosz K., Schjonning P., Christensen B. T., and Elmholt S. (1997) Phospholipid fatty acid profiles and C availability in wet-stable macro-aggregates from conventionally and organically farmed soils. *Geoderma* **78**(3–4), 181–196.

Pohlman A. A. and McColl J. G. (1988) Soluble organics from forest litter and their role in metal dissolution. *Soil Sci. Soc. Am. J.* **52**(1), 265–271.

Poirier N., Derenne S., Rouzaud J. N., Largeau C., Mariotti A., Balesdent J., and Maquet J. (2000) Chemical structure and sources of the macromolecular, resistant, organic fraction isolated from a forest soil (Lacadee, south-west France). *Org. Geochem.* **31**(9), 813–827.

Poirier N., Derenne S., Balesdent J., Rouzaud J. N., Mariotti A., and Largeau C. (2002) Abundance and composition of the refractory organic fraction of an ancient, tropical soil (Pointe Noire, Congo). *Org. Geochem.* **33**(3), 383–391.

Pongracic S., Kirschbaum M. U. F., and Raison R. J. (1997) Comparison of soda lime and infrared gas analysis techniques for *in situ* measurement of forest soil respiration. *Can. J. Forest Res.* **27**(11), 1890–1895.

Post W. M., Emanuel W. R., Zinke P. J., and Stangenberger A. G. (1982) Soil carbon pools and world life zones. *Nature* **298**, 156–159.

Prescott C. E. (1995) Does nitrogen availability control rates of litter decomposition in forests? *Plant Soil* **168–169**(0), 83–88.

Prescott C. E., Corbin J. P., and Parkinson D. (1992) Immobilization and availability of nitrogen and phosphorus in the forest floors of fertilized Rocky Mountain coniferous forests. *Plant Soil* **143**(1), 1–10.

Prescott C. E., Taylor B. R., Parsons W. F. J., Durall D. M., and Parkinson D. (1993) Nutrient release from decomposing litter in Rocky-Mountain coniferous forests-influence of nutrient availability. *Can. J. Forest Res.: Revue Canadienne De Recherche Forestiere* **23**(8), 1576–1586.

Preston C. M., Trofymow J. A., Sayer B. G., and Niu J. N. (1998) C-13 nuclear magnetic resonance spectroscopy with cross-polarization and magic-angle spinning investigation of the proximate-analysis fractions used to assess litter quality in decomposition studies. *Can. J. Botany: Rev. Can. Botanique* **75**(9), 1601–1613.

Puget P., Chenu C., and Balesdent J. (2000) Dynamics of soil organic matter associated with particle-size fractions of water-stable aggregates. *Euro. J. Soil Sci.* **51**(4), 595–605.

Qualls R. G. and Haines B. L. (1991) Geochemistry of dissolved organic nutrients in water percolating through a forest ecosystem. *Soil Sci. Soc. Am. J.* **55**(4), 1112–1123.

Qualls R. G., Haines B. L., and Swank W. T. (1991) Fluxes of dissolved organic nutrients and humic substances in a deciduous forest. *Ecology (Tempe)* **72**(1), 254–266.

Quideau S. A., Graham R. C., Chadwick O. A., and Wood H. B. (1998) Organic carbon sequestration under chaparral and pine after four decades of soil development. *Geoderma* **83**(3–4), 227–242.

Quideau S. A., Anderson M. A., Graham R. C., Chadwick O. A., and Trumbore S. E. (2000) Soil organic matter processes: characterization by C-13 NMR and C-14 measurements. *Forest Ecol. Manage.* **138**(1–3), 19–27.

Quideau S. A., Chadwick O. A., Trumbore S. E., Johnson-Maynard J. L., Graham R. C., and Anderson M. A. (2001) Vegetation control on soil organic matter dynamics. *Org. Geochem.* **32**(2), 247–252.

Raich J. W. and Schlesinger W. H. (1992) The global carbon dioxide flux in soil respiration and its relationship to vegetation and climate. *Tellus Ser. B: Chem. Phys. Meteorol.* **44**(2), 81–99.

Ramsh M. V., Saha B. C., Mathupala S. P., Podkovyrov S., and Zeikus J. G. (1994) Biodegradation of starch and alpha-glycan polmyers. In *Biochemistry of Microbial Degradation* (ed. C. Ratledge). Kluwer Academic.

Rastin N., Schlechte G., Huettermann A., and Rosenplaenter K. (1990) Seasonal fluctuation of some biological and biochemical soil factors and their dependence on certain soil factors on the upper and lower slope of a spruce forest. *Soil Biol. Biochem.* **22**(8), 1049–1062.

Ratledge C. (1994) *Biochemistry of Microbial Degradation*. Kluwer Academic.

Reddy M. V. (1995a) Litter Arthropods. In *Soil Organisms and Litter Decomposition in the Tropics* (ed. M. V. Reddy). Westview Press, pp. 113–140.

Reddy M. V. (1995b) *Soil Organisms and Litter Decomposition in the Tropics*. Westview Press.

Reddy M. V. and Venkataiah B. (1989) Influence of microarthropod abundance and climatic factors on weight loss and mineral nutrient contents of Eucalyptus leaf litter during decomposition. *Biol. Fertility Soils* **8**(4), 319–324.

Regina I. S. (2001) Litter fall, decomposition and nutrient release in three semi-arid forests of the Duero basin, Spain. *Forestry* **74**(4), 347–358.

Rice J. A. (2001) Humin. *Soil Sci.* **166**(11), 848–857.

Rihani M., Cancela Da Fonseca J. P., and Kiffer E. (1995a) Decomposition of beech leaf litter by microflora and mesofauna: II. Food preferences and action of oribatid mites on different substrates. *Euro. J. Soil Biol.* **31**(2), 67–79.

Rihani M., Kiffer E., and Botton B. (1995b) Decomposition of beech leaf litter microflora and mesofauna: I. In vitro action of white-rot fungi on beech leaves and foliar components. *Euro. J. Soil Biol.* **31**(2), 57–66.

Rogers H. J., Perkins H. R., and Ward J. B. (1980) *Microbial Cell Walls and Membranes*. Chapman and Hall.

Rosenbrock P., Buscot F., and Munuch J. C. (1995) Fungal succession and changes in the fungal degradation potential during the initial stage of litter decomposition in a black alder forest (*Alnus glutinosa* (L.) Gaertn.). *Euro. J. Soil Biol.* **31**(1), 1–11.

Rustad L. E. (1994) Element dynamics along a decay continuum in a red spruce ecosystem in Maine, USA. *Ecology (Tempe)* **85**(4), 867–879.

Rutigliano F. A., Alfani A., Bellini L., and De Santo A. V. (1998) Nutrient dynamics in decaying leaves of Fagus sylvatica L. and needles of Abies alba Mill. *Biol. Fertility Soils* **27**(2), 119–126.

Ryan M. G., Melillo J. M., and Ricca A. (1990) A comparison of methods for determining proximate carbon fractions of forest litter. *Can. J. Forest Res.: J. Canadien De La Recherche Forestiere* **20**(2), 166–171.

Saiz-Jimenez C. (1994) Analytical pyrolysis of humic substances—pitfalls, limitations, and possible solutions. *Environ. Sci. Technol.* **28**(11), 1773–1780.

Saiz-Jimenez C. (1996) The chemical structure of humic substances: recent advances. In *Humic Substances in Terrestrial Ecosystems* (ed. A. Piccolo). Elsevier, pp. 1–44.

Sallih Z. and Bottner P. (1988) Effect of wheat (Triticum aestivum) roots on mineralization rates of soil organic matter. *Biol. Fertility Soils* **7**(1), 67–70.

Sanderman J., Amundson R., and Baldocchi D. D. (2003) Application of eddy covariance measurements to the temperature dependence of soil organic matter mean residence times. *Global Biogeochem. Cycles*, **17**(2), 1061.

Savin M. C., Gorres J. H., Neher D. A., and Amador J. A. (2001) Uncoupling of carbon and nitrogen mineralization: role of microbivorous nematodes. *Soil Biol. Biochem.* **33**(11), 1463–1472.

Scheu S. (1991) Mucus excretion and carbon turnover of endogeic earthworms. *Biol. Fertility Soils* **12**(3), 217–220.

Scheu S. (1992) Decomposition of lignin in soil microcompartments: a methodical study with three different carbon-14-labelled lignin substrates. *Biol. Fertility Soils* **13**(3), 160–164.

Scheu S. and Parkinson D. (1994) Changes in bacterial and fungal biomass C, bacterial and fungal biovolume and ergosterol content after drying, remoistening and incubation of different layers of cool temperate forest soils. *Soil Biol. Biochem.* **26**(11), 1515–1525.

Scheu S. and Parkinson D. (1995) Successional changes in microbial biomass, respiration and nutrient status during litter decomposition in an aspen and pine forest. *Biol. Fertility Soils* **19**(4), 327–332.

Schimel D. S., Coleman D. C., and Horton K. A. (1985) Soil organic matter dynamics in paired rangeland and cropland toposequences in North Dakota. *Geoderma* **36**, 201–214.

Schimel J. (2001) Biogeochemical models: implicit versus explicit microbiology. In *Global Biogeochemical Cylces in the Climate System* (eds. E. D. Schulze, M. Heimann, S. Harrison, E. A. Holland, J. Lloyd, I. C. Prentice, and D. Schimel). Academic Press, pp. 177–183.

Schlesinger W. H. (1985) Decomposition of chaparral shrub foliage. *Ecology (Tempe)* **66**(4), 1353–1359.

Schlesinger W. H. (1990) Evidence from Chronosequence Studies for a low carbon-storage potential of soils. *Nature (London)* **348**(6298), 232–234.

Schlesinger W. H. (1991) *Biogeochemistry: An Analysis of Global Change*. Academic Press.

Schlesinger W. H. and Hasey M. M. (1981) Decomposition of chaparral shrub foliage: losses of organic and inorganic constituents from deciduous and evergreen leaves. *Ecology* **62**(3), 762–774.

Schmidt M. W. I., Knicker H., Hatcher P. G., and Koegel-Knabner I. (1997) Improvement of C-13 and N-15 CPMAS NMR spectra of bulk soils, particle size fractions and organic material by treatment with 10% hydrofluoric acid. *Euro. J. Soil Sci.* **48**(2), 319–328.

Schnitzer M. (1990) Selected methods for the characterization of soil humic substances. In *Humic Substances in Soil and Crop Sciences: Selected Readings* (eds. P. MacCarthy, C. E. Clapp, R. L. Malcolm, and P. R. Bloom). American Society of Agronomy, pp. 65–89.

Schnitzer M. and Schuppli P. (1989) The extraction of organic matter from selected soils and particle size fractions with 0.5 M sodium hydroxide and 0.1 M tetrasodium pyrophosphate solutions. *Can. J. Soil Sci.* **69**(2), 253–262.

Schwarz W. H. (2001) The cellulosome and cellulose degradation by anaerobic bacteria. *Appl. Microbiol. Biotechnol.* **56**(5–6), 634–649.

Scott N. A., Cole C. V., Elliott E. T., and Huffman S. A. (1996) Soil textural control on decomposition and soil organic matter dynamics. *Soil Sci. Soc. Am. J.* **60**(4), 1102–1109.

Seastedt T. R. (1984) The role of microarthropods in decomposition and mineralization processes. *Annu. Rev. Entomol.* **29**, 25–46.

Seastedt T. R., Parton W. W., and Ojima D. S. (1992) Mass loss and nitrogen dynamics of decaying litter of grassland: the apparent low nitrogen immobilization potential of root detritus. *Can. J. Botany* **70**(2), 384–391.

Sharma R. and Sharma S. B. (1995) Nematodes and tropical litter decomposition. In *Soil Organisms and Litter Decomposition in the Tropics* (ed. M. V. Reddy). Westview Press, pp. 75–88.

Shaw P. J. A. (1992) Fungi, fungivores and fungal foodwebs. In *The Fungal Community—Its Organization and Role in the Ecosystem* (eds. G. C. Carroll and D. T. Wicklow). Dekker, pp. 295–310.

Shipitalo M. J. and Protz R. (1989) Chemistry and micromorphology of aggregation in earthworm casts. *Geoderma* **45**(3–4), 357–374.

Silver W. L. and Miya R. K. (2001) Global patterns in root decomposition: comparisons of climate and litter quality effects. *Oecologia* **V129**(N3), 407–419.

Silver W. L., Neff J., McGroddy M., Veldkamp E., Keller M., and Cosme R. (2000) Effects of soil texture on belowground carbon and nutrient storage in a lowland Amazonian forest ecosystem. *Ecosystems* **V3**(N2), 193–209.

Sims P. L. and Singh J. S. (1978a) The structure and function of tem western North American grasslands: I. Abiotic and vegetational characteristics. *J. Ecol.* **66**, 251–285.

Sims P. L. and Singh J. S. (1978b) The structure and function of ten western North American grasslands: III. Net primary production, turnover and efficiencies of energy capture and water use. *J. Ecol.* **66**, 573–597.

Six J., Guggenberger G., Paustian K., Haumaier L., Elliott E. T., and Zech W. (2001) Sources and composition of soil organic matter fractions between and within soil aggregates. *Euro. J. Soil Sci.* **52**(4), 607–618.

Skjemstad J. O., Dalal R. C., and Barron P. F. (1986) Spectroscopic investigations of cultivation effects on organic matter of vertisols. *Soil Sci. Soc. Am. J.* **50**(2), 354–359.

Skjemstad J. O., Clarke P., Taylor J. A., Oades J. M., and Newman R. H. (1994) The Removal of magnetic materials from surface soils—a solid state C-13 Cp/Mas Nmr study. *Austral. J. Soil Res.* **32**(6), 1215–1229.

Skjemstad J. O., Clarke P., Taylor J. A., Oades J. M., and McClure S. G. (1996) The chemistry and nature of protected carbon in soil. *Austral. J. Soil Res.* **34**(2), 251–271.

Sohlenius B. (1980) Abundance, biomass, and contribution to energy flow by soil nematodes in terrestrial ecosystems. *Oikos* **34**, 186–194.

Sollins P., Homann P., and Caldwell B. A. (1996) Stabilization and destabilization of soil organic matter: mechanisms and controls. *Geoderma* **74**(1–2), 65–105.

Sørensen L. H. (1981) Carbon–nitrogen relationships during the humification of cellulose in soils containing different amounts of clay. *Soil Biol. Biochem.* **13**, 313–321.

Sparling G. S., Cheshire M. V., and Mundie C. M. (1982) Effect of barley plants on the decomposition of 14C-labelled soil organic matter. *J. Soil Sci.* **33**, 89–100.

St. John T. V. (1980) Influences of litterbags on growth of fungal vegetative structures. *Oecologia* **46**, 130–132.

Staaf H. and Berg B. (1982) Accumulation and release of plant nutrients in decomposing Scots pine needle litter—long-term decomposition in a Scots pine forest: 2. *Can. J. Botany: Revue Canadienne De Botanique* **60**(8), 1561–1568.

Stevenson F. J. (1994) *Humus Chemistry: Genesis, Composition, Reactions*. Wiley.

Sutton D. D. and Lampen J. O. (1962) Localization of sucrose and maltose fermenting systems in *Saccharomyces cerevisiae*. *Biochim. Biophys. Acta* **56**, 303–312.

Swift M. J., Heal O. W., and Anderson J. M. (1979) *Decomposition in Terrestrial Ecosystems*. University of California Press.

Swift M. J., Russel-Smith A., and Perfect T. J. (1981) Decomposition and mineral nutrient dynamics of plant litter in a regenerating bush-fallow in the sub-humid tropics. *J. Ecol.* **69**, 981–995.

Sylvia D. M., Fuhrmann J. J., Hartel P. G., and Zuberer D. A. (1999) *Principles and Applications of Soil Microbiology*. Prentice-Hall.

Tanaka M., Sugi Y., Tanaka S., Mishima Y., and Hamada R. (1978) Animal population, biomass and production. Soil invertebrates. In *Biological Production in a Warm Temperate Evergreen Oak Forest of Japan* (eds. T. Kira, Y. Ono, and T. Hosokawa). University of Tokyo Press, pp. 147–163.

Taylor B. R. and Parkinson D. (1988a) Aspen and pine leaf litter decomposition in laboratory microcosms: 2. Interactions of temperature and moisture level. *Can. J. Botany: Rev. Can. Botanique* **66**(10), 1966–1973.

Taylor B. R. and Parkinson D. (1988b) Patterns of water-absorption and leaching in pine and aspen leaf litter. *Soil Biol. Biochem.* **20**(2), 257–258.

Taylor B. R., Parkinson D., and Parsons W. F. J. (1989) Nitrogen and lignin content as predictors of litter decay rates: a microcosm test. *Ecology (Tempe)* **70**(1), 97–104.

Taylor B. R., Prescott C. E., Parsons W. J. F., and Parkinson D. (1991) Substrate control of litter decomposition of four Rocky Mountain coniferous forests. *Can. J. Botany* **69**(10), 2242–2250.

Taylor J. P., Wilson B., Mills M. S., and Burns R. G. (2002) Comparison of microbial numbers and enzymatic activities in surface soils and subsoils using various techniques. *Soil Biol. Biochem.* **34**(3), 387–401.

Tenney F. G. and Waksman S. A. (1929) Composition of natural organic materials and their decomposition in the soil: IV. The nature and rapidity of decomposition of the various organic complexes in different plant materials, under aerobic conditions. *Soil Sci.* **28**, 55–84.

Teplitski M., Robinson J. B., and Bauer W. D. (2000) Plants secrete substances that mimic bacterial *N*-acyl homoserine lactone signal activities and affect population density-dependent behaviors in associated bacteria. *Molecul. Plant–Microbe Interact.* **13**(6), 637–648.

Thomsen I. K., Schjonning P., Jensen B., Kristensen K., and Christensen B. T. (1999) Turnover of organic matter in differently textured soils: II. Microbial activity as influenced by soil water regimes. *Geoderma* **89**(3–4), 199–218.

Tian G., Kang B. T., and Brussaard L. (1992a) Biological effects of plant residues with contrasting chemical—compositions under humid tropical conditions—decomposition and nutrient release. *Soil Biol. Biochem.* **24**(10), 1051–1060.

Tian G., Kang B. T., and Brussaard L. (1992b) Effects of chemical-composition on N, Ca, and Mg release during incubation of leaves from selected agroforestry and fallow plant-species. *Biogeochemistry* **16**(2), 103–119.

Tice K. R., Graham R. C., and Wood H. B. (1996) Transformations of 2:1 phyllosilicates in 41-year-old soils under oak and pine. *Geoderma* **70**(1), 49–62.

Tiessen H., Chacon P., and Cuevas E. (1994) Phosphorus and nitrogen status in soils and vegetation along a toposequences of dystrophic rainforests on the upper Rio Negro. *Oecologia (Berlin)* **99**(1–2), 145–150.

Tingey D. T., Phillips D. L., and Johnson M. G. (2000) Elevated CO_2 and conifer roots: effects on growth, life span and turnover. *New Phytol.* **147**(1), 87–103.

Tisdall J. M. (1994) Possible role of soil microorganisms in aggregation in soils. *Plant Soil* **159**, 115–121.

Tisdall J. M. and Oades J. M. (1982) Organic matter and water stable aggregates in soils. *J. Soil Sci.* **33**, 141–161.

Tisdall J. M., Smith S. E., and Rengasamy P. (1997) Aggregation of soil by fungal hyphae. *Austral. J. Soil Res.* **35**(1), 55–60.

Tiwari S. C. and Mishra R. R. (1993) Fungal abundance and diversity in earthworm casts and in uningested soil. *Biol. Fertility Soils* **16**(2), 131–134.

Tokumasu S. (1998) Fungal successions on pine needles fallen at different seasons: the succession of interior colonizers. *Mycoscience* **39**(4), 409–416.

Tokunaga T. K., Wan J., Firestone M. K., Hazen T. C., Schwartz E., Sutton S. R., and Newville M. (2001) Chromium diffusion and reduction in soil aggregates. *Environ. Sci. Technol.* **35**(15), 3169–3174.

Torn M. S., Trumbore S. E., Chadwick O. A., Vitousek P. M., and Hendricks D. M. (1997) Mineral control of soil organic carbon storage and turnover. *Nature (London)* **389**(6647), 170–173.

Torn M. S., Vitousek P. M., and Trumbore S. E. (in review) The influence of nutrient availability on soil organic matter turnover estimated by incubations and radiocarbon modeling. *Ecology*.

Torsvik V., Sorheim R., and Goksoyr J. (1996) Total bacterial diversity in soil and sediment communities: a review. *J. Indust. Microbiol.* **17**(3–4), 170–178.

Townsend A. R., Vitousek P. M., Desmarais D. J., and Tharpe A. (1997) Soil carbon pool structure and temperature sensitivity inferred using CO-2 and 13CO-2 incubation fluxes from five Hawaiian soils. *Biogeochemistry (Dordrecht)* **38**(1), 1–17.

Traore O., Groleau-Renaud V., Plantureux S., Tubeileh A., and Boeuf-Tremblay V. (2000) Effect of root mucilage and modelled root exudates on soil structure. *Euro. J. Soil Sci.* **51**(4), 575–581.

Trigo D. and Lavelle P. (1993) Changes in respiration rate and some physicochemical properties of soil during gut transit through Allolobophora molleri (Lumbricidae, Oligochaeta). *Biol. Fertility Soils* **15**(3), 185–188.

Trumbore S. E. (1993) Comparison of carbon dynamics in tropical and temperate soils using radiocarbon measurements. *Global Biogeochem. Cycles* **7**(2), 275–290.

Trumbore S. E., Chadwick O. A., and Amundson R. (1996) Rapid exchange between soil carbon and atmospheric carbon dioxide driven by temperature change. *Science* **272**(5260), 393–396.

Tunlid A. and White D. C. (1992) Biochemical analyses of biomass, community structure, nutritional status, and metabolic activity of microbial communities in soil. In *Soil Biochemistry* (ed. G. Stotzky). Dekker, vol. 7, pp. 229–262.

Ugolini F. C., Dawson H., and Zachara J. (1977) Direct evidence of particle migration in soil solution of a podzol. *Science* **198**(4317), 603–605.

Ulery A. L., Graham R. C., Chadwick O. A., and Wood H. B. (1995) Decade-scale changes of soil carbon, nitrogen and exchangeable cations under chaparral and pine. *Geoderma* **65**(1–2), 121–134.

Uren N. C. and Reisenauer H. M. (1988) The role of root exudates in nutrient acquisition. In *Advances in Plant Nutrition* (eds. P. B. Tinker and A. Lauchli). Praeger, pp. 79–114.

Van Bergen P. F., Nott C. J., Bull I. D., Poulton P. R., and Evershed R. P. (1998) Organic geochemical studies of soils from the Rothamsted Classical Experiments: IV. Preliminary results from a study of the effect of soil pH on organic matter decay. *Org. Geochem.* **29**(5–7), 1779–1795.

Van Gestel M., Merckx R., and Vlassak K. (1993) Microbial biomass responses to soil drying and rewetting: the fate of fast- and slow-growing microorganisms in soils from different climates. *Soil Biol. Biochem.* **25**(1), 109–123.

Van Veen J. A. and Kuikman P. J. (1990) Soil structural aspects of decomposition of organic matter by microorganisms. *Biogeochemistry (Dordrecht)* **11**(3), 213–234.

Van Veen J. A., Ladd J. N., and Amato M. (1985) Turnover of carbon and nitrogen through the microbial biomass in a sandy loam and a clay soil incubated with uniformly carbon-14-labeled glucose and nitrogen-15 ammonium sulfate under different moisture regimes. *Soil Biol. Biochem.* **17**(6), 747–756.

Van Vuuren M. M. I., Berendse F., and De Visser W. (1993) Species and site differences in the decomposition of litters and roots from wet heathlands. *Can. J. Botany* **71**(1), 167–173.

Veeman W. S. (1998) Nuclear magnetic resonance, a simple introduction to the principles and applications. *Geoderma* **80**(3–4), 225–242.

Vitousek P. M. and Sanford R. L., Jr. (1986) Nutrient cycling in moist tropical forest. In *Annual Review of Ecology and Systematics*, Annual Reviews (ed. R. F. Johnston) vol. 17, pp. 137–168.

Vitousek P. M., Turner D. R., Parton W. J., and Sanford R. L. (1994) Litter decomposition on the Mauna Loa environmental matrix, Hawaii: patterns, mechanisms, and models. *Ecology (Tempe)* **75**(2), 418–429.

Vogt K. (1991) Carbon budgets of temperate forest ecosystems. *Tree Physiol.* **9**(1–2), 69–86.

Wagner G. H. and Wolf D. C. (1999) Carbon transformations and soil organic matter formation. In *Principles and Applications of Soil Microbiology* (eds. D. M. Sylvia, J. J. Fuhrmann, P. G. Hartel, and D. A. Zuberer). Prentice Hall, pp. 218–258.

Waksman S. A. (1932) *Humus.* Williams and Wilkins.

Waksman S. A. and Tenney F. G. (1927) The composition of natural organic materials and their decomposition in the soil: I. Methods of quantitative analysis of plant materials. *Soil Sci.* **24**, 275–283.

Waksman S. A. and Tenney F. G. (1928) Composition of natural organic materials and their decomposition in the soil: III. The influence of nature of plant upon the rapidity of its decomposition. *Soil Sci.* **27**, 155–171.

Wallwork J. A. (1970) *Ecology of Soil Animals.* McGraw-Hill, London.

Walse C., Berg B., and Sverdrup H. (1998) Review and synthesis of experimental data on organic matter decomposition with respect to the effect of temperature, moisture, and acidity. *Environ. Rev.* **6**(1), 25–40.

Ward D. M., Bateson M. M., Weller R., and Ruff-Roberts A. L. (1992) Ribosomal RNA analysis of microorganisms as they occur in nature. In *Advances in Microbial Ecology* (ed. K. C. Marshall). Plenum, pp. 219–286.

Wasilewska L. (1991) The role of nematode in the process of elemental recycling on drained fen differentiated by peat origin. *Polish Ecol. Stud.* **17**, 191–201.

Weil R. R. (1992) Inside the heart of sustainable farming. *The New Farm January*, 43–48.

West A. W., Grant W. D., and Sparling G. P. (1987) Use of ergosterol, diaminopimelic acid and glucosamine contents of soils to monitor changes in microbial populations. *Soil Biol. Biochem.* **19**(5), 607–612.

West A. W., Sparling G. P., Feltham C. W., and Reynolds J. (1992) Microbial activity and survival in soils dried at different rates. *Austral. J. Soil Res.* **30**(2), 209–222.

Whipps J. M. and Lynch J. M. (1985) Energy losses by the plant in rhizodeposition. *Ann. Proc. Phytochem. Soc. Euro.* **26**, 59–71.

White D. L., Haines B. L., and Boring L. R. (1988) Litter decomposition in southern appalachian black locust and pine hardwood stands–litter quality and nitrogen dynamics. *Can. J. Forest Res.: Revue Canadienne De Recherche Forestiere* **18**(1), 54–63.

White G. (1789) *The Natural History and Antiquities of Selborne, in the County of Southampton: With Engravings, and an Appendix.* Printed by T. Bensley, for B. White and son.

Whitford W. G. (1989) Abiotic controls on the functional structure of soil food webs. *Biol. Fertility Soils* **8**(1), 1–6.

Whitford W. G., Meentemyer V., Seastedt T. R., Cromack K., Jr., Crossley D. A., Santos P., Todd R. L., and Waide J. B. (1981) Exceptions to the AET model: deserts and clear-cut forest. *Ecology* **62**(1), 275–277.

Whitford W. G., Steinberger Y., and Ettershank G. (1982) Contributions of subterranean termites to the economy of Chihuahuan desert ecosystems. *Oecologia* **55**, 298–302.

Whitman W. B., Coleman D. C., and Wiebe W. J. (1998) Prokaryotes: the unseen majority. *Proceedings of the National Academy of Sciences of the United States of America* **95**(12), 6578–6583.

Wilson M. A. (1989) Solid-state nuclear magnetic resonance spectroscopy of humic substances: basic concepts and techniques. In *Humic Substances II: in Search of Structure* (eds. M. H. B. Hayes, P. MacCarthy, R. L. Malcolm, and M. J. Swift). Wiley, pp. 309–338.

Wilson M. A. (1990) Application of nuclear magnetic resonance spectroscopy to organic matter in whole soils. In *Humic Substances in Soil and Crop Sciences: Selected Readings* (eds. P. MacCarthy, C. E. Clapp, R. L. Malcolm, and P. R. Bloom). American Society of Agronomy, pp. 221–260.

Wilson M. A., Pugmire R., Zilm K. W., Goh K. M., Heng S., and Grant D. (1981) Cross-polarization 13C NMR spectroscopy with "magic angle" spinning characterizes organic matter in whole soils. *Nature* **294**, 648–650.

Witkamp M. and Crossley D. A. (1966) The role of microarthropods and microflora in the breakdown of white oak litter. *Pedobiologia* **6**, 293–303.

Witkamp M. and Olson J. S. (1963) Breakdown of confined and nonconfined oak litter. *Oikos* **14**, 138–147.

Woese C. R. (1987) Bacterial Evolution. *Microbiol. Rev.* **51**(2), 221–271.

Wolters V. (2000) Invertebrate control of soil organic matter stability. *Biol. Fertility Soils* **31**(1), 1–19.

Wood T. M. and Garcia-Campayo V. (1994) Enzymes and mechanisms involved in microbial cellulolysis. In *Biochemistry of Microbial Degradation* (ed. C. Ratledge). Kluwer Academic, pp. 197–231.

Xu M. and Qi Y. (2001) Spatial and seasonal variations of Q(10) determined by soil respiration measurements at a Sierra Nevadan forest. *Global Biogeochem. Cycles* **15**(3), 687–696.

Yu. Z. S., Dahlgren R. A., and Northup R. R. (1999) Evolution of soil properties and plant communities along an extreme edaphic gradient. *Euro. J. Soil Biol.* **35**(1), 31–38.

Zak J. C., Willig M. R., Moorhead D. L., and Wildman H. G. (1994) Functional diversity of microbial communities: a quantitative approach. *Soil Biol. Biochem.* **26**(9), 1101–1108.

Zarda B., Hahn D., Chatzinotas A., Schoenhuber W., Neef A., Amann R. I., and Zeyer J. (1997) Analysis of bacterial community structure in bulk soil by *in situ* hybridization. *Arch. Microbiol.* **168**(3), 185–192.

Zech W., Johansson M. B., Haumaier L., and Malcolm R. L. (1987) Cpmas carbon-13 Nmr and Ir spectra of spruce and pine litter and of the Klason lignin fraction at different stages of decomposition. *Zeitschrift fuer Pflanzenernaehrung und Bodenkunde* **150**(4), 262–265.

Zech W., Ziegler F., Koegel-Knabner I., and Haumaier L. (1992) Humic substances distribution and transformation in forest soils. *Sci. Total Environ.* 117–118, 155–174.

Zech W., Senesi N., Guggenberger G., Kaiser K., Lehmann J., Miano T. M., Miltner A., and Schroth G. (1997) Factors controlling humification and mineralization of soil organic matter in the tropics. *Geoderma* **79**(1–4), 117–161.

Zhang H. and Schrader S. (1993) Earthworm effects on selected physical and chemical properties of soil aggregates. *Biol. Fertility Soils* **15**(3), 229–234.

Zogg G. P., Zak D. R., Ringelberg D. B., MacDonald N. W., Pregitzer K. S., and White D. C. (1997) Compositional and functional shifts in microbial communities due to soil warming. *Soil Sci. Soc. Am. J.* **61**(2), 475–481.

8.08
Anaerobic Metabolism: Linkages to Trace Gases and Aerobic Processes

J. P. Megonigal
Smithsonian Environmental Research Center, Edgewater, MD, USA
M. E. Hines
University of Massachusetts Lowell, MA, USA
and
P. T. Visscher
University of Connecticut Avery Point, Groton, CT, USA

8.08.1 OVERVIEW OF LIFE IN THE ABSENCE OF O_2

8.08.1.1 Introduction

Life evolved and flourished in the absence of molecular oxygen (O_2). As the O_2 content of the atmosphere rose to the present level of 21% beginning about two billion years ago, anaerobic metabolism was gradually supplanted by aerobic metabolism. Anaerobic environments have persisted on Earth despite the transformation to an oxidized state because of the combined influence of water and organic matter. Molecular oxygen diffuses about 10^4 times more slowly through water than air, and organic matter supports a large biotic O_2 demand that consumes the supply faster than it is replaced by diffusion. Such conditions exist in wetlands, rivers, estuaries, coastal marine sediments, aquifers, anoxic water columns, sewage digesters, landfills, the intestinal tracts of animals, and the rumen of herbivores. Anaerobic microsites are also embedded in oxic environments such as upland soils and marine water columns. Appreciable rates of aerobic respiration are restricted to areas that are in direct contact with air or those inhabited by organisms that produce O_2.

Rising atmospheric O_2 reduced the global area of anaerobic habitat, but enhanced the overall rate of anaerobic metabolism (at least on an area basis) by increasing the supply of electron donors and acceptors. Organic carbon production increased dramatically, as did oxidized forms of nitrogen, manganese, iron, sulfur, and many other elements. In contemporary anaerobic ecosystems, nearly all of the reducing power is derived from photosynthesis, and most of it eventually returns to O_2, the most electronegative electron acceptor that is abundant. This photosynthetically driven redox gradient has been thoroughly exploited by aerobic and anaerobic microorganisms for metabolism. The same is true of hydrothermal vents (Tunnicliffe, 1992) and some deep subsurface environments (Chapelle *et al.*, 2002), where thermal energy is the ultimate source of the reducing power.

Although anaerobic habitats are currently a small fraction of Earth's surface area, they have a profound influence on the biogeochemistry of the planet. This is evident from the observation that the O_2 and CH_4 content of Earth's atmosphere are in extreme disequilibrium (Sagan *et al.*, 1993). The combination of high aerobic primary production and anoxic sediments provided the large deposits of fossil fuels that have become vital and contentious sources of energy for modern industrialized societies. Anaerobic metabolism is responsible for the abundance of N_2 in the atmosphere; otherwise N_2-fixing bacteria would have consumed most of the N_2 pool long ago (Schlesinger, 1997). Anaerobic microorganisms are common symbionts of termites, cattle, and many other animals, where they aid digestion. Nutrient and pollutant chemistry are strongly modified by the reduced conditions that prevail in wetland and aquatic ecosystems.

This review of anaerobic metabolism emphasizes aerobic oxidation, because the two processes cannot be separated in a complete treatment of the topic. It is process oriented and highlights the fascinating microorganisms that mediate anaerobic biogeochemistry. We begin this review with a brief discussion of CO_2 assimilation by autotrophs, the source of most of the reducing power on Earth, and then consider the biological processes that harness this potential energy. Energy liberation begins with the decomposition of organic macromolecules to relatively simple compounds, which are simplified further by fermentation. Methanogenesis is considered next because CH_4 is a product of acetate fermentation, and thus completes the catabolism of organic matter, particularly in the absence of inorganic electron acceptors. Finally, the organisms that use nitrogen, manganese, iron, and sulfur for terminal electron acceptors are considered in order of decreasing free-energy yield of the reactions.

8.08.1.2 Overview of Anaerobic Metabolism

Microorganisms derive energy by transferring electrons from an external electron source or *donor* to an external electron sink or *terminal electron acceptor*. Organic electron donors vary from monomers that support fermentation to simple compounds such as acetate and CH_4. The common inorganic electron donors are molecular hydrogen (H_2), ammonium (NH_4^+), manganous manganese (Mn(II)), ferrous iron (Fe(II)), and hydrogen sulfide (H_2S). Energy is harnessed by shuttling electrons through transport chains within a cell until a final transfer is made to a terminal electron acceptor. The common terminal electron acceptors are nitrate (NO_3^-), manganic manganese Mn(IV), ferric iron (Fe(III)), sulfate (SO_4^{2-}), and carbon dioxide (CO_2) (Table 1).

Anaerobic organisms often have the capacity to reduce two or more terminal electron acceptors. In many cases, these alternative reactions do not support growth, as with the fermenting bacteria that reduce Fe(III) (Lovley, 2000b). In other cases, the ability to use multiple electron acceptors is presumably an adaptation for remaining active in an environment where the supply of specific electron acceptors is variable. For example, denitrification permits normally aerobic bacteria to respire in the absence of O_2, albeit at a slower rate.

Table 1 Thermodynamic sequence for reduction of inorganic substances by organic matter[a].

Reaction	*Eh* (V)	ΔG^b
Reduction of O_2		
$O_2 + 4H^+ + 4e^- \rightleftarrows 2H_2O$	0.812	−29.9
Reduction of NO_3^-		
$NO_3^- + 6H^+ + 6e^- \rightleftarrows N_2 + 3H_2O$	0.747	−28.4
Reduction of Mn^{4+} to Mn^{2+}		
$MnO_2 + 4H^+ + 2e^- \rightleftarrows Mn^{2+} + 2H_2O$	0.526	−23.3
Reduction of Fe^{3+} to Fe^{2+}		
$Fe(OH)_3 + 3H^+ + e^- \rightleftarrows Fe^{2+} + 3H_2O$	−0.047	−10.1
Reduction of SO_4^{2-} to H_2S		
$SO_4^{2-} + 10H^+ + 8e^- \rightleftarrows H_2S + 4H_2O$	−0.221	−5.9
Reduction of CO_2 to CH_4		
$CO_2 + 8H^+ + 8e^- \rightleftarrows CH_4 + 2H_2O$	−0.244	−5.6

Source: Schlesinger (1997).
[a] Units are kcal mol^{-1} e^{-1} assuming coupling to the oxidation reaction $\frac{1}{4}CH_2O + \frac{1}{4}H_2O \rightarrow \frac{1}{4}CO_2 + H^+ + e^-$. [b] $\Delta G = -RT \ln (K)$; pH 7.0, 25 °C.

The most fundamental difference between aerobic and anaerobic metabolism is energy yield. Oxidation of glucose yields ~2,900 kJ mol^{-1} under aerobic conditions (Equation (1)) compared to ~400 kJ mol^{-1} under typical methanogenic conditions (Equation (2)):

$$C_6H_{12}O_6 + 6O_2 \rightarrow 6CO_2 + 6H_2O \quad (1)$$

$$C_6H_{12}O_6 \rightarrow 3CO_2 + 3CH_4 \quad (2)$$

The high-energy yield from aerobic metabolism permits a single organism to completely oxidize complex organic compounds to CO_2. In comparison, no single anaerobic microorganism can completely degrade organic polymers to CO_2 and H_2O (Fenchel and Finlay, 1995), and most have highly specialized substrate demands. Under anaerobic conditions, the mineralization of organic carbon to CO_2 is a multistep process that involves a consortium of organisms, each of which conserves a fraction of the potential-free energy that was available in the original organic carbon substrate. As a result, anaerobic organisms are adapted to conserve quantities of energy that often approach the theoretical minimum (20 kJ mol^{-1}) required for metabolism (Valentine, 2001).

The first step in anaerobic decomposition is the breakdown of complex organic molecules to simple molecules such as sugars (Section 8.08.3.1). Next, these are fermented to even simpler molecules such as acetate and H_2 (Section 8.08.3.2). In the final step, fermentation products serve as electron donors for the reduction of inorganic compounds. The low-energy yield of fermentation, SO_4^{2-} reduction, methanogenesis and most other anaerobic pathways make them particularly sensitive to the concentrations of substrates and products. Because fermentation is inhibited by its end products, the process is dependent on end product consumption by non-fermentative organisms. This thermodynamic dependence among anaerobic microorganisms, in which several species must function together in order to consume a single substrate, is known as *syntrophy*, which is a type of mutualism. Mutualism is a particularly important ecological interaction that determines the structure and function of anaerobic communities.

Competition is another important ecological mechanism that affects the structure and function of anaerobic microbial communities. Competition for fermentation products, such as H_2 and acetate, is particularly keen, because they usually limit microbial activity in anaerobic ecosystems. The outcome of competition for these substrates favors those pathways with the highest thermodynamic yield: NO_3^- reduction > Mn(IV) reduction > Fe(III) reduction > SO_4^{2-} reduction > HCO_3^- reduction (i.e., methanogenesis) (Table 1). This hierarchy also tends to apply to the expression of metabolic pathways within a single microorganism that can reduce multiple electron acceptors (Nealson and Myers, 1992). The presence of O_2 suppresses all the anaerobic metabolic pathways primarily because it is a toxin, but also because it has the highest energy yield of all the common terminal electron acceptors. To a first approximation, a single metabolic pathway dominates anaerobic carbon cycling until it is limited by the availability of electron acceptors. However, it is not uncommon for the pathways to coexist because of spatial variation in the abundance of terminal electron acceptors. Coexistence may also occur because the supply of a *competitive* electron donor (e.g., H_2) is large and nonlimiting, or because there is a supply of a *noncompetitive* electron donor that can be used by some organisms, but not others (e.g., Oremland *et al.*, 1982). The outcome of competition for electron donors establishes the redox potential (Yao and Conrad, 1999), and not vice versa as is often suggested.

The thermodynamic yield of the various anaerobic pathways is related to their affinity for substrates. Sulfate-reducing bacteria outcompete methanogens by maintaining H_2 concentrations below the threshold for the process to be thermodynamically feasible (Lovley *et al.*, 1982), and Fe(III) reducers do the same to SO_4^{2-} reducers. In fact, the equilibrium concentration of H_2 can be used to predict the dominant metabolic pathways in a given environment (Lovley and Goodwin, 1988; Lovley *et al.*, 1994a).

8.08.1.3 Anaerobic–Aerobic Interface Habitats

Spatial variation in the abundance of electron donors and acceptors explains large-scale and small-scale patterns of anaerobic metabolism. Sulfate reduction dominates anaerobic carbon metabolism on about two-thirds of the planet because of the high abundance of SO_4^{2-} in seawater (Capone and Kiene, 1988). Fe(III) reduction is important in all anaerobic ecosystems with mineral-dominated soils or sediments, regardless of whether they are marine or freshwater (Thamdrup, 2000). Methanogenesis is important in freshwater environments generally, and it dominates the anaerobic carbon metabolism of bogs, fens, and other wetlands that exist on organic (i.e., peat) soils.

The contribution of the various anaerobic metabolic pathways to carbon metabolism varies temporally and spatially due to changes in the abundance of electron donors and acceptors (Figure 1). Organic carbon is most abundant at the surface of soils and sediments where detritus is deposited, most of which is derived from aerobic photosynthesis. The aerobic zone is also the source of most terminal electron acceptors, some of which diffuse into the anaerobic zone from the atmosphere or water column (e.g., O_2 and SO_4^{2-} in marine ecosystems); others are regenerated at the aerobic–anaerobic interface due to oxidation of NH_4^+, Fe(II), Mn(II), or H_2S. Regeneration at the aerobic–anaerobic interface can supply a large fraction of the terminal electron acceptors consumed in anaerobic metabolism.

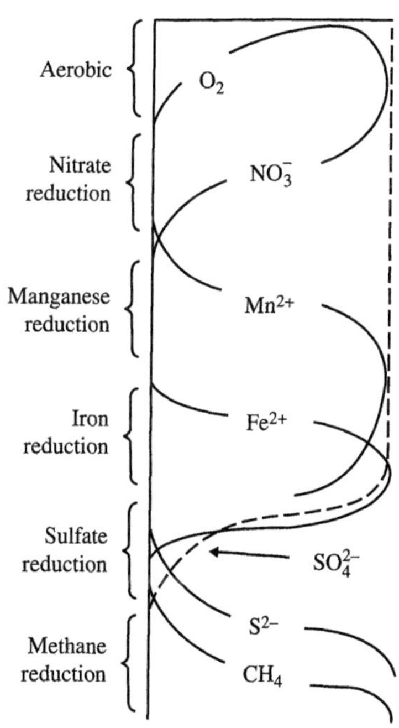

Figure 1 Vertical biogeochemical zones in sediments. The top is the sediment–water interface. Processes on the left represent the use of various electron acceptors (respirations) during the degradation of organic matter. Plots on the right represent the chemical profiles most widely used to delineate the vertical extent of each zone. Rotating the figure 90° to the left shows the sequence of electron acceptors used over time (x-axis) if a sample of oxic sediment were enclosed and allowed to become anaerobic over time.

Oxidant regeneration is stimulated dramatically by the presence of animals and plants. The burrowing activity of animals (i.e., *bioturbation*) is a particularly important agent of Fe(III) and Mn(IV) regeneration in marine sediments and salt marshes (Thamdrup, 2000; Kostka *et al.*, 2000c; Gribsholt *et al.*, 2003; Gribsholt and Kristensen, 2002). Wetland plants promote regeneration by serving as conduits of O_2 infusion deep into the soil profile. Such plants tolerate flooding partly by supplying O_2 to their root system, where some of it leaks into the soil due to *radial* O_2 *loss*. In the absence of physical mixing, burrowing, or radial O_2 loss, the depth of O_2 penetration into saturated soils and sediments is a few millimeters. The influence of plants and animals on anaerobic metabolism is similar in the sense that both effectively increase the aerobic–anaerobic surface area (Mayer *et al.*, 1995; Armstrong, 1964). In addition, plants are a source of labile organic carbon compounds that fuel anaerobic metabolism.

Competition for fermentation products produces a succession of dominant metabolic pathways as distance from the source of electron donors and acceptors increases. Aerobic metabolism dominates the surface of sediments, and methanogenesis the deeper depths, as suggested by their free energy yield (Table 1). There is often very little overlap between each zone, suggesting nearly complete exclusion of one group by another. The same pattern is observed with distance from the surface of a root or burrow, or with distance downstream from an organic pollutant source in rivers and aquifers.

8.08.1.4 Syntax of Metabolism

In ecological and biogeochemical studies, any organism can be described by three basic attributes that define a "feeding" (Greek *trophos*) niche: energy source, electron donor source, and carbon source (Table 2). Organisms that use light energy are *phototrophic*, while those that use chemical energy are *chemotrophic*. Organisms that use inorganic electron donors are *lithotrophic*, while those that use organic electron donors are *organotrophic*. Finally, *autotrophic* organisms assimilate CO_2 for biosynthesis, while *heterotrophic* organisms assimilate organic carbon (e.g., Lwoff *et al.*, 1946; Barton *et al.*, 1991). The modifiers are linked in the order energy–electrons–carbon. For example, plant photosynthesis is technically *photolithoautotrophy*. It is not surprising that

Table 2 Classification terms for microbial metabolism. Each term is composed of modifiers for the source of energy (chemical versus light), the source of electrons (inorganic versus organic), and the source of carbon (inorganic versus organic). The modifiers are linked in the order energy–electrons–carbon. Metabolisms for which organic carbon is both the carbon and energy source are abbreviated as "heterotrophy."

Energy and carbon source	*Electron source*	
	*Inorganic (**litho**trophy)*	*Organic (**organo**trophy)*
I. Chemical energy (***chemo***trophy)		
Carbon source organic (***hetero***trophy)	Chemolithoheterotrophy	Chemoheterotrophy
Carbon source inorganic (***auto***trophy)	Chemolithoautotrophy	Chemoorganoautotrophy
II. Light energy (***photo***trophy)		
Carbon source organic (***hetero***trophy)	Photolithoheterotrophy	Photoheterotrophy
Carbon source inorganic (***auto***trophy)	Photolithoautotrophy	Photoorganoautotrophy

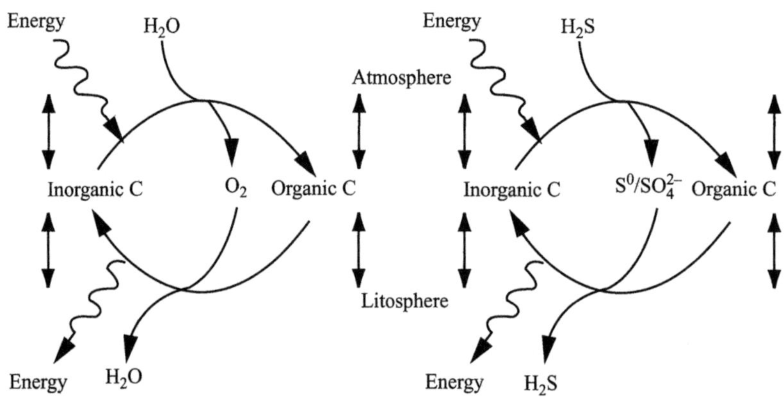

Figure 2 A diagram of the biological carbon cycle. The conversion from inorganic to organic carbon requires light or chemical energy and an electron donor (e.g., H_2O, H_2S, Fe(II)), and is the process of autotrophy. The reverse reaction, in which organic carbon is oxidized to CO_2, releases energy while reducing an electron acceptor (e.g., O_2, SO_4^{2-}/S^0, Fe(III)). This part of the cycle is referred to as heterotrophy.

some organisms do not fit neatly into this classification scheme. *Mixotrophic* organisms alternate between two or more distinct types of metabolism, such as autotrophy and heterotrophy.

In practice, microbiologists use abbreviated terms because certain combinations of physiological traits commonly occur together (Syliva *et al.*, 1998). Because most organisms that metabolize organic compounds can extract electrons, energy, and carbon from them; they are referred to simply as *heterotrophs*. Chemolithoautotrophs are often abbreviated as *chemolithotrophs*, *lithotrophs*, or *autotrophs* because lithotrophic organisms are usually also autotrophic

Microorganisms are also described by more specific substrate requirements or environmental preferences. For example, organisms that require H_2 are *hydrogenotrophic* (i.e., H_2-feeding) and those that thrive in very cold environments are *psychrophilic* (i.e., "cold-loving"). The most important aerobic organisms in oxic–anoxic interfaces are *microaerophiles*, which thrive at very low O_2 concentrations.

8.08.2 AUTOTROPHIC METABOLISM

From a microbial point of view, the carbon cycle is merely an energy cycle (Figure 2). Reduction of CO_2 through a variety of biochemical pathways produces organic carbon, thereby changing the oxidation state of carbon from +IV to between +III and −IV. The main source of energy that drives this process on Earth is quantum energy or light, but there are ecosystems on Earth that are entirely dependent on chemolithoautotrophy. Both forms of autotrophic metabolism are possible under anaerobic conditions. Here, we briefly consider the energy sources that are the ultimate source of reducing power for anaerobic respiration.

8.08.2.1 Phototroph (Photolithoautotrophy) Diversity and Metabolism

Perhaps the most familiar photosynthetic prokaryotes are the cyanobacteria (formerly blue-green algae). These organisms use the Calvin cycle to assimilate CO_2, releasing O_2 as a waste product in the process (i.e., oxygenic photolithoautotrophy; Table 3). Cyanobacteria include single-cell and filamentous pelagic and benthic species, and they are found across a range of environmental conditions from frigid Antarctic dry valleys to mildly hot springs. The upper temperature limit of photosynthesis is 70–74 °C. The physiological versatility of this group is evident in the capacity of some members to fix N_2, to derive energy from fermentation and to use HS^- as an electron donor (through PS I) for photosynthesis. Cyanobacteria are dominant species in microbial mats and stromatolites and are believed to be responsible for the development of an oxidized (O_2-containing) atmosphere 2.2–2.0 Gyr ago (Des Marais, 2000). In addition, N_2-fixing cyanobacteria are able to produce copious amounts of H_2 during the dark period, which may have had important implications for the development of an O_2-rich atmosphere (Hoehler *et al.*, 2001).

Many phototrophs do not produce O_2 as a waste product. Such *anoxygenic phototrophs* are comprised of purple sulfur, purple nonsulfur, green sulfur, and green nonsulfur bacteria. Although purple sulfur bacteria are typically found in anoxic zones of lakes and sediments, many are capable of photosynthesis under oxic conditions (Van Gemerden, 1993). Most fix N_2 and store S^0 intra- or extracellularly, and some are capable of chemolithoautotrophic growth. Extreme halophilic, sulfidic, and mildly thermophilic environments harbor purple sulfur bacteria. In contrast, purple nonsulfur bacteria typically thrive in less sulfidic, organic rich habitats and are metabolically more diverse. For example, some are photoheterotrophs, which use light as an energy source, but also require organic precursors to synthesize a portion of their organic compounds. Purple bacteria use the Calvin cycle for autotrophic CO_2 fixation (Table 3). In contrast, the green sulfur bacteria are strict anaerobes that store S^0 only outside the cell and predominantly use the reversed citric acid cycle for carbon fixation. Some use ferrous iron as their photosynthetic electron donor (Heising *et al.*, 1999; Widdel *et al.*, 1993). Green nonsulfur bacteria typically grow photoheterotrophically, but also use H_2 or H_2S during autotrophic growth. The upper temperature of anoxygenic photosynthesis is 70–73 °C.

During photosynthesis, the energy from light quanta is converted into chemical energy that can be used to drive biochemical reactions (Warburg and Negelein, 1923). Photosynthetic organisms use a variety of light-harvesting systems, which collectively cover most of the visible spectrum (Stolz, 1991; Samsonoff and MacColl, 2001). Eukaryotic phototrophic organisms contain either Chlorophyll a (Chl*a*) or Chl*b*, which have their maximum absorption at 680 nm and 660 nm, respectively. Among the prokaryotes, aerobic organisms such as cyanobacteria and prochlorophytes typically contain Chl*a*, while anoxygenic phototrophs have a wide variety of chlorophyll

Table 3 Autotrophic reduction–oxidation reactions coupled to CO_2 assimilation. The most common carbon assimilation pathways for each type of metabolism are listed, with alternative modes of CO_2 fixation provided in parentheses. The free energy yield of the reaction is provided for pathways that do not require light.

Metabolism	*Reaction*[a]	$\Delta G^{0\prime}$ (kJ)	*C assimilation pathway*[b]
H_2 oxidation	$H_2 + 0.5O_2 \rightarrow H_2O$	−237	Calvin (RTCA)
CO oxidation	$CO + 0.5O_2 \rightarrow CO_2$	−257	Calvin
S oxidation	$HS^- + 2O_2 \rightarrow SO_4^{2-} + H^+$	−798	Calvin (AcetylCoA)
S oxidation	$5HS^- + 8NO_3^- + 3H^+ \rightarrow 5SO_4^{2-} + 4N_2 + 4H_2O$	−3,723	Calvin
NH_4^+ oxidation	$NH_4^+ + 1.5O_2 \rightarrow NO_2^- + 2H_2O$	−287	Calvin (PEP)
NO_2^- oxidation	$NO_2^- + 0.5O_2 \rightarrow NO_3^-$	−74	Calvin
Fe(II) oxidation	$Fe^{2+} + H^+ + 0.25O_2 \rightarrow Fe^{3+} + 0.5H_2O$	−33	Calvin
Mn(II) oxidation	$Mn^{2+}0.5O_2 + H_2O \rightarrow MnO_2 + 2H^+$	−68	Calvin
Anaerobic phototrophic Fe(II) oxidation (and denitrification)	$4FeCO_3 + 10H_2O + h\nu \rightarrow 4Fe(OH)_3 + (CH_2O) + 3HCO_3^-$		Calvin
Acetogenesis	$4H_2 + 2HCO_3^- + H^+ \rightarrow CH_3COO^- + 4H_2O$	−105	AcetylCoA
Hydrogenotrophic methanogenesis	$4H_2 + HCO_3^- \rightarrow CH_4 + 3H_2O$	−136	AcetylCoA
Oxygenic phototrophy	$CO_2 + H_2O + h\nu \rightarrow (CH_2O) + O_2$		Calvin
Anoxygenic phototrophy	$2CO_2 + H_2S + 2H_2O + h\nu \rightarrow 2(CH_2O) + SO_4^{2-} + H^+$		Calvin (RTCA, HPC)

[a] $h\nu$ = photon energy that is required to drive the reaction (hence no ΔG reported). [b] Calvin = Calvin cycle, RTCA = reductive or reversed tri-carboxylic acid cycle, PEP = Phosphoenolpyruvate carboxylation, and AcetylCoA = acetyl-CoA pathway, HPC = hydroxy-priopionate cycle.

pigments. The purple bacteria contain either bacteriochlorophyll a Bchl*a* (805 nm and 830–890 nm) or BChl*b* (835–850; 1,020–1,040 nm); green bacteria contain BChl*c* (745–755 nm), BChl*d* (705–740 nm) or BChl*e* (719–726 nm); and heliobacteria contain BChl*g* (670 nm and 788 nm). BChl*a* is also found in some aerobic, heterotrophic phototrophs (e.g., erythrobacters; Yurkov and Van Gemerden, 1993) and anaerobic, heterotrophic phototrophs (e.g., some purple nonsulfur bacteria). A variety of antenna pigments, such as carotenoids and phycobiliproteins, cover the remaining windows in the visible spectrum.

Oxygenic photosynthesis requires two photosystems with different standard potentials and reaction center chlorophylls (Figure 3): PS II ($E^0 = +1.0$ V; P680), which uses H_2O as electron donor, and PS I ($E^0 = +0.3$ V; P700). In contrast, anoxygenic photosynthesis utilizes only a single photosystem: P870 ($E^0 = +0.5$ V) in purple bacteria (Figure 3), P840 ($E^0 = +0.3$ V) in green bacteria and P798 ($E^0 = +0.2$ V) in heliobacteria. Electron donors for anoxygenic phototrophs include a range of reduced S-compounds, H_2 and Fe(II). Oxygenic photosynthesis involves noncyclic electron transfer, light-driven energy generation and light-driven reducing power, whereas in anoxygenic photosynthesis, electron transfer is cyclic and only energy generation is driven by light. In hypersaline environments, bacteriorhodopsin-based phototrophy is found in some halobacteria (Hartman *et al.*, 1980) that can grow in brine (32%, or 5.5 M NaCl). Phototrophic growth in these red-pigmented (bacterioruberin, an antenna pigment) microbes does not involve chlorophyll and takes place under microaerophilic conditions. The phototrophic growth potential is extremely limited. Maximum bacteriorhodopsin absorption is at 570 nm.

8.08.2.2 Chemotroph (Chemolithoautotrophy) Diversity and Metabolism

Inorganic redox reactions provide an alternative to using light as a source of energy and reducing equivalents to assimilate CO_2 (Table 3). Electron donors for chemolithotrophy include H_2 (Bowien and Schlegel, 1981), carbon monoxide (CO_2) (Shiba *et al.*, 1985), H_2S and other reduced sulfur compounds, NH_4^+, and other reduced nitrogen compounds, Fe(II), and Mn(II). It is likely that other reduced elements (e.g., As(IV), Cr(III), Sb(III), Se(-II)/Se(0), U(-IV)) also function as reductants (Battaglia-Brunet *et al.*, 2002;

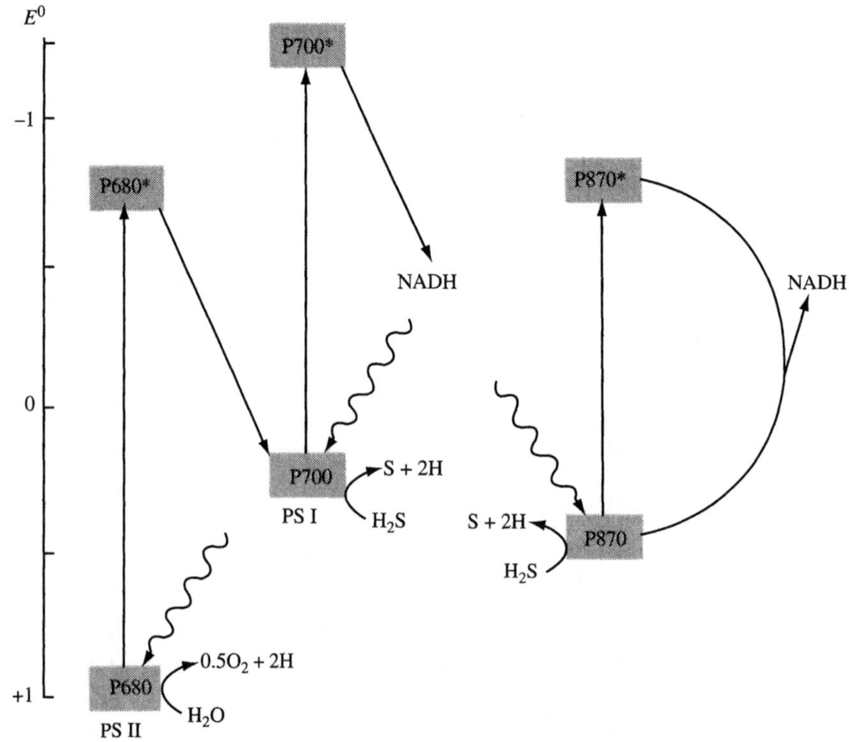

Figure 3 A schematic representation of photosystems and associated standard potentials involved in oxygenic photosynthesis in cyanobacteria (left panel) and anoxygenic photosynthesis (right panel). The flow of electrons in the former is linear, while in the latter it is cyclic.

Dowdle and Oremland, 1998; Ehrlich, 1999, 2002). Several of these pathways are discussed here in the context of the element cycles they influence, including acetogenesis (Section 8.08.3.2.1), hydrogenotrophic methanogenesis (Section 8.08.4.2), NH_4^+ oxidation (Section 8.08.5.3.3), Fe(II) oxidation (Section 8.08.6.5), and sulfur oxidation (Section 8.08.7.9).

8.08.2.3 Pathways of CO_2 Fixation

All microbes, including chemoorganoheterotrophs, possess some ability to engage in reversible carboxylation (i.e., CO_2–C assimilation into an organic compound) and decarboxylation reactions, some of which lead to the incorporation of a significant amount of CO_2 (Wood, 1985). Here, we briefly consider the biochemical pathways that photo- and chemolithotrophic bacteria deploy in order to produce the majority of their biomass. The four major CO_2-fixing pathways are the Calvin cycle, the acetyl-CoA pathway, the reductive tricarboxylic acid (TCA) cycle, and the 3-hydroxypriopionate cycle.

The Calvin cycle or reductive pentose phosphate cycle occurs in all green plants and many microorganisms. Carboxylation is catalyzed by ribulose-bis-phosphate carboxylase–oxygenase (Rubisco). Rubisco also functions as an oxygenase during photorespiration, but its affinity for O_2 is 20–80 times lower than for CO_2. As with most enzymes, Rubisco has a preference for lighter stable isotopes and CO_2 fixation results in the depletion of ^{13}C ($\delta^{13}C$) ranging from $-10‰$ to $-20‰$.

The acetyl-CoA or Ljungdahl–Wood pathway is found in anaerobes, including methanogenic bacteria, acetogenic bacteria, and autotrophic sulfate reducers. Two parallel pathways fix CO_2, one of which results in an enzyme-bound carbonyl group, the other one in an enzyme-bound methyl group. Combining the two ultimately yields the acetyl-CoA pathway. The key enzyme in this pathway is carbon monoxide dehydrogenase. In methanogens, the biosynthesis proceeds via the acetyl-CoA pathway as well. Isotope fractionation in the acetyl-CoA pathway yields $\delta^{13}C$ values ranging from -20 to $-40‰$.

The reductive tricarboxylic acid cycle is basically the reverse of the oxidative tricarboxylic acid cycle that heterotrophs use to generate reducing equivalents (NADH, $FADH_2$) that function as electron donors for energy generation. The 3-hydroxyproprionate cycle was found in the green nonsulfur bacterium *Chloroflexus* (Strauss and Fuchs, 1993) and more recently also in some autotrophic Archea (Mendez *et al.*, 1999). In this pathway, 3-hydroxyproprionate is a key intermediate.

Many other carboxylation reactions exist (Barton *et al.*, 1991). For example, in methylotrophic bacteria, formaldehyde and CO_2 are combined to produce acetyl-CoA in the serine or hydroxypyruvate pathway. In contrast, the ribulose monophosphate cycle, which is another methylotrophic pathway of formaldehyde fixation, does not involve carboxylation steps. In addition to those described above, commonly found carboxylation reactions include those of pyruvate or phosphoenol pyruvate. In view of several relatively recent discoveries of novel CO_2 assimilation pathways (e.g., the hydoxypropionate cycle and anaerobic ammonium oxidation) and growing interest in deep-subsurface microbiology, novel pathways of CO_2 incorporation may be discovered in the near future.

8.08.3 DECOMPOSITION AND FERMENTATION

Most knowledge regarding the anaerobic decomposition of organic materials in depositional environments has been derived from studies of terminal processes in the microbial food web, particularly NO_3^-, Fe(III), Mn(IV), and SO_4^{2-} reduction, and methanogenesis. These processes are far easier to study than those responsible for polymer degradation. Because the degradation products of labile compounds eventually pass through a terminal step, it is generally considered that terminal decomposition exerts a major control on the decomposition pathway by regulating intermediary metabolism and the rate of degradation (Conrad, 1999). However, it is clear that the overall rate of organic matter decomposition is limited by the rate at which polymeric materials are depolymerized. Much information about decomposition can be provided by studies limited solely to intermediary or terminal metabolic events.

Organic matter deposited in sedimentary or wetland habitats is composed of a complex mixture of biopolymers. Some of these compounds, such as proteins, carbohydrates, and lipids are easily degraded by microorganisms (i.e., labile), while other compounds, such as lignin and hemicellulose, are resistant to decomposition (i.e., recalcitrant). Biopolymers are degraded in a multistep process. First, microorganisms simplify polymers to monomers such as amino acids, fatty acids, and monosaccharides (Figure 4). The monomers are further mineralized to CO_2, or to a combination of CO_2 and CH_4.

Under aerobic conditions, the conversion of monomers to fully mineralized products is rather simple because O_2-respiring bacteria can degrade monomers completely to CO_2. However, under anaerobic conditions this process requires

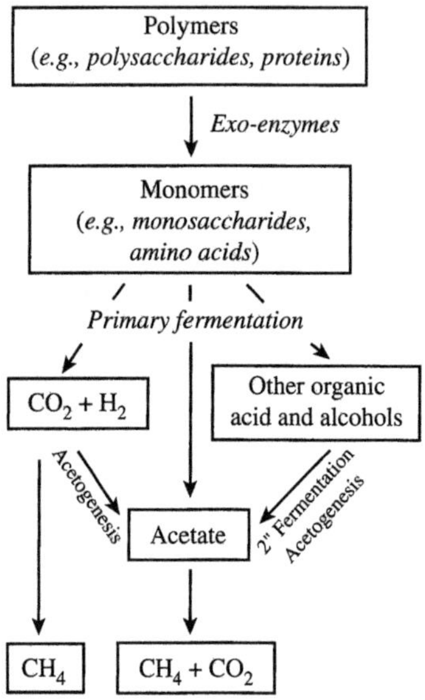

Figure 4 Metabolic scheme for the degradation of complex organic matter, culminating in methanogenesis. Polymers are cleaved via extracellular or cell-surface associated enzymes to monomers that are fermented to organic products, H_2 and CO_2. Methane is formed primarily from the oxidation of H_2 coupled to CO_2 reduction or by the fermentation of acetate. Acetate is formed by primary fermentation, acetogenesis from H_2/CO_2, and from secondary fermentation of primary fermentation products.

a consortium of bacteria that degrade monomers in a series of steps. The first step is primary fermentation to low molecular weight products such as alcohols and volatile fatty acids. Next, primary fermentation products are either mineralized to CO_2 and CH_4, or they undergo secondary fermentation to smaller volatile fatty acids. Finally, the secondary fermentation products are mineralized by respiratory organisms using inorganic terminal electron acceptors, a process that yields CO_2, or CO_2 and CH_4. Secondary fermentation is prevalent under methanogenic conditions.

8.08.3.1 Polymer Degradation

The quality of organic matter degraded in anaerobic environments varies depending on its origin with a gradient of structural complexity occurring from phytoplankton (labile) to vascular plants (recalcitrant) (Wetzel, 1992). Vascular plant detritus is resistant to decomposition because of an abundance of high-molecular-weight structural compounds such as lignocelluloses and complex polysaccharides (Benner *et al.*, 1991), while phytoplankton cells are composed largely of carbohydrates (Benner *et al.*, 1992). Cowie and Hedges (1993) noted that the reactivity of organic matter in sediments was generally amino acids > neutral sugars > total organic carbon > lignin. This pattern was independent of the O_2 content and bioturbation of sediments.

Seasonal variations in decomposition are affected strongly by temperature, but temporal changes in organic matter deposition also affect the degradation rates in sediments. Evidence of this is the observation that carbon mineralization varies strongly across seasons in sediments that are uniformly cold year-around because of seasonal variation in plankton production (Schulz and Conrad, 1995). It is clear that polymer breakdown is often the rate limiting step in organic matter degradation (Glissmann and Conrad, 2002; Reineke, 2001; Wu *et al.*, 2001).

Mineralization of organic macromolecules is initiated by extracellular enzymes because bacteria are unable to hydrolyze substrates that are much larger than about 600 Da (Weiss *et al.*, 1991). Not all bacteria are capable of synthesizing these enzymes, as is often the case with those responsible for terminal decomposition and some intermediary metabolisms. As a result, these terminal organisms depend heavily on the activities of other bacteria for substrates. It is clear that polymer hydrolysis occurs since these compounds are required to support microbial activities in sediments, but some studies have failed to detect polymer hydrolysis potentials sufficient to support *in situ* rates of metabolism (Arnosti, 1998). Such studies underscore the difficulties of examining hydrolytic processes.

The approaches used in studies of polymer hydrolysis include additions of intact plant material to incubating sediments (Battersby and Brown, 1982; Wainwright, 1981), inhibition of carbohydrate consumption with toluene and measurement of carbohydrate increases (Boschker *et al.*, 1995), and the use of labeled materials. The latter approach usually involves estimating the hydrolytic potential of microbial communities by amending sediments with fluorescently labeled precursors that serve as proxy organic macromolecules. A particularly common fluorophore is methylumbelliferyl (MUF) attached to a monomer such as a monosaccharide or an amino acid (Boetius and Lochte, 1994; Boschker and Cappenberg, 1994; Hoppe, 1983; King, 1986; Meyer-Reil, 1986). When these labels are attached to a monomer, they may fail to discern the activities of true extracellular enzymes because they are small enough to enter the bacterial periplasmic space. Thus, labeled compounds may be poor proxies of large-molecular-weight materials (Arnosti, 1998; Martinez and Azam, 1993). In some cases, hydrolytic potential has been assayed using whole polymers that were fluorescently labeled (Arnosti, 1996) or specific polymers have been introduced into growing cultures of isolated bacteria (Reichardt, 1988).

8.08.3.1.1 Polysaccharides

Studies of polymer use have often focused on the hydrolysis of polysaccharides because carbohydrates make up a large portion of phytoplankton biomass (Parsons *et al.*, 1961) and cellulose is the main polysaccharide in terrestrial ecosystems (Watanabe *et al.*, 1993; Glissman and Conrad, 2002). In general, hydrolytic activity in sediments decreases rapidly with depth in parallel to declining organic matter reactivity and general bacterial metabolic activity (King, 1986). However, Arnosti (1998) found that the potential for hydrolysis of algal-derived polysaccharides, such as pullulan and laminarin, was uniformly rapid throughout the upper ~11 cm of Arctic sediments. Additions of metabolic inhibitors failed to completely arrest hydrolytic activity, indicating that a significant portion of hydrolytic enzymes in sediments are either free in pore waters or attached to particles and active in lieu of bacterial metabolism (Arnosti, 1998; Boschker *et al.*, 1995). Although hydrolytic rate optima have been shown to correlate with environmental factors such as pH or salinity (King, 1986), they often exhibit temperature optima that greatly exceed ambient temperatures (King, 1986; Mayer, 1989; Reichardt, 1988). For example, extracellular enzyme activity in sea-ice bacterial communities exhibited temperature optima near 15 °C and a psychrophilic isolate yielded a protease extract with an optimum activity at 20 °C (Huston *et al.*, 2000).

Relatively slow decomposition in anaerobic environments may partly be due to depressed extracellular enzyme activity caused by low pH, low O_2 concentrations, or other factors (Kang and Freeman, 1999). Freeman *et al.* (2001) determined that depressed extracellular enzyme activity in a peatland soil was due to the high content of phenolic compounds, and concluded that phenols were not degraded because of an O_2 limitation on phenol oxidase activity. This led them to speculate that the vast pool of organic carbon currently sequestered in peatland soils (one-third of all soil carbon) is under the control of a single enzymatic "latch," the activity of which could increase dramatically if drought or drainage were to increase O_2 availability. In other cases, extracellular hydrolase activities may not be affected by environmental factors such as O_2, sulfide, or iron chemistry (King, 1986), and do not necessarily follow substrate concentrations or microbial biomass on an annual basis (Mayer, 1989). Because microbial activity is consistently affected by such factors, these findings reflect the extracellular nature of the hydrolytic enzymes.

8.08.3.1.2 Lignin

Organic matter deposited in near-shore marine and most freshwater habitats is composed primarily (~75% by weight) of lignocellulose, a complex mixture of lignin and the polysaccharides cellulose and hemicellulose (Benner *et al.*, 1985). Lignin is a unique phenolic polymer of nonrepeating units that makes up 25–30% of the biopolymers in vascular plants and is second only to cellulose as the most abundant organic carbon source in the biosphere. It is highly resistant to microbial degradation (Kawakami, 1989) and its association with cellulose and hemicellulose polysaccharides imparts degradation resistance to these polymers as well (Crawford, 1981). Hence, lignin is widely distributed in depositional environments such as soils and peats (Hedges and Oades, 1997; Miyajima *et al.*, 1997; Tsutsuki *et al.*, 1994; Yavitt *et al.*, 1997), riverine sediments (Hedges *et al.*, 1986, 2000; Meyers *et al.*, 1995), and coastal marine sediments (Dittmar and Lara, 2001; Gough *et al.*, 1993; Hedges *et al.*, 1997; Miltner and Emeis, 2001). Lignin's aromatic character makes it the major source of naturally occurring aromatic compounds.

The degradation of lignocelluloses in detritus-based ecosystems like wetlands is crucial to maintain carbon balance since macrophytes are composed of 50–80% lignocellulose (Maccubbin and Hodson, 1980). Early studies suggested that the lignin polymer was essentially inert in the absence of oxygen (Hackett *et al.*, 1977). However, it has since been shown that the complex lignin polymer can undergo anaerobic degradation (Benner *et al.*, 1985, 1986), but the process can be rather slow and tends to yield unmetabolized products (Colberg and Young, 1982; Young and Frazer, 1987). Anaerobic loss of lignin polymers is ~3–30% as complete as aerobic degradation of the same polymers (Benner *et al.*, 1984); the cellulose component is more labile than the lignin component.

Both fungi and bacteria degrade lignocelluloses. Fungi tend to dominate decomposition in upland soils (Orth *et al.*, 1993; Witkamp and Ausmus, 1976), whereas bacteria dominate in most aquatic environments (Benner *et al.*, 1986). The latter authors noted that bacteria were responsible for the degradation of lignin and associated polysaccharides in marine wetlands. Eukaryotes also contributed significantly to lignin degradation in an acidic freshwater wetland, and eukaryotes degraded both lignin and polysaccharides in a slightly alkaline freshwater wetland. The polysaccharide component is usually decomposed several fold more rapidly than lignin, and the lignocellulose of herbaceous plants decomposes more readily than that of woody plants (Benner *et al.*, 1985). The aerobic catabolism of lignin by fungi and filamentous bacteria utilizes ligninolytic peroxidases (Black and Reddy, 1991; Eriksson *et al.*, 1990), that often require manganese to be active (Brown *et al.*, 1990; Gettemy *et al.*, 1998). Less is known of the enzymes involved in the anaerobic decomposition process.

Under anaerobic conditions, lignin oligomers can be depolymerized to monomers, and lignin monomers can be mineralized to CO_2 (Colberg, 1988; Colberg and Young, 1982; Young and Frazer, 1987). The degradation of the lignin polymers releases aromatic subunits, and many studies have examined the anaerobic pathways by which these monomers are used. In fact, lignin monomers have been used often as lignin model compounds (Pareek *et al.*, 2001; Phelps and Young, 1997). Although it is clear that aromatic rings are readily cleaved aerobically by dioxygenase and peroxidase enzymes, these reactions do not occur anaerobically; aromatic rings are cleaved in the absence of O_2 via a reductive mechanism whereby hydrogenation of the aromatic ring nucleus results in a cyclohexane derivative, the ring of which is then opened (Reineke, 2001; Young and Frazer, 1987). Lignin monomers generally contain hydroxyl, methoxyl, and/or carboxyl groups that are removed from aromatic rings prior to reduction and cleavage of the ring. The O-demethylation (demethyoxylation) of phenylmethylethers has received considerable attention since the methyl product can serve as a C_1 substrate (i.e., not C–C bonds) for bacterial growth (Evans and Fuchs, 1988; Young and Frazer, 1987). Many of the O-demethylating strains of bacteria are acetogenic (Section 8.08.3.2.1) (Frazer, 1995; Kreft and Schink, 1993; Kreft and Schink, 1997; Küsel *et al.*, 2000; Wu *et al.*, 1988), but several other types of anaerobic and facultatively anaerobic bacteria have the ability to demethoxylate aromatic rings, including sulfate reducers, nitrate reducers, fermentative bacteria, and other strains involved in anaerobic syntrophy (Section 8.08.3.2.2) (Cocaign *et al.*, 1991; Krumholz and Bryant, 1985; Liu and Suflita, 1993; Mountfort *et al.*, 1988; Phelps and Young, 1997; Young and Frazer, 1987). Some pure cultures are capable of removing a variety of functional groups from aromatic rings (Küsel *et al.*, 2000). In many instances, the aromatic ring is not cleaved after these groups are removed (Young and Frazer, 1987). During O-demethyoxylation, the released methyl group can be used to methylate sulfide, which leads to the production of the gases methane thiol and dimethylsulfide; the remaining aromatic compound is not further degraded (Bak *et al.*, 1992; Finster *et al.*, 1990, 1992). Bacteria capable of degrading plant-derived aromatic compounds such as ferulic and syringic acids are often capable of degrading xenobiotic aromatics such as chlorinated pollutants.

8.08.3.2 Fermentation

Fermentation is a metabolic process in which organic compounds serve as both electron donors and acceptors. Primary fermentation is the exergonic breakdown of glucose and other monomers to products such as alcohols, fatty acids, H_2, and CO_2. Secondary fermentation further converts these primary products to acetate and other low-molecular weight organic acids (Figure 4). Fermentation differs dramatically from anaerobic respiration because it occurs inside the cell, and energy is generated by organic matter dismutation and substrate-level phosphorylation. By comparison, aerobic and anaerobic respiration requires an external electron acceptor and they proceed by oxidative phosphorylation via electron transport. Fermentation generates very little energy relative to aerobic or anaerobic respiration. However, it is a key component of the anaerobic mineralization process because most nonfermentative organisms cannot use the typical monomers released during polymer hydrolysis. As a result, fermenters can greatly outnumber bacteria dependent on terminal respiration processes in some environments. For example, in SO_4^{2-}-reducing sediments, the SO_4^{2-} reducers typically account for ~5% of the all bacteria present, while most bacteria are involved in polymer hydrolysis and fermentation (Devereux *et al.*, 1996). The situation is different in anaerobic wetland soils, where plant roots release acids and alcohols that can be used directly by SO_4^{2-} reducers. In this case, the SO_4^{2-} reducer populations on roots can account for over 30% of the total microbial community because fermentation is not required (Hines *et al.*, 1999; Rooney-Varga *et al.*, 1997).

In pure culture, fermenting bacteria consume a large variety of organic compounds. The list includes sugars, amino acids, purines, pyrimidines, some aromatics, acetylene, and a broad range of organic acids (C_1–C_{18}). They produce a wide spectrum of fermentation products, but C_1–C_{18} acids and alcohols, H_2, and CO_2 are most prevalent (Schink and Stams, 2002). In methanogenic habitats, all fatty acids longer than two carbons, branched-chain and aromatic fatty acids, and all alcohols longer than one carbon require secondary fermentation prior to use by methanogenic bacteria (Schmitz *et al.*, 2001). Hence, secondary fermentation is essential for complete mineralization to occur during methanogenesis. However, SO_4^{2-}- and metal-reducing bacteria are more metabolically versatile than methanogens and are capable of mineralizing most primary fermentation products directly. Organic mineralization can proceed via a two-step process in metal oxide and SO_4^{2-}-rich environments like marine sediments where primary fermenters hydrolyze polymers and ferment monomers while metal and SO_4^{2-} reducers utilize the fermentation products (Widdel, 1988). However, if fatty acid oxidizing SO_4^{2-} reducers are absent, then mineralization of fatty acids is possible via secondary fermentation and the use of H_2 by SO_4^{2-}-reducing bacteria (Monetti and Scranton, 1992). The SO_4^{2-}-reducing community does exhibit some other cooperative activities (Section 8.08.7.2.2).

8.08.3.2.1 Acetogenesis

Besides the "classical" fermentation described above, organic monomers can also be degraded to acetate via acetogenesis. This term is somewhat nebulous and has been used to describe any reaction leading to acetate, including fermentation. However, a widely accepted definition holds that acetogens are obligately anaerobic bacteria that use the acetyl-CoA pathway both for the reductive synthesis of acetyl-CoA from CO_2, and as a terminal electron-accepting and energy conserving process (Drake, 1994). If acetate is usually the sole reduced end product, the acetogen is considered to be a homoacetogen. The acetyl-CoA pathway is responsible for the anaerobic dark fixation of CO_2 into organic compounds, and most acetogens are able to use this pathway to reduce CO_2 during the oxidation of H_2 to form acetate autotrophically. In a typical acetate fermentation reaction, glucose is converted to two moles of acetate and two moles of CO_2. However, acetogens can reduce the two CO_2 molecules to acetate yielding a total of three moles of acetate. It is theoretically possible for acetogenic catabolism of glucose to acetate, followed by its degradation to CH_4 and CO_2, to be a major degradative pathway in anaerobic environments because it is highly favorable thermodynamically. However, this joint pathway does not seem to be important *in situ*, and the role of acetogenesis in methanogenic habitats remains unclear (Conrad, 1999).

Acetogenic bacteria are a very diverse group and display a wide range of catabolic capabilities including the utilization of sugars, acids, alcohols, lignin-derived aromatic methoxyl groups, C_1 and C_2 compounds, CO, and methyl halides (Drake, 1994). In addition, acetogens have been isolated that are capable of reverse acetogenesis where acetate is converted to CO_2 and H_2 (Zinder and Koch, 1984). The acetyl-CoA pathway is also present in some SO_4^{2-}-reducing and methanogenic bacteria and is involved in acetate oxidation, disproportionation (i.e., CH_4 production from acetate), autotrophic fixation of CO_2, CO oxidation, and the assimilation of C_1 compounds (Fuchs, 1994).

The autotrophic formation of acetate from H_2/CO_2 can directly compete with autotrophic methanogenesis. Methane formation from H_2/CO_2 is thermodynamically more favorable than acetogenesis, but the latter pathway is dominant in some habitats. Küsel and Drake (1994) noted that flooded soils accumulated acetate for several months without becoming methanogenic. Acetate also accumulates transiently in sediments and wetland peat soils prior to its use by methanogens, suggesting that acetoclastic methanogens grow more slowly than H_2/CO_2 utilizers or are more sensitive to O_2 (Avery *et al.*, 1999; Crill and Martens, 1986; Sansone and Martens, 1982; Shannon and White, 1996). In addition, it appears that acetate accumulates all season in colder high latitude wetlands (Duddleston *et al.*, 2002; Hines *et al.*, 2001), but it is unknown what portion of this acetate is due to fermentation or acetogenesis. In rice paddy soils, up to 40% of the fatty acids can be derived from CO_2 reduction, including virtually all of the acetate formed (Conrad and Klose, 2000). The bacteria physically associated with rice roots exhibited diverse anaerobic metabolisms, including the coexistence of acetogenesis and methanogenesis from CO_2 reduction (Conrad and Klose, 1999; Liesack *et al.*, 2000). This apparent relief of competition may have been due to the abundance of substrates released by roots (Conrad and Klose, 1999; Liesack *et al.*, 2000).

8.08.3.2.2 Syntrophy and interspecies hydrogen transfer

Many anaerobic microorganisms require secondary fermentation products such as acetate for electron donors. However, the ability of bacteria to perform secondary fermentation can be limited if H_2 (a metabolic waste product of the process) accumulates, and causes the reaction to become endergonic. For example, fermentation of organic acids and alcohols such as butyrate, propionate and ethanol does not yield sufficient energy to support growth if the H_2 concentration rises above 10^{-4} atm. Yet, these fermentation reactions are generally exergonic *in situ*, because H_2 is continually removed by H_2-consuming bacteria (Table 4). As a result, H_2 has a short turnover time and usually occurs at very low concentrations (Conrad *et al.*, 1986,1989) even though it is an common intermediate in metabolism (Conrad, 1999). The exchange of H_2 makes these organisms metabolic partners, and allows primary fermentation products to be completely mineralized in anaerobic habitats. Such *syntrophic* relationships (Biebl and Pfennig, 1978) are a symbiotic cooperation between two metabolically different bacteria that depend on each other for the degradation of a substrate, typically for energetic

Table 4 Examples of reactions occurring in methanogenic environments illustrating the effect on energy yield of the consumption of fermentation products. Maintenance of low reactant concentrations allows secondary fermentation reactions that are endergonic under standard conditions to be exergonic (negative ΔG).

Reaction	*Free-energy change* (kJ)	
	$\Delta G^{0,a}$	ΔG^{b}
Glucose + $4H_2O \rightarrow 2\,acetate^- + 2HCO_3^- + 4H^+ + 4H_2$	−207	−319
Glucose + $2H_2O \rightarrow butyrate^- + 2HCO_3^- + 3H^+ + 2H_2$	−135	−284
$Butyrate^- + 2H_2O \rightarrow 2\,acetate^- + H^+ + 2H_2$	+48.2	−17.6
$Propionate^- + 3H_2O \rightarrow acetate^- + HCO_3^- + H^+ + H_2$	+76.2	−5.5
2 Ethanol + $2H_2O \rightarrow 2\,acetate^- + 2H^+ + 4H_2$	+19.4	−37
Benzoate + $6H_2O \rightarrow 3\,acetate^- + 2H^+ + CO_2 + 3H_2$	+47	−18

Source: Zinder (1984).
[a] Standard conditions: solutes, 1 M; gases, 1 atm. [b] Concentrations of reactants typical of anaerobic habitats: fatty acids, 1 mM; glucose, 10 μM; CH_4, 0.6 atm; H_2, 10^{-4} atm; HCO_3^-; 20 mM.

reasons. In many cases, H_2 is the compound transferred to the terminal organism and this process is known as *interspecies* H_2 *transfer.* Although syntrophy and interspecies H_2 transfer are synonymous in many instances, it should be noted that H_2 transfer is a mechanism of electron transfer and other electron carriers can be involved. Because formate is transferred between organisms similarly to H_2, the term *interspecies* H_2 *transfer* often refers to both formate and H_2 transfer (Boone *et al.*, 1989; Thiele and Zeikus, 1988). However, in some instances formate transfer is insignificant compared to H_2 transfer and vice versa (Schmidt and Ahring, 1995). In addition, electron transfer can occur via other electron carriers such as acetate (Dong *et al.*, 1994), or amino acids such as cysteine (Cord-Ruwisch *et al.*, 1998).

The classical example of interspecies H_2 transfer was recognized when a presumably pure ethanol-consuming methanogenic culture, *Methanobacillus omelianskii* (Barker, 1940), was found to actually be a co-culture of an ethanol utilizing syntroph (S Strain) and an H_2-consuming methanogen (Strain M.o.H.) (Bryant *et al.*, 1967):

Strain S :

$$2CH_3CH_2OH + 2H_2O \rightarrow 2CH_3COO^- + 2H^+ + 4H_2 \qquad \Delta G_0' = +19\ \text{kJ mol}^{-1}$$

Strain M.o.H. :

$$4H_2 + CO_2 \rightarrow CH_4 + 2H_2O \qquad \Delta G_0' = -131\ \text{kJ mol}^{-1}$$

Co-culture :

$$CH_3CH_2OH + CO_2 \rightarrow 2CH_3COO^- + 2H^+ + CH_4 \qquad \Delta G_0' = -112\ \text{kJ mol}^{-1}$$

The bacterium responsible for ethanol fermentation will not grow using ethanol without the H_2-scavenging methanogen because the reaction is endergonic at the H_2 concentrations encountered. However, the reaction is exergonic if the H_2 partial pressure is maintained at $<10^{-3}$ atm by the methanogen with a net co-culture.

Primary fermenting bacteria can also profit from the activities of H_2-consuming partners at the end of the degradation chain. The maintenance of low H_2 partial pressures allows primary fermentation to favor the production of H_2 and more oxidized end products like acetate and CO_2. This permits fermentation to be more efficient in general, and leads to additional ATP synthesis (Thauer *et al.*, 1977; Schink and Stams, 2002; Schmitz *et al.*, 2001). The H_2-utilizing population is the primary regulator in the degradation process because its presence dictates which pathways are energetically capable of proceeding (Bryant, 1979; Zeikus, 1977). In an active anaerobic community, the presence of a microbial H_2 sink allows electron and carbon flow to bypass much of the secondary fermentation process, except for the use of long-chained or branched intermediates produced during lipid and amino acid fermentations (Schink, 1997). Therefore, it is conceivable that secondary fermentation could be a rather small portion of anaerobic metabolism if primary fermentation yielded only acetate as the organic product. However, many natural habitats are not as microbially active as those encountered in culture, and secondary fermentation tends to be an important, but rather poorly understood, component of natural anaerobic environments (Conrad, 1999).

Since its discovery around the mid-1960s, a variety of interspecies H_2 transfer reactions have been elucidated. Early studies focused primarily on processes involving methanogenic bacteria because this association was the first described. The small amount of energy available in methanogenic metabolism effectively forces the bacteria into symbiotic relationships that neither partner can function without (Schink, 1997). Although the original ethanol-utilizing partner (S Strain) of the co-culture comprising "*M. omelianskii*" was lost, other syntrophic ethanol-oxidizing bacteria have been isolated (Ben-Bassat *et al.*, 1981; Schink, 1984). The suite of H_2-releasing compounds used by syntrophs includes alcohols (Bryant *et al.*, 1967; Eichler and Schink, 1986; Schink *et al.*, 1985), fatty acids (McInerney *et al.*, 1981, 1979; Schink, 1985b; Schink and Friedrich, 1994; Zinder and Koch, 1984), aromatic compounds (Dolfing and Tiedje, 1991; Elshahed *et al.*, 2001; Knoll and Winter, 1989; Mountfort *et al.*, 1984), glycolic acid (Friedrich *et al.*, 1991), and amino acids (Nagase and Matsuo, 1982; Nanninga and Gottschal, 1985; Stams and Hansen, 1984; Winter *et al.*, 1987).

Many syntrophic microorganisms can grow in the absence of a H_2-utilizing partner by using slightly more oxidized substrates in a dismutation fermentation reaction (Eichler and Schink, 1986; Elshahed and McInerney, 2001a; Schink, 1997). For example, a bacterium that can only degrade propionate when in co-culture with an H_2-consuming partner can be grown in pure culture on pyruvate (Wallrabenstein *et al.*, 1994), and an ethanol-utilizing syntroph can be grown alone on acetaldehyde analogs like acetylene (Schink, 1985a). The need for physical or chemical removal of H_2 allows for only limited growth of pure culture syntrophs (Mountfort and Kaspar, 1986).

Because the degradation of fatty acids to acetate and H_2 is more endergonic than ethanol oxidation, the H_2 partial pressure must be ~1 atm lower for fatty acids to be degraded under methanogenic conditions than for ethanol degradation (Schink, 1997; Wallrabenstein *et al.*, 1994). The energy yield of fatty acid oxidation is improved when both acetate and H_2 are being removed, thereby "pulling" the reactions to completion. This has been demonstrated in tricultures that degraded either butyrate (Ahring and Westermann, 1988) or propionate (Dong *et al.*, 1994). Fatty acid degradation by a fermenter was enhanced when one methanogenic strain consumed acetate while the other consumed H_2. Thus, the combination of low acetate and low H_2 concentrations improved degradation efficiency via syntrophy. It should be mentioned that acetate can also be consumed *indirectly* by methanogens. This occurs when acetate is first converted to H_2 and CO_2, both of which are then consumed by a hydrogenotrophic methanogen (Zinder and Koch, 1984).

All known propionate oxidizers are also capable of reducing SO_4^{2-}, and the biochemical components may include part of the SO_4^{2-}-reducing apparatus (Schink, 1997). Syntrophic propionate use occurs when SO_4^{2-} is limiting and involves the transfer of H_2 to a methanogenic bacterium. Microscopic studies have shown that syntrophic associations between fatty acid-oxidizers and methanogens can produce structured or layered microcolonies composed of propionate utilizers and H_2-consuming methanogens, and these structures can be surrounded by acetate-utilizing methanogens and other bacteria (Harmsen *et al.*, 1996b). Finding an organized juxtaposition of bacteria is not surprising considering the physically close association required for the transfer of products between syntrophic partners. Further work has shown that acetate and H_2-consuming methanogens tend to segregate into micro-clusters as well (Gonzalez- Gil *et al.*, 2001; Rocheleau *et al.*, 1999).

Anaerobic bacteria can ferment a wide range of amino acids and many do so using interspecies H_2 transfer in association with H_2 utilizers. The classic mode of amino acid fermentation is the Strickland reaction in which a pair of amino acids is consumed, one acting as an oxidant and the other as a reductant. For example, a single bacterium can oxidize alanine to acetate and NH_4^+ while reducing glycine to acetate and NH_4^+:

$$\begin{aligned}&CH_3CH(NH_3^+)COO^-(\text{alanine})\\&\quad + 2CH_2(NH_3^+)COO^-(\text{glycine})\\&\quad\quad \rightarrow 3NH_4^+ + 3CH_3COO^- + CO_2\end{aligned}$$

However, this reaction can be uncoupled in which one bacterium oxidizes alanine to acetate, CO_2, NH_4^+, and H_2, and a second bacterium uses the H_2 to reduce glycine to acetate and NH_4^+:

$$\begin{aligned}&CH_3CH(NH_3^+)COO^- + 2H_2O \rightarrow CH_3COO^-\\&\quad + NH_4^+ + CO_2 + 2H_2\\&\qquad\qquad \Delta G_0{}' = +2.7\ \text{kJ mol}^{-1}\end{aligned}$$

$$\begin{aligned}&CH_2(NH_3^+)COO^- + H_2 \rightarrow CH_3COO^- + NH_4^+\\&\qquad\qquad \Delta G_0{}' = -78\ \text{kJ mol}^{-1}\end{aligned}$$

Adding these equations, we obtain

$$\begin{aligned}&CH_3CH(NH_3^+)COO^- + 2CH_2(NH_3^+)COO^-\\&\quad + 2H_2O \rightarrow 3CH_3COO^- + 3NH_4^+ + CO_2\\&\qquad\qquad \Delta G_0{}' = -153\ \text{kJ mol}^{-1}\end{aligned}$$

A single bacterium has been isolated that can conduct all three reactions (Zindel *et al.*, 1988). However, H_2 and acetate generated from alanine oxidation can be consumed by other partners such as methanogens (Nagase and Matsuo, 1982), SO_4^{2-} reducers (Nanninga and Gottschal, 1985), or acetogens (Zindel *et al.*, 1988). The Strickland reaction is energetically favorable over the syntrophic fermentation of amino acids (Schink, 1997) and the former may dominate in habitats rich in amino acids. However, syntrophic pathways may be important where amino acid concentrations are low and methanogenesis is active. A wide variety of bacteria are capable of fermenting various amino acids syntrophically using methanogens as H_2 scavengers with acetate, propionate, and butyrate as common end products (Baena *et al.*, 2000, 1999; Meijer *et al.*, 1999; Stams and Hansen, 1984; Wildenauer and Winter, 1986; Winter *et al.*, 1987). The ratio of organic products formed is a function of the H_2 partial pressure, which underscores the importance of the H_2 scavenger in controlling electron and carbon flow (Stams and Hansen, 1984).

The role of the H_2 scavenger in syntrophic relationships can be accomplished by a variety of H_2-utilizing anaerobes. Methanogens have received the most attention due to the importance of syntrophy in methanogenic habitats since CH_4-producing bacteria are extremely limited in the scope of electron donors used. Anaerobic, SO_4^{2-}-dependent CH_4 oxidation is the most recent example of microbial syntrophy between a methanogen-like organism and a SO_4^{2-} reducer (Section 8.08.4.5). However, this role can be replaced by several other physiologic groups including SO_4^{2-}-, S^0-, metal-, nitrate-, glycine-, and fumarate-reducing, and acetogenic bacteria (Schink, 1997). The ability of an H_2-scavenging reaction to compete for transferred electrons depends on the redox potential of the terminal electron acceptor with nitrate and fumarate $> SO_4^{2-} > CO_2/CH_4 > CO_2$/acetate (Cord-Ruwisch *et al.*, 1988).

Respiring bacteria resort to syntrophy to decompose low molecular weight compounds when exogenous oxidant levels are low. The most widely studied are SO_4^{2-} reducers that utilize fermentation processes coupled to H_2 transfer to methanogens when SO_4^{2-} is depleted (Bryant *et al.*, 1977; Cord-Ruwisch *et al.*, 1986; Harmsen *et al.*, 1996a; van Kuijk and Stams, 1995; Wallrabenstein *et al.*, 1995). Interestingly, SO_4^{2-}-reducing bacteria can fill an opposite role in which H_2 produced during the degradation of methanol and acetate by a methanogenic bacterium is consumed by a SO_4^{2-}-reducing bacterium during SO_4^{2-} reduction (Phelps *et al.*, 1985). This SO_4^{2-}-dependent interspecies H_2 transfer results in a more complete oxidation of organic carbon substrates yielding more CO_2 and less CH_4 than the methanogen would produce alone (Achtnich *et al.*, 1995b). Hence, although SO_4^{2-} reducers and CH_4 producers often compete for substrates, this competition is circumvented in the absence of SO_4^{2-} when fermenting SO_4^{2-} reducers rely on methanogens as H_2 sinks, or when methanogens use SO_4^{2-} reducers as H_2 sinks. Studies using sediment slurries revealed a dual role for SO_4^{2-} reducers in which they act as direct consumers of fatty acids as well as affecting fatty acid degradation by removing H_2 during syntrophy with other fatty acid fermenting species (Monetti and Scranton, 1992). Metal-reducing bacteria such as *Geobacter* sp. are also capable of degrading fatty acids like acetate by interspecies H_2 transfer to nitrate reducing bacteria (Cord-Ruwisch *et al.*, 1998). It appears that the transfer of electrons to the nitrate-reducing partner by the *Geobacter* sp. occurs using cysteine as the electron-transferring agent (Kaden *et al.*, 2002).

Fermenting bacteria can also transfer electrons to oxidized humic acids, allowing for the formation of more oxidized fermentation products than those produced in the absence of humics (Benz *et al.*, 1998). It was shown previously that Fe(III)-respiring bacteria can transfer reducing equivalents from acetate to various humic acid preparations including the quinoid model compound 2,4-anthraquinone disulfonate (AQDS) (Lovley *et al.*, 1996b, 1998). Humic acids act catalytically as electron shuttles by transferring electrons to oxidized iron. It has also been shown that fermenting bacteria can transfer electrons to Fe(III) in this manner, but it appears that these bacteria do not conserve energy via electron transport like iron-respiring bacteria do (Benz *et al.*, 1998). However, the electron sink serves a similar function as a syntrophic partner. For example, propionate fermentation is endergonic in the absence of an H_2-consuming partner, but propionate can be fermented exergonically to acetate without a bacterial partner when electrons are transferred to humic acids (Benz *et al.*, 1998).

8.08.4 METHANE

Methanogenesis is the final step in the anaerobic degradation of organic carbon. The principal steps performed by methanogens are fermentation of acetate to CO_2 and CH_4, and oxidation of H_2 to H_2O. In both cases, the waste product, CH_4, still holds potential energy in the form of reducing equivalents that can support additional anaerobic metabolism. Thus, the true end point of anaerobic organic matter degradation in some ecosystems is the anaerobic oxidation of CH_4 to CO_2 (Section 8.08.4.5). In the presence of O_2, methane is a source of energy for CH_4-oxidizing bacteria (i.e., methanotrophs), some of which are symbionts of deep-sea mussels (Childress *et al.*, 1986; Fiala–Médioni *et al.*, 2002; Kochevar *et al.*, 1992).

The many environmental and economic issues that involve CH_4 have stimulated research in all aspects of methane cycling. Here, we consider the processes that influence CH_4 production and oxidation. The literature on CH_4 emissions is considered only briefly in this article (Section 8.08.4.7).

8.08.4.1 Methane in the Environment

Human activity has increased the atmospheric CH_4 concentration from ~0.7 $\mu L\,L^{-1}$ to 1.8 $\mu L\,L^{-1}$ (ppmv) since about the 1850s, primarily by stimulating methanogenesis in soils and sediments (Lelieveld *et al.*, 1998; see below). Methane concentrations are currently double the highest level recorded in a 420,000-year ice core (Petit *et al.*, 1999), and it accounts for 20% of human-induced radiative forcing (Prather *et al.*, 2001). Rising CH_4 concentrations during periods of increased solar insolation and interglacial warming have substantially amplified global warming in the past (Petit *et al.*, 1999), and there is concern that this will occur in the future. In the past, changes in the atmospheric CH_4 content were accompanied by changes in the area of northern and tropical wetlands (Blunier *et al.*, 1995; Chappellaz *et al.*, 1993). Because CH_4 hydrate deposits may have been a large and sudden source of atmospheric CH_4 that contributed to past climate change (Nisbet, 1992; Thorpe *et al.*, 1996), the current stability of these vast CH_4 reservoirs is a topic of considerable importance (Kvenvolden, 1999; Wood *et al.*, 2002).

About 70% of the current CH_4 sources are anthropogenic, with roughly equal contributions from fossil fuel-related industries, waste management systems, and enteric fermentation associated with raising livestock (Table 5). Of the natural sources, wetlands are 70% of the total and therefore a major research focus. About 60% of natural wetland sources and most rice paddies occur in the tropical latitudes. Another 35% are in northern latitudes, and 10% are in mid-latitudes. The combined contribution of natural and managed wetlands to global CH_4 emissions is ~32% at present, and perhaps 70% of all sources before the industrial revolution (Lelieveld *et al.*, 1998). Estuaries are <9% of ocean CH_4 sources (Middelburg *et al.*, 2002).

The dynamics of CH_4 in the atmosphere are quite different than CO_2. It is a reactive gas and participates in atmospheric chemical reactions that influence the concentrations of NO, NO_2, CO, and O_3 (Crutzen, 1995). On a molar basis, CH_4 is 3–22 times stronger as a greenhouse gas than CO_2, depending on the period of time over which its impact is considered (Rodhe, 1990; Whiting and Chanton, 2001). Methane concentrations are more responsive than CO_2 to changes in sources or sinks because of a far shorter atmospheric residence time for CH_4 (~10 yr versus >100 yr). These characteristics have inspired the recommendation that efforts to slow the pace of global warming should focus initially on abating CH_4 emissions (Hansen *et al.*, 2000; Fuglestvedt *et al.*, 2000).

Methane is an important compound economically. In the form of natural gas, it is a clean-burning energy source and the major carbon substrate for alkane formation in economic gas reservoirs (Sherwood Lollar *et al.*, 2002). Methane represents an unwelcome loss of metabolic energy from flatulent livestock (Johnson *et al.*, 1993). Although methanogens do not directly degrade organic pollutants, they participate in the process by consuming fermentation products, thereby maintaining an environment that is thermodynamically conducive to the anaerobic degradation of multi-carbon compounds (Lovley, 2000a; Weiner and Lovley, 1998; Section 8.08.3.2.2). It has been demonstrated that methanogens play this role in the degradation of alkanes, a group of particularly stable hydrocarbons that are abundant in fossil fuel deposits (Anderson and Lovley, 2000; Zengler *et al.*, 1999). Aerobic bacteria that depend entirely on CH_4 for their carbon and energy (i.e., methanotrophs) directly degrade organic pollutants such as trichloroethylene (TCE) (R. S. Hanson and T. E. Hanson, 1996).

Table 5 Estimated sources and sinks of methane in the atmosphere in units of 10^{12} g CH_4 yr^{-1}.

	Range	*Likely*
Sources		
Natural		
Wetlands		
Tropics	30–80	65
Northern latitude	20–60	40
Others	5–15	10
Termites	10–50	20
Ocean	5–50	10
Freshwater	1–25	5
Geological	5–15	10
Total		160
Anthropogenic		
Fossil fuel related		
Coal mines	15–45	30
Natural gas	25–50	40
Petroleum industry	5–30	15
Coal combustion	5–30	15
Waste management system		
Landfills	20–70	40
Animal waste	20–30	25
Domestic sewage treatment	15–80	25
Enteric fermentation	65–100	85
Biomass burning	20–80	40
Rice paddies	20–100	60
Total		375
Total Sources		535
Sinks		
Reaction with OH	330–560	445
Removal in stratosphere	25–55	40
Removal by soils	15–45	30
Total Sinks		515
Atmospheric increase	30–35	30

Source : Schlesinger (1997).

8.08.4.2 Methanogen Diversity and Metabolism

Methanogens are strict anaerobes that produce CH_4 as a waste product of energy metabolism. Several characteristics set the methanogens apart from most other microorganisms. They are the largest and most diverse group in the Archea, which is phylogenetically distinct from the other two domains of life, the Bacteria and Eukaryota. Many members of the Archea grow at extremes of temperature, pH, or salinity, although the distribution of methanogens is fairly cosmopolitan. Methanogens have a number of unique coenzymes (Wolfe, 1996) and differ from Bacteria in the construction of their cell walls, a characteristic that makes them insensitive to penicillin and other antibiotics. The methanogens and other Archea can be identified by an electron carrier, F_{420}, that autofluoresces at 420 nm under UV light (Edwards and McBride, 1975). Many methanogens lack cytochromes and other features of electron transport chains such as quinones (Fenchel and Finlay, 1995). It has been suggested that a quinone-like role in electron transfer may be filled by phenazine compounds in the *Methanosarcinales* (Abken *et al.*, 1998; Deppenmeier *et al.*, 1999). In a detailed taxonomic treatment, Boone *et al.* (1993) defined 26 genera and 74 species of methanogens.

Despite a great deal of taxonomic diversity, the methanogens use a limited variety of simple energy sources compared to the other major forms of anaerobic metabolism (Zinder, 1993). The compounds that support energy conservation for growth are H_2, acetate, formate, some alcohols, and methylated compounds. The most important of these are generally H_2 and acetate. About 73% of methanogenic species consume H_2 (Garcia *et al.*, 2000):

$$4H_2 + CO_2 \rightarrow CH_4 + 2H_2O \qquad (3)$$

Hydrogenotrophic methanogenesis (also known as *CO_2/H_2 reduction* or *H_2-dependent methanogenesis*) is a chemoautotrophic process in which H_2 is the source of both energy and electrons, and CO_2 is often both an electron sink and the source of cellular carbon. Some hydrogenotrophic methanogens require an additional organic carbon source for growth (Vogels *et al.*, 1988). CO_2/H_2 reduction requires a 4:1 molar ratio of H_2 to CO_2, yet H_2 is typically at nM concentrations while CO_2 is at mM concentrations in natural systems. Thus, substrate limitation of hydrogenotrophic methanogenesis must always be caused by a lack of the electron donor, H_2. Roughly 45% of all hydrogenotrophic methanogens can substitute formate for H_2 in reaction (3) (Garcia *et al.*, 2000; Thiele and Zeikus, 1988).

Whereas hydrogenotrophy is wide spread among the methanogens, acetotrophy (also known as *acetate fermentation* or *acetoclastic methanogenesis*) is restricted to just two genera, the *Methanosarcina* and *Methanosaeta* (formerly *Methanothrix*), which comprise ~10% of methanogenic species:

$$CH_3COOH \rightarrow CO_2 + CH_4 \qquad (4)$$

The *Methanosarcina* use a wide variety of substrates and have high potential growth rates, but their affinity for acetate is low (Jetten *et al.*, 1992). In contrast, the *Methanosaeta* specialize in using acetate and have a high affinity for the substrate, but their potential growth rate is low. Despite its limited taxonomic distribution, acetotrophy is the dominant methanogenic pathway in many ecosystems. Acetotrophic methanogenesis can be considered a special case of methylotrophy whereby a portion of the substrate molecule is oxidized to CO_2, while a methyl group on the same molecule is reduced to CH_4 (Fenchel and Finlay, 1995; Hornibrook *et al.*, 2000; Pine and Barker, 1956). In this case, the methyl group is the electron donor and the carboxyl group is the electron acceptor.

About 26% of methanogenic species can use methylated substrates other than acetate, such as methanol, methylated amines, and methylated sulfur compounds (Hippe *et al.*, 1979; Kiene, 1991b). Because the process does not require an external electron acceptor, methylotrophy is a type of fermentation. The contribution of these organisms to CH_4 production in natural ecosystems appears to be minor, but they may contribute substantially to the metabolism of methylated sulfur compounds. Scholten *et al.* (2003) provided thermodynamic constraints on the feasibility of such reactions.

Some methanogens are dependent on a single substrate, such as acetate or methanol, while others are able to grow on two or more alternative substrates. All known formate-oxidizing methanogens are also hydrogenotrophs (Garcia *et al.*, 2000), and at least one methanogen also grows by fermenting pyruvate (Bock and Schönheit, 1995). Members of the *Methanosaetaceae* use acetate but not H_2, whereas the *Methanosarcinaceae* can use both substrates. Both genera fall in the order *Methanosarcinales*.

Methanogens metabolize several substrates that do not support growth. *Methanosarcina barkeri* was able to lower the redox potential by dissimilatory reduction of Fe(III) until the redox potential reached 50 mV, at which point methanogenesis began (Fetzer and Conrad, 1993). In a survey of five methanogenic species drawn from a wide range of phylogenetic and physiological types, all the hydrogenotrophs could reduce Fe(III) (Bond and Lovley, 2002). The ability to grow on Fe(III) was not investigated. Carbon monoxide is converted to CH_4, but the physiological role this substrate plays is uncertain (Vogels *et al.*, 1988). Rich and King (1999) measured maximum potential uptake velocities of 1–2 nmol CO cm^{-3} sediment h^{-1} in anaerobic soils, but the response to amendments of SO_4^{2-}, Fe(III), and the methanogen inhibitor bromoethanesulfonic acid (BES), suggested that no more than 30% of the oxidation activity was due to methanogens. Methanogens degrade chlorinated pollutants and have been used for bioremediation (Fathepure and Boyd, 1988; Mikesell and Boyd, 1990).

Methanogens grow at temperatures ranging from 4 °C to 100 °C, salinities from freshwater to brine, and pH from 3 to 9. Most grow optimally at temperatures ≥30 °C, but thermophilic methanogens have temperature optima near 100 °C, and a few isolates are adapted to frigid conditions (Franzmann *et al.*, 1997). In a survey of 68 methanogenic species, most species grew best in a pH range from 6 to 8, and none could grow at pH <5.6 (Garcia *et al.*, 2000). Observations of CH_4 production in acidic environments suggests that there are uncultured methanogens that can grow at pH <5.6 (Walker, 1998).

8.08.4.3 Regulation of Methanogenesis

Methane production is regulated mainly by O_2 concentration, pH, temperature, salinity, organic substrates, and nutrient availability. The amount of CH_4 emitted from wetlands is further influenced by a large number of factors that are not addressed in this review, including plant physiology, community composition, and hydrology. In addition to their direct effects on methanogen physiology, these factors influence the process indirectly by regulating the flow of methanogenic substrates from fermenting and syntrophic microorganisms. Methanogenesis can be limited by any single link in the chain of reactions that begins with detrital inputs.

8.08.4.3.1 O_2 and oxidant inhibition

Methanogenesis is inhibited by O_2. This is evident from field studies that show no overlap in the depth distributions of O_2 penetration in soils or sediments and net CH_4 production. Inhibition by O_2 is one reason that water table depth in soils is often a strong predictor of CH_4 emissions (Kettunen *et al.*, 1999; Roulet and Moore, 1995; Sundh *et al.*, 1995), the other reason being its influence on CH_4 oxidation.

Methanogens are not necessarily as oxidant sensitive as it was once believed. Some methanogens are fairly tolerant of O_2 (Kiener and Leisinger, 1983) and have adaptations such as superoxide dismutase (Kirby *et al.*, 1981). Methane production by *Methanosarcina barkeri* starts when the redox potential drops below 50 mV (Fetzer and Conrad, 1993), and CH_4 production has been observed in rice paddy soils at redox potentials >200 mV (Peters and Conrad, 1996; Yao and Conrad, 1999). Despite the fact that they do not form spores or other resting stages, methanogens can survive for long periods of time in largely dry and oxic soils. Mayer and Conrad (1990) observed a rapid increase in CH_4 production within 25 d of flooding an upland agricultural soil and a forest soil. Viable methanogen populations survived in an oxic, air-dried paddy soil for at least 2 yr (Ueki *et al.*, 1997). von Fischer and Hedin (2002) recently developed a ^{13}C pool dilution method that gives a sensitive estimate of *gross* CH_4 production, and found that CH_4 is in fact produced in many dry, oxic soils (mean = 0.15 mg CH_4–C m^{-2} d^{-1}). However, they stopped short of attributing this activity to methanogens because of reports that CH_4 is a minor metabolic product in some eubacteria such as *Clostridia* (Rimbault *et al.*, 1988). Methanogens may be able to survive in small anaerobic microsites imbedded in dry soils, or they may be protected from O_2 by reactive soil minerals. For example, the presence of FeS_2 (pyrite) in paddy soils was shown to enhance the survival of methanogens exposed to O_2 (Fetzer *et al.*, 1993).

The lack of CH_4 production in the presence of O_2 *in situ* may be due to a combination of factors, of which O_2 toxicity is just one. For example, methanogens were more sensitive to desiccation than O_2 exposure in a paddy soil (Fetzer *et al.*, 1993). The oxidized products of denitrification, NO and N_2O, have a toxic effect on methanogens similar to that of O_2 (Klüber and Conrad, 1998). A negative correlation between redox potential and CH_4 production is often observed in the absence of O_2, but this probably reflects competition between methanogens and their competitors for reductants, not a physiological requirement for a certain redox potential.

8.08.4.3.2 Nutrients and pH

Laboratory incubations of wetland soils with nitrogen and phosphorus amendments have shown either no effect or an inhibitory effect on methanogenesis (Bodelier *et al.*, 2000a,b; Bridgham and Richardson, 1992; Wang and Lewis, 1992). A low rate of phosphate supply to rice roots stimulated CH_4 emission (Lu *et al.*, 1999), while phosphate concentrations ≥20 mM specifically inhibited acetotrophic methanogenesis (Conrad *et al.*, 2000).

Methanogens require a somewhat unique suite of micronutrients that include nickel, cobalt, iron, and sodium (Jarrell and Kalmokoff, 1988). Methanogenesis was stimulated by molybdenum, nickel, boron, iron, zinc, vanadium, and cobalt in a rice paddy soil (Banik *et al.*, 1996), and by a cocktail of nickel, cobalt, and iron in *Sphagnum*-derived peat (Basiliko and Yavitt, 2001). The micronutrient cocktail did not stimulate CO_2 production in the same soils, suggesting that methanogens, rather than fermenters, were directly limited by trace elements. Freshwater methanogens required at least 1 mM Na^+ to drive ATP formation by an Na^+/K^+ pump (Kaesler and Schönheit, 1989). Trace element availability could limit methanogenesis in peatlands that are isolated from groundwater inputs and sea salt deposition. The latter effect could be important in bogs in the interior of continents.

Most methanogenic communities seem to be dominated by neutrophilic species. Some acidic peats have responded to an increase in pH with higher CH_4 production (Dunfield *et al.*, 1993; Valentine *et al.*, 1994), while other peats have not (Bridgham and Richardson, 1992). A substantial portion of the acetate pool may not be available to methanogens at low pH because it is prevented from dissociating (Fukuzaki *et al.*, 1990).

8.08.4.3.3 Temperature

Methanogenesis is often more sensitive to temperature than other biological processes, which typically double in rate with a 10 °C increase in temperature (i.e., $Q_{10} = 2$). A compilation of temperature-response studies using wetlands soils reported an average Q_{10} of 4.1 for CH_4 production and 1.9 for aerobic CH_4 oxidation (Segers, 1998). van Hulzen *et al.* (1999) argued that such high Q_{10} values are often caused by the changing availability of alternative electron acceptors or methanogenic substrates over the course of an experiment. The time required for methanogen competitors (e.g., Fe(III) reducers) to deplete the pool of alternative electron acceptors decreases with increasing temperature. Thus, if CH_4 production is measured cumulatively over a fixed period of time, it can appear to increase at a $Q_{10} > 2$ even though the underlying processes increased at a far lower Q_{10}. This would explain an apparent correlation between Q_{10} and the ratio of CO_2 to CH_4 in anaerobic incubations (Figure 5). It is likely that temperature-sensitive steps during fermentation or acetogenesis contribute to the temperature sensitivity of CH_4 production. It is more difficult to explain high CH_4 Q_{10} values in some pure methanogen cultures (Segers, 1998). Methanogenesis can also be insensitive to temperature if fermentation is limited by other factors such as carbon quality (Yavitt *et al.*, 1988).

Because methanogenesis is severely suppressed by low temperature, CH_4 emissions are often

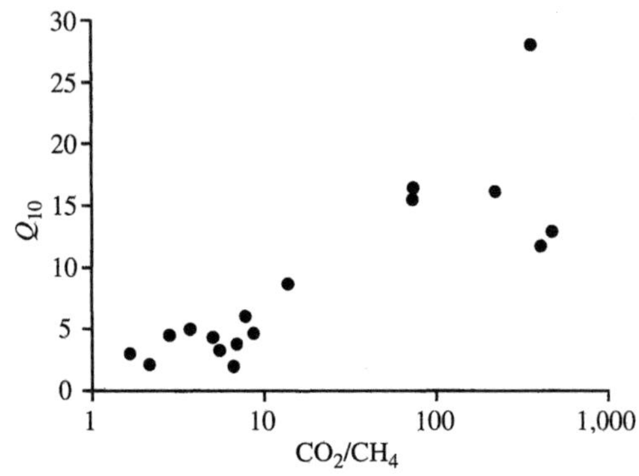

Figure 5 The relationship between the temperature sensitivity (i.e., Q_{10}) of CH_4 production and the ratio CO_2 to CH_4 produced in anaerobic incubations (van Hulzen *et al.*, 1999) (reproduced by permission of Elsevier from *Soil Biol. Biochem.* **1999**, *31*, 1919–1929).

assumed to be negligible during the winter. This is clearly not the case in peatlands where winter CH_4 emissions contribute 2–21% of annual fluxes (Alm *et al.*, 1999; Dise, 1992; Melloh and Crill, 1996). Zimov *et al.* (1997) estimated that 75% of CH_4 emissions from Siberian lakes is emitted in the winter.

8.08.4.3.4 Carbon quantity and quality

The quantity and quality (i.e., chemical composition) of organic carbon compounds is a master variable regulating methanogenesis. Organic carbon fermentation is the ultimate source of both the electron donors and the electron acceptors required by the two major methanogenic pathways (Section 8.08.4.2). There are inorganic sources of CO_2, but this substrate is abundant compared to H_2 and does not limit hydrogenotrophic methanogenesis. Organic carbon also exerts a strong influence on another key regulator of methanogenesis, namely the supply of inorganic compounds that are toxic to methanogens (O_2, NO_3^-) or more energetically favorable as electron acceptors for competing organisms (NO_3^-, Fe(III), humic acids, SO_4^{2-}). Methanogenesis is most vigorous when the consumption rate of alternative electron acceptors exceeds their supply rate. Because the organisms that consume these alternative substrates are heterotrophic and usually carbon limited, such conditions are most likely to develop in systems with a high supply of labile organic carbon compounds (Section 8.08.8).

A wide variety of evidence suggests that carbon availability limits methanogenesis *in situ*. The fact that methanogenesis is often inhibited by the presence of alternative electron acceptors such as Fe(III) and SO_4^{2-} is evidence of competition for fermentation products and thus widespread carbon limitation of the process (Section 8.08.8).

Methanogenesis is stimulated by plant detritus amendments in rice paddy soils (Dannenberg and Conrad, 1999; Inubushi *et al.*, 1997; Kludze and DeLaune, 1995; Tanji *et al.*, 2003) and peat soils (Valentine *et al.*, 1994), and by amendments of H_2 (Bridgham and Richardson, 1992; Yavitt *et al.*, 1987;Yavitt and Lang, 1990) and dissolved organic carbon (Lu *et al.*, 2000b). Methane *emissions* are stimulated by organic amendments (Watanabe *et al.*, 1995; Bronson *et al.*, 1997), but this effect may be due partly to lower CH_4 oxidation caused by more rapid heterotrophic O_2 consumption at the soil surface.

The chemical composition of organic compounds (i.e., carbon quality) influences methanogenesis by regulating the production of fermentation products (Section 8.08.3). This explains two common observations about changes in methanogenesis with increasing depth in freshwater wetland soil profiles: (i) rates of potential CH_4 production decline with depth below the aerobic zone (Clymo and Pearce, 1995; Kettunen *et al.*, 1999; Megonigal and Schlesinger, 2002; Moore and Dalva, 1997; Sundh *et al.*, 1994; Updergraff *et al.*, 1995; Valentine *et al.*, 1994; Yavitt *et al.*, 1988, 1990) and (ii) methanogenesis is increasingly dependent on hydrogenotrophy (Hornibrook *et al.*, 2000). These depth-dependent patterns reflect a combination of changes in the soil organic carbon quality, and increasing distance from labile carbon substrates deposited near the surface. The mineralization of fresh organic matter is thought to favor acetotrophic methanogenesis (Schoell, 1988; Sugimoto and Wada, 1993). In bogs, common measures of carbon quality, such as lignin content, lignin : N ratio and C : N ratio, show soil organic matter becoming progressively older, more decomposed and recalcitrant with depth (Yavitt and Lang, 1990; Updergraff *et al.*, 1995; Valentine *et al.*, 1994). In a fen, 90% of the CH_4 produced from soil organic matter mineralization came from particles with a diameter >2 mm, which represents recent plant detritus (van den Pol-van Dasselaar and Oenema, 1999). Because the >2 mm soil organic carbon fraction decreased rapidly with depth, 70% of the total CH_4 production occurred in the top 5 cm of the soil profile. In other cases, the depth-dependent decline in potential CH_4 production does not correlate to changes in the quality of the soil organic carbon pool. For example, potential CH_4 production, but not the quality of soil organic matter, decreased strongly with depth in two northern fens (Valentine *et al.*, 1994). Here, the methanogens were apparently using labile carbon compounds that were produced in the root zone and were less abundant with increasing distance from the soil surface. This is consistent with $^{14}CH_4$ measurements that showed microbial respiration in fens, but not in bogs, were based on recently fixed photosynthates that originated in the root zone, and were transported by groundwater through the soil profile (Chasar *et al.*, 2000a).

Soil organic matter quality reflects the chemical characteristics of the dominant plant species, and it is quite poor in peatlands dominated by *Sphagnum* moss. *Sphagnum* tissue has an exceedingly low nitrogen content (Aerts and de Caluwe, 1999; Hobbie, 1996; Johnson and Damman, 1991), which is one reason for lower CH_4 emissions from northern bogs than fens (Bridgham *et al.*, 1995; Dise *et al.*, 1993; Moore and Knowles, 1990). Although *Sphagnum* is abundant in both ecosystem types, fens also support a variety of vascular plant species with comparatively high tissue quality. The dominance of *Sphagnum* is itself a reflection of low nutrient availability, and is ultimately caused by the geomorphologic and hydrologic characteristics of the landscape.

Organic carbon indirectly influences methanogenesis by governing the rate that alternative electron acceptors are consumed. A combination of carbon content and alternative electron acceptor availability explained 90% of the variation in CH_4 production in 10 rice paddy soils (Gaunt *et al.*, 1997). The sediments at Cape Lookout Bight support exceptionally high rates of CH_4 production for a marine sediment because organic carbon inputs are rapid enough to deplete the porewater SO_4^{2-} pool before exhausting the labile organic carbon pool (Martens and Klump, 1984). High sedimentation rates also contribute to SO_4^{2-} depletion at Cape Lookout Bight by increasing the diffusion path length, thus slowing SO_4^{2-} resupply from the water column. Salt marshes tend to have higher CH_4 emissions than marine sediments per unit area (Bartlett *et al.*, 1987) because they receive considerable organic inputs from plants, thus relieving the substrate competition between methanogens and other microorganisms. Blair (1998) demonstrated the effectiveness of combining the labile carbon flux rate and oxidant availability (e.g., O_2 and SO_4^{2-}) into a ratio for modeling the proportion of organic carbon metabolized to CH_4 in marine sediments.

8.08.4.3.5 Plants as carbon sources

Plants and phytoplankton are the most abundant and labile organic carbon sources in ecosystems, supplying photosynthates either in the form of exudates or fresh detritus. Several isotope tracer studies have demonstrated a tight coupling between plant photosynthesis and methanogenesis in pot studies designed to represent rice paddies (Lu *et al.*, 2002; Dannenburg and Conrad, 1999; Minoda and Kimura, 1994; Minoda and Kimura, 1996), freshwater marshes (Megonigal *et al.*, 1999), and arctic tundra (King and Reeburgh, 2002; King *et al.*, 2002). A full cycle of CO_2

assimilation by plants, release into soils, and emission as CH_4 requires as little as 2 h, and up to 6% of the assimilated CO_2 is emitted as CH_4. Photosynthate-derived carbon was estimated to contribute 10–100% of the CH_4–C in rice paddies depending on the growth stage of the crop (Minoda and Kimura, 1994). Lu *et al.* (2000a,b) found that differences in the root exudation rates of three rice cultivars corresponded to variation in dissolved organic carbon in the rhizosphere and CH_4 emissions. It should be noted that isotopic labeling studies alone cannot determine whether there is an *energetic* link between the plants and microbes (Megonigal *et al.*, 1999).

In most instances, it is difficult to apply a radiocarbon label *in situ* because of environmental regulations (for an exception see Wieder and Yavitt, 1994). However, a ^{14}C label was applied worldwide during aboveground nuclear bomb testing, which peaked in 1965. Radiocarbon dates indicate that a significant amount of the pore-water CH_4 that currently exists in peatlands (Chanton *et al.*, 1995; Charman *et al.*, 1994; Martens *et al.*, 1992b), and an aquifer (Harvey *et al.*, 2002) was produced since 1965. In peatlands, pore-water CH_4–C was significantly younger than the bulk soil carbon (Aravena *et al.*, 1993), indicating that recently assimilated organic carbon compounds are carried downward into the soil profile by advection. An interesting exception to this pattern was reported in Siberian lakes where the radiocarbon age of CH_4 indicated that 68–100% of the CH_4–C had been assimilated during the Pleistocene (Zimov *et al.*, 1997). However, the contribution of this "old" carbon fell to 23–46% in the summer with inputs of more recent carbon sources.

There are several additional lines of evidence for a link between methanogenesis and photosynthesis. One is the observation that CH_4 emissions are often strongly related to primary production. Such a relationship was first suggested by Whiting and Chanton (1993), who reported a positive correlation between CH_4 emissions and net ecosystem exchange of CO_2 across North American wetlands distributed from the subarctic to the subtropics (Figure 6(a)). Net ecosystem exchange (NEE) is the difference between gross primary production (GPP) and the respiration of plants (R_p) and heterotrophic organisms (R_h):

$$NEE = GPP - R_p - R_h \quad (5)$$

Similar relationships have been reported for many individual peatland ecosystems (Alm *et al.*, 1997; Chanton *et al.*, 1995; Whiting and Chanton, 1992; Whiting *et al.*, 1991). Methanogenesis accounted for 4% of NEE when averaged across several of these studies (Bellisario *et al.*, 1999). NEE is not a direct measure of photosynthetic activity because it includes R_h, but other studies have shown that the relationship holds when photosynthesis is considered independently. Bridgham *et al.* (2001) observed a positive correlation between GPP and CH_4 emissions in bog and fen mesocosms (Figure 6(b)). Differences in CH_4 emissions that were caused by manipulating temperature and nitrogen availability could be explained by changes in photosynthesis. The effects of plant community type (i.e., bog or fen) and water

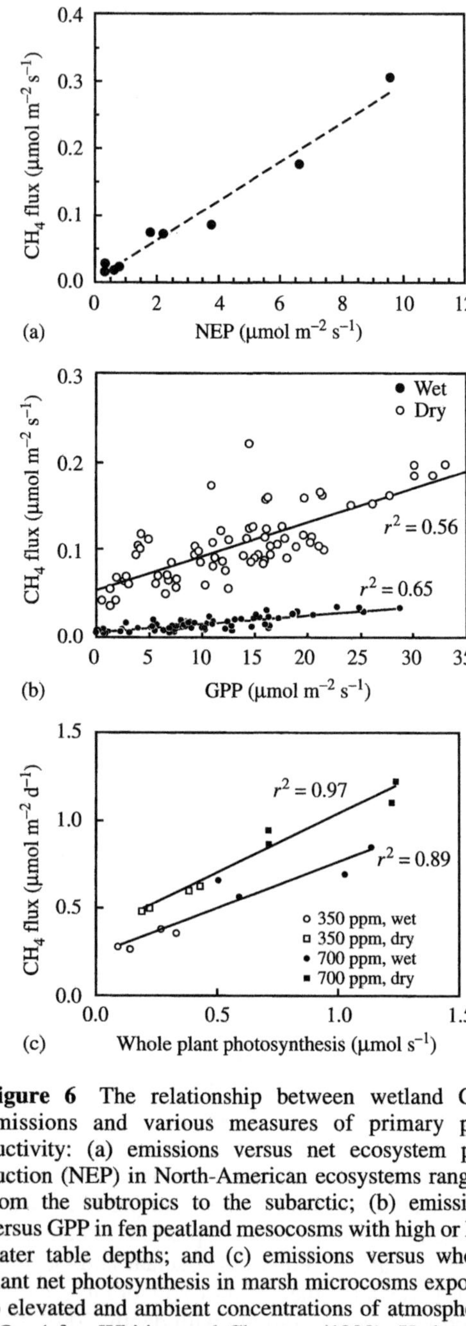

Figure 6 The relationship between wetland CH_4 emissions and various measures of primary productivity: (a) emissions versus net ecosystem production (NEP) in North-American ecosystems ranging from the subtropics to the subarctic; (b) emissions versus GPP in fen peatland mesocosms with high or low water table depths; and (c) emissions versus whole-plant net photosynthesis in marsh microcosms exposed to elevated and ambient concentrations of atmospheric CO_2. (after Whiting and Chanton, (1993); Updergraff *et al.*, (2001); and Vann and Megonigal (2003), respectively).

table depth needed to be accounted for separately. Vann and Megonigal (2003) found that elevated atmospheric CO_2 stimulated CH_4 flux in direct proportion to net photosynthesis in a greenhouse study (Figure 6(c)), but again the regression relationships varied by plant species and water table depth.

A final line of evidence for a direct link between methanogenesis and photosynthesis is a report that CH_4 emissions from a group of peatlands were correlated to $\delta^{13}C$-CH_4 (Chanton *et al.*, 1995). The positive slope of this relationship indicated that the highest fluxes occurred at sites where labile carbon was relatively abundant, which favors acetotrophic methanogenesis and ^{13}C enrichment (Section 8.08.4.4.1). Similar observations have been reported from other types of wetland and aquatic ecosystems (Chanton and Martens, 1988; Martens *et al.*, 1986; Tyler *et al.*, 1994). Collectively, these studies suggest that wetland methanogens depend on labile, high-quality organic carbon supplied by plants in the form of root exudates or detritus.

8.08.4.4 Contributions of Acetotrophy versus Hydrogenotrophy

Acetotrophy and hydrogenotrophy are the dominant methanogenic pathways *in situ*, although trimethylamines and methylated sulfur species may be important in some marine sediments (Oremland *et al.*, 1982). Because acetotrophic methanogenesis is associated with high rates of total CH_4 production, it is useful to understand the factors that govern the relative dominance of these two pathways. The theoretical contribution of hydrogenotrophic methanogenesis to CH_4 production during anaerobic degradation of carbohydrates is 33%, and this value is often observed in freshwater ecosystems (see Conrad (1999) for a partial compilation). However, the relative contributions of the two pathways can vary considerably across ecosystems, seasons, and depths. An early generalization was that acetotrophic methanogenesis is dominant in freshwater ecosystems, while hydrogenotrophic methanogenesis is dominant in marine systems (Whiticar, 1999). This is a reasonable first approximation because SO_4^{2-} reducers readily outcompete methanogens for the limited acetate supply in marine systems. However, there are examples of freshwater systems dominated by hydrogenotrophic methanogenesis (Lansdown *et al.*, 1992), marine systems with a relatively large contribution from acetotrophic methanogenesis (Martens *et al.*, 1986), and wide variation in time and space within a given site (Martens *et al.*, 1986; Sugimoto and Wada, 1993).

8.08.4.4.1 *Organic carbon availability*

The availability of labile organic carbon is perhaps a more appropriate basis than salinity on which to generalize about the relative importance of the two primary methanogenic pathways. Freshwater wetlands that are dominated by acetotrophic methanogenesis also support high rates of primary productivity, and thus high rates of organic carbon mineralization. Labile carbon pools are typically high in rice paddy soils, which are highly productive and dominated (>50%) by acetotrophic methanogenesis (Conrad, 1999). By comparison, *Sphagnum* bog peatlands have exceptionally low nutrient availability, low primary productivity, and highly recalcitrant soil organic matter pools. Isotopic evidence and direct rate measurements indicate that these systems are dominated by hydrogenotrophic methanogenesis (Bellisario *et al.*, 1999; Chasar *et al.*, 2000a; Hines *et al.*, 2001; Lansdown *et al.*, 1992). Despite deep accumulations of nearly pure organic soils in bogs, the size of the labile organic carbon pool is small because of poor carbon quality (Section 8.08.4.3.4). Whereas bogs are isolated from groundwater, fen peatlands receive substantial groundwater inputs and are characterized by comparatively high nutrient availability, herbaceous plant biomass, labile carbon availability, CH_4 emissions, and acetotrophic methanogenesis. This is particularly evident at the soil surface.

The size of the labile carbon pool is an important predictor of methanogenic pathways in marine sediments (Blair, 1998). Sediments with low carbon availability are dominated by hydrogenotrophic methanogenesis, while carbon-rich systems such as Cape Lookout Bight support relatively more acetotrophic methanogenesis (Martens *et al.*, 1986). The labile portion of the organic carbon pool in marine sediments is determined by several factors such as chemical composition, adsorption by mineral surfaces, and burial efficiency (Hedges and Keil, 1995). Poor organic carbon quality is expected in coastal sediments that receive large inputs of terrestrially derived, highly degraded particulate organic matter from rivers (Hedges *et al.*, 1994; Martens *et al.*, 1992a).

Stable isotopes provide a nondestructive alternative to radiolabeled substrates for inferring the relative importance of methanogenic pathways. Whiticar *et al.* (1986) recognized that acetotrophic methanogenesis yields CH_4 that is ^{13}C-enriched ($\delta^{13}C = -65‰$ to $-50‰$) compared to hydrogenotrophy ($\delta^{13}C = -110‰$ to $-60‰$). The reason for this difference is a larger $^{12}C/^{13}C$ fractionation during CH_4 production by hydrogenotrophy ($\alpha_c = 1.055-1.090$) than acetotrophy ($\alpha_c = 1.04-1.055$). Similarly, CH_4 from acetotrophy is deuterium-depleted

($\delta D = -400‰$ to $-250‰$) compared to hydrogenotrophy ($\delta D = -250‰$ to $-170‰$). Subsequent studies have suggested somewhat broader ranges for these fractionation factors (Tyler, 1991). Plots of δD versus $\delta^{13}C$ can be used to infer the extent of aerobic or anaerobic oxidation of CH_4 (Alperin *et al.*, 1988; Coleman *et al.*, 1981). If oxidation has a substantial influence on the CH_4 pool, the relationship between δD and $\delta^{13}C$ is positive. In the absence of oxidation, and provided that the relative contributions of the methanogenic pathways varies in time or space, the relationship between δD and $\delta^{13}C$ is negative. In such cases, the variability in $\delta^{13}C$-CH_4 can often be interpreted as a shift in the contribution of acetotrophy and hydrogenotrophy to total methanogenesis (Burke *et al.*, 1988, 1992). There are several factors other than the methanogenic pathways and oxidation that can affect isotope fractionation factors (Happell *et al.*, 1993; Avery and Martens, 1999; Bergamaschi, 1997; Tyler, 1992), and these make it difficult to interpret stable isotope data in terms of absolute rates. Additional fractionations occur when CH_4 passes through plants, which must be accounted for when interpreting the stable isotope ratios of CH_4 emitted from wetlands (Chanton *et al.*, 1999a,b; 2002; Chanton and Dacey, 1991; Harden and Chanton, 1994).

Plots of $\delta^{13}C$-CH_4 versus $\delta^{13}C$-$\sum CO_2$ in peatland soils typically indicate a progressive transition from acetotrophic methanogenesis at the surface to hydrogenotrophic methanogenesis at depth (Chasar *et al.*, 2000b; Hornibrook *et al.*, 1997; 2000), that coincides with increasingly recalcitrant soil organic matter pools and distance from plant-derived labile carbon at the soil surface (Section 8.08.4.3.4). In a survey of the δD and $\delta^{13}C$ content of CH_4 in gas bubbles collected from lake sediments in western Alaska, there was a larger contribution from acetotrophic methanogenesis at vegetated sites than those that were unvegetated (Martens *et al.*, 1992b). Miyajima *et al.* (1997) reported that ^{13}C-CH_4 enrichment increased in proportion to increasing leaf litter decomposition rate; decomposition rate was in turn negatively related to lignin content (Figure 7). Thus, this study made a direct link between carbon quality and the relative contributions of hydrogenotrophic and acetotrophic methanogenesis to overall CH_4 production. This study was also notable because it was a rare investigation of anaerobic metabolism in a *tropical* peatland. In a variety of shelf and slope marine sediments, $\delta^{13}C$-CH_4 is positively related to the flux of labile organic carbon to the sediment surface (Boehme *et al.*, 1996) and to the rate of pore-water SO_4^{2-} depletion (Figure 8). This relationship indicates that acetotrophic methanogenesis makes a relatively large contribution to CH_4 production in marine sediments when the labile carbon flux is high enough to deplete pools of O_2, Mn(IV), Fe(III) and SO_4^{2-}. However, the relationship did not hold for deep-sea sediments (Blair, 1998).

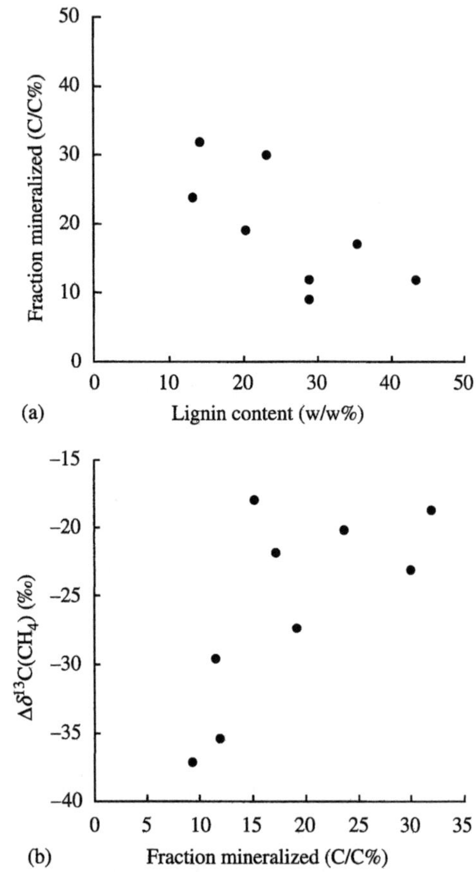

Figure 7 Influence of (a) lignin content on leaf litter decomposition rates, and (b) leaf litter decomposition rates on methanogenic pathway as reflected in the $\delta^{13}C$ of CH_4. In (b), the *y*-axis is the difference in $\delta^{13}C$-CH_4 and $\delta^{13}C$ of total mineralized carbon ($CH_4 + CO_2$) (Miyajima *et al.*, 1997) (reproduced by permission of Elsevier from *Geochim. Cosmochim. Acta*, **1997**, *61*, 3739–3751).

The contribution of acetotrophy to methanogenesis can be quite stable in time (Avery and Martens, 1999; Burke *et al.*, 1992; Hines *et al.*, 2001), or it can vary strongly with the seasons. A mid-latitude bog was dominated by hydrogenotrophic methanogenesis for 10 months of the year, then switched to acetotrophic methanogenesis (Avery *et al.*, 2002). Because the switch was accompanied by a large increase in CH_4 emissions, acetotrophy contributed ~50% of total annual emissions. Seasonal shifts in methanogenic pathways were observed in several peatlands located at mid-latitudes (Kelley *et al.*, 1992; Lansdown *et al.*, 1992; Shannon and White, 1996). By comparison, no such switch occurred in a group of bogs and fens located at high latitudes (Duddleston *et al.*, 2002; Hines *et al.*, 2001).

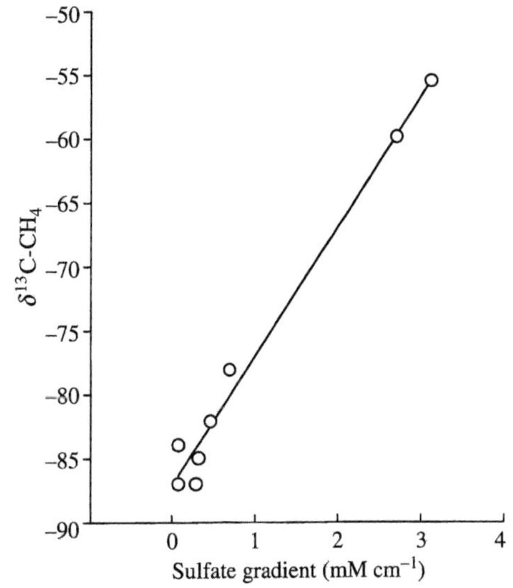

Figure 8 The relationship between that rate at which SO_4^{2-} is consumed down-core in shallow marine sediments and the $\delta^{13}C\text{-}CH_4$. The r^2 fit of the regression line is 0.98 (Blair, 1998) (reproduced by permission of Elsevier from *Chem. Geol.*, **1998**, *152*, 139–150).

These sites were completely dominated by hydrogenotrophic methanogenesis despite an accumulation of acetate in the pore water. Incubating these soils at 24 °C for 5 months did not trigger acetotrophic CH_4 production, even though acetate continued to accumulate during the period. Apparently, the methanogens in these systems are limited by factors other than temperature, such as microbial community composition, pH, or perhaps trace nutrient availability (Section 8.08.4.3.2).

Chapelle *et al.* (2002) recently described a subsurface microbial community in which hydrogenotrophic methanogens were >90% of the 16S ribosomal DNA sequences and geothermal H_2 was the primary energy source. Since geothermal H_2 is likely to be an abundant energy source for microbial metabolism on other planets, hydrogenotrophy may be an important pathway of anaerobic metabolism elsewhere in the universe.

8.08.4.4.2 Temperature

Temperature is one of several factors that influence the contributions of acetotrophic and hydrogenotrophic methanogenesis to overall CH_4 production. The most comprehensive work on this topic has been done by Conrad and colleagues in rice paddy soils and lake sediments. They have repeatedly observed that the contribution from hydrogenotrophic methanogenesis declines at low temperatures, while the contribution of acetotrophic methanogenesis shows the opposite pattern (Figure 9(a); Chin and Conrad, 1995; Chin *et al.*, 1999a; Conrad *et al.*, 1987; Fey and Conrad, 2000; Schultz and Conrad, 1996; Schultz *et al.*, 1997; Yao and Conrad, 1999). Although changes in the relative contributions of the two pathways coincide with shifts in the structure of methanogenic communities (Chin *et al.*, 1999b; Fey and Conrad, 2000), they are probably not a direct response of methanogens to temperature. Rather, the methanogens appear to be responding to changes in substrate availability (Figure 9(b)), suggesting that the actual source of the observed temperature limitation was organisms that produce or consume H_2 and acetate. Indeed, H_2 amendments stimulated hydrogenotrophic methanogenesis at low (15 °C) temperature in a paddy soil, but not cellulose amendments, which would have required fermentation to methanogenic substrates (Schultz *et al.*, 1997). This result is consistent with temperature limitation of syntrophic bacteria, which produce H_2 by fermentation (Section 8.08.3.2.2). It is more difficult to explain an increase in acetate concentrations at low temperatures. One proposal is that low temperature favors the fermentation of multicarbon substrates directly to acetate (Conrad, 1996; Fey and Conrad, 2000), but this has not been demonstrated. Acetogenesis was shown to be <10% of the acetate sources at low temperature (Rothfuss and Conrad, 1993; Thebrath *et al.*, 1992). This body of work demonstrates that building a mechanistic understanding of the temperature responses of methanogens will require comprehensive studies that include fermenting, syntrophic, and homoacetogenic bacteria.

Physiological differences in the two methanogenic genera that can use acetate may influence temperature-driven changes in methanogenic pathways (reviewed by Liesack *et al.*, 2000). The members of the *Methanosaetaceae* that have been isolated to date use acetate exclusively and have a relatively low threshold for the substrate (typically <100 μM), while members of the *Methanosarcinaceae* use both acetate and H_2/CO_2, and have a relatively high threshold for acetate (typically between 200 and 1200 μM) (Jetten *et al.*, 1992; Großkopf *et al.*, 1998). The relative abundance of the two genera in paddy soils appears to be a function of both temperature and acetate concentration (reviewed in Liesack *et al.*, 2000). When acetate concentrations were above the threshold of both genera, the *Methanosaeta* were dominant, presumably because they were more tolerant of suboptimal (15 °C) temperature than the *Methanosarcina* (Chin *et al.*, 1999a,b). The *Methanosarcina* became dominant at high (30 °C) temperature, which is consistent with the fact that they can also use H_2. The reverse of this pattern was observed when acetate levels were maintained below the threshold concentration for both genera. In this case, the *Methanosaeta* were dominant at the high

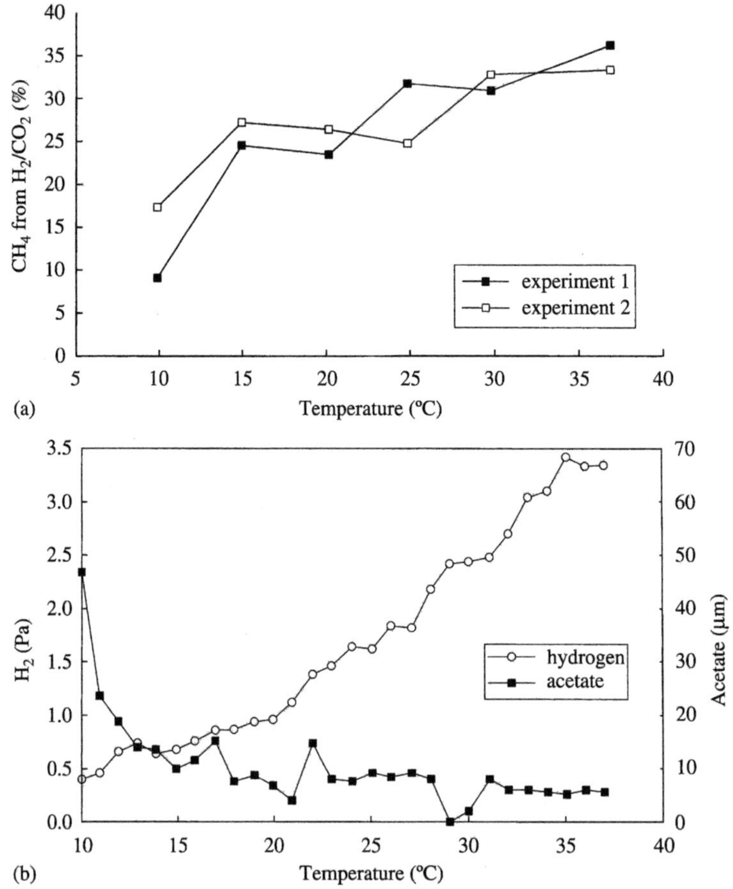

Figure 9 The influence of temperature on: (a) methanogenic pathways and (b) methanogenic substrates (after Fey and Conrad, 2000).

temperature because acetate concentrations were below the threshold concentration of the *Methanosarcina* (Fey and Conrad, 2000).

The extensive work on methanogenic pathways in profundal lake sediments and rice paddy soils does not appear to apply to a broader sample of anaerobic ecosystems. The proportions of CH_4 produced from hydrogenotrophic and acetotrophic methanogenesis did not vary seasonally in a temperate tidal freshwater estuarine sediment (Avery and Martens, 1999; Avery *et al.*, 1999) or a northern peatland soil (Hines *et al.*, 2001) despite changes of 25 °C or more. Temperature alone could not explain transient changes in the methanogenic pathways in a mid-latitude (45° N) peatland (Avery *et al.*, 1999, 2002). Clearly, detailed mechanistic studies on temperature regulation of fermentation and methanogenesis need to be done in a wider variety of anaerobic ecosystems (e.g., Kotsyurbenko *et al.*, 1996).

8.08.4.5 Anaerobic Methane Oxidation

Anaerobic CH_4 oxidation in marine sediments was first observed around the mid-1970s (Barnes and Goldberg, 1976; Reeburgh, 1976; Martens and Berner, 1977). The process was inferred from pore-water SO_4^{2-} and CH_4 profiles because it imparted the CH_4 profile with a strong "concave up" shape (Martens and Berner, 1977). Further geochemical evidence for this process included conversion of $^{14}CH_4$ to $^{14}CO_2$ under anoxic conditions, and a zone of relatively ^{13}C-enriched CH_4 that coincided with a zone of ^{13}C-depleted CO_2 (citations in Valentine, 2002). Although anaerobic CH_4 oxidation was shown to be microbially mediated long ago (Reeburgh, 1982), recent advances in molecular biology and stable isotope techniques have provided compelling new biological insights on the process. Nevertheless, there are no pure cultures or co-cultures of organisms that oxidize CH_4 anaerobically, and several competing hypotheses about the metabolic pathways involved are still considered viable explanations of the existing data (reviewed by Hinrichs and Boetius, 2002; Valentine, 2002).

Three primary mechanisms have been offered to explain anaerobic CH_4 oxidation in marine sediments. The process may be carried out by a single organism that couples methanotrophy to SO_4^{2-} reduction. Anaerobic methane oxidation coupled to sulfate reduction is thermodynamically favorable under conditions found in marine sediments (Martens and Berner, 1977):

$$CH_4 + SO_4^{2-} \rightarrow HS - +HCO_3^- + H_2O \quad (6)$$

A second possibility yields the same net reaction but is driven by a syntrophic relationship between an organism that oxidizes CH_4 to CO_2/H_2 and an SO_4^{2-}-reducing bacterium that consumes H_2, thereby maintaining a thermodynamically favorable environment for the reaction (Hoehler *et al.*, 1994). In this case, anaerobic CH_4 oxidation is analogous to hydrogenotrophic methanogenesis in reverse:

$$CH_4 + 2H_2O \rightarrow CO_2 + 4H_2 \quad (7)$$

$$SO_4^{2-} + 4H_2 + H^+ \rightarrow HS^- + HCO_3^- + 4H_2O \quad (8)$$

The ability to run metabolic pathways in forward or reverse, thereby swapping end products for substrates, has been documented previously in anaerobic organisms. For example, the homoacetogen nicknamed *Reversibacter* oxidizes acetate to CO_2/H_2 when the H_2 concentration is low, and vice versa (Lee and Zinder, 1988; Zinder and Koch, 1984). However, attempts to reverse methanogenesis in pure cultures by varying the H_2 concentration have not been successful (Valentine *et al.*, 2000). Thermodynamic and kinetic considerations suggest that acetogenic methanogenesis running in reverse is feasible only in high-CH_4 environments such as CH_4 seeps, as are pathways that require interspecies transfer of acetate (Valentine, 2002). Likewise, pathways the require interspecies transfer of H_2, formate or methanol are not favorable for anaerobic CH_4 oxidation under high- or low-CH_4 conditions (Sørensen *et al.*, 2001). Nauhaus *et al.* (2002) were unable to stimulate SO_4^{2-} reduction in sediments collected from a CH_4 seep by adding H_2, formate, acetate, or methanol.

A modification of the reverse methanogenesis hypothesis was offered by Valentine and Reeburg (2000). This reaction would yield twice the amount of energy normally associated with anaerobic CH_4 oxidation (reaction (6)), making it more thermodynamically favorable:

$$2CH_4 + 2H_2O \rightarrow CH_3COOH + 4H_2 \quad (9)$$

In this case, the methanogens operating in reverse produce acetate and H_2, which are then consumed by SO_4^{2-}-reducing bacteria:

$$SO_4^{2-} + 4H_2 \rightarrow S^{2-} + 4H_2O \quad (10)$$

$$SO_4^{2-} + CH_3COOH \rightarrow H_2S + 2HCO_3^- \quad (11)$$

All of these mechanisms conform to the observation that CH_4 is consumed and S^{2-} is produced in a molar ratio of 1 : 1 in marine sediments (Nauhaus *et al.*, 2002).

The fact that CH_4 is highly depleted in ^{13}C has proven useful for constraining the nature of the microorganisms that oxidize methane anaerobically. Because the $\delta^{13}C$ of CH_4 ranges from about −50‰ to −110‰ PDB (Whiticar *et al.*, 1986), the fate of CH_4-derived carbon can be traced through food webs by means of its distinctive isotopic ratio. Such work has tended to support some form of reverse methanogenesis. Most isotope studies have been done in sediments overlaying CH_4 hydrate deposits or seeps where the high porewater CH_4 concentration favors rapid anaerobic CH_4 oxidation. At such sites, Archea-specific lipids are abundant and highly depleted in ^{13}C (often <−100‰), indicating they could only be derived from CH_4 (Bian *et al.*, 2001; Elvert *et al.*, 1999, 2000; Hinrichs *et al.*, 1999, 2000; Orphan *et al.*, 2001a; Peckmann *et al.*, 1999; Pancost *et al.*, 2000, 2001; Thiel *et al.*, 1999, 2001). Interestingly, fatty acids that are characteristic of SO_4^{2-}-reducing bacteria are also highly ^{13}C-depleted (up to −75‰) suggesting that SO_4^{2-} reducers are consuming an organic intermediate produced by methane oxidizers (Boetius *et al.*, 2000b; Hinrichs *et al.*, 1999; Orphan *et al.*, 2001a,b; Thiel *et al.*, 2001). If so, this is evidence that anaerobic CH_4 oxidation is performed by an Archea–Bacteria consortium.

Perhaps the most striking evidence of a consortium between Archea and Bacteria is their spatial arrangement in hydrate seep sediments. Using fluorescent phylogenetic stains, Boetius *et al.* (2000b) observed clusters of ~100 Archeal cells surrounded by a 1–2 cell-thick shell of sulfate-reducing bacteria (Figure 10). Orphan *et al.* (2001b) further showed that highly depleted $\delta^{13}C$

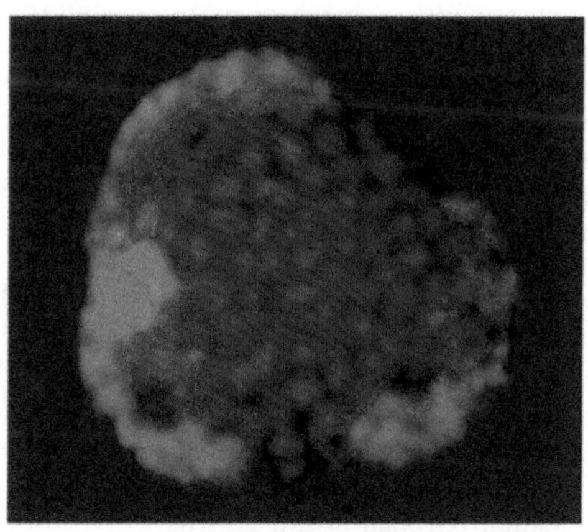

Figure 10 *In situ* identification of microbial aggregates consisting of Archea (red) in the center and SO_4^{2-}-reducing bacteria (green) on the periphery. Microorganisms were labeled with rRNA-targeted oligionucleotide probes. (source Boetius *et al.*, 2000b).

ratios coincided with the Archeal core of these aggregates. Collectively, the evidence supports a syntrophic model of anaerobic CH_4 oxidation in which methanogen-like Archea oxidize CH_4, and SO_4^{2-}-reducing bacteria consume end products, thereby making the reaction thermodynamically favorable. It remains to be determined whether the Archeal member is a methanogen capable of both CH_4 production and oxidation, or an obligate methanotroph. Highly ^{13}C-depleted Archeal lipids have also been observed in solitary organisms, suggesting a microbial consortium is not necessarily a requirement for anaerobic CH_4 oxidation (Orphan *et al.*, 2002).

Little is known about the organisms that oxidize methane anaerobically except for the broad outlines of their phylogeny. Sequences of 16S rRNA cloned from several CH_4 seep sediments are dominated by two groups of previously unknown Archeal genes, ANME-1 and ANME-2 (Boetius *et al.*, 2000b; Hinrichs *et al.*, 1999; Orphan *et al.*, 2001a). The ANME-1 group typically occurs as single filaments or monospecific mats, and does not appear require a bacterial partner (Orphan *et al.*, 2002). They have been recovered from hydrothermal vents (Takai and Horikoshi, 1999; Teske *et al.*, 2002), methane hydrates (Lanoil *et al.*, 2001), and shallow marine sediments (Thomsen *et al.*, 2001). Massive ANME-1 "reefs" were recently discovered in the Black Sea (Michaelis *et al.*, 2002). The cores of anaerobic methane-oxidizing aggregates have so far been composed of ANME-2 type Archea. This group is closely related to known methanogens in the *Methanosarcinales*, an order with members that ferment acetate to CH_4, but may also oxidize small amounts of acetate to CH_4 (Zehnder and Brock, 1979). The proposition that some members of the anaerobic consortium are capable of oxidizing CH_4 to acetate and H_2 (reaction (9)) is attractive because it provides a mechanism to explain the presence of highly ^{13}C-depleted lipids in SO_4-reducing bacteria (Valentine and Reeburg, 2000). It is known that acetate is consumed by members of the same SO_4-reducing group found in CH_4 seep sediments (i.e., *Desulfosarcinales*).

Judging by the microbial diversity of methane seeps, anaerobic CH_4 oxidation may not be fully explained by any single mechanism proposed so far. Many of the SRB associated with anaerobic CH_4 oxidation environments are unknown and unique (Thomsen *et al.*, 2001), and biomarker data suggest a great deal of taxonomic diversity among both Archea and Bacteria in sediments that support anaerobic CH_4 oxidation (Orphan *et al.*, 2001a; Thiel *et al.*, 2001). Specific inhibitors intended for methanogens, SO_4^{2-} reducers, and acetate oxidizers have sometimes influenced rates of anaerobic CH_4 oxidation (Hoehler *et al.*, 1994) and had no effect at other times (Alperin and Reeburgh, 1985). Anaerobic CH_4 oxidation occurs in freshwater environments where SO_4^{2-} concentrations are low (Smith *et al.*, 1991; Miura *et al.*, 1992; Murase and Kimura, 1994a,b; Smemo and Yavitt, 2000; Grossman *et al.*, 2002), suggesting that electron acceptors other than SO_4^{2-} may play a role in the process. Thermodynamic considerations suggest that anaerobic CH_4 oxidation pathways may vary between sites as a function of CH_4 partial pressure (Valentine, 2002). Further progress will undoubtedly require isolating or enriching the responsible organisms.

Very little information exists about the ecological conditions that govern anaerobic CH_4 oxidation. Manipulation of a CH_4 seep sediment demonstrated a broad temperature optimum of 4–16 °C, and rates that varied predictably with temperature (Nauhaus *et al.*, 2002). The process is widespread in marine sediments and water columns provided there is a sufficient supply of labile organic carbon to permit substantial methanogenesis (Section 8.0.8.4.3.4), or some other CH_4 source exists. Anaerobic CH_4 oxidation is the primary sink for CH_4 in the world oceans (D'Hondt *et al.*, 2002), and it accounts for a large fraction of the SO_4^{2-} reduced in some marine sediments (Table 6), particularly in the narrow zone where CH_4 and SO_4^{2-} profiles intersect. Sulfate reduction rates as high as 140 mmol m^{-2} d^{-1} have been reported above decomposing CH_4 hydrate deposits (Boetius *et al.*, 2000b), suggesting that anaerobic oxidation significantly reduces CH_4 flux to the water column from these vast carbon reservoirs (Gornitz and Fung, 1994).

8.08.4.6 Aerobic Methane Oxidation

Methane is subject to aerobic oxidation by methanotrophic bacteria when it diffuses across an anoxic–oxic interface before escaping to the atmosphere (King, 1992):

$$CH_4 + 2O_2 \rightarrow CO_2 + 2H_2O \qquad (12)$$

Methanotrophs occur at the oxic–anoxic interface of methanogenic habitats, in symbiotic association with animals (Kochevar *et al.*, 1992), and *inside* wetland plants (Bosse and Frenzel, 1997). Although methanotrophs dominate aerobic CH_4 oxidation, NH_4-oxidizing bacteria may account for a small amount of the CH_4 oxidation activity in soils and sediments (Bodelier and Frenzel, 1999).

8.08.4.6.1 Methanotroph diversity

Methanotrophs are obligate aerobes that use CH_4 as a sole carbon and energy source (reviewed by R. S. Hanson and T. E. Hanson, 1996). They are a subset of the methylotrophic bacteria, all of which oxidize compounds lacking C–C bonds

Table 6 Selected estimates of the proportion of SO_4^{2-} reduction in marine sediments that is mediated by anaerobic CH_4 oxidation.

Site	*Peak contribution to SO_4^{2-} reduction (%)*	*Depth-integrated contribution to SO_4^{2-} reduction (%)*	*Citation*
Aarhus Bay, Denmark	47–52	9	Thomsen *et al.* (2001)
Kattegat, Denmark	61	10	Iverson and Jørgensen (1985)
Skagerrak	89	10	Iverson and Jørgensen (1985)
Upwelling Zone, Namibia	100		Niewöhner *et al.* (1998)
Amazon Fan Sediment	50–85		Burns (1998)
Norsminde Fjorde	10–30		Hansen *et al.* (1998)
Big Soda Lake, Nevada		2	Iversen *et al.* (1987)
Kysing Fjord		<0.1	Iversen and Blackburn (1981)
Hydrate Ridge, Oregon[a]	100	100	Boetius *et al.* (2000b)

[a] Inferred by comparing SRR above decomposing CH_4 hydrates to nearby nonhydrate sites.

(i.e., C_1 compounds) (Murrell and Kelley, 1993). The ability to metabolize C_1 compounds is a feature that methylotrophs have in common with methanogens. Indeed, the two groups share many homologous genes for C_1 metabolism, despite the large evolutionary distance between the *Archea* and the *Proteobacteria* (Chistoserdova *et al.*, 1998).

Methanotrophic species are separated into two types that differ with respect to phylogeny, ultrastructure, lipid composition, biochemistry, and physiology. The type I methanotrophs belong to the family *Methylococcaceae* in the gamma proteobacteria; the type II methanotrophs are in the family *Methylocystaceae* in the alpha proteobacteria. There is growing evidence that the two groups also differ ecologically. Amaral and Knowles (1995) used opposing gradients of CH_4 and O_2 to determine that a type I methanotroph preferred a somewhat lower CH_4 concentration than a type II methanotroph. An often-cited ecological distinction between the two groups is that only the type II methanotrophs are N_2 fixers (R .S. Hanson and T. E. Hanson, 1996). This is consistent with a report that a type II methanotroph out-competed a type I methanotroph in a nitrogen-limited chemostat culture (Graham *et al.*, 1993), and in a field study where NH_4^+ fertilization increased the proportion of type I methanotrophs in planted rice paddy soils (Bodelier *et al.*, 2000b). In the latter study, the abundance of type II methanotrophs did not change in response to nitrogen fertilization, suggesting that only the type I methanotrophs were nitrogen limited. However, a survey of several type I and type II methanotrophic strains for N_2-fixation genes (*nif*H) and nitrogenase activity demonstrated that the capacity for N_2 fixation occurs in both groups (Auman *et al.*, 2001). Thus, it is no longer valid to assume that type I methanotrophs lack the capacity for N_2 fixation. The extent to which nitrogenase activity is expressed by type I methanotrophs *in situ* remains to be determined.

All known methanotrophs have a low affinity for CH_4 ($K_m > 1$ μM) and none can maintain growth on CH_4 at atmospheric concentration (for a compilation of kinetic data see Conrad, 1996). Although there are no pure cultures of "high affinity" CH_4 oxidizers, these organisms are known to exist in upland soils where they account for ~10% of the annual global CH_4 sink (Table 5). High-affinity CH_4 oxidizers evidently occur in wetland soils as well because there have been reports of net CH_4 consumption from the atmosphere in response to falling water table depth (Happell and Chanton, 1993; Harriss *et al.*, 1982; Pulliam, 1993; Roulet *et al.*, 1993; Shannon and White, 1994), low temperature (Megonigal and Schlesinger, 2002), and low soil organic matter (Giani *et al.*, 1996). The capacity for both types of oxidation kinetics in saturated soils and sediments has been assumed to be due to the presence of mixed methanotrophic populations (Bender and Conrad, 1992). This was confirmed in a recent study that demonstrated the power of combining lipid biomarker analysis with isotope pulse-labeling (Boschker *et al.*, 1998). Bull *et al.* (2000) pulse-labeled a forest soil with $^{13}CH_4$, and followed its incorporation into phospholipid fatty acids (PLFAs). At the soil surface, where ambient CH_4 levels prevail, the label was incorporated into novel organisms with PLFAs similar to type II methanotrophs. A buried soil horizon with higher CH_4 levels (100 μL L^{-1}) indicated the presence of both type I and type II methanotrophs. Another promising technique is "stable-isotope probing" in which a microbial community is exposed to an isotopically labeled substrate (e.g., $^{13}CH_4$), thereby changing the mass of organisms that assimilate the substrate. The labeled and unlabeled microorganisms are separated by density-gradient centrifugation and DNA sequenced (Radajewski *et al.*, 2000). This technique demonstrated that novel proteobacteria contributed to high-affinity CH_4 oxidation in a peat soil (Morris *et al.*, 2002).

Until recently, all cultured methanotrophs required a pH $\geq$5 and most were neutrophilic (R. S. Hanson and T. E. Hanson, 1996). Yet, CH_4 oxidation at pH <5 has been demonstrated in acidic *Sphagnum*-bogs (Born *et al.*, 1990; Dunfield *et al.*, 1993; Fechner and Hemond, 1992; Yavitt *et al.*, 1990). Relatively recently, acidophilic methanotrophic bacteria have been isolated and two species have been described with an optimum activity between pH 4.5 and 5.5 (Dedysh *et al.*, 1998a,b, 2000, 2002). This demonstrates that some methanotrophs are adapted to the ambient pH conditions of their environment.

The relative abundance of various methanotrophic species or types has been investigated in peatlands, rice paddies, lakes and elsewhere with PLFA analyses, community DNA analyses (McDonald and Murrell, 1997), and fluorescent *in situ* hybridization (FISH). A suite of FISH probes has been developed for methanotrophs with a range of specificity for types, genera and species (Amann *et al.*, 1990b; Bourne *et al.*, 2000; Dedysh *et al.*, 2001; Eller *et al.*, 2001; Gulledge *et al.*, 2001; Rosselló-Mora *et al.*, 1995). The use of such probes in an acidic northern peatland showed that <1% of the methanotrophic community was type I. Although the remainder could tolerate low pH, <5% were known acidophiles (Dedysh *et al.*, 2003). A single genus, *Methylocystis*, contributed 60–95% of all detectable methanotrophic cells.

8.08.4.6.2 *Regulation of methanotrophy*

Methane oxidation rates can be limited by O_2, CH_4, or nitrogen. Limitation by O_2 is most likely to occur in continuously flooded or submerged systems that are strongly methanogenic. Methane oxidation is limited by O_2 flux to the sediment surface in lakes (Frenzel *et al.*, 1990) and flooded wetlands (King, 1990a,b; King *et al.*, 1990). For this reason, the presence of algal mats or submerged aquatic vegetation can cause diurnal variations in CH_4 oxidation, with the highest rates during daylight due to photosynthetic O_2 production (King, 1990b). Heilman and Carlton (2001) reported a particularly dramatic example of diel variation in CH_4 emissions driven by O_2 release from submerged aquatic macrophytes. In this case, the sediment was a net CH_4 sink in daylight and a CH_4 source at night. However, they concluded that O_2 was inhibiting methanogenesis rather than stimulating methanotrophy. Methanotrophy itself can account for a substantial fraction of the O_2 demand in some systems; it accounted for 30% of O_2 consumption in Lake Erie (Adams *et al.*, 1982) and >9% of O_2 consumption in Lake Constance (Frenzel *et al.*, 1990).

Root-associated methanotrophy has been demonstrated to be O_2-limited in the rhizosphere of emergent aquatic macrophytes such as *Phragmites australis* (van der Nat and Middelburg, 1998), *Sparganium eurycarpum* (King, 1996a), and *Typha latifolia* (Lombardi *et al.*, 1997). Plant species with high rates of root O_2 loss also support high rates of rhizosphere CH_4 oxidation (Calhoun and King, 1997), further suggesting the process is generally O_2 limited. Differences in root O_2 loss rates among plant species should cause species-specific differences in rhizosphere CH_4 oxidation rates if the process is O_2 limited. Indeed, soil-free root CH_4 oxidation potentials vary among plant species (Sorrell *et al.*, 2002). Because >75% of the CH_4 efflux from wetlands may pass through plants (Chanton and Dacey, 1991; King, 1996a; Shannon and White, 1994), O_2 limitation in the rhizosphere should translate into O_2 limitation on an ecosystem basis.

By comparison with uncultivated plants, the case for O_2 limitation in the rice rhizosphere is not so clear. Linear correlations between rhizosphere CH_4 oxidation and CH_4 concentration suggested that oxidation is CH_4-limited (Bosse and Frenzel, 1998; Gilbert and Frenzel, 1995; Gilbert and Frenzel, 1998). This result is consistent with a study by Denier van der Gon and Neue (1996) in which the atmospheric O_2 concentration was doubled to 40% around rice shoots, but CH_4 oxidation increased by only 20%. However, kinetic and theoretical considerations have also suggested that O_2 limits rhizosphere CH_4 oxidation in rice (Bosse and Frenzel, 1997; van Bodegom *et al.*, 2001).

To the extent that O_2 limitation exists, O_2 competition is an important factor regulating CH_4 oxidation efficiency. Based on kinetic modeling, van Bodegom *et al.* (2001) concluded that methanotrophs were only able to out compete heterotrophic bacteria for O_2 when O_2 concentrations were <10 μM. The competitive advantage of the methanotrophs at low O_2 concentrations was due to their high affinity for O_2. There are other microaerophilic bacteria with a high affinity for O_2 (e.g., Fe(II)-oxidizing bacteria, Section 8.08.6.5.2) that compete with methanotrophs for O_2, as well as abiotic oxidation reactions involving Fe(II) or H_2S. The continued development of simulation models, such as the one offered by van Bodegom *et al.* (2001), will help integrate the effects of the multiple factors affecting rhizosphere CH_4 oxidation.

Methane availability is likely to limit oxidation in systems that are weakly methanogenic, such as marine sediments or dry-end wetlands that have exposed soils and subsurface water tables. Methane-limited methanotrophy was recently demonstrated in a pair of tidal forested wetlands that were aerobic in the top 5–10 cm (Megonigal and Schlesinger, 2002; Figure 11). Indirect evidence for CH_4-limited methanotrophy in peatland ecosystems is the observation that potential CH_4 oxidation tends to peak near the water table boundary, where CH_4 concentrations are high and O_2 concentrations are low (Kettunen *et al.*, 1999; Sundh *et al.*, 1995; Sundh *et al.*, 1994). Such a relationship should be common in bog-type peatlands that are typically dry at the soil surface, have relatively low potential CH_4 production rates, and are dominated by nonvascular plants. Methanotrophy in bogs can also be limited by high rates of diffusivity through the soil profile due to low bulk density (Freeman *et al.*, 2002).

Methane oxidation can be limited by nitrogen due to enzyme-level inhibition or unmet methanotroph nitrogen demand. The oxidation of CH_4 by methanotrophic bacteria and NH_4^+ by ammonium-oxidizing bacteria is initiated by similar enzymes—methane monooxygenase (MMO) and ammonium monooxygenase (AMO). The membrane form of MMO is evolutionarily related to AMO (Holmes *et al.*, 1995); there is also a soluble form of MMO. MMO and AMO fortuitously oxidize a variety of additional compounds (Hanson and Hanson, 1996), including NH_4^+ (MMO) and CH_4 (AMO). Bodelier and Frenzel (1999) determined that the contribution of nitrifiers to CH_4 oxidation was negligible in rice paddy microcosms; however, methanotrophs contributed substantially to NH_4^+ oxidation. Competition between NH_4^+ and CH_4 for the active site on MMO is one explanation for the common observation that nitrogenous fertilizers inhibit CH_4 oxidation (Steudler *et al.*, 1989; Conrad and Rothfuss, 1991; Crill *et al.*, 1994; King, 1990a), and it is consistent with fact that inhibition often varies with the relative concentrations of CH_4 and NH_4^+ (Bosse *et al.*, 1993; Boeckx and Van Cleemput, 1996; Dunfield and Knowles, 1995; King and Schnell, 1994a,b; Schnell and King, 1994; van der Nat *et al.*, 1997). A second explanation for depressed CH_4 oxidation is NO_2^- toxicity (King and Schnell, 1994b; Kravchenko, 1999). Both of these factors should be relatively unimportant in anaerobic environments because CH_4 concentrations are high and nitrification rates are low due to limited O_2 availability. Indeed, nitrogen fertilization unexpectedly stimulated CH_4 oxidation in rice paddy soils (Bodelier *et al.*, 2000b), where both CH_4 concentrations and plant nitrogen demand are typically high. These authors used a combination of radioisotope labeling and PFLA analysis to determine that type I methanotrophs were nitrogen-limited in the absence of fertilizer application. Using the same technique, Nold *et al.* (1999) observed that NH_4^+ additions decreased CH_4 incorporation into methanotroph lipids in a lake sediment. Although CH_4 oxidation is clearly influenced by fertilization, the variety of different responses that have been observed and the number of mechanisms that have been proposed for these effects makes it difficult to generalize about the process (Gulledge *et al.*, 1997).

In a recent compilation of the wetland CH_4 oxidation literature, Segers (1998) concluded that pH is not an important factor governing aerobic CH_4 oxidation, salt concentrations of ~40 mM inhibit the process, and aerobic CH_4 oxidation responds to temperature with Q_{10} = ~2. The Q_{10} for CH_4 oxidation can be <2 when there is a phase-transfer limitation on CH_4 diffusion (e.g., King and Adamsen, 1992).

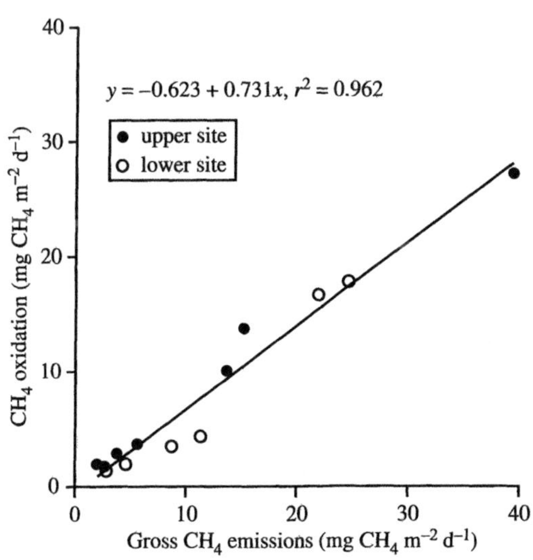

Figure 11 Relationship between rates of gross CH_4 emission and CH_4 oxidation measured over a 13-month period in two tidal freshwater wetlands (source Megonigal and Schlesinger, 2002).

8.08.4.6.3 Methane oxidation efficiency

The amount of CH_4 emitted to the atmosphere is just a fraction of that produced because of methanotrophy. Globally, it accounts for ~80% of gross CH_4 production (Reeburgh, 1996; Reeburgh *et al.*, 1993). The wetland CH_4 oxidation sink has been estimated at 40% (King, 1996b) to 70% (Reeburgh *et al.*, 1993) of gross CH_4 production, or roughly 100–400 Tg yr^{-1}. Methane oxidation efficiency is generally higher in wetland forests than marshes, perhaps because forests occupy drier positions on the landscape and develop a relatively deep oxic zone at the soil surface (Megonigal and Schlesinger, 2002). It follows that wetland forests are more likely than marshes to be CH_4 limited. In some cases, CH_4 oxidation efficiency is 100% and the site becomes a net CH_4 sink (Harriss *et al.*, 1982).

The path taken by a CH_4 molecule en route to the atmosphere affects the chance it will be oxidized. Oxidation is perhaps most likely when CH_4 diffuses across a soil or sediment surface because of the potential for a wide aerobic zone, and a correspondingly long residence time in the aerobic zone. One of the highest CH_4 oxidation rates that has been reported was measured in a landfill soil that had high pore space CH_4 concentrations overlaid by >1.5 m of aerobic topsoil (Whalen *et al.*, 1990). In most cases, the potential for CH_4 oxidation in the rhizosphere will be limited by a narrow aerobic zone of perhaps 1–2 mm, or an aerobic zone may be absent altogether (e.g., King *et al.*, 1990; Roura-Carol and Freeman, 1999). However, there is a notable exception in which O_2 from roots can penetrate a few centimeters into the soil (Pedersen *et al.*, 1995). Methane oxidation efficiency in the rhizosphere is likely to be lower than at the soil surface because the residence time of CH_4 in the rhizosphere is relatively short. For example, King (1996a) determined that 43% of the CH_4 crossing the soil surface was oxidized compared to 27% of the CH_4 crossing the rhizosphere in a freshwater marsh. Methane bubbles (i.e., ebullition) bypass the CH_4 oxidation zone altogether. Ebullition occurs when the CH_4 partial pressure exceeds a critical threshold. It is a significant process in aquatic sediments (Chanton and Martens, 1988; Martens and Klump, 1980; Zimov *et al.*, 1997) and is generally under-appreciated as a source of CH_4 emissions from wetlands (Romanowicz *et al.*, 1993). Because plants efficiently evacuate CH_4 to the atmosphere by passive diffusion or mass flow through stems (Chanton and Dacey, 1991), they improve CH_4 oxidation efficiency by decreasing ebullition (Bosse and Frenzel, 1998; Grünfeld and Brix, 1999). Understanding the proportion of CH_4 efflux that occurs by these three pathways is essential for modeling the response of natural wetlands, rice paddies and landfills to management or climate change (Bogner *et al.*, 2000).

A variety of techniques have been used to estimate the proportion of gross production that is consumed by CH_4 oxidation (i.e., CH_4 oxidation efficiency), most of which involve some degree of manipulation. A common approach is to measure methanotrophy in the presence and absence of O_2. However, this approach will underestimate the process if it simultaneously stimulates CH_4 production. An alternative approach is to apply a specific inhibitor of methanotrophy. This technique was used by de Bont *et al.* (1978) to estimate methanotrophy in the rice rhizosphere. They applied acetylene gas to the shoot, and allowed it to diffuse through the stem, along with O_2, into the root zone where it blocked methanotrophy. Acetylene has the unfortunate property of causing irreversible inhibition, but it is far less expensive and equally effective as some alternative inhibitors (King, 1996a). Methylfluoride (CH_3F) is an attractive inhibitor because the effects are reversible, so that it can be used many times on the same site (Epp and Chanton, 1993; Oremland and Culbertson, 1992a,b). However, CH_3F inhibits acetotrophic methanogenesis and may therefore underestimate CH_4 oxidation. This is not necessarily a problem when the gas is applied to shoots (King, 1996a; Megonigal and Schlesinger, 2002), but the concentration and duration of application should be considered beforehand (Lombardi *et al.*, 1997). Recently, difluoromethane was shown to reversibly inhibit CH_4 oxidation without severely affecting hydrogenotrophic or acetotrophic methanogenesis (Miller *et al.*, 1998). Stable isotope ratios offer a nonintrusive means of estimating CH_4 oxidation (e.g., Popp and Chanton, 1999).

8.08.4.7 Wetland Methane Emissions and Global Change

The influence of global-scale changes in climate, nutrient availability and land use on CH_4 efflux from freshwater wetlands is of much current interest because they are one-third of all CH_4 sources to the atmosphere. Methane emissions from wetlands *per se* are not considered in detail here because the topic falls outside the primary scope of this review and many excellent reviews already exist (Matthews and Fung, 1987; Aselmann and Crutzen, 1989; Bouwman, 1991; Neue, 1993; Bartlett and Harriss, 1993; Bubier and Moore, 1994; Bridgham *et al.*, 1995; Denier van der Gon and Neue, 1995; Vourlitis and Oechel, 1997; Minami and Takata, 1997; Le Mer and Roger, 2001). In a compilation of 48 published studies of wetland CH_4 emissions (Le Mer and Roger, 2001), median rates were 0.43 kg CH_4 ha^{-1} d^{-1} for peatlands, 0.72 kg CH_4 ha^{-1} d^{-1} for other nonagriculture freshwater wetlands, and 1.0 kg CH_4 ha^{-1} d^{-1} for rice paddies. This ranking matches that for primary production and is consistent with the large body of evidence that suggests methanogenesis is closely coupled to the photosynthetic carbon supply (Sections 8.08.4.3.4 and 8.08.4.3.5). In this section, we briefly consider the influence of global change on CH_4 emissions.

Rising atmospheric CO_2 will influence CH_4 emissions from wetlands even in the absence of changes in climate. Temperate and tropical wetland ecosystems, including rice paddies, consistently respond to elevated CO_2 with an increase in photosynthesis and CH_4 emissions (Dacey *et al.*, 1994; Megonigal and Schlesinger, 1997; Ziska *et al.*, 1998), although one exception has been reported (Schrope *et al.*, 1999). One study reported a strong linear relationship ($r^2 = 0.87-0.97$) between photosynthesis and CH_4 emissions, suggesting that the elevated CO_2 response is due to an increase in the supply of labile carbon compounds (Vann and Megonigal, 2003). By comparison, the responses of northern peatland ecosystems have been inconsistent, with several studies reporting negligible effects on photosynthesis, primary production or CH_4 emissions (Berendse *et al.*, 2001; Hutchin *et al.*, 1995; Hoosbeek *et al.*, 2001; Kang *et al.*, 2001; Saarnio *et al.*, 1998, 2000). A likely reason for the absence of a photosynthetic response to elevated CO_2 in these systems is nitrogen limitation. In a study of tussock tundra, Oechel *et al.* (1994) found that the elevated CO_2 treatment did not enhance photosynthesis for more than a few months unless it was crossed with a 4 °C increase in air temperature. They suggested that warming relieved a severe nutrient limitation on photosynthesis by increasing nitrogen mineralization. Thus, the effects of elevated CO_2 alone are expected to interact with nitrogen deposition and temperature. The influence of elevated CO_2 may be particularly important in tropical wetlands in which nitrogen mineralization and methanogenesis are not temperature-limited. These systems account for about 60% of all CH_4 emitted from natural wetlands (Table 5).

Temperature changes are anticipated to be the largest at latitudes above 45° N that hold 57% of all wetland areas and 33% of the global soil carbon pool. Northern hemisphere wetlands contribute about 10% of all CH_4 emissions (Table 5). Tropical areas emit amounts of CH_4 comparable to northern-hemisphere wetlands (Aselmann and Crutzen, 1989; Matthews and Fung, 1987). This region is also expected to warm in the coming decades (Kattenberg *et al.*, 1996), but efforts to quantify the contribution of southern hemisphere peatlands to global carbon storage and fluxes are just beginning (Thompson *et al.*, 2001). Increasing temperature stimulates the decomposition and fermentation of soil organic matter (Section 8.08.4.3.3), and many studies have reported temperature-dependent increases in CH_4 emissions as a result (Bartlett and Harriss, 1993). However, the most important impacts of rising temperature on CH_4 emissions are likely to be realized through changes in the species composition and metabolic activity of plants (Schimel, 1995). A temperature-induced increase in nitrogen mineralization is expected to stimulate plant productivity in northern bogs, and simultaneously lengthen the growing season (Myneni *et al.*, 1997). Verville *et al.* (1998) concluded that plant community composition had a larger effect on CH_4 emissions than direct changes in temperature in wet tussock tundra. A similar conclusion was reached in a study of global change variables in a fen peatland (Granberg *et al.*, 2001). These results are consistent with evidence that the influence of plants on substrate availability (Section 8.08.4.3.5), CH_4 ventilation, and methanotrophy (Section 8.08.4.6.3) vary strongly among plant species.

The effects of global warming on hydrology are relatively uncertain compared to temperature. The long-term effects of a falling water table will be a decrease in CH_4 emissions (e.g., Nykänen *et al.*, 1998) caused by lower CH_4 production, lower CH_4 oxidation or both. Freeman *et al.* (2002) concluded that reduced CH_4 emissions following a simulated drought in a peatland were due mainly to its effect on methanogenesis, and only secondarily to methanotrophy. They proposed that CH_4 diffusion through the unsaturated soil surface was too rapid to permit an increase in CH_4 oxidation efficiency; this is plausible in peat soils because they typically have a low bulk density at the surface, but is unlikely to apply to mineral soil wetlands unless they are on sands. Methane emissions often recover slowly following drought because of reduced methanogen populations and regeneration of alternative electron acceptors.

Human activity has substantially increased the hydrologic import of oxidized nitrogen and sulfur into wetland ecosystems. Emissions of sulfur are expected to double over the next 50 years (Rodhe, 1999), and anthropogenic N_2 fixation will also increase dramatically (Section 8.08.5.1). This raises the possibility that methanogenesis will be suppressed due to substrate competition between methanogens and denitrifiers or SO_4^{2-} reducers. It is well known that methanogenesis is low in SO_4^{2-}-rich marine sediments, and SO_4^{2-} amendments to rice paddies suppress CH_4 production and emissions. However, even the low doses of SO_4^{2-} in acid rain can suppress methanogenesis (Dise and Verry, 2001). Gauci *et al.* (2002) suggested that 1990 levels of SO_4^{2-} deposition depressed CH_4 emissions from northern wetlands by 5–17%. The effects of nitrogen fertilization on CH_4 emissions are variable and depend on the integrated responses of plants, methanogens and methanotrophs, which can offset one another (Bodelier *et al.*, 2000b). Several studies have shown a net increase in CH_4 emissions with nitrogen fertilization (Banik *et al.*, 1996; Lindau *et al.*, 1991).

It is important to test how various factors that force global changes interact because the results may be nonadditive or counterintuitive. Granberg *et al.* (2001) evaluated CH_4 emissions from a fen in response to crossed treatments of elevated air temperature, nitrogen deposition, and sulfur deposition. They found that SO_4^{2-} additions depressed CH_4 emissions at ambient temperature, but not at elevated temperature. The reason for this unexpected result was that pore-water SO_4^{2-} concentrations did not increase at elevated temperature, presumably because of stronger SO_4^{2-} sinks such as plant uptake.

Global changes in climate and nutrient availability are expected to alter the existing source–sink balance of several radiatively active trace gases including CO_2, CH_4, and N_2O. Elevated atmospheric CO_2 and falling water tables have the potential to increase carbon sequestration in biomass if forests replace herbaceous communities (Trettin and Jurgensen, 2003). However, an increase in primary production will translate into an increase in CH_4 emissions due to improved substrate supply (Whiting and Chanton, 2001). Falling water tables will decrease CH_4 emissions, but may increase N_2O emissions (Martikainen *et al.*, 1995). At present, it is uncertain whether feedbacks between climate change and wetland metabolism will exacerbate or mitigate radiative forcing. This can be determined only by accounting for changes in the current fluxes of CO_2, CH_4, and N_2O, allowing for differences in their global warming potential (e.g., Liikanen *et al.*, 2002).

8.08.5 NITROGEN

All forms of anaerobic metabolism are influenced directly or indirectly by O_2, but this is particularly true of nitrogen metabolism. Many of the organisms that anaerobically transform nitrogen perform best under aerobic conditions, resorting to anaerobic metabolism in order to cope with low O_2 availability (Tiedje, 1988). This physiological versatility is favored by the fact that nitrogen oxides can replace O_2 as terminal electron acceptors with only a small loss of energy yield (Table 1), and it is the reason that nitrogen-based anaerobic metabolism is so cosmopolitan. Denitrification occurs in even the driest ecosystems on Earth (Peterjohn, 1991). Whereas most of this review is concerned with saturated soils and sediments as venues for anaerobic metabolism, the section on nitrogen includes work in upland soils. Here we focus on anaerobic metabolism involving nitrogen, but it should be recognized that aerobic metabolism is important to the physiological ecology of many of the organisms concerned.

Research on nitrogen-based anaerobic metabolism has addressed denitrification to the exclusion of several other pathways. Recent advances make it clear that some less-studied pathways make a larger contribution to NO_3^- consumption or N_2 production than previously believed (Zehr and Ward, 2002). Furthermore, completely new metabolic pathways have been discovered in the past decade that will require a re-evaluation of nitrogen sinks in many ecosystems.

Although processes are the focus of this review, there have been significant strides made in quantifying N_2 fixation, denitrification and other nitrogen transformations at large scales (e.g., van Breemen *et al.*, 2002). There exist several recent compilations and reviews of nitrogen oxide emissions (Olivier *et al.*, 1998) and denitrification rates (Barton *et al.*, 1999; Herbert, 1999), including one focused on studies that used the ^{15}N isotope pairing technique (Steingruber *et al.*, 2001).

8.08.5.1 Nitrogen in the Environment

Microbial transformations of nitrogen have a dramatic influence on the structure and function of ecosystems because nitrogen availability often establishes the upper limit of productivity, particularly in terrestrial and marine ecosystems. Widespread nitrogen limitation is somewhat ironic because the atmosphere is a vast reservoir of N_2, yet no eukaryote can use nitrogen in this form. Plants require nitrogen in the form of NO_3^- (nitrate), NH_4^+ (ammonium) or organic-N, and most animals ultimately derive nitrogen from plants (except perhaps rock-eating snails (Jones and Shachak, 1990)!). These compounds have single or double covalent bonds that require far less energy to sever than the triple bonds of N_2, and they are referred to collectively as *fixed*, *combined* or *biologically available* nitrogen. Prokaryotes are unique in their ability to convert N_2 to fixed nitrogen (i.e., perform N_2 fixation), and they participate in nearly every other important nitrogen transformation (Figure 12).

Among several factors that favor persistent nitrogen limitation is the fact that prokaryotes also convert fixed nitrogen back to N_2 in the process of denitrification (Vitousek and Howarth, 1991), a dominant form of anaerobic microbial metabolism. The global inventory of fixed nitrogen is negligible compared to the atmospheric N_2 reservoir, because denitrification has roughly balanced N_2 fixation over eons. Denitrification is also a source of nitrous oxide (N_2O) and nitric oxide (NO), which are far less benign gases. Nitrous oxide is a greenhouse gas that is ~300 times more effective at radiative forcing than CO_2 on a mole basis (Ramaswamy *et al.*, 2001), and accounts for ~6% of the radiative forcing since ca. 1750. Because its lifetime in the atmosphere is very long, N_2O mixes in to the stratosphere where it promotes O_3 destruction. Nitric oxide is an ingredient in smog and promotes tropospheric O_3 production, which is a danger to human health. Emissions of N_2O and NO have increased in the

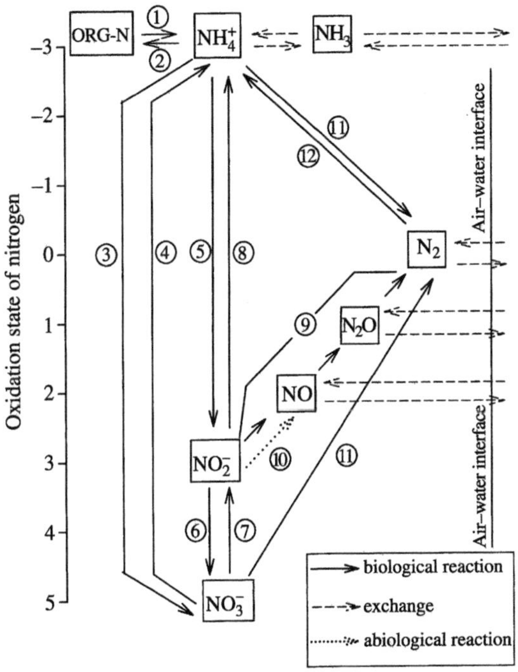

Figure 12 Major reduction–oxidation reactions involving nitrogen. The reactions are numbered as follows: (1) mineralization, (2) ammonium assimilation, (3) nitrification, (4) assimilatory or dissimilatory nitrate reduction, (5) ammonium oxidation, (6) nitrite oxidation, (7) assimilatory or dissimilatory nitrate reduction, (8) assimilatory or dissimilatory nitrite reduction, (9) denitrification, (10) chemodenitrification, (11) anaerobic ammonium oxidation, and (12) dinitrogen fixation (after Capone, 1991) (reproduced by permission of ASM Press from *Microbial Production and Consumption of Greenhouse Gases: Methane, Nitrogen Oxides, and Halomethanes*, **1991**).

past decades from aerobic and anaerobic sources due to the widespread use of fertilizers and fossil fuel combustion.

Human activity has roughly doubled the rate of N_2 fixation (Vitousek *et al.*, 1997), and thereby enhanced microbial nitrogen transformations (Galloway *et al.*, 1995). The enhanced metabolism of N_2-fixing bacteria through cultivation of legumes was an intended outcome of this activity, but many other effects have been unintended. Transport of nitrogen through water, air, and sediments has increased the productivity of adjacent ecosystems, resulting in terrestrial and aquatic eutrophication (Meyer-Reil and Köster, 2000). For example, nitrogen export from the Mississippi River basin has been linked to the expansion of anaerobic sediments in the Gulf of Mexico (Rabalais *et al.*, 2002). Elevated NO_3^- levels are a direct human health threat in newborn infants and adults deficient in glucose–phosphate dehydrogenase (Comly, 1945; Payne, 1981). Denitrification is an effective means of removing fixed nitrogen from sewage effluents and is used widely in wastewater treatment facilities. Wetland and aquatic ecosystems often provide the same service for the cost of protecting these areas as a resource (Bowden, 1987; Groffman, 1994).

8.08.5.2 Nitrogen Fixation

In a sense, all biological N_2 fixation is anaerobic because the nitrogenase enzyme is strongly inhibited by O_2. As a result, many aerobic diazotrophs (N_2-fixing microorganisms) have specialized structures (i.e., heterocysts) for keeping the site of N_2 fixation O_2-free. In the absence of such structures, N_2 fixation varies with O_2 concentrations, often peaking at night when oxygenic photosynthesis is absent (Bebout *et al.*, 1987; Currin *et al.*, 1996).

Dinitrogen fixation is an energetically expensive reaction and is inhibited by micromolar levels of NH_4^+, which requires comparatively little energy to assimilate. The energetic demands of the process is one reason why many diazotrophs are either photosynthetic themselves or occur in symbiotic relationships with plants. However, diazotrophy is certainly not restricted to these groups; it occurs widely among heterotrophic and chemoautotrophic microorganisms, and both aerobes and anaerobes (Lovell *et al.*, 2001).

A provocative link between diazotrophy and methanogenesis was recently proposed by Hoehler *et al.* (2001), who reported that cyanobacteria (formerly blue-green algae) in a coastal mudflat were a "hot spot" of H_2 production, presumably due to a side-reaction of the nitrogenase enzyme system. In some cases, H_2 accumulated underneath the mats to levels that were high enough to support methanogenesis, despite >50 mM SO_4^{2-}. However, the more important link to anaerobic metabolism was the possibility that H_2 production by diazotrophs promoted the oxidation of the primordial Earth. The loss of diazotroph-produced hydrogen to space may have promoted the accumulation of oxidants, thereby initiating the transition from an anaerobic to an aerobic planet before there were significant amounts of atmospheric O_2.

8.08.5.3 Respiratory Denitrification

The term *denitrification* is often used in the general literature to describe any process that converts nitrogen oxides (NO_3^- or NO_2^-) to reduced nitrogen gases (N_2O or N_2). In the context of microbial metabolism the term is used more narrowly to describe a specific respiratory pathway. Denitrification is the most common form of anaerobic respiration based on nitrogen. Energy is conserved by coupling electron transport phosphorylation to the reduction of nitrogen oxides located outside the cell. Because nitrogen is not assimilated into the cell, the process is *dissimilatory*. Respiratory denitrification is more energetically favorable than Fe(III) reduction, SO_4^{2-} reduction or methanogenesis (Table 1), and it tends to be the dominant form of anaerobic carbon metabolism when NO_3^- or NO_2^- are available. Many microorganisms also reduce nitrogen oxides without conserving energy for growth, in which case the process is nonrespiratory. Nonrespiratory nitrogen oxide reduction may be assimilatory (i.e., assimilated into the cell) or dissimilatory. Dissimilatory NO_3^- reduction to NH_4^+ is a potentially important sink for NO_3^- in most ecosystems (Section 8.08.5.4).

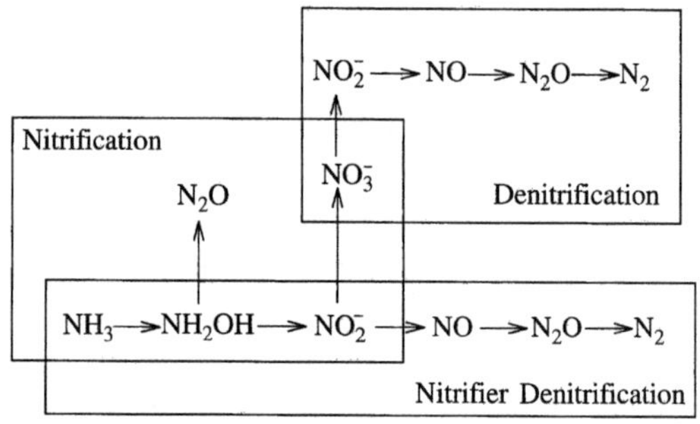

Figure 13 Relationships between three pathways of inorganic nitrogen oxidation and reduction (Wrage *et al.*, 2001) (reproduced by permission of Elsevier from *Soil Biol. Biochem.* **2001**, *33*, 1723–1732).

8.08.5.3.1 Denitrifier diversity and metabolism

Denitrifying bacteria are aerobes that substitute NO_3^- (or NO_2^-) for O_2 as the terminal electron acceptor when there is little or no O_2 available (Payne, 1981; Firestone, 1982). Aerobic respiration yields more free energy than NO_3^- respiration, and it is favored metabolically because O_2 inhibits key denitrification enzymes. Although some organisms can denitrify at O_2 concentrations up to 80% of air saturation (Robertson and Kuenen, 1991), the process is most rapid *in situ* under anaerobic conditions provided there is a supply of NO_3^-. The critical O_2 concentration where denitrifiers switch to mostly anaerobic respiration is roughly $\leq 10\ \mu mol\ L^{-1}$ (Seitzinger, 1988; Tiedje, 1988). Denitrification activity survives well in aerobic soils, even without new enzyme synthesis or cell growth (Smith and Parsons, 1985). It persists in habitats that lack O_2 or NO_3^- presumably because denitrifiers can maintain themselves with a low level of fermentation (Jørgensen and Tiedje, 1993).

Denitrifying bacteria use all three energy sources available to bacteria including organic carbon compounds (organotrophs), inorganic compounds (lithotrophs), and light (phototrophs). The dominant populations of denitrifiers appear to be organotrophs such as *Pseudomonas* and *Alcaligenes*. Others are able to ferment or use H_2, sulfur, or NH_4^+ as energy sources. Some denitrifiers are also N_2 fixers that grow in association with plants, including members of the genus *Rhizobium* (Tiedje, 1988), although it is not clear if many of these strains are capable of growing solely on NO_3^- as an electron acceptor. Many denitrifiers are chemolithotrophs, including the obligate chemolithotroph *Nitrosomonas europaea* (Poth and Focht, 1985; Poth, 1986), which is better known as a *nitrifying* microorganism.

In its most familiar form, denitrification requires organic carbon for the electron donor and NO_3^- or NO_2^- for the terminal electron acceptor. A variety of intermediate compounds are produced as the terminal electron acceptor is reduced stepwise to N_2 (Figure 13), including the solute nitrite (NO_2^-), and the gases nitric oxide (NO) and nitrous oxide (N_2O) (Ye *et al.*, 1994). With a few exceptions, a single organism can complete the entire series of reductions from NO_3^- to N_2 (Tiedje, 1988). Nonrespiratory denitrifiers tend to produce N_2O rather than N_2 (Tiedje, 1988). Because these intermediate compounds are required for metabolism, it is reasonable to assume that denitrifiers can also consume them. Consumption of external NO and N_2O creates many variations on the basic denitrification scheme. Nitrite, NO, and N_2O can serve as terminal electron acceptors rather than intermediates, and they can also replace N_2 as terminal products. It was recently observed that nitrogen dioxide gas (NO_2) is a terminal electron acceptor leading to N_2 production by the nitrifying bacterium *Nitrosomonas eutrophica* (Schmidt and Bock, 1997).

Respiratory denitrification is widespread among various physiologic and taxonomic groups of bacteria (Tiedje, 1988; Zumft, 1997), and may be considered the most versatile form of anaerobic metabolism. The only prokaryotic group that does not include denitrifying members is the *Enterobacteriaceae* (Zumft, 1997). Interestingly, most members of this group are capable of dissimilatory nitrate reduction to ammonium (Tiedje, 1994), a process that competes with denitrification for NO_3^-. With a few exceptions (Zumft, 1997), denitrification is absent among the gram-positive bacteria and obligate anaerobes.

The use of 16S rDNA gene sequences to detect the presence of denitrifying bacteria in microbial communities is limited by the high phylogenetic diversity of this group. Priemé *et al.* (2002) characterized the heterogeneity of gene fragments coding for two types of nitrate reductase (*nirK* and *nirS*) in an upland soil and a wetland soil.

They concluded that both soils, but particularly the wetland soil, had a high richness of *nir* genes, most of which have not yet been found in cultivated denitrifiers. This contrasts with a similar survey by Rösch *et al.* (2002) in which denitrification was not a genetic trait of most of the uncultured bacteria in a hardwood forest soil.

Fusarium oxysporum and a few other fungi have the capacity to denitrify NO_3^- or NO_2^- to N_2O (Shoun and Tanimoto, 1991; Usuda *et al.*, 1995). Their denitrifying enzyme system is coupled to the mitochondrial electron transport chain where it produces ATP (Kobayashi *et al.*, 1996). They do not grow under strictly anaerobic conditions, but require a minimal amount of O_2 (Zhou *et al.*, 2001). *F. oxysporum* can also produce N_2 using NO_2 as an electron acceptor (Tanimoto *et al.*, 1992). Zhou *et al.* (2002) have reported that *F. oxysporum* is capable of at least three types of metabolism depending on O_2 availability. Aerobic respiration is favored when O_2 is abundant, denitrification occurs under microaerobic conditions, and NO_3^- is reduced to NH_4^+ during fermentation under anoxic conditions. There have been virtually no field studies of NO_3-based metabolism by these organisms. However, Laughlin and Stevens (2002) selectively inhibited fungal activity using cycloheximide and observed a decrease in N_2O fluxes of 89% in a grassland soil; the antibiotic streptomycin decreased N_2O flux by 23%. Because fungi represent a large portion of the microbial biomass in soils (Ruzicka *et al.*, 2000), they may be an important source of N_2O in some terrestrial ecosystems.

8.08.5.3.2 Regulation of denitrification

Foremost on the long list of factors that regulate denitrification are those that influence the availability of O_2, NO_3^-, and organic carbon, the primary substrates that are metabolized by denitrifying bacteria. Some factors (e.g., nitrification rate) influence substrate pools directly, while other factors (e.g., soil water content) influence the transfer of substrates to sites where denitrification occurs. These factors have been reviewed previously (Cornwell *et al.*, 1999; Herbert, 1999; Seitzinger, 1988). Although it is convenient to consider the factors that regulate denitrification individually, the ability to predict *in situ* rates requires an understanding of their complex interactions (Parton *et al.*, 1996; Strong and Fillery, 2002).

Molecular oxygen (O_2) is the physiologically preferred terminal electron acceptor for denitrifying bacteria, and its presence represses denitrification enzyme synthesis and activity. Although some microorganisms are capable of performing denitrification in the presence of O_2, it occurs only under anaerobic or microaerobic conditions *in situ*. The presence of water is nearly a prerequisite for denitrification activity because it slows O_2 diffusion by a factor of 10^4 compared to air. In upland soils, abundant air-filled pore spaces permit rapid gas exchange and O_2 concentrations are typically 21% to depths of 1 m or more (Megonigal *et al.*, 1993). Water reduces the O_2 supply by blocking a fraction of the pores, thereby increasing the effective distance (i.e., tortuosity) an O_2 molecule travels to a given microsite (Renault and Sierra, 1994; Figure 14). As the soil water content increases, some sites are blocked completely so that there is no air-filled pathway to the atmosphere. This explains why denitrification rates in soils are often positively related to water content (Drury *et al.*, 1992; Groffman and Tiedje, 1991). For example, denitrification enzyme activity in a temperate hardwood forest was higher in wet years than dry years (Bohlen *et al.*, 2001).

Soil texture is a good predictor of denitrification rates at the landscape scale (Groffman, 1991), in part because it captures the interaction between water content and soil porosity with respect to gas and solute diffusion path length. At a given soil water content, the small pores found in clayey soils are more likely to be blocked than the relatively large pores found in loam and sand soils. Soil texture also influences temporal variability in soil water content because it largely establishes the water infiltration rate and water holding capacity (de Klein and van Logtestijn, 1996; Sexstone *et al.*, 1985). A particularly useful expression for soil water content is *percent water-filled porosity* (Williams *et al.*, 1992), which is the ratio of volumetric water content to total soil porosity. Based on a compilation of denitrification studies in agricultural and forest soils, Barton *et al.* (1999) determined that denitrification increases dramatically above 65% water-filled porosity on average, with higher values for sandy soils (74–83%) than clayey soils (50–74%).

Carbon availability largely controls O_2 consumption rates, either directly by fueling aerobic heterotrophic respiration, or indirectly by supporting the anaerobic production of reductants, such as Fe(II), that subsequently react with O_2. Even small amounts of organic carbon can produce anaerobic conditions in soils that are partially or completely saturated. In upland soils, anaerobic sites occur inside soil aggregates. Højberg *et al.* (1994) used O_2 and N_2O microsensors to demonstrate simultaneous O_2 consumption and denitrification in the surface of soil aggregates (Figure 15). Anaerobic microsites are particularly pronounced near sources of organic carbon such as roots or detritus. For example, Parkin (1987) found that a single leaf constituting 1% of the soil mass supported

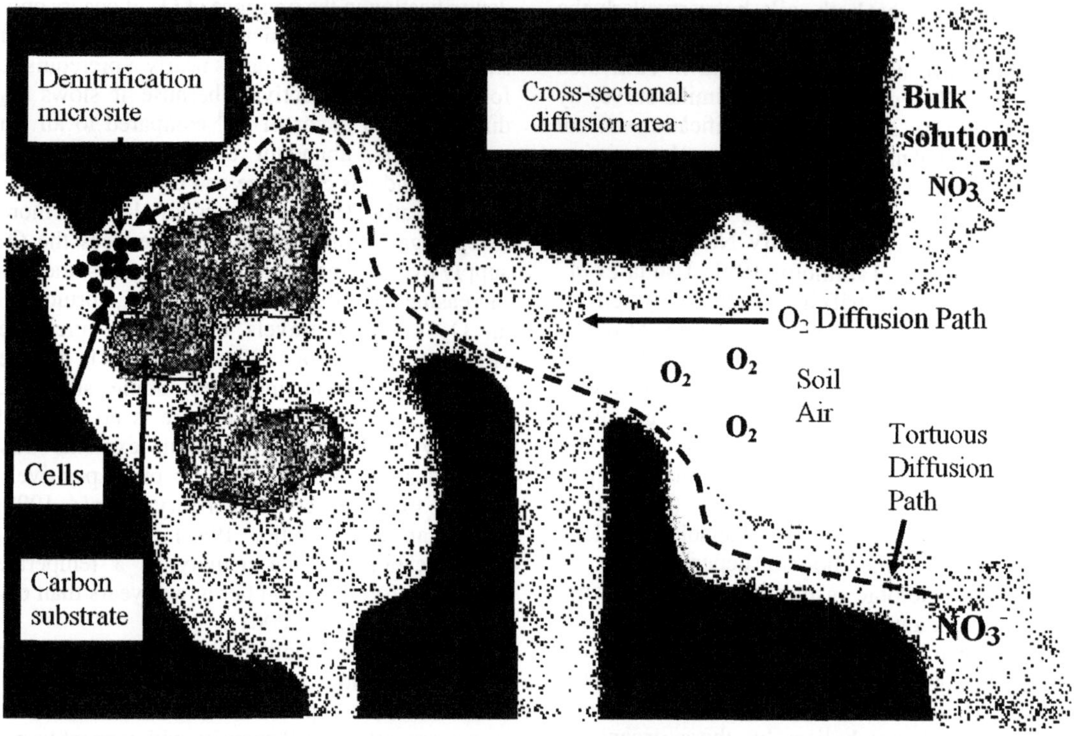

Figure 14 The physical and chemical factors that regulate substrate diffusion to denitrifying bacteria (Strong and Fillery, 2002) (reproduced by permission of Elsevier from *Soil Biol. Biochem.* **2002**, *34*, 945–954).

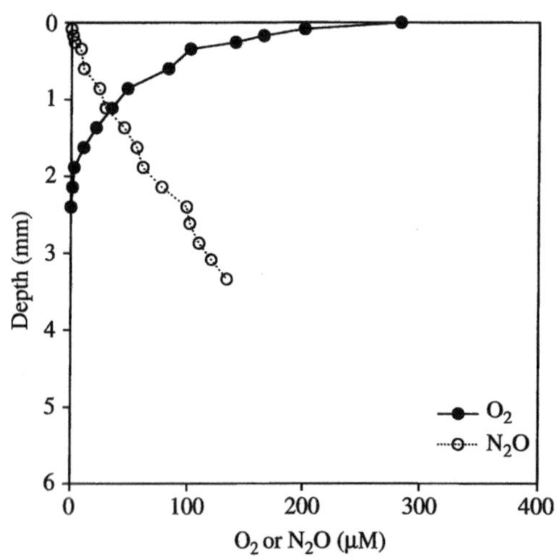

Figure 15 Profiles of O_2 and N_2O with depth from the surface of a soil aggregate. The profiles were determined with microelectrodes (source Højberg *et al.*, 1994).

85% of the denitrification. Such "hot-spots" are one explanation for the notoriously high spatial variability in N_2O production measured in upland soils (Parkin, 1987) and, to a lesser extent, in estuarine sediments (Kana *et al.*, 1998). Although microsites in upland soils are a small fraction of the total soil volume, they contribute perhaps 70% of global N_2O emissions (Conrad, 1996).

Some forms of organic carbon support higher denitrification rates than others due to differences in carbon quality. For example, fresh pine needles supported higher denitrification rates than senescent pine needles in a riparian wetland (Schipper *et al.*, 1994), presumably because the former had more labile compounds such as soluble carbohydrates. van Mooy *et al.* (2002) interpreted depth-dependent changes in the amino acid content of sinking particulate organic carbon in the Pacific Ocean as evidence that denitrifiers preferred to metabolize nitrogen-rich amino acids. If this is the case, calculations of denitrification in the eastern tropical North Pacific Ocean may be 9% higher than previous estimates that were based on typical Redfield ratios.

Denitrification can be limited by carbon availability when O_2 is absent and NO_3^- is abundant. Additions of glucose stimulated denitrification in 11 of 13 agricultural soils that were presumably fertilized (Drury *et al.*, 1991). Similar observations have been made in water columns (Brettar and Rheinheimer, 1992), marine sediments (Slater and Capone, 1987), river sediments (Bradley *et al.*, 1995), aquifers (Smith and Duff, 1988; Obenhuber and Lowrance, 1991), wastewater treatment wetlands (Ingersoll and Baker, 1998), and forested wetlands (DeLaune *et al.*, 1996). Tiedje (1988) proposed that the major influence of carbon on *in situ* denitrification is to promote anaerobic conditions.

Due to a high demand for nitrogen by all organisms, the NO_3^- pool in upland ecosystems is commonly small. In wetland and aquatic ecosystems, NO_3^- availability may be limited further by anaerobic conditions and low nitrification rates. Nitrate addition studies have demonstrated that NO_3^- availability limits denitrification in a variety of aquatic ecosystems (Lohse *et al.*, 1993; Nowicki, 1994). In upland soils, NO_3^- limitation can be an indirect effect of low soil water content, which increases the length of the diffusion pathway between aerobic sites where nitrification takes place and anaerobic microsites (Figure 14). Thus, low soil water content suppresses denitrification in upland soils by simultaneously increasing the O_2 supply and decreasing the NO_3^- supply. Seitzinger (1994) concluded that NO_3^--limitation of denitrification in eight riparian wetlands was an indirect response to limited organic carbon availability. Organic carbon governed mineralization rates and, therefore, the NH_4^+ supply to nitrifying bacteria. The relationship between NO_3^- concentration and denitrification rate often approximates a Michaelis–Menten function. Half-saturation concentrations generally range from 27 μM to 53 μM for stirred marine sediments (Seitzinger, 1988); the values for intact soils can be 30-fold higher because of diffusion limitation (Schipper *et al.*, 1993; Strong and Fillery, 2002).

8.08.5.3.3 Nitrification–denitrification coupling

Microbial metabolism and anthropogenic activity are the primary sources of NO_3^- used by denitrifying bacteria. Nitrate is a waste product of *chemoautotrophic nitrification*, a series of two dissimilatory oxidation reactions, each performed by a distinct group of bacteria. In the first step, NH_4^+ is oxidized to NO_2^- by *Nitrosomonas europaea* and other "Nitroso-" genera including *Nitrosococcus* and *Nitrosospira*. Although $\geq 95\%$ of the total $NH_3 + NH_4^+$ pool is in the form of NH_4^+ at $pH \leq 8$, these organisms are also known as *ammonia oxidizers* because NH_3 is used at the enzymatic level. In the second step, NO_2^- is oxidized to NO_3^- by nitrite-oxidizing bacteria belonging to "Nitro-" genera such as *Nitrobacter* and *Nitrospira*. The two genera constitute the *Nitrobacteriaceae* (Buchanan, 1917). Ammonium is also oxidized by microorganisms that specialize in other types of metabolism. The most important of these are methanotrophic bacteria (Section 8.08.4.6.1), which were estimated to mediate 44–85% of the NH_4^+ oxidation in a rice paddy soil (Bodelier and Frenzel, 1999). In addition, methanotrophic bacteria produce and consume NO (Ren *et al.*, 2000), a common nitrification product (Section 8.08.5.3.5). Heterotrophic oxidation of NH_4^+ or organic-N to NO_2^- and NO_3^- has been observed in fungi, but bacteria are generally considered the dominant *in situ* source of NO_3^- in most ecosystems. The factors regulating nitrification in marine sediments were reviewed by Henriksen and Kemp (1988).

Because nitrification occurs only under aerobic (or microaerobic) conditions and denitrification under anaerobic conditions, the two processes are spatially separated. However, if the sites where these processes occur are sufficiently close together, NO_3^- transport and consumption are very rapid and the overall process is considered to be *coupled nitrification-denitrification*. On the basis of a literature review, Seitzinger (1988) concluded that nitrification is generally the major source of NO_3^- for denitrification in river, lake, and coastal sediments. The same is likely to be true of nonagricultural soils that are largely dependent on mineralization and atmospheric deposition for fixed nitrogen.

Denitrification can be uncoupled from nitrification when NO_3^- is delivered to soils and sediments from outside sources such as overlying water, fertilizers, and atmospheric deposition. Fertilizers are a source of NO_3^- to groundwater (Spalding and Exner, 1993), which in turn is a source of NO_3^- to riparian forests (Hill, 1996), tidal marshes (Tobias *et al.*, 2001a), and estuarine sediments (Capone and Bautista, 1985). Surface water runoff carries NO_3^- directly to rivers, lakes, estuaries and oceans (Jordan and Weller, 1996; Joye and Paerl, 1993). Because of high NO_3^- loading, denitrification rates in estuarine sediments are often directly proportional to the NO_3^- concentration in the overlying water column (Kana *et al.*, 1998; Nielsen *et al.*, 1995: Pelegrí *et al.*, 1994), indicating that denitrification is controlled by the NO_3^- diffusion rate across the aerobic sediment surface.

The response of denitrification to certain environmental factors varies according to how strongly the process is coupled to nitrification. An increase in O_2 penetration into sediments can stimulate coupled nitrification–denitrification by enhancing NO_3^- availability (Rysgaard *et al.*, 1994), but simultaneously depress water column-supported denitrification because of the increased distance that NO_3^- must diffuse to reach the anaerobic layer (Cornwell *et al.*, 1999). Risgaard-Petersen *et al.* (1994) reported that a photosynthesis-induced increase in O_2 penetration into an estuarine sediment doubled the coupled nitrification–denitrification rate, but reduced the amount of water column-supported denitrification by 50%. Because the contribution of coupled nitrification–denitrification to total denitrification at this site was small (~16%), the net effect of deeper O_2 penetration was to reduce denitrification. An increase in O_2 penetration in North Sea sediments had an entirely different outcome (i.e., higher overall denitrification) because the sediment, rather than the water column, was the dominant NO_3^- source supporting denitrification (Lohse *et al.*, 1993).

8.08.5.3.4 Animals and plants

Animals and plants influence denitrification in wetland and aquatic ecosystems by altering the availability of O_2, NO_3^-, and carbon. In organic-rich sites, nitrification rates tend to be limited by the depth of O_2 penetration (Kemp *et al.*, 1990). Benthic invertebrates enhance pools of O_2 and NO_3^- by digging burrows and irrigating them with overlying water (Pelegrí *et al.*, 1994). Macrofaunal tubes and burrow walls support higher potential rates of nitrification than nearby surface sediments, and contribute >25% of benthic nitrification is some coastal ecosystems (Blackburn and Henriksen, 1983; Kristensen *et al.*, 1985). The degree to which infauna enhance potential nitrification varies among worm species due to variations in irrigation behavior (Mayer *et al.*, 1995). Nitrification and denitrification are coupled in burrow wall sediments just as they are in surface sediments (Sayama and Kurihara, 1983). Animals also enhance denitrification by concentrating organic matter into faecal pellets, thereby creating "hot-spots" of O_2 demand.

Aquatic plants influence sediments in much the same manner as animals. Reddy *et al.* (1989) applied $^{15}N\text{-}NH_4^+$ to the saturated root zone of three emergent aquatic macrophytes and recovered ~25% of the label as $^{15}N_2$ over 18 d. Since $^{15}N_2$ was not recovered from the unplanted controls, it was apparent that coupled nitrification–denitrification was stimulated by the rhizosphere. Similar observations have been reported for submerged aquatic plants (Caffrey and Kemp, 1991; Christensen and Sørensen, 1986) and microalgae (Law *et al.*, 1993). Risgaard-Petersen and Jensen (1997) reported that denitrification was enhanced in the presence of a submerged aquatic macrophyte by six-fold. This impressive effect was due in part to the fact that O_2 was released well below the sediment surface, creating a zone of nitrification capped above and below by zones of denitrification. The result was a highly efficient coupling between the two processes. The influence of plants on anaerobic metabolism is apparently species- or site-specific because studies of other marine angiosperms have found no effects on denitrification (Rysgaard *et al.*, 1996). In addition to introducing O_2, plants introduce organic carbon in the form of root exudates (Lynch and Whipps, 1990) and detritus, and compete with microorganisms for NH_4^+ (Lin *et al.*, 2002). In upland systems, plants are more likely to have an overall negative effect on denitrification due to competition for NH_4^+ and NO_3^-, and consumption of water (Tiedje, 1988).

8.08.5.3.5 N_2O and NO fluxes

Classical coupled nitrification–denitrification requires NH_4^+, organic carbon, aerobic conditions, and anaerobic conditions. It involves three distinct populations of microorganisms, some of which are heterotrophic and others autotrophic. As a result, regulation of the process is rather complex and the relative proportions of NO_3^-, NO, N_2O, and N_2 as end products varies widely with environmental and ecological conditions.

Several factors can cause the reduction of NO_3^- to N_2 to be incomplete, resulting in the pooling of NO and N_2O. Denitrification and nitrification enzymes are inhibited by H_2S (Joye and Hollibaugh, 1995) and O_2. The sensitivity of denitrifying enzymes to O_2 inhibition is inversely proportional to the oxidation state of the nitrogen substrate (Figure 12), increasing in the order: NO_3^- reductase < NO_2^- reductase < NO reductase < N_2O reductase (Dendooven and Anderson, 1994; McKenney *et al.*, 1994). Pooling can be caused by an excess of NO_3^- relative to organic carbon, or an imbalance in the kinetics of the various steps. Some organisms lack key enzymes in the sequence and release NO or N_2O as waste products (Tiedje, 1982). Reduction to N_2 appears to be favored over N_2O at circumneutral pH (Šimek *et al.*, 2002).

Nitrogenous gas emission from soils varies strongly with soil water content (Williams *et al.*, 1992). The water content at which efflux from soils peaks generally increases in the order: NO > N_2O > N_2. Yang and Meixner (1997) reported that NO fluxes peaked at ~20% water-filled pore space in a grassland soil, which is in qualitative agreement with other studies (Otter *et al.*, 1999; Potter *et al.*, 1996). Nitrous oxide emissions peak at higher levels of soil moisture than NO emissions (Drury *et al.*, 1992), and N_2 emissions are highest in saturated soils. This pattern is caused by differences in the O_2 sensitivity of denitrifying enzymes, and by the influence of soil water content on gas diffusion. High soil water content restricts the diffusion of gases and enhances the diffusion of solutes (Section 8.08.5.3.2). Because nitrifying bacteria require both a gas (O_2) and a solute (NH_4^+), the optimal availability of substrates occurs where soils are wet, but not saturated (Williams *et al.*, 1992). Such conditions favor N_2O production because nitrification and denitrification can be simultaneously producing N_2O (Stevens *et al.*, 1997). The influence of soil water on NO and N_2O

emissions is complicated by that fact that these gases are also consumed by microorganisms. Water promotes microbial consumption of NO and N_2O by restricting diffusion to the atmosphere, thus increasing their residence time in the soil (Davidson, 1992; Skiba *et al.*, 1997).

It is difficult to distinguish between nitrification and denitrification as sources of NO and N_2O. Selective inhibition of nitrification by 10 Pa C_2H_2 (acetylene) and denitrification by 10 kPa C_2H_2 (Davidson *et al.*, 1986) has widely been used, but it suffers from a number of problems. These include the possibility of C_2H_2-insensitive denitrifiers (Dalsgaard and Bak, 1992), and NO scavenging by O_2 (Bollmann and Conrad, 1997; McKenney *et al.*, 1997). Techniques based on ^{15}N or ^{13}N tracing present different problems (Boast *et al.*, 1988; Arah, 1997), but are favored in more recent studies. The contribution of nitrification to NO and N_2O production is highest under the conditions that favor nitrification, namely, moderate O_2 partial pressure and high NH_4^+ concentrations (Avrahami *et al.*, 2002). Nitrification is generally considered to be the dominant source of NO upland in soils, but there are studies that suggest the opposite (reviewed by Ludwig *et al.*, 2001). The dominant source may vary with environmental factors such as pH (Remde and Conrad, 1991).

Nitric oxide and N_2O are also produced abiotically by chemical decomposition of NO_2^- (Hooper and Terry, 1979). The reaction is favorable at low pH and yields mainly NO (van Cleemput and Baert, 1984), although N_2O can also be produced (Martikainen and De Boer, 1993). Abiotic production of NO and N_2O is assumed to be relatively unimportant in most ecosystems (e.g., Webster and Hopkins, 1996).

8.08.5.4 Dissimilatory Nitrate Reduction to Ammonium (DNRA)

Nitrate reduction studies have focused overwhelmingly on denitrification at the expense of other NO_3^- sinks such as dissimulatory NO_3^- reduction to NH_4^+ (also known as DNRA or *nitrate ammonification*). The ecological implications of reducing NO_3^- to NH_4^+, versus N_2 are vastly different because NH_4^+ is more readily retained in the ecosystem, and it is a form that is readily assimilated by biota. Thus, DNRA contributes to eutrophication by reducing the quantity of fixed nitrogen that is returned to the atmosphere as N_2.

There are two pathways by which NO_3^- can be reduced to NH_4^+. In assimilatory NO_3^- reduction, NO_3^--N is taken up by microorganisms or plants, reduced to NH_4^+, and assimilated into organic nitrogen compounds such as amino acids. The process is inhibited by low concentrations ($0.1\ \mu L\ L^{-1}$) of NH_4^+ or organic nitrogen (Rice and Tiedje, 1989), and not regulated by O_2. These factors have the opposite effect on dissimilatory NO_3^- reduction to NH_4^+. Assimilatory NO_3^- reduction by microorganisms is usually assumed to be negligible due to inhibition by NH_4^+.

8.08.5.4.1 Physiology and diversity of DNRA bacteria

Dissimilatory nitrate reduction to ammonium is an anaerobic pathway that is insensitive to NH_4^+ and yields energy. The first step of the process is termed *nitrate respiration* because it is coupled to electron transport phosphorylation that generates ATP:

$$NO_3^- + H_2 \rightarrow NO_2^- + H_2O \qquad (13)$$

Nitrate respiration is widely found among microorganisms, some of which further reduce NO_2^- to NH_4^+:

$$NO_2^- + 3H_2 + 2H^+ \rightarrow NH_4^+ + 2H_2O \qquad (14)$$

This second step is not coupled to energy production except in a few species that are not expected to be abundant *in situ*, such as *Campylobacter*, *Deulfovibrio*, and *Wolinella* (Tiedje, 1988). The overall DNRA reaction transfers $8e^-$ and yields 600 kJ mol^{-1} NO_3^- (Tiedje, 1994):

$$NO_3^- + 4H_2 + 2H^+ \rightarrow NH_4^+ + 3H_2O \qquad (15)$$

The second step in which NO_2^- is reduced to NH_4^+ is ecologically significant because it ensures that the nitrogen can be retained in the ecosystem for plant uptake, microbial assimilation or adsorption to cation exchange sites; otherwise, NO_2^- would accumulate and be subjected to denitrification. Because NO_2^- reduction to NH_4^+ does not normally yield energy, it presumably has other physiological or ecological advantages. Perhaps its most likely role is to serve as an electron sink for the regeneration of NADH from $NADH_2$ (Bonin, 1996).

DNRA bacteria can be aerobic, facultatively anaerobic or obligately anaerobic. As a group they are unlike the denitrifying bacteria in that most species are fermentative (Bonin, 1996; Tiedje, 1988). DNRA bacteria are abundant in aerobic soils and other environments that do not favor DNRA activity *per se*. This suggests that they can compete with other fermentative bacteria or aerobes for carbon substrates (Tiedje, 1988). Much less is understood about the diversity and physiology of DNRA bacteria than denitrifiers, despite the fact that they are sometimes the larger of the two dissimilatory NO_3^- sinks (Table 7).

Table 7 Examples of the contribution of dissimilatory nitrate reduction to ammonium (DNRA) to total dissimulatory nitrate reduction in various freshwater, marine, and terrestrial ecosystems. The ranges are inclusive of all sites and treatments reported for which appropriate data could be drawn, including those that manipulated substrate levels.

Ecosystem	*%DNRA*[a]	*Method*[b]	*Citation*
Marine sediment	79–93	ABT + estimate of DNRA	Bonin (1996)
Marine sediment	18–97	^{15}N assay + ABT	Bonin *et al.* (1998)
Marine sediment	98	^{15}N assay + ABT	Gilbert *et al.* (1997)
Marine sediment	33	^{15}N assay	Goeyens *et al.* (1987)
Marine sediment	46–68	^{15}N assay + ABT	Sørensen (1978)
Marine sediment	0–18	^{15}N assay + ABT	Kaspar *et al.* (1985)
Estuarine sediment	5–30	^{15}N assay + ABT	Kaspar (1983)
Estuarine sediment	0–85	^{15}N assay	Christensen *et al.* (2000)
Estuarine sediment	73–82	ABT + estimate of DNRA	Jørgensen and Sørensen (1985)
Estuarine sediment	66–75	ABT + estimate of DNRA	Jørgensen and Sørensen (1988)
Estuarine sediment	6–75	^{15}N assay + ABT	Jørgensen (1989)
Estuarine sediment	10–61	^{15}N assay	Koike and Hattori (1978)
Estuarine sediment	2–3	^{15}N assay	Pelegrí *et al.* (1994)
Mangrove soil	3–100	^{15}N assay	Riviera-Monroy (1995)
Brackish marsh soil	7–70	^{15}N assay	Tobias *et al.* (2001a)
Brackish marsh soil	1–7	^{15}N assay	Tobias *et al.* (2001b)
River sediment[c]	6–10	^{15}N assay	Kelso *et al.* (1997)
Aquifer sediment	22–60	^{15}N assay	Bengtsson and Annadotter (1989)
Rice paddy soil	1–56	^{15}N assay	Buresh and Patrick (1978)
Rice paddy soils	4–12	^{15}N assay	Yin *et al.* (2002)
Boreal forest	40–93	^{15}N assay + ABT	Bengtsson and Bergwall (2000)
Calcareous clay soil	4–19	^{15}N assay	Fazzolari *et al.* (1998)
Clay loam soil	1–2	^{15}N assay	Chen *et al.* (1995)
Silt loam soil	1–77	^{15}N assay + ABT	deCatanzaro *et al.* (1987)
Silt loam soil[c]	11–30	^{15}N assay	Stanford *et al.* (1975)
Wet tropical forest	71–83	^{15}N assay	Silver *et al.* (2001)

[a] Percent DNRA calculated as (NO_3^- reduction to NH_4)/(NO_3^- reduction to $NH_4 + N_2 + N_2O$). [b] ABT is the acetylene block technique.
[c] Used last reported incubation time point.

8.08.5.4.2 DNRA versus denitrification

DNRA and denitrification occur simultaneously and can be expected to compete for carbon and NO_3^-. Tiedje *et al.* (1982) proposed that the partitioning of NO_3^- to N_2 versus NH_4^+ is a function of the carbon : NO_3^- ratio. They reasoned that a combination of abundant electron donors (i.e., carbon) and limited electron acceptors (i.e., NO_3^-) should favor organisms that use electron acceptors most efficiently. In this case, DNRA has a competitive advantage because it transfers 8 moles of electrons per mole of NO_3^- reduced, while denitrification transfers five moles of electrons. Nitrite respiration alone transfers two moles of electrons. Thus, denitrifying bacteria should out compete DNRA bacteria for organic carbon when carbon is limiting. Several studies support the generalization that high labile carbon availability and/or low NO_3^- availability favor DNRA over denitrification (Bonin, 1996; Fazzolari *et al.*, 1998; King and Nedwell, 1985; Nijburg *et al.*, 1997; Yin *et al.*, 2002). In wetlands, the presence of emergent aquatic macrophytes appeared to stimulate DRNA activity (Nijburg *et al.*, 1997; Nijburg and Laanbroek, 1997), perhaps because of increased carbon availability due to root exudates. Nonetheless, competition between the two groups of organisms has not been adequately studied and there are exceptions to this generalization that suggest the influence of other factors.

A series of laboratory and field studies showed that low temperatures favor denitrifying bacteria while high temperatures favor DNRA bacteria in temperate salt marsh and estuarine sediments of the Colne estuary, UK (King and Nedwell, 1984; Ogilvie *et al.*, 1997a,b). It is not yet clear whether this effect can be generalized to other systems. Field estimates of NO_3^- partitioning to the two processes in a North Sea estuary were best explained when it was assumed that DNRA activity was favored by *both* high and low temperatures (Kelly-Gerreyn *et al.*, 2001). Temperature effects may explain why the relative importance of DNRA versus denitrification decreased from May to October in a temperate tidal marsh despite a three-fold increase in dissolved carbon concentrations (Tobias *et al.*, 2001a).

DNRA activity has been assumed to be important only under highly reduced conditions

($E_h = -200$ mV; Buresh and Patrick, 1981). However, there is growing evidence that DNRA can be important in relatively oxidized environments. DNRA accounted for ~75% of NO_3^- reduction in a humid tropical rainforest where the soil O_2 content was 15% (Silver *et al.*, 2001). DNRA activity was stimulated by the amphipod *Corophium volutator* (Pelegrí *et al.*, 1994), an organism that increases redox potential through burrowing. Fazzolari *et al.* (1998) suggested that DNRA bacteria are less O_2 sensitive than denitrifying bacteria. Clearly, highly reduced conditions are not a prerequisite for DNRA to be an important NO_3^- sink.

DNRA has an abiotic equivalent reaction that can proceed at rates comparable to the biotic reaction in the presence of "green rust" (Hansen *et al.*, 1996) or trace metals such as Cu(II) (Ottley *et al.*, 1997). Green rusts are Fe(II)–Fe(III) precipitates that form in nonacid, Fe(II)-rich soils and sediments (Hansen *et al.*, 1994). A similar reaction may have made a significant contribution to the NH_4^+ inventory on the prebiotic Earth (Summers and Chang, 1993).

In the past, DNRA has been considered to be inconsequential in nonmarine ecosystems (Tiedje, 1988). There is now evidence that the process can be an important, and even dominant, NO_3^- sink in a wide variety of freshwater aquatic and terrestrial systems (Table 7). Strong relationships between DRNA activity, carbon : NO_3^- ratio, O_2 and temperature provide a basis for formulating hypotheses on the range of conditions where DRNA activity should be important. Many aspects of the microbial ecology of DNRA bacteria are ripe for study, such as the factors that determine the outcome of competition with denitrifying bacteria. It is likely that some of the NO_3^- consumption attributed to denitrifying bacteria is actually due to NO_3^--respiring bacteria that do not express DNRA activity. Because the two groups differ in key aspects of their physiology, this difference in NO_3^- reduction pathway could have implications for understanding NO_3^--dependent anaerobic carbon metabolism.

8.08.5.5 Alternative Pathways to N_2 Production

Alternative pathways for denitrification have been proposed that do not involve the classical enzyme systems, but their existence has only recently been confirmed or investigated in detail. In the absence of O_2, a number of elements and compounds have the potential to oxidize NH_4^+, including Mn(II), MnO_2, NO_3^-, NO_2^- and NO_2. Because anaerobic oxidation of NH_4^+ directly to N_2 bypasses the multiple oxidation steps required for coupled nitrification–denitrification (Figure 12), the substrate and environmental controls on anaerobic NH_4^+ oxidation are likely to be quite different from those for the classical sequence of reactions. The contribution of these processes to overall denitrification is largely unknown, but the limited evidence suggests they may be quantitatively important in both natural ecosystems and wastewater treatment processes.

8.08.5.5.1 Anammox

Ammonium oxidation linked to NO_2^- (nitrite) reduction was first recognized in a wastewater treatment system and patented under the process name *anammox* for *ana*erobic *ammo*nium *ox*idation (Mulder *et al.*, 1995). The discovery was motivated by observations of simultaneous NH_4^+ and NO_3^- losses balanced by N_2 production under anaerobic conditions. The existence of a "missing" chemolithotrophic organism capable of coupling NH_4^+ oxidation to either NO_3^- or NO_2^- reduction had been predicted nearly two decades earlier based on thermodynamic considerations (Broda, 1977). The anammox reaction was initially proposed to involve NO_3^- (Mulder *et al.*, 1995):

$$5NH_4^+ + 3NO_3^- \rightarrow 4N_2 + 9H_2O + 2H^+ \quad (16)$$

However, it was subsequently determined that NO_2^- is the oxidant in anammox (van de Graaf *et al.*, 1995):

$$NO_2^- + NH_4^+ \rightarrow N_2 + 2H_2O \quad (17)$$

Reactions (16) and (17) were distinguished using an ^{15}N isotope pairing technique in which the NH_4^+ pool was enriched with ^{15}N while the NO_3^- and NO_2^- pools remained dominated by ^{14}N. With the NH_4^+ pool dominated by ^{15}N, reaction (16) yields 75% $^{29}N_2$ (i.e., $^{14}N-^{15}N$) and 25% $^{30}N_2$, whereas reaction (17) yields 100% $^{29}N_2$ (van de Graaf *et al.*, 1995, 1997). The anammox process also produces small amounts of NO_3^- that are thought to provide reducing equivalents for CO_2 assimilation (Jetten *et al.*, 1999).

The microorganisms responsible for anaerobic NH_4^+ oxidation have proven extremely difficult to isolate and no pure cultures exist. However, two organisms have been enriched from wastewater treatment plants that perform the anammox reaction (Egli *et al.*, 2001; Strous *et al.*, 1999a; Toh *et al.*, 2002), and aspects of their physiology have been described (Strous *et al.*, 1999b). Both the archetype strain "*Candidatus* Brocadia anammoxidans" and its relative "*Candidatus* Kuenenia stuttgartiensis" are deeply branching members of the order *Planctomycetales*, a major division of the Bacteria. Members of this division share several unusual features such as internal compartmentalization similar to the Eukaryotes

(Lindsay *et al.*, 2001). To add further intrigue to these organisms, the NH_4^+ oxidizing members of the order have a specialized compartment, the anammoxosome, composed of "ladderane" lipids that have never before been observed in nature (Sinninghe Damsté *et al.*, 2002). The anammoxosome lipid membrane is highly impermeable and may be specialized to protect the cell from toxic intermediates generated by the anammox process. These include hydrazine (N_2H_4), an active ingredient in rocket fuel (DeLong, 2002), and hydroxalamine (NH_2OH). The organisms are chemolithoautotrophic and grow exceptionally slowly in enrichment culture, dividing once every nine days under optimal conditions (Egli *et al.*, 2001; Strous *et al.*, 1999a). The function of a highly impermeable anammoxosome membrane may be to maintain an electrochemical gradient in spite of low rates of catabolism (Sinninghe Damsté *et al.*, 2002). Anammox bacteria are reversibly inhibited by O_2 concentrations as low as 2 μM and NO_2^- concentrations between 5 mM and 10 mM (Jetten *et al.*, 1999).

The nature of interactions between anammox bacteria and other microorganisms in the nitrogen cycle is a matter of speculation (Schmidt *et al.*, 2002). If anammox bacteria consume NO_2^- supplied directly by autotrophic NH_3-oxidizers, they are probably competing with NO_2^--oxidizers for a limited substrate (i.e., NO_2^-). In this case, N_2 production is bypassing the NO_3^- link between nitrifiers and denitrifiers in the classical nitrogen cycle (Zehr and Ward, 2002). It is clear that NO_3^--N is readily consumed in the anammox process (Mulder *et al.*, 1995; Dalsgaard and Thamdrup, 2002), although it is first reduced to NO_2^-. Nitrification is presumably the NO_3^- source in this case, and anammox is coupled to nitrification in a manner analogous to coupled nitrification–denitrification. A key difference between anammox and denitrification is that anammox produces twice the amount of N_2 per mole of NO_3^-. Thus, anammox may enhance N_2 production in soils and sediments where the supply of oxidized nitrogen substrates is limited by nitrification rates.

Few studies have quantified the contribution of anammox to nitrogen cycling, and most of these were done in wastewater reactors under conditions that do not exist in most ecosystems (e.g., Dong and Tollner, 2003). The contribution of anaerobic NH_4^+ oxidation to overall N_2 production in natural systems is largely unknown aside from a pair of studies in marine sediments (but see very recent contributions from Dalsgaard *et al.* (2003) and Kuypers *et al.* (2003). In one study, sediments were taken from three sites across the Baltic–North Sea transition that differed in sediment organic carbon content (Thamdrup and Dalsgaard, 2002). Isotope pairing showed that up to 67% of total N_2 production was due to anaerobic NH_4^+ oxidation. Moreover, a 1:1 stoichiometry between consumption of $^{15}NH_4^+$ and $^{15}NO_3^-$, and nearly 100% production of $^{29}N_2$, suggested that it was an anammox-like process. A subsequent study confirmed that that anaerobic NH_4^+ oxidation required NO_2^- rather than NO_3^- (Dalsgaard and Thamdrup, 2002).

In the Baltic–North Sea transition study of Thamdrup and Dalsgaard (2002), anammox contributed 33–67% of total N_2 production at continental shelf sites, but just 2% of total N_2 production at the site in a coastal bay (Figure 16). The difference between the bay site and the closest continental shelf site was due primarily to an increase in the denitrification rate, rather than a decrease in anaerobic oxidation. Denitrification declined in parallel with declining sediment organic carbon content from the coastal bay seaward. Absolute rates of anaerobic NH_4^+ oxidation were in the range 30–99 μM d^{-1}, and were lowest at the site where the percent contribution was highest. This pattern is consistent with evidence that anammox organisms are autotrophic, whereas denitrifiers are heterotrophic.

Anaerobic NH_4^+ oxidation may be most significant in ecosystems where denitrification is limited by carbon availability rather than NH_4^+, NO_3^-, or the presence of O_2. This may include large areas of the continental shelves and slopes (Seitzinger and Giblin, 1996), O_2-deficient waters such as the Black Sea, and eutrophic soils and sediments. Anammox-specific 16S rRNA gene sequences (Schmidt *et al.*, 2002) and lipids (Sinninghe Damsté *et al.*, 2002), and ^{15}N isotope pairing provide a suite of powerful techniques for surveying ecosystems for anammox bacteria and anaerobic NH_4^+ oxidation activity.

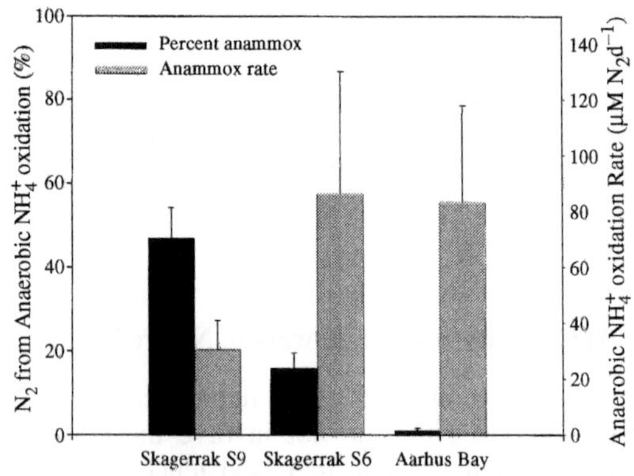

Figure 16 Summary of experiments from three sites on the continental shelf that quantified the absolute rates of anaerobic ammonium oxidation (i.e., anammox) and its contribution to total N_2 production (i.e., anammox + denitrification). The sites constitute a transect with Aarhus Bay closest to shore and Skagerrak S9 furthest from shore. (after Dalsgaard and Thamdrup, 2002).

8.08.5.5.2 *Nitrifier denitrification*

NH_3-oxidizing bacteria are typically characterized as a NO_2^- source for other bacteria that ultimately produce NO_3^- (Henriksen and Kemp, 1988; Schlesinger, 1997). However, it is important to recognize the metabolic versatility of NH_3-oxidizers (i.e., nitrifiers) and to consider the other roles they may play in nitrogen cycling. Some nitrifying bacteria produce N_2 from NH_4^+ using O_2 or NO_2 (nitrogen dioxide) as oxidants. The process has been named *nitrifier denitrification* to indicate that it involves autotrophic NH_3-oxidizers with an enzyme system similar to that of the heterotrophic denitrifying bacteria (Poth and Focht, 1985; Wrage *et al.*, 2001). In fact, there may be an evolutionary linkage between denitrifying NH_3-oxidizers and denitrifiers, as suggested by a high degree of similarity in their nitrite reductase gene sequences (Casciotti and Ward, 2001). During nitrifier denitrification, NH_3 is oxidized to NO_2^-, as in typical nitrification, then reduced to NO, N_2O, or N_2 (Figure 13). Thus, the process couples NH_3 oxidation and denitrification within a single organism, such as *Nitrosomonas europaea* or *N. eutropha* (Ritchie and Nicholas, 1972; Bock *et al.*, 1995).

The most recent metabolic pathway described in *Nitrosomonas* is anaerobic NH_3 oxidation in which O_2 is replaced by nitrogen dioxide (NO_2) or its dimeric form, N_2O_4:

$$NH_3 + N_2O_4 \rightarrow NO_2^- + 2NO + 3H^+ \quad (18)$$

This pathway has been described in *N. eutropha* and a few other related species (Schmidt and Bock, 1997; Schmidt *et al.*, 2002). It is superficially similar to aerobic NH_3 oxidation in that hydroxylamine is an intermediate, NO_2^- is a product, and a portion of the NO_2^- may be reduced further N_2:

$$NO_2^- + 4H^+ \rightarrow 0.5N_2 + 2H_2O \quad (19)$$

Schmidt and Bock (1997) demonstrated that anaerobic nitrifier denitrification supports growth.

The anaerobic and aerobic nitrifier denitrification pathways differ in that NO is an end product under anaerobic conditions rather than an intermediate compound. In addition, nitrogen dioxide-dependent NH_3 oxidation by *N. eutropha* does not require ammonium monooxygenase (Schmidt *et al.*, 2002), demonstrating that the two pathways are enzymatically different. In the absence of NH_3, *N. eutropha* can use H_2 or simple organic compounds as electron donors (Abeliovich and Vonhak, 1992; Bock *et al.*, 1995). In contrast to the anammox process, which is strictly anaerobic, O_2 does not inhibit NO_2-dependent NH_3 oxidation and N_2 production can occur even under aerobic conditions (Zart and Bock, 1998). However, Shrestha *et al.* (2002) observed N_2 production only under conditions of low O_2 concentration or anaerobiosis in *N. europaea*.

Very few studies have attempted to establish whether these alternative metabolic pathways in NH_3-oxidizing bacteria contribute to N_2O or N_2 production *in situ*. Webster and Hopkins (1996) estimated that nitrifier denitrification contributed 29% of the N_2O produced in a dry soil and 3% in a wet soil, but an earlier study concluded that their contribution to N_2O production was negligible (Robertson and Tiedje, 1987). In a review of the topic, Wrage *et al.* (2001) proposed the highest contributions from nitrifying denitrifiers are likely to occur under conditions of low carbon and nitrogen content. The contribution of nitrifier denitrification to anaerobic metabolism and N_2 production will ultimately be limited by the low availability of NO_2 (nitrogen dioxide) in anoxic soils and sediments.

Autotrophic microorganisms are of interest in wastewater treatment because they can denitrify without an organic carbon supplement (Jetten *et al.*, 1999; Verstraete and Philips, 1998). The microbial ecology of these artificial systems is better understood than natural systems at present. The anammox bacterium *B. anammoxidans* and the nitrifier *N. eutropha* were able to coexist in a laboratory-scale reactor (Schmidt *et al.*, 2002). Indeed, the specific anaerobic NH_3 oxidation activity of *N. eutropha* was 10 times higher than in co-culture than pure culture. Rather than the two groups producing N_2 simultaneously, their metabolism is more likely to be coupled in a nitrification–denitrification reaction that bypasses nitrite-oxidizing bacteria (Schmidt *et al.*, 2002). Wastewater reactors that favor the anammox process contain aerobic and anaerobic NH_3-oxidizers, but not nitrite oxidizers such as *Nitrobacter*. This suggests that the nitrite oxidizers cannot compete effectively with the NH_3 oxidizers for O_2, nor can anaerobic NH_4^+ oxidizers compete with NH_3 oxidizers for NO_2^- (Schmidt *et al.*, 2002). The discovery of anammox-like metabolism in marine sediments (Thamdrup and Dalsgaard, 2002) raises many ecological questions about the nature of interactions between these groups of organisms in natural ecosystems.

8.08.5.5.3 *Abiotic and autotrophic denitrification*

Denitrification can be supported by electron donors other than organic carbon such as Fe(II), Mn(II) and H_2S, and they can proceed by both abiotic and biotic pathways. Abiotic reduction of NO_3^- to N_2 coupled to Fe(II) oxidation may occur at low rates in the pH range 2–7 (Postma, 1990):

$$10Fe^{2+} + 2NO_3^- + 14H_2O \rightarrow 10FeOOH + N_2 + 18H^+ \quad (20)$$

The reaction is quite sensitive to temperature (Sørensen and Thorling, 1991), and is catalyzed by freshly precipitated Fe(III) oxides and Cu^{2+} (Buresh and Moraghan, 1976; Postma, 1990). In contrast to the abiotic reaction, chemolithoautotrophic denitrification coupled to NO_3^- proceeds rapidly under the low temperature, circumneutral pH conditions that are typical of the Earth's surface (Weber *et al.*, 2001) (Section 8.08.6.5.2).

Two pathways have been proposed that form N_2 from abiotic reactions between fixed nitrogen and mangenese, and both are thermodynamically feasible in typical marine pore water. Reduction of NO_3^- by Mn^{2+} (i.e., dissolved Mn(II)) was proposed by Aller (1990):

$$2NO_3^- + 5Mn^{2+} + 4H_2O \rightarrow N_2 + 5MnO_2 + 8H^+ \quad (21)$$

Because Mn^{2+} reacts rapidly with O_2, reaction (21) may be restricted to anoxic zones in manganese-rich sediments. Although this reaction is similar to a microbially mediated reaction involving Fe(II) (reaction (26)), there is no evidence at present to suggest it is performed biotically. Luther *et al.* (1997) proposed a mechanism for producing N_2 from NH_4^+ that should be favored in aerobic zones where MnO_2 is rapidly regenerated from Mn^{2+}:

$$2NH_3 + 3MnO_2 + 6H^+ \rightarrow N_2 + 3Mn^{2+} + 6H_2O \quad (22)$$

The reaction was demonstrated in anaerobic marine sediments amended with freshly synthesized MnO_2 and NH_4^+ (Luther *et al.*, 1997). However, it could not be detected with a sensitive ^{15}N-labeling technique in an unamended, manganese-rich sediment (Thamdrup and Dalsgaard, 2000). The reaction may be favorable only with certain forms of MnO_2 (Hulth *et al.*, 1999) or high levels of NH_4^+.

Chemoautotrophic denitrification coupled to H_2S, S^0, or $S_2O_3^{2-}$ occurs in some bacteria of the genus *Thiobacillus* such as *T. denitrificans* (Hoor, 1981; Section 8.08.7.9). Nitrate reduction by such a pathway increased with FeS additions and followed Michaelis–Menten kinetics in a marine sediment (Garcia-Gil and Golterman, 1993).

8.08.6 IRON AND MANGANESE

Iron and manganese oxides are the most abundant components of Earth's surface that can serve as anaerobic terminal electron acceptors in microbial metabolism, yet it was recognized only recently that microorganisms play a key role their cycling. Despite early reports that suggested biological Fe(III) reduction was important in wet rice paddy soils (Kamura *et al.*, 1963; Takai *et al.*, 1963a,b), the process was thought to be either unimportant or dominated by abiotic mechanisms (Lovley, 2000c). Similarly, biological Fe(II) oxidation was assumed to be negligible except under acidic (pH ≤ 3) conditions. It is now clear that microorganisms mediate both Fe(III) reduction and Fe(II) oxidation across a broad range of pH, temperature, and salinity conditions.

Dissimilatory Fe(III) reduction is the dominant form of anaerobic carbon metabolism in many ecosystems, and may be one of the earliest forms of microbial metabolism to evolve on Earth or elsewhere (Lovley, 2000c). In contrast to metabolism based on oxygen, nitrogen, or carbon terminal electron acceptors, metabolisms involving iron, manganese, and sulfur (Section 8.08.7) are regulated strongly by both geochemical and biochemical phenomena. Geochemistry is fundamental to understanding the use of metals in anaerobic metabolism, and several of the most intriguing questions about iron and manganese bacteria concern their interactions with mineral surfaces or their influence on geochemistry. Previous reviews have addressed many of the salient interactions between iron cycling and other biogeochemical processes (Figure 17).

Because Mn(IV) reduction generally makes a small contribution to carbon metabolism, it is not considered in detail in this review. However, there are many similarities between the cycles of the two metals that we allude to in the text. In fact, many Fe(III)-reducing bacteria are more aptly described as metal-reducing bacteria because they also reduce Mn(IV) and a variety of other metals. A recent and excellent review of both Fe(III) and Mn(IV) cycling was provided by Thamdrup (2000), and Tebo *et al.* (1997) reviewed microbial Mn(IV) oxidation.

8.08.6.1 Iron and Manganese in the Environment

Fe(III) and Mn(IV) respiration influences the cycling of many other elements that are of concern to environmental scientists. Fe(III)- and Mn(IV)-reducers interfere with the metabolism of SO_4^{2-}-reducers and methanogens by competing for organic carbon. Because anaerobic metabolism is often organic carbon limited and Fe(III) reduction yields more free energy than methanogenesis, Fe(III)-reducing bacteria suppress freshwater emissions of CH_4, an important greenhouse gas (Section 8.08.4). Roden and Wetzel (1996) concluded that CH_4 emissions from a freshwater wetland were reduced by 70% due to Fe(III) cycling.

Some Fe(III)-reducing bacteria oxidize organic pollutants (Anderson *et al.*, 1998; Heider *et al.*, 1999; Lovley, 2000a; Lovley and Anderson, 2000; Gibson and Harwood, 2002).

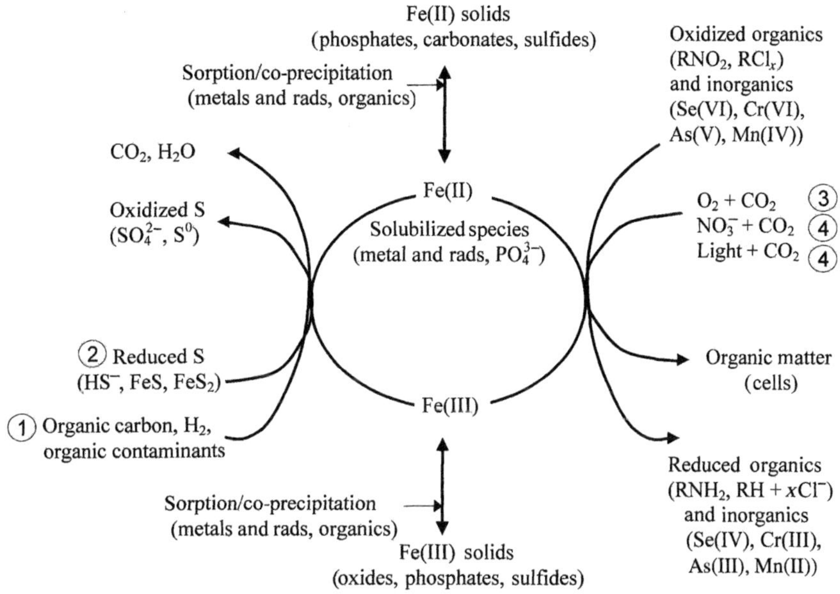

Figure 17 Substrates and processes coupled to Fe reduction–oxidation. Circled numbers refer to recent reviews of the role of microbial processes in various iron transformations. (1) Lovley and Anderson (2000), Thamdrup (2000), Johnson (1998), Blake and Johnson (2000), Küsel *et al.* (1999), and Peine *et al.* (2000). (2) Thamdrup *et al.*, (1993), Lovely and Phillips (1994a, Schipper and Jørgensen (2002), Blake and Johnson (2000), Pronk and Johnson (1992). (3) Emerson (2000), Johnson (1998), Blake and Johnson (2000), Edwards *et al.* (2000b), and Roden *et al.* (in press). (4) Straub *et al.* (2001). (after Tebo and He, 1999; Roden *et al.*, in press).

Benzene degradation is stimulated in the laboratory by substances that enhance Fe(III) reduction (Lovley *et al.*, 1994b, 1996b), and benzene-degrading sediments are enriched in *Geobacter* species (Anderson *et al.*, 1998; Rooney-Varga *et al.*, 1999), the only genus of microorganism known to couple Fe(III) reduction to the oxidation of aromatic compounds. Because petroleum-contaminated sediments are often anaerobic, Fe(III)-reducing bacteria are attractive candidates for bioremediation.

Some Fe(III)-reducing organisms, particularly members of the *Geobacteraceae*, have the ability to reduce heavy metal pollutants such as U(VI) (Holmes *et al.*, 2002). Because reduction changes uranium from a soluble to an insoluble form, such organisms may be used to remove uranium from polluted waters (Lovley, 1997; Lovley and Phillips, 1992a). Fe(III)-reducing bacteria can reduce a long list of metals including gold, silver, chromium, cobalt, selenium, and technetium (Lovley, 1997).

Iron and manganese have a strong influence on the availability of trace metal pollutants through precipitation–dissolution reactions (Burdige, 1993; Cornell and Schwertmann, 1996). Trace metals form surface complexes or co-precipitate with Fe(III) and Mn(IV) oxides, and they are released upon Fe(III) and Mn(IV) reduction (Zachara *et al.*, 2001). For example, processes that oxidize Fe(II) retain arsenic in sediments (Raven *et al.*, 1998; Senn and Hemond, 2002), and vice versa (Cummings *et al.*, 1999; Harvey *et al.*, 2002; Zachara *et al.*, 2001). Fe(III) oxides complex and co-precipitate phosphorus (Gunnars *et al.*, 2002; Bjerrum and Canfield, 2002), perhaps the biosphere's ultimate limiting nutrient (Tyrrell, 1999), and P is released upon Fe(III) reduction (Smolders *et al.*, 2001).

8.08.6.2 Iron and Manganese Geochemistry

Iron and manganese total 5% of the continental crust, with iron contributing 98% and manganese the remainder (Weaver and Tarney, 1984). They are subject to rapid changes in redox state mediated by both geochemical and biological processes. Iron atoms in near-surface Earth environments cycle between an oxidized or *ferric* state, Fe(III), and a reduced or *ferrous* state, Fe(II). Manganese exists in three redox states: Mn(II), Mn(III), and Mn(IV). In this review, the abbreviation *Mn(IV)* is understood to represent Mn(IV) and Mn(III).

Oxidized iron in equilibrium with even the most unstable of iron minerals has a concentration of about 10^{-8} M in seawater at pH 8 (Stumm and Morgan, 1981), and oxidized forms of manganese are nearly insoluble at neutral pH. Due to their poor solubility, these elements are conserved in soils and sediments derived from rock weathering. The solubility of Fe(III) and Mn(IV) is greatly enhanced when complexed by ligands (Lovley and Woodward, 1996; Luther *et al.*, 1996; Stone, 1997). Contrary to the assumption that most dissolved iron is Fe(II), pore-water concentrations of Fe(III) can be comparable to Fe(II) (Ratering and Schnell, 2000), presumably due to organic complexation (Luther *et al.*, 1996; Tallefert *et al.*, 2000). Thus, iron cycling is partially mediated by the production and degradation of organic chelators.

Iron and manganese solubility increases dramatically upon reduction, yet most of the Fe(II) and Mn(II) in sediments at any given moment is not in a soluble form (Heron and Christensen, 1995). Dissolved forms of both metals sorb strongly onto cation exchange sites. Fe(II) precipitates as a variety of secondary minerals, the most common of which are iron monosulfide (FeS), pyrite (FeS_2), siderite ($FeCO_3$), vivianite ($Fe_3(PO_4)_2$), and magnetite ($Fe^{II}Fe_2^{III}O_4$), a magnetic, mixed valance mineral. Sulfide minerals dominate estuarine and marine sediments (Section 8.08.7), and are stable in neutral to weakly acidic habitats in the absence of O_2. Siderite formation is favored in water with carbonate alkalinity >1 mM (King, 1998), and it is the most abundant form of Fe(II) in sediments dominated by Fe(III) reduction or methanogenesis (Coleman *et al.*, 1993). These minerals are also produced by Fe(III)-reducing bacteria during ferrihydrite reduction (Zachara *et al.*, 2002). In contrast to Fe(II), Mn(II) is quite stable in the presence of O_2 at circumneutral pH (<8) and therefore tends to accumulate in oxic environments. Mn(II) may precipitate as $MnCO_3$ (van Cappellen *et al.*, 1998), but most is either adsorbed, dissolved or organically complexed (Burdige, 1993).

About 35% of the iron and 75% of the manganese in soils and sediments is in the form of free oxides (Canfield, 1997; Cornell and Schwertmann, 1996; Thamdrup, 2000). The remainder occurs as a minor constituent of silicate minerals. The lattice structure of Fe(III) oxide minerals varies widely. Freshly oxidized Fe(III) precipitates rapidly as ferrihydrite ($Fe(OH)_3$), a reddish-brown, amorphous, poorly crystalline mineral. Ferrihydrite is the dominant product of Fe(II) oxidation whether it occurs by abiotic oxidation, aerobic microbial oxidation, or anaerobic microbial oxidation (Straub *et al.*, 1998). Over a period of weeks to months, amorphous ferrihydrite crystals undergo diagenesis to yield well-ordered, strongly crystalline, stable minerals such as hematite(α-Fe_2O_3) and goethite (α-FeOOH) (Cornell and Schwertmann, 1996).

Fe(III)-reducing bacteria growing on ferrihydrite can produce extracellular fine-grained magnetite (Fe_3O_4) (Lovley *et al.*, 1987; Fredrickson *et al.*, 1998). Recently, it was discovered that *Shewanella oneidensis* (formerly *S. putrefaciens*) produced and deposited an unidentified iron mineral intracellularly (Glasauer *et al.*, 2002). Intracellular deposits of magnetite were previously known only from magnetotactic bacteria and a few higher organisms (Bazylinski and Moskowitz, 1997).

8.08.6.3 Microbial Reduction of Iron and Manganese

Microorganisms that reduce extracellular Fe(III) or Mn(III, IV) to support metabolism or growth (i.e., dissimilatory Fe(III)- or Mn (IV)-reducing bacteria) can be classified into two broad physiological categories. The most important of these for anaerobic carbon metabolism are the organisms that use metals as their primary terminal electron acceptor for the partial or complete oxidation of organic compounds or H_2, thereby conserving energy via Fe(III) or Mn(III, IV) respiration. The second group are fermenting bacteria that channel a small portion (<5%) of their organic-carbon-derived electron transport to Fe(III) or Mn(IV) reduction, thereby using metals as nonrespiratory electron sinks (Lovley, 1987, 1997). Most work on microbial metal reduction has focused on iron because it is abundant and interacts strongly with other elements, whether in an oxidized or reduced state. However, it is recognized that metal-reducing microorganisms are often capable of using several alternative elements as terminal electron acceptors, and likewise, a given element can be reduced by a wide variety of microorganisms.

8.08.6.3.1 Metabolic diversity in Fe(III)- and Mn(IV)-reducing organisms

The capacity for dissimilatory Fe(III) and Mn(IV) reduction is widely distributed among the subdivisions of the *Bacteria*, including all subclasses of the *Proteobacteria* (Coates *et al.*, 2001) and some hyperthermophiles. Most of the known species of Fe(III)-respiring bacteria are in the delta subclass of the *Proteobacteria*. This subclass includes the *Geobacteraceae* family, which is populated entirely by Fe(III)-reducing bacteria, including the well-studied species *Geobacter metallireducens* (Lovley *et al.*, 1993a). Metal reduction has also been reported in hyperthermophiles in the *Archea* (Vargas *et al.*, 1998). More than 40 isolates that couple anaerobic growth to dissimilatory Fe(III) respiration have been characterized (Coates *et al.*, 2001).

The energy sources used by Fe(III)- and Mn(IV)-respiring organisms include H_2 and acetate, which are the same primary substrates that support methanogenesis and SO_4^{2-} reduction. Unlike the methanogens, some metal-respiring microorganisms also oxidize multicarbon organic compounds such as lactate, pyruvate, long-chain fatty acids, aromatic compounds, amino acids and glucose (Küsel *et al.*, 1999; Lovley, 2000b). These compounds can be completely oxidized to CO_2, or they can be fermented to acetate. Some fermentative organisms cannot grow solely on Fe(III), but reduce small quantities of Fe(III) as a nonrespiratory electron sink (Lovley, 1987, 2000b).

Complete oxidation of organic carbon to CO_2 is found in three of the four genera in the *Geobacteraceae* family (Lovley, 2000c), and phylogenetically distinct genera such as *Geovibrio* and *Deferribacter* (Greene *et al.*, 1997). The gamma and epsilon subclasses of the *Proteobacteria* tend to oxidize a more limited range of organic acids, and often carbon oxidation is incomplete (Lovley, 2000b). A well-studied member of this group is *Shewanella oneidensis* (formerly *S. putrefaciens*), one of the first described nonfermentative Fe(III)-reducers (Myers and Nealson, 1988; Nealson and Myers, 1992), and notable because it is also a facultative anaerobe (Kieft *et al.*, 1999). The ability to use H_2 for an energy source is common among incompletely oxidizing Fe(III)-reducers, and hyperthermophilic Fe(III)-reducers in the *Archea* and *Bacteria* (Slobodkin *et al.*, 2001; Vargas *et al.*, 1998). Relatively few Fe(III)-reducing bacteria outside these groups are known to oxidize H_2 (Coates *et al.*, 1999). However, some species of *Geobacter* can use both H_2 and acetate (Caccavo *et al.*, 1994, 1996).

Most Fe(III)-reducing organisms have the capacity to reduce Mn(IV) and at least one other common oxidant such as O_2, NO_3^-, SO_4^{2-}, elemental sulfur (S^0), or humic substances (Thamdrup, 2000; Senko and Stolz, 2001). In contrast to the small penalty in free energy that occurs when denitrifying bacteria switch from O_2 to NO_3^--respiration, growth on Fe(III) by the facultative anaerobe *S. oneidensis* is far inferior to that on O_2 (Kostka *et al.*, 2002a). Humic substances may prove to be the most common alternative electron acceptor for metal-reducing bacteria, particularly in freshwater environments where S^0 is not abundant (Lovley *et al.*, 1996a, 1998). There are several SO_4^{2-}-reducing bacteria that can also reduce Fe(III) and U(VI), although they generally cannot support growth on transition metals alone (Lovley *et al.*, 1993b). In fact, *Desulfovibrio desulfuricans* will simultaneously reduce SO_4^{2-} and Fe(III), or SO_4^{2-} and U(VI), if H_2 is available (Lovley and Phillips, 1992b; Coleman *et al.*, 1993).

Relatively little is known about the abundance of Fe(III)-reducing species *in situ*. Attempts to quantify dissimilatory Fe(III)-reducing bacteria in aquifer sediments suggest that they are dominated by the genera *Geobacter* or *Geothrix* (Anderson *et al.*, 1998; Rooney-Varga *et al.*, 1999; Snoeyenbos-West *et al.*, 2000). More ecosystems will need to be surveyed before generalizations are made about the dominant groups elsewhere (e.g., Rosselló-Mora *et al.*, 1995).

8.08.6.3.2 Physical mechanisms for accessing oxides

A fundamental difference between Fe(III) and Mn(IV) reduction and the other major pathways of anaerobic metabolism is that the oxidants are sparingly soluble solids. In order to access the terminal electron acceptor a microorganism must either: (i) make physical contact with an oxide surface, (ii) reduce an intermediate compound that can shuttle electrons from the cell to a distant oxide surface, or (iii) use the metal in a dissolved or chelated form. All three adaptations are observed among Fe(III)- and Mn(IV)-respiring organisms. Metal-respiring organisms do not necessarily require direct physical contact with oxides as some early studies suggested (Lovley and Phillips, 1986; Tugel *et al.*, 1986), but such contact may be common. Evidence of this is the observation that Fe(III) reductase activity tends to be localized in the outer membrane of both *G. sulfurreducens* and *S. oneidensis* (Beliaev and Saffarini, 1998; Gaspard *et al.*, 1998; C. R. Myers and J. M. Myers, 1992). Using atomic force microscopy, Lower *et al.* (2001) measured the adhesion strength of *S. oneidensis* cells to mineral surfaces and found that it increased rapidly upon exposure to goethite under anaerobic conditions. There was some evidence for a mobile iron reductase in the outer cell membrane that facilitated electron transport from the cell to the oxide surface. Grantham *et al.* (1997) reported that sites where *S. oneidensis* was observed adhering to Fe(III) oxide surfaces corresponded to pits caused by Fe(III) dissolution. The shape of the pits suggested that the organisms were more mobile under anaerobic conditions than aerobic conditions, an adaptation that may enhance their access to fresh Fe(III) oxide surfaces. A similar adaptation was observed in *G. metallireducens* which expressed flagella and was chemotaxic toward Fe(II) and Mn(II) when grown on insoluble oxides, but not when grown on soluble Fe(III) or Mn(IV) (Childers *et al.*, 2002). This adaptation had gone unnoticed because most culture work had previously been based on complexed, soluble Fe(III) media, highlighting the importance of culturing metal-reducing organisms on substrates that are common *in situ* such as ferrihydrite or goethite. More species must be tested to determine if chemotaxis is unique to

Fe(III)-reducing organisms that do not produce soluble electron shuttling compounds.

8.08.6.4 Factors that Regulate Fe(III) Reduction

Because Fe(III) respiration is the most recent major pathway of anaerobic metabolism to be investigated, there has been relatively little work on the processes that regulate it *in situ*. Rather, most advances have concerned the mechanisms by which microorganisms access Fe(III) oxides and interact with the physicochemical processes that govern Fe(III) oxide dissolution.

8.08.6.4.1 Electron shuttling compounds

Humic substances are effective electron shuttling compounds that are ubiquitous in soils and sediments (Stevenson, 1994). The redox-active moieties in these compounds are primarily quinones and related aromatic reductants (Lovley and Blunt-Harris, 1999; Scott *et al.*, 1998; Stone *et al.*, 1994). Because purified humic acids are expensive and it is difficult to evaluate their redox state, the synthetic quinone 2,6-anthraquinone disulphonate (AQDS) has been used as a model compound for studying the potential for humic acids to serve as terminal electron acceptors (Lovley *et al.*, 1996a). AQDS may not be an appropriate substitute for native humic acids in studies of reaction kinetics (Nurmi and Tratnyek, 2002), but it has proved to be useful for isolating humic-reducing microorganisms from a variety of ecosystems including lakes, wetlands and estuaries (Coates *et al.*, 1998). It appears that most, if not all, Fe(III)-reducing bacteria can obtain energy by the reduction of AQDS or humic substances (Lovley *et al.*, 2000). The strains tested to date include members of the *Geobacteraceae* family and hyperthermophiles. The metabolism is present in fermentative and nonfermentative organisms, and may be particularly widespread among thermophilic organisms (Lovley, 2000c).

There are several reasons why the ability to reduce humic substances could be an advantage to metal-respiring microorganisms. Higher rates of reduction should be possible with humic acids and other soluble oxidants than with solid-phase oxidants that require the organism to continually establish physical contact in order to access fresh oxide surfaces. Humic substances are ubiquitous and accumulate to levels that could make them significant alternative electron acceptors in some ecosystems. As with other anaerobic electron acceptors, the capacity to support respiration is enhanced by processes that regenerate humic substances in an oxidized form. Reduced humics and hydroquinones (i.e., reduced quinones) are rapidly oxidized by Fe(III) and Mn(IV), making them available again for microbial respiration. Some humic substances may be more efficient than microbial cells at accessing physical locations on metal oxides because they are relatively small (Zachara *et al.*, 1998). Finally, because enzymatically reduced humics may pass electrons to Fe(III), Mn(IV), U(VI), or other oxidants (Stone, 1987), an organism can indirectly access several potential "secondary" electron acceptors simultaneously rather than just one (Figure 18). The ability to shuttle electrons between microorganisms and inorganic oxidants explains why even low concentrations of AQDS (5 μmol kg^{-1}) and humics can stimulate Fe(III) reduction in sediments (Nevin and Lovley, 2000).

An actual contribution of humic substances to metal oxide reduction in natural systems has not been demonstrated, and there are processes such as adsorption or decomposition that could limit their effectiveness. Kostka *et al.* (2002a) observed that AQDS additions elicited a larger increase in Fe(III) reduction by *S. oneidensis* growing on ferrihydrite than smectite clay minerals. This suggests that the influence of humic substances may depend on soil or sediment mineralogy. Nevertheless, there is ample evidence to suggest that a portion of the anaerobic metabolism that was previously attributed to direct enzymatic Fe(III) and Mn(IV) reduction was actually nonenzymatic reduction by microbially reduced humic substances.

Variability in the effectiveness of humic substances as electron shuttling compounds is expected due to differences in their chemical structure. For example, the electron-accepting capacity of humic substances from three distinct sources varied nearly 700-fold in the following order: soils > sediments > dissolved river-borne (Scott *et al.*, 1998; Figure 19). Microorganisms are a source of quinone moities and other electron shuttling compounds. For example, *S. oneidensis*

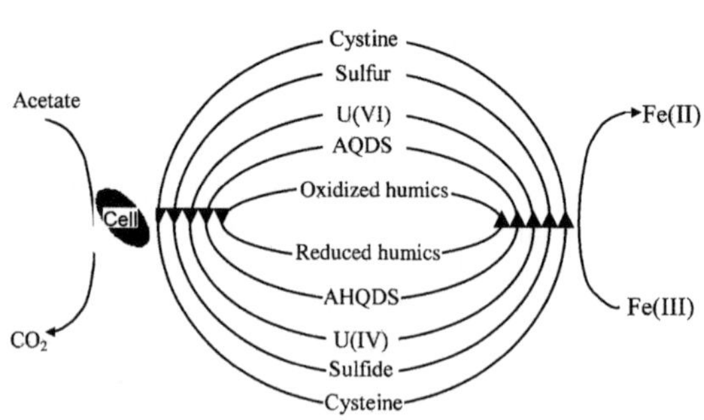

Figure 18 The variety of electron shuttles that promote Fe(III) reduction. (after Nevin and Lovley, 2000).

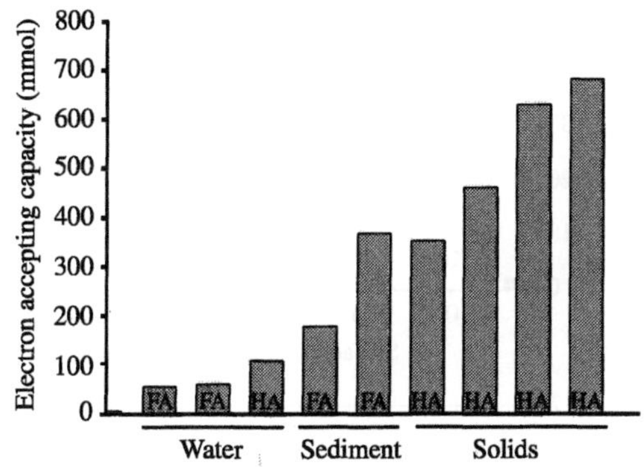

Figure 19 Ecosystem-level differences in the ability of humic substances to accept electrons. Samples were taken from the water column of freshwater lakes and rivers, aquatic sediments and soils (FA = fulvic acid, HA = humic acid) (after Scott *et al.*, 1998).

excretes an unidentified quinone-based compound that can reduce ferrihydrite and MnO_2 (Newman and Kolter, 2000), and *G. sulfurreducens* excretes a *c*-type cytochrome with similar capability (Seeliger *et al.*, 1998). These studies raise the prospect that some organisms can synthesize their own electron shuttle compounds if the supply of oxidants is limiting.

8.08.6.4.2 Fe(III) chelators

There are a variety of humic and nonhumic organic chelating agents that enhance the dissolution and solubility of metals (Stone, 1997; Luther *et al.*, 1992). Some forms of chelated Fe(III) are more rapidly reduced by bacteria than insoluble Fe(III) oxides (Lovley, 1991). Amending aquifer sediments with the synthetic chelator nitrilotriacetic acid (NTA) increased dissolved Fe(III) and stimulated the enzymatic reduction of Fe(III)-oxides (Lovley and Woodward, 1996). The increase in Fe(III) reduction was attributed to enhanced availability of dissolved Fe(III) rather than a stimulation of growth caused by trace elements or metabolizable organic carbon. However, the effect could also be explained by NTA complexation of Fe(II), which would prevent Fe(II)-poisoning of the mineral surface (see next paragraph). Other complexing agents such as hydroxamate have no effect on Fe(III) dissolution rates (Holmén *et al.*, 1999). As with humic acids, there is very little evidence that chelators substantially influence microbial iron metabolism *in situ*, and there are reasons that model chelators such as NTA are not good surrogates for those in natural systems (Straub *et al.*, 2001). The stability of ligand-Fe(III) complexes influence Fe(III) reduction rates because the cell must be able to out compete the chelator for Fe(III). In a study with *S. oneidensis*, the stability of soluble ligand-Fe(III) complexes was inversely related to Fe(III) reduction rates (Haas and DiChristina, 2002). Of the chelators tested, the highest Fe(III) reduction rates were produced by citrate, which is a weakly complexing ligand similar to most naturally occurring organic chelators.

Organic ligands also form complexes with Fe(II) (Luther *et al.*, 1996) and may enhance Fe(III) dissolution by removing the Fe(II) that coats and passively blocks access to fresh Fe(III) oxides surfaces. Urrutia *et al.* (1999) considered this mechanism and concluded that it was more important than Fe(III) complexation. Bacterial cells adsorbed to mineral surfaces also inhibit Fe(III) reduction rates (Urrutia *et al.*, 1999).

8.08.6.4.3 Mineral reactivity

Metal oxides in soils and sediments are mixtures of poorly crystalline, strongly crystalline and silicate-bound minerals. Differences in the intrinsic reactivity of these broad categories is sufficient to explain much of the variation in Fe(II) and Mn(II) reduction rates observed in nature, which spans several orders of magnitude. Mixtures of Fe(III) oxides are typically separated by exposing them to a sequence of chemical extractions involving increasingly strong solutions of acids, ligands, and reductants. For example, the poorly crystalline fraction can be extracted with a combination of aerobic and anaerobic oxalate solutions (Thamdrup and Canfield, 1996) or a dilute solution of HCl (0.5 M), while dithionate–citrate–bicarbonate is used to extract the strongly crystalline fraction. A limitation of such techniques is that the fractions are operationally defined and not mineral specific. Efforts have been made to calibrate wet chemical extraction protocols against minerals of known composition (Canfield, 1989; Haese *et al.*, 1997; Kostka and Luther, 1994). Nevertheless, no extraction technique is entirely discriminating for a specific metal phase or mineral (Zachara *et al.*, 2002). Recognizing the wide range of reactivities represented in a single operationally defined Fe(III) oxide pool, Postma (1993) proposed characterizing reactivity with a single continuous extraction. The "reactivity continuum" method allows straightforward comparisons of the bulk reactivity of pure and mixed iron (hydr)oxide pools (Figure 20).

The susceptibility of metal oxides to reduction and dissolution depends on mineralogy, crystallinity, surface area, the effectiveness of reducing and chelating agents, and microbial activity. Early culture studies with Fe(III)-respiring bacteria demonstrated that Fe(III) reduction rates vary with mineral form or crystallinity

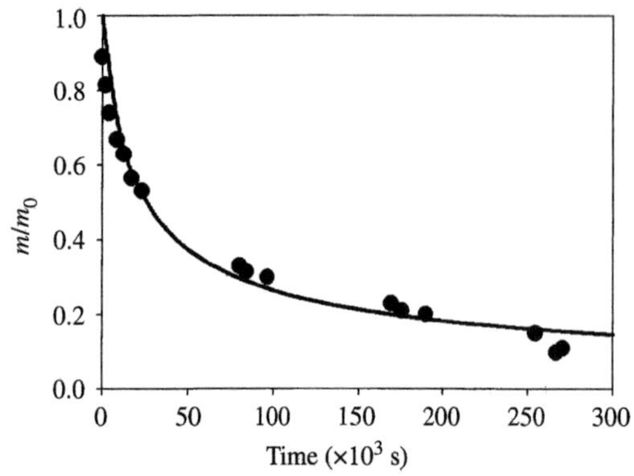

Figure 20 Application of the reactivity continuum method to Fe(III) oxide pools in oxidized aquifer sediments. The ratio m/m_0 is the fraction of undissolved Fe-oxide. The slope of the curve decreases with time because reactive minerals are increasingly depleted leaving behind relatively more recalcitrant forms (source Postma, 1993).

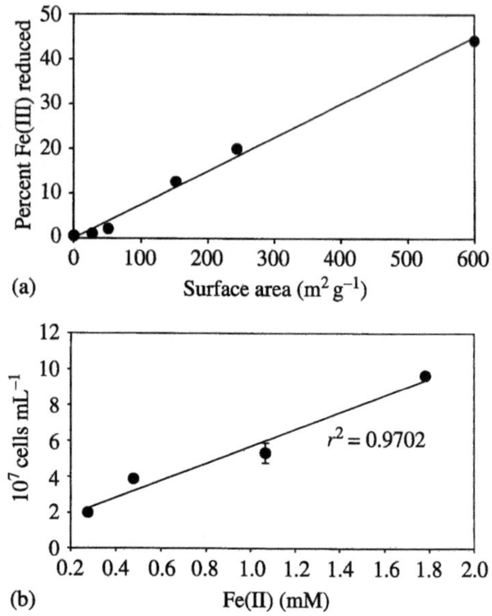

Figure 21 Relationship of Fe(III)-reducing bacteria activity and growth to oxide surface area. (a) Percent Fe(III) reduced as a function of oxide surface area. Surface area corresponded to different mineral types and included hematite, goethite, and ferrihydrite. (b) The density of *Shewanella oneidensis* cells as a function of the amount of structural Fe(III) reduction to Fe(II) in smectite clay, a strongly crystalline, high-surface-area Fe mineral. Differences in Fe(II) content reflect different amounts of clay particles inoculated into a minimal basal media (after Roden and Zachara, 1996 and Kostka *et al.*, 2002a, respectively).

(Lovley and Phillips, 1986). Minerals such as ferrihydrite and lepidocrocite (γ-FeOOH) are generally reduced more rapidly than relatively stable minerals such as goethite and hematite (Postma, 1993). Amorphous manganese oxides such as vernadite are more easily reduced than strongly crystalline forms such as pyrolusite, but the overall influence of crystallinity on reduction kinetics appears to be weaker for manganese and iron oxides (Burdige *et al.*, 1992).

It has now become evident that the primary reason for mineral-related differences in reactivity is not thermodynamic stability, but the fact that amorphous minerals have a far higher surface area than crystalline minerals (Figure 21). Kostka *et al.* (2002a) reported that Fe(III)-reducers grew nearly as well on smectite clay as on ferrihydrite, even though smectite clay is considered a crystalline mineral. Although smectite clay is more highly crystalline than ferrihydrite, the two minerals have a comparable surface area (~700 $m^2\ g^{-1}$) (Schwertmann and Cornell, 1991). As a result, Fe(III)-reducing bacteria were able to remove 20–50% of the total clay-bound iron and thereby alter the physical and chemical characteristics of these ubiquitous minerals (Kostka *et al.*, 1999a,c). Although Fe(III) reduction rates are reduced by one to two orders of magnitude in the presence of minerals with low- versus high surface area, there is ample evidence demonstrating that dissimilatory Fe(III)-reducing bacteria can gain energy for growth using goethite, hematite, and magnetite (Kostka and Nealson, 1995; Roden and Zachara, 1996), in some cases consuming up to 30% of the oxide-bound Fe(III). Fe(III)-rich clays and other highly crystalline forms account for much of the Fe(III) mass in soils and sediments (Stucki *et al.*, 1996), and may therefore support a substantial portion of the planet's anaerobic microbial carbon metabolism.

The reactivity of Fe(III) and Mn(IV) minerals is often highest immediately after they precipitate following Fe(II) or Mn(IV) oxidation. Because metal reduction is usually limited by electron acceptor availability (or co-limited with organic carbon), ecosystem processes that favor Fe(II) and Mn(II) oxidation also promote Fe(III) and Mn(IV) reduction. Regeneration by O_2 occurs slowly in rivers, lakes, and oceans because O_2 diffusion in water is exceedingly slow and sediments have a high biological oxygen demand. Oxidation is much faster when the sediments are mixed by currents or bioturbation (Aller, 1990). Periodic drawdown of the water table in intertidal sediments and wetlands rapidly introduces O_2 and accelerates metal oxidation. Wetland plants enhance metal oxidation by introducing O_2 directly in the soil from porous root systems (Kostka and Luther, 1995), a process known as radial oxygen loss, and by removing soil water by transpiration (Dacey and Howes, 1984). Metal oxides are also regenerated anaerobically through a variety of enzymatic and nonenzymatic mechanisms coupled to redox cycles of sulfur, iron, and manganese. Because all of these processes can occur in wetland soils, a larger proportion of the total iron pool may be reactive in wetlands than nonwetland ecosystems (Weiss, 2002). Thus, wetlands may be "hot spots" of iron cycling.

8.08.6.4.4 Abiotic versus biotic reduction

Iron and manganese differ from other common terminal electron acceptors such as O_2, NO_3^-, SO_4^{2-}, and HCO_3^- in that they are subject to rapid nonenzymatic reduction. For example, Fe(II) reduces Mn(IV) according the generalized reaction (Lovley and Phillips, 1988; Postma, 1985):

$$2Fe(II) + Mn(IV) \rightarrow 2Fe(III) + Mn(II) \quad (23)$$

This reaction makes manganese the ultimate electron sink for a portion of the organic carbon consumed by dissimilatory Fe(III)-reducing bacteria, and it reduces diffusive losses of iron from sediments. Nonenzymatic Mn(IV) reduction by Fe(II) can be significant in high-Mn(IV) sediments (Aller, 1990), but its global significance is ultimately limited by the low content of manganese in the Earth's crust compared to iron.

Nonenzymatic reduction by sulfides is widely considered to be the primary mechanism for Fe(III) and Mn(IV) reduction in systems where SO_4^{2-} is abundant and sulfate reduction dominates anaerobic metabolism (Aller and Rude, 1988; Burdige and Nealson, 1986; Jacobson, 1994; Kostka and Luther, 1995; Postma, 1985; Pyzik and Sommer, 1981). Yet, Fe(III)-reducing bacteria are abundant in marine soils and sediments (Lowe *et al.*, 2000; Kostka *et al.*, 2002c), and several workers have noted that Fe(II) continues to accumulate when H_2S production is blocked by molybdate, apparently because of enzymatic Fe(III) reduction (Canfield, 1989; Canfield *et al.*, 1993b; Hines *et al.*, 1997; Jacobson, 1994; Joye *et al.*, 1996; Kostka *et al.*, 2002c; Lovley and Phillips, 1987; Sørensen, 1982).

Organic compounds have the potential to abiotically reduce Fe(III) and Mn(IV) (Luther *et al.*, 1992; Stone, 1987). Phenols and a variety of other aromatic compounds reduce Fe(III) rapidly at acidic pH, but slowly at circum neutral pH (LaKind and Stone, 1989). Humics can reduce Fe(III) effectively at circumneutral pH and they are abundant in soils and sediments. Because humics and other organic compounds often serve as electron shuttles between metal-reducing bacteria and metal oxides (Lovley *et al.*, 1996a), it may be difficult to separate microbial and nonmicrobial sources of electrons. Finally, aerobic photoreduction of Fe(III) has been observed in freshwater and marine environments (Barbeau *et al.*, 2001; Emmenegger *et al.*, 2001), but it is unknown to what degree this process contributes to nonenzymatic Fe(III) reduction in water columns or sediments.

The relative contributions of enzymatic and nonenzymatic pathways to Fe(III) reduction in marine environments is difficult to measure and highly variable. Because nonenzymatic reduction is rapid enough to compete with microorganisms for Fe(III)-oxide substrates (Thamdrup, 2000), nonenzymatic reduction should be favored over enzymatic reduction in environments where sulfides are abundant and reactive Fe(III) oxides are scarce. In such sediments, biological reduction may be restricted to microsites where sulfide levels are low and poorly-crystalline Fe(III) is abundant (Canfield, 1989). However, H_2S does not accumulate in the porewater of many marine environments (i.e. they are not *sulfidic*) because it reacts with metal oxides. In sediments where sulfide is produced but does not accumulate, enzymatic Fe(III) reduction rates are often substantial. Stoichiometric considerations suggest that enzymatic reduction should be a large part of the overall Fe(III) reduction in such sediments (Thamdrup, personal communication). For example, the oxidation of one mole of organic carbon to CO_2 can support the reduction of either 4 moles of Fe(III) to Fe(II) or 0.5 moles of SO_4^{2-} to S^{2-}. Assuming that SO_4^{2-} reduction produces H_2S, the H_2S could oxidize as little as 1/3 moles of Fe(III) (i.e. $3H_2S + 2FeOOH \rightarrow 2FeS + S$). In this case, the relative contributions of enzymatic and nonenzymatic processes to Fe(III) reduction would be comparable even if Fe(III) reduction contributed just 10% of total organic carbon oxidation. It is necessary to quantify the proportion of total Fe(III) reduction contributed by Fe(III)-reducing bacteria in order to understand the ecology of these ubiquitous organisms. However, from the perspective of overall microbial metabolism, it should be noted that Fe(III) reduction is ultimately a result of microbial respiration regardless of the mechanism. That is, Fe(III) serves indirectly as a terminal electron acceptor for SO_4^{2-}-reducing bacteria when Fe(III) undergoes non-enzymatic reduction by H_2S.

8.08.6.4.5 Separating enzymatic and nonenzymatic Fe(III) reduction

A common approach to estimating the contribution of microorganisms to Fe(III) and Mn(IV) reduction based on carbon mass balance. Total anaerobic microbial respiration is estimated by measuring the carbon mineralization rate (i.e., $\sum CO_2 + CH_4$ production). The carbon that was respired by SO_4^{2-}-reducing bacteria and methanogens is subtracted from the total, and the difference is assumed to be the contribution of Fe(III)- and Mn(IV)-reducing bacteria (Canfield *et al.*, 1993b; Thamdrup, 2000). Typically, the contribution of aerobic metabolism to carbon mineralization is excluded by design, denitrification is assumed to be negligible because of limited NO_3^- availability, and it is assumed that non-methanogenic fermentation does not completely oxidize organic carbon to CO_2. The approach is attractive in marine systems because methanogenesis is negligible, which simplifies the calculation, and SO_4^{2-} reduction can be measured accurately using a ^{35}S radioisotope technique (Jørgensen, 1978a,b,c). Sulfate reduction measurements in salt marsh soils are more problematic because severed roots can introduce fermentable organic carbon compounds, and spatial variability introduces error into the difference calculation.

A second method for separating enzymatic and nonenzymatic Fe(III) reduction by H_2S is to block SO_4^{2-} reduction with molybdate (MoO_4^{2-}). The technique has been used effectively to demonstrate the importance of enzymatic reduction in marine and freshwater sediments (Section 8.08.6.4.4). As with all inhibitor techniques, there is the possibility that molybdate additions directly or indirectly affect processes other than SO_4^{2-} reduction. For example, it could overestimate biotic Fe(III) reduction if the enzymatic process was stimulated by a cessation of competition with H_2S for Fe(III) substrates, or underestimate if SO_4^{2-} reduction was not completely blocked. Despite these *potential* limitations, the molybdate method produces patterns that are consistent with other types of geochemical data, and it is therefore widely used.

Perhaps the most elegant method for separating enzymatic and nonenzymatic Fe(III) reduction would exploit a difference in natural Fe-isotope fractionation. However, such a method has proved to be elusive. Experiments with the Fe(III)-reducing bacterium *Shewanella algae* yielded a 1.3‰ fractionation between the ferrihydrite substrate and soluble Fe(II), suggesting that variations in the natural abundance $\delta^{56}Fe$ could be interpreted biologically (Beard *et al.*, 1999). Unfortunately, even larger iron fractionations (up to 3.6‰) are possible by nonbiological mechanisms (Anbar *et al.*, 2000). The $^{16}O/^{18}O$ ratio of siderite ($FeCO_3$) was similar in the presence and absence of Fe(III)-reducing bacteria, suggesting that oxygen isotopes will not be a useful signature of enzymatic Fe(III) reduction (2001). However, the isotopic composition of biogenic iron minerals could prove to be useful as paleothermometers. Temperature-dependent fractionation of $\delta^{18}O$ in biogenic siderite occurred in cultures of Fe(III)-reducing bacteria (Zhang *et al.*, 1997; Zhang *et al.*, 2001), and in the intracellular magnetite produced by magnetotactic bacteria (Mandernack *et al.*, 1999). Radiolabeling the iron oxide pool to more precisely measure Fe(III) reduction rates proved unsuccessful because of rapid isotope exchange between the Fe(III) and Fe(II) pools (Roden and Lovley, 1993).

8.08.6.4.6 Fe(III) reduction in ecosystems

Field studies of enzymatic Fe(III) reduction are scarce, but they generally indicate that the availability of reactive Fe(III) and organic carbon govern rates. Thamdrup (2000) reported a positive relationship between the content of poorly crystalline Fe(III) in marine sediments and the fraction of carbon oxidation mediated by Fe(III)-reducing bacteria (Figure 22). The range of values indicates that Fe(III) reduction can account for up to 90% of the carbon oxidation in some sediments. Roden and Wetzel (2002) found a similar relationship in a freshwater marsh, also with a contribution from Fe(III) reduction of between 20% to 90% to the overall anaerobic carbon metabolism. It is noteworthy that Fe(III) reduction was never Fe(III)-saturated in either study, suggesting that Fe(III) availability may limit Fe(III) reduction within the range of labile Fe(III) concentrations that are generally found in soils and aquatic sediments. By analogy to denitrification, SO_4^{2-} reduction and methanogenesis, organic carbon availability is expected to also limit Fe(III) reduction. Roden and Wetzel (2002) reported a strong linear relationship ($r^2 = 0.99$) between microbial respiration (a measure of carbon availability) and initial Fe(III) reduction rate, and proposed that overall Fe(III) reduction was regulated by both carbon and Fe(III) availability according to the equation:

$$R_{Fe(III)} = \alpha b R_{oc} Fe(III)_{reac} \quad (24)$$

where $R_{Fe(III)}$ and R_{oc} are rates of Fe(III) reduction and organic carbon decomposition, respectively. $Fe(III)_{reac}$ is the concentration of reactive Fe(III), α is the stoichiometric ratio of Fe(III) atoms reduced per carbon atoms oxidized (4 : 1 in this case), and b is a rate constant. Thus, for any given initial amount of reactive Fe(III), organic carbon

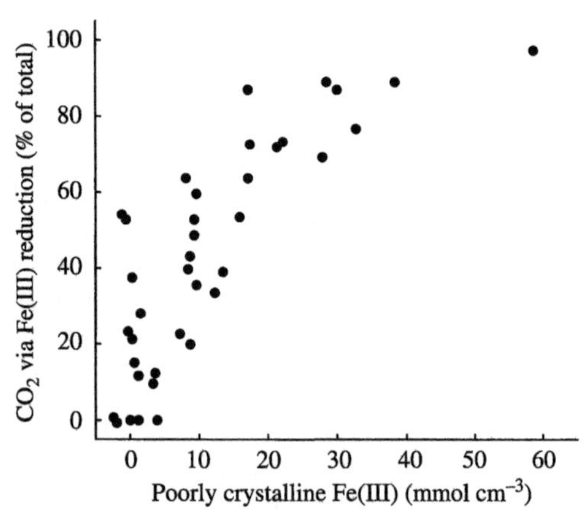

Figure 22 The fraction of carbon metabolism due to Fe(III) reduction in marine sediments as a function of the poorly crystalline Fe(III) content (after Thamdrup, 2000).

availability regulates Fe(III) reduction rates until the reactive Fe(III) pool has been exhausted.

8.08.6.5 Microbial Oxidation of Iron and Manganese

Fe(II) and Mn(II) oxidation regenerate high-quality (i.e., poorly crystalline) substrates for Fe(III) and Mn(IV)-reducing microorganisms. In contrast to other terminal electron acceptors, these elements rapidly precipitate upon oxidation and settle so that they are efficiently recycled in the ecosystem. Oxidation occurs under both aerobic and anaerobic conditions, and it can be biologically mediated or autocatalytic in either case. Mechanisms for chemical oxidation of Fe(II) under anaerobic conditions include reaction with Mn(IV) (Postma, 1985) and NO_2^- (Hansen *et al.*, 1994; Weber *et al.*, 2001). Organisms that oxidize Fe(II) and Mn(II) to support growth must compete with chemical oxidation for substrates.

8.08.6.5.1 Energetics of Fe(II) and Mn(II) oxidation

Microbial growth based on Fe(II) oxidation proceeds readily at acidic pH (<4) because Fe^{2+} (i.e., aqueous Fe(II)) is stable (Patrick and Henderson, 1981). In contrast, biological Fe(II) oxidation at circumneutral pH has been assumed to be unimportant because autocatalytic oxidation is rapid, and the reaction produces relatively little free energy (Straub *et al.*, 2001). However, neutralophilic Fe(II)-oxidizing bacteria growing on Fe(II) may actually be able to generate more free energy than acidophilic bacteria when one considers the forms of Fe(II) and Fe(III) in natural ecosystems. At a pH of 6 or 7, Fe(II) will often be in the form of $FeCO_3$, and Fe(III) will be in the form of an insoluble amorphous hydroxide, such as $Fe(OH)_3$ or FeOOH, which removes the product from the solution (Emerson, 2000). Under these conditions, Fe(II) oxidation coupled to O_2 respiration can generate substantial free energy.

Because Mn(II) is stable at pH <8, significant contributions from microorganisms to Mn(II) oxidation has been relatively easy to demonstrate, and a number of Mn(II) oxidizing bacteria have been isolated. Microbial oxidation is considered the primary mechanism of Mn(II) oxidation in circumneutral freshwater (Ghiorse, 1984; Nealson *et al.*, 1988). The sheathed bacterium *Leptothrix discophora* is perhaps the most-studied species of Mn(II)-oxidizing bacteria, and rate laws have been developed to describe Mn(II) oxidation as a function of pH, temperature, dissolved O_2, and Cu concentration (Zhang *et al.*, 2002). *Bacillus* sp. strain SG-1 has also been well studied. This species and related strains have been shown to oxidize Mn(II) even while in a spore stage (Francis and Tebo, 2002). All of the known organisms that perform Mn(II) oxidation are heterotrophic, and the process has not yet been linked to chemolithoautotrophy. Further details on Mn(II) oxidation are provided by Tebo *et al.* (1997) and Emerson (2000).

8.08.6.5.2 Anaerobic Fe(II) oxidation

Anaerobic Fe(II) oxidation is one of the most recently recognized categories of anaerobic metabolism. The first report of such metabolism described anaerobic, phototrophic organisms that required only Fe(II), CO_2, and light for growth (Widdel *et al.*, 1993):

$$4Fe^{2+} + CO_2 + 11H_2O \rightarrow 4Fe(OH)_3 + (CH_2O) + 8H^+ \quad (25)$$

Since then, several such bacteria have been isolated from freshwater and marine environments (Ehrenreich and Widdel, 1994; Heising and Schink, 1998; Heising *et al.*, 1999; Straub *et al.*, 1999). One implication of this metabolism is that microorganisms may have contributed to the banded iron formations, which resulted from widespread Fe(III)-oxide deposition in oceans at a time when there was very little atmospheric O_2 (Kump, 1993). Little is known about the physiology or ecology of these organisms (Straub *et al.*, 2001), but most strains can couple Fe(II) oxidation to NO_3^- reduction.

Even more recent was the first report of anaerobic Fe(II) oxidation coupled to NO_3^- reduction (Straub *et al.*, 1996; Hafenbradl *et al.*, 1996):

$$10FeCO_3 + 2NO_3^- + 24H_2O \rightarrow 10Fe(OH)_3 + N_2 + 10HCO_3^- + 8H^+ \quad (26)$$

Although NO_3^- reduction to NH_4^+ is thermodynamically feasible, all known strains produce primarily N_2 (Straub *et al.*, 2001), making this a novel metabolic pathway for N_2 production (Section 8.08.5.5.3). Cultures from a variety of environments grow solely on aqueous or solid-phase Fe(II) and CO_2 (i.e., they are chemolithoautotrophs), while others require Fe(II), CO_2 and an organic co-substrate such as acetate (i.e., they are putative mixotrophs). Organisms capable of anaerobic Fe(II) oxidation accounted for up to 58% of the total cultivatable denitrifying community in the profundal sediments of a lake (Hauck *et al.*, 2001), but not more than 0.8% of denitrifiers in other aquatic sediments (Straub and Buchholz-Cleven, 1998). Freshwater organisms can couple NO_3^- reduction to the oxidation of microbially reduced goethite and other solid-phase Fe(II) (Weber *et al.*, 2001), and enrichment cultures from marine sediments oxidize FeS, but not FeS_2 (Schippers and Jørgensen, 2002). The contribution of these organisms to N_2 production and Fe(II) oxidation activity in ecosystems has not been evaluated. It has been suggested that their role in Fe(II) oxidation is ultimately limited by NO_3^- availability, which is typically low due to NO_3^- assimilation by plants and microorganisms. However, many of these organisms are facultative anaerobes and can use O_2 as an alternative electron acceptor for heterotropic growth. The capacity to switch between O_2 and NO_3^- as terminal electron acceptors may be a common trait among chemoautotrophic bacteria that link Fe(II) oxidation to N_2 production, just as it is among strictly heterotrophic denitrifying bacteria.

8.08.6.5.3 Aerobic Fe(II) oxidation

Iron oxidation occurs at the interface of aerobic and anaerobic environments according to the following generalized reactions:

$$Fe^{2+} + \tfrac{1}{4}O_2 + H^+ \rightarrow Fe^{3+} + \tfrac{1}{2}H_2O \quad (27)$$

$$Fe^{3+} + 3H_2O \rightarrow Fe(OH)_3(s) + 3H^+ \quad (28)$$

The relative contributions of enzymatic versus nonenzymatic oxidation to overall Fe(II) oxidation rates is influenced by a variety of factors such as pH, concentrations of O_2 and Fe(II), and the rate of Fe(II) delivery (Neubauer *et al.*, 2002). At pH <4, the nonenzymatic oxidation of Fe(II) by O_2 is very slow, and microbial activity can increase oxidation rates by a factor $>10^6$ (Singer and Stumm, 1970). The practical implications of this observation for the abatement of acid mine drainage has motivated a great deal of research on chemolithotrophic, acidophilic prokaryotes such as *Acidothiobacillus ferrooxidans* (formerly *Thiobacillus ferrooxidans*) and *Leptospirillum ferrooxidans* (Nordstrom and Southam, 1997). The influence of such organisms on the pH of mine drainage may vary, depending on the niche they occupy in an acid mine waste stream. *L. ferrooxidans* tolerates extremely low pH (<1.0) conditions typical of solutions in contact with pyrite ores, and it may be responsible for initiating a series of reactions that ultimately enhance pyrite dissolution and generates acidity (Schrenk *et al.*, 1998). Archea in the order *Thermoplasmales* are also abundant at such sites and presumably contribute to Fe(II) oxidation (Edwards *et al.*, 2000b). Contrary to its presumed role of enhancing pyrite dissolution, it has been shown that *A. ferriooxidans* actually occurs in somewhat less acidic environments where it would have little influence on pyrite dissolution (Schrenk *et al.*, 1998). In fact, *A. ferriooxidans* may serve a beneficial role by reducing aqueous loads of Fe(II) and other metals. As these studies illustrate, there is much insight to be gained from investigating the ecology of microorganisms *in situ*.

The only neutrophilic Fe(II)-oxidizer that had been cultured in the laboratory previous to the 1990s was *Gallionella ferruginea*, a microaerobe that forms a helical stalk (Emerson, 2000). *G. ferruginea* is a member of the beta-Proteobacteria, and is capable of chemolithoautotrophic or mixotrophic growth (Hallbeck and Pederson, 1991; Hallbeck *et al.*, 1993). A number of novel strains of chemolithoautotrophic Fe(II)-oxidizing bacteria have been isolated from groundwater (Emerson and Moyer, 1997) and hydrothermal vents (Emerson and Moyer, 2002). These strains are unicellular and do not form extracellular stalks. They are members of the gamma-*Proteobacteria* subclass in the family *Xanthomonaceae*. Another recent isolate of lithotrophic Fe(II)-oxidizing bacteria, strain TW-2, was recovered from a freshwater wetland and determined to be in the beta-*Proteobacteria* (Sobolev and Roden, 2003).

It is often assumed that Fe(II) oxidation is a nonenzymatic process at the circumneutral pH conditions that prevail in most ecosystems because autocatalytic oxidation is quite rapid. This may be the case when O_2 levels are high due to physical mixing, bioturbation, or rapid changes in water table depth. But when the position of the oxic–anoxic interface is stable, opposing diffusion gradients of Fe(II) and O_2 intersect to form microaerobic zones ($[O_2] <0.5\%$), the ideal niche for organisms that require both reduced inorganic elements and O_2 for respiration. The half-life of nonenzymatic Fe^{2+} (i.e., dissolved Fe(II)) oxidation at low O_2 concentration is up to 300-fold longer than at high O_2 concentration (Roden *et al.*, in press). A long Fe^{2+} half-life should favor microbial oxidation. Indeed, chemolithoautotrophic bacteria competed successfully with nonenzymatic processes for O_2 and Fe(II) when grown in pure culture at circumneutral pH (Emerson and Revsbech, 1994).

Pure culture studies have alluded to some of the factors that may regulate enzymatic Fe(II) oxidation in complex natural environments such as sediments, soils, and the rhizosphere. Neubauer *et al.* (2002), used microcosms fed with environmentally relevant concentrations of O_2 and Fe(II) to investigate the metabolism of an Fe(II)-oxidizing strain isolated from the wetland rhizosphere. They found that both biotic and abiotic Fe(II) oxidation increased linearly ($r^2 \geq 0.90$) with the rate of Fe(II) addition (Figure 23). Since the experimental Fe(II) addition rate approximated *in situ* Fe(III)-diffusion rates in freshwater

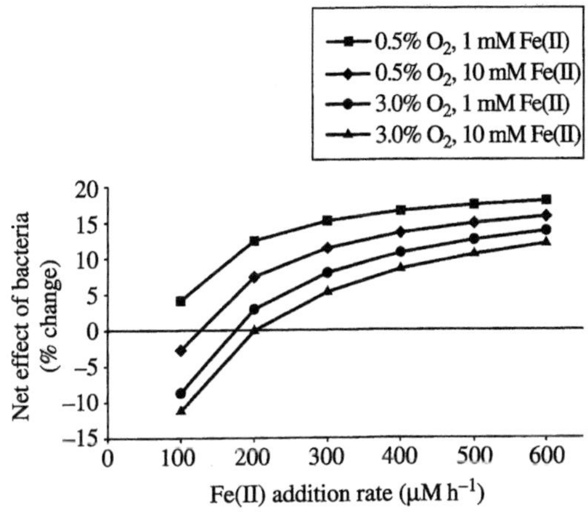

Figure 23 The influence of a chemolithoautotrophic bacterium on Fe(II) oxidation rates as a function of Fe(II) addition rate (*x*-axis) as determined in stirred microcosms. The effects varied depending on the O_2 concentration and the total Fe(II) oxide content in the microcosm. The bacterium was isolated from the rhizosphere of wetland plants (reproduced from Neubauer *et al.*, 2002).

wetland soils, these data suggested that microbial Fe(II) oxidation in wetlands is Fe(II)-limited. The bacteria fared best in competition with nonenzymatic oxidation when the total iron that had accumulated in the chambers was low, suggesting that competition is mediated in part by factors such as particle size distribution, texture, and mineralogy. Each of these factors influences the abundance and nature of surfaces that can adsorb Fe(II) and coordinate oxidation reactions. Finally, the presence of Fe(II)-oxidizing bacteria actually slowed the rate of overall Fe(II) oxidation in some circumstances. This presumably was caused by the inhibition of nonenzymatic Fe(II) oxidation via chelation and stabilization of aqueous Fe(II), perhaps by exopolymers or other organic extracellular molecules that are produced by Fe(II)-oxidizing bacteria (Emerson and Moyer, 1997). Similar compounds have been proposed to chelate Fe(III), allowing it to diffuse into anaerobic zones after microbial oxidation (Sobolev and Roden, 2001).

There is currently no satisfactory way to separate enzymatic from nonenzymatic Fe(II) oxidation *in situ*. However, pure culture studies have shown that bacteria mediated up to 90% of Fe(II) oxidation (Emerson and Revsbech, 1994; Neubauer *et al.*, 2002; Sobolev and Roden, 2001). In contrast to these studies, van Bodegom *et al.* (2001) reported that microbial Fe(II) oxidation was insignificant compared to nonenzymatic oxidation in a model rice system. However, their experiment included periods of vigorous sample agitation that may have artificially favored abiotic Fe(II) oxidation (Neubauer *et al.*, 2002). Indirect evidence that aerobic, chemolithotrophic Fe(II)-oxidizing bacteria contribute to Fe(III) deposition in circumneutral environments is the observation that they are ubiquitous and can account for 1% or more of the total microbial population (Weiss *et al.*, 2003).

8.08.6.6 Iron Cycling

Nealson (1983) articulated the first detailed description of iron cycling in which microbial metabolism was a central theme. Research since then has confirmed the more speculative elcments of a microbial iron cycle, and contributed new insights on the processes that link Fe(III) reduction and Fe(II) oxidation in a "ferrous wheel," particularly in wetland soils. Microbial iron cycling promotes organic carbon oxidation via Fe(III) reduction, and it suppresses the metabolism of sulfate reducers and methanogens that do not compete effectively with Fe(III)-reducing bacteria for carbon substrates (Section 8.08.1). Many of the new developments in this area have come from the work of Roden and colleagues, who investigated a freshwater wetland where nonenzymatic Fe(III) reduction by sulfides was relatively unimportant (see synthesis by Roden *et al.*, in press). They proposed a model in which Fe(II)-oxidizing and Fe(III)-reducing bacteria are spatially organized into aggregates reminiscent of those that mediate anaerobic methane oxidation (Figure 10; Section 8.08.4.5). The aggregates have an Fe(III)-coated particle at the center, surrounded by a mass of Fe(III)-reducing bacteria, and then an outermost layer of Fe(II)-oxidizing bacteria (Sobolev and Roden, 2002). The Fe(II)-oxidizing bacteria enhance Fe(III) reduction by two mechanisms: (i) they generate a steep O_2 gradient that maintains anaerobic conditions inside the aggregate, and (ii) they produce compounds that chelate Fe(III) so that it can diffuse back into the anoxic center of the aggregate where it will be re-reduced to Fe(II). The Fe(II) produced by Fe(III)-reducing bacteria diffuses toward the periphery of the aggregate, completing the iron cycle. The end result would be the oxidation of organic carbon without the net consumption or depletion of Fe(III) oxides. Certain details of this model are supported by experimental evidence (Roden *et al.*, in press), but this model should be considered speculative. It is interesting to further speculate that these aggregates might also include bacteria that are capable of linking anaerobic Fe(II) oxidation to denitrification, with the NO_3^- provided by nitrifying bacteria at the periphery.

Iron cycling may also be favored by processes that operate at larger spatial and temporal scales than those proposed by Roden *et al.* (in press).

In wetland soils, iron cycling is promoted by several factors related to the presence of plants (Kostka and Luther, 1995; Roden and Wetzel, 1996; Frenzel *et al.,* 1999). Roots introduce O_2 into anaerobic soils (Schütz *et al.*, 1991; Hines, 1991; Holmer *et al.*, 2002), which supports biotic and abiotic Fe(II) oxidation (Neubauer *et al.*, 2002), and the deposition of poorly-crystalline Fe(III) oxides (Mendelssohn *et al.*, 1995). Due to repeated annual cycles of root growth and senescence, non-rhizosphere soils (i.e. mm or more from a root surface) in wetlands are enriched in poorly-crystalline Fe(III) oxides compared to sediments that lack plants altogether (Weiss, 2002). The combination of abundant poorly-crystalline Fe(III) minerals and labile organic carbon from roots favors rapid Fe(III) reduction during periods of anoxia. Because variations in plant development and photosynthetic activity can cause the rate of O_2 release by roots to vary hourly, daily and seasonally (Risgaard-Petersen and Jensen, 1997), parts of the root system may support net deposition of poorly-crystalline Fe(III) oxides at times when O_2 is released, then switch to net Fe(III)-reduction when O_2 is absent. Indeed, Fe(III)-reducing and Fe(II)-oxidizing bacteria occur on the same 1-cm lengths of plant roots (Weiss *et al.*, 2003). Burrowing animals influence sediments in much the same manner as roots (Kostka *et al.,* 2002c; Gribsholt *et al.,* 2003), and water table fluctuations can cause sediments to shift rapidly between reducing and oxidizing conditions. Although the processes discussed here have been demonstrated, their individual contributions to regulating Fe(III) cycling at an ecosystem level is largely unknown.

8.08.7 SULFUR

Sulfur (S) is an important element biochemically and geochemically. It is the fourteenth most abundant element in the Earth's crust and is ~1% of the dry mass of organisms where it serves many structural and enzymatic functions. Sulfur acts as a significant electron donor and acceptor in many bacterial metabolisms (Jørgensen, 1988). Microbial sulfur transformations are closely linked with the carbon cycle; sulfur reduction coupled to organic carbon oxidation is a major mineralization pathway in anoxic habitats, and autotrophic sulfur oxidation can occur aerobically and anaerobically. Sulfur compounds are often highly reactive, which results in a tight coupling of the oxidative and reductive portions of the biological sulfur cycle, particularly at the redoxcline where sulfur cycling can be extremely rapid (Figure 24). Although sulfur cycling occurs in terrestrial and freshwater aquatic environments, it is most prominent in marine ecosystems due to

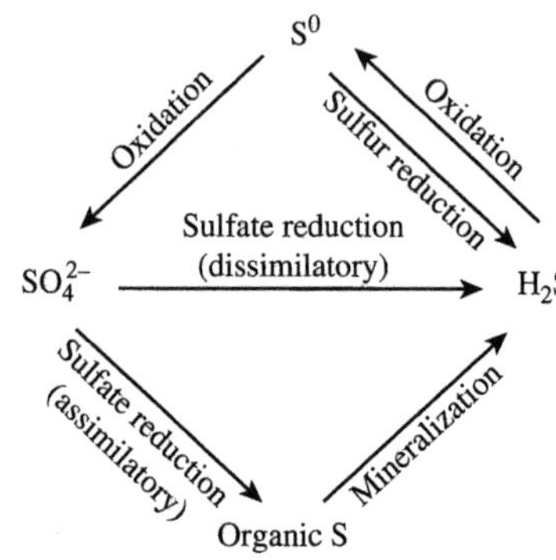

Figure 24 A simplified biological redox cycle for sulfur.

the abundance of sulfate (SO_4^{2-}) in seawater. Many specialized, small-scale environments, including hypersaline and geothermal (hot spring and hydrothermal vent) ecosystems also thrive on microbial sulfur transformations (e.g., Baas Becking, 1934; Ehrlich, 2002). In marine environments, a large fraction (up to 80%) of the organic carbon is respired through SO_4^{2-} reduction (Canfield *et al.*, 1993a, b), so copious amounts of sulfide are available for geochemical and microbial transformations.

8.08.7.1 Sulfur Geochemistry

The average crustal abundance of sulfur is 260 $\mu g\ g^{-1}$ and most on Earth is present as metal sulfide, gypsum, anhydrite, and dissolved sulfate. Sulfur can be found in a range of valence states from the highly reduced sulfide (−2) to the most oxidized form in SO_4^{2-} (+6). There are several intermediate valence forms of sulfur that can serve as both electron donors and acceptors for bacteria depending on environmental conditions, the most important being elemental sulfur and thiosulfate (Odom and Singleton, 1993). Sulfate is highly soluble and is the second most abundant anion in freshwater (after bicarbonate) and in seawater (after chloride) with the ocean acting as a major reservoir of SO_4^{2-}. Sulfate is also found in evaporite deposits, primarily as gypsum ($CaSO_4{\cdot}2H_2O$). Gypsum dissolution is a source of SO_4^{2-} for microorganisms (Machel, 2001). Sulfate-containing minerals are common oxidation products of sulfide minerals and can include anhydrous sulfates such as barite ($BaSO_4$) and hydroxylated sulfates such as alunite ($KAl_3(SO_4)_2(OH)_6$) and jarosite ($KFe_3(SO_4)_2(OH)_6$) (Deer *et al.*, 1992). Elemental sulfur (S^0) is formed hydrothermally and as an oxidation product of sulfide weathering. However, S^0 is also a common product of sulfide oxidation by bacteria and can serve as an electron acceptor for bacterial respiration (Jørgensen, 1982a).

Reduced S(II), or *sulfide*, is present in a variety of forms, most of which are solids. Dissolved sulfide exists as bisulfide ion (HS^-) at neutral pH, sulfide ion (S^{2-}) at alkaline pH, and hydrogen sulfide (H_2S) at low pH. H_2S is the only form of sulfide that is volatile and it imparts a characteristic "rotten egg" smell to sediments at low tide. Dissolve sulfides react strongly with base and transition metal ions to form insoluble sulfide minerals, the most prominent of which are iron sulfides that make anoxic sediments black. Freshly precipitated iron sulfides are amorphous, but tend to crystallize rather quickly into other acid-soluble authigenic minerals such as Mackinawite (tetragonal FeS) and Greigite (cubic Fe_3S_4). Many metals form insoluble sulfide minerals, which result in economically important deposits, and as many as 95 sulfide minerals appear in standard lists including the monosulfides galena (PbS), covellite (CuS), and cinnabar (HgS). Economically important metal sulfide deposits may be precipitated from hydrothermal solutions in vein-type or replacement deposits, but some of the largest zinc, copper, and lead deposits were formed from diagenetic reactions in sedimentary basins from biogenic sulfide. The acid-soluble forms of iron-containing sulfides convert to disulfide pyrite (cubic FeS_2) by the addition of elemental sulfur. Pyrite is not acid soluble and accumulates as the major end product of sulfur diagenesis in reducing sediments (Goldhaber and Kaplan, 1975), a process that occurs slowly over several years. Pyrite can also be formed rapidly when iron monosulfide levels are undersaturated, as is usually the case in salt marsh sediments (Howarth, 1979; Howarth and Merkel, 1984). Disulfides can also incorporate other elements to form minerals such as arsenopyrite (FeAsS) and molybdenite (MoS_2).

8.08.7.2 Microbial Reduction of Sulfate

8.08.7.2.1 Overview of sulfate reduction

Biological SO_4^{2-} reduction is an ancient process and evolved earlier than ~3.5 Gyr ago (Shen *et al.*, 2001). Sulfate-reducing bacteria are anaerobes that are situated at the terminus of the anaerobic food web where they act as an important cog in the sulfur and carbon cycles. Sulfate-reducing bacteria liberate CO_2 and S(II) as the primary end products of organic matter mineralization. They gain energy by coupling the oxidation of organic compounds or H_2 to SO_4^{2-} reduction (Figure 25). The process is also termed *dissimilatory* SO_4^{2-} *reduction* to differentiate it from *assimilatory* SO_4^{2-} *reduction*, which produces reduced sulfur for biosynthesis. Assimilatory SO_4^{2-} reduction is common among organisms and does not lead to the excretion of sulfide. The presence of H_2S is

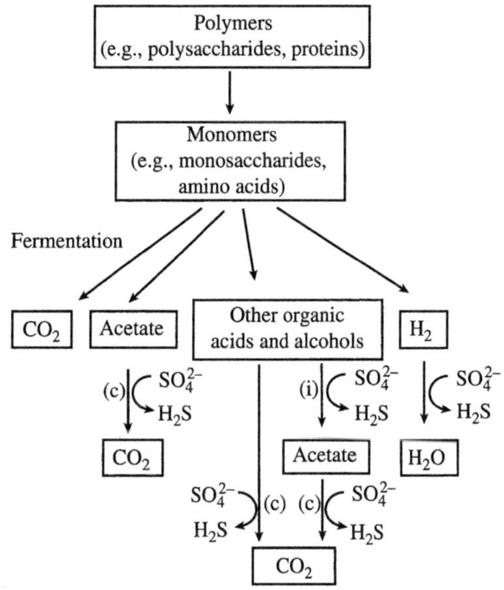

Figure 25 Anaerobic decomposition with sulfate reduction as the terminal step. Fermentation leads to several possible products including low molecular weight organic acids and alcohols and hydrogen and carbon dioxide. Incomplete oxidizers (i) produce acetate as an end product, whereas complete oxidizers (c) mineralize organic compounds, including acetate, to carbon dioxide.

noted by its characteristic foul smell, black ferrous sulfide precipitates, or white patches of S^0 as an oxidation product. Dissimilatory SO_4^{2-} reduction can account for half or more of the total organic carbon mineralization in many environments (Canfield *et al.*, 1993a; Jørgensen, 1982b). It has been estimated that ~5×10^{12} kg yr^{-1} of SO_4^{2-}-S is reduced by bacteria globally, with greater than 95% the activity occurring in the ocean (Skyring, 1987). Most of the SO_4^{2-} reduction in marine ecosystems occurs in coastal regions that receive high inputs of organic material. However, SO_4^{2-} reduction in freshwater environments can account for a significant portion of anaerobic mineralization processes, and in some instances can be the dominant pathway (Bak and Pfennig, 1991; Holmer and Storkholm, 2001; Lovley and Klug, 1983a; Urban *et al.*, 1994).

Besides its importance in the degradation of organic compounds, the reduced sulfur produced during SO_4^{2-} reduction is important geochemically since it is highly reactive and involved in the precipitation of highly insoluble metal sulfides and the accumulation of dissolved sulfide. The reduced products can be readily reoxidized under oxidizing conditions and can act as substrates for autotrophic bacteria including phototrophs, thus completing a dynamic sulfur cycle. The dissolved and solid-phase products of SO_4^{2-} reduction are responsible for consuming a significant portion of the O_2 that diffuses into sediments, and this O_2 sink can be equal to or greater than the amount of O_2 consumed by aerobic bacterial respiration (Jørgensen, 1977). In typical coastal marine sediments, it is possible that 90% of the reduced sulfur produced during sulfate reduction each year is recycled back to sulfate. The remaining sulfide is buried as a metal sulfide and eventually transformed into FeS_2 (Jørgensen, 1982a).

The primary role of SO_4^{2-} reducers in the carbon cycle is the mineralization of relatively small organic substrates to CO_2. Although the group as a whole contains a diverse array of metabolic capabilities, SO_4^{2-} reducers do not generally degrade polymers. An exception is the thermophilic Archeal species *Archeoglobus* (Stetter, 1988; Stetter *et al.*, 1987), and few other species seem to be capable of degrading compounds more complicated than simple monomers or fermentation products. The energy gain from dissimilatory SO_4^{2-} reduction is relatively low. The free-energy yield ($\Delta G'$) of the complete oxidation of acetate or lactate to CO_2 is -48 kJ or -128 kJ, respectively, whereas acetate or lactate oxidation with O_2 yields -844 kJ or -1323 kJ, respectively. However, the sulfide produced by these reactions can act as an energy source for other bacteria, especially photoautotrophic and chemoautotrophic bacteria in aerobic surface sediments (Jørgensen, 1988).

8.08.7.2.2 Metabolic diversity

(i) *Complete and incomplete organic carbon oxidation.* Physiologically, SO_4^{2-} reduction is divided between complete and incomplete oxidation processes. Incomplete oxidizers utilize a variety of substrates, many of which are incompletely degraded to acetate. For example:

$$2\,\text{lactate} + SO_4^{2-} \rightarrow 2\,\text{acetate} + 2CO_2 + 2H_2O + S^{2-} \quad (29)$$

or

$$4\,\text{propionate} + 3O_4^{2-} \rightarrow 4\,\text{acetate} + 4CO_2 + 2H_2O + 3S^{2-} \quad (30)$$

Some incomplete oxidizers are also capable of fermenting substrates without SO_4^{2-} respiration, for example:

$$3\,\text{lactate} \rightarrow 2\,\text{propionate} + \text{acetate} + CO_2 \quad (31)$$

However, some strains of SO_4^{2-} reducers are capable of fermentations that yield H_2 as an end product that must be removed quickly by a bacterial partner. It was noted early that strains produced acetate as an end product, so SO_4^{2-} reducers were regarded as fermenting bacteria for decades. It was not until the 1950s that the respiratory nature of SO_4^{2-} reduction was elucidated (Ishimoto *et al.*, 1954; Postgate, 1954). All SO_4^{2-} reducers isolated prior to about 1977 were incomplete oxidizers, including most members of the genus *Desulfovibrio*, the first of which was isolated over 100 yr ago (Voordouw, 1995). Although no acetate-consuming isolates existed prior to the late 1970s, it was postulated for many years that SO_4^{2-}-reducing bacteria were capable of utilizing acetate as an electron donor (Widdel, 1988).

The first acetate utilizer isolated was a gram-positive, sporulating bacterium of the genus *Desulfotomaculum* (Widdel and Pfennig, 1977). Since then, several species of acetate-utilizing gram-negative bacteria have been recovered representing a suite of metabolic capabilities and representing new genera. Complete oxidation entails the production of CO_2 as a major end product and includes the use of acetate, for example:

$$\text{acetate} + SO_4^{2-} \rightarrow 2CO_2 + H_2O + S^{2-} \quad (32)$$

Although physiologically distinct in many ways, the incomplete and complete oxidizing SO_4^{2-}-reducing bacteria can coexist, with the former supplying acetate to the latter in some instances.

Sulfate-reducing bacteria can engage in interspecies H_2 transfer, in which H_2 equivalents are transferred to other bacteria in lieu of SO_4^{2-} respiration. For example:

Desulfovibrio

$$2\,\text{lactate} + 4H_2O \rightarrow 2\,\text{acetate} + 2CO_2 + 4H_2 \quad (33)$$

Hydrogenotrophic methanogen

$$CO_2 + 4H_2 \rightarrow CH_4 + 2H_2O \quad (34)$$

In essence, SO_4^{2-} reducers in this case are fermenting bacteria that depend on H_2-consuming methanogens to maintain a low H_2 partial pressure, thereby making fermentation thermodynamically feasible. In addition, methanogenic bacteria can consume the acetate produced by the SO_4^{2-} reducers.

(ii) *Electron donors.* Energy sources used by SO_4^{2-}-reducing bacteria are similar to those utilized by metal-reducing and methanogenic bacteria. It is generally accepted that SO_4^{2-}-reducing bacteria oxidize fermentation products such as fatty acids, alcohols, and H_2 (Christensen, 1984; Parkes *et al.*, 1989; Sørensen *et al.*, 1981), but this list is incomplete and is continually growing as new metabolisms are discovered. Sulfate reducers can also metabolize chemolithotrophically (autotrophically) when utilizing H_2 as an electron donor. Since most of the substrates used by SO_4^{2-} reducers are provided by a variety of other bacteria, these compounds represent rapidly recycled intermediates that accumulate quickly when SO_4^{2-} reduction is inhibited by the use of SO_4^{2-} analogues such as the group IV oxyanions molybdate and selenate (Oremland and Capone, 1988; Smith and Klug, 1981; Taylor and Oremland, 1979).

Besides the common fermentation products listed above, SO_4^{2-}-reducing bacteria have been shown to consume a variety of other substrates including xenobiotics and other aromatic compounds (Bolliger *et al.*, 2001; Elshahed and McInerney, 2001b; Kniemeyer *et al.*, 2003; Kuever *et al.*, 2001; Lovley *et al.*, 1995), carbohydrates such as fructose and sucrose (Sass *et al.*, 2002), amino acids (Burdige, 1989; Coleman, 1960), alkanes and alkenes up to C_{20} (Aeckersberg *et al.*, 1991, 1998), phosphite (Schink *et al.*, 2002), aldehydes (Tasaki *et al.*, 1992), dicarboxylic acids (Postgate, 1984), glycolate (Friedrich and Schink, 1995), methylated nitrogen and sulfur compounds (Finster *et al.*, 1997; Heijthuijsen and Hansen, 1989; Kiene, 1988; van der Maarel *et al.*, 1996a), acetone (Platen *et al.*, 1990), and sulfonates such as taurine and cysteate (Visscher *et al.*, 1999).

Sulfate-reducing bacteria are able to use a variety of organic compounds as both electron acceptors and electron donors. For example, *Desulfovibrio* species can ferment fumarate or malate, but will reduce these species to succinate if an additional electron donor is available (Miller and Wakerley, 1966).

(iii) *Electron acceptors.* Although the reduction of SO_4^{2-} is considered to be the classic role of SO_4^{2-}-reducing bacteria, these organisms are capable of utilizing a wide variety of inorganic sulfur compounds as electron acceptors including sulfite, bisulfite, metabisulfite, dithionite, tetrathionate, thiosulfate, dimethylsulfoxide, sulfur dioxide, and elemental sulfur (S^0) (Fitz and Cypionka, 1990). Although some species can grow with S^0 as an electron acceptor, the growth of many species is inhibited in the presence of S^0 (Bak and Widdel, 1986; Widdel and Pfennig, 1982), presumably due to increases in redox potential (Rabus *et al.*, 2000).

Nitrate and NO_2^- reduction are rather widespread in SO_4^{2-} reducers (Keith and Herbert, 1983; McCready *et al.*, 1983; Mitchell *et al.*, 1986; Moura *et al.*, 1997; Seitz and Cypionka, 1986), and in some cases NO_3^- is preferred over SO_4^{2-} (Seitz and Cypionka, 1986). The bisulfite reductase enzyme in SO_4^{2-} reducers displays some activity toward NO_2^-, which partially explains the ubiquity of NO_2^- reduction. However, it appears that specific NO_2^- reductases are also present (Liu and Peck Jr., 1981). Unlike denitrifying bacteria that reduce NO_3^- to N_2, the end product of NO_3^- reduction in SO_4^{2-} reducers is NH_4^+ (Widdel and Pfennig, 1982). Thus, SO_4^{2-} reducers contribute to DNRA (Section 8.08.5.4).

Iron reduction has been observed in SO_4^{2-}-reducing bacteria (Bale *et al.*, 1997; Knoblauch *et al.*, 1999b; Lovley *et al.*, 1993b), but this ability is more ubiquitous in the non-SO_4^{2-}-reducing members of the δ Proteobacteria (Kostka *et al.*, 2002a; Nielsen *et al.*, 2002), and cell growth has not been observed. Sulfate reducers are capable of reducing a variety of other metals such as uranium, chromium, technetium, and gold, but like iron, no growth occurs (Lovley and Phillips, 1992b, 1994b; Lovley *et al.*, 1993b). The reduction of arsenate to arsenite supports growth of SO_4^{2-}-reducing bacteria and, in some cases, arsenate is the preferred electron acceptor (Macy *et al.*, 2000; Newman *et al.*, 1997a,b; Stolz and Oremland, 1999). Sulfate reducers can reduce selenate (Stolz and Oremland, 1999) and Mn(IV) (Tebo and Obraztsova, 1998).

Other entries on the list nonsulfur inorganic electron acceptors used by SO_4^{2-} reducers are carbonate, which is reduced to acetate (Klemps *et al.*, 1985), and O_2 (Dilling and Cypionka, 1990). Oxygen reduction is a relatively common feature of SO_4^{2-}-reducing bacteria (Cypionka, 2000; Dannenberg *et al.*, 1992). However, aerobic growth in pure cultures is poor or absent and it appears that O_2 reduction, despite being enzymatic, is primarily an O_2 removal mechanism (Cypionka, 2000). Many of the SO_4^{2-}-reducing bacteria isolated from oxic environments belong to the genus *Desulfovibrio*, although other genera dominate in some instances (Krekeler *et al.*, 1997; Sass *et al.*, 1997; Wieringa *et al.*, 2000). Organic compounds that can serve as electron acceptors for SO_4^{2-} reduction include malate, aspartate, cysteine, sulfonates, pyruvate, acrylate, and oxidized glutathione (Rabus *et al.*, 2000).

Sulfate-reducing bacteria couple the reductive dehalogenation of aromatic compounds to growth (DeWeerd *et al.*, 1990; Dolfing and Tiedje, 1987), and both chlorinated benzoates and bromophenols are used as electron acceptors (Boyle *et al.*, 1999; Mohn and Tiedje, 1990). Separate populations of SO_4^{2-} reducers can be either sources (indirectly) or sinks for acrylate. Acrylate is formed during the breakdown of the marine osmolyte dimethylsulfoniopropionate, a process that is catalyzed by SO_4^{2-} reducers in sediments, and then reduced by other SO_4^{2-}-reducing species (van der Maarel *et al.*, 1998, 1996d).

8.08.7.3 Taxonomic Considerations

Sulfate-reducing bacteria are a complex physiologic group and classifying them has traditionally required the consideration of several properties, the most important of which are motility, cell shape, the guanine plus cytosine content of DNA, the presence of desulfoviridin and cytochromes, growth temperature, use of various electron donors, and the ability to conduct complete or incomplete oxidation. New analysis of ribosomal RNA (rRNA) sequences has allowed for a more thorough organization of the SO_4^{2-}-reducing bacteria into four major groups: gram-negative mesophilic, gram-positive spore forming, thermophilic bacterial, and thermophilic Archeal (Castro *et al.*, 2000). The gram-negative mesophilic group of SO_4^{2-} reducers is placed within the delta (δ) subdivision of the *Proteobacteria* and includes two major families, the *Desulfovibrionaceae* and *Desulfobacteriaceae* although many genera fall outside of these families. The *Desulfovibrionaceae* includes the genera *Desulfovibrio* and *Desulfomicrobium*, and this family appears to be rather phylogenetically distinct (Devereux *et al.*, 1990). The *Desulfobacteriaceae* family is much less distinct and includes several genera (perhaps >20), including many of the complete oxidizing species (Castro *et al.*, 2000; Widdel and Bak, 1992). The gram-positive spore-forming SO_4^{2-} reducers constitute primarily members of the genus *Desulfotomaculum*.

Prior to the late 1970s, only two genera of SO_4^{2-}-reducing bacteria were known, *Desulfovibrio* and *Desulfotomaculum* (Widdel, 1988; Widdel and Bak, 1992). The desulfovibrios have received the most attention because they are relatively easily isolated from the environment and are not difficult to maintain in laboratory culture. They were originally described as gram-negative bacteria that are curved rods, do not produce spores, and utilize primarily H_2 and lactate as electron donors (Postgate and Campbell, 1966). However, the group has been expanded to include members that are capable of using several other electron donors and electron acceptors, the latter including NO_3^-, O_2, and metal oxides (Barton *et al.*, 1983; Cypionka, 2000; Lovley *et al.*, 1993b; Moura *et al.*, 1997). They are the classic examples of SO_4^{2-} reducers that conduct incomplete metabolism with acetate as an important end product. Phylogenetic analyses have shown the *desulfovibrios* to be sufficiently diverse and distinct to warrant placement within their own family (Devereux *et al.*, 1990). *Desulfovibrio* species are routinely isolated from marine sediments and probably are an important component of SO_4^{2-} reduction in the sea. The *desulfovibrios* are the only group observed to enter into syntrophic relationships with methanogenic bacteria (Fiebig and Gottschalk, 1983; Pankhania *et al.*, 1988).

Desulfotomaculum species are gram-positive, rod shaped, spore-forming SO_4^{2-}-reducing bacteria that as a group display wide phylogenetic diversity (Rabus *et al.*, 2000). They are also quite phylogenetically distinct from all other SO_4^{2-}-reducing bacteria. The genus exhibits a great nutritional versatility comparable to that of non-spore-forming sulfate reducers, including the use of H_2, alcohols, fatty acids, other aliphatic mono-carboxylic or dicarboxylic acids, alanine, hexoses, or phenyl-substituted organic acids as electron donors for dissimilatory SO_4^{2-} reduction (Widdel and Pfennig, 1999). *Desulfotomaculum* species are not considered to be particularly important in marine sediments compared to the non-spore-forming SO_4^{2-} reducers. However, they are easily isolated from organic-rich sediments and they possess metabolic capabilities that are known to be important in sediments. Their ability to form spores gives them a competitive advantage in some habitats and they are found to dominate environments like rice paddy sediments that undergo wetting–drying cycles that could harm nonsporing cells (Rabus *et al.*, 2000). *Dtm. acetoxidans* was the first SO_4^{2-} reducer isolated that was capable of acetate oxidation (Widdel and Pfennig, 1977).

Since the early 1980s, many new lineages of SO_4^{2-} reducers have been isolated and described. The phylogeny of SO_4^{2-} reducers has expanded to include the family *Desulfobacteriaceae* (Widdel and Bak, 1992), which includes many new genera that are capable of complete and/or incomplete oxidation; the genera *Desulfobacter*, *Desulfobacterium*, and *Desulfococcus* to name a few. Members of the filamentous, gliding SO_4^{2-}-reducing genus *Desulfonema* also appear to be common inhabitants of organic-rich sediments, especially within sharp redox gradients (Fukui *et al.*, 1999). Sulfate reducers as a whole are phylogenetically distinct from other bacteria, which has led to the discovery of signature DNA sequences of ribosomal subunit genes that have been used as oligonucleotide probes and polymerase chain reaction primers for the detection, determination of relative abundance, and microscopic visualization of members of the group (Amann *et al.*, 1990a,b; Daly *et al.*, 2000; Devereux *et al.*, 1992; Devereux and Stahl, 1993). Probes specific for SO_4^{2-}-reducing bacteria have been applied to marine water columns (Ramsing *et al.*, 1996; Teske *et al.*, 1996), biofilms and bacterial mats (Fukui *et al.*, 1999; Minz *et al.*, 1999a; Ramsing *et al.*, 1993; Santegoeds *et al.*, 1999), and sediments (Hines *et al.*, 1999; LlobetBrossa *et al.*, 1998; Rooney-Varga *et al.*, 1997; Sahm *et al.*, 1999b; Sass *et al.*, 1998). In addition, reverse sample genome probing (Voordouw *et al.*, 1991), hydrogenase (Wawer *et al.*, 1997) and dissimilatory sulfite-reductase genes (Minz *et al.*, 1999b; Wagner *et al.*, 1998), and denaturing gradient gel electrophoresis (Okabe *et al.*, 2002) have been used to study SO_4^{2-} reducers in nature.

8.08.7.4 Sulfate-reducing Populations

The population composition of SO_4^{2-}-reducing bacteria has been investigated using culturing, biochemical, and genetic methods. In general, estimates of abundance using viable counting methods such as colony counts on solid media or growth in liquid media after serial dilutions (most probable number (MPN) methods), are 10^2–$10^5\ ml^{-1}$, which appear low when considered in terms of the rate of *in situ* SO_4^{2-} reduction and culture estimates of rates per cell. This discrepancy is undoubtedly due to the inability of viable counting techniques to recover the majority of bacteria present. However, MPN techniques yield more realistic estimates of SO_4^{2-} reducer abundance (10^6–$10^8\ ml^{-1}$) when applied to organic-rich habitats such as marine microbial mats (Visscher *et al.*, 1992; Ramsing *et al.*, 1993) and salt marsh sediments (Hines *et al.*, 1999).

Due to the complex nature of the anaerobic bacterial food web, it is generally held that SO_4^{2-} reducers account for only ~5% of the total bacteria present despite their important role at the end of the food web (Devereux *et al.*, 1996; Li *et al.*, 1999; LlobetBrossa *et al.*, 1998). *In situ* hybridization techniques that use fluorescent oligonucleotide probes to visualize individual cells of specific bacteria groups have yielded SO_4^{2-}-reducing bacteria counts in marine sediments as high as $3 \times 10^7\ ml^{-1}$, which represented up to 6% of the total *Bacteria* (LlobetBrossa *et al.*, 1998). Hybridizations of bulk sedimentary RNA with probes also demonstrated that ~1–6% of the total bacteria present in sediments were SO_4^{2-}-reducing (Devereux *et al.*, 1996), although this technique demonstrated that 20% of the prokaryotes in a subtidal sediment were likely SO_4^{2-}-reducing (Sahm *et al.*, 1999b). Sulfate reducers in salt marsh sediments can account for >30% of the total bacteria (Hines *et al.*, 1999). This high percentage is likely due to the fact that marsh grasses exude organic substrates from roots that can be used directly by SO_4^{2-} reducers, thus circumventing the need for fermenting bacteria.

A wide variety of SO_4^{2-}-reducing bacteria are found within sediments. Enumerations of bacteria using MPN methods supplemented with specific substrates for SO_4^{2-}-reducing groups have shown the presence of bacteria able to use many substrates including lactate, ethanol, acetate, malate, and propionate (Laanbroek and Pfennig, 1981). Lactate and acetate utilizers often outnumber other groups illustrating the importance of both incomplete and complete oxidizing SO_4^{2-} reducers in sediments. Sulfate-reducing marine sediments display an abundance of even chain bacterial phospholipid fatty acids indicative of the presence of acetate-utilizing bacteria of the genus *Desulfobacter* (Parkes *et al.*, 1993). Using MPN methods, acetate-utilizing sulfate reducers were most abundant in a marine microbial mat (Visscher *et al.*, 1992; Teske *et al.*, 1998), while ethanol utilizers greatly outnumbered acetate utilizers in a salt marsh sediment (Hines *et al.*, 1999).

Molecular analyses have furthered the description of SO_4^{2-}-reducing groups in depositional environments. Both whole-cell *in situ* hybridizations and hybridizations using RNA extracted from sediments have been employed to investigate the diversity of SO_4^{2-} reducers (Ramsing *et al.*, 1993; Rooney-Varga *et al.*, 1997). Incomplete oxidizing groups of sulfate reducers seem to dominate in marine sediments, primarily members of the *Desulfovibrionaceae*, but also *Desulfobulbus* species are abundant. However, the distribution of groups varies with depth and among habitats. For example, members of the *Desulfobacteriaceae* and *Desulfovibrionaceae* were equally abundant in the upper 2.0 cm of sediments, but the incomplete oxidizers (*Desulfovibrionaceae* species) dominated at greater depths (Devereux *et al.*, 1996). Complete oxidizers were essentially absent from Arctic Ocean sediments (Sahm *et al.*, 1999a), yet they dominated sediments inhabited by salt marsh grasses (Hines *et al.*, 1999). Complete oxidizers, i.e., *Desulfobacteriaceae* species accounted for over 20% of the total recovered RNA in some instances in the marsh, while *Desulfovibrionaceae* species were a small fraction in all cases (Hines *et al.*, 1999; Rooney-Varga *et al.*, 1997). Members of the *Desulfobacteriaceae* are metabolically diverse and may be well suited for environments that undergo widely changing seasonal redox changes like those encountered in salt marsh sediments and at the sediment surface. The presence of high numbers of *Desulfobulbus* species in marsh sediments (Hines *et al.*, 1999) may reflect the ability of members of the genus to conduct sulfur disproportionation reactions, which would be stimulated by plant-mediated redox cycling within the rhizosphere and the production of intermediate redox states of sulfur.

8.08.7.5 Factors Regulating Sulfate Reduction Activity

8.08.7.5.1 Sulfate-reducing activity

Knowledge of the role of SO_4^{2-} reduction in sediments has increased greatly since the introduction of the use of ^{35}S as a tracer for determining sulfate reduction rates (Jørgensen and Fenchel, 1974; Sorokin, 1962). At first, rates were estimated from the incorporation of ^{35}S into dissolved sulfide (H_2S) and acid volatile sulfides (iron monosulfides, FeS), which were thought to be the only significant products (Jørgensen, 1978a). However, it was determined that in some habitats, such as salt marsh sediments and near the oxic/anoxic boundary, a significant fraction of the reduced sulfur produced during SO_4^{2-} reduction is rapidly converted to pyrite (FeS_2), a compound that was previously thought to be produced at rates on the order of several years (Howarth, 1979). Methods now routinely include a reduction step using reduced chromium under acidic conditions to incorporate all major reduced inorganic sulfur species that may form during SO_4^{2-} reduction (Westrich, 1983; Fossing and Jørgensen, 1989; Meier *et al.*, 2000). In highly organic soils and sediments, it is usually necessary to recover the carbon-bonded ^{35}S label (Wieder and Lang, 1988). Sulfate reduction rates can also be determined from losses of SO_4^{2-} in incubated sediments and from mathematical models that describe the loss of SO_4^{2-} with sediment depth in terms of sedimentation and diffusion (Jørgensen, 1978a,b,c). Depth increases in reduced sulfur compounds and the abundance of SO_4^{2-}-reducing bacteria are useful as comparative indicators of the SO_4^{2-} reduction process, but they are poor indicators of actual rates of activity.

The rate of organic matter input (i.e., sedimentation) and the availability of SO_4^{2-} control the rate of SO_4^{2-} reduction in sediments. Sulfate is rarely limiting in marine systems except in brackish estuarine waters and with depth in sediments where SO_4^{2-} has been depleted. Sulfate reduction rates in sediments can span several orders of magnitude, but typical near-shore rates in the upper 5–10 cm of marine sediments are often 50–500 nmol cm^{-3} d^{-1} (Skyring, 1987). Rates in sediment with unusually rapid sedimentation rates can be reach 2,000 nmol ml^{-1} d^{-1} (Crill and Martens, 1987). Sulfate reduction in salt marshes and microbial mats can reach rates as high as 4,000 nmol ml^{-1} d^{-1} and 14,000 nmol ml^{-1} - d^{-1}, respectively (Canfield and Des Marais, 1991; Hines *et al.*, 1999). Anaerobic CH_4 oxidation supports a large fraction of the SO_4^{2-} reduction in some marine sediments (Table 6).

8.08.7.5.2 *Temperature*

Sulfate reduction occurs over a wide range (0–110 °C) of temperatures (Castro *et al.*, 2000; Elsgaard *et al.*, 1994; Jørgensen *et al.*, 1992; Knoblauch and Jorgensen, 1999; Knoblauch *et al.*, 1999a; Kostka *et al.*, 1999b; Sievert and Kuever, 2000; Stetter *et al.*, 1993). Above ~115 °C, SO_4^{2-} reduction is believed to occur only by thermochemical reactions (Machel, 2001). Like other biological processes, biological SO_4^{2-} reduction is affected strongly by temperature. Seasonal changes in SO_4^{2-} reduction activity often follow temperature well, except for lags in activity due to the time required to remove oxygen and other competing electron acceptors (Crill and Martens, 1987; Hines *et al.*, 1982; Jørgensen, 1977). In general, rates of activity can vary by factors of <5 to >30 between winter and summer, and these changes have profound influence on redox conditions and the accumulation of reduced species (Jørgensen, 1977).

8.08.7.5.3 *Carbon*

Sulfate reduction is controlled strongly by the quantity and quality of organic matter present. In marine sediments, there is a strong relationship between organic sedimentation rate and SO_4^{2-} reduction rate (Goldhaber and Kaplan, 1975) and it is generally held that SO_4^{2-} reduction is limited by organic matter availability except when SO_4^{2-} concentration is quite low (Boudreau and Westrich, 1984; Dornblaser *et al.*, 1994; Westrich and Berner, 1984). Although SO_4^{2-} reducers usually consume a relatively narrow suite of organic compounds, the tight coupling of SO_4^{2-} reduction with degradative processes prior to SO_4^{2-} reduction results in a stoichiometric relationship between the degradation of complex organic matter and the reduction of SO_4^{2-} (Richards, 1965). In general, there is a 2 : 1 molar relationship between labile carbon deposited into the SO_4^{2-} reduction zone and SO_4^{2-} reduced (Thamdrup and Canfield, 1996). The quantity of nitrogen and phosphorus mineralized during SO_4^{2-} reduction can be predicted using C/N/P ratios of organic matter and SO_4^{2-} reduction stoichiometry (Hines and Lyons, 1982; Martens *et al.*, 1978). Sulfate reduction activity responds rapidly to increased inputs of organic material with the seasonal deposition of spring phytoplankton blooms resulting in significant increases in activity in lake (Hadas and Pinkas, 1995) and ocean sediments (Boetius *et al.*, 2000a). Although it is clear that low-molecular-weight fatty acids are important substrates for SO_4^{2-}-reducing bacteria, different groups of SO_4^{2-} reducers may consume different acids (Boschker *et al.*, 2001).

Some electron donors are more readily metabolized by methanogens than SO_4^{2-} reducers and these "noncompetitive" substrates can allow methanogenesis to occur in the presence of active SO_4^{2-} reduction (Oremland *et al.*, 1982). Examples of these types of substrates include C_1 compounds like methylated amines, methylated sulfur compounds (e.g., dimethylsulfide and methane thiol), and methanol. Methylated nitrogen and sulfur compounds are common degradation products of osmoregulating compounds found in certain marine algae and salt marsh grasses, and their use by methanogens partially explains the occurrence of significant concentrations of CH_4 in SO_4^{2-}-containing estuarine and salt marsh sediments (Dacey *et al.*, 1987; Kiene, 1996b). In fact, the degradation of the osmoregulant glycine betaine in marine sediments produces acetate and trimethylamine, the former of which is consumed by sulfate-reducing bacteria, while the latter is consumed by methanogens (King, 1984). Methanol is a degradation product of plant structural components such as pectin.

8.08.7.5.4 Sulfate and molecular oxygen concentrations

The SO_4^{2-} concentration in sediments affects SO_4^{2-} reduction only when concentrations are quite low. The reduction of SO_4^{2-} in marine sediments appears to be zero-order with respect to SO_4^{2-} to concentrations of ~2 mM (Boudreau and Westrich, 1984; Goldhaber and Kaplan, 1974). In freshwaters, SO_4^{2-} concentrations must be much lower before they limit SO_4^{2-} reduction (Bak and Pfennig, 1991; Lovley and Klug, 1983b; Sinke *et al.*, 1992). Because freshwater contains very little SO_4^{2-} compared to seawater, the importance of SO_4^{2-} reduction in sediments increases in an estuary as the salinity increases (Capone and Kiene, 1988). Therefore, the vertical extent of the SO_4^{2-} reduction zone increases substantially as more SO_4^{2-} becomes available, while the methanogenic zone is "pushed" deeper into the sediment and its contribution to carbon mineralization decreases in importance.

Sulfate-reducing bacteria are generally anaerobic, but recent studies have shown that some varieties are capable of O_2 use, are able to withstand several hours of full aeration, and are common inhabitants of oxic regions of microbial mats (Cypionka, 2000; Krekeler *et al.*, 1997). Sulfate reducers withstand O_2 stress better in the presence of a co-metabolizing bacterium that presumably consumes O_2 (Gottschalk and Szwezyk, 1985). However, studies of co-cultures of a SO_4^{2-} reducer and a facultative anaerobe suggested the occurrence of O_2-dependent growth by the SO_4^{2-} reducer in the absence of SO_4^{2-} (Sigalevich *et al.*, 2000a,b; Sigalevich and Cohen, 2000).

Sulfate-reducing populations within active sediments, like microbial mats, can exhibit a bimodal distribution with two distinct maxima of differing populations, one within surficial oxic layers and another in deeper anoxic sediments (Minz *et al.*, 1999a; Ramsing *et al.*, 1993; Risatti *et al.*, 1994; Sass *et al.*, 1997). Sulfate reducers within or near oxic regions can benefit from labile organic compounds released by photosynthetic bacteria (Canfield and Des Marais, 1991; Teske *et al.*, 1998; Visscher *et al.*, 1992) or vascular plants (Scheid and Stubner, 2001; Blaabjerg and Finster, 1998; Hines *et al.*, 1999). Indeed, rates of SO_4^{2-} reduction measured in oxic regions of microbial mats can equal or exceed those noted in deeper anoxic lamina (Canfield and Des Marais, 1991; Frund and Cohen, 1992; Visscher *et al.*, 1992). The complete oxidizing and gliding SO_4^{2-} reducer, *Desulfonema*, is a common inhabitant of the oxic–anoxic interface in microbial mats (Fukui *et al.*, 1999; Risatti *et al.*, 1994; Teske *et al.*, 1998), and is also abundant within the partially oxygenated rhizosphere of salt marsh plants (Rooney-Varga *et al.*, 1997). Filamentous morphology, aggregate formation, and diurnal migration are all adaptations that allow this species to thrive in rapidly changing conditions.

8.08.7.6 Microbial Reduction of Sulfur

The microbial reduction of elemental sulfur (S^0) was not discovered as early as the reduction of SO_4^{2-}. The first report of the sole use of S^0 as a terminal electron acceptor for growth was reported well into the twentieth century (Pelsh, 1936), and the first pure culture (*Desulfuromonas acetoxidans*) was isolated four decades later (Pfennig and Biebl, 1976). *D. acetoxidans* was discovered as a partner living syntrophically with an anaerobic phototroph that oxidized sulfide to S^0 and provided organic electrons donors. Hence, a complete anaerobic sulfur cycle was maintained. Several other S^0 reducers were isolated soon thereafter, which couple the reduction of S^0 with the oxidation of compounds like acetate, formate, lactate, fumarate, aspartate, dimethylsulfoxide, and H_2, (Widdel, 1988). Some SO_4^{2-}-reducing bacteria possess the ability to reduce S^0 (Biebl and Pfennig, 1977; He *et al.*, 1986), but this is relatively rare. Certain Fe(III)-reducing bacteria also possess the ability to reduce S^0, and species have been isolated that are able to use a variety of other electrons acceptors such as NO_3^-, Mn(IV), thiosulfate, and O_2 (Caccavo *et al.*, 1994; Myers and Nealson, 1988; Roden and Lovley, 1993). Dissimilatory S^0 reduction also occurs in bacteria that are known to grow aerobically at normal O_2 tension (Balashova, 1985; Lovley *et al.*, 1989; Myers and Nealson, 1988). The ability to use organic disulfide molecules, such as cysteine or glutathione, as terminal electron acceptors appears to be restricted to certain members of the S^0-reducing bacteria (Pfennig and Biebl, 1976). Among the S^0-reducing bacteria, several species are able to generate ATP during the reduction of S^0 (Hedderich *et al.*, 1999). However, many members of the Archea are strictly fermentative S^0 reducers that use S^0 reduction as an electron sink in lieu of respiratory S^0 reduction (Stetter, 1996). Since S^0 is highly insoluble, it is likely that many S^0 reducers utilize polysulfide as an electron acceptor, or at least as an intermediate in S^0 reduction (Hedderich *et al.*, 1999). In general, the habitat of sulfur reducers is similar to that of SO_4^{2-} reducers, so ecologically they coexist. However, in most instances SO_4^{2-} reducers produce more sulfide.

Several thermophilic microorganisms are capable of sulfur reduction (Bonch-Osmolovskaya *et al.*, 1990; Huber *et al.*, 1992; L'Haridon *et al.*, 1998) and this capability was employed in the early isolation of thermophilic Archea (Stetter, 1996), some of which are capable of aerobic growth using sulfur as an electron donor and anaerobic growth using sulfur as an electron acceptor (Segerer *et al.*, 1985). In fact, most of the sulfur-reducing bacteria known belong to the Archeal domain (Hedderich *et al.*, 1999). In the presence of sulfur, methanogenic Archea, especially thermophilic strains, reduce sulfur to H_2S, while CH_4 formation is highly curtailed (Stetter and Gaag, 1983). Sulfur reduction is common in microorganisms situated within deep branches of the phylogenetic tree suggesting that sulfur reduction is a very ancient process (Stetter, 1997; Woese, 1987).

A *Desulfuromonas* sp. was the first eubacterium to be classified using rRNA techniques and it was found to affiliate with the δ-subclass of Proteobacteria and closely with the completely oxidizing SO_4^{2-}-reducing bacteria (Fowler *et al.*, 1986; Liesack and Finster, 1994). However, sulfur reducing eubacteria are quite diverse and are found within other subclasses of the Proteobacteria, many of which are not related phylogenetically to SO_4^{2-} reducers (Lau *et al.*, 1987; Schleifer and Ludwig, 1989). In general, the sulfur reducers as a group still require proper classification (Widdel and Pfennig, 1999).

8.08.7.7 Disproportionation

It has been demonstrated that SO_4^{2-}-reducing bacteria have the ability to conduct inorganic fermentations of sulfur compounds, or *sulfur disproportionation* (Bak and Cypionka, 1987). The ability to disproportionate thiosulfate, sulfite and elemental sulfur (S^0) has been found in many SO_4^{2-}-reducing genera (Kramer and Cypionka, 1989), and it occurs as follows:

Thiosulfate

$$S_2O_3^{2-} + H_2O \rightarrow SO_4^{2-} + HS^- + H^+ \quad (35)$$

Elemental sulfur

$$4S^0 + 4H_2O \rightarrow SO_4^{2-} + 3HS^- + 5H^+ \quad (36)$$

Sulfite

$$4SO_3^{2-} + H^+ \rightarrow 3SO_4^{2-} + HS^- \quad (37)$$

Pure cultures have been isolated that are capable of all these reactions including an isolate that is an obligate disproportionator and does not reduce SO_4^{2-} (Finster *et al.*, 1998).

Disproportionation reactions are important in sediments because they provide a mechanism for anaerobic sulfur cycling (Jørgensen, 1990a), and SO_4^{2-} reducers capable of this type of metabolism are numerically abundant (Bak and Pfennig, 1987; Jørgensen and Bak, 1991). Thiosulfate appears to be a key component of the sulfur cycle in sediments because it can be oxidized, reduced, and disproportionated. Sulfate-reducing bacteria are key components of this cycling since they are capable of reducing SO_4^{2-} and thiosulfate, and they are able to disproportionate thiosulfate. A "thiosulfate shunt" seems to exist in sediments where thiosulfate couples both reductive and oxidative pathways of the sulfur cycle (Jørgensen, 1990b) (Figure 26). Thiosulfate can be the main product of sulfide oxidation in reducing sediments, and much of this can be disproportionated with the remainder either oxidized to SO_4^{2-} or reduced to sulfide. Hence, most of the sulfide produced during SO_4^{2-} reduction may be ultimately oxidized to SO_4^{2-} after passing through a thiosulfate intermediate.

Although S^0 disproportionation is important in the sulfur cycle, it is not an energetically favorable reaction under standard conditions (Bak and Cypionka, 1987) and some have argued that it is not a major mode of metabolism in terms of bacterial energetics or growth (Jørgensen and Bak, 1991). However, bacteria are able to grow via this process when metal oxides are present to rapidly remove sulfide (Lovley and Phillips, 1994a; Thamdrup *et al.*, 1993). Such sulfide scavenging by sedimentary metal oxides may allow the

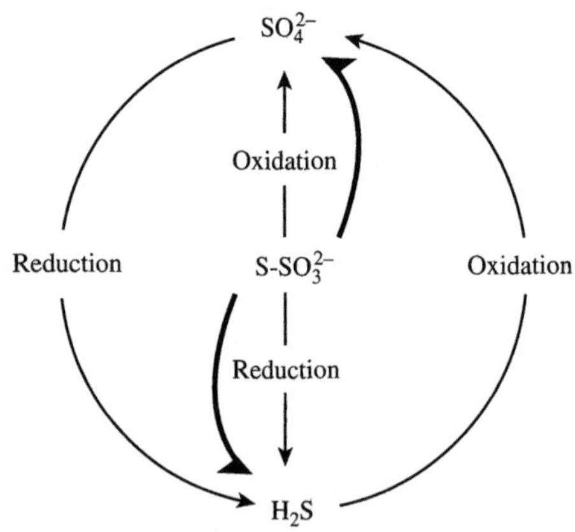

Figure 26 Thiosulfate disproportionation (heavy lines). Typical oxidation (aerobic) and reduction (anaerobic) reactions are included (light lines). A combination of these reactions leads to a thiosulfate shunt that allows for the anaerobic production of intermediate redox states of S in anoxic sediments.

process to act as a common metabolic transformation (Finster *et al.*, 1998).

8.08.7.8 Sulfur Gases

Gaseous sulfur compounds are produced and consumed by microorganisms and they play a significant role in the global sulfur cycle because they connect the terrestrial, freshwater, and marine environments via the atmosphere, and they affect atmospheric chemistry and physics. The most important volatile sulfur compounds are H_2S, dimethylsulfide (DMS), methane thiol (MeSH), carbonyl sulfide (OCS), carbon disulfide (CS_2), and dimethyl disulfide (DMDS) (Bates *et al.*, 1992; Hines, 1996; Kiene, 1996a; Lomans *et al.*, 2002b). Reduced sulfur compounds are subject to chemical and photochemical oxidation in the atmosphere, which yield acidic products that contribute to acid precipitation, and aerosol particles that directly attenuate incoming solar radiation or lead to cloud condensation nuclei, both of which affect the global radiative balance (Charlson *et al.*, 1987; Kelly and Baker, 1990; Panter and Penzhorn, 1980). Early work predicted that H_2S produced by dissimilatory SO_4^{2-} reduction was the primary biogenic sulfur gas emitted into the atmosphere (Bremner and Steele, 1978; Natusch and Slatt, 1978). However, although the biogenic production of organosulfur gases has been known since the 1930s (Haas, 1935), it was only since the 1970s that the importance of these gases has been recognized (Graedel, 1979; Lovelock *et al.*, 1972; Rodhe and Isaksen, 1980). It is now clear that DMS emissions constitute the bulk of the sulfur that enters the atmosphere each year (~75%, primarily from oceanic sources). See Chapter 8.14 of this volume for details on the role of sulfur gases in the global sulfur cycle.

8.08.7.8.1 Hydrogen sulfide

The liberation of H_2S is controlled not only by the rate of its production by SO_4^{2-}-reducing bacteria, but also by its pH-dependent speciation, its tendency to rapidly precipitate as metal sulfides, and its rapid chemical and biological oxidation. Only the protonated species (H_2S) is volatile, and at neutral pH, most inorganic sulfide is present as bisulfide ion (HS^-), whereas sulfide (S^{2-}) dominates under alkaline conditions. These three species are known collectively as ΣH_2S. Hence, the escape of sulfide should be enhanced at low pH. Sulfate reduction is most dominant in marine sediments and this is where the highest emissions of gaseous H_2S occur (Hines, 1996). However, DeLaune *et al.* (2002) reported higher emissions of H_2S from brackish marshes than from true salt marshes. In most cases, despite the fact that dissolved sulfide concentrations can be very high (mM) (Hines *et al.*, 1989; King *et al.*, 1982), only a small portion of the gross production of sulfide escapes to the atmosphere (Jørgensen and Okholm-Hansen, 1985; Kristensen *et al.*, 2000). Bodenbender *et al.* (1999) reported H_2S fluxes from intertidal marine sediments that were up to 2.6×10^4 times less than the rate of SO_4^{2-} reduction. Release of H_2S is usually highest at night, because during the day, photosynthetic microorganisms at the sediment surface either directly consume sulfide or increase the penetration of the O_2 into sediments that enhances chemical oxidation (Bodenbender *et al.*, 1999; Castro and Dierberg, 1987; Hansen *et al.*, 1978; Jørgensen and Okholm-Hansen, 1985). Sulfide emissions from intertidal sediments can increase during flooding due to tidal pumping (Jørgensen and Okholm-Hansen, 1985). Vegetated sediments tend to release less H_2S due to the oxidation of sulfides by O_2 released from roots (Hines, 1996). In general, sulfureta-like sediments, i.e., highly reducing, unvegetated, and sulfide-rich environments, allow for a significant loss of H_2S to the atmosphere since O_2 penetration is minimal and metals capable of precipitating sulfide are already removed. However, these environments are the exception rather than the rule.

8.08.7.8.2 Methylsulfides

Numerous studies of DMS formation and consumption have appeared since the recognition of the importance of this compound in marine biogeochemistry and atmospheric chemistry. Indeed, DMS accounts for ~90% of the biogenic sulfur emissions from the marine environment (Andreae and Crutzen, 1997). A link has been proposed among DMS production and flux, the atmospheric oxidation of DMS, and the subsequent formation of SO_4^{2-} aerosols and cloud condensation nuclei (Charlson *et al.*, 1987). In this model, increasing global temperatures lead to enhanced production of DMS, which attenuates further warming by radiative backscatter from aerosols and reflection of radiation from increased cloud cover. Although most work to date has been conducted on the cycling of DMS in seawater, considerable effort has also been invested in understanding the biogeochemistry of DMS in wetlands and sediments (Kiene, 1996b).

Several compounds can serve as precursors of DMS including dimethylsulfonium compounds such as dimethylsulfoniopropionate (DMSP), S-containing amino acids, MeSH, dimethylsulfoxide (DMSO), and methoxylated aromatic compounds, which are capable of methylating inorganic sulfide and MeSH (Bak *et al.*, 1992; Finster *et al.*, 1990; Kadota and Ishida, 1972; Kiene and Taylor, 1988a,b; Kreft and Schink, 1993; Lomans *et al.*, 2002b). Methionine is ubiquitous in organisms and has been shown to be a precursor of several sulfur gases including DMS (Bremner and Steele, 1978; Kiene and Capone, 1988; Kiene and Visscher, 1987b; Zhang *et al.*, 2000; Zinder and Brock, 1978c). Methionine degradation produces MeSH, which can be subsequently methylated to DMS (Kiene and Capone, 1988; Kiene and Visscher, 1987b). S-methyl-cysteine can also be degraded to MeSH, whereas S-methyl-methionine degrades primarily to DMS (Kiene and Capone, 1988). DMSO reduction by bacteria also produces DMS (Griebler, 1997; Zinder and Brock, 1978a; Zinder and Brock, 1978b) and several SO_4^{2-}-reducing bacteria with this ability have been isolated (Bale *et al.*, 1997; Jonkers *et al.*, 1996; van der Maarel *et al.*, 1998).

DMSP is a major source of DMS in marine systems (Kiene, 1996a). DMSP is produced by marine micro- and macroalgae and halophytic plants where it acts as an osmoregulant or as part of an antioxidant system (Sunda *et al.*, 2002; Yoch, 2002). Although many phytoplankton contain a DMSP lyase enzyme, bacteria are thought to be the primary degraders of DMSP in seawater (Kiene, 1992). Fungi associated with decaying salt marsh plants also possess a DMSP lyase (Bacic *et al.*, 1998). Cleavage of DMSP can lead to the production of DMS and acrylate, but DMSP can also be demethylated to 3-methylmercaptopropionate, which can be demethiolated to MeSH or demethylated to 3-mercaptopropionate (Kiene and Taylor, 1988a,b; van der Maarel *et al.*, 1996c; Visscher and Taylor, 1994). Members of the *Proteobacteria*, especially the α-group, appear to be important DMS producers in the sea (Gonzalez *et al.*, 1999). However, DMS producers belonging to the β-, δ-, and γ-subdivisions have been identified in seawater (de Souza and Yoch, 1995; Jonkers *et al.*, 1998; Ledyard *et al.*, 1993; van der Maarel *et al.*, 1996b) and members of the γ-subdivision dominated DMS-producing isolates from a salt marsh (Ansede *et al.*, 2001). It has been hypothesized for aerobic seawater that the pathway used for DMSP degradation is controlled by DMSP concentrations and the demand for sulfur, and that DMS formation is usually favored at high DMSP concentrations (Kiene *et al.*, 2000).

Both direct cleavage of DMSP to DMS/acrylate and DMSP demethylation pathways have been observed in anoxic sediments and oceanic waters (Kiene and Taylor, 1988b; Yoch, 2002). However, unlike oxic environments, the demethylation of DMSP in anoxic sediments appears to require two organisms, one to demethylate DMSP to 3-methiolpropionate, followed by another to form MeSH (van der Maarel *et al.*, 1995, 1993). Sulfate-reducing bacteria have been isolated that are capable of cleaving DMSP to DMS and acrylate (van der Maarel *et al.*, 1996d) as well as demethylating DMSP (van der Maarel *et al.*, 1996c). A SO_4^{2-}-reducing bacterium has been isolated that cleaves DMSP and then reduces the liberated acrylic acid (van der Maarel *et al.*, 1996d). Acrylate and SO_4^{2-} reduction occurred simultaneously, but SO_4^{2-} reduction yielded more growth.

DMS (and MeSH) concentrations in freshwater sediments are considerably lower than in marine habitats, primarily due to the low concentrations of DMSP in freshwater plants (Bechard and Rayburn, 1979). However, some small ponds and wetlands exhibit DMS levels similar to those in seawater (3–25 nM) (Kiene and Hines, 1995; Nriagu *et al.*, 1987), and DMS emissions from freshwater habitats can be significant (Hines, 1996; Kelly and Smith, 1990). It has been known for sometime that freshwater bacteria have the ability to form DMS from DMSP, and an obligate anaerobe with this ability was isolated early (Wagner and Stadtman, 1962). Freshwater sediments are capable of producing DMS from DMSP amendments even though these sediments may be nearly DMS-free (Yoch *et al.*, 2001). Enrichments of these sediments with DMSP led to the isolation of a suite of gram-positive DMS-producing bacteria related to *Rhodococcus* spp.

Most studies indicate that the primary path of DMS formation in freshwater sediments is via the methylation of inorganic sulfide to MeSH, which is subsequently methylated to DMS (Bak *et al.*, 1992; Drotar *et al.*, 1987a; Finster *et al.*, 1990; Kiene and Hines, 1995; Lomans *et al.*, 2001a, 1997). Thiol methyltransferase enzymes that catalyze this methylation are relatively widespread in microorganisms including bacteria and protozoa (Drotar *et al.*, 1987a,b; Drotar and Fall, 1985). Methoxylated aromatic compounds, which are degradation products of lignin, have been shown to be important as methyl donors for the methylation of sulfide and MeSH (Finster *et al.*, 1990; Kiene and Hines, 1995), and several anaerobic bacteria have been isolated that can perform this sulfide-mediated *O*-demethylation reaction (Bak *et al.*, 1992; Lomans *et al.*, 2001a; Mechichi *et al.*, 1999). The hydroxylated aromatic residue that remains after demethylation is degraded to acetate or acetate and butyrate (Kreft and Schink, 1993). This differs from the more common pathway for acetate production in which homoacetogenic bacteria transfer methyl groups from methoxylated aromatic compounds to CO to produce acetate (Bache and Pfennig, 1981).

DMS and other volatile organo-sulfur compounds rarely accumulate in sediments, because they are consumed by microorganisms as rapidly as they are produced (Lomans *et al.*, 2002a,b). Several bacteria are able to degrade DMS and MeSH aerobically, but most of those isolated are *Thiobacillus*, *Methylophaga* or *Hyphomicrobium* spp. (Cho *et al.*, 1991; de Bont *et al.*, 1981; de Zwart *et al.*, 1996; Sivelä and Sundmann, 1975; Smith and Kelly, 1988a; Visscher *et al.*, 1991). Many bacteria are capable of oxidizing DMS to DMSO in the presence of an additional carbon source (Zhang *et al.*, 1991).

Zinder and Brock (1978d) were the first to report the anaerobic conversion of DMS and MeSH to CO_2 and CH_4 in freshwater lake sediments and sewage sludge. Since then, this process has been reported to occur in many anoxic environments including salt marsh and mangrove sediments, hypersaline lakes, and a variety of freshwater sediments (Jonkers *et al.*, 2000; Kiene, 1988; Kiene *et al.*, 1986; Lomans *et al.*, 1999a, 2001b; Lyimo *et al.*, 2000, 2002). Degradation of DMS and MeSH in anaerobic environments is catalyzed primarily by methanogenic, SO_4^{2-}-reducing, NO_3^--reducing, and phototrophic bacteria (Kiene, 1988, 1991a; Kiene *et al.*, 1986; Liu *et al.*, 1990; Lomans *et al.*, 2002b; Lyimo *et al.*, 2000; Oremland *et al.*, 1989; Tanimoto and Bak, 1994; Visscher *et al.*, 1995; Zeyer *et al.*, 1987). Despite the ubiquity of CH_4 formation from DMS and MeSH degradation, SO_4^{2-}-reducing bacteria appear to dominate degradation in marine sediments (Kiene, 1996a) and SO_4^{2-}-reducing bacteria are known to degrade DMS in freshwater environments as well. Methanogens and SO_4^{2-}-reducing bacteria appear to compete for DMS, but methanogens out-compete SO_4^{2-} reducers at high DMS concentrations even when SO_4^{2-} is abundant (Lomans *et al.*, 2002b). The role of SO_4^{2-} reduction in DMS degradation is based primarily on the observed decrease in DMS consumption when SO_4^{2-} reduction is inhibited by molybdate or tungstate (Kiene and Visscher, 1987a). However, a pure culture of a DMS-utilizing SO_4^{2-} reducer has been isolated from a thermophilic digester (Tanimoto and Bak, 1994). The energetic contribution of transformation of methylated sulfur compounds to anaerobic bacteria remains unclear, but the thermodynamics of the use these compounds by methanogens and SO_4^{2-} reducers has recently been evaluated (Scholten *et al.*, 2003). A pure culture of a DMS-degrading denitrifying bacterium has also been isolated (Visscher and Taylor, 1993b).

Several methanogens capable of using DMS have been isolated from marine and hypersaline environments (Finster *et al.*, 1992; Kiene *et al.*, 1986; Oremland *et al.*, 1989; Rajagopal and Daniels, 1986), and a freshwater methanogen with this capability has also been isolated and represents a new genus (Lomans *et al.*, 1999b). The degradation of DMS and MeSH appears to occur similarly to other C_1 compounds such as methanol and methylated amines (Ferry, 1999), but the consumption of methylated sulfur compounds appears to be due to separate inducible enzymes (Ni and Boone, 1993). The methyltransferase responsible for the use of DMS in *Methanosarcina barkeri* is distinct from those used for acetate, methanol, or methylated amines (Tallant and Krzycki, 1997; Tallant *et al.*, 2001). The methanogenic degradation of two moles of DMS yields three moles of CH_4 and one of CO_2. One mole of methyl groups is oxidized to CO_2, which generates reducing equivalents for the reduction of three moles of methyl groups to CH_4 (Finster *et al.*, 1992; Kiene *et al.*, 1986). Complete degradation of methylated sulfides to CO_2 by methanogens may occur when H_2 is removed by hydrogenotrophic bacteria such as SO_4^{2-} or NO_3^- reducers (interspecies H_2 transfer) (Lomans *et al.*, 1999c). Methanogens in oligotrophic peat in northern bogs do not consume DMS, which accumulates during incubations (Kiene and Hines, 1995). It was shown that acetate accumulates in these peats as well (Hines *et al.*, 2001).

8.08.7.8.3 Carbonyl sulfide and carbon disulfide

Bacteria are also involved in the formation and degradation of other volatile sulfur compounds such as carbonyl sulfide (OCS) and carbon disulfide (CS_2). Much less is known about the bacterially mediated transformations of these compounds compared to the methylated sulfur species, especially with respect to the role of anaerobic bacteria. They both constitute a rather minor portion of the global flux of biogenic sulfur gases to the atmosphere (Bates *et al.*, 1992; Hines, 1996). Carbonyl sulfide is a long-lived species in the atmosphere and is the most abundant gaseous form of S, whereas CS_2 has a low atmospheric abundance. Both species appear to form photochemically in aquatic systems, but both also can be formed biologically from a variety of organosulfur precursors including S-containing amino acids (Banwart and Bremner, 1975; Bremner and Steele, 1978) and more unusual sulfur compounds like djenkolic acid ($CH_2[SCH_2CH(CH_2)COOH]_2$) and lanthionine ($S[CH_2CH(NH_2)COOH]_2$) (Piluk *et al.*, 1998). In general, these gases are formed anaerobically in organic-rich environments, but the mechanisms of formation are unclear in most instances (Bremner and Steele, 1978; Kiene, 1996a; Wakeham *et al.*, 1984). Both OCS and CS_2 are toxic to certain bacteria (Borjesson, 2001; Bremner and Bundy, 1874; Seefeldt *et al.*, 1995) and their release from roots has been implicated as a plant defense mechanism (Kanda and Tsuruta, 1995). Both OCS and CS_2 are consumed aerobically, primarily by bacteria involved in oxidation of inorganic sulfur compounds (Kelly and Baker, 1990; Kelly *et al.*, 1994). They are also consumed in anaerobic sediments (Zinder and Brock, 1978a), but the mechanism for this consumption is unresolved. CS_2 can be oxidized to OCS and H_2S anaerobically (Smith and Kelly, 1988b) and denitrifying bacteria have been isolated that are capable of degrading CS_2 anaerobically (Jordan *et al.*, 1997).

8.08.7.9 Microbial Oxidation of Sulfur

Sulfide and a wide variety of additional organic and inorganic reduced sulfur compounds can be used as electron donors for microbial redox reactions. In addition to its importance in "natural" depositional and hot spring environments, sulfur oxidation plays a significant role in sewage sludge, paper- and wood-mill effluents, coal desulfurization, and leaching of ores and minerals. Sulfide-oxidizing microbes remove noxious sulfides (organic and inorganic) and interact with minerals through oxidation of metal sulfides (e.g., FeS, FeS_2), deposition of elemental sulfur (S^0), precipitation of gypsum and dissolution of calcium carbonate (Baas Becking, 1934; Bos and Kuenen, 1990; Edwards *et al.*, 2000a; Ehrlich, 2002; Lens and Kuenen, 2001; Visscher and Van Gemerden, 1993).

The Gibbs free energy yield of carbon oxidation with SO_4^{2-} as a terminal electron acceptor is about 4–10 times lower than with oxygen. However, sulfide, the product of SO_4^{2-} reduction is an excellent electron donor; the reaction of which with oxygen:

$$HS^- + 2O_2 \rightarrow SO_4^{2-} + H^+ \qquad (38)$$

has a $\Delta G^{0\prime}$ of $-798\ \mathrm{kJ\ mol^{-1}}$. Alternatively, anaerobic oxidation using NO_3^-, which couples the sulfur and nitrogen cycles, provides ample energy as well:

$$5HS^- + 8NO_3^- \rightarrow 5SO_4^{2-} + 4N_2 + 3OH^- + H_2O \qquad (39)$$

with a $\Delta G^{0\prime}$ of $-744\ \mathrm{kJ\ mol^{-1}}\ H_2S$. Chemolithotrophic microbes that derive their energy in this fashion are collectively referred to as colorless sulfur bacteria (CSB) in contrast to a physiologically distinct group of pigmented organisms (e.g., purple-sulfur bacteria (PSB)). The latter couple the oxidation of sulfur to photoautotrophic CO_2 fixation (Van Niel, 1931, 1941):

$$HS^- + 2CO_2 + 2H_2O \rightarrow SO_4^{2-} + 2[CH_2O] + H^+$$

HS^- and certain other reduced sulfur-compounds can serve as electron donors for photosystem (PS) I in anoxygenic photolithotrophs (purple and green bacteria) and a few cyanobacteria species. The electrons of HS^- are donated to PS I (Section 8.08.2.1) and utilized to generate reducing biochemical equivalents (NAD(P)H) and ATP. In contrast to oxygenic photolithoautotrophs, the oxidation product of photosynthesis is SO_4^{2-} instead of O_2. In both CSB and phototrophic sulfur bacteria, much of the energy generated from sulfur oxidation is used to assimilate CO_2 into cell material.

Other reduced sulfur compounds that play a significant role in oxidation–reduction reactions in the environment include inorganic compounds such as FeS, FeS_2, $S_2O_3^{2-}$, S_x^{2-}(polysulfides), $S_yO_6^{2-}$ (polythionates), and S^0; organic compounds include methylsulfides (CH_3SH, $(CH_3)_2S$). Clearly, in sulfidic environments, reduced sulfur compounds represent an important source of energy for chemo- and photolithoautotrophic sulfur oxidizers. However, sulfur oxidizers must compete with a variety of biotic and abiotic reactions for terminal electron acceptors.

8.08.7.9.1 Colorless sulfur bacteria

CSB gain energy through chemolithotrophic sulfur-oxidation using O_2 or NO_3^- as terminal electron acceptors. CSB include a wide variety of morphologically and phylogenetically distinct groups (e.g., filamentous *Beggiatoa* and *Thioploca* spp. and single-cell *Thiobacillus* and *Thiomicrospira* spp.). The filamentous CSB have a conspicuous morphology, with cell sizes that allow observation with the naked eye. For example, *Thiomargarita namibiensis* has a cell diameter of up to 0.75 mm (Schulz *et al.*, 1999; Schulz and Jorgensen, 2001) and filaments of certain *Thioploca* species may be as long 70 mm (Jørgensen and Gallardo, 1999). Not surprisingly, Winogradsky (1887) made his early observations of "chemosynthetic" sulfur oxidation in *Beggiatoa* spp. It has been claimed that the *Thioploca-Beggiatoa*-dominated mats in the upwelling area off the coast of Chile embody the largest microbial ecosystem, estimated $10^4\ \mathrm{km}^2$ (Jørgensen and Gallardo, 1999), although it is possible that other such systems exist (Gallardo *et al.*, 1998; Namsaraev *et al.*, 1994). However, the single-celled CSB are abundant in most sulfur-containing ecosystems and are metabolically diverse (Kelly, 1982, 1988; Kelly *et al.*, 1997; Kuenen and Beudeker, 1982). Thus, they can be considered to be at least equally important from a global perspective.

8.08.7.9.2 Single-cell colorless sulfur bacteria

Single-celled, rod-shaped microbes that derive energy from oxidation of reduced sulfur-compounds were traditionally classified as *Thiobacillus* spp. (Vishniac and Santer, 1957), whereas smaller spirillum-shaped species were *Thiomicrospira* (Timmer-Ten Hoor, 1975). Based on molecular phylogeny, the validity of early classification has recently been challenged and the phylogeny of single-celled CSB is currently under revision. Furthermore, physiological studies revealed the need for a careful description of the products of sulfide-oxidation (Gottschall and Kuenen, 1980; Tuttle and Jannasch, 1977; Vishniac and Santer, 1957). Many heterotrophic microorganisms that seem capable of oxidizing HS^- or $S_2O_3^{2-}$ produce polythionates ($S_yO_6^{2-}$) rather than SO_4^{2-} as reaction products (Tuttle and Jannasch, 1977, 1979). Some heterotrophs have been found to oxidize SO_3^{2-} and other reduced sulfur compounds to SO_4^{2-} (Sorokin and Lysenko, 1993), but since thiosulfate utilization is generally not tested as a metabolic trait, it is possible that many more heterotrophs possess the capacity to oxidize sulfur-compounds. Although these heterotrophs are not considered to be CSB, their role in HS^- oxidation in the environment may be significant (Sorokin and Lysenko, 1993; Tuttle and Jannasch, 1979).

Single-celled CSB oxidize a wide variety of inorganic and some organic sulfur-compounds, mixtures of which are typically present in the environment (Luther and Church, 1988; Luther *et al.*, 1999; Visscher and Van Gemerden, 1991; Zopfi *et al.*, 2001). In mixtures of sulfur-compounds, CSB have a strong preference for HS^- (Kuenen and Beudeker, 1982), and under oxygen-limitation, S^0, polythionates and polysulfides are the oxidation products in species that do not utilize NO_3^- (Van den Ende and Van Gemerden, 1993). S^0 deposited outside the cell, consists primarily of long-chained polythionates that form micelle-like structures (Steudel *et al.*, 1987). This "zero-valent" sulfur contains hydrophilic, charged sulfonate groups on the outside, which explains how growth on highly insoluble "elemental" sulfur is possible. The enzymes involved in the sulfide oxidation pathway, which are well characterized (Kelly, 1982; Kelly *et al.*, 1997), include two that oxidize sulfite to sulfate: sulfite oxidase and reverse adenine phosphosulfate reductase (Kappler *et al.*, 2001). The latter is highly efficient in energy conservation and couples oxidation of sulfite directly to ATP generation. This could in part explain the observation that two groups of thiobacilli can be distinguished based on the amount of energy required for autotrophic growth (Kelly, 1982, 1988). The first group has a low growth yield and oxidizes ~4 mol HS^- per mole of CO_2 fixed into structural cell material. Members include *Thiobacillus neapolitanus*, *T. thiooxidans*, *T. versutus*. The second group has a high growth yield and requires ~2 mol of HS^- oxidized per mole CO_2 fixed. Members of the second group include *T. thioparus*, *T. aquaesulis* and *T. denitrificans*. Autotrophic thiobacilli deploy the Calvin cycle for CO_2 fixation and some have cytoplasmic inclusions that contain crystalline Rubisco (Beudeker *et al.*, 1980). Many CSB are not obligate chemolithoautotrophs, but are either facultative autotrophs (e.g., *T. pantotropha, T.*(or *Paracoccus*) *denitrificans*, *T.* (or *Paracoccus*) *versutus*) or so-called mixotrophs (chemolithoorganotrophs; e.g., *T. novellus*) that use sulfur-compounds as electron donors but use organic carbon for biomass.

Single-celled CSB are found in a variety of environments, both terrestrial and aquatic, ranging from freshwater to hypersaline, with pH values from <1 to >10, and displaying psychrophilic, mesophilic, and thermophilic temperature responses. Several species are symbionts in hydrothermal vent invertebrates (e.g., the tube worm *Riftia pachyptilia*, the bivalve *Calyptogena magnifica*, the mussel *Bathymodiolus puteoserpentis;* Cavanaugh, 1994; Cavanaugh *et al.*, 1981; Southward *et al.*, 2001) as well as in bivalves (e.g., *Thyasira* sp.) and oligochetes and clams in estuarine and other environments (Dubilier *et al.*, 1999; Krueger *et al.*, 1996; Wood and Kelly, 1989).

Thiobacilli are typically present in high densities, especially in marine sediments near the oxycline, where densities of 10^6-10^9 cells cm^{-3} have been reported (Gonzalez *et al.*, 1999; Sorokin, 1972; Teske *et al.*, 1996; Visscher *et al.*, 1991, 1992, 1995). Similar densities could be expected in sewage outfall and wastewater treatment facilities, where high rates of SR or of HS^- loading can be expected (Lens and Kuenen, 2001). Some CSB species thrive in extreme environments. For example, *T. ferrooxidans* is commonly found in acid mine drainage where it oxidizes Fe(II) to Fe(III), Cu(I) to Cu(II), and HS^- to SO_4^{2-}. This organism is capable of pyrite oxidation and plays an important role in leaching of ores (e.g., copper, nickel, and uranium) and desulfurization of coal (Bos and Kuenen, 1990). *T. thioparus* is versatile in organic carbon utilization and oxidizes MeSH and DMS (Kanagawa and Kelly, 1986; Smith and Kelly, 1988a; Visscher and Taylor, 1993b; Visscher and Van Gemerden, 1991). Another *Thiobacillus* sp., resembling *T. thioparus*, is able to degrade DMS with NO_3^- instead of O_2 and can also oxidize the thiol group of a range of alkylthiols to sulfate (Visscher and Taylor, 1993a).

The genus *Thiomicrospira* contains three known species, all of which are chemolithoautotrophs. These organisms thrive under microaerophilic conditions, and one strain can couple sulfide and thiosulfate oxidation to NO_3^-

reduction (Timmer-Ten Hoor, 1975). Another exceptional genus of sulfide-oxidizing Eubacteria is *Achromatium*, a facultative autotroph that stores S^0 and calcite intracellularly (Head *et al.*, 2000) and is found in freshwater, brackish, and marine environments. The three known species that belong to this genus all have mixotrophic growth capabilities. The calcite deposits make this organism one of the largest free-living single-celled prokaryotic cells known (estimated biovolume of up to 8×10^4 μm^3 $cell^{-1}$) (Schulz and Jorgensen, 2001). The role of the calcite is obscure; it could perhaps act as a buffer in the cytoplasm when SO_4^{2-} is produced, or provide CO_2 for autotrophic growth. The latter scenario is somewhat unlikely, since precipitation of $CaCO_3$ yields H^+, and CSB are actually believed to dissolve this mineral when growing autotrophically (Visscher *et al.*, 1998).

The Archeabacteria contain two genera of sulfur-oxidizing organisms, both of which are hyperthermophiles. *Sulfolobus* is an aerobic sulfur-oxidizer, which is also capable of using Fe^{2+} and organic carbon (facultative chemolithoautotroph that grows chemoorganoheterotrophically). *Acidianus* resembles *Sulfolobus*, but can grow both in the presence and absence of O_2. Under oxic conditions it oxidizes S^0 to SO_4^{2-}, while under anoxic conditions, S^0 is used as electron acceptor and sulfide is produced.

8.08.7.9.3 Filamentous colorless sulfur bacteria

The filamentous CSB include the genera *Thioploca*, *Thiothrix*, and *Beggiatoa* (Larkin and Strohl, 1983; Schulz *et al.*, 1999), which are relatively closely grouped phylogenetically. *Beggiatoa* spp. chemolithoautotrophs were the first to be described (Winogradsky, 1887), but later investigations demonstrated that they might have a chemolithoorganotrophic (or mixotrophic) lifestyle. These organisms are found in freshwater and marine habitats at the interface of O_2 and HS^- (Schulz and Jorgensen, 2001).

Beggiatoa stores "elemental" sulfur inside the cell, which gives the filaments a white appearance. Massive blooms of this organism in sulfidic hot springs, salt marshes, and marine sediments make this gliding organism conspicuous (Larkin and Strohl, 1983) when it forms veils or mats. Hydrothermal vents at the deep seafloor support sulfide-based ecosystems with unusually large *Beggiatoa* spp. (Jannasch *et al.*, 1989) in addition to unicellular CSB in trophosomes of *Riftia* spp. (Cavanaugh, 1994). In some *Beggiatoa* spp., the sulfide can be oxidized with O_2 or NO_3^- as an electron acceptor, and N_2 is the product of nitrate reduction (Sweerts *et al.*, 1990).

Thioploca spp. are sheathed filamentous CSB, that resemble *Beggiatoa* spp. in many characteristics. They thrive in the O_2-minimum zone in marine and freshwater mud and seem to accumulate NO_3^- in their vacuoles (as do some *Beggiatoa* spp.), which enables these organisms to thrive under conditions where HS^- and O_2 or NO_3^- are not present simultaneously. The NO_3^- concentration in the vacuole can reach 0.5 M, or more than 100 times that outside the cell (Jørgensen and Gallardo, 1999). However, under anoxic conditions, NO_3^- is reduced to NH_4^+ and not N_2 (Otte *et al.*, 1999). In addition to a positive tactic response to O_2, *Thioploca* spp. also move towards NO_3^- and low concentrations of sulfide; microaerophilic conditions are preferred. *Thioploca* sheaths may contain SO_4^{2-}-reducing bacteria (*Desulfonema* spp.; (Jørgensen and Gallardo, 1999)) that may provide sufficient HS^-. Filaments that have characteristic tapered ends can live outside the sheath, and S^0 is stored intracellularly. *Thioploca* spp. have a mixotrophic lifestyle and, based on cell diameter, four size categories can be distinguished: two small groups (2.5–5 and 12–20 μm), a larger group (30–43 μm), and a group with gigantic cells (125 μm). Although the mats off the Chilean and Peruvian coast are quite extensive (see above), *Thioploca* spp. have been found at other offshore locations including the Arabian Sea, the Benguela Current, and the Mediterranean Sea, as well as in lakes (Lake Erie, Lake Baikal, Lake Biwa) (Gallardo *et al.*, 1998; Jørgensen and Gallardo, 1999; Namsaraev *et al.*, 1994; Nishino *et al.*, 1998).

8.08.7.9.4 Anoxyphototrophic bacteria

Purple and green sulfur bacteria use reduced sulfur compounds (e.g., HS^-, S^0, S_x^{2-}, $S_2O_3^{2-}$) as electron donors for photosynthesis. Purple and green nonsulfur bacteria seem to have limited capabilities to use these donors as well. Interestingly, some PSB have the ability to oxidize sulfide chemolithotrophically in the presence of O_2 (De Wit and Van Gemerden, 1987). For example, although *Thiocapsa roseopersicina* can use HS^- phototrophically, in the presence of O_2, it requires anoxia for synthesis of photopigments (De Wit and Van Gemerden, 1987). If these conditions are not met, *T. roseopersicina* will ultimately be nonpigmented and resort to chemotrophic sulfide oxidation for energy requirements. Immediately following even a brief period of anoxia, the reverse shift from chemolithotrophy to photolithotrophy occurs.

A phylogenetic link between PSB and thiobacilli has been suggested, in which the PSB (and/or other phototrophic sulfur bacteria) are considered to be the ancestors of certain thiobacilli. Blooms of phototrophic bacteria are observed in lakes (e.g., Guerrero *et al.*, 1985; Overmann, 1997)), microbial mats (Pierson *et al.*, 1987; Van Gemerden, 1993) and in salt marshes (Baas Becking, 1934). Purple nonsulfur bacteria (PNSB) can use a wide variety of organic carbon compounds (Madigan, 1988) and typically thrive in habitats where sulfur and organic carbon loading are high (e.g., sewage effluents). Some PSB and PNSB can use organic sulfur compounds, either as electron donors (Visscher and Taylor, 1993c; Visscher and Van Gemerden, 1991) or as electron donors and/or carbon sources (Chien *et al.*, 1999).

8.08.7.9.5 *Ecological aspects of sulfide oxidation*

As the phylogeny, physiology, and biochemistry of sulfide-oxidizing microbes is under revision, the ecological aspects and biogeochemical role of these organisms in carbon, sulfur, and nitrogen cycles is anticipated to change as well. For example, as outlined above, the capability to oxidize sulfide or thiosulfate is not one of the "standard tests" that is performed when characterizing organisms. This may well explain the scattered distribution of single-celled CSB in the phylogenetic trees (although the sequence of 16S rRNA and metabolic traits are not synonymous). In addition to the abundance of sulfur-oxidizers, it is important to acknowledge the environmental characteristics, especially those related to O_2 and HS^- (and other sulfur-compounds; Jørgensen *et al.*, 1979; Luther *et al.*, 1991 ; Visscher and Van den Ende, 1994; Visscher and Van Gemerden, 1993; Zopfi *et al.*, 2001).

The expanding role of cultured organisms that were not considered to be involved in sulfur transformations and some novel insights in associations of various sulfur bacteria were recently reviewed (Overmann and van Gemerden, 2000). One example of interactions among sulfide-oxidizers is that of CSB and PSB. When PSB (*Thiocapsa roseopersina*) and CSB (*Thiobacillus thioparus*) compete for sulfide under alternating oxic–anoxic conditions, PSB outcompete CSB despite a better affinity for sulfides in the latter group of organisms (Visscher *et al.*, 1992). The storage of S^0 intracellularly and the capability to use light provide a competitive advantage. Similarly, other investigators hinted that a combination of environmental factors, including light quality and quantity and sulfide and oxygen concentrations, need to be considered when evaluating the competitive position between CSB and PSB (Jørgensen and Des Marais, 1986). In further investigations, where the oxygen/sulfide ratios were varied, the PSB were only able to coexist with the CSB when the ratio was less than 1.6 (indicating O_2-limitation), in which case the PSB used partially oxidized sulfur compounds excreted by the CSB (Van den Ende *et al.*, 1996).

Although several physiological and ecological aspects of sulfur-oxidation have been outlined above, others exist, including the fermentation of sulfur compounds (Bak and Cypionka, 1987), the use of HS^- or $S_2O_3^{2-}$ as electron donor in certain cyanobacteria (e.g., De Wit and Van Gemerden, 1987; Pringault and GarciaPichel, 2000), and the coexistence of CSB and SRB under atmospheric O_2 concentrations (Van den Ende and Van Gemerden, 1993). The need for an improved understanding of ecophysiological functioning of microbes and their distribution as well as biogeochemical transformations in the sulfur cycle are still unambiguous despite 125 years of research.

8.08.8 COUPLED ANAEROBIC ELEMENT CYCLES

Although it is pragmatic to consider the element cycles independently, an understanding of how elements and organisms interact is required to interpret the relative contributions of the various pathways to anaerobic metabolism in nature. This concluding section elaborates on these interactions and briefly reviews the relative contribution of the major anaerobic pathways to carbon metabolism in ecosystems.

8.08.8.1 Evidence of Competitive Interactions

The outcome of competition between microorganisms for electron donors can be predicted from thermodynamic theory (Section 8.08.1.2; Table 1; Zehnder and Stumm, 1988), and these predictions are generally consistent with empirical data. Temporal succession of the microbial metabolic pathways that dominate respiration occurs upon the flooding of an oxidized soil or sediment (Figure 1). Not surprisingly, most examples of temporal succession in anaerobic respiration processes have come from wetland soils, which are subject to cycles of flooding and exposure (Turner and Patrick, 1968; Ponnamperuma, 1972; Achtnich *et al.*, 1995a; Yao *et al.*, 1999). However, the same pattern is observed in sediments and even upland soils (Peters and Conrad, 1996).

Yao *et al.* (1999) recognized three phases of reduction following the flooding of rice paddy soil. In the *reduction phase* (phase I), the inorganic electron acceptors were sequentially consumed and CO_2 emissions were highest; in phase II, CH_4 production became dominant and peaked, and in phase III there was a steady-state ratio of CH_4 and CO_2. Peak CH_4 production during phase II was highly correlated ($r^2 = 0.79$) with the ratio of nitrogen to inorganic electron acceptors, reflecting the dual influence of carbon quality and competition on methanogenesis. Methane production is favored by a high nitrogen content, because it is indicative of a large labile carbon pool (Section 8.08.4.3.4), which is consumed preferentially by nonmethanogenic bacteria until the inorganic electron acceptor pools are exhausted. Thus, electron flow through methanogenesis is highest when the labile carbon pool is large and the inorganic electron acceptor pools are small. The length of phase I increases with the inorganic electron acceptor pool size, and decreases with a decrease in the labile pool size or an increase in temperature (van Hulzen *et al.*, 1999).

Spatial zonation develops due the progressive consumption of terminal electron acceptors from their source to points downgradient. Such zonation occurs as a function of depth in systems that are continually inundated such as aquatic sediments. It also occurs downstream of an organic carbon point source in rivers and aquifers (Lovley *et al.*, 1994a).

A final source of observations that indicate competition for electron donors comes from direct manipulation of soils or sediments. Addition of 30 g ferrihydrite per kg soil reduced CH_4 emission from a paddy soil by 84% (Jäckel and Schnell, 2000), and similar responses have been reported following a fall in the water table (Krüger *et al.*, 2001). Freeman *et al.* (1994) attributed a decline in CH_4 emissions following drought to inhibition by SO_4^{2-} reduction, which is consistent with the results of direct SO_4^{2-} addition studies (Gauci *et al.*, 2002).

8.08.8.2 Mechanisms of Competition

Molecular hydrogen (H_2) is the most abundant product of fermentation and the most common electron donor in terminal anaerobic metabolism. Extremely low (nanamolar level) H_2 concentrations in anaerobic sediments are evidence of keen competition for H_2. Metabolic pathways that yield relatively large amounts of free energy (i.e., those with more negative ΔG values) tend to be associated with low pore-water H_2 concentrations (Table 1; Figure 27). Thus, SO_4^{2-}-reducers inhibit H_2-dependent methanogens by reducing the H_2 concentration to a value below the threshold at which CH_4 production is thermodynamically feasible (Lovley *et al.*, 1982); Fe(III) reducers have the same influence on SO_4^{2-} reducers, and so forth according to the free-energy yield of the processes when operating near chemical equilibrium (Lovley and Goodwin, 1988; Postma and Jakobsen, 1996).

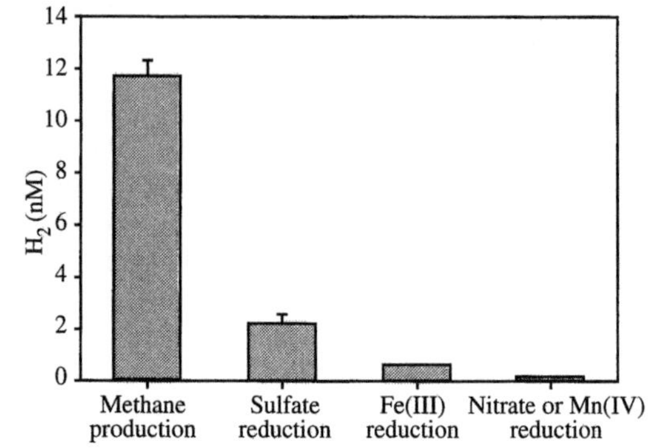

Figure 27 Steady-state H_2 concentrations in sediments with different dominant terminal electron accepting processes (Lovely and Goodwin, 1988) (reproduced by permission of Elsevier from *Geochim. Cosmochim. Acta.* **1988**, *52*, 2993–3003).

The link between bioenergetics and H_2 offers a nondestructive method of determining the dominant terminal electron accepting process *in situ* (Hoehler *et al.*, 1998). Such an approach is most effective when the influence of temperature, pH, mineral reactivity and other factors are taken into account by calculating the free energy of the system. Hoehler *et al.* (1998) observed that the ΔG value for a given terminal respiration process remained constant when temperature and pH were manipulated, and suggested that it reflected the need for microorganisms to operate at their energetic limit in order to successfully compete for H_2. The theoretical minimum energy yield that can support life (i.e., the *biological energy quantum*) is one-third ATP per round of metabolism or about -20 kJ mol^{-1} (Schink, 1997). This limit agrees at least roughly with *in situ* observations. The low energy yields of fermentation, methanogenesis, and most other forms of anaerobic metabolism suggest that anaerobic organisms function at a level of near-starvation (Valentine, 2001).

Thermodynamic considerations suggest that competition for acetate should also favor the metabolic pathway with the highest free-energy yield. However, acetate concentrations have been reported to change in response to a shift in the dominant metabolic pathway in some cases but not others (Achtnich *et al.*, 1995b; Chidthaisong and Conrad, 2000; Sigren *et al.*, 1997). Thermodynamic control of acetate concentrations may be superseded by kinetic effects due to its slow diffusion rate.

A more recent addition to the list of terminal electron acceptors is humic substances (Section 8.08.6.4.1). Cervantes *et al.* (2000) demonstrated that a humic acid analogue (AQDS) inhibited methanogenesis due to a combination of toxic and competitive effects. The thermodynamic yield of AQDS reduction was estimated to be more favorable than SO_4^{2-} reduction or methanogenesis, but less favorable than Fe(III) reduction. Because humic substances can quickly transfer electrons to other terminal acceptors such as Fe(III), they are rapidly recycled and could be important in anaerobic metabolism despite typically low concentrations.

8.08.8.3 Exceptions

There are many instances when the succession of anaerobic microbial pathways or segregation in space varies from the classical pattern (Section 8.08.8.1). There is a tendency for overlap between pairs of terminal electron acceptors with similar free-energy yields (e.g., NO_3^- and Mn; Fe and SO_4^{2-}). These exceptions can often be explained by the absence of competition for electron donors (Oremland *et al.*, 1982; Crill and Martens, 1986; Holmer and Kristensen, 1994; Chidthaisong and Conrad, 2000; Mitterer *et al.*, 2001; McGuire *et al.*, 2002). Small amounts of CH_4 are produced in rice paddy soils immediately upon the onset of anoxia, despite the presence of NO_3^-, Fe(III) and SO_4^{2-}, and a high redox potential (Roy *et al.*, 1997;Yao and Conrad, 1999). The initial burst of methanogenic activity coincides with a peak in H_2 concentrations, suggesting that H_2 production from fermentation can exceed demand for a period of time (Yao and Conrad, 1999). Once Fe(III) reduction and SO_4^{2-} reduction draw down H_2 to a level below the threshold required for methanogenesis, CH_4 production ceases. A seemingly uncommon reason for the absence of competition between terminal electron accepting processes is that the organisms require different electron donor substrates (Section 8.08.7.5.3).

The order of competing terminal electron accepting processes can vary with any number of factors that influence the thermodynamics of the system. One factor that must be considered in ecosystems with mineral sediments or soils is the composition of Fe(III) and Mn(IV) minerals. The typical sequence of Fe(III) reduction before SO_4^{2-} reduction can be reversed with a change in the abundance of labile Fe(III) minerals such as ferrihydrite (Postma and Jakobsen, 1996). This is one explanation for the common observation that the zones of Fe(III) reduction and SO_4^{2-} reduction overlap in marine sediments (Boesen and Postma, 1988; Canfield, 1989; Canfield *et al.*, 1993b; Goldhaber *et al.*, 1977; Jakobsen and Postma, 1994). Postma and Jakobsen (1996) predicted that the overlap between Fe(III) reduction and SO_4^{2-} reduction should increase as Fe(III) oxide stability (or surface area) increases.

Spatial heterogeneity is a likely explanation for the coexistence of competing terminal electron accepting processes. It is well established that denitrifying bacteria are active in anaerobic microsites imbedded in upland soils (Section 8.08.5.3.2; Figure 14), and there is evidence of methanogenic activity, the most O_2 sensitive of anaerobic metabolisms, in upland soils as well (von Fischer and Hedin, 2002). Redox potentials are notoriously variable in anaerobic sediments, suggesting that redox microsites occur in the total absence of O_2. Spatial variability has been proposed to explain deviations from the expected suppression of SO_4^{2-} reduction by Fe(III) reduction (Hoehler *et al.*, 1998). It is now possible to observe such small-scale variability in microbial populations using molecular techniques (e.g., Boetius *et al.*, 2000b; Orphan *et al.*, 2001b) and element-specific microelectrodes, including a recently developed system for measuring iron speciation (e.g., Luther *et al.*, 2001).

8.08.8.4 Noncompetitive Interactions

In many cases, the segregation of terminal electron accepting processes is due to factors other than competition. The ability of NO_3^- to suppress Fe(III) reduction (DiChristina, 1992) and methanogenesis has been shown to be due in part to inhibition by denitrification intermediates (Klüber and Conrad, 1998, p. 331; Roy and Conrad, 1999). Concentrations of $<100\ \mu M\ NO_3^-$, 1–2 μM NO and <1 mM N_2O have been reported to completely inhibit hydrogenotrophic methanogenesis (Balderston and Payne, 1976; Klüber and Conrad, 1998), while somewhat higher concentrations are necessary to inhibit acetogenic methanogenesis (Clarens *et al.*, 1998). Chidthaisong and Conrad (2000) reported that NO_3^- amendments inhibited glucose turnover in a paddy soil, which indicates that nitrogen oxide toxicity may have affected the methanogens indirectly by inhibiting fermentation.

8.08.8.5 Contributions to Carbon Metabolism

The factors that influence the flux of energy through aerobic–anaerobic interface ecosystems have been the focus of this review chapter. Here we consider the net effect of these influences on carbon metabolism in marine and freshwater ecosystems. Thamdrup (2000) recently compiled studies that reported the relative contributions of O_2 reduction, Fe(III) reduction, and SO_4^{2-} reduction to carbon metabolism in marine ecosystems ($n = 16$). On average, the dominant pathway was SO_4^{2-} reduction ($62 \pm 17\%$, $\bar{x} \pm$ SD). Aerobic respiration and Fe(III) respiration contributed equally to carbon metabolism ($18 \pm 10\%$ and $17 \pm 15\%$, respectively). Compared to previous compilations, ~50% of the amount attributed to O_2 reduction is now credited to Fe(III) reduction, while the contribution of SO_4^{2-} reduction is unchanged (Thamdrup, 2000). Fe(III) reduction can also be the dominant carbon oxidation pathway in salt marsh sediments (Kostka *et al.*, 2002c; Gribsholt *et al.*, 2003).

There have been remarkably few attempts to determine the contribution of Fe(III) reduction to anaerobic carbon metabolism in freshwater ecosystems where it may be the dominant pathway. The most extensive such study examined 16 rice paddy soils collected from China, Italy, and the Philippines (Yao *et al.*, 1999). Fe(III) reduction was 58–79% of carbon metabolism during the reduction phase (Section 8.08.8.1), with most of the remainder attributed to methanogenesis. Fe(III) reduction contributed ~70% of the anaerobic metabolism in a *Juncus effusus* marsh (Roden and Wetzel, 1996).

There is growing evidence that plant activity causes carbon metabolism to shift away from methanogenesis to Fe(III) reduction. Roden and Wetzel (1996) reported that the contribution of methanogenesis to anaerobic carbon metabolism shifted from 69% in the absence of plants to <30% in their presence. Similar observations have been reported for freshwater marshes dominated by *Scirpus lacustris* and *Phragmites australis* (van der Nat and Middelburg, 1998) and a rice paddy soil (Frenzel *et al.*, 1999). The mechanism for this effect is an abundance of poorly crystalline Fe(III) in the rhizosphere compared to the bulk soil (Kostka and Luther 1995; Weiss, 2002), and perhaps a larger labile organic carbon supply. Fe(III) and Mn(IV) respiration are subordinate terminal electron-accepting processes in saturated soils and sediments in the absence of a mechanism for their regeneration to oxides. Plants, bioturbation, or physical mixing have this effect. In salt marshes, bioturbation by crabs may be more important to Fe(III) regeneration than radial O_2 loss from roots in some cases (Kostka *et al.*, 2002c), but not in others (Gribsholt *et al.*, 2003). In marine sediments, there is very little Fe(III) or Mn(IV) reduction in the absence of mixing (Thamdrup, 2000).

8.08.8.6 Concluding Remarks

Research since 1990 has greatly expanded our understanding of anaerobic metabolism. Novel microorganisms have been discovered, such as those performing anaerobic ammonium oxidation to N_2 (anammox). These organisms were predicted to exist based on thermodynamic considerations, but have since been shown to contribute substantially to N_2 production in some marine sediments. Organisms that were known to exist for sometime have been shown to perform unexpected types of metabolism, such as nitrifying bacteria that are capable of denitrification and Fe(III)-reducing bacteria that produce CH_4. The reduction of Fe(III) was thought to be primarily an abiotic process, but it is now understood to account for much of the anaerobic carbon metabolism in many freshwater ecosystems. Microorganisms appear to play a larger role in Fe(II) oxidation at circumneutral pH that previously believed. Progress has been aided by the development of molecular and stable isotope techniques that yield detailed descriptions of microbial communities. Such techniques have recently provided strong support for the hypothesis that anaerobic CH_4 oxidation is achieved by a syntrophic relationship between *Archea* and *Bacteria*. The significance of these new microorganisms and metabolic pathways for element cycling *in situ* remains to be determined, but it is likely to lead to important revisions of the element budgets for carbon, nitrogen, manganese, iron, and sulfur, and the mechanisms that regulate their cycling. The recent downward revision of aerobic carbon mineralization rates in marine sediments that followed the discovery of respiratory Fe(III) reducing bacteria (Section 8.08.8.4) is a poignant example of how important the field of environmental microbiology has become for advancing our knowledge of biogeochemistry. Likewise, geochemistry has made significant contributions to environmental microbiology by discovering the activity of anaerobic CH_4 oxidizing bacteria and the anammox process. More exciting discoveries can be expected as these fields continue to become integrated.

ACKNOWLEDGMENTS

The authors gratefully acknowledge the people who reviewed portions of the manuscript: Dave Emerson, Joel Kostka, Scott Neubauer, Robin Sutka, Bo Thamdrup, and Johanna Weiss. They thank Laura Lipps for assisting in every phase of assembling the manuscript and Mei Mei Chong for producing most of figures. Support for this activity was provided by NSF (JPM, MEH, PTV), DOE (JPM) and NASA's Astrobiology Institute (PTV).

REFERENCES

Abeliovich A. and Vonhak A. (1992) Anaerobic metabolism of *Nitrosomonas europaea*. *Arch. Microbiol.* **158**, 267–270.

Abken H. J., Tietze M., Brodersen J., Baumer S., Beifuss U., and Deppenmeier U. (1998) Isolation and characterization of methanophenazine and function of phenazines in membrane-bound electron transport of *Methanosarcian* mazei Göl. *J. Bacteriol.* **180**, 2027–2032.

Achtnich C., Bak F., and Conrad R. (1995a) Competition for electron donors among nitrate reducers, ferric iron reducers, sulfate reducers, and methanogens in anoxic paddy soil. *Biol. Fertility Soils* **19**, 65–75.

Achtnich C., Schuhmann A., Wind T., and Conrad R. (1995b) Role of interspecies H_2 transfer to sulfate and ferric iron-reducing bacteria in acetate consumption in anoxic paddy soil. *FEMS Microbiol. Ecol.* **16**, 61–69.

Adams D. D., Matisoff G., and Snodgrass W. J. (1982) Flux of reduced chemical constituents (Fe^{2+}, Mn^{2+}, NH_4^+, and CH_4) and sediment oxygen demand in Lake Erie. *Hydrobiologia* **92**, 405–414.

Aeckersberg F., Bak F., and Widdel F. (1991) Anaerobic oxidation of saturated hydrocarbons to CO_2 by a new type of sulfate-reducing bacterium. *Arch. Microbiol.* **156**, 5–14.

Aeckersberg F., Rainey F. A., and Widdel F. (1998) Growth, natural relationships, cellular fatty acids and metabolic adaptation of sulfate-reducing bacteria that utilize long-chain alkanes under anoxic conditions. *Arch. Microbiol.* **170**, 361–369.

Aerts R. and de Caluwe H. (1999) Nitrogen deposition effects on carbon dioxide and methane emissions from temperate peatland soils. *Oikos* **84**, 44–54.

Ahring B. K. and Westermann P. (1988) Product inhibition of butyrate metabolism by acetate and hydrogen in a thermophilic coculture. *Appl. Environ. Microbiol.*, **54**.

Aller R. C. (1990) Bioturbation and manganese cycling in hemipelagic sediments. *Phil. Trans. Roy. Soc. London A* **331**, 51–68.

Aller R. C. and Rude P. D. (1988) Complete oxidation of solid phase sulfides by manganese and bacteria in anoxic marine sediments. *Geochim. Cosmochim. Acta* **52**, 751–765.

Alm J., Talanov A., Saarnio S., Silvola J., Ikkonen E., Aaltonen H., Hykänen H., and Martikainen P. J. (1997) Reconstruction of the carbon balance for microsites in a boreal oligotrophic pine fen, Finland. *Oecologia* **110**, 423–431.

Alm J., Saarnio S., Nykänen H., Silbola J., and Martikainen P. J. (1999) Winter CO_2, CH_4, and N_2O fluxes on some natural and drained boreal peatlands. *Biogeochemistry* **44**, 163–186.

Alperin M. J. and Reeburgh W. S. (1985) Inhibition experiments on anaerobic methane oxidation. *Appl. Environ. Microbiol.* **50**, 940–945.

Alperin M. J., Reeburgh W. S., and Whiticar M. J. (1988) Carbon and hydrogen isotope fractionation resulting from anaerobic methane oxidation. *Global Biogeochem. Cycles* **2**, 279–288.

Amann R. I., Binder B., Chisholm S. W., Olsen R., Devereux R., and Stahl D. A. (1990a) Combination of 16S rRNA-targeted oligonucleotide probes with flow cytometry for analyzing mixed microbial populations. *Appl. Environ. Microbiol.* **56**, 1919–1925.

Amann R. I., Krumholz L., and Stahl D. A. (1990b) Fluorescent-oligonucleotide probing of whole cells for determinative, phylogenetic, and environmental studies in microbiology. *J. Bacteriol.* **172**, 762–770.

Amaral J. A. and Knowles R. (1995) Growth of methanotrophs in methane and oxygen counter gradients. *FEMS Microbiol. Lett.* **126**, 215–220.

Anbar A. D., Roe J. E., Barling J., and Nealson K. H. (2000) Nonbiological fractionation of iron isotopes. *Science* **288**, 126–128.

Anderson R. T. and Lovley D. R. (2000) Hexadecane decay by methanogenesis. *Nature* **404**, 722–723.

Anderson R. T., Rooney-Varga J. N., Gaw C. V., and Lovley D. R. (1998) Anaerobic benzene oxidation in the Fe(III) reduction zone of petroleum-contaminated aquifers. *Environ. Sci. Technol.* **32**, 1222–1229.

Andreae M. O. and Crutzen P. J. (1997) Atmospheric aerosols: biogeochemical sources and role in atmospheric chemistry. *Science* **276**, 1052–1058.

Ansede J. H., Friedman R., and Yoch D. C. (2001) Phylogenetic analysis of culturable dimethyl sulfide-producing bacteria from a Spartina-dominated salt marsh and estuarine water. *Appl. Environ. Microbiol.* **67**, 1210–1217.

Arah J. R. M. (1997) Apportioning nitrous oxide fluxes between nitrification and denitrification using gas-phase mass spectrometry. *Soil Biol. Biochem.* **29**, 1295–1299.

Aravena R., Warner B. G., Charman D. J., Belyea L. R., Mathur S. P., and Dinel H. (1993) Carbon isotopic composition of deep carbon gases in an ombrogenous peatland, northwestern Ontario, Canada. *Radiocarbon* **35**, 271–276.

Armstrong W. (1964) Oxygen diffusion from the roots of some British bog plants. *Nature* **204**, 801–802.

Arnosti C. (1996) A new method for measuring polysaccharide hydrolysis rates in marine environments. *Org. Geochem.* **25**, 105–115.

Arnosti C. (1998) Rapid potential rates of extracellular enzymatic hydrolysis in Arctic sediments. *Limnol. Oceanogr.* **43**, 315–324.

Aselmann I. and Crutzen P. J. (1989) Global distribution of natural freshwater wetlands and rice paddies, their net primary productivity, seasonality and possible methane emissions. *J. Atmos. Chem.* **8**, 307–358.

Auman A. J., Speake C. C., and Lidstrom M. E. (2001) *nif*H sequences and nitrogen fixation in type I and type II methanotrophs. *Appl. Environ. Microbiol.* **67**, 4009–4016.

Avery G. B., Jr. and Martens C. S. (1999) Controls on the stable carbon isotopic composition of biogenic methane produced in a tidal freshwater estuarine sediment. *Geochim. Cosmochim. Acta* **63**, 1075–1082.

Avery G. B., Jr., Shannon R. D., White J. R., Martens C. S., and Alperin M. J. (1999) Effect of seasonal changes in the pathways of methanogenesis on the $\delta^{13}C$ values of pore water methane in a Michigan peatland. *Global Biogeochem. Cycles* **13**, 475–484.

Avery G. B., Jr., Shannon R. D., White J. R., Martens C. S., and Alperin M. J. (2002) Controls on methane production in a tidal freshwater estuary and a peatland: methane production via acetate fermentation and CO_2 reduction. *Biogeochemistry* **62**, 19–37.

Avrahami S., Conrad R., and Braker G. (2002) Effect of soil ammonium concentration on N_20 release and the community structure of ammonia oxidizers and denitrifiers. *Appl. Environ. Microbiol.* **68**, 5685–5692.

Baas Becking L. G. M. (1934) Geobiology. Stockum & Zoon, NV, The Netherlands, 263pp.

Bache R. and Pfennig N. (1981) Selective isolation of *Acetobacterium woodii* on methoxylated aromatic acids and determination of growth yields. *Arch. Microbiol.* **130**, 255–261.

Bacic M. K., Newell S. Y., and Yoch D. C. (1998) Release of dimethylsulfide from dimethylsulfonio-propionate by plant-associated salt marsh fungi. *Appl. Environ. Microbiol.* **64**, 1484–1489.

Baena S., Fardeau M.-L., Ollivier B., Labat M., Thomas P., Garcia J.-L., and Patel B. K. C. (1999) *Aminomonas paucivorans* gen. nov., sp. nov., a mesophilic, anaerobic, amino-acid-utilizing bacterium. *Int. J. Systemat. Bacteriol.* **49**, 975–982.

Baena S., Fardeau M.-L., Labat M., Ollivier B., Garcia J.-L., and Patel B. K. C. (2000) *Aminobacterium mobile*, sp. nov., a new anaerobic amino-acid degrading bacterium. *Int. J. Systemat. Bacteriol.* **50**, 259–264.

Bak F. and Cypionka H. (1987) A novel type of energy metabolism involving fermentation of inorganic sulphur compounds. *Nature* **326**, 891–892.

Bak F. and Pfennig N. (1987) Chemolithotrophic growth of *Desulfovibrio sulfodismutans* sp. nov. by disproportionation of inorganic sulfur compouds. *Arch. Microbiol.* **147**, 184–189.

Bak F. and Pfennig N. (1991) Microbial sulfate reduction in littoral sediment of Lake Constance. *FEMS Microbiol. Ecol.* **85**, 31–42.

Bak F. and Widdel F. (1986) Anaerobic degradation of phenol and phenol derivatives by *Desulfobacterium phenolicum* sp. nov. *Arch. Microbiol.* **146**, 177–180.

Bak F., Finster K., and Rothfuss F. (1992) Formation of dimethylsulfide and methanethiol from methoxylated aromatic compounds and inorganic sulfide by newly isolated anaerobic bacteria. *Arch. Microbiol.* **157**, 529–534.

Balashova V. V. (1985) The use of molecular sulfur as an agent oxidizing hydrogen by the facultative anaerobic *Pseudomonas* strain. *Mikrobiologiya, USSR* **51**, 324–326.

Balderston W. L. and Payne W. J. (1976) Inhibition of methanogenesis in salt marsh sediments and whole-cell suspensions of methanogenic bacteria by nitrogenous oxides. *Appl. Environ. Microbiol.* **32**, 264–269.

Bale S. J., Goodman K., Rochelle P. A., Marchesi J. R., Fry J. C., Weightman A. J., and Parkes R. J. (1997) *Desulfovibrio profundus* sp. nov., a novel barophilic sulfate-reducing bacterium from deep sediment layers in the Japan Sea. *Int. J. Systemat. Bacteriol.* **47**, 515–521.

Banik S., Sen M., and Sen S. P. (1996) Effects of inorganic fertilizers and micronutrients on methane production from wetland rice (*Oryza sativa* L.). *Biol. Fertility Soils* **21**, 319–322.

Banwart W. L. and Bremner J. M. (1975) Formation of volatile sulfur compounds by microbial decomposition of sulfur-containing amino acids in soils. *Soil Biol. Biochem.* **7**, 359–364.

Barbeau K., Rue E. L., Bruland K. W., and Butler A. (2001) Photochemical cycling of iron in the surface ocean mediated by microbial iron(III)-binding ligands. *Nature* **413**, 409–413.

Barker H. A. (1940) Studies upon the methane fermentation: IV. The isolation and culture of *Methanobacterium omelianskii*. *Antonie Van Leeuwenhoek* **6**, 201–220.

Barnes R. O. and Goldberg E. D. (1976) Methane production and consumption in anoxic marine sediments. *Geology* **20**, 962–970.

Bartlett K. B. and Harriss R. C. (1993) Review and assessment of methane emissions from wetlands. *Chemosphere* **26**, 261–320.

Bartlett K. B., Bartlett D. S., Harriss R. C., and Sebacher D. I. (1987) Methane emissions along a salt marsh salinity gradient. *Biogeochemistry* **4**, 183–202.

Barton L. L., LeGall J., Odom J. M., and Peck H. D. J. (1983) Energy coupling to nitrite respiration in the sulfate-reducing bacterium *Desulfovibrio gigas*. *J. Bacteriol.* **153**, 867–871.

Barton L. L., Shively J. M., and Lascelles J. (1991) Autotrophs: variations and versatilities. In *Variations in Autotrophic Life* (eds. J. Shively and L. L. Barton). Academic Press, London, UK, pp. 1–23.

Barton L., McLay C. D. A., Schipper L. A., and Smith C. T. (1999) Annual denitrification rates in agricultural and forest soils: a review. *Austral. J. Soil Res.* **37**, 1073–1093.

Basiliko N. and Yavitt J. B. (2001) Influence of Ni, Co, Fe, and Na additions on methane production in *Sphagnum*-dominated northern American peatlands. *Biogeochemistry* **52**, 133–153.

Bates T. S., Lamb B. K., Guenther A., Dignon J., and Stoiber R. E. (1992) Sulfur emissions to the atmosphere from natural sources. *J. Atmos. Chem.* **14**, 315–337.

Battaglia-Brunet F., Dictor M. C., Garrido F., Crouzet C., Morin D., Dekeyser K., Clarens M., and Baranger P. (2002) An arsenic(III)-oxidizing bacterial population: selection, characterization, and performance in reactors. *J. Appl. Microbiol.* **93**, 656–667.

Battersby N. S. and Brown C. M. (1982) Microbial activity in organically enriched marine sediments. In *Sediment Microbiology* (eds. D. B. Nedwell and C. M. Brown). Academic Press, New York, pp. 147–170.

Bazylinski D. A. and Moskowitz B. M. (1997) Microbial biomineralization of magnetic minerals: microbiology, magnetism, and environmental significance. In *Geomicrobiology: Interactions between Microbes and Minerals*, Rev. Mineral. 35, Washington, DC, pp. 181–223.

Beard B. L., Johnson C. M., Cox L., Sun H., Nealson K. H., and Aguilar C. (1999) Iron isotope biosignatures. *Science* **285**, 1889–1891.

Bebout B. M., Paerl H. W., Crocker H. W., and Prufert L. E. (1987) Diel interactions of oxygenic photosynthesis and N_2 fixation (acetylene reduction) in a marine microbial mat community. *Appl. Environ. Microbiol.* **53**, 2353–2362.

Bechard M. J. and Rayburn W. R. (1979) Volatile organic sulfides from freshwater algae. *J. Phycol.* **15**, 379–383.

Beliaev A. and Saffarini D. A. (1998) *Shewanella putrefaciens mtr B* encodes an outer membrane protein required for Fe(III) and Mn(IV) reduction. *J. Bacteriol.* **180**, 6292.

Bellisario L. M., Bubier J. L., Moore T. R., and Chanton J. P. (1999) Controls on CH_4 emissions from a northern peatland. *Global Biogeochem. Cycles* **13**, 81–91.

Ben-Bassat A., Lamed R., and Zeikus J. G. (1981) Ethanol production by thermophilic bacteria: metabolic control of end product formation in *Thermoanaerobium brockii*. *J. Bacteriol.* **146**, pp. 192–199.

Bender M. and Conrad R. (1992) Kinetics of CH_4 oxidation in oxic soils exposed to ambient air or high CH_4 mixing ratios. *FEMS Microbiol. Ecol.* **101**, 261–270.

Bengtsson G. and Annadotter H. (1989) Nitrate reduction in a groundwater microcosm determined by 15N gas chromatography-mass spectrometry. *Appl. Environ. Microbiol.* **55**, 2861–2870.

Bengtsson G. and Bergwall C. (2000) Fate of ^{15}N labelled nitrate and ammonium in a fertilized forest soil. *Soil Biol. Biochem.* **32**, 545–557.

Benner R., Maccubbin A. A., and Hodson R. E. (1984) Anaerobic biodegradation of the lignin and polysaccharide components of lignocellulose and synthetic lignin by sediment microflora. *Appl. Environ. Microbiol.* **47**, 998–1004.

Benner R., Moran M. A., and Hodson R. E. (1985) Effects of pH and plant source on lignocellulose biodegradation rates in two wetland ecosystems, the Okefenokee Swamp and a Georgia salt marsh. *Limnol. Oceanogr.* **30**, 489–499.

Benner R., Moran M. A., and Hodson R. E. (1986) Biogeochemical cycling of lignocellulosic carbon in marine and freshwater ecosystems: relative contributions of prokaryotes and eukaryotes. *Limnol. Oceanogr.* **31**, 89–100.

Benner R., Fogel M. L., and Sprague E. K. (1991) Diagenesis of belowground biomass of *Spartina alterniflora* in salt-marsh sediments. *Limnol. Oceanogr.* **36**, 1358–1374.

Benner R., Pakulski J. D., Mccarthy M., Hedges J. I., and Hatcher P. G. (1992) Bulk chemical characteristics of dissolved organic matter in the ocean. *Science* **255**, 1561–1564.

Benz M., Schink B., and Brune A. (1998) Humic acid reduction by *Propionibacterium freudenreichii* and other fermenting bacteria. *Appl. Environ. Microbiol.* **64**, 4507–4512.

Berendse F., van Breeman N., Rydin H., Buttler A., Heijmans M., Hoosbeek M. R., Lee J. A., Mitchell E., Saarinen T., Vasander H., and Wallen B. (2001) Raised atmospheric CO_2 levels and increased N deposition cause shifts in plant species composition and production in Sphagnum bogs. *Global Change Biol.* **7**, 591–598.

Bergamaschi P. (1997) Seasonal variations of stable hydrogen and carbon isotope ratios in methane from a Chinese rice paddy. *J. Geophys. Res.* **102**, 25383–25393.

Beudeker R. F., Cannon G. C., Kuenen J. G., and Shively J. M. (1980) Relations between D-ribulose-1,5-bisphosphate carboxylase, carboxysomes, and CO_2-fixing capacity in the obligately chemolithotroph *Thiobacillus neapolitanus* grown under different limitations in the chemostat. *Arch. Microbiol.* **124**, 185–189.

Bian L., Hinrichs K.-U., Xie T., Brassell S. C., Iversen N., Fossing H., Jorgensen B. B., and Hayes J. M. (2001) Algal and archaeal polyisoprenoids in a recent marine sediment: molecular isotopic evidence for anaerobic oxidation of methane. *Geochem. Geophys. Geosys.* **2** paper number 2000GC000112.

Biebl H. and Pfennig N. (1977) Growth of sulfate-reducing bacteria with sulfur as electron acceptor. *Arch. Microbiol.* **112**, 115–117.

Biebl H. and Pfennig N. (1978) Growth yields of green sulfur bacteria in mixed cultures with sulfur and sulfate-reducing bacteria. *Arch. Microbiol.* **117**, 9–16.

Bjerrum C. J. and Canfield D. E. (2002) Ocean productivity before about 1.9 Gyr ago limited by phosphorus adsorption onto iron oxides. *Nature* **417**, 159–162.

Blaabjerg V. and Finster K. (1998) Sulphate reduction associated with roots and rhizomes of the marine macrophyte Zostera marina. *Aquat. Microbiol. Ecol.* **15**, 311–314.

Black A. K. and Reddy C. A. (1991) Cloning and characterization of a lignin peroxidase gene from the white-rot fungus *Trametes versicolor*. *Biochem. Biophys. Res. Commun.* **179**, 428–435.

Blackburn T. H. and Henriksen K. (1983) Nitrogen cycling in different types of sediments from Danish waters. *Limnol. Oceanogr.* **28**, 477–493.

Blair N. (1998) The $d^{13}C$ of biogenic methane in marine sediments: the influence of C_{org} deposition rate. *Chem. Geol.* **152**, 139–150.

Blake R. and Johnson D. B. (2000) Phylogenetic and biochemical diversity among acidophilic bacteria that respire on iron. In *Environmental Metal-microbe Interactions* (ed. D. R. Lovley). ASM Press, Washington, DC, pp. 53–78.

Blunier T., Chappellaz J., Schwander J., Stauffer B., and Raynaud D. (1995) Variations in atmospheric methane concentration during the Holocene epoch. *Nature* **374**, 46–49.

Boast C. W., Mulvaney R. L., and Baveye P. (1988) Evaluation of nitrogen-15 tracer techniques for direct measurement of denitrification in soil: 1. Theory. *Soil Sci. Soc. Am. J.* **52**, 1317–1322.

Bock A. K. and Schönheit P. (1995) Growth of *Methanosarcina barkeri* (Fusaro) under nonmethanogenic conditions by the fermentation of pyruvate to acetate: ATP synthesis via the mechanism of substrate level phosphorylation. *J. Bacteriol.* **177**, 2002–2007.

Bock E., Schmidt I., Stüven R., and Zart D. (1995) Nitrogen loss caused by denitrifying *Nitrosomonas* cells using ammonium or hydrogen as electron donors and nitrite as electron acceptors. *Arch. Microbiol.* **163**, 16–20.

Bodelier P. L. E. and Frenzel P. (1999) Contribution of methanotrophic and nitrifying bacteria to CH_4 and NH_4^+ oxidation in the rhizosphere of rice plants as determined by new methods of discrimination. *Appl. Environ. Microbiol.* **65**, 1826–1833.

Bodelier P. L. E., Hahn A. P., Arth I. R., and Frenzel P. (2000a) Effects of ammonium-based fertilisation on microbial processes involved in methane emission from soils planted with rice. *Biogeochemistry* **51**, 225–257.

Bodelier P. L. E., Roslev P., Henckel T., and Frenzel P. (2000b) Stimulation by ammonium-based fertilizers of methane oxidation in soil around rice roots. *Nature* **403**, 421–424.

Bodenbender J., Wassmann R., Papen H., and Rennenberg H. (1999) Temporal and spatial variation of sulfur-gas-transfer between coastal marine sediments and the atmosphere. *Atmos. Environ.* **33**, 3487–3502.

Boeckx P. and Van Cleemput O. (1996) Methane oxidation in a neutral landfill cover soils: influence of moisture content, temperature, and nitrogen-turnover. *J. Environ. Qual.* **25**, 178–183.

Boehme S. E., Blair N. E., Chanton J. P., and Martens C. S. (1996) A mass balance of ^{13}C and ^{12}C in an organic-rich methane-producing marine sediment. *Geochim. Cosmochim. Acta* **60**, 3835–3848.

Boesen C. and Postma D. (1988) Pyrite formation in anoxic environments of the Baltic. *Am. J. Sci.* **288**, 575–603.

Boetius A. and Lochte K. (1994) Regulation of microbial enzymatic degradation of organic matter in Deep-Sea sediments. *Mar. Ecol. Prog. Ser.* **104**, 299–307.

Boetius A., Ferdelman T., and Lochte K. (2000a) Bacterial activity in sediments of the deep Arabian Sea in relation to vertical flux. *Deep-Sea Res. II: Top. Stud. Oceanogr.* **47**, 2835–2875.

Boetius A., Ravenschlag K., Schubert C. J., Rickert D., Widdel F., Gieseke A., Amann R., Jørgensen B. B., Witte U., and Pfannkuche O. (2000b) A marine microbial consortium apparently mediating anaerobic oxidation of methane. *Nature* **407**, 623–626.

Bogner J. E., Sass R. L., and Walter B. P. (2000) Model comparisons of methane oxidation across a management gradient: wetlands, rice production systems, and landfill. *Global Biogeochem. Cycles* **14**, 1021–1033.

Bohlen P. J., Groffman P. M., Driscoll C. T., Fahey T. J., and Siccama T. G. (2001) Plant-soil-microbial interactions in a northern hardwood forest. *Ecology* **82**, 965–978.

Bolliger C., Kleikemper J., Zeyer J., Schroth M. H., and Bernasconi S. M. (2001) Sulfur isotope fractionation during microbial sulfate reduction by toluene-degrading bacteria. *Geochim. Cosmochim. Acta* **65**, 3289–3298.

Bollmann A. and Conrad R. (1997) Acetylene blockage technique leads to underestimation of denitrification rates in oxic soils due to scavenging on intermediate nitric oxide. *Soil Biol. Biochem.* **29**, 1067–1077.

Bonch-Osmolovskaya E. A., Sokolova T. G., Kostrikina N. A., and Zavarzin G. A. (1990) *Desulfurella acetivorans gen. nov.* and *sp. nov.* a new thermophilic sulfur-reducing eubacterium. *Arch. Microbiol.* **153**, 151–155.

Bond D. R. and Lovley D. R. (2002) Reduction of Fe(III) oxide by methanogens in the presence and absence of extracellular quinones. *Environ. Microbiol.* **4**, 115–124.

Bonin P. (1996) Anaerobic nitrate reduction to ammonium in two strains isolated from coastal marine sediment: a dissimilatory pathway. *FEMS Microbiol. Ecol.* **19**, 27–38.

Bonin P., Omnes P., and Chalamet A. (1998) Simultaneous occurrence of denitrification and nitrate ammonification in sediments of the French Mediterranean coast. *Hydrobiologia* **389**, 169–182.

Boone D. R., Johnson R. L., and Liu Y. (1989) Diffusion of the interspecies electron carriers H_2 and formate in methanogenic ecosystems and its implications in the measurement of K_m for H_2 or formate uptake. *Appl. Environ. Microbiol.* **55**, 1735–1741.

Boone D. R., Whitman W. B., and Rouvičre P. (1993) Diversity and taxonomy of methanogens. In *Methanogenesis: Ecology, Physiology, Biochemistry and Genetics* (ed. J. G. Jerry). Chapman and Hall, New York, pp. 35–80.

Borjesson G. (2001) Inhibition of methane oxidation by volatile sulfur compounds (CH3SH and CS2) in landfill cover soils. *Waste Manage. Res.* **19**, 314–319.

Born M., Dörr H., and Levin I. (1990) Methane consumption in aerated soils of the temperate zone. *Tellus* **42B**, 2–8.

Bos J. P. and Kuenen J. G. (1990) Microbial treatment of coal. In *Microbial Mineral Recovery* (eds. H. L. and C. Brierley). McGraw Hill, New York, Ehrlich, pp. 344–377.

Boschker H. T. S. and Cappenberg T. E. (1994) A sensitive method using 4-methylumbelliferyl-β-cellobiose as a substrate to measure (1,4)-β-glucanase activity in sediments. *Appl. Environ. Microbiol.* **60**, 3592–3596.

Boschker H. T. S., Bertilsson S. A., Dekkers E. M. J., and Cappenberg T. E. (1995) An inhibitor-based method to measure initial decomposition of naturally occurring polysaccharides in sediments. *Appl. Environ. Microbiol.* **61**, 2186–2192.

Boschker H. T. S., deGraaf W., Koster M., MeyerReil L. A., and Cappenberg T. E. (2001) Bacterial populations and processes involved in acetate and propionate consumption in anoxic brackish sediment. *FEMS Microbiol. Ecol.* **35**, 97–103.

Boschker H. T. S., Nold, S. C., Wellsbury P., Bos D., de Graaf W., Pel R., Parkes R. J., and Cappenberg T. E. (1998) Direct linking of microbial populations to specific biogeochemical processes by ^{13}C-labelling of biomarkers. *Nature* **392**, 801–803.

Bosse U. and Frenzel P. (1997) Activity and distribution of methane-oxidizing bacteria in flooded rice soil microcosms and in rice plants (*Oryza sativa*). *Appl. Environ. Microbiol.* **63**, 1199–1207.

Bosse U. and Frenzel P. (1998) Methane emissions from rice microcosms: the balance of production, accumulation and oxidation. *Biogeochemistry* **41**, 199–214.

Bosse U., Frenzel P., and Conrad R. (1993) Inhibition of methane oxidation by ammonium in the surface layer of a littoral sediment. *FEMS Microbiol. Ecol.* **13**, 123–134.

Boudreau B. P. and Westrich J. T. (1984) The dependence of bacterial sulfate reduction on sulfate concentrations in marine sediments. *Geochim. Cosmochim. Acta* **48**, 2503–2516.

Bourne D. G., Holmes A. J., Iversen N., and Murrell J. C. (2000) Fluorescent oligonucleotide rDNA probes for specific detection of methane oxidising bacteria. *FEMS Microbiol. Ecol.* **31**, 29–38.

Bouwman A. F. (1991) Agronomic aspects of wetland rice cultivation and associated emissions. *Biogeochemistry* **15**, 65–88.

Bowden W. B. (1987) The biogeochemistry of nitrogen in freshwater wetlands. *Biogeochemistry* **4**, 313–348.

Bowien B. and Schlegel H. G. (1981) Physiology and biochemistry of aerobic hydrogen-oxidizing bacteria. *Ann. Rev. Microbiol.* **35**, 405–452.

Boyle A. W., Phelps C. D., and Young L. Y. (1999) Isolation from estuarine sediments of a *Desulfovibrio* strain which can grow on lactate coupled to the reductive dehalogenation of 2, 4,6-tribromophenol. *Appl. Environ. Microbiol.* **65**, 1133–1140.

Bradley P. M., McMahon P. B., and Chapelle F. H. (1995) Effects of carbon and nitrate on denitrification in bottom sediments of an effluent-dominated river. *Water Resour. Res.* **31**, 1063–1068.

Bremner J. M. and Bundy L. G. (1874) Inhibition of nitrification in soils by volatile sulfur compounds. *Soil. Biol. Biochem.* **6**, 161–165.

Bremner J. M. and Steele C. G. (1978) Role of microorganisms in the atmospheric sulfur cycle. In *Advances in Microbial Ecology*. (ed. M. Alexander) Plenum, New York, pp. 155–201.

Brettar I. and Rheinheimer G. (1992) Influence of carbon availability on denitrification in the central Baltic Sea. *Limnol. Oceanogr.* **37**, 1146–1163.

Bridgham S. D. and Richardson C. J. (1992) Mechanisms controlling soil respiration (CO_2 and CH_4) in southern peatlands. *Soil Biol. Biochem.* **24**, 1089–1099.

Bridgham S. D., Johnston C. A., Pastor J., and Updegraff K. (1995) Potential feedbacks of northern wetlands on climate change. *Bioscience* **45**, 262–274.

Bridgham S. D., Updergraff K., and Pastor J. (2001) A comparison of nutrient availability indices along an ombrotrophic-minerotrophic gradient in Minnesota wetlands. *Soil Sci. Soc. Am. J.* **65**, 259–269.

Broda E. (1977) Two kinds of lithotrophs missing in nature. *Z. Allg. Mikrobiol.* **17**, 491–493.

Bronson K. F., Neue H.-U., Singh U., and Abao E. B., Jr. (1997) Automated chamber measurements of methane and nitrous oxide flux in a flooded rice soil: I. Residue, nitrogen, and water management. *Soil Sci. Soc. Am. J.* **61**, 981–987.

Brown J. A., Glenn J. K., and Gold M. H. (1990) Manganese regulates expression of manganese peroxidase by *Phanerochaete chrysosporium*. *J. Bacteriol.* **172**, 3125–3130.

Bryant M. P. (1979) Microbial methane production-theoretical aspects. *J. Animal Sci.* **48**, 193–210.

Bryant M. P., Wolin E. A., Wolin M. J., and Wolfe R. S. (1967) *Methanobacillus omelianskii*, a symbiotic association of two species of bacteria. *Arch. Microbiol.* **59**, 20–31.

Bryant M. P., Campbell L. L., Reddy C. A., and Crabill M. R. (1977) Growth of *Desulfovibrio* in lactate or ethanol media low in sulfate in association with H_2-utilizing methanogenic bacteria. *Appl. Environ. Microbiol.* **33**, 1162–1169.

Bubier J. L. and Moore T. R. (1994) An ecological perspective on methane emissions from northern wetlands. *Tree* **9**, 460–464.

Buchanan R. E. (1917) Studies of the nomenclature and classification of the bacteria: III. The families of the Eubacteriales. *J. Bacteriol.* **2**, 347–350.

Bull I. D., Parekh N. R., Hall G. H., Ineson§ P., and Evershed R. P. (2000) Detection and classification of atmospheric methane oxidizing bacteria in soil. *Nature* **405**, 175–178.

Burdige D. J. (1989) The effects of sediment slurrying on microbial processes, and the role of amino acids as substrates for sulfate reduction in anoxic marine sediments. *Biogeochemistry* **8**, 1–23.

Burdige D. J. (1993) The biogeochemistry of manganese and iron reduction in marine sediments. *Earth Sci. Rev.* **35**, 249–284.

Burdige D. J. and Nealson K. H. (1986) Chemical and microbiological studies of sulfide-mediated manganese reduction. *Geomicrobiol. J.* **4**, 361–387.

Burdige D. J., Dhakar S. P., and Nealson K. H. (1992) Effects of manganese oxide mineralogy on microbial and chemical manganese reduction. *Geomicrobiol. J.* **10**, 27–48.

Buresh R. J. and Moraghan J. T. (1976) Chemical reduction of nitrate by ferrous iron. *J. Environ. Qual.* **5**, 320–325.

Buresh R. J. and Patrick W. H., Jr. (1978) Nitrate reduction to ammonium in anaerobic soil. *Soil Sci. Soc. Am. J.* **42**, 913–918.

Buresh R. J. and Patrick W. H., Jr. (1981) Nitrate reduction to ammonium and organic nitrogen in an estuarine sediment. *Soil. Biol. Biochem.* **13**, 279–283.

Burke R. A., Martens C. S., and Sackett W. S. (1988) Seasonal variations of D/H and 13C/12C ratios of microbial methane in surface sediments. *Nature* **332**, 829–830.

Burke R. A., Jr., Barber T. M., and Sackett W. M. (1992) Seasonal variations of stable hydrogen and carbon isotope ratios of methane in subtropical freshwater sediments. *Global Biogeochem. Cycles* **6**, 125–138.

Burns D. A. (1998) Retention of NO_3^- in an upland stream environment: a mass balance approach. *Biogeochemistry* **40**, 73–96.

Caccavo F., Lonergan D. J., Lovley D. R., Davis M., Stolz J. F., and Mcinerney M. J. (1994) *Geobacter sulfurreducens sp. nov.*, a hydrogen- and acetate-oxidizing dissimilatory metal-reducing microorganism. *Appl. Environ. Microbiol.* **60**, 3752–3759.

Caffrey J. M. and Kemp W. M. (1991) Seasonal and spatial patterns of oxygen production, respiration and root-rhizome release in *Potomogeton perfoliatus* L. and *Zostera marina* L. *Aquat. Bot.* **40**, 109–128.

Calhoun A. and King G. M. (1997) Regulation of root-associated methanotrophy by oxygen availability in the rhizosphere of two aquatic macrophytes. *Appl. Environ. Microbiol.* **63**, 3051–3058.

Canfield D. E. (1989) Reactive iron in marine sediments. *Geochim. Cosmochim. Acta* **53**, 619–632.

Canfield D. E. (1997) The geochemistry of river particulates from the continental USA: major elements. *Geochim. Cosmochim. Acta* **61**, 3349–3365.

Canfield D. E. and Des Marais D. J. (1991) Aerobic sulfate reduction in microbial mats. *Science* **251**, 1471–1473.

Canfield D. E., Jørgensen B. B., Fossing H., Glud R., Gundersen J., Ramsing N. B., Thamdrup B., Hansen J. W., Nielsen L. P., and Hall P. O. J. (1993a) Pathways of organic carbon oxidation in three continental margin sediments. *Mar. Geol.*, **113**.

Canfield D. E., Thamdrup B., and Hansen J. W. (1993b) The anaerobic degradation of organic matter in Danish coastal sediments: Iron reduction, manganese reduction, and sulfate reduction. *Geochim. Cosmochim. Acta* **57**, 3867–3883.

Capone D. G. (1991) Aspects of the marine nitrogen cycle with relevance to the dynamics of nitrous and nitric oxide.

In *Microbial Production and Consumption of Greenhouse Gases: Methane, Nitrous Oxides, and Halomethanes* (eds. J. E. Rogers and W. B. Whitman). American Society for Microbiology, Washington, DC, pp. 255–275.

Capone D. G. and Bautista M. F. (1985) A groundwater source of nitrate in nearshore marine sediments. *Nature* **313**, 214–216.

Capone D. C. and Kiene R. P. (1988) Comparison of microbial dynamics in marine and freshwater sediments: contrasts in anaerobic carbon catabolism. *Limnol. Oceanogr.* **33**, 725–749.

Casciotti K. L. and Ward B. B. (2001) Dissimilatory nitrite reductase genes form autotrophic ammonia-oxidizing bacteria. *Appl. Environ. Microbiol.* **67**, 2213–2221.

Castro H. F., Williams N. H., and Ogram A. (2000) Phylogeny of sulfate-reducing bacteria. *FEMS Microbiol. Ecol.* **31**, 1–9.

Castro M. S. and Dierberg F. E. (1987) Biogenic hydrogen sulfide emissions from selected Florida wetlands. *Water Air Soil Pollut.* **33**, 1–13.

Cavanaugh C. M. (1994) Microbial symbiosis: patterns of diversity in the marine environment. *Am. Zool.* **34**, 79–89.

Cavanaugh C. M., Gardiner S. L., Jones M. L., Jannasch H. W., and Waterbury J. B. (1981) Prokaryotic cells in the hydrothermal vent tube worm *Riftia pachyptila* Jones: possible chemoautotrophic symbionts. *Science* **213**, 340–342.

Cervantes F. J., van der Velde S., Lettinga G., and Field J. A. (2000) Competition between methanogenesis and quinone respiration for ecologically important substrates in anaerobic consortia. *FEMS Microbiol. Ecol.* **34**, 161–171.

Chanton J. P. and Dacey J. W. H. (1991) Effects of vegetation on methane flux, reservoirs, and carbon isotopic composition. In *Trace Gas Emissions by Plants* (eds. T. D. Sharkey, E. A. Holland, and H. A. Mooney). Academic Press, San Diego, pp. 65–92.

Chanton J. P. and Martens C. S. (1988) Seasonal variations in ebullitive flux and carbon isotopic composition of methane in a tidal freshwater estuary. *Global Biogeochem. Cycles* **2**, 289–298.

Chanton J. P., Martens C. S., Kelley C. A., Crill P. M., and Showers W. J. (1992a) Methane transport mechanisms and isotopic fractionation in emergent macrophytes of an Alaskan tundra lake. *J. Geophys. Res.* **97**(D15), 16681–16688.

Chanton J. P., Whiting G. J., Showers W. J., and Crill P. M. (1992b) Methane flux from Peltandra virginica: stable isotope tracing and chamber effects. *Global Biogeochem. Cycles* **6**, 15–31.

Chanton J. P., Bauer J. E., Glaser P. A., Siegel D. I., Kelley C. A., Tyler S. C., Romanowicz E. H., and Lazrus A. (1995) Radiocarbon evidence for the substrates supporting methane formation within northern Minnesota peatlands. *Geochim. Cosmochim. Acta* **59**, 3663–3668.

Chanton J. P., Arkebauer T. J., Harden H. S., and Verma S. B. (2002) Diel variation in lacunal CH_4 and CO_2 concentration and $\delta^{13}C$ in *Phragmites australis*. *Biogeochemistry* **59**, 287–301.

Chapelle F. H., O'Neill K., Bradley P. M., Methé B. A., Ciufo S. A., Knobel L. L., and Lovley D. R. (2002) A hydrogen-based subsurface microbial community dominated by methanogens. *Nature* **415**, 312–315.

Chappellaz J. A., Fung I. Y., and Thompson A. M. (1993) The atmospheric CH_4 increase since the last glacial maximum (1). Source estimates. *Tellus* **45B**, 228–241.

Charlson R. J., Lovelock J. E., Andreae M. O., and Warren S. G. (1987) Oceanic phytoplankton, atmospheric sulphur, cloud albedo, and climate. *Nature* **326**, 655–661.

Charman D. J., Aravena R., and Warner B. G. (1994) Carbon dynamics in a forested peatland in north-eastern Ontario, Canada. *J. Ecol.* **82**, 55–62.

Chasar L. S., Chanton J. P., Glaser P. H., and Siegel D. I. (2000a) Methane concentration and stable isotope distribution as evidence of rhizospheric processes: comparison of a fen and bog in the glacial Lake Agassiz Peatland complex. *Ann. Bot.* **86**, 655–663.

Chasar L. S., Chanton J. P., Glaser P. H., Siegel D. I., and Rivers J. S. (2000b) Radiocarbon and stable carbon isotopic evidence for transport and transformation of dissolved organic carbon, dissolved inorganic carbon, and CH_4 in a northern Minnesota peatland. *Global Biogeochem. Cycles* **14**, 1095–1108.

Chen D. L., Chalk P. M., and Freney J. R. (1995) Distribution of reduced products of ^{15}N-labelled nitrate in anaerobic soils. *Soil Biol. Biochem.* **27**, 1539–1545.

Chidthaisong A. and Conrad R. (2000) Turnover of glucose and acetate coupled to reduction of nitrate, ferric iron and sulfate and to methanogenesis in anoxic rice field soil. *FEMS Microbiol. Ecol.* **31**, 73–86.

Chien C. C., Leadbetter E. R., and Godchaux W. (1999) *Rhodococcus* spp. utilize taurine (2-aminoethanesulfonate) as sole source of carbon, energy, nitrogen, and sulfur for aerobic respiratory growth. *FEMS Microbiol. Lett.* **176**, 333–337.

Childers S. E., Ciufo S., and Lovley D. R. (2002) *Geobacter metallireducens* accesses insoluble Fe(III) oxide by chemotaxis. *Nature* **416**, 767–769.

Childress J. J., Fisher C. R., Brooks J. M., Kennicutt M. C., II, and Bidigare R. A. A. E. (1986) A methanotrophic marine molluscan (Bivalvia, Mytilidae) symbiosis: mussels fueled by gas. *Science* **233**, 1306–1308.

Chin K. and Conrad R. (1995) Intermediary metabolism in methanogenic paddy soil and the influence of temperature. *FEMS Microbiol. Ecol.* **18**, 85–102.

Chin K., Lukow T., and Conrad R. (1999a) Effect of temperature on structure and function of the methanogenic archaeal community in an anoxic rice field soil. *Appl. Environ. Microbiol.* **65**, 2341–2349.

Chin K., Lukow T., Stubner S., and Conrad R. (1999b) Structure and function of the methanogenic archaeal community in stable cellulose-degrading enrichment cultures at two different temperatures (15 and 30 °C). *FEMS Microbiol. Ecol.* **30**, 313–326.

Chistoserdova L., Vorholt J. A., Thauer R. K., and Lidstrom M. E. (1998) C_1 transfer enzymes and coenzymes linking methylotrophic bacteria and methanogenic archaea. *Science* **281**, 99–102.

Cho K. S., Hirai M., and Shoda M. (1991) Degradation characteristics of hydrogen sulfide, methanethiol, dimethyl sulfide, and dimethyl disulfide by *Thiobacillus thioparus* DW44 isolated from peat biofilter. *J. Ferment. Bioeng.* **71**, 384–389.

Christensen D. (1984) Determination of substrates oxidized by sulfate reduction in intact cores of marine sediments. *Limnol. Oceanogr.* **29**, 189–192.

Christensen P. B. and Sørensen J. (1986) Temporal variation of denitrification in plant-covered, littoral sediment from Lake Hampen, Denmark. *Appl. Environ. Microbiol.* **51**, 1174–1179.

Christensen P. B., Rysgaard S., Sloth N. P., Dalsgaard T., and Schwaerter S. (2000) Sediment mineralization, nutrient fluxes, denitrification, and dissimilatory nitrate reduction to ammonium in an estuarine fjord with sea cage trout farms. *Aquat. Microbial. Ecol.* **21**, 73–84.

Clarens M., Bernet N., Delgenes J. P., and Moletta R. (1998) Effects of nitrogen oxides and denitrification by *Pseudomonas stutzeri* on acetotrophic methanogenesis by *Methanosarcina mazei*. *FEMS Microbiol. Ecol.* **25**, 271–276.

Clymo R. S. and Pearce D. M. E. (1995) Methane and carbon dioxide production in, transport through, and efflux from a peatland. *Phils. Trans., Phys. Sci. Eng.* **351**, 249–259.

Coates J. D., Phillips E. J. P., Lonergan D. J., Jenter H., and Lovley D. R. (1996) Isolation of *Geobacter* species from diverse sedimentary environments. *Appl. Environ. Microbiol.* **62**, 1531–1536.

Coates J. D., Ellis D. J., Blunt-Harris E. L., Gaw C. V., Roden E. E., and Lovley D. R. (1998) Recovery of humic-reducing bacteria from a diversity of environments. *Appl. Environ. Microbiol.* **64**, 1504–1509.

Coates J. D., Ellis D. J., Gaw C. V., and Lovley D. R. (1999) Geothrix fermentans gen. nov., sp. nov., a novel Fe(III)-reducing bacterium from a hydrocarbon-contaminated aquifer. *Int. J. Systemat. Bacteriol.* **49**, 1615–1622.

Coates J. D., Bhupathiraju V. K. A. L. A., McInerney M. J., and Lovley D. R. (2001) *Geobacter hydrogenophilus, Geobacter chapellei,* and *Geobacter grbiciae*, three new, strictly anaerobic, dissimulatory Fe(III)-reducers. *Int. J. Syst. Evol. Microbiol.* **51**, 581–588.

Cocaign M., Wilberg E., and Lindley N. D. (1991) Sequential demethoxylation reactions during methyltrophic growth of methoxylated aromatic substrates with *Eubacterium limosum. Arch. Microbiol.* **155**, 496–499.

Colberg P. J. (1988) Anaerobic microbial degradation of cellulose, lignin, oligonols and monomeric lignin derivatives. In *Biology of Anaerobic Microorganisms* (ed. A. J. G. Zehnder). Wiley, New York, pp. 333–372.

Colberg P. J. and Young L. Y. (1982) Microbial degradation of lignin-derived compounds under anaerobic conditions. *Can. J. Microbiol.* **28**, 886–889.

Coleman D. D., Risatti J. B., and Schoell M. (1981) Fractionation of carbon and hydrogen isotopes by methane-oxidizing bacteria. *Geochim. Cosmochim. Acta* **45**, 1033–1037.

Coleman G. S. (1960) A sulfate-reducing bacterium from the sheep rumen. *J. Gen. Microbiol.* **22**, 423–436.

Coleman M. L., Hedrick D. B., Lovley D. R., White D. C., and Pye K. (1993) Reduction of Fe(III) in sediments by sulphate-reducing bacteria. *Nature* **361**, 436–438.

Comly H. H. (1945) Cyanosis in infants caused by nitrate in well water. *J. Am. Med. Assoc.* **129**, 112–116.

Conrad R. (1996) Soil microorganisms as controllers of atmospheric trace gases (H_2, CO, CH_4, N_2O, and NO). *Microbiol. Rev.* **60**, 609–640.

Conrad R. (1999) Contribution of hydrogen to methane production and control of hydrogen concentrations in methanogenic soils and sediments. *FEMS Microbiol. Ecol.* **28**, 193–202.

Conrad R. and Klose M. (1999) Anaerobic conversion of carbon dioxide to methane acetate and propionate on washed rice roots. *FEMS Microbiol. Ecol.* **30**, 147–155.

Conrad R. and Klose M. (2000) Selective inhibition of reactions involved in methanogenesis and fatty acid production on rice roots. *FEMS Microbiol. Ecol.* **34**, 27–34.

Conrad R. and Rothfuss F. (1991) Methane oxidation in the soil surface layer of a flooded rice field and the effect of ammonium. *Biol. Fertility Soils* **12**, 28–32.

Conrad R., Schink B., and Phelps T. J. (1986) Thermodynamics of H_2-producing and H_2-consuming metabolic reactions in diverse methanogenic environments under *in situ* conditions. *FEMS Microbiol. Ecol.* **38**, 353–360.

Conrad R., Schütz H., and Babbel M. (1987) Temperature limitation of hydrogen turnover and methanogenesis in anoxic paddy soil. *FEMS Microbiol. Ecol.* **45**, 281–289.

Conrad R., Mayer H. P., and Wüst M. (1989) Temporal change of gas metabolism by hydrogen-syntrophic methanogenic bacterial associations in anoxic paddy soil. *FEMS Microbiol. Ecol.* **62**, 265–274.

Conrad R., Klose M., and Claus P. (2000) Phosphate inhibits acetotrophic methanogenesis on rice roots. *Appl. Environ. Microbiol.* **66**, 828–831.

Cord-Ruwisch R., Ollivier B., and Garcia J. L. (1986) Fructose degradation by *Desulfovibrio* sp. in pure culture and in coculture with *Methanospirillum hungatei. Current Microbiol.* **13**, 285–289.

Cord-Ruwisch R., Seitz H. J., and Conrad R. (1988) The capacity of hydrogenotrophic anaerobic bacteria to compete for traces of hydrogen depends on the redox potential of the terminal electron acceptor. *Arch. Microbiol.* **149**, 350–357.

Cord-Ruwisch R., Lovley D. R., and Schink B. (1998) Growth of *Geobacter sulfurreducens* with acetate in syntrophic cooperation with hydrogen-oxidizing anaerobic partners. *Appl. Environ. Microbiol.* **64**, 2232–2236.

Cornell R. M. and Schwertmann U. (1996) *The Iron Oxides: Structure, Properties, Reactions, Occurrences, and Uses.* VCH, Weinheim, Germany.

Cornwell J. C., Kemp W. M., and Kana T. M. (1999) Denitrification in coastal ecosystems: methods, environmental controls, and ecosystem level controls, a review. *Aquat. Ecol.* **33**, 41–54.

Cowie G. L. and Hedges J. I. (1993) A comparison of organic matter sources, diagenesis and preservation in oxic and anoxic coastal sites. *Chem. Geol.* **107**, 447–451.

Crawford R. L. (1981) *Lignin Biodegradation and Transformation.* Wiley, New York.

Crill P. M. and Martens C. S. (1986) Methane production from bicarbonate and acetate in an anoxic marine sediment. *Geochim. Cosmochim. Acta* **50**, 2089–2097.

Crill P. M. and Martens C. S. (1987) Biogeochemical cycling in an organic-rich coastal marine basin: 6. Temporal and spatial variations in sulfate reduction rates. *Geochim. Cosmochim. Acta* **51**, 1175–1186.

Crill P. M., Martikainen P. J., Nykänen H., and Silvola J. (1994) Temperature and N fertilization effects on methane oxidation in a drained peatland soil. *Soil Biol. Biochem.* **26**, 1331–1339.

Crutzen P. J. (1995) On the role of CH_4 in atmospheric chemistry: sources, sinks and possible reductions in anthropogenic sources. *Ambio* **24**, 52–55.

Cummings D. E., Caccavo F., Jr., Fendorf S., and Rosenzweig R. F. (1999) Arsenic mobilization by the dissimilatory Fe(III)-reducing bacterium *Shewanella alga* BrY. *Environ. Sci. Technol.* **33**, 723–729.

Currin C. A., Joye S. B., and Paerl H. W. (1996) Diel rates of N_2-fixation and denitrification in a transplanted *Spartina alterniflora* marsh: implications for N-flux dynamics. *Estuar. Coast. Shelf Sci.* **42**, 597–616.

Cypionka H. (2000) Oxygen respiration by *Desulfovibrio* species. *Ann. Rev. Microbiol.* **54**, 827–848.

Dacey J. W. H. and Howes B. L. (1984) Water uptake by roots controls water table movement and sediment oxidation in short *Spartina* marsh. *Science* **224**, 487–489.

Dacey J. W., King G. M., and Wakeham S. G. (1987) Factors controlling emission of dimethylsulfide from salt marshes. *Nature* **330**, 643–645.

Dacey J. W. H., Drake B. G., and Klug M. J. (1994) Stimulation of methane emission by carbon dioxide enrichment of marsh vegetation. *Nature* **370**, 47–49.

Dalsgaard T. and Bak F. (1992) Effect of acetylene on nitrous oxide reduction and sulfide oxidation in batch and gradient cultures of *Thiobacillus denitrificans*. *Appl. Environ. Microbiol.* **58**, 1601–1608.

Dalsgaard T. and Thamdrup B. (2002) Factors controlling anaerobic ammonium oxidation with nitrite in marine sediments. *Appl. Environ. Microbiol.* **68**, 3802–3808.

Dalsgaard T., Canfield D. E., Petersen J., Thamdrup B., and Acuña-González J. (2003) N_2 production by the anammox reaction in the anoxic water column of Golfo Dulce, Costa Rica. *Nature* **422**, 606–608.

Daly K., Sharp R. J., and McCarthy A. J. (2000) Development of oligonucleotide probes and PCR primers for detecting phylogenetic subgroups of sulfate-reducing bacteria. *Microbiology* **146**, 1693–1705.

Dannenberg S. and Conrad R. (1999) Effect of rice plants on methane production and rhizospheric metabolism in paddy soil. *Biogeochemistry* **45**, 53–71.

Dannenberg S., Kroder M., Dilling W., and Cypionka H. (1992) Oxidation of H_2, organic compounds and inorganic sulfur compounds coupled to reduction of O_2 or nitrate by sulfate-reducing bacteria. *Arch. Microbiol.* **158**, 93–99.

Davidson E. A. (1992) Sources of nitric oxide and nitrous oxide following wetting of dry soil. *Soil Sci. Soc. Am. J.* **56**, 95–102.

Davidson E. A., Swank W. T., and Perry T. O. (1986) Distinguishing between nitrification and denitrification as sources of gaseous nitrogen production in soil. *Appl. Environ. Microbiol.* **52**, 1280–1286.

de Bont J. A. M., Lee K. K., and Bouldin D. F. (1978) Bacterial oxidation of methane in a rice paddy. In *Environmental Role of Nitrogen-fixing Blue-green Algae and Asymbiotic Bacteria. Ecol. Bull. (Stockholm)* **26**, 91–96.

de Bont J. A. M., van Dijken J. P., and Harder W. (1981) Dimethyl sulphoxide and dimethyl sulphide as a carbon, sulphur, and energy source for growth of *Hyphomicrobium* S. *J. Gen. Microbiol.* **127**, 315–323.

deCatanzaro J. B., Beauchamp E. G., and Drury C. F. (1987) Denitrification vs. dissimilatory nitrate reduction in soil with alfalfa, straw, glucose and sulfide treatments. *Soil. Biol. Biochem.* **19**, 583–587.

Dedysh S. N., Panikov N. S., Liesack W., Groβkopf R., Zhou J., and Tiedje J. M. (1998a) Isolation of acidophilic methane-oxidizing bacteria from northern peat wetlands. *Science* **282**, 281–284.

Dedysh S. N., Panikov N. S., and Tiedje J. M. (1998b) Acidophilic methanotrophic communities from *Sphagnum* peat bogs. *Appl. Environ. Microbiol.* **64**, 922–929.

Dedysh S. N., Liesack W., Khmelenina V. N., Suzina N. E., Trotsenko Y. A., Semrau J. D., Bares A. M., Panikov N. S., and Tiedje J. M. (2000) *Methylocella palustris* gen. nov., sp. nov., a new methane-oxidizing acidophilic bacterium from peat bogs representing a novel sub-type of serine pathway methanotrophs. *Int. J. Syst. Evol. Microbiol.* **50**, 955–969.

Dedysh S. N., Derakshani M., and Liesack W. (2001) Detection and enumeration of methanotrophs in acidic *Sphagnum* peat by 16S rRNA flurorescence *in situ* hybridization, including the use of newly developed oligonucleotide probes for *Methylocella palustris*. *Appl. Environ. Microbiol.* **67**, 4850–4857.

Dedysh S. N., Khmelenina V. N., Suzina N. E., Trotsenko Y. A., Semrau J. D., Liesack W., and Tiedje J. M. (2002) *Methylocapsa acidiphila* gen. nov., sp. nov., a novel methane-oxidizing and dinitrogen-fixing acidophilic bacterium from *Sphagnum* bog. *Int. J. Syst. Evol. Microbiol.* **52**, 251–261.

Dedysh S. N., Dunfield P. F., Derakshani M., Stubner S., Heyer J., and Liesack W. (2003) Differential detection of type II methanotrophic bacteria in acidic peatlands using newly developed 16S rRNA-targeted fluorescent oligonucleotide probes. *FEMS Microbiol. Ecol.* **43**, 299–308.

Deer W., Howie R., and Sussman J. (1992) *An Introduction to the Rock-forming Minerals*. Longman, Horlow, Essex, UK.

de Klein C. A. M. and van Logtestijn R. S. P. (1996) Denitrification in grassland soils in The Netherlands in relation to irrigation, nitrogen application, soil water content and soil temperature. *Soil Biol. Biochem.* **28**, 231–237.

DeLaune R. D., Boar R. R., Lindau C. W., and Kleiss B. A. (1996) Denitrification in bottomland hardwood wetland soils of the Cache River. *Wetlands* **16**, 309–320.

DeLaune R. D., Devai I., and Lindau C. W. (2002) Flux of reduced sulfur gases along a salinity gradient in Louisiana coastal marshes. *Estuar. Coast. Shelf Sci.* **54**, 1003–1011.

DeLong E. F. (2002) All in the packaging. *Nature* **419**, 676–677.

Dendooven L. and Anderson J. M. (1994) Dynamics of reduction enzymes involved in the denitrification process in pasture soil. *Soil Biol. Biochem.* **26**, 1501–1506.

Denier van der Gon H. A. C. and Neue H.-U. (1995) Influence of organic matter incorporation on the methane emission from a rice field. *Global Biogeochem. Cycles* **9**, 11–22.

Denier van der Gon H. A. C. and Neue H.-U. (1996) Oxidation of methane in the rhizosphere of rice plants. *Biol. Fertility Soils* **22**, 359–366.

Deppenmeier U., Lienard T., and Gottschalk G. (1999) Novel reactions involved in energy conservation by methanogenic archaea. *Federat. Euro. Biochem. Soc. Lett.* **457**, 291–297.

Des Marais D. J. (2000) Evolution—when did photosynthesis emerge on earth? *Science* **289**, 1703–1705.

de Souza M. P. and Yoch D. C. (1995) Purification and characterization of dimethylsulfoniopropionate lyase from an alcaligenes-like dimethyl sulfide-producing marine isolate. *Appl. Environ. Microbiol.* **61**, 21–26.

Devereux R. and Stahl D. A. (1993) Phylogeny of sulfate-reducing bacteria and a perspective for analyzing their natural communities. In *Sulfate-reducing Bacteria: Contemporary Perspectives* (eds. J. M. Odom and R. Singleton). Springer, New York, pp. 131–160.

Devereux R., He S.-H., Doyle C. L., Orkland S., Stahl D. A., LeGall J., and Whitman W. B. (1990) Diversity and origin of *Desulfovibrio* species: phylogenetic definition of a family. *J. Bacteriol.* **172**, 3609–3619.

Devereux R., Kane M. D., Winfrey J., and Stahl D. A. (1992) Genus- and group-specific hybridization probes for determinative and environmental studies of sulfate-reducing bacteria. *Syst. Appl. Microbiol.* **15**, 149–601.

Devereux R., Hines M. E., and Stahl D. A. (1996) S-cycling: characterization of natural communities of sulfate-reducing bacteria by 16S rRNA sequence comparisons. *Microbiol. Ecol.* **32**, 283–292.

DeWeerd K. A., Mandelco L., Tanner R. S., Woese C. R., and Suflita J. M. (1990) *Desulfomonile iedjei* gen. nov. and sp. nov., a novel anaerobic, dehalogenating, sulfate-reducing bacterium. *Arch. Microbiol.* **154**, 23–30.

De Wit R. and Van Gemerden H. (1987) Chemolithotrophic growth of the purple sulfur bacterium *Thiocapsa roseopersicina*. *FEMS Microbiol. Ecol.* **45**, 117–126.

de Zwart J. M. M., Nelisse P. N., and Kuenen J. G. (1996) Isolation and characterization of *Methylophaga sulfidovorans* sp. nov.: an obligately methylotrophic, aerobic, dimethylsulfide oxidizing bacterium from a microbial mat. *FEMS Microbiol. Ecol.* **20**, 261–270.

D'Hondt S., Rutherford S., and Spivack A. J. (2002) Metabolic activity of subsurface life in deep-sea sediments. *Science* **295**, 2067–2070.

DiChristina T. J. (1992) Effects of nitrate and nitrite on dissimilatory iron reduction by *Shewanella putrefaciens* 200. *J. Bacteriol.* **174**, 1891–1896.

Dilling W. and Cypionka H. (1990) Aerobic respiration in sulfate-reducing bacteria. *FEMS Microbiol. Lett.* **71**, 123–128.

Dise N. B. (1992) Winter fluxes of methane from Minnesota peatlands. *Biogeochemistry* **17**, 17–83.

Dise N. B. and Verry E. S. (2001) Supression of peatland methane emission by cumulative sulfate desposition in simulated acid rain. *Biogeochemistry* **53**, 143–160.

Dise N. E., Gorham E., and Verry E. S. (1993) Environmental factors influencing the seasonal release of methane from peatlands in northern Minnesota. *J. Geophys. Res.* **98**, 10583–10594.

Dittmar T. and Lara R. J. (2001) Molecular evidence for lignin degradation in sulfate-reducing mangrove sediments (Amazonia, Brazil). *Geochim. Cosmochim. Acta* **65**, 1417–1428.

Dolfing J. and Tiedje J. M. (1987) Growth yield increase linked to reductive dechlorination in a defined 3-chlorobenzoate degrading methanogenic coculture. *Arch. Microbiol.* **149**, 102–105.

Dolfing J. and Tiedje J. M. (1991) Kinetics of two complementary hydrogen sink reactions in a defined 3-chlorobenzoate degrading methanogenic co-culture. *FEMS Microbiol. Ecol.* **86**, 25–32.

Dong X. and Tollner E. W. (2003) Evaluation of anammox and denitrification during anaerobic digestion of poultry manure. *Bioresour. Technol.* **86**, 139–145.

Dong X., Plugge C. M., and Stams A. J. M. (1994) Anaerobic degradation of propionate by a mesophilic acetogenic bacterium in coculture and triculture with different methanogens. *Appl. Environ. Microbiol.* **60**, 2834–2838.

Dornblaser M., Giblin A. E., Fry B., and Peterson B. J. (1994) Effects of sulfate concentration in the overlying water on sulfate reduction and sulfur storage in lake sediments. *Biogeochemistry* **24**, 129–144.

Dowdle P. R. and Oremland R. S. (1998) Microbial oxidation of elemental selenium in soil slurries and bacterial cultures. *Environ. Sci. Technol.* **32**, 3749–3755.

Drake H. L. (1994) Acetogenesis, acetogenic bacteria, and the acetyl-CoA "Wood/Ljungdahl" pathway: past and current perspectives. In *Acetogenesis* (ed. H. L. Drake). Chapman and Hall, New York, pp. 3–60.

Drotar A. G. A., Burton J., Tavernier J. E., and Fall R. (1987a) Widespread occurrence of bacterial thiol methyltransferases and the biogenic emission of methylated sulfur gases. *Appl. Environ. Microbiol.* **53**, 1626–1644.

Drotar A. M. and Fall R. (1985) Methylation of xenobiotic thiols by *Euglena gracilis*: characterization of cytoplasmic thiol methyltransferase. *Plant Cell Physiol.* **26**, 847–854.

Drotar A. M., Fall L. R., Mishalanie E. A., Tavernier J. E., and Fall R. (1987b) Enzymatic methylation of sulfide, selenide, and organic thiols by *Tetrahymena thermophila*. *Appl. Environ. Microbiol.* **53**, 2111–2118.

Drury C. F., McKenney D. J., and Findlay W. I. (1991) Relationships between denitrification, microbial biomass and indigenous soil properties. *Soil Biol. Biochem.* **23**, 751–755.

Drury C. F., McKenney D. J., and Findlay W. I. (1992) Nitric oxide and nitrous oxide production from soil: water and oxygen effects. *Soil Sci. Soc. Am. J.* **56**, 766–770.

Dubilier N., Amann R., Erseus C., Muyzer G., Park S. Y., Giere O., and Cavanaugh C. M. (1999) Phylogenetic diversity of bacterial endosymbionts in the gutless marine oligochete Olavius loisae (Annelida). *Mar. Ecol. Prog. Ser.* **178**, 271–280.

Duddleston K. N., Kinney M. A., Kiene R. P., and Hines M. E. (2002) Anaerobic microbial biogeochemistry in a northern bog: acetate a dominant metabolic end product. *Global Biogeochem. Cycles* **16**(11), 1–9.

Dunfield P. and Knowles R. (1995) Kinetics of inhibition of methane oxidation by nitrate, nitrite, and ammonium in a Humisol. *Appl. Environ. Microbiol.* **61**, 3129–3135.

Dunfield P., Knowles R., Dumont R., and Moore T. R. (1993) Methane production and consumption in temperate and subarctic soils: response to temperature and pH. *Soil Biol. Biochem.* **25**, 321–326.

Edwards K. J., Bond P. L., and Banfield J. F. (2000a) Characteristics of attachment and growth of *Thiobacillus caldus* on sulphide minerals: a chemotactic response to sulphur minerals? *Environ. Microbiol.* **2**, 324–332.

Edwards K. J., Bond P. L., Gihring T. M., and Banfield J. F. (2000b) An archaeal iron-oxidizing extreme acidophile important in acid mine drainage. *Science* **287**, 1796–1799.

Edwards T. and McBride B. C. (1975) A new method for the isolation and identification of methanogenic bacteria. *Appl. Microbiol.* **29**, 540–545.

Egli K., Fanger U., Alvarez P. J. J., Siegrist H., van der Meer J. R., and Zehnder A. J. B. (2001) Enrichment and characterization of an anammox bacterium from a rotating biological contactor treating ammonium-rich leachate. *Arch. Microbiol.* **175**, 198–207.

Ehrenreich A. and Widdel F. (1994) Anaerobic oxidation of ferrous iron by purple bacteria, a new type of phototrophic metabolism. *Appl. Environ. Microbiol.* **60**, 4517–4526.

Ehrlich H. L. (1999) Microbes as geologic agents: their role in mineral formation. *Geomicrobiol. J.* **16**, 135–153.

Ehrlich H. L. (2002) *Geomicrobiology* 4th edn. Dekker, New York.

Eichler B. and Schink B. (1986) Fermentation of primary alcohols and diols, and pure culture of syntrophically alcohol-oxidizing anaerobes. *Arch. Microbiol.* **143**, 60–66.

Eller G., Stubner S., and Frenzel P. (2001) Group-specific 16S rRNA targeted probes for the detection of type I and type II methanotrophs by fluorescence *in situ* hybridisation. *FEMS Microbiol. Lett.* **198**, 91–97.

Elsgaard L., Jannasch H. W., Isaksen M. F., Jorgensen B. B., and Alayse A. M. (1994) Microbial sulfate reduction in deep-sea sediments at the Guaymas Basin hydrothermal vent area: influence of temperature and substrates. *Geochim. Cosmochim. Acta* **58**, 3335–3343.

Elshahed M. S. and McInerney M. J. (2001a) Benzoate fermentation by the anaerobic bacterium *Syntrophus aciditrophicus* in the absence of hydrogen-using microorganisms. *Appl. Environ. Microbiol.* **67**, 5520–5525.

Elshahed M. S. and McInerney M. J. (2001b) Is interspecies hydrogen transfer needed for toluene degradation under sulfate-reducing conditions? *FEMS Microbiol. Ecol.* **35**, 163–169.

Elshahed M. S., Bhupathiraju V. K., Wofford N. Q., Nanny M. A., and McInerney M. J. (2001) Metabolism of benzoate, cyclohex-1-ene carboxylate, and cyclohexane carboxylate by "*Syntrophus aciditrophicus*" strain SB in syntrophic association with H_2 using microorganisms. *Appl. Environ. Microbiol.* **67**, 1728–1738.

Elvert M., Suess E., and Whiticar M. J. (1999) Anaerobic methane oxidation associated with marine gas hydrates: superlight C-isotopes from saturated and unsaturated C-20 and C-25 irregular isoprenoids. *Naturwissenschaften* **86**, 295–300.

Elvert M., Suess E., Greinert J., and Whiticar M. J. (2000) Archaea mediating anaerobic methane oxidation in deep-sea sediments at cold seeps of the eastern Aleutian subduction zone. *Org. Geochem.* **31**, 1175–1187.

Emerson D. (2000) Microbial oxidation of Fe(II) and Mn(II) at circumneutral pH. In *Environmental Microbe-metal Interactions* (ed. D. R. Lovley). ASM Press, Washington, DC, pp. 31–52.

Emerson D. and Moyer C. (1997) Isolation and characterization of novel iron-oxidizing bacteria that grow at circumneutral pH. *Appl. Environ. Microbiol.* **63**, 4784–4792.

Emerson D. and Moyer C. L. (2002) Neutrophilic Fe-oxidizing bacteria are abundant at the Loihi seamount hydrothermal vents and play a major role in Fe oxide deposition. *Appl. Environ. Microbiol.* **68**, 3085–3093.

Emerson D. and Revsbech N. P. (1994) Investigation of an iron-oxidizing microbial mat community located near Aarhus, Denmark: field studies. *Appl. Environ. Microbiol.* **60**, 4022–4031.

Emmenegger L., Schönenberger R., Sigg L., and Sulzberger B. (2001) Light-induced redox cycling of iron in circumneutral lakes. *Limnol. Oceanogr.* **46**, 49–61.

Epp M. A. and Chanton J. P. (1993) Rhizospheric methane oxidation determined via the methylfluoride inhibition technique. *J. Geophys. Res.* **98**, 18413–18422.

Eriksson K. E., Blanchette R. A., and Ander P. (1990) *Microbial and Enzymatic Degradation of Wood and Wood Components*. Springer, Berlin.

Evans W. C. and Fuchs G. (1988) Anaerobic degradation of aromatic compounds. *Ann. Rev. Microbiol.* **4**, 289–317.

Fathepure B. Z. and Boyd S. A. (1988) Reductive dechlorination of perchloroethylene and the role of methanogens. *FEMS Microbiol. Lett.*, **49**, 149–156.

Fazzolari E., Mariotti A., and Germon J. C. (1990) Nitrate reduction to ammonia: a dissimilatory process in *Enterobacter amnigenus*. *Can. J. Microbiol.* **36**, 779–785.

Fazzolari É., Nicolardot B., and Germon J. C. (1998) Simultaneous effects of increasing levels of glucose and oxygen partial pressures on denitrification and dissimilatory nitrate reduction to ammonium in repacked soil cores. *Euro. J. Soil Biol.* **34**, 47–52.

Fechner E. J. and Hemond H. F. (1992) Methane transport and oxidation in the unsaturated zone of a *Sphagnum* peatland. *Global Biogeochem. Cycles* **6**, 33–44.

Fenchel T. and Finlay B. J. (1995) *Ecology and Evolution in Anoxic Worlds*. Oxford University Press, New York.

Ferry J. G. (1999) Enzymology of one-carbon metabolism in methanogenic pathways. *FEMS Microbiol. Rev.* **23**, 13–38.

Fetzer S. and Conrad R. (1993) Effect of redox potential on methanogenesis by *Methanosarcina barkeri*. *Arch. Microbiol.* **160**, 108–113.

Fetzer S., Bak F., and Conrad R. (1993) Sensitivity of methanogenic bacteria from paddy soil to oxygen and desiccation. *FEMS Microbiol. Ecol.* **12**, 107–115.

Fey A. and Conrad R. (2000) Effect of temperature on carbon and electron flow and on the archaeal community in methanogenic rice field soil. *Appl. Environ. Microbiol.* **66**, 4790–4797.

Fiala-Médioni A., McKiness Z. P., Dando P., Boulegue J., Mariotti A., Alayse-Danet A. M., Robinson J. J., and Cavanaugh C. M. (2002) Ultrastructural, biochemical, and immunological characterization of two populations of the mytilid mussel *Bathymodiolus azoricus* from the Mid-Atlantic Ridge: evidence for a dual symbiosis. *Mar. Biol.* **141**, 1035–1043.

Fiebig K. and Gottschalk G. (1983) Methanogenesis from choline by a coculture of *Desulfovibrio* sp. and Methanosarcina barkeri. *Appl. Environ. Microbiol.* **45**, 161–168.

Finster K., King G. M., and Bak F. (1990) Formation of methyl mercaptan and dimethylsulfide from methoxylated aromatic compounds in anoxic marine and freshwater sediments. *FEMS Microbiol. Ecol.* **74**, 295–302.

Finster K., Tanimoto Y., and Bak F. (1992) Fermentation of methanethiol and dimethylsulfide by a newly isolated methanogenic bacterium. *Arch. Microbiol.* **157**, 425–430.

Finster K., Liesack W., and Tindall B. J. (1997) *Desulfospira joergensenii*, gen. nov., sp. nov., a new sulfate-reducing isolated from marine surface sediment. *Syst. Appl. Microbiol.* **20**, 201–208.

Finster K., Liesack W., and Thamdrup B. (1998) Elemental sulfur and thiosulfate disproportionation by Desulfocapsa sulfoexigens sp. nov., a new anaerobic bacterium isolated from marine surface sediment. *Appl. Environ. Microbiol.* **64**, 119–125.

Firestone M. K. (1982) Biological denitrification. In *Nitrogen in Agriculture* (ed. F. J. Stevenson). American Society of Agronomy, Madison, Wisconsin, pp. 289–326.

Fitz R. M. and Cypionka H. (1990) Formation of thiosulfate and trithionate during sulfite reduction by washed cells of *Desulfovibrio desulfuricans*. *Arch. Microbiol.* **154**, 400–406.

Fossing H. and Jørgensen B. B. (1989) Measurement of bacterial sulfate reduction in sediments: evaluation of a single step chromium reduction. *Biogeochem.* **8**, 205–222.

Fowler V. J., Widdel F., Pfennig N., and Wese C. R. (1986) Phylogenetic relationships of sulfate- and sulfur-reducing eubacteria. *Syst. Appl. Microbiol.* **8**, 32–41.

Francis C. A. and Tebo B. M. (2002) Enzymatic manganese(II) oxidation by metabolically dormant spores of diverse *Bacillus* species. *Appl. Environ. Microbiol.* **68**, 874–880.

Franzmann P. D., Liu Y., Balkwill D. L., Aldrich H. C., Conway de Macario E., and Boone D. R. (1997) *Methanogenium frigidum* sp. nov., a psychrophilic, H_2 using methanogen from Ace Lake Antarctica. *Int. J. Systemat. Bacteriol.* **47**, 1068–1072.

Frazer A. C. (1995) O-demethylation and other transformations of aromatic compounds by acetogenic bacteria. In *Acetogenesis* (ed. H. L. Drake). Chapman and Hall, New York, pp. 445–483.

Fredrickson J. K., Zachara J. M., Kennedy D. W., Dong H., Onstott T. C., Hinman N. W., and Li S.-M. (1998) Biogenic iron mineralization accompanying the dissimilatory reduction of hydrous ferric oxide by a groundwater bacterium. *Geochim. Cosmochim. Acta* **62**, 3239–3257.

Freeman C., Hudson J., Lock M. A., Reynolds B., and Swanson C. (1994) A possible role of sulphate in the suppression of wetland methane fluxes following drought. *Soil Biol. Biochem.* **26**, 1439–1442.

Freeman C., Ostle N., and Kang H. (2001) An enzymatic 'latch' on a global carbon store. *Nature* **409**, 149.

Freeman C., Nevison G. B., Kang H., Hughes S., Reynolds B., and Hudson J. A. (2002) Contrasted effects of simulated drought on the production and oxidation of methane in a mid-Wales wetland. *Soil Biol. Biochem.* **34**, 61–67.

Frenzel P., Thebrath B., and Conrad R. (1990) Oxidation of methane in the oxic surface layer of a deep lake sediment (Lake Constance). *FEMS Microbiol. Ecol.* **73**, 149–158.

Frenzel P., Bosse U., and Janssen P. H. (1999) Rice roots and methanogenesis in a paddy soil: ferric iron as an alternative electron acceptor in the rooted soil. *Soil Biol. Biochem.* **31**, 421–430.

Friedrich M. and Schink B. (1995) Isolation and characterization of a desulforubidin-containing sulfate-reducing bacterium growing with glycolate. *Arch. Microbiol.* **164**, 271–279.

Friedrich M., Laderer U., and Schink B. (1991) Fermentative degradation of glycolic acid by defined syntrophic cocultures. *Arch. Microbiol.* **156**, 398–404.

Frund C. and Cohen Y. (1992) Diurnal cycles of sulfate reduction under oxic conditions in cyanobacterial mats. *Appl. Environ. Microbiol.* **58**, 70–77.

Fuchs G. (1994) Variations of the acetyl-CoA pathway in diversely realted microorganisms that are not acetogens. In *Acetogenesis* (ed. H. L. Drake). Chapman and Hall, New York, pp. 507–520.

Fuglestvedt J. S., Berntsen T. K., Godal O., and Skodvin T. (2000) Climate implications of GWP-based reductions in greenhouse gas emissions. *Geophys. Res. Lett.* **27**, 409–412.

Fukui M., Teske A., Aßmus B., Muyzer G., and Widdel F. (1999) Physiology, phylogenetic relationships, and ecology of filamentous sulfate-reducing bacteria (genus *Desulfonema*). *Arch. Microbiol.* **172**, 193–203.

Fukuzaki S., Nishio N., and Nagai S. (1990) Kinetics of the methanogenic fermentation of acetate. *Appl. Environ. Microbiol.* **56**, 3158–3163.

Gallardo V. A., Klingelhoeffer E., Arntz W., and Graco M. (1998) First report of the bacterium *Thioploca* in the Benguela ecosystem off Namibia. *J. Mar. Biol. Assoc. UK* **78**, 1007–1010.

Galloway J. N., Schlesinger W. H., Levy H. I., Micheals A., and Schoor J. L. (1995) Nitrogen fixation: anthropogenic enhancement-environment response. *Global Biogeochem. Cycles* **9**, 235–252.

Garcia J.-L., Patel B. K. C., and Ollivier B. (2000) Taxonomic, phylogenetic, and ecological diversity of methanogenic *Archaea*. *Anaerobe* **6**, 205–226.

Garcia-Gil L. J. and Golterman H. L. (1993) Kinetics of FeS-mediated denitrification in sediments from the Camargue (Rhone delta, southern France). *FEMS Microbiol. Ecol.* **13**, 85–92.

Gaspard S., Vazquez F., and Holliger C. (1998) Localization and solubilization of the Iron(III) reductase of *Geobacter sulfurreducens*. *Appl. Environ. Microbiol.* **64**, 3188–3194.

Gauci V., Dise N., and Fowler D. (2002) Controls on suppression of methane flux from a peat bog subjected to simulated acid rain sulfate deposition. *Global Biogeochem. Cycles* **16**, 4-1–4-12.

Gaunt J. L., Neue H. U., Bragais J., Grant I. F., and Giller K. E. (1997) Soil characteristics that regulate soil reduction and methane production in wetland rice soils. *Soil Sci. Soc. Am. J.* **61**, 1526–1531.

Gettemy J. M., Ma B., Alic M., and Gold M. H. (1998) Reverse transcription-PCR analysis of the regulation of the manganese peroxidase gene family. *Appl. Environ. Microbiol.* **64**, 569–574.

Ghiorse W. C. (1984) Biology of iron- and manganese-depositing bacteria. *Ann. Rev. Microbiol.* **38**, 515–550.

Giani L., Dittrich K., Marstfeld-Hartmann A., and Peters G. (1996) Methanogenesis in salt marsh soils of the North Sea coast of Germany. *Euro. J. Soil Sci.* **47**, 175–182.

Gibson J. and Harwood C. S. (2002) Metabolic diversity in aromatic compound utilization by anaerobic microbes. *Ann. Rev. Microbiol.* **56**, 345–369.

Gilbert B. and Frenzel P. (1995) Methanotrophic bacteria in the rhizosphere of rice microcosms and their effect on porewater methane concentration and methane emission. *Biol. Fertility Soils* **20**, 93–100.

Gilbert B. and Frenzel P. (1998) Rice roots and CH_4 oxidation: the activity of bacteria, their distribution and the microenvironment. *Soil Biol. Biochem.* **30**, 1903–1916.

Gilbert B., Souchu P., Bianchi M., and Bonin P. (1997) Influence of shellfish farming activities on nitrification, nitrate reduction to ammonium and denitrification at the water-sediment interface of the Thau lagoon, France. *Mar. Ecol. Prog. Ser.* **151**, 143–153.

Glasauer S., Langley S., and Beveridge T. J. (2002) Intracellular iron minerals in a dissimulatory iron-reducing bacterium. *Science* **295**, 117–119.

Glissmann K. and Conrad R. (2002) Saccharolytic activity and its role as a limiting step in methane formation during the anaerobic degradation of rice straw in rice paddy soil. *Biol. Fertility Soils* **35**, 62–67.

Goeyens L., de Vries R. T. P., Bakker J. F., and Helder W. (1987) An experiment on the relative importance of denitrification, nitrate reduction and ammonification in coastal marine sediment. *Neth. J. Sea Res.* **21**, 171–175.

Goldhaber M. B. and Kaplan I. R. (1974) The sulfur cycle. In *The Sea* (ed. E. D. Goldberg). Wiley, New York, vol. 5, pp. 569–665.

Goldhaber M. B. and Kaplan I. R. (1975) Controls and consequences of sulfate reduction in recent marine sediments. *Soil Sci.* **119**, 42–55.

Goldhaber M. B., Aller R. C., Cochran J. K., Rosenfeld J. K., Martens C. S., and Berner R. A. (1977) Sulfate reduction, diffusion, and bioturbation in Long Island Sound sediments: report of the FOAM group. *Am. J. Sci.* **277**, 193–237.

Gonzalez J. M., Kiene R. P., and Moran M. A. (1999) Transformation of sulfur compounds by an abundant lineage of marine bacteria in the alpha-subclass of the class Proteobacteria. *Appl. Environ. Microbiol.* **65**, 3810–3819.

Gonzalez-Gil G., Lens P. N. L., Van Aelst A., Van As H., Versprille A. I., and Lettinga G. (2001) Cluster structure of anaerobic aggregates of an expanded granular sludge bed reactor. *Appl. Environ. Microbiol.* **67**, 3683–3692.

Gornitz V. and Fung I. (1994) Potential distribution of methane hydrates in the world's oceans. *Global Biogeochem. Cycles* **8**, 335–347.

Gottschalk J. and Szwezyk R. (1985) Growth of a facultative anaerobe under oxygen-limiting conditions in pure culture and in co-culture with a sulfate-reducing bacterium. *FEMS Microbiol. Ecol.* **31**, 159–170.

Gottschall J. C. and Kuenen J. G. (1980) Selective enrichment of facultatively chemolithotrophic thiobacilli and related organisms in continuous culture. *FEMS Microbiol. Lett.* **7**, 241–247.

Gough M. A., Fauzi R., Mantoura C., and Preston M. (1993) Terrestrial plant biopolymers in marine sediments. *Geochim. Cosmochim. Acta* **57**, 945–964.

Graedel T. (1979) Reduced sulfur emissions from the oceans. *Geophys. Res. Lett.* **6**, 329–331.

Graham D. W., Chaudhary J. A., Hanson R. S., and Arnold R. G. (1993) Factors affecting competition between type I and type II methanotrophs in two-organism, continuous-flow reactors. *Microbiol. Ecol.* **25**, 1–17.

Granberg G., Sundh I., Svensson B. H., and Nilsson M. (2001) Effects of temperature, and nitrogen and sulfur deposition, on methane emission from a boreal mire. *Ecology* **82**, 1982–1998.

Grantham M. C., Dove P. M., and DiChristina T. J. (1997) Microbially catalyzed dissolution of iron and aluminum oxyhydroxide mineral surface coatings. *Geochim. Cosmochim. Acta* **61**, 4467–4477.

Greene A. C., Patel B. K. C., and Sheehy A. J. (1997) Defferribacter thermophilus gen. nov., sp. nov., a novel thermophilic manganese- and iron-reducing bacterium isolated from a petroleum reservoir. *Int. J. Systemat. Bacteriol.* **47**, 505–509.

Griebler C. (1997) Dimethylsulfoxide (DMSO) reduction: a new approach to determine microbial activity in freshwater sediments. *J. Microbiol. Meth.* **29**, 31–40.

Gribsholt B. and Kristensen E. (2002) Effects of bioturbation and plant roots on salt marsh biogeochemistry: a mesocosm study. *Mar. Ecol. Prog. Ser.* **241**, 71–87.

Gribsholt B., Kostka J. E., and Kristensen E. (2003) Impact of fiddler crabs and plant roots on sediment biogeochemistry in a Georgia saltmarsh. *Mar. Ecol. Prog. Ser.* (in press).

Großkopf R., Stubner S., and Liesack W. (1998) Novel euryarchaeotal lineages detected on rice roots and in the anoxic bulk soil of flooded rice microcosms. *Appl. Environ. Microbiol.* **64**, 4983–4989.

Groffman P. M. (1991) Ecology of nitrification and denitrification in soil evaluated as scales relevant to atmospheric chemistry. In *Microbial Production and Consumption of Greenhouse Gases: Methane, Nitrogen Oxides, and Halomethanes* (eds. J. E. Rogers and W. B. Whitman). American Society of Microbiology, Washington, DC, pp. 201–217.

Groffman P. M. (1994) Denitrification in freshwater wetlands. *Curr. Topics Wetland Biogeochem.* **1**, 15–35.

Groffman P. M. and Tiedje J. M. (1991) Relationships between denitrification, CO_2 production and air-filled porosity in soils of different texture and drainage. *Soil Biol. Biochem.* **23**, 299–302.

Grossman E. L., Cifuentes L. A., and Cozzarelli I. M. (2002) Anaerobic methane oxidation in a landfill-leachate plume. *Environ. Sci. Technol.* **36**, 2436–2442.

Grünfeld S. and Brix H. (1999) Methanogenesis and methane emissions: effects of water table, substrate type and presence of *Phragmites australis*. *Aquat. Bot.* **64**, 63–75.

Guerrero R., Montesinos E., Pedros-Alio C., Esteve I., Mas J., Van Gemerden H., Hofman P. A. G., and Bakker J. F. (1985) Phototrophic sulfur bacteria in two Spanish Lakes: vertical distributions and limiting factors. *Limnol. Oceanogr.* **30**, 919–931.

Gulledge J., Doyle A. P., and Schimel J. P. (1997) Different NH_4^+-inhibition patterns of soil CH_4 consumption: a result of distinct CH_4-oxidizer populations across sites? *Soil Biol. Biochem.* **29**, 13–21.

Gulledge J., Ahmad A., Steudler P. A., Pomerantz W. J., and Cavanaugh C. M. (2001) Family- and genus-level 16S rRNA-targeted oligonucleotide probes for ecological studies of methanotrophic bacteria. *Appl. Environ. Microbiol.* **67**, 4726–4733.

Gunnars A., Blomqvist S., Johansson B., and Andersson C. (2002) Formation of Fe(III) oxyhydroxide colloids in freshwater and brackish seawater, with incorporation of phosphate and calcium. *Geochim. Cosmochim. Acta* **66**, 445–458.

Haas J. R. and DiChristina T. J. (2002) Effects of Fe(III) chemical speciation on dissimulatory Fe(III) reduction by *Shewanella putrefaciens*. *Environ. Sci. Technol.* **36**, 373–380.

Haas P. (1935) CLVII. The liberation of methyl sulphide by seaweed. *Biochem. J.* **29**, 1297–1298.

Hackett W. F., Connors W. J., Kirk T. K., and Zeikus J. G. (1977) Microbial decomposition of synthetic ^{14}C-labled lignins in nature: lignin biodegration in a variety of natural materials. *Appl. Environ. Microbiol.* **33**, 43–51.

Hadas O. and Pinkas R. (1995) Sulfate reduction processes in sediments at different sites in Lake Kinneret, Israel. *Microbial Ecol.* **30**, 55–66.

Haese R. R., Wallmann K., Dahmke A., Kretamann U., Müller P. J., and Schulz H. D. (1997) Iron species determination to investigate early diagenetic reactivity in marine sediments. *Geochim. Cosmochim. Acta* **61**, 63–72.

Hafenbradl D., Keller M., Dirmeier R., Rachel R., Roßnagel P., Burggraf S., Huber H., and Stetter K. O. (1996) *Ferroglobus placidus* gen. nov., sp. nov. a novel hyperthermophilic archaeum that oxidizes Fe^{2+} at neutral pH under anoxic conditions. *Arch. Microbiol.* **166**, 308–314.

Hallbeck L. and Pederson K. (1991) Autotrophic and mixotrophic growth of *Gallionella ferruginea*. *J. Gen. Microbiol.* **137**, 2657–2661.

Hallbeck L., Ståhl F., and Pederson K. (1993) Phylogeny and phenotypic characterization of the stalk-forming and iron-oxidizing bacterium *Gallionella ferruginea*. *J. Gen. Microbiol.* **139**, 1531–1535.

Hansen H. C. B., Børggaard O. K., and Sørensen J. (1994) Evaluation of the free energy of formation of Fe(II)–Fe(III) hydroxide-sulphate (green rust) and its reduction of nitrite. *Geochim. Cosmochim. Acta* **12**, 2599–2608.

Hansen H. C. B., Koch C. B., Nancke-Krogh H. B. O. K., and Sørensen J. (1996) Abiotic nitrate reduction to ammonium: key role of green rust. *Environ. Sci. Technol.* **30**, 2053–2056.

Hansen J. E., Sato M., Lacis A. R. R., Tegen I., and Matthews E. (1998a) Climate forcings in the industrial era. *Proc. Natl. Acad. Sci. USA* **95**, 12753–12758.

Hansen J., Sato M., Ruedy R., Lacis A., and Oinas V. (2000) Global warming in the twenty-first century: an alternative scenario. *Proc. Natl. Acad. Sci. USA* **97**, 9875–9880.

Hansen L. B., Finster K., Fossing H., and Iversen N. (1998b) Anaerobic methane oxidation in sulfate depleted sediments: effects of sulfate and molybdate additions. *Aquat. Microbiol. Ecol.* **14**, 195–204.

Hansen M. H., Ingvorsen K., and Jørgensen B. B. (1978) Mechanisms of hydrogen sulfide release from coastal marine sediments to the atmosphere. *Limnol. Oceanogr.* **23**, 68–76.

Hanson R. S. and Hanson T. E. (1996) Methanotrophic bacteria. *Microbiol. Rev.* **60**, 439–471.

Happell J. D. and Chanton J. P. (1993) Carbon remineralization in a North Florida swamp forest: effects of water level on the pathways and rates of soil organic matter decomposition. *Global Biogeochem. Cycles* **7**, 475–490.

Happell J. D., Chanton J. P., Whiting G. J., and Showers W. J. (1993) Stable isotopes as tracers of methane dynamics in Everglades marshes with and without active populations of methane oxidizing bacteria. *J. Geophys. Res.* **98**, 14771–14782.

Harden H. S. and Chanton J. P. (1994) Locus of methane release and mass-dependent fractionation from two wetland macrophytes. *Limnol. Oceanogr.* **39**, 148–154.

Harmsen H. J. M., Akkermans A. D. L., Stams A. J. M., and De Vos W. M. (1996a) Population dynamics of propionate-oxidizing bacteria under methanogenic and sulfidogenic conditions in anaerobic granular sludge. *Appl. Environ. Microbiol.* **62**, 2163–2168.

Harmsen H. J. M., Kengen H. M. P., Akkermans A. D. L., Stams A. J. M., and De Vos W. M. (1996b) Detection and localization of syntrophic propionate-oxidizing bacteria in granular sludge by *in situ* hybridization using 16S rRNA-based oligonucleotide probes. *Appl. Environ. Microbiol.* **62**, 1656–1663.

Harriss R. C., Sebacher D. I., and Day F. P., Jr. (1982) Methane flux in the Great Dismal Swamp. *Nature* **297**, 673–674.

Hartman R., Stickinger H. D., and Oesterhelt D. (1980) Anaerobic growth of halobacteria. *Proc. Natl. Acad. Sci. USA* **77**, 3812–3825.

Harvey C. F., Swartz C. H., Badruzzaman A. B. M., Keon-Blute N., Yu W., Ali M. A., Jay J., Beckie R., Niedan V., Brabander D., Oates P. M., Ashfaque K. N., Islam S., Hemond H. F., and Ahmed M. F. (2002) Arsenic mobility and groundwater extraction in Bangladesh. *Science* **298**, 1602–1606.

Hauck S., Benz M., Brune A., and Schink B. (2001) Ferrous iron oxidation by denitrifying bacteria in profundal sediments of a deep lake (Lake Constance). *FEMS Microbiol. Ecol.* **37**, 127–134.

He S. H., DerVartanian D. V., and LeGall J. (1986) Isolation of fumarate reductase from *Desulfovibrio multispirans*, a sulfate-reducing bacterium. *Biochem. Biophys. Res. Commun.* **135**, 1000–1007.

Head I. M., Gray N. D., Babenzien H. D., and Glockner F. O. (2000) Uncultured giant sulfur bacteria of the genus Achromatium. *FEMS Microbiol. Ecol.* **33**, 171–180.

Hedderich R., Klimmek O., Kroger A., Dirmeier R., Keller M., and Stetter K. O. (1999) Anaerobic respiration with elemental sulfur and with disulfides. *FEMS Microbiol. Rev.* **22**, 353–381.

Hedges J. I. and Keil R. G. (1995) Sedimentary organic matter preservation: an assessment and speculative synthesis. *Mar. Chem.* **49**, 81–115.

Hedges J. I. and Oades J. M. (1997) Comparative organic geochemistries of soils and marine sediments. *Org. Geochem.* **27**, 319–361.

Hedges J. I., Clark W. A., Quay P. D., Richey J. E., Devol A. L., and Santos U. M. (1986) Compositions and fluxes of particulate organic material in the Amazon River. *Limnol. Oceanogr.* **31**, 717–738.

Hedges J. I., Cowie G. L., Richey J. E., Quay P. D., Benner R., Strom M., and Forsberg B. R. (1994) Origins and processing of organic matter in the Amazon River as indicated by carbohydrates and amino acids. *Limnol. Oceanogr.* **39**, 743–761.

Hedges J. I., Keil R. G., and Benner R. (1997) What happens to terrestrial organic matter in the ocean? *Org. Geochem.* **27**, 195–212.

Hedges J. I., Mayorga E., Tsamakis E., McClain M. E., Aufdenkampe A., Quay P., Richey J. E., Benner R., Opsahl S., Black B., Pimentel T., Quintanilla J., and Maurice L. (2000) Organic matter in Bolivian tributaries of the Amazon River: a comparison to the lower mainstream. *Limnol. Oceanogr.* **45**, 1449–1466.

Heider J., Spormann A. M., Beller H. B., and Widdel F. (1999) Anaerobic bacterial metabolism of hydrocarbons. *FEMS Microbiol. Rev.* **22**, 459–473.

Heijthuijsen J. H. F. G. and Hansen T. A. (1989) Betaine Fermentation and Oxidation by Marine *Desulfuromonas* Strains. *Appl. Environ. Microbiol.* **55**, 965–969.

Heilman M. A. and Carlton R. G. (2001) Methane oxidation associated with submersed vascular macrophytes and its impact on plant diffusive mathane flux. *Biogeochemistry* **52**, 207–224.

Heising S. and Schink B. (1998) Phototrophic oxidation of ferrous iron by a *Rhodomicrobium vannielii* strain. *Microbiology* **144**, 2263–2269.

Heising S., Richter L., Ludwig W., and Schink B. (1999) *Chlorobium ferrooxidans* sp. nov., a phototrophic green sulfur bacterium that oxides ferrous iron in coculture with a *Geospirillum* sp. strain. *Arch. Microbiol.* **172**, 116–124.

Henriksen K. and Kemp W. M. (1988) Nitrification in estuarine and coastal marine sediments. In *Nitrogen Cycling in Coastal Marine Environments* (eds. T. H. Blackburn and J. Sorensen). Wiley, New York, pp. 207–249.

Herbert R. A. (1999) Nitrogen cycling in coastal marine ecosystems. *FEMS Microbiol. Rev.* **23**, 563–590.

Heron G. and Christensen T. H. (1995) Impact of sediment-bound iron on redox buffering in a landfill leachate polluted aquifer (Vejen, Denmark). *Environ. Sci. Technol.* **29**, 187–192.

Hill A. R. (1996) Nitrate removal in stream riparian zones. *J. Environ. Qual.* **25**, 743–755.

Hines M. E. (1991) The role of certain infauna and vascular plants in the mediation of redox reactions in marine sediments. In *Diversity of Environmental Geochemistry* (ed. J. Berthelin). Elsevier, Amsterdam, The Netherlands, pp. 275–286.

Hines M. E. (1996) Emissions of sulfur gases from wetlands. In *Cycling of Reduced Gases in the Hydrosphere*. (eds. D. D. Adams, P. M. Crill, and S. P. Seitzinger). Mitt. Int. Verein. Limnol. vol. 25, pp. 153–161.

Hines M. E. and Lyons W. B. (1982) Biogeochemistry of nearshore Bermuda sediments: I. Sulfate reduction rates and nutrient generation. *Mar. Ecol. Prog. Ser.* **8**, 87–94.

Hines M. E., Orem W. H., Lyons W. B., and Jones G. E. (1982) Microbial activity and bioturbation-induced oscillations in pore water chemistry of estuarine sediments in spring. *Nature* **299**, 433–435.

Hines M. E., Knollmeyer S. L., and Tugel J. B. (1989) Sulfate reduction and other sedimentary biogeochemistry in a northern New England salt marsh. *Limnol. Oceanogr.* **34**, 578–590.

Hines M. E., Faganeli J., and Planinc R. (1997) Sedimentary anaerobic microbial biogeochemistry in the Gulf of Trieste, northern Adriatic Sea: influences of bottom water oxygen depletion. *Biogeochemistry* **39**, 65–86.

Hines M. E., Evans R. S., Genthner B. R. S., Willis S. G., Friedman S., Rooney-Varga J. N., and Devereux R. (1999) Molecular phylogenetic and biogeochemical studies of sulfate-reducing bacteria in the rhizosphere of *Spartina alterniflora*. *Appl. Environ. Microbiol.* **65**, 2209–2216.

Hines M. E., Duddleston K. H., and Kiene R. P. (2001) Carbon flow to acetate and C_1 compounds in northern wetlands. *Geophys. Res. Lett.* **28**, 4251–4254.

Hinrichs K.-U. and Boetius A. (2002) The anaerobic oxidation of methane: new insights in microbial ecology and biogeochemistry. In *Ocean Margin Systems* (eds. G. Wefer, D. Billet, D. Hebbeln, B. B. Jorgensen, M. Schlueter, and T. V. Weering). Springer, Heidelberg, Germany, pp. 457–477.

Hinrichs K.-U., DeLong E. F., Hayes J. M., Sylva S. P., and Brewer P. G. (1999) Methane-consuming archaebacteria in marine sediments. *Nature* **398**, 802–805.

Hinrichs K.-U., Summons R. E., Orphan V., Sylva S. P., and Hayes J. M. (2000) Molecular and isotopic analysis of anaerobic methane-oxidizing communities in marine sediments. *Org. Geochem.* **31**, 1685–1701.

Hippe H. D., Caspari D., Feibig K., and Gottschalk G. (1979) Utilization of trimethylamine and other N-methyl compounds for growth and methane formation by *Methanosarcina barkeri*. *Proc. Natl. Acad. Sci. USA* **76**, 494–498.

Hobbie S. E. (1996) Temperature and plant species control over litter decomposition in Alaskan tundra. *Ecol. Monogr.* **66**, 503–522.

Hoehler T. M., Alperin M. J., Albert D. B., and Martens C. S. (1994) Field and laboratory studies of methane oxidation in an anoxic marine sediment: evidence for a methanogen-sulfate reducer consortium. *Global Biogeochem. Cycles* **8**, 451–463.

Hoehler T. M., Alperin M. J., Albert D. B., and Martens C. S. (1998) Thermodynamic control on hydrogen concentrations in anoxic sediments. *Geochim. Cosmochim. Acta* **62**, 1745–1756.

Hoehler T. M., Bebout B. M., and Des Marais D. J. (2001) The role of microbial mats in the production of reduced gases on the early Earth. *Nature* **412**, 324–328.

Højberg O., Revsbech N. P., and Tiedje J. M. (1994) Denitrification in soil aggregates analyzed with microsensors for nitrous oxide and oxygen. *Soil Sci. Soc. Am. J.* **58**, 1691–1698.

Holmén B. A., Sison J. D., Nelson D. C., and Casey W. H. (1999) Hydroxamate siderophores, cell growth and Fe(III) cycling in two anaerobic iron oxide media containing *Geobacter metallireducens*. *Geochim. Cosmochim. Acta* **63**, 227–239.

Holmer M. and Kristensen E. (1994) Coexistence of sulfate reduction and methane production in an organic-rich sediment. *Mar. Ecol. Prog. Ser.* **107**, 177–184.

Holmer M. and Storkholm P. (2001) Sulphate reduction and sulphur cycling in lake sediments: a review. *Freshwater Biol.* **46**, 431–451.

Holmer M., Gribsholt B., and Kristensen E. (2002) Effects of sea level rise on growth of *Spartina anglica* and oxygen dynamics in rhizosphere and salt marsh sediments. *Mar. Ecol. Prog. Ser.* **225**, 197–204.

Holmes A. J., Costello A., Lidstrom M. E., and Murrell J. C. (1995) Evidence that particulate methane monooygenase and ammonia monooxygenase may be evolutionarily related. *FEMS Microbiol. Lett.* **132**, 203–208.

Holmes D. E., Finneran K. T., O'Neil R. A., and Lovley D. R. (2002) Enrichment of members of the family *Geobacteraceae* associated with stimulation of dissimilatory metal reduction in uranium-contaminated aquifer sediments. *Appl. Environ. Microbiol.* **68**, 2300–2306.

Hooper A. B. and Terry K. R. (1979) Hydroxylamine oxidoreductase of *Nitrosomonas*: production of nitric oxide from hydroxylamine. *Biochim. Biophys. Acta* **571**, 12–20.

Hoor A. T. (1981) Cell yield and bioenergetics of *Thiomicrospira denitrificans* compared with *Thiobacilllus denitrificans*. *Antonie Van Leeuwenhoek* **47**, 231–243.

Hoosbeek M. R., van Breeman N., Berendse F., Brosvernier P., Vasander H., and Wallén B. (2001) Limited effect of increased atmospheric CO_2 concentration on ombrotrophic bog vegetation. *New Phytologist* **150**, 459–463.

Hoppe H. G. (1983) Significance of exoenzyme activities in the ecology of brackish water: measurments by means of methylumbelliferyl-substrates. *Mar. Ecol. Prog. Ser.* **11**, 299–308.

Hornibrook E. R. C., Longstaffe F. J., and Fyfe W. S. (1997) Spatial distribution of microbial methane production pathways in temperate zone wetland soils: stable carbon and hydrogen isotope evidence. *Geochim. Cosmochim. Acta* **61**, 745–753.

Hornibrook E. R. C., Longstaffe F. J., and Fyfe W. S. (2000) Evolution of stable carbon isotope compositions for methane and carbon dioxide in freshwater wetlands and other anaerobic environments. *Geochim. Cosmochim. Acta* **64**, 1013–1027.

Howarth R. W. (1979) Pyrite: its rapid formation in a salt marsh and its importance in ecosystem metabolism. *Science* **203**, 49–51.

Howarth R. W. and Merkel S. (1984) Pyrite formation and the measurement of sulfate reduction in salt marsh sediments. *Limnol. Oceanogr.* **29**, 598–608.

Huber R., Wilharm T., Huber D., Trincone A., Burggraf S., König H., Rachel R., Rockinger I., Fricke H., and Stetter K. O. (1992) *Aquifex pyrophilus* gen. nov., sp. nov., represents a novel group of marine hyperthermophilic hydrogen-oxidizing bacteria. *Syst. Appl. Microbiol.* **15**, 340–351.

Hulth S., Aller R. C., and Gilbert F. (1999) Coupled anoxic nitrification/manganese reduction in marine sediments. *Geochim. Cosmochim. Acta* **63**, 49–66.

Huston A. L., Krieger-Brockett B. B., and Deming J. W. (2000) Remarkably low temperature optima for extracellular enzyme activity from Arctic bacteria and sea ice. *Environ. Microbiol.* **2**, 383–388.

Hutchin P. R., Press M. C., Lee J. A., and Ashenden T. W. (1995) Elevated concentrations of CO_2 may double methane emissions from mires. *Global Change Biol.* **1**, 125–128.

Ingersoll T. L. and Baker L. A. (1998) Nitrate removal in wetland microcosms. *Water Res.* **32**, 677–684.

Inubushi K., Hori K., Matsumoto S., and Wada H. (1997) Anaerobic decomposition of organic carbon in paddy soil in relation to methane emission to the atmosphere. *Water Sci. Technol.* **36**, 523–530.

Ishimoto M., Koyama J., and Nagai Y. (1954) Biochemical studies on sulfate-reducing bacteria: IV. The cyctochrome system of sulfate-reducing bacteria. *J. Biochem.* **41**, 763–770.

Iversen N. and Blackburn T. H. (1981) Seasonal rates of methane oxidation in anoxic marine sediments. *Appl. Environ. Microbiol.* **41**, 1295–1300.

Iverson N. and Jørgensen B. B. (1985) Anaerobic methane oxidation at the sulfate-methane transition in marine sediments from the Kattegat and Skagerrak (Denmark). *Limnol. Oceanogr.* **30**, 944–955.

Iverson N., Oremland R. S., and Klug M. J. (1987) Big Soda Lake (Nevada): 3. Pelagic methanogenesis and anaerobic methane oxidation. *Limnol. Oceanogr.* **32**, 804–814.

Jäckel U. and Schnell S. (2000) Suppression of methane emission from rice paddies by ferric iron fertilization. *Soil Biol. Biochem.* **32**, 1811–1814.

Jacobson M. E. (1994) Chemical and biological mobilization of Fe(III) in marsh sediments. *Biogeochemistry* **25**, 41–60.

Jakobsen R. and Postma D. (1994) *In situ* rates of sulfate reduction in an aquifer (Rømø, Denmark) and implications for the reactivity of organic matter. *Geology* **23**, 1103–1106.

Jannasch H. W., Nelson D. C., and Wirsen C. O. (1989) Massive natural occurrence of unusually large bacteria (*Beggiatoa* sp.) at a hydrothermal deep-sea site. *Nature* **342**, 834–836.

Jarrell K. F. and Kalmokoff M. L. (1988) Nutritional requirements of the methanogenic archeabacteria. *Can. J. Microbiol.* **34**, 557–576.

Jetten M. S. M., Stams A. J. M., and Zehnder A. J. B. (1992) Methanogenesis from acetate: a comparison of the acetate metabolism in *Methanothrix soehngenii* and *Methanosarcina* spp. *FEMS Microbiol. Rev.* **88**, 181–198.

Jetten M. S. M., Strous M., van de Pas-Schoonen K. T., Schalk J., van Dongen U. G. J. M., van de Graaf A. A., Logemann S., Muyzer G., van Loosdrecht M. C. M., and Kuenen J. G. (1999) The anaerobic oxidation of ammonium. *FEMS Microbiol. Rev.* **22**, 421–437.

Johnson D. B. (1998) Biodiversity and ecology of acidophilic microorganisms. *FEMS Microbiol. Ecol.* **11**, 63–70.

Johnson D. E., Hill T. M., Ward G. M., Johnson K. A., Branine M. E., Carmean B. R., and Lodman D. W. (1993) Ruminants and other animals. In *Atmospheric Methane: Sources, Sinks, and Role* (ed. M. A. K. Khalil). Springer, Berlin, pp. 199–229.

Johnson L. C. and Damman A. W. (1991) Species-controlled *Sphagnum* decay on a south Swedish raised bog. *Oikos* **61**, 234–242.

Jones C. G. and Shachak M. (1990) Fertilization of the desert soil by rock-eating snails. *Nature* **346**, 839–841.

Jonkers H. M., Van der Maarel M., Van Gemerden H., and Hansen T. A. (1996) Dimethylsulfoxide reduction by marine sulfate-reducing bacteria. *FEMS Microbiol. Lett.* **136**, 283–287.

Jonkers H. M., deBruin S., and vanGemerden H. (1998) Turnover of dimethylsulfoniopropionate (DMSP) by the purple sulfur bacterium *Thiocapsa roseopersicina* M11: ecological implications. *FEMS Microbiol. Ecol.* **27**, 281–290.

Jonkers H. M., vanBergeijk S. A., and vanGemerden H. (2000) Microbial production and consumption of dimethyl sulfide (DMS) in a sea grass (*Zostera noltii*)-dominated marine intertidal sediment ecosystem (Bassin d'Arcachon, France). *FEMS Microbiol. Ecol.* **31**, 163–172.

Jordan S. L., McDonald I. R., Kraczkiewicz-Dowjat A. J., Kelly D. P., Rainey F. A., Murrell J. C., and Wood A. P. (1997) Autotrophic growth on carbon disulfide is a property of novel strains of *Paracoccus denitrificans*. *Arch. Microbiol.* **168**, 225–236.

Jordan T. E. and Weller D. E. (1996) Human contributions to terrestrial nitrogen flux. *Bioscience* **46**, 655–664.

Jørgensen B. B. (1977) The sulfur cycle of a coastal marine sediment. *Limnol. Oceanogr.* **22**, 814–832.

Jørgensen B. B. (1978a) A comparison of methods for the quantification of bacterial sulfate reduction in coastal marine sediments: I. Measurements with radiotracer techniques. *Geomicrobiol. J.* **1**, 11–27.

Jørgensen B. B. (1978b) A comparison of methods for the quantification of bacterial sulfate reduction in coastal marine sediments: II. Calculations from mathematical models. *Geomicrobiol. J.* **1**, 29–51.

Jørgensen B. B. (1978c) A comparison of methods for the quantification of bacterial sulfate reduction in coastal marine sediments: III. Estimation from chemical and bacteriological field data. *Geomicrobiol. J.* **1**, 49–64.

Jørgensen B. B. (1982a) Ecology of the bacteria of the sulphur cycle with special reference to anoxic-oxic interface environments. *Phil. Trans. Roy. Soc. London* **298**, 543–561.

Jørgensen B. B. (1982b) Mineralization of organic matter in the sea bed: the role of sulphate reduction. *Nature* **296**, 643–645.

Jørgensen B. B. (1988) Ecology of the sulphur cycle: oxidative pathways in sediments. In *The Nitrogen and Sulphur Cycles* (eds. J. A. Cole and S. J. Ferguson). Cambridge University Press, Cambridge, pp. 31–63.

Jørgensen K. S. (1989) Annual pattern of denitrification and nitrate ammonification in estuarine sediment. *Appl. Environ. Microbiol.* **55**, 1841–1847.

Jørgensen B. B. (1990a) The sulfur cycle of freshwater sediments: role of thiosulfate. *Limnol. Oceanogr.* **35**, 1329–1342.

Jørgensen B. B. (1990b) A thiosulfate shunt in the sulfur cycle of marine sediments. *Science* **249**, 152–154.

Jørgensen B. B. and Bak F. (1991) Pathways and microbiology of thiosulfate transformations and sulfate reduction in a marine sediment (Kattegat, Denmark). *Appl. Environ. Microbiol.* **57**, 847–856.

Jørgensen B. B. and Des Marais D. J. (1986) Competition for sulfide among colorless and purple sulfur bacteria. *FEMS Microbiol. Ecol.* **38**, 179–186.

Jørgensen B. B. and Fenchel T. (1974) The sulfur cycle of a marine sediment model system. *Mar. Biol.* **24**, 189–201.

Jørgensen B. B. and Gallardo V. A. (1999) Thioploca spp: filamentous sulfur bacteria with nitrate vacuoles. *FEMS Microbiol. Ecol.* **28**, 301–313.

Jørgensen B. B. and Okholm-Hansen B. (1985) Emissions of biogenic sulfur gases from a Danish estuary. *Atmos. Environ.* **19**, 1737–1749.

Jørgensen B. B. and Sørensen J. (1985) Seasonal cycles of O_2, NO_3^-, and SO_4^{2-} reduction in estuarine sediments: the significance of an NO_3^- reduction maximum in spring. *Mar. Ecol. Prog. Ser.* **24**, 65–74.

Jørgensen B. B. and Sørensen J. (1988) Two annual maxima of nitrate reduction and denitrification in estuarine sediment (Norsminde Fjord, Denmark). *Mar. Ecol. Prog. Ser.* **48**, 147–154.

Jørgensen B. B., Revsbech N. P., Blackburn T. H., and Cohen Y. (1979) Diurnal cycle of oxygen and sulfide microgradients and microbial photosynthesis in a cyanobacterial microbial mat sediment. *Appl. Environ. Microbiol.* **38**, 46–58.

Jørgensen B. B., Isaksen M. F., and Jannasch H. W. (1992) Bacterial sulfate reduction above 100-degrees-C in deep-sea hydrothermal vent sediments. *Science* **258**, 1756–1757.

Jørgensen K. S. and Tiedje J. M. (1993) Survival of denitrifiers in nitrate-free, anaerobic environments. *Appl. Environ. Microbiol.* **59**, 3297–3305.

Joye S. B. and Paerl H. W. (1993) Contemporaneous nitrogen fixation and denitrification in intertidal microbial mats: rapid response to runoff events. *Mar. Ecol. Prog. Ser.* **94**, 267–274.

Joye S. B. and Hollibaugh J. T. (1995) Influence of sulfide inhibition of nitrification on nitrogen regeneration in sediments. *Science* **270**, 623–625.

Joye S. B., Mazzotta M. L., and Hollibaugh J. T. (1996) Community metabolism in microbial mats: the occurrence of biologically-mediated iron and manganese reduction. *Estuar. Coast. Shelf Sci.* **43**, 747–766.

Kaden J., Galushko A. S., and Schink B. (2002) Cysteine-mediated electron transfer in syntrophic acetate oxidation by co-cultures of *Geobacter sulfurreducens* and *Wolinella succinogenes*. *Arch. Microbiol.* **178**, 53–58.

Kadota H. and Ishida Y. (1972) Production of volatile sulfur compounds by microorganisms. *Rev. Microbiol.* **26**, 127–138.

Kaesler B. and Schönheit P. (1989) The sodium cycle in methanogenesis. *Euro. J. Biochem.* **186**, 309–316.

Kamura T., Takai Y., and Ishikawa K. (1963) Microbial reduction mechanism of ferric iron in paddy soils (Part 1). *Soil Sci. Plant Nutr.* **9**, 171–175.

Kana T. M., Sullivan M. B., Cornwell J. C., and Groszkowski K. M. (1998) Denitrification in estuarine sediments determined by membrane inlet mass spectrometry. *Limnol. Oceanogr.* **43**, 334–339.

Kanagawa T. and Kelly D. P. (1986) Breakdown of dimethyl sulphide by mixed cultures and by *Thiobacillus thioparus*. *FEMS Microbiol. Lett.* **34**, 13–19.

Kanda K.-i. and Tsuruta H. (1995) Emissions of sulfur gases from various types of terrestrial higher plants. *Soil Sci. Plant Nutr. (Tokyo, Japan)* **41**, 321–328.

Kang H. and Freeman C. (1999) Phosphatase and arylsulphatase activities in wetlands soils: annual variation and controlling factors. *Soil Biol. Biochem.* **31**, 449–454.

Kang H., Freeman C., and Ashendon T. W. (2001) Effects of elevated CO_2 on fen peat biogeochemistry. *Sci. Tot. Environ.* **279**, 45–50.

Kappler U., Friedrich C. G., Truper H. G., and Dahl C. (2001) Evidence for two pathways of thiosulfate oxidation in *Starkeya novella* (formerly *Thiobacillus novellus*). *Arch. Microbiol.* **175**, 102–111.

Kaspar H. F. (1983) Denitrification, nitrate reduction to ammonium, and inorganic nitrogen pools in intertidal sediments. *Mar. Biol.* **74**, 133–139.

Kaspar H. F., Asher R. A., and Boyer I. C. (1985) Microbial nitrogen transformations in sediments and inorganic nitrogen fluxes across the sediment/water interface on the South Island West Coast, New Zealand. *Estuar. Coast. Shelf Sci.* **21**, 245–255.

Kattenberg A., Giorgi F., Grassl H., Meehl G. A., Mitchell J. F. B., Stouffer R. J., Tokioka T., Weaver A. J., and Wigley T. M. L. (1996) Climate models-projections of future climate. In *Climate Change 1995* (eds. J. T. Houghton, L. G. Meira, B. A. Callander, N. Harris, A. Kattenberg, and K. Maskell). Cambridge University Press, Cambridge, pp. 285–357.

Kawakami H. (1989) Degradation of lignin-related aromatics and lignins by several pseudomonas. In *Lignin Biogedradation: Microbiology, Chemistry, and Potential Applications* (eds. K. T. Kirk, H. Takayoshi, and C. Hou-moin). CRC Press, Boca Raton, FL, vol. 2.

Keith S. M. and Herbert R. A. (1983) Dissimilatory nitrate reduction by a stain of *Desulfovibrio desulfuricans*. *FEMS Microbiol. Lett.* **18**, 55–59.

Kelley C. A., Dise N. B., and Martens C. S. (1992) Temporal variations in the stable carbon isotopic composition of methane emitted from Minnesota peatlands. *Global Biogeochem. Cycles* **6**, 263–269.

Kelly D. P. (1982) Biochemistry of chemolithotrophic oxidation of inorganic sulphur. *Phil. Trans. Roy. Soc. London B* **298**, 499–528.

Kelly D. P. (1988) Oxidation of sulphur compounds. In *The Nitrogen and Sulphur Cycles*, Soc. Gen. Microbiol. Symp. (eds. J. A. Cole and S. J. Ferguson). Cambridge University Press, Cambridge, vol. 42, pp. 65–98.

Kelly D. P. and Baker S. C. (1990) The organosulphur cycle: aerobic and anaerobic processes leading to turnover of C_1-sulphur compounds. *FEMS Microbiol. Rev.* **87**, 241–246.

Kelly D. P. and Smith N. A. (1990) Organic sulfur compounds in the environment: biogeochemistry, microbiology, and ecological aspects. *Adv. Microb. Ecol.* **11**, 345–385.

Kelly D. P., Wood A. P., Jordan S. L., Padden A. N., Gorlenko V. M., and Dubinina G. A. (1994) Biological production and consumption of gaseous organic sulfur compounds. *Biochem. Soc. Trans.* **22**, 1011–1015.

Kelly D. P., Shergill J. K., Lu W.-P., and Wood A. P. (1997) Oxidative metabolism of inorganic sulfur compounds by bacteria. *Antonie Van Leeuwenhoek* **71**, 95–107.

Kelly-Gerreyn B. A., Trimmer M., and Hydes D. J. (2001) A diagenetic model discriminating denitrification and dissimilatory nitrate reduction to ammonium in a temperate estuarine sediment. *Mar. Ecol. Prog. Ser.* **220**, 33–46.

Kelso B. H. L., Smith R. V., Laughlin R. J., and Lennox S. D. (1997) Dissimilatory nitrate reduction in anaerobic sediments leading to river nitrite accumulation. *Appl. Environ. Microbiol.* **63**, 4679–4685.

Kemp W. M., Sampou P., Caffrey J., Mayer M., Henriksen K., and Boynton W. R. (1990) Ammonium recycling versus denitrification in Chesapeake Bay sediments. *Limnol. Oceanogr.* **35**, 1545–1563.

Kettunen A., Kaitala V., Lehtinen A., Lohila A., Alm J., Silvola J., and Martikainen P. J. (1999) Methane production and oxidation potentials in relation to water table fluctuations in two boreal mires. *Soil Biol. Biochem.* **31**, 1741–1749.

Kieft T. L., Fredrickson J. K., Onstott T. C., Gorby Y. A., Kostandarithes H. M., Bailey T. J., Kennedy D. W., Li S. W., Plymale A. E., Spadoni C. M., and Gray M. S. (1999) Dissimilatory reduction of Fe(III) and other electron acceptors by a *Thermus* isolate. *Appl. Environ. Microbiol.* **65**, 1214–1221.

Kiene R. P. (1988) Dimethyl sulfide metabolism in salt marsh sediments. *FEMS Microbiol. Ecol.* **53**, 71–78.

Kiene R. P. (1991a) Evidence for the biological turnover of thiols in anoxic marine sediments. *Biogeochemistry* **13**, 117–135.

Kiene R. P. (1991b) Production and consumption of methane in aquatic systems. In *Microbial Production and Consumption of Greenhouse Gases: Methane, Nitrogen Oxides, and Halomethanes* (eds. J. E. Rogers and W. B. Whitman). American Society for Microbiology, Washington, DC, pp. 111–146.

Kiene R. P. (1992) Dynamics of dimethyl sulfide and dimethylsulfoniopropionate in oceanic water samples. *Mar. Chem.* **37**, 29–52.

Kiene R. P. (1996a) Microbial cycling of organosulfur gases in marine and freshwater environments. In *Cycling of Reduced Gases in the Hydrosphere.* (eds. D. D. Adams, P. M. Crill, and S. P. Seitzinger) Mitt. Int. Verein. Limnol. 25, pp. 137–151.

Kiene R. P. (1996b) Microbiological controls on the emissions of dimethylsulfide from wetlands and the ocean. In *Microbiology of Atmospheric Trace Gases.* NATO ASI Series 1 (eds. J. C. Murrell and D. P. Kelly). Springer, Berlin, vol. 39, pp. 205–225.

Kiene R. P. and Capone D. G. (1988) Microbial transformations of methylated sulfur compounds in anoxic salt marsh sediments. *Microbiol. Ecol.* **15**, 275–291.

Kiene R. P. and Hines M. E. (1995) Microbial formation of dimethyl sulfide in anoxic *Sphagnum* peat. *Appl. Environ. Microbiol.* **61**, 2720–2726.

Kiene R. P. and Taylor B. F. (1988a) Biotransformations of organosulphur compounds in sediments via 3-mercaptopropionate. *Nature* **332**, 148–150.

Kiene R. P. and Taylor B. F. (1988b) Demethylation of dimethylsulfonioproprionate and production of thiols in anoxic marine sediments. *Appl. Environ. Microbiol.* **54**, 2208–2212.

Kiene R. P. and Visscher P. T. (1987a) Production and fate of methylated sulfur compounds from methionine and dimethyl-sulfoniopropionate in anoxic salt marsh sediments. *Appl. Environ. Microbiol.* **53**, 2426–2434.

Kiene R. P. and Visscher P. T. (1987b) Production and fate of methylated sulfur compounds from methionine and dimethylsulfoniopropionate in anoxic salt marsh sediments. *Appl. Environ. Microbiol.* **53**, 2426–2434.

Kiene R. P., Oremland R. S., Catena A., Miller L. G., and Capone D. G. (1986) Metabolism of reduced methylated sulfur compounds in anaerobic sediments and by a pure

culture of an estuarine methanogen. *Appl. Environ. Microbiol.* **52**, 1037–1045.
Kiene R. P., Linn L. J., and Bruton J. A. (2000) New and important roles for DMSP in marine microbial communities. *J. Sea Res.* **43**, 209–224.
Kiener A. and Leisinger T. (1983) Oxygen sensitivity of methanogenic bacteria. *Syst. Appl. Microbiol.* **4**, 305–312.
King D. and Nedwell D. B. (1984) Changes in nitrate-reducing community of an anaerobic salt marsh sediment in response to seasonal selection by temperature. *J. Gen. Microbiol.* **130**, 2935–2941.
King D. and Nedwell D. B. (1985) The influence of nitrate concentration upon the end-products of nitrate dissimilation by bacteria in anaerobic salt marsh sediments. *FEMS Microbiol. Ecol.* **31**, 23–28.
King D. W. (1998) Role of carbonate speciation on the oxidation rate of Fe(II) in aquatic systems. *Environ. Sci. Technol.* **32**, 2997–3003.
King G. M. (1984) Utilization of hydrogen, acetate, and "noncompetitive" substrates by methanogenic bacteria in marine sediments. *Geomicrobiol. J.* **3**, 275–306.
King G. M. (1986) Characterization of B-glucosidase activity in intertidal marine sediments. *Appl. Environ. Microbiol.* **51**, 373–380.
King G. M. (1990a) Dynamics and controls of methane oxidation in a Danish wetland sediment. *FEMS Microbiol. Ecol.* **74**, 309–324.
King G. M. (1990b) Regulation by light of methane emissions from a wetland. *Nature* **345**, 513–515.
King G. M. (1992) Ecological aspects of methane oxidation, a key determinant of global methane dynamics. *Adv. Microb. Ecol.* **12**, 431–468.
King G. M. (1996a) *In situ* analyses of methane oxidation associated with the roots and rhizomes of a bur reed, *Sparganium eurycarpum*, in a Maine wetland. *Appl. Environ. Microbiol.* **62**, 4548–4555.
King G. M. (1996b) Physiological limitations of methanotrophic activity *in situ*. In *Microbiology of Atmospheric Trace Gases* (eds. J. C. Murrell and D. P. Kelly). Springer, Berlin, pp. 17–32.
King G. M. and Adamsen A. P. S. (1992) Effects of temperature on methane consumption in a forest soil and in pure cultures of the methanotroph *Methylomonas rubra*. *Appl. Environ. Microbiol.* **58**, 2758–2763.
King G. M. and Schnell S. (1994a) Ammonium and nitrite inhibition of methane oxidation by *Methylobacter albus* BG8 and *Methylosinus trichosporium* OB3b at low methane concentrations. *Appl. Environ. Microbiol.* **60**, 3508–3513.
King G. M. and Schnell S. (1994b) Effect of increasing atmospheric methane concentration on ammonium inhibition of soil methane consumption. *Nature* **370**, 282–284.
King G. M., Klug M. J., Wiegert R. G., and Chalmers A. G. (1982) Relation of soil water movement and sulfide concentration to *Spartina alterniflora* production in a Georgia salt marsh. *Science* **218**, 61–63.
King G. M., Roslev P., and Skovgaard H. (1990) Distribution and rate of methane oxidation in sediments of the Florida Everglades. *Appl. Environ. Microbiol.* **56**, 2902–2911.
King J. Y. and Reeburgh W. S. (2002) A pulse-labeling experiment to determine the contribution of recent plant photosynthates to net methane emission in arctic wet sedge tundra. *Soil Biol. Biochem.* **34**, 173–180.
King J. Y., Reeburgh W. S., Thielder K. K., Kling G. W., Loya W. M., Johnson L. C., and Nadelhoffer K. J. (2002) Pulse-labeling studies of carbon cycling in Arctic tundra ecosystems: the contribution of photosynthates to methane emission. *Global Biogeochem. Cycles* **1062**, 2002 doi: 10.1029/2001GB001456.
Kirby T. W., Lancaster J. R., Jr., and Fridovich I. (1981) Isolation and characterization of the iron-containing superoxide dismutase of *Methanobacterium bryantii*. *Arch. Biochem. Biophys.* **210**, 140–148.
Klemps R., Cypionka H., Widdel F., and Pfennig N. (1985) Growth with hydrogen, and further physiological characteristics of *Desulfotomaculum* species. *Arch. Microbiol.* **143**, 203–208.
Klüber H. D. and Conrad R. (1998) Effects of nitrate, nitrite, NO and N_2O on methanogenesis and other redox processes in anoxic rice field soil. *FEMS Microbiol. Ecol.* **25**, 301–318.
Kludze H. K. and DeLaune R. D. (1995) Straw application effects on methane and oxygen exchange and growth in rice. *Soil Sci. Soc. Am. J.* **59**, 824–830.
Kniemeyer O., Fischer T., Wilkes H., Glockner F. O., and Widdel F. (2003) Anaerobic degradation of ethylbenzene by a new type of marine sulfate-reducing bacterium. *Appl. Environ. Microbiol.* **69**, 760–768.
Knoblauch C. and Jorgensen B. B. (1999) Effect of temperature on sulphate reduction, growth rate, and growth yield in five psychrophilic sulphate-reducing bacteria from Arctic sediments. *Environ Microbiol* **1**, 457–467.
Knoblauch C., Jorgensen B. B., and Harder J. (1999a) Community size and metabolic rates of psychrophilic sulfate-reducing bacteria in Arctic marine sediments. *Appl. Environ. Microbiol.* **65**, 4230–4233.
Knoblauch C., Sahm K., and Jørgensen B. B. (1999b) Psychrophilic sulfate-reducing bacteria isolated from permanently cold arctic marine sediments: description of *Desulfofrigus oceanense* gen. nov., sp. nov., *Desulfofrigus fragile* sp. nov., *Desulfofaba gelida* gen. nov., sp. nov., *Desulfotalea psychrophila* gen. nov., sp. nov., and *Desulfotalea actica* sp. nov. *Int. J. Syst. Evol. Microbiol.* **49**, 1631–1643.
Knoll G. and Winter J. (1989) Degradation of phenol via carboxylation to benzoate by a defined, obligate syntrophic consortium of anaerobic bacteria. *Appl. Microbiol. Biotech.* **30**, 318–324.
Kobayashi M., Matsuo Y., Tanimoto A., Suzuki S., Maruo F., and Shoun H. (1996) Denitrification, a novel type of respiratory metabolism in fungal mitochondrion. *J. Biol. Chem.* **271**, 16263–16267.
Kochevar R. E., Childress J. J., Fisher C. R., and Minnich E. (1992) The methane mussel: roles of symbiont and host in the metabolic utilization of methane. *Mar. Biol.* **112**, 389–401.
Koike I. and Hattori A. (1978) Denitrification and ammonia formation in anaerobic coastal sediments. *Appl. Environ. Microbiol.* **35**, 278–282.
Kostka J. E. and Luther G. W. I. (1994) Partitioning and speciation of solid phase iron in salt marsh sediments. *Geochim. Cosmochim. Acta* **7**, 1701–1710.
Kostka J. E. and Luther G. W. I. (1995) Seasonal cycling of Fe in salt marsh sediments. *Biogeochemistry* **29**, 159–181.
Kostka J. E. and Nealson K. H. (1995) Dissolution and reduction of magnetite by bacteria. *Environ. Sci. Technol.* **10**, 2535–2539.
Kostka J. E., Haefele E., Viehweger R., and Stucki J. W. (1999a) Respiration and dissolution of iron(III)-containing clay minerals by bacteria. *Environ. Sci. Technol.* **33**, 3127–3133.
Kostka J. E., Thamdrup B., Glud R. N., and Canfield D. E. (1999b) Rates and pathways of carbon oxidation in permanently cold Arctic sediments. *Mar. Ecol. Prog. Ser.* **180**, 7–21.
Kostka J. E., Wu J., Nealson K. H., and Stucki J. W. (1999c) The impact of structural Fe(III) reduction by bacteria on the surface chemistry of smectite clay minerals. *Geochim. Cosmochim. Acta* **63**, 3703–3705.
Kostka J. E., Dalton D. D., Skelton H., Dollhopf S., and Stucki J. W. (2002a) Growth of Iron(III)-reducing bacteria on clay minerals as the sole electron acceptor and comparison of growth yields on a variety of oxidized iron forms. *Appl. Environ. Microbiol.* **68**, 6256–6262.
Kostka J. E., Roychoudhury A., and van Cappellen P. (2002b) Rates and controls of anaerobic microbial respiration across

spatial and temporal gradients in salt marsh sediments. *Biogeochemistry* **60**, 49–76.

Kostka J. E., Gribsholt B., Petrie E., Dalton D., Skelton H., and Kristensen E. (2002c) The rates and pathways of carbon oxidation in bioturbated salt marsh sediments. *Limnol. Oceanogr.* **47**, 230–240.

Kotsyurbenko O. R., Nozhevnikova A. N., Soloviova T. I., and Zavarzin G. A. (1996) Methanogenesis at low temperature by microflora of tundra wetland soil. *Antonie Leeuwenhoek* **69**, 75–86.

Kramer M. and Cypionka H. (1989) Sulfate formation via ATP sulfurylase in thiosulfate- and sulfite-disproportionating bacteria. *Arch. Microbiol.* **151**, 232–237.

Kravchenko I. K. (1999) The inhibiting effect of ammonium on the activity of the methanotrophic microbial community of a raised *Sphagnum* bog in West Siberia. *Microbiology* **68**, 203–208.

Kreft J. U. and Schink B. (1993) Demethylation and degradation of phenylmethylethers by the sulfide-methylating homoacetogenic bacterium strain TMBS4. *Arch. Microbiol.* **159**, 308–315.

Kreft J. U. and Schink B. (1997) Specificity of O-demethylation in extracts of the homoacetogenic Holophaga foetida and demethylation kinetics measured by a coupled photometric assay. *Arch. Microbiol.* **167**, 363–368.

Krekeler D., Sigalevich P., Teske A., Cohen Y., and Cypionka H. (1997) A sulfate-reducing bacterium from the oxic layer of a microbial mat from Solar Lake (Sinai), *Desulfovibrio oxyclinae* sp. nov. *Arch. Microbiol.* **167**, 369–375.

Kristensen E., Jensen M. H., and Andersen T. K. (1985) The impact of polychaete (*Nereis virens* Sars.) on nitrification and denitrification in estuarine sediments. *J. Exp. Mar. Biol. Ecol.* **85**, 75–91.

Kristensen E., Rennenberg H., Jensen K. M., Bodenbender J., and Jensen M. H. (2000) Sulfur cycling of intertidal Wadden Sea sediments (Konigshafen, Island of Sylt, Germany): sulfate reduction and sulfur gas emission. *J. Sea Res.* **43**, 93–104.

Krueger D. M., Dubilier N., and Cavanaugh C. M. (1996) Chemoautotrophic symbiosis in the tropical clam *Solemya occidentalis* (Bivalvia: Protobranchia): ultrastructural and phylogenetic analysis. *Mar. Biol.* **126**, 55–64.

Krüger M., Frenzel P., and Conrad R. (2001) Microbial processes influencing methane emission from rice fields. *Global Change Biol.* **7**, 49–63.

Krumholz L. R. and Bryant M. P. (1985) *Clostridium pfennigii* sp. nov., uses methyoxyl groups of monobenzenoid and produces butyrate. *Int. J. Syst Bacteriol.* **35**, 454–456.

Kuenen J. G. and Beudeker R. F. (1982) Microbiology of thiobacilli and other sulphur-oxidizing autotrophs, mixotrophs, and heterotrophs. *Phil. Trans. Royal Soc. London B* **298**, 473–497.

Kuever J., Konneke M., Galushko A., and Drzyzga O. (2001) Reclassification of *Desulfobacterium phenolicum* as *Desulfobacula phenolica* comb. nov. and description of strain Sax (T) as *Desulfotignum balticum* gen. nov., sp. nov. *Int. J. Syst. Evol. Microbiol.* **51**, 171–177.

Kump L. (1993) Bacteria forge a new link. *Nature* **362**, 790–791.

Kuypers M. M. M., Sliekers A. O., Lavik G., Schmid M., Jørgensen B. G., Kuenen J. G., Sinninghe Damsté J. S., Strous M., and Jetten M. S. M. (2003) Anaerobic ammonium oxidation by anammox bacteria in the Black Sea. *Nature* **422**, 608–611.

Küsel K. and Drake H. L. (1994) Acetate synthesis in soil from a Bavarian beech forest. *Appl. Environ. Microbiol.* **60**, 1370–1373.

Küsel K., Dorsch T., Acker G., and Stackebrandt E. (1999) Microbial reduction of Fe(III) in acidic sediments: isolation of *Acidiphilium cryptum* JF-5 capable of coupling the reduction of Fe(III) to the oxidation of glucose. *Appl. Environ. Microbiol.* **65**, 3633–3640.

Küsel K., Dorsch T., Acker G., Stackebrandt E., and Drake H. L. (2000) *Clostridium scatologenes* strain SL1 isolated as an acetogenic bacterium from acidic sediments. *Int. J. Syst. Evol. Microbiol.* **50**, 537–546.

Kvenvolden K. A. (1999) Potential effects of gas hydrate on human welfare. *Proc. Natl. Acad. Sci. USA* **96**, 3420–3426.

Laanbroek H. J. and Pfennig N. (1981) Oxidation of short-chain fatty acids by sulfate-reducing bacteria in freshwater and marine sediments. *Arch. Microbiol.* **128**, 330–335.

LaKind J. and Stone A. T. (1989) Reductive dissolution of goethite by phenolic reductants. *Geochim. Cosmochim. Acta* **53**, 961–971.

Lanoil B. D., Sassen R., La Duc M. T., Sweet S. T., and Nealson K. H. (2001) Bacteria and Archaea physically associated with Gulf of Mexico gas hydrates. *Appl. Environ. Microbiol.* **67**, 5143–5153.

Lansdown J. M., Quay P. D., and King S. L. (1992) CH_4 production via CO_2 reduction in a temperate bog: a source of ^{13}C-depleted CH_4. *Geochim. Cosmochim. Acta* **56**, 3493–3503.

Larkin J. M. and Strohl W. R. (1983) *Beggiatoa*, *Thiothrix*, and *Thioploca*. *Ann. Rev. Microbiol.* **37**, 341–367.

Lau P. P., DeBrunner-Vossbrinck B., Dunn B., Miotto K., MacDonell M. T., Rollins D. M., Pillidge C. J., Hespell R. B., Colwell R. R., Sogin M. L., and Fox G. E. (1987) Phylogenetic diversity and position of the genus *Campylobacter*. *Syst. Appl. Microbiol.* **9**, 231–238.

Laughlin R. J. and Stevens R. J. (2002) Evidence for fungal dominance of denitrification and codenitrification in a grassland soil. *Soil Sci. Soc. Am. J.* **66**, 1540–1548.

Law C. S., Rees A. P., and Owens N. J. P. (1993) Nitrous oxide production by estuarine epiphyton. *Limnol. Oceanogr.* **38**, 435–441.

Ledyard K. M., DeLong E. F., and Dacey J. W. H. (1993) Characterization of a DMSP-degrading bacterial isolate from the Sargasso Sea. *Arch. Microbiol.* **160**, 312–318.

Lee M. J. and Zinder S. H. (1988) Isolation and characterization of a thermophilic bacterium which oxidizes acetate in syntrophic association with a methanogen and which grows acetogenically on H_2–CO_2. *Appl. Environ. Microbiol.* **54**, 124–129.

Lelieveld J., Crutzen P. J., and Dentener F. J. (1998) Changing concentration, lifetime and climate forcing of atmospheric methane. *Tellus B* **50**, 128–150.

Le Mer J. and Roger P. (2001) Production, oxidation, emission, and consumption of methane by soils: a review. *Euro. J. Soil Biol.* **37**, 25–50.

Lens P. N. L. and Kuenen J. G. (2001) The biological sulfur cycle: novel opportunities for environmental biotechnology. *Water Sci. Technol.* **44**, 57–66.

L'Haridon S., Cilia V., Messner P., Raguénès G., Gambacorta A., Sleytr U. B., Prieur D., and Jeanthon C. (1998) *Desulfurobacterium thermolithotrophum* gen. nov., sp. nov., a novel autotrophic, sulphur-reducing bacterium isolated from a deep-sea hydrothermal vent. *Int. J. Systemat. Bacteriol.* **48**, 701–711.

Li J. H., Hayashi H., Purdy K. J., and Takii S. (1999) Seasonal changes in ribosomal RNA of sulfate-reducing bacteria and sulfate-reducing activity in a freshwater lake sediment. *FEMS Microbiol. Ecol.* **28**, 31–39.

Liesack W. and Finster K. (1994) Phylogenetic analysis of five strains of gram-negative, obligately anaerobic, sulfur-reducing bacteria and description of *Desulfuromusa* gen. nov., including *Desulfuromusa kysingii* sp. nov., *Desulfuromusa bakii* sp. nov. and *Desulfuromusa succinoxidans* sp. nov. *Int. J. Systemat. Bacteriol.* **44**, 753–758.

Liesack W., Schnell S., and Revsbech N. P. (2000) Microbiology of flooded rice paddies. *FEMS Microbiol. Ecol.* **24**, 625–645.

Liikanen A., Flöjt L., and Martikainen P. (2002) Gas dynamics in Eutrophic lake sediments affected by oxygen, nitrate, sulfate. *J. Environ. Qual.* **31**, 338–349.

Lin Y. F., Jing S. R., Wang T. W., and Lee D. Y. (2002) Effects of macrophytes and external carbon sources on nitrate removal from groundwater in constructed wetlands. *Environ. Pollut.* **119**, 413–420.

Lindau C. W., Bollich P. K., DeLaune R. D., Patrick W. H., Jr., and Law V. J. (1991) Effects of urea fertilizer and environmental factors on CH_4 emissions from a Louisiana, USA rice field. *Plant Soil* **136**, 195–203.

Lindsay M. R., Webb R. I., Strous M., Jetten M. S. M., Butler M. K., Forde R. J., and Fuerst J. A. (2001) Cell compartmentalization in planctomycetes: novel types of structural organization for the bacterial cell. *Arch. Microbiol.* **175**, 413–429.

Liu C. L. and Peck H. D., Jr. (1981) Comparative bioenergetics of sulfate reduction in *Desulfovibrio* and *Desulfotomaculum*. *J. Bacteriol.* **145**, 966–973.

Liu S. and Suflita J. M. (1993) H_2–CO_2-dependent anaerobic O-demethylation activity in subsurface sediments and by an isolated bacterium. *Appl. Environ. Microbiol.* **59**, 1325–1331.

Liu Y., Boone D. R., and Choy C. (1990) *Methanohalophilus oregonense* sp. nov., a methylotrophic methanogen from an alkaline, saline aquifer. *Int. J. Systemat. Bacteriol.* **40**, 111–116.

LlobetBrossa E., RosselloMora R., and Amann R. (1998) Microbial community composition of Wadden Sea sediments as revealed by fluorescence *in situ* hybridization. *Appl. Environ. Microbiol.* **64**, 2691–2696.

Lohse L., Malschaert J. F. P., Slomp C. P., Helder W., and van Raaphorst W. (1993) Nitrogen cycling in North Sea sediments: interaction of denitrification and nitrification in offshore and coastal areas. *Mar. Ecol. Prog. Ser.* **101**, 283–296.

Lomans B. P., Smolders A. J. P., Intven L. M., Pol A., Op Den Camp H. J. M., and Van Der Drif C. (1997) Formation of dimethyl sulfide and methanethiol in anoxic freshwater sediments. *Appl. Environ. Microbiol.* **63**, 4741–4747.

Lomans B. P., denCamp H. J. M. O., Pol A., and Vogels G. D. (1999a) Anaerobic versus aerobic degradation of dimethyl sulfide and methanethiol in anoxic freshwater sediments. *Appl. Environ. Microbiol.* **65**, 438–443.

Lomans B. P., Maas R., Luderer R., denCamp H. J. M. O., Pol A., vanderDrift C., and Vogels G. D. (1999b) Isolation and characterization of *Methanomethylovorans hollandica* gen. nov., sp nov., isolated from freshwater sediment, a methylotrophic methanogen able to grow on dimethyl sulfide and methanethiol. *Appl. Environ. Microbiol.* **65**, 3641–3650.

Lomans B. P., Op den Camp H. J. M., Pol A., van der Drift C., and Vogels G. D. (1999c) Role of methanogens and other bacteria in degradation of dimethyl sulfide and methanethiol in anoxic freshwater sediments. *Appl. Environ. Microbiol.* **65**, 2116–2121.

Lomans B. P., Leijdekkers P., Wesselink J., Bakkes P., Pol A., van der Drift C., and Op den Camp H. J. (2001a) Obligate sulfide-dependent degradation of methoxylated aromatic compounds and formation of methanethiol and dimethyl sulfide by a freshwater sediment isolate, *Parasporobacterium paucivorans* gen. nov., sp. nov. *Appl. Environ. Microbiol.* **67**, 4017–4023.

Lomans B. P., Luderer R., Steenbakkers P., Pol A., vanderDrift A. P., Vogels G. D., and denCamp H. J. M. O. (2001b) Microbial populations involved in cycling of dimethyl sulfide and methanethiol in freshwater sediments. *Appl. Environ. Microbiol.* **67**, 1044–1051.

Lomans B., Pol A., and Op Den Camp H. J. M. (2002a) Microbial cycling of volatile organic sulfur compounds in anoxic environments. *Water Sci. Technol.* **45**, 55–60.

Lomans B., van der Drift C., Pol A., and Op Den Camp H. J. M. (2002b) Microbial cycling of volatile organic sulfur compounds. *Cell. Mol. Life Sci.* **59**, 575–588.

Lombardi J. E., Epp M. A., and Chanton J. P. (1997) Investigation of the methyl fluoride technique for determining rhizospheric methane oxidation. *Biogeochemistry* **36**, 153–172.

Lovell C. R., Friez M. J., Longshore J. W., and Bagwell C. E. (2001) Recovery and phylogenetic analysis of *nifH* sequences from diazotrophic bacteria associated with dead aboveground biomass of *Spartina alterniflora*. *Appl. Environ. Microbiol.* **67**, 5308–5314.

Lovelock J. E., Maggs R. J., and Rasmussen R. A. (1972) Atmospheric dimethyl sulphide and the natural sulphur cycle. *Nature* **237**, 452–453.

Lovley D. R. (1987) Organic matter mineralization with the reduction of ferric iron: a review. *Geomicrobiol. J.* **5**, 375–399.

Lovley D. R. (1991) Dissimilatory Fe(III) and Mn(IV) reduction. *Microbiol. Rev.* **55**, 259–287.

Lovley D. R. (1997) Microbial Fe(III) reduction in subsurface environments. *FEMS Microbiol. Rev.* **20**, 305–313.

Lovley D. R. (2000a) Anaerobic benzene degradation. *Biodegradation* **11**, 107–116.

Lovley D. R. (2000b) Fe(III)- and Mn(IV)-reducing prokaryotes. In *The Prokaryotes* (eds. M. Dworkin, S. Falow, E. Rosenberg, K.-H. Schleifer, and E. Stackebrandt). Springer, New York.

Lovley D. R. (2000c) Fe(III) and Mn(IV) reduction. In *Environmental Microbe-metal Interactions* (ed. D. R. Lovley). ASM Press, Washington, DC, pp. 3–30.

Lovley D. R. and Anderson R. T. (2000) Influence of dissimilatory metal reduction on fate of organic and metal contaminants in the subsurface. *Hydrogeol. J.* **8**, 77–88.

Lovley D. R. and Blunt-Harris E. L. (1999) Role of humic-bound iron as an electron transfer agent in dissimilatory Fe(III) reduction. *Appl. Environ. Microbiol.* **65**, 4252–4254.

Lovley D. R. and Goodwin S. (1988) Hydrogen concentrations as an indicator of the predominant terminal electron-accepting reactions in aquatic sediments. *Geochim. Cosmochim. Acta* **52**, 2993–3003.

Lovley D. R. and Klug M. J. (1983a) Methanogenesis from methanol and methylamines and acetogenesis from hydrogen and carbon dioxide in the sediments of a eutrophic lake. *Appl. Environ. Microbiol.* **45**, 1310–1315.

Lovley D. R. and Klug M. J. (1983b) Sulfate reducers can outcompete methanogens at freshwater sulfate concentrations. *Appl. Environ. Microbiol.* **65**, 438–443.

Lovley D. R. and Phillips E. J. P. (1986) Availability of ferric iron for microbial reduction in bottom sediments of the freshwater tidal Potomac river. *Appl. Environ. Microbiol.* **52**, 751–757.

Lovley D. R. and Phillips E. J. P. (1987) Rapid assay for microbially reducible ferric iron in aquatic sediments. *Appl. Environ. Microbiol.* **53**, 1536–1540.

Lovley D. R. and Phillips E. J. P. (1988) Manganese inhibition of microbial iron reduction in anaerobic sediments. *Geomicrobiol. J.* **6**, 145–155.

Lovley D. R. and Phillips E. J. P. (1992a) Bioremediation of uranium contamination with enzymatic uranium reduction. *Environ. Sci. Technol.* **26**, 2228–2234.

Lovley D. R. and Phillips E. J. P. (1992b) Reduction of uranium by *Desulfovibrio desulfuricans*. *Appl. Environ. Microbiol.* **58**, 850–856.

Lovley D. R. and Phillips E. J. P. (1994a) Novel processes for anaerobic sulfate production from elemental sulfur by sulfate-reducing bacteria. *Appl. Environ. Microbiol.* **60**, 2394–2399.

Lovley D. R. and Phillips E. J. P. (1994b) Reduction of chromate by *Desulfovibrio vulgaris* and its c(3) cytochrome. *Appl. Environ. Microbiol.* **60**, 726–728.

Lovley D. R. and Woodward J. C. (1996) Mechanisms for chelator stimulation of microbial Fe(III)-oxide reduction. *Chem. Geol.* **132**, 19–24.

Lovley D. R., Dwyer D. F., and Klug M. J. (1982) Kinetic analysis of competition between sulfate reducers and methanogens for hydrogen in sediments. *Appl. Environ. Microbiol.* **82**, 1373–1379.

Lovley D. R., Stolz J. F., Nord G. L., Jr., and Phillips E. J. P. (1987) Anaerobic production of magnetite by a dissimilatory iron-reducing microorganism. *Nature* **330**, 252–254.

Lovley D. R., Phillips E. J. P., and Lonergan D. J. (1989) Hydrogen and formate oxidation coupled to the dissimilatory reduction of iron or manganese by *Alteromonas putrefaciens*. *Appl. Environ. Microbiol.* **55**, 700–706.

Lovley D. R., Giovannoni S. J., White D. C., Champine J. E., Phillips E. J. P., Gorby Y. A., and Goodwin S. (1993a) *Geobacter metallireducens* gen. nov. sp. nov., a microorganism capable of coupling the complete oxidation of organic matter to the reduction of iron and other metals. *Arch. Microbiol. (Berlin)* **159**, 336–344.

Lovley D. R., Roden E. E., Phillips E. J. P., and Woodward J. C. (1993b) Enzymatic iron and uranium reduction by sulfate-reducing bacteria. *Mar. Geol.* **113**, 41–53.

Lovley D. R., Chapelle F. H., and Woodward J. C. (1994a) Use of dissolved H_2 concentrations to determine distribution of microbially catalyzed redox reactions in anoxic groundwater. *Environ. Sci. Technol.* **28**, 1205–1210.

Lovley D. R., Woodward J. C., and Chapelle F. H. (1994b) Stimulated anoxic biodegradation of aromatic hydrocarbons using Fe(III) ligands. *Nature* **370**, 128–131.

Lovley D. R., Coates J. D., Woodward J. C., and Phillips E. J. P. (1995) Benzene oxidation coupled to sulfate reduction. *Appl. Environ. Microbiol.* **61**, 953–958.

Lovley D. R., Coates J. D., Blunt-Harris E. L., Phillips E. J. P., and Woodward J. C. (1996a) Humic substances as electron acceptors for microbial respiration. *Nature* **382**, 445–448.

Lovley D. R., Woodward J. C., and Chapelle F. H. (1996b) Rapid anaerobic benzene oxidation with a variety of chelated Fe(III) forms. *Appl. Environ. Microbiol.* **62**, 288–291.

Lovley D. R., Fraga J. L., Blunt-Harris E. L., Hayes L. A., Phillips E. J. P., and Coates J. D. (1998) Humic substances as a mediator for microbially catalyzed metal reduction. *Acta Hydrochim. Hydrobiol.* **26**, 152–157.

Lovley D. R., Kashefi K., Vargas M., Tor J. M., and Blunt-Harris E. L. (2000) Reduction of humic substances and Fe(III) by hyperthermophilic microorganisms. *Chem. Geol.* **169**, 289–298.

Lowe K. L., Dichristina T. J., Roychoudhury A. N., and van Cappellen P. (2000) Microbiological and geochemical characterization of microbial Fe(III) reduction in salt marsh sediments. *Geomicrobiol. J.* **17**, 163–178.

Lower S. K., Hochella M. F., Jr., and Beveridge T. J. (2001) Bacterial recognition of mineral surfaces: nanoscale interactions between Shewanella and α-FeOOH. *Science* **292**, 1360–1363.

Lu Y., Wassmann R., Neue H. U., and Huang C. (1999) Impact of phosphorus supply on root exudation, aerenchyma formation and methane emission of rice plants. *Biogeochemistry* **47**, 202–218.

Lu Y., Wassmann R., Neue H.-U., and Huang C. (2000a) Dissolved organic carbon and methane emissions from a rice paddy fertilized with ammonium and nitrate. *J. Environ. Qual.* **29**, 1733–1740.

Lu Y., Wassmann R., Neue H.-U., and Huang C. (2000b) Dynamics of dissolved organic carbon and methane emissions in a flooded rice soil. *Soil Sci. Soc. Am. J.* **64**, 2011–2017.

Lu Y., Watanabe A., and Kimura M. (2002) Contribution of plant-derived carbon to soil microbial biomass dynamics in a paddy rice microcosm. *Biol. Fertility Soils* **36**, 136–142.

Ludwig J., Meixner F. X., Vogel B., and Förstner J. (2001) Soil-air exchange of nitric oxide: an overview of processes, environmental factors, and modeling studies. *Biogeochemistry* **52**, 225–257.

Luther G. W., III, and Church T. M. (1988) Seasonal cycling of sulfur and iron in porewaters of a Delaware salt marsh. *Mar. Chem.* **23**, 295–309.

Luther G. W., III, Church T. M., and Powell D. (1991) Sulfur speciation and sulfide oxidation in the water column of the Black Sea. *Deep-Sea Res. Part A: Oceanogr. Res* **38**, S1121–S1137.

Luther G. W., III, Kostka J. E., Church T. M., Sulzberger B., and Stumm W. (1992) Seasonal iron cycling in a salt marsh sedimentary environment: the importance of ligand complexes with Fe(II) and Fe(III) in the dissolution of Fe(III) minerals and pyrite, respectively. *Mar. Chem.* **40**, 81–103.

Luther G. W., III, Shellenbarger P. A., and Brendel P. J. (1996) Dissolved organic Fe(III) and Fe(II) complexes in salt marsh porewaters. *Geochim. Cosmochim. Acta* **60**, 951–960.

Luther G. W., III, Sundby B., Lewis B. L., Brendel P. J., and Silverberg N. (1997) Interactions of manganese with the nitrogen cycle: alternative pathways to dinitrogen. *Geochim. Cosmochim. Acta* **61**, 4043–4052.

Luther G. W., III, Church T. M., and Powell D. (1999) Sulfur speciation and sulfide oxidation in the water column of the Black Sea. *Deep-Sea Res.* **38**(suppl. 2), S1121–S1137.

Luther G. W., III, Rozan T. F., Taillefert M., Nuzzio D. B., Di Meo C., Shank T. M., Lutz R. A., and Cary S. C. (2001) Chemical speciation drives hydrothermal vent ecology. *Nature* **410**, 813–816.

Lwoff A., van Niel C. B., Ryan T. F., and Tatum E. L. (1946) Nomenclature of nutritional types of microorganisms. *Cold Spring Harbor Symp. Quantit. Biol.* **11**, 302–303.

Lyimo T. J., Pol A., den Camp H., Harhangi H. R., and Vogels G. D. (2000) *Methanosarcina semesiae* sp. nov., a dimethylsulfide-utilizing methanogen from mangrove sediment. *Int. J. Syst. Evol. Microbiol.* **50**, 171–178.

Lyimo T. J., Pol A., and denCamp H. J. M. O. (2002) Sulfate reduction and methanogenesis in sediments of Mtoni mangrove forest, Tanzania. *Ambio* **31**, 614–616.

Lynch J. M. and Whipps J. M. (1990) Substrate flow in the rhizosphere. *Plant Soil* **129**, 1–10.

Maccubbin A. A. and Hodson R. E. (1980) Mineralization of detrital lignocelluloses by salt marsh sediment microflora. *Appl. Environ. Microbiol.* **40**, 735–740.

Machel H. G. (2001) Bacterial and thermochemical sulfate reduction in diagenetic settings: old and new insights. *Sedim. Geol.* **140**, 143–175.

Macy J. M., Santini J. M., Pauling B. V., O'Neill A. H., and Sly L. I. (2000) Two new arsenate/sulfate-reducing bacteria: mechanism of arsenate reduction. *Arch. Microbiol.* **173**, 49–57.

Madigan M. T. (1988) Microbiology, physiology, and ecology of phototrophic bacteria. In *Biology of Anaerobic Microorganisms* (ed. A. J. B. Zehnder). Wiley, New York, pp. 39–111.

Mandernack K. W., Bazylinski D. A., Shanks W. C., and Bullen T. D. (1999) Oxygen and iron isotopes studies of magnetite produced by magnetotactic bacteria. *Science* **285**, 1892–1895.

Martens C. S. and Berner R. A. (1977) Interstitial water chemistry of anoxic Long Island Sound sediments: 1. Dissolved gases. *Limnol. Oceanogr.* **22**, 10–25.

Martens C. S. and Klump J. V. (1980) Biogeochemical cycling in an organic-rich coastal marine basin: 1. Methane sediment-water exchange processes. *Geochim. Cosmochim. Acta* **44**, 471–490.

Martens C. S. and Klump J. V. (1984) Biogeochemical cycling in an organic-rich coastal marine basin: 4. An organic carbon budget for sediments dominated by sulfate reduction and methanogenesis. *Geochim. Cosmochim. Acta* **48**, 1987–2004.

Martens C. S., Berner R. A., and Rosenfeld J. K. (1978) Interstitial water chemistry of anoxic Long Island Sound sediments: 2. Nutrient regeneration and phosphate removal. *Limnol. Oceanogr.* **23**, 605–717.

Martens C. S., Blair N. E., Green C. D., and Des Marais D. J. (1986) Seasonal variations in the stable carbon isotope signature of biogenic methane in a coastal sediment. *Science* **233**, 1300–1303.

Martens C. S., Haddad R. I., and Chanton J. P. (1992a) Organic matter accumulation, remineralization, and burial in an anoxic coastal sediment. In *Organic Matter: Productivity, Accumulation, and Preservation in Recent and Ancient Sediments* (eds. J. K. Whelan and J. W. Farrington). Columbia University Press, New York, pp. 82–98.

Martens C. S., Kelley C. A., Chanton J. P., and Showers W. J. (1992b) Carbon and hydrogen isotopic characterization of methane from wetlands and lakes of the Yukon-Kushokwim Delta, Western Alaska. *J. Geophys. Res.* **97**, 16689–16701.

Martikainen P. and De Boer W. (1993) Nitrous oxide production and nitrification in acidic soil from a Dutch coniferous forest. *Soil Biol. Biochem.* **25**, 343–347.

Martikainen P. J., Nykänen H., Alm J., and Silvola J. (1995) Change in fluxes of carbon dioxide, methane, and nitrous oxide due to forest drainage of mire sites of different trophy. *Plant Soil* **168-169**, 571–577.

Martinez J. and Azam F. (1993) Periplasmic aminopeptidase and alkaline phosphatase activities in a marine bacterium: implications for substrate processing in the sea. *Mar. Ecol. Prog. Ser.* **92**, 89–97.

Matthews E. and Fung I. (1987) Methane emission from natural wetlands: global distribution, area, and environmental characteristics of sources. *Global Biogeochem. Cycles* **1**, 61–86.

Mayer H. P. and Conrad R. (1990) Factors influencing the population of methanogenic bacteria and the initiation of methane production upon flooding of paddy soil. *FEMS Microbiol. Ecol.* **73**, 103–112.

Mayer L. M. (1989) Extracellular proteolytic enzyme activity in sediments of an intertidal mudflat. *Limnol. Oceanogr.* **34**, 973–981.

Mayer M. S., Schaffner L., and Kemp W. M. (1995) Nitrification potentials of benthic macrofaunal tubes and burrow walls: effects of sediment NH_4^+ and animal irrigation behavior. *Mar. Ecol. Prog. Ser.* **121**, 157–169.

McCready R. G. L., Gould W. D., and Cook F. D. (1983) Respiratory nitrate reduction by *Desulfovibrio* sp. *Arch. Microbiol.* **135**, 182–185.

McDonald I. R. and Murrell J. C. (1997) The particulate methane monooxygenase gene pmoA and its use as a functional gene probe for methanotrophs. *FEMS Microbiol. Lett.* **156**, 205–210.

McGuire J. T., Long D. T., Klug M. J., Haack S. K., and Hyndman D. W. (2002) Evaluating behavior of oxygen, nitrate, and sulfate during recharge and quantifying reduction rates in a contaminated aquifer. *Environ. Sci. Technol.* **36**, 2693–2700.

McInerney M. J., Bryant M. P., and Pfennig N. (1979) Anaerobic bacterium that degrades fatty acids in syntrophic association with methanogens. *Arch. Microbiol.* **122**, 129–135.

McInerney M. J., Bryant M. P., Hespell R. B., and Costerton J. W. (1981) *Syntrophomonas wolfei* gen. nov. sp. nov., an anaerobic, syntrophic, fatty acid-oxidizing bacterium. *Appl. Environ. Microbiol.* **41**, 1029–1039.

McKenney D. J., Drury C. F., Findlay W. I., Mutus B., McDonnell T., and Gajala C. (1994) Kinetics of denitrification by *Pseudomonas fluorescens*: oxygen effects. *Soil Biol. Biochem.* **26**, 901–908.

McKenney D. J., Drury C. F., and Wang S. W. (1997) Reaction of nitric oxide with acetylene and oxygen: implications for denitrification assays. *Soil Sci. Soc. Am. J.* **61**, 1370–1375.

Mechichi T., Labat M., Garcia J.-L., Thomas P., and Patel B. K. C. (1999) *Sporobacterium olearium* gen. nov. sp. nov., a new methanethiol-producing bacterium that degrades aromatic compounds, isolated from an olive mill wastewater treatment digester. *Int. J. Systemat. Bacteriol.* **49**, 1741–1748.

Megonigal J. P. and Schlesinger W. H. (1997) Enhanced CH_4 emissions from a wetland soil exposed to elevated CO_2. *Biogeochemistry* **37**, 77–88.

Megonigal J. P. and Schlesinger W. H. (2002) Methane production and oxidation in a tidal freshwater swamp. *Global Biogeochem. Cycles* **16**, 1062 doi: 10.1029/2001GB001594.

Megonigal J. P., Patrick W. H., Jr., and Faulkner S. P. (1993) Wetland identification in seasonally flooded forest soils: soil morphology and redox dynamics. *Soil Sci. Soc. Am. J.* **57**, 140–149.

Megonigal J. P., Whalen S. C., Tissue D. T., Bovard B. D., Albert D. B., and Allen A. S. (1999) A plant-soil-atmosphere microcosm for tracing radiocarbon from photosynthesis through methanogenesis. *Soil Sci. Soc. Am. J.* **63**, 665–671.

Meier J., Voigt A., and Babenzien H. D. (2000) A comparison of S-35–SO42-radiotracer techniques to determine sulphate reduction rates in laminated sediments. *J. Microbiol. Meth.* **41**, 9–18.

Meijer W., Nienhuis-Kuiper M. E., and Hansen T. A. (1999) Fermentative bacteria from estuarine mud: phylogenetic position of *Acidaminobacter hydrogenoformans* and description of a new type of gram-negative, propionigenic bacterium as *Propionibacter pelophilus* gen. nov., sp. nov. *Int. J. Systemat. Bacteriol.* **49**, 1039–1044.

Melloh R. A. and Crill P. M. (1996) Winter methane dynamics in a temperate peatland. *Global Biogeochem. Cycles* **10**, 247–254.

Mendelssohn, I. A., Kleiss, B. A., and Wakeley, J. S. (1995) Factors controlling the formation of oxidized root channels: a review. *Wetlands* **15**, 37–46.

Mendez C., Bauer A., Huber H., Gad'on N., Stetter K. O., and Fuchs G. (1999) Presence of acetyl coenzyme A (CoA) carboxylase and propionyl-CoA carboxylase in autotrophic *Crenarchaeota* and indication for operation of a 3-hydroxypropionate cycle in autotrophic carbon fixation. *J. Bacteriol.* **181**, 1088–1098.

Meyer-Reil L. A. (1986) Measurements of hydrolytic activity and incorporation of dissolved organic substrates by microogranisms in marine sediments. *Mar. Ecol. Prog. Ser.* **31**, 143–149.

Meyer-Reil L. and Köster M. (2000) Eutrophication of marine waters: effects of benthic microbial communities. *Mar. Pollut. Bull.* **41**, 255–263.

Meyers P. A., Leenheer M. J., and Bourbonniere R. A. (1995) Diagenesis of vascular plant organic matter components during burial in lake sediments. *Aquat. Geochem.* **1**, 35–52.

Michaelis W., Seifert R., Nauhaus K., Treude T., Thiel V., Blumenberg M., Knittel K., Gieseke A., Peterknecht K., Pape T., Boetius A., Amann R., Jørgensen B. B., Widdel F., Peckmann J., Pimenov N. V., and Gulin M. B. (2002) Microbial reefs in the Black Sea fueled by anaerobic oxidation of methane. *Science* **297**, 1013–1015.

Middelburg J. J., Nieuwenhuize J., Iversen N., Högh N., de Wilde H., Helder W., Seifert R., and Christof O. (2002) Methane distribution in European tidal estuaries. *Biogeochemistry* **59**, 95–119.

Mikesell M. D. and Boyd S. A. (1990) Dechlorination of chloroform by *Methanosarcina* strains. *Appl. Environ. Microbiol.* **56**, 1198–1201.

Miller J. D. A. and Wakerley D. S. (1966) Growth of sulphate-reducing bacteria by fumarate dismutation. *J. Gen. Microbiol.* **43**, 101–107.

Miller L. G., Sasson C., and Oremland R. S. (1998) Difluoromethane, a new and improved inhibitor of methanotrophy. *Appl. Environ. Microbiol.* **64**, 4357–4362.

Miltner A. and Emeis K. C. (2001) Terrestrial organic matter in surface sediments of the Baltic Sea, Northwest Europe, as determined by CuO oxidation. *Geochim. Cosmochim. Acta* **65**, 1285–1299.

Minami K. and Takata K. (1997) Atmospheric methane: sources, sinks, and strategies for reducing agricultural emissions. *Water Sci. Technol.* **36**, 509–516.

Minoda T. and Kimura M. (1994) Contribution of photosynthesized carbon to the methane emitted from paddy fields. *Geophys. Res. Lett.* **21**, 2007–2010.

Minoda T. and Kimura M. (1996) Photosynthates as dominant source of CH_4 and CO_2 in soil water and CH_4 emitted to the atmosphere from paddy fields. *J. Geophys. Res.* **101**, 21091–21097.

Minz D., Fishbain S., Green S. J., Muyzer G., Cohen Y., Rittmann B. E., and Stahl D. A. (1999a) Unexpected population distribution in a microbial mat community: sulfate-reducing bacteria localized to the highly oxic chemocline in contrast to a eukaryotic preference for anoxia. *Appl. Environ. Microbiol.* **65**, 4659–4665.

Minz D., Flax J. L., Green S. J., Muyzer G., Cohen Y., Wagner M., Rittmann B. E., and Stahl D. A. (1999b) Diversity of sulfate-reducing bacteria in oxic and anoxic regions of a microbial mat characterized by comparative analysis of dissimilatory sulfite reductase genes. *Appl. Environ. Microbiol.* **65**, 4666–4671.

Mitchell G. J., Jones J. G., and Cole J. A. (1986) Distribution and regulation of nitrate and nitrite reduction by *Desulfovibrio* and *Desulfotomaculum* species. *Arch. Microbiol.* **144**, 35–40.

Mitterer R. M., Wortman U. G., Logan G. A., Feary D. A., Hine A. C., Malone M. J., Goodfriend G. A., and Swart P. K. (2001) Co-generation of hydrogen sulfide and methane in marine carbonate sediments. *Geophys. Res. Lett.* **28**, 3931–3934.

Miura Y., Watanabe A., Murase J., and Kimura M. (1992) Methane production and its fate in paddy fields: II. Oxidation of methane and its coupled ferric oxide reduction in subsoil. *Soil Sci. Plant Nutr.* **38**, 673–679.

Miyajima T., Wada E., Hanba Y. T., and Vijarnsorn P. (1997) Anaerobic mineralization of indigenous organic matters and methanogenesis in tropical wetland soils. *Geochim. Cosmochim. Acta* **61**, 3739–3751.

Mohn W. W. and Tiedje J. M. (1990) Strain DCB-1 conserves energy for growth from reductive dechlorination coupled to formate oxidation. *Arch. Microbiol.* **153**, 267–271.

Monetti M. A. and Scranton M. I. (1992) Fatty acid oxidation in anoxic marine sediments: the importance of hydrogen sensitive reactions. *Biogeochemistry* **17**, 23–47.

Moore T. R. and Knowles R. (1990) Methane emissions from fen, bog, and swamp peatlands. *Biogeochemistry* **11**, 45–61.

Moore T. R. and Dalva M. (1997) Methane and carbon dioxide exchange potentials of peat soils on aerobic and anaerobic laboratory incubations. *Soil Biol. Biochem.* **29**, 1157–1164.

Morris S. A., Radajewski S., Willison T. W., and Murrell J. C. (2002) Identification of the functionally active methanotroph population in a peat soil microcosm by stable-isotope probing. *Appl. Environ. Microbiol.* **68**, 1446–1453.

Mountfort D. O. and Kaspar H. F. (1986) Palladium-mediated hydrogenation of unsaturated hydrocarbons with hydrogen gas released during anaerovic cellulose degradation. *Appl. Environ. Microbiol.* **52**, 744–750.

Mountfort D. O., Brulla W. J., Krumholz L. R., and Bryant M. P. (1984) *Syntrophus buswellii* gen. nov., sp. nov.: a benzoate catabolizer from methanogenic ecosystems. *Int. J. Systemat. Bacteriol.* **34**, 216–217.

Mountfort D. O., Grant W. D., Clake R., and Asher R. A. (1988) *Eubacterium callanderii* sp. nov. that demethoxylates O-methoxylated aromatic acids to volatile fatty acids. *Int. J. Systemat. Bacteriol.* **38**, 254–258.

Moura I., Bursakov S., Costa C., and Moura J. J. G. (1997) Nitrate and nitrite utilization in sulfate-reducing bacteria. *Anaerobe* **3**, 279–290.

Mulder A., van de Graaf A. A., Robertson L. A., and Kuenen J. G. (1995) Anaerobic ammonium oxidation discovered in a denitrifying fluidized bed reactor. *FEMS Microbiol. Ecol.* **16**, 177–184.

Murase J. and Kimura M. (1994a) Methane production and its fate in paddy fields: 6. Anaerobic oxidation of methane in plow layer soil. *Soil Sci. Plant Nutr.* **40**, 505–514.

Murase J. and Kimura M. (1994b) Methane production and its fate in paddy fields: 7. Electron acceptors responsible for anaerobic methane oxidation. *Soil Sci. Plant Nutr.* **40**, 647–654.

Murrell J. C. and Kelley D. P. (eds.) (1993) *Microbial Growth on C_1 Compounds*. Intercept Press, Andover, UK.

Myers C. R. and Nealson K. H. (1988) Bacterial manganese reduction and growth with manganese oxide as the sole electron acceptor. *Science* **240**, 1319–1321.

Myers C. R. and Myers J. M. (1992) Localization of cytochromes to the outer membrane of anaerobically grown *Shewanella putrefaciens*. *J. Bacteriol.* **174**, 3429–3438.

Myneni R. B., Keeling C. D., Tucker C. J., Asrar G., and Nemani R. R. (1997) Increased plant growth in the northern high latitudes from 1981 to 1991. *Nature* **386**, 698–702.

Nagase M. and Matsuo T. (1982) Interaction between amino-acid degrading bacteria and methanogenic bacteria in anaerobic digestion. *Biotechnol. Bioeng.* **24**, 2227–2239.

Namsaraev B. B., Dulov L. E., Dubinina G. A., Zemskaya T. I., Granina L. Z., and Karabanov E. V. (1994) Bacterial synthesis and destruction of organic matter in microbial mats of Lake Baikal. *Microbiology* **63**, 193–197.

Nanninga H. J. and Gottschal J. C. (1985) Amino acid fermentation and hydrogen transfer in mixed cultures. *FEMS Microbiol. Ecol.* **31**, 261–269.

Natusch D. and Slatt B. (1978) Hydrogen sulfide as an air pollutant. In *Air Pollution Control* (ed. W. Strauss). Wiley-Interscience, New York, pp. 459–518.

Nauhaus K., Boetius A., Kruger M., and Widdel F. (2002) *In vitro* demonstration of anaerobic oxidation of methane coupled to sulphate reduction in sediment from a marine gas hydrate area. *Environ. Microbiol.* **4**, 296–305.

Nealson K. H. (1983) The microbial iron cycle. In *Microbial Geochemistry* (ed. W. E. Krumbein). Blackwell, Boston, pp. 159–190.

Nealson K. H., Tebo B. M., and Rosson R. A. (1988) Occurrence and mechanisms of microbial oxidation of manganese. *Adv. Appl. Microbiol.* **33**, 279–318.

Nealson K. H. and Myers C. R. (1992) Microbial reduction of manganese and iron: new approaches to carbon cycling. *Appl. Environ. Microbiol.* **58**, 439–443.

Neubauer S. C., Emerson D., and Megonigal J. P. (2002) Life at the energetic edge: kinetics of circumneutral iron oxidation by lithotrophic iron-oxidizing bacteria isolated from the wetland-plant rhizophere. *Appl. Environ. Microbiol.* **68**, 3988–3995.

Neue H. H. (1993) Methane emission from rice fields. *Bioscience* **43**, 466–474.

Nevin K. P. and Lovley D. R. (2000) Potential for nonenzymatic reduction of Fe(III) via electron shuttling in subsurface sediments. *Environ. Sci. Technol.* **34**, 2472–2478.

Newman D. K., Beveridge T. J., and Morel F. M. M. (1997a) Precipitation of arsenic trisulfide by Desulfotomaculum auripigmentum. *Appl. Environ. Microbiol.* **63**, 2022–2028.

Newman D. K., Kennedy E., Coates J. D., Ahmann D., Ellis D. J., *et al.* (1997b) Dissimilatory arsenate and sulfate reduction in *Desulfotomaculum auripidmentum* sp. nov. *Arch. Microbiol.* **168**, 380–388.

Newman D. K. and Kolter R. (2000) A role for excreted quinones in extracellular electron transfer. *Nature* **405**, 94–97.

Ni S. and Boone D. R. (1993) Catabolism of dimethyl sulfide and methanethiol by methylotrophic methanogens. In *Biogeochemistry of Global Change* (ed. R. S. Oremland). Chapman and Hall, New York, pp. 796–810.

Nielsen J. L., Juretschko S., Wagner M., and Nielsen P. H. (2002) Abundance and phylogenetic affiliation of iron reducers in activated sludge as assessed by fluorescence *in situ* hybridization and microautoradiography. *Appl. Environ. Microbiol.* **68**, 4629–4636.

Nielsen K., Nielsen L. P., and Rasmussen P. (1995) Estuarine nitrogen retention independently estimated by denitrification rate and mass balance methods: a study of Norsminde Fjord, Denmark. *Mar. Ecol. Prog. Ser.* **119**, 275–283.

Niewöhner C., Hensen C., Kasten S., Zabel M., and Schulz H. D. (1998) Deep sulfate reduction completely mediated by anaerobic methane oxidation in sediments of the upwelling area off Namibia. *Geochim. Cosmochim. Acta* **62**, 455–464.

Nijburg J. W. and Laanbroek H. J. (1997) The influence of *Glyceria maxima* and nitrate input on the composition and nitrate metabolism of the dissimilatory nitrate-reducing bacteria community. *FEMS Microbiol. Ecol.* **22**, 57–63.

Nijburg J. W., Coolen M. J. L., Gerards S., Klein Gunnewiek P. J. A., and Laanbroek H. J. (1997) Effects of nitrate availability and the presence of *Glyceria maxima* on the composition and activity of the dissimilatory nitrate-reducing bacteria community. *Appl. Environ. Microbiol.* **63**, 931–937.

Nisbet E. G. (1992) Sources of atmospheric CH_4 in early postglacial time. *J. Geophys. Res.* **97**, 12859–12867.

Nishino M., Fukui M., and Nakajima T. (1998) Dense mats of Thioploca, gliding filamentous sulfur bacteria in Lake Biwa, Japan. *Water Res.* **32**, 953–957.

Nishio T., Koike I., and Hatori A. (1983) Estimates of denitrification and nitrification in coastal and estuarine sediments. *Appl. Environ. Microbiol.* **45**, 444–450.

Nordstrom D. K. and Southam G. (1997) Geomicrobiology of sulfide mineral oxidation. In *Geomicrobiology: Interactions between Microbes and Minerals.* (eds. J. F. Banfield and K. H. Nealson). Mineralogical Society of America, Washington, DC, vol. 35, 361–390.

Nold S. C., Boschker H. T. S., Pel R., and Laanbroek H. J. (1999) Ammonium addition inhibits ^{13}C-methane incorporation into methanotroph membrane lipids in a freshwater sediment. *FEMS Microbiol. Ecol.* **29**, 81–89.

Nowicki B. L. (1994) The effect of temperature, oxygen, salinity, and nutrient enrichment on estuarine denitrification rates measured with a modified nitrogen gas flux technique. *Estuar. Coast. Shelf Sci.* **38**, 137–156.

Nriagu J. O., Holdway D. A., and Coker R. D. (1987) Biogenic sulfur and the acidity of rainfall in remote areas of Canada. *Science* **237**, 1189–1192.

Nurmi J. T. and Tratnyek P. G. (2002) Electrochemical properties of natural organic matter (NOM), fractions of NOM, and model bieogeochemical electron shuttles. *Environ. Sci. Technol.* **36**, 617–624.

Nykänen H., Alm J., Silvola J., Tolonen K., and Martikainen P. J. (1998) Methane fluxes on boreal peatlands of different fertility and the effect of long-term experimental lowering of the water table on flux rates. *Global Biogeochem. Cycles* **12**, 53–69.

Obenhuber D. C. and Lowrance R. (1991) Reduction of nitrate in aquifer microcosms by carbon additions. *J. Environ. Qual.* **20**, 255–258.

Odom J. M. and Singleton R., Jr. (1993) *The Sulfate-reducing Bacteria: Contemporary Perspectives*. Springer, New York.

Oechel W. C., Cowles S., Grulke N., Hastings S. J., Lawrence B., Prudhomme T., Riechers G., Strain B., Tissue D., and Vourlitis G. (1994) Transient nature of CO_2 fertilization in Arctic tundra. *Nature* **371**, 500–502.

Ogilvie B. G., Nedwell D. B., Harrison R., Robinson A., and Sage A. (1997a) High nitrate, muddy estuaries as nitrogen sinks: the nitrogen budget of the River Colne estuary (UK). *Mar. Ecol. Prog. Ser.* **150**, 217–228.

Ogilvie B. G., Rutter M., and Nedwell D. B. (1997b) Selection by temperature of nitrate-reducing bacteria from estuarine sediments: species composition and competition for nitrate. *FEMS Microbiol. Ecol.* **23**, 11–22.

Okabe S., Santegoeds C. M., Watanabe Y., and deBeer D. (2002) Successional development of sulfate-reducing bacterial populations and their activities in an activated sludge immobilized agar gel film. *Biotechnol. Bioeng.* **78**, 119–130.

Olivier J. G. J., Bouwman A. F., Van der Hoek K. W., and Berdowski J. J. M. (1998) Global air emission inventories for anthropogenic sources of NO_x, NH_3, and N_2O in 1990. *Environ. Pollut.* **102**(S1), 135–148.

Oremland R. S. and Capone D. G. (1988) Use of "specific" inhibitors in biogeochemistry and microbial ecology. *Adv. Microb. Ecol.* **10**, 285–383.

Oremland R. S. and Culbertson C. W. (1992a) Evaluation of methyl fluoride and dimethyl ether as inhibitors of aerobic methane oxidation. *Appl. Environ. Microbiol.* **58**, 883–892.

Oremland R. S. and Culbertson C. W. (1992b) Importance of methane-oxidizing bacteria in the methane budget as revealed by the use of a specific inhibitor. *Nature* **356**, 421–423.

Oremland R. S., Marsh L. M., and Polcin S. (1982) Methane production and simultaneous sulfate reduction in anoxic salt marsh sediments. *Nature* **296**, 143–145.

Oremland R. S., Kiene R. P., Whiticar M. J., and Boone D. R. (1989) Description of an estuarine methyltrophic methanogen which grows on dimethylsulfide. *Appl. Environ. Microbiol.* **55**, 994–1002.

Orphan V. J., Hinrichs K.-U., Ussler W. P. C. K., III, Taylor L. T., Sylva S. P., Hayes J. M., and DeLong E. F. (2001a) Comparative analysis of methane-oxidizing archaea and sulfate-reducing bacteria in anoxic marine sediments. *Appl. Environ. Microbiol.* **67**, 1922–1934.

Orphan V. J., House C. H., Hinrichs K. U., McKeegan K. D., and DeLong E. F. (2001b) Methane-consuming archaea revealed by directly coupled isotopic and phylogenetic analysis. *Science* **293**, 484–487.

Orphan V. J., House C. H., Hinrichs K.-U., McKeegan K. D., and DeLong E. F. (2002) Multiple archaeal groups mediate methane oxidation in anoxic cold seep sediments. *Proc. Natl. Acad. Sci. USA* **99**, 7663–7668.

Orth A. B., Roye D. J., and Tien D. M. (1993) Ubiquity of lignin degrading peroxidases among various wood degrading fungi. *Appl. Environ. Microbiol.* **59**, 4017–4023.

Otte S., Kuenen G., Nielsen L. P., Paerl H. W., Zopfi J., Schulz H. N., Teske A., Strotmann B., Gallardo V. A., and Jørgensen B. B. (1999) Nitrogen, carbon, and sulfur metabolism in natural *Thioplaca* samples. *Appl. Environ. Microbiol.* **65**, 3148–3157.

Otter L. B., Yang W. X., Scholes M. C., and Meixner F. X. (1999) Nitric oxide emissions from a southern African savanna. *J. Geophys. Res.* **104**(D15), 18471–18485.

Ottley C. J., Davison W., and Edmunds W. M. (1997) Chemical catalysis of nitrate reduction by iron(II). *Geochim. Cosmochim. Acta* **61**, 1819–1828.

Overmann J. (1997) Mahoney Lake: a case study of the ecological significance of phototrophic sulfur bacteria. *Adv. Microb. Ecol.* **15**, 251–288.

Overmann J. and van Gemerden H. (2000) Microbial interactions involving sulfur bacteria: implications for the ecology and evolution of bacterial communities. *FEMS Microbiol. Rev.* **24**, 591–599.

Pancost R. D., Damste J. S. S., deLint S., vanderMaarel M. J. E. C., and Gottschal J. C. (2000) Biomarker evidence for widespread anaerobic methane oxidation in Mediterranean sediments by a consortium of methanogenic archaea and bacteria. *Appl. Environ. Microbiol.* **66**, 1126–1132.

Pancost R. D., Hopmans E. C., and Sinninghe Damsté J. S. (2001) Archaeal lipids in Mediterranean cold seeps: molecular proxies for anaerobic methane oxidation. *Geochim. Cosmochim. Acta* **65**, 1611–1627.

Pankhania I. P., Spormann A. M., and Thauer R. K. (1988) Lactate conversion to acetate, CO_2 and H_2 in cell suspensions of *Desulfovibrio vulgaris* (Marburg): indications for the involvement of an energy driven reaction. *Arch. Microbiol.* **150**, 26–31.

Panter R. and Penzhorn R. (1980) Alkyl sulfonic acids in the atmosphere. *Atmos. Environ.* **14**, 149–151.

Pareek S., Azuma J. I., Matsui S., and Shimizu Y. (2001) Degradation of lignin and lignin model compound under sulfate-reducing condition. *Water Sci. Technol.* **44**, 351–358.

Parkes R., Gibson G., Mueller-Harvey I., Buckingham W., and Herbert R. A. (1989) Determination of the substrates for sulfphate-reducing bacteria within marine and estuarine sediments with different rates of sulphate reduction. *J. Gen. Microbiol.* **135**, 175–187.

Parkes R. J., Dowling N. J. E., White D. C., Herbert R. A., and Gibson G. R. (1993) Characterization of sulphate-reducing bacterial populations within marine and estuarine sediments with different rates of sulphate reduction. *FEMS Microbiol. Ecol.* **102**, 235–250.

Parkin T. B. (1987) Soil microsites as a source of denitrification variability. *Soil Sci. Soc. Am. J.* **51**, 1194–1199.

Parsons T. R., Stephens K., and Strickland J. D. H. (1961) On the chemical composition of eleven species of marine phytoplankton. *J. Fish. Res. Board Can.* **18**, 1001–1016.

Parton W. J., Mosier A. R., Ojima D. S., Valentine D. W., Schimel D. S., Weier K., and Kulmala A. E. (1996) Generalized model for N_2 and N_2O production from nitrification and denitrification. *Global Biogeochem. Cycles* **10**, 401–412.

Patrick W. H., Jr. and Henderson R. E. (1981) Reduction and reoxidation cycles of manganese and iron in flooded soil and in water solution. *Soil Sci. Soc. Am. J.* **45**, 855–859.

Payne W. J. (1981) *Denitrification*. Wiley, New York.

Peckmann J., Thiel V., Michaelis W., Clari P., Gaillard C., Martire L., and Reitner L. (1999) Cold seep deposits of Beauvoisin (Oxfordian; southeastern France) and Marmorito (Miocene; northern Italy): microbially induced authigenic carbonates. *Int. J. Earth Sci.* **88**, 60–75.

Pedersen O., Sand-Jensen K., and Revsbech N. P. (1995) Diel pulses of O_2 and CO_2 in sandy lake sediments inhabited by *Lobelia dortmanna*. *Ecology* **76**, 1536–1545.

Peine A., Tritschler A., Küsel K., and Peiffer S. (2000) Electron flow in an iron-rich acidic sediment–evidence for an acidity-driven iron cycle. *Limnol. Oceanogr.* **45**, 1077–1087.

Pelegrí S. P., Nielsen L. P., and Blackburn T. H. (1994) Denitrification in estuarine sediment stimulated by the irrigation activity of the amphipod *Corophium volutator*. *Mar. Ecol. Prog. Ser.* **105**, 285–290.

Pelsh A. D. (1936) About new autotrophic hydrogenthiobacteria (in Russian). *Trudy Solyanoi Laboratorii, vypusk M.-L., Izdatelstvo AN SSSR* **5**, 109–126.

Peterjohn W. T. (1991) Denitrification: enzyme content and activity in desert soils. *Soil Biol. Biochem.* **23**, 845–855.

Peters V. and Conrad R. (1996) Sequential reduction processes and initiation of CH_4 production upon flooding of oxic upland soils. *Soil Biol. Biochem.* **28**, 371–382.

Petit J. R., Jouzel J., Raynaud D., Barkov N. I., Barnola J. M., Vasile I., Bender M., Chappellaz J., Davis, Delaygue G., Delmotte M., Kotlyakov V. M., Legrand M., Lipenkov V. Y., Lorius C., Pepin L., Ritz C., Saltzman E., and Stievenard M. (1999) Climate and atmospheric history of the past 420,000 years from the Vostok ice core, Antarctica. *Nature* **399**, 429–436.

Pfennig N. and Biebl H. (1976) *Desulfuromonas acetoxidans* gen. nov. and sp. nov., a new anaerobic, sulfur-reducing, acetate-oxidizing bacterium. *Arch. Microbiol.* **110**, 3–12.

Phelps C. D. and Young L. Y. (1997) Microbial metabolism of the plant phenolic compounds ferulic and syringic acids under three anaerobic conditions. *Microbial. Ecol.* **33**, 206–215.

Phelps T. J., Conrad R., and Zeikus J. G. (1985) Sulfate-dependent interspecies H_2 transfer between Methanosarcina barkeri and Desulfovibrio vulgaris during coculture metabolism of acetate or methanol. *Appl. Environ. Microbiol.* **50**, 589–594.

Pierson B. K., Oesterle A., and Murphy G. L. (1987) Pigments, light penetration, and photosynthetic activity in the multi-layered microbial mats of Great Sippewissett Salt Marsh, Massachusetts. *FEMS Microbiol. Ecol.* **45**, 365–376.

Piluk J., Hartel P. G., and Haines B. L. (1998) Production of carbon disulfide (CS_2) from L-djenkolic acid in the roots of *Mimosa pudica* L. *Plant Soil* **200**, 27–32.

Pine M. J. and Barker H. A. (1956) Studies on methane fermentation: XII. The pathway of hydrogen in the acetate fermentation. *J. Bacteriol.* **71**, 644–648.

Platen H., Temmes A., and Schink B. (1990) Anaerobic degradation of acetone by *Desulfococcus biacutus* sp. nov. *Arch. Microbiol.* **154**, 335–361.

Ponnamperuma F. N. (1972) The chemistry of submerged soils. In *Advances in Agronomy* (ed. N. C. Brady). Academic Press, New York, pp. 29–96.

Popp T. J. and Chanton J. P. (1999) Methane stable isotope distribution at a Carex dominated fen in north central Alberta. *Global Biogeochem. Cycles* **13**, 1063–1077.

Postgate J. (1954) Presence of cytochrome in an obligate anaerobe. *Biochem. J.* **56**, 11–12.

Postgate J. R. (1984) *The Sulphate-reducing Bacteria*. Cambridge University Press, Cambridge.

Postgate J. R. and Campbell L. L. (1966) Classification of Desulfovibrio species, the nonsporulating sulfate-reducing bacteria. *Bacteriol. Rev.* **30**, 732–738.

Postma D. (1985) Concentration of Mn and separation from Fe in sediments: I. Kinetics and stoichiometry of the reaction between birnessite and dissolved Fe(II) at 10 °C. *Geochim. Cosmochim. Acta* **49**, 1023–1033.

Postma D. (1990) Kinetics of nitrate reduction in a sandy aquifer. *Geochim. Cosmochim. Acta* **54**, 903–908.

Postma D. (1993) The reactivity of iron oxides in sediments: a kinetic approach. *Geochim. Cosmochim. Acta* **57**, 5027–5034.

Postma D. and Jakobsen R. (1996) Redox zonation: equilibrium constraints on the Fe(III)/SO_4–reduction interface. *Geochim. Cosmochim. Acta* **60**, 3169–3175.

Poth M. (1986) Dinitrogen production from nitrite by a *Nitrosomonas* isolate. *Appl. Environ. Microbiol.* **52**, 957–959.

Poth M. and Focht D. D. (1985) ^{15}N kinetic analysis of N_2O production by *Nitrosomonas europaea*: an examination of nitrifier denitrification. *Appl. Environ. Microbiol.* **49**, 1134–1141.

Potter C. S., Matson P. A., Vitousek P. M., and Davidson E. A. (1996) Process modeling of control on nitrogen trace gas emissions from soils worldwide. *J. Geophys. Res.* **101**, 1361–1377.

Prather M., Ehhalt D., Dentener F., Derwent R., Dlugokencky E., Holland E., Isaksen I., Katima J., Kirchhoff V., Matson P., Midgley P., and Wang (2001) Atmospheric chemistry and greenhouse gases. In *Climate Change 2001: the Scientific Basis. Contribution of Working Group I to the Third Assessment Report of the Intergovernmental Panel on Climate Change* (eds. J. T. Houghton, Y. Ding, D. J. Griggs, M. Noguer, P. J. van der Linden, X. Dai, K. Maskell, and C. A. Johnson). Cambridge University Press, Cambridge, pp. 236–287.

Priemé A., Braker G., and Tiedje J. M. (2002) Diversity of nitrite reductase (*nirK* and *nirS*) gene fragments in forested upland and wetland soils. *Appl. Environ. Microbiol.* **68**, 1893–1900.

Pringault O. and GarciaPichel F. (2000) Monitoring of oxygenic and anoxygenic photosynthesis in a unicyanobacterial biofilm, grown in benthic gradient chamber. *FEMS Microbiol. Ecol.* **33**, 251–258.

Pronk J. T. and Johnson D. B. (1992) Oxidation and reduction of iron by acidophilic bacteria. *Geomicrobiol. J.* **10**, 153–171.

Pulliam W. M. (1993) Carbon dioxide and methane exports from a southeastern floodplain swamp. *Ecol. Monogr.* **63**, 29–53.

Pyzik A. J. and Sommer S. E. (1981) Sedimentary iron monosulfides: kinetics and mechanism of formation. *Geochim. Cosmochim. Acta* **45**, 687–698.

Rabalais N. N., Turner R. E., Dortch Q., Justic D., Bierman V. J., Jr., and Wiseman W. J. (2002) Nutrient-enhanced productivity in the northern Gulf of Mexico: past, present, and future. *Hydrobiologia* **475/476**, 39–63.

Rabus R., Hansen T., and Widdel F. (2000) Dissimilatory sulfate- and sulfur-reducing prokaryotes. In *The Prokaryotes: an Evolving Electronic Resource for the Microbiological Community* (ed. M. Dworkin, *et al.*). 3rd edn, release 3.3. Springer, New York., http://link.springer-ny.com/link/service/books/10125/.

Radajewski S., Ineson P., Parekh N. R., and Murrell J. C. (2000) Stable-isotope probing as a tool in microbial ecology. *Nature* **403**, 646–649.

Rajagopal B. S. and Daniels L. (1986) Investigations of mercaptans, organic sulfides, and inorganic sulfur sources for the growth of methanogenic bacteria. *Curr. Microbiol.* **14**, 137–144.

Ramaswamy V., Boucher O., Haigh J., Hauglustaine D., Haywood J., Myhre G., Nakajima T., Shi G. Y., and Solomon S. (2001) Radiative forcing of climate change. In *Climate Change 2001: the Scientific Basis. Contribution of Working Group I to the Third Assessment Report of the Intergovernmental Panel on Climate Change* (eds. J. T. Houghton, Y. Ding, D. J. Griggs, M. Noguer, P. J. van der Linden, X. Dai, K. Maskell, and C. A. Johnson). Cambridge University Press, Cambridge, pp. 236–287.

Ramsing N. B., Kuhl M., and Jørgensen B. B. (1993) Distribution of sulfate-reducing bacteria, O_2 and H_2S in photosynthetic biofilms determined by oligonucleotide probes and microelectrodes. *Appl. Environ. Microbiol.* **59**, 3840–3849.

Ramsing N. B., Fossing H., Ferdelman T. G., Andersen F., and Thamdrup B. (1996) Distribution of bacterial populations in a stratified fjord (Mariager Fjord, Denmark) quantified by *in situ* hybridization and related to chemical gradients in the water column. *Appl. Environ. Microbiol.* **62**, 1391–1404.

Ratering S. and Schnell S. (2000) Localization of iron-reducing activity in paddy soil by profile studies. *Biogeochemistry* **48**, 341–365.

Raven K. P., Jain A., and Loeppert R. H. (1998) Arsenite and arsenate adsorption on ferrihydrite: kinetics, equilibrium, and adsorption envelopes. *Environ. Sci. Technol.* **32**, 344–349.

Reddy K. R., Patrick W. H., Jr., and Lindau C. W. (1989) Nitrification-denitrification at the plant root-sediment interface in wetlands. *Limnol. Oceanogr.* **34**, 1004–1024.

Reeburgh W. S. (1976) Methane consumption in Cariaco Trench waters and sediments. *Earth Planet. Sci. Lett.* **28**, 337–344.

Reeburgh W. S. (1982) A major sink and flux control for methane in marine sediments: anaerobic consumption. In *The Dynamics of the Ocean Floor* (eds. K. A. Fanning and F. T. Manheim). D. C. Heath and Co, Lexington, MA, pp. 203–218.

Reeburgh W. S. (1996) 'Soft spots' in the global methane budget. In *Microbial Growth on C-1 Compounds* (eds. M. E. Lidstrom and F. R. Tabita). Kluwer, Dordrecht, The Netherlands, pp. 335–342.

Reeburgh W. S., Whalen S. C., and Alperin M. J. (1993) The role of methanotrophy in the global methane budget. In *Microbial Growth on C-1 Compounds* (eds. J. C. Murrell and D. P. Kelly). Intercept Press, Andover, UK, pp. 1–14.

Reichardt W. (1988) Impact of the Antarctic benthic fauna on the enrichment of biopolymer degrading pshychrotrophic bacteria. *Microb. Ecol.* **15**, 311–321.

Reineke W. (2001) Aerobic and anaerobic biodegradation potentials of microorganisms. In *Biodegradation and Persistence* (ed. B. Beek). Springer, Berlin, vol. 2, pp. 1–161.

Remde A. and Conrad R. (1991) Role of nitrification and denitrification for NO metabolism in soil. *Biogeochemistry* **12**, 189–205.

Ren T., Roy R., and Knowles R. (2000) Production and consumption of nitric oxide by three methanotrophic bacteria. *Appl. Environ. Microbiol.* **66**, 3891–3897.

Renault P. and Sierra J. (1994) Modeling oxygen diffusion in aggregated soils: II. Anaerobiosis in topsoil layers. *Soil Sci. Soc. Am. J.* **58**, 1023–1030.

Rice C. W. and Tiedje J. M. (1989) Regulation of nitrate assimilation by ammonium in soils and in isolated soil microorganisms. *Soil Biol. Biochem.* **21**, 597–602.

Rich J. J. and King G. M. (1999) Carbon monoxide consumption and production by wetland peats. *FEMS Microbiol. Ecol.* **28**, 215–224.

Richards F. A. (1965) Anoxic basins and fjords. In *Chemical Oceanography* (eds. J. P. Riley and G. Skirrow). Academic Press, New York, vol. 1, pp. 611–645.

Rimbault A., Niel P., Virelizier H., Darbord J.-C., and Leluan G. (1988) L-Methionine, a precursor of trace methane in some proteolytic Clostridia. *Appl. Environ. Microbiol.* **54**, 1581–1586.

Risatti J. B., Chapman W. C., and Stahl D. A. (1994) Community structure of a microbial mat: the phylogenetic dimension. *Proc. Natl. Acad. Sci. USA* **91**, 10173–10177.

Risgaard-Petersen N. S. and Jensen K. (1997) Nitrification and denitrification in the rhizosphere of the aquatic macrophyte *Lobelia dortmanna* L. *Limnol. Oceanogr.* **42**, 529–537.

Risgaard-Petersen N. S., Rysgaard S., Nielsen L. P., and Revsbech N. P. (1994) Diurnal variation of denitrification and nitrification in sediments colonized by benthic microphytes. *Limnol. Oceanogr.* **39**, 573–579.

Ritchie G. A. F. and Nicholas D. J. D. (1972) Identification of nitrous oxide produced by oxidative and reductive processes in *Nitrosomonas europaea*. *Biochem. J.* **126**, 1181–1191.

Riviera-Monroy V. H., Twilley R. R., Boustany R. G., Day J. W., Vera-Herrera F., and Ramirez M. C. (1995) Direct denitrification in mangrove sediments in Terminos Lagoon, Mexico. *Mar. Ecol. Prog. Ser.* **126**, 97–109.

Robertson G. P. and Tiedje J. M. (1987) Nitrous oxide sources in aerobic soils: nitrification, denitrification and other biological processes. *Soil Biol. Biochem.* **19**, 187–193.

Robertson L. A. and Kuenen J. G. (1991) Physiology of nitrifying and denitrifying bacteria. In *Microbial Production and Consumption of Greenhouse Gases: Methane Nitrogen Oxides and Halomethanes* (eds. J. E. Rogers and W. B. Whitman). American Society for Microbiology, Washington, DC, pp. 189–199.

Rocheleau S., Greer C. W., Lawrence J. R., Cantin C., Laramee L., and Guiot S. R. (1999) Differentiation of *Methanosaeta concilii* and *Methanosarcina barkeri* in anaerobic mesophilic granular sludge by fluorescent *in situ* hybridization and confocal scanning laser microscopy. *Appl. Environ. Microbiol.* **65**, 2222–2229.

Roden E. E. and Lovley D. R. (1993) Dissilatory Fe(III) reduction by the marine microorganism Desulfuromonas acetooxidans. *Appl. Environ. Microbiol.* **59**, 734–742.

Roden E. E. and Wetzel R. G. (1996) Organic carbon oxidation and suppression of methane production by microbial Fe(III) oxide reduction in vegetated freshwater wetland sediments. *Limnol. Oceanogr.* **41**, 1733–1748.

Roden E. E. and Wetzel R. G. (2002) Kinetics of microbial Fe(III) oxide reduction in freshwater wetland sediments. *Limnol. Oceanogr.* **47**, 198–211.

Roden E. E. and Zachara J. M. (1996) Microbial reduction of crystalline iron(III) oxides: influence of oxide surface area and potential for cell growth. *Environ. Sci. Technol.* **30**, 1618–1628.

Roden E. E., Sobolev D., Glazer B., and Luther G. W. (in press) New insights into the biogeochemical cycling of iron in circumneutral sedimentary environments: potential for a rapid microscale bacterial Fe redox cycle at the aerobic-anaerobic interface. In *Iron in the Natural Environment: Biogeochemistry, Microbial Diversity, and Bioremediation* (eds. J. D. Coates and C. Zhang). Kluwer, (in press).

Rodhe H. (1990) A comparison of the contribution of various gases to the greenhouse effect. *Science* **248**, 1217–1219.

Rodhe H. (1999) Human impact on the atmosphere sulfur balance. *Tellus A–B*, 110–122.

Rodhe H. and Isaksen I. (1980) Global distribution of sulfur compounds in treh troposphere estimated in a height/latitude transport model. *J. Geophys. Res.* **85**, 7401–7409.

Romanowicz E. A., Siegel D. I., and Glaser P. H. (1993) Hydraulic reversals and episodic methane emissions during drought cycles in mires. *Geology* **21**, 231–234.

Rooney-Varga J. N., Devereux R., Evans R. S., and Hines M. E. (1997) Seasonal changes in the relative abundance of uncultivated sulfate-reducing bacteria in a salt marsh sediment and rhizosphere of *Spartina alterniflora*. *Appl. Environ. Microbiol.* **63**, 3895–3901.

Rooney-Varga J. N., Anderson R. T., Fraga J. L., Ringelberg D., and Lovley D. R. (1999) Microbial communities associated with anaerobic benzene mineralization in a petroleum-contaminated aquifer. *Appl. Environ. Microbiol.* **65**, 3056–3063.

Rösch C., Mergel A., and Bothe H. (2002) Biodiversity of denitrifying bacteria and dinitrogen-fixing bacteria in an acid forest soil. *Appl. Environ. Microbiol.* **68**, 3818–3829.

Rosselló-Mora R. A., Ludwig W., Kempfer P., Amann R., and Schleifer K. H. (1995) *Ferrimonas balearica* gen. nov., spec. nov., a new marine facultative Fe(III)-reducing bacterium. *Syst. Appl. Microbiol.* **18**, 196–202.

Rothfuss F. and Conrad R. (1993) Vertical profiles of CH_4 concentrations, dissolved substrates and processes involved in CH_4 production in a flooded Italian rice field. *Biogeochemistry* **18**, 137–152.

Roulet N. T. and Moore T. R. (1995) The effect of forestry drainage practices on the emissions of methane from northern peatlands. *Can. J. Forest Res.* **25**, 491–499.

Roulet N. T., Ash R., Quinton W., and Moore T. (1993) Methane flux from a drained northern peatlands: effect of a persistent water table lowering on flux. *Global Biogeochem. Cycles* **7**, 749–769.

Roura-Carol M. and Freeman C. (1999) Methane release from peat soils: effects of Sphagnum and Juncus. *Soil Biol. Biochem.* **31**, 323–325.

Roy R. and Conrad R. (1999) Effect of methanogenic precursors (acetate, hydrogen, and propionate) on the suppression of methane production by nitrate in anoxic rice field soil. *FEMS Microbiol. Ecol.* **28**, 49–61.

Roy R., Klüber H. D., and Conrad R. (1997) Early initiation of methane production in anoxic rice soil despite the presence of oxidants. *FEMS Microbiol. Ecol.* **24**, 311–320.

Ruzicka S. D., Edgerton D., Norman M., and Hill T. (2000) The utility of ergosterol as a bioindicator of fungi in temperate soils. *Soil Biol. Biochem.* **32**, 989–1005.

Rysgaard S., Risgaard-Petersen N., Sloth N. P., Jensen K., and Nielsen L. P. (1994) Oxygen regulation of nitrification and denitrification in sediments. *Limnol. Oceanogr.* **39**, 1643–1652.

Rysgaard S., Risgaard-Petersen N., and Sloth N. P. (1996) Nitrification, denitrification and nitrate ammonification in two coastal lagoons in southern France. In *Coastal Lagoon Eutrophication and Anaerobic Processes* (eds. P. Caumette, J. Castel, and R. Herbert). Kluwer, Dordrecht, The Netherlands, pp. 133–144.

Saarnio S., Alm J., Martikainen P. J., and Silvola J. (1998) Effects of raised CO_2 on potential CH_4 production and oxidation in, and CH_4 emission from, a boreal mire. *J. Ecol.* **86**, 261–268.

Saarnio S., Saarinen T., Vasander H., and Silvola J. (2000) A moderate increase in the annual CH_4 efflux by raised CO_2 or NH_4NO_3 supply in a boreal oligotrophic mire. *Global Change Biol.* **6**, 137–144.

Sagan C., Thompson W. R., Carlson R., Gurnett D., and Hord C. (1993) A search for life on Earth from the Galileo spacecraft. *Nature* **365**, 715–721.

Sahm K., Knoblauch C., and Amann R. (1999a) Phylogenetic affiliation and quantification of psychrophilic sulfate-reducing isolates in marine Arctic sediments. *Appl. Environ. Microbiol.* **65**, 3976–3981.

Sahm K., MacGregor B. J., Jorgensen B. B., and Stahl D. A. (1999b) Sulphate reduction and vertical distribution of sulphate-reducing bacteria quantified by rRNA slot-blot hybridization in a coastal marine sediment. *Environ. Microbiol.* **1**, 65–74.

Samsonoff W. A. and MacColl R. (2001) Biliproteins and phycobilisomes from cyanobacteria and red algae at the extremes of habitat. *Arch. Microbiol.* **176**, 400–405.

Sansone F. J. and Martens C. S. (1982) Volatile fatty acid cycling in organic-rich marine sediments. *Geochim. Cosmochim. Acta* **45**, 101–121.

Santegoeds C. M., Damgaard L. R., Hesselink C., Zopfi J., Lens P., Muyzer G., and DeBeer D. (1999) Distribution of sulfate-reducing and methanogenic bacteria in anaerobic aggregates determined by microsensor and molecular analyses. *Appl. Environ. Microbiol.* **65**, 4618–4629.

Sass A., Rutters H., Cypionka H., and Sass H. (2002) *Desulfobulbus mediterraneus* sp. nov., a sulfate-reducing bacterium growing on mono- and disaccharides. *Arch. Microbiol.* **177**, 468–474.

Sass H., Cypionka H., and Babenzien H. D. (1997) Vertical distribution of sulfate-reducing bacteria at the oxic-anoxic interface in sediments of the oligotrophic Lake Stechlin. *FEMS Microbiol. Ecol.* **22**, 245–255.

Sass H., Wieringa E., Cypionka H., Babenzien H. D., and Overmann J. (1998) High genetic and physiological diversity of sulfate-reducing bacteria isolated from an oligotrophic lake sediment. *Arch. Microbiol.* **170**, 243–251.

Sayama M. and Kurihara Y. (1983) Relationship between activity of the Polychaetes annelid *Neanthes japonica* (Izuka) and nitrification–denitrification processes in the sediments. *J. Exp. Mar. Biol. Ecol.* **72**, 233–241.

Scheid D. and Stubner S. (2001) Structure and diversity of gram-negative sulfate-reducing bacteria on rice roots. *FEMS Microbiol. Ecol.* **36**, 175–183.

Schimel J. P. (1995) Plant transport and methane production as controls on methane flux from arctic wet meadow tundra. *Biogeochemistry* **28**, 183–200.

Schink B. (1984) Fermentation of 2.3-butanediol by *Pelobacter carbinolicus* sp. nov., and *Pelobacter propionicus*, sp. nov., and evidence for propionate formation from C_2 compounds. *Arch. Microbiol.* **137**, 33–41.

Schink B. (1985a) Fermentation of acetylene by an obligate anaerobe, *Pelobacter acetylenicus*, sp. nov. *Arch. Microbiol.* **142**, 295–301.

Schink B. (1985b) Mechanism and kinetics of succinate and propionate degradation in anoxic freshwater sediments and sewage slugde. *J. Can. Microbiol.* **131**, 643–650.

Schink B. (1997) Energetics of syntrophic cooperation in methanogenic degradation. *Microbiol. Mol. Biol. Rev.* **61**, 262–280.

Schink B. and Friedrich M. (1994) Energetics of syntrophic fatty acid oxidation. *FEMS Microbiol. Rev.* **15**, 85–94.

Schink B. and Stams A. J. M. (2002) Syntrophism among prokaryotes. In *The Prokaryotes: an Evolving Electronic Resource for the Microbiological Community* (ed. M. Dworkin, *et al.*). 3rd edn, release 3.8. Springer, New York, http://link.springer-ny.com/link/service/books/10125/.

Schink B., Phelps T. J., Eichler B., and Zeikus J. G. (1985) Comparison of ethanol degradation pathways in anoxic freshwater environments. *J. Gen. Microbiol.* **131**, 651–660.

Schink B., Thiemann V., Laue H., and Friedrich M. W. (2002) *Desulfotignum phosphitoxidans* sp. nov., a new marine sulfate reducer that oxidizes phosphite to phosphate. *Arch. Microbiol.* **177**, 381–391.

Schipper L. A., Cooper A. B., Harfoot C. G., and Dyck W. J. (1993) Regulators of denitrification in an organic riparian soil. *Soil Biol. Biochem.* **25**, 925–933.

Schipper L. A., Harfoot C. G., McFarlane P. N., and Cooper A. B. (1994) Anaerobic decomposition and denitrification during plant decomposition in an organic soil. *J. Environ. Qual.* **23**, 923–928.

Schippers A. and Jørgensen B. B. (2002) Biogeochemistry of pyrite and iron sulfide oxidation in marine sediments. *Geochim. Cosmochim. Acta* **66**, 85–92.

Schleifer K. H. and Ludwig W. (1989) Phylogenetic relationships among bacteria. In *The Hierarchy of Life* (eds. B. Fernholm, K. Bremer, and H. Jörnvall). Elsevier, Amsterdam, pp. 103–117.

Schlesinger W. H. (1997) *Biogeochemistry: an Analysis of Global Change*. Academic Press, San Diego.

Schmidt I. and Bock E. (1997) Anaerobic ammonia oxidation with nitrogen dioxide by *Nitrosomonas eutropha*. *Arch. Microbiol.* **167**, 106–111.

Schmidt I., Sliekers O., Schmid M., Cirpus I., Strous M., Bock E., Kuenen J. G., and Jetten M. S. M. (2002) Aerobic and anaerobic ammonia oxidizing bacteria—competitors or natural partners? *FEMS Microbiol. Ecol.* **39**, 175–181.

Schmidt J. E. and Ahring B. K. (1995) Interspecies electron transfer during propionate and butyrate degradation in mesophilic, granular sludge. *Appl. Environ. Microbiol.* **61**, 2765–2767.

Schmitz R. A., Daniel R., Deppenmeir U., and Gottschalk G. (2001) The anerobic way of life. In *The Prokaryotes: An Evolving Electronic Resource for the Microbiological Community* (eds. M. Dworkin, *et al.*). 3rd edn., release 3.5. Springer, New York, http://link.springer-ny.com/link/service/books/10125/.

Schnell S. and King G. M. (1994) Mechanistic analysis of ammonium inhibition of atmospheric methane consumption in forest soils. *Appl. Environ. Microbiol.* **60**, 3514–3521.

Schoell M. (1988) Multiple origins for methane in the earth. *Chem. Geol.* **71**, 1–10.

Scholten J. C. M., Murrell J. C., and Kelly D. P. (2003) Growth of sulfate-reducing bacteria and methanogenic archaea with methylated sulfur compounds: a commentary on the thermodynamic aspects. *Arch. Microbiol.* **179**, 135–144.

Schrenk M. O., Edwards K. J., Goodman R. M., Hamers R. J., and Banfield J. F. (1998) Distribution of *Thiobacillus ferrooxidans* and *Leptospirillum ferroxidans*: implications for generation of acid mine drainage. *Science* **279**, 1519–1522.

Schrope M. K., Chanton J. P., Allen L. H., and Baker J. T. (1999) Effect of CO_2 enrichment and elevated temperature on methane emissions from rice, *Oryza sativa*. *Global Change Biol.* **5**, 587–599.

Schultz S. and Conrad R. (1996) Influence of temperature on pathways to methane production in the permanently cold profundal sediment of Lake Constance. *FEMS Microbiol. Ecol.* **20**, 1–14.

Schultz S., Matsuyama H., and Conrad R. (1997) Temperature dependence of methane production from different precursors in a profundal sediment (Lake Constance). *FEMS Microbiol. Ecol.* **22**, 207–213.

Schulz H. N. and Jorgensen B. B. (2001) Big bacteria. *Ann. Rev. Microbiol.* **55**, 105–137.

Schulz H. N., Brinkhoff T., Ferdelamn T. G., Hernandez M., Teske A., and Jørgensen B. B. (1999) Dense populations of a giant sulfur bacterium in Namibian Shelf sediments. *Science* **284**, 493–495.

Schulz S. and Conrad R. (1995) Effect of algal deposition on acetate and methane concentrations in the profundal sediment of a deep lake (Lake Constance). *FEMS Microbiol. Ecol.* **16**, 251–259.

Schütz H., Schröder P., and Rennenberg H. (1991) Role of plants in regulating the methane flux to the atmosphere. In *Trace Gas Emissions by Plants* (eds. T. D. Sharkey, E. A. Holland, and H. A. Mooney). Academic Press, pp. 29–63.

Schwertmann U. and Cornell R. M. (1991) *Iron Oxides in the Laboratory*. VCH, Weinheim, Germany.

Scott D. T., McKnight D. M., Blunt-Harris E. L., Kolesar S. E., and Lovley D. R. (1998) Quinone moieties act as electron acceptors in the reduction of humic substances by humic-reducing microorganisms. *Environ. Sci. Technol.* **32**, 2984–2989.

Seefeldt L. C., Rasche M. E., and Ensign S. A. (1995) Carbonyl sulfide and carbon dioxide as new substrates, and carbon disulfide as a new inhibitor, of nitrogenase. *Biochemistry* **34**, 5382–5389.

Seeliger S., Cord-Ruwisch R., and Schink B. (1998) A periplasmatic and extracellular c-type cytochrome of *Geobacter sulfurreducens* acts as a ferric iron reductase and as an electron carrier to other acceptors or to partner bacteria. *J. Bacteriol.* **180**, 3686–3691.

Segerer A., Stetter K. O., and Klink F. (1985) Two contrary modes of chemolithotrophy in the same archaebacterium. *Nature* **313**, 787–789.

Segers R. (1998) Methane production and methane consumption: a review of processes underlying wetland methane fluxes. *Biogeochemistry* **41**, 23–51.

Seitz H. J. and Cypionka H. (1986) Chemolithotrophic growth of Desulfovibrio desulfuricans with hydrogen coupled to ammonification of nitrate and nitrite. *Arch. Microbiol.* **146**, 63–67.

Seitzinger S. P. (1988) Denitrification in freshwater and coastal marine ecosystems: ecological and geochemical significance. *Limnol. Oceanogr.* **33**, 702–724.

Seitzinger S. P. (1994) Linkages between organic matter mineralization and denitrification in eight riparian wetlands. *Biogeochemistry* **25**, 19–39.

Seitzinger S. P. and Giblin A. E. (1996) Estimating denitrification in North Atlantic continental shelf sediments. *Biogeochemistry* **35**, 235–260.

Seitzinger S. P., Nielsen L. P., Caffrey J., and Christensen P. B. (1993) Denitrification measurements in aquatic sediments: a comparison of three methods. *Biogeochemistry* **23**, 147–167.

Senko J. M. and Stolz J. F. (2001) Evidence for iron-dependent nitrate respiration in the dissimulatory iron-reducing bacterium Geobacter metallireducens. *Appl. Environ. Microbiol.* **67**, 3750–3752.

Senn D. B. and Hemond H. F. (2002) Nitrate controls on iron and arsenic in an urban lake. *Science* **296**, 2373–2376.

Sexstone A. J., Parkin T. P., and Tiedje J. M. (1985) Temporal response of soil denitrification to rainfall and irrigation. *Soil Sci. Soc. Am. J.* **49**, 99–103.

Shannon R. D. and White J. R. (1994) A three-year study of controls on methane emissions from two Michigan peatlands. *Biogeochemistry* **27**, 35–60.

Shannon R. D. and White J. R. (1996) The effects of spatial and temporal variations in acetate and sulfate on methane cycling in two Michigan peatlands. *Limnol. Oceanogr.* **41**, 435–443.

Shen Y. A., Buick R., and Canfield D. E. (2001) Isotopic evidence for microbial sulphate reduction in the early Archaean era. *Nature* **410**, 77–81.

Sherwood Lollar B., Westgate T. D., Ward J. A., Slater G. F., and Lacrampe-Couloume G. (2002) Abiogenic formation of alkanes in the Earth's crust as a minor source for global hydrocarbon reservoirs. *Nature* **416**, 522–524.

Shiba H., Kawasumi T., Igarashi Y., and Minoda Y. (1985) The CO_2 assimilation via the reductive tricarboxylic acid cycle in an obligately autotrophic aerobic hydrogen-oxidizing bacterium, *Hydrogenobacter thermophilus*. *Arch. Microbiol.* **141**, 198–203.

Shoun H. and Tanimoto T. (1991) Denitrification by the fungus *Fusarium oxysporum* and involvement of cytochrome P-450 in the respiratory nitrite reduction. *J. Biol. Chem.* **266**, 11078–11082.

Shrestha N. K., Hadano S., Kamachi T., and Okura I. (2002) Dinitrogen production from ammonia by *Nitrosomonas europaea*. *Appl. Catal. A: General* **237**, 33–39.

Sievert S. M. and Kuever J. (2000) *Desulfacinum hydrothermale* sp. nov., a thermophilic, sulfate-reducing bacterium from geothermally heated sediments near Milos Island (Greece). *Int. J. Syst. Evol. Microbiol.* **50**, 1239–1246.

Sigalevich P. and Cohen Y. (2000) Oxygen-dependent growth of the sulfate-reducing bacterium *Desulfovibrio oxyclinae* in coculture with *Marinobacter* sp. strain MB in an aerated sulfate-depleted chemostat. *Appl. Environ. Microbiol.* **66** 5019 + .

Sigalevich P., Baev M. V., Teske A., and Cohen Y. (2000a) Sulfate reduction and possible aerobic metabolism of the sulfate-reducing bacterium *Desulfovibrio oxyclinae* in a chemostat coculture with *Marinobacter* sp. strain MB under exposure to increasing oxygen concentrations. *Appl. Environ. Microbiol.* **66** 5013 + .

Sigalevich P., Meshorer E., Helman Y., and Cohen Y. (2000b) Transition from anaerobic to aerobic growth conditions for the sulfate-reducing bacterium *Desulfovibrio oxyclinae* results in flocculation. *Appl. Environ. Microbiol.* **66** 5005 + .

Sigren L. K., Byrd G. T., Fisher F. M., and Sass R. L. (1997) Comparison of soil acetate concentrations and methane production, transport, and emission in two rice cultivars. *Global Biogeochem. Cycles* **11**, 1–14.

Silver W. L., Herman D. J., and Firestone M. K. (2001) Dissimilatory nitrate reduction to ammonium in upland tropical forest soils. *Ecology* **82**, 2410–2416.

Šimek M., Jísová L., and Hopkins D. W. (2002) What is the so-called optimum pH for denitrification in soil? *Soil Biol. Biochem.* **34**, 1227–1234.

Singer P. C. and Stumm W. (1970) Acid mine drainage: the rate-determining step. *Science* **167**, 1121–1123.

Sinke A. J. C., Cornelese A. A., Cappenberg T. E., and Zehnder A. J. B. (1992) Seasonal variation in sulfate reduction and methanogenesis in peaty sediments of eutrophic Lake Loosdrecht, The Netherlands. *Biogeochemistry* **16**, 43–61.

Sinninghe Damsté J. S., Strous M., Rijpstra W. I. C., Hopmans E. C., Geenevasen J. A. J., van Duin A. C. T., van Niftrik L. A., and Jetten M. S. M. (2002) Linearly concatenated cyclobutane lipids from a dense bacterial membrane. *Nature* **419**, 708–712.

Sivelä S. and Sundmann V. (1975) Demonstration of Thiobacillus-type bacteria, which ultilize methyl sulphides. *Arch. Microbiol.* **103**, 303–304.

Skiba U., Fowler D., and Smith K. A. (1997) Nitric oxide emissions from agricultural soils in temperate and tropical climates: sources, controls and mitigation options. *Nutr. Cycl. Agroecosyst.* **48**, 139–153.

Skyring G. W. (1987) Sulfate reduction in coastal ecosystems. *Geomicrobiol. J.* **5**, 295–374.

Slater J. M. and Capone D. G. (1987) Denitrification in aquifer soils and nearshore marine sediments influenced by groundwater nitrate. *Appl. Environ. Microbiol.* **53**, 1292–1297.

Slobodkin A., Campbell B., Cary S. C., Bonch-Osmolovskaya E., and Jeanthon C. (2001) Evidence for the presence of thermophilic Fe(III)-reducing microorganisms in deep-sea hydrothermal vents at 13°N (East Pacific Rise). *FEMS Microbiol. Ecol.* **36**, 235–243.

Smemo K. and Yavitt J. B. (2000) Evidence for anaerobic methane oxidation in freshwater peatlands. *EOS Trans., AGU* **81**(48) Fall Meeting Supplement, B71C-02.

Smith M. S. and Parsons L. L. (1985) Persistence of denitrifying enzyme activity in dried soils. *Appl. Environ. Microbiol.* **49**, 316–320.

Smith N. A. and Kelly D. P. (1988a) Isolation and physiological characterization of autotrophic sulphur bacteria oxidizing dimethyl disulphide as sole source of energy. *J. Gen. Microbiol.* **134**, 1407–1417.

Smith N. A. and Kelly D. P. (1988b) Oxidation of carbon disulphide as the sole source of energy for the autotrophic growth of Thiobacillus thioparus strain TK-m. *J. Gen. Microbiol.* **134**, 3041–3048.

Smith R. L. and Duff J. H. (1988) Denitrification in a sand and gravel aquifer. *Appl. Environ. Microbiol.* **54**, 1071–1078.

Smith R. L. and Klug M. J. (1981) Electron donors utilized by sulfate-reducing bacteria in eutrophic lake sediments. *Appl. Environ. Microbiol.* **42**, 116–121.

Smith R. L., Howes B. L., Garabedian S. P. (1991) In situ measurement of methane oxidation in groundwater by using natural-gradient tracer tests. *Appl. Environ. Microbiol.* **57**, 1997–2004.

Smolders A. J. P., Lamers L. P. M., Moonen M., Zwaga K., and Roelofs J. G. M. (2001) Controlling phosphate release form phosphate-enriched sediments by adding various iron compounds. *Biogeochemistry* **54**, 219–228.

Snoeyenbos-West O. L., Nevin K. P., and Lovley D. R. (2000) Stimulation of dissimulatory Fe(III) reduction results in a predominance of Geobacter species in a variety of sandy aquifers. *Microb. Ecol.* **39**, 153–167.

Sobolev D. and Roden E. E. (2001) Suboxic deposition of ferric iron by bacteria in oppposing gradients of Fe(II) and oxygen at circumneutral pH. *Appl. Environ. Microbiol.* **67**, 1328–1334.

Sobolev D. and Roden E. E. (2002) Evidence for rapid microscale bacterial redox cycling of iron in circumneutral environments. *Antonie Van Leeuwenhoek* **81**, 587–597.

Sobolev D. and Roden E. E. (2003) Characterization of a neutrophilic, chemolithotrophic Fe(II)-oxidizing β-proteobacterium from freshwater wetland sediments. *Geomicrobiol. J.* (in press).

Sørensen J. (1978) Capacity for denitrification and reduction of nitrate to ammonia in a coastal marine sediment. *Appl. Environ. Microbiol.* **35**, 301–305.

Sørensen J. (1982) Reduction of ferric iron in anaerobic, marine sediment and interaction with reduction of nitrate and sulfate. *Appl. Environ. Microbiol.* **43**, 319–324.

Sørensen J. and Thorling L. (1991) Stimulation by lepidocrocite (Upsilon-FeOOH) of Fe(II)-dependent nitrite reduction. *Geochim. Cosmochim. Acta* **55**, 1289–1294.

Sørensen J., Christensen D., and Jørgensen B. B. (1981) Volatile fatty acids and hydrogen as substrates for sulfate-reducing bacteria. *Appl. Environ. Microbiol.* **42**, 5–11.

Sørensen K. B., Finster K., and Ramsing N. B. (2001) Thermodynamic and kinetic requirements in anaerobic methane oxidizing consortia exclude hydrogen, acetate, and methanol as possible electron shuttles. *Microbial. Ecol.* **42**, 1–10.

Sorokin D. Y. and Lysenko A. M. (1993) Heterotrophic bacteria from the Black Sea oxidizing reduced sulfur compounds to sulfate. *Microbiology (English Translation)* **62**, 594–602.

Sorokin Y. I. (1962) Experimental investigations of bacterial sulfate reduction in the Black Sea using ^{35}S. *Mikrobiologiya* **31**, 329–335.

Sorokin Y. I. (1972) The bacterial population and the process of hydrogen sulfide oxidation in the Black Sea. *J. Cons. Int. Explor. Mer.* **34**, 423–454.

Sorrell B. K., Downes M. T., and Stanger C. L. (2002) Methanotrophic bacteria and their activity on submerged aquatic macrophytes. *Aquat. Bot.* **72**, 107–119.

Southward E. C., Gebruk A., Kennedy H., Southward A. J., and Chevaldonne P. (2001) Different energy sources for three symbiont-dependent bivalve molluscs at the Logatchev hydrothermal site (Mid-Atlantic Ridge). *Mar Biol. Assoc. UK* **81**, 655–661.

Spalding R. F. and Exner M. E. (1993) Occurrence of nitrate in groundwater–a review. *J. Environ. Qual.* **22**, 393–402.

Stams A. J. M. and Hansen T. A. (1984) Fermentation of glutamate and other compounds by *Acidaminobacter hydrogenoformans* gen. nov. sp. nov., an obligate anaerobe isolated from black mud. Studies with pure cultures and mixed cultures with sulfate-reducing and methanogenic bacteria. *Arch. Microbiol.* **137**, 329–337.

Stanford G., Legg J. O., Dzienia S., and Simpson E. C. (1975) Denitrification and associated nitrogen transformations in soils. *Soil Sci.* **120**, 147–152.

Steingruber S. M., Friedrich J., Gächter R., and Wehrli B. (2001) Measurement of denitrification in sediments with the ^{15}N isotope pairing technique. *Appl. Environ. Microbiol.* **67**, 3771–3778.

Stetter K. O. (1988) *Archaeoglobus fulgidus gen. nov.*, sp. nov.: a new taxon of extremely thermophilic archaebacteria. *Syst. Appl. Microbiol.* **10**, 171–173.

Stetter K. O. (1996) Hyperthermophilic prokaryotes. *FEMS Microbiol. Rev.* **18**, 149–158.

Stetter K. O. (1997) Primative archaea and bacteria in the cycles of sulfur and nitrogen near the temperature limit of life. In *Progress in Microbial Ecology. Proceedings of the 7th International Symposium on Microbial Ecology, Santos, Sao Paulo, Brazil* (ed. M. T. Martins). SBM/COME pp. 55–61.

Stetter K. O. and Gaag G. (1983) Reduction of molecular sulphur by methanogenic archaea. *Nature* **305**, 309–311.

Stetter K. O., Lauerer G., Thomm M., and Neuner A. (1987) Isolation of extremely thermophilic sulfate reducers: evidence for a novel brance of archaebacteria. *Science* **236**, 822–824.

Stetter K. O., Huber R., Bloechl E., Kurr M., Eden R. D., Fielder M., Cash H., and Vance I. (1993) Hyperthermophilic archaea are thriving in deep North Sea and Alaskan oil reservoirs. *Nature* **365**, 743–745.

Steudel R., Holdt G., Goebel T., and Hazeu W. (1987) Chromatographic separation of higher polythionates SnO_6^{2-} ($n = 3...22$) and their detection in cultures of Thiobacillus ferrooxidans: molecular composition of bacterial sulfur secretions. *Angew. Chemie. Int. Ed. Engl.* **26**, 151–153.

Steudler P. A., Bowden R. D., Melillo J. M., and Aber J. D. (1989) Influence of nitrogen fertilization on methane uptake in temperate forest soils. *Nature* **341**, 314–316.

Stevens R. J., Laughlin R. J. B. L. C., Arah J. R. M., and Hood R. C. (1997) Measuring the contributions of nitrification and denitrification to the flux of nitrous oxide from soil. *Soil Biol. Biochem.* **29**, 139–151.

Stevenson F. J. (1994) *Humus Chemistry*. Wiley, New York.

Stolz J. F. (1991) The ecology of phototrophic bacteria. In *Structure of Phototrophic Prokaryotes* (ed. J. F. Stolz). CRC Press, Boca Raton, FL, pp. 105–123.

Stolz J. F. and Oremland R. S. (1999) Bacterial respiration of arsenic and selenium. *FEMS Microbiol. Rev.* **23**, 615–627.

Stone A. T. (1987) Microbial metabolites and the reductive dissolution of manganese oxides. *Geochim. Cosmochim. Acta* **51**, 919–925.

Stone A. T. (1997) Reactions of extracellular organic ligands with dissolved metal ions and mineral surfaces. *Rev. Mineral.* **35**, 309–344.

Stone A. T., Godtfredsen K. L., and Deng B. (1994) Sources and reactivity of reductants encountered in aquatic environments. In *Chemistry of Aquatic Systems: Local and Global Perspectives* (eds. G. Bidoglio and W. Stumm). Kluwer, Boston, pp. 337–374.

Straub K. L., Benz M., Schlink B., and Widdel F. (1996) Anaerobic, nitrate-dependent microbial oxidation of ferrous iron. *Appl. Environ. Microbiol.* **62**, 1458–1460.

Straub K. L. and Buchholz-Cleven B. E. E. (1998) Enumeration and detection of anaerobic ferrous iron-oxidizing, nitrate-reducing bacteria from diverse European sediments. *Appl. Environ. Microbiol.* **64**, 4846–4856.

Straub K. L., Hanzlik M., and Buchholz-Cleven B. E. E. (1998) The use of biologically produced ferrihydrite for the isolation of novel iron-reducing bacteria. *Syst. Appl. Microbiol.* **21**, 442–449.

Straub K. L., Rainey F. A., and Widdel F. (1999) *Rhodovulum iodosum* sp. nov. and *Rhodovulum robiginosum* sp. nov., two new marine phototrophic ferrous-iron-oxidizing purple bacteria. *Int. J. Systemat. Bacteriol.* **49**, 729–735.

Straub K. L., Benz M., and Schink B. (2001) Iron metabolism in anoxic environments at near neutral pH. *FEMS Microbiol. Ecol.* **34**, 181–186.

Strauss G. and Fuchs G. (1993) Enzymes of a novel autotrophic CO_2 fixation pathway in the phototrophic bacterium *Chloroflexus aurantiacus*, the 3-hydroxypropionate cycle. *Euro. J. Biochem.* **215**, 633–643.

Strong D. T. and Fillery I. R. P. (2002) Denitrification response to nitrate concentrations in sandy soils. *Soil Biol. Biochem.* **34**, 945–954.

Strous M., Fuerst J. A., Kramer E. H. M., Logemann S., Muyzer G., van de Pas-Schoonen K. T., Webb R., Kuenen J. G., and Jetten M. S. M. (1999a) Missing lithotroph identified as new planctomycete. *Nature* **400**, 446–449.

Strous M., Kuenen J. G., and Jetten M. S. M. (1999b) Key physiology of anaerobic ammonium oxidation. *Appl. Environ. Microbiol.* **65**, 3248–3250.

Stucki J. W., Bailey G. W., and Gan H. (1996) Oxidation-reduction mechanisms in iron-bearing phyllosilicates. *Appl. Clay Sci.* **10**, 417–430.

Stumm W. and Morgan J. J. (1981) *Aquatic Chemistry*. Wiley, New York.

Sugimoto A. and Wada E. (1993) Carbon isotopic composition of bacterial methane in a soil incubation experiment: contributions of acetate and CO_2/H_2. *Geochim. Cosmochim. Acta* **57**, 4015–4027.

Summers D. P. and Chang S. (1993) Prebiotic ammonia from reduction of nitrite by iron(II) on the early Earth. *Nature* **365**, 630–633.

Sunda W., Kieber D. J., Kiene R. P., and Huntsman S. (2002) An antioxidant function for DMSP and DMS in marine algae. *Nature* **418**, 317–320.

Sundh I., Nilsson M., Granberg G., and Svensson B. H. (1994) Depth distribution of microbial production and oxidation of methane in northern boreal peatlands. *Microb. Ecol.* **27**, 253–265.

Sundh I., Mikkelä C., Nilsson M., and Svensson B. H. (1995) Potential aerobic methane oxiation in *Sphagnum*-dominated peatland—controlling factors and relation to methane emission. *Soil Biol. Biochem.* **27**, 829–837.

Sweerts J. P. R. A., De Beer D., Nielsen L. P., Verdouw H., Van den Heuvel J. C., Cohen Y., and Cappenberg T. E. (1990) Denitrification by sulphur oxidizing *Beggiatoa* spp. mats on freshwater sediments. *Nature* **344**, 762–763.

Sylvia D. M., Fuhrmann J. J., Hartel P. G., and Zuberer D. A. (1998) *Principles and Applications of Soil Microbiology*. Prentice-Hall.

Takai K. and Horikoshi K. (1999) Genetic diversity of archaea in deep-sea hydrothermal vent environments. *Genetics* **152**, 1285–1297.

Takai Y., Koyama T., and Kamura T. (1963a) Microbial metabolism in reduction process of paddy soils (Part 2). *Soil Sci. Plant Nutr. (Tokyo)* **9**, 176–185.

Takai Y., Koyama T., and Kamura T. (1963b) Microbial metabolism in reduction process of paddy soils (Part 3). *Soil Sci. Plant Nutr. (Tokyo)* **9**, 207–211.

Tallant T. C. and Krzycki J. A. (1997) Methiol: coenzyme M methyltransferase from *Methanosarcina barkeri*, an enzyme of methanogenesis from dimethylsulfide and methylmercaptopropionate. *J. Bacteriol.* **179**, 6902–6911.

Tallant T. C., Paul L., and Krzycki J. A. (2001) The MtsA subunit of the methylthiol: coenzyme M methyltransferase of *Methanosarcina barkeri* catalyses both half-reactions of corrinoid-dependent dimethylsulfide: coenzyme M methyl transfer. *J. Biol. Chem.* **276**, 4485–4493.

Taillefert M., Bono A. B., and Luther G. W., III (2000) Reactivity of freshly formed Fe(III) in synthetic solutions and (pore)waters: voltammetric evidence of an aging process. *Environ. Sci. Technol.* **34**, 2169–2177.

Tanimoto T., Hatano K., Kim D., Uchiyama H., and Shoun H. (1992) Co-denitrification by the denitrifying system of the fungus Fusarium oxysporum. *FEMS Microbiol. Lett.* **93**, 177–180.

Tanimoto Y. and Bak F. (1994) Anaerobic degradation of methylmercaptan and dimethyl sulfide by newly isolated thermophilic sulfate-reducing bacteria. *Appl. Environ. Microbiol.* **60**, 2450–2455.

Tanji K. K., Gao S., Scardaci S. C., and Chow A. T. (2003) Characterizing redox status of paddy soils with incorporated rice straw. *Geoderma* **114**, 333–354.

Tasaki M., Kamagata Y., Nakamura K., and Mikami E. (1992) Propionate formation from alcohols or aldehydes by

Desulfobulbus propionicus in the absence of sulfate. *J. Ferment. Bioeng.* **73**, 329–331.

Taylor B. F. and Oremland R. S. (1979) Depletion of adenosine triphosphate in *Desulfovibrio* by oxyanions of Group VI elements. *Curr. Microbiol.* **3**, 101–103.

Tebo B. M. and He L. M. (1999) Microbially mediated oxidation precipitation reactions. In *Mineral-Water Interfacial Reactions* (eds. D. L. Sparks and T. J. Grundl). American Chemical Society, Washington, DC, pp. 393–414.

Tebo B. M. and Obraztsova A. Y. (1998) Sulfate-reducing bacterium grows with Cr(VI), U(VI), Mn(IV), and Fe(III) as electron acceptors. *FEMS Microbiol. Lett.* **162**, 193–198.

Tebo B. M., Ghiorse W. C., van Wassbergen L. G., Siering P. L., and Caspi R. (1997) Bacterially-mediated mineral formation: insights into manganese(II) oxidation from molecular genetic and biochemical studies. *Rev. Mineral.* **35**, 225–266.

Teske A., Wawer C., Muyzer G., and Ramsing N. B. (1996) Distribution of sulfate-reducing bacteria in a stratified fjord (Mariager Fjord, Denmark) as evaluated by most-probable-number counts and denaturing gradient gel electrophoresis of PCR-amplified ribosomal DNA fragments. *Appl. Environ. Microbiol.* **62**, 1405–1415.

Teske A., Kuver J., Jorgensen B. B., Cohen Y., Ramsing N. B., Habicht K., and Fukui M. (1998) Sulfate-reducing bacteria and their activities in cyanobacterial mats of Solar Lake (Sinai, Egypt). *Appl. Environ. Microbiol.* **64**, 2943–2951.

Teske A., Hinrichs K.-U., Edgcomb V., Gomez A. V., Kysela D., Sylva S. P., Sogin M. L., and Jannasch H. W. (2002) Microbial diversity of hydrothermal sediments in the Guaymas Basin: evidence for anaerobic methanotrophic communities. *Appl. Environ. Microbiol.* **68**, 1994–2007.

Thamdrup B. (2000) Bacterial manganese and iron reduction in aquatic sediments. *Adv. Microb. Ecol.* **16**, 41–84.

Thamdrup B. and Canfield D. E. (1996) Pathways of carbon oxidation in continental margin sediments off central Chile. *Limnol. Oceanogr.* **41**, 1629–1650.

Thamdrup B. and Canfield D. E. (2000) Benthic respiration in aquatic sediments. In *Methods in Ecosystem Science* (eds. O. E. Sala, R. B. Jackson, H. A. Mooney, and R. W. Howarth) Springer, New York, pp. 86–103.

Thamdrup B. and Dalsgaard T. (2000) The fate of ammonium in anoxic manganese oxide-rich sediment. *Geochim. Cosmochim. Acta* **64**, 4157–4164.

Thamdrup B. and Dalsgaard T. (2002) Production of N_2 through anaerobic ammonium oxidation coupled to nitrate reduction in marine sediments. *Appl. Environ. Microbiol.* **68**, 1312–1318.

Thamdrup B., Finster K., Hansen J. W., and Bak F. (1993) Bacterial disproportionation of elemental sulfur coupled to chemical reduction of iron or manganese. *Appl. Environ. Microbiol.* **59**, 101–108.

Thamdrup B., Rosselő-Mora R., and Amman R. (2000) Microbial manganese and sulfate reduction in Black Sea shelf sediments. *Appl. Environ. Microbiol.* **66**, 2888–2897.

Thauer R. K., Jungermann K., and Dekker K. (1977) Energy conservation in chemotrophic anaerobic bacteria. *Bacteriol. Rev.* **41**, 100–180.

Thebrath B., Mayer H. P., and Conrad R. (1992) Bicarbonate-dependent production and methanogenic consumption of acetate in anoxic paddy soil. *FEMS Microbiol. Lett.* **86**, 295–302.

Thiel V., Peckmann J., Seifert R., Wehrung P., Reitner J., and Michaelis W. (1999) Highly isotopically depleted isoprenoids: molecular markers for ancient methane venting. *Geochim. Cosmochim. Acta* **63**, 3959–3966.

Thiel V., Peckmann J., Richnow H. H., Luth U., Reitner J., and Michaelis W. (2001) Molecular signals for anaerobic methane oxidation in Black Sea seep carbonates and a microbial mat. *Mar. Chem.* **73**, 97–112.

Thiele J. H. and Zeikus J. G. (1988) Control of interspecies electron flow during anaerobic digestion: significance of formate transfer versus hydrogen transfer during syntrophic methanogenesis in flocs. *Appl. Environ. Microbiol.* **54**, 20–29.

Thompson A. M., Witte J. C., Hudson R. D., Guo H., Herman J. R., and Fujiwara M. (2001) Tropical tropospheric ozone and biomass burning. *Science* **291**, 2128–2132.

Thomsen T. R., Finster K., and Ramsing N. B. (2001) Biogeochemical and molecular signatures of anaerobic methane oxidation in a marine sediment. *Appl. Environ. Microbiol.* **67**, 1646–1656.

Thorpe R. B., Law K. S., Bekki S., and Pyle J. A. (1996) Is methane-driven deglaciation consistent with the ice-core record? *J. Geophys. Res.* **101**, 28627–28635.

Tiedje J. M. (1982) Denitrification. In *Methods of Soil Analysis, Part 2. Agronomy Monograph No. 9* (eds. A. L. Page, R. H. Miller, and D. R. Keeney). American Society of Agronomy, Madison, Wisconsin, pp. 1011–1026.

Tiedje J. M. (1988) Ecology of denitrification and dissimilatory nitrate reduction to ammonium. In *Biology of Anaerobic Microorganisms* (ed. A. J. B. Zehnder). Wiley, New York, pp. 179–244.

Tiedje J. M. (1994) Denitrifiers. In *Methods of Soil Analysis, Part 2* (eds. R. W. Weaver, *et al.*). Soil Science Society of America, Madison, Wisconsin, pp. 245–267.

Tiedje J. M., Sexstone A. J., Myrold D. D., and Robinson J. A. (1982) Denitrification: ecological niches, competition and survival. *Antonie Van Leeuwenhoek* **48**, 569–583.

Timmer-Ten Hoor A. (1975) A new type of thiosulphate oxidizing, nitrate reducing microorganism: Thiomicrospira denitrificans. *Neth. J. Sea Res.* **9**, 351–353.

Tobias C. R., Anderson I. C., Canuel E. A., and Macko S. A. (2001a) Nitrogen cycling through a fringing marsh-aquifer ecotone. *Mar. Ecol. Prog. Ser.* **210**, 25–39.

Tobias C. R., Macko S. A., Anderson I. C., Canuel E. A., and Harvey J. W. (2001b) Tracking the fate of a high concentration groundwater nitrate plume through a fringing marsh: a combined groundwater tracer and in situ isotope enrichment study. *Limnol. Oceanogr.* **46**, 1977–1989.

Toh S. K., Webb R. I., and Ashbolt N. J. (2002) Enrichment of autotrophic anaerobic ammonium-oxidizing consortia from various wastewaters. *Microb. Ecol.* **43**, 154–167.

Trettin C. C. and Jurgensen M. F. (2003) Carbon cycling in wetland forest soils. In *The Potential of US Forest Soils to Sequester Carbon and Mitigate the Greenhouse Effect* (eds. J. M. Kimble, L. S. Heath, R. A. Birdsey, and R. Lal). CRC Press, Boca Raton, FL, pp. 311–331.

Trimmer M., Nedwell D. B., Sivyer D. B., and Malcolm S. J. (1998) Nitrogen fluxes through the lower estuary of the river Great Ouse, England: the role of bottom sediments. *Mar. Ecol. Prog. Ser.* **163**, 109–124.

Tsutsuki K., Esaki I., and Kuwatsuka S. (1994) CuO-oxidation products of peat as a key to the analysis of the paleo-environmental changes in a wetland. *Soil Sci. Plant Nutr.* **40**, 107–116.

Tugel J. B., Hines M. B., and Jones G. E. (1986) Microbial iron reduction by enrichment cultures isolated from estuarine sediments. *Appl. Environ. Microbiol.* **52**, 1167–1172.

Tunnicliffe V. (1992) Hydrothermal-vent communities of the deep sea. *Am. Sci.* **80**, 336–349.

Turner F. T. and Patrick W. H., Jr. (1968) Chemical changes in waterlogged soils as a result of oxygen depletion. *Trans. 9th Cong. Soil Sci.* **4**, 53–65.

Tuttle J. H. and Jannasch H. W. (1977) Thiosulfate stimulation of dark assimilation of carbon dioxide in shallow marine environments. *Microb. Ecol.* **4**, 9–25.

Tuttle J. H. and Jannasch H. W. (1979) Microbial dark assimilation of CO_2 in the Cariaco Trench. *Limnol. Oceanogr.* **24**, 747–753.

Tyler S. C. (1991) The global methane budget. In *Microbial Production of Greenhouse Gases: Methane, Nitrogen Oxides, and Halomethanes* (eds. J. E. Rogers and W. B. Whitman). American Society of Microbiology, Washington, DC, pp. 7–38.

Tyler S. C. (1992) Kinetic isotope effects and their use in studying atmospheric trace species. In *Isotope Effects in Gas Phase Chemistry* (ed. J. Kaye). American Chemical Society, Washington, DC, pp. 390–408.

Tyler S. C., Brailsford G. W., Yagi K., Minami K., and Cicerone R. J. (1994) Seasonal variations in methane flux and $d^{13}CH_4$ values for rice paddies in Japan and their implications. *Global Biogeochem. Cycles* **8**, 1–12.

Tyrrell T. (1999) The relative influences of nitrogen and phosphorus on oceanic primary production. *Nature* **400**, 525–531.

Ueki A., Ono K., Tsuchiya A., and Ueki K. (1997) Survival of methanogens in air-dried paddy field soil and their heat tolerance. *Water Sci. Technol.* **36**, 517–522.

Updergraff K., Pastor J., Bridgham S. D., and Johnston C. A. (1995) Environmental and substrate controls over carbon and nitrogen mineralization in northern wetlands. *Ecol. Appl.* **5**, 151–163.

Updergraff K., Bridgham S. D., Pastor J., and Weishampel P. H. C. (2001) Ecosystem respiration response to warming and water-table manipulations in peatland mesocosms. *Ecol. Appl.* **11**, 311–326.

Urban N. R., Brezonik P. L., Baker L. A., and Sherman L. A. (1994) Sulfate reduction and diffusion in sediments of Little Rock Lake, Wisconsin. *Limnol. Oceanogr.* **39**, 797–815.

Urrutia M. M., Roden E. E., and Zachara J. M. (1999) Influence of aqueous and solid-phase Fe(II) complexants on microbial reduction of crystalline iron(III) oxides. *Environ. Sci. Technol.* **33**, 4022–4028.

Usuda K., Toritsuka N., Matsuo Y., Kim D. H., and Shoun H. (1995) Denitrification by the fungus *Cylindrocarpon tonkinense*: anaerobic cell growth and two isozyme forms of cytochrome P-450nor. *Appl. Environ. Microbiol.* **61**, 883–889.

Valentine D. L. (2001) Thermodynamic ecology of hydrogen-based syntrophy. In *Symbiosis: Mechanisms and Model Systems* (ed. J. Seckback). Kluwer, Dordrecht, The Netherlands, pp. 147–161.

Valentine D. L. (2002) Biogeochemistry and microbial ecology of methane oxidation in anoxic environments: a review. *Antonie Van Leeuwenhoek* **81**, 271–282.

Valentine D. L. and Reeburg W. S. (2000) New perspectives on anaerobic methane oxidation. *Environ. Microbiol.* **2**, 477–484.

Valentine D. W., Holland E. A., and Schimel D. S. (1994) Ecosystem and physiological controls over methane production in northern wetlands. *J. Geophys. Res.* **99**, 1563–1571.

Valentine D. L., Blanton D. C., and Reeburgh W. S. (2000) Hydrogen production by methanogens under low-hydrogen conditions. *Arch. Microbiol.* **174**, 415–421.

van Bodegom P., Stams F., Mollema L., Boeke S., and Leffelaar P. (2001) Methane oxidation and the competition for oxygen in the rice rhizosphere. *Appl. Environ. Microbiol.* **67**, 3586–3597.

van Breemen N., Boyer E. W., Goodale C. L., Jaworski N. A., Paustian K., Seitzinger S. P., Lajtha K., Mayer B., van Dam D., Howarth R. W., Nadelhoffer K. J., Eve M., and Billen G. (2002) Where did all the nitrogen go? Fate of nitrogen inputs to large watersheds in the northeastern USA. *Biogeochemistry* **57/58**, 267–293.

van Cappellen P., viollier E., Roychoudhury A., Clark L., Ingall E., Lowe K., and DiChristina T. (1998) Biogeochemical cycles of manganese and iron at the oxic-anoxic transition of a stratified marine basin (Orca Basin, Gulf of Mexico). *Environ. Sci. Technol.* **32**, 2931–2939.

van Cleemput O. and Baert L. (1984) Nitrite: a key compound in N loss processes under acid conditions? *Plant Soil* **76**, 233–241.

van de Graaf A. A., Mulder A., de Bruijn P., Jetten M. S. M., Roberston L. A., and Kuenen J. G. (1995) Anaerobic oxidation of ammonium is a biologically mediated process. *Appl. Environ. Microbiol.* **61**, 1246–1251.

van de Graaf A. A., de Bruijn P., Roberston L. A., Jetten M. S. M., and Kuenen J. G. (1997) Metabolic pathway of anaerobic ammonium oxidation on the basis of ^{15}N studies in a fluidized bed reactor. *Microbiology* **143**, 2415–2421.

Van den Ende F. P. and Van Gemerden H. (1993) Sulfide oxidation under oxygen limitation by a *Thiobacillus thioparus* isolated from a marine microbial mat. *FEMS Microbiol. Ecol.* **13**, 69–78.

Van den Ende F. P., Laverman A. M., and Van Gemerden H. (1996) Coexistence of aerobic chemotrophic and anaerobic phototrophic sulfur bacteria under oxygen limitation. *FEMS Microbiol. Ecol.* **19**, 141–151.

van den Pol-van Dasselaar A. and Oenema O. (1999) Methane production and carbon mineralisation of size and density fractions of peat soils. *Soil Biol. Biochem.* **31**, 877–886.

van der Maarel M., Jansen M., and Hansen T. A. (1995) Methanogenic conversion of 3-S-methylmercaptopropionate to 3-mercaptopropionate. *Appl. Environ. Microbiol.* **61**, 48–51.

van der Maarel M., Jansen M., Haanstra R., Meijer W. G., and Hansen T. A. (1996a) Demethylation of dimethylsulfoniopropionate to 3-S-methylmercaptopropionate by marine sulfate-reducing bacteria. *Appl. Environ. Microbiol.* **62**, 3978–3984.

van der Maarel M. J. E. C., Quist P., Dijkhuizen L., and Hansen T. (1993) Degradation of dimethylsulfoniopropionate to 3-S-methylmercaptopropionate by a marine *Desulfobacterium* strain. *Arch. Microbiol.* **60**, 411–412.

van der Maarel M. J. E. C., Aukema W., and Hansen T. A. (1996b) Purification and characterization of a dimethylsulfoniopropionate cleaving enzyme from *Desulfovibrio acrylicus. FEMS Microbiol. Lett.* **143**, 241–245.

van der Maarel M. J. E. C., Jansen M., Haanstra R., Meijer W. G., and Hansen T. A. (1996c) Demethylation of dimethylsulfoniopropionate to 3-S-methylmercaptopropionate by marine sulfate-reducing bacteria. *Appl. Environ. Microbiol.* **62**, 3978–3984.

van der Maarel M. J. E. C., van Bergeijk S., van Werkhoven A., Laverman A. M., Meijer W. G., Stam W., and Hansen T. (1996d) Cleavage of dimethylsulfonioproionate and reduction of acrylate by *Desulfvibrio acrylicus* sp. nov. *Arch. Microbiol.* **166**, 109–115.

van der Maarel M. J. E. C., Jansen M., Jonkers H. M., and Hansen T. A. (1998) Demethylation and cleavage of dimethylsulfoniopropionate and reduction of dimethyl sulfoxide by sulfate-reducing bacteria. *Geomicrobiol. J.* **15**, 37–44.

van der Nat F. J. W. A. and Middelburg J. J. (1998) Seasonal variation in methane oxidation by the rhizosphere of *Phragmites australis* and *Scirpus lacustris. Aquat. Bot.* **61**, 95–110.

van der Nat F. J. W. A., Brouwer J. F. C., Middelburg J. J., and Laanbroek H. J. (1997) Spatial distribution and inhibition by ammonium of methane oxidation in intertidal freshwater marshes. *Appl. Environ. Microbiol.* **63**, 4734–4740.

Van Gemerden H. (1993) Microbial mats: a joint venture. *Mar. Geol.* **113**, 3–25.

van Hulzen J. B., Segers R., van Bodegom P. M., and Leffelaar P. A. (1999) Temperature effects on soil methane production: an explanation for observed variability. *Soil Biol. Biochem.* **31**, 1919–1929.

van Kuijk B. L. M. and Stams A. J. M. (1995) Sulfate reduction by a syntrophic proionate-oxidizing bacterium. *Antonie Van Leeuwenhoek* **68**, 293–296.

van Mooy B. A. S., Keil R. G., and Devol A. H. (2002) Impact of suboxia on sinking particulate organic carbon: enhanced carbon flux and preferential degradation of amino acids via denitrification. *Geochim. Cosmochim. Acta* **66**, 457–465.

Vann C. D. and Megonigal J. P. (2003) Elevated CO_2 and water depth regulation of methane emissions: comparison of woody and non-woody wetland plant species. *Biogeochemistry* **63**, 117–134.

Van Niel C. B. (1931) On the morphology and physiology of the purple and green sulphur bacteria. *Arch. Microbiol.* **3**, 1–112.

Van Niel C. B. (1941) The bacterial photosyntheses and their importance for the general problem of photosynthesis. *Adv. Enzymol.* **1**, 263–328.

Vargas M., Kashefi K., Blunt-Harris, and Lovley D. R. (1998) Microbiological evidence for Fe(III) reduction on early Earth. *Nature* **395**, 65–67.

Verstraete W. and Philips S. (1998) Nitrification-denitrification processes and technologies in new contexts. *Environ. Pollut.* **102**(S1), 717–726.

Verville J. H., Hobbie S. E., Chapin F. S., and Hooper D. U. (1998) Response of tundra CH_4 and CO_2 flux to manipulation of temperature and vegetation. *Biogeochemistry* **41**, 215–235.

Vishniac W. and Santer M. (1957) The thiobacilli. *Bacteriol. Rev.* **21**, 195–213.

Visscher P. T. and Van Gemerden H. (1991) Photoautotrophic growth of *Thiocapsa roseopersicina* on dimethyl sulfide. *FEMS Microbiol. Lett.* **81**, 247–250.

Visscher P. T. and Taylor B. F. (1993a) Aerobic and anaerobic degradation of a range of alkyl sulfides by a denitrifying marine bacterium. *Appl. Environ. Microbiol.* **59**, 4083–4089.

Visscher P. T. and Taylor B. F. (1993b) A new mechanism for the aerobic catabolism of dimethyl sulfide. *Appl. Environ. Microbiol.* **59**, 3784–3789.

Visscher P. T. and Taylor B. F. (1993c) Organic thiols as organolithotrophic substrates for growth of phototrophic bacteria. *Appl. Environ. Microbiol.* **59**, 93–96.

Visscher P. T. and Taylor B. F. (1994) Demethylation of dimethylsulfoniopropionate to 3-mercaptopropionate by an aerobic marine bacterium. *Appl. Environ. Microbiol.* **60**, 4617–4619.

Visscher P. T. and Van den Ende F. (1994) Diel and spatial fluctuations of sulfur transformations. In *Microbial Mats. Structure, Development, and Environmental Significance* (eds. L. J. Stal and P. Caumette). Springer, Berlin, pp. 353–360.

Visscher P. T. and Van Gemerden H. (1993) Sulfur cycling in laminated marine ecosystems. In *Biogeochemistry of Global Change: Radiatively Active Trace Gases* (ed. R. S. Oremland). Chapman and Hall, New York, pp. 672–693.

Visscher P. T., Quist P., and Gemerden H. (1991) Methylated sulfur compounds in microbial mats: *In situ* concentrations and metabolism by a colorless sulfur bacterium. *Appl. Environ. Microbiol.* **57**, 1758–1763.

Visscher P. T., Vandenende F. P., Schaub B. E. M., and Vangemerden H. (1992) Competition between anoxygenic phototrophic bacteria and colorless sulfur bacteria in a microbial mat. *FEMS Microbiol. Ecol.* **101**, 51–58.

Visscher P. T., Taylor B. F., and Kiene R. P. (1995) Microbial consumption of dimethyl sulfide and methanethiol in coastal marine sediments. *FEMS Microbiol. Ecol.* **18**, 145–154.

Visscher P. T., Macintyre I. G., Thompson J. A., Jr., Reid R. P., Bebout B. M., and Hoeft S. E. (1998) Formation of lithified micritic laminae in modern marine stromatolites (Bahamas): the role of sulfur cycling. *Am. Mineral.* **83**, 1482–1493.

Visscher P. T., Gritzer R. F., and Leadbetter E. R. (1999) Low-molecular-weight sulfonates, a major substrate for sulfate reducers in marine microbial mats. *Appl. Environ. Microbiol.* **65**, 3272–3278.

Vitousek P. M. and Howarth R. W. (1991) Nitrogen limitation on land and in the sea: how can it occur? *Biogeochemistry* **13**, 87–116.

Vitousek P. M., Aber J. D., Howarth R. W., Likens G. E., Matson P. A., Schindler D. W., Schlesinger W. H., and Tilman G. D. (1997) Human alterations of the global nitrogen cycle: causes and consequences. *Ecol. Appl.* **7**, 737–750.

Vogels G. D., Keltjens J. T., and van der Drift C. (1988) Biochemistry of methane production. In *Biology of Anaerobic Microorganisms* (ed. A. J. B. Zehnder). Wiley, New York.

von Fischer J. C. and Hedin L. O. (2002) Separating methane production and consumption with a field-based isotope pool dilution technique. *Global Biogeochem. Cycles* **16**(8), 1–13.

Voordouw G. (1995) Mini review—the genus *Desulfovibrio*: the centennial. *Appl. Environ. Microbiol.* **61**, 2813–2819.

Voordouw G., Voordouw J. K., Karkhoffschweizer R. R., Fedorak P. M., and Westlake D. W. S. (1991) Reverse sample genome probing, a new technique for identification of bacteria in environmental samples by DNA hybridization, and its application to the identification of sulfate-reducing bacteria in oil field samples. *Appl. Environ. Microbiol.* **57**, 3070–3078.

Vourlitis G. L. and Oechel W. C. (1997) The role of northern ecosystems in the global methane budget. In *Global Change and Arctic Terrestrial Ecosystems* (eds. W. C. Oechel, T. Callaghan, T. Gilmanov, J. I. Holten, B. Maxwell, U. Molau, and B. Sveinbjörnsson). Springer, New York, pp. 266–289.

Wagner C. and Stadtman E. R. (1962) Bacterial fermentation of dimethyl-B-propiothetin. *Arch. Biochem. Biophys.* **98**, 331–336.

Wagner M., Roger A. J., Flax J. L., Brusseau G. A., and Stahl D. A. (1998) Phylogeny of dissimilatory sulfite reductases supports an early origin of sulfate respiration. *J. Bacteriol.* **180**, 2975–2982.

Wainwright M. (1981) Assay and properties of alginate lyase and 1.3-β-glucanase in intertidal sands. *Plant Soil* **59**, 83–89.

Wakeham S. G., Howes B. L., and Dacey J. W. H. (1984) Dimethyl sulphide in a stratified coastal salt pond. *Nature* **310**, 770–772.

Walker D. A. a. o. (1998) Energy and trace-gas fluxes across a soil pH boundary in the Arctic. *Nature* **394**, 469–472.

Wallrabenstein C., Hauschild E., and Schink B. (1994) Pure culture and cytological properties of *Syntrophobacter wolinii. FEMS Microbiol. Lett.* **123**, 249–254.

Wallrabenstein C., Hauschild E., and Schink B. (1995) *Syntrophobacter pfennigii* sp. nov., new syntrophically propionate-oxidizing anaerobe growing in pure culture with propionate and sulfate. *Arch. Microbiol.* **131**, 360–365.

Wang K. and Lewis T. J. (1992) Geothermal evidence from Canada for a cold period before recent climatic warming. *Science* **256**, 1003–1005.

Warburg O. and Negelein E. (1923) Über den Energieumsatz der Kohlensäureassimilation. *Z. Phys. Chem.* **102**, 235–266.

Watanabe A., Katoh K., and Kimura M. (1993) Effect of rice straw application on CH_4 emission from paddy fields: 2. Contribution of organic constituents in rice straw. *Soil Sci. Plant Nutr.* **39**, 707–712.

Watanabe A., Satoh Y., and Kimura M. (1995) Estimation of the increase in CH_4 emission from paddy soils by rice straw application. *Plant Soil* **173**, 225–231.

Wawer C., Jetten M. S. M., and Muyzer G. (1997) Genetic diversity and expression of the [NiFe] hydrogenase large-subunit gene of *Desulfovibrio* spp. in environmental samples. *Appl. Environ. Microbiol.* **63**, 4360–4369.

Weaver B. L. and Tarney J. (1984) Empirical approach to estimating the composition of the continental crust. *Nature* **310**, 575–577.

Weber K. A., Picardal F. W., and Roden E. E. (2001) Microbially catalyzed nitrate-dependent oxidation of biogenic solid-phase Fe(II) compounds. *Environ. Sci. Technol.* **35**, 1644–1650.

Webster E. A. and Hopkins D. W. (1996) Contributions from different microbial processes to N_2O emission from soil under different moisture regimes. *Biol. Fertility Soils* **22**, 331–335.

Weiner J. M. and Lovley D. R. (1998) Rapid benzene degradation in methanogenic sediments from a petroleum-contaminated aquifer. *Appl. Environ. Microbiol.* **64**, 1937–1939.

Weiss J. V. (2002) Microbially-mediated iron cycling in the rhizosphere of wetland plants. PhD Dissertation, George Mason University.

Weiss J. V., Emerson D. E., Backer S. M., and Megonigal J. P. (2003) Enumeration of Fe(II)-oxidizing and Fe(III)-reducing bacteria in the root zone of wetland plants: implications for a rhizosphere iron cycle. *Biogeochemistry* **64**, 77–96.

Weiss M. S., Abele U., Weckesser J., Welte W., Schiltz E., and Schulz G. E. (1991) Molecular architecture and electrostatic properties of a bacterial porin. *Science* **254**, 1627–1630.

Westrich J. T. (1983) The consequences and control of bacterial sulfate reduction in marine sediments. PhD Dissertation, Yale University.

Westrich J. T. and Berner R. A. (1984) The role of sedimentary organic matter in bacterial sulfate reduction: the G model tested. *Limnol. Oceanogr.* **29**, 236–249.

Wetzel R. G. (1992) Gradient-dominated ecosystems: sources and regulatory functions of dissolved organic matter in freshwater systems. *Hydrobiologia* **229**, 181–198.

Whalen S. C., Reeburgh W. S., and Sandbeck K. A. (1990) Rapid methane oxidation in a landfill cover soil. *Appl. Environ. Microbiol.* **56**, 3405–3411.

Whiticar M. J. (1999) Carbon and hydrogen isotope systematics of bacterial formation and oxidation of methane. *Chem. Geol.* **161**, 291–314.

Whiticar M. J., Faber E., and Schoell M. (1986) Biogenic methane formation in marine and freshwater environments: CO_2 reduction vs. acetate fermentation: isotope evidence. *Geochim. Cosmochim. Acta* **50**, 693–709.

Whiting G. J. and Chanton J. P. (1992) Plant-dependent CH_4 emission in a subarctic Canadian fen. *Global Biogeochem. Cycles* **6**, 225–231.

Whiting G. J. and Chanton J. P. (1993) Primary production control of methane emission from wetlands. *Nature* **364**, 794–795.

Whiting G. J. and Chanton J. P. (2001) Greenhouse carbon balance of wetlands: methane emission versus carbon sequestration. *Tellus* **53B**, 521–528.

Whiting G. J., Chanton J. P., Bartlett D. S., and Happell J. D. (1991) Relationships between CH_4 emission, biomass, and CO_2 exchange in a subtropical grassland. *J. Geophys. Res.* **96**, 13067–13071.

Widdel F. (1988) Microbiology and ecology of sulfate- and sulfur-reducing bacteria. In *Biology of Anaerobic Microorganisms* (ed. A. J. B. Zehnder). Wiley, New York, pp. 469–585.

Widdel F. and Bak F. (1992) Gram-negative mesophilic sulfate-reducing bacteria. In *The Procaryotes* (eds. A. Balows, H. G. Trüper, M. Dworkin, W. Harder, and K.-H. Schleifer). Springer, New York, pp. 583–624.

Widdel F. and Pfennig N. (1977) A new anaerobic, sproring, acetate-oxidizing, sulfate-reducing bacterium, *Desulfotomaculum (emend.) acetoxidans. Arch. Microbiol.* **112**, 119–122.

Widdel F. and Pfennig N. (1982) Studies on dissimilatory suflate-reducing bacteria that decompose fatty acids: II. Incomplete oxidation of propionate by *Desulfobulbus propionicus* gen. nov., sp. nov. *Arch. Microbiol.* **131**, 360–365.

Widdel F. and Pfennig N. (1999) The genus *Desulfuromonas* and other gram-negative sulfur-reducing eubacteria. In *The Prokaryotes: an Evolving Electronic Resource for the Microbiological Community* (eds. M. Dworkin, *et al.*). 3rd edn, release 3.0. Springer, New York, http://link.springer-ny.com/link/service/books/10125/.

Widdel F., Schnell S., Heising S., Ehrenreich A., Assmus B., and Schink B. (1993) Ferrous iron oxidation by anoxygenic phototrophic bacteria. *Nature* **362**, 834–836.

Wieder R. K. and Lang G. E. (1988) Cycling of inorganic and organic sulfur in peat from Big Run Bog, West Virginia. *Biogeochemistry* **5**, 221–242.

Wieder R. K. and Yavitt J. B. (1994) Peatlands and global climate change: insights from comparative studies of sites situated along a latitudinal gradient. *Wetlands* **14**, 229–238.

Wieringa E. B. A., Overmann J., and Cypionka H. (2000) Detection of abundant sulphate-reducing bacteria in marine oxic sediment layers by a combined cultivation and molecular approach. *Environ. Microbiol.* **2**, 417–427.

Wildenauer F. X. and Winter J. (1986) Fermentation of isoleucine and arginine by pure and syntrophic cultures of *Clostridium sporogenes. FEMS Microbiol. Ecol.* **38**, 373–379.

Williams E. J., Hutchinson G. L., and Fehsenfeld F. C. (1992) NO_x and N_2O emissions from soil. *Global Biogeochem. Cycles* **6**, 351–388.

Winogradsky S. (1887) Über Schwefelbakterien. *Botanische Zeitung* **45**, 489–523.

Winter J., Schindler F., and Wildenauer F. X. (1987) Fermentation of alanine and glycine by pure and syntrophic cultures of *Clostridium sporogenes. FEMS Microbiol. Ecol.* **45**, 153–161.

Witkamp M. and Ausmus B. S. (1976) Processes in decomposition and nutrient transfer in forest systems. In *The Role of Terrestrial and Aquatic Organisms in Decomposition Processes* (eds. J. M. Anderson and A. Macfadyen). Blackwell, Edinburg, pp. 75–396.

Woese C. R. (1987) Bacterial evolution. *Microbiol. Rev.* **51**, 221–271.

Wolfe R. S. (1996) 1776–1996: Alessandro Volta's combustible air. *ASM News* **62**, 529–534.

Wood A. P. and Kelly D. P. (1989) Isolation and physiological characterization of *Thiobacillus thyasyris sp. nov.*, a novel marine facultative autotroph and putative symbiont of *Thyasira flexuosa. Arch. Microbiol.* **152**, 160–162.

Wood H. G. (1985) Then and now. *Ann. Rev. Biochem.* **54**, 1–41.

Wood W. T., Gettrust J. F., Chapman N. R., Spence G. D., and Hyndman R. D. (2002) Decreased stability of methane hydrates in marine sediments owing to phase-boundary roughness. *Nature* **420**, 656–660.

Wrage N., Velthof G. L. V., Beusichem M. L., and Oenema O. (2001) Role of nitrifier denitrification in the production of nitrous oxide. *Soil Biol. Biochem.* **33**, 1723–1732.

Wu X. L., Chin K. J., Stubner S., and Conrad R. (2001) Functional patterns and temperature response of cellulose-fermenting microbial cultures containing different methanogenic communities. *Appl. Microbiol. Biotechnol.* **56**, 212–219.

Wu Z., Daniel S. L., Hsu T., and Drake H. L. (1988) Characterization of a CO-dependent O-demethylating enzyme system from the acetogen *Clostridium thermoaceticum. J. Bacteriol.* **170**, 5705–5708.

Yang W. X. and Meixner F. X. (1997) Laboratory studies on the release of nitric oxide from sub-tropical grassland soils: the effect of soil temperature and moisture. In *Gaseous Nitrogen Emissions from Grasslands* (eds. S. C. Jarvis and B. F. Pain). CAB International, Wallingford, UK, pp. 67–71.

Yao H. and Conrad R. (1999) Thermodynamics of methane production in different rice paddy soils from China, the Philippines and Italy. *Soil Biol. Biochem.* **31**, 463–473.

Yao H., Conrad R., Wassmann R., and Neue H. U. (1999) Effect of soil characteristics on sequential reduction and methane production in sixteen rice paddy soils from China, the Philippines, and Italy. *Biogeochemistry* **47**, 267–293.

Yavitt J. B. and Lang G. E. (1990) Methane production in contrasting wetland sites: response to organic-chemical components of peat and to sulfate reduction. *Geomicrobiol. J.* **8**, 27–46.

Yavitt J. B., Lang G. E., and Wieder R. K. (1987) Control of carbon mineralization to CH_4 and CO_2 in anaerobic, Sphagnum-derived peat from Big Run Bog, West Virginia. *Biogeochemistry* **4**, 141–157.

Yavitt J. B., Lang G. E., and Downey D. M. (1988) Potential methane production and methane oxidation rates in the peatland ecosystems of the Appalachian mountains, United States. *Global Biogeochem. Cycles* **2**, 253–268.

Yavitt J. B., Downey D. M., Lancaster E., and Lang G. E. (1990) Methane consumption in decomposing *Sphagnum*-derived peat. *Soil Biol. Biochem.* **22**, 441–447.

Yavitt J. B., Williams C. J., and Wieder R. K. (1997) Production of methane and carbon dioxide in peatland ecosystems across north America: effects of temperature, aeration, and organic chemistry of peat. *Geomicrobiol. J.* **14**, 299–316.

Ye R. W., Averill B. A., and Tiedje J. M. (1994) Denitrification: production and consumption of nitric oxide. *Appl. Environ. Microbiol.* **60**, 1053–1058.

Yin S. X., Chen D., Chen L. M., and Edis R. (2002) Dissimilatory nitrate reduction to ammonium and responsible microorganisms in two Chinese and Australian paddy soils. *Soil Biol. Biochem.* **34**, 1131–1137.

Yoch D. C. (2002) Dimethylsulfoniopropionate: its sources, role in the marine food web, and biological degradation to dimethylsulfide. *Appl. Environ. Microbiol.* **68**, 5804–5815.

Yoch D. C., Carraway R. H., Friedman R., and Kulkarni N. (2001) Dimethylsulfide (DMS) production from dimethylsulfoniopropionate by freshwater river sediments: phylogeny of gram-positive DMS-producing isolates. *FEMS Microbiol. Ecol.* **37**, 31–37.

Young L. Y. and Frazer A. C. (1987) The fate of lignin and lignin-derived compounds in anaerobic environments. *Geomicrobiol. J.* **5**, 261–293.

Yurkov V. and Van Gemerden H. (1993) Impact of light/dark regime on growth rate, biomass formation and bacteriochlorophyll synthesis in *Erythromicrobium hydrolyticum*. *Arch. Microbiol.* **159**, 84–89.

Zachara J. M., Fredrickson J. K., Li S. M., Kennedy D. W., Smith S. C., and Gassman P. L. (1998) Bacterial reduction of crystalline Fe^{3+} oxides in single phase suspensions and subsurface materials. *Am. Mineral.* **83**, 1426–1443.

Zachara J. M., Fredrickson J. K., Smith S. C., and Gassman P. L. (2001) Solubilization of Fe(III) oxide-bound trace metals by a dissimulatory Fe(III) reducing bacterium. *Geochim. Cosmochim. Acta* **65**, 75–93.

Zachara J. M., Kukkadapu R. K., Fredrickson J. M., Gorby Y. A., and Smith S. C. (2002) Biomineralization of poorly crystalline Fe(III) oxides by dissimulatory metal reducing bacteria (DMRB). *Geomicrobiol. J.* **19**, 179–207.

Zart D. and Bock E. (1998) High rates of aerobic nitrification and denitrification by *Nitrosomonas eutropha* grown in a fermentor with complete biomass retention in the presence of gaseous NO_2 or NO. *Arch. Microbiol.* **169**, 282–286.

Zehnder A. J. and Brock T. D. (1979) Methane formation and methane oxidation by methanogenic bacteria. *J. Bacteriol.* **137**, 420–432.

Zehnder A. J. B. and Stumm W. (1988) Geochemistry and biogeochemistry of anaerobic habitats. In *Biology of Anaerobic Microorganisms* (ed. A. J. B. Zehnder). Wiley, New York.

Zehr J. P. and Ward B. B. (2002) Nitrogen cycling in the ocean: new perspectives on processes and paradigms. *Appl. Environ. Microbiol.* **68**, 1015–1024.

Zeikus J. G. (1977) The biology of methanogenic bacteria. *Bacteriol. Rev.* **41**, 514–541.

Zengler K., Richnow H. H., Rosselló-Mora R., Michaelis W., and Widdel F. (1999) Methane formation from long-chain alkanes by anaerobic microorganisms. *Nature* **401**, 266–269.

Zeyer J., Eicher P., Wakeham S. G., and Schwarzenbach R. P. (1987) Oxidation of dimethyl sulfide to dimethyl sulfoxide by phototrophic purple bacteria. *Appl. Environ. Microbiol.* **53**, 2026–2076.

Zhang C. L., Liu S., Phelps T. J., Cole D. R., Horita J., Fortier S. M., Elless M., and Valley J. W. (1997) Physiochemical, mineralogical, and isotopic characterization of magnetite-rich iron oxides formed by thermophilic iron-reducing bacteria. *Geochim. Cosmochim. Acta* **61**, 4621–4632.

Zhang C. L., Horita J., Cole D. R., Zhou J., Lovley D. R., and Phelps T. J. (2001) Temperature-dependent oxygen and carbon isotope fractionations of biogenic siderite. *Geochim. Cosmochim. Acta* **65**, 2257–2271.

Zhang J., Zhou Z., Nie Y., Zhang J., Xi S., and Yang Z. (2000) Factors affecting volatile sulfur gas discharge from decomposition of methionine in paddy soil. *Huanjing Kexue* **21**, 37–41.

Zhang J., Lion L. W., Nelson Y. M., Shuler M. L., and Ghiorse W. C. (2002) Kinetics of Mn(II) oxidation by *Leptothrix discophora* SS1. *Geochim. Cosmochim. Acta* **65**, 773–781.

Zhang L., Kuniyoshi I., Hirai M., and Shoda M. (1991) Oxidation of dimethylsulfide by *Pseudomonas acidovorans* DMR-11 isolated from a peat biofilter. *Biotechol. Lett.* **13**, 223–228.

Zhou Z., Takaya N., Sakairi M. A. C., and Shoun H. (2001) Oxygen requirement for denitrification by the fungus *Fusarium oxysporum*. *Arch. Microbiol.* **175**, 19–25.

Zhou Z., Takaya N., Nakamura A., Yamaguchi M., Takeo K., and Shoun H. (2002) Ammonia fermentation, a novel anoxic metabolism of nitrate by fungi. *J. Biol. Chem.* **277**, 1892–1896.

Zimov S. A., Voropaev Y. V., Semiletov I. P., Davidov S. P., Prosiannikov S. F., Chapin F. S. I., Chapin M. C., Trumbore S., and Tyler S. (1997) North Siberian lakes: a methane source fueled by pleistocene carbon. *Science* **277**, 800–802.

Zindel U., Freudenberg W., Rieth M., Andreesen J. R., Schnell J., and Widdel F. (1988) *Eubacterium acidaminophilum* sp. nov., a versatile amino acid-degrading anaerobe producing or utilizing H_2 or formate: description and enzymatic studies. *Arch. Microbiol.* **150**, 254–266.

Zinder S. H. (1984) Microbiology of anaerobic conversion of organic wastes to methane: recent developments. *Am. Soc. Microbiol. News* **50**, 294–298.

Zinder S. H. (1993) Physiological ecology of methanogens. In *Methanogensis: Ecology, Physiology, Biochemistry and Genetics* (ed. J. G. Ferry). Chapman and Hall, New York, pp. 128–206.

Zinder S. H. and Brock T. D. (1978a) Dimethyl sulfoxide as an electron acceptor for anaerobic growth. *Arch. Microbiol.* **116**, 35–40.

Zinder S. H. and Brock T. D. (1978b) Dimethyl sulfoxide reduction by microorganisms. *J. Gen. Microbiol.* **105**, 335–342.

Zinder S. H. and Brock T. D. (1978c) Methane, carbon dioxide, and hydrogen sulfide production from the terminal methiol group of methionine by anaerobic lake sediments. *Appl. Environ. Microbiol.* **35**, 344–352.

Zinder S. H. and Brock T. D. (1978d) Production of methane and carbon dioxide from methane thiol and dimethyl sulfide by anaerobic lake sediments. *Nature* **273**, 226–228.

Zinder S. H. and Koch M. (1984) Non-aceticlastic methanogenesis from acetate: acetate oxidation by a thermophilic syntrophic culture. *Arch. Microbiol.* **138**, 263–272.

Ziska L. H., Moya T. B., Wassmann R., Namuco O. S., Lantin R. S., Aduna J. B., Abao E., Jr., Bronson K. F., Neue H. S., and Olszyk D. (1998) Long-term growth at elevated carbon dioxide stimulates methane emission in tropical paddy rice. *Global Change Biol.* **4**, 657–665.

Zopfi J., Ferdelman T. G., Jorgensen B. B., Teske A., and Thamdrup B. (2001) Influence of water column dynamics on sulfide oxidation and other major biogeochemical processes in the chemocline of Mariager Fjord (Denmark). *Mar. Chem.* **74**, 29–51.

Zumft W. G. (1997) Cell biology and molecular basis of denitrification. *Microbiol. Mol. Biol. Rev.* **61**, 533–616.

8.09
The Geologic History of the Carbon Cycle

E. T. Sundquist and K. Visser
US Geological Survey, Woods Hole, MA, USA

8.09.1 INTRODUCTION

Geologists, like other scientists, tend to view the global carbon cycle through the lens of their particular training and experience. The study of Earth's history requires a view both humbled by the knowledge of past global transformations and emboldened by the imagination of details not seen in the fragments of the rock record. In studying the past behavior of the carbon cycle, geologists are both amazed by unexpected discoveries and reassured by the extent to which "the present is the key to the past." Understanding the present-day carbon cycle has become a matter of societal urgency because of concerns about the effects of human activities on atmospheric chemistry and global climate. This public limelight has had far-reaching consequences for research on the geologic history of the carbon cycle as well as for studies of its present and future. The burgeoning new "interdiscipline" of biogeochemistry claims among its adherents many geologists as well as biologists, chemists, and other scientists. The pace of discovery demands that studies of the geologic history of the carbon cycle cannot be isolated from the context of present and future events.

This chapter describes the behavior of the carbon cycle prior to human influence. It describes events and processes that extend back through geologic time and include the exchange of carbon between the Earth's surface and the long-term reservoirs in the lithosphere. Chapter 8.10 emphasizes carbon exchanges that are important over years to decades, with a focus on relatively recent human influences and prospects for change during the coming century. Chapter 4.03 presents an overview of the biogeochemistry of methane, again with emphasis on relatively recent events. In these chapters as well as in the present chapter, relationships between the carbon cycle and global climate are a central concern. Together, these chapters provide an overview of how our knowledge of the present-day carbon cycle can be applied both to contemporary issues and to the record of the past. Similarly, these chapters collectively reflect the collaborative efforts of biogeochemists to utilize information about past variations in the carbon cycle to understand both Earth's history and modern changes.

This chapter begins with an overview of the carbon exchanges and processes that control the variations observed in the geologic record of the carbon cycle. Then examples of past carbon-cycle change are described, beginning with the most recent variations seen in cores drilled from glaciers and the sea floor, and concluding with the distant transformations inferred from the rock record of the Precambrian. Throughout this treatment, three themes are prominent. One is that different processes control carbon cycling over different timescales. A second theme is that relatively "abrupt" changes have played a central role in the evolution of the carbon cycle throughout Earth's history. The third theme is that the geologic cycling of carbon over all timescales passes through the atmosphere and the hydrosphere, and "it is this common course that unites the entire carbon cycle and allows even its most remote constituents to influence our environment and biosphere" (Des Marais, 2001).

The description of geologic events in this chapter includes examples from a broad span of the geologic record, but does not distribute attention in proportion to the distribution of geologic time in Earth's history. Readers will note, in particular, a disproportional emphasis on the Quaternary period, the most recent but briefest of geologic periods. The reason for this emphasis is twofold. First, the quality of the available Quaternary record of carbon-cycle change is far better than that available for earlier geologic periods. Second, the Quaternary record reveals a particularly illuminating array of details about interactions among the atmosphere, the biosphere, and the hydrosphere—the subset of the carbon cycle that must be understood in order to comprehend carbon cycling over nearly all times and timescales.

8.09.2 MODES OF CARBON-CYCLE CHANGE

8.09.2.1 The Carbon Cycle over Geologic Timescales

Throughout Earth's history, the principal forms of carbon in the atmosphere have been carbon dioxide (CO_2) and methane (CH_4). These gases have played crucial, yet distinct, roles in the development of life forms and the alteration of Earth's surface environment. Carbon dioxide is a principal medium of photosynthesis, metabolism, and organic decomposition. Through its transformation in weathering and carbonate precipitation, it supplies a major portion of the Earth's sedimentary rocks and contributes to the cycling of volatiles through the lithosphere. Methane represents the anaerobic side of carbon cycling, from its importance in microbial metabolism to its release from organic matter trapped in rocks and sediments. While the major transfers of mass in the carbon cycle are usually associated with the cycling of carbon dioxide through the atmosphere, methane may have played a more important role in the past, and may be a more sensitive indicator of changes in Earth's processes. Together, carbon dioxide and methane are the primary compounds through which carbon cycling over all timescales has influenced the Earth's surface.

Figures 1 and 2 depict the carbon reservoirs and fluxes that affect atmospheric CO_2 and CH_4, respectively, over geologic timescales. Estimates are shown for both the most recent glacial period and the Late Holocene Epoch prior to human influence. Numerical values in these figures are shown in the units most commonly used in the literature: petagrams carbon (Pg C, 10^{15} g carbon) for the transfer and storage of carbon and carbon dioxide, and teragrams methane (Tg CH_4, 10^{12} g methane) for the transfer and storage of methane. (For direct comparison between CO_2 and CH_4, Table 1(c) includes estimates of CH_4 fluxes and reservoirs in the units for carbon transfer and storage (Pg C).) In Figures 1 and 2, the vertical alignment of each reservoir represents the approximate timescale over which it may significantly influence the atmosphere. The array of processes that can affect atmospheric CO_2 and CH_4, and the spectrum of timescales over which these processes act, comprise a principal topic of this chapter.

8.09.2.1.1 The "carbon dioxide" carbon cycle

Estimates of many Late Holocene carbon fluxes shown in Figure 1 are derived from values given by Houghton (see Chapter 8.10) and by Sarmiento and Gruber (2002) for the contemporary carbon cycle, adjusted for the estimates of human

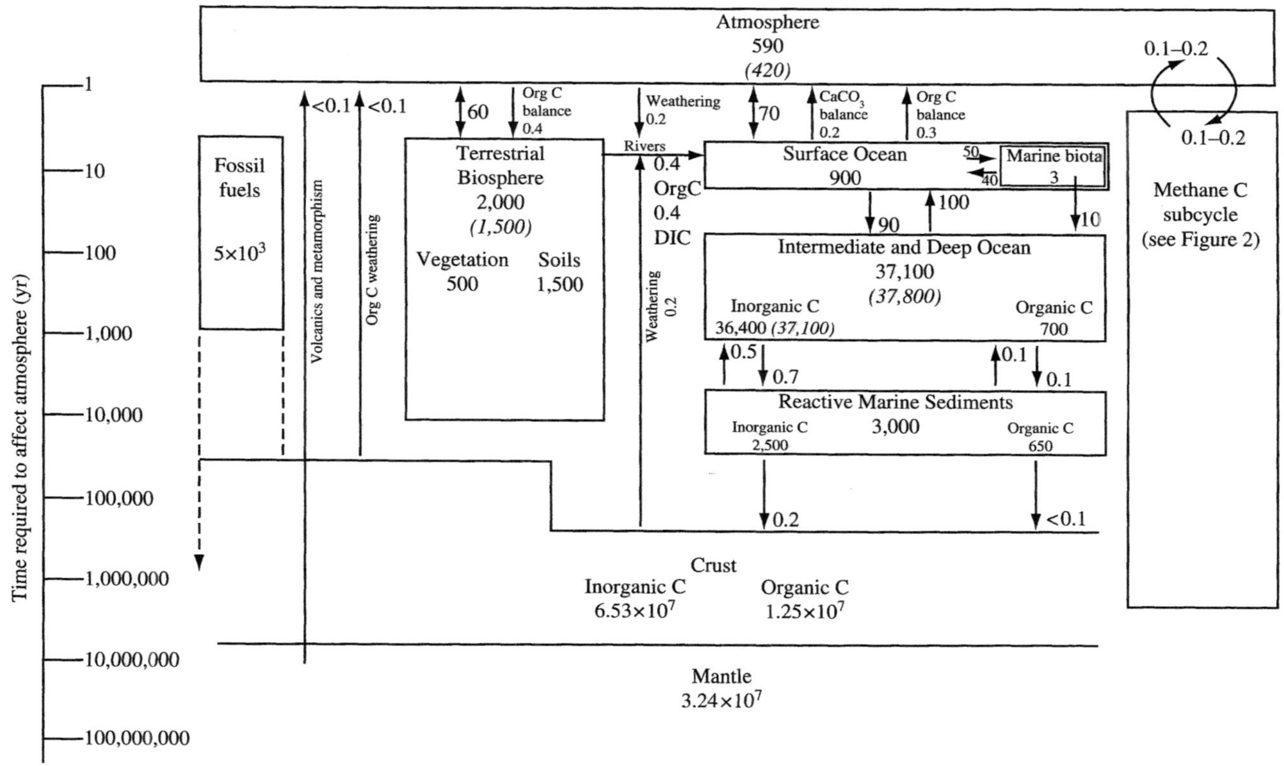

Figure 1 Reservoirs (Pg C; boxes) and fluxes (Pg C yr^{-1}; arrows) of the pre-industrial global carbon cycle. Values for glacial periods, where available, are shown in parentheses. Values for the glacial and pre-industrial (1750 AD) atmospheric carbon reservoir were calculated using CO_2 concentrations of 200 ppmv and 278 ppmv, respectively, and the equation M_a (g) $= (P_a \times 10^{-6}) \times (12.01/28.97) \times (5.12 \times 10^{21})$, where P_a is the atmospheric CO_2 concentration in ppmv, 12.01 is the atomic weight of carbon, 28.97 is the effective atmospheric molecular weight, and 5.12×10^{21} g is the mass of the (dry) atmosphere. See Section 8.09.2.1.1 for further derivation of estimates for glacial reservoirs. The vertical bar on the left shows the approximate time (in years) necessary for the different reservoirs to affect the atmosphere. The data for the figure are primarily from Chapter 8.10, Li (2000), Sarmiento and Gruber (2002), Sundquist (1985), and Sundquist (1993) (see also Table 1(a)). Atmospheric "balance" fluxes are shown to indicate the small net atmospheric exchange required to maintain a steady state with respect to sedimentation of organic carbon and calcium carbonate. The terrestrial biosphere and oceanic reservoir values are rounded to the nearest 100 Pg C, total reactive marine sediments to the nearest 1,000 Pg C, and all other reservoirs to the number of significant figures shown, according to the references cited in Table 1(a). All fluxes are rounded to one significant figure.

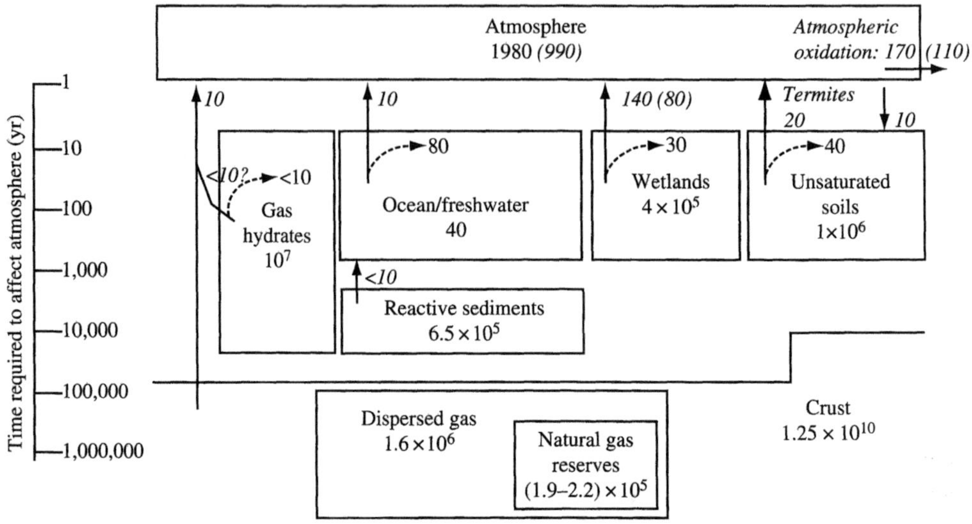

Figure 2 Reservoirs (Tg CH_4 or Tg C) and fluxes (Tg CH_4 yr^{-1}) of the natural, pre-industrial methane carbon subcycle. Values for glacial periods, where available, are shown in parentheses. Values for reactive sediments, wetlands, unsaturated soils, and crustal reservoirs represent organic carbon that might be converted to methane, and are all given as Tg C. Values for the glacial and pre-industrial (1750 AD) atmospheric reservoir were calculated using concentrations of 350 ppbv and 700 ppbv, respectively, and the equation $M_a\,(g) = (P_a \times 10^{-9}) \times (16.04/28.97) \times (5.12 \times 10^{21})$, where P_a is the atmospheric CH_4 concentration in ppbv, 16.04 is the molecular weight of methane, 28.97 is the effective atmospheric molecular weight, and 5.12×10^{21} g is the mass of the (dry) atmosphere. See Section 8.09.2.1.2 for further derivation of glacial flux estimates. The vertical bar on the left shows the approximate time (in years) necessary for the different reservoirs to affect the atmosphere. Estimates of methane consumption within reservoirs, from Chapter 4.03, are shown as dashed arrows (see Section 8.09.2.1.2). Gross production of methane within a reservoir can be calculated by adding the flux to the atmosphere and the consumption value. The flux and consumption values are rounded to the nearest 10 Tg CH_4 or Tg C. See Table 1(c) for the data sources.

influence since ~1750 AD. The adjustments for human influence are based on relatively well-known estimates of the historical cumulative increase in atmospheric CO_2 (190 Pg C), and the corresponding uptake of CO_2 by the oceans (155 Pg C). These estimates are calculated from the estimates in Chapter 8.10 of the Treatise for the period 1850–2000 AD, corrected to the period 1750–2000 AD by adding 15 Pg C to the cumulative atmospheric increase (assuming a global mean atmospheric CO_2 concentration of 285 ppmv in 1850) and 15 Pg C to the cumulative oceanic increase. As described in Chapter 8.10, these estimates can be used with the estimated cumulative CO_2 production from burning fossil fuels (275 Pg C) to infer the net cumulative change in the amount of carbon stored in land plants and soils. This change is not applied to the estimates shown in Figure 1 because it is too small (<50 Pg C) to be considered significant relative to probable uncertainties in the total size of the reservoirs. The adjustments for human influence can be compared to uncertainties suggested by the range of recent estimates in reservoir sizes and fluxes in Table 1(a). Table 1(b) shows details of recent estimates of carbon in fossil fuel resources.

Figure 1 and Table 1(a) include estimates of the "geologic" components of the carbon cycle (weathering, river transport, sedimentation, volcanic/metamorphic emissions, and the rock and sediment reservoirs), so called because they are generally more important over longer timescales. Some sediment processes (the burial and remineralization of organic matter and calcium carbonate) may affect the atmosphere over timescales as short as a few thousand years. The sediment carbon reservoirs associated with these timescales are depicted in Figure 1 and Table 1(a) as "reactive sediment" reservoirs (Sundquist, 1985). Although the estimated magnitude of these reservoirs is somewhat arbitrary and process-dependent (see Broecker and Takahashi, 1977; Sundquist, 1985), they are large enough to affect ocean–atmosphere chemistry significantly through such processes as carbonate dissolution and organic carbon diagenesis.

Another large and potentially reactive sediment carbon reservoir is the pool of methane hydrates, shown as part of the "methane carbon subcycle" in Figure 2. Methane is produced most commonly by anoxic bacterial metabolism of CO_2 or of organic substrates derived ultimately from photosynthesis of CO_2. In the presence of oxygen or other electron acceptors containing oxygen, methane is oxidized to CO_2. Thus the cycling of methane is depicted in Figure 1 as a subcycle of the "carbon dioxide" carbon cycle. The methane carbon

Table 1(a) Representative range of recent estimates for natural, pre-industrial reservoirs and geologic fluxes of carbon (see Sundquist (1985) for a summary of older published estimates).

Reservoir	*Size* (Pg C)	*References*
Atmosphere	590	See Figure 1 caption
Oceans	$(3.71-3.90) \times 10^4$	Sundquist, 1985, 1993; see Chapter 8.10
Surface layer—inorganic	700–900	Sundquist, 1985; Sarmiento and Gruber, 2002; see Chapter 8.10
Deep layer—inorganic	$(3.56-3.80) \times 10^4$	Sundquist, 1993; Sarmiento and Gruber, 2002; see Chapter 8.10
Total organic	685–700	Sharp, 1997; Hansell and Carlson, 1998 (depths >1,000 m); Doval and Hansell, 2000
Aquatic biosphere	1–3	Falkowski *et al.*, 2000; Sarmiento and Gruber, 2002
Terrestrial biosphere and soils	$2.0-2.3 \times 10^3$	Sundquist, 1993; Sarmiento and Gruber, 2002; see Chapter 8.10
Vegetation	500–600	Sundquist, 1993; see Chapter 8.10
Soil	$(1.5-1.7) \times 10^3$	See Chapter 8.10
Reactive marine sediments	3,000	Sundquist, 1985
Inorganic	2,500	
Organic	650	
Crust	$(7.78-9.0) \times 10^7$	Holland, 1978; Li, 2000
Sedimentary carbonates	6.53×10^7	Li, 2000
Organic carbon	1.25×10^7	Li, 2000
Mantle	3.24×10^8	Des Marais, 2001
Fossil fuel reserves and resources	$(4.22-5.68) \times 10^3$	See Table 1(b) for references
Oil (conv. and unconv.)	636–842	
Natural gas (conv. and unconv.)	483–564	
Coal	$(3.10-4.27) \times 10^3$	

Flux	*Size* ($Pg\ C\ yr^{-1}$)	*References*
Carbonate burial	0.13–0.38	Berner *et al.*, 1983; Berner and Berner, 1987; Meybeck, 1987; Drever *et al.*, 1988; Milliman, 1993; Wollast, 1994
Organic carbon burial	0.05–0.13	Lein, 1984; Berner, 1982; Berner and Raiswell, 1983; Dobrovolsky, 1994; Schlesinger, 1997
Rivers (dissolved inorganic carbon)	0.39–0.44	Berner *et al.*, 1983; Berner and Berner, 1987; Meybeck, 1987; Drever *et al.*, 1988
Rivers (total organic carbon)	0.30–0.41	Schlesinger and Melack, 1981; Meybeck, 1981; Meybeck, 1988; Degens *et al.*, 1991
Rivers-dissolved organic carbon	0.21–0.22	Meybeck, 1981; Meybeck, 1988; Spitzy and Leenheer, 1991
Rivers-particular organic carbon	0.17–0.30	Meybeck, 1981; Milliman *et al.*, 1984; Ittekot, 1988; Meybeck, 1988
Volcanism	0.04–0.10	Gerlach, 1991; Kerrick *et al.*, 1995; Arthur, 2000; Kerrick, 2001; Morner and Etiope, 2002
Mantle exchange	0.022–0.07	Sano and Williams, 1996; Marty and Tolstikhin, 1998; Des Marais, 2001

Table 1(b) Global fossil fuel energy reserves and resources, in exajoules (EJ), and the equivalent carbon content in petagrams carbon (Pg C). Fossil fuel reserves are identified deposits that can be recovered using current technology under existing economic conditions. Resources are defined as "the occurrences of material in recognizable form" (WEC, 2000) but not extractable under current economic or technological conditions. Resources are essentially the amount of fossil fuels of "foreseeable economic interest." The differences between reserve and resource estimates illustrate the wide range of inherent uncertainties (geological, economic, and technological) in the resource estimates. "Additional occurrences" refers to fossil fuels that are not believed to be potentially recoverable (Moomaw, 2001). Unconventional occurrences differ from conventional occurrences by either "the nature of existence (being solid rather than liquid for oil) or the geological location (coal bed methane or clathrates)." Unconventional occurrences include oil shale, tar sands, coalbed methane, clathrates (Moomaw, 2001).

	EJ[c]	Pg C[d]
Reserves		
Oil—conventional[1,2,3,4,5,6]	5,908–6,759	111–127
Oil—unconventional[1,2]	6,624–8,639	124–162
Natural gas—conventional[1,2,3,4,5,6]	5,058–6,311	73–91
Natural gas—unconventional[1,2]	8,102–8,594	117–125
Coal[1,2,4,6]	28,825–41,985	499–1,094
Resources		
Oil—conventional[1,2,3,5,a]	6,490–14,055	122–264
Oil—unconventional[1,2,a]	14,860–15,397	279–289
Natural gas—conventional[1,2,3,5,b]	9,355–12,488	136–181
Natural gas—unconventional[1,2,b]	10,787–11,458	157–167
Coal[1,2]	100,352–125,059	2,605–3,177
Additional occurrences		
Oil—unconventional[1,2]	61,008–85,004	1,145–1,596
Natural gas—unconventional[1,2]	15,979–17,904	232–260
Coal[1,2]	120,986–134,280	3,122–3,411

References: [1]Nakicenovic *et al.*, 1998; [2]Moomaw 2001 (IPCC); [3]USGS World Energy Assessment, 2000; [4]WEC, 2000; [5]Masters *et al.*, 1994; [6]Oil and Gas Journal, 2001.
[a] For oil resources, the USGS World Energy Assessment (2000) estimated that there is a 5% probability that 16,466 EJ is recoverable, a 50% probability of 8,872 EJ and a 95% probability of 4,283 EJ. The mean values given in the assessment are used in the ranges shown above. [b] For natural gas resources, the USGS World Energy Assessment (2000) estimated that there is a 5% probability that 15,207 EJ is recoverable, a 50% probability of 8,057 EJ and a 95% probability of 3,946 EJ. The mean values given in the assessment are used in the ranges shown above. [c] The CDIAC "Common Energy Unit Conversion Factors Table 4" (O'Hara, 1990; http://cdiac.esd.ornl.gov/pns/convert.html) was used to convert all energy units to Exajoules. [d] The energy units were converted to equivalent carbon content (Pg C) using the methods detailed in Sundquist (1985) as well as the conversion factor for the average carbon content of coal [1 TJ = 25.4 mt carbon (ORNL Bioenergy Feedstock Development Programs, 2003—http://bioenergy.ornl.gov/papers/misc/energy_conv.html)] in cases where the breakdown of coal rank was not available.

subcycle is further described in Section 8.09.2.1.2. It is important to note that, during the early history of the Earth, the processes now associated with the cycling of CO_2 may have emerged initially as a subcycle of the more prevalent cycling of methane (see Section 8.09.5).

The amount of CO_2 in the glacial atmosphere can be calculated accurately from the concentration of CO_2 in bubbles of air trapped in ice that formed during the most recent glacial period. (These measurements are further described in Section 8.09.3.1.) The glacial atmospheric CO_2 concentration also provides a way to estimate the dissolved inorganic carbon (DIC) content of the glacial ocean surface mixed layer. DIC exists primarily in four forms: dissolved CO_2, carbonic acid (H_2CO_3), bicarbonate ions (HCO_3^-), and carbonate ions (CO_3^{2-}). The ionic forms may combine with other ions to form complex ions in solutions such as seawater. Chemical reaction among the dissolved inorganic carbon species is rapid, and they occur in proportions that can be calculated from well-known thermodynamic equilibrium relationships. CO_2 exchange between the atmosphere and the mixed layer is so rapid that, for global estimates like those shown in Figure 1, the ocean surface layer can be considered to be near chemical equilibrium with the atmosphere (Sundquist and Plummer, 1981). The glacial-to-interglacial increase in ocean surface DIC, corresponding to the atmospheric CO_2 increase (170 Pg C) shown in Figure 1, is estimated to have been ~30 Pg C (Sundquist, 1993).

Estimates of other carbon reservoirs and fluxes during the most recent glacial period depend largely on estimates of the amount of carbon stored in the terrestrial biosphere. Because the transfer of carbon over timescales of thousands of years is limited largely to translocation among the atmosphere, the biosphere, and the oceans, a loss or gain of carbon in one of these reservoirs can be used as a good approximation for a corresponding gain or loss in the others. (see Section 8.09.2.2 for discussion of this "rule" and its possible exceptions.) Estimates of the glacial terrestrial biosphere are referenced to the Late Holocene or present-day terrestrial biosphere, either by reconstruction of ecological changes or by inference from the shift in the isotopic composition of oceanic dissolved carbon. There is a

Table 1(c) Representative range of recent estimates for natural (pre-industrial) fluxes and reservoirs of methane.

Reservoir	*Size* (Tg CH_4 or Tg C)	*Size* (*Pg C*)	*References*
Atmosphere	1980 Tg CH_4	1.5	See Figure 2 caption
Oceans	22–65 Tg CH_4	0.016–0.049	Holmes *et al.*, 2000; Kelley and Jeffrey, 2002
Wetlands	$(2.2–4.9) \times 10^5$ Tg C	224–489	Gorham, 1991; Botch *et al.*, 1995; Lappalainen, 1996; Schlesinger, 1997, after Schlesinger, 1977; Clymo *et al.*, 1998
Reactive marine sediments (org. C)	6.5×10^5 Tg C	650	Sundquist, 1985
Non-wetland soils	9.7×10^5 Tg C	970	Schlesinger, 1997, after Schlesinger, 1977
Geological sources			
Crust	1.25×10^{10} Tg C	1.25×10^7	Li, 2000
Hydrates	$5 \times 10^5 - 2.4 \times 10^7$ Tg CH_4	$4 \times 10^2 - 1.8 \times 10^4$	Kvenvolden and Lorenson, 2001
Dispersed gas in sedimentary basins	1.6×10^6 Tg CH_4	1.2×10^3	Hunt, 1996
Natural gas reserves (part of dispersed gas in sed. basins)	$(2.5–2.9) \times 10^5$ Tg CH_4	190–216	See Table 1(b) for references

Flux	*Size* (Tg CH_4 yr^{-1})	*Size*[a] (Pg C yr^{-1})	*References*
Oceans	0.4–20	0–0.01	Ehhalt, 1974; Watson *et al.*, 1990; Fung *et al.*, 1991; Chappellaz *et al.*, 1993b; Lambert and Schmidt, 1993; Bange *et al.*, 1994; Bates *et al.*, 1996; Lelieveld *et al.*, 1998; Holmes *et al.*, 2000; Kelley and Jeffrey, 2002
Marine sediments	0.4–12.2	0–0.01	Judd, 2000
Wetlands	92–260	0.07–0.19	Fung *et al.*, 1991; Bartlett and Harriss, 1993; Chappellaz *et al.*, 1993b; Cao *et al.*, 1996; Hein *et al.*, 1997; Matthews, 2000; Walter *et al.*, 2001
Termites	2–22	0–0.02	Cicerone and Oremland, 1988; Chappellaz *et al.*, 1993b; Sanderson, 1996; Sugimoto *et al.*, 1998
Wild fires	2	0	Levine *et al.*, 2000
Geological sources (*Hydrates, volcanoes, natural gas seeps, geothermal*)	5–65	0–0.05	Chappellaz *et al.*, 1993b; Hovland, 1993; Lacroix, 1993; Hornafius *et al.*, 1999; Judd, 2000, after Lacroix, 1993; Etiope and Klusman, 2002; Judd *et al.* 2002; see Chapter 4.03
Total source	159–290	0.12–0.22	McElroy, 1989; Chappellaz *et al.*, 1993b; Thompson *et al.*, 1993; Martinerie *et al.*, 1995; Etheridge *et al.*, 1998; Brook *et al.*, 2000

[a] Fluxes less than 0.01 Pg C yr^{-1} are rounded to 0.

broad consensus that the amount of carbon stored in vegetation and soils was smaller during the most recent glacial period than during the Holocene Epoch (Shackleton, 1977; Crowley, 1995). Paleoecological reconstructions, based on data from soils and sediments (see, e.g., Adams *et al.*, 1990; Adams and Faure, 1998) or on models of glacial climate/vegetation relationships (see, e.g., Prentice *et al.*, 1993; Otto *et al.*, 2002), yield estimates of glacial carbon storage ranging from a few hundred to more than one thousand Pg C less than carbon storage in Holocene plants and soils. This broad range of estimates reflects diverse sources of uncertainty, including limitations in paleoclimatic and paleoecological data, climate models, and assumptions about effects of atmospheric CO_2, interactions between climate and vegetation, and amounts of carbon stored in various biomes and coastal environments (Prentice and Fung, 1990; Van Campo *et al.*, 1993; Francois *et al.*, 1999).

The carbon isotope record of marine sediments provides an important constraint on glacial/interglacial changes in carbon storage by the terrestrial biosphere. During photosynthesis, plants preferentially assimilate $^{12}CO_2$, leaving the atmosphere relatively enriched in $^{13}CO_2$. Carbon assimilated during photosynthesis by land plants may be depleted in ^{13}C by ~3–25‰ relative to the inorganic carbon source. (Carbon isotope compositions are denoted by per mil (‰) differences in the ratio of ^{13}C to ^{12}C relative to a standard.) The extent of photosynthetic ^{13}C depletion varies primarily because plants may use several different biochemical pathways to assimilate carbon. The ribulose-1,5-bisphosphate carboxylation, or "C_3," pathway (also known as the Calvin cycle) is the most common and reduces the $^{13}C:{}^{12}C$ ratio by ~15 to 25‰ (O'Leary, 1988). The phosphenolpyruvate carboxylation, or "C_4," pathway is observed most commonly in corn and other grasses. Although the C_4 pathway operates in close association with ribulose-1,5-bisphosphate carboxylation, the resulting reduction in the $^{13}C:{}^{12}C$ ratio is only ~3–8‰ (Deleens *et al.*, 1983). Because the C_3 pathway is the most common among land plants, this pathway probably determined the average $^{13}C:{}^{12}C$ ratio of the atmospheric CO_2 transferred to and from vegetation and soils during glacial/interglacial transitions.

Exchange of atmospheric CO_2 with the oceans extends the isotopic fractionation effect of land plants to the ocean DIC reservoir. Over timescales of ocean mixing, the carbon isotope effect of land plants is diluted but nevertheless measurable in the larger oceanic DIC reservoir. A significant global change in the amount of photosynthesized carbon stored in plants and soils would be expected to change the isotopic composition of carbon in atmospheric CO_2 and oceanic DIC. The carbon isotope compositions of both atmospheric CO_2 and oceanic DIC are also affected by other factors, most notably sea-surface gas exchange, chemical reactions among DIC species, and photosynthesis by marine organisms. For example, photosynthesis by marine organisms reduces the $^{13}C:{}^{12}C$ ratio by ~10–30‰ (Deines, 1980). Marine photosynthesis is widely believed to occur primarily via the C_3 biochemical pathway, but effects of C_4 photosynthesis by diatoms may account for part of the relatively wide range of carbon isotope ratios observed in planktonic ecosystems (Reinfelder *et al.*, 2000). Glacial/interglacial changes in the factors affecting oceanic isotope fractionation must be quantified in order to discriminate the carbon isotopic signature of changes in the terrestrial biosphere (Hofmann *et al.*, 1999).

The isotopic fractionation effects of marine photosynthesis, gas exchange, and chemical reactions are most pronounced in atmospheric CO_2 and ocean surface DIC. Deep-ocean DIC is relatively unaffected by these factors. Thus, the carbon isotope effects of changes in the terrestrial biosphere are best seen in the fossils of organisms that form their shells in the deep ocean. Glacial/interglacial isotopic changes in oceanic DIC are recorded in the calcareous shells of organisms deposited in sediments, and isotopic changes in atmospheric CO_2 are recorded in ice cores. These records generally indicate a glacial-to-interglacial increase in the $^{13}CO_2:{}^{12}CO_2$ ratio of oceanic DIC by ~0.3–0.5‰ (Sarnthein *et al.*, 1988; Curry *et al.*, 1988; Duplessy *et al.*, 1988; Ku and Luo, 1992; Crowley, 1995), and a corresponding increase in atmospheric CO_2 by ~0.1–0.5‰ (Leuenberger *et al.*, 1992; Smith *et al.*, 1999). Mass balance calculations based on these isotopic shifts imply a glacial-to-interglacial expansion of ~400–800 Pg C in the amount of organic carbon stored in plants and soils (Sundquist, 1993; Bird *et al.*, 1996). Sundquist (1993) suggested a range of 450–750 Pg C, while Bird *et al.* (1996) suggested a range of 300–700 Pg C based on somewhat different assumptions. Here we derive an estimated range of 400–800 Pg C from the mass balance equations of Bird *et al.* (1996) applied to the range of atmospheric and oceanic carbon isotope shifts given in the text, and to a range of −22‰ to −25‰ (relative to the common PDB standard) for the ^{13}C content of the carbon taken up by the biosphere during deglaciation. These calculations do not take into account the possible effect of increased methane production and oxidation during the period of deglaciation (Maslin and Thomas, 2003); see Section 8.09.3.3.3).

The procedures described above provide a reasonable basis for estimating the redistribution of carbon storage among the atmosphere, terrestrial biosphere, and oceans during the most recent glacial period. The mass balance of glacial carbon losses and gains, relative to Holocene values, can be tallied as follows:

Glacial carbon losses:	
Organic carbon in plants and soils	400–800 Pg C
Atmosphere	170
Ocean surface DIC	30
Total losses	600–1,000 Pg C
Glacial carbon gains:	
Deep-ocean DIC	600–1,000 Pg C

Although these calculations are used to derive the estimates of glacial carbon reservoirs shown in Figure 1, glacial carbon fluxes are much less certain and are not shown.

8.09.2.1.2 The "methane" carbon subcycle

Estimates of Late Holocene CH_4 fluxes and reservoirs are shown in Figure 2. Many of the flux estimates in this figure are derived from the values for natural sources presented by Reeburgh (see Chapter 4.03) and Lelieveld *et al.* (1998). Estimates of the carbon reservoirs that contribute to methane cycling have been added. The wetland soil and sediment organic carbon reservoirs estimated in Figure 2 are also represented in the soil and sediment reservoirs depicted in Figure 1, because these reservoirs may yield both CO_2 and CH_4 depending on environmental conditions. Uncertainties in the values shown in Figure 2 can be inferred from the ranges of estimates listed in Table 1(c).

Global fluxes of CH_4 to and from the atmosphere are generally smaller than those of CO_2 (Tables 1(a) and 1(c) and Figures 1 and 2). Although CH_4 may be less significant in exchange of carbon mass, atmospheric CH_4 can be an important indicator of changes in carbon cycling that do not entail large transfers of carbon. Moreover, atmospheric CH_4 has direct importance for the carbon cycle due to its disproportionate influence on climate.

A molecule of CH_4 absorbs more than 20 times as much long-wave radiation as a molecule of CO_2 (Ramaswamy *et al.*, 2001). In considering the climatic effects of different gases, their radiative properties are often compared using calculated indices such as the greenhouse warming potential (GWP; Shine *et al.*, 1990), which is defined as the time-integrated radiative effect following an instantaneous injection of a particular gas relative to the effect of an injection of an equal mass of CO_2. The GWP is referenced to specific integration times in order to adjust for the different time trajectories of instantaneous trace gas injections. The GWP of CH_4 is 7 times that of CO_2 over a 500 yr period (Ramaswamy *et al.*, 2001). The increase in atmospheric CH_4 that occurred during the time between the most recent glacial period and the Late Holocene Epoch (Figure 2) had a direct radiative effect that was approximately one-fifth that of the corresponding increase in atmospheric CO_2 (Figure 1) (Petit *et al.*, 1999). The relative impact of methane's radiative properties on the global greenhouse effect is diminished by its low concentration and short lifetime in the atmosphere, but CH_4 clearly plays an important role in the complex interactions between the carbon cycle and the climate system.

As depicted in Figure 2 and detailed in Chapter 4.03, the rate of CH_4 release to the oceans and atmosphere is only a fraction of the rate at which CH_4 is produced in anoxic environments. In wetlands and marine sediments, zones of CH_4 production are frequently overlain by or interspersed with zones where aerobic or anaerobic methanotrophic microbes are active consumers of CH_4 (Reeburgh, 1976; Scranton and Brewer, 1978; Yavitt *et al.*, 1988; Whalen and Reeburgh, 1990; Moosavi *et al.*, 1996; R. S. Hanson and T. E. Hanson, 1996; Orphan *et al.*, 2001; Hinrichs and Boetius, 2002). Although some CH_4 may bypass oxidation in these zones by transport via bubbles or vascular plants (Dacey and Klug, 1979; Bartlett *et al.*, 1988), significant quantities of CH_4 are consumed in these layers and therefore are not released to the atmosphere and oceans (Galchenko *et al.*, 1989; Oremland and Culbertson, 1992; Reeburgh *et al.*, 1993). This process is difficult to quantify (see Chapter 4.03), but it is probably a significant component of the "methane" carbon subcycle (King, 1992). Several studies suggest that additional CH_4 may be produced in anoxic microsites within oxic soils and ocean environments, which are on the whole net consumers of atmospheric CH_4 (Yavitt *et al.*, 1995; Holmes *et al.*, 2000; von Fischer and Hedin, 2002). Thus the extent and magnitude of the methane carbon subcycle may be greater than suggested by the fluxes and reservoirs shown in Figure 2.

Surprisingly, CH_4 has an atmospheric lifetime longer than that of CO_2. "Atmospheric lifetime" is calculated here as the ratio of abundance in the atmosphere to the sum of annual sources or removals, assuming a steady state. Using the fluxes represented in Figure 1, CO_2 would have an atmospheric lifetime of ~4–5 yr. However, this calculation does not reflect the full extent of cycling of CO_2 through plants because it considers only net primary production and heterotrophic respiration. When the cycling of CO_2 via gross primary productivity and autotrophic respiration is taken into account, the exchange of CO_2 between the atmosphere and the terrestrial biosphere is approximately doubled, and the atmospheric lifetime of CO_2 is reduced to ~3 yr. The atmospheric lifetime of CH_4 is calculated from the fluxes depicted in Figure 2 to be ~11 yr. For the glacial values shown in Figure 2, the atmospheric lifetime of CH_4 is decreased to ~8 yr. Because the atmospheric oxidation of CH_4 is coupled to the abundance of OH and other reactants, its atmospheric lifetime may vary (Prinn *et al.*, 1983; Lelieveld *et al.*, 1998; Breas *et al.*, 2002).

CO_2 is removed from the atmosphere by exchange with the biosphere, the oceans, and the lithosphere. The primary mechanism for removal of atmospheric CH_4 is chemical oxidation in the atmosphere without further exchange. Thus, while atmospheric CO_2 is controlled by exchange with other components of the carbon cycle, the concentration of CH_4 in the atmosphere reflects a balance between its rate of supply to the atmosphere and its oxidation in the atmosphere. This fundamental difference in control mechanisms affects the time dependence of responses of atmospheric CO_2 and CH_4 to relatively abrupt perturbations. The response of CH_4 will occur rapidly through effects on its rate of oxidation in the atmosphere, whereas the response of CO_2 will be mediated by the more complex array of processes that govern its rate of exchange with other large carbon reservoirs and the rates of response of those reservoirs. Thus, even though CH_4 has a longer atmospheric lifetime than CO_2, the response of atmospheric CH_4 to perturbations will tend to be more rapid.

Methane is oxidized primarily in the troposphere by reactions involving the hydroxyl radical (OH). Methane is the most abundant hydrocarbon species in the atmosphere, and its oxidation affects atmospheric levels of other important reactive species, including formaldehyde (CH_2O), carbon monoxide (CO), and ozone (O_3) (Wuebbles and Hayhoe, 2002). The chemistry of these reactions is well known, and the rate of atmospheric CH_4 oxidation can be calculated from the temperature and concentrations of the reactants, primarily CH_4 and OH (Prinn *et al.*, 1987). Tropospheric OH concentrations are difficult to measure directly, but they are reasonably well constrained by observations of other reactive trace gases (Thompson, 1992; Martinerie *et al.*, 1995; Prinn *et al.*, 1995; Prinn *et al.*, 2001). Thus, rates of tropospheric CH_4 oxidation can be estimated from knowledge of atmospheric CH_4 concentrations. And because tropospheric oxidation is the primary process by which CH_4 is removed from the atmosphere, the estimated rate of CH_4 oxidation provides a basis for approximating the total rate of supply of CH_4 to the atmosphere from all sources at steady state (see Section 8.09.2.2) (Cicerone and Oremland, 1988).

Atmospheric CH_4 concentrations during the Late Holocene Epoch and the most recent glacial period are known from analyses of air bubbles in ice cores (see Section 8.09.3). These concentrations provide the basis for calculating the rates of atmospheric CH_4 oxidation and total rates of atmospheric CH_4 supply shown in Figure 2 (Chappellaz *et al.*, 1993b; Martinerie *et al.*, 1995; Lelieveld *et al.*, 1998). Although these overall budgets are reasonably well determined, the relative contributions of individual sources contributing to the atmospheric CH_4 supply are less well known (Cicerone and Oremland, 1988). Likewise, the sizes of the reservoirs other than atmospheric CH_4 can only be approximated.

Methane hydrates comprise the largest CH_4 reservoir in the methane carbon subcycle. Methane hydrates are formed when abundant dissolved methane accumulates under specific conditions of cold temperature and high pressure. These conditions commonly occur in marine sediments below water depths of a few hundred meters, and in continental sediments at high latitudes (Figure 3). When environmental conditions in these locations change, methane hydrates may become unstable and yield large quantities of dissolved and gaseous methane. The sensitivity of methane hydrates to changing environmental conditions implies that they may have played a role in past carbon-cycle changes (see Sections 8.09.3.3 and 8.09.4.4).

8.09.2.2 Timescales of Carbon-cycle Change

In studying the many factors that contribute to carbon-cycle change over geologic timescales, geochemists commonly observe that the relative importance of various processes and reservoirs depends on the timescale under consideration. The time axes shown in Figures 1 and 2 portray approximate timescales over which various fluxes and reservoirs may influence the atmosphere. With important exceptions (see below), these timescales can be used to categorize components of the carbon cycle into time-related frames of reference (Sundquist, 1986). Over relatively short timescales (up to ~10^3 yr), the most common variations in atmospheric CO_2 and CH_4 involve exchange with only the terrestrial biosphere and the oceans. Over somewhat longer timescales (~10^3–10^5 yr), the frame of reference must be expanded to include "reactive" carbon in the uppermost layers of marine sediments. Over timescales of millions of years and longer, the frame of reference must include carbon in the Earth's crust.

These frames of reference define fundamentally different modes of carbon-cycle change. Over the

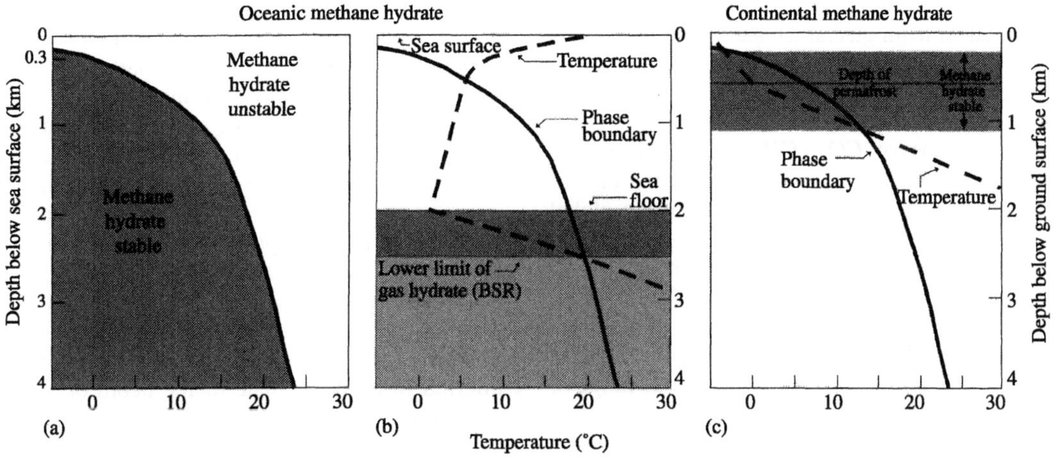

Figure 3 (a) Phase boundary and stability fields for methane hydrate. The vertical axis represents pressure, increasing downward and plotted as equivalent depth (km) assuming hydrostatic pressure. (b) The same phase boundary as shown in (a) in a marine environment, with seafloor inserted at 2 km, and a typical marine temperature profile through the water column and seafloor (dashed line). The dark gray area denotes the vertical extent of methane hydrate under the specified conditions (after Dillon, 2001, figure 2). (c) The phase boundary of methane hydrate in a high-latitude continental environment, plotted with a typical temperature (geothermal) gradient (dashed line) and showing depths assuming a hydrostatic pressure gradient. The dark gray area denotes the vertical extent of methane hydrate stability, under the specified conditions (after Kvenvolden and Lorenson, 2001, figure 1).

timescales of Quaternary glacial/interglacial cycles, changes in atmospheric CO_2 and CH_4 reflect primarily the redistribution of carbon among the atmosphere, biosphere, and oceans, with important contributions from reactive sediments. Over timescales of millions of years and longer, atmospheric CO_2 and CH_4 are controlled not only by carbon cycling at the Earth's surface, but also by the balance between long-term releases from the Earth's crust and carbon burial in sediments. With respect to the ocean–atmosphere–biosphere system, these relatively short-term and long-term carbon-cycle modes can be characterized respectively by internal redistributions and external exchange. Both modes of change operate simultaneously, and sorting out their effects is one of the most challenging problems in understanding the geologic evidence for carbon-cycle change (see, e.g., Sundquist (1991)).

The dynamic behavior of the carbon cycle and other complex systems may tend toward conditions of no change or "steady state" when exchanges are balanced by feedback loops. For example, model simulations of historical and projected effects of anthropogenic CO_2 and CH_4 emissions are usually based on an assumed carbon-cycle steady state before the onset of human influence. It is important to understand that the concept of steady state refers to an approximate condition within the context of a particular time-dependent frame of reference. Sundquist (1985) examined this problem rigorously using eigenanalysis of a hierarchy of carbon-cycle box models in which boxes were mathematically added and "lumped" to span a broad range of timescales. This analysis demonstrated that the nature and accuracy of any geochemical steady-state approximation depends on the timescale of interest and the defined frame of reference for that timescale. For example, the particular steady-state approximation used in projecting atmospheric CO_2 and CH_4 concentrations for the next century might not be appropriate for projections over timescales of hundreds to thousands of years (Sundquist, 1990b).

Recent studies of the geologic history of the carbon cycle have revealed many surprising examples of relatively abrupt change that cannot be characterized by the time-scale relationships depicted in Figures 1 and 2. While time-dependent frames of reference and steady-state approximations have provided a basis for many advances in understanding the carbon cycle, it is clear that these concepts are not adequate to guide studies of events such as bolide impacts or methane hydrate outbursts. A rigorous theoretical treatment of abrupt events is beyond the scope of this chapter. We offer two examples and some general comments in Section 8.09.4.4.

8.09.3 THE QUATERNARY RECORD OF CARBON-CYCLE CHANGE

The study of the Quaternary history of the global carbon cycle is powerfully constrained by the analysis of air trapped in ice from the continental ice sheets of Greenland and Antarctica. Cores taken from glacial ice and firn have yielded samples of air ranging in age from a few decades to more than 400,000 yr. The ice cores provide direct evidence of past variations in atmospheric concentrations of CO_2, CH_4, other greenhouse gases, aerosols, and dust. These variations in atmospheric chemistry reflect variations in global carbon cycling. The ice cores also offer a record of climate change that can be closely correlated with the record of carbon-cycle variations. The ice-core record from ice sheets provides compelling evidence for the intimate association between climate and the carbon cycle over a broad range of timescales.

8.09.3.1 Analysis of CO_2 and CH_4 in Ice Cores

The analysis of ice-core gas chemistry presents a very challenging technical problem. The gas samples are extremely small (a 10 g sample of ice might yield less than 1 cm^3 of trapped air (Schwander and Stauffer, 1984)), and they are very susceptible to alteration during extraction and analysis. Because CO_2 might dissolve in melted ice or condensed water vapor, the extraction of samples of this gas must be conducted by crushing the ice in cold and completely dry conditions (Delmas *et al.*, 1980). Methane is generally not extracted with CO_2 because friction during the dry crushing process may produce CH_4 (Stauffer *et al.*, 1985). Fortunately, CH_4 is less soluble in water than CO_2, so the extraction of samples for CH_4 can be accomplished by melting the sample (see, e.g., Rasmussen and Khalil (1984) or by special grinding techniques (Etheridge *et al.*, 1988)). Other sampling and analytical difficulties include the effects of breakdown of clathrates, which are the principal form of *in situ* gas storage in the deepest and coldest ice (Neftel *et al.*, 1983). Techniques to address these problems have been pioneered by the ice-core laboratories of the University of Bern in Switzerland, the Laboratoire de Glaciologie et de Géophysique de l'Environment in Grenoble, France, and the Division of Atmospheric Research at CSIRO in Australia (see the many references to work from these laboratories in Table 2).

Ice cores can be dated by counting annual layers, modeling ice accumulation and flow, and correlating recognizable events with other datable records (see, e.g., Dansgaard *et al.*, 1969; Johnsen *et al.*, 1972; Hammer *et al.*, 1978). However, the entrapped air is significantly younger than the surrounding ice, because the air is not completely isolated from the atmosphere until burial to depths that may exceed 100 m (Schwander and Stauffer, 1984). An additional problem arises because the process of complete bubble enclosure (coinciding with the conversion of granular firn to solid ice) occurs gradually during ice accumulation. Thus, the bubbles of air trapped at any particular depth are not only younger than the surrounding ice, but they also represent a range of ages reflecting the range of depths and times of bubble enclosure. Ice-core gas samples are generally too small, and many are too old, for radiocarbon dating of the enclosed carbon compounds. Thus the dating of ice-core gas records is very difficult.

Ice-core gas samples are typically dated by applying an age-difference correction relative to the age of the enclosing ice, based on the offset inferred from an estimate of the depth of bubble enclosure. An age range is also assigned based on the estimated rate of bubble enclosure during ice accumulation. The dating of gas samples therefore depends on assumptions about the physical properties of firn and ice during periods of past accumulation. Fortunately, these properties can be estimated from models confirmed by observations in modern firn layers (Schwander and Stauffer, 1984) and extrapolated to past climatic conditions inferred from the ice record itself (Schwander *et al.*, 1997). Corrections and uncertainties are larger for ice that accumulated more slowly, and thus they vary according to changes in accumulation conditions from location to location and through time (Schwander *et al.*, 1997). For example, the age difference between the ice and the air entrapped during the most recent 200 years at the Siple Station (in Ellsworth Land, West Antarctica) is estimated to be 80–85 yr (Neftel *et al.*, 1985), whereas the corresponding age difference for the more slowly accumulating ice at the Dome C site (on the polar plateau in East Antarctica) approaches 5,500 yr for samples from the last glacial maximum (LGM) (~18–24 ka) (Monnin *et al.*, 2001). The range of gas ages at any given depth likewise depends on the accumulation rate of the ice. Where accumulation rates are relatively low, the range is generally 5–10% of the age difference between the ice and the enclosed air. The relative range may be higher in ice that accumulates more rapidly. For ice cores that provide records of relatively recent trends, estimates of the gas age difference correction and age range can be confirmed by comparison with historical atmospheric measurements (see, e.g., Levchenko *et al.*, 1996). However, these estimates are less certain for longer ice-core records. Uncertainties in the dating of ice-core gases are an important concern in efforts to correlate the ice record of carbon-cycle changes with other variations recorded in the ice itself.

Some ice-core gas samples do not represent old air because they have undergone postdepositional alteration. In glaciated regions where seasonal melting occurs, the frozen and buried melt layers are enriched in CO_2 (Neftel *et al.*, 1983).

Table 2(a) Carbon dioxide data from polar ice cores. Carbon dioxide concentrations and carbon isotope values (in per mil relative to the PDB standard) are listed for the LGM (~18–24 ka), the Holocene Epoch (1–10 ka), and the most recent pre-industrial part of the Holocene Epoch (1000–1800 AD). Key references for the CO_2 data from each core site are listed in the right-hand column. The superscript letters in the data table indicate the references from which the indicated values were derived. Values for Vostok (LGM and Holocene) were calculated for this table from the numerical data cited.

Site	*Location*	*Data type*	*LGM*	*Holocene*	*Pre-industrial Holocene*	*References*
Vostok	78° 28′ S, 106° 48′ E	CO_2	187.7 ± 2–3 ppmv[a]	266.2 ± 2–3 ppmv[a]		Lorius *et al.*, 1985; Barnola *et al.*, 1987; Barnola *et al.*, 1991; Jouzel *et al.*, 1993; Fischer *et al.*, 1999; Petit *et al*, 1999[a]
Law Dome	66° 46′ S, 112° 48′ E	CO_2			275–284 (±1.2) ppmv (1000–~1800) with lower levels during 1550–1800 AD[a]; 279.2 ppmv (1006 AD)[b]	Pearman *et al.*, 1986; Etheridge *et al.*, 1988; Etheridge *et al.*, 1996[a]; Francey *et al.*, 1999[b]; Joos *et al.*, 1999; Gillett *et al.*, 2000
		$\delta^{13}CO_2$			−6.44 ± 0.013‰ (1006 AD)	Francey *et al.*, 1999
Taylor Dome	77° 48′ S, 158° 43′ E	CO_2	186–190 ppmv[a,d]	~275 ppmv ± 7.5 (avg Hol.)[b]; 268 ± 1 ppmv (10.5 ka); 260 ± 1 ppmv (8.2 ka); slow increase to 285 ppmv (1 ka)[c]		Smith *et al.*, 1999[a]; Fischer *et al.*, 1999[b]; Indermuhle *et al.*, 1999[c]; Indermuhle *et al.*, 2000[d]
		$\delta^{13}CO_2$	−7.5 to −7.0‰ (18–16.5 ka)[a]	−6.4 ± 1.6‰ (avg)[a]; −6.6‰ (11 ka), −6.3‰ (8 ka), slow decrease to −6.1‰ (1 ka)[b] (uncertainty: ±0.085‰)	−6.5‰[b]	Smith *et al.*, 1999[a]; Indermuhle *et al.*, 1999[b]
Siple Dome	75° 55′ S, 83° 55′ W	CO_2			280 ± 5 ppmv (1750); 279 ± 3 ppmv (1734–1756 & 1754–1776)[a]; 276.8 ppmv (1744)[b]	Neftel *et al.*, 1985[a]; Friedli *et al.*, 1986[b]
		$\delta^{13}CO_2$			−6.48 ± 0.15‰ (1744); −6.41‰ (avg of 3 samples from before 1800)	Friedli *et al.*, 1986
Byrd	80° 01′ S, 119° 31′ W	CO_2	~200 ppmv[a]	unreliable Holocene values, no results from 8.7–2.2 ka (brittle zone)[b]; ~280 ppmv (with 40 ppmv fluctuations-early Hol)[a]		Berner *et al.*, 1980; Neftel *et al.*, 1988[a]; Stauffer *et al.*, 1998; Indermuhle *et al.*, 1999[b]

(continued)

Table 2(a) (continued).

Site	*Location*	*Data type*	*LGM*	*Holocene*	*Pre-industrial Holocene*	*References*
		$\delta^{13}CO_2$	−6.84 ± 0.12‰	early Holocene: −6.65 ± 13‰; avg: −6.78‰ (excluding samples from brittle zone)	−6.52 ± 0.12‰	Leuenberger *et al.*, 1992
South Pole	90° S, 0°	CO_2			283 ± 5 ppmv (avg 430–770 yr BP)[b]; 278 ± 3 ppmv (1660–1880)[a]; 281 ± 3 ppmv (1450–1670)[a]; 279 ± 4 ppmv (950–1170)[a]	Neftel *et al.*, 1985[a]; Friedli *et al.*, 1986[b]; Siegenthaler *et al.*, 1988
		$\delta^{13}CO_2$			−6.69 ± 0.8‰ (avg. 430–770 yr BP) (uncertainty: 0.22‰)[a]; −6.34 ± 0.3‰ (500–1000 yr BP) (uncertainty: 0.22‰)[b]	Friedli *et al.*, 1984; Stauffer and Oeschger, 1985[b]; Siegenthaler *et al.*, 1988
Dome C	75° 06′ S, 123° 24′ E	CO_2	188 ppmv[a]	265 ppmv (Early Holocene): variation during Hol: 260–280 ppmv[a]		Lorius *et al.*, 1979; Delmas *et al.*, 1980; Monnin *et al.*, 2001[a]; Fluckiger *et al.*, 2002
D47	67° 23′ S, 154° 03′ E	CO_2		240 ppmv (6.0 ka) to 270 ppmv (6.8 ka): slow increase to 280 ppmv	273.2 ppmv (914 AD)	Barnola *et al.*, 1995
D57	68° 11′ S, 137° 33′ E	CO_2			284.8 ppmv (1310 AD)	Barnola *et al.*, 1995; Raynaud and Barnola, 1985
GISP2	72° 36′ N, 38° 30′ W	CO_2	~195 ppmv[b]	275 ± 9 ppmv (3.1–1.9 ka)[b]	280 ± 5 ppmv[a]	Wahlen *et al.*, 1991[a]; Smith *et al.*, 1997[b]
GRIP	72° 35′ N, 37° 38′ W	CO_2	205 ± 19 ppmv[b]		280–287.4 ppmv (946 AD)[a]	Barnola *et al.*, 1995[a]; Anklin *et al.*, 1995; Anklin *et al.*, 1997[b]
Camp Century	77° 12′ N, 61° 06′ W	CO_2	~200 ppmv			Berner *et al.*, 1980

Table 2(b) Methane data from polar ice cores, and Mt. Logan ice core (Yukon Territory, Canada). Methane concentrations are listed for the LGM (~18–24 ka), the Holocene Epoch (1–10 ka), and the most recent pre-industrial part of the Holocene Epoch (1000–1800 AD). Key references for the CH_4 data from each core site are listed in the right-hand column. The superscript letters in the data table indicate the references from which the indicated values were derived. Values for Vostok (LGM and Holocene) and Dome C (LGM and Holocene) were calculated for this table from the numerical data cited.

Site	*Location*	*LGM*	*Holocene*	*Pre-industrial Holocene*	*References*
Vostok	78°28′ S, 106°48′ E	369 ± 20 ppbv[a]	620.6 ± 20 ppbv[a]		Chappellaz *et al.*, 1990; Jouzel *et al.*, 1993; Blunier *et al.*, 1998; Petit *et al.*, 1999[a]
Law Dome	66°46′ S, 112°48′ E			avg: 695 ppb, with ~40 ppbv variation on century timescale (1000–1800 AD)	Pearman *et al.*, 1986; Etheridge *et al.*, 1992; Etheridge *et al.*, 1998
Taylor Dome	77°48′ S, 158°43′ E	377 ± 7 ppbv[a]	534 ± 11 ppbv (5–7 ka) (1 data point)[a]		Steig *et al.*, 1998; Brook *et al.*, 2000[a]
Dome C	75°06′ S, 123°24′ E	367 ± 10 ppbv[a]	673 ppbv (9–11.1 ka); Variation during Hol: 562–674 ppbv[a]	653 ppbv (0.5–1 ka)[b]	Monnin *et al.*, 2001[a]; Fluckiger *et al.*, 2002[b]
Siple Dome	75°55′ S, 83°55′ W			780 ± 90 ppbv (1771)	Stauffer *et al.*, 1985
Byrd	80°01′ S, 119°31′ W	350 ± 25 ppbv[a]		650 ± 25 ppbv[a]	Rasmussen and Khalil, 1984; Stauffer *et al.*, 1988[a]; Chappellaz *et al.*, 1997; Blunier *et al.*, 1998; Dallenbach *et al.*, 2000
South Pole	90° S, 0°			840 ± 60 ppbv (1630)	Stauffer *et al.*, 1985
GISP2	72°36′ N, 38°30′ W	395 ± 5 ppbv[b]	~730 ppbv (early Holocene); drop at 8.2 ka to mid-Hol values of ~575 ppbv; slow rise beg at 3 ka[a]		Brook *et al.*, 1996[a]; Chappellaz *et al.*, 1997; Sowers *et al.*, 1997; Brook *et al.*, 1999; Brook *et al.*, 2000[b]; Blunier and Brook, 2001
GRIP	72°35′ N, 37°38′ W	362 ± 2.6 ppbv[b]	654 ppbv (avg); 617 ± 2 (2.5–5 ka), 608 ± 5 (5–7 ka), 718 ± 3 (9.5–11.5 ka)[a]	710 ± 2 (0.25–1 ka)[a]	Raynaud *et al.*, 1988; Blunier *et al.*, 1993; Chappellaz *et al.*, 1993a; Blunier *et al.*, 1995; as corrected by Chappellaz *et al.*, 1997[a]; Blunier *et al.*, 1998; Dallenbach *et al.*, 2000[b]
Camp Century	77°12′ N, 61°06′ W			700 ± 30 ppbv	Rasmussen and Khalil, 1984
Dye 3	65°12′ N, 43°48′ W	350 ± 40 ppbv[a]		650 ± 30 ppbv[a]	Craig and Chou, 1982; Stauffer *et al.*, 1988[a]
Mt. Logan (Yukon Territory)	60°35′ N, 140°35′ W			750 ± 60 ppbv (1802)	Dibb *et al.*, 1993

High CO_2 concentrations may also be caused by chemical reactions among other constituents buried in the firn and ice. For example, many features of the CO_2 record in ice cores from Greenland are probably not reliable because of reactions between carbonate dust and trace acidic compounds in the ice (Delmas, 1993; Anklin *et al.*, 1995). Oxidation of organic compounds may likewise produce CO_2 that contaminates the CO_2 in entrapped air (Haan and Raynaud, 1998; Tschumi and Stauffer, 2000). *In situ* bacterial activity is another possible source of contamination (Sowers, 2001).

Other postdepositional changes may be caused by physical fractionation of gases in the thick and porous firn layer. Gravitational fractionation causes gases with higher molecular weights to be enriched at the depths of bubble enclosure. Although this process has a relatively minor effect (less than 1%) on concentrations of gases like CO_2 and CH_4 in air, it can significantly alter the more subtle concentration differences among species that differ only in their isotopic compositions (Craig *et al.*, 1988; Schwander, 1989). Similarly, thermal fractionation among different isotopic species may occur under conditions when a change in temperature at the ice surface is large and persistent enough to cause a temperature gradient in the firn (Severinghaus *et al.*, 1998). Additional fractionation may occur due to molecular size exclusion during bubble enclosure (Craig *et al.*, 1988). These effects must be taken into account in interpreting the stable isotopic compositions of ice-core gas samples.

Because of the possibility of postdepositional alteration, and the other problems relating to extraction, analysis, and dating described above, ice-core records are most compelling when they can be shown to agree across different locations, accumulation conditions, and analytical procedures. These conditions have been demonstrated for many ice-core gas records. For example, Table 2 shows the agreement among records of CO_2 and CH_4 concentrations during the LGM (~18–24 ka) and the Holocene Epoch. The agreement among some ice-core gas records is so substantial that more detailed time-dependent features have proven to be an important tool in the correlation of variations in other properties of the ice cores. Millennial variations in atmospheric methane concentrations, recorded in ice cores from both Antarctica and Greenland, provide a basis for detailed correlation between ice-core records from these locations (Steig *et al.*, 1998; Blunier *et al.*, 1998; Blunier *et al.*, 1999; Blunier and Brook, 2001). More gradual variations in the isotopic composition of atmospheric oxygen are likewise reflected in gases extracted from ice cores in both Greenland and Antarctica (Sowers and Bender, 1995). Because variations in the isotopic composition of atmospheric oxygen are controlled mainly by the isotopic composition of oxygen in seawater, the isotopic record of oxygen in ice-core air provides not only a tool for correlation among ice-core gas records, but also a robust means of correlation with the widely used marine oxygen isotope record measured in marine sediments (Shackleton, 2000).

8.09.3.2 Holocene Carbon-cycle Variations

Recent anthropogenic influences on greenhouse gases are clearly documented in ice-core records. Several ice cores from Antarctica provide strong evidence that the atmospheric concentration of CO_2 before the onset of human influence was 280 ± 5 ppmv, and ice cores from both Antarctica and Greenland document CH_4 levels of 650–730 ppbv before human influence (see Table 2 and references therein). Modern (2000 AD) atmospheric concentrations of CO_2 and CH_4 are near 370 ppmv and 1,800 ppmv, respectively (Blasing and Jones, 2002), reflecting the rapid increases due to anthropogenic sources during the last two centuries. The ice-core record also shows that the carbon isotope ratio of atmospheric CO_2, expressed as $\delta^{13}CO_2$, was $\sim -6.4‰$ before human influence (Friedli *et al.*, 1986; Francey *et al.*, 1999), compared to its present value near $-7.9‰$ (Francey *et al.*, 1999), reflecting the influence of ^{13}C-depleted anthropogenic CO_2 added to the atmosphere. ($\delta^{13}CO_2$ denotes the $^{13}CO_2 : {}^{12}CO_2$ ratio difference in per mil relative to the PDB standard. See Section 8.09.2.1.1.) The effects of human activities on atmospheric CO_2 and CH_4 are discussed in other chapters of this treatise (see Chapter 4.03 and Chapter 8.10). These recent trends provide an important test of the geologic ice-core record, because the ice-core gas measurements can be closely matched for recent decades with records from firn air and atmospheric measurements (Figures 4(a) and (b)). Measurements of carbon dioxide and its carbon isotopes in firn and in the uppermost sections of ice cores taken at the Siple Station and Law Dome sites in Antarctica show close agreement with the record of direct atmospheric measurements (Neftel *et al.*, 1985; Pearman *et al.*, 1986; Friedli *et al.*, 1986; Etheridge *et al.*, 1996; Francey *et al.*, 1999). Extrapolated trends in ice-core methane measurements at these sites (Stauffer *et al.*, 1985; Etheridge *et al.*, 1992) and at the Dye 3 site in Greenland (Craig and Chou, 1982) are likewise consistent with recent atmospheric measurements. These observations provide powerful support for the validity of ice-core gas records extending back in time.

Models of the modern global carbon cycle typically assume a pre-disturbance steady state in

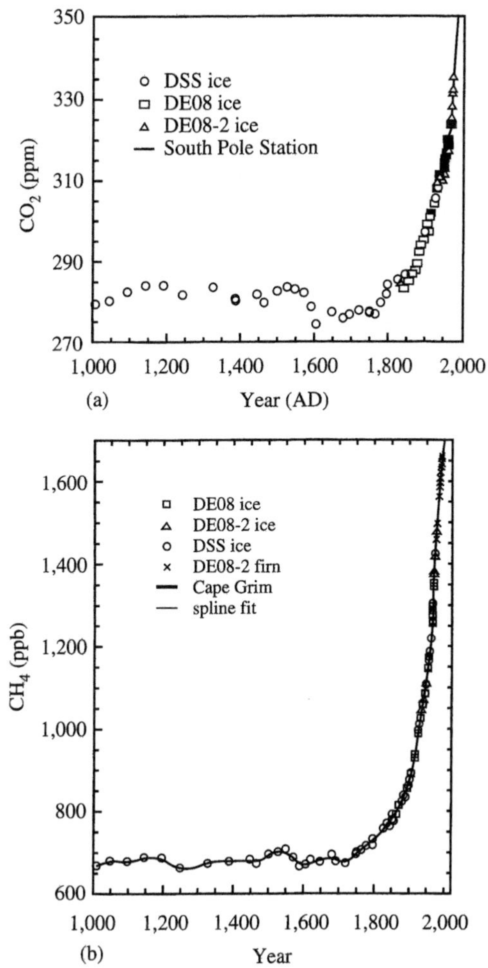

Figure 4 (a) Carbon dioxide concentrations over the last 1,000 yrs derived from three ice cores from Law Dome, Antarctica, and from air samples collected at South Pole Station since the late 1950s (Etheridge *et al.* (1996), reproduced by permission of the American Geophysical Union from *J. Geophys. Res.*, **1996**, *101*, 4115–4128 (Figure 4); the air sample data are from Keeling, 1991). (b) Methane concentrations over the last 1,000 yr derived from three Antarctic ice cores, the Antarctic firn, and archived air samples collected since 1978 at Cape Grim, Tasmania, Australia (Etheridge *et al.* (1998); reproduced by permission of the American Geophysical Union from *J. Geophys. Res.*, **1998**, *103*, 15979–15993 (figure 2); the air sample data are from Langenfelds *et al.*, 1996).

calculations of historical and future human influences on atmospheric greenhouse gases. The ice-core gas record provides a basis for testing this steady-state assumption. The record shows that atmospheric levels of both CO_2 and CH_4 varied during the most recent 10,000 years prior to human influence. The documented Holocene variations, while small and slow compared to the recent anthropogenic changes in atmospheric chemistry, provide important information about interactions among greenhouse gases, climate, and the global carbon cycle. Implications for the assumption of steady state in models are discussed below and in Section 8.09.2.2.

Subtle variations in atmospheric CO_2 during the most recent millennium, first hinted in some of the earliest Antarctic ice-core measurements (Raynaud and Barnola, 1985; Stauffer *et al.*, 1988; Etheridge *et al.*, 1988), have been confirmed by more detailed and precise analyses (Figure 5). CO_2 concentrations appear to have risen to near 285 ppmv in the thirteenth century AD, then decreased to ~275 ppmv during the sixteenth century AD, and then increased again during the eighteenth century AD (Barnola *et al.*, 1995; Etheridge *et al.*, 1996; Indermuhle *et al.*, 1999). Some of the earliest ice-core CH_4 measurements also suggested a period of decreased concentrations during the sixteenth and seventeenth centuries (Rasmussen and Khalil, 1984; Khalil and Rasmussen, 1989). CH_4 measurements in ice from the Law Dome site in Antarctica show small variations that parallel the CO_2 trends over the most recent millennium, including a decrease of ~40 ppbv coinciding with the CO_2 decrease in the sixteenth and seventeenth centuries (Etheridge *et al.*, 1998). The decrease in atmospheric CO_2 and CH_4 during the sixteenth and seventeenth centuries coincides with a maximum in $\delta^{13}CO_2$ at ~− 6.3‰ (Francey *et al.*, 1999). Because variations in terrestrial sources and sinks are the most likely cause of the CH_4 and carbon isotope trends, they are probably associated with changes in terrestrial carbon cycling, although simultaneous oceanic changes cannot be ruled out (Trudinger *et al.*, 1999; Joos *et al.*, 1999).

Times of slightly increased and decreased CO_2 and CH_4 during the most recent millennium appear to coincide roughly with the times of slightly warmer and cooler climate known as the "Medieval Warm Period" and the "Little Ice Age." The extent and significance of these climatic variations are somewhat unclear (see, e.g., Bradley (1999) and references therein), and the implied changes in climate forcing by the greenhouse gases are very small (Etheridge *et al.*, 1998). Nevertheless, these coincidences of greenhouse gas and climate trends are consistent with an understanding of close coupling between climate and carbon cycling that emerges from observations over a very broad range of timescales.

The likelihood of a "Little Ice Age" carbon-cycle perturbation poses a particular concern for the assumption of pre-anthropogenic steady state in models of the modern carbon cycle. Part of the increase in concentrations of CO_2 and CH_4 during recent centuries may be due to trends associated with the natural climate anomaly. This association may seriously affect estimates of eighteenth- and early nineteenth-century human impacts on the

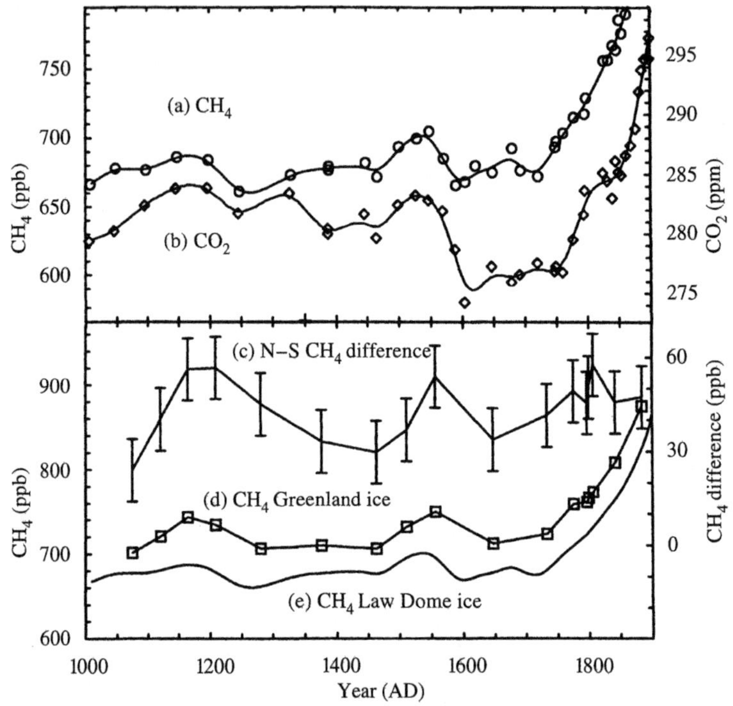

Figure 5 Variability of (a) methane (using the same dataset as that of Figure 4(b)) and (b) carbon dioxide (using the same dataset as that of Figure 4(a)) concentrations over the past 1,000 yr. (c) The interpolar difference in methane, which is the difference between the Greenland ice-core record (d) and the Law Dome methane record (e, same as curve a) for corresponding years. Estimated 1σ uncertainty for the interpolar difference is 10 ppb (Etheridge *et al.* (1998); reproduced by permission of the American Geophysical Union from *J. Geophys. Res.*, **1998**, *103*, 15979–15993 (figure 3)).

terrestrial biosphere based on greenhouse gas trends. More importantly, it may imply the need for climate and carbon cycling to be coupled in models of the modern carbon cycle (Enting, 1992). The maximum natural rates of change in atmospheric CO_2 and CH_4 during the most recent millennium were at least an order of magnitude smaller than current anthropogenic rates of change, and the natural variations were much less persistent than the increasing trends of the most recent two centuries. Thus the primary concern raised by natural carbon-cycle variations during the most recent millennium is not that they may be significant relative to anthropogenic changes, but that they suggest a degree of global carbon/climate coupling that is not well represented in current models of present-day and potential future carbon cycling.

Coupling between variations in carbon cycling and climate is also a prominent theme in efforts to understand the ice-core gas records extending back through the most recent 10,000 years. The records of CO_2, $\delta^{13}CO_2$, and CH_4 for this period show variations that are larger and more complex than those of the most recent millennium. Measurements from the Taylor Dome and Dome C sites provide evidence that atmospheric CO_2 concentrations decreased from ~270 ppmv at 10.5 ka to 260 ppmv at 8 ka and then gradually increased to values near 285 ppmv at 1 ka (Figure 6; (Indermuhle *et al.*, 1999; Fluckiger *et al.*, 2002)). The Holocene $\delta^{13}CO_2$ record from Taylor Dome is much less detailed and precise, and cannot be confirmed by overlap with the $\delta^{13}CO_2$ record of the most recent millennium (Francey *et al.*, 1999), but it appears to show a maximum near −6.3‰ between 7 ka and 8 ka and a minimum near −6.6‰ between 2 ka and 3 ka (Indermuhle *et al.*, 1999). Measurements in ice cores from both Greenland and Antarctica show that atmospheric methane concentrations also varied significantly during the Holocene Epoch (Figure 6). Concentrations decreased from levels near 700 ppbv at 10 ka to levels less than 600 ppbv at 5 ka, followed by a gradual return to values near 700 ppbv at 1 ka (Blunier *et al.*, 1995; Chappellaz *et al.*, 1997; Fluckiger *et al.*, 2002).

Carbon-cycle responses to Holocene climate change have been hypothesized to explain the atmospheric CO_2 and CH_4 trends observed in the ice-core record of this period. The early Holocene expansion of terrestrial vegetation and soils in areas that were previously glaciated has been suggested as the cause of the early Holocene decrease in CO_2 (Indermuhle *et al.*, 1999). Likewise, the increase in atmospheric CO_2 between 8 ka and 1 ka has been attributed to a release of biospheric carbon caused by a global trend during

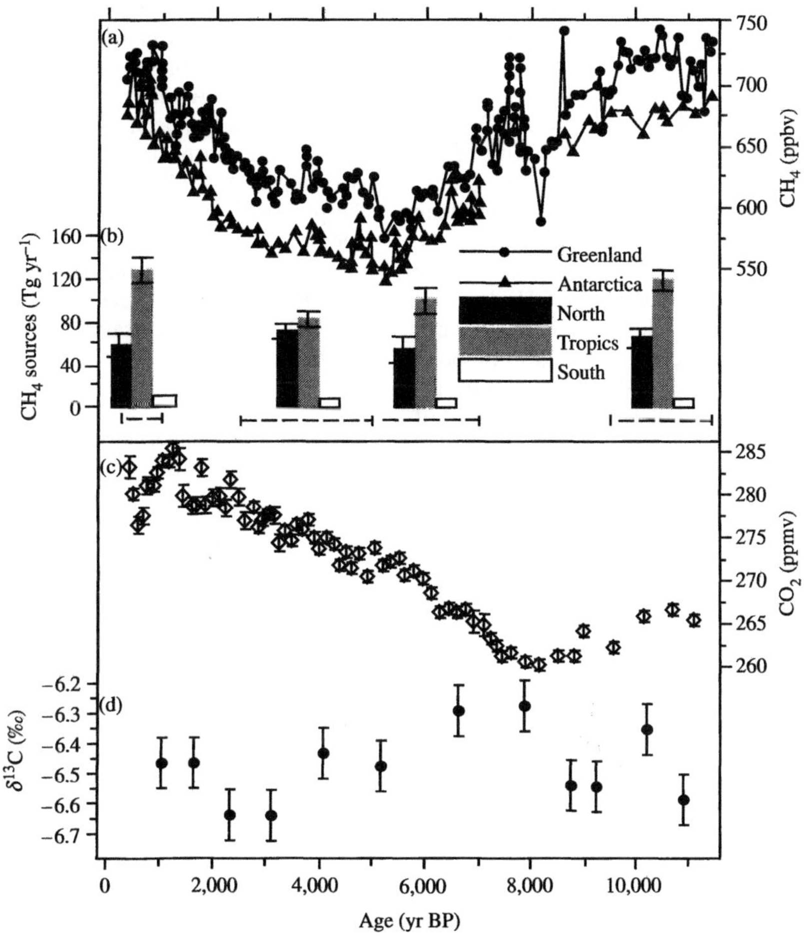

Figure 6 The Holocene ice-core record of the variability of methane, carbon dioxide, and $\delta^{13}CO_2$. (a) The methane record from ice cores in Greenland and Antarctica, showing significant variability including variability in the interpolar gradient (Chappellaz *et al.*, 1997). (b) The results of a three-box model used to infer the latitudinal distribution sources of methane at four different time spans during the Holocene Epoch (Chappellaz *et al.*, 1997). (c, d) The carbon dioxide concentration and its isotopic composition from Taylor Dome, Antarctica (Indermuhle *et al.*, 1999). Note the dissimilarities among the trends in this figure (Raynaud *et al.* (2000); reproduced by permission of Elsevier from *Quat. Sci. Rev.*, **2000**, *19*, 9–17 (figure 1)).

that period toward cooler and drier conditions (Indermuhle *et al.*, 1999). In contrast, one explanation for the Late Holocene increase in atmospheric CH_4 is an expansion of boreal wetland source areas (Blunier *et al.*, 1995; Chappellaz *et al.*, 1997; Velichko *et al.*, 1998). (Some investigators have suggested that the Late Holocene increase in atmospheric CH_4 may be partly due to emissions from the onset of early human rice cultivation (Subak, 1994; Chappellaz *et al.*, 1997; Ruddiman and Thomson, 2001)). Oceanic responses to climate change have also been suggested to explain the Holocene CO_2 and $\delta^{13}CO_2$ trends (Indermuhle *et al.*, 1999; Broecker *et al.*, 2001). This variety of hypotheses about Holocene carbon cycling is not surprising, because the Holocene variations in CO_2, $\delta^{13}CO_2$, and CH_4 are clearly dissimilar (Figure 6). Moreover, as discussed in Section 8.09.2.1.2, significant changes in atmospheric CH_4 can be caused by changes in specific sources and sinks that do not necessarily require the larger translocations of carbon needed to change atmospheric CO_2 and $\delta^{13}CO_2$. It seems likely that a variety of processes, reflecting the regional and global complexity of both the carbon cycle and the climate system, contributed to the trends observed in the ice-core record of Holocene CO_2 and CH_4.

Important clues to these processes come from analysis of details available in the Holocene ice-core CH_4 record. The nature of CH_4 cycling through the atmosphere (see Section 8.09.2.1.2) assures that any significant change in CH_4 sources will quickly affect atmospheric CH_4 concentrations. Although the atmospheric lifetime of CO_2 is shorter than that of CH_4, a change in CO_2 sources must be much larger (in terms of carbon transfer) to affect the larger mass of atmospheric

CO_2, and the response of atmospheric CO_2 will be attenuated and prolonged by its exchange with the ocean surface and the terrestrial biosphere (see Section 8.09.2.1.1). Thus, atmospheric CH_4 concentrations are more susceptible than CO_2 concentrations to rapid variations resulting from changes in sources. Abrupt changes and "spikes" are more common in the ice-core CH_4 record than in the ice-core CO_2 record. For example, the Greenland GRIP and GISP2 ice cores (see references in Table 2) record a sharp CH_4 decrease at 8.2 ka (see Figure 6). This feature corresponds to a widespread climatic event linked to outburst flooding from the melting of the Laurentide ice sheet (Barber *et al.*, 1999). The ice-core record of abrupt changes in atmospheric CH_4 is an important source of information about links between climate and carbon-cycle changes.

Whereas the oxidation of CH_4 occurs in the atmosphere worldwide (see Section 8.09.2.1.2), the sources of CH_4 are not distributed uniformly. The dominant natural sources of CH_4 are wetlands, which occur today mainly in the tropics and the northern hemisphere. The CH_4 produced by northern wetland sources is oxidized during its transit southward in the atmosphere, resulting in a north-to-south decreasing atmospheric CH_4 concentration gradient, well documented in ice-core CH_4 measurements of air from Greenland and Antarctica (Rasmussen and Khalil, 1984; Nakazawa *et al.*, 1993). The gradient appears to have varied during the most recent millennium, with atmospheric CH_4 concentrations over Greenland exceeding those over Antarctica by $24-58 \pm 10$ ppbv (Etheridge *et al.*, 1998) (Figure 5). During the most recent 10,000 years, the difference between Greenland and Antarctica CH_4 concentrations ranged from 33 ± 7 ppbv during the middle Holocene Epoch (~7–5 ka) to values as high as 50 ± 3 ppbv during the Late Holocene Epoch (~5–2.5 ka) and 44 ± 4 ppbv during the Early Holocene Epoch (11.5–9.5 ka) (Chappellaz *et al.*, 1997; Figure 6). The ice core record of the atmospheric CH_4 gradient implies that northern wetlands (as well as tropical wetlands) have been an important CH_4 source throughout the Epoch. The variations in the interhemispheric CH_4 gradient, together with the trends in CH_4 concentrations described above and shown in Figure 6, have been used to infer changes in the magnitude and relative importance of tropical versus northern wetland CH_4 sources (Chappellaz *et al.*, 1997; Etheridge *et al.*, 1998). Recent anthropogenic sources of CH_4, which occur predominantly in the northern hemisphere, have caused the modern north-to-south CH_4 concentration gradient to be greater than the Holocene gradient by a factor of about three (Chappellaz *et al.*, 1997). Anthropogenic sources of CO_2 have similarly caused a present-day north-to-south atmospheric CO_2 gradient of several ppmv (Pearman *et al.*, 1983; Fung *et al.*, 1983; Enting and Mansbridge, 1991) (see Chapter 8.10). A small south-to-north decreasing CO_2 gradient has been hypothesized for the Late Holocene atmosphere prior to human influence (Keeling *et al.*, 1989; Taylor and Orr, 2000), but the proposed gradient is difficult to verify in ice-core records because it requires correction for postdepositional artifacts in Greenland ice, and its magnitude is close to the measurement precision for CO_2 in ice-core samples (but see Barnola (1999)).

8.09.3.3 Glacial/interglacial Carbon-cycle Variations

The ice coring effort at Vostok Station in East Antarctica required several decades and a remarkable combination of international (Russian, French, and American) technological and analytical expertise. The resulting record (Figure 7; Jouzel *et al.*, 1987; Barnola *et al.*, 1987; Chappellaz *et al.*, 1990; Jouzel *et al.*, 1993; Petit *et al.*, 1999) extends continuously to depths exceeding 3 km and provides detailed evidence for the close relationship between climate and carbon cycling throughout the four most recent glacial/interglacial cycles. The Vostok ice-core data reveal the range of natural variations in atmospheric CO_2 and CH_4 during this period, and demonstrate that present-day levels of CO_2 and CH_4 are much higher than they have been at any time during the most recent 420,000 years (Petit *et al.*, 1999). The Vostok record is a preeminent constraint on analyses and hypotheses concerning variations in climate and the carbon cycle over glacial/interglacial timescales.

The most conspicuous feature of the Vostok data (Figure 7) is the close similarity among trends in CO_2, CH_4, and local temperature (calculated from the ratio of deuterium to hydrogen in the ice). These trends all show a pattern of relatively rapid onsets of warm (interglacial) conditions, followed by more gradual transitions to cool (glacial) conditions, repeated in cycles ~100 kyr long. This "sawtoothed" pattern resembles characteristics long observed in paleoclimate records from marine sediments worldwide (see, e.g., Emiliani, 1966). The low CO_2 and CH_4 concentrations observed in ice that formed during glacial periods confirmed earlier measurements in ice from the most recent glacial period collected at other sites in both Greenland and Antarctica (Berner *et al.*, 1980; Delmas *et al.*, 1980; Stauffer *et al.*, 1988). At times of minimum glacial temperatures, CO_2 concentrations dropped to values near 180 ppmv, and then rapidly rose to values of 280–300 ppmv during interglacials. Similarly, glacial periods coincided with minimum CH_4 concentrations of

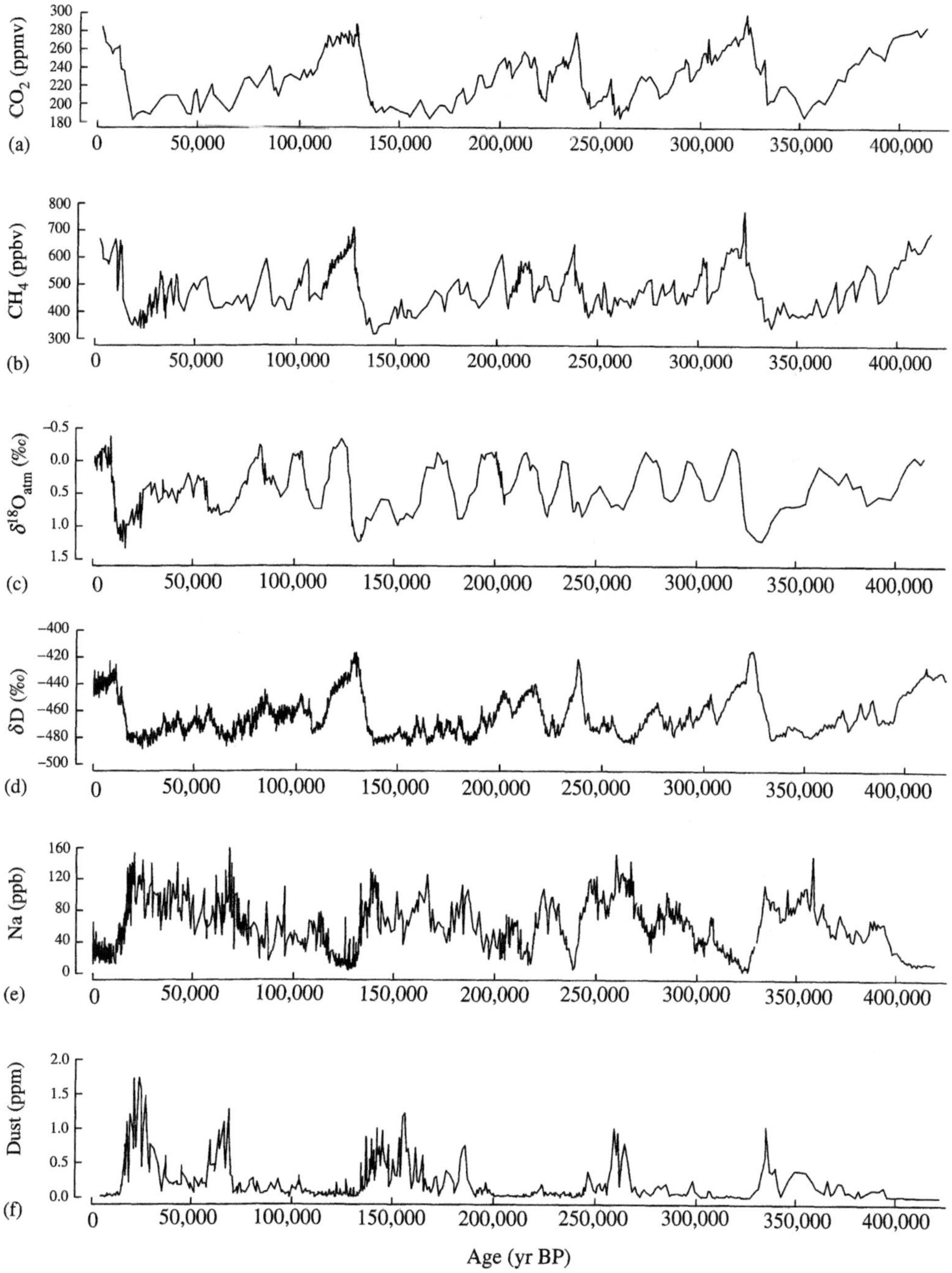

Figure 7 The Vostok ice-core record (Petit *et al.*, 1999). (a)–(c) Carbon dioxide, methane, and oxygen isotope ratios (expressed as $\delta^{18}O$) of oxygen in air extracted from the ice. (d) Hydrogen isotope ratios (expressed as δ^2H, or δD) in the ice. This record is a proxy for local temperature. (e) Sodium in the ice, a measure of sea salt aerosol deposition. (f) Dust in the ice, a measure of continental aerosol deposition. (a)–(c) are plotted against the estimated age of the ice air; (d)–(f) are plotted against the estimated age of the ice.

320–350 ppbv, and CH_4 concentrations rose to maximum values of 650–770 ppbv during interglacials. The Vostok record revealed that maximum and minimum temperature, CO_2, and CH_4 levels were similar during each of the four most recent glacial cycles, including the Holocene Epoch prior to human influence. Other detailed features of the CO_2, CH_4, and isotopic temperature variations appear to be related: statistical correlation of the Vostok data yields overall r^2 values exceeding 0.7 for both CO_2 and CH_4 with respect to the ice isotopic composition, which is a proxy for local temperatures (Petit *et al.*, 1999).

The correlation among these records is so high that local or regional causes—rather than global conditions—might be considered to explain such a close correspondence. However, there is excellent agreement among glacial/interglacial ice-core CH_4 records from Antarctica and Greenland (Brook *et al.*, 1996; Brook *et al.*, 2000), and among glacial/interglacial CO_2 records from diverse locations in Antarctica (see, e.g., Fischer *et al.*, 1999). Unfortunately, due to higher levels of reactive impurities in Greenland ice (see Section 8.09.3.1), there is no ice-core CO_2 record from Greenland that can be compared in detail to the Antarctic CO_2 measurements. But the Antarctic CO_2 measurements are much less susceptible to effects of impurities (Anklin *et al.*, 1997). The agreement among CH_4 measurements in the same ice cores suggests strongly that the Antarctic ice-core CO_2 record, like the CH_4 record, can be considered a record of global glacial/interglacial changes in atmospheric composition. Likewise, the Vostok deuterium temperature record and other Antarctic and Greenland records of stable isotopes in ice can be linked to well-known global climate variations by direct matching of obvious similarities to marine sediment isotope records (Shackleton *et al.*, 2000; Shackleton, 2001) and by correlations between oxygen isotope variations in the ice-core air and in marine sediments (Bender *et al.*, 1999; Shackleton, 2000).

The Southern Ocean might be an important regional influence on both temperature and CO_2 in Antarctica. But if this influence contributes to the close correlation of temperature and CO_2 in the Vostok record, then the Southern Ocean is likely part of a complex interaction among processes at high southern latitudes that plays an important role in both global climate and the global carbon cycle (Petit *et al.*, 1999). There is no known local or regional CH_4 source near Antarctica; thus, the close correlation of temperature and CH_4 at Vostok seems to imply an even broader scope of interconnected global processes. The Vostok ice-core record provides compelling evidence that climate and carbon-cycle variations were closely interactive throughout the Late Pleistocene period.

This understanding of Pleistocene climate/carbon coupling is illustrated by the fact that many paleoclimatologists conclude that the variable greenhouse effect of atmospheric CO_2 and CH_4 has contributed significantly to glacial/interglacial climate change, while many geochemists consider climate change to have been an essential driver for glacial/interglacial variations in the carbon-cycle processes that control atmospheric CO_2 and CH_4.

8.09.3.3.1 Carbon-cycle influences on glacial/interglacial climate

Paleoclimatologists have analyzed the timing of Pleistocene climate cycles and confirmed statistical similarities to the periodicities of cyclic changes in orbital parameters that control the seasonal and latitudinal distribution of solar radiation (insolation) reaching the Earth's surface (Broecker and van Donk, 1970; Shackleton and Opdyke, 1976; Imbrie *et al.*, 1992; Imbrie *et al.*, 1993). The influence of these orbital parameters on global climate was hypothesized and calculated by Milutin Milankovitch (1879–1958), for whom the identified periodicities have come to be called "Milankovitch cycles." Although these periodicities (predominantly at 100,000 yr, 41,000 yr, 23,000 yr, and 19,000 yr) have been identified in many paleoclimate records, there is still no consensus concerning the exact mechanisms by which variations in the Earth's orbital configuration affect climate. Mathematical models and other analyses of global climate suggest strongly that the insolation changes directly attributed to orbital causes are not sufficient to explain the magnitude and the relative importance of the periodicities observed in differences between glacial and interglacial climates (Berger, 1979, 1988; Rind *et al.*, 1989; Imbrie *et al.*, 1993). Thus, the study of Pleistocene climate requires a focus on the processes that might have been capable of amplifying the subtle effects of orbital changes.

Paleoclimatologists have identified several possible amplifying feedback in the differences observed between glacial and interglacial climatic conditions. One important feedback is the "ice–albedo" feedback, which is caused by changes in the reflection of solar radiation back into space by expanding and retreating glaciers. The changing greenhouse effect is another likely amplifying feedback, through variations in the retention of heat in the atmosphere caused by rising and falling concentrations of greenhouse gases, including CO_2 and CH_4 (Broecker, 1982; Manabe and Broccoli, 1985).

No single feedback mechanism appears to be sufficient to explain the full magnitude or character of the climatic amplification between Late Pleistocene glacial and interglacial conditions. Variations in heat absorption due to greenhouse gases might have accounted for about half of the total glacial–interglacial difference in radiative forcing (Lorius *et al.*, 1990). The glacial–interglacial difference in the greenhouse effect is attributed primarily to CO_2, which contributed ~5 times as much as CH_4 and N_2O combined to the change in radiative forcing (Petit *et al.*, 1999). Thus, the variable greenhouse effect is viewed as an important contributing feedback mechanism in the glacial–interglacial climate system, but the influence of greenhouse gases is viewed as operating in a very complex association with other important feedback (Hewitt and Mitchell, 1997; Felzer *et al.*, 1998; Berger *et al.*, 1998; Petit *et al.*, 1999; Yoshimori *et al.*, 2001).

In addition to contributing to the global temperature differences between glacial and interglacial conditions, the greenhouse gas feedback might interact in complex ways with other aspects of Pleistocene climate variability. For example, one of the major mysteries of Pleistocene climate is the unexplained appearance, beginning ~700 ka, of the pronounced 100 kyr cycles that dominate paleoclimate records such as the Vostok data (Shackleton and Opdyke, 1976). Because the magnitude of the orbital insolation effects with 100 kyr periodicity is relatively weak in comparison to the effects with other periodicities, the 100 kyr cycles would be expected to be less conspicuous than those at the other orbital periodicities (as evident in paleoclimate records prior to 700 ka). Long-term feedback processes have been suggested as a way to explain the enhancement of the longer cycles. Several studies have suggested that the variable greenhouse effect of CO_2 or CH_4 might provide the needed long-term feedback because of the long response times associated with some components of the carbon cycle (see Section 8.09.2) (Pisias, 1984; Saltzman, 1987; Shackleton, 2000; Kennett *et al.*, 2003). Alternatively, others have suggested that the changing character of Late Pleistocene climate variability, including the unexplained appearance of pronounced 100 kyr periodicity, might be a response of shifting climate thresholds associated with a long-term decrease in atmospheric CO_2 since the Pliocene Epoch (Berger *et al.*, 1999).

8.09.3.3.2 *Climate influences on glacial/interglacial carbon cycling*

Geochemists have intensely debated the processes that caused atmospheric CO_2 and CH_4 concentrations to change between glacial and interglacial periods. The broad scope of this debate reflects the many ways that climate change can influence the global carbon cycle.

The influence of glacial/interglacial climate on the terrestrial carbon cycle is seen most dramatically in the ice-core CH_4 record. Because rates of atmospheric CH_4 oxidation during glacial periods were probably not drastically different from those during interglacials (see Section 8.09.2.1.2), differences between glacial and interglacial atmospheric CH_4 levels are attributed primarily to variations in the principal natural CH_4 source, wetlands. Wetlands occur primarily in tropical regions, where they are sensitive to variations in monsoonal rainfall, and in northern midlatitude and boreal regions, where they are sensitive to the changing extent of glacial ice and associated conditions of temperature and water balance (Matthews and Fung, 1987; Bubier and Moore, 1994). It is very likely that the distribution and extent of these wetland CH_4 sources were affected by differences between glacial and interglacial climate (Prell and Kutzbach, 1987; Khalil and Rasmussen, 1989; Chappellaz *et al.*, 1990; Petit-Maire *et al.*, 1991; Crowley, 1991; Chappellaz *et al.*, 1993a,b; Velichko *et al.*, 1998). The ice-core record of atmospheric CH_4 can be used to infer changes not only in the magnitude of wetland CH_4 sources, but also in their geographic distribution (see Section 8.09.3.2). The CH_4 concentration gradient between Greenland and Antarctica during the most recent glacial period ranged from ~40 ppbv during relatively warm periods to near zero during the coldest part of the last glacial period (Brook *et al.*, 2000; Dallenbach *et al.*, 2000). Thus, northern CH_4 sources appear to have been significant during all but the coldest glacial climates, and northern wetlands may have persisted throughout most of the last glacial period.

Climatic influences are apparent not only in the record of differences between glacial and interglacial atmospheric CH_4 concentrations and budgets, but also in more frequent variations that occurred over millennial timescales extending throughout the most recent glacial period (Figure 8). These variations correlate very closely with the abrupt interstadial warming events known as Dansgaard–Oeschger events, documented in the stable isotope record in Greenland ice cores (Oeschger *et al.*, 1984; Dansgaard *et al.*, 1984; Chappellaz *et al.*, 1993a; Brook *et al.*, 1996). The millennial-scale CH_4 variations, which range in magnitude from ~50 ppbv to ~300 ppbv, can be readily correlated in ice cores from both Greenland and Antarctica. In addition to providing a global record of carbon-cycle response to climate change, they have been used as a correlation tool to demonstrate synchronies and asynchronies between the Dansgaard–Oeschger events and corresponding variations in Antarctic climate (Steig *et al.*, 1998; Blunier *et al.*, 1998; Blunier and Brook, 2001).

The correlation of millennial climate fluctuations in Greenland and Antarctica is very important in efforts to understand the source of atmospheric CO_2 variations over millennial timescales during the most recent glacial period (Figure 9). These variations, on the order of 20 ppmv, are not as conspicuous as the millennial-scale CH_4 variations, and they appear to have occurred only in association with the largest and longest Dansgaard–Oeschger events (Stauffer *et al.*, 1998). Larger glacial CO_2 variations, in excess of 50 ppmv, were observed to correlate with Dansgaard–Oeschger events in ice cores from Greenland (Stauffer *et al.*, 1984). However, these features are now definitively interpreted as artifacts of postdepositional chemical reactions within the ice (Delmas, 1993; Tschumi and Stauffer, 2000). The smaller glacial CO_2

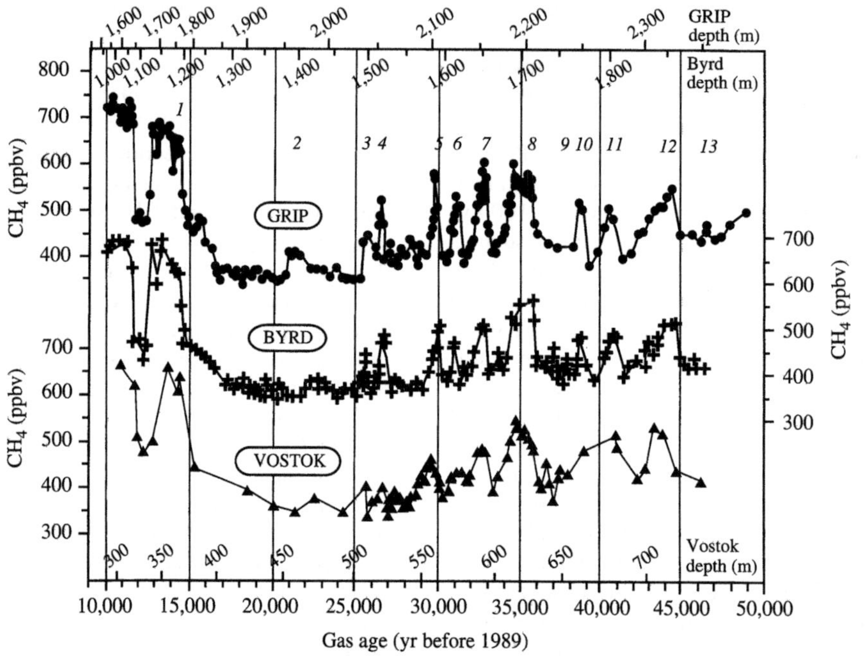

Figure 8 Methane variations during the most recent glacial period as recorded in ice cores from Greenland (GRIP) and Antarctica (Byrd, Vostok) (Blunier *et al.*, 1998; Stauffer *et al.*, 1998). The records were correlated by statistical matching of the variations shown (Blunier *et al.* (1999); reproduced by permission of Kluwer Academic/Plenum Publishers from *Reconstructing Ocean History: A Window into the Future,* **1999**, pp. 121–138 (figure 1)).

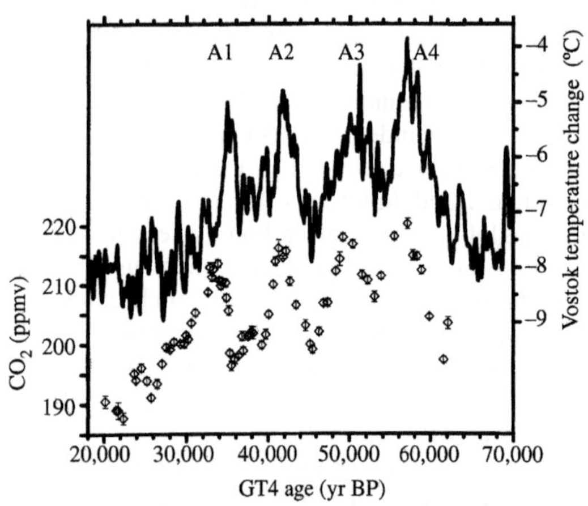

Figure 9 Correlation of Antarctic temperature variations (solid line, calculated from hydrogen isotope ratios in the ice at Vostok) and CO_2 variations in air extracted from the Taylor Dome ice core. The temperature curve is a running mean of the data from Petit *et al.* (1999) (Indermuhle *et al.* (2000); reproduced by permission of the American Geophysical Union from *Geophys. Res. Lett.*, **2000**, *27*, 735–738 (figure 2)).

variations seen in Antarctic ice cores do not appear to be affected by postdepositional reactions (Stauffer *et al.*, 1998). Detailed correlation, based on the more clearly defined millennial-scale CH_4 variations, suggests that the glacial variations in atmospheric CO_2 more closely paralleled the less abrupt and out-of-phase temperature variations recorded in Antarctic ice cores (Blunier *et al.*, 1999; Indermuhle *et al.*, 2000). Given these subtle differences in sensitivity and phasing, atmospheric CO_2 and CH_4 appear to have been associated with different components of glacial climate variability at millennial timescales. Similar differences are observed in the record of CO_2 and CH_4 variations during the Holocene Epoch (see Section 8.09.3.2).

These differences do not obscure the obvious similarities between the larger CO_2 and CH_4 trends that characterize the longer glacial/interglacial cycles (Figure 7). Because the differences between glacial and interglacial CH_4 levels are attributed to changes in terrestrial sources, it seems likely that the parallel changes in atmospheric CO_2 might also reflect changes in the terrestrial biosphere. Unfortunately, the changes that most likely controlled glacial/interglacial CH_4 sources would be expected to affect CO_2 sources in the opposite manner. Wetlands—particularly peatlands—are known to store far more carbon per unit area than the soils that form under drier climatic conditions (Schlesinger, 1997), and there is some evidence of a direct relationship between the intensity of CH_4 emissions and the rate of carbon production and storage among individual wetland ecosystems (Whiting and Chanton, 1993). Thus, an expansion of wetlands and their CH_4 emissions during interglacial times would be associated with a decrease in soil CO_2 emissions. Indeed, as we have seen in Section 8.09.2.1.1, most estimates of climatic effects on terrestrial carbon storage (by wetlands and other ecosystems) suggest that the global land surface stored significantly less carbon during the most recent glacial period than during the Holocene Epoch. While these terrestrial trends might help to explain glacial/interglacial differences in atmospheric CH_4, they require that we look elsewhere in order to understand the glacial/interglacial CO_2 budget.

As discussed in Section 8.09.2.1.1, the deep ocean is the largest and most probable reservoir capable of large transfers of carbon over the timescales of glacial/interglacial transitions. All of the processes that account for deep-ocean carbon storage are susceptible to effects of climate change. These processes can be summarized in three general categories that can be described as the "solubility pump," the "soft-tissue pump," and the "carbonate pump" (Volk and Hoffert, 1985).

The oceanic solubility pump stores CO_2 in the deep ocean by simple dissolution of atmosphere CO_2. As deep-ocean water forms by sinking from the surface, it carries with it the CO_2 dissolved by gas exchange with the atmosphere. Because deep water forms where the ocean surface is coldest, and because the solubility of CO_2 in seawater is highest at low temperatures, the amount of CO_2 dissolved in sinking deep water tends to be greater than the average amount of CO_2 dissolved in global ocean surface waters. The solubility pump is sensitive to the conditions that affect local gas exchange at the sites of deep-water formation. Climate affects all of these conditions, including temperature, salinity, wind, sea state, sea-ice cover, planktonic metabolism, and prevailing currents and vertical mixing patterns. The effects of climate change on the solubility pump have long comprised a central topic in studies of glacial/interglacial CO_2 change (Newell *et al.*, 1978), and the importance of these effects continues to be a topic of intense debate (see, e.g., Bacastow (1996) and Toggweiler *et al.* (2003)).

The oceanic soft-tissue pump stores carbon in the deep ocean by the sinking of organic debris produced by microorganisms at the ocean surface. Much of the sinking organic material is oxidized in the deep ocean, yielding CO_2 and smaller quantities of nitrate, phosphate, and other nutrients. Like the solubility pump, the soft-tissue pump is sensitive to climate change. The assimilation of organic matter by photosynthesis at the ocean surface depends on temperature, salinity, and the availability of light and nutrients. Nitrate and phosphate are supplied to the ocean surface primarily by the upward mixing of deeper waters, where their concentration ratio is nearly the same as in the sinking organic material from which they are derived. Thus, the soft-tissue pump is sensitive not only to direct effects of climate such as light and temperature at the ocean surface, but also to more complex effects such as the impact of climate on ocean mixing and the cycling of nutrients. Changes in oceanic phosphate levels may have occurred as a result of the glacial/interglacial exposure and flooding of continental shelves (and their "reactive sediment" reservoir; see Figure 1) during changes in sea level (Broecker, 1982). Changes in nitrate concentrations may have resulted from glacial/interglacial variations in ocean mixing and other conditions that affect rates of oceanic nitrogen fixation and denitrification (McElroy, 1983; Altabet *et al.*, 1999). Changes in the availability of iron and other trace nutrients may have been caused by glacial/interglacial alterations in the windborne transport of continental dust to ocean areas (Martin *et al.*, 1990; Martin, 1990). The oceanic soft-tissue pump encompasses a wide array of potential climate sensitivities because it links deep-ocean carbon storage to the oceanic cycles of nitrogen, phosphorus, and other nutrients.

The oceanic carbonate pump removes carbon from the ocean surface by the production of calcium carbonate ($CaCO_3$) in the shells, tests, and skeletal frameworks of various marine organisms. Some of this $CaCO_3$ settles as debris into the deep ocean, where much is dissolved and added to the deep-ocean DIC reservoir. The $CaCO_3$ that does not dissolve is buried in sediments, where it may persist to comprise a major component of the carbon cycle over timescales of millions of years. The carbonate pump differs from the solubility and soft-tissue pumps in three fundamental ways. First, unlike the solubility and soft-tissue pumps, the carbonate pump transfers significant quantities of alkalinity as well as carbon from the ocean surface to the deep sea. The result of this "alkalinity pump" is to partially offset the effect of deep-ocean carbon storage on atmospheric CO_2. This offset results from the effect of alkalinity removal on the chemical equilibrium between dissolved CO_2 and the other dissolved inorganic carbon species in ocean surface waters. For every mole of $CaCO_3$ removed from the ocean surface, ~0.6 mol of CO_2 are released to the atmosphere (Sundquist, 1993). Because many of the organisms that produce $CaCO_3$ in the ocean surface are restricted to warmer waters and/or shallow shelf substrates, the alkalinity pump is sensitive to climate and sea-level change.

Second, unlike the solubility and soft-tissue pumps, the carbonate pump supplies a large reservoir of "reactive sediments" (Figure 1) that almost certainly affected glacial/interglacial ocean carbon storage by contributing large net removals or additions through changes in $CaCO_3$ production and dissolution. The glacial/interglacial exposure and flooding of warm shallow-water shelf areas probably affected rates of $CaCO_3$ production and dissolution in coral reefs and other carbonate-rich sedimentary environments (Berger, 1982; Berger and Keir, 1984). Glacial/interglacial fluctuations in the extent of deep-sea $CaCO_3$ dissolution are one of the most widespread and conspicuous features in the Quaternary marine sediment record (Broecker, 1971; Peterson and Prell, 1985; Farrell and Prell, 1989). The influence of these fluctuations on atmospheric CO_2 probably occurred primarily through their effect on oceanic alkalinity. For example, a decrease in deep-sea $CaCO_3$ dissolution tends to decrease deep-ocean alkalinity as well as DIC. When these changes are mixed upward to the ocean surface, atmospheric CO_2 will increase in a manner identical to the chemical equilibrium effect of removing $CaCO_3$ from the ocean surface. Conversely, an increase in deep-sea dissolution will tend to increase deep-ocean alkalinity and DIC, and every mole of dissolved $CaCO_3$ that consequently reaches the ocean surface will cause ~0.6 mol of CO_2 to be absorbed from the atmosphere (Sundquist, 1993).

The third characteristic that distinguishes the carbonate pump is the fact that it is metered by the chemical equilibrium relationship between seawater and the carbonate minerals (primarily calcite) in the reactive sediment pool. Changes in the DIC and/or alkalinity of deep-ocean waters may cause changes in carbonate dissolution through their effect on the state of saturation of the water with respect to the carbonate minerals calcite and aragonite. Although carbonate dissolution is also affected by the diagenesis of organic matter in sediment pore waters (Archer and Maier-Reimer, 1994), chemical equilibrium appears to be the primary control on the distribution of carbonate dissolution in the deep sea (Broecker and Takahashi, 1978; Plummer and Sundquist, 1982). The response time of this process is on the order of 5–10 kyr (Sundquist, 1990a; Archer *et al.*, 1997), providing a mechanism for maintaining the overall balance between the oceanic supply and removals of both alkalinity and DIC over longer timescales (see Section 8.09.4.1.1). Thus, the carbonate pump is an important mechanism not only in the glacial/interglacial storage of oceanic carbon, but also in the carbon cycle over much longer timescales (Garrels *et al.*, 1976; Sundquist, 1991; Archer *et al.*, 1997).

The oceanic carbon pumps (solubility, soft-tissue, and carbonate) are inherently linked. The amount of CO_2 dissolved in ocean surface waters is controlled not only by its equilibrium solubility, but also by the relative rates of gas exchange, ocean mixing, and soft-tissue and carbonate production. Nutrients that are necessary for soft-tissue assimilation arrive at the ocean surface with excess dissolved CO_2 from deeper waters, and these nutrients nourish carbonate as well as soft-tissue production. Carbonate production may be affected by dissolved CO_2 concentrations (Riebesell *et al.*, 2000) and by competition from production by non-carbonate-producing organisms (Nozaki and Yamamoto, 2001). The "rain ratio" of $CaCO_3$ to organic carbon debris reaching the sea floor is a very important factor controlling the magnitude and distribution of deep-sea $CaCO_3$ dissolution (Archer, 1991; Archer and Maier-Reimer, 1994). Any change in deep-ocean carbon storage will be accompanied by a $CaCO_3$ dissolution response within 5–10 kyr (see above). Efforts to explain the role of the oceans in controlling glacial/interglacial CO_2 levels must address all of these interconnections in a manner consistent with the evidence available in ice and marine sediment cores. Progress in these important endeavors is detailed elsewhere in this treatise (see Chapters 6.10, 6.18, and 6.19).

8.09.3.3.3 Carbon/climate interactions at glacial terminations

Given the many ways in which glacial/interglacial climate change have probably affected the carbon cycle, and given the likely importance of changes in atmospheric CO_2 and CH_4 as amplifiers of climate change, it is very difficult to separate cause from effect in the evolution of Quaternary climate and carbon cycling. Some of the most important evidence for mechanisms of change comes from detailed analysis of the sequence of events that occurred during the terminations of glacial periods. These events are the most rapid and dramatic variations in the pattern that characterizes Late Quaternary climate and carbon-cycle records. They are not only one of the most important aspects of glacial/interglacial change, but also among the best opportunities for detailed global correlation using the methods described in Section 8.09.3.1.

Analysis of glacial terminations in ice-core data has suggested that the onset of Antarctic warming may have preceded the onset of rising CO_2 levels by a few hundred years (Fischer *et al.*, 1999). However, there is a general consensus that this difference is too small to be distinguished from errors due to sampling and dating the ice cores (see Section 8.09.3.1), and that the initial increases in CO_2, CH_4 and Antarctic temperatures occurred simultaneously within the available dating resolution (Petit *et al.*, 1999; Monnin *et al.*, 2001). More significant timing differences have provided a basis for describing a general sequence of events that characterize at least the two most recent termination episodes (Broecker and Henderson, 1998; Petit *et al.*, 1999). First, the initiation of the termination is characterized by a dramatic decrease in the delivery of continental dust to Antarctica and nearby oceans. Second, atmospheric CO_2 and CH_4 rise in synchrony with Antarctic temperatures. Finally, several thousand years later, the rising temperatures and greenhouse gas concentrations are joined by the oxygen isotope trends in air and seawater that indicate melting of the continental ice sheets primarily in the northern hemisphere.

These events occurred within the context of very complex and still controversial relationships between northern and southern hemisphere climate trends and the influence of Earth's variable orbital configuration (see, e.g., Steig *et al.*, 1998; Bender *et al.*, 1999; Henderson and Slowey, 2000; Alley *et al.*, 2002). Asynchronies between northern and southern climate are particularly (perhaps uniquely) apparent in the record of the most recent glacial termination. As shown in Figure 10, these asynchronies affected atmospheric CO_2 and CH_4 (Monnin *et al.*, 2001). The Antarctic warming trend was interrupted by the Antarctic Cold Reversal, which appears to have coincided with a temporary 2 kyr cessation in the rise of atmospheric CO_2 concentrations. Atmospheric CH_4 levels, on the other hand, increased dramatically during this period, then fell precipitously to near-glacial values for ~1.5 kyr, and then rose to Holocene values just as suddenly. These CH_4 trends, observed in ice-core records from both Antarctica and Greenland, correlate closely with the warm Bølling/Allerød event followed by the cold Younger Dryas event in the Greenland ice-core isotope record (Blunier *et al.*, 1997). Thus, the CH_4 record during the most recent glacial termination seems to reflect a strong influence by climate variations that are most conspicuous in the northern hemisphere (Dallenbach *et al.*, 2000;

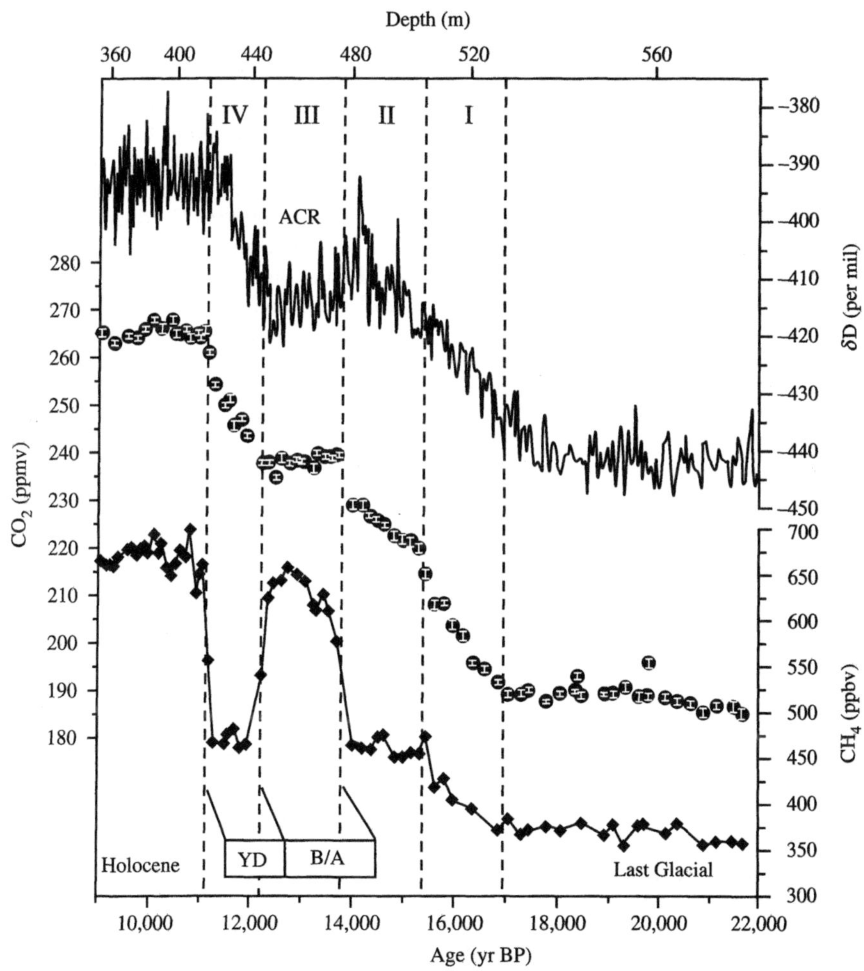

Figure 10 Events during the termination of the most recent glacial period, as recorded in the Dome C ice core. The top curve (δD) is a proxy for local temperatures, showing the Antarctic Cold Reversal (ACR). The solid circles represent CO_2 (means and error bars of six samples each). The diamonds represent CH_4 (uncertainty is 10 ppb). The depth scale at the top refers only to the CO_2 and CH_4 records. Bars indicate the Younger Dryas (YD) and Bolling/Allerod (B/A) events, which are recorded in the isotope ratios of Greenland ice cores (Monnin *et al.* (2001); reproduced by permission of the American Association for the Advancement of Science from *Science,* **2001**, *291*, 112–114 (figure 1)).

Brook *et al.*, 2000). The CO_2 record of this period more closely resembles the record of Antarctic climate, supporting suggestions that the Southern Ocean played a key role in the oceanic variations needed to account for changes in atmospheric CO_2 and terrestrial carbon storage (Francois *et al.*, 1997; Blunier *et al.*, 1997; Broecker and Henderson, 1998; Toggweiler, 1999; Petit *et al.*, 1999; Stephens and Keeling, 2000). Although more specific mechanisms continue to elude explanation, the importance of CO_2 and CH_4 as climate amplifiers seems to be reinforced by the observation that the melting of continental ice sheets—and hence the ice–albedo amplifying feedback—was not widespread until relatively late in the sequence of glacial termination events.

One of the most remarkable features of the carbon cycle during the Quaternary period is the rapidity of many global atmospheric CH_4 changes observed in the ice-core record. Using an elegant analysis of the ice-core evidence for transient thermal diffusion in nitrogen isotopes, Severinghaus *et al.* (1998) were able to delineate the relative timing of the CH_4 and temperature increases at the end of the Younger Dryas event without the ambiguities inherent in the age difference between the ice and the enclosed air (see Section 8.09.3.1). They concluded that atmospheric CH_4 began to rise within 0–30 yr following the sudden temperature increase. The speed of these changes has been cited in support of the hypothesis that climatically important changes in atmospheric CH_4 throughout the Quaternary period have been caused by sudden release from the very large volume of methane hydrates stored in marine and continental sediments (Nisbet, 1990, 1992; Kennett *et al.*, 2003) (see Section 8.09.2.1.2). Others have argued for rapid changes in CH_4 release from wetlands and a less significant role for CH_4 as a driver of climate change (Brook *et al.*, 1999; Severinghaus and Brook, 1999; Brook *et al.*, 2000). The disparity of these views, and the intensity of the debate, serves to remind us that our understanding of carbon-cycle behavior during the Quaternary period is still quite limited, despite the wealth of detailed information available in the record of ice and sediment cores.

8.09.4 THE PHANEROZOIC RECORD OF CARBON-CYCLE CHANGE

We have emphasized in Section 8.09.3 that the carbon-cycle variations observed on glacial/interglacial time scales during the Quaternary period can be understood in terms of the redistribution of carbon among the atmosphere, oceans, biosphere, and reactive sediments. Over timescales of millions of years and longer, many carbon-cycle trends cannot be explained by similar redistributions among Earth-surface reservoirs. In the pre-Quaternary past, although Earth-surface reallocations likely caused some degree of carbon-cycle variability, the geologic record reveals more substantial changes that appear to have required the influence of imbalances in exchange between the Earth's surface carbon reservoirs and the rocks of the Earth's crust. In this section we describe the types of crustal exchange processes that have been identified as causes of gradual geologic carbon-cycle change.

8.09.4.1 Mechanisms of Gradual Geologic Carbon-cycle Change

As shown in Figures 1 and 2, the amount of carbon in the Earth's crust vastly exceeds the amount stored in the atmosphere, biosphere, and oceans combined. A persistent imbalance in the exchange of crustal carbon could, in principle, cause a drastic depletion or buildup of carbon at the Earth's surface (Holland, 1978). Therefore, the study of gradual geologic carbon-cycle change involves seeking both the potential causes of change and the feedback mechanisms that might limit the extent of change (Berner and Caldeira, 1997; Berner, 1999). Feedback mechanisms have been identified in the balances between carbonate weathering and sedimentation, between silicate weathering and metamorphic decarbonation, and between organic carbon production and oxidation.

8.09.4.1.1 The carbonate weathering-sedimentation cycle

The most abundant anion delivered by rivers to the oceans is bicarbonate ion (HCO_3^-), and most of the bicarbonate alkalinity in rivers comes from the weathering of carbonate rocks (Meybeck, 1987). The chemical weathering of limestones and dolostones by dissolved CO_2 can be represented by the reactions for dissolution of calcite and dolomite:

$$CaCO_3 + CO_2 + H_2O \rightarrow Ca^{2+} + 2HCO_3^- \quad (1a)$$

$$CaMg(CO_3)_2 + 2CO_2 + 2H_2O \rightarrow Ca^{2+} + Mg^{2+} + 4HCO_3^- \quad (1b)$$

In the oceans, reef and planktonic organisms precipitate calcium carbonate (both calcite and aragonite), which comprises a major component of marine sediments. The precipitation of calcium carbonate is essentially the reverse of reaction (1a) above:

$$Ca^{2+} + 2HCO_3^- \rightarrow CaCO_3 + CO_2 + H_2O \quad (2a)$$

Although dolomite is abundant in Proterozoic and Paleozoic rocks, its relative contribution to more recent carbonate sediments is significantly less (Daly, 1909), and it comprises only ~10% of modern carbonate sediments (Holland and Zimmerman, 2000). Thus, the precipitation of dolomite,

$$Ca^{2+} + Mg^{2+} + 4HCO_3^- \rightarrow CaMg(CO_3)_2 + 2CO_2 + 2H_2O \quad (2b)$$

is not as significant today as it has been during some of the geologic past. The weathering of abundant dolostones and the lack of comparable Mg-carbonate sedimentation during the last 40 Myr may have contributed to a non-steady-state increase in the oceanic concentration of Mg^{2+} (Zimmerman, 2000; Horita *et al.*, 2002). But as described in Section 8.09.3.3.2, the oceanic carbonate pump should have responded rapidly to any imbalance in oceanic bicarbonate associated with the Mg^{2+} imbalance. The mechanism of this response would have been an enhancement of $CaCO_3$ removal brought about effectively by diminished carbonate dissolution. The rate of reaction (2a) changes effectively to match any change in the sum of the rates of reactions (1a) and (1b). Simple stoichiometry requires that this condition returns as much CO_2 to the ocean/atmosphere system via reaction (2a) as the amount of CO_2 consumed in reactions (1a) and (1b). Changes in atmospheric CO_2 might occur in association with the response of the carbonate pump to an imbalance between carbonate weathering and sedimentation, but these changes would involve the reactive sediments of the Earth-surface system (see Figure 1 and Table 1(a)) rather than changes in the exchange of crustal carbon (Sundquist, 1991).

Interestingly, the Late Tertiary increase in oceanic Mg^{2+} concentrations appears to have been accompanied by a decrease in oceanic Ca^{2+} concentrations (Horita *et al.*, 2002), and there is strong evidence that the rate of present-day $CaCO_3$ sedimentation significantly exceeds the rate of Ca^{2+} input to the oceans (Milliman, 1993). However, interpretation of these trends must consider not only the balance between carbonate weathering and sedimentation, but also the importance of changes in the silicate–carbonate weathering–decarbonation cycle (see Section 8.09.4.1.2), variations in hydrothermal and cation exchange reactions involving mid-ocean ridge basalts and sediments, and fluctuations in the relative magnitudes of shallow-water and deep-water carbonate sedimentation (Spencer and Hardie, 1990; Hardie, 1996; Holland and Zimmerman, 2000). A full discussion of dolomite weathering and sedimentation is beyond the scope of this chapter. The Tertiary trends described above may be related to the proliferation of planktonic calcareous organisms and the consequent shift of $CaCO_3$ deposition to the deep sea (Holland and Zimmerman, 2000); thus, they may not be analogous to earlier Phanerozoic trends in carbonate mineralogy and oceanic cation concentrations inferred from the record of nonskeletal carbonates and fluid inclusions (Sandberg, 1983, 1985; Horita *et al.*, 2002). The relationships described above demonstrate that steady-state feedback in the long-term cycling of carbon must be viewed in the context of significant non-steady-state trends in the chemistry of major dissolved oceanic cations. The carbonate weathering-sedimentation cycle is sometimes ignored or trivialized in treatments of the geological carbon cycle, because reactions (1) and (2) do not appear to have an impact on long-term atmospheric CO_2 trends. It is important to remember that the weathering and sedimentation of carbonate minerals are, like other carbon-cycle feedback systems, coupled to the global cycling of other materials.

8.09.4.1.2 The silicate-carbonate weathering-decarbonation cycle

Although the chemical weathering of silicate minerals contributes less than the weathering of carbonates to the cycling of materials from the land surface to ocean waters and sediments, silicate weathering presents a much more complex challenge to geochemists seeking to discern the feedback mechanisms that assure a balance in crustal carbon exchange. This balance can be illustrated by the following cyclic reactions:

Weathering:

$$3H_2O + 2CO_2 + CaSiO_3 \rightarrow Ca^{2+} + 2HCO_3^- + Si(OH)_4 \quad (3a)$$

$$3H_2O + 2CO_2 + MgSiO_3 \rightarrow Mg^{2+} + 2HCO_3^- + Si(OH)_4 \quad (3b)$$

Sedimentation:

$$Ca^{2+} + 2HCO_3^- + Si(OH)_4 \rightarrow CaCO_3 + SiO_2 + 3H_2O + CO_2 \quad (4a)$$

$$Mg^{2+} + 2HCO_3^- + Si(OH)_4 \rightarrow MgCO_3 + SiO_2 + 3H_2O + CO_2 \quad (4b)$$

Decarbonation:

$$CaCO_3 + SiO_2 + 3H_2O + CO_2 \rightarrow 3H_2O + 2CO_2 + CaSiO_3 \quad (5a)$$

$$MgCO_3 + SiO_2 + 3H_2O + CO_2 \rightarrow 3H_2O + 2CO_2 + MgSiO_3 \quad (5b)$$

Reactions (3)–(5) are an oversimplification of the wide variety of contributing minerals and reactions, but they can be used to illustrate the most important aspects of the silicate–carbonate weathering–decarbonation cycle. Like carbonate weathering, silicate weathering consumes CO_2 and yields cations and bicarbonate ions that are delivered by rivers to the oceans. However, the sedimentation of carbonate minerals releases only some of the CO_2 consumed during weathering. The remaining CO_2 is released by decarbonation reactions that occur during burial, subduction, and other tectonic processes.

Arrhenius (1896) and Chamberlin (1898) proposed feedback linkages among tectonic activity, atmospheric CO_2, chemical weathering, and carbonate deposition. Budyko and Ronov (1979) were the first to use silicate–carbonate feedback linkages to estimate past atmospheric CO_2 concentrations. They observed a correlation between volcanic and carbonate rock abundances in the sediments of the Russian platform, and they calculated Phanerozoic atmospheric CO_2 concentrations by assuming a simple proportionality to rates of carbonate deposition. Walker *et al.* (1981) contributed a very important extension to this idea, proposing that the feedback mechanism is mediated by the climatic effect of atmospheric CO_2. According to this hypothesis, higher atmospheric CO_2 would increase the greenhouse effect, raising global temperatures and silicate weathering reaction rates, and providing a negative feedback to the CO_2 increase. Conversely, an increase in silicate weathering would lead to a decrease in atmospheric CO_2 and global temperatures, providing a negative feedback to the weathering increase. The CO_2–climate feedback was adopted as a central concept in the time-dependent model of Berner *et al.* (1983), popularly known as the BLAG model (see Section 8.09.4.2).

Much debate has been devoted to the question of which of the reactions (3)–(5) "control" atmospheric CO_2. The BLAG model (Berner *et al.*, 1983; Berner, 1991) suggested that rates of silicate weathering (reaction (3)) have adjusted essentially instantaneously to maintain a steady state with respect to rates of CO_2 production from decarbonation (reaction (5)). In this view, the rate-limiting step in the cycle is, therefore, the tectonically induced production of CO_2. Others have emphasized the importance of continental uplift as a factor that could affect atmospheric CO_2 through changes in weathering without necessarily changing the rate of decarbonation CO_2 supply (Raymo *et al.*, 1988; Edmond *et al.*, 1995). The two views are not incompatible because steady-state weathering feedback can incorporate effects of uplift (Sundquist, 1991; Berner, 1991). There is now a broad consensus that both weathering and tectonism are critical to the relationship between atmospheric CO_2 and the silicate–carbonate cycle (see, e.g., Berner, 1994; Bickle, 1996; Berner and Kothavala, 2001; Ruddiman, 1997; Wallmann, 2001a).

It is very difficult to infer present-day values for the carbon fluxes and relationships relevant to the silicate–carbonate weathering–decarbonation cycle (see Table 1(a)), and it is even more daunting to estimate how these interactions might have varied in the geologic past (Kump *et al.*, 2000; Boucot and Gray, 2001). The difficulty is further complicated by the suggestion that low-temperature hydrothermal reactions may precipitate significant quantities of $CaCO_3$ from seawater circulating through the upper ocean crust (Staudigel *et al.*, 1990; Caldeira, 1995; Alt and Teagle, 1999). The overall stoichiometry of these "sea-floor weathering" reactions is very complex, but they appear to be a process of global if enigmatic significance to the long-term oceanic carbon and alkalinity balance (Wallmann, 2001a). Thus reactions (3)–(5) must be considered illustrative not only in the sense that they are idealized representations of more complicated reactions, but also in the sense that they may not represent all of the basic silicate–carbonate reaction types that contribute exchange of carbon between the Earth's crust and surface reservoirs.

8.09.4.1.3 *The organic carbon production-consumption-oxidation cycle*

Since the Precambrian, oxygenic photosynthesis and aerobic respiration have been the dominant reactions in the cycling of organic carbon (Des Marais *et al.*, 1992) (see Section 8.09.5). The production of organic matter by oxygenic photosynthesis can be illustrated by the production of glucose:

$$6CO_2 + 6H_2O \rightarrow C_6H_{12}O_6 + 6O_2 \qquad (6)$$

The transformation of glucose and other carbohydrates into the vast array of compounds that are buried and further transformed in sediments is beyond the scope of this chapter. The fundamental relationship in reaction (6) is the production of oxygen that accompanies the production of organic carbon. The aerobic cycling of organic carbon is completed by oxidative consumption, which can be illustrated by the reverse of reaction (6):

$$C_6H_{12}O_6 + 6O_2 \rightarrow 6CO_2 + 6H_2O \qquad (7)$$

Reactions (6) and (7) demonstrate that the cycling of organic carbon is inherently associated with the cycling of oxygen. This relationship has two very important consequences. First, a net excess of organic matter production relative to oxidation is accompanied by net production of oxygen.

Conversely, a net excess of oxidation is accompanied by net consumption of oxygen. Thus, as long as reactions (6) and (7) have dominated the cycling of organic carbon, net burial of organic matter has been associated with production of atmospheric oxygen over timescales of millions of years and longer (Garrels and Lerman, 1981; Shackleton, 1987; Kump *et al.*, 1991; Des Marais *et al.*, 1992). Second, through its linkage to the cycling of oxygen, the geological carbon cycle has also been intimately connected to the biogeochemical cycles of sulfur (see below), phosphorus, iron, nitrogen, and other elements that are cycled in forms that are sensitive to the oxidation state of their environment (Ingall and Van Cappellen, 1990; Van Cappellen and Ingall, 1996; Petsch and Berner, 1998; Berner, 1999).

In keeping with our perspective that the modern cycling of methane can be viewed as a subcycle of the CO_2 carbon cycle, we represent microbial methane production by the fermentation of glucose:

$$C_6H_{12}O_6 \rightarrow 3CH_4 + 3CO_2 \quad (8)$$

and we represent methane oxidation by the stoichiometrically consistent reaction:

$$3CH_4 + 6O_2 \rightarrow 6H_2O + 3CO_2 \quad (9)$$

The sum of reactions (8) and (9) is identical to reaction (7), demonstrating the treatment of methane production and oxidation as a subcycle.

Under anaerobic conditions, microbial consumption of methane may occur by reactions such as:

$$3CH_4 + 3SO_4^{2-} \rightarrow 3HCO_3^- + 3HS^- + 3H_2O \quad (10)$$

Reaction (10) illustrates an important connection between the cycling of carbon and sulfur. The bisulfide produced by this kind of reaction is often precipitated and buried in marine sediments as iron sulfide (Berner, 1982). Over geological timescales, the removal of iron sulfide from the Earth-surface sulfur cycle affects the oxygen cycle in a manner analogous to the net removal (burial) of organic carbon from the Earth-surface carbon cycle. The burial of reduced forms of carbon and sulfur allows oxygen to accumulate in the atmosphere (Garrels and Lerman, 1984; Berner, 1987; Berner, 2001). Thus, the anaerobic consumption of methane may have played an important part in the geologic history of atmospheric oxygen.

Global trends in the organic carbon production-consumption-oxidation cycle can be discerned from the carbon isotope record of marine carbonates. Because the formation of organic matter selectively assimilates ^{12}C (see Section 8.09.2.1.1), changes in the relative proportions of net removals or additions of organic and inorganic carbon are reflected in the ratio of ^{13}C to ^{12}C in the carbon of the oceans and atmosphere. Although the carbon isotope ratios of marine carbonate shells are affected by environmental and biological variables as well as by the isotopic composition of oceanic carbon, the record from fossil shells shows significant trends through the Phanerozoic Eon that most likely represent global changes in the ratio of ^{13}C to ^{12}C in oceanic DIC (Figure 11). More generally, the marine carbonate record is characterized by $^{13}C:^{12}C$ ratios that are consistently greater than the corresponding ratio of the carbon released to the Earth's surface environment from the mantle (Schidlowski, 2001). This relationship implies that the burial of carbonates has been accompanied by burial of ^{13}C-depleted organic matter. If mantle carbon reaches the Earth's surface with a $\delta^{13}C$ value of −5‰ (Deines, 1992), and if carbonate and organic carbon are buried with $\delta^{13}C$ values of 0‰ and −25‰, respectively, then the steady-state ratio of organic carbon burial to carbonate carbon burial can be calculated by isotopic mass balance to be 1 : 4 (Schidlowski, 2001). Variations such as those depicted in Figure 11 probably reflect transient changes in the relative rates of organic carbon and carbonate burial (Schidlowski and Junge, 1981; Berner and Raiswell, 1983; Garrels and Lerman, 1984; Berner, 1987), as well as possible changes in the degree of fractionation of carbon isotopes due to changes in ocean/atmosphere chemistry and other environmental factors (Hayes *et al.*, 1999; Berner *et al.*, 2000; Berner, 2001). This record is an important constraint on models of the geologic carbon cycle (see Section 8.09.4.2).

The fractionation of carbon isotopes between ^{13}C-depleted organic matter and ^{13}C-enriched carbonates is also observed in Precambrian sediments extending back to the Archean Eon (Schidlowski *et al.*, 1983; Schidlowski, 2001). Prior to the early Proterozoic Eon, the cycling of organic carbon probably occurred through reactions and life forms that did not require significant levels of atmospheric oxygen (Holland, 1994). These early modes of organic carbon cycling and their relationships to ocean/atmosphere chemistry are discussed briefly in Section 8.09.5 and more extensively in Holland (2003).

8.09.4.2 Model Simulations of Gradual Geologic Carbon-cycle Change

The mechanisms of gradual geologic carbon-cycle change are complex and poorly understood. They span a broad range of timescales, and they include many poorly understood relationships and feedback. They present challenges to our understanding of the cycling of many other elements in addition to carbon. Mathematical modeling is an

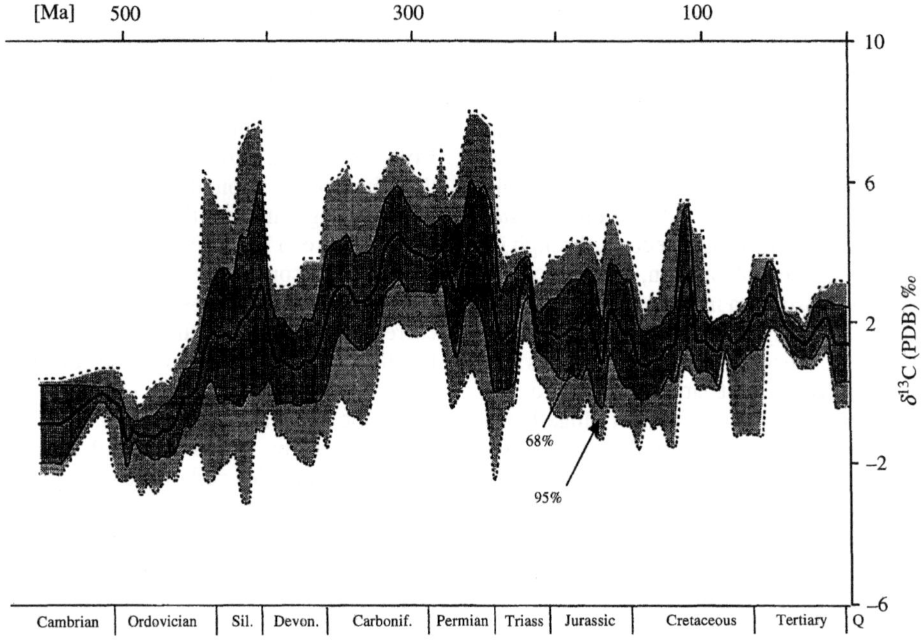

Figure 11 Phanerozoic $\delta^{13}C$ trends in marine carbonates. The central curve represents running means calculated using 20 Ma window and 5 Ma forward step. The shaded areas show relative uncertainties (Veizer *et al.* (1999); reproduced by permission of Elsevier from *Chem. Geol.*, **1999**, *161*, 59–88 (figure 10)).

essential tool in the quest for this understanding. Models provide a crucial device for quantifying relationships, formulating and testing hypotheses, and comparing simulated results to the evidence of the geologic record. Biogeochemists use models of the global carbon cycle as an iterative self-teaching tool, reconfiguring and reformulating as new ideas and evidence emerge.

The most widely cited and applied example of this iterative modeling approach is the work of Berner and coworkers on the development of the GEOCARB models (Berner *et al.*, 1983; Lasaga *et al.*, 1985; Berner, 1990, 1991, 1994, Berner and Kothavala, 2001). This effort, which began with the publication of the BLAG model twenty years before the writing of this chapter, is distinguished by its prolonged capacity to test innovative ideas and to incorporate evolving insights (including many initiated by the criticisms of colleagues). The core of the GEOCARB model is a simple steady-state ocean-atmosphere carbon mass balance expression:

$$F_{wc} + F_{mc} + F_{wg} + F_{mg} = F_{bc} + F_{bg} \quad (11)$$

where F_{wc} is the rate of weathering of carbonates (reaction (1)), F_{mc} is the production of CO_2 by decarbonation (reaction (5)), F_{wg} is the rate of production of CO_2 by oxidative weathering of sedimentary organic matter (reaction (7)), F_{mg} is the rate of production of CO_2 by the metamorphism of sedimentary organic matter (reactions (7)–(9)), F_{bc} is the rate of burial of sedimentary carbonates (reactions (2) and (4)), and F_{bg} is the rate of burial of sedimentary organic material (reaction (6)). Each of these terms is defined in the model as a function of variables that represent the estimated effects of subduction and tectonics (for F_{mc} and F_{mg}) and of land area, relief, climate, vegetation, and atmospheric CO_2 (for the weathering fluxes). The rate of CO_2 consumption during weathering of silicate minerals is expressed by the equation:

$$F_{ws} = F_{bc} - F_{wc} \quad (12)$$

which simply represents the steady-state approximation that the rate of carbonate burial will be equal to the sum of the rates of bicarbonate delivered to the oceans from weathering of silicate and carbonate minerals. Using a model configured to represent the time-dependent behavior of ocean chemistry and reactive sediment interactions, Sundquist (1991) confirmed that carbonate sedimentation does indeed respond very quickly (within a few thousand years) to balance changes in weathering, and determined that the overall response time of weathering to changes in the terms in Equations (11) and (12) was on the order of a few hundred thousand years. Thus the steady-state approximation of these equations appears to be reasonable for the timescales to which GEOCARB has been applied.

The calculation of past CO_2 levels using GEOCARB begins with the iterative solution of equations (11) and (12) and a parallel mass balance equation for carbon isotopes to yield fluxes that are consistent with the carbonate carbon isotope record (Figure 11) and a variety of other sources of information for the factors that determine relative variations of the individual fluxes through time. Further iteration is then applied to derive atmospheric CO_2 concentrations consistent with the steady-state weathering fluxes. GEOCARB results have shown the same general pattern through more than ten years of modification and refinement (Figure 12). The calculated CO_2 trend shows very high levels in the Early Paleozoic Era, a significant decline during the Devonian and Carboniferous periods, high values during the Early Mesozoic Era, and a decline beginning at ~170 Ma toward the relatively low values of the Cenozoic Era.

Like all geochemical simulations requiring broad assumptions and extrapolations, the GEOCARB results are not immune to criticism. Some have argued that the model does not adequately represent the effects of pre-Devonian weathering (Boucot and Gray, 2001), and others have taken issue with the inference that atmospheric CO_2 plays a significant role in the climatic history of the Phanerozoic Eon (see, e.g., Cowling, 1999; Veizer *et al.*, 2000; Wallmann (2001b). While a number of independent modeling efforts have yielded results similar to those of GEOCARB (see, e.g., Tajika, 1998; Wallmann, 2001a), others have not (see, e.g., Rothman, 2002). There is general agreement that all models of Phanerozoic carbon cycling are poorly constrained by the incomplete and circumstantial nature of the geologic evidence. The ongoing assessment and evolution of the GEOCARB model exemplifies the effective use of what information is available.

8.09.4.3 Geologic Evidence for Phanerozoic Atmospheric CO_2 Concentrations

Figure 12 shows some of the geologic evidence to which the GEOCARB model results are compared. The general Phanerozoic pattern of atmospheric CO_2 (Figure 12, panel a) and derived radiative forcing (panel b) appears to correspond to the pattern of continental glaciation episodes (panel d) in a manner consistent with a major effect of CO_2 on global climate. However, this relationship does not appear to be robust in comparing the CO_2 and derived radiative forcing to trends in low-latitude temperatures derived from the oxygen isotope ratios of marine calcite fossils (panel c). These discrepancies remain unresolved. It seems likely that the record of Phanerozoic climate and carbon cycling, like the more detailed record of Quaternary climate and carbon cycling described in Section 8.09.3, reflects the influence on climate of many complex factors including—but not limited to—atmospheric CO_2.

Figure 12 also displays some of the proxy data to which model simulations of atmospheric CO_2 can be more directly compared. These proxy data have recently been reviewed by Royer *et al.* (2001) and Beerling and Royer (2002). All of the proxy measures are limited by significant uncertainties.

The carbon isotope ratios of pedogenic carbonates have been used to infer atmospheric CO_2 concentrations from calculations based on a model for steady-state diffusive mixing with soil-respired CO_2 in the soil profile (Cerling, 1991; Cerling, 1992; Ekart *et al.*, 1999; Ghosh *et al.*, 2001). Significant sources of uncertainty in this approach include the dependence of soil CO_2 diffusion on temperature and moisture, the contribution of C_3 versus C_4 plants to respired CO_2, and the somewhat arbitrary choice of values for the mole fraction of respired CO_2 at depth in the soil.

The extent of carbon isotope fractionation during photosynthesis is sensitive to atmospheric CO_2 concentrations. This sensitivity has been used to estimate atmospheric CO_2 concentrations from fossil marine and terrestrial plant residues (Rau *et al.*, 1989; Popp *et al.*, 1989; Marino and McElroy, 1991; Freeman and Hayes, 1992; Pagani *et al.*, 1999a,b; Grocke, 2002; Beerling and Royer, 2002; Pagani, 2002). Uncertainties in this method include (for marine plants) variations in the dissolved CO_2 concentration of ocean surface waters and effects of temperature, salinity, and cell size and growth rate; and (for land plants) variations in canopy CO_2 concentrations and effects of temperature and leaf-to-air vapor pressure. Analysis of alkenones extracted from marine plankton appears to show some promise of narrowing uncertainties but yields disparate estimates of Miocene CO_2 levels (Freeman and Hayes, 1992; Pagani, 2002). Analysis of fossil land plants has demonstrated some encouraging correlations with carbon isotope anomalies recorded in marine carbonates (see, e.g., Koch *et al.*, 1992), but efforts to reconstruct past atmospheric CO_2 levels are much more tentative (Beerling and Royer, 2002).

Land plants take up CO_2 through leaf pores called stomates. C_3 plants appear to regulate the number of stomates per unit of leaf area (stomatal density) to maximize CO_2 uptake while minimizing water loss through the stomates. The stomatal density of many modern plants has been shown to be inversely related to ambient CO_2 concentrations (Woodward, 1987; Beerling and Royer, 2002), and this relationship has been applied to analysis of stomatal densities of fossil plant leaves (Beerling, 1993; McElwain and Chaloner, 1995; Rundgren and Beerling, 1999; Retallack, 2001, 2002; Beerling and Royer, 2002). Although there is impressive support for this approach in its

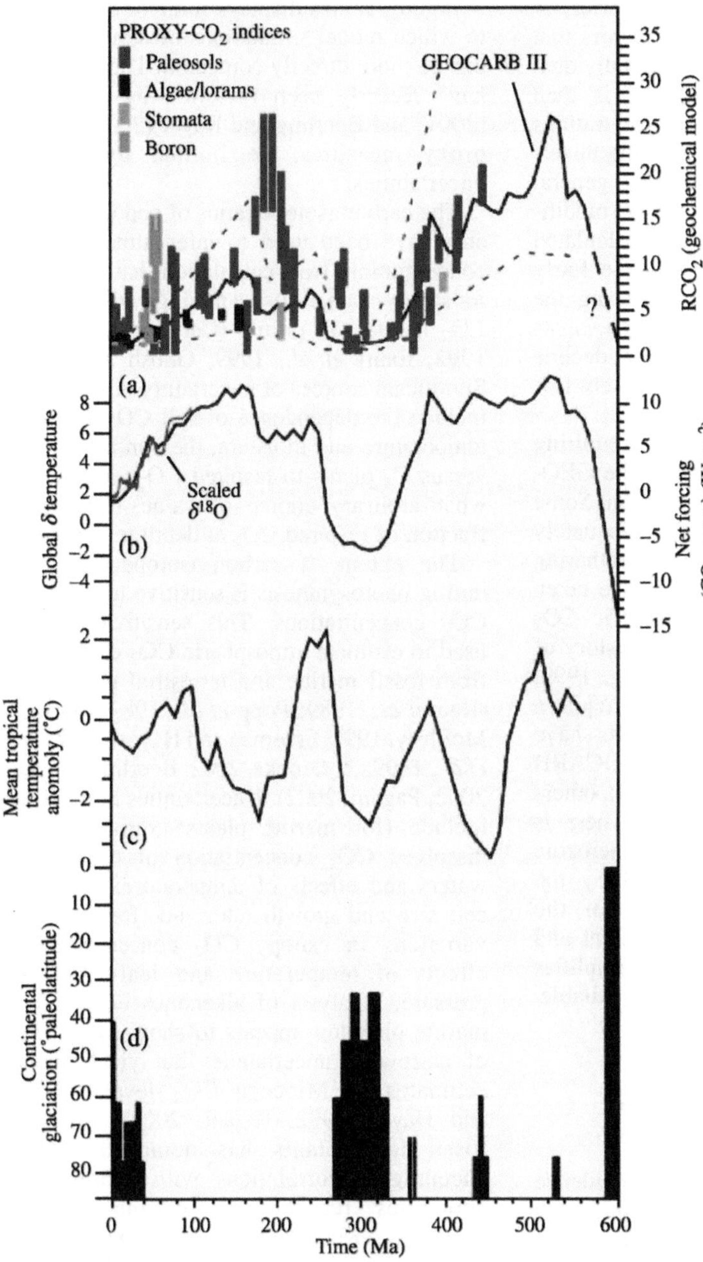

Figure 12 Phanerozoic model results (Berner and Kothavala, 2001) compared to climate indicators. Crowley and Berner (2001) describe this figure as follows: "(a) Comparison of CO_2 concentrations from the GEOCARB III model (Berner and Kothavala, 2001) with a compilation of proxy-CO_2 evidence (vertical bars; Royer *et al.*, 2001). Dashed lines: estimates of uncertainty in the geochemical model values (Berner and Kothavala, 2001). Solid line: conjectured extension to the Late Neoproterozoic (~590–600 Ma). R_{CO_2}, ratio of CO_2 levels with respect to the present (300 ppm). Other carbon-cycle models (Tajika, 1998; Wallmann, 2001b) for the past 150 Myr are in general agreement with the results from this model. (b) Radiative forcing for CO_2 calculated from (Kiehl and Dickinson, 1987) and corrected for changing luminosity (Crowley and Baum, 1993) after adjusting for an assumed 30% planetary albedo. Deep-sea oxygen isotope data over the past 100 Ma (Douglas and Woodruff, 1981; Miller *et al.*, 1987) have been scaled to global temperature variations according to Crowley (2000). (c) Oxygen isotope-based low-latitude paleotemperatures (Veizer *et al.*, 2000). (d) Glaciological data for continental-scale ice sheets modified from Crowley (2000) and Crowell (1999) based on many sources. The duration of the Late Neoproterozoic glaciation is a subject of considerable debate" (Crowley and Berner (2001); reproduced by permission of the American Association for the Advancement of Science from *Science*, **2001**, *292*, 870–872 (figure 1)).

validation for subtle Holocene CO_2 trends (Rundgren and Beerling, 1999), its application to a larger range of CO_2 variations and plant species remains uncertain.

The ratio of the boron isotopes ^{11}B and ^{10}B is known to depend on ambient pH in the boron incorporated in the carbonate shells of marine foraminifera (Sanyal *et al.*, 1996; Sanyal *et al.*, 2001). The use of boron isotopes as an indicator for seawater paleo-pH has been extended to the calculation of past CO_2 concentrations in ocean surface waters and in the atmosphere (Spivack, 1993; Sanyal *et al.*, 1997; Pearson and Palmer, 1999, 2000; Sanyal and Bijma, 1999; Palmer and Pearson, 2003). Sources of uncertainty in these estimates include the fractionation of boron isotopes during incorporation in carbonate shells, effects of diagenesis, the assumptions needed to calculate CO_2 concentrations from pH, and the influence of changing boron isotope ratios in ambient seawater (Lemarchand *et al.*, 2000; Lemarchand *et al.*, 2002). The latter problem is especially serious for estimates based on samples older than the 15 Myr residence time of boron in the oceans.

Given the discrepancies and uncertainties enumerated above, it is clear that the study of Phanerozoic carbon-cycle change does not enjoy the luxury of a paleo-CO_2 "gold standard" analogous to the ice-core records of the Late Quaternary period. Measurements that reflect the global mass balance of carbon, such as the carbon isotope record of marine carbonates (Figure 11), remain the most powerful kind of evidence for analysis of gradual geologic carbon-cycle change. This approach requires geochemical models to quantify and test hypothesized relationships among global fluxes and reservoirs.

8.09.4.4 Abrupt Carbon-cycle Change

Like the concept of steady state described in Section 8.09.2.2, the concept of "abrupt" change requires the context of a particular timescale and frame of reference. In Section 8.09.3, we described examples of abrupt change in the context of the Late Quaternary ice-core record. Abrupt changes in the context of the Phanerozoic rock record might be viewed as leisurely transitions from the perspective of Quaternary timescales. A formal treatment of abrupt change is beyond the scope of this chapter. However, we describe two examples that seem to qualify as "abrupt" changes in the carbon cycle by any definition.

About 55 Myr ago, a brief period known as the Late Paleocene thermal maximum (Zachos *et al.*, 1993) is defined in marine and terrestrial records by a sudden increase in proxy temperature estimates and a very large decrease in the ratio of ^{13}C to ^{12}C (Figure 13). Most of the carbon isotope shift appears to have occurred within a period of a few thousand years (Norris and Rohl, 1999), suggesting that the cause might have been a redistribution of carbon from the terrestrial biosphere to the oceans and atmosphere. But the magnitude of the shift, ~−2.5‰ to −3‰, would have required the transfer of an amount of terrestrial organic carbon equivalent at least to virtually the entire modern terrestrial biosphere,

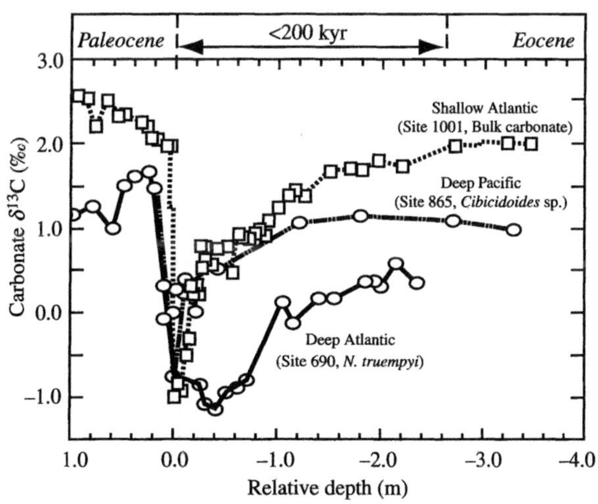

Figure 13 High-resolution carbon isotope records showing the anomaly corresponding to the Late Paleocene Thermal Maximum. The data are from three widely separated Ocean Drilling Program cores (Kennett and Stott, 1991; Bralower *et al.*, 1995; Bralower *et al.*, 1997) plotted on a common depth scale with the $\delta^{13}C$ minimum at 0.0 m (Dickens (2001); reproduced by permission of the American Geophysical Union from *Natural Gas Hydrates: Occurrence, Distribution, and Detection,* **2001**, pp. 19–38 (figure 1)).

including soils (Dickens *et al.*, 1995). Instead, the most likely cause is a sudden transfer of methane hydrate to the ocean–atmosphere system (Dickens *et al.*, 1995, 1997; Dickens, 2000). The extremely depleted ratio of ^{13}C to ^{12}C in methane means that only nearly one-third as much methane carbon must be transferred to account for the carbon isotope shift. Even so, the amount estimated to have caused the Late Paleocene isotope shift is on the order of 1,000–2,000 Pg C. The precise mechanism of this methane release is not clear, but the hypothesized release appears to be consistent with the evidence for abrupt global warming and (assuming that the methane was rapidly oxidized to CO_2) for $CaCO_3$ dissolution (Thomas and Shackleton, 1996). The carbon isotope anomaly seems to have decayed somewhat exponentially over ~150 kyr (Figure 13), which is comparable to the modern residence time of oceanic carbon with respect to sediment burial (see Figure 1). From evidence for an episode of widespread biogenic barium coincident with the carbon isotope anomaly, Bains *et al.* (2000) suggested that oceanic productivity increased in response to a warming-induced enhancement of continental weathering and delivery of nutrients to the oceans. The Late Paleocene thermal maximum seems to provide not only an example of an abrupt event, but also an opportunity to document consequent feedback mechanisms ranging from climate change to carbonate dissolution to continental weathering and oceanic productivity.

The Cretaceous–Tertiary boundary, ~65 Myr ago, is defined by a bolide impact event putatively associated with the Chicxulub Crater in Mexico (Alvarez *et al.*, 1980). Among the cataclysmic environmental effects of this event, a large but uncertain quantity of CO_2 was likely injected into the ocean–atmosphere system from the volatilization of $CaCO_3$ rocks at the impact site (O'Keefe and Ahrens, 1989; Pope *et al.*, 1997; Pierazzo *et al.*, 1998). Not surprisingly, a large simultaneous release of methane hydrate has also been hypothesized (Max *et al.*, 1999). Although the effects of these carbon transfers are greatly complicated by the wide range of biological and geochemical consequences of the impact, they provide another opportunity to document and analyze carbon-cycle feedback mechanisms.

An understanding of geologically abrupt change requires not only evidence of the abrupt event itself, but also analysis of the sequence of subsequent responses. These responses may "cascade" through a wide range of timescales, reflecting and perhaps shedding new light on many of the relationships between processes and timescales that characterize more gradual change. Further geological evidence is needed to determine the extent to which carbon-cycle change has occurred by abrupt rather than gradual transitions. If abrupt change turns out to be a significant mode of long-term carbon-cycle evolution, new theoretical understandings may be needed to accommodate more pronounced effects of non-steady-state relationships (see, e.g., Sundquist, 1991).

8.09.5 THE PRECAMBRIAN RECORD OF CARBON-CYCLE CHANGE

Many aspects of the Precambrian history of the carbon cycle are summarized in Holland (2003). Although the carbon isotope record suggests the common existence of ^{13}C-depleted organic carbon of biological origin by 3 Ga, the atmosphere contained very little oxygen and the carbon cycle did not begin to resemble its modern form until the Late Archean Eon and the transition to the Proterozoic Eon (Des Marais *et al.*, 1992; Kasting, 1993; Des Marais, 2001). The earliest forms of life appear to have been based on fundamentally different modes of carbon cycling in which organic synthesis and metabolism occurred without the production and consumption of oxygen. Anoxygenic photosynthesis is observed in several forms of bacteria today, and the antiquity of these forms is suggested by analyses based on genomic sequencing and the molecular structure and function of key protein complexes (Mathis, 1990; Schubert *et al.*, 1998; Rhee *et al.*, 1998; Xiong *et al.*, 2000; Hedges *et al.*, 2001). Early forms of life may have cycled carbon through redox reactions involving hydrogen, sulfur, iron, or non-biotic organic compounds (see, e.g., Walker, 1987; Widdel *et al.*, 1993; MacLeod *et al.*, 1994; Russell and Hall, 1997; Pace, 1997; Rasmussen, 2000; Kelley *et al.*, 2001).

The origin of oxygenic photosynthesis is attributed to the cyanobacteria, which appear to have evolved hundreds of millions of years before the rise in atmospheric oxygen (Buick, 1992; Brocks *et al.*, 1999; Summons *et al.*, 1999; Des Marais, 2000). During the period of transition to oxygenated conditions, the oxygen produced by photosynthesis was probably consumed by reaction with reduced compounds except in particular oxygenated environments (Cloud Jr., 1968; Walker *et al.*, 1983; Kasting, 1987, 1993). The quantitative pace of the rise in atmospheric oxygen is a topic of lively debate (Holland, 1994, 2003; Ohmoto, 1996; Holland and Rye, 1997; Lasaga and Ohmoto, 2002). The expansion of oxygenated conditions may have been affected by the evolving differentiation of the Earth's mantle and crust, changes in the oxidation state of volcanic gases, episodes of tectonic activity, hydrogen escape to space from the upper atmosphere, and the increasing availability of stable cratonic platforms for more productive life forms and more active subaerial weathering (Walker *et al.*, 1983; Des Marais *et al.*, 1992; Kasting *et al.*, 1993; Kasting, 1993, 2001; Hoehler *et al.*, 2001; Catling *et al.*, 2001; Holland, 2002). The variety of processes affecting atmospheric oxygen must also have affected the cycling of carbon, and the atmosphere may have contained both CO_2 and CH_4 at this time (Kasting, 1993; Hayes, 1994; Rye *et al.*, 1995), but see Kasting *et al.* (1983). The cycling of CH_4 by microbial production and consumption may have been so important that this period has been dubbed the "Age of Methanotrophs" (Hayes, 1994). High atmospheric concentrations of CO_2 and/or CH_4 were probably necessary in order to offset the lowered luminosity of the Sun during this period (Sagan and Mullen, 1972; Owen *et al.*, 1979; Kuhn and Kasting, 1983; Kasting *et al.*, 1988; Rye *et al.*, 1995).

As the oxidation state of the Earth's surface gradually changed, the anaerobic oxidation of methane via reaction (10) may have been an important transitional pathway (Hinrichs and Boetius, 2002). The first significant increase in atmospheric oxygen (the so-called "Great Oxidation Event") has been inferred from a positive anomaly in the carbon isotope ratios of marine carbonates deposited between 2.22 Ga and 2.06 Ga (Karhu and Holland, 1996). This shift is interpreted as an effect of increased organic carbon burial, leading to more atmospheric oxygen as discussed in Section 8.09.4.1.3. The timing of this carbon and oxygen event is generally supported in the sediment record by the earliest appearance of redbeds (oxidized subaerial sediments) and by the disappearance of banded iron formations and detrital pyrite and uraninite (Cloud Jr., 1968; Walker *et al.*, 1983; Holland, 1994). Further support has been documented in the disappearance of sediment sulfur isotope signatures that could only have originated through mass-independent fractionation in a low-oxygen atmosphere (Farquhar *et al.*, 2000).

A second positive carbon isotope excursion is apparent in marine carbonates formed during the Neoproterozoic. This feature is punctuated by large oscillations associated with glaciogenic sediments that have been attributed to so-called "snowball Earth" conditions (Hoffman *et al.*, 1998; Hoffman and Schrag, 2002). Declining concentrations of both CO_2 and CH_4 have been hypothesized as causes of the onset of intense global glaciation during this time, and increasing CO_2 has been hypothesized as a cause of subsequent deglaciation (Hoffman and Schrag, 2002; Pavlov *et al.*, 2003).

8.09.6 CONCLUSIONS

It is clear that the carbon cycle and global climate are linked in many ways throughout the history of the Earth. But it is equally apparent that the complex interactions evident in the geologic record defy simple attribution of cause and effect. Collection of more data and further analysis and modeling will continue to improve our understanding of these interactions, which are now so important to the near-term relationship between human activities and the global environment.

Until recently, the paradigm for analysis of the geologic carbon cycle has been to focus on particular processes and carbon stores that are thought to be relevant to chosen timescales of interest (see, e.g., Sundquist, 1985; Sundquist, 1986). This paradigm has provided a platform for development of a range of models, each appropriate to particular timescales. But the geologic record is revealing a growing array of "abrupt" events that cannot be analyzed in this way. New models are needed with capabilities to span a broad range of timescales, from the abruptness of the events to the prolonged cascade of subsequent effects. The need for these models is urgent because the effects of current human activities must be viewed as a geologically abrupt event by any standard.

ACKNOWLEDGMENTS

Work on this chapter was supported by the National Research Program of the US Geological Survey. We are grateful to Dick Holland, Skee Houghton, Clive Oppenheimer, Bill Reeburgh, and Roberta Rudnick for providing preprints of their chapters for this treatise, enabling us to benefit from and complement their work. We also thank Richie S. Williams and Dick Holland for thoughtful and constructive comments. Finally, in this twentieth anniversary year of the BLAG publication, it is fitting for a paper on the geologic history of the carbon cycle to take special note of Bob Berner's prolonged and insightful contributions. In that spirit, we add that this chapter benefited from the collaborative musical inspiration of Miles Davis, Cannonball Adderly, Paul Chambers, James Cobb, John Coltrane, Bill Evans, and Wynton Kelly.

REFERENCES

Adams J. M. and Faure H. (1998) A new estimate of changing carbon storage on land since the last glacial maximum, based on global land ecosystem reconstruction. *Global Planet. Change* **16–17**, 3–24.

Adams J. M., Faure H., Faure-Denard L., McGlade J. M., and Woodward F. I. (1990) Increases in terrestrial carbon storage from the last glacial maximum to the present. *Nature* **348**, 711–714.

Alley R. B., Brook E. J., and Anandakrishnan S. (2002) A northern lead in the orbital band: north-south phasing of ice-age events. *Quat. Sci. Rev.* **21**, 431–441.

Alt J. C. and Teagle D. A. H. (1999) The uptake of carbon during alteration of oceanic crust. *Geochim. Cosmochim. Acta* **63**, 1527–1536.

Altabet M. A., Murray D. W., and Prell W. L. (1999) Climatically linked oscillations in Arabian Sea denitrification over the past 1 my: implications for the marine N cycle. *Paleoceanography* **14**, 732–743.

Alvarez L. A., Alvarez W., Asaro F., and Michel H. V. (1980) Extraterrestrial cause for the Cretaceous–Tertiary extinction. *Science* **208**, 1095–1108.

Anklin M., Barnola J.-M., Schwander J., Stauffer B., and Raynaud D. (1995) Processes affecting the CO_2 concentrations measured in Greenland ice. *Tellus* **47B**, 461–470.

Anklin M., Schwander J., Stauffer B., Tschumi J., Fuchs A., Barnola J. M., and Raynaud D. (1997) CO_2 record between 40 and 8 kyr BP from the Greenland Ice Core Project ice core. *J. Geophys. Res.-Oceans* **102**, 26539–26545.

Archer D. (1991) Modeling the calcite lysocline. *J. Geophys. Res.-Oceans* **96**, 17037–17050.

Archer D. and Maier-Reimer E. (1994) Effect of deep-sea sedimentary calcite preservation on atmospheric CO_2 concentration. *Nature* **367**, 260–263.

Archer D., Kheshgi H., and Maier-Reimer E. (1997) Multiple time-scales for neutralization of fossil fuel CO_2. *Geophys. Res. Lett.* **24**, 405–408.

Arrhenius S. (1896) On the influence of carbonic acid in the air upon the temperature of the ground. *Phil. Mag. J. Sci.* **41**, 237–276.

Arthur M. A. (2000) Volcanic contributions to the carbon and sulfur geochemical cycles and global change. In *Encyclopedia of Volcanoes* (eds. H. Sigurdsson, B. F. Houghton, S. R. McNutt, H. Rymer, and J. Stix). Academic Press, San Diego, pp. 1045–1056.

Bacastow R. B. (1996) The effect of temperature change of the warm surface waters of the oceans on atmospheric CO_2. *Global Biogeochem. Cycles* **10**, 319–333.

Bains S., Norris R. D., Corfield R. M., and Faul K. L. (2000) Termination of global warmth at the Palaeocene/Eocene boundary through productivity feedback. *Nature* **407**, 171–174.

Bange H. W., Bartell U. H., Rapsomanikis S., and Andreae M. O. (1994) Methane in the Baltic and North Seas and a reassessment of the marine emissions of methane. *Global Biogeochem. Cycles* **8**, 465–480.

Barber D. C., Dyke A., Hillaire-Marcel C., Jennings A. E., Andrews J. T., Kerwin M. W., Bilodeau G., McNeely R., Southon J., Morehead M. D., and Gagnon J.-M. (1999) Forcing of the cold event of 8,200 years ago by catastrophic drainage of Laurentide lakes. *Nature* **400**, 344–348.

Barnola J. M. (1999) Status of the atmospheric CO_2 reconstruction from ice cores analyses. *Tellus* **51B**, 151–155.

Barnola J. M., Raynaud D., Korotkevich Y. S., and Lorius C. (1987) Vostok ice core provides 160,000-year record of atmospheric CO_2. *Nature* **329**, 408–414.

Barnola J. M., Pimienta P., Raynaud D., and Korotkevich Y. S. (1991) CO_2–climate relationship as deduced from the Vostok ice core: a reexamination based on new measurements and on a re-evaluation of the air dating. *Tellus* **43B**, 83–90.

Barnola J. M., Anklin M., Porcheron J., Raynaud D., Schwander J., and Stauffer B. (1995) CO_2 evolution during the last millennium as recorded by Antarctic and Greenland ice. *Tellus* **47B**, 264–272.

Bartlett K. B. and Harriss R. C. (1993) Review and assessment of methane emissions from wetlands. *Chemosphere* **26**, 261–320.

Bartlett K. B., Crill P. M., Sebacher D. I., Harriss R. C., Wilson J. O., and Melack J. M. (1988) Methane flux from the central Amazonian floodplain. *J. Geophys. Res.-Atmos.* **93**, 1571–1582.

Bates T. S., Kelley K. C., Johnson J. E., and Gammon R. H. (1996) A reevaluation of the open ocean source of methane to the atmosphere. *J. Geophys. Res.-Atmos.* **101**, 6953–6961.

Beerling D. J. (1993) Changes in the stomatal density of *Betula nana* leaves in response to increases in atmospheric carbon dioxide concentration since the late-glacial. *Spec. Pap. Palaeontol.* **49**, 181–187.

Beerling D. J. and Royer D. L. (2002) Fossil plants as indicators of the Phanerozoic global carbon cycle. *Ann. Rev. Earth. Planet. Sci.* **30**, 527–556.

Bender M., Malaize B., Orchardo J., Sowers T., and Jouzel J. (1999) High precision correlations of Greenland and Antarctic ice core records over the last 100 kyr. In *Mechanisms of Global Climate Change at Millennial Time Scales: Geophysical Monograph 112* (eds. P. U. Clark, R. S. Webb, and L. D. Keigwin). American Geophysical Union, Washington, DC, pp. 149–164.

Berger A. (1988) Milankovitch theory and climate. *Rev. Geophys.* **26**, 624–657.

Berger A., Loutre M. F., and Gallee H. (1998) Sensitivity of the LLN climate model to the astronomical and CO_2 forcings over the last 200 ky. *Clim. Dyn.* **14**, 615–629.

Berger A., Li X. S., and Loutre M. F. (1999) Modeling northern hemisphere ice volume over the last 3 Ma. *Quat. Sci. Rev.* **18**, 1–11.

Berger A. L. (1979) Insolation signatures of Quaternary climatic change. *Il Nuovo Cimento* **2C**, 63–87.

Berger W. H. (1982) Increase of carbon dioxide in the atmosphere during deglaciation: the coral reef hypothesis. *Naturwissenschaften* **69**, 87–88.

Berger W. H. and Keir R. S. (1984) Glacial-Holocene changes in atmospheric CO_2 and the deep-sea record. In *Climate Processes and Climate Sensitivity: Geophysical Monograph 29* (eds. J. E. Hansen and T. Takahashi). American Geophysical Union, Washington, DC, pp. 337–351.

Berner E. K. and Berner R. A. (1987) *The Global Water Cycle.* Prentice-Hall, Englewood Cliffs.

Berner R. A. (1982) Burial of organic carbon and pyrite sulfur in the modern ocean: its geochemical and environmental significance. *Am. J. Sci.* **282**, 451–473.

Berner R. A. (1987) Models for carbon and sulfur cycles and atmospheric oxygen: application to Paleozoic geologic history. *Am. J. Sci.* **287**, 177–196.

Berner R. A. (1990) Atmospheric carbon dioxide levels over Phanerozoic time. *Science* **249**, 1382–1386.

Berner R. A. (1991) A model for atmospheric CO_2 over Phanerozoic time. *Am. J. Sci.* **291**, 339–376.

Berner R. A. (1994) Geocarb II: a revised model of atmospheric CO_2 over Phanerozoic time. *Am. J. Sci.* **294**, 56–91.

Berner R. A. (1999) A new look at the long-term carbon cycle. *GSA Today* **9**, 1–6.

Berner R. A. (2001) Modeling atmospheric O_2 over Phanerozoic time. *Geochim. Cosmochim. Acta* **65**, 685–694.

Berner R. A. and Caldeira K. (1997) The need for mass balance and feedback in the geochemical carbon cycle. *Geology* **25**, 955–956.

Berner R. A. and Kothavala Z. (2001) Geocarb III: a revised model of atmospheric CO_2 over Phanerozoic time. *Am. J. Sci.* **301**, 182–204.

Berner R. A. and Raiswell R. (1983) Burial of organic carbon and pyrite sulfur in sediments over Phanerozoic time: a new theory. *Geochim. Cosmochim. Acta* **47**, 855–862.

Berner R. A., Lasaga A. C., and Garrels R. M. (1983) The carbonate-silicate geochemical cycle and its effect on atmospheric carbon dioxide over the past 100 million years. *Am. J. Sci.* **283**, 641–683.

Berner R. A., Petsch S. T., Lake J. A., Beerling D. J., Popp B. N., Lane R. S., Laws E. A., Westley M. B., Cassar N., Woodward F. I., and Quick W. P. (2000) Isotope fractionation and atmospheric oxygen: implications for Phanerozoic O_2 evolution. *Science* **287**, 1630–1633.

Berner W., Oeschger H., and Stauffer B. (1980) Information on the CO_2 cycle from ice core studies. *Radiocarbon* **22**, 227–235.

Bickle M. J. (1996) Metamorphic decarbonation, silicate weathering and the long-term carbon cycle. *Terra Nova* **8**, 270–276.

Bird M. I., Lloyd J., and Farquhar G. D. (1996) Terrestrial carbon-storage from the last glacial maximum to the present. *Chemosphere* **33**, 1675–1685.

Blasing T. J. and Jones S. (2002) *Current Greenhouse Gas Concentrations. Trends: A Compendium of Data on Global Change*. Carbon Dioxide Information Analysis Center, US Department of Energy, Oak Ridge, TN. http://cdiac.esd.ornl.gov/pns/current_ghg.html

Blunier T. and Brook E. J. (2001) Timing of millennial-scale climate change in Antarctica and Greenland during the last glacial period. *Science* **291**, 109–112.

Blunier T., Chappellaz J. A., Schwander J., Barnola J.-M., Desperts T., Stauffer B., and Raynaud D. (1993) Atmospheric methane, record from a Greenland ice core over the last 1000 years. *Geophys. Res. Lett.* **20**, 2219–2222.

Blunier T., Chappellaz J., Schwander J., Stauffer B., and Raynaud D. (1995) Variations in the atmospheric methane concentration during the Holocene epoch. *Nature* **374**, 46–49.

Blunier T., Schwander J., Stauffer B., Stocker T., Dallenbach A., Indermuhle A., Tschumi J., Chappellaz J., Raynaud D., and Barnola J.-M. (1997) Timing of Antarctic Cold Reversal and the atmospheric CO_2 increase with respect to the Younger Dryas event. *Geophys. Res. Lett.* **24**, 2683–2686.

Blunier T., Chappellaz J., Schwander J., Dallenbach A., Stauffer B., Stocker T. F., Raynaud D., Jouzel J., Clausen H. B., Hammer C. U., and Johnsen S. J. (1998) Asynchrony of Antarctic and Greenland climate change during the last glacial period. *Nature* **394**, 739–743.

Blunier T., Stocker T. F., Chappellaz J., and Raynaud D. (1999) Phase lag of Antarctic and Greenland temperature in the last glacial and link between CO_2 variations and Heinrich events. In *Reconstructing ocean history: A window into the future (Proceedings of the Sixth International Conference on Paleoceanography)* (eds. F. Abrantes and A. Mix). Kluwer Academic/Plenum, New York, pp. 121–138.

Botch M. S., Kobak K. I., Vinson T. S., and Kolchugina T. P. (1995) Carbon pools and accumulation in peatlands of the former Soviet Union. *Global Biogeochem. Cycles* **9**, 37–46.

Boucot A. J. and Gray J. (2001) A critique of Phanerozoic climatic models involving changes in the CO_2 content of the atmosphere. *Earth Sci. Rev.* **56**, 1–159.

Bradley R. S. (1999) *Paleoclimatology: Reconstructing Climates of the Quaternary*. Harcourt Academic Press, New York.

Bralower T. J., Zachos J. C., Thomas E., Parrow M., Paull C. K., Kelly D. C., Premoli Silva I., Sliter W. V., and Lohmann K. C. (1995) Late Paleocene to Eocene paleoceanography of the equatorial Pacific Ocean: stable isotopes recorded at Ocean Drilling Program Site 865, Allison Guyot. *Paleoceanography* **10**, 841–865.

Bralower T. J., Thomas D. J., Zachos J. C., Hirschmann M. M., Rohl U., Sigurdsson H., Thomas E., and Whitney D. L. (1997) High-resolution records of the late Palaeocene thermal maximum and circum-Caribbean volcanism: is there a causal link? *Geology* **25**, 963–966.

Breas O., Guillou C., Reniero F., and Wada E. (2002) The global methane cycle: isotopes and mixing ratios, sources and sinks. *Isotopes Environ. Health. Stud.* **37**, 257–379.

Brocks J. J., Logan G. A., Buick R., and Summons R. E. (1999) Archean molecular fossils and the early rise of eukaryotes. *Science* **285**, 1033–1036.

Broecker W. S. (1971) Calcite accumulation rates and glacial to interglacial changes in ocean mixing. In *The Late Cenozoic Glacial Ages* (ed. K. K. Turekian). Yale University Press, New Haven, CT, pp. 239–265.

Broecker W. S. (1982) Ocean chemistry during glacial time. *Geochim. Cosmochim. Acta* **46**, 1689–1705.

Broecker W. S. and Henderson G. M. (1998) The sequence of events surrounding Termination II and their implications for the cause of glacial-interglacial CO_2 changes. *Paleoceanography* **13**, 352–364.

Broecker W. S. and Takahashi T. (1977) Neutralization of fossil fuel CO_2 by marine calcium carbonate. In *The Fate of Fossil Fuel CO_2 in the oceans* (eds. N. R. Andersen and A. Malahoff). Plenum, New York, pp. 213–242.

Broecker W. S. and Takahashi T. (1978) The relationship between lysocline depth and *in situ* carbonate ion concentration. *Deep Sea Res.* **25**, 65–95.

Broecker W. S. and van Donk J. (1970) Insolation changes, ice volumes, and the O^{18} record in deep-sea cores. *Rev. Geophys. Space Phys.* **8**, 169–198.

Broecker W. S., Lynch-Stieglitz J., Clark E., Hajdas I., and Bonani G. (2001) What caused the atmosphere's CO_2 content to rise during the last 8000 years? *Geochem. Geophys. Geosys.* **2**, 2001GC000177.

Brook E. J., Sowers T., and Orchardo J. (1996) Rapid variations in atmospheric methane concentration during the past 110,000 years. *Science* **273**, 1087–1091.

Brook E. J., Severinghaus J., Harder S., and Bender M. (1999) Atmospheric methane and millennial scale climate change. In *Mechanisms of Global Climate Change at Millennial Time Scales: Geophysical Monograph 112* (eds. P. U. Clark, R. S. Webb, and L. D. Keigwin). American Geophysical Union, Washington, DC, pp. 165–175.

Brook E. J., Harder S., Severinghaus J., Steig E. J., and Sucher C. M. (2000) On the origin and timing of rapid changes in atmospheric methane during the last glacial period. *Global Biogeochem. Cycles* **14**, 559–572.

Bubier J. L. and Moore T. R. (1994) An ecological perspective on methane emissions from northern wetlands. *Trends Ecol. Evol.* **9**, 460–464.

Budyko M. I. and Ronov A. B. (1979) Chemical evolution of the atmosphere in the Phanerozoic. *Geochem. Int.* **16**, 1–9.

Buick R. (1992) The antiquity of oxygenic photosynthesis: evidence from stromatolites in sulphate-deficient Archaean lakes. *Science* **255**, 74–77.

Caldeira K. (1995) Long-term control of atmospheric carbon dioxide: low-temperature seafloor alteration or terrestrial silicate-rock weathering? *Am. J. Sci.* **295**, 1077–1114.

Cao M., Marshall S., and Gregson K. (1996) Global carbon exchange and methane emissions from natural wetlands: application of a process-based model. *J. Geophys. Res.-Atmos.* **101**, 14399–14414.

Catling D. C., Zahnle K. J., and McKay C. P. (2001) Biogenic methane, hydrogen escape, and the irreversible oxidation of early Earth. *Science* **293**, 839–843.

Cerling T. E. (1992) Use of carbon isotopes in paleosols as an indicator of the pCO_2 of the paleoatmosphere. *Global Biogeochem. Cycles* **6**, 307–314.

Cerling T. H. (1991) Carbon dioxide in the atmosphere: evidence from Cenozoic and Mesozoic paleosols. *Am. J. Sci.* **291**, 377–400.

Chamberlin T. C. (1898) The influence of great epochs of limestone formation upon the constitution of the atmosphere. *J. Geol.* **6**, 609–621.

Chappellaz J., Barnola J. M., Raynaud D., Korotkevich Y. S., and Lorius C. (1990) Ice-core record of atmospheric methane over the past 160,000 years. *Nature* **345**, 127–131.

Chappellaz J., Blunier T., Raynaud D., Barnola J.-M., Schwander J., and Stauffer B. (1993a) Synchronous changes in atmospheric CH_4 and Greenland climate between 40 and 8 kyr BP. *Nature* **366**, 443–445.

Chappellaz J. A., Fung I. Y., and Thompson A. M. (1993b) The atmospheric CH_4 increase since the Last Glacial Maximum: 1. Source estimates. *Tellus* **45B**, 228–241.

Chappellaz J., Blunier T., Kints S., Dallenbach A., Barnola J.-M., Schwander J., Raynaud D., and Stauffer B. (1997) Changes in the atmospheric CH_4 gradient between Greenland and Antarctica during the Holocene. *J. Geophys. Res.-Atmos.* **102**, 15987–15997.

Cicerone R. J. and Oremland R. S. (1988) Biogeochemical aspects of atmospheric methane. *Global Biogeochem. Cycles* **2**, 299–327.

Cloud P. E., Jr. (1968) Atmospheric and hydrospheric evolution on the primitive Earth. *Science* **160**, 729–736.

Clymo R. S., Turunen J., and Tolonen K. (1998) Carbon accumulation in peatland. *Oikos* **81**, 368–388.

Cowling S. A. (1999) Plants and temperature-CO_2 uncoupling. *Science* **285**, 1500–1501.

Craig H. and Chou C. C. (1982) Methane: the record in polar ice cores. *Geophys. Res. Lett.* **9**, 1221–1224.

Craig H., Horibe Y., and Sowers T. (1988) Gravitational separation of gases and isotopes in polar ice caps. *Science* **242**, 1675–1678.

Crowell J. C. (1999) *Pre-Mesozoic ice Ages: Their Bearing on understanding the Climate System: GSA Memoir 192*. Geological Society of America, Boulder.

Crowley T. J. (1991) Ice-age methane variations. *Nature* **353**, 122–123.

Crowley T. J. (1995) Ice age terrestrial carbon changes revisited. *Global Biogeochem. Cycles* **9**, 377–389.

Crowley T. J. (2000) Carbon dioxide and Phanerozoic climate: an overview. In *Warm Climates in Earth History* (eds. B. T. Huber, K. G. MacLeod, and S. L. Wing). Cambridge University Press, Cambridge, pp. 425–444.

Crowley T. J. and Baum S. K. (1993) Effect of decreased solar luminosity on late Precambrian ice extent. *J. Geophys. Res.-Atmos.* **98**, 16723–16732.

Crowley T. J. and Berner R. A. (2001) CO_2 and climate change. *Science* **292**, 870–872.

Curry W. B., Duplessy J. C., Labeyrie L. D., and Shackleton N. J. (1988) Changes in the distribution of $\delta^{13}C$ of deep water $\sum CO_2$ between the last glaciation and the Holocene. *Paleoceanography* **3**, 317–341.

Dacey J. W. H. and Klug M. J. (1979) Methane efflux from lake sediments through water lillies. *Science* **203**, 1253–1255.

Dallenbach A., Blunier T., Fluckiger J., Stauffer B., Chappellaz J., and Raynaud D. (2000) Changes in the atmospheric CH_4 gradient between Greenland and Antarctica during the last glacial and the transition to the Holocene. *Geophys. Res. Lett.* **27**, 1005–1008.

Daly R. A. (1909) First calcareous fossils and the evolution of the limestones. *Geol. Soc. Am. Bull.* **20**, 155–170.

Dansgaard W., Johnsen S. J., Moller J., and Langway C. C., Jr. (1969) One thousand centuries of climatic record from Camp Century on the Greenland ice sheet. *Science* **166**, 377–381.

Dansgaard W., Johnsen S. J., Clausen H. B., Dahl-Jensen D., Gundestrup N., Hammer C. U., and Oeschger H. (1984) North Atlantic climatic oscillations revealed by deep Greenland ice cores. In *Climate Processes and Climate Sensitivity: Geophysical Monograph 29* (eds. J. E. Hansen and T. Takahashi). American Geophysical Union, Washington, DC, pp. 288–298.

Degens E. T., Kempe S., and Richey J. E. (1991) Summary: biogeochemistry of major world rivers. In *SCOPE 42: Biogeochemistry of Major World Rivers* (eds. E. T. Degens, S. Kempe, and J. E. Richey). Wiley, New York, pp. 323–347.

Deines P. (1980) The isotopic composition of reduced organic carbon In *Handbook of Environmental Isotope Geochemistry* (eds. P. Fritz and J. Ch. Fontes). Elsevier, New York, vol. 1, pp. 329–406.

Deines P. (1992) Mantle carbon: concentration, mode of occurrence, and isotopic composition. In *Early Organic Evolution: Implications for Mineral and Energy Resources* (eds. M. Schidlowski, S. Golubic, M. M. Kimberley, D. M. McKirdy, and P. A. Trudinger). Springer, Berlin, pp. 133–146.

Deleens E., Ferhi A., and Queiroz O. (1983) Carbon isotope fractionation by plants using the C4 pathway. *Physiol. Veg.* **21**, 897–905.

Delmas R. J. (1993) A natural artefact in Greenland ice-core CO_2 measurements. *Tellus* **45B**, 391–396.

Delmas R., Ascencio J.-M., and Legrand P. (1980) Polar ice evidence that atmospheric CO_2 20,000 yr BP was 50% of present. *Nature* **284**, 155–157.

Des Marais D. J. (2000) Evolution: when did photosynthesis emerge on Earth? *Science* **289**, 1703–1705.

Des Marais D. J. (2001) Isotopic evolution of the biogeochemical carbon cycle during the Precambrian. *Rev. Mineral. Geochem.* **43**, 555–578.

Des Marais D. J., Strauss H., Summons R. E., and Hayes J. M. (1992) Carbon isotope evidence for the stepwise oxidation of the Proterozoic environment. *Nature* **359**, 605–609.

Dibb J. E., Rasmussen R. A., Mayewski P. A., and Holdsworth G. (1993) Northern Hemisphere concentrations of methane and nitrous oxide since 1800: results from the Mt. Logan and 20D ice cores. *Chemosphere* **27**, 2413–2423.

Dickens G. R. (2000) Methane oxidation during the Late Palaeocene Thermal Maximum. *Bull. Soc. Geol. France* **171**, 37–49.

Dickens G. R. (2001) Modeling the global carbon cycle with a gas hydrate capacitor: significance for the Latest Paleocene Thermal Maximum. In *Natural Gas Hydrates: Occurrence, Distribution, and Detection: Geophysical Monograph 124* (eds. C. K. Paull and W. P. Dillon). American Geophysical Union, Washington, DC, pp. 19–38.

Dickens G. R., Oneil J. R., Rea D. K., and Owen R. M. (1995) Dissociation of oceanic methane hydrate as a cause of the carbon isotope excursion at the end of the Paleocene. *Paleoceanography* **10**, 965–971.

Dickens G. R., Castillo M. M., and Walker J. C. G. (1997) A blast of gas in the latest Paleocene: simulating first-order effects of massive dissociation of oceanic methane hydrate. *Geology* **25**, 259–262.

Dillon W. P. (2001) Gas hydrate in the ocean environment. In *Encyclopedia of Physical Science and Technology* (ed. R. Meyers). Academic Press, San Diego, vol. 6, pp. 473–486.

Dobrovolsky V. V. (1994) *Biogeochemistry of the World's Land*. CRC Press, Boca Raton, Florida.

Douglas R. G. and Woodruff F. (1981). Deep sea benthic foraminifera. In *The Sea* (ed. C. Emiliani). Wiley, New York, vol. 7, pp. 1233–1327.

Doval M. D. and Hansell D. A. (2000) Organic carbon and apparent oxygen utilization in the western South Pacific and the central Indian Oceans. *Mar. Chem.* **68**, 249–264.

Drever J. I., Li Y.-H., and Maynard J. B. (1988) Geochemical cycles: the continental crust and the oceans. In *Chemical Cycles in the Evolution of the Earth* (eds. C. B. Gregor, R. M. Garrels, F. T. Mackenzie, and J. B. Maynard). Wiley, New York, pp. 17–53.

Duplessy J. C., Shackleton N. J., Fairbanks R. G., Labeyrie L., Oppo D., and Kallel N. (1988) Deepwater source variations during the last climatic cycle and their impact on the global deepwater circulation. *Paleoceanography* **3**, 343–360.

Edmond J. M., Palmer M. R., Measures C. I., Grant B., and Stallard R. F. (1995) The fluvial geochemistry and denudation rate of the Guayana Shield in Venezuela, Colombia, and Brazil. *Geochim. Cosmochim. Acta* **59**, 3301–3325.

Ehhalt D. H. (1974) The atmospheric cycle of methane. *Tellus* **26**, 58–70.

Ekart D. D., Cerling T. E., Montanez I. P., and Tabor N. J. (1999) A 400 million year carbon isotope record of pedogenic carbonate: implications for paleoatmospheric carbon dioxide. *Am. J. Sci.* **299**, 805–827.

Emiliani C. (1966) Paleotemperature analysis of Caribbean cores P6304-8 and P6304-9 and generalized temperature curve for the past 425,000 years. *J. Geol.* **74**, 109–126.

Enting I. G. (1992) The incompatibility of ice-core CO_2 data with reconstructions of biotic CO_2 sources (II): the influence of CO_2-fertilised growth. *Tellus* **44B**, 23–32.

Enting I. G. and Mansbridge J. V. (1991) Latitudinal distribution of sources and sinks of CO_2: results of an inversion study. *Tellus* **43B**, 156–170.

Etheridge D. M., Pearman G. I., and de Silva F. (1988) Atmospheric trace-gas variations as revealed by air trapped in an ice core from Law Dome, Antarctica. *Ann. Glaciol.* **10**, 28–33.

Etheridge D. M., Pearman G. I., and Fraser P. J. (1992) Changes in tropospheric methane between 1841 and 1978 from a high accumulation-rate Antarctic ice core. *Tellus* **44B**, 282–294.

Etheridge D. M., Steele L. P., Langenfelds R. L., Francey R. J., Barnola J.-M., and Morgan V. I. (1996) Natural and anthropogenic changes in atmospheric CO_2 over the last 1000 years from air in Antarctic ice and firn. *J. Geophys. Res.-Atmos.* **101**, 4115–4128.

Etheridge D. M., Steele L. P., Francey R. J., and Langenfelds R. L. (1998) Atmospheric methane between 1000 AD and present: evidence of anthropogenic emissions and climatic variability. *J. Geophys. Res.-Atmos.* **103**, 15979–15993.

Etiope G. and Klusman R. W. (2002) Geologic emissions of methane to the atmosphere. *Chemosphere* **49**, 777–789.

Falkowski P., Scholes R. J., Boyle E., Canadell J., Canfield D., Elser J., Gruber N., Hibbard K., Hogberg P., Linder S., Mackenzie F. T., Moore I. B., Pedersen T., Rosenthal Y., Seitzinger S., Smetacek V., and Steffen W. (2000) The global carbon cycle: a test of our knowledge of Earth as a system. *Science* **290**, 291–296.

Farquhar J., Bao H. M., and Thiemens M. (2000) Atmospheric influence of Earth's earliest sulfur cycle. *Science* **289**, 756–758.

Farrell J. W. and Prell W. L. (1989) Climatic change and $CaCO_3$ preservation: an 800,000 year bathymetric reconstruction from the central equatorial Pacific Ocean. *Paleoceanography* **4**, 447–466.

Felzer B., Webb T., and Oglesby R. J. (1998) The impact of ice sheets, CO_2, and orbital insolation on late Quaternary climates: sensitivity experiments with a general circulation model. *Quat. Sci. Rev.* **17**, 507–534.

Fischer H., Wahlen M., Smith J., Mastroianni D., and Deck B. (1999) Ice core records of atmospheric CO_2 around the last three glacial terminations. *Science* **283**, 1712–1714.

Fluckiger J., Monnin E., Stauffer B., Schwander J., Stocker T. F., Chappellaz J., Raynaud D., and Barnola J.-M. (2002) High-resolution Holocene N_2O ice core record and its relationship with CH_4 and CO_2. *Global Biogeochem. Cycles* **16**, 10.29/2001GB001417.

Francey R. J., Allison C. E., Etheridge D. M., Trudinger C. M., Enting I. G., Leuenberger M., Langenfelds R. L., Michel E., and Steele L. P. (1999) A 1000-year high precision record of $\delta^{13}C$ in atmospheric CO_2. *Tellus* **51B**, 170–193.

Francois L. M., Godderis Y., Warnant P., Ramstein G., de Noblet N., and Lorenz S. (1999) Carbon stocks and isotopic budgets of the terrestrial biosphere at mid-Holocene and last glacial maximum times. *Chem. Geol.* **159**, 163–189.

Francois R., Altabet M. A., Yu E.-F., Sigman D. M., Bacon M. P., Frank M., Bohrmann G., Bareille G., and Labeyrie L. D. (1997) Contribution of Southern Ocean surface-water stratification to low atmospheric CO_2 concentrations during the last glacial period. *Nature* **389**, 929–935.

Freeman K. H. and Hayes J. M. (1992) Fractionation of carbon isotopes by phytoplankton and estimates of ancient CO_2 levels. *Global Biogeochem. Cycles* **6**, 185–198.

Friedli H., Moor E., Oeschger H., Siegenthaler U., and Stauffer B. (1984) $^{13}C/^{12}C$ ratios in CO_2 extracted from Antarctic ice. *Geophys. Res. Lett.* **11**, 1145–1148.

Friedli H., Lotscher H., Oeschger H., Siegenthaler U., and Stauffer B. (1986) Ice core record of the $^{13}C/^{12}C$ ratio of atmospheric CO_2 in the past two centuries. *Nature* **324**, 237–238.

Fung I., Prentice K., Matthews E., Lerner J., and Russell G. (1983) Three-dimensional tracer model of atmospheric CO_2: response to seasonal exchanges with the terrestrial biosphere. *J. Geophys. Res.-Oceans* **88**, 1281–1294.

Fung I., John J., Lerner J., Matthews E., Prather M., Steele L. P., and Fraser P. J. (1991) Three-dimensional model synthesis of the global methane cycle. *J. Geophys. Res.-Atmos.* **96**, 13033–13065.

Galchenko V. F., Lein A., and Ivanov M. (1989) Biological sinks of methane. In *Exchange of Trace Gases between Terrestrial Ecosystems and the Atmosphere* (eds. M. O. Andreae and D. S. Schimel). Wiley, New York, pp. 59–71.

Garrels R. M. and Lerman A. (1981) Phanerozoic cycles of sedimentary carbon and sulfur. *Proc. Natl. Acad. Sci.* **78**, 4652–4656.

Garrels R. M. and Lerman A. (1984) Coupling of the sedimentary sulfur and carbon cycles-an improved model. *Am. J. Sci.* **284**, 989–1007.

Garrels R. M., Lerman A., and Mackenzie F. T. (1976) Controls of atmospheric O_2 and CO_2: past, present, and future. *Am. Sci.* **64**, 306–315.

Gerlach T. M. (1991) Present-day CO_2 emissions from volcanoes. *EOS, Trans., AGU* **72**, 249, 254–255.

Ghosh P., Ghosh P., and Bhattacharya S. K. (2001) CO_2 levels in the Late Palaeozoic and Mesozoic atmosphere from soil carbonate and organic matter, Satpura basin, Central India. *Palaeogeogr. Palaeoclimatol. Palaeoecol.* **170**, 219–236.

Gillett R. W., van Ommen T. D., Jackson A. V., and Ayers G. P. (2000) Formaldehyde and peroxide concentrations in Law Dome (Antarctica) firn and ice cores. *J. Glaciol.* **46**, 15–19.

Gorham E. (1991) Northern peatlands: role in the carbon cycle and probable responses to climatic warming. *Ecol. Appl.* **1**, 182–195.

Grocke D. R. (2002) The carbon isotope composition of ancient CO_2 based on higher-plant organic matter. *Phil. Trans. Roy. Soc. London Ser. A: Math. Phys. Eng. Sci.* **360**, 633–658.

Haan D. and Raynaud D. (1998) Ice core record of CO variations during the two last millennia: atmospheric implications and chemical interactions within the Greenland ice. *Tellus* **50B**, 253–262.

Hammer C. U., Clausen H. B., Dansgaard W., Gundestrup N., Johnsen S. J., and Reeh N. (1978) Dating of Greenland ice cores by flow models, isotopes, volcanic debris, and continental dust. *J. Glaciol.* **20**, 3–26.

Hansell D. A. and Carlson C. A. (1998) Deep-ocean gradients in the concentration of dissolved organic carbon. *Nature* **395**, 263–266.

Hanson R. S. and Hanson T. E. (1996) Methanotrophic bacteria. *Microbiol. Rev.* **60**, 439–471.

Hardie L. A. (1996) Secular variation in seawater chemistry: an explanation for the coupled secular variation in the mineralogies of marine limestones and potash evaporites over the past 600 m. y. *Geology* **24**, 279–283.

Hayes J. M. (1994) Global methanotrophy at the Archean-Proterzoic transition. In *Early Life on Earth: Nobel Symposium No. 84* (ed. S. Bengtson). Columbia University Press, New York, pp. 220–236.

Hayes J. M., Strauss H., and Kaufman A. J. (1999) The abundance of ^{13}C in marine organic matter and isotopic fractionation in the global biogeochemical cycle of carbon during the past 800 Ma. *Chem. Geol.* **161**, 103–125.

Hedges S. B., Chen H., Kumar S., Wang D. Y. C., Thompson A. S., and Watanabe H. (2001) A genomic timescale for the origin of eukaryotes. *BMC Evol. Biol.* **1**, article 4: (1471-2148/1/4).

Hein R., Crutzen P. J., and Heimann M. (1997) An inverse modeling approach to investigate the global atmospheric methane cycle. *Global Biogeochem. Cycles* **11**, 43–76.

Henderson G. M. and Slowey N. C. (2000) Evidence from U–Th dating against Northern Hemisphere forcing of the penultimate deglaciation. *Nature* **404**, 61–66.

Hewitt C. D. and Mitchell J. F. B. (1997) Radiative forcing and response of a GCM to ice age boundary conditions: cloud feedback and climate sensitivity. *Clim. Dyn.* **13**, 821–834.

Hinrichs K.-U. and Boetius A. (2002) The anaerobic oxidation of methane: new insights in microbial ecology and biogeochemistry. In *Ocean Margin Systems* (eds. G. Wefer, D. Billett, D. Hebbeln, B. B. Jorgensen, M. Schluter, and T. van Weering). Springer, New York, pp. 457–477.

Hoehler T. M., Bebout B. M. and Des Marais D. J.. (2001). The role of microbial mats in the production of reduced gases on the early Earth. *Nature* **412**, 324–327.

Hoffman P. F. and Schrag D. P. (2002) The snowball Earth hypothesis: testing the limits of global change. *Terra Nova* **14**, 129–155.

Hoffman P. F., Kaufman A. J., Halverson G. P., and Schrag D. P. (1998) A Neoproterozoic snowball earth. *Science* **281**, 1342–1346.

Hofmann M., Broecker W. S., and Lynch-Stieglitz J. (1999) Influence of a $[CO_{2(aq)}]$ dependent biological C-isotope fractionation on glacial C-13/C-12 ratios in the ocean. *Global Biogeochem. Cycles* **13**, 873–883.

Holland H. D. (1978) *Chemistry of the Atmosphere and Oceans*. Wiley, New York.

Holland H. D. (1994) Early Proterozoic atmospheric change. In *Early Life on Earth: Nobel Symposium No. 84* (ed. S. Bengtson). Columbia University Press, New York, pp. 237–244.

Holland H. D. (2002) Volcanic gases, black smokers, and the Great Oxidation Event. *Geochim. Cosmochim. Acta* **66**, 3811–3826.

Holland H. D. (2003) Discussion of the article by A. C. Lasaga and H. Omoto on "The oxygen geochemical cycle: dynamics and stability," *Geochim. Cosmochim. Acta* **66**, 361–381, 2002. *Geochim. Cosmochim. Acta* **67**, 787–789.

Holland H. D. and Rye R. (1997) Evidence in pre-2.2 Ga paleosols for the early evolution of atmospheric oxygen and terrestrial biota: comment and reply. *Geology* **25**, 857–859.

Holland H. D. and Zimmerman H. (2000) The dolomite problem revisited. *Int. Geol. Rev.* **42**, 481–490.

Holmes M. E., Sansone F. J., Rust T. M., and Popp B. N. (2000) Methane production, consumption, and air-sea exchange in the open ocean: an evaluation based on carbon isotopic ratios. *Global Biogeochem. Cycles* **14**, 1–10.

Horita J., Zimmermann H., and Holland H. D. (2002) Chemical evolution of seawater during the Phanerozoic: implications from the record of marine evaporites. *Geochim. Cosmochim. Acta* **66**, 3733–3756.

Hornafius J. S., Quigley D., and Luyendyk B. P. (1999) The world's most spectacular marine hydrocarbon seeps (Coal Oil Point, Santa Barbara Channel, California): quantification of emissions. *J. Geophys. Res.-Oceans* **104**, 20703–20711.

Hovland M., Judd A. G., and Burke J. R. A. (1993) The global flux of methane from shallow submarine sediments. *Chemosphere* **26**, 559–578.

Hunt J. M. (1996) *Petroleum Geochemistry and Geology*. W. H. Freeman, New York.

Imbrie J., Boyle E. A., Clemens S. C., Duffy A., Howard W. R., Kukla G., Kutzbach J., Martinson D. G., McIntyre A., Mix A. C., Molfino B., Morley J. J., Peterson L. C., Pisias N. G., Prell W. G., Raymo M. E., Shackleton N. J., and Toggweiler J. R. (1992) On the structure and origin of major glaciation cycles: 1. Linear responses to Milankovitch forcing. *Paleoceanography* **7**, 701–738.

Imbrie J., Berger A., Boyle E. A., Clemens S. C., Duffy A., Howard W. R., Kukla G., Kutzbach J., Martinson D. G., McIntyre A., Mix A. C., Molfino B., Morley J. J., Peterson L. C., Pisias N. G., Prell W. G., Raymo M. E., Shackleton N. J., and Toggweiler J. R. (1993) On the structure and origin of major glaciation cycles: 2. The 100,000-year cycle. *Paleoceanography* **8**, 699–735.

Indermuhle A., Stocker T. F., Joos F., Fischer H., Smith H. J., Wahlen M., Deck B., Mastroianni D., Tschumi J., Blunier T., Meyer R., and Stauffer B. (1999) Holocene carbon-cycle dynamics based on CO_2 trapped in ice at Taylor Dome, Antarctica. *Nature* **398**, 121–126.

Indermuhle A., Monnin E., Stauffer B., Stocker T. F., and Wahlen M. (2000) Atmospheric CO_2 concentration from 60 to 20 kyr BP from the Taylor Dome ice core, Antarctica. *Geophys. Res. Lett.* **27**, 735–738.

Ingall E. D. and Van Cappellen P. (1990) Relation between sedimentation rate and burial of organic phosphorus and organic carbon in marine sediments. *Geochim. Cosmochim. Acta* **54**, 373–386.

Ittekot V. (1988) Global trends in the nature of organic matter in river suspensions. *Nature* **332**, 436–438.

Johnsen S. J., Dansgaard W., Clausen H. B., and Langway C. C., Jr. (1972) Oxygen isotope profiles through the Antarctic and Greenland ice sheets. *Nature* **235**, 429–434.

Joos F., Meyer R., Bruno M., and Leuenberger M. (1999) The variability in the carbon sinks as reconstructed for the last 1000 years. *Geophys. Res. Lett.* **26**, 1437–1440.

Jouzel J., Lorius C., Petit J. R., Genthon C., Barkov N. I., Kotlyakov V. M., and Petrov V. M. (1987) Vostok ice core: a continuous isotope temperature record over the last climatic cycle (160,000 years). *Nature* **329**, 403–408.

Jouzel J., Barkov N. I., Barnola J.-M., Bender M., Chappellaz J., Genthon C., Kotlyakov V. M., Lipenkov V., Lorius C., Petit J.-R., Raynaud D., Raisbeck G., Ritz C., Sowers T., Sievenard M., Yiou F., and Yiou P. (1993) Extending the Vostok ice-core record of paleoclimate to the penultimate glacial period. *Nature* **364**, 407–412.

Judd A. G. (2000) Geological sources of methane. In *Atmospheric Methane: Its Role in the Global Environment* (ed. M. A. K. Khalil). Springer, New York, pp. 280–303.

Judd A. G., Hovland M., Dimitrov L. I., Garcia Gil S. G., and Jukes V. (2002) The geological methane budget at continental margins and its influence on climate change. *Geofluids* **2**, 109–126.

Karhu J. A. and Holland H. D. (1996) Carbon isotopes and the rise of atmospheric oxygen. *Geology* **24**, 867–870.

Kasting J. F. (1987) Theoretical constraints on oxygen and carbon dioxide concentrations in the Precambrian atmosphere. *Precamb. Res.* **34**, 205–228.

Kasting J. F. (1993) Earth's early atmosphere. *Science* **259**, 920–926.

Kasting J. F. (2001) The rise of atmospheric oxygen. *Science* **293**, 819–820.

Kasting J. F., Zahnle K. J., and Walker J. C. G. (1983) Photochemistry of methane in the Earth's early atmosphere. *Precamb. Res.* **20**, 121–148.

Kasting J. F., Toon O. B., and Pollack J. B. (1988) How climate evolved on the terrestrial planets. *Sci. Am.* **256**, 90–97.

Kasting J. F., Eggler D. H., and Raeburn S. P. (1993) Mantle redox evolution and the case for a reduced Archean atmosphere. *J. Geol.* **101**, 245–257.

Keeling C. D. (1991) *Atmospheric CO_2—modern record, South Pole*. Trends: A Compendium of Data on Global Change. Carbon Dioxide Information Analysis Center, US Department of Energy, Oak Ridge, TN.

Keeling C. D., Bacastow R. B., Carter A. F., Piper S. C., Whorf T. P., Heimann M., Mook W. G., and Roeloffzen H. (1989) A three-dimensional model of atmospheric CO_2 transport based on observed winds: 1. Analysis of observational data. In *Aspects of Climate Variability in the Pacific and the Western Americas: Geophysical Monograph 55* (ed. D. H. Peterson). American Geophysical Union, Washington, DC, pp. 165–236.

Kelley C. A. and Jeffrey W. H. (2002) Dissolved methane concentration profiles and air-sea fluxes from 41°S to 27°N. *Global Biogeochem. Cycles* **16**, 10.1029/2001GB001809.

Kelley D. S., Karson J. A., Blackman D. K., Fruh-Green G. L., Butterfield D. A., Lilley M. D., Olson E. J., Schrenk M. O., Roe K. K., Lebon G. T., Rivizzigno P., and and the AT3-60 Shipboard Party (2001) An off-axis hydrothermal vent field near the Mid-Atlantic Ridge at 30N. *Nature* **412**, 145–149.

Kennett J. P. and Stott L. D. (1991) Abrupt deep sea warming, paleoceanographic changes and benthic extinctions at the end of the Palaeocene. *Nature* **353**, 319–322.

Kennett J. P., Cannariato K. G., Hendy I. L., and Behl R. J. (2003) *Methane Hydrates in Quaternary Climate Change: The Clathrate Gun Hypothesis*. American Geophysical Union, Washington, DC.

Kerrick D. M. (2001) Present and past nonanthropogenic CO_2 degassing from the solid earth. *Rev. Geophys.* **39**, 565–585.

Kerrick D. M., McKibben M. A., Seward T. M., and Caldeira K. (1995) Convective hydrothermal CO_2 emission from high heat flow regions. *Chem. Geol.* **121**, 285–293.

Khalil M. A. K. and Rasmussen R. A. (1989) Climate-induced feedback for the global cycles of methane and nitrous oxide. *Tellus* **41B**, 554–559.

Kiehl J. T. and Dickinson R. E. (1987) A study of the radiative effects of enhanced atmospheric CO_2 and CH_4 on early Earth surface temperatures. *J. Geophys. Res.-Atmos.* **92**, 2991–2998.

King G. M. (1992) Ecological aspects of methane oxidation, a key determinant of global methane dynamics. In *Advances in Microbial Ecology* (ed. K. C. Marshall). Plenum, New York, vol. 12, pp. 431–468.

Koch P. L., Zachos J. C., and Gingerich P. D. (1992) Correlation between isotope records in marine and continental carbon reservoirs near the Paleocene/Eocene boundary. *Nature* **358**, 319–322.

Ku T.-L. and Luo S. (1992) Carbon isotopic variations on glacial-to-interglacial time scales in the ocean: modeling and implications. *Paleoceanography* **7**, 543–562.

Kuhn W. R. and Kasting J. F. (1983) Effects of increased CO_2 concentrations on surface temperature of the early Earth. *Nature* **301**, 53–55.

Kump L. R., Kasting J. F., and Robinson J. M. (1991) Atmospheric oxygen variation through geologic time–introduction. *Palaeogeogr. Palaeoclimatol. Palaeoecol.* **97**, 1–3.

Kump L. R., Brantley S. L., and Arthur M. A. (2000) Chemical weathering, atmospheric CO_2, and climate. *Ann. Rev. Earth. Planet. Sci.* **28**, 611–667.

Kvenvolden K. A. and Lorenson T. D. (2001) The global occurrence of natural gas hydrate. In *Natural Gas Hydrates: Occurrence, Distribution, and Detection: Geophysical Monograph 124* (eds. C. K. Paull and W. P. Dillon). American Geophysical Union, Washington, DC, pp. 3–18.

Lacroix A. V. (1993) Unaccounted-for sources and isotopically enriched methane and their contribution to the emissions inventory: a review synthesis. *Chemosphere* **26**, 507–557.

Lambert G. and Schmidt S. (1993) Reevaluation of the oceanic flux of methane: uncertainties and long term variations. *Chemosphere* **26**, 579–589.

Langenfelds R. L., Fraser P. J., Francey R. J., Steele L. P., Porter L. W., and Allison C. E. (1996) The Cape Grim air archive: the first seventeen years, 1978–1995. In *Baseline Atmospheric Program Australia* (ed. R. J. Francey *et al.*). Bureau of Meteorology and CSIRO Division of Atmospheric Research, Melbourne, Australia, pp. 53–70.

Lappalainen E. (1996) General review on world peatland and peat resources. In *Global Peat Resources* (ed. E. Lappalainen). International Peat Society, Finland, pp. 53–56.

Lasaga A. C. and Ohmoto H. (2002) The oxygen geochemical cycle: dynamics and stability. *Geochim. Cosmochim. Acta* **66**, 361–381.

Lasaga A. C., Berner R. A., and Garrels R. M. (1985) An improved geochemical model of atmospheric CO_2 fluctuations over the past 100 million years. In *The Carbon Cycle and Atmospheric CO_2: Natural Variations Archean to Present: Geophysical Monograph 32* (eds. E. T. Sundquist and W. S. Broecker). American Geophysical Union, Washington, DC, pp. 397–411.

Lein A. Y. (1984) Anaerobic consumption of organic matter in modern marine sediments. *Nature* **312**, 148–150.

Lelieveld J., Crutzen P. J., and Dentener F. J. (1998) Changing concentration, lifetime, and climate forcing of atmospheric methane. *Tellus* **50B**, 128–150.

Lemarchand D., Gaillardet J., Lewin E., and Allegre C. J. (2000) The influence of rivers on marine boron isotopes and implications for reconstructing past ocean pH. *Nature* **408**, 951–954.

Lemarchand D., Gaillardet J., Lewin E., and Allegre C. J. (2002) Boron isotope systematics in large rivers: implications for the marine boron budget and paleo-pH reconstruction over the Cenozoic. *Chem. Geol.* **190**, 123–140.

Leuenberger M., Siegenthaler U., and Langway C. C. (1992) Carbon isotope composition of atmospheric CO_2 during the last ice age from an Antarctic ice core. *Nature* **357**, 450–488.

Levchenko V. A., Francey R. J., Etheridge D. M., Tuniz C., Head J., Morgan V. I., Lawson E., and Jacobsen G. (1996) The ^{14}C "bomb spike" determines the age spread and age of CO_2 in Law Dome firn and ice. *Geophys. Res. Lett.* **23**, 3345–3348.

Levine J. S., Cofer W. R., and Pinto J. P. (2000) Biomass burning. In *Atmospheric Methane: Its Role in the Global Environment* (ed. M. A. K. Khalil). Springer, New York, pp. 190–201.

Li Y.-H. (2000) *A compendium of geochemistry: from solar nebula to the human brain*. Princeton University Press, Princeton.

Lorius C., Merlivat L., Jouzel J., and Pourchet M. (1979) A 30, 000-yr isotope climatic record from Antarctic ice. *Nature* **280**, 644–648.

Lorius C., Jouzel J., Ritz C., Merlivat L., Barkov N. I., Korotkevich Y. S., and Kotlyakov V. M. (1985) A 150,000-year climatic record from Antarctic ice. *Nature* **316**, 591–596.

Lorius C., Jouzel J., Raynaud D., Hansen J., and Le Treut H. (1990) The ice-core record: climate sensitivity and future greenhouse warming. *Nature* **347**, 139–145.

MacLeod G., McKeown C., Hall H. J., and Russell M. J. (1994) Hydrothermal and oceanic pH conditions of possible relevance to the origin of life. *Orig. Life Evol. Biosph.* **23**, 19–41.

Manabe S. and Broccoli A. J. (1985) A comparison of climate model sensitivity with data from the last glacial maximum. *J. Atmos. Sci.* **42**, 2643–2651.

Marino B. D. and McElroy M. B. (1991) Isotopic composition of atmospheric CO_2 inferred from carbon in C4 plant cellulose. *Nature* **349**, 127–131.

Martin J. H. (1990) Glacial-interglacial CO_2 change: the iron hypothesis. *Paleoceanography* **5**, 1–13.

Martin J. H., Fitzwater S. E., and Gordon R. M. (1990) Iron deficiency limits phytoplankton growth in Antarctic waters. *Global Biogeochem. Cycles* **4**, 5–12.

Martinerie P., Brasseur G. P., and Granier C. (1995) The chemical composition of ancient atmospheres: a model study constrained by ice core data. *J. Geophys. Res.-Atmos.* **100**, 14291–14304.

Marty B. and Tolstikhin I. N. (1998) CO_2 fluxes from mid-ocean ridges, arcs, and plumes. *Chem. Geol.* **145**, 233–248.

Maslin M. A. and Thomas E. (2003) Balancing the deglacial global carbon budget: the hydrate factor. *Quat. Sci. Rev.* **22**, 1729–1736.

Masters C. D., Attanasi E. D., and Root D. H. (1994) USGS World Petroleum Assessment and Analysis. In *Proceedings of the 14th World Petroleum Congress*. Wiley, New York.

Mathis P. (1990) Compared structure of plant and bacterial photosynthetic reaction centers. Evolutionary implications. *Biochim. Biophys. Acta (BBA): Bioenergetics* **1018**(2–3), 163–167.

Matthews E. (2000) Wetlands. In *Atmospheric Methane: Its Role in the Global Environment* (ed. M. A. K. Khalil). Springer, New York, pp. 202–233.

Matthews E. and Fung I. (1987) Methane emission from natural wetlands: global distribution, area, and environmental characteristics of sources. *Global Biogeochem. Cycles* **1**, 61–86.

Max M. D., Dillon W. P., Nishimura C., and Hurdle B. G. (1999) Sea-floor methane blow-out and global firestorm at the K–T boundary. *Geo. Mar. Lett.* **18**, 285–291.

McElroy M. B. (1983) Marine biological controls on atmospheric CO_2 and climate. *Nature* **302**, 328–329.

McElroy M. B. (1989) Studies of polar ice: insights for atmospheric chemistry. In *The Environmental Record in Glaciers and Ice Sheets* (eds. H. Oeschgers and Langway, Jr.). Wiley, New York, pp. 363–377.

McElwain J. C. and Chaloner W. G. (1995) Stomatal density and index of fossil plants track atmospheric carbon dioxide in the Paleozoic. *Ann. Bot.* **76**, 389–395.

Meybeck M. (1981) River transport of organic carbon to the ocean. In *Flux of Organic Carbon by Rivers to the Oceans*. Carbon Dioxide Effects Research and Assessment Program (eds. G. E. Likens and F. T. Mackenzie). US DOE, Washington, DC, pp. 219–269.

Meybeck M. (1987) Global chemical weathering of surficial rocks estimated from river dissolved loads. *Am. J. Sci.* **288**, 401–428.

Meybeck M. (1988) How to establish and use world budgets of riverine materials. In *Physical and Chemical Weathering in Geochemical Cycles* (eds. A. Lerman and M. Meybeck). Kluwer Academic, Dordrecht, pp. 247–272.

Miller K. G., Fairbanks R. G., and Mountain G. S. (1987) Tertiary oxygen isotope synthesis, sea level history, and continental margin erosion history. *Paleoceanography* **2**, 1–19.

Milliman J. D. (1993) Production and accumulation of calcium carbonate in the ocean: budget of a nonsteady state. *Global Biogeochem. Cycles* **7**, 927–957.

Milliman J. D., Quinchun X., and Zuosheng Y. (1984) Transfer of particulate organic carbon and nitrogen from the Changjiang River to the ocean. *Am. J. Sci.* **284**, 824–834.

Monnin E., Indermuhle A., Dallenbach A., Fluckiger J., Stauffer B., Stocker T. F., Raynaud D., and Barnola J. M. (2001) Atmospheric CO_2 concentrations over the last glacial termination. *Science* **291**, 112–114.

Moomaw W. R., Moreira J. R., Blok K., Greene D. L., Gregory K., Jaszay T., Kashiwagi T., Levine M., McFarland M., Prasad N. S., Price L., Rogner H.-H., Sims R., Zhou F., and Zhou P. (2001) Technological and economic potential of greenhouse gas emissions reduction. In *Climate Change 2001: Mitigation. Contribution of Working Group III to the Third Assessment Report of the Intergovernmental Panel on Climate Change* (eds. B. Metz, O. Davidson, R. Swart, and J. Pan). Cambridge University Press, Cambridge, pp. 171–299.

Moosavi S. C., Crill P. M., Pullman E. R., Funk D. W., and Peterson K. M. (1996) Controls on CH_4 flux from an Alaskan boreal wetland. *Global Biogeochem. Cycles* **10**, 287–296.

Morner N.-A. and Etiope G. (2002) Carbon degassing from the lithosphere. *Global Planet. Change* **33**, 185–203.

Nakazawa T., Machida T., Tanaka M., Fujii Y., Aoki S., and Watanabe O. (1993) Differences of the atmospheric CH_4 concentration between the Arctic and Antarctic regions in pre-Industrial/pre-Agricultural era. *Geophys. Res. Lett.* **20**, 943–946.

Nakicenovic N., Grubler A., and McDona A. (1998) *Global Energy Perspectives*. Cambridge University Press, New York.

Neftel A., Oeschger H., Schwander J., and Stauffer B. (1983) Carbon dioxide concentration in bubbles of natural cold ice. *J. Phys. Chem.* **87**, 4116–4120.

Neftel A., Moor E., Oeschger H., and Stauffer B. (1985) Evidence from polar ice cores for the increase in atmospheric CO_2 in the past two centuries. *Nature* **315**, 45–47.

Neftel A., Oeschger H., Staffelbach T., and Stauffer B. (1988) CO_2 record in the Byrd ice core 50,000–5,000 years BP. *Nature* **331**, 609–611.

Newell R. E., Navato A. R., and Hsiung J. (1978) Long-term global sea surface temperature fluctuations and their possible influence on atmospheric CO_2 concentrations. *Pure Appl. Geophys.* **116**, 351–371.

Nisbet E. G. (1990) The end of the ice age. *Can. J. Earth Sci.* **27**, 148–157.

Nisbet E. G. (1992) Sources of atmospheric CH_4 in early postglacial time. *J. Geophys. Res.-Atmos.* **97**, 12859–12867.

Norris R. D. and Rohl U. (1999) Carbon cycling and chronology of climate warming during the Palaeocene/Eocene transition. *Nature* **401**, 775–778.

Nozaki Y. and Yamamoto Y. (2001) Radium 228 based nitrate fluxes in the eastern Indian Ocean and the South China Sea and a silicon-induced "alkalinity pump" hypothesis. *Global Biogeochem. Cycles* **15**, 555–567.

O'Hara F., Jr. (1990) *CDIAC Glossary: Carbon Dioxide and Climate*. ORNL/CDIAC-39. Carbon Dioxide Information and Analysis Center, Oak Ridge National Laboratory, Oak Ridge, TN. http://cdiac.esd.ornl.gov/pns/convert.html

O'Keefe J. D. and Ahrens T. J. (1989) Impact production of CO_2 by the Cretaceous/Tertiary extinction bolide and the resultant heating of the Earth. *Nature* **338**, 247–249.

O'Leary M. H. (1988) Carbon isotopes in photosynthesis. *BioScience* **38**, 328–335.

Oeschger H., Beer J., Siegenthaler U., and Stauffer B. (1984) Late glacial climate history from ice cores. In *Climate Processes and Climate Sensitivity: Geophysical Monograph 29* (eds. J. E. Hansen and T. Takahashi). American Geophysical Union, Washington, DC, pp. 299–306.

Ohmoto H. (1996) Evidence in pre-2.2 Ga paleosols for the early evolution of atmospheric oxygen and terrestrial biota. *Geology* **24**, 1135–1138.

Oil and Gas Journal (2001) Worldwide look at reserves and production. *Oil Gas J.* 126–127.

Oremland R. S. and Culbertson C. W. (1992) Importance of methane-oxidizing bacteria in the methane budget as revealed by the use of a specific inhibitor. *Nature* **356**, 421–423.

ORNL Bioenergy Feedstock Development Programs (2003) *Bioenergy conversion factors*. http://bioenergy.ornl.gov/papers/misc/energy_conv.html

Orphan V. J., House C. H., Hinrichs K.-U., McKeegan K. D., and DeLong E. F. (2001) Methane-consuming Archaea revealed by directly coupled isotopic and phylogenetic analysis. *Science* **293**, 484–487.

Otto D., Rasse D., Kaplan J., Warnant P., and Francois L. (2002) Biospheric carbon stocks reconstructed at the last glacial maximum: comparison between general circulation models using prescribed and computed sea surface temperatures. *Global Planet. Change* **33**, 117–138.

Owen T., Cess R. D., and Ramanathan V. (1979) Enhanced CO_2 greenhouse to compensate for reduced solar luminosity on early Earth. *Nature* **277**, 640–642.

Pace N. R. (1997) A molecular view of microbial diversity and the biosphere. *Science* **276**, 734–774.

Pagani M. (2002) The alkenone-CO_2 proxy and ancient atmospheric carbon dioxide. *Phil. Trans. Roy. Soc. London Ser. A: Math. Phys. Eng. Sci.* **360**, 609–632.

Pagani M., Arthur M. A., and Freeman K. H. (1999a) Miocene evolution of atmospheric carbon dioxide. *Paleoceanography* **14**, 273–292.

Pagani M., Freeman K. H., and Arthur M. A. (1999b) Late Miocene atmospheric CO_2 concentrations and the expansion of C-4 grasses. *Science* **285**, 876–879.

Palmer M. R. and Pearson P. N. (2003) A 23,000-year record of surface water pH and pCO_2 in the western equatorial Pacific Ocean. *Science* **300**.

Pavlov A. A., Hurtgen M. T., Kasting J. F., and Arthur M. A. (2003) Methane-rich Proterozoic atmosphere? *Geology* **31**, 87–90.

Pearman G. I., Hyson P., and Fraser P. J. (1983) The global distribution of atmospheric carbon dioxide: 1. Aspects of observations and modeling. *J. Geophys. Res.-Oceans* **88**, 3581–3590.

Pearman G. I., Etheridge D., de Silva F., and Fraser P. J. (1986) Evidence of changing concentrations of atmospheric CO_2, N_2O and CH_4 from air bubbles in Antarctic ice. *Nature* **320**, 248–250.

Pearson P. N. and Palmer M. R. (1999) Middle Eocene seawater pH and atmospheric carbon dioxide concentrations. *Science* **284**, 1824–1826.

Pearson P. N. and Palmer M. R. (2000) Atmospheric carbon dioxide concentrations over the past 60 million years. *Nature* **406**, 695–699.

Peterson L. C. and Prell W. L. (1985) Carbonate dissolution in recent sediments of the eastern equatorial Indian Ocean: preservation patterns and carbonate loss above the lysocline. *Mar. Geol.* **64**, 259–290.

Petit J. R., Jouzel J., Raynaud D., Barkov N. I., Barnola J. M., Basile I., Bender M., Chappellaz J., Davis M., Delaygue G., Delmotte M., Kotlyakov V. M., Legrand M., Lipenkov V. Y., Lorius C., Pepin L., Ritz C., Saltzman E., and Stievenard M. (1999) Climate and atmospheric history of the past 420,000 years from the Vostok ice core, Antarctica. *Nature* **399**, 429–436.

Petit-Maire N., Fontugne M., and Rouland C. (1991) Atmospheric methane ratio and environmental changes in the Sahara and Sahel during the last 130 kyrs. *Palaeogeogr. Palaeoclimatol. Palaeoecol.* **86**, 197–204.

Petsch S. T. and Berner R. A. (1998) Coupling the geochemical cycles of C, P, Fe, and S: the effect on atmospheric O_2 and the isotopic records of carbon and sulfur. *Am. J. Sci.* **298**, 246–262.

Pierazzo E., Kring D. A., and Melosh H. J. (1998) Hydrocode simulation of the Chicxulub impact event and the production of climatically active gases. *J. Geophys. Res.-Planet* **103**, 28607–28625.

Pisias N. G. S. N. J. (1984) Modeling the global climate response to orbital forcing and atmospheric carbon dioxide changes. *Nature* **310**, 757–759.

Plummer L. N. and Sundquist E. T. (1982) Total individual ion activity coefficients of calcium and carbonate in seawater at 25°C and 35‰ salinity, and implications to the agreement between apparent and thermodynamic constants of calcite and aragonite. *Geochim. Cosmochim. Acta* **46**, 247–258.

Pope K. O., Baines K. H., Ocampo A. C., and Ivanov B. A. (1997) Energy, volatile production, and climatic effects of the Chicxulub Cretaceous/Tertiary impact. *J. Geophys. Res.-Planet* **102**, 21645–21664.

Popp B. N., Takigiku R., Hayes J. M., Louda J. W., and Baker E. W. (1989) The post-Paleozoic chronology and mechanism of ^{13}C depletion in primary marine organic matter. *Am. J. Sci.* **289**, 436–454.

Prell W. L. and Kutzbach J. E. (1987) Monsoon variability over the past 150,000 years. *J. Geophys. Res.-Atmos.* **92**, 8411–8425.

Prentice I., Sykes M., Lautenschlager M., Harrison S., Denissenko O., and Bartlein P. (1993) Modelling global vegetation patterns and terrestrial carbon storage at the last glacial maximum. *Global Ecol. Biogeogr. Lett.* **3**, 67–76.

Prentice K. C. and Fung I. Y. (1990) The sensitivity of terrestrial carbon storage to climate change. *Nature* **346**, 48–51.

Prinn R., Cunnold D., Rasmussen R., Simmonds P., Alyea F., Crawford A., Fraser P., and Rosen R. (1987) Atmospheric trends in methylchloroform and the global average for the hydroxyl radical. *Science* **238**, 945–950.

Prinn R. G., Simmonds P. G., Rasmussen R. A., Rosen R. D., Alyea F. N., Cardelino C. A., Crawford A. J., Cunnold D. M., Fraser P. J., and Lovelock J. E. (1983) The atmospheric lifetime experiment: 1. Introduction, instrumentation, and overview. *J. Geophys. Res.-Oceans* **88**, 8353–8367.

Prinn R. G., Weiss R. F., Miller B. R., Huang J., Alyea F. N., Cunnold D. M., Fraser P. J., Hartley D. E., and Simmonds P. G. (1995) Atmospheric trends and lifetime of CH_3CCl_3 and global OH concentrations. *Science* **269**, 187–192.

Prinn R. G., Huang J., Weiss R. F., Cunnold D. M., Fraser P. J., Simmonds P. G., McCulloch A., Harth C., Salameh P., O'Doherty S., Wang R. H. J., Porter L., and Miller B. R. (2001) Evidence for substantial variations of atmospheric hydroxyl radicals in the past two decades. *Science* **292**, 1882–1888.

Ramaswamy V., Boucher O., Haigh J., Hauglustaine D., Haywood J., Myhre G., Nakajima T., Shi G. Y., Solomon S., Betts R., Charlson R., Chuang C., Daniel J. S., Del Genio A., van Dorland R., Feichter J., Fuglestvedt J., de. F. Forster P. M., Ghan S. J., Jones A., Kiehl J. T., Koch D., Land C., Lean J., Lohmann U., Minschwaner K., Penner J. E., Roberts D. L., Rodhe H., Roelofs G. J., Rotstayn L. D., Schneider T. L., Schumann U., Schwartz S. E., Schwarzkopf M. D., Shine K. P., Smith S., Stevenson D. S., Stordal F., Tegen I., Zhang Y., and Jones A. (2001) Radiative forcing of climate change. In *Climate Change 2001: The Scientific Basis. Contribution of Working Group I to the Third Assessment Report of the Intergovernmental Panel on Climate Change* (eds. J. T. Houghton, Y. Ding, D. J. Griggs, M. Noguer, P. J. van der Linden, X. Dai, K. Maskell, and C. A. Johnson). Cambridge University Press, Cambridge, pp. 350–416.

Rasmussen B. (2000) Filamentous microfossils in a 3,235-million-year-old volcanogenic massive sulphide deposit. *Nature* **405**, 676–679.

Rasmussen R. A. and Khalil M. A. K. (1984) Atmospheric methane in the recent and ancient atmospheres: concentrations, trends, and interhemispheric gradient. *J. Geophys. Res.-Atmos.* **89**, 11599–11605.

Rau G. H., Takahashi T., and Des Marais D. J. (1989) Latitudinal variations in plankton $\delta^{13}C$: implications for CO_2 and productivity in past oceans. *Nature* **341**, 516–518.

Raymo M. E., Ruddiman W. F., and Froelich P. N. (1988) Influence of late Cenozoic mountain building on ocean geochemical cycles. *Geology* **16**, 649–653.

Raynaud D. and Barnola J. M. (1985) An Antarctic ice core reveals atmospheric CO_2 variations over the past few centuries. *Nature* **315**, 309–311.

Raynaud D., Chappellaz J., Barnola J. M., Korotkevich Y. S., and Lorius C. (1988) Climatic and CH_4 cycle implications of glacial-interglacial CH_4 change in the Vostok ice core. *Nature* **333**, 655–657.

Raynaud D., Barnola J. M., Chappellaz J., Blunier T., Indermuhle A., and Stauffer B. (2000) The ice record of greenhouse gases: a view in the context of future changes. *Quat. Sci. Rev.* **19**, 9–17.

Reeburgh W. S. (1976) Methane consumption in Cariaco Trench waters and sediments. *Earth Planet. Sci. Lett.* **28**, 337–344.

Reeburgh W. S., Whalen S. C., and Alperin M. J. (1993) The role of methylotrophy in the global methane budget. In *Microbial Growth on C1 Compounds* (eds. J. C. Murrell and D. P. Kelley). Intercept Press, Andover, UK, pp. 1–14.

Reinfelder J. R., Kraepiel A. M., and Morel F. M. M. (2000) Unicellular C_4 photosynthesis in a marine diatom. *Nature* **407**, 996–999.

Retallack G. J. (2001) A 300-million-year record of atmospheric carbon dioxide from fossil plant cuticles. *Nature* **411**, 287–290.

Retallack G. J. (2002) Carbon dioxide and climate over the past 300 Myr. *Phil. Trans. Roy. Soc. London Ser. A: Math. Phys. Eng. Sci.* **360**, 659–673.

Rhee K.-H., Morris E. P., Barber J., and Kuhlbrandt W. (1998) Three-dimensional structure of the plant photosystem II reaction centre at 8A resolution. *Nature* **396**, 283–286.

Riebesell U., Zondervan I., Rost B., Tortell P. D., Aeebe R. E., and Morel F. M. M. (2000) Reduced calcification of marine plankton in response to increased atmospheric CO_2. *Nature* **407**, 364–367.

Rind D., Peteet D., and Kukla G. (1989) Can Milankovitch orbital variations initiate the growth of ice sheets in a general circulation model? *J. Geophys. Res.-Atmos.* **94**, 12851–12871.

Rothman D. H. (2002) Atmospheric carbon dioxide levels for the last 500 million years. *Proc. Natl. Acad. Sci.* **99**, 4167–4171.

Royer D. L., Berner R. A., and Beerling D. J. (2001) Phanerozoic atmospheric CO_2 change: evaluating geochemical and paleobiological approaches. *Earth Sci. Rev.* **54**, 349–392.

Ruddiman W. F. (1997) *Tectonic Uplift and Climate Change.* Plenum, New York.

Ruddiman W. F. and Thomson J. S. (2001) The case for human causes of increased atmospheric CH_4. *Quat. Sci. Rev.* **20**, 1769–1777.

Rundgren M. and Beerling D. (1999) A Holocene CO_2 record from the stomatal index of subfossil *Salix herbacea L.* leaves from northern Sweden. *Holocene* **9**, 509–513.

Russell M. J. and Hall A. J. (1997) The emergence of life from iron monosulphide bubbles at a submarine hydrothermal redox and pH front. *J. Geol. Soc.* **154**, 377–402.

Rye R., Kuo P. H., and Holland H. D. (1995) Atmospheric carbon dioxide concentrations before 2.2 billion years ago. *Nature* **378**, 603–605.

Sagan C. and Mullen G. (1972) Earth and Mars: evolution of atmospheres and surface temperatures. *Science* **177**, 52–56.

Saltzman B. (1987) Carbon dioxide and the $\delta^{18}O$ record of late-Quaternary climatic change: a global model. *Clim. Dyn.* **1**, 77–85.

Sandberg P. A. (1983) An oscillating trend in Phanerozoic non-skeletal carbonate mineralogy. *Nature* **305**, 19–22.

Sandberg P. A. (1985) Nonskeletal aragonite and pCO_2 in the Phanerozoic and Proterozoic. In *The Carbon Cycle and Atmospheric CO_2: Natural Variations Archean to Present: Geophysical Monograph 32* (eds. E. T. Sundquist and W. S. Broecker). American Geophysical Union, Washington, DC, pp. 585–594.

Sanderson M. G. (1996) Biomass of termites and their emissions of methane and carbon dioxide: a global database. *Global Biogeochem. Cycles* **10**, 543–557.

Sano Y. and Williams S. N. (1996) Fluxes of mantle and subducted carbon along convergent plate boundaries. *Geophys. Res. Lett.* **23**, 2749–2752.

Sanyal A. and Bijma J. (1999) A comparative study of the northwest Africa and eastern equatorial Pacific upwelling zones as sources of CO during glacial periods based on boron isotope paleo-pH estimation. *Paleoceanography* **14**, 753–759.

Sanyal A., Hemming N. G., Broecker W. S., Lea D. W., Spero H. J., and Hanson G. N. (1996) Oceanic pH control on the boron isotopic composition of foraminifera: evidence from culture experiments. *Paleoceanography* **11**, 513–517.

Sanyal A., Hemming N. G., Broecker W. S., and Hanson G. (1997) Changes in pH in the eastern equatorial Pacific across stage 5-6 boundary based on boron isotopes in foraminifera. *Global Biogeochem. Cycles* **11**, 125–133.

Sanyal A., Bijma J., Spero H., and Lea D. W. (2001) Empirical relationship between pH and the boron isotope composition of *Globigerinoides sacculifer*: implications for the boron isotope paleo-pH proxy. *Paleoceanography* **16**, 515–519.

Sarmiento J. L. and Gruber N. (2002) Sinks for anthropogenic carbon. *Phys. Today* **5**, 30–36.

Sarnthein M., Winn K., Duplessy J. C., and Fontugne M. R. (1988) Global variations of surface ocean productivity in low and mid latitudes: influence on CO_2 reservoirs of the deep ocean and atmosphere during the last 21,000 years. *Paleoceanography* **3**, 361–399.

Schidlowski M. (2001) Carbon isotopes as biogeochemical recorders of life over 3.8 Ga of Earth history: evolution of a concept. *Precamb. Res.* **106**, 117–134.

Schidlowski M. and Junge C. E. (1981) Coupling among the terrestrial sulfur, carbon and oxygen cycles: numerical modeling based on revised Phanerozoic carbon isotope record. *Geochim. Cosmochim. Acta* **45**, 589–594.

Schidlowski M., Hayes J. M., and Kaplan I. R. (1983) Isotopic inferences of ancient biochemistries: carbon, sulfur, hydrogen, nitrogen. In *Earth's Earliest Biosphere: Its Origin and Evolution* (ed. J. W. Schopf). Princeton University Press, Princeton, NJ, pp. 149–186.

Schlesinger W. H. (1977) Carbon balance in terrestrial detritus. *Ann. Rev. Ecol. Sys.* **8**, 51–81.

Schlesinger W. H. (1997) *Biogeochemistry: An Analysis of Global Change.* Academic Press, New York.

Schlesinger W. S. and Melack J. M. (1981) Transport of organic carbon in the world's rivers. *Tellus* **33B**, 172–187.

Schubert W.-D., Klukas O., Saenger W., Witt H. T., Fromme P., and Kraub N. (1998) A common ancestor for oxygenic and anoxygenic photosynthetic systems: a comparison based on the structural model of photosystem I. *J. Mol. Biol.* **280**, 297–314.

Schwander J. (1989) The transformation of snow to ice and the occlusion of gases. In *The Environmental Record in Glaciers and Ice Sheets* (eds. H. Oeschger and C. C. Langway). Wiley, New York, pp. 53–67.

Schwander J. and Stauffer B. (1984) Age difference between polar ice and the air trapped in its bubbles. *Nature* **311**, 45–47.

Schwander J., Sowers T., Barnola J. M., Blunier T., Fuchs A., and Malaize B. (1997) Age scale of the air in the Summit ice—implication for glacial-interglacial temperature change. *J. Geophys. Res.-Atmos.* **102**, 19483–19493.

Scranton M. I. and Brewer P. G. (1978) Consumption of dissolved methane in the deep ocean. *Limnol. Oceanogr.* **23**, 1207–1213.

Severinghaus J. P. and Brook E. J. (1999) Abrupt climate change at the end of the last glacial period inferred from trapped air in polar ice. *Science* **286**, 930–934.

Severinghaus J. P., Sowers T., Brook E. J., Alley R. B., and Bender M. L. (1998) Timing of abrupt climate change at the end of the Younger Dryas interval from thermally fractionated gases in polar ice. *Nature* **391**, 141–146.

Shackleton N. J. (1977) Carbon-13 in *Uvigerina*: tropical rainforest history and the equatorial Pacific carbonate dissolution cycles. In *The Fate of Fossil Fuel CO_2 in the Oceans* (eds. N. R. Andersen and A. Malahoff). Plenum, New York, pp. 401–427.

Shackleton N. J. (1987) The carbon isotope record of the Cenozoic: history of organic carbon burial and of oxygen in the ocean and atmosphere. In *Marine Petroleum Source Rocks: Geological Society Special Publication 26* (eds. J. Brooks and A. J. Fleet). Blackwell, Boston, pp. 423–434.

Shackleton N. J. (2000) The 100,000-year ice-age cycle identified and found to lag temperature, carbon dioxide, and orbital eccentricity. *Science* **289**, 1897–1902.

Shackleton N. J. (2001) Climate change across the hemispheres. *Science* **291**, 58–59.

Shackleton N. J. and Opdyke N. D. (1976) Oxygen isotope and paleomagnetic stratigraphy of Pacific core V28-239 late Pliocene to latest Pleistocene. In *Investigation of Late Quaternary Paleoceanography and Paleoclimatology: Geological Society of America Memoir 145* (eds. R. M. Cline and J. D. Hays). Geological Society of America, Boulder, CO, pp. 449–464.

Shackleton N. J., Hall M. A., and Vincent E. (2000) Phase relationships between millennial-scale events 64,000–24,000 years ago. *Paleoceanography* **15**, 565–569.

Sharp J. H. (1997) Marine dissolved organic carbon: are the older values correct? *Mar. Chem.* **56**, 265–277.

Shine K. P., Derwent R. G., Wuebbles D. J., and Morcrette J.-J. (1990) Radiative forcing of climate. In *Climate Change: The IPCC Scientific Assessment* (eds. J. T. Houghton, G. J. Jenkins, and J. J. Ephraums). Cambridge University Press, New York, pp. 45–68.

Siegenthaler U., Friedli H., Loetscher H., Moor E., Neftel A., Oeschger H., and Stauffer B. (1988) Stable-isotope ratios and concentration of CO_2 in air from polar ice cores. *Ann. Glaciol.* **10**, 151–156.

Smith H. J., Wahlen M., Mastroianni D., and Taylor K. C. (1997) The CO_2 concentration of air trapped in GISP2 ice from the last glacial maximum-Holocene transition. *Geophys. Res. Lett.* **24**, 1–4.

Smith H. J., Fischer H., Wahlen M., Mastroianni D., and Deck B. (1999) Dual modes of the carbon cycle since the last glacial maximum. *Nature* **400**, 248–250.

Sowers T. (2001) N_2O record spanning the penultimate deglaciation from the Vostok ice core. *J. Geophys. Res.-Atmos.* **106**, 31903–31914.

Sowers T. and Bender M. (1995) Climate records covering the last deglaciation. *Science* **269**, 210–214.

Sowers T., Brook E., Etheridge D., Blunier T., Fuchs A., Leuenberger M., Chappellaz J., Barnola J. M., Wahlen M., Deck B., and Weyhenmeyer C. (1997) An interlaboratory comparison of techniques for extracting and analyzing trapped gases in ice cores. *J. Geophys. Res.-Oceans* **102**, 26527–26538.

Spencer R. J. and Hardie L. A. (1990) Control of seawater composition by mixing of river waters and mid-ocean ridge hydrothermal brines. In *Fluid-mineral interactions: a tribute to H. P. Eugster: Geochemical Society Special Publication 2* (eds. R. J. Spencer and I.-M. Chou). Geochemical Society, San Antonio, pp. 409–419.

Spitzy A. and Leenheer J. (1991) Dissolved organic carbon in rivers. In *SCOPE 42: Biogeochemistry of Major World Rivers* (eds. E. T. Degens, S. Kempe, and J. E. Richey). Wiley, New York, pp. 213–232.

Spivack A. J. (1993) Foraminiferal boron isotope ratios as a proxy for surface ocean pH over the past 21 Myr. *Nature* **363**, 149–151.

Staudigel H. R., Hart S. R., Schmincke H. U., and Smith B. M. (1990) Cretaceous ocean crust at DSDP Sites 417 and 418: carbon uptake from weathering versus loss by magmatic outgassing. *Geochim. Cosmochim. Acta* **53**, 3091–3094.

Stauffer B. and Oeschger H. (1985) Gaseous components in the atmosphere and the historic record revealed by ice cores. *Ann. Glaciol.* **7**, 54–60.

Stauffer B., Hofer H., Oeschger H., Schwander J., and Siegenthaler U. (1984) Atmospheric CO_2 concentration during the last glaciation. *Ann. Glaciol.* **5**, 160–164.

Stauffer B., Fischer G., Neftel A., and Oeschger H. (1985) Increase of atmospheric methane recorded in Antarctic ice core. *Science* **229**, 1386–1388.

Stauffer B., Lochbronner E., Oeschger H., and Schwander J. (1988) Methane concentration in the glacial atmosphere was only half that of the pre-industrial Holocene. *Nature* **332**, 812–814.

Stauffer B., Blunier T., Dallenbach A., Indermuhle A., Schwander J., Stocker T. F., Tschumi J., Chappellaz J., Raynaud D., Hammer C. U., and Clausen H. B. (1998) Atmospheric CO_2 concentration and millennial-scale climate change during the last glacial period. *Nature* **392**, 59–62.

Steig E. J., Brook E. J., White J. W. C., Sucher C. M., Bender M. L., Lehman S. J., Morse D. L., Waddington E. D., and Clow G. D. (1998) Synchronous climate changes in Antarctica and the North Atlantic. *Science* **282**, 92–95.

Stephens B. B. and Keeling R. F. (2000) The influence of Antarctic sea ice on glacial-interglacial CO_2 variations. *Nature* **404**, 171–174.

Subak S. (1994) Methane from the House of Tudor and Ming Dynasty: anthropogenic emissions in the sixteenth century. *Chemosphere* **29**, 843–854.

Sugimoto A., Inoue T., Kirtibutr N., and Abe T. (1998) Methane oxidation by termite mounds estimated by the carbon isotopic composition of methane. *Global Biogeochem. Cycles* **12**, 595–605.

Summons R. E., Jahnke L. L., Hope J. M., and Logan G. A. (1999) 2-Methylhopanoids as biomarkers for cyanobacterial oxygenic photosynthesis. *Nature* **400**, 554–557.

Sundquist E. T. (1985) Geological perspectives on carbon dioxide and the carbon cycle. In *The Carbon Cycle and Atmospheric CO_2: Natural Variations Archean to Present: Geophysical Monograph 32* (eds. E. T. Sundquist and W. S. Broecker). American Geophysical Union, Washington, DC, pp. 5–60.

Sundquist E. T. (1986) Geologic analogs: their value and limitations in carbon dioxide research. In *The Changing Carbon Cycle, A Global Analysis* (eds. J. R. Trabalka and D. E. Reichle). Springer, New York, pp. 371–402.

Sundquist E. T. (1990a) Influence of deep-sea benthic processes on atmospheric CO_2. *Phil. Trans. Roy. Soc. London Ser. A: Math. Phys. Eng. Sci.* **331**, 155–165.

Sundquist E. T. (1990b) Long-term aspects of future atmospheric CO_2 and sea-level changes. *Sea Level Change: Natl. Res. Council*, 193–207.

Sundquist E. T. (1991) Steady- and non-steady-state carbonate-silicate controls on atmospheric CO_2. *Quat. Sci. Rev.* **10**, 283–296.

Sundquist E. T. (1993) The global carbon dioxide budget. *Science* **259**, 934–941.

Sundquist E. T. and Plummer L. N. (1981) Carbon dioxide in the ocean surface layer: some modeling considerations. In *Carbon Cycle Modelling* (ed. B. Bolin). Wiley, New York, pp. 259–270.

Tajika E. (1998) Climate change during the last 150 million years: reconstruction from a carbon cycle model. *Earth Planet. Sci. Lett.* **160**, 695–707.

Taylor J. A. and Orr J. C. (2000) The natural latitudinal distribution of atmospheric CO_2. *Global Planet. Change* **26**, 375–386.

Thomas E. and Shackleton N. J. (1996) The Paleocene-Eocene benthic foraminiferal extinction and stable isotope anomalies. In *Correlations of the early Paleogene in Northwest Europe: Geological Society Special Publication 101* (eds. R. O. Knox, R. M. Corfield, and R. E. Dunay). Geological Society of London, London, pp. 401–411.

Thompson A. M. (1992) The oxidizing capacity of the Earth's atmosphere: probable past and future changes. *Science* **256**, 1157–1165.

Thompson A. M., Chappellaz J. A., Fung I. Y., and Kucsera T. L. (1993) The atmospheric CH4 increase since the Last Glacial Maximum: 2. Interactions with oxidants. *Tellus* **45B**, 242–257.

Toggweiler J. R. (1999) Variation of atmospheric CO_2 by ventilation of the ocean's deepest water. *Paleoceanography* **14**, 571–588.

Toggweiler J. R., Gnanadesikan A., Carson S., Murnane R., and Sarmiento J. L. (2003) Representation of the carbon cycle in box models and GCMs: 1. Solubility pump. *Global Biogeochem. Cycles* **17**, 1026, doi:10.1029/2001GB001401.

Trudinger C. M., Enting I. G., Francey R. J., Etheridge D. M., and Rayner P. J. (1999) Long-term variability in the global carbon cycle inferred from a high-precision CO_2 and delta C-13 ice-core record. *Tellus* **51B**, 233–248.

Tschumi J. and Stauffer B. (2000) Reconstructing past atmospheric CO_2 concentration based on ice-core analyses: open questions due to *in situ* production of CO_2 in the ice. *J. Glaciol.* **46**, 45–53.

US Geological Survey World Energy Assessment Team (2000) *World Petroleum Assessment, 2000.* USGS Digital Data Series 60.

Van Campo E., Guiot J., and Peng C. (1993) A data-based re-appraisal of the terrestrial carbon budget at the last glacial maximum. *Global Planet. Change* **8**, 189–201.

Van Cappellen P. and Ingall E. D. (1996) Redox stabilization of the atmosphere and oceans by phosphorus-limited marine productivity. *Science* **271**, 493–496.

Veizer J., Ala D., Azmy K., Bruckschen P., Buhl D., Bruhn F., Carden G. A. F., Diener A., Ebneth S., and Godderis Y. (1999) $^{87}Sr/^{86}Sr$, $\delta^{13}C$ and $\delta^{18}O$ evolution of Phanerozoic seawater. *Chem. Geol.* **161**, 59–88.

Veizer J., Godderis Y., and Francois L. M. (2000) Evidence for decoupling of atmospheric CO_2 and global climate during the Phanerozoic eon. *Nature* **408**, 698–701.

Velichko A. A., Kremenetski C. V., Borisova O. K., Zelikson E. M., Nechaev V. P., and Faure H. (1998) Estimates of methane emission during the last 125,000 years in Northern Eurasia. *Global Planet. Change* **16–17**, 159–180.

Volk T. and Hoffert M. I. (1985) Ocean carbon pumps: analysis of relative strengths and efficiencies in ocean-driven atmospheric CO_2 changes. In *The Carbon Cycle and Atmospheric CO_2: Natural Variations Archean to Present: Geophysical Monograph 32* (eds. E. T. Sundquist and W. S. Broecker). American Geophysical Union, Washington, DC, pp. 99–110.

von Fischer J. C. and Hedin L. O. (2002) Separating methane production and consumption with a field-based isotope pool dilution technique. *Global Biogeochem. Cycles* **16**, 10.1029/2001GB001448.

Wahlen M. D., Allen D., and Deck B. (1991) Initial measurements of CO_2 concentrations (1530-1940 AD) in air occluded in the GISP2 ice core from central Greenland. *Geophys. Res. Lett.* **18**, 1457–1460.

Walker J. C. G. (1987) Was the Archean biosphere upside down? *Nature* **329**, 710–712.

Walker J. C. G., Hays P. B., and Kasting J. F. (1981) A negative feedback mechanism for the long-term stabilization of Earth's surface temperature. *J. Geophys. Res.-Oceans* **86**, 9776–9782.

Walker J. C. G., Klein C., Schidlowski M., Schopf J. W., Stevenson D. J., and Walter M. R. (1983) Environmental evolution of the Archean-Early Proterozoic Earth. In *Earth's Earliest Biosphere: Its Origin and Evolution* (ed. J. W. Schopf). Princeton University Press, Princeton, NJ, pp. 260–290.

Wallmann K. (2001a) Controls on the Cretaceous and Cenozoic evolution of seawater composition, atmospheric CO_2 and climate. *Geochim. Cosmochim. Acta* **65**, 3005–3025.

Wallmann K. (2001b) The geological water cycle and the evolution of marine delta O-18 values. *Geochim. Cosmochim. Acta* **65**, 2469–2485.

Walter B. P., Heimann M., and Matthews E. (2001) Modeling modern methane emissions from natural wetlands: 1. Model description and results. *J. Geophys. Res.-Atmos.* **106**, 34189–34206.

Watson R. T., Rodhe H., Oeschger H., and Siegenthaler U. (1990) Greenhouse gases and aerosols. In *Climate Change: The IPCC Scientific Assessment* (eds. J. T. Houghton, G. J. Jenkins, and J. J. Ephraums). Cambridge University Press, Cambridge, pp. 1–40.

Whalen S. C. and Reeburgh W. S. (1990) Consumption of atmospheric methane by tundra soils. *Nature* **346**, 160–162.

Whiting G. J. and Chanton J. P. (1993) Primary production control of methane emission from wetlands. *Nature* **364**, 794–795.

Widdel F., Schnell S., Heising S., Ehrenreich A., Assmus B., and Schink B. (1993) Ferrous iron oxidation by anoxygenic phototrophic bacteria. *Nature* **362**, 834–836.

Wollast R. (1994) The relative importance of biomineralization and dissolution of $CaCO_3$ in the global carbon cycle. In *Past and Present Biomineralization Processes* (ed. F. Doumenge). Musee Oceanographie, Monaco, pp. 13–35.

Woodward F. I. (1987) Stomatal numbers are sensitive to increases in CO_2 from pre-industrial levels. *Nature* **327**, 617–618.

World Energy Council (WEC) (2000) *Survey of Energy Resources*. World Energy Council, London.

Wuebbles D. J. and Hayhoe K. (2002) Atmospheric methane and global change. *Earth Sci. Rev.* **57**, 177–210.

Xiong J., Fischer W. M., Inoue K., Nakahara M., and Bauer C. E. (2000) Molecular evidence for the early evolution of photosynthesis. *Science* **289**, 1724–1730.

Yavitt J. B., Lang G. E., and Downey D. M. (1988) Potential methane production and methane oxidation rates in peatland ecosystems of the Appalachian Mountains, United States. *Global Biogeochem. Cycles* **2**, 253–268.

Yavitt J. B., Fahey T. J., and Simmons J. A. (1995) Methane and carbon dioxide dynamics in a northern hardwood ecosystem. *Soil. Sci. Soc. Am. J.* **59**, 796–804.

Yoshimori M., Weaver A. J., Marshall S. J., and Clarke G. K. C. (2001) Glacial termination: sensitivity to orbital and CO_2 forcing in a coupled climate system model. *Clim. Dyn.* **17**, 571–588.

Zachos J. C., Lohmann K. C., Walker J. C. G., and Wise S. W. (1993) Abrupt climate change and transient climates during the Paleogene: a marine perspective. *J. Geol.* **101**, 191–213.

Zimmerman H. (2000) Tertiary seawater chemistry–implications from fluid inclusions in primary marine halite. *Am. J. Sci.* **300**, 723–767.

8.10
The Contemporary Carbon Cycle

R. A. Houghton
Woods Hole Research Center, MA, USA

8.10.1 INTRODUCTION

The global carbon cycle refers to the exchanges of carbon within and between four major reservoirs: the atmosphere, the oceans, land, and fossil fuels. Carbon may be transferred from one reservoir to another in seconds (e.g., the fixation of atmospheric CO_2 into sugar through photosynthesis) or over millennia (e.g., the accumulation of fossil carbon (coal, oil, gas) through deposition and diagenesis of organic matter). This chapter emphasizes the exchanges that are important over years to decades and includes those occurring over the scale of months to a few centuries. The focus will be on the years 1980–2000 but our considerations will broadly include the years ~1850–2100. Chapter 8.09, deals with longer-term processes that involve rates of carbon exchange that are small on an annual timescale (weathering, vulcanism, sedimentation, and diagenesis).

The carbon cycle is important for at least three reasons. First, carbon forms the structure of all life on the planet, making up ~50% of the dry weight of living things. Second, the cycling of carbon approximates the flows of energy around the Earth, the metabolism of natural, human, and industrial systems. Plants transform radiant energy into chemical energy in the form of sugars, starches, and other forms of organic matter; this energy, whether in living organisms or dead organic matter, supports food chains in natural ecosystems as well as human ecosystems, not the least of which are industrial societies habituated (addicted?) to fossil forms of energy for heating, transportation, and generation of electricity. The increased use of fossil fuels has led to a third reason for interest in the carbon cycle. Carbon, in the form of carbon dioxide (CO_2) and methane (CH_4), forms two of the most important greenhouse gases. These gases contribute to a natural greenhouse effect that has kept the planet warm enough to evolve and support life (without the greenhouse effect the Earth's average temperature would be −33 °C). Additions of greenhouse gases to the atmosphere from industrial activity, however, are increasing the concentrations of these gases, enhancing the greenhouse effect, and starting to warm the Earth.

The rate and extent of the warming depend, in part, on the global carbon cycle. If the rate at which the oceans remove CO_2 from the atmosphere were faster, e.g., concentrations of CO_2 would have increased less over the last century. If the processes removing carbon from the atmosphere and storing it on land were to diminish, concentrations of CO_2 would increase more rapidly than projected on the basis of recent history. The processes responsible for adding carbon to, and withdrawing it from, the atmosphere are not well enough understood to predict future levels of CO_2 with great accuracy. These processes are a part of the global carbon cycle.

Some of the processes that add carbon to the atmosphere or remove it, such as the combustion of fossil fuels and the establishment of tree plantations, are under direct human control. Others, such as the accumulation of carbon in the oceans or on land as a result of changes in global climate (i.e., feedbacks between the global carbon cycle and climate), are not under direct human control except through controlling rates of greenhouse gas emissions and, hence, climatic change. Because CO_2 has been more important than all of the other greenhouse gases under human control, combined, and is expected to continue so in the future, understanding the global carbon cycle is a vital part of managing global climate.

This chapter addresses, first, the reservoirs and natural flows of carbon on the earth. It then addresses the sources of carbon to the atmosphere from human uses of land and energy and the sinks of carbon on land and in the oceans that have kept the atmospheric accumulation of CO_2 lower than it would otherwise have been. The chapter describes changes in the distribution of carbon among the atmosphere, oceans, and terrestrial ecosystems over the past 150 years as a result of human-induced emissions of carbon. The processes responsible for sinks of carbon on land and in the sea are reviewed from the perspective of feedbacks, and the chapter concludes with some prospects for the future.

Earlier comprehensive summaries of the global carbon cycle include studies by Bolin *et al.* (1979, 1986), Woodwell and Pecan (1973), Bolin (1981), NRC (1983), Sundquist and Broecker (1985), and Trabalka (1985). More recently, the Intergovernmental Panel on Climate Change (IPCC) has summarized information on the carbon cycle in the context of climate change (Watson *et al.*, 1990; Schimel *et al.*, 1996; Prentice *et al.*, 2001). The basic aspects of the global carbon cycle have been understood for decades, but other aspects, such as the partitioning of the carbon sink between land and ocean, are being re-evaluated continuously with new data and analyses. The rate at which new publications revise estimates of these carbon sinks and re-evaluate the mechanisms that control the magnitude of the sinks suggests that portions of this review will be out of date by the time of publication.

8.10.2 MAJOR RESERVOIRS AND NATURAL FLUXES OF CARBON

8.10.2.1 Reservoirs

The contemporary global carbon cycle is shown in simplified form in Figure 1. The four major reservoirs important in the time frame of decades to centuries are the atmosphere, oceans, reserves of fossil fuels, and terrestrial ecosystems, including vegetation and soils. The world's oceans contain ~50 times more carbon than either the atmosphere or the world's terrestrial vegetation, and thus shifts in the abundance of carbon among the major reservoirs will have a much greater significance for the terrestrial biota and for the atmosphere than they will for the oceans.

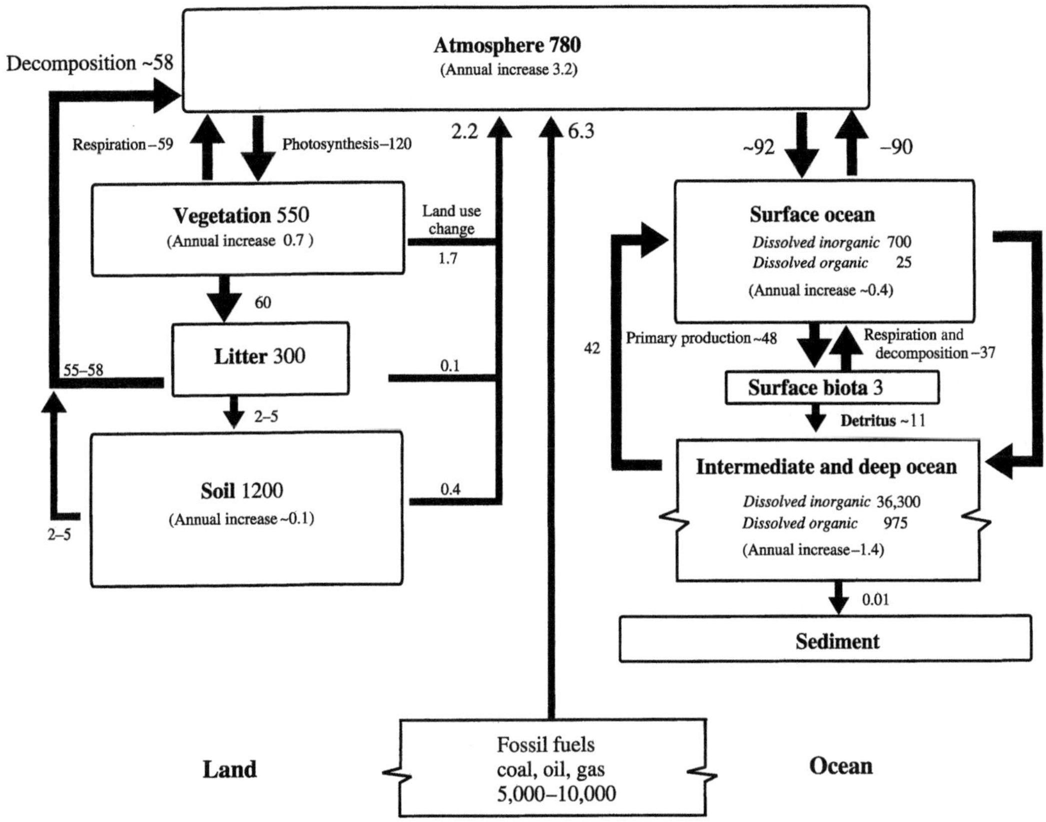

Figure 1 The contemporary global carbon cycle. Units are Pg C or Pg C yr^{-1}.

8.10.2.1.1 The atmosphere

Most of the atmosphere is made up of either nitrogen (78%) or oxygen (21%). In contrast, the concentration of CO_2 in the atmosphere is only ~0.04%. The concentrations of CO_2 in air can be measured to within one tenth of 0.1 ppmv, or 0.00001%. In the year 2000 the globally averaged concentration was ~0.0368%, or 368 ppmv, equivalent to ~780 Pg C (1 Pg = 1 petagram = 10^{15} g = 10^9 t) (Table 1).

The atmosphere is completely mixed in about a year, so any monitoring station free of local contamination will show approximately the same year-to-year increase in CO_2. There are at least 77 stations worldwide, where weekly flask samples of air are collected, analyzed for CO_2 and other constituents, and where the resulting data are integrated into a consistent global data set (Masarie and Tans, 1995; Cooperative Atmospheric Data Integration Project—Carbon Dioxide, 1997). The stations generally show the same year-to-year increase in concentration but vary with respect to absolute concentration, seasonal variability, and other characteristics useful for investigating the global circulation of carbon.

Most of the carbon in the atmosphere is CO_2, but small amounts of carbon exist in concentrations of CH_4, carbon monoxide (CO), and non-methane hydrocarbons. These trace gases are important because they modify the chemical and/or the radiative properties of the Earth's atmosphere. Methane is present at ~1.7 ppm, two orders of magnitude more dilute than CO_2. Methane is a reduced form of carbon, is much less stable than

Table 1 Stocks and flows of carbon.

Carbon stocks (Pg C)	
Atmosphere	780
Land	2,000
Vegetation	500
Soil	1,500
Ocean	39,000
Surface	700
Deep	38,000
Fossil fuel reserves	10,000
Annual flows (Pg C yr^{-1})	
Atmosphere-oceans	90
Atmosphere-land	120
Net annual exchanges (Pg C yr^{-1})	
Fossil fuels	6
Land-use change	2
Atmospheric increase	3
Oceanic uptake	2
Other terrestrial uptake	3

CO_2, and has an average residence time in the atmosphere of 5–10 years. Carbon monoxide has an atmospheric residence time of only a few months. Its low concentration, ~0.1 ppm, and its short residence time result from its chemical reactivity with OH radicals. Carbon monoxide is not a greenhouse gas, but its chemical reactivity affects the abundances of ozone and methane which are greenhouse gases. Non-methane hydrocarbons, another unstable form of carbon in the atmosphere, are present in even smaller concentrations. The oxidation of these biogenic trace gases is believed to be a major source of atmospheric CO, and, hence, these non-methane hydrocarbons also affect indirectly the Earth's radiative balance.

8.10.2.1.2 Terrestrial ecosystems: vegetation and soils

Carbon accounts for only ~0.27% of the mass of elements in the Earth's crust (Kempe, 1979), yet it is the basis for life on Earth. The amount of carbon contained in the living vegetation of terrestrial ecosystems (550 ± 100 Pg) is somewhat less than that present in the atmosphere (780 Pg). Soils contain 2–3 times that amount (1,500–2,000 Pg C) in the top meter (Table 2) and as much as 2,300 Pg in the top 3 m (Jobbágy and Jackson, 2000). Most terrestrial carbon is stored in the vegetation and soils of the world's forests. Forests cover ~30% of the land surface and hold ~75% of the living organic carbon. When soils are included in the inventory, forests hold almost half of the carbon of the world's terrestrial ecosystems. The soils of woodlands, grasslands, tundra, wetlands, and agricultural lands store most of the rest of the terrestrial organic carbon.

8.10.2.1.3 The oceans

The total amount of dissolved inorganic carbon (DIC) in the world's oceans is ~3.7×10^4 Pg, and the amount of organic carbon is ~1,000 Pg. Thus, the world's oceans contain ~50 times more carbon than the atmosphere and 70 times more than the world's terrestrial vegetation. Most of this oceanic carbon is in intermediate and deep waters; only 700–1,000 Pg C are in the surface layers of the ocean, that part of the ocean in direct contact with the atmosphere. There are also 6,000 Pg C in reactive ocean sediments (Sundquist, 1986), but the turnover of sediments is slow, and they are not generally considered as part of the active, or short-term, carbon cycle, although they are important in determining the long-term concentration of CO_2 in the atmosphere and oceans.

Carbon dioxide behaves unlike other gases in the ocean. Most gases are not very soluble in water and are predominantly in the atmosphere. For example, only ~1% of the world's oxygen is in the oceans; 99% exists in the atmosphere. Because of the chemistry of seawater, however, the distribution of carbon between air and sea is reversed: 98.5% of the carbon in the ocean–atmosphere systems is in the sea. Although this inorganic carbon is dissolved, less than 1% of it is in the form of dissolved CO_2 (p_{CO_2}); most of the inorganic carbon is in the form of bicarbonate and carbonate ions (Table 3).

About 1,000 Pg C in the oceans (out of the total of 3.8×10^4 Pg) is organic carbon. Carbon in living organisms amounts to ~3 Pg in the sea, in comparison to ~550 Pg on land. The mass of animal life in the oceans is almost the same as on land, however, pointing to the very different trophic structures in the two environments.

Table 2 Area, carbon in living biomass, and net primary productivity of major terrestrial biomes.

Biome	*Area* (10^9 ha)		*Global carbon stocks* (Pg C)						*Carbon stocks* (Mg C ha^{-1})				*NPP* (Pg C yr^{-1})	
	WBGU	*MRS*	*WGBU*			*MRS IGBP*			*WBGU*		*MRS IGPB*		*Ajtay*	*MRS*
			Plants	*Soil*	*Total*	*Plants*	*Soil*	*Total*	*Plants*	*Soil*	*Plants*	*Soil*		
Tropical forests	17.6	17.5	212	216	428	340	214	553	120	123	194	122	13.7	21.9
Temperate forests	1.04	1.04	59	100	159	139	153	292	57	96	134	147	6.5	8.1
Boreal forests	1.37	1.37	88	471	559	57	338	395	64	344	42	247	3.2	2.6
Tropical savannas and grasslands	2.25	2.76	66	264	330	79	247	326	29	117	29	90	17.7	14.9
Temperate grasslands and shrublands	1.25	1.78	9	295	304	23	176	199	7	236	13	99	5.3	7.0
Deserts and semi-deserts	4.55	2.77	8	191	199	10	159	169	2	42	4	57	1.4	3.5
Tundra	0.95	0.56	6	121	127	2	115	117	6	127	4	206	1.0	0.5
Croplands	1.60	1.35	3	128	131	4	165	169	2	80	3	122	6.8	4.1
Wetlands	0.35		15	225	240				43	643			4.3	
Total	15.12	14.93	466	2,011	2,477	654	1,567	2,221					59.9	62.6

Source: Prentice *et al.* (2001).

Table 3 The distribution of 1,000 CO_2 molecules in the atmosphere–ocean.

Atmosphere	15
Ocean	985
CO_2	5
HCO_3^-	875
CO_3^{2-}	105
Total	1,000

Source: Sarmiento (1993).

The ocean's plants are microscopic. They have a high productivity, but the production does not accumulate. Most is either grazed or decomposed in the surface waters. Only a fraction (~25%) sinks into the deeper ocean. In contrast, terrestrial plants accumulate large amounts of carbon in long-lasting structures (trees). The distribution of organic carbon between living and dead forms of carbon is also very different on land and in the sea. The ratio is ~1 : 3 on land and ~1 : 300 in the sea.

8.10.2.1.4 Fossil fuels

The common sources of energy used by industrial societies are another form of organic matter, so-called fossil fuels. Coal, oil, and natural gas are the residuals of organic matter formed millions of years ago by green plants. The material escaped oxidation, became buried in the Earth, and over time was transformed to a (fossil) form. The energy stored in the chemical bonds of fossil fuels is released during combustion just as the energy stored in carbohydrates, proteins, and fats is released during respiration.

The difference between the two forms of organic matter (fossil and nonfossil), from the perspective of the global carbon cycle, is the rate at which they are cycled. The annual rate of formation of fossil carbon is at least 1,000 times slower than rates of photosynthesis and respiration. The formation of fossil fuels is part of a carbon cycle that operates over millions of years, and the processes that govern the behavior of this long-term system (sedimentation, weathering, vulcanism, seafloor spreading) are much slower from those that govern the behavior of the short-term system. Sedimentation of organic and inorganic carbon in the sea, e.g., is ~0.2 Pg C yr^{-1}. In contrast, hundreds of petagrams of carbon are cycled annually among the reservoirs of the short-term, or active, carbon cycle. This short-term system operates over periods of seconds to centuries. When young (nonfossil) organic matter is added to or removed from the atmosphere, the total amount of carbon in the active system is unchanged. It is merely redistributed among reservoirs. When fossil fuels are oxidized, however, the CO_2 released represents a net increase in the amount of carbon in the active system.

The amount of carbon stored in recoverable reserves of coal, oil, and gas is estimated to be 5,000–10,000 Pg C, larger than any other reservoir except the deep sea, and ~10 times the carbon content of the atmosphere. Until ~1850s this reservoir of carbon was not a significant part of the short-term cycle of carbon. The industrial revolution changed that.

8.10.2.2 The Natural Flows of Carbon

Carbon dioxide is chemically stable and has an average residence time in the atmosphere of about four years before it enters either the oceans or terrestrial ecosystems.

8.10.2.2.1 Between land and atmosphere

The inorganic form of carbon in the atmosphere (CO_2) is fixed into organic matter by green plants using energy from the Sun in the process of photosynthesis, as follows:

$$6CO_2 + 6H_2O \leftrightharpoons C_6H_{12}O_6 + 6O_2$$

The reduction of CO_2 to glucose ($C_6H_{12}O_6$) stores some of the Sun's energy in the chemical bonds of the organic matter formed. Glucose, cellulose, carbohydrates, protein, and fats are all forms of organic matter, or reduced carbon. They all embody energy and are nearly all derived ultimately from photosynthesis.

The reaction above also goes in the opposite direction during the oxidation of organic matter. Oxidation occurs during the two, seemingly dissimilar but chemically identical, processes of respiration and combustion. During either process the chemical energy stored in organic matter is released. Respiration is the biotic process that yields energy from organic matter, energy required for growth and maintenance. All living organisms oxidize organic matter; only plants and some microbes are capable of reducing CO_2 to produce organic matter.

Approximately 45–50% of the dry weight of organic matter is carbon. The organic carbon of terrestrial ecosystems exists in many forms, including living leaves and roots, animals, microbes, wood, decaying leaves, and soil humus. The turnover of these materials varies from less than one year to more than 1,000 years. In terms of carbon, the world's terrestrial biota is almost entirely vegetation; animals (including humans) account for less than 0.1% of the carbon in living organisms.

Each year the atmosphere exchanges ~120 Pg C with terrestrial ecosystems through photosynthesis and respiration (Figure 1 and Table 1). The uptake of carbon through photosynthesis is gross primary production (GPP). At least half of this production is respired by the plants, themselves (autotrophic respiration (Rs_a)), leaving a net primary production (NPP) of ~60 Pg C yr^{-1}. Recent estimates of global terrestrial NPP vary between 56.4 Pg C $yr^{-1}$and 62.6 Pg C yr^{-1} (Ajtay *et al.*, 1979; Field *et al.*, 1998; Saugier *et al.*, 2001). The annual production of organic matter is what fuels the nonplant world, providing food, feed, fiber, and fuel for natural and human systems. Thus, most of the NPP is consumed by animals or respired by decomposer organisms in the soil (heterotrophic respiration (Rs_h)). A smaller amount (~4 Pg C yr^{-1} globally) is oxidized through fires. The sum of autotrophic and heterotrophic respiration is total respiration or ecosystem respiration (Rs_e). In steady state the net flux of carbon between terrestrial ecosystems and the atmosphere (net ecosystem production (NEP)) is approximately zero, but year-to-year variations in photosynthesis and respiration (including fires) may depart from this long-term balance by as much as 5–6 Pg C yr^{-1}. The annual global exchanges may be summarized as follows:

$$\text{NPP} = \text{GPP} - Rs_a$$
$$(\sim 60 = 120 - 60\ \text{Pg C yr}^{-1})$$

$$\text{NEP} = \text{GPP} - Rs_a - Rs_h$$
$$(\sim 0 = 120 - 60 - 60\ \text{Pg C yr}^{-1})$$

$$\text{NEP} = \text{NPP} - Rs_h$$
$$(\sim 0 = 60 - 60\ \text{Pg C yr}^{-1})$$

Photosynthesis and respiration are not evenly distributed either in space or over the course of a year. About half of terrestrial photosynthesis occurs in the tropics where the conditions are generally favorable for growth, and where a large proportion of the Earth's land area exists (Table 2). Direct evidence for the importance of terrestrial metabolism (photosynthesis and respiration) can be seen in the effect it has on the atmospheric concentration of CO_2 (Figure 2(a)). The most striking feature of the figure is the regular sawtooth pattern. This pattern repeats itself annually. The cause of the oscillation is the metabolism of terrestrial ecosystems. The highest concentrations occur at the end of each winter, following the season in which respiration has exceeded photosynthesis and thereby caused a net release of CO_2 to the atmosphere. Lowest concentrations occur at the end of each summer, following the season in which photosynthesis has exceeded respiration and drawn CO_2 out of the atmosphere. The latitudinal variability in the amplitude of this oscillation suggests that it is driven largely by northern temperate and boreal ecosystems: the highest amplitudes (up to ~16 ppmv) are in the northern hemisphere with the largest land area. The phase of the amplitude is reversed in the southern hemisphere, corresponding to seasonal terrestrial metabolism there. Despite the high rates of production and respiration in the tropics, equatorial regions are thought to contribute little to this oscillation. Although there is a strong seasonality in precipitation throughout much of the tropics, the seasonal changes in moisture affect photosynthesis and respiration almost equally and thus the two processes remain largely in phase with little or no net flux of CO_2.

8.10.2.2.2 Between oceans and atmosphere

There is ~50 times more carbon in the ocean than in the atmosphere, and it is the amount of DIC in the ocean that determines the atmospheric concentration of CO_2. In the long term (millennia) the most important process determining the exchanges of carbon between the oceans and the atmosphere is the chemical equilibrium of dissolved CO_2, bicarbonate, and carbonate in the ocean. The rate at which the oceans take up or release carbon is slow on a century timescale, however, because of lags in circulation and changes in the availability of calcium ions. The carbon chemistry of seawater is discussed in more detail in the next section.

Two additional processes besides carbon chemistry keep the atmospheric CO_2 lower than it otherwise would be. One process is referred to as the solubility pump and the other as the biological pump. The solubility pump is based on the fact that CO_2 is more soluble in cold waters. In the ocean, CO_2 is ~2 times more soluble in the cold mid-depth and deep waters than it is in the warm surface waters near the equator. Because sinking of cold surface waters in Arctic and Antarctic regions forms these mid-depth and deep waters, the formation of these waters with high CO_2 keeps the CO_2 concentration of the atmosphere lower than the average concentration of surface waters.

The biological pump also transfers surface carbon to the intermediate and deep ocean. Not all of the organic matter produced by phytoplankton is respired in the surface waters where it is produced; some sinks out of the photic zone to deeper water. Eventually, this organic matter is decomposed at depth and reaches the surface again through ocean circulation. The net effect of the sinking of organic matter is to enrich the deeper waters relative to surface waters and thus to reduce the CO_2 concentration of the atmosphere. Marine photosynthesis and the sinking of organic matter out of the surface water are estimated to keep the concentration of CO_2 in air ~30% of what it would be in their absence.

Together the two pumps keep the DIC concentration of the surface waters ~10% lower

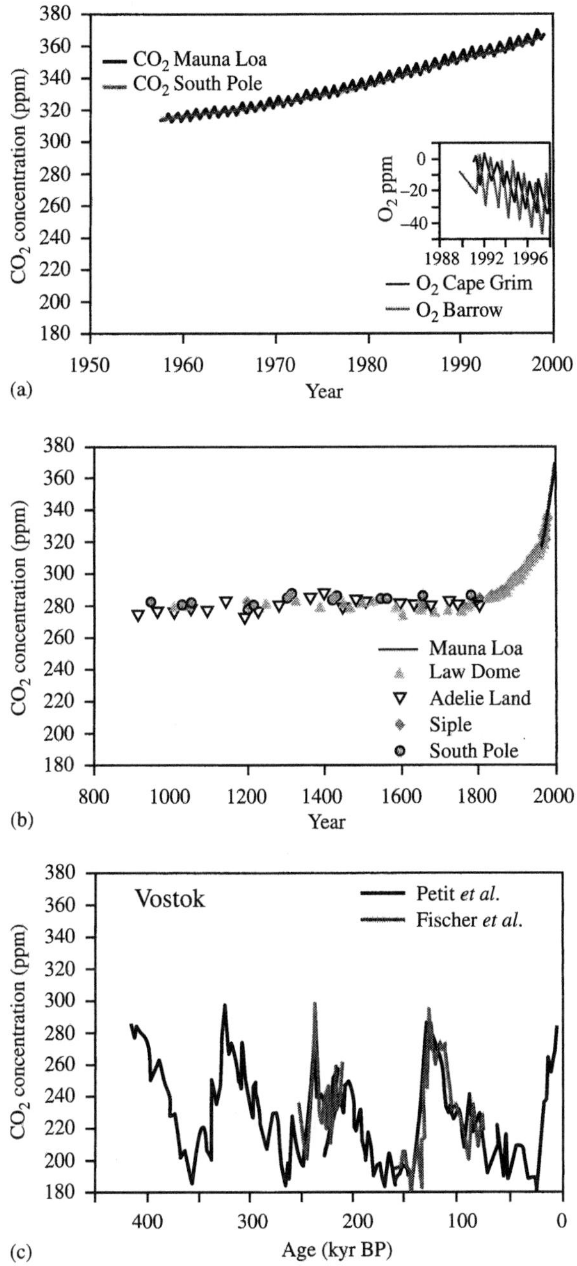

Figure 2 Concentration of CO_2 in the atmosphere: (a) over the last 42 years, (b) over the last 1,000 years, and (c) over the last $\sim 4 \times 10^5$ years (Prentice *et al.*, 2001) (reproduced by permission of Intergovernmental Panel on Climate Change from *Climate Change 2001: The Scientific Basis*, **2001**, pp. 183–237).

than at depth. Ocean models that simulate both carbon chemistry and oceanic circulation show that the concentration of CO_2 in the atmosphere (280 ppmv pre-industrially) would have been 720 ppmv if both pumps were turned off (Sarmiento, 1993).

There is another biological pump, called the carbonate pump, but its effect in reducing the concentration of CO_2 in the atmosphere is small. Some forms of phytoplankton have $CaCO_3$ shells that, in sinking, transfer carbon from the surface to deeper water, just as the biological pump transfers organic carbon to depth. The precipitation of $CaCO_3$ in the surface waters, however, increases the partial pressure of CO_2, and the evasion of this CO_2 to the atmosphere offsets the sinking of carbonate carbon.

Although ocean chemistry determines the CO_2 concentration of the atmosphere in the long term and the solubility and biological pumps act to modify this long-term equilibrium, short-term exchanges of carbon between ocean and atmosphere result from the diffusion of CO_2 across the air–sea interface. The diffusive exchanges transfer ~90 Pg C yr^{-1} across the air–sea interface in both directions (Figure 1). The transfer has been estimated by two different methods. One method is based on the fact that the transfer rate of naturally produced ^{14}C into the oceans should balance the decay of ^{14}C within the oceans. Both the production rate of ^{14}C in the atmosphere and the inventory of ^{14}C in the oceans are known with enough certainty to yield an average rate of transfer of ~100 Pg C yr^{-1}, into and out of the ocean.

The second method is based on the amount of radon gas in the surface ocean. Radon gas is generated by the decay of ^{226}Ra. The concentration of the parent ^{226}Ra and its half-life allow calculation of the expected radon gas concentration in the surface water. The observed concentration is ~70% of expected, so 30% of the radon must be transferred to the atmosphere during its mean lifetime of six days. Correcting for differences in the diffusivity of radon and CO_2 allows an estimation of the transfer rate for CO_2. The transfer rates given by the ^{14}C method and the radon method agree within ~10%.

The net exchange of CO_2 across the air–sea interface varies latitudinally, largely as a function of the partial pressure of CO_2 in surface waters, which, in turn, is affected by temperature, upwelling or downwelling, and biological production. Cold, high-latitude waters take up carbon, while warm, lower-latitude waters tend to release carbon (outgassing of CO_2 from tropical gyres). Although the latitudinal pattern in net exchange is consistent with temperature, the dominant reason for the exchange is upwelling (in the tropics) and downwelling, or deep-water formation (at high latitudes).

The annual rate of photosynthesis in the world oceans is estimated to be ~48 Pg C (Table 4) (Longhurst *et al.*, 1995). About 25% of the primary production sinks from the photic zone to deeper water (Falkowski *et al.*, 1998; Laws *et al.*, 2000). The gross flows of carbon between the surface ocean and the intermediate and deep ocean are estimated to be ~40 Pg C yr^{-1}, in part from the sinking of organic production (11 Pg C yr^{-1}) and in part from physical mixing (33 Pg C yr^{-1}) (Figure 1).

Table 4 Annual net primary production of the ocean.

Domain or ecosystem	*NPP* (Pg C yr^{-1})
Trade winds domain (tropical and subtropical)	13.0
Westerly winds domain (temperate)	16.3
Polar domain	6.4
Coastal domain	10.7
Salt marshes, estuaries, and macrophytes	1.2
Coral reefs	0.7
Total	48.3

Source: Longhurst *et al.* (1995).

8.10.2.2.3 Between land and oceans

Most of the carbon taken up or lost by terrestrial ecosystems and the ocean is exchanged with the atmosphere, but a small flux of carbon from land to the ocean bypasses the atmosphere. The river input of inorganic carbon to the oceans (0.4 Pg C yr^{-1}) is almost balanced in steady state by a loss of carbon to carbonate sediments (0.2 Pg C yr^{-1}) and a release of CO_2 to the atmosphere (0.1 Pg C yr^{-1}) (Sarmiento and Sundquist, 1992). The riverine flux of organic carbon is 0.3–0.5 Pg C yr^{-1}, and thus, the total flux from land to sea is 0.4–0.7 Pg C yr^{-1}.

8.10.3 CHANGES IN THE STOCKS AND FLUXES OF CARBON AS A RESULT OF HUMAN ACTIVITIES

8.10.3.1 Changes Over the Period 1850–2000

8.10.3.1.1 Emissions of carbon from combustion of fossil fuels

The CO_2 released annually from the combustion of fossil fuels (coal, oil, and gas) is calculated from records of fuel production compiled internationally (Marland *et al.*, 1998). Emissions of CO_2 from the production of cement and gas flaring add small amounts to the total industrial emissions, which have generally increased exponentially since ~1750. Temporary interruptions in the trend occurred during the two World Wars, following the increase in oil prices in 1973 and 1979, and following the collapse of the former Soviet Union in 1992 (Figure 3). Between 1751 and 2000, the total emissions of carbon are estimated to have been ~275 Pg C, essentially all of it since 1860. Annual emissions averaged 5.4 Pg C yr^{-1} during the 1980s and 6.3 Pg C yr^{-1} during the 1990s. Estimates are thought to be known globally to within 20% before 1950 and to within 6% since 1950 (Keeling, 1973; Andres *et al.*, 1999).

The proportions of coal, oil, and gas production have changed through time. Coal was the major contributor to atmospheric CO_2 until the late 1960s, when the production of oil first exceeded that of coal. Rates of oil and gas consumption grew rapidly until 1973. After that their relative rates of growth declined dramatically, such that emissions of carbon from coal were, again, as large as those from oil during the

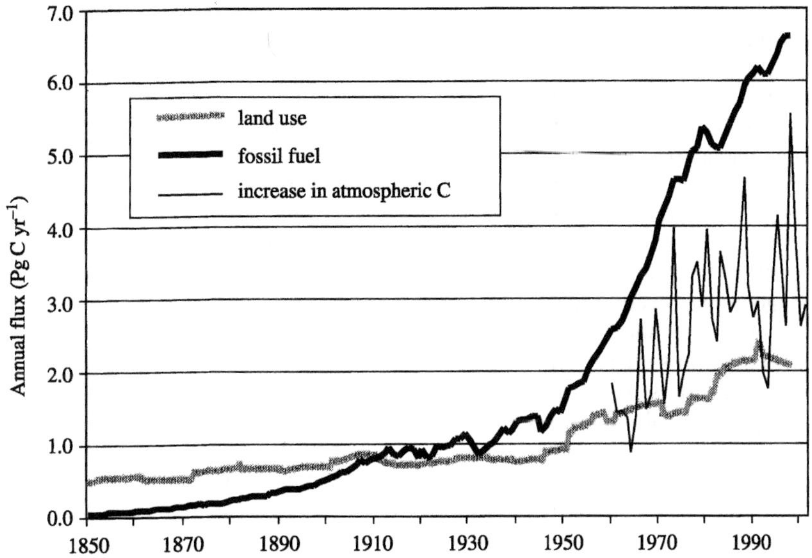

Figure 3 Annual emissions of carbon from combustion of fossil fuels and from changes in land use, and the annual increase in atmospheric CO_2 (in Pg C) since ~1750 (interannual variation in the growth rate of atmospheric CO_2 is greater than variation in emissions).

second half of the 1980s and in the last years of the twentieth century.

The relative contributions of different world regions to the annual emissions of fossil fuel carbon have also changed. In 1925, the US, Western Europe, Japan, and Australia were responsible for ~88% of the world's fossil fuel CO_2 emissions. By 1950 the fraction contributed by these countries had decreased to 71%, and by 1980 to 48%. The annual rate of growth in the use of fossil fuels in developed countries varied between 0.5% and 1.4% in the 1970s. In contrast, the annual rate of growth in developing nations was 6.3% during this period. The share of the world's total fossil fuel used by the developing countries has grown from 6% in 1925, to 10% in 1950, to ~20% in 1980. By 2020, the developing world may be using more than half of the world's fossil fuels annually (Goldemberg *et al.*, 1985). They may then be the major source of both fossil fuel and terrestrial CO_2 to the atmosphere (Section 8.10.3.1.5).

Annual emissions of CO_2 from fossil fuel combustion are small relative to the natural flows of carbon through terrestrial photosynthesis and respiration (~120 Pg C yr^{-1}) and relative to the gross exchanges between oceans and atmosphere (~90 Pg C yr^{-1}) (Figure 1). Nevertheless, these anthropogenic emissions are the major contributor to increasing concentrations of CO_2 in the atmosphere. They represent a transfer of carbon from the slow carbon cycle (see Chapter 8.09) to the active carbon cycle.

8.10.3.1.2 The increase in atmospheric CO_2

Numerous measurements of atmospheric CO_2 concentrations were made in the nineteenth century (Fraser *et al.*, 1986), and Callendar (1938) estimated from these early measurements that the amount of CO_2 had increased by 6% between 1900 and 1935. Because of geographical and seasonal variations in the concentrations of CO_2, however, no reliable measure of the rate of increase was possible until after 1957 when the first continuous monitoring of CO_2 concentrations was begun at Mauna Loa, Hawaii, and at the South Pole (Keeling *et al.*, 2001). In 1958 the average concentration of CO_2 in air at Mauna Loa was ~315 ppm. In the year 2000 the concentration had reached ~368 ppm, yielding an average rate of increase of ~1 ppm yr^{-1} since 1958. However, in recent decades the rate of increase in the atmosphere has been ~1.5 ppm yr^{-1} (~3 Pg C yr^{-1}).

During the early 1980s, scientists developed instruments that could measure the concentration of atmospheric CO_2 in bubbles of air trapped in glacial ice. Ice cores from Greenland and Antarctica show that the pre-industrial concentration of CO_2 was between 275 ppm and 285 ppm (Neftel *et al.*, 1985; Raynaud and Barnola, 1985; Etheridge *et al.*, 1996) (Figure 2(b)). The increase between 1700 and 2000, therefore, has been ~85 ppm, equivalent to ~175 Pg C, or 30% of the pre-industrial level.

Over the last 1,000 years the concentration of CO_2 in the atmosphere has varied by less than 10 ppmv (Figure 2(b)). However, over the last 4.2×10^5 years (four glacial cycles), the concentration of CO_2 has consistently varied from ~180 ppm during glacial periods to ~280 ppm during interglacial periods (Figure 2(c)). The correlation between CO_2 concentration and the surface temperature of the Earth is evidence for the greenhouse effect of CO_2, first advanced almost a century ago by the Swedish climatologist Arrhenius (1896). As a greenhouse gas, CO_2 is more transparent to the Sun's energy entering the Earth's atmosphere than it is to the re-radiated heat energy leaving the Earth. Higher concentrations of CO_2 in the atmosphere cause a warmer Earth and lower concentrations a cooler one. There have been abrupt changes in global temperature that were not associated with a change in CO_2 concentrations (Smith *et al.*, 1999), but never in the last 4.2×10^5 years have concentrations of CO_2 changed without a discernible change in temperature (Falkowski *et al.*, 2000). The glacial–interglacial difference of 100 ppm corresponds to a temperature difference of ~10 °C. The change reflects temperature changes in the upper troposphere and in the region of the ice core (Vostoc, Antarctica) and may not represent a global average. Today's CO_2 concentration of 368 ppm represents a large departure from the last 4.2×10^5 years, although the expected increase in temperature has not yet occurred.

It is impossible to say that the increase in atmospheric CO_2 is entirely the result of human activities, but the evidence is compelling. First, the known sources of carbon are more than adequate to explain the observed increase in the atmosphere. Balancing the global carbon budget requires additional carbon sinks, not an unexplained source of carbon (see Section 8.10.3.1.4). Since 1850, ~275 Pg C have been released from the combustion of fossil fuels and another 155 Pg C were released as a result of net changes in land use, i.e., from the net effects of deforestation and reforestation (Section 8.10.3.1.5). The observed increase in atmospheric carbon was only 175 Pg C (40% of total emissions) over this 150-year period (Table 5).

Table 5 The global carbon budget for the period 1850 to 2000 (units are Pg C).

Fossil fuel emissions	275
Atmospheric increase	175
Oceanic uptake	140
Net terrestrial source	40
Land-use net source	155
Residual terrestrial sink	115

Second, for several thousand years preceding 1850 (approximately the start of the industrial revolution), the concentration of CO_2 varied by less than 10 ppmv (Etheridge *et al.*, 1996) (Figure 2(b)). Since 1850, concentrations have increased by 85 ppmv (~30%). The timing of the increase is coincident with the annual emissions of carbon from combustion of fossil fuels and the net emissions from land-use change (Figure 3).

Third, the latitudinal gradient in CO_2 concentrations is highest at northern mid-latitudes, consistent with the fact that most of the emissions of fossil fuel are located in northern mid-latitudes. Although atmospheric transport is rapid, the signal of fossil fuel combustion is discernible.

Fourth, the rate of increase of carbon in the atmosphere and the distribution of carbon isotopes and other biogeochemical tracers are consistent with scientific understanding of the sources and sinks of carbon from fossil fuels, land, and the oceans. For example, while the concentration of CO_2 has increased over the period 1850–2000, the ^{14}C content of the CO_2 has decreased. The decrease is what would be expected if the CO_2 added to the system were fossil carbon depleted in ^{14}C through radioactive decay.

Concentrations of other carbon containing gases have also increased in the last two centuries. The increase in the concentration of CH_4 has been more than 100% in the last 100 years, from background levels of less than 0.8 ppm to a value of ~1.75 ppm in 2000 (Prather and Ehhalt, 2001). The temporal pattern of the increase is similar to that of CO_2. There was no apparent trend for the 1,000 years before 1700. Between 1700 and 1900 the annual rate of increase was ~1.5 ppbv, accelerating to 15 ppb yr^{-1} in the 1980s. Since 1985, however, the annual growth rate of CH_4 (unlike CO_2) has declined. The concentration is still increasing, but not as rapidly. It is unclear whether sources have declined or whether atmospheric sinks have increased.

Methane is released from anaerobic environments, such as the sediments of wetlands, peatlands, and rice paddies and the guts of ruminants. The major sources of increased CH_4 concentrations are uncertain but are thought to include the expansion of paddy rice, the increase in the world's population of ruminants, and leaks from drilling and transport of CH_4 (Prather and Ehhalt, 2001). Atmospheric CH_4 budgets are more difficult to construct than CO_2 budgets, because increased concentrations of CH_4 occur not only from increased sources from the Earth's surface but from decreased destruction (by OH radicals) in the atmosphere as well. The increase in atmospheric CH_4 has been more significant for the greenhouse effect than it has for the carbon budget. The doubling of CH_4 concentrations since 1700 has amounted to only ~1 ppm, in comparison to the CO_2 increase of almost 90 ppm. Alternatively, CH_4 is, molecule for molecule, ~15 times more effective than CO_2 as a greenhouse gas. Its atmospheric lifetime is only 8–10 years, however.

Carbon monoxide is not a greenhouse gas, but its chemical effects on the OH radical affect the destruction of CH_4 and the formation of ozone. Because the concentration of CO is low and its lifetime is short, its atmospheric budget is less well understood than budgets for CO_2 and CH_4. Nevertheless, CO seems to have been increasing in the atmosphere until the late 1980s (Prather and Ehhalt, 2001). Its contribution to the carbon cycle is very small.

8.10.3.1.3 Net uptake of carbon by the oceans

As discussed above, the chemistry of carbon in seawater is such that less than 1% of the carbon exists as dissolved CO_2. More than 99% of the DIC exists as bicarbonate and carbonate anions (Table 3). The chemical equilibrium among these three forms of DIC is responsible for the high solubility of CO_2 in the oceans. It also sets up a buffer for changes in oceanic carbon. The buffer factor (or Revelle factor), ξ, is defined as follows:

$$\xi = \frac{\Delta p_{CO_2}/p_{CO_2}}{\Delta\Sigma CO_2/\Sigma CO_2}$$

where p_{CO_2} is the partial pressure of CO_2 (the atmospheric concentration of CO_2 at equilibrium with that of seawater), ΣCO_2 is total inorganic carbon (DIC), and Δ refers to the change in the variable. The buffer factor varies with temperature, but globally averages ~10. It indicates that p_{CO_2} is sensitive to small changes in DIC: a change in the partial pressure of CO_2 (p_{CO_2}) is ~10 times the change in total CO_2. The significance of this is that the storage capacity of the ocean for excess atmospheric CO_2 is a factor of ~10 lower than might be expected by comparing reservoir sizes (Table 1). The oceans will not take up 98% of the carbon released through human activity, but only ~85% of it. The increase in atmospheric CO_2 concentration by ~30% since 1850s has been associated with a change of only ~3% in DIC of the surface waters. The other important aspect of the buffer factor is that it increases as DIC increases. The ocean will become increasingly resistant to taking up carbon (see Section 8.10.4.2.1).

Although the oceans determine the concentration of CO_2 in the atmosphere in the long term, in the short term, lags introduced by other processes besides chemistry allow a temporary disequilibrium. Two processes that delay the transfer of anthropogenic carbon into the ocean are: (i) the transfer of CO_2 across the air–sea interface and (ii) the mixing of water masses within the sea. The rate of transfer of CO_2 across the air–sea interface was discussed above (Section 8.10.2.2.2). This transfer is believed to have reduced the oceanic absorption of CO_2 by ~10% (Broecker *et al.*, 1979).

The more important process in slowing the oceanic uptake of CO_2 is the rate of vertical mixing within the oceans. The mixing of ocean waters is determined from measured profiles of natural ^{14}C, bomb-produced ^{14}C, bomb-produced tritium, and other tracers. Profiles of these tracers were obtained during extensive oceanographic surveys: one called Geochemical Ocean Sections (GEOSECS) carried out between 1972 and 1978), a second called Transient Tracers in the Ocean (TTO) carried out in 1981, and a third called the Joint Global Ocean Flux Study (JGOSFS) carried out in the 1990s. The surveys measured profiles of carbon, oxygen, radioisotopes, and other tracers along transects in the Atlantic and Pacific Oceans. The differences between the profiles over time have been used to calculate directly the penetration of anthropogenic CO_2 into the oceans (e.g., Gruber *et al.*, 1996, described below). As of 1980, the oceans are thought to have absorbed only ~40% of the emissions (20–47%, depending on the model used; Bolin, 1986).

Direct measurement of changes in the amount of carbon in the world's oceans is difficult for two reasons: first, the oceans are not mixed as rapidly as the atmosphere, so that spatial and temporal heterogeneity is large; and, second, the background concentration of dissolved carbon in seawater is large relative to the change, so measurement of the change requires very accurate methods. Nevertheless, direct measurement of the uptake of anthropogenic carbon is possible, in theory if not practically, by two approaches. The first approach is based on measurement of changes in the oceanic inventory of carbon and the second is based on measurement of the transfer of CO_2 across the air–sea interface.

Measurement of an increase in oceanic carbon is complicated by the background concentration and the natural variability of carbon concentrations in seawater. The total uptake of anthropogenic carbon in the surface waters of the ocean is calculated by models to have been ~40 $\mu mol\ kg^{-1}$ of water. Annual changes would, of course, be much smaller than 40 $\mu mol\ kg^{-1}$, as would the increase in DIC concentrations in deeper waters, where less anthropogenic carbon has penetrated. By comparison, the background concentration of DIC in surface waters is 2,000 $\mu mol\ kg^{-1}$. Furthermore, the seasonal variability at one site off Bermuda was 30 $\mu mol\ kg^{-1}$. Against this background and variability, direct measurement of change is a challenge. Analytical techniques add to uncertainties, although current techniques are capable of a precision of 1.5 μmol kg^{-1} within a laboratory and 4 μmol kg^{-1} between laboratories (Sarmiento, 1993).

A second method for directly measuring carbon uptake by the oceans, measurement of the air–sea exchange, is also made difficult by spatial and temporal variability. The approach measures the concentration of CO_2 in the air and in the surface mixed layer. The difference defines the gradient, which, together with a model that relates the exchange coefficient to wind speed, enables the rate of exchange to be calculated. An average air–sea difference (gradient) of 8 ppm, globally, is equivalent to an oceanic uptake of 2 Pg C yr^{-1} (Sarmiento, 1993), but the natural variability is greater than 10 ppm. Furthermore, the gas transfer coefficient is also uncertain within a factor of 2 (Broecker, 2001).

Because of the difficulty in measuring either changes in the ocean's inventory of carbon or the exchange of carbon across the air–sea interface, the uptake of anthropogenic carbon by the oceans is calculated with models that simulate the chemistry of carbon in seawater, the air–sea exchanges of CO_2, and oceanic circulation.

Ocean carbon models. Models of the ocean carbon cycle include three processes that affect the uptake and redistribution of carbon within the ocean: the air–sea transfer of CO_2, the chemistry of CO_2 in seawater, and the circulation or mixing of the ocean's water masses.

Three tracers have been used to constrain models. One tracer is CO_2 itself. The difference between current distribution of CO_2 in the ocean and the distribution expected without anthropogenic emissions yields an estimate of oceanic uptake (Gruber *et al.*, 1996). The approach is based on changes that occur in the chemistry of seawater as it ages. With age, the organic matter present in surface waters decays, increasing the concentration of CO_2 and various nutrients, and decreasing the concentration of O_2. The hard parts ($CaCO_3$) of marine organisms also decay with time, increasing the alkalinity of the water. From data on the concentrations of CO_2, O_2, and alkalinity throughout the oceans, it is theoretically possible to calculate the increased abundance of carbon in the ocean as a result of the increased concentration in the atmosphere. The approach is based on the assumption that the surface waters were in equilibrium with the atmosphere when they sank, or, at least, that the extent of disequilibrium is known. The approach is sensitive to seasonal variation in the CO_2 concentration in these surface waters.

A second tracer is bomb ^{14}C. The distribution of bomb ^{14}C in the oceans (Broecker *et al.*, 1995), together with an estimate of the transfer of $^{14}CO_2$ across the air–sea interface (Wanninkhof, 1992) (taking into account the fact that $^{14}CO_2$ equilibrates ~10 times more slowly than CO_2 across this interface), yields a constraint on uptake. A third constraint is based on the penetration of CFCs into the oceans (Orr and Dutay, 1999; McNeil *et al.*, 2003).

Ocean carbon models calculate changes in the oceanic carbon inventory. When these changes, together with changes in the atmospheric carbon inventory (from atmospheric and ice core CO_2 data), are subtracted from the emissions of carbon from fossil fuels, the result is an estimate of the net annual terrestrial flux of carbon.

Most current models of the ocean reproduce the major features of oceanic carbon: the vertical gradient in DIC, the seasonal and latitudinal patterns of p_{CO_2} in surface waters, and the interannual variability in p_{CO_2} observed during El Niños (Prentice *et al.*, 2001). However, ocean models do not capture the spatial distribution of ^{14}C at depth (Orr *et al.*, 2001), and they do not show an interhemispheric transport of carbon that is suggested from atmospheric CO_2 measurements (Stephens *et al.*, 1998). The models also have a tight biological coupling between carbon and nutrients, which seems not to have existed in the past and may not exist in the future. The issue is addressed below in Section 8.10.4.2.2.

8.10.3.1.4 Land: net exchange of carbon between terrestrial ecosystems and the atmosphere

Direct measurement of change in the amount of carbon held in the world's vegetation and soils may be more difficult than measurement of change in the oceans, because the land surface is not mixed. Not only are the background levels high (~550 Pg C in vegetation and ~1,500 in soils), but the spatial heterogeneity is greater on land than in the ocean. Thus, measurement of annual changes even as large as 3 Pg C yr^{-1}, in background levels 100 times greater, would require a very large sampling approach. Change may be measured over short intervals of a year or so in individual ecosystems by measuring fluxes of carbon, as, e.g., with the eddy flux technique (Goulden *et al.*, 1996), but, again, the results must be scaled up from 1 km^2 to the ecosystem, landscape, region, and globe.

Global changes in terrestrial carbon were initially estimated by difference, i.e., by estimates of change in the other three reservoirs. Because the global mass of carbon is conserved, when three terms of the global carbon budget are known, the fourth can be determined by difference. For the period 1850–2000, three of the terms (275 Pg C released from fossil fuels, 175 Pg C accumulated in the atmosphere, and 140 Pg C taken up by the oceans) define a net terrestrial uptake of 40 Pg C (Table 5). Temporal variations in these terrestrial sources and sinks can also be determined through inverse calculations with ocean carbon models (see Section 8.10.3.1.3). In inverse mode, models calculate the annual sources and sinks of carbon (output) necessary to produce observed concentrations of CO_2 in the atmosphere (input). Then, subtracting known fossil fuel sources from the calculated sources and sinks yields a residual flux of carbon, presumably terrestrial, because the other terms have been accounted for (the atmosphere and fossil fuels directly, the oceans indirectly). One such inverse calculation or deconvolution (Joos *et al.*, 1999b) suggests that terrestrial ecosystems were a net source of carbon until ~1940 and then became a small net sink. Only in the early 1990s was the net terrestrial sink greater than 0.5 Pg C yr^{-1} (Figure 4).

8.10.3.1.5 Land: changes in land use

At least a portion of terrestrial sources and sinks can be determined more directly from the large changes in vegetation and soil carbon that result from changes in land use, such as the conversion of forests to cleared lands. Changes in the use of land affect the amount of carbon stored in vegetation and soils and, hence, affect the flux of carbon between land and the atmosphere. The amount of carbon released to the atmosphere or accumulated on land depends not only on the magnitude and types of changes in land use, but also on the amounts of carbon held in different ecosystems. For example, the conversion of grassland to pasture may release no carbon to the atmosphere because the stocks of carbon are unchanged. The net release or accumulation of carbon also depends on time lags introduced by the rates of decay of organic matter, the rates of oxidation of wood products, and the rates of regrowth of forests following harvest or following abandonment of agriculture land. Calculation of the net terrestrial flux of carbon requires knowledge of these rates in different ecosystems under different types of land use. Because there are several important forms of land use and many types of ecosystems in different parts of the world, and because short-term variations in the magnitude of the flux are important, computation of the annual flux requires a computer model.

Changes in terrestrial carbon calculated from changes in land use. Bookkeeping models (Houghton *et al.*, 1983; Hall and Uhlig, 1991; Houghton and Hackler, 1995) have been used to calculate net sources and sinks of carbon resulting from land-use change in all the world's regions.

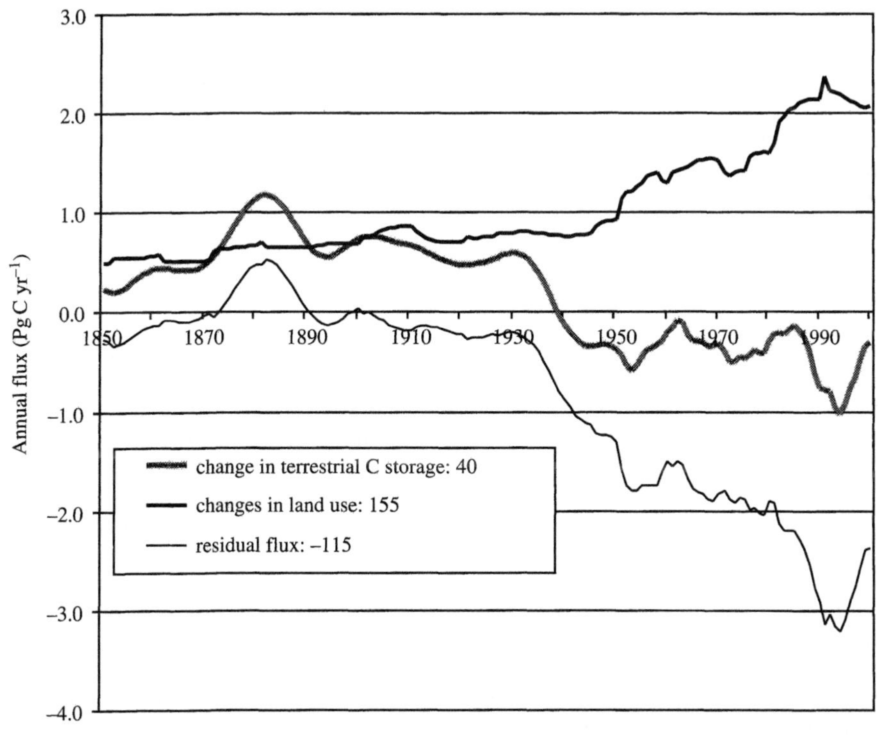

Figure 4 The net annual flux of carbon to or from terrestrial ecosystems (from inverse calculations with an ocean model (Joos *et al.*, 1999b), the flux of carbon from changes in land use (from Houghton, 2003), and the difference between the net flux and the flux from land-use change (i.e., the residual terrestrial sink). Positive values indicate a source of carbon from land and negative values indicate a terrestrial sink.

Calculations are based on two types of data: rates of land-use change and per hectare changes in carbon stocks that follow a change in land use. Changes in land use are defined broadly to include the clearing of lands for cultivation and pastures, the abandonment of these agricultural lands, the harvest of wood, reforestation, afforestation, and shifting cultivation. Some analyses have included wildfire because active policies of fire exclusion and fire suppression have affected carbon storage (Houghton *et al.*, 1999).

Bookkeeping models used to calculate fluxes of carbon from changes in land use track the carbon in living vegetation, dead plant material, wood products, and soils for each hectare of land cultivated, harvested, or reforested. Rates of land-use change are generally obtained from agricultural and forestry statistics, historical accounts, and national handbooks. Carbon stocks and changes in them following disturbance and growth are obtained from field studies. The data and assumptions used in the calculations are more fully documented in Houghton (1999) and Houghton and Hackler (2001).

The calculated flux is not the net flux of carbon between terrestrial ecosystems and the atmosphere, because the analysis does not consider ecosystems undisturbed by direct human activity. Rates of decay and rates of regrowth are defined in the model for different types of ecosystems and different types of land-use change, but they do not vary through time in response to changes in climate or concentrations of CO_2. The processes explicitly included in the model are the ecological processes of disturbance and recovery, not the physiological processes of photosynthesis and respiration.

The worldwide trend in land use over the last 300 years has been to reduce the area of forests, increase the area of agricultural lands, and, therefore, reduce the amount of carbon on land. Although some changes in land use increase the carbon stored on land, the net change for the 150-year period 1850–2000 is estimated to have released 156 Pg C (Houghton, 2003). An independent comparison of 1990 land cover with maps of natural vegetation suggests that another 58–75 Pg C (or ~30% of the total loss) were lost before 1850 (DeFries *et al.*, 1999).

The net annual fluxes of carbon to the atmosphere from terrestrial ecosystems (and fossil fuels) are shown in Figure 3. The estimates of the net flux from land before 1800 are relatively less reliable, because early estimates of land-use change are often incomplete. However, the absolute errors for the early years are small because the fluxes themselves were small. There were no worldwide economic or cultural developments in the eighteenth century that would have caused changes in land use of the magnitude that began in the nineteenth century and accelerated to the present day. The net annual biotic flux of carbon to the atmosphere before 1800 was probably less than 0.5 Pg and probably less than 1 Pg C until ~1950.

It was not until the middle of the last century that the annual emissions of carbon from combustion of fossil fuels exceeded the net terrestrial source from land-use change. Since then the fossil fuel contribution has predominated, although both fluxes have accelerated in recent decades with the intensification of industrial activity and the expansion of agricultural area.

The major releases of terrestrial carbon result from the oxidation of vegetation and soils associated with the expansion of cultivated land. The harvest of forests for fuelwood and timber is less important because the release of carbon to the atmosphere from the oxidation of wood products is likely to be balanced by the storage of carbon in regrowing forests. The balance will occur only as long as the forests harvested are allowed to regrow, however. If wood harvest leads to permanent deforestation, the process will release carbon to the atmosphere.

In recent decades the net release of carbon from changes in land use has been almost entirely from the tropics, while the emissions of CO_2 from fossil fuels were almost entirely from outside the tropics. The highest biotic releases were not always from tropical countries. The release of terrestrial carbon from the tropics is a relatively recent phenomenon, post-1945. In the nineteenth century the major sources were from the industrialized regions—North America, Europe, and the Soviet Union—and from those regions with the greatest numbers of people—South Asia and China.

8.10.3.1.6 *Land: a residual flux of carbon*

The amount of carbon calculated to have been released from changes in land use since the early 1850s (156 Pg C) (Houghton, 2003) is much larger than the amount calculated to have been released using inverse calculations with global carbon models (40 Pg C) (Joos *et al.*, 1999b) (Section 8.10.3.1.4). Moreover, the net source of CO_2 from changes in land use has generally increased over the past century, while the inversion approach suggests, on the contrary, that the largest releases of carbon from land were before 1930, and that since 1940 terrestrial ecosystems have been a small net sink (Figure 4).

The difference between these two estimates is greater than the errors in either one or both of the analyses, and might indicate a flux of carbon from processes not related to land-use change. The approach based on land-use change includes only the sources and sinks of carbon directly attributable to human activity; ecosystems not directly modified by human activity are left out of the analysis (assumed neither to accumulate nor release carbon). The approach based on inverse analyses with atmospheric data, in contrast, includes all ecosystems and all processes affecting carbon storage. It yields a net terrestrial flux of carbon. The difference between the two approaches thus suggests a generally increasing terrestrial sink for carbon attributable to factors other than land-use change. Ecosystems not directly cut or cleared could be accumulating or releasing carbon in response to small variations in climate, to increased concentrations of CO_2 in air, to increased availability of nitrogen or other nutrients, or to increased levels of toxins in air and soil resulting from industrialization. It is also possible that management practices not considered in analyses of land-use change may have increased the storage of carbon on lands that have been affected by land-use change. These possibilities will be discussed in more detail below (Section 8.10.4.1). Interestingly, the two estimates (land-use change and inverse modeling) are generally in agreement before 1935 (Figure 4), suggesting that before that date the net flux of carbon from terrestrial ecosystems was largely the result of changes in land use. Only after 1935 have changes in land use underestimated the net terrestrial carbon sink. By the mid-1990s this annual residual sink had grown to ~3 Pg C yr^{-1}.

8.10.3.2 Changes Over the Period 1980–2000

The period 1980–2000 deserves special attention not because the carbon cycle is qualitatively different over this period, but because scientists have been able to understand it better. Since 1980 new types of measurements and sophisticated methods of analysis have enabled better estimates of the uptake of carbon by the world's oceans and terrestrial ecosystems. The following section addresses the results of these analyses, first at the global level, and then at a regional level. Attention focuses on the two outstanding questions that have concerned scientists investigating the global carbon cycle since the first carbon budgets were constructed in the late 1960s (SCEP, 1970): (i) How much of the carbon released to the atmosphere from combustion of fossil fuels and changes in land use is taken up by the oceans and by terrestrial ecosystems? (ii) What are the mechanisms responsible for the uptake of carbon? The mechanisms for a carbon sink in terrestrial ecosystems have received considerable attention, in part because different mechanisms have different implications for future rates of CO_2 growth (and hence future global warming).

The previous section addressed the major reservoirs of the global carbon cycle, one at a time. This section addresses the methods used to determine changes in the amount of carbon held on land and in the sea, the two reservoirs for which changes in carbon are less well known. In contrast, the atmospheric increase in CO_2 and the emissions from fossil fuels are well documented. The order in which methods are presented is arbitrary. To set the stage, top-down (i.e., atmospherically based) approaches are described first, followed by bottom–up (ground-based) approaches (Table 6). Although the results of different methods often differ, the methods are not entirely comparable. Rather, they are complementary, and discrepancies sometimes suggest mechanisms responsible for transfers of carbon (Houghton, 2003; House *et al.*, 2003). The results from each method are presented first, and then they are added to an accumulating picture of the global carbon cycle. Again, the emphasis is on, first, the fluxes of carbon to and from terrestrial ecosystems and the ocean and, second, the mechanisms responsible for the terrestrial carbon sink.

Table 6 Characteristics of methods use to estimate terrestrial sinks.

	Geographic limitations	*Temporal resolution*	*Attribution of mechanism(s)*	*Precision*
Inverse modeling: oceanic data	No geographic resolution	Annual	No	Moderate
Land-use models	Data limitations in some regions	Annual	Yes	Moderate
Inverse modeling: atmospheric data	Poor in tropics	Monthly to annual	No	High: North–South Low: East–West
Forest inventories	Nearly nonexistent in the tropics	5–10 years	Yes (age classes)	High for biomass; variable for soil carbon
CO_2 flux	Site specific (a few km^2); difficult to scale up	Hourly to annual	No	Some problems with windless conditions
Physiologically based models	None	Hourly to annual	Yes	Variable; difficult to validate

8.10.3.2.1 The global carbon budget

(i) Inferring changes in terrestrial and oceanic carbon from atmospheric concentrations of CO_2 and O_2. According to the most recent assessment of climate change by the IPCC, the world's terrestrial ecosystems were a net sink averaging close to zero (0.2 Pg C yr^{-1}) during the 1980s and a significantly larger sink (1.4 Pg C yr^{-1}) during the 1990s (Prentice *et al.*, 2001). The large increase during the 1990s is difficult to explain. Surprisingly, the oceanic uptake of carbon was greater in the 1980s than the 1990s. The reverse would have been expected because atmospheric concentrations of CO_2 were higher in the 1990s. The estimates of terrestrial and oceanic uptake were based on changes in atmospheric CO_2 and O_2 and contained a small adjustment for the outgassing of O_2 from the oceans.

One approach for distinguishing terrestrial from oceanic sinks of carbon is based on atmospheric concentrations of CO_2 and O_2. CO_2 is released and O_2 taken up when fossil fuels are burned and when forests are burned. On land, CO_2 and O_2 are tightly coupled. In the oceans they are not, because O_2 is not very soluble in seawater. Thus, CO_2 is taken up by the oceans without any change in the atmospheric concentration of O_2. Because of this differential response of oceans and land, changes in atmospheric O_2 relative to CO_2 can be used to distinguish between oceanic and terrestrial sinks of carbon (Keeling and Shertz, 1992; Keeling *et al.*, 1996b; Battle *et al.*, 2000). Over intervals as short as a few years, slight variations in the seasonality of oceanic production and decay may appear as a change in oceanic O_2, but these variations cancel out over many years, making the method robust over multiyear intervals (Battle *et al.*, 2000).

Figure 5 shows how the method works. The individual points show average annual global CO_2/O_2 concentrations over the years 1990–2000.

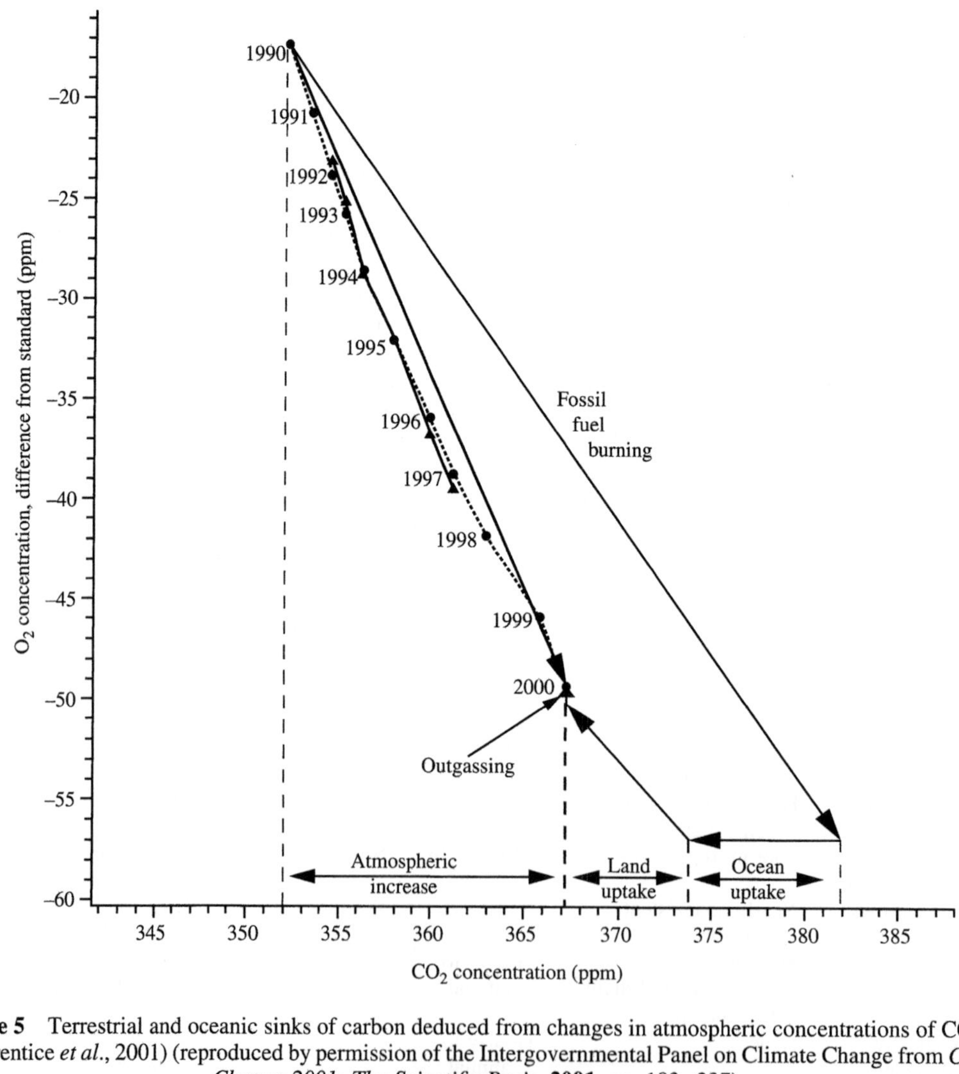

Figure 5 Terrestrial and oceanic sinks of carbon deduced from changes in atmospheric concentrations of CO_2 and O_2 (Prentice *et al.*, 2001) (reproduced by permission of the Intergovernmental Panel on Climate Change from *Climate Change 2001: The Scientific Basis*, **2001**, pp. 183–237).

Changes in the concentrations expected from fossil fuel combustion (approximately 1 : 1) during this interval are drawn, starting in 1990. The departure of these two sets of data confirms that carbon has accumulated somewhere besides the atmosphere. The oceans are assumed not to be changing with respect to O_2, so the line for the oceanic sink is horizontal. The line for the terrestrial sink is approximately parallel to the line for fossil fuel, and drawn through 2000. The intersection of the terrestrial and the oceanic lines thus defines the terrestrial and oceanic sinks. According to the IPCC (Prentice *et al.*, 2001), these sinks averaged 1.4 Pg C yr^{-1} and 1.7 Pg C yr^{-1}, respectively, for the 1990s. The estimate also included a small correction for outgassing of O_2 from the ocean (in effect, recognizing that the ocean is not neutral with respect to O_2).

Recent analyses suggest that such outgassing is significantly larger than initially estimated (Bopp *et al.*, 2002; Keeling and Garcia, 2002; Plattner *et al.*, 2002). The observed decadal variability in ocean temperatures (Levitus *et al.*, 2000) suggests a warming-caused reduction in the transport rate of O_2 to deeper waters and, hence, an increased outgassing of O_2. The direct effect of the warming on O_2 solubility is estimated to have accounted for only 10% of the loss of O_2 (Plattner *et al.*, 2002). The revised estimates of O_2 outgassing change the partitioning of the carbon sink between land and ocean. The revision increases the oceanic carbon sink of the 1990s relative to that of the 1980s (average sinks of 1.7 Pg C yr^{-1} and 2.4 Pg C yr^{-1}, respectively, for the 1980s and 1990s). The revised estimates are more consistent with estimates from ocean models (Orr, 2000) and from analyses based on $^{13}C/^{12}C$ ratios of atmospheric CO_2 (Joos *et al.*, 1999b; Keeling *et al.*, 2001). The revised estimate for land (a net sink of 0.7 Pg C yr^{-1} during the 1990s) (Table 7) is half of that given by the IPCC (Prentice *et al.*, 2001). The decadal change in the terrestrial sink is also much smaller (from 0.4 Pg C yr^{-1} to 0.7 Pg C yr^{-1} instead of from 0.2 Pg C yr^{-1} to 1.4 Pg C yr^{-1}).

(ii) Sources and sinks inferred from inverse modeling with atmospheric transport models and atmospheric concentrations of CO_2, $^{13}CO_2$, and O_2. A second top-down method for determining oceanic and terrestrial sinks is based on spatial and temporal variations in concentrations of atmospheric CO_2 obtained through a network of flask air samples (Masarie and Tans, 1995; Cooperative Atmospheric Data Integration Project—Carbon Dioxide, 1997). Together with models of atmospheric transport, these variations are used to infer the geographic distribution of sources and sinks of carbon through a technique called inverse modeling.

Variations in the carbon isotope of CO_2 may also be used to distinguish terrestrial sources and sinks from oceanic ones. The ^{13}C isotope is slightly heavier than the ^{12}C isotope and is discriminated against during photosynthesis. Thus, trees have a lighter isotopic ratio (−22 ppt to −27 ppt) than does air (−7 ppt) (ratios are expressed relative to a standard). The burning of forests (and fossil fuels) releases a disproportionate share of the lighter isotope, reducing the isotopic ratio of $^{13}C/^{12}C$ in air. In contrast, diffusion of CO_2 across the air–sea interface does not result in appreciable discrimination, so variations in the isotopic composition of CO_2 suggest terrestrial and fossil fuels fluxes of carbon, rather than oceanic.

Spatial and temporal variations in the concentrations of CO_2, $^{13}CO_2$, and O_2 are used with models of atmospheric transport to infer (through inverse calculations) sources and sinks of carbon at the Earth's surface. The results are dependent upon the model of atmospheric transport (Figure 6; Ciais *et al.*, 2000).

The interpretation of variations in ^{13}C is complicated. One complication results from isotopic disequilibria in carbon pools (Battle *et al.*, 2000). Disequilibria occur because the $\delta^{13}C$ taken up by plants, e.g., is representative of the $\delta^{13}C$ currently in the atmosphere (allowing for discrimination), but the $\delta^{13}C$ of CO_2 released through decay represents not the $\delta^{13}C$ of the current atmosphere but of an atmosphere several decades ago. As long as the $\delta^{13}C$ of the atmosphere is changing, the $\delta^{13}C$ in pools will reflect a mixture of earlier and current conditions. Uncertainties in the turnover of various carbon pools add uncertainty to interpretation of the $\delta^{13}C$ signal.

Table 7 The global carbon budget (Pg C yr^{-1}).

	1980s	*1990s*
Fossil fuel emissions[a]	5.4 ± 0.3	6.3 ± 0.4
Atmospheric increase[a]	3.3 ± 0.1	3.2 ± 0.2
Oceanic uptake[b]	−1.7 + 0.6 (−1.9 ± 0.6)	−2.4 + 0.7 (−1.7 ± 0.5)
Net terrestrial flux[b]	−0.4 + 0.7 (−0.2 ± 0.7)	−0.7 + 0.8 (−1.4 ± 0.7)
Land-use change[c]	2.0 ± 0.8	2.2 ± 0.8
Residual terrestrial flux	−2.4 + 1.1 (−2.2 ± 1.1)	−2.9 + 1.1 (−3.6 ± 1.1)

[a] Source: Prentice *et al.* (2001). [b] Source: Plattner *et al.* (2002) (values in parentheses are from Prentice *et al.*, 2001). [c] Houghton (2003).

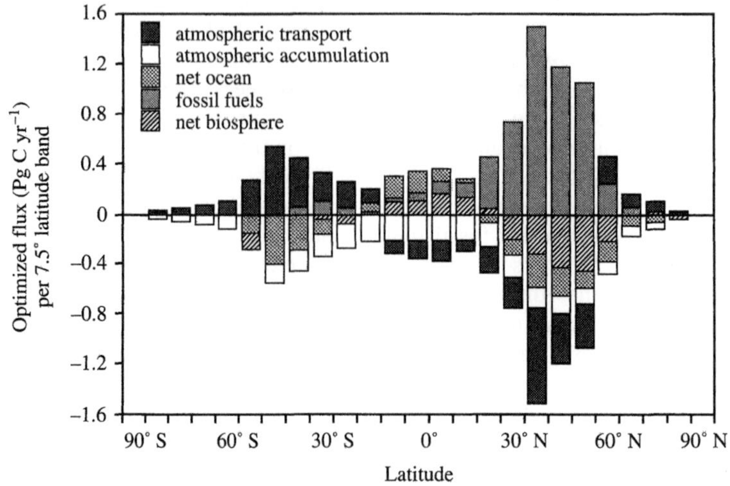

Figure 6 Terrestrial and oceanic sources and sinks of carbon inferred from inverse calculations with an atmospheric transport model and spatial and temporal variations in CO_2 concentrations. The net fluxes inferred over each region have been averaged into 7.5°-wide latitude strips (Ciais *et al.*, 2000) (reproduced by permission of the Ecological Society of America from *Ecol. Appl.*, **2000**, *10*, 1574–1589).

Another complication results from unknown year-to-year variations in the photosynthesis of C_3 and C_4 plants (because these two types of plants discriminate differently against the heavier isotope). C_4 plants discriminate less than C_3 plants and leave a signal that looks oceanic, thus confounding the separation of land and ocean exchanges. These uncertainties of the $\delta^{13}C$ approach are most troublesome over long periods (Battle *et al.*, 2000); the approach is more reliable for reconstructing interannual variations in sources and sinks of carbon.

An important distinction exists between global approaches (e.g., O_2, above) and regional inverse approaches, such as implemented with ^{13}C. In the global top-down approach, changes in terrestrial or oceanic carbon *storage* are calculated. In contrast, the regional inverse method yields *fluxes* of carbon between the land or ocean surface and the atmosphere. These fluxes of carbon include both natural and anthropogenic components. Horizontal exchange between regions must be taken into account to estimate changes in storage. For example, the fluxes will not accurately reflect changes in the amount of carbon on land or in the sea if some of the carbon fixed by terrestrial plants is transported by rivers to the ocean and respired there (Sarmiento and Sundquist, 1992; Tans *et al.*, 1995; Aumont *et al.*, 2001).

An example of inverse calculations is the analysis by Tans *et al.* (1990). The concentration of CO_2 near the Earth's surface is ~3 ppm higher over the northern mid-latitudes than over the southern hemisphere. The "bulge" in concentration over northern mid-latitudes is consistent with the emissions of carbon from fossil fuel combustion at these latitudes. The extent of the bulge is also affected by the rate of atmospheric mixing. High rates of mixing would dilute the bulge; low rates would enhance it. By using the latitudinal gradient in CO_2 and the latitudinal distribution of fossil fuel emissions, together with a model of atmospheric transport, Tans *et al.* (1990) determined that the bulge was smaller than expected on the basis of atmospheric transport alone. Thus, carbon is being removed from the atmosphere by the land and oceans at northern mid-latitudes. Tans *et al.* estimated removal rates averaging between 2.4 Pg C yr^{-1} and 3.5 Pg C yr^{-1} for the years 1981–1987. From p_{CO_2} measurements in surface waters, Tans *et al.* calculated that the northern mid-latitude oceans were taking up only 0.2–0.4 Pg C yr^{-1}, and thus, by difference, northern mid-latitude lands were responsible for the rest, a sink of 2.0–3.4 Pg C yr^{-1}. The range resulted from uncertainties in atmospheric transport and the limited distribution of CO_2 sampling stations. Almost no stations exist over tropical continents. Thus, Tans *et al.* (1990) could not constrain the magnitude of a tropical land source or sink, but they could determine the magnitude of the northern sink relative to a tropical source. A large tropical source, as might be expected from deforestation, implied a large northern sink; a small tropical source implied a smaller northern sink.

The analysis by Tans *et al.* (1990) caused quite a stir because their estimate for oceanic uptake was only 0.3–0.8 Pg C yr^{-1}, while analyses based on ocean models yielded estimates of 2.0 ± 0.8 Pg C yr^{-1}. The discrepancy was subsequently reconciled (Sarmiento and Sundquist, 1992) by accounting for the effect of skin temperature on the calculated air–sea exchange, the effect of atmospheric transport and oxidation of CO on the carbon budget, and the effect of riverine transport of carbon on changes in carbon storage (see below). All of the adjustments increased the estimated oceanic uptake of carbon to values obtained by ocean models and lowered the estimate of the mid-latitude terrestrial sink.

Similar inverse approaches, using not only CO_2 concentrations but also spatial variations in O_2 and $^{13}CO_2$ to distinguish oceanic from terrestrial fluxes, have been carried out by several groups since 1990. An intercomparison of 16 atmospheric transport models (the TransCom 3 project) by Gurney *et al.* (2002) suggests average oceanic and terrestrial sinks of 1.3 Pg C yr^{-1} and 1.4 Pg C yr^{-1}, respectively, for the period 1992–1996.

The mean global terrestrial sink of 1.4 Pg C yr^{-1} for the years 1992–1996 is higher than that obtained from changes in O_2 and CO_2 (0.7 Pg C yr^{-1}) (Plattner *et al.*, 2002). However, the estimate from inverse modeling has to be adjusted to account for terrestrial sources and sinks of carbon that are not "seen" by the atmosphere. For example, the fluxes inferred from atmospheric data will not accurately reflect changes in the amount of carbon on land or in the sea if some of the carbon fixed by terrestrial plants or used in weathering minerals is transported by rivers to the ocean and respired and released to the atmosphere there. Under such circumstances, the atmosphere sees a terrestrial sink and an oceanic source, while the storage of carbon on land and in the sea may not have changed. Several studies have tried to adjust atmospherically based carbon budgets by accounting for the river transport of carbon. Sarmiento and Sundquist (1992) estimated a pre-industrial net export by rivers of 0.4–0.7 Pg C yr^{-1}, balanced by a net terrestrial uptake of carbon through photosynthesis and weathering. Aumont *et al.* (2001) obtained a global estimate of 0.6 Pg C yr^{-1}. Adjusting the net terrestrial sink obtained through inverse calculations (1.4 Pg C yr^{-1}) by 0.6 Pg C yr^{-1} yields a result (0.8 Pg C yr^{-1}) similar to the estimate obtained through changes in the concentrations of O_2 and CO_2 (Table 8). The two top-down methods based on atmospheric measurements yield similar global estimates of a net terrestrial sink (~0.7 (±0.8) Pg C yr^{-1} for the 1990s).

(iii) Land-use change. Another method, independent of those based on atmospheric data and models, that has been used to estimate terrestrial sources and sinks of carbon, globally, is a method based on changes in land use (see Section 8.10.3.1.5). This is a ground-based or bottom-up approach. Changes in land use suggest that deforestation, reforestation, cultivation, and logging were responsible for a carbon source, globally, that averaged 2.0 Pg C yr^{-1} during the 1980s and 2.2 Pg C yr^{-1} during the 1990s (Houghton, 2003). The approach includes emissions of carbon from the decay of dead plant material, soil, and wood products and sinks of carbon in regrowing ecosystems, including both vegetation and soil. Analyses account for delayed sources and sinks of carbon that result from decay and regrowth following a change in land use.

Other recent analyses of land-use change give results that bound the results of this summary, although differences in the processes and regions included make comparisons somewhat misleading. An estimate by Fearnside (2000) of a 2.4 Pg C yr^{-1} source includes only the tropics. A source of 0.8 Pg C yr^{-1} estimated by McGuire *et al.* (2001) includes changes in global cropland area but does not include either the harvest of wood or the clearing of forests for pastures, both of which contributed to the net global source. The average annual release of carbon attributed by Houghton (2003) to changes in the area of croplands (1.2 Pg C yr^{-1} for the 1980s) is higher than the estimate found by McGuire *et al.* (0.8 Pg C yr^{-1}).

The calculated *source* of 2.2 (±0.8) Pg C yr^{-1} for the 1990s (Houghton, 2003) is very different from the global net terrestrial *sink* determined from top-down analyses (0.7 Pg C yr^{-1}) (Table 8). Are the methods biased? Biases in the inverse calculations may be in either direction. Because of the "rectifier effect" (the seasonal covariance

Table 8 Estimates of the annual terrestrial flux of carbon (Pg C yr^{-1}) in the 1990s according to different methods. Negative values indicate a terrestrial sink.

	O_2 and CO_2	*Inverse calculations CO_2, $^{13}CO_2$, O_2*	*Forest inventories*	*Land-use change*
Globe	−0.7 (±0.8)[a]	−0.8 (±0.8)[b]		2.2 (±0.6)[c]
Northern mid-latitudes		−2.1 (±0.8)[d]	−0.6 to −1.3[e]	−0.03 (±0.5)[c]
Tropics		1.5 (±1.2)[f]	−0.6 (±0.3)[g]	0.5 to 3.0[h]

[a] Plattner *et al.* (2002). [b] −1.4 (±0.8) from Gurney *et al.* (2002) reduced by 0.6 to account for river transport (Aumont *et al.*, 2001). [c] Houghton, 2003. [d] −2.4 from Gurney *et al.* (2002) reduced by 0.3 to account for river transport (Aumont *et al.*, 2001). [e] −0.65 in forests (Goodale *et al.*, 2002) and another 0.0–0.65 assumed for nonforests (see text). [f] 1.2 from Gurney *et al.* (2002) increased by 0.3 to account for river transport (Aumont *et al.*, 2001). [g] Undisturbed forests: −0.6 from Phillips *et al.* (1998) (challenged by Clark, 2002). [h] 0.9 (range 0.5–1.4) from DeFries *et al.* (2002) 1.3 from Achard *et al.* (2002) adjusted for soils and degradation (see text) 2.2 (±0.8) from Houghton (2003). 2.4 from Fearnside (2000).

between the terrestrial carbon flux and atmospheric transport), inverse calculations are thought to underestimate the magnitude of a northern mid-latitude sink (Denning *et al.*, 1995). However, if the near-surface concentrations of atmospheric CO_2 in northern mid-latitude regions are naturally lower than those in the southern hemisphere, the apparent sink in the north may not be anthropogenic, as usually assumed. Rather, the anthropogenic sink would be less than 0.5 Pg C yr^{-1} (Taylor and Orr, 2000).

In contrast to the unknown bias of atmospheric methods, analyses based on land-use change are deliberately biased. These analyses consider only those changes in terrestrial carbon resulting directly from human activity (conversion and modification of terrestrial ecosystems). There may be other sources and sinks of carbon not related to land-use change (such as caused by CO_2 fertilization, changes in climate, or management) that are captured by other methods but ignored in analyses of land-use change. In other words, the flux of carbon from changes in land use is not necessarily the same as the net terrestrial flux from all terrestrial processes.

If the net terrestrial flux of carbon during the 1990s was 0.7 Pg C yr^{-1}, and 2.2 Pg C yr^{-1} were emitted as a result of changes in land use, then 2.9 Pg C yr^{-1} must have accumulated on land for reasons not related to land-use change. This residual terrestrial sink was discussed above (Table 7 and Figure 4). That the residual terrestrial sink exists at all suggests that processes other than land-use change are affecting the storage of carbon on land. Recall, however, that the residual sink is calculated by difference; if the emissions from land-use change are overestimated, the residual sink will also be high.

8.10.3.2.2 Regional distribution of sources and sinks of carbon: the northern mid-latitudes

Insights into the magnitude of carbon sources and sinks and the mechanisms responsible for the residual terrestrial carbon sink may be obtained from a consideration of tropical and extratropical regions separately. Inverse calculations show the tropics to be a moderate source, largely oceanic as a result of CO_2 outgassing in upwelling regions. Some of the tropical source is also terrestrial. Estimates vary greatly depending on the models of atmospheric transport and the years included in the analyses. The net global oceanic sink of 1.3 Pg C yr^{-1} for the period 1992–1996 is distributed in northern (1.2 Pg C yr^{-1}) and southern oceans (0.8 Pg C yr^{-1}), with a net source from tropical gyres (0.5 Pg C yr^{-1}) (Gurney *et al.*, 2002).

The net terrestrial sink of ~0.7 Pg C yr^{-1} is not evenly distributed either. The comparison by Gurney *et al.* (2002) showed net terrestrial sinks of 2.4 ± 0.8 Pg C yr^{-1} and 0.2 Pg C yr^{-1} for northern and southern mid-latitude lands, respectively, offset to some degree by a net tropical land source of 1.2 ± 1.2 Pg C yr^{-1}. Errors are larger for the tropics than the nontropics because of the lack of sampling stations and the more complex atmospheric circulation there.

River transport and subsequent oceanic release of terrestrial carbon are thought to overestimate the magnitude of the atmospherically derived northern terrestrial sink by 0.3 Pg C yr^{-1} and underestimate the tropical source (or overestimate its sink) by the same magnitude (Aumont *et al.*, 2001). Thus, the northern terrestrial sink becomes 2.1 Pg C yr^{-1}, while the tropical terrestrial source becomes 1.5 Pg C yr^{-1} (Table 8).

Inverse calculations have also been used to infer east–west differences in the distribution of sources and sinks of carbon. Such calculations are more difficult because east–west gradients in CO_2 concentration are an order of magnitude smaller than north–south gradients. Some estimates placed most of the northern sink in North America (Fan *et al.*, 1998); others placed most of it in Eurasia (Bousquet *et al.*, 1999a,b). More recent analyses suggest a sink in both North America and Eurasia, roughly in proportion to land area (Schimel *et al.*, 2001; Gurney *et al.*, 2002). The analyses also suggest that higher-latitude boreal forests are small sources rather than sinks of carbon during some years.

The types of land use determining fluxes of carbon are substantially different inside and outside the tropics (Table 9). As of early 2000s, the fluxes of carbon to and from northern lands are dominated by rotational processes, e.g., logging and subsequent regrowth. Changes in the area of forests are small. The losses of carbon from decay of wood products and slash (woody debris generated as a result of harvest) are largely offset by the accumulation of carbon in regrowing forests (reforestation and regrowth following harvest). Thus, the net flux of carbon from changes in land use is small: a source of 0.06 Pg C yr^{-1} during the 1980s changing to a sink of 0.02 Pg C yr^{-1} during the 1990s. Both the US and Europe are estimated to have been carbon sinks as a result of land-use change.

Inferring changes in terrestrial carbon storage from analysis of forest inventories. An independent estimate of carbon sources and sinks in northern mid-latitudinal lands may be obtained from forest inventories. Most countries in the northern mid-latitudes conduct periodic inventories of the growing stocks in forests. Sampling is designed to yield estimates of total growing stocks (volumes of merchantable wood) that are

Table 9 Estimates of the annual sources (+) and sinks (−) of carbon resulting from different types of land-use change and management during the 1990s (Pg C yr^{-1}).

Activity	*Tropical regions*	*Temperate and boreal zones*	*Globe*
Deforestation	2.110[a]	0.130	2.240
Afforestation	−0.100	−0.080[b]	−0.190
Reforestation (agricultural abandonment)	0[a]	−0.060	−0.060
Harvest/management	0.190	0.120	0.310
Products	0.200	0.390	0.590
Slash	0.420	0.420	0.840
Regrowth	−0.430	−0.690	−1.120
Fire suppression[c]	0	−0.030	−0.030
Nonforests			
Agricultural soils[d]	0	0.020	0.020
Woody encroachment[c]	0	−0.060	−0.060
Total	2.200	0.040	2.240

[a] Only the net effect of shifting cultivation is included here. The gross fluxes from repeated clearing and abandonment are not included. [b] Areas of plantation forests are not generally reported in developed countries. This estimates includes only China's plantations. [c] Probably an underestimate. The estimate is for the US only, and similar values may apply in South America, Australia, and elsewhere. [d] These values include loss of soil carbon resulting from cultivation of new lands; they do not include accumulations of carbon that may have resulted from recent agricultural practices.

estimated with 95% confidence to within 1–5% (Powell *et al.*, 1993; Köhl and Päivinen, 1997; Shvidenko and Nilsson, 1997). Because annual changes due to growth and mortality are small relative to the total stocks, estimates of wood volumes are relatively less precise. A study in the southeastern US determined that regional growing stocks (m^3) were known with 95% confidence to within 1.1%, while changes in the stocks ($m^3\ yr^{-1}$) were known to within 39.7% (Phillips *et al.*, 2000). Allometric regressions are used to convert growing stocks (the wood contained in the boles of trees) to carbon, including all parts of the tree (roots, stumps, branches, and foliage as well as bole), nonmerchantable and small trees and nontree vegetation. Other measurements provide estimates of the carbon in the forest floor (litter) and soil. The precision of the estimates for these other pools of carbon is less than that for the growing stocks. An uncertainty analysis for 140×10^6 ha of US forests suggested an uncertainty of 0.028 Pg C yr^{-1} (Heath and Smith, 2000). The strength of forest inventories is that they provide direct estimates of wood volumes on more than one million plots throughout northern mid-latitude forests, often inventoried on 5–10-year repeat cycles. Some inventories also provide estimates of growth rates and estimates of mortality from various causes, i.e., fires, insects, and harvests. One recent synthesis of these forest inventories, after converting wood volumes to total biomass and accounting for the fate of harvested products and changes in pools of woody debris, forest floor, and soils, found a net northern mid-latitude terrestrial sink of between 0.6 Pg C yr^{-1} and 0.7 Pg C yr^{-1} for the years around 1990 (Goodale *et al.*, 2002). The estimate is ~30% of the sink inferred from atmospheric data corrected for river transport (Table 8). Some of the difference may be explained if nonforest ecosystems throughout the region are also accumulating carbon. Inventories of nonforest lands are generally lacking, but in the US, at least, nonforests are estimated to account for 40–70% of the net terrestrial carbon sink (Houghton *et al.*, 1999; Pacala *et al.*, 2001).

It is also possible that the accumulation of carbon below ground, not directly measured in forest inventories, was underestimated and thus might account for the difference in estimates. However, the few studies that have measured the accumulation of carbon in forest soils have consistently found soils to account for only a small fraction (5–15%) of measured ecosystem sinks (Gaudinski *et al.*, 2000; Barford *et al.*, 2001; Schlesinger and Lichter, 2001). Thus, despite the fact that the world's soils hold 2–3 times more carbon than biomass, there is no evidence, as of early 2000s, that they account for much of a terrestrial sink.

The discrepancy between estimates obtained from forest inventories and inverse calculations might also be explained by differences in the dates of measurements. The northern sink of 2.1 Pg C yr^{-1} from Gurney *et al.* (−2.4 + 0.3 for riverine transport) is for 1992–1996 and would probably have been lower (and closer to the forest inventory-based estimate) if averaged over the entire decade (see other estimates in Prentice *et al.*, 2001). Top-down measurements based on atmospheric data are sensitive to large year-to-year variations in the growth rate of CO_2 concentrations.

Both forest inventories and inverse calculations with atmospheric data show terrestrial ecosystems to be a significant carbon sink, while changes in land use show a sink near zero. Either the analyses of land-use change are incomplete, or other mechanisms besides land-use change must be responsible for the observed sink, or some combination of both. With respect to the difference between forest inventories and land-use change, a regional comparison suggests that the recovery of forests from land-use change (abandoned farmlands, logging, fire suppression) may either overestimate or underestimate the sinks measured in forest inventories (Table 10). In Canada and Russia, the carbon sink calculated for forests recovering from harvests (land-use change) is greater than the measured sink. The difference could be error, but it is consistent with the fact that fires and insect damage increased in these regions during the 1980s and thus converted some of the boreal forests from sinks to sources (Kurz and Apps, 1999). These sources would not be counted in the analysis of land-use change, because natural disturbances were ignored. In time, recovery from these natural disturbances will increase the sink above that calculated on the basis of harvests alone, but as of early 2000s the sources from fire and insect damage exceed the net flux associated with harvest and regrowth.

In the three other regions (Table 10), changes in land use yield a sink that is smaller than measured in forest inventories. If the results are not simply a reflection of error, the failure of past changes in land use to explain the measured sink suggests that factors not considered in the analysis have enhanced the storage of carbon in the forests of the US, Europe, and China. Such factors include past natural disturbances, more subtle forms of management than recovery from harvest and agricultural abandonment (and fire suppression in the US), and environmental changes that may have enhanced forest growth. It is unclear whether the differences between estimates (changes in land use and forest inventories) are real or the result of errors and omissions. The differences are small, generally less than 0.1 Pg C yr^{-1} in any region. The likely errors and omissions in analyses of land-use change include uncertain rates of forest growth, natural disturbances, and many types of forest management (Spiecker *et al.*, 1996).

8.10.3.2.3 Regional distribution of sources and sinks of carbon: the tropics

How do different methods compare in the tropics? Inverse calculations show that tropical lands were a net source of carbon, 1.2 ± 1.2 Pg C yr^{-1} for the period 1992–1996 (Gurney *et al.*, 2002). Accounting for the effects of rivers (Aumont *et al.*, 2001) suggests a source of 1.5 (±1.2) Pg C yr^{-1} (Table 8).

Forest inventories for large areas of the tropics are rare, although repeated measurements of permanent plots throughout the tropics suggest that undisturbed tropical forests are accumulating carbon, at least in the neotropics (Phillips *et al.*, 1998). The number of such plots was too small in tropical African or Asian forests to demonstrate a change in carbon accumulation, but assuming the plots in the neotropics are representative of undisturbed forests in that region suggests a sink of 0.62 (±0.30) Pg C yr^{-1} for mature humid neotropical forests (Phillips *et al.*, 1998). The finding of a net sink has been challenged, however, on the basis of systematic errors in measurement. Clark (2002) notes that many of the measurements of diameter included buttresses and other protuberances, while the allometric regressions used to estimate biomass were based on above-buttress relationships. Furthermore, these stem protuberances display disproportionate rates of radial growth. Finally, some of the plots were on floodplains where primary forests accumulate carbon. When plots with buttresses were excluded (and when recent floodplain (secondary) forests were excluded as well), the net increment was not statistically different from zero (Clark, 2002). Phillips *et al.* (2002) counter that the errors are minor, but the results remain contentious.

Thus, the two methods most powerful in constraining the northern net sink (inverse analyses and forest inventories) are weak or lacking in the tropics (Table 14), and the carbon balance of the tropics is less certain.

Direct measurement of CO_2 flux. The flux of CO_2 between an ecosystem and the atmosphere can be calculated directly by measuring the

Table 10 Annual net changes in the living vegetation of forests (Pg C yr^{-1}) in northern mid-latitude regions around the year 1990. Negative values indicate an increase in carbon stocks (i.e., a terrestrial sink).

Region	*Land-use change*[a]	*Forest inventory*[b]	*Sink from land-use change relative to inventoried sink*
Canada	−0.025	0.040	0.065 (larger)
Russia	−0.055	0.040	0.095 (larger)
USA	−0.035	−0.110	0.075 (smaller)
China	0.075	−0.040	0.115 (smaller)
Europe	−0.020	−0.090	0.070 (smaller)
Total	−0.060	−0.160	0.100 (smaller)

[a] Houghton (2003). [b] From Goodale *et al.* (2002).

covariance between concentrations of CO_2 and vertical wind speed (Goulden *et al.*, 1996). The approach is being applied at ~150 sites in North America, South America, Asia, and Europe. The advantage of the approach is that it includes an integrated measure for the whole ecosystem, not only the wood or the soil. The method is ideal for determining the short-term response of ecosystems to diurnal, seasonal, and interannual variations of such variables as temperature, soil moisture, and cloudiness. If measurements are made over an entire year or over a significant number of days in each season, an annual carbon balance can be determined. The results of such measured fluxes have been demonstrated in at least one ecosystem to be in agreement with independent measurements of change in the major components of the ecosystem (Barford, 2001).

As NEP is often small relative to the gross fluxes of photosynthesis and ecosystem respiration, the net flux is sometimes less than the error of measurement. More important than error is bias, and the approach is vulnerable to bias because both the fluxes of CO_2 and the micrometeorological conditions are systematically different day and night. Wind speeds below 17 cm s^{-1} in a temperate zone forest, e.g., resulted in an underestimate of nighttime respiration (Barford *et al.*, 2001). A similar relationship between nighttime wind speed and respiration in forests in the Brazilian Amazon suggests that the assumption that lateral transport is unimportant may have been invalid (Miller *et al.*, in press).

Although the approach works well where micrometeorological conditions are met, the footprint for the measured flux is generally less than 1 km^2, and it is difficult to extrapolate the measured flux to large regions. Accurate extrapolations require a distribution of tower sites representative of different flux patches, but such patches are difficult to determine *a priori*. The simple extrapolation of an annual sink of 1 Mg C ha yr^{-1} (based on 55 days of measurement) for a tropical forest in Brazil to all moist forests in the Brazilian Amazon gave an estimated sink of ~1 Pg C yr^{-1} (Grace *et al.*, 1995). In contrast, a more sophisticated extrapolation based on a spatial model of CO_2 flux showed a basin-wide estimate averaging only 0.3 Pg C yr^{-1} (Tian *et al.*, 1998). The modeled flux agreed with the measured flux in the location of the site; spatial differences resulted from variations in modeled soil moisture throughout the basin.

Initially, support for an accumulation of carbon in undisturbed tropical forests came from measurements of CO_2 flux by eddy correlation (Grace *et al.*, 1995; Malhi *et al.*, 1998). Results showed large sinks of carbon in undisturbed forests, that, if scaled up to the entire tropics, yielded sinks in the range of 3.9–10.3 Pg C yr^{-1} (Malhi *et al.*, 2001), much larger than the sources of carbon from deforestation. Tropical lands seemed to be a large net carbon sink. Recent analyses raise doubts about these initial results.

When flux measurements are corrected for calm conditions, the net carbon balance may be nearly neutral. One of the studies in an old-growth forest in the Tapajós National Forest, Pará, Brazil, showed a small net CO_2 source (Saleska *et al.*, in press). The results in that forest were supported by measurements of biomass (forest inventory) (Rice *et al.*, in press). Living trees were accumulating carbon, but the decay of downed wood released more, for a small net source. Both fluxes suggest that the stand was recovering from a disturbance several years earlier.

The observation that the rivers and streams of the Amazon are a strong source for CO_2 (Richey *et al.*, 2002) may help balance the large sinks measured in some upland sites. However, the riverine source is included in inverse calculations based on atmospheric data and does not change those estimates of a net terrestrial source (Gurney *et al.*, 2002).

Changes in land use in the tropics are clearly a source of carbon to the atmosphere, although the magnitude is uncertain (Detwiler and Hall, 1988; Fearnside, 2000; Houghton, 1999, 2003). The tropics are characterized by high rates of deforestation, and this conversion of forests to nonforests involves a large loss of carbon. Although rotational processes of land use, such as logging, are just as common in the tropics as in temperate zones (even more so because shifting cultivation is common in the tropics), the sinks of carbon in regrowing forests are dwarfed in the tropics by the large releases of carbon resulting from permanent deforestation.

Comparisons of results from different methods (Table 8) suggest at least two, mutually exclusive, interpretations for the net terrestrial source of carbon from the tropics. One interpretation is that a large release of carbon from land-use change (Fearnside, 2000; Houghton, 2003) is partially offset by a large sink in undisturbed forests (Malhi *et al.*, 1998; Phillips *et al.*, 1998, 2002). The other interpretation is that the source from deforestation is smaller (see below), and that the net flux from undisturbed forests is nearly zero (Rice *et al.*, in press; Saleska *et al.*, in press). Under the first interpretation, some sort of growth enhancement (or past natural disturbance) is required to explain the large current sink in undisturbed forests. Under the second, the entire net flux of carbon may be explained by changes in land use, but the source from land-use change is smaller than estimated by Fearnside (2000) or Houghton (2003).

A third possibility, that the net tropical source from land is larger than indicated by inverse calculations (uncertain in the tropics), is constrained by the magnitude of the net sink in northern mid-latitudes. The latitudinal gradient in CO_2 concentrations constrains the difference between the northern sink and tropical source more than it constrains the absolute fluxes. The tropical source can only be larger than indicated by inverse calculations if the northern mid-latitude sink is also larger. As discussed above, the northern mid-latitude sink is thought to be in the range of 1–2.6 $Pg\,C\,yr^{-1}$, but the estimates are based on the assumption that the pre-industrial north–south gradient in CO_2 concentrations was zero (similar concentrations at all latitudes). No data exist for the pre-industrial north–south gradient in CO_2 concentrations, but following Keeling *et al.* (1989), Tayor and Orr extrapolated the current CO_2 gradient to a zero fossil fuel release and found a negative gradient (lower concentrations in the north). They interpreted this negative gradient as the pre-industrial gradient, and their interpretation would suggest a northern sink larger than generally believed. In contrast, Conway and Tans (1999) interpret the extrapolated zero fossil fuel gradient as representing the current sources and sinks of carbon in response to fossil fuel emissions and other human activities, such as present and past land-use change. Most investigators of the carbon cycle favor this interpretation.

The second interpretation of existing estimates (a modest source of carbon from deforestation and little or no sink in undisturbed forests) is supported by satellite-based estimates of tropical deforestation. The high estimates of Fearnside (2000) and Houghton (2003) were based on rates of deforestation reported by the FAO (2001). If these rates of deforestation are high, the estimates of the carbon source are also high. Two new studies of tropical deforestation (Achard *et al.*, 2002; DeFries *et al.*, 2002) report lower rates than the FAO and lower emissions of carbon than Fearnside or Houghton. The study by Achard *et al.* (2002) found rates 23% lower than the FAO for the 1990s (Table 11). Their analysis used high resolution satellite data over a 6.5% sample of tropical humid forests, stratified by "deforestation hot-spot areas" defined by experts. In addition to observing 5.8×10^6 ha of outright deforestation in the tropical humid forests, Achard *et al.* also observed 2.3×10^6 ha of degradation. Their estimated carbon flux, including changes in the area of dry forests as well as humid ones, was 0.96 $Pg\,C\,yr^{-1}$. The estimate is probably low because it did not include the losses of soil carbon that often occur with cultivation or the losses of carbon from degradation (reduction of biomass within forests). Soils and degradation accounted for 12% and 26%, respectively, of Houghton's (2003) estimated flux of carbon for tropical Asia and America and would yield a total flux of 1.3 $Pg\,C\,yr^{-1}$ if the same percentages were applied to the estimate by Archard *et al.*

A second estimate of tropical deforestation (DeFries *et al.*, 2002) was based on coarse resolution satellite data (8 km), calibrated with high-resolution satellite data to identify percent tree cover and to account for small clearings that would be missed with the coarse resolution data. The results yielded estimates of deforestation that were, on average, 54% lower than those reported by the FAO (Table 11). According to DeFries *et al.*, the estimated net flux of carbon for the 1990s was 0.9 (range 0.5–1.4) $Pg\,C\,yr^{-1}$.

If the tropical deforestation rates obtained by Archard *et al.* and DeFries *et al.* were similar, there would be little doubt that the FAO estimates are high. However, the estimates are as different from each other as they are from those of the FAO (Table 11). Absolute differences between the two studies are difficult to evaluate because Achard *et al.* considered only humid tropical forests, whereas DeFries *et al.* considered all tropical forests. The greatest differences are in tropical Africa, where the percent tree cover mapped by DeFries *et al.* is most unreliable because of the large areas of savanna. Both studies suggest that the FAO estimates of tropical deforestation are high, but the rates are still in question (Fearnside and Laurance, 2003; Eva *et al.*, 2003). The tropical emissions of carbon estimated by the two studies (after adjustments for degradation and soils) are

Table 11 Annual rate of change in tropical forest area[a] for the 1990s.

	Tropical humid forests			*All tropical forests*		
	FAO (2001) (10^6 ha yr^{-1})	*Achard et al. (2002)*		*FAO (2001)* (10^6 ha yr^{-1})	*DeFries et al. (2002)*	
		10^6 ha yr^{-1}	*% lower than FAO*		10^6 ha yr^{-1}	*% lower than FAO*
America	2.7	2.2	18	4.4	3.179	28
Asia	2.5	2.0	20	2.4	2.008	16
Africa	1.2	0.7	42	5.2	0.376	93
All tropics	6.4	4.9	23	12.0	5.563	54

[a] The net change in forest area is not the rate of deforestation but, rather, the rate of deforestation minus the rate of afforestation.

about half of Houghton's estimate: 1.3 Pg C yr^{-1} and 0.9 Pg C yr^{-1}, as opposed to 2.2 Pg C yr^{-1} (Table 8).

8.10.3.2.4 Summary: synthesis of the results of different methods

Top-down methods show consistently that terrestrial ecosystems, globally, were a small net sink in the 1980s and 1990s. The sink was in northern mid-latitudes, partially offset by a tropical source. The northern sink was distributed over both North America and Eurasia roughly in proportion to land area. The magnitudes of terrestrial sinks obtained through inverse calculations are larger (or the sources smaller) than those obtained from bottom-up analyses (land-use change and forest inventories). Is there a bias in the atmospheric analyses? Or are there sinks not included in the bottom-up analyses?

For the northern mid-latitudes, when estimates of change in nonforests (poorly known) are added to the results of forest inventories, the net sink barely overlaps with estimates determined from inverse calculations. Changes in land use yield smaller estimates of a sink. It is not clear how much of the discrepancy is the result of omissions of management practices and natural disturbances from analyses of land-use change, and how much is the result of environmentally enhanced rates of tree growth. In other words, how much of the carbon sink in forests can be explained by age structure (i.e., previous disturbances and management), and how much by enhanced rates of carbon storage? The question is important for predicting future concentrations of atmospheric CO_2 (see below).

In the tropics, the uncertainties are similar but also greater because inverse calculations are more poorly constrained and because forest inventories are lacking. Existing evidence suggests two possibilities. Either large emissions of carbon from land-use change are somewhat offset by large carbon sinks in undisturbed forests, or lower releases of carbon from land-use change explain the entire net terrestrial flux, with essentially no requirement for an additional sink. The first alternative (large sources and large sinks) is most consistent with the argument that factors other than land-use change are responsible for observed carbon sinks (i.e., management or environmentally enhanced rates of growth). The second alternative is most consistent with the findings of Caspersen *et al.* (2000) that there is little enhanced growth. Overall, in both northern and tropical regions changes in land use exert a dominant influence on the flux of carbon, and it is unclear whether other factors have been important in either region. These conclusions question the assumption used in predictions of climatic change, the assumption that the current terrestrial carbon sink will increase in the future (see below).

8.10.4 MECHANISMS THOUGHT TO BE RESPONSIBLE FOR CURRENT SINKS OF CARBON

8.10.4.1 Terrestrial Mechanisms

Distinguishing between regrowth and enhanced growth in the current terrestrial sink is important. If regrowth is dominant, the current sink may be expected to diminish as forests age (Hurtt *et al.*, 2002). If enhanced growth is important, the magnitude of the carbon sink may be expected to increase in the future. Carbon cycle models used to calculate future concentrations of atmospheric CO_2 from emissions scenarios assume the latter (that the current terrestrial sink will increase) (Prentice *et al.*, 2001). These calculated concentrations are then used in general circulation models to project future rates of climatic change. If the current terrestrial sink is largely the result of regrowth, rather than enhanced growth, future projections of climate may underestimate the extent and rate of climatic change.

The issue of enhanced growth versus regrowth can be illustrated with studies from the US. Houghton *et al.* (1999) estimated a terrestrial carbon sink of 0.15–0.35 Pg C yr^{-1} for the US, attributable to changes in land use. Pacala *et al.* (2001) revised the estimate upwards by including additional processes, but in so doing they included sinks not necessarily resulting from land-use change. Their estimate for the uptake of carbon by forests, e.g., was the uptake measured by forest inventories. The measured uptake might result from previous land use (regrowth), but it might also result from environmentally enhanced growth, e.g., CO_2 fertilization (Figure 7). If all of the accumulation of carbon in US forests were the result of recovery from past land-use practices (i.e., no enhanced growth), then the measured uptake should equal the flux calculated on the basis of land-use change. The residual flux would be zero. The study by Caspersen *et al.* (2000) suggests that such an attribution is warranted because they found that 98% of forest growth in five US states could be attributed to regrowth rather than enhanced growth. However, the analysis by Houghton *et al.* (1999) found that past changes in land use accounted for only 20–30% of the observed accumulation of carbon in trees. The uptake calculated for forests recovering from agricultural abandonment, fire suppression, and earlier harvests was only 20–30% of the uptake measured by forest inventories (~40% if the uptake attributed to woodland "thickening" (0.26 Pg C yr^{-1}; Houghton, 2003) is included (Table 12)). The results are inconsistent with

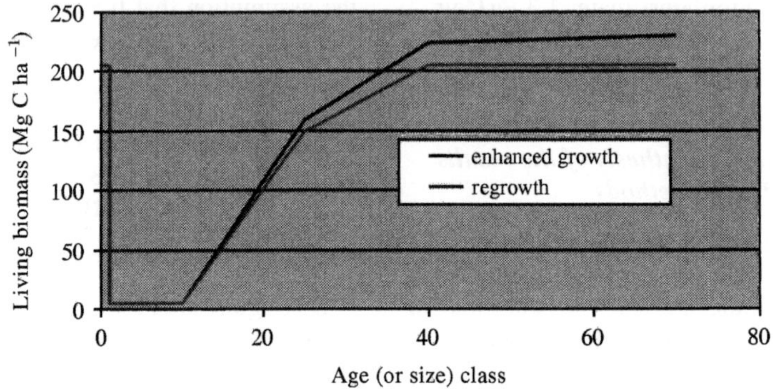

Figure 7 Idealized curves showing the difference between enhanced growth and regrowth in the accumulation of carbon in forest biomass.

Table 12 Estimated rates of carbon accumulation in the US (Pg C yr^{-1} in 1990).

	Pacala et al.[a] *(2001)*		*Houghton*[b] *et al. (1999)*	*Houghton*[b] *(2003)*	*Goodale et al. (2002)*
	Low	*High*			
Forest trees	−0.11	−0.15	−0.072[c]	−0.046[d]	−0.11
Other forest organic matter	−0.03	−0.15	0.010	0.010	−0.11
Cropland soils	0.00	−0.04	−0.138	0.00	NE
Woody encroachment	−0.12	−0.13	−0.122	−0.061	NE
Wood products	−0.03	− 0.07	−0.027	−0.027	−0.06
Sediments	−0.01	−0.04	NE	NE	NE
Total sink	−0.30	−0.58	−0.35	−0.11	−0.28
% of total sink neither in forests nor wood products	43%	36%	74%	55%	NE

NE is "not estimated". Negative values indicate an accumulation of carbon on land.
[a] Pacala *et al.* (2001) also included the import/export imbalance of food and wood products and river exports. As these would create corresponding sources outside the US, they are ignored here. [b] Includes only the direct effects of human activity (i.e., land-use change and some management). [c] 0.020 Pg C yr^{-1} in forests and 0.052 Pg C yr^{-1} in the thickening of western pine woodlands as a result of early fire suppression. [d] 0.020 Pg C yr^{-1} in forests and 0.026 Pg C yr^{-1} in the thickening of western pine woodlands as a result of early fire suppression.

those of Caspersen *et al.* (2000). Houghton's analysis requires a significant growth enhancement to account for the observed accumulation of carbon in trees; the analysis by Caspersen *et al.* suggests little enhancement.

Both analyses merit closer scrutiny. Joos *et al.* (2002) have pointed out, e.g., that the relationship between forest age and wood volume (or biomass) is too variable to constrain the enhancement of growth to between 0.001% and 0.01% per year, as Caspersen *et al.* claimed. An enhancement of 0.1% per year fits the data as well. Furthermore, even a small enhancement of 0.1% per year in NPP yields a significant sink (~2 Pg C yr^{-1}) if it applies globally (Joos *et al.*, 2002). Thus, Caspersen *et al.* may have underestimated the sink attributable to enhanced growth.

However, Houghton's analysis of land-use change (Houghton *et al.*, 1999; Houghton, 2003) most likely underestimates the sink attributable to regrowth. Houghton did not consider forest management practices other than harvest and subsequent regrowth. Nor did he include natural disturbances, which in boreal forests are more important than logging in determining the current age structure and, hence, rate of carbon accumulation (Kurz and Apps, 1999). Forests might now be recovering from an earlier disturbance. A third reason why the sink may have been underestimated is that Houghton used net changes in agricultural area to obtain rates of agricultural abandonment. In contrast, rates of clearing and abandonment are often simultaneous and thus create larger areas of regrowing forests than would be predicted from net changes in agricultural area. It is unclear how much of the carbon sink in the US can be attributed to changes in land use and management, and how much can be attributed to enhanced rates of growth.

The mechanisms responsible for the current terrestrial sink fall into two broad categories (Table 13 and Figure 7): (i) enhanced growth from physiological or metabolic factors that affect rates of photosynthesis, respiration, growth, and decay and (ii) regrowth from past disturbances, changes in land use, or management, affecting the mortality of forest stands, the age structure of forests, and hence their rates of carbon accumulation. What evidence do we have that

Table 13 Proposed mechanisms for terrestrial carbon sinks.[a]

Metabolic or physiological mechanisms
CO_2 fertilization
N fertilization
Tropospheric ozone, acid deposition
Changes in climate (temperature, moisture)
Ecosystem mechanisms
Large-scale regrowth of forests following human disturbance (includes recovery from logging and agricultural abandonment)[b]
Large-scale regrowth of forests following natural disturbance[b]
Fire suppression and woody encroachment[b]
Decreased deforestation[b]
Improved agricultural practices[b]
Erosion and re-deposition of sediment
Wood products and landfills[b]

[a] Some of these mechanisms enhance growth; some reduce decomposition. In some cases these same mechanisms may also yield sources of carbon to the atmosphere. [b] Mechanisms included in analyses of land-use change (although not necessarily in all regions).

these mechanisms are important? Consider, first, enhanced rates of growth.

8.10.4.1.1 Physiological or metabolic factors that enhance rates of growth and carbon accumulation

CO_2 fertilization. Numerous reviews of the direct and indirect effects of CO_2 on photosynthesis and plant growth have appeared in the literature (Curtis, 1996; Koch and Mooney, 1996; Mooney *et al.*, 1999; Körner, 2000), and only a very brief review is given here. Horticulturalists have long known that annual plants respond to higher levels of CO_2 with increased rates of growth, and the concentration of CO_2 in greenhouses is often deliberately increased to make use of this effect. Similarly, experiments have shown that most C_3 plants (all trees, most crops, and vegetation from cold regions) respond to elevated concentrations of CO_2 with increased rates of photosynthesis and increased rates of growth.

Despite the observed stimulative effects of CO_2 on photosynthesis and plant growth, it is not clear that the effects will result in an increased storage of carbon in the world's ecosystems. One reason is that the measured effects of CO_2 have generally been short term, while over longer intervals the effects are often reduced or absent. For example, plants often acclimate to higher concentrations of CO_2 so that their rates of photosynthesis and growth return to the rates observed before the concentration was raised (Tissue and Oechel, 1987; Oren *et al.*, 2001).

Another reason why the experimental results may not apply to ecosystems is that most

Table 14 Increases observed for a 100% increase in CO_2 concentrations.

Increased rates
60% increase in *photosynthesis* of young trees
33% average increase in net primary productivity (*NPP*) of crops
25% increase in *NPP* of a young pine forest
Increased stocks
14% average increase in *biomass* of grasslands and crops
~0% increase in the carbon content of mature forests

experiments with elevated CO_2 have been conducted with crops, annual plants, or tree seedlings. The few studies conducted at higher levels of integration or complexity, such as with mature trees and whole ecosystems, including soils as well as vegetation, suggest much reduced responses. Table 14 summarizes the results of experiments at different levels of integration. Arranged in this way (from biochemical processes to ecosystem processes), observations suggest that as the level of complexity, or the number of interacting processes, increases, the effects of CO_2 fertilization are reduced. This dampening of effects across ever-increasing levels of complexity has been noted since scientists first began to consider the effects of CO_2 on carbon storage (Lemon, 1977).

In other words, a CO_2-enhanced increase in photosynthesis is many steps removed from an increase in carbon storage. An increase in NPP is expected to lead to increased carbon storage until the carbon lost from the detritus pool comes into a new equilibrium with the higher input of NPP. But, if the increased NPP is largely labile (easily decomposed), then it may be decomposed rapidly with little net carbon storage (Davidson and Hirsch, 2001). Results from a loblolly pine forest in North Carolina suggest a very small increase in carbon storage. Elevated CO_2 increased litter production (with a turnover time of about three years) but did not increase carbon accumulation deeper in the soil layer (Schlesinger and Lichter, 2001). Alternatively, the observation that microbes seemed to switch from old organic matter to new organic matter after CO_2 fertilization of a grassland suggests that the loss of carbon may be delayed in older, more refractory pools of soil organic matter (Cardon *et al.*, 2001).

The central question is whether natural ecosystems will accumulate carbon as a result of elevated CO_2, and whether the accumulation will persist. Few CO_2 fertilization experiments have been carried out for more than a few years in whole ecosystems, but where they have, the results generally show an initial CO_2-induced increment in biomass that diminishes after a few years. The diminution of the initial CO_2-induced effect occurred after two years in an arctic tundra (Oechel *et al.*, 1994) and after three years in a rapidly growing loblolly pine forest (Oren *et al.*, 2001). Other forests may behave differently, but the North Carolina forest was chosen in part because CO_2 fertilization, if it occurs anywhere, is likely to occur in a rapidly growing forest. The longest CO_2 fertilization experiment, in a brackish wetland on the Chesapeake Bay, has shown an enhanced net uptake of carbon even after 12 years, but the expected accumulation of carbon at the site has not been observed (Drake *et al.*, 1996).

Nitrogen fertilization. Human activity has increased the abundance of biologically active forms of nitrogen (NO_x and NH_4), largely through the production of fertilizers, the cultivation of legumes that fix atmospheric nitrogen, and the use of internal combustion engines. Because the availability of nitrogen is thought to limit NPP in temperate-zone ecosystems, the addition of nitrogen through human activities is expected to increase NPP and, hence, terrestrial carbon storage (Peterson and Melillo, 1985; Schimel *et al.*, 1996; Holland *et al.*, 1997). Based on stoichiometric relations between carbon and nitrogen, many physiologically based models predict that added nitrogen should lead to an accumulation of carbon in biomass. But the extent to which this accumulation occurs in nature is unclear. Adding nitrogen to forests does increase NPP (Bergh *et al.*, 1999). It may also modify soil organic matter and increase its residence time (Fog, 1988; Bryant *et al.*, 1998). But nitrogen deposited in an ecosystem may also be immobilized in soils (Nadelhoffer *et al.*, 1999) or lost from the ecosystem, becoming largely unavailable in either case (Davidson, 1995).

There is also evidence that additions of nitrogen above some level may saturate the ecosystem, causing: (i) increased nitrification and nitrate leaching, with associated acidification of soils and surface waters; (ii) cation depletion and nutrient imbalances; and (iii) reduced productivity (Aber *et al.*, 1998; Fenn *et al.*, 1998). Experimental nitrogen additions have had varied effects on wood production and growing stocks. Woody biomass production increased in response to nitrogen additions to two New England hardwood sites, although increased mortality at one site led to a net decrease in the stock of woody biomass (Magill *et al.*, 1997, 2000). Several studies have shown that chronic exposure to elevated nitrogen inputs can inhibit forest growth, especially in evergreen species (Tamm *et al.*, 1995; Makipaa, 1995). Fertilization decreased rates of wood production in high-elevation spruce-fir in Vermont (McNulty *et al.*, 1996) and in a heavily fertilized red pine plantation (Magill *et al.*, 2000). The long-term effects of nitrogen deposition on forest production and carbon balance remain uncertain. Furthermore, because much of the nitrogen deposited on land is in the form of acid precipitation, it is difficult to distinguish the fertilization effects of nitrogen from the adverse effects of acidity (see below).

Atmospheric chemistry. Other factors besides nitrogen saturation may have negative effects on NPP, thus reducing the uptake of carbon in ecosystems and perhaps changing them from sinks to sources of carbon. Two factors that have received attention are tropospheric ozone and sulfur (acid rain). Experimental studies show leaf injury and reduced growth in crops and trees exposed to ozone. At the level of the ecosystem, elevated levels of ozone have been associated with reduced forest growth in North America (Mclaughlin and Percy, 2000) and Europe (Braun *et al.*, 2000). Acidification of soil as a result of deposition of NO_3^- and SO_4^{2-} in precipitation depletes the soils of available plant nutrients (Ca^{2+}, Mg^{2+}, K^+), increases the mobility and toxicity of aluminum, and increases the amount of nitrogen and sulfur stored in forest soils (Driscoll *et al.*, 2001). The loss of plant nutrients raises concerns about the long-term health and productivity of forests in the northeastern US, Europe, and southern China.

Although the effects of tropospheric ozone and sulfur generally reduce NPP, their actual or potential effects on carbon stocks are not known. The pollutants could potentially increase carbon stocks if they reduce decomposition of organic matter more than they reduce NPP.

Climatic variability and climatic change. Year-to-year differences in the growth rate of CO_2 in the atmosphere are large (Figure 3). The annual rate of increase ranged from 1.9 Pg C in 1992 to 6.0 Pg C in 1998 (Prentice *et al.*, 2001; see also Conway *et al.*, 1994). In 1998 the net global sink (ocean and land) was nearly 0 Pg C, while the average combined sink in the previous eight years was $\sim$3.5 Pg C yr^{-1} (Tans *et al.*, 2001). The terrestrial sink is generally twice as variable as the oceanic sink (Bousquet *et al.*, 2000). This temporal variability in terrestrial fluxes is probably caused by the effect of climate on carbon pools with short lifetimes (foliage, plant litter, soil microbes) through variations in photosynthesis, respiration, and possibly fire (Schimel *et al.*, 2001). Measurements in terrestrial ecosystems suggest that respiration, rather than photosynthesis, is the major contributor to variability (Valentini *et al.*, 2000). Annual respiration was almost twice as variable as photosynthesis over a five-year period in the Harvard Forest (Goulden *et al.*, 1996). Respiration is also more sensitive than photosynthesis to changes in both temperature and moisture. For example, during a dry year at the Harvard Forest, both photosynthesis and respiration were reduced, but the reduction in respiration was greater, yielding a greater than average net uptake of carbon for the year (Goulden *et al.*, 1996). A tropical forest in the Brazilian Amazon behaved similarly (Saleska *et al.*, in press).

The greater sensitivity of respiration to climatic variations is also observed at the global scale. An analysis of satellite data over the US, together with an ecosystem model, shows that the variability in NPP is considerably less than the variability in the growth rate of atmospheric CO_2 inferred from inverse modeling, suggesting that the cause of the year-to-year variability in carbon fluxes is largely from varying rates of respiration rather than photosynthesis (Hicke *et al.*, 2002). Also, global NPP was remarkably constant over the three-year transition from El Niño to La Niña (Behrenfeld *et al.*, 2001). Myneni *et al.* (1995) found a positive correlation between annual "greenness," derived from satellites, and the growth rate of CO_2. Greener years presumably had more photosynthesis and higher GPP, but they also had proportionately more respiration, thus yielding a net release of carbon from land (or reduced uptake) despite increased greenness.

Climatic factors influence terrestrial carbon storage through effects on photosynthesis, respiration, growth, and decay. However, prediction of future terrestrial sinks resulting from climate change requires an understanding of not only plant and microbial physiology, but the regional aspects of future climate change, as well. The important aspects of climate are: (i) temperature, including the length of the growing season; (ii) moisture; and (iii) solar radiation and clouds. Although year-to-year variations in the growth rate of CO_2 are probably the result of terrestrial responses to climatic variability, longer-term changes in carbon storage involve acclimation and other physiological adjustments that generally reduce short-term responses.

In cold ecosystems, such as those in high latitudes (tundra and taiga), an increase in temperature might be expected to increase NPP and, perhaps, carbon storage (although the effects might be indirect through increased rates of nitrogen mineralization; Jarvis and Linder, 2000). Satellite records of "greenness" over the boreal zone and temperate Europe show a lengthening of the growing season (Myneni *et al.*, 1997), suggesting greater growth and carbon storage. Measurements of CO_2 flux in these ecosystems do not consistently show a net uptake of carbon in response to warm temperatures (Oechel *et al.*, 1993; Goulden *et al.*, 1998), however, presumably because warmer soils release more carbon than plants take up. Increased temperatures in boreal forests may also reduce plant growth if the higher temperatures are associated with drier conditions (Barber *et al.*, 2000; Lloyd and Fastie, 2002). The same is true in the tropics, especially as the risk of fires increases with drought (Nepstad *et al.*, 1999; Page *et al.*, 2002). A warming-enhanced increase in rates of respiration and decay may already have begun to release carbon to the atmosphere (Woodwell, 1983; Raich and Schlesinger, 1992; Houghton *et al.*, 1998).

The results of short-term experiments may be misleading, however, because of acclimation or because the more easily decomposed material is respired rapidly. The long-term, or equilibrium, effects of climate on carbon storage can be inferred from the fact that cool, wet habitats store more carbon in soils than hot, dry habitats (Post *et al.*, 1982). The transient effects of climatic change on carbon storage, however, are difficult to predict, in large part because of uncertainty in predicting regional and temporal changes in temperature and moisture (extremes as well as means) and rates of climatic change, but also from incomplete understanding of how such changes affect fires, disease, pests, and species migration rates.

In the short term of seasons to a few years, variations in terrestrial carbon storage are most likely driven by variations in climate (temperature, moisture, light, length of growing season). Carbon dioxide fertilization and nitrogen deposition, in contrast, are unlikely to change abruptly. Interannual variations in the emissions of carbon from land-use change are also likely to be small (<0.2 Pg C yr^{-1}) because socioeconomic changes in different regions generally offset each other, and because the releases and uptake of carbon associated with a land-use change lag the change in land use itself and thus spread the emissions over time (Houghton, 2000). Figure 8 shows the annual net emissions of carbon from deforestation and reforestation in the Brazilian Amazon relative to the annual fluxes observed in the growth rates of trees and modeled on the basis of physiological responses to climatic variation. Clearly, metabolic responses to climatic variations are more important in the short term than interannual variations in rates of land-use change.

Understanding short-term variations in atmospheric CO_2 may not be adequate for predicting longer-term trends, however. Organisms and populations acclimate and adapt in ways that generally diminish short-term responses. Just as increased rates of photosynthesis in response to elevated levels of CO_2 often, but not always, decline within months or years (Tissue and Oechel, 1987), the same diminished response has been observed for higher temperatures (Luo *et al.*, 2001). Thus, over decades and centuries the factors most important in influencing concentrations of atmospheric CO_2 (fossil fuel emissions, land-use change, oceanic uptake) are probably different from those factors important in determining the short-term variations in

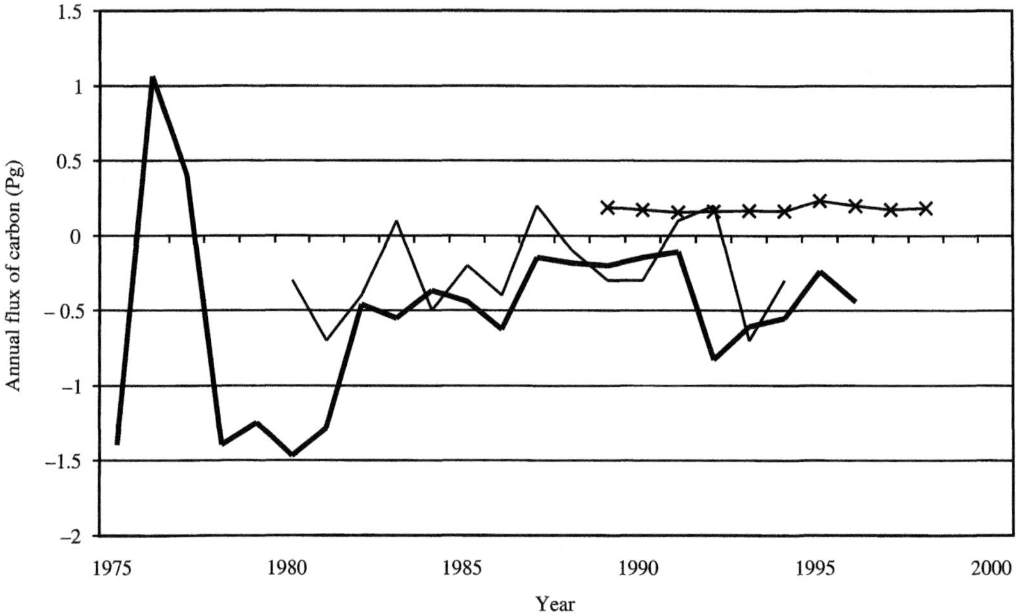

Figure 8 Net annual sources (+) and sinks (−) of carbon for the Brazilian Amazon, as determined by three different methods: (×) land-use change (Houghton *et al.*, 2000) (reproduced by permission of the American Geophysical Union from *J. Geophys. Res.*, **2000**, *105*, 20121–20130); (**—**) tree growth (Phillips *et al.*, 1998); and (—) modeled ecosystem metabolism (Tian *et al.*, 1998).

atmospheric CO_2 (Houghton, 2000). Long-term changes in climate, as opposed to climatic variability, may eventually lead to long-term changes in carbon storage, but probably not at the rates suggested by short-term experiments.

One further observation is discussed here. Over the last decades the amplitude of the seasonal oscillation of CO_2 concentration increased by ~20% at Mauna Loa, Hawaii, and by ~40% at Point Barrow, Alaska (Keeling *et al.*, 1996a). This winter–summer oscillation in concentrations seems to be largely the result of terrestrial metabolism in northern mid-latitudes. The increase in amplitude suggests that the rate of processing of carbon may be increasing. Increased rates of summer photosynthesis, increased rates of winter respiration, or both would increase the amplitude of the oscillation, but it is difficult to ascertain which has contributed most. Furthermore, the increase in the amplitude does not, by itself, indicate an increasing terrestrial sink. In fact, the increase in amplitude is too large to be attributed to CO_2 fertilization or to a temperature-caused increase in winter respiration (Houghton, 1987; Randerson *et al.*, 1997). It is consistent with the observation that growing seasons have been starting earlier over the last decades (Randerson *et al.*, 1999). The trend has been observed in the temperature data, in decreasing snow cover (Folland and Karl, 2001), and in the satellite record of vegetation activity (Myneni *et al.*, 1997).

Synergies among physiological "mechanisms." The factors influencing carbon storage often interact nonadditively. For example, higher concentrations of CO_2 in air enable plants to acquire the same amount of carbon with a smaller loss of water through their leaves. This increased water-use efficiency reduces the effects of drought. Higher levels of CO_2 may also alleviate other stresses of plants, such as temperature and ozone. The observation that NPP is increased relatively more in "low productivity" years suggests that the indirect effects of CO_2 in ameliorating stress may be more important than the direct effects of CO_2 on photosynthesis (Luo *et al.*, 1999).

Another example of synergistic effects is the observation that the combination of nitrogen fertilizer and elevated CO_2 concentration may have a greater effect on the growth of biomass in a growing forest than the expected additive effect (Oren *et al.*, 2001). The relative increase was greater in a nutritionally poor site. The synergy between nitrogen and CO_2 was different in a grassland, however (Hu *et al.*, 2001). There, elevated CO_2 increased plant uptake of nitrogen, increased NPP, and increased the carbon available for microbes; but it reduced microbial decomposition, presumably because the utilization of nitrogen by plants reduced its availability for microbes. The net effect of the reduced decomposition was an increase in the accumulation of carbon in soil.

Relatively few experiments have included more than one environmental variable at a time. A recent experiment involving combinations of four variables shows the importance of such work.

Shaw *et al.* (2003) exposed an annual grassland community in California to increased temperature, precipitation, nitrogen deposition, and atmospheric CO_2 concentration. Alone, each of the treatments increased NPP in the third year of treatment. Across all multifactor treatments, however, elevated CO_2 decreased the positive effects of the other treatments. That is, elevated CO_2 increased productivity under "poor" growing conditions, but reduced it under favorable growing conditions. The most likely explanation is that some soil nutrient became limiting, either because of increased microbial activity or decreased root allocation (Shaw *et al.*, 2003).

The expense of such multifactor experiments has led scientists to use process-based ecosystem models (see the discussion of "terrestrial carbon models" below) to predict the response of terrestrial ecosystems to future climates. When predicting the effects of CO_2 alone, six global biogeochemical models showed a global terrestrial sink that began in the early part of the twentieth century and increased (with one exception) towards the year 2100 (Cramer *et al.*, 2001). The maximum sink varied from $\sim$4 Pg C yr^{-1} to $\sim$10 Pg C yr^{-1}. Adding changes in climate (predicted by the Hadley Centre) to these models reduced the future sink (with one exception), and in one case reduced the sink to zero near the year 2100.

Terrestrial carbon models. A number of ecosystem models have been developed to calculate gross and net fluxes of carbon from environmentally induced changes in plant or microbial metabolism, such as photosynthesis, plant respiration, decomposition, and heterotrophic respiration (Cramer *et al.*, 2001; McGuire *et al.*, 2001). For example, six global models yielded net terrestrial sinks of carbon ranging between 1.5 Pg C yr^{-1} and 4.0 Pg C yr^{-1} for the year 2000 (Cramer *et al.*, 2001). The differences among models became larger as environmental conditions departed from existing conditions. The magnitude of the terrestrial carbon flux projected for the year 2100 varied between a source of 0.5 Pg C yr^{-1} and a sink of 7 Pg C yr^{-1}. Other physiologically based models, including the effects of climate on plant distribution as well as growth, projected a net source from land as tropical forests were replaced with savannas (White *et al.*, 1999; Cox *et al.*, 2000).

The advantage of such models is that they allow the effects of different mechanisms to be distinguished. However, they may not include all of the important processes affecting changes in carbon stocks. To date, e.g., few process-based terrestrial models have included changes in land use.

Although some processes, such as photosynthesis, are well enough understood for predicting responses to multiple factors, other processes, such as biomass allocation, phenology, and the replacement of one species by another, are not. Even if the physiological mechanisms and their interactions were well understood and incorporated into the models, other nonphysiological factors that affect carbon storage (e.g., fires, storms, insects, and disease) are not considered in the present generation of models. Furthermore, the factors influencing short-term changes in terrestrial carbon storage may not be the ones responsible for long-term changes (Houghton, 2000) (see next section). The variability among model predictions suggests that they are not reliable enough to demonstrate the mechanisms responsible for the current modest terrestrial sink (Cramer *et al.*, 2001; Knorr and Heimann, 2001).

8.10.4.1.2 *Demographic or disturbance mechanisms*

Terrestrial sinks also result from the recovery (growth) of ecosystems disturbed in the past. The processes responsible for regrowth include physiological and metabolic processes, but they also involve higher-order or more integrated processes, such as succession, growth, and aging. Forests accumulate carbon as they grow. Regrowth is initiated either by disturbances or by the planting of trees on open land. Disturbances may be either natural (insects, disease, some fires) or human induced (management and changes in land use, including fire management). Climatic effects—e.g., droughts, storms, or fires—thus affect terrestrial carbon storage not only through physiological or metabolic effects on plant growth and respiration, but also through effects on stand demography and growth.

In some regions of the world—e.g., the US and Europe—past changes in land use are responsible for an existing sink (Houghton *et al.*, 1999; Caspersen *et al.*, 2000; Houghton, 2003). Processes include the accumulation of carbon in forests as a result of fire suppression, the growth of forests on lands abandoned from agriculture, and the growth of forests earlier harvested. In tropical regions carbon accumulates in forests that are in the fallow period of shifting cultivation. All regions, even countries with high rates of deforestation, have sinks of carbon in recovering forests, but often these sinks are offset by large emissions (Table 9). The sinks in tropical regions as a result of logging are nearly the same in magnitude as those outside the tropics.

Sinks of carbon are not limited to forests. Some analyses of the US (Houghton *et al.*, 1999; Pacala *et al.*, 2001) show that a number of processes in nonforest ecosystems may also be responsible for carbon sinks. Processes include the encroachment of woody vegetation into formerly herbaceous ecosystems, the accumulation of carbon in agricultural soils as a result of conservation tillage or other practices, exportation of wood and food, and the riverine export of carbon from land to the sea (Table 12). At least a portion of these last two processes (import/export of food and wood and river export) represents an export of carbon from the US (an apparent sink) but not a global sink because these exports presumably become sources somewhere else (either in ocean waters or in another country).

Which terrestrial mechanisms are important? Until recently, the most common explanations for the residual carbon sink in the 1980s and 1990s were factors that affect the physiology of plants and microbes: CO_2 fertilization, nitrogen deposition, and climatic variability (see Table 13). Several findings have started to shift the explanation to include management practices and disturbances that affect the age structure or demography of ecosystems. For example, the suggestion that CO_2 fertilization may be less important in forests than in short-term greenhouse experiments (Oren *et al.*, 2001) was discussed above. Second, physiological models quantifying the effects of CO_2 fertilization and climate change on the growth of US forests could account for only a small fraction of the carbon accumulation observed in those forests (Schimel *et al.*, 2000). The authors acknowledged that past changes in land use were likely to be important. Third, and most importantly, 98% of recent accumulations of carbon in US forests can be explained on the basis of the age structure of trees without requiring growth enhancement due to CO_2 or nitrogen fertilization (Caspersen *et al.*, 2000). Either the physiological effects of CO_2, nitrogen, and climate have been unimportant or their effects have been offset by unknown influences. Finally, the estimates of sinks in the US (Houghton *et al.*, 1999; Pacala *et al.*, 2001; Table 12) are based, to a large extent, on changes in land use and management, and not on physiological models of plant and soil metabolism.

To date, investigations of these two different classes of mechanisms have been largely independent. The effects of changing environmental conditions have been largely ignored in analyses of land-use change (see Section 8.10.3.1.5), and physiological models have generally ignored changes in land use (see Section 8.10.4.1.1).

As of early 2000s, the importance of different mechanisms in explaining known terrestrial carbon sinks remains unclear. Management and past disturbances seem to be the dominant mechanisms for a sink in mid-latitudes, but they are unlikely to explain a large carbon sink in the tropics (if one exists). Recovery from past disturbances is unlikely to explain a large carbon sink in the tropics, because both the area of forests and the stocks of carbon within forests have been declining. Rates of human-caused disturbance have been accelerating. Clearly there are tropical forests recovering from natural disturbances, but there is no evidence that the frequency of disturbances changed during the last century, and thus no evidence to suggest that the sink in recovering forests is larger or smaller today than in previous centuries. The lack of systematic forest inventories over large areas in the tropics precludes a more definitive test of where forests are accumulating carbon and where they are losing it.

Enhanced rates of plant growth cannot be ruled out as an explanation for apparent sinks in either the tropics or mid-latitude lands, but it is possible that the current sink is entirely the result of recovery from earlier disturbances, anthropogenic and natural.

How will the magnitude of the current terrestrial sink change in the future? Identifying the mechanisms responsible for past and current carbon sinks is important because some mechanisms are more likely than others to persist into the future. As discussed above, physiologically based models predict that CO_2 fertilization will increase the global terrestrial sink over the next 100 years (Cramer *et al.*, 2001). Including the effects of projected climate change reduces the magnitude of projected sinks in many models but turns the current sink into a future global source in models that include the longer-term effects of climate on plant distribution (White *et al.*, 1999; Cox *et al.*, 2000). Thus, although increased levels of CO_2 are thought to increase carbon storage in forests, the effect of warmer temperatures may replace forests with savannas and grasslands, and, in the process, release carbon to the atmosphere. Future changes in natural systems are difficult to predict.

To the extent the current terrestrial sink is a result of regrowth (changes in age structure), the future terrestrial sink is more constrained. First, the net effect of continued land-use change is likely to release carbon, rather than store it. Second, forests that might have accumulated carbon in recent decades (whatever the cause) will cease to function as sinks if they are turned into croplands. Third, the current sink in regrowing forests will diminish as forests mature (Hurtt *et al.*, 2002).

Despite the recent evidence that changes in land use are more important in explaining the current terrestrial carbon sink than physiological responses to environmental changes in CO_2, nitrogen, or climate, most projections of future rates of climatic change are based on the assumption that the current terrestrial sink will not only continue, but grow in proportion to concentrations of CO_2. Positive biotic feedbacks and changes in land use are not included in the general circulation models (GCMs) used to predict future rates of climate change. The GCMs include physical feedbacks such as water vapor, clouds, snow, and polar ice, but not biotic feedbacks (Woodwell and Mackenzie, 1995). Thus, unless negative feedbacks in the biosphere become more important in the future, through physiological or other processes, these climate projections underestimate the rate and extent of climatic change. If the terrestrial sink were to diminish in the next decades, concentrations of CO_2 by the year 2100 might be hundreds of ppm higher than commonly projected.

8.10.4.2 Oceanic Mechanisms

8.10.4.2.1 Physical and chemical mechanisms

Increasing the concentration of CO_2 in the atmosphere is expected to affect the rate of oceanic uptake of carbon through at least eight mechanisms, half of them physical or chemical, and half of them biological. Most of the mechanisms reduce the short-term uptake of carbon by the oceans.

The buffer factor. The oceanic buffer factor (or Revelle factor), by which the concentration of CO_2 in the atmosphere is determined, increases as the concentration of CO_2 increases. The buffer factor is discussed above in Section 8.10.3.1.3. Here, it is sufficient to describe the chemical equation for the dissolution of CO_2 in seawater.

$$2HCO_3 + Ca \rightleftharpoons CaCO_3 + CO_2 + H_2O$$

Every molecule of CO_2 entering the oceans consumes a molecule of carbonate as the CO_2 is converted to bicarbonate. Thus, as CO_2 enters the ocean, the concentration of carbonate ions decreases, and further additions of CO_2 remain as dissolved CO_2 rather than being converted to HCO_3^-. The ocean becomes less effective in taking up additional CO_2. The effect is large. The change in DIC for a 100 ppm increase above 280 ppm (pre-industrial) was 40% larger than a 100 ppm increase would be today. The change in DIC for a 100 ppm increase above 750 ppm will be 60% lower than it would be today (Prentice *et al.*, 2001). Thus, the fraction of added CO_2 going into the ocean decreases and the fraction remaining in the atmosphere increases as concentrations continue to increase.

Warming. The solubility of CO_2 in seawater decreases with temperature. Raising the ocean temperature 1 °C increases the equilibrium p_{CO_2} in seawater by 10–20 ppm, thus increasing the atmospheric concentration by that much as well. This mechanism is a positive feedback to a global warming.

Vertical mixing and stratification. If the warming of the oceans takes place in the surface layers first, the warming would be expected to increase the stability of the water column. As discussed in Section 8.10.3.1.3, the bottleneck for oceanic uptake of CO_2 is largely the rate at which the surface oceans exchange CO_2 with the intermediate and deeper waters. Greater stability of the water column, as a result of warming, might constrict this bottleneck further. Similarly, if the warming of the Earth's surface is greater at the poles than at the equator, the latitudinal gradient in surface ocean temperature will be reduced; and because that thermal gradient plays a role in the intensity of atmospheric mixing, a smaller gradient might be expected to subdue mixing and increase stagnation. Alternatively, the increased intensity of the hydrologic cycle expected for a warmer Earth will probably increase the intensity of storms and might, thereby, increase oceanic mixing. Interactions between oceanic stability and biological production might also change the ocean's carbon cycle, with consequences for the oceanic uptake of carbon that are difficult to predict (Sarmiento *et al.*, 1998; Matear and Hirst, 1999).

One aspect of the ocean's circulation that seems particularly vulnerable to climate change is the thermohaline circulation, which is related to the formation of North Atlantic Deep Water (NADW). Increased warming of surface waters may intensify the hydrologic cycle, leading to a reduced salinity in the sea surface at high latitudes, a reduction (even collapse) of NADW formation, reduction in the surface-to-deep transport of anthropogenic carbon, and thus a higher rate of CO_2 growth in the atmosphere. In a model simulation, modest rates of warming reduced the rate of oceanic uptake of carbon, but the reduced uptake was largely compensated by changes in the marine biological cycle (Joos *et al.*, 1999a). For higher rates of global warming, however, the NADW formation collapsed and the concentration of CO_2 in the atmosphere was 22% (and global temperature 0.6 °C) higher than expected in the absence of this feedback.

Rate of CO_2 emission. High rates of CO_2 emissions will increase the atmosphere–ocean gradient in CO_2 concentrations. Although this gradient drives the uptake of carbon by surface waters, if the rate of CO_2 emissions is greater than the rate of CO_2 uptake, the fraction of emitted CO_2 remaining in the atmosphere will be higher. Under the business-as-usual scenario for future CO_2 emissions, rates of emissions increase by more than a factor of 3, from approximately 6 Pg C yr^{-1} in the 1990s to 20 Pg C yr^{-1} by the end of the twenty-first century.

8.10.4.2.2 Biological feedback/processes

Changes in biological processes may offset some of the physical and chemical effects described above (Sarmiento *et al.*, 1998; Joos *et al.*, 1999a), but the understanding of these processes is incomplete, and the net effects far from predictable. Potential effects fall into four categories (Falkowski *et al.*, 1998).

(i) Addition of nutrients limiting primary production. Nutrient enrichment experiments and observations of nutrient distributions throughout the oceans suggest that marine primary productivity is often limited by the availability of fixed inorganic nitrogen. As most of the nitrogen for marine production comes from upwelling, physical changes in ocean circulation might also affect oceanic primary production and, hence, the biological pump. Some nitrogen is made available through nitrogen fixation, however, and some is lost through denitrification, both of which are biological processes, limited by trace nutrients and the concentration of oxygen. The two processes are not coupled, however, and differential changes in either one would affect the inventory of fixed nitrogen in the ocean.

(ii) Enhanced utilization of nutrients. One of the mysteries of ocean biology today is the observation of "high nutrient, low chlorophyll (HNLC) regions." That is, why does primary production in major regions of the surface ocean stop before all of the available nitrogen and phosphorous have been used up? It is possible that grazing pressures keep phytoplankton populations from consuming the available nitrogen and phosphorous, and any reduction in grazing pressures might increase the export of organic matter from the surface. Another possibility that has received considerable attention is that iron may limit production (Martin, 1990). In fact, deliberate iron fertilization of the ocean has received serious attention as a way of reducing atmospheric CO_2 (see Section 8.10.5.2, below). Iron might also become more available naturally as a result of increased human eutrophication of coastal waters, or it might be less available as a result of a warmer (more strongly stratified) ocean or reduced transport of dust (Falkowski *et al.*, 1998). The aeolian transport of iron in dust is a major source of iron for the open ocean, and dust could either increase or decrease in the future, depending on changes in the distribution of precipitation.

(iii) Changes in the elemental ratios of organic matter in the ocean. The elemental ratio of C:N:P in marine organic particles has long been recognized as conservative (Falkowski *et al.*, 1998). The extent to which the ratios can depart from observed concentrations is not known, yet variations could reduce the limitation of nitrogen and thus act in the same manner as the addition of nitrogen in affecting production, export, and thus oceanic uptake of CO_2.

(iv) Increases in the organic carbon/carbonate ratio of export production. The biological and carbonate pumps are described above (Section 8.10.2.2.2). Both pumps transport carbon out of the surface waters, and the subsequent decay at depth is responsible for the higher concentration of carbon in the intermediate and deep ocean. The formation of carbonate shells in the surface waters has the additional effect of increasing the p_{CO_2} in these waters, thus negating the export of the carbonate shells out of the surface. Any increase in the organic carbon/ carbonate ratio of export production would enhance the efficiency of the biological pump.

8.10.5 THE FUTURE: DELIBERATE SEQUESTERING OF CARBON (OR REDUCTION OF SOURCES)

Section 8.10.4 addressed the factors thought to be influencing current terrestrial and oceanic sinks, and how they might change in the future. It is possible, of course, that CO_2 fertilization will become more important in the future as concentrations of CO_2 increase. Multiyear, whole ecosystems experiments with elevated CO_2 do not uniformly support this possibility, but higher concentrations of CO_2, together with nitrogen deposition or increases in moisture, might yet be important. Rather than wait for a more definitive answer, a more cautious approach to the future, besides reducing emissions of CO_2, would consider strategies for withdrawing carbon from the atmosphere through management. Three general options for sequestering carbon have received attention: terrestrial, oceanic, and geological management.

8.10.5.1 Terrestrial

Even if CO_2 fertilization and other environment effects turn out to be unimportant in enhancing terrestrial carbon storage, terrestrial sinks can still be counted on to offset carbon emissions or to reduce atmospheric concentrations of CO_2. Increasing the amount of carbon held on land might be achieved through at least six management options (Houghton, 1996; Kohlmaier *et al.*, 1998): (i) a reduction in the rate of deforestation (current rates of deforestation in the tropics are responsible for an annual release of 1–2 Pg C (Section 8.10.3.2.3); (ii) an increase in the area of forests (afforestation); (iii) an increase in the stocks of carbon within existing forests; (iv) an increase in the use of wood (including increased efficiency of wood harvest and use); (v) the substitution of wood fuels for fossil fuels; and (vi) the substitution of wood for more energy-intensive materials, such as aluminum, concrete, and steel. Estimates of the amount of carbon that might be sequestered on land over the 55-year period 1995–2050 range between 60 Pg C and 87 Pg C (1–2 Pg C yr^{-1} on average)

(Brown, 1996). Additional carbon might also be sequestered in agricultural soils through conservation tillage and other agricultural management practices, and in grassland soils (Sampson and Scholes, 2000). An optimistic assessment, considering all types of ecosystems over the Earth, estimated a potential for storing 5–10 Pg C yr^{-1} over a period of 25–50 years (DOE, 1999).

The amount of carbon potentially sequestered is small relative to projected emissions of CO_2 from business-as-usual energy practices, and thus the terrestrial options for sequestering carbon should be viewed as temporary, "buying time" for the development and implementation of longer-lasting measures for reducing fossil fuel emissions (Watson *et al.*, 2000).

8.10.5.2 Oceanic

Schemes for increasing the storage of carbon in the oceans include stimulation of primary production with iron fertilization and direct injection of CO_2 at depth. As pointed out in Section 8.10.4.2.2., there are large areas of the ocean with high nutrient, low chlorophyll, concentrations. One explanation is that marine production is limited by the micronutrient iron. Adding iron to these regions might thus increase the ocean's biological pump, thereby reducing atmospheric CO_2 (Martin, 1990; Falkowski *et al.*, 1998). Mesoscale fertilization experiments have been carried out (Boyd *et al.*, 2000), but the effects of large-scale iron fertilization of the ocean are not known (Chisholm, 2000).

The direct injection of concentrated CO_2 (probably in liquid form) below the thermocline or on the seafloor might sequester carbon for hundreds of years (Herzog *et al.*, 2000). The gas might be dissolved within the water column or held in solid, ice-like CO_2 hydrates. The possibility is receiving attention in several national and international experiments (DOE, 1999). Large uncertainties exist in understanding the formation and stability of CO_2 hydrates, the effect of the concentrated CO_2 on ocean ecosystems, and the permanence of the sequestration.

8.10.5.3 Geologic

CO_2 may be able to be sequestered in geological formations, such as active and depleted oil and gas reservoirs, coalbeds, and deep saline aquifers. Such formations are widespread and have the potential to sequester large amounts of CO_2 (Herzog *et al.*, 2000). A model project is underway in the North Sea off the coast of Norway. The Sleipner offshore oil and natural gas field contains a gas mixture of natural gas and CO_2 (9%). Because the Norwegian government taxes emissions of CO_2 in excess of 2.5%, companies have the incentive to separate CO_2 from the natural gas and pump it into an aquifer 1,000 m under the sea. Although the potential for sequestering carbon in geological formations is large, technical and economic aspects of an operational program require considerable research.

8.10.6 CONCLUSION

We are conducting a great geochemical experiment, unlike anything in human history and unlikely to be repeated again on Earth. "Within a few centuries we are returning to the atmosphere and oceans the concentrated organic carbon stored in sedimentary rocks over hundreds of millions of years" (Revelle and Suess, 1957). During the last 150 years (~1850–2000), there has been a 30% increase in the amount of carbon in the atmosphere. Although most of this carbon has come from the combustion of fossil fuels, an estimated 150–160 Pg C have been lost during this time from terrestrial ecosystems as a result of human management (another 58–75 Pg C were lost before 1850). The global carbon balance suggests that other terrestrial ecosystems have accumulated ~115 Pg C since about 1930, at a steadily increasing rate. The annual net fluxes of carbon appear small relative to the sizes of the reservoirs, but the fluxes have been accelerating. Fifty percent of the carbon mobilized over the last 300 years (~1700–2000) was mobilized in the last 30–40 of these years (Houghton and Skole, 1990) (Figure 3). The major drivers of the geochemical experiment are reasonably well known. However, the results are uncertain, and there is no control. Furthermore, the experiment would take a long time to stop (or reverse) if the results turned out to be deleterious.

In an attempt to put some bounds on the experiment, in 1992 the nations of the world adopted the United Nations Framework Convention on Climate Change, which has as its objective "stabilization of greenhouse gas concentrations in the atmosphere at a level that would prevent dangerous anthropogenic interference with the climate system" (UNFCCC, 1992). The Convention's soft commitment suggested that the emissions of greenhouse gases from industrial nations in 2000 be no higher than the emissions in 1990. This commitment has been achieved, although more by accident than as a result of deliberate changes in policy. The "stabilization" resulted from reduced emissions from Russia, as a result of economic downturn, balanced by increased emissions almost everywhere else. In the US, e.g., emissions were 18% higher in the year 2000 than they had been in 1990. The near-zero increase in industrial nations' emissions between 1990 and 2000 does not suggest that the stabilization will last.

Ironically, even if the annual rate of global emissions were to be stabilized, concentrations of the gases would continue to increase. Stabilization of concentrations at early 2000's levels, e.g., would require reductions of 60% or more in the emission of long-lived gases, such as CO_2. The 5% average reduction in 1990 emissions by 2010, agreed to by the industrialized countries in the Kyoto Protocol (higher than 5% for the participating countries now that the US is no longer participating), falls far short of stabilizing atmospheric concentrations. Such a stabilization will require nothing less than a switch from fossil fuels to renewable forms of energy (solar, wind, hydropower, biomass), a switch that would have salubrious economic, political, security, and health consequences quite apart from limiting climatic change. Nevertheless, the geophysical experiment seems likely to continue for at least the near future, matched by a sociopolitical experiment of similar proportions, dealing with the consequences of either mitigation or not enough mitigation.

REFERENCES

Aber J. D., McDowell W. H., Nadelhoffer K. J., Magill A. H., Berntson G., Kamakea M., McNulty S., Currie W., Rustad L., and Fernandex I. (1998) Nitrogen saturation in temperate forest ecosystems. *BioScience* **48**, 921–934.

Achard F., Eva H. D., Stibig H.-J., Mayaux P., Gallego J., Richards T., and Malingreau J.-P. (2002) Determination of deforestation rates of the world's humid tropical forests. *Science* **297**, 999–1002.

Ajtay G. L., Ketner P., and Duvigneaud P. (1979) Terrestrial primary production and phytomass. In *The Global Carbon Cycle* (eds. B. Bolin, E. T. Degens, S. Kempe, and P. Ketner). Wiley, New York, pp. 129–182.

Andres R. J., Fielding D. J., Marland G., Boden T. A., Kumar N., and Kearney A. T. (1999) Carbon dioxide emissions from fossil-fuel use, 1751–1950. *Tellus* **51B**, 759–765.

Arrhenius S. (1896) On the influence of carbonic acid in the air upon the temperature of the ground. *Phil. Magazine J. Sci.* **41**, 237–276.

Aumont O., Orr J. C., Monfray P., Ludwig W., Amiotte-Suchet P., and Probst J.-L. (2001) Riverine-driven interhemispheric transport of carbon. *Global Biogeochem. Cycles* **15**, 393–405.

Barber V., Juday G. P., and Finney B. (2000) Reduced growth of Alaskan white spruce in the twentieth century from temperature-induced drought stress. *Nature* **405**, 668–673.

Barford C. C., Wofsy S. C., Goulden M. L., Munger J. W., Hammond Pyle E., Urbanski S. P., Hutyr L., Saleska S. R., Fitzjarrald D., and Moore K. (2001) Factors controlling long- and short-term sequestration of atmospheric CO_2 in a mid-latitude forest. *Science* **294**, 1688–1691.

Battle M., Bender M., Tans P. P., White J. W. C., Ellis J. T., Conway T., and Francey R. J. (2000) Global carbon sinks and their variability, inferred from atmospheric O_2 and $\delta^{13}C$. *Science* **287**, 2467–2470.

Behrenfeld M. J., Randerson J. T., McClain C. R., Feldman G. C., Los S. O., Tucker C. J., Falkowski P. G., Field C. B., Frouin R., Esaias W. W., Kolber D. D., and Pollack N. H. (2001) Biospheric primary production during an ENSO transition. *Science* **291**, 2594–2597.

Bergh J., Linder S., Lundmark T., and Elfving B. (1999) The effect of water and nutrient availability on the productivity of Norway spruce in northern and southern Sweden. *Forest Ecol. Manage.* **119**, 51–62.

Bolin B. (ed.) (1981) *Carbon Cycle Modelling*. Wiley, New York.

Bolin B. (1986) How much CO_2 will remain in the atmosphere? In *The Greenhouse Effect, Climatic Change, and Ecosystems* (eds. B. Bolin, B. R. Doos, J. Jager, and R. A. Warrick). Wiley, Chichester, England, pp. 93–155.

Bolin B., Degens E. T., Kempe S., and Ketner P. (eds.) (1979) *The Global Carbon Cycle*. Wiley, New York.

Bolin B., Doos B. R., Jager J., and Warrick R. A. (1986) *The Greenhouse Effect, Climatic Change, and Ecosystems*. Wiley, New York.

Bopp L., Le Quéré C., Heimann M., and Manning A. C. (2002) Climate-induced oceanic oxygen fluxes: implications for the contemporary carbon budget. *Global Biogeochem. Cycles* **16**, doi: 10.1029/2001GB001445.

Bousquet P., Ciais P., Peylin P., Ramonet M., and Monfray P. (1999a) Inverse modeling of annual atmospheric CO_2 sources and sinks: 1. Method and control inversion. *J. Geophys. Res.* **104**, 26161–26178.

Bousquet P., Peylin P., Ciais P., Ramonet M., and Monfray P. (1999b) Inverse modeling of annual atmospheric CO_2 sources and sinks: 2. Sensitivity study. *J. Geophys. Res.* **104**, 26179–26193.

Bousquet P., Peylin P., Ciais P., Le Quérè C., Friedlingstein P., and Tans P. P. (2000) Regional changes in carbon dioxide fluxes of land and oceans since 1980. *Science* **290**, 1342–1346.

Boyd P. W., Watson A. J., Cliff S., Law C. S., Abraham E. R., Trulli T., Murdoch R., Bakker D. C. E., Bowie A. R., Buesseler K. O., Chang H., Charette M., Croot P., Downing K., Frew R., Gall M., Hadfield M., Hall J., Harvey M., Jameson G., Laroche J., Liddicoat M., Ling R., Maldonado M.T., Mckay R.M., Nodder S., Pickmere S., Pridmore R., Rintoul S., Safi K., Sutton P., Strzepek R. T., Tanneberger K., Turner S., Waite A., and Zeldis J. (2000) A mesoscale phytoplankton bloom in the polar Southern Ocean stimulated by iron fertilization. *Nature* **407**, 695–702.

Braun S., Rihm B., Schindler C., and Fluckiger W. (2000) Growth of mature beech in relation to ozone and nitrogen deposition: an epidemiological approach. *Water Air Soil Pollut.* **116**, 356–364.

Broecker W. S. (2001) A Ewing Symposium on the contemporary carbon cycle. *Global Biogeochem. Cycles* **15**, 1031–1032.

Broecker W. S., Takahashi T., Simpson H. H., and Peng T.-H. (1979) Fate of fossil fuel carbon dioxide and the global carbon budget. *Science* **206**, 409–418.

Broecker W. S., Sutherland S., Smethie W., Peng T. H., and Ostlund G. (1995) Oceanic radiocarbon: separation of the natural and bomb components. *Global Biogeochem. Cycles* **9**, 263–288.

Brown S. (1996) Management of forests for mitigation of greenhouse gas emissions. In *Climatic Change 1995. Impacts, Adaptations and Mitigation of Climate Change: Scientific-technical Analyses* (eds. J. T. Houghton, G. J. Jenkins, and J. J. Ephraums). Cambridge University Press, Cambridge, pp. 773–797.

Bryant D. M., Holland E. A., Seastedt T. R., and Walker M. D. (1998) Analysis of litter decomposition in an alpine tundra. *Canadian J. Botany* **76**, 1295–1304.

Callendar G. S. (1938) The artificial production of carbon dioxide and its influence on temperature. *Quarterly J. Roy. Meteorol. Soc.* **64**, 223–240.

Cardon Z. G., Hungate B. A., Cambardella C. A., Chapin F. S., Field C. B., Holland E. A., and Mooney H. A. (2001) Contrasting effects of elevated CO_2 on old and new soil carbon pools. *Soil Biol. Biochem.* **33**, 365–373.

Caspersen J. P., Pacala S. W., Jenkins J. C., Hurtt G. C., Moorcroft P. R., and Birdsey R. A. (2000) Contributions of land-use history to carbon accumulation in US forests. *Science* **290**, 1148–1151.

Chisholm S. W. (2000) Stirring times in the Southern Ocean. *Nature* **407**, 685–687.

Ciais P., Peylin P., and Bousquet P. (2000) Regional biospheric carbon fluxes as inferred from atmospheric CO_2 measurements. *Ecol. Appl.* **10**, 1574–1589.

Clark D. (2002) Are tropical forests an important carbon sink? Reanalysis of the long-term plot data. *Ecol. Appl.* **12**, 3–7.

Conway T. J. and Tans P. P. (1999) Development of the CO_2 latitude gradient in recent decades. *Global Biogeochem. Cycles* **13**, 821–826.

Conway T. J., Tans P. P., Waterman L. S., Thoning K. W., Kitzis D. R., Masarie K. A., and Zhang N. (1994) Evidence for interannual variability of the carbon cycle from the National Oceanic and Atmospheric Administration/Climate Monitoring and Diagnostics Laboratory Global Air Sampling Network. *J. Geophys. Res.* **99**, 22831–22855.

Cooperative Atmospheric Data Integration Project—Carbon Dioxide (1997) *GLOBALVIEW—CO_2*. National Oceanic and Atmospheric Administration, Boulder, CO (CD-ROM).

Cox P. M., Betts R. A., Jones C. D., Spall S. A., and Totterdell I. J. (2000) Acceleration of global warming due to carbon-cycle feedbacks in a coupled climate model. *Nature* **408**, 184–187.

Cramer W., Bondeau A., Woodward F. I., Prentice I. C., Betts R. A., Brovkin V., Cox P. M., Fisher V., Foley J. A., Friend A. D., Kucharik C., Lomas M. R., Ramankutty N., Sitch S., Smith B., White A., and Young Molling C. (2001) Global response of terrestrial ecosystem structure and function to CO_2 and climate change: results from six dynamic global vegetation models. *Global Change Biol.* **7**, 357–373.

Curtis P. S. (1996) A meta-analysis of leaf gas exchange and nitrogen in trees grown under elevated carbon dioxide. *Plant Cell Environ.* **19**, 127–137.

Davidson E. A. (1995) Linkages between carbon and nitrogen cycling and their implications for storage of carbon in terrestrial ecosystems. In *Biotic Feedbacks in the Global Climatic System: Will the Warming Feed the Warming?* (eds. G. M. Woodwell and F. T. Mackenzie). Oxford University Press, New York, pp. 219–230.

Davidson E. A. and Hirsch A. I. (2001) Fertile forest experiments. *Nature* **411**, 431–433.

DeFries R. S., Field C. B., Fung I., Collatz G. J., and Bounoua L. (1999) Combining satellite data and biogeochemical models to estimate global effects of human-induced land cover change on carbon emissions and primary productivity. *Global Biogeochem. Cycles* **13**, 803–815.

DeFries R. S., Houghton R. A., Hansen M. C., Field C. B., Skole D., and Townshend J. (2002) Carbon emissions from tropical deforestation and regrowth based on satellite observations for the 1980s and 90s. *Proc. Natl. Acad. Sci.* **99**, 14256–14261.

Denning A. S., Fung I. Y., and Randall D. A. (1995) Latitudinal gradient of atmospheric CO_2 due to seasonal exchange with land biota. *Nature* **376**, 240–243.

Department of Energy (DOE) (1999) *Carbon Sequestration Research and Development*. National Technical Information Service, Springfield, Virginia (www.ornl.gov/carbon_sequestration/).

Detwiler R. P. and Hall C. A. S. (1988) Tropical forests and the global carbon cycle. *Science* **239**, 42–47.

Drake B. G., Muche M. S., Peresta G., Gonzalez-Meler M. A., and Matamala R. (1996) Acclimation of photosynthesis, respiration and ecosystem carbon flux of a wetland on Chesapeake Bay, Maryland to elevated atmospheric CO_2 concentration. *Plant Soil* **187**, 111–118.

Driscoll C. T., Lawrence G. B., Bulger A. J., Butler T. J., Cronan C. S., Eagar C., Lambert K. F., Likens G. E., Stoddard J. L., and Weathers K. C. (2001) Acidic deposition in the northeastern United States: sources and inputs, ecosystem effects, and management strategies. *BioScience* **51**, 180–198.

Etheridge D. M., Steele L. P., Langenfelds R. L., Francey R. J., Barnola J. M., and Morgan V. I. (1996) Natural and anthropogenic changes in atmospheric CO_2 over the last 1000 years from air in Antarctic ice and firn. *J. Geophys. Res.* **101**, 4115–4128.

Eva H. D., Achard F., Stibig H. J., and Mayaux P. (2003) Response to comment on Achard *et al.* (2002). *Science* **299**, 1015b.

Falkowski P. G., Barber R. T., and Smetacek V. (1998) Biogeochemical controls and feedbacks on ocean primary production. *Science* **281**, 200–206.

Falkowski P., Scholes R. J., Boyle E., Canadell J., Canfield D., Elser J., Gruber N., Hibbard K., Högberg P., Linder S., Mackenzie F. T., Moore B., Pedersen T., Rosenthal Y., Seitzinger S., Smetacek V., and Steffen W. (2000) The global carbon cycle: a test of our knowledge of earth as a system. *Science* **290**, 291–296.

Fan S., Gloor M., Mahlman J., Pacala S., Sarmiento J., Takahashi T., and Tans P. (1998) A large terrestrial carbon sink in North America implied by atmospheric and oceanic CO_2 data and models. *Science* **282**, 442–446.

FAO (2001) *Global Forest Resources Assessment 2000*. Main Report, FAO Forestry Paper 140, Rome.

Fearnside P. M. (2000) Global warming and tropical land-use change: greenhouse gas emissions from biomass burning, decomposition and soils in forest conversion, shifting cultivation and secondary vegetation. *Climat. Change* **46**, 115–158.

Fearnside P. M. and Laurance W. F. (2003) Comment on Achard *et al.* (2002). *Science* **299**, 1015a.

Fenn M. E., Poth M. A., Aber J. D., Baron J. S., Bormann B. T., Johnson D. W., Lemly A. D., McNulty S. G., Ryan D. F., and Stottlemyer R. (1998) Nitrogen excess in North American ecosystems: predisposing factors, ecosystem responses and management strategies. *Ecol. Appl.* **8**, 706–733.

Field C. B., Behrenfeld M. J., Randerson J. T., and Falkowski P. G. (1998) Primary production of the biosphere: integrating terrestrial and oceanic components. *Science* **281**, 237–240.

Fog K. (1988) The effect of added nitrogen on the rate of decomposition of organic matter. *Biol. Rev. Cambridge Phil. Soc.* **63**, 433–462.

Folland C. K. and Karl T. R. (2001) Observed climate variability and change. In *Climate Change 2001: The Scientific Basis. Contribution of Working Group I to the 3rd Assessment Report of the Intergovernmental Panel on Climate Change* (eds. J. T. Houghton, Y. Ding, D. J. Griggs, M. Noguer, P. J. van der Linden, X. Dai, K. Maskell, and C. A. Johnson). Cambridge University Press, Cambridge, UK and New York, pp. 99–181.

Fraser P. J., Elliott W. P., and Waterman L. S. (1986) Atmospheric CO_2 record from direct chemical measurements during the 19th century. In *The Changing Carbon Cycle. A Global Analysis* (eds. J. R. Trabalka and D. E. Reichle). Springer, New York, pp. 66–88.

Gaudinski J. B., Trumbore S. E., Davidson E. A., and Zheng S. (2000) Soil carbon cycling in a temperate forest: radiocarbon-based estimates of residence times, sequestration rates and partitioning of fluxes. *Biogeochemistry* **51**, 33–69.

Goldemberg J., Johansson T. B., Reddy A. K. N., and Williams R. H. (1985) An end-use oriented global energy strategy. *Ann. Rev. Energy* **10**, 613–688.

Goodale C. L., Apps M. J., Birdsey R. A., Field C. B., Heath L. S., Houghton R. A., Jenkins J. C., Kohlmaier G. H., Kurz W., Liu S., Nabuurs G.-J., Nilsson S., and Shvidenko A. Z. (2002) Forest carbon sinks in the northern hemisphere. *Ecol. Appl.* **12**, 891–899.

Goulden M. L., Munger J. W., Fan S.-M., Daube B. C., and Wofsy S. C. (1996) Exchange of carbon dioxide by a deciduous forest: response to interannual climate variability. *Science* **271**, 1576–1578.

Goulden M. L., Wofsy S. C., Harden J. W., Trumbore S. E., Crill P. M., Gower S. T., Fries T., Daube B. C., Fau S., Sulton D. J., Bazzaz A., and Munger J. W. (1998) Sensitivity of boreal forest carbon balance to soil thaw. *Science* **279**, 214–217.

Grace J., Lloyd J., McIntyre J., Miranda A. C., Meir P., Miranda H. S., Nobre C., Moncrieff J., Massheder J., Malhi Y., Wright I., and Gash J. (1995) Carbon dioxide uptake by an undisturbed tropical rain forest in southwest Amazonia, 1992 to 1993. *Science* **270**, 778–780.

Gruber N., Sarmiento J. L., and Stocker T. F. (1996) An improved method for detecting anthropogenic CO_2 in the oceans. *Global Biogeochem. Cycles* **10**, 809–837.

Gurney K. R., Law R. M., Denning A. S., Rayner P. J., Baker D., Bousquet P., Bruhwiler L., Chen Y.-H., Ciais P., Fan S., Fung I. Y., Gloor M., Heimann M., Higuchi K., John J., Maki T., Maksyutov S., Masarie K., Peylin P., Prather M., Pak B. C., Randerson J., Sarmiento J., Taguchi S., Takahashi T., and Yuen C.-W. (2002) Towards robust regional estimates of CO_2 sources and sinks using atmospheric transport models. *Nature* **415**, 626–630.

Hall C. A. S. and Uhlig J. (1991) Refining estimates of carbon released from tropical land-use change. *Canadian J. Forest Res.* **21**, 118–131.

Heath L. S. and Smith J. E. (2000) An assessment of uncertainty in forest carbon budget projections. *Environ. Sci. Policy* **3**, 73–82.

Herzog H., Eliasson B., and Kaarstad O. (2000) Capturing greenhouse gases. *Sci. Am.* **282**(2), 72–79.

Hicke J. A., Asner G. P., Randerson J. T., Tucker C., Los S., Birdsey R., Jenkins J. C., Field C., and Holland E. (2002) Satellite-derived increases in net primary productivity across North America, 1982–1998. *Geophys. Res. Lett.*, 10.1029/2001GL013578.

Holland E. A., Braswell B. H., Lamarque J.-F., Townsend A., Sulzman J., Muller J.-F., Dentener F., Brasseur G., Levy H., Penner J. E., and Roelots G.-J. (1997) Variations in the predicted spatial distribution of atmospheric nitrogen deposition and their impact on carbon uptake by terrestrial ecosystems. *J. Geophys. Res.* **102**, 15849–15866.

Houghton R. A. (1987) Biotic changes consistent with the increased seasonal amplitude of atmospheric CO_2 concentrations. *J. Geophys. Res.* **92**, 4223–4230.

Houghton R. A. (1996) Converting terrestrial ecosystems from sources to sinks of carbon. *Ambio* **25**, 267–272.

Houghton R. A. (1999) The annual net flux of carbon to the atmosphere from changes in land use 1850–1990. *Tellus* **51B**, 298–313.

Houghton R. A. (2000) Interannual variability in the global carbon cycle. *J. Geophys. Res.* **105**, 20121–20130.

Houghton R. A. (2003) Revised estimates of the annual net flux of carbon to the atmosphere from changes in land use and land management 1850–2000. *Tellus* **55B**, 378–390.

Houghton R. A. and Hackler J. L. (1995) *Continental Scale Estimates of the Biotic Carbon Flux from Land Cover Change: 1850–1980.* ORNL/CDIAC-79, NDP-050, Oak Ridge National Laboratory, Oak Ridge, TN, 144pp.

Houghton R. A. and Hackler J. L. (2001) *Carbon Flux to the Atmosphere from Land-use Changes: 1850–1990.* ORNL/CDIAC-131, NDP-050/R1, US Department of Energy, Oak Ridge National Laboratory, Carbon Dioxide Information Analysis Center, Oak Ridge, TN.

Houghton R. A. and Skole D. L. (1990) Carbon. In *The Earth as Transformed by Human Action* (eds. B. L. Turner, W. C. Clark, R. W. Kates, J. F. Richards, J. T. Mathews, and W. B. Meyer). Cambridge University Press, Cambridge, pp. 393–408.

Houghton R. A., Hobbie J. E., Melillo J. M., Moore B., Peterson B. J., Shaver G. R., and Woodwell G. M. (1983) Changes in the carbon content of terrestrial biota and soils between 1860 and 1980: a net release of CO_2 to the atmosphere. *Ecol. Monogr.* **53**, 235–262.

Houghton R. A., Davidson E. A., and Woodwell G. M. (1998) Missing sinks, feedbacks, and understanding the role of terrestrial ecosystems in the global carbon balance. *Global Biogeochem. Cycles* **12**, 25–34.

Houghton R. A., Hackler J. L., and Lawrence K. T. (1999) The US carbon budget: contributions from land-use change. *Science* **285**, 574–578.

Houghton R. A., Skole D. L., Nobre C. A., Hackler J. L., Lawrence K. T., and Chomentowski W. H. (2000) Annual fluxes of carbon from deforestation and regrowth in the Brazilian Amazon. *Nature* **403**, 301–304.

House J. I., Prentice I. C., Ramankutty N., Houghton R. A., and Heimann M. (2003) Reconciling apparent inconsistencies in estimates of terrestrial CO_2 sources and sinks. *Tellus* **(55B)**, 345–363.

Hu S., Chapin F. S., Firestone M. K., Field C. B., and Chiariello N. R. (2001) Nitrogen limitation of microbial decomposition in a grassland under elevated CO_2. *Nature* **409**, 188–191.

Hurtt G. C., Pacala S. W., Moorcroft P. R., Caspersen J., Shevliakova E., Houghton R. A., and Moore B., III. (2002) Projecting the future of the US carbon sink. *Proc. Natl. Acad. Sci.* **99**, 1389–1394.

Jarvis P. and Linder S. (2000) Constraints to growth of boreal forests. *Nature* **405**, 904–905.

Jobbágy E. G. and Jackson R. B. (2000) The vertical distribution of soil organic carbon and its relation to climate and vegetation. *Ecol. Appl.* **10**, 423–436.

Joos F., Plattner G.-K., Stocker T. F., Marchal O., and Schmittner A. (1999a) Global warming and marine carbon cycle feedbacks on future atmospheric CO_2. *Science* **284**, 464–467.

Joos F., Meyer R., Bruno M., and Leuenberger M. (1999b) The variability in the carbon sinks as reconstructed for the last 1000 years. *Geophys. Res. Lett.* **26**, 1437–1440.

Joos F., Prentice I. C., and House J. I. (2002) Growth enhancement due to global atmospheric change as predicted by terrestrial ecosystem models: consistent with US forest inventory data. *Global Change Biol.* **8**, 299–303.

Keeling C. D. (1973) Industrial production of carbon dioxide from fossil fuels and limestone. *Tellus* **25**, 174–198.

Keeling C. D., Bacastow R. B., Carter A. F., Piper S. C., Whorf T. P., Heimann M., Mook W. G., and Roeloffzen H. (1989) A three-dimensional model of atmospheric CO_2 transport based on observed winds: 1. Analysis of observational data. In *Aspects of Climate Variability in the Pacific and the Western Americas.* Geophysical Monograph 55 (ed. D. H. Peterson). American Geophysical Union, Washington, DC, pp. 165–236.

Keeling C. D., Chin J. F. S., and Whorf T. P. (1996a) Increased activity of northern vegetation inferred from atmospheric CO_2 observations. *Nature* **382**, 146–149.

Keeling C. D., Piper S. C., Bacastow R. B., Wahlen M., Whorf T. P., Heimann M., and Meijer H. A. (2001) *Exchanges of Atmospheric CO_2 and $^{13}CO_2$ with the Terrestrial Biosphere and Oceans from 1978 to 2000: I. Global Aspects.* Scripps Institution of Oceanography, Technical Report SIO Reference Series, No. 01-06 (Revised from SIO Reference Series, No. 00-21), San Diego.

Keeling R. F. and Garcia H. (2002) The change in oceanic O_2 inventory associated with recent global warming. *Proc. US Natl. Acad. Sci.* **99**, 7848–7853.

Keeling R. F. and Shertz S. R. (1992) Seasonal and interannual variations in atmospheric oxygen and implications for the global carbon cycle. *Nature* **358**, 723–727.

Keeling R. F., Piper S. C., and Heimann M. (1996b) Global and hemispheric CO_2 sinks deduced from changes in atmospheric O_2 concentration. *Nature* **381**, 218–221.

Kempe S. (1979) Carbon in the rock cycle. In *The Global Carbon Cycle* (eds. B. Bolin, E. T. Degens, S. Kempe, and P. Ketner). Wiley, New York, pp. 343–377.

Knorr W. and Heimann M. (2001) Uncertainties in global terrestrial biosphere modeling: 1. A comprehensive sensitivity analysis with a new photosynthesis and energy balance scheme. *Global Biogeochem. Cycles* **15**, 207–225.

Koch G. W. and Mooney H. A. (1996) Response of terrestrial ecosystems to elevated CO_2: a synthesis and summary. In *Carbon Dioxide and Terrestrial Ecosystems* (eds. G. W. Koch and H. A. Mooney). Academic Press, San Diego, pp. 415–429.

Köhl M. and Päivinen R. (1997). *Study on European Forestry Information and Communication System.* Office for Official Publications of the European Communities, Luxembourg, Volumes 1 and 2, 1328pp.

Kohlmaier G. H., Weber M., and Houghton R. A. (eds.) (1998) *Carbon Dioxide Mitigation in Forestry and Wood Industry.* Springer, Berlin.

Körner C. (2000) Biosphere responses to CO_2-enrichment. *Ecol. Appl.* **10**, 1590–1619.

Kurz W. A. and Apps M. J. (1999) A 70-year retrospective analysis of carbon fluxes in the Canadian forest sector. *Ecol. Appl.* **9**, 526–547.

Laws E. A., Falkowski P. G., Smith W. O., Ducklow H., and McCarthy J. J. (2000) Temperature effects on export production in the open ocean. *Global Biogeochem. Cycles* **14**, 1231–1246.

Lemon E. (1977) The land's response to more carbon dioxide. In *The Fate of Fossil Fuel CO_2 in the Oceans* (eds. N. R. Andersen and A. Malahoff). Plenum Press, New York, pp. 97–130.

Levitus S., Antonov J. I., Boyer T. P., and Stephens C. (2000) Warming of the world ocean. *Science* **287**, 2225–2229.

Lloyd A. H. and Fastie C. L. (2002) Spatial and temporal variability in the growth and climate response of treeline trees in Alaska. *Climatic Change* **52**, 481–509.

Longhurst A., Sathyendranath S., Platt T., and Caverhill C. (1995) An estimate of global primary production in the ocean from satellite radiometer data. *J. Plankton Res.* **17**, 1245–1271.

Luo Y. Q., Reynolds J., and Wang Y. P. (1999) A search for predictive understanding of plant responses to elevated $[CO_2]$. *Global Change Biol.* **5**, 143–156.

Luo Y., Wan S., Hui D., and Wallace L. L. (2001) Acclimatization of soil respiration to warming in a tall grass prairie. *Nature* **413**, 622–625.

Magill A. H., Aber J. D., Hendricks J. J., Bowden R. D., Melillo J. M., and Steudler P. A. (1997) Biogeochemical response of forest ecosystems to simulated chronic nitrogen deposition. *Ecol. Appl.* **7**, 402–415.

Magill A., Aber J., Berntson G., McDowell W., Nadelhoffer K., Melillo J., and Steudler P. (2000) Long-term nitrogen additions and nitrogen saturation in two temperate forests. *Ecosystems* **3**, 238–253.

Makipaa R. (1995) Effect of nitrogen input on carbon accumulation of boreal forest soils and ground vegetation. *Forest Ecol. Manage.* **79**, 217–226.

Malhi Y., Nobre A. D., Grace J., Kruijt B., Pereira M. G. P., Culf A., and Scott S. (1998) Carbon dioxide transfer over a central Amazonian rain forest. *J. Geophys. Res.* **103**, 31593–31612.

Malhi Y., Phillips O., Kruijt B., and Grace J. (2001) The magnitude of the carbon sink in intact tropical forests: results from recent field studies. In *6th International Carbon Dioxide Conference, Extended Abstracts.* Tohoku University, Sendai, Japan, pp. 360–363.

Marland G., Andres R. J., Boden T. A., and Johnston C. (1998) *Global, Regional and National CO_2 Emission Estimates from Fossil Fuel Burning, Cement Production, and Gas Flaring: 1751–1995* (revised January 1998). ORNL/CDIAC NDP-030/R8, http://cdiac.esd.ornl.gov/ndps/ndp030.html

Martin J. H. (1990) Glacial–interglacial CO_2 change: the iron hypothesis. *Paleoceanography* **5**, 1–13.

Masarie K. A. and Tans P. P. (1995) Extension and integration of atmospheric carbon dioxide data into a globally consistent measurement record. *J. Geophys. Res.* **100**, 11593–11610.

Matear R. J. and Hirst A. C. (1999) Climate change feedback on the future oceanic CO_2 uptake. *Tellus* **51B**, 722–733.

McGuire A. D., Sitch S., Clein J. S., Dargaville R., Esser G., Foley J., Heimann M., Joos F., Kaplan J., Kicklighter D. W., Meier R. A., Melillo J. M., Moore B., Prentice I. C., Ramankutty N., Reichenau T., Schloss A., Tian H., Williams L. J., and Wittenberg U. (2001) Carbon balance of the terrestrial biosphere in the twentieth century: Analyses of CO_2, climate and land use effects with four process-based ecosystem models. *Global Biogeochem. Cycles* **15**, 183–206.

Mclaughlin S. and Percy K. (2000) Forest health in North America: some perspectives on actual and potential roles of climate and air pollution. *Water Air Soil Pollut.* **116**, 151–197.

McNeil B. I., Matear R. J., Key R. M., Bullister J. L., and Sarmiento J. L. (2003) Anthropogenic CO_2 uptake by the ocean based on the global chlorofluorocarbon data set. *Science* **299**, 235–239.

McNulty S. G., Aber J. D., and Newman S. D. (1996) Nitrogen saturation in a high elevation spruce-fir stand. *Forest Ecol. Manage.* **84**, 109–121.

Miller S. D., Goulden M. L., Menton M. C., da Rocha H. R., Freitas H. C., Figueira A. M., and Sousa C. A. D. Tower-based and biometry-based measurements of tropical forest carbon balance. *Ecol. Appl.* (in press).

Mooney H. A., Canadell J., Chapin F. S., Ehleringer J., Körner C., McMurtrie R., Parton W. J., Pitelka L., and Schulze E.-D. (1999) Ecosystem physiology responses to global change. In *Implications of Global Change for Natural and Managed Ecosystems: A Synthesis of GCTE and Related Research* (ed. B. H. Walker, W. L. Steffen, J. Canadel, and J. S. I. Ingram). Cambridge University Press, Cambridge, pp. 141–189.

Myneni R. B., Los S. O., and Asrar G. (1995) Potential gross primary productivity of terrestrial vegetation from 1982–1990. *Geophys. Res. Lett.* **22**, 2617–2620.

Myneni R. B., Keeling C. D., Tucker C. J., Asrar G., and Nemani R. R. (1997) Increased plant growth in the northern high latitudes from 1981 to 1991. *Nature* **386**, 698–702.

Nadelhoffer K. J., Emmett B. A., Gundersen P., Kjønaas O. J., Koopmans C. J., Schleppi P., Teitema A., and Wright R. F. (1999) Nitrogen deposition makes a minor contribution to carbon sequestration in temperate forests. *Nature* **398**, 145–148.

National Research Council (NRC) (1983) *Changing Climate.* National Academy Press, Washington, DC.

Neftel A., Moor E., Oeschger H., and Stauffer B. (1985) Evidence from polar ice cores for the increase in atmospheric CO_2 in the past two centuries. *Nature* **315**, 45–47.

Nepstad D. C., Verissimo A., Alencar A., Nobre C., Lima E., Lefebvre P., Schlesinger P., Potter C., Moutinho P., Mendoza E., Cochrane M., and Brooks V. (1999) Large-scale impoverishment of Amazonian forests by logging and fire. *Nature* **398**, 505–508.

Oechel W. C., Hastings S. J., Vourlitis G., Jenkins M., Riechers G., and Grulke N. (1993) Recent change of arctic tundra ecosystems from a net carbon dioxide sink to a source. *Nature* **361**, 520–523.

Oechel W. C., Cowles S., Grulke N., Hastings S. J., Lawrence B., Prudhomme T., Riechers G., Strain B., Tissue D., and Vourlitis G. (1994) Transient nature of CO_2 fertilization in Arctic tundra. *Nature* **371**, 500–503.

Oren R., Ellsworth D. S., Johnsen K. H., Phillips N., Ewers B. E., Maier C., Schäfer K. V. R., McCarthy H., Hendrey G., McNulty S. G., and Katul G. G. (2001) Soil fertility limits carbon sequestration by forest ecosystems in a CO_2-enriched atmosphere. *Nature* **411**, 469–472.

Orr J. C. (2000) OCMIP carbon analysis gets underway. *Research GAIM* **3**(2), 4–5.

Orr J. C. and Dutay J.-C. (1999) OCMIP mid-project workshop. *Res. GAIM Newslett.* **3**, 4–5.

Orr J., Maier-Reimer E., Mikolajewicz U., Monfray P., Sarmiento J. L., Toggweiler J. R., Taylor N. K., Palmer J., Gruber N., Sabine C. L., Le Quéré C., Key R. M., and Boutin J. (2001) Estimates of anthropogenic carbon uptake from four 3-D global ocean models. *Global Biogeochem. Cycles* **15**, 43–60.

Pacala S. W., Hurtt G. C., Baker D., Peylin P., Houghton R. A., Birdsey R. A., Heath L., Sundquist E. T., Stallard R. F.,

Ciais P., Moorcroft P., Caspersen J. P., Shevliakova E., Moore B., Kohlmaier G., Holland E., Gloor M., Harmon M. E., Fan S.-M., Sarmiento J. L., Goodale C. L., Schimel D., and Field C. B. (2001) Consistent land- and atmosphere-based US carbon sink estimates. *Science* **292**, 2316–2320.

Page S. E., Siegert F., Rieley L. O., Boehm H.-D. V., Jaya A., and Limin S. (2002) The amount of carbon released from peat and forest fires in Indonesia during 1997. *Nature* **420**, 61–65.

Peterson B. J. and Melillo J. M. (1985) The potential storage of carbon by eutrophication of the biosphere. *Tellus* **37B**, 117–127.

Phillips D. L., Brown S. L., Schroeder P. E., and Birdsey R. A. (2000) Toward error analysis of large-scale forest carbon budgets. *Global Ecol. Biogeogr.* **9**, 305–313.

Phillips O. L., Malhi Y., Higuchi N., Laurance W. F., Núñez P. V., Vásquez R. M., Laurance S. G., Ferreira L. V., Stern M., Brown S., and Grace J. (1998) Changes in the carbon balance of tropical forests: evidence from land-term plots. *Science* **282**, 439–442.

Phillips O. L., Malhi Y., Vinceti B., Baker T., Lewis S. L., Higuchi N., Laurance W. F., Vargas P. N., Martinez R. V., Laurance S., Ferreira L. V., Stern M., Brown S., and Grace J. (2002) Changes in growth of tropical forests: evaluating potential biases. *Ecol. Appl.* **12**, 576–587.

Plattner G.-K., Joos F., and Stocker T. F. (2002) Revision of the global carbon budget due to changing air-sea oxygen fluxes. *Global Biogeochem. Cycles* **16**(4), 1096, doi:10.1029/2001GB001746.

Post W. M., Emanuel W. R., Zinke P. J., and Stangenberger A. G. (1982) Soil carbon pools and world life zones. *Nature* **298**, 156–159.

Powell D. S., Faulkner J. L., Darr D. R., Zhu Z., and MacCleery D. W. (1993) Forest resources of the US, 1992. General Technical Report RM-234,. USDA Forest Service, Rocky Mountain Forest and Range Experiment Station, Fort Collins, CO.

Prather M. and Ehhalt D. (2001) Atmospheric chemistry and greenhouse gases. In *Climate Change 2001: The Scientific Basis. Contribution of Working Group I to the 3rd Assessment Report of the Intergovernmental Panel on Climate Change* (eds. J. T. Houghton, Y. Ding, D. J. Griggs, M. Noguer, P. J. van der Linden, X. Dai, K. Maskell, and C. A. Johnson). Cambridge University Press, Cambridge, UK and New York, pp. 239–287.

Prentice I. C., Farquhar G. D., Fasham M. J. R., Goulden M. L., Heimann M., Jaramillo V. J., Kheshgi H. S., Le Quéré C., Scholes R. J., and Wallace D. W. R. (2001) The carbon cycle and atmospheric carbon dioxide. In *Climate Change 2001: The Scientific Basis. Contribution of Working Group I to the 3rd Assessment Report of the Intergovernmental Panel on Climate Change* (eds. J. T. Houghton, Y. Ding, D. J. Griggs, M. Noguer, P. J. van der Linden, X. Dai, K. Maskell, and C. A. Johnson). Cambridge University Press, Cambridge, UK and New York, pp. 183–237.

Raich J. W. and Schlesinger W. H. (1992) The global carbon dioxide flux in soil respiration and its relationship to vegetation and climate. *Tellus* **44B**, 81–99.

Randerson J. T., Thompson M. V., Conway T. J., Fung I. Y., and Field C. B. (1997) The contribution of terrestrial sources and sinks to trends in the seasonal cycle of atmospheric carbon dioxide. *Global Biogeochem. Cycles* **11**, 535–560.

Randerson J. T., Field C. B., Fung I. Y., and Tans P. P. (1999) Increases in early season ecosystem uptake explain recent changes in the seasonal cycle of atmospheric CO_2 at high northern latitudes. *Geophys. Res. Lett.* **26**, 2765–2768.

Raynaud D. and Barnola J. M. (1985) An Antarctic ice core reveals atmospheric CO_2 variations over the past few centuries. *Nature* **315**, 309–311.

Revelle R. and Suess H. E. (1957) Carbon dioxide exchange between atmosphere and ocean and the question of an increase of atmospheric CO_2 during the past decades. *Tellus* **9**, 18–27.

Rice A. H., Pyle E. H., Saleska S. R., Hutyra L., de Camargo P. B., Portilho K., Marques D. F., and Wofsy S. C. Carbon balance and vegetation dynamics in an old-growth Amazonian forest. *Ecol. Appl.* (in press).

Richey J. E., Melack J. M., Aufdenkampe A. K., Ballester V. M., and Hess L. L. (2002) Outgassing from Amazonian rivers and wetlands as a large tropical source of atmospheric CO_2. *Nature* **416**, 617–620.

Saleska S. R., Miller S. D., Matross D. M., Goulden M. L., Wofsy S. C., da Rocha H., de Camargo P. B., Crill P. M., Daube B. C., Freitas C., Hutyra L., Keller M., Kirchhoff V., Menton M., Munger J. W., Pyle E. H., Rice A. H., and Silva H. Carbon fluxes in old-growth Amazonian rainforests: unexpected seasonality and disturbance-induced net carbon loss (in press).

Sampson R. N. and Scholes R. J. (2000) Additional human-induced activities—article 3.4. In *Land Use, Land-use Change, and Forestry. A Special Report of the IPCC* (eds. R. T. Watson, I. R. Noble, B. Bolin, N. H. Ravindranath, D. J. Verardo, and D. J. Dokken). Cambridge University Press, New York, pp. 181–281.

Sarmiento J. L. (1993) Ocean carbon cycle. *Chem. Eng. News* **71**, 30–43.

Sarmiento J. L. and Sundquist E. T. (1992) Revised budget for the oceanic uptake of anthropogenic carbon dioxide. *Nature* **356**, 589–593.

Sarmiento J. L., Hughes T. M. C., Stouffer R. J., and Manabe S. (1998) Simulated response of the ocean carbon cycle to anthropogenic climate warming. *Nature* **393**, 245–249.

Saugier B., Roy J., and Mooney H. A. (2001) Estimations of global terrestrial productivity: converging toward a single number? In *Terrestrial Global Productivity* (eds. J. Roy, B. Saugier, and H. A. Mooney). Academic Press, San Diego, California, pp. 543–557.

SCEP (Study of Critical Environmental Problems) (1970) *Man's Impact on the Global Environment.* The MIT Press, Cambridge, Massachusetts.

Schimel D. S., Alves D., Enting I., Heimann M., Joos F., Raynaud D., and Wigley T. (1996) CO_2 and the carbon cycle. In *Climate Change 1995* (eds. J. T. Houghton, L. G. M. Filho, B. A. Callendar, N. Harris, A. Kattenberg, and K. Maskell). Cambridge University Press, Cambridge, pp. 76–86.

Schimel D., Melillo J., Tian H., McGuire A. D., Kicklighter D., Kittel T., Rosenbloom N., Running S., Thornton P., Ojima D., Parton W., Kelly R., Sykes M., Neilson R., and Rizzo B. (2000) Contribution of increasing CO_2 and climate to carbon storage by ecosystems in the United States. *Science* **287**, 2004–2006.

Schimel D. S., House J. I., Hibbard K. A., Bousquet P., Ciais P., Peylin P., Braswell B. H., Apps M. J., Baker D., Bondeau A., Canadell J., Churkina G., Cramer W., Denning A. S., Field C. B., Friedlingstein P., Goodale C., Heimann M., Houghton R. A., Melillo J. M., Moore B., III, Murdiyarso D, Noble I., Pacala S. W., Prentice I. C., Raupach M. R., Rayner P. J., Scholes R. J., Steffen W. L., and Wirth C. (2001) Recent patterns and mechanisms of carbon exchange by terrestrial ecosystems. *Nature* **414**, 169–172.

Schlesinger W. H. and Lichter J. (2001) Limited carbon storage in soil and litter of experimental forest plots under increased atmospheric CO_2. *Nature* **411**, 466–469.

Shaw M. R., Zavaleta E. S., Chiariello N. R., Cleland E. E., Mooney H. A., and Field C. B. (2003) Grassland responses to global environmetal changes suppressed by elevated CO_2. *Science* **298**, 1987–1990.

Shvidenko A. Z. and Nilsson S. (1997) Are the Russian forests disappearing? *Unasylva* **48**, 57–64.

Smith H. J., Fischer H., Wahlen M., Mastroianni D., and Deck B. (1999) *Nature* **400**, 248–250.

Spiecker H., Mielikainen K., Kohl M., Skovsgaard J. (eds.) (1996) *Growth Trends in European Forest—Studies from 12 Countries.* Springer, Berlin.

Stephens B. B., Keeling R. F., Heimann M., Six K. D., Murnane R., and Caldeira K. (1998) Testing global ocean carbon cycle models using measurements of atmospheric O_2 and CO_2 concentration. *Global Biogeochem. Cycles* **12**, 213–230.

Sundquist E. T. (1986) Geologic analogs: their value and limitation in carbon dioxide research. In *The Changing Carbon Cycle. A Global Analysis* (eds. J. R. Trabalka and D. E. Reichle). Springer, New York, pp. 371–402.

Sundquist E. T. and Broecker W. S. (eds.) (1985) *The Carbon Cycle and Atmospheric CO_2: Natural Variations Archean to Present*, Geophysical Monograph 32. American Geophysical Union, Washington, DC.

Tamm C. O., Aronsson A., and Popovic B. (1995) Nitrogen saturation in a long-term forest experiment with annual additions of nitrogen. *Water Air Soil Pollut.* **85**, 1683–1688.

Tans P. P., Fung I. Y., and Takahashi T. (1990) Observational constraints on the global atmospheric CO_2 budget. *Science* **247**, 1431–1438.

Tans P. P., Fung I. Y., and Enting I. G. (1995) Storage versus flux budgets: the terrestrial uptake of CO_2 during the 1980s. In *Biotic Feedbacks in the Global Climatic System.* Will the Warming Feed the Warming (eds. G. M. Woodwell and F. T. Mackenzie). Oxford University Press, New York, pp. 351–366.

Tans P. P., Bakwin P. S., Bruhwiler L., Conway T. J., Dlugokencky E. J., Guenther D. W., Kitzis D. R., Lang P. M., Masarie K. A., Miller J. B., Novelli P. C., Thoning K. W., Vaughn B. H., White J. W. C., and Zhao C. (2001) Carbon cycle. In *Climate Monitoring and Diagnostics Laboratory Summary Report No. 25 1998–1999* (eds. R. C. Schnell, D. B. King, and R. M. Rosson). NOAA, Boulder, CO, pp. 24–46.

Taylor J. A. and Orr J. C. (2000) The natural latitudinal distribution of atmospheric CO_2. *Global Planet. Change* **26**, 375–386.

Tian H., Melillo J. M., Kicklighter D. W., McGuire A. D., Helfrich J. V. K., Moore B., and Vorosmarty C. J. (1998) Effect of interannual climate variability on carbon storage in Amazonian ecosystems. *Nature* **396**, 664–667.

Tissue D. T. and Oechel W. C. (1987) Response of *Eriophorum vaginatum* to elevated CO_2 and temperature in the Alaskan tussock tundra. *Ecology* **68**, 401–410.

Trabalka J. R. (ed.) (1985) *Atmospheric Carbon Dioxide and the Global Carbon Cycle.* DOE/ER-0239, US Department of Energy, Washington, DC.

UNFCCC (1992) *Text of the United Nations Framework Convention on Climate Change* (UNEP/WMO Information Unit on Climate Change.), Geneva, Switzerland, 29pp.

Valentini R., Matteucci G., Dolman A. J., Schulze E.-D., Rebmann C., Moors E. J., Granier A., Gross P., Jensen N. O., Pilegaard K., Lindroth A., Grelle A., Bernhofer C., Grünwald T., Aubinet M., Ceulemans R., Kowalski A. S., Vesala T., Rannik Ü., Berbigier P., Loustau D., Gudmundsson J., Thorgeirsson H., Ibrom A., Morgenstern K., Clement R., Moncrieff J., Montagnani L., Minerbi S., and Jarvis P. G. (2000) Respiration as the main determinant of European forests carbon balance. *Nature* **404**, 861–865.

Wanninkhof R. (1992) Relationship between wind-speed and gas-exchange over the ocean. *J. Geophys. Res.* **97**, 7373–7382.

Watson R. T., Rodhe H., Oeschger H., and Siegenthaler U. (1990) Greenhouse gases and aerosols. In *Climate Change*, The IPCC Scientific Assessment (eds. J. T. Houghton, G. J. Jenkins, and J. J. Ephraums). Cambridge University Press, Cambridge, pp. 1–40.

Watson R. T., Noble I. R., Bolin B., Ravindranath N. H., Verardo D. J., and Dokken D. J. (eds.) (2000) *Land Use, Land-Use Change, and Forestry.* A Special Report of the IPCC, Cambridge University Press, New York.

White A., Cannell M. G. R., and Friend A. D. (1999) Climate change impacts on ecosystems and the terrestrial carbon sink: a new assessment. *Global Environ. Change* **9**, S21–S30.

Woodwell G. M. (1983) Biotic effects on the concentration of atmospheric carbon dioxide: a review and projection. In *Changing Climate.* National Academy Press, Washington, DC, pp. 216–241.

Woodwell G. M. and Mackenzie F. T. (eds.) (1995) *Biotic Feedbacks in the Global Climatic System. Will the Warming Feed the Warming?* Oxford University Press, New York.

Woodwell G. M. and Pecan E. V. (eds.) (1973) *Carbon and the Biosphere*, US Atomic Energy Commission, Symposium Series 30. National Technical Information Service, Springfield, Virginia.

8.11
The Global Oxygen Cycle

S. T. Petsch
University of Massachusetts, Amherst, MA, USA

8.11.1 INTRODUCTION

One of the key defining features of Earth as a planet that houses an active and diverse biology is the presence of free molecular oxygen (O_2) in the atmosphere. Biological, chemical, and physical processes interacting on and beneath the Earth's surface determine the concentration of O_2 and variations in O_2 distribution, both temporal and spatial. In the present-day Earth system, the process that releases O_2 to the atmosphere (photosynthesis) and the processes that consume

O_2 (aerobic respiration, sulfide mineral oxidation, oxidation of reduced volcanic gases) result in large fluxes of O_2 to and from the atmosphere. Even relatively small changes in O_2 production and consumption have the potential to generate large shifts in atmospheric O_2 concentration within geologically short periods of time. Yet all available evidence supports the conclusion that stasis in O_2 variation is a significant feature of the Earth's atmosphere over wide spans of the geologic past. Study of the oxygen cycle is therefore important because, while an equable O_2 atmosphere is central to life as we know it, our understanding of exactly why O_2 concentrations remain nearly constant over large spans of geologic time is very limited.

This chapter begins with a review of distribution of O_2 among various reservoirs on Earth's surface: air, sea, and other natural waters. The key factors that affect the concentration of O_2 in the atmosphere and surface waters are next considered, focusing on photosynthesis as the major process generating free O_2 and various biological and abiotic processes that consume O_2. The chapter ends with a synopsis of current models on the evolution of an oxygenated atmosphere through 4.5 billion years of Earth's history, including geochemical evidence constraining ancient O_2 concentrations and numerical models of atmospheric evolution.

8.11.2 DISTRIBUTION OF O_2 AMONG EARTH SURFACE RESERVOIRS

8.11.2.1 The Atmosphere

The partial pressure of oxygen in the present-day Earth's atmosphere is ~0.21 bar, corresponding to a total mass of $\sim 34 \times 10^{18}$ mol O_2 (0.20946 bar (force/area) multiplied by the surface area of the Earth (5.1×10^{14} m^2), divided by average gravitational acceleration g (9.8 m s^{-2}) and the formula weight for O_2 (32 g mol^{-1}) yields $\sim 34 \times 10^{18}$ mol O_2). There is a nearly uniform mixture of the main atmospheric gases (N_2, O_2, Ar) from the Earth's surface up to ~80 km altitude (including the troposphere, stratosphere, and mesosphere), because turbulent mixing dominates over molecular diffusion at these altitudes. Because atmospheric pressure (and thus gas molecule density) decreases exponentially with altitude, the bulk of molecular oxygen in the atmosphere is concentrated within several kilometers of Earth's surface. Above this, in the thermosphere, gases become separated based on their densities. Molecular oxygen is photodissociated by UV radiation to form atomic oxygen (O), which is the major form of oxygen above ~120 km altitude.

Approximately 21% O_2 in the atmosphere represents an average composition. In spite of well-developed turbulent mixing in the lower atmosphere, seasonal latitudinal variations in O_2 concentration of ±15 ppm have been recorded. These seasonal variations are most pronounced at high latitudes, where seasonal cycles of primary production and respiration are strongest (Keeling and Shertz, 1992). In the northern hemisphere, the seasonal variations are anticorrelated with atmospheric p_{CO_2}; summers are dominated by high O_2 (and high inferred net photosynthesis), while winters are dominated by lower O_2. In addition, there has been a measurable long-term decline in atmospheric O_2 concentration of $\sim 10^{14}$ mol yr^{-1}, attributed to oxidation of fossil fuels. This decrease has been detected in both long-term atmospheric monitoring stations (Keeling and Shertz, 1992, Figure 1(a)) and in atmospheric gases trapped in Antarctic firn ice bubbles (Figure 1(b)). The polar ice core records extend the range of direct monitoring of atmospheric composition to show that a decline in atmospheric O_2 linked to oxidation of fossil fuels has been occurring since the Industrial Revolution (Bender *et al.*, 1994b; Battle *et al.*, 1996).

8.11.2.2 The Oceans

Air-saturated water has a dissolved O_2 concentration dependent on temperature, the Henry's law constant k_H, and ionic strength. In pure water at 0 °C, O_2 saturation is 450 μM; at 25 °C, saturation falls to 270 μM. Other solutes reduce O_2 solubility, such that at normal seawater salinities, O_2 saturation is reduced by ~25%. Seawater is, of course, rarely at perfect O_2 saturation. Active photosynthesis may locally increase O_2 production rates, resulting in supersaturation of O_2 and degassing to the atmosphere. Alternately, aerobic respiration below the sea surface can consume dissolved O_2 and lead to severe O_2-depletion or even anoxia.

Lateral and vertical gradients in dissolved O_2 concentration in seawater reflect balances between O_2 inputs from air–sea gas exchange, biological processes of O_2 production and consumption, and advection of water masses. In general terms, the concentration of O_2 with depth in the open ocean follows the general structures described in Figure 2. Seawater is saturated to supersaturated with O_2 in the surface mixed layer (~0–60 m water depth). Air–sea gas exchange and trapping of bubbles ensures constant dissolution of atmospheric O_2. Because gas solubility is temperature dependent, O_2 concentrations are greater in colder high-latitude surface waters than in waters near the equator. Oxygen concentrations in surface waters also vary strongly with

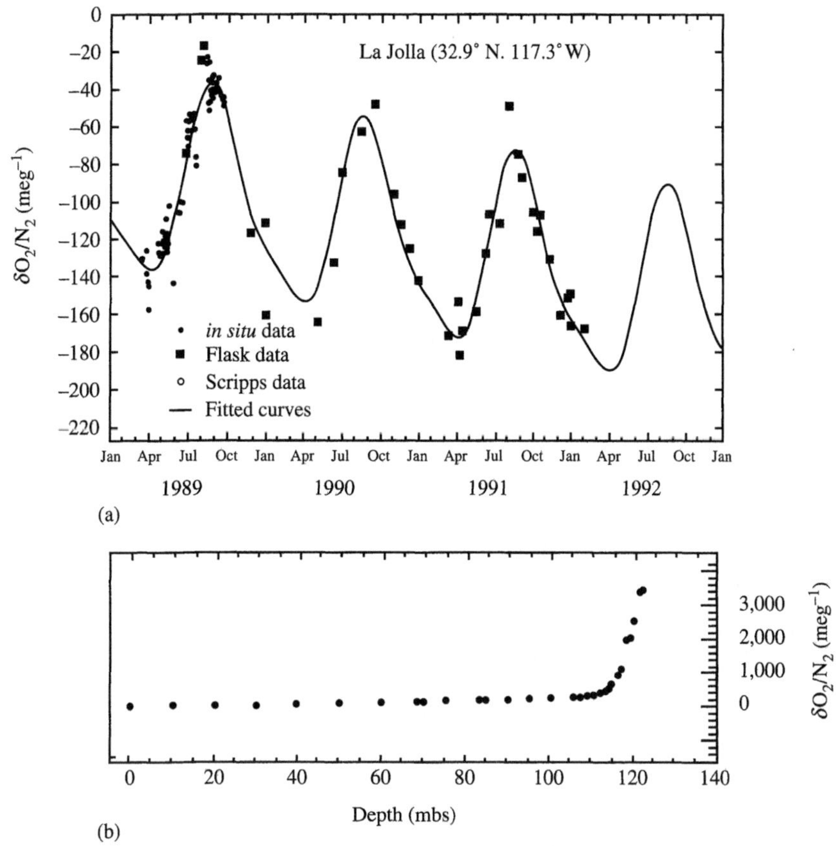

Figure 1 (a) Interannual variability of atmospheric O_2/N_2 ratio, measured at La Jolla, California. 1 ppm O_2 is equivalent to ~4.8 meg^{-1} (source Keeling and Shertz, 1992). (b) Variability in atmospheric O_2/N_2 ratio measured in firn ice at the South Pole, as a function of depth in meters below surface (mbs). The gentle rise in O_2/N_2 between 40 m and 100 m reflects a loss of atmospheric O_2 during the last several centuries due to fossil fuel burning. Deeper than 100 m, selective effusion of O_2 out of closing bubbles into firn air artificially boosts O_2/N_2 ratios (source Battle *et al.*, 1996).

season, especially in high productivity waters. Supersaturation is strongest in spring and summer (time of greatest productivity and strongest water column stratification) when warming of surface layers creates a shallow density gradient that inhibits vertical mixing. Photosynthetic O_2 production exceeds consumption and exchange, and supersaturation can develop. O_2 concentrations drop below the surface mixed layer to form O_2 minimum zones (OMZs) in many ocean basins. O_2 minima form where biological consumption of O_2 exceeds resupply through advection and diffusion. The depth and thickness of O_2 minima vary among ocean basins. In the North Atlantic, the OMZ extends several hundred meters. O_2 concentrations fall from an average of ~300 μM in the surface mixed layer to ~160 μM at 800 m depth. In the North Pacific, however, the O_2 minimum extends deeper, and O_2 concentrations fall to <100 μM. Along the edges of ocean basins, where OMZs impinge on the seafloor, aerobic respiration is restricted, sediments are anoxic at or near the sediment–water interface, and burial of organic matter in sediments may be enhanced. Below the oxygen minima zones in the open ocean, O_2 concentrations gradually increase again from 2000 m to the seafloor. This increase in O_2 results from the slow progress of global thermohaline circulation. Cold, air-saturated seawater sinks to the ocean depths at high-latitudes in the Atlantic, advecting in O_2-rich waters below the O_2 minimum there. Advection of O_2-rich deep water from the Atlantic through the Indian Ocean into the Pacific is the source of O_2 in deep Pacific waters. However, biological utilization of this deep-water O_2 occurs along the entire path from the North Atlantic to the Pacific. For this reason, O_2 concentrations in deep Atlantic water are slightly greater (~200 μM) than in the deep Pacific (~150 μM).

In some regions, dissolved O_2 concentration falls to zero. In these regions, restricted water circulation and ample organic matter supply result in biological utilization of oxygen at a rate that exceeds O_2 resupply through advection and diffusion. Many of these are temporary zones of

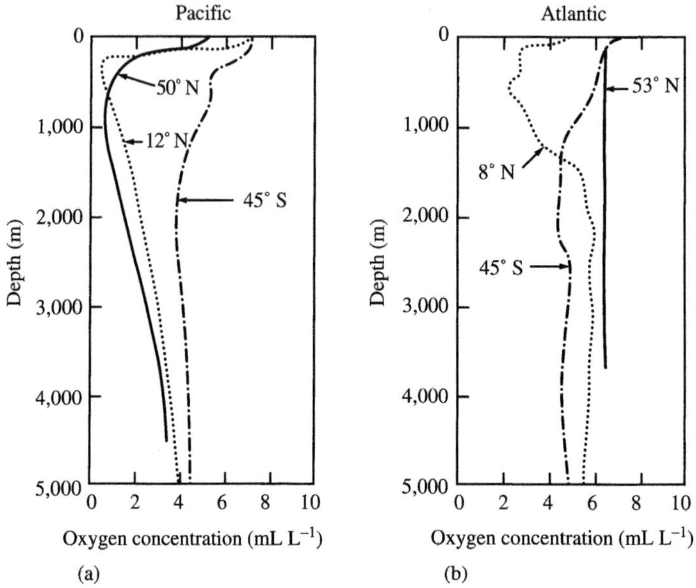

Figure 2 Depth profiles of oxygen concentration dissolved in seawater for several latitudes in the Pacific (a) and Atlantic (b). Broad trends of saturation or supersaturation at the surface, high dissolved oxygen demand at mid-depths, and replenishment of O_2 through lateral advection of recharged deep water are revealed, although regional influences of productivity and intermediate and deep-water heterotropy are also seen (source Ingmanson and Wallace, 1989).

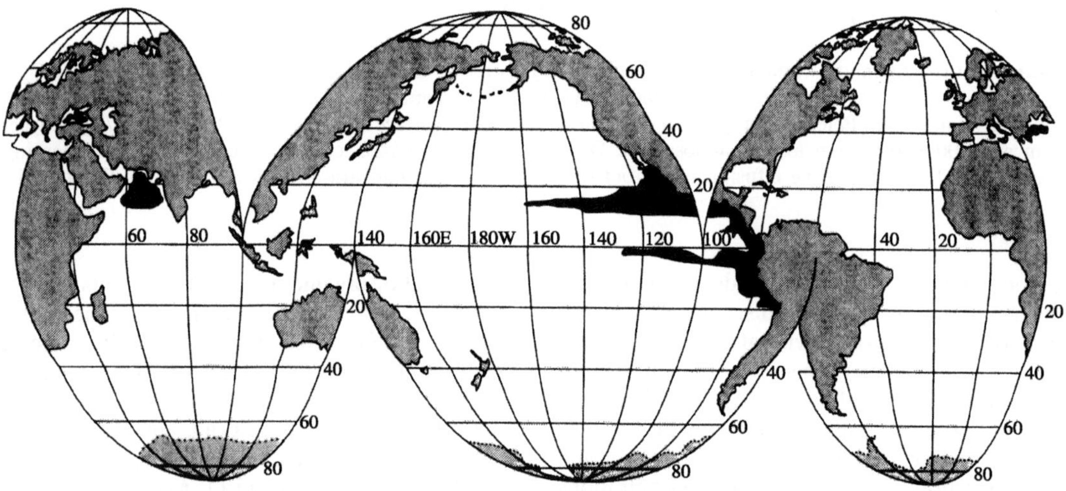

Figure 3 Map detailing locations of extensive and permanent oxygen-deficient intermediate and deep waters (Deuser, 1975) (reproduced by permission of Elsevier from *Chemical Oceanography* **1975**, p.3).

anoxia that form in coastal regions during summer, when warming facilitates greatest water column stratification, and primary production and organic matter supply are high. Such O_2-depletion is now common in the Chesapeake Bay and Schelde estuaries, off the mouth of the Mississippi River and other coastal settings. However, there are several regions of the world oceans where stratification and anoxia are more permanent features (Figure 3). These include narrow, deep, and silled coastal fjords, larger restricted basins (e.g., the Black Sea, Cariaco Basin, and the chain of basins along the southern California Borderlands). Lastly, several regions of the open ocean are also associated with strong O_2-depletion. These regions (the equatorial Pacific along Central and South America and the Arabian Sea) are associated with deep-water upwelling, high rates of surface water primary productivity, and high dissolved oxygen demand in intermediate waters.

Oxygen concentrations in the pore fluids of sediments are controlled by a balance of entrainment of overlying fluids during sediment deposition, diffusive exchange between the sediment and the water column, and biological utilization. In marine sediments, there is a good correlation between the rate of organic matter supply and the depth of O_2 penetration in the sediment (Hartnett *et al.*, 1998). In coastal sediments and on the continental shelf, burial of organic matter is sufficiently rapid to deplete the sediment of oxygen within millimeters to centimeters of the sediment–water interface. In deeper abyssal sediments, where organic matter delivery is greatly reduced, O_2 may penetrate several meters into the sediment before being entirely consumed by respiration.

There is close coupling between surface water and atmospheric O_2 concentrations and air–sea gas exchange fluxes (Figure 4). High rates of marine primary productivity result in net outgassing of O_2 from the oceans to the atmosphere in spring and summer, and net ingassing of O_2 during fall and winter. These patterns of air–sea O_2 transfer relate to latitude and season: outgassing of O_2 during northern hemisphere high productivity months (April through August) are accompanied by simultaneous ingassing in southern latitudes when and where the productivity is lowest (Najjar and Keeling, 2000). Low-latitude ocean surface waters show very little net air–sea O_2 exchange and minimal change in outgassing or ingassing over an annual cycle.

8.11.2.3 Freshwater Environments

Oxygen concentrations in flowing freshwater environments closely match air-saturated values, due to turbulent mixing and entrainment of air bubbles. In static water bodies, however, O_2-depletion can develop much like in the oceans. This is particularly apparent in some ice-covered lakes, where inhibited gas exchange and wintertime respiration can result in O_2-depletion and fish kills. High productivity during spring and summer in shallow turbid aquatic environments can result in extremely sharp gradients from strong O_2 supersaturation at the surface to near O_2-depletion within a few meters of the surface. The high concentration of labile dissolved and particulate organic matter in many freshwater environments leads to rapid O_2-depletion where advective resupply is limited. High rates of O_2 consumption have been measured in many temperature and tropical rivers.

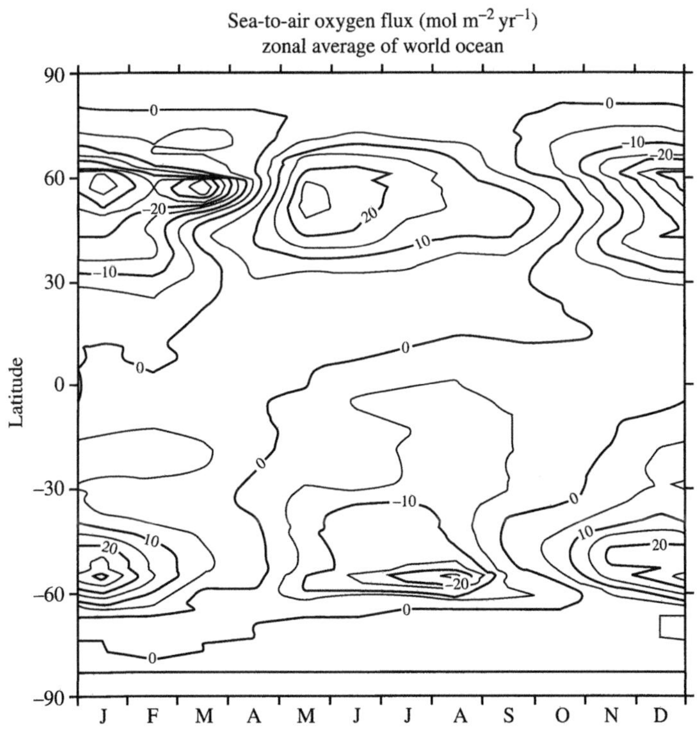

Figure 4 Zonal average monthly sea-to-air oxygen flux for world oceans. Outgassing and ingassing are concentrated at mid- to high latitudes. Outgassing of oxygen is strongest when primary production rates are greatest; ingassing is at a maximum during net respiration. These patterns oscillate during an annual cycle from northern to southern hemisphere (source Najjar and Keeling, 2000).

8.11.2.4 Soils and Groundwaters

In soil waters, oxygen concentrations depend on gas diffusion through soil pore spaces, infiltration and advection of rainwater and groundwater, air–gas exchange, and respiration of soil organic matter (see review by Hinkle, 1994). In organic-matter-rich temperate soils, dissolved O_2 concentrations are reduced, but not entirely depleted. Thus, many temperate shallow groundwaters contain some dissolved oxygen. Deeper groundwaters, and water-saturated soils and wetlands, generally contain little dissolved O_2. High-latitude mineral soils and groundwaters contain more dissolved O_2 (due to lower temperature and lesser amounts of soil organic matter and biological O_2 demand). Dry tropical soils are oxidized to great depths, with dissolved O_2 concentration less than air saturated, but not anoxic. Wet tropical forests, however, may experience significant O_2-depletion as rapid oxidation of leaf litter and humus occurs near the soil surface. Soil permeability also influences O_2 content, with more clay-rich soils exhibiting lower O_2 concentrations.

In certain environments, localized anomalously low concentrations of soil O_2 have been used by exploration geologists to indicate the presence of a large body of chemically reduced metal sulfides in the subsurface. Oxidation of sulfide minerals during weathering and soil formation draws down soil gas p_{O_2} below regional average. Oxidation of sulfide minerals generates solid and aqueous-phase oxidation products (i.e., sulfate anion and ferric oxyhydroxides in the case of pyrite oxidation). In some instances, the volume of gaseous O_2 consumed during mineral oxidation generates a mild negative pressure gradient, drawing air into soils above sites of sulfide mineral oxidation (Lovell, 2000).

8.11.3 MECHANISMS OF O_2 PRODUCTION

8.11.3.1 Photosynthesis

The major mechanism by which molecular oxygen is produced on Earth is through the biological process of photosynthesis. Photosynthesis occurs in higher plants, the eukaryotic protists collectively called algae, and in two groups of prokaryotes: the cyanobacteria and the prochlorophytes. In simplest terms, photosynthesis is the harnessing of light energy to chemically reduce carbon dioxide to simple organic compounds (e.g., glucose). The overall reaction (Equation (1)) for photosynthesis shows carbon dioxide and water reacting to produce oxygen and carbohydrate:

$$6CO_2 + 6H_2O \rightarrow 6O_2 + C_6H_{12}O_6 \quad (1)$$

Photosynthesis is actually a two-stage process, with each stage broken into a cascade of chemical reactions (Figure 5). In the light reactions of photosynthesis, light energy is converted to chemical energy that is used to dissociate water to yield oxygen and hydrogen and to form the reductant NADPH from $NADP^+$.

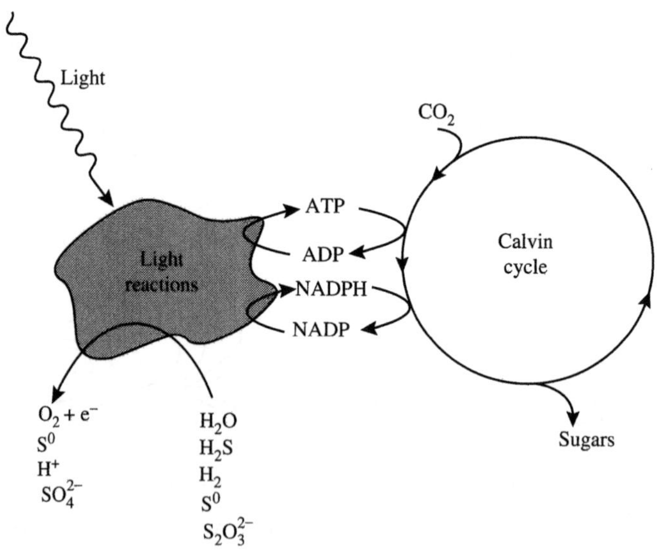

Figure 5 The two stages of photosynthesis: light reaction and the Calvin cycle. During oxygenic photosynthesis, H_2O is used as an electron source. Organisms capable of anoxygenic photosynthesis can use a variety of other electron sources (H_2S, H_2, S^0, $S_2O_3^{2-}$) during the light reactions, and do not liberate free O_2. Energy in the form of ATP and reducing power in the form of NADPH are produced by the light reactions, and subsequently used in the Calvin cycle to deliver electrons to CO_2 to produce sugars.

The next stage of photosynthesis, the Calvin cycle, uses NADPH to reduce CO_2 to phosphoglyceraldehyde, the precursor for a variety of metabolic pathways, including glucose synthesis. In higher plants and algae, the Calvin cycle operates in special organelles called chloroplasts. However, in bacteria the Calvin cycle occurs throughout the cytosol. The enzyme ribulose-1,5-biphosphate carboxylase (rubisco) catalyzes reduction of CO_2 to phosphoglycerate, which is carried through a chain of reactions that consume ATP and NADPH and eventually yield phosphoglyceraldehyde. Most higher plants are termed C_3 plants, because the first stable intermediate formed during the carbon cycle is a three-carbon compound. Several thousand species of plant, spread among at least 17 families including the grasses, precede the carbon cycle with a CO_2-concentrating mechanism which delivers a four-carbon compound to the site of the Calvin cycle and rubisco. These are the C_4 plants. This four-carbon compound breaks down inside the chloroplasts, supplying CO_2 for rubisco and the Calvin cycle. The C_4-concentrating mechanisms is an advantage in hot and dry environments where leaf stomata are partially closed, and internal leaf CO_2 concentrations are too low for rubisco to efficiently capture CO_2. Other plants, called CAM plants, which have adapted to dry climates utilize another CO_2-concentrating mechanisms by closing stomata during the day and concentrating CO_2 at night. All higher plants, however, produce O_2 and NADPH from splitting water, and use the Calvin cycle to produce carbohydrates. Some prokaryotes use mechanisms other than the Calvin cycle to fix CO_2 (i.e., the acetyl-CoA pathway or the reductive tricarboxylic acid pathway), but none of these organisms is involved in oxygenic photosynthesis.

Global net primary production estimates have been derived from variations in the abundance and isotopic composition of atmospheric O_2. These estimates range from 23×10^{15} mol yr^{-1} (Keeling and Shertz, 1992) to 26×10^{15} mol yr^{-1}, distributed between 14×10^{15} mol yr^{-1} O_2 production from terrestrial primary production and 12×10^{15} mol yr^{-1} from marine primary production (Bender *et al.*, 1994a). It is estimated that ~50% of all photosynthetic fixation of CO_2 occurs in marine surface waters. Collectively, free-floating photosynthetic microorganisms are called phytoplankton. These include the algal eukaryotes (dinoflagellates, diatoms, and the red, green, brown, and golden algae), various species of cyanobacteria (*Synechococcus* and *Trichodesmium*), and the common prochlorophyte *Prochlorococcus*. Using the stoichiometry of the photosynthesis reaction, this equates to half of all global photosynthetic oxygen production resulting from marine primary production. Satellite-based measurements of seasonal and yearly average chlorophyll abundance (for marine systems) and vegetation greenness (for terrestrial ecosystems) can be applied to models that estimate net primary productivity, CO_2 fixation, and O_2 production. In the oceans, there are significant regional and seasonal variations in photosynthesis that result from limitations by light, nutrients, and temperature. Yearly averages of marine chlorophyll abundance show concentrated primary production at high-latitudes in the North Atlantic, North Pacific, and coastal Antarctica, in regions of seawater upwelling off of the west coasts of Africa, South America, and the Arabian Sea, and along the Southern Subtropical Convergence. At mid- and high latitudes, marine productivity is strongly seasonal, with primary production concentrated in spring and summer. At low latitudes, marine primary production is lower and varies little with season or region. On land, primary production also exhibits strong regional and seasonal patterns. Primary production rates (in g C m^{-2} yr^{-1}) are greatest year-round in the tropics. Tropical forests in South America, Africa, and Southeast Asia are the most productive ecosystems on Earth. Mid-latitude temperate forests and high-latitude boreal forests are also highly productive, with a strong seasonal cycle of greatest production in spring and summer. Deserts (concentrated at ~30° N and S) and polar regions are less productive. These features of seasonal variability in primary production on land and in the oceans are clearly seen in the seasonal variations in atmospheric O_2 (Figure 4).

Transfer of O_2 between the atmosphere and surface seawater is controlled by air–sea gas exchange. Dissolved gas concentrations trend towards thermodynamic equilibrium, but other factors may complicate dissolved O_2 concentrations. Degassing of supersaturated waters can only occur at the very surface of the water. Thus, in regions of high primary production, concentrations of O_2 can accumulate in excess of the rate of O_2 degassing. In calm seas, where the air–sea interface is a smooth surface, gas exchange is very limited. As seas become more rough, and especially during storms, gas exchange is greatly enhanced. This is in part because of entrainment of bubbles dispersed in seawater and water droplets entrained in air, which provide much more surface for dissolution or degassing. Also, gas exchange depends on diffusion across a boundary layer. According to Fick's law, the diffusive flux depends on both the concentration gradient (degree of super- or undersaturation) and the thickness of the boundary layer. Empirically it is observed that the boundary layer thickness decreases with increasing wind speed, thus enhancing diffusion and gas exchange during high winds.

8.11.3.2 Photolysis of Water

In the upper atmosphere today, a small amount of O_2 is produced through photolysis of water vapor. This process is the sole source of O_2 to the atmospheres on the icy moons of Jupiter (Ganymede and Europa), where trace concentrations of O_2 have been detected (Vidal *et al.*, 1997). Water vapor photolysis may also have been the source of O_2 to the early Earth before the evolution of oxygenic photosynthesis. However, the oxygen formed by photolysis would have been through reactions with methane and carbon monoxide, preventing any accumulation in the atmosphere (Kasting *et al.*, 2001).

8.11.4 MECHANISMS OF O_2 CONSUMPTION

8.11.4.1 Aerobic Cellular Respiration

In simple terms, aerobic respiration is the oxidation of organic substrates with oxygen to yield chemical energy in the form of ATP and NADH. In eukaryotic cells, the respiration pathway follows three steps (Figure 6). Glycolysis occurs throughout the cytosol, splitting glucose into pyruvic acid and yielding some ATP. Glycolysis does not directly require free O_2, and thus occurs among aerobic and anaerobic organisms. The Krebs citric acid (tricarboxylic acid) cycle and oxidative phosphorylation are localized within the mitochondria of eukaryotes, and along the cell membranes of prokaryotes. The Krebs cycle completes the oxidation of pyruvate to CO_2 from glycolysis, and together glycolysis and the Krebs cycle provide chemical energy and reductants (in the form of ATP, NADH, and $FADH_2$) for the third step—oxidative phosphorylation. Oxidative phosphorylation involves the transfer of electrons from NADH and $FADH_2$ through a cascade of electron carrying compounds to molecular oxygen. Compounds used in the electron transport chain of oxidative phosphorylation include a variety of flavoproteins, quinones, Fe–S proteins, and cytochromes. Transfer of electrons from NADH to O_2 releases considerable energy, which is used to generate a proton gradient across the mitochondrial membrane and fuel significant ATP synthesis. In part, this gradient is created by reduction of O_2 to H_2O as the last step of oxidative phosphorylation. While eukaryotes and many prokaryotes use the Krebs cycle to oxidize pyruvate to CO_2, there are other pathways as well. For example, prokaryotes use the glyoxalate cycle to metabolize fatty acids. Several aerobic prokaryotes also can use the Entner–Doudoroff pathway in place of normal glycolysis. This reaction still produces pyruvate, but yields less energy in the form of ATP and NADH.

Glycolysis, the Krebs cycle, and oxidative phosphorylation are found in all eukaryotes (animals, plants, and fungi) and many of the aerobic prokaryotes. The purpose of these reaction pathways is to oxidize carbohydrates with O_2, yielding CO_2, H_2O, and chemical energy in the form of ATP. While macrofauna generally require a minimum of ~0.05–0.1 bar (~10 μM) O_2 to survive, many prokaryotic microaerophilic organisms can survive and thrive at much lower O_2 concentrations. Because most biologically mediated oxidation processes occur through the activity of aerotolerant microorganisms, it is unlikely that a strict coupling between limited atmospheric O_2 concentration and limited global respiration rates could exist.

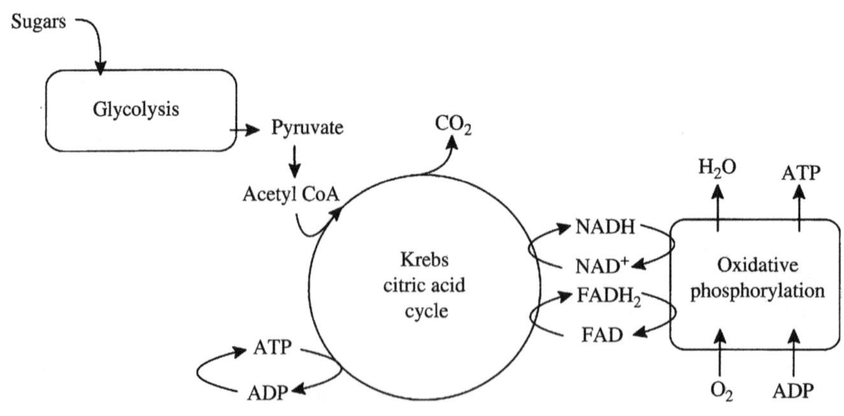

Figure 6 The three components of aerobic respiration: glycolysis, the Krebs cycle, and oxidative phosphorylation. Sugars are used to generate energy in the form of ATP during glycolysis. The product of glycolyis, pyruvate, is converted to acetyl-CoA, and enters the Krebs cycle. CO_2, stored energy as ATP, and stored reducing power as NADH and $FADH_2$ are generated in the Krebs cycle. O_2 is only directly consumed during oxidative phosphorylation to generate ATP as the final component of aerobic respiration.

8.11.4.2 Photorespiration

The active site of rubisco, the key enzyme involved in photosynthesis, can accept either CO_2 or O_2. Thus, O_2 is a competitive inhibitor of photosynthesis. This process is known as photorespiration, and involves addition of O_2 to ribulose-biphosphate. Products of this reaction enter a metabolic pathway that eventually produces CO_2. Unlike cellular respiration, photorespiration generates no ATP, but it does consume O_2. In some plants, as much as 50% of the carbon fixed by the Calvin cycle is respired through photorespiration. Photorespiration is enhanced in hot, dry environments when plant cells close stomata to slow water loss, CO_2 is depleted and O_2 accumulates. Photorespiration does not occur in prokaryotes, because of the much lower relative concentration of O_2 versus CO_2 in water compared with air.

8.11.4.3 C_1 Metabolism

Beyond metabolism of carbohydrates, there are several other biological processes common in prokaryotes that consume oxygen. For example, methylotrophic organisms can metabolize C_1 compounds such as methane, methanol, formaldehyde, and formate, as in

$$CH_4+NADH+H^+ + O_2 \rightarrow CH_3OH+NAD^+ + H_2O$$

$$CH_3OH+PQQ \rightarrow CH_2O+PQQH_2$$

$$CH_2O+NAD^+ + H_2O \rightarrow HCOOH+NADH+H^+$$

$$HCOOH+NAD^+ \rightarrow CO_2+NADH+H^+$$

These compounds are common in soils and sediments as the products of anaerobic fermentation reactions. Metabolism of these compounds can directly consume O_2 (through monooxygenase enzymes) or indirectly, through formation of NADH which is shuttled into oxidative phosphorylation and the electron transport chain. Oxidative metabolism of C_1 compounds is an important microbial process in soils and sediments, consuming the methane produced by methanogenesis.

8.11.4.4 Inorganic Metabolism

Chemolithotrophic microorganisms are those that oxidize inorganic compounds rather than organic substrates as a source of energy and electrons. Many of the chemolithotrophs are also autotrophs, meaning they reduce CO_2 to generate cellular carbon in addition to oxidizing inorganic compounds. In these organisms, CO_2 fixation is not tied to O_2 production by the stoichiometry of photosynthesis. Hydrogen-oxidizing bacteria occur wherever both O_2 and H_2 are available. While some H_2 produced by fermentation in anoxic environments (deep soils and sediments) may escape upward into aerobic environments, most H_2 utilized by hydrogen-oxidizing bacteria derives from nitrogen fixation associated with nitrogen-fixing plants and cyanobacteria. Nitrifying bacteria are obligate autotrophs that oxidize ammonia. Ammonia is produced in many environments during fermentation of nitrogen compounds and by dissimilatory nitrate reduction. Nitrifying bacteria are common at oxic–anoxic interfaces in soils, sediments, and the water column. Other chemolithoautotrophic bacteria can oxidize nitrite. Non-photosynthetic bacteria that can oxidize reduced sulfur compounds form a diverse group. Some are acidophiles, associated with sulfide mineral oxidation and tolerant of extremely low pH. Others are neutrophilic and occur in many marine sediments. These organisms can utilize a wide range of sulfur compounds produced in anaerobic environments, including H_2S, thiosulfate, polythionates, polysulfide, elemental sulfur, and sulfite. Many of the sulfur-oxidizing bacteria are also autotrophs. Some bacteria can live as chemoautotrophs through the oxidation of ferrous-iron. Some of these are acidophiles growing during mining and weathering of sulfide minerals. However, neutrophilic iron-oxidizing bacteria have also been detected associated with the metal sulfide plumes and precipitates at mid-ocean ridges. Other redox-sensitive metals that can provide a substrate for oxidation include manganese, copper, uranium, arsenic, and chromium. As a group, chemolithoautotrophic microorganisms represent a substantial flux of O_2 consumption and CO_2 fixation in many common marine and terrestrial environments. Because the net reaction of chemolithoautotrophy involves both O_2 and CO_2 reduction, primary production resulting from chemoautotrophs has a very different O_2/CO_2 stoichiometry than does photosynthesis. Chemolithoautotrophs use the electron flow from reduced substrates (metals, H_2, and reduced sulfur) to O_2 to generate ATP and NAD(P)H, which in turn are use for CO_2-fixation. It is believed that most aerobic chemoautotrophs utilize the Calvin cycle for CO_2-fixation.

8.11.4.5 Macroscale Patterns of Aerobic Respiration

On a global scale, much biological O_2 consumption is concentrated where O_2 is abundant.

This includes surficial terrestrial ecosystems and marine surface waters. A large fraction of terrestrial primary production is consumed by aerobic degradation mechanisms. Although most of this is through aerobic respiration, some fraction of aerobic degradation of organic matter depends on anaerobic breakdown of larger biomolecules into smaller C_1 compounds, which, if transported into aerobic zones of soil or sediment, can be degraded by aerobic C_1 metabolizing microorganisms. Partially degraded terrestrial primary production can be incorporated into soils, which slowly are degraded and eroded. Research has shown that soil organic matter (OM) can be preserved for up to several millennia, and riverine export of aged terrestrial OM may be a significant source of dissolved and particulate organic carbon to the oceans.

Aside from select restricted basins and specialized environments, most of the marine water column is oxygenated. Thus, aerobic respiration dominates in open water settings. Sediment trap and particle flux studies have shown that substantial fractions of marine primary production are completely degraded (remineralized) prior to deposition at the sea floor, and thus by implication, aerobic respiration generates an O_2 consumption demand nearly equal to the release of oxygen associated with photosynthesis. The bulk of marine aerobic respiration occurs within the water column. This is because diffusion and mixing (and thus O_2 resupply) are much greater in open water than through sediment pore fluids, and because substrates for aerobic respiration (and thus O_2 demand) are much less concentrated in the water column than in the sediments. O_2 consumption during aerobic respiration in the water column, coupled with movement of deep-water masses from O_2-charged sites of deep-water formation, generates the vertical and lateral profiles of dissolved O_2 concentration observed in seawater.

In most marine settings, O_2 does not penetrate very far into the sediment. Under regions of high primary productivity and limited water column mixing, even if the water column is oxygenated, O_2 may penetrate 1 mm or less, limiting the amount of aerobic respiration that occurs in the sediment. In bioturbated coastal sediments, O_2 penetration is facilitated by the recharging of pore fluids through organisms that pump overlying water into burrows, reaching several centimeters into the sediment in places. However, patterns of oxic and anoxic sediment exhibit a great degree of spatial and temporal complexity as a result of spotty burrow distributions, radial diffusion of O_2 from burrows, and continual excavation and infilling through time. Conversely, in deep-sea (pelagic) sediments, where organic matter delivery is minor and waters are cold and charged with O_2 from sites of deep-water formation, O_2 may penetrate uniformly 1 m or more into the sediment.

8.11.4.6 Volcanic Gases

Gases emitted from active volcanoes and fumaroles are charged with reduced gases, including CO, H_2, SO_2, H_2S, and CH_4. During explosive volcanic eruptions, these gases are ejected high into the atmosphere along with H_2O, CO_2, and volcanic ash. Even the relatively gentle eruption of low-silicon, low-viscosity-shield volcanoes is associated with the release of reduced volcanic gases. Similarly, reduced gases are released dissolved in waters associated with hot springs and geysers. Oxidation of reduced volcanic gases occurs in the atmosphere, in natural waters, and on the surfaces of minerals. This is predominantly an abiotic process, although many chemolithoautotrophs have colonized the walls and channels of hot springs and fumaroles, catalyzing the oxidation of reduced gases with O_2. Much of biological diversity in hyperthermophilic environments consists of prokaryotes employing these unusual metabolic types.

Recent estimates of global average volcanic gas emissions suggest that volcanic sulfur emissions range, $0.1-1\times10^{12}$ mol S yr^{-1}, is nearly equally distributed between SO_2 and H_2S (Halmer *et al.*, 2002; Arthur, 2000). This range agrees well with the estimates used by both Holland (2002) and Lasaga and Ohmoto (2002) for average volcanic S emissions through geologic time. Other reduced gas emissions (CO, CH_4, and H_2) are estimated to be similar in magnitude (Mörner and Etiope, 2002; Arthur, 2000; Delmelle and Stix, 2000). All of these gases have very short residence times in the atmosphere, revealing that emission and oxidative consumption of these gases are closely coupled, and that O_2 consumption through volcanic gas oxidation is very efficient.

8.11.4.7 Mineral Oxidation

During uplift and erosion of the Earth's continents, rocks containing chemically reduced minerals become exposed to the oxidizing conditions of the atmosphere. Common rock-forming minerals susceptible to oxidation include olivine, Fe^{2+}-bearing pyroxenes and amphiboles, metal sulfides, and graphite. Ferrous-iron oxidation is a common feature of soil formation. Iron oxides derived from oxidation of Fe^{2+} in the parent rock accumulate in the B horizon of temperate soils, and extensive laterites consisting of iron and aluminum oxides develop in tropical soils to many meters of depth. Where erosion rates are high, iron-bearing silicate minerals may be transported short distances in rivers; however, iron oxidation is so efficient that very few sediments show deposition of clastic ferrous-iron minerals. Sulfide minerals are extremely susceptible to oxidation, often being completely weathered from near-surface rocks. Oxidation of sulfide minerals generates appreciable acidity, and in areas where mining has brought sulfide minerals in contact with the atmosphere or O_2-charged rainwater, low-pH discharge has become a serious environmental problem. Although ferrous silicates and sulfide minerals such as pyrite will oxidize under sterile conditions, a growing body of evidence suggests that in many natural environments, iron and sulfur oxidation is mediated by chemolithoautotrophic microorganisms. Prokaryotes with chemolithoautotrophic metabolic pathways have been isolated from many environments where iron and sulfur oxidation occurs. The amount of O_2 consumed annually through oxidation of Fe^{2+} and sulfur-bearing minerals is not known, but based on sulfur isotope mass balance constraints is on the order of $(0.1-1)\times 10^{12}$ mol yr^{-1} each for iron and sulfur, similar in magnitude to the flux of reduced gases from volcanism.

8.11.4.8 Hydrothermal Vents

The spreading of lithospheric plates along mid-ocean ridges is associated with much undersea volcanic eruptions and release of chemically reduced metal sulfides. Volcanic gases released by subaerial volcanoes are also generated by submarine eruptions, contributing dissolved reduced gases to seawater. Extrusive lava flows generate pillow basalts, which are composed in part of ferrous-iron silicates. Within concentrated zones of hydrothermal fluid flow, fracture-filling and massive sulfide minerals precipitate within the pillow lavas, large chimneys grow from the seafloor by rapid precipitation of Fe–Cu–Zn sulfides, and metal sulfide-rich plumes of high-temperature "black smoke" are released into seawater. In cooler zones, more gradual emanations of metal and sulfide rich fluid diffuse upward through the pillow basalts and slowly mix with seawater. Convective cells develop, in which cold seawater is drawn down into pillow lavas off the ridge axis to replace the water released at the hydrothermal vents. Reaction of seawater with basalt serves to alter the basalt. Some sulfide is liberated from the basalt, entrained seawater sulfate precipitates as anhydrite or is reduced to sulfide, and Fe^{2+} and other metals in basalt are replaced with seawater-derived magnesium. Oxygen dissolved in seawater is consumed during alteration of seafloor basalts. Altered basalts containing oxidized iron-mineral can extend as much as 500 m below the seafloor. Much more O_2 is consumed during oxidation of black smoker and chimney sulfides. Chimney sulfide may be only partially oxidized prior to transport off-axis through spreading and burial by sediments. However, black smoker metal-rich fluids are fairly rapidly oxidized in seawater, forming insoluble iron and manganese oxides, which slowly settle out on the seafloor, generating metalliferous sediments.

Because of the diffuse nature of reduced species in hydrothermal fluids, it is not known what role marine chemolithoautotrophic microorganisms may play in O_2 consumption associated with fluid plumes. Certainly such organisms live within the walls and rubble of cooler chimneys and basalts undergoing seafloor weathering. Because of the great length of mid-ocean ridges throughout the oceans, and the abundance of pyrite and other metal sulfides associated with these ridges, colonization and oxidation of metal sulfides by chemolithoautotrophic organisms may form an unrecognized source of primary production associated with consumption, not net release, of O_2.

8.11.4.9 Iron and Sulfur Oxidation at the Oxic–Anoxic Transition

In restricted marine basins underlying highly productive surface waters, where consumption of O_2 through aerobic respiration near the surface allows anoxia to at least episodically extend beyond the sediment into the water column, oxidation of reduced sulfur and iron may occur when O_2 is present. Within the Black Sea and many anoxic coastal fjords, the transition from oxic to anoxic environments occurs within the water column. At other locations, such as the modern Peru Shelf and coastal California basins, the transition occurs right at the sediment–water interface. In these environments, small concentrations of sulfide and O_2 may coexist within a very narrow band where sulfide oxidation occurs. In such environments, appreciable sulfur recycling may occur, with processes of sulfate reduction, sulfide oxidation, and sulfur disproportionation acting within millimeters of each other. These reactions are highly mediated by microorganisms, as evidenced by extensive mats of chemolithoautotrophic sulfur-oxidizing *Beggiatoa* and *Thioploca* found where the oxic–anoxic interface and seafloor coincide.

Similary, in freshwater and non-sulfidic brackish environments with strong O_2 demand, dissolved ferrous-iron may accumulate in groundwaters and anoxic bottom waters. Significant iron oxidation will occur where the water table outcrops with the land surface (i.e., groundwater outflow into a stream), or in lakes and estuaries, at the oxic–anoxic transition within the water column. Insoluble iron oxides precipitate and settle down to the sediment. Recycling of iron may occur if sufficient organic matter exists for ferric-iron reduction to ferrous-iron.

The net effect of iron and sulfur recycling on atmospheric O_2 is difficult to constrain. In most cases, oxidation of sulfur or iron consumes O_2 (there are some anaerobic chemolithoautotrophic microorganisms that can oxidize reduced substrates using nitrate or sulfate as the electron acceptor). Reduction of sulfate or ferric-iron are almost entirely biological processes, coupled with the oxidation of organic matter; sulfate and ferric-iron reduction individually have no effect on O_2. However, the major source of organic substrates for sulfate and ferric-iron reduction is ultimately biomass derived from photosynthetic organisms, which is associated with O_2 generation. The net change derived from summing the three processes (S^{2-}- or Fe^{2+}-oxidation, SO_4^{2-}- or Fe^{3+}- reduction, and photosynthesis), is net gain of organic matter with no net production or consumption of O_2.

8.11.4.10 Abiotic Organic Matter Oxidation

Aerobic respiration is the main means by which O_2 is consumed on the Earth. This pathway occurs throughout most Earth-surface environments: soils and aquatic systems, marine surface waters supersaturated with O_2, within the water column and the upper zones of sediments. However, reduced carbon materials are also reacted with O_2 in a variety of environments where biological activity has not been demonstrated. Among these are photo-oxidation of dissolved and particulate organic matter and fossil fuels, fires from burning vegetation and fossil fuels, and atmospheric methane oxidation. Olefins (organic compounds containing double bonds) are susceptible to oxidation in the presence of transition metapls, ozone, UV light, or gamma radiation. Low-molecular-weight oxidized organic degradation products form from oxidation reactions, which in turn may provide organic substrates for aerobic respiration or C_1 metabolism. Fires are of course high-temperature combustion of organic materials with O_2. Research on fires has shown that O_2 concentration has a strong influence on initiation and maintenance of fires (Watson, 1978; Lenton and Watson, 2000). Although the exact relationship between p_{O_2} and initiation of fire in real terrestrial forest communities is debated (see Robinson, 1989, 1991), it is generally agreed that, at low p_{O_2}, fires cannot be started even on dry wood, although smoldering fires with inefficient oxidation can be maintained. At high p_{O_2}, even wet wood can support flame, and fires are easily initiated with a spark discharge such as lightning.

Although most methane on Earth is oxidized during slow gradual transport upwards through soils and sediments, at select environments there is direct injection of methane into the atmosphere. Methane reacts with O_2 in the presence of light or metal surface catalysts. This reaction is fast enough, and the amount of atmospheric O_2 large enough, that significant concentrations of atmospheric CH_4 are unlikely to accumulate. Catastrophic calving of submarine, CH_4-rich hydrates during the geologic past may have liberated large quantities of methane to the atmosphere. However, isotopic evidence suggests that this methane was oxidized and consumed in a geologically short span of time.

8.11.5 GLOBAL OXYGEN BUDGETS AND THE GLOBAL OXYGEN CYCLE

One window into the global budget of oxygen is the variation in O_2 concentration normalized to N_2. O_2/N_2 ratios reflect changes in atmospheric O_2 abundance, because N_2 concentration is assumed to be invariant through time. Over an annual cycle, $\delta(O_2/N_2)$ can vary by 100 per meg or more, especially at high-latitude sites (Keeling and Shertz, 1992). These variations reflect latitudinal variations in net O_2 production and consumption related to seasonal high productivity during summer. Observations of O_2/N_2 variation have been expanded beyond direct observation (limited to the past several decades) to records of atmospheric composition as trapped in Antarctic firn ice and recent ice cores (Battle *et al.*, 1996; Sowers *et al.*, 1989; Bender *et al.*, 1985, 1994b). These records reflect a slow gradual decrease in atmospheric O_2 abundance over historical times, attributed to the release and oxidation of fossil fuels.

As yet, details of the fluxes involved in the processes that generate and consume molecular oxygen are too poorly constrained to establish a balanced O_2 budget. A summary of the processes believed to dominate controls on atmospheric O_2, and reasonable best guesses for the magnitude of these fluxes, if available, are shown in Figure 7 (from Keeling *et al.*, 1993).

8.11.6 ATMOSPHERIC O_2 THROUGHOUT EARTH'S HISTORY

8.11.6.1 Early Models

Starting with Cloud (1976), two key geologic formations have been invoked to constrain the

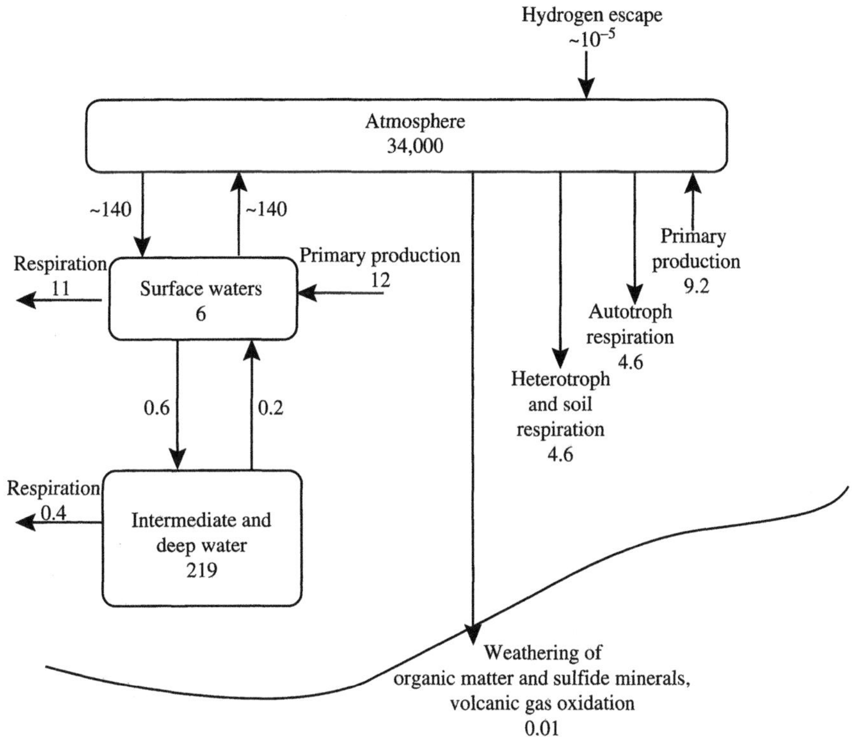

Figure 7 Global budget for molecular oxygen, including gas and dissolved O_2 reservoirs (sources Keeling *et al.*, 1993; Bender *et al.*, 1994a,b).

history of oxygenation of the atmosphere: banded iron formations (BIFs) and red beds (Figure 8). BIFs are chemical sediments containing very little detritus, and consist of silica laminae interbedded with layers of alternately high and low ratios of ferric- to ferrous-iron. As chemical sediments, BIFs imply the direct precipitation of ferrous-iron from the water column. A ferrous-iron-rich ocean requires anoxia, which by implication requires an O_2-free atmosphere and an anaerobic world. However, the ferric-iron layers in BIFs do reflect consumption of molecular O_2 (oxidation of ferrous-iron) at a rate much greater than supply of O_2 through prebiotic H_2O photolysis. Thus, BIFs may also record the evolution of oxygenic photosynthesis and at least localized elevated dissolved O_2 concentrations (Walker, 1979). Red beds are sandy sedimentary rocks rich in coatings, cements, and particles of ferric-iron. Red beds form during and after sediment deposition, and thus require that both the atmosphere and groundwater are oxidizing. The occurrence of the oldest red beds (~2.0 Ga) coincides nearly with the disappearance of BIFs (Walker, 1979), suggesting that some threshold of atmospheric O_2 concentration was reached at this time. Although the general concept of a low-O_2 atmosphere before ~2.0 Ga and accumulated O_2 since that time has been agreed on for several decades, the details and texture of oxygenation of the Earth's atmosphere are still being debated.

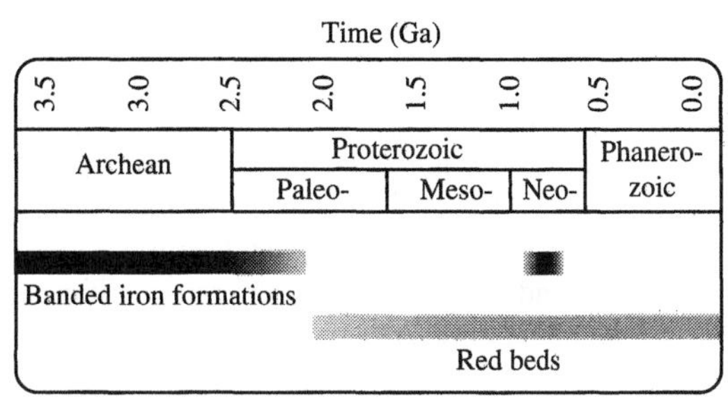

Figure 8 Archean distribution of banded iron formations, with short reoccurrence associated with widespread glaciation in the Neoproterozoic, and the Proterozoic and Phanerozoic distribution of sedimentary rocks containing ferric-iron cements (red beds). The end of banded iron formation and beginning of red bed deposition at ~2.2 Ga has been taken as evidence for a major oxygenation event in Earth's atmosphere.

Geochemists and cosmochemists initially looked to models of planetary formation and comparison with other terrestrial planets to understand the earliest composition of Earth's atmosphere. During planetary accretion and core formation, volatile components were liberated from a molten and slowly convecting mixture of silicates, metals, and trapped gases. The gravitational field of Earth was sufficient to retain most of the gases released from the interior. These include CH_4, H_2O, N_2, NH_3, and H_2S. Much H_2 and He released from the interior escaped Earth's gravitational field into space; only massive planets such as Jupiter, Saturn, Neptune, and Uranus have retained an H_2–He rich atmosphere. Photolysis of H_2O, NH_3, and H_2S produced free O_2, N_2, and S, respectively. O_2 was rapidly consumed by oxidation of CH_4 and H_2S to form CO_2, CO, and SO_2. High partial pressures of CO_2 and CH_4 maintained a strong greenhouse effect and warm average Earth surface temperature (~90 °C), in spite of much lower solar luminosity. Recognizing that the early Earth contained an atmospheric substantially richer in strong greenhouse gases compared with the modern world provided a resolution to subfreezing average Earth surface temperatures predicted for the early Earth due to reduced solar luminosity (Kasting *et al.*, 2001, 1983; Kiehl and Dickinson, 1987).

Liquid water on the early planet Earth allowed a hydrologic cycle and silicate mineral weathering to develop. Fairly quickly, much of the atmosphere's CO_2 was reacted with silicates to produce a bicarbonate-buffer ocean, while CH_4 was rapidly consumed by oxygen produced through photolysis of H_2O. Early microorganisms (and many of the most primitive organisms in existence today) used inorganic substrates to derive energy, and thrive at the high temperatures expected to be widespread during Earth's early history. These organisms include Archea that oxidize H_2 using elemental sulfur. Once photolysis and CH_4 oxidation generated sufficient p_{CO_2}, methanogenic Archea may have evolved. These organisms reduce CO_2 to CH_4 using H_2. However, sustainable life on the planet is unlikely to have developed during the first several hundred million years of Earth's history, due to large and frequent bolide impacts that would have sterilized the entire Earth's surface prior to ~3.8 Ga (Sleep *et al.*, 1989; Sleep and Zahnle, 1998; Sleep *et al.*, 2001; Wilde *et al.*, 2001), although recent work by Valley *et al.* (2002) suggests a cool early Earth that continually supported liquid water as early as 4.4 Ga.

Much of present understanding of the earliest evolution of Earth's atmosphere can trace descent from Walker (1979) and references therein. The prebiological atmosphere (before the origin of life) was controlled principally by the composition of gases emitted from volcanoes. Emission of H_2 in volcanic gases has contributed to net oxidation of the planet through time. This is achieved through several mechanisms. Simplest is hydrodynamic escape of H_2 from Earth's gravity. Because H_2 is a strongly reducing gas, loss of H_2 from the Earth equates to loss of reducing power or net increase in whole Earth oxidation state (Walker, 1979). Today, gravitational escape of H_2 and thus increase in oxidation state is minor, because little if any H_2 manages to reach the upper levels of the atmosphere without oxidizing. Early in Earth's history, sources of H_2 included volcanic gases and water vapor photolysis. The small flux of O_2 produced by photolysis was rapidly consumed by reaction with ferrous-iron and sulfide, contributing to oxidation of the crust.

Today, volcanic gases are fairly oxidized, consisting mainly of H_2O and CO_2, with smaller amounts of H_2, CO, and CH_4. The oxidation state of volcanic gases derives in part from the oxidation state of the Earth's mantle. Mantle oxygen fugacity today is at or near the quartz-fayalite-magnesite buffer (QFM), as is the f_{O_2} of eruptive volcanic gases. Using whole-rock and spinel chromium abundance from volcanogenic rocks through time, Delano (2001) has argued that the average oxidation state of the Earth's mantle was set very early in Earth's history (~3.6–3.9 Ga) to f_{O_2} at or near the QFM buffer. Magmas with this oxidation state release volcanic gases rich in H_2O, CO_2, and SO_2, rather than more reducing gases. Thus, throughout much of Earth's history, volcanic gases contributing to the atmosphere have been fairly oxidized. More reduced magma compositions have been detected in diamond-bearing assemblages likely Hadean in age (>4.0 Ga) (Haggerty and Toft, 1985). The increase in mantle oxidation within several hundred million years of early Earth's history reveals very rapid "mantle + crust" overturn and mixing at this time, coupled with subduction and reaction of the mantle with hydrated and oxidized crustal minerals (generated from reaction with O_2 produced through H_2O photolysis).

8.11.6.2 The Archean

8.11.6.2.1 Constraints on the O_2 content of the Archean atmosphere

Several lines of geochemical evidence support low to negligible concentrations of atmospheric O_2 during the Archean and earliest Proterozoic, when oxygenic photosynthesis may have evolved. The presence of pyrite and uraninite in detrital Archean sediments reveals that the atmosphere in the earliest Archean contained no free O_2 (Cloud, 1972). Although Archean-age detrital pyrites from South Africa may be hydrothermal in origin, Australian sediments of the Pilbara craton (3.25–2.75 Ga) contain rounded grains of pyrite, uraninite, and siderite (Rasmussen and Buick, 1999), cited as evidence for an anoxic atmosphere at this time. Although disputed (Ohmoto, 1999), it is difficult to explain detrital minerals that are extremely susceptible to dissolution and oxidation under oxidizing condition unless the atmosphere of the Archean was essentially devoid of O_2.

Archean paleosols provide other geochemical evidence suggesting formation under reducing conditions. For example, the 2.75 Ga Mount Roe #2 paleosol of Western Australia contains up to 0.10% organic carbon with isotope ratios between −33‰ and −55‰ (Rye and Holland, 2000). These isotope ratios suggest that methanogenesis and methanotrophy were important pathways of carbon cycling in these soils. For modern soils in which the bulk organic matter is strongly ^{13}C-depleted (<−40‰), the methane fueling methanotrophy must be derived from somewhere outside the soil, because reasonable rates of fermentation and methanogenesis cannot supply enough CH_4. By extension, Rye and Holland (2000) argue that these soils formed under an atmosphere rich in CH_4, with any O_2 consumed during aerobic methanotrophy having been supplied by localized limited populations of oxygenic photoautotrophs. Other paleosol studies have used lack of cerium oxidation during soil formation as an indicator of atmospheric anoxia in the Archean (Murakami *et al.*, 2001). In a broader survey of Archean and Proterozoic paleosols, Rye and Holland (1998) observe that all examined paleosols older than 2.4 Ga indicate loss of iron during weathering and soil formation. This chemical feature is consistent with soil development under an atmosphere containing $<10^{-4}$ atm O_2 (1 atm = 1.01325×10^5 Pa), although some research has suggested that anoxic soil development in the Archean does not necessarily require an anoxic atmosphere (Ohmoto, 1996).

Other evidence for low Archean atmospheric oxygen concentrations come from studies of mass-independent sulfur isotope fractionation. Photochemical oxidation of volcanic sulfur species, in contrast with aqueous-phase oxidation and dissolution that characterizes the modern sulfur cycle, may have been the major source of sulfate to seawater in the Archean (Farquhar *et al.*, 2002; Farquhar *et al.*, 2000). Distinct shifts in $\delta^{33}S$ and $\delta^{34}S$ in sulfide and sulfate from Archean rocks occurred between 2.4–2.0 Ga, consistent with a shift from an O_2-free early atmosphere in which SO_2 photochemistry could dominate among seawater sulfate sources to an O_2-rich later atmosphere in which oxidative weathering of sulfide minerals predominated over photochemistry as the major source of seawater sulfate. Sulfur isotope heterogeneities detected in sulfide inclusions in diamonds also are believed to derive from photochemical SO_2 oxidation in an O_2-free atmosphere at 2.9 Ga (Farquhar *et al.*, 2002). Not only do these isotope ratios require an O_2-free atmosphere, but they also imply significant contact between the mantle, crust, and atmosphere as recently as 2.9 Ga.

Nitrogen and sulfur isotope ratios in Archean sedimentary rocks also indicate limited or negligible atmospheric O_2 concentrations (Figure 9). Under an O_2-free environment, nitrogen could only exist as N_2 and reduced forms (NH_3, etc.). Any nitrate or nitrite produced by photolysis would be quickly reduced, likely with Fe^{2+}. If nitrate is not available, then denitrification (reduction of nitrate to free N_2) cannot occur. Denitrification is associated with a substantial nitrogen-isotope discrimination, generating N_2 that is substantially depleted in ^{15}N relative to the NO_3^- source. In the modern system, this results in seawater nitrate (and organic matter) that is ^{15}N-enriched relative to air. Nitrogen in kerogen from Archean sedimentary rocks is not enriched in ^{15}N (as is found in all kerogen nitrogen from Proterozoic age to the present), but instead is depleted relative to modern atmospheric N_2 by several ‰ (Beaumont and Robert, 1999). This is consistent with an Archean nitrogen cycle in which no nitrate and no free O_2 was available, and nitrogen cycling was limited to N_2-fixation, mineralization and ammonia volatilization. Bacterial sulfate reduction is associated with a significant isotope discrimination, producing sulfide that is depleted in ^{34}S relative to substrate sulfate. The magnitude of sulfur isotope fractionation during sulfate reduction depends in part on available sulfate concentrations. Very limited differences in sulfur isotopic ratios among Archean sedimentary sulfide and sulfate minerals (<2‰ $\delta^{34}S$) indicate only minor isotope fractionation during sulfate reduction in the Archean oceans (Canfield *et al.*, 2000; Habicht *et al.*, 2002), best explained by extremely low SO_4^{2-} concentrations (<200 μM in contrast with modern concentrations of ~28 mM) (Habicht *et al.*, 2002). The limited supply of sulfate in the Archean ocean suggests that the major source of sulfate to Archean seawater was volcanic gas, because oxidative weathering of sulfide minerals could not occur under an O_2-free atmosphere. The limited sulfate concentration would have suppressed the activity of sulfate-reducing bacteria and facilitated methanogenesis. However, by ~2.7 Ga, sedimentary sulfides that are ^{34}S-depleted relative to sulfate are detected, suggesting at least sulfate reduction, and by implication sources of sulfate to seawater through sulfide oxidation, may have developed (Canfield *et al.*, 2000).

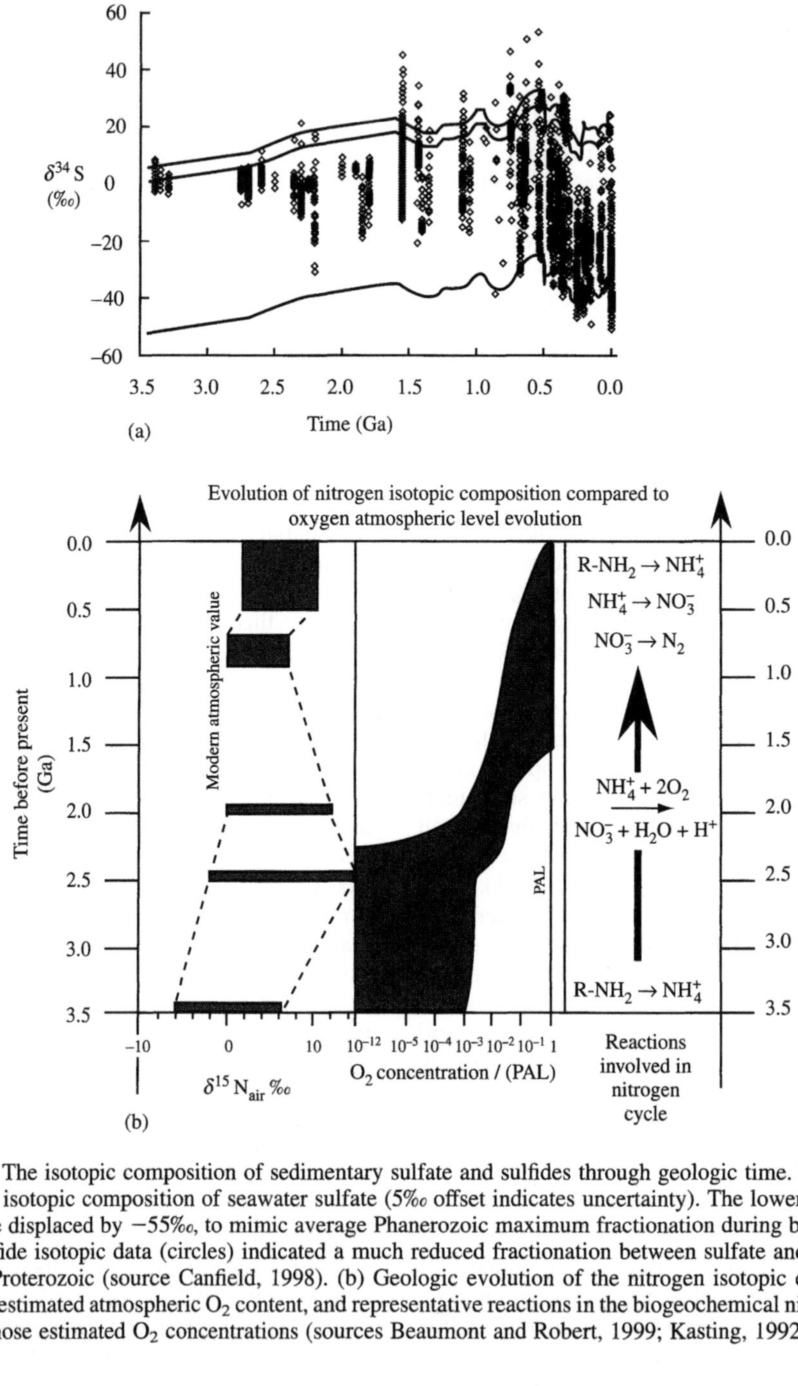

Figure 9 (a) The isotopic composition of sedimentary sulfate and sulfides through geologic time. The two upper lines show the isotopic composition of seawater sulfate (5‰ offset indicates uncertainty). The lower line indicated δ^{34}S of sulfate displaced by −55‰, to mimic average Phanerozoic maximum fractionation during bacterial sulfate reduction. Sulfide isotopic data (circles) indicated a much reduced fractionation between sulfate and sulfide in the Archean and Proterozoic (source Canfield, 1998). (b) Geologic evolution of the nitrogen isotopic composition of kerogen (left), estimated atmospheric O_2 content, and representative reactions in the biogeochemical nitrogen cycle at those estimated O_2 concentrations (sources Beaumont and Robert, 1999; Kasting, 1992).

8.11.6.2.2 The evolution of oxygenic photosynthesis

In the early Archean, methanogenesis (reaction of $H_2 + CO_2$ to yield CH_4) was likely a significant component of total primary production. O_2 concentrations in the atmosphere were suppressed due to limited O_2 production and rapid consumption with iron, sulfur, and reduced gases, while CH_4 concentrations were likely very high (Kasting *et al.*, 2001). The high methane abundance is calculated to have generated a hydrocarbon-rich smog that could screen UV light and protect early life in the absence of O_2 and ozone (Kasting *et al.*, 2001). Other means of protection from UV damage in the O_2-free Archean include biomineralization of cyanobacteria within UV-shielded iron–silica sinters (Phoenix *et al.*, 2001).

Biological evolution may have contributed to early Archean oxidation of the Earth. Catling *et al.*, (2001) have recognized that CO_2 fixation associated with early photosynthesis may have been coupled with active fermentation and methanogenesis. Prior to any accumulation of atmospheric O_2, CH_4 may have been a large component of Earth's atmosphere. A high flux of biogenic methane is supported by coupled ecosystem–climate models of the early Earth (Pavlov *et al.*, 2000; Kasting *et al.*, 2001) as a supplement to Earth's greenhouse warming under reduced solar luminosity in the Archean. CH_4 in the upper atmosphere is consumed by UV light to yield hydrogen (favored form of hydrogen in the upper atmosphere), which escapes to space. Because H_2 escape leads to net oxidation, biological productivity and methanogenesis result in slow oxidation of the planet. Net oxidation may be in the form of direct O_2 accumulation (if CO_2 fixation is associated with oxygenic photosynthesis) or indirectly (if CO_2 fixation occurs via anoxygenic photosynthesis or anaerobic chemoautotrophy) through production of oxidized iron or sulfur minerals in the crust, which upon subduction are mixed with other crustal rocks and increase the crustal oxidation state. One system that may represent a model of early Archean biological productivity consists of microbial mats found in hypersaline coastal ponds. In these mats, cyanobacteria produce H_2 and CO that can be used as substrates by associated chemoautotrophs, and significant CH_4 fluxes out of the mats have been measured (Hoehler *et al.*, 2001). Thus, mats may represent communities of oxygenic photosynthesis, chemoautotrophy, and methanogenesis occurring in close physical proximity. Such communities would have contributed to elevated atmospheric CH_4 and the irreversible escape of hydrogen from the Archean atmosphere, contributing to oxidation of the early Earth.

The earliest photosynthetic communities may have contributed to oxidation of the early Earth through production of oxidized crustal minerals without requiring production of O_2. Crustal rocks today are more oxidized than the mantle, with compositions ranging between the QFM and hematite-magnesite f_{O_2} buffers. This is best explained as an irreversible oxidation of the crust associated with methanogenesis and hydrogen escape (Catling *et al.*, 2001). It is unlikely that O_2 produced by early photosynthesis ever directly entered the atmosphere.

Maintaining low O_2 concentrations in the atmosphere while generating oxidized crustal rocks requires oxidation mechanisms that do not involve O_2. Among these may be serpentinization of seafloor basalts (Kasting and Siefert, 2002). During seafloor weathering, ferrous-iron is released to form ferric oxyhydroxides. Using H_2O or CO_2 as the ferrous-iron oxidant, H_2 and CH_4 are generated. This H_2 gas (either produced directly, or indirectly through UV decomposition of CH_4) can escape the atmosphere, resulting in net oxidation of the crust and accumulation of oxidized crustal rocks. BIF formation in the Archean may be related to dissolved Fe^{2+} that was oxidized in shallow-water settings associated with local oxygenic photosynthesis or chemoautotrophy (Kasting and Siefert, 2002). After the biological innovation of oxygenic photosynthesis evolved, there was still a several hundred million year gap until O_2 began to accumulate in the atmosphere, because O_2 could only accumulate once the supply of reduced gases (CO and CH_4) and ferrous-iron fell below rates of photosynthetic O_2 supply.

8.11.6.2.3 Carbon isotope effects associated with photosynthesis

The main compound responsible for harvesting light energy to produce NADPH and splitting water to form O_2 is chlorophyll. There are several different structural variants of chlorophyll, including chlorophyll *a*, chlorophyll *b*, and bacteriochlorophylls *a–e* and *g*. Each of these shows optimum excitation at a different wavelength of light. All oxygenic photosynthetic organisms utilize chlorophyll *a* and/or chlorophyll *b*. Other non-oxygenic photoautotrophic microorganisms employ a diverse range of chlorophylls.

The earliest evidence for evolution of oxygenic photosynthesis comes from carbon isotopic signatures preserved in Archean rocks (Figure 10). CO_2 fixation through the Calvin cycle is associated with a significant carbon isotope discrimination, such that organic matter produced through CO_2 fixation is depleted in ^{13}C relative to ^{12}C by several per mil. In a closed or semi-closed system (up to and including the whole ocean–atmosphere system), isotope discrimination during fixation of CO_2 then results in a slight enrichment in the $^{13}C/^{12}C$ ratio of CO_2 not taken up during photosynthesis. Thus, a biosignature of CO_2 fixation is an enrichment in $^{13}C/^{12}C$ ratio in atmospheric CO_2 and seawater bicarbonate over a whole-Earth averaged isotope ratio. When carbonate minerals precipitate from seawater, they record the seawater isotope value. Enrichment of the $^{13}C/^{12}C$ ratio in early Archean carbonate minerals, and by extension seawater bicarbonate, is taken as early evidence for CO_2 fixation. Although anoxygenic photoautotrophs and chemoautotrophs fix CO_2 without generating O_2, these groups either do not employ the Calvin cycle (many anoxygenic photoautotrophs use the acetyl-CoA or reverse tricarboxylic acid pathways) or require O_2 (most chemolithoautotrophs). Thus, the most likely group of organisms responsible for this isotope effect are oxygenic photoautotrophs such as cyanobacteria. Kerogen and graphite that is isotopically depleted in ^{13}C is a common and continuous feature of the sedimentary record, extending as far back as ~3.8 Ga to the Isua

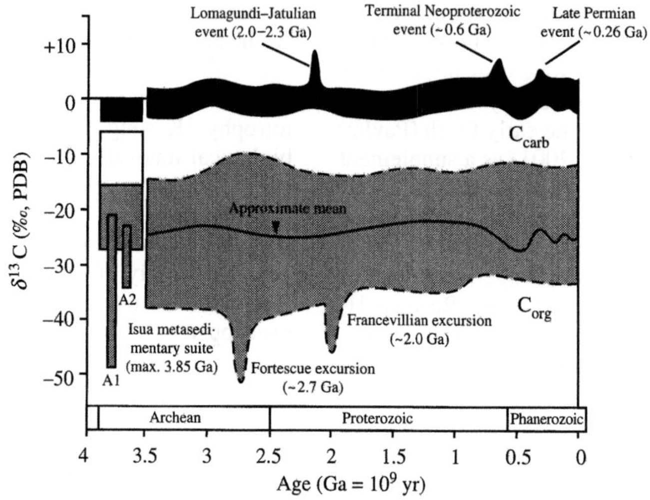

Figure 10 Isotopic composition of carbonates and organic carbon in sedimentary and metasedimentary rocks through geologic time. The negative excursions in organic matter $\delta^{13}C$ at 2.7 Ga and 2.0 Ga may relate to extensive methanogenesis as a mechanism of carbon fixation (Schidlowski, 2001) (reproduced by permission of Elsevier from *Precanb. Res.* **2001**, *106*, 117–134).

Supracrustal Suite of Greenland (Schidlowski, 1988, 2001; Nutman *et al.*, 1997). Although some isotopically depleted graphite in the metasediments from the Isua Suite may derive from abiotic hydrothermal processes (van Zuilen *et al.*, 2002), rocks interpreted as metamorphosed turbidite deposits retain ^{13}C-depleted graphite believed to be biological in origin. Moreover, the isotopic distance between coeval carbonate and organic matter (in the form of kerogen) can be used to estimate biological productivity through time. With a few exceptions, these isotope mass balance estimates reveal that, since Archean times, global scale partitioning between inorganic and organic carbon, and thus global productivity and carbon burial, have not varied greatly over nearly 4 Gyr of Earth's history (Schidlowski, 1988, 2001). Approximately 25% of crustal carbon burial is in the form of organic carbon, and the remainder is inorganic carbonate minerals. This estimate derives from the mass- and isotope-balance equation:

$$\delta^{13}C_{avg} = f\delta^{13}C_{organic\ matter} + (1-f) \times \delta^{13}C_{carbonate} \qquad (2)$$

where $\delta^{13}C_{avg}$ is the average isotopic composition of crustal carbon entering the oceans from continental weathering and primordial carbon emitted from volcanoes, $\delta^{13}C_{organic\ matter}$ is the average isotopic composition of sedimentary organic matter, $\delta^{13}C_{carbonate}$ is the average isotopic composition of carbonate sediments, and f is the fraction of carbon buried as organic matter in sediments.

The observation that the proportion of carbon buried in sediments as organic matter versus carbonate has not varied throughout geologic time raises several intriguing issues. First, biogeochemical cycling of carbon exhibits remarkable constancy across 4 Gyr of biological evolution, in spite of large-scale innovations in primary production and respiration (including anoxygenic and oxygenic photosynthesis, chemoautotrophy, sulfate reduction, methanotrophy, and aerobic respiration). Thus, with a few notable exceptions expressed in the carbonate isotope record, burial flux ratios between organic matter and carbonate have remained constant, in spite of varying dominance of different modes of carbon fixation and respiration through time, not to mention other possible controls on organic matter burial and preservation commonly invoked for Phanerozoic systems (anoxia of bottom waters, sedimentation rate, selective preservation, or cumulative oxygen exposure time). Second, the constancy of organic matter versus carbonate burial through time reveals that throughout geologic time, the relative contributions of various sources and sinks of carbon to the "ocean + atmosphere" system have remained constant. In other words, to maintain a constant carbonate isotopic composition through time, not only must the relative proportion of organic matter versus carbonate burial have remained nearly constant, but the relative intensity of organic matter versus carbonate weathering also must have remained nearly constant. In the earliest stages of Earth's history, when continents were small and sedimentary rocks were sparse, inputs of carbon from continental weathering may have been small relative to volcanic inputs. However, the several billion year sedimentary record of rocks rich in carbonate minerals and organic matter suggests that continental weathering must have formed a significant contribution to total oceanic carbon inputs fairly early in the Archean. Intriguingly, this indicates that oxidative weathering of ancient sedimentary rocks may have been active even in the Archean, prior to accumulation of O_2 in the atmosphere. This runs counter to traditional interpretations of geochemical carbon–oxygen cycling, in which organic matter burial is equated to O_2 production and organic matter weathering is equated to O_2 consumption.

8.11.6.2.4 Evidence for oxygenic photosynthesis in the Archean

Fossil evidence for photosynthetic organisms from the same time period can be traced to the existence of stromatolites (Schopf, 1992, 1993; Schopf *et al.*, 2002; Hofmann *et al.*, 1999), although evidence for these early oxygenic photosynthetic communities is debated (Buick, 1990; Brasier *et al.*, 2002). Stromatolites are laminated sediments consistently of alternating organic matter-rich and organic matter-lean layers; the organic matter-rich layers are largely composed of filamentous cyanobacteria. Stromatolites and similar mat-forming cyanobacterial crusts occur today in select restricted shallow marine environments. Fossil evidence from many locales in Archean rocks (Greenland, Australia, South Africa) and Proterozoic rocks (Canada, Australia, South America) coupled with carbon isotope geochemistry provides indirect evidence for the evolution of oxygenic photosynthesis early in Earth's history (~3.5 Ga).

Other evidence comes from molecular fossils. All cyanobacteria today are characterized by the presence of 2α-methylhopanes in their cell membranes. Brocks *et al.*, (1999) have demonstrated the existence of this taxon-specific biomarker in 2.7 Ga Archean rocks of the Pilbara Craton in NW Australia. Also in these rocks is found a homologous series of C_{27} to C_{30} steranes. Steranes today derive from sterols mainly produced by organisms in the domain Eukarya. Eukaryotes are obligate aerobes that require molecular O_2, and thus Brocks and colleagues argue that the presence of these compounds provides strong evidence for both oxygenic photosynthesis and at least localized utilization of accumulated O_2. Of course, coexistence of two traits within a biological lineage today reveals nothing about which evolved first. It is uncertain whether 2α-methylhopane lipid biosynthesis preceded oxygenic photosynthesis among cyanobacteria. Sterol synthesis certainly occurs in modern prokaryotes (including some anaerobes), but no existing lineage produces the distribution of steranes found in the Brocks *et al.*, (1999) study except eukaryotes.

8.11.6.3 The Proterozoic Atmosphere

8.11.6.3.1 Oxygenation of the Proterozoic atmosphere

Although there is evidence for the evolution of oxygenic photosynthesis several hundred million years before the Huronian glaciation (e.g., Brocks *et al.*, 1999) or earlier (Schidlowski, 1988, 2001), high fluxes of UV light reacting with O_2 derived from the earliest photosynthetic organisms would have created dangerous reactive oxygen species that severely suppressed widespread development of large populations of these oxygenic photoautotrophs. Cyanobacteria, as photoautotrophs, need be exposed to visible light and have evolved several defense mechanisms to protect cell contents and repair damage. However, two key metabolic pathways (oxygenic photosynthesis and nitrogen fixation) are very sensitive to UV damage. Indisputably cyanobacterial fossil occur in 1,000 Ma rocks, with putative fossils occurring 2,500 Ma and possibly older (Schopf, 1992). Sediment mat-forming cyanobacteria and stromatolites are least ambiguous and oldest. Terrestrial encrusting cyanobacteria are only known in the Phanerozoic. Planktonic forms are not known for the Archean and early Proterozoic. They may not have existed, or they may not be preserved. Molecular evolution, specifically coding for proteins that build gas vesicles necessary for planktonic life, are homologous and conserved in all cyanobacteria. Thus, perhaps planktonic cyanobacteria existed throughout Earth's history since the late Archean.

Today, ozone forms in the stratosphere by reaction of O_2 with UV light. This effectively screens much incoming UV radiation. Prior to the accumulation of atmospheric O_2, no ozone could form, and thus the UV flux to the Earth's surface would be much greater (with harmful effects on DNA and proteins, which adsorb and are altered by UV). A significant ozone shield could develop at $\sim 10^{-2}$ PAL O_2 (Kasting, 1987). However, a fainter young Sun would have emitted somewhat lower UV, mediating the lack of ozone. Although seawater today adsorbs most UV light by 6–25 m (1% transmittance cutoff), seawater with abundant dissolved Fe^{2+} may have provided an effective UV screen in the Archean (Olson and Pierson, 1986). Also, waters rich in humic materials, such as modern coastal oceans, are nearly UV opaque. If Archean seawater contained DOM, this could adsorb some UV. Iron oxidation and BIF at ~2.5–1.9 Ga would have removed the UV screen in seawater. Thus, a significant UV stress may have developed at this time. This would be mediated coincidentally by accumulation of atmospheric O_2.

In the early Archean before oxygenic photosynthesis evolved, cyanobacteria were limited.

Planktonic forms were inhibited, and limited by dissolved iron content of water, and existence of stratified, UV-screen refuges. Sedimentary mat-forming and stromatolite-forming communities were much more abundant. Iron-oxide precipitation and deposition of screen enzymes may have created UV-free colonies under a shield even in shallow waters.

Advent of oxygenic photosynthesis in the Archean generated small oxygen oases (where dissolved O_2 could accumulate in the water) within an overall O_2-free atmosphere waters containing oxygenic photoautotrophs might have reached 10% air saturation (Kasting, 1992). At this time, both unscreened UV radiation and O_2 may have coincided within the water column and sediments. This would lead to increased UV stress for cyanobacteria. At the same time, precipitation of iron from the water would make the environment even more UV-transparent. To survive, cyanobacteria would need to evolve and optimize defense and repair mechanisms for UV damage (Garcia-Pichel, 1998). Perhaps this explains the ~500 Myr gap between origin of cyanobacteria and accumulation of O_2 in the atmosphere. For example, the synthesis of scytonemin (a compound found exclusively in cyanobacteria) requires molecular oxygen (implying evolution in an oxic environment); it optimally screens UV-a, the form of UV radiation only abundant in an oxygenated atmosphere.

Oxygenation of the atmosphere at ~2.3–2.0 Ga may derive from at least three separate causes. First, discussed earlier, is the titration of O_2 with iron, sulfur, and reduced gases. The other is global rates of photosynthesis and organic carbon burial in sediments. If the rate of atmospheric O_2 supply (oxygenic photosynthesis) exceeds all mechanisms of O_2 consumption (respiration, chemoautotrophy, reduced mineral oxidation, etc.), then O_2 can accumulate in the atmosphere. One means by which this can be evaluated is through seawater carbonate $\delta^{13}C$. Because biological CO_2 fixation is associated with significant carbon isotope discrimination, the magnitude of carbon fixation is indicated by the isotopic composition of seawater carbonate. At times of more carbon fixation and burial in sediments, relatively more ^{12}C is removed from the atmosphere + ocean inorganic carbon pool than is supplied through respiration, organic matter oxidation, and carbonate mineral dissolution. Because carbon fixation is dominated by oxygenic photosynthesis (at least since the late Archean), periods of greater carbon fixation and burial of organic matter in sediments (observed as elevated seawater carbonate $\delta^{13}C$) are equated to periods of elevated O_2 production through oxygenic photosynthesis. The early Proterozoic Lomagundi event (~2.3–2.0 Ga) is recorded in the sediment record as a prolonged period of elevated seawater carbonate $\delta^{13}C$, with carbonate $\delta^{13}C$ values reaching nearly 10‰ in several sections around the world (Schidlowski, 2001). This represents an extended period of time (perhaps several hundred million years) during which removal of carbon from the "ocean + atmosphere" system as organic matter greatly exceeded supply. By implication, release of O_2 through photosynthesis was greatly accelerated during this time.

A third mechanism for oxygenation of the atmosphere at ~2.3 Ga relies on the slow, gradual oxidation of the Earth's crust. Irreversible H_2 escape and basalt-seawater reactions led to a gradual increase in the amount of oxidized and hydrated minerals contained in the Earth's crust and subducted in subduction zones throughout the Archean. Gradually this influenced the oxidation state of volcanic gases derived in part from subducted crustal rocks. Thus, although mantle oxygen fugacity may not have changed since the early Archean, crustal and volcanic gas oxygen fugacity slowly increased as the abundance of oxidized and hydrated crust increased (Holland, 2002; Kasting *et al.*, 1993; Kump *et al.*, 2001). Although slow to develop, Holland (2002) estimates that an increase in f_{O_2} of less than 1 log unit is all that would have been required for transition from an anoxic to an oxic atmosphere, assuming rates of oxygenic photosynthesis consistent with modern systems and the sediment isotope record. Once a threshold volcanic gas f_{O_2} had been reached, O_2 began to accumulate. Oxidative weathering of sulfides released large amounts of sulfate into seawater, facilitating bacterial sulfate reduction.

There are several lines of geochemical evidence that suggest a rise in oxygenation of the atmosphere ~2.3–2.0 Ga, beyond the coincident last occurrence of BIFs (with one late Proterozoic exception) and first occurrence of red beds recognized decades ago, and carbon isotopic evidence suggesting ample burial of sedimentary organic matter (Karhu and Holland, 1996; Bekker *et al.*, 2001; Buick *et al.*, 1998). The Huronian glaciation (~2.3 Ga) is the oldest known glacial episode recorded in the sedimentary record. One interpretation of this glaciation is that the cooler climate was a direct result of the rise of photosynthetically derived O_2. The rise of O_2 scavenged and reacted with the previously high atmospheric concentration of CH_4. Methane concentrations dropped, and the less-effective greenhouse gas product CO_2 could not maintain equable surface temperatures. Kasting *et al.*, (1983) estimate that a rise in p_{O_2} above $\sim 10^{-4}$ atm resulted in loss of atmospheric CH_4 and onset of glaciation.

Paleosols have also provided evidence for a change in atmosphere oxygenation at some time between 2.3–2.0 Ga. Evidence from rare earth element enrichment patterns and U/Th fractionation suggests a rise in O_2 to ~0.005 bar by the time of formation of the Flin Flon paleosol of Manitoba, Canada, 1.85 Ga (Pan and Stauffer, 2000; Holland *et al.*, 1989; Rye and Holland, 1998). Rye and Holland (1998) examined several early Proterozoic paleosols and observed that negligible iron loss is a consistent feature from soils of Proterozoic age through the present. These authors estimate that a minimum p_{O_2} of >0.03 atm is required to retain iron during soil formation, and thus atmospheric O_2 concentration has been 0.03 or greater since the early Paleozoic. However, re-evaluation of a paleosol crucial to the argument of iron depletion during soil formation under anoxia, the Hekpoort paleosol dated at 2.2 Ga, has revealed that the iron depletion detected by previous researchers may in fact be the lower zone of a normal oxidized lateritic soil. Upper sections of the paleosol that are not depleted in iron have been eroded away in the exposure examined by Rye and Holland (1998), but have been found in drill core sections. The depletion of iron and occurrence of ferrous-iron minerals in the lower sections of this paleosol have been reinterpreted by Beukes *et al.*, (2002) to indicate an abundant soil surface biomass at the time of deposition that decomposed to generate reducing conditions and iron mobilization during the wet season, and precipitation of iron oxides during the dry season.

Sulfur isotope studies have also provided insights into the transition from Archean low p_{O_2} to higher values in the Proterozoic. In the same studies that revealed extremely low Archean ocean sulfate concentrations, it was found that by ~2.2 Ga, isotopic compositions of sedimentary sulfates and sulfides indicate bacterial sulfate reduction under more elevated seawater sulfate concentrations compared with the sulfate-poor Archean (Habicht *et al.*, 2002; Canfield *et al.*, 2000). As described above, nitrogen isotope ratios in sedimentary kerogens show a large and permanent shift at ~2.0 Ga, consistent with denitrification, significant seawater nitrate concentrations, and thus available atmospheric O_2.

Prior to ~2.2 Ga, low seawater sulfate concentrations would have limited precipitation and subduction of sulfate-bearing minerals. This would maintain a lower oxidation state in volcanic gases derived in part from recycled crust (Holland, 2002). Thus, even while the oxidation state of the mantle has remained constant since ~4.0 Ga (Delano, 2001), the crust and volcanic gases derived from subduction of the crust could only achieve an increase in oxidation state once seawater sulfate concentrations increased.

A strong model for oxygenation of the atmosphere has developed based largely on the sulfur isotope record and innovations in microbial metabolism. The classical interpretation of the disappearance of BIFs relates to the rise of atmospheric O_2, oxygenation of the oceans, and removal of dissolved ferrous-iron by oxidation. However, another interpretation has developed, based largely on the evolving Proterozoic sulfur isotope record. During the oxygenation of the atmosphere at ~2.3–2.0 Ga, the oceans may not have become oxidized, but instead remained anoxic and became strongly sulfidic as well (Anbar and Knoll, 2002; Canfield, 1998 and references therein). Prior to ~2.3 Ga, the oceans were anoxic but not sulfidic. Ferrous-iron was abundant, as was manganese, because both are very soluble in anoxic, sulfide-free waters. The high concentration of dissolved iron and manganese facilitated nitrogen fixation by early cyanobacteria, such that available nitrogen was abundant, and phosphorus became the nutrient limiting biological productivity (Anbar and Knoll, 2002). Oxygenation of the atmosphere at ~2.3 Ga led to increased oxidative weathering of sulfide minerals on the continents and increased sulfate concentration in seawater. Bacterial sulfate reduction generated ample sulfide, and in spite of limited mixing, the deep oceans would have remained anoxic and now also sulfidic as long as p_{O_2} remained below ~0.07 atm (Canfield, 1998), assuming reasonable rates of primary production. Both iron and manganese form insoluble sulfides, and thus were effectively scavenged from seawater once the oceans became sulfidic. Thus, the Proterozoic oxygenation led to significant changes in global oxygen balances, with the atmosphere and ocean mixed layer becoming mildly oxygenated (probably <0.01 atm O_2), and the deep oceans becoming strongly sulfidic in direct response to the rise of atmospheric O_2.

In addition to increased oxygen fugacity of volcanic gases, and innovations in biological productivity to include oxygenic photosynthesis, the oxygenation of the Proterozoic atmosphere may be related to large-scale tectonic cycles. There are several periods of maximum deposition of sedimentary rocks rich in organic matter through geologic time. These are ~2.7 Ga, 2.2 Ga, 1.9 Ga, and 0.6 Ga (Condie *et al.*, 2001). The increased deposition of black shales at 2.7 Ga and 1.9 Ga are associated with superplume events: highly elevated rates of seafloor volcanism, oceanic crust formation. Superplumes lead to increased burial of both organic matter and carbonates (through transgression, increased atmospheric CO_2, accelerated weathering, and nutrient fluxes to the oceans), with no net effect on carbonate isotopic composition. Thus, periods of increased absolute rates of organic matter burial and O_2 production may be masked by a lack of carbon isotopic signature. Breakup of supercontinents may be related to the black shale depositional events at 2.2 Ga and 0.6 Ga. Breakup of supercontinents may lead to more sediment accommodation space on continental shelves, as well as accelerated continental weathering and delivery of nutrients to seawater, fertilizing primary production and increasing organic matter burial. These supercontinent breakup events at 2.2 Ga and 0.6 Ga are clearly observed on the carbonate isotopic record as increases in relative burial of organic matter versus carbonate. Modeling efforts examining the evolution of the carbon, sulfur, and strontium isotope records have shown that gradual growth of the continents during the Archean and early Proterozoic may in fact play a very large role in controlling the onset of oxygenation of the atmosphere, and that biological innovation may not be directly coupled to atmospheric evolution (Godderís and Veizer, 2000).

8.11.6.3.2 Atmospheric O_2 during the Mesoproterozoic

The sulfur-isotope-based model of Canfield and colleagues and the implications for limiting nutrient distribution proposed by Anbar and Knoll (2002) suggest that for nearly thousand million years (~2.2–1.2 Ga), oxygenation of the atmosphere above p_{O_2} ~0.01 atm was held in stasis. Although oxygenic photosynthesis was active, and an atmospheric ozone shield had developed to protect surface-dwelling organisms from UV radiation, much of the deep ocean was still anoxic and sulfidic. Removal of iron and manganese from seawater as sulfides generated severe nitrogen stress for marine communities, suggesting that productivity may have been limited throughout the entire Mesoproterozoic. The sluggish but consistent primary productivity and organic carbon burial through this time is seen in the carbonate isotope record. For several hundred million years, carbonate isotopic composition varied by no more than ±2‰, revealing very little change in the relative carbonate/organic matter burial in marine sediments. Anbar and Knoll (2002) suggest that this indicates a decoupling of the link between tectonic events and primary production, because although variations in tectonic activity (and associated changes in sedimentation, generation of restricted basins, and continental weathering) occurred during the Mesoproterozoic, these are not observed in the carbonate isotope record. This decoupling is a natural result of a shift in the source of the biological limiting nutrient from phosphorus (derived from continental weathering) to nitrogen (limited by N_2 fixation). Furthermore, the isotopic composition of carbonates throughout the Mesoproterozoic is 1–2‰ depleted relative to average carbonates from the early and late Proterozoic and Phanerozoic. This is consistent with a decrease in the relative proportion of carbon buried as organic matter versus carbonate during this time. It appears that after initial oxygenation of the atmosphere from completely anoxic to low p_{O_2} in the early Proterozoic, further oxygenation was halted for several hundred million years.

Global-scale reinvigoration of primary production, organic matter burial and oxygenation of the atmosphere may be observed in the latest Mesoproterozoic. Shifts in carbonate $\delta^{13}C$ of up to 4‰ are observed in sections around the globe at ~1.3 Ga (Bartley *et al.*, 2001; Kah *et al.*, 2001). These positive carbon isotope excursions are associated with the formation of the Rodinian supercontinent, which led to increased continental margin length, orogenesis, and greater sedimentation (Bartley *et al.*, 2001). The increased organic matter burial and atmospheric oxygenation associated with this isotope excursion may be related to the first occurrence of laterally extensive $CaSO_4$ evaporites (Kah *et al.*, 2001). Although the rapid 10‰ increase in evaporate $\delta^{34}S$ across these sections is taken to indicate a much reduced seawater sulfate reservoir with much more rapid turnover times than found in the modern ocean, the isotope fractionation between evaporate sulfates and sedimentary sulfides indicates that the oceans were not sulfate-limiting. Atmospheric O_2 was of sufficient concentration to supply ample sulfate for bacterial sulfate reduction throughout the Mesoproterozoic.

8.11.6.3.3 Neoproterozoic atmospheric O_2

Elevated carbonate isotopic compositions (~4‰) and strong isotope fractionation between sulfate and biogenic sulfide minerals through the early Neoproterozoic indicates a period of several hundred million years of elevated biological productivity and organic matter burial. This may relate in part to the oxygenation of the atmosphere, ammonia oxidation, and increased seawater nitrate availability. Furthermore, oxidative weathering of molybdenum-bearing sulfide minerals, and greater oxygenation of the oceans increased availability of molybdenum necessary for cyanobacterial pathways of N_2-fixation (Anbar and Knoll, 2002). Much of the beginning of the Neoproterozoic, like the late Mesoproterozoic, saw gradual increases in oxygenation of the atmosphere, and possibly of the surface oceans. Limits on the oxygenation of the atmosphere are provided by sulfur isotopic and molecular evidence for the evolution of sulfide oxidation and sulfur disproportionation. The increase in sulfate–sulfide isotope fractionation in the Neoproterozoic (~1.0–0.6 Ga) reflects a shift in sulfur cycling from simple one-step reduction of sulfate to sulfide, to a system in which sulfide was oxidized to sulfur intermediates such as thiosulfate or elemental sulfur, which in turn were disproportionated into sulfate and sulfide (Canfield and Teske, 1996). Sulfur disproportionation is associated with significant isotope effects, generating sulfide that is substantially more ^{34}S-depleted than can be achieved through one-step sulfate reduction. The sulfur isotope record reveals that an increase in sulfate–sulfide isotope fractionation occurred between 1.0 Ga and 0.6 Ga, consistent with evolution of sulfide-oxidation at this time as derived from molecular clock based divergence of non-photosynthetic sulfide-oxidizing bacteria (Canfield and Teske, 1996). These authors estimated that innovation of bacterial sulfide oxidation occurred when much of the coastal shelf sediment (<200 m water depth) was exposed to water with 13–46 μM O_2, which corresponds to 0.01–0.03 atm p_{O_2}. Thus, after the initial oxygenation of the atmosphere in the early Proterozoic (~2.3–2.0 Ga) to ~0.01 atm, p_{O_2} was constrained to this level until a second oxygenation in the late Proterozoic (~1.0–0.6 Ga) to 0.03 atm or greater.

During the last few hundred million years of the Proterozoic (~0.7–0.5 Ga), fragmentation of the Rodinian supercontinent was associated with at least two widespread glacial episodes (Hoffman *et al.*, 1998; Hoffman and Schrag, 2002). Neoproterozoic glacial deposits are found in Canada, Namibia, Australia, and other locations worldwide (Evans, 2000). In some of these, paleomagnetic evidence suggests that glaciation extended completely from pole to equator (Sumner, 1997; see also Evans (2000) and references therein). The "Snowball Earth", as these events have come to be known, is associated with extreme fluctuations in carbonate isotopic stratigraphy, reoccurrence of BIFs after an ~1.0 Ga hiatus, and precipitation of enigmatic, massive cap carbonate sediments immediately overlying the glacial deposits (Figure 11). It has been proposed that the particular configuration of continents in the latest Neoproterozoic, with land masses localized within the middle–low latitudes, lead to cooling of climate through several mechanisms. These include higher albedo in the subtropics (Kirschvink, 1992), increased silicate weathering as the bulk of continents were located in the warm tropics, resulting in a drawdown of atmospheric CO_2

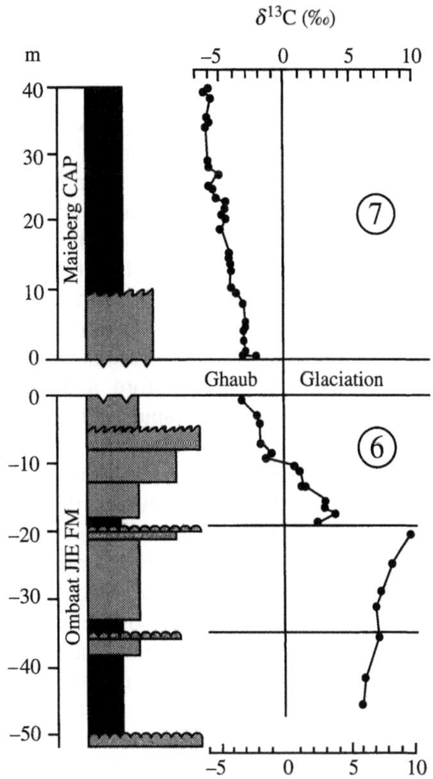

Figure 11 Composite section of carbonate isotope stratigraphy before and after the Ghaub glaciation from Namibia. Enriched carbonates prior to glaciation indicate active primary production and organic matter burial. Successive isotopic depletion indicates a shut down in primary production as the world became covered in ice (source Hoffman *et al.*, 1998).

(Hoffman and Schrag, 2002), possibly accelerated through high rates of OM burial, and reduced meridional Hadley cell heat transport, because tropical air masses were drier due to increased continentality (Hoffman and Schrag, 2002). Growth of initially polar ice caps would have created a positive ice–albedo feedback such that greater than half of Earth's surface became ice covered, global-scale growth of sea ice became inevitable (Pollard and Kasting, 2001; Baum and Crowley, 2001). Kirschvink (1992) recognized that escape from the Snowball Earth becomes possible because during extreme glaciation, the continental hydrologic cycle would be shut down. He proposed that sinks for atmospheric CO_2, namely, photosynthesis and silicated weathering, would have been eliminated during the glaciation. Because volcanic degassing continued during the glaciation, CO_2 in the atmosphere could rise to very high concentrations. It was estimated that an increase in p_{CO_2} to 0.12 bar would be sufficient to induce a strong enough greenhouse effect to begin warming the planet and melting the ice (Caldeira and Kasting, 1992). Once meltback began, it would subsequently accelerate through the positive ice–albedo feedback. Intensely warm temperatures would follow quickly after the glaciation, until accelerated silicate weathering under a reinvigorated hydrologic cycle could consume excess CO_2 and restore p_{CO_2} to more equable values. Carbonate mineral precipitation was inhibited during the global glaciation, so seawater became enriched in hydrothermally derived cations. At the end of glaciation, mixing of cation-rich seawater with high alkalinity surface runoff under warming climates led to the precipitation of massive cap carbonates.

Geochemical signatures before and after the Snowball episodes reveal rapid and short-lived changes in global biogeochemical cycles that impacted O_2 concentrations in the atmosphere and oceans. Prior to the Snowball events, the Neoproterozoic ocean experienced strong primary production and organic matter burial, as seen in the carbon isotopic composition of Neoproterozoic seawater in which positive isotope excursions reached 10‰ in some sections (Kaufman and Knoll, 1995). These excursions are roughly coincident with increased oxygenation of the atmosphere in the Neoproterozoic (Canfield and Teske, 1996; Des Marais *et al.*, 1992). Although sulfur isotopic evidence suggests that much of the ocean may have become oxygenated during the Neoproterozoic, it is very likely that this was a temporary phenomenon. The global-scale glaciations of the late Neoproterozoic would have driven the oceans to complete anoxia through several mechanisms (Kirschvink, 1992). Ice cover inhibited air–sea gas exchange, and thus surface waters and sites of deep-water formation were cut off from atmospheric O_2 supplies. Intense oxidation of organic matter in the water column and sediment would have quickly consumed any available dissolved O_2. Extreme positive sulfur-isotope excursions associated with snowball-succession deposits reflect nearly quantitative, closed-system sulfate reduction in the oceans during glaciation (Hurtgen *et al.*, 2002). As seawater sulfate concentrations were reduced, hydrothermal inputs of ferrous-iron exceeded sulfide supply, allowing BIFs to form (Hoffman and Schrag, 2002). Closed-system sulfate reduction and iron formations require ocean anoxia throughout the water column, although restricted oxygenic photosynthesis beneath thin tropical sea ice may be responsible for ferrous-iron oxidation and precipitation (Hoffman and Schrag, 2002).

Once global ice cover was achieved, it is estimated to have lasted several million years, based on the amount of time required to accumulate 0.12 bar CO_2 at modern rates of volcanic CO_2 emission (Caldeira and Kasting, 1992). This must have placed extreme stress on eukaryotes and other organisms dependent on aerobic respiration. Obviously, oxygenated refugia must have existed, because the rise of many eukaryotes predates the Neoproterozoic Snowball events (Butterfield and Rainbird, 1998; Porter and Knoll, 2000). Nonetheless, carbon isotopic evidence leading up to and through the intense glacial intervals suggests that overall biological productivity was severely repressed during Snowball events. Carbonate isotope compositions fall from extremely enriched values to depletions of ~(−6‰) at glacial climax (Hoffman *et al.*, 1998; Kaufman *et al.*, 1991; Kaufman and Knoll, 1995). These carbonate isotope values reflect primordial (volcanogenic) carbon inputs, and indicate that effectively zero organic matter was buried in sediments at this time. Sustained lack of organic matter burial and limited if any oxygenic photosynthesis under the glacial ice for several million years would have maintained an anoxic ocean, depleted in sulfate, with low sulfide concentrations due to iron scavenging, and extremely high alkalinity and hydrothermally-derived ion concentrations. Above the ice, the initially mildly oxygenated atmosphere would slowly lose its O_2. Although the hydrologic cycles were suppressed, and thus oxidative weathering of sulfide minerals and organic matter exposed on the continents was inhibited, the emission of reduced volcanic gases would have been enough to fully consume 0.01 bar O_2 within several hundred thousand years. Thus, it is hypothesized that during Snowball events, the Earth's atmosphere returned to pre-Proterozoic anoxic conditions for at least several million years. Gradual increases in carbonate $\delta^{13}C$ after glaciation and deposition of cap carbonates suggests slow restored increases in primary productivity and organic carbon burial.

Carbon isotopic evidence suggests that reoxygenation, at least of seawater, was extremely gradual throughout the remainder of the Neoproterozoic. Organic matter through the terminal Proterozoic derives largely from bacterial heterotrophs, particularly sulfate reducing bacteria, as opposed to primary producers (Logan *et al.*, 1995). These authors suggested that throughout the terminal Neoproterozoic, anaerobic heterotrophy dominated by sulfate reduction was active throughout the water column, and O_2 penetration from surface waters into the deep ocean was inhibited. Shallow-water oxygen-deficient environments became widespread at the Precambrian–Cambrian boundary (Kimura and Watanabe, 2001), corresponding to negative carbonate $\delta^{13}C$ excursions and significant biological evolution from Ediacaran-type metazoans to emergence of modern metazoan phyla in the Cambrian.

8.11.6.4 Phanerozoic Atmospheric O_2

8.11.6.4.1 Constraints on Phanerozoic O_2 variation

Oxygenation of the atmosphere during the latest Neoproterozoic led to a fairly stable and well-oxygenated atmosphere that has persisted through the present day. Although direct measurement or a quantifiable proxy for Phanerozoic paleo-p_{O_2} concentrations have not been reported, multiple lines of evidence point to upper and lower limits on the concentration of O_2 in the atmosphere during the past several hundred million years. Often cited is the nearly continuous record of charcoal in sedimentary rocks since the evolution of terrestrial plants some 350 Ma. The presence of charcoal indicates forest fires throughout much of the Phanerozoic, which are unlikely to have occurred below $p_{O_2} = 0.17$ atm (Cope and Chaloner, 1985). Combustion and sustained fire are difficult to achieve at lower p_{O_2}. The existence of terrestrial plants themselves also provides a crude upper bound on p_{O_2} for two reasons. Above the compensation point of p_{O_2}/p_{CO_2} ratios, photorespiration outcompetes photosynthesis, and plants experience zero or negative net growth (Tolbert *et al.*, 1995). Although plants have developed various adaptive strategies to accommodate low atmospheric p_{CO_2} or aridity (i.e. C_4 and CAM plants), net terrestrial photosynthesis and growth of terrestrial ecosystems are effectively inhibited if p_{O_2} rises too high. This upper bound is difficult to exactly constrain, but is likely to be ~0.3–0.35 atm. Woody tissue is also extremely susceptible to combustion at high p_{O_2}, even if the tissue is wet. Thus, terrestrial ecosystems would be unlikely above some upper limit of p_{O_2} because very frequent reoccurrence of wildfires would effectively wipe out terrestrial plant communities, and there is no evidence of this occurring during the Phanerozoic on a global scale. What this upper p_{O_2} limit is, however, remains disputed. Early experiments used combustion of paper under varying humidity and p_{O_2} (Watson, 1978) but paper may not be the most appropriate analog for inception of fires in woody tissue with greater moisture content and thermal thickness (Robinson, 1989; Berner *et al.*, 2002). Nonetheless, the persistence of terrestrial plant communities from the middle Paleozoic through the present does imply that p_{O_2} concentrations have not risen too high during the past 350 Myr.

Prior to evolution of land plants, even circumstantial constraints on early Paleozoic p_{O_2} are difficult to obtain. Invertebrate metazoans, which have a continous fossil record from the Cambrian on, require some minimum amount of dissolved oxygen to support their aerobic metabolism. The absolute minimum dissolved O_2 concentration able to support aerobic metazoans varies from species to species, and is probably impossible to reconstruct for extinct lineages, but modern infaunal and epifaunal metazoans can accommodate dissolved O_2 dropping to the tens of micromolar in concentration. If these concentrations are extrapolated to equilibrium of the atmosphere with well-mixed cold surface waters, they correspond to ~0.05 atm p_{O_2}. Although local and regional anoxia occurred in the oceans at particular episodes through Phanerozoic time, the continued presence of aerobic metazoans suggests that widespread or total ocean anoxia did not occur during the past ~600 Myr, and that p_{O_2} has been maintained at or above ~5% O_2 since the early Paleozoic.

8.11.6.4.2 Evidence for variations in Phanerozoic O_2

Several researchers have explored possible links between the final oxygenation of the atmosphere and explosion of metazoan diversity at the Precambrian–Cambrian boundary (McMenamin and McMenamin, 1990; Gilbert, 1996).While the Cambrian explosion may not record the origin of these phyla in time, this boundary does record the development of large size and hard skeletons required for fossilization (Thomas, 1997). Indeed, molecular clocks for diverse metazoan lineages trace the origin of these phyla to ~400 Myr before the Cambrian explosion (Doolittle *et al.*, 1996; Wray *et al.*, 1996). The ability of lineages to develop fossilizable hardparts may be linked with increasing oxygenation during the latest Precambrian. Large body size requires elevated p_{O_2} and dissolved O_2 concentrations, so the diffusion can supply O_2 to internal tissues. Large body size provides several advantages, and may have evolved rather quickly during the earliest Cambrian (Gould, 1995), but large size also requires greater structural support. The synthesis of collagen (a ubiquitous structural protein among all metazoans and possible precursor to inorganic structural components such as carbonate or phosphate biominerals) requires elevated O_2. The threshold for collagen biosynthesis, and associated skeletonization and development of large body size, could not occur until p_{O_2} reached some critical threshold some time in the latest Precambrian (Thomas, 1997).

There is evidence to suggest that the late Paleozoic was a time of very elevated p_{O_2}, to concentrations substantially greater than observed in the modern atmosphere. Coals from the Carboniferous and Permian contain a greater abundance of fusain, a product of woody tissue combustion and charring, than observed for any period of the subsequent geologic past (Robinson, 1989, 1991; Glasspool, 2000), suggesting more abundant forest fires and by implication possibly higher p_{O_2} at this time. Less ambiguous is the biological innovation of gigantism at this time among diverse arthropod lineages (Graham *et al.*, 1995). All arthropods rely on tracheal networks for diffusion of O_2 to support their metabolism; active pumping of O_2 through a vascular system as found in vertebrates does not occur. This sets upper limits on body size for a given p_{O_2} concentration. In comparison with arthropod communities of the Carboniferous and Permian, modern terrestrial arthropods are rather small. Dragonflies at the time reached 70 cm wingspan, mayflies reached 45 cm wingpans, millipedes reached 1 m in length. Even amphibians, which depend in part on diffusion of O_2 through their skin for aerobic respiration, reached gigantic size at this time. These large body sizes could not be supported by today's 21% O_2 atmosphere, and instead require elevated p_{O_2} between 350 Ma and 250 Ma. Insect taxa that were giants in the Carboniferous do not survive past the Permian, suggesting that declining p_{O_2} concentrations in the Permian and Mesozoic led to extinction (Graham *et al.*, 1995; Dudley, 1998).

Increases in atmospheric O_2 affect organisms in several ways. Increased O_2 concentration facilitates aerobic respiration, while elevated p_{O_2} against a constant p_{N_2} increases total atmospheric pressure, with associated changes in atmospheric gas density and viscosity (Dudley, 1998, 2000). In tandem these effects may have played a strong role in the innovation of insect flight in the Carboniferous (Dudley, 1998, 2000), with secondary peaks in the evolution of flight among birds, bats and other insect lineages corresponding to times of high O_2 in the late Mesozoic (Dudley, 1998).

In spite of elevated p_{O_2} in the late Paleozoic, leading up to this time was an extended period of water column anoxia and enhanced burial of organic matter in the Devonian. Widespread deposition of black shales, fine-grained laminated sedimentary rocks rich in organic matter, during the Devonian indicate at least partial stratification of several ocean basins around the globe, with oxygen deficiency throughout the water column. One example of this is from the Holy Cross Mountains of Poland, from which particular molecular markers for green sulfur bacteria have been isolated (Joachimski *et al.*, 2001). These organisms are obligately anaerobic chemophotoautotrophs, and indicate that in the Devonian basin of central Europe, anoxia extended upwards through the water column well into the photic zone. Other black shales that indicate at least episodic anoxia and enhanced organic matter burial during the Devonian are found at several sites around the world, including the Exshaw Shale of Alberta (Canada), the Bakken Shale of the Williston Basin (Canada/USA), the Woodford Shale in Oklahoma (USA), and the many Devonian black shales of the Illinois, Michigan, and Appalachian Basins (USA). Widespread burial of organic matter in the late Devonian has been linked to increased fertilization of the surface waters through accelerated continental weathering due to the rise of terrestrial plant communities (Algeo and Scheckler, 1998). High rates of photosynthesis with relatively small rates of global respiration led to accumulation of organic matter in marine sediments, and the beginnings of a pulse of atmospheric hyperoxia that extended through the Carboniferous into the Permian. Coalescence of continental fragments to form the Pangean supercontinent at this time led to widespread circulation-restricted basins that facilitated organic matter burial and net oxygen release, and later, to extensive infilling to generate near-shore swamps containing terrestrial vegetation which was often buried to form coal deposits during the rapidly fluctuating sea levels of the Carboniferous and Permian.

The largest Phanerozoic extinction occurred at the end of the Permian (~250 Ma). A noticeable decrease in the burial of organic matter in marine sediments across the Permian–Triassic boundary may be associated with a global decline in primary productivity, and thus, with atmospheric p_{O_2}. The gigantic terrestrial insect lineages, thought to require elevated p_{O_2}, do not survive across this boundary, further suggesting a global drop in p_{O_2}, and the sedimentary and sulfur isotope records indicate an overall increase in sulfate reduction and burial of pyritic shales (Berner, 2002; Beerling and Berner, 2000). Although a long-duration deep-sea anoxic event has been proposed as a cause for the Permian mass extinction, there are competing models to explain exactly how this might have occurred. Hotinski *et al.*, (2001) has shown that while stagnation of the water column to generate deep-water anoxia might at first seem attractive, global thermohaline stagnation would starve the oceans of nutrients, extremely limiting primary productivity, and thus shutting down dissolved O_2 demand in the deep oceans. Large negative excursions in carbon, sulfur, and strontium isotopes during the late Permian may indicate stagnation and reduced ventilation of seawater for extended periods, coupled with large-scale overturn of anoxic waters. Furthermore, sluggish thermohaline circulation at this time could derive from a warmer global climate and warmer water at the sites of high-latitude deep-water formation (Hotinski *et al.*, 2001). The late Permian paleogeography of one supercontinent (Pangea) and one superocean (Panthalassa) was very different from the arrangement of continents and oceans on the modern Earth. Coupled with elevated p_{CO_2} at the time (Berner, 1994; Berner and Kothavala, 2001), GCM models predict warmer climate, weaker wind stress, and low equator to pole temperature gradients. Although polar deep-water formation still occurred, bringing O_2 from the atmosphere to the deep oceans, anoxia was likely to develop at mid-ocean depths (Zhang *et al.*, 2001), and thermohaline circulation oscillations between thermally versus salinity driven modes of circulation were likely to develop. During salinity driven modes, enhanced bottom-water formation in warm, salty low-latitude regions would limit oxygenation of the deep ocean. Thus, although sustained periods of anoxia are unlikely to have developed during the late Permian, reduced oxygenation of deep water through sluggish thermohaline circulation, coupled with episodic anoxia driven by low-latitude warm salty bottom-water formation, may have led to reoccurring episodes of extensive ocean anoxia over period of several million years.

Other researchers have invoked extraterrestrial causes for the End-Permian extinction and anoxia. Fullerenes (cage-like hydrocarbons that effectively trap gases during formation and heating) have been detected in late Permian sediments from southern China. The noble gas complement in these fullerenes indicates an extraterrestrial origin, which has been interpreted by Kaiho *et al.*, (2001) to indicate an unrecognized bolide impact at the Permian–Triassic boundary. The abrupt decrease in $\delta^{34}S$ across this boundary (from 20‰ to 5‰) implies an enormous and rapid release of ^{34}S-depleted sulfur into the ocean–atmosphere system. These authors propose that volatilized bolide- and mantle-derived sulfur (~0‰) oxidized in air, consumed atmospheric and dissolved O_2, and generated severe oxygen and acid stress in the oceans. Isotope mass balance estimates require $\sim 10^{19}$ mol sulfur to be released, consuming a similar mass of oxygen. 10^{19} mol O_2 represents some 10–40% of the total available inventory of atmospheric and dissolved O_2 at this time, removal of which led to immediate anoxia, as these authors propose.

Other episodes of deep ocean anoxia and extensive burial of organic matter are known from the Jurassic, Cretaceous, Miocene, and Pleistocene, although these have not been linked to changes in atmospheric O_2 and instead serve as examples of the decoupling between atmospheric and deep-ocean O_2 concentrations through much of the geologic past. Widespread Jurassic black shale facies in northern Europe (Posidonien Schiefer, Jet Rock, and Kimmeridge Clay) were deposited in a restricted basin on a shallow continental margin. Strong monsoonal circulation led to extensive freshwater discharge and a low-salinity cap on basin waters during summer, and intense evaporation and antiestuarine circulation during winter (Röhl *et al.*, 2001), both of which contributed to water column anoxia and black shale deposition. Several oceanic anoxic events (OAEs) are recognized from the Cretaceous in all major ocean basins, suggesting possible global deep-ocean anoxia. Molecular markers of green sulfur bacteria, indicating photic zone anoxia, have been detected from Cenomanian–Turonian boundary section OAE sediments from the North Atlantic (Sinninghe Damsté and Köster, 1998). The presence of these markers (namely, isorenieratene, a diaromatic carotenoid accessory pigment used during anoxygenic photosynthesis) indicates that the North Atlantic was anoxic and euxinic from the base of the photic zone (~50–150 m) down to the sediment. High concentrations of trace metals scavenged by sulfide and an absence of bioturbation further confirm anoxia throughout the water column. Because mid-Cretaceous oceans were not highly productive, accelerated dissolved O_2 demand from high rates of respiration and primary production cannot be the prime cause of these OAEs. Most likely, the warm climate of the Cretaceous led to low O_2 bottom waters generated at warm, high salinity regions of low-latitude oceans. External forcing, perhaps through Milankovic-related precession-driven changes in monsoon intensity and strength, influenced the rate of salinity-driven deep-water formation, ocean basin oxygenation, and OAE formation (Wortmann *et al.*, 1999). Sapropels are organic matter-rich layers common to late Cenozoic sediments of the eastern Mediterranean. They are formed through a combination of increased primary production in surface waters, and increased organic matter preservation in the sediment likely to be associated with changes in ventilation and oxygenation of the deep eastern Mediterranean basins (Stratford *et al.*, 2000). The well-developed OMZ located off the coast of Southern California today may have been more extensive in the past. Variations in climate affecting intensity of upwelling and primary production, coupled with tectonic activity altering the depth of basins and height of sills along the California coast, have generated a series of anoxia-facies organic-matter-rich sediments along the west coast of North America, beginning with the Monterey Shale and continuing through to the modern sediments deposited in the Santa Barbara and Santa Monica basins.

As shown by the modeling efforts of Hotinski *et al.*, (2001) and Zhang *et al.*, (2001), extensive deep ocean anoxia is difficult to achieve for extended periods of geologic time during the Phanerozoic when p_{O_2} were at or near modern levels. Thus, while localized anoxic basins are common, special conditions are required to generate widespread, whole ocean anoxia such as observed in the Cretaceous. Deep-water formation in highly saline low-latitude waters was likely in the geologic past when climates were warmer and equator to pole heat gradients were reduced. Low-latitude deep-water formation has a significant effect on deep-water oxygenation, not entirely due to the lower O_2 solubility in warmer waters, but also the increased efficiency of nutrient use and recycling in low-latitude surface waters (Herbert and Sarmiento, 1991). If phytoplankton were 100% efficient at using and recycling nutrients, even with modern high-latitude modes of cold deep-water formation, the deep oceans would likely become anoxic.

8.11.6.4.3 Numerical models of Phanerozoic oxygen concentration

Although photosynthesis is the ultimate source of O_2 to the atmosphere, in reality photosynthesis and aerobic respiration rates are very closely coupled. If they were not, major imbalances in atmospheric CO_2, O_2, and carbon isotopes would result. Only a small fraction of primary production (from photosynthesis) escapes respiration in the water column or sediment to become buried in deep sediments and ultimately sedimentary rocks. This flux of buried organic matter is in effect "net photosynthesis", or total photosynthesis minus respiration. Thus, while over timescales of days to months, dissolved and atmospheric O_2 may respond to relative rates of photosynthesis or respiration, on longer timescales it is burial of organic matter in sediments (the "net photosynthesis") that matters. Averaged over hundreds of years or longer, burial of organic matter equates to release of O_2 into the "atmosphere + ocean" system:

Photosynthesis:

$$6CO_2 + 6H_2O \rightarrow C_6H_{12}O_6 + 6O_2 \qquad (3)$$

Respiration:

$$C_6H_{12}O_6 + 6O_2 + 6CO_2 + 6H_2O \qquad (4)$$

If for every 1,000 rounds of photosynthesis there are 999 rounds of respiration, the net result is one round of organic matter produced by photosynthesis that is not consumed by aerobic respiration. The burial flux of organic matter in sediments represents this lack of respiration, and as such is a net flux of O_2 to the atmosphere. In geochemists' shorthand, we represent this by the reaction

Burial of organic matter in sediments:

$$CO_2 + H_2O \rightarrow O_2 + \text{“}CH_2O\text{”} \qquad (5)$$

where "CH_2O" is not formaldehyde, or even any specific carbohydrate, but instead represents sedimentary organic matter. Given the elemental composition of most organic matter in sediments and sedimentary rocks, a more reduced organic matter composition might be more appropriate, i.e., $C_{10}H_{12}O$, which would imply release of 12.5 mol O_2 for every mole of CO eventually buried as organic matter. However, the simplified stoichiometry of (5) is applied for most geochemical models of C–S–O cycling.

If burial of organic matter equates to O_2 release to the atmosphere over long timescales, then the oxidative weathering of ancient organic matter in sedimentary rocks equates to O_2 consumption. This process has been called "georespiration" by some authors (Keller and Bacon, 1998). It can be represented by Equation (6), the reverse of (5):

Weathering of organic matter from rocks:

$$O_2 + \text{“}CH_2O\text{”} \rightarrow CO_2 + H_2O \qquad (6)$$

Both Equations (3) and (4) contain terms for addition and removal of O_2 from the atmosphere. Thus, if we can reconstruct the rates of burial and weathering of OM into/out of sedimentary rocks through time, we can begin to quantify sources and sinks for atmospheric O_2. The physical manifestation of this equation is the reaction of organic matter with O_2 during the weathering and erosion of sedimentary rocks. This is most clearly seen in the investigation of the changes in OM abundance and composition in weathering profiles developed on black shales (Petsch *et al.*, 2000, 2001).

In addition to the C–O system, the coupled C–S–O system has a strong impact on atmospheric oxygen (Garrels and Lerman, 1984; Kump and Garrels, 1986; Holland, 1978, 1984). This is through the bacterial reduction of sulfate to sulfide using organic carbon substrates as electron donors. During bacterial sulfate reduction (BSR), OM is oxidized and sulfide is produced from sulfate. Thus, BSR provides a means of resupplying oxidized carbon to the "ocean + atmosphere" system without consuming O_2. The net reaction for BSR shows that for every 15 mol of OM consumed, 8 mol sulfate and 4 mol ferric-iron are also reduced to form 4 mol of pyrite (FeS_2):

$$4\,Fe(OH)_3 + 8\,SO_4^{2-} + 15\,CH_2O \rightarrow 4\,FeS + 15\,HCO_3 + 13\,H_2O + (OH) \qquad (7)$$

Oxidation of sulfate using organic substrates as electron donors provides a means of restoring inorganic carbon to the ocean + atmosphere system without consuming free O_2. Every 4 mol of pyrite derived from BSR buried in sediments represents 15 mol of O_2 produced by photosynthesis to generate organic matter that will not be consumed through aerobic respiration. In effect, pyrite burial equates to net release of O_2 to the atmosphere, as shown by (8), obtained by the addition of Equation (7) to (5):

$$4Fe(OH)_3 + 8SO_4^{2-} + 15CH_2O \rightarrow 4FeS_2 + 15HCO_3^- + 13H_2O + (OH)^-$$

$$+15CO_2 + 15H_2O \rightarrow 15O_2 + 15\text{“}CH_2O\text{”}$$

$$4Fe(OH)_3 + 8SO_4^{2-} + 15CO_2 + 2H_2O \rightarrow 4FeS_2 + 15HCO_3^- + (OH)^- + 15O_2 \quad (8)$$

Oxidative weathering of sedimentary sulfide minerals during exposure and erosion on the continents results in consumption of O_2 (Equation (9):

$$4FeS_2 + 15O_2 + 14H_2O \rightarrow 4Fe(OH)_3 + 8SO_4^{2-} + 16H^+ \quad (9)$$

Just as for the C–O geochemical system, if we can reconstruct the rates of burial and oxidative weathering of sedimentary sulfide minerals through geologic time, we can use these to estimate additional sources and sinks for atmospheric O_2 beyond organic matter burial and weathering.

In total, then, the general approach taken in modeling efforts of understanding Phanerozoic O_2 variability is to catalog the total sources and sinks for atmospheric O_2, render these in the form of a rate of change equation in a box model, and integrate the changing O_2 mass through time implied by changes in sources and sinks:

$$\begin{aligned} dM_{O_2}/dt &= \sum F_{O_2} \text{ into the atmosphere} - \sum F_{O_2} \text{out of the atmosphere} \\ &= F_{\text{burial of organic matter}} + (15/8)F_{\text{burial of pyrite}} \\ &\quad - F_{\text{weathering of organic matter}} - (15/8)F_{\text{weathering of pyrite}} \end{aligned} \quad (10)$$

One approach to estimate burial and weathering fluxes of organic matter and sedimentary sulfides through time uses changes in the relative abundance of various sedimentary rock types estimated over Phanerozoic time. Some sedimentary rocks are typically rich in both organic matter and pyrite. These are typically marine shales. In contrast, coal basin sediments contain much organic carbon, but very low amounts of sedimentary sulfides. Non-marine coarse-grained clastic sediments contain very little of either organic matter or sedimentary sulfides. Berner and Canfield (1989) simplified global sedimentation through time into one of three categories: marine shales + sandstone, coal basin sediments, and non-marine clastic sediments. Using rock abundance estimates derived from the data of Ronov and others (Budyko *et al.*, 1987; Ronov, 1976), these authors estimated burial rates for organic matter and pyrite as a function of time for the past ~600 Myr (Figure 12). Weathering rates for sedimentary organic matter and pyrite were calculated as first order dependents on the total mass of sedimentary organic matter or pyrite, respectively. Although highly simplified, this model provided several new insights into global-scale coupling C–S–O geochemistry. First, the broad-scale features of Phanerozoic O_2 evolution were established. O_2 concentrations in the atmosphere were low in the early Paleozoic, rising to some elevated p_{O_2} levels during the Carboniferous and Permian (probably to a concentration substantially greater than today's 0.21 bar), and then falling through the Mesozoic and Cenozoic to more modern values. This model confirmed the suspicion that in contrast with the Precambrian, Phanerozoic O_2 evolution was a story of relative stability through time, with no great excursions in p_{O_2}. Second, by linking C–S–O cycles with sediments and specifically sedimentation rates, this model helped fortify the idea that a strong control on organic matter burial rates globally, and thus ultimately on release of O_2 to the atmosphere, may be rates of sedimentation in near-shore environments. These authors extended this idea to propose that the close linkage between sedimentation and erosion (i.e., the fact that global rates of sedimentation are matched nearly exactly to global rates of sediment production—in other words, erosion) may in fact be a stabilizing influence on atmospheric O_2 fluctuations. If higher sedimentation rates result in greater burial of organic matter and pyrite, and greater release of O_2 to the atmosphere, at the same time there will be greater rates of erosion on the continents, some of which will involve oxidative weathering of ancient organic matter and/or pyrite.

The other principal approach towards modeling the Phanerozoic evolution of atmospheric O_2 rests on the isotope systematics of the carbon and sulfur geochemical cycles. The significant isotope discriminations associated with biological fixation of CO_2 to generate biomass and with bacterial reduction of sulfate to sulfide have been mentioned several times previously in this chapter. Given a set of simplifications of the exogenic cycles of carbon, sulfur and oxygen, these isotopic discriminations and the isotopic composition of seawater through time ($\delta^{13}C$, $\delta^{34}S$) can be used to estimate global rates of burial and weathering of organic matter, sedimentary carbonates, pyrite, and evaporative sulfates (Figure 13).

(i) The first required simplification is that the total mass of exogenic carbon is constant through time (carbon in the oceans, atmosphere, and sedimentary rocks). This of course neglects inputs of carbon and sulfur from volcanic activity and metamorphic degassing, and outputs into the mantle at subduction zones. However, if these fluxes into and out of the exogenic cycle are small enough (or have no effect on bulk crustal carbon or sulfur isotopic composition), then this simplification may be acceptable.

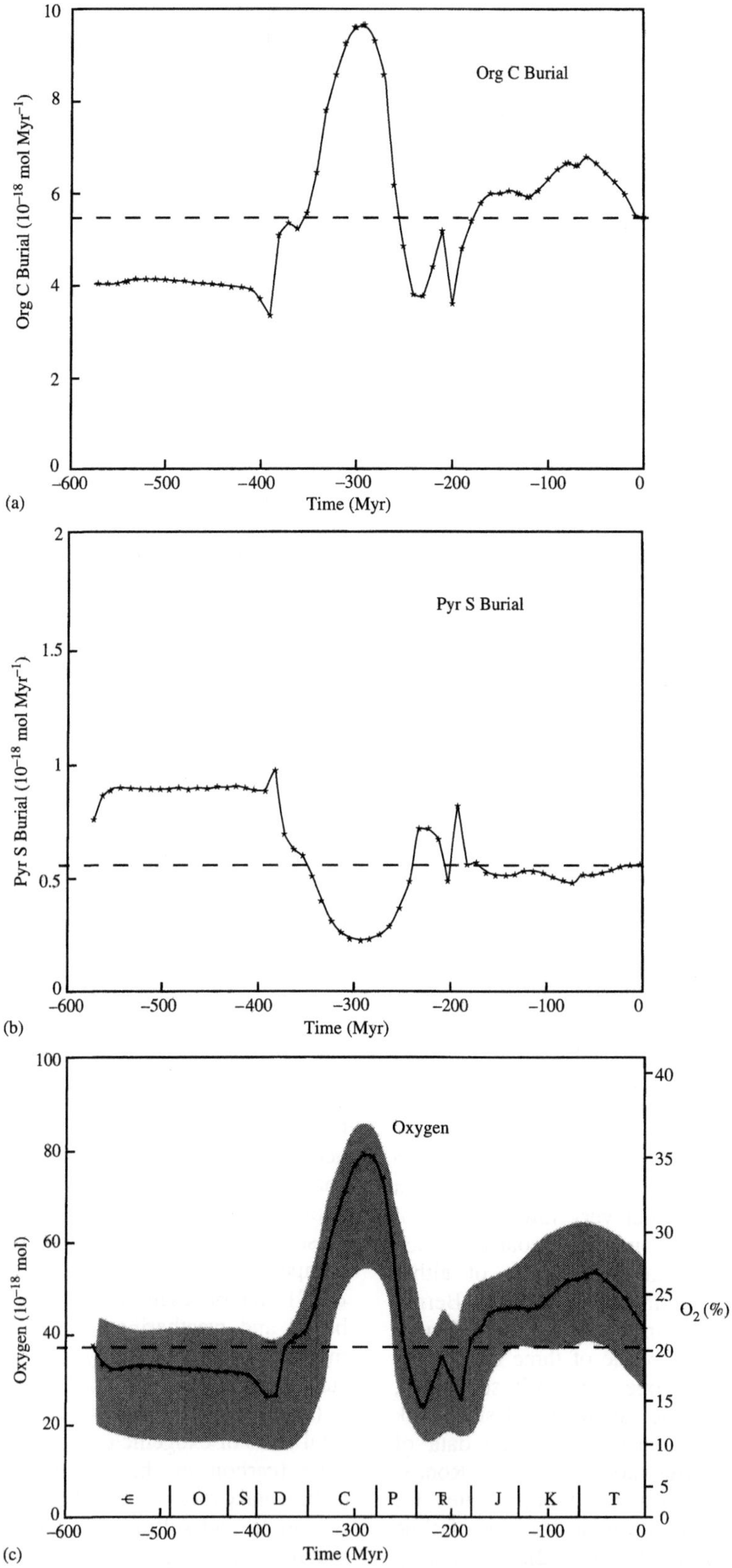

Figure 12 Estimated organic carbon burial (a), pyrite sulfur burial (b), and atmospheric oxygen concentrations (c) through Phanerozoic time, derived from estimates of rock abundance and their relative organic carbon and sulfide content (source Berner and Canfield, 1989).

(ii) The second simplification is that the total mass of carbon and sulfur dissolved in seawater plus the small reservoir of atmospheric carbon and sulfur gases remains constant through time. For carbon, this may be a realistic simplification. Dissolved inorganic carbon in seawater is strongly buffered by carbonate mineral precipitation and dissolution, and thus it is unlikely that extensive regions of the world ocean could have become significantly enriched or depleted in inorganic carbon during the geologic past. For sulfur, this assumption may not be completely accurate. Much of the interpretation of sulfur isotope records (with implications for atmospheric O_2 evolution) in the Precambrian depend on varying, but generally low dissolved sulfate concentrations. Unlike carbonate, there is no great buffering reaction maintaining stable sulfate concentrations in seawater. And while the sources of sulfate to seawater have likely varied only minimally, with changes in the sulfate flux to seawater increasing or decreasing smoothly through time as the result of broad-scale tectonic activity and changes in bulk continental weathering rates, removal of sulfate through excessive BSR or rapid evaporate formation may be much more episodic through time, possibly resulting in fairly extensive shifts in seawater sulfate concentration, even during the Phanerozoic. Nonetheless, using this simplification allows us to establish that for C–S–O geochemical models, total weathering fluxes for carbon and sulfur must equal total burial fluxes for carbon and sulfur, respectively.

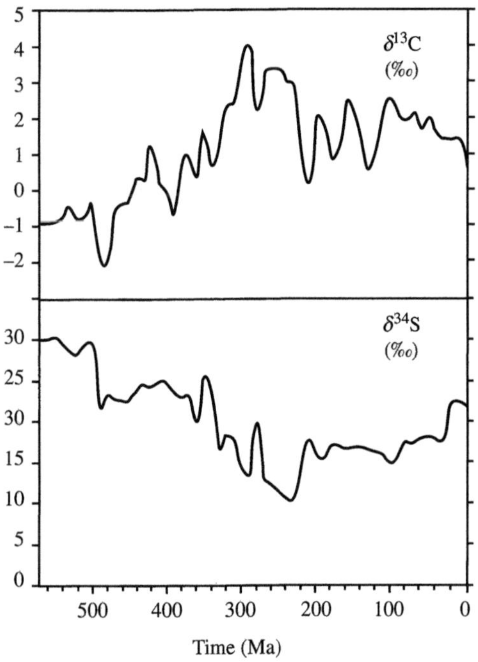

Figure 13 Globally averaged isotopic composition of carbonates ($\delta^{13}C$) and sulfates ($\delta^{34}S$) through Phanerozoic time (source Lindh, 1983).

Referring to Figure 14, we can see that these simplifications allow us to say that

$$dM_{oc}/dt = F_{wg} + F_{wc} - F_{bg} - F_{bc} = 0 \quad (11)$$

$$dM_{os}/dt = F_{ws} + F_{wp} - F_{bs} - F_{bp} = 0 \quad (12)$$

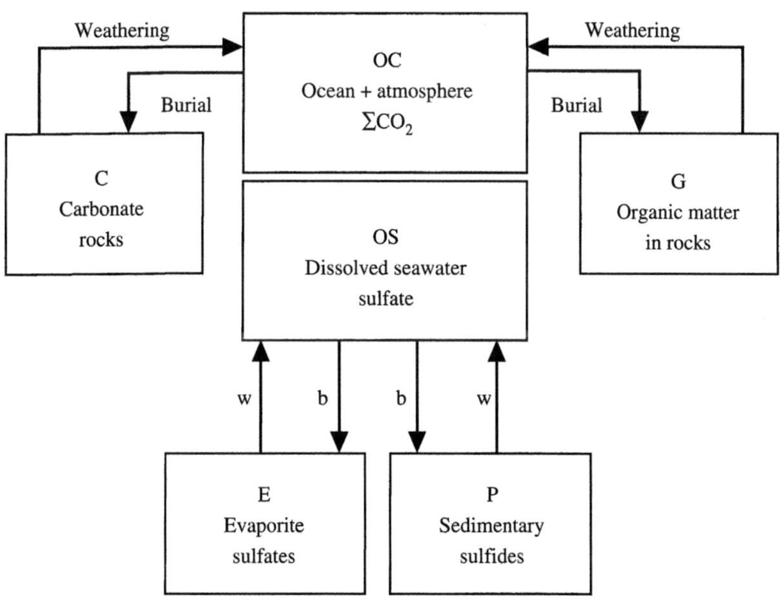

Figure 14 The simplified geochemical cycles of carbon and sulfur, including burial and weathering of sedimentary carbonates, organic matter, evaporites, and sulfides. The relative fluxes of burial and weathering of organic matter and sulfide minerals plays a strong role in controlling the concentration of atmospheric O_2.

The full rate equations for each reservoir mass in Figure 14 are as follows:

$$dM_c/dt = F_{bc} - F_{wc} \quad (13)$$

$$dM_g/dt = F_{bg} - F_{wg} \quad (14)$$

$$dM_s/dt = F_{bs} - F_{ws} \quad (15)$$

$$dM_p/dt = F_{bp} - F_{wp} \quad (16)$$

$$dM_{O_2}/dt = F_{bg} - F_{wg} + 15/8(F_{bp} - F_{wp}) \quad (17)$$

This system of equations has four unknowns: two burial fluxes and two weathering fluxes.

(iii) At this point, a third simplification of the carbon and sulfur systems is often applied to the weathering fluxes of sedimentary rocks. As a first approach, it is not unreasonable to guess that the rate of weathering of a given type of rock relates in some sense to the total mass of that rock type available on Earth's surface. If that relation is assumed to be first order with respect to rock mass, an artificial weathering rate constant for each rock reservoir can be derived. Such constants have been derived by assuming that the weathering rate equation has the form $F_{wi} = k_i M_i$. If we can establish the mass of a sedimentary rock reservoir i, and also the average global river flux to the oceans due to weathering of reservoir i, then k_i can easily be calculated. For example, if the total global mass of carbonate in sedimentary rock is 5000×10^{18} mol C, and annually there are 20×10^{12} mol C discharged from rivers to the oceans from carbonate rock weathering, then $k_{carbonate}$ becomes $(20 \times 10^{12}$ mol C yr$^{-1})$/ $(5,000 \times 10^{18}$ mol C), i.e., 4×10^{-3} Myr^{-1}. These simple first-order weathering rate constants have been calculated for each sedimentary rock reservoir in the C–S–O cycle, derived entirely from estimated preanthropogenic carbon and sulfur fluxes from continental weathering. Lack of a true phenomenological relationship relating microscale and outcrop-scale rock weathering reactions to regional- and global-scale carbon and sulfur fluxes remains one of the primary weaknesses limiting the accuracy of numerical models of the coupled C–S–O geochemical cycles.

If weathering fluxes from eroding sedimentary rocks are independently established, such as through use of mass-dependent weathering fluxes, then all of the weathering fluxes in our system of equations above become effectively known. This leaves only the burial fluxes as unknowns in solving the evolution of the C–S–O system.

The exact isotope discrimination that occurs during photosynthesis and during BSR is dependent on many factors. For carbon, these include cell growth rate, geometry and nutrient availability (Rau *et al.*, 1989, 1992), CO_2 availability species-specific effects, modes of CO_2 sequestration (i.e., C_3, C_4, and CAM plants). For sulfur, these can include species–specific effects, the degree of closed-system behavior and sulfate concentration, sulfur oxidation and disproportionation. However, as a simplification in geochemical modeling of the C–S–O cycles, variability in carbon and sulfur isotopic fractionation is limited. In the simplest case, fractionations are constant through all time for all environments. As a result, for example, the isotopic composition of organic matter buried at a given time is set at a constant 25‰ depletion relative to seawater dissolved carbonate at that time, and pyrite isotopic composition is set at a constant 35‰ depletion relative to seawater sulfate. Of course, in reality, α_c and α_s (the isotopic discriminations assigned between inorganic–organic carbon and sulfate–sulfide, respectively) vary greatly in both time and space. Regardless of how α_c and α_s are set, however, once fractionations have been defined, the mass balance equations given in (13)–(17) above can be supplemented with isotope mass balances as well.

Based on the first simplification listed above, the exogenic cycles of carbon and sulfur are regarded as closed systems. As such, the bulk isotopic composition of average exogenic carbon and sulfur do not vary through time. We can write a rate equation for the rate of change in (mass × isotopic composition) of each reservoir in Figure 14 to reflect this isotope mass balance. For example:

$$\frac{d(\delta_{oc}M_{oc})}{dt} \equiv \frac{\delta_{oc}dM_{oc}}{dt} + \frac{M_{oc}d\delta_{oc}}{dt} = \delta_c F_{wc} + \delta_c F_{wg} - \delta_{oc}F_{bg} - (\delta_{oc} - \alpha_c)F_{bg} \quad (18)$$

Using the simplification that $dM_{oc}/dt = 0$, Equation (18) reduces to an equation relating the organic matter burial flux F_{bg} in terms of other known entities:

$$F_{bg} = \left(\frac{1}{\alpha_c}\right)\left[M_{oc}\left(\frac{d\delta_{oc}}{dt}\right) + F_{wc}(\delta_{oc} - \delta_c) + F_{wg}(\delta_{oc} - \delta_g)\right] \quad (19)$$

Using (11) above,

$$F_{bc} = F_{wg} + F_{wc} - F_{bg} \quad (20)$$

A similar pair of equations can be written for the sulfur system:

$$F_{bp} = \left(\frac{1}{\alpha_s}\right)\left[M_{os}\left(\frac{d\delta_{os}}{dt}\right) + F_{ws}(\delta_{os} - \delta_s) + F_{wp}(\delta_{os} - \delta_p)\right] \quad (21)$$

$$F_{bs} = F_{wp} + F_{ws} - F_{bp} \quad (22)$$

Full rate equations for the rate of change in sedimentary rock reservoir isotopic composition can be written as

$$\frac{d\delta_c}{dt} = \frac{F_{bc}(\delta_{oc} - \delta_c)}{M_c} \quad (23)$$

$$\frac{d\delta_g}{dt} = \frac{F_{bg}(\delta_{oc} - \alpha_c - \delta_g)}{M_g} \quad (24)$$

$$\frac{d\delta_s}{dt} = \frac{F_{bw}(\delta_{os} - \delta_s)}{M_s} \quad (25)$$

$$\frac{d\delta_p}{dt} = \frac{F_{bp}(\delta_{os} - \alpha_s - \delta_p)}{M_p} \quad (26)$$

as well as the rate of change in the mass of O_2. The isotopic composition of seawater carbonate and sulfate through time comes directly from the sedimentary rock record. However, it is uncertain how to average globally as well as over substantial period of geologic time. The rates of change in seawater carbonate and sulfate isotopic compositions are fairly small terms, and have been left out of many modeling efforts directed at the geochemical C–S–O system; however, for completeness sake these terms are included here.

One implication of the isotope-driven modeling approach to understanding the C–S–O coupled cycle is that burial fluxes of organic matter and pyrite (which are the sources of atmospheric O_2 in these models) are nearly proportional to seawater isotopic composition of inorganic carbon and sulfate. Recalling the equation for organic matter burial above, it is noted that because sedimentary carbonate and organic matter mass are nearly constant through time, weathering fluxes do not vary greatly through time. Likewise, the average isotopic composition of carbonates and organic matter are also nearly constant, as is the mass of dissolved carbonate and the rate of change in dissolved inorganic carbon isotopic composition. Assuming that F_{wc}, F_{wg}, M_{oc}, $d\delta_{oc}/dt$, δ_c, and δ_g are constant or nearly constant, F_{bg} becomes linearly proportional to the isotopic composition of seawater dissolved inorganic carbon. The same rationale applies for the sulfur system, with pyrite burial becoming linearly proportional to the isotopic composition of seawater sulfate. Although these relationships are not strictly true, even within the constraints of the model simplifications, these relationships provide a useful guide for evaluating changes in organic matter and pyrite burial fluxes and the impact these have on atmospheric O_2, simply by examining the isotopic records of marine carbonate and sulfate.

Early efforts to model the coupled C–S–O cycles yielded important information. The work of Garrels and Lerman (1984) showed that the exogenic C and S cycles can be treated as closed systems over at least Phanerozoic time, without exchange between sedimentary rocks and the deep "crust + mantle." Furthermore, over timescales of millions of years, the carbon and sulfur cycles were seen to be closely coupled, with increase in sedimentary organic carbon mass matched by loss of sedimentary pyrite (and vice versa). Other models explored the dynamics of the C–S–O system. One important advance was promoted by Kump and Garrels (1986). In these authors' model, a steady-state C–S–O system was generated and perturbed by artificially increasing rates of organic matter burial. These authors tracked the shifts in seawater carbon and sulfur isotopic composition that resulted, and compared these results with the true sedimentary record. Importantly, these authors recognized that although there is a general inverse relationship between seawater $\delta^{13}C$ and $\delta^{34}S$, the exact path along an isotope–isotope plot through time is not a straight line. Instead, because of the vastly different residence times of sulfate versus carbonate in seawater, any changes in the C–S–O system are first expressed through shifts in carbon isotopes, then sulfur isotopes (Figure 15). The authors also pointed out large-scale divisions in C–S isotope coupling through Phanerozoic time. In the Paleozoic, organic matter and pyrite burial were closely coupled (largely because the same types of depositional environment favor burial of marine organic matter and pyrite). During this time, seawater carbonate and sulfur isotopes co-varied positively, indicating concomitant increases (or decreases) in burial of organic matter and pyrite. During the late Paleozoic, Mesozoic, and Cenozoic, terrestrial depositional environments became important settings for burial of organic matter. Because pyrite formation and burial in terrestrial environments is extremely limited, organic matter and pyrite burial became decoupled at this time, and seawater carbonate and sulfur isotopes co-varied negatively, indicating close matching of increased sedimentary organic matter with decreased sedimentary pyrite (and vice versa), perhaps suggesting a net balance in O_2 production and consumption, and maintenance of nearly constant, equable p_{O_2} throughout much of the latter half of the Phanerozoic. The model of Berner (1987) introduced the concept of rapid recycling: the effort to numerically represent the observation that younger sedimentary rocks are more likely to be eroded and weathered than are

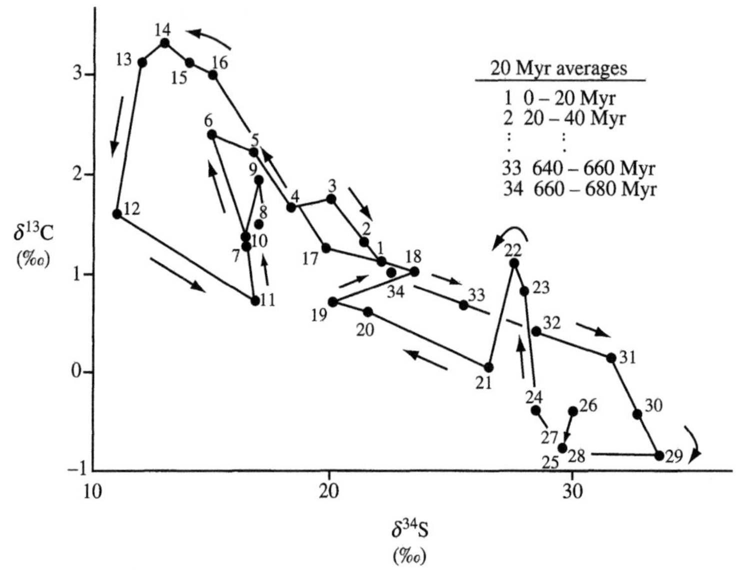

Figure 15 Twenty-million-year average values of seawater $\delta^{34}S$ plotted against concomitant carbonate $\delta^{13}C$ for the last 700 Myr (source Kump and Garrels, 1986).

older sedimentary rocks. Because young sedimentary rocks are likely to be isotopically distinct from older rocks (because they are recording any recent shifts in seawater carbon or sulfur isotopic composition), restoring that isotopically distinct carbon or sulfur more quickly back into seawater provides a type of negative feedback, dampening excessively large or small burial fluxes required for isotope mass balance. This negative feedback serves to reduce calculated fluctuations in organic matter and pyrite burial rates, which in turn reduce fluctuations in release of O_2 to the atmosphere. Results from this study also predict large increases in OM burial fluxes during the Permocarboniferous (~300 Ma) above values present earlier in the Paleozoic. This increase, likely associated with production and burial of refractory terrigenous organic matter (less easily degraded than OM produced by marine organisms), led to elevated concentrations of O_2 ~300 Ma.

One flaw with efforts to model the evolution of Phanerozoic O_2 using the carbon and sulfur isotope records is that unreasonably large fluctuations in organic matter and pyrite burial fluxes (with coincident fluctuations in O_2 production rates) would result. Attempts to model the whole Phanerozoic generated unreasonably low and high O_2 concentrations for several times in the Phanerozoic (Lasaga, 1989), and applications of what seemed to be a realistic feedback based on reality (weathering rates of sedimentary organic matter and pyrite dependent on the concentration of O_2) were shown to actually become positive feedbacks in the isotope-driven C–S–O models (Berner, 1987; Lasaga, 1989).

Phosphorus is a key nutrient limiting primary productivity in many marine environments. If phosphorus supply is increased, primary production and perhaps organic matter burial will also increase. Degradation and remineralization of OM during transit from surface waters into sediments liberates phosphorus, but most of this is quickly scavenged by adsorption onto the surfaces of iron oxyhydroxides. However, work by Ingall and Jahnke (1994) and Van Cappellen and Ingall (1996) has shown that phosphorus recycling and release into seawater is enhanced under low O_2 or anoxic conditions. This relationship provides a strong negative feedback between primary production, bottom water anoxia, and atmospheric O_2. As atmospheric O_2 rises, phosphorus scavenging on ferric-iron is enhanced, phosphorus recycling back into surface waters is reduced, primary production rates are reduced, and O_2 declines. If O_2 concentrations were to fall, phosphorus scavenging onto ferric-iron would be inhibited, phosphorus recycling back into surface waters would be accelerated, fueling increased primary production and O_2 release to the atmosphere. Van Cappellen and Ingall (1996) applied these ideas to a mathematical model of the C–P–Fe–O cycle to show how O_2 concentrations could be stabilized by phosphorus recycling rates.

Petsch and Berner (1998) expanded the model of Van Cappellen and Ingall (1996) to include the sulfur system, as well as carbon and sulfur isotope effects. This study examined the response of the C–S–O–Fe–P system, and in particular carbon and sulfur isotope ratios, to perturbation in global ocean overturn rates, changes in continental weathering, and shifts in the locus of organic matter burial from marine to terrestrial depocenters. Confirming the idea promoted by Kump and Garrels (1986), these authors showed that perturbations of the exogenic C–S–O cycle result in shifts in seawater carbon and sulfur isotopic composition similar in amplitude and duration to observed isotope excursions in the sedimentary record.

Other proposed feedbacks stabilizing the concentration of atmospheric O_2 over Phanerozoic time include a fire-regulated PO_4 feedback (Kump, 1988, 1989). Terrestrial primary production requires much less phosphate per mole CO_2 fixed during photosynthesis than marine primary production. Thus, for a given supply of PO_4, much more CO_2 can be fixed as biomass and O_2 released from photosynthesis on land versus in the oceans. If terrestrial production proceeds too rapidly, however, p_{O_2} levels may rise slightly and lead to increased forest fires. Highly weatherable, PO_4-rich ash would then be delivered through weathering and erosion to the oceans. Primary production in the oceans would lead to less CO_2 fixed and O_2 released per mole of PO_4.

Hydrothermal reactions between seawater and young oceanic crust have been proposed as an influence on atmospheric O_2 (Walker, 1986; Carpenter and Lohmann, 1999; Hansen and Wallmann, 2002). While specific periods of oceanic anoxia may be associated with accelerated hydrothermal release of mantle sulfide (i.e., the Mid-Cretaceous, see Sinninghe-Damsté and Köster, 1998), long-term sulfur and carbon isotope mass balance precludes substantial inputs of mantle sulfur to the Earth's surface of a different net oxidation state and mass flux than what is subducted at convergent margins (Petsch, 1999; Holland, 2002).

One recent advance in the study of isotope-driven models of the coupled C–S–O cycles is re-evaluation of isotope fractionations. Hayes *et al.*, (1999) published a compilation of the isotopic composition of inorganic and organic carbon for the past 800 Myr. One feature of this dual record is a distinct shift in the net isotopic distance between carbonate and organic carbon, occurring during the past ~100 million years. When carbon isotope distance is compared to estimates of Cenozoic and Mesozoic p_{O_2}, it becomes apparent that there may be some relationship between isotopic fractionation associated with organic matter production and burial and the concentration of O_2 in the atmosphere. The physiological underpinning behind this proposed relationship rests on competition between photosynthesis and photorespiration in the cells of photosynthetic organisms. Because O_2 is a competitive inhibitor of CO_2 for attachment to the active site of Rubisco, as ambient O_2 concentrations rise relative to CO_2, so will rates of photorespiration. Photorespiration is a net consumptive process for plants; previously fixed carbon is consumed with O_2 to produce CO_2 and energy. CO_2 produced through photorespiration may diffuse out of the cell, but it is also likely to be taken up (again) for photosynthesis. Thus, in cells undergoing fairly high rates of photorespiration in addition to photosynthesis, a significant fraction of total CO_2 available for photosynthesis derives from oxidized, previously fixed organic carbon. The effect of this on cellular carbon isotopic composition is that because each round of photosynthesis results in ^{13}C-depletion in cellular carbon relative to CO_2, cells with high rates of photorespiration will contain more ^{13}C-depleted CO_2 and thus will produce more ^{13}C-depleted organic matter.

In controlled-growth experiments using both higher plants and single-celled marine photosynthetic algae, a relationship between ambient O_2 concentration and net isotope discrimination has been observed (Figure 16) (Berner *et al.*, 2000; Beerling *et al.*, 2002). The functional form of this relationship has been expressed in several ways. The simplest is to allow isotope discrimination to vary linearly with changing atmospheric O_2 mass: $\alpha_c = 25 \times (M_{O_2}/38)$. More complicated relationships have also been derived, based on curve-fitting the available experimental data on isotopic fractionation as a function of $[O_2]$. O_2-dependent isotopic fractionation during photosynthesis has provided the first mathematically robust isotope-driven model of the C–S–O cycle consistent with geologic observations (Berner *et al.*, 2000). Results of this model show that allowing isotope fractionation to respond to changes in ambient O_2

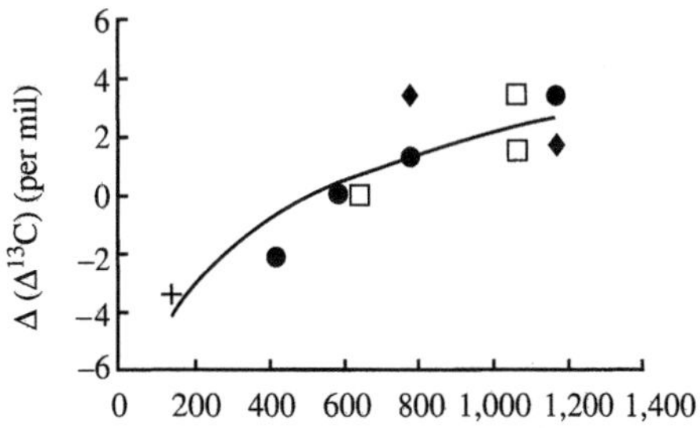

Figure 16 Relationship between change in $\Delta(\Delta^{13}C)$ of vascular land plants determined experimentally in response to growth under different O_2/CO_2 atmospheric mixing ratios. $\Delta(\Delta^{13}C)$ is the change in carbon isotope fractionation relative to fractionation for the controls at present day conditions (21% O_2, 0.036% CO_2). The solid line shows the nonlinear curve fitted to the data, given by $\Delta(\Delta^{13}C) = -19.94 + 3.195 \times \ln(O_2/CO_2)$; (●) *Phaseolus vulgaris*; (◆) *Sinapis alba*; (□) from Berner *et al.* (2000); (+) from Berry *et al.* (1972) (Beerling *et al.*, 2002) (reproduced by permission of Elsevier from *Geochem. Cosmochim. Acta* **2002**, *66*, 3757–3767).

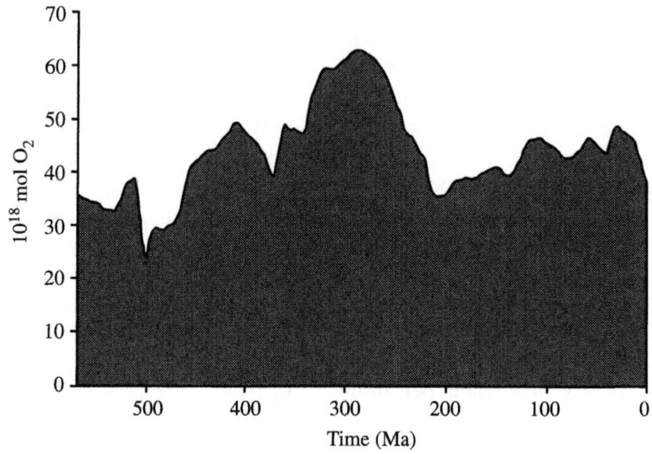

Figure 17 Evolution of the mass of atmospheric O_2 through Phanerozoic time, estimated using an isotope mass balance described in Equations (11)–(26). The model employs the isotope date of Figure 13, and includes new advances in understanding regarding dependence of carbon isotope discrimination during photosynthesis and sulfur isotope discrimination during sulfur disproportionation and organic sulfur formation. The system of coupled differential equations were integrated using an implicit fourth-order Kaps–Rentrop numerical integration algorithm appropriate for this stiff set of equations (sources Petsch, 2000; Berner *et al.*, 2000).

provides a strong negative feedback dampening excessive increases or decreases in organic matter burial rates. Rates of organic matter burial in this model are no longer simply dependent on seawater carbonate $\delta^{13}C$, but now also vary with $1/\alpha_c$. As fractionation becomes greater (through elevated O_2), less of an increase in organic matter burial rates is required to achieve the observed increase in seawater $\delta^{13}C$ than if α_c were constant.

The same mathematical argument can be applied to sulfate–sulfide isotope fractionation during BSR. As O_2 concentrations increase, so does sulfur isotope fractionation, resulting in a strong negative feedback on pyrite burial rates. This is consistent with the broad-scale changes in sulfur isotope dynamics across the Proterozoic, reflecting a large increase in $\Delta^{34}S$ (between sulfate and sulfide) when atmospheric O_2 concentrations were great enough to facilitate bacterial sulfide oxidation and sulfur disproportionation. Perhaps during the Phanerozoic, when O_2 concentrations were greater, sulfur recycling (sulfate to sulfide through BSR, sulfur oxidation, and sulfur disproportionation) was increased, resulting in greater net isotopic distance between sulfate and sulfide. Another means of changing net sulfur isotope discrimination in response to O_2 may be the distribution of reduced sulfur between sulfide minerals and organic matter-associated sulfides. Work by Werne *et al.*, 2000, 2003) has shown that organic sulfur is consistently ~10 ‰ enriched in ^{34}S relative to associated pyrite. This is believed to result from different times and locations of organic sulfur versus pyrite formation. While pyrite may form in shallow sediments or even anoxic portions of the water column, reflecting extreme sulfur isotope depletion due to several cycles of BSR, sulfide oxidation and sulfur disproportionation, organic matter is sulfurized within the sediments. Closed, or nearly closed, system behavior of BSR in the sediments results in late-stage sulfide (the source of sulfur in sedimentary organic matter) to be more enriched compared with pyrite in the same sediments. It is known that burial of sulfide as organic sulfur is facilitated in low O_2 or anoxic waters. If lower atmospheric O_2 in the past encouraged development of more extensive anoxic basins and increased burial of sulfide as organic sulfur instead of pyrite, the 10‰ offset between pyrite and organic sulfur would become effectively a change in net sulfur fractionation in response to O_2.

Applying these newly recognized modifications of carbon and sulfur isotope discrimination in response to changing O_2 availability has allowed development of new numerical models of the evolution of the coupled C–S–O systems and variability of Phanerozoic atmospheric O_2 concentration (Figure 17).

8.11.7 CONCLUSIONS

Molecular oxygen is generated and consumed by a wide range of processes. The net cycling of O_2 is influenced by physical, chemical, and most importantly, biological processes acting on and beneath the Earth's surface. The exact distribution of O_2 concentrations depends on the specific interplay of these processes in time and space. Large inroads have been made towards

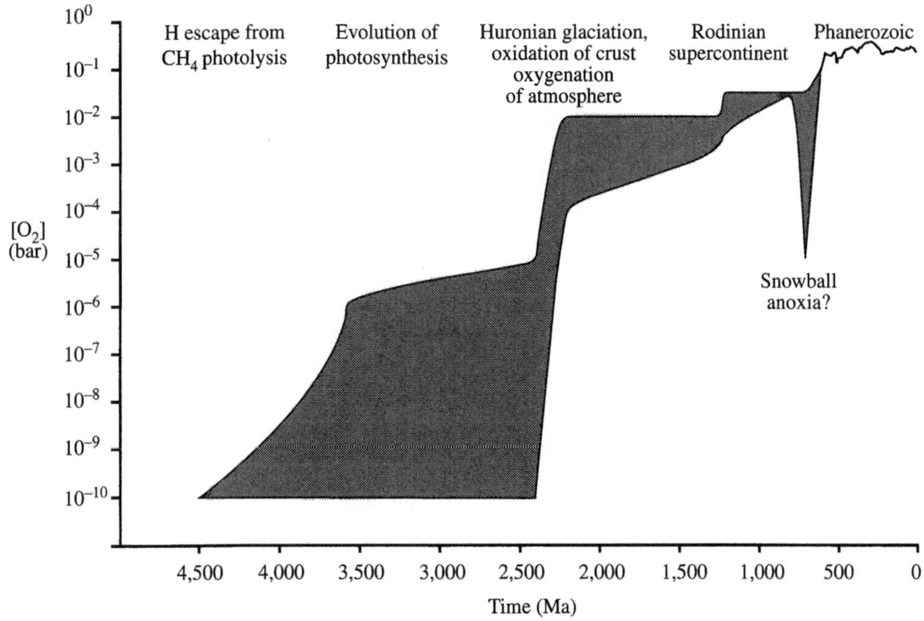

Figure 18 Composite estimate of the evolution of atmospheric oxygen through 4.5 Gyr of Earth's history. Irreversible oxidation of the Earth resulted from CH_4 photolysis and hydrogen escape during early Earth's history. Evolution of oxygenic photosynthesis preceded substantial oxygenation of the atmosphere by several hundred million years. Relative stasis in atmospheric p_{O_2} typified much of the Proterozoic, with a possible pulse of oxygenation associated with formation of the Rodinian supercontinent in the Late Mesoproterozoic, and possible return to anoxia associated with snowball glaciation in the Neoproterozoic (sources Catling *et al.*, 2001; Kasting, 1992; Rye and Holland, 1998; Petsch, 2000; Berner *et al.*, 2000).

understanding the processes that control the concentration of atmospheric O_2, especially regarding O_2 as a component of coupled biogeochemical cycles of many elements, including carbon, sulfur, nitrogen, phosphorus, iron, and others.

Earth's modern oxygenated atmosphere is the product of over four billion years of its history (Figure 18). The early anoxic atmosphere was slowly oxidized (although not oxygenated) as the result of slow H_2 escape. Evolution of oxygenic photosynthesis accelerated the oxidation of Earth's crust and atmosphere, such that by ~2.2 Ga a small but significant concentration of O_2 was likely present in Earth's atmosphere. Limited primary production and oxygen production compared with the flux of reduced volcanic gases maintained this low p_{O_2} atmosphere for over one billion years until the Neoproterozoic. Rapid oscillations in Earth's carbon and sulfur cycles associated with global Snowball glaciation may also have expression in a return to atmospheric anoxia at this time, but subsequent to the late Proterozoic isotope excursions, oxygenation of the atmosphere to near-modern concentrations developed such that by the Precambrian–Cambrian boundary, O_2 concentrations were high enough to support widespread skeletonized metazoans. Phanerozoic seawater and atmospheric O_2 concentrations have fluctuated in response to tectonic forcings, generating regional-scale anoxia in ocean basins at certain times when biological productivity and ocean circulation facilitate anoxic conditions, but in the atmosphere, O_2 concentrations have remained within ~0.05–0.35 bar p_{O_2} for the past ~600 Myr.

Several outstanding unresolved gaps in our understanding remain, in spite of a well-developed understanding of the general features of the evolution of atmospheric O_2 through time. These gaps represent potentially meaningful directions for future research, including:

(i) assessing the global importance of mineral oxidation as a mechanism of O_2 consumption;

(ii) the flux of reduced gases from volcanoes, metamorphism, and diffuse mantle/lithosphere degassing;

(iii) the true dependence of organic matter oxidation on availability of O_2, in light of the great abundance of microaerophilic and anaerobic microorganisms utilizing carbon respiration as a metabolic pathway, carbon isotopic evidence suggesting continual and essentially constant organic matter oxidation as part of the sedimentary rock cycle during the entire past four billion years, and the inefficiency of organic matter oxidation during continental weathering;

(iv) stasis in the oxygenation of the atmosphere during the Proterozoic;

(v) contrasting biochemical, fossil, and molecular evidence for the antiquity of the innovation of oxygenic photosynthesis; and

(vi) evaluating the relative strength of biological productivity versus chemical evolution of the Earth's crust and mantle in controlling the early stages of oxygenation of the atmosphere.

Thus, study of the global biogeochemical cycle of oxygen, the component of our atmosphere integral and crucial for life as we know it, remains a fruitful direction for Earth science research.

REFERENCES

Algeo T. J. and Scheckler S. E. (1998) Terrestrial-marine teleconnections in the Devonian: links between the evolution of land plants, weathering processes, and marine anoxic events. *Phil. Trans. Roy. Soc. London B* **353**, 113–128.

Anbar A. D. and Knoll A. H. (2002) Proterozoic ocean chemistry and evolution: a bioinorganic bridge? *Science* **297**, 1137–1142.

Arthur M. A. (2000) Volcanic contributions to the carbon and sulfur geochemical cycles and global change. In *Encyclopedia of Volcanoes* (eds. H. Sigurdsson, B. F. Houghton, S. R. McNutt, H. Rymer, and J. Stix). Academic Press, San Diego, pp. 1045–1056.

Bartley J. K., Semikhatov M. A., Kaufman A. J., Knoll A. H., Pope M. C., and Jacobsen S. B. (2001) Global events across the Mesoproterozoic–Neoproterozoic boundary: C and Sr isotopic evidence from Siberia. *Precamb. Res.* **111**, 165–202.

Battle M., Bender M., Sowers R., Tans P. P., Butler J. H., Elkins J. W., Ellis J. T., Conway T., Zhang N., Pang P., and Clarke A. D. (1996) Atmospheric gas concentrations over the past century measured in air from firn at the South Pole. *Nature* **383**, 231–235.

Baum S. K. and Crowley T. J. (2001) GCM response to late Precambrian (~590 Ma) ice-covered continents. *Geophys. Res. Lett.* **28**, 583–586.

Beaumont V. and Robert F. (1999) Nitrogen isotope ratios of kerogens in Precambrian cherts: a record of the evolution of atmospheric chemistry? *Precamb. Res.* **96**, 63–82.

Beerling D. J. and Berner R. A. (2000) Impact of a Permo–Carboniferous high O_2 event on the terrestrial carbon cycle. *Proc. Natl. Acad. Sci.* **97**, 12428–12432.

Beerling D. J., Lake J. A., Berner R. A., Hickey L. J., Taylor D. W. and Royer D. L. (2002) Carbon isotope evidence implying high O_2/CO_2 ratios in the Permo–Carboniferous atmosphere. *Geochim. Cosmochim. Acta* **66**, 3757–3767.

Bekker A., Kaufman A. J., Karhu J. A., Beukes N. J., Quinten S. D., Coetzee L. L., and Kenneth A. E. (2001) Chemostratigraphy of the Paleoproterozoic Duitschland Formation, South Africa. Implications for coupled climate change and carbon cycling. *Am. J. Sci.* **301**, 261–285.

Bender M., Labeyrie L. D., Raynaud D., and Lorius C. (1985) Isotopic composition of atmospheric O_2 in ice linked with deglaciation and global primary productivity. *Nature* **318**, 349–352.

Bender M., Sowers T., and Labeyrie L. (1994a) The Dole effect and its variations during the last 130,000 years as measured in the Vostok ice core. *Global Biogeochem. Cycle* **8**, 363–376.

Bender M. L., Sowers T., Barnola J.-M., and Chappellaz J. (1994b) Changes in the O_2/N_2 ratio of the atmosphere during recent decades reflected in the composition of air in the firn at Vostok Station, Antarctica. *Geophys. Res. Lett.* **21**, 189–192.

Berner R. A. (1987) Models for carbon and sulfur cycles and atmospheric oxygen: application to Paleozoic geologic history. *Am. J. Sci.* **287**, 177–196.

Berner R. A. (1994) GEOCARB II: a revised model of atmospheric CO_2 over Phanerozoic time. *Am. J. Sci.* **294**, 56–91.

Berner R. A. (2002) Examination of hypotheses for the Permo–Triassic boundary extinction by carbon cycle modeling. *Proc. Natl. Acad. Sci.* **99**, 4172–4177.

Berner R. A. and Canfield D. E. (1989) A new model for atmospheric oxygen over Phanerozoic time. *Am. J. Sci.* **289**, 333–361.

Berner R. A. and Kothavala Z. (2001) GEOCARB III: a revised model of atmospheric CO_2 over Phanerozoic time. *Am. J. Sci.* **301**, 182–204.

Berner R. A., Petsch S. T., Lake J. A., Beerling D. J., Popp B. N., Lane R. S., Laws E. A., Westley M. B., Cassar N., Woodward F. I., and Quick W. P. (2000) Isotope fractionation and atmospheric oxygen: Implications for Phanerozoic O_2 evolution. *Science* **287**, 1630–1633.

Berry J. A., Troughton J. H., and Björkman O. (1972) Effect of oxygen concentration during growth on carbon isotope discrimination in C_3 and C_4 species of *Atriplex*. *Carnegie Inst. Yearbook* **71**, 158–161.

Beukes N. J., Dorland H., Gutzmer J., Nedachi M., and Ohmoto H. (2002) Tropical laterites, life on land, and the history of atmospheric oxygen in the Paleoproterozoic. *Geology* **30**, 491–494.

Brasier M. D., Green O. R., Jephcoat A. P., Kleppe A. K., van Kranendonk M. J., Lindsay J. F., Steele A., and Grassineau N. V. (2002) Questioning the evidence for Earth's earliest fossils. *Nature* **416**, 76–81.

Brocks J. J., Logan G. A., Buick R., and Summons R. E. (1999) Archean molecular fossils and the early rise of eukaryotes. *Science* **285**, 1033–1036.

Budyko M. I., Ronov A. B., and Yanshin A. L. (1987) *History of the Earth's Atmosphere*. Springer, Berlin.

Buick R. (1990) Microfossil recognition in Archean rocks: an appraisal of spheroids and filaments from a 3500 M. Y. old chert-barite unit at North Pole, Western Australia. *Palaios* **5**, 441–459.

Buick I. S., Uken R., Gibson R. L., and Wallmach T. (1998) High-$\delta^{13}C$ Paleoproterozoic carbonates from the Transvaal Supergroup, South Africa. *Geology* **26**, 875–878.

Butterfield N. J. and Rainbird R. H. (1998) Diverse organic-walled fossils, including possible dinoflagellates, from early Neoproterozoic of arctic Canada. *Geology* **26**, 963–966.

Caldeira K., and Kasting J. F. (1992) Susceptibility of the early Earth to irreversible glaciation caused by carbon dioxide clouds. *Nature* **359**, 226–228.

Canfield D. E. (1998) A new model for Proterozoic ocean chemistry. *Nature* **396**, 450–453.

Canfield D. E. and Teske A. (1996) Late Proterozoic rise in atmospheric oxygen concentration inferred from phylogenetic and sulphur-isotope studies. *Nature* **382**, 127–132.

Canfield D. E., Habicht K. S., and Thamdrup B. (2000) The Archean sulfur cycle and the early history of atmospheric oxygen. *Science* **288**, 658–661.

Carpenter S. J. and Lohmann K. C. (1999) Carbon isotope ratios of Phanerozoic marine cements: re-evaluating global carbon and sulfur systems. *Geochim. Cosmochim. Acta* **61**, 4831–4846.

Catling D. C., Zahnle K. J., and McKay C. P. (2001) Biogenic methane, hydrogen escape, and the irreversible oxidation of the early Earth. *Science* **293**, 839–843.

Cloud P. (1972) A working model of the primitive Earth. *Am. J. Sci.* **272**, 537–548.

Cloud P. E. (1976) Beginnings of biospheric evolution and their biogeochemical consequences. *Paleobiol.* **2**, 351–387.

Condie K. C., Des Marais D. J., and Abbott D. (2001) Precambrian superplumes and supercontinents: a record in black shales, carbon isotopes, and paleoclimates? *Precamb. Res.* **106**, 239–260.

Cope M. J. and Chaloner W. G. (1985) Wildfire: an interaction of biological and physical processes. In *Geological Factors and the Evolution of Plants* (ed. B. H. Tiffney). Yale University Press, New Haven, CT, pp. 257–277.

Delano J. W. (2001) Redox history of the Earth's interior since ~3900 Ma: implications for prebiotic molecules. *Origins Life Evol. Biosphere* **31**, 311–341.

Delmelle P. and Stix J. (2000) Volcanic gases. In *Encyclopedia of Volcanoes* (eds. H. Sigurdsson, B. F. Houghton, S. R. McNutt, H. Rymer, and J. Stix). Academic Press, San Diego, pp. 803–816.

Des Marais D. J., Strauss H., Summons R. E., and Hayes J. M. (1992) Carbon isotope evidence for stepwise oxidation of the Proterozoic environment. *Nature* **359**, 605–609.

Deuser W. G. (1975) *Chemical Oceanography*. Academic Press, Orlando, FL, p. 3.

Doolittle R. F., Feng D. F., Tsang S., Cho G., and Little E. (1996) Determining divergence times of the major kingdoms of living organisms with a protein clock. *Science* **271**, 470–477.

Dudley R. (1998) Atmospheric oxygen, giant Paleozoic insects and the evolution of aerial locomotor performance. *J. Exp. Biol.* **201**, 1043–1050.

Dudley R. (2000) The evolutionary physiology of animal flight: Paleobiological and present perspectives. *Ann. Rev. Phys.* **62**, 135–155.

Evans D. A. D. (2000) Stratigraphic, geochronological and paleomagnetic constraints upon the Neoproterozoic climatic paradox. *Am. J. Sci.* **300**, 347–433.

Farquhar J., Bao H., and Thiemens M. (2000) Atmospheric influence of Earth's earliest sulfur cycle. *Science* **289**, 756–758.

Farquhar J., Wing B. A., McKeegan K. D., Harris J. W., Cartigny P., and Thiemens M. H. (2002) Mass-independent sulfur of inclusions in diamond and sulfur recycling on early Earth. *Science* **297**, 2369–2372.

Garcia-Pichel F. (1998) Solar ultraviolet and the evolutionary history of cyanobacteria. *Origins Life Evol. Biosphere* **28**, 321–347.

Garrels R. M. and Lerman A. (1984) Coupling of the sedimentary sulfur and carbon cycles—an improved model. *Am. J. Sci.* **284**, 989–1007.

Gilbert D. L. (1996) Evolutionary aspects of atmospheric oxygen and organisms. In *Environmental Physiology:* 2 (eds. M. J. Fregly, and C. M. Blatteis). Oxford University Press, Oxford, UK, pp. 1059–1094.

Glasspool I. (2000) A major fire event recorded in the mesofossils and petrology of the late Permian, Lower Whybrow coal seam, Sydney Basin, Australia. *Palaeogeogr. Palaeoclimat. Palaeoecol.* **164**, 373–396.

Godderís Y. and Veizer J. (2000) Tectonic control of chemical and isotopic composition of ancient oceans: the impact of continental growth. *Am. J. Sci.* **300**, 434–461.

Gould S. J. (1995) Of it and not above it. *Nature* **377**, 681–682.

Graham J. B., Dudley R., Anguilar N., and Gans C. (1995) Implications of the late Paleozoic oxygen pulse for physiology and evolution. *Nature* **375**, 117–120.

Habicht K. S., Gade M., Thamdrup B., Berg P., and Canfield D. E. (2002) Calibration of sulfate levels in the Archean ocean. *Science* **298**, 2372–2374.

Haggerty S. E. and Toft P. B. (1985) Native iron in the continental lower crust: petrological and geophysical implications. *Science* **229**, 647–649.

Halmer M. M., Schmincke H.-U., and Graf H.-F. (2002) The annual volcanic gas input into the atmosphere, in particular into the stratosphere: a global data set for the past 100 years. *J. Volcanol. Geotherm. Res.* **115**, 511–528.

Hansen K. W. and Wallmann K. (2003) Cretaceous and Cenozoic evolution of seawater composition, atmospheric O_2 and CO_2: a model perspective. *Am. J. Sci.* **303**, 94–148.

Hartnett H. E., Keil R. G., Hedges J. I., and Devol A. H. (1998) Influence of oxygen exposure time on organic carbon preservation in continental margine sediments. *Nature* **391**, 572–574.

Hayes J. M., Strauss H., and Kaufman A. J. (1999) The abundance of ^{13}C in marine organic matter and isotopic fractionation in the global biogeochemical cycle of carbon during the past 800 Myr. *Chem. Geol.* **161**, 103–125.

Herbert T. D. and Sarmiento J. L. (1991) Ocean nutrient distribution and oxygenation: limits on the formation of warm saline bottom water over the past 91 m.y. *Geology* **19**, 702–705.

Hinkle M. E. (1994) Environmental conditions affecting the concentrations of He, CO_2, O_2 and N_2 in soil gases. *Appl. Geochem.* **9**, 53–63.

Hoehler T. M., Bebout B. M., and Des Marais D. J. (2001) The role of microbial mats in the production of reduced gases on the early Earth. *Nature* **412**, 324–327.

Hoffman P. F. and Schrag D. P. (2002) The snowball Earth hypothesis: testing the limits of global change. *Terra Nova* **14**, 129–155.

Hoffman P. F., Kaufman A. J., Halverson G. P., and Schrag D. P. (1998) A Neoproterozoic snowball Earth. *Science* **281**, 1342–1346.

Hofmann H. J., Gery K., Hickman A. H., and Thorpe R. I. (1999) Origin of 2.45 Ga coniform stromatolites in Warrawoona Group, Western Australia. *Geol. Soc. Am. Bull.* **111**, 1256–1262.

Holland H. D. (1978) *The Chemistry of the Atmosphere and Oceans*. Wiley, New York.

Holland H. D. (1984) *The Chemical Evolution of the Atmosphere and Oceans*. Princeton University Press, Princeton, NJ.

Holland H. D. (2002) Volcanic gases, black smokers and the great oxidation event. *Geochim. Cosmochim. Acta* **66**, 3811–3826.

Holland H. D., Feakes C. R., and Zbinden E. A. (1989) The Flin Flon paleosol and the composition of the atmosphere 1.8 BYBP. *Am. J. Sci.* **289**, 362–389.

Hotinski R. M., Bice K. L., Kump L. R., Najjar R. G., and Arthur M. A. (2001) Ocean stagnation and End-Permian anoxia. *Geology* **29**, 7–10.

Hurtgen M. T., Arthur M. A., Suits N. S., and Kaufman A. J. (2002) The sulfur isotopic composition of Neoproterozoic seawater sulfate: implications for a snowball Earth? *Earth Planet. Sci. Lett.* **203**, 413–429.

Ingall E. and Jahnke R. (1994) Evidence for enhanced phosphorus regeneration from marine sediments overlain by oxygen depleted waters. *Geochim. Cosmochim. Acta* **58**, 2571–2575.

Ingmanson D. E. and Wallace W. J. (1989) *Oceanography: An Introduction*. Wadworth, Belmont, CA, pp. 99.

Joachimski M. M., Ostertag-Henning C., Pancost R. D., Strauss H., Freeman K. H., Littke R., Sinninghe-Damsté J. D., and Racki G. (2001) Water column anoxia, enhanced productivity and concomitant changes in $\delta^{13}C$ and $\delta^{34}S$ across the Frasnian–Famennian boundary (Kowala—Holy Cross Mountains/Poland). *Chem. Geol.* **175**, 109–131.

Kah L. C., Lyons T. W., and Chesley J. T. (2001) Geochemistry of a 1.2 Ga carbonate-evaporite succession, northern Baffin and Bylot Islands: implications for Mesoproterozoic marine evolution. *Precamb. Res.* **111**, 203–234.

Kaiho K., Kajiwara Y., Nakano T., Miura Y., Kawahata H., Tazaki K., Ueshima M., Chen Z., and Shi G. (2001) End-Permian catastrophe by a bolide impact: evidence of a gigantic release of sulfur from the mantle. *Geology* **29**, 815–818.

Karhu J. and Holland H. D. (1996) Carbon isotopes and the rise of atmospheric oxygen. *Geology* **24**, 867–870.

Kasting J. F. (1987) Theoretical constraints on oxygen and carbon dioxide concentrations in the Precambrian atmosphere. *Precamb. Res.* **34**, 205–229.

Kasting J. F. (1992) Models relating to Proterozoic Atmospheric and Oceanic Chemistry. In *The Proterozoic Biosphere* (eds. J. W. Schopf, and C. Klein). Cambridge University Press, Cambridge, pp. 1185–1187.

Kasting J. F. and Siefert J. L. (2002) Life and the evolution of Earth's atmosphere. *Science* **296**, 1066–1068.

Kasting J. F., Zahnle K. J., and Walker J. C. G. (1983) Photochemistry of methane in the Earth's early atmosphere. *Precamb. Res.* **20**, 121–148.

Kasting J. F., Eggler D. H., and Raeburn S. P. (1993) Mantle redox evolution and the oxidation state of the Archean atmosphere. *J. Geol.* **101**, 245–257.

Kasting J. F., Pavlov A. A., and Siefert J. L. (2001) A coupled ecosystem-climate model for predicting the methane concentration in the Archean atmosphere. *Orig. Life Evol. Biosph.* **31**, 271–285.

Kaufman A. J. and Knoll A. H. (1995) Neoproterozoic variations in the C-isotopic composition of seawater: stratigraphic and biogeochemical implications. *Precamb. Res.* **73**, 27–49.

Kaufman A. J., Hayes J. M., Knoll A. H., and Germs G. J. B. (1991) Isotopic composition of carbonates and organic carbon from upper Proterozoic successions in Namibia: stratigraphic variation and the effects of diagenesis and metamorphism. *Precamb. Res.* **49**, 301–327.

Keeling R. F. and Shertz S. R. (1992) Seasonal and interannual variations in atmospheric oxygen and implications for the carbon cycle. *Nature* **358**, 723–727.

Keeling R. F., Najjar R. P., Bender M. L., and Tans P. P. (1993) What atmospheric oxygen measurements can tell us about the global carbon cycle. *Global Biogeochem. Cycles* **7**, 37–67.

Keller C. K. and Bacon D. H. (1998) Soil respiration and georespiration distinguished by transport analyses of vadose CO_2, $^{13}CO_2$, and $^{14}CO_2$. *Global Biogeochem. Cycles* **12**, 361–372.

Kiehl J. T. and Dickinson R. E. (1987) A study of the radiative effects of enhanced atmospheric CO_2 and CH_4 on early Earth surface temperatures. *J. Geophys. Res.* **92**, 2991–2998.

Kimura H. and Watanabe Y. (2001) Oceanic anoxia at the Precambrian–Cambrian boundary. *Geology* **29**, 995–998.

Kirschvink J. L. (1992) Late Proterozoic low-latitude global glaciation: the snowball earth. In *The Proterozoic Biosphere* (eds. J. W. Schopf, and C. Klein). Cambridge University Press, Cambridge, pp. 51–52.

Kump L. R. (1988) Terrestrial feedback in atmospheric oxygen regulation by fire and phosphorus. *Nature* **335**, 152–154.

Kump L. R. (1989) Chemical stability of the atmosphere and ocean. *Palaeogeogr. Palaeoclimat. Paleoecol.* **75**, 123–136.

Kump L. R. and Garrels R. M. (1986) Modeling atmospheric O_2 in the global sedimentary redox cycle. *Am. J. Sci.* **286**, 337–360.

Kump L. R., Kasting J. F., and Barley M. E. (2001) Rise of atmospheric oxygen and the upside down Archean mantle. *Geochem. Geophys. Geosys.* **2** No. 2000 GC 000114.

Lasaga A. C. (1989) A new approach to isotopic modeling of the variation in atmospheric oxygen through the Phanerozoic. *Am. J. Sci.* **289**, 411–435.

Lasaga A. C. and Ohmoto H. (2002) The oxygen geochemical cycles: dynamics and stability. *Geochim. Cosmochim. Acta* **66**, 361–381.

Lenton T. M. and Watson A. J. (2000) Redfield revisited 2. What regulates the oxygen content of the atmosphere? *Global Biogeochem. Cycles* **14**, 249–268.

Lindh T. B. (1983) Temporal variations in ^{13}C, ^{34}S and global sedimentation during the Phanerozoic. MS Thesis, University of Miami.

Logan G. A., Hayes J. M., Hleshima G. B., and Summons R. E. (1995) Terminal Proterozoic reorganization of biogeochemical cycles. *Nature* **376**, 53–56.

Lovell J. S. (2000) Oxygen and carbon dioxide in soil air. In *Handbook of Exploration Geochemistry,* Geochemical Remote Sensing of the Subsurface (ed. M. Hale). Elsevier, Amsterdam, vol. 7, pp. 451–469.

McMenamin M. A. S. and McMenamin D. L. S. (1990) *The Emergence of Animals—The Cambrian Breakthrough.* Columbia University Press, New York.

Mörner N.-A. and Etiope G. (2002) Carbon degassing from the lithosphere. *Global. Planet. Change* **33**, 185–203.

Murakami T., Utsunomiya S., Imazu Y., and Prasad N. (2001) Direct evidence of late Archean to early Proterozoic anoxic atmosphere from a product of 2.5 Ga old weathering. *Earth Planet. Sci. Lett.* **184**, 523–528.

Najjar R. G. and Keeling R. F. (2000) Mean annual cycle of the air–sea oxygen flux: a global view. *Global Biogeochem. Cycles* **14**, 573–584.

Nutman A. P., Mojzsis S. J., and Friend C. R. L. (1997) Recognition of ≧3850 Ma water-lain sediments in West Greenland and their significance for the early Archean Earth. *Geochim. Cosmochim. Acta* **61**, 2475–2484.

Ohmoto H. (1996) Evidence in pre-2.2 Ga paleosols for the early evolution of atmospheric oxygen and terrestrial biota. *Geology* **24**, 1135–1138.

Ohmoto H. (1999) Redox state of the Archean atmosphere: evidence from detrital heavy minerals in ca. 3250–2750 Ma sandstones from the Pilbara Craton, Australia: comment and reply. *Geology* **27**, 1151–1152.

Olson J. M. and Pierson B. K. (1986) Photosynthesis 3.5 thousand million years ago. *Photosynth. Res.* **9**, 251–259.

Pan Y. and Stauffer M. R. (2000) Cerium anomaly and Th/U fractionation in the 1.85 Ga Flin Flon Paleosol: clues from REE- and U-rich accessory minerals and implications for paleoatmospheric reconstruction. *Am. Mineral.* **85**, 898–911.

Pavlov A. A., Kasting J. F., Brown L. L., Rages K. A., and Freedman R. (2000) Greenhouse warming by CH_4 in the atmosphere of early Earth. *J. Geophys. Res.* **105**, 11981–11990.

Petsch S. T. (1999) Comment on Carpenter and Lohmann (1999). *Geochim Cosmochim Acta* **63**, 307–310.

Petsch S. T. (2000) A study on the weathering of organic matter in black shales and implications for the geochemical cycles of carbon and oxygen. PhD Dissertation. Yale University.

Petsch S. T. and Berner R. A. (1998) Coupling the geochemical cycles of C, P, Fe, and S: the effect on atmospheric O_2 and the isotopic records of carbon and sulfur. *Am. J. Sci.* **298**, 246–262.

Petsch S. T., Berner R. A., and Eglinton T. I. (2000) A field study of the chemical weathering of ancient sedimentary organic matter. *Org. Geochem.* **31**, 475–487.

Petsch S. T., Smernik R. J., Eglinton T. I., and Oades J. M. (2001) A solid state ^{13}C-NMR study of kerogen degradation during black shale weathering. *Geochim. Cosmochim. Acta* **65**, 1867–1882.

Phoenix V. R., Konhauser K. O., Adams D. G., and Bottrell S. H. (2001) Role of biomineralization as an ultraviolet shield: implications for Archean life. *Geology* **29**, 823–826.

Pollard D. and Kasting J. K. (2001) Coupled GCM-ice sheet simulations of Sturtian (750–720 Ma) glaciation: when in the snowball-earth cycle can tropical glaciation occur? *EOS* **82**, S8.

Porter S. M. and Knoll A. H. (2000) Testate amoebae in the Neoproterozoic Era: evidence from vase-shaped microfossils in the Chuar Group, Grand Canyon. *Paleobiology* **26**, 360–385.

Rasmussen B. and Buick R. (1999) Redox state of the Archean atmosphere: evidence from detrital heavy minerals in ca. 3250–2750 Ma sandstones from the Pilbara Craton, Australia. *Geology* **27**, 115–118.

Rau G. H., Takahashi T., and Des Marais D. J. (1989) Latitudinal variations in plankton $\delta^{13}C$: implications for CO_2 and productivity in past oceans. *Nature* **341**, 516–518.

Rau G. H., Takahashi T., Des Marais D. J., Repeta D. J., and Martin J. H. (1992) The relationship between $d^{13}C$ of organic matter and $[CO_{2(aq)}]$ in ocean surface water: Data from a JGOFS site in the northeast Atlantic Ocean and a model. *Geochim. Cosmochim. Acta* **56**, 1413–1419.

Robinson J. M. (1989) Phanerozoic O_2 variation, fire, and terrestrial ecology. *Palaeogeogr. Palaeoclimat. Palaeoecol. (Global Planet Change)* **75**, 223–240.

Robinson J. M. (1991) Phanerozoic atmospheric reconstructions: a terrestrial perspective. *Palaeogeogr. Palaeoclimat. Palaeoecol.* **97**, 51–62.

Röhl H.-J., Schmid-Röhl A., Oschmann W., Frimmel A., and Schwark L. (2001) The Posidonia Shale (Lower Toarcian) of SW-Germany: an oxygen-depleted ecosystem controlled by sea level and paleoclimate. *Palaeogeogr. Palaeoclimat. Palaeocol.* **165**, 27–52.

Ronov A. B. (1976) Global carbon geochemistry, volcanism, carbonate accumulation, and life. *Geochem. Int.* **13**, 172–195.

Rye R. and Holland H. D. (1998) Paleosols and the evolution of atmospheric oxygen: a critical review. *Am. J. Sci.* **298**, 621–672.

Rye R. and Holland H. D. (2000) Life associated with a 2.76 Ga ephemeral pond? Evidence from Mount Roe #2 paleosol. *Geology* **28**, 483–486.

Schidlowski M. (1988) A 3,800-million year isotopic record of life from carbon in sedimentary rocks. *Nature* **333**, 313–318.

Schidlowski M. (2001) Carbon isotopes as biogeochemical recorders of life over 3.8 Ga of Earth history: evolution of a concept. *Precamb. Res.* **106**, 117–134.

Schopf J. W. (1992) Paleobiology of the Archean. In *The Proterozoic Biosphere* (eds. J. W. Schopf, and C. Klein). Cambridge University Press, pp. 25–39.

Schopf J. W. (1993) Microfossils of the early Archean Apex Chert: new evidence of the antiquity of life. *Science* **260**, 640–646.

Schopf J. W., Kudryavtvev A. B., Agresti D. G., Wdowiak T. J., and Czaja A. D. (2002) Laser-Raman imagery of Earth's earliest fossils. *Nature* **416**, 73–76.

Sinninghe-Damsté J. S. and Köster J. (1998) A euxinic southern North Atlantic Ocean during the Cenomanian/Turonian oceanic anoxic event. *Earth Planet. Sci. Lett.* **158**, 165–173.

Sleep N. H. and Zahnle K. (1998) Refugia from asteroid impacts on early Mars and the early Earth. *J. Geophys. Res. E—Planets.* **103**, 28529–28544.

Sleep N. H., Zahnle K. J., Kasting J. F., and Morowitz H. J. (1989) Annihilation of ecosystems by large asteroid impacts on the early earth. *Nature* **342**, 139–142.

Sleep N. H., Zahnle K. J., and Neuhoff P. S. (2001) Initiation of clement surface conditions on the earliest Earth. *Proc. Natl. Acad. Sci.* **98**, 3666–3672.

Sowers T., Bender M., and Raynaud D. (1989) Elemental and isotopic composition of occluded O_2 and N_2 in polar ice. *J. Geophys. Res.* **94**, 5137–5150.

Stratford K., Williams R. G., and Myers P. G. (2000) Impact of the circulation on sapropel formation in the eastern Mediterranean. *Global Biogeochem. Cycles* **14**, 683–695.

Sumner D. Y. (1997) Carbonate precipitation and oxygen stratification in late Archean seawater as deduced from facies and stratigraphy of the Gamohaan and Frisco Formations, Transvaal Supergroup, South Africa. *Am. J. Sci.* **297**, 455–487.

Thomas A. L. R. (1997) The breath of life—did increased oxygen levels trigger the Cambrian Explosion? *Trends Ecol. Evol.* **12**, 44–45.

Tolbert N. E., Benker C., and Beck E. (1995) The oxygen and carbon dioxide compensation points of C_3 plants: possible role in regulating atmospheric oxygen. *Proc. Natl. Acad. Sci.* **92**, 11230–11233.

Valley J. W., Peck W. H., King E. M., and Wilde S. A. (2002) A cool early Earth. *Geology* **30**, 351–354.

Van Cappellen P. and Ingall E. D. (1996) Redox stabilization of the atmosphere and oceans by phosphorus-limited marine productivity. *Science* **271**, 493–496.

van Zuilen M., Lepland A., and Arrhenius G. (2002) Reassessing the evidence for the earliest traces of life. *Nature* **418**, 627–630.

Vidal R. A., Bahr D., Baragiola R. A., and Peters M. (1997) Oxygen on Ganymede: laboratory studies. *Science* **276**, 1839–1842.

Walker J. C. G. (1979) The early history of oxygen and ozone in the atmosphere. *Pure Appl. Geophys.* **117**, 498–512.

Walker J. C. G. (1986) Global geochemical cycles of carbon, sulfur, and oxygen. *Mar. Geol.* **70**, 159–174.

Watson A. J. (1978) *Consequences for the Biosphere of Grassland and Forest Fires.* PhD Dissertation. Reading University, UK.

Werne J. P., Hollander D. J., Behrens A., Schaeffer P., Albrecht P., and Sinninghe-Damsté J. S. (2000) Timing of early diagenetic sulfurization of organic matter: a precursor-product relationship in Holocene sediments of the anoxic Cariaco Basin, Venezuela. *Geochim. Cosmochim. Acta* **64**, 1741–1751.

Werne J. P., Lyons T. W., Hollander D. J., Formolo M., and Sinninghe-Damsté J. S. (2003) Reduced sulfur in euxinic sediments of the Cariaco Basin: sulfur isotope constraints on organic sulfur formation. *Chem. Geol.* **195**, 159–179.

Wilde S. A., Valley J. W., Peck W. H., and Graham C. M. (2001) Evidence from detrital zircons for the existence of continental crust and oceans on the Earth 4.4 Gyr ago. *Nature* **409**, 175–178.

Wortmann U. G., Hesse R., and Zacher W. (1999) Major-element analysis of cyclic black shales: paleoceanographic implications for the early Cretaceous deep western Tethys. *Paleoceanography* **14**, 525–541.

Wray G. A., Levinton J. S., and Shapiro L. H. (1996) Molecular evidence for deep Precambrian divergences among Metazoan phyla. *Science* **274**, 568–573.

Zhang R., Follows M. J., Grotzinger J. P., and Marshall J. (2001) Could the late Permian deep ocean have been anoxic? *Paleoceanography* **16**, 317–329.

8.12
The Global Nitrogen Cycle

J. N. Galloway
University of Virginia, Charlottesville, VA, USA

8.12.1 INTRODUCTION

Once upon a time nitrogen did not exist. Today it does. In the intervening time the universe was formed, nitrogen was created, the Earth came into existence, and its atmosphere and oceans were formed! In this analysis of the Earth's nitrogen cycle, I start with an overview of these important events relative to nitrogen and then move on to the more traditional analysis of the nitrogen cycle itself and the role of humans in its alteration.

The universe is ~15 Gyr old. Even after its formation, there was still a period when nitrogen did not exist. It took ~300 thousand years after the big bang for the Universe to cool enough to create atoms; hydrogen and helium formed first. Nitrogen was formed in the stars through the process of nucleosynthesis. When a star's helium mass becomes great enough to reach the necessary pressure and temperature, helium begins to fuse into still heavier elements, including nitrogen.

Approximately 10 Gyr elapsed before Earth was formed (~4.5 Ga (billion years ago)) by the accumulation of pre-assembled materials in a multistage process. Assuming that N_2 was the predominate nitrogen species in these materials and given that the temperature of space is −270 °C, N_2 was probably a solid when the Earth was formed since its boiling point (b.p.) and melting point (m.p.) are −196 °C and −210 °C, respectively. Towards the end of the accumulation period, temperatures were probably high enough for significant melting of some of the accumulated material. The volcanic gases emitted by the resulting volcanism strongly influenced the surface environment. Nitrogen was converted from a solid to a gas and emitted as N_2. Carbon and sulfur were probably emitted as CO and H_2S (Holland, 1984). N_2 is still the most common nitrogen volcanic gas emitted today at a rate of ~2 Tg N yr^{-1} (Jaffee, 1992).

Once emitted, the gases either remained in the atmosphere or were deposited to the Earth's surface, thus continuing the process of biogeochemical cycling. The rate of transfer depended on the reactivity of the emitted material. At the lower extreme of reactivity are the noble gases, neon and argon. Most neon and argon emitted during the degassing of the newly formed Earth is still in the atmosphere, and essentially none has been transferred to the hydrosphere or crust. At the other extreme are carbon and sulfur. Over 99% of the carbon and sulfur emitted during degassing are no longer in the atmosphere, but reside in the hydrosphere or the crust. Nitrogen is intermediate. Of the $\sim 6 \times 10^6$ Tg N in the atmosphere, hydrosphere, and crust, ~2/3 is in the atmosphere as N_2 with most of the remainder in the crust. The atmosphere is a large nitrogen reservoir primarily, because the triple bond of the N_2 molecule requires a significant amount of energy to break. In the early atmosphere, the only sources of such energy were solar radiation and electrical discharges.

At this point we had an earth with mostly N_2 and devoid of life. How did we get to an earth with mostly N_2 and teeming with life? First, N_2 had to be converted into reactive N (Nr). (The term reactive nitrogen (Nr) includes all biologically active, photochemically reactive, and radiatively active nitrogen compounds in the atmosphere and biosphere of the Earth. Thus, Nr includes inorganic reduced forms of nitrogen (e.g., NH_3 and NH_4^+), inorganic oxidized forms (e.g., NO_x, HNO_3, N_2O, and NO_3^-), and organic compounds (e.g., urea, amines, and proteins).) The early atmosphere was reducing and had limited NH_3. However, NH_3 was a necessary ingredient in forming early organic matter. One possibility for NH_3 generation was the cycling of seawater through volcanics (Holland, 1984). Under such a process, NH_3 could then be released to the atmosphere where, when combined with CH_4, H_2, H_2O, and electrical energy, organic molecules including amino acids could be formed (Miller, 1953). In essence, electrical discharges and UV radiation can convert mixtures of reduced gases into mixtures of organic molecules that can then be deposited to land surfaces and oceans (Holland, 1984).

To recap, Earth was formed at 4.5 Ga, water condensed at 4 Ga, and organic molecules were formed thereafter. By 3.5 Ga simple organisms (prokaryotes) were able to survive without O_2 and produced NH_3. At about the same time, the first organisms that could create O_2 in photosynthesis (e.g., cyanobacteria) evolved. It was not until 1.5–2.0 Ga that O_2 began to build up in the atmosphere. Up to this time, the O_2 had been consumed by chemical reactions (e.g., iron oxidation). By 0.5 Ga the O_2 concentration of the atmosphere reached the same value found today. As the concentration of O_2 built up, so did the possibility that NO could be formed in the atmosphere during electrical discharges from the reaction of N_2 and O_2.

Today we have an atmosphere with N_2 and there is energy to produce some NO (reaction of N_2 and O_2). Precipitation can transfer Nr to the Earth's surface. Electrical discharges can create nitrogen-containing organic molecules. Simple cells evolved ~3.5 Ga and, over the succeeding years, more complicated forms of life have evolved, including humans. Nature formed nitrogen and created life. By what route did that "life" discover nitrogen?

To address this question, we now jump from 3.5 Ga to $\sim 2.3 \times 10^{-7}$ Ga. In the 1770s, three scientists—Carl Wilhelm Scheele (Sweden), Daniel Rutherford (Scotland), and Antoine Lavosier (France)—independently discovered the existence of nitrogen. They performed experiments in which an unreactive gas was produced. In 1790, Jean Antoine Claude Chaptal formally named the gas *nitrogène*. This discovery marked the beginning of our understanding of nitrogen and its role in Earth systems.

By the beginning of the second half of the nineteenth century, it was known that nitrogen is a common element in plant and animal tissues, that it is indispensable for plant growth, that there is constant cycling between organic and inorganic compounds, and that it is an effective fertilizer. However, the source of nitrogen was still uncertain. Lightning and atmospheric deposition were thought to be the most important sources. Although the existence of biological nitrogen fixation (BNF) was unknown at that time, in 1838 Boussingault demonstrated that legumes restore Nr to the soil and that somehow they create Nr directly. It took almost 50 more years to solve the puzzle. In 1888, Herman Hellriegel (1831–1895) and Hermann Wilfarth (1853–1904) published their work on microbial communities. They noted that microorganisms associated with legumes have the ability to assimilate atmospheric N_2 (Smil, 2001). They also said that it was necessary for a symbiotic relationship to exist between legumes and microorganisms.

Other important processes that drive the cycle were elucidated in the nineteenth century. In the late 1870s, Theophile Scholesing proved the bacterial origins of nitrification. About a decade later, Serfei Nikolaevich Winogradsky isolated the two nitrifers—*Nitrosomonas* and *Nitrobacter*—and showed that the species of the former genus oxidize ammonia to nitrite and that the species of the latter genus convert nitrite to nitrate. Then in 1885, Ulysse Gayon isolated cultures of two bacteria that convert nitrate to N_2. Although there are only two bacterial genera that can convert N_2 to Nr, several can convert Nr back to N_2, most notably *Pseudomonas*, *Bacillus*, and *Alcaligenes* (Smil, 2001).

By the end of the nineteenth century, humans had discovered nitrogen and the essential components of the nitrogen cycle. In other words, they then knew that some microorganisms convert N_2 to NH_4^+, other microorganisms convert NH_4^+ to NO_3^-, and yet a third class of microorganisms convert NO_3^- back to N_2, thus completing the cycle.

The following sections of this chapter examine the biogeochemical reactions of Nr, the distribution of Nr in Earth's reservoirs, and the exchanges between the reservoirs. This chapter then discusses Nr creation by natural and anthropogenic processes and nitrogen budgets for the global land mass and for continents and oceans using Galloway and Cowling (2002) and material from Cory Cleveland (University of Colorado) and Douglas Capone (University of Southern California) from a paper in review in Biogeochemistry (Galloway *et al.*, 2003a). This chapter also presents an overview of the consequences of Nr accumulation in the environment (using Galloway *et al.* (2003b) as a primary reference) and then concludes with estimates of minima and maxima Nr creation rates in 2050.

8.12.2 BIOGEOCHEMICAL REACTIONS

8.12.2.1 The Initial Reaction: Nr Creation

In the formation of Earth, most nitrogen was probably in the form of N_2, the most stable molecule and thus the reservoir from which all other nitrogen compounds are formed. It is thus fitting to begin this section on chemical reactions with N_2—with both atoms of the molecule representing the Adam and Eve of nitrogen. N_2 is converted to Nr by either converting N_2 to NH_3 or by converting N_2 to NO. High energy is required to break the triple bond of the N_2 molecule—226 kcal mol^{-1}. In the case of both NH_3 and NO formation, a natural process and an anthropogenic process form Nr from N_2. The next few paragraphs review these processes; a later section discusses rates of Nr formation as a function of time and process.

BNF is a microbially mediated process that occurs in several types of bacteria and blue-green algae. This process uses the enzyme nitrogenase in an anaerobic environment to convert N_2 to NH_3. The microbes can be free-living or in a symbiotic association with the roots of higher plants. Legumes are the best-known example of this type of relationship (Schlesinger, 1997; Mackenzie, 1998):

$$2N_2 + 6H_2O \rightarrow 4NH_3 + 3O_2$$

As will be explained in more detail later, BNF can occur in both unmanaged and managed ecosystems. In the former, natural ecosystems produce Nr. In the latter, the cultivation of legumes enhances BNF and the cultivation of some crops (e.g., wetland rice) creates the necessary anaerobic environment to promote BNF.

Over geologic history, most Nr has been formed by BNF. However, in the last half of the twentieth century, the Haber–Bosch process has replaced BNF as the dominant terrestrial process creating Nr. The Haber–Bosch process was invented and developed commercially in the early 1900s. It uses high temperature and pressure with a metallic catalyst to produce NH_3:

$$N_2 + 3H_2 \rightarrow 2NH_3$$

This process was used extensively during World War I to produce munitions and, since the early 1950s, has become the world's largest source of nitrogenous fertilizer (Smil, 2001). N_2 is taken from the atmosphere, whereas H_2 is produced from a fossil fuel, usually natural gas.

Two processes create Nr in the form of NO from the oxidation of N_2. The natural process is lightning; the anthropogenic process is fossil-fuel combustion. As previously mentioned, at an early point in Earth's history, lightning was an important process in creating Nr from N_2. Although globally much less important now, N_2 is still converted to NO by electrical discharges:

$$N_2 + O_2 + \text{electrical energy} \rightarrow 2NO$$

Lightning formation of NO is most important in areas of deep convective activities such as occurring in tropical continental regions.

The high temperatures and pressures found during fossil-fuel combustion provide the energy to convert N_2 to NO through reaction with O_2:

$$N_2 + O_2 + \text{fossil energy} \rightarrow 2NO$$

Note that, during fossil-fuel combustion, NO can also be formed from the oxidation of fossil organic nitrogen, which is primarily found in coal. This is technically not a creation of new Nr but rather the mobilization of Nr that has been sequestered for millions of years (Socolow, 1999).

Thus, independent of the process, N_2 is transformed to either NO in the atmosphere or a combustion chamber or to NH_3 in an organism or a factory. The rest of this section examines the fate of NO and NH_3 (and their reaction products) in the atmosphere, terrestrial ecosystems, aquatic ecosystems, and agro-ecosystems.

Except in the atmosphere (where nitrogen chemistry is essentially a series of oxidation reactions), the basic structure of nitrogen reactions is similar in all reservoirs. For soils, freshwaters, coastal waters, and oceans, the chemical reactions of nitrogen are generally self-contained sequences of oxidation and reduction reactions driven by microbial activity and environmental conditions. The central framework of the reactions is constructed through the processes of nitrogen fixation, assimilation, nitrification, decomposition, ammonification, and denitrification (Figure 1). This framework provides nitrogen for amino acid and protein synthesis by primary producers followed perhaps by consumption of the protein by a secondary producer. This framework also provides a mechanism to convert the amino acids/protein back to inorganic nitrogen following excretion from, or the death of, the primary or secondary producer.

8.12.2.2 Atmosphere

The atmospheric nitrogen cycle is the simplest cycle, because the direct influence of biota is limited and thus chemical and physical processes primarily control transformations. In addition, the cycle of oxidized inorganic nitrogen (NO_y) is, for the most part, decoupled from the cycle of reduced inorganic nitrogen (NH_x). As discussed in the next section, nitrogen cycles in terrestrial and aquatic ecosystems are substantially more complex because of the microbially mediated nitrogen transformations that occur (Figure 1).

The atmospheric chemistry of nitrogen can be divided into four groupings of nitrogen species

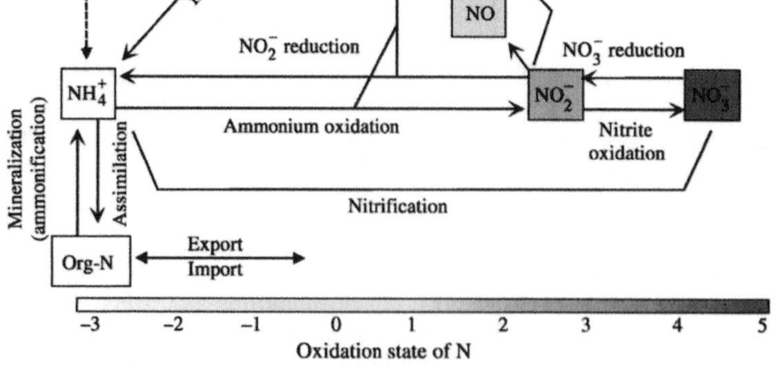

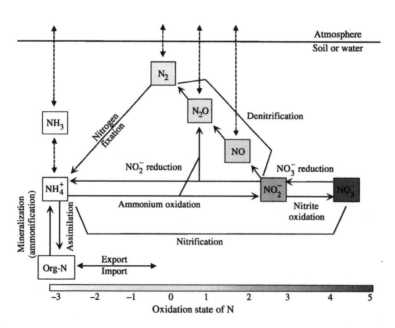

Redrawn from Karl

Figure 1 The processes of nitrogen fixation, assimilation, nitrification, decomposition, ammonification, and denitrification (after Karl, 2002).

that are generally independent of each other—inorganic reduced nitrogen, inorganic oxidized nitrogen, organic reduced nitrogen, and organic oxidized nitrogen. Reactions are generally within the groupings. For each grouping, I first discuss the emitted species and then its fate in the atmosphere.

8.12.2.2.1 Inorganic reduced nitrogen

There are two species in this grouping—ammonia (NH_3) and ammonium (NH_4^+) with one valence state (−III) (Table 1). The primary species emitted to the atmosphere is NH_3 produced during organic matter decomposition and emitted when the partial pressure in the soil, water, or plant is greater than the partial pressure in the atmosphere. It is the most common atmospheric gaseous base and, once in the atmosphere, can be converted to an aerosol in an acid–base reaction with a gas (e.g., HNO_3) or aerosol (e.g., H_2SO_4):

$$NH_{3(g)} + HNO_{3(g)} \rightarrow NH_4NO_{3(s)}$$

$$NH_{3(g)} + H_2SO_{4(s)} \rightarrow NH_4HSO_{4(s)}$$

$$NH_{3(g)} + NH_4HSO_{4(s)} \rightarrow (NH_4)_2SO_{4(s)}$$

All species are readily removed by atmospheric deposition. NH_3 is primarily removed by dry deposition (often close to its source). Aerosol NH_4^+ is primarily removed by wet deposition; in fact, the hydroscopic aerosol is a cloud-condensation nuclei. If NH_x (NH_3 and NH_4^+) is lifted above the planetary boundary layer (PBL), it can be transported large distances (1,000 km or more). Emissions in one location can impact receptors far downwind.

8.12.2.2.2 Inorganic oxidized nitrogen

This grouping has many species and valence states (Table 1). Most oxidized nitrogen species in the atmosphere are part of NO_y, which includes $NO + NO_2 + HNO_3 + PAN$ and other trace oxidized species. Within this group, $NO + NO_2$ are referred to as NO_x. All these species are relatively reactive and most have lifetimes of minutes to days in the atmosphere; some (e.g., PAN) can have longer lifetimes. NO is the most commonly emitted species and has several sources that are usually of two types. As discussed earlier, the first is from the conversion of N_2 to NO by lightning or the high-temperature combustion of fossil fuels. The second is the conversion of one Nr species to NO generally through fire or microbial activity. Once in the atmosphere, NO is quickly oxidized to NO_2 that is then oxidized to HNO_3 and then potentially reacts with NH_3 to form an aerosol:

$$NO + O_3 \rightarrow NO_2 + O_2$$

$$NO_2 + OH \rightarrow HNO_3$$

$$HNO_3 + NH_3 \rightarrow NH_4NO_3$$

However, in this process a significant cycle involving hydrocarbons and ozone has important implications for the oxidation capacity of the atmosphere and human and ecosystem health.

The oxidized inorganic nitrogen species not included in NO_y is N_2O. It is produced during nitrification and denitrification (Figure 1). It has an atmospheric residence time of ~100 yr and, because of this, emissions are globally distributed. Because of its stability, no significant chemical reactions take place in the troposphere but, once in the stratosphere, it is converted to NO by UV radiation:

$$N_2O + O(^1D) \rightarrow 2NO$$

The NO produced will then destroy stratospheric ozone in a reaction that regenerates it:

$$NO + O_3 \rightarrow NO_2 + O_2$$

$$O_3 \rightarrow O + O_2$$

$$NO_2 + O \rightarrow NO + O_2$$

The net reaction is

$$2O_3 \rightarrow 3O_2$$

Increasing concentrations of atmospheric N_2O contribute to two environmental issues of the day. In the troposphere it contributes to the greenhouse potential; in the stratosphere it contributes to ozone destruction.

Table 1 Valence state of N species.

Grouping	*Valence/oxidation state*	*Species*
Inorganic oxidized N	5	NO_3^-, HNO_3
	4	NO_2
	3	NO_2^-
	2	NO
Diatomic N	0	N_2
Inorganic reduced N	−3	NH_3, NH_4^+
Organic reduced N	−3	$R{-}NH_2$

8.12.2.2.3 Reduced organic nitrogen

Atmospheric organic nitrogen can occur as bacteria, particulate matter, and soluble species (e.g., amines) (Neff *et al.*, 2002). These materials are emitted to the atmosphere through low-temperature processes such as turbulence and high-temperature processes like biomass burning. The concentrations of bacteria range from ~10 bacteria m^{-3} in the marine environment to >1,000 bacteria m^{-3} in urban environments (Neff *et al.*, 2002). Atmospheric reactions involving bacteria are limited. Particulate organic nitrogen is composed of organic debris and soil matter. Again, there is probably limited atmospheric chemistry of this material. Conversely, soluble reduced organic nitrogen compounds (e.g., urea, free amino acids, and other methylated amines) are quite reactive in the atmosphere. Many species can react quickly with HO and organic nitrates and thus are not transported far from their emission points (Neff *et al.*, 2002).

8.12.2.2.4 *Oxidized organic nitrogen*

Atmospheric oxidized organic nitrogen species are generally formed in the atmosphere as the end products of reactions of hydrocarbons with NO_x (Neff *et al.*, 2002). Hydrocarbons can form organic radicals (RO, RO_2) through reaction with light, OH, or ozone. The resulting species can react with NO_2 to form $RONO_2$.

8.12.2.3 Biosphere

The biosphere is defined as the terrestrial and aquatic ecosystems, including the oceans. As mentioned above, microbial processes have a strong, in many cases controlling, influence on the biogeochemistry of nitrogen in these systems. These processes themselves constitute a cycle (Figure 1). The individual processes are briefly defined below.

Nitrogen fixation is the processes by which N_2 is converted to any nitrogen compound where nitrogen has a nonzero oxidation state. Historically, the most common process has been the biologically driven reduction of N_2 to NH_3 or NH_4^+. However, currently anthropogenically enhanced nitrogen fixation dominates on continents (see Section 8.12.4).

Ammonia assimilation is the uptake of NH_3 or NH_4^+ by an organism into its biomass in the form of an organic nitrogen compound. For organisms that can directly assimilate reduced inorganic nitrogen, this is an efficient process to incorporate nitrogen into biomass.

Nitrification is the aerobic process by which microorganisms oxidize ammonium to nitrate and derive energy. It is the combination of two bacterial processes: one group of organisms oxidizes ammonia to nitrite (e.g., *Nitrosomonas*), after which a different group oxidizes nitrite to nitrate (e.g., *Nitrobactor*):

$$2NH_4^+ + 3O_2 \rightarrow 2NO_2^- + 2H_2O + 4H$$

$$2NO_2^- + O_2 \rightarrow 2NO_3^-$$

Assimilatory nitrate reduction is the uptake of NO_3^- by an organism and incorporation as biomass through nitrate reduction. It is an important process, because it allows the mobile nitrate ion to be transported to a receptor which can then reduce it to ammonia for subsequent uptake. It is an important input of nitrogen for many plants and organisms.

Ammonification is the primary process that converts reduced organic nitrogen ($R{-}NH_2$) to reduced inorganic nitrogen (NH_4^+) through the action of microorganisms. This is part of the general process of decomposition where heterotrophic microbes use organic matter for energy and in the process convert organic N to NH_4^+.

Denitrification is the reduction of NO_3^- to any gaseous nitrogen species, normally N_2.

$$5CH_2O + 4H^+ + 4NO_3^- \rightarrow 2N_2 + 5CO_2 + 7H_2O$$

It is an anaerobic process and requires nitrate (NH_x that can be nitrified) and organic matter. Microorganisms use nitrate as an oxidant to obtain energy from organic matter. It is prevalent in waterlogged soils and is the primary process that converts Nr back to N_2.

8.12.3 NITROGEN RESERVOIRS AND THEIR EXCHANGES

The reservoirs of the Earth contain $\sim 5 \times 10^{21}$ g N, 80% of which is in the atmosphere. Sedimentary rocks contain almost all the remainder with just a trace (<1% of total) in oceans and living and dead organic matter (Table 2) (Mackenzie, 1998).

The chemical form of the nitrogen depends on the reservoir. In the atmosphere, except for trace amounts of N_2O, NO_y, NH_x, and organic N, it occurs as N_2. In oceans and soils, it primarily occurs as organic nitrogen, nitrate, and ammonium.

Table 2 N amounts in global reservoirs (Tg N yr^{-1}).

Reservoirs	*Amount*	*Percentage of total*
Atmosphere, N_2	3,950,000,000	79.5
Sedimentary rocks	999,600,000	20.1
Ocean		
N_2	20,000,000	0.4
NO_3^-	570,000	0.0
Soil organics	190,000	0.0
Land biota	10,000	0.0
Marine biota	500	0.0

Source: Mackenzie (1998) except ocean, N_2 from Schlesigner (1997).

Atmospheric and hydrologic transport provide the primary paths for exchanges between reservoirs.

8.12.3.1 Land to Atmosphere

Nitrogen is transferred to the atmosphere by low- and high-temperature processes. The high-temperature processes are biomass combustion and fossil-fuel combustion; the low-temperature processes are volatilization of gases from soils and waters and turbulent injection of particulate matter into the atmosphere. The gases are generated primarily as a result of microbial activity (e.g., nitrification, denitrification, and ammonification).

8.12.3.2 Ocean to Atmosphere

When the partial pressure of gas in seawater is greater than the partial pressure of gas in the atmosphere, nitrogen can be emitted from the ocean through volatilization of NH_3, N_2, and N_2O. Nitrogen can also be transferred to the atmosphere via aerosol formed by breaking waves or bubbles.

8.12.3.3 Atmosphere to Surface

Nitrogen is deposited to land and ocean surfaces by wet and dry deposition. Wet deposition includes rain, snow, and hail; the nitrogen can be inorganic or organic. Dry deposition includes gases and aerosols. The two most important dry-deposited species are gaseous HNO_3 and NH_3. Aerosol dry deposition of nitrate and ammonium does occur, but the fluxes are generally small relative to gaseous deposition.

8.12.3.4 Land to Ocean

Both inorganic and organic nitrogen compounds are transferred from continents to the coastal ocean via discharge of rivers and groundwater. The former can be soluble (e.g., NO_3^-) or particulate (NH_4^+ adsorbed on surfaces). The latter can also be soluble or particulate. Most nitrogen is transported as either NO_3^- or particulate organic nitrogen (Seitzinger *et al.*, 2002).

8.12.4 Nr CREATION

8.12.4.1 Introduction

As discussed earlier, N_2 is converted to Nr by four basic processes—lightning, BNF, combustion, and the Haber–Bosch process. This section details these processes and concludes with an estimate of the trends in Nr creation from 1860 to 2000.

8.12.4.2 Lightning—Natural

High temperatures in lightning strikes produce NO in the atmosphere from molecular oxygen and nitrogen, primarily over tropical continents. Subsequently this NO is oxidized to NO_2 and then to HNO_3 and quickly (i.e., days) introduces Nr into ecosystems through wet and dry deposition. More current estimates of Nr creation by lightning range between $3\,Tg\,N\,yr^{-1}$ and $10\,Tg\,N\,yr^{-1}$ (Ehhalt *et al.*, 2001). In this analysis I have used a global estimate of $5.4\,Tg\,N\,yr^{-1}$ (Lelieveld and Dentener, 2000). As will be seen below, although this number is small relative to terrestrial BNF, it can be important for regions that do not have other significant Nr sources. It is also important, because it creates NO_x high in the free troposphere as opposed to NO_x emitted at the Earth's surface. As a result it has a longer atmospheric residence time and is more likely to contribute to tropospheric O_3 formation, which significantly impacts the oxidizing capacity of the atmosphere (E. Holland, personal communication).

8.12.4.3 Terrestrial BNF—Natural

Before human activity, BNF was the most important process in converting N_2 to Nr. However, quantifying BNF is difficult primarily because of the uncertainty and variability in the existing estimates of BNF rates at the plot scale. In addition, as noted in Cleveland *et al.* (1999), for many large areas where BNF is likely to be important, particularly in the tropical regions of Asia, Africa, and South America, there are virtually no data on natural terrestrial rates of BNF. In a recent compilation of rates of natural BNF by Cleveland *et al.* (1999), symbiotic BNF rates for several biome types are based on a few published rates of symbiotic BNF at the plot scale within each particular biome. For example, based on the few estimates of symbiotic BNF available for tropical rain forests, published estimates indicate that natural BNF in these systems annually represents ~24% of the total global, natural terrestrial BNF (Cleveland *et al.*, 1999).

Difficulties notwithstanding, earlier estimates of BNF in terrestrial ecosystems range from $\sim 30\,Tg\,N\,yr^{-1}$ to $200\,Tg\,N\,yr^{-1}$ (e.g., Delwiche, 1970; Söderlund and Rosswall, 1982; Stedman and Shetter, 1983; Paul and Clark, 1997; Schlesinger, 1997). Most studies merely present BNF estimates as "global values" and, at best, are broken into a few very broad components (e.g., "forest," "grassland," and "other"; e.g., Paul and Clark (1997)). However, such coarse divisions average enormous land areas that contain significant variation in both BNF data sets and biome types, thus diminishing their usefulness. Many studies also lack a list of data sources from which their estimates are derived (e.g., Söderland and Rosswall, 1982; Stedman and Shetter, 1983; Paul and Clark, 1997; Schlesinger, 1997). In contrast, Cleveland *et al.* (1999) provide a range of estimates of BNF in natural ecosystems from 100 Tg N yr^{-1} to 290 Tg N yr^{-1} (with a "best estimate" of 195 Tg N yr^{-1}). These estimates are based on published, data-based rates of BNF in natural ecosystems and differ only in the percentage of cover estimates of symbiotic nitrogen fixers used to scale plot-level estimates to the biome scale.

Although the data-based estimates of Cleveland *et al.* (1999) provide more of a documented, constrained range of terrestrial BNF, there are several compelling reasons to believe that an estimate in the lower portion of the range is more realistic than the higher estimates (Galloway *et al.*, 2003a; C. Cleveland, personal communication). First, rate estimates of BNF in the literature are inherently biased since investigations of BNF are frequently carried out in plot-level studies in areas where BNF is most likely to be important. For example, rates of BNF in temperate forests are based on estimates that include nitrogen inputs from alder and black locust (Cleveland *et al.*, 1999). Although the rates of BNF may be very high within stands dominated by these species (Boring and Swank, 1984; Binkley *et al.*, 1994), these species are certainly not dominant in temperate forests as a whole (Johnson and Mayeux, 1990). Similarly, although the species with high rates of BNF are often common in early successional forests (Vitousek, 1994), they are often rare in mature or late successional forests, especially in the temperate zone (Gorham *et al.*, 1979; Boring and Swank, 1984; Blundon and Dale, 1990). Literature-derived estimates based on reported coverage of nitrogen-fixing species are thus inflated by these inherent biases (Galloway *et al.*, 2003a; C. Cleveland, personal communication).

Therefore, based on the above, it is reasonable to assume that annual global BNF contributed between 100 Tg N yr^{-1} and 290 Tg N yr^{-1} to natural terrestrial ecosystems before large-scale human disturbances. However, because of the biases inherent in plot-scale studies of nitrogen-fixation rates, the true rate of BNF is probably at the lower end of this range (C. Cleveland, personal communication). Asner *et al.* (2001) use actual evapotranspiration (ET) values generated in the model Terraflux together with the strong, positive relationship between ET and nitrogen fixation (Cleveland *et al.*, 1999) to generate a new, single estimate of BNF before any large-scale human disturbance. This analysis is still based on the relationship between ET and BNF, but uses rates of BNF calculated using the low percentage cover values of symbiotic nitrogen fixers over the landscape (i.e., 5%; Cleveland *et al.*, 1999). This analysis suggests that within the range of 100–290 Tg N yr^{-1}, natural BNF in terrestrial ecosystems contributes 128 Tg N yr^{-1}. This value is supported by an analysis comparing BNF to nitrogen requirement (by type of biome). Using the Cleveland *et al.* (1999) relationship between ET and BNF, a global nitrogen-fixation value of 128 Tg N yr^{-1} would suggest an average of ~15% of the nitrogen requirement across all biome types is met via BNF; higher estimates of BNF would imply that at least 30% of the nitrogen requirement across all biomes is met via natural BNF (Asner *et al.*, 2001).

This estimate (128 Tg N yr^{-1}) represents potential BNF before large-scale human disturbances and does not account for decreases in BNF from land-use change or from any other physical, chemical, or biological factor. To estimate natural terrestrial BNF for 1890 and the early 1990s, BNF is scaled to the extent of altered land at those two times. Of the 1.15×10^4 Mha of natural vegetated land (Mackenzie, 1998), Houghton and Hackler (2002) estimate that in 1860 and 1995, 760 Mha and 2,400 Mha, respectively, had been altered by human action (e.g., cultivation, conversion of forests to pastures). Therefore, in this analysis of BNF in the terrestrial landscape, 128 Tg N yr^{-1} is fixed before landscape alteration of the natural world, 120 Tg N yr^{-1} for 1860, and 107 Tg N yr^{-1} for the present world (Galloway *et al.*, 2003a). These values are similar to ones of Galloway and Cowling (2002) (which also relied on work of C. Cleveland) for 1890 and 1990, and are also used in this comparison (see below).

8.12.4.4 Anthropogenic

8.12.4.4.1 Introduction

As discussed above, in the prehuman world, two processes—BNF and lightning—generated enough energy to break the N_2 triple bond. The former process uses metabolic energy to create Nr on purpose and the latter process uses electrical energy to produce Nr by accident. Humans developed one new process (Haber–Bosch) and enhanced a natural process (BNF) to create Nr on purpose to sustain food production. Humans also create Nr by accident by fossil fuel combustion.

8.12.4.4.2 Food production

Humans not only need nitrogen to survive but they need it in the form of amino acids because they are not able to synthesize amino acids from inorganic nitrogen. Early hunter gatherers met their nitrogen requirements by consuming naturally occurring amino acids in plants (which can make their own organic nitrogen compounds (e.g., amino acids)) and animals (which eat plants or other animals to get organic nitrogen compounds). Once people established settled communities and started using and reusing the same land, they had to find ways to introduce the harvested Nr back into the earth.

The history of Nr additions can be divided into four stages—recycling of organic matter (e.g., crop residues, manure), C-BNF (e.g., legumes), importing existing Nr (e.g., guano), and creation of mineral nitrogen (e.g., Haber–Bosch). The first two stages are not distinctly separate. Recycling has probably been practiced since the advent of agriculture, and legumes were certainly early candidates for cultivation. Archeological evidence points to legume cultivation over 6.5 ka (Smith, 1995) and perhaps as early as 1.2×10^4 yr BP (Kislev and Baryosef, 1988). Soybeans have been cultivated in China for at least 3,100 yr (Wang, 1987). Rice cultivation, which began in Asia perhaps as early as 7,000 yr ago (Wittwer *et al.*, 1987), also resulted in anthropogenic-induced creation of Nr by creating anaerobic environments where BNF could occur. There are, however, limits to recycling and C-BNF; by the early nineteenth century additional nitrogen sources were required. Guano imported from arid tropical and subtropical islands and from South America became an early source of new Nr with nitrogen contents up to ~30 times higher than most manures (Smil, 2001). From about 1830 to the 1890s, guano was the only source of additional nitrogen. However, by the late 1890s many were concerned that there was not enough nitrogen to provide food for the world's growing population. This concern was highlighted by Sir William Crookes (1832–1919), President of the British Association of Science. In a September 1898 speech he warned that "all civilized nations stand in deadly peril of not having enough to eat" (Crookes, 1871) because of the lack of nitrogen. Crookes went on to encourage chemists to learn how to convert atmospheric N_2 to NH_3 (Smil, 2001). As discussed below, in a very short 15 years, the Haber–Bosch process was developed exactly for that purpose. However, before that happened, agriculture used two other industrial processes to create nitrogen for fertilizer. When coal is burned or heated in the absence of air, part of the organic nitrogen in the fuel is converted to NH_3 (coke ovens, Table 3). On a global basis, by 1890 trace amounts of NH_3 were captured for agriculture; by 1935 it had overtaken guano and sodium nitrate as the primary nitrogen-fertilizer source (Table 3). Another industrial process was the synthesis of cyanamide—the first method of creating Nr from N_2. In this process, calcium carbide reacts with N_2 to form $CaCN_2$, which is then combined with superheated steam to produce $CaCO_3$ and NH_3. Although this process did create usable Nr, it required a large amount of energy. By the late 1920s, the traditional nitrogen sources, recycling crop residues and manure, were replaced by sodium nitrate, coke-oven gas, and cyanamide synthesis. By 1929, the total amount

Table 3 Guano and Chilean nitrate extraction: ammonium sulfate, cyanimide, calcium production from 1850 to 2000 ($Tg\,N\,yr^{-1}$).

	$NaNO_3$	*Guano*	*Coke oven*	*$CaCN_2$*	*Electric arc*	*Haber–Bosch*	*Total*
1850	0.01						0.01
1860	0.01	0.07					0.08
1870	0.03	0.07					0.10
1880	0.05	0.03					0.08
1890	0.13	0.02					0.15
1900	0.22	0.02	0.12				0.36
1905	0.25	0.01	0.13				0.39
1910	0.36	0.01	0.23	0.01			0.61
1913	0.41	0.01	0.27	0.03	0.01		0.73
1920	0.41	0.01	0.29	0.07	0.02	0.15	0.95
1929	0.51	0.01	0.43	0.26	0.02	0.93	2.15
1935	0.18	0.01	0.37	0.23		1.30	2.09
1940	0.20	0.01	0.45	0.29		2.15	3.10
1950	0.27		0.50	0.31		3.70	4.78
1960	0.20		0.95	0.30		9.54	10.99
1970	0.12		0.95	0.30		30.23	31.60
1980	0.09		0.97	0.25		59.29	60.60
1990	0.12		0.55	0.11		76.32	77.10
2000	0.12		0.37	0.08		85.13	85.70

Source: Smil (2001).

of nitrogen produced by these processes was 1.2 Tg N.

There was, of course, another source of new Nr—the reaction of N_2 and H_2 to produce NH_3. It has long been realized that NH_3 cannot be formed at ambient temperature and pressure. Indeed, the first experiments using higher temperatures began in 1788 (Smil, 2001). It took more than another century for the experiments to bear fruit. In 1904, Fritz Haber began his experiments on ammonia synthesis. By 1909 he had developed an efficient process to create NH_3 from N_2 and H_2 at the laboratory scale using high temperature, high pressure, and a catalyst. For this he received the Nobel prize in Chemistry in 1920. The path from a laboratory synthesis to a factory synthesis took four years. In 1913, based on the work of Carl Bosch, the first major NH_3-generating plant became operational in Germany. In honor of his work, Bosch received the Nobel prize in Chemistry in 1932. For the first few years, most production went for munitions to sustain Germany during World War I. It was not until the early 1930s that the Haber–Bosch process became the primary source of NH_3 for agricultural use.

In summary, from the advent of agriculture to about 1850, the sole source of nitrogen for food production was BNF naturally occurring in the soil, cultivation-induced BNF and nitrogen added by recycling of organic matter (stage 1). From 1850 to 1890, an additional source was guano and sodium nitrate deposits (stage 2). From 1890 to 1930, the coke oven and $CaCN_2$ processes accounted for up to 40% of the nitrogen fertilizer produced (stage 3). After 1930, the Haber–Bosch process became the dominant source of fertilizer nitrogen (stage 4). Of the early processes, the only ones that are still important sources for some regions are C-BNF and organic matter recycling.

8.12.4.4.3 Energy production

Combustion of fossil fuels creates energy on purpose and several waste gases by accident, including NO, which is created in the combustion chamber by the reaction of N_2 and O_2. In addition, nitrogen fixed by plants hundreds of millions of years ago (fossil nitrogen) has the potential to become reactive again when fossil fuels are extracted and burned. More than 1% by weight of coal, for example, is nitrogen fixed in geological times. In all, ~50 Tg N yr^{-1} of fossil nitrogen are removed from belowground today, primarily in coal but also in crude oil. When fossil fuel is burned, fossil nitrogen either joins the global pool of Nr in the form of NO_x emissions or is changed back to nonreactive N_2. In a typical coal-fired plant without pollution control, less than half the fossil nitrogen becomes NO_x emissions, a fraction that is much smaller with pollution control (Galloway *et al.*, 2002; Ayres *et al.*, 1994).

8.12.4.5 Nr Creation Rates from 1860 to 2000

Nr creation rates (Galloway *et al.*, 2003b) are related to global population trends from 1860 to 2000 in Figure 2. For 1920, 1930, and 1940, global anthropogenic fertilizer production (Smil, 2001) was made equivalent to total Nr production by the Haber–Bosch process. For 1950 onwards Nr creation rates by Haber–Bosch were obtained from USGS minerals (Kramer, 1999). The rate in 1900 is estimated to be ~15 Tg N yr^{-1} (V. Smil,

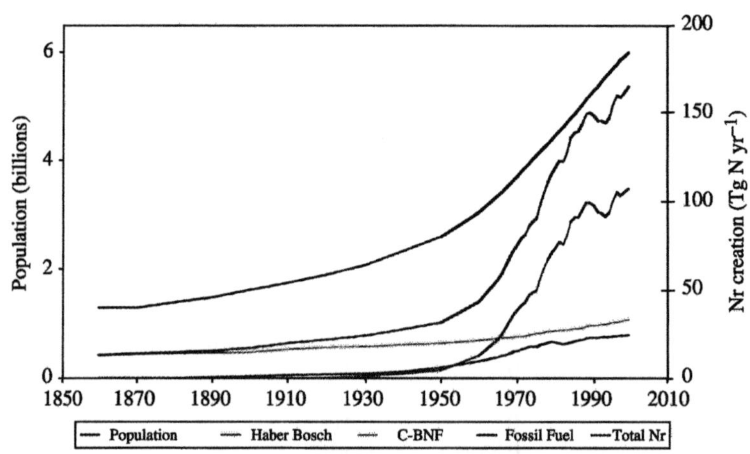

Figure 2 The (purple line) global population from 1860 to 2000 (left axis: population in billions; right axis: Nr creation in Tg N yr^{-1}) showing (green line) Nr creation via the Haber–Bosch process, including production of NH_3 for nonfertilizer purposes; (blue line) Nr creation from cultivation of legumes, rice and sugar cane; (brown line) Nr creation from fossil-fuel combustion; and (red line) the sum created by these three processes (source Galloway *et al.*, 2003b).

personal communication). Nr creation rates for 1860, 1870, 1880, and 1890 were estimated from population data using the 1900 data on population and Nr creation. For the period 1961–1999, Nr creation rates were calculated from crop-specific data on harvested areas (FAO, 2000) and fixation rates (Smil, 1999). Decadal data from 1910 to 1950 were interpolated between 1900 and 1961. The data from 1860 to 1990 are from a compilation from Elisabeth Holland and are based on Holland and Lamarque (1997), Müller (1992), and Keeling (1993). These data agree well with those published by van Aardenne *et al.* (2001) for the decadal time steps from 1890 to 1990. The data for 1991–2000 are estimated by scaling NO_x emissions to increases in fossil-fuel combustion over the same period.

The global rate of increase in Nr creation by humans was relatively slow from 1860 to 1960. Since 1960, however, the rate of increase has accelerated sharply (Figure 2). Cultivation-induced Nr creation increased from ~15 Tg N yr^{-1} in 1860 to ~33 Tg N yr^{-1} in 2000. Nr creation by fossil-fuel combustion increased from ~0.3 Tg N yr^{-1} in 1860 to ~25 Tg N yr^{-1} in 2000. Nr creation from the Haber–Bosch process went from zero before 1910 to >100 Tg N yr^{-1} in 2000 with ~85% used in the production of fertilizers. Thus, between 1860 and 2000, the anthropogenic Nr creation rate increased from ~15 Tg N yr^{-1} to ~165 Tg N yr^{-1} with ~5 times more Nr coming from food production than from energy production and use (Galloway *et al.*, 2002).

8.12.5 GLOBAL TERRESTRIAL NITROGEN BUDGETS

8.12.5.1 Introduction

Over the last several decades, there have been numerous compilations of the global nitrogen cycle (Table 4). Most cover the basic fluxes—BNF and denitrification in terrestrial and marine

Table 4 Global creation and distribution rates of Nr (Tg N yr^{-1}).

	~1970 (Delwiche, 1970)	*1970 (Svensson and Söderlund, 1976)*	*~1980 (Rosswall, 1983[a])*	*1990 (Galloway et al., 1995)*	*1990s (Schlesinger, 1997)*
Natural Nr creation					
Terrestrial BNF	30	140	44–200	90–130	100
Marine BNF	10	30–130	1–130	40–200	15
Total lightning	7.6	?	0.5–30	3	5
Anthropogenic Nr creation					
Haber–Bosch	30	36	60	78	80
BNF, cultivation	14	89		43	40
Fossil-fuel combustion		19	10–20	21	24
Total terrestrial	74	194		255	249
Total global	174	274		375	264
Atmospheric emission					
NO_x, fossil-fuel combustion		19	10–20	21	24
NO_x, other		21–89	0–90	14.5	24
Terrestrial NH_3		113–244	36–250	53	62
Marine NH_3		0		13	13
Total emissions		253		102	123
Atmospheric deposition					
Terrestrial NO_y		32–83	110–240	26.5	30
Marine NO_y		11–33		12.3	14[b]
Terrestrial NH_x		91–186	40–116	52	40
Marine NH_x		19–50		17	16[b]
Organic N			10–100		
Total deposition		253	173–496	110	100
Riverine flux to coast	30	13–24	13–40	76[c]	36[d]
Denitrification[e]					
Continental N_2O		16–69	16–69	9.1	11.7
Marine N_2O		20–80	9–90	2	4
Continental N_2	43	91–92	43–390	130–290	13–233
Marine N_2	40	5–99	0–330	150–180	110
Total denitrification	83	236		386	249

[a] Deposition values for land plus ocean; NH_3 emissions include marine. [b] Wet. [c] Total. [d] Dissolved. [e] N_2O emissions are included with the realization that N_2O is also produced during nitrification.

environments, anthropogenic Nr creation, Nr emissions and deposition, and riverine discharge. A better understanding of the anthropogenic processes that create Nr has led to more precise estimates, and spatially defined databases have led to a better understanding of how nitrogen fluxes vary. Although the current estimates reflect definite improvements, there is still substantial uncertainty about these fluxes.

One of the first global-scale nitrogen cycles was created by Delwiche (1970) (Table 4). He noted that industrial fixation (Haber–Bosch) was the flux with the highest confidence and that all other fluxes "could well be off by a factor of 10." He also stated, and perhaps was the first to do so, that anthropogenic Nr was accumulating in the environment, because denitrification was not keeping up with increased Nr created by humans. Shortly thereafter, Svensson and Söderlund (1976) published a budget based on 1970 data. Their budget was more complete (atmospheric emission and deposition were included). Rosswall (1992) used a number of sources to estimate global-scale nitrogen fluxes for ~1980. In the last several years, several papers have addressed the nitrogen cycle on the global scale (e.g., Ayres *et al.*, 1994; Mackenzie, 1994; Galloway *et al.*, 1995; Vitousek *et al.*, 1997; Galloway, 1998; Seitzinger and Kroeze, 1998; Galloway and Cowling, 2002).

To illustrate the impact of this significant increase in the rate of Nr creation on the global nitrogen cycle, Table 5 contrasts Nr creation and distribution in 1890 with those in 1990 using Galloway and Cowling (2002) as the primary source. Because there was limited Nr created by human activities in 1890, it is an appropriate point to begin, examining the nitrogen cycle. Although the global population was ~25% its population at the turn of twenty-first century, the world was primarily agrarian and produced only 2% of the energy and 10% of the grain produced today. Most energy (75%) was provided by biomass fuels; coal provided most of the rest (Smil, 1994). Little petroleum or natural gas was produced and what was produced was of little consequence when compared to the global supply of energy and the creation of Nr as NO_x through combustion.

Table 5 Global Nr creation and distribution in 1890 and 1990 ($Tg\,N\,yr^{-1}$).

	1890	*1990*
Natural Nr creation		
BNF, terrestrial	100	89
BNF, marine	120	120
Total lightning	5	5
Anthropogenic Nr creation		
Haber–Bosch	0	85
BNF, cultivation	15	33
Fossil-fuel combustion	0.6	21
Nr creation		
Total terrestrial	121	233
Total global	241	353
Atmospheric emission		
NO_x, fossil-fuel combustion	0.6	21
NO_x, other	6.2	13.0
NH_3, terrestrial	8.7	43.0
NH_3, marine	8	8
Total emissions	24	85
Atmospheric deposition		
NO_y, terrestrial	8	33
NO_y, marine	5	13
NH_x, terrestrial	8	43
NH_x, marine	12	14
Total deposition	33	103
Riverine flux to coast (DIN)	5	20

Source: Galloway and Cowling (2002). N_2O emissions are included here with the realization that N_2O is also produced during nitrification; deposition values are for land plus ocean.

8.12.5.2 Nr Creation

Fossil-fuel combustion created only ~0.6 $Tg\,N\,yr^{-1}$ in 1890 through production of NO_x (Table 5). Crop production was primarily sustained by recycling crop residue and manure on the same land where food was raised. Since the Haber–Bosch process had not yet been invented, the only new Nr created by human activities was by legume and rice cultivation (the latter promotes Nr creation because rice cultivation creates an anaerobic environment that enhances nitrogen fixation). Although estimates are not available for 1890, Smil (1999) estimates that in 1900 cultivation-induced Nr creation was ~15 $Tg\,N\,yr^{-1}$. Additional Nr was mined from guano (~0.02 $Tg\,N\,yr^{-1}$) and nitrate deposits (~0.13 $Tg\,N\,yr^{-1}$) (Smil, 2000).

Thus, in 1890 the total anthropogenic Nr creation rate was ~15 $Tg\,N\,yr^{-1}$, almost all of which was from food production. In contrast, the natural rate of Nr creation was ~220 $Tg\,N\,yr^{-1}$. Terrestrial ecosystems created ~100 $Tg\,N\,yr^{-1}$ and marine ecosystems created ~120 $Tg\,N\,yr^{-1}$ (within a range of 87–156 $Tg\,N\,yr^{-1}$) (D. Capone, personal communication). An additional ~5 $Tg\,N\,yr^{-1}$ was fixed by lightning. Thus, globally human activities created ~6% of the total Nr fixed and ~13% when only terrestrial systems are considered.

One century later the world's population had increased by a factor of ~3.5, from ~1.5 billion to ~5.3 billion, but the global food and energy production increased approximately sevenfold and 90-fold, respectively. Just as was the case in 1890, in 1990 (and now) food production accounts for most new Nr created. The largest change since 1890 has been the magnitude of Nr created by humans. Smil (1999) estimates that in the mid-1990s cultivation-induced Nr production was $\sim 33\ \mathrm{Tg\,N\,yr^{-1}}$. The Haber–Bosch process, which did not exist in 1890, created an additional $\sim 85\ \mathrm{Tg\,N\,yr^{-1}}$ in 1990, mostly for fertilizer ($\sim 78\ \mathrm{Tg\,N\,yr^{-1}}$) and the remainder in support of industrial activities such as the manufacture of synthetic fibers, refrigerants, explosives, rocket fuels, nitroparaffins, etc.

From 1890 to 1990, energy production for much of the world was transformed from a biofuel to a fossil-fuel economy. The increase in energy production by fossil fuels resulted in increased NO_x emissions—from $\sim 0.6\ \mathrm{Tg\,N\,yr^{-1}}$ in 1890 to $\sim 21\ \mathrm{Tg\,N\,yr^{-1}}$ in 1990. By 1990 over 90% of the energy produced created new Nr. There was substantial atmospheric dispersal. Thus, in 1990 Nr created by anthropogenic activities was $\sim 140\ \mathrm{Tg\,N\,yr^{-1}}$, an almost ninefold increase over 1890 even though there was only a ~3.5-fold increase in global population. With the increase in Nr creation by human activities came a decrease in natural terrestrial nitrogen fixation (from $\sim 100\ \mathrm{Tg\,N\,yr^{-1}}$ to $\sim 89\ \mathrm{Tg\,N\,yr^{-1}}$) because of the conversion of natural grasslands and forests to croplands, etc. (C. Cleveland, personal communication).

8.12.5.3 Nr Distribution

What is the fate of anthropogenic Nr? The immediate fate for the three anthropogenic sources is clear: NO_x from fossil-fuel combustion is emitted directly into the atmosphere; $R{-}NH_2$ from rice and legume cultivation is incorporated into biomass; NH_3 from the Haber–Bosch process is converted primarily into commercial fertilizer applied to agro-ecosystems to produce food. However, little fertilizer nitrogen actually enters the human mouth in the form of food; in fact, most created Nr is released to environmental systems (Smil, 1999).

In 1890 both the creation and fate of Nr were dominated by natural processes (Figure 3(a)). Only limited Nr was transferred via atmospheric and hydrologic pathways compared to the amount of Nr created. For terrestrial systems, of the $\sim 115\ \mathrm{Tg\,Nr\ yr^{-1}}$ created, only about $\sim 15\ \mathrm{Tg\,N\,yr^{-1}}$ were emitted to the atmosphere as either NH_3 or NO_x. There was limited connection between terrestrial and marine

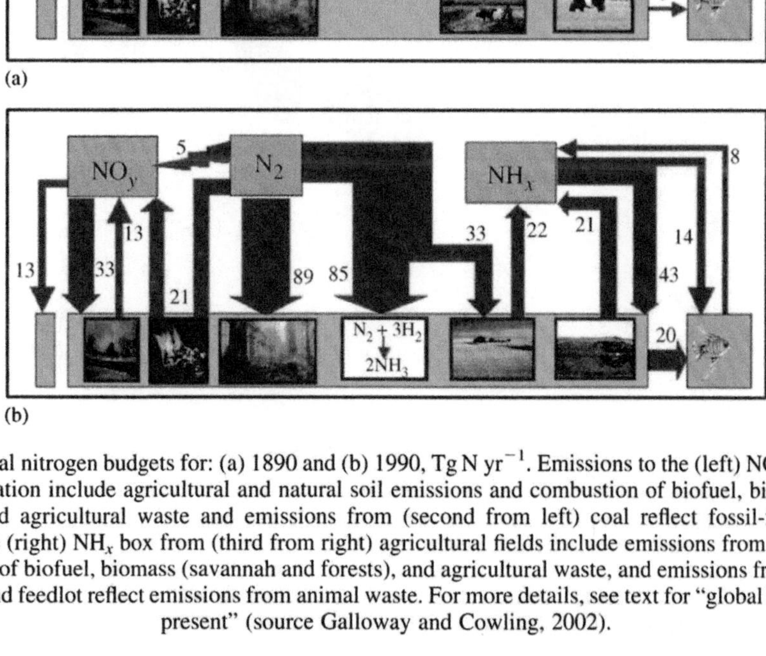

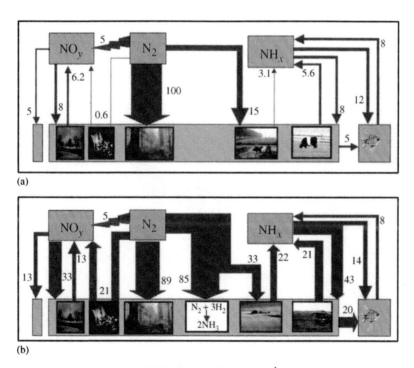

Figure 3 Global nitrogen budgets for: (a) 1890 and (b) 1990, $\mathrm{Tg\,N\,yr^{-1}}$. Emissions to the (left) NO_y box from (first from left) vegetation include agricultural and natural soil emissions and combustion of biofuel, biomass (savannah and forests), and agricultural waste and emissions from (second from left) coal reflect fossil-fuel combustion. Emissions to the (right) NH_x box from (third from right) agricultural fields include emissions from agricultural land and combustion of biofuel, biomass (savannah and forests), and agricultural waste, and emissions from (second from right) the cow and feedlot reflect emissions from animal waste. For more details, see text for "global N cycle: past and present" (source Galloway and Cowling, 2002).

ecosystems; only about 5 Tg N yr^{-1} of dissolved inorganic nitrogen were transferred via rivers into coastal ecosystems in 1890 and only about 17 Tg N yr^{-1} were deposited to the ocean surface.

By contrast, in 1990 when creation of Nr was dominated by human activities (Figure 3(b)), Nr distribution changed significantly. Increased food production caused NH_3 emissions to increase from ~9 Tg N yr^{-1} to ~43 Tg N yr^{-1}, and NO_x emissions increased from ~7 Tg N yr^{-1} to ~34 Tg N yr^{-1} because of increased energy and food production. These increased emissions resulted in widespread distribution of Nr to downwind ecosystems (Figure 4). The transfer of Nr to marine systems also increased. By 1990 riverine fluxes of dissolved inorganic nitrogen to the coastal ocean had increased to 20 Tg N yr^{-1}, and atmospheric nitrogen deposition to marine regions had increased to 27 Tg N yr^{-1}. Although evidence suggests that most riverine nitrogen is denitrified in coastal and shelf environments (Seitzinger and Giblin, 1996), most atmospheric flux is deposited directly to the open ocean, with part of the 27 Tg N yr^{-1} deposited to coastal ocean and shelf regions with significant ecological consequences (Rabalais, 2002).

Another Nr species emitted to the atmosphere, N_2O, bears mention. Produced by nitrification and denitrification, natural emissions of N_2O are ~9.6 Tg N yr^{-1} primarily from the oceans and tropical soils (Mosier *et al.*, 1998). As compiled in Ehhalt *et al.* (2001), several recent estimates of anthropogenic N_2O emissions for the 1990s range from 4.1 Tg N yr^{-1} to 8.1 Tg N yr^{-1}; the most recent estimate is 6.9 Tg N yr^{-1} (Ehhalt *et al.*, 2001). N_2O accumulates either in the troposphere (current estimate is 3.8 Tg N yr^{-1}) or is destroyed in the stratosphere.

8.12.5.4 Nr Conversion to N_2

Another key component missing from Figure 3 is the ultimate fate of the ~140 Tg N yr^{-1} Nr created by human action in 1990. On a global basis, Nr created by human action is either accumulated (stored) or denitrified. Several recent studies estimate denitrification in land and associated freshwaters relative to inputs for large regions. At the scale of continents, denitrification is estimated as 40% of Nr inputs in Europe (van Egmond *et al.*, 2002) and ~30% in Asia (Zheng *et al.*, 2002). At the scale of large regions, denitrification estimates as a percentage of Nr inputs include 33% for land areas draining to the North Atlantic Ocean (Howarth *et al.*, 1996) and 37% for land areas draining to the Yellow-Bohai Seas (Bashkin *et al.*, 2002). Country-scale estimates of the percentage of Nr inputs denitrified to N_2 in soils and waters include ~40% for the Netherlands (Kroeze *et al.*, 2003), 23% for the USA (Howarth *et al.*, 2002a), 15% for China

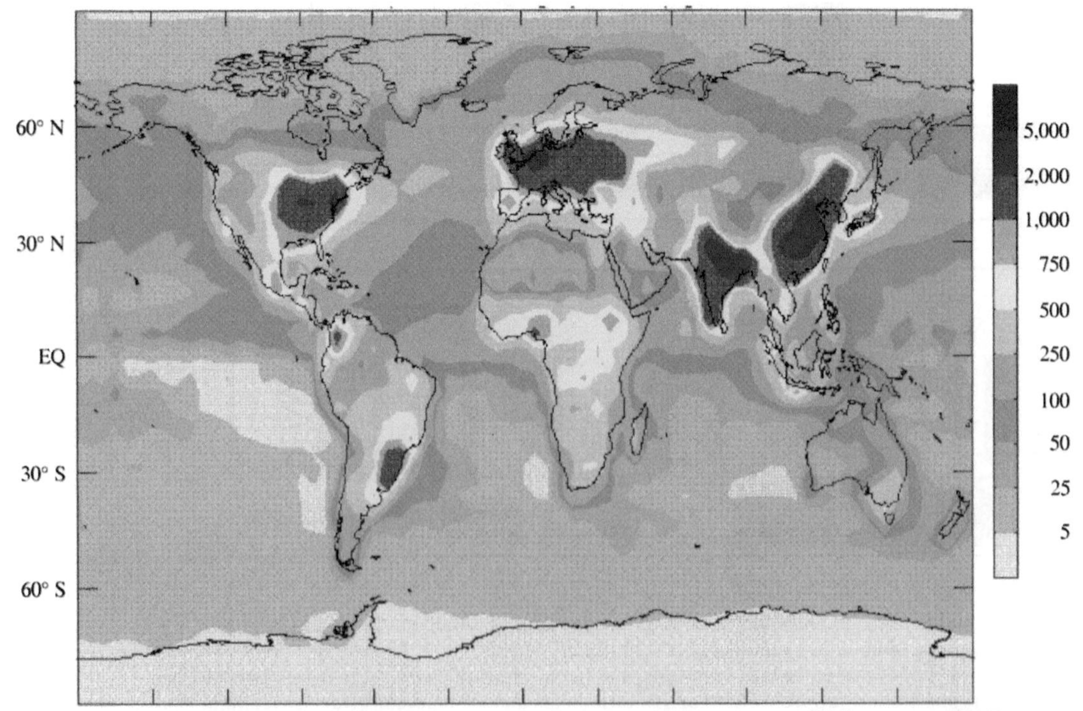

Figure 4 Global atmospheric deposition of Nr to the oceans and continents of the Earth in 1993 (mg N m^{-2} yr^{-1}) (sources F. J. Dentener, personal communication; Lelieveld and Dentener, 2000).

(Xing and Zhu, 2002), and 16% for the Republic of Korea (Bashkin *et al.*, 2002).

On a watershed scale, van Breemen *et al.* (2002) estimate that 47% of total Nr inputs to the collective area of 16 large watersheds in the northeastern US are converted back to N_2: 35% in soils and 12% in rivers. Moreover, within the Mississippi River watershed, Burkart and Stoner (2001) divide the basin into six large sub-basins and conclude that soil denitrification losses of N_2 range from a maximum of ~10% of total inputs in the Upper Mississippi region to less than 2% in the Tennessee and Arkansas/Red regions. Goolsby *et al.* (1999) estimate denitrification within soils of the entire Mississippi–Atchafalaya watershed to be ~8% but do not quantify additional denitrification losses of Nr inputs in river systems. Relative to inputs, landscape-scale estimates of Nr denitrified to N_2 in terrestrial systems and associated freshwaters are quite variable reflecting major differences in the amount of nitrogen inputs available to be denitrified and in environmental conditions that promote this process. All the above-mentioned studies reporting denitrification at regional scales claim a high degree of uncertainty in their estimates, highlighting the fact that our knowledge of such landscape-level rates of denitrification is quite poor. The few estimates that do exist are subject to enormous uncertainties and must often be derived as the residual after all other terms in a regional nitrogen budget are estimated, terms which themselves are often difficult to constrain (e.g., Erisman *et al.*, 2001; van Breemen *et al.*, 2002).

The wetland/stream/river/estuary/shelf region provides a continuum with substantial capacity for denitrification. Nitrate is commonly found, there is abundant organic matter, and sediments and suspended particulate microsites offer anoxic environments. In this section we discuss denitrification in the stream/river/estuary/shelf continuum. Although there are several specific studies of denitrification in wetlands, the role of wetlands in Nr removal at the watershed scale needs to be better understood.

Seitzinger *et al.* (2002) estimate that, of the Nr that enters the stream/river systems that drain 16 large watersheds in the eastern US, 30–70% can be removed within the stream/river network, primarily by denitrification. In an independent analysis on the same watersheds, van Breemen *et al.* (2002) estimated by difference that the lower end of the Seitzinger *et al.* (2002) range is the more likely estimate. For Nr that enters estuaries, 10–80% can be denitrified, depending primarily on the residence time and depth of water in the estuary (Seitzinger, 1988; Nixon *et al.*, 1996).

Of the Nr that enters the continental shelf environment of the North Atlantic Ocean from continents, Seitzinger and Giblin (1996) estimate that >80% is denitrified. The extent of the denitrification depends, in part, on the size of the continental shelf. The shelf denitrification potential is large on the south and east coasts of Asia and the east coasts of South America and North America. These are also the regions where riverine Nr inputs are the largest. Although most riverine/estuary Nr that enters the shelf region is denitrified, total denitrification in the shelf region is larger than that supplied from the continent and thus Nr advection from the open ocean is required.

The uncertainties about these estimates of denitrification on large scales are great enough that the relative importance of denitrification versus storage is unknown. However, the calculations do support the findings that denitrification is important in all portions of the stream/river/estuary/shelf continuum and that, on a global basis, the continuum is a permanent sink for Nr created by human action.

In summary, the approximate doubling of new Nr added to the world between 1890 and 1990 had a substantial impact on the global nitrogen cycle (Figure 3):

- atmospheric emissions of NO_x increased from ~14 Tg N yr^{-1} to 39 Tg N yr^{-1};
- atmospheric emissions of NH_3 increased from ~17 Tg N yr^{-1} to ~50 Tg N yr^{-1};
- terrestrial N_2O emissions increased from 8 Tg N yr^{-1} to 12 Tg N yr^{-1};
- riverine discharge of DIN to the coastal zone increased from 5 Tg N yr^{-1} to 20 Tg N yr^{-1};
- deposition of inorganic nitrogen to continents increased from 16 Tg N yr^{-1} to 76 Tg N yr^{-1}; and
- deposition of inorganic nitrogen to oceans increased from 17 Tg N yr^{-1} to 27 Tg N yr^{-1}.

In 1990, human action dominated the addition of new Nr into the landscape. Substantial added Nr is redistributed throughout environmental reservoirs via atmospheric and riverine transport with, as will be demonstrated in a later section, substantial and long-lasting consequences.

8.12.6 GLOBAL MARINE NITROGEN BUDGET

For the most part, nitrogen cycles within the ocean have the same microbial processes as in the soil (Figure 1). BNF is the dominant source of new nitrogen to the ocean, with other smaller contributions coming from atmospheric deposition and riverine runoff. Ammonification, nitrification, uptake, and decomposition are all critical components. There are two primary removal mechanisms—denitrification and burial in marine sediments. The former is by far the most important.

The pool of Nr in the oceans is ~570 Gt (as $N-NO_3$) (Table 2). Over long periods of time, the inputs and outputs of Nr (and thus the pool size) have to balance; however, over the short term (e.g., centuries), they are probably not in balance given the different controls on the two major fluxes, BNF and denitrification.

Because many ocean regions are nitrogen limited, once Nr is added, it is rapidly used by phytoplankton in the euphotic zone. Zooplankton graze on the phytoplankton. Nr can recycle within the euphotic zone or can be transferred to the subeuphotic zone as total organic nitrogen (TON) where most is mineralized to nitrate and some is deposited to the sediments.

Estimates of marine BNF rates have varied substantially over the last few decades. In the 1970 and early 1980s, estimates ranged from 1 to 130 (Delwiche, 1970; Svensson and Söderlund, 1976). As recently as the late 1980s and early 1990s, global marine BNF rate was ~10–15 Tg N yr^{-1} (Capone and Carpenter, 1982; Walsh *et al.*, 1988; Carpenter and Romans, 1991). However, the more recent estimates range from 100 Tg N yr^{-1} to 200 Tg N yr^{-1} (Karl *et al.*, 2002). In support of this higher range, a recent assessment of marine BNF (D. Capone, personal communication; Galloway *et al.*, 2003a) gives a global range of 87–156 Tg N yr^{-1} (mostly in pelagic regions) and states that all ocean basins are important contributors (Table 6).

The other Nr sources to the ocean are riverine injection and atmospheric deposition. Over the last few decades, riverine inputs to the ocean have ranged from 13 Tg N yr^{-1} to 76 Tg N yr^{-1} with the more recent estimates ~40–50 Tg N yr^{-1} (Table 4). These inputs to coastal regions have significantly altered the associated ecosystems (NRC, 2000; Rabalais, 2002).

Using an empirical model that calculates total nitrogen riverine inputs from the inputs of new Nr to a watershed (BNF, nitrogen fertilizer consumption, cultivation-induced BNF, and atmospheric deposition of NO_y from fossil-fuel combustion), Galloway *et al.*, (2003a) quantify riverine Nr export from world regions for 1990. They find that ~59 Tg N yr^{-1} is discharged via riverine export, with ~11 Tg N yr^{-1} transported to inlands and drylands and ~48 Tg N yr^{-1} transported to coastal waters. Asia has the highest rate of Nr transported to inland waters/drylands (5.1 Tg N yr^{-1}) and North America the lowest. Interestingly, about twice as much Nr is transported to the inland waters/drylands of Oceania than is transported to the coast. A comparison of Nr riverine transport to coasts shows that the highest transport rate (~16.7 Tg N yr^{-1}) is in Asia, with transport in all other regions except Oceania in the range of 6–9 Tg N yr^{-1}.

Atmospheric deposition provides ~27 Tg N yr^{-1} of total NO_y and NH_x deposition (Figure 3). The deposition of organic nitrogen is potentially important. On a global basis, Neff *et al.* (2002) estimate that atmospheric deposition of organic nitrogen ranges from 10 Tg N yr^{-1} to 50 Tg N yr^{-1}, with highest rates (for marine regions) in the coastal zone. Atmospheric deposition to coastal and shelf regions (defined by the 200 m depth line) is ~8 Tg N yr^{-1} compared to ~25 Tg N yr^{-1} to the open ocean. Nr inputs to the ocean are balanced primarily by denitrification and, to a much lesser extent, by sediment burial.

River discharge, atmospheric deposition, and BNF introduce Nr to the coastal oceans. Riverine injection is the most important even though, in some regions, atmospheric nitrogen deposition can be as much as 40% of total inputs and has a measurable effect on coastal productivity (Paerl *et al.*, 2002). However, since most Nr that enters the coastal and shelf region is denitrified, there is little continental-based Nr transferred to the open ocean (Nixon *et al.*, 1996; Seitzinger and Giblin, 1996). Although open oceans are essentially chemically decoupled from continents, they are connected through the atmosphere, which deposits ~27 Tg N yr^{-1} to the global ocean.

8.12.7 REGIONAL NITROGEN BUDGETS

An analysis of regional nitrogen budgets is important as it illustrates the differences in Nr creation and distribution as a function of level of development and geographic location. In addition, the short atmospheric lifetimes of NO_x (and most of its reaction products), and NH_x (one day to one week, depending on total burden and altitude) mean that these chemical species vary substantially in both space and time. Atmospheric transport and dynamics, nitrogen emissions, chemical processing, and removal mechanism (dry versus wet deposition) all interact to alter the spatial distribution of reactive nitrogen. Dividing the Earth's land surface by geopolitical region offers a convenient mechanism for examining this spatial variation.

There are several recent studies of the nitrogen cycle on regional scale: Asia—Galloway (2000), Bashkin *et al.* (2002), and Zheng *et al.* (2002); North Atlantic Ocean and watershed—Galloway *et al.* (1996) and Howarth *et al.* (1996); oceans—Karl (1999) and Capone (2001); Europe—van Egmond *et al.* (2002); and the US—Howarth *et al.* (2002a). To give an appreciation of how the regional budgets vary, this section contrasts Nr creation for several regions of the world: Africa, Asia, Europe, Latin America, North America, and Oceania, as defined by the United Nations Food and Agriculture Organization (FAO, 2000).

Table 6 N fixation and denitrification by ocean basin (Tg N yr^{-1}).

	N_2 *fixation*		*High N_2 fixation*		*Denitrification*		*High denitrification*	
North Atlantic								
Pelagic	12	Capone (pers. com.)	42	Gruber and Sarmiento (1997)	0		3	Codispoti *et al.* (2001)
Shelf	0.38	Howarth *et al.* (1988)	3.7	Capone (1983)	15	Christensen *et al.* (1987)	75	Devol (1991)
Deep sediments	0		0		1.2	Capone (pers. com.)	2.4	Bender *et al.* (1977)
South Atlantic								
Pelagic	5	Capone (pers. com.)	21		0		2	Codispoti *et al.* (2001)
Shelf	0.38	Howarth *et al.* (1988)	3.7	Capone (1983)	15	Christensen *et al.* (1987)	75	Devol (1991)
Deep sediments	0		0		1.2	Capone (pers. com.)	2.4	Bender *et al.* (1977)
North Pacific								
Pelagic	29	Capone (pers. com.)	35	Deutsch *et al.* (2001)	22	Deutsch *et al.* (2001)	40	Codispoti *et al.* (2001)
Shelf	0.28	Howarth *et al.* (1988)	2.8	Capone (1983)	14	Christensen *et al.* (1987)	70	Devol (1991)
Deep sediments	0		0		2.1	Capone (pers. com.)	4.2	Bender *et al.* (1977)
South Pacific								
Pelagic	19	Capone (pers. com.)	24	Deutsch *et al.* (2001)	26	Deutsch *et al.* (2001)	40	Codispoti *et al.* (2001)
Shelf	0.28	Howarth *et al.* (1988)	2.8	Capone (1983)	7	Christensen *et al.* (1987)	35	Devol (1991)
Deep sediments	0		0		2.1	Capone (pers. com.)	4.2	Bender *et al.* (1977)
Indian Ocean								
Pelagic	20	Capone (pers. com.)	19	Bange *et al.* (2000)	33	Bange *et al.* (2000)	65	Codispoti *et al.* (2001)
Shelf	0.19	Howarth *et al.* (1988)	1.9	Capone (1983)	6.4	Christensen *et al.* (1987)	32	Devol (1991)
Deep sediments	0		0		1.8	Capone (pers. com.)	3.6	Bender *et al.* (1977)
All pelagic	85		141		81		150	
All shelves	1.5		14.9		57		287	
All deep sediments	0		0		8.4		17	
Total	87		156		147		454	

Source: Galloway *et al.* (2003b).

Table 7 Nr creation rates for various regions of the world in mid-1990s (Tg N yr^{-1}).

World regions	*Fertilizer production*	*Cultivation*	*Combustion*	*Net import/export*	*Total*
Africa	2.5	1.8	0.8	0.2	5.3
Asia	40.1	13.7	6.4	8.7	68.9
Europe + FSU	21.6	3.9	6.6	−5.6	26.5
Latin America	3.2	5.0	1.4	−0.2	9.4
North America	18.3	6.0	7.4	−3.3	28.4
Oceania	0.4	1.1	0.4	0.3	2.2
World	~86	~30	~23	0.1	~140

Source: Galloway and Cowling (2002).

For the mid-1990s in each region, Galloway and Cowling (2002) estimate the amount of Nr created by fertilizer production, fossil-fuel combustion (F. J. Dentener, personal communication), and rice/legume cultivation (FAO, 2000). Most fertilizer was produced in Asia (40.1 Tg N yr^{-1}), Europe (21.6 Tg N yr^{-1}), and North America (18.3 Tg N yr^{-1}), with smaller amounts in Latin America, Africa, and Oceania (3.2 Tg N yr^{-1}, 2.5 Tg N yr^{-1}, and 0.4 Tg N yr^{-1}, respectively). The other Nr creation process involved in food production is cultivation-induced BNF. Smil (1999) estimates that ~33 Tg N yr^{-1} were globally produced by cultivation in the mid-1990s. The regional breakdown for these BNF rates from cultivation of seed legumes, rice, and sugar cane is 13.7 Tg N yr^{-1} for Asia; ~6.0 Tg N yr^{-1}, ~3.9 Tg N yr^{-1}, and ~5.0 Tg N yr^{-1} for North America, Europe plus the former Soviet Union, and Latin America, respectively; and ~<2 Tg N yr^{-1} for Africa and Oceania. Most Nr creation by fossil-fuel combustion was in North America, Europe plus the former Soviet Union, and Asia (~7.4 Tg N yr^{-1}, ~6.6 Tg N yr^{-1}, and ~6.4 Tg N yr^{-1}, respectively). Latin America, Africa, and Oceania by comparison had more modest Nr creation rates from fossil-fuel combustion with each being <1.5 Tg N yr^{-1} (Table 7).

In addition to the regional creation of Nr, Nr is also exchanged among regions. Exports of nitrogen-containing fertilizer, plant material (e.g., grain), and meat from one region can be a source of nitrogen for another region. As reported in Galloway and Cowling (2002), fertilizer is the commodity most often exchanged between regions. In 1995, the global production of nitrogen fertilizers was ~86 Tg N yr^{-1}. Of this amount, ~24.9 Tg N yr^{-1} was exported to other regions. Over half of the exports were from Europe (~13.2 Tg N yr^{-1}). Other regions with significant exports were Asia (~10.7 Tg N yr^{-1}) and North America (~5.2 Tg N yr^{-1}). The primary receiving regions were Asia (~7.6 Tg N yr^{-1}) and Europe (Tg N yr^{-1}). Thus, although ~30% of the fertilizer nitrogen produced was exported, the only region that had a large net loss was Europe (~6.6 Tg N yr^{-1}); the largest net gain of any region was for Asia (~6.4 Tg N yr^{-1}).

Table 8 Regional per capita Nr creation during food and energy production in mid-1990s (kg N $person^{-1}$ yr^{-1}).

World regions	*Food*	*Energy*	*Total*
Africa	5.7	1.1	6.8
Asia	15	1.8	17
Europe + FSA	35	9.1	44
Latin America	16	2.8	19
North America	80	24	100
Oceania	50	13	63
World	20	3.9	24

Source: Galloway and Cowling (2002).

The next most frequently exchanged commodity is plant material, mostly cereal grains. Asia, Europe, and Africa had net gains in nitrogen-containing plant material (~2.2 Tg N yr^{-1}, ~1.0 Tg N yr^{-1}, ~0.5 Tg N yr^{-1}, respectively), whereas North America, Latin America, and Oceania had net losses (~2.8 Tg N yr^{-1}, ~0.8 Tg N yr^{-1}, and ~0.2 Tg N yr^{-1}, respectively). Net meat exchange was <0.1 Tg N yr^{-1} for each region. Summing all three categories, Asia gained the most Nr: ~8.7 Tg N yr^{-1}. Although Oceania and Africa also gained nitrogen, the gains were small (~0.3 Tg N yr^{-1} and ~0.2 Tg N yr^{-1}, respectively). Europe and North America had net losses of ~5.6 Tg N yr^{-1} and ~3.3 Tg N yr^{-1}, respectively. Latin America had a small net loss. When these net import/exports are added to the Nr creation in each region (Table 7), North America, with ~5% of the world's population, was responsible for creating ~20% of the world's Nr. Africa, with ~13% of the world's population, was responsible for creating ~6% of the world's Nr.

Given these regional differences, it is also interesting to express Nr creation and use on a per capita basis (calculated from Table 7 (not including net import/export) and FAO population data (2000)) to illustrate the average amount of Nr mobilized per person, by region (Table 8). At one extreme North America mobilized ~100 kg N $person^{-1}$ yr^{-1}, and at the other extreme Africa mobilized about an order of magnitude less or ~6.8 kg N $person^{-1}$ yr^{-1}. In all regions, food production was larger than energy production, and the primary causes for the regional differences were the amounts of nitrogen mobilized per capita in producing food. In North America, it was ~80 kg N $person^{-1}$ yr^{-1}, and in Africa it was ~6 kg N $person^{-1}$ yr^{-1}. These values show the regions responsible for Nr creation in excess of what is needed for human body sustenance (~2 kg N $capita^{-1}$ yr^{-1}).

8.12.8 CONSEQUENCES

8.12.8.1 Introduction

Much has been written about the impacts of increased Nr creation and its accumulation in environmental reservoirs. The largest impact is, of course, that much of the world's population is sustained by the increased food production made possible by the Haber–Bosch process and by cultivation-induced BNF (Smil, 2000). However, as noted in Galloway *et al.* (2003b), there are also substantial negative impacts.

- Nr increase leads to the production of tropospheric ozone and aerosols that induce serious respiratory diseases, cancer, and cardiac diseases in humans (Pope *et al.* 1995; J. R. Follett and R. F. Follett, 2001; Wolfe and Patz, 2002).
- Nr increases and then decreases forest and grassland productivity wherever atmospheric Nr deposition has increased significantly and critical thresholds have been exceeded. Nr additions probably also decrease biodiversity in many natural habitats (Aber *et al.*, 1995).
- Nr is responsible (together with sulfur) for the acidification and loss in biodiversity of lakes and streams in many regions of the world (Vitousek *et al.*, 1997).
- Nr is responsible for eutrophication, hypoxia, loss of biodiversity, and habitat degradation in coastal ecosystems. Nr is now considered the biggest pollution problem in coastal waters (e.g., Howarth *et al.*, 2000; NRC, 2000; Rabalais, 2002).
- Nr contributes to global climate change and stratospheric ozone depletion, both of which impact human and ecosystem health (e.g., Cowling *et al.*, 1998).

Although most of these nitrogen problems are being studied separately, research is increasingly indicating that these Nr-related problems are linked. For example, the same nitrogen that produces urban air pollution can also contribute to water pollution. These linkages, referred to as the nitrogen cascade, are discussed in detail in Galloway *et al.* (2003b). This section summarizes that discussion. The nitrogen cascade is defined as the sequential transfer of Nr through environmental systems, which results in environmental changes as Nr moves through or is temporarily stored within each system.

Two scenarios in Figure 5 illustrate the nitrogen cascade. The first example shows the fate of

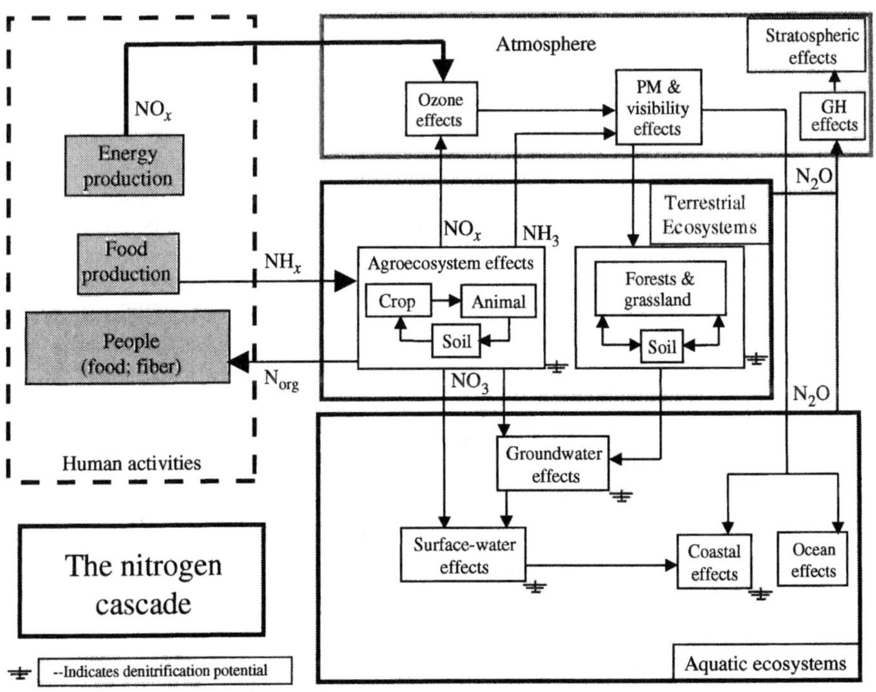

Figure 5 The nitrogen cascade illustrates the sequential effects a single atom of N can have in various reservoirs after it has been converted from a nonreactive to a reactive form (source Galloway *et al.*, 2003b).

NO_x produced during fossil fuel combustion. In the sequence, an atom of nitrogen mobilized as NO_x in the atmosphere first increases ozone concentrations, then decreases atmospheric visibility and increases concentrations of small particles, and then increases precipitation acidity. After that same nitrogen atom is deposited into the terrestrial ecosystem; it may increase soil acidity (if a base cation is lost from the system), decrease biodiversity, and either increase or decrease ecosystem productivity. If discharged to the aquatic ecosystem, the nitrogen atom can increase surface-water acidity in mountain streams and lakes. Following transport to the coast, nitrogen atom can countribute to coastal eutrophication. If the nitrogen atom is converted to N_2O and emitted back to the atmosphere, it can first increase greenhouse warming potential and then decrease stratospheric ozone.

Example 2 illustrates a similar cascade of effects of Nr from food production. In this case, atmospheric N_2 is converted to NH_3 in the Haber–Bosch process. The NH_3 is primarily used to produce fertilizer. About half the Nr fertilizer applied to global agro-ecosystems is incorporated into crops harvested from fields and used for human food and livestock feed (Smil, 1999, 2001). The other half is transferred to the atmosphere as NH_3, NO, N_2O, or N_2, lost to the aquatic ecosystems, primarily as nitrate or accumulate in the soil nitrogen pool. Once transferred downstream or downwind, the nitrogen atom becomes part of the cascade. As Figure 5 illustrates, Nr can enter the cascade at different places depending on its chemical form. An important characteristic of the cascade is that, once it starts, the source of the Nr (e.g., fossil-fuel combustion and fertilizer production) becomes irrelevant. Nr species can be rapidly interconverted from one Nr to another. Thus, the critical step is the *formation* of Nr.

The simplified conceptual overview of the cascade shown in Figure 5 does not cover some of its important aspects. The internal cycling of Nr within each step of the cascade is ignored; long-term Nr storage is not accounted for and neither are all the multiple pathways that Nr follows as it flows from one step to another. In a recent paper, Galloway *et al.* (2003b) cover these aspects for the atmosphere (troposphere and stratosphere), for terrestrial ecosystems (agro-ecosystems, forests, and grasslands), and for aquatic ecosystems (groundwater, wetlands, streams, lakes, rivers, and marine coastal regions). They evaluate the potential for accumulation and cycling of Nr within the system; the loss of Nr through conversion to N_2 by denitrification; the transfer of Nr to other systems; and the effects of Nr within the system. Galloway *et al.* (2003b) conclude that the cascade of Nr from one system to another will be enhanced if there is a limited potential for Nr to be accumulated or for the loss of N_2 through denitrification within a given system thereby increasing the potential for transfer to the next system. There will be a lag in the cascade if there is a large potential for accumulation within a system. The cascade will be decreased if there is a large potential for denitrification to N_2 within a system. Their analyses are summarized in Table 9, and the remainder of this section presents an overview of their findings (Galloway *et al.*, 2003b).

8.12.8.2 Atmosphere

The atmosphere receives Nr mainly as air emissions of NO_x, NH_3, and N_2O from aquatic and terrestrial ecosystems and of NO_x from the combustion of biomass or fossil fuels. NO_x and NH_3 (and their reaction products), can accumulate in the troposphere on a regional scale. However, because of their short residence times in the atmosphere and lack of potential for formation of N_2 by denitrification, almost all Nr emitted as NO_x and NH_3 is transferred back to the Earth's surface within hours to days. There is also an internal cascade of effects. (NO increases the potential first for ozone and then for aerosol formation.) Except for N_2O, there is very limited potential for the long-term storage of Nr (and thus limited lag time), but there are significant effects from Nr while it remains in the atmosphere. There is no potential for denitrification back to N_2 within the troposphere, and a large potential for Nr transfer to the next receptors—terrestrial and aquatic ecosystems.

8.12.8.3 Terrestrial Ecosystems

Intensively managed agro-ecosystems and, even more so, confined animal feeding operations (CAFOs) are the technical means by which most human needs and dietary preferences are met. About 75% of the Nr created around the world by humans is added to agro-ecosystems to sustain food production. About 70% of that is from the Haber–Bosch process and ~30% from C-BNF (Figure 2). Smil (2001) estimates that ~40% of all the people alive today owe their life to the production and wide use of fertilizers produced by the Haber–Bosch process.

Most Nr applied to agro-ecosystems is transferred to other systems along the nitrogen cascade; a much smaller portion is denitrified to N_2 (Table 9). Nitrogen-use efficiency in major grain and animal production systems can be improved through the collaboration of ecologists, agronomists, soil scientists, agricultural economists, and politicians. Actual fertilizer nitrogen-use efficiency, nitrogen losses, and loss pathways in major cropping and animal systems must be more

Table 9 Characteristics of different systems relevant to the nitrogen cascade.

System	*Accumulation potential*	*Transfer potential*	*N_2 production potential*	*Links to systems down the cascade*	*Effects*
Atmosphere	Low	Very high	None	All but groundwater	Human and ecosystem health, climate change
Agroecosystems	Low to moderate	Very high	Low to moderate	All	Human and ecosystem health, climate change
Forests	High	Moderate to high in places	Low	All	Biodiversity, net primary productivity, mortality, groundwater
Grasslands	High	Moderate to high in places	Low	All	Biodiversity, net primary productivity, groundwater
Groundwater	Moderate	Moderate	Moderate	Surface water, atmosphere	Human and ecosystem health, climate change
Wetlands, streams, lakes, and rivers	Low	Very high	Moderate to high	Atmosphere marine costal systems	Biodiversity, ecological structure, fish
Marine coastal regions	Low to moderate	Moderate	High	Atmosphere	Biodiversity, ecological structure, fish, harmful algal blooms

Source: Galloway *et al.* (2003b).

accurately measured so that we can: (i) identify opportunities for increased nitrogen-use efficiency through improved crop and soil management; (ii) quantify nitrogen-loss pathways in major food crops, including animal feeding operations; and (iii) improve human understanding of local, regional, and global nitrogen balances and nitrogen losses from major cropping and animal systems.

The residence times (and lag times) of Nr in forests can be years to centuries depending on forest history, forest type, and Nr inputs. The effects of Nr accumulation in forests are numerous, most relate to changes in forest and microbial productivity and function. There is significant potential for Nr to be transferred to the atmosphere as NO and N_2O and especially to surface waters as NO_3^- once Nr additions or availability exceed biotic requirements. Relative to inputs in high Nr deposition areas, there is a limited potential for Nr to be removed from the cascade through N_2 formation.

Unmanaged grasslands receive most of their Nr from BNF and atmospheric deposition, with the latter being much more important where deposition rates are large. As with forests, temperate grasslands are potentially major storage reservoirs and short- to long-term sinks within the nitrogen cascade.

Grasslands managed for animal production (e.g., cattle) are much more leaky with respect to loss of added Nr. The addition of fertilizer and/or grazing animals increases the amount of Nr available for loss, especially via the atmosphere (e.g., NH_3 (Sommer and Hutchings, 1997) and N_2O (Fowler *et al.*, 1997)). Thus, the effective residence times for managed ecosystems are potentially less than those for unmanaged systems.

8.12.8.4 Aquatic Ecosystems

Leaching from agro-ecosystems is the primary source of anthropogenic Nr for groundwater, although in some regions human waste disposal can also be important (Puckett *et al.*, 1999; Nolan and Stoner, 2000). Nitrate is the most common Nr species (Burkart and Stoner, 2001). Nitrogen is lost from groundwater both from denitrification to N_2 and from losses of Nr to surface waters and the atmosphere but the relative mix of these fates is site dependent, as is the residence time of Nr in groundwater reservoirs.

The effects of elevated Nr in groundwater do pose a significant human-health risk through contaminated drinking water. In the human body, nitrate is converted to nitrite, which can cause methemoglobinemia through interference with the ability of hemoglobin to take up O_2. Methemoglobinemia usually occurs after water with high concentrations of nitrate has been consumed; infants and people on kidney dialysis are particularly susceptible (J. R. Follett and R. F. Follett, 2001). Other potential effects associated with elevated concentrations of nitrate in drinking water include respiratory infection, alteration of thyroid metabolism, and cancers induced by conversion of nitrate to N-nitroso compounds in the body (J. R. Follett and R. F. Follett, 2001; Wolfe and Patz, 2002).

Groundwater systems are accumulating Nr, and the effects of contaminated groundwater can be significant. Nr is lost from groundwater through denitrification, from advection of nitrate to surface waters, and from conversion to gaseous Nr forms that diffuse or flow to the atmosphere. Just as there is a lag in Nr release in forests because of the nitrogen-saturation phenomena (Aber *et al.*, 2003), there is also a lag for Nr release from groundwaters to surface waters but this is highly variable among sites (Groffman *et al.*, 1998; Puckett and Cowdery, 2002). At some time, the Nr introduced into the groundwater will be transferred to surface water if the Nr is not permanently stored in groundwater or denitrified to N_2.

Surface freshwater ecosystems consist of wetlands (e.g., bogs, fens, marshes, swamps, prairie potholes, etc.), streams, lakes (and artificial reservoirs), and rivers. Surface freshwater ecosystems receive most of their Nr from their associated watersheds, from atmospheric deposition, and from BNF within the system. There is limited potential for Nr to accumulate within surface-water ecosystems, because the residence time of Nr within surface waters, like the water itself, is very brief. Residence times may be relatively longer in the sediments associated with wetlands and some larger lakes but are still short when compared to terrestrial ecosystems or the oceans.

However, many headwater streams and lakes are now in highly disturbed landscapes and thus have high nitrate concentrations; this can lead to eutrophication problems either locally or further downstream. In addition, for water in headwater streams and lakes draining poorly buffered soils, increased nitrate concentrations can acidify streams and impact biota.

The potential for Nr accumulation in streams, lakes, rivers, and associated wetlands is small (Table 9). Although changes in Nr inputs to surface waters may significantly alter the internal cycles in these systems, globally such changes are merely extremely efficient avenues for propagating the effects of the nitrogen cascade from higher to lower components. Even though wetlands may delay or prevent this transfer locally (and denitrification may short-circuit some of the Nr transport along the way), surface freshwater ecosystems essentially move Nr from the mountains to the sea and ensure that perturbations at one point in the cascade quickly lead to changes elsewhere.

Most Nr in coastal ecosystems (e.g., estuaries) is from river water and groundwater; direct atmospheric deposition is an important source in some systems, and inputs from the ocean are important in others. These inputs have increased several-fold as a consequence of human activities (Meybeck, 1982; Howarth *et al.*, 1996; Seitzinger and Kroeze, 1998; NRC, 2000; Howarth *et al.*, 2002b). Because of the dynamic nature of coastal ecosystems, there is limited potential for Nr accumulation. In addition, although the potential for Nr to be transferred to the continental shelf regions is large because of the high rates of denitrification (mostly as N_2), transport to the shelf is limited, and the Nr that is transferred is mostly converted to N_2 before being transported to the open ocean.

Although Nr has a short residence time in coastal ecosystems (compared to terrestrial ecosystems), the time that it does spend there can have a profound impact on the ecosystem. Primary production in most coastal rivers, bays, and seas of the temperate zone is limited by Nr supplies (Vitousek and Howarth, 1991; Nixon *et al.*, 1996; NRC, 2000). In the USA, the increased Nr flux is now viewed as the most serious pollution problem in coastal waters (Howarth *et al.*, 2000; NRC, 2000; Rabalais *et al.*, 2002). One-third of the nation's coastal rivers and bays are severely degraded, and another one-third is moderately degraded from nutrient over enrichment (Bricker *et al.*, 1999).

An obvious consequence of increasing Nr inputs to coastal waters over the past few decades has been an increase in the size of water masses that are anoxic (completely devoid of oxygen) or hypoxic (concentrations of oxygen less than 2–3 $mg\ L^{-1}$). These "dead zones" can be found in the Gulf of Mexico, Chesapeake Bay, Long Island Sound, Florida Bay, the Baltic Sea, the Adriatic Sea, and many other coastal areas (Diaz and Rosenberg, 1995; NRC, 2000).

Other major effects of Nr increases in coastal regions include the loss of seagrass beds and macro-algal beds and changes in coral reefs (Lapointe and O'Connell, 1989; Valiela *et al.*, 1997; NRC, 2000; Howarth *et al.*, 2000). Thus, Nr inputs to coastal ecosystems have increased significantly over the last few decades. Although most Nr is eventually denitrified to N_2 within the coastal ecosystems and its associated shelf, significant and widespread impacts on various ecosystem components and on human health remain (Table 9).

8.12.9 FUTURE

In 1995 the rate of global creation of Nr from human activities was $\sim 140\ Tg\ N\ yr^{-1}$, which equates to an average per capita Nr creation rate of ~24 kg N person^{-1} yr^{-1}: 20 kg N person^{-1} yr^{-1} attributable to food production and 4 kg N person^{-1} yr^{-1} attributable to energy production. These figures varied greatly in different areas. In Africa, the per capita Nr creation rate in the mid-1990s was 6.8 kg N yr^{-1} (5.7 kg N capita^{-1} yr^{-1} for food production and 1.1 kg N capita^{-1} yr^{-1} for energy production) and in North America it was ~100 kg N capita^{-1} yr^{-1} of which ~80 kg N capita^{-1} yr^{-1} was used to sustain food production (Table 8).

What will it be in 2050? The United Nations (1999) estimates that the world population will be ~8.9 billion. Some of the numerous estimates of future energy and food production scenarios use sophisticated scenario development (e.g., Houghton *et al.*, 2001). For this analysis, I use a simple calculation to set a range of Nr creation rates using projections for population increases and two assumptions of per capita Nr creation rate. To estimate the lower end of the possible range of Nr creation in 2050, I assume that, although the population increases, the Nr per capita creation rate by region will be the same in 2050 as it was in the mid-1990s (Table 8). Multiplying 2050 population rates by 1995 Nr creation rates gives a global total of ~190 Tg N yr^{-1} created (compared to ~140 Tg N yr^{-1} in 1990), with most of the increase occurring (by definition) in those regions where population increases are the largest (e.g., Asia) (Figure 6). To estimate an upper limit of Nr creation in 2050, I assume that all the world's peoples will have Nr creation rates the same as those in North America in 1990 (100 kg N capita^{-1} yr^{-1}). The resulting global total is ~960 Tg N yr^{-1} for 2050 or about a factor of 7 greater than the 1990 rate. Latin America, Africa, and Asia show the largest relative increases and Asia has the largest absolute increase (Figure 6).

The likelihood that the 2050 Nr creation rates at either end of the range will occur is slight. In all probability per capita Nr creation rates will rise to some degree, especially in those areas that will experience the largest population increases; therefore, the lower estimate is probably indeed low. Although the actual level reached in 2050 will probably be larger than the low estimate, it is also very likely to be less (much less?) than the high estimate. This, however, does not preclude the eventuality of the high estimate being reached sometime after 2050, after population growth has leveled off but during a prolonged period of growth in per capita Nr creation.

Given the concerns about Nr, it is perhaps unlikely that the high estimate will be reached. Whatever the final maximum Nr creation rate turns out to be, however, will depend greatly on how the world manages its use of nitrogen for food production and its control of nitrogen in energy production.

There is ample opportunity for a creative management strategy for Nr that will optimize food and energy production as well as environmental health. Several papers in the special issue (**31**) of *Ambio* devoted to the plenary papers of the *Second International Nitrogen Conference* address the issue of Nr management to maximize food and energy production while protecting the health of people and ecosystems: management of energy producing systems (Moomaw, 2002; Bradley and Jones, 2002) and management of food production systems (Cassman *et al.*, 2002; Fixen and West, 2002; Oenema and Pietrzak, 2002; Roy *et al.*, 2002; Smil, 2002). Other recent publications focus more generally on policy issues related to nitrogen (Erisman *et al.*, 2001; Mosier *et al.*, 2001; Melillo and Cowling, 2002).

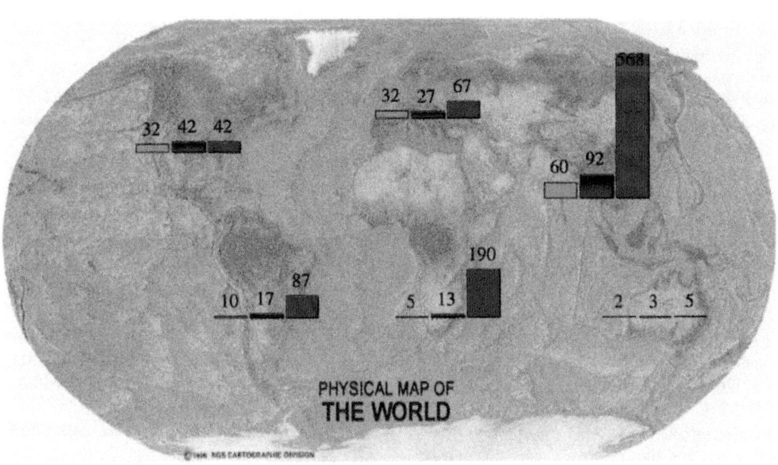

Figure 6 Continental Nr creation rates for (left column) 1995 and low (center column) and high (right column) estimates for 2050, Tg N yr^{-1}.

8.12.10 SUMMARY

The global cycle of nitrogen has changed over the course of a century from one dominated by natural processes to one dominated by anthropogenic processes. The increased abundance of Nr has positive and negative effects that can occur in sequence as Nr cascades through the environment. A critical topic for research is the fate of anthropogenic Nr. Where is it going? How much is stored? How much is denitrified? A broader question that needs to be answered is how can society optimize nitrogen's beneficial role in sustainable food production and minimize nitrogen's negative effects on human health and the environment resulting from food and energy production.

These critical issues will require a new way of examining how people manage the environment and use the Earth's resources.

ACKNOWLEDGMENTS

I thank Aaron Mills and Fred Mackenzie for discussions on various aspects of this paper. I thank Mary Scott Kaiser for putting the paper into the correct format and for editing the manuscript. I also thank The Ecosystems Center of the Marine Biological Laboratory and the Woods Hole Oceanographic Institution for providing a sabbatical home to write this paper and to the University of Virginia for the Sesquicentennial Fellowship.

REFERENCES

Aber J. D., Magill A., McNulty S. G., Boone R. D., Nadelhoffer K. J., Downs M., and Hallett R. (1995) Forest biogeochemistry and primary production altered by nitrogen saturation. *Water Air Soil Pollut.* **85**, 1665–1670.

Aber J. D., Goodale C. L., Ollinger S. V., Smith M. -L., Magill A. H., Martin M. E., Hallett R. A., and Stoddard J. L. (2003) Is nitrogen deposition altering the nitrogen status of northeastern forests?. *Bioscience* **53**, 375–389.

Asner G. P., Townsend A. R., Riley W. J., Matson P. A., Neff J. C., and Cleveland C. C. (2001) Physical and biogeochemical controls over terrestrial ecosystem responses to nitrogen deposition. *Biogeochemistry* **54**, 1–39.

Ayres R. U., Schlesinger W. H., and Socolow R. H. (1994) Human impacts on the carbon and nitrogen cycles. In *Industrial Ecology and Global Change* (eds. R. H. Socolow, C. Andrews, R. Berkhout, and V. Thomas). Cambridge University Press, pp. 121–155.

Bange H., Rixen T., Johansen A., Siefert R., Ramesh R., Ittekkot V., Hoffmann M., and Andreae M. (2000) A revised nitrogen budget for the Arabian Sea. *Global Biogeochem. Cycles* **14**, 1283–1297.

Bashkin V. N., Park S. U., Choi M. S., and Lee C. B. (2002) Nitrogen budgets for the Republic of Korea and the Yellow Sea region. *Biogeochemistry* **57**, 387–403.

Bender M., Ganning K. A., Froelich P. M., Heath G. R., and Maynard V. (1977) Interstitial nitrate profiles and oxidation of sedimentary organic matter in the Eastern Equatorial Atlantic. *Science* **198**, 605–609.

Binkley D., Cromack K., Jr., and Baker D. D. (1994) Nitrogen fixation by red alder: biology, rates, and controls. In *The Biology and Management of Red Alder* (eds. D. E. Hibbs, D. S. DeBell, and R. F. Tarrant). Oregon State University Press, Corvallis, pp. 57–72.

Blundon D. J. and Dale M. R. T. (1990) Denitrogen fixation (acetylene reduction) in primary succession near Mount Robson, British Columbia, Canada. *Arct. Alp. Res.* 255–263.

Boring L. R. and Swank W. T. (1984) The role of black locust (*Robinia pseudo-acacia*) in forest succession. *J. Ecol.* **72**, 749–766.

Boyer E. W., Goodale C. L., Jaworski N. A., and Howarth R. W. (2002) Anthropogenic nitrogen sources and relationships to riverine nitrogen export in the Northeastern USA. *Biogeochemical* **57**, 137–169.

Bradley M. J. and Jones B. M. (2002) Reducing global NO_x emissions: promoting the development of advanced energy and transportation technologies. *Ambio* **31**, 141–149.

Bricker S. B., Clement C. G., Pirhalla D. E., Orland S. P., and Farrow D. G. G. (1999) *National Estuarine Eutrophication Assessment: A Summary of Conditions, Historical Trends, and Future Outlook.* US Department of Commerce, National Oceanic and Atmospheric Administration, National Ocean Service, Silver Spring, MD.

Burkart M. R. and Stoner J. D. (2001) Nitrogen in groundwater associated with agricultural systems. In *Nitrogen in the Environment: Sources, Problems and Management* (eds. R. F. Follett and J. L. Hatfield). Elsevier, New York, pp. 123–145.

Capone D. G. (1983) Benthic nitrogen fixation. In *Nitrogen in the Marine Environment* (eds. E. J. Carpenter and D. G. Capone). Elsevier, New York, pp. 105–137.

Capone D. (2001) Marine nitrogen fixation: what's the fuss? *Current Opinions Microbiol.* **4**, 341–348.

Capone D. G. and Carpenter E. J. (1982) Nitrogen fixation in the marine environment. *Science* **217**, 1140–1142.

Carpenter E. J. and Romans K. (1991) Major role of the cyanobacterium tricodesmium in nutrient cycling in the North Atlantic Ocean. *Science* **254**, 1356–1358.

Cassman K. G., Dobermann A. D., and Walters D. (2002) Agro-ecosystems, nitrogen-use efficiency, and nitrogen management. *Ambio* **31**, 132–140.

Christensen J. P., Murray J. W., Devol A. H., and Codispoti L. A. (1987) Denitrification in continental shelf sediments has a major impact on the oceanic nitrogen budget. *Global Biogeochem. Cycles* **1**, 97–116.

Cleveland C. C., Townsend A. R., Schimel D. S., Fisher H., Howarth R. W., Hedin L. O., Perakis S. S., Latty E. F., Von Fischer J. C., Elseroad A., and Wasson M. F. (1999) Global patterns of terrestrial biological nitrogen (N_2) fixation in natural ecosystems. *Global Biogeochem. Cycles* **13**, 623–645.

Codispoti L., Brandes J., Christensen J., Devol A., Naqvi S., Paerl H., and Yoshinari T. (2001) The oceanic fixed nitrogen and nitrous oxide budgets: moving targets as we enter the anthropocene? *Scientia Marina* **65**(suppl. 2), 85–105.

Cowling E., Erisman J. W., Smeulders S. M., Holman S. C., and Nicholson B. M. (1998) Optimizing air quality management in Europe and North America: justification for integrated management of both oxidized and reduced forms of nitrogen. *Environ. Pollut.* **102**, 599–608.

Crookes W. Sir (1871) *Select Methods in Chemical Analysis (Chiefly inorganic).* Longmans, Green, London. (microform. Landmarks of Science, Readex Microprint, 1969, Micro-opaque, New York).

Delwiche C. C. (1970) The nitrogen cycle. *Sci. Am.* **223**, 137–146.

Deutsch C., Gruber N., Key R. M., Sarmiento J. L., and Ganachaud A. (2001) Denitrification and N_2 fixation in the Pacific Ocean. *Global Biogeochem. Cycles* **15**, 483–506.

Devol A. H. (1991) Direct measurement of nitrogen gas fluxes from continental shelf sediments. *Nature* **349**, 319–322.

Diaz R. and Rosenberg R. (1995) Marine benthic hypoxia: a review of its ecological effects and the behavioral responses of benthic macrofauna. *Oceanogr. Mar. Biol. Ann. Rev.* **33**, 245–303.

Ehhalt D., Prather M., Dentener F., Derwent R., Dlugokencky E., Holland E., Isaksen I., Katima J., Kirchhoff V., Matson P., Midgley P., and Wang M. (2001). Atmospheric chemistry and greenhouse gases. In *Climate Change 2001: The Scientific Basis* (eds. J. T. Houghton, Y. Ding, D. J. Griggs, M. Noguer, P. J. van der Linden, X. Da, K. Maskell, and C. A. Johnson). Third Assessment Report, Intergovernmental Panel on Climate change, Cambridge University Press, New York, pp. 239–288 (http://www.grida.no/climate/ipcc_tar/wg1/index.htm).

Erisman J. W., de Vries W., Kros H., Oenema O., van der Eerden L., and Smeulders S. (2001) An outlook for a national integrated nitrogen policy. *Environ. Sci. Policy* **4**, 87–95.

FAO (2000) *FAO Statistical Databases*. http://apps.fao.org

Fixen P. E. and West F. B. (2002) Nitrogen fertilizers meeting the challenge. *Ambio* **31**, 169–176.

Follett J. R. and Follett R. F. (2001) Utilization and metabolism of nitrogen by humans. In *Nitrogen in the Environment: Sources, Problems and Management* (eds. R. Follett and J. L. Hatfield). Elsevier Science, New York, pp. 65–92.

Fowler D., Skiba U., and Hargreaves K. J. (1997) Emissions of nitrous oxide from grasslands. In *Gaseous Nitrogen Emissions from Grasslands* (eds. S. C. Jarvis and B. F. Pain). CAB International, Wallingford, UK, pp. 147–164.

Galloway J. N. (1998) The global nitrogen cycle: changes and consequences. *Environ. Pollut.* **102**(S1), 15–24.

Galloway J. N. (2000) Nitrogen mobilization in Asia. *Nutr. Cycl. Agroecosyst.* **57**, 1–12.

Galloway J. N. and Cowling E. B. (2002) Reactive nitrogen and the world: 200 years of change. *Ambio* **31**, 64–71.

Galloway J. N., Schlesinger W. H., Levy H., II, Michaels A., and Schnoor J. L. (1995) Nitrogen fixation: anthropogenic enhancement—environmental response. *Global Biogeochem. Cycles* **9**, 235–252.

Galloway J. N., Howarth R. W., Michaels A. F., Nixon S. W., Prospero J. M., and Dentener F. J. (1996) Nitrogen and phosphorus budgets of the North Atlantic ocean and its watershed. *Biogeochemistry* **35**, 3–25.

Galloway J. N., Cowling E. B., Seitzinger S. J., and Socolow R. (2002) Reactive nitrogen: too much of a good thing? *Ambio* **31**, 60–63.

Galloway J. N., Dentener F. J., Capone D. G., Boyer E. W., Howarth R. W., Seitzinger S. P., Asner G. P., Cleveland C., Green P., Holland E., Karl D. M., Michaels A F., Porter J. H., Townsend A., and Vörösmary C. (2003a) Nitrogen cycles: past, present, and future. *Biogeochemistry* (submitted).

Galloway J. N., Aber J. D., Erisman J. W., Seitzinger S. P., Howarth R. W., Cowling E. B., and Cosby B. J. (2003b) The nitrogen cascade. *Bioscience* **53**, 1–16.

Goolsby D. A., Battaglin W. A., Lawrence G. B., Artz R. S., Aulenbach B. T., Hooper R. P., Keeney D. R., and Stensland G. J. (1999) Flux and sources of nutrients in the Mississippi–Atchafalaya River Basin: Topic 3. *Report for the Integrated Assessment on Hypoxia in the Gulf of Mexico*. Coastal Ocean Program Decision Analysis Series No. 17, US Dept. of Commerce, National Oceanic and Atmospheric Administration, National Ocean Service, Silver Spring, MD.

Gorham E., Vitousek P. M., and Reiners W. A. (1979) The regulation of chemical budgets over the course of terrestrial ecosystem succession. *Ann. Rev. Ecol. Sys.* **10**, 53–84.

Groffman P. M., Gold A. J., and Jacinthe P. A. (1998) Nitrous oxide production in riparian zones and groundwater. *Nutr. Cycl. Agroecosyst.* **52**, 179–186.

Gruber N. and Sarmiento J. (1997) Global patterns of marine nitrogen fixation and denitrification. *Global Biogeochem. Cycles* **11**, 235–266.

Holland H. D. (1984) *The Chemical Evolution of the Atmosphere and Oceans*. Princeton University, NJ.

Holland E. A. and Lamarque J.-F. (1997) Bio-atmospheric coupling of the nitrogen cycle through NO_x emissions and NO_y deposition. *Nutr. Cycl. Agroecosyst.* **48**, 7–24.

Houghton R. A. and Hackler J. L. (2002) Carbon flux to the atmosphere from land-use changes. In *Trends: A Compendium of Data on Global Change*. Carbon Dioxide Information Analysis Center, Oak Ridge National Laboratory, US Department of Energy, Oak Ridge, TN.

Houghton J. T., Ding Y., Griggs D. J., Noguer M., van der Linden P. J., Da X., Maskell, K., and Johnson C. A. (eds.) (2001) *Climate Change 2001: The Scientific Basis*, Third Assessment Report, Intergovernmental Panel on Climate Change. Cambridge University Press, Cambridge, New York (http://www.grida.no/climate/ipcc_tar/wg1/index.htm).

Howarth R. W., Marino R., and Cole J. J. (1988) Nitrogen fixation in freshwater, estuarine, and marine ecosystems: 2. Biogeochemical controls. *Limnol. Oceanogr.* **33**, 688–701.

Howarth R. W., Billen G., Swaney D., Townsend A., Jaworski N., Lajtha K., Downing J. A., Elmgren R., Caraco N., Jordan T., Berendse F., Freney J., Kudeyarov V., Murdoch P., and Zhao-Liang Z. (1996) Regional nitrogen budgets and riverine N and P fluxes for the drainages to the North Atlantic Ocean: natural and human influences. *Biogeochemistry* **35**, 75–139.

Howarth R. W., Anderson D., Cloern J., Elfring C., Hopkinson C., Lapointe B., Malone T., Marcus N., McGlathery K., Sharpley A., and Walker D. (eds.) (2000) *Nutrient Pollution of Coastal Rivers, Bays, and Seas. Issues in Ecology 7*. Ecological Society of America, Washington, DC. (http://esa.sdsc.edu).

Howarth R. W., Boyer E., Pabich W., and Galloway J. N. (2002a) Nitrogen use in the United States from 1961–2000, and estimates of potential future trends. *Ambio* **31**, 88–96.

Howarth R. W., Sharpley A. W., and Walker D. (2002b) Sources of nutrient pollution to coastal waters in the United States: implications for achieving coastal water quality goals. *Estuaries* **25**, 656–676.

Jaffee D. A. (1992) The global nitrogen cycle. In *Global Biogeochemical Cycles* (eds. S. S. Butcher, G. H. Orians, R. J. Charlson, and G. V. Wolfe). Academic Press, London, pp. 263–284.

Johnson H. B. and Mayeux H. S. (1990) *Prosopis glandulosa* and the nitrogen balance of rangelands: extent and occurrence of nodulation. *Oecologia* **84**, 176–185.

Karl D. M. (1999) A sea of change: biogeochemical variability in the North Pacific Subtropical Gyre. *Ecosystems* **2**, 181–214.

Karl D. M. (2002) Nutrient dynamics in the deep blue sea. *Trends Microbiol.* **10**, 410–418.

Karl D. M., Michaels A., Berman B., Capone D., Carpenter E., Letelier R., Lipschultz F., Paerl H., Sigman D., and Stal L. (2002) Denitrogen fixation in the world's oceans. *Biogeochemistry* **57**, 47–52.

Keeling C. D. (1993) Global historical CO_2 emissions. In *Trends '93: A Compendium of Data on Global Change* (eds. T. A. Boden, D. P. Kaiser, R. J. Sepanski, and F. W. Stoss). ORNL/CDIAC-65. Carbon Dioxide Information Analysis Center, Oak Ridge National Laboratory, Oak Ridge, TN, pp. 501–504.

Kislev M. E. and Baryosef O. (1988) The legumes—the earliest domesticated plants in the Near East. *Current Anthropol.* **29**, 175–179.

Kramer D. A. (1999) *Minerals Yearbook: Nitrogen*. US Geological Survey Minerals Information (http://minerals.usgs.gov/minerals/pubs/commodity/nitrogen/).

Kroeze C., Aerts R., van Breemen N., van Dam D., van der Hoek K., Hofschreuder P., Hoosbeek M., de Klein J., Kros H., van Oene H., Oenema O., Tietema A., van der Veeren R., and de Vries W. (2003) Uncertainties in the fate of nitrogen: I. An overview of sources of uncertainty illustrated with a Dutch case study. *Nutr. Cycl. Agroecosyst.* **66**, 43–69.

Lapointe B. E. and O'Connell J. D. (1989) Nutrient-enhanced productivity of *Cladophroa prolifera* in Harrington Sounds, Bermuda: eutrophication of a confined, phosphorus-limited marine ecosystem. *Est. Coast Shelf Sci.* **28**, 347–360.

Lelieveld J. and Dentener F. (2000) What controls tropospheric ozone? *J. Geophys. Res.* **105**, 3531–3551.

Mackenzie F. T. (1994) Global climatic change: climatically important biogenic gases and feedbacks. In *Biotic Feedbacks in the Global Climatic System: Will the Warming Feed the Warming* (eds. G. M. Woodwell and F. T. Mackenzie). Oxford University Press, New York, pp. 22–46.

Mackenzie F. T. (1998) *Our Changing Planet: An Introduction to Earth System Science and Global Environmental Change*, 2nd edn. Prentice Hall, Upper Saddle River, NJ.

Melillo J. M. and Cowling E. B. (2002) Reactive nitrogen and public policies for environmental protection. *Ambio* **31**, 150–158.

Meybeck M. (1982) Carbon, nitrogen, and phosphorus transport by world rivers. *Am. J. Sci.* **282**, 401–450.

Miller S. L. (1953) A production of amino acids under possible primitive Earth conditions. *Science* **117**, 529–529.

Moomaw W. R. (2002) Energy, industry and nitrogen: strategies for reducing reactive nitrogen emissions. *Ambio* **31**, 184–189.

Mosier A., Kroeze C., Nevison C., Oenema O., Seitzinger S., and van Cleemput O. (1998) Closing the global atmospheric N_2O budget: nitrous oxide emissions through the agricultural nitrogen cycle. *Nutr. Cycl. Agroecosyst.* **52**, 225–248.

Mosier A. R., Bleken M. A., Chaiwanakupt P., Ellis E. C., Freney J. R., Howarth R. B., Matson P. A., Minami K., Naylor R., Weeks K., and Zhu Z.-L. (2001) Policy implications of human accelerated nitrogen cycling. *Biogeochemistry* **52**, 281–320.

Müller J. (1992) Geographical distribution and seasonal variation of surface emissions and deposition velocities of atmospheric trace gases. *J. Geophys. Res.* **97**, 3787–3804.

Neff J. C., Holland E. A., Dentener F. J., McDowell W. H., and Russell K. M. (2002) The origin, composition and rates of organic nitrogen deposition: a missing piece of the nitrogen cycle? *Biogeochemistry* **57**, 99–136.

Nixon S. W., Ammerman J. W., Atkinson L. P., Berounsky V. M., Billen G., Boicourt W. C., Boynton W. R., Church T. M., DiToro D. M., Elmgren R., Garber J. H., Giblin A. E., Jahnke R. A., Owens N. J. P., Pilson M. E. Q., and Seitzinger S. P. (1996) The fate of nitrogen and phosphorus at the land-sea margin of the North Atlantic Ocean. *Biogeochemistry* **35**, 141–180.

Nolan B. T. and Stoner J. D. (2000) Nutrients in groundwaters of the conterminous United States, 1992–1995. *Environ. Sci. Tech.* **34**, 1156–1165.

NRC (National Research Council) (2000) *Clean Coastal Waters: Understanding and Reducing the Effects of Nutrient Pollution*. National Academy Press, Washington, DC.

Oenema O. and Pietrzak S. (2002) Nutrient management in food production: achieving agronomic and environmental targets. *Ambio* **31**, 132–140.

Paerl H. W., Dennis E. L., and Whitall D. D. (2002) Atmospheric deposition of nitrogen: implications for nutrient over-enrichment of coastal waters. *Estuaries* **25**, 677–693.

Paul E. A. and Clark F. E. (1997) *Soil Microbiology and Biochemistry*. Elsevier/Academic Press, New York.

Pope C. A., III, Thun M. J., Namboodiri M. M., Dockery D. W., Evans J. S., Speizer F. E., and Heath C. W., Jr. (1995) Particulate air pollution as a predictor of mortality in a prospective study of US adults. *Am. J. Resp. Crit. Care Med.* **151**, 669–674.

Puckett L. J. and Cowdery T. K. (2002) Transport and fate of nitrate in a glacial outwash aquifer in relation to groundwater age, land use practices and redox processes. *J. Environ. Qual.* **31**, 782–796.

Puckett L. J., Cowdery T. K., Lorenz D. L., and Stoner J. D. (1999) Estimation of nitrate contamination of an agro-ecosystem outwash aquifer using a nitrogen mass-balance budget. *J. Environ. Qual.* **28**, 2015–2025.

Rabalais N. (2002) Nitrogen in aquatic ecosystems. *Ambio* **31**, 102–112.

Rabalais N. N., Turner R. E., and Scavia D. (2002) Beyond science into policy: gulf of mexico hypoxia and the mississippi river. *Bioscience* **52**, 129–142.

Rosswall T. (1983) The nitrogen cycle. In *The major Biogechemical Cycles and Their Interactions* (eds. B. Bolin and R. B. Cook). SCOPE, Wiley, Chichester, Vol. 21, pp. 46-50.

Rosswall T. (1992) The international geosphere-biosphere program—a study of global change (IGBP). *Environ. Geol. Water Sci.* **20**, 77–78.

Roy R. N., Misra R. V., and Montanez A. (2002) Reduced reliance on mineral nitrogen, yet more food. *Ambio* **31**, 177–183.

Schlesinger W. H. (1997) *Biogeochemistry: An Analysis of Global Change*. Elsevier, New York.

Seitzinger S. P. (1988) Denitrification in freshwater and coastal marine ecosystems: ecological and geochemical importance. *Limnol. Oceanogr.* **33**, 702–724.

Seitzinger S. P. and Giblin A. E. (1996) Estimating denitrification in North Atlantic continental shelf sediments. *Biogeochemistry* **35**, 235–260.

Seitzinger S. P. and Kroeze C. (1998) Global distribution of nitrous oxide production and N inputs in freshwater and coastal marine ecosystems. *Global Biogeochem. Cycles* **12**, 93–113.

Seitzinger S. P., Styles R. V., Boyer E., Alexander R. B., Billen G., Howarth R., Mayer B., and van Breemen N. (2002) Nitrogen retention in rivers: model development and application to watersheds in the Northeastern USA. *Biogeochemistry* **57**, 199–237.

Smil V. (1994) *Energy in World History*. Westview Press, Boulder, CO.

Smil V. (1999) Nitrogen in crop production: an account of global flows. *Global Biogeochem. Cycles* **13**, 647–662.

Smil V. (2000) *Feeding the World*. MIT Press, Cambridge.

Smil V. (2001) *Enriching the Earth*. MIT Press, Cambridge.

Smil V. (2002) Nitrogen and food production: proteins for human diets. *Ambio* **31**, 126–131.

Smith B. D. (1995) *The Emergence of Agriculture*. Scientific American, New York.

Socolow R. (1999) Nitrogen management and the future of food: lessons from the management of energy and carbon. *Proc. Natl. Acad. Sci. USA* **96**, 6001–6008.

Söderlund R. and Rosswall T. (1982) The nitrogen cycles: Topic 1. Report for the integrated assessment on hypoxia in the Gulf of Mexico. In *NOAA Coastal Ocean Program Decision Analysis Series No. 17*. NOAA, Silver Spring, MD.

Sommer S. G. and Hutchings N. J. (1997) Components of ammonia volatilization from cattle and sheep production. In *Gaseous Nitrogen Emissions from Grasslands* (eds. S. C. Jarvis and B. F. Pain). CAB International, UK, pp. 79–93.

Stedman D. H. and Shetter R. (1983) The global budget of atmospheric nitrogen species. In *Trace Atmospheric Constituents: Properties, Transformations and Fates* (ed. S. S. Schwartz). Wiley, New York, pp. 411–454.

Svensson B. H. and Söderlund R. (eds.) (1976) Nitrogen, phosphorus and sulphur—global cycles. Scientific Committee on Problems of the Environment (SCOPE) Report on Biogeochemical Cycles, Royal Swedish Academy of Science.

United Nations (1999) *World Population Prospects: the 1998 Revision*. United Nations, New York.

Valiela I., McClelland J., Hauxwell J., Behr P. J., Hersh D., and Foreman K. (1997) Macroalgal blooms in shallow estuaries: controls and ecophysiological and ecosystem consequences. *Limnol. Oceanogr.* **42**, 1105–1118.

van Aardenne J. A., Dentener F. J., Klijn Goldewijk C. G. M., Lelieveld J., and Olivier J. G. J. (2001) A 1°–1° resolution data set of historical anthropogenic trace gas emissions for the period 1890–1990. *Global Biogeochem. Cycles* **15**, 909–928.

van Breemen N., Boyer E. W., Goodale C. L., Jaworski N. A., Paustian K., Seitzinger S., Lajtha L. K., Mayer B., Van Dam D., Howarth R. W., Nadelhoffer K. J., Eve M., and Billen G. (2002) Where did all the nitrogen go? Fate of nitrogen inputs to large watersheds in the Northeastern USA. *Biogeochemistry* **57**, 267–293.

van Egmond N. D., Bresser A. H. M., and Bouwman A. F. (2002) The European nitrogen case. *Ambio* **31**, 72–78.

Vitousek P. M. (1994) Potential nitrogen fixation during primary succession in Hawaii volcanoes national park. *Biotropica* **26**, 234–240.

Vitousek P. M. and Howarth R. W. (1991) Nitrogen limitation on land and in the sea. How can it occur? *Biogeochemistry* **13**, 87–115.

Vitousek P. M., Aber J. D., Howarth R. W., Likens G. E., Matson P. A., Schindler D., Schlesinger W. H., and Tilman G. D. (1997) *Human Alteration of the Global Nitrogen Cycle: Causes and Consequences, Issues in Ecology 1*. Ecological Society of America, Washington, DC, (http://esa.sdsc.edu).

Walsh J. J., Dieterle D. A., and Meyers M. B. (1988) A simulation analysis of the fate of phytoplankton within the mid-Atlantic Bight. *Cont. Shelf Res.* **8**, 757–787.

Wang L. (1987) Soybeans—the miracle bean of China. In *Feeding a Billion: Frontiers of Chinese Agriculture)* (eds. S. Wittwer, L. Wang, and Y. Yu). Michigan State University Press, East Lansing, MI.

Wittwer S., Wang L., and Yu Y. (1987) *Feeding a Billion: Frontiers of Chinese Agriculture*. Michigan State University Press, East Lansing.

Wolfe A. and Patz J. A. (2002) Nitrogen and human health: direct and indirect impacts. *Ambio* **31**, 120–125.

Xing G. X. and Zhu Z. L. (2002) Regional nitrogen budgets for China and its major watersheds. *Biogeochemistry* **57**, 405–427.

Zheng X., Fu C., Xu X., Yan X., Chen G., Han S., Huang Y., and Hu F. (2002) The Asian nitrogen case. *Ambio* **31**, 79–87.

8.13
The Global Phosphorus Cycle

K. C. Ruttenberg

University of Hawaii at Manoa, Honolulu, HI, USA

8.13.1 INTRODUCTION

Phosphorus is an essential nutrient for all life forms. It is a key player in fundamental biochemical reactions (Westheimer, 1987) involving genetic material (DNA, RNA) and energy transfer (ATP), and in structural support of organisms provided by membranes (phospholipids) and bone (the biomineral hydroxyapatite). Photosynthetic organisms utilize dissolved phosphorus, carbon, and other essential nutrients to build their tissues using energy from the Sun. Biological productivity is contingent upon the availability of phosphorus to these simple organisms that constitute the base of the food web in both terrestrial and aquatic systems. (For reviews of P-utilization, P-biochemicals, and pathways in aquatic plants, see Fogg (1973), Bieleski and Ferguson (1983), and Cembella *et al.* (1984a, 1984b).)

Phosphorus locked up in bedrock, soils, and sediments is not directly available to organisms. Conversion of unavailable forms to dissolved orthophosphate, which can be directly assimilated, occurs through geochemical and biochemical reactions at various stages in the global phosphorus cycle. Production of biomass fueled by P-bioavailability results in the deposition of organic matter in soils and sediments, where it acts as a source of fuel and nutrients to microbial communities. Microbial activity in soils and sediments, in turn, strongly influences the concentration and chemical form of phosphorus incorporated into the geological record.

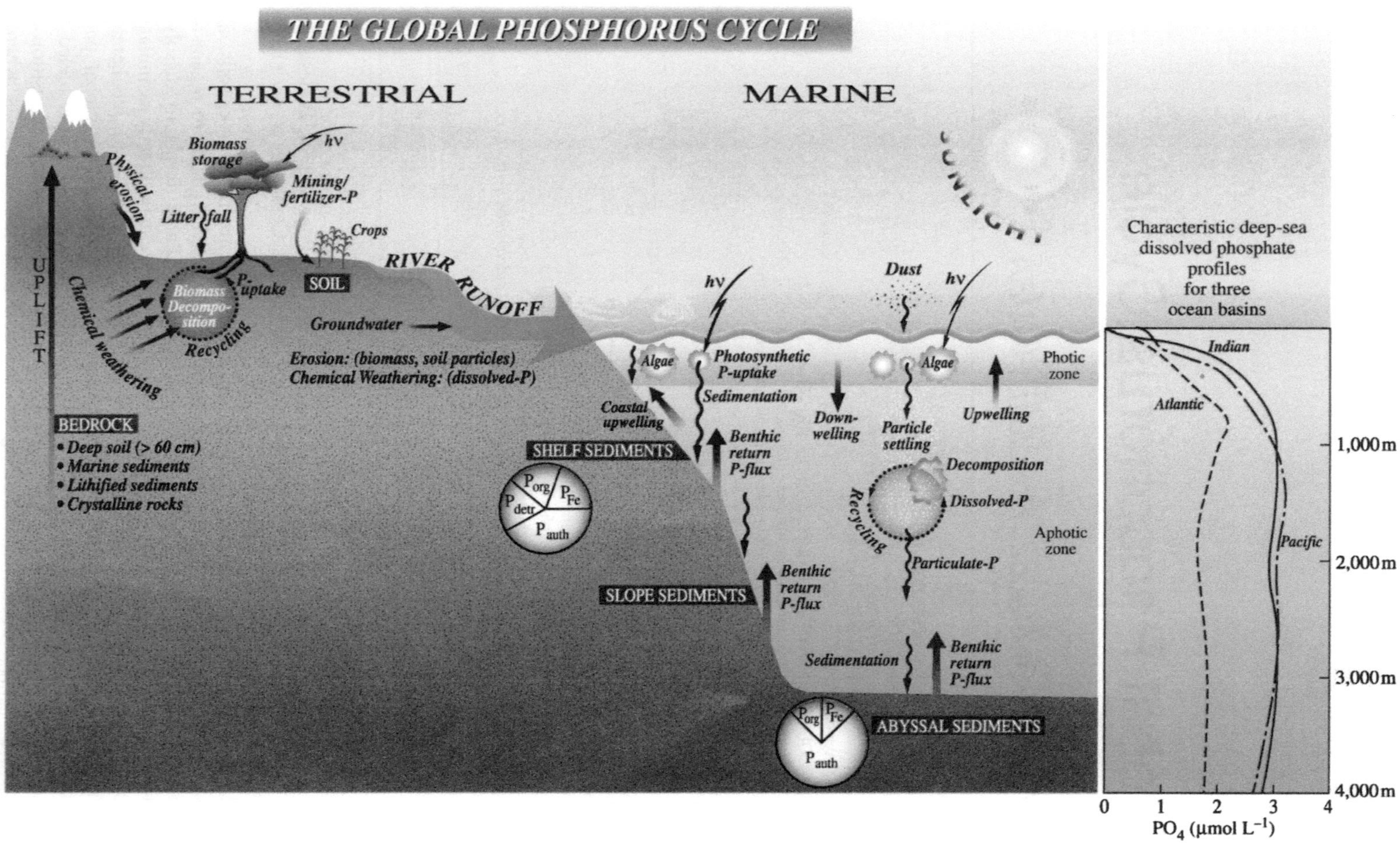

THE GLOBAL PHOSPHORUS CYCLE
TERRESTRIAL
MARINE
UPLIFT
Physical erosion
Biomass storage
hν
Mining/ fertilizer-P
Litter fall
Crops
SOIL
P-uptake
Biomass Decomposition
Recycling
Chemical weathering
RIVER RUNOFF
Groundwater
Erosion: (biomass, soil particles)
Chemical Weathering: (dissolved-P)
BEDROCK
• Deep soil (> 60 cm)
• Marine sediments
• Lithified sediments
• Crystalline rocks
SUNLIGHT
hν
Dust
hν
Algae
Photosynthetic P-uptake
Algae
Photic zone
Sedimentation
Coastal upwelling
Down-welling
Particle settling
Upwelling
Benthic return P-flux
SHELF SEDIMENTS
P org
P Fe
P detr
P auth
Decomposition
Dissolved-P
Recycling
Aphotic zone
Particulate-P
Benthic return P-flux
SLOPE SEDIMENTS
Sedimentation
Benthic return P-flux
P org
P Fe
P auth
ABYSSAL SEDIMENTS
Characteristic deep-sea dissolved phosphate profiles for three ocean basins
Indian
Atlantic
Pacific
1,000m
2,000m
3,000m
4,000m
0
1
2
3
4
PO4 (μmol L−1)

The global phosphorus cycle has four major components: (i) tectonic uplift and exposure of phosphorus-bearing rocks to the forces of weathering; (ii) physical erosion and chemical weathering of rocks producing soils and providing dissolved and particulate phosphorus to rivers; (iii) riverine transport of phosphorus to lakes and the ocean; and (iv) sedimentation of phosphorus associated with organic and mineral matter and burial in sediments (Figure 1). The cycle begins anew with uplift of sediments into the weathering regime.

This chapter begins with a brief overview of the various components of the global phosphorus cycle. Estimates of the mass of important phosphorus reservoirs, transport rates (fluxes) between reservoirs, and residence times are given in Tables 1 and 2. As is clear from the large uncertainties associated with these estimates of reservoir size and flux, many aspects of the global phosphorus cycle remain poorly understood. Following the overview, various aspects of the global phosphorus cycle will be examined in more depth, including a discussion of the most pressing research questions currently being posed, and research efforts presently underway to address these questions.

8.13.2 THE GLOBAL PHOSPHORUS CYCLE: OVERVIEW

8.13.2.1 The Terrestrial Phosphorus Cycle

In terrestrial systems phosphorus resides in three pools: bedrock, soil, and living organisms (biomass) (Table 1). Weathering of continental bedrock is the principal source of phosphorus to the soils that support continental vegetation (F_{12}); atmospheric deposition is relatively unimportant (F_{82}). Phosphorus is weathered from bedrock by dissolution of phosphorus-bearing minerals such as apatite ($Ca_{10}(PO_4)_6(OH,F,Cl)_2$), the most abundant primary P-mineral in crustal rocks. Weathering reactions are driven by exposure of minerals to naturally occurring acids derived mainly from microbial activity (e.g., Cosgrove, 1977; Frossard *et al.*, 1995). Phosphate solubilized during weathering is available for uptake by terrestrial plants, and is returned to the soil by decay of litterfall (e.g., Likens *et al.*, 1977).

Soil solution phosphate is maintained at low levels as a result of P-sorption by various soil constituents, particularly ferric iron and aluminum oxyhydroxides. Sorption is considered the most important process controlling terrestrial P-bioavailability (Lajtha and Harrison, 1995). Plants have different physiological strategies for obtaining P-despite low soil solution concentrations. For example, some plants can increase root volume and surface area to optimize uptake potential. Alternatively, plant roots and/or associated fungi can produce (i) chelating compounds that solubilize ferric iron, aluminum, and calcium-bound phosphorus, (ii) enzymes and/or (iii) acids in the root vicinity, to solubilize phosphate (Lajtha and Harrison, 1995). Plants also minimize P-loss by resorbing much of their phosphorus prior to litterfall, and by efficient recycling from fallen litter. In extremely unfertile soils (e.g., in tropical rain forests) P-recycling is so efficient that topsoil contains virtually no phosphorus; it is all tied up in biomass (e.g., Vitousek *et al.*, 1997).

Systematic changes in the total amount and chemical form of phosphorus occur during soil development. In initial stages, phosphorus is present mainly as primary minerals such as apatite. In mid-stage soils, the reservoir of primary apatite is diminished; less-soluble secondary minerals and organic-P make up an increasing fraction of soil phosphorus. Late in soil development, phosphorus is partitioned mainly between refractory minerals and organic-P (Figure 2). This classic model articulated by Walker and Syers (1976) has been validated in

Figure 1 The major reservoirs and fluxes of the global phosphorus cycle are illustrated (see Tables 1 and 2, and text). The oceanic photic zone, idealized in the cartoon, is typically thinner in coastal environments due to turbidity from continental terrigenous input, and deepens as the water column clarifies with distance away from the continental margins. The distribution of phosphorus among different chemical/mineral forms in marine sediments is given in the pie diagrams, where the abbreviations used are: organic phosphorus (P_{org}), iron-bound phosphorus (P_{Fe}), detrital apatite (P_{detr}), authigenic/biogenic apatite (P_{auth}). The P_{org}, P_{Fe}, and P_{auth} reservoirs represent potentially reactive-P pools (see text and Tables 2 and 3 for discussion), whereas the P_{detr} pool reflects mainly detrital apatite weathered off the continents and passively deposited in marine sediments (note that P_{detr} is not an important sedimentary phosphorus component in abyssal sediments, far from continents). Continental margin P-speciation data were compiled from Louchouarn *et al.* (1997), and Ruttenberg and Berner (1993). Abyssal sediment P-speciation data were compiled from Filippelli and Delaney (1996), and Ruttenberg (1990). The "global phosphorus cycle" cartoon is from Ruttenberg (2002). The vertical water column phosphate distributions typically observed in the three ocean basins are shown in the panel to the right of the "global phosphorus cycle" cartoon, and are from Sverdrup *et al.* (1942).

Table 1 Major reservoirs active in the global phosphorus cycle and associated residence times.

Reservoir #	*Reservoir description*	*Reservoir size* (mole P × 10^{12})	*Reference*	*Residence time* τ (yr)
R1	Sediments (crustal rocks and soil >60 cm deep and marine sediments)	$0.27\times10^8-1.3\times10^8$	b, a = c = d	$42-201\times10^6$
R2	Land (≈total soil <60 cm deep: organic + inorganic)	3,100–6,450	b, a = c = d	425–2,311
R3	Land biota	83.9–96.8	b, a = c = d	13–48
R4	Surface ocean, 0–300 m (total dissolved P)	87.4	a = c	2.46–4.39
R5	Deep sea, 300–3300 m (total dissolved P)	2,810	a = c & d	1,502
R6	Oceanic biota	1.61–4.45	b & d, a = c & d	0.044–0.217 (16–78 d)
R7	Minable P	323–645	a = c, b & d	718–1,654
R8	Atmospheric P	0.0009	b = c = d	0.009 (80 h)

(1) Ranges are reported for those reservoirs for which a consensus on a single best estimated reservoir size does not exist. Maximum and minimum estimates found in a survey of the literature are reported. References cited before the comma refer to the first (lowest) estimate, those after the comma refer to the second (higher) estimate. References that give identical values are designated by an equality sign, references giving similar values are indicated by an ampersand. As indicated by the wide ranges reported for some reservoirs, all calculations of reservoir size have associated with them a large degree of uncertainty. Methods of calculation, underlying assumptions, and sources of error are given in the references cited.

(2) Residence times are calculated by dividing the concentration of phosphorus contained in a given reservoir by the sum of fluxes out of the reservoir. Where ranges are reported for reservoir size and flux, maximum and minimum residence time values are given; these ranges reflect the uncertainties inherent in reservoir size and flux estimates. Fluxes used to calculate residence times for each reservoir are as follows: R1 (F_{12}), R2 ($F_{23}+F_{28}+F_{24(d)}+F_{24(p)}$), R3 ($F_{32}$), R4 ($F_{45}+F_{46}$), R5 ($F_{54}$), R6 ($F_{64}+F_{65}$), R7 ($F_{72}$), R8 ($F_{82}+F_{84}$). Flux estimates are given in Table 2. The residence time of R5 is decreased to 1,492 yr by inclusion of the scavenged flux of deep-sea phosphate at hydrothermal MOR systems, mostly onto ferric oxide and oxyhydroxide phases (Wheat *et al.* (1996).

(3) Estimates for the partitioning of the oceanic reservoir between dissolved inorganic phosphorus and particulate phosphorus are given in references b and d as follows: (2,581–2,600) × 10^{12} mol dissolved inorganic phosphorus (b, d) and (20–21) × 10^{12} mol particulate phosphorus (d, b).

(4) The residence times estimated for the minable phosphorus reservoir reflect estimates of current mining rates; if mining activity increases or diminishes the residence time will change accordingly.

[a] Lerman *et al.* (1975) [b] Richey (1983) [c] Jahnke (1992) [d] Mackenzie *et al.* (1993).

Table 2 Fluxes between the major phosphorus reservoirs.

Flux #	*Description of flux*	*Flux* (moles P × 10^{12} yr^{-1})	*References and comments*
Reservoir fluxes:			
F_{12}	rocks/sediments → soils (erosion/weathering, soil accumulation)	0.645	a = c&d
F_{21}	soils → rocks/sediments (deep burial, lithification)	0.301–0.603	d, a = c
F_{23}	soils → land biota	2.03–6.45	a = c, b&d
F_{32}	land biota → soils	2.03–6.45	a = c, b&d
$F_{24(d)}$	soil → surface ocean (river total dissolved P flux)	0.032–0.058	e, a = c; ca. >50% of TDP is DOP (e)
$F_{24(p)}$	soil → surface ocean (river particulate P flux)	0.59–0.65	d, e; ca. 40% of RSPM-P is org. P (e); it is estimated that between 25% and 45% is reactive once it enters the ocean (f).
F_{46}	surface ocean → oceanic biota	19.35–35	b, d; a = c = 33.5, b reports upper limit of 32.3; d reports lower limit of 28.2
F_{64}	oceanic biota → surface ocean	19.35–35	b,d; a&c = 32.2, b reports upper limit of 32.3; d reports lower limit of 28.2
F_{65}	oceanic biota → deep sea (particulate rain)	1.13–1.35	d, a = c
F_{45}	surface ocean → deep sea (downwelling)	0.581	a = c
F_{54}	deep sea → surface ocean (upwelling)	1.87	a = c
F_{42}	surface ocean → land (fisheries)	0.01	d
F_{72}	minable P → land (soil)	0.39–0.45	a = c = d, b
F_{28}	land (soil) → atmosphere	0.14	b = c = d
F_{82}	atmosphere → land (soil)	0.1	b = c = d
F_{48}	surface ocean → atmosphere	0.01	b = c = d
F_{84}	atmosphere → surface ocean	0.02–0.05	c, b; d gives 0.04; ca. 30% of atmospheric aerosol-P is soluble (g)
Sub-reservoir fluxes:			
marine sediments			
sF_{ms}	marine sediment accumulation (total)	0.265–0.280	i, j; for higher estimate (j), use of sediment P-concentration below the diagenesis zone implicitly accounts for P-loss via benthic remineralization flux and yields pre-anthropogenic net burial flux. For estimates of reactive-P burial see note (j).

(continued)

Table 2 (continued).

Flux #	*Description of flux*	*Flux* (moles P × 10^{12} yr^{-1})	*References and comments*
sF_{cs}	continental margin ocean sediments → burial	0.150–0.223	j, i; values reported reflect total-P, reactive-P burial constitutes from 40% to 75% of total-P (h). These values reflect pre-agricultural fluxes, modern value estimated as 0.33 (d).
sF_{as}	abyssal (deep sea) sediments →burial	0.042–0.130	i, j; a = c gives a value of 0.055. It is estimated that 90–100% of this flux is reactive-P (h). These values reflect pre-agricultural fluxes, modern value estimates range from 0.32 (d) to 0.419 (b).
sF_{cbf}	coastal sediments → coastal waters (remineralization, benthic flux)	0.51–0.84	d, k; these values reflect pre-agricultural fluxes, modern value estimated as 1.21 with uncertainties ±40% (k).
sF_{abf}	abyssal sediments → deep sea (remineralization, benthic flux)	0.41	k; this value reflects pre-agricultural fluxes, modern value estimated as 0.52, uncertainty ±30% (k)

(1) Reservoir fluxes (*F*) represent the P-flux between reservoirs #R1–R8 defined in Table 1. The sub-reservoir fluxes (sF) refer to the flux of phosphorus into the marine sediment portion of reservoir #1 via sediment burial, and the flux of diagenetically mobilized phosphorus out of marine sediments via benthic return flux. These sub-fluxes have been calculated as described in references h–k. Note that the large magnitude of these sub-fluxes relative to those into and out of reservoir #1 as a whole, and the short oceanic-P residence time they imply (Tables 1 and 5), highlight the dynamic nature of the marine phosphorus cycle.
(2) Ranges are reported where consensus on a single best estimate does not exist. References cited before the comma refer to the first (lower) estimate, those after the comma refer to the second (higher) estimate. References that give identical values are designated by an equality sign, references giving similar values are indicated by an ampersand. Maximum and minimum estimates found in a survey of the literature are reported. In some cases this range subsumes ranges reported in the primary references. As indicated by the wide ranges reported, all flux calculations have associated with them a large degree of uncertainty. Methods of calculation, underlying assumptions, and sources of error are given in the references cited.
[a] Lerman, *et al.* (1975) [b] Richey (1983) [c] Jahnke (1992) [d] Mackenzie *et al.* (1993) [e] Meybeck (1982)
[f] The range of riverine suspended particulate matter that may be solubilized once it enters the marine realm (e.g., the so-called "reactive-P") is derived from three sources. Colman and Holland (2000) estimate that 45% may be reactive, based on RSPM-P compositional data from a number of rivers and estimated burial efficiency of this material in marine sediments. Berner and Rao (1994) and Ruttenberg and Canfield (1994) estimate that 35% and 31% of RSPM-P is released upon entering the ocean, based on comparison of RSPM-P and adjacent deltaic surface sediment phosphorus in the Amazon and Mississippi systems, respectively. Lower estimates have been published (8%: Ramirez and Rose (1992); 18%: Froelich (1988); 18%: Compton *et al.* (2000). Higher estimates have also been published (69%: Howarth *et al.* (1995).
Howarth *et al.* (1995) also estimate the total flux of riverine particulate phosphorus to the oceans at 0.23×10^{12} moles P yr^{-1}, an estimate likely too low because it uses the suspended sediment flux from Milliman and Meade (1983), which does not include the high sediment flux rivers from tropical mountainous terranes (Milliman and Syvitski (1992). [g] Duce *et al.* (1991) [h] Ruttenberg (1993) [i] Howarth *et al.* (1995) [j] P-burial flux estimates as reported in Ruttenberg (1993) modified using pre-agricultural sediment fluxes updated by Colman and Holland (2000). Using these total P burial fluxes and the ranges of likely reactive P given in the table, the best estimate for reactive P burial flux in the oceans lies between $(0.177–0.242)\times10^{12}$ moles P yr^{-1}. Other estimates of whole-ocean reactive P burial fluxes range from, at the low end: 0.032 to 0.081×10^{12} moles P yr^{-1} (Compton *et al.* (2000), and 0.09×10^{12} moles P yr^{-1} (Wheat *et al.* (1996); to values more comparable to those derived from the table above (0.21×10^{12} moles P yr^{-1}: Filippelli and Delaney (1996). [k] Colman and Holland (2000).

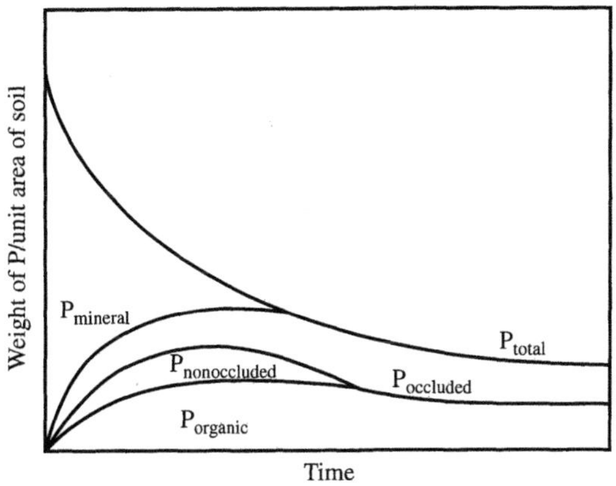

Figure 2 The fate of phosphorus during soil formation can be viewed as the progressive dissolution of primary mineral phosphorus (dominantly apatite), some of which is lost from the system by leaching (decrease in P_{total}), and some of which is reincorporated into nonoccluded, occluded, and organic fractions within the soil. Nonoccluded phosphorus is defined as phosphate sorbed to surfaces of hydrous oxides of iron and aluminum, and calcium carbonate. Occluded phosphorus refers to phosphorus present within the mineral matrix of discrete mineral phases. The initial buildup in organic phosphorus results from organic matter return to soil from vegetation supported by the soil. The subsequent decline in $P_{organic}$ is due to progressive mineralization and soil leaching. The timescale over which these transformations occur depends upon the initial soil composition, topographic, and climatic factors (after Walker and Syers, 1976).

numerous subsequent works (e.g., Smeck, 1985; Lajtha and Schlesinger, 1988; Crews *et al.*, 1995; Schlesinger *et al.*, 1998; Filippelli and Souch, 1999; Chadwick *et al.*, 1999).

8.13.2.2 Transport of Phosphorus from Continents to the Ocean

Phosphorus is transferred from the continental to the oceanic reservoir primarily by rivers (F_{24}). Deposition of atmospheric aerosols (F_{84}) is a minor flux. Groundwater seepage to the coastal ocean is a potentially important but poorly documented flux. Riverine phosphorus derives from weathered continental rocks and soils. Because phosphorus is particle reactive, most riverine-P is associated with particulate matter. By most estimates, well over 90% of the phosphorus delivered by rivers to the ocean is as particulate-P ($F_{24(p)}$). Dissolved phosphorus in rivers occurs in both inorganic and organic forms. The scant data on dissolved organic P suggest that it may account for 50% or more of dissolved riverine-P (Meybeck, 1982). The chemical form of phosphorus associated with riverine particles is variable and depends upon the drainage basin geology, extent of weathering of the substrate, and on the nature of the river itself. Available data suggest that ~20–40% of phosphorus in suspended particulate matter is organic; inorganic forms are partitioned mainly between ferric oxyhydroxides and apatite (Lucotte and d'Anglejan, 1983; Lebo, 1991; Berner and Rao, 1994; Ruttenberg and Canfield, 1994). Aluminum oxyhydroxides and clays may also be significant carriers of phosphorus (Lebo, 1991).

The fate of phosphorus entering the ocean via rivers is variable. Dissolved phosphorus in estuaries at the continent–ocean interface typically displays nonconservative behavior. Both negative and positive deviations from conservative mixing can occur, sometimes changing seasonally within the same estuary (Froelich *et al.*, 1982; Fox *et al.*, 1985, 1986, 1987). Net removal of phosphorus in estuaries is typically driven by flocculation of humic–iron complexes and biological uptake (e.g., Sholkovitz *et al.*, 1978; Fox, 1990). Net P-release is due to a combination of desorption from freshwater particles entering high ionic strength marine waters, and flux of diagenetically mobilized phosphorus from benthic sediments (e.g., Chase and Sayles, 1980; Nixon, 1981; Nixon and Pilson, 1984; Fox *et al.*, 1986; Berner and Rao, 1994; Chambers *et al.*, 1995; Conley *et al.*, 1995). Accurate estimates of bioavailable riverine-P flux to the ocean must take into account, in addition to dissolved forms, the fraction of riverine particulate-P released to solution upon entering the ocean (Tables 2 and 3).

8.13.2.2.1 *Human impacts on the global phosphorus cycle*

The mining of phosphate rock (mostly from terrestrially emplaced marine phosphorite deposits) for use as agricultural fertilizer has increased dramatically in the latter half of this century (F_{72}). In addition to fertilizer use, deforestation, increased cultivation, urban and industrial waste disposal all have enhanced phosphorus transport from terrestrial to aquatic systems, often with deleterious results. For example, elevated phosphorus concentrations in rivers resulting from these activities have resulted in eutrophication in some lakes and coastal areas, stimulating nuisance algal blooms and promoting hypoxic or anoxic conditions harmful or lethal to natural populations (e.g., Caraco, 1995; Fisher *et al.*, 1995; Melack, 1995).

Increased erosion due to forest clear-cutting and widespread cultivation has increased riverine suspended matter concentrations, and thus increased the riverine particulate-P flux. Dams, in contrast, decrease sediment loads in rivers and therefore diminish the phosphorus flux to the oceans. However, increased erosion below dams

Table 3 Revised oceanic phosphorus input fluxes, removal fluxes, and estimated oceanic residence time.

	Flux description[a]	*Flux* (moles P $\times 10^{12}$ yr^{-1})	*Residence time*[e] (yr)
Input fluxes:			
F_{84}	atmosphere → surface ocean	0.02–0.05	
$F_{24(d)}$	soil → surface ocean (river dissolved P flux)[b]	0.032–0.058	
$F_{24(p)}$	soil → surface ocean (river particulate P flux)[b]	0.59–0.65	
	Minimum reactive-P input flux	0.245	12,000
	Maximum reactive-P input flux	0.301	10,000
Removal fluxes:			
sF_{cs}	Best estimate of total-P burial in continental margin marine sediments (Table 2, note j)[c]	0.150	
sF_{as}	Best estimate of total-P burial in abyssal marine sediments (Table 2, note j)[c]	0.130	
	Minimum estimate of reactive-P burial in marine sediments[d]	0.177	17,000
	Maximum estimate of reactive-P burial in marine sediments[d]	0.242	12,000

[a] All fluxes are from Table 2. [b] As noted in Table 2, 30% of atmospheric aerosol-P (Duce *et al.*, 1991) and 25–45% of the river particulate flux (see note (f) in Table 2) is believed to be mobilized upon entering the ocean. The reactive-P input flux was calculated as the sum of $0.3(F_{84}) + F_{24(d)} + 0.35(F_{24(p)})$, where the mean value of the fraction of riverine particulate phosphorus flux estimated as reactive-P (35%) was used. Reactive-P is defined as that which passes through the dissolved oceanic P-reservoir, and thus is available for biological uptake. [c] These estimates are favored by the author, and reflect the minimum sF_{cs} and maximum sF_{as} fluxes given in Table 2. Because the reactive-P content of continental margin and abyssal sediments differs (see Table 2 and note d, below), these fluxes must be listed separately in order to calculate the whole-ocean reactive-P burial flux. See note (j) in Table 2 for other published estimates of reactive-P burial flux. [d] As noted in Table 2, between 40% and 75% of phosphorus buried in continental margin sediments is potentially reactive, and 90% to 100% of phosphorus buried in abyssal sediments is potentially reactive. The reactive-P fraction of the total sedimentary P-reservoir represents that which may have passed through the dissolved state in oceanic waters, and thus represents a true P-sink from the ocean. The minimum reactive-P burial flux was calculated as the sum of $0.4(sF_{cs}) + 0.9\ (sF_{as})$; the maximum reactive-P burial flux was calculated as the sum of $0.75(sF_{cs}) + 1(sF_{as})$. Both the flux estimates and the percent reactive-P estimates have associated with them large uncertainties. [e] Residence time estimates are calculated as the oceanic phosphorus inventory (reservoirs #4 and 5 (Table 1) = 3×10^{15} moles P) divided by the minimum and maximum input and removal fluxes.

and diagenetic mobilization of phosphorus in sediments trapped behind dams moderates this effect. The overall effect has been a 50% to threefold increase in riverine-P flux to the oceans above pre-agricultural levels (Melack, 1995; Howarth *et al.*, 1995).

8.13.2.3 The Marine Phosphorus Cycle

Phosphorus in its simplest form, dissolved orthophosphate, is taken up by photosynthetic organisms at the base of the marine food web. The marine phosphorus (P) cycle is linked to the marine carbon (C) and nitrogen (N) cycles through the photosynthetic fixation of these elements by the microscopic floating marine plants, or phytoplankton, which form the base of the marine food web. In the seminal work of Alfred Redfield (1958; Redfield *et al.*, 1963), it was recognized that marine phytoplankton, on average, have a C : N : P molar ratio of 106C : 16N : 1P. This ratio can vary due to such factors as nutrient availability and nutritive state of the phytoplankton, but the relative consistency of this ratio is striking. Further, Redfield (1958) and Redfield *et al.* (1963) pointed out that the ratio of inorganic dissolved C : N : P in seawater (as HCO_3^-, NO_3^-, HPO_4^{2-}) is 1,000 : 15 : 1, when contrasted with the average oceanic phytoplankton C : N : P, suggests that limitation of marine phytoplankton is poised closely between N- and P-limitation. When phosphate is exhausted, however, organisms may utilize more complex forms by converting them to orthophosphate via enzymatic and microbiological reactions (e.g., Karl and Bjorkman, 2000; Sections 8.13.3.3.3 and 8.13.3.3.4); the same is true for nitrogen. Thus, the simple, though elegant formulation of Redfield is now believed to be substantially more complex, given the role of dissolved organic nutrients in supporting marine biological productivity. In the open ocean most phosphorus associated with biogenic particles is recycled within the upper water column. Efficient stripping of phosphate from surface waters by photosynthesis combined with buildup at depth due to respiration of biogenic particles results in the classic oceanic dissolved nutrient profile. The progressive accumulation of respiration-derived phosphate at depth along the deep-water circulation trajectory results in higher phosphate concentrations in Pacific Ocean deep waters at the end of the trajectory than in the North Atlantic where deep water originates (Figure 1).

The sole means of phosphorus removal from the oceans is burial with marine sediments

(Froelich *et al.*, 1982; Ruttenberg, 1993; Delaney, 1998). The P-flux to shelf and slope sediments is larger than the P-flux to the deep sea (Table 2) for several reasons. Coastal waters receive continentally derived nutrients via rivers (including P, N, Si, Fe), which stimulate high rates of primary productivity relative to the deep sea, and result in a higher flux of organic matter to continental margin sediments. Organic matter is an important, perhaps primary carrier of phosphorus to marine sediments. Due to the shorter water column in coastal waters, less respiration occurs prior to deposition. The larger flux of marine organic-P to margin sediments is accompanied by a larger direct terrigenous flux of particulate phosphorus (organic and inorganic), and higher sedimentation rates overall (e.g., Filippelli, 1997a,b). These factors combine to enhance retention of sedimentary phosphorus. During high sea-level stands the sedimentary phosphorus reservoir on continental margins expands, increasing the P-removal flux and therefore shortening the oceanic-P residence time (Ruttenberg, 1993).

Terrigenous-dominated shelf and slope (hemipelagic) sediments and abyssal (pelagic) sediments have distinct P-distributions. While both are dominated by authigenic Ca–P (P_{auth}: mostly authigenic apatite), this reservoir comprises a larger fraction of total phosphorus in pelagic sediments. The remaining phosphorus in hemipelagic sediments is partitioned between ferric iron bound-P (P_{Fe}: mostly oxyhydroxides), detrital apatite (P_{detr}), and organic-P (P_{org}); in pelagic sediments detrital apatite is unimportant, most likely due to limited transport of this heavy mineral beyond the near-shore. Certain coastal environments characterized by extremely high, upwelling-driven biological productivity and low terrigenous input are enriched in authigenic apatite; these are proto-phosphorite deposits (see Section 8.13.3.3.2 for further discussion). A unique process contributing to the pelagic sedimentary Fe–P reservoir is sorptive removal of phosphate onto ferric oxyhydroxides in mid-ocean ridge (MOR) hydrothermal systems (e.g., Wheat *et al.*, 1996; see Section 8.13.3.3.2).

Mobilization of sedimentary phosphorus by microbial activity during diagenesis causes dissolved phosphate buildup in sediment pore waters, promoting benthic efflux of phosphate to bottom waters or incorporation in secondary authigenic minerals. The combined benthic flux from coastal (sF_{cbf}) and abyssal (sF_{abf}) sediments is estimated to exceed the total riverine-P flux ($F_{24(d+p)}$) to the ocean. Reprecipitation of diagenetically mobilized phosphorus in secondary phases significantly enhances phosphorus burial efficiency, impeding return of phosphate to the water column (see Section 8.13.3.3.2). Both processes impact the marine phosphorus cycle by affecting the primary productivity potential of surface waters.

8.13.3 PHOSPHORUS BIOGEOCHEMISTRY AND CYCLING: CURRENT RESEARCH

Many compelling lines of research are being pursued that seek to understand the biogeochemical controls on cycling of the essential nutrient, phosphorus, in terrestrial and aquatic systems; the research field is broad and deep. Limited space dictates that the summary of research presented in this chapter cannot be comprehensive. The more lengthy section on marine systems reflects the research bias of the author, and is not meant to convey any hierarchy of importance of marine research over terrestrial soil or freshwater studies. In fact, it can be argued from the standpoint of societal relevance that studies on terrestrial systems have higher significance, because anthropogenic impacts on terrestrial environments historically have been more severe. This is unfortunately changing, however, as the impact of humans on the terrestrial environment is bleeding over into the coastal ocean (e.g., Mackenzie *et al.*, 2002; Rabalais *et al.*, 1996; and references therein).

8.13.3.1 Phosphorus Cycling in Terrestrial Ecosystems and Soils

The process of sorptive binding of phosphate by iron (Fe)- and aluminum (Al)-oxide and oxyhydroxide phases is of enormous importance in terrestrial ecosystems, and has been extensively studied in soil science. Its importance lies in the fact that phosphorus can be a limiting nutrient in terrestrial ecosystems, and sorptive removal of natural or fertilizer phosphorus can impact the health and production level of crops and forests (e.g., Barrow, 1983; Guzman *et al.*, 1994; Frossard *et al.*, 1995). Unlike the Al-oxides/oxyhydroxides, Fe-oxides/oxyhydroxies are subject to reductive dissolution under anoxic conditions, and thus redox conditions can play an important role in soil P-bioavailability (e.g., Miller *et al.*, 2001). Fe- and Al-complexes with organic matter also can be important sinks for phosphate, as can clays, but the latter are less efficient scavengers of phosphate than the oxyhydroxides. Mineralogy and morphology of oxyhydroxides also exert control on relative efficiency of P-sorption; e.g., goethite is a more efficient substrate for P-sorption than is hematite (Torrent *et al.*, 1990; Filho and Torrent, 1993). Dissolved organic phosphorus compounds can also be sorptively removed from soil solution by various solid-phase components, including clays, oxides, and oxyhydroxides (see Frossard *et al.* (1995) for a review of soil reactions involving phosphorus).

One unique ecosystem that has yielded an unparalleled natural laboratory for the study of soil and terrestrial ecosystem development in general, and the evolution of phosphorus and other nutrient bioavailability in particular, is the Hawaiian Island chain in the central subtropical Pacific Ocean (Vitousek *et al.*, 1997). These volcanic islands all have a soil substrate of basaltic rock with an essentially identical initial chemistry, are similar in climate, and are arrayed along an age transect from young, active volcanoes in the southeast to the oldest islands in the northwest. This system thus offers the opportunity for contemporaneous study of soil and ecosystem development processes that have occurred on thousand-year timescales, by comparing soil chemistry, nutrient availability, and ecosystem development on each of the islands, which range in age from 0.3 kyr to 4,100 kyr (Vitousek *et al.*, 1997).

During chemical weathering of bedrock and soil formation, rock-derived plant nutrients such as phosphorus, calcium, magnesium, and potassium are leached from parent rock and soil and are eventually lost from the system, or in the case of phosphorus, residual phosphorus is both lost and converted to unavailable forms (Figure 2), resulting ultimately in phosphorus limitation of the ecosystem. In their study of the Hawaiian Islands, Chadwick *et al.* (1999) validate this classical model of soil development, and determine that phosphorus acts as the "master regulator" of biological productivity in the most weathered soils, in agreement with existing conceptual models for terrestrial biogeochemistry. They also illuminate interesting variations on this classical theme. For example, as the soil evolved from newly formed to extensively aged substrate, nutrient limitation of the ecosystem evolved as well, from a system initially co-limited by phosphorus and nitrogen (when rock-derived nutrients calcium, magnesium, and potassium are plentiful, there has not been adequate time for soil accumulation of atmospherically derived nitrogen, and phosphorus is in bio-unavailable forms), to intermediate soils where cation availability is lower but nitrogen and phosphorus are more plentiful, and finally to highly weathered soils where cations and P-bioavailability are low but there is plentiful nitrogen. By evaluating nutrient sources to these Hawaiian soils, Chadwick *et al.* (1999) determine that in end-stage soils, atmospheric sources of cations from seasalt aerosols and phosphorus from atmospheric dust deposition are the dominant nutrient sources. By the oldest site on the soil chronosequence, phosphorus provided by Asian dust is substantially larger than that provided by the parent rock and, in the absence of this distal source of phosphorus, these ecosystems would be far more severely P-limited (Chadwick *et al.*, 1999).

8.13.3.2 Phosphorus Cycling in Terrestrial Aquatic Systems: Lakes, Rivers and Estuaries

8.13.3.2.1 Biogeochemistry and cycling of phosphorus in lakes

Owing to their finite boundaries, lakes provide a more tractable venue as natural laboratories for mechanistic studies than do the oceans, and have therefore been the site of many elegant and informative studies on mechanisms of phosphorus cycling in aquatic systems (e.g., Schindler, 1970; Hecky and Kilham, 1988). Several concepts and processes of current intense interest in oceanography were first defined and examined in lakes, including (i) redox-driven coupled Fe–P cycling and benthic phosphorus efflux from sediments (Einsele, 1936a,b; Mortimer, 1941, 1942; and more recently, Böstrom *et al.*, 1988); (ii) the potential for direct biological control of benthic phosphorus efflux or uptake driven by shifts in redox conditions (Carlton and Wetzel, 1988; Gächter and Meyer, 1993; Gunnars and Blomqvist, 1997); and (iii) critical assessment of analytically defined "orthophosphate" (also known as soluble reactive phosphorus, or SRP) concentrations, as opposed to "true orthophosphate" concentrations (Rigler, 1966, 1968; and more recently Hudson *et al.*, 2000; see below). Other aspects P-biogeochemistry of keen interest in current oceanographic research have previously been extensively studied in lakes, including (i) definitive demonstration of phosphorus limitation of photosynthetic primary productivity (Schindler, 1970; Hecky and Kilham, 1988; Tarapchak and Nawelajko, 1986; however, see review by Fisher *et al.* (1995) summarizing studies of N-limited, or nitrogen and phosphorus co-limited lakes), (ii) bioavailability and composition of dissolved organic phosphorus (DOP) (Herbes *et al.*, 1975; Lean and Nawelajko, 1976; and more recently Nanny and Minear, 1997); (iii) the development and use of sequential extraction methods for sedimentary P-speciation (e.g., Williams *et al.*, 1976; Böstrom and Petterson, 1982); and (iv) phosphate mineral authigenesis (e.g., Emerson, 1976; Emerson and Widmer, 1978; Williams *et al.*, 1976). Research on a number of these topics is summarized in the remainder of this section.

There is disagreement about the relative importance of an indirect versus a direct microbial role in benthic phosphate flux as triggered by changes in redox state of bottom waters and sediments. The classical view is that microbial reduction and re-oxidation/re-precipitation of ferric oxyhydroxides controls P-release versus P-retention by sediments by virtue of the high sorption affinity of ferric oxyhydroxides for phosphate (Einsele, 1936, 1938; Mortimer, 1941, 1942); an indirect microbial effect on benthic P-flux. While this mechanism clearly is important (e.g., Gunnars and Blomqvist, 1997), it has been shown that bacteria in surficial sediments directly take up and release phosphate in response to changes in redox state (Carlton and Wetzel, 1988; Gächter and Meyer, 1993), presumably due to redox-triggered changes in physiology. The relative importance of these two pathways for phosphorus cycling at the sediment–water interface is unclear. Further, whether the relative importance of these two mechanisms is different in lacustrine versus marine systems (Gunnars and Blomqvist, 1997) is an important unresolved question.

In an effort to systematize differences in the absolute magnitude of benthic phosphate efflux in freshwater versus marine systems, Caraco *et al.* (1989) argue that more efficient benthic P-release occurs in lake relative to marine sediments as a direct consequence of the presence of higher sulfate in seawater, and that redox conditions exert secondary control. This argument is overly simplistic, however, because redox conditions control production of sulfide from sulfate, and it is the removal of ferrous iron from solution into insoluble ferrous sulfides that decouples the iron and phosphorus cycles (e.g., Golterman, 1995a,b,c; Rozen *et al.*, 2002). Thus, the presence of sulfate is a necessary but not sufficient criterion to account for differences in benthic P-cycling in marine versus freshwater systems; redox conditions are an equally crucial factor.

The importance of redox effects on coupled iron–phosphorus cycling in freshwater systems has been the subject of study in applied environmental science, where phosphate removal from eutrophic natural waters and wastewaters, by sorption onto Fe-oxyhydroxide phases, has been explored as a remediation measure. Phosphate also has a pronounced tendency to sorb onto Al-oxyhydroxides, and these phases have been used in remediation of phosphate overenriched aquatic systems, as well (e.g., Leckie and Stumm, 1970).

Recently, Hudson *et al.* (2000) reported the smallest phosphate concentrations to date for any aquatic system (27 pM) in a study using a new, steady-state bioassay technique for estimating orthophosphate concentrations. In phosphate-limited aquatic systems, accurate determination of orthophosphate is critical because it is the only form of phosphorus that can be directly assimilated by primary producers (e.g., Cembella *et al.*, 1984a,b). The standard phosphomolybdate blue method for phosphate determination falls short of this goal, and is widely thought to overestimate "true" orthophosphate concentrations due to reagent-promoted hydrolysis of DOP (e.g., Bentzen and Taylor, 1991; Rigler, 1966, 1968; however, see Monaghan and Ruttenberg (1999)). The findings of Hudson *et al.* (2000) suggest that "true" orthophosphate levels are 2–3 orders of magnitude lower than previously thought, equivalent to levels observed for micronutrients such as dissolved iron and zinc. Implications of this study include the possibility of diffusion-limited phosphate uptake, a situation favoring small organisms with large surface-to-volume ratios. This would further imply that bacteria may be net sinks for phosphate, rather than efficient remineralizers, as is the commonly held view, and may mean that phytoplankton have to adopt other competitive P-uptake strategies such as utilization of DOP (Karl, 2000). In another analytical development, Field and Sherrell (2003) have recently directly quantified the lowest ever measured concentrations of total dissolved phosphorus (and a number of other trace elements) in Lake Superior using a new ICP-MS method. Their method is unable to distinguish orthophosphate from other phosphorus forms, however, without application of prior separation techniques.

Alkaline phosphatase (APase) has been widely used in lake studies to demonstrate physiological phosphate stress, or P-limitation, of the bulk lake phytoplankton community (e.g., Healey and Hendzel, 1980; Cembella *et al.*, 1984a,b). A recent development in phytoplankton nutrient physiology research involves the use of a new, cell-specific labeling method for monitoring physiological phosphate-stress in phytoplankton and bacteria at the single-cell level (e.g., Carlsson and Caron, 2001; Rengefors *et al.*, 2001, 2003). In this method, when a phosphate group is cleaved from the ELF-97® (Molecular Probes) fluorescent substrate, the remaining molecule precipitates near the site of enzyme activity, thus fluorescently tagging cells that are expressing APase. The importance of this new method is that it monitors phytoplankton physiology at the single-cell level permitting, for example, the determination of differential phosphate stress among taxa, and even within taxa under the same environmental conditions (e.g., Rengefors *et al.*, 2003). This sort of cell-specific resolution is not possible with bulk APase assays, and the insights into phytoplankton nutrition made possible by the new ELF method have already revealed, and promise to continue to reveal previously inaccessible information on the nutrient physiological ecology of aquatic systems.

In summary, current research on P-cycling in lakes illustrates well that continued development and improvement of analytical methods is critical to addressing many long-standing unresolved questions about biogeochemical P-cycling. This is true not just for lakes, but for other Earth surface environments, as well.

8.13.3.2.2 Biogeochemistry and cycling of phosphorus in rivers and estuaries

Rivers and streams are the major conduits of phosphorus transfer to the oceans and to many lakes (groundwaters are likely an important but unquantified dissolved phosphorus transport medium). Rivers themselves function as ecosystems, and biogeochemical processes that occur during riverine transport can modify the form of phosphorus *en route*, with consequences for its chemical reactivity and biological availability once it reaches the recipient water body. Because of its extreme particle reactivity, most phosphorus in rivers is associated with particulate matter, dominantly through sorption processes (Froelich, 1988). A substantial body of work has established that, in pristine rivers with high enough turbidity to minimize autotrophic biological phosphorus uptake, dissolved inorganic phosphorus (DIP) levels are set by sorption equilibrium with suspended sediments; this controlling mechanism is known as the phosphate buffer mechanism (see Froelich (1988) and Fox (1993) for a summary of research on this topic). Through controlled laboratory experiments monitoring P-uptake and release from synthetic ferric oxyhydroxides and natural riverine suspended sediments, under a range of riverine and estuarine conditions, Fox (1989, 1993) has more explicitly described the P-buffering mechanism as a thermodynamic equilibrium between DIP and a solid-solution of ferric phosphate–hydroxide in suspended colloids and sediments. This mechanism appears to be valid for a wide number of turbid rivers with low calcium levels, including the world's three largest rivers: the Amazon, Zaire, and Orinoco (Fox, 1993), suggesting a globally significant mechanism. In calcium-rich rivers such as the Mississippi, in contrast, DIP solubility appears to be controlled by equilibrium with a calcium phosphate mineral (Fox *et al.*, 1985, 1987). The impetus for investigating the P-buffering phenomenon has been the recognition that (i) phosphate buffering by suspended river sediments can maintain immediately bioavailable DIP at near-constant levels in rivers, streams, and estuaries; (ii) suspended sediments can act as a large source of potentially bioavailable phosphorus; and (iii) P-sorption onto soil and suspended sediments can effectively sequester pollutive-P (e.g., derived from fertilizer, waste water, etc.) in forms that are not immediately bioavailable, thereby reducing the effect of excess P-loading into terrestrial and coastal aquatic ecosystems.

The first comprehensive, systematic analysis of phosphorus transport by the world's rivers was compiled by Meybeck (1982). In this compilation, DIP, dissolved organic phosphorus (DOP), and particulate inorganic and organic phosphorus forms are separately reported for a wide range of river systems, recognizing that these different forms have different reactivities both geochemically and biologically. More recent global riverine-P flux estimates include Froelich *et al.* (1982), GESAMP (1987), Howarth *et al.* (1995), Colman and Holland (2000) and Smith *et al.* (2003). These compilations usually quantify phosphate concentrations and fluxes separately for pre-agricultural (pristine, or natural) and modern rivers.

An important use of global river flux estimates is in construction of element budgets for the oceans. Rivers are by far the dominant phosphorus source to the ocean (Table 2), and the oceanic phosphorus budget has been formulated by balancing riverine inputs against phosphorus burial with sediments by numerous groups (e.g., Froelich *et al.*, 1982; Colman and Holland, 2000; Compton *et al.*, 2000; Table 3, this chapter). An important factor in evaluating riverine P-input to the ocean, from the standpoint of accurately quantifying bioavailable-P input and balancing the marine phosphorus budget, is that some fraction of river suspended sediment releases its phosphorus to seawater. The magnitude of this "releasable-P" has been estimated as anywhere from 25% to 45% of the riverine suspended sediment flux (see Table 2, note (f), and Table 3). Detailed P-speciation studies in two major world rivers, the Amazon (Berner and Rao, 1994) and the Mississippi (Ruttenberg and Canfield, 1994), indicate that the phases responsible for riverine suspended sediment P-release in the coastal ocean are ferric oxyhydroxide associated phosphorus (see also Chase and Sayles, 1980) and particulate organic phosphorus. Important unanswered questions remain, including:

(i) To what extent are the Mississippi and Amazon representative of other rivers regarding the magnitude of P-release? Although these are globally significant rivers in terms of dissolved inputs, they do not have the high sediment yields of rivers in mountainous tropical regions (Milliman and Syvitski, 1992), for which there is scant phosphorus data.

(ii) To what extent are the Mississippi and Amazon representative of other rivers regarding the phases responsible for P-release? For example, within the Chesapeake Bay estuary, organic phosphorus does not appear to be an important source of P-release (Conley *et al.*, 1995). Until P-speciation data exists for more coupled river/shelf systems, it is not possible to evaluate the relative importance of different forms of particulate-P as sources of bioavailable-P to the ocean.

(iii) What is the locus of and mechanisms for P-release from these (or other) phases at the continent-ocean interface? In particular, what are the relative importance of desorption from suspended sediments in the water column versus diagenetic remobilization in and release from benthic sediments?

(iv) At what point along the riverine/estuarine/marine salinity trajectory does P-release occur?

As the interface between rivers and the coastal ocean, estuaries are important sites of biogeochemical modification of phosphorus, as noted above (see also Section 8.13.2.2). Although not strictly terrestrial, the geography of many estuaries is often such that it is difficult to draw a clear boundary delineating where the terrestrial zone ends, and the oceanic zone begins. The nature of estuarine P-transformations has been a topic of interest since the 1950s, due to the importance of riverine transport in the global phosphorus cycle, and more recently because of recognition that human activity impacts the health of coastal zones. Recently published studies focused on resolving possible mechanisms of estuarine P-transformations, their locus of occurrence, and projected impacts on the adjacent and/or global ocean include studies on the Amazon shelf (DeMaster and Aller, 2001), Chesapeake Bay (Conley *et al.*, 1995); the northern Gulf of Mexico (Shiller, 1993; Rabalais *et al.*, 1996); and the estuaries fringing the North Atlantic Ocean (Nixon *et al.*, 1996; Galloway *et al.*, 1996; Howarth *et al.*, 1996); these recently published studies are a rich source of references to earlier work.

Elevated input of phosphorus and other nutrients via rivers to lakes and the coastal ocean has resulted in "cultural eutrophication" in many water bodies, a process that can lead to excessive accumulation of autotrophic (e.g., algal, macrophyte) biomass and ecosystem shifts to undesirable algal species. When excess algal biomass accumulates in bottom sediments, it stimulates heterotrophic activity that consumes oxygen, sometimes leading to development of anoxic sediments and bottom waters, with adverse effects for higher trophic levels (see discussions in Fisher *et al.*, 1995; Melack, 1995; Howarth *et al.*, 1995; Caraco, 1995; Nixon, 1995; Rabalais *et al.*, 1996; Richardson and Jørgensen, 1996). Recognizing the vulnerability of natural water systems to anthropogenic nutrient over-enrichment, and that integration of high-quality scientific data is required to formulate successful remediation strategies, a variety of national and international groups have banded together at various times to synthesize data from regional and/or global studies to provide a sound basis for management of aquatic resources (e.g., see summary of such efforts in Fisher *et al.*, 1995). One recent example is "The Land Ocean Interactions in the Coastal Zone (LOICZ) Project," a core project of the International Geosphere–Biosphere Program (IGBP), whose stated goals include a determination of nutrient fluxes between land and sea (with emphasis on carbon, phosphorus, and nitrogen), and an assessment of how coastal systems respond to varying terrestrial inputs of nutrients (Gordon *et al.*, 1996; Smith, 2001).

8.13.3.3 Biogeochemistry and Cycling of Phosphorus in the Modern Ocean

Areas of active research on the modern oceanic phosphorus cycle range from inquiries into the molecular composition of the dissolved organic phosphorus pool, to determinations of the global scale and distribution of authigenic phosphate minerals, to the perpetually plaguing question of whether phosphorus limits marine primary productivity and, if so, on what time and space scales. The next series of sections summarize first, an historic overview of the way in which our understanding of the oceanic phosphorus cycle has evolved over the past two decades. Next, recent and ongoing work on some of the most intriguing questions about the character and functioning of the oceanic phosphorus cycle are summarized.

8.13.3.3.1 Historical perspective: the marine phosphorus budget

The current vision of the marine phosphorus cycle differs substantially from that which prevailed as of the early 1980s. These changes have been driven by methodological developments, which have made new observations possible, as well as by challenges made to accepted paradigms as new studies have worked to reconcile new data with old, and sometimes entrenched, views.

The first comprehensive global marine phosphorus budget took the approach of quantifying phosphorus removal from the ocean by characterizing P burial rates in different depositional environments (Froelich *et al.*, 1982). These researchers took the important approach of separately quantifying P-burial rates for different depositional environments, and different sediment types, recognizing that the processes dominating in different environments would be distinct, and therefore subject to different controlling factors. This early incarnation of the marine phosphorus budget focused almost exclusively on the deep sea; the only ocean margin data included were from areas characterized by upwelling circulation and phosphorite formation, as the latter were recognized as hot-spots for phosphorus burial.

Notably, upwelling margins represent only 2% of the total marginal area of the ocean (Ganeshram *et al.*, 2002; Berner, 1982).

The emphasis on the pelagic realm was the result of the prevailing bias at the time, in both research and funding, toward blue water oceanography, whereas continental margins received far less research attention. The pelagic budget described by Froelich *et al.* (1982) was viewed as self-consistent, because the P-removal flux with sediments was balanced by the dissolved phosphate input flux with rivers, with a residence time of 80 kyr, similar to the canonical value of 100 kyr, the accepted residence time estimate at that time (Broecker and Peng, 1982).

Adopting the budgetary model of Froelich *et al.* (1982), Ruttenberg (1993) presented a revised budget, this time taking into account the importance of continental margins, *in toto*, in the global marine P-cycle. Inclusion of the margins was motivated by a growing recognition of continental margins as extremely important depocenters for organic matter (Berner, 1982), and the knowledge that organic matter is one of the most, if not the most important vector for delivery of phosphorus to the seabed. An additional motivator for inclusion of continental margins was the recognition that the early diagenetic regime in organic-matter-rich margin sediments of all types, and not exclusively those underlying upwelling regimes, make these depositional environments likely places for authigenic carbonate fluorapatite (CFA) formation.

Identification of CFA in nonupwelling environments required the use of new, and indirect methods of detection, because dilution of this authigenic phase by the large burden of terrigenous material in many continental margin settings makes identification by direct methods, such as XRD, impossible. Ruttenberg (1992) adapted existing methods for sequential extraction of phosphorus and trace metals from a variety of disciplines, including soil science, limnology, and marine geochemistry, resulting in the SEDEX method, which is able to separately quantify CFA. Application of a coupled SEDEX–pore water approach to continental margin sediments revealed that formation of authigenic CFA is not restricted to margin environments characterized by upwelling (Ruttenberg and Berner, 1993, and others, see Section 8.13.3.3.2). Inclusion of continental margins in the global marine phosphorus budget increased the P-removal flux by 2–6 times (depending upon which burial flux estimates are used), due to high rates of burial in ocean margin sediments of organic phosphorus, authigenic CFA, and iron-bound phosphorus, the latter particularly important in deltaic marginal environments (Table 3; Ruttenberg, 1993).

More recent studies have further refined the estimated burial fluxes of phosphorus in the global marine phosphorus budget (Table 4), including better estimates of P-removal with iron oxyhydroxides at MORs (Wheat *et al.*, 1996), inclusion of burial fluxes for authigenic rare earth element- and thorium-phosphates (Rasmussen, 2000), phosphates buried in association with hydroxyapatite from fish bones, scales, and teeth (using a modified SEDEX method to separately quantify hydroxyapatite as distinct from CFA: Schenau *et al.*, 2000); and taking into account the return benthic flux of phosphate out of sediments (Colman and Holland, 2000). These and other studies (Compton *et al.*, 2000; Filippelli and Delaney, 1996) concur with Ruttenberg (1993) that the earlier pelagic-focused budget of Froelich *et al.* (1982) underestimated global ocean P-burial fluxes, and therefore overestimated P-residence times (Table 3).

The higher P-burial rate estimates that these diverse groups have converged upon set up an imbalance in the global marine phosphorus budget when contrasted with the riverine dissolved P-input rate. This imbalance can be reconciled if some fraction of the phosphorus associated with riverine particulate matter is solubilized upon entry into the ocean (Table 2). Estimates of the quantity of phosphorus that might be liberated upon delivery from rivers to the oceans, or "releasable-P," have been made in several studies (Ruttenberg, 1993; Colman and Holland, 2000; Ruttenberg and Canfield, 1994; Berner and Rao, 1994; Compton *et al.*, 2000; Howarth *et al.*, 1995; Ramirez and Rose, 1992; Froelich, 1988). Estimates made on the basis of P-inputs that include this "releasable" riverine particulate-P yield residence times that fall within the range of residence time estimates derived from P-burial fluxes (Table 3). Despite the large uncertainties associated with these numbers, as evidenced by the maximum and minimum values derived from both input and removal fluxes (Table 3), these updated residence times are all significantly shorter than the canonical value of 100 kyr. Residence times on the order of 10–17 kyr make feasible a role for phosphorus in perturbations of the ocean–atmosphere CO_2 reservoir on the timescale of glacial–interglacial climate change.

8.13.3.3.2 *Diagenesis and burial of phosphorus in marine sediments*

The sources of particulate phosphorus to the seabed include detrital inorganic and organic material transported by rivers to the ocean, biogenic material produced in the marine water column that sinks to the seabed, and atmospheric

Table 4 Geochemical partitioning of reactive-P burial fluxes.

Phosphorus reservoir	*Phosphorus burial flux* (10^{10} mol yr^{-1})	*Method of determination*
Organic-P	1.1	Delaney, 1998 (calculated from C_{org} burial rate and P_{org}/C_{org} ratios)
	1.5	Froelich *et al.*, 1982 (calculated from C_{org} burial rate and P_{org}/C_{org} ratios)
	1.6	Froelich, 1984 (calculated from C_{org} burial rate and P_{org}/C_{org} ratios)
	2.0	Mach *et al.*, 1987 (calculated from C_{org} burial rate and P_{org}/C_{org} ratios)
	4.1	Ruttenberg, 1993 (measured P_{org} via SEDEX method and estimated sediment delivery flux to the oceans)
		Comments: Variability in first four estimates results from different C_{org} burial fluxes and different P_{org}/C_{org} ratios chosen for estimated P_{org} burial fluxes (see Delaney, 1998 for summary).
Authigenic CFA, Biogenic HAP, $CaCO_3$-P	0.4	Froelich *et al.*, 1982, 1988 (based on diagenetic modeling of sediment pore water)
	8.0	Filippelli and Delaney, 1996 (assuming 80–90% of P_{total} measured via SEDEX method is P_{CFA} combined with P_{total} accumulation rates)
	9.1 (2.2)	Ruttenberg, 1993 (measured P_{CFA} via SEDEX method and estimated sediment delivery flux to the oceans; value in parentheses is minimum estimate, see comments)
		Comments: Froelich *et al.* estimates are for upwelling, classical phosphogenic provinces only. Ruttenberg's (1993) maximum estimate assumes all phosphorus measured in step III of the SEDEX method is truly authigenic. The minimum estimate assumes only the portion observed to increase above the concentration in the shallowest sediment interval is truly authigenic, accounting for the possibility that there may be a nonauthigenic background component to this reservoir (see Ruttenberg and Berner (1993) for an expanded discussion). HAP = hydroxyapatite (fish bones, teeth, scales).
Ferric iron-bound P:		
Hydrothermal MOR Processes:		
High-temperature ridge axis	0.01	Wheat *et al.* (1996)
Low-temperature ridge flank	0.65	Wheat *et al.* (1996) basalt seawater reactions during convective circulation of seawater in sediments and crust of flanks of MORs
Hydrothermal plume scavenging	0.77	Wheat *et al.* (1996); Feeley *et al.* (1994)
Total hydrothermal	1.43	
Non-hydrothermal scavenging onto Fe-oxyhydroxides:	1.5	Froelich *et al.*, 1982 (in the Froelich *et al.* (1982) study, this quantity was attributed to burial with $CaCO_3$, determined by dissolving foram and coccolith tests from deep-sea cores. The work of Sherwood *et al.* (1987) and Palmer (1985) demonstrated conclusively that phosphorus associated with $CaCO_3$ tests in the deep sea is nearly all associated with Fe-oxyhydroxide coatings on the tests. This quantity is therefore more accurately attributable to phosphorus burial with reactive Fe-oxyhydroxide phases (see also Ruttenberg, 1993)
	4.0 (0.4)	Ruttenberg, 1993 (measured P_{Fe} via SEDEX method and estimated sediment delivery flux to the oceans; value in parentheses is minimum estimate, see comments)

(continued)

Table 4 (continued).

Phosphorus reservoir	*Phosphorus burial flux* (10^{10} mol yr^{-1})	*Method of determination*
		Comments: Ruttenberg's (1993) maximum estimate assumes all phosphorus measured in step II of the SEDEX method is truly reactive-P. The bulk of the phosphorus reported in Ruttenberg (1993) was observed in deltaic sediments, however, and the possibility exists that some portion of this reservoir is detrital, and therefore does not represent phosphorus removed from seawater. Thus, the minimum estimate assumes that all of the deltaic P_{Fe} is detrital (this is an extreme, and likely unrealistic minimum end-member)
Loosely sorbed P	1.3	Ruttenberg, 1993 (measured P_{exchg} via SEDEX method and estimated sediment delivery flux to the oceans)
Calcium carbonate	<0.009	Delaney, 1998 (calculated from Holocene $CaCO_3$ burial flux and maximum foraminiferal $CaCO_3$-P content from Palmer, 1985) *Comments:* Any phosphorus associated with $CaCO_3$ tests, not surface metal oxyhydroxide coatings, is included in the quantity measured by step III of the SEDEX method; this estimate serves as a measure of the fraction of the step III, or the burial flux of authigenic apatite + biogenic apatite + $CaCO_3$, that can be accounted for by burial of $CaCO_3$.
REE phosphates and aluminum-phosphates	6.56	Rasmussen, 2000 (determined for ancient marine sandstones via microscopy, where sedimentary textures are the primary evidence use to argue that they are authigenic, and therefore representative of a reactive-P burial flux. *Comments:* These phases have yet to be observed in modern marine sediments, and their role as reactive phosphorus sinks in the ocean has yet to be verified. Verification of authigenic REE–P and Al–P formation in the modern ocean would strengthen arguments for the authigenic nature of these phases in ancient sediments.

After Ruttenberg (1993), Delaney (1998), and Compton *et al.* (2000).

dust that becomes entrained with sinking particulate material and is thus transported to the seabed. Once at the sediment–water interface, refractory phosphorus phases, such as detrital apatites and other P-containing refractory crystalline minerals, are simply passively buried. Particulate phases that are reactive in the early diagenetic environment are subject to a number of biogeochemical processes that affect the extent to which they are retained in the sediment, and the form in which they are ultimately buried and become part of the sedimentary record. These processes include microbial breakdown of organic phosphorus and production of dissolved inorganic and organic phosphorus, uptake of phosphate via sorption, uptake of phosphate during formation of secondary phosphate minerals (the so-called authigenic minerals, because they form in place), and benthic efflux of dissolved phosphorus from sediments to bottom waters (Figure 3). Studies focusing on these and a number of other important early diagenetic processes are discussed in subsequent subsections.

Because phosphorus is an essential nutrient, processes controlling the extent of P-retention versus P-release from sediments are important regulators of biological cycling and thus impact the workings of the global carbon cycle. This effect can be manifest on short timescales, where reactive particulate phases release phosphorus after deposition that is then returned to the water column where it is available for biological uptake. On longer, geologic timescales, the ultimate form in which phosphorus is buried will affect its susceptibility to weathering once it is uplifted into the weathering regime, and therefore its propensity for being rendered bioavailable in the next tectonic cycle. These links between early diagenesis of phosphorus, global ocean biological productivity, and the global carbon cycle, have been the motivation for much recent work on the biogeochemistry of phosphorus in modern marine sediments.

Sedimentary organic phosphorus: composition and reactivity. Organic phosphorus (P_{org}) is the primary vector of phosphorus delivery to marine sediments, and constitutes an important fraction (~25–30%) of total phosphorus buried in marine sediments (Froelich *et al.*, 1982; Ruttenberg, 1993; Colman and Holland, 2000). Despite its importance to the total marine sedimentary phosphorus

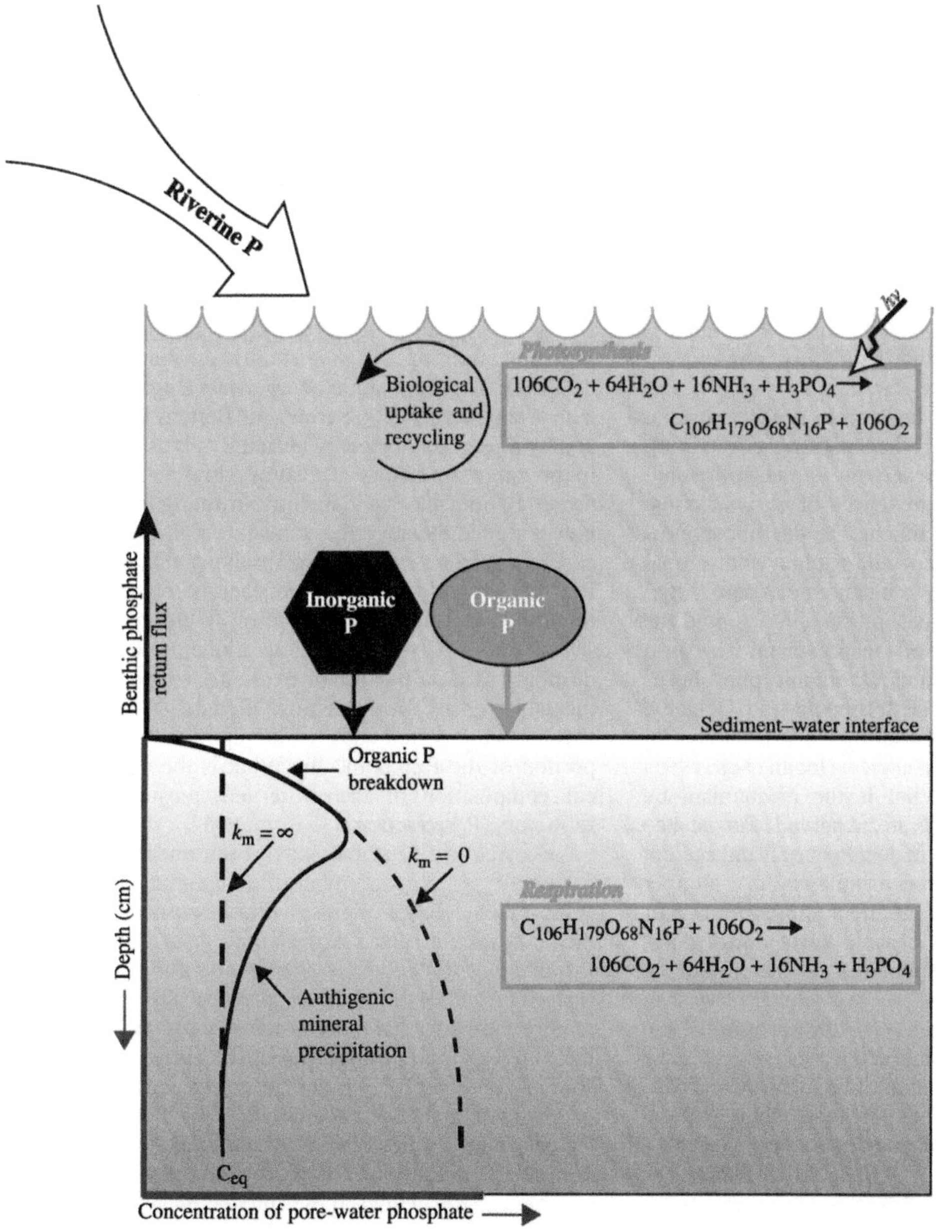

Figure 3 Processes important during early diagenetic transformations of phosphorus in marine sediments are illustrated. Sources of phosphorus to the sediment–water interface include allocthonous river-borne inorganic and organic phosphorus, and autocthonous, biogenic phosphorus formed through photosynthesis and subsequent food-web processes. Once delivered to the sediment–water interface, organic phosphorus is subject to breakdown via microbial respiration, a process often called "mineralization" because it transforms organic matter into its inorganic, "mineral" constituents, such as phosphate, nitrate, and carbon dioxide (dissolved organic phosphorus, nitrogen, and carbon, are also products of respiration, although these products are not shown). A representative equation for oxygenic respiration is given as an example, but a well-documented sequence of electron acceptors are utilized by microbial communities, typically in order of decreasing metabolic energy yield, to affect respiration (nitrate, oxides of iron and manganese, sulfate; however, for exceptions to this strict hierarchy of oxidants, see Canfield, 1993; Aller, 1994; Hulth *et al.*, 1999; Anschutz *et al.*, 2000). All of these respiration reactions result in a buildup of phosphate and other metabolites in pore waters. Schematic of pore-water profiles for the general situation of steady-state phosphate diagenesis, with organic matter as the sole source of phosphate to pore waters, is after Berner (1980). (Another important source of phosphate to pore waters, not depicted in this cartoon, is release of sorbed phosphate from host Fe-oxyhydrixides when these phases are buried into suboxic and anoxic zones within the sediment (see text for discussion).) Once released to pore waters, phosphate can escape from sediments via diffusional transport, resuspension, or irrigation by benthos. An important process for retention of pore-water phosphate within sediments is secondary authigenic mineral formation. The dashed profiles illustrate phosphate profile shapes encountered when

inventory, advances in understanding the compositional make-up of sedimentary organic phosphorus have been fairly limited, due in large part to the analytical difficulties associated with characterizing its molecular forms. The principal analytical challenge is that only a small fraction of sedimentary organic phosphorus can be separated into individual organic phosphorus compounds for identification and analysis; the bulk of the pool (in most cases well over 90%) is intimately associated with high-molecular-weight (HMW) bulk organic matter (Laarkamp, 2000), and is inaccessible to most existing analytical methodologies.

One of the most intriguing, and persistent, questions concerning the marine sedimentary organic phosphorus reservoir is the following: Given that the ultimate source of organic phosphorus to marine sediments is the phosphorus biochemicals contained within plant and animal tissues, and that these biochemicals are high-energy compounds that presumably should be labile after the death of the organism, how is it that significant quantities of organic phosphorus persist in deeply buried (Filippelli and Delaney, 1996; Delaney and Anderson, 1997; Tamburini *et al.*, 2003) and even in ancient (Ingall *et al.*, 1993) marine sediments? What is the mechanism by which these compounds are preserved? Persistence of organic phosphorus in deeply buried and ancient marine sediments has been explained in a number of ways, including, (i) preferential preservation of inherently refractory organic phosphorus compounds such as phosphonates or inositol phosphates (Froelich *et al.*, 1982; Ingall and Van Cappellen, 1990; Suzumura and Kamatami, 1995), and (ii) presence of bacterial biomass or derivative compounds (Froelich *et al.*, 1982; Ingall and Van Cappellen, 1990; Ruttenberg and Goñi, 1997a). Without insight into the composition of sedimentary P_{org}, it is not possible to conclusively determine controls on its relative reactivity or lability during early diagenesis, and thus to understand preservation mechanisms.

Most information about sedimentary organic phosphorus derives from studies focusing on the size of the total P_{org} pool and bulk organic C : P ratios in sediments (Filipek and Owen, 1981; Froelich *et al.*, 1982; Krom and Berner, 1981; Ingall and Van Cappellen, 1990; Ingall *et al.*, 1993; Morse and Cook, 1978; Reimers *et al.*, 1996; Ruttenberg, 1993; Ruttenberg and Goñi, 1997a,b; Anderson *et al.*, 2001) without examining distribution among specific P_{org} compounds or compound classes. The size of the bulk P_{org} pool in marine sediments typically decreases with depth, indicating partial mineralization during early diagenesis (Krom and Berner, 1981; Morse and Cook, 1978; Filippelli and Delaney, 1996; Reimers *et al.*, 1996; Ruttenberg and Berner, 1993; Slomp *et al.*, 1996a,b; Ruttenberg and Goñi, 1997a; Shenau *et al.*, 2000; Filippelli, 2001; Van der Zee *et al.*, 2002; see also Figure 9(a), this chapter). Deeper in the sediments, mineralization of P_{org} slows to undetectable levels. The initial rapid mineralization of P_{org} is usually attributed to destruction of more labile components (Ingall and Van Cappellen, 1990; Krom and Berner, 1981), and by inference the deeply buried P_{org} is assumed to be more refractory. In some slow sediment accumulation rate sites such as in the deep sea, however, sedimentary P_{org} concentration profiles can be invariant with depth, implying that even P_{org} at the sediment water interface is refractory toward microbial mineralization (Ruttenberg, 1990). Bulk P_{org} concentrations alone, however, do not provide a means for explicitly supporting these inferences. Key questions include: (i) What specific P_{org} compounds make up the "labile" portion of the P_{org} pool? (ii) What is the chemical composition of the preserved, presumably refractory, P_{org} fraction?

Less frequently, studies have been undertaken to quantify a limited number of compounds (e.g., ATP, DNA, phospholipids) within marine sediments (Crave *et al.*, 1986; Harvey *et al.*, 1986; White *et al.*, 1979). The relative rarity of the latter types of studies is in large part because they require substantially more work than studies of the bulk P_{org} pool. Furthermore, the aim of such studies is usually distinct from the goal of understanding the sedimentary P_{org} pool *in toto*. Rather, these studies typically seek to understand biomass distribution or microbial activity in sediments, for which these biochemicals may serve as proxies.

Suzumura and Kamatani (1995) pursued a study of the fate of inositol phosphates in sediments, another specific class of compounds that can be isolated and quantified, with the quite different aim of evaluating whether these terrestrial plant-derived P_{org} compounds might be refractory in the marine environment. Their results suggest that these compounds are minor constituents of total P_{org} in marine sediments, and that they are vulnerable to microbial breakdown during early

(i) organic matter breakdown is the dominant process, and there is no precipitation ($k_m = 0$: exponential increase of pore-water phosphate with increasing depth), and (ii) there is very rapid precipitation (k_m = infinity: vertical gradient at a concentration (C_{eq}) in equilibrium with the authigenic phase). The intermediate case is given in the solid curve, where a reversal of the initially exponentially increasing pore-water gradient is observed, indicating removal of phosphate from pore waters during phosphate mineral authigenesis.

diagenesis and do not persist to depth. Thus, these compounds do not provide an explanation for persistent preservation of P_{org} in marine sediments.

Although studies of specific P_{org} compounds such as these have expanded our understanding of the sedimentary P_{org} pool, their narrow focus on compounds which make up at most a few percent of the total P_{org} pool contribute little to our understanding of the forces driving bulk P_{org} trends in marine sediments.

Early work undertaken to examine the composition of the bulk sedimentary P_{org} pool separately quantified phosphorus associated with humic and fulvic acids in marine sediments (Nissenbaum, 1979). Results of this study suggested that phosphorus is preferentially mineralized during the diagenetic transition from fulvic to humic acid. However, because the distinction between organic and inorganic phosphorus in these fractions was not made, it is unclear whether the trends reported accurately reflect changes in P_{org}.

A decade after Nissenbaum's (1979) work, new advances in our understanding of the bulk sedimentary P_{org} pool began to be made with the application of phosphorus-31 nuclear magnetic resonance spectroscopy (^{31}P-NMR) to marine sediments. ^{31}P-NMR is currently the most promising tool for characterizing P_{org} in sediments. Application of ^{31}P-NMR to the insoluble "protokerogen" fraction of marine sediments has revealed the presence of phosphonates (Ingall *et al.*, 1990; Laarkamp, 2000; Ruttenberg and Laarkamp, 2000). Phosphonates were originally viewed as promising candidates for compounds that might make up the refractory sedimentary P_{org} pool because their structure (a direct carbon–phosphate bond) was thought to render them more stable than organic phosphates (Froelich *et al.*, 1982; Ingall *et al.*, 1990; Ingall and Van Cappellen, 1990). However, recent work using solution phase ^{31}P-NMR coupled with a new organic phosphorus sequential extraction method (Laarkamp, 2000) has shown that phosphonate esters are equally, if not more, vulnerable to microbial breakdown during early diagenesis than phosphate esters. Thus, the direct C–P bond in phosphonates does not appear to render these compounds more resistant to microbial respiration in marine sediments (Laarkamp, 2000; Ruttenberg and Laarkamp, 2000). The nature of the "refractory" organic phosphorus that escapes breakdown during early diagenesis, substantial quantities of which make it into the rock record, and can be quantified as P_{org} in ancient shales (Ingall *et al.*, 1993; Laarkamp, 2000), thus remains an open question.

Authigenic Carbonate Fluorapatite (CFA): Modern Phosphorites. Carbonate fluorapatite (CFA), or francolite, is the dominant phosphatic mineral in phosphorite deposits. Phosphorites are marine sedimentary deposits containing greater than 5 wt.% and up to 40 wt.% P_2O_5 (McKelvey, 1967; Riggs, 1979; Cook, 1984); a lower threshold of 15–20% P_2O_5 is commonly cited (Bentor, 1980; Jarvis *et al.*, 1994). These high concentrations are in contrast to most sedimentary rocks and sea-floor sediments, which contain less than 0.3 wt.% P_2O_5 (0.13 wt.% P) (Riggs, 1979). The high phosphorus concentrations of phosphorites place them in the category of economic ore deposits, and they are actively mined for P used predominantly in fertilizer (e.g., Fisher *et al.*, 1995; Melack, 1995; and see Section 8.13.2.2.1).

CFA is a substituted form of fluorapatite ($Ca_{10}(PO_4)_6F_2$), with a variety of cations substituting for calcium, and anions substituting for both phosphate and fluoride, respectively. Apatites have an extremely accommodating crystal lattice, and their chemical composition takes on the characteristics of the precipitating fluid (McConnell, 1973; McClellan, 1980; Kolodny, 1981; Jarvis *et al.*, 1994). Marine authigenic CFA, which forms within marine sediments, thus incorporates aspects of the chemistry of the interstitial pore fluids of the sediment within which it forms. There have been numerous general chemical formulae proposed to capture the possible permutations of the chemical composition of marine authigenic CFA (e.g., Jahnke, 1992). A fairly comprehensive general formula has been proposed by Jarvis *et al.* (1994): $(Ca_{10\text{-}a\text{-}b}\ Na_a\ Mg_b\ (PO_4)_{6\text{-}x}(CO_3)_{x\text{-}y\text{-}z}(CO_3{\cdot}F)_y(SO_4)_zF_2)$. The most important distinction between pure fluorapatite, typically formed by igneous processes, and CFA, is the presence of carbonate within the apatite crystal lattice. Disagreement exists about whether the carbonate substitutes for phosphate or for fluoride (LeGeros *et al.*, 1969; McConnell, 1973; McClellan, 1980; Kolodny, 1981), but regardless of its position within the lattice, its presence acts to distort the lattice and increase its solubility (LeGeros, 1965; LeGeros *et al.*, 1967; McConnell, 1973; McClellan, 1980; Jahnke, 1984). It is this difference in solubility that forms the basis for the ability to separately quantify CFA and detrital fluorapatite by the SEDEX method. Once out of the early diagenetic environment, with increasing age, and particularly upon metamorphism or uplift and exposure to subaerial weathering, CFA has a tendency to lose its marine-derived substituents and recrystallize to the thermodynamically more stable unsubstituted fluorapatite (Jarvis *et al.*, 1994; Lucas *et al.*, 1980; McArthur, 1980; McClellan, 1980).

Phosphorites are generally thought to be the result of postdepositional concentration of CFA from a primary sediment deposit, which contained more disseminated CFA, by secondary physical processes such as winnowing and reworking. The classical depositional environments in which these deposits form have two main characteristics: high marine organic matter (and therefore P_{org}) flux to sediments and low detrital input. Such environments are found in the eastern margins of the oceans, where wind-driven upwelling of nutrient-rich deep waters sustains a highly productive biological community in surface waters, which in turn provides a rich source of organic matter, and thus organically bound phosphorus, to underlying sediments. The continents adjacent to margins characterized by wind-driven upwelling tend to be arid, such that transport of continental detrital material to the adjacent ocean by runoff is relatively low, minimizing potential dilution of authigenic CFA within the sediment column. These two factors predispose the sediments to formation and concentration of CFA at the high levels necessary for formation of a phosphorite deposit (Baturin, 1983; Cook and McElhinny, 1979; Bentor, 1980; Kolodny, 1981; Burnett *et al.*, 1983; Föllmi, 1995a,b). Other factors have been cited as potentially important for phosphorite formation (Table 5), but there is not consensus on many of these.

Interestingly, although archaic phosphorites had been well documented in the geologic record and on the seafloor since the mid-1800s (Bentor, 1980; Föllmi, 1995a,b), until recently it was widely held that phosphorites were not forming in the modern ocean (Kolodny, 1969; Kolodny and Kaplan, 1970). This lent a certain mystique to these ancient deposits, in that it was unclear how the oceans of the past might have differed from the present day ocean, such that deposition of these hugely concentrated phosphorus deposits was favored. This paradox was de-mystified in the 1970s, with the application of uranium-series dating to seafloor phosphorites in sediments from the Peru upwelling zone, which demonstrated that contemporary phosphorite formation was indeed occurring (Baturin *et al.*, 1972; Veeh *et al.*, 1973; Burnett, 1977; Burnett *et al.*, 1980). Later detailed studies of pore water and solid phase chemistry provided corroborating evidence for contemporary formation of CFA in these sediments (Froelich *et al.*, 1988; Glenn and Arthur, 1988). Burnett *et al.* (2000) have pursued a number of radiochemical studies of phosphorite nodules and crusts in order to document rates and modes of precipitation, and find growth rates of 2–9 mm kyr^{-1} in response to downward diffusion of phosphate from pore-water phosphate maxima present just below the sediment–water interface. Postprecipitation winnowing causes re-exposure of the so-called phosphorite "proto-crusts" at the sediment surface, where erosion into the more common and widespread phosphatic hardgrounds, conglomerates, and nodules found in sediments of the Peru shelf occurs. Modern-day phosphorites are now also known to form on the Namibian shelf (Thompson *et al.*, 1984; Baturin, 2000, and references therein), in sediments adjacent to Baja California (Jahnke *et al.*, 1983; Schuffert *et al.*, 1994), on the Oman margin (Schenau and De Lange, 2000), and on the eastern Australian margin (Heggie *et al.*, 1990; O'Brien *et al.*, 1990).

In contemporary Peruvian, Baja Californian, and eastern Australian margin sediments, the confluence of pore-water gradients in phosphate and fluoride concentration with sedimentary phosphorite layers are consistent with active CFA formation, and careful work in these environments has provided insight into the process of modern-day phosphorite formation in these classical phosphorite environments. For example, Schuffert *et al.* (1994), working in sediments underlying the upwelling regime off the west coast of Baja California, Mexico, observed downward decreasing pore-water concentration gradients of both phosphate and fluoride, reflecting removal of these ions from pore water as they are incorporated into authigenic CFA (Figure 4). Coincidence of inflections in these pore-water gradients with the occurrence of XRD-identified CFA layers in the solid phase (stippled bands in Figure 4) is conclusive evidence for CFA formation in these sediments. In all but one core examined, the coupled pore-water and solid-phase data suggest contemporary formation of CFA in the uppermost one or two layers, while deeper layers reflect relict episodes of CFA formation, presumably that were active when these layers were located closer to the sediment–water interface (Schuffert *et al.*, 1994). As further confirmation of removal of phosphate from pore

Table 5 Summary of factors leading to phosphorite formation.

Factor	*Reference*
Large supply of particulate organic matter	1, 2, 3, 4, 5, 6
Warm temperatures	1, 2
Associated with interglacials, high sea level	1, 7
High salinity	2
Elevated pH	1, 2, 8
Low accumulation rate of inorganic material	2, 4
Associated with extinctions, major evolutionary events	3, 9
Associated with the boundaries of the oxygen minimum zone	1, 4
High pore water calcium:magnesium ratio	1, 10

After Jahnke *et al.* (1983).
References: (1) Burnett (1977); (2) Gulbrandsen (1969); (3) Piper and Codispoti (1975); (4) Manheim *et al.* (1975); (5) Baturin and Bezrukov (1979); (6) Van Cappellen and Berner (1988); (7) Riggs (1984); (8) Reimers *et al.* (1996); (9) Cook and Shergold (1984); (10) Van Cappellen and Berner (1991). See also Cook *et al.* (1990) for a succinct summary of models for phosphogenesis.

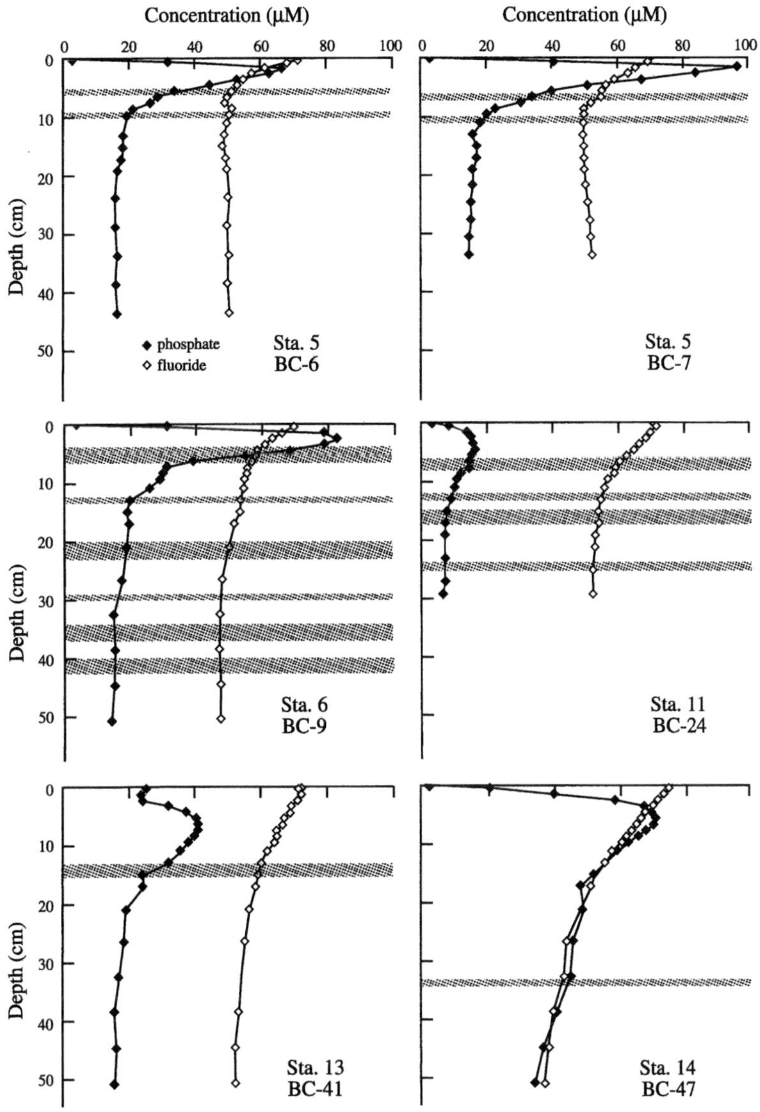

Figure 4 Pore-water profiles showing coupled removal of dissolved phosphate and fluoride from pore waters with depth in sediments, suggesting active growth of CFA in these Baja California sediments. Stippled bands indicate position and width of discrete phosphorite layers, as detected visually and confirmed by X-ray diffraction. Downward extent of concentration gradients indicates that CFA can precipitate simultaneously in two or more phosphorite layers (after Schuffert *et al.*, 1994).

water to form CFA, Schuffert *et al.* (1994) executed a stoichiometric nutrient regeneration model to predict pore-water phosphate profiles from pore-water ammonium and total alkalinity gradients, and contrasted these predicted profiles to the empirically observed phosphate profile (Figure 5). The results of this model reveal that the subsurface maximum in phosphate, which is unpredicted by the stoichiometric nutrient regeneration model, must be the result of input from some other process than "normal" organic matter decay, a finding highlighted for this region previously by Van Cappellen and Berner (1988). One likely explanation proposed by Schuffert *et al.* (1994) is that it is linked to the diagenetic redox cycling of iron oxyhydroxides, as has also been suggested in other studies (Shaffer, 1986; Heggie *et al.*, 1990; Sundby *et al.*, 1992; Jarvis *et al.*, 1994; Slomp *et al.*, 1996; see also Section 8.13.3.3.2). A schematic of the coupled cycling of iron and phosphorus, and the proposed link to CFA (or francolite) formation during early diagenesis, shows the sequence of events and their distribution relative to different redox zones within a sediment (Figure 6). Fluoride is also scavenged by *in situ* formed iron oxyhydroxides (Ruttenberg and Canfield, 1988), and thus the iron redox cycle in sediments acts as a concentrating mechanism for these CFA substituents which, when released to pore water upon reduction of

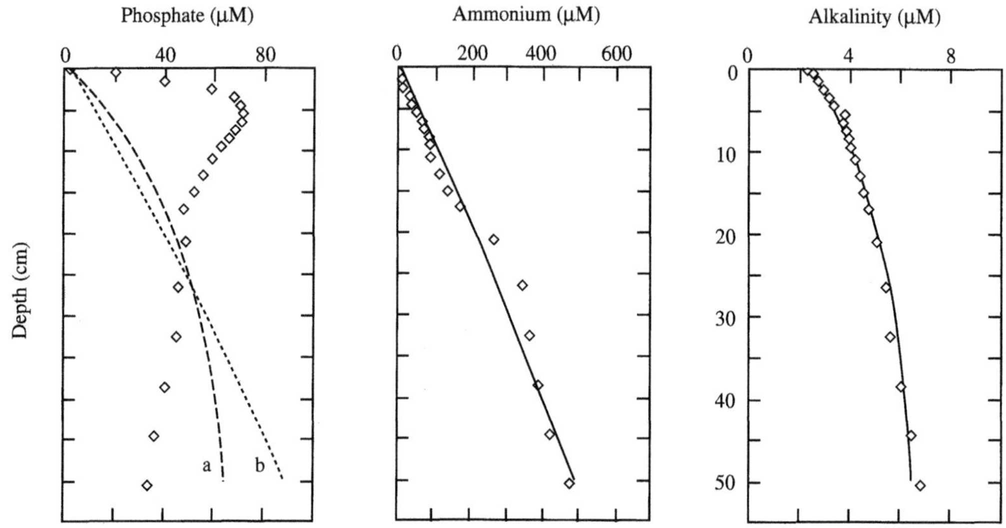

Figure 5 Measured and model-predicted pore-water profiles of dissolved nutrients for one of the Baja California cores shown in Figure 4: Station 14 (BC-47), after Schuffert *et al.*, 1994. Data appear as filled diamonds. Solid curves for alkalinity and ammonium represent empirical fits generated from a steady-state diagenetic model for stoichiometric nutrient regeneration from microbial breakdown of organic matter (see Schuffert *et al.*, 1994, for details). Dashed curves for pore-water phosphate represent model predictions for phosphate concentration gradients derived from the curve fits shown for the (a) alkalinity and (b) ammonium data, assuming organic matter is the sole source of phosphate, and no authigenic CFA formation. The overprediction of pore-water phosphate at depth suggests that the model requires a removal term, e.g., authigenic CFA formation. The underprediction at the surface implies another source of phosphorus to pore waters in addition to organic matter decay (see text for discussion).

the iron oxyhydroxide substrate at depth, provide a spike of phosphate and fluoride which may be instrumental in triggering CFA formation (Heggie *et al.*, 1990; Ruttenberg and Berner, 1993; Jarvis *et al.*, 1994; Reimers *et al.*, 1996; Slomp *et al.*, 1996a).

The over-prediction of pore-water phosphate at depth (Figure 5) is likely due to phosphate removal of organic-matter-derived phosphate into authigenic CFA, a finding that is in accordance with other, similar applications of stoichiometric nutrient regeneration models to document the process of authigenic CFA formation (Ruttenberg and Berner, 1993; Slomp *et al.*, 1996a). A further conclusion of the modeling work on pore-water data was an estimate of authigenic CFA precipitation rates for this site, which agree well with rates estimated for the contemporary Peruvian margin phosphorites (Froelich *et al.*, 1988), and for several large phosphorites of the recent geologic past, now exposed on land (Froelich *et al.*, 1988; Filippelli and Delaney, 1992).

Disseminated Authigenic Carbonate Fluorapatite. Once thought to be a phase that formed only in limited areas characterized by specific and stringent depositional environmental conditions (Table 5; see previous section), evidence accumulated during the 1990s strongly suggests that formation of authigenic CFA is a widespread phenomenon in the oceans (Figure 7). As mentioned previously, in most continental margin areas and in pelagic environments, detection of authigenic CFA against a large burden of detrital sediment cannot be accomplished through direct mineralogical analysis by XRD or SEM, the standard methods used for its detection and quantification in phosphorite and protophosphorite deposits (e.g., Schuffert *et al.*, 1990; Schuffert *et al.*, 1994; Baturin, 2000). Those studies during the 1990s that have documented CFA in nonphosphorite sediments have used the SEDEX method (Ruttenberg, 1992), a sequential extraction method designed expressly to quantify CFA as distinct from detrital igneous and metamorphic apatites. The most robust of these studies have used pore-water profiles in combination with the SEDEX method, the so-called "coupled pore-water–SEDEX approach," which allows the application of four diagnostic indicators of CFA formation (Ruttenberg and Berner, 1993): (i) downward decreasing pore-water fluoride gradients, (ii) decoupled pore-water phosphate and ammonium gradients where, assuming stoichiometric nutrient regeneration in sediments, deficits in phosphate relative to ammonium indicate phosphate removal (see Figure 5), (iii) calculated saturation state of CFA, and (iv) identification of CFA in sediments via the SEDEX method. These diagnostic indicators, although indirect, are the only available tools for identifying authigenic CFA when it is diluted in sediments with a high burden of terrigenous detrital material.

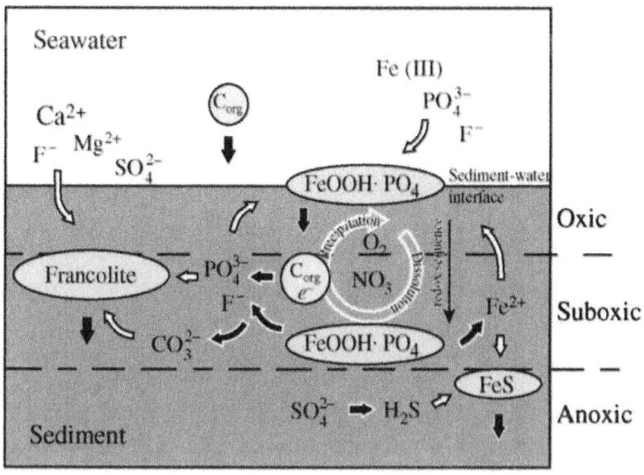

Figure 6 Schematic diagram of the coupled iron and phosphate cycles in during early diagenesis in marine sediments. Light gray ovals and circles represent solid phases, black arrows are solid-phase fluxes. White-outlined black arrows indicate reactions, white arrows are diffusion pathways. Ferric oxyhydroxides (FeOOH) precipitated in the water column and at the sediment–water interface scavenge phosphate (PO_4^{3-}) and some fluoride (F^-) from seawater. During burial and mixing, microbial respiration of organic matter utilizes a sequence of electron acceptors in order of decreasing thermodynamic advantage. Oxygen is used first, followed by nitrate and nitrite, manganese- and iron-oxyhydroxides, and sulfate. Phosphate is liberated to pore waters upon decomposition of organic matter, and reductive dissolution of FeOOH liberates Fe^{2+}, PO_4^{3-}, and F^-, resulting in increases in concentrations of these ions in pore waters (e.g., see phosphate profile in Figure 3 and model curves in Figure 4). If concentration levels are sufficient to exceed saturation with respect to authigenic CFA (denoted as francolite in figure), this phase will precipitate out of solution, sometimes first as a precursor phase that then recrystallizes to CFA proper. Excess phosphate diffuses up towards the sediment–water interface, where it is readsorbed by FeOOH. Ferrous iron (Fe^{2+}) diffuses both downwards to be precipitated with sulfide as FeS in the anoxic zone of sediments, and upwards to be re-oxidized in the oxic zone, where it is reprecipitated as FeOOH. The Fe-redox cycle provides an effective means of trapping phosphate in sediments, and can promote the precipitation of CFA (after Jarvis *et al.*, 1994).

The SEDEX method (Ruttenberg, 1992) is a multistep selective, sequential extraction method that separately quantifies five distinct sedimentary phosphorus reservoirs on the basis of their chemical reactivity to a sequence of solvents: (i) loosely sorbed, or exchangeable phosphorus, (ii) ferric-iron-bound phosphorus, (iii) authigenic carbonate fluorapatite, biogenic hydroxyapatite, plus $CaCO_3$-bound phosphorus, (iv) detrital apatite, and (v) organic phosphorus (Figure 8). The method was standardized for application in marine sediments using analogues for marine phosphorus phases. There have been several modifications proposed to the method that have aimed to streamline the procedure (Anderson and Delaney, 2000; Slomp *et al.*, 1996a,b; Berner and Rao, 1994; Ruttenberg and Ogawa, 2002), but the sequence of solvents in all of these remains fundamentally the same. Exceptions to the streamlining approach are Jensen *et al.* (1998), who expanded the SEDEX method by combining it with another sequential extraction method permitting separation of seven P-reservoirs, and Vink *et al.* (1997), who inserted an additional step between steps I and II of the SEDEX method to remove labile organic phosphorus using the surfactant sodium dodecyl sulfate (SDS). The SDS step was standardized in the same way that the original SEDEX method was standardized (Ruttenberg, 1992), and it appears that it is efficient and specific for labile organic phosphorus. The merit of this approach is that it separately quantifies labile organic phosphorus that might otherwise be lost in subsequent, more aggressive steps of the sequence. The potential benefits of these expanded methods must, however, be weighed against the burden of extending an already arduous analytical method.

The most salient feature of the SEDEX scheme is the ability to separately quantify authigenic CFA as distinct from detrital igneous and metamorphic apatite, as the former represents an active sink for phosphorus from the ocean, whereas the latter does not. Schenau and De Lange (2000) recently proposed adding a series of NH_4Cl extractions to the SEDEX method in order to separately quantify biogenic hydroxyapatite (fish bones, teeth, and scales), as distinct from authigenic apatite. This modification is important for studies wishing to quantify fish debris in sediments, but provides no additional information on removal of reactive phosphorus from the ocean, since the hydroxyapatite reservoir is quantified in step III of the original SEDEX scheme, along with CFA and $CaCO_3$–P.

One of the most important outcomes of application of the SEDEX method in studies of early diagenesis of phosphorus in marine sediments has been identification of modern-day disseminated CFA formation in sediments from nonupwelling regimes (Figure 7). This finding is important because it requires a revision of thinking regarding the role of phosphogenesis and phosphorites in the oceans. That is, there is nothing glaringly unique about the classical phosphorite-forming environment with regard to its propensity for authigenic CFA formation; such conditions potentially exist anywhere that pore-water concentrations of constituent ions satisfy the thermodynamic condition of saturation or supersaturation with respect to CFA. Rather, the uniqueness of phosphorite forming environments has to do with secondary processes that act to concentrate CFA, and/or with low sedimentation rates, which minimize dilution of CFA. The finding of disseminated CFA in a wide range of

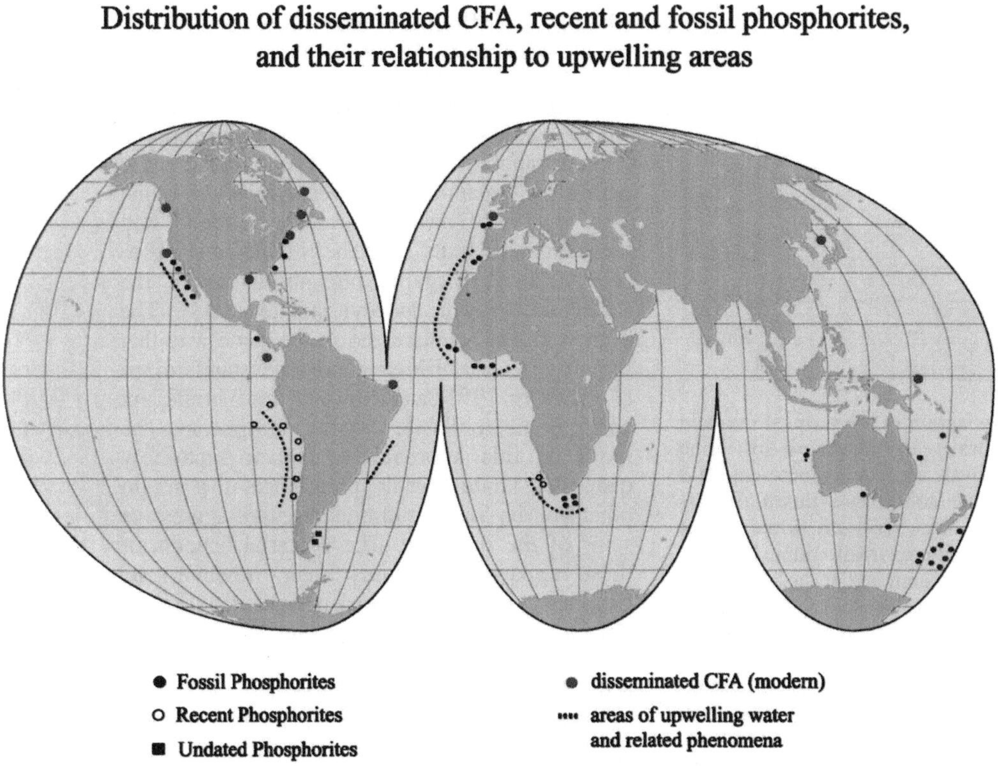

Figure 7 Locations of disseminated (nonphosphorite) authigenic CFA occurrence, as identified using the SEDEX method, as well as locations of fossil, recent, and undated phosphorites. Note that most phosphorites are located in continental margin areas characterized by upwelling, a process whereby nutrient-rich deep waters are advected to the surface causing high biological productivity and a resulting large flux of organic matter to underlying sediments. Sites of disseminated CFA, in contrast, are not restricted to these classical phosphorite-forming environments. Disseminated CFA data are from Cha, 2002 (East Sea between Korea and Japan); Delaney and Anderson, 1997 (Ceara Rise); Filippelli, 2001 (Saanich Inlet); Filippelli and Delaney, 1996 (eastern and western equatorial Pacific); Kim *et al.*, 1999 and Reimers *et al.*, 1996 (California Borderland Basins); Louchouarn *et al.*, 1997 (Gulf of St. Lawrence); Lucotte *et al.*, 1994 (Labrador Sea); Ruttenberg and Berner, 1993 (Long Island Sound and Gulf of Mexico); Slomp *et al.*, 1996a (North Atlantic continental platform; Van der Zee *et al.*, 2002 (Iberian margin in the NE Atlantic). See text for more detailed discussion of selected studies. Figure is modified after Kolodny (1981), by addition of disseminated CFA locales; see Kolodny (1981) for discussion of phosphorite locales.

geographically distinct, nonclassical phosphorite-forming environments has an even more important ramification for the global marine phosphorus cycle, however, in that CFA acts as a permanent sink for reactive phosphorus from the oceans. Thus, the discovery that formation of this phase is not restricted to upwelling environments substantially increases burial rate estimates, and thus decreases the residence time estimate, of phosphorus in the ocean (Tables 3 and 4).

The source of phosphorus for incorporation into CFA in the early diagenetic regime is solid-phase phosphorus liberated either by microbial mineralization of organic matter, or by release of phosphorus associated with iron oxyhydroxides upon reduction of the iron oxyhydroxide substrate once it is transported into suboxic or anoxic zones within sediments. Examination of SEDEX-generated phosphorus profiles from a number of studies show roughly mirror-image profiles of the reservoirs representing phosphorus source: organic matter (organic phosphorus) and/or iron oxyhydroxides (iron-bound phosphorus), and phosphorus sink (CFA). Examples of such co-varying profiles are shown in Figure 9, in which authigenic CFA is forming at the expense of organic-P in Mississippi Delta sediments (Figure 9(a), after Ruttenberg and Berner, 1993), whereas CFA is forming at the expense of iron-bound P in sediments from the North Atlantic Platform (Figure 9(b), after Slomp *et al.*, 1996a,b). These mirror-image profiles illustrate the transfer of phosphorus from initially deposited P-reservoirs to CFA, the authigenic phase that is secondarily formed in the sediment. Note that iron oxyhydroxides may also be authigenic, forming *in situ* in surficial oxidized regions of the sediment and subsequently liberating associated phosphate to pore water upon reduction, once buried below the redox boundary (Figure 6). Thus, iron-bound

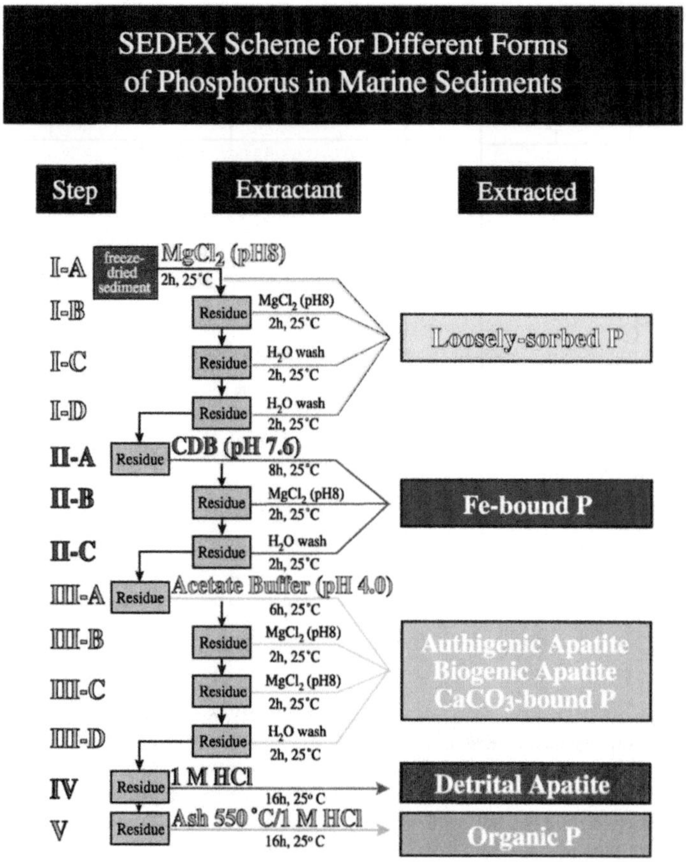

Figure 8 Sequence of extractants and extraction conditions that make up the SEDEX sequential extraction method for quantifying different forms of phosphorus in marine sediments (after Ruttenberg, 1992).

phosphorus may play an intermediate role in the transfer of phosphorus from organic matter to CFA in some depositional environments (see following discussion). The process of transfer from one phosphorus reservoir to another as a result of early diagenetic reactions has been termed "sink-switching" (Ruttenberg and Berner, 1993; Ruttenberg, 1993). This process can greatly enhance the retention of phosphorus by sediments (Ruttenberg and Berner, 1993; Slomp *et al.*, 1996a,b; Louchouarn *et al.*, 1997; Filippelli, 2001), retaining phosphorus that would otherwise have been diffusively lost from sediments, and thus plays an important role in governing phosphorus burial rates and therefore residence time. Enhanced phosphorus retention in sediments by the mechanism of sink-switching of phosphorus from a phase that is unstable in the early diagenetic environment (labile organic phosphorus and ferric oxyhydroxides), to one that is stable (CFA), also impacts the global carbon cycle in that it permanently removes otherwise bioavailable phosphorus from the ocean.

An interesting variant to the sink-switching scheme is found in carbonate sediments hosting seagrass beds, where pore-water fluoride and SEDEX analyses indicate CFA formation, but CFA does not accumulate because it is redissolved in the rhizosphere, where it provides an important source of phosphorus to otherwise P-limited seagrasses (Jensen *et al.*, 1998). Fluoride removal from pore water in carbonate sediments had been recognized in earlier studies and CFA was presumed to be the fluoride sink (Berner, 1974; Gaudette and Lyons, 1980). Rude and Aller (1991) cast some uncertainty on this conclusion, however, as they demonstrate that fluoride can be mobilized/immobilized by carbonate phases other than CFA.

Because the SEDEX scheme, like all sequential extraction schemes, is operationally defined, it is important to obtain corroborating evidence for the identity of the separately quantified sedimentary phosphorus reservoirs whenever possible. This can be accomplished by analyzing pore-water chemistry and other solid-phase components of the host sediments, and by linking depth profiles of various solutes and components to SEDEX phosphorus profiles. Some of the most elegant and comprehensive work of this type has been done by

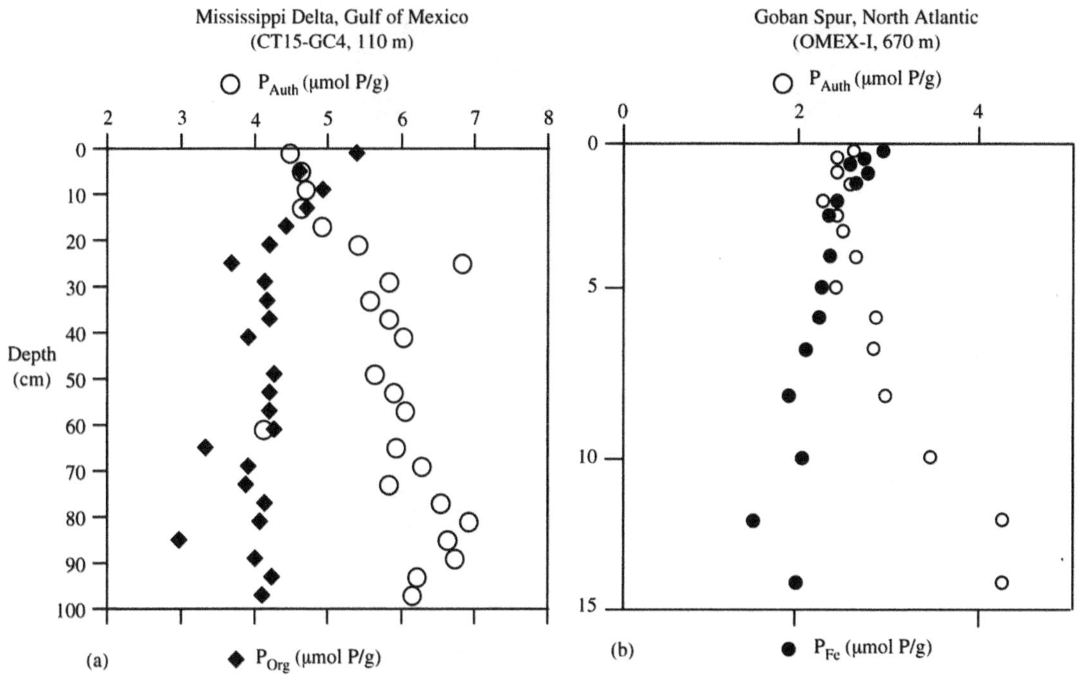

Figure 9 Sink-switching: Mirror-image SEDEX-generated phosphorus profiles from (a) the Mississippi Delta, showing formation of authigenic CFA at the expense of organic-P (after Ruttenberg and Berner, 1993), and (b) the Goban Spur on the North Atlantic Platform, showing CFA is forming at the expense of iron bound-P (after Slomp *et al.*, 1996a,b). These mirror-image profiles illustrate the transfer of phosphorus from initially deposited P-reservoirs to CFA, the authigenic phase that is secondarily formed in the sediment. Sink-switching is an important mechanism by which reactive phosphorus is retained in marine sediments.

Slomp *et al.* (1996a,b) in a study targeting early phosphorus diagenesis in North Atlantic continental platform sediments. In this study, application of a diagenetic model to pore-water and SEDEX-derived solid-phase phosphorus profiles indicates that CFA is forming at the expense of phosphorus bound to iron oxyhydroxides (Figure 9(b)). The model indicates that the iron-bound phosphorus forms *in situ* as phosphate released from decomposing organic matter is sorbed onto authigenic iron oxyhydroxides precipitating in the surficial oxidized layer of sediment. Once transported below the redox boundary by bioturbation, reduction of the host iron oxyhydroxides releases the iron-bound phosphorus, which is then incorporated into precipitating CFA. Slomp *et al.* (1996a,b) hypothesize that this mechanism, in which iron-bound phosphorus plays a key role in early diagenetic CFA formation, may be particularly important in sediments with low sedimentation rates where the most rapid organic matter mineralization takes place near the sediment–water interface. In such environments, sorption onto authigenic iron oxyhydroxides traps organic-matter-derived phosphate that would otherwise be diffusively lost from the sediments. Slomp *et al.* (1996a,b) propose that such a mechanism could also be at work in another low-sedimentation environment, the Labrador Sea, where SEDEX phosphorus data indicate formation of CFA at the expense of iron-bound phosphorus (Lucotte *et al.*, 1994). In more rapidly accumulating sediments, organic phosphorus may be a more important direct source of phosphorus for CFA formation (e.g., Long Island Sound: Ruttenberg and Berner, 1993; see Figure 9(a), the Oman Margin: Shenau *et al.*, 2000), while in still other sediments organic phosphorus and iron-bound phosphorus may both act as sources of phosphorus for CFA formation (the Mississippi Delta: Ruttenberg and Berner, 1993; the Gulf of St. Lawrence: Louchouarn *et al.*, 1997).

Experimental studies of authigenic apatite precipitation. Mechanisms and rates of authigenic apatite formation in the early diagenetic environment are difficult to resolve, because of the wide variety of biological, chemical, and physical factors that can affect its formation. Experimental studies of apatite formation under controlled conditions have provided important information for placing constraints on modes and rates of CFA authigenesis. Examples of such studies include those of Ames (1959), who documented nucleation of CFA on calcium carbonate; Gulbrandsen *et al.* (1984), who documented rates of CFA formation in seawater; Jahnke (1984), who evaluated the effect of carbonate substitution on CFA solubility; Van Cappellen and Berner (1989, 1991), who focused on the idealized, noncarbonate-containing fluorapatite, to resolve dependence of growth rate on solution supersaturation, pH, temperature, and dissolved magnesium (believed to inhibit CFA formation: Martens and Harriss, 1970), and nucleation processes.

One process that has been explored experimentally, and invoked using field evidence, is the role of microbes in phosphogenesis. An unresolved question is whether this role is direct or indirect (see Krajewski *et al.*, 1994, for a review). The indirect role of microbial activity, that is microbial breakdown of organic matter and reduction of Fe-oxyhydroxides with subsequent pore-water phosphate buildup, is well documented and well accepted. Whether microbes play a direct role in CFA formation remains controversial. Petrographic and SEM evidence, showing phosphatized microbial remains in rocks and sediments, has been used to argue for microbes as both active and passive players in CFA formation (e.g., O'Brien *et al.*, 1981; Soudry and Champetier, 1983; Abed and Fakhouri, 1990; Lamboy, 1990). Experimental studies have also been devised to argue for a direct microbial role in apatite precipitation (e.g., Lucas and Prévôt, 1985).

Other Authigenic Phosphate Minerals. Although CFA has received the lion's share of attention in studies of authigenic phosphate minerals in marine sediments, there are other authigenic phosphate mineral phases that have intriguing diagenetic pathways, and may be important sinks for reactive phosphate from the oceans. Once such phase is authigenic vivianite, a hydrous ferrous phosphate with the chemical formula $Fe_3(PO_4)_2 \cdot 8H_2O$. The presence of this phase in marine sediments was first suspected as the result of pore-water studies of early diagenesis in a particularly organic-matter-rich, anoxic site in Long Island Sound, CT, USA: Sachem's Head (Martens *et al.*, 1978; Ruttenberg, 1991), and has also been observed in Amazon Fan sediments (Ruttenberg and Goñi 1997a,b). Sediments at both these sites display the classic phosphate profile observed in situations of authigenic phosphate mineral (Figure 3): an initial buildup in pore-water phosphate due largely to organic matter remineralization, and the reversal to a negative gradient at depth, reflecting removal of phosphate to the solid phase (Figure 10). The Sachem site is extremely anoxic, as evidenced by a rapid decline of sulfate just below the sediment–water interface to complete disappearance of sulfate by ~45 cm depth (Martens *et al.*, 1978). Pore-water-dissolved ferrous iron is undetectable until all of the sulfate has been converted into sulfide and precipitated out as ferrous sulfide, at which point ferrous iron is allowed to buildup in

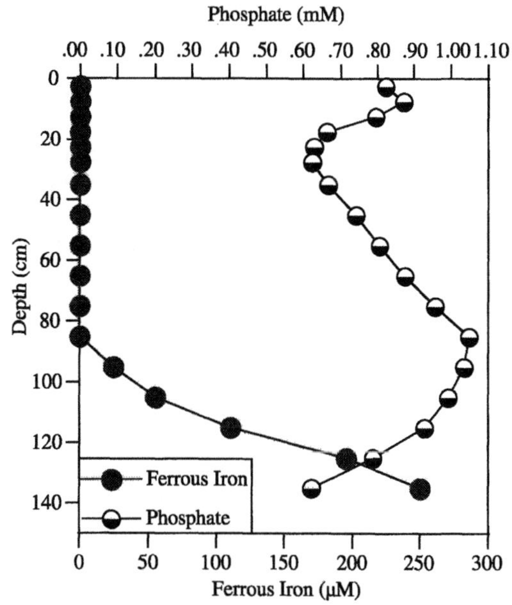

Figure 10 Authigenic vivianite formation: Pore-water-dissolved phosphate and ferrous iron profiles from Sachem's Head, Long Island Sound (phosphate data from Ruttenberg, unpubl., iron data from Canfield, unpubl.). Similar profiles were observed by Martens *et al.* (1978). Removal of phosphate coincident with build up of ferrous iron in pore waters, as well as saturation state calculations, are suggestive of authigenic vivianite formation. Shallow subsurface phosphate maximum is likely due to non-steady-state deposition of organic matter, possibly after a phytoplankton bloom.

pore waters. Coincident with the appearance of ferrous iron in pore waters, pore-water phosphate concentrations begin to drop, indicating probable removal into vivianite (Figure 10). In Sachem's Head and Amazon Fan sediments, sulfide is depleted above the zone of iron and phosphate uptake, due to a high burden of solid-phase iron in these sediments. After the point of sulfide depletion, some of the ferrous iron that builds up in pore water is taken out of solution by formation of authigenic vivianite. This is supported by saturation state arguments (Martens *et al.*, 1978; Berner, 1990; Ruttenberg, 1991), and in the case of Amazon Fan Sediments, nodules of vivianite were recovered providing visible confirmation of processes deduced from indirect pore-water evidence (Ruttenberg and Goñi, 1997a,b). The prevalence of vivianite as an early diagenetic authigenic phosphate phase in the oceans has not been estimated, but it should be restricted to environments with a heavy burden of reactive iron oxyhydroxides, such that they are not completely consumed by iron sulfide formation and may supply ferrous iron for vivianite formation after sulfide has been completely consumed. Deltaic marine environments are prime candidates for such a regime, and if vivianite is an important mechanism for phosphate removal in such environments, it is possible that it could contribute significantly to phosphate removal from the oceans.

Both vivianite and CFA are authigenic minerals that have been detected in muddy sediments of the ocean's margins, and the type of processes involved in their formation, e.g., organic matter diagenesis and redox cycling causing a buildup of pore-water phosphate to the extent that supersaturation with respect to authigenic phosphate minerals form, have been envisaged as most important, if not exclusively important, in muddy sediments with high organic matter content. However, in a very different kind of environment, a different set of authigenic phosphate minerals have been detected and quantified, which may form an important and until recently unrecognized sink for reactive marine phosphorus. These are the authigenic aluminophosphates, often rare earth element (REE)-enriched, that have been found in marine sandstones through exacting scanning electron microscope analyses (Rasmussen, 1996, 2000; see Table 4). The dominant mineral in this assemblage is florencite ($(REE)Al_3(PO_4)_2(OH)_6$), with minor amounts of crandallite ($CaAl_3(PO_4)_2(OH)5H_2O$), gorceixite ($BaAl_3(PO_4)_2(OH)5H_2O$), and xenotime (YPO_4). These minerals appear to be ubiquitous, but volumetrically minor in the sandstones studied; diagenetic textures and the presence of structural sulfate have been cited as evidence of their authigenic early diagenetic nature (Rasmussen, 2000). They have escaped detection in earlier studies because of their small crystal size (<10 μm), and the fact that they are insoluble in the extractants commonly used to quantify marine authigenic phosphate minerals (Rasmussen, 2000). Studies conducted thus far have focused on sandstones of Early Cretaceous age or older (Rasmussen, 2000), so the formation of these phases in the modern ocean remains to be documented. However, based on the calculations of Rasmussen (2000), the phosphorus burial flux associated with these authigenic phases may exceed that of authigenic CFA, and thus impact estimates of residence time to an extent that would require a reassessment of the implications of an even shorter phosphorus residence time than has been proposed (Table 3).

Sedimentary organic carbon to organic phosphorus ($(C:P)_{org}$) ratios. Sediments are the repository of marine organic matter produced in overlying waters (authocthonous organic matter), as well as organic matter transported from the continents, dominantly through riverine transport, with minor atmospheric flux (allocthonous organic matter) (Figure 1, Table 2). The carbon and phosphorus cycles in both marine and terrestrial systems are linked through their coupled uptake during photosynthesis, as illustrated by the following two equations, representative of photosynthetic fixation of carbon and nutrients at sea, by marine phytoplankton:

$$106CO_2 + 64H_2O + 16NH_3 + H_3PO_4 + h\nu \rightarrow C_{106}H_{179}O_{68}N_{16}P + 106O_2 \qquad (1)$$

and on land, by terrestrial plants:

$$830CO_2 + 600H_2O + 9NH_3 + H3PO_4 + h\nu \rightarrow C_{830}H_{1230}O_{604}N_9P + 830O_2 \qquad (2)$$

As a result of the coupled C- and P-uptake, the $(C:P)_{org}$ ratio of organic matter preserved in marine sediments can, in theory, be used to make inferences about the coupled carbon and phosphorus cycles.

Since the early 1980s, there has been a focused effort to define and understand organic carbon to organic phosphorus ratios (hereafter $(C:P)_{org}$) in marine sediments. The motivation for this work stems from two research objectives. First, and foremost, by virtue of the coupled uptake of phosphorus and carbon during marine photosynthesis, and due to the fact that over long timescales it is likely that phosphorus is the limiting nutrient for oceanic biological productivity (Holland, 1978; Broecker and Peng, 1982; Codispoti, 1989; Smith, 1984; also see Section 8.13.3.3.4), various researchers have used the sedimentary $(C:P)_{org}$ to hind-cast levels of marine biological productivity during different periods of Earth's history (Holland, 1984; Sarmiento and Toggweiler, 1984; Delaney and Boyle, 1988; Delaney and Filippelli, 1994; Van Cappellen and Ingall, 1994a,-b; Delaney, 1998). Some of these research efforts are summarized in later sections of this chapter (see Section 8.13.3.4). A second objective has been to use the distinct $(C:P)_{org}$ ratios of marine and terrestrial organic matter, in much the way that $(C:N)_{org}$ ratios traditionally have been used (e.g., Ruttenberg and Goñi, 1997a,b), to determine sources of organic matter in marine sediments. In other words, these ratios can potentially be used to evaluate the relative partitioning of the sedimentary organic matter pool between marine and terrestrial organic matter sources.

In order to make sense of observed trends in sedimentary $(C:P)_{org}$ ratios, and to evaluate the utility of this parameter in hind-casting past ocean productivity or in partitioning sedimentary organic matter as to source, it is important to understand the sources of organic matter to sediments, and the processes that modify the $(C:P)_{org}$ ratio while organic matter is in transit to sediments, and then during its burial history. The $(C:P)_{org}$ ratios of marine and terrestrial organic matter are distinct. This ratio in marine phytoplankton hovers closely around the classical

Redfield ratio of 106C : 1P, and whereas deviations from this ratio due to such things as phytoplankton nutritional status and environmental factors have been observed (e.g., Goldman *et al.*, 1979; Goldman, 1986), the adherence to the canonical 106 : 1 ratio by marine phytoplankton, first pointed out by Redfield (1958; see also Redfield *et al.*, 1963) is truly remarkable. In contrast, the $(C:P)_{org}$ ratio observed in terrestrial plants can vary quite widely, but is always substantially enriched in carbon relative to phosphorus. This C-enrichment is illustrated by the 830C : 1P ratio given in Equation (2), which represents a reasonable average for terrestrial, vascular plants. The high $(C:P)_{org}$ ratios of terrestrial vascular plant tissue are due to the dominance in these tissues of cellulose, a polymer of glucose, whereas phosphorus-containing biochemicals are a relatively minor component. Even considering the variability that has been observed for both phytoplankton and terrestrial plants, the distinction between the $(C:P)_{org}$ ratios of marine versus terrestrial organic matter is robust.

An overview of previous studies on $(C:P)_{org}$ ratios has been compiled recently (Anderson *et al.*, 2001), which nicely summarizes the various observations and interpretations of different workers in the field (Table 6). All of the studies summarized in Table 6 report observing $(C:P)_{org}$ ratios greater than the Redfield ratio, particularly in organic-rich, continental margin sediments. Most of these studies interpret the higher ratios, which often increase with depth below the sediment–water interface, as a reflection of preferential regeneration of phosphorus relative to carbon during microbial mineralization of marine organic matter. This has been a well-accepted concept in the field of early diagenesis for some time (e.g., Berner, 1980; Ingall and Van Cappellen, 1990), and has been understood to reflect the more labile nature of P-biochemicals relative to most nonphosphorus containing organic carbon compounds. A second rationale for this observation has been that, because phosphorus is an essential and potentially limiting nutrient, it is preferentially targeted for remineralization to support subsequent biological productivity. The paradigm of preferential regeneration of phosphorus relative to carbon suggests that progressively more extensive mineralization should lead to ever higher $(C:P)_{org}$ ratios. However, in contradiction to this paradigm, several studies have observed sedimentary $(C:P)_{org}$ ratios close to the Redfield ratio in pelagic sediments (Froelich *et al.*, 1982; Ingall and Van Cappellen, 1990; Ruttenberg, 1990) and in iron-dominated Amazon Shelf sediments (Ruttenberg and Goñi, 1997a), both of which are sites of intensely remineralized organic matter. In the pelagic environment the high degree of organic matter degradation is due to low sediment accumulation rates and long oxygen exposure times, whereas in Amazon Shelf sediments it is due to prolonged oxygen exposure due to repeated resuspension into the water column, under conditions that have been likened to a fluidized bed reactor (Aller, 1998). This observation does not fit into the paradigm of preferential regeneration of phosphorus relative to carbon with progressively more extensive mineralization. The explanations variously given for this unexpected observation are either that sedimentary organic matter at these sites is enriched in refractory organic phosphorus compounds, left behind after extensive remineralization of more labile organic carbon, or that these low ratios reflect living sedimentary bacterial communities or their residua (Froelich *et al.*, 1982; Ingall and Van Cappellen, 1990; Ruttenberg and Goñi, 1997a).

Ingall and Van Cappellen (1990) systematized the range of $(C:P)_{org}$ ratios observed in different depositional environments by relating them to sediment accumulation rate (Figure 11). Sediment accumulation rate has been recognized as a robust proxy for extent of organic carbon degradation (e.g., Henrichs and Reeburgh, 1987; Berner, 1989; Canfield, 1989), and thus it is reasonable to expect that it may also be a proxy for extent of P_{org} degradation. The rationale given for the observed relationship between sedimentary $(C:P)_{org}$ ratios and sediment accumulation rate (Figure 11) is as follows. In the highest sedimentation rate region, $(C:P)_{org}$ ratios approximate the Redfield ratio because organic matter has very little time to decompose before it is buried into the zone of less efficient, anoxic diagenesis. Therefore, minimal P_{org} is regenerated preferentially to C_{org}, and the organic matter buried has a $(C:P)_{org}$ ratio closely approximating the Redfield ratio for fresh phytoplankton. In the lowest sediment accumulation rate region, the near-Redfield $(C:P)_{org}$ ratios are explained as either residual organic matter rich in refractory organic phosphorus compounds, bacterial biomass, or both. In the mid-range of sediment accumulation rates the high $(C:P)_{org}$ ratios suggest preferential P_{org} regeneration relative to C_{org} during incomplete mineralization of organic matter. The sediment accumulation rate framework appears to work well for medium to high sedimentation rates. The explanation of low $(C:P)_{org}$ ratios in low sediment accumulation regions is substantially weaker, however. Attempts to identify the nature of refractory organic phosphorus compounds that can withstand the most intense remineralization regimes have not been successful (e.g., Section 8.13.3.3.2 cites studies in which the two most promising candidates for refractory P_{org} are discounted: phosphonates—Laarkamp, 2000; Laarkamp and Ruttenberg, 2000; Ruttenberg and Laarkamp, 2000; and inositol phosphates—Suzumura and Kamatani, 1996).

Table 6 Overview of Previous Studies on (C:P)organic Ratios in Marine Sediments (after Anderson *et al.*, 2001).

#	*References*	*Geographic region*	*Water depth range* (m)	*Observations and interpretations*
1	Hartmann *et al.* (1973)	Northwest African continental slope	2037–2066	$(C:P)_{org}$ increased with increasing sediment depth to a ratio of ~265 at depths >90 cm. Preferential decomposition of P_{org} compounds relative to C_{org}.
2	Hartmann *et al.* (1976)	Northwest African continental slope	~600–3700 (also on shelf core)	$(C:P)_{org}$ increased with increasing sediment depth; higher $(C:P)_{org}$ in rapidly depositing sediments (>480 for >11 cm kyr^{-1}) than in slowly depositing sediments (<240 for <6 cm kyr^{-1}). Preferential decomposition of P_{org} compounds during early diagenesis; formation of authigenic P compounds.
3	Filipek and Owen (1981)	Gulf of Mexico	30 and 112	$(C:P)_{org}$ increased with increasing sediment depth to a mean ratio of 207 in shallow and 259 in deep station. Preferential decomposition of P_{org} compounds; formation of authigenic P compounds.
4	Krom and Berner (1981)	Long Island Sound	9	$(C:P)_{org}$ constant (320) with increasing sediment depth; mobilization of P via metal oxide dissolution within zone of bioturbation only. C_{org} delivered to sediment depleted in P_{org} relative to Redfield Ratio for marine phytoplankton.
5	Suess (1981)	Peru continental margin	180–645	$(C:P)_{org}$ > Redfield Ratio at all sediment depths. Fish debris dissolution significant source of interstitial dissolved phosphate
6	Froelich *et al.* (1982)	Northwest African continental margin; west African continental margin, French equatorial Atlantic; eastern tropical Pacific (MANOP sites M and H); western subtropical Atlantic, east North American continental slope and rise; Gulf of Mexico; Santa Barbara Basin, Long Island Sound; Peru continental margin; central Pacific gyre (includes data sets from references 1–5)	Continental margin to open ocean water depths	P_{org} concentrations in marine sediments relatively constant; $(C:P)_{org}$ > Redfield Ratio in organic rich sediments, <Redfield Ratio in organic carbon poor sediments (<1 wt% C_{org}). Preferential decomposition of P_{org} compounds; organic carbon poor sediments enriched in stable, P-rich moieties.

7	Mach *et al.* (1987)	Data set from Froelich *et al.* (1982) plus additional Peru sites	Continental margin to open ocean water depths	P_{org} concentrations in marine sediments linearly related to C_{org}; $(C:P)_{org}$ > Redfield Ratio. Earlier conclusions from same data set were incorrect because of problems with analytical detection limits and difference methods.
8	Ingall and Van Cappellen (1990)	Amazon Delta, eastern equatorial Atlantic; eastern equatorial Pacific; Long Island Sound; Mississippi Delta; Northwest African continental margin; Pacific deep sea; Peru continental margin; Santa Barbara Basin; western Gulf of Mexico (includes data sets from references 2, 3, 5, and 6; from Sholkovitz, (1973), Froelich *et al.*, (1988), Ruttenberg, (1990), and their own data)	Continental margin to open ocean water depths	$(C:P)_{org}$ varies with sedimentation rate: <200 at sedimentation rates <2 and >1000 cm kyr^{-1}, and up to 600 at intermediate sedimentation rates. Preferential decomposition of C_{org} relative to P_{org} occurs during oxic respiration, leaving behind P_{org} enriched organic matter, whereas preferential decomposition of P_{org} during incomplete degradation of organic matter under anoxic conditions leaves organic matter depleted in P_{org} relative to C_{org}.
9	Ruttenberg and Berner (1993)	Long Island Sound; Mississippi Delta	9 and 110	$(C:P)_{org}$ > Redfield Ratio, accompanied by transformation of P_{org} to carbonate fluorapatite. This "sink switching" results in enhanced P-retention by sediments.
10	Ingall *et al.* (1993)	Various shale sequences, both bioturbated and laminated, especially from the Camp Run Member of the New Albany Shale (Late Devonian-Early Mississippian age)		$(C:P)_{org}$ ~ 150 for bioturbated shales; ~3900 for laminated shales.
				High $(C:P)_{org}$ in anoxic sediments results from limited bacterial storage of P; extensive P-regeneration, and enhanced C_{org} preservation.
11	Calvert *et al.* (1996)	Reexamination of Camp Run Member of the New Albany Shale		Sediment sources and water depths were different during deposition of bioturbated and laminated shale sequences, making comparison of the two difficult.

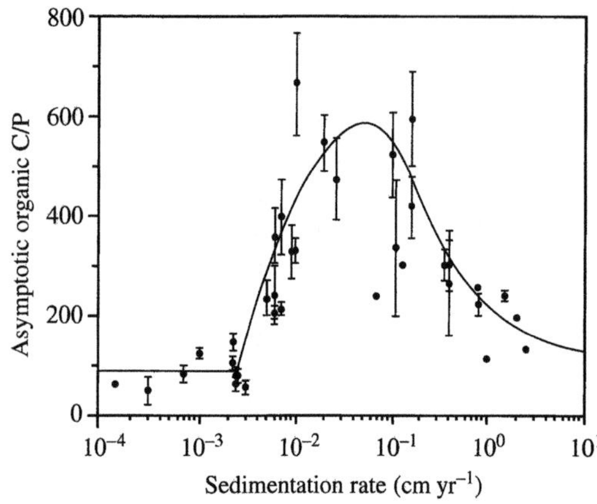

Figure 11 Calculated asymptotic organic C : P ratios plotted as a function of sedimentation rate. Designation of these $(C:P)_{org}$ ratios as "asymptotic" derives from the fact that they represent the composition of organic matter buried below the depth at which the $(C:P)_{org}$ increases, and so reflect the composition of buried (preserved) organic matter. Error bars represent the standard deviation of the average asymptotic $(C:P)_{org}$ values, or the absolute range of values where their number was less than or equal to 3. Solid curve is the model-predicted asymptotic $(C:P)_{org}$ ratio versus sedimentation rate. See text for discussion (after Ingall and Van Cappellen, 1990).

The possibility of sediment enriched in bacteria or their remains is a viable explanation, but such an enrichment in bacteria has not been conclusively demonstrated. Another weak point in the Ingall and Van Cappellen study, and in most other studies of $(C:P)_{org}$ in marine sediments, is the failure to recognize explicitly that, particularly in continental margin settings, terrestrial organic matter may significantly affect the observed $(C:P)_{org}$ ratios in marine sediments. In these environments, one should expect the sedimentary $(C:P)_{org}$ ratios to reflect a mixture of marine and terrestrial organic matter (e.g., Ruttenberg and Goñi, 1997a, Ruttenberg and Goñi, 1997b). In sites where the observed sedimentary $(C:P)_{org}$ ratios exceed the Redfield ratio, for example, it is difficult to evaluate whether these high ratios reflect preferential phosphorus regeneration from autocthonous marine phytodetritus, the presence of allocthonous, high-$(C:P)_{org}$ ratio terrestrial organic matter, or both. Thus, there remain many open questions about what, in fact, sedimentary $(C:P)_{org}$ ratios represent. As a result, it remains unclear how best to interpret marine sedimentary $(C:P)_{org}$ ratios, which are a complex reflection of source material, depositional environment, extent of degradation, and degradation pathway.

In a departure from the work summarized in the foregoing paragraphs, Anderson *et al.* (2001) argue that the parameter of interest for reconstructing marine paleoproductivity is not the sedimentary $(C:P)_{org}$ ratio, but the ratio of organic carbon to reactive phosphorus (hereafter, $C_{organic}:P_{reactive}$), where $P_{reactive}$ is defined (as it is in Table 2, note (f)), as the sum of SEDEX-quantified organic-, authigenic-, loosely sorbed-, and iron-bound phosphorus. The rationale for preferred use of the $C_{organic}:P_{reactive}$ ratio is that, because of diagenetic sink-switching (see Section 8.13.3.3.2), the phosphorus contained in the SEDEX-quantified authigenic-, loosely sorbed-, and iron-bound phosphorus pools derived originally from P_{org}. Thus, the quantity $P_{reactive}$ is a reflection of the P_{org} originally deposited with sediments. Two potential weaknesses of this model are that (i) it implies 100% P-retention by sediments, which is likely not a good assumption as, except in the most rapidly accumulating sediments, it is expected that some remineralized P_{org} will diffuse out of sediments; and (ii) it does not take into account the sizeable authigenic-P "background" that has been found in most SEDEX studies (e.g., Ruttenberg and Berner, 1993; Slomp *et al.*, 1996a,b; Louchouarn *et al.*, 1997). If this background pool is not truly authigenic phosphorus, i.e., if it is a detrital phase that was passively deposited and did not derive its phosphorus from organic matter (e.g., see discussion in Ruttenberg and Berner (1993)), then the $C_{organic}:P_{reactive}$ ratio will be systematically offset. Despite these potential weaknesses, Anderson *et al.* (2001) make an interesting and compelling case for use of the $C_{organic}:P_{reactive}$ ratio in lieu of the traditional $(C:P)_{org}$ ratio. Further, their results lead them to question whether sediment accumulation rate and state of anoxia exert strong control on sedimentary $(C:P)_{org}$ ratios, as has been argued in earlier studies (e.g., Ingall and Van Cappellen, 1990; Ingall *et al.*, 1993). There remain substantial questions about what, precisely, the sedimentary $(C:P)_{org}$ ratios, or $C_{organic}:P_{reactive}$ ratios represent. The degree of uncertainty in results of models employing these ratios to infer past variations in coupled carbon and phosphorus cycling will continue to challenge scientists, until these questions can be further resolved.

Coupled Iron-phosphorus Cycling. The affinity of phosphate for sorptive association with ferric oxide and oxyhydroxide phases, well documented in soil and freshwater systems (see Sections 8.13.3.1 and 8.13.3.2), is also a well-studied process in marine systems. Three distinct marine environments where coupled iron–phosphorus cycling has been identified as an important process are MOR systems, estuaries, and continental margin sediments. The purely physicochemical process of sorption is essentially the same in these three distinct environments, where an initial, rapid surface sorption phase is followed, given enough time, by a redistribution of adsorbed phosphate into the interior of iron oxyhydroxides by solid-state diffusion (Bolan *et al.*, 1985;

Froelich, 1988). The quantity of phosphate sorbed depends upon the nature of the ferric iron phase, with less crystalline phases having higher surface area and therefore more sorption capacity (Bolan *et al.*, 1985; Ruttenberg, 1992). The specific characteristics of the different environments, however, impact the timescale and the pattern of this coupled cycling, and the ultimate fate of the iron and phosphorus that are entrained in the coupled cycling process. Each of these environments will be discussed in turn.

Sorptive removal of phosphate by iron oxyhydroxides in hydrothermal MOR environments was first documented in a study of the East Pacific Rise MOR system by Berner (1973), who recognized that these environments could constitute a significant sink for phosphate from the ocean. Subsequent work has expanded on this initial study, focusing on phosphate scavenging onto volcanogenic ferric oxyhydroxides in water column hydrothermal plumes above MORs (Feely *et al.*, 1990; Rudnicki and Elderfield, 1993), and removal in ridge-axis and ridge-flank hydrothermal systems at different MOR systems throughout the world's oceans (Froelich *et al.*, 1977; Wheat *et al.*, 1996). The more recent studies confirmed earlier suggestions of phosphate uptake by ridge-flank basalts, which had been inferred from highly correlated phosphate and ferric iron concentrations in progressively altered seafloor basalts (Hart, 1970; Thompson, 1983). By the latest estimates (Wheat *et al.*, 1996), MOR ridge-flank systems and water column hydrothermal plume particles dominate phosphorus removal at MOR systems, removing approximately 22% and 27%, respectively, of the preindustrial dissolved riverine phosphorus flux into the oceans. Plume particles form when ferrous iron exhaled from MOR vents encounters the cold, oxidized oceanic waters above the vents, and precipitates out as amorphous ferric oxyhydroxide particles (Figure 12). Phosphate is scavenged onto these volcanogenic ferric oxyhydroxides by co-precipitation and surface sorption. As hydrothermal plumes reach a state of neutral buoyancy and are advected away from the ridge axis, fallout from these plumes drapes the seafloor beneath with reactive ferric oxyhydroxides and their associated phosphorus. Scavenging of phosphorus in ridge-flanks occurs via basalt–seawater reactions promoted by hydrothermally driven seawater circulation through oceanic crust (Figure 12). Whereas phosphate removal with hydrothermal plume particles and in un-sedimented segments of MOR systems clearly occurs by sorption onto volcanogenic ferric oxyhydroxides, some component of phosphate removal in ridge-flank sediments may be due to authigenic apatite formation. Pore-water phosphate and fluoride gradients in sediments deposited on ridge flanks at some sites are consistent with authigenic apatite (CFA) formation (Wheat *et al.*, 1996), but the relative importance of CFA versus sorption onto Fe-oxyhydroxides as removal mechanisms for phosphate in MOR systems has not been determined. The magnitude of phosphorus removal at MOR systems suggests that any changes in hydrothermal activity at MOR in the past or in the future could perturb the magnitude of this sink, and thus have significant impact on the global phosphorus cycle (Wheat *et al.*, 1996).

The mixing of seawater and freshwater in estuaries causes coagulation and flocculation of metal-organic rich colloidal material (Sholkovitz *et al.*, 1978; Fox, 1990), and estuarine water

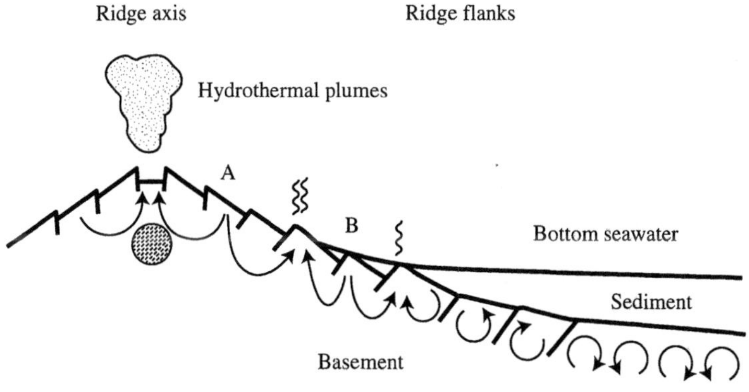

Figure 12 Conceptual model of seawater circulation through oceanic crust on the flanks of MOR systems (after Wheat *et al.*, 1996). Circulation cell A illustrates the case where bottom seawater enters basement directly through faults and basaltic outcrops, and there is no direct contact with sediment, while circulation cell B illustrates the case where seawater downwells through the sediment into basement and interacts with sediment along its flow trajectory. Phosphate is removed in both instances via sorption onto ferric oxyhydroxides. At sediment covered ridge-flank segments, additional P-removal as authigenic CFA may occur as a result of diagenesis and sink-switching within the sediments. In addition, ferric oxyhydroxides precipitating out of hydrothermal plumes, formed as hot hydrothermal water is vented from the ridge axis, scavenge phosphate from seawater (see text for discussion).

columns often have a high concentration of suspended sediments due to riverine input of particulates, and due to resuspension of bottom sediments. It has been shown both in laboratory and field studies that suspended sediments influence the concentration of dissolved phosphorus in a process that has been termed "phosphate buffering" (e.g., Carritt and Goodgal, 1954; Pomeroy *et al.*, 1965; Chase and Sayles, 1980; Lucotte and d'Anglejan, 1983; Fox *et al.*, 1985, 1986; Fox, 1989; Lebo, 1991; see Section 8.13.3.2.2). This buffering mechanism causes dissolved phosphate levels to remain relatively constant, and has been attributed to solid-phase sorption. It is widely believed that ferric oxyhydroxides are the prime substrates for phosphate sorption in estuaries (e.g., Fox, 1989). The overall effect on phosphorus in estuaries, however, is complicated by other process such as biological uptake, and periodic seabed storage and diagenesis followed by resuspension, such that the role of estuaries as sources or sinks of phosphorus to the oceans is variable, and cannot be generalized (Froelich *et al.*, 1982; Kaul and Froelich, 1996; Vink *et al.*, 1997).

In marine continental margin sediments, coupled cycling of iron and phosphorus is a dynamic phenomenon. In river-dominated margins, detrital iron-bound phosphorus of variable reactivity is delivered to marine sediments (Berner and Rao, 1994; Ruttenberg and Canfield, 1994). Whether these phases participate in redox cycling during subsequent diagenesis depends upon the lability of the iron mineral substrate. Within sediments (as shown schematically in Figure 6), a portion of the pore-water phosphate derived from organic matter mineralization that does not escape to overlying waters via diffusion may be sorbed to detrital or authigenic iron oxyhydroxides within the oxidized zone of the sediment (e.g., Krom and Berner, 1980; Sundby *et al.*, 1992; Slomp and Van Raaphorst, 1993; Jensen *et al.*, 1995; Slomp *et al.*, 1996a,b). Once buried into the reduced zone of sediments, the ferric oxyhydroxide substrate is subject to reductive solubilization, and concomitant phosphate release to pore waters. Both the dissolved ferrous iron and phosphate can then diffuse upward into the oxidized zone, where the iron is subject to reoxidation, and the phosphate to renewed scavenging by the freshly precipitated ferric oxyhydroxide (Heggie *et al.*, 1990; Sundby *et al.*, 1992; Jensen *et al.*, 1995; Slomp *et al.*, 1996a,b). Alternatively, the phosphate released within the reduced zone of sediments may be taken up by authigenic minerals (Van Cappellen and Berner, 1988; Heggie *et al.*, 1990; O'Brien *et al.*, 1990; Slomp *et al.*, 1996a,b; Reimers *et al.*, 1996).

The oxidized zone can be of variable thickness, depending on a host of environmental parameters, including sediment accumulation rate, organic matter concentration and reactivity, bottom-water oxygen, infaunal activity, and physical resuspension processes. Many of these parameters vary seasonally, and as such, the coupled cycling of iron and phosphorus can have a distinct seasonality (Lijklema, 1980; Aller, 1980; Jensen *et al.*, 1995; Colman and Holland, 2000). In summer the oxidized zone will decrease due to increased organic matter loading and enhanced microbial activity at warmer summer temperatures. In winter, the oxidized zone will expand due to a reversal of these conditions. These parameters can also vary spatially due, for example, to variability in bottom-water topography or in overlying water conditions that can affect organic matter productivity and subsequent organic loading of underlying sediments. This spatial and temporal variability affects another important phenomenon in the marine phosphorus cycle, that of the return phosphate flux out of marine sediments into overlying waters, or the benthic phosphate flux.

Benthic Return Flux of Phosphorus From the Seabed. A number of diagenetic processes act to enrich pore-water phosphate concentrations above bottom water levels and, as a result, can lead to an appreciable benthic return flux of phosphate from the seabed to overlying bottom waters. These processes, discussed in earlier sections of this chapter (see Figures 1, 3, and 6) include (i) microbial respiration of organic matter in sediments; (ii) microbial reduction and solubilization of ferric oxyhydroxides with subsequent release of associated phosphate (iron-bound phosphorus (P_{Fe})); and (iii) abiotic reduction of ferric oxyhydroxides by H_2S and liberation of P_{Fe} (Krom and Berner, 1981). Although not explicitly demonstrated in marine sediments, several studies in freshwater systems also suggest that bacteria can liberate phosphate directly to pore waters in response to physiological cues tied to redox changes in the sedimentary environment (Shapiro, 1967; Gächter *et al.*, 1988; Gächter and Meyers, 1993; Carlton and Wetzel, 1988; however, see Gunnars and Blomqvist (1997)). The most extreme contrasts between pore-water and bottom-water phosphate concentrations occur in organic-rich, anoxic marine sediments, typically in continental margin environments, where phosphate concentrations can reach levels of 100–500 μM phosphate in the upper 10 cm below the sediment–water interface, in contrast to bottom water phosphate concentrations of 2–3 μm (e.g., Krom and Berner, 1981; Klump and Martens, 1987; Ruttenberg and Berner, 1993; McManus *et al.*, 1997; Reimers *et al.*, 1996).

The magnitude of phosphate benthic flux can be estimated from the pore-water gradient, using Fick's first law of diffusion (e.g., Krom and

Berner, 1981; Klump and Martens, 1981; Sundby *et al.*, 1992) corrected for bioturbation, bioirrigation, and abiotic mixing/irrigation where necessary (e.g., Berner, 1980; Aller, 1980, 1982; Klump and Martens, 1981), or by mass balance calculations (e.g., Krom and Berner, 1981; Klump and Martens, 1987). Alternatively, the benthic flux out of sediments can be measured directly during incubations of retrieved cores, or by deploying flux chambers to make *in situ* benthic flux measurements (e.g., Aller, 1980; Ingall and Jahnke, 1994, 1997; Jensen *et al.*, 1995; Klump and Martens, 1987; McManus *et al.*, 1997). Studies in which directly measured phosphate benthic fluxes are compared to the flux calculated from pore-water profiles often find that benthic fluxes obtained by these two approaches disagree (e.g., Krom and Berner, 1981; Klump and Martens, 1981, 1987; McManus *et al.*, 1997). Cases in which the pore-water gradient-derived flux is high relative to the directly measured benthic flux are typically explained by removal processes that occur at or near the sediment–water interface. These processes thus impede escape of pore-water phosphate from the seabed, but do not perturb the pore-water profile. Likely processes include phosphate sorption onto iron oxyhydroxides in the surficial oxidized layer of sediments, and biological uptake by interfacial microbial mats (Klump and Martens, 1981, 1987; Jensen *et al.*, 1995; Colman and Holland, 2000). Instances in which the pore-water gradient-derived flux is low relative to the directly measured benthic flux have been explained by rapid decomposition of highly labile organic matter right at the sediment–water interface, such that phosphate is liberated directly to bottom water and does not contribute to the pore-water buildup (e.g., Krom and Berner, 1981). While the *in situ* measurement approach may provide more accurate flux estimates, this approach is much less commonly taken because it requires specialized equipment, and is more time consuming (benthic chambers must be deployed for time periods of one to several days; e.g., McManus *et al.*, 1997; Berelson *et al.*, 1996) than core retrieval.

A number of studies have highlighted the role of bottom water and sediment redox state on the benthic return flux of phosphate (Klump and Martens, 1981, 1987; Sundby *et al.*, 1992; Ingall and Jahnke, 1994, 1997; Jensen *et al.*, 1995; McManus *et al.*, 1997; Colman and Holland, 2000). When there is sufficient oxygen in bottom waters to support formation of iron oxyhydroxides in surficial sediments, these can act as a trap for pore-water phosphate, diminishing or eliminating the benthic phosphate return flux (see previous discussion). Phosphate released from decomposing organic matter below the redox boundary is transported down the concentration gradient into the oxic surficial zone of sediments, where it can be sorptively scavenged onto authigenic ferric oxyhydroxides (e.g., Figure 6). This may happen multiple times prior to ultimate burial (e.g., Jensen *et al.*, 1995). The development of an oxic surface layer in sediments can be a seasonal phenomenon, and the depth of the redox boundary below the sediment–water interface can likewise vary seasonally, in response to seasonal changes in temperature, temperature-sensitive metabolic processes, and organic matter flux to sediments (Aller, 1980; Klump and Martens, 1981, 1987; Sundby *et al.*, 1992; Jensen *et al.*, 1995; Colman and Holland, 2000).

Sorptive removal of phosphate from pore waters decouples phosphate from carbon during organic matter remineralization. Ingall and Jahnke (1994, 1997) have made a strong case for elemental fractionation of phosphorus from carbon during sediment diagenesis, and argue that phosphorus burial efficiency (relative to carbon) is reduced in low-oxygen continental margin environments, relative to environments with higher bottom-water oxygen. Drawing on a significantly larger data set, McManus *et al.* (1997) find that the picture is not necessarily as simple as that painted by Ingall and Jahnke (1994, 1997), and that some environments with low bottom-water oxygen do not show reduced phosphorus burial efficiency. Both of these studies utilized *in situ* benthic phosphate flux chamber data. In a synthesis of almost 200 pore-water phosphate profiles from the literature, Colman and Holland (2000) argue convincingly that the variability observed in the Ingall and Jahnke (1994, 1997) and McManus *et al.* (1997) studies could well be explained by seasonal variability. This argument is consistent with studies that have focused on measuring *in situ* benthic phosphorus flux on a seasonal basis (e.g., Aller, 1980; Klump and Martens, 1981, 1987), and have documented substantial differences in magnitude and direction of benthic flux as a function of season.

Mechanisms proposed to explain the relationship between bottom-water anoxia and phosphate benthic flux are (i) coupled iron–phosphorus cycling (Aller, 1980; Klump and Martens, 1981, 1987), and/or (ii) bacterial release of phosphate under anoxic conditions (Ingall and Jahnke, 1994, 1997, citing the work of Gächter *et al.*, 1988; Gächter and Meyers, 1993, in freshwater systems). It is likely that both mechanisms are operant, and both contribute to variability in benthic phosphate return flux and fractionation of phosphorus from carbon during organic matter mineralization (Colman and Holland, 2000). Drawing on their large and geographically diverse data set, Colman and Holland (2000) argue convincingly that the benthic phosphate return flux is a quantitatively important term in the marine phosphorus budget. The link between the benthic phosphate return flux and oxidation state of bottom water/sediments, and the coupled cycling of phosphorus, iron, and carbon that results, has been proposed as a negative feedback that has stabilized atmospheric oxygen throughout the Phanerozoic (Betts and Holland, 1991; Colman and Holland, 1994, 2000; Holland, 1994; Colman *et al.*, 1997; see Section 8.13.3.4.2 for further discussion).

8.13.3.3.3 Phosphorus in the oceanic water column: composition and cycling

Dissolved inorganic phosphorus (DIP). Uncombined dissolved inorganic phosphorus exists as three ionic species in natural waters, corresponding to the conjugate bases of weak, triprotic phosphoric acid (H_3PO_4). In seawater of pH 8, the dominant species is HPO_4^{2-}(87%), over PO_4^{3-} (12%), and $H_2PO_4^-$ (1%), (Kester and Pytkowicz, 1967). Since PO_4^{3-} is the ionic species taken up into authigenic minerals, it is necessary to have the apparent dissociation constants to evaluate the activity of this species for use in solubility calculations. In a series of experiments designed to evaluate the apparent dissociation constants of phosphoric acid in natural waters, Kester and Pytkowicz (1967) compared the distribution of phosphate species in freshwater, artificial seawater, and a sodium chloride solution of ionic strength identical to the artificial seawater (Figure 13). This approach permitted distinction between nonspecific effects (e.g., those due to ionic strength only) and specific interactions (e.g., those resulting from interactions such as ion pair and complex formation). The shift in phosphate ionic species distribution from freshwater (Figure 13(a)) to the 0.68 M NaCl solution (Figure 13(b)) is attributable to ionic strength effects, whereas the shift observed from NaCl to the artificial seawater solution (Figure 13(c)) is brought about by specific interactions between phosphate species and seawater dissolved cations. Kester and Pytkowicz (1967) estimated that 99.6% of the PO_4^{3-} and 44% of the HPO_4^{2-} species are complexed by cations other than Na^+, dominantly by Mg^{2+} and Ca^{2+}. For the that portion of the DIP pool present as cation-complexes, Atlas *et al.* (1976) estimate that various Mg-phosphate complexes dominate (43%), followed by Na- (15%) and Ca-phosphate (12%) complexes (Figure 14).

Polyphosphate, another form of DIP, is a polymer of phosphate that accumulates intercellularly in phosphate-replete phytoplankton and bacteria (e.g., Solórzano and Strickland, 1968, and citations therein). Attempts to measure polyphosphate in seawater have been limited. Solórzano and Strickland (1968) found barely detectable levels in two coastal studies, and speculate that any polyphosphate formed by marine organisms that is liberated to seawater is likely rapidly reused

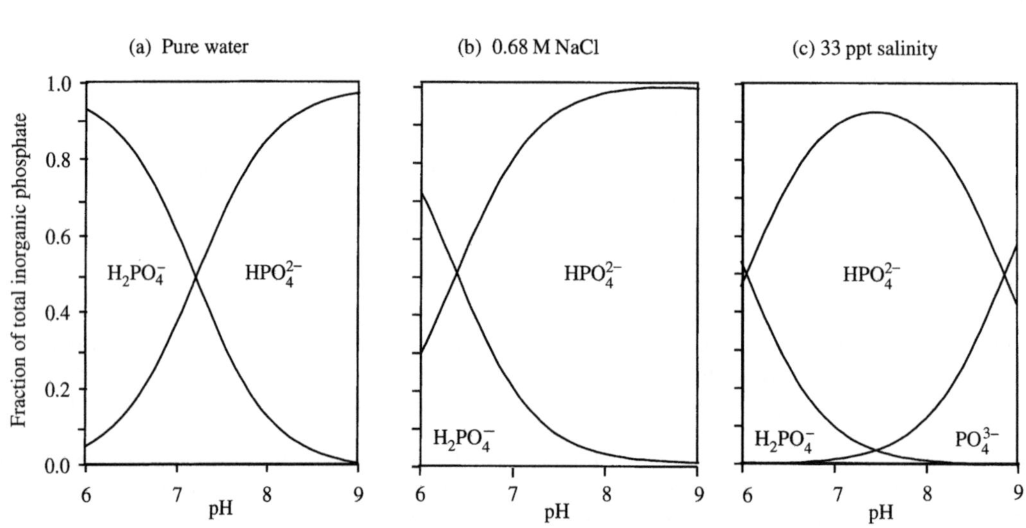

$K_1=7.52\times10^{-3}$ $K_2=6.23\times10^{-8}$ $K_3=4.80\times10^{-13}$

$K_1=2.83\times10^{-2}$ $K_2=0.41\times10^{-6}$ $K_3=0.10\times10^{-10}$

$K_1=2.35\times10^{-2}$ $K_2=0.88\times10^{-6}$ $K_3=1.37\times10^{-9}$

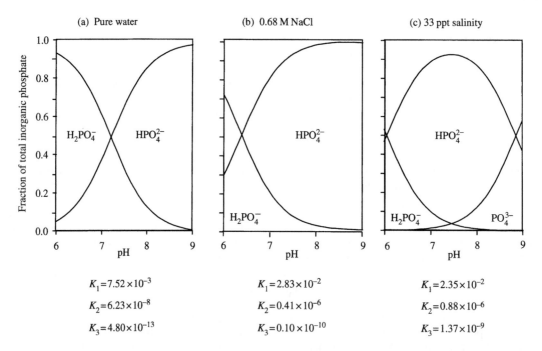

Figure 13 Distribution of phosphoric acid species as a function of pH, and dissociation constants, in (a) pure water; (b) 0.68 M NaCl; and (c) artificial seawater of salinity 33 ppt (after Kester and Pytkowicz, 1967).

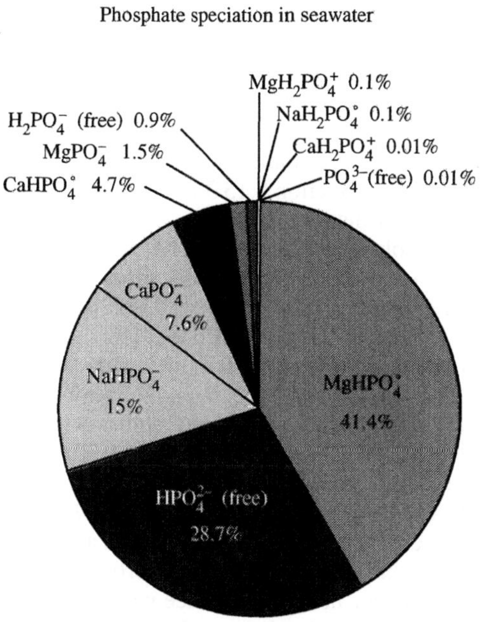

Figure 14 Calculated speciation of inorganic phosphate in seawater at 20 °C, 34.8 ppt salinity, and pH 8 (after Atlas *et al.*, 1976).

by phytoplankton and bacteria, owing to its high degree of bioavailability, and therefore its accumulation in seawater is highly unlikely.

The fundamental characteristics of the oceanic distribution of phosphate were summarized earlier (Section 8.13.2.3, and Figure 1). Levitus *et al.* (1993) present average nutrient distributions in the global ocean, compiled using data available through the National Oceanographic Data Center. This compilation clearly shows the general pattern of phosphate depletion in surface waters due to biological uptake (except in upwelling and high-latitude regions), and increased concentrations with depth due to regeneration from sinking organic matter. Conkright *et al.* (2000) present a similar compilation, taking into account seasonal variability. A clear limitation of these sorts of global compilations are the unavoidable data gaps that exist due to incomplete sampling coverage by oceanographic surveys. For surface waters, which are of interest because of the link between surface phosphate concentration, biological productivity, and the carbon cycle, this problem rapidly becomes insurmountable because of the high degree of variability on seasonal, and shorter, timescales.

With the advent of satellite remote sensing of surface seawater properties, attempts are now being made to use remotely sensed parameters to estimate surface-ocean nutrient (including phosphate) concentrations and distributions. This approach, although it has its own set of limitations, is one that can potentially redress the "data-gap" problem of oceanographic surveys. As a recent example, Kamykowski *et al.* (2002) used the relationship between sea surface temperature and nutrient concentrations to generate date-specific estimates of nutrient levels in the world ocean. They use a derived parameter termed the "nutrient depletion temperature," defined as the temperature above which nutrients are analytically undetectable, to estimate multinutrient availability in the world ocean. The output of this analysis was then linked to satellite-derived chlorophyll-*a* distributions to infer phytoplankton cell-size and taxonomic composition, with the aim of evaluating the impact of available nutrient (phosphate, nitrate, silicate, and iron) levels on phytoplankton community structure. There are other similar approaches being developed (see Kamykowski *et al.*, 2002 for citations), with the overall goal of determining global biogeochemical cycling of phosphate and other bioactive elements, and their impact on the global carbon cycle.

Dissolved organic phosphorus (DOP). Despite early studies indicating the importance of dissolved organic phosphorus (DOP) as a phosphorus source to marine primary producers (e.g., Redfield *et al.*, 1937; Ketchum *et al.*, 1955; Butler *et al.*, 1979; Jackson and Williams, 1985; Rivkin and Swift, 1985; Orrett and Karl, 1987), until recently many assessments of nutrient inventories and phytoplankton nutrition have ignored dissolved organic nutrient reservoirs. Two factors account for past lack of routine DOP data collection: (i) an historical bias in oceanographic nutrient studies favoring nitrogen as the limiting nutrient (e.g., Hecky and Kilham, 1988), relegating phosphate to a role of lesser importance, and (ii) the absence of a widely accepted, quick and convenient method of DOP measurement. Since the 1990s, there has been a growing appreciation of DOP as a significant player in phytoplankton nutrition (e.g., Björkman and Karl, 2003; Wu *et al.*, 2000; Abell *et al.*, 2000; Loh and Bauer, 2000; Monaghan and Ruttenberg, 1999). Accompanying this renewed interest, there have been a number of re-evaluations and improvements in analytical methods (e.g., Karl and Tien, 1992; Thomson-Bulldis and Karl, 1998; Monaghan and Ruttenberg, 1999). This section begins with a brief overview on current methods and recent method developments, followed by a sampling of recent work, highlighting unresolved research questions (see Karl and Bjorkman (2002) for a recent, comprehensive review of DOP, including an historical perspective).

Most available methods determine DOP as the difference between total dissolved phosphorus (TDP) and DIP, where TDP and DIP are quantified on separate splits of a single sample of filtered seawater. DIP is that portion of dissolved phosphorus that readily forms the phosphomolybdate blue complex (e.g., Koroleff, 1983). This operationally defined fraction is sometimes referred to as "soluble reactive phosphorus (SRP)," since it is suspected that some easily hydrolysable DOP compounds may be converted to orthophosphate upon contact with the acidic colorimetric reagents (e.g., Rigler, 1968; Hudson *et al.*, 2000), although experimental acid hydrolysis studies of standard DOP compounds suggest that this effect may be small for many DOP compounds (Monaghan and Ruttenberg, 1999; however, see Baldwin, 1998). TDP is determined on a second split of filtered sample after oxidation/hydrolysis, which converts combined forms of dissolved phosphorus (DOP, polyphosphate) to orthophosphate. Concentration is then determined colorimetrically, as for DIP. Three oxidation/hydrolysis methods have been commonly used: high-temperature combustion/acid hydrolysis, persulfate oxidation, and UV oxidation. Although persulfate oxidation appears to be the most widely used method, Monaghan and Ruttenberg (1999) report inefficient recovery of phospholipids by this method and prefer the high-temperature combustion/hydrolysis method, which has the highest TDP recoveries of all DOP compound classes. In 1990, Ridal and Moore found that UV-oxidation showed poor recovery of nucleotide polyphosphates, and suggest combining the UV- and persulfate-oxidation methods for more complete DOP recovery (for recent reviews of TDP methods see Monaghan and Ruttenberg, 1999, Karl and Björkman, 2002, Mitchell and Baldwin, 2003). Aside from questions about the operationally defined nature of DIP, and hence the accuracy of DIP and DOP concentration determinations, another feature that plagues efforts to quantify DOP is the fact that it is a product of a "difference" method. When TDP and DIP concentrations are similar, a common occurrence in natural waters, and thus DOP is obtained by subtracting two similar numbers, the associated error can be quite large. This is not an uncommon problem in geochemistry, a field often limited by analytical capabilities, where the so-called "difference methods" are quite common.

The MAGIC (magnesium induced co-precipitation) method (Karl and Tien, 1992; Thomson-Bulldis and Karl, 1998) is currently finding wide usage in studies of dissolved phosphorus cycling in surface waters of oligotrophic oceanic regions, where dissolved phosphorus can be at or below detection limits of standard methods. This method calls for adding base to seawater, inducing precipitation of brucite ($Mg(OH)_2$), which effectively strips out dissolved phosphorus via co-precipitation. The $Mg(OH)_2$ pellet is retrieved after centrifugation, and redissolved for colorimetric phosphate determination of DIP. As with the previously described difference methods, a pellet from a separate split is subjected to oxidation/hydrolysis for determination of TDP, and DOP is derived by difference. This method has also been adapted for use in concentrating DOP for isotopic studies of phosphate oxygen in seawater (see Section 8.13.3.3.3; Colman, 2002).

DOP can make up a significant fraction of the total dissolved phosphorus (TDP) pool in surface waters, in some cases seasonally surpassing levels of DIP (Butler *et al.*, 1979; Orrett and Karl, 1987; Cotner *et al.*, 1987; Karl and Tien, 1997). Karl and Björkman (2002) have compiled data from 23 studies from a variety of geographic locations, and report that marine DOP concentrations range from undetectable to as high as 99% of TDP. This compilation indicates that concentrations are highest in surface waters (37–99%TDP), while the lowest concentrations are observed at depth. In view of the extremely high proportions of DOP that can occur in marine surface waters, the locus of phytoplanktonic photosynthesis, a convincing case can be made for the importance of evaluating DOP bioavailability to marine primary producers.

The DOP pool includes phosphate esters (C–O–P bonded compounds, both mono- and diesters), phosphonates (C–P bonded compounds), and P associated with HMW organic matter, such as humic and fulvic acids (Table 7). Most DOP compounds are not available for direct uptake by phytoplankton or bacteria (e.g., Cembella *et al.*, 1984a,b). DOP can be rendered bioavailable by enzymatic hydrolytic production of orthophosphate, a form that can be directly assimilated (Ammerman, 1991). The question of bioavailability is essential for evaluating coupled C–N–P cycling and associated issues such as nutrient limitation, coupled ocean–atmosphere CO_2 dynamics, and climate change (through the nutrient–CO_2 connection, see Section 8.13.3.4.2), and coupled ocean–atmosphere O_2 dynamics (through the O_2–Fe–P–C connection, see Section 8.13.3.4.2). The fact that different DOP compounds display variable enzyme susceptibility, and will therefore be of variable ecological significance, is the primary driving force behind studies of DOP composition in seawater. Studies on the molecular composition of DOP are, however, even less numerous than those on the bulk DOP pool. The paucity of information on DOP in natural waters is due again to analytical difficulties associated with analyzing low DOP levels that are typical of natural waters, a factor that is exacerbated in seawater where any measurements must be made against the background of a complex salt matrix. The situation of analytical roadblocks to detailed study of DOP molecular composition is not dissimilar to that described for solid-phase organic phorphorus (see Section 8.13.3.3.2). A number of approaches are

Table 7 Selected organic phosphorus compounds identified or likely to be present in seawater.

Compound	*Chemical formula* (molecular weight)	*P* (% by weight)	*Molar C :N : P*
Monophosphate esters			
Ribose-5-phosphoric acid (R-5-P)	$C_5H_{11}O_8P$ (230.12)	13.5	5:_:1
Phospho(enol)pyruvic acid (PEP)	$C_3H_5O_6P$ (168)	18.5	3:_:1
Glyceraldehyde 3-phosphoric acid (G-3-P)	$C_3H_7O_6P$ (170.1)	18.2	3:_:1
Glycerphosphoric acid (gly-3-P)	$C_3H_9O_6P$ (172.1)	18.0	3:_:1
Creatine phosphoric acid (CP)	$C_4H_{10}N_3O_5P$ (211.1)	14.7	4:3:1
Glucose-6-phosphoric acid (glu-6-P)	$C_6H_{13}O_9P$ (260.14)	11.9	6:_:1
Ribulose-1,5-bisphosphoric acid (RuBP)	$C_5H_6O_{11}P_2$ (304)	20.4	2.5:_:1
Fructose-1,6-diphosphoric acid (F-1,6-DP)	$C_6H_{14}O_{12}P_2$ (340.1)	18.2	3:_:1
Phosphoserine (PS)	$C_3H_8NO_6P$ (185.1)	16.7	3:1:1
Nucleotides and derivatives			
Adenosine 5′-triphosphoric acid (ATP)	$C_{10}H_{16}N_5O_{13}P_3$ (507.2)	18.3	3.3:1.7:1
Uridylic acid (UMP)	$C_9H_{13}N_2O_9P$ (324.19)	9.6	9:2:1
Uridine-diphosphate-glucose (UDPG)	$C_{15}H_{24}N_2O_{17}P_2$ (566.3)	10.9	7.5:1:1
Guanosine 5′-diphosphate-3′-diphosphate (ppGpp)	$C_{10}H_{17}N_5O_{17}P_4$ (603)	20.6	2.5:1.5:1
Pyridoxal 5-monophosphoric acid (PyMP)	$C_8H_{10}NO_6P$ (247.2)	12.5	8:1:1
Nicotinamide adenine dinucleotide phosphate (NADP)	$C_{22}H_{28}N_2O_{14}\ N_6P_2$ (662)	9.4	11:3:1
Ribonucleic acid (RNA)	Variable	~9.2	~9.5:4:1
Deoxyribonucleic acid (DNA)	Variable	~9.5	~10:4:1
Inositohexaphosphoric acid, or phytic acid (PA)	$C_6H_{18}O_{24}P_6$ (660.1)	28.2	1:_:1
Vitamins			
Thiamine pyrophosphate (vitamin B_1)	$C_{12}H_{19}N_4O_7\ P_2S$ (425)	14.6	6:2:1
Riboflavin 5′-phosphate (vitamin B_2-P)	$C_{17}H_{21}N_4O_9\ P$ (456.3)	6.8	17:4:1
Cyanocobalamin (vitamin B_{12})	$C_{63}H_{88}CoN_{14}O_{14}P$ (1355.42)	2.3	63:14:1
Phosphonates			
Methylphosphonic acid (MPn)	CH_5O_3P (96)	32.3	1:_:1
Phosphonoformic acid (FPn)	CH_3O_5P (126)	24.6	1:_:1
2-aminoethylphosphonic acid (2-AEPn)	$C_2H_8NO_4P$ (141)	22.0	2:1:1
Other compounds/compound classes			
Marine fulvic acid (FA)[a]	Variable	0.4–0.8	80–100:_:1
Marine humic acid (HA)[a]	Variable	0.1–0.2	>300:_:1
Phospholipids (PL)	Variable	≤0.4	~40:1:1
Malathion (Mal)	$C_9H_{16}O_5PS$ (267)	11.6	9:_:1
"Redfield" phytoplankton	Variable	1–3	106:16:1

After Karl and Björkman (2002).
[a] Marine HA and FA are operationally defined fractions, and their composition may be variable (values are from Nissenbaum, 1979). Phosphate associated with HA and FA may be organically bound. Alternatively, it may be inorganic orthophosphate linked to HA and/or FA through metal bridges (Laarkamp, 2000).

currently in use to determine DOP composition and bioavailability, and have provided valuable insights.

^{31}P-NMR spectroscopy applied to the HMW fraction of DOP, concentrated from seawater by ultrafiltration (Clark *et al.*, 1999; Kolowith *et al.*, 2001), has revealed a striking uniformity in HMW-DOP composition at different geographical locations, regardless of water depth or proximity to shore. One unanticipated result of this work was the large proportion of phosphonates (~25%) present in HMW-DOP, the remainder being composed of phosphoesters. These authors hypothesize that the high proportion of phosphonates result from selective preservation of these compounds over time, based on the assumption that phosphonates are less easily degradable than organophosphates (however, see discussion in Section 8.13.3.3.2). Another intriguing result of these ^{31}P-NMR studies was the observation of a sixfold decrease in HMW-DOP concentration from surface to 4,000 m water depth in abyssal sites, with no indication of preferential mineralization of phosphoesters over phosphonates. The latter finding is apparently at odds with the initial hypothesis of preferential preservation of phosphonates in the HMW-DOP pool. Finally, Clark *et al.* (1998) note a shift in the molar C : N : P ratios of the HMW-DOP pool with depth, shifting from 247C : 15N : 1P in surface waters, to 321 : 19 : 1 at 375 m, and finally to 539 : 30 : 1 at 4,000 m water depth, indicating preferential mineralization of phosphorus from HMW-DOP, relative to nitrogen and carbon. Several drawbacks to this methodology in its current state of development include: (i) the necessity of preconcentration (typically by ultrafiltration) to meet detection limit constraints, with possible distortion of *in situ* composition resulting from preconcentration treatments (e.g., Nanny and Minear, 1997; Bauer *et al.*, 1996); (ii) inaccessibility of lower-molecular-weight, possibly more bioavailable, DOP. The HMW (>1 nm) fraction of the DOP pool, which is amenable to preconcentration via ultrafiltration and subsequent NMR analysis, represents only one-third of the total DOP pool; the other two-thirds of this pool, made up of smaller molecular weight DOP compounds, remain outside our window of current analytical accessibility. Despite these drawbacks, ^{31}P-NMR is an exciting and promising tool for the study of DOP, and has provided insights that were not possible with previously available methods.

Other methods that have been used for DOP characterization are briefly summarized here (see Karl and Björkman (2002) for a more detailed discussion of these approaches, including case studies; see also review by Mitchell and Baldwin (2003)). Fractionation of DOP according to molecular weight, the most recent incarnation of which uses tangential flow ultrafiltration (e.g., Nanny *et al.*, 1994; Nanny and Minear, 1997), is often coupled to other diagnostic analytical methods, such as ^{31}P-NMR (see above), or enzyme hydrolysis characterization (Suzumura *et al.*, 1998). Enzyme hydrolysis studies add specific enzymes to natural waters to achieve partial determination of its DOP composition (e.g., Kobori and Taga, 1979; McKelvie *et al.*, 1995; Suzumura *et al.*, 1998). This type of approach is also used to determine the potential bioavailability of the DOP pool (e.g., DeFlaun *et al.*, 1987). Karl and Yanagi (1997) utilized a partial photochemical oxidation technique to gain insights into the composition of oceanic DOP. A number of specific compounds have also been directly measured in seawater (Table 7) using a variety of methodologies, including nucleic acids, ATP and related nucleotides, phospholipids, and vitamins. As it is the quality of bioavailability that is of prime ecological significance, a number of studies have used radiophosphorus uptake studies to determine bioavailability of model DOP compounds (Björkman and Karl, 1994) and *in situ* marine DOP (Björkman *et al.*, 2000; Björkman and Karl, 2003; Bossard and Karl, 1986).

Water column C : P ratios. The concept of preferential regeneration of phosphorus relative to carbon during respiration of organic matter in marine sediments (see Section 8.13.3.3.2) has also been applied to particulate $(C : P)_{org}$ ratios in the marine water column. If it is true, as many sediment studies suggest, that phosphorus is preferentially regenerated during microbial respiration in the early diagenetic environment, it stands to reason that this process would have already begun in the water column. In a series of phytoplankton decomposition experiments, Grill and Richards (1976) demonstrated that phosphorus was regenerated preferentially to nitrogen. Consistent with these experimental findings, a number of sediment trap studies have observed increasing marine particulate $(C : P)_{org}$ ratios with water column depth, to ratios elevated above the Redfield ratio (e.g., Copin-Montegut and Copin-Montegut, 1972; Knauer *et al.*, 1979; Liebezeit, 1991; Honjo and Manganini, 1993). Most of these studies did not account for the effect of mineralization during particulate storage in the sediment trap, and therefore could represent overestimates of the preferential phosphorus regeneration effect (Peng and Broecker, 1987; also discussed in Delaney, 1998 and Anderson *et al.*, 2001). However, Honjo and Manganini (1993) corrected particulate phosphorus concentrations for in-trap mineralization, and the trend of preferential P-mineralization was still observed. In line with these results for water column particulate matter, Clark *et al.* (1998) and Loh and Bauer (2000) have found evidence for preferential phosphorus regeneration (relative to nitrogen and carbon) in HMW- and bulk-DOP, respectively, in the form of increasing ratios with increasing water depth.

In direct contradiction to the suggestion of preferential phosphorus remineralization derived from water column particulates and DOP, outlined above, regression analyses of dissolved nutrient ratios along isopycnals do not find evidence for preferential regeneration of phosphorus relative to carbon in the water column (e.g., Peng and Broecker, 1987). Anderson and Sarmiento concluded that organic matter flux to the sediment water interface must be dominated by "fast-flux particles," such as fecal pellets, and must not differ substantially from the near-Redfield remineralization ratios observed in their isopycnal analysis. They go on to caution that their results may not be applicable to high-latitude regions, to the ocean above 400 m (the upper depth limit of their analysis), or to short time- or length-scales. This disclaimer is significant, given that most organic matter in the modern ocean is buried in relatively shallow, continental margin sediments (Berner, 1982). It is also important to note that each of the studies using hydrographic data to model C : N : P ratios have imbedded in them a series of assumptions, and different approaches can lead to different results. For example, Thomas (2002) conducts a regression analysis that indicates increasing C : P ratios with water depth in the North Atlantic, consistent with sediment trap data (Honjo and Manganini, 1993), but in disagreement with other regression analysis studies. The discrepancies that exist between studies of modeled versus measured C : P ratios indicate that we have not yet arrived at the "truth" about the extent of preferential P-mineralization of sinking organic matter in the marine water column. Reconciling these conflicting lines of argument is an important priority for achieving accurate coupled models of carbon, nitrogen, and phosphorus.

Cosmogenic ^{32}P and ^{33}P as tracers of phosphorus cycling in surface waters. There are two radioactive isotopes of phosphorus, ^{32}P (half-life = 14.3 d) and ^{33}P (half-life = 25.3 d). Both have been widely used in the study of biologically mediated phosphorus cycling in aquatic systems. Until very recently, these experiments have been conducted by artificially introducing radiophosphorus into laboratory incubations, or far more rarely, by direct introduction into natural waters under controlled circumstances. Such experiments necessarily involve significant perturbation of the system, which can complicate interpretation of results. Recent advances in phosphorus sampling and radioisotope measurement have made it possible to use naturally produced ^{32}P and ^{33}P as *in situ* tracers of phosphorus cycling in surface waters (Lal and Lee, 1988; Lee *et al.*, 1991; Waser and Bacon, 1995; Waser *et al.*, 1996; Benitez-Nelson and Buesseler, 1998, 1999; Benitez-Nelson and Karl, 2002). These advances have permitted studies of net P-recycling in the absence of experimental perturbation caused by addition of artificially introduced radiophosphorus.

^{32}P and ^{33}P are naturally produced in the atmosphere by cosmic ray interactions with atmospheric argon nuclei. They are then quickly scavenged onto aerosol particles, and delivered to the ocean surface predominantly in rain. The ratio of $^{33}P/^{32}P$ introduced to the oceans by rainfall remains relatively constant, despite the fact that absolute concentrations can vary from one precipitation event to another. Once the dissolved phosphorus is incorporated into a given surface water phosphorus pool (e.g., uptake by phytoplankton or bacteria, grazing of phytoplankton or bacteria by zooplankton, or abiotic sorption), the $^{33}P/^{32}P$ ratio will increase in a systematic way as a given pool ages. This increase in the $^{33}P/^{32}P$ ratio with time results from the different half-lives of the two phosphorus radioisotopes. By measuring the $^{33}P/^{32}P$ ratio in rain and in different marine phosphorus pools (e.g., DIP, DOP (sometimes called soluble nonreactive phosphorus, or SNP), particulate-P of various size classes corresponding to different levels in the food chain), the net age of phosphorus in any of these reservoirs can be determined (Table 8). New insights into P-cycling in oceanic surface waters derived from recent work using the cosmogenically produced $^{33}P/^{32}P$ ratio include (refer to citations in Table 8): (i) turnover rates of DIP in coastal and oligotrophic oceanic surface waters range from 1 d to 20 d; (ii) variable turnover rates in the DOP pool range from <1 week to >100 d, suggesting differences in either the demand for DOP, or the lability of DOP toward enzymatic breakdown; (iii) in the Gulf of Maine, DOP turnover times vary seasonally, increasing from 28 d in July to >100 d in August, suggesting that the DOP pool may evolve compositionally during the growing seasons; (iv) highly variable TDP residence times

Table 8 Turnover rates of [a]DIP and [b]DOP in surface seawater.

Phosphorus pool	*P turnover rate*		
	Coastal	*Open ocean*	*References*
DIP	<1 h to 10 d (>1,000 d in Bedford Basin)	Weeks to months	1, 2, 3, 4, 5, 6, 7, 8, 9, 10, 11
Total DOP	3 d to >90 d	20–300 d	10, 11, 12, 13, 14, 15, 16, 22
Bioavailable DOP (model compounds)	2–30 d	1–4 d	9, 17, 18, 19
Microplankton (<1 μm)	>1–3 d	NA	11
Phytoplankton (>1 μm)	<1–8 d	<1 week	11, 20
Macrozooplankton (>280 μm)	14–40 d	30–80 d	11, 14, 15, 20, 21

After Benifez-Nelson (2000).

References: (1) Pomeroy (1960); (2) Duerden (1973); (3) Taft *et al.* (1975); (4) Harrison *et al.* (1977); (5) Perry and Eppley (1981); (6) Smith *et al.* (1985); (7) Sorokin (1985); (8) Harrison and Harris (1986); (9) Björkman and Karl (1994); (10) Björkman *et al.* (2000); (11) Benitez-Nelson and Buesseler (1999); (12) Jackson and Williams (1985); (13) Orrett and Karl (1987); (14) Lal and Lee (1988); (15) Lee *et al.* (1992); (16) Karl and Yanagi (1997); (17) Ammerman and Azam (1985); (18) Nawrocki and Karl (1989); (19) Björkman and Karl (2001); (20) Waser *et al.* (1996); (21) Lee *et al.* (1991); (22) Benitez-Nelson and Karl (2002).

[a] DIP is equivalent to the soluble reactive P (SRP) pool, which may include some phosphate derived from hydrolysis of DOP (e.g., see Monaghan, and Ruttenberg, 1998). [b] DOP is equivalent to the soluble nonreactive P (SNP) pool which may include dissolved inorganic polyphosphates (e.g., see Karl and Yanagi, 1997.

in the North Pacific Subtropical Gyre suggest dynamic phosphorus utilization over short timescales of weeks to months, and correlation of apparent TDP ages with C-fixation rates suggests preferential removal of younger, presumably more labile TDP; (v) comparison of the $^{33}P/^{32}P$ ratio in different particulate size classes indicates that the age of phosphorus generally increases at successive levels in the food chain; and (vi) under some circumstances, the $^{33}P/^{32}P$ ratio can reveal which dissolved pool is being ingested by a particular size class of organisms. Utilization of this new tool highlights the dynamic nature of P-cycling in surface waters by revealing the rapid rates and temporal variability of P-turnover. It further stands to provide new insights into ecosystem nutrient dynamics by revealing, for example, that (i) low phosphorus concentrations can support high primary productivity through rapid turnover rates, and (ii) there is preferential utilization of particular dissolved phosphorus pools by certain classes of organisms.

Oxygen isotopes of phosphate in seawater. Use of the oxygen isotopic composition of phosphate in biogenic hydroxyapatite (bones, teeth) as a paleotemperature and climate indicator was pioneered by Longinelli (1966), and has since been fairly widely and successfully applied (Kolodny *et al.*, 1983; Shemesh *et al.*, 1983; Luz *et al.*, 1984). The relationship between phosphate oxygen isotopic fractionation and temperature originally proposed (Longinelli and Nuti, 1973) has been confirmed by later work (Blake *et al.*, 1997; Colman, 2002).

Phosphorus has only one stable isotope (^{31}P) and occurs almost exclusively as orthophosphate (PO_4) under Earth surface conditions. The phosphorus-oxygen bond in phosphate is highly resistant to nonenzymatic oxygen isotope exchange reactions, but when phosphate is metabolized by living organisms, that is, when organic phosphorus compounds are first biosynthesized and subsequently enzymatically hydrolyzed in the intracellular environment, oxygen isotopic exchange is rapid and extensive (Blake *et al.*, 1997, 1998; Paytan *et al.*, 2002). Such exchange results in temperature-dependent fractionations between phosphate and ambient water (Colman, 2002). This property renders phosphate oxygen isotopes useful as indicators of present or past metabolic activity of organisms, and allows distinction of biotic from abiotic processes operating in the cycling of phosphorus through the environment.

New methods for the isolation and purification of inorganic phosphate (P_i) from natural waters (Colman *et al.*, 2000; Colman, 2002) have permitted phosphate–oxygen isotopic ($\delta^{18}O-P_i$) analysis of dissolved seawater inorganic phosphate as a tracer of phosphate source, water mass mixing, and biological productivity (Colman, 2002). In this study, Colman (2002) demonstrates that microbial phosphate cycling imprints the $\delta^{18}O-P_i$ signature of major aquatic phosphate reservoirs. For example, on the basis of residence time constraints, Colman (2002) demonstrates that microbial cycling of P_i in Long Island Sound (USA) estuary is rapid enough to overprint, on a time scale of weeks, the distinctly different $\delta^{18}O-P_i$ value of the major water source to the estuary (Connecticut River). Biological P_i turnover appears to keep pace with significant temperature shifts (3–9°C cooling) experienced by Long Island Sound surface waters during the study period, leading Colman (2002) to conclude that turnover rates could be even more rapid than implied on the basis of residence time constraints alone. This observation is particularly interesting, given that Long Island Sound is generally considered to be strongly nitrogen limited (Colman, 2002).

Another example of insights gained into microbial P_i cycling through application of $\delta^{18}O-P_i$ analysis derives from the observation (Colman, 2002) that deepwater ocean samples from two oligotrophic oceanic regions (the Subtropical North Pacific Gyre, and the Sargasso Sea in the North Atlantic) are offset from equilibrium values. The direction of offset (to lower values) implies that some fraction of the P_i reservoir at depth is regenerated extracellularly from sinking particulate organic matter, resulting in only partial equilibration of the regenerated P_i flux (Colman, 2002).

The $\delta^{18}O-P_i$ system has also recently been applied to phosphates associated with ferric iron oxyhydroxide precipitates in submarine ocean ridge sediments (Blake *et al.*, 2000, 2001). The $\delta^{18}O-P_i$ signature of phosphate associated with these authigenic Fe-oxyhydroxide precipitates indicates microbial phosphate turnover at elevated temperatures. The latter observation suggests that phosphate oxygen isotopes may be useful biomarkers for fossil hydrothermal vent systems. On the basis of this work, Blake *et al.* (2001) also hypothesize that authigenic phases extant on other planets may retain imprints of primitive biospheres, in the form of detectable and diagnostic $\delta^{18}O-P_i$ composition, imparted by biochemical, enzymatic processes.

8.13.3.3.4 Phosphorus limitation of marine primary photosynthetic production

In terrestrial soils and in the euphotic zone of lakes and the ocean the concentration of dissolved orthophosphate is typically low. When bioavailable phosphorus is exhausted prior to more abundant nutrients, it limits the amount of sustainable biological productivity. This is the ecological principle often referred to as Liebig's (1840) Law of the Miniminum, first established for terrestrial plants and then adapted to phytoplankton growth (Blackman, 1905), in which the standing phytoplankton stock is limited by the substance least available relative to the amount required for synthesis of healthy biomass. Phosphorus limitation in lakes is widely accepted (e.g., Hecky and Kilham, 1988), and terrestrial ecosystems are often phosphorus-limited (e.g., Lajtha and Harrison, 1995; Chadwick *et al.*, 1999). In the oceans, however, phosphorus limitation is the subject of controversy and debate (e.g., Smith, 1984; Hecky and Kilham, 1988; Codispoti, 1989; Falkowski, 1997; Palenik and Dyhrman, 1998).

The prevailing paradigm among geochemists is that phosphorus is most probably the limiting nutrient on long, geologic time-scales (e.g., Redfield, 1958; Holland, 1978; Broecker and Peng, 1982; Smith and Mackenzie, 1987; Kump, 1988). According to this paradigm, while there is an abundant reservoir of nitrogen (gaseous N_2) in the atmosphere that can be rendered bioavailable by N-fixing photosynthetic organisms, phosphorus supply to the ocean is limited to that weathered off the continents and delivered by rivers, with some minor atmospheric input. As a consequence of continental weathering control on phosphorus supply to the oceans (and ultimately, tectonic control, see Section 8.13.3.4.1) phosphorus availability limits net primary production on geological timescales (however, see Codispoti (1989), Falkowski (1997), and Lenton and Watson (2000), for counter-arguments that favor nitrogen limitation).

In the modern ocean, in contrast, the prevailing view has been that nitrogen, and not phosphorus, is the limiting nutrient for marine primary productivity (Codispoti, 1989; Tyrrell, 1999). This paradigm was emplaced in the 1970s, largely due to the work of Ryther and Dunstan (1971), who showed that nitrate stimulated phytoplankton growth off the coast of Long Island, NY, whereas phosphate did not. This view has been widely supported by studies using large oceanographic data sets (e.g., GEOSECS, TTO, WOCE), which show that when nitrate and phosphate concentrations are plotted against one another for the world's oceans, nitrate drops to undetectable limits first, with residual phosphate present; or alternatively that the nitrate: phosphate ratio for much of the world's oceans is equal to or slightly lower than the Redfield ratio (e.g., Tyrrell and Law, 1997; Tyrrell, 1999).

Several factors not considered by the N-limitation paradigm have been receiving attention recently. For example, the recognition that phytoplankton can utilize dissolved organic nutrients, both phosphorus (DOP) and nitrogen, render questionable any conclusions drawn on the basis of inorganic nutrients alone (e.g., Jackson and Williams, 1985; Karl *et al.*, 1997; Palenik and Dyhrman, 1998; Karl and Björkman, 2002). Further, there is a growing appreciation for the heterogeneity of the ocean, with different regions potentially experiencing different limiting nutrient conditions (e.g., Palenik and Dyhrman, 1998), sometimes variable on decadal (e.g., Karl, 1999) or shorter timescales (e.g., Monaghan and Ruttenberg, 1999; Dyhrman and Palenik, 1999). Consideration of these and other factors suggest that the nutrient limitation debate has been too simplistic.

The dearth of studies on marine phosphorus biogeochemistry prior to the early 1980s can in part be explained by the entrenched view of marine scientists that phosphorus was the "lesser important" of the essential nutrients, since nitrogen and not phosphorus was considered to be *the* limiting nutrient. This view is currently being challenged in a number of different venues, by a number of different scientists, and a growing consensus is emerging that P-limitation may be important not just on geologic time scales, but in the modern ocean, as well. The remainder of this section highlights the growing body of literature that present evidence for phosphorus limitation of primary productivity in some marine systems.

In the oligotrophic gyres of both the western North Atlantic and subtropical North Pacific, evidence in the form of dissolved N : P ratios has been used to argue convincingly that these systems are currently P-limited (e.g., Ammerman *et al.*, 2003; Cavender-Barres *et al.*, 2001; Karl *et al.*, 2001; Wu *et al.*, 2000). The N : P ratio of phytoplankton under nutrient sufficient conditions is 16N : 1P (the Redfield ratio). A positive deviation from this ratio indicates probable phosphorus limitation, while a negative deviation indicates probable nitrogen limitation. In the North Pacific at the Hawaiian Ocean Time Series (HOT) site, there has been a shift since the 1988 inception of the time series to N : P ratios exceeding the Redfield ratio in both particulate and surface ocean dissolved nitrogen and phosphorus (Figure 15). Coincident with this shift has been an increase in the prevalence of the N-fixing cyanobacterium *Trichodesmium* (Table 9). Currently, it appears that the supply of new nitrogen has shifted from a limiting flux of upwelled nitrate from below the euphotic zone to an unlimited pool of atmospheric N_2 rendered bioavailable by the action of nitrogen-fixers. This shift is believed to result from climatic changes that promote water column stratification, a condition that selects for N_2-fixing microorganisms, thus driving the system to P-limitation. A similar situation exists in the subtropical Sargasso Sea at the Bermuda OceanTime Series (BATS) site,

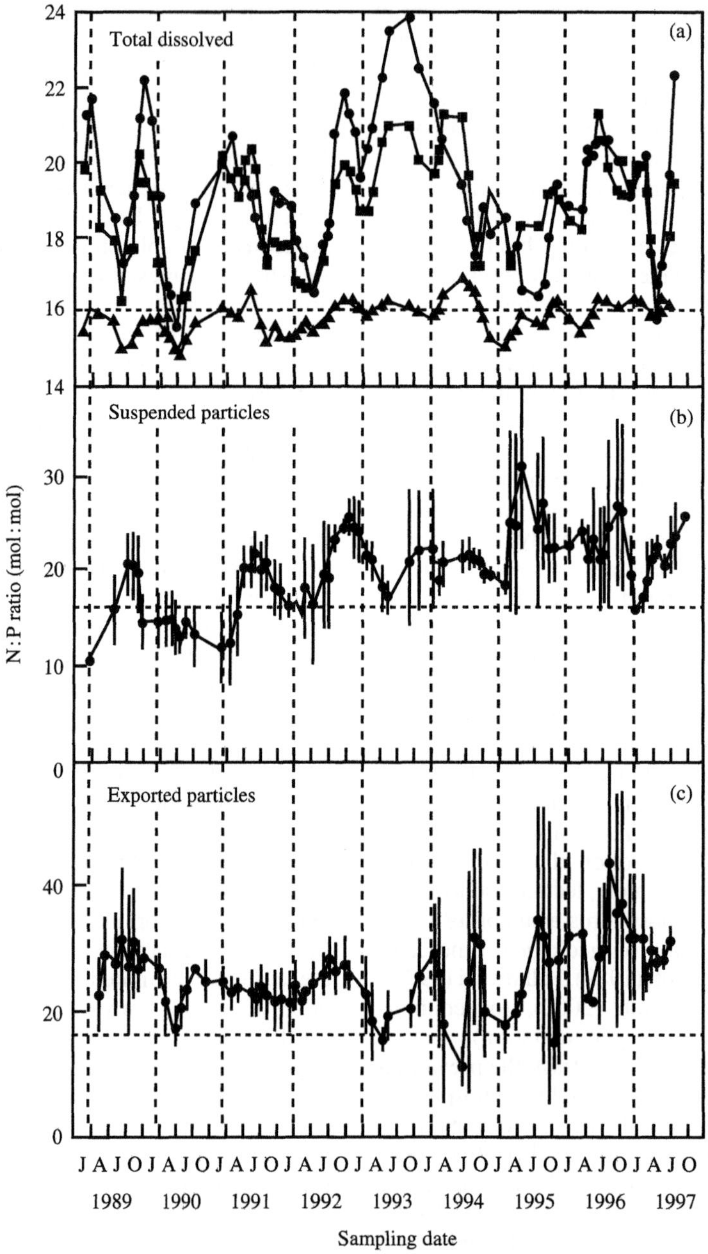

Figure 15 Time-series molar N : P ratios in (a) the dissolved pool, (b) suspended particulate matter, and (c) exported particulate matter from the HOT time-series site at station ALOHA in the subtropical North Pacific near Hawaii. Panel (a) shows the three-point running mean N:P ratios for 0–100 m (circles), 100–200 m (squares), and 200–500 m (triangles). Panel (b) shows the three-point running mean (±1 SD) for the average suspended particulate matter in the upper water column (0–100 m). Panel (c) shows the three-point running mean (±1 SD) for the average N : P ratio of sediment trap-collected particulate matter at 150 m. The Redfield ratio (N : P = 16) is represented by a dashed line in all three panels. Particulate and upper water column dissolved pools show an increasing N : P ratio throughout the time-series, with a preponderance of values in excess of the Redfield ratio (after Karl *et al.*, 1997).

where currently the dissolved phosphorus concentrations (especially DIP) are significantly lower than at the HOT site, indicating even more severe P-limitation (Table 9; see also Ammerman *et al.*, 2003).

A number of coastal systems also display evidence of P-limitation, sometimes shifting seasonally from nitrogen to phosphorus limitation in concert with changes in environmental features such as upwelling and river runoff. On the Louisiana Shelf in the Gulf of Mexico (MacRae *et al.*, 1994; Pakulski *et al.*, 2000), the Eel River Shelf of northern California (USA) (Monaghan and Ruttenberg, 1999), the upper Chesapeake Bay

Table 9 Parameters affecting nutrient limitation: comparison between North Atlantic and North Pacific gyres.

Parameter	*Sargasso Sea*	*Pacific HOT site*
DIP(nM)	0.48 ± 0.27[a]	9–40[b]
TDN (nM)	4512 ± 430	5680 ± 620[b]
TDP (nM)	75 ± 42	222 ± 14[b]
TDN:TDP	60 ± 7	26 ± 3[b]
N_2-fixation rate (mmol N m^{-2} yr^{-1})	72[c]	31–51[c]

After Wu *et al.* (2000).
[a] Average DIP between 26° and 31° N in Sargasso Sea surface waters in March 1998. [b] North Pacific near Hawaii at station ALOHA (the HOT site) during 1991–1997. [c] See Wu *et al.* (2000) for method of calculation or measurement.

(USA) (Fisher *et al.*, 1992; Malone *et al.*, 1996), and regions of the Baltic Sea (Granéli *et al.*, 1990) surface water column dissolved inorganic N : P ratios indicate seasonal phosphorus limitation. The suggestion of P-limitation is reinforced in the Louisiana and Eel River Shelf studies by the occurrence of alkaline phosphatase activity (MacRae *et al.*, 1994; Monaghan and Ruttenberg, 1999), an enzyme induced only under conditions of physiologically stressful low phosphate concentrations that can be P-limiting (e.g., Ammerman, 1991; Dyhrman and Palenik, 1999). Alkaline phosphatase has also been observed seasonally in Narragansett Bay (Dyhrman and Palenik, 1999). In this latter study, a novel probe was utilized that permits evaluation of phytoplankton P physiology at the single-cell level, by fluorescently labeling the site of alkaline phosphatase activity (see Section 8.13.3.2.1, discussion of the ELF® probe). Although these coastal sites are recipients of anthropogenically-derived nutrients (N and P) that stimulate primary productivity above 'natural' levels, the processes that result in shifts in the limiting nutrient are not necessarily related to anthropogenic effects. Some other oceanic sites where P limitation of primary productivity has been documented include the Mediterranean Sea (Krom *et al.*, 1991), Florida Bay (USA) (Fourqurean *et al.*, 1993), Northeastern Gulf of Mexico (Myers and Iverson, 1981); the Chesapeake Bay (Fisher *et al.*, 1992, 1995; Malone *et al.*, 1996) Shark Bay (Australia) (Smith and Atkinson, 1984), Trondheims fjord, Norway (Myklestad and Sakshaug, 1983) and Oslofjord, Norway (Paasche and Erga, 1988), the Baltic Sea (Granéli *et al.*, 1990).

One key question that studies of N- and P-limitation must address before meaningful conclusions may be drawn about P- versus N-limitation of marine primary productivity, is the extent to which the dissolved organic nutrient pools are accessible to phytoplankton. In brief, this is the question of bioavailability. Many studies of nutrient cycling and nutrient limitation do not include measurement of these quantitatively important nutrient pools (e.g., Downing, 1997), even though there is indisputable evidence that some portion of the DOP and DON pools are bioavailable (e.g., Rivkin and Swift, 1980; Björkman and Karl, 1994; Karl and Björkman, 2002). Dissolved inorganic N : P ratios and the presence of alkaline phosphatase (APase) activity are two common diagnostic parameters for inferring P-limitation in field studies. The occurrence of high N : P ratios and APase activity imply that phytoplankton are stressed by low DIP levels. It is important to note that P-stress will progress to P-limitation only if the phytoplankton physiological response to P-stress, e.g., synthesis of APase, fails to relieve the P-stress. Progression to P-limitation thus is linked to DOP bioavailability. An important direction for future research is to characterize the DOP (and DON) pools at the molecular level, and to evaluate what fraction of these are bioavailable. The analytical challenge of identifying the molecular composition of the DOP pool is significant. As discussed in Section 8.13.3.3.3, recent advances in ^{31}P-NMR spectroscopy have permitted a first look at the high molecular weight (>1 nm) fraction of the DOP pool, but smaller, possibly more bioavailable DOP compounds remain outside the window of current analytical accessibility.

8.13.3.3.5 The oceanic residence time of phosphorus

As phosphorus is considered the most likely limiting nutrient on geologic timescales (e.g., Redfield, 1958; Broecker and Peng, 1982; Holland, 1978, 1984; Smith, 1984; Howarth *et al.*, 1995; see Section 8.13.3.3.4), an accurate determination of its oceanic residence time is crucial to understanding how levels of primary productivity may have varied in the Earth's past. Residence time provides a means of evaluating how rapidly the oceanic phosphorus inventory may have changed in response to variations in either input (e.g., continental weathering, dust flux) or output (e.g., burial with sediments). In its role as limiting nutrient, phosphorus will dictate the amount of surface–ocean net primary productivity, and hence atmospheric CO_2 draw-down that will occur by photosynthetic biomass production. It has been suggested that this so-called "nutrient–CO_2"connection may link the oceanic phosphorus cycle to climate change due to reductions or increases in the atmospheric greenhouse gas inventory (Broecker, 1982a,b; Boyle, 1990; see Section 8.13.3.4.2). Oceanic phosphorus residence time and response time (the inverse of residence time) will dictate the timescales over which such a P-induced climate effect may operate.

Since the 1990s there have been several re-evaluations of the marine phosphorus cycle in the literature, reflecting changes in our understanding of the identities and magnitudes of important P-sources and P-sinks. Newly identified marine P-sinks include disseminated authigenic carbonate fluorapatite (CFA) in sediments in nonupwelling environments (see Section 8.13.3.3.2), authigenic (REE)-aluminophosphates in sandstones (Rasmussen, 2000; Section 8.13.3.3.2), and hydrothermal P-scavenging via basalt–seawater interaction on ridge-flanks of MOR systems (Wheat *et al.*, 1996; Section 8.13.3.3.2). Continental margins in general are quantitatively important sinks for organic and ferric iron-bound phosphorus, as well. When newly calculated P-burial fluxes in continental margins, including the newly identified CFA and aluminophosphate phosphorus (dominantly Al-REE-P) sinks are combined with older estimates of P-burial fluxes in the deep sea, the overall burial flux results in a much shorter residence time than the canonical value of 100 kyr found in most text books (Table 3). This shorter residence time suggests that the oceanic P-cycle is subject to perturbations on shorter timescales than has previously been believed.

The revised, larger burial flux cannot be balanced by the dissolved riverine input alone. However, when the fraction of riverine particulate-P that is believed to be released upon entering the marine realm is taken into account, the possibility of a balance between inputs and outputs becomes more feasible. Residence times estimated on the basis of P-inputs that include this "releasable" riverine particulate-P fall within the range of residence time estimates derived from P-burial fluxes (Table 3). Despite the large uncertainties associated with these numbers, as evidenced by the maximum and minimum values derived from both input and removal fluxes, these updated residence times are all significantly shorter than the canonical value of 100 kyr. Residence times on the order of 10–17 kyr make P-perturbations of the ocean–atmosphere CO_2 reservoir on the timescale of glacial–interglacial climate change feasible. It is interesting to speculate about the fate of P sequestered in continental margin sediments during high sea level stands once sea level drops, as it does during glacial times. As the phosphorus in margin sediments becomes subjected to subaereal weathering and erosion as sea level lowers, if it is remobilized in bioavailable forms, it could augment productivity-driven CO_2 draw-down and enhance glacial cooling trends (Compton *et al.*, 1993; Ruttenberg, 1993).

8.13.3.4 Phosphorus Cycling Over Long, Geologic Timescales

8.13.3.4.1 The role of tectonics in the global phosphorus cycle

Given the well-studied role of tectonics in carbon cycling on timescales of tens to hundreds of millions of years (e.g., Berner *et al.*, 1983), it has recently been recognized that, owing to the linkage of carbon and phosphorus cycles through photosynthetic uptake (see discussion in Sections 8.13.2.3 and 8.13.3.3.2), tectonics likely play an important role in the global phosphorus cycle over these long timescales, as well (Guidry *et al.*, 2000). In this formulation, the balance between subduction of phosphorus bound up in marine sediments and underlying crust, and creation of new crystalline rock, sets the mass of exogenic phosphorus.

In brief, P-delivery rate to the ocean is controlled by continental weathering. Unlike carbon and nitrogen, there is no gaseous phosphorus phase of any importance to supplement the continental weathering flux. (Note that the reduced phosphorus gas phosphine (PH_3) has been found in trace quantities in highly reducing media (Dévai *et al.*, 1988; Gassman, 1994), but is an insignificant reservoir at current Earth surface conditions (Burford and Bremner, 1972; Schink and Friedrich, 2000), unless it is a more important component of reduced volcanic gases that has been recognized.) P-removal rate from the ocean is controlled by burial with sediments. The primary removal mechanism is production, deposition, and burial of organic matter in sediments. Although some phosphorus is lost from sediments during diagenesis, and some is converted from organic to inorganic phases through sink-switching (see Section 8.13.3.3.2), the phosphorus that remains within the sediment pile accumulating on the sea bottom is eventually recycled into the mantle, along with accompanying bioactive elements (e.g., carbon, nitrogen), upon subduction of the oceanic plate. Marine sedimentary organic matter undergoes metamorphism at the elevated temperatures and pressures experienced by the subducting plate, causing volatilization of organic carbon and nitrogen, with subsequent release as CO_2 and N_2. Because there is no quantitatively important phosphorus gas phase that forms under metamorphic conditions, sedimentary organic phosphorus on the subducting oceanic plate is likely incorporated into crystalline apatite during subduction-zone metamorphism. The net result is that subducted organic phosphorus cannot be returned to the Earth's surface at the same rate as subducted organic carbon and nitrogen, and thus phosphorus is decoupled from carbon and nitrogen cycles during subduction and metamorphism. Furthermore, once exposed on land through tectonic uplift processes, the phosphorus residing in crystalline rocks formed during subduction and metamorphism must be chemically weathered in order to supply bioavailable phosphorus to the ocean (Guidry *et al.*, 2000).

Thus, if phosphorus is the limiting nutrient for the ocean over geologic timescales, as has been the paradigm of geochemists (see Section 8.13.3.3.4 for summary of arguments for and against the paradigm of P-limitation), return of phosphorus to the Earth's surface via crystalline rock production, sediment uplift, and subsequent chemical weathering controls the amount of phosphorus returning to the ocean, and thus controls oceanic productivity over geologic timescales (Guidry *et al.*, 2000). In this way, Guidry *et al.* (2000) argue convincingly that tectonics may play the ultimate role in controlling the exogenic phosphorus mass, resulting in long-term P-limited productivity in the ocean.

8.13.3.4.2 Links to other biogeochemical cycles on long, geologic timescales

The nutrient–CO_2 connection. The biogeochemical cycles of phosphorus and carbon are linked through photosynthetic uptake and release during respiration (Section 8.13.3.3.2). During times of elevated marine biological productivity, enhanced uptake of surface water CO_2 by photosynthetic organisms results in increased CO_2 invasion from the atmosphere, which persists until the supply of the least abundant nutrient is exhausted. On geologic timescales, phosphorus is likely to function as the limiting nutrient and thus play a role in atmospheric CO_2 regulation by limiting CO_2 draw-down by oceanic photosynthetic activity. Because atmospheric CO_2 exerts a greenhouse warming effect, changes in phosphorus (or other nutrient) inventories can impact atmospheric CO_2 levels and play a role in global climate change. Several studies have argued that oceanic phosphorus has exerted just such an effect on climate, triggering or enhancing climate change on glacial–interglacial timescales (Broecker, 1982a,b; Boyle, 1990; Ganeshram *et al.*, 2002; Tamburini *et al.*, 2003).

In order to estimate paleoceanographic nutrient levels, and thereby explore links between nutrient variability, atmospheric CO_2 and climate change, a range of nutrient proxies have been developed. The ratio of cadmium (Cd) to calcium (Ca) in benthic foraminifera has been used as a proxy for dissolved phosphate, based on the observation that dissolved cadmium and phosphate concentrations are linearly correlated in modern oceanic waters (Boyle, 1988). While the basis of the Cd : PO_4 relationship is not understood, from either a biochemical or geochemical standpoint (e.g., Cullen *et al.*, 2003), the Cd : Ca ratio in benthic foraminifera has found wide application as a proxy for dissolved phosphate in the paleo-ocean (see summary in Boyle, 1990). Important questions about the reliability of Cd : Ca as a phosphate proxy remain to be addressed. For example, since it appears that benthic phosphate flux from the seabed is a globally important source of phosphorus to the ocean (Colman and Holland, 2000), it is critical to examine the relative behavior of phosphate and cadmium during early diagenesis. Uncoupling of cadmium and phosphate during early diagenesis, with implications for uncoupled benthic flux of these elements, could negatively impact the reliability of the cadmium : phosphorus proxy (e.g., McCorkle and Klinkhammer, 1991; Rosenthal *et al.*, 1995a). It has been argued that the decoupling, although it has been documented, has minimal effects on the oceanic budgets of cadmium and phosphate (Rosenthal *et al.*, 1995b). Another consideration is, while cadmium may be a good proxy for phosphate, if dissolved organic phosphorus is important in phytoplankton nutrition, as seems likely (see Sections 8.13.3.3.3 and 8.13.3.3.4), the utility of the Cd : PO_4 proxy for assessing nutrient-constraints on levels of oceanic primary production may be diminished. This latter point suggests that it may be interesting to examine the relationship between cadmium and TDP, although it is still unclear what fraction of the DOP pool is bioavailable. Finally, it appears that phytoplankton Cd : PO_4 ratios vary as a function of environmental parameters that affect phytoplankton growth rates, suggesting that extrapolation from Cd : Ca ratios to surface phosphate concentrations may not be a straightforward endeavor (Cullen *et al.*, 2003).

The phosphorus–iron–oxygen connection. Phosphorus and oxygen cycles are linked through the redox chemistry of iron. Ferrous iron is unstable at the Earth's surface in the presence of oxygen, and oxidizes to form ferric iron oxyhydroxide precipitates, which are extremely efficient scavengers of dissolved phosphate (see Sections 8.13.3.1, 8.13.3.2 and 8.13.3.3.2). Resupply of phosphate to surface waters, where it can fertilize biological productivity, is reduced when oceanic bottom waters are well-oxygenated due to scavenging of phosphate by ferric oxyhydroxides. In contrast, during times in Earth's history when oxygen was not abundant in the atmosphere (Precambrian), and when expanses of the deep ocean were anoxic (e.g., Cretaceous), the potential for a larger oceanic dissolved phosphate inventory could have been realized due to the reduced importance of sequestering with ferric oxyhydroxides. This iron–phosphorus–oxygen coupling produces a negative feedback, which may have kept atmospheric O_2 within equable levels throughout the Phanerozoic. The feedback between atmospheric oxygen and phosphorus availability has been recognized in a number of studies (Kump, 1988; Berner and Canfield, 1989; Holland, 1990, 1994; Betts and Holland, 1991; Herbert and Sarmiento, 1991; Colman and Holland, 1994; Van Cappellen and Ingall, 1994a,b; Colman *et al.*, 1997; Petsch and Berner, 1998).

In some of the most recent work on this topic, Van Cappellen and Ingall formalize and explore this feedback through mass balance calculations using a coupled model of carbon, phosphorus, oxygen, and iron biogeochemical cycles, and conclude that oceanic P-cycling plays a determining role in long-term stabilization of atmospheric oxygen. Colman *et al.* (1997) take exception to some of the assumptions made in Van Cappellen and Ingall's work, and demonstrate that explicit inclusion of weathering rates, not included in the original model, markedly impacts the oceanic bioavailable P-inventory and affects the model's prediction of O_2 levels. Petsch and Berner (1998) incorporate the weathering rate effect demonstrated by Colman *et al.* (1997) into a revised coupled $C-O_2-Fe-P$ model. Colman *et al.* (1997) also point out that questions remain concerning preferential P-regeneration from sedimentary organic matter under anoxic conditions, and this is a core condition of the negative feedback in Van Cappellen and Ingall's model. Further, according to subsequent work (Anderson *et al.*, 2001), it is unclear that sedimentary organic C : P ratios are a robust indicator of organic-P burial in any case, since diagenetic sink-switching entraps some portion of regenerated phosphorus from organic matter into secondary inorganic phases (see discussion in Section 8.13.3.3.2).

8.13.3.4.3 *Phosphorus in paleoceanography: P-burial as a Proxy for weathering, paleoproductivity, and climate change*

A number of studies have used records of phosphorus burial in marine sediments as direct indicators of weathering, delivery of bioavailable phosphorus to the oceans, paleoproductivity, and resultant impacts on climate through the nutrient–CO_2 connection (Filippelli and Delaney, 1994; Delaney and Filippelli, 1994; Filippelli, 1997a,b; Föllmi, 1995a,b). This is a challenging undertaking, as it is difficult to extrapolate from regional studies to global-scale phenomena, and many uncertainties remain about the geological, chemical, and biological controls on fluxes of phosphorus into and out of the ocean (Ruttenberg, 1994; this chapter). In a more regional study, Slomp *et al.* (2002) use P-speciation of sediments from Mediterranean cores to infer depositional conditions that existed during ancient sapropel formation. In another type of paleoceanographic study, Tamburini *et al.* (2003) document sediment P-speciation in a North Atlantic sediment core in which they are able to correlate phosphorus geochemistry with Heinrich events (ice-rafting of terrestrially derived detrital material to the open ocean during glacial times). In this latter study, P-speciation is used to infer bottom water redox conditions and make inferences about water column stratification induced by enhanced freshwater input coincident with the Heinrich events. Variations in the global phosphorus cycle on a number of different timescales are reviewed by Compton *et al.* (2000), ranging from glacial–interglacial time periods to timescales greater than 1 Myr.

8.13.3.4.4 *Ancient phosphorites*

The episodic occurrence of phosphorite formation as implied from the episodic abundance of giant phosphorite deposits throughout the Phanerozoic rock record (Figure 16), has motivated questions about perturbations to the global phosphorus cycle throughout geologic time, and the implications of such perturbations on other global cycles of other bio-elements, and climate (Cook and McElhinney, 1979; Cook *et al.*, 1990; Arthur and Jenkyns, 1981; Föllmi, 1990; Donnelly *et al.*, 1990). The huge variations in P-removal implied by these enormous phosphorite deposits imply extreme perturbations to the global ocean. Among the paleoceanographic conditions that have been linked to the genesis of giant phosphorite deposits are ocean anoxic events (Cook and McElhinny, 1979; Arthur and Jenkyns, 1981), sea-level change (Arthur and Jenkyns, 1981; Sheldon, 1980; Riggs, 1984; Riggs *et al.*, 1985), and plate tectonics (Cook and McElhinny, 1979; Sheldon, 1980; Compton *et al.*, 2000). Several studies have suggested a link between phosphorite formation and evolution (Cook and Shergold, 1984; Donnelly *et al.*, 1990). These and other studies are reviewed by Cook *et al.* (1990). No master variable has been identified that can account for the formation of these giant phosphorite deposits, and it may be that the confluence of different factors were important at different times. There is still uncertainty about whether these giant phosphorite deposits represent increased input of reactive phosphorus to the ocean, or whether they instead represent regionally focused phosphorus removal resulting from changes in ocean circulation (e.g., large, sustained upwelling sites) and concentration by later reworking (Compton *et al.*, 2000). With the recent identification of disseminated CFA formation in nonphosphorite forming environments (see Section 8.13.3.3.2), it is interesting to speculate how superimposition of

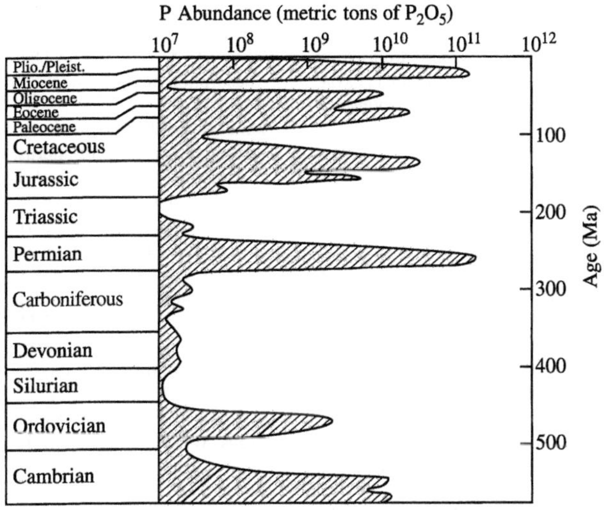

Figure 16 The abundance of phosphate in sedimentary rocks as a function of geologic age show episodicity of giant phosphorite deposition (after Cook and McElhinny, 1979).

P-burial as disseminated CFA onto the occurrence of phosphorus in phosphorites throughout geologic time (Figure 16) would change perceptions of the global impact of phosphogenic events. Would the high P-abundances during periods of phosphorite formation be enhanced by addition of disseminated CFA burial in nonphosphorite environments, thus augmenting their suggestion of extreme conditions of oceanic P-removal? Or would disseminated CFA burial during nonphosphorite forming time periods minimize the valleys perceptible between eras of phosphorite giant deposition?

8.13.4 SUMMARY

The global cycle of phosphorus is truly a biogeochemical cycle, owing to the involvement of phosphorus in both biochemical and geochemical reactions and pathways. There have been marked advances since the 1990s on numerous fronts of phosphorus research, resulting from application of new methods, as well as rethinking of old assumptions and paradigms. An oceanic phosphorus residence time on the order of 10–20 kyr, a factor of 5–10 shorter than previously cited values, casts phosphorus in the role of a potential player in climate change on glacial–interglacial timescales through the nutrient–CO_2 connection. This possibility is bolstered by findings in a number of recent studies that phosphorus does function as the limiting nutrient in some modern oceanic settings. Both oxygen isotopes in phosphate ($\delta^{18}O$–PO_4) and *in situ* produced radiophosphorus isotopes (^{33}P and ^{32}P) are providing new insights into how phosphorus is cycled through metabolic pathways in the marine environment. ^{31}P-NMR and a new probe for DOP-hydrolyzing enzymes are illuminating the composition of DOP and its importance for phytoplankton nutrition. Finally, new ideas about global phosphorus cycling on long, geologic timescales include a possible role for phosphorus in regulating atmospheric oxygen levels via the coupled iron–phosphorus–oxygen cycles, and the potential role of tectonics in setting the exogenic mass of phosphorus. The interplay of new findings in each of these areas, and others touched upon in this review chapter, are providing us with a fresh look at the global phosphorus cycle, one which is sure to evolve further as these and other new areas are explored in more depth by future studies.

REFERENCES

Abed A. M. and Fakhouri K. (1990) Role of microbial processes in the genesis of Jordanian Upper Cretaceous phosphorites. In *Phosphorite Research and Development*. Geol. Soc. Spec. Publ. No. 52 (eds. A. J. G. Notholt and I. Jarvis), pp. 193–203.

Abell J., Emerson S., and Renaud P. (2000) Distributions of TOP, TON, and TOC in the North Pacific Subtropical Gyre: implications for nutrient supply in the surface ocean and remineralization the upper thermocline. *J. Mar. Res.* **58**, 203–222.

Aller R. C. (1980). Diagenetic processes near the sediment–water interface of Long Island Sound: I. Decomposition and nutrient element geochemistry (S, N, P). In *Advances in Geophysics*, Academic Press, vol. 22, pp. 237–350.

Aller R. C. (1982) The effects of macrobenthos on chemical properties of marine sediment and overlying water. In *Animal–Sediment Relations* (eds. P. L. McCall and J. M. S. Tevesz). Plenum, pp. 53–102.

Aller R. C. (1994) Bioturbation and remineralization of sedimentary organic matter: effects of redox oscillation. *Chem. Geol.* **114**, 331–345.

Aller R. C. (1998) Mobile deltaic and continental shelf muds as fluidized bed reactors. *Mar. Chem.* **61**, 143–155.

Ames L. L., Jr. (1959) The genesis of carbonate apatites. *Econ. Geol.* **54**, 829–841.

Ammerman J. W. (1991) Role of ecto-phosphohydrolases in phosphorus regeneration in estuarine and coastal ecosystems. In *Microbial Enzymes in Aquatic Environments* (ed. R. J. Chróst). Springer, New York, pp. 165–186.

Ammerman J. W. and Azam F. (1985) Bacterial 5'-nucleotidase activity in estuarine and coastal marine waters: role in phosphorus regeneration. *Limnol. Oceanogr.* **36**, 1437–1447.

Ammerman J. W., Hood R. R., Case D. A., and Cotner J. B. (2003) Phosphorus deficiency in the Atlantic: an emerging paradigm in Oceanography. *EOS, Trans., AGU* **84**(18) 165, 170.

Anderson L. D. and Delaney M. L. (2000) Sequential extraction and analysis of phosphorus in marine sediments: streamlining of the SEDEX procedure. *Limnol. Oceanogr.* **45**, 509–515.

Anderson L. D., Delaney M. L., and Faul K. L. (2001) Carbon to phosphorus ratios in sediments: implications for nutrient cycling. *Global Biogeochem. Cycles* **15**, 65–80.

Anschutz P., Sundby B., Lefrancois L., Luther G., and Mucci A. (2000) Interactions between metal oxides and species of nitrogen andiodine in bioturbated marine sediments. *Geochim. Cosmochim. Acta* **64**, 2751–2763.

Arthur M. A. and Jenkyns H. C. (1981) Phosphorites and paleoceanography. *Oceanoglog. Acta* (special publication), 83–96.

Atlas E., Culbertson C., and Pytkowicz R. M. (1976) Phosphate association with Na^{+}, Ca^{++}, Mg^{++} in seawater. *Mar. Chem.* **4**, 243.

Baldwin D. S. (1998) Reactive "organic" phosphorus revisited. *Water Res.* **32**, 2265–2270.

Barrow N. J. (1983) On the reversibility of phosphate sorption by soils. *J. Soil Sci.* **34**, 751–758.

Baturin G. H. (1983) Some unique sedimentological and geochemical features of deposits in coastal upwelling regions. In *Coastal Upwelling – Its Sediment Record,* part B. NATO Conference Series IV, Marine Sciences 10B (eds. J. Theide and E. Suess). Plenum, pp. 11–27.

Baturin G. H. (2000) Formation and evolution of phosphorite grains and nodules on the Namibian shelf, from recent to Pleistocene. In *Marine Authigenesis: From Global to Microbial.* Spec. Publ. 66 (eds. C. R. Glenn, L. Prevot-Lucas, and J. Lucas). SEPM, pp. 185–199.

Baturin G. H., Merkulova K. I., and Chalov P. I. (1972) Radiometric evidence for recent formation of phosphatic nodules in marine shelf sediments. *Mar. Geol.* **13**, M37–M41.

Bauer J. E., Ruttenberg K. C., Wolgast D. M., Monaghan E., and Schrope M. K. (1996) Cross-flow filtration of dissolved and colloidal nitrogen and phosphorus in seawater: results from an intercomparison study. *Mar. Chem.* **55**, 33–52.

Benitez-Nelson C. R. (2000) The biogeochemical cycling of phosphorus in marine systems. *Earth-Sci. Rev.* **51**, 109–135.

Benitez-Nelson C. R. and Buesseler K. O. (1998) Measurement of cosmogenic ^{32}P and ^{33}P activities in rainwater and seawater. *Anal. Chem.* **70**, 502–505.

Benitez-Nelson C. R. and Buesseler K. O. (1999) Temporal variability of inorganic and organic phosphorus turnover rates in the coastal ocean. *Nature* **398**, 64–72.

Benitez-Nelson C. R. and Karl D. M. (2002) Phosphorus cycling in the North Pacific Subtropical Gyre using cosmogenic ^{32}P and ^{33}P. *Limnol. Oceanogr.* **47**, 762–770.

Bentor Y. K. (1980) Phosphorites—the unsolved problems. In *Marine Phosphorites*, Spec. Publ. 29 (ed. Y. K. Bentor). SEPM, pp. 3–18.

Bentzen E. and Taylor W. D. (1991) Estimating Michaelis–Menton parameters and lake water phosphate by the Rigler bioassay: importance of fitting technique, plankton size, and substrate range. *Can. J. Fish. Aquat. Sci.* **48**, 73–83.

Berelson W. M., McManus J., Coale K. H., Johnson K. S., Kilgore T., Burdige D., and Pilskaln C. (1996) Biogenic matter diagenesis on the sea floor: a comparison between two continental margin transects. *J. Mar. Res.* **54**, 731–762.

Berner R. A. (1973) Phosphate removal from sea water by adsorption on volcanogenic ferric oxides. *Earth Planet. Sci. Lett.* **18**, 77–86.

Berner R. A. (1974) Kinetic models for the early diagenesis of nitrogen, sulfur, phosphorus, and silicon in anoxic marine sediments. In *The Sea* (ed. E. D. Goldberg). Wiley, vol. 5, pp. 427–450.

Berner R. A. (1980) *Early Diagenesis: A Theoretical Approach.* Princeton University Press, NJ, 241pp.

Berner R. A. (1982) Burial of organic carbon and pyrite sulfur in the modern oceans: its geochemical and environmental significance. *Am. J. Sci.* **282**, 451–473.

Berner R. A. (1990) Diagenesis of phosphorus in sediments from non-upwelling areas. In *Phosphate Deposits of the World Vol. 3: Neogene to Modern Phosphorites.* International Geological Correlation Project 156: Phosphorites (eds. W. C. Burnett and S. R. Riggs). Cambridge University Press, Cambridge, pp. 27–32.

Berner R. A. and Canfield D. E. (1989) A new model for atmospheric oxygen over phanerozoic time. *Am. J. Sci.* **289**, 333–361.

Berner R. A. and Rao J.-L. (1994) Phosphorus in sediments of the Amazon River and estuary: implications for the global flux of phosphorus to the sea. *Geochim. Cosmochim. Acta* **58**, 2333–2339.

Berner R. A., Lasaga A. C., and Garrels R. M. (1983) The carbonate silicate geochemical cycle and its effect on atmospheric carbon dioxide over the past 100 million years. *Am. J. Sci.* **283**, 641–683.

Betts J. H. and Holland H. D. (1991) The oxygen content of ocean bottom waters, the burial efficiency of organic carbon, and the regulation of atmospheric oxygen. *Palaeogeogr. Palaeoclimatol. Palaeoecol.* (Global and Planetary Change Section) **97**, 5–18.

Bieleski R. L. and Ferguson I. B. (1983) Physiology and metabolism of phosphate and its compounds. In *Inorganic Plant Nutrition, Encyclopedia of Plant Physiology*, New Series (eds. A. Länchli and R. L. Bieleski) vol. 15A.

Björkman K. and Karl D. M. (1994) Bioavailability of inorganic and organic P compounds to natural assemblages of microorganisms in Hawaiian coastal waters. *Mar. Ecol. Prog. Ser.* **111**, 265–273.

Björkman K. and Karl D. M. (2001) A novel method for the measurement of dissolved adenosine and juanosine triphosphate in aquatic habitats: applications to marine microbial ecology. *J. Microbiol. Meth.* **47**, 159–167.

Björkman K. and Karl D. M. (2003) Bioavailability of dissolved organic phosphorus in the euphotic zone at Station ALOHA, North Pacific Subtropical Gyre. *Limnol. Oceanogr.* **48**, 1049–1057.

Björkman K., Thomson-Bulldis A. L., and Karl D. M. (2000) Phosphorus dynamics in the North Pacific Subtropical Gyre. *Aquat. Microbial. Ecol.* **22**, 185–198.

Blackman F. F. (1905) Optima and limiting factors. *Ann. Bot.* **19**, 281–295.

Blake R. E., O'Neil J. R., and Garcia G. A. (1997) Oxygen isotope systematics of biologically mediated reactions of phosphate: I. Microbial degradation of organophosphorus compounds. *Geochim. Cosmochim. Acta* **61**, 4411–4422.

Blake R. E., O'Neil J. R., and Garcia G. A. (1998) Effects of microbial activity on the $\delta^{18}O$ of dissolved inorganic phosphate and textural features of synthetic apatites. *Am. Mineral.* **83**, 1516–1531.

Blake R. E., Alt J. C., Moreira N. F., and Martini A. M. (2000) $\delta^{18}O$ of PO_4: a geochemical fingerprint for biological activity. *EOS, Trans., AGU* **81**(48).

Blake R. E., Alt J. C., and Martini A. M. (2001) Oxygen isotope ratios of PO_4: an inorganic indicator of enzymatic activity and P metabolism and a new biomarker in the search for life. *Proc. Natl. Acad. Sci.* **98**, 2148–2153.

Bolan N. S., Barrow N. J., and Posner A. M. (1985) Describing the effect of time on sorption of phosphate by iron and aluminum hydroxyoxides. *J. Soil Sci.* **36**, 187–197.

Bossard P. and Karl D. M. (1986) The direct measurement of ATP and adenine nucleotide pool turnover in microorganisms: a new method for environmental assessment of metabolism, energy flux and phosphorus dynamics. *J. Plankton Res.* **8**, 1–13.

Böstrom B. and Petterson K. (1982) Different patterns of phosphorus release from lake sediments in laboratory experiments. *Hydrobiologia* **92**, 415–429.

Böstrom B., Andersen J. M., Fleischer S., and Jansson M. (1988) Exchange of phosphorus across the sediment–water interface. *Hydrobiologia* **170**, 229–244.

Boyle E. A. (1988) Cadmium: chemical tracer of deep-water paleoceanography. *Paleoceanogrsphy* **3**, 471–489.

Boyle E. A. (1990) Quaternary deepwater paleoceanography. *Science* **249**, 863–870.

Broecker W. S. (1982a) Ocean chemistry during glacial time. *Geochim. Cosmochim. Acta* **46**, 1689–1705.

Broecker W. S. (1982b) *Prog. Oceanogr.* **11**, 151.

Broecker W. S. and Peng T.-H. (1982) *Tracers in the Sea.* Eldigio Press, Palisades, NY, 690pp.

Burford J. R. and Bremner J. M. (1972) Is phosphate reduced to phosphine in waterlogged soils? *Soil Biol. Biochem.* **4**, 489–495.

Burnett W. C. (1977) Geochemistry and origin of phosphorite deposits from off Peru and Chile. *Geol. Soc. Am. Bull.* **88**, 813–823.

Burnett W. C., Veeeh H. H., and Soutar A. (1980) U-series, oceanographic and sedimentary evidence in support of recent formation of phosphate nodules off Peru. In *Marine Phosphorites.* Spec. Publ. 29 (ed. Y. K. Bentor). SEPM, pp. 61–71.

Burnett W. C., Roe K. K., and Piper D. Z. (1983) Upwelling and phosphorite formation in the ocean. In *Coastal Upwelling: Part A* (eds. E. Suess and J. Thiede). Plenum, New York, vol. 4, pp. 377–397.

Burnett W. C., Glenn C. R., Yeh C. C., Schultz M., Chanton J., and Kashgarian M. (2000) U-Series, 14C, and stable isotope studies of recent phosphatic "protocrusts" from the Peru margin. In *Marine Authigenesis: From Global to Microbial*, Spec. Publ. 66 (eds. C. R. Glenn, L. Prevot-Lucas, and J. Lucas). SEPM, pp. 163–183.

Butler E. I., Knox S., and Liddicoat M. I. (1979) The relationship between inorganic and organic nutrients in seawater. *J. Mar. Biol. Assoc. UK* **59**, 239–250.

Canfield D. E. (1993) Organic matter oxidation in marine sediments. In *Interactions of C, N, P, and S Biogeochemical Cycles.* NATO-ARW (eds. R. Wollast, L. Chou, and F. Mackenzie). Springer, Berlin, pp. 333–363.

Caraco N. F. (1995) Influence of human populations on P transfers to aquatic systems: a regional scale study using large rivers. In *Phosphorus in the Global Environment.* SCOPE, chap. 14 (ed. H. Tiessen). Wiley, pp. 235–244.

Caraco N. F., Cole J. J., and Likens G. E. (1989) Evidence for suphate-controlled phosphorus release from sediments of aquatic systems. *Nature* **341**, 316–318.

Carlsson P. and Caron D. A. (2001) Seasonal variation of phosphorus limitation of bacterial growth in a small lake. *Limnol. Oceanogr.* **46**, 108–120.

Carlton R. G. and Wetzel R. G. (1988) Phosphorus flux from lake sediments: effect of epipelagic algal oxygen production. *Limnol. Oceanogr.* **33**, 562–570.

Carritt D. E. and Goodgal S. (1954) Sorption reactions and some ecological implications. *Deep-Sea Res.* **1**, 224–243.

Cavender-Barres K. K., Karl D. M., and Chisholm S. W. (2001) Nutrient gradients in the western North Atlantic Ocean: relationship to microbial community structure and comparison to patterns in the Pacific Ocean. *Deep-Sea Res. I.* **48**, 2373–2395.

Cembella A. D., Antia N. J., and Harrison P. J. (1984a) The utilization of inorganic and organic phosphorus compounds as nutrients by eukaryotic microalgae: a multidisciplinary perspective. Part 1. *CRC Crit. Rev. Microbiol.* **10**, 317–391.

Cembella A. D., Antia N. J., and Harrison P. J. (1984b) The utilization of inorganic and organic phosphorus compounds as nutrients by eukaryotic microalgae: a multidisciplinary perspective. Part 2. *CRC Crit. Rev. Microbiol.* **11**, 13–81.

Cha H.-J. (2002) Geochemistry of surface sediments and diagenetic redistribution of phosphorus in the southwestern East Sea. PhD Thesis, Seoul National University, 191pp.

Chadwick O. A., Derry L. A., Vitousek P. M., Buebert B. J., and Hedin L. O. (1999) Changing sources of nutrients during four million years of ecosystem development. *Nature* **397**, 491–497.

Chambers R. M., Fourqurean J. W., Hollibaugh J. T., and Vink S. M. (1995) Importance of terrestrially derived particulate phosphorus to phosphorus dynamics in a west coast estuary. *Estuaries* **18**(3), 518–526.

Chase E. M. and Sayles F. L. (1980) Phosphorus in suspended sediments of the Amazon River. *Estuar. Coast. Mar. Sci.* **11**, 383–391.

Clark L. L., Ingall E. D., and Benner R. (1998) Marine phosphorus is selectively mineralized. *Nature* **393**, 426.

Clark L. L., Ingall E. D., and Benner R. (1999) Marine organic phosphorus cycling: novel insights from nuclear magnetic resonance. *Am. J. Sci.* **299**, 724–737.

Codispoti L. A. (1989) Phosphorus vs. nitrogen limitation of new and export production. In *Productivity of the Ocean: Present and Past* (eds. W. H. Berger, V. S. Smetacek, and G. Wefer). Wiley, New York, pp. 377–394.

Colman A. S. (2002) The oxygen isotope composition of dissolved inorganic phosphate and the marine phosphorus cycle. PhD. Thesis, Yale University, 230pp.

Colman A. S. and Holland (1994) Benthic phosphorus regeneration: the global diffusive flux of phosphorus from marine sediments into the oceans. *EOS, Trans., AGU* **75**, 96.

Colman A. S. and Holland H. D. (2000) The global diagenetic flux of phosphorus from marine sediments to the oceans: redox sensitivity and the control of atmospheric oxygen levels. In *Marine Authigenesis: From Global to Microbial.* Spec. Publ. #66 (eds. C. R. Glenn, L. Prévôt-Lucas, and J. Lucas). SEPM, pp. 53–75.

Colman A. S., Holland H. D., and Mackenzie F. T. (1997) Redox stabilization of the atmosphere and oceans by phosphorus limited marine productivity: discussion and reply. *Science* **276**, 406–408.

Colman A. S., Karl D. M., Fogel M. L., and Blake R. E. (2000) A new technique for the measurement of phosphate oxygen isotopes of dissolved inorganic phosphate in natural waters. *EOS, Trans., AGU*, F176.

Compton J. S., Hodell D. A., Garrido J. R., and Mallinson D. J. (1993) Origin and age of phosphorite from the south-central Florida Platform: relation of phosphogenesis to sea-level fluctuations and $\delta^{13}C$ excursions. *Geochim. Cosmochim. Acta* **57**, 131–146.

Compton J., Mallinson D., Glenn C. R., Filippelli G., Föllmi K., Shields G., and Zanin Y. (2000) Variations in the global phosphorus cycle. In *Marine Authigenesis: From Global to Microbial.* Spec. Publ. #66 (eds. C. R. Glenn, L. Prévôt-Lucas, and J. Lucas). SEPM, pp. 21–33.

Conkright M. E., Gegg W., and Levitus S. (2000) Seasonal cycle of phosphate in the open ocean. *Deep Sea Res., Part I.* **47**, 159–175.

Conley D. J., Smith W. M., Cornwell J. C., and Fisher T. R. (1995) Transformation of particle-bound phosphorus at the land–sea interface. *Estuar. Coast. Shelf Sci.* **40**, 161–176.

Cook P. J. (1984) Spatial and temporal controls on the formation of phosphate deposits—a review. In *Phosphate Minerals* (eds. P. B. Moore and J. O. Nriagu). Springer, New York, pp. 242–274.

Cook P. J. and McElhinny M. W. (1979) A re-evaluation of the spatial and temporal distribution of sedimentary phosphorite deposits in light of plate tectonics. *Econ. Geol.* **74**, 315–330.

Cook P. J. and Shergold J. H. (1984) Phosphorus, phosphorites, and skeletal evolution at the Precambrian–Cambrian boundary. *Nature* **308**, 231–236.

Cook P. J., Shergold J. H., Burnett W. C., and Riggs S. R. (1990) Phosphorite research: a historical overview. In *Phosphorite Research and Development*. Geol. Soc. Spec. Publ. No. 52 (eds. A. J. G. Notholt and I. Jarvis), pp. 1–22.

Copin-Montegut C. and Copin-Montegut G. (1972) Chemical analysis of suspended particulate matter collected in the northeast Atlantic. *Deep-Sea Res.* **19**, 445–452.

Cosgrove D. J. (1977) Microbial transformations in the phosphorus cycle. *Adv. Microb. Ecol.* **1**, 95–135.

Cotner J. B., Ammerman J. W., Peele E. R., and Bentzen E. (1987) Phosphorus-limited bacterioplankton growth in the Sargasso Sea. *Aquat. Microb. Ecol.* **13**, 141–149.

Crave D. B., Jahnke R. A., and Carlucci A. F. (1986) Fine-scale vertical distributions of microbial biomass and activity in California Borderland Basin sediments. *Deep-Sea Res.* **33**, 379–390.

Crews T. E., Kitayama K., Fownes J. H., Riley R. H., Herbert D. A., Dombois D. M., and Vitousek P. M. (1995) Changes in soil phosphorus fractions and ecosystem dynamics across a long chronosequence in Hawaii. *Ecology* **76**, 1407–1423.

Cullen J. T., Chase Z., Coale K. H., Fitzwater S. E., and Sherrell R. M. (2003) Effect of iron limitation on the cadmium to phosphorus ratio of natural phytoplankton assemblages from the Southern Ocean. *Limnol. Oceanogr.* **48**(3), 1079–1087.

DeFlaun M. F., Paul J. H., and Jeffrey W. H. (1987) Distribution and molecular weight of dissolved DNA in subtropical estuarine and oceanic environments. *Mar. Ecol. Prog. Ser.* **38**, 65–73.

Delaney M. L. (1998) Phosphorus accumulation in marine sediments and the oceanic phosphorus cycle. *Global Biogeochem. Cycles* **12**, 563–672.

Delaney M. L. and Anderson L. D. (1997) Phosphorus geochemistry in Ceara Rise sediments. *Proc. ODP Sci. Results* **154**, 475–482.

Delaney M. L. and Filippelli G. M. (1994) An apparent contradiction in the role of phosphorus in Cenozoic chemical mass balances for the world ocean. *Paleoceanography* **9**(4), 513–527.

DeMaster D. J. and Aller R. C. (2001) Biogeochemical processes on the Amazon Shelf: changes in dissolved and particulate fluxes during river/ocean mixing. In *The Biogeochemistry of the Amazon Basin* (eds. M. E. McClain, R. L. Victoria, and J. E. Richey). Oxford Press, pp. 328–357.

Dévai I., Felföldy L., Wittner H., and Plósz S. (1988) Detection of phosphine: new aspects of the phosphorus cycle in the hydrosphere. *Nature* **333**, 343–345.

Donnelly T. H., Shergold J. H., Southgate P. N., and Barnes C. J. (1990) Events leading to global phosphogenesis around the Proterozoic/Cambrian boundary. In *Phosphorite Research and Development*. Geol. Soc. Spec. Publ. No. 52 (eds. A. J. G. Notholt and I. Jarvis), pp. 273–287.

Downing J. A. (1997) Marine nitrogen: phosphorus stoichiometry and the global N:P cycle. *Biogeochemistry* **37**, 237–252.

Duce R. A., Liss P. S., Merrill J. T., Atlans E. L., Buat-Menard P., Hicks B. B., Miller J. M., Prospero J. M., Arimoto R., Church T. M., Ellis W., Galloway J. N., Hansen L., Jickells T. D., Knap A. H., Reinhardt K. H., Schneider B., Soudine A., Tokos J. J., Tsunogai S., Wollast R., and Zhou M. (1991) The atmospheric input of trace species to the world ocean. *Global Biogeochem. Cycles* **5**, 193–259.

Dyhrman S. T. and Palenik B. (1999) Phosphate stress in cultures and field populations of the dinoflagellate Prorocentrum minimum detected by a single-cell alkaline phosphatase assay. *Appl. Environ. Microbiol.* **65**, 3205–3212.

Duerden C. F. (1973) Aspects of phytoplankton production and phosphate exchange in Bedford Basin, Nova Scotia. PhD Thesis, Dalhousie University, Halifax.

Einsele W. (1936a) Über die beziehungen des eisenkreislaufs zum phosphatkreislauf im eutrofen see. *Arch. Hydrobiol.* **29**, 664–686.

Einsele W. (1936b) Über chemische und kolloidchemische vorgange in eisen-phosphat-systemen unter limnochemischen und limnogeologischen gesichtspunkten. *Arch. Hydrobiol.* **33**, 361–387.

Emerson S. (1976) Early diagenesis in anaerobic lake sediments: thermodynamic and kinetic factors controlling the formation of iron phosphate. *Geochim. Cosmochim Acta* **42**, 1307–1316.

Emerson S. and Widmer G. (1978) Early diagenesis in anaerobic lake sediments: 2. Chemical equilibria in interstitial waters. *Geochim. Cosmochim Acta* **40**, 925–934.

Falkowski P. G. (1997) Evolution of the nitrogen cycle and its influence on the biological sequestration of CO_2 in the ocean. *Nature* **387**, 272–275.

Field M. P. and Sherrell R. M. (2003) Direct determination of ultra-trace levels of metals in fresh water using desolvating micronebulization and HR-ICP-MS: application to Lake Superior waters. *J. Anal. At. Spectrom.* **18**, 254–259.

Filho M. V., de Mesquita, and Torrent J. (1993) Phosphate sorption as related to mineralogy of a hydrosequence of soils from the Cerrado region (Brazil). *Geoderma* **58**, 107–123.

Filipek L. H. and Owen R. M. (1981) Diagenetic controls on phosphorus in outer continental shelf sediments from the Gulf of Mexico. *Chem. Geol.* **33**, 181–204.

Filippelli G. M. (1997a) Controls on phosphorus concentration and accumulation in marine sediments. *Mar. Geol.* **139**, 231–240.

Filippelli G. M. (1997b) Intensification of the Asian monsoon and a chemical weathering event in the late Miocene–early Pliocene: implications for late Neogene climate change. *Geology* **25**, 27–30.

Filippelli G. M. (2001) Carbon and phosphorus cycling in anoxic sediments of the Saanich Inlet. *British Columbia. Mar. Geol.* **174**, 307–321.

Filippelli G. M. and Delaney M. L. (1992) Similar phosphorus fluxes in ancient phosphorite deposits and a modern phosphogenic environment. *Geology* **20**, 709–712.

Filippelli G. M. and Delaney M. L. (1994) The oceanic phosphorus cycle and continental weathering during the Neogene. *Paleoceanography* **9**(5), 643–652.

Filippelli G. M. and Delaney M. L. (1996) Phosphorus geochemistry of equatorial Pacific sediments. *Geochim. Cosmochim. Acta* **60**, 1479–1495.

Filippelli G. M. and Souch C. (1999) Effects of climate and landscape development on the terrestrial phosphorus cycle. *Geology* **27**, 171–174.

Fisher T., Peele E., Ammerman J., and Harding L. (1992) Nutrient limitation of phytoplankton in Chesapeake Bay. *Mar. Ecol. Prog. Ser.* **82**, 51–63.

Fisher T. R., Melack J. M., Grobbelaar J. U., and Howarth R. W. (1995) Nutrient limitation of phytoplankton and eutrophication of inland, estuarine, and marine waters. In *Phosphorus in the Global Environment*. SCOPE, chap. 18 (ed. H. Tiessen). Wiley, pp. 301–322.

Fogg G. E. (1973) Phosphorus in primary aquatic plants. *Water Res.* **7**, 77–91.

Föllmi K. B. (1990) Condensation and phosphogenesis: example of the Helvetic mid-Cretaceous (northern Tethyan margin). In *Phosphorite Research and Development*. Geol. Soc. Spec. Publ. No. 52 (eds. A. J. G. Notholt and I. Jarvis), pp. 237–252.

Föllmi K. B. (1995a) The phosphorus cycle, phosphogenesis and marine phosphate-rich deposits. *Earth-Sci. Rev.* **40**, 55–124.

Föllmi K. B. (1995b) A 160 m.y. record of marine sedimentary phosphorus burial: coupling of climate and continental weathering under greenhouse and icehouse conditions. *Geology* **23**, 859–862.

Fourqurean J. S., Jones R. D., and Zieman J. C. (1993) Processes influencing water column nutrient characteristics and phosphorus limitation of phytoplankton biomass in Florida Bay, FL, USA: inferences from spatial distributions. *Estuar. Coast. Shelf Sci.* **36**, 295–314.

Fox L. E. (1989) A model for inorganic phosphate concentrations in river waters. *Geochim. Cosmochim. Acta* **53**, 417–428.

Fox L. E. (1990) Geochemistry of dissolved phosphate in the Sepik River Estuary, Papua, New Guinea. *Geochim. Cosmochim. Acta* **54**, 1019–1024.

Fox L. E. (1993) The chemistry of aquatic phosphate: inorganic processes in rivers. *Hydrobiologia* **253**, 1–16.

Fox L. E., Sager S. L., and Wofsy S. C. (1985) Factors controlling the concentrations of soluble phosphorus in the Mississippi estuary. *Limnol. Oceanogr.* **30**, 826–832.

Fox L. E., Sager S. L., and Wofsy S. C. (1986) A chemical survey of the Mississippi estuary. *Estuaries* **10**, 1–12.

Fox L. E., Lipschult F., Kerkhof L., and Wofsy S. C. (1987) The chemical control of soluble phosphorus in the Amazon estuary. *Geochim. Cosmochim. Acta* **50**, 783–794.

Froelich P. N. (1988) Kinetic control of dissolved phosphate in natural rivers and estuaries: a primer on the phosphate buffer mechanism. *Limnol. Oceanogr.* **33**, 649–668.

Froelich P. N., Bender M. L., and Heath G. R. (1977) Phosphorus accumulation rates in metalliferous sediments on the East Pacific Rise. *Earth Planet. Sci. Lett.* **34**, 351–359.

Froelich P. N., Bender M. L., Luedtke N. A., Heath G. R., and Devries T. (1982) The marine phosphorus cycle. *Am. J. Sci.* **282**, 474–511.

Froelich P. N., Arthur M. A., Burnett W. C., Deakin M., Hensley V., Jahnke R., Kaul L., Kim K. H., Soutar A., and Vathakanon C. (1988) Early diagenesis of organic matter in Peru continental margin sediments: phosphorite precipitation. *Mar. Geol.* **80**, 309–346.

Frossard E., Brossard M., Hedley M. J., and Meterell A. (1995) Reactions controlling the cycling of P in soils. In *Phosphorus in the Global Environment*. SCOPE 54, chap. 7 (ed. H. Tiessen). Wiley, New York, pp. 107–137.

Gächter R. and Meyers J. S. (1993) The role of microorganisms in sediment phosphorus dynamics in relation to mobilization and fixation of phosphorus. *Hydrobiologia* **253**, 103–121.

Gächter R., Meyer J. S., and Mares A. (1988) Contribution of bacteria to release and fixation of phosphorus in lake sediments. *Limnol. Oceanogr.* **33**, 1542–1558.

Galloway J. N., Howarth R. W., Michaels A. F., Nixon S. W., Prospero J. M., and Dentener F. J. (1996) Nitrogen and phosphorus budgets of the North Atlantic Ocean and its watershed. *Biogeochemistry* **35**, 3–25.

Ganeshram R. S., Pedersen T. F., Calvert S. E., and Francois R. (2002) Reduced nitrogen fixation in the glacial ocean inferred from changes in marine nitrogen and phosphorus inventories. *Nature* **415**, 156–159.

Gassman G. (1994) Phosphine in the fluvial and marine hydrosphere. *Mar. Chem.* **45**, 197–205.

Gaudette H. E. and Lyons W. B. (1980) Phosphoate geochemistry in nearshore carbonate sediments: a suggestion of apatite formation. In *Marine Phosphorites: A Symposium*. Spec. Publ. 29 (ed. Y. Bentor). SEPM, pp. 215–225.

GESAMP (1987) Land/sea boundary flux of contaminants: contributions from rivers. Report and studies #32 of the IMCO/FAO/UNESCO/WHOM/WHO/IAEA/UNEP Joint Group of Experts on the Scientific Aspects of Marine Pollution. UNESCO, Paris.

Glenn C. R. and Arthur M. A. (1988) Petrology and major element geochemistry of Peru margin phosphorites and associated diagenetic minerals: authigenesis in modern organic-rich sediments. *Mar. Geol.* **80**, 231–267.

Goldman J. C. (1986) On phytoplankton growth rates and particulate C:N:P ratios at low light. *Limnol. Oceanogr.* **31**, 1358–1361.

Goldman J. C., McCarthy J. J., and Peavey D. G. (1979) Growth rate influence on the chemical composition of phytoplankton in oceanic waters. *Nature* **279**, 210–215.

Golterman H. L. (1995a) The role of the ferric hydroxide–phosphate–sulphide system in sediment water exchange. *Hydrobiologia* **297**, 43–54.

Golterman H. L. (1995b) The labyrinth of nutrient cycles and buffers in wetlands: results based on research in the Camargue (Southern France). *Hydrobiologia* **315**, 39–58.

Golterman H. L. (1995c) Theoretical aspects of the adsorption of *ortho*-phosphate onto ironhydroxide. *Hydrobiologia* **315**, 59–68.

Gordon D. C., Jr., Boudreau P. R., Mann K. H., Ong J.-E., Silvert W. L., Smith S. V., Wattayakorn G., Wull F., and Yanagi T. (1996) LOICZ Biogeochemical modeling guidelines. In *LOICZ Reports and Studies* No. 5, LOICZ/RandS/95-5, vi +96pp. LOICZ, Texel, The Netherlands.

Granéli E., Wallström K., Larsson U., Granéli W., and Elmgren E. (1990) Nutrient limitation of primary productionin the Baltic Sea area. *Ambio* **19**, 142–151.

Grill E. and Richards F. A. (1976) Nutrient regeneration from phytoplankton decomposing in seawater. *J. Mar. Res.* **22**, 51–59.

Guidry M. W., Mackenzie F. T., and Arvidson R. S. (2000) Role of tectonics in phosphorus distribution and cycling. In *Marine Authigenesis: From Global to Microbial*. Spec. Publ. #66 (eds. C. R. Glenn, L. Prévôt-Lucas, and J. Lucas). SEPM, pp. 35–51.

Gulbrandsen R. A., Roberson C. E., and Neil S. T. (1984) Time and the crystallization of apatite in seawater. *Geochim. Cosmochim. Acta* **48**, 213–218.

Gunnars A. and Blomqvist S. (1997) Phosphate exchange across the sediment water interface when shifting from anoxic to oxic conditions—an experimental comparison of freshwater and brackish-marine systems. *Biogeochemistry* **37**, 203–226.

Guzman G., Alcantara E., Barrón M., and Torrent J. (1994) Phytoavailability of phosphate adsorbed on ferrihydrite, hematite and goethite. *Plant Soil* **159**, 219–225.

Harrison W. G. and Harris L. R. (1986) Isotope-dilution and its effects on measurements of nitrogen and phosphorus uptake by oceanic microplankton. *Mar. Ecol. Prog. Ser.* **27**, 253–261.

Harrison W. G., Azam F., Renger E. H., and Eppley R. W. (1977) Some experiments on phosphate assimilation by coastal marine plankton. *Mar. Biol.* **40**, 9–18.

Hart R. (1970) Chemical exchange between sea water and deep ocean basalts. *Earth Planet. Sci. Lett.* **9**, 269–279.

Harvey H. R., Fallon R. D., and Patton J. S. (1986) The effect of organic matter and oxygen on the degradation of bacterial membrane lipids in marine sediments. *Geochim. Cosmochim. Acta* **50**, 795–804.

Healey F. P. and Hendzel L. L. (1980) Physiological indicators of nutrient deficiency in lake phytoplankton. *Can. J. Fish. Aquat. Sci.* **37**, 442–453.

Hecky R. E. and Kilham P. (1988) Nutrient limitation of phytoplankton in freshwater and marine environments: a review of recent evidence on the effects of enrichment. *Limnol. Oceanogr.* **33**, 796–822.

Heggie D. T., Skyring G. W., O'Brien, Reimers C., Herczeg A., Moriarty D. J. W., Burnett W. C., and Milnes A. R. (1990) Organic carbon cycling and modern phosphorite formation on the east Australian continental margin: an overview. In *Phosphorite Research and Development*. Geol. Soc. Spec. Publ. 52 (eds. A. G. J. Notholt and I. Jarvis), pp. 87–117.

Herbert T. D. and Sarmiento J. L. (1991) Ocean nutrient distribution and oxygenation: limits on the formation of warm saline bottom water over the past 91 m.y. *Geology* **19**, 702–705.

Herbes S. E., Allen H. E., and Mancy K. H. (1975) Enzymatic characterization of soluble organic phosphorus in lake water. *Science* **187**, 432.

Holland H. D. (1978) *The Chemistry of the Atmosphere and Oceans*. Wiley, New York, 351pp.

Holland H. D. (1984) *The Chemical Evolution of the Atmosphere and Oceans*. Princeton University Press, Princeton, NJ, 582pp.

Holland H. D. (1990) The origins of breathable air. *Nature* **347**, 17.

Holland H. D. (1994) The phosphate–oxygen connection. *EOS, Trans., AGU* **75**, 96.

Honjo S. and Manganini S. (1993) Annual biogenic particle fluxes to the interior of the North Atlantic Ocean. *Deep-Sea Res.* **40**, 587–607.

Howarth R. W., Jensen H. S., Marino R., and Postma H. (1995) Transport to and processing of P in near-shore and oceanic waters. In *Phosphorus in the Global Environment*. SCOPE 54, chap. 19 (ed. H. Tiessen). Wiley, Chichester, pp. 323–345, (see other chapters in this volume for additional information of P-cycling).

Howarth R. W., Billen G., Swaney D., Townsend A., Jaworski N., Lajtha K., Downing J. A., Elmgren R., Caraco N., Jordan T., Berendse F., Freney J., Kudeyarov V., Murdoch P., and Zhao-Liang Z. (1996) Regional nitrogen budgets and riverine N and P fluxes for the drainages to the North Atlantic Ocean: natural and human influences. *Biogeochemistry* **35**, 75–139.

Hudson J. J., Taylor W. D., and Schindler D. W. (2000) Phosphate concentration in lakes. *Nature* **406**, 54–56.

Hulth S., Aller R. C., and Gilbert F. (1999) Coupled anoxic nitrification/manganese reduction in marine sediments. *Geochim. Cosmochim. Acta* **63**, 49–66.

Ingall E. D. and Jahnke R. A. (1994) Evidence for enhanced phosphorus regeneration from marine sediments overlain by oxygen depleted waters. *Geochim. Cosmochim. Acta* **58**, 2571–2575.

Ingall E. D. and Jahnke R. A. (1997) Influence of water-column anoxia on the elemental fractionation of carbon and phosphorus during sediment diagenesis. *Mar. Geol.* **139**, 219–229.

Ingall E. D. and Van Cappellen P. (1990) Relation between sedimentation rate and burial of organic phosphorus and organic carbon in marine sediments. *Geochim. Cosmochim. Acta* **54**, 373–386.

Ingall E. D., Schroeder P. A., and Berner R. A. (1990) The nature of organic phosphorus in marine sediments: new insights from ^{31}P-NMR. *Geochim. Cosmochim. Acta* **54**, 2617–2620.

Ingall E. D., Bustin R. M., and Van Cappellen P. (1993) Influence of water column anoxia on the burial and preservation of carbon and phosphorus in marine shales. *Geochim. Cosmochim. Acta* **57**, 303–316.

Jackson G. A. and Williams P. M. (1985) Importance of dissolved organic nitrogen and phosphorus to biological nutrient cycling. *Deep-Sea Res.* **32**, 223–235.

Jahnke R. A. (1984) The synthesis and solubility of carbonate fluorapatite. *Am. J. Sci.* **284**, 58–78.

Jahnke R. A. (1992) The phosphorus cycle. In *Global Geochemical Cycles*. chap. 14 (eds. S. S. Butcher, R. J. Charlson, G. H. Orians, and G. V. Wolff). Academic, San Diego, pp. 301–315.

Jahnke R. A., Emerson S. R., Kim K. H., and Burnett W. C. (1983) The present day formation of apatite in Mexican continental margin sediments. *Geochim. Cosmochim. Acta* **47**, 259–266.

Jarvis I., Burnett W. C., Nathan Y., Almbaydin F., Attia K. M., Castro L. N., Flicoteaux R., Hilmy M. E., and Husain V. (1994) Phosphorite geochemistry: state-of-the-art and environmental concerns. In *Concepts and Controversies in Phosphogenesis*. Eclogae Geol. Helv. 87 (ed. K. B. Follmi) pp. 643–700.

Jensen H. S., Mortensen P. B., Andersen F. O., Rasmussen E., and Jensen A. (1995) Phosphorus cycling in a coastal marine sediment, Aarhus Bay, Denmark. *Limnol. Oceanogr.* **40**, 908–917.

Jensen H. S., McGlathery K. J., Marino R., and Howarth R. W. (1998) Forms and availabiliity of sediment phosphorus in carbonate sand of Bermuda seagrass beds. *Limnol. Oceanogr.* **43**, 799–810.

Kamykowski D., Zentara S.-J., Morrison J. M., and Switzer A. C. (2002) Dynamic patterns of nitrate, phosphate, silicate, and iron availability and phytoplankton community composition from remote sensing data. *Global Biogeochem. Cycles* **16**(4), 1077 doi:10.1029/2001GB001640.

Karl D. M. (1999) A sea of change: biogeochemical variability in the North Pacific Subtropical Gyre. *Ecosystems* **2**, 181–214.

Karl D. M. (2000) Phosphorus, the staff of life. *Nature* **406**, 32–33.

Karl D. M. and Björkman K. M. (2002) Dynamics of DOP. In *Biogeochemistry of Marine Organic Matter* (eds. D. Hansell and C. Carlson), pp. 249–366.

Karl D. M. and Tien G. (1992) MAGIC: a sensitive and precise method for measuring dissolved phosphorus in aquatic environments. *Limnol. Oceanogr.* **37**(1), 77–96.

Karl D. M. and Tien G. (1997) Temporal variability in dissolved phosphorus concentrations in the subtropical North Pacific Ocean. *Mar. Chem.* **56**, 77–96.

Karl D. M. and Yanagi K. (1997) Partial characterization of the dissolved organic phosphorus pool in the oligotrophic North Pacific Ocean. *Limnol. Oceanogr.* **4**, 1398–1405.

Karl D. M., Letelier R., Tupas L., Dore J., Christian J., and Hebel D. (1997) The role of nitrogen fixation in the biogeochemical cycling in the subtropical North Pacific Ocean. *Nature* **388**, 533–538.

Karl D. M., Bjorkman K. M., Dore J. E., Fujieki L., Hebel D. V., Houlihan T., Letelier R. M., and Tupas L. (2001) Ecological nitrogen-to-phosphorus stoichiometry at station ALOHA. *Deep-Sea Res. I.* **48**, 1529–1566.

Kaul L. W. and Froelich P. N. (1996) Modeling estuarine nutrient geochemistry in a simple system. *Geochim. Cosmochim. Acta* **48**, 1417–1433.

Kester and Pytkowicz (1967) Determination of the apparent dissociation constants of phosphoric acid in seawater. *Limnol. Oceanogr.* **12**, 243–252.

Ketchum B. H., Corwin N., and Keen D. J. (1955) The significance of organic phosphorus determinations in ocean waters. *Deep-Sea Res.* **2**, 172–181.

Kim, *et al.* (1999) *Geochim. Cosmochim. Acta* **63**, 3485–8477.

Klump J. V. and Martens C. S. (1981) Biogeochemical cycling in an organic-rich coastal marine basin: II. Nutrient sediment-water exchange processes. *Geochim. Cosmochim. Acta* **45**, 101–121.

Klump J. V. and Martens C. S. (1987) Biogeochemical cycling in an organic-rich coastal marine basin: 5. Sedimentary nitrogen and phosphorus budgets based upon kinetic models, mass balances, and the stoichiometry of nutrient regeneration. *Geochim. Cosmochim. Acta* **51**, 1161–1173.

Knauer G. A., Martin J. H., and Bruland K. W. (1979) Fluxes of particulate carbon, nitrogen, and phosphorus in the upper water column of the northeast Pacific. *Deep-Sea Res.* **26A**, 97–108.

Kobori H. and Taga N. (1979) Phosphatase activity and its role in the mineralization of phosphorus in coastal seawater. *J. Exp. Mar. Biol. Ecol.* **36**, 23–39.

Kolodny Y. (1969) Are marine phosphorites forming today? *Nature* **224**, 1017–1019.

Kolodny Y. (1981) Phosphorites. In *The Sea* (ed. C. Emiliani). Wiley, vol. 7, pp. 981–1023.

Kolodny Y. and Kaplan I. R. (1970) Uranium isotopes in sea floor phosphorites. *Geochim. Cosmochim. Acta* **34**, 3–24.

Kolodny Y., Luz B., and Navon O. (1983) Oxygen isotope variations in phosphate of biogenic apatite: I. Fish bone apatite—rechecking the rules of the game. *Earth Planet. Sci. Lett.* **64**, 398–404.

Kolowith L. C., Ingall E. D., and Benner R. (2001) Composition and cycling of marine organic phosphorus. *Limnol. Oceanogr.* **46**, 309–320.

Koroleff F. (1983) Determination of phosphorus. In *Methods of Seawater Analysis*, Weinheim, Verlag Chemie.

Krajewski K. P., Van Cappellen P., Trichet J., Kuhn O., Lucas J., Martín-Algarra A., Prévôt L., Tweari V. C., Gaspar L., Knight R. I., and Lamboy M. (1994) Biological processes and apatite formation in sedimentary environments. *Eclogae Geol. Helv.* **87**, 701–745.

Krom M. D. and Berner R. A. (1980) Adsorption of phosphate in anoxic marine sediments. *Limnol. Oceanogr.* **25**, 797–806.

Krom M. D. and Berner R. A. (1981) The diagenesis of phosphorus in a nearshore marine sediment. *Geochim. Cosmochim. Acta* **45**, 207–216.

Krom M. D., Kress N., Brenner S., and Gordon L. (1991) Phosphorus limitation of primary production in the eastern Mediterranean. *Limnol. Oceanogr.* **36**, 424–432.

Kump L. (1988) Terrestrial feedback in atmospheric oxygen regulation by fire and phosphorus. *Nature* **335**, 152–154.

Laarkamp K. L. (2000) Organic phosphorus in marine sediments: chemical structure, diagenetic alteration, and mechanisms of preservation. PhD Thesis, MIT-WHOI Joint Program in Oceanography, 286pp.

Laarkamp K. L. and Ruttenberg K. C. (2000) Assessment of the contribution of biomass to total sediment organic P, and the preservation potential of primary P biochemicals. *EOS, Trans., AGU* (12/2000).

Lajtha K. and Harrison A. F. (1995) Strategies of phosphorus acquisition and conservation by plant species and communities. In *Phosphorus in the Global Environment.* SCOPE 54, chap. 8 (ed. H. Tiessen). Wiley, New York, pp. 139–147.

Lajtha K. and Schlesinger W. H. (1988) The biogeochemistry of phosphorus cycling and phosphorus availability along a desert soil chronosequence. *Ecology* **69**, 24–39.

Lal D. and Lee T. (1988) Cosmogenic ^{32}P and ^{33}P used as tracers to study phosphorus recycling in the upper ocean. *Nature* **333**, 752–754.

Lamboy M. (1990) Microbial mediation in phosphatogenesis: new data from the Cretaceous phosphatic chalks of northern France. In *Phosphorite Reserarch and Development.* Geol. Soc. Spec. Publ. No. 52 (eds. A. J. G. Notholt and I. Jarvis), pp. 157–167.

Lean D. R. S. and Nawelajko C. (1976) Phosphate exchange and organic phosphorus excretion by freshwater algae. *J. Fish. Board. Can.* **33**, 1312.

Lebo M. (1991) Particle-bound phosphorus along an urbanized coastal plain estuary. *Mar. Chem.* **34**, 225–246.

Leckie J. and Stumm W. (1970) In *Water Quality Improvement by Physical and Chemical Processes* (eds. E. F. Gloyna and W. W. Eckenfelder). University of Texas Press, Austin.

Lee T., Barg E., and Lal D. (1991) Studies of vertical mixing in the Southern California Bight with cosmogenic radionuclides ^{32}P and ^{7}Be. *Limnol. Oceanogr.* **36**, 1044–1053.

Lee T., Barg E., and Lal D. (1992) Techniques for extraction of dissolved inorganic and organic phosphorus from large volumes of seawater. *Anal. Chim. Acta* **260**, 113–121.

LeGeros R. Z. (1965) Effect of carbonate on the lattice parameters of apatite. *Nature* **206**, 403–404.

LeGeros R. Z., LeGeroz J. P., Trautz O. R., and Shirra W. P. (1967) Apatite crystallites: effects of carbonate ion on morphology. *Science* **155**, 1409–1411.

LeGeros R. Z., Trautz O. R., Klein E., and LeGeros J. P. (1969) Two types of carbonate substitution in the apatite structure. *Experientia* **25**, 5–7.

Lenton T. M. and Watson A. J. (2000) Redfield Revisited: 1. Regulation of nitrate, phosphate, and oxygen in the ocean. *Global Biogeochem. Cycles* **14**, 225–248.

Lerman A., Mackenzie F. T., and Garrels R. M. (1975) Modeling of geochemical cycles: phosphorus as an example. *Geol. Soc. Am. Memoirs.* **142**, 205–217.

Levitus S., Conkright M. W., Reid J. I., Najjar R., and Mantyla A. (1993) Distribution of nitrate, phosphate, and silicate in the world oceans. **31**, 245–273.

Liebig J. Von (1840) *Chemistry and its Application to Agriculture and Physiology.* Taylor and Walton, London, 4th edn. 352pp.

Liebezeit G. (1991) Analytical phosphorus fractionation of sediment trap material. *Mar. Chem.* **33**, 61–69.

Lijklema L. (1980) Interaction of orthophosphate with iron (III) and aluminum hydroxide. *Environ. Sci. Tech.* **14**, 537–541.

Likens G. E., Bormann F. H., Pierce R. S., Eaton J. S., and Johnson N. M. (1977) *Biogeochemistry of a Forested Ecosystem.* Springer, New York.

Loh A. N. and Bauer J. E. (2000) Distributions, partitioning, and fluxes of dissolved and particulate organic C, N, and P in the eastern north Pacific and Southern Oceans. *Deep-Sea Res. I.* **4**, 2287–2316.

Longinelli A. (1966) Ratios of oxygen-18: oxygen-16 in phosphate from living and fossil marine organisms. *Nature* **21**, 923–927.

Longinelli A. and Nuti S. (1973) Revised phosphate-water isotopic temperature scale. *Earth Planet. Science Lett.* **19**, 373–376.

Louchouarn P., Lucotte M., Duchemin E., and de Vernal A. (1997) Early diagenetic processes in recent sediments of the Gulf of St. Lawrence: phosphorus, carbon and iron burial rates. *Mar. Geol.* **139**, 181–200.

Lucas J. and Prévôt L. (1985) The synthesis of apatite by bacterial activity: mechanism. *Sci. Geol. Mem. (Strasbourg)* **77**, 83–92.

Lucas J., Flicoteaux R., Nathan U., Prévôt L., and Shahar Y. (1980) Different aspects of phosphorite weathering. In *Marine Phosphorites.* Spec. Publ. 29 (ed. Y. K. Bentor). SEPM, pp. 41–51.

Lucotte M. and D'Anglejan B. (1983) Forms of phosphorus and phosphorus–iron relationships in the suspended matter of the St. Lawrence Estuary. *Can. J. Earth Sci.* **20**, 1880–1890.

Lucotte M., Mucci A., and Hillairemarcel A. (1994) Early diagenetic processes in deep Labrador Sea sediments: reactive and nonreactive iron and phosphorus. *Can. J. Earth Sci.* **31**, 14–27.

Luz B., Kolodny Y., and Horowitz M. (1984) Fractionation of oxygen isotopes between mammalian bone-phosphate and environmental drinking water. *Geochim. Cosmochim. Acta* **63**, 855–862.

Mach D. M., Ramirez A., and Holland H. D. (1987) Organic phosphorus and carbon in marine sediments. *Am. J. Sci.* **278**, 429–441.

Mackenzie F. T., Ver L. M., Sabine C., Lane M., and Lerman A. (1993) C, N, P, S Global biogeochemical cycles and modeling of global change. In *Interactions of C, N, P and S Biogeochemical Cycles and Global Change.* NATO ASI Series 1 (eds. R. Wollast, F. T. Mackenzie, and L. Chou). Springer, Berlin, vol. 4, pp. 1–61.

Mackenzie F. T., Ver L. M., and Lerman A. (2002) Century-scale nitrogen and phosphorus controls of the carbon cycle. *Chem. Geol.* **190**, 13–32.

MacRae M., Glover W., Ammerman J., Sada R., and Ruvalcaba B. (1994) Seasonal phosphorus deficiency in the Mississippi River Plume: unusually large aerial extent during the record flood of 1993. *EOS, Trans., AGU* **75**, 30.

Malone T. C., Conley D. J., Fisher T. R., Glibert P. M., and Harding L. Wl. (1996) Scales of nutrient-limited phytoplankton productivity in the Chesapeake Bay. *Estuaries* **19**, 371–385.

Martens C. S. and Harriss R. C. (1970) Inhibition of apatite precipitation in the marine environment by magnesium ions. *Geochim. Cosmochim. Acta* **34**, 621–625.

Martens C. S., Berner R. A., and Rosenfeld J. K. (1978) Interstitial water chemistry of anoxic Long Island Sound Sediments: 2. Nutrient regeneration and phosphate removal. *Limnol. Oceanogr.* **23**, 605–617.

McArthur J. M. (1980) Post-depositional alteration of the carbonate-fluorapatite phase of Morocccan Phosphates. In *Marine Phosphorites*. Spec. Publ. 29 (ed. Y. K. Bentor). SEPM, pp. 53–60.

McClellan G. H. (1980) Mineralogy of carbonate fluorapatites. *J. Geol. Soc. London* **137**, 675–681.

McConnel D. (1973) *Apatite, its Crystal Chemistry, Mineralogy, Utilization and Geologic and Biologic Occurrences*. Springer, New York, 111pp.

McCorkle D. C. and Klinkhammer G. P. (1991) Pore-water cadmium geochemistry and the pore-water cadmium: $\delta^{13}C$ relationship. *Geochim. Cosmochim. Acta* **55**, 161–168.

McKelvie I. D., Hart B. T., Cardwell T. J., and Cattrall R. W. (1995) Use of immobilized 3-phytase and flow injection for the determination of phosphorus species in natural waters. *Anal. Chim. Acta* **316**, 277–289.

McKelvey V. E. (1967) Phosphoate deposits. *US Geol. Surv. Bull.* **1252-D**, 1–22.

McManus J., Berelson W. M., Coale K. H., Johnson K. S., and Kilgore T. E. (1997) Phosphorus regeneration in continental margin sediments. *Geochim. Cosmochim. Acta* **61**, 2891–2907.

Melack J. M. (1995) Transport and transformations of P in fluvial and lacustrine ecosystems. In *Phosphorus in the Global Environment*. SCOPE, chap. 15 (ed. H. Tiessen). Wiley, pp. 245–254.

Meybeck M. (1982) Carbon, nitrogen, and phosphorus transport by world rivers. *Am. J. Sci.* **282**, 401–450.

Miller A. J., Schuur E. A. G., and Chadwick O. A. (2001) Redox control of phosphorus pools in Hawaiian montane forest soils. *Geoderma* **102**, 219–237.

Milliman J. D. and Meade R. H. (1983) World-wide delivery of river sediment to the oceans. *J. Geol.* **91**, 1–21.

Milliman J. D. and Syvitski J. P. M. (1992) Geomorphic/tectonic control of sediment discharge to the ocean: the importance of small mountainous rivers. *J. Geol.* **100**, 525–544.

Mitchell A. M. and Baldwin D. S. (2003) Organic phosphorus in the aquatic environment: speciation, transformations, and interactions with nutrient cycles. In *Organic Phosphorus in the Environment*. chap. 14 (ed. B. Turner) (in press).

Monaghan E. J. and Ruttenberg K. C. (1999) Dissolved organic phosphorus in the coastal ocean: reassessment of available methods and seasonal phosphorus profiles from the Eel River Shelf. *Limnol. Oceanogr.* **44**, 1702–1714.

Morse J. W. and Cook N. (1978) The distribution and form of phosphorus in North Atlantic Ocean deep-sea and continental slope sediments. *Limnol. Oceangr.* **23**, 825–830.

Mortimer C. H. (1941) The exchange of dissolved substances between mud and water in lakes: I and II. *J. Ecol.* **29**, 280–329.

Mortimer C. H. (1942) The exchange of dissolved substances between mud and water in lakes: III and IV. *J. Ecol.* **30**, 147–201.

Myers V. B. and Iverson R. I. (1981) Phosphorus and nitrogen limited phytoplankton productivity in northeastern Gulf of Mexico coastal estuaries. In *Estuaries and Nutrients* (eds. B. J. Nielson and L. E. Cronin). Humana, Clifton, NJ, pp. 569–582.

Myklestad S. and Sakshaug E. (1983) Alkaline phosphatase activity of Skeletonema costatum populations in the Trondheims fjord. *J. Plankton Res.* **5**, 558–565.

Nanny M. A. and Minear R. A. (1997) Characterization of soluble unreactive phosphorus using ^{31}P nuclear magnetic resonance spectroscopy. *Mar. Geol.* **130**, 77–94.

Nanny M. A., Kim S., Gadomski J. E., and Minear R. A. (1994) Aquatic soluble unreactive phosphorus: concentration by ultrafiltration and reverse osmosis membranes. *Water Res.* **28**, 1355–1365.

Nawrocki M. P. and Karl D. M. (1989) Dissolved ATP turnover in the Bransfield Strait, Antarctica during the spring bloom. *Mar. Ecol. Prog. Ser.* **57**, 35–44.

Nissenbaum A. (1979) Phosphorus in marine and non-marine humic substances. *Geochim. Cosmochim. Acta* **43**, 1973–1978.

Nixon S. W. (1981) Remineralization and nutrient cycling in coastal marine ecosystems. In *Estuaries and Nutrients* (eds. B. J. Neilson and L. E. Cronin). Humana, Clifton, NJ, pp. 111–138.

Nixon S. W. (1995) Coastal marine eutrophication: a definition, social causes, and future concerns. *Ophelia* **41**, 199–220.

Nixon S. W. and Pilson M. E. (1984) Estuarine total system metabolism and organic exchange calculated from nutrient ratios: an example from Narragansett Bay. In *The Estuary as a Filter* (ed. V. S. Kennedy). Academic Press, Orlando, pp. 261–290.

Nixon S. W., Ammerman J. W., Atkinson L. P., Berounsky V. M., Billen G., Boicourt W. C., Boynton W. R., Church T. M., Ditoro D. M., Elmgren R., Garber J. H., Giblin A. E., Jahnke R. A., Owens N. J. P., Pilson M. E. Q., and Seitzinger S. P. (1996) The fate of nitrogen and phosphorus at the land-sea interface. *Biogeochemistry* **35**, 141–180.

O'Brien G. W., Harris J. R., Milnes A. R., and Veeh H. H. (1981) Bacterial origin of East Australian continental margin phosphorites. *Nature* **294**, 442–444.

O'Brien G. W., Milnes A. R., Veeh H. H., Heggie D. T., Riggs S. R., Cullen D. J., Marshall J. F., and Cook P. J. (1990) Sedimentation dynamics and redox iron-cycling: controlling factors for the apatite-glauconite association on the east Australian continental margin. In *Phosphorite Research and Development*. Geol. Soc. Spec. Publ. 52 (eds. A. G. J. Notholt and I. Jarvis), pp. 61–86.

Orrett K. and Karl D. M. (1987) Dissolved organic phosphorus production in surface waters. *Limnol. Oceanogr.* **32**, 383–395.

Paasche E. and Erga S. R. (1988) Phosphorus and nitrogen limitation of phytopllantkonin the inner Oslofjord (Norway). *Sarsia* **73**, 229–243.

Pakulski J. D., Benner R., Whitledge T., Amon R., Eadie B., Cifuentes L., Ammerman J., and Stockwell D. (2000) Microbial metabolism and nutrient cycling in the Mississippi and Atchafalaya River Plumes. *Estuar. Coast. Shelf Sci.* **50**, 173–184.

Palenik B. and Dyhrman S. T. (1998) Recent progress in understanding the regulation of marine primary productivity by phosphorus. In *Phosphorus in Plant Biology: Regulatory Roles in Molecular, Cellular, Organismic, and Ecosystem Processes* (eds. J. P. Lynch and J. Deikman). American Society of Plant Physiologists, pp. 26–38.

Paytan A., Kolodny Y., Neori A., and Luz B. (2002) Rapid biologically mediated oxygen isotope exchange between water and phosphate. *Global Biogeochem. Cycles* **16**, 10.1029/2001GB001430.

Peng T.-H. and Broecker W. S. (1987) C/P ratios in marine detritus. *Global Biogeochem. Cycles* **1**, 155–161.

Perry M. J. and Eppley R. W. (1981) Phosphate uptake by phytoplankton in the central North Pacific Ocean. *Deep-Sea Res.* **28**, 39–49.

Petsch S. T. and Berner R. A. (1998) Coupling the geochemical cycles of C, P, Fe and S: the effect on atmospheric O2 and the isotopic records of carbon and sulfur. *Am J. Sci.* **298**, 246–262.

Pomeroy L. R. (1960) Residence time of dissolved phosphate in natural waters. *Science* **131**, 1731–1732.

Pomeroy L. R., Smith E. E., and Grant C. M. (1965) The exchange of phosphate between estuarine water and sediments. *Limnol. Oceanogr.* **10**, 167–172.

Rabalais N. N., Turner R. E., Justic D., Dortch Q., Wiseman W. J., and Sen Gupta B. (1996) Nutrient changes in the

Mississippi River and system responses on the adjacent continental shelf. *Estuaries* **19**, 386–407.

Ramirez A. J. and Rose A. W. (1992) Analytical geochemistry of organic phosphorus and its correlation with organic carbon in marine and fluvial sediments and soils. *Am. J. Sci.* **292**, 421–454.

Rasmussen B. (1996) Early-diagenetic REE-phosphate minerals (florencite, gorceixite, crandallite and xenotime) in marine sandstones: a major sink for oceanic phosphorus. *Am. J. Sci.* **296**, 601–632.

Rasmussen B. (2000) The impact of early-diagenetic aluminophosphate precipitation on the oceanic phosphorus budget. In *Marine Authigenesis: From Global to Microbial.* Spec. Publ. 66 (eds. C. R. Glenn, L. Prevot-Lucas, and J. Lucas). SEPM, pp. 89–101.

Redfield A. C. (1958) The biological control of chemical factors in the environment. *Am. Sci.* **46**, 205–222.

Redfield A. C., Smith H. P., and Ketchum B. (1937) The cycle of organic phosphorus in the Gulf of Maine. *Biol. Bull.* **73**, 421–443.

Redfield A. C., Ketchum B. H., and Richards F. A. (1963) The influence of organisms on the composition of sea water. In *The Sea* (ed. M. N. Hill). Interscience, New York, Vol. 2, pp. 26–77.

Reimers C. E., Ruttenberg K. C., Canfield D. E., Christiansen M. B., and Martin J. B. (1996) Pore water pH and authigenic phases formed in the uppermost sediments of the Santa Barbara Basin. *Beochim. Cosmochim. Acta* **60**(21), 4037–4057.

Rengefors K., Petterson K., Blenckner T., and Anderson D. M. (2001) Species-specific alkaline phosphatase activity in freshwater spring phytoplankton: application of a novel method. *J. Plankton Res.* **23**, 443–453.

Rengefors K., Ruttenberg K., Haupert C., Howes B., and Taylor C. (2003) Experimental investigation of taxon-specific response of alkaline phosphatase activity in natural freshwater phytoplankton. *Limnol. Oceanogr.* **48**, 1167–1175.

Ryther J. H. and Dunstan W. M. (1971) Nitrogen, phosphorus, and eutrophication in the coastal marine environment. *Science* **171**, 1008–1013.

Richardson K. and Jørgensen B. B. (1996) Eutrophication: definition, history, and effects. In *Eutrophication in Coastal Marine Ecosystems Coastal and Estuarine Studies.* Am. Geophys. Union Publ. (eds. B. B. Jørgensen and K. Richardson) vol. 52, pp. 1–20.

Richey J. E. (1983) The phosphorus cycle. In *The Major Biogeochemical Cycles and Their Interactions.* SCOPE **21** (eds. B. Bolin and R. B. Cook). Wiley, Chichester, pp. 51–56.

Riggs S. R. (1979) Phosphorite sedimentation in Florida—a model phosphogenic system. *Econ. Geol.* **74**, 285–314.

Riggs S. R. (1984) Paleoceanographic model of Neogene phosphorite deposition US Atlantic continental margin. *Science* **223**, 123–131.

Riggs S. R., Snyder S. W., Hine A. C., Ellington M. D., and Mallette P. M. (1985) Geological framework of phosphate resources in Onslow Bay. *North Carolina continental shelf. Econ. Geol.* **80**, 716–738.

Rigler F. H. (1966) Radiobiological analysis of inorganic phosphorus in lakewater. *Vehr. Int. Verein. Theor. Angew. Limnol.* **16**, 465–470.

Rigler F. H. (1968) Further observations inconsistent with the hypothesis that the molybdenum blue method measured orthophosphate in lake water. *Limnol. Oceanogr.* **13**, 7–13.

Rivkin R. B. and Swift E. (1980) Characterization of alkaline phosphatase and organic phosphorus utilization in the oceanic dinoflagellate *Pyrocystis noctiluca. Mar. Biol.* **61**, 1–8.

Rivkin R. B. and Swift E. (1985) Phosphorus metabolism of oceanic dinoflagellates: phosphate uptake, chemical composition, and growth of *Pyrocystis noctiluca. Mar. Biol.* **88**, 189–198.

Rosenthal Y., Lam P., Boyle E. A., and Thompson J. (1995a) Authigenic cadmium enrichments in suboxic sediments: precipitation and postdepositional mobility. *Earth Planet. Sci. Lett.* **132**, 99–111.

Rosenthal Y., Boyle E. A., Labeyrie L., and Oppo D. (1995b) Glacial enrichments of authigenic Cd and U in Subantarctic sediments: a climatic control on the elements' oceanic budget? *Paleoceanography* **10**(3), 395–414.

Rozen T. F., Taillefert M., Trouwborst R. E., Glazer B. T., Ma S., Herszage J., Valdes L. M., Price K. S., and Luther G. W. (2002) Iron-sulfur-phosphorus cycling in the sediments of a shallow coastal bay: implications for sediment nutrient release and benthic macroalgal blooms. *Limnol. Oceanogr.* **47**, 1346–1354.

Rude P. D. and Aller R. C. (1991) Fluorine mobility during early diagenesis of carbonate sediment: an indicator of mineral transformations. *Geochim. Cosmochim. Acta* **55**, 2491–2509.

Rudnicki M. D. and Elderfield H. (1993) A chemical model of the buoyant and neutrally buoyant plume above the TAG vent field, 26 degrees N, Mid-Atlantic Ridge. *Geochim. Cosmochim. Acta* **57**, 2939–2957.

Ruttenberg K. C. (1990) Diagenesis and burial of phosphorus in marine sediments: implications for the marine phosphorus budget. PhD Thesis, Yale University 375pp.

Ruttenberg K. C. (1991) The role of bottom sediments in the aquatic phosphorus cycle. In *Proceedings of the National Workshop on Phosphorus in Australian Freshwaters.* Occas. Paper No. 03/93, Charles Sturt University Environmental and Analytical Labs, Wagga Wagga, NSW, Australia.

Ruttenberg K. C. (1993) Reassessment of the oceanic residence time of phosphorus. *Chem. Geol.* **107**, 405–409.

Ruttenberg K. C. (1994) Proxy paradox for P-prediction. *Nature* **372**, 224–225.

Ruttenberg K. C. (2002) The global phosphorus cycle. In *The Encyclopedia of Global Change* (eds. A. S. Goudie and D. J. Cuff). Oxford University Press, vol. 2, pp. 241–245.

Ruttenberg K. C. and Berner R. A. (1993) Authigenic apatite formation and burial in sediments from non-upwelling continental margins. *Geochim. Cosmoshim. Acta* **57**, 991–1007.

Ruttenberg K. C. and Canfield D. E. (1988) Fluoride adsorption onto ferric oxyhydroxides in sediments. *EOS, Trans., AGU* **69**, 1235.

Ruttenberg K. C. and Canfield D. E. (1994) Chemical distribution of phosphorus in suspended particulate matter from twelve North American Rivers: evidence for bioavailability of particulate-P. *EOS, Trans., AGU* **75**(3), 110.

Ruttenberg K. C. and Goñi M. A. (1997a) Phosphorus distribution, elemental ratios, and stable carbon isotopic composition of arctic, temperate, and tropical coastal sediments: tools for characterizing bulk sedimentary organic matter. *Mar. Geol.* **139**, 123–145.

Ruttenberg K. C. and Goñi M. A. (1997b) Depth trends in phosphorus distribution and C:N:P ratios of organic matter in Amazon Fan sediments: indices of organic matter source and burial history. In *Proc. Ocean Drilling Program.* Sci. Res. (eds. R. D. Flood, D. J. W. Piper, A. Klaus, and L. C. Peterson), vol. 155, pp. 505–517.

Ruttenberg K. C. and Laarkamp K. L. (2000) Diagenetic trends in molecular-level organic phosphorus composition in Santa Barbara Basin sediments: insights into mechanisms of preservation. *EOS, Trans., AGU.* Fall Meet. (12/2000).

Ruttenberg K. C. and Ogawa N. O. (2002) A high through-put solid-phase extraction manifold (SPEM) for separating and quantifying different forms of phosphorus and iron in particulate matter and sediments, GES-6 Meeting, May 2002, Honolulu, HI.

Schenau S. J. and De Lange G. J. (2000) Phosphogenesis and active phosphorite formation in sediments from the Arabian Sea oxygen minimum zone. *Mar. Geol.* **169**, 1–20.

Schenau S. J., Slomp C. P., and De Lange G. J. (2000) A novel chemical method to quantify fish debris in marine sediments. *Limnol. Oceanogr.* **45**, 963–971.

Schindler D. W. (1970) Evolution of phosphorus limitation in lakes. *Science* **195**, 260–262.

Schink B. and Friedrich M. (2000) Phosphite oxidation by sulfate reduction. *Nature*, 406.

Schlesinger W. H., Bruijnzeel L. A., Bush M. B., Klein E. M., Mace K. A., Raikes J. A., and Whittaker R. J. (1998) The biogeochemistry of ph9osphorus after the first century of soil development on Rakata Island, Krakatau, Indonesia. *Biogeochemistry* **40**, 37–55.

Schuffert J. D., Kastner M., Emanuele G., and Jahnke R. A. (1990) Carbonate ion substitution in francolite: a new equation. *Geochim. Cosmochim. Acta* **54**, 2323–2328.

Schuffert J. D., Jahnke R. A., Kastner M., Leather J., Sturz A., and Wing M. R. (1994) Rates of formation of modern phosphorite off western Mexico. *Geochim. Cosmochim. Acta* **58**, 5001–5010.

Shaffer G. (1986) Phosphate pumps and shuttles in the Black Sea. *Nature* **321**, 515–517.

Shapiro J. (1967) Induced rapid release and uptake of phosphate by microorganisms. *Science* **155**, 1269–1271.

Sheldon R. P. (1980) Episodicity of phosphate deposition and deep ocean circulation—a hypothesis. In *Marine Phosphorites*. Spec. Publ. 29 (ed. Y. K. Bentor). SEPM, pp. 239–247.

Shemesh A., Kolodny Y., and Luz B. (1983) Oxygen isotope variations in phosphate of biogenic apatites: II. Phosphate rocks. *Earth Planet. Sci. Lett.* **64**, 405–416.

Shiller A. M. (1993) Comparison of nutrient and trace element distributions in the delta and shelf outflow regions of the Mississippi and Atchafalaya Rivers. *Estuaries* **16**(3A), 541–546.

Sholkovitz E. R., Boyle E. A., and Price N. B. (1978) The removal of dissolved humic acids and iron during estuarine mixing. *Earth Planet. Sci. Lett.* **40**, 130–136.

Slomp C. P. and Van Raaphorst W. (1993) Phosphate adsorption in oxidized marine sediments. *Chem. Geol.* **107**, 477–480.

Slomp C. P., Epping E. H. G., Helder W., and Van Raaphorst W. (1996a) A key role for iron-bound phosphorus in authigenic apatite formation in North Atlantic continental platform sediments. *J. Mar. Res.* **54**, 1179–1205.

Slomp C. P., Van der Gaast S. J., and Van Raaphorst W. (1996b) Phosphorus binding by poorly crystalline iron oxides in North Sea sediments. *Mar. Chem.* **52**, 55–73.

Slomp C. P., Thomson J., and De Lange G. J. (2002) Enhanced regeneration of phosphorus during formation of the most recent eastern Mediterranean sapropel (S1). *Geochim. Cosmochim. Acta* **66**(7), 1171–1184.

Smeck N. E. (1985) Phosphorus dynamics in soils and landscapes. *Geoderma* **36**, 185–199.

Smith R. E., Harrison W. G., and Harris L. (1985) Phosphorus exchange in marine microplankton communities near Hawaii. *Mar. Biol.* **86**, 75–84.

Smith S. V. (1984) Phosphorus versus nitrogen limitation in the marine environment. *Limnol. Oceanogr.* **29**, 1149–1160.

Smith S. V. (2001) Carbon–nitrogen–phosphorus fluxes in the coastal zone: the LOICZ approach to global assessment, and scaling issues with available data. LOICZ Newsletter, December 2001, no. 21, pp 1–3.

Smith S. V. and Atkinson M. J. (1984) Phosphorus limitation of net production in a confined aquatic ecosystem. *Nature* **307**, 626–627.

Smith S. V. and Mackenzie F. T. (1987) The ocean as a net heterotrophic system. *Global Biogeochem. Cycles* **1**, 187–198.

Smith S. V., Swaney D. P., Talaue-McManus L., Bartley J. D., Sandhei P. T., McLaughlin C. J., Dupra V. C., Crossland C. J., Buddemeier R. W., Maxwell B. A., and Wulff F. (2003) Humans, hydrology, and the distribution of inorganic nutrient loading to the ocean. *Bioscience* **53**(3), 235–245.

Solórzano L. and Strickland J. D. H. (1968) Polyphosphate in seawater. *Limnol. Oceanogr.* **13**, 515–518.

Sorokin Y. I. (1985) Phosphorus metabolism in planktonic communities of the eastern tropical Pacific Ocean. *Mar. Ecol. Prog. Ser.* **27**, 87–97.

Soudry D. and Champetier Y. (1983) Microbial processes in the Negev phosphorites (southern Israel). *Sedimentology* **30**, 411–423.

Sundby B., Gobeil C., Silverberg N., and Mucci A. (1992) The phosphorus cycle in coastal marine sediments. *Limnol. Oceanogr.* **37**, 1129–1145.

Suzumura M. and Kamatami A. (1995) Origin and distribution of inositol hexaphosphate in estuarine and coastal sediments. *Limnol. Oceanogr.* **40**, 1254–1261.

Suzumura M., Ishikawa K., and Ogawa H. (1998) Characterization of dissolved organic phosphorus using ultrafiltration and phosphohydrolytic enzymes. *Limnol. Oceanogr.* **43**, 1553–1564.

Sverdrup H. V., Johnson M. W., and Fleming R. H. (1942) *The Oceans, their Physics, Chemistry and General Biology*. Prentice Hall, New York, 1087pp.

Taft J. L., Taylor W. R., and McCarthy J. J. (1975) Uptake and release of phosphorus by phytoplankton in the Chesapeake Bay, USA. *Mar. Biol.* **33**, 21–32.

Tamburini F., Föllmi K. B., Adatte T., Bernasconi S. M., and Steinmann P. (2003) Sedimentary phosphorus record from the Oman margin: new evidence of high productivity during glacial periods. *Paleoceanography* **18**(1), 1015 doi:10.1029/2000PA000616.

Tarapchak S. J. and Nawelajko C. (1986) Synopsis: phosphorus-plankton dynamics symposium. *Can. J. Fish. Aquat. Sci.* **43**, 416–419.

Thomas H. (2002) Remineralization ratios of carbon, nutrients, and oxygen in the North Atlantic Ocean: a field data based assessment. *Global Biogeochem. Cycles* **16**(3), 1051 doi: 10.1029/2001 GB001452.

Thompson G. (1983) Basalt-seawater interaction. In *Hydrothermal Processes at Seafloor Spreading Centers* (ed. P. A. Rona, *et al.*). Plenum, pp. 225–278.

Thompson J., Calvert S. E., Mukherjee S., Burnett W. C., and Bremner J. M. (1984) Further studies of the nature, composition and ages of contemporary phosphorite from the Namibian Shelf. *Earth Planet. Sci. Lett.* **69**, 341–353.

Thomson-Bulldis A. and Karl D. (1998) Application of a novel method for phosphorus determination in the oligotrophic North Pacific Ocean. *Limnol. Oceanogr.* **43**(7), 1565–1577.

Torrent J., Barrón V., and Schwertmann U. (1990) Phosphate adsorption and desorption by goethites differeng in crystal morphology. *Soil Sci. Soc. Am. J.* **54**(4), 1007–1012.

Tyrrell T. (1999) The relative influences of nitrogen and phosphorus on oceanic productivity. *Nature* **400**, 525–531.

Tyrrell T. and Law C. S. (1997) Low nitrate: phosphate ratios in the global ocean. *Nature* **387**, 793–796.

Van Cappellen P. and Berner R. A. (1988) A mathematical model for the early diagenesis of phosphorus and fluorine in marine sediments: apatite precipitation. *Am. J. Sci.* **288**, 289–333.

Van Cappellen P. and Berner R. A. (1989) Marine apatite precipitation. In *Water-Rock Interactions* (ed. D. Miles). Balkema, pp. 707–710.

Van Cappellen P. and Berner R. A. (1991) Fluorapatite crystal growth from modified seawater solutions. *Geochim. Cosmochim. Acta* **55**, 1219–1234.

Van Cappellen P. and Ingall E. D. (1994a) Benthic phosphorus regeneration, net primary production, and ocean anoxia: a model of the coupled marine biogeochemical cycles of carbon and phosphorus. *Paleoceanography* **9**(5), 677–692.

Van Cappellen P. and Ingall E. D. (1994b) Redox stabilization of the atmosphere by phosphorus-limited marine productivity. *Science* **271**, 493–496.

Van der Zee C., Slomp C. P., and van Raaphorst W. (2002) Authigenic P formation and reactive P burial in sediments of the Nazare canyon on the Iberian margin (NE Atlantic). *Mar. Geol.* **185**, 379–392.

Veeh H. H., Burnett W. C., and Soutar A. (1973) Contemporary phosphorites on the continental margin of Peru. *Science* **181**, 844–845.

Vink S., Chambers R. M., and Smith S. V. (1997) Distribution of phosphorus in sediments from Tomales Bay, California. *Mar. Geol.* **139**, 157–179.

Vitousek P. M., Chadwick O. A., Crews T. E., Fownes J. H., Hendricks D. M., and Herbert D. (1997) Soil and ecosystem development across the Hawaiian Islands. *GSA Today* **7**(9), 1–8.

Walker T. W. and Syers J. K. (1976) The fate of phosphorus during pedogenesis. *Geoderma* **15**, 1–19.

Waser N. A. D. and Bacon M. P. (1995) Wet deposition fluxes of cosmogenic ^{32}P and ^{33}P and variations in the $^{33}P/^{32}P$ ratios at Bermuda. *Earth Planet. Sci. Lett.* **133**, 71–80.

Waser N. A. D., Bacon M. P., and Michaels A. F. (1996) Natural activities of ^{32}P and ^{33}P and the $^{33}P/^{32}P$ ratio in suspended particulate matter and plankton in the Sargasso Sea. *Deep-Sea Res. II.* **43**, 421–436.

Westheimer F. H. (1987) Why nature chose phosphates. *Science* **235**, 1173–1178.

Wheat C. G., Feely R. A., and Mottl M. J. (1996) Phosphate removal by oceanic hydrothermal processes: an update of the phosphorus budget of the oceans. *Geochim. Cosmochim. Acta* **60**, 3593–3608.

White D. C., Davis W. M., Nickels J. S., King J. D., and Bobbie R. J. (1979) Determination of the sedimentary microbial biomass by extractable lipid phosphate. *Oecologia* **40**, 51–62.

Williams J. D. H., Jaquet J.-M., and Thomas R. L. (1976) Forms of phosphorus in the surficial sediments of Lake Erie. *J. Fish Res. Board Can.* **33**, 413–429.

Wu J.-F., Sunda W., Boyle E. A., and Karl D. M. (2000) Phosphate depletion in the western North Atlantic Ocean. *Science* **289**, 759–762.

8.14
The Global Sulfur Cycle

P. Brimblecombe

University of East Anglia, Norwich, UK

8.14.1 ELEMENTARY ISSUES

8.14.1.1 History

Sulfur is one of the elements from among the small group of elements known from ancient times. Homer described it as a disinfectant, and in *In Fasti* (IV, 739–740) Ovid wrote: "caerulei fiant puro de sulpure fumi, tactaque fumanti sulpure balet ovis...," which explains how the blue smoke from burning pure sulfur made the sheep bleat. Less poetical Roman writers gave more detail and distinguished among the many forms of elemental sulfur, which could be mined from volcanic regions. It was used in trade and craft activities such as cleaning of wool. In addition to the native form, sulfur is also widely found as sulfate and sulfide minerals. It was important to alchemists, because they were able to produce sulfuric acid by heating the mineral, green vitriol, $FeSO_4{\cdot}7H_2O$ and then condensing the acid from the vapor:

$$FeSO_4{\cdot}7H_2O_{(s)} \rightarrow H_2SO_{4(l)} + FeO_{(s)} + 6H_2O_{(g)}$$

Sulfuric acid rose to become such an important industrial chemical that the demand for sulfur in the production of the acid has sometimes been considered an indicator of national wealth.

Early geochemists, such as Victor Moritz Goldschmidt (1888–1947), recognized the difficulties associated with the geochemistry of a highly mobile and biologically active element such as sulfur. Fortunately, the composition of various reservoirs was known by the beginning of the twentieth century and carefully arranged analyses in terms of reservoirs: the crust, waters, and the atmosphere. The collected data were compiled and clearly arranged by Frank W. Clarke in his *US Geological Survey (USGS) Bulletins: The Data of Geochemistry* from 1908. Soon after Waldemar Lindgren, who had also been with the USGS and became head of the Massachusetts Institute of Technology's Department of Geology, gave the "story of sulfur" in an address on the concentration and circulation of the elements (Lindgren, 1923). The story is essentially a sulfur cycle with the notion of the volcanic release of reduced sulfur and oxidation and the transport as soluble sulfates to the sea. He recognized that sulfate in the oceans would rapidly have dominated over chloride were it not for biologically mediated reduction. Later Conway (1942) speculated on the importance of oceanic hydrogen sulfide as a source to the sulfur cycle. The idea of elemental cycles formed a central part of Rankama and Sahama's (1950) *Geochemistry*. There were many early sulfur cycles drawn up including those of Robinson and Robins, Granat *et al.*, Kellogg *et al.*, and Friend *et al.* that formed the basis of more recent cycles (e.g., Brimblecombe *et al.*, 1989; Rodhe, 1999) (Figure 1).

8.14.1.2 Isotopes

There are four stable isotopes of sulfur as listed in Table 1. The isotopic abundances vary slightly and this is frequently used to distinguish the source of the element. Because measurement of absolute isotope abundance is difficult, relative isotopic ratios are measured by comparison with the abundance of the natural isotopes in a standard sample. The Canyon Diablo meteorite has been used as a standard for sulfur isotopes.

There are nine known radioactive isotopes and six are listed in Table 2. Sulfur-35 has the longest half-life and is produced by cosmogenic synthesis in the upper atmosphere: cosmogenic S-35 (Tanaka and Turekian, 1991) is sufficiently long lived to be useful in determining overall removal and transformation rates of SO_2 from the atmosphere and an estimated dry deposition flux to total flux ratio is ~0.20 in the eastern US (Turekian and Tanaka, 1992).

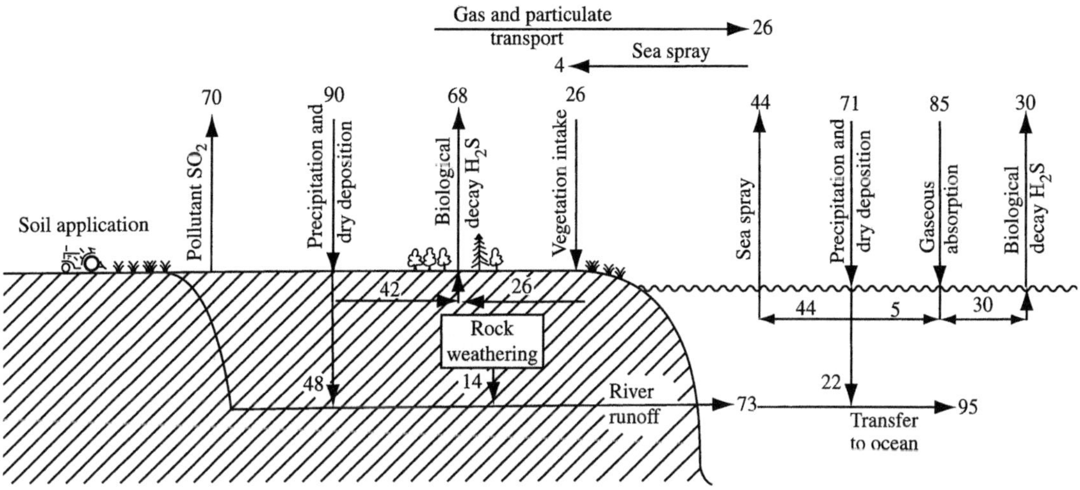

Figure 1 A generalized geochemical cycle for sulfur of the early 1970s. Note the large emissions of hydrogen sulfide from the land and oceans and that volcanic sulfur emissions are neglected (units: Tg (s) a^{-1}).

Table 1 Stable isotopes of sulfur.

Isotope	*Mass*	*Abundance* (%)
S-32	31.97207	94.93
S-33	32.971456	0.76
S-34	33.967886	4.29
S-36	35.96708	0.02

Table 2 Radioactive isotopes of sulfur.

Isotope	*Half-life*	*Principal radiation* (MeV)
S-29	0.19 s	p
S-30	1.4 s	β^+ 5.09, β^+ 4.2, γ 0.687
S-31	2.7 s	β^+ 4.42, γ 1.27
S-35	88 d	β^- 0.167
S-37	5.06 min	β^- 4.7, β^- 1.6, γ 3.09
S-38	2.87 h	β^- 3.0, β^- 1.1, γ 1.27

8.14.1.3 Allotropes

In 1772 Lavoisier proved that the sulfur is an elementary substance, which was not necessarily obvious given that it is characterized by a number of allotropes (i.e., different forms). Sulfur has more allotropes than any element because of the readiness to form S—S bonds (Table 3). These bonds can be varied both in terms of length and angle, so open and cyclic allotropes of S_n are known where n ranges between 2 and 20. The familiar form of yellow orthorhombic sulfur (α-sulfur) is a cyclic crown S_8 ring. Two other S_8 ring forms are known the β-orthorhombic and the γ-orthorhombic found at higher temperatures. Engel in 1891 prepared a rhombohedral form ε-sulfur, which was ultimately shown to be an S_6 ring. The S_6 and S_8 rings have equally spaced sulfur atoms, which are essentially equivalent. This is not true of the S_7 form, which is found in four crystalline modifications. Here the interatomic distances vary between 199.3 pm and 218.1 pm. This latter distance is exceptionally large compared to the 2.037–2.066 pm typical of other sulfur rings.

8.14.1.4 Vapor Pressure

Sulfur vapor is a relatively volatile element and the vapor contains polyatomic species over the range S_2–S_{10}, with S_7 the main form at high temperatures (see Figure 2). The strong sulfur–sulfur double bond in S_2 (422 kJ mol^{-1}) means that monatomic sulfur is found only at very high temperatures (>2,200 °C).

Molten sulfur is known from volcanic lakes (Oppenheimer and Stevenson, 1989). The elemental liquid is a complex material. Elemental sulfur melts at ~160 °C giving a yellow liquid, which becomes brown and increasingly viscous as the temperature rises in the range 160–195 °C, which is interpreted as a product of polymerization into forms that can contain more than 2×10^5 sulfur atoms. As temperature increases above this, the chain length (and thus viscosity) decreases to ~100 by 600 °C. If molten sulfur is cooled rapidly by pouring into water, it condenses into *plastic sulfur* that can be stretched as long fibers, which appear to be helical chains of sulfur atoms with ~3.5 atoms for each turn of the helix (Cotton *et al.*, 1999).

8.14.1.5 Chemistry

Sulfur is very reactive and will combine directly with many elements. It will equilibrate with hydrogen to form hydrogen sulfide at elevated

Table 3 Sulfur allotropes.

Allotrope	*Color*	*Density* ($g\ cm^{-3}$)	*m.p.* (°C)	*Interatomic* (distance/pm)
S_2 (gas)	Blue-violet		*s*	188.9
S_3 (gas)	Cherry red		*s*	
S_6	Orange red	2.209	$d > 50$	205.7
S_7	Yellow	2.183(@ − 110°)	$d \sim 39$	199.3–218.1
αS_8	Yellow	2.069	112.8	203.7
βS_8	Yellow	1.94–2.01	119.6	204.5
γS_8	Light yellow	2.19	106.8	
S_9	Intense yellow		$s <$ rt	
S_{10}	Yellow green	2.103(@ − 110°)	$d > 0$	205.6
S_{12}	Pale yellow	2.036	148	205.3
S_{18}	Lemon yellow	2.090	128	205.9
S_{20}	Pale yellow	2.016	124	204.7

Source: Greenwood and Earnshaw (1997). d decomposition, s stable at high temperature, rt room temperature.

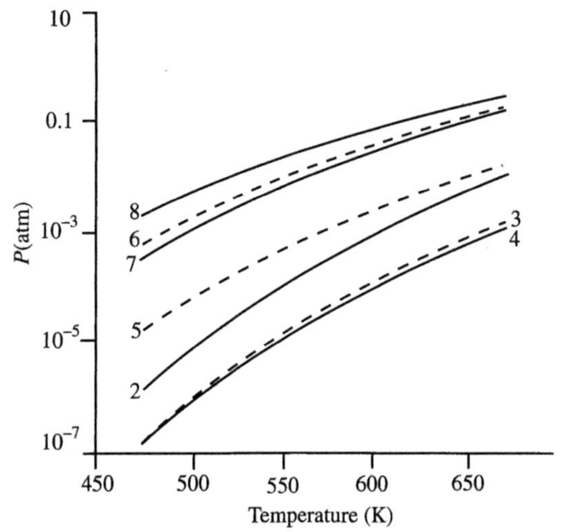

Figure 2 Partial pressure of sulfur polymers as a function of temperature. The numbers refer to the number of atoms in the polymer.

temperatures and will burn in oxygen and fluorine to form sulfur dioxide and sulfur hexafluoride. Reactions with chlorine and bromine are also known along with reactions with many metals even at low temperatures where it can be observed tarnishing copper, silver, and mercury.

Hydrogen sulfide is the only stable sulfane and is a poisonous gas with the smell of rotten eggs. Although the smell is noticeable at 0.02 ppm, it can rapidly anesthetize the nose; so higher toxic concentrations can sometimes pass unnoticed. Much less is known about prolonged exposure to low concentrations, but there are associations with *persistent neurobehavioral dysfunction* (Kilburn and Warshaw, 1995). The gas occurs widely in volcanic regions and as the product of protein degradation. It is slightly soluble in water (K_H, 0.1 mol L^{-1} atm^{-1} at 25 °C) and dissociates as a dibasic weak acid:

$$H_2S \rightleftharpoons H^+ + HS^-$$

$$HS^- \rightleftharpoons H^+ + S^{2-}$$

The first dissociation constant takes the value $10^{-6.88 \pm 0.02}$ at 20 °C, but the second has given greater problems. Some have argued for a value close to $\sim 10^{-17}$ (Licht *et al.*, 1991; Migdisov *et al.*, 2002), while others justify the use of $10^{-14 \pm 0.2}$ at 25 °C for calculations at low ionic strength (D. J. Phillips and S. L. Phillips, 2000).

Polysulfanes with the general formula H_2S_n ($n = 2-8$) are reactive oils in their pure state. Heating aqueous sulfide solutions yields polysulfide ions S_n^{2-} most typically with $n = 3$ or 4. The presence of sulfanes has been used to explain the high concentrations of sulfur compounds in some hydrothermal fluids (Migdisov and Bychkov, 1998).

Metal sulfides often have unusual noninteger stoichiometries and may be polymorphic and alloy like. The alkali metal sulfides have antifluorite structures. The alkaline earth and less basic metals adopt an NaCl-type structure. Many of the later transition metal sulfides show an increasing tendency to covalency. Iron, cobalt, and nickel sulfides adopt a nickel arsenide structure, with each metal atom surrounded octahedrally by six sulfur atoms (Figure 3). Disulfides are also common, among which the best known is pyrite (FeS_2); but this is characterized by a nonstoichiometric tendency, which leads to a 50–55% presence of sulfur, while retaining the nickel arsenide structure. There are more complex metal sulfides that adopt chain ring and sheet structures (Cotton *et al.*, 1999). Nonmetals often yield polymers that involve sulfur bridges.

Many oxides of sulfur are known, and although SO_2 and SO_3 are the most common, there are a range of cyclic oxides (e.g., S_nO ($n = 5-10$), S_6O_2, S_7O_2), and some thermally unstable acyclic oxides such as S_2O_2, S_2O, and SO. The oxides S_2O and SO appear to be components of terrestrial

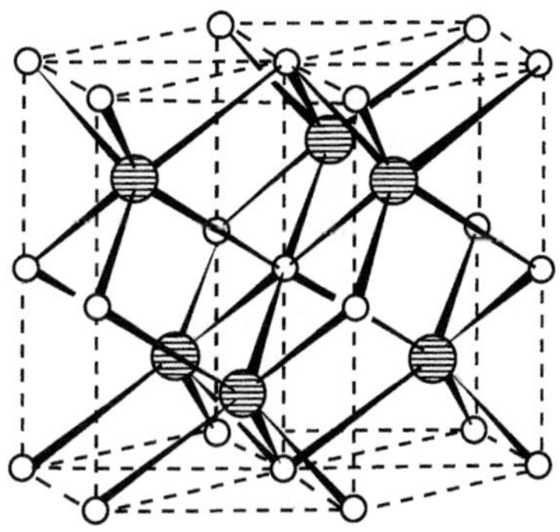

Figure 3 The nickel arsenide structure (reproduced by permission of Oxford University Press from *Structural Inorganic Chemistry*, **1945**, p. 387).

volcanic gases. More recently, there has been considerable interest in volcanogenic sulfur monoxide on the Jovian moon, Io (Russell and Kivelson, 2001; Zolotov and Fegley, 1998).

Sulfur dioxide (SO_2) is typically produced when sulfur burns in air or sulfide minerals are heated. It has been one of the most characteristic pollutants produced during the combustion of high sulfur-rich fuels. It is toxic with a characteristic choking odor. It liquefies very readily at $-10\ °C$ and has often been used as a nonaqueous solvent. It has been used as a bleaching agent and a preservative, and frequently found in wines. It is soluble in water, and is a weak dibasic acid, which has considerable relevance to its atmospheric chemistry.

Sulfur trioxide (SO_3) is an industrially important compound key to the production of sulfuric acid. It tends to polymeric forms both in the solid and liquid states. As a gas, the molecules have a planar triangular structure in which the sulfur atom has a high affinity for electrons. This explains its action as a strong Lewis acid towards bases that it does not oxidize. It can thus crystallize complexes with pyridine or trimethylamine. It has a very strong affinity for water and hence rapidly associates with water in the environment.

Sulfur dioxide and trioxide dissolve to give sulfurous and sulfuric acid. These acids are widely present in aqueous systems, particularly in the atmosphere. It is not possible to prepare pure sulfurous acid, but sulfuric acid can be obtained at high concentration. Sulfurous acid undergoes a range of reactions in atmospheric droplets, which are relevant to its removal from the atmosphere. Many of these reactions oxidize it to the stronger sulfuric acid, which has been a key component of acid rain.

Sulfuric acid is a dibasic acid and although regarded as a strong acid, the second dissociation constant is rather modest ($\sim 0.01\ mol\ L^{-1}$, although care must be taken to consider the nonideality of electrolyte solutions when undertaking calculations of the position of this equilibrium):

$$HSO_4^- \rightleftharpoons H^+ + SO_4^{2-}$$

The acid is a weak oxidizing agent, but has a strong affinity for water meaning that it chars carbohydrates. Although the acid is typically dilute in the environment, it may reach high concentrations (many $mol\ kg^{-1}$) in atmospheric droplets. At the low temperatures found in the stratosphere, these can crystallize as a range of hydrates.

Sulfates are widely known in the environment and a few such as those of calcium, strontium, barium, and lead are insoluble. The sulfates of magnesium and the alkali metals are soluble and so are typically found in brines and salt deposits. Sulfites are rarer because of the ease of oxidation, but calcium sulfite, perhaps as the hemihydrate, can be detected on the surface of building stone in urban atmospheres containing traces of sulfur dioxide (Gobbi *et al.*, 1998).

A wide range of oxyacids is possible with more than one sulfur atom. Although the acids are not always very stable, the oxyanions are often more easily prepared. Many of these are likely intermediates in the oxidation of aqueous sulfites and occur at low concentration in water droplets in the atmosphere. Higher concentrations of sulfur oxyanions such as $S_4O_6^{2-}$, $S_5O_6^{2-}$, and $S_6O_6^{2-}$ can be found in mineralized acid-sulfate waters of volcanic crater lakes (e.g., Sriwana *et al.*, 2000).

Sulfur can form halogen compounds and nitrides. The nitrogen compounds are found as interstellar molecules and chlorides postulated on the Jovian moon, Io, but these compounds do not appear to be a significant part of biogeochemical cycles on the Earth.

Compounds of sulfur with carbon having the general formula C_xS_y are characterized through a series of linear polycarbon sulfides C_nS ($n = 2–9$), some of which can be detected in interstellar clouds. Carbon disulfide (CS_2) occurs in the atmosphere and oceans of the Earth. Carbonyl sulfide (OCS) is also well known in nature and is one of many polycarbon oxide sulfides represented by the formula OC_nS where $n < 6$.

Many organosulfur compounds occur in the environment. Divalent sulfur as thiols (e.g., CH_3SH) or thioethers (e.g., CH_3SCH_3) is analogous to alcohols and ethers. These may be present as polysulfides, e.g., the dimethylpolysulfides ($CH_3S_nCH_3$, $n = 2–4$) detected in some water bodies (Gun *et al.*, 2000). Simpler volatile compounds, most typically the methyl derivatives, are present in the atmosphere. Thiols are much more acidic than the corresponding alcohols, e.g., phenylthiol has a pK_a of 6.5. Tri- and tetracoordinated organosulfur compounds are known in the form of sulfoxides and sulfones. Proteins often contain sulfur present in the amino acids methionine, cystine, and cysteine. Organosulfur compounds are common components of fossil fuels appearing as aliphatic or aromatic thiols or sulfides and disulfides in addition to the heterocyclic combinations as thiophenes or dibenzothiophene. Thioethers may be found in lower-grade coals (Wawrzynkiewicz and Sablik, 2002).

Oxo compounds such as alkyl sulfates and sulfonates are used as detergents in large quantities, while simple hydroxysulfonates are formed through the reactions of sulfur dioxide and aldehydes in rain and cloud water. Oxidation processes, most particularly in soils, seem to generate a range of sulfenic, sulfinic, and sulfonic acids. Sulfenic acids take on the general formula RSOH (R $\neq$ H) and the sulfinic acids RS(=O)OH. Methyl sulfenic acid (CH_3SOH) is probably produced in the atmosphere through the oxidation of dimethyl disulfide (DMDS) (Barnes *et al.*, 1994) and the parent sulfinic acid may occur on the Jovian moon, Europa (Carlson *et al.*, 2002). The sulfonates are rather more widely known with methanesulfonic acid (CH_3SO_3H), being an important oxidation product of compounds such as dimethyl sulfide (DMS) in the Earth's atmosphere.

8.14.2 ABUNDANCE OF SULFUR AND EARLY HISTORY

Sulfur is the 10th most abundant element in the Sun, and has less than half the abundance of silicon. Its abundance in the Earth's crust is very much less, indicative of a mobile and reactive element as seen in Table 4, perhaps most easily seen by comparison with silicon. Note that nitrogen, whose cycle is often compared with sulfur, is also much depleted in the crust, but is unreactive and not so mineralized effectively.

Table 4 Average abundances of some elements in the cosmos and continental crust. This is best interpreted by comparing elements relative to silicon which take about the same values in this representation.

Element	*Cosmos relative hydrogen atoms*	*Continental crust (%)*
Hydrogen	*1,000,000*	
Helium	*140,000*	
Carbon	*300*	0.2
Nitrogen	*91*	0.006
Oxygen	*680*	*47.2*
Aluminum	1.9	7.96
Sulfur	9.5	0.07
Silicon	25	28.8
Calcium	1.7	3.85
Magnesium	29	2.2
Iron	8	4.32

Source: Wedepohl (1995).

8.14.2.1 Sulfur in the Cosmos

The universe is largely composed of hydrogen and it is in the older stars that heavier elements are found. This offers one of the central clues to the origin of the elements. The evolution of stars is interpreted with the aid of the Hertzsprung–Russell diagram. Stars that lie in the main sequence burn hydrogen into helium although the rate of this process varies with mass: heavier stars burning up more quickly perhaps in only a few million years for the brightest. The sun will take billions of years to evolve before becoming a cooler red giant, shedding its outer layers and ending up as a white dwarf. Heavier stars evolve into super giants, which synthesize heavier elements in their interiors. They may end up exploding as supernovae, forming more elements and distributing these through the cosmos. Interstellar sulfur isotopes are synthesized by oxygen burning in massive stars. This derives from a sequence

$$^{1}H + ^{1}H = ^{2}H + e^{+} + \nu$$

$$^{1}H + ^{2}H = ^{3}He + \gamma$$

$$^{3}He + ^{3}He = ^{4}He + ^{1}H + ^{1}H$$

A small proportion of the helium reacts by PPII and PPIII processes that yield ^{7}Be and ^{7}Li and some ^{8}B. In addition to the PP chains, there are CNO cycles. These are more important in heavier stars and are started by small catalytic amounts of ^{12}C, and so represent an important source of ^{13}C, ^{14}N, ^{15}N, and ^{17}O. Hydrogen burning yields helium in the stellar core, and helium burning can give rise to both ^{12}C and ^{16}O. Once helium is exhausted, carbon and oxygen burning takes place and it is at this stage the sulfur element forms

$$^{16}O + ^{16}O = ^{31}S + n \text{ and } ^{16}O + ^{16}O = ^{32}S + \gamma$$

An equilibrium phase is possible at very high temperatures (3×10^9 K) which would lead to high abundances of elements around the same atomic number as iron, and less of lighter elements such as sulfur. However, this situation exists only in the core, and in the surrounding shells substantial amounts of other elements such as sulfur will be present.

There are a number of estimates of the isotopic abundance of sulfur in the cosmos from Galactic cosmic rays (Thayer, 1997) and interstellar clouds (Mauersberger *et al.*, 1996). Measurements from the *Ulysses High Energy Telescope* resolve the individual isotopes of sulfur (S-32, S-33, and S-34), giving a measured ratio of Galactic cosmic-ray S-34/S-32 in the heliosphere is 24.2 $\pm$ 2.7% and a ratio for S-33/S-32 in the heliosphere as 19.0 $\pm$ 2.4%. The Galactic source composition for S-34/S-32 has been found to be 6.2%. The source abundance of S-33 is estimated as 2.6 ± 2.4%, which is higher than, but consistent with, the solar system abundance of 0.8% (Thayer, 1997). Enrichment of S-33 in ureilite meteorites may derive from heterogeneity within the presolar nebula (Farquhar *et al.*, 2000). Molecules of CS in interstellar space suggest the abundance ratio S-34/S-36 to be 115(±17). This is smaller than in the solar system ratio of 200 and Mauersberger *et al.* (1996) argue that it supports the idea that S-36 is, unlike the other stable sulfur isotopes, a purely secondary nucleus that is produced by s-process nucleosynthesis in massive stars.

Sulfur-containing molecules have been found in clouds in interstellar space: CS, C_2S, C_3S, SiS, SO, SO^+, H_2S, HCS^+, NS, OCS, SO_2, H_2CS, HNCS among them. Although sulfur is a very abundant and important element, it appears to be depleted in cold dense molecular clouds with embedded young stellar objects (Keller *et al.*, 2002). In dense molecular clouds solid SO_2 accounts for only 0.6–6% of the cosmic sulfur abundance (Boogert *et al.*, 1997). Keller *et al.* (2002) have determined the infrared spectra of FeS grains from primitive meteorites in the laboratory and shown a broad similarity with grains in interplanetary dust, which show an FeS peak, suggesting that the missing sulfur appears to reside as sulfides in solid grains.

8.14.2.2 Condensation, Accretion, and Evolution

The simplest models for the composition of the planets presume that the differences between them can be explained in terms of an equilibrium condensation. At the highest temperatures a sequence of mixed oxides of calcium, titanium, and aluminum would be found (>1,400 K). This would be followed, at lower temperatures, by metal and silicate fractions. At temperatures somewhat greater than 600 K alkali metals enter the silicate phase along with sulfur, which combines with iron at ~650 K to form triolite

$$Fe + H_2S \rightleftharpoons FeS + H_2$$

Hydration reactions become possible at lower temperatures with the potential to form serpentine and talc. Although such models for the accretion of planets have significant explanatory power, the process was doubtless more complex. Many scientists have sought explanations involving the late heterogeneous accretion of infalling meteoritic materials. These would have brought in large amount of more volatile substances to planetary bodies.

After accretion there is further planetary evolution, especially through increases in temperature from the decay of long-lived radioactive elements. Planets, such as the Earth, thus become layered either through accretion or differential melting processes. The heating deep inside the planet, such as the Earth, causes outgassing of volatile materials especially water from the dehydration of tremolite:

$$Ca_2Mg_5Si_8O_{22}(OH)_2 \rightleftharpoons 2CaMgSi_2O_6 + MgSiO_3 + SiO_2 + H_2O$$

or sulfur gases:

$$FeS + 2FeO \rightleftharpoons 3Fe + SO_2$$

$$FeS + CO \rightleftharpoons Fe + OCS$$

$$FeS + H_2O \rightleftharpoons FeO + H_2S$$

along with hydrogen, nitrogen, ammonia, and traces of hydrogen fluoride and chloride. The heating was usually intense enough to cause melting. As there was insufficient oxygen available, it became associated with magnesium, silicon, and aluminum, while elements such as iron and nickel remained in an unoxidized state. Gravitational settling meant that these heavy materials moved towards the center of the Earth. Iron sulfide would also be heavy enough to settle. Thus although the Earth as a whole may be almost 2% sulfur by weight, the crust is much depleted at 700 ppm.

As dense minerals sank and formed the Earth's mantle, silicate-rich magma floated on the surface. It is this that became the crust, which represents the outermost layer of rocks forming the solid Earth. It is distinguished from the underlying mantle rocks by its composition and density. The silica-rich continental crust includes the oldest rocks as "shields" or "cratons," which in the extreme may be ~4 Gyr old. Crust under the oceans is only ~5 km thick compared to continental crust, which can be up to 65 km thick under mountains. The oceanic crust has a remarkably uniform composition (49 ± 2% SiO_2) and thickness (7 ± 1 km). The ocean floor is the most dynamic part of the Earth's surface, such that no part of the oceanic crust is more than 200 Myr old. It is constantly renewed through seafloor spreading at mid-ocean ridges.

8.14.2.3 Sulfur on the Early Earth

Sulfur on the early Earth has been studied, because it gives insight into the oxidation state of the early atmosphere ocean system (Canfield and Raiswell, 1999). The occurrence of evaporitic sulfate from the 3.5 Ga as in the Warrawoona Group of Western Australia suggests that sulfate must have been present in elevated concentrations at least at some sites. Although now present as barite, the original precipitation occurred as gypsum. These barites and early sulfates precipitate from the ocean. Evidence from isotope ratios in Early Archean sedimentary sulfides suggest that seawater sulfate concentrations could be as low as 1 mM (Canfield *et al.*, 2000). More recent work of Habicht *et al.* (2002) suggests even lower oceanic Archean sulfate concentrations of <200 μM, which is less than one-hundredth of present marine sulfate levels. They argue that such low sulfate concentrations were maintained by volcanic sulfur dioxide and were so low that they severely suppressed sulfate reduction rates allowing for a carbon cycle dominated by methanogenesis (Habicht *et al.*, 2002).

Studies of the sulfide inclusions in diamonds have given some indications of sulfur cycling on the Archean earth. These inclusions often appear to be enriched in ^{33}S, which has been interpreted as indicative of photolysis of volcanic SO_2 at wavelengths <193 nm. This photolysis yields elemental sulfur enriched in ^{33}S, while the oceanic sulfate was depleted in ^{33}S (see Figure 4). The elemental sulfide becomes incorporated into sulfides and can be subducted, and included in diamonds. By contrast seawater has sulfates depleted in ^{33}S and gave Archean barites that are also depleted in the ^{33}S. While such studies offer an indication of the sulfur cycle in the Archean, they can also offer insight into the nature of mantle convection through time (Farquhar *et al.*, 2002).

Some of the earliest organisms on the Earth utilized sulfur compounds particularly through anoxogenic photosynthesis. Wächtershauser and co-workers have argued that inorganic reactions involving S–S links have a reducing potential that is strong enough to drive a reductive metabolism based on carbon dioxide fixation (e.g., Hafenbradl *et al.*, 1995):

$$FeS + H_2S \rightarrow FeS_2 + H_2$$

The sulfur-isotope signal in sedimentary record of the early Earth provides evidence for the early evolution of life. Sulfate-reducing bacteria deplete the resultant sulfides in ^{34}S. The isotopic record of ^{34}S supports this, although some evidence now suggests that seawater sulfate concentrations were so low that they severely suppressed sulfate reduction rates allowed for a carbon cycle dominated by methanogenesis (Habicht *et al.*, 2002).

Modern taxonomy based on sequences in rRNA indicates three domains of living organisms: Eucarya (includes fungi, plants, and animals), Bacteria, and Archea (prokaryotes with no nuclei or organelles that are similar to bacteria, but with distinguishing ether linked membrane lipids). Both Bacteria and Archea include organisms capable of reducing elemental sulfur and it is likely that this was one of the most primitive early metabolisms. Such sulfur may have been generated on an anoxic Earth via the reaction between SO_2 and H_2S from fumaroles (Canfield and Raiswell, 1999).

The ability to undertake anoxygenic photosynthesis is widespread among bacteria. Its presence among green nonsulfur and sulfur bacteria and the fact that this uses a single photosystem to utilize solar energy indicate that this is an ancient process, much simpler than that required in oxygenic photosynthesis. Nevertheless, much of the ability to undertake oxygenic photosynthesis is probably associated with the production of abundant organic material. During oxidation this meant the production of large amounts of reduced compounds such as H_2S and Fe^{2+}. This would have led to large increase in the range of environments offering potential for anoxygenic photosynthesis (Canfield and Raiswell, 1999).

An important consequence of anoxygenic photosynthesis is the production of large amounts

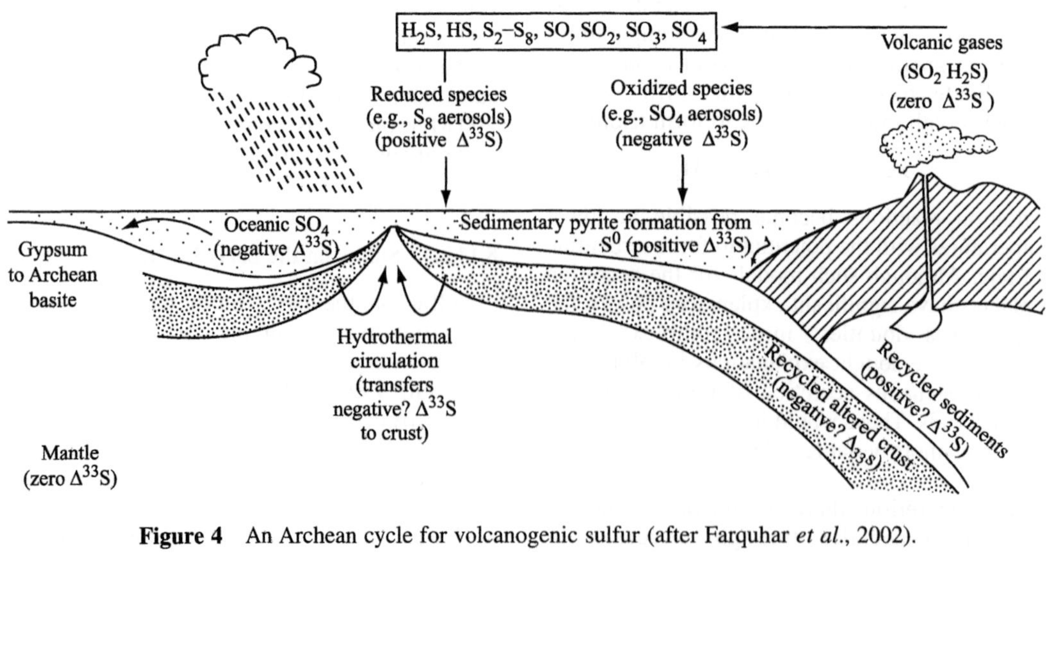

Figure 4 An Archean cycle for volcanogenic sulfur (after Farquhar *et al.*, 2002).

of sulfate which can be used as an electron acceptor in the remineralization of organic carbon by sulfate-reducing bacteria. Thus, it is not surprising that sulfate-reducing bacteria are found almost as deep in the evolutionary SSU rRNA record as the earliest anoxygenic photosynthetic bacteria (Canfield and Raiswell, 1999). This is important because the photosynthesis represented an energy resource that is orders of magnitude larger than that available from oxidation–reduction reactions associated with weathering and perhaps more importantly submarine hydrothermal activity. Hydrothermal sources deliver $(0.13-1.1)\times 10^{12}$ mol yr^{-1} of S(−II), Fe^{2+}, Mn^{2+}, H_2, and CH_4 on a global basis, sufficient to allow microorganisms capable of using hydrothermal energy to produce only limited amounts of organic carbon ($\sim(0.2-2.0)\times 10^{12}$ mol C yr^{-1}) (Des Marais, 2000).

8.14.3 OCCURRENCE OF SULFUR

8.14.3.1 Elemental Sulfur

Elemental sulfur is found in many parts of the world. Such native sulfur can be detected by the faint odor of sulfur dioxide gas or more often the small amount of hydrogen sulfide from reaction with traces of water. Although there are some deposits related to volcanic sublimates, the largest deposits are due to sulfate reduction. For example, deeper in the Gulf Coast of the US, salt containing gypsum as an impurity has dissolved, thus leaving the insoluble gypsum. A range of biologically mediated processes intervened. First hydrogen sulfide was produced possibly by reaction of methane with the gypsum. The hydrogen sulfide reacted with this gypsum to form free sulfur, which became trapped within limestone beds.

8.14.3.2 Sulfides

There are widespread deposits of sedimentary sulfides, mostly as the iron sulfides from the reduction of sulfate in seawater. Although pyrite (FeS_2) is an important source of sulfur in extractive mining, the deposits most utilized are often of volcanogenic origin where the deposits may consist of as much as 90% pyrite. Although most frequently found associated with iron, sulfur can also be found as calcopyrite ($CuFeS_2$), galena (PbS, lead sulfide), and cinnabar (HgS, mercury sulfide).

Extensive sulfide deposits were associated with volcanism in the earliest stages of the Earth's history. It is possible to identify at least four short epochs of massive sulfide deposition: (i) 2.69–2.72 Ga; (ii) 1.77–1.90 Ga; (iii) the Devonian–Early Carboniferous; and (iv) the Cambrian–Early Ordovician. The Devonian–Early Carboniferous is the most important period and the Cambrian–Early Ordovician less noteworthy. These processes are cyclic in nature, following tectonic cycles that include the convergence of continental masses and the formation and subsequent breakup of supercontinents. As each cycle proceeds, we see the appearance of different sequences within massive sulfide deposits (Eremin *et al.*, 2002).

8.14.3.3 Evaporites

Evaporites are typically found in arid regions. The salts deposited in desert basins are often sulfates, chloride, and carbonates of sodium and calcium with smaller amounts of potassium and magnesium. Saline lakes often have a composition that resembles seawater, although many have a greater abundance of carbonate. When both sodium chloride and calcium sulfate have crystallized, the remaining water may contain substantial amounts of magnesium, potassium and bromide. The dead seawater contains substantial bromide concentrations (5,900 ppm) and although it has high concentrations of calcium (1.58×10^4 ppm), sulfate concentrations are low (540 ppm) because of the low solubility of calcium sulfate.

The most important salt deposits have formed through the evaporation of seawater. As this evaporation occurs, calcium carbonate is the first mineral to precipitate, although there is insufficient carbonate to give thick deposits. Calcium sulfate deposits in substantial quantities under continued evaporation as either gypsum ($CaSO_4{\cdot}2H_2O$) or anhydrite ($CaSO_4$). After about a half of the calcium sulfate has been deposited, anhydrite becomes the stable phase. When seawater has evaporated to 10%, its original volume halite (NaCl) starts to separate. Ultimately, polyhalite ($K_2Ca_2Mg(SO_4)_4{\cdot}2H_2O$) begins to crystallize. The volume has to be reduced almost a 100-fold before salts of potassium and magnesium separate. Other sulfate minerals from such systems include magnesium sulfate (Epsom salt and kyeserite $MgSO_4{\cdot}H_2O$), barium sulfate (barite), strontium sulfate (celestite), and sodium sulfate (glauberite).

8.14.3.4 The Geological History of Sulfur

The sulfur cycle over geological periods can be represented by the simplified cycle shown in Figure 5(a). A central feature of this cycle is marine sulfate, which may be crystallized

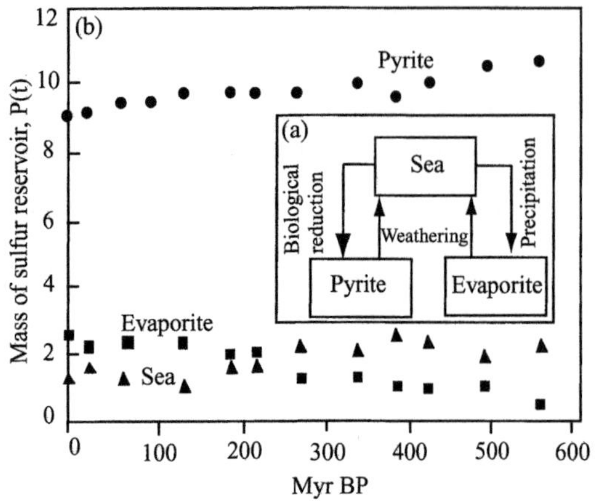

Figure 5 (a) (Inset) Simplified box model of the sulfur cycle. (b) Mass of sulfur in reservoirs in Pt (10^{21} g) (source Holser *et al.*, 1989).

as sulfate evaporites or biologically reduced to sulfidic sediments (Holser *et al.*, 1989). In the long term this sulfide becomes fixed by reaction with iron in a process that can be represented as

$$8SO_4^{2-} + 2Fe_2O_3 + 8H_2O + 15C_{(org)} \rightarrow 4FeS_2 + 16OH^- + 15CO_2$$

The reduction is typically limited by the availability of organic carbon and often occurs in shallow waters at continental margins. Thus, global sulfide production would be dependent on the availability of biological productive areas over geological time. Sulfur-isotope data can be used to constrain simple models of the sulfur cycle over geological time and establish the size of the reservoirs as shown in Figure 5(b).

The crystallization of sulfate as evaporite deposits requires specific and relatively short-lived geographical features such as: sabkha shorelines, lagoons, and deep basins of cratonic, Mediterranean or rift valley origin, where brines can concentrate and precipitate. This suggests that sulfate removal from the oceans would be episodic over geological time, although there are problems with an imperfect preservation of deposits. Numerical models for paleozoic global evaporite deposition suggest that there were major changes in the evaporite accumulations (as in Figure 6) and rate of evaporite deposition related to alteration in the flux of sulfate to and from the ocean. These can be the result of the height distribution of the continental mass and its geographic location. Most evaporite deposition appears in the 5–45° latitude range and as seen in Figure 6 this changes significantly over geological time. The removal of large quantities of sulfate from the oceans also has implications for ocean chemistry, which would have to change, i.e., oceanic sulfate is not in steady state. Currently evaporative environments are relatively rare, as a result of a *low stand* disposition of the continents. This means that sulfate is accumulating in the oceans (Railsback, 1992).

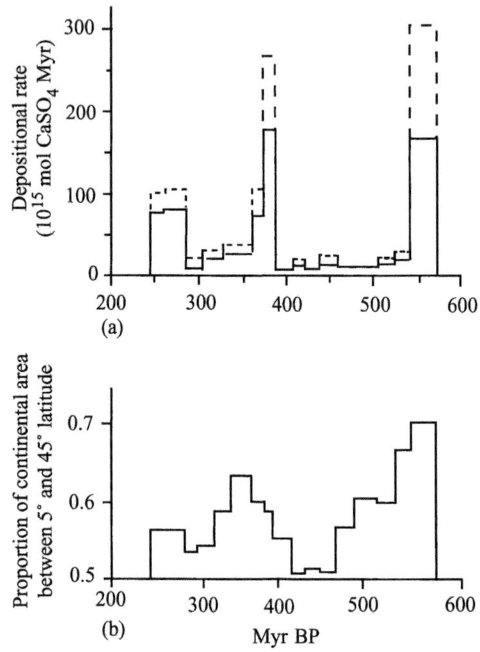

Figure 6 (a) Global deposition rates for evaporites over time. (b) The proportion of the Earth's total continental area located between paleo-latitudes 5° and 45° (after Railsback, 1992).

The changing fluxes in various parts of the sulfur cycle over time imply that we have to consider the potential for significant changes in seawater composition over geological time. Although it is often assumed that in the Phanerozoic period (the last 590 Myr) seawater composition became stable, there is evidence that there were significant fluctuations in some components. Fluid inclusions in Silurian (395–430 Myr ago) suggest that the Silurian ocean had lower concentrations of Mg^{2+}, Na^+, and SO_4^{2-}, and much higher concentrations of Ca^{2+} relative to the present-day ocean composition. Concentration of sulfate was probably 11 mM then compared with 29 mM as we see nowadays. Although this could be caused by localized changes, it is also possible that global scale dolomitization of carbonate sediments or changing inputs of mid-ocean ridge brines could be responsible (Brennan and Lowenstein, 2002).

8.14.3.5 Utilization and Extraction of Sulfur Minerals

Elemental sulfur is widely known and has been mined since the earliest times with archeological sites in the United Arab Emirates, Yangmingshan National Park in Taiwan and from Etruscan activities in Sicily. In some parts of the world laborers still carry the sulfur out of volcanoes by hand. Sulfur in addition to being a vital ingredient of gunpowder was important to the overall industrialization of Europe. Historically the most important source was from Sicily, which has been active for more than two thousand years. Towns such as Ravanusa became notable producers during the Spanish Bourbon rule (after 1735). Italian output peaked at 0.57 Mt in 1905 declining from that point to just a few kt by the 1970s.

The United States is now the major sulfur producer, accounting for 20% of world production. The most important sources in the US are from Louisiana and Texas, and other major producers are Japan, Canada, China, Russia, and Mexico. World sulfur production (and apparent consumption) peaked at nearly 60 Mt in 1989 and declined by almost 14% to 52.8 Mt in 1993 (Figure 7). There was a partial recovery from this time and future growth is expected.

One of the most important developments in sulfur mining was the development of the Frasch process which allowed it to be extracted from salt domes, particularly along the US Gulf Coast. Herman Frasch became involved in oil and sulfur mining and invented a process which allows liquid sulfur to be recovered with injection wells. Water at 330 °C is injected through wells into formations. The molten sulfur is extracted in a very pure form. It is an efficient method for mining sulfur, but the process can also produce saline wastewaters, increases in pH levels, and a high concentration of dissolved salts such as sodium chloride (TWRI, 1986).

Sulfur may also be recovered from H_2S in natural gas where it can be precipitated. Natural gas can contain up to 28% hydrogen sulfide gas. A range of approaches are possible, often known as Claus processes to oxidize the H_2S, often recovering more than 95% of the sulfur

$$1\tfrac{1}{2}O_2 + H_2S \rightarrow H_2O + SO_2$$

$$2H_2S + SO_2 \rightarrow 3S + 2H_2O$$

More than 90% of world sulfur consumption is used in the production of sulfuric acid, much of which goes to the fertilizer industry. Smaller amounts of sulfur are used in the manufacture of gunpowder, matches, phosphate, insecticides, fungicides, medicines, wood, and paper products, and in vulcanizing rubber. Despite slight uncertainties in sulfur demand in the 1990s, its use is still predicted to grow.

Sulfide ores are frequently associated with a range of important metals, most notably copper, lead, nickel, zinc, silver, and gold. This means the mining operations to extract these metals can mobilize large amounts of sulfur.

8.14.4 CHEMISTRY OF VOLCANOGENIC SULFUR

8.14.4.1 Deep-sea Vents

Deep-sea vents have only been studied in the late twentieth century, so there is still much to learn in terms of their global contribution because of their inaccessibility. However, they can affect global fluxes, and some estimates would suggest that warm ridge-flank sites may remove each year, as much as 35% of the riverine flux of sulfur to the oceans (Wheat and Mottl, 2000). The hydrothermal vents are locally important sources of sulfide-containing materials. The black smokers yield polymetal sulfides, that will

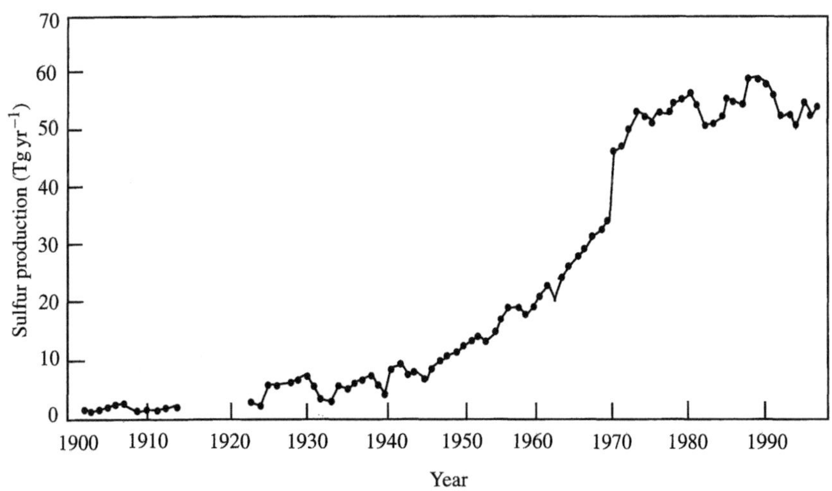

Figure 7 World sulfur production (source http://pubs.usgs.gov/of/of01-197/html/app5.htm).

oxidize to sulfur and ultimately sulfates. The reduced sulfur is also utilized by the ecological communities that develop close to the vents (Jannasch, 1989).

8.14.4.2 Aerial Emissions

Volcanoes represent a very large source of sulfur gases to the atmosphere. Mafic magmas such as basalts are likely to allow sulfur to be released in large quantities. However, despite the low solubility and diffusivity of sulfur in the silicic magmas typical of explosive eruptions large quantities of sulfur are inevitably released. The presence of fluids may account for higher than expected sulfur yields and indicates a redox control of sulfur degassing in silicic magmas. This means that different eruptions can have very different efficiencies in terms of sulfur release. Very large eruptions that involve cool silicic magmas may yield little sulfur. Krakatoa in 1883 gave 1.2 km^3 of eruptive magma, with an estimated petrological yield of up to 7 Tg in reasonable agreement with ice core data, which suggests 10–18 Tg. By contrast, the Taupo eruption of 177 AD in New Zealand gave more than 35 km^3 of magma, but perhaps only 0.13 Tg of sulfur (Scaillet *et al.*, 1998).

There are numerous calculations of the multi-component chemical equilibria in gas–solid–liquid systems within volcanoes (e.g., for Mt. St. Helens see: Symonds and Reed, 1993). These typically show the importance of SO_2 at high temperatures (see Figure 8) and a transition to a dominance of H_2S at lower temperatures in systems such as

$$SO_2 + 3H_2 \rightarrow H_2S + 2H_2O$$

where entropy changes favor fewer molecules of gas.

Much volcanism is a sporadic process and on a year-to-year basis the emissions can vary.

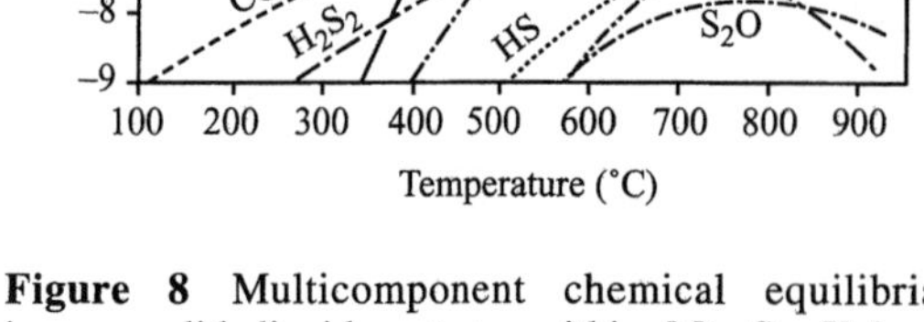

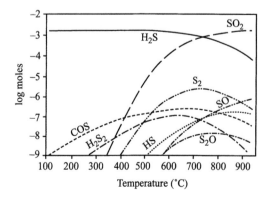

Figure 8 Multicomponent chemical equilibria in gas–solid–liquid systems within Mt. St. Helens (source Symonds and Reed, 1993).

There are also difficult problems in balancing the contributions from the more explosive processes against that from fumaroles that are more continuous. Explosive processes also place the sulfur high up in the atmosphere. There are also inputs as particulate sulfur and sulfate along with small amounts of carbon disulfide and OCS of volcanic origin, but these probably amount to no more than a Tg(S) a^{-1} (Andres and Kasgnoc, 1998). There are lesser amounts of S_2, S_3, H_2S_2, S_2O and a range of sulfides of arsenic, lead, and antimony.

There are a variety of estimates of the total volcanic source of sulfur. A recent one considered continuous and sporadic volcanoes (Andres and Kasgnoc, 1998). It is necessary to recognize that measurements do not include all volcanic eruptions, so assumptions have to be made about the distribution of emissions among the smaller sources. A time-averaged subaerial volcanic sulfur (S) emission rate of 10.4 Tg(S) is probably a conservative estimate of which ~65% is as SO_2 along with substantial amounts of hydrogen sulfide. The high variability of emissions is illustrated in Figure 9 as established by the TOMS emission group from satellite measurements.

8.14.4.3 Fumaroles

Fumaroles represent a gentler and more continuous source of sulfur. The sources can be dispersed and quite small, so the total emissions from this source are not easy to estimate. Some of them are dominated by H_2S. The sulfur gases, SO_2, H_2S, S_8, have been found in a range of fumaroles (Montegrossi *et al.*, 2001). Although present S_8 remains a minor component several orders of magnitude below SO_2 and H_2S. The production of sulfuric acid through aerial oxidation of sulfur(IV) is the most familiar process but it can readily be produced by disproportionation in fumarolic systems (Kusakabe *et al.*, 2000):

$$3SO_2 + 3H_2O \rightarrow 2HSO_4^- + S + 2H^+$$

$$4SO_2 + 4H_2O \rightarrow 3HSO_4^- + H_2S + 3H^+$$

It is this process that can be responsible for large $\delta^{34}S_{HSO_4}$ values in crater lakes.

8.14.4.4 Crater Lakes

High concentrations of acids and sulfur compounds in crater lakes give rise to a complex chemistry within the waters (Sriwana *et al.*, 2000). Reactions between hydrogen sulfide and sulfur dioxide lead to the formation of polythionates, with $S_4O_6^{2-}$, $S_5O_6^{2-}$, and $S_6O_6^{2-}$ being the most typical. The initial ratio of SO_2/H_2S controls

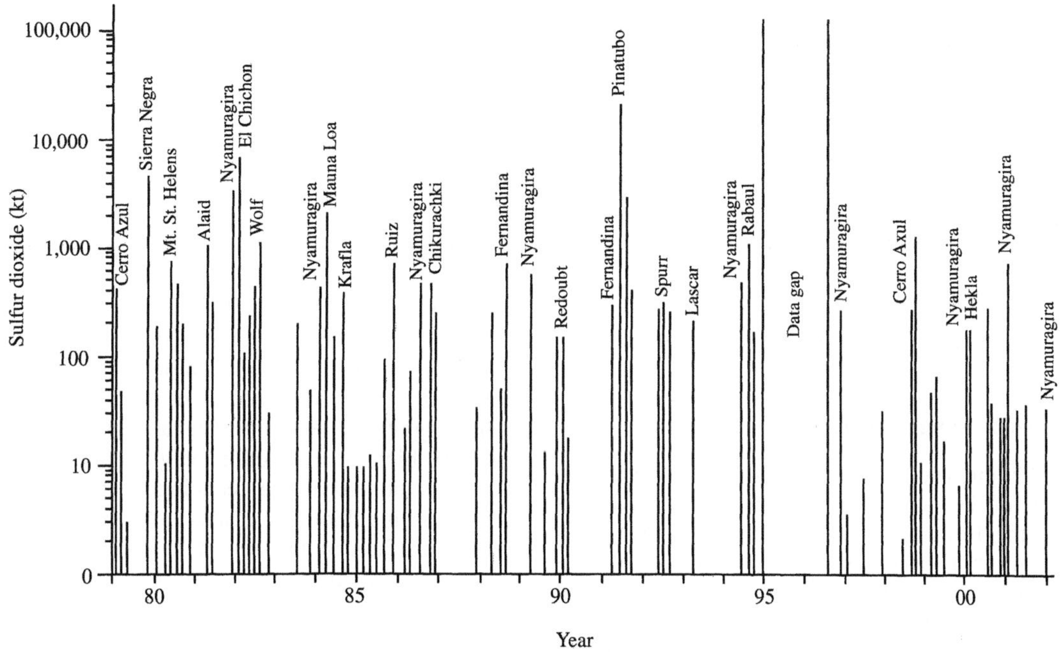

Figure 9 Volcanic sulfur emissions as established by the TOMS emission group (http://skye.gsfc.nasa.gov/).

the relative abundance of the oxyanions. The anion $S_6O_6^{2-}$ is always at the lowest concentrations, but the ratio $SO_2/H_2S > 0.07$ favors $S_4O_6^{2-} > S_5O_6^{2}$. Polythionate concentrations vary up to a few hundred milligrams per liter and seem to be a useful, indicator of changes in the subaqueous fumarolic activity with volcanic lakes corresponding to the activity of the volcano (Takano *et al.*, 1994). The polythionates are destroyed by increases in SO_2 input (as bisulfite):

$$(3x - 7)HSO_3^- + S_xO_6^{2-} \rightarrow (2x - 3)SO_4^{2-} + (2x - 4)S + (x - 1)H^+ + (x - 3)H_2O$$

Lakes of liquid sulfur at 116 °C have been found in the crater of the Poas volcano in Costa Rica. These form after evaporation of water from crater lakes and the underlying sulfur deposits melt and are remobilized into lakes (Oppenheimer and Stevenson, 1989).

8.14.4.5 Impacts of Emissions on Local Environments

The volcanoes of Hawaii emit large quantities of fumarolic sulfur dioxide which causes widespread damage to human health and vegetation downwind. These emissions often take the appearance of widespread hazes and are locally known as vogs. The primarily component is of sulfuric acid and sulfate formed through oxidation of the sulfur dioxide. The vog particles also contain trace elements selenium, mercury, arsenic, and iridium (USGS, 2000). There is evidence that volcanic SO_2 and HF affect the diversity of plant communities downwind and damage crops (Delmelle *et al.*, 2002). Emissions of SO_2 and to a lesser extent HCl are responsible for substantial fluxes of acidity to soils downwind. Such acid deposition can cause extreme acid loading to local ecosystems (Delmelle *et al.*, 2001). Hydrogen sulfide concentrations can also be important in some volcanic areas (Siegel *et al.*, 1990), where the odors can be strong and increase corrosion of metals, especially those related to electrical switches and discolor lead-based paints.

8.14.5 BIOCHEMISTRY OF SULFUR

8.14.5.1 Origin of Life

An important stage in the origin of life is the abiotic synthesis of important biological molecules. Interest has traditionally focused on the production of amino acids following the work of Stanley L. Miller, and Harold C. Urey, who created laboratory atmospheres consisting of methane, ammonia, and hydrogen that could be forced to react to give the amino acid glycine. Currently there is a view that RNA might have established what is now called the RNA world in which RNA catalyzed all the reactions to replicate proteins. Such chemistry has focused on nitrogen.

The discovery of deep-sea vents has led to speculation that they could be involved in the origin of life and here a unique sulfur chemistry could come into play (Wachtershauser, 2000). In particular, there is much interest in the potential of iron and nickel sulfides to act as catalysts in a range of reactions that can lead to keto acids such as pyruvic acids. Quite complex amino acids seem to be easily assembled in the presence of iron sulfides, e.g., the conversion of phenyl pyruvate formation of phenylalanine in a reaction where CO_2 seems to act as a catalyst (Hafenbradl *et al.*, 1995):

$$RCOCOOH + 2FeS + NH_3 + 2H^+ \rightarrow RCH(NH_2)COOH + FeS_2 + Fe^{2+} + H_2O$$

8.14.5.2 Sulfur Biomolecules

Although we have discussed the low oxidation state of the early atmosphere and its implications for the form of sulfur in organisms, today this chemistry is largely restricted to anoxic environments. Here sulfur is available as hydrogen sulfide and can be incorporated into amino acids such as cysteine and methionine and then into proteins. The thiolate group RS^- of cysteine of the thioether of methionine can act as bases or ligands for transition metals such as iron, zinc, molybdenum, and copper.

In modern organisms operating under oxic conditions this is more difficult, because sulfur arrives as sulfate and this has to be bound and reduced so that the sulfur can be utilized biologically (see Figure 10). Initially sulfate can be bound to ATP (adenosine trisphosphate) as APS (phospho-adenosine monophosphate sulfate). In the cell's Golgi apparatus it can be converted to sulfate polysaccharides. Alternatively, further reactions with ATP can lead to PAPS which can then be reduced to sulfide through sulfite using a molybdenum in the initial step and then a haem/Fe_nS_n sulfite reductase (Frausto da Silva and Williams, 2001).

Much sulfur is incorporated into proteins as the thiolate or thioether of the amino acids, cysteine, and methionine. There are also more fundamental roles for thiolate in redox reactions and in enzymes and in controls in the cytoplasm involving glutathione and thioredoxin. Glutathione is a tripeptide made up of the amino acids gamma-glutamic acid, cysteine, and glycine. The primary biological function of glutathione is to act as a nonenzymatic reducing agent to help keep cysteine thiol side chains in a reduced state on the surface of proteins. Glutathione confers protection by maintaining redox potential and communicating the redox balance between metabolic pathways at a redox potential of ~ -0.1 V. Glutathione is also involved in reductive synthesis with the enzyme thioredoxin in processes that involve redox reactions of –S–S– bridges. The –S–S– bridge is also important as a cross-linking unit in extracellular proteins. This cross-linking may have become important once the atmosphere gained its oxygen. Additionally, glutathione is utilized to prevent oxidative stress in cells by trapping free radicals that can damage DNA and RNA. There is a correlation with the speed of aging and the

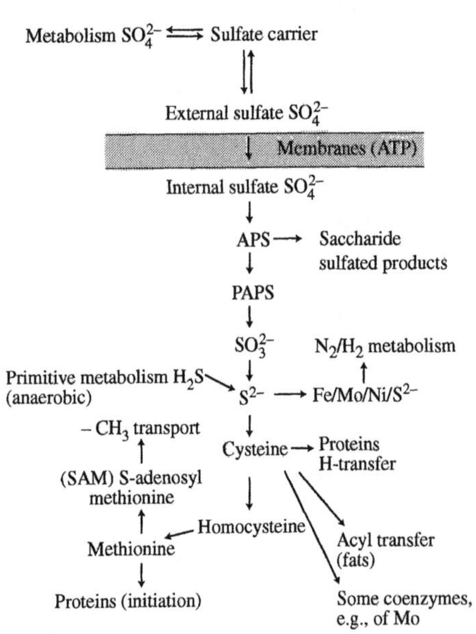

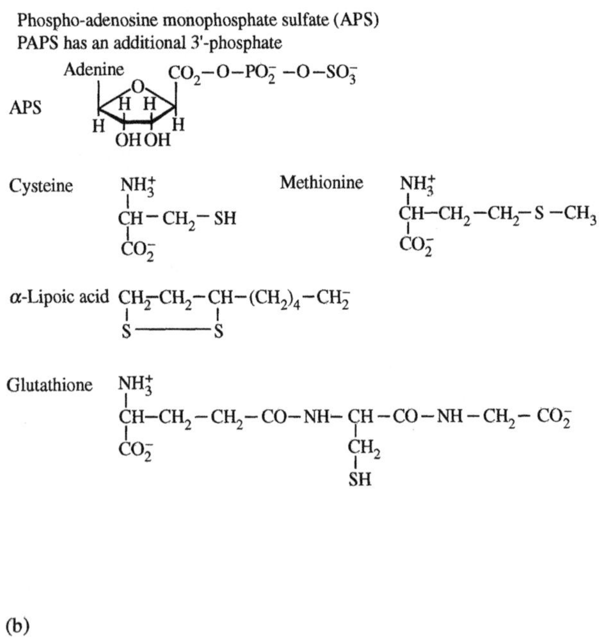

Figure 10 Scheme for the uptake and incorporation of sulfate into cells and the formulas of some biologically important sulfur-containing molecules (source Frausto da Silva and Williams, 2001).

reduction of glutathione concentrations in intracellular fluids.

The biological importance of sulfur means that modern agriculture frequently confronts the high demand of some crops and livestock for sulfur. In Europe as the deposit of pollutant sulfur dioxide has decreased, sulfur deficiency has been more widely recognized especially in wheat, cereals, and rape-seed (Blake-Kalff *et al.*, 2001). Some animals have a particularly high demand for sulfur, and under domestication this can increase still further. It is evident in sheep where the amount of sulfur that sheep can obtain through its diet fundamentally limits wool growth. Transgenic lupin seeds have been shown to provide higher amounts of methionine and cysteine in a protein that is both rich in sulfur amino acids and stable in the sheep's rumen leading to increased wool yield (White *et al.*, 2001).

8.14.5.3 Uptake of Sulfur

A wide variety of forms of sulfur are used by microorganisms. Sulfur dioxide can readily deposit in wet leaf surfaces and through the stomata. Purple nonsulfur bacteria can utilize thiosulfate and sulfide (Sinha and Banerjee, 1997). Lichen and leafy vegetation can also act remove sulfur compounds from the atmosphere. OCS removal is the best known and probably represents a dominant sink for tropospheric OCS. The fate of the oxidation products of DMS has not always been clear. There have been some biochemical studies that suggest that they may degrade biologically. Facultatively methylotrophic species of Hyphomicrobium and Arthrobacter seem to be able to produce enzymes necessary for a reductive/oxidative pathway for dimethylsulfoxide and dimethylsulfone (although the former appears to be more generally utilizable). Methanesulfonic acid is stable to photochemical decomposition in the atmosphere, and its fate on land has been puzzling. A range of terrestrial methylotropic bacteria that appear common in soils can mineralize MSA to carbon dioxide and sulfate (Kelly and Murrell, 1999).

8.14.6 SULFUR IN SEAWATER

Seawater is rich in sulfur and, while containing high concentrations of inorganic sulfur, it also has an elaborate organic chemistry, driven largely by biological activities.

8.14.6.1 Sulfate

Most of the sulfur in seawater present as sulfate. At a salinity of 35‰ sulfate is found at $2.712\ \mathrm{g\ L^{-1}}$, and the ratio of sulfate chloride mass in normal seawater is 0.14. This ratio is fairly constant in the oceans. Nevertheless, it can be much lower in deep brines of the type found in the southern end of the Red Sea possibly because of clays in sediments acting as semi-permeable membranes. Even though sulfuric acid has a modest second dissociation constant, it is sufficient to ensure that the bisulfate ion is at very low concentrations in seawater.

The double charge on the sulfate ion means that it can associate with other cations, and typically seawater has been argued to have more than 30% of the sulfate associated with sodium and magnesium ion pairs. The magnesium sulfate chemistry of seawater has considerable importance in underwater acoustics. Although sound absorption coefficients are dependent on pH at frequencies below 1 kHz, they are dependent upon $MgSO_4$ over the range 10–100 kHz (Brewer *et al.*, 1995). Some models have assigned and association constant of $10^{5.09}$ to the important and powerful association between magnesium and sulfate (Millero and Hawke, 1992). However, ion association models are complex and difficult to use, so recent models of seawater and other electrolyte systems in the environment rely increasingly on the formalism of Pitzer, which treats the electrolyte in terms of ion association (Clegg and Whitfield, 1991).

Seaspray can be blown directly from the sea surface during high winds, but the particles of salt from this process tend to be large and short lived in the atmosphere so do not travel far inland. Finer particles are produced during bubble bursting. The very finest arise when the cap of the surfacing bubble shatters giving film drops that dry into salt crystals of 5–50 pg. Along with these are larger jet drops that emerge as the bubble cavity collapses and give particles of ~0.15 μg. Such processes lead to the production of around a 10–30 Pg of seasalt each year, of which some 260–770 Tg would be sulfur in the form of sulfate. High chloride and sulfate concentrations are found in coastal rainwater, but perhaps only 10% of the solutes generated through the action of wind or through bubble bursting are deposited over land, as most falls over the oceans (Warneck, 1999).

The sulfate that derives from seasalt is called seasalt sulfate. It is important to distinguish it from sulfates that are found in the marine aerosol from the oxidation of gaseous sulfur compounds, which are known as non-seasalt sulfate.

8.14.6.2 Hydrogen Sulfide

Hydrogen sulfide is now recognized as occurring widely throughout the oceans (Cutter *et al.*, 1999). About 12% of the total dissolved sulfide is free and the rest is largely complexed to metals in typical North Atlantic water (Cutter *et al.*, 1999). Hydrogen sulfide has long been known to occur in anoxic regions, but in surface waters of the open ocean it was thought to oxidize rapidly. Hydrogen sulfide is produced from the hydrolysis of OCS (Elliott *et al.*, 1987; Elliott *et al.*, 1989). However, studies of the North Atlantic show that the rate of total sulfide production from hydrolysis and through atmospheric input is some 20 times less than sulfide removal via oxidation and that bound to sinking particulate materials. This suggests additional sources. The similarity between the depth distribution of total sulfide and chlorophyll suggests that there is a biological involvement in the production of total sulfide, perhaps by phytoplankton in the open ocean (Radfordknoery and Cutter, 1994).

The low concentrations of hydrogen sulfide found in the marine atmosphere suggest that there might be a significant flux of the gas from the oceans, but most studies conclude that it is a rather insignificant flux (Shooter, 1999). There is also the potential for the sulfate reduction to occur on particles in the ocean (Cutter and Krahforst, 1988) and some may arise from deep-sea vents, but this source is restricted to water close to the vent (Shooter, 1999).

Hydrogen sulfide can be oxidized in less than an hour in seawater. This removal can be through oxidation by oxygen or iodate. There is a possibility of oxidation, by hydrogen peroxide, but it is probably a minor pathway (Radfordknoery and Cutter, 1994). Photo-oxidation is also possible (Pos *et al.*, 1998), along with oxidation by Fe(III) oxide particles. This latter process is dependent on the way in which the particle forms and on pH with a maximum near 6.5. The Fe(III) oxide route gives mostly elemental sulfur as a product, which may have implications for pyrite formation (Yao and Millero, 1996).

There are much greater production rates of hydrogen sulfide in some anoxic basins. The Black Sea is the largest of these where sulfide production in the water column is estimated to be 30–50 Tg(S) yr^{-1}, taking place between depths of 500 m and 2,000 m. This may amount to as much as one-third of the global sedimentary budget (120–140 Tg(S) yr^{-1}). There is also pyrite formation in the water column of the Black Sea, through reaction with iron although this is relatively small compared with oxidation at the oxic anoxic interface (Neretin *et al.*, 2001).

8.14.6.3 OCS and Carbon Disulfide

Other sulfides seem readily produced by photochemical processes in seawater along with a parallel photoproduction of carbon monoxide (Pos *et al.*, 1998). Colored dissolved organic material is frequently involved in such processes. Laboratory irradiation has confirmed that cysteine and cystine are efficient precursors of CS_2 and that OH radicals are likely to be important intermediates (Xie *et al.*, 1998), but it is also likely that some of these sulfur gases are produced directly by biological processes. Ocean waters appear to be supersaturated in carbon disulfide and capable of yielding ~0.1 Tg(S) yr^{-1} (Xie and Moore, 1999).

8.14.6.4 Organosulfides

The discovery of substantial amounts of volatile organosulfides in the oceans was one of the major additions to the sulfur cycle in the second half of the twentieth century. The largest flux of reduced sulfur to the atmosphere from the oceans is as DMS. The importance of this compound that was largely unknown in nature until the 1970s was revealed by Lovelock *et al.* (1972) as a potential explanation for the imbalance in the sulfur cycle. Over time it has become clear that this process has important implications to the atmosphere and offers a source of sulfate to form cloud condensation nuclei.

The production of DMS is driven by the activity of microbiological organisms in the surface waters of the ocean (see Figure 11). The precursor to DMS is dimethylsulfoniopropionate (DMSP) produced within cells, but its concentration can vary over five orders of magnitude. Prymnesiophytes and some dinoflagellates are strong producers of DMSP. It is usually considered as an algal osmolyte globally distributed in the marine euphotic zone, where it represents a major form of reduced sulfur in seawater. It is typically partitioned between particulates (33%; range 6–85%;); dissolved nonvolatile degradation products (46%; range 21–74%); and in a volatile form (9%; range 2–21%) (Kiene and Linn, 2000). The natural turnover of dissolved DMSP results in the conversion of a small fraction (although large amount) to reduced sulfur gases. DMSP is released from cells during senescence, zooplankton grazing or viral attack. Once in the oceans it undergoes cleavage, which is usually microbiologically mediated or catalyzed by the enzyme DMSP lyase to DMS and acrylic acid (Kiene, 1990; Ledyard and Dacey, 1996; Malin and Kirst, 1997):

$$(CH_3)_2S^+CH_2CH_2COO^- + OH^- \\ \rightarrow CH_3SCH_3 + CH_2CHCOO^- + H_2O$$

The precursor DMSP is often present at concentrations an order of magnitude higher than DMS. These concentrations are so high that DMSP could support 1–13% of the bacterial carbon demand in surface waters, making it a

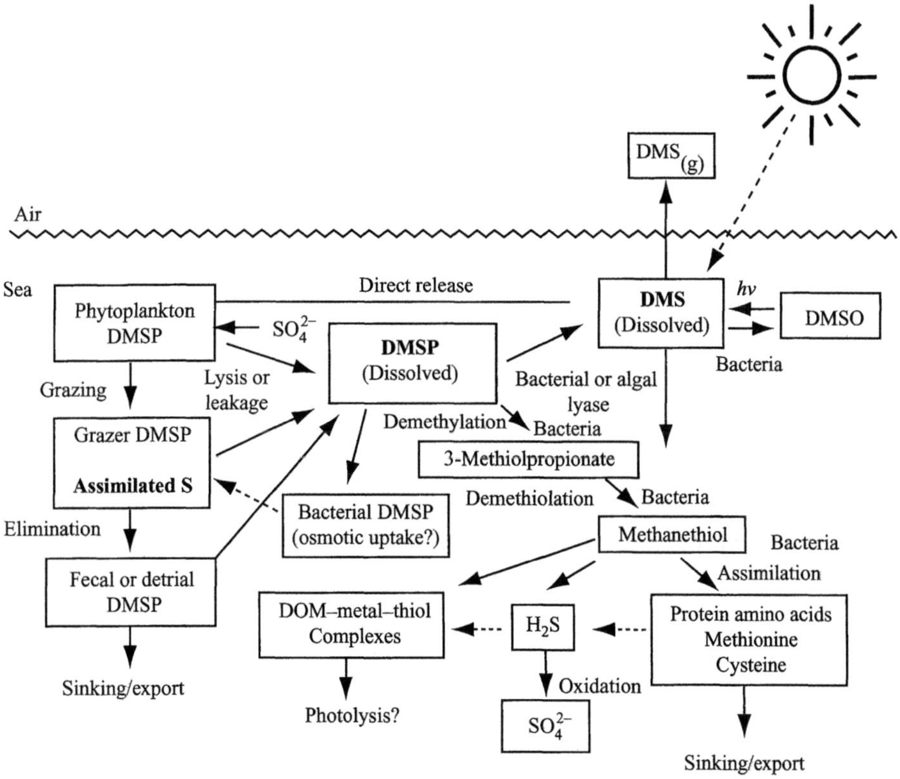

Figure 11 The biogeochemical cycles of DMSP and DMS in surface waters of the oceans (source Kiene *et al.*, 2000).

significant substrate for bacterioplankton (Kiene *et al.*, 2000). It is likely that some 90% of the DMSP in seawater is converted to methanethiol (MeSH) via 3-methiolpropionate (Kiene and Linn, 2000). The MeSH provides sulfur for incorporation into bacterial proteins, such as methionine. If the thiol is not assimilated, it is likely to react to give dissolved nonvolatile compounds, such as sulfate and DOM–metal–MeSH complexes. High production rates of MeSH in seawater (3–90 nM d^{-1}) offer the potential for the thiol to affect metal availability of trace metals and their chemistry in seawater (Kiene *et al.*, 2000).

Maximum concentrations of DMS in the oceans are found, at or a few meters below, the surface (as high as micrograms per liter under bloom conditions), which fall rapidly at depths where phytoplanktonic species are no longer active. In Arctic and Antarctic waters *Phaeocystis* sp. is responsible for enhanced DMS in the oceans particularly at the final stages of production of a bloom.

Questions are still asked about the role of DMS, which is sometimes seen, along with DMSP as having an antioxidant function in marine algae because the breakdown products of DMSP (i.e., DMS, acrylate, dimethylsulfoxide, and methane sulfinic acid) readily scavenge hydroxyl radicals and other reactive oxygen species (Sunda *et al.*, 2002).

Once produced the DMS is relatively volatile and is evaded efficiently from seawater. However, the notion that the polar seas represent an underestimated source of DMS has been questioned as the concentrations of DMS in polar waters are not always high in the spring blooms (Bouillon *et al.*, 2002).

Although large quantities of DMS are transferred across the air–sea interface into the atmosphere, it can also be chemically transformed in seawater. The presence of relatively large amounts of dimethylsulfoxide (DMSO) suggested the oxidation of DMS as a likely source. Andreae (1980) proposed bacterial oxidation, while Brimblecombe and Shooter (1986) have shown that DMS is susceptible to photo-oxidation by visible light and thermal oxidation at much slower rates (Shooter and Brimblecombe, 1989). However, more recent work has indicated that DMSO is unlikely to be the only oxidation product (Hatton, 2002; Kieber *et al.*, 1996). As DMS is transparent to sunlight, photo-oxidation requires the presence of naturally occurring photosensitizers such as chlorophyll derivatives, fulvic and humic acids, or bile-type pigments. Laboratory measurements of Kieber *et al.* (1996) have indicated that the photo-oxidation of DMS is wavelength dependent and can take place in the UV and also in the range 380–460 nm. More recently, Hatton (2002) made shipboard measurements that suggested increased oxidation rates in the UV. It is not certain what fraction of DMS in the ocean is oxidized, but it may well be close to the amounts that are lost via evasion to the atmosphere.

DMS may be mixed downwards and contribute to the budget in the deep oceans. Concentrations here are low, but the water volumes are large. Transport to the deep ocean is taken to be via sinking or transport and at these depths there may be a very slow chemical oxidation to DMSO (Shooter and Brimblecombe, 1989).

Dimethylsulfoxide is found at higher concentrations than DMS in the oceans. Although oxidation of DMS has usually been seen as the main source, more recently it has become clear that phytoplankton can also biosynthesize DMSO directly (Bouillon *et al.*, 2002). Incubation experiments suggest that the biological cycling of DMSO involves both its production and consumption in seawater (Simo *et al.*, 2000). Dimethylsulfoxide is found at highest concentrations where biological activity is high, but it appears to be long lived compared with DMS. Thus, it is better mixed within the water column and found at significantly higher concentrations than DMS in deep waters (Hatton *et al.*, 1998).

These marine sources of reduced sulfur gases can be important as a source of sulfur to the continents. The gypsum accumulations of the hyper-arid Central Namib Desert seem to be mainly derived from non-seasalt sulfur, in particular oxidation products of marine DMS (Eckardt and Spiro, 1999).

8.14.6.5 Coastal Marshes

The salt and brackish waters found in coastal areas are also large potential sources of reduced sulfur compounds, given the high level of biological activity in such zones. The major emissions are: H_2S, OCS, DMS, and DMDS. It has been difficult to estimate the global contributions to this source because of the high degree of variability in measured fluxes from these environments, such that total flux can span four orders of magnitude. The sea–air flux of OCS from a unit area of the estuarine waters, Chesapeake Bay, are over 50 times greater than those typical of the open ocean. Nevertheless, it has not been possible to identify definite seasonal trends in the fluxes. High concentrations of OCS were found in pore waters from estuarine sediments, which were at a maximum in the summer months. Microbial sulfate reduction in sediments contributes to the estuarine budget of OCS (Zhang *et al.*, 1998). A range of factors, such as temperature and biological activity, drive the variability in emission flux, but work at the coastal margins suggests that salinity is a very important control (DeLaune *et al.*, 2002).

The types of sulfur compounds change significantly along salinity gradients in coastal marshes with emissions from saltmarsh sites mostly as DMS. Saltmarshes also showed the highest sulfur emission fluxes followed by brackish marshes, with freshwater marshes having the lowest emissions. Brackish marshes gave hydrogen sulfide, while freshwaters were predominantly a source of OCS (see Figure 12). The low emissions of hydrogen sulfide from saltmarshes were attributed to iron sulfide formation (DeLaune *et al.*, 2002).

Interstitial waters of intertidal mud flats also show high concentrations (up to 1 mM) of reduced sulfur. There are wide variations in these concentrations with both season and depth indicating that annelids on tidal flats can be exposed to large concentrations of hydrogen sulfide (Thiermann *et al.*, 1996). Hydrogen sulfide emissions are much enhanced where there are significant discharges of industrial and domestic wastewaters into estuarine areas. Limited oxygen inputs and warm climate lead to optimal medium for anaerobic processes and the release of large concentrations of H_2S that affect nearby areas (Muezzinoglu *et al.*, 2000).

8.14.7 SURFACE AND GROUNDWATERS

Inorganic sulfur compounds are at a much lower concentration in surface waters than in the ocean. Nevertheless, rivers move a large amount of dissolved sulfate to the sea each year (Meybeck, 1987) and industrial activities and agriculture have added much to this flux. Typically estimates are perhaps a little less than a $100\,Tg(S)\,yr^{-1}$, with additional loads of the same magnitude as those from industrial and agricultural sources. There is also some

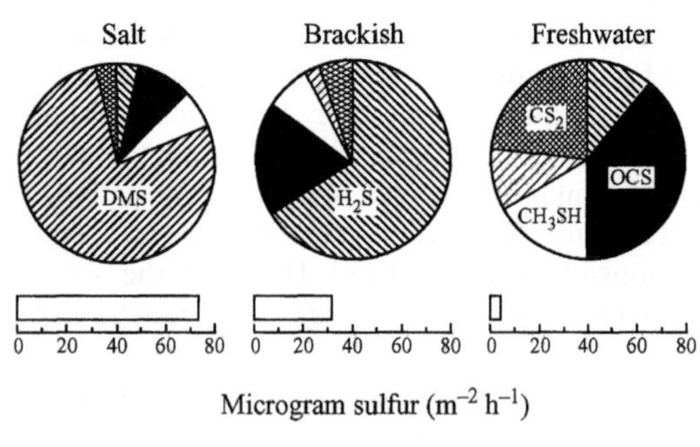

Figure 12 The relative amounts of reduced sulfur species released from various water bodies, with the total emission rate shown as histograms (source DeLaune *et al.*, 2002).

100 Tg(S) yr^{-1} transported as suspended particulate matter (Brimblecombe *et al.*, 1989).

There are a number of natural sources of sulfate in rivers. In volcanic regions the sulfate in river and lake water can be derived from volcanic waters enriched in sulfur (Robinson and Bottrell, 1997). The oxidation of pyritic glaciofluvial sediments and bedrock sulfides may be an important source of sulfate in aquifers (Robinson and Bottrell, 1997; Sidle *et al.*, 2000). Chemical attack on acid-insoluble metal sulfides FeS_2, MoS_2, and WS_2 by Fe(III) hexahydrate ions generates thiosulfate. Other metal sulfides are attacked by Fe(III) ions resulting in the formation of elemental sulfur via intermediary polysulfides. Elemental sulfur and sulfate are typically the only products of FeS oxidation, whereas FeS_2 was oxidized to a variety of sulfur compounds, including intermediates such as thiosulfate, trithionate, tetrathionate, and pentathionate. Sulfur and the sulfur oxyanions are ultimately oxidized to sulfuric acid (Schippers and Jorgensen, 2001).

In extreme cases the oxidation of sulfidic floodplain sediments can cause pulses of acids in rivers during floods after long dry periods. The acids can bring large amounts of dissolved aluminum and iron into streams. Such processes can be enhanced as a result of attempts to modify drainage and mitigate floods (Sammut *et al.*, 1996).

Polysulfates, most typically thiosulfates, are found in anoxic riverine sediments. Additionally inorganic polysulfides H_2S_n ($n = 1-5$) can be present even in oxygen-rich aquatic systems. The polysulfide ions can readily be converted to OCS or perhaps carbon disulfide under sunlight irradiation (Gun *et al.*, 2000) or via:

$$H^+ + CO + S_2^{2-} \rightarrow HS^- + OCS$$

As seen in the discussion of coastal marshes, even freshwater lakes contain a range of volatile sulfur compounds. Seasonally varying concentrations of DMS can sometimes be in the same range as in seawater, i.e., a few nmol L^{-1}. DMDS and methanthiol are also found (Gröne, 1997). DMS is the dominant organosulfur species found in lakes, along with OCS, MSH, DMDS, and CS_2. The concentrations of volatile organosulfur compounds found in lakes are dependent on their sulfate content. Hypersaline lakes with sulfate concentrations greater than 20 g SO_4^{2-} L^{-1} yield volatile organosulfur concentrations several orders of magnitude higher than typical freshwater lakes. The flux to the atmosphere from such water bodies can rival that from the ocean. However, even where the sulfate content is exceeding low there still appears to be mechanisms for the production of these compounds (Richards *et al.*, 1994). In some saline lakes dimethylpolysulfides ($CH_3S_nCH_3$, $n = 2-4$) are found probably as a result of biological methylation of polysulfides (Gun *et al.*, 2000).

8.14.8 MARINE SEDIMENTS

The oxidation of organic matter in anaerobic sediments can utilize a number of species as oxidants, of which sulfate is the most important. In seawater at pH values close to 8, sulfate-reducing bacteria metabolize organic matter according to the following simplified equation:

$$2CH_2O_{(s)} + SO_4^{2-}{}_{(aq)} \rightarrow 2HCO_{(aq)}^{3-} + HS_{(aq)}^- + H_{(aq)}^+$$

This process is widespread in marine sediments but is most important at the continental margin sediments, where organic matter accumulation is fastest. Sulfate reduction can occur even meters below the sediment/water interface as long as seawater sulfate can diffuse, or be pumped by bioturbation induced through the actvities of sediment-dwelling organisms. The reaction yields the bisulfide anion (HS^-), much of which diffuses upward and is reoxidized to SO_4^{2-} in oxygenated seawater closer to the sediment surface. However, ~10% of the HS^- precipitates soluble Fe(II) to yield iron monosulfide:

$$Fe_{(aq)}^{2+} + HS_{(aq)}^- \rightarrow FeS_{(s)} + H_{(aq)}^+$$

This iron monosulfides is then converted to pyrite (FeS_2) (Butler and Rickard, 2000). At Eh below −250 mV and at pH around 6, the conversion of FeS to FeS_2 occurs through the oxidation of dissolved FeS by hydrogen sulfide:

$$FeS_{(aq)} + H_2S_{(aq)} \rightarrow FeS_{2(s)} + H_{2(g)}$$

Under milder reducing conditions, the oxidation can use polysulfides, polythionates, or thiosulfate anions, e.g.,

$$FeS_{(s)} + S_2O_3^{2-}{}_{(aq)} \rightarrow FeS_{2(s)} + SO_3^{2-}{}_{(aq)}$$

The sulfite (SO_3^{2-}) anion can undergo reoxidation by oxygen closer to the sediment surface to produce polysulfur anions or even sulfate. Elemental sulfur may be found in marine sediments as a product of the oxidation of iron sulfides (Schippers and Jorgensen, 2001). The incorporation of pyrite and organosulfides, formed as a by-product of sulfate reduction in marine sediments, is a major sink for seawater SO_4^{2-}. Over geological time we have seen that this is an important process in controlling the concentrations of seawater sulfate.

An important consequence of this reduction, in terms of the human use of fossil fuels, is the incorporation of this sulfide into the formation of high sulfur coal. In addition to the sulfide as iron sulfides, organic sulfur in the coals is formed through reaction of reduced of organic materials with sulfur species. This takes place in the early stages of diagenesis by reaction of H_2S or polysulfides ($HS^-_{(x)}$) with humic substances formed by bacterial decomposition of the peat. Organic sulfur species are typically found in coals as thiols, sulfides, disulfides, and thiophenes (see Figure 13). The thiophenic fraction of organic sulfur increases with the carbon content of coals. Alkylkated thiophenes are converted to alkylated benzo(b)thiophenes and dibenzothiophenes via cyclization and subsequent aromatization reactions with coalification (Damste and Deleeuw, 1992; Wawrzynkiewicz and Sablik, 2002).

The importance of organic sulfur to the global sulfur cycle has only recently been recognized, with the awareness that organic sulfur is the second largest pool of reduced sulfur in sediments after pyrite. The incorporation of reduced inorganic sulfur into organic matter represents a significant mechanism for the preservation of functionalized organic compounds, such as thianes, thiolanes, and thiophenes but the timing of this process is not currently known (Werne *et al.*, 2000).

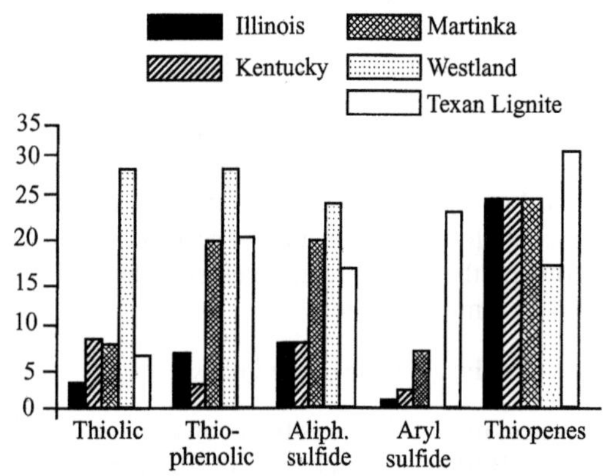

Figure 13 The relative amount of various organic forms of sulfur in some coals (source Attar, 1979).

8.14.9 SOILS AND VEGETATION

Sulfur in soils is present in both inorganic and organic forms. In calcareous soils some 90% is present in the organic form. Inorganic sulfate is present in solution, adsorbed, and insoluble forms. It is this form that is typically taken up by plants. In some soils such as tundra, there are low sulfate concentrations in the water. Changes in recent years in Europe have raised interest in sulfate as a nutrient, because increasingly soils are found to be sulfur deficient. Coarse textured soils typically low in organic matter tend to lose sulfate through leaching (Havlin *et al.*, 1999). Some of the sulfate in soils comes from the atmosphere (or added as fertilizers), but it is also biologically mineralized from organic sulfur, via sulfides (see Figure 14).

Long-term studies of the nutrient balance at the Hubbard Brook Experimental Forest show that much of the SO_4^{2-} entering via atmospheric deposition passes through vegetation and microbial biomass before being released to the soil solution and stream water. Gaseous emission loss of sulfur is probably small. The residence time for S in the soil was determined to be ~9 yr (Likens *et al.*, 2002).

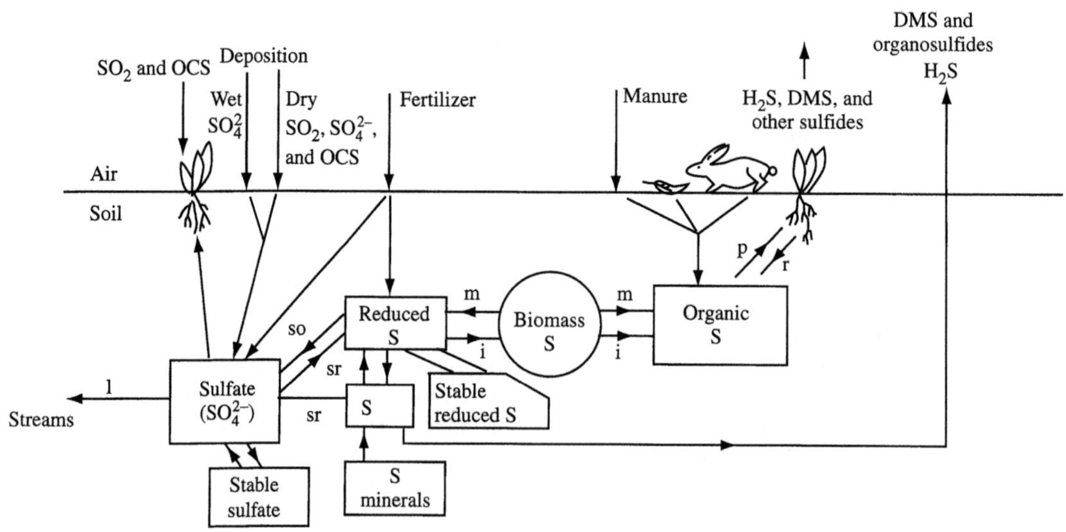

Figure 14 The cycle of sulfur in the soil (abbreviations: i, immobilization; m, mineralization; p, plant uptake; r, root exudation and turnover; so, oxidation; l, leaching; sr, reduction (after Havlin *et al.*, 1999).

The sulfur in noncalcareous soils occurs mostly as organic sulfur. The typical C/N/S ratio (on a weight basis) is 120/10/1.4 and the N/S ratio remains within a fairly narrow range 6–8.1. Organic sulfur is present in three forms: (i) HI-reducible, (ii) carbon-bonded, and (iii) residual sulfur. The HI-reducable sulfur can be reduced to H_2S with hydroiodic acid and the sulfur, is largely present as esters and ethers with C–O–S linkages. These would include arylsulfates, phenolic sulfates and sulfated polysaccharized and lipids, and represent about half the organic sulfur. Carbon bonded sulfur is found as the amino acids methionine and cystine and some 10–20% of the organic sulfur is found in this form. It also includes sulfoxides, sulfones, and sulfenic, sulfinic and sulfonic acids (Havlin *et al.*, 1999). Sulfur is additionally associated with humic and fulvic acids in a range of oxidized and reduced forms (Morra *et al.*, 1997). After conversion to inorganic sulfide, the sulfur can be oxidized quite rapidly via elemental sulfur and the thiosulfate ion through to a range of polythionates, which are ultimately converted to sulfite and thence oxidized to sulfate.

Volatilization of sulfur from soils occurs mainly as organosulfides, with DMS usually accounting for at least half (Havlin *et al.*, 1999). The emission of reduced sulfur compounds from soils and most particularly wetlands has been seen as an important source of these compounds to the global budget (Hines, 1992). Table 5 shows the biochemical origin of volatile sulfides from soils. Vegetation is also a source of trace sulfur gases and lichens notably emit H_2S and DMS (Gries *et al.*, 1994).

Sulfate reduction can take place in waterlogged soils, especially environments such as paddy fields. Soils with high sulfate content, particularly as a result of input of seawater, can over time accumulate large concentrations of sulfide through sulfate reduction. When these soils are exposed to air, they can become acidic through the production of sulfuric acid. This can be a serious problem to agriculture, but may also lead to the acidification of rivers.

Table 5 The biochemical origin of volatile sulfides from soils under anaerobic conditions through the microbial degradation of organic matter.

Volatile	*Biochemical precursor*
H_2S	Proteins, polypeptides, cystine, cysteine, glutadione
CH_3SH	Methionine, methionine sulfoxide, methionine sulfone, S-methyl cysteine
CH_3SCH_3	As for CH_3SH plus homocysteine
CH_3SSCH_3	As for CH_3SH
CS_2	Cystine, cysteine, homocysteine, lanthionine, djencolic acid
OCS	Lanthionine, djencolic acid

Adapted from Bremner and Steele by Warneck (1999).

Dusts with large amounts of sulfate, most notably as gypsum, are often driven into the air under windy conditions. In drier regions, particularly from the surfaces of dry lake beds, the fluxes can be high. The exposed areas change dramatically over geological time.

8.14.10 TROPOSPHERE

8.14.10.1 Atmospheric Budget of Sulfur Compounds

The reservoirs treated above contain most of the mass of sulfur. The amount in the atmosphere is small compared with these reservoirs (see Figure 15). However, this does not mean that the chemistry of sulfur in the atmosphere is of proportionately small importance. The residence times of compounds in the atmosphere are frequently short. The atmosphere combines great mobility with active chemistry and the part of this biogeochemical cycles is often easy to observe because of the transparency of the gas phase. Interest in atmospheric sulfur chemistry has been further stimulated by the magnitude of anthropogenic sulfur emissions. Global anthropogenic sulfur dioxide emissions were of the same magnitude as natural emissions through the twentieth century. Although similar arguments can be made about anthropogenic contributions to the nitrogen cycle, it was the importance of emissions from high sulfur coals that have caused problems during a long period of human industrialization. On a regional scale, the oxidation of sulfur dioxide to sulfuric acid makes it the main solute in acid rain. Studies of acidification led to great improvements in the scientific understanding of the liquid phase chemistry of rain from the 1950s onwards. By comparison other aspects of liquid-phase atmospheric chemistry (such as the chemistry of water-soluble organic compounds) tended to be much neglected until the very end of the twentieth century.

The focus on oxidized forms of sulfur also meant that comparatively little was known about the more reduced sulfur gases until recently. The presence of hydrogen sulfide had been evident in volcanic regions and on coastal marshes because of its strong smell, but the detection of atmospheric organosulfides had to await the development of gas chromatographic methods and sensitive detectors.

8.14.10.2 Hydrogen Sulfide

The fluxes of hydrogen sulfide to the atmosphere have not been easy to establish with

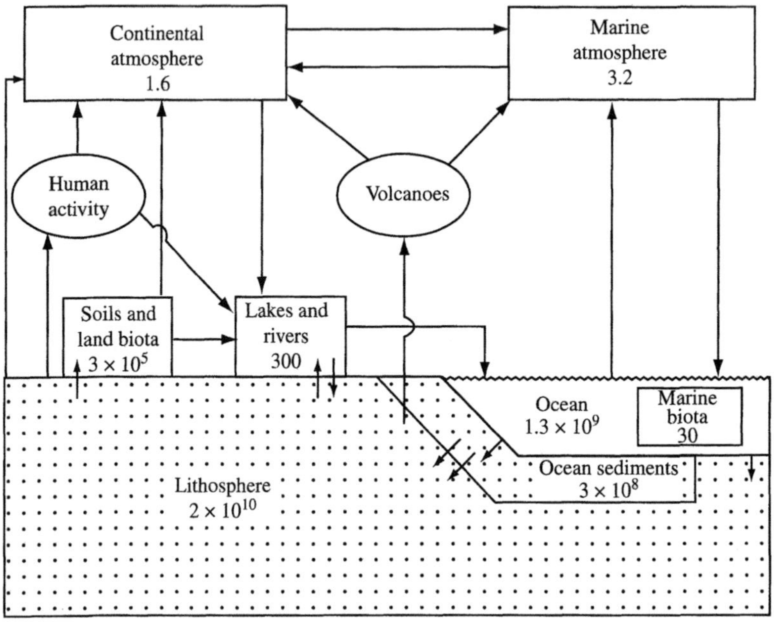

Figure 15 Major reservoirs and burdens of sulfur as Tg(S) (source Charlson *et al.*, 1992).

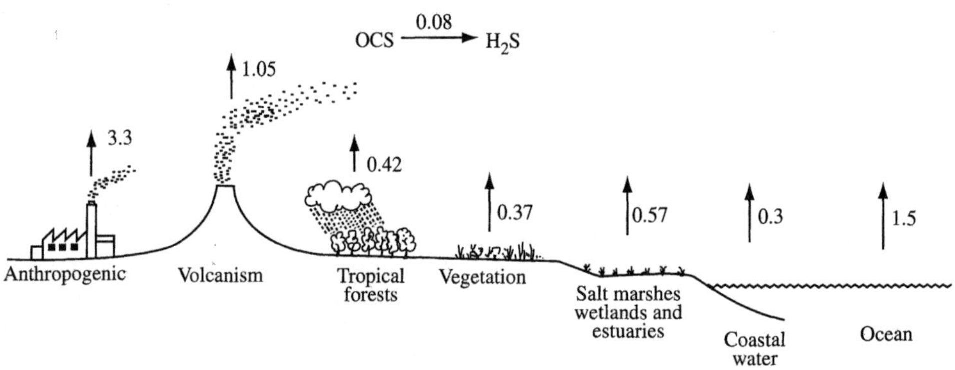

Figure 16 Sources and sinks of atmospheric hydrogen sulfide (source Watts, 2000).

estimates ranging from 16 Tg a^{-1} to 60 Tg a^{-1} (Watts, 2000). There is a poor correlation between concentrations of hydrogen sulfide in the marine atmosphere and both biological emissions and radon concentrations, which does little to give a clearer idea of sources. The estimates made by Watts (2000) are shown in Figure 16.

In remote air hydrogen sulfide is typically in the 30–100 pmol mol^{-1} range. Close to sources such as tidal mud flats it can be higher, and in volcanic regions it can be higher still. In Sulfur Bay near Rotorua New Zealand, average concentrations between 2,000–4,000 ppb were found exposing both human and wildlife (Siegel *et al.*, 1986). The gas is oxidized in air probably through hydroxyl radical attack:

$$OH + H_2S \rightarrow HS + H_2O$$

The HS radical reacts with either oxygen or ozone, the latter giving the HSO radical. The oxidation processes ultimately yield SO_2. The typical lifetime for atmospheric H_2S is ~3 d. Global sources and sinks of H_2S are estimated as 7.72 ± 1.25 Tg a^{-1} and 8.50 ± 2.80 Tg a^{-1}, respectively, with an imbalance that was indefinable (Watts, 2000).

8.14.10.3 Carbonyl Sulfide

There is a range of sources for atmospheric COS and these seem fairly evenly balanced (see Figure 17). This coupled with its long residence time mean that it has a constant concentration throughout the atmosphere. It is typically found at concentrations close to 500 ppt

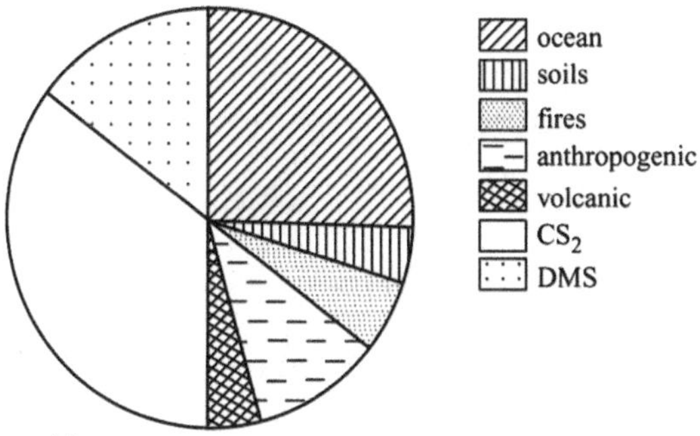

(a)

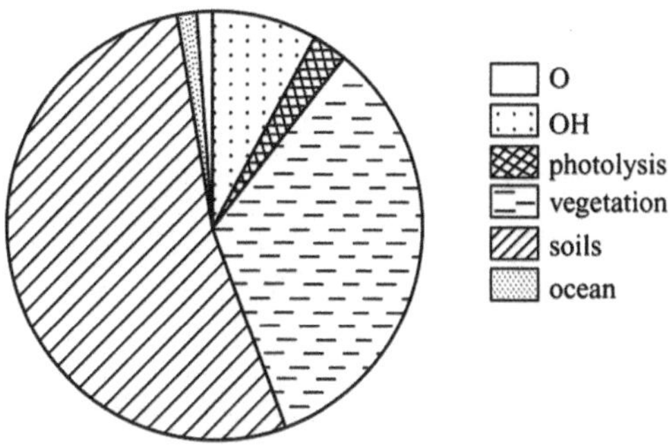

(b)

Figure 17 Production ($1.31 \pm 0.25\ Tg\ yr^{-1}$) and loss ($1.66 \pm 0.79\ Tg\ yr^{-1}$) processes for OCS (source Watts, 2000).

in remote and unpolluted air. There has been some debate as to whether the oceans are a source or a sink for OCS, but in the analysis of Watts (2000) the oceans are taken as a source. The long lifetime of ~25 yr is the result of a slow rate of reaction with the OH radical. This can proceed via addition to form the OCS–OH adduct, but it may decompose to the original reactants. Another possibility is that it reacts:

$$OH + OCS \rightarrow HS + CO_2$$

The long lifetime means that a significant fraction (see Figure 17) of the OCS is transferred to the stratosphere. Uptake by vegetation and deposition to soils are important loss processes, although the large uptake by oxic soils is particularly uncertain (Kjellstrom, 1998; Watts, 2000).

8.14.10.4 Carbon Disulfide

Carbon disulfide has a much shorter lifetime than OCS (~7 d), so is necessarily far more variable in concentration. At ground level, concentrations are typically in the range 15–30 pmol mol^{-1}, while in the unpolluted free troposphere concentrations are probably ~1–6 pmol mol^{-1}. Oxidation of carbon disulfide proceeds:

$$OH + CS_2 + M \rightarrow HOCS_2$$

There are many ways that the adduct $HOCS_2$ can react with oxygen, one of which is an important source of OCS:

$$HOCS_2 + O_2 \rightarrow HOSO + OCS$$

or

$$HOCS_2 + O_2 \rightarrow H + SO_2 + OCS$$

Estimates of sources and sinks of CS_2 are $0.66 \pm 0.19\ Tg\ yr^{-1}$ and $1.01 \pm 0.45\ Tg\ yr^{-1}$ (Watts, 2000). The amounts that are destroyed through oxidation are probably similar to the amount that is lost to oxic soils (Watts, 2000).

8.14.10.5 Dimethyl Sulfide

In the early 1970s there was some concern that the sulfur cycle did not balance and that there needed to be additional global sources. When James Lovelock discovered DMS in the atmosphere of remote Ireland, this seemed a fine candidate to balance the sulfur cycle. The production of DMS in the oceans and its subsequent oxidation have been extensively investigated, particularly because of its role in producing sulfuric acid droplets that can act as an important cloud condensation nuclei in the remote marine atmosphere.

A large number of measurements of seawater DMS concentrations are available and these have been extensively used to model the flux of DMS to the atmosphere. It is not always easy to get agreement between various models and data sets. The work of Kettle *et al.* (1999) gives rather higher fluxes of DMS and shows a larger seasonal variation especially from the Southern Ocean than that of Maier-Reimer (http://www.mpimet.mpg.de/en/depts/bgc). Utilizing the data of Kettle and Andreae (2000) shows a latitudinal band of high DMS concentrations over the oceans at 60° S. Figure 18 shows global fluxes modeled by Maier-Reimer using the Kettle *et al.* data and yielding a total annual flux of 25 Tg(S) yr^{-1}. In the past estimates of marine DMS have been as high as 30–50 Tg(S) yr^{-1}, but increasingly somewhat lower values are favored and it may well be in the range 10–20 Tg(S) yr^{-1}.

Free tropospheric concentrations are low, probably less than 3 pmol mol^{-1} and its residence time is short probably ~2 d. It is much higher in surface air close to areas of active production. Here concentrations can average 450 pmol mol^{-1} with a diurnal cycle (amplitude 85 pmol mol^{-1}) showing a decreased concentration during the day because of reactions with OH in the sunlit atmosphere (Yvon *et al.*, 1996):

$$OH + CH_3SCH_3 \rightarrow CH_3SCH_2 + H_2O$$

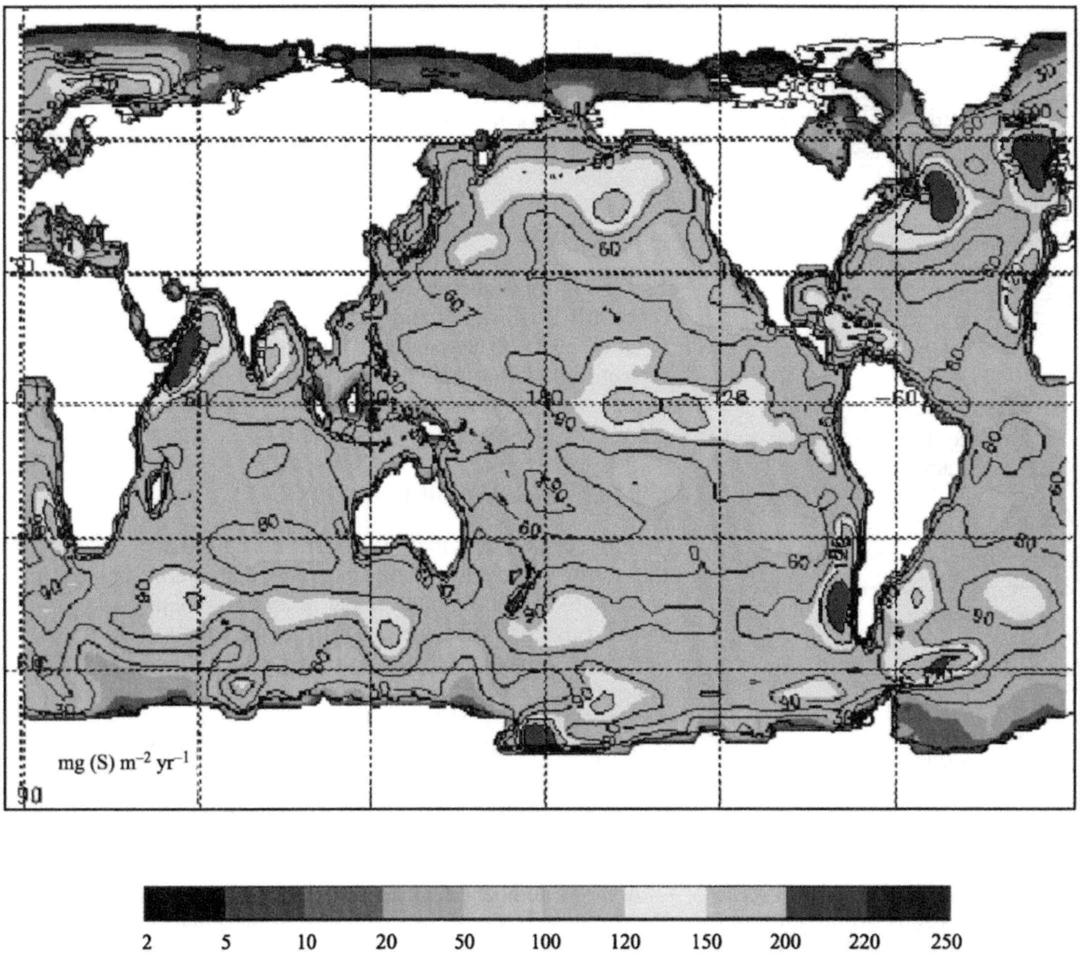

Figure 18 Global fluxes of marine DMS to the atmosphere modeled by Maier-Reimer using the Kettle *et al.* data (http://www.mpimet.mpg.de/en/depts/bgc).

or via addition

$$OH + CH_3SCH_3 + M \rightarrow CH_3S(OH)CH_3$$

Some of this addition can lead onto dimethylsulfoxide, which can be detected in the atmosphere (Sciare *et al.*, 2000):

$$CH_3S(OH)CH_3 + O_2 \rightarrow CH_3SOCH_3 + HO_2$$

DMS can also react with the halogens and their oxides to yield a range of products that include DMSO.

8.14.10.6 Dimethylsulfoxide and Methanesulfonic Acid

Dimethylsulfoxide and methanesulfonic acid are two of the most important organic oxidation products of DMS. It is not entirely clear how methanesulfonic acid, $CH_3S(O)(O)(OH)$, forms, but methanesulfinic acid, $CH_3S(O)(OH)CH_3$ has been reported during oxidation in OH–DMS systems. Further addition of OH to methanesulfinic acid, followed by reaction with oxygen, can yield methanesulfonic acid. At lower temperatures found in the Arctic there are a wide variety of oxidation products of DMS that include the MSA, DMS, and dimethylsulfone, $CH_3S(O)(O)CH_3$.

Although DMS is only moderately soluble in rainwater, there has been some interest in the oxidation in the liquid phase, where modeling suggests that the multiphase reactions can be important. Ozone seems to be the most important oxidant, where there is a predicted lifetime for DMS of a few days in clouds. The oxidation reactions offer the potential for these heterogeneous processes to yield more soluble oxidized sulfur compounds such as DMSO and $DMSO_2$ (Betterton, 1992; Campolongo *et al.*, 1999).

The concentrations of DMSO have been measured in the southern Indian Ocean with mixing ratios range from 0.3 ppt to 5.8 ppt.

Typically concentrations of DMSO in the air are ~1–2% of the DMS concentrations in air (Jourdain and Legrand, 2001). There is a seasonal cycle with a minimum in winter and a maximum in summer similar to that observed for atmospheric DMS (Ayers *et al.*, 1991). There is also a diurnal cycle for DMSO with maximum values around 09:00 and minimum ones during night, which implies OH reactions with DMS as an important source (Sciare *et al.*, 2000). Being soluble DMSO is also found in rainwater with concentrations from 7.0 nM to 369 nM and a seasonal maximum in the summer following much the same pattern as DMS in the atmosphere (Sciare *et al.*, 1998).

Dimethylsulfoxide can readily be removed onto particles and there it can undergo an efficient oxidation through to methanesulfonate. This adds a significant pathway to the gas-phase production of methanesulfonic acid, which is present largely in the submicron aerosol fraction. Peak summer concentrations are 0.6 ± 0.3 nmol m^{-3} and at times this can amount to almost a quarter of the non-seasalt sulfate in the remote marine atmosphere (Jourdain and Legrand, 2001).

Dimethylsulfoxide is also oxidized to SO_2, which can then be converted onto non-seasalt sulfate. The yield probably ranges from 50% to100% in the tropics with the potential for it to be somewhat lower, perhaps 20–40% in mid-latitudes (de Bruyn *et al.*, 2002).

8.14.10.7 Sulfur

Elemental sulfur has been found in marine air (e.g., Atlas, 1991) although it does not seem to occur in continental air. The sources of this are not known, but there are hints that it is of biochemical origin and could arise from either the reduction of sulfates or the oxidation of hydrogen sulfide. Given the strong tendency of sulfur to form polyatomic molecules, it is unlikely to be present in a monatomic form.

8.14.10.8 Sulfur Dioxide

Volcanoes and to a small extent biomass burning represent the major natural primary sources of sulfur dioxide to the atmosphere. Further sulfur dioxide is produced through the oxidation of sulfides, in the atmosphere. In addition to the natural sources there is a very large anthropogenic source that arises from fossil fuel combustion that is comparable in magnitude to the natural sources.

Concentrations in the remote atmosphere are low. In the middle to upper troposphere SO_2 concentrations range between 20 ppt and 100 ppt. Sulfur dioxide concentrations decrease with altitude (see Figure 19), suggesting ground level sources and oxidation to sulfate within the atmosphere. This means that the sulfate concentrations decline more slowly with altitude than those of SO_2. Over the oceans there is little change in SO_2 concentrations with height and these are typically ~40 ppt. The most obvious source for marine SO_2 is the oxidation of reduced sulfur compounds such as DMS, although long-range transport from the continents must also contribute, most particularly in the upper troposphere (Warneck, 1999).

Sulfur dioxide can undergo homogeneous oxidation through attack by the OH radical:

$$SO_2 + OH \rightarrow HOSO_2$$

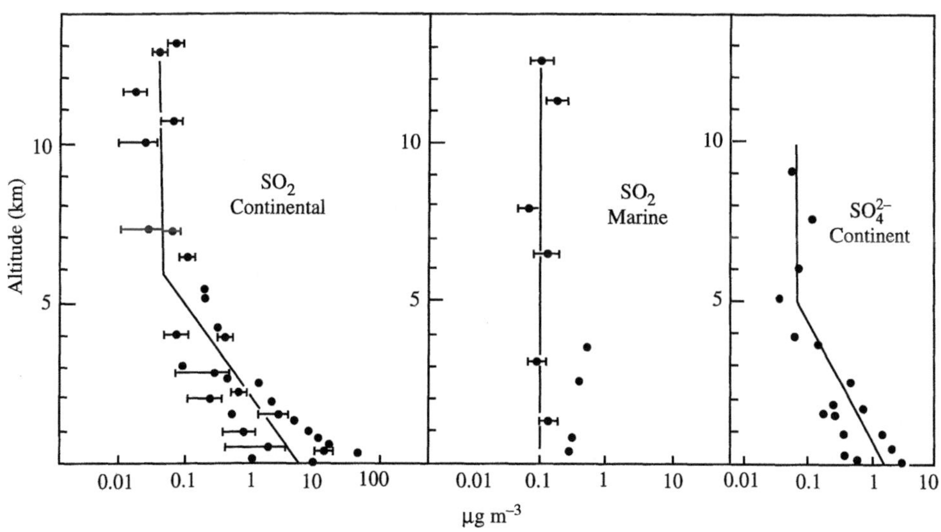

Figure 19 Vertical distribution of sulfur dioxide and sulfate in the remote continental and marine atmosphere (Atlantic) (source Warneck, 1999).

$$HOSO_2 \rightarrow SO_3 + HO_2$$

The sulfur trioxide reacts very rapidly with water to form a $H_2O{\cdot}SO_3$ complex that reacts with a further water to form the aquated sulfuric acid molecule. It is also likely that SO_3 reacts with water in aerosols. Although there are possibilities that SO_3 could react with ammonia to form sulfamic acid, most of the trioxide will be converted to sulfuric acid (Findlayson-Pitts and Pitts, 2000).

The heterogeneous oxidation can be significant, but in the presence of liquid water there is, as we will see, an important heterogeneous process for the production of sulfate from SO_2. Sulfur dioxide is not especially soluble in water, but subsequent equilibria increase the partitioning into cloud water:

$$SO_{2(g)} + H_2O \rightleftharpoons H_2SO_{3(aq)}$$

$$H_2SO_{3(aq)} \rightleftharpoons H^+_{(aq)} + HSO^-_{3(aq)}$$

$$HSO^-_{3(aq)} \rightleftharpoons H^+_{(aq)} + SO^{2-}_{3(aq)}$$

$$K_H = mH_2SO_3/pSO_2,\ 5.4\ \text{mol L}^{-1}\ \text{atm}^{-1}\ \text{at } 15\ ^\circ\text{C}$$

$$K' = mH^+ mHSO_3^-/mH_2SO_3,\ 0.027\ \text{mol L}^{-1}\ \text{at } 15\ ^\circ\text{C}$$

$$K'' = mH^+ SO^{2-}_{3(aq)}/mHSO_3^- \sim 10^{-7}\ \text{mol L}^{-1}\ \text{at } 15\ ^\circ\text{C}$$

Strictly, SO_2 dissolves in water as $(SO_2)_{aq}$ with little forming sulfurous acid, $H_2SO_{3(aq)}$, but it is usual to neglect the distinctions between these two species. The ionization equilibria are typically fast and in the case of the hydration of aqueous SO_2 the hydration reaction proceeds with rate a constant of $3.4 \times 10^6\ s^{-1}$ which allows the formation of the bisulfite anion to be exceedingly rapid. Although $H_2SO_{3(aq)}$ is a dibasic acid, the second dissociation constant is so small that the bisulfite anion (HSO_3^-) dominates as the subsequent dissociation to the sulfite ion $SO^{2-}_{3(aq)}$ would not be important except in the most alkaline of solutions. At around pH 5.4 in a typical cloud with a gram of liquid water in each cubic meter, SO_2 will partition equally into both phases, because of the hydrolysis reactions.

The dissolution of sulfur dioxide in water at the Earth's surface, most particularly seawater, which is alkaline, represents an important sink. Land surfaces, especially those covered by vegetation, also represent a removal process as SO_2 is dry deposited to these surfaces. Although we use the term dry deposition, the vegetated surfaces act as if they are wet and gas exchange takes place with effectively wet surfaces on or inside leaves, via access through the stomata. Such dry deposition is most significant where SO_2 concentrations are high (polluted regions).

The other removal process, wet deposition, removes sulfur from the atmosphere as sulfates in rain. This would be the fate of sulfuric acid produced via the homogeneous oxidation of SO_2, but oxidation also proceeds within droplets. Aqueous sulfur dioxide is oxidized only slowly by dissolved oxygen, but the production of sulfuric acid, which is much stronger, leads to acidification

$$\tfrac{1}{2}O_2 + HSO_3^- \rightarrow H^+ + SO_4^{2-}$$

This process can be catalyzed by iron, manganese, and other transition metals in the atmosphere, e.g.,

$$M(III)(OH)_n + HSO_3^- \rightarrow M(II)(OH)_{n-1} + SO_3^- + H_2O$$

This initiates a radical chain that leads to sulfate production:

$$SO_3^- + O_2 \rightarrow SO_5^-$$

$$SO_5^- + SO_3^{2-} \rightarrow SO_4^- + SO_4^{2-}$$

$$SO_4^- + SO_3^{2-} \rightarrow SO_3^- + SO_4^{2-}$$

In cloud droplets in remote regions the metal concentrations are likely to be low. Here more typically the reaction proceeds with oxidants such as dissolved hydrogen peroxide (or other atmospheric peroxides) and ozone. Hydrogen peroxide is an especially important droplet phase oxidant, because the gas is very soluble in water so can dissolve from the atmosphere. Additionally, it is readily produced within droplets in the atmosphere via photochemical processes. Oxidation by hydrogen peroxide is also significant, because the reaction is faster in acidic solutions, which means that the oxidation process does not become much slower as droplets become more acidic with the production of sulfuric acid. This oxidation can be represented as

$$ROOH + HSO_3^- \rightleftharpoons ROOSO_2^- + H_2O$$

$$ROOSO_2^- \rightarrow ROSO_3^-$$

$$ROSO_3^- + H_2O \rightarrow ROH + SO_4^{2-} + H^+$$

Ozone although abundant has relatively low solubility in water and the oxidation reaction does not increase in rate at low pH values. However, the oxidation does appear to increase with ionic strength. The overall reaction can be written as

$$HSO_3^- + O_3 \rightarrow H^+ + SO_4^{2-} + O_2$$

and may proceed via the generation of aqueous OH radicals which can react with HSO_3^-:

$$OH + HSO_3^- \rightarrow SO_3^- + H_2O$$

The SO_3 radical can enter into the radical chains previously described.

A range of other oxidants dissolve or are produced in cloud droplets, such as is peroxynitric acid, which contributes to the production of sulfate:

$$HOONO_2 + HSO_3^- \rightarrow 2H^+ + NO_3^- + SO_4^{2-}$$

The relative importance of the various heterogeneous oxidation pathways depends on pH. At pH values below ~4.5 the hydrogen peroxide pathway typically dominates. In urban areas hydrogen peroxide may not be abundant enough to be the most important oxidant. Here transition metal catalysts can enhance the rates considerably, especially if there are alkaline materials from fly ash or ammonia to neutralize the growing acidity of droplet phases, which otherwise limits SO_2 solubility.

When we consider of the heterogeneous oxidation of SO_2, it has to include not only the oxidation of the SO_2 within the droplet phase, but the transfer of further SO_2 into the droplet and the overall depletion of the gas in the air mass as a whole. In general the overall oxidation in the remote atmosphere is rather slow and takes 2–4 d, but under some conditions it can be much faster. The depletion rates of sulfur dioxide in volcanic plumes can sometimes be very fast with residence times of as little as 15 min in moist plumes, where catalytic mechanisms similar to urban air masses probably occur.

There are a number of other reactions of SO_2 in solution most notably the formation of complexes with soluble aldehydes that also partition effectively into droplets. Formaldehyde, acetaldehyde, glyoxal, and methylglyoxal are the most common of these reactive aldehydes:

$$HCHO + SO_2 = CH_2(OH)SO_3^-$$

The formation of such complexes can result in an order of magnitude increase in the overall solubility of SO_2 in droplets, but some of the aldehyde reactions with dissolved SO_2 can be relatively slow so they may not compete with oxidation to sulfate (Findlayson-Pitts and Pitts, 2000).

8.14.10.9 Aerosol Sulfates and Climate

Sulfate is the ultimate product of the oxidation of SO_2. We might expect small concentrations of other species, such as dithionates that are intermediates in the oxidation process, although these have not been detected. Concentrations of sulfate particles in the remote atmosphere are typically at a few nanomoles per cubic meter. In the marine atmosphere the sulfate is found both in coarse particles where it derives from seasalt and in fine particles around a micron in diameter as sulfuric acid. This non-seasalt sulfate in the smaller particle sizes is an important cloud condensation nucleus (CCN) in the atmosphere.

Over the oceans much of the non-seasalt sulfate derived from the biological production of DMS in seawater (Ayers and Gillett, 2000), which has suggested a biological coupling between climate and living organisms. This coupling has been seen as support (Charlson *et al.*, 1987) for James Lovelock's *Gaia Hypothesis*. This is a popular concept that goes well beyond accepting that there are biological impacts on the geochemical cycling. It sees the atmosphere as an extension of the biosphere, such that it becomes regulated in a similar way to homeostasis within cells. The regulation of the surface temperature of the Earth has been very important for life. It is possible to argue that enhanced DMS from marine sources would lead to sulfate particles and thus to more CCN. This would be expected to increase the number of cloud droplets and cloud albedo so more light would be reflected into space leading to a cooler Earth. If the Earth cooled there could be a gradual reduction in the biological activity of the oceans and thence less clouds and lower albedo. Such a feedback offers the potential for temperature regulation. The notion has clear attractions, but is difficult to assess and raises, in addition to scientific questions, a range of much more philosophical ones. These include some about the nature of hypotheses through to concerns that we might be assigning altruism to ecosystems.

The role that DMS plays in producing more clouds has offered the possibility for humans to control climate. In the Southern Ocean where iron is a limiting nutrient it would be possible to add large quantities of iron to fertilize the ocean and thus increase DMS emissions and hence cloudiness thus inducing lower temperatures (Watson and Liss, 1998). Experiments show that iron can indeed fertilize the ocean and increase phytoplanktonic activity, but it would be risky to undertake such an experiment with our only planet.

In continental air sulfate tends to be associated with finer particles, and as ammonia is more likely to be present in the air this can neutralize the sulfuric acid with the formation of ammonium sulfate- or bisulfate-containing particles over land. Sulfuric acid can displace chloride from seasalt aerosols and represent a source of hydrogen chloride:

$$H_2SO_4 + Cl^- \rightarrow HCl_{(g)} + HSO_4^-$$

8.14.10.10 Deposition

We have already seen that sulfur dioxide can be removed from the atmosphere in a number of ways. It is soluble in water, especially at the higher pH values associated with seawater.

The rapid absorption of the gas into water means that it is readily transferred into the oceans or other water bodied. Vegetation also acts as an important sink for sulfur dioxide. This can be particularly efficient when the vegetation is wet, such as when it is covered by dew or rainwater. However, even when the vegetation is not wet, sulfur dioxide can enter the leaves through the stomata and enhance deposition. This process is called dry deposition because although it involves water, the water is not in the atmosphere.

The alternative process, wet deposition, deposits the sulfur in rain or other forms of precipitation. Here it is largely as sulfate, which has been incorporated from aerosols or through oxidation of dissolved sulfur dioxide. Sulfate particles can also be dry deposited to the Earth's surface.

The magnitude of the various processes varies with locality. Over vegetated continental areas typically a fifth of the sulfur is dry-deposited as SO_2. In areas of high SO_2 concentrations much higher amounts can be deposited. Where sulfate concentrations in particulate material are high, dry deposition rates can be greater. The balance of wet and dry deposition of sulfate particles over the ocean is uncertain, while some authors suggest dry deposition dominates others favor wet deposition (Warneck, 1999).

8.14.11 ANTHROPOGENIC IMPACTS ON THE SULFUR CYCLE

Human activities have vastly affected the sulfur cycle (Brimblecombe *et al.*, 1989). The sulfur released from combustion of fossil fuels for example, exceeds the average natural releases into the atmosphere. Thus sulfur has long been seen as a pollutant central to the acid rain debate of the 1980s. However human progress has had other impacts on the cycle.

8.14.11.1 Combustion Emissions

Sulfur emissions from the combustion of high sulfur coals has been a problem from the thirteenth century when the fossil fuel began to be used in London after the depletion of nearby wood supplies. The intensity of coal use increased reaching its peak within the early twentieth century in Europe and North America. Although the use of coal has declined in these areas, the late twentieth century saw a profound increase in coal use in developing countries, most notably in Asia (see Figure 20). Here emissions have continued to grow with the enormous pressures for industrial development, although changing patterns of fuel use here may lead to decreased emissions in the twenty-first century.

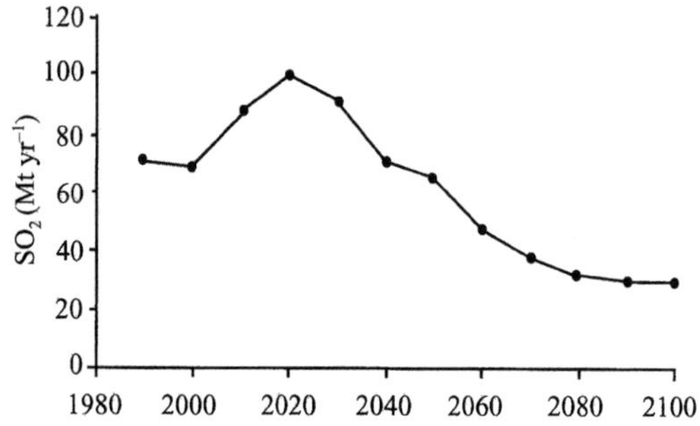

Figure 20 Predicted sulfur emissions from anthropogenic sources (http://sres.ciesin.org).

The combustion of fuels leads to release of SO_2 in a simple, but effective oxidation:

$$S + O_{2(g)} \rightarrow SO_{2(g)}$$

Many refining and extractive processes release large amounts of air pollution containing SO_2. For example, sulfide ores have been roasted in the past with uncontrolled emissions

$$Ni_2S_3 + 4O_2 \rightarrow 2NiO + 3SO_2$$

This sulfur dioxide often destroyed large tracts of vegetation downwind from smelters, such as those at Sudbury in Canada. However, changes in the processes and the construction of a very tall chimney stack have lessened the problems.

The decline in sulfur emissions in Europe and North America has come as part of a shift away from coal as a fuel in all but extremely large industrial plants. Sulfur in coal is about half as pyrites, which is relatively easy to remove, but the rest is organically bound which makes it difficult to remove at an economic rate. Improved controls on stack emissions increasingly rely on the treatment of exhaust gases. In the past this was sometimes by scrubbing the exhaust gas with water to dissolve the SO_2, but the late twentieth century saw a range of well-developed methods. The use of lime (calcium hydroxide) or limestone (calcium carbonate) slurries to absorb sulfur dioxide is widely adopted. The main product, calcium sulfate, is notionally not seen as an environmental problem by-product, although it can be contaminated with trace metals. The process is also hampered by the large amounts of lime that can be required. Regenerative desulfurization processes such as the Wellman-Lord procedure absorb SO_2 into sodium sulfite solutions converting them to sodium bisulfite. The SO_2 is later degassed and can be used as a feedstock for the production of sulfuric acid, for example.

Sulfur is also found in petroleum in organic forms. It can occur at high concentration in some residual oils. This sulfur can be removed by catalytic hydrodesulfurization, but it leads to fuels that tend to become waxy at low temperature. In vehicles catalytic converters have been used to remove nitrogen oxides, carbon monoxide, and hydrocarbons from exhaust streams. However under fuel rich driving cycles (i.e., lots of accelerating and decelerating), hydrogen gas is produced in the exhaust. Three-way catalysts containing cerium dioxide store sulfur from the exhaust stream, under driving conditions, as cerium sulfate. This can at other times be reduced by hydrogen gas to form hydrogen sulfide, which creates a noticeable odor, where traffic is heavy (Watts and Roberts, 1999).

8.14.11.2 Organosulfur Gases

Although the natural sources of organosulfur compounds are best known, there are a number of anthropogenic sources of this class of air pollutant. These are often released from sulfur-rich wastewaters and sewage sludges where DMDS is of particular concern because of its odor problems. MSH and DMS can cause similar problems along with hydrogen sulfide, OCS and carbon disulfide.

OCS is found as a major sulfur compound in anodic gases of commercial aluminum smelters. Studies suggest a specific OCS emission of 1–7 $kg\,t^{-1}$ (Al). In 1993 aluminum production was responsible for between 0.02 Tg(S) yr^{-1} and 0.14 Tg(S) yr^{-1} of OCS emissions, which is only a small fraction of the annual global budget (Harnisch *et al.*, 1995). Other sources of OCS are coal combustion 0.019 Tg(S) yr^{-1}, industry 0.001 Tg(S) yr^{-1}, the wear on tyres 0.04 Tg(S) $yr^{-1}$and the combustion products from automobiles 0.002 Tg(S) yr^{-1} (Kjellstrom, 1998). Carbon disulfide also has a range of anthropogenic sources such as chemical production 0.26 Tg(S) yr^{-1}, industry 0.002 Tg(S) yr^{-1}, aluminum production 0.004 Tg(S) yr^{-1}, and combustion products from automobiles 0.0002 Tg(S) yr^{-1} (Kjellstrom, 1998).

There are more complex organosulfur compounds associated with the aerosols. Even relatively volatile compounds such as thiophene can bind to particulate materials (Huggins *et al.*, 2000). The emissions from sulfur-containing fuels may be characterized by the presence of dibenzothiophene and benzonaphthothiophene. Coal combustion (notably lignite in an atmospheric fluidized bed) yields a range of heterocyclic compounds containing oxygen, nitrogen, or sulfur, often with three or four rings. The types of sulfur compounds found include: dibenzothiophene, methyldibenzothiophene, dibenzothieno(3,2-b)(1)benzothiophene, benzo(b)napthothiophene, and benzo(2,3)phenanthro-(4,5-bcd) (Stefanova *et al.*, 2002).

Water-soluble organic compounds in urban atmospheric particles can also contain organosulfur compounds. Methanesulfonic acid and hydroxymethanesulfonic acid have been found as the major organosulfur compounds in urban aerosols, most particularly in particles with the diameter range of 0.43–1.1 μm. Monomethyl hydrogen sulfate has also been detected on urban particles from localities where no oil or coal power plant exist (Suzuki *et al.*, 2001).

8.14.11.3 Acid Rain

Rain is naturally acidic because of the weakly acidic carbon dioxide, but the oxidation of sulfur dioxide leads to the much stronger sulfuric acid. The large-scale use of coal caused this to become a problem in nineteenth-century Europe. Remote observations of polluted black rain in Scotland and Scandinavia were made in the late nineteenth century. Some of the most significant early work came from Norway with work of the geologist Waldemar Brøgger and later Amund Helland, who reported on the loss of fish stocks from the 1890s, possibly because of acidification. Modern studies by Knut Dahl and Haakon Torgersen's in the 1920s and 1930s confirmed these observations and the importance of adding lime to reduce the acidity of streams. Rainfall composition was determined by agricultural networks set up to monitor the flux of nutrients to crops in the late nineteenth century. In parallel this also established the impact of combustion processes in the rural environment. The monitoring work recommenced after World War II and gave the information that established the modern picture of acidification of precipitation by the late 1960s. Despite this acid rain did not emerge as a global environmental issue until the 1980s, with concerns over the impact of acidification to the ecosystems across Europe and North America.

The hemispheric changes in precipitation chemistry have been reflected in the records left in high latitude snow, particularly in Greenland. Here the pre-industrial snow shows about equal input of sulfur from marine sources and industrial emissions. From the late nineteenth century, there is evidence of increased anthropogenic impact (Patris *et al.*, 2002), with much of the sulfate arising from North America (Goto-Azuma and Koerner, 2001). The anthropogenic sulfate is sufficient to displace hydrogen chloride from seasalt aerosols such that increasing inputs of anthropogenic HCl can be found in alpine and Arctic ice core records of the twentieth century (Legrand *et al.*, 2002).

The declines in sulfur emissions from the North Atlantic sector have made it is easy for politicians to believe that the acid rain problem has gone away. It is true, emissions here are lower, with a shift away from high sulfur fuels, most particularly coal. In parallel, the amount of sulfur deposited in rain Europe and North America has declined. Indeed, the decline in some parts of the UK and Germany has been so large that crops such as oats and oil seed rape can suffer from sulfur deficiency.

However, the decreases in deposited sulfur are not always matched by equivalent improvements in the amount of acid brought down in rain. Declining sulfur emissions have not always been accompanied by declines in the emission of the nitrogen oxides, which give rise to atmospheric nitric acid in precipitation. Differences in the mode of oxidation that forms nitric acid (a homogeneous oxidation) mean that nitric acid rain has a different geographical distribution to that of sulfuric acid rain. This distribution of acidity can easily change. Furthermore, calcium was once more abundant in dusts and available to neutralize some of the acidity. In recent decades, the amount of alkaline particulate material has declined, perhaps because there is less dust from unsealed roads and less grit from industry and power generation.

However, not all that has been learnt about acid rain in temperate regions is easily applicable in the Asian context. Research and regulation needs to face a different acid rain problem here. Entirely new ecosystem will be confronted by acidic deposition. We have to recognize some of the novel factors along the Asia-Pacific rim have been present for many centuries. Kosa dust and forest fires are a feature of the region, although we know only a little of their history. Alkaline dust offers the potential to buffer acids in rainfall. Forest fires can produce acids, but can also generate large amounts of alkaline ash that disperses along with the acids. However, such neutralization processes are not always well understood. The greater prevalence of acid rain into the tropics, where soils are often deeply weathered, makes available new routes for mobilizing toxic metals within ecosystems.

8.14.11.4 Water and Soil Pollutants

Agriculture and industrial activities have been responsible for doubling the load of sulfate in rivers and the concentrations can be much enhanced in rivers that pass through regions with high anthropogenic activity. In addition, industrial releases sometimes have catastrophic impacts. Examples of such a water pollution incident have occurred when toxic water and tailings from pyrite mining are accidentally released into river basins. In April 1998, releases from a pyrite mine in Aznalcollar, southern Spain, spilled into the Agrio and Guadiamar River Basin affecting some 40 km^2. Rapid oxidation meant an increase in concentration and the solubilization of the zinc, cadmium and copper (Simon *et al.*, 2001).

Linear alkylbenzene sulfonates (from detergents) are found in the range 0–5.0 mg kg^{-1} in both freshwater and marine sediments. Higher levels occasionally found in sediments are associated with untreated sewage effluent (Cavalli *et al.*, 2000).

Benzothiazoles from vulcanization of rubber tires are found so enriched in some situations that they are used as a marker for street run-off water. Thiocyanatomethylthiobenzothiazole used as a fungicide in wood protection and antifouling paints may degrade into benzothiazoles and methylthiobenzothiazole is used in vulcanization processes, so these materials find their way into rivers and ultimately estuaries. Bester *et al.* (1997) determined the presence of benzothiazole and methylthiobenzothiazole in estuarine and marine waters. Their concentrations range from 0.04 ng L^{-1} to 1.37 ng L^{-1}. Methylthiobenzothiazole and benzothiazole vary from 0.25 ng L^{-1} to 2.7 ng L^{-1} in the North Sea, while 55 ng L^{-1} methylthiobenzothiazole was found in the Elbe River.

Soils are often sulfur deficient, so sulfur can be added as a fertilizer in agricultural activities. Sulfur can be mobilized by agriculture and rivers readily polluted through the use of sulfur-rich fertilizers (Robinson and Bottrell, 1997). Although this can be added as sulfate, it is readily moved into the soil profile by rain or irrigation. It can also be applied as elemental sulfur, ammonium thiosulfate, and ammonium polysulfide. Ultimately, this fertilizer is likely to appear in runoff waters. Sulfur-containing pesticides now represent a significant input of sulfur to agricultural soils (Killham, 1994) and can pollute runoff waters as mentioned in the section above.

The widespread agricultural use of pesticides increases their concentration in river waters. Many of these contain sulfur, e.g., herbicides thiobencarb (S-4-chlorobenzyl diethylthiocarbamate) isoprothiolane (diisopropyl 1,3-dithiolan-2-ylidenemalonate), diazinon (O–O-diethyl 0-2-isopropyl-6-methylpyrimidin-4-yl) and fenitrothion (O,O-dimethyl O-4-nitro-m-toryl phosphorothioate), fungicide (isoprothiolane) and two insecticides (diazinon and fenitrothion). The concentrations of these pesticides typically show strong seasonal patterns which reflect their use (Sudo *et al.*, 2002).

Thiobencarb controls broad leaf weeds mostly in rice (so much has been used in Japan), lettuce, celery, and endive (894000lbs were used in California in 1997). Thiobencarb may be found at concentrations lower than expected from its application rate and sorption on soil particles (Sudo *et al.*, 2002) and reactions that degrade it. When thiobenzcarb is present at higher concentrations (>1 ppb), a bitter taste in drinking water can be associated with the oxidation product thiobencarb sulfoxide. Other degradation products include 2-hydroxythiobenzcarb (Fan and Alexeeff, 2000) along with 4-chlorobenzaldehyde. The pesticides thiobencarb and ethiofencarb undergo direct photolysis in aqueous solution although thiobencarb yields 4-chlorobenzaldehyde and ethiofencarb gives the expected sulfoxide (Vialaton and Richard, 2000). The sulfur-containing pesticides fenthion and the pollutant from fossil fuels dibenzothiophene can undergo photodegradation typically to the sulfones (Huang and Mabury, 2000a,b).

Windblown sulfates, typically as gypsum, are found in arid regions. Water utilization or agricultural changes can enhance the production of such dusts. The decreasing water volume in the Aral Sea has much enhanced the production of sulfate dusts over recent decades that has become associated with severe environmental problems (O'Hara *et al.*, 2000).

8.14.11.5 Coastal Pollution

Because seawater has high sulfate concentrations, the release of sulfate to coastal waters is not a major problem. However, when organic loads in the discharge waters are heavy, the sulfate can be reduced with a problem of bad odors. A range of larger organosulfides can be found in seawater. Dibenzothiophene is representative of a group of sulfur-containing heterocyclic organic compounds that are common as a form of organosulfur in oils. The relative stability of the dibenzothiophenes has made them useful markers in the weathering of crude oils (Barakat *et al.*, 2001). Dibenzothiophene has been found as a common constituent of coastal sediments with concentrations up to several hundred ng g^{-1}. Sediments of the South China Sea show a concentration range of 11–66 ng g^{-1} dry sediment. Dibenzothiophene content was higher nearshore relative to offshore sediments, but clay and organic carbon contents appear as two prime factors controlling the sediment dibenzothiophene levels (Yang *et al.*, 1998). The source of this sedimentary dibenzothiophene is generally seen as the deposition of particles from combustion sources. However, it is also possible from crude oil pollution or terrestrial runoff as river sediments can be high in dibenzothiophene (West *et al.*, 1988).

8.14.12 SULFUR IN UPPER ATMOSPHERES

8.14.12.1 Radiation Balance and Sulfate Particles

Sulfuric acid is an important component of upper planetary atmospheres. The background sulfur found in the Earth's stratosphere derives from OCS that is resistant to attack by the OH radical in the troposphere. This means that there is a flux of OCS to the stratosphere of 0.03–0.06 Tg(S) yr^{-1}. Although resistant to oxidation in the lower atmosphere, there is the potential for photochemistry or chemistry involving atomic oxygen in the stratosphere (Warneck, 1999):

$$OCS + h\nu \rightarrow S + CO$$

$$O + OCS \rightarrow SO + CO$$

Subsequently,

$$S + O_2 \rightarrow SO + O$$

$$SO + O_2 \rightarrow O + SO_2$$

The sulfur dioxide will ultimately be converted to sulfuric acid through addition of OH much as was shown for the troposphere. Oxidative processes lead to sulfuric acid, which has such a strong affinity for water that even under conditions where its abundance is low solution droplets form, mostly through the condensation onto small nuclei in equatorial regions. The stratospheric sulfate aerosol has a size range 0.1–0.3 μm and is found between the tropopause and ~30 km. This is often called the Junge layer after Christian Junge, who discovered it in the1960s.

During periods of intense volcanic activity large quantities of SO_2 can be injected into the stratosphere, increasing the concentration of sulfate aerosols. In 1991 Mt. Pinatubo, in the Philippines, injected some 20 Tg of SO_2 into the stratosphere. Under normal conditions aerosol sulfate concentrations are 1–10 particles cm^{-3}, although after eruptions this can rise by as much as two orders of magnitude. Peak sulfate levels in the Junge layer can increase from around 0.1 μg m^{-3} to 40 μg m^{-3}. The sulfate layer appears to take about six months to form through the slow oxidation of SO_2 into a sulfate aerosol.

The volcanic particles, including those formed from sulfuric acid droplets, intercept incoming solar radiation. The sulfuric acid particles have a greater effect that the larger volcanic ash particles fall out more quickly. This absorption of radiation by stratospheric particles warms the stratosphere and cools the troposphere. After major eruptions this can amount to a 1° increase in the middle troposphere and changes in surface climate for a few years afterwards. Tropospheric cooling after an eruption is mitigated in the northern hemisphere winter, because tropical eruptions can induce a stronger polar vortex, with a stronger jet stream producing characteristic stationary wave pattern within the tropospheric circulation. Thus, we can find warmer northern hemisphere continents in these winters. This indirect advective effect on winter climate is stronger than the radiative cooling effect, which can dominate at lower latitudes and in the summer months. Volcanic effects play a significant role in interdecadal climate change on longer timescales (Robock, 2000).

8.14.12.2 Ozone

The sulfate droplets also have an important role in stratospheric chemistry. In the lower stratosphere the relative humidity is also low, so sulfuric acid concentrations can be high (65–80% H_2SO_4 mass fraction). In the higher stratosphere where water is more abundant or under conditions where the airmass cools, the sulfuric acid becomes less concentrated (around 30% H_2SO_4 mass fraction). The larger amount of water and lower temperatures allows other components of the stratosphere, most importantly nitric acid and ultimately hydrogen chloride to dissolve in the droplets. Further cooling allows various hydrates to crystallize out, although it is also possible for these solutions to remain as liquid droplets below the frost point.

These sulfate aerosols provide sites for an extensive heterogeneous chemistry and have stimulated interest in the role of polar stratospheric clouds in the depletion of ozone at high latitudes. This depletion processes have been enhanced by the increased amount of halogens introduced into the stratosphere from halogens containing compounds such as CFCs/freons. One way to view the heterogeneous chemistry in the sulfate aerosol is to consider it in terms of separating of the chlorine and the nitrogen species of the stratosphere. Chlorine and nitrogen can be found combined as chlorine nitrate, $ClONO_2$, limiting the destruction of ozone by chlorine. Heterogeneous processes within the cloud particles allow chlorine to distribute into the gas phase with nitrogen compounds as nitric acid in the solid phase, e.g.,

$$ClONO_{2(g)} + HCl_{(s)} \rightarrow Cl_{2(g)} + HNO_{3(s)}$$

This can be followed by gas-phase processes, such as photolysis:

$$Cl_{2(g)} + h\nu \rightarrow 2Cl_{(g)}$$

and a sequence of reactions:

$$2Cl_{(g)} + 2O_{3(g)} \rightarrow 2ClO_{(g)} + 2O_{2(g)}$$

$$2ClO_{(g)} + M \rightarrow Cl_2O_{2(g)} + M$$

$$Cl_2O_{2(g)} + h\nu \rightarrow ClO_{2(g)} + Cl_{(g)}$$

$$ClO_{2(g)} + M \rightarrow Cl_{(g)} + O_{2(g)} + M$$

These four equations sum

$$2O_{3(g)} \rightarrow 3O_{2(g)}$$

showing ozone destruction. One can see that the cloud chemistry of the sulfate aerosol can be important in depleting ozone. The cloud particles separate the chlorine and the nitrogen species and additionally some of the reactions proceed faster at the low temperatures found over the poles. It should also be noted that the formation of $Cl_2O_{2(g)}$ is second order so potentially sensitive to chlorine concentration.

Such heterogeneous processes are not only important in polar regions. During the Pinatubo eruption the presence of enhanced numbers of sulfate particles caused a 10–15% loss of ozone after about a year at 40° N. This gradually recovered over the following years (Warneck, 1999).

8.14.12.3 Aircraft

Aircraft are also an important source of sulfur in the stratosphere. Although the sulfur content of aviation fuels is rather low (typical sulfur content ~400 ppm by mass), the current subsonic fleet injects some 0.02 Tg(S) yr^{-1}compared with the OCS-derived input of 0.03–0.06 Tg(S) yr^{-1}. This is a significant fraction of the background flux, although considerably smaller than the input from large volcanic eruptions. Aircraft also emit sulfur compounds in the troposphere, but here there are other larger anthropogenic sources at these lower altitudes. Nevertheless, the aircraft emissions are very important because only a small fraction of sulfur sources at the surface of the Earth reach the upper troposphere. Such transfer to high altitudes depends on deep mid-latitude and tropical convection processes, while aircraft emissions occur at high altitude and thus do not require vertical transport. In addition, although these emissions are often in remote areas, they tend to concentrate along well-used flight paths (Fahey and Schumann, 1999).

The sulfur emitted from aircraft assists the formation of contrails, by increasing the number of ice particles and decreasing particle size. Contrails can develop into more extensive contrail cirrus clouds in air masses that are supersaturated with respect to ice and here the ice particles can grow through the uptake of water vapor. Aircraft can act as a kind of trigger for cloud formation. Currently, line-shaped contrails are more frequent over North America, the North Atlantic, and Europe. Above Europe these clouds amount to 0.5% of the daytime coverage. These contrails induce a radiative forcing that increases upper temperatures, while decreasing that at the surface. Although aviation will increase in the future, some of its effects on cirrus formation could decrease through reduced sulfur and soot emissions.

8.14.13 PLANETS AND MOONS

Although this chapter focuses on the Earth's biogeochemistry, it is instructive to consider what examining sulfur cycling on other bodies in the solar system reveals. The atmospheric chemistry is most studied because of the comparative effectiveness of spectroscopic methods of planetary observation.

8.14.13.1 Venus

Studies of Venus and most particularly the fly-by missions of early spacecraft gave an impression of a hostile environment with very high surface temperatures (735 K). Infrared observations indicated the presence of sulfur dioxide in the upper cloud layers. The sulfur dioxide (SO_2) content of atmosphere observed at the cloud tops (~50 km) varies through time and appears to have decreased by 50-fold since 1978 (Bezard *et al.*, 1993). At altitudes in the range 35–45 km region concentrations are ~130 ppm and may be more stable. They are seen as a likely tracer of Venusian volcanism. The yellow color of the visual images has been attributed to suspended sulfur particles in the upper atmosphere. There has been a long debate over sulfuric acid droplets in the atmosphere of Venus, but they may be less likely than once thought. OCS can be produced in the atmosphere:

$$SO_2 + 3CO = OCS + 2CO_2$$

with nominal concentrations of 5 ppm at the surface of Venus (Hong and Fegley, 1997), but equilibrium might not be attained.

8.14.13.2 Jupiter

Sulfur on Jupiter is better understood since the impact of the cometary fragment *Shoemaker-Levy 9* into the Jovian atmosphere in July 1994. The impact forced a great plume of material from deep in the atmosphere up some 3,000 km. This plume was characterized by the presence of ammonia, sulfur, ammonium hydrosulfide, and helium, and led to observations of hydrogen sulfide on Jupiter. The impacts left dark spots in the atmosphere, which gave strong signatures from sulfur-bearing compounds such as diatomic sulfur (S_2), carbon disulfide (CS_2), and hydrogen sulfide (H_2S). Modeling studies have examined the post-impact sulfur photochemistry in the Jovian atmosphere (Moses *et al.*, 1995). They suggest that the sulfur polymerizes to S-8 in the first few days after an impact. Other important sulfur reservoirs are CS, whose abundance increased markedly with time. There is also the potential for sulfur to react with hydrocarbons to give thioformaldehyde or, depending on reaction rates, possibly organosulfides such as DMS. There is also the potential for NS to be significant as its loss rates may be small.

8.14.13.3 Io

The active volcanoes of the Jovian moon Io release large quantities of sulfur and other materials (Spencer and Schneider, 1996) that recover the surface at a rapid rate and maintain a tenuous atmosphere. This sulfur is largely as sulfur dioxide, which is also found as condensate that covers some three-quarters of the surface. However, sulfur is also found as elemental sulfur, with perhaps traces of hydrogen sulfide (Russell and Kivelson, 2001; Zolotov and Fegley, 1998). Low pressures in the atmosphere of Io mean that sulfur can remain in seemingly exotic forms such as sulfur monoxide (SO), which has been calculated to have an SO/SO_2 ratio of 3–10% (Zolotov and Fegley, 1998). Others suggest that OSOSO, and its cation, are likely present in the Io's atmosphere (Cacace *et al.*, 2001).

8.14.13.4 Europa

Recent observations from the Galileo's near infrared mapping spectrometer have suggested the presence of enhanced concentrations of sulfuric acid on the trailing side of the Jovian moon Europa. This face of Europa is struck by sulfur ions coming from Jupiter's innermost moon Io. A dark surface material, which spatially correlates with the sulfuric acid concentration, is identified as radiolytically altered sulfur polymers. Radiolysis of the surface by magnetospheric plasma bombardment continuously cycles sulfur between three forms: sulfuric acid, sulfur dioxide, and sulfur polymers, with sulfuric acid being ~50 times as abundant as the other forms. Sulfur species continually undergo interconversion in radiolytic cycles as shown in Figure 21 These maintain a dynamic equilibrium between sulfur polymers S_n, sulfur dioxide SO_2, hydrogen sulfide H_2S, and $H_2SO_4{\cdot}nH_2O$ (Carlson *et al.*, 2002).

8.14.14 CONCLUSIONS

The Earth's sulfur cycle (Figure 22) transfers enormous amounts of this biologically important element through various reservoirs each year. Sulfur has a wide range of oxidation states and shows the ability to form a large number of oxides and oxyanions, many of which are found in the environment. It also has the potential to form polymeric species with a significant number of sulfur atoms. Although the nitrogen cycle is critical to the biosphere and agricultural production, it may be fair to argue that it does not show quite the same range of polymerism.

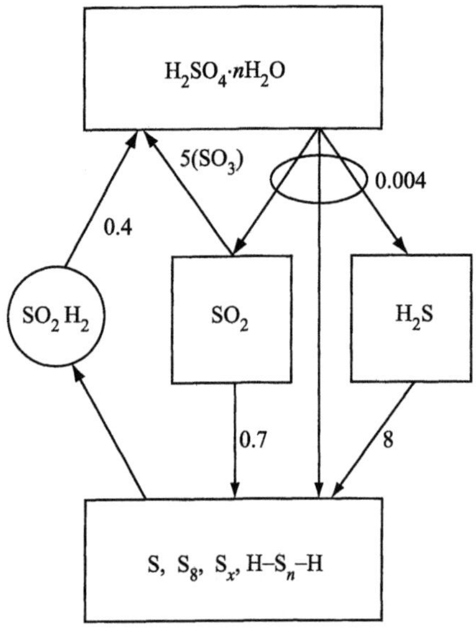

Figure 21 Radiolytic sulfur cycling on the Jovian moon Europa with reaction rates marked numerically as reaction efficiencies. The transient sulfinic acid is circled as an intermediate (source Carlson *et al.*, 2002).

The insolubility of iron sulfides offers the potential for the burial and mineralization of large amounts of sulfur in sediments. However, sulfur shares with nitrogen a complex organic chemistry, but organosulfur compounds have not always been widely studied. Nevertheless the discovery of large fluxes of DMS to the atmosphere has revealed just how important organosulfur compounds can be.

The role that DMS plays in the formation of cloud condensation nuclei has implications for climate that has provoked a debate and interest that has gone well beyond scientists and has been seen as support for the Gaia Hypothesis.

As with the nitrogen cycle human inputs of sulfur have altered the cycles in profound ways. Human activities, although growing and dependent on economic cycles, are relatively constant compared with the high variability of some natural emissions such as explosive volcanism. Ever since the beginning of the twentieth century, global industrialization has had a remarkable effect on the sulfur cycle, perhaps most notably in the production of acid rain. There have also been increases in the release of sulfur into rivers from mining, industrialization, and particularly agriculture. Wind-blown dusts have also been enhanced usually through loss of ground cover or in some very notable incidents, the removal water from inland seas.

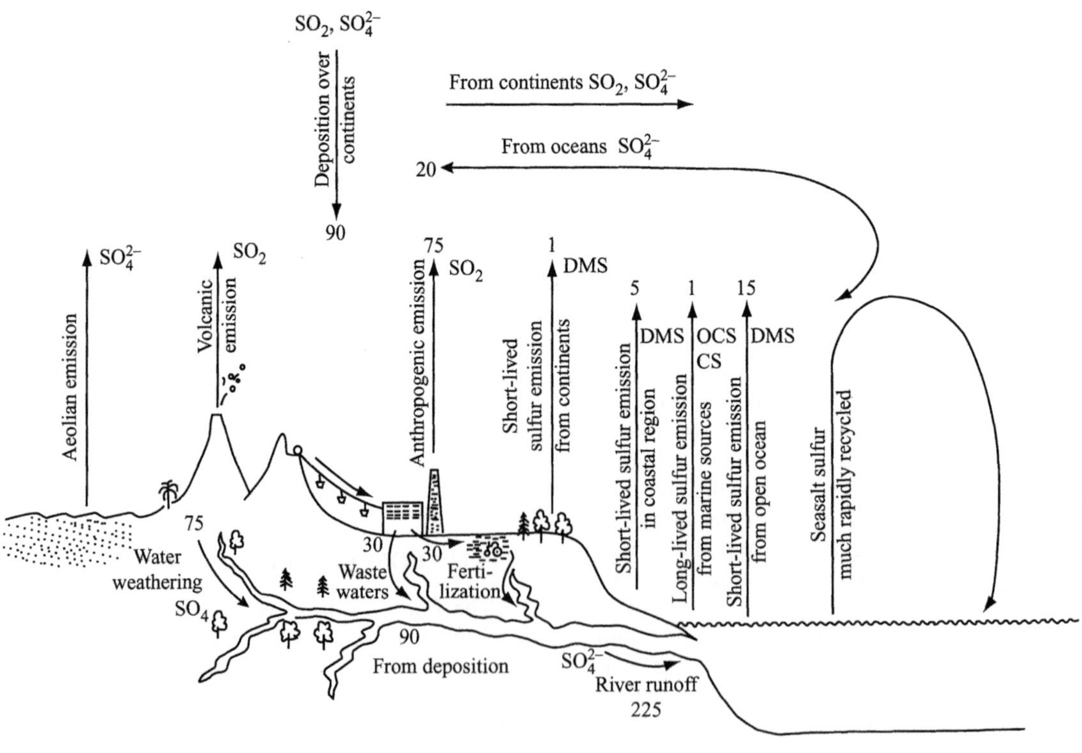

Figure 22 Global cycle showing key fluxes at Tg(S) yr^{-1}. The formula for the most significant components are marked against each flux.

There will be changes in the future. Human emissions of sulfur dioxide to the atmosphere are likely to reach a maximum in the early twenty-first century. The biogeochemical cycle of sulfur seems set to undergo further change, so our retained interest is bound to unlock more of its secrets.

REFERENCES

Andres R. J. and Kasgnoc A. D. (1998) A time-averaged inventory of subaerial volcanic sulfur emissions. *J. Geophys. Res.: Atmos.* **103**(D19), 25251–25261.

Atlas E. (1991) Observation of possible elemental sulfur in the marine atmosphere and speculation on its origin. *Atmos. Environ. Part a: General Topics* **25**(12), 2701–2705.

Attar A. (1979) Evaluate sulfur in coal. *Hydrocarb. Process.* **58**, 175–179.

Ayers G. P. and Gillett R. W. (2000) DMS and its oxidation products in the remote marine atmosphere: implications for climate and atmospheric chemistry. *J. Sea Res.* **43**(3–4), 275–286.

Ayers G. P., Ivey J. P., and Gillett R. W. (1991) Coherence between seasonal cycles of dimethyl sulfide, methanesulfonate and sulfate in marine air. *Nature* **349**(6308), 404–406.

Barakat A. O., Qian Y. R., Kim M., and Kennicutt M. C. (2001) Chemical characterization of naturally weathered oil residues in arid terrestrial environment in Al-Alamein, Egypt. *Environ. Int.* **27**(4), 291–310.

Barnes I., Becker K. H., and Mihalopoulos N. (1994) An FTIR product study of the photooxidation of dimethyl disulfide. *J. Atmos. Chem.* **18**(3), 267–289.

Bester K., Huhnerfuss H., Lange W., and Theobald N. (1997) Results of non-target screening of lipophilic organic pollutants in the German Bight: I. Benzothiazoles. *Sci. Tot. Environ.* **207**(2–3), 111–118.

Betterton E. A. (1992) Oxidation of alkyl sulfides by aqueous peroxymonosulfate. *Environ. Sci. Technol.* **26**(3), 527–532.

Bezard B., Debergh C., Fegley B., Maillard J. P., Crisp D., Owen T., Pollack J. B., and Grinspoon D. (1993) The abundance of sulfur-dioxide below the clouds of Venus. *Geophys. Res. Lett.* **20**(15), 1587–1590.

Blake-Kalff M. M. A., Zhao F. J., Hawkesford M. J., and McGrath S. P. (2001) Using plant analysis to predict yield losses caused by sulphur deficiency. *Ann. Appl. Biol.* **138**(1), 123–127.

Boogert A. C. A., Schutte W. A., Helmich F. P., Tielens A. G. G. M., and Wooden D. H. (1997) Infrared observations and laboratory simulations of interstellar CH_4 and SO_2. *Astron. Astrophys.* **317**, 929–94.

Bouillon R. C., Lee P. A., de Mora S. J., Levasseur M., and Lovejoy C. (2002) Vernal distribution of dimethylsulphide, dimethylsulphoniopropionate, and dimethylsulphoxide in the North Water in 1998. *Deep-Sea Res. II: Top. Stud. Oceanogr.* **49**(22–23), 5171–5189.

Brennan S. T. and Lowenstein T. K. (2002) The major-ion composition of Silurian seawater. *Geochim. Cosmochim. Acta* **66**(15), 2683–2700.

Brewer P. G., Glover D. M., Goyet C., and Shafer D. K. (1995) The pH of the North-Atlantic Ocean—improvements to the global-model for sound-sbsorption in seawater. *J. Geophys. Res.: Oceans* **100**(C5), 8761–8776.

Brimblecombe P. and Shooter D. (1986) Photooxidation of dimethylsulfide in aqueous-solution. *Mar. Chem.* **19**(4), 343–353.

Brimblecombe P., Hammer C., Rodhe H., Ryaboshapko A., and Boutron C. F. (1989) Human influence on the sulfur cycle. In *Evolution of the Global Biogeochemical Sulphur Cycle* (eds. P. Brimblecombe and A. Y. Lein). Wiley, Chichester, vol. 39, pp. 77–121.

Butler I. B. and Rickard D. (2000) Framboidal pyrite formation via the oxidation of iron(II) monosulfide by hydrogen sulphide. *Geochim. Cosmochim. Acta* **64**(15), 2665–2672.

Cacace F., Cipollini R., de Petris G., Rosi M., and Troiani A. (2001) A new sulfur oxide, OSOSO, and its cation, likely present in the Io's atmosphere: detection and characterization by mass spectrometric and theoretical methods. *J. Am. Chem. Soc.* **123**(3), 478–484.

Campolongo F., Saltelli A., Jensen N. R., Wilson J., and Hjorth J. (1999) The role of multiphase chemistry in the oxidation of dimethylsulphide (DMS): a latitude dependent analysis. *J. Atmos. Chem.* **32**(3), 327–356.

Canfield D. E. and Raiswell R. (1999) The evolution of the sulfur cycle. *Am. J. Sci.* **299**(7–9), 697–723.

Canfield D. E., Habicht K. S., and Thamdrup B. (2000) The Archean sulfur cycle and the early history of atmospheric oxygen. *Science* **288**(5466), 658–661.

Carlson R. W., Anderson M. S., Johnson R. E., Schulman M. B., and Yavrouian A. H. (2002) Sulfuric acid production on Europa: the radiolysis of sulfur in water ice. *Icarus* **157**(2), 456–463.

Cavalli L., Cassani G., Vigano L., Pravettoni S., Nucci G., Lazzarin M., and Zatta A. (2000) Surfactants in sediments. *Tenside Surfactants Detergents* **37**(5), 282–288.

Charlson R. J., Lovelock J. E., Andreae M. O., and Warren S. G. (1987) Oceanic phytoplankton, atmospheric sulfur, cloud albedo and climate. *Nature* **326**(6114), 655–661.

Charlson R. J., Anderson T. L., and McDuff R. E. (1992) The sulfur cycle. In *Global Biogeochem. Cycles* (eds. S. S. Butcher, R. J. Charlson, G. H. Orians, and G. V. Wolfe). Academic Press, London, pp. 285–300.

Clegg S. L. and Whitfield M. (1991) Activity coefficients in natural waters. In *Activity Coefficients in Electrolyte Solutions* (ed. K. S. Pitzer). CRC Press, vol. 102A, pp. 279–434.

Conway E. J. (1942) Mean geochemical data in relation to oceanic evolution. *Proc. Roy. Irish Acad.* **48**, 119–159.

Cotton F. A., Wilkinson G., Murillo C. A., and Bochmann M. (1999) *Advanced Inorganic Chemistry*. Wiley, New York.

Cutter G. A. and Krahforst C. F. (1988) Sulfide in Surface waters of the western Atlantic-Ocean. *Geophys. Res. Lett.* **15**(12), 1393–1396.

Cutter G. A., Walsh R. S., and de Echols C. S. (1999) Production and speciation of hydrogen sulfide in surface waters of the high latitude North Atlantic Ocean. *Deep-Sea Res. II: Top. Stud. Oceanogr.* **46**(5), 991–1010.

Damste J. S. S. and Deleeuw J. W. (1992) Organically bound sulfur in coal—a molecular approach. *Fuel Process. Technol.* **30**(2), 109–178.

de Bruyn W. J., Harvey M., Cainey J. M., and Saltzman E. S. (2002) DMS and SO_2 at Baring Head, New Zealand: implications for the yield of SO_2 from DMS. *J. Atmos. Chem.* **41**(2), 189–209.

DeLaune R. D., Devai I., and Lindau C. W. (2002) Flux of reduced sulfur gases along a salinity gradient in Louisiana coastal marshes. *Estuar. Coast. Shelf Sci.* **54**(6), 1003–1011.

Delmelle P., Stix J., Bourque C. P. A., Baxter P. J., Garcia-Alvarez J., and Barquero J. (2001) Dry deposition and heavy acid loading in the vicinity of Masaya volcano, a major sulfur and chlorine source in Nicaragua. *Environ. Sci. Technol.* **35**(7), 1289–1293.

Delmelle P., Stix J., Baxter P. J., Garcia-Alvarez J., and Barquero J. (2002) Atmospheric dispersion, environmental effects and potential health hazard associated with the low-altitude gas plume of Masaya volcano, Nicaragua. *Bull. Volcanol.* **64**(6), 423–434.

Des Marais D. J. (2000) Evolution—when did photosynthesis emerge on earth? *Science* **289**(5485), 1703–1705.

Eckardt F. D. and Spiro B. (1999) The origin of sulphur in gypsum and dissolved sulphate in the Central Namib Desert, Namibia. *Sedim. Geol.* **123**(3–4), 255–273.

Elliott S., Lu E., and Rowland F. S. (1987) Carbonyl sulfide hydrolysis as a source of hydrogen-sulfide in open ocean seawater. *Geophys. Res. Lett.* **14**(2), 131–134.

Elliott S., Lu E., and Rowland F. S. (1989) Rates and mechanisms for the hydrolysis of carbonyl sulfide in natural-waters. *Environ. Sci. Technol.* **23**(4), 458–461.

Eremin N. I., Dergachev A. L., Pozdnyakova N. V., and Sergeeva N. E. (2002) Epochs of volcanic-hosted massive sulfide ore formation in the Earth's history. *Geol. Ore Dep.* **44**(4), 227–241.

Fahey D. W. and Schumann U. (1999) Aviation produced aerosol and cloudiness. In *Aviation and the Global Atmosphere* (eds. J. E. Penner, D. H. Lister, D. J. Griggs, D. J. Dokken, and M. McFarland). Cambridge University Press, Cambridge, pp. 65–120.

Fan A. M. and Alexeeff G. V. (2000) *Thiobenzcarb*. California Environmental Protection Agency, Sacromento.

Farquhar J., Jackson T. L., and Thiemens M. H. (2000) A S-33 enrichment in ureilite meteorites: evidence for a nebular sulfur component. *Geochim. Cosmochim. Acta* **64**(10), 1819–1825.

Farquhar J., Wing B. A., McKeegan K. D., Harris J. W., Cartigny P., and Thiemens M. H. (2002) Mass-independent sulfur of inclusions in diamond and sulfur recycling on early earth. *Science* **298**(5602), 2369–2372.

Findlayson-Pitts B. J. and Pitts J. N. (2000) *Chemistry of the Upper and Lower Atmosphere*. Academic Press, San Diego.

Frausto da Silva J. J. R. and Williams R. J. P. (2001) *The Biological Chemistry of the Elements*. Oxford University Press, Oxford.

Gobbi G., Zappia G., and Sabbioni C. (1998) Sulphite quantification on damaged stones and mortars. *Atmos. Environ.* **32**(4), 783–789.

Goto-Azuma K. and Koerner R. M. (2001) Ice core studies of anthropogenic sulfate and nitrate trends in the Arctic. *J. Geophys. Res.: Atmos.* **106**(D5), 4959–4969.

Greenwood N. N. and Earnshaw A. (1997) *Chemistry of the Elements*. Heinemann, Oxford.

Gries C., Nash T. H., and Kesselmeier J. (1994) Exchange of reduced sulfur gases between lichens and the atmosphere. *Biogeochemistry* **26**(1), 25–39.

Gröne T. (1997) Volatile organic sulfur species in three North Italian lakes: seasonality, possible sources and flux to the atmosphere. *Mem. Ist. ital. Idrobiol.* **56**, 77–94.

Gun J., Goifman A., Shkrob I., Kamyshny A., Ginzburg B., Hadas O., Dor I., Modestov A. D., and Lev O. (2000) Formation of polysulfides in an oxygen rich freshwater lake and their role in the production of volatile sulfur compounds in aquatic systems. *Environ. Sci. Technol.* **34**(22), 4741–4746.

Habicht K. S., Gade M., Thamdrup B., Berg P., and Canfield D. E. (2002) Calibration of sulfate levels in the Archean Ocean. *Science* **298**(5602), 2372–2374.

Hafenbradl D., Keller M., Wachtershauser G., and Stetter K. O. (1995) Primordial amino-acids by reductive amination of alpha-oxo acids in conjunction with the oxidative formation of pyrite. *Tetrahedron Lett.* **36**(29), 5179–5182.

Harnisch J., Borchers R., Fabian P., and Kourtidis K. (1995) Aluminium production as a source of atmospheric carbonyl sulfide (COS). *Environ. Sci. Pollut. Res.* **2**(3), 161–162.

Hatton A. D. (2002) Influence of photochemistry on the marine biogeochemical cycle of dimethylsulphide in the northern North Sea. *Deep-Sea Res. II.*

Hatton A. D., Turner S. M., Malin G., and Liss P. S. (1998) Dimethylsulphoxide and other biogenic sulphur compounds in the Galapagos Plume. *Deep-Sea Res. II: Top. Stud. Oceanogr.* **45**(6), 1043–1053.

Havlin J. L., D B. J., Tisdale S. L., and Nelson W. L. (1999) *Soil Fertility and Fertilizers*. Prentice-Hall, Upper Saddle River, NJ.

Hines M. E. (1992) Emissions of biogenic sulfur gases from Alaskan tundra. *J. Geophys. Res.: Atmos.* **97**(D15), 16703–16707.

Holser W. T., Maynard J. B., and Cruikshank K. M. (1989) Modelling the natural cycle of sulphur through phanerozoic time. In *Evolution of the Global Biogeochemical Sulphur Cycle* (eds. P. Brimblecombe and A. Y. Lein). Wiley, Chichester, vol. 39, pp. 21–56.

Hong Y. and Fegley B. (1997) Formation of carbonyl sulfide (OCS) from carbon monoxide and sulfur vapor and applications to Venus. *Icarus* **130**(2), 495–504.

Huang J. and Mabury S. A. (2000a) The role of carbonate radical in limiting the persistence of sulfur-containing chemicals in sunlit natural waters. *Chemosphere* **41**(11), 1775–1782.

Huang J. P. and Mabury S. A. (2000b) Hydrolysis kinetics of fenthion and its metabolites in buffered aqueous media. *J. Agri. Food Chem.* **48**(6), 2582–2588.

Huggins F. E., Shah N., Huffman G. P., and Robertson J. D. (2000) XAFS spectroscopic characterization of elements in combustion ash and fine particulate matter. *Fuel Process. Technol.* **65**, 203–218.

Jannasch H. W. (1989) Sulphur emission and transformations at deep sea hydrothermal vents. In *Evolution of the Global Biogeochemical Sulphur Cycle* (eds. P. Brimblecombe and A. Y. Lein). Wiley, Chichester, vol. 39, pp. 181–190.

Jourdain B. and Legrand M. (2001) Seasonal variations of atmospheric dimethylsulfide, dimethylsulfoxide, sulfur dioxide, methanesulfonate, and non-seasalt sulfate aerosols at Dumont d'Urville (coastal Antarctica) (December 1998 to July 1999). *J. Geophys. Res.: Atmos.* **106**(D13), 14391–14408.

Keller L. P., Hony S., Bradley J. P., Molster F. J., Waters L., Bouwman J., de Koter A., Brownlee D. E., Flynn G. J., Henning T., and Mutschke H. (2002) Identification of iron sulphide grains in protoplanetary disks. *Nature* **417**(6885), 148–150.

Kelly D. P. and Murrell J. C. (1999) Microbial metabolism of methanesulfonic acid. *Arch. Microbiol.* **172**(6), 341–348.

Kettle A. J. and Andreae M. O. (2000) Flux of dimethylsulfide from the oceans: a comparison of updated data seas and flux models. *J. Geophys. Res. Atmos.* **105**(D22), 26793–26808.

Kettle A. J., Andreae M. O., Amouroux D., Andreae T. W., Bates T. S., Berresheim H., Bingemer H., Boniforti R., Curran M. A. J., DiTullio G. R., Helas G., Jones G. B., Keller M. D., Kiene R. P., Leck C., Levasseur M., Malin G., Maspero M., Matrai P., McTaggart A. R., Mihalopoulos N., Nguyen B. C., Novo A., Putaud J. P., Rapsomanikis S., Roberts G., Schebeske G., Sharma S., Simo R., Staubes R., Turner S., Uher G. (1999) A global database of sea surface dimethylsulfide (DMS) measurements and a procedure to predict sea surface DMS as a function of latitude, longitude, and month. *Global Biogeochem. Cycles* **13**, 399–444.

Kieber D. J., Jiao J., Kiene R. P., and Bates T. S. (1996) Impact of dimethylsulfide photochemistry on methyl sulfur cycling in the equatorial Pacific Ocean. *J. Geophys. Res.* **101**, 3715–3722.

Kiene R. P. (1990) Dimethyl sulfide production from dimethylsufoniopropionate in coastal seawater samples and bacterial cultures. *Appl. Environ. Microbiol.* **56**, 3292–3297.

Kiene R. P. and Linn L. J. (2000) The fate of dissolved dimethylsulfoniopropionate (DMSP) in seawater: tracer studies using S-35-DMSP. *Geochim. Cosmochim. Acta* **64**(16), 2797–2810.

Kiene R. P., Linn L. J., and Bruton J. A. (2000) New and important roles for DMSP in marine microbial communities. *J. Sea Res.* **43**(3–4), 209–224.

Kilburn K. H. and Warshaw R. H. (1995) Hydrogen-sulfide and reduced-sulfur gases adversely affect neurophysiological functions. *Toxicol. Ind. Health* **11**(2), 185–197.

Killham K. (1994) *Soil Ecology*. Cambridge University Press, Cambridge.

Kjellstrom E. (1998) A three-dimensional global model study of carbonyl sulfide in the troposphere and the lower stratosphere. *J. Atmos. Chem.* **29**(2), 151–177.

Kusakabe M., Komoda Y., Takano B., and Abiko T. (2000) Sulfur isotopic effects in the disproportionation reaction of sulfur dioxide in hydrothermal fluids: implications for the delta S-34 variations of dissolved bisulfate and elemental sulfur from active crater lakes. *J. Volcanol. Geotherm. Res.* **97**(1–4), 287–307.

Ledyard K. M. and Dacey J. W. H. (1996) Micronial cycling of DMSP and DMS in coastal and oligotrophic seawater. *Limonol. Oceanogr.* **41**, 33–40.

Legrand M., Preunkert S., Wagenbach D., and Fischer H. (2002) Seasonally resolved Alpine and Greenland ice core records of anthropogenic HCl emissions over the 20th century. *J. Geophys. Res.: Atmos.* **107**(D12), article no. 4139.

Licht S., Longo K., Peramunage D., and Forouzan F. (1991) Conductometric analysis of the 2nd acid dissociation-constant of H2S in highly concentrated aqueous-media. *J. Electroanal. Chem.* **318**, 111–129.

Likens G. E., Driscoll C. T., Buso D. C., Mitchell M. J., Lovett G. M., Bailey S. W., Siccama T. G., Reiners W. A., and Alewell C. (2002) The biogeochemistry of sulfur at Hubbard Brook. *Biogeochemistry* **60**(3), 235–316.

Lindgren W. (1923) Concentration and circulation of the elements from the standpoint of *Econ. Geol.* **18**, 419–442.

Lovelock J. E., Maggs R. J., and Rasmussen R. A. (1972) Atmospheric dimethyl sulfide and the natural sulfur cycle. *Nature* **237**, 452–453.

Malin G. and Kirst G. O. (1997) Algal production of dimethyl sulfide and its atmospheric role. *J. Phycol.* **33**, 889–896.

Mauersberger R., Henkel C., Langer N., and Chin Y. N. (1996) Interstellar S-36: a probe of s-process nucleosynthesis. *Astron. Astrophys.* **313**, L1–L4.

Meybeck M. (1987) Global chemical-weathering of surficial rocks estimated from river dissolved loads. *Am. J. Sci.* **287**(5), 401–428.

Migdisov A. A. and Bychkov A. Y. (1998) The behaviour of metals and sulphur during the formation of hydrothermal mercury-antimony-arsenic mineralization, Uzon caldera, Kamchatka, Russia. *J. Volcanol. Geotherm. Res.* **84**, 153–171.

Migdisov A. A., Williams-Jones A. E., Lakshtanov L. Z., and Alekhin Y. V. (2002) Estimates of the second dissociation constant of H_2S from the surface sulfidation of crystalline sulfur. *Geochim. Cosmochim. Acta* **66**, 1713–1725.

Millero F. J. and Hawke D. J. (1992) Ionic interactions of divalent metals in natural-waters. *Mar. Chem.* **40**(1–2), 19–48.

Montegrossi G., Tassi F., Vaselli O., Buccianti A., and Garofalo K. (2001) Sulfur species in volcanic gases. *Anal. Chem.* **73**(15), 3709–3715.

Morra M. J., Fendorf S. E., and Brown P. D. (1997) Speciation of sulfur in humic and fulvic acids using X-ray absorption near-edge structure (XANES) spectroscopy. *Geochim. Cosmochim. Acta* **61**(3), 683–688.

Moses J. I., Allen M., and Gladstone G. R. (1995) Post-Sl9 sulfur photochemistry on Jupiter. *Geophys. Res. Lett.* **22**(12), 1597–1600.

Muezzinoglu A., Sponza D., Koken I., Alparslan N., Akyarli A., and Ozture N. (2000) Hydrogen sulfide and odor control in Izmir Bay. *Water Air Soil Pollut.* **123**(1–4), 245–257.

Neretin L. N., Volkov II, Bottcher M. E., and Grinenko V. A. (2001) A sulfur budget for the Black Sea anoxic zone. *Deep-Sea Res.: I. Oceanogr. Res. Pap.* **48**(12), 2569–2593.

O'Hara S. L., Wiggs G. F. S., Mamedov B., Davidson G., and Hubbard R. B. (2000) Exposure to airborne dust contaminated with pesticide in the Aral Sea region. *Lancet* **355**(9204), 627–628.

Oppenheimer C. and Stevenson D. (1989) Liquid sulfur lakes at Poas volcano. *Nature* **342**(6251), 790–793.

Patris N., Delmas R. J., Legrand M., De Angelis M., Ferron F. A., Stievenard M., and Jouzel J. (2002) First sulfur isotope measurements in central Greenland ice cores along the preindustrial and industrial periods. *J. Geophys. Res.: Atmos.* **107**(D11), article no. 4115.

Phillips D. J. and Phillips S. L. (2000) High temperature dissociation constants of HS and the standard thermodynamic values for S2. *J. Chem. Eng. Data* **45**, 981–987.

Pos W. H., Riemer D. D., and Zika R. G. (1998) Carbonyl sulfide (OCS) and carbon monoxide (CO) in natural waters: evidence of a coupled production pathway. *Mar. Chem.* **62**(1–2), 89–101.

Radfordknoery J. and Cutter G. A. (1994) Biogeochemistry of dissolved hydrogen-sulfide species and carbonyl sulfide in the western North-Atlantic Ocean. *Geochim. Cosmochim. Acta* **58**(24), 5421–5431.

Railsback L. B. (1992) A geological numerical-model for paleozoic global evaporite deposition. *J. Geol.* **100**(3), 261–277.

Rankama K. and Sahama T. G. (1950) *Geochemistry*. University of Chicago Press, Chicago.

Richards S. R., Rudd J. W. M., and Kelly C. A. (1994) Organic volatile sulfur in lakes ranging in sulfate and dissolved salt concentration over 5 orders of magnitude. *Limnol. Oceanogr.* **39**(3), 562–572.

Robinson B. W. and Bottrell S. H. (1997) Discrimination of sulfur sources in pristine and polluted New Zealand river catchments using stable isotopes. *Appl. Geochem.* **12**(3), 305–319.

Robock A. (2000) Volcanic eruptions and climate. *Rev. Geophys.* **38**(2), 191–219.

Rodhe H. (1999) Human impact on the atmospheric sulfur balance. *Tellus Ser. a: Dyn. Meteorol. Oceanogr.* **51**(1), 110–122.

Russell C. T. and Kivelson M. G. (2001) Evidence for sulfur dioxide, sulfur monoxide, and hydrogen sulfide in the Io exosphere. *J. Geophys. Res.: Planets* **106**, 33267–33272.

Sammut J., White I., and Melville M. D. (1996) Acidification of an estuarine tributary in eastern Australia due to drainage of acid sulfate soils. *Mar. Freshwater Res.* **47**(5), 669–684.

Scaillet B., Clemente B., Evans B. W., and Pichavant M. (1998) Redox control of sulfur degassing in silicic magmas. *J. Geophys. Res.: Solid Earth* **103**(B10), 23937–23949.

Schippers A. and Jorgensen B. B. (2001) Oxidation of pyrite and iron sulfide by manganese dioxide in marine sediments. *Geochim. Cosmochim. Acta* **65**(6), 915–922.

Sciare J., Baboukas E., Hancy R., Mihalopoulos N., and Nguyen B. C. (1998) Seasonal variation of dimethylsulfoxide in rainwater at Amsterdam island in the southern Indian Ocean: implications on the biogenic sulfur cycle. *J. Atmos. Chem.* **30**(2), 229–240.

Sciare J., Kanakidou M., and Mihalopoulos N. (2000) Diurnal and seasonal variation of atmospheric dimethylsulfoxide at Amsterdam Island in the southern Indian Ocean. *J. Geophys. Res.: Atmos.* **105**(D13), 17257–17265.

Shooter D. (1999) Sources and sinks of oceanic hydrogen sulfide—an overview. *Atmos. Environ.* **33**(21), 3467–3472.

Shooter D. and Brimblecombe P. (1989) Dimethylsulfide oxidation in the Ocean. *Deep-Sea Res. a: Oceanogr. Res. Pap.* **36**(4), 577–585.

Sidle W. C., Roose D. L., and Shanklin D. R. (2000) Isotopic evidence for naturally occurring sulfate pollution of ponds in the Kankakee River Basin, Illinois-Indiana. *J. Environ. Qual.* **29**(5), 1594–1603.

Siegel B. Z., Nachbarhapai M., and Siegel S. M. (1990) The Contribution of sulfate to rainfall pH around Kilauea volcano, Hawaii. *Water Air Soil Pollut.* **52**(3–4), 227–235.

Siegel S. M., Penny P., Siegel B. Z., and Penny D. (1986) Atmospheric hydrogen-sulfide levels at the Sulfur Bay wildlife area, Lake Rotorua, New-Zealand. *Water Air Soil Pollut.* **28**(3–4), 385–391.

Simo R., Pedros-Alio C., Malin G., and Grimalt J. O. (2000) Biological turnover of DMS, DMSP, and DMSO in contrasting open-sea waters. *Mar. Ecol.: Prog. Ser.* **203**, 1–11.

Simon M., Martin F., Ortiz I., Garcia I., Fernandez J., Fernandez E., Dorronsoro C., and Aguilar J. (2001) Soil pollution by oxidation of tailings from toxic spill of a pyrite mine. *Sci. Tot. Environ.* **279**(1–3), 63–74.

Sinha S. N. and Banerjee R. D. (1997) Ecological role of thiosulfate and sulfide utilizing purple nonsulfur bacteria of a riverine ecosystem. *FEMS Microbiol. Ecol.* **24**(3), 211–220.

Spencer J. R. and Schneider N. M. (1996) Io on the eve of the Galileo mission. *Ann. Rev. Earth Planet. Sci.* **24**, 125–190.

Sriwana T., van Bergen M. J., Varekamp J. C., Sumarti S., Takano B., van Os B. J. H., and Leng M. J. (2000) Geochemistry of the acid Kawah Putih lake, Patuha volcano, West Java, Indonesia. *J. Volcanol. Geotherm. Res.* **97**(1–4), 77–104.

Stefanova M., Marinov S. P., Mastral A. M., Callen M. S., and Garcia T. (2002) Emission of oxygen, sulphur and nitrogen containing heterocyclic polyaromatic compounds from lignite combustion. *Fuel Process. Technol.* **77**, 89–94.

Sudo M., Kunimatsu T., and Okubo T. (2002) Concentration and loading of pesticide residues in Lake Biwa Basin (Japan). *Water Res.* **36**(1), 315–329.

Sunda W., Kieber D. J., Kiene R. P., and Huntsman S. (2002) An antioxidant function for DMSP and DMS in marine algae. *Nature* **418**(6895), 317–320.

Suzuki Y., Kawakami M., and Akasaka K. (2001) H-1 NMR application for characterizing water-soluble organic compounds in urban atmospheric particles. *Environ. Sci. Technol.* **35**(13), 2656–2664.

Symonds R. B. and Reed M. H. (1993) Calculation of multicomponent chemical-equilibria in gas-solid- liquid systems- calculation methods, thermochemical data, and applications to studies of high-temperature volcanic gases with examples from Mount St-Helens. *Am. J. Sci.* **293**(8), 758–864.

Takano B., Ohsawa S., and Glover R. B. (1994) Surveillance of Ruapehu Crater Lake, New-Zealand, by aqueous polythionates. *J. Volcanol. Geotherm. Res.* **60**(1), 29–57.

Tanaka N. and Turekian K. K. (1991) Use of Cosmogenic S-35 to determine the rates of removal of atmospheric SO_2. *Nature* **352**(6332), 226–228.

Thayer M. R. (1997) An investigation into sulfur isotopes in the galactic cosmic rays. *Astrophys. J.* **482**, 792–795.

Thiermann F., Niemeyer A. S., and Giere O. (1996) Variations in the sulfide regime and the distribution of macrofauna in an intertidal flat in the North Sea. *Helgolander Meeresuntersuchungen* **50**(1), 87–104.

Turekian K. K. and Tanaka N. (1992) The use of atmospheric cosmogenic S-35 and Be-7 in determining depositional fluxes of SO_2. *Geophys. Res. Lett.* **19**(17), 1767–1770.

TWRI (1986) *Threats to Groundwater Quality*. Texas Water Resources Institute, College Station, TX.

USGS (2000) *Volcanic Air Pollution—a Hazard in Hawaii*. US Geological Survey, Hawaii Volcanoes National Park.

Vialaton D. and Richard C. (2000) Direct photolyses of thiobencarb and ethiofencarb in aqueous phase. *J. Photochem. Photobiol. a: Chem.* **136**(3), 169–174.

Wachtershauser G. (2000) Life as we don't know it. *Science* **289**(5483), 1307–1308.

Warneck P. (1999) *Chemistry of the Natural Atmosphere*. Academic Press, San Diego.

Watson A. J. and Liss P. S. (1998) Marine biological controls on climate via the carbon and sulphur geochemical cycles. *Phil. Trans. Roy. Soc. London Ser. B: Biol. Sci.* **353**(1365), 41–51.

Watts S. F. (2000) The mass budgets of carbonyl sulfide, dimethyl sulfide, carbon disulfide and hydrogen sulfide. *Atmos. Environ.* **34**(5), 761–779.

Watts S. F. and Roberts C. N. (1999) Hydrogen sulfide from car catalytic converters. *Atmos. Environ.* **33**(1), 169–170.

Wawrzynkiewicz W. and Sablik J. (2002) Organic sulphur in the hard coal of the stratigraphic layers of the Upper Silesian coal basin. *Fuel* **81**(14), 1889–1895.

Wedepohl K. H. (1995) The composition of the continental-crust. *Geochim. Cosmochim. Acta* **59**(7), 1217–1232.

Werne J. P., Hollander D. J., Behrens A., Schaeffer P., Albrecht P., and Damste J. S. S. (2000) Timing of early diagenetic sulfurization of organic matter: a precursor-product relationship in Holocene sediments of the anoxic Cariaco Basin, Venezuela. *Geochim. Cosmochim. Acta* **64**(10), 1741–1751.

West W. R., Smith P. A., Booth G. M., and Lee M. L. (1988) Isolation and detection of genotoxic components in a Black River sediment. *Environ. Sci. Technol.* **22**(2), 224–228.

Wheat C. G. and Mottl M. J. (2000) Composition of pore and spring waters from Baby Bare: global implications of geochemical fluxes from a ridge flank hydrothermal system. *Geochim. Cosmochim. Acta* **64**(4), 629–642.

White C. L., Tabe L. M., Dove H., Hamblin J., Young P., Phillips N., Taylor R., Gulati S., Ashes J., and Higgins T. J. V. (2001) Increased efficiency of wool growth and live weight gain in Merino sheep fed transgenic lupin seed containing sunflower albumin. *J. Sci. Food Agri.* **81**(1), 147–154.

Xie H. X. and Moore R. M. (1999) Carbon disulfide in the North Atlantic and Pacific Oceans. *J. Geophys. Res.: Oceans* **104**(C3), 5393–5402.

Xie H. X., Moore R. M., and Miller W. L. (1998) Photochemical production of carbon disulphide in seawater. *J. Geophys. Res.: Oceans* **103**(C3), 5635–5644.

Yang G. P., Liu X. L., and Zhang J. W. (1998) Distribution of dibenzothiophene in the sediments of the South China Sea. *Environ. Pollut.* **101**(3), 405–414.

Yao W. S. and Millero F. J. (1996) Oxidation of hydrogen sulfide by hydrous Fe(III) oxides in seawater. *Mar. Chem.* **52**(1), 1–16.

Yvon S. A., Saltzman E. S., Cooper D. J., Bates T. S., and Thompson A. M. (1996) Atmospheric sulfur cycling in the tropical Pacific marine boundary layer (12 degrees S, 135 degrees W): a comparison of field data and model results:1. Dimethylsulfide. *J. Geophys. Res.: Atmos.* **101**(D3), 6899–6909.

Zhang L., Walsh R. S., and Cutter G. A. (1998) Estuarine cycling of carbonyl sulfide: production and sea-air flux. *Mar. Chem.* **61**(3–4), 127–142.

Zolotov M. Y. and Fegley B. (1998) Volcanic production of sulfur monoxide (SO) on lo. *Icarus* **132**, 431–434.

Subject Index

The index is in letter-by-letter order, whereby hyphens and spaces within index headings are ignored in the alphabetization (e.g. Arabian–Nubian Shield precedes Arabian Sea). Terms in parentheses are excluded from the initial alphabetization. In line with normal materials science practice, compound names are not inverted but are filed under substituent prefixes.

The index is arranged in set-out style, with a maximum of three levels of heading. Location references refer to the page number. Major discussion of a subject is indicated by bold page numbers. Page numbers suffixed by *f* or *t* refer to figures or tables.